MEDICAL TOXICOLOGY OF NATURAL SUBSTANCES

MEDICAL TOXICOLOGY OF NATURAL SUBSTANCES

Foods, Fungi, Medicinal Herbs, Plants, and Venomous Animals

DONALD G. BARCELOUX, MD, FAACT, FACMT, FACEP

Clinical professor of medicine, department of emergency medicine
David Geffen school of medicine
University of California at Los Angeles
Los Angeles, California

Senior partner, CEP America
Emeryville, California

Staff physician, department of emergency medicine
Pomona valley hospital medicine center
Pomona, California

A JOHN WILEY & SONS, INC., PUBLICATION

Published by John Wiley & Sons, Inc., Hoboken, New Jersey
Published simultaneously in Canada

Library of Congress Cataloging-in-Publication Data:

Barceloux, Donald G.
Medical toxicology of natural substances : foods, fungi, medicinal herbs, plants, and venomous animals / Donald G. Barceloux.
p. ; cm.
Includes index.
ISBN 978-0-471-72761-3 (cloth)
1. Natural products–Toxicology. I. Title.
[DNLM: 1. Plants, Toxic. 2. Animals, Poisonous. 3. Food Poisoning. 4. Furgi. 5. Medicinal Herbs. WD 500 B242m 2008]
RA1211.B285 2008
615.9—dc22

2007051402

Printed in the United States of America

10 9 8 7 6 5 4 3 2 1

To my wife, Kimberly, your loving support and encouragement created the environment to write this book, and your patience allowed me to focus on this book away from what I enjoy most—being with you.

To my son, Colin, your computer expertise has been immensely helpful solving the technical issues in this book, and I admire the application of those skills to the development of your own company, Bookrenter.com.

To my daughter, Shannon, sharing your knowledge for the treatment of animals and your passion for becoming a veterinarian is a great joy for me.

CONTENTS

FOREWORD

Men and women live in a miraculous world, surrounded by natural beauty, diverse environmental conditions and habitats, and evolutionary marvels. Within this intricate assembly of flora and fauna, many plant and animals have evolved that are more than passive inhabitants of this planet. They are endowed with substances both offensive and defensive, namely, potent toxins capable of slowly poisoning or rapidly subduing very large animals, including humans. In addition, modern man utilizes plant and animal products and extracts for commercial, medical, religious, and other purposes. These exposures range from naturopathic cures to a casual encounters with cactus spines during a wilderness expedition, from a diver's encounters with the needles of sea urchins to covert politically motivated assassinations utilizing ricin from castor beans. Wild mushroom foragers grow old only if they are not too bold, while amateur aquarists who reach into their saltwater tanks learn about the toxicity under the cover of lionfish plumage.

We can never learn or know too much about how best to deal with natural toxins, whether we seek to eliminate them from our immediate food supply or treat acutely intoxicated victims. Furthermore, from the understanding of syndromes and therapies, we are offered insights into their possible therapeutic value. No matter what the ultimate fate of man, the seeds, spores, fangs, and venom glands will survive. While the relationship of toxins to humans may not be always be characterized as symbiotic, there will remain a coexistence that is predicted, yet always in part unpredictable.

The medical toxicology of natural substances is predicated upon their existence, which will diminish as humans continue to erode and consume their environment. Until then, we should continue to catalogue, record, evaluate, and teach. Medical practitioners and toxicologists should accept the responsibility to perpetuate these traditions because the wisdom of indigenous shamans is being lost as rapidly as the rainforests in which they reside.

What we do not seek to protect may soon disappear. It is my fervent wish that this book not only serve as a superb medical reference for those who seek to cure, but that all who read it are inspired to preserve the landscapes and seascapes that support the origination of everything that is natural and sustainable upon this Earth.

Paul S. Auerbach, MD, MS, FACEP, FAWM

Division of Emergency Medicine, Department of Surgery,
Stanford University School of Medicine
Stanford, California

PREFACE

Medical Toxicology of Natural Substances: Foods, Fungi, Medicinal Herbs, Plants, and Venomous Animals is designed to provide in-depth, evidence-based coverage of the most important natural toxins. This book is the first of a four volume series in Medical Toxicology, which will include drugs of abuse and psychoactive plants, occupational and environmental exposures, and pharmaceutical overdoses. Scientific knowledge in the field of Medical Toxicology increased considerably since I co-authored the First Edition of *Medical Toxicology: Diagnosis and Treatment of Human Poisoning* with the late Matthew J. Ellenhorn, published in 1988. That book was designed as an authoritative, concise volume for the immediate treatment of poisoning including natural toxins, pharmaceutical agents, and occupational exposures. In the last 20 years, sufficient interest has developed in natural toxins, food contamination, medicinal herbs, chronic occupational exposures, and bioterrorism to justify coverage of the field of Medical Toxicology by a book series rather than a single volume. *Medical Toxicology of Natural Substances: Foods, Fungi, Medicinal Herbs, Plants, and Venomous Animals* retains the consistent, formatted style I developed for the First Edition of *Medical Toxicology: Diagnosis and Treatment of Human Poisoning*. Once the reader is familiar with the templates used in my book series, the consistency of the organization allows the reader quickly to locate the appropriate information necessary for informed decisions regarding the sources, effect, regulation, recognition, and management of toxic exposures.

Conversions for length and temperature in metric and imperial systems are provided to ease the use of this book by an international readership, whereas the metric system for mass and concentrations are retained to limit any confusion about doses in the United States.

The following provides organizational details on the material under the headings for each toxin:

History provides interesting historical facts involving the use and recognized effects of the toxicity of natural substances.

Identifying Characteristics and **Botanical Descriptions** helps the reader identify the characteristics and geographic distribution of the specific toxin (e.g., venomous animal, toxic plant).

Exposure discusses the sources, uses, and regulation of exposures to these toxins.

Principal Toxins identifies the main toxins in the natural substances and provides data on the chemistry, structure, and physical properties of the toxin that are important for the reader's understanding of the clinical response to the toxin. This part discusses the basic science and the composition of the toxins along with factors that affect the delivery of the toxin including discussions of the venom apparatus and seasonal variation in the locations of the toxins. Additionally, this part covers the biochemical and pathophysiological basis for the toxic responses.

Dose Response covers data on the lethality and clinical effects of the toxin both in animals and in humans as well as factors that affect the potency of the toxin. The emphasis is on dose-related effects, but important adverse and idiosyncratic reactions are also discussed.

Toxicokinetics discusses the disposition of the principal toxins in the body including the distribution, absorption, and elimination of the principal toxins. The emphasis is on human data, but animal data may be included when human data are sparse.

Clinical Response provides data on the clinical features of poisoning following exposure to the toxin including the onset, duration, and type of clinical effects.

Diagnostic Testing covers information important to the interpretation of the clinical significance of the laboratory data. This section includes current laboratory methods to determine the presence of the toxin, effects of storage, biomarkers of exposure in blood, urine, and postmortem material, and the laboratory abnormalities detected by imaging studies and ancillary tests.

Treatment provides details on current methods to treat the poisoning including information important for first responders, life-threatening problems associated with the poisoning, the use of antidotes, and measures of supportive care.

Medical Toxicology of Natural Substances: Foods, Fungi, Medicinal Herbs, Plants, and Venomous Animals focuses on scientifically validated facts about specific toxins based on clinical experience and the medical literature. References are documented to validate the statements and to provide sources for further inquiry. The interdisciplinary, evidence-based approach is designed to reach beyond clinical settings to increase the scientific understanding of those in associated fields (analytical laboratories, universities, regulatory agencies, coroner's offices) involved with decisions regarding toxic exposures. My hope is that increased scientific communication between the fields aligned with Medical Toxicology will inspire more inquiry into the pathophysiology, clinical effects, biomarkers, treatment, and prevention of toxic exposures.

Donald G. Barceloux, MD

June 12, 2008

ACKNOWLEDGMENTS

I deeply respect the expertise and appreciate the dedication of the following people for their help in improving the quality, depth, and accuracy of my book, *Medical Toxicology of Natural Substances: Foods, Fungi, Medicinal Herbs, Plants, and Venomous Animals*.

Bob Esposito, Senior Editor, John Wiley & Sons
His belief in the importance of my Book Series to the medical literature and his guidance made this book a reality.

Kate McKay, Text Editor
Her outstanding editorial comments, meticulous style, and judgment substantially improved the consistency and clarity of the text.

Ed Wickland
His support and astute advice on the medical publishing field was an essential part of the placement of this book series.

Donna Seger, MD, Managing Editor
Her management skills and resources for the arduous task of developing this book were an invaluable contribution.

Rob Palmer, PhD, Illustration Editor and Analytical Toxicology Reviewer
His breadth of analytical knowledge, clinical judgment, and artistic expertise were important contributions to the interdisciplinary approach of this book.

Cyrus Rangan, MD
His scholarly chapters on food contamination, food additives, and bacteria were vital contributions to the content of this book.

Rune Midtgaard
The sharing of his exquisite photographs from the professional gallery at Nature's Window (www.natureswindow.dk) was extremely generous.

Review Panel
The critical reviews and clinical insights of the Review Panel were an essential part of the scientific basis of this book.

CEP America
The support and shared clinical experiences of my medical colleagues have been indispensable to the completion of my book: Wes Curry, President; James Kim, Medical Director; Ken Moore, former Medical Director; Ken Nakamoto, Vice President of Medical Affairs; Phil Piccinni, Regional Medical Director; Emergency Department physicians, Mark Boettger, Ivan Schatz, Richard Dorosh, Greg Burke, Greg Murphy, Howard Friedman, Marc Eckstein, Matt Janssen, Geoffrey Pableo, John Lee, Tom Umemoto, Ali Jamehdor, Ludwig Cibelli, Brian Rhee, Vicki Shook, Hanne Rechtschaffen, Susan Salazar, Tom Edholm; and physician assistants, Anne Castle, Glenn deGuzman, Jaison Fraizer, Erin Merchant, Janet Nakamura, Frank Pastor, and Erik Smith.

Pomona Valley Hospital Medical Center
I admire the incredible hard work and dedication of the Emergency Department nurses and support staff in the care of a community challenged by difficult medical and social problems.

UCLA Emergency Department Colleagues
I appreciate all those who shared their expertise and clinical experience at UCLA Toxicology Rounds especially Marshall T. Morgan, Director of the Emergency Medicine Center and David A. Talan, Chairman, Department of Emergency Medicine, Olive View-UCLA Medical Center.

UCLA/Pomona Librarians
The writing of this book required the review of thousands of references and the technical assistance of the UCLA Biomedical and Pomona Valley Hospital Medical Center libraries, particularly Glenn Sato, Joseph Babi, and Debra Klein.

REVIEW PANEL

Joseph M. Betz, PhD
Director, Analytical Methods and Reference Materials Program, Office of Dietary Supplements, National Institutes of Health, Bethesda, MD

Jude T. McNally, RPh, DABAT
Managing Director, Arizona Poison & Drug Information Center, College of Pharmacy, University of Arizona, Tucson, AZ

Robert B. Palmer, PhD, DABAT
Toxicology Associates, Prof LLC, Assistant Clinical Professor, Department of Surgery, University of Colorado School of Medicine, Faculty Attending Toxicologist, Rocky Mountain Poison and Drug Center, Denver, CO

Sylvie Rapior, PhD
Professor of Botany and Mycology, University of Montpellier, Montpellier, France

Donna L. Seger, MD, FAACT, FACEP
Associate Professor of Medicine and Emergency Medicine, Vanderbilt University Medical Center, Medical Director, Tennessee Poison Center, Nashville, TN

Daniel L. Sudakin, MD, MPH, FACMT, FACOEM
Associate Professor, Department of Environmental and Molecular Toxicology, Oregon State University, Corvallis, OR

Richard S. Vetter
Research Associate, Department of Entomology, University of California, Riverside, Riverside, CA

Gary S Wasserman, DO, FAAP, FAACT, FACMT
Chief of the Section of Medical Toxicology, Children's Mercy Hospitals & Clinics, Professor of Pediatrics, University of Missouri-Kansas City School of Medicine, Assistant Medical Director of the University of Kansas Hospital Poison Control Center, Kansas City, MO

Aruna Weerasooriya, PhD, FLS
Taxonomist and Manager, Medicinal Plants Garden, The University of Mississippi, University, MS

Suzanne R. White, MD, FACMT
Medical Director, Children's Hospital of Michigan Regional Poison Control Center, Dayanandan Professor and Chair, Emergency Medicine, Wayne State University School of Medicine, Detroit, MI

Saralyn R. Williams, M.D., FACEP, FACMT
Associate Professor of Medicine and Emergency Medicine, Vanderbilt University, Nashville, TN

PART 1

FOODBORNE and MICROBIAL TOXINS

PART 1 FOODBORNE and MICROBIAL TOXINS

I Chemical Contamination and Additives

Chapter 1

FOOD CONTAMINATION

CYRUS RANGAN, MD, FAAP

Records of outbreaks of human illness caused by toxic contaminants in foods appeared at least several centuries BC, when mad honey poisoning was associated with an illness among troops under the command of the Greek historian and mercenary, Xenophon.[1] One of the first recorded foodborne outbreaks of ergotism occurred in Limoges, France during the Capetian Dynasty in 994 AD.[2] The ingestion of rye bread contaminated with ergot alkaloids during this epidemic caused the deaths of approximately 20,000–50,000 victims. Numerous episodes of ergotism have occurred throughout history. Although ergotism is a possible explanation for the bizarre behavior that occurred before and during the Salem Witch Trials of 1692, there is no definite evidence that ergotism was a contributing factor.[3] In the 1950s, mining operations in the Toyama region of Japan released cadmium into the Jinzu River. The use of this water for drinking and for irrigation of nearby rice fields resulted in a disease called *itai-itai* (translation: "ouch-ouch"), manifest by osteoporosis and renal dysfunction primarily in middle-aged women.[4] In the same decade, numerous neonates born near Minamata Bay, Japan, developed birth defects and neurological abnormalities after pregnant women were exposed to seafood contaminated by methyl mercury released into the bay from a local factory. A methyl mercury-based fungicide caused an outbreak of mercury poisoning in Iraq in 1971 after grain seeds treated with the fungicide were inadvertently used for food manufacturing instead of planting.[5] Prominent outbreaks of illnesses associated with chemical contamination of cooking oils include tri-*ortho*-cresyl phosphate-induced neuropathy (Morocco, 1959),[6] yusho ("rice oil disease," Japan, 1968), yu-cheng ("oil disease," Taiwan, 1979), toxic oil syndrome (Spain, 1981), and epidemic dropsy (India, 1998). Some contaminants are unavoidable in food manufacturing. In the United States, the Food and Drug Administration (FDA) imposes "Current Good Manufacturing Practices" (CGMP) on food manufacturers. These mandatory codes enable the FDA to cite food products as *unfit* if an unavoidable contaminant poses a risk of harm by violating a standard or action level for that unavoidable contaminant (e.g., aflatoxin). Food products are considered *adulterated* when concentrations of avoidable contaminants (e.g., pesticides) exceed established standards, sometimes prompting food recalls after sale and distribution.

Metal contaminants such as lead, mercury, arsenic, and cadmium come from factory emissions, mining operations, and metal-containing industrial products used in food production. Methyl mercury found in commercially sold seafood is deemed an unavoidable contaminant because contamination preexists in the raw material; therefore, the contamination does not result from food processing or distribution. Fish and shellfish acquire methyl mercury primarily from microorganisms that methylate environmental inorganic mercury compounds released primarily from industrial sources.

Medical Toxicology of Natural Substances, by Donald G. Barceloux, MD

YUSHO and YU-CHENG

HISTORY

The first known case of yusho (rice oil disease) involved a 3-year-old girl in northern Kyushu, Japan, who had an acute onset of an acneiform rash (chloracne) in June, 1968. Her family members, followed by other familial clusters, presented to a single clinic with complaints of acneiform rash, hyperpigmentation, and eye discharge over the next 2 months. By January 1969, 325 cases were reported. After a small minority of patients initially identified rice oil as the causative agent of yusho, Kyushu University convened the *Study Group for Yusho* to investigate yusho; about 2,000 afflicted patients were subsequently identified. The clinical features of yusho included fatigue, headache, cough, abdominal pain, peripheral numbness, hepatomegaly, irregular menstrual cycles, nail deformities, and hypersecretion of sebaceous glands. A field survey of canned rice oil associated the disease with the use of "K Rice Oil" produced or shipped by the K Company on February 5–6, 1968.[7] The yu-cheng epidemic involved over 2,000 individuals in Taiwan in 1979, when an accidental leakage of thermal exchange fluid resulted in the contamination of rice-bran oil with polychlorinated biphenyls (PCBs), dibenzofurans (PCDFs), and quaterphenyls (PCQs).[8] The clinical features of yu-cheng and yusho were similar.

EXPOSURE

Source

Polychlorinated biphenyls (PCBs) and polychlorinated dibenzofurans (PCDFs) are thermal heat exchanger compounds used in food processing machinery. Leakage of these compounds into rice oils during manufacturing led to the yusho and yu-cheng outbreaks.

Yusho

Epidemiological studies revealed that 95.7% ($p < 0.01$) of surveyed patients recalled consumption of rice oil from K Company in Western Japan.[7] A case-control study revealed rice oil as the only associated etiologic factor, and a cohort study demonstrated a 64% risk of yusho in K rice oil consumers compared with no risk for nonexposed individuals.[7] Food engineers confirmed the leakage of dielectric thermal exchange fluid (Kanechlor 400) containing PCBs into the rice oil. This contaminant contained PCB compounds, primarily tetra-chlorinated biphenyls. In 1969, the Study Group initially concluded that PCBs caused yusho.[9] However, a lack of similar symptoms (besides chloracne) in PCB workers who had significantly higher tissue burdens (mean blood PCB level: 45 ppb) contradicted this conclusion. Furthermore, the dermatological lesions could not be reproduced in animals following the oral administration of PCB compounds or by Kanechlor 400, and the severity of the clinical features of yusho did not correlate to serum concentrations of PCB compounds. Therefore, other compounds (e.g., polychlorinated dibenzofurans) in the adulterated rice oil probably contributed to the development of yusho.[10,11]

Yu-cheng

As with the yusho incident, the suspected causative agents of yu-cheng were PCDFs rather than PCBs.[12] Contamination of the cooking oil occurred when PCBs used for the indirect heating of rice-bran oil leaked into the cooking oil. Repeated heating of the partially degraded PCBs produced PCDFs, as well as polychlorinated terphenyl and polychlorinated quaterphenyl compounds.[13]

Food Processing

High temperatures (>200 °C) in dielectric thermal exchange fluid during the deodorization step of oil refining contributed to the development of yusho and yu-cheng by degrading PCBs in the contaminated rice oil to PCDFs, PCDDs (polychlorinated dibenzo dioxins), and PCQs (polychlorinated quaterphenyls).[14]

DOSE RESPONSE

Exposure to toxic contaminants in the rice oil from the yusho and yu-cheng epidemics was assessed by recording the lot numbers of purchased oil containers and comparison of the volume of oil purchased to the volume of oil remaining in the containers retrieved from affected households. Consumption of the contaminated rice oil by household members was estimated by proportional distribution to each family member. Positive relationships were observed between estimated individual oil consumption and incidences of yusho and yu-cheng.[15,16] The mean concentrations of PCBs, polychlorinated quaterphenyls (PCQs), and PCDFs in five samples of contaminated cooking oil from the yu-cheng outbreak were 62 ppm, 20 ppm, and 0.14 ppm, respectively.[17] The congeners of these compounds were similar in the cooking oils from these two outbreaks, but ye-cheng cooking oil samples contained about 10% of the concentrations of these compounds found in cooking oil from the yusho incident along with three to four times lower PCQs/PCBs and PCDFs/PCBs ratios.

A cross-sectional study of 79 patients with documented yusho demonstrated a dose-response relationship between estimated consumption of contaminated rice oil and the symptoms of extremity numbness, coughing, expectoration, and the sensation of "elevated teeth."[18] These symptoms were evaluated by self-administered questionnaires. Symptoms failing to demonstrate a dose-response relationship to the estimated ingestion of contaminated rice oil included fatigue, eye discharge, fever, headache, dizziness, abdominal pain, swollen joints, menstrual irregularities, and alopecia. The estimated mean total intake of PCBs and PCDFs by yusho patients was about 633 mg and 3.4 mg, respectively, compared with 973 mg and 3.84 mg, respectively, for yu-cheng patients.[19]

CLINICAL RESPONSE

Animal studies confirm a strong association between high concentrations of PCBs (i.e., at least 60% chlorination) in diets and the incidence of hepatic carcinomas. However, no human studies have confirmed an association between PCB exposure and cancer. Therefore, PCBs are listed as probable human carcinogens by the International Agency for Research on Cancer (IARC) and the US Environmental Protection Agency (EPA).[20]

Yusho

Clinical features of this illness included fatigue, headache, cough, abdominal pain, peripheral numbness, hepatomegaly, irregular menstrual cycles, nail deformities, and sebaceous gland hypersecretion. The most common symptoms were eye discharge, hyperpigmentation (skin, mucous membranes, nails), acneform lesions, and weakness.[19] Although the severity of these symptoms has decreased since the epidemic, follow-up studies of yusho survivors indicated that these symptoms persisted at least through 1993.[21] The chloracne resolved relatively rapidly in children, but hyperpigmentation and hypertrichosis remained in some patients. Thirteen children born to yusho-affected mothers exhibited gray-brown skin discoloration at birth ("black babies," "cola-colored babies"), but the discoloration spontaneously disappeared after several weeks. These babies exhibited no other symptoms consistent with yusho.[22]

Yu-cheng

The clinical features of yu-cheng and yusho are similar. Clinical findings include chloracne, hyperpigmentation, edema, weakness, vomiting, diarrhea, and hepatomegaly. The acneform eruptions were open comedones, papules, and pustules with dark heads distributed on the axilla, extremities, and external genitalia.[23] Abnormalities in children of yu-cheng patients included low birth weights,[24] prematurity, neurobehavioral changes such as delayed autonomic maturity,[25] normal menarche with shortened menstrual cycles,[26] abnormal reflexes, dysfunctions in visual recognition memory,[27] and decreased intelligence scores.[28] A 24-year follow-up study of yu-cheng victims demonstrated increased mortality from chronic liver disease and cirrhosis in men, but not in women.[29] There was an increased incidence of systemic lupus erythematosus in exposed women in the later years. The mortality rates for cancers were similar between the exposed group and the background population.

DIAGNOSTIC TESTING

Analytical Methods

Gel permeation chromatography and high resolution gas chromatography/high resolution mass spectrometry detect and differentiate PCBs and PCDFs in oil samples and in human samples.[30] Methods that utilize high performance liquid chromatography/mass spectroscopy (HPLC/MS) or high performance liquid chromatography/tandem mass spectrometry (HPLC/MS/MS) reliably detect the di-oleyl-phenyl amino propanediol ester (OOPAP) and other acylated phenyl amino propanediol derivatives (PAPs) in cooking oils.[31] Methods to detect OOPAP in humans are not available.

Biomarkers

Rice oil from the yu-cheng epidemic contained approximately 3,000 ppm total PCBs. The mean blood concentration of PCBs in patients with chloracne and hyperpigmentation was in the range of approximately 5 ppb. There was a linear correlation between the severity of skin lesions and the total PCB concentrations in blood samples. Studies of human and animal subjects indicate that PCB concentrations in the range of 10–25 ppm cause similar skin abnormalities. Analysis of PCBs, PCDFs, and PCQs in contaminated rice-oil samples collected from factory cafeterias, school cafeterias, and the families of patients with yu-cheng ranged from 53–99 ppm, 0.18–0.40 ppm, and 25–53 ppm, respectively.[32]

Although the specific compound associated with yusho or yu-cheng remains unknown, PCDDs are more appropriate biological biomarkers for the severity of yusho and yu-cheng than other polychlorinated hydrocarbons because the presence of PCDDs in blood samples indicates exposure to these compounds or parent compounds as PCDDs do not occur in nature.[33] These compounds persist in the blood for years.[34] Clearance of PCDFs and PCBs in humans is nonlinear with faster elimination rates at higher concentrations. In blood samples from 3 yu-cheng patients, the whole blood elimination half-life of two persistent toxic congeners, 2,3,4,7,8-pentachlorodibenzofuran (PnCDF) and 1,2,3,4,7,8-hexachlorodibenzofuran (HxCDF) was approximately $2\frac{1}{2}$ years.[35] The calculated blood elimination half-lives of the same PCDF congeners in yusho patients were more variable with median values near 10 years. In a follow-up study of 359 patients with yusho, the serum concentrations of PnCDF and PCBs remained significantly elevated over 35 years after the incident.[36] The mean blood concentration of PnCDF in exposed patients was 177.50 pg/g lipids compared with 15.2 ± 8.9 pg/g lipids in the blood of healthy controls. The blood PnCDF concentration in these patients correlated to some clinical symptoms of yusho including acneform eruption, comedones, oral pigmentation, constipation, numbness in the extremities, and body weight loss. Follow-up studies indicate that the blood elimination half-lives are shorter (i.e., about 1–5 years) for PCB congeners than PCDF congeners. Follow-up studies of blood samples from yusho patients 34 years after the incident indicate that PCDFs contribute about 65% of the remaining total toxic equivalents of dioxins (non-*ortho* PCBs, mono-*ortho*-PCBs, PCDDs, PCDFs).[37] In particular, the mean concentration of some PCDF congeners were substantially higher than controls including 1,2,3,6,7,8-HxCDF (3.9 times), 1,2,3,4,7,8-HxCDF (12 times), and PnCDF (11.3 times). The blood samples from 165 yu-cheng patients collected 9–18 months after the onset of poisoning contained 10–720 ppb PCBs with a mean value of 38 ppb.[8]

Abnormalities

Yusho and yu-cheng patients occasionally had mild elevation of serum hepatic aminotransferase concentrations, but results of most laboratory studies were within normal ranges.[38] Hepatorenal function is usually normal.

TREATMENT

The treatment for yusho and yu-cheng is supportive. Dermatologic changes may persist for several months, and topical or systemic medications typically do not alter the time-course of these lesions. Yusho and yu-cheng patients require long-term follow-up for the development of liver dysfunction and malignancy.[38]

TOXIC OIL SYNDROME

HISTORY

In the early 1980s, Spanish laws banned the importation of rapeseed oil for human consumption to protect the Spanish olive oil market.[39] These laws required the denaturation of imported rapeseed oil with aniline, methylene blue, or castor oil as a means to make this imported oil unfit for human consumption. During late 1980 and early 1981, a substantial amount of aniline-denatured rapeseed oil imported from France was fraudulently diverted for human consumption through Catalonia. These oils were diluted with other edible oils, refined, and resold through a network of distributors and itinerant salesmen. Although these oils contained aniline, the toxic oil syndrome was apparently not associated with the aniline.

During the first quarter of 1981, an oil distributor based in the center of Madrid named RAELCA entered the illicit oil sales market with mixed aniline-denatured rapeseed oil. In March 1981, RAELCA purchased five lots of aniline-denatured rapeseed oil from two French food oil companies, and they shipped three lots to the ITH oil refinery in Seville.[40] The other two lots went to the Danesa Bau refinery in Madrid. Although the denatured oils were intended for industrial use, roadside vendors sold the oils as "olive oil" to residents over a 2-month period.[41] Most of the cases of toxic oil syndrome appeared in the period from May to August 1981. Figure 1.1 displays the number of cases of toxic oil syndrome diagnosed in 1981. Epidemiological studies concluded that the ITH oil refinery was the point source for the epidemic.[42]

EXPOSURE

Source

The exact causal agent in the contaminated oil involved with toxic oil syndrome remains unknown, in part, because of the many potential toxins in denatured aniline compounds associated with this illness. In a study of toxic oil syndrome-associated oils, fractionation of

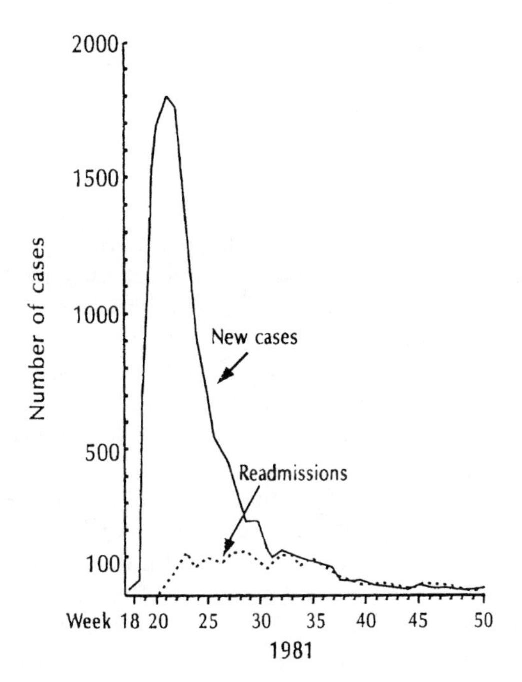

FIGURE 1.1. Onset of cases of toxic oil syndrome in Spain during 1981. From Philen RM, et al. Toxic oil syndrome and eosinophilia-myalgia syndrome: May 8–10, 1991, World Health Organization Meeting Report. Semin Arthritis Rheum 1993;23:106. Reprinted with permission from Elsevier.

the oils by high performance liquid with ultraviolet detection (HPLC/UV) and analysis by high performance liquid chromatography/atmospheric pressure chemical ionization/ tandem mass spectrometry (HPLC/APCI/MS/MS) demonstrated 115 aniline derivatives from nine aniline-related families of chemicals.[43] Toxic oil syndrome is a unique combination of vasculitis, thrombosis, and immunologic changes (e.g., T-lymphocyte activation, cytokine release) that differs from the toxicity previously associated with refinery products, additives, or contaminants.[44] The full spectrum of toxic oil syndrome has not been reproduced in experimental animals.[39] The two suspected groups of toxic compounds are fatty acid anilides and amino-propanediol derivatives (fatty acid mono- and di-esters).[45] Analysis of contaminated oil samples indicates that pentachlorophenol and pentachloroanisole were not etiological agents of toxic oil syndrome.[46]

Food Processing

In contrast to other illicit rapeseed oil distributors, RAELCA mixed the denatured rapeseed oil with other oils *after* the refining process. All other illicit refiners of denatured rapeseed oil in Catalonia mixed the denatured and edible oils *before* refining the adulterated oils. The processing of the fraudulently diverted rapeseed oil probably contributed to the formation of toxic compounds in the oil. High temperatures during deodorization catalyzed the reaction of 2% aniline with triglycerides in the rapeseed oil associated with toxic oil syndrome.[47] In experimental studies, the yield of 3-(*N*-phenylamino)-1,2-propanediol (PAP) esters in toxic oil syndrome is highest at 250–300 °C (~480–570 °F), similar to those temperatures achieved during the deodorizing step of oil refining.[48] These studies indicate that the heating of denatured oil samples stored for 3 weeks yields higher concentrations of potentially toxic PAP esters compared with samples stored for 1 week. Distillations times did not significantly affect the formation of fatty acid anilide compounds. The development of toxic oil syndrome following the ingestion of denatured rapeseed oil stored for about 1 year suggests that the toxic compounds in these contaminated cooking oils are stable for at least 1 year.[49]

DOSE RESPONSE

Dose response in toxic oil syndrome was evaluated by a case control study known as the "Toxi-Epi Study," which followed sales and distribution chains of the fraudulently distributed rapeseed oils using the chemical markers, oleyl-anilide and di-oleyl-3-phenylamino-1,2-propanediol (OOPAP). The content of these compounds varied several fold in different samples of contaminated cooking oil. Cooking oil containers with a characteristic shape purchased by residents were traced back to the ITH oil refinery in Seville in southern Spain. Analytical studies showed a linear statistical correlation between the concentration of the chemical markers detected in the refinery's oil and the log of the odds ratio for dose-response.[50] Rapeseed oil concentrations of oleyl-anilide and OOPAP traced to this plant were 1900 ppm and 150 ppm, respectively.

CLINICAL RESPONSE

The first cases of toxic oil syndrome were reported on May 1, 1981 in central and northwestern Spain. Initial epidemiological studies demonstrated clustering in incidence and mortality for toxic oil syndrome in households distributed along transportation routes throughout the affected regions. In July 1981, Spanish health officials announced that a fraudulently distributed, industrial oil sold as edible cooking oil was the etiology of this illness, initially termed pneumonic paralytic eosinophilic syndrome.[51] The Spanish government subsequently initiated a consumer exchange program for olive oil, and they stockpiled the contaminated oil for further study.[52] In 1983, the World Health Organization named the illness toxic oil syndrome.[53] The official census originally consisted of records of patients with clinically suspected toxic oil syndrome that did not necessarily fulfill the case definition of the 1981 Spanish

Clinical Commission. A 1987 review of these records by the World Health Organization Regional Office for Europe Scientific Committee for the Investigation of the Toxic Oil Syndrome indicated that about 20,000 people were affected with over 10,000 hospitalizations.[54] Although during the first few years about 300 patients died of toxic oil syndrome, the overall mortality of the cohort with toxic oil syndrome was not elevated after the first year of the epidemic when compared with the general Spanish population.[55]

Toxic oil syndrome is a progressive multisystemic disease with three distinct clinical phases (acute, intermediate, chronic). The average latency period was about 4 to 7 days after the ingestion of the contaminated oil with a maximum of about 10 days.[56] The basic underlying pathological lesion is a nonnecrotizing vasculitis with associated thrombotic events. Common initial symptoms of toxic oil syndrome included fever, cough, dyspnea, and chest pain. Other acute symptoms included urticarial rash, pruritus, abdominal cramping, and headache. The differential diagnosis for toxic oil syndrome includes other autoimmune disorders such as eosinophilia-myalgia syndrome, eosinophilic fasciitis, systemic sclerosis, scleroderma, and systemic lupus erythematosus.[57] Porphyria cutanea tarda did not occur in these patients.

Acute (Pulmonary)

The characteristic manifestation of the acute phase is the development of noncardiogenic pulmonary edema with dyspnea, alveolar-interstitial infiltrates with or without pleural effusions, peripheral eosinophilia, fever, and rash. Most patients recovered from this early pneumonic phase of the illness. Deaths during this phase occurred from respiratory insufficiency initially from degeneration of type I and type II pneumocytes and later from thromboembolic complications.[58] During the acute phase (i.e., first 2 months), the primary lesion occurs in the endothelium of multiple organs with the exception of the central nervous system.[59] Severe myalgias and muscle cramps occur at the end of the acute phase.[39]

Intermediate (Thrombotic)

About 60% of the patients with the acute phase progress to the intermediate phase that involved the development of sensory peripheral neuropathy along with intense myalgias, dermal induration, and weight loss. More serious complications during this phase began about 2 months after the onset of illness, including pulmonary hypertension and thromboembolism of large vessels. In severe cases, histological examination demonstrated proliferation of the intimal lining of the vessels along with fibrosis and thrombosis.

Chronic (Neuromuscular)

Approximately 2 months after the onset of the intermediate phase, the chronic phase of toxic oil syndrome began, characterized by sclerodermiform changes, motor neuropathy, musculoskeletal contractures, muscle wasting, myalgias, muscle cramps, weight loss, limited joint mobility, peripheral eosinophilia, hepatomegaly, pulmonary hypertension, and Sjögren syndrome.[60] Some patients developed this chronic phase without experiencing the acute pneumonic phase.[58] The most common persistent symptoms were cough and dyspnea with about 20% of affected patients developing reduction in the carbon monoxide diffusing capacity.[61] Mortality in the chronic phase was primarily due to infectious complications of respiratory insufficiency secondary to neuromuscular weakness, thromboembolism, or pulmonary hypertension with cor pulmonale.[62,63] A minority of toxic oil syndrome patients developed severe pulmonary arterial hypertension during exercise over 20 years after exposure.[64] Long-term follow-up studies of survivors suggest reduction in the quality of life of these patients with elevated rates of depression (odds ratio [OR] = 9.66), functional disabilities (OR = 4.74), and psychosocial disabilities (OR = 2.82).[65]

Mortality was high during the first year, with a significant decline in mortality rates in subsequent years.[55] Most of the decline in mortality was observed in elderly populations; however, the mortality rate in women <40 years of age increased as a result of complications from pulmonary hypertension.[66] Concurrent clustering of incidence and mortality in toxic oil syndrome suggested that a genetic predisposition determines the severity of toxic oil syndrome. Linkage mapping of the human genome revealed increased mortality in patients with a chromosome 6-associated risk factor for the HLA-DR2 phenotype.[67] Studies in enzyme mechanics implicate a role for impaired hepatic acetylation in mediating individual susceptibilities to toxic oil syndrome.[68]

DIAGNOSTIC TESTING

Analytical Methods

Analytical methods to identify and quantify OOPAP and other PAPs in contaminated cooking oil samples include high performance liquid chromatography/atmospheric pressure ionization/tandem mass spectrometry (HPLC/API/MS/MS) and HPLC/MS.[31,69]

Biomarkers

The analysis of contaminated oil was complicated by poor identification markers on cooking oil bottles during the recall process. Two case control studies suggested a dose-related association between the presence of three fatty acid anilide compounds (oleyl, linoleyl, palmityl) and the risk of developing toxic oil syndrome.[50,70] Of these fatty acid anilide compounds, oleyl anilide occurred in the highest concentration. Subsequent analyses suggested that 3-(*N*-phenylamino)-1,2-propanediol (DEPAP), a by-product of the same reaction of aniline with triglycerides, is an equally sensitive and a more specific biomarker of toxic oil syndrome than fatty anilide compounds.[71] Animal studies suggest that the liver converts fatty acid mono- and diesters of 3-(*N*-phenylamino)propane-1,2-diol to 3-(4′-hydroxyphenylamino)-propane-1,2-diol, which generates the electrophilic metabolite quinoneimine intermediate-2 (QI-2).[72] However, the specific chemical causing toxic oil syndrome has not been identified.

Abnormalities

Laboratory changes associated with toxic oil syndrome include peripheral eosinophilia, hypertriglyceridemia, and coagulation disorders in patients with liver involvement. Analytical studies of toxic oil syndrome patients demonstrate eosinophilia and a high concentration of mRNA for T-helper-2 cytokines, IL-4 and IL-5, in the lungs.[73] Antemortem sera from fatal cases of toxic oil syndrome also contained elevated serum IL-2R and total IgE concentrations along with a high frequency of HLA-DR2 on chromosome 6.[74] The sinus and atrioventricular nodes may exhibit dense fibrosis, hemorrhages, or cystic degeneration similar to findings in scleroderma and systemic lupus erythematosus. Coronary arteries may exhibit focal fibromuscular dysplasia and cystic myointimal degeneration with embolization. Histopathological analyses show lymphocytic inflammatory lesions of coronary arteries and the cardiac conduction system, similar to the findings in eosinophilia-myalgia syndrome. Cardiac lesions associated with eosinophilia-myalgia syndrome, however, are distinguished by cytotoxic T-cells directed against cardiac neural structures and sinus nodal myocytes, whereas toxic oil syndrome cardiac lesions are characterized by a prominence of B cells and T-helper cells.[75]

Chest x-ray demonstrates an interstitial-alveolar pattern with progression to acute respiratory distress syndrome (adult respiratory distress syndrome or ARDS) in the acute phase of toxic oil syndrome.[76] Eosinophilia is a universal finding in toxic oil syndrome patients.[77] Hypoxemia, respiratory alkalosis, and an increased alveolar-arterial oxygen gradient (A-a gradient) as determined by arterial blood gases are common in patients with severe toxic oil syndrome-induced noncardiogenic pulmonary edema during the acute phase.[78]

TREATMENT

During the acute phase, respiratory compromise from ARDS is the most serious complications of toxic oil syndrome. Patients with toxic oil syndrome are also at increased risk of cardiovascular disease that may manifest several years after exposure. Echocardiography, cholesterol screening, and weight management are effective tools for the screening and detection of cardiovascular sequelae in patients with toxic oil syndrome.[79,80] Long-term neuromuscular and articular complaints are prominent, and abnormalities are treated symptomatically.[81] Patients should be monitored for the prominent risk factors most closely associated with early mortality: female <40 years old, liver disease, pulmonary hypertension, frequent pulmonary infections, motor neuropathy, and eosinophilia.[82]

EPIDEMIC DROPSY

HISTORY

Although the first of several epidemics of epidemic dropsy was documented in 1877 in Calcutta, the most prominent epidemic dropsy epidemic occurred in Delhi, India during August, 1998.[83] The Indian Ministry of Health received over 2,552 reports of poisoning with 65 deaths from August to October 1998. The incidence of epidemic dropsy was higher in lower socioeconomic groups, probably as a result of purchasing less expensive, loosely packaged mustard oil from roadside vendors. Epidemic dropsy (argemone toxicity) is characterized by the pathological accumulation of lymph throughout the body along with gastrointestinal distress, erythrocyanosis, sarcoid-like skin lesions, myalgias, paresthesias, and painful edema of the lower extremities.

EXPOSURE

Source

Argemone oil (Katkar oil) mixed with mustard oil and ghee (clarified butter) caused epidemic dropsy outbreaks in India, South Africa, and Nepal. *Argemone mexicana* L. (Mexican prickly poppy, Satyanashi [translation: *devastating*], Papaveraceae) is an invasive weed with a yellow flower similar to the mustard flower (*Brassica nigra* (L.) W.D.J. Koch). Indian mustard oil merchants commonly remark to public health officials that argemone seeds are inadvertently harvested with mustard seeds because they thrive in similar climates, with potentially overlapping harvests (Mustard—February/March; Argemone—April/May).[84,85] However, the canopy-like structure of the mustard plant does not permit concurrent growth of other plants in the same fields; therefore, mustard seeds are generally harvested in February, leading to the widely accepted theory that adulterations of mustard oil with argemone oil are intentional.

Argemone oil contains two toxic alkaloids: dihydrosanguinarine (CAS RN: 3606-45-9, $C_{20}H_{15}NO_4$) and much smaller amounts of sanguinarine (CAS RN: 2447-54-3, $C_{20}H_{14}NO_4$). Although the concentration of dihydrosanguinarine in argemone oil is greater than sanguinaria, the toxicity of the latter compound is greater.[86] The names originate from the bloodroot (*Sanguinaria canadensis* L.), which is the original source of this alkaloid.[87] These alkaloids are benzophenanthridine derivatives of glycosides that cause increased permeability of blood vessels, particularly in the heart, liver, eyes, gastrointestinal tract, and kidneys.[88] Figure 1.2 and Figure 1.3 demonstrate the chemical structure of dihydrosanguinarine and sanguinarine, respectively.

The exact mechanism of toxicity of epidemic dropsy is unknown. Sanguinarine and dihydrosanguinarine accumulate in the gastrointestinal tract, serum, and tissues after binding tightly to plasma proteins.[89] *Potentially* these toxic alkaloids cause damage by binding to Na^+-K^+-ATPase,[90] inactivating hepatic cytochrome P-450 enzymes,[91] depleting hepatic glutathione,[92] and disrupting carbohydrate metabolism. The latter effect inhibits active transport of glucose across intestinal villi, resulting in increased glycogenolysis and the formation of excess glucose-1-phosphate, pyruvate, and lactate.[93] These processes lead to the formation of reactive oxygen species, causing oxidation of plasma proteins and lipids.[94]

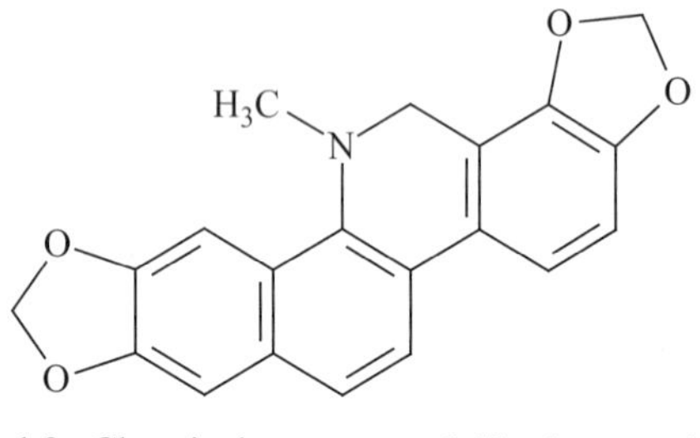

FIGURE 1.2. Chemical structure of dihydrosanguinarine.

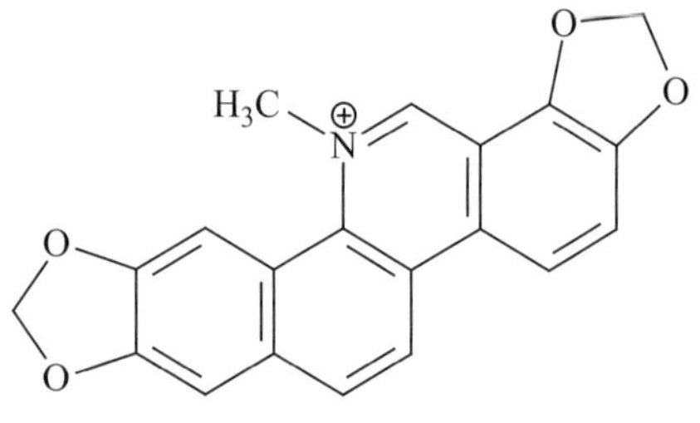

FIGURE 1.3. Chemical structure of sanguinarine.

Food Processing

Mustard seeds are differentiated from the seeds of *Argemone mexicana* L. by physical appearance: mustards seeds are smooth and round; argemone seeds are creased with spiked edges. Batches of mustard seeds are visually inspected for argemone seeds during processing with the argemone seed discarded upon detection. The practice of sorting is painstaking and cost-prohibitive, as 1% adulteration of mustard oil by argemone oil is enough to cause symptoms.[95] The early harvesting of mustard avoids the contamination of mustard oil by argemone seeds. The toxic alkaloids in argemone oil are heat-stable to 240 °C (~460 °F). Consequently, consumers in endemic areas are advised to heat mustard oils to 240 °C for at least 15 minutes to deactivate the toxic alkaloids present in argemone seeds.[96]

DOSE RESPONSE

As little as 1% adulteration of cooking oil with argemone oil is probably sufficient to cause clinical toxicity.[86] Because sanguinarine and dihydrosanguinarine accumulate in human tissue, toxicity can result from subacute, low-dose exposure.

CLINICAL RESPONSE

The clinical features of epidemic dropsy begin approximately 1–3 weeks after consumption of contaminated oil with a variety of symptoms including nausea, vomiting, diarrhea, bloating, anorexia, hyperplastic lesions in the mouth and other mucous membranes, erythematous rash with bluish mottling (erythrocyanosis), nodular sarcoid-like skin lesions or telangiectasias,[97] bilateral glaucomatous findings,[98] anemia, muscle tenderness, numbness, tingling, and painful bilateral pitting edema, particularly of the lower extremities.[99,100] Capillary leakage and congestion of gut mucosa epithelia cause watery gastrointestinal symptoms with occasional hematemesis, melena, or hematochezia. The ophthalmologic features of epidemic dropsy occur relatively late in the course of the disease. Ocular signs include increased ocular pressure and glaucoma from protein accumulation in the aqueous humor. Visual complications include subconjunctival hemorrhage, superficial retinal hemorrhages, retinal venous dilatation and tortuosity, subhyaloid hemorrhages, macular and papillary edema, central retinal vein occlusion, visual field defects, and rarely permanent visual defects from optic nerve damage.[101–103] In a case series of 230 documented cases of epidemic dropsy, the incidence of intraocular pressures exceeding 22 mmHg was about 11% with most elevations of intraocular pressure returning to normal values within 12 weeks.[104] Inflammation of the anterior segment is usually absent.

Renal insufficiency with hypoalbuminemia and proteinuria may occur in serious cases of epidemic dropsy, resulting in acute renal failure.[100,105] Bleeding occurs from sarcoidal lesions and telangiectasias on mucous membranes and gastrointestinal epithelial mucosa.[106,107] Noncardiogenic pulmonary edema and, rarely, cardiac failure develop in severe cases with hypoxemia, respiratory alkalosis, high-output cardiac failure, tachycardia, dyspnea, and increased A-a gradient. Death results primarily from cardiac arrest secondary to congestive heart failure, noncardiogenic pulmonary edema, or pericardial fluid accumulation.[108] Postmortem histological findings include venous congestion, extramedullary hematopoiesis, focal hemorrhages, proliferation of capillary endothelial cells, dilation of hepatic sinusoids, and mononuclear infiltration of the central veins in the liver.[86]

DIAGNOSTIC TESTING

Analytical Methods

During an epidemic, samples of adulterated mustard oil are screened for sanguinarine by the addition of ferric chloride to the sample.[109] The appearance of a red-brown precipitate suggests the presence of ≥0.25% argemone oil in the sample. The presence of sanguinarine in these samples is confirmed by HPLC with limits of detection in the range of 0.001% argemone oil in the sample.[110]

Biomarkers

In a study of 45 patients with epidemic dropsy during an outbreak in New Delhi, sanguinarine was detected in eight urine samples collected within 2–3 weeks of onset of dropsy with concentrations ranging between 0.4 and 3.6 μg/100 mL. Three of 18 serum samples in the same group were positive for sanguinarine with concentrations of 1.2, 1.6, and 3.6 μg/100 mL.[111]

Abnormalities

Epidemic dropsy patients typically have hypoalbuminemia, hypocalcemia, proteinuria, and azotemia, similar to nephrotic syndrome along with reduced concentrations of tocopherol and retinol secondary to depletion of antioxidants by reactive oxygen species.[112] Chest x-ray may reveal signs of pulmonary edema and right ventricular cardiomegaly. Case series suggest that the dyspnea associated with epidemic dropsy usually results from a restrictive ventilatory defect with reduced carbon monoxide diffusing capacity rather than a cardiomyopathy.[113] The electrocardiogram may demonstrate ST-T wave changes and premature ventricular contractions.[96] Normochromic, normocytic anemia is a common finding despite expected hemoconcentration from intravascular fluid loss.[107]

TREATMENT

Treatment for epidemic dropsy is supportive. Inpatient treatment includes compression stockings, protein-rich diet, correction of symptomatic hypocalcemia, and albumin infusions as indicated by the condition of the patient. Deficiencies in tocopherol and retinol may be corrected with supplementation, but there are inadequate clinical data to determine the clinical efficacy of the administration of antioxidants to patients with epidemic dropsy.[114] Serial ophthalmologic examinations are necessary for several weeks to evaluate for the onset of glaucoma, which occurs in up to 11% of cases.[115] Topical beta-receptor antagonists (timolol, levobunolol), $alpha_2$ receptor antagonists (brimonidine, iopidine) and parasympathetics (pilocarpine) may reduce intraocular pressures. Surgical intervention with trabeculectomy or laser trabeculoplasty may be necessary in severe cases. Patients should be followed for several weeks for signs of cardiac decompensation, cardiogenic and noncardiogenic pulmonary edema, and pericardial effusion. Digitalis and diuretics may improve the outcome of cardiovascular manifestations depending on the cause of reduced cardiac output (e.g., pericardial effusion or right-sided heart failure). Full recovery generally occurs within 3 weeks to 3 months, with a 5% mortality rate.[116]

EOSINOPHILIA-MYALGIA SYNDROME

HISTORY

In 1980, Sternberg et al. reported a scleroderma-like illness with eosinophilia in a patient treated with L-5-

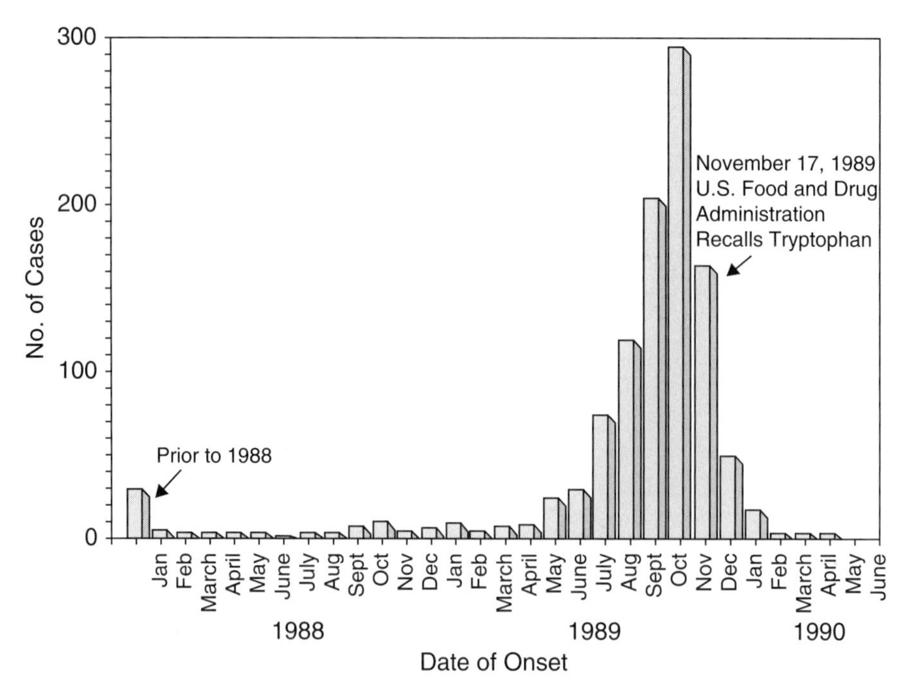

FIGURE 1.4. Onset of cases of eosinophilia-myalgia syndrome, 1988–1990. From Philen RM, et al. Toxic oil syndrome and eosinophilia-myalgia syndrome: May 8–10, 1991, World Health Organization Meeting Report. Semin Arthritis Rheum 1993;23:108. Reprinted with permission from Elsevier.

hydroxytryptophan and carbidopa.[117] In the fall of 1989, the worldwide outbreak (United States, Canada, Germany, United Kingdom) of a disease with clinical features (eosinophilia, myalgias) similar to toxic oil syndrome appeared, primarily in the United States.[118] This disease was subsequently called eosinophilia-myalgia syndrome. The first reported cases of myalgia and eosinophilia associated with L-tryptophan use occurred in three patients from New Mexico.[119] The disease subsequently affected at least 1,500 people with about 30–40 deaths. Figure 1.4 displays the onset of cases of eosinophilia-myalgia syndrome during the 1989–1990 epidemic.

Typically, patients developed intense myalgias, peripheral eosinophilia, and dermatological lesions. The onset of eosinophilia-myalgia syndrome was associated *mostly* with the ingestion of contaminated L-tryptophan manufactured by a single Japanese manufacturer (Showa Denko K.K., Tokyo, Japan).[120] Case reviews of patients who ingested L-tryptophan produced by other manufacturers (e.g., Optimax®, Merck & Co., Inc., Whitehouse Station, NJ) did not support the diagnosis of eosinophilia-myalgia syndrome in these patients.[121] Since 1991, the current surveillance system has detected a few cases of eosinophilia-myalgia syndrome. The occurrence of nontryptophan-related cases of eosinophilic-myalgia syndrome now is similar to the rate before the 1989–1990 epidemic.[122]

EXPOSURE

Source

To date, analysis of case-associated L-tryptophan from Showa Denko revealed the following six contaminants: peak AAA, (undefined); peak E, (1,1′-ethylidenebis[L-tryptophan]); peak 200, [2-(3-indolylmethyl)-L-tryptophan], peak C, (3a-hydroxy-1,2,3,3a,8,8a-hexahydropyrrolo-[2-3b]-indole-2-carboxylic acid); peak FF, [2-(2-hydroxy-indoline)-L-tryptophan]; and peak UV-5, [3-(*N*-phenylamino)-L-alanine].[123,124] The concentrations of these compounds in the L-tryptophan preparations were very low, and these concentrations did not exceed purity specification for the United States Pharmacopeia. Analysis of aniline-contaminated rapeseed oil from the toxic oil syndrome suggested a *possible* link between toxic oil syndrome and eosinophilia-myalgia syndrome, based on the structural similarity between the latter contaminant [3-(*N*-phenylamino)-L-alanine] and 3-(*N*-phenylamino)-1,2-propanediol in contaminated rapeseed oil.[125] Rodent studies indicate that metabolism of 3-(*N*-phenylamino)-1,2-propanediol produces 3-(*N*-phenylamino)-L-alanine.[126,127] *In vitro* studies suggest that the biotransformation of 3-(*N*-phenylamino)-L-alanine by human liver microsomes produces a toxic metabolite (4-aminophenol) similar to the metabolism of the toxic biomarker (3-(*N*-phenylamino)-1,2-propanediol) associated with toxic oil syndrome.[128] Potentially, the formation of this toxic metabolite causes the release of hazardous carbonyl species. However, animal studies have not reproduced eosinophilia-myalgia syndrome following administration of implicated tryptophan lots or case-associated contaminants (peak E, peak UV-5).[129] Consequently, the identities of the responsible contaminant or contaminants remain unconfirmed. Isolated case reports have associated eosinophilia-myalgia syndrome with the ingestion of 5-hydroxytryptophan and lysine unrelated to the L-tryptophan associated with eosinophilia-myalgia syndrome.[130,131] The inability to identify specific contaminants in cases of eosinophilia-myalgia syndrome and the lack of dose-response relationships between L-tryptophan and eosinophilia-myalgia syndrome suggests that the cause of eosinophilia-myalgia syndrome is multifactorial.[132] The cause of eosinophilia-myalgia syndrome remains controversial.[133]

Food Processing

The association of eosinophilia-myalgia syndrome with the ingestion of L-tryptophan from a single Japanese company suggests that contamination of the L-tryptophan products occurred during the fermentation process with genetically modified strains of *Bacillus amyloliquefaciens*.[134] This manufacturer purified and isolated the tryptophan from the fermentation broth by using ion exchange resins followed by processing through an activated charcoal column prior to crystallization of the product. Before the epidemic of eosinophilia-myalgia syndrome, the fermentation and purification processes underwent several modifications including a reduction in the amount of activated charcoal and a change to *B. amyloliquefaciens* strain V.[135]

DOSE RESPONSE

The risk of developing eosinophilia-myalgia syndrome increased with increased consumption of L-tryptophan Showa Denko products, suggesting a dose-response effect. In a cohort of 157 individuals from a psychiatric practice using Showa Denko L-tryptophan, the number of patients developing definite eosinophilia-myalgia syndrome increased from 13% in persons using 250 mg–1500 mg/daily to 50% in persons receiving >4,000 mg/daily.[136]

CLINICAL RESPONSE

Following the index cases of eosinophilia-myalgia syndrome in New Mexico, the US Centers for Disease Control (CDC) established a voluntary national surveillance system to monitor the course of eosinophilia-myalgia syndrome. The CDC defined a case of eosinophilia-myalgia syndrome as follows: 1) peripheral eosinophil count ≥1,000 cells/mm^3, 2) generalized myalgias severe enough to disrupt the patient's usual daily activities, and 3) absence of any infection or neoplasm that accounts for the patient's symptoms.[137] Based on this relatively specific case definition, two case-control studies linked the ingestion of Showa Denko L-tryptophan with the development of eosinophilia-myalgia syndrome.[137,138] Although most authors accept the causal link between contaminated L-tryptophan and eosinophilia-myalgia syndrome, these studies have been criticized for methodological flaws including diagnostic, recall and reporting biases, bias in the inclusion and exclusion of cases and controls, inequalities between cases and controls, failure to ensure that L-tryptophan exposure preceded the illness, and failure to exclude other illness.[139,140]

The onset of eosinophilia-myalgia syndrome occurs over days to weeks, primarily in Caucasian, middle-aged women. Clinical features associated with eosinophilia-myalgia syndrome include diffuse myalgias, fatigue, headache, skin lesions (peau d'orange, fasciitis, erythematous, maculopapular rash), sicca syndrome, and sensorimotor neuropathy. These effects are similar to, but less intense than the clinical features associated with the toxic oil syndrome. For most patients, the acute phase begins with the abrupt onset of intense myalgias and peripheral blood eosinophilia along with variable degrees of weakness, edema, cough, dyspnea, paresthesias, and induration. The fingers and toes are spared, and the myalgias and weakness typically involved the proximal muscles. After a few weeks to several months, a chronic phase develops, manifest by persistent myalgias, peripheral neuropathy, cognitive dysfunction, and varying degrees of dermal sclerosis. Follow-up studies of patients with eosinophilia-myalgia syndrome suggest that the number and severity of symptoms diminishes with time.[141] Although most patients remain symptomatic, a few patients develop new symptoms one year after onset of the disease.[142] Histological features of eosinophilia-myalgia syndrome include increased collagen deposition and perivascular accumulation of eosinophils, plasma cells, and lymphocytes in affected tissue. The differential diagnosis of eosinophilia-myalgia syndrome includes relatively rare diseases, such as eosinophilic myositis, Churg–Strauss syndrome, Loeffler's syndrome, hypereosinophilic syndrome, eosinophilic gastroenteritis, and toxic oil syndrome.

DIAGNOSTIC TESTING

Analytical Methods

Methods to detect and quantify contaminants in L-tryptophan preparations include high performance liquid chromatography (HPLC),[143] reversed phase HPLC (RP/HPLC),[144] and RP/HPLC with online ultraviolet detection and mass spectrometry.[145] The limit of detection of 1,1′-ethylidenebis(L-tryptophan) using RP/HPLC was 0.6 µg/g.

Biomarkers

Although at least six contaminants are associated with eosinophilia-myalgia syndrome, a specific contaminant responsible for eosinophilia-myalgia syndrome has not been identified. Therefore, to date there are no unique biomarkers for this disease.

Abnormalities

Case reports associated elevated serum aldolase, increased serum hepatic aminotransferases, and less frequently elevated serum creatine kinase concentrations with eosinophilia-myalgia syndrome. Antinuclear antibodies with a speckled pattern are occasionally detected. Marked peripheral eosinophilia may occur in the absence of symptoms.

TREATMENT

Treatment is supportive.

References

1. Biberoglu S, Biberoglu K, Komsuoglu G. Mad honey. JAMA 1988;259:1943.
2. Head T, Landes R. The peace of God: social violence and religious response in France around the year 1000. Speculum 1994;69:163–169.
3. Woolf A. Witchcraft or mycotoxin? The Salem witch trials. J Toxicol Clin Toxicol 2000;38:457–460.
4. Emmerson BT. "Ouch-ouch" disease: The osteomalacia of cadmium nephropathy. Ann Intern Med 1970; 73:854–855.
5. Bakir F, Damluji SF, Amin-Zaki L, Murtadha M, Khalidi A, Al-Rawi NY, et al. Methyl mercury poisoning in Iraq. Science 1973;181:230–241.
6. Smith HV, Spalding JM. Outbreak of paralysis in Morocco due to ortho-cresyl phosphate poisoning. Lancet 1959;2:1019–1021.
7. Kuratsune M, Yoshimura T, Matsuzaka J, Yamaguchi A. Epidemiologic study of polychlorinated biphenyls. Environ Health Perspect 1972;1:119–128.
8. Chen PH, Wong CK, Rappe C, Nygren M. Polychlorinated biphenyls, dibenzofurans and quaterphenyls in toxic rice-bran oil and in the blood and tissues of patients with PCB poisoning (Yu-cheng) in Taiwan. Environ Health Perspect 1985;59:59–65.
9. Kuratsune M, Yoshimura H, Hori Y, Okumura M, Matsuda Y. Yusho—a human disaster caused by PCB and related compounds. Fukuoka: Kyushu University Press; 1996.
10. Yoshimura T. Yusho in Japan. Ind Health 2003; 41:139–148.
11. Kunita N, Kashimoto T, Miyata H, Fukushima S, Hori S, Obana H. Causal agents of yusho. Prog Clin Biol Res 1984;137:45–58.
12. Kashimoto T, Miyata H, Fukushima S, Kunita N, Ohi G, Tung TC. PCBs, PCQs and PCDFs in the blood of Yusho and Yu-cheng patients. Environ Health Perspect 1985;59:73–78.
13. Guo YL, Lambert HG, Hsu C-C, Hsu MM. Yucheng: health effects of prenatal exposure to polychlorinated biphenyls and dibenzofurans. Int Arch Occup Environ Health 2004;77:153–158.
14. Asahi M. Clinical features and pathogenesis of Yusho (PCB poisoning). J Univ Occup Environ Health 1993;15:1–11.
15. Yoshimura T. Epidemiological analysis of "yusho" patients with special reference to sex, age, clinical grades and oil consumption. Fukuoka Acta Medica 1971;62:104–108.
16. Hayabuchi H, Yoshimura T, Kuratsune M. Consumption of toxic rice oil by Yusho patients and its relation to the clinical response and latent period. Food Cosmet Toxicol 1979;17:455–461.
17. Miyata H, Fukushima S, Kashimoto T, Kunita N. PCBs, PCQs and PCDFs in tissues of yusho and yu-cheng patients. Environ Health Perspect 1985;59:67–72.
18. Yoshimura T, Hayabuchi H. Relationship between the amount of rice oil ingested by patients with yusho and their subjective symptoms. Environ Health Perspect 1985;59:47–51.
19. Masuda Y. Health status of Japanese and Taiwanese after exposure to contaminated rice oil. Environ Health Perspect 1985;60:321–325.
20. Norback DH, Weltman RH. Polychlorinated biphenyl induction of hepatocellular carcinoma in the Sprague-Dawley rat. Environ Health Perspect 1985; 60:97–105.
21. Hirota Y, Tokunaga S, Kataoka K, Shinohara S. Symptoms and blood PCB level among chronic yusho patients, twenty-five years after outbreak. Fukuoka Igaku Zasshi 1997;88:220–225.
22. Yoshimura T. Epidemiological study on yusho babies born to mothers who had consumed oil contaminated by PCB. Fukuoka Acta Medica 1974;65:74–80.

23. Lu YC, Wong PN. Dermatological, medical, and laboratory findings of patients in Taiwan and their treatments. Am J Ind Med 1984;5:81–115.
24. Fein GG, Jacobson JL, Jacobson SW, Schwartz PM, Dowler JK. Prenatal exposure to polychlorinated biphenyls: effects on birth size and gestational age. J Pediatrics 1984;105:315–320.
25. Gladen BC, Rogan WJ. Decrements on six month and one-year Bayley scores and prenatal polychlorinated biphenyls (PCB) exposure. Am J Epidemiol 1988;128: 912.
26. Yang C-Y, Yu M-L, Guo M-L, Lai T-J, Hsu C-C, Lambert G, Guo YL. The endocrine and reproductive function of the female yu-cheng adolescents prenatally exposed to PCBs/PCDFs. Chemosphere 2005;61:355–360.
27. Jacobson SW, Fein GG, Jacobson JL, Schwartz PM, Dowler JK. The effect of intrauterine PCB exposure on visual recognition memory. Child Dev 1985;56:853–860.
28. Rogan WJ, Gladen BC, Hung KL, Koong SL, Shih LY, Taylor JS, et al. Congenital poisoning by polychlorinated biphenyls and their contaminants in Taiwan. Science 1988;241:334–336.
29. Tsai PC, Ko YC, Huang W, Liu HS, Guo YL. Increased liver and lupus mortalities in 24-year follow-up of the Taiwanese people highly exposed to polychlorinated biphenyls and dibenzofurans. Sci Total Environ 2007;374:216–222.
30. Hori T, Tobiishi K, Ashizuka Y, Nakagawa R, Todaka T, Hirakawa H, Iida T. Congener specific determination of PCBs in human blood using gel permeation chromatography (GPC) and high resolution gas chromatography/high resolution mass spectrometry (HRGC/HRMS). Fukuoka Igaku Zasshi 2005;96:220–226.
31. Calaf RE, Pena J, Paytubi S, Blount BC, Posada de La Paz M, Gelpi E, et al. Determination of aniline derivatives in oils related to the toxic oil syndrome by atmospheric pressure ionization-tandem mass spectrometry. Anal Chem 2001;73:3828–3837.
32. Chen PH, Luo ML, Wong CK, Chen CJ. Polychlorinated biphenyls, dibenzofurans, and quaterphenyls in the toxic rice-bran oil and PCBs in the blood of patients with PCB poisoning in Taiwan. Am J Ind Med 1984;5:133–145.
33. Nagayama J, Masuda Y, Kuratsune M. Determination of polychlorinated dibenzofurans in tissues of patients with Yusho. Food Cosmet Toxicol 1977;15:195–198.
34. Lung S-C, Guo Y-L, Chang H-Y. Serum concentrations and profiles of polychlorinated biphenyls in Taiwan Yu-cheng victims twenty years after the incident. Environ Pollut 2005;136:71–79.
35. Ryan JJ, Levesque D, Panopio LG, Sun WF, Masuda Y, Kuroki H. Elimination of polychlorinated dibenzofurans (PCDFs) and polychlorinated biphenyls (PCBs) from human blood in the yusho and yu-cheng rice oil poisonings. Arch Environ Contam Toxicol 1993;24:504–512.
36. Imamura T, Kanagawa Y, Matsumoto S, Tajima B, Uenotsuchi T, Shibata S, Furue M. Relationship between clinical features and blood levels of pentachlorodibenzofuran in patients with yusho. Environ Toxicol 2007;22:124–131.
37. Todaka T, Hirakawa H, Hori T, Tobiishi K, Iida T, Furue M. Concentrations of polychlorinated dibenzo-*p*-dioxins, polychlorinated dibenzofurans, and non-*ortho* and mono-*ortho* polychlorinated biphenyls in the blood of Yusho patients. Chemosphere 2007;66:1983–1989.
38. Masuda Y, Yoshimura H. Polychlorinated biphenyls and dibenzofurans in patients with yusho and their toxicological significance: a review. Am J Ind Med 1984;5:31–44.
39. Posada de la Paz M, Philen RM, Abaitua Borda I. Toxic oil syndrome: The perspective after 20 years. Epidemiol Rev 2001;23:231–247.
40. Posada de la Paz M, Philen RM, Abaitua Borda I, Bernert JT Jr, Gancedo JC, DuClos PJ, Kilbourne EM. Manufacturing processes at two French rapeseed oil companies: possible relationships to toxic oil syndrome in Spain. Food Chem Toxicol 1991;29:797–803.
41. Rigau-Perez JG, Perez-Alvarez L, Duenas-Castro S, Choi K, Thacker SB, Germain JL, et al. Epidemiologic investigation of an oil-associated pneumonic paralytic eosinophilic syndrome in Spain. Am J Epidemiol 1984;119:250–260.
42. Posada M, Castro M, Kilbourne EM, Diaz de Rojas F, Abaitua I, Tabuenca JM, et al. Toxic-oil syndrome: case reports associated with the ITH oil refinery in Seville. Food Chem Toxicol 1987;25:87–90.
43. Reig N, Calaf RE, Messeguer A, Morató A, Escabros J, Gelpí E, Abian J. LC-MS ion maps for the characterization of aniline derivatives of fatty acids and triglycerides in laboratory-denatured rapeseed oil. J Mass Spectrom 2007;42:527–541.
44. Hard GC. Short-term adverse effects in humans of ingested mineral oils, their additives and possible contaminants—a review. Hum Exp Toxicol 2000;19: 158–172.
45. Closa D, Folch E, Calaf RE, Abian J, Rosello-Catafau J, Gelpi E. Absorption and effects of 3-(*N*-phenylamino)-1,2-propanediol esters in relation to toxic oil syndrome. Lipids 2001;36:1125–1133.
46. Guitart R, Norgaard L, Abaitua Borda I, Posada de la Paz M, Abian J, Gelpi E. Pentachlorophenol and pentachloroanisole in oil samples associated with the toxic oil syndrome. Bull Environ Contam Toxicol 1999;62:1–7.
47. Morato A, Escabros J, Manich A, Reig N, Castano Y, Abian J, Messeguer A. On the generation and outcome of 3-(*N*-phenylamino)propane-1,2-diol derivatives in deodorized model oils related to toxic oil syndrome. Chem Res Toxicol 2005;18:665–674.
48. Ruiz-Mendez MV, Posada de la Paz M, Abian J, Calaf RE, Blount B, Castro-Molero N, et al. Storage time and deodorization temperature influence the formation of aniline-derived compounds in denatured rapeseed oils. Food Chem Toxicol 2001;39:91–96.

49. Posada de la Paz M, Abaitua Borda I, Kilbourne EM, Tabuenca Oliver JM, Diaz de Rojas F, Castro Garcia M. Late cases of toxic oil syndrome: evidence that the aetiological agent persisted in oil stored for up to one year. Food Chem Toxicol 1989;27:517–521.
50. Kilbourne EM, Bernert JT Jr, Posada de la Paz M, Hill RH Jr, Abaitua Borda I, Kilbourne BW, et al. Chemical correlates of pathogenicity of oils related to the toxic oil syndrome epidemic in Spain. Am J Epidemiol 1988;127:1210–1227.
51. Doll R. The aetiology of the Spanish Toxic Syndrome: interpretation of the epidemiological evidence. SPA/CEH 502. Copenhagen, Denmark: World Health Organization; 1985.
52. Tabuenca JM. Toxic-allergic syndrome caused by ingestion of rapeseed oil denatured with aniline. Lancet 1981;2:567–568.
53. Grandjean P, Tarkowsky S. Review of investigations and findings. In: World Health Organization: Toxic Oil Syndrome: Mass food poisoning in Spain. 1st ed. Copenhagen, Denmark: WHO, Regional Office for Europe; 1984:3–16.
54. World Health Organization. Report on the 21st Meeting of the WHO/FIS Scientific Committee for the Toxic Oil Syndrome. Copenhagen, Denmark: WHO; 1997.
55. Abaitua Borda I, Philen RM, Posada de la Paz M, Gomez de la Camara A, Diez Ruiz-Navarro M, Gimenez Ribota O, et al. Toxic oil syndrome mortality: the first 13 years. Int J Epidemiol 1998;27:1057–1063.
56. Diaz de Rojas F, Castro Garcia M, Abaitua Borda I, Alonso Gordo JM, Posada de la Paz M, Kilbourne EM, Tabuenca Oliver JM. The association of oil ingestion with toxic oil syndrome in two convents. Am J Epidemiol 1987;125:907–911.
57. Kaufman LD, Krupp LB. Eosinophilia-myalgia syndrome, toxic-oil syndrome, and diffuse fasciitis with eosinophilia. Curr Opin Rheumatol 1995;7:560–567.
58. Kilbourne EM, Rigau-Perez JG, Heath CW Jr, Zack MM, Falk H, Martin-Marcos M, et al. Clinical epidemiology of toxic-oil syndrome. Manifestations of a new illness. N Engl J Med 1983;309:1408–1414.
59. Martinez-Tello FJ, Navas-Palacios JJ, Ricoy JR, Gil-Martin R, Conde-Zurita JM, Colina-Ruiz Delgado F, et al. Pathology of a new toxic syndrome caused by ingestion of adulterated oil in Spain. Virchows Arch A Pathol Anat Histol 1982;397:261–285.
60. Alonso-Ruiz A, Zea-Mendoza AC, Salazar-Vallinas JM, Rocamora-Ripoll A, Beltran-Gutierrez J. Toxic oil syndrome: a syndrome with features overlapping those of various forms of scleroderma. Semin Arthritis Rheum 1986;15:200–212.
61. Escribano PM, Diaz de Atauri MJ, Gomez Sanchez MA. Persistence of respiratory abnormalities four years after the onset of toxic oil syndrome. Chest 1991;100:336–339.
62. James TN, Posada de la Paz M, Abaitua Borda I, Gomez Sanchez A, Martinez Tello FJ, Soldevilla LB. Histologic abnormalities of large and small coronary arteries, neural structures, and the conduction system of the heart found in postmortem studies of individuals dying from the toxic oil syndrome. Am Heart J 1991;121:803–815.
63. Abaitua Borda I, Kilbourne EM, Posada de la Paz M, Diez Ruiz-Navarro M, Gabriel Sanchez R, Falk H. Mortality among people affected by toxic oil syndrome. Int J Epidemiol 1993;22:1077–1084.
64. Tello de Meneses R, Gomez de la Camara A, Nogales-Moran MA, Escribano-Subias P, Barainca-Oyague MT, Gomez-Sanchez MA, et al. Pulmonary hypertension during exercise in toxic oil syndrome. Med Clin Barc 2005;19:685–688.
65. Martin-Arribas MC, De Andres Copa P, De La Paz MP. Quality of life, disability and handicap in patients with toxic oil syndrome. J Adv Nurs 2005;50:595–604.
66. Gomez de la Camara A, Posada de la Paz M, Abaitua Borda I, Barainca Oyague MT, Abraira Santos V, Ruiz-Navarro MD, et al. Heath status measurement in toxic oil syndrome. J Clin Epidemiol 1998;51:867–873.
67. Cardaba B, Gallardo S, Del Pozo V, Ezendam J, Izquierdo M, Martin C, et al. DR2 antigens are associated with severity of disease in toxic oil syndrome (TOS). Tissue Antigens 2000;55:110–117.
68. ladona MG, Izquierdo-Martinez M, Posada de La Paz MP, De La Torre R, Ampurdanes C, Segura J, et al. Pharmacogenetic profile of xenobiotic enzyme metabolism in survivors of the Spanish toxic oil syndrome. Environ Health Perspect 2001;109:369–375.
69. Hill RH Jr, Schurz HH, Posada de la Paz M, Abaitua Borda I, Philen RM, Kilbourne EM, et al. Possible etiologic agents for toxic oil syndrome: fatty acid esters of 3-(*N*-phenylamino)-1,2-propanediol. Arch Environ Contam Toxicol 1995;28:259–264.
70. Posada de la Paz M, Philen RM, Abaitua Borda I, Diez Ruiz-Navarro M, Abraira Santos V, Pozo Rodriguez F, et al. Factors associated with pathogenicity of oils related to the toxic oil syndrome epidemic in Spain. Epidemiology 1994;5:404–409.
71. Posada de la Paz M, Philen RM, Schurz H, Hill RH Jr, Gimenez Ribota O, Gomez de la Camara A, et al. Epidemiologic evidence for a new class of compounds associated with toxic oil syndrome. Epidemiology 1999;10:130–134.
72. Martinez-Cabot A, Morato A, Messeguer A. Synthesis and stability studies of the glutathione and *N*-acetylcysteine adducts of an iminoquinone reactive intermediate generated in the biotransformation of 3-(*N*-phenylamino)propane-1,2-diol: implications for toxic oil syndrome. Chem Res Toxicol 2005;18:1721–1728.
73. Gallardo S, Cardaba B, del Pozo V, De Andres B, Cortegano I, Jurado A, et al. Study of apoptosis in human lymphocytes by toxic substances implicated in toxic oil syndrome. Toxicol 1997;118:71–82.
74. Gallardo S, Cardaba B, Posada M, del Pozo V, Messeguer A, David CS, et al. Toxic oil syndrome: genetic restriction

and immunomodulatory effects due to adulterated oils in a model of HLA transgenic mice. Toxicol Lett 2005;159:173–181.

75. Hayashi T, James TN. Immunohistochemical analysis of lymphocytes in postmortem study of the heart from fatal cases of the eosinophilia myalgia syndrome and of the toxic oil syndrome. Am Heart J 1994;127:1298–1308.
76. Esteban A, Guerra L, Ruiz-Santana S, Fernandez A, Fernandez-Segoviano P. ARDS due to ingestion of denatured rapeseed oil. Chest 1983;84:166–169.
77. Gelpi E, de la Paz MP, Terracini B, Abaitua I, De La Camara AG, Kilbourne EM, et al. WHO/CISAT Scientific Committee for the Toxic Oil Syndrome, Centro de Investigacion para el Sindrome del Aceite Toxico: The Spanish toxic oil syndrome 20 years after its onset: a multidisciplinary review of scientific knowledge. Environ Health Perspect 2002;110:457–464.
78. Gomez-Sanchez MA, Mestre de Juan MJ, Gomez-Pajuelo C, Lopez JI, Diaz de Atauri MJ, Martinez-Tello FJ. Pulmonary hypertension due to toxic oil syndrome. A clinicopathologic study. Chest 1989;95:325–331.
79. Gomez de la Camara A, Gomez Mateos MA, Ferrando Vivas P, Barianca Oyague MT, Abiatua Borda I, Posada de la Paz M. Prevalence of cardiovascular risk factors in a cohort affected by the toxic oil syndrome. Med Clin Barc 2003;121:405–407. [Spanish]
80. Plaza Cano MM, Garcia de Albeniz Martinez XA, De Andres Copa P, Braun Saro B, Suarez Alvarez C, Garcia De Aguinaga ML, et al. Diagnostic performance of echocardiography in the follow-up of patients with toxic oil syndrome. Rev Esp Cardiol 2003;56:1195–1201.
81. Kaufman LD, Krupp LB. Eosinophilia-myalgia syndrome, toxic-oil syndrome, and diffuse fasciitis with eosinophilia. Curr Opin Rheumatol 1995;7:560–567.
82. Sanchez-Porro Valades P, Posada de la Paz M, de Andres Copa P, Gimenez Ribota O, Abaitua Borda I. Toxic oil syndrome: survival in the whole cohort between 1981 and 1995. J Clin Epidemiol 2003;56:701–708.
83. Lyon IB. Textbook of medical jurisprudence of India. 1st ed. Calcutta, India: Thacker, Spink & Company; 1889:214.
84. Gomber S, Daral TS, Sharma PP, Faridi MM. Resurgence of epidemic dropsy. Indian Pediatr 1997;34:953–957.
85. Khanna SK, Upreti KK, Singh GB. Trend of adulteration of mustard oil in rural markets of Uttar Pradesh. Sci Cult 1986;52:149–156.
86. Verma SK, Dev G, Tyagi AK, Goomber S, Jain GV. Argemone mexicana poisoning: autopsy findings of two cases. Forensic Sci Int 2001;115:135–141.
87. Newton SM, Lau C, Gurcha SS, Besra GS, Wright CW. The evaluation of forty-three plant species for *in vitro* antimycobacterial activities; isolation of active constituents from *Psoralea corylifolia* and *Sanguinaria canadensis*. J Ethnopharmacol 2002;79:57–67.
88. Chakravarty NK, Chaudhuri RN, Werner G. Observations on vascular changes produced by *Argemone* alkaloids. Indian J Med Res 1955;43:107–112.
89. Tandon S, Das M, Khanna SK. Biometabolic elimination and organ retention profile of *Argemone* alkaloid, sanguinarine, in rats and guinea pigs. Drug Metab Disp 1993;21:194–197.
90. Pitts BJR, Meyerson LR. Inhibition of Na^{+}-K^{+}-ATPase activity and ouabain binding of sanguinarine. Drug Dev Res 1981;1:43–47.
91. Upreti KK, Das M, Khanna SK. Biochemical toxicology of argemone oil: effect on hepatic cytochrome P-450 and xenobiotic metabolizing enzymes. J Appl Toxicol 1991;11:203–209.
92. Upreti KK, Das M, Kumar A, Singh GB, Khanna SK. Biochemical toxicology of argemone oil. IV. Short-term oral feeding response in rats. Toxicol 1989;58:285–298.
93. Tandon S, Das M, Khanna SK. Effect of sanguinarine on the transport of essential nutrients in an everted gut sac model: role of Na^{+}-K^{+}-ATPase. Nat Toxins 1993;1:235–240.
94. Das M, Babu K, Reddy NP, Srivastava LM. Oxidative damage of plasma proteins and lipids in epidemic dropsy patients: alterations in antioxidant status. Biochim Biophys Acta 2005;1722:209–217.
95. Lal RB, Dasgupta AC. Investigation into epidemiology of epidemic dropsy X: a note on an outbreak of epidemic dropsy associated with the use of mustard oil pressed from adulterated seed. Indian J Med Res 1941;29:157–163.
96. Das M, Khanna SK. Clinicoepidemiological, toxicological, and safety evaluation studies on argemone oil. Crit Rev Toxicol 1997;27:273–297.
97. Kar HK, Jain RK, Sharma PK, Gautam RK, Kumar P, Bhardwaj M. Epidemic dropsy: a study of cutaneous manifestations with histopathological correlation. Indian J Venereol Leprol 2001;67:178–179.
98. Sood NN, Sachdev MS, Verma LK. Pressure dynamics and treatment modalities of epidemic dropsy glaucoma. Glaucoma 1990;12:20–27.
99. Sachdev HP, Sachdev MS, Verma L, Sood NN, Moonis M. Electrophysiological studies of the eye, peripheral nerves and muscles in epidemic dropsy. J Trop Med Hyg 1989;92:412–415.
100. Sharma BD, Bhatia V, Rahtee M, Kumar R, Mukharjee A. Epidemic dropsy: observations on pathophysiology and clinical features during the Delhi epidemic of 1998. Trop Doct 2002;32:70–75.
101. Singh K, Singh MJ, Das JC. Visual field defects in epidemic dropsy. Clin Toxicol 2006;44:159–163.
102. Singh NP, Anuradha S, Dhanwal DK, Singh K, Prakash A, Madan K, Agarwal SK. Epidemic dropsy—a clinical study of the Delhi outbreak. J Assoc Physicians India 2000;48:877–880.
103. Sachdev MS, Sood NN, Verma LK, Gupta SK, Jaffery NF. Pathogenesis of epidemic dropsy glaucoma. Arch Ophthalmol 1988;106:1221–1223.
104. Malik KP, Dadeya S, Gupta VS, Sharan P, Guliani, Dhawan M. Pattern of intraocular pressure in epidemic dropsy in India. Trop Doct 2004;34:161–162.

105. Prakash J. Acute renal failure in epidemic dropsy. Renal Failure 1999;21:707–711.
106. Tandon RK, Tandon HD, Nayak NC, Tandon BN. Liver in epidemic dropsy. Ind J Med Res 1976;64:1064–1069.
107. Senegupta PC, Napier LE. Haematological changes in epidemic dropsy. Ind J Med Res 1940;28:197–203.
108. Chopra RN, Basu VP. Cardiovascular manifestations of epidemic dropsy and their treatment. Ind Med Gaz 1980;65:546–550.
109. Singh GB, Khanna SK. Argemone mexicana strikes again. Sci Rep 1983;20:108–110.
110. Husain S, Narsimha R, Rao RN. Separation, identification and determination of sanguinarine in argemone and other adulterated edible oils by reversed-phase high-performance liquid chromatography. J Chromatogr Anal 1999;863:123–126.
111. Tandon RK, Singh DS, Arora RR, Lal P, Tandon BN. Epidemic dropsy in New Delhi. Am J Clin Nutr 1975;28:883–887.
112. Das M, Babu K, Reddy NP, Srivastava LM. Oxidative damage of plasma proteins and lipids in epidemic dropsy patients: alterations in antioxidant status. Biochim Biophys Acta 2005;1722:209–217.
113. Khilnani GC, Trikha I, Malhotra OP, Prabhakaran D, Pande JN. Pulmonary functions in patients with epidemic dropsy. Indian J Chest Dis Allied Sci 2003;45:25–29.
114. Upreti KK, Das M, Khanna SK. Role of antioxidants and scavengers on argemone oil induced toxicity. Arch Environ Contam Toxicol 1991;20:531–537.
115. Singh K, Singh MJ, Das JC. Visual field defects in epidemic dropsy. Clin Toxicol 2006;44:159–163.
116. Sainani GS Epidemic dropsy. In: Ahuja MM, ed: Progress in clinical medicine in India. New Dehli, India: Arnold Heinemann Publishers, 1976:92–104.
117. Sternberg EM, Van Woert MH, Young SN, Magnussen I, Baker H, Gauthier S, Osterland CK. Development of a scleroderma-like illness during therapy with L-5-hydroxytryptophan and carbidopa. N Engl J Med 1980;303:782–787.
118. Swygert LA, Maes EF, Sewell LE, Miller L, Falk H, Kilbourne EM. Eosinophilia-myalgia syndrome. Results of national surveillance. JAMA 1990;264:1698–1703.
119. Hertzman PA, Blevins WL, Mayer J, Greenfield B, Ting M, Gleich GJ. Association of the eosinophilia-myalgia syndrome with the ingestion of tryptophan. N Engl J Med 1990;322:869–873.
120. Kilbourne EM, Philen RM, Kamb ML, Falk H. Tryptophan produced by Showa Denko and epidemic eosinophilia-myalgia syndrome. J Rheumatol 1996; 23(suppl 46):81–88.
121. McKeon P, Swanwick G, Manley P. L-Tryptophan and the eosinophilia-myalgia syndrome: a clinical and laboratory study. Acta Psychiatr Scand 1993;90:451–454.
122. Sullivan EA, Staehling N, Philen RM. Eosinophilia–myalgia syndrome among the non-L-tryptophan users and pre-epidemic cases. J Rheumatol 1996;23: 1784–1787.
123. Naylor S, Williamson BL, Johnson KL, Gleich GJ. Structural characterization of case-associated contaminants peak C and FF in L-tryptophan implicated in eosinophilia-myalgia syndrome. Adv Exp Med Biol 1999;467:453–460.
124. Hill RH Jr, Caudill SP, Philen RM, Bailey SL, Flanders WD, Driskell WJ. Contaminants in L-tryptophan associated with eosinophilia myalgia syndrome. Arch Environ Contam Toxicol 1993;25:134–142.
125. Philen RM, Posada M. Toxic oil syndrome and eosinophilia-myalgia syndrome: May 8–10, 1991, World Health Organization Meeting Report. Semin Arthritis Rheum 1993;23:104–124.
126. Philen RM, Hill RH Jr, Flanders WD, Caudill SP, Needham L, Sewell L, et al. Tryptophan contaminants associated with eosinophilia-myalgia syndrome. Am J Epidemiol 1993;138:154–159.
127. Mayeno AN, Benson LM, Naylor S, Colberg-Beers M, Puchalski JT, Gleich GJ. Biotransformation of 3-(phenylamino)-1,2-propanediol to 3-(phenylamino)alanine: a chemical link between toxic oil syndrome and eosinophilia-myalgia syndrome. Chem Res Toxicol 1995; 8:911–916.
128. Martinez-Cabot A, Messeguer A. Generation of quinoneimine intermediates in the bioactivation of 3(*N*-phenylamino)alanine (PAA) by human liver microsomes: a potential link between eosinophilia-myalgia syndrome and toxic oil syndrome. Chem Res Toxicol 2007; 20:1556–1562.
129. Sato F, Hagiwara Y, Kawase Y. Subchronic toxicity of 3-phenylamino alanine, an impurity in L-tryptophan reported to be associated with eosinophilia-myalgia syndrome. Arch Toxicol 1995;69:444–449.
130. Patmas MA. Eosinophilia-myalgia syndrome not associated with L-tryptophan. N J Med 1992;89:285–286.
131. Michelson D, Page SW, Casey R, Trucksess MW, Love LA, Milstien S, et al. An eosinophilia-myalgia syndrome related disorder associated with exposure to L-5-hydroxytryptophan. J Rheumatol 1994;21:2261–2265.
132. Martin RW, Duffy J, Engel AG, Lie JT, Bowles CA, Moyer TP, Gleich GJ. The clinical spectrum of the eosinophilia-myalgia syndrome associated with L-tryptophan ingestion. Clinical features in 20 patients and aspects of pathophysiology. Ann Intern Med 1990;113:124–134.
133. Smith MJ, Garrett RH. A heretofore undisclosed crux of eosinophilia-myalgia syndrome: compromised histamine degradation. Inflamm Res 2005;54:435–450.
134. Slutsker L, Hoesly FC, Miller L, Williams LP, Watson JC, Fleming DW. Eosinophilia-myalgia syndrome associated with exposure to tryptophan from a single manufacturer. JAMA 1990;264:213–217.
135. Belongia EA, Hedberg CW, Gleich GJ, White KE, Mayeno AN, Loegering DA. An investigation of the

cause of the eosinophilia-myalgia syndrome associated with tryptophan use. N Engl J Med 1990;323:357–365.

136. Kamb ML, Murphy JJ, Jones JL, Caston JC, Nederlof K, Horney LF. Eosinophilia-myalgia syndrome in L-tryptophan-exposed patients. JAMA 1992;267:77–82.
137. Centers for Disease Control. Eosinophilia-myalgia syndrome and L-tryptophan-containing products—New Mexico, Minnesota, Oregon, New York, 1989. MMWR Morb Mortal Wkly Rep 1989;38:785–788.
138. Eidson M, Philen RM, Sewell CM, Voorhees R, Kilbourne EM. L-Tryptophan and eosinophilia-myalgia syndrome in New Mexico. Lancet 1990;335:645–648.
139. Shapiro S. Epidemiologic studies of the association of L-tryptophan with the eosinophilia-myalgia syndrome: a critique. J Rheumatol 1996;23(suppl 46):44–59.
140. Horwitz RI, Daniels SR. Bias or biology: evaluating the epidemiologic studies of L-tryptophan and the eosinophilia-myalgia syndrome. J Rheumatol 1996;23(suppl 46):60–72.
141. Sullivan EA, Kamb ML, Jones JL, Meyer P, Philen RM, Falk H, Sinks T. The natural history of eosinophilia-myalgia syndrome in a tryptophan-exposed cohort in South Carolina. Arch Intern Med 1996;156:973–979.
142. Pincus T. Eosinophilia-myalgia syndrome: patient status 2–4 years after onset. J Rheumatol Suppl 1996; 46:19–24.
143. Meyer K, Moller A, Karg C, Steinhart H. Formation and determination of aldehyde-addition products of tryptophan. Adv Exp Med Biol 1996;398:711–717.
144. Trucksess MW, Thomas FS, Page SW. High-performance liquid chromatographic determination of 1,1′-ethylidenebis(L-tryptophan) in L-tryptophan preparations. J Pharm Sci 1994;83:720–722.
145. Williamson BL, Tomlinson AJ, Hurth KM, Posada de la Paz M, Gleich GJ, Naylor S. Rapid HPLC screening method for contaminants found in implicated L-tryptophan associated with eosinophilia myalgia syndrome and adulterated rapeseed oil associated with toxic oil syndrome. Biomed Chromatogr 1998;12:255–261.

Chapter 2

FOOD ADDITIVES and SENSITIVITIES

CYRUS RANGAN, MD, FAAP

HISTORY

For centuries, food additives have been used for flavoring, coloring, and extension of the useful shelf-life of food, as well as the promotion of food safety. President Theodore Roosevelt signed into law the original Food and Drugs Act in 1906.[1] This law prohibited interstate commerce in mislabeled or adulterated drugs, foods, and drinks. The Gould Amendment of 1913 required the plain and conspicuous marking of weight, numerical count, or measure on the outside of food packages. Reorganization of the US Bureau of Chemistry in 1927 resulted in the formation of the Food, Drug, and Insecticide Administration (later shortened to the Food and Drug Administration [FDA]) as a regulatory agency. The passage of the Food, Drug & Cosmetics Act of 1938 established mechanisms for providing safe tolerances for unavoidable toxic substances in food, and this act authorized standards on the identity and quality of foods. In 1940, the FDA transferred from the Department of Agriculture to the Federal Security Agency, and Walter G. Campbell became the first Commissioner of Food and Drugs.

The FDA regulates and monitors administrative, chemical, and toxicological information on food additives through a program called Priority-based Assessment of Food Additives (PAFA). This program categorizes additives in an informational database called Everything Added to Food in the United States (EAFUS) based on available toxicological information, the uses of substances in food production, and the presence of banned substances.[2] A subset of EAFUS is a list of ingredients generally recognized as safe (GRAS). Substances that are GRAS do not require FDA authorization for use as food additives.[3] Conversely, all substances introduced into the marketplace after 1958 with evidence of potential harm to humans or animals undergo FDA scrutiny and investigation. Substances prohibited from addition to food are listed in Table 2.1.

The Food Additives Amendment (1958) and the Color Additives Amendment (1960) included the Delaney clause, stating, "No additive shall be deemed to be safe if it is found to induce cancer when ingested by man or laboratory animals or if it is found, after tests which are appropriate for the evaluation of the safety of food additives, to induce cancer in man or animals." The Delaney clause authorizes the FDA to permit continued use of certain additives on a provisional basis during safety evaluations. Safety of food additives undergoes continual reexamination by the FDA. Food additives are divided into the following five broad categories according to their function: 1) taste enhancers, 2) antioxidants, 3) preservatives, 4) stabilizers & emulsifiers, and 5) coloring agents. There are over 3,000 additives approved for use in US food, including about 1,800 taste enhancers.

EXPOSURE

Environment

Natural additives derived from plant and mineral sources include caramel color, fruit juice, chlorophyll,

TABLE 2.1. Banned Food Additives

Banned Additive (Year)	Purpose	Source/Description	Toxicity
Calamus (1968)	Taste enhancer	β-Asarone contained in oil extract of *Acorus calamus*	Carcinogenic
Chlorofluorocarbon propellants (1978)	Preservative	Aliphatic halogenated hydrocarbons	Hepatotoxic, Cardiotoxic
Cinnamyl anthranilate (1985)	Taste enhancer	Ester of cinnamyl alcohol and anthranilic acid ($C_{16}H_{15}NO_2$)	Carcinogenic
Cobaltous salts (1966)	Stabilizer	Inorganic salts ($CoCl_2$, $CoSO_4$)	Cardiotoxic
Coumarin (1954)	Unavoidable ingredient of tonka beans	1,2-benzopyrone ($C_9H_6O_2$)	Hepatotoxic
Cyclamate (1969)	Taste enhancer	Salts of cyclohexane sulfamic acid ($C_6H_{12}NO_3S$)	Carcinogenic
Diethylpyrocarbonate (1972)	Stabilizer	Pyrocarbonic acid diethyl ester ($C_6H_{10}O_5$)	Carcinogenic
Dulcin (1950)	Taste enhancer	4-ethoxyphenylurea ($C_9H_{12}N_2O_2$)	Carcinogenic
Monochloroacetic acid (1941)	Preservative	Chloroacetic acid ($C_2H_3ClO_2$)	Carcinogenic, Neurotoxic, Cardiotoxic
Nordihydroguaiaretic acid (1968)	Antioxidant	4,4′-(2,3-dimethyltetramethylene) dipyrocatechol ($C_{18}H_{22}O_4$)	Hepatotoxic
P-4000 (1950)	Taste enhancer	5-nitro-2-propoxyaniline ($C_9H_{12}N_2O_3$)	Carcinogenic
Safrole (1960)	Taste enhancer	4-allyl-1,2-methylenedioxy-benzene ($C_{10}H_{10}O_2$)	Carcinogenic
Thiourea (1977)	Preservative	Thiocarbamide (CH_4N_2S)	Carcinogenic

annatto extracts B-G (*cis*-bixin in seeds of the tropical lipstick tree, *Bixa orellana* L.), turmeric yellow or curcumin (solvent extract of ground rhizomes of *Curcuma longa* L.), quillaia extracts (glycosides of quillaic acid from aqueous extracts of milled inner bark of *Quillaia saponaria* Molina), dried algae meal, beet juice, and paprika.[4] More recently introduced natural substances include LE399 α-amylase from *Bacillus licheniformis* (Weigmann) Chester and laccase from *Myceliophthora thermophila* (Apinis) Oorschot. The former genetically modified enzyme is used in the production of nutritive sweeteners in syrups and alcoholic beverages, whereas the latter enzyme is used for the oxidation of off-flavor phenolic compounds (e.g., *trans*-2-nonenal) in the brewing of beer.[5] Xylanase is a hydrolytic enzyme from a genetically modified strain of *Thermomyces lanuginosus* Tsikl. that improves the handling and stability of dough. Most of these enzymes are destroyed during food processing.[5] Almost all synthetic food additives are derivatives of sulfonated aromatic amines, organic acids and salts, or inorganic acids and salts.

Animal

Gelatin, a common additive, is derived from the hooves of livestock.[6] Carminic acid is a red dye extracted from the dried bodies of pregnant female cochineal (*Dactylopius coccus* Costa), which are scale-type insects farmed on cactus plantations in Peru and the Canary Islands. The bitter Italian liquor, Campari, contains carminic acid.[7]

Food

Some food additives (e.g., nitrates) occur naturally. Leafy vegetables represent almost 100% of the human daily intake of nitrates. The content of naturally occurring nitrates is not regulated by the FDA, but calculations on the cumulative estimated daily intake of nitrates include nitrates from natural sources.[8] Notame is a flavor enhancer that is a dipeptide methyl ester with a potency of about 30–60 times the sweetness of aspartame.[5]

DOSE RESPONSE

Human exposure to food additives is evaluated by the acceptable daily intake (ADI) and estimated daily intake (EDI). ADI is the federal standard for maximum allowable dietary exposure to a food additive. ADI is an index derived from the no-observed-adverse-effect level (NOAEL) from lifetime toxicological studies in animals.

NOAEL is divided by 100 to compute the ADI. EDI is a cumulative index derived from both the concentration of additives in various foods and the average daily intake of those foods. ADI and EDI are computed for every food additive, and food additive concentrations are adjusted to reflect the addition of new additives and foods. For FDA approval, the sum of all EDI's of a food additive from all sources must not exceed the ADI. These indices are utilized primarily for the evaluation of chronic effects, such as mutagenesis and carcinogenesis.[9,10] The ADI for monosodium glutamate (MSG) is 120 mg/kg/day. MSG may be associated with acute symptoms in sensitive individuals at lower doses. A randomized-controlled trial found a 33% occurrence of headache, numbness, tingling, and weakness following the ingestion of single doses of 2.5 g MSG.[11] The ADI for nitrates is 3.64 mg/kg/day. The vast majority (99%) of nitrate intake comes from leafy vegetables, with less than 1% from food additives. The EDI for nitrates is less than 70% of the ADI.[12] This safety window may explain the lack of association between nitrates and cases of human cancer. Idiosyncratic responses to food additives (e.g., methemoglobinemia) are nonimmunological and dose-independent.

CLINICAL RESPONSE

As much as 2% of the US population develops sensitivity or other adverse reactions to food additives.[13] In children with asthma, the incidence of these responses is as high as 10%.[14] Immune-mediated hypersensitivity reactions and idiosyncratic reactions comprise the majority of adverse responses to food additives. In general, there are no safety concerns for the use of alicyclic, alicyclic-fused, and aromatic-fused ring lactones as flavoring agents including tuberose lactone, mintlactone, dihydromintlactone, sclareolide, octahydrocoumarin, 3-propylidenephthalide, 3-butylidenephthalide, dihydrocoumarin, and 6-methylcoumarin.[5]

Taste Enhancers

Taste is comprised of the rudimentary sensations of sweet, sour, bitter, and salty, whereas flavor is the combined sensation of taste and the olfactory perception of food. Methods to enhance the taste of food include the addition of salts, spices, and sweeteners to food.

Monosodium Glutamate

The most closely studied taste enhancer is MSG, the sodium salt of glutamic acid. MSG enhances the savory flavors imparted by glutamic acid, which occurs naturally in proteinaceous foods (e.g., meats, seafood, stews, soups, sauces). Toxicity was first reported in 1968 as the "Chinese Restaurant Syndrome," which was characterized by the triad of palpitations, generalized weakness, and sensory numbness originating at the nape of the neck with radiation to the arms and back.[15] Controlled clinical trials have not confirmed this response to MSG, suggesting that MSG may cause symptoms only in select populations (e.g., asthmatic or atopic patients).[16,17] Short-term responses to MSG occurred in one double-blind trial of healthy patients ingesting 5 grams of MSG on an empty stomach, and in patients with severe asthma. Transient symptoms (MSG symptom complex) associated with MSG consumption included numbness, burning sensation, tingling, facial tightness, chest pain, headache, nausea, palpitations, drowsiness, and weakness.[11] These effects may result from MSG-induced stimulation of peripheral nerve receptors in the esophagus.[16] Although some epidemiological studies demonstrated correlations between MSG and adverse responses in severely asthmatic patients, a single-blind, placebo-controlled challenge study did not detect symptoms of wheezing or reduction in forced expiratory volume in 100 asthmatic patients ingesting 2.5 mg MSG.[18] Long-term health effects do not usually occur. MSG remains on the list of GRAS ingredients, but the US FDA requires manufacturers to list MSG and related compounds (monopotassium glutamate, monoammonium glutamate) on the food label. MSG was banned from the production of infant foods because of the occurrence of irreversible retinal lesions in neonatal rodents.[19]

Aspartame

Aspartame is a synthetic dipeptide formed by the reaction of L-aspartic acid with L-phenylalanine methyl ester. The gastrointestinal tract hydrolyzes aspartame to aspartyl phenylalanine and methanol. Further hydrolysis of aspartyl phenylalanine to aspartic acid and the essential amino acid, phenylalanine, produces a risk for patients with the homozygous gene for phenylketonuria (PKU). Healthy individuals hydroxylate phenylalanine to tyrosine after absorption by the gastrointestinal tract. Patients with PKU lack phenylalanine hydroxylase, and thus they can not hydroxylate phenylalanine, resulting in the accumulation of phenylalanine in the blood.[20] Acute hyperphenylalaninemia causes vomiting, agitation, a mousy odor to the urine, and a rash consistent with atopic dermatitis.[21] Chronic hyperphenylalaninemia produces hypertonia, hyperreflexia, craniofacial abnormalities, growth retardation, central nervous system dysfunction, mental retardation, and seizures. Chronic findings may be due to long-term disruption of glutamatergic synaptic transmission.[22] Individuals with the heterozygous gene for PKU do not usually develop the clinical effects of PKU following exposure to aspartame.[23]

Saccharin

Saccharin (CAS RN: 81-07-2) is a synthetic product with 500 times the sweetness of sugar. Figure 2.1 displays the chemical structure of saccharin. In 1981, the National Toxicology Program (NTP) listed saccharin as a likely human carcinogen, based on laboratory studies demonstrating an increased incidence of bladder cancer in male rats. However, female rats and mice of either gender were not similarly affected. Continued marketing of saccharin was permitted with the provision that packages disclose the recognition of carcinogenicity in laboratory animals. In 1997, a report was issued by the NTP that bladder cancer incidence was explained by the tortuous anatomy of male rat ureters and the possible presence of contaminants in the original study. Furthermore, numerous studies documented the lack of association of human cancer to saccharin exposure.[24] Saccharin was subsequently delisted as a cause of human cancer.

Sodium Cyclamate

The synthetic sweetener, sodium cyclamate, was originally designated in the list of GRAS ingredients. The liver converts sodium cyclamate to cyclohexylamine, a well-documented cause of bladder cancer in both male and female rats. Although the extrapolated human dose equivalent for sodium cyclamate would be impossible to achieve, the additive was banned in 1969 because of strongly supportive evidence for carcinogenesis in animals.[25] Subsequent epidemiological data did not support an increased risk of bladder cancer in individuals using cyclamate or saccharin.[25] The International Agency for Research on Cancer lists cyclamates as compounds (Group 3) for which the evidence of carcinogenicity is inadequate in humans and inadequate or limited in experimental animals.

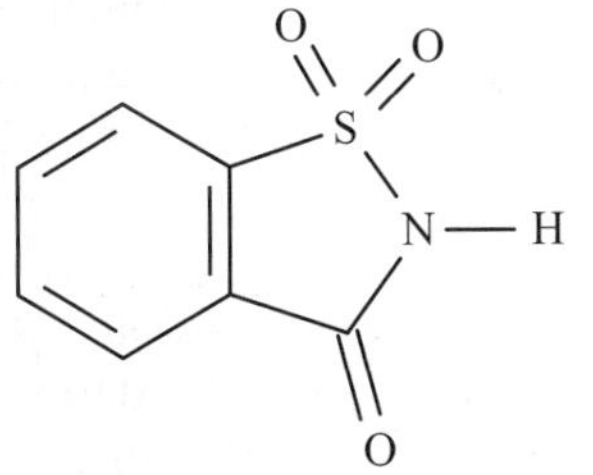

O O S N — H O

FIGURE 2.1. Chemical structure of saccharin.

Diacetyl and Bronchiolitis Obliterans

A case report of eight workers from a single microwave popcorn plant associated the developed bronchiolitis obliterans with exposure to particulates and a range of organic vapors from flavoring agents.[26] Air monitoring of this plant indicated that the predominant compounds were acetic acid and ketones (diacetyl, methyl ethyl ketone, acetoin, 2-nonanone). Additional individual case reports associated the exposure of workers at flavor-manufacturing plants with the development of bronchiolitis obliterans.[27] Subsequent cross-sectional studies confirmed the occurrence of bronchiolitis obliterans among workers from chemical production plants that produce diacetyl.[28] In this plant, the strength of association between exposure to diacetyl and severe fixed pulmonary obstruction was about 10-fold. Pulmonary abnormalities of this illness included fixed airflow obstruction, air trapping, reduced FEV_1, decreased FEV_1 to FVC ratio, and increased total lung capacity (TLC). Decline in FEV_1 exceeding 8% (>330 mL) within 6 to 12 months suggests the development of obstructive pulmonary disease.[29]

Diacetyl (CAS RN: 431-03-8, 2,3-butanedione) is a water-soluble, yellowish liquid that smells like butter and is a natural constituent of many foods. Figure 2.2 displays the chemical structure of diacetyl. This compound was the predominant ketone in air samples from this plant. The low boiling point (88 °C/190 °F) and relatively low vapor pressure (56.8 mmHg at 25 °C/77 °F) of diacetyl increase the risk of airborne exposure, but this compound has good odor and irritant warning properties. Animal studies suggest that this compound can cause airway epithelial injury, although the contribution of other compounds (e.g., acetoin, acetaldehyde) in the ambient air remains unresolved.[30] The granulocytic inflammatory infiltrates associated with diacetyl-associated bronchiolitis obliterans differs from the lymphocytic infiltrates commonly associated with posttransplant bronchiolitis obliterans.[31]

Antioxidants

Antioxidant food additives protect oily and fatty foods from spoilage by inhibiting lipid peroxidation and preventing the disintegration of lipid-soluble vitamins. Common antioxidant food additives in cereals, marga-

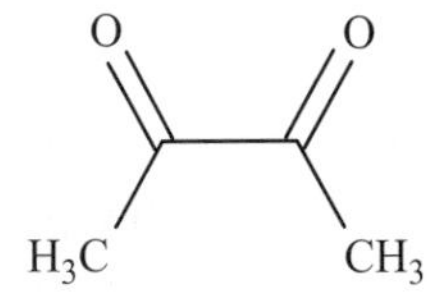

O O H_3C CH_3

FIGURE 2.2. Chemical structure of diacetyl.

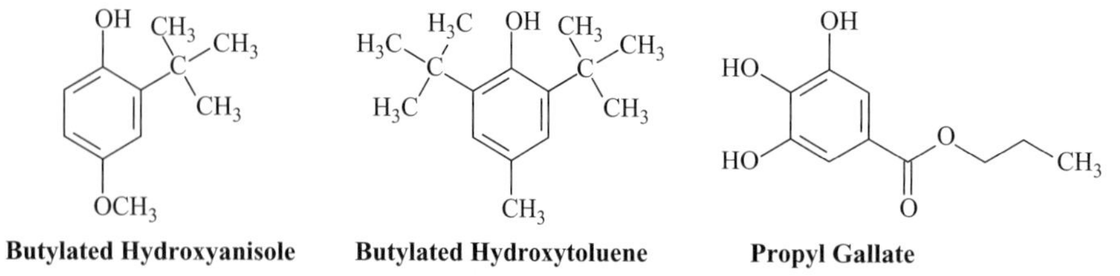

Butylated Hydroxyanisole **Butylated Hydroxytoluene** **Propyl Gallate**

FIGURE 2.3. Chemical structures of butylated hydroxyanisole, butylated hydroxytoluene, and propyl gallate.

rine, cooking oils and shortening, and dry mixes include ascorbic acid (vitamin C), tocopherol (vitamin E), butylated hydroxyanisole (CAS RN: 25013-16-5, BHA), butylated hydroxytoluene (CAS RN: 128-37-0, BHT), and propyl gallate (CAS RN: 121-79-9). Figure 2.3 displays the chemical structures of butylated hydroxyanisole, butylated hydroxytoluene, and propyl gallate. Antioxidants are generally nontoxic in the amounts used in food production practices.[32] Paradoxically, excessive application of antioxidants to food may promote lipid peroxidation in copper and iron-containing foods such as meats.[33] BHA and BHT are synthetic phenol-derivatives that have been associated with exacerbations of urticaria in patients with chronic urticaria.[34] High doses of BHT that far exceed the dose of BHT in food can cause human toxicity. A case report associated the onset of acute gastroenteritis (epigastric pain, vomiting, weakness, confusion, syncope) with the ingestion of 4 g BHT,[35] while the ingestion of 80 g BHT produced lightheadedness, slurred speech, unsteady gait, and lethargy. These patients recovered without sequelae.[36] Reports of carcinogenicity from BHA and BHT free-radical metabolites are limited to rodent squamous cells of the forestomach, an organ that does not have a human counterpart.[37] The International Agency for Research on Cancer (IARC) lists butylated hydroxyanisole as an animal carcinogen without evidence of carcinogenicity in humans (group 2B). Similarly, the US National Toxicology (NTP) Program lists this compound as reasonably anticipated to be a human carcinogen. Neither the IARC nor the US NTP lists butylated hydroxytoluene as a human or animal carcinogen.

At extremely high doses (i.e., 10,000 mg/kg feed), the oral administration of propyl gallate to rats causes growth retardation, anemia, hepatorenal dysplasia, and forestomach hyperplasia.[38] The ADI for propyl gallate is 0.2 mg/kg body weight as established by the US FDA. Neither the IARC nor the US NTP lists propyl gallate as a human carcinogen.

Preservatives

Preservatives enhance food safety and extend shelf-life by limiting viral, bacterial, and fungal growth. They are added to a wide variety of foods, especially those with high carbohydrate content (e.g., beverages, marmalades, jellies, jams, desserts, dairy products, breads, dough, cured meats and fish, fermented fruits and vegetables).

Benzoic Acid

Benzoic acid and sodium benzoate are most suitable for naturally acidic foods, fruit juices, and soft drinks. Benzoic acid occurs naturally in many plant and animal products (e.g., milk) at concentrations up to about 40 mg/kg, whereas maximum concentrations of benzoic acid and sodium benzoate in food as preservatives range up to 2,000 mg/kg.[39] Sodium benzoate is conjugated with glycine to form hippurate, which the kidney rapidly excretes in the urine. The administration of sodium benzoate is a treatment for inborn errors of the urea cycle. The dose of sodium benzoate for this disease is 250 mg/kg body weight up to 500 mg/kg daily for several years.[40] The most common adverse responses to this treatment are anorexia and vomiting.[41] However, adverse responses from the use of these doses of sodium benzoate are difficult to separate from the effects of the disease. Table 2 lists recommendations for the storage of selected foods.[42]

Many preservatives, such as sorbates, hydrogen peroxide, benzoic acid (CAS RN: 65-85-0), and sodium benzoate (CAS RN: 532-32-1) cause animal toxicities in extremely high doses; human toxicities primarily involve rare case reports of mild hypersensitivity reactions (urticaria, pruritus).[43,44] Under special circumstances, small amounts of benzene can form from preservative systems comprised of sodium benzoate and citric acid. The decarboxylation of benzoic acid produces measurable amounts of benzene in the presence of transition metals, acid, and heat.[45]

Sulfites

Sulfites, bisulfites, and metabisulfites occur naturally in fermented beverages, such as beer and wine. Similar to food additives, sulfite derivatives are added primarily to packaged fruits, vegetables, and seafood to prevent discoloration. Sulfite-induced hypersensitivity is the most well-established adverse response to a food additive. Four to 8% of asthmatic patients develop sensitivity to sulfites.[46] The hepatic enzyme, sulfite oxidase, catalyzes the conversion of sulfites to inorganic sulfates. Some

TABLE 2.2. Food Storage Recommendations

Food	Fresh (Refrigerated)	Frozen
Beef		
Organs (Liver, Heart)	1 day	1–2 months
Ground	1–2 days	2–3 months
Chops	2–3 days	6–9 months
Steaks	2–3 days	6–9 months
Sausage	2–3 days	1–2 months
Roast	2–4 days	6–12 months
Lunch meats	4–6 days	
Chicken		
Meat	2–3 days	12 months
Liver	1–2 days	3 months
Eggs, fresh	4–5 weeks	Not recommended
Eggs, hard-boiled	1 week	Not recommended
Pork		
Chops	2–3 days	2–3 months
Ground	1–2 days	1–2 months
Roasts	2–4 days	3–6 months
Gravy		1 month
Bread		
Home-made		1 month
Commercial		2–3 months
Butter	1–2 weeks	
Cheese		
Cottage cheese	10–30 days	
Cream cheese (open)	2 weeks	
Processed cheeses	34 weeks	
Ice Cream		2–4 months
Lettuce		
Washed, drained	3–5 days	
Refrigerator crisper	5–7 days	
Mayonnaise	2–3 months in pantry, then refrigerate when open	
Milk	5 days	
Ketchup	12 months in pantry, then refrigerate when open	
Vegetables		
Fresh	1–7 days	
Frozen		8 months

Source: Adapted from Ref 42.

sulfite-sensitive patients lack sulfite oxidase, possibly leading to an accumulation of sulfite and subsequent conversion to sulfur dioxide, a known pulmonary irritant.[47] Some sulfite-sensitive patients have a positive skin test to sulfite, suggesting a relationship with an IgE-mediated response in selected patients. The clinical presentation is an anaphylactic or anaphylactoid reaction with acute bronchospasm. Sulfite sensitivity occurs more commonly in patients with asthma, but sulfite-induced symptoms may develop in patients without documented asthma or other allergy-related disorders. Symptoms include urticaria, angioedema, and IgE-mediated anaphylaxis.[48] Although the FDA banned the use of sulfites on raw foods, sulfites are permitted in processed foods. The presence of sulfites must be declared on food labels if the concentration exceeds 10 ppm, even though sulfite concentrations below 10 ppm in food may trigger reactions.[49] Sulfites are not permitted as additives to foods (enriched flour, produce) high in thiamine (vitamin B_1) because sulfites degrade thiamine.[50]

Nitrates and Nitrites

Nitrates and nitrites are added to hot dogs and other meat products as a method to prevent the growth of *Clostridium botulinum*. Nitrates are reduced to nitrites by enzymes in the mouth and gastrointestinal tract. Nitrites are known oxidative stressors that induce methemoglobinemia in a dose independent fashion.[51] Cases of methemoglobinemia from food exposures occur primarily in infants ingesting home-grown vegetables.[52] Nitrites combine with amine groups to form *N*-nitrosamines in foods that subsequently decompose into free-radical alkyl carbonium ions at physiologic pH. These radicals are mutagenic and carcinogenic in rodents, but studies do not support human carcinogenicity.[12] Although a case control study suggested that salivary nitrate and nitrite concentrations are higher in patients with oral cancer than in controls,[53] epidemiological studies to date have not demonstrated a statistically significant association between dietary exposure to nitrates or nitrites and gastric[54] or colon cancers.[55] In the absence of a definite causal link between nitrate or nitrate exposure and human cancer, the concern over *Clostridium botulinum* control supersedes proposals to reduce nitrate and nitrite additives in food production.[12]

Propionic Acid

Propionic acid is an antifungal agent in floury foods, such as cakes and pastries. The administration of feed containing 4% propionic acid to rats causes an excess of forestomach tumors, when administered during long-term studies.[56] However, the IARC and the US NTP do not list propionic acid as a human carcinogen.

Antibiotics

Antimicrobials including tetracycline, penicillin, and erythromycin are consumed by food-producing animals (e.g., cattle and poultry) to control pathogenic organisms during food production and to promote growth. This practice results in potential exposure of humans to antibiotic residues in meat, milk, and eggs; however, this level of exposure does not cause human toxicity.[57] A broader, potential risk to human health is the promotion of antibiotic-resistant bacteria by administration of low doses of antibiotics to animals. These bacteria can be transmitted to consumers in contaminated meat as well as in soil, air, and water, when animal manure is added to agricultural fields.[58] Resistant bacterial strains include food borne *Shigella dysenteriae, Escherichia coli, Campylobacter* spp., and *Salmonella* spp. In 1998, 5 patients were hospitalized during an outbreak of *Salmonella typhimurium* DT 104 following ingestion of contaminated pork in Denmark. The same strain of *Salmonella* was detected at a Danish slaughterhouse one month prior to the outbreak, based on molecular fingerprinting. The strain was resistant to ampicillin, chloramphenicol, streptomycin, sulfonamides, and tetracycline. Subsequently, the outbreak spread to 22 other victims, including a hospital employee and a previously hospitalized patient at the same facility.[59] Similar outbreaks have occurred in Canada and the United States.[60]

Stabilizers and Emulsifiers

Food stabilizers and emulsifiers include lecithin, gelatins, corn starch, waxes, gums, propylene glycol, and cation scavengers such as ethylenediaminetetraacetic acid (EDTA). These agents prevent the natural propensity of food to separate, thin, melt, and precipitate over time. Gums such as guar, carrageenan, cellulose, and xanthan are frequently used in small amounts to stabilize cheeses and to thicken salad dressings and ice cream.[61] In Quebec City, Canada a syndrome of "Beer Drinker's Cardiomyopathy" was described in 1966. Epidemiological analysis of this epidemic revealed numerous beer-drinking patients with congestive heart failure and pericardial effusion. The cause was multifactorial, but the addition of cobalt as a foam stabilizer by microbreweries probably contributed to the cardiotoxicity.[62]

Guar gum is a hydrophilic polysaccharide composed of galactose and mannose (galactomannan). The source of this non-cellulosic, dietary fiber is the endosperm of the Indian clusterbean [*Cyamopsis tetragonoloba* (L.) Taub.]. This bean is an animal food crop in India that contains about 20% guar gum by dry weight.[63] In the presence of water, guar gum forms a highly viscous gel that undergoes fermentation in the colon to short-chain fatty acids. The viscosity of this gel depends on the amount, concentration, and molecular weight of the guar gum. Guar gum is not a uniform product because of variability in the amount of galactomannan cross-linking between different plant sources. Although there is a lack of clinical data, potential uses for this dietary fiber include a dietary supplement for weight reduction, type 2 diabetes, and hypercholesterolemia.[64] Adverse effects result primarily from increased intraluminal pressures and the formation of gas following fermentation. These adverse effects include nausea, diarrhea, and flatulence.[65] Rarely, case reports associate the use of guar gum-containing products with anaphylaxis,[66] occupational asthma,[67] and allergic rhinitis.[68]

Although guar gum-containing tablets are intended to produce a gelatinous mass in the stomach that causes early satiety,[69] numerous case reports associate the use of these tablets with esophageal obstruction, particularly in patients with preexisting esophageal disease (e.g., strictures).[70,71] The use of guar gum in diet pills has been banned in several countries including the United States and Australia as a result of the association of esophageal and small bowel obstruction with the use of these tablets.[72] There is no clear evidence that the long-term use of guar gum causes the permanent weight reduction.[65]

Coloring Agents

Coloring agents are derived from multiple animal, vegetable, and mineral sources, and approximately 10% of all food consumed in the United States contain these compounds.[73] Coloring agents and dyes were commonplace long before experimental testing was performed to document safety. Food producers in the 1800s often used pigments containing heavy metals including arsenic, lead, mercury, thallium, and antimony.[74] Coloring agents are ineligible to appear on the GRAS list; thus, all coloring agents must obtain premarket approval from the FDA before their use in foods. The FDA names synthetic dyes by a food, drug, and cosmetic designation ("FD&C"), followed by a color, and either a number or letter. In the 1950s, the FDA received numerous reports involving the association of candies and popcorn colored with excessive amounts of FD&C Orange No. 1 with acute gastrointestinal distress in children. As a result, the Food, Drug, and Cosmetics Act of 1938 was amended by the Color Additives Amendment of 1960. This legislation allows the FDA to evaluate safety profiles of color additives and limits of all synthetic dyes before marketing. Each manufactured lot or batch of food color must be submitted to the FDA for certification because many food colors are synthesized from highly reactive azo compounds. This process ensures that manufactured lots of color do not contain nonreacted intermediates or toxic side-products from the synthesis of the food colors. FDA chemists evaluate every manufactured batch of color additives for impurities and contaminants that may be carried from their organic sources into the final product.

Each year, the FDA certifies almost 12 million pounds of synthetic color additives and dyes.[73] FDA-certifiable synthetic coloring agents include FD&C Blue No. 1, FD&C Blue No. 2, FD&C Green No. 3, FD&C Red No. 3, FD&C Red No. 40, FD&C Yellow No. 5, FD&C Yellow No. 6, Orange B, and Citrus Red No. 2. Most dyes are sulfonated derivatives of aromatic amines. Sulfonation confers a highly polar moiety to the additive, rendering it poorly absorbable by the gastrointestinal tract. As a result, acute toxicities are not generally observed unless the dyes are present in very large amounts.[75] Concern over dyes and human health effects began in 1959 when reports associated the occurrence of asthma, hypersensitivity, and urticaria with the synthetic aniline dye tartrazine (FD&C Yellow No. 5).[76] Reported symptoms also included migraine, blurred vision, itching, and rhinitis. A double-blind study documented an association between tartrazine and asthma exacerbation in young children,[77] but experimental trials on tartrazine have not conclusively identified a specific clinical syndrome following exposure tartrazine.[78] Since then, dyes have undergone intense evaluation.

Safrole, which occurs naturally in minimal quantities in cinnamon and other spices, was banned as a food additive in 1960 after hepatic and esophageal cancers developed in long-term studies of rats fed safrole. This compound is an alkylbenzene-derivative similar to thujone, which is found in absinthe and oil of wormwood. A reactive metabolite of safrole, 1-hydroxy sulfate ester, is the suspected carcinogen. In 1976, the FDA banned the sale of sassafras tea because of continued detection of safrole as an unavoidable harmful ingredient. Today, safrole-free extracts from sassafras are used as food additives for filé powder and root beer.[79] Amaranth (FD&C Red No. 2) was banned for its association with an increased incidence in embryotoxicity as well as renal pelvis hyperplasia and calcification.[80,81] Lipid-soluble formulations of erythrosine (FD&C Red No. 3) were banned because of the increased incidence of thyroid tumors in male rats fed this dye. Water-soluble derivatives of erythrosine are permitted in foods and oral medications because the gastrointestinal absorption of these compounds is very low.[82,83] FD&C Red No. 4, Green No. 1, Green No. 2, and Violet No. 1 were also banned because of concerns over animal carcinogenesis.[84] As a result of possible carcinogenicity, the use of Orange Number B is restricted to hot dog and sausage casings in concentrations up to 150 ppm of Orange B per food weight. Twenty color additives (e.g., caramel color, fruit juice, chlorophyll, turmeric, dried algae meal, beet juice, paprika) were exempted from FDA certification by the Color Additives Amendment of 1960 because they are extracted from natural sources.[85] Other exemptions include synthetic versions of naturally based compounds (e.g. beta-carotene).[86] Despite these exemptions, natural color additives may still produce adverse responses.[87] Carminic acid (red), annatto (orange), and saffron (yellow) are noncertified dyes that have been associated with IgE-mediated anaphylaxis, urticaria, and angioedema.[88,89]

DIAGNOSTIC TESTING

Sensitivity to food additives is a potential adverse response in patients with sensitivity to multiple, unrelated foods and with negative diagnostic workups for food protein allergies. Meticulous inspection of food labels is the initial approach to detect food sensitivities, but some ingredients are occasionally listed under alternate names. Recent FDA food recall notices are an important source of information. Systematic elimination of foods from the diet may assist the clinician in determining the offending food or additive.

Analytical Methods

Laboratory screening for quantitative IgE antibodies is a nonspecific method to support the presence of food sensitivity. Skin tests are available for coloring agents, sulfites, and other preservatives.[90] Double-blind, placebo-controlled provocative challenges with individual food additives may determine the presence of sensitivity, although challenge doses of most ingredients are not standardized. Provocative challenge tests are frequently used for confirmation of sulfite sensitivity.[91] Food may be analyzed for the presence of any additive; however, this practice is generally limited to epidemiological studies. Methemoglobin concentration is assessed by direct co-oximetry rather than the pulse oximetry used in most clinical settings because the latter method is unable to distinguish methemoglobin from oxyhemoglobin. There are various AOAC International (Association of Analytical Chemists International, Gaithersburg, MD; http://www.eoma.aoac.org/) official methods of analysis for intermediates and reaction side-products in specific FDA-approved food colors. For example, AOAC Method 968.17 is a gas chromatographic method for butylated hydroxytoluene and butylated hydroxyanisole in cereals. Cyclamates are detectable in beverages by solid-phase extraction and capillary electrophoresis with indirect ultraviolet detection, and the limits of detection are in the range of 10 μg/g.[92]

Biomarkers

For PKU patients, the long-term goal is to control diet to maintain blood phenylalanine concentrations between

2 and 10 mg/dL to prevent neurologic and neuromuscular sequelae.[93] The vast majority of other sensitivities and adverse responses to food additives are idiosyncratic and dose independent. Therefore, biomarkers for concentrations of additives in humans are not generally used for clinical purposes.

Abnormalities

Normal results of food additive skin tests or IgE concentrations do not rule out sensitivity.[90] Blood phenylalanine concentrations may exceed 100 mg/dL in acute exacerbations of PKU, although phenylalanine concentrations do not consistently correlate to the severity of toxicity. Methemoglobin concentration is reported as a percentage of total hemoglobin. Correlation of methemoglobin concentrations to cyanosis depends on the total amount of hemoglobin present.

TREATMENT

Patients with confirmed sensitivities to food additives should avoid foods with these additives under the guidance of an allergist and a clinical dietician. Newborns are screened for PKU routinely in every US state, and families with affected newborns are counseled to avoid excess intake of foods containing phenylalanine and related compounds.[94] Patients with sensitivities require education regarding the potential for unlisted ingredients and proper interpretation of food labels, and these patients should wear medical alert identification tags at all times. Antihistamines are adequate to treat mild hypersensitivity symptoms such as itching, urticaria, and rhinitis. Certain patients with mild, predictable reactions may benefit from daily prophylaxis or pretreatment with antihistamines to tolerate offending food ingredients.[95] Patients with recalcitrant or persistent symptoms may benefit from corticosteroid therapy under the guidance of an allergist. Bronchodilators are used to treat bronchospasm and related pulmonary symptoms. Asthmatic patients with food sensitivity should carry rescue inhalers at all times; patients with history of severe responses may also need to carry rapid self-injectable epinephrine devices.[96] Decontamination of a patient by cleansing the lips and hands for residual deposits of the offending food is a reasonable approach, but there are inadequate epidemiologic data to demonstrate that this practice improves clinical outcome. Anaphylaxis and anaphylactoid reactions are treated by standard advanced life support protocols.[97] Methemoglobinemia is treated with supportive care, supplemental oxygen, and methylene blue in moderate-to-severe cases. Desensitizing immunotherapy has not been evaluated for food additive sensitivity.

References

1. Food and Drug Administration (FDA) (2005). Milestones in U.S. food and drug law history. Available at http://www.fda.gov/opacom/backgrounders/miles.html. Accessed February 17, 2008.
2. Food and Drug Administration (FDA). (2008). EAFUS: a food additive database. Available at http://www.cfsan.fda.gov/~dms/eafus.html. Accessed February 17, 2008.
3. Food and Drug Administration (FDA). (2008). Numerical listing of GRAS notices. Available at http://www.cfsan.fda.gov/~rdb/opa-gras.html. Accessed February 17, 2008.
4. Cheetham PS. Bioprocesses for the manufacture of ingredients for foods and cosmetics. Adv Biochem Eng Biotechnol 2004;86:83–158.
5. Joint FAO/WHO Expert Committee on Food Additives. Evaluation of certain food additives and contaminants. World Health Organ Tech Rep Ser 2004;922:1–176.
6. Djagny VB, Wang Z, Xu S. Gelatin: a valuable protein for food and pharmaceutical industries. Crit Rev Food Sci Nutr 2001;41:481–492.
7. Lucas CD, Hallagan JB, Taylor SL. The role of natural color additives in food allergy. Adv Food Nutr Res 2001;43:195–216.
8. McKnight GM, Duncan CW, Leifert C, Golden MH. Dietary nitrate in man: friend or foe? Br J Nutr 1999;81:349–358.
9. Renwick AG. Acceptable daily intake and the regulation of intense sweeteners. Food Addit Contam 1990; 7:463–475.
10. Truhaut R. The concept of the acceptable daily intake: an historical review. Food Addit Contam 1991;8:151–162.
11. Yang WH, Drouin MA, Herbert M, Mao Y, Karsh J. The monosodium glutamate symptom complex: assessment in double-blind, placebo-controlled, randomized study. J Aller Clin Immunol 1997;99:757–762.
12. Eichholzer M, Gutzwiller F. Dietary nitrates, nitrites, and N-nitroso compounds and cancer risk: a review of the epidemiologic evidence. Nutr Rev 1998;56:95–105
13. Madsen C: Prevalence of food additive intolerance. Hum Exp Toxicol 1994;13:393–399.
14. Fuglsang G, Madsen G, Halken S, Jorgensen S, Ostergaard PA, Osterballe O. Adverse reactions to food additives in children with atopic symptoms. Allergy 1994;49: 31–37.
15. Kwok RH. Chinese-restaurant syndrome. N Engl J Med 1968;278:796.
16. Geha RS, Beiser A, Ren C, Patterson R, Greenberger PA, Grammer LC, et al. Review of alleged reaction to monosodium glutamate and outcome of a multicenter double-blind placebo-controlled study. J Nutr 2000 (45 Suppl);130:1058S–1062S.
17. Tarasoff L, Kelly MF. Monosodium L-glutamate: a double-blind study and review. Food Chem Toxicol 1993; 31:1019–1035.

18. Woessner KM, Simon RA, Stevenson DD. Monosodium glutamate sensitivity in asthma. J Allergy Clin Immunol 1999;104:305–310.
19. Stricker-Krongrad A, Burlet C, Beck B. Behavioral deficits in monosodium glutamate rats: specific changes in structure and behavior. Life Sci 1998;62:2127–2132.
20. Erlandsen H, Patch MG, Gamez A, Straub M, Stevens RC. Structural studies on phenylalanine hydroxylase and implications toward understanding and treating phenylketonuria. Pediatrics 2003;112:1557–1565.
21. Northam EA. Neuropsychological and psychosocial correlates of endocrine and metabolic disorders—a review. J Pediatr Endocrinol Metab 2004;17:5–15.
22. Glushakov AV, Glushakova O, Varshney M, Bajpai LK, Sumners C, Laipis PJ, et al. Long-term changes in glutamatergic synaptic transmission in phenylketonuria. Brain 2005;128:300–307.
23. Butchko HH, Stargel WW, Comer CP, Mayhew DA, Benninger C, Blackburn GL, et al. Aspartame: review of safety. Regul Toxicol Pharmacol 2002;35:S1–S93.
24. Weihrauch MR, Diehl V. Artificial sweeteners—do they bear a carcinogenic risk? Ann Oncol 2004;15:1460–1465.
25. Ahmed FE, Thomas DB. Assessment of the carcinogenicity of the non-nutritive sweetener cyclamate. Crit Rev Toxicol 1992;22:81–118.
26. Kullman G, Boylstein R, Jones W, Piacitelli C, Pendergrass S, Kreiss K. Characterization of respiratory exposure at a microwave popcorn plant with cases of bronchiolitis obliterans. J Occup Environ Hyg 2005;2:169–178.
27. Centers for Disease Control and Prevention (CDC). Fixed obstructive lung disease among workers in the flavor-manufacturing industry—California, 2004–2007. MMWR Morb Mortal Wkly Rep 2007;56:389–393.
28. van Rooy FG, Rooyackers JM, Prokop M, Houba R, Smit LA, Heederik DJ. Bronchiolitis obliterans syndrome in chemical workers producing diacetyl for food flavorings. Am J Respir Crit Care Med 2007;176:498-504.
29. Kreiss K. Occupational bronchiolitis obliterans masquerading as COPD. Am J Respir Crit Care Med 2007; 176:427-429.
30. Hubbs AF, Battelli LA, Goldsmith WT, Porter DW, Frazer D, Friend S, et al. Necrosis of nasal and airway epithelium in rats inhaling vapors of artificial butter flavoring. Toxicol Appl Pharmacol 2002;185:128–135.
31. Harber P, Saechao K, Boomus C. Diacetyl-induced lung disease. Toxicol Rev 2006;25:261–272.
32. Daniel JW. Metabolic aspects of antioxidants and preservatives. Xenobiotica 1986;16:1073–1078.
33. Reddy AC, Lokesh BR. Studies on spice principles as antioxidants in the inhibition of lipid peroxidation of rat liver microsomes. Mol Cell Biochem 1992;111:117–124.
34. Goodman DL, McDonnell JT, Nelson HS, Vaughan TR, Weber RW. Chronic urticaria exacerbated by the antioxidant food preservatives, butylated hydroxyanisole (BHA) and butylated hydroxytoluene (BHT). J Allergy Clin Immunol 1990;86:570–575.
35. Shlian DM, Goldstone J. Toxicity of butylated hydroxytoluene. N Engl J Med 1986;314:648–649.
36. Grogan WA. Toxicity from BHT ingestion. West J Med 1986;145:245–246.
37. Altmann HJ, Grunow W, Mohr U, Richter-Reichhelm HB, Wester PW. Effects of BHA and related phenols on the forestomach of rats. Food Chem Toxicol 1986;24:1183–1188.
38. van der Heijden CA, Janssen PJ, Strik JJ.: Toxicology of gallates: a review and evaluation. Food Chem Toxicol 1986;24:1067–1070.
39. International Programme on Chemical Safety. Concise international chemical assessment document No. 26 benzoic acid and sodium benzoate. Geneva: World Health Organization; 2000.
40. Feillet F, Leonard JV. Alternative pathway therapy for urea cycle disorders. J Inherit Metab Dis 1998;21(suppl 1):101–111.
41. Kubota K, Ishizaki T. Effect of single oral dose of sodium benzoate on ureagenesis in healthy men and two patients with late onset citrullinaemia. Eur J Clin Pharmacol 1993;45:465–468.
42. Miller RW. The ABC's of food storage. FDA Consumer 1986;20(2):12–14.
43. Parke DX, Lewis DF. Safety aspects of food preservatives. Food Addit Contam 1992;9:561–577.
44. Walker R. Toxicology of sorbic acid and sorbates. Food Addit Contam 1990;7:671–676.
45. Gardner LK, Lawrence GD. Benzene production from decarboxylation of benzoic acid in the presence of ascorbic acid and a transition-metal catalyst. J Agric Food Chem 1993;41:693–695.
46. Taylor SL, Bush RK, Selner JC, Nordlee JA, Wiener MB, Holden K, et al. Sensitivity to sulfited foods among sulfite-sensitive subjects with asthma. J Allergy Clin Immunol 1988;81:1159–1167.
47. Gunnison AF, Jacobsen DW. Sulfite hypersensitivity. a critical review. CRC Crit Rev Toxicol 1987;17:185–214.
48. Sokol WN, Hydick IB. Nasal congestion, urticaria, and angioedema caused by an IgE-mediated reaction to sodium metabisulfites. Ann Allergy 1990;65:233–237.
49. Taylor SL, Hefle SL. Ingredient and labeling issues associated with allergenic foods. Allergy 2001;56:65–69.
50. Stammati A, Zanetti C, Pizzoferrato L, Quattrucci E, Tranquilli GB. *In vitro* model for the evaluation of toxicity and antinutritional effects of sulphites. Food Addit Contam 1992;9:551–560.
51. Chan TY. Food-borne nitrates and nitrites as a cause of methemoglobinemia. Southeast Asian J Trop Med Public Health 1996;27:189–192.
52. Sanchez-Echaniz J, Benito-Fernandez J, Mintegui-Raso S. Methemoglobinemia and consumption of vegetables in infants. Pediatrics 2001;107:1024–1028.
53. Badawi AF, Hosny G, el-Hadary M, Mostafa MH. Salivary nitrate, nitrate and nitrite reductase activity in relation to risk of oral cancer in Egypt. Dis Markers 1998;14:91–97.

54. van Leon AJ, Botterweck AA, Goldbohm RA, Brants HA, Van Klaveren JD, Van den Brandt PA. Intake of nitrate and nitrite and the risk of gastric cancer: a prospective cohort study. Br J Cancer 1998;78:129–135.
55. Knekt P, Jarvinen R, Dich J, Halkulinen T. Risk of colorectal and other gastro-intestinal cancers after exposure to nitrate, nitrite and N-nitroso compounds: a follow-up study. In J Cancer 1999;80:552–856.
56. Harrison PT. Propionic acid and the phenomenon of rodent forestomach tumorigenesis: a review. Food Chem Toxicol 1992;30:333–340.
57. Lathers CM. Clinical pharmacology of antimicrobial use in humans and animals. J Clin Pharmacol 2002;42:587–600.
58. White DG, Zhao S, Singh R, McDermott PF. Antimicrobial resistance among gram-negative foodborne bacterial pathogens associated with foods of animal origin. Foodborne Pathog Dis 2004;1:137–152.
59. Molbak K, Baggesen DL, Aarestrup FM, Ebbesen JM, Engberg J, Frydendahl K, et al. An outbreak of multidrug-resistant, quinolone-resistant *Salmonella enterica* serotype *typhimurium* DT104. N Engl J Med 1999;341:1420–1425.
60. Helms M, Ethelberg S, Molbak K. DT104 Study Group: International *Salmonella typhimurium* DT104 infections, 1992–2001. Emerg Infect Dis 2005;11:859–867.
61. Wilde P, Mackie A, Husband F, Gunning P, Morris V. Proteins and emulsifiers at liquid interfaces. Adv Colloid Interface Sci 2004;108–109:63–71.
62. Seghizzi P, D'Adda F, Borleri D, Barbic F, Mosconi G. Cobalt myocardiopathy. a critical review of literature. Sci Total Environ 1994;150:105–109.
63. Singh SP, Misra BK. Lipids of guar seed meal (*Cyamopsis tetragonoloba* L. Taub.). J Agric Food Chem 1981; 29:907–909.
64. Todd PA, Benfield P, Goa KL. Guar gum a review of its pharmacological properties, and use as a dietary adjunct in hypercholesterolaemia. Drugs 1990;39:917–928.
65. Pittler MH, Ernst E. Guar gum for body weight reduction: meta-analysis of randomized trials. Am J Med 2001; 110:724–730.
66. Roesch A, Haegele T, Vogt T, Babilas P, Landthaler M, Szeimies R-M. Severe contact urticaria to guar gum included as gelling agent in a local anaesthetic. Contact Dermatitis 2005;52:307–308.
67. Lagier F, Cartier A, Somer J, Dolovich J, Malo JL. Occupational asthma caused by guar gum. J Allergy Clin Immunol 1990;85:735–790.
68. Kanerva L, Tupasela O, Jolanki R, Vaheri E, Estlander T, Keskinen H. Occupational allergic rhinitis from guar gum. Clin Allergy 1988;18:245–252.
69. Pittler MH, Schmidt K, Ernst E. Adverse events of herbal food supplements for body weight reduction: systematic review. Obes Rev 2005;6:93–111.
70. Opper FH, Isaacs KL, Warshauer DM. Esophageal obstruction with a dietary fiber product designed for weight reduction. J Clin Gastroenterol 1990;12:667–669.
71. Seidner DL, Roberts IM, Smith MS. Esophageal obstruction after ingestion of a fiber-containing diet pill. Gastroenterology 1990;99:1820–1822.
72. Lewis JH. Esophageal and small bowel obstruction from guar gum-containing "diet pills": analysis of 26 cases reported to the Food and Drug Administration. Am J Gastroenterol 1992;87:1424–1428.
73. Green S. Research activities in toxicology and toxicological requirements of the FDA for food and color additives. Biomed Environ Sci 1988;1:424–430.
74. Drake JJ. Food colours–harmless aesthetics or epicurean luxuries? Toxicology 1975;5:3–42.
75. Chung KT, Cerniglia CE. Mutagenicity of azo dyes: structure-activity relationships. Mutat Res 1992;277:201–20.
76. Lockey SD. Allergic reactions due o FD&C yellow no. 5 tartrazine, an aniline dye used as a coloring and identifying agent in various studies. Ann Allergy 1959;17:719–725.
77. Hariparsad D, Wilson N, Dixon C, Silverman M. Oral tartrazine challenge in childhood asthma: effect on bronchial reactivity. Clin Allergy 1984;14:81–85.
78. Ram FS, Ardern KD. Tartrazine exclusion for allergic asthma. Cochrane Database Syst Rev 2001;4:CD000460.
79. Ioannides C, Delaforge M, Parke DV. Safrole: its metabolism, carcinogenicity and interactions with cytochrome P-450. Food Cosmet Toxicol 1981;19:657–666
80. Collins TF, Keeler HV, Black TN, Ruggles DI. Long-term effects of dietary amaranth in rats. I. Effects on reproduction. Toxicology 1975;3:115–128.
81. Clode SA, Hooson J, Grant D, Butler WH. Long-term toxicity study of amaranth in rats using animals exposed in utero. Food Chem Toxicol 1987;25:937–946.
82. Lakdawalla AA, Netrawali MS. Mutagenicity, comutagenicity, and antimutagenicity of erythrosine (FD and C red 3), a food dye, in the Ames/Salmonella assay. Mutat Res 1988;204:131–139.
83. Borzelleca JF, Capen CC, Hallagan JB. Lifetime toxicity/carcinogenicity study of FD & C Red No. 3 (erythrosine) in mice. Food Chem Toxicol 1987;25:735–737.
84. Palmer S, Mathews RA. The role of non-nutritive dietary constituents in carcinogenesis. Surg Clin North Am 1986;66:891–915.
85. Hallagan JB, Allen DC, Borzelleca JF. The safety and regulatory status of food, drug and cosmetics colour additives exempt from certification. Food Chem Toxicol 1995;33:515–528.
86. Vandamme EJ. Production of vitamins, coenzymes and related biochemicals by biotechnological processes. J Chem Technol Biotechnol 1992;53:313–327.
87. Lucas CD, Hallagan JB, Taylor SL. The role of natural color additives in food allergy. Adv Food Nutr Res 2001;43:195–216.
88. DiCello MC, Myc A, Baker JR Jr, Baldwin JL. Anaphylaxis after ingestion of carmine colored foods: two case reports and a review of the literature. Allergy Asthma Proc 1999;20:377–382.

89. Moneret-Vautrin DA, Morisset M, Lemerdy P, Croizier A, Kanny G. Food allergy and IgE sensitization caused by spices: CICBAA data (based on 589 cases of food allergy). Allerg Immunol (Paris) 2002;34:135–140.
90. Gordon BR. Approaches to testing for food and chemical sensitivities. Otolaryngol Clin North Am 2003;36:917–940.
91. Simon RA. Sulfite challenge for the diagnosis of sensitivity. Allergy Proc 1989;10:357–362.
92. Horie M, Ishikawa F, Oishi M, Shindo T, Yasui A, Ito K. Rapid determination of cyclamate in foods by solid-phase extraction and capillary electrophoresis. J Chromatogr A 2007;1154:423–428.
93. National Institutes of Health Consensus Development Panel. National Institutes of Health Consensus Development Conference Statement: phenylketonuria: screening and management, October 16–18, 2000. Pediatrics 2001;108:972–982.
94. Collins JE, Leonard JV. The dietary management of inborn errors of metabolism. Hum Nutr Appl Nutr 1985;39: 255–272.
95. Fogg MI, Spergel JM. Management of food allergies. Expert Opin Pharmacother 2003;4:1025–1037.
96. Sampson HA. Food anaphylaxis. Br Med Bull 2000;56: 925–935.
97. Tang AW. A practical guide to anaphylaxis. Am Fam Physician 2003;68:1325–1332.

II Staples and Spices

Chapter 3

AKEE FRUIT and JAMAICAN VOMITING SICKNESS (*Blighia sapida* Köenig)

HISTORY

Blighia sapida Köenig is a native plant of West Africa that was introduced to Jamaica in 1776 by Thomas Clark.[1] The infamous Captain Bligh took the breadfruit tree to the West Indies.[2] Toxicity has been associated with the consumption of unripe akee fruit since the late 19th century.[3] Between 1880 and 1955, a variety of case reports associated the ingestion of akee fruit with vomiting, generalized weakness, altered consciousness, and death including an epidemic of 151 cases and 32 deaths during 1954 near Montego Bay, Jamaica.[1] Hassal et al first isolated the two toxic constituents (hypoglycin A and B) from the arils and seeds of unripe akee in 1954.[4] In 1976, Tanaka et al confirmed the ingestion of hypoglycin A from unripe akee fruit as the cause of Jamaican vomiting sickness.[5]

BOTANICAL DESCRIPTION

Common Name: Akee tree

Scientific Name: *Blighia sapida* Köenig

Botanical Family: Sapindaceae (soapberry)

Physical Description: The akee tree is a tropical evergreen tree with broad, pinnate leaves. The tree bears an oblong-shaped, yellow-red fruit that is about 10 cm (4 in.) wide and weighs 100 g. As the akee fruit ripens, the color of the fruit changes from green to yellow, to yellow-red, and then to red. When ripe, the fruit splits longitudinally into three sections to reveal glassy black seeds in each section surrounded by a thick oily aril.

Distribution and Ecology: The akee tree is a native species of West Africa, which was introduced into Jamaica and has naturalized in southern California and Florida, as well as the Antilles, and Central America. Consumption of the akee fruit is uncommon outside these areas with the exception of the illicit importation of this fruit into countries with a ban on akee fruit. Figure 3.1 displays a specimen of the akee fruit.

EXPOSURE

The ripe fleshy akee aril is a food staple among poor Jamaican and West African children. The akee is the national fruit of Jamaica. Although akee fruit is banned from commerce in the United States, canned akee fruit is available in the Caribbean Islands.[6] To prevent toxicity, the seeds and membrane of the akee fruit must be carefully removed and the aril thoroughly washed before consumption.

PRINCIPAL TOXINS

Physiochemical Properties

Ingestion of unripe akee fruit causes Jamaican vomiting sickness. The unripe fruit contain a water-soluble toxin, hypoglycin A (L-α-aminomethylenecyclopropylpropionic acid) and the less toxic hypoglycin B. The latter compound is the γ-glutamyl conjugate of hypoglycin A. Unripe akee fruit also contain glutamate analogs that are carboxycyclopropylglycine compounds.[7] Cooking *unripe* akee fruit does not destroy the toxins in these

FIGURE 3.1. Akee fruit. Photograph courtesy of Stephan Russ. See color insert.

fruit, whereas cooking effectively eliminates the toxicity of the *ripe* fruit by leaching hypoglycin A from the ripe fruit.[6,8]

Poisonous Parts

The akee fruit and the white aril surrounding the seeds are edible only when the fruit spontaneously opens (i.e., stage 8 of the 10-stage maturity scale assigned to akee fruit).[9] This maturity scale ranges from stage 1 (blossom development with very small fruit) to stage 10 (the onset of spoilage with decay of the aril). Full maturity (i.e., stage 8) of the akee fruit occurs when the red or yellow pod is completely open with the seed and aril clearly exposed. The concentration of hypoglycin A is substantially higher in unripe than ripe akee fruit, and the hypoglycin A content of the aril decreases substantially as the fruit ripens. In uncooked samples of akee fruit, the hypoglycin A content in unripe and ripe fruit was approximately 124 ± 7 mg/100 g fresh weight and about 6 ± 1 mg/100 g fresh weight, respectively.[10] In another study, the concentration of hypoglycin A decreased from 711 mg/100 g in raw, unripe fruit to below the limits of detection (1.2 mg/100 g) in the ripe aril, as measured by ion-exchange chromatography.[11] The concentration of hypoglycin A in the membrane and aril is similar as the fruit matures, but the membrane contains detectable amounts (i.e., <40 ppm) of hypoglycin A even at the edible stage.[9] The ripe aril does not usually contain detectable amounts of hypoglycin B.

The concentration of hypoglycin A remains relatively constant in the seeds until maturation. The husk and seeds of the fresh akee fruit retain significant concentrations of hypoglycin A as the fruit matures.[12] The seeds contain several times higher concentrations of hypoglycin A than unripe akee aril. In a study of akee fruit from a Florida subtropical horticulture research station, the hypoglycin A content decreased from 939 mg/100 g in unripe seeds to 269 mg/100 g in ripe seeds, whereas the hypoglycin A content in the husk remained near 40 mg/100 g.[11] The hypoglycin A content in canned akee fruit is variable. In six samples of commercially canned akee fruit, the hypoglycin A content in the drained solid or edible portion of the six samples ranged between 11 mg and 66 mg with a mean of 35.6 mg.[13] The mean weight of the solid portion of the six samples was approximately 378 ± 13 g. The liquid portion of the samples accounted for about one-third of the weight of the contents, and the mean hypoglycin A content in the liquid portion of the six cans was 20.4 g with a range between 7.8 g and 38.7 g. Akee oil is the lipid fraction of the unripe akee aril and the yield of akee oil is >60% wet weight.

Mechanism of Toxicity

Metabolism of hypoglycin A produces methylenecyclopropylacetic acid, which reduces several cofactors (e.g., coenzyme A, carnitine) essential to the β-oxidation of long-chain fatty acids and inhibits the transport of long-chain fatty acids into the mitochondria.[14,15] Consequently, fatty acids conjugated with carnitines (e.g., octanoylcarnitine) accumulate in the serum and urine. Accumulation of short-chain fatty acids (e.g., propionic, *n*-butyric, isovaleric acid) in the serum results from the suppression of short-chain acyl-CoA dehydrogenases and β-oxidation. Long-chain fatty acids undergo omega-oxidation by mixed function oxidases in the endoplasmic reticulum of the liver as a result of the impairment of β-oxidation, and accumulation of short- and medium-chain dicarboxylic acids (e.g., glutaric, ethylmalonic acid) occurs in the urine. The reduction in fatty acid metabolism causes an increased use of glucose, and the blockade of the substrate for hepatic gluconeogenesis causes hypoglycemia after the depletion of NADH and hepatic glycogen stores. However, hypoglycemia does not explain the entire clinical picture because vomiting and central nervous system (CNS) depression occur during intoxication without the presence of hypoglycemia.[16] Although the increased concentration of short-chain fatty acids (e.g., isovaleric, 1-methylbutyric acids) may contribute to the metabolic acidosis present during severe akee poisoning, the concentrations of these fatty acids are too low to cause lethargy and coma.[5] The potential involvement of glutamate analogs (i.e., carboxycyclopropylglycine compounds) may account for some of the CNS effects present during Jamaican vomiting sickness.[17]

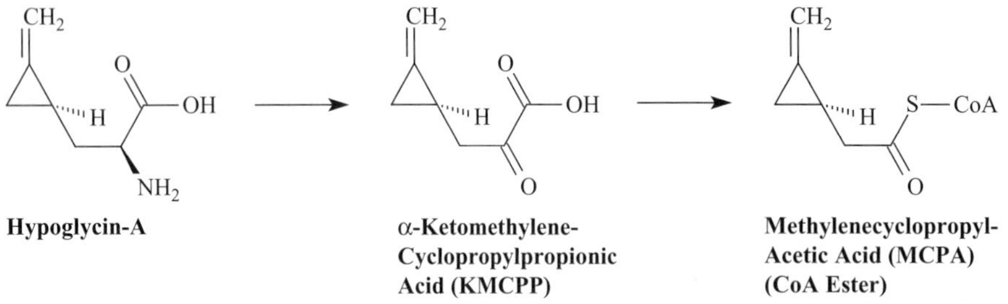

Hypoglycin-A → α-Ketomethylene-Cyclopropylpropionic Acid (KMCPP) → Methylenecyclopropyl-Acetic Acid (MCPA) (CoA Ester)

FIGURE 3.2. Metabolism of hypoglycin to the toxic metabolite.

Pathophysiology

The intraperitoneal administration of the lipid portion of unripe akee fruit (i.e., akee oil) to rats produces petechial hemorrhages and perivascular mononuclear infiltrates in the lungs, probably as a result of a hypersensitivity reaction.[18] Similar reactions in humans following the consumption of akee fruit have not been documented in the medical literature. Findings upon postmortem examination of fatal akee fruit intoxications include fatty metamorphosis of the liver and kidney, depletion of liver glycogen, diffuse petechial hemorrhages, and generalized hyperemia of the internal organs.[2] Pathological changes in the liver are similar to the histological abnormalities associated with Reye's syndrome and impila toxicity.[19]

Toxicokinetics

The metabolic pathway for the biotransformation of hypoglycin A and other branched-chain amino acids is similar. Metabolism of hypoglycin A produces the toxic metabolite, methylenecyclopropylacetic acid (MCPA-CoA), which inhibits the oxidation of long-chain fatty acids leading to hypoglycemia. α-Ketomethylenecyclopropylpropionic acid is an intermediate metabolite of hypoglycin A as demonstrated by Figure 3.2. The kidneys excrete un-reacted MCPA-CoA as methylenecyclopropyl-acetyl-glycine.[20] Urine specimens from two children with akee fruit poisoning contained substantial amounts of methylenecyclopropylacetic acid as well as the medium-chain dicarboxylic acids, 2-ethylmalonate, adipate, and glutarate.[21]

DOSE RESPONSE

The ingestion of 12 and 24 uncooked, raw akee fruit by two adults produced vomiting and drowsiness that progressed to mild hypoglycemia and coma.[8] Within 2–3 days after ingestion, respiratory depression required endotracheal intubation for respiratory support, but within about one week recovery was complete. In a study of male and female rats, the acute toxic dose for male and female rats was 231.19 ± 62.5 5 mg hypoglycin A/kg body weight and 215.99 ± 63.33 mg hypoglycin A/kg body weight, respectively.[22] This study used the World Health Organization definition of acute akee toxicity (impairment of movement, raised hairs, arched backs, 10% loss of body weight, and black feces). The maximum tolerated daily dose of hypoglycin A over the 30-day study was 1.50 ± 0.07 mg hypoglycin A/kg body weight.

CLINICAL RESPONSE

The syndrome of Jamaican vomiting sickness is similar to Reye's syndrome and is characterized by gastrointestinal distress, hypoglycemia, and CNS depression.[23] Symptoms typically develop within 6–48 hours after ingestion of unripe akee fruit with symptoms occasionally occurring within 2 hours of ingestion.[24] Hypotonia and seizures occur commonly in severe cases of Jamaican vomiting sickness, particularly in young children. Nausea, vomiting, and abdominal pain are usually, but not always, the initial symptoms. These gastrointestinal symptoms appear within several hours after ingestion of unripe akee fruit.[25] Other initial symptoms of akee poisoning include headache, thirst, and paresthesia.[6] Lethargy, hypotonia, hypothermia, convulsions, and coma occur after a brief quiescent period followed in some cases by death up to 48 hours after ingestion. Fever usually is absent and death is more common in children than in adults.[26] A case report suggests that chronic exposure to unripe akee fruit produces cholestatic jaundice characterized by jaundice, pruritus, diarrhea, and right upper quadrant abdominal pain.[27] However, the liver biopsy demonstrated cholestasis and centrilobular necrosis, which are atypical of histological changes associated with Jamaican vomiting sickness. Rarely, case reports associate the development of hypersensitivity reactions (urticaria, angioedema, chest tightness, brief syncope) with the ingestion of akee fruit by patients with known histories of food allergies or anaphylaxis.[28]

DIAGNOSTIC TESTING

Analytical Methods

Methods for the determination of hypoglycin A content of akee fruit include ion exchange chromatography (IEC)[13] and reversed phase high performance liquid chromatography with UV detection at 254 nm (RP/HPLC/UV).[29] The precision and accuracy of HPLC/UV is sufficient to determine hypoglycin A concentrations in a range between 50 µg/g and 400 µg/g akee fruit, whereas the limit of detection for IEC is about 5 µg/mL.

Biomarkers

The metabolism of hypoglycin A is rapid, and therefore this compound is not usually detectable in biological samples from affected patients. The presence of the metabolite of hypoglycin A (methylenecyclopropylacetic acid) is a biomarker of exposure to akee fruit.[5] The inhibition of acetyl CoA enzymes increases serum carboxylic acids and the renal excretion of dicarboxylic acids with 5–10 carbons (propionic, isobutyric, *n*-butyric, isovaleric, *n*-hexanoic, glutaric, adipic, suberic, sebacic). Adipic acid and lactic acid occur commonly in urine samples of patients poisoned with unripe akee fruit.[8] Although the hypoglycin A concentration is usually nondetectable in postmortem samples from patients dying of Jamaican vomiting sickness, urinary concentrations of organic acids are typically elevated including adipic acid, glutaric acid, and suberic acid.[24] Increased urine concentrations of glutaric and ethylmalonic acids occur only after ingestion of hypoglycin A and a rare genetic disorder (type II glutaric aciduria).[30] Additionally, the concentrations of some carnitine compounds are elevated in serum and urine samples from affected patients, particularly octanoylcarnitine and the metabolite, hexanoylcarnitine.[24]

Abnormalities

In addition to severe hypoglycemia, serious intoxication laboratory abnormalities include metabolic acidosis, electrolyte imbalance (e.g., hypokalemia), and laboratory evidence of dehydration (e.g., elevated serum BUN to creatinine ratio).

TREATMENT

Decontamination measures are usually unnecessary because of the prominence of vomiting and the late presentation (i.e., >1 hour after ingestion). All symptomatic patients should receive measurements of serum glucose, serum electrolytes, and arterial pH. The presence of any degree of obtundation mandates immediate bedside analysis of the serum glucose concentration and the intravenous administration of glucose as needed. The serum glucose should be monitored serially, particularly when the mental status deteriorates, along with serum hepatic aminotransferases and serum creatinine. Treatment is primarily supportive with restoration of fluid, electrolyte, glucose, and pH balance. Experimental studies suggest that glycine[31–32] and methylene blue[33] are potential treatments for akee fruit poisoning, but there are no clinical data to support the efficacy of these agents during akee fruit poisoning, particularly when administered more than several hours after ingestion. An *in vitro* study of rat hepatocytes did not support the use of carnitine as an antidote for the inhibition of multiple acyl-CoA dehydrogenase enzymes following the consumption of unripe akee fruit.[34]

References

1. Feng PC. Hypoglycin–from ackee: a review. West Indian Med J 1969;18:238–243.
2. Hill KR. The vomiting sickness of Jamaica. A review. West Indian Med J 1952;1:243–264.
3. Scott H. On the vomiting sickness of Jamaica. Ann Trop Med Parasitol 1916;10:1–78.
4. Hassall CH, Reyle K. Hypoglycin A,B: biologically active polypeptides from *Blighia sapida*. Nature 1954; 173:356–357.
5. Tanaka K, Kean EA, Johnson B. Jamaican vomiting sickness: Biochemical investigation of two cases. N Engl J Med 1976;295:461–467.
6. McTague JA, Forney R Jr. Jamaican vomiting sickness in Toledo, Ohio. Ann Emerg Med 1994;23:1116–1118.
7. Natalini B, Capodiferro V, De Luca C, Espinal R. Isolation of pure (2S,1′S,2′S)-2-(2′-carboxycyclopropyl)glycine from *Blighia sapida* (akee). J Chromatogr A 2000;873: 283–286.
8. Golden KD, Kean EA, Terry SI. Jamaican vomiting sickness: a study of two adult cases. Clin Chim Acta 1984;142:293–298.
9. Brown M, Bates RP, McGowan C, Cornell JA. Influence of fruit maturity on the hypoglycin A level in ackee (*Blighia sapida*). J Food Safety 1992;12:167–177.
10. Golden KD, Williams OJ, Bailey-Shaw Y. High-performance liquid chromatographic analysis of amino acids in ackee fruit with emphasis on the toxic amino acid hypoglycin A. J Chromatogr Sci 2003;40:441–446.
11. Chase GW Jr, Landen WO Jr, Soliman AG. Hypoglycin A content in the aril, seeds, and husks of ackee fruit at various stages of ripeness. J Assoc Off Anal Chem 1990;73:318–319.

12. Manchester KL. Biochemistry of hypoglycin. FEBS Lett 1974;23;40(suppl):S133–S139.
13. Chase GW Jr, Landen WO Jr, Gelbaum LT, Soliman AG. Ion-exchange chromatographic determination of hypoglycin A in canned ackee fruit. J Assoc Off Anal Chem 1989;72:374–377.
14. Bressler R, Corredor C, Brendel K. Hypoglycin and hypoglycinlike compounds. Pharmacol Rev 1969;21:105–130.
15. Wenz A, Thorpe C, Ghisla S. Inactivation of general acyl-CoA dehydrogenase from pig kidney by a metabolite of hypoglycin A. J Biol Chem 1981;256:9809–9812.
16. Bressler R. The unripe akee-forbidden fruit. N Engl J Med 1976;295:500–501.
17. Kean EA. Commentary on a review on the mechanism of ackee-induced vomiting sickness. West Indian Med J 1988;3:139–142.
18. Singh P, Gardner M, Poddar S, Choo-Kang E, Coard K, Rickards E. Toxic effects of ackee oil (*Blighia sapida* L) following subacute administration to rats. West Indian Med J 1992;41:23–26.
19. Hautekeete ML, Degott C, Benhamou JP. Microvesicular steatosis of the liver. Acta Clin Belgica 1990;45:311–326.
20. Tanaka K, Isselbacher KJ, Shih V. Isovaleric and α-methylbutyric acidemias induced by hypoglycin A: mechanism of Jamaican vomiting sickness. Science 1972;175: 69–71.
21. Tanaka K, Ikeda Y. Hypoglycin and Jamaican vomiting sickness. Prog Clin Biol Res 1990;321:167–184.
22. Blake OA, Bennink MR, Jackson JC. Ackee (*Blighia sapida*) hypoglycin A toxicity: dose response assessment in laboratory rats. Food Chem Toxicol 2006;44:207–213.
23. Moya J. Ackee (*Blighia sapida*) poisoning in the Northern Province, Haiti, 2001. Epidemiol Bull 2001;22:8–9.
24. Joskow R, Belson M, Vesper H, Backer L, Rubin C. Ackee fruit poisoning: an outbreak investigation in Haiti 2000–2001, and review of the literature. Clin Toxicol 2006;44:267–273.
25. Jellife DB, Stuart KL. Acute toxic hypoglycemia in the vomiting sickness of Jamaica. Br Med J 1954;1: 75–77.
26. Meda HA, Diallo B, Buchet J-P, Lison D, Barennes H, Ouangre A, Sanou M, et al. Epidemic of fatal encephalopathy in preschool children in Burkina Faso and consumption of unripe ackee (*Blighia sapida*) fruit. Lancet 1999;35;536–540.
27. Larson J, Vender R, Camuto P. Cholestatic jaundice due to ackee fruit poisoning. Am J Gastroenterol 1994; 89:1577–1578.
28. Lebo DB, Ditto AM, Boxer MB, Grammer LC, Bonagura VR, Roberts M. Anaphylaxis to ackee fruit. J Allergy Clin Immunol 1996;98:997–998.
29. Ware GM. Method validation study of hypoglycin A determination in ackee fruit. J AOAC Int 2002;85:933–937.
30. Shimizu N, Yamaguchi S, Orii T. A study of urinary metabolites in patients with dicarboxylic aciduria for differential diagnosis. Acta Paediatr Japan 1994;36:139–145.
31. Al-Bassam SS, Sherratt HS. The antagonism of the toxicity of hypoglycin by glycine. Biochem Pharmacol 1981; 30:2817–2824.
32. Krieger I, Tanaka K. Therapeutic effects of glycine in isovaleric acidemia. Pediatr Res 1976;10:25–29.
33. Barennes H, Valea I, Boudat AM, Idle JR, Nagot N. Early glucose and methylene blue are effective against unripe ackee apple (*Blighia sapida*) poisoning in mice. Food Chem Toxicol 2004;42:809–815.
34. Lieu YK, Hsu BY, Price WA, Corkey BE, Stanley CA. Carnitine effects on coenzyme A profiles in rat liver with hypoglycin inhibition of multiple dehydrogenases. Am J Physiol 1997;272:E359–E366.

Chapter 4

CINNAMON
(*Cinnamomum* Species)

HISTORY

Cinnamon has been used as a spice for thousands of years; several references to it are found in the Bible. In Egypt, cinnamon was a spice used in embalming fluid. In Ayurvedic medicine, cinnamon bark was used as an antiemetic, antidiarrheal, antiflatulent, and general stimulant. The Portuguese found cinnamon trees growing in Sri Lanka (Ceylon) during the early 16th century; they subsequently imported cinnamon to Europe during the 16th and 17th centuries. The Dutch occupied Sri Lanka in the mid-17th century until the British captured the island in 1796. The East India Company then became the main exporter of cinnamon to Europe. The Dutch cultivated cinnamon in Java, and the exports of Ceylon cinnamon decreased as a result of heavy export duties. Nevertheless, Sri Lanka is the only regular supplier of cinnamon bark and leaf oils. The food industry prefers Ceylon cinnamon, but pharmaceutical manufacturers use both oils from Ceylon cinnamon (cinnamon oil) and from Chinese cinnamon (cassia oil) interchangeably. China is the main exporter of cassia oil.

BOTANICAL DESCRIPTION

The genus *Cinnamomum* comprises over 250 aromatic evergreen trees and shrubs, primarily located in Asia and Australia.

Common Name: Cinnamon, Ceylon cinnamon

Scientific Name: *Cinnamomum verum* J. Presl (*Cinnamomum zeylanicum* Blume, *Laurus cinnamomum* L.)

Botanical Family: Lauraceae (laurel family)

Physical Description: Large evergreen trees with young branches that are smooth and brown. The leaves are opposite, leathery, ovate to broadly ovate with three (and rarely, five) prominent veins. Young leaves are reddish and later turn dark green. Small, pale yellow flowers are borne in axillary or terminal panicles. The fruit is a fleshy, ovoid drupe, which contains one seed and turns dark purple or black when ripe.

Distribution and Ecology: Ceylon cinnamon is an indigenous tree of Sri Lanka and southwestern India.

Common Name: Cassia, Chinese cinnamon, false cinnamon, cassia lignea, bastard cinnamon, cassia bark, cassia-bark tree, Chinese cassia, Saigon cinnamon

Scientific Name: *Cinnamomum aromaticum* Nees (*Cinnamomum cassia* Nees ex Blume, *Cinnamomum cassia* J. Presl)

Botanical Family: Lauraceae (laurel family)

Physical Description: This slender, evergreen tree grows up to 65 ft (20 meters) high. Young branches are smooth and brown. The leaves are subopposite, slender, lanceolate or oblanceolate with three prominent veins. These leaves are reddish when young and dark green when mature. The small, white flowers are borne in axillary or terminal panicles. The fruit is a green, fleshy, globose drupe, which contains one seed and turns dark purple or black when mature. This fruit is similar in size to a small olive.

Medical Toxicology of Natural Substances, by Donald G. Barceloux, MD

Distribution and Ecology: Chinese cinnamon is an indigenous bush in the mountains of southern China, and cultivation for cassia oil production now occurs primarily in the southern Chinese provinces of Guangxi (Kwangsi) and Guangdong (Kwangtung). Other similar species occur on the Indonesian islands of Sumatra and Java [*C. burmannii* (Nees & T. Nees) Blume, Indonesia cassia], in Vietnam (*C. loureiroi* Nees, Vietnamese cassia), and in India and Nepal [*C. tamala* (Buch.-Ham.) Nees & Eberm., Indian cassia]. Commercial cultivation of these cassia species occurs in all of these areas.

EXPOSURE

Sources

Cinnamon leaf and bark are spices and sources of cinnamon oil, primarily from the *Cinnamomum verum* J. Presl (Ceylon cinnamon). However, most of products are extracted from cinnamon trees cultivated at altitudes up to about 1500 ft (460 meters). Cortex cinnamomi is the dried inner bark of shoots from *Cinnamomum verum* J. Presl or the stripped trunk bark of *Cinnamomum cassia* Blume. *Cinnamomum aromaticum* Nees (Chinese cinnamon) is the main source of internationally traded cassia oil, which is a distillate from a mixture of leaves, twigs, and bark fragments. Limited commercial production of cassia oil occurs in other regions including Indonesia [*C. burmannii* (Nees & T. Nees) Blume], Vietnam (*Cinnamomum loureiroi* Nees), and India [*C. tamala* (Buch.-Ham.) Nees & Eberm.].

Commercial production of cinnamon usually begins about 3–4 years after planting and pruning of the stems to force the growth of young stems. These stems are cut during the rainy season to facilitate the peeling of the bark. After scraping the outer skin of the bark, workers rub off the bark with a brass block. The bark is then split from end to end with a special round knife. The long strips of bark are then formed into the familiar compound quills (cinnamon) or hollow quills (cassia) of the spices that are stuffed with fragments of bark and then dried. Bleaching with sulfur dioxide disinfects the material and imparts a golden hue to the quills.

Uses

Traditional

The herbal use of cinnamon includes application as an astringent, germicide, and antispasmodic. Cinnamon was one of the early treatments for chronic bronchitis.[1] Other traditional uses include the treatment of impotence, frigidity, dyspnea, inflammation of the eye, leukorrhea, vaginitis, rheumatism, and neuralgia, as well as wounds and toothaches.[2]

Current

Cinnamon *bark* oil has a delicate aroma along with a sweet, pungent taste that results in use primarily as a flavoring in dental and pharmaceutical preparations, seasonings, sauces, baked goods, drinks, and tobacco. Investigational uses of cinnamon bark include use as a hypoglycemic and cholesterol lowering agent,[3] promotion of wound healing,[4] antimicrobial agent,[5] and an antiinflammatory compound.[6] The flavoring agent cinnamaldehyde in cinnamon oil is added to toothpaste to mask the taste of pyrophosphate, which is an unpleasant tasting compound that inhibits plaque calcification by interrupting the conversion of amorphous calcium phosphate to hydroxyapatite. The skin sensitizing properties of cinnamon bark oil limits use of this essential oil in cosmetic and other topical products.[7] Cinnamon *leaf* oil has a warm, spicy, and somewhat harsh odor that lacks the smooth consistency of cinnamon bark oil. Major uses of this essential oil include as flavoring in seasonings and a fragrance in soaps and insecticides. Cinnamon leaf oil is also a source for eugenol. Cassia oil is used medicinally as a carminative, antidiarrheal, antimicrobial, and antiemetic. Traditional uses for dried cassia include digestive complaints such as flatulence, colic, dyspepsia, diarrhea, and nausea, as well as colds, influenza, fevers, arthritis, and rheumatism. The major commercial uses of cassia oil are the flavoring of cola-type drinks and, to a lesser extent, bakery goods, sauces, confectionery products, and liquors. Like cinnamon bark oil, the use of cassia oil in topical preparations is limited by its skin-sensitizing properties. Cassia oil is a constituent of tiger balm. Smoking cessation products frequently involve the use of candies, chewing gum, or fresheners that often contain cinnamon flavoring. A few clinical studies suggest that cinnamon supplementation may lower blood glucose concentrations in patients with diabetes type 2; however, the sample sizes of these studies were small and there are inadequate data to recommend the regular use of cinnamon in diabetic patients.[8]

Regulatory Status

In the United States, cinnamon has GRAS (generally recognized as safe) status as a food additive.

PRINCIPAL TOXINS

Chemical Composition

Volatile oils are distilled products from the bark, leaves, flowers, or buds of *Cinnamomum* species, and the chemical composition of these oils varies depending on the part of the plant used for the distillation process. Cinnamon bark and leaf oil are steam distillation products obtained from the inner bark and leaves, respectively, of Ceylon cinnamon (*Cinnamomum verum*).[9] Other sources of cinnamon include Japanese cinnamon, Java cinnamon, and Taiwan cinnamon. Most of the chemical constituents of the essential oils from cinnamon are monoterpenes, sesquiterpenes, and related oxygen derivatives of these two types of compounds. The major monoterpene hydrocarbons in volatile components of cinnamon extracts are α-pinene, camphene, and limonene.[10] The main constituent of cinnamon *bark* oil is cinnamaldehyde (Figure 4.1), whereas eugenol is the main constituent (i.e., about 81–85%) of cinnamon leaf oil.[11] A commercial sample of essential oils from *Cinnamomum verum* contained approximately 63% cinnamaldehyde, 8% limonene, 7% eugenol, 5.5% cinnamaldehyde propylene, and <1–2% of a variety of terpenoid compounds (α-pinene, camphene) as measured by gas chromatography/mass spectrometry.[5] Cinnamon *leaf* oil contains a variety of constituents including eugenol (CAS RN: 97-53-0, $C_{10}H_{12}O_2$) and cinnamaldehyde (CAS RN: 104-55-2, C_9H_8O), which is a local mucous and dermal membrane irritant. In a study of the essential oils from leaves of *Cinnamomum osmophloeum* (Taiwan cinnamon), terpenoid compounds accounted for approximately 90% of the chemical compounds with 1,8-cineole, spathulenol, santolina triene, and caryophyllene oxide being the most common compounds.[6]

The essential oils from leaves of *Cinnamomum* species accounts for about 0.5% dry weight. Analysis of a steam-distilled volatile oil from cinnamon fruit stalks yielded 27 compounds with cinnamyl acetate (36.59%) and caryophyllene (22.36%) being the major components.[12] Analysis of the hydro-distilled volatile oil from buds of *Cinnamomum verum* (*C. zeylanicum*) yielded terpene hydrocarbons (78%) and oxygenated terpenoids (9%) with the sesquiterpenes, α-bergamotene (27%) and α-copaene (23%), being the most common compounds.[13] Minor compounds included α-humulene, α-muurolene, and δ-cadinenes. The volatile oil of the buds contains more monoterpene and sesquiterpene compounds than oils from the flowers and fruits, whereas the concentration of *trans*-cinnamyl acetate is much higher in the volatile oils from flowers and fruit than from the buds. A study of cinnamon essential oil from *C. verum* grown in Madagascar indicated that the major constituent was *trans*-cinnamaldehyde (41.3%).[14] Cassia oil is extracted from the leaves, bark, twigs, and stalks of *C. cassia* by steam distillation.

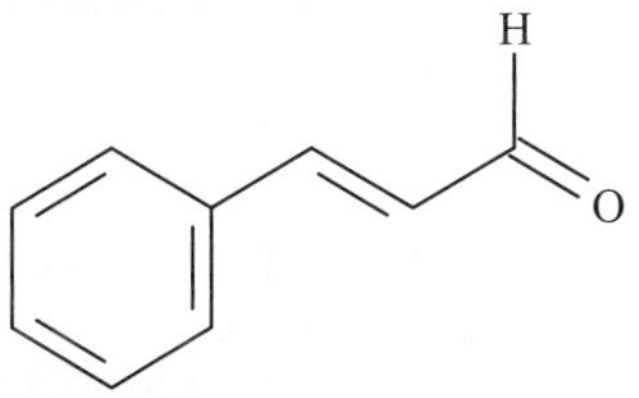

H
O

FIGURE 4.1. Chemical structure of cinnamaldehyde.

Physicochemical Properties

Cinnamon leaf oil has a fragrant odor and a very pungent taste. Experimental studies suggest that cinnamon has some mosquito larvicidal activity[15] and some antibacterial properties when added to certain food products.[16] However, there is no conclusive evidence that the antimicrobial properties of cinnamon produce clinically efficacious results.[17] Table 4.1 lists some of the physical properties of cinnamaldehyde.

DOSE RESPONSE

Medicinal Use

The average daily dose of the crude drug (cortex cinnamomi) is 2–4g or 0.05–0.2g of the essential oil.

Toxicity

The ingestion of an estimated 60mL of cinnamon oil was associated with burning sensation in the gastrointestinal tract along with lethargy, double vision, vomiting, and lightheadedness.[18] The symptoms resolved spontaneously within 5 hours with no complications.

TABLE 4.1. Some Physical Properties of Cinnamaldehyde

Physical Property	Value
Melting point	−7.50°C
Boiling point	246°C
log P (octanol-water)	1.9
Water solubility	1420mg/L (25°C)
Henry's Law Constant	1.60E-06atm m^3/mole (25°C)
Atmospheric OH rate constant	3.79E-11 cm^3 molecule/second (25°C)

TOXICOKINETICS

There are few human toxicokinetic data for the constituents in cinnamon oil. Rodent studies indicate that the major metabolic pathway of *o*-methoxycinnamaldehyde involves oxidation to the corresponding cinnamic and phenylpropionic acids with subsequent urinary excretion of benzoic and hippuric acids.[19]

CLINICAL RESPONSE

Most case reports of toxicity from cinnamon oil involve local irritation and allergic reactions to cinnamon oil as a constituent of personal hygiene (toilet soaps, mouthwash, toothpaste, perfumes, mud baths),[20] beverages (colas, vermouth, bitters), or baking products.[21–23] Allergic reactions include contact dermatitis, perioral dermatitis, cheilitis, stomatitis, gingivitis, glossitis, chronic lichenoid mucositis, contact urticaria,[24] and rarely immediate hypersensitivity reactions (asthma, urticaria).[25,26] Clinical manifestations of intraoral reactions include pain, swelling, erythema, ulcerations, fissures, vesicles, and white patches.[27] These reactions are local, and distal skin involvement is rare.[28] Occupational allergic contact dermatitis from spices is rare, and typically involves the hands.[29] More common food sensitizers include carrot, cucumber, tomato, melon, fish, potato, orange, green pepper, onion, red cherry, and garlic. Cinnamon oil contains local mucous membrane irritants such as cinnamaldehyde and cinnamic acid. Prolonged skin contact (48 hours) from a cinnamon oil spill produced superficial partial-thickness burns.[30] Chronic use of cinnamon-flavored gum can produce submucosal inflammation and alteration of the surface epidermis resembling oral leukoplakia, manifest on biopsy by acanthosis, hyperkeratosis, parakeratosis, plasma cell infiltration, fibrosis of the lamina, and focal atypia.[31] These changes are not pathognomonic for cinnamon-induced mucositis. The differential diagnosis of chronic mucositis associated with hypersensitivity to cinnamon includes local trauma, smokeless tobacco keratosis (snuff dipper's lesion), hyperkeratosis, lichen planus, lupus erythematosus, candidiasis, premalignant lesions, lichenoid mucositis, and carcinoma.

The severity of the local mucosal reaction depends on the duration of cinnamon-gum chewing. In contrast to the diffuse gingival reaction associated with cinnamon-flavored toothpaste, oral lesions associated with gum chewing occur on the lateral border of the free tongue or adjacent buccal mucosa. Sequelae of contact dermatitis associated with cinnamon include desquamation and hyperpigmentation.[32] Although a case report associated the development of squamous cell carcinoma of the tongue with prolonged use of cinnamon gum,[33] the International Agency for Research on Cancer (IARC) and the US National Toxicology Program do not list cinnamon as a potential carcinogen.

School-aged children abuse cinnamon oil by sucking on toothpicks or fingers dipped in the oil.[34] Reported effects include facial flushing, sensation of warmth, and intraoral hyperesthesias. Although nausea and abdominal pain may occur, systemic symptoms do not usually result from this type of exposure. Ingestion of cinnamon oil may cause central nervous system depression, predisposing the patient to aspiration pneumonia.

DIAGNOSTIC TESTING

Analytical methods for the identification of the constituents of cinnamon oil include high performance liquid chromatography[35] and gas chromatography/mass spectrometry.[10,36]

TREATMENT

Treatment is supportive. Patients with leukoplakia after chronic use of cinnamon-containing gums should be observed for resolution after cessation of use. Persistent oral lesions should be evaluated for oral cancer.

References

1. Ziment I. History of the treatment of chronic bronchitis. Respiration 1991;58(Suppl 1):37–42.
2. World Health Organization. WHO monographs on selected medicinal plants. Vol. 1. Geneva: World Health Organization; 1999.
3. Khan A, Khattak KN, Safdar M, Anderson RA, Khan MM. Cinnamon improves glucose and lipids of people with type 2 diabetes. Diabetes Care 2003;26:3215–3218.
4. Kamath JV, Rana AC, Chowdhury AR. Pro-healing effect of *Cinnamomum zeylanicum* bark. Phytother Res 2003; 17:970–972.
5. Simic A, Sokovic MD, Ristic M, Grujic-Jovanovic S, Vukojevic J, Marin PD. The chemical composition of some Lauraceae essential oils and their antifungal activities. Phytother Res 2004;18:713–717.
6. Chao LK, Hua K-F, Hsu H-Y, Cheng S-S, Liu J-Y, Chang S-T. Study on the antiinflammatory activity of essential oil from leaves of *Cinnamomum osmophloeum*. J Agric Food Chem 2005;53:7274–7278.
7. Hartmann K, Hunzelmann N. Allergic contact dermatitis from cinnamon as an odour-neutralizing agent in shoe insoles. Contact Dermatitis 2004;50:253–254.
8. Pham AQ, Kourlas H, Pham DQ. Cinnamon supplementation in patients with type 2 diabetes mellitus. Pharmacotherapy 2007;27:595–599.

9. Mino Y, Ota N. Inorganic chemical approaches to pharmacognosy. V. X-ray fluorescence spectrometric studies on the inorganic constituents of crude drugs. (3). On the cinnamoni cortex. Chem Pharm Bull (Tokyo) 1990; 38:709–713.
10. Miyazawa M, Hashimoto Y, Taniguchi Y, Kubota K. Headspace constituents of the tree remain of *Cinnamomum camphora*. Nat Prod Lett 2001;15:63–69.
11. Mallavarapu GR, Ramesh S, Chandrasekhara RS, Rajeswara Rao BR, Kaul PN, Battacharya AK. Investigation of the essential oil of cinnamon leaf grown at Bangalore and Hyderabad. Flavour Fragrance J 1995;10; 239–242.
12. Jayaprakasha GK, Rao LJ, Sakariah KK. Volatile constituents from *Cinnamomum zeylanicum* fruit stalks and their antioxidant activities. J Agric Food Chem 2003;51: 4344–4348.
13. Jayaprakasha GK, Rao LJ, Sakariah KK. Chemical composition of volatile oil from *Cinnamomum zeylanicum* buds. Z Naturforsch 2003;57:990–993.
14. Mollenbeck S, Konig T, Schreier P, Schwab W, Rajanonarivony J, Ranariveol L. Chemical composition and analysis of enantiomers of essential oils from Madagascar. Flavour Fragrance J 1997;12:63–69.
15. Cheng SS, Liu JY, Tsai KH, Chen WJ, Chang ST. Chemical composition and mosquito larvicidal activity of essential oils from leaves of different *Cinnamomum osmophloeum* provenances. J Agric Food Chem 2004;52:4395–4400.
16. Yuste J, Fung DY. Inactivation of *Salmonella typhimurium* and *Escherichia coli* O157:H7 in apple juice by a combination of nisin and cinnamon. J Food Prot 2004;67:371–377.
17. Martin KW, Ernst E. Herbal medicines for treatment of bacterial infections: a review of controlled clinical trials. J Antimicrob Chemother 2003;51:241–246.
18. Pilapil VR. Toxic manifestations of cinnamon oil ingestion in a child. Clin Pediatr 1989;28:276.
19. Samuelsen OB, Brenna J, Solheim E, Scheline RR. Metabolism of the cinnamon constituent *o*-methoxycinnamaldehyde in the rat. Xenobiotica 1986;16:845–852.
20. Garcia-Abujeta JL, de Larramendi CH, Berna JP, Palomino EM. Mud bath dermatitis due to cinnamon oil. Contact Dermatitis 2005;52:234.
21. Calnan CD. Cinnamon dermatitis from an ointment. Contact Dermatitis 1976;2;167–170.
22. Sainio EL, Kanerva L. Contact allergens in toothpastes and a review of their hypersensitivity. Contact Dermatitis 1995;33:100–105.
23. Nixon R. Cinnamon allergy in a baker. Australas J Dermatol 1995;36:4.
24. Kirton V. Contact urticaria and cinnamic aldehyde. Contact Dermatitis 1978;4:374–375.
25. Lamey PJ, Lewis MA, Rees TD, Fowler C, Binnie WH, Forsyth A. Sensitivity reaction to the cinnamonaldehyde component of toothpaste. Br Dent J 1990;168:115–118.
26. Uragoda CG. Asthma and other symptoms in cinnamon workers. Br J Ind Med 1984;41:224–227.
27. Jacob SE, Steele T. Tongue erosions and diet cola. Ear Nose Throat J 2007;86:232–233.
28. Drake TE, Maibach HI. Allergic contact dermatitis and stomatitis caused by a cinnamic aldehyde-flavored toothpaste. Arch Dermatol 1976;112:202–203.
29. Kanerva L, Estlander T, Jolanki R. Occupational allergic contact dermatitis from spices. Contact Dermatitis 1996; 35:157–162.
30. Sparks T. Cinnamon oil burn. West J Med 1985;142:835.
31. Mihail RC. Oral leukoplakia caused by cinnamon food allergy. J Otolaryngol 1992;21:366–367.
32. Garcia-Abujeta JL, de Larramendi CH, Berna JP, Palomino EM. Mud bath dermatitis due to cinnamon oil. Contact Dermatitis 2005;52:234.
33. Westra WH, McMurray JS, Califano J, Flint PW, Corio RL. Squamous cell carcinoma of the tongue associated with cinnamon gum use: a case report. Head Neck 1998; 20:430–433.
34. Perry PA, Dean BS, Krenzelok EP. Cinnamon oil abuse by adolescents. Vet Hum Toxicol 1990;32:162–164.
35. Sagara K, Oshima T, Yoshida T, Tong YY, Zhang G, Chen YH. Determination of *Cinnamomi cortex* by high-performance liquid chromatography. J Chromatogr 1987;409:365–370.
36. Jayaprakasha GK, Jagan Mohan Rao L, Sakariah KK. Chemical composition of the flower oil of *Cinnamomum zeylanicum* Blume. J Agric Food Chem 2000;48: 4294–4295.

Chapter 5

CYANOGENIC FOODS
(Cassava, Fruit Kernels, and Cycad Seeds)

HISTORY

Cassava is a staple in Africa that contains toxic concentrations of cyanogenic glycosides when improperly processed. Occasionally, outbreaks of cyanide poisoning occur following the consumption of bitter cassava, particularly during periods of drought.[1]

BOTANICAL DESCRIPTION

The distribution of cyanogenic glycosides is widespread throughout the plant kingdom, occurring in at least 2,500 plants species, primarily in the Compositae, Fabaceae, Linaceae, and Rosaceae families.[2] Of these plants, clinically important human toxicity probably occurs only following the ingestion of cassava. Prunasin occurs in at least six plant families (Myoporaceae, Myrtaceae, Polypodiaceae, Rosaceae, Saxifragaceae, Scrophulariaceae).

Cassava

Common Name: Cassava

Scientific Name: *Manihot esculenta* Crantz

Botanical Family: Euphorbiaceae (Spurge)

Physical Description: Cassava is a perennial, woody shrub with large, brown, tapered tubers that are staples in many poor countries. This shrub or small tree reaches about 1.25–5 m (4–16 ft) in height. The leaves are glabrous and deeply parted with spatulate to linear-lanceolate acuminate lobes up to 15 cm (6 in.) long. The flowers form panicles.

Distribution and Ecology: This plant is a native species of Central and South America that was spread by 17th century explorers to Africa and later to the Indian subcontinent. Cassava is widely cultivated in subtropical and tropical areas for the high starch content of the tubers.

Fruit Kernels (*Prunus spp.*)

Common Name: Apple

Scientific Name: *Malus domestica* Borkh.

Botanical Family: Rosaceae (rose)

Physical Description: This deciduous fruit tree forms a broad crown and reaches up to 12 m (39 ft) in height. The leaves are simple and alternate with a serrated tip and petiole up to 5 cm (2 in.) long. The white flower appears in spring with five petals. The fruit matures in autumn with five, star-shaped carpels containing one or two seeds.

Distribution and Ecology: The apple is a native plant of the mountains in Central Asia, and now numerous cultivars of the apple are widely distributed throughout the world.

Common Name: Peach

Scientific Name: *Prunus persica* (L.) Batsch.

Botanical Family: Rosaceae (rose)

Physical Description: The deciduous tree reaches up to 10 m (33 ft) in height with broad, lanceolate leaves up to 15 cm (6 in.) long. In early spring, the solitary, pink flowers appear before the leaves with five petals. The drupe (stone fruit) contains a single, large seed with a hard, corrugated cover.

Medical Toxicology of Natural Substances, by Donald G. Barceloux, MD

Distribution and Ecology: This peach tree is a native species of China, which is now widely cultivated in temperate climates. However, the chilling requirements and limited cold tolerance of the fruit restricts the areas of production to places other than tropical and subtropical climates.

Cycad Seeds

Cycas plants (e.g., *Cycas revoluta* Thunb. or sago palm) belong to the order Cycadales, which contains approximately 11 genera of tropical and subtropical plants that produce terminal oblong cones containing orange-yellow seeds. This slow growing fern-like tree is a native species of southern Japan, but this plant has been naturalized throughout temperate and tropical habitats.

EXPOSURE

Cyanogenic glycosides in economically important plants include linamarin in *Manihot esculenta* Crantz (cassava), *Linum usitatissimum* L. (flax), and *Trifolium repens* L. (white clover); lotaustralin in *Lotus corniculatus* L. (bird's foot trefoil); dhurrin in *Sorghum* spp.; and amygdalin in Rosaceae species (peach, apricot, apple seeds).[2] Medicinal uses of fruits from the genus *Prunus* include the treatment of headache, hypertension, diarrhea, and constipation.[3] Cassava tuber is an important food source with high carbohydrate and low protein content for up to 500 million people in tropical Africa and Latin America.[4] Although millet, rice, and potatoes are usually preferred food sources, cassava root is a cheap source of energy, particularly during difficult economic times. Gari is a grit derived from the cassava root by grating, dehydrating, and fermenting the root pulp and then roasting the resultant mash. Eba is a paste formed from the addition of boiling water to gari. The Cycas plant has a long history of use as a food and medicine.[5] In traditional Chinese medicine, cycad seeds are used to treat hypertension, musculoskeletal disorders, gastrointestinal distress, cough, and amenorrhea.

PRINCIPAL TOXINS

Structure and Properties

Cyanogenic glycosides are secondary plant metabolites composed of an α-hydroxynitrile aglycone and a sugar moiety (e.g., D-glucose). Many of these cyanogenic glycosides are structurally similar to prunasin including amygdalin (Rosaceae), sambunigrin (Caprifoliaceae, Mimosaceae, Oleaceae), and vicianin (Fabaceae, Polypodiaceae) as demonstrated in Table 5.1. The biosynthetic precursors of cyanogenic glycosides in these plants are various L-amino acids that are hydroxylated to *N*-hydroxylamino acids, aldoximes and then to nitriles. Linamarin and lotaustralin are derivatives of valine, isoleucine and leucine, whereas prunasin, amygdalin, and sambunigrin are derivatives of phenylalanine.[6] Dhurrin is a tyrosine derivative. Hydroxylation of nitriles to α-hydroxynitrile followed by glycosylation forms the cyanogenic glycosides. The initial metabolism of cyanogenic glycosides involves the cleavage of the carbohydrate moiety by one or more β-glycosidases that yields the corresponding α-hydroxynitrile.[7] Hydroxynitrile compounds may decompose either spontaneously or enzymically in the presence of a hydroxynitrile lyase to produce hydrogen cyanide and an aldehyde or ketone. The separation of cyanogenic glycosides and the degrading enzymes in the plant tissues prevents the formation of significant amounts of cyanide until the physical disruption of the plant compartments.

Fruit Kernels

The seeds of stone fruits from *Prunus* species (e.g., cherries, peaches, apricots) contain up to 6% amygdalin on a fresh weight basis, whereas the cyanogenic glucoside content of the pulp ranges from about 0.001–0.01%.[8] Consequently, canned stone fruits can contain small amounts (i.e., up to 4 mg/kg) of hydrogen cyanide. Amygdalin is the cyanogenic diglucoside, D-mandelonitrile-β-D-gentiobioside, present in the kernels of many stone fruits. Amygdalin hydrolase and prunasin hydrolase are two β-glycosidases that catalyze the hydrolysis of amygdalin in a stepwise reaction by removing two glucose residues and releasing benzaldehyde and hydrogen cyanide.[9] These enzymes are released following the crushing of moistened, amygdalin hydrolase-containing kernels. Amygdalin hydrolase removes the terminal glucose of amygdalin and forms the monoglucoside, prunasin. Certain human intestinal bacteria may also possess glycosidase activity, so that ingestion of amygdalin in the absence of this plant enzyme can cause cyanide toxicity. *In vitro* studies indicate that the addition of β-glucuronidase to amygdalin does not liberate hydrogen cyanide.[10] In addition to amygdalin and prunasin, peach seeds also contain several minor cyanogenic glycosides including amygdalinic acid, mandelic acid β-D-glucopyranoside, benzyl β-gentiobioside, and benzyl β-D-glucopyranoside.[11]

Cassava

Linamarin and the methyl derivative, lotaustralin, are the two main cyanogenic glycoside compounds in cassava root-derived foods, while the content of hydrogen cyanide and cyanohydrins are generally low.[12,13]

TABLE 5.1. Chemical Structure of Common Cyanogenic Glucosides

Base Structure	Substituent R	Glycoside	Sugar	Occurrence
Glycosides with aromatic substituents				
H, O— Glycoside, C, R, CN	Phenyl	Prunasin	D-Glucose	*Prunus* spp.
	Phenyl	Amygdalin	Gentiobiose	*Prunus* spp.
	Phenyl	Lucumin	Primeverose	*Lucuma* spp.
	Phenyl	Vicianin	Vicianose	*Vicia* spp.
	Phenyl	Sambunigrin	D-Glucose	*Sambucus* spp.
	p-Hydroxyphenyl	Dhurrin	D-Glucose	*Sorghum* spp.
	p-Hydroxyphenyl	Taxiphyllin	D-Glucose	*Taxus* spp.
	p-Hydroxyphenyl	Zierin	D-Glucose	*Zieria* spp.
	p-Glucosyloxyphenyl	Proteacin	D-Glucose	*Macadamia* spp.
Glycosides with a free α-hydroxynitrile				
		p-Glucosyloxymandelonitrile		*Nandina* spp.
Glycosides with Aliphatic substituents				
R, O— Glucose, C, R', CN	R = R′ = CH_3	Linamarin	D-Glucose	*Linum* spp. *Trifolium* spp.
	R = CH_3 R′ = CH_2CH_3	Lotaustralin	D-Glucose	*Loyus* spp. *Maniholt* spp.
O— Glucose, R=C, CN	R = $CH(CH_3)_2$	Acaciapetalin	D-Glucose	*Acacia* spp.
	R = $HCO_2CH{=}CH(CO_2HCH_2)C$	Triglochinin	D-Glucose	*Triglochin* spp.
O— Glucose, CN, R, R'	R = R′ = H	Deidaclin Tetraphyllin A	D-Glucose D-Glucose	*Deidamia* spp. *Tetrapathaea* spp.
	R = OH; R′ = H R = OH; R′ = H	Tetraphyllin B Gynocardin	D-Glucose D-Glucose	*Tetrapathaea* spp. *Gynocardia* spp. *Pangium* spp.

Source: Adapted from Ref 2.

These two cyanogenic glycosides are hydrolyzed to cyanohydrins, which undergo nonenzymatic conversion to hydrogen cyanide via hydrolysis by linamarinase.[4] Processing (e.g., sun drying, soaking, and boiling) removes about 80–95% of the cyanogenic compounds present in traditional African foods (gari, lafun, fufu) derived from the cassava root.[14] Disruption of the root tissue releases an endogenous enzyme, linamarinase, which hydrolyses linamarin to acetone cyanohydrins in the presence of high moisture content and low temperatures. After the formation of cyanohydrins, heat and low moisture content facilitate the spontaneous conversion of cyanohydrins to volatile hydrogen cyanide.[15] Inefficient processing of cassava root results in the persistence of cyanohydrins in the cassava meal. The storage of cassava in a deep freeze for 4 months after the slicing of the cassava in a food processor resulted in the hydrolysis of almost all of the linamarin in the cassava, whereas much smaller amounts of linamarin were lost during storage under the same conditions when the cassava was cut into 20-g slices.[16] Storage of gari or eba for 4 weeks at room temperature resulted in the loss of about one-half of the original linamarin content of roasted cassava that was fermented for at least 4 days prior to storage.[17]

CYCAD SEEDS

Cycad seeds contain the cyanogenic glycosides, cycasin and neocycasin. Cycasin and neocycasin A, B, C, and E (methylazoxymethanol β-D-glycoside) are unique toxins present in cycad species. These azoxyglucosides are glycosides of the same aglycone, methylazoxymethanol.[18] Cyanide production results from the hydrolysis of the glycoside to D-glucose and hydroxynitrile by the enzyme β-glycosidase. Hydroxynitrile forms hydrogen cyanide and the corresponding aldehyde or ketone following either spontaneous hydrolysis or catalytic conversion by hydroxynitrile lyase. However, the liberation of cyanide from this process is a minor pathway compared to the formation of nitrogen gas, formaldehyde, and methanol. Food processing removes some of the cyanide by inactivation of β-glycosidase or removal of the cyanogenic glycosides by exposure to β-glycosidases.

Poisonous Parts

The lima bean (*Phaseolus lunatus* L.) contains low concentrations of cyanogenic glycosides, and clinical toxicity in humans following the ingestion of these fruits is not well documented.[19] A sample of dry, mature, raw lima beans contained a cyanide content of about 38 mg/g that decreased to nondetectable concentrations after autoclaving or toasting for 20 minutes, as measured by colorimetry after acid hydrolysis.[20] The total amount of cyanide-containing compounds is substantially higher in young leaves than in older leaves. In an experimental study, the cyanogenic potential (i.e., total amount of cyanide-containing compounds), ranged from 0.7 μmol HCN/g leaf fresh weight in mature leaves of one strain of lima beans to 82.0 μmol HCN/g leaf fresh weight in young leaves of another strain.[21]

Sorghum (e.g., *Sorghum bicolor* L.) contains the cyanogenic glucoside, dhurrin, which releases hydrogen cyanide as a result of the activity of the heterotetrameric enzyme, hydroxynitrile lyase.[22] Flaxseed is an oilseed that contains about 35–45% oil along with cyanogenic glycosides, primarily linustatin and neolinustatin.[23] Flaxseed is a desirable food product because of the high content of α-linolenic acid and dietary fiber. The cyanide content of 10 cultivars of flaxseed ranged from 124–196 μg/g, as analyzed by high performance liquid chromatography with electrochemical detection.[24] Boiling water or solvent extraction removes 90–100% of the HCN content of flaxseeds.[25] The highest cyanogenic glucoside concentrations occur in extracted flaxseed mucilage, but baking muffins with flaxseed flour (i.e., 150 g flaxseed/kg) at 230°C (446°F) for 15–18 minutes eliminates the cyanogenic glucosides.[26] These compounds also are not detectable in processed flaxseed oil.

FRUIT KERNELS

Apricot, peach, and apple kernels as well as the stems, leaves, and roots of many species from the rose (Rosaceae) family contain cyanogenic glycosides, such as amygdalin. In a review of several studies on the HCN content of fruit kernels, the approximate amount of HCN releases were as follows: apricot, <0.05–4 mg/g; peach, 0.4–2.6 mg/g; and apple, 0.6 mg/g.[10] The cyanide content of apricot kernels (*Prunus armeniaca*) varies widely depending on a variety of poorly defined factors including cultivation practices, variety, and moisture content.[27] In several studies, the cyanide content of apricot kernels ranged from <0.05 mg/g to about 4 mg/g material compared with about 56 mg HCN/g amygdalin.[28,29] A study of pulverized apricot kernels demonstrated average cyanide yields of approximately 3 mg/g ground kernels, as measured by a simple colorimetric method.[30] In general, the hydrogen cyanide content is lower in peach kernels than in bitter almond or apricot kernels. Analysis of two peach kernels in one study demonstrated an average hydrogen cyanide content of approximately 0.45 mg/g of material as measured by a colorimetric method.[31] In another study, the hydrogen cyanide content of peach and apricot kernels was 2.6 mg/g and 2.92 mg/g, respectively, compared with 55.9 mg HCN/g amygdalin.[10] Various factors account for some the differences in cyanide content of peach kernels between these studies including variety, moisture content, physical properties used for analytical methods (e.g., chromatography or spectroscopy), and expression of results (e.g., HCN or glycoside). The cyanide content of apple (*Malus domestica*) seeds is low (0.6 mg/g) compared with bitter almond and apricot kernels, and toxicity from the ingestion of these seeds is unusual.[10]

CASSAVA

Cassava root typically contains <100 mg HCN equivalent/kg wet weight, but some "bitter" cultivars contain up to 500 mg HCN equivalent/kg wet weight. However, the amount of cyanide in foods (e.g., gari) derived from the cassava root depends on the processing methods.[32] All parts of the cassava plant contain cyanogenic glycosides with cyanogen content in tubers varying substantially (e.g., 15–400 mg HCN equivalent/kg fresh weight).[14] The main cyanogen in cassava is linamarin (i.e., >90%) with minor amounts of the closely related compound, lotaustralin.[16] Most cassava cultivars contain roots with cyanogenic glucoside content below 100 mg/kg wet weight, although some bitter cultivars may contain cyanogenic glycoside content up to 500 mg/kg wet weight.[2] Processing and fermenting of the cassava root for 5 days in water removes most (i.e., about 90%) of the cyanogenic glycosides and converts the processed root to the storage form (e.g., gari). Studies of children eating meals from fried and boiled cassava root during normal times indicate that the exposure to cyanide from this food source is relatively low.[33] Acute intoxication from the ingestion of cassava root occurs primarily during periods of drought or intensive commercialization that results in inadequate processing of the cassava root.[34] The leaves and buds contain cyanogenic glycosides that may produce cyanide poisoning, especially if the leaves are boiled as tea. Intact cyanogenic plants do not release hydrogen cyanide because the substrates and the degrading enzymes occur in separate intracellular compartments.[2]

Mechanism of Toxicity

The release of hydrogen cyanide from cyanogenic precursors in certain plants is a constitutive plant defense against tissue disruption by herbivores. The effectiveness of this defense mechanism results primarily from the endogenous release of hydrogen cyanide within the gastrointestinal tract of the herbivore rather than the release of this gas during the feeding process.[35] The production of cyanide from cyanogenic plants depends both on the biosynthesis of cyanogenic glycosides and the coexistence of degrading enzymes localized in the apoplast. The generation of cyanide from cyanogenic glycosides is a two-step process involving deglycosilation and cleavage of the molecule. The latter process is regulated by β-glucosidase and α-hydroxynitrile lyase.[2] For example, the generation of cyanide from linamarin in cassava root involves the deglycosilation of linamarin by a β-glucosidase (linamarase) to acetone cyanohydrin, and the subsequent cleavage of acetone cyanohydrin to acetone and hydrocyanic acid by α-hydroxynitrile lyase. The latter reaction can occur spontaneously as a result of the instability of the cyanohydrin.

DOSE RESPONSE

Fruit Kernels

Ingestion of 20–40 apricot kernels by an adult female resulted in acute cyanide toxicity with disorientation that responded promptly to the administration of sodium nitrite and thiosulfate.[29] The chewing of approximately 30 apricot kernels by a 41-year-old woman produced generalized weakness, dyspnea, diaphoresis, and dysphagia within 20 minutes, which subsequently responded to intravenous sodium nitrite and sodium thiosulfate.[36] A man and a woman developed vomiting, headache, flushing, diaphoresis, and lightheadedness about 1 hour after consuming a milkshake containing 48 pulverized apricot kernels.[37] The symptoms rapidly resolved without sequelae.

Cassava

Experimental studies indicate that the consumption of cassava does not usually cause a significant increase in cyanide exposure. In a study of nonsmoking subjects, the ingestion of 1–4 kg fresh boiled cassava over 2 days did not produce a significant increase in urinary thiocyanate concentrations.[38] In an experimental study, volunteers ingested a meal of gari containing cyanohydrin equivalent to 128 μmol of cyanide.[39] Analysis of plasma from these volunteers demonstrated small amounts of cyanide that suggests that cyanide may accumulate in plasma following chronic ingestion of cassava.

TOXICOKINETICS

Absorption/Distribution

There are few data on the toxicokinetics of amygdalin or prunasin in humans. Limited animal studies suggest that the gastrointestinal tract poorly absorbs (i.e., <3%) oral doses of amygdalin, whereas the oral bioavailability of prunasin is about 50%.[40] In animal models, the occurrence of peak whole blood cyanide concentrations after oral administration of amygdalin and linamarin depends on the rate of enzymatic hydrolysis. A study of female golden Syrian hamsters indicated that peak whole blood cyanide concentrations following the oral administration of amygdalin and linamarin occurred 1 hour and 3 hours, respectively, after administration.[41] The protein binding of amygdalin is minimal.

Biotransformation/Elimination

The biotransformation of cyanogenic glycosides begins with the cleavage of the carbohydrate moiety by one or more β-glycosidases that form α-hydroxynitrile. Depending on the plant species, there are two distinct pathways that involve 1) the hydrolysis of cyanogenic disaccharides by the stepwise removal of the sugar residues, and 2) hydrolysis of the aglycone-disaccharide bond producing mandelonitrile and the corresponding disaccharides.[42] Figure 5.1 demonstrates these two pathways. Hydrolysis of amygdalin produces the active metabolite, prunasin (D-mandelonitrile-β-D-glucoside). Animal studies indicate that the elimination of amygdalin is rapid with most of the absorbed dose (62–96%) excreted in the urine during the first 24 hours after ingestion as unchanged drug.[43,44] The hepatic excretion of amygdalin is relatively small compared with renal excretion.[45] Following intravenous administration of 500 mg amygdalin

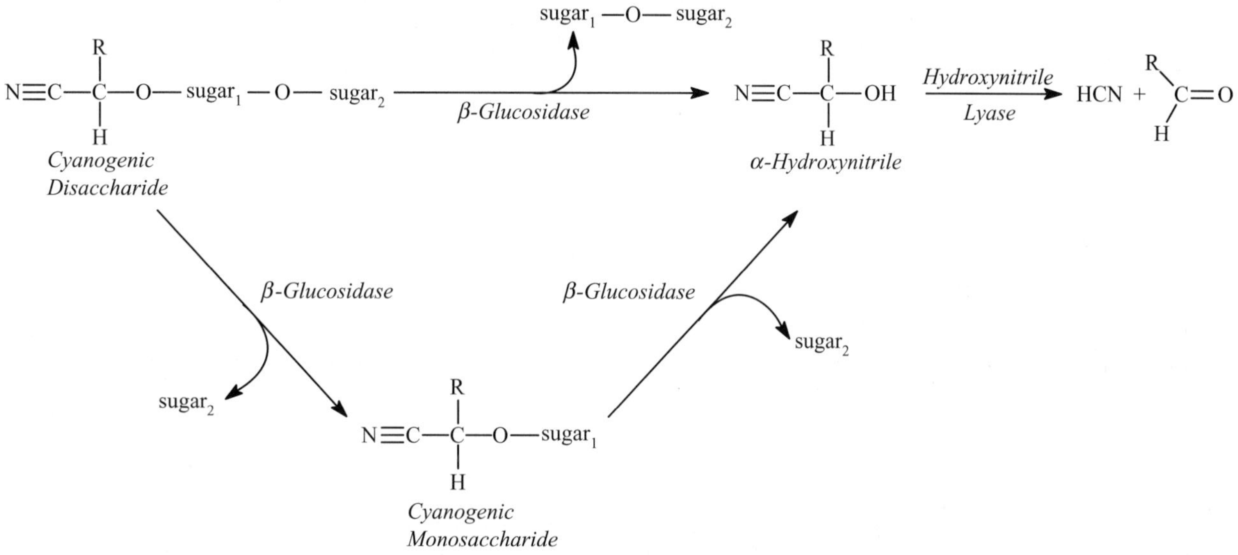

sugar$_1$—O—sugar$_2$

N≡C—C(R)(H)—O—sugar$_1$—O—sugar$_2$ → (*β-Glucosidase*) → N≡C—C(R)(H)—OH → (*Hydroxynitrile Lyase*) → HCN + R(H)C=O

Cyanogenic Disaccharide

α-Hydroxynitrile

β-Glucosidase

sugar$_2$

β-Glucosidase

sugar$_2$

N≡C—C(R)(H)—O—sugar$_1$

Cyanogenic Monosaccharide

FIGURE 5.1. Biotransformation pathways of cyanogenic disaccharides. From Poulton JE. Localization and catabolism of cyanogenic glycosides. Ciba Found Symp 1988;140:69. Copyright John Wiley & Sons Limited. Reprinted with permission.

to cancer patients, the mean elimination half-life of amygdalin was approximately 2 hours with a mean plasma clearance of about 100 mL/min.[46]

Frequent intake of cassava root-derived foods (e.g., gari) substantially increases the plasma concentrations of thiocyanate, the major metabolic product of cyanide, despite the low concentrations of cyanide in the processed foods. Limited pharmacokinetic data indicate that approximately one-half of the linamarin is converted to cyanide and subsequently to thiocyanate following ingestion of linamarin-containing foods.[47] About one-fourth of the ingested linamarin appears in the urine unchanged, and the rest of the linamarin is metabolized to other compounds. The rise in the plasma cyanide concentration following ingestion of a single meal of well-processed cassava root is small (i.e., doubling of baseline concentrations) with a return to baseline plasma cyanide concentrations averaging about 7 ± 2 hours.[39]

CLINICAL RESPONSE

Acute

The clinical features associated with the ingestion of cyanogenic glycosides mimics cyanide poisoning, and the severity of the intoxication correlates to the dose of active cyanogenic compounds. The onset of vomiting, abdominal pain, weakness, dyspnea, diaphoresis, and lightheadedness followed by convulsions, stupor, disorientation, hypotension, metabolic acidosis, coma, respiratory failure, and cardiovascular collapse develops after exposure to high doses of cyanogenic glycoside-containing foods[48,49] or from the ingestion of laetrile.[50,51] Symptoms typically start within one half hour after consumption and progress quickly. A case series associated the ingestion of wild apricot seed ingestions by Turkish children with the sudden onset of vomiting and crying followed by fainting, lethargy, and coma.[52]

In a retrospective review of charts from 21 patients, the ingestion of cooked cycad seeds was associated with severe vomiting.[18] Other symptoms present in less than one-half of these patients included abdominal pain, headache, dizziness, weakness, tachycardia, and diarrhea. All patients tested for blood cyanide concentrations had elevated blood cyanide concentrations, but the blood cyanide concentrations were below values (0.5–1 mg/L) associated with serious toxicity. All symptoms resolved within 24 hours. Although case reports associated the ingestion of seeds and other plant material from *Cycas* species with fatalities,[53] forensic analysis of these cases did not include documentation of cyanide exposure. The ingestion of yam beans [*Pachyrhizus erosus* (L.)Urb.] causes an acute intoxication similar to cyanide poisoning including metabolic acidosis, nausea, vomiting, perioral paresthesias, and coma.[54] However, this bean contains rotenone rather than cyanide. Rotenone blocks nicotinamide adenine dinucleotide CoQ reductase (complex I) in the first step of mitochondrial electron transport.[55]

Chronic

Tropical Ataxic Polyneuropathy

Tropical ataxic polyneuropathy is a sporadic and endemic disease that causes gait ataxia, sensory polyneuropathy, optic atrophy, and neurosensory deafness. This disease is uncommon in patients under 40 years of age, and women are affected more often than men.[56] The endemic form occurs primarily in communities in southwestern Nigeria. Epidemiological studies in the 1950s and 1960s associated this disease with the high intake of cassava.[57] Although recent studies confirm the relatively high consumption of cassava foods in Nigerian communities with endemic ataxic polyneuropathy,[58] consumption of large quantities of cassava occurs in Nigerian communities without this disease.[59] A nested case-control study of cases of endemic ataxic polyneuropathy did not detect a significant difference in the intake of cassava or the plasma thiocyanate concentrations between cases and controls.[60]

Konzo

Konzo (i.e., "tired legs") is a permanent spastic upper motor neuron disorder occurring in rural areas of Mozambique, Tanzania, Central African Republic, Cameroon, and the Democratic Republic of Congo. The abrupt onset of this spastic paraparesis or tetraparesis is associated with the consumption of cassava, particularly in children and women of childbearing age with low protein intake (i.e., low sulfur amino acids).[61] About one-half of patients with konzo also have an optic neuropathy manifest by decreased visual acuity, alteration of the visual fields, and/or pallor of the optic disk.[62] The most common eye complaint in patients with konzo is transient blurred vision that is not usually associated with progressive loss of vision. In contrast to optic effects (i.e., amblyopia) associated with tropical ataxic polyneuropathy, visual field defects and pallor of the optic disks are the most common eye abnormalities associated with konzo.[63] Visual evoked potentials are abnormal in approximately half of the patients with konzo, manifest by delayed latency and reduced amplitudes. A small number of patients have pendular nystagmus consisting of involuntary, lateral, and medial eye movements with equal velocity in both directions. The severity of the ophthalmological complaints does not correlate to the severity of spastic paresis.[62] The differential diagnosis includes epidemic optic neuropathy in Tanzania, Nigeria, and Cuba, tobacco amblyopia, Leber's hereditary optic neuropathy, and vitamin B deficiency (e.g., thiamine, cobalamin).

Western Pacific Amyotrophic Lateral Sclerosis/Parkinsonism-Dementia Complex

Western Pacific amyotrophic lateral sclerosis/parkinsonism-dementia complex (WP ALS-PD) is a prototypic neurodegenerative disorder with symptoms similar to amyotrophic lateral sclerosis, Parkinson's disease, and Alzheimer's disease. The incidence of this neurodegenerative disease is 50–100 times more common in the Chamorro people of Guam than other Pacific populations.[64] The nonprotein amino acid β-methylamino-L-alanine (BMAA) in the Guam ecosystem is the most likely neurotoxic compound rather than a cyanogenic compound.

Hypersensitivity

Food allergies may develop following the consumption of the skin of certain fruits (apricot, plum, cherry, peach), particularly in patients with allergic rhinitis.[65]

DIAGNOSTIC TESTING

Analytical Methods

Methods to detect amygdalin include thin layer chromatography (TLC), high performance liquid chromatography (HPLC), and gas chromatography mass spectrometry (GC/MS).[66] Amygdalin and prunasin are quantifiable in biological samples (urine, plasma) by HPLC on a reversed phase column with detection at 215 nm and confirmation by mass spectrometry.[67] Methods for detecting hydrogen cyanide in biological samples include spectrometry, electrochemical analysis with cyanide ion-selective electrode, GC, GC/MS, and HPLC/MS.[68,69] The latter two methods are widely used methods for the quantitation of cyanide after headspace extraction or derivatization. More recent and rapid methods of detecting cyanide in biological specimens include capillary gas chromatography with cryogenic oven trapping (limit of detection [LOD]: 2 ng/mL),[70] capillary electrophoresis with fluorescence detection (LOD 0.1 ng/mL),[71] automated headspace GC (LOD: 13.8 μg/L),[72] and isotope-dilution GC-MS (LOD: 0.003-0.007 μg/g).[73]

Biomarkers

There are few data correlating blood cyanide concentrations to the ingestion of specific amounts of cyanogenic plants. In the presence of a compatible history, the presence of elevated blood cyanide concentrations confirms exposure to cyanogenic plants, but the results are not usually available to guide clinical management.

Abnormalities

Cyanide poisoning inevitably produces a metabolic acidosis that results in an increased serum lactate concentration from anaerobic metabolism.[74] The lack of metabolic acidosis in a symptomatic patient suggests a diagnosis other than cyanide poisoning.

TREATMENT

Treatment of poisoning from exposure to cyanogenic plants is similar to the treatment of cyanide poisoning. Decontamination measures are not usually necessary because of the limited absorptive capacity of activated charcoal to cyanide. Mild to moderate ingestions usually require only supportive care. Indications for the use of an antidote (nitrites/thiosulfate or hydroxocobalamin) include alteration in mental status, severe acidosis, continuous seizures, and refractory hypotension. The administration of the cyanide antidote kit to children requires the appropriate adjustments for weight and hemoglobin content to avoid methemoglobinemia. Clinical features that suggest an appropriate response to nitrites and thiosulfate or hydroxocobalamin include the following: normalization of vital signs (blood pressure, pulse), normal mentation, cessation of seizure activity, spontaneous respirations, and resolution of metabolic acidosis.[75] The presence of persistent metabolic acidosis may require a continuous infusion of sodium thiosulfate (2 g/h) for 24 hours.[36] Survival has occurred in patients with cyanide poisoning including laetrile-induced, severe cyanide intoxication (coma, metabolic acidosis, profound hypotension requiring aggressive support with vasopressors) following only supportive care.[76]

References

1. Cliff J, Coutinho J. Acute intoxication form newly-introduced cassava during drought in Mozambique. Trop Doctor 1995;25:193.
2. Vetter J. Plant cyanogenic glycosides. Toxicon 2000; 38:11–36.
3. Pieroni A. Medicinal plants and food medicines in the folk traditions of the upper Lucca Province, Italy. J Ethnopharmacol 2000;70:235–273.
4. Aregheore EM, Agunbiade OO. The toxic effects of casava (*Manihot esculenta Grantz*) diets on humans: a review. Vet Human Toxicol 1991;33:274–275.
5. Hall WT. Cycad (zamia) poisoning in Australia. Aust Vet J 1987;64:149–151.
6. Seigler DS. Isolation and characterization of naturally occurring cyanogenic compounds. Phytochemistry 1975;14: 9–29.
7. Poulton JE. Localization and catabolism of cyanogenic glycosides. Ciba Found Symp 1988;140:67–91.
8. Voldrich M, Kyzlink V. Cyanogenesis in canned stone fruits. J Food Sci 1992;57:161–162, 189.
9. Haisman DR, Knight DJ. The enzymic hydrolysis of amygdalin. Biochem J 1967;103:528–534.
10. Holzbecher MD, Moss MA, Ellenberger HA. The cyanide content of laetrile preparations, apricot, peach and apple seeds. Clin Toxicol 1984;22:341–347.
11. Fuuda T, Ito H, Mukainaka T, Tokuda H, Nishino H, Yoshida T. Anti-tumor promoting effect of glycosides from *Prunus persica* seeds. Biol Pharm Bull 2003; 26:271–273.
12. Butter GW. The distribution of the cyanoglycoside linamarin and lotaustralin in higher plants. Phytochemistry 1965;4:127–131.
13. Mlingi NL, Bainbridge ZA, Poulter NH, Rosling H. Critical stages in cyanogen removal during cassava processing in southern Tanzania. Food Chem 1995;53:29–33.
14. Padmaja G. Cyanide detoxification in cassava for food and feed uses. Crit Rev Food Sci Nutr 1995;35:299–339.
15. Tylleskar T, Banea M, Bikangi N, Cooke RD, Poulter NH, Rosling H. Cassava cyanogens and konzo, an upper motoneuron disease found in Africa. Lancet 1992; 339:208–211.
16. Bradbury JH, Egan SV, Lynch MJ. Analysis of cyanide in cassava using acid hydrolysis of cyanogenic glucosides. J Sci Food Agric 1991;55:277–290.
17. Onabolu AO, Oluwole OS, Bokanga M. Loss of residual cyanogens in a cassava food during short-term storage. Int J Food Sci Nutr 2002;53:343–349.
18. Chang S-S, Chan Y-L, Wu M-L, Deng J-F, Chiu F-F, Chen J-C, et al. Acute cycad seed poisoning in Taiwan. J Toxicol Clin Toxicol 2004;42:49–54.
19. Montgomery RD. The medical significance of cyanogen in plant foodstuffs. Am J Clin Nutr 1965;17:103–113.
20. Adeparusi EO. Effect of processing on the nutrients and anti-nutrients of lima bean (*Phaseolus lunatus* L.) flour. Nahrung 2001;45:94–96.
21. Ballhorn DJ, Lieberei R, Ganzhorn JU. Plant cyanogenesis of *Phaseolus lunatus* and its relevance for herbivore plant interaction: the importance of quantitative data. J Chem Ecol 2005;31:1445–1473.
22. Wajant H, Mundry K-W, Pfizenmaier K. Molecular cloning of hydroxynitrile lyase from *Sorghum bicolor* (L.). Homologies to serine carboxypeptidases. Plant Mol Biol 1994;26:735–746.
23. Oomah BD, Mazza G, Kenaschuk EO. Cyanogenic compounds in flaxseed. J Agric Food Chem 1992;40: 1346–1348.
24. Chadha RK, Lawrence JF, Ratnayake WM. Ion chromatographic determination of cyanide released from flaxseed under autohydrolysis conditions. Food Addit Contam 1995;12:527–533.
25. Hall C III, Tulbek MC, Xu Y. Flaxseed. Adv Food Nutr Res 2006;51:1–97.
26. Cunnane SC, Ganguli S, Menard C, Liede AC, Hamadeh MJ, Zhen-Yu C, et al. High a-linolenic acid flaxseed (*Linum usitatissimum*): some nutritional properties in humans. Br J Nutr 1993;69:443–453.
27. Stoewsand GS, Anderson JL, Lamb RC. Cyanide content of apricot kernels. J Food Sci 1975;40:1107.
28. Miller KW, Anderson JL, Stoewsand GS. Amygdalin metabolism and effect on reproduction of rats fed apricot kernels. J Toxicol Environ Health 1981;7:457–467.

29. Rubino MJ, Davidoff F. Cyanide poisoning from apricot seeds. JAMA 1979;241:359.
30. Egli KL. Colorimetric determination of cyanide liberated from apricot kernels. J AOAC 1977;60:954–956.
31. Machel AR, Dorsett CI. Cyanide analysis of peaches. Econ Bot 1970;24:51–52.
32. Bainbridge Z, Harding S, French L, Kapinga R, Westby A. A study of the role of tissue disruption in the removal of cyanogens during cassava root processing. Food Chem 1998;62:291–297.
33. Mlingi N, Abrahamsson M, Yuen J, Gebre-Medhin M, Rosling H. Low cyanide exposure from consumption of cassava in Dar es Salaam, Tanzania. Nat Toxins 1998;6;6:67–72.
34. Mlingi N, Poulter N, Rosling H. An outbreak of acute intoxication from consumption of insufficiently processed cassava in Tanzania. Nutr Res 1992;12;677–687.
35. Ballhorn, DJ, Heil M, Lieberei R. Phenotypic plasticity of cyanogenesis in lima bean *Phaseolus lunatus*—activity and activation of β-glucosidase. J Chem Ecol 2006; 32:261–275.
36. Suchard JR, Wallace KL, Gerkin RD. Acute cyanide toxicity caused by apricot kernel ingestion. Ann Emerg Med 1998;32:742–744.
37. Townsend WA, Boni B. Cyanide poisoning from ingestion of apricot kernels. MMWR 1975;24:427.
38. Hernández T, Lundquist P, Oliveira L, Pérez Cristiá R, Rodriguez E, Rosling H. Fate in humans of dietary intake of cyanogenic glycosides from roots of sweet cassava consumed in Cuba. Nat Toxins 1995;3;114–117.
39. Oluvole OSA, Onabolu AO, Sowunmi A. Exposure to cyanide following a meal of cassava food. Toxicol Lett 2002;135:19–23.
40. Rauws AG, Olling M, Timmerman A. The pharmacokinetics of prunasin, a metabolite of amygdalin. J Toxicol Clin Toxicol 1982–1983;19:851–856.
41. Frakes RA, Sharma RP, Willhite CC. Comparative metabolism of linamarin and amygdalin in hamsters. Food Chem Toxicol 1986;24:417–420.
42. Poulton JE. Localization and catabolism of cyanogenic glycosides. Ciba Found Symp 1988;140:67–91.
43. Moertel CG, Ames MM, Kovach JS, Moyer TP, Rubin JR, Tinker JH. A pharmacologic and toxicological study of amygdalin. JAMA 1981;245:591–594.
44. Ames MM, Kovach JS, Flora KP. Initial pharmacologic studies of amygdalin (laetrile) in man. Res Commun Chem Pathol Pharmacol 1978;22:175–185.
45. Rauws AG, Olling M, Timmerman A. The pharmacokinetics of amygdalin. Arch Toxicol 1982;49:311–319.
46. Ames MM, Moyer TP, Kovach JS, Moertel CG, Rubin J. Pharmacology of amygdalin (laetrile) in cancer patients. Cancer Chemother Pharmacol 1981;6:51–57.
47. Carlsson L, Mlingi N, Juma A, Ronquist G, Rosling H. Metabolic fates in humans of linamarin in cassava flour ingested as stiff porridge. Food Chem Toxicol 1999; 37:307–312.
48. Akintonwa A, Tunwashe OL. Fatal cyanide poisoning from cassava-based meal. Hum Exp Toxicol 1992;11: 47–49.
49. Espinoza OB, Perez M, Ramirez MS. Bitter cassava poisoning in eight children: a case report. Vet Hum Toxicol 1992;34:65.
50. Braico KT, Humbert JR, Terplan KL, Lehotay JM. Laetrile intoxication. Report of a fatal case. N Engl J Med 1979;300:238–240.
51. Beamer WC, Shealy RM, Prough DS. Acute cyanide poisoning from laetrile ingestion. Ann Emerg Med 1983;12:449–451.
52. Sayre JW, Kaymakcalan S. Cyanide poisoning of apricot seeds among children in central Turkey. N Engl J Med 1964;270:1113–1115.
53. Hirono I, Kachi H, Kato T. A survey of acute toxicity of cycads and mortality rate from cancer in the Miyako Islands, Okinawa. Acta Pathol Japan 1970;20: 327–337.
54. Hung Y-M, Hung S-Y, Olson KR, Chou K-J, Lin S-L, Chung H-M, et al. Yam bean seed poisoning mimicking cyanide intoxication. Intern Med J 2007;37:130–132.
55. Narongchai P, Narongchai S, Thampituk S. The first fatal case of yam bean and rotenone toxicity in Thailand. J Med Assoc Thai 2005;88:984–987.
56. Thomas PK, Plant GT, Baxter P, Bates C, Santiago Luis R. An epidemic of optic neuropathy and painful sensory neuropathy in Cuba: clinical aspects. J Neurol 1995;242: 629–638.
57. Osuntokun BO. Epidemiology of tropical nutritional neuropathy in Nigerians. Trans R Soc Trop Med Hyg 1971;65:454–479.
58. Onabolu AO, Oluwole OS, Bokanga M, Rosling H. Ecological variation of intake of cassava food and dietary cyanide load in Nigerian communities. Pub Health Nutr 2001;4:871–876.
59. Oluwole OS, Onabolu AO, Cotgreave IA, Rosling H, Persson A, Link H. Low prevalence of ataxic polyneuropathy in a community with high exposure to cyanide from cassava foods. J Neurol 2002;249:1034–1040.
60. Oluwole OS, Onabolu AO, Cotgreave IA, Rosling H, Persson A, Link H. Incidence of endemic ataxic polyneuropathy and its relation to exposure to cyanide in a Nigerian community. J Neurol Nerosurg Psychiatry 2003; 74:1417–1422.
61. Tylleskar T, Legue FD, Peterson S, Kpizingui E, Stecker P. Konzo in the Central African Republic. Neurology 1994;44:959–961.
62. Mwanza J-C, Tshala-Katumbay D, Tylleskar T. Neuro-ophthalmologic manifestations of konzo. Environ Toxicol Pharmacol 2005;19:491–496.
63. Mwanza J-CK, Tashala-Katumbay D, Kayembe KL, Eeg-Olofsson KE, Tylleskar T. Neuro-ophthalmologic findings in konzo, an upper motor neuron disorder in Africa. Eur J Ophthalmol 2003;13:383–389.

64. Cox PA, Sacks OW. Cycad neurotoxins, consumption of flying foxes, and ALS-PDC disease in Guam. Neurology 2002;58:956–959.
65. Lleonart R, Cistero A, Carreira J, Batista A, del Prado JM. Food allergy: identification of the major IgE-binding component of peach (*Prunus persica*). Ann Allergy 1992; 69:128–130.
66. Balkon J. Methodology for the detection and measurement of amygdalin in tissues and fluids. J Anal Toxicol 1982;6:244–246.
67. Rauws AG, Gramberg LG, Olling M. Determination of amygdalin and its major metabolite prunasin in plasma and urine by high pressure liquid chromatography. Pharm Wkly Sci 1982;4:172–175.
68. Tracqui A, Raul JS, Geraut A, Berthelon L, Ludes B. Determination of blood cyanide by HPLC-MS. J Anal Toxicol 2002;26:144–148.
69. Kage S, Nagata T, Kudo K. Determination of cyanide and thiocyanate in blood by gas chromatography and gas chromatography-mass spectrometry. J Chromatogr B 1996;675:27–32.
70. Ishii A, Seno H, Watanabe-Suzuki K, Suzuki O. Determination of cyanide in whole blood by capillary gas chromatography with cryogenic oven trapping. Anal Chem 1998;70:4873–4876.
71. Chinaka S, Tanaka S, Takayama N, Tsuji N, Takou S, Ueda K. High-sensitivity analysis of cyanide by capillary electrophoresis with fluorescence detection. Anal Sci 2001; 17:649–652.
72. Calafat AM, Stanfill SB. Rapid quantitation of cyanide in whole blood by automated headspace gas chromatography. J Chromatogr 2002;772:131–137.
73. Murphy KE, Schantz MM, Butler TA, Genner BA Jr, Wood LJ, Turk GC. Determination of cyanide in blood by isotope-dilution gas chromatography-mass spectrometry. Clin Chem 2006;52:458–467.
74. Ruangkanchanasetr S, Wananukul V, Suwanjutha S. Cyanide poisoning, 2 cases report and treatment review. J Med Assoc Thai 1999;82(suppl 1):S162–S166.
75. Hall AH, Linden CH, Kulig KW, Rumack BH. Cyanide poisoning from laetrile ingestion: Role of nitrite therapy. Pediatrics 1986;78:269–272.
76. O'Brien B, Quigg C, Leong T. Severe cyanide toxicity from "vitamin supplements". Eur J Emerg Med 2005;12: 257–258.

Chapter 6

CYCAD SEEDS and CHRONIC NEUROLOGIC DISEASE (*Cycas* Species)

HISTORY

Neurodegenerative diseases are disorders of the central nervous system that result from the selective and premature atrophy of functionally related neurons without obvious causes, such as slow viruses. The most prevalent neurodegenerative human diseases are Alzheimer's dementia (AD), Parkinson's disease, and amyotrophic lateral sclerosis (ALS). Between 1942 and 1944, the Japanese Army occupied Guam, and the Chamorros (indigenous peoples of Guam) relied primarily on cycad as one of the few available food sources. In the 1950s, the occurrence of ALS in the Chamorro people in the Mariana Islands (Guam, Rota) markedly exceeded the prevalence of this disease in the United States and other developed countries. Further investigations revealed that other neurodegenerative diseases (Alzheimer's-like dementia, Parkinson's disease) also exceeded the prevalence of these diseases in developed countries. The term, Western Pacific amyotrophic lateral sclerosis-parkinsonism-dementia complex (WP ALS-PD), was applied to this set of neurodegenerative diseases. Excess rates of this disease were also found in the Auyu and Jakai people of West New Guinea and Japanese residents of the Kii Peninsula (Honshu Island).[1,2] A prospective, follow-up study of 899 Chamorros in the Mariana Islands detected 23 new cases of parkinsonism-dementia and 5 new cases of ALS between 1968 and 1983.[3] Of 23 variables selected during baseline examinations in 1968, only the preference for traditional Chamorro food was significantly associated with an increased risk of parkinsonism-dementia. The decline of WP ALS-PD after World War II along with the absence of recognizable transmissible or heritable factors resulted in a search for neurotoxins in traditional foods (e.g., cycad flour) and medicines. Early studies suggested that β-*N*-methylamino-L-alanine (L-BMAA), an amino acid present in food sources derived from the seeds of the cycad tree (*Cycas circinalis*), was a potential slow neurotoxin.[4] This compound is a *N*-methyl-D-aspartate (NMDA)-receptor agonist at high concentrations and a non-NMDA-receptor agonist at low concentrations that is structurally related to, but much less potent than the neurotoxin associated with lathyrism (β-*N*-oxalyl-amino-L-alanine).[5] However, experimental models using L-BMAA have not reproduced WP ALS-PD, and the consumption of large amounts of processed (i.e., washed, fermented) cycad seeds does not necessarily produce the disease.[6] Although the incidence of neurodegenerative disease has declined and the median age of onset has increased since the 1950s along with a reduction in the consumption of cycad seeds, neurodegenerative diseases continue to significantly impact the health of the Chamorro people of Guam.[7]

BOTANICAL DESCRIPTION

The Cycadaceae (cycad) family contains about 45 primitive species of seed-bearing plants that probably dominated the world's vegetation during the Jurassic and early Cretaceous periods.[8] Although the cycad tree has fronds and an appearance similar to the sago palm

Medical Toxicology of Natural Substances, by Donald G. Barceloux, MD

(*Metroxylon sagu* Rottb.) from the Arecaceae (palm) family, the cycad tree (*Cycas circinalis* L.) is not a true palm like the true sago palm. Because the cycad tree has fronds, a popular term for this tree is the false sago palm. The genus, *Cycas*, contains other species including *C. revoluta* Thunb. (sago palm) and C. *rumphii* (Queen sago).

Common Name: Cycad tree, false sago palm, queen sago

Scientific Name: *Cycas circinalis* L.

Botanical Family: Cycadaceae

Physical Description: A perennial evergreen, tree-like herb with pinnate foliage and pollen cones. The mature seeds are pale reddish yellow.

Distribution and Ecology: Species in the genus are widely distributed throughout the world from Madagascar and the east coast of Africa to northern Australia, the Mariana Islands, and southern Japan. In southern Florida, the Gulf Coast, and warm sections of California, *C. revoluta* and *C. circinalis* are used as ornamental plants.

Common Name: Sago palm, king sago palm, cycad, bread tree (Africa)

Scientific Name: *Cycas revoluta* Thunb.

Botanical Family: Cycadaceae

Physical Description: A slow growing fern-like tree that reaches up to 3 m (10 ft), usually as a thick, shaggy, solitary trunk topped by a terminal crown of stiff, deep glossy green, feathery palm-like pinnate fronds 1–2 m (3–7 ft) long. The shiny leaflets are tipped with spines and curved at the margins. The brown, petal-less flowers are clustered together, and form an orange fruit.

Distribution and Ecology: This cycad species is a native of Southern Japan, but has been naturalized throughout temperate and tropical habitats including Taiwan.

Common Name: Queen sago

Scientific Name: *Cycas rumphii* Miq.

Physical Description: The bright green, glossy leaves are about the same size as *C. circinalis* with 150–200 leaflets. The fusiform pollen cones are yellow to pale brown. The flatten-ovoid seed are 45 mm (2 in.) long with a spongy endocarp present.

Distribution and Ecology: This plant is a native species of Indonesia that inhabits closed woodlands or forests near shorelines in S. Borneo, N.E. Java, Maluku Papua, and Sulawesi. Figure 6.1 displays a female specimen of *C. rumphii* with seeds on the edges of the leaves.

FIGURE 6.1. Female specimen of *Cycas rumphii.* Miq. © 2005 Rune Midtgaard. See color insert.

EXPOSURE

In traditional Chinese medicine, the seeds of *Cycas revoluta* are used as treatments for gastrointestinal disturbances, musculoskeletal pain, cough, and hypertension.

PRINCIPAL TOXINS

Unwashed cycad flour contains cycasin (methylazoxymethanol *beta*-D-glucoside), an amino sugar with a toxic aglycone called MAM (methylazoxymethanol). This compound is unique to the cycad family, and cycasin is a weak cyanogen compound (pseudocyanogen). In a study of 17 samples of traditionally processed cycad flours obtained from Chamorro residents of Guam and the adjacent island of Rota, the content of β-*N*-methylamino-L-alanine (BMAA) and methylazoxymethanol *beta*-D-glucoside (cycasin) ranged from nondetectable to 18.39 μg/g (M = 5.44 ± 1.56 μg/g) and 0.004–75.93 μg/g (M = 12.45 ± 5.0 μg/g), respectively, as measured by high performance liquid chromatography (HPLC).[9] Animal tests suggest that MAM is a hepatic carcinogen and animal teratogen, but neuropathology in animals does not correlate to the abnormalities associated with the WP ALS-PD complex.[10,11] Washing, boiling, fermenting, and storage of cycad flour removes substantial amounts of toxins (cycasin, MAM, L-BMAA) as well as β-glycosidase enzymes.[12] Unwashed cycad flour also contains the neurotoxin, β-methylamino-L-alanine (L-BMAA), which is a *N*-methyl-D-aspartate (NMDA) agonist that produces neurodegeneration following some routes of administration. However, washing of cycad flour substantially reduces the L-BMAA content, and the residual amount (0.005%) of L-BMAA after washing may not be sufficient to cause neurodegenerative disease.[13] In animal studies, L-BMAA is substantially less neurotoxic than the structurally related excitatory plant amino acid, β-*N*-oxalylamino L-alanine (L-BOAA), which is associated with lathyrism.[14] Animal studies suggest that another neurotoxin (e.g., β-sitosterol-β-D-glucoside) or a toxic metabolite is a causal factor in WP ALS-PD complex.[15,16] Other nonprotein amino acids present in cycad flour include β-alanine, α-aminoadipic acid, γ-aminobutyric acid, and pyroglutamic acid.[17]

DOSE RESPONSE

The ingestion of 1–30 seeds from *C. revoluta* was associated with severe gastrointestinal distress and other nonspecific symptoms that resolved within 24 hours.[18] These patients did not develop clinically significant respiratory depression or hypotension.

TOXICOKINETICS

Cycasin is a pseudocyanogenic compound that releases minor amounts of hydrogen cyanide following the hydrolysis of cycasin by bacterial β-glycosidases in the alkaline environment of the small intestines. Most of the aglycone (methylazoxymethanol) formed by this reaction is degraded to nitrogen gas, methanol, and formaldehyde, particularly in the acidic conditions of the stomach. In contrast to cycasin, the unstable cyanohydrin (α-hydroxynitrile) formed from cyanogenic compounds (amygdalin, prunasin) releases substantial amounts of hydrogen cyanide following hydrolysis of these compounds.

CLINICAL RESPONSE

In domestic animals and humans, ingestion of cycad seeds causes the following two distinct syndromes: 1) an acute gastrointestinal illness sometimes involving the liver, and 2) a chronic neurological disorder characterized by ataxia.

Acute

The ingestion of improperly or unprocessed cycad seeds can cause severe gastrointestinal distress manifest by nausea, vomiting, diarrhea, headache, and lightheadedness. Although historical accounts suggest that the acute ingestion of cycad seeds can cause loss of consciousness, hepatitis, and death,[19] the role of cycad seeds in the development of these serious complications is unclear because of the lack of direct evaluation of the affected patients. In a retrospective chart review conducted at the Poison Control Center in Taiwan, 21 cases of cycad seed poisoning were identified.[18] Severe vomiting was the most prominent symptom with a mean onset about 3 hours after ingestion of the cycad seeds. Other symptoms included abdominal pain, headache, dizziness, weakness, tachycardia, diarrhea, and hypertension. Although blood cyanide and thiocyanate concentrations were elevated in the 6 patients tested, these concentrations did not reach toxic concentrations.

Chronic

WP ALS-PD complex is a chronic neurodegenerative disease prevalent in small populations in the Mariana Islands, West New Guinea, and the Japanese island of Honshu. The onset of the disease and the progression of the disease are variable with both rapid and slow (i.e., years to decades) onset after exposure to suspected causal factors (cycad seeds).[20] Characteristics of ALS include progressive weakness of the extremities, spasticity, fasciculations, muscle atrophy, and bulbar dysfunction. The parkinsonism-dementia complex includes bradykinesia, resting tremor, decreased blink reflex, loss of fine motor movements, muscle rigidity, dysphagia, dysarthria, incontinence, upper- and lower-motor neurologic signs with progression to flexion-contractures, and progressive dementia. Pathological changes involve the presence of neurofibrillary tangles in association with prominent degeneration of the neurons in the upper and lower motor neurons, hippocampus, substantia nigra, and medial basal forebrain.[21]

The suspected link is the presence of symbiotic, free-living cyanobacteria in the coralloid roots of cycad trees that produce BMAA. BMAA is concentrated in the cycad seeds and also in flying foxes (*Pteropus mariannus* Desmarest) that forage on these seeds.[22] Flying foxes are a delicacy for the indigenous Chamorro people, who eat the fox after boiling the whole fox in coconut cream. A case-control study of patients with dementia, mild cognitive impairment, or parkinsonism-dementia complex on Guam demonstrated an increased risk for patients exposed to traditional cycad-derived food (fadang), particularly for men.[23] The adjusted odds ratio for eating fadang in young adulthood was 2.87 (95% CI: 1.48–5.56). Although constituents in unprocessed flour are carcinogenic in long term animal studies,[24] there are inadequate animal and human data to indicate that the consumption of processed cycad flour increases the incidence of cancer.[25]

DIAGNOSTIC TESTING

Analytical methods to detect the presence of BMAA in plant and animal tissue in the subpicomole range include gas chromatography mass spectrometry (GC/MS)[17] and reversed phase HPLC method with precolumn derivatization by 9-fluorenylmethyl chloroformate.[26] Based on an electrophysiology study of 29 Chamorro patients with neurodegenerative diseases, a peripheral neuropathy can occur in these patients. Three patients with parkinsonism-dementia out of 16 patients with neurodegenerative disease had mild reduction in central nerve conduction manifest by mildly abnormal tibial somatosensory evoked potentials (SEPs).[27]

TREATMENT

Treatment is supportive. Acute ingestion of unprocessed cycad seeds produces gastroenteritis, and patients should be evaluated for fluid and electrolyte balance depending on the severity of symptoms. Decontamination measures are unnecessary.

References

1. Spencer PS, Ohta M, Palmer VS. Cycad use and motor neurone disease in Kii peninsula of Japan. Lancet 1987;2(8573):1462–1463.
2. Gajdusek DC, Salazar AM. Amyotrophic lateral sclerosis and parkinsonian syndromes in high incidence among the Auyu and Jakai people of West New Guinea. Neurology 1982;32:107–126.
3. Reed D, Labarthe D, Chen KM, Stallones R. A cohort study of amyotrophic lateral sclerosis and parkinsonism-dementia on Guam and Rota. Am J Epidemiol 1987;125:92–100.
4. Spencer PS, Nunn PB, Hugon J, Ludolph AC, Ross SM, Roy RD, Robertson RC. Guam amyotrophic lateral sclerosis-Parkinsonism-dementia linked to a plant excitant neurotoxin. Science 1987;237:517–522.
5. Allen CN, Spencer PS, Carpenter DO. Beta-*N*-methylamino-L-alanine in the presence of bicarbonate is an agonist at non-*N*-methyl-D-aspartate-type receptors. Neuroscience 1993;54:567–574.
6. Spencer PS, Allen CN, Kisby GE, Ludolph AC. On the etiology and pathogenesis of chemically induced neurodegenerative disorders. Neurobiol Aging 1994;15:265–267.
7. Waring SC, Esteban-Santillan C, Reed DM, Craig UK, Labarthe DR, Petersen RC, Kurland LT. Incidence of amyotrophic lateral sclerosis and of the parkinsonism-dementia complex of Guam, 1950–1989. Neuroepidemiology 2004;23:192–200.
8. Burrows GE, Tyrl RJ. Toxic plants of North America. Ames, IO: Iowa State University Press; 2001.
9. Kisby GE, Ellison M, Spencer PS. Content of the neurotoxins cycasin (methylazoxymethanol beta-D-glucoside) and BMAA (beta-*N*-methylamino-L-alanine) in cycad flour prepared by Guam Chamorros. Neurology 1992; 42:1336–1340.
10. Lazueur GL, Mickelsen O, Whiting MG, Kurland LT. Carcinogenic properties of nuts from *Cycas circinalis* L. indigenous to Guam. J Natl Cancer Inst 1963;31:919–951.
11. Smith DW. Mutagenicity of cycasin aglycone (methylazoxymethanol), a naturally occurring carcinogen. Science 1966;152:1273–1274.
12. Dastur DK, Palekar RS. Effect of boiling and storing on cycasin content of *Cycas circinalis* L. Nature 1966; 210:841–843.
13. Duncan MW, Steele JC, Kopin IJ, Markey SP. 2-Amino-3-(methylamino)-propanoic acid (BMAA) in cycad flour: an unlikely cause of amyotrophic lateral sclerosis and parkinsonism-dementia of Guam. Neurology 1990;40:767–772.
14. Pai KS, Shankar SK, Ravindranath V. Billionfold difference in the toxic potencies of two excitatory plant amino acids, L-BOAA and L-BMAA: biochemical and morphological studies using mouse brain slices. Neurosci Res 1993;17:241–248.
15. Khabazian I, Bains JS, Williams DE, Cheung J, Wilson JM, Pasqualotto BA, et al. Isolation of various forms of sterol β-D-glucoside from the seed of *Cycas circinalis*: neurotoxicity and implications of ALS-parkinsonism dementia complex. J Neurochem 2002;82:516–528.
16. Wilson JM, Khabazian I, Wong MC, Seyedalikhani A, Bains JS, Pasqualotto BA, et al. Behavioral and neurological correlates of ALS-Parkinsonism dementia complex in adult mice fed washed cycad flour. Neuromolecular Med 2002;1:207–221.
17. Oh C-H, Brownson DM, Marby TJ. Screening for non-protein amino acids in seeds of the Guam cycad, *Cycas circinalis*, by an improved GC-MS method. Planta Med 1995;61:66–70.
18. Chang S-S, Chan Y-L, Wu M-L, Deng J-F, Chiu T-F, Chen J-C, et al: Acute *Cycas* seed poisoning in Taiwan. J Toxicol Clin Toxicol 2004;42:49–54.
19. Hirono I, Kachi H, Kato T. A survey of acute toxicity of cycads and mortality rate from cancer in the Miyako Islands, Okinawa. Acta Pathol Jpn 1970;20:327–337.
20. Kurland LT. Amyotrophic lateral sclerosis and Parkinson's disease complex on Guam linked to an environmental neurotoxin. Trends Neurosci 1988;11:51–54.
21. Hirano A, Malamud N, Elizan TS, Kurland LT. Amyotrophic lateral sclerosis and Parkinsonism-dementia complex on Guam. Further pathologic studies. Arch Neurol 1966;15:35–51.
22. Cox PA, Banack SA, Murch SJ. Biomagnification of cyanobacterial neurotoxins and neurodegenerative disease

among the Chamorro people of Guam. Proc Natl Acad Sci U S A 2003;100:13380–13383.

23. Borenstein AR, Mortimer JA, Schofield E, Wu Y, Salmon DP, Gamst A, et al. Cycad exposure and risk of dementia, MCI, and PDC in the Chamorro population of Guam. Neurology 2007;68:1764–1771.
24. Sieber SM, Correa P, Dalgard DW, McIntire KR, Adamson RH. Carcinogenicity and hepatotoxicity of cycasin and its aglycone methylazoxymethanol acetate in nonhuman primates. J Natl Cancer Inst 1980;65:177–189.
25. Yang MG, Mickelsen O, Campbell ME, Laqueur GL, Keresztesy JC. Cycad flour used by Guamanians: effects produced in rats by long-term feeding. J Nutr 1966;90:153–156.
26. Kisby GE, Roy DN, Spencer PS. Determination of β-*N*-methylamino-L-alanine (BMAA) in plant (*Cycas circinalis* L.) and animal tissue by precolumn derivatization with 9-fluorenylmethyl chloroformate (FMOC) and reversed-phase high-performance liquid chromatography. J Neurosci Methods 1988;26:45–54.
27. Ahlskog JE, Litchy WJ, Peterson RC, Waring SC, Esteban-Santillan C, Chen KM, et al. Guamanian neurodegenerative disease: electrophysiologic findings. J Neurol Sci 1999;166:28–35.

Chapter 7

DJENKOL BEAN
[*Archidendron jiringa* (Jack) I. C. Nielsen]

BOTANICAL DESCRIPTION

Common Name: Djenkol tree (Indonesia), krakos (Cambodia), jering (Malaysia), niang-yai (Thailand), jenkol, genkol, yiniking, yi-ring, ma-niang, cha-niang, niang, kra-niang

Scientific Name: *Archidendron jiringa* (Jack) I. C. Nielsen [*Abarema jiringa* Kosterm, *Albizzia jiringa* (Jack) Kurz, *Inga jiringa* (Jack) Jack, *Mimosa jiringa* Jack, *Pithecellobium jiringa* (Jack) Prain, *Pithecellobium lobatum* Benth., *Zygia jiringa* (Jack) Kosterm]

Botanical Family: Mimosaceae (Mimosa)

Physical Description: The djenkol bean has a broad, round shape and reddish color. The beans grow in large, dark purple pods (3–9 beans/pod) on the djenkol tree, which grows to 25 m (82 ft) in height. These beans resemble a large, flattened horse chestnut, and the crushed bean emits a faint sulfurous odor.

EXPOSURE

Djenkol beans are a local delicacy in Indonesia, Malaysia, southern Thailand, and Myanmar. The beans are commercially available in markets during most of the year and are consumed raw, roasted, or fried.

PRINCIPAL TOXINS

Chemical Composition

The djenkol bean contains the sulfur-containing amino acid, djenkol acid (CAS RN: 498-59-9, $C_{11}H_{23}N_3S_3O_6$) in the range of about 0.3–1.3 g/100 g wet weight.[1] This compound is a cysteine thioacetal of formaldehyde as displayed in Figure 7.1.

Physiochemical Properties

Boiling, frying, or roasting the djenkol bean does not necessarily prevent toxicity as a case report temporally associated the development of acute renal failure with the ingestion of fried djenkol beans.[2] Typically, the ingestion of djenkol beans produces a pungent smell in the breath and urine of affected individuals.[3]

Mechanism of Toxicity

The mechanism of renal failure has not been clearly established. Animal studies demonstrate variable amounts of acute tubular necrosis with occasional glomerular involvement after the oral administration of djenkolic acid extracted from djenkol beans with 70% ethanol and water.[1] The peak incidence of djenkolism

Medical Toxicology of Natural Substances, by Donald G. Barceloux, MD

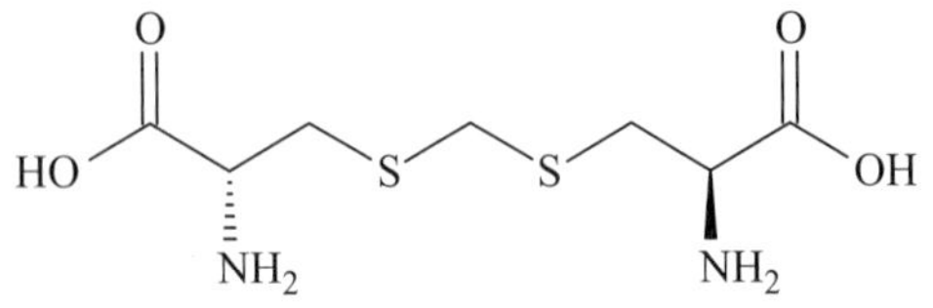

FIGURE 7.1. Chemical structure of djenkol acid.

occurs in the rainy season (September–January) corresponding to the bloom time of the djenkol tree.[4]

DOSE RESPONSE

Most people eat djenkol beans without ill effects, and there is no clear dose-response effect for the development of djenkolism. Although the occurrence of renal failure is frequently associated with the consumption of raw beans with low fluid intake,[5] the development of djenkolism does not depend on the method of preparation or the age of the fruit. The dose of djenkol beans associated with acute renal failure is highly variable (1–20 beans).[6] The development of djenkol bean poisoning can occur after numerous symptom-free meals of djenkol beans, and the attack rate can be low in large families consuming similar amounts of djenkol beans. Prior poisoning with djenkol beans does not confer immunity or hypersensitivity to subsequent meals of djenkol beans.

TOXICOKINETICS

There are inadequate data to determine the toxicokinetics of djenkol acid in humans.

CLINICAL RESPONSE

Acute Effects

The clinical features of djenkolism range from asymptomatic hematuria, dysuria, and abdominal pain to gross hematuria, severe flank pain, and acute renal failure. Mild djenkolism develops within 2–12 hours after ingesting djenkol beans, manifest by hypogastric cramps, flank pain, vomiting, diarrhea, dysuria, hematuria, and opaque ("milky") urine. Case reports associate acute djenkolism with the development of acute renal failure, manifest by nausea, vomiting, bilateral flank pain, gross hematuria, hypertension, and transient oliguria.[2,7] Phase-contrast microscopy of urinary sediment and renal biopsy reported in this case report was consistent with acute tubular necrosis rather than glomerular disease. There was no evidence of ureteral obstruction. Symptoms usually resolve within 3 days with supportive care.[6]

Rare case reports associate the consumption of djenkol beans with the development of urethral stones, obstruction of the ureters, and urinary retention.[8]

Chronic Effects

In a study of 609 school children aged 7–11 years in an urban area of Thailand, 78% of the children had a history of djenkol bean consumption with 38% ingesting these beans within 24 hours of the interview.[9] The presence of hematuria was about 4-fold (crude odds ratio = 3.7) higher in children with a history of djenkol bean consumption compared with children not consuming djenkol beans. There was no significant difference in the presence of crystalluria or pyuria between these two groups.

DIAGNOSTIC TESTING

Urinary abnormalities in cases of renal failure associated with the ingestion of djenkol beans include hematuria, proteinuria, and needlelike crystals of djenkolic acid. Crystals of djenkol acid are usually, but not always, present in the urine of patients with djenkolism. Persistent or recurrent microscopic hematuria in children consuming djenkol beans is not usually associated with clinical effects, but there are few data on the long- term effects of hematuria secondary to chronic djenkol bean consumption.[9] Renal biopsies of patients with djenkolism typically demonstrate evidence of acute tubular necrosis with little interstitial nephritis, ureteral obstruction, or glomerular lesions.[2]

TREATMENT

Treatment is supportive with careful monitoring of renal function. Generous fluid intake and alkalinization potentially increase the dissolution of djenkol crystals (pK_a = 5.1), although there are few clinical data to confirm the efficacy of this treatment. Decontamination measures are unnecessary. Most patients recover without sequelae.

References

1. Areekul S, Kirdudom P, Chaovanapricha K. Studies on djenkol bean poisoning (djenkolism) in experimental animals. Southeast Asian J Trop Med Pub Health 1976;7:551–558.

2. Segasothy M, Swaminathan M, Kong NC, Bennett WM. Djenkol bean poisoning (djenkolism): an unusual cause of acute renal failure. Am J Kidney Dis 1995;25:63–66.
3. Wong JS, Ong T-A, Chua H-H, Tan C. Acute anuric renal failure following jering bean ingestion. Asian J Surg 2007;30:80–81.
4. Suharjono, Sadatun. "Djengkol" intoxication in children. Paediatr Indones 1968;8:20–29.
5. Jha V, Chugh KS. Nephropathy associated with animal, plant, and chemical toxins in the tropics. Semin Nephrol 2003;23:49–65.
6. Reimann HA, Sukaton RU. Djenkol bean poisoning (djenkolism): a cause of hematuria and anuria. Am J Med Sci 1956;232:172–174.
7. H'ng PK, Nayar SK, Lau WM, Segasothy M. Acute renal failure following jering ingestion. Singapore Med J 1991;32:148–149.
8. Areekul S, Muangman V, Bohkerd C, Saenghirun C. Djenkol bean as a cause of urolithiasis. Southeast Asian Trop Med Pub Health 1978;9:427–432.
9. Vachvanichsanong P, Lebel L. Djenkol beans as a cause of hematuria in children. Nephron 1997;76:39–42.

Chapter 8

GRASS PEA and NEUROLATHYRISM (*Lathyrus sativus* L.)

HISTORY

Carbonized fossils indicate that the grass pea was domesticated in the Near East, and then cultivated with cereals in the Balkans since the Neolithic period.[1] Human lathyrism is a spastic lower-extremity paraplegia that has been associated with the ingestion of grass peas since the time of Hippocrates.[2] The Duke of Wurttenberg prohibited the baking and consumption of bread containing grass peas in Germany during the late 17th century.[3] In 1873, Cantani identified this neurological disease as lathyrism.[4] Epidemics of neurolathyrism usually occur during times of food shortages, such as drought, flooding, or wars.[5,6] Epidemics of neurolathyrism occurred during the Spanish Civil War (1936–1939), in China, Bangladesh, and India during the 1970s, and in Ethiopia during the 1970s and 1990s.[7]

BOTANICAL DESCRIPTION

Although *Lathyrus sativus* is the species commonly associated with neurolathyrism, other *Lathyrus* species also contain neurotoxins including *Lathyrus cicera* L. (red pea) and *Lathyrus clymenum* L. There are more than 100 *Lathyrus* species in the United States. The chick pea (*Cicer arietinum* L.) is a nontoxic legume that does not produce neurolathyrism following chronic consumption.

Common Name: Grass pea, white pea, chickling pea, vetchling, khasari (India), guaya (Ethiopia), shan li dou (China), pois carrè (France)

Scientific Name: *Lathyrus sativus* L.

Botanical Family: *Fabaceae* (pea)

Physical Description: A perennial vine that grows up to 1 m (about 3 ft) in length. The pinnately compound leaves usually have two, small linear-lanceolate leaflets with the upper leaflets being tendrils.

Distribution and Ecology: *Lathyrus sativus* is a vine introduced into the United States from Europe with distribution limited to Minnesota, California, Oregon, Maryland, and Massachusetts. This legume grows as a weed in cultivated land, particularly in moist areas with full sun. However, this plant is relatively drought tolerant and widely cultivated in Central, South, and East Europe, the Mediterranean, Southeast Asia, and Africa.

EXPOSURE

The grass pea is an inexpensive, drought-tolerant food staple for parts of Africa (e.g., Ethiopia) and Asia (e.g., Bangladesh, India, and Nepal) that are susceptible to drought. Advantageous characteristics of grass pea include tolerance to extreme water and poor soil conditions, resistance to pests, high protein content in seeds, and nitrogen fixation. Under normal conditions this substance is a small part of the diet, but during periods of food shortage the consumption of grass peas increases dramatically because of the ability of *Lathyrus sativus* to survive under harsh environmental conditions.[8] The Indian staple, ghotu, is prepared by cooking a mixture of grass pea and rice to form a stiff porridge. In Ethio-

Medical Toxicology of Natural Substances, by Donald G. Barceloux, MD

pia, chapati and kitta are unleavened breads formed from grass pea seeds. The latter bread may contain high concentrations of neurotoxins.[9] This legume is used as flour in cattle feed.

PRINCIPAL TOXINS

Structure and Properties

The spastic paraparesis of neurolathyrism results from the ingestion of peas and foliage from drought-tolerant *Lathyrus sativus* that contain the neuroexcitatory amino acid, 3-*N*-oxalyl-L-2,3-diaminopropanoic acid (β-ODAP).[10] A synonym for β-ODAP is L-2-amino-3-oxalylaminopropanoic acid (β-oxalylaminoalanine). β-ODAP is a structural analog of glutamine.[11] The α-isomer represents only about 5% of the total quantity of ODAP, but heat and storage at neutral pH causes increased formation of α-ODAP.[12] Prolonged cooking decreases the amount of β-ODAP in grass pea seeds by converting this neurotoxin to the less toxic isomer, α-ODAP.[13] Animal studies indicate that the α-isomer possesses relatively low neurotoxicity compared with β-ODAP, and the conversion of β-ODAP to α-ODAP during food processing reduces the toxicity of chickling peas.[14] In addition to β-ODAP, the seedlings of *L. sativus* also contain other potential neurotoxins [e.g., α-amino-γ-(isoxazolin-5-on-2-yl)-butyric acid].[15] β-ODAP is water soluble, and boiling grass peas for 1 hour followed by decanting the water removes about 70% of the β-ODAP from the pea.[16]

Poisonous Parts

The average concentration of β-ODAP in ripe seeds from *L. sativus* is usually <1%, depending on the variety, season, and growing conditions.[8] The mean β-ODAP content in dry seeds purchased in an Ethiopian market was 0.74% wet weight as measured by flow injection assay.[17] In general, seedlings contain the highest concentrations of neurotoxins compared with ripe seeds.[18] Analysis of 3-day-old seedlings demonstrated β-ODAP concentrations ranging from 0.34–0.79 mg/g dry weight in low-toxin lines to 5.5–7.16 mg/g dry weight in wild types.[19] In a study of 63 samples of grass pea seeds used in food preparations at an Ethiopian village, the β-ODAP concentration ranged from 282–810 mg/100 g dry seeds with a mean of 455 mg/100 g.[9] The highest concentrations of β-ODAP occur in bread and roasted forms of grass pea foods, whereas boiled snacks and flour derived from grass pea seeds steeped in boiling water contained relatively low β-ODAP concentrations. Detoxification of grass pea seeds requires the removal of the husk and boiling of the dehusked seeds for several hours in hot water. The supernatant contains about 70–80% of the neurotoxin originally present in the grass pea seeds.[9] This water is discarded. The scarcity of water during a drought increases the risk that the grass pea seeds will not be properly detoxified. A simulated kitchen experiment suggested that steeping grass pea seeds in large volumes of water for 3 minutes along with the decanting of excess water removes approximately 30% of the β-ODAP content in the grass pea seeds.[9] Grass pea contains a 24 kDa allergenic protein, but the exact structure of this allergen has not been delineated.[20]

DOSE RESPONSE

The development of neurolathyrism depends on the daily dose of grass peas, the duration of consumption, climatic conditions, and physical condition. Generally, the development of neurolathyrism requires the ingestion of grass peas for 2–3 months as one-third to one-half of the total dietary intake.[2] Cases of neurolathyrism usually occur only during food-shortages because this disease does not appear when grass pea is a minor component of a varied diet.[13] The attack rate and severity of neurolathyrism is variable, even in the same family consuming similar diets. The daily consumption of rations consisting of 400 g of grass pea (*Lathyrus sativus*) boiled in water and 200 g baked bread (80% barley, 20% chopped straw) for 3 months was associated with various degrees of neurolathyrism in about 70% of Jewish prisoners in a German forced labor camp.[21] Ukrainian and Russian inmates survived on a diet of 200 g/day of grass peas for 3–6 months until the amount of grass peas increased to 400 g/day. Neurological signs develop in primates 3–10 months after the initiation of a grass pea diet containing 1.1–1.5 g/kg body weight.[22]

Mechanism of Toxicity

The human variety of lathyrism is a chronic neurologic disease called neurolathyrism that results from the excessive ingestion of the peas from *Lathyrus sativus*.[23] Duplication of this disease in animals is difficult. Animals fed peas from *Lathyrus* species containing β-aminopropionitrile develop skeletal deformities and aortic aneurysms rather than neurological disorders. This animal disease affects primarily collagen tissue and is called osteolathyrism. Neurolathyrism and the minor bone changes seen in humans probably result from the neurotoxin β-*N*-oxalylamino-L-alanine (L-3-oxalylamino-2-aminopropionic acid), which is structurally similar to the putative neurotransmitter glutamate.[24] Experimentally, this neurotoxin is a competitive agonist of the 3-hydroxy-5-methylisoxazole-4-propionic acid (AMPA) subtype of the glutamate receptor.[22] These inotropic quisqualate receptors are one of the postsynaptic neuronal receptors that bind the putative excitatory neurotransmitter glutamate. Persistent depolarization causes failure of energy-dependent cellular homeostasis and neuronal death. Animal studies suggest that the neurotoxicity induced by L-β-ODAP is partially mediated by the activation of group I metabotropic glutamatergic receptors by indirect mechanisms.[25] Pathological changes in patients with neurolathyrism are usually mild. The prominent abnormalities involve the symmetrical axonal degeneration of crossed and uncrossed pyramidal tracts in the lumbar region along with mild degeneration of the spinocerebellar tracts.[26] Light microscopy typically demonstrates degeneration of the corticospinal pathways, (e.g., lateral pyramidal tracts) distal to the mid-thoracic level with preservation of anterior horn cells in the spinal cord.[21]

TOXICOKINETICS

The low incidence of neurolathyrism in humans after excessive consumption of the pulse, *Lathyrus sativus*, even under severe drought and famine conditions, suggests that humans usually metabolize the major neurotoxic amino acid (β-*N*-oxalyl-L-α,β-diaminopropionic acid or β-ODAP) to a nontoxic metabolite. In contrast to animal studies, the urinary excretion of unchanged β-ODAP in humans is low.[27] The difference in toxicokinetics may explain the low incidence of neurolathyrism in humans compared with susceptible animal species.

CLINICAL RESPONSE

Central Nervous System

The ingestion of parts including peas from the genus *Lathyrus* produces a progressive, chronic intoxication rather than acute toxicity. The disease is irreversible, but neurolathyrism rarely progresses after the cessation of further consumption of grass peas. The clinical pattern of neurolathyrism involves varying spasticity of the lower extremities along with relatively mild muscle weakness. The spasticity of the lower extremities causes neurological effects ranging from ataxic gait to contractures that prevent ambulation. Increased muscle tone in the legs (thigh extensors, thigh adductors, gastrocnemius) along with muscle weakness produce inversion of the feet and a lurching, cross-legged gait.[28] Evidence of a peripheral neuropathy does not usually occur unless the patient has severe neurolathyrism. Superficial sensation, rectal sphincter tone, cranial nerve function, upper extremity muscle tone, and mentation are usually normal. In a follow-up study of 800 concentration camp victims with neurolathyrism, the number of former prisoners with spastic paresis of the lower legs, upper and lower motor lesions, and peripheral neuropathies were 80%, 15%, and 5%, respectively.[21] During the acute stage of the disease, young men (10–20 years) were most affected.

Neurolathyrism is rare in well-nourished adults and postmenopausal women.[29] There are no well-defined, predisposing factors for the development of neurolathyrism. Other features of neurolathyrism include gastrointestinal distress, urinary frequency and urgency, sexual dysfunction, and neurasthenic symptoms (insomnia, thirst, short-term memory loss).[30]

Skeletal System

Abnormalities outside the nervous system are uncommon. Some skeletal bony abnormalities were reported, but these bony abnormalities probably resulted from secondary stresses produced by the neurologic lesions.[31] In a study of 500 patients with neurolathyrism in Bangladesh, 60 patients had skeletal abnormalities and arthralgias.[32] Examination of these 60 patients demonstrated the characteristic features of osteolathyrism (e.g., failure of fusion in vertebral and iliac epiphyses) in only 2 patients.

Hypersensitivity Reactions

Occasional case reports associate exposure to *Lathyrus sativus* flour with allergic responses including occupational asthma.[33] A 42-year-old man developed rhinorrhea, pruritus, tearing, wheezing, and dyspnea after occupational exposure to saw dust that contained flour from *L. sativa*.[34] He had a history of seasonal rhinitis and a family history of atopy. Skin prick testing and bronchial provocation tests were positive following exposure to *L. sativus* flour.

DIAGNOSTIC TESTING

Analytical Methods

Methods to quantitate β-ODAP in solid samples include spectrophotometry, reverse phase high performance liquid chromatography (RP/HPLC),[35] capillary zone electrophoresis,[36] and liquid chromatography with refractory index and bioelectrochemical detection.[37] The limit of detection for β-ODAP using the latter method was 2 μM. The use of precolumn derivatization with 1-fluoro-2,4-dinitrobenzene and RP/HPLC with UV detection (360 nm) allows the detection β-ODAP in the range of 10 pmol.[38]

Abnormalities

Electrophysiological changes associated with neurolathyrism are usually minor and nonspecific. In a study of eight former Wapniarka concentration camp prisoners with neurolathyrism, all nerve conduction studies including F-wave latencies were normal.[39] However, several of the patients had electromyographic evidence of α-motor neuron disease in the lumber spinal cord that may have resulted from chronic muscle denervation. In an electrophysiological study of 17 patients with endemic neurolathyrism (mean duration 15.6 years with a range of 2–30 years), 14 patients demonstrated slowing of the central motor conduction to the tibialis anterior muscle.[40] Slowing of peripheral motor nerve conduction of the upper extremity was present in only 4 patients.

TREATMENT

Management of grass pea ingestions involves supportive care. Decontamination is not usually necessary for ingestions by children.

References

1. Kislev ME. Origin of the cultivation of *Lathyrus sativus* and *L. cicera* (Fabaceae). Econ Bot 1989;43:262–270.
2. Cohn DF, Kislev ME. Neurolathyrism. Adler Museum Bull 1987;13:5–8.
3. Cohn DF, Streifler M. Neurolathyrism. Historical notes. Koroth 1978;7:147–152.
4. Cantani A. Latirismo (Lathyrismus). Morgagni 1873; 15:745–765.
5. Gebreab T, Wolde Gabriel Z, Maffi M, Ahmed Z, Ayele T, Fanta H. Neurolathyrism. A review and report of an epidemic. Ethiop Med J 1978;16:1–11.
6. Kessler A. Lathyrismus. Psychiatrie Neurologie 1947;113: 345–376.
7. Getahun H, Lambein F, Vanhoorne M, Van der Stuyft P. Pattern and associated factors of the neurolathyrism epidemic in Ethiopia. Trop Med Int Health 2002;7:118–124.
8. Spencer PS, Ludolph AC, Kisby GE. Neurologic diseases associated with use of plant components with toxic potential. Environ Res 1993;62:106–113.
9. Tekle-Haimanot T, Abegaz BM, Wuhib E, Kassina A, Kidane Y, Kebede N, et al. Pattern of *Lathyrus sativus* (grass pea) consumption and β-*N*-oxalyl-α,β-diaminopropionic acid (β-ODAP) content of food samples in the lathyrism endemic region of northwest Ethiopia. Nutr Res 1993;13:1113–1126.
10. Haque A, Hossain M, Khan JK, Kuo YH, Lambein F, De Reuck J. New findings and symptomatic treatment for neurolathyrism, a motor neuron disease occurring in North West Bangladesh. Paraplegia 1994;32:193–195.
11. De Bruyn A, Van Haver D, Lambein F, Abegaz BM. Chemical properties of the natural neurotoxin of *Lathyrus sativus* 3-*N*-oxalyl-2,3-diamino-propanoic acid (beta-ODAP), its nontoxic 2-*N*-oxalyl isomer, and its hydrolysis product 2,3-diamino-propanoic acid (DAPRO) by 1H- and 13C-NMR spectroscopy. Nat Toxins 1993;1:328–340.
12. Khan JK, Kebede N, Kuo YH, Lambein F, De Bruyn A. Analysis of the neurotoxin β-ODAP and its α-isomer by precolumn derivatization with phenylisothiocyanate. Anal Biochem 1993;208:237–240.
13. Yan Z-Y, Spencer PS, Li Z-X, Liang Y-M, Wang Y-F, Wang C-Y, Li F-M. *Lathyrus sativus* (grass pea) and its neurotoxin ODAP. Phytochemistry 2006;67:107–121.
14. Chase RA, Pearson S, Nunn PB, Lantos PL. Comparative toxicities of alpha- and beta-*N*-oxalyl-L-alpha, beta-diaminopropionic acids to rat spinal cord. Neurosci Lett 1985;55:89–94.
15. Lambein F, Khan JK, Kuo YH. Free amino acids and toxins in *Lathyrus sativus* seedlings. Planta Med 1992; 58:380–381.
16. Jha K. Effect of the boiling and decanting method of Khesari (*Lathyrus sativus*) detoxification on changes in selected nutrients. Arch Latinoam Nutr 1987;37:101–107.
17. Moges G, Johansson G. Flow injection assay for the neurotoxin β-ODAP using an immobilized glutamate oxidase reactor with prereactors to eliminate glutamate interferences. Anal Chem 1994;66:3834–3839.
18. Ressler C. Neurotoxic amino acids of certain species of *Lathyrus* and vetch (*Vicia sativa*). Fed Proc 1964;23: 135–1353.
19. Lambein F, Khan JK, Kuo YH, Campbell CG, Briggs CJ. Toxins in the seedlings of some varieties of grass pea (*Lathyrus sativus*). Nat Toxins 1993;1:246–249.
20. Qureshi IA, Sethi DK, Salunke DM. Purification, identification and preliminary crystallographic studies of an allergenic protein from *Lathyrus sativus*. Acta Crystallograph Sect F Struct Biol Cryst Commun 2006;62:869–872.
21. Streifler M, Cohn PF. Chronic central nervous system toxicity of the chickling pea (*Lathyrus sat*ivus). Clin Toxicol 1981;18:1513–1517.
22. Spencer PS. Food toxins, AMPA receptors, and motor neuron diseases. Drug Metab Rev 1999;31:561–587.
23. Weaver AL, Spittell JA. Lathyrism. Mayo Clin Proc 1964;39:485–489.
24. Mehta T, Zarghami NS, Cusick PK, Parker AJ, Haskell BE. Tissue distribution and metabolism of the *Lathyrus sativus* neurotoxin, L-3-oxalylamino-2-aminopropionic acid in the squirrel monkey. J Neurochem 1976;27:1327–1331.
25. Kusama-Eguchi K, Kusama T, Suda A, Masuko T, Yamamoto M, Ikegami F, Igarashi K, Kuo YH, Lambein F, Watanabe K. Partial involvement of group I metabotropic glutamate receptors in the neurotoxicity of 3-*N*-oxalyl-L-2,3-diaminopropanoic acid (L-beta-ODAP). Biol Pharm Bull 2004;27:1052–1058.
26. Sachdev S, Sachdev JC, Puri D. Morphological study in a case of lathyrism. J Indian Med Assoc 1969;52:320–322.

27. Pratap Rudra MP, Singh MR, Junaid MA, Jyothi P, Rao SL. Metabolism of dietary ODAP in humans may be responsible for the low incidence of neurolathyrism. Clin Biochem 2004;37:318–322.

28. Ludolph AC, Hugon J, Dwivedi MP, Schaumburg HH, Spencer PS. Studies on the aetiology and pathogenesis of motor neuron diseases. 1. Lathyrism: clinical findings in established cases. Brain 1987;110:149–165.

29. Haque A, Hosain M, Wouters G, Lambein F. Epidemiological study of lathyrism in northwestern districts of Bangladesh. Neuroepidemiology 1996;15:83–91.

30. Spencer PS, Roy DN, Ludolph A, Hugon J, Dwivedi MP, Schaumburg HH. Lathyrism: Evidence for role of the neuroexcitatory amino acid BOAA. Lancet 1986;2: 1066–1067.

31. Paissios CS, Demopoulos T. Human lathyrism: A clinical and skeletal study. Clin Orthop 1982;23:236–249.

32. Haque A, Hossain M, Lambein F, Bell EA. Evidence of osteolathyrism among patients suffering from neurolathyrism in Bangladesh. Nat Toxins 1997;5:43–46.

33. Valdivieso R, Quirce S, Sainz T. Bronchial asthma caused by *Lathyrus sativus* flour. Allergy 1988;43:536–539.

34. Anton Girones M, de la Hoz Caballer B, Munoz Martin T, Cuevas Agustin M, Sanchez-Cano M. Occupational rhinoconjunctivitis and asthma by exposure to *Lathyrus sativus* flour. Allergol Immunopathol 2005;33:326–328.

35. Chen X, Wang F, Chen Q, Qin XC, Li Z. Analysis of neurotoxin 3-N-oxalyl-L-2,3-diaminopropionic acid and its alpha-isomer in *Lathyrus sativus* by high-performance liquid chromatography with 6-aminoquinolyl-N-hydroxysuccinimidyl carbamate (AQC) derivatization. J Agric Food Chem 2000;48:3383–3386.

36. Zhao L, Chen X, Hu Z, Li Q, Chen Q, Li Z. Analysis of beta-*N*-oxalyl-L-alpha, beta-diaminopropionic acid and homoarginine in *Lathyrus sativus* by capillary zone electrophoresis. J Chromatogr A 1999;857:295–302.

37. Yigzaw Y, Larsson N, Gorton L, Ruzgas T, Solomon T. Liquid chromatographic determination of total and beta-*N*-oxalyl-L-alpha, beta-diaminopropionic acid in *Lathyrus sativus* seeds using both refractive index and bioelectrochemical detection. J Chromatogr A 2001;929:13–21.

38. Wang F, Chen X, Chen Q, Qin X, Li Z. Determination of neurotoxin 3-N-oxalyl-2,3-diaminopropionic acid and non-protein amino acids in *Lathyrus sativus* by precolumn derivatization with 1-fluoro-2,4-dinitrobenzene. J Chromatogr A 2000;883:113–118.

39. Drory VE, Rabey MJ, Cohn DF. Electrophysiologic features inpatients with chronic neurolathyrism. Acta Neurol Scand 1992;85;401–403.

40. Misra UK, Sharma RP. Peripheral and central conduction studies in neurolathyrism. J Neurol Neurosurg Psychiatr 1994;57:572–577.

Chapter 9

NUTMEG
(*Myristica fragrans* Houtt.)

HISTORY

Nutmeg probably was imported into Europe during the 12th century by Arab merchants. For many years, this spice has been used as an aromatic stimulant, abortifacient, antiflatulent and as a means to induce menses.[1] The Portuguese discovered the nutmeg tree on the Banda Islands of Indonesia (Spice Islands) in 1512. In 1576, de Lobel reported the first case of nutmeg intoxication in a pregnant English woman, who ingested 10–12 nutmeg nuts.[2] Beginning in the 17th century, the Dutch controlled the Spice Islands, and they monopolized the spice trade until the British obtained nutmeg seedlings from the Banda Islands at the end of the 18th century. References to the central nervous system effects of nutmeg appeared in the first part of the 19th century when Purkinje developed lethargy after consuming three nutmeg nuts.[1] Because of the alleged hallucinogenic and euphoria-inducing properties of nutmeg, abuse of this spice has occurred for many years, particularly in persons with limited access to other drugs. The autobiography of Malcolm X contains accounts of his use of nutmeg while incarcerated.[3]

BOTANICAL DESCRIPTION

Common Name: Nutmeg

Scientific Name: *Myristica fragrans* Houtt.

Botanical Family: Myristicaceae (nutmeg)

Physical Description: This aromatic evergreen tree grows 9–12m (30–39ft) high with spreading branches and a yellow fleshy fruit similar in appearance to an apricot or peach. The ripe fruit splits to expose a single glossy brown nut enclosed by a scarlet aril. The tree produces fruit year round, but the harvest usually occurs in April and November.

Distribution and Ecology: The nutmeg tree is indigenous to the Maluku Province of Indonesia, formerly known as the Spice Islands. The nutmeg tree is grown commercially on the Caribbean islands of Grenada and Trinidad as well as in Central and East Java.

EXPOSURE

Traditional uses of nutmeg include the treatment of rheumatism, cholera, psychosis, stomach cramps, nausea, diarrhea, flatulence, and anxiety in addition to use as an aphrodisiac and an abortifacient. Nutmeg is not used as a narcotic or hallucinogen in traditional Indonesian culture.[4] Although a case report associated the resolution of diarrhea secondary to thyroid medullary carcinoma with the use of nutmeg,[5] there are no approved medical indications for this spice. Nutmeg is a substitute psychotomimetic substance of abuse, particularly for adolescents, students, and prisoners with limited access to other psychotomimetic agents. The unpleasant taste, large doses required for effect, frequent adverse effects, and relative lack of potency as a hallucinogen limit the abuse of nutmeg.[6]

The mature fruit from the nutmeg tree contains a 2.5cm (1in.) nut with a bright red, fleshy covering (scarlet aril). Nutmeg is the nut, whereas the dried scarlet aril is called mace. The nut and mace are dried

Medical Toxicology of Natural Substances, by Donald G. Barceloux, MD

and processed separately. Mace is also a spice, and this spice has been used as an aphrodisiac by physicians in the Near East. There are relatively few reports of poisoning following the ingestion of mace.

PRINCIPAL TOXINS

Structure and Properties

The concentration of active components depends on the botanical source, environmental conditions, storage, and analytical methods. The two oils of nutmeg are fixed oil (expressed oil, nutmeg butter) and essential oil. The fixed oil is an orange, butter-like material obtained by applying heat and hydraulic pressure to nutmeg. This oil contains primarily trimyristin, and the product has no culinary value. Essential oil of nutmeg is a steam distillate that appears as a pale-yellow, nearly colorless liquid with the characteristic odor of spice. Table 9.1 lists some properties of myristicin, which is the main psychoactive constituent of nutmeg. This compound is structurally similar to kawain and related psychoactive constituents of kava (*Piper methysticum* Forst.). Myristicin is also the major component of the aromatic ether fraction of essential oil of mace.[7] Figure 9.1 displays the chemical structure of myristicin.

Poisonous Parts

This essential oil contains from 5–15% volatile oils comprised of about 80% monoterpenes (α-pinene, sabinene, 1,8-*p*-methadiene, β-pinene, 1,4-*p*-menthadiene, camphene), 5% monoterpene alcohols, an aromatic ether fraction, and miscellaneous compounds.[7] The aromatic fraction contains myristicin, elemicin, safrole, and minor constituents (methyleugenol, eugenol, isoeugenol, toluene). The essential oil of nutmeg contains variable amounts (0.1–18%) of methyleugenol, which is a potentially genotoxic compound with DNA-binding potency similar to safrole.[8] Myristicin, elemicin, and safrole account for the majority (85–95%) of the compounds in the aromatic fraction, and myristicin represents about 4–12% of the compounds present in the essential oil.[9] Myristicin is present in plants from the carrot (Umbelliferae) family including dill, celery, parsnip, parsley, and carrot. This compound is also a minor constituent of oil of black pepper (*Piper nigrum*). The estimated intake of myristicin from these sources by the general population is a few mg daily.[10]

TABLE 9.1. Identifying Characteristics and Properties of Myristicin

Property	Value
CAS Number	607-91-0
Chemical name	1-allyl-5-methoxy-3,4-methylenedioxybenzene
Synonyms	Myristicin, methoxysafrole
Empirical formula	$C_{11}H_{12}O_3$
Molecular weight	192.22
Appearance	Colorless oil

FIGURE 9.1. Chemical structure of myristicin.

Mechanism of Toxicity

Structural similarities of myristicin to classical hallucinogenic compounds (e.g., mescaline) suggest that myristicin may act as a serotonin receptor agonist and hallucinogenic compound. However, the acute toxicity of myristicin is relatively low. Although myristicin comprises the largest fraction (i.e., about 4–8%) of compounds in the aromatic fraction of nutmeg, human studies with myristicin have not duplicated the effects of nutmeg intoxication on the central nervous system.[17] In rodent studies, myristicin and elemicin impair coordination and decreased motor activity.[11] Safrole, eugenol, and isoeugenol do not have similar behavioral effects is these animal studies.

DOSE RESPONSE

One grated nutmeg is approximately one tablespoon and weighs about 6–7 g with typical recreational doses ranging from 5–30 g.[7] The administration of 6 g nutmeg to students did not significantly alter performance on neuropsychological tests.[12] Symptoms do not usually develop following the ingestion of <10 g nutmeg.[7] The ingestion of two whole nutmegs caused moderate toxicity;[1] the ingestion of an estimated dose of 18 g of finely ground nutmeg powder resulted in prolonged periods of obtundation.[13] The ingestion of 25–28 g of nutmeg powder produced tachycardia, anxiety, miosis, paresthesias, palpitations, anticholinergic signs (mydriasis, difficulty voiding, tachycardia), and paranoid behavior without hallucinations.[14,15] The ingestion of an estimated dose of 37 g nutmeg blended in a milkshake was associated with tachycardia, palpitations, drowsiness, nausea, dry mouth, anxiety, restlessness, and agitation without hallucinations.[16] Variation in toxicity may result from the loss of essential oils from the ground nutmeg.

The ingestion of a single dose 400 mg myristicin by 10 subjects was associated with alertness, a feeling of irresponsibility, and euphoria in 2 subjects as well as an unpleasant reaction (anxiety, fear, tremor, nausea, tachycardia) in 2 subjects.[17] Onset of symptoms occurred about 1–2 hours after ingestion and resolved within 24 hours. This dose of myristicin is equivalent to the amount of myristicin present in 40 g of nutmeg.

TOXICOKINETICS

In the 1970s, an *in vitro* study suggested that psychoactive amphetamine derivatives (e.g., 3-methoxy-4,5-methylenedioxyamphetamine, MMDA) are potential metabolites of myristicin.[18] Although the alkenebenzene derivatives, myristicin, elemicin, safrole, are extensively metabolized following ingestion, subsequent studies have not confirmed the presence of amphetamine metabolites. Limited toxicokinetic studies indicate that these alkenebenzene undergo hydroxylation of the side chain, and elemicin undergoes *O*-demethylation prior to hydroxylation. Metabolites detected in the urine of rats fed these compounds include *O*-demethyl elemicin, *O*-demethyl dihydroxy elemicin, demethylenyl myristicin, dihydroxy myristicin, and demethylenyl safrole.[19] *In vivo* animal studies suggest that enzymatic hydrolysis of myristicin produces at least two metabolites (5-allyl-1-methoxy-2,3-dihydroxybenzene, 1′-hydroxymyristicin) that are conjugated and excreted in the urine.[20,21] Amphetamine derivatives were not detected in these two studies.

CLINICAL RESPONSE

The clinical features of nutmeg intoxication resemble belladonna (anticholinergic) toxicity manifest by facial flushing, tachycardia, hypertension, dry mouth, blurred vision, and delirium.[22,23] However, mydriasis is uncommon. Symptoms usually begin about 3–6 hours after ingestion and resolve by 24–36 hours.[6,7] Initially, these symptoms include nausea, vomiting, abdominal pain, chest pain, restlessness, agitation, tremor, ataxia, nystagmus, vertigo, and a feeling of doom. These symptoms occur with alternating periods of lethargy and delirium.[13,24] Other effects of nutmeg intoxication documented in case reports include the sensation of warmth and coldness of the extremities, distortion of space and colors, auditory and tactile hallucinations, headache, and generalized weakness.[25] Patients usually recover without sequelae.[1] The medical literature does not contain any fatalities solely related to nutmeg intoxication since the first decade of the 20th century.[2] There are few data on the reproductive effects of nutmeg. A 29-year old woman, who developed signs of nutmeg intoxication at 30 weeks gestation, delivered a healthy infant at term.[22]

The processing of nutmeg does not usually produce irritative symptoms in spice workers.[26] However, allergic contact dermatitis may develop in workers sensitized to allergens (e.g., isoprenyl myristate) in nutmeg.[27] Occupational asthma may also occur in workers sensitized to spices including mace.[28]

DIAGNOSTIC TESTING

Chromatographic methods easily separate marijuana and the components of nutmeg and mace.[29] Human metabolites of the constituents in nutmeg are detectable by the use of gas chromatography mass spectrometry after acid hydrolysis, liquid-liquid extraction of analytes, and microwave-assisted acetylation of extracted analytes.[19] These metabolites include *O*-demethyl elemicin, *O*-demethyl dihydroxy elemicin, demethylenyl myristicin, dihydroxy myristicin, and demethylenyl safrole. Developing biomarkers for nutmeg intoxication is complicated by limited data on the toxic constituents of nutmeg. Six hours after ingestion 14–21 g nutmeg powder, a 16-year-old adolescent developed tachycardia, drowsiness, dry mouth, warm skin, and mydriasis.[30] The myristicin concentration in a blood sample drawn 8 hours after ingestion was 2 µg/mL. The postmortem blood from a 55-year-old woman contained 4 µg/mL of myristicin and 0.072 µg/mL of flunitrazepam.[30] Death was attributed to the combination of the two substances. However, myristicin does not necessarily account for the toxic effects of nutmeg, and the contribution of myristicin and nutmeg to this death remains speculative. Routine laboratory tests during nutmeg intoxication are usually normal.[24] The ingestion of nutmeg does not produce positive urine drug screens.[14]

TREATMENT

Management is usually supportive. The presence of hemodynamic instability suggests the presence of other toxins or illnesses. Decontamination measures are usually unnecessary because of the presence of vomiting or delayed contact (i.e., >1–2 hours after ingestion) with the health care facility. There are no clinical data to guide management of nutmeg intoxication. The use of standard antiemetics (prochlorperazine, trimethobenzamide, odansetron, metoclopramide) and intravenous fluids may be required to treat protracted nausea and vomiting. Sedatives (diazepam, haloperidol) should be used with caution because of alternating periods of delirium and obtundation during nutmeg intoxication.

References

1. Green RC. Nutmeg poisoning. JAMA 1959;171:1342–1344.
2. Cushny AR. Nutmeg poisoning. Proc Roy Soc Med 1908;1:39–44.
3. Malcolm X, Haley A. The autobiography of Malcolm X. New York: Grove Press Inc; 1964.
4. van Gils C, Cox PA. Ethnobotany of nutmeg in the Spice Islands. J Ethnopharmacol 1994;42:117–124.
5. Barrowman JA, Bennett A, Hillenbrand P, Rolles K, Pollock DJ, Wright JT. Diarrhoea in thyroid medullary carcinoma: role of prostaglandins and therapeutic effect of nutmeg. Br Med J 1975;3:11–12.
6. Sangalli BC, Chiang W. Toxicology of nutmeg abuse. Clin Toxicol 2000;38:671–678.
7. Forrest JE, Heacock RA. Nutmeg and mace. The psychotropic species from *Myristica fragrans*. Lloydia 1972;35:440–449.
8. De Vincenzi M, Silano M, Stacchini P, Scazzocchio B. Constituents of aromatic plants: I. Methyleugenol. Fitoterapia 2000;71:216–221.
9. Farnsworth NR. Hallucinogenic plants. Science 1968;162:1086–1092.
10. Hallstrom H, Thuvander A. Toxicological evaluation of myristicin. Nat Toxins 1997;5:186–192.
11. Cesario de Mello A, Carlini EA. Behavioral observations on compounds found in nutmeg. Psychopharmacologia 1973;31:349–363.
12. Beattie RT. Nutmeg as a psychoactive agent. Br J Addict Alcohol Other Drugs 1968;63:105–109.
13. McCord JA, Jervey LP. Nutmeg (myristicin) poisoning. A case report. SC Med Assoc J 1962;58:436–439.
14. Abernethy MK, Becker LB. Acute nutmeg intoxication. Am J Emerg Med 1992;10:429–430.
15. McKenna A, Nordt SP, Ryan J. Acute nutmeg poisoning. Eur J Emerg Med 2004;11:240–241.
16. Demetriades AK, Wallman PD, McGuiness A, Gavalas MC. Low cost, high risk: accidental nutmeg intoxication. Emerg Med J 2005;22:223–225.
17. Truitt EB, Callaway E, Braude MC, Krantz JC. The pharmacology of myristicin: a contribution to the psychopharmacology of nutmeg. J Neurophsychiatr 1961;2:205–210.
18. Braun U, Kalbhen DA. Evidence for the biogenic formation of amphetamine derivatives from components of nutmeg. Pharmacology 1973;9:312–316.
19. Beyer J, Ehlers D, Maurer HH. Abuse of nutmeg (*Myristica fragrans* Houtt.): studies on the metabolism and the toxicologic detection of its ingredients elemicin, myristicin, and safrole in rat and human urine using gas chromatography/mass spectrometry. Ther Drug Monit 2006;28: 568–575.
20. Oswald EO, Fishbein L, Corbett BJ, Walker MP. Urinary excretion of tertiary amino methoxy methylene-dioxy propiophenones as metabolites of myristicin in the rat and the guinea pig. Biochim Biophys Acta 1971;244:322–328.
21. Lee HS, Jeong TC, Kim JH. *In vitro* and *in vivo* metabolism of myristicin in the rat. J Chromatogr B Biomed Sci Appl 1998;705:367–372.
22. Lavy G. Nutmeg intoxication in pregnancy a case report. J Reprod Med 1987;32:63–64.
23. Payne RB. Nutmeg intoxication. N Engl Med J 1973;269:36–38.
24. Painter JC, Shanor SP, Winek CL. Nutmeg poisoning—a case report. Clin Toxicol 1971;4:1–4.
25. Fras 1, Friedman JJ. Hallucinogenic effects of nutmeg in adolescents. NY State J Med 1969;69:463–465.
26. Uragoda CG. Symptoms in spice workers. J Trop Med Hyg 1992;95:136–139.
27. van den Akker TW, Roesyanto-Mahadi ID, van Toorenenbergen AW, van Joost T. Contact allergy to spices. Contact Dermatitis 1990;22:267–272.
28. Sastre J, Olmo M, Novalvos A, Ibanez D, Lahoz C. Occupational asthma due to different spices. Allergy 1996; 51:117–120.
29. Forrest JE, Heacock RA. A chromatographic comparison of the constituents of nutmeg and mace (*Myristica fragrans* Houtt.) with those of marihuana and hashish (*Cannabis sativa* L.). J Chromatogr 1974;89:113–117.
30. Stein U, Greyer H, Hentschel H. Nutmeg (myristicin) poisoning—report on a fatal case and a series of cases recorded by a poison information centre. Forensic Sci Int 2001;118:87–90.

Chapter 10

PEPPER and CAPSAICIN (*Capsicum* and *Piper* Species)

HISTORY

Mexican Indians probably used chili peppers long before the birth of Christ.[1] The archaeologist, R. S. Macneish, found pepper seeds dating from about 7500 BC in Mexico.[2] The physician, Chauca, on the second voyage of Columbus to the New World noted the culinary and medicinal use of chili peppers by Native Americans. Irritant smoke from the burning of chili peppers was a weapon used by Native Americans against invaders. Chemical investigations into the ingredients in chili peppers began in the early 19th century. In 1846, Thresh crystallized the active component in chili peppers, and he named the substance capsaicin.[3] Nelson and Dawson determined the chemical structure of capsaicin in the early 20th century.[4,5]

BOTANICAL DESCRIPTION

There are a variety of *Capsicum* peppers beside the common *C. annuum* L. (cayenne pepper, chili pepper) including *C. baccatum* L. (aji, aji amarillo, locoto), *C. chinense* Jacq. (aji dulce, habanero, rica red), and *C. pubescens* Ruiz & Pavón (rocoto, manzano).

Common Name: chili pepper, red pepper, Christmas pepper, Tabasco pepper, cayenne pepper

Scientific Name: *Capsicum annuum* L. (*Capsicum frutescens*)

Botanical Family: Solanaceae

Physical Description: This annual plant has densely branched stems that reach up to 1.5 m (5 ft) in length. Single, white flowers produce small fruit in the fall. The fruit changes from white to shades of yellow, orange, and red as the fruit matures.

Distribution and Ecology: This pepper is a native species of South America that is now cultivated worldwide in warm, dry climates.

Common Name: Black pepper (ripe), white pepper (unripe)

Scientific Name: *Piper nigrum* L.

Botanical Family: Piperaceae (pepper)

There are a large number of pepper species in the genus *Piper* including *P. auritum* Kunth (Vera Cruz pepper, Mexican pepperleaf), *P. jacquemontianum* Kunth (Caracas pepper), *P. glabrescens* (Miq.) C. DC. (Guyanese pepper), *P. hispidum* Sw. (Jamaican pepper), *P. longum* L. (Indian long pepper), *P. marginatum* Jacq. (marigold pepper), and *P. retrofractum* Vahl (Javanese long pepper).

Physical Description: This perennial vine has smooth, round, woody, branched stems that reach up to 4 m (13 ft) in length. The leaves are broad-ovate, smooth, and dark green; and sheath-like petioles attach the leaves to joints in the branches. The small, white flowers produce globular, sessile drupes, which turn red on maturity.

Distribution and Ecology: This pepper plant is a native species of India and Sri Lanka. It is cultivated widely in the tropical areas of China, India, Indonesia, and the Philippines. These plants do not tolerate frost.

Medical Toxicology of Natural Substances, by Donald G. Barceloux, MD

EXPOSURE

Chili peppers are the pungent fruits of various species of the genus *Capsicum* that are grown as condiments, spices, ornamentals, and pharmaceutical preparations (i.e., counterirritant in external analgesic preparations). Some dog-repellent products and pepper sprays contain up to 10% oleoresin capsicum as well as other capsaicinoid compounds.[6] Pepper spray is a chemical irritant that has been available commercially since the 1970s as an aerosol to temporarily disable assailants. Increasing the concentration of capsaicin prolongs the irritant effects of the spray rather than increasing the initial response.[7] Capsaicin-containing creams are used for the treatment of painful conditions, such as psoriasis, rheumatoid arthritis, diabetic neuropathy, postherpetic neuralgia, postmastectomy syndrome, cluster headache, and reflex sympathetic dystrophy. For the treatment of these conditions, the exact role of capsaicin compared with other analgesics remains undefined.[8] An evidence-based review of the efficacy of capsaicin in chronic pain suggests that topical capsaicin is useful only for a small number of patients who do not respond to conventional analgesics.[9] Capsaicin is not a traditional counterirritant like oleoresin capsicum, which contains only trace amounts of capsaicin.

PRINCIPAL TOXINS

Structure and Properties

CHILI PEPPERS AND CAPSAICIN

Capsaicin (trans-8-methyl-*N*-vanillyn-6-noneamide, CAS RN: 404-86-4) is the major alkaloid responsible for the mucosal irritant properties of plant species from the genus *Capsicum*.[10] The molecular weight of capsaicin is 305.4 Da, and the molecular formula is $C_{18}H_{27}NO_3$. Other compounds in chili pepper include homologous branched- and straight-chain alkyl vanillylamides (i.e., capsaicinoid compounds) such as dihydrocapsaicin, nordihydrocapsaicin, homocapsaicin, and homodihydrocapsaicin.[11] These compounds also possess pungent properties. Nonivamide (CAS RN: 2444-46-4) is a chemical structurally similar to capsaicin with a molecular formula of $C_{17}H_{27}NO_3$. Although sometimes termed "synthetic capsaicin," nonivamide possesses slightly less pungent properties compared with capsaicin.[12] Civamide is a synthetic chemical (CAS RN: 25775-90-0) that is the *cis*-isomer of capsaicin.

Capsaicin is soluble in ethanol, acetone, and fatty oils, but this compound is insoluble in cold water. The concentrations of these capsaicinoid compounds in chili peppers are much smaller (i.e., 0.1–1%) than capsaicin.[12] The aroma from the roasting of red pepper seed oil results from the formation of pyrazine compounds (e.g., 2,5-dimethylpyrazine).[13] Aflatoxins may contaminate seeds of black pepper.[14]

OLEORESIN CAPSICUM

Oleoresin capsicum is a complex mixture of a variety of branched- and straight-chained alkyl vanillylamides with capsaicin being the principal constituent.[15] There are over 100 compounds in this mixture, which is extracted from dried, ripe fruit of chili peppers (*Capsicum annuum*). The actual composition depends on the environmental conditions, season, maturity of the fruit, and the extraction process. Depending on the variety of chili pepper, oleoresin capsicum contains about 0.01–1% capsaicinoids (vanillylamides) on a dry mass basis.[16] The approximate distribution of these capsaicinoid compounds is as follows: capsaicin, 70%; dihydrocapsaicin, 20%; norhydrocapsaicin, 7%, homocapsaicin, 1%, and monodihydrocapsaicin, 1%. Commercial pepper sprays with oleoresin capsicum contain a variety of natural and synthetic capsaicinoid compounds depending on the formulation and source. Brands of pepper spray intended for sale to the general public contains ≤1% oleoresin capsicum; law enforcement products typically contain 5–10% oleoresin capsicum. Capsaicin is usually the active ingredient of these formulations along with other capsaicinoid compounds (dihydrocapsaicin, nordihydrocapsaicin, nonivamide). Analysis of two samples of commercial pepper spray yielded a total of 10.1 mg/mL and 19.8 mg/mL capsaicinoid compounds with capsaicin accounting for 3.6 mg/mL and 8.5 mg/mL of the content, respectively.[6] The other major capsaicinoid compound in the spray was dihydrocapsaicin, which possesses pungent properties similar to capsaicin.

BLACK PEPPER

The active ingredient in black pepper is piperine (1-piperoylpiperidine), the amide of the alkaloid piperidine, and the concentration of this compound varies substantially between *Piper* species.[17] Piperine accounts for about one-half of the nonvolatile ether extract of black pepper.[18] Safrole is a minor constituent of *P. nigrum* at an approximate concentration of 0.1%.[19]

Poisonous Parts

Capsicum is the dried, ripe fruit of *Capsicum annuum*, and capsaicin is the purified chemical that is the active ingredient in capsicum. Capsaicin content increases with maturation of the fruit.[20] In general, the capsaicin content of chili peppers is <1% dry weight. Capsicum oleoresin is a thick, dark reddish-brown liquid concentrate produced by the extraction of fruits with volatile solvents. This liquid contains over 100 chemicals (e.g., alcohols, carbonyls, carboxylic acids, esters, pyrazine compounds, terpenes) in the volatile component of the resin. However, the capsaicin content of the oleoresin is highly variable, and the oleoresin does not uniformly mediate substance P release.[12] This neuropeptide causes vasodilation due to the release of nitric oxide from the endothelium.

Mechanism of Toxicity

The medical effects of capsaicin result form the initial period of intense excitation of primary afferent sensory nerves followed by a prolonged period of relative resistance to nociceptive chemical stimuli. Capsaicin reduces the transmission of painful impulses from the periphery to the central nervous system by depleting substance P in sensory neurons, particularly in unmyelinated C-type and some myelinated A delta-type nerve fibers.[10] Physiologically, capsaicin reduces vasodilation and heat pain thresholds, leading to increased pain sensitivity.[21] Repeat exposure to capsaicin causes desensitization to the stimulatory effect on sensory nerves.[22]

In volunteer studies, both red pepper and black pepper increase parietal and pepsin secretions, potassium loss, and the exfoliation of gastric cells.[23] The clinical consequences of these changes are not known. Although there is substantial interindividual variation, these studies suggest that gastrointestinal hemorrhage may occur after the ingestion of black or red peppers. However, there are inadequate clinical data to determine the effect of the chronic consumption of peppers on the gastric mucosa.

DOSE RESPONSE

Chili peppers are edible and nontoxic in the usual doses.

TOXICOKINETICS

There are limited data on the toxicokinetics of capsaicin and capsaicinoid compounds in humans. Analysis of breast milk indicates that piperine is excreted in breast milk following the ingestion of black pepper.[24]

CLINICAL RESPONSE

Chili (Cayenne) Pepper

Ingestion

Contact with products containing capsaicin produces local irritation and lacrimation. A case series of five contestants in a jalapeno pepper-eating contest associated the ingestion of large quantities of jalapeno peppers with lacrimation, rhinorrhea, dysuria, and burning discomfort on defecation within 24 hours.[25] The eponym "jalaproctitis" was given to this syndrome, which also occurs in children abused with jalapeno peppers.[26]

The potential metabolism of capsaicinoid compounds to highly reactive intermediates (phenoxyl radicals, arene oxides) and the high incidence of gastric cancer in populations consuming relative large amounts of jalapeno peppers suggests the possibility of genotoxicity and/or carcinogenicity following the chronic consumption of chili peppers.[27] A population-based case-control study in Mexico City compared the risk of gastric cancer in chili pepper consumers and nonconsumers.[28] The odds ratio for gastric cancer was increased (OR = 5.49, 95% CI = 2.72–11.06) in chili pepper consumers compared with age- and sex-adjusted controls (nonconsumers). The odds ratio for gastric cancer demonstrated a dose-response relationship for the amount of chili peppers consumed, but not for the frequency of consumption. Animal studies do not consistently demonstrate carcinogenic effects following chronic consumption of chili peppers. Beginning at the age of 6 weeks, 0.03125% capsaicin was administered in a semisynthetic powdered diet for the lifespan of Swiss mice.[29] Although there was an increased incidence of benign polypoid adenomas in the cecum, the rate of malignant tumors in the treated and control groups was similar.

Dermal Exposure

The most common symptoms reported to the American Association of Poison Control Centers Toxic Exposure Surveillance System after reported exposure to *Capsicum annum* were mucous membrane irritation (oral, dermal, ocular, throat), rash (erythema, hives), and pruritus.[30] In this case series, exposure to chili peppers was based on historical reports, and there was no independent confirmation. Case reports associate exposure to dried red chili peppers (*Capsicum annum* cultivar) with severe burning of abraded fingertips. The eponym "Hunan hand" was given to this syndrome, which results from the effect of capsaicin on dermal pain fibers.[21,31] Rare case reports associate dermal exposure to jalapeno peppers with acute febrile neutrophilic dermatosis

(Sweet's syndrome) characterized by painful erythematous papules, plaques, pustules, and hemorrhagic bullae on the hands.[32]

INHALATION

The most common symptoms reported to poison control centers following contact with pepper sprays involve irritation of the eyes, skin, and upper respiratory tract.[33] Contact with pepper spray produces local inflammation, erythema, and pain in the skin and conjunctiva with transient reduction in visual acuity resulting from secondary tearing. Symptoms usually resolve within one-half hour to 2 hours after decontamination.[34] Although exposure to capsaicin causes cough and a transient increase in airway resistance, experimental studies and case series do not indicate that patients with bronchial hypersensitivity are substantially more susceptible to the irritant effects of pepper spray than patients without bronchial hypersensitivity.[34,35] Experimental studies demonstrate that some asthmatic patients may develop a transient decrease in expiratory flow rates (e.g., FEV_1), but there is no direct correlation between methacholine and capsaicin responses in the bronchioles.[36] There is little clinical evidence that the use of standard doses of oleoresin capsicum spray significantly reduces oxygen saturation or ventilation in healthy subjects. A controlled clinical trial involving 35 healthy subjects sprayed with oleoresin capsicum did not detect significant changes in the mean percent predicted forced vital capacity (FVC), percent predicted forced expiratory volume in 1 second (FEV_1), oxygen saturation, or end-tidal CO_2 concentration, when compared with population norms.[37] There is a progressive mild restriction of pulmonary function with changes in position from sitting to supine, prone, and restraint positions. In this study, the use of the prone maximum restraint position reduced the FVC and FEV_1 by approximately 15% compared with the sitting position, but there was no statistically significant difference in these parameters between the use of oleoresin capsicum and controls in the restraint position. Furthermore, there was no clinically significant impairment of ventilation.

Serious complications following exposure to pepper spray are rare,[38] and case reports usually involve the direct aspiration of the spray. An 11-year-old child directly inhaled a jet spray from a pressurized container containing capsaicin spray.[39] He recovered after developing subglottic obstruction of the trachea and bilateral pulmonary infiltrates that required intubation. Signs of clinically significant airway edema included drooling, expiratory wheezing, stridor, and use of accessory muscles. The accidental discharge of pepper spray next to a 4-week-old infant produced life-threatening respiratory distress followed by the development of copious bronchial secretions, pneumonia, and progressive respiratory failure.[40] The infant survived after receiving extracorporeal membrane oxygenation. Corneal abrasions may complicate the use of pepper spray.[41] Case reports document the temporal association between the use of pepper spray and death during police custody.[7] However, the causal role of pepper spray in these deaths is unclear because of the multiple etiologies of "in-custody" deaths and the lack of significant changes in respiratory parameters following use of the pepper spray.

OCCUPATIONAL EXPOSURE

The inhalation of capsaicin in red pepper produces coughing in a dose-dependent relationship.[42] A cross-sectional study of workers exposed to capsaicin detected a statistically significant increase in complaints of cough in capsaicin-exposed workers compared with nonexposed controls.[43] There was no significant difference in measures of pulmonary function (FEV_1, FVC) between the two groups.

Black Pepper

Aspiration of black pepper is a rare cause of respiratory distress, and case reports indicate that the aspiration of large amounts of black pepper can produce airway obstruction and hypoxia. The aspiration of black pepper by a child with a history of pica resulted in respiratory distress, asphyxia, and cardiopulmonary arrest.[44] The child died 4 days later when respiratory support was withdrawn. Several case reports document child abuse associated with the fatal, intrapharyngeal administration and aspiration of black pepper.[45,46] A 21-month-old infant developed respiratory distress and thick, copious, bronchial secretions after the accidental aspiration of an unknown amount of pepper.[47] She survived after intensive care that included intubation and tracheostomy.

DIAGNOSTIC TESTING

UV spectrophotometric methods are relatively specific tests for the presence of the active ingredient in black pepper, piperine. However, UV spectrophotometric methods provide slightly higher piperine content compared with reverse phase high performance liquid chromatography (HPLC).[18] Gas chromatographic methods are available to separate capsaicinoid compounds, such as nonivamide.[48] Following separation of capsaicinoid compounds in blood samples by reversed phase HPLC and identification by tandem mass spectrometry, the limit of detection (LOD) and lower limit of quantitation

(LLOQ) were 0.25 mg/mL and 1.0 mg/mL, respectively, for the major capsaicinoid compounds (capsaicin, dihydrocapsaicin, nonivamide).[49]

TREATMENT

Management of symptoms associated with exposure to capsaicin is supportive. Decontamination involves cleansing of the eyes and affected skin with water (i.e., at least 1–2 L saline). Cases of aspiration may require bronchoscopy and pulmonary lavage. There are limited clinical data regarding the optimal treatment of the paraesthesias associated with peppers. Options for the treatment of the dermal pain associated with capsaicin include immersion in milk, cold water, topical lidocaine gel, vegetable oil, or vinegar (5% acetic acid).[50,51] Although immersion in cold water initially provided better relief of pain associated with peeling green chili (*C. annum*), *complete* immersion in vegetable oil provides greater long-term relief than cold water as measured by a visual analog scale.[52] Prolonged immersion (i.e., several hours) may be necessary for severe cases. A pilot study on the use of magnesium-aluminum hydroxide suspension (Maalox®, Novartis, East Hanover, NJ) suggested that the immediate application of this compound reduces pain associated with commercial defensive spray containing 10% capsaicin during the first 30 minutes, but pain scores did not differ between the treatment and saline control group after 60 minutes postexposure.[53] Rodent studies suggest that bismuth subsalicylate can reduce the dyspepsia and gastric pain associated with the consumption of peppers.[54] Corneal abrasions may complicate the use of pepper spray. If a corneal abrasion cannot be excluded clinically after decontamination measures, the cornea should be examined with fluorescein staining by a Wood's lamp or slit lamp.

References

1. Govindarajan VS. Capsicum—production, technology, chemistry, and quality. I. History, botany, cultivation and primary processing. CRC Crit Rev Food Sci Nutr 1985;22:109–176.
2. Fett DD. Botanical briefs: *Capsicum* peppers. Cutis 2003;72:21–23.
3. Thresh LT. Isolation of capsaicin. Pharm J 1846;6:941–942.
4. Nelson EK, Dawson LE. The constitution of capsaicin—the pungent principle of *Capsicum*, III. J Am Chem Soc 1923;45:2179–2181.
5. Nelson EK. The constitution of capsaicin—the pungent principle of *Capsicum*. J Am Chem Soc 1919;41:1115–1121.
6. Hass JS, Whipple RE, Grant PM, Andresen BD. Chemical and elemental comparison of two formulations of oleoresin capsicum. Sci Justice 1997;37:15–24.
7. Steffee CH, Lantz PE, Flannagan LM, Thompson RL, Jason DR. Oleoresin capsicum (pepper) spray and "in-custody deaths". Am J Forensic Med Pathol 1995;16:185–192.
8. Hautkappe M, Roizen MF, Toledano A, Roth S, Jeffries JA, Ostermeier AM. Review of the effectiveness of capsaicin for painful cutaneous disorders and neural dysfunction. Clin J Pain 1998;14:97–106.
9. Mason L, Moore RA, Derry S, Edwards JE, McQuay HJ. Systematic review of topical capsaicin for the treatment of chronic pain. BMJ 2004;328:991.
10. Buck SH, Burks TF. The neuropharmacology of capsaicin: review of some recent observations. Pharmacol Rev 1985;38:179–226.
11. Masada Y, Hasimoto K, Inoue T, Suzuki M. Analysis of the pungent principles of *Capsicum annum* by combined gas chromatography-mass spectrometry. J Food Sci 1971;36:858–860.
12. Cordell GA, Araujo OE. Capsaicin: identification, nomenclature, and pharmacotherapy. Ann Pharmacother 1993;27:330–336.
13. Jung MY, Bock JY, Baik SO, Lee JH, Lee TK. Effects of roasting on pyrazine contents and oxidative stability of red pepper seed oil prior to its extraction. J Agric Food Chem 1999;47:1700–1704.
14. Roy AK, Sinha KK, Chourasia HK. Aflatoxin contamination of some common drug plants. Appl Environ Microbiol 1988;54:842–843.
15. Olajos EJ, Salem H. Riot control agents: pharmacology, toxicology, biochemistry and chemistry. J Appl Toxicol 2001;21:355–391.
16. Govindarajan VS, Sathyanarayana MN. Capsicum–production, technology, chemistry, and quality. Part V. Impact on physiology, pharmacology, nutrition, and metabolism; structure, pungency, pain, and desensitization sequences. Crit Rev Food Sci Nutr 1991;29:435–474.
17. Nakatani N, Inatani R, Ohta H, Nishioka A. Chemical constituents of peppers (*Piper* spp.) and application to food preservation: naturally occurring antioxidative compounds. Environ Health Perspect 1986;67:135–142.
18. Rathnawathie M, Buckle KA. Determination of piperine in pepper (*Piper nigrum*) using high-performance liquid chromatography. J Chromatogr 1983;264:316–320.
19. Farag SE, Abo-Zeid M. Degradation of the natural mutagenic compound safrole in spices by cooking and irradiation. Nahrung 1997;41:359–361.
20. Balbaa SI, Karawya MS, Girgis AN. The capsaicin content of capsicum fruits at different stages of maturity. Lloydia 1968;31:272–274.
21. Weinberg RB. Hunan hand. N Engl J Med 1981;305:1020.
22. Carpenter SE, Lynn B. Vascular and sensory responses of human skin to mild injury after topical treatment with capsaicin. Br J Pharmacol 1981;73:755–758.

23. Myers BM, Smith JL, Graham DY. Effect of red pepper and black pepper on the stomach. Am J Gastroenterol 1987;82:211–214.
24. Khachik F, Spangler CJ, Smith JC Jr, Canfield LM, Steck A, Pfander H. Identification, quantification, and relative concentrations of carotenoids and their metabolites in human milk and serum. Anal Chem 1997;69:1873–1881.
25. Diehl AK, Bauer RL. Jalaproctitis. N Engl J Med 1978;299:1137–1138.
26. Tominack RL, Spyker DA. Capsicum and capsaicin—a review: case report of the use of hot peppers in child abuse. Clin Toxicol 1987;25:591–601.
27. Lopez-Carrillo L, Lopez-Cervantes M, Robles-Diaz G, Ramirez-Espitia A, Mohar-Betancourt A, Meneses-Garcia A, et al. Capsaicin consumption, *Helicobacter pylori* positivity and gastric cancer in Mexico. Int J Cancer 2003;106:277–282.
28. Lopez-Carrillo L, Hernandez Avila M, Dubrow R. Chili pepper consumption and gastric cancer in Mexico: a case-control study. Am J Epidemiol 1994;139:263–271.
29. Toth B, Gannett P. Carcinogenicity of lifelong administration of capsaicin of hot pepper in mice. In Vivo 1992;6:59–63.
30. Mrvos R, Krenzelok EP, Jacobsen TD. Toxidromes associated with the most common plant ingestions. Vet Hum Toxicol 2001;43:366–369.
31. Williams SR, Clark RF, Dunford JV. Contact dermatitis associated with capsaicin: Hunan hand syndrome. Ann Emerg Med 1995;25:713–715.
32. Greer JM, Rosen T, Tschen JA. Sweet's syndrome with an exogenous cause. Cutis 1993;51:112–114.
33. Forrester MB, Stanley SK. The epidemiology of pepper spray exposures reported in Texas in 1998–2002. Vet Hum Toxicol 2003;45:327–330.
34. Watson WA, Stremel KR, Westdorp EJ. Oleoresin capsicum (Cap-Stun) toxicity from aerosol exposure. Ann Pharmacother 1996;30:733–735.
35. Fuller RW. Pharmacology of inhaled capsaicin in humans. Respir Med 1991;85(Suppl A):31–34.
36. Hathaway TJ, Higenbottam TW, Morrison JFJ, Clelland CA, Wallwork J. Effects of inhaled capsaicin in heart-lung transplant patients and asthmatic subjects. Am Rev Respir Dis 1993;148:1233–1237.
37. Chan TC, Vilke GM, Clausen J, Clark RF, Schmidt P, Snowden T, Neuman T. The effect of oleoresin capsicum "pepper" spray inhalation on respiratory function. J Forensic Sci 2002;47:299–304.
38. Smith J, Greaves I. The use of chemical incapacitant sprays: a review. J Trauma 2002;52:595–600.
39. Winograd HL. Acute croup in an older child. An unusual toxic origin. Clin Pediatr 1977;16:884–887.
40. Billmire DF, Vinocur C, Ginda M, Robinson NB, Panitch H, Friss H, et al. Pepper-spray-induced respiratory failure treated with extracorporeal membrane oxygenation. Pediatr 1996;98:961–963.
41. Brown L, Takeuchi D, Challoner K. Corneal abrasions associated with pepper spray exposure. Am J Emerg Med 2000;18:271–272.
42. Choudry NB, Fuller RW, Pride NB. Sensitivity of the human cough reflex: effect of inflammatory medicators prostaglandin E_2, bradykinin, and histamine. Am Rev Respir Dis 1989;140;137–141.
43. Blanc P, Liu D, Juarez C, Boushey HA: Cough in hot pepper workers. Chest 1991;99:27–32.
44. Sheahan K, Page DV, Kemper T, Suarez. Childhood sudden death secondary to accidental aspiration of black pepper. Am J Forensic Med Pathol 1988;9:51–53.
45. Adelson L. Homicide by pepper. J Forensic Sci 1964; 9:391–395.
46. Cohle SD. Homicidal asphyxia by pepper aspiration. J Forensic Sci 1986;31:1475–1478.
47. Flintoff WM, Poushter DL. Aspiration of black pepper. A case report. Arch Otolaryngol 1974;100:375–376.
48. Jerenitsch J, Leinmuller R. Quantification of nonylic acid vanillylamide and other capsaicinoids in the pungent principle of Capsicum fruits and preparation by gas-liquid chromatography on glass capillary columns. J Chromatogr 1980;189:389–397.
49. Reilly CA, Crouch DJ, Yost GS, Fatah AA. Determination of capsaicin, nonivamide, and dihydrocapsaicin in blood and tissue by liquid chromatography-tandem mass spectrometry. J Anal Toxicol 2002;26:313–319.
50. Nasrawi CW, Pangborn RM. Temporal effectiveness of mouth-rinsing on capsaicin mouth-burn. Physiol Behav 1990;47:617–623.
51. Vogl TP. Treatment of Hunan hand. N Engl J Med 1982;306:178.
52. Jones LA, Tandberg D, Troutman WG. Household treatment for "chile burns" of the hands. Clin Toxicol 1987;25:483–491.
53. Lee DC, Ryan JR. Magnesium-aluminum hydroxide suspension for the treatment of dermal capsaicin exposures. Acad Emerg Med 2003;10;688–690.
54. Lichtenberger LM, Romero JJ, Carryl OR, Illich PA, Walters ET. Effect of pepper and bismuth subsalicylate on gastric pain and surface hydrophobicity in the rat. Aliment Pharmacol Ther 1998;12:483–490.

Chapter 11

POTATOES, TOMATOES, and SOLANINE TOXICITY (*Solanum tuberosum* L., *Solanum lycopersicum* L.)

HISTORY

The Inca people first cultivated the potato in the Altiplano of the High Andes of Chile during prehistoric times.[1] Spanish explorers brought the potato to Europe near the end of the 16th century. Although initially considered "peasant food," the cultivation of the potato spread across Europe including continental Europe, Britain, and Ireland. By 1780, the potato became a staple crop in Ireland. Although cultivation of the potato contributed to the rapid expansion of the Irish population between 1750 and 1850, a famine began in the late 1840s as a result of a fungal-induced potato blight [*Phytophthora infestans* (Mont.) de Bary]. During The Great Hunger in Ireland between 1845 and 1852, approximately one million deaths occurred along with the migration of 1.5 million people.[1] The creation of new transgenic potato cultivars with increased resistance against phytopathogens and improved composition of nutrients is an area of active research.

Although the toxicity of plants (e.g., deadly nightshade, henbane) from the nightshade (Solanaceae) family has been known since ancient times, toxicity from members of the *Solanum* genus (e.g., potato, tomato) has been documented only in modern times. Solanine intoxication is relatively rare considering the frequent use of potatoes and tomatoes as food staples. Outbreaks of solanine poisoning occur primarily when toxic glycoalkaloid concentrations increase substantially in commercial potatoes. Most sporadic cases of solanine toxicity involve children ingesting poisonous wild plant parts.[2]

BOTANICAL DESCRIPTION

Sweet potatoes [*Ipomoea batatas* (L.) Lam.] are a seasonal crop grown in tropical and subtropical regions, primarily for human consumption of the fresh or dried tubers. This South American native is a member of the morning glory family (Convolvulaceae) rather than the potato family. Sweet potatoes are widely cultivated in warm areas throughout the world as annual herbaceous plants. Three closely related, cultivated species of eggplant include *Solanum macrocarpon* L. (gboma eggplant), *Solanum aethiopicum* L. (scarlet eggplant, Ethiopian nightshade), and *Solanum melongena* L. (brinjal or aubergine eggplant).[3]

Common Name: Garden tomato

Scientific Name: *Solanum lycopersicum* L.

Botanical Family: Solanaceae (potato)

Physical Description: This herbaceous perennial plant produces a large berry with 2–12 locules that contain many small seeds. Most varieties of tomatoes are red when mature.

Distribution and Ecology: The tomato plant was originally cultivated in Mexico and Peru. European explorers brought seeds to Europe, and the tomato is now widely grown in temperate climates throughout the world.

Common Name: Eggplant, aubergine, brinjal

Scientific Name: *Solanum melongena* L.

Botanical Family: Solanaceae (potato)

Physical Description: This hairy, erect, annual, herbaceous plant reaches up to 1 m (about 3 ft) in height. The branched stems contain alternate, petiolate leaves with an ovate-elliptic shape. The flowers have a lobed calyx and a tubular, violet corolla. The fruit is a large ellipsoid berry that contains many brown seeds and turns purple when mature.

Distribution and Ecology: This perennial, frost-tender plant requires sunny and moist conditions. This species of eggplant is cultivated worldwide. Major areas of cultivation include China, Turkey, Japan, Mexico, Greece, Egypt, and Syria.

Common Name: Irish potato

Scientific Name: *Solanum tuberosum* L.

Botanical Family: Solanaceae (potato)

Physical Description: These annual, sprawling plants produce lavender to white flowers with yellow stamen. The relatively weak stems reach up to about 1 m (about 3 ft) in length with long pinnate leaves and ovate leaflets.

Distribution and Ecology: The potato is widely cultivated in temperate regions for the edible underground tubers.

EXPOSURE

The genus *Solanum* contains a number of food-producing species including eggplant (*S. melongena*), potatoes (*S. tuberosum*), and tomatoes (*S. lycopersicum*). African societies use the cholinesterase-lowering activity of solanine and chaconine in potatoes for the traditional treatment of human immunodeficiency virus (HIV) infections. Some communities (e.g., Bangladesh immigrants in the UK) use tomato leaves as a food source without obvious toxicity.

PRINCIPAL TOXINS

Structure and Properties

Glycoalkaloids are natural plant glycosides that contain nitrogen in a steroidal structure (aglycone) and a carbohydrate side chain at the 3-OH position. The hexacyclic alkaloid aglycones are derived from cholesterol. Species of the genus *Solanum* that contain glycoalkaloids include the potato (α-solanine, α-chaconine), the tomato (α-tomatine, dehydrotomatine), and the eggplant (solasonine, solamargine).[4] Table 11.1 lists the basic glycoalkaloids and the corresponding steroid skeleton (aglycone) for potatoes, tomatoes, and eggplants. The glycoalkaloid content varies substantially between different cultivars depending on the strain and postharvest conditions (light, mechanical injury, storage). Potatoes may also contain small amounts of the glycoalkaloid hydrolysis products (β- and γ-chaconines, β- and γ-solanines) and solanidine.[5] Food processing (cooking, baking, frying) does not significantly alter the glycoalkaloid content. Boiling removed <3.5% of the main glycoalkaloids in potatoes, whereas microwaving decreases the concentrations of these compounds about 15%.[5] Significant degradation of these glycoalkaloids begins at temperatures above 170 °C (338 °F), and deep-frying at 150 °C (302 °F) does not significantly alter the concentrations of these glycoalkaloids. Heating potatoes to 210 °C (410 °F) for 10 minutes reduces the α-chaconine and α-solanine concentrations about 40%.[5] Potatoes and eggplants also contain water-soluble nortropane alkaloids, such as calystegine A_3 and calystegine B_2.[6] There are few data on the human toxicity of these nortropane alkaloids.

TABLE 11.1. Glycoalkaloid Structure of Food Products from Genus Solanum

Plant	Scientific Name	Aglycone Skeleton	Glycoalkaloid
Potato	*Solanum tuberosum*	Solanidine	α-chaconine, α-solanine
Tomato	*Solanum lycopersicum*	Tomatidine	α-tomatine, dehydrotomatine
Eggplant	*Solanum melongena*	Solasodine	Solamargine, solasonine

Potatoes

Glycoalkaloids occur in all parts of potato plants including tubers, roots, sprouts, and leaves. In general, glycoalkaloids are α-compounds, whereas enzymatic or acid hydrolysis of the trisaccharide side chains of these glycoalkaloids produces β- and γ-compounds. The two major glycoalkaloids, α-solanine and α-chaconine, share the same aglycone (solanidine), but these two glycoalkaloids differ with respect to the composition of the sugar side-chain. α-Solanine (CAS RN: 20562-02-1, $C_{45}H_{73}NO_{15}$) is a tri-glycoside compound composed of galactose, glucose, and rhamnose, whereas α-chaconine is composed of glucose and two rhamnose moieties. Accumulation of glycoalkaloids stops after processing, but home processing (baking, boiling, frying, microwaving) does not eliminate significant amounts of glycoalkaloids from the potato.[7] Solanine is practically insoluble in water. Table 11.2 lists some of the physical properties of solanine.

TABLE 11.2. Some Physical Properties of Solanine

Physical Property	Value
Melting point	285 °C[a]
Log P (octanol-water)	2.000
Water solubility	1.380 mg/L (25 °C)
Vapor pressure	1.67E-34 mmHg (25 °C)
Henry's Law Constant	1.22E-31 atm m^3/mole (25 °C)
Atmospheric OH rate constant	3.72E-10 cm^3 molecule/second (25 °C)

[a]Decomposes at this temperature.

TOMATOES

Tomatidine is the basic aglycone for the glycoalkaloids, α-tomatine and dehydrotomatine, in the tomato plant. These glycoalkaloids probably aid the defense of the tomato plant against bacteria, fungi, viruses, and insects.

EGGPLANT

Eggplant fruit contains a relatively large amount of antioxidant phenolic compounds, primarily hydroxycinnamic acid conjugates such as chlorogenic and dicaffeoylquinic acids.[8]

Poisonous Parts

Although the sweet potatoes contain low concentrations of a potentially toxic furanoterpenoid (ipomearone),[9] consumption of sweet potatoes has not been associated with human toxicity.

POTATO

Although glycoalkaloids are present throughout the potato plant, the highest concentrations occur in the foliage, blossoms, periderm, cortex, and areas of high metabolic activity (e.g., eyes, green skin, stems, sprouts) followed by the peel and the tuber. α-Chaconine and α-solanine represent up to 95% of the glycoalkaloid content in potato tubers, and the general term solanine typically describes the glycoalkaloids present in *Solanum tuberosum*.[10] The glycoalkaloid concentration varies due to environmental and genetic factors. Ripe fruits contain the least amount of solanine. Factors that increase the solanine concentrations include physical injury to the plant, specific species (e.g., potato cultivar Lenape), physiological stress, immaturity (i.e., green potato), low storage temperature, and storage in bright lighting.[11]

The typical ratio of α-chaconine to α-solanine in the tuber is about 60:40. The glycoalkaloid content of peels is substantially higher than tubers, particularly α-chaconine. In a study of Mexican potatoes, the peel always contained higher total glycoalkaloid concentrations than the tuber, and α-chaconine accounted for 65–71% of the total content of glycoalkaloids.[12] Potato leaves also contain higher concentrations of glycoalkaloids than the tubers, particularly α-chaconine. In a study of UK potatoes, the leaves contained 0.06–55.7 mg α-chaconine/100 g and 0.64–22.6 mg α-solanine/100 g.[13] Unprocessed tubers contained 0.3–0.63 mg α-chaconine/100 g and 0.05–0.65 mg α-solanine/100 g.

Normally, potato tubers (*Solanum tuberosum*) contain low concentrations of toxic glycoalkaloids (e.g., solanine) unless adverse storage conditions or cultivation methods increase the solanine content. In a study of Mexican potatoes, the glycoalkaloid content of boiled peeled potatoes ranged from nondetectable to 9 mg/100 g fresh weight.[12] Because of the bitter taste of the alkaloids, solanine poisoning is rare except in times of food shortages when stressed or green potatoes are consumed. Normal food processing methods (baking, frying, broiling, microwaving) do not remove substantial amounts of glycoalkaloids from potatoes.[14] In a study of 20 commercial potato products, the α-chaconine and α-solanine content was lowest in canned boiled potatoes (0.04–0.08 mg/100 g and 0.04–0.06 mg/100 g, respectively) and highest in deep fat fried peels (93.1–97.9 mg/100 g and 46.1–48.0 mg/100 g, respectively).[15] The α-chaconine and α-solanine concentrations of individual commercial potato products were similar. The boiling of sprouted potatoes allows the diffusion of glycoalkaloids from the sprouts to the tuber and therefore increases the glycoalkaloid content of the tuber.[16]

TOMATO

Tomato leaves and vines contain glycoalkaloids similar to solanine, but toxic glycoalkaloids are not usually detectable in the fruit. As listed in Table 11.3, the glycoalkaloid content varies between plant part and the type of glycoalkaloid. Approximately 72% of the original amount of α-tomatine remained after the preparation of Southern fried green tomatoes using traditional cooking methods.[17] In a study of three tomato plants, the ratio of mean concentrations of α-tomatine to dehydrotomatine ranged from 4.7 in the stems and flowers to 10.9 in green fruit as measured by high performance liquid chromatography (HPLC) with UV detection.[18] There were trace amounts of glycoalkaloid compounds in the roots and nondetectable concentrations in mature fruit.

TABLE 11.3. Glycoalkaloid Content in Various Parts of the Tomato Plant

Tomato Part	Dehydrotomatine[a]	α-Tomatine[a]
Fruit (green)	1498 ± 49	16,285 ± 112
Fruit (red)	ND	ND
Flowers	1023 ± 3	4,825 ± 191
Calyxes	370 ± 17	2,870 ± 129
Leaves	304 ± 15	1,847 ± 112
Stems	331 ± 4	1,547 ± 32
Roots	Trace	Trace

Source: Data from Ref 18.
[a]Mean ± standard deviation of three samples expressed as μg/g fresh weight; ND, nondetectable.

Mechanism of Toxicity

In vitro studies indicate that both α-solanine and α-chaconine are reversible inhibitors of human plasma cholinesterase (butyrylcholinesterase).[19] In a study of rabbits, intraperitoneal doses of solanine caused mild to moderate inhibition of both specific and nonspecific cholinesterases.[20] The relevance of plasma cholinesterase inhibition to human solanine toxicity is unclear because solanine toxicity in not classically associated with a cholinergic syndrome. In experimental animals, α-solanine and α-chaconine also demonstrate cytotoxic properties.[13,21]

DOSE RESPONSE

In animal models, the toxicity of solanine depends on the dose, species, and route of administration with the parenteral route much more toxic than the oral route of exposure.[22] Volunteer studies using taste panels suggest that a bitter taste and burning sensation in the mouth occurs when the glycoalkaloid content of potatoes exceeds 14 mg/100 g and 22 mg/100 g fresh weight, respectively.[23,24] The presence of total glycoalkaloid content exceeding 60 mg/100 g fresh weight produces strong bitterness and burning.[25] The total glycoalkaloid content of most commercial potatoes does not usually exceed 10 mg/100 g. The US Food & Drug Administration (FDA) limits solanine content in potatoes to ≤20 mg/100 g potatoes.

TOXICOKINETICS

Animal studies indicate that the absorption of solanine from the gastrointestinal tract is poor, and solanine distributes in the highest concentration to the spleen with progressively lower concentrations in the kidney, liver, lung, fat, heart, brain, and blood.[26] In a study of seven healthy volunteers, the ingestion of mashed potatoes containing glycoalkaloid concentrations of 1 mg/kg (α-solanine:α-chaconine ratio = 41:59) produced peak serum α-solanine and α-chaconine concentrations at approximately 5 ± 1 hours and 6 ± 1 hours, respectively.[27] The biological half-lives of α-solanine and α-chaconine in serum samples were about 11 hours and 19 hours, respectively. The stomach hydrolyzes solanine to the less toxic glycoalkaloid, solanidine. Elimination occurs rapidly in the feces and to a lesser extent in urine. Animal studies indicate that the gastrointestinal tract also poorly absorbs α-chaconine with excretion occurring rapidly in the feces, probably by biliary excretion.[28] The elimination of α-chaconine and α-solanine is similar with the major metabolite being the aglycone, solanidine. Figure 11.1 demonstrates the degradation of major glycoalkaloids in the potato.

CLINICAL RESPONSE

Clinical features of solanine poisoning include gastrointestinal and neurologic symptoms, particularly vomiting, headache, and flushing. The glycoalkaloid content of young leaves is substantially higher than the tubers;[13] therefore, ingestion of the above-ground portion of these plants can cause gastrointestinal distress. The largest series of solanine poisoning involved an English day school where 78 schoolboys developed diarrhea and vomiting after eating potatoes stored since the summer term.[29] Symptoms began 7–19 hours after ingestion with vomiting, diarrhea, anorexia, and malaise. Of the 78 boys, 17 were admitted to the hospital. Other symptoms included fever (88%), altered mental status (drowsiness, confusion, delirium) (82%), restlessness (47%), headache (29%), and hallucinations (23%). Three boys were seriously ill with hypotension, tachycardia, and stupor out of proportion to fluid and electrolyte imbalance. These boys were discharged 6–11 days after admission, and they had nonspecific symptoms and visual blurring for several weeks after release from the hospital.

Fatalities from solanine poisoning are not well documented in the modern medical literature. Deaths have been associated with consumption of toxic potatoes, but those reports involved malnourished patients who may not have received adequate care.[30] Headache, abdominal pain, vomiting, thirst, restlessness, and apathy preceded death, but no convulsions or fever were reported. There is no conclusive data that constituents of potatoes or tomatoes are teratogenic to humans.[31] In studies of hamsters, acute maternal toxicity limited the administration of dosages high enough to induce statistically significant levels of terata in litters receiving α-chaconine and α-solanine.[32] However, the administration of toxic doses of the aglycone solanidine and the derivative

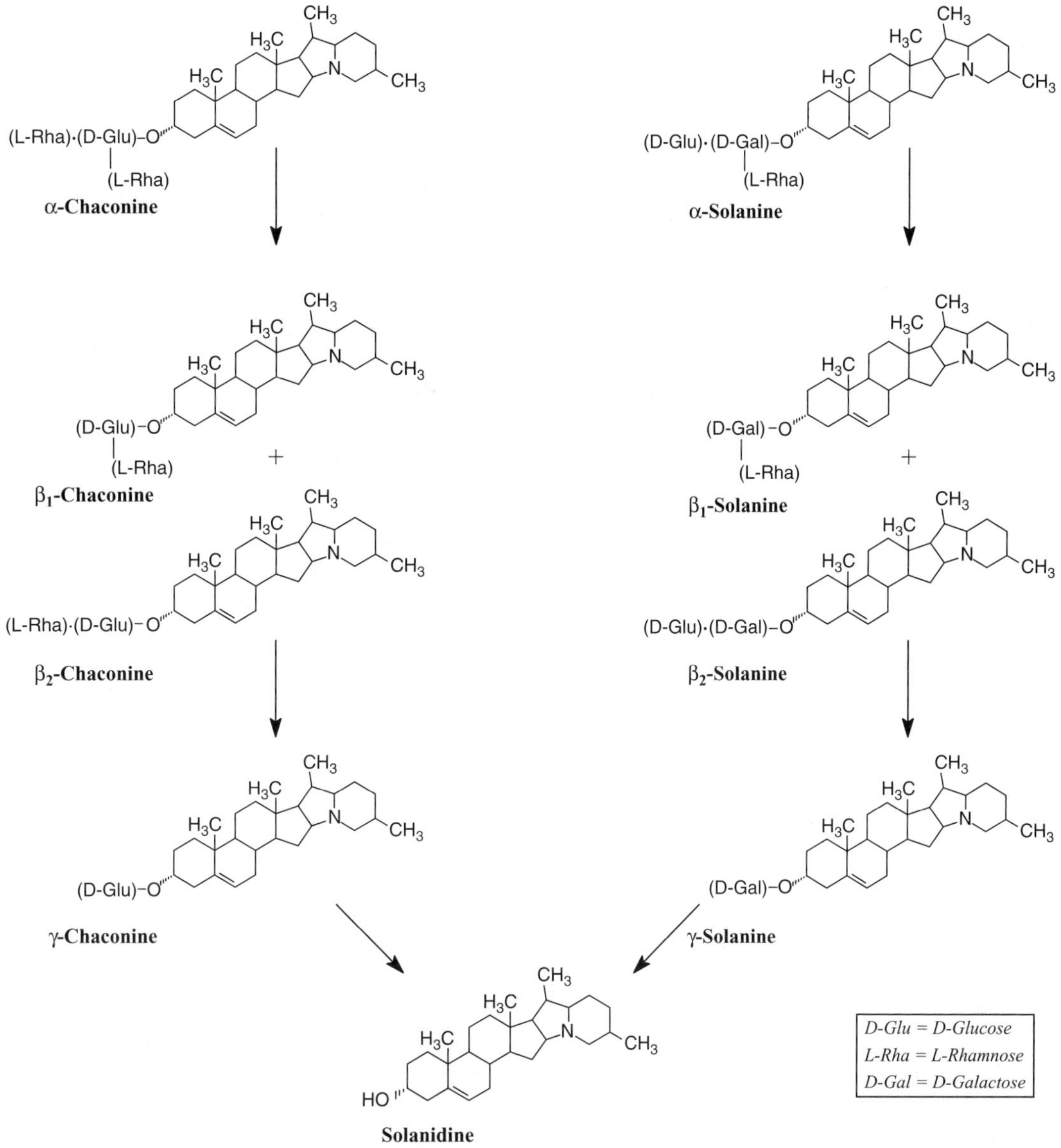

(L-Rha)·(D-Glu)–O
(L-Rha)
α-Chaconine

(D-Glu)·(D-Gal)–O
(L-Rha)
α-Solanine

(D-Glu)–O
(L-Rha)
β_1-Chaconine

(D-Gal)–O
(L-Rha)
β_1-Solanine

(L-Rha)·(D-Glu)–O
β_2-Chaconine

(D-Glu)·(D-Gal)–O
β_2-Solanine

(D-Glu)–O
γ-Chaconine

(D-Gal)–O
γ-Solanine

HO
Solanidine

D-Glu = D-Glucose
L-Rha = L-Rhamnose
D-Gal = D-Galactose

FIGURE 11.1. Hydrolysis of the trisaccharide side chains of potato glycoalkaloids α-chaconine and α-solanine to the aglycone solanidine. Adapted from Friedman M, et al. Postharvest changes in glycoalkaloid content of potatoes. Adv Exp Med Biol 1999;459:124.

solanidine *N*-oxide produced a statistically significant ($p < 0.005$) increase in the incidence of malformations. Although teas brewed from the tomato plant have been associated with solanine-like poisoning in some reference books,[33] there are few data that indicate tomatoes cause toxicity.

DIAGNOSTIC TESTING

Methods for the detection and quantitation of glycoalkaloids include gas chromatography,[34] high performance liquid chromatography (HPLC),[35,36] HPLC with electrospray ionization and tandem mass spectrometry,[37] and gas chromatography mass spectrometry.[38] Enzyme-linked immunosorbent assay (ELISA) methods allow the rapid determination of glycoalkaloid content in plant parts from the potato including the tuber.[39] HPLC is commonly used for the determination of glycoalkaloid content of fresh and processed potatoes, including the concentration of individual glycoalkaloids in various plant parts. The limits of detection for individual glycoalkaloids in serum are about 0.3 ng/mL as determined by HPLC.[27] The AOAC International (Association of Analytical Chemists International, Gaithersburg, MD; http://www.eoma.aoac.org/) official method of analysis for glycoalkaloids in potato tubers involves

HPLC with ultraviolet detection (HPLC/UV).[40] The analytical range for this method is 10–200 mg α-solanine/kg and 20–250 mg α-chaconine/kg. Following the ingestion of approximately 0.41 mg α-solanine/kg body weight and 0.59 mg α-chaconine/kg body weight, the mean peak concentration of these glycoalkaloids were about 8 ng/mL and 14 ng/mL, respectively.[27] In general, glycoalkaloid content of potatoes should not exceed 200 mg/kg fresh potatoes.

TREATMENT

The treatment of solanine poisoning is entirely supportive. Most patients with solanine poisoning develop vomiting and diarrhea; therefore, decontamination measures are not usually necessary. There are no clinical or experimental data to indicate that decontamination measures improve outcome during solanine poisoning. Dehydration and electrolyte imbalance are the most common serious complications of solanine poisoning, and patients with substantial vomiting and diarrhea should be evaluated for the presence of these complications. Treatment involves standard measures to correct fluid and electrolyte imbalance. For those seriously ill patients who do not respond to fluid replacement, cardiac monitoring and vasopressors may be required. Those patients with neurologic abnormalities or serious electrolyte imbalance should be observed until these complications resolve.

References

1. Lee MR. The Solanaceae: foods and poisons. J R Coll Physicians Edinb 2006;36:162–169.
2. Edwards RO Jr. Poisoning from plant ingestions. J Fla Med Assoc 1965;52:875–881.
3. van Eck J, Snyder A. Eggplant (*Solanum melongena* L.). Meth Mol Biol 2006;343:439–447.
4. Jadhav SJ, Sharma RP, Salunkhe DK. Naturally occurring toxic alkaloids in foods. CRC Crit Rev Toxicol 1981;9:21–104.
5. Friedman M. Potato glycoalkaloids and metabolites: roles in the plant and in the diet. J Agric Food Chem 2006;54:8655–8681.
6. Friedman M, Roitman JN, Kozukue N. Glycoalkaloid and calystegine contents of eight potato cultivars. J Agric Food Chem 2003;51:2964–2973.
7. Friedman M, McDonald GM. Postharvest changes in glycoalkaloid content of potatoes. Adv Exp Med Biol 1999;459:121–143.
8. Whitaker BD, Stommel JR. Distribution of hydroxycinnamic acid conjugates in fruit of commercial eggplant (*Solanum melongena* L.) cultivars. J Agric Food Chem 2003;51:3448–3454.
9. Coxon DJ, Curtis RF, Howard B. Ipomearone, a toxic furanoterpenoid in sweet potatoes (*Ipomea batatas*) in the United Kingdom. Food Cosmet Toxicol 1975;13:87–90.
10. Slanina P. Solanine (glycoalkaloids) in potatoes: toxicological evaluation. Food Chem Toxicol 1990;28:759–761.
11. Dimenstein L, Lisker N, Kedar N, Lefy D. Changes in the content of steroidal glycoalkaloids in potato tubers grown in the field and in the greenhouse under different conditions of light, temperature and daylength. Physiol Mol Plant Pathol 1997;50:391–402.
12. Sotelo A, Serrano B. High-performance liquid chromatographic determination of the glycoalkaloids α-solanine and α-chaconine in 12 commercial varieties of Mexican potato. J Agric Food Chem 2000;48:2472–2475.
13. Phillips BJ, Hughes JA, Phillips JC, Walters DG, Anderson D, Tahourdin CS. A study of the toxic hazard that might be associated with the consumption of green potato tops. Food Chem Toxicol 1996;34:439–448.
14. Takagi K, Toyoda M, Fuiyama Y, Saito Y. Effect of cooking on the contents of α-chaconine and α-solanine of potatoes. J Food Hyg Soc Japan 1990;31:67–73.
15. Bushway RJ, Ponnampalam R. α-Chaconine and α-solanine content of potato products and their stability during several modes of cooking. J Agric Food Chem 1981;29:814–817.
16. Gonmori K, Shindo S. Effect of cooking on the concentration of solanine in potato. Res Pract Forensic Med 1985;28;91–93.
17. Friedman M, Levin CE. α-Tomatine content in tomatoes and tomato products determined by HPLC with pulsed amperometric detection. J Agric Food Chem 1995;43: 1507–1511.
18. Kozukue N, Han JS, Lee KR, Friedman M. Dehydrotomatine and alpha-tomatine content in tomato fruits and vegetative plant tissues. J Agric Food Chem 2004;52: 2079–2083.
19. Nigg HN, Ramos LE, Graham EM, Sterling J, Brown S, Cornell JA. Inhibition of human plasma and serum butyrylcholinesterase (EC3.1.1.8) by α-chaconine and α-solanine. Fund Appl Toxicol 1996;33:272–281.
20. Patil BC, Sharma RP, Salunkhe DK, Salunkhe K. Evaluation of solanine toxicity. Food Cosmet Toxicol 1972;10: 395–398.
21. Roddick JG, Rijnenberg AL, Osman SF. Synergistic interaction between potato glycoalkaloids α-solanine and α-chaconine in relation to destabilization of cell membranes. J Chem Ecol 1988;14:889–902.
22. Dalvi RR, Bowie WC. Toxicology of solanine: An overview. Vet Hum Toxicol 1983;25:13–15.
23. Johns T, Keen SL. Taste evaluation of potato glycoalkaloids by the Ayamara: a case study in human chemical ecology. Hum Ecol 1986;14;437–452.
24. Zitnak A, Filadelfi MA. Estimation of taste thresholds of three potato glycoalkaloids. J Can Inst Food Sci Technol 1985;18:337–339.

25. Sinden SL, Deahl KL, Aulenbach BB. Effect of glycoalkaloids and phenolics on potato flavor. J Food Sci 1976;41: 520–523.
26. Nishie K, Gumbmann MR, Keyl AC. Pharmacology of solanine. Toxicol Appl Pharmacol 1971;19:81–92.
27. Hellenas K-E, Nyman A, Slanina P, Loof L, Gabrielsson J. Determination of potato glycoalkaloids and their aglycone in blood serum by high-performance liquid chromatography. J Chromatogr Biomed Appl 1992;573:69–78.
28. Norred WP, Nishie K, Osman SF. Excretion, distribution and metabolic fate of 3H-alpha-chaconine. Res Commun Chem Pathol Pharmacol 1976;13:161–171.
29. McMillan M, Thompson JC. An outbreak of suspected solanine poisoning in school boys: An examination of criteria of solanine poisoning. Q J Med 1979;48:227–243.
30. Hansen AA. Two fatal cases of potato poisoning. Science 1925;61:348–349.
31. Chaube S, Swinyard CA. Teratological and toxicological studies of alkaloidal and phenolic compounds from *Solanum tuberosum* L. Toxicol Appl Pharmacol 1976;36: 227–237.
32. Gaffield W, Keeler RF. Induction of terata in hamsters by solanidane alkaloids derived from *Solanum tuberosum*. Chem Res Toxicol 1996;9:426–433.
33. Hardin JW, Arena JM. Human poisoning from native and cultivated plants. 2nd ed. Durham, NC: Duke University Press; 1974:140.
34. Bushway RJ, McGann DF, Bushway A. Gas chromatographic method for the determination of solanidine and tis application to a study of fed-milk transfer in the cow. J Agric Food Chem 1984;32:548–551.
35. Sotelo A, Serrano B. High-performance liquid chromatographic determination of the glycoalkaloids alpha-solanine and alpha-chaconine in 12 commercial varieties of Mexican potato. J Agric Food Chem 2000;48:2472–2475.
36. Kodamatani H, Saito K, Niina N, Yamazaki S, Tanaka Y. Simple and sensitive method for determination of glycoalkaloids in potato tubers by high-performance liquid chromatography with chemiluminescence detection. J Chromatogr A 2005;1100:26–31.
37. Cataldi TR, Lelario F, Bufo SA. Analysis of tomato glycoalkaloids by liquid chromatography coupled with electro-spray ionization tandem mass spectrometry. Rapid Commun Mass Spectrometry 2005;19:3103–3110.
38. van Gelder WM, Tuinstra LG, Van Der Greef J, Scheffer JJ. Characterization of novel steroidal alkaloids from tubers of *Solanum* species by combined gas chromatography-mass spectrometry. Implications for potato breeding. J Chromatogr 1989;482:13–22.
39. Friedman M, Bautista F, Stanker LH, Larkin KA. Analysis of potato glycoalkaloids by a new ELISA kit. J Agric Food Chem 1998;46:5097–5102.
40. Horwitz W, editor. Official methods of analysis of AOAC International. 17th ed. Gaithersburg, MD: AOAC International; 2000, Official method 997.13.

Chapter 12

RHUBARB and OXALOSIS (*Rheum* Species)

HISTORY

Although the petioles of rhubarb leaves have been a food source in modern times, the dried root or rhizome has been used as a medicinal herb since the third millennium BC. Chinese rhubarb is traditional herb used as a purgative and bacteriocidal agent for dysentery.[1] In the 16th and 17th centuries, Chinese rhubarb was coveted as a medicinal remedy that was superior to European varieties. During this time, Russian sources (the Commerce Collegium in St. Petersburg, the Rhubarb Commission on the Mongolian border) and the British East India Company supplied Europe with large quantities of rhubarb.[2] This popular laxative was the "All Bran of the Age of Reason." Although the Russian explorer, Nikolai Mikhailovich Przheval'skii, obtained rhubarb seeds from the major Chinese production center (Sining) in the late 1800s, European growing conditions could not duplicate the luxuriant growth of rhubarb cultivated in China. In the early 1900s, the anthraquinone compounds responsible for the purgative action of rhubarb were separated from powdered rhubarb root at the Wellcome Research Laboratories (Beckenham, Kent, UK). Consumption of rhubarb leaves as a food substitute for spinach was encouraged in England during World War I until several deaths were attributed to the ingestion of cooked leaves.[3]

BOTANICAL DESCRIPTION

Species in the oxalate-containing genus *Rheum* include *Rheum officinale* Baillon (Chinese rhubarb), *Rheum palmatum* L. (Turkey rhubarb), *Rheum rhabarbarum* L. (garden rhubarb), and *Rheum rhaponticum* L. (false rhubarb). Other plant species that contain substantial quantities of oxalate include *Halogeton glomeratus* (Bieb.) C. A. Mey. (barilla, saltlover), *Oxalis caerulea* (Small) R. Knuth (blue woodsorrel), *Oxalis corniculata L.* (creeping oxalis, creeping woodsorrel, yellow oxalis), *Portulaca oleracea* L. (common purslane, duckweed), *Rumex acetosa* L. (garden sorrel), *Rumex crispus* L. (curley dock, yellow dock),[4] and *Tetragonia tetragonioides* (Pallas) Kuntze (New Zealand spinach).

Common Name: Garden rhubarb

Scientific Name: *Rheum rhabarbarum* L.

Botanical Family: Polygonaceae (buckwheat)

Physical Description: A garden vegetable plant with large, leathery, heart-shaped leaves and a reddish color.

Distribution and Ecology: *R. rhabarbarum* L. is an introduced, perennial plant that inhabits Alaska, the Rocky Mountain States, Georgia, and the northeastern United States.

EXPOSURE

The leaf petioles (stalks) are used as a vegetable and a constituent of pies. The dried roots have been used medicinally as a cathartic and as a tonic.[5,6] In traditional Chinese medicine, official rhubarb (dahuang) is a purgative and detoxicant used to treat fever, constipation, cancer, abdominal distention and obstruction, jaundice, hematemesis, appendicitis, amenorrhea, skin lesions,

Medical Toxicology of Natural Substances, by Donald G. Barceloux, MD

food poisoning, inflammation, hypertension, and renal failure.[7,8] The official Chinese herbal medicine contains dried rhizome and root of *R. palmatum* L., *R. tanguticum* Maxim. ex Balf., and *R. officinale* Baillon. The Japanese version of Rhei Rhizoma (rhubarb) also contains *R. coreanum* Nakai.

PRINCIPAL TOXINS

Structure and Properties

Salts of oxalic acid are the main toxic constituents in rhubarb. This chemical is the simplest dicarboxylic acid $[(COOH)_2]$ and oxalic acid forms soluble (iron, lithium, potassium, sodium) and insoluble (calcium, magnesium) salts. Boiling spinach for 1 minute removes only about 10% of the insoluble oxalate salts compared with 47% of the soluble oxalate salts.[9] Cooking does not substantially alter the oxalate concentration in rhubarb. Roots of some species of rhubarb (*R. palmatum* L.) contain up to 2% anthraquinone glycoside derivatives similar in structure to the sennosides, which also possess laxative properties.[7] The major active components of the herbs derived from rhubarb (e.g., dahuang) are hydroxyanthraquinone compounds including rhein, emodin, aloe-emodin, and chrysophanol.[10] Figure 12.1 displays the structure of some hydroxyanthraquinone compounds in herbs derived from the Chinese rhubarb (*Rheum officinale*). Emodin is a naturally occurring anthraquinone present in the rhubarb as well as the roots and bark of numerous plants of the genus *Rhamnus* including buckthorn (*Rhamnus cathartica* L.), senna (*Senna alexandrina* Mill.), cascara (*Frangula purshiana* (DC) J.G. Cooper), and aloe (*Aloe ferox* Mill.). This hydroxyanthraquinone compound is a common constituent of herb-based stimulant laxatives that produce mild nephrotoxicity in some rodents following the administration of very high doses (i.e., about 1 g emodin/kg body weight daily for 2 years).[11]

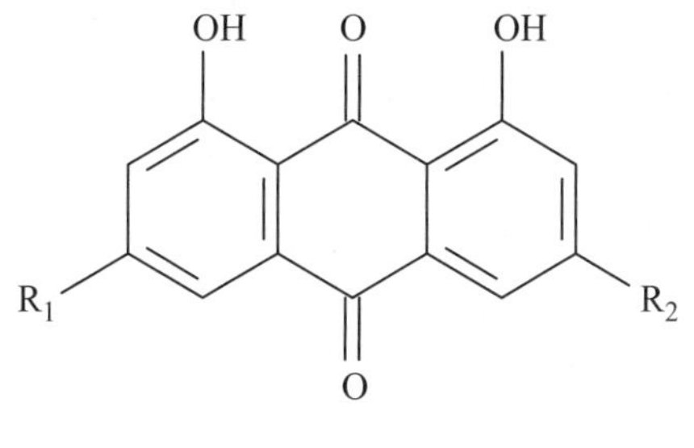

OH O OH

R_1 R_2

O

	R_1	R_2
Rhein	H	COOH
Emodin	CH_3	OH
Aloe-Emodin	OH	CH_2OH
Chrysophanol	H	CH_3
Physcion	CH_3	OCH_3

FIGURE 12.1. Hydroxyanthraquinone compounds in rhubarb.

Poisonous Parts

Oxalate occurs in most plant tissues and the amount depends on growing conditions, season, and plant part. The average Western diet contains about 50–150 mg oxalate/d with vegetarian diets somewhat higher (i.e., 150 mg/d).[12] Spinach contains oxalate compounds divided into approximately 55% soluble oxalate and 45% calcium oxalate, whereas rhubarb contains primarily calcium oxalate.[13] The most toxic part of the rhubarb plant is the leaf, which contains <1% fresh weight soluble oxalates.[14] The petioles of rhubarb leaves are edible because of the low oxalate content. Food sources beside rhubarb that contain substantial amounts of oxalate include sugar beets, chocolate, coconut, peanuts, spinach, strawberries, wheat bran, and tea.[15,16] Table 12.1 lists the oxalate content of some common foods. The major bioactive constituents of herbal preparations of rhubarb are phenolic compounds (sennosides, anthraquinone

TABLE 12.1 Oxalate Content of Some Foods

Food Group	Food	Soluble Oxalate[a]	Total Oxalate[a]
Fruits			
	Apple	0.3–1.8	0.4–5.8
	Banana	0.1–2.2	0.5–23.9
	Orange	0.2	1.8
	Strawberry	0.6–1.9	1.5–4.3
Vegetables			
	Asparagus[b]	0.5–1.1	1.8–3.1
	Beans	1.5	13.9
	Potatoes[b]	8.8–18.9	8.8–35.3
	Rhubarb	380	570–1900
	Spinach[b]	33.3–168	100–627
	Tomato	2.5–4.5	3.7–13.7
Bread			
	Rye	0.9	
	White	4.9–8.6	
Drinks/Sweets			
	Beer	1.7–1.8	
	Black Tea	2.5–6.2	
	Chocolate	7.1	
	Cocoa		154–980
	Coffee	0.5–0.7	
	Coke	0.05	
	Iced Tea	0.46–1.72	0.35–2.05

Source: Adapted from Ref 18.
[a]mg oxalate/100 g food.
[b]Cooked.

glycosides, glucose gallates, naphthalenes, catechins).[17] These concentrations and types of phenolic compounds (e.g., sennosides, anthraquinone glycosides) vary between different species of *Rheum*.

Mechanism of Toxicity

The ingestion of rhubarb causes oxalate poisoning that is associated with local corrosive effects and renal damage from the excretion of oxalate crystals.[5] The formation of large amounts of calcium oxalate crystals in the body may produce hypocalcemia as well as renal dysfunction, renal calculi, and electrolyte imbalance.[18] The deposition of calcium oxalate crystals in the myocardium of patients with renal failure may occur without the presence of oxalate poisoning.[19] Pathological changes associated with fatal oxalate ingestions include corrosive effects on the gastrointestinal tract, cloudy swelling and hyalin degeneration of the renal tubules, glomerulosclerosis, and birefringent crystals in vascular walls throughout the body.[20]

DOSE RESPONSE

The ingestion of small amounts of plant parts containing oxalates usually causes only mild gastrointestinal irritation. Serious intoxication from the ingestion of rhubarb leaf is not well-documented in the modern medical literature. Fatalities associated with the ingestion of rhubarb probably involved other agents, toxins, or etiologies. The intravenous injection of 20mg sodium oxalate/kg produced profound hypocalcemia and cardiac arrest.[21]

TOXICOKINETICS

Gastrointestinal absorption of oxalate occurs in the small intestine by active transport and by passive diffusion along the small intestine and the colon. The bioavailability of oxalate is low (<2–6% as total oxalate) and varies with plant species.[15] In a study of volunteers, the average bioavailability of oxalate in sugar beets and spinach was 0.7% and 4.5%, respectively.[22] The mean bioavailability of a solution of sodium (soluble) oxalate in the same study was 6.2%. Endogenous intermediary metabolism in humans produces oxalate as a result of glycine, ascorbic acid, and glycolate metabolism. As end-products of metabolism, oxalates are excreted by the kidney rather than metabolized. Most renal excretion of oxalate absorbed from the diet occurs within 8–12 hours.[13] In a study of 12 volunteers, the elimination half-life of the hydroxyanthraquinone compound, rhein, derived from the Chinese rhubarb was 3.38 ± 0.35 hours.[23]

CLINICAL RESPONSE

Most casual exposures of children to the rhubarb plant produce mild gastrointestinal symptoms (vomiting, diarrhea) that resolve within a few hours.[24] The initial symptoms of oxalate intoxication result from irritation of the oropharynx and the gastrointestinal tract manifest by sore throat, dysphagia, nausea, vomiting, anorexia, diarrhea, abdominal pain, and occasionally hematemesis. Symptoms typically begin within 2–12 hours after ingestion.[14] Serious oxalate toxicity produces renal dysfunction and electrolyte imbalance that develops after signs of gastrointestinal distress. The presence of paresthesias, tetany, hyperreflexia, muscle twitches, and muscle cramps suggest hypocalcemia. Seizures may complicate the clinical course as a result of hypocalcemia.[25] Other sources of oxalosis include excessive oxalate formation as a result pyridoxine deficiency,[26] primary hyperoxaluria with glycolic aciduria or L-glyceric aciduria,[27] increased oxalate intake from the diet, toxins (ethylene glycol, glycolate), renal failure, and fungal infections.[28,29] Rarely, case reports associate exposure to rhubarb with the development of vesiculobullous, photosensitivity dermatitis.[30] Another case report associated acute renal failure, atrophy of the renal tubules, and interstitial renal fibrosis with the chronic ingestion of the nonsteroidal antiinflammatory drug, diclofenac, and a slimming pill that contained anthraquinone derivatives extracted from Rhizoma Rhei (rhubarb).[31]

DIAGNOSTIC TESTING

Analytical Methods

Methods for detecting anthraquinone derivatives in rhubarb include thin layer chromatography,[32] capillary electrophoresis,[33] high performance liquid chromatography (HPLC),[34] high speed counter-current chromatography,[10] high performance liquid chromatography with ultraviolet detection and mass spectrometry,[35] and HPLC coupled with electrospray ionization mass spectrometry.[17] HPLC with mobile phase gradient conditions and UV detection (280nm) can detect at least 30 compounds in rhubarb at detection limits ranging from 0.05–2μg/mL.[36] These compounds include anthraquinones, anthraquinone glucosides, dianthrones, phenylbutanones, stilbenes, galloylglucoses, acylglucoses, flavan-3-ols, procyanidins, and gallic acid.

Biomarkers

Normal values of oxalate in the serum range from approximately 60–230μg/dL.[37] The concentration of urinary oxalate depends on urine volume and electrolytes as well as diet. In a study of 20 men with recurrent calcium containing kidney stones, the upper ranges of urine oxalate concentrations for daily urine volumes of 1, 2, and 3 liters were 46 mg, 55 mg, and 63 mg, respectively.[38] Women excreted slightly lower daily amounts of urine oxalate.

Abnormalities

The definition of hyperoxaluria is the urinary excretion of >0.5 mmol oxalate/1.73 m^2 body surface area daily. Electrolyte imbalance (e.g., hypocalcemia) complicates serious cases of oxalate poisoning. Laboratory evidence of renal dysfunction includes the presence of anuria, oliguria, proteinuria, hematuria, and oxaluria. Microscopic examination of urine sediment may demonstrate red blood cells, leukocytes, or birefringent crystals of calcium oxalate as either needle-like or envelope-shaped dimorphic forms.

TREATMENT

Decontamination

Decontamination is not usually necessary following the ingestion of oxalate-containing plant parts. Symptomatic treatment for pharyngeal irritation includes demulcents (milk, chipped ice) and antihistamines. For serious oxalate ingestions, lavage with 0.15% calcium hydroxide (lime water) to precipitate insoluble calcium oxalate in the gastrointestinal tract was a previously recommended treatment for patients presenting to a health care facility within one hour of ingestion, but there are no clinical data to substantiate the efficacy or safety of this therapeutic modality.

Supplemental Care

For symptomatic patients, generous fluid replacement (>2 L/m^2/d) should be administered to promote the excretion of calcium oxalate crystals from the kidney tubules. The treatment of patients with *chronic* secondary oxalosis includes the use of sodium citrate (0.15 g/kg/d) to inhibit calcium oxalate crystallization in the kidneys,[39] but there are few clinical data on the use of sodium citrate following the *acute* ingestion of calcium oxalate-containing plants. Symptomatic patients should receive a complete blood count, serum electrolytes including serum calcium, and kidney function tests (serum creatinine and BUN [blood urea nitrogen], urinalysis). Kidney function may deteriorate over the first week and careful management of fluid and electrolyte balance may be necessary. Intravenous calcium is not usually required to reverse the oxalate-induced hypocalcemia unless the patient is symptomatic. The initial adult treatment for symptomatic hypocalcemia is 10 mL of 10% calcium gluconate intravenously with cardiac monitoring over a 10-minute period and repeated if symptoms, signs, and electrocardiogram (ECG) evidence (e.g., shortened QT interval) persist. Hemodialysis easily removes oxalate from the blood with most elimination within the first 2 hours.[40]

References

1. Butler AR. The Fifth Haldane Tait Lecture. The coming of rhubarb. Rep Proc Scott Soc Hist Med 1994;96:52–55.
2. Foust CM. Mysteries of rhubarb: Chinese medicinal rhubarb through the ages. Pharm His 1994;36:155–159.
3. Robb HF. Death from rhubarb leaves due to oxalic acid poisoning. JAMA 1919;73:627–662.
4. Farre M, Xirgu J, Salgado A, Peracaula R, Reig R, Sanz P. Fatal oxalic acid poisoning from sorrel soup. Lancet 1989;2:1524.
5. Jacobziner H, Raybin HW. Rhubarb poisoning. NY J Med 1962;62:1676–1678.
6. Harima S, Matsuda H, Kuo M. Study of various rhubarbs regarding the cathartic effect and endotoxin-induced disseminated intravascular coagulation. Biol Pharm Bull 1994;17:1522–1525.
7. Wojcikowski K, Johnson DW, Gobe G. Medicinal herbal extracts — renal friend or foe: Part two: Herbal extracts with potential renal benefits. Nephrol 2004;9:400–405.
8. Peigen X, Liyi H, Liwei W. Ethnopharmacologic study of Chinese rhubarb. J Ethnopharmacol 1984;10:275–293.
9. Ohkawa H. Gas chromatographic determination of oxalic acid in foods. J Assoc Off Anal Chem 1985;68:108–111.
10. Liu R, Li A, Sun A. Preparative isolation and purification of hydroxyanthraquinones and cinnamic acid from the Chinese medicinal herb *Rheum officinale* Baill. by high-speed counter-current chromatography. J Chromatogr A 2004;1052:217–221.
11. National Toxicology Program. NTP toxicology and carcinogenesis studies of EMODIN (CAS NO. 518-82-1) feed studies in F344/N rats and B6C3F1 mice. Natl Toxicol Program Tech Rep Ser 2001;493:1–278.
12. Siener R, Ebert D, Nicolay C, Hesse A. Dietary risk factors for hyperoxaluria in calcium oxalate stone formers. Kidney Int 2003;63:1037–1043.
13. Prenen JAC, Boer P, Dorhout Mees EJ. Absorption kinetics of oxalate from oxalate-rich food in man. Am J Clin Nutr 1984;40:1007–1010.
14. Tallqvist H, Vaananen I. Death of a child from oxalic acid poisoning due to eating rhubarb leaves. Ann Paediatr Fenn 1960;6:144–147.
15. Massey LK, Roman-Smith H, Sutton RA. Effect of dietary oxalate and calcium on urinary oxalate and risk of forma-

tion of calcium oxalate kidney stones. J Am Diet Assoc 1993;93:901–906.

16. Finch AM, Kasidas GP, Rose GA. Urine composition in normal subjects after oral ingestion of oxalate-rich foods. Clin Sci (London) 1981;60:411–418.
17. Ye M, Han J, Chen H, Zheng J, Guo D. Analysis of phenolic compounds in rhubarbs using liquid chromatography coupled with electrospray ionization mass spectrometry. J Am Soc Mass Spectrom 2007;18: 82–91.
18. Hoppe B, Leumann E, von Unruh G, Laube N, Hesse A. Diagnostic and therapeutic approaches in patients with secondary hyperoxaluria. Front Biosci 2003;8:e437–e443.
19. Salyer WR, Hutchins GM. Cardiac lesions in secondary oxalosis. Arch Intern Med 1974;134:250–252.
20. Sanz P, Reig R. Clinical and pathological findings in fatal plant oxalosis. Am J Forensic Med Pathol 1992;13: 342–345.
21. Dvorackova I. [Fatal poisoning following intravenous administration of sodium oxalate]. Arch Toxikol 1966;22:63–67 [German].
22. Hanson CF, Frankos VH, Thompson WO. Bioavailability of oxalic acid from spinach, sugar beet fibre and a solution of sodium oxalate consumed by female volunteers. Food Chem Toxicol 1989;27:181–184.
23. Lee J-H, Kim JM, Kim C. Pharmacokinetic analysis of rhein in *Rheum undulatum* L. J Ethnopharmacol 2003;84:5–9.
24. Lamminpää A, Kinos M. Plant poisonings in children. Hum Exp Toxicol 1996;15:245–249.
25. Kaliiala H, Kauste O. Ingestion of rhubarb leaves as cause of oxalic acid poisoning. Ann Paediatr Fenn 1964;10: 228–231.
26. Williams HE, Smith LH Jr. Disorders of oxalate metabolism. Am J Med 1968;45:715–735.
27. Kuiper JJ. Initial manifestation of primary hyperoxaluria type I in adults: recognition, diagnosis, and management. West J Med 1996;164:42–53.
28. Nime FA, Hutchins GM. Oxalosis caused by *Aspergillus* infection. Johns Hopkins Med J 1973;133:183–194.
29. Weiner ES, Hutchins GM. Localized endotracheal oxalosis probably secondary to aspiration of rhubarb. Arch Intern Med 1979;139:602.
30. Diffey BL, Lawlor EF, Hindson TC. Photoallergic contact dermatitis to rhubarb wine. Photodermatology 1984;1: 43–44.
31. Kwan TH, Tong MK, Leung KT, Lai CK, Poon WT, Chan YW, et al. Acute renal failure associated with prolonged intake of slimming pills containing anthraquinones. Hong Kong Med J 2006;12:394–397.
32. Zhang HX, Liu MC. Separation procedures for the pharmacologically active components of rhubarb. J Chromatogr B Analyt Technol Biomed Life Sci 2004;812: 175–181.
33. Koyama J, Morita I, Kobayashi N. Simultaneous determination of anthraquinones in rhubarb by high-performance liquid chromatography and capillary electrophoresis. J Chromatogr A 2007;1145:183–189.
34. Zhang YZ, Lu YH, Wei DZ, Chou GX, Zhu EY. Preparative isolation and purification of hydroxyanthraquinones from *Rheum tanguticum* Maxim. on normal phase silica gel: using a Flash Master Personal system. Prep Biochem Biotechnol 2007;37:185–193.
35. Lin CC, Wu CI, Lin TC, Sheu SJ: Determination of 19 rhubarb constituents by high-performance liquid chromatography-ultraviolet-mass spectrometry. J Sep Sci 2006;29: 2584–2593.
36. Komatsu K, Nagayama Y, Tanaka K, Ling Y, Basnet P, Meselhy MR. Development of a high performance liquid chromatographic method for systematic quantitative analysis of chemical constituents in rhubarb. Chem Pharm Bull 2006;54:941–947.
37. Hodgkinson A, Zarembski PM. Oxalic acid metabolism in man: a review. Calcif Tissue Res 1968;2:115–132.
38. Oreopoulos DG, Husdan H, Leung M, Reid AD, Rapoport A. Urine oxalic acid: relation to urine flow. Ann Intern Med 1976;85:617–618.
39. Leumann E, Hoppe B, Neuhaus T, Blau N. Efficacy of oral citrate administration in primary hyperoxaluria. Nephrol Dial Transplant 1995;10(8 Suppl):14–16.
40. Langman CB. The optimal approach to the patient with oxalosis. Adv Renal Replace Ther 2001;8:214–222.

III Microbes

A Bacteria

Chapter 13

BACILLUS CEREUS

CYRUS RANGAN, MD, FAAP

HISTORY

Bacillus cereus Frankland and Frankland (*B. cereus*) is one of over 100 species belonging to the genus *Bacillus*, which contains other foodborne species (*B. anthracis, B. thuringiensis, B. mycoides*, and *B. pseudomycoides*).[1] *B. cereus* is usually a gram-positive aerobic spore-forming bacterium that causes two distinct gastrointestinal illnesses, the emetic syndrome and the diarrheal syndrome. *B. cereus* can become a gram-negative or a facultative anaerobe with age while maintaining toxin-producing properties.[2] In 1955, the first reported cases of *B. cereus*-induced food poisoning occurred in Scandinavia, when four outbreaks involving over 600 patients were linked to consumption of vanilla sauce contaminated by *B. cereus*.[3] Although food poisoning from *B. cereus* is well known, cases are vastly underreported or diagnosed as presumed viral infections, secondary to a short and generally mild course of afebrile gastrointestinal illness that usually resolves spontaneously. Detection of *B. cereus* during food poisoning episodes requires analysis of stool, vomitus, and food specifically for this bacterium. In an outbreak during 1985, 11 of 36 customers of a Japanese restaurant developed acute gastrointestinal symptoms. Stool and vomitus specimens were positive for *B. cereus*, which was also found in hibachi steak served to these patrons.[4] In 1993, 14 of 67 staff and children at two Virginia daycare centers developed gastrointestinal symptoms after consuming a catered lunch from a Chinese restaurant. *B. cereus* was isolated from the vomitus of one child and from leftover chicken fried rice, which was held at room temperature for several hours after initial cooking and reheating.[5]

EXPOSURE

Source

ENVIRONMENT

Bacillus cereus is ubiquitous in water and soil. Uptake of spores and the bacterium by rice plants occurs in rice fields contaminated by *B. cereus*. Transfer of *B. cereus* to food occurs not only from soil and fertilizers during harvesting and postharvest handling, but also from contamination during processing, shipping, and marketing.[6] Although this bacterium frequently contaminates vegetables, the occurrence of diarrheal illness secondary to the ingestion of *B. cereus* is relatively uncommon.

ANIMAL

Bacillus cereus can colonize the gastrointestinal tract of some invertebrates (e.g., worms, flying insect larvae), resulting in the exposure of cows to this bacteria. *B. cereus* spores may contaminate dairy feed and provide a source for the contamination of cow milk by *B. cereus*.[7] The life cycle of this bacterium may contribute to the occurrence of *B. cereus* in the milk from cows and goats.[8] Some humans transiently carry *B. cereus* in the gut without symptoms (14–43%), as a result of the ingestion of nonpathogenic concentrations of the bacterium from food.[9,10]

Medical Toxicology of Natural Substances, by Donald G. Barceloux, MD

Food

Bacillus cereus can contaminate most foods, but the development of *Bacillus cereus*-induced food poisoning requires high bacterial counts. Foods frequently associated with high *B. cereus* content include rice (particularly fried rice), cereals, pasta, corn, cornstarch, beef, poultry, pasteurized dairy products,[11] infant formulas, meats, fish, vegetables, and soups.[12,13] The diarrheal type is typically associated with vegetables, sauces, soups, meat, and milk products, whereas the emetic type is usually associated with rice, pasta, noodles, and pastry.[14] Bacterial growth is enhanced in proteinaceous environments. The preparation of fried rice provides an ideal milieu for bacterial growth. Rice is typically cooked in advance, and then cooled at room temperature when spores begin to germinate. Subsequent stir-frying does not generate sufficient heat to destroy toxins formed during the storage of precooked rice. *In vitro* studies indicate that quantities of emetic-type toxin sufficient to cause human illness can form at room temperature (20 °C/68 °F) within 12–16 hours after inoculation of rice with *B. cereus*.[15] These studies also indicate that the addition of vinegar, mayonnaise, or catsup to foods contaminated with *B. cereus* inhibits both bacterial proliferation and toxin formation. When the rice is briefly reheated or flash-fried, the addition of proteinaceous foods (e.g., chicken, beef, eggs) augments the growth of *B. cereus*.[16,17] Although bacterial growth occurs in these proteinaceous foods, experimental studies suggest that the production of emetic-type toxin is relatively low in eggs, milk, and meat compared with rice.[15] Consequently, these proteinaceous foods are a source of contamination for improperly refrigerated rice, but rice remains the most common source for *B. cereus*-associated food poisoning.

Food Processing

Proper refrigeration does not eliminate *B. cereus* because of the ability of these bacteria to survive low temperatures (i.e., 5 °C–8 °C/41 °F–46 °F).[18] These psychrotrophic organisms multiply optimally at 68 °F (20 °C), but these organisms remain viable at temperatures from 32 °F–104 °F (0 °C–40 °C). Accordingly, room temperature (72 °F/22 °C) provides a favorable environment for *B. cereus* growth, while refrigeration (35 °F/2 °C) and reheating do not always inhibit growth or kill these organisms.[19] The US Department of Agriculture and the European Food Safety Authority do not have specific regulations on *B. cereus* in foodstuffs.[20] The ability to control *B. cereus* in commercial food processing is limited by a number of factors including the ubiquity of the organism, the resistances of spores to commercial disinfectant processes, and the lack of easily detectable contamination by smell or sight.[21] Vegetative cells can form films on stainless steel equipment and containers, conferring even higher resistance to disinfectants and heat.[22] Chlorine, chlorine dioxides, and FIT powder (i.e., a combination of glycerin, starch, trisodium phosphate, sodium lauryl sulfate, and grapefruit oil) are effective disinfectants against *B. cereus* vegetative cells and spores.[23] Pasteurization is an effective means of killing vegetative cells, but the pasteurization process does not kill all *B. cereus* spores.[24] Each vegetative cell is capable of producing one spore. The spore core, or protoplast, is surrounded by a cortex and three protein coats that protect the core from environmental stresses, such as heat, radiation, chemical disinfectants, and desiccants.[2] Spores from *B. cereus* withstand high temperatures during cooking and pasteurization, allowing them to germinate as food cools. Therefore, the storage of precooked food for later serving requires immediate refrigeration after cooking to reduce germination of *B. cereus* spores. Food should be reheated at high temperatures (>140 °F/60 °C,) to kill new vegetative cells, and food should be served immediately.[6] Microwave ovens cook food from the outside-in, so a thermometer should be placed into the center of the food to confirm adequate reheating temperatures. Although proper reheating can kill vegetative cells, the spores and preformed toxins can survive normal food preparation methods.[25] Use of separate utensils and containers for cooked and uncooked food may help prevent recontamination of cooked food, and food handlers must practice strict personal hygiene and receive proper food safety training.[26,27]

TOXINS

Physiochemical Properties

Distinct toxins are responsible for two forms of gastrointestinal illness caused by *B. cereus*. The emetic-type of illness results from the formation of the dodecadepsipeptide toxin, cereulide, whereas the diarrheal form of the illness probably involves several enterotoxins.[28]

Emetic Toxin

Cereulide is a preformed, 1165-Da cyclic dodecadepsipeptide that is heat stable to 259 °F (126 °C).[29] The production of cereulide by *B. cereus* depends on several factors including the type of the food, temperature, pH, preparation methods (e.g., aeration), and the specific strain of *B. cereus*.[30,31] The chemical structure of cereulide includes a ring-shaped structure consisting of three repeats of a four-amino acid sequence (leucine, alanine, valine, valine). Cereulide is chemically related to valinomycin, a potassium-selective ionophore antibiotic produced by *Streptomyces fulvissimus* (Jensen) Waksman and Henrici. *S. fulvissimus* is a filamentous or rod-shaped microorganism of the order *Actinomycetales*, which also occurs in soil.[32] Cereulide is resistant to pepsin and stable in the pH range between 2 and 11. This latter property accounts for the ability of cereulide to resist degradation in the human stomach.[33]

Diarrheal Toxin

Bacillus cereus produces at least four heat-labile enterotoxins capable of producing diarrheal illness. These enterotoxins include two protein complexes (hemolysin BL, nonhemolytic enterotoxin) and two enterotoxic proteins (enterotoxin T, cytotoxin K).[34] Hemolysin BL (HBL), the diarrheal toxin, is a 40-kDa, three-component protein that is heat labile and pH stable between 4 and 11.[35] A minority of *B. cereus* strains produce a nonhemolytic enterotoxin (Nhe) that also causes the diarrheal syndrome.[36] The hemolytic notation of HBL refers to the hemolysis induced by this toxin in sheep red blood cells *in vitro*; however, there is no direct evidence of a similar effect in humans.[37] Whether a particular strain of *B. cereus* will elaborate cereulide or HBL depends in part on the following flagellar (H) serotype: cereulide production, types 1, 5, 8, 12, and 19; HBL production, types 1, 2, 3, 8, 10, 12, and 19. Although some *B. cereus* strains are capable of producing both toxins, only one toxin is usually elaborated in disease states.[1] Some strains of *Bacillus* species other than *B. cereus* can produce emetic toxins or enterotoxins including *B. subtilis*, *B. mojavensis*, *B. pumilus*, and *B. fusiformis*.[38]

Mechanism of Toxicity

Both spores and vegetative cells are found in many foods. Cooking destroys vegetative cells, but spores survive to germinate when food is cooled to room temperature. Emetic (cereulide) or diarrheal (HBL) enterotoxins are subsequently elaborated when food is improperly stored and reheating fails to destroy these toxins. Following ingestion, these enterotoxins produce either the emetic syndrome or the diarrheal syndrome. The mechanism and site of action of cereulide is unknown. High biliary concentrations of cereulide suggest that this compound undergoes enterohepatic recirculation. Rapid onset of illness and the lack of a diarrheal component suggest that the stomach is the site of action.[39] Animal studies indicate that cereulide disrupts mitochondrial membrane potential and impairs mitochondrial fatty acid metabolism, resulting in a dose-dependent increase of small fatty droplets and massive degeneration of hepatocytes following high doses.

Hemolysin BL consists of three polypeptide components: B, L_1, and L_2. The B component forms pores in cell membranes of the intestinal mucosa, allowing the L_1 and L_2 components to lyse cells and activate adenylate cyclase, resulting in altered tissue permeability, tissue damage, and fluid secretion.[40] *B. cereus* also produces a series of phospholipases that have an affinity for phosphatidylcholine and sphingomyelinase on intestinal cell membranes.[41] These enzymes play a role in mucosal tissue degradation, and are structurally similar to alpha-toxin produced by *Clostridium perfringens*.[42]

DOSE RESPONSE

The bacterial dose required for the development of the emetic syndrome has not been established because the number of bacteria required to produce sufficient preformed toxin to cause symptoms is not known. Animal studies suggest 9–12.9 µg/kg of cereulide enterotoxin are required to induce emesis.[43,44] The bacterial dose for the diarrheal syndrome is accepted as $>10^5$ CFU/g of food substance. This dose is based on epidemiological calculations of data obtained from confirmed *B. cereus* outbreaks with diarrheal syndrome (range: 5×10^4–10^{11} CFU/g of food substance).[45]

CLINICAL RESPONSE

Bacillus cereus causes the following two forms of food poisoning: a heat-stable toxin-induced emetic form that is similar to *Staphylococcus aureus* food poisoning, and heat-labile enterotoxin-induced diarrheal form that produces symptoms similar to *Clostridium perfringens* food poisoning. Nausea, abdominal cramping, bloating, and vomiting characterize the emetic syndrome. Symptoms occur 30 minutes to 6 hours after ingestion and resolve within 6–24 hours.[1] Mild diarrhea is occasionally associated with the emetic form of gastrointestinal illness. The diarrheal syndrome is characterized by mild nausea, occasional vomiting, tenesmus, and profuse, nonbloody diarrhea. Symptoms occur 8–16 hours after ingestion and persist 12–24 hours. Case series suggest that prolonged diarrhea occasionally occurs after *B. cereus* infections. In a 1993 case series of 139 reports of *B. cereus*-induced diarrheal disease, 34% of patients developed diarrhea outside of the usual range of 6–24 hours, and 23% of patients reported fever based on a postoutbreak survey.[46] Death from either syndrome is extremely rare, and person-to-person spread does not occur.

Rarely, case reports document the development of serious extra gastrointestinal infections caused by *B. cereus*, including meningitis in an immunocompromised patient,[47] sepsis in children with acute leukemia,[48] bacteremia in neonates,[49] and fulminant necrotizing fascitis in a healthy child following a puncture wound from a tree branch.[50] A case report associated the development of fulminant liver failure and rhabdomyolysis in a 17-year-old boy with high concentrations of *B. cereus* emetic toxin (cereulide).[51] Postmortem examination of other fatal cases of *B. cereus* poisoning demonstrated liver steatosis,[52] but these patients did not develop fulminant hepatic failure and the cause of this abnormality remains unclear. Postmortem examination of this boy revealed a diffuse microvesicular steatosis and midzonal necrosis, consistent with impaired β-oxidation of liver mitochondria. The differential diagnosis for microvesicular steatosis includes hypoglycin (ackee fruit, renta yams), neem (margosa) oil, aflatoxin, valproic acid, aspirin, tetracycline, dideoxynucleoside drugs (zidovudine, didanosine, lamivudine), fialuridine, fatty liver of pregnancy, and Reye's syndrome.

DIAGNOSTIC TESTING

Analytical Methods

In the healthcare setting, routine analytical testing for *B. cereus* or the associated toxins is not clinically useful, as symptoms are self-limited and short in duration. Official methods of analysis are available from AOAC International (Association of Analytical Chemists International, Gaithersburg, MD; http://www.eoma.aoac.org/) for the enumeration and confirmation of *Bacillus cereus* in foods[53] and for the differentiation of members of the *Bacillus cereus* group.[54] In epidemiological studies, isolation of *B. cereus* bacteria from stool, vomit, or leftover food is performed on mannitol-egg yolk-polymyxin agar, followed by serotyping of vegetative cells.[55] Because asymptomatic people can transiently shed *B. cereus*, simple identification in clinical specimens is not sufficient to confirm *B. cereus* foodborne poisoning unless both clinical and food specimens contain $>10^5$ CFU/g of an identical serotype.[56] Food sampling in outbreaks is performed on representative samples of 50 grams each on different parts of the food in question. If transport of food to a laboratory is necessary, raw or cooked food should be packed in insulated containers at temperatures ≥42 °F (≥5 °C).[57] Dried food may be transported without refrigeration. Three methods for the detection of emetic toxin include cytotoxicity assay, high performance liquid chromatography (HPLC) with ion trap mass spectrometry, and a sperm-based bioassay.[14] Cereulide enterotoxin is identified by a HEp-2 vacuolization assay, in which the toxin uncouples oxidative phosphorylation in HEp-2 mitochondria. Visualization of mitochondrial swelling and measurement of oxygen utilization indicates a positive test for cereulide. Methods for confirmation and quantitation of cereulide involve HPLC with ion trap mass spectrometry.[58] The limit of detection with this method was 10 pg cereulide per injection. Parts of the tripartite hemolysin BL enterotoxin can be identified via latex agglutination assay and immunoassay, but these tests are limited by their inability to identify the component B, L_1, and L_2 peptides simultaneously. All three components are necessary for HBL to be bioactive.[59] Hemolysin BL and cereulide can also be confirmed using ileal loop assays in rhesus monkeys and rabbits, but these tests are labor-intensive and expensive. Consequently, these tests are reserved for epidemiologic studies of outbreak that require definite identification.

Biomarkers

Postmortem analysis of a case associated with cereulide toxicity and fatal fulminant hepatic failure demonstrated the accumulation of cereulide in the small intestine, liver, and bile as well as in a cooking pan used by the patient.[60] Hemolysin has been identified only in fecal specimens from cases of foodborne poisoning associated with *B. cereus*.

Abnormalities

Both the emetic syndrome and the diarrheal syndrome can produce severe serum electrolyte abnormalities, with a profile similar to that seen in dehydration from gastrointestinal fluid losses. Severe dehydration involves hypernatremia (serum Na >150 mEq/L) or hyponatremia (serum Na <130 mEq/L); hypokalemia (serum K <3.0 mEq/L); hypochloremia, and elevated urine specific gravity (SpG >1.020). Low serum bicarbonate results from lactic acid production or elevated bicarbonate secondary to contraction alkalosis. The short duration of illness limits the severity of fluid and electrolyte imbalances, which usually resolve spontaneously.[61]

TREATMENT

Treatment includes rehydration and bowel rest for both the emetic and the diarrheal syndromes. Oral rehydration is usually sufficient, but intravenous fluid replacement and hospitalization may be necessary in cases of severe dehydration or severe electrolyte abnormalities, particularly in young children or elderly patients.[62,63] Anti-emetics may alleviate symptoms in the emetic syndrome. Antimotility agents (e.g., loperamide, diphenoxylate) may curtail symptoms in the diarrheal syndrome, but these agents should be administered cautiously. The short duration of both syndromes generally limits the necessity of these medications. There are no clinical trials that demonstrate an improvement in clinical outcome, avoidance of hospitalization, or shortened duration of symptoms following the use of either antiemetics or antidiarrheal agents. *B. cereus* produces beta-lactamases, generating resistance to penicillins and cephalosporins, but *B. cereus* is usually susceptible to aminoglycosides, sulfonamides, vancomycin, erythromycin, tetracycline, ciprofloxacin, and clindamycin.[64,65] Although antimicrobials are beneficial in extra gastrointestinal infections of *B. cereus* (e.g., septicemia, endophthalmitis, keratitis, endocarditis, meningitis, pneumonia, wound infections), they are not routinely used in cases of *B. cereus* foodborne poisoning, even when symptoms are prolonged.

References

1. Kramer JM, Gilbert RJ. *Bacillus cereus* and other *Bacillus* species. In: Doyle MP, editor. Foodborne bacterial pathogens. New York: Marcel Dekker; 1989:21–70.
2. Sneath PH. Endospore-forming gram-positive rods and cocci. In: Sneath PH, Mair NS, Sharpe E, Holt JG, editors. Bergey's manual of systematic bacteriology. Baltimore: Williams & Wilkins; 1986:1104–1105.
3. Hauge S. Food poisoning caused by aerobic spore forming bacilli. J Appl Bacteriol 1955;18:591–595.
4. Centers for Disease Conrol (CDC). *Bacillus cereus*–Maine. MMWR Morb Mortal Wkly Rep 1986;35:408–410.
5. Centers for Disease Conrol (CDC). *Bacillus cereus* food poisoning associated with fried rice at two child day care centers–Virginia, 1993. MMWR Morb Mortal Wkly Rep 1994;43:177–178.
6. Andersson A, Ronner U, Granum PE. What problems does the food industry have with the spore forming pathogen's *Bacillus cereus* and *Clostridium perfringens*? Int J Food Microbiol 1995;28:145–155.
7. Magnusson M, Christiansson A, Svensson B. *Bacillus cereus* spores during housing of dairy cows: factors affecting contamination of raw milk. J Dairy Sci 2007;90: 2745–2754.
8. Margulis L, Jorgensen JZ, Dolan S, Kolchinsky R, Rainey FA, Lo SC. The *Arthromitus* stage of *Bacillus cereus*: intestinal symbionts of animals. Proc Natl Acad Sci USA 1998;95:1236–1241.
9. Ghosh AC. Prevalence of *Bacillus cereus* in the faeces of healthy adults. J Hyg 1978;80:233–236.
10. Turnbull PC, Kramer JM. Intestinal carriage of *Bacillus cereus*: faecal isolation studies in three population groups. J Hyg 1985;95:629–638.
11. Te Giffel MC, Beumer RR, Granum PE, Rombouts FM. Isolation and characterization of *Bacillus cereus* from pasteurized milk in household refrigerators in the Netherlands. Int J Food Microbiol 1997;34:307–318.
12. Granum PE, Brynestad S, Kramer JM. Analysis of enterotoxin production by *Bacillus cereus* from dairy products, food poisoning incidents, and non-gastrointestinal infections. Int J Food Microbiol 1993;17:269–279.
13. Rusul G, Yaacob NH. Prevalence of *Bacillus cereus* in selected foods and detection of enterotoxin using TECRA-VIA and BCET-RPLA. Int J Food Microbiol 1995;25: 131–139.
14. Ehling-Schulz M, Fricker M, Scherer S. *Bacillus cereus*, the causative agent of an emetic type of foodborne illness. Mol Nutr Food Res 2004;48:479–487.
15. Agata N, Ohta M, Yokoyama K. Production of *Bacillus cereus* emetic toxin (cereulide) in various foods. Int J Food Microbiol 2002;73:23–27.
16. Pan TM, Wang TK, Lee CL, Chien SW, Horng C-B. Foodborne disease outbreaks due to bacteria in Taiwan 1986–1995. J Clin Microbiol 1997;35:1260–1262.
17. Nichols GL, Little CL, Mithani V, de Louvois J. The microbiological quality of cooked rice from restaurants and take-away premises in the United Kingdom. J Food Prot 1999;62:877–882.
18. Jaquette CB, Beuchat LR. Combined effects of pH, nisin, and temperature on growth and survival of psychrotrophic *Bacillus cereus*. J Food Prot 1998;61:563–570.
19. Valero M, Fernandez PS, Salmeron MC. Influence of pH and temperature on growth of *Bacillus cereus* in vegetable substrates. Int J Food Microbiol 2003;82:71–79.
20. McCabe-Sellers BJ, Beattie SE. Food safety: emerging trends in foodborne illness surveillance and prevention. J Am Diet Assoc 2004;104:1708–1717.
21. Kotiranta A, Lounatmaa K, Haapasalo M. Epidemiology and pathogenesis of *Bacillus cereus* infection. Microbes Infect 2000;2:189–198.
22. Lindsay D, Brozel VS, Mostert JF, von Holy A. Physiology of dairy-associated *Bacilli* spp. over a wide pH range. In J Food Microbiol 2000;54:49–62.
23. Beuchat LR, Pettigrew CA, Tremblay ME, Roselle BJ, Scouten AJ. Lethality of chlorine, chlorine dioxide, and a commercial fruit and vegetable sanitizer to vegetative cells and spores of *Bacillus cereus* and spores of *Bacillus thuringiensis*. J Ind Microbiol Biotechnol 2005;32: 301–308.
24. Novak JS, Call J, Tomasula P, Luchansky JB. An assessment of pasteurization treatment of water, media, and milk with respect to *Bacillus* spores. J Food Prot. 2005;68: 751–757.
25. Fermanian C, Fremy JM, Lahellac C. *Bacillus cereus* pathogenicity: a review. J Rapid Methods Automation Microbiol 1993;2:83–114.
26. Reij MW, Den Aantrekker ED. ILSI Europe Risk Analysis in Microbiology Task Force: Recontamination as a source of pathogens in processed foods. Int J Food Microbiol 2004;91:1–11.
27. Redmond EC, Griffith CJ. Consumer food handling in the home: a review of food safety studies. J Food Prot 2003;66:130–161.

28. Ghelardi E, Celandroni F, Salvetti S, Barsotti C, Baggiani A, Senesi S. Identification and characterization of toxigenic *Bacillus cereus* isolates responsible for two food-poisoning outbreaks. FEMS Microbiol Lett 2002;208: 129–134.
29. Turnbull PC, Kramer JM, Melling J. Bacillus. In: Wilson GS, Miles AS, editors. Tropley and Wilson's principles of bacteriology, virology, and immunity. London: Edward Arnold; 1990:188–210.
30. Carlin F, Fricker M, Pielaat A, Heisterkamp S, Shaheen R, Salkinoja Salonen M, et al. Emetic toxin-producing strains of *Bacillus cereus* show distinct characteristics within the *Bacillus cereus* group. Int J Food Microbiol 2006;109: 132–138.
31. Rajkovic A, Uyttendaele M, Ombregt SA, Jaaskelainen E, Salkinoja-Salonen M, Debevere J. Influence of type of food on the kinetics and overall production of *Bacillus cereus* emetic toxin. J Food Prot 2006;69:847–852.
32. Granum PE, Lund T. *Bacillus cereus* and its food poisoning toxins. FEMS Microbiol Lett 1997;157:223–228.
33. Taylor JM, Sutherland AD, Aidoo KE, Logan NA. Heat-stable toxin production by strains of *Bacillus cereus, Bacillus firmus, Bacillus megaterium, Bacillus simplex* and *Bacillus licheniformis*. FEMS Microbiol Lett 2005;242: 313–317.
34. Lund T, De Buyser ML, Granum PE. A new cytotoxin from *Bacillus cereus* that may cause necrotic enteritis. Mol Microbiol 2000;38:254–261.
35. Garcia-Arribas ML, Kramer JM. The effect of glucose, starch, and pH on growth, enterotoxin and haemolysin production by strains of *Bacillus cereus* associated with food poisoning and non-gastrointestinal infection. Int J Food Microbiol 1990;11:21–33.
36. Lund T, Granum PE. Characterisation of a non-haemolytic enterotoxin complex from *Bacillus cereus* isolated after a foodborne outbreak. FEMS Microbiol Lett 1996;141:151–156.
37. Beecher DJ, Wong AC. Tripartite hemolysin BL from *Bacillus cereus*. Hemolytic analysis of component interactions and a model for its characteristic paradoxical zone phenomenon. J Biol Chem 1997;272:233–239.
38. From C, Pukall R, Schumann P, Hormazabal V, Granum PE. Toxin-producing ability among *Bacillus* spp. outside the *Bacillus cereus* group. Appl Environ Microbiol 2005;71:1178–1183.
39. Yokoyama K, Ito M, Agata N, Isobe M, Shibayama K, Horii T, Ohta M. Pathological effect of synthetic cereulide, an emetic toxin of *Bacillus cereus*, is reversible in mice. FEMS Immunol Med Microbiol 1999;24:115–120.
40. Beecher DJ, Macmillan JD. Characterization of the components of hemolysin BL from *Bacillus cereus*. Infect Immun 1991;59:1778–1784.
41. Beecher DJ, Wong AC. Cooperative, synergistic and antagonistic haemolytic interactions between haemolysin BL, phosphatidylcholine phospholipase C and sphingomyelinase from *Bacillus cereus*. Microbiology 2000;146: 3033–3039.
42. Leslie D, Fairweather N, Pickard D, Dougan G, Kehoe M. Phospholipase C and haemolytic activities of *Clostridium perfringens* alpha-toxin cloned in Escherichia coli: sequence and homology with a *Bacillus cereus* phospholipase C. Mol Microbiol 1989;3:383–392.
43. Agata N, Ohta M, Mori M, Isobe M. A novel dodecadepsipeptide, cereulide, is an emetic toxin of *Bacillus cereus*. FEMS Microbiol Lett 1995;129:17–20.
44. Shinagawa K, Konuma H, Sekita H, Sugii S. Emesis of rhesus monkeys induced by intragastric administration with the HEp-2 vacuolation factor (cereulide) produced by *Bacillus cereus*. FEMS Microbiol Lett 1995;130:87–90.
45. Granum PE, Lund T. *Bacillus cereus* and its food poisoning toxins. FEMS Microbiol Lett 1997;157:223–228.
46. Luby S, Jones J, Dowda H, Kramer J, Horan J. A large outbreak of gastroenteritis caused by diarrheal toxin-producing *Bacillus cereus*. J Infect Dis 1993;167:1452–1455.
47. Haase R, Sauer H, Dagwadordsch U, Foell J, Lieser U. Successful treatment of *Bacillus cereus* meningitis following allogenic stem cell transplantation. Pediatr Transplant 2005;9:338–341.
48. Arnaout MK, Tamburro RF, Bodner SM, Sandlund JT, Rivera GK, Pui CH, Ribeiro RC. *Bacillus cereus* causing fulminant sepsis and hemolysis in two patients with acute leukemia. J Pediatr Hematol Oncol 1999;21:431–435.
49. Hilliard NJ, Schelonka RL, Waites KB. *Bacillus cereus* bacteremia in a preterm neonate. J Clin Microbiol 2003;41: 3441–3444.
50. Darbar A, Harris IA, Gosbell IB. Necrotizing infection due to *Bacillus cereus* mimicking gas gangrene following penetrating trauma. J Orthop Trauma 2005;19:353–355.
51. Mahler H, Pasi A, Kramer JM, Schulte P, Scoging AC, Bar W, Krahenbuhl S. Fulminant liver failure in association with the emetic toxin of *Bacillus cereus*. N Engl J Med 1997;336:1142–1148.
52. Takabe F, Oya M. An autopsy case of food poisoning associated with *Bacillus cereus*. Forensic Sci 1976;7: 97–101.
53. Horwitz W, editor. Official methods of analysis of AOAC International. 17th ed. Gaithersburg, MD: AOAC International; 2000, official method 980.31.
54. Horwitz, W, editor. Official methods of analysis of AOAC International. 17th ed. Gaithersburg, MD: AOAC International; 2000, official method 983.26.
55. Bennett RW, Belay N. Bacillus cereus. In: Downs FP, Ito K, editors. Compendium of methods for the microbiological examination of foods. Washington, DC: American Public Health Association; 2001:311–316.
56. Kramer JM, Turnbull PC, Munshi G, Gilbert RJ. Identification and characterization of *Bacillus cereus* and other Bacillus species associated with foods and food poisoning. In Corry EL, Roberts D, Skinner FA, editors. Isolation and identification methods for food poisoning organisms. London: Academic Press; 1982:261–286.
57. Andrews WH, Hammack TJ. Food sampling and preparation of sample homogenate. In: U.S. Food and Drug

Administration: Bacteriological Analytical Manual. Washington, DC: U.S. Food and Drug Administration; 2003: 1–10.

58. Haggblom MM, Apetroaie C, Andersson MA, Salkinoja-Salonen MS. Quantitative analysis of cereulide, the emetic toxin of *Bacillus cereus*, produced under various conditions. Appl Environ Microbiol 2002;68:2479–2483.
59. Beecher DJ, Wong AC. Tripartite haemolysin BL. isolation and characterization of two distinct homologous sets of components from a single *Bacillus cereus* isolate. Microbiology 2000;146:1371–1380.
60. Mahler H, Pasi A, Kramer JM, Schulte P, Scoging AC, Bar W, Krahenbuhl S. Fulminant liver failure in association with the emetic toxin of *Bacillus cereus*. N Engl J Med 1997;336:1142–1148.
61. Guerrant RL, Van Gilder T, Steiner TS, Thielman NM, Slutsker L, Tauxe RV, et al and the Infectious Diseases Society of America. Practice guidelines for the management of infectious diarrhea. Clin Infect Dis 2001;32: 331–351.
62. King CK, Glass R, Bresee JS, Duggan C. Centers for Disease Control and Prevention: Managing acute gastroenteritis among children: oral rehydration, maintenance, and nutritional therapy. MMWR Morb Mortal Wkly Rep Recomm Rep; 2003;52:1–16.
63. Smith JL. Foodborne illness in the elderly. J Food Prot 1998;61:1229–1239.
64. Gigantelli JW, Torres Gomez J, Osato MS. *In vitro* susceptibilities of ocular *Bacillus cereus* isolates to clindamycin, gentamicin, and vancomycin alone or in combination. Antimicrob Agents Chemother 1991;35:201–202.
65. Dubouix A, Bonnet E, Alvarez M, Bensafi H, Archambaud M, Chaminade B, Chabanon G, Marty N. *Bacillus cereus* infections in traumatology-orthopaedics department: retrospective investigation and improvement of healthcare practices. J Infect 2005;50:22–30.

Chapter 14

CAMPYLOBACTER JEJUNI

CYRUS RANGAN, MD, FAAP

HISTORY

Campylobacter jejuni (Jones et al.) Véron and Chatelain (*C. jejuni*) is a gram-negative, catalase-positive, oxidase-positive, nitrate-positive, S-shaped, flagellated, motile rod that was first identified by Theodor Escherich in 1886 as a rare cause of infant cholera in Germany.[1] Until case reports of *Campylobacter* bacteremia in humans in the 1950s, *Campylobacter* was considered an animal pathogen.[2] Originally, *Campylobacter* species were categorized as a *Vibrio* species;[3] however, *Campylobacter* became a unique genus in the 1970s following the discovery that multiplication of these bacteria required a microaerophilic (5% oxygen) environment and these bacteria could not metabolize carbohydrates or lipids.[4] In the 1970s, *Campylobacter* was recognized as a common cause of human gastroenteritis.[5] Although numerous other species in Campylobacteraceae have been implicated in gastroenteritis (e.g., *C. upsaliensis*, *C. lari*, *C. hyointestinalis* subsp. *hyointestinalis*, *C. coli*), *C. jejuni* causes over 90% of reported cases.[6] *C. jejuni* is the most commonly reported cause of bacterial gastroenteritis in the United States, with an estimated 2 million cases per year in the United States and 4 million cases per year worldwide.[7] *C. jejuni* is a frequent cause of traveler's diarrhea.[8]

EXPOSURE

Source

ENVIRONMENT

Water runoff containing animal feces can contaminate nonchlorinated water sources (e.g., streams and ponds) with *C. jejuni*.[9] This organism can survive for extended periods in still waters by living inside waterborne protozoa.[10]

ANIMAL

Healthy humans do not carry *C. jejuni*, but a variety of animals are carriers of *C. jejuni* including asymptomatic wild animals (water fowl, rodents, flying insects),[11] food animals (poultry, cattle),[12] and domestic animals (dogs, cats, birds).[13] *C. jejuni* is commensal in healthy cattle, particularly during the summer. Healthy rodents frequently excrete *Campylobacter* species in their stool including *C. jejuni*.[14] *C. jejuni* thrives at 37 °C–42 °C (99 °F–108 °F), and therefore an optimal reservoir for this bacterium is the chicken, which has a mean body temperature of 41 °C (106 °F).[15] Shellfish caught in contaminated waters may also carry *C. jejuni*.[16] Contact with live animals colonized with *C. jejuni* is a well-documented source of human exposure, especially in young children.[17]

FOOD

Raw milk, untreated drinking water, and undercooked meat (e.g., chicken) are the most common sources of foodborne illness in humans.[18,19] Undercooked pork, beef, and sausages are also documented food sources.[20] *C. jejuni* frequently contaminates raw chicken. Surveys of retail chickens demonstrate 20–100% contamination rates by *C. jejuni*.[21] *C. jejuni* does not proliferate in food when the ambient temperature is below 37 °C (99 °F) or when ambient oxygen is present at atmospheric conditions. Thus, human infections are sporadic, rarely resulting in outbreaks.[22]

Medical Toxicology of Natural Substances, by Donald G. Barceloux, MD

Food Processing

Pasteurizing milk and chlorinating drinking water control *C. jejuni* growth.[23] Proper food handling during preparation and cooking food to >170 °C (338 °F) effectively eliminate *C. jejuni* from consumer products.[24] Screening and isolation of *C. jejuni* in food processing is difficult because the bacteria are usually present in relatively low numbers.

TOXINS

Physiochemical Properties

Campylobacter jejuni proliferates at temperatures from 37 °C–42 °C (99 °F–108 °F).[25] *In vitro* studies indicate that growth and chemotaxis of *Campylobacter* strains are greater at 37 °C (98.6 °F) than at 42 °C (107.6 °F), and these properties suggest that the physiological temperature of humans (37.0 °C/98.6 °F) is potentially more favorable for the expression of virulence than for birds (body temperature, 42.0 °C/107.6 °F).[26] *C. jejuni* is microaerophilic, requiring 3–5% oxygen and 2–10% carbon dioxide for optimal growth.[27] *C. jejuni* is relatively fragile, and these bacteria are sensitive to numerous environmental stresses, including 21% atmospheric oxygen, drying, freezing, heating, disinfectants, and pH < 5.[28] Survival of *C. jejuni* outside the gut is relatively poor, secondary to aerobic stress induced by ambient oxygen concentrations (i.e., 21% oxygen). However, the production of iron-regulated alkyl hydroperoxide reductase (*AhpC*)[29] and a superoxide dismutase (*SodB*)[30] by *C. jejuni* provides some protection from stressful oxidative and aerobic conditions. When stressed, *C. jejuni* absorbs amino acids rapidly from its environment and assumes a dormant coccoidal form, limiting its ability to grow while maintaining its outer membrane. In this state, the organism remains intact, but this form does not grow in these conditions or in culture media. This unique state of *Campylobacter* is known as viable-but-not-culturable (VNC).[31] Once exposed to a microaerophilic environment, VNC organisms resume growth.[32] *C. jejuni* is a highly motile organism with a characteristic S-shape and polar flagella. These organisms do not metabolize lipids or oxidize carbohydrates.[33] Isolation of antimicrobial-resistant *C. jejuni* has evolved concurrently with the administration of ciprofloxacin and enrofloxacin to farm chickens. Since 1995 when the use of quinolones in poultry farms began, human cases of resistant campylobacteriosis have increased annually.[34] Developing countries have higher rates of antimicrobial-resistant *C. jejuni* compared with more developed countries.[35] Antibiotic-resistant isolates demonstrate a genetic mutation that may additionally confer improved fitness for survival because rates of quinolone-resistant strains outpace rates of nonresistant bacteria in food animals, even when antibiotics are not administered.[36]

Mechanism of Toxicity

Darting, corkscrew-like motility allows *C. jejuni* to pass through the acid milieu of the stomach and to colonize the colon and distal ileum.[37] Conditions of decreased gastric acidity enhance survivability of *C. jejuni*.[38] Chemotaxis, adherence to mucosa, and invasion of *C. jejuni* into mucosal cells are mediated by flagella[39] and two cell surface proteins, CadF and PEB1.[40] *Campylobacter* possesses lipopolysaccharide and lipo-oligosaccharide endotoxins, which enhance adherence, increase invasiveness, and improve survival in human serum during bacteremia.[41] *C. jejuni* is vacuolized by intestinal epithelial cells. During vacuolization, *C. jejuni* resists lysosomal degradation via catalase production and anti-oxidant activities of *AhpC* and *SodB*.[42] *C. jejuni* then translocates to the basolateral surface of the epithelial cell and undergoes exocytosis.[43] Subsequently, *C. jejuni* proliferates in the lamina propria and mesenteric lymph nodes, resulting in dissemination and extraintestinal infections (e.g., sepsis, meningitis, mesenteric adenitis).[44] Destruction of intestinal epithelial cells is provoked by a heat-labile, cytolethal-distending toxin composed of three subunits: A, B, and C. Cytolethal-distending toxin subunit B attacks intestinal endothelial cell DNA, and recruits protein kinases to block cells from entering into mitosis.[45] *C. jejuni* induces a local inflammatory reaction mediated by interleukin-8, leukotriene B4, prostaglandin E2, and cyclic AMP. Endothelial cells die, resulting in watery or bloody diarrhea with fecal leukocytes.[46] Guillain-Barré syndrome, an acute autoimmune demyelinating disease of peripheral nerve bundles, is a sequela of *C. jejuni* infection mediated by antigenic similarities between the terminal tetrasaccharide of the *C. jejuni* lipo-oligosaccharide endotoxin and the GM_1 ganglioside on the membrane of peripheral nerves.[47,48]

DOSE RESPONSE

Human feeding studies suggest that 800 *C. jejuni* organisms are required to cause illness in healthy individuals, whereas 400–500 bacteria may cause illness in more susceptible individuals (e.g., young children, elderly individuals, immunocompromised patients).[49,50] Approximately 200 annual fatalities occur in the United States as a result of *C. jejuni* infections, primarily from septicemia in immunocompromised patients.[51] The severity of the infection depends on the strain of *C. jejuni* as well as host characteristics.[52,53] Watery diarrhea typically occurs in individuals in developing countries following *C. jejuni* infection, whereas individuals in developed countries usually have bloody diarrhea with fever. This dichotomy results from increased immunity in individuals in developing countries secondary to repeated exposures to *C. jejuni* during childhood and throughout life. When infected in developing countries, travelers acquire symptoms that are similar to symptoms observed in their countries of origin.[54]

CLINICAL RESPONSE

Course

Symptoms of *C. jejuni*-induced gastroenteritis include watery or bloody diarrhea, cramps, nausea, myalgias, and fever, similar to the clinical responses to *Shigella* spp., *Salmonella* spp., enteroinvasive *Escherichia coli, E. coli* 0157:H7, *Yersinia enterocolitica*, and *Vibrio parahaemolyticus*.[51] In a study of 877 episodes of bloody diarrhea (>3 stools over 24 hours) presenting to a US emergency department, 168 enteropathogens (30.6% of stools samples) were identified.[55] The most common pathogens were *Shigella* spp. (15.3%), *Campylobacter* spp. (6.2%), *Salmonella* spp. (5.8%), and Shiga toxin-producing *E. coli* (2.6%). Abdominal pain, fever, headache, lethargy, and anorexia may precede diarrhea for several hours. Tenderness to palpation in the right lower quadrant may accompany abdominal pain, resulting in a clinical picture that mimics appendicitis before the onset of diarrhea.[56] Appearance of blood in stools may be occult or grossly bloody.[57] Incubation time is approximately 2–5 days after exposure.[58] Symptoms typically resolve within 3–10 days after the onset of illness, with recrudescence in approximately 25% of untreated cases.[58] Diarrhea persists for 1–2 weeks in 20–30% of cases, and longer than 2 weeks in about 5–10% of cases.[58] Adults may have severe inflammatory diarrhea, resembling Crohn's disease, ulcerative colitis, or pseudomembranous enterocolitis.[51] Young children usually have frequent, watery, secretory diarrhea, often without blood.[59] *Campylobacter*-induced abdominal pain and tenderness with bloody diarrhea in young infants may mimic intussusception.[56]

The annual incidence of *C. jejuni* infections in infants in the US population is approximately 14 per 100,000.[60] This rate decreases to 4/100,000 during adolescence, but annual incidence increases to 8/100,000 in young adults. Older adults incur the illness at a rate of 3/100,000. Population susceptibility and low threshold for seeking medical attention cause the higher disease rate in infants; poor dietary and food-handling practices cause the upswing during young adulthood. Breast-fed infants have lower rates of infection with *C. jejuni* than bottle-fed infants. However, the incidence of this illness in infants increases after weaning.[61]

Complications

Transient bacteremia is relatively common, but severe complications (e.g., septic shock, meningitis, urinary tract infection, cholecystitis) occur in <0.1% of reported cases of campylobacteriosis, most commonly in elderly and in immunocompromised patients.[51] *C. jejuni*-induced septic abortion is a rare complication.[62] The human fatality rate from *C. jejuni* infection is less than 0.1%, and occurs primarily from sepsis in immunocompromised patients.[51] Prominent postinfectious manifestations include reactive arthritis and Guillain-Barré syndrome. Reactive arthritis occurs in 1% of *Campylobacter* cases, typically 7–10 days after the onset of diarrhea. Symptoms of this reactive arthritis are usually monoarticular and confined to the knee, but inflammation may involve any large joint.[63] The appearance of reactive arthritis has a 60% association with human lymphocyte antigen B27 (HLA-B27).[64] However, there is some variation in the association between spondyloarthropathies and various HLA-B27 subtypes (e.g., HLA-B*2706 among Southeast Asian populations and HLA-B*2709 among Sardinians lack an association).[65] Guillain-Barré syndrome is an autoimmune disorder with ascending paralysis and respiratory failure. Some patients demonstrate the Miller-Fisher variant, accompanied by ophthalmoplegia, areflexia, and cerebellar ataxia.[66] One in 3,000 cases of *C. jejuni* is associated with Guillain-Barré syndrome, and *C. jejuni* may be responsible for up to 40% of all Guillain-Barré syndrome cases.[67] *Campylobacter*-associated Guillain-Barré syndrome is frequently very severe with a 20% rate of residual disabilities and a 2–3% mortality rate.[68] Rare postinfectious sequelae of *C. jejuni* infection include infectious myocarditis,[69] erythema nodosum, IgA-nephropathy, and hemolytic-uremic syndrome.[70]

DIAGNOSTIC TESTING

Analytical Methods

Phase-contrast microscopy and darkfield microscopy demonstrates S-shaped organisms with darting, corkscrew-like motility.[37] Gram stain reveals thin, curved, gram-negative rods that are difficult to distinguish from *Vibrio* spp.[71] Direct plating of stool specimens is an inexpensive method to culture *C. jejuni*, but sensitivity is relatively low due to increased prevalence of VNC or stressed organisms with exposure to atmospheric oxygen.[32] Under aerobic conditions, *C. jejuni* remains intact, but these organisms are unable to grow in culture. Specimens may be pre-enriched to decrease the prevalence of VNC organisms by raising ambient temperature to 37°C–42°C (99°F–108°F).[72] Isolation requires an antibiotic-containing enrichment broth, special antibiotic-containing agar plates (i.e., Campy-BAP, containing vancomycin, trimethoprim, and polymyxin), and a microaerophilic atmosphere with 5% oxygen and an elevated concentration of carbon dioxide (10%).[73] These conditions are necessary to reduce the growth of competing organisms. Because of these requirements, most clinical laboratories do not routinely isolate *Campylobacter* species, and the isolation of these organisms occurs only after specific requests. These methods require several days to a week to complete. Stool specimens should be submitted for culture in refrigerated, airtight containers to minimize contact with atmospheric oxygen.[74] Rectal swabs in bacteria transport media are also adequate. Stools may remain positive for *C. jejuni* for several weeks with or without antimicrobial therapy.[75]

The identity of *C. jejuni* isolates are confirmed by phenotyping or latex-agglutination.[76] Enzyme-linked immunosorbent assay (ELISA) detects *C. jejuni* with 96% sensitivity and 99% specificity.[77] Hybrid polymerase chain reaction (PCR)-ELISA assay is the most rapid (<24 hours) method to identify *C. jejuni* with high (99.7%) sensitivity and specificity (99%).[78,79] Serologic methods are available, but these methods are reserved for epidemiological investigations of outbreaks rather than clinical management.[80] There are no official AOAC International (Association of Analytical Chemists International, Gaithersburg, MD; http://www.eoma.aoac.org/) methods for the detection of *Campylobacter* species in food.

Biomarkers

Bacteremia is confirmed by positive routine culture for *Campylobacter* species in blood samples. Routine analytical methods are not available for the detection of the toxins associated with *C. jejuni*.

Abnormalities

Stools frequently contain gross or occult blood. Light microscopy reveals fecal leukocytes in approximately 75% of cases.[81] Complete blood count typically reveals a normal white blood cell count, occasionally with a left shift; however, the white blood cell count is usually elevated in *Campylobacter* septicemia.[51] Mild elevations in erythrocyte sedimentation rate, C-reactive protein, and alanine aminotransferase are often observed in patients with reactive arthritis, Reiter syndrome, and Guillain-Barré syndrome.[25] Severe dehydration involves hypernatremia (Na > 150mEq/L) or hyponatremia (Na < 130mEq/L); hypokalemia (K < 3.0mEq/L); hypochloremia, or elevated urine specific gravity (SpG > 1.020). Low serum bicarbonate results from lactic acid production or elevated bicarbonate secondary to contraction alkalosis.

TREATMENT

Treatment primarily involves bowel rest and fluid and electrolyte replacement. Oral rehydration is usually sufficient, but intravenous fluid replacement and hospitalization may be necessary in cases of severe dehydration, severe electrolyte abnormalities, or severe anemia and blood loss, particularly in young children or elderly patients.[82,83] Most *C. jejuni* infections are self-limited, and consequently the treatment of *C. jejuni* infections with antibiotics is appropriate only in selected cases. Early antimicrobial treatment may benefit patients with persistent fevers, bloody diarrhea, or more than eight stools per 24-hour period; patients with symptoms persisting for more than one week; pregnant patients; and immunocompromised patients.[84] *C. jejuni* demonstrates high rates of resistance to penicillins, most cephalosporins, tetracycline, and metronidazole.[85] Treatment with erythromycin for 5–7 days reduces illness duration by 5–10 days and attenuates subsequent fecal shedding of bacteria. Mild resistance to macrolide antibiotics exists, and the resistance to newer (e.g., azithromycin, clarithromycin) and to older (e.g., erythromycin) macrolide antibiotics is similar.[86,87] Antimicrobials are most beneficial when administered early in the course of illness. However, antimicrobial therapy typically begins after the return of *C. jejuni*-positive stool cultures, thus limiting the practical benefit of antimicrobial therapy in *C. jejuni* infections.[88] *C. jejuni* exhibits a significant rate of resistance to quinolones, especially in populations of developing countries.[89] Resistance may prolong illness and increase the severity of infection, especially in bacteremic patients, and the use of quinolones in susceptible strains may enhance the survival of the organism.

Loperamide may reduce the severity of symptoms in adult patients, but antidiarrheal agents theoretically increase the risk of systemic bacterial invasion when used alone. Therefore, loperamide is recommended only with concurrent administration of an antibiotic.[90] Patients with sepsis, meningitis, or other severe systemic infections should be treated intravenously with a third-generation cephalosporin and gentamicin, pending laboratory results of antimicrobial susceptibilities.[91] Patients with Guillain-Barré syndrome require intensive care monitoring, and should be observed closely for the development of respiratory failure. Some Guillain-Barré syndrome patients benefit from the administration of intravenous immunoglobulin or plasmapheresis. Survivors of Guillain-Barré syndrome may require long-term physical therapy and rehabilitation.[92]

References

1. Kist M. [Who discovered *Campylobacter jejuni/coli*? A review of hitherto disregarded literature.] Zentralbl Bakteriol Mikrobiol Hyg 1986;261:177–186. [German]
2. King EO. Human infections with *Vibrio fetus* and a closely related vibrio. J Infect Dis 1957;101:119–128.
3. Jones FS, Orcutt M, Little RB. *Vibrios* (*Vibrio jejuni*, n. sp.) associated with intestinal disorders of cows and calves. J Exp Med 1931;53:853–864.
4. Dekeyser P, Gossuin-Detrain M, Butzler JP, Sternon J. Acute enteritis due to related vibrio: first positive stool cultures. J Infect Dis 1972;125:390–392.
5. Butzler JP, Skirrow MB. *Campylobacter* enteritis. Clin Gastroenterol 1979;8:737–765.
6. Butzler JP. *Campylobacter*, from obscurity to celebrity. Clin Microbiol Infect 2004;10:868–876.
7. Centers for Disease Control (CDC). Foodborne diseases active surveillance network, 1996. JAMA 1997;227:1344–1345.
8. Gallardo F, Gascon J, Ruiz J, Corachan M, Jimenez de Anta M, Vila J. *Campylobacter jejuni* as a cause of traveler's diarrhea: clinical features and antimicrobial susceptibility. J Travel Med 1998;5:23–26.
9. Center for Disease Control (CDC). *Campylobacter* enteritis-New Zealand, 1990. MMWR Morb Mortal Wkly Rep 1991;40:116–117.
10. Snelling WJ, McKenna JP, Lecky DM, Dooley JS. Survival of *Campylobacter jejuni* in waterborne protozoa. Appl Environ Microbiol 2005;71:5560–5571.
11. Meerburg BG, Jacobs-Reitsma WF, Wagenaar JA, Kijlstra A. Presence of *Salmonella* and *Campylobacter* spp. in wild small mammals on organic farms. Appl Environ Microbiol 2006;72:960–962.
12. Doyle MP, Erickson MC. Reducing the carriage of foodborne pathogens in livestock and poultry. Poult Sci 2006;85:960–973.
13. Altekruse SF, Hunt JM, Tollefson LK, Madden JM. Food and animal sources of human *Campylobacter jejuni* infection. J Am Vet Med Assoc 1994;204:57–61.
14. Fernie DS, Park RW. The isolation and nature of campylobacters (microaerophilic vibrios) from laboratory and wild rodents. J Med Microbiol 1977;10:325–329.
15. Sahin O, Morishita TY, Zhang Q. *Campylobacter* colonization in poultry: sources of infection and modes of transmission. Anim Health Res Rev 2002;3:95–105.
16. Wilson IG, Moore JE. Presence of *Salmonella* spp. and *Campylobacter* spp. in shellfish. Epidemiol Infect 1996;116:147–153.
17. Amieva MR. Important bacterial gastrointestinal pathogens in children: a pathogenesis perspective. Pediatr Clin North Am 2005;52:749–777.
18. Wood RC, MacDonald KL, Osterholm MT. *Campylobacter* enteritis outbreaks associated with drinking raw milk during youth activities. A 10-year review of outbreaks in the United States. JAMA 1992;268:3228–3230.
19. Deming MS, Tauxe RV, Blake PA, Dixon SE, Fowler BS, Jones TS, et al. *Campylobacter* enteritis at a university: transmission from eating chicken and from cats. Am J Epidemiol 1987;126:526–534.
20. Zhao C, Ge B, De Villena J, Sudler R, Yeh E, Zhao S, et al. Prevalence of *Campylobacter* spp., *Escherichia coli*, and *Salmonella* serovars in retail chicken, turkey, pork, and beef from the Greater Washington, D.C. area. Appl Environ Microbiol 2001;67:5431–5436.
21. Meldrum RJ, Smith RM, Wilson IG. Three-year surveillance program examining the prevalence of *Campylobacter* and *Salmonella* in whole retail raw chicken. J Food Prot 2006;69:928–931.
22. Centers for Disease Control and Prevention (CDC). Preliminary FoodNet data on the incidence of infection with pathogens transmitted commonly through food-10 States, United States, 2005. MMWR Morb Mortal Wkly Rep 2006;55:392–395.
23. Peterson MC. *Campylobacter jejuni* enteritis associated with consumption of raw milk. J Environ Health 2003;65:20–26.
24. Christensen BB, Rosenquist H, Sommer HM, Nielsen NL, Fagt S, Andersen NL, et al. A model of hygiene practices and consumption patterns in the consumer phase. Risk Anal 2005;25:49–60.
25. Bereswill S, Kist M. Recent developments in *Campylobacter* pathogenesis. Curr Opin Infect Dis 2003;16:487–491.
26. Khanna MR, Bhavsar SP, Kapadnis BP. Effect of temperature on growth and chemotactic behaviour of *Campylobacter jejuni*. Lett Appl Microbiol 2006;43:84–90.
27. Bolton FJ, Coates D. A study of the oxygen and carbon dioxide requirements of thermophilic campylobacters. J Clin Pathol 1983;36:829–834.
28. Blaser MJ, Hardesty HL, Powers B, Wang WL. Survival of *Campylobacter fetus* subsp. *jejuni* in biological milieus. J Clin Microbiol 1980;27:309–313.
29. Baillon ML, van Vliet AH, Ketley JM, Constantinidou C, Penn CW. An iron-regulated alkyl hydroperoxide reductase (AhpC) confers aerotolerance and oxidative stress resistance to the microaerophilic pathogen *Campylobacter jejuni*. J Bacteriol 1999;181:4798–4804.
30. Pesci EC, Cottle DL, Pickett CL. Genetic, enzymatic, and pathogenic studies of the iron superoxide dismutase of *Campylobacter jejuni*. Infect Immun 1994;62:2687–2694.
31. Rollins DM, Colwell RR. Viable but non-culturable stage of *Campylobacter jejuni* and its role in survival in the natural aquatic environment. Appl Environ Microbiol 1986;52:531–538.

32. Tholozan JL, Cappelier JM, Tissier JP, Delattre G, Federighi M. Physiological characterization of viable-but-non-culturable *Campylobacter jejuni* cells. Appl Environ Microbiol 1999;65:1110–1116.
33. Griffiths PL, Park RW. Campylobacters associated with human diarrhoeal disease. J Appl Bacteriol 1990;69:281–301.
34. Hein I, Schneck C, Knogler M, Feierl G, Plesss P, Kofer J, et al. *Campylobacter jejuni* isolated from poultry and humans in Styria, Austria: epidemiology and ciprofloxacin resistance. Epidemiol Infect 2003;130:377–386.
35. Aarestrup FM, Engberg J. Antimicrobial resistance of thermophilic *Campylobacter*. Vet Res 2001;32:311–321.
36. Zhang Q, Lin J, Pereira S. Fluoroquinolone-resistant *Campylobacter* in animal reservoirs: dynamics of development, resistance mechanisms and ecological fitness. Anim Health Res Rev 2003;4:63–71.
37. Ferrero RL, Lee A. Motility of *Campylobacter jejuni* in a viscous environment: comparison with conventional rod-shaped bacteria. J Gen Microbiol 1988;134:53–59.
38. Wagner S, Schuler A, Gebel M, Freise J, Schmidt FW. *Campylobacter pylori* and acid secretion. Lancet 1989;2: 562.
39. Grant CC, Konkel ME, Cieplak W Jr, Tompkins LS. Role of flagella in adherence, internalization, and translocation of *Campylobacter jejuni* in nonpolarized and polarized epithelial cell cultures. Infect Immun 1993;61:1764–1771.
40. De Melo MA, Pechere JC. Identification of *Campylobacter jejuni* surface proteins that bind to Eucaryotic cells *in vitro*. Infect Immun 1990;58:1749–1756.
41. McSweegan E, Walker RI. Identification and characterization of two *Campylobacter jejuni* adhesins for cellular and mucous substrates. Infect Immun 1986;53:141–148.
42. De Melo MA, Gabbiani G, Pechere JC. Cellular events and intracellular survival of *Campylobacter jejuni* during infection of HEp-2 cells. Infect Immunol 1989;57:2214–2222.
43. Konkel ME, Mead DJ, Hayes SF, Cieplak W Jr. Translocation of *Campylobacter jejuni* across human polarized epithelial cell monolayer cultures. J Infect Dis 1992;166:308–315.
44. Duffy MC, Benson JB, Rubin SJ. Mucosal invasion in *Campylobacter* enteritis. Am J Clin Pathol 1980;73:706–708.
45. Ceelen LM, Decostere A, Ducatelle R, Haesebrouck F. Cytolethal distending toxin generates cell death by inducing a bottleneck in the cell cycle. Microbiol Res 2006;161: 109–120.
46. Everest PH, Cole AT, Hawkey CJ, Knutton S, Goossens H, Butzler JP, et al. Roles of leukotriene B4, prostaglandin E2, and cyclic AMP in *Campylobacter jejuni*-induced intestinal fluid secretion. Infect Immunol 1993;61:4885–4887.
47. Vriesendorp FJ, Mishu B, Blaser MJ, Koski CL. Serum antibodies to GM1, GD1b, peripheral nerve myelin, and *Campylobacter jejuni* in patients with Guillain-Barré syndrome and controls: correlation and prognosis. Ann Neurol 1993;34:130–135.
48. Yuki N, Odaka M. Ganglioside mimicry as a cause of Guillain-Barré syndrome. Curr Opin Neurol 2005;18:557–561.
49. Black RE, Levine MM, Clements ML, Hughes TP, Blaser MJ. Experimental *Campylobacter jejuni* infection in humans. J Infect Dis 1988;157:472-479.
50. Robinson DA. Infective dose of *Campylobacter jejuni* in milk. Br Med J (Clin Res Ed) 1981;282:1584.
51. Peterson MC. Clinical aspects of *Campylobacter jejuni* infections in adults. West J Med 1994;161:148–152.
52. Schonberg-Norio D, Sarna S, Hanninen ML, Katila ML, Kaukoranta SS, Rautelin H. Strain and host characteristics of *Campylobacter jejuni* infections in Finland. Clin Microbiol Infect 2006;12:754–760.
53. Hofreuter D, Tsai J, Watson RO, Novik V, Altman B, Benitez M, et al. Unique features of a highly pathogenic *Campylobacter jejuni* strain. Infect Immun 2006;74:4694–4707.
54. Coker AO, Isokpehi RD, Thomas BN, Amisu KO, Obi CL. Human campylobacteriosis in developing countries. Emerg Infect Dis 2002;8:237–244.
55. Talan DA, Moran GJ, Newdow M, Ong S, Mower WR, Nakase JT, et al. Etiology of bloody diarrhea among patients presenting to United States emergency departments: prevalence of *Escherichia coli* O157:H7 and other enteropathogens. Clin Infect Dis 2001;32;573–580.
56. Blakelock RT, Beasley SW. Infection and the gut. Semin Pediatr Surg 2003;12:265–274.
57. Ina K, Kusugami K, Ohta M. Bacterial hemorrhagic enterocolitis. J Gastroenterol 2003;38:111–120.
58. Allos BM. *Campylobacter jejuni* Infections: update on emerging issues and trends. Clin Infect Dis 2001;32: 1201–1206.
59. Crushell E, Harty S, Sharif F, Bourke B. Enteric *Campylobacter*: purging its secrets? Pediatr Res 2004;55:3–12.
60. Slutsker L, Altekruse SF, Swerdlow DL. Foodborne diseases. Emerging pathogens and trends. Infect Dis Clin North Am 1998;12:199–216.
61. Ruiz-Palacios GM, Cervantes LE, Ramos P, Chavez-Munguia B, Newburg DS. *Campylobacter jejuni* binds intestinal H(O) antigen (Fuc alpha 1, 2Gal beta 1, 4GlcNAc), and fucosyloligosaccharides of human milk inhibit its binding and infection. J Biol Chem 2003;278: 14112–14120.
62. Simor AE, Karmali MA, Jadavji T, Roscoe M. Abortion and perinatal sepsis associated with *Campylobacter* infection. Rev Infect Dis 1986;8:397–402.
63. Hill Gaston JS, Lillicrap MS. Arthritis associated with enteric infection. Best Pract Res Clin Rheumatol 2003; 17:219–239.
64. Yu D, Kuipers JG. Role of bacteria and HLA-B27 in the pathogenesis of reactive arthritis. Rheum Dis Clin North Am 2003;29:21–36.
65. Khan MA, Ball EJ. Genetic aspects of ankylosing spondylitis. Best Pract Res Clin Rheumatol 2002;16:675–690.

66. Hahn AF. Guillain-Barré syndrome. Lancet 1998;352:635–641.

67. Allos BM. *Campylobacter jejuni* infection as a cause of the Guillain-Barré syndrome. Infect Dis Clin North Am 1998;12:173–184.

68. Blaser MJ. Epidemiologic and clinical features of *Campylobacter jejuni* infections. J Infect Dis 1997;176:103–105.

69. Pena LA, Fishbein MC. Fatal myocarditis related to *Campylobacter jejuni* infection: a case report. Cardiovasc Pathol 2007;16:119–121.

70. Sillero M, Almirall J. *Campylobacter jejuni* and hemolytic-uremic syndrome. Nephron 1999;82:363–364.

71. Wang H, Murdoch DR. Detection of *Campylobacter* species in faecal samples by direct Gram stain microscopy. Pathology 2004;36:343–344.

72. Humphrey TJ. An appraisal of the efficacy of pre-enrichment for the isolation of *Campylobacter jejuni* from water and food. J Appl Bacteriol 1989;66:119–126.

73. On SL. Identification methods for campylobacters, helicobacters, and related organisms. Clin Microbiol Rev 1996;9:405–422.

74. Nachamkin I. *Campylobacter* and *Arcobacter*. In: Murray PR, Baron EJ, Pfaller MA, Tenover FC, Yolken RH, editors. Manual of clinical microbiology. 6th ed. Washington, DC: ASM Press; 1995:483–491.

75. Stephen J. Pathogenesis of infectious diarrhea. Can J Gastroenterol 2001;15:669–683.

76. Scotter SL, Humphrey TJ, Henley A. Methods for the detection of thermotolerant campylobacters in foods: results of an inter-laboratory study. J Appl Bacteriol 1993;74:155–163.

77. Sails AD, Fox AJ, Bolton FJ, Wareing DR, Greenway DL, Borrow R. Development of a PCR ELISA assay for the identification of *Campylobacter jejuni* and *campylobacter coli*. Mol Cell Probes 2001;15:291–300.

78. Olsen JE, Aabo S, Hill W, Notermans S, Wernars K, Granum PE, et al. Probes and polymerase chain reaction for detection of food-borne bacterial pathogens. Int J Food Microbiol 1995;28:1–78.

79. Hong Y, Berrang ME, Liu T, Hofacre CL, Sanchez S, Wang L, et al. Rapid detection of *Campylobacter coli, C. jejuni*, and *Salmonella enterica* on poultry carcasses by using PCR-enzyme-linked immunosorbent assay. Appl Environ Microbiol 2003;69:3492–3499.

80. Siragusa GR, Line JE, Brooks LL, Hutchinson T, Laster JD, Apple RO. Serological methods and selective agars to enumerate *Campylobacter* from broiler carcasses: data from inter- and intra-laboratory analyses. J Food Prot 2004;67:901–907.

81. Huicho L, Sanchez D, Contreras M, Paredes M, Murga H, Chinchay L, et al. Occult blood and fecal leukocytes as screening tests in childhood infectious diarrhea: an old problem revisited. Pediatr Infect Dis J 1993;12:474–477.

82. King CK, Glass R, Bresee JS, Duggan C, Centers for Disease Control and Prevention (CDC). Managing acute gastroenteritis among children: oral rehydration, maintenance, and nutritional therapy. MMWR Recomm Rep 2003;52:1–16.

83. Smith JL. Foodborne illness in the elderly. J Food Prot 1998;61:1229–1239.

84. Adachi JA, Ostrosky-Zeichner L, DuPont HL, Ericsson CD. Empirical antimicrobial therapy for traveler's diarrhea. Clin Infect Dis 2000;31:1079–1083.

85. Schonberg-Norio D, Hanninen ML, Katila ML, Kaukoranta SS, Koskela M, Eerola E, et al. Activities of telithromycin, erythromycin, fluoroquinolones, and doxycycline against *Campylobacter* strains isolated from Finnish subjects. Antimicrob Agents Chemother 2006;50:1086–1088.

86. Gibreel A, Taylor DE. Macrolide resistance in *Campylobacter jejuni* and *Campylobacter coli*. J Antimicrob Chemother 2006;58:243–255.

87. Tarlow MJ, Block SL, Harris J, Kolokathis A. Future indications for macrolides. Pediatr Infect Dis J 1997;16:457–462.

88. DuPont HL. Travellers' diarrhoea: contemporary approaches to therapy and prevention. Drugs 2006;66:303–314.

89. Jesse TW, Englen MD, Pittenger-Alley LG, Fedorka-Cray PJ. Two distinct mutations in gyrA lead to ciprofloxacin and nalidixic acid resistance in *Campylobacter coli* and *Campylobacter jejuni* isolated from chickens and beef cattle. J Appl Microbiol 2006;100:682–688.

90. Taylor DN, Sanchez JL, Candler W, Thornton S, McQueen C, Echeverria P. Treatment of travelers' diarrhea: ciprofloxacin plus loperamide compared with ciprofloxacin alone. A placebo-controlled, randomized trial. Ann Intern Med 1991;114:731–734.

91. Reimer LG, Wilson ML, Weinstein MP. Update on detection of bacteremia and fungemia. Clin Microbiol Rev 1997;10:444–465.

92. Cosi V, Versino M. Guillain-Barré syndrome. Neurol Sci 2006;27:47–51.

Chapter 15

CLOSTRIDIUM BOTULINUM (Botulism)

CYRUS RANGAN, MD, FAAP

HISTORY

Clostridium botulinum (van Ermengem) Bergey et al. (*C. botulinum*) is a gram-positive, anaerobic, spore-forming, motile bacillus. The German physician, Justinus Kerner, first described botulism (botulus [*Latin*]: "sausage") in 1817, when 230 Germans fell ill from eating home-fermented blood sausage.[1] Belgian bacteriologist Emile Pierre van Ermengem isolated *C. botulinum* in 1895 after he traced an outbreak of botulism to smoked ham consumed at a local funeral.[2] American biochemist Edward J. Schantz isolated botulinum neurotoxin in 1944.[3] In the United States each year, the Centers for Disease Control (CDC) report approximately 100 cases of botulism (0.1 per 1 million people) divided into the following categories: infant botulism, 72% (50–60% in the state of California); foodborne botulism, 25%; and wound botulism, 3%. The botulism fatality rate dropped from 60% (1899–1949) to 15.5% (1950–1996), primarily as a result of early clinical recognition and advances in treatment. Since the 1990s, mortality has been <2% for all toxin types.[4]

Since September 11, 2001, public health programs have raised awareness for terrorism-associated releases of botulinum toxin into food, water supplies, and air.[5] The Japanese, Soviet, British, Canadian, and American militaries investigated botulinum toxin as a biologic agent beginning in World War II. The infamous Japanese Unit 731 fed botulinum toxin to prisoners in Manchuria during experiments with this potential biological agent.[6] The Soviets exposed animals to botulinum toxin during experiments on Vozrozhdeiye Island in the Aral Sea. In the early 1990s, the Japanese cult, Aum Shinrikyo, released a *C. botulinum* preparation in Japan, but the attempt was unsuccessful. Prior to the first Gulf War, botulinum toxin was the main weaponizable biological agent in the Iraqi arsenal.[7]

EXPOSURE

Source

Botulinum toxin is produced by all strains of *Clostridium botulinum* and certain strains of three other *Clostridia* species (*C. argentinense* Suen, *C. baratii* corrig. (Prévot) Holdeman and Moore, and *C. butyricum* Prazmowski). For practical purposes, all of these toxin-producing strains are referenced under the single heading of *C. botulinum*, and designated by seven distinct neurotoxin serotypes (A–G). Although the neurotoxins of these seven types are structurally different, these neurotoxins all produce similar clinical effects. All of these species are associated with human disease except *C. argentinense*, which causes neurotoxicity only in animals. Strains of *C. botulinum* produce one and occasionally two of the seven types of neurotoxins.

ENVIRONMENT AND DISTRIBUTION

Clostridium botulinum is found worldwide in soil and in both freshwater and saltwater aquatic sediments, where dormant spores remain viable for several years.[8] Windy conditions and disturbed soils (e.g., construction sites) increase the risk of exposure to *C. botulinum*.[9] Vacuum

Medical Toxicology of Natural Substances, by Donald G. Barceloux, MD

cleaners may emit or resuspend botulinum spores into indoor air spaces, but the clinical significance of this exposure is unclear.[10,11] Human cases are caused primarily by *C. botulinum* types A, B, and E. Type A accounts for 86% of *C. botulinum* cases west of the Mississippi River; whereas type B accounts for 60% of *C. botulinum* cases east of the Mississippi River.[4] Type E causes 25% of foodborne botulism cases and 16% of all botulism cases in the United States, primarily in Alaska and the Great Lakes.[12] *C. botulinum* types C and D cause botulism in animals. *C. botulinum* type F is a rare cause of foodborne and infant botulism.[13,14] *C. botulinum* type G has been implicated, but not confirmed as a cause of sporadic deaths in Argentina and Switzerland.[15]

Animal

The feces of asymptomatic grazing animals may harbor *C. botulinum* spores.[16,17] *C. botulinum* type *E* is found naturally in marine life in colder climates.[18,19] Fermented preparations of beaver, fish heads, fish eggs, whale, and seal have been associated with botulism,[20] but the potential for the introduction of this bacterium during the food preparation process limits conclusions regarding the source of the botulism.

Food

Home-preserved or home-fermented vegetables, fruits, and canned meats account for 65% of reported foodborne cases of botulism.[12] Commercial foods account for approximately 7% of foodborne cases. About 18% of foodborne cases do not have a traceable food source. The traceable sources in most US outbreak investigations are stored vegetables, such as asparagus, beets, carrots, carrot juice, corn, green beans, and mushrooms.[12] Canned fruits, corn syrup, and honey harbor significantly fewer vegetative bacterial cells and less botulinum toxin because *C. botulinum* grows more slowly in high-sugar environments; however, 15% to 25% of honey products harbor high counts of botulinum spores (especially type B).[21] Spores may germinate readily in the relatively alkaline and low oxygen environment in the gastrointestinal system of an infant.[22] The source of the inoculum is unknown in approximately 85% of cases of infant botulism with the ingestion of honey confirmed in only 15% of cases.[4] The CDC recommends that honey not be given to infants younger than 12 months old. Improperly prepared home-canned baby food is a potential source of botulism in older infants.[23] The source of botulism is frequently unknown in breastfed children <2 months of age, and predisposing factors (e.g., constipation, preexisting intestinal flora, acidic food) may be more important than food sources.[24] A bulging or dented container may indicate gas production from *C. botulinum* in canned foods.[12,25] Since 1950, sporadic outbreaks of botulism have been traced to fried onions,[26] venison jerky,[27] poultry stuffing, baked potatoes,[28] potato salad,[29] hash browns,[30] hazelnut yogurt,[31] home-prepared fermented tofu,[32] and garlic-in-oil infusions.[33]

Medical

BOTOX® (Allergan Inc., Irvine, CA) is a crystalline botulinum toxin type A formulated for cosmetic purposes, such as the treatment of strabismus, blepharospasm, glabellar lines, primary axillary hyperhidrosis, and cervical dystonia. Myobloc® and Neurobloc® (Solstice Neuroscience, Malvern, PA) contain botulinum toxin B that is approved for the treatment of cervical dystonia.

Food Processing

Clostridium botulinum growth can be controlled by high acidity, high salinity, high concentration of dissolved sugar, high oxygen concentrations, low moisture, and the use of preservatives.[34] Canned vegetables are low acid foods that may not be sufficiently heated to kill spores, resulting in an oxygen-poor medium that allows botulinum spores to germinate and to produce botulinum toxin. Pressure canners can be used; 0.2 M salt solutions and preservatives may be added to home-canned vegetables to prevent *C. botulinum* growth. Canned tomatoes and tomato sauce produce sufficient acid to prevent growth. The US Food and Drug Administration (FDA) requires all commercially canned foods to undergo cooking at 121 °C under pressure for 3 minutes.[35] All home-canned foods should be boiled for at least 20 minutes before serving to eliminate the heat-labile, preformed botulinum toxin.[36] Nitrates and nitrites prevent *C. botulinum* growth in meats and sausages smoked at low temperatures.[37] Citrate is added to canned fruits to chelate divalent cations essential for *C. botulinum* growth.[37] Other effective means to control food contamination include the following: proper cooking and refrigeration, purification of water, pasteurization of dairy products, sanitation of food preparation surfaces with at least 1% bleach solution and 20-minute contact time, and boiling of all canned foods to remove preformed botulinum toxin.[38]

TOXINS

Physiochemical Properties

Clostridium botulinum types A, B, and F are proteolytic organisms with chromosomally encoded toxins. *C. botu-*

linum type E strains are saccharolytic bacteria with a bacteriophage-encoded toxin. *C. botulinum* type G has a plasmid-encoded toxin. Botulinum neurotoxins differ slightly in chemical makeup and mechanisms of toxicity; however, clinical manifestations of all types of botulism are similar.[39] In the United States, types A and B cause the majority of infant, foodborne, and wound botulism cases.[4] Optimum temperatures for growth of *C. botulinum* are 78 °F–95 °F (25 °C–35 °C) for proteolytic types A and B, and 38 °F–77 °F (3 °C–25 °C) for the cold-tolerant, saccharolytic type E.[34,40] Bacteria proliferate more slowly at 118 °F (48 °C), but do not die until heated to 212 °F (100 °C).[41] Spores must be heated for several hours at 212 °F (100 °C) before complete inactivation.[42] Botulinum toxin is synthesized as one of three distinct progenitor toxin complexes: 19s (900 kDa), 16s (500 kDa), or 12s (300 kDa).[43] Synthesis of assorted progenitor toxin complexes depends on antigenic type. Each progenitor toxin is composed of a nonneurotoxic component and a unique 150 kDa active neurotoxin.[44] The nonneurotoxic component of each progenitor toxin contains hemagglutinins and other proteins that protect the 150 kDa neurotoxin from acid and protease degradation in the gastrointestinal tract and blood.[45] The 150 kDa neurotoxin consists of a 100 kDa heavy chain and a 50 kDa light chain linked by a disulfide bond, which must remain intact before the neurotoxin enters nerve cells. The heavy chain binds to neuronal membrane receptors; the light chain is a zinc-dependent metalloprotease that mediates neurotoxicity.[46] See Table 15.1 for properties and characteristics of *C. botulinum*.

Mechanism of Toxicity

Foodborne botulism is caused by consuming preformed neurotoxins of *C. botulinum* in foods.[4] Infant botulism is a tandem infection-intoxication, beginning with ingestion of *C. botulinum* spores, gastrointestinal tract proliferation, and neurotoxin production *in vivo*.[11] Wound botulism is caused by neurotoxin production from wounds infected with *C. botulinum*.[47,48] Botulinum toxin binds with moderate affinity to cells in the gastrointestinal tract. The nonneurotoxic portion of the botulinum progenitor toxin binds to receptors on the apical surfaces of gastric and epithelial cells. These cells transcytose the progenitor toxin-receptor complex to the basolateral surface, where the complex enters the lymphatic circulation.[49] Following aerosolized exposure, progenitor toxin crosses the epithelial cells of the lung

TABLE 15.1. Characteristics of *Clostridium Botulinum* Antigenic Types

	A	B	C	D	E	F	G
Infant	50%	50%	–	–	–	<1%	–
Foodborne	50%	25%	–	–	25%	<1%	–
Wound	80%	20%	–	–	–	–	–
Progenitor toxin	19s,16s,12s	16s,12s	16s,12s	16s,12	12s	12s	16s
Gastrointestinal symptoms	+	+	+ (Animals)	+ (Animals)	++	+/–	Unknown
Neuronal target protein	SNAP-25	VAMP	SNAP-25, syntaxin	VAMP	VAMP	SNAP-25	SNAP-25
Geographic predominance	Western US[a]	Eastern US[b]	US wetlands		Pacific Northwest, Great Lakes	California, Denmark	Argentina
Common food sources	Home-processed vegetables, fruits, meats; rare commercial foods		Fish (Disease occurs in fish-eating animals.)		Alaska Native foods	Meat, seafood, jerky	Unknown
Cell inactivation	100 °C (212 °F) (>25 minutes) pH < 5.2				100 °C (212 °F) (>1 minute) pH < 4.8	100 °C (212 °F) (>25 minutes) pH < 5.2	
Spore inactivation	100 °C (212 °F), >2 hours, pH < 4.6 Spore germination slows at 60 °C (140 °F), but may resume several hours after cooling.						
Toxin inactivation	80 °C (176 °F), >15 minutes						

[a]US states with no reported foodborne botulism: Delaware, New Hampshire, South Carolina, Vermont.
[b]US states with no reported infant botulism: Maine, Rhode Island, South Carolina, Minnesota.

in a similar manner.[5] The nonneurotoxic component of progenitor toxin protects the 150 kDa neurotoxin from serum proteases during transit through the blood and lymphatic systems.[50] The neurotoxin binds preferentially to presynaptic cholinergic neurons at ganglionic synapses, postganglionic parasympathetic synapses, and neuromuscular junctions.[51,52] The 50 kDa C-terminal of the neurotoxin heavy chain binds to protein and carbohydrate moieties on the outer presynaptic membrane: synaptotagmins I and II, and polysialogangliosides GD1b and GT1b.[53,54] The 50 kDa N-terminal of the neurotoxin heavy chain subsequently translocates through the neuronal membrane, culminating in endocytosis of neurotoxin-receptor complex by the presynaptic neuron.[55] The low acidity of the endosome severs the neurotoxin disulfide bond via proteolysis to separate the 100 kDa heavy chain from the 50 kDa light chain; the light chain then penetrates the endosomal membrane to enter the cytosol.[56] The light chain prevents the docking of acetylcholine-containing synaptosomes to a multimeric transport complex on the presynaptic membrane (soluble *N*-ethylmaleimide-sensitive factor attachment protein receptor [SNARE] complex). The SNARE complex mediates the release of acetylcholine into the synapse.[57] Types A, C, and E light chains disrupt the SNARE complex by hydrolyzing unique locations on a 25 kDa synaptosomal-associated protein (SNAP-25).[58] Types B D, F, and G light chains disrupt the acetylcholine-containing synaptosome by hydrolyzing unique locations on a vesicle-associated membrane protein complex (VAMP/synaptobrevin II).[59] Type C hydrolyzes an additional SNARE protein called syntaxin.[60] Each of these mechanisms results in irreversible inhibition of acetylcholine exocytosis, resulting in clinical findings consistent with acetylcholine deficiency in the cranial nerves, parasympathetic nervous system, and neuromuscular junctions.[61] Botulinum toxin does not cross the blood–brain barrier. Consequently, central nervous system effects are generally absent and deep tendon reflexes remain intact.[62] Nausea, vomiting, abdominal discomfort, and constipation arise from paralytic ileus inflicted by the neurotoxin.[63] Brief diarrhea may occur early in the course of infant or foodborne botulism, sometimes as a result of the effects of coexisting gastrointestinal microorganisms or toxins in the offending food.[64] Saccharolytic enzymes of *C. botulinum* type E cause more prominent gastrointestinal symptoms before neurotoxicity.[65–67]

DOSE RESPONSE

Botulinum toxin is one of the most potent toxins known. The lethal dose of botulinum toxin following ingestion is approximately 1 μg/kg, whereas the lethal dose following inhalation probably is substantially less (i.e., about 100 times).[68] The typical dose of BOTOX® is 4–20 units with a therapeutic range up to 300 units. Based on primate studies, the approximate lethal parenteral dose in humans is about 40 U/kg with each unit being the estimated median lethal dose (LD_{50}) for mice.[69] A case report associated the development of botulism with the intradermal administration of high doses of unlicensed botulinum toxin A secondary to a dilution error.[70] This error resulted in the intradermal injection of up to 2,857 times the estimated human lethal dose of botulinum toxin. The patient required prolonged ventilation and physical rehabilitation.

Infants are particularly susceptible to spore-derived botulism because of the ineffective production of clostridial spore-destroying stomach and bile acids along with a relative deficiency of protective gastrointestinal bacterial flora to compete for colonization.[71] Formula-fed infants are relatively protected from spore-derived botulism compared with solely breastfed infants, presumably because of gut colonization with enterococci and coliforms earlier in life.[72,73] After infants begin eating solid food, increased stomach acid production and the increase in gastrointestinal flora tend to resist bacterial colonization.[71] Risk factors for older children and adults include inflammatory bowel disease, achlorhydria, acid-reducing medications (e.g., proton pump inhibitors, antacids), and diminished gastrointestinal flora from recent antibiotic usage.[74,75]

CLINICAL RESPONSE

Rare case reports associate limited paresis and dysphagia with the intradermal and intramuscular use of botulinum toxin A for medicinal purposes.[76,77]

Foodborne Botulism

Foodborne botulism occurs after ingestion of preformed toxin. Onset of symptoms typically occur 6–36 hours after exposure; however, a delay of up to 10 days is rare but may occur.[78,79] All adult forms of botulism manifest similar symptoms. Early symptoms of foodborne botulism include nausea, vomiting, abdominal discomfort, and constipation. These nonspecific symptoms may not initially prompt the patient to seek medical attention, and occasionally botulism presents with gastrointestinal symptoms and minimal cranial nerve dysfunction.[80] Speech disturbances including dysarthria and dysphonia are very common along with dysphagia and mild anticholinergic symptoms (dry mouth, dilated pupils, decreased bowel and bladder motility). A symmetrical progressive descending paralysis begins with bulbar paralysis including paralysis of accommodation, diplopia, lateral rectus palsy (strabismus), oculomotor palsy (ptosis), and paralysis of respiratory muscles (intercostal, diaphragm). An acute generalized flaccid paralysis follows, but sensory deficits and central nervous system involvement is absent.[4] The clinical features of botulism progress over 1–3 days. This pattern of paralysis mimics the Miller-Fisher variant of Guillain-Barré syndrome. However, the clinical presentation of this disease often involves sensory paresthesias, absent deep tendon reflexes, and elevated cerebrospinal fluid (CSF) protein, whereas the clinical features of botulism include normal sensation, diminished or normal deep tendon reflexes, and normal CSF protein.[81,82] The combination of early bulbar findings, descending paralysis including large skeletal muscle groups, normal mental status, and normal temperature is highly suggestive of botulism. The presence of fever does not exclude the diagnosis of botulism.[83] Although the anticholinergic syndrome and botulism share many peripheral findings, the characteristic absence of changes in mental status, memory, temperature, and heart rate indicates botulism.[84] Vital signs are typically normal with the exception of low oxygen saturation secondary to respiratory insufficiency. Occasionally, paresthesias or asymmetric limb weakness occurs.[85] Differential diagnosis for botulism includes anticholinergic syndrome, Guillain-Barré syndrome, encephalitis, meningitis, hypermagnesemia, multiple sclerosis, myasthenia gravis, poliomyelitis, stroke, and tick paralysis.[86–88] Table 15.2 illustrates the clinical and diagnostic features that differentiate botulism from similar conditions.

Infant Botulism

Infant botulism affects over 100 infants per year in the United States; 95% of infant cases occur between the ages of 3 weeks and 6 months of age at presentation (median: 13 weeks).[72] Infant botulism occurs after spore ingestion, germination, and production of botulinum toxin in the gut. Symptoms develop 3–30 days after exposure. The first symptom of infant botulism is usually constipation, followed gradually by poor head control, feeding difficulties with poor sucking and swallowing, weak cry, and generalized weakness or floppiness.[89] Constipation typically precedes symptoms of weakness by several weeks. A brief period of diarrhea (19%) may precede constipation (73%).[72] Classic bulbar signs may be subtle and difficult to recognize in infants; however, diminished tracking (i.e., ophthalmoplegia) and loss of smile or facial grimace suggest bulbar involvement.[72] Seizures are rarely associated with infant botulism. Infants with botulism may be dehydrated from poor oral intake; tachycardia may accompany infant botulism secondary to dehydration in contrast to the bradycardia associated with adult botulism.[72] Dry mouth, decreased urination, and hypoactive bowel sounds also occur. The differential diagnosis includes failure-to-thrive, septicemia, congenital and acquired hypotonic disorders, Guillain-Barré syndrome, myasthenia gravis, and poliomyelitis.[72] Older children and adults with risk factors for gut colonization can develop botulism from spore ingestion.[90] Overall, botulism mortality rate since 1950 is about 2% with the case fatality rate <1% for infants hospitalized with botulism.[4] Botulism is part of the differential diagnosis of sudden infant death syndrome (SIDS),[91] but the causal role of botulism in SIDS is unclear.[92]

Wound Botulism

Wound botulism occurs after botulinum toxin is absorbed from traumatic wounds colonized with

TABLE 15.2. Distinguishing Clinical and Diagnostic Features in the Differential Diagnosis of Botulism*

Disease	Distinguishing Clinical and Diagnostic Features
Botulism	Normal temperature, normal mental status, normal heart rate, diminished or normal reflexes, normal CSF/CT/MRI, descending generalized flaccid paralysis, negative edrophonium test. EMG: Normal sensory/motor nerve conduction; brief small, abundant potentials
Antimuscarinic syndrome	Fever, altered mental status, tachycardia, intact skeletal muscle strength
Encephalitis/meningitis	Fever, altered mental status, abnormal CSF
Guillain-Barré syndrome	Sensory paresthesias, absent deep tendon reflexes, and elevated CSF protein, ascending paralysis (except Miller-Fisher variant)
Hypermagnesemia	High magnesium level
Multiple sclerosis	Abnormal CSF, MRI, nerve conduction studies
Myasthenia gravis	Positive edrophonium test, acetylcholine receptor antibodies, response to pyridostigmine
Poliomyelitis	Muscle pain and tenderness EMG: Motor axon dysfunction, sharp waves, fibrillation
Stroke	Abnormal CT/MRI, focal paralysis
Tetanus	Rigid/spastic paralysis
Tick paralysis	Ascending paralysis, identification of tick

*CSF indicates cerebrospinal fluid; CT, computed tomography; EMG, electromyogram; MRI, magnetic resonance imaging.

C. botulinum. The incubation period is 4–14 days. Clinical findings are similar to foodborne and infant botulism;[93,94] the diagnosis is often delayed in drug users presenting with early complaints consistent with bulbar palsy.[95] Fever may accompany wound botulism secondary to wound infection from other microorganisms.[93] Wound botulism is an occasional complication of illicit drug use, particularly following "skin popping" with Mexican black tar heroin.[47] The differential diagnosis includes Eaton-Lambert syndrome, hypermagnesemia, aminoglycoside toxicity, Miller-Fisher variant of Guillain-Barré syndrome, diphtheria in unimmunized patients, and myasthenia gravis.[96] Tetanus primarily affects the central nervous system and produces spasticity along with cranial nerve deficits. Tick paralysis produces an ascending rather than a descending paralysis, and poliomyelitis causes a more asymmetric ascending paralysis, primarily in the lower extremities.

Inhalational Botulism

Aerosolized exposure to botulinum toxin does not occur naturally, and airborne botulism is highly suspicious of an intentional release during terrorist acts. The clinical syndrome is probably similar to foodborne botulism without prodromal gastrointestinal symptoms. Inhalational botulism developed in three laboratory technicians beginning 3 days after performing an autopsy on a laboratory animal with botulism. Clinical features included dysphagia, ocular paresis, rotatory nystagmus, dysarthria, dilated pupils, ataxia, and generalized weakness. All 3 patients recovered after 2 weeks of treatment with botulinum antitoxin. Suspected events should be reported to the CDC or local public health departments immediately.[5]

DIAGNOSTIC TESTING

Analytical Methods

Culture of *C. botulinum* bacteria from patient specimens confirms the diagnosis of botulism; however, bacteria are rarely recovered.[97] Routine laboratory diagnosis of botulism is based on the detection of botulinum neurotoxin in the serum, feces, gastric contents, wound swabs of the patient, or in the food source, if applicable. Specimens should be obtained as early as possible after presentation, and transported in anaerobic media under refrigerated conditions (4°C/39°F). Neurotoxin detection is based on a mouse lethality assay. Mice are injected intraperitoneally with specimen extracts, alone or mixed with monovalent antitoxins (A, B, C, D, E, and F). Mice are observed for death over a 4-day period; however, death usually occurs within 24 hours with neurotoxin-positive specimens. Assays are conducted at public health laboratories and the CDC.[21] An enzyme-linked immunosorbent (ELISA) assay is an alternative to the mouse bioassay, but this immunoassay is available primarily in research laboratories.[98] More rapid molecular and biochemical methods for detecting botulinum toxin currently are being developed, such as multiplex polymerase chain reaction (PCR) assay.[99,100] AOAC International (Association of Analytical Chemists International, Gaithersburg, MD; http://www.eoma.aoac.org/) official method 977.26 is applicable to the determination of *C. botulinum* and associated toxins in foods.[101]

Biomarkers

There are no human biomarkers for monitoring or surveillance of *C. botulinum*.

Abnormalities

Electrolytes, complete blood counts, CSF fluid analysis, and edrophonium tests are normal in patients with botulism. Documentation of negative studies helps to narrow the differential diagnosis to botulism, and health officials may utilize this information in an individual case to determine the potential benefit of botulinum antitoxin. The electromyogram (EMG) typically shows brief, small, abundant action potentials (BSAPs) that increase in amplitude under repetitive stimulation at 50Hz; nerve conduction remains intact.[102] Severe dehydration results in hypernatremia (serum Na > 150mEq/L) or hyponatremia (serum Na < 130mEq/L), hypokalemia (serum K < 3.0mEq/L), hypochloremia, elevated urine specific gravity (SpG > 1.020), and/or hyperphosphatemia. Low serum bicarbonate results from lactic acid production or elevated serum bicarbonate secondary to contraction alkalosis.

TREATMENT

Gastrointestinal symptoms are mild and self-limited. Affected patients may experience significant anorexia, necessitating intravenous fluid replacement and hospitalization in cases of significant dehydration or severe electrolyte abnormalities, particularly in young children or elderly patients.[103,104] Supportive care of botulism includes stabilization of airway, breathing, and circulation. Cases with bulbar findings usually progress to respiratory failure, and require mechanical ventilation.[4] Nutrition is supported with tube feedings and rarely by total parenteral nutrition.[72] Activated charcoal may be beneficial when patients present early, but there are no clinical data to confirm the efficacy of activated charcoal for the treatment of botulism.[105] Guanidine enhances acetylcholine release from botulinum-poisoned nerve terminals *in vitro*; however, guanidine has no clear benefit *in vivo*, and is not recommended. *C. botulinum* is susceptible to a number of antibiotics (e.g., aminoglycosides, clindamycin);[106] however, antibiotics are not recommended because additional toxin is released from bacterial lysis following the administration of antibiotics.[78] The CDC offers 24-hour telephone consultation to discuss suspected cases of botulism and the appropriateness of botulinum antitoxin (day phone: 404-639-2206, night phone: 404-639-2888). The clinician may also report suspected cases to regional or state health officers, infection control officers, or poison control centers.

Treatment with antitoxin is determined through direct clinical discussion with local, state, or CDC health officials. Laboratory testing may take several hours or days to complete; treatment should be given as early as possible during the clinical course, and should occur before laboratory confirmation in cases with high suspicion.[107] Botulinum antitoxin halts the progression of illness by binding circulating, unbound toxin; toxin bound to nerve tissue is not affected by antitoxin.[108] Two primary botulinum antitoxins are available. Trivalent (A, B, E) botulinum antitoxin contains intact IgG antibodies derived from horses. This antitoxin is available from state and local health departments via the CDC. A single vial is administered after skin testing for sensitivity.[107] Horse-derived antitoxin is not recommended for infant botulism secondary to risk of allergic reactions.[109] Infant botulism may be treated with human botulism immune globulin (BabyBIG®), available through the California Department of Health Care Services (Sacramento, CA).[110] BabyBIG® is dosed at 1 mL/kg (50 mg/kg) as a single intravenous infusion. The package insert should be reviewed by the clinician for updates in dosage and administration. When used within 3 days of onset of symptoms, BabyBIG® reduces mean hospital stay from 5.7 weeks to 2.6 weeks, with significant reductions in the length of stay in the intensive care unit along with decreased duration of mechanical ventilation and of tube feedings.[111–113] Botulinum antitoxins are housed in CDC satellite pharmacies at major airports to facilitate the transportation of affected patients. Patients require continued supportive care after antitoxin administration until respiratory drive is sufficient to maintain ventilation and oxygenation without mechanical support. Many patients require extensive physical and occupational therapy during recovery. Cholinergic neurons regenerate terminal axons and new synapses over several weeks to months in successfully treated cases.[114] The US Army developed a broad heptavalent (A, B, C, D, E, F, G) botulinum antitoxin, derived from equine IgG antibodies digested with pepsin to cleave most of the Fc portion of the antibody, leaving the $F(ab')_2$ portion. This preparation is less immunogenic, and is effective against all known strains of botulism when not contraindicated. The US Department of Health and Human Services commissioned the Cangene Corporation (Winnipeg, Manitoba, Canada) to develop 200,000 doses of heptavalent botulinum antitoxin for delivery into the Strategic National Stockpile from 2007–2012.[115]

References

1. Erbguth FJ, Naumann M. On the first systematic descriptions of botulism and botulinum toxin by Justinus Kerner (1786–1862). J Hist Neurosci 2000;9:218–220.
2. van Ermengem E. Classics in infectious diseases. A new anaerobic bacillus and its relation to botulism. E. van Ermengem. Originally published as "Ueber einen neuen anaeroben Bacillus und seine Beziehungen zum Botulismus" in Zeitschrift fur Hygiene und Infektionskrankheiten 26: 1–56, 1897. Rev Infect Dis 1979;1:701–719.
3. Schantz EJ, Sugiyama H. Toxic proteins produced by *Clostridium botulinum*. J Agric Food Chem 1974;22:26–30.
4. Shapiro RL, Hatheway C, Swerdlow DL. Botulism in the United States: a clinical and epidemiologic review. Ann Intern Med 1998;129:221–228.
5. Los Angeles County Department of Public Health, Emergency Medical Services Agency. Terrorism agent information and treatment guidelines for clinicians and hospitals, Los Angeles: County of Los Angeles Public Health; July 2006.
6. Arnon SS, Schechter R, Inglesby TV, Henderson DA, Bartlett JG, Ascher MS, et al. Botulinum toxin as a biological weapon: medical and public health management. JAMA 2001;285:1059–1070.
7. Zilinskas RA. Iraq's biological weapons. The past as future? JAMA 1997;278:418–424.
8. Haagsma J. Pathogenic anaerobic bacteria and the environment. Rev Sci Tech 1991;10:749–764.
9. Istre GR, Compton R, Novotny T, Young JE, Hatheway CL, Hopkins RS. Infant botulism: three cases in a small town. Am J Dis Child 1986;140:1013–1014.
10. Nevas M, Lindstrom M, Virtanen A, Hielm S, Kuusi M, Arnon SS, et al. Infant botulism acquired from household dust presenting as sudden infant death syndrome. J Clin Microbiol 2005;43:511–513.
11. Thompson JA, Glasgow LA, Warpinski JR. Infant botulism: Clinical spectrum and epidemiology. Pediatrics 1980;66:936–942.
12. Sobel J, Tucker N, Sulka A, McLaughlin J, Maslanka S. Foodborne botulism in the United States, 1990–2000. Emerg Infect Dis 2004;10:1606–1611.

13. Gupta A, Sumner CJ, Castor M, Maslanka S, Sobel J. Adult botulism type F in the United States, 1981–2002. Neurology 2005;65:1694–1700.
14. Barash JR, Tang TW, Arnon SS. First case of infant botulism caused by *Clostridium baratii* type F in California. J Clin Microbiol 2005;43:4280–4282.
15. Sonnabend WF, Sonnabend UP, Krech T. Isolation of *Clostridium botulinum* type G from Swiss soil specimens by using sequential steps in an identification scheme. Appl Environ Microbiol 1987;53:1880–1884.
16. Braun U. [Botulism in cattle.] Schweiz Arch Tierheilkd 2006;148:331–339. [German]
17. Dahlenborg M, Borch E, Radstrom P. Prevalence of *Clostridium botulinum* types B, E and F in faecal samples from Swedish cattle. Int J Food Microbiol 2003;82:105–110.
18. Chiou LA, Hennessy TW, Horn A, Carter G, Butler JC. Botulism among Alaska natives in the Bristol Bay area of southwest Alaska: a survey of knowledge, attitudes, and practices related to fermented foods known to cause botulism. Int J Circumpolar Health 2002;61:50–60.
19. McLaughlin JB. Botulism type E outbreak associated with eating a beached whale, Alaska. Emerg Infect Dis 2004;10:1685–1687.
20. Shaffer N, Wainwright RB, Meddaugh JP, Tauxe RV. Botulism among Alaska Natives the role of changing food preparation and consumption practices. West J Med 1990;153:390–393.
21. Arnon SS, Midura TF, Damus K. Honey and other environmental risk factors for infant botulism. J Pediatr 1979;94:331–336.
22. Tanzi MG, Gabay MP. Association between honey consumption and infant botulism. Pharmacotherapy 2002;22:1479–1483.
23. Armada M, Love S, Barrett E, Monroe J, Peery D, Sobel J. Foodborne botulism in a six-month-old infant caused by home-canned baby food. Ann Emerg Med 2003;42: 226–229.
24. Spika JS, Shaffer N, Hargrett-Bean N, Collin S, MacDonald KL, Blake PA. Risk factors for infant botulism in the United States. Am J Dis Child 1989;143:828–832.
25. Spalter RM. Botulism: a danger of home canning. J Ky Med Assoc 1965;63:507–509.
26. MacDonald KL, Spengler RF, Hatheway CL, Hargrett NT, Cohen ML. Type A botulism from sauteed onions. Clinical and epidemiologic observations. JAMA 1985;253: 1275–1278.
27. Midura TF, Nygaard GS, Wood RM, Bodily HL. *Clostridium botulinum* type F: isolation from venison jerky. Appl Microbiol 1972;24:165–167.
28. Angulo FJ, Getz J, Taylor JP, Hendricks KA, Hatheway CL, Barth SS, et al. A large outbreak of botulism: the hazardous baked potato. J Infect Dis 1998;178: 172–177.
29. Seals JE, Snyder JD, Edell TA, Hatheway CL, Johnson CJ, Swanson RC, et al. Restaurant-associated type A botulism: transmission by potato salad. Am J Epidemiol 1981;113:436–444.
30. Ferreira JL, Eliasberg SJ, Harrison MA, Edmonds P. Detection of preformed type A botulinal toxin in hash brown potatoes by using the mouse bioasssay and a modified ELISA test. J AOAC Int 2001;84: 1460–1464.
31. O'Mahony M, Mitchell E, Gilbert RJ, Hutchinson DN, Begg NT, Rodhouse JC, et al. An outbreak of foodborne botulism associated with contaminated hazelnut yoghurt. Epidemiol Infect 1990;104:389–395.
32. Meyers H, Inami G, Rosenberg J MD, Mohle-Boetani J, Vugia D. Brief Report: Foodborne botulism from home-prepared fermented tofu—California, 2006. MMWR Morb Mortal Wkly Rep 2007;56:96–97.
33. Morse DL, Pickard LK, Guzewich JJ, Devine BD, Shayegani M. Garlic-in-oil associated botulism: episode leads to product modification. Am J Public Health 1990;80:1372–1373.
34. Graham AF, Mason DR, Peck MW. Inhibitory effect of combinations of heat treatment, pH, and sodium chloride on a growth from spores of nonproteolytic *Clostridium botulinum* at refrigeration temperature. Appl Environ Microbiol 1996;62:2664–2668.
35. Centers for Disease Control and Prevention (CDC). Botulism Associated with Commercial Carrot Juice—Georgia and Florida, September 2006. MMWR Morb Mortal Wkly Rep 2006;55;1098–1099.
36. Licciardello JJ, Nickerson JTR, Ribich CA, Goldblith SA. Thermal inactivation of type E botulinum toxin. Appl Microbiol 1967;15:249–256.
37. Wong DM, Young-Perkins KE, Merson RL. Factors influencing *Clostridium botulinum* spore germination, outgrowth, and toxin formation in acidified media. Appl Environ Microbiol 1988;54:1446–1450.
38. Christensen BB, Rosenquist H, Sommer HM, Nielsen NL, Fagt S, Andersen NL, et al. A model of hygiene practices and consumption patterns in the consumer phase. Risk Anal 2005;25:49–60.
39. Collins MD, East AK. Phylogeny and taxonomy of the food-borne pathogen *Clostridium botulinum* and its neurotoxins. J Appl Microbiol 1998;84:5–17.
40. Anniballi F, Fenicia L, Franciosa G, Aureli P. Influence of pH and temperature on the growth of and toxin production by neurotoxigenic strains of *Clostridium butyricum* type E. J Food Prot 2002;65:1267–1270.
41. Skinner GE, Larkin JW. Conservative prediction of time to *Clostridium botulinum* toxin formation for use with time-temperature indicators to ensure the safety of foods. J Food Prot 1998;61:1154–1160.
42. Grecz N, Arvay LH. Effect of temperature on spore germination and vegetative cell growth of *Clostridium botulinum*. Appl Environ Microbiol 1982;43:331–337.
43. Yoneda S, Shimazawa M, Kato M, Nonoyama A, Torii Y, Nishino H, et al. Comparison of the therapeutic indexes of different molecular forms of botulinum toxin type A. Eur J Pharmacol 2005;508:223–229.

44. Inoue K, Sobhany M, Transue TR, Oguma K, Pedersen LC, Negishi M. Structural analysis by x-ray crystallography and calorimetry of a haemagglutinin component (HA1) of the progenitor toxin from *Clostridium botulinum*. Microbiology 2003;149:3361–3370.
45. Sharma SK, Singh BR. Hemagglutinin binding mediated protection of botulinum neurotoxin from proteolysis. J Nat Toxins 1998;7:239–253.
46. Schiavo G, Rossetto O, Benfenati F, Poulain B, Montecucco C. Tetanus and botulinum neurotoxins are zinc proteases specific for components of the neuroexocytosis apparatus. Ann NY Acad Sci 1994;710:65–75.
47. Passaro DJ, Werner SB, McGee J, Mac Kenzie WR, Vugia DJ. Wound botulism associated with black tar heroin among injecting drug users. JAMA 1998;279:859–863.
48. Centers for Disease Control and Prevention (CDC). Wound botulism among black tar heroin users—Washington, 2003. MMWR Morb Mortal Wkly Rep 2003; 52:885–886.
49. Ahsan CR, Hajnoczky G, Maksymowych AB, Simpson LL. Visualization of binding and transcytosis of botulinum toxin by human intestinal epithelial cells. J Pharmacol Exp Ther 2005;315:1028–1035.
50. Fujinaga Y. Transport of bacterial toxins into target cells: pathways followed by cholera toxin and botulinum progenitor toxin. J Biochem 2006;140:155–160.
51. Jin R, Rummel A, Binz T, Brunger AT. Botulinum neurotoxin B recognizes its protein receptor with high affinity and specificity. Nature 2006;444:1092–1095.
52. Lalli G, Bohnert S, Deinhardt K, Verastegui C, Schiavo G. The journey of tetanus and botulinum neurotoxins in neurons. Trends Microbiol 2003;11:431–437.
53. Dong M, Richards DA, Goodnough MC, Tepp WH, Johnson EA, Chapman ER. Synaptotagmins I and II mediate entry of botulinum neurotoxin B into cells. J Cell Biol 2003;162:1293–1303.
54. Schengrund CL, DasGupta BR, Ringler NJ. Binding of botulinum and tetanus neurotoxins to ganglioside, GT1b and derivatives thereof. J Neurochem 1991;57:1024–1032.
55. Poulain B, Tauc L, Maisey EA, Wadsworth JD, Mohan PM, Dolly JO. Neurotransmitter release is blocked intracellularly by botulinum neurotoxin, and this requires uptake of both toxin polypeptides by a process mediated by the larger chain. Proc Natl Acad Sci USA 1988;85:4090–4094.
56. Simpson LL, Maksymowych AB, Park JB, Bora RS. The role of the interchain disulfide bond in governing the pharmacological actions of botulinum toxin. J Pharmacol Exp Ther 2004;308:857–864.
57. Humeau Y, Doussau F, Grant NJ, Poulain B. How botulinum and tetanus neurotoxins block neurotransmitter release. Biochimie 2000;82:427–446.
58. Zhou L, de Paiva A, Liu D, Aoki R, Dolly JO. Expression and purification of the light chain of botulinum neurotoxin A: a single mutation abolishes its cleavage of SNAP-25 and neurotoxicity after reconstitution with the heavy chain. Biochemistry 1995;34:15175–15181.
59. Adler M, Shafer HF, Manley HA, Hackley BE Jr, Nicholson JD, Keller JE, et al. A capillary electrophoresis technique for evaluating botulinum neurotoxin B light chain activity. J Protein Chem 2003;22:441–448.
60. Blasi J, Chapman ER, Yamasaki S, Binz T, Niemann H, Jahn R. Botulinum neurotoxin C1 blocks neurotransmitter release by means of cleaving HPC-1/syntaxin. EMBO J 1993;12:4821–4828.
61. Simpson LL. Identification of the major steps in botulinum toxin action. Ann Rev Pharmacol Toxicol 2004;44: 167–193.
62. Horowitz BZ. Botulinum toxin. Crit Care Clin 2005;21: 825–839.
63. Critchley EM, Mitchell JD. Human botulism. Br J Hosp Med 1990;43:290–292.
64. Hurst DL, Marsh WW. Early severe infantile botulism. J Pediatr 1993;122:909–911.
65. Badhey H, Cleri DJ, D'Amato RF, Vernaleo JR, Veinni V, Tessler J, et al. Two fatal cases of type E adult foodborne botulism with early symptoms and terminal neurologic signs. J Clin Microbiol 1986;23:616–618.
66. Weber JT, Hibbs RG Jr, Darwish A, Mishu B, Corwin AL, Rakha M, et al. A massive outbreak of type E botulism associated with traditional salted fish in Cairo. J Infect Dis 1993;167:451–454.
67. Fenicia L, Franciosa G, Pourshaban M, Aureli P. Intestinal toxemia botulism in two young people, caused by *Clostridium butyricum* type E. Clin Infect Dis 1999;29: 1381–1387.
68. Schantz EJ, Johnson EA. Properties and use of botulinum toxin and other microbial neurotoxins in medicine. Microbiol Rev 1992;56:80–99.
69. Herrero BA, Ecklung AE, Streett CS, Ford DF, King JK. Experimental botulism in monkeys—a clinical pathological study. Exp Mol Pathol 1967;6:84–95.
70. Chertow Ds, Tan E, Maslanka SE, Schlte J, Bresnitz EA, Weisman RS, et al. Botulism in 4 adults following cosmetic injections with an unlicensed, highly concentrated botulinum preparation. JAMA 2006;296: 2476–2479.
71. Hentges DJ. The intestinal flora and infant botulism. Rev Infect Dis 1979;1:668–673.
72. Cox N, Hinkle R. Infant botulism. Am Fam Physician 2002;65:1388–1392.
73. Giacoia GP, Catz C, Yaffe SJ. Environmental hazards in milk and infant nutrition. Clin Obstet Gynecol 1983;26: 458–466.
74. Critchley EM. A comparison of human and animal botulism: a review. J R Soc Med 1991;84:295–298.
75. Caya JG, Agni R, Miller JE. *Clostridium botulinum* and the clinical laboratorian: a detailed review of botulism, including biological warfare ramifications of botulinum toxin. Arch Pathol Lab Med 2004;128:653–662.

76. Tugnoli V, Eleopra R, Quatrale R, Capone JG, Sensi M, Gastaldo E. Botulism-like syndrome after botulinum toxin type A injections for focal hyperhidrosis. Br J Dermatol 2002;147:808–809.
77. Comella CL, Tanner CM, DeFoor-Hill L, Smith C. Dysphagia after botulinum toxin injections for spasmodic torticollis: clinical and radiologic findings. Neurology 1992;42:1307–1310.
78. Woodruff BA, Griffin PM, McCroskey LM, Smart JF, Wainwright RB, Bryant RG, et al. Clinical and laboratory comparison of botulism from toxin types A, B, and E in the United States, 1975–1988. J Infect Dis 1992;166: 1281–1286.
79. St. Louis ME, Peck SH, Bowering D, Morgan GB, Blatherwick J, Banerjee S, et al. Botulism from chopped garlic: delayed recognition of a major outbreak. Ann Intern Med 1988;108:363–368.
80. Sobel J, Malavet M, John S. Outbreak of clinically mild botulism type E illness from home-salted fish in patients presenting with predominantly gastrointestinal symptoms. Clin Infect Dis 2007;45:e14–e16.
81. Spies JM. Cranial and peripheral neuropathies. Med J Aust 2001;174:598–604.
82. Van der Meche FG, Van Doorn PA, Meulstee J, Jennekens FG; GBS-consensus group of the Dutch Neuromuscular Research Support Centre. Diagnostic and classification criteria for the Guillain-Barré syndrome. Eur Neurol 2001;45:133–139.
83. Critchley EM, Hayes PJ, Isaacs PE. Outbreak of botulism in Northwest England and Wales, June, 1989. Lancet 1989;2(8667):849–853.
84. Goldfrank L, Flomenbaum N, Lewin N, Weisman R, Howland MA, Kaul B. Anticholinergic poisoning. J Toxicol Clin Toxicol 1982;19:17–25.
85. Hughes JM, Blumenthal JR, Merson MH, Lombard GL, Dowell VR Jr, Gangarosa EJ. Clinical features of types A and B food-borne botulism. Ann Intern Med 1981;95:442–445.
86. Hart RG, Kanter MC. Acute autonomic neuropathy. Two cases and a clinical review. Arch Intern Med 1990;150:2373–2376.
87. Campbell WW Jr, Swift TR. Differential diagnosis of acute weakness. South Med J 1981;74:1371–1375.
88. Shelton GD. Myasthenia gravis and disorders of neuromuscular transmission. Vet Clin North Am Small Anim Pract 2002;32:189–206.
89. Schreiner MS, Field E, Ruddy R. Infant botulism: a review of 12 years' experience at the Children's Hospital of Philadelphia. Pediatrics 1991;87:159–165.
90. Li LY, Kelkar P, Exconde RE, Day J, Parry GJ. Adult-onset "infant" botulism: an unusual cause of weakness in the intensive care unit. Neurology 1999;53:891.
91. Bohnel H, Behrens S, Loch P, Lube K, Gessler F. Is there a link between infant botulism and sudden infant death? Bacteriological results obtained in Central Germany. Eur J Pediatr 2001;160:623–628.
92. Byard RW, Moore L, Bourne AJ, Lawrence AJ, Goldwater PN. *Clostridium botulinum* and sudden infant death syndrome: a 10 year prospective study. J Paediatr Child Health 1992;28;156–157.
93. Merson MH, Dowell VR. Epidemiologic, clinical and laboratory aspects of wound botulism. N Engl J Med 1973;289:1005–1010.
94. Akbulut D, Dennis J, Gent M, Grant KA, Hope V, Ohai C, et al. Wound botulism in injectors of drugs: upsurge in cases in England during 2004. Euro Surveill 2005;10: 172–174.
95. Cooper JG, Spilke CE, Denton M, Jamieson S. *Clostridium botulinum*: an increasing complication of heroin misuse. J Emerg Med 2005;12:251–252.
96. Burningham MD, Walter FG, Mechem C, Haber J, Ekins RB. Wound botulism. Ann Emerg Med 1994;24:1184–1187.
97. Dowell VR Jr, McCroskey LM, Hatheway CL, Lombard GL, Hughes JM, Merson MH. Coproexamination for botulinal toxin and *Clostridium botulinum*. JAMA 1977;238:1829–1832.
98. Ferreira JL, Eliasberg SJ, Edmonds P, Harrison MA. Comparison of the mouse bioassay and enzyme-linked immunosorbent assay procedures for the detection of type A botulinal toxin in food. J Food Prot 2004;67:203–206.
99. Lindstrom M, Keto R, Markkula A, Nevas M, Hielm S, Korkeala H. Multiplex PCR assay for detection and identification of *Clostridium botulinum* types A, B, E, and F in food and fecal material. Appl Environ Microbiol 2001;67:5694–5699.
100. Sharma SK, Whiting RC. Methods for detection of *Clostridium botulinum* toxin in foods. J Food Prot 2005;68: 1256–1263.
101. Horowitz W, editor. Official methods of analysis of AOAC International. 18th ed. Gaithersburg, MD: AOAC International; 2006.
102. Jones HR Jr, Darras BT. Acute care pediatric electromyography. Muscle Nerve 2000;9(Suppl):53–62.
103. King CK, Glass R, Bresee JS, Duggan C, Centers for Disease Control and Prevention. Managing acute gastroenteritis among children: oral rehydration, maintenance, and nutritional therapy. MMWR Morb Mortal Wkly Rep 2003;52:1–16.
104. Smith JL. Foodborne illness in the elderly. J Food Prot 1998;61:1229–1239.
105. Gomez HF, Johnson R, Guven H, McKinney P, Phillips S, Judson F, et al. Adsorption of botulinum toxin to activated charcoal with a mouse bioassay. Ann Emerg Med 1995;25:818–822.
106. Swenson JM, Thornsberry C, McCroskey LM, Hatheway CL, Dowell VR. Susceptibility of *Clostridium botulinum* to thirteen antimicrobial agents. Antimicrob Agents Chemother 1980;18:13–19.
107. Tacket CO, Shandera WX, Mann JM, Hargrett NT, Blake PA. Equine antitoxin use and other factors that predict

outcome in type A foodborne botulism. Am J Med 1984;76:794–798.

108. Sugiyama H. *Clostridium botulinum* neurotoxin. Microbiol Rev 1980;44:419–448.

109. Black RE, Gunn RA. Hypersensitivity reactions associated with botulinal antitoxin. Am J Med 1980;69:567–570.

110. Francisco AM, Arnon SS. Clinical mimics of infant botulism. Pediatrics 2007;119:826–828.

111. Thompson JA, Filloux FM, Van Orman CB, Swoboda K, Peterson P, Firth SD, et al. Infant botulism in the age of botulism immune globulin. Neurology 2005;64:2029–2032.

112. Arnon SS, Schechter R, Maslanka SE, Jewell NP, Hatheway CL. Human botulism immune globulin for the treatment of infant botulism. N Engl J Med 2006; 354:462–471.

113. U.S. Food and Drug Administration. Summary Basis of Approval, California Department of Health Services Botulism Immune Globulin Intravenous (Human) BabyBIG®. Rockville, MD: U.S. Food and Drug Administration; 2003 Oct 10:1–13.

114. Meunier FA, Schiavo G, Molgo J. Botulinum neurotoxins: from paralysis to recovery of functional neuromuscular transmission. J Physiol Paris 2002;96:105–113.

115. Department of Health and Human Services (DHHS). (2006). HHS awards bioshield contract for botulism antitoxin. Available at http://www.dhhs.gov/news/press/2006pres/20060601.html. Accessed March 18, 2008.

Chapter 16

CLOSTRIDIUM PERFRINGENS

CYRUS RANGAN, MD, FAAP

HISTORY

Clostridium perfringens (Veillon & Zuber) Hauduroy et al. (*C. perfringens*) is a strict anaerobic, gram-positive, spore-forming bacterium that causes a variety of human and veterinary diseases. In 1892, William H. Welch and George Nuttall discovered these bacteria in gangrenous wounds. At that time, they designated the bacteria as *Bacillus aerogenes capsulatus*. Foodborne disease from *C. perfringens* was first described by L.S. McClung in 1945.[1] Although *C. perfringens* is largely recognized as the chief cause of clostridial myonecrosis (gas gangrene), it is the third most commonly reported cause of human foodborne gastrointestinal disease in the United States after *Salmonella* spp. and *Staphylococcus aureus* with 10–20 reported annual outbreaks in the United States.[2] *C. perfringens* is serotyped A–E, based on distinct combinations of major and minor clostridial toxins (e.g., alpha, beta, epsilon, iota). Epsilon toxin is a bioterrorism select agent. Serotypes A and C are associated with human disease, whereas serotypes B, D, and E are associated with veterinary diseases. All five serotypes elaborate α-toxin (a variant of phospholipase C), which is the toxin responsible for clostridial myonecrosis. About 5% of serotype A isolates also elaborate *Clostridial perfringens* enterotoxin, which mediates the vast majority of foodborne diseases from *C. perfringens*.[3] Serotype C produces β-toxin that causes rare cases of human necrotic enteritis. In Germany and Denmark during WWII, outbreaks of this necrotic form of enteritis occurred in soldiers, who were given large meals after days to weeks of starvation. The disease was named "Darmbrand."[4] In the early 1960s in Papua New Guinea, outbreaks of enteritis necroticans occurred in natives after large pig feasts lasting several days. In the native dialect, this disease was called "Pigbel." In both settings, *C. perfringens* type C was isolated in peritoneal fluid and food.[5] Most reported foodborne outbreaks associated with *C. perfringens* are mild diseases caused by *C. perfringens* type A. The term "cafeteria diarrhea" has been applied to food poisoning outbreaks from *C. perfringens* because these outbreaks often occur following consumption of meals prepared in schools, prisons, hospitals, and cruise ships.

EXPOSURE

Source

ENVIRONMENT

Clostridial perfringens spores are ubiquitous in soil and sediment, especially in areas that receive runoff from animal or human feces.[6,7] This bacteria is part of the normal intestinal flora of both humans and animals. However, only a few strains carry the specific *C. perfringens* enterotoxin gene encoded for the production of the enterotoxin, which causes food poisoning.

ANIMAL

Clostridial perfringens colonizes and sporulates in the intestines of asymptomatic humans and various animals including poultry, cattle, and domestic pets.[8] These bacteria may also contaminate poultry feeds. A necrotic

Medical Toxicology of Natural Substances, by Donald G. Barceloux, MD

enteritis and a subclinical form of *C. perfringens* infections occur in poultry as a result of colonization by α-toxin producing type A strains and, to a lesser extent, by α- and β-toxin producing type C strains. Predisposing factors to the development of these diseases in poultry include coccidial-induced mucosal damage and diets high in indigestible, water soluble, nonstarch polysaccharides. Lactose supplementation and reduction of rye in the diets of poultry result in significant reductions in colonization by *C. perfringens*.[9]

Food

Undercooked or inadequately stored meats, poultry, meat stocks, stews, and gravies constitute the vast majority of food sources responsible for outbreaks of *C. perfringens*-induced food poisoning.[10] However, any food including nonanimal protein (e.g., spinach boiled with fried bean curd) that undergoes slow cooling or inadequate reheating (<65 °C/149 °F) is a potential source of human exposure to *Clostridium perfringens*.[11,12] In a US survey of about 900 raw, retail foods unassociated with any food poisoning outbreaks, approximately 1.4% of the foods contained *C. perfringens* isolates carrying the enterotoxin gene.[13] The development of *C. perfringens*-induced cases of food poisoning typically require high bacterial counts with isolates that contain the enterotoxin gene (*cpe*).[14]

Food Processing

Sufficient refrigeration and adequate reheating of meat-containing dishes stored after cooking are the best methods for preventing germination of spores.[15,16] The addition of table salt (NaCl) to uncooked meats (e.g., ham, beef) restricts growth and sporulation of *C. perfringens*.[17] Growth is also inhibited by the addition of organic acids (e.g., sodium citrate, sodium lactate) to vacuum-packaged foods.[18] Proper disinfection and the use of separate utensils and containers for cooked and uncooked food may help prevent cross-contamination of cooked food.[19,20]

TOXINS

Physiochemical Properties

Type A Clostridia Enterotoxin

Clostridial perfringens is a gram-positive, spore-forming anaerobic bacterial rod that germinates at a wide variety of temperatures (12 °C–50 °C/54 °F–122 °F) and pH (5.0–9.0). Spores survive domestic cooking (>60 °C/140 °F) and freezing (<4 °C/39 °F) temperatures. *C. perfringens* enterotoxin (CPE) is a 35 kDa heat-labile protein stored by *C. perfringens* type A in paracrystalline inclusion bodies during sporulation.[21] Upon lysis of the mother cell in 8–24 hours, both CPE and the spore are released.[22] The link between spore formation and CPE production is well established, and enterotoxin-negative spore-forming strains of *C. perfringens* do not cause food poisoning.[24] The gene (*cpe*), which encodes CPE, lies on a chromosome rather than a plasmid; therefore, *C. perfringens*-associated food poisoning usually requires a relatively large inoculation to produce enough CPE to cause disease. Although most cases of *C. perfringens*-induced food poisoning involve heat-resistant (i.e., 100 °C/212 °F for >10 minutes), *cpe*-chromosomal-borne strains, occasionally heat-sensitive, *cpe*-plasmid-borne strains cause outbreaks of food poisoning.[23]

In cases of sporadic (nonfood-related poisoning) diarrhea from *C. perfringens*, CPE is encoded on a plasmid that transfers to normal gastrointestinal flora for subsequent transcription.[24] Consequently, only a few organisms are required to cause disease.[25] Only a small percentage of *C. perfringens* isolates are positive for the *cpe* gene. *C. perfringens* type C elaborates a plasmid-encoded β-toxin that is easily degraded by trypsin produced by the pancreas. In contrast to the plasmid-encoded β-toxin, the presence of trypsin enhances CPE activity.[26] All *C. perfringens* serotypes elaborate α-toxin that is structurally similar to the enterotoxic phospholipases produced by *Bacillus cereus*.[27]

Epsilon Enterotoxin and Serotype B–E Clostridial Perfringens Toxins

Table 16.1 displays the serotypes of *C. perfringens* isolates and the associated diseases and toxins. Epsilon enterotoxin is a 33-kDa protein produced by serotypes B and D. These strains produce fatal hemorrhagic enteritis in lambs and goats.[28] This clostridial toxin is the third most lethal of all clostridial toxin, ranking behind only tetanus and botulinum neurotoxins. Epsilon toxin is considered a potential bioterrorism agent because of the ability of this toxin to produce diffuse edema, renal damage and neurotoxicity.[29] This toxin is synthesized as an inactive proto-toxin that requires proteolytic activation by intestinal proteases (trypsin, α-chymotrypsin) prior to systemic absorption.

Mechanism of Toxicity

Both spores and vegetative cells of *C. perfringens* are found in many foods. Cooking destroys vegetative cells, but spores survive to germinate when food is cooled to room temperature. *C. perfringens* type A produces *C. perfringens* enterotoxin that ultimately erodes the intestinal mucosa by disrupting epithelial tight junctions. The

TABLE 16.1. Diseases and Toxins Associated with Types of *Clostridial Perfringens*

Type	Major Toxins[a]	Diseases
A	α	Human myonecrosis (gas gangrene), necrotic enteritis in fowls, enterotoxemias of cattle and lambs, mild necrotizing enteritis of piglets
	α, CPE	Human food poisoning, human nonfoodborne gastrointestinal disease, veterinary diarrhea
	α, CPE, β_2	Human nonfoodborne gastrointestinal disease
	α, β_2	Porcine enteritis
B	α, β, ε	Dysentery in newborn lambs, chronic enteritis in older lambs, hemorrhagic enteritis in neonatal calves and foals, hemorrhagic enterotoxemias in sheep, biodefense concerns
C	α, β	Human necrotizing enteritis (Pigbel, Darmbrand), enteritis in hemorrhagic/necrotic enterotoxemias in neonatal pigs, lambs, calves goats, and foals, acute enterotoxemia (struck) in adult sheep
	α, β, β_2	Porcine enteritis
D	α, ε	Enterotoxemia in lambs (pulpy kidney disease) and calves, enterocolitis in neonatal and adult goats and cattle, biodefense concerns
E	α, ί	Canine, bovine, porcine enteritis

Source: From Ref 29.
[a]Major *C. perfringens* toxins include alpha (α), beta (β), epsilon (ε), and iota (ί), and enterotoxin (CPE).

lysis of the vegetative cells during sporulation releases *C. perfringens* enterotoxin into the intestines. The enteric toxins of *C. perfringens* are all single polypeptides of modest size (25–35 kDa) that form pores or channels in plasma membranes of host cells.[30] Structure and function of tight junctions are maintained by membrane-spanning proteins called claudins and occludins. *C. perfringens* enterotoxin (CPE) first induces shortening of the intestinal villi and desquamation of the epithelial lining, primarily in the ileum. Then CPE binds preferentially to claudins 3, 4, 6, 7, 8, and 14 (but not to claudins 1, 2, 5, or 10) on the apical membrane of epithelial cells. This specific interaction produces a large disruptive protein complex that stimulates the influx of calcium into the cell.[31] Calcium binds and activates calmodulin and calpain, releasing intracellular stores of calcium and the subsequent depolarization of mitochondrial membranes.[32] Mitochondrial cytochrome C is released into the cytosol, where an apoptosome (i.e., a moiety that mediates apoptosis) forms with the large CPE protein complex. The apoptosome activates intracellular caspase-3, which enters the cell nucleus and cleaves DNA, resulting in cell damage.[33] This damage exposes the basolateral membrane of the epithelial cell that houses the occludins. *C. perfringens* enterotoxin binds to these occludins, thus finalizing the compromised integrity of epithelial tight junctions. Major alterations in cell permeability ensue, as water, proteins, and fragmented epithelial cells pour into the intestinal lumen.[34] Mucosal tissue degradation by CPE is aided by α-toxin-mediated binding to phosphatidylcholine and sphingomyelinase on intestinal cell membranes. CPE is structurally similar to the phospholipases produced by *Bacillus cereus*.[27] *C. perfringens* type C β-toxins are internalized by epithelial cells, where they oligomerize to form pores in the basolateral membrane and induce cell death.[35] There are limited data suggesting that β-toxins produce neurotoxic effects, resulting in damage to preganglionic sympathetic fibers and subsequent vasoconstriction.[36] A chronic low-protein diet (e.g., sweet potatoes rich in heat-stable trypsin inhibitors) predisposes individuals to the toxic effects of β-toxin.[37]

DOSE RESPONSE

The onset of gastrointestinal symptoms requires significant bacterial replication in the gastrointestinal tract, characterized by stool colony counts of $\geq 10^6$ CFU/g. In a human volunteer study, an inoculum of a minimum 5×10^8 viable vegetative *C. perfringens* bacteria were necessary to produce diarrheal symptoms in 100% of subjects.[14]

CLINICAL RESPONSE

Clostridial perfringens food poisoning is characterized by abdominal cramping and diarrhea with or without blood after an incubation period of 8–24 hours. Fewer than 20% of patients experience fever or vomiting. Symptoms usually persist for 24 hours.[38] The presentation is clinically similar to the diarrheal syndrome caused by enterotoxigenic strains of *B. cereus*.[39] Case reports indicate that potential fatal acute necrotizing enterocolitis can occur following foodborne CPE-positive *C. perfringens* type A infection.[40,41] Food poisoning from *C. perfringens* serotype C (pigbel) is characterized by bloody diarrhea, fever, and rare cases of bowel ischemia, perforation, and peritonitis.[42,43]

DIAGNOSTIC TESTING

Analytical Methods

The symptoms associated with *C. perfringens* are usually self-limiting, and routine analytical testing for *C. perfringens* is unnecessary. Verification of these bacteria as the source of food poisoning requires identification beyond the species level because of the ubiquitous presence of these bacteria in humans, animals, soil, and surface water. Bacterial strain typing methods aid the epidemiological investigations of the sources, limit the spread of the food poisoning outbreak, and supply information on the contamination of food for the improvement of food distribution. For epidemiological purposes, traditional phenotypical methods (biotyping, serotyping and phage typing of isolates, antimicrobial susceptibility testing) provide insufficient information to track the source of a food poisoning epidemic in contrast to molecular methods. During epidemiological studies, isolation of *C. perfringens* strains from stool, vomit, or leftover food is performed by molecular methods including hydrophobic grid membrane filter-colony hybridization (HGMF-CH), pulse-field gel electrophoresis (PFGE), ribotyping,[44] and plasmid gene profiling.[45,46] The toxins of *C. perfringens* can also be identified by latex agglutination and by enzyme-linked immunosorbent assay (ELISA).[47] Bacteria can be analyzed for the presence of the *cpe* gene by polymerase chain reaction (PCR) and Western blot if the toxin is not identified directly.[23,48] In contrast to serological methods, PCR-based assays do not require the isolates to sporulate *in vitro* before detection. However, PCR identifies *C. perfringens* isolates as *potentially* enterotoxigenic because some isolates carry a silent, unexpressed *cpe* gene.

Biomarkers

Confirmation of *C. perfringens*-induced food poisoning includes direct assay for CPE by reversed passive latex agglutination test,[49] detection of $\geq 10^5$ *cpe*-positive colony count/g in food, $\geq 10^6$ *cpe*-positive colony count/g in stool, and type-to-type serotype confirmation of stool to food.[14] The presence of all four criteria provides definite proof of *C. perfringens* food poisoning. However, sufficient testing material may not be present to meet all four criteria, and a presumptive diagnosis of *C. perfringens* food poisoning can be made with fewer criteria.

Abnormalities

Clostridial perfringens food poisoning can produce severe serum electrolyte abnormalities including hypernatremia (serum Na > 150 mEq/L), hyponatremia (serum Na < 130 mEq/L), hypokalemia (serum K < 3.0 mEq/L), and hypochloremia, as well as elevated urine specific gravity (SpG > 1.020). Low serum bicarbonate results from diarrhea-induced, chloride-responsive, non-anion gap metabolic acidosis. Other laboratory abnormalities include lactic acid production or elevated bicarbonate secondary to contraction alkalosis. The short duration of disease induced by *C. perfringens* limits the severity of fluid and electrolyte imbalances that usually resolve spontaneously.[50]

TREATMENT

Treatment is primarily supportive including rehydration and bowel rest for the diarrheal syndrome. Oral rehydration is usually sufficient, but intravenous fluid replacement and hospitalization may be necessary in cases of severe dehydration or severe electrolyte abnormalities, particularly in young children or elderly patients.[51,52] There are no clinical trials that demonstrate an improvement in clinical outcome, avoidance of hospitalization, or shortened duration of symptoms following the use of either antiemetics or antidiarrheal agents. The short duration of symptoms usually limits the need for these medications; however, some patients may experience symptomatic relief with these agents when symptoms are severe. Although antimicrobials are beneficial in cases of clostridial myonecrosis, these drugs are not routinely used in most cases of *C. perfringens* foodborne poisoning, even when symptoms are prolonged.[35] Enteritis necroticans should be evaluated with prompt surgical consultation and aggressive supportive care to maintain fluid and electrolyte balance.[42]

References

1. Smyser CF, Cleverdon RC, Kulp WL. The use of mice in a modified Welch-Nuttall test for the isolation of *Clostridium perfringens*. Am J Clin Pathol 1952;22:1036–1037.
2. McCabe-Sellers BJ, Beattie SE. Food safety: emerging trends in foodborne illness surveillance and prevention. J Am Diet Assoc 2004;104:1708–1717.
3. Daube G, Simon P, Limbourg B, Manteca C, Mainil J, Kaeckenbeeck A. Hybridization of 2,659 *Clostridium perfringens* isolates with gene probes for seven toxins (alpha, beta, epsilon, iota, theta, mu, and enterotoxin) and for sialidase. Am J Vet Res 1996;57:496–501.
4. Tevene AM, Dari P. On an unusual observation of "Darmbrand" type acute segmental enteritis. Minerva Chir 1960;15:1215–1221.
5. Shepherd A. Clinical features and operative treatment of pigbel–enteritis necroticans. P N G Med J 1979;22:18–23.

6. Fujioka RS. Monitoring coastal marine waters for spore-forming bacteria of faecal and soil origin to determine point from non-point source pollution. Water Sci Technol 2001;44:181–188.
7. Thurston-Enriquez JA, Gilley JE, Eghball B. Microbial quality of runoff following land application of cattle manure and swine slurry. J Water Health 2005;3:157–171.
8. Van Immerseel F, De Buck J, Pasmans F, Huyghebaert G, Haesebrouck F, Ducatelle R. *Clostridium perfringens* in poultry: an emerging threat for animal and public health. Avian Pathol 2004;33:537–549.
9. Takeda T, Fukata T, Miyamoto T, Sasai K, Baba E, Arakawa A. The effects of dietary lactose and rye on cecal colonization of *Clostridium perfringens* in chicks. Avian Dis 1995;39:375–381.
10. Ochiai H, Ohtsu T, Tsuda T, Kagawa H, Kawashita T, Takao S, et al. *Clostridium perfringens* foodborne outbreak due to braised chop suey supplied by chafing dish. Acta Med Okayama 2005;59:27–32.
11. Miwa N, Masuda T, Terai K, Kawamura A, Otani K, Miyamoto H. Bacteriological investigation of an outbreak of *Clostridium perfringens* food poisoning caused by Japanese food without animal protein. Int J Food Microbiol 1999;49:103–106.
12. Taormina PJ, Bartholomew GW, Dorsa WJ. Incidence of *Clostridium perfringens* in commercially produced cured raw meat product mixtures and behavior in cooked products during chilling and refrigerated storage. J Food Prot 2003;66:72–81.
13. Wen Q, McClane BA. Detection of enterotoxigenic *Clostridium perfringens* type A isolates in American retail foods. Appl Environ Microbiol 2004;70:2685–2691.
14. Ackermann G, Thomalla S, Ackermann F, Schaumann R, Rodloff AC, Ruf BR. Prevalence and characteristics of bacteria and host factors in an outbreak situation of antibiotic-associated diarrhoea. J Med Microbiol 2005;54:149–153.
15. Andersson A, Ronner U, Granum PE. What problems does the food industry have with the spore-forming pathogens *Bacillus cereus* and *Clostridium perfringens*? Int J Food Microbiol 1995;28:145–155.
16. Kalinowski RM, Tompkin RB, Bodnaruk PW, Pruett WP Jr. Impact of cooking, cooling, and subsequent refrigeration on the growth or survival of *Clostridium perfringens* in cooked meat and poultry products. J Food Prot 2003;66:1227–1232.
17. Zaika LL. Influence of NaCl content and cooling rate on outgrowth of *Clostridium perfringens* spores in cooked ham and beef. J Food Prot 2003;66:1599–1603.
18. Sabah JR, Thippareddi H, Marsden JL, Fung DY. Use of organic acids for the control of *Clostridium perfringens* in cooked vacuum-packaged restructured roast beef during an alternative cooling procedure. J Food Prot 2003;66:1408–1412.
19. Reij MW, Den Aantrekker ED; ILSI Europe Risk Analysis in Microbiology Task Force. Recontamination as a source of pathogens in processed foods. Int J Food Microbiol 2004;91:1–11.
20. Redmond EC, Griffith CJ. Consumer food handling in the home: a review of food safety studies. J Food Prot 2003;66:130–161.
21. Loffler A, Labbe R. Characterization of a parasporal inclusion body from sporulating, enterotoxin-positive *Clostridium perfringens* type A. J Bacteriol 1986;165:542–548.
22. Tseng W, Labbe RG. Characteristics of a sporulation stimulating factor from *Clostridium perfringens* type A. Lett Appl Microbiol 2000;30:254–257.
23. Nakamura M, Kato A, Tanaka D, Gyobu Y, Higaki S, Karasawa T, et al. PCR identification of the plasmid-borne enterotoxin gene (cpe) in *Clostridium perfringens* strains isolated from food poisoning outbreaks. Int J Med Microbiol 2004;294:261–265.
24. Lukinmaa S, Takkunen E, Siitonen A. Molecular epidemiology of *Clostridium perfringens* related to food-borne outbreaks of disease in Finland from 1984 to 1999. Appl Environ Microbiol 2002;68:3744–3749.
25. Cornillot E, Saint-Joanis B, Daube G, Katayama S, Granum PE, Canard B, et al. The enterotoxin gene (cpe) of *Clostridium perfringens* can be chromosomal or plasmid-borne. Mol Microbiol 1995;15:639–647.
26. Granum PE, Whitaker JR, Skjelkvale R. Trypsin activation of enterotoxin from *Clostridium perfringens* type A: fragmentation and some physicochemical properties. Biochim Biophys Acta 1981;668:325–332.
27. Leslie D, Fairweather N, Pickard D, Dougan G, Kehoe M. Phospholipase C and haemolytic activities of *Clostridium perfringens* alpha-toxin cloned in *Escherichia coli*: sequence and homology with a *Bacillus cereus* phospholipase C. Mol Microbiol 1989;3:383–392.
28. Songer JG. Clostridial enteric diseases of domestic animals. Clin Microbiol Rev 1996;9:216–234.
29. Smedley JG III, Fisher DJ, Sayeed S, Chakrabarti G, McClane BA. The enteric toxins of *Clostridium perfringens*. Rev Physiol Biochem Pharmacol 2004;152:183–204.
30. Smedley JG, Uzal FA, McClane BA. Identification of a prepore large-complex stage in the mechanism of action of *Clostridium perfringens* enterotoxin. Infect Immun 2007;75:2381–2390.
31. Sonoda N, Furuse M, Sasaki H, Yonemura S, Katahira J, Horiguchi Y, et al. *Clostridium perfringens* enterotoxin fragment removes specific claudins from tight junction strands: Evidence for direct involvement of claudins in tight junction barrier. J Cell Biol 1999;147:195–204.
32. Chakrabarti G, McClane BA. The importance of calcium influx, calpain and calmodulin for the activation of CaCo-2 cell death pathways by *Clostridium perfringens* enterotoxin. Cell Microbiol 2005;7:129–146.
33. Chakrabarti G, Zhou X, McClane BA. Death pathways activated in CaCo-2 cells by *Clostridium perfringens* enterotoxin. Infect Immun 2003;71:4260–4270.

34. McClane BA. The complex interactions between *Clostridium perfringens* enterotoxin and epithelial tight junctions. Toxicon 2001;39:1781–1791.
35. Tweten RK. *Clostridium perfringens* beta toxin and *Clostridium septicum* alpha toxin: their mechanisms and possible role in pathogenesis. Vet Microbiol 2001;82:1–9.
36. Sakurai J, Nagahama M, Ochi N. Major toxins of *Clostridium perfringens*. J Toxicol Toxin Rev 1997;16:195–214.
37. Sakurai J, Nagahama M. *Clostridium perfringens* beta-toxin: characterization and action. Toxin Rev 2006;25:89–108.
38. Meer RR, Songer JG, Park DL. Human disease associated with *Clostridium perfringens* enterotoxin. Rev Environ Contam Toxicol 1997;150:75–94.
39. Laohachai KN, Bahadi R, Hardo MB, Hardo PG, Kourie JI. The role of bacterial and non-bacterial toxins in the induction of changes in membrane transport: implications for diarrhea. Toxicon 2003;42:687–707.
40. Mandrella B. A recent outbreak of necrotizing enteritis in eastern Sri Lanka. Trop Doct 2007;37:52–54.
41. Sobel J, Mixter CG, Kolhe P, Gupta A, Guarner J, Zaki S, et al. Necrotizing enterocolitis associated with *clostridium perfringens* type A in previously healthy North American adults. J Am Coll Surg 2005;201:48–56.
42. Clarke LE, Diekmann-Guiroy B, McNamee W, Java DJ Jr, Weiss SM. Enteritis necroticans with midgut necrosis caused by *Clostridium perfringens*. Arch Surg 1994;129:557–560.
43. Zhao W, Daroca PJ Jr, Crawford BE. Clostridial enteritis necroticans versus secondary clostridial infection superimposed upon ischemic bowel disease. J La State Med Soc 2002;154:251–255.
44. Schalch B, Sperner B, Eisgruber H, Stolle A. Molecular methods for the analysis of *Clostridium perfringens* relevant to food hygiene. FEMS Immunol Med Microbiol 1999;24:281–286.
45. Heikinheimo A, Lindstrom M, Korkeala H. Enumeration and isolation of cpe-positive *Clostridium perfringens* spores from feces. J Clin Microbiol 2004;42:3992–3997.
46. Lukinmaa S, Nakari UM, Eklund M, Siitonen A. Application of molecular genetic methods in diagnostics and epidemiology of food-borne bacterial pathogens. APMIS 2004;112:908–929.
47. Petit L, Gibert M, Popoff MR. *Clostridium perfringens*: toxinotype and genotype. Trends Microbiol 1999;7:104–110.
48. Wen Q, Miyamoto K, McClane BA. Development of a duplex PCR genotyping assay for distinguishing *Clostridium perfringens* Type A isolates carrying chromosomal enterotoxin (*cpe*) genes from those carrying plasmid-borne enterotoxin (*cpe*) genes. J Clin Microbiol 2003;41:1494–1498.
49. Kato N, Kato H. Human diseases caused by exotoxins produced by anaerobes and their rapid detection. Rinsho Biseibutshu Jinsoku Shindan Kenkyukai Shi 1998;9:97–104.
50. Guerrant RL, Van Gilder T, Steiner TS, Thielman NM, Slutsker L, Tauxe RV, et al. Infectious Diseases Society of America: Practice guidelines for the management of infectious diarrhea. Clin Infect Dis 2001;32:331–351.
51. King CK, Glass R, Bresee JS, Duggan C, Centers for Disease Control and Prevention (CDC). Managing acute gastroenteritis among children: oral rehydration, maintenance, and nutritional therapy. MMWR Recomm Rep 2003;52:1–16.
52. Smith JL. Foodborne illness in the elderly. J Food Prot 1998;61:1229–1239.

Chapter 17

ESCHERICHIA COLI

CYRUS RANGAN, MD, FAAP

HISTORY

Escherichia coli (Migula) Castellani and Chalmers (*E. coli*) are lactose-fermenting, gram-negative, facultative anaerobic rods, first isolated and characterized by Austrian pediatrician and bacteriologist Theodor Escherich in 1885.[1] Since that time, over 170 serogroups of *E. coli* have been identified that are distinguished by the expression of O (somatic), H (flagellar), and/or K (capsular) antigens. Over the last part of the 20th century, several groups of pathogenic *E. coli* have been identified. Of these serotypes, 43 enterohemorrhagic serotypes of *E. coli* (EHEC) are associated with bloody diarrhea.[2] *E. coli* O157:H7 has emerged as the most notorious enterohemorrhagic *E. coli* serotype based on the threat of these infections to public health. The first recognized outbreak of *E. coli* O157:H7 occurred in 1982, when an outbreak of food poisoning ("hamburger diarrhea") resulted from the contamination of hamburger with this *E. coli* serotype.[3] In the United States, *E. coli* O157:H7 remains among the most common organism isolated from patients with bloody stool specimens.[4] Enteropathogenic *E. coli* were the first disease-causing strains of *E. coli* identified, and these strains of *E. coli* remain a common cause of diarrhea in children ("infant diarrhea"), particularly in developing nations. During the 1960s, enterotoxigenic strains of *E. coli* were recognized as a cause of "traveler's diarrhea." Enterotoxigenic strains of *E. coli* remain a major cause of travel-associated diarrhea (Asia, Mexico, Latin America, Africa), as well as gastrointestinal illness in military settings. Recently, enteroaggregative strains of *E. coli* have been associated with persistent watery diarrhea in developing countries. Each year, 73,000 cases of pathogenic *E. coli* illness are reported worldwide, resulting in over 60 deaths.[5]

EXPOSURE

Classification

Enterovirulent strains of *E. coli* are important causes of foodborne illness, and there are four major groups: enterohemorrhagic (EHEC), enteropathogenic (EPEC), enteroinvasive (EIEC), and enterotoxigenic (ETEC). In addition, there are two new enterovirulent groups that include enteroaggregative (EAggEC), and diffusely adhering (DAEC) *E. coli*. In contrast to other pathogenic *E. coli* groups, the toxicity of EIEC is not limited to the intestinal mucosal surface.

Source

ENVIRONMENT

Worldwide, the most common environmental sources of *E. coli* infections are drinking water, feces, and sewage. In particular, public swimming pools and lakes are frequent sources of community exposure to *E. coli*.[6] Environmental contamination may persist for up to 10 months, resulting in latent infections.[7] Chlorination adequately minimizes the risk of *E. coli* in drinking water.[8]

ANIMAL

Escherichia coli survive in the intestines of various animal reservoirs including rabbits,[9] sheep,[10] healthy cattle, deer,[11] and humans.[12] These reservoirs potentially include both ruminants (cattle, sheep, goats, buffalo, deer, elk, giraffes, camels) and monogastric animals (humans, pigs, chickens, rabbits). Food and water sources may be contaminated directly by human or animal feces. Well-fed cattle with high volatile fatty acid diets, low intestinal pH, and low lactate concentrations usually have lower *E. coli* colony counts than fasted, poorly fed cattle.[13] In animal models, the consumption of corn contaminated by the mycotoxin, fumonisin B_1, is associated with increased intestinal colonization of *E. coli*.[14] Concomitant ingestion of *E. coli* and ochratoxin increases mortality in chickens.[15] Cattle are resistant to the effects of *E. coli*-derived toxins as a result of the difference in expression of toxin receptors between humans and animal reservoirs.[16,17]

FOOD

Humans contract *E. coli* from raw food, processed food, and beverages. Person-to-person transmission during outbreaks of *E. coli* infections is common, particularly in day care centers and nursing homes.[18] Foods left outdoors for extended periods of time are particularly vulnerable to *E. coli* contamination. Attendance at agricultural fairs is an epidemiological risk factor, and may contribute significantly to summer peak incidence of *E. coli* exposures in the United States.[19] In addition to food, *E. coli* contaminates herbal dietary supplements, when bacterial contamination results from inappropriate manufacturing methods.[20] The distribution of dietary supplements in the United States is regulated by the Dietary Supplement Health and Education Act of 1994. Good manufacturing practices have been established for the formulation of these herbal dietary supplements including a zero tolerance for the presence of pathogens.[21] Infections from *E. coli* O157:H7 strains occur sporadically in the United States, and occasionally outbreaks result from the ingestion of contaminated food (e.g., raw spinach).[22]

Although ground beef is a common source of EHEC, other food sources include potatoes, alfalfa sprouts, unpasteurized fruit juice, dry-cured salami, deer jerky, lettuce, game meat, cheese curds, artichokes, raw unpasteurized cow's milk, hard taco shells, coleslaw, fruit salads, and drinking water.[23–26] Typical foods associated with contamination by EPEC and EIEC are not well defined. Potential sources of contamination by these *E. coli* serotypes include raw meats, unpasteurized milk, imported cheese, guacamole, and contaminated water, or any other food source exposed to human or animal feces. Infantile disease associated with these serotypes suggests the use of *E. coli*-contaminated water for the mixing of bottled baby formula.[27] Human feces is the primary source of ETEC; potential sources include dairy products, semisoft cheeses, raw fruits and vegetables, tabbouleh, parsley, raw seafood, meat, poultry, and untreated water. Outbreaks of ETEC in the United States have occurred at a Mexican restaurant, a hospital cafeteria, and on a cruise ship due to ETEC-contaminated food and water.[27,28]

Food Processing

All phases of food handling and processing permit the potential entry of *E. coli* into the food supply. *E. coli* does not survive in temperatures above 155 °F (68 °C) during cooking, and *E. coli* is sensitive to virtually all commercially available alkaline cleaners in the food industry. However, exposure to a cleaner confers some cross-protection to heat.[29] Contamination of produce occurs following contact with cattle manure-based fertilizers, improperly sanitized equipment, or inadequately cleaned produce. Cattle feed derived from chicken manure must be properly stored for 4–8 weeks, during which time naturally produced temperatures of 160 °F–170 °F (71 °C–77 °C) destroy bacteria including *E. coli*. Irrigation water increases the risk of contamination when recycled water is used on fertilized fields. Because *E. coli* can thrive on both food and on unclean food handling equipment, public health regulations require food handlers to maintain short fingernails and to scrub the nails with soap and nailbrush.[30] Proper hygiene during harvesting, slaughtering, transportation, storage, and packaging are extremely important to reduce the risk of contamination. High risk of *E. coli* contamination occurs during certain food processes, such as the reworking of ground beef when leftover beef is stored until the next day's production run. This process also reduces the ability to trace sources of *E. coli* when outbreaks of food poisoning occur. Although irradiation is an approved method for infection control in chicken and produce, this process is not widely used as a result of issues related to consumer concern and acceptance. *E. coli* cross-contamination occurs also during food transportation and storage, especially if vehicles or facilities are used alternately for both fertilizer and food. Refrigeration can slow the rate of growth of *E. coli*, but this process does not kill bacteria or bacterial toxins, and EHEC survives freezing in ground beef. Therefore, refrigeration does not alter previous *E. coli* contamination.

Escherichia coli O157:H7 contains a genetic sequence that controls intrinsic resistance to multiple structurally and functionally unrelated antibiotics added during food processing.[31] The hemolytic, antimicrobial peptide dermaseptin is optimally bactericidal to *E. coli* at high NaCl concentrations, low pH, and temperature extremes. Consequently, dermaseptin is a useful food preservative for multiple phases of food processing.[32] Nontoxic, probiotic *E. coli* strains produce colicin, a natural antibacterial compound that has the potential to compete for pathogenic *E. coli* in cattle reservoirs.[33] Experimental *in vitro* evidence suggests the efficacy of the combination of nisin and cinnamon to inactivate *E. coli* in homemade apple juice.[34]

TOXINS

Physiochemical Properties

Enterohemorrhagic *Escherichia Coli*

Bacteria-derived effector molecules and a special mechanism called the type-III secretion system (TTSS) determine the virulence of EHEC. Diffusely adhering *E. coli* also possess a TTSS mechanism, making it a potentially virulent form of *E. coli*.[35] EHEC and EPEC genomes contain a 35-kbp pathogenicity island called the locus of enterocyte effacement (LEE), which codes for three protein sets: adhesive intimin on the outer bacterial membrane;[36] *E. coli* secretory proteins (*EspA*, *EspB*, and *EspD*);[37] and translocated intimin receptor (*Tir*). These three proteins interact to anchor bacteria to intestinal epithelial cells.[38] EHEC contain additional loci encoding enterotoxins called Shiga toxins (*Stx1* and *Stx2*), which are nearly identical to the enterotoxins of *Shigella flexneri*.[39,40] The structure (A subunit linked to five B subunits) and functions of these Shiga toxins and cholera toxin are similar. EHEC also produce a colonic acid slime polysaccharide to combat extremes in physiologic stress such as pH and temperature,[41] adding the inherent ability of *E. coli* to survive gastric acid.[42]

Enteropathogenic *Escherichia Coli*

EPEC possess an identical TTSS, secreting the same series of proteins that enable it to anchor to intestinal epithelial cells. However, EPEC do not elaborate toxins.

Enteroinvasive *Escherichia Coli*

EIEC contain an invasiveness plasmid that is identical to a plasmid found in *Shigella* species. This genetic material encodes for proteins that permit epithelial invasion, bacterial proliferation, and ultimate destruction of epithelial cells. EIEC probably developed from a hybrid strain of *E. coli* and *Shigella* species, though it does not elaborate Shiga enterotoxins.[43]

Enterotoxigenic *Escherichia Coli*

ETEC exhibit specialized pili and fimbriae termed colonization-factor antigens (CFAs), comprised of alpha helical arrangements of lectin adhesins that enable the bacteria to colonize epithelial cell surfaces in the small intestine. The ETEC genome codes for two distinct polypeptides, a heat-stable enterotoxin (ST), and a heat-labile enterotoxin (LT), which are secreted via a type-II secretion system. ETEC strains may synthesize one or both of these toxins. LT bears a 77% genotypic resemblance to *Vibrio cholera* toxin and functions similarly.[44]

Mechanism of Toxicity

Enterohemorrhagic *Escherichia Coli*

EHEC induce attachment and effacement (A/E) lesions in the intestine, followed by the elaboration of Shiga toxins that cause hemorrhagic diarrhea. EHEC initially cause A/E lesions in the terminal ileum and colon. These lesions are characterized by rapid colonization and binding to the intestinal epithelium, followed by a rearrangement of the actin scaffolding of intestinal villi cells.[45,46] EHEC may initially communicate with host cells via an auto-inducer protein that cross-communicates with epinephrine secreted from intestinal cells.[47] Phentolamine, propanolol, and haloperidol interfere with the action of intestinal norepinephrine and dopamine, and diminish adherence of EHEC to cecal epithelial cells *in vitro*.[48] EHEC make first contact with epithelial cells via fimbriae on the outer bacterial membrane; these fimbriae also mediate bacteria aggregation and formation of microcolonies.[49,50] Advanced postnatal age increases adherence, due to progressive postnatal development of gut receptors for these fimbriae.[51] The type-III secretory system releases secretory proteins, beginning with *EspA*, which forms a filamentous, needle-like chute. *EspB* is exported down the chute to the surface of the intestinal epithelial cell, where it punches pores into the cell membrane. Translocated intimin receptor (*Tir*) travels down the chute and settles into the pores. *Tir* then anchors adhesive intimin that coats the outer membranes of EHEC to complete the process of the attachment.[52] Adherence stimulates host cell mediators to promote the release of stored intracellular calcium, which stimulates the enzyme villin to depolymerize the scaffold of cytoplasmic actin filaments supporting the cytoskeleton of the intestinal villi cells. Free actin microfilaments are then recruited, and they accumulate directly beneath bacteria (F-actin aggregation). In this location, they are cross-linked by the enzymes α-actinin, talin, erzin, and villin to form pedestals, completing the process of effacement and destruction of microvilli.[53] EHEC also injects cyclomodulin into host cells, halting progression to mitosis and slowing down epithelial cell turnover, which is an important human defense mechanism against gastrointestinal infections. High HCO_3 concentration increases production of intimin, *EspA*, *EspB*, and *Tir*, explaining the predilection of EHEC for the distal portions of the intestinal tract.[54]

The B subunit of the toxin binds with high affinity to Gb3 receptors on epithelial and endothelial cells. Once attachment and effacement occur, Shiga toxins bind to the membrane glycolipid globotriaosylceramide, cross the intestinal epithelial barrier, and injure host cells via inhibition of protein synthesis, stimulation of prothrombotic messages, and induction of apoptosis. After endocytosis, the intact toxin travels to the endoplasmic reticulum via the Golgi apparatus, where cleavage of the A1 and A2 subunits occurs.[55] Release of the A1 subunit into the cytosol causes the deactivation of ribosomal RNA on the 60S ribosomal subunit. Cellular protein machinery thereby shuts down, leading to cell death, edema, hemorrhage, and subsequent bloody diarrhea.[56] EHEC also suppresses mucosal inflammation and proliferation of intraepithelial lymphocytes.[57] Shiga toxins stimulate production of proinflammatory mediators IL-1-B, IL-6, TNF-alpha, gamma-interferon, and von Willebrand factor (vWF); and suppress antiinflammatory cytokines IL-2 and IL-10. Hemolytic-uremic syndrome (HUS) is an acquired Coombs-negative hemolytic anemia with thrombocytopenia and renal failure. Following at least 6 months of recovery from hemolytic-uremic syndrome, these patients demonstrate a similar cytokine profile, suggesting a possible immunogenetic predisposition to this syndrome.[58]

Hemolytic-uremic syndrome occurs as a result of the translocation of Shiga toxins into circulatory system and the transport of these toxins by leukocytes to capillary endothelium of the renal vasculature, where they bind to renal globotriaosylceramide and are endocytosed.[39] In patients with hemolytic-uremic syndrome, plasminogen activator (vWF) and platelet aggregating factor are released from the renal endothelium, forming platelet-fibrin-hyaline thrombi at the injured sites. Platelets are consumed, and circulating erythrocytes are deformed and fragmented as they move through the occluded microvasculature. The spleen then removes the damaged erythrocytes from the circulation. Children have Gb3 receptors in multiple extraintestinal loci, namely the glomerulus. Shiga toxin binds to both glomeruli and renal tubules in toddlers, but only to renal tubules in the elderly, suggesting that the binding of *Stx* to glomeruli in children may be responsible for the appearance of hemolytic-uremic syndrome.[59] Although EHEC is clearly associated with hemolytic-uremic syndrome, the evidence linking Shiga toxin to the pathology associated with HUS remains largely indirect.

Thrombotic thrombocytopenic purpura (TTP) is a potential sequela of EHEC infection in adults. This illness shares some of the clinical features (e.g., fever, altered mental status) of hemolytic-uremic syndrome. The causes of these two illnesses are uncertain, but both begin with a similar pattern of vascular endothelial injury, resulting in microvascular lesions, platelet-fibrin-hyaline microthrombi, occluded arterioles and capillaries, and a consumptive thrombocytopenia. Microthrombi occlude the vasculature of multiple organs, including the skin, intestines, skeletal muscle, pancreas, spleen, adrenal glands, heart, and most prominently in the central nervous system, kidney, and liver.

Enteropathogenic Escherichia Coli

EPEC induce similar attachment and effacement pathology seen in EHEC in the terminal ileum and colon, but EPEC do not elaborate toxins. Instead, they induce changes in Na^+/H^+ isoform activity, leading to the accumulation of intraluminal sodium and water, and subsequent watery diarrhea.[60]

Enteroinvasive Escherichia Coli

EIEC physically invade intestinal epithelial cells of the colon, causing bloody, *Shigella*-like dysentery. A large 140–MDa plasmid confers *Shigella*-like dysentery, similar to the plasmid associated with *Shigella* gastroenteritis. Micro ulcers are produced in the wake of cell death, liberating erythrocytes, leucocytes, mucus, cell debris, and bacterial progeny. After cell death, EIEC invades superficial tissue, but neutrophils and macrophages quickly phagocytose organisms, preventing systemic dissemination. Extrusion and regeneration of colonic epithelial calls limit the disease process to 5–7 days.[43]

Enterotoxigenic Escherichia Coli

ETEC causes a secretory diarrhea mediated by bacterial endotoxins. ST and/or LT enterotoxins are elaborated after bacterial colonization on intestinal epithelial cells. LT activates adenyl cyclase, leading to increased cAMP, activation of cAMP-dependent protein kinases, and phosphorylation of the cystic fibrosis transmembrane conductance regulator. ST directly stimulates guanylate cyclase on brush border membranes of the jejunum and ileum, increasing cellular cGMP. Chloride ion flow is subsequently reversed, leading to an accumulation of chloride, sodium, and water in the intestinal lumen.[44,45]

Enteroaggregative Escherichia Coli

EAggEC exhibits a "stacked brick" adherence pattern on intestinal epithelial cells in the colon without invasion. EAggEC stool cultures are positive for fecal lactoferrin, which supports the notion that EAggEC infection is associated with an intestinal inflammatory response.[61] EAggEC produces heat-stable and heat-labile toxins that induce watery, intestinal secretions.[62]

DOSE RESPONSE

There are limited data on the inoculation size of various *E. coli* groups necessary to cause toxicity based primarily on basic research done with *Shigella* species. EHEC and EIEC require as little as 10 organisms to cause severe clinical effects, based on their functional similarities to *Shigella* spp. in epidemiological studies.[63] EPEC requires at least 1 million organisms, and ETEC requires 100 million to 10 billion organisms to elaborate sufficient toxin to produce clinical effects.[64] Minimum inoculations of EAggEC and DAEC are not known.

CLINICAL RESPONSE

Enterohemorrhagic *Escherichia Coli*

Enterohemorrhagic *E. coli* is often used synonymously with *E. coli* serotype O157:H7. EHEC is a common cause of bacterial diarrhea, particularly in children under four years of age. The typical presentation of EHEC involves the initial onset of watery diarrhea followed by the development of severe, frequently bloody stools and abdominal cramps with mild or no fever. Infection with this strain is usually associated with hemorrhagic colitis despite the absence of bloody diarrhea in approximately 25% of recognized cases.[65] Rare case reports associate infection by *E. coli* serotype O157:H7 with ischemic pseudomembranous colitis.[66] The differential diagnosis for EHEC includes gastrointestinal hemorrhage, acute mesenteric infarction, intussusception, inflammatory bowel disease, intestinal polyps, Meckel's diverticulum, and infectious enteritis (*Campylobacter*, *Salmonella*, *Shigella*). An asymptomatic carrier state can occur following the resolution of the gastrointestinal illness.

Hemolytic-Uremic Syndrome

Children with gastrointestinal exposure to EHEC may develop hemolytic-uremic syndrome, but this illness is rare in adults. In EHEC-induced hemolytic-uremic syndrome, 10% of cases do not develop initial gastrointestinal symptoms.[67] After an incubation period of 3–5 days, patients typically have an initial 3–4-day phase of watery diarrhea, dehydration, and severe cramps, followed by bloody diarrhea.[67] Vomiting and low-grade fevers occur occasionally. Gastrointestinal symptoms are self-limited, and usually resolve 1–14 days after the onset of symptoms. However, patients can shed EHEC for 1–2 weeks after symptoms resolve. In young children, 10–15% of cases progress to hemolytic-uremic syndrome, marked by the triad of microangiopathic thrombocytopenia, hemolytic anemia, and oliguric renal failure within 2 weeks of onset of diarrhea.[68] This illness is the most common cause of acute renal failure in children in the United States with up to 50% of these cases requiring dialysis. The mortality rate associated with hemolytic-uremic syndrome with intensive care is 3–5%.[68] In a meta-analysis of 3,476 patients (aged 1 month–18 years at the time of recruitment) in 49 studies, death or permanent end-stage renal disease (ESRD) ranged from 0–30% with a pooled incidence of 12% (95% CI: 10–15%).[69] Approximately 25% of survivors demonstrate long-term renal sequelae (proteinuria, decreased glomerular filtration rate). Eight to ten percent of patients with hemolytic-uremic syndrome develop complications including biliary lithiasis, colonic strictures, abdominal pain, hypertension, seizures, blindness, paralysis, and short-gut syndrome.[67,70] The clinical presentation, endoscopic findings, and histopathology of mucosal biopsies are identical to that of ischemic colitis. The differential diagnosis of hemolytic-uremic syndrome includes other infectious causes (non-O157 enterohemorrhagic *E. coli*, *Streptococcus pneumoniae*, *Salmonella* serovar Typhi, *Proteus* spp., and *Shigella* spp.), hemotoxic drugs (mitomycin, cyclosporin, tacrolimus), high-dose total body irradiation, and familial hemolytic-uremic syndrome associated with coagulation abnormalities.[71]

Thrombotic Thrombocytopenic Purpura

Case reports have associated EHEC with the development of thrombotic thrombocytopenic purpura (TTP) in adults.[72] The clinical presentation of TTP includes fever (60%), impaired renal function, and neurologic dysfunction (36%), as well as purpuric rash with thrombocytopenia and microangiopathic hemolytic anemia. However, not all signs are apparent on presentation. Splenomegaly, jaundice, and hypertension may also occur. Severe cases may proceed to develop seizures, hemiplegia, paresthesias, arthralgias, and abdominal pain. Mortality in the elderly following *E. coli*-induced TTP approaches 50%.

Enteropathogenic *Escherichia Coli*

EPEC causes mild, watery diarrhea after an incubation period of 9–24 hours.[40,73] Symptoms typically resolve spontaneously within 1–7 days.[74] Prolonged diarrhea (i.e., <14 days) can occur in bottle-fed infants, with breast-feeding providing some protection.

Enteroinvasive *Escherichia Coli*

EIEC classically causes *Shigella*-like dysentery, but most often cause watery diarrhea followed by scant mucus-bloody stools 12–72 hours after ingestion.[65] Cramps, diarrhea, vomiting, fever, chills, tenesmus, and malaise are sometimes present, further mimicking shigellosis. Symptoms typically resolve spontaneously within 5–7 days.[43,73] An asymptomatic carrier state can occur following EIEC infection.

Enterotoxigenic *Escherichia Coli*

ETEC causes high-volume, cholera-like, watery diarrhea 14–72 hours after ingestion. Cramps are more common than in cholera. Diarrhea may be accompanied by low-grade fever, nausea, malaise, vomiting, headache, muscle aches, bloating, and loss of appetite. Symptoms are similar to a viral flu-like illness, and these symptoms typically resolve within 3–4 days.[75]

Enteroaggregative *Escherichia Coli*

EAggEC causes a persistent watery, secretory diarrhea in adults and children, lasting 14 days or longer. The incubation period lasts approximately 20–48 hours.[76]

DIAGNOSTIC TESTING

Analytical Methods

In food-processing practices, a selective differential medium called hemorrhagic colitis agar isolates EHEC from foods. This culturing process is limited, however, by the extended period of time necessary to identify pathogenic organisms. Newer techniques are being developed to accelerate identification times. Mulitplex polymerase chain reaction (PCR) detects multiple strains of *E. coli* and other pathogens in produce within 30 hours.[77] A multianalyte antibody-based dipstick prototype is under development, which can provide immediate detection of EHEC foods.[78] Genotyping with pulsed-field gel electrophoresis (PFGE) verifies slaughter hygiene in cattle.[79] Biomagnetic separation (BMS) uses paramagnetic microspheres coated with ligands that have a special affinity to specific microbes.[80] A novel test uses a silicone microcantilever that is coated with antibodies, and the resulting structure bends upon recognition of antibodies to *E. coli*.[81]

The self-limited course of most *E. coli*-related gastrointestinal diseases often limits the necessity of laboratory confirmation in the healthcare setting. Rapid assays for Shiga toxin-producing strains include enzyme-linked immunosorbent assay (ELISA) and enzyme-immunoassay (EIA). Because of false-positive results from these assays, confirmation and serotyping is necessary by stool culture and sorbitol-MacConkey agar analysis.[82] Cases of severe bloody diarrhea or dysentery may warrant testing for *E. coli* O157:H7, both for public health implications and for differentiation from other important bacterial causes of diarrhea such as *Salmonella* and *Shigella* spp. Currently, the US Center for Disease Control (CDC) recommends culturing all bloody stools with screening for EHEC. AOAC International (Association of Analytical Chemists International, Gaithersburg, MD; http://www.eoma.aoac.org/) has at least 23 official methods of analysis for qualitative and quantitative determination of *E. coli* in various matrices (water, various foods) and for identifying serotype O157.[83] Table 17.1 lists some official methods for the analysis of *E. coli* in various matrices.

TABLE 17.1. Some Official Methods of Analysis of *Escherichia Coli* in Various Matrices as Listed by the AOAC International

Method	Matrix	AOAC Number
Microbiological	Fish/cooked seafood, meat and meat products/frozen cooked meat products, nuts and nut products/tree nut meats	966.23
Coliform and *Escherichia coli* counts in foods	Foods	991.14
Total coliforms in *Escherichia coli* in water	Water	991.15
Escherichia coli enterotoxins	Mouse/suckling mouse, mouse/mouse adrenal cell	984.35
Escherichia coli O157:H7 in selected foods	Meat and meat products/cooked beef, eggs and egg products/liquid eggs, meat and meat products, poultry products	996.09, 996.10, 2000.13, 2005.04
Enterotoxigenic *Escherichia coli*	Media	986.34

Enterohemorrhagic *Escherichia Coli*

Laboratory detection of *E. coli* O157:H7 is important for public health concerns to differentiate this *E. coli* strain from *Shigella* and for the monitoring of the development of hemolytic-uremic syndrome. The isolation of E. coli O157:H7 requires special media, and subtyping with PFGE allows the tracking of related infections for determination of the source of the infection. In suspected clinical cases of *E. coli* O157:H7 infection, all laboratory requests should ask specifically for *E. coli* O157:H7 detection because the isolation of this strain is not routine for all *E. coli* strains. All *E. coli* grow on lactose enriched MacConkey agar, forming pink colonies. Substitution of sorbitol for lactose singles out EHEC into nonpink islands because this strain does not ferment sorbitol. Tests are 92% positive in the first week of illness, but 33% positive thereafter, illustrating the benefit of early testing.[84] These colonies are then tested for O157 lipopolysaccharide antigen side chain by latex agglutination. Positive specimens are often referred to public health laboratories for H7 antigen detection.[85,86] Antibodies to *E. coli* secretory proteins are produced in the acute phase of EHEC infection, particularly in response to *EspB*. Detection of these antibodies in serum may be useful when fecal strains have not been isolated.[38,87] Multiplex PCR can rapidly identify *E. coli* O157:H7 in 24 hours, using antibodies to genes encoding for intimin and Shiga toxins. This method also differentiates O157:H7 from non O157:H7 isolates.[88] Multiplex PCR may be the broader alternative to conventional PCR restriction fragment length polymorphism because different genes can be amplified together in the same reaction. Thus, simultaneous detection of EHEC, EPEC, ETEC, and EIEC is possible.[89]

Enteropathogenic *Escherichia Coli*

Clinical laboratories do not routinely test for EPEC, though it can be identified by serologic and cultural assays from stool. A fluorescent-actin stain of *E. coli* isolates detects actin accumulation associated with EPEC A/E lesions.[90] PCR or multiplex PCR allow precise detection with close to 100% sensitivity.[91]

Enteroinvasive *Escherichia Coli*

Clinical laboratories do not routinely test for EIEC, but it can be detected by genetic probes for the enteroinvasiveness plasmid (similar to identification of *Shigella* spp.) or by demonstration of enteroinvasiveness in tissue culture (positive Seleny test). A novel monoclonal antibody blot immune assay can identify and differentiate EIEC from *Shigella* spp. quickly, thus enabling early decision-making on antibiotic use.[92] PCR is significantly more sensitive than culture for the detection of EIEC with a sensitivity of 96% and 54%, respectively.[93]

Enterotoxigenic *Escherichia Coli*

Clinical laboratories do not routinely test for ETEC, but the evaluation of foodborne outbreaks of gastroenteritis may require special testing. ETEC stool cultures show large numbers of fimbriae-expressing *E. coli*. Serotyping follows the isolation of *E. coli* by culture. PCR for genes coding enterotoxins and/or fimbriae improves the sensitivity of culture results. More advanced testing involves the inoculation of the isolates into tissue culture of mouse adrenal cells or Chinese hamster ovarian cells. The stimulation of adenylate cyclase by the heat-labile toxin (LT) produces positive results. ELISA can also detect the LT, and DNA probes can detect the gene sequence.[94]

Biomarkers

Biomarkers are not routinely used for the identification of specific toxins in clinical cases. Biomarkers are often used to provide reliable information on identification and classification of homologous clones for research purposes.[95]

Abnormalities

Fecal leucocytes commonly occur with EIEC and EHEC, but not generally with ETEC, EPEC, or EAggEC. Significant anemia (Hgb < 10) can develop during *E. coli*-induced hemorrhagic colitis secondary to blood loss. All strains can produce severe serum electrolyte abnormalities with a profile similar to that seen in dehydration from gastrointestinal fluid losses. Severe dehydration involves hypernatremia (serum Na > 150 mEq/L) or hyponatremia (serum Na < 130 mEq/L); hypokalemia (serum K < 3.0 mEq/L); hypochloremia, elevated urine specific gravity (SpG > 1.020); low serum bicarbonate secondary to lactic acid production, or elevated serum bicarbonate secondary to contraction alkalosis. Progression of laboratory abnormalities often requires inpatient hospital admission, particularly if the patient is not able to tolerate adequate oral fluid intake.

Both hemolytic-uremic syndrome and thrombotic thrombocytopenic purpura (TTP) are primarily clinical diagnoses, but several laboratory parameters help substantiate the diagnoses. In hemolytic-uremic syndrome, a complete blood count typically reveals low hemoglobin and hematocrit, resulting from acute, mechanical (Coombs-negative) intravascular hemolysis along with increased reticulocyte count, serum aspartate aminotransferase, serum alanine aminotransferase, serum indirect bilirubin, and serum lactate dehydrogenase. Increased mean platelet volume reflects reduced platelet survival time secondary to increased platelet consumption. Giant platelets may also develop. Platelet count is usually less than 60,000 per mL. The hallmark of hemolytic-uremic syndrome is the presence of deformed, fragmented, or helmet-shaped red blood cells (i.e., schistocytes) on peripheral blood smear. Consumptive coagulopathy does not occur; therefore, coagulation parameters are within normal limits including prothrombin time (PT), activated partial thromboplastin time (aPTT), D-dimer, and fibrinogen. This feature differentiates both hemolytic-uremic syndrome and TTP from disseminated intravascular coagulation (DIC). The leucocyte count is sometimes mildly elevated, but rarely greater than 20,000 per mL. Serum globulin, complement, and albumin determinations are decreased.[67,96] Urine, if available, may contain protein, erythrocytes, hemoglobin, leucocytes, and casts.

As platelet microthrombi occlude the glomeruli, renal failure ensues with marked elevations in blood urea nitrogen (BUN) and serum creatinine. Urinalysis in oliguric patients typically demonstrates leucocytes, erythrocytes, protein, hemoglobin, and cellular casts.[67] Endoscopy, though rarely necessary, shows a pattern identical to ulcerative colitis, and abdominal ultrasound shows bowel wall edema and ascites. Thrombotic thrombocytopenic purpura causes a hemopathologic profile similar to hemolytic-uremic syndrome.

TREATMENT

Treatment of gastrointestinal manifestations of all strains of *E. coli* includes maintenance of fluid and electrolyte balance along with bowel rest. Laboratory confirmation of *E. coli* spp. is not usually available for several hours to days after clinical presentation. Therefore, supportive care should be directed at correction of dehydration; replacement of ongoing losses from stools and emesis; and blood transfusion if indicated. The use of antimotility agents, opioids, and anticholinergics should be avoided because these drugs may potentiate the pathogenicity of *E. coli* and predispose the patient to the development of hemolytic-uremic syndrome. Based on clinical similarities, suspected EIEC cases often cannot be differentiated clinically from shigellosis, and clinical experience suggests that the use of antimotility agents is contraindicated.[97]

Enterohemorrhagic *Escherichia Coli*

Although there are only limited data, antimicrobials are not usually recommended in cases of EHEC because of the perceived risks of increased Shiga-toxin production, the worsening renal injury, and the progression to hemolytic-uremic syndrome.[98] Previous clinical data on the treatment of EHEC with antibiotics suggest that antibiotic treatment had no statistically significant effect on the progression of symptoms, the excretion of fecal pathogen, or the incidence of hemolytic-uremic syndrome.[99] However, more recent prospective studies demonstrated an increased risk (RR = 14.3, 95% CI: 2.9–70.7) of this disease with antibiotic treatment in children with *E. coli* O157:H7.[98] Because the similarity in onset of bloody diarrhea during both *Shigella* and *E. coli* O157:H7 infections, antibiotic therapy with a second- or third-generation cephalosporin (e.g., ceftriaxone) is a potential option until stool cultures return. However, health departments typically serotype EHEC, and thus the definitive diagnosis is often delayed. Oral Shiga toxin-binding agents do not reduce the severity of the gastrointestinal illness in pediatric patients.[100]

In hemolytic-uremic syndrome, supportive treatment of hypertension, renal failure, and other complications is recommended. Fluid overload must be avoided, and early dialysis is often necessary for acute renal failure. Platelet transfusions should be avoided unless the patient has active bleeding because of the association between platelet transfusion and exacerbation of renal damage. Plasma exchange (plasmapheresis combined with fresh-frozen plasma replacement) is an option for severe cases of hemolytic-uremic syndrome. However, there are inadequate data to confirm the ability of plasma exchange to alter clinical outcome. Secretory immunoglobulin A (sIgA) prevents the attachment of enteropathogens to gut epithelium in newborn infants. Breast milk colostrum contains sIgA towards *EspA*, *EspB*, intimin, and Shiga toxins. Anti-*EspA* sIgA is produced in high quantities, offering a potential target for immunization and anti-toxin strategies in EHEC and EPEC.[101–104]

In TTP, aggressive supportive care of anemia and fluid imbalance is important, and fluid replacement with fresh frozen plasma may be necessary. Although corticosteroids frequently are administered, there are inadequate data to determine whether the use of corticosteroids alters clinical outcome. Case reports suggest that the use of antiplatelet drugs improves renal function, but there are inadequate data at the present time to support the efficacy of this treatment, and these agents may increase the risk of hemorrhage.[105] Platelet infusion is contraindicated because of an association with worsening platelet aggregation seen in postmortem examination of the CNS of these patients. Some patients benefit from splenectomy in severe cases. Refractory cases have been treated with vincristine, with an unknown mechanism of action.[106] Hemodialysis may be required for renal failure. The use of plasmapheresis for *E. coli*-associated TTP has not altered clinical outcome in pregnant patients or in patients with cancers or HIV infection.[106]

Enteropathogenic *Escherichia Coli*

Treatment of EPEC is entirely supportive. The use of antibiotics during EPEC infection has not improved clinical outcome.[107]

Enteroinvasive *Escherichia Coli*

EIEC is a self-limited disease that does not require antimicrobial therapy, although antibiotics are often administered because of suspected *Shigella* infection. Empiric antibiotic use did not worsen the clinical syndrome of EIEC in case reports.[108] Supportive treatment recommendations are identical to those proposed for shigellosis, and the initial use of antibiotics is reasonable, based on the clinical similarities between EIEC and *Shigella*-induced dysentery. When laboratory evaluation rules out *Shigella* species, antibiotics may be discontinued. No controlled trials have examined the efficacy of antimicrobial therapy in EIEC infections, but case reports suggest the use of trimethoprim-sulfamethoxazole, fluoroquinolones, and ciprofloxacin shortens the duration of disease. The lack of consistent efficacy is likely due to significant antibiotic resistance.[109]

Enterotoxigenic *Escherichia Coli*

ETEC responds very well to supportive care, and most cases do not require hospitalization. Antibiotics are not generally recommended, but fluoroquinolones, tetracycline, and trimethoprim-sulfamethoxazole have reduced the duration of symptoms in limited trials. Significant resistance to antibiotics is likely responsible for the lack of consistent therapeutic efficacy.[110] Breast milk confers passive immunity to infants via immunoglobulins, and breast milk also contains receptor molecules that neutralize *E. coli* toxins and colonization-factor antigens. Bismuth salicylate may provide symptomatic relief of both cramping and stool frequency.[45]

References

1. Finkelstein H. Theodor Escherich. Deutsche Medizinische Wochenschrift 1911;37:604–605.
2. Riemann HP, Oliver DO. *Escherichia coli* O157:H7. Vet Clin North Am Food Anim Pract 1998;14:41–48.
3. Riley LW, Remis RS, Helgerson SD, McGee HB, Wells JG, Davis BR, et al. Hemorrhagic colitis associated with a rare *E. coli* serotype. N Engl J Med 1983; 308:681–685.
4. Slutsker L, Ries AA, Greene KD, Wells JG, Hutwagner L, Griffin PM. *Escherichia coli* O157:H7 diarrhea in the United States: clinical and epidemiological features. Ann Intern Med 1997;126:505–513.
5. Oldfield EC. Emerging foodborne pathogens: keeping your patients and your families safe. Rev Gastroenterol Disord 2001;1:177–186.
6. Ackman D, Marks S, Mack P, Caldwell M, Root T, Birkhead G. Swimming-associated hemorrhagic colitis due to *Escherichia coli* O157:H7 infection: evidence of prolonged contamination of a fresh water lake. Epidemiol Infect 1997;119:1–8.
7. Varma JK, Greene KD, Reller ME, DeLong SM, Trottier J, Nowicki SF, et al. An outbreak of *Escherichia coli* O157 infection following exposure to a contaminated building. JAMA 2003;290:2709–2712.
8. Water Quality Disinfection Committee. Survey of water utility disinfection practices. J Am Water Works Assoc 1992;84:121–128.
9. Garcia A, Fox JG. The rabbit as a new reservoir host of enterohemorrhagic *Escherichia coli*. Emerg Infect Dis 2003;9:1592–1597.
10. Rey J, Blanco JE, Blanco M, Mora A, Dahbi G, Alonso JM, et al. Serotypes, phage types and virulence genes of Shiga-producing *Escherichia coli* isolated from sheep in Spain. Vet Microbiol 2003;94:47–56.
11. Keene WE, Sazie E, Kok J, Rice DH, Hancock DD, Balan VK, Zhao T, Doyle MP. An outbreak of *Escherichia coli* O157:H7 infections traced to jerky made from deer meat. JAMA 1997;277:1229–1231.
12. Albihn A, Eriksson E, Wallen C, Aspan A. Verotoxinogenic *Escherichia coli* (VTEC) O157:H7—a nationwide Swedish survey of bovine faeces. Acta Vet Scand 2003;44:43–52.
13. Cobbold RN, Desmarchelier PM. *In vitro* studies on the colonization of bovine colonic mucosa by Shiga-toxigenic *Escherichia coli* (STEC). Epidemiol Infect 2004; 132:87–94.
14. Oswald IP, Desautels C, Laffitte J, Fournout S, Peres SY, Odin M, et al. Mycotoxin Fumonisin B-1 increases intestinal colonization by pathogenic *Escherichia coli* in pigs. App Environ Microbiol 2003;69:5870–5874.
15. Kumar A, Jindal N, Shukla CL, Pal Y, Ledoux DR, Rottinghaus GE. Effect of ochratoxin A on *Escherichia coli*-challenged broiler chicks. Avian Diseases 2003 47:415–424.
16. Hoey DE, Currie C, Else RW, Nutikka A, Lingwood CA, Gally DL, et al. Expression of receptors for verotoxin 1 from *Escherichia coli* O157 on bovine intestinal epithelium. J Med Microbiol 2002;51:143–149.
17. Kang SJ, Ryu SJ, Chae JS, Aloe SK, Woo GJ, Lee JH. Occurrence and characteristics of enterohemorrhagic

Escherichia coli O157 in calves associated with diarrhea. Vet Microbiol 2004;98:323–328.

18. Bell BP, Griffin PM, Lozano P, Christie DL, Kobayashi JM, Tarr PI. Predictors of hemolytic uremic syndrome in children during a large outbreak of *Escherichia coli* O157:H7 infection. Pediatrics 1997;100:E12–E17.
19. Crump JA, Braden CR, Dey ME, Hoekstra RM, Rickelman-Apsia JM, Baldwin DA, et al. Outbreaks of *Escherichia coli* O157 infections at multiple county agricultural fairs: a hazard of mixing cattle, concession stands, and children. Epidemiol Infect 2003;131:1055–1062.
20. Czech E, Kneifel W, Kopp B. Microbiological status of commercially available medicinal herbal drugs—a screening study. Planta Med 2001;67:263–269.
21. Hoffman FA. Regulation of dietary supplements in the United States: understanding the Dietary Supplement Health and Education Act. Clin Obstet Gynecol 2001;44:780–788.
22. Centers for Disease Control (CDC). Ongoing multistate outbreak of *Escherichia coli* serotype O157:H7 infections associated with consumption of fresh spinach—United States, September 2006. MMWR Morb Mortal Wkly Rep 2006;55:1045–146.
23. Allerberger F, Friedrich AW, Grif K, Dierich MP, Dornbusch HJ, Mache CJ, et al. Hemolytic-uremic syndrome associated with enterohemorrhagic *Escherichia coli* O26: H infection and consumption of unpasteurized cow's milk. In J Infect Dis 2003;7:42–45.
24. Sanz S, Gimenez M, Olarte C. Survival and growth of Listeria monocytogenes and enterohemorrhagic *Escherichia coli* O157:H7 in minimally processed artichokes. J Food Prot 2003;66:2203–2209.
25. Solomon EB, Pang HJ, Matthews KR. Persistence of *Escherichia coli* O157:H7 on lettuce plants following spray irrigation with contaminated water. J Food Prot 2003;66:2198–2202.
26. Su C, Brandt LJ. *Escherichia coli* O157:H7 infection in humans. Ann Intern Med 1995;123:698–714.
27. Olsvik O, Wasteson Y, Lund A, Hornes E. Pathogenic *Escherichia coli* found in food. Int J Food Microbiol 1991;12:103–113.
28. Naimi TS, Wicklund JH, Olsen SJ, Krause G, Wells JG, Bartjus JM, et al. Concurrent outbreaks of *Shigella sonnei* and enterotoxigenic *Escherichia coli* infections associate with parsley: implications for surveillance and control of foodborne illness. J Food Prot 2003;66:535–541.
29. Sharma M, Beuchat LR. Sensitivity of *Escherichia coli* O157:H7 to commercially available alkaline cleaners and subsequent resistance to heat and sanitizers. Appl Environ Microbiol 2004;70:1795–1803.
30. Lin CM, Wu FM, Kim HK, Doyle MP, Michael BS, Williams LK. A comparison of hand washing techniques to remove *Escherichia coli* and caliciviruses under natural or artificial fingernails. J Food Prot 2003; 66:2296–2301.
31. Golding SS, Matthews KR. Intrinsic mechanism decreases susceptibility of *Escherichia coli* O157:H7 to multiple antibiotics. J Food Prot 2004;67:34–39.
32. Yaron S, Rydlo T, Shachar D, Mor A. Activity of dermaseptin K4-S4 against foodborne pathogens. Peptides 2003;24:1815–1821.
33. Schamberger GP, Diez-Gonzalez F. Characterization of colicinogenic *Escherichia coli* strains inhibitory to enterohemorrhagic *Escherichia coli*. J Food Prot 2004; 67:486–492.
34. Yuste J, Fung DY. Inactivation of Salmonella typhimurium and *Escherichia coli* O157:H7 in apple juice by a combination of nisin and cinnamon. J Food Prot 2004;67: 371–377.
35. Kyaw CM, De Araujo CR, Lima MR, Gondim EG, Brigido MM, Giugliano LG. Evidence for the presence of a type III secretion system in diffusely adhering *Escherichia coli* (DAEC). Infect Genet Evol 2003;3:111–117.
36. Elliott SJ, Wainwright LA, McDaniel TK, Jarvis KG, Deng YK, Lai LC, et al. The complete sequence of the locus of enterocyte effacement (LEE) from enteropathogenic *Escherichia coli* E2348/69. Mol Microbiol 1998; 28:1–4.
37. Torres AG, Kaper JB. Multiple elements controlling adherence of enterohemorrhagic *Escherichia coli* O157: H7 to HeLa cells. Infect Immun 2003;71:4985–4995.
38. McDaniel TK, Kaper JB. A cloned pathogenicity island from enteropathogenic *Escherichia coli* confers the attaching and effacing phenotype on *E. coli* K-12. Mol Microbiol 1997;23:399–407.
39. Karmali MA. Infection by Shiga toxin-producing *Escherichia coli*: an overview. Mol Biotechnol 2004;26: 117–122.
40. Stephan R, Borel N, Zeifel C, Blanco M, Blanco JE. First isolation and further classification of enteropathogenic *Escherichia coli* (EPEC) O157:H45 strains from cattle. BMC Microbiol 2004;4:10–15.
41. Lee SM, Chen J. Survival of *Escherichia coli* O157:H7 in set yogurt as influenced by the production of an exopolysaccharide, colonic acid. J Food Prot 2004;67:252–255.
42. Smith JL. The role of gastric acid in preventing foodborne disease and how bacteria overcome acid conditions. J Food Prot 2003;66:1292–1303.
43. Baldassarri L, Caprioli A, Donelli G. Adherence to and penetration of cultured cells by an invasive strain of *Escherichia coli*: an ultrastructural study. Microbiologica 1987;10:317–323.
44. Tauschek M, Gorrell RJ, Strugnell RA, Robins-Browne RM. Identification of a protein secretory pathway for the secretion of heat-labile enterotoxin by an enterotoxigenic strain of *Escherichia coli*. Proc Natl Acad Sci USA 2002;99:7066–7071.
45. Farthing MJ. Bugs and the gut: an unstable marriage. Best Pract Res Clin Gastroenterol 2004;18:233–239.
46. Moon HW, Whipp SC, Argenzio RA, Levine MM, Gianella RA. Attaching and effacing activities of rabbit

and human enteropathogenic *Escherichia coli* in pig and rabbit intestines. Infect Immunol 1983;41:1340–1351.

47. Sperandio V, Torres AG, Jarvis B, Nataro JP, Kaper JB. Bacteria-host communication: the language of hormones. Proc Natl Acad Sci USA 2003;100:8951–8956.
48. Chen C, Brown DR, Xie Y, Green BT, Lyte M. Catecholamines modulate *Escherichia coli* O157:H7 adherence to murine cecal mucosa. Shock 2003;20:183–188.
49. Giron JA, Ho AS, Schoolnik GK. An inducible bundle-forming pilus of enteropathogenic *Escherichia coli*. Science 1991;254:710–713.
50. Newton HJ, Sloan J, Bennett-Wood V, Adams LM, Robins-Browne RM, Hartland EL. Contribution of long polar fimbriae to the virulence of rabbit-specific enteropathogenic *Escherichia coli*. Infect Immun 2004;72: 1230–1239.
51. Ashkenazi S, May L, LaRocco M, Lopez EL, Cleary TG. The effect of postnatal age on the adherence of enterohemorrhagic *Escherichia coli* to rabbit intestinal cells. Pediatr Res 1991;29:14–19.
52. Roe AJ, Hoey DE, Gally DL. Regulaton, secretion, and activity of type III-secreted proteins of enterohemorrhagic *Escherichia coli* O157. Biochem Soc Trans 2003;31:98–103.
53. Campellone KG, Leong JM. Tails of two Tirs: actin pedestal formation by enteropathogenic *Escherichia coli* and enterohemorrhagic *Escherichia coli* O157:H7. Curr Opin Microbiol 2003;6:82–90.
54. Abe H, Tatsuno I, Tobe T, Okutani A, Sasakawa C. Bicarbonate ion stimulates the expression of locus of enterocyte effacement-encoded genes in enterohemorrhagic *Escherichia coli* O157:H7. Infect Immunol 2002;70: 3500–3509.
55. Salzman M, Madsen JM, Greenberg MI. Toxins: bacterial and marine toxins. Clin Lab Med 2006;26:397–419.
56. Karch H. The role of virulence factors in enterohemorrhagic *Escherichia coli* (EHEC)—associated hemolytic-uremic syndrome. Semin Thromb Hemost 2001;27: 207–213.
57. Menge C, Stamm I, Van Dieman PM, Sopp P, Baljer G, Wallis TS, et al. Phenotypic and functional characterization of intraepithelial lymphocytes in a bovine ligated intestinal loop model of enterohaemorrhagic *Escherichia coli* infection. J Med Microbiol 2004;53:573–579.
58. Westerholt S, Pieper A-K, Griebel M, Volk HD, Hartung T, Oberhoffer R. Characterization of the cytokine immune response in children who have experienced an episode of typical hemolytic-uremic syndrome. Clin Diagn Lab Immunol 2003;10:1090–1095.
59. Chaisri U, Nagata M, Kurazono H, Horie H, Tongtawe P, Hayashi H, et al. Localization of Shiga toxins of enterohaemorrhagic *Escherichia coli* in kidneys of paediatric and geriatric patients with fatal haemolytic uraemic syndrome. Microb Pathog 2001;31:59–67.
60. Hecht G, Hodges K, Gill RK, Kear F, Tyagi S, Malakooti J, et al. Differential regulation of Na+/H+ exchange isoform activities by enteropathogenic *Escherichia coli* in human intestinal epithelial cells. Am J Physiol Gastrointest Liver Physiol 2004;287:370–378.
61. Beuchat LR, Scouten AJ. Viability of acid-adapted *Escherichia coli* O157:H7 in ground beef treated with acidic calcium sulfate. J Food Prot 2004;67:591–595.
62. Savarino SJ, Fasano A, Watson J, Martin BM, Levine MM, Guandlini S, et al. Enteroaggregative *Escherichia coli* heat-stable enterotoxin 1 represents another subfamily of *E. coli* heat-stable toxin. Proc Natl Acad Sci USA 1993;90:3093–3097.
63. Begue RE, Mehta DI, Blecker U. *Escherichia coli* and the hemolytic-uremic syndrome. South Med J 1998;91: 798–804.
64. Clarke SC. Diarrhoeagenic *Escherichia coli*—an emerging problem? Diagn Microbiol Infect Dis 2001;41: 93–98.
65. Gill CJ, Hamer DH. Foodborne illnesses. Curr Treat Options Gastroenterol 2001;4:23–38.
66. Kendrick JB, Risbano M, Groshong SD, Frankel SK. A rare presentation of ischemic pseudomembranous colitis due to *Escherichia coli* O157:H7. Clin Infect Dis 2007;45:217–219.
67. Siegler RL. The hemolytic uremic syndrome. Pediatr Clin North Am 1995;42:1505–1529.
68. Griffin PM, Tauxe RV. The epidemiology of infections caused by *Escherichia coli* O157:H7, other enterohemorrhagic *E. coli* and the associated hemolytic uremic syndrome. Epidemiol Rev 1991;13:60–98.
69. Garg AX, Suri RS, Barrowman N, Rehman F, Matsell D, Rosas-Arellano MP. Long-term renal prognosis of diarrhea-associated hemolytic uremic syndrome: a systematic review, meta-analysis, and meta-regression. JAMA 2003;290:1360–1370.
70. Cordovez A, Prado V, Maggi L, Cordero J, Martinez J, Misraji A, et al. Enterohemorrhagic *Escherichia coli* associated with hemolytic-uremic syndrome in Chilean children. J Clin Microbiol 1992;30:2153–2157.
71. Pickering LK, Obrig TG, Stapleton FB. Hemolytic-uremic syndrome and enterohemorrhagic *Escherichia coli*. Pediatr Infect Dis J 1994;13:459–476.
72. Cohen MB, Giannella RA. Hemorrhagic colitis associated with *Escherichia coli* O157:H7. Adv Intern Med 1992;37:173–195.
73. Gilligan PH. *Escherichia coli*. EAEC, EHEC, EIEC, ETEC. Clin Lab Med 1999;19:505–21.
74. Clarke SC, Haigh RD, Freestone PP, Williams PH. Enteropathogenic *Escherichia coli* infection: history and clinical aspects. Br J Biomed Sci 2002;59:123–127.
75. Bouckenooghe AR, DuPont HL, Jiang ZD, Adachi J, Mathewson JJ, Verenkar MP, et al. Markers of enteric inflammation in enteroaggregative *Escherichia coli* diarrhea in travelers. Am J Trop Med Hyg 2000;62:711–713.
76. Huang DB, Okhuysen PC, Jiang ZD, DuPont HL. Enteroaggregative *Escherichia coli*: an emerging enteric pathogen. Am J Gastroenterol 2004;99:383–389.

77. Li Y, Mustapha A. Simultaneous detection of *Escherichia coli* O157:H7, Salmonella, and *Shigella* in apple cider and produce by a multiplex PCR. J Food Prot 2004;67: 27–33.
78. Aldus CF, Van Amerongen A, Ariens RM, Peck MW, Wichers JH, Wyatt GM. Principles of some novel rapid dipstick methods for detection and characterization of verotoxigenic *Escherichia coli*. J Appl Microbiol 2003;95:380–389.
79. Zweifel C, Zychowska MA, Stephan R. Prevalence and characteristics of Shiga toxin-producing *Escherichia coli, Salmonella* spp. and *Campylobacter* spp. isolated from slaughtered sheep in Switzerland. Int J Food Microbiol 2004;92:45–53.
80. Deponte S, Steingroewer J, Loser C, Boschke E, Bley T. Biomagnetic separation of *Escherichia coli* by use of anion-exchange beads: measurement and modeling of the kinetics of cell-bead interactions. Anal Bioanal Chem 2004;379:419–426.
81. Zhang J, Ji HF. An anti Escherichia coli O157:H7 antibody-immobilized microcantilever for the detection of *Escherichia coli (E. coli)*. Anal Sci 2004;20:585–587.
82. Centers for Disease Control and Prevention (CDC). University outbreak of calicivirus infection mistakenly attributed to Shiga toxin-producing *Escherichia coli* O157:H7—Virginia, 2000. MMWR Morb Mortal Wkly Rep 2001;50:489–491.
83. Horowitz W, editor. Official Methods of Analysis of AOAC International. 18th ed. Gaithersburg, MD: AOAC International; 2006.
84. Tarr PI, Neill MA, Clausen CR, Watkins SL, Christie DL, Hickman RO. *Escherichia coli* O157:H7 and the hemolytic uremic syndrome: importance of early cultures in establishing the etiology. J Infect Dis 1990;162:553–556.
85. Carroll KC, Adamson K, Korgenski K, Croft A, Hankmeier R, Daly J, et al. Comparison of a commercial reversed passive latex agglutination assay to an enzyme immunoassay for the detection of Shiga toxin-producing *Escherichia coli*. Eur J Clin Microbiol Infect Dis 2003;22:689–692.
86. March SB, Ratnam S. Latex agglutination test for detection of *Escherichia coli* serotype O157. J Clin Microbiol 1989; 27:1675–1677.
87. Tsutsumi R, Ichinohe N, Shimooki O, Obata F, Takahashi K, Inada K, et al. Homologous and heterologous antibody responses to lipopolysaccharide after enterohemorrhagic *Escherichia coli* infection. Microbiol Immunol 2004;48:27–38.
88. Sharma VK, Dean-Nystrom EA. Detection of enterohemorrhagic *Escherichia coli* O157:H7 by using a multiplex real-time PCR assay for genes encoding intimin and Shiga toxins. Vet Microbiol 2003;93:247–260.
89. Rahman H. Multiplex PCR for the detection of *Stx* genes of *Escherichia coli*. Indian J Med Res 2002;115:251–254.
90. Knutton S, Phillips AD, Smith HR. Screening for enteropathogenic *Escherichia coli* in infants with diarrhea by the fluorescent-actin staining test. Infect Immun 1991;59:365–371.
91. Aranda KR, Fagundes-Neto U, Scaletsky IC. Evaluation of multiplex PCRs for diagnosis of infection with diarrheagenic *Escherichia coli* and *Shigella* spp. J Clin Microbiol 2004;42:5849–5853.
92. Szakal DD, Schneider G, Pal T. A colony blot immune assay to identify enteroinvasive *Escherichia coli* and *Shigella* in stool samples. Diagn Microbiol Infect Dis 2003;45:165–171
93. Dutta S, Chatterjee A, Dutta P, Rajendran K, Roy S, Pramanik KC, et al. Sensitivity and performance characteristics of a direct PCR with stool samples in comparison to conventional techniques for diagnosis of *Shigella* and enteroinvasive *Escherichia coli* infection in children with acute diarrhoea in Calcutta, India. J Med Microbiol 2001;50:667–674.
94. Amar CF, East C, Maclure E, McLaughlin J, Jenkins C, Duncanson P, et al. Blinded application of microscopy, bacteriological culture, immunoassays and PCR to detect gastrointestinal pathogens from faecal samples of patients with community-acquired diarrhoea. Eur J Clin Microbiol Infect Dis 2004;23:529–534.
95. Xu W, Chen Q, Wang H. Application of biomarkers in homologous clone identification of enterotoxigenic *Escherichia coli*. Zhonghua Yu Fang Yi Xue Za Zhi 1998;32:22–24.
96. Bouckenooghe AR, Jiang ZD, De La Cabada FJ, Ericsson CD, DuPont HL. Enterotoxigenic *Escherichia coli* as cause of diarrhea among Mexican adults and US travelers in Mexico. J Travel Med 2002;9:137–140.
97. Bell BP, Goldoft M, Griffin PM, Davis MA, Gordon DC, Tarr PI, et al. A multistate outbreak of *Escherichia coli* O157:H7-associated bloody diarrhea and hemolytic-uremic syndrome from hamburgers. JAMA 1994; 272:1349–1353.
98. Wong CS, Jelacic S, Habeeb RL, Watkins SL, Tarr PI. The risk of the hemolytic-uremic syndrome after antibiotic treatment of *Escherichia coli* O157:H7 infections. N Engl J Med 2000;342:1930–1936.
99. Proulx F, Turgeon JP, Delage G, LaFleur L, Chicoine L. Randomized, controlled trial of antibiotic therapy for *Escherichia coli* O157:H7 enteritis. J Pediatr 1992;121: 299–303.
100. Trachtman H, Cnaan A, Christen E, Gibbs K, Zhao S, Acheson DW, et al. Effect of an oral Shiga toxin-binding agent on diarrhea-associated hemolytic uremic syndrome in children: a randomized controlled trial. JAMA 2003;290:1337–1344.
101. Funatogawa K, Ide T, Kirikae F, Saruta K, Nakano M, Kirikae T. Use of immunoglobulin enriched bovine colostrums against oral challenge with enterohemorrhagic *Escherichia coli* O157:H7 in mice. Microbiol Immunol 2002;46:761–766.
102. Mukherjee J, Chios K, Fishwild D, Hudson D, O'Donnell S, Rich SM, et al. Human *Stx*2-specific monoclonal anti-

bodies prevent systemic complications of *Escherichia coli* O157:H7 infection. Infect Immunol 2002;70:612–619.

103. Noguera-Obenza M, Ochoa TJ, Gomez HF, et al. Human milk secretory antibodies against attaching and effacing *Escherichia coli* antigens. Emerg Infect Dis 2003;9: 545–551.
104. Van de Perre P. Transfer of antibody via mother's milk. Vaccine 2003;21:3374–3376.
105. Yarranton H, Machin SJ. An update on the pathogenesis and management of acquired thrombotic thrombocytopenic purpura. Curr Opin Neurol 2003;16:367–373.
106. Ziman A, Mitri M, Klapper E, Pepkowitz SH, Goldfinger D. Combination vincristine and plasma exchange as initial therapy in patients with thrombotic thrombocytopenic purpura: one institution's experience and review of the literature. Transfusion 2005;45:41–49.
107. Mani R, Udgaonkar U, Pawar S. Study of enteropathogenic *Escherichia coli* (EPEC) diarrhoea in children. Indian J Pathol Microbiol 2003;46:118–120.
108. Nijssen S, Florijn A, Bonten MJ, Schmitz FJ, Verhoef J, Fluit AC. Beta-lactam susceptibilities and prevalence of ESBL-producing isolates among more than 5000 European Enterobacteriaceae isolates. Int J Antimicrob Agents 2004;24:585–591.
109. DuPont HL. Guidelines on acute infectious diarrhea in adults. Am J Gastroenterol 1997;92:1962–1975.
110. Maynard C, Fairbrother JM, Bekal S, Sanschagrin F, Levesque RC, Brousseau R, et al. Antimicrobial resistance genes in enterotoxigenic *Escherichia coli* O149:K91 isolates obtained over a 23-year period from pigs. Antimicrob Agents Chemother 2003;47:3214–3221.

Chapter 18

LISTERIA MONOCYTOGENES

Cyrus Rangan, MD, FAAP

HISTORY

Listeria monocytogenes (Murray et al.) Pirie (*L. monocytogenes*) is a gram-positive, facultatively anaerobic, catalase-positive, motile coccobacillus. This organism was first cultured in 1921 by French microbiologists Jérôme Dumont and Louis Cotoni at L'Institut Pasteur de Paris, France, and named for British surgeon Lord Lister.[1] In 1926, Murray et al. first reported the occurrence of listeriosis in rabbits.[2] The first documented outbreak of foodborne listeriosis occurred in 23 patients at a Boston hospital in 1979, and definitive proof of the foodborne transmission of human listeriosis was first reported in 1983.[3] Most exposures of healthy individuals to *L. monocytogenes* do not produce symptoms, and up to 10% of healthy individuals harbor *L. monocytogenes* in the gastrointestinal tract. In the United States, approximately 2,500 cases of listeriosis are confirmed annually with a mortality rate of 20–30%. This mortality rate is the highest mortality rate of all foodborne bacterial infections.[4]

EXPOSURE

Source

Environment

Listeria species are tolerant to extreme salt, temperature, and pH. These organisms are ubiquitous in soil, water, fertilizers, and effluents worldwide. These organisms proliferate readily on decomposing vegetation.[5,6] *Listeria* species occur in the feces of 1–5% of healthy individuals, particularly nonpathogenic species (e.g., *L. innocua*).[7] However, there are inadequate data to distinguish between a carrier state and transient passage of these bacteria due to environmental exposures.

Animal

Listeria monocytogenes is widely distributed in nature including domestic and wild animals, water soil, and plant materials. Humans, ruminant animals, and fowl carry *L. monocytogenes* asymptomatically.[8] However, most strains in animals are not pathogenic in humans.[9] Pregnant females may transmit infections vertically to fetuses or neonates.[10] Healthcare providers may prevent rare nosocomial transmission of *L. monocytogenes* in the neonatal inpatient setting via hand washing and utilization of 70% ethanol antiseptic gels.[11]

Food

Listeria monocytogenes occurs in a wide variety of foods including prepared, ready-to-eat, refrigerated foods, (e.g., turkey deli meats, sliced ham, pâtés, hot dogs, sausages), soft cheeses, coleslaw, undercooked chicken, leafy vegetables (lettuce, carrots), and dairy products (e.g., contamination of pasteurized milk).[12–15] Although *L. monocytogenes* is present in meat, poultry, diary, and vegetables, contamination levels are typically low and human listeriosis is relative rare.[16] In a study of Italian

Medical Toxicology of Natural Substances, by Donald G. Barceloux, MD

TABLE 18.1. *Listeria Monocytogenes* Serogroups Identified in 167 Strains Isolated from Retail Foods in Florida and Greater Washington, DC

Serotypes	% Serotypes
1/2b, 3b	41
4b, 4d, 4e	32
1/2a, 3a	26
1/2c, 3c	1.2
4a, 4c	0.6

Source: From Ref 20.

dry sausage, 22.7% of the samples contained detectable amounts of *L. monocytogenes* (1/2c, 1/2a, 1/2b, 4b), but the contamination levels were below 10 CFU/g.[17] Shelf-stable forms of these foods harbor substantially lower rates of *L. monocytogenes*. The bacteria proliferate particularly well at the pH range of unpasteurized soft cheeses (pH 5.2–6.4).[18] There are at least 13 serotypes of *L. monocytogenes* with serotype 1/2a being the most prevalent serotype in food.[19] Table 18.1 lists the serogroups found in 167 isolates of *L. monocytogenes* from retail foods including ready-to-eat meats, raw chicken carcasses, and fresh produce.[20]

Although *L. monocytogenes* is ubiquitous, most human outbreaks of foodborne listeriosis involve a small number of closely related strains, particularly two clonal groups (epidemic clone [EC] I and II) of serotype 4b.[21,22] The genomic diversity is higher among the strains with ECII-like characteristics than among those strains carrying the ECI-specific genetic markers.[23]

Food Processing

The ability of *L. monocyto*genes to proliferate under relatively dry, refrigerated conditions enhances contamination of food by these bacteria.[24] The most effective means to control *L. monocytogenes* include the following: proper cooking and refrigeration; pasteurization of dairy products; separation of cooked and uncooked foods; sanitation of food preparation surfaces with at least 1% bleach solution; and washing of vegetables.[25] *L. monocytogenes* survives in environments of up to 20% NaCl. Hence, refrigeration or the addition of salt does not effectively control the growth of *L. monocytogenes*.[26] Freezing temperatures (<–20°C/–4°F) and high pressures effectively limit *L. monocytogenes* growth.[27,28] Cooking foods to 120°C –170°C (248°F–338°F) also kills this bacteria; however, a significant portion of *L. monocytogenes* exposures result from the consumption of refrigerated, ready-to-eat foods that do not require cooking.[29]

TOXINS

Physiochemical Properties

The genus *Listeria* contains eight species (*L. agrestis, L. grayi, L. innocua, L. ivanovii, L. monocytogenes, L. murrayi, L. seeligeri, L. welshimeri*), which are gram-positive, nonspore-forming, rod-shaped, facultative anaerobes. Only *L. monocytogenes* and *L. ivanovii* are pathogenic with the latter species being primarily an animal and, rarely, a human pathogen.[30] *L. monocytogenes* is morphologically indistinguishable from other *Listeria* species, and these bacteria are divided into at least 13 serovars based on unique combinations of somatic (O) and flagellar (H) antigens. Serovars 4b, 1/2b, and 1/2a are the main serovars causing human disease with the latter two serovars being responsible primarily for sporadic cases of human listeriosis. Other serovars of *L. monocytogenes* cause disease in animals or are nonpathogenic.[31] *L. monocytogenes* grows over a wide temperature range (1°–45°C/34°F–113°F), and thus proliferates under refrigeration (0°C–4°C/32°F–39°F), at room temperature (20°C–25°C/68°F–77°F), and at human body temperature (37.0°C/98.6°F). This organism thrives in high salinity conditions, (e.g., bile salts in the intestines). *L. monocytogenes* grows at pH 4.4 to 6.0 with survival at pH 3.6 to 9.5.[32] Gastric acid kills *L. monocytogenes*, and patients ingesting medication causing decreased gastric acidity may be at higher risk for *L. monocytogenes* infection.[33,34]

Mechanism of Toxicity

Listeria monocytogenes is a facultative intracellular bacterium, which can invade and multiply in both phagocytic and nonphagocytic cells. This pathogen has the ability to cross intestinal, placental, and blood–brain barriers, producing local infections. The entry of this pathogen into human cells results from a species-specific interaction of a surface protein, internalin, with the human receptor, E-cadherin. *L. monocytogenes* can synthesize flagella at 20°C–25°C (68°F–77°F), but not at human body temperature (37.0°C/98.6°F); thus, *L. monocytogenes* motility relies on a characteristic tumbling motion that counters the flow of peristalsis, permitting prolonged transit through the gastrointestinal lumen.[35,36] Several factors are responsible for the infection of human cells by *L. monocytogenes* including the following: 1) internalin (InlA, InlB) causes the internalization of the bacteria into the cell, 2) listeriolysin along with two phospholipases (PlcA, PlcB) allow the escape of the bacteria from phagocytes, and 3) ActA mediates the actin-based intracellular and extracellular movement of the pathogen. The prominence of factors (e.g., internalin in serotype 4b) accounts for the prevalence of this serotype in human listeriosis.[37] *L. monocytogenes* adheres to enterocytes and intestinal macrophages via D-galactose residues on the bacteria surface.[38] Internalin protein, AnlA, on the bacterial cell wall binds to the transmembrane glycoprotein E-cadherin on the mucosal surface of polarized epithelial cells. Connected directly to the actin cytoskeleton, E-cadherins are associated with two adjacent surface proteins, myosin VIIa and vezatin that mediate enterocytic phagocytosis of *L. monocytogenes*.[39,40] Another invasion protein, InlB, uses the signalling pathway of the hepatocyte growth factor receptor, Met, to induce endocytosis of the receptor and the internalization of the bacteria into nonpolarized cells (e.g., hepatocytes, fibroblasts, epithelioid cells).[41] *L. monocytogenes* escapes the highly oxidative and acidic environment of the phagosome by producing catalase, superoxide dismutase, and two hemolysins (listeriolysin O, phospholipase C). These hemolysins lyse the phagosome and allow *L. monocytogenes* to proliferate in the cytosol.[42,43] The listeriolysin-encoding gene (hlyA) occurs only in virulent strains of *L. monocytogenes*.[44] After lysis of the primary vacuoles, *L. monocytogenes* grows and multiplies in the cytosol. The intracellular mobility and cell-to-cell spread of this bacteria depends on another *L. monocytogenes* cell wall surface protein, ActA. This 67kDa protein reorganizes the enterocytic actin cytoskeleton into rope-like polymers, which punch pocket-like filopods along the enterocytic inner membrane.[45] These Act A-bound actin polymers thrust microcolonies of *L. monocytogenes* into the filopods, thus propelling the filopods into adjacent enterocytes.[46] Adjacent enterocytes subsequently engulf protruding filopods, thus permitting *L. monocytogenes* to pass safely into new enterocytes, where they continue to proliferate while evading host immune defenses.[47] Because epithelial cells, M cells, and macrophages are rarely destroyed by *L. monocytogenes*, enteric macrophages readily transport *L. monocytogenes* to the liver and spleen.[48] *L. monocytogenes* subsequently induces phagocytosis in parenchymal cells of the liver and spleen, where they proliferate further before disseminating to other locations, particularly the central nervous system.[49] Bacterial uptake into brain parenchymal cells and the meninges causes infection of the central nervous system. T-cells, natural killer cells, interferon-gamma, and interleukin-18 prevent extraintestinal manifestations of *L. monocytogenes* in most healthy individuals.[50] In utero, *L. monocytogenes* invades placental endothelial cells in similar fashion, resulting in premature labor, fetal death, or perinatal listeriosis.[10] Both underdeveloped fetal immune system and two-way transfer of *L. monocytogenes* between the placenta and maternal organs increase the fetal risk of listerosis during pregnancy.[51]

DOSE RESPONSE

The current US standard for *L. monocytogenes* content in food is zero, but the European Union and Canada allow 100CFU/g for ready-to-eat foods unable to support the growth of these bacteria. The infective dose of *L. monocytogenes* in healthy persons is unknown, and these individuals do not typically manifest clinical symptoms.[52] The estimated average median dose of these bacteria in healthy individuals, who developed fever and diarrhea after ingesting contaminated chocolate milk, was 2.9×10^{11} CFU.[53] In susceptible populations, fewer than 1,000 organisms may cause gastrointestinal or systemic disease. High-risk populations include pregnant woman, fetuses, neonates, elderly individuals, immunocompromised patients, steroid-dependent patients, and patients with decreased gastric acidity. Pregnant women in the third trimester of pregnancy are infected at a rate 20 times that of the general population, secondary to impaired cell-mediated immunity.[10] Immunocompromised patients are infected at a rate 300 times that of the general population.[52] The overall mortality rate of listerosis is 20–30%, but this rate reaches 60% in patients with central nervous system disease, and 50% in cases of newborn listeriosis.[4,54]

CLINICAL RESPONSE

Listeria monocytogenes incubates for 1–2 days before gastrointestinal symptoms occur. A latency period of 3 days to 10 weeks occurs before systemic listeriosis appears. In healthy individuals exposed to foods containing high concentrations of *L. monocytogenes*, classic foodborne listeriosis begins with an asymptomatic infection or by a mild, self-limited flu-like illness manifest by fever, malaise, nausea, vomiting, diarrhea, and myalgias.[55] The gastroenteritis occurs without symptoms of bacteremia or more serious infections. However, in susceptible patients, systemic invasion leads to septicemia and/or meningoencephalitis with clinical effects including headache, photophobia, stiff neck, ataxia, and alteration of mental status.[56] Seizures occur in 25% of cases with nervous system involvement.[57] Pregnant woman may spread *L. monocytogenes* vertically after asymptomatic or symptomatic infection, inducing premature labor, miscarriage, stillbirth, and neonatal listeriosis. Neonates may become infected via multiple routes including in utero, contact with vaginal fluid during birth, contact with maternal shedding up to 10 days after delivery, and following consumption of contaminated breast milk.[10] Neonates may be asymptomatic for up to 8 weeks before developing systemic manifestations, including septicemia, meningitis, disseminated granulomatosis, or pneumonia.[58] Meningitis, meningoencephalitis, and rare cases of brain abscess and endocarditis occur following *L. monocytogenes* infections, primarily in elderly and immunocompromised patients.[59,60] Infected or colonized patients may shed bacteria in feces for several months.[61]

DIAGNOSTIC TESTING

Analytical Methods

Biochemical methods of identifying *L. monocytogenes* involve isolation with selective and enrichment media, followed by Gram stain and multiple biochemical tests. *L. monocytogenes* is gram-positive; however, Gram staining alone is only 60% specific for *L. monocytogenes*. These bacteria may be coccoidal, resembling *Streptococcus*, or rod-shaped, resembling *Corynebacteria*.[57] The characteristic features of *L. monocytogenes* are a tumbling motion under light microscopy similar to *Vibrio* species, the production of hemolysins (listeriolysin O, phospholipase C), and the presence of flagellae at room temperature.[35,36] Laboratories must be specifically advised to evaluate for *L. monocytogenes* in suspected specimens because conventional methods for the detection of foodborne bacterial pathogens rely on the use of selective microbiological media (e.g., FDA enrichment broth). Specimens from blood, cerebrospinal fluid, aborted fetuses, or placentas must be cold-enriched at 4°C (39°F) for 28–72 hours before culturing on chromogenic agar, where colonies display a turquoise hue (sensitivity 85.7%, and specificity 100%).[62] *Corynebacteria* also grow under similar laboratory conditions. Stool cultures are not utilized to detect listerosis because of the high rates of asymptomatic carriage of *L. monocytogenes* species.[61] Cultured colonies of *L. monocytogenes* exhibit a weak zone of β-hemolysis after streaking onto sheep blood agar.[63]

DNA-based methods allow the amplification of specific genetic signals from a few cells, and therefore these analytical methods are much more sensitive and specific than the use of selective microbiological media.[44] Polymerase chain reaction (PCR) is the basis of many nucleic acid-based detection systems. This method detects both live and dead strains of *L. monocytogenes*, and PCR for this bacteria yields 99% specificity and 100% sensitivity.[64] PCR after immunomagnetic separation is a sensitive method for detecting *L. monocytogenes* in food samples at concentrations >100 CFU/g, but the detection of lower concentrations of these bacteria requires the use of enrichment broth.[65] Subtyping procedures are useful for tracking individual strains responsible for outbreaks of listeriosis. Analytical methods for subtyping include phenotypic subtyping (serotyping, phage typing, multilocus enzyme electrophoresis, esterase typing) and genotypic subtyping (pulsed-field gel electrophoresis, ribotyping, PCR-based subtyping techniques, amplified fragment length polymorphism, PCR-restriction fragment length polymorphism, repetitive element PCR, DNA sequencing-based subtyping techniques, such as multilocus sequence typing). These subtyping techniques are reserved for the investigation of epidemics.[66]

There are 13 different AOAC International (Association of Analytical Chemists International, Gaithersburg, MD; http://www.eoma.aoac.org/) official methods of analysis for *L. monocytogenes* in foods.[67]

1. Methods 992.18: A qualitative biochemical identification method for *L. monocytogenes* species in foods
2. Method 992.19: A qualitative biochemical identification method for *L. monocytogenes* species in foods
3. Method 993.09: A qualitative colorimetric DNA hybridization method for dairy products, seafoods, and meats. (The rest of the official methods utilize commercially available test kits)
4. Method 993.12: A qualitative selective enrichment and isolation method for *L. monocytogenes* in milk and dairy products
5. Method 994.03: A qualitative colorimetric monoclonal enzyme-linked immunosorbent assay method for determination of *L. monocytogenes* in dairy products, seafoods, and meats
6. Method 995.22: A qualitative colorimetric polyclonal enzyme immunoassay screening method for *L. monocytogenes* in foods
7. Method 996.14: A polyclonal enzyme immunoassay for *L. monocytogenes* and related species in selected foods, including dairy products, vegetables, fruits and fruit products, red meats, pork, pasta, cacao bean and its products, and eggs and egg products
8. Method 997.03: A qualitative visual immunoprecipitate assay for *L. monocytogenes* and related species in selected foods, including dairy products, vegetables, fruits and fruit products, red meats, pork, pasta, fish, and eggs and egg products
9. Method 999.06: A qualitative enzyme-linked immunofluorescent assay for *L. monocytogenes* in foods
10. Method 2002.09: A qualitative colorimetric polyclonal enzyme immunoassay screening method for *L. monocytogenes* in foods
11. Method 2003.12: A qualitative PCR method for *L. monocytogenes* in foods

12. Method 2004.02: A quantitative instrumental method for determination of *L. monocytogenes* in foods
13. Method 2004.06: An automated enzyme-linked fluorescent immunoassay for the detection of *L. monocytogenes* in dairy products, vegetables, seafood, raw meats and poultry, and processed meats and poultry

Biomarkers

Maternal fever around the time of delivery is a nonspecific sign of threatened perinatal bacterial infection. There is no biomarker for screening for *L. monocytogenes* during pregnancy; however, gram-positive stains of amniotic fluid in febrile patients with preterm labor may suggest *L. monocytogenes* infection.[68]

Abnormalities

Listeria monocytogenes elicits a marked monocytosis in animals, but rarely in humans.[57] Cerebrospinal fluid demonstrates a pleocytosis with moderately elevated protein and normal-to-low glucose.[69] Severe dehydration involves hypernatremia (serum Na > 150 mEq/L) or hyponatremia (serum Na < 130 mEq/L), hypokalemia (serum K < 3.0 mEq/L), hypochloremia, elevated urine specific gravity (SpG > 1.020), and hyperphosphatemia. Low serum bicarbonate results from lactic acid production or elevated serum bicarbonate secondary to contraction alkalosis.

TREATMENT

Treatment of *L. monocytogenes* infection includes bowel rest and rehydration with fluid and electrolytes replacement. Oral replacement may be sufficient, but affected patients may experience significant anorexia, necessitating intravenous fluid replacement and hospitalization in cases of significant dehydration or severe electrolyte abnormalities, particularly in young children or elderly patients.[70,71] Most gastrointestinal effects and associated flu-like symptoms rarely require hospitalization, and these effects resolve spontaneously in 1–2 days.[72] In cases of severe systemic listeriosis, prompt, early antimicrobial treatment is paramount. Virtually all human pathogenic *L. monocytogenes* organisms are susceptible to penicillin and its close derivatives (e.g., ampicillin, amoxicillin). Patients with confirmed or suspected listeriosis with extraintestinal manifestation should receive penicillin or a derivative until antimicrobial susceptibilities have been determined.[73] Therapy should continue for 4–6 weeks; however, patients with brain abscess or endocarditis may require a longer course of therapy.[74] Piperacillin, azlocillin, carbenicillin, and ticarcillin demonstrate intermediate antimicrobial activity against *L. monocytogenes*.[75] Penicillin-allergic patients may be placed on erythromycin or clarithromycin.[76] For immunocompromised patients, imipenem is a useful agent with a historically lower minimum inhibitory concentration than penicillin and broader coverage against other opportunistic organisms that threaten such patients.[77] Vancomycin is another alternative for immunocompromised patients; however, resistant *L. monocytogenes* strains are increasingly documented.[78] *L. monocytogenes* is resistant to cephalosporins, secondary to a wealth of penicillin-binding protein 3 to which cephalosporins bind poorly.[79] *L. monocytogenes* is also resistant to nalidixic acid and ofloxacin, and this organism is intermediately susceptible to levofloxacin, ciprofloxacin, and chloramphenicol.[80] Bacteriostatic activity of trimethoprim-sulfamethoxazole is not effective *in vivo*, secondary to rapid intracellular proliferation of *L. monocytogenes*.[80] Gentamicin is highly bacteriocidal, but this antibiotic is ineffective as a single agent because of the rapid transport of gentamicin into cellular lysosomes, where inactivation and limited contact with *L. monocytogenes* in the cytosol occurs. However, ampicillin plus gentamicin is a proven synergistic combination.[81]

References

1. Dumont J, Cotoni L. Bacille semblable au bacille du rouget du porc rencontré dans le liquide céphalo- rachidien d'un méningitique. Ann Inst Pasteur 1921;35:625–633.
2. Murray EG, Webb RA, Swann RA. A disease of rabbits characterised by a large mononuclear leucocytosis, caused by a hitherto undescribed bacillus *Bacterium monocytogenes* (n.sp.). J Pathol Bacteriol 1926;29:407–439.
3. Schlech WF 3rd, Lavigne PM, Bortolussi RA, Allen AC, Haldane EV, Wort AJ, et al. Epidemic listeriosis—evidence for transmission by food. N Engl J Med 1983;308:203–206.
4. Gross E, Slauson S. Update on emerging infections: news from the Centers for Disease Control and Prevention. Preliminary FoodNet Data on the incidence of infection with pathogens transmitted commonly through food—selected sites, United States, 2003. Ann Emerg Med 2004;44:532–536.
5. Watkins J, Sleath KP. Isolation and enumeration of *L. monocytogenes* from sewage, sewage sludge and river water. J Appl Bacteriol 1981;50:1–9.
6. Weis J, Seeliger HP. Incidence of *L. monocytogenes* in nature. Appl Microbiol 1975;30:29–32.
7. Hof H. *L. monocytogenes*: a causative agent of gastroenteritis? Eur J Clin Microbiol Infect Dis 2001;20:369–373.

8. McCarthy SA. *L. monocytogenes* in the environment. In: Miller AJ, Smith JL, Somkuti GA, editors. Foodborne listeriosis. New York: Elsevier; 1990:25–29.
9. Kalorey DR, Kurkure NV, Warke SR, Rawool DB, Malik SV, Barbuddhe SB. Isolation of pathogenic *L. monocytogenes* in faeces of wild animals in captivity. Compar Immunol Micorobiol Infect Dis 2006;29:295–300.
10. Mylonakis E, Paliou M, Hohmann EL. Listeriosis during pregnancy: a case series and review of 222 cases. Medicine (Baltimore) 2002;81:260–269.
11. Graham JC, Lanser S, Bignardi G, Pedler S, Hollyoak V. Hospital-acquired listeriosis. J Hosp Infect 2002;51:136–139.
12. Farber JM, Peterkin PI. *L. monocytogenes*, a food-borne pathogen. Microbiol Rev 1991;55:476–511.
13. McLauchlin J, Greenwood MH, Pini PN. The occurrence of *L. monocytogenes* in cheese from a manufacturer associated with a case of listeriosis. Int J Food Microbiol 1990;10:255–262.
14. McLauchlin J, Hall SM, Velani SK, Gilbert RJ. Human listeriosis and pâté: a possible association. Br Med J 1991;303:773–775.
15. Pinner RW, Schuchat A, Swaminathan B, Hayes PS, Deaver KA, Weaver RE, et al; the *L. monocytogenes* Study Group. Role of foods in sporadic listeriosis. II. Microbiologic and epidemiologic investigation. JAMA 1992;267:2046–2050.
16. Schuchat A, Deaver KA, Wenger JD, Plikaytis BD, Mascola L, Pinner RW, et al. Role of foods in sporadic listeriosis. 1. Case-control study of dietary risk factors. JAMA 1992;267:2041–2045.
17. Gianfranceschi M, Gattuso A, Fiore A, D'Ottavio MC, Casale M, Palumbo A, Aureli P. Survival of *L. monocytogenes* in uncooked Italian dry sausage (salami). J Food Protect 2006;69:1533–1538.
18. Pintado CM, Oliveira A, Pampulha ME, Ferreira MA. Prevalence and characterization of *L. monocytogenes* isolated from soft cheese. Food Microbiol 2005;22:79–85.
19. Raybourne RB. Virulence testing of *L. monocytogenes*. J AOAC Int 2002;85:516–523.
20. Zhang Y, Yeh E, Hall G, Cripe J, Bhagwat AA, Meng J. Characterization of *L. monocytogenes* isolated from retail foods. Int J Food Microbiol 2007;113:47–53.
21. Kathariou S, Graves L, Buchrieser C, Glaser P, Siletzky RM, Swaminathan B. Involvement of closely related strains of a new clonal group of L. monocytogenes in the 1998–99 and 2002 multistate outbreaks of foodborne listeriosis in the United States. Foodborne Pathog Dis 2006;3:292–302.
22. Yildirim S, Lin W, Hitchins AD, Jaykus LA, Altermann E, Klaenhammer TR, Kathariou S. Epidemic clone I-specific genetic markers in strains of *L. monocytogenes* serotype 4b from foods. Appl Environ Microbiol 2004;70:4158–4164.
23. Franciosa G, Scalfaro C, Maugliani A, Floridi F, Gattuso A, Hodzic S, Aureli P. Distribution of epidemic clonal genetic markers among Listeria monocytogenes 4b isolates. J Food Prot 2007;70:574–581.
24. Koutsoumanis KP, Sofos JN. Effect of inoculum size on the combined temperature, pH and Aw limits for growth of *L. monocytogenes*. Int J Food Microbiol 2005;104:83–91.
25. Gravani R. Incidence and control of *L. monocytogenes* in food-processing facilities. In: Ryser ET, Marth EH, editors. *L. monocytogenes*, listeriosis, and food safety. 2nd eds. New York: Marcel Dekker Inc.; 1999:657–709.
26. Lammerding AM, Doyle MP. Stability of *L. Monocytogenes* To non-thermal processing conditions. In: Miller AJ, Smith JL, Somkuti GA, editors. Foodborne listeriosis. New York: Elsevier; 1990:195–202.
27. Morales P, Calzada J, Nunez M. Effect of high-pressure treatment on the survival of *L. monocytogenes* Scott A in sliced vacuum-packaged Iberian and Serrano cured hams. J Food Protect 2006;69:2539–2543.
28. Yoon KS, Burnette CN, Abou-Zeid KA, Whiting RC. Control of growth and survival of *L. monocytogenes* on smoked salmon by combined potassium lactate and sodium diacetate and freezing stress during refrigeration and frozen storage. J Food Prot 2004;67:2465–2471.
29. Houben JH, Eckenhausen F. Surface pasteurization of vacuum-sealed precooked ready-to-eat meat products. J Food Prot 2006;69:459–468.
30. Low JC, Donachie W. A review of *L. monocytogenes* and listeriosis. Vet J 1997;153:9–29.
31. Heymer B, Wirsing von Konig CH, Finger H, Hof H, Emmerling P. Histomorphology of experimental listeriosis. Infection 1988;16:106–111.
32. Datta AR, Kothary MH. Effects of glucose, growth temperature, and pH on listeriolysin O production in L. monocytogenes. Appl Environ Microbiol 1993;59:3495–3497.
33. Glupczynski Y. Prolonged gastric acid suppression therapy: a significant risk factor for *L. monocytogenes* infection? Eur J Gastroenterol Hepatol 1996;8:1063–1036.
34. Merrell DS, Camilli A. Acid tolerance of gastrointestinal pathogens. Curr Opin Microbiol 2002;5:51–55.
35. Williams T, Joseph B, Beier D, Goebel W, Kuhn M. Response regulator DegU of *L. monocytogenes* regulates the expression of flagella-specific genes. FEMS Microbiol Lett 2005;252:287–298.
36. Galsworthy SB, Girdler S, Koval SF. Chemotaxis in *L. monocytogenes*. Acta Microbiol Hung 1990;37:81–85.
37. Lecuit M. Understanding how *L. monocytogenes* targets and crosses host barriers. Clin Microbiol Infect 2005;11:430–436.
38. Tresse O, Lebret V, Benezech T, Faille C. Comparative evaluation of adhesion, surface properties, and surface protein composition of *L. monocytogenes* strains after cultivation at constant pH of 5 and 7. J Appl Microbiol 2006;101:53–62.
39. Seveau S, Bierne H, Giroux S, Prevost MC, Cossart P. Role of lipid rafts in E-cadherin- and HGF-R/Met-mediated

entry of *L. monocytogenes* into host cells. J Cell Biol 2004;166:743–753.

40. Sousa S, Cabanes D, El-Amraoui A, Petit C, Lecuit M, Cossart P. Unconventional myosin VIIa and vezatin, two proteins crucial for *L. monocytogenes* entry into epithelial cells. J Cell Sci 2004;117:2121–2130.
41. Pizarro-Cerda J, Cossart P. Subversion of cellular functions by *L. monocytogenes*. J Pathol 2006;208:215–223.
42. Kayal S, Charbit A. Listeriolysin O: a key protein of *L. monocytogenes* with multiple functions. FEMS Microbiol Rev 2006;30:514–529.
43. Poussin MA, Goldfine H. Involvement of *L. monocytogenes* phosphatidylinositol-specific phospholipase C and host protein kinase C in permeabilization of the macrophage phagosome.Infect Immun 2005;73:4410–4413.
44. Churchill RL, Lee H, Hall JC. Detection of *L. monocytogenes* and the toxin listeriolysin O in food. J Microbiol Methods 2006;64:141–170.
45. Wisniewski JM, Bielecki J. Polymerizer-mediated intracellular movement. Pol J Microbiol 2004;53:35–38.
46. Zeile WL, Zhang F, Dickinson RB, Purich DL. *L. monocytogenes*'s right-handed helical rocket-tail trajectories: mechanistic implications for force generation in actin-based motility. Cell Motil Cytoskeleton 2005;60:121–128.
47. Rafelski SM, Theriot JA. Mechanism of polarization of *L. monocytogenes* surface protein ActA. Mol Microbiol 2006;59:1262–1279.
48. Berg RD. Bacterial translocation from the gastrointestinal tract. Trends Microbiol 1995;3:149–154.
49. Wells CL. Relationship between intestinal microecology and the translocation of intestinal bacteria. Antonie Van Leeuwenhoek 1990;58:87–93.
50. Berg RE, Crossley E, Murray S, Forman J. Relative contributions of NK and CD8 T cells to IFN-gamma mediated innate immune protection against *L. monocytogenes*. J Immunol 2005;175:1751–1757.
51. Bakardjiev AI, theriot JA, Portnoy DA. *L. Monocytogenes* traffics from maternal organs to the placenta and back. PLoS Pathog 2006;2:e66.
52. Skogberg K, Syrjanen J, Jahkola M. Clinical presentation and outcome of listeriosis in patients with and without immunosuppressive therapy. Clin Infect Dis 1992;14:815–821.
53. Dalton CB, Austin CC, Sobel J, Hayes PS, Bibb WF, Graves LM, et al. An outbreak of gastroenteritis and fever due to *L. monocytogenes* in milk. N Engl J Med 1997;336:100–105.
54. Cherubin CE, Appleman MD, Heseltine PN. Epidemiological spectrum and current treatment of listeriosis. Rev Infect Dis 1991;13:1108–1114.
55. Ooi ST, Lorber B. Gastroenteritis due to *L. monocytogenes*. Clin Infect Dis 2005;40:1327–1332.
56. Pigrau C, Almirante B, Pahissa A. Clinical presentation and outcome in cases of listeriosis. Clin Infect Dis 1993;17:143–144.
57. Mylonakis E, Hohmann EL, Calderwood SB. Central nervous system infection with *L. monocytogenes*. 33 years' experience at a general hospital and review of 776 episodes from the literature. Medicine (Baltimore) 1998;77:313–336.
58. Posfay-Barbe KM, Wald ER. Listeriosis. Pediatr Rev 2004;25:151–159.
59. Brouwer MC, van de Beek D, Heckenberg SG, Spanjaard L, de Gans J. Community-acquired *L. monocytogenes* meningitis In adults. Clin Infect Dis 2006;43:1233–1238.
60. Uldry PA, Kuntzer T, Bogousslavsky J. Early symptoms and outcome of *L. monocytogenes* rhombencephalitis: 14 adult cases. J Neurol 1993;240:235–242.
61. Grif K, Patscheider G, Dierich MP, Allerberger F. Incidence of fecal carriage of *L. monocytogenes* in three healthy volunteers: a one-year prospective stool survey. Eur J Clin Microbiol Infect Dis 2003;22:16–20.
62. Restaino L, Frampton EW, Irbe RM, Schabert G, Spitz H. Isolation and detection of *L. monocytogenes* using fluorogenic and chromogenic substrates for phosphatidylinositol-specific phospholipase C. J Food Prot 1999;62:244–251.
63. Parrisius J, Bhakdi S, Roth M, Tranum-Jensen J, Goebel W, Seeliger HP. Production of listeriolysin by beta-hemolytic strains of *L. monocytogenes*. Infect Immun 1986;51:314–319.
64. Greisen K, Loeffelholz M, Purohit A, Leong D. PCR primers and probes for the 16S rRNA gene of most species of pathogenic bacteria, including bacteria found in cerebrospinal fluid. J Clin Microbiol 1994;32:335–351.
65. Ueda S, Maruyama T, Kuwabara Y. Detection of *L. monocytogenes* from food samples by PCR after IMS-plating. Biocontrol Sci 2006;11:129–134.
66. Liu D. Identification, subtyping and virulence determination of *L. monocytogenes*, an important foodborne pathogen. J Med Microbiol 2006;55:645–659.
67. Horowitz W, editor. Official methods of analysis of AOAC International. 18th ed. Gaithersburg, MD: AOAC International; 2006.
68. Mazor M, Froimovich M, Lazer S, Maymon E, Glezerman M. *L. monocytogenes*. The role of transabdominal amniocentesis in febrile patients with preterm labor. Arch Gynecol Obstet 1992;252:109–112.
69. Doganay M. Listeriosis: clinical presentation. FEMS Immunol Med Microbiol 2003;35:173–175.
70. King CK, Glass R, Bresee JS, Duggan C. Centers for Disease Control and Prevention. Managing acute gastroenteritis among children: oral rehydration, maintenance, and nutritional therapy. MMWR Morb Mortal Wkly Rep 2003;52:1–16.
71. Smith JL. Foodborne illness in the elderly. J Food Prot 1998;61:1229–1239.
72. Gellin BG, Broome CV. Listeriosis. JAMA 1989;261:1313–1320.

73. Charpentier E, Gerbaud G, Jacquet C, Rocourt J, Courvalin P. Incidence of antibiotic resistance in *L. monocytogenes* species. J Infect Dis 1995;172:277–821.
74. Temple ME, Nahata MC. Treatment of listeriosis. Ann Pharmacother 2000;34:656–661.
75. McCracken GH Jr, Nelson JD, Thomas ML. Discrepancy between carbenicillin and ampicillin activities against enterococci and *L. monocytogenes*. Antimicrob Agents Chemother 1973;3:343–349.
76. Bauer J, Hof H, Nichterlein T, Wuenscher M. Comparative activities of macrolide derivatives on murine listeriosis. Zentralbl Bakteriol 1993;278:112–119.
77. Kim KS. *In vitro* and *in vivo* studies of imipenem-cilastatin alone and in combination with gentamicin against *L. monocytogenes*. Antimicrob Agents Chemother 1986;29: 289–293.
78. Krcmery V, Sefton A. Vancomycin resistance in gram-positive bacteria other than *Enterococcus* spp. Int J Antimicrob Agents 2000;14:99–105.
79. Tenover FC. Mechanisms of antimicrobial resistance in bacteria. Am J Infect Control 2006;34:3–10.
80. Boisivon A, Guiomar C, Carbon C. *In vitro* bactericidal activity of amoxicillin, gentamicin, rifampicin, ciprofloxacin and trimethoprim-sulfamethoxazole alone or in combination against *L. monocytogenes*. Eur J Clin Microbiol Infect Dis 1990;9:206–209.
81. Hof H. Listeriosis: therapeutic options. FEMS Immunol Med Microbiol 2003;35:203–205.

Chapter 19

SALMONELLA

Cyrus Rangan, MD, FAAP

HISTORY

During the late 1800s to the mid-1900s, *Salmonella* serovar Typhi (*Salmonella* Typhi) was the leading cause of salmonellosis in the United States, but by the 1950s improvements in sanitation nearly eliminated typhoid fever. Now nontyphoidal *Salmonella* infections are more common than typhoid.[1] Antibiotics were introduced into the practice of veterinary medicine during the 1940s and into animal husbandry as feed additives to promote faster growth of livestock and domestic fowl during the 1950s.[2] In England and Wales during the 1960s, drug-resistant strains of *Salmonella* (*Salmonella typhimurium* phage type DT 29, DT204, DT 193, DT203c) appeared in cattle.[3] During the 1980s and 1990s, *Salmonella* serovar Enteritidis (*Salmonella* Enteritidis) emerged as an important contaminant of raw and undercooked chicken eggs.[4] The incidence rate for isolates of this serovar peaked in 1996 at 3.6 per 100,000 persons, and the rate declined at the end of the 1990s to about 2 isolates per 100,000 persons.[5] During this period changes in meat- and poultry-processing mandated by the Pathogen Reduction and Hazard Analysis and Critical Control Points (HACCP) rule of the US Department of Agriculture occurred. The fraction of *Salmonella* serovar Typhimurium (*Salmonella* Typhimurium) isolates with multidrug resistance increased substantially in the 1990s, primarily as a result of the epidemic spread of DT104.[6] By 1997, *Salmonella* Typhimurium again replaced *Salmonella* Enteritidis as the most common serovar in *Salmonella* enterocolitis.[7] Rodents are an important animal reservoir for *Salmonella* Enteritidis. During the late 1800s and early 1900s, *Salmonella* Enteritidis was used as a rodenticide, including use during outbreaks of *Yersinia pestis* (Lehmann and Neumann) van Loghem. This use was banned in the United States during the 1920s and in the United Kingdom and Denmark during the 1960s.[8] *Salmonella*-based rodenticides are still used in some countries with the phage type being similar to past strains of *Salmonella* that caused human illness and fatalities.[9]

EXPOSURE

Nomenclature

Officially, the *Approved Lists of Bacterial Names* identifies the genus *Salmonella* as containing the following five species: *Salmonella arizonae, Salmonella choleraesuis* (type species of the genus), *Salmonella enteritidis, Salmonella typhi* and *Salmonella typhimurium*.[10] However, DNA relatedness studies indicate that strains from all five of these species form a single DNA hybridization group with seven phenotypically different subgroups.[11] Based on Judicial Opinion 80 from the Judicial Commission of the International Committee on the Systematics of Prokaryotes and recent publications, the most common nomenclature for the genus *Salmonella* consists of the following names: *Salmonella bongori*, a relatively new species, *Salmonella subterranea*, and *Salmonella enterica* including *Salmonella enterica* subsp. *arizonae, Salmonella enterica* subsp. *diarizonae, Salmonella enterica* subsp. *enterica, Salmonella enterica* subsp. *houtenae, Salmonella enterica* subsp. *indica*, and *Salmonella enterica* subsp. *salamae*.[12] The subspecies, *S. enterica* subsp. *enterica* (synonym: *Salmonella choleraesuis* subsp.

Medical Toxicology of Natural Substances, by Donald G. Barceloux, MD

choleraesuis) contains over 99% of human isolates.[13] Each of the subspecies of *Salmonella enterica* (*Salmonella choleraesuis*) has serovars. Serovars are not species and therefore the serovar names are not italicized. To avoid confusion between serovars and species, the names of serovars are printed in Roman type starting with a capital letter. Although the names of some serovars denote a syndrome or host specificity (e.g. *Salmonella* Typhi, *Salmonella* Abortusequi), most recent serovars are named for the geographical origin of the first strain of the serovar (e.g., London, Panama). Consequently, simplified nomenclature for *Salmonella enterica* subsp. *enterica* serovar Typhi is *Salmonella* Typhi (typhoid fever). There are over 2,500 *Salmonella* serovars.[14] According to the International Code of Nomenclature of Bacteria (1990 Revision), the term serovar is preferred to the term serotype.[15] The shortened nomenclature prohibits abbreviation of the name of the genus (*Salmonella* = *S.*) because the abbreviation of a name of a genus is only authorized when followed by the name of a species.

Source

The vast majority of cases of salmonellosis result from animal to human transmission via food, and *Salmonella* strains remain the most common cause of food poisoning among confirmed isolates of nine common enteric pathogens in the United States followed by *Campylobacter* and *Shigella*.[16] Most of the identified cases of *Salmonella* involve sporadic outbreaks of strains of *Salmonella enterica* subsp. *enterica* in single individuals rather than in several persons sharing a common meal.[17] Common serovars involved in human disease include *Salmonella* Choleraesuis (serovar rather than species), *Salmonella* Derby, *Salmonella* Enteritidis, *Salmonella* Infantis, *Salmonella* Muenchen, *Salmonella* Panama, *Salmonella* Typhimurium, and *Salmonella* Virchow. Other less common serovars include *Salmonella* Newport and *Salmonella* Heidelberg. The serovars *Salmonella* Typhi (typhoid fever) and *Salmonella* Paratyphi are highly adapted to humans and these serovars do not have other natural hosts, whereas *Salmonella* Typhimurium and *Salmonella* Enteritidis invade a broad range of hosts. Rodents are an important reservoir of *Salmonella* Enteritidis, particularly in poultry farms.

Animal

Salmonella enterica (*S. choleraesuis*) is a ubiquitous organism, particularly in the intestines of kittens, hedgehogs, birds (pet chickens, ducklings), pet rodents (hamsters, mice, rats),[18,19] and reptiles (lizards, snakes, turtles, crocodiles, kangaroo).[20–22] Pet ducklings are a source of *Salmonella* enterocolitis (*Salmonella* Hadar),[23] and small pet turtles harbor rare *Salmonella* serovars (e.g., *Salmonella* Pomona) pathogenic to humans, especially infants and children <5 years of age.[24,25] Approximately 7% of the cases of salmonellosis are attributed to pet reptile or amphibian contacts, and these infections usually involve relatively rare *Salmonella* serovars (e.g., Abaetetuba, Carrau, Chameleon, Ealing, Flint, Marina, Java, Kintambo, Poona, Stanley, Wassenaar).[26,27] Common serovars in food processing facilities include the following: broiler plants—*Salmonella* Kentucky, *Salmonella* Heidelberg; turkey plants—*Salmonella* Hadar, *Salmonella* Heidelberg; and swine plants—*Salmonella* Derby, *Salmonella* Typhimurium.[28] Although egg farms contain a variety of *Salmonella* serovars (*Salmonella* Pullorum, *Salmonella* Gallinarum, *Salmonella* Heidelberg), *Salmonella* Enteritidis is the only human pathogen that routinely contaminates chicken eggs.[29]

Food

Foods contaminated by *Salmonella* strains usually appear and taste normal. Sources of salmonellosis include contamination or cross-contamination of raw or undercooked eggs,[30,31] alfalfa sprouts (*Salmonella enterica* subsp. *enterica* serovars Newport, Stanley, Kottbus),[32–34] cantaloupe,[35] peanut butter,[36] undercooked meat (beef, pork),[37,38] poultry products,[39] unpasteurized milk[41,41] and dairy products (e.g., unpasteurized ice cream, cheese).[42] Shell eggs are an important source of *Salmonella* Enteritidis-induced enterocolitis.[43,44] Potential sources of salmonellosis from food containing raw eggs include homemade eggnog or ice cream, hollandaise sauce, cold deserts, and Caesar salad dressing. Commercial eggnog does not contain clinically significant numbers of *Salmonella* species because of the use of pasteurized eggs. *Salmonella* species occur most often on the shell of the egg as a result of fecal contamination. Internal contamination of the egg occurs from transmission of the bacteria through the shell (e.g., damp or contaminated environment) or transovarially during the formation of the yolk in the ovary or oviduct of an infected hen.[17] The ingestion of unpasteurized milk is a major risk factor for the acquisition of *Salmonella* Dublin infection.[45] Freshly cooked eggs do not usually cause *Salmonella* enterocolitis because the infective dose is low, unless the yolk is inadequately heated. Cross-contamination of food (spices, shellfish) and food contact surfaces by poultry carcasses[46] or infected food handlers[47] may cause outbreaks of *Salmonella* food poisoning. Other less common sources of Salmonella include lettuce, uncooked tomatoes,[48] cantaloupe (*Salmonella* Poona, *Salmonella* Saphra),[49] and cilantro.[50]

Food Processing

Although temperatures as low as 60°C (140°F) kill *Salmonella* bacteria in free egg fluid within a few minutes, light cooking does not eliminate *Salmonella* from the interior of the egg. Cooking must completely coagulate both the white and yolk to ensure the destruction of viable bacteria.[51] Antibiotic resistance in strains of *Salmonella* is chromosomally mediated, and proper cooking techniques (e.g., pasteurization) destroy these strains of *Salmonella* in a manner similar to that for less resistant strains.[52] *Salmonella* can survive refrigeration temperatures in an acidic environment, such as unpasteurized orange juice, for sufficient time to cause enterocolitis.[53] Table 19.1 lists recommendations for storage, cooking, and serving techniques that reduce the incidence of salmonellosis associated with the consumption of eggs.

TOXINS

Physiochemical Properties

Large numbers of *Salmonella* are usually required to cause invasion of the large- and small-bowel mucosa.[54] *Salmonella* species excrete a heat-labile enterotoxin that directly stimulates the adenylate cyclase-cAMP system causing secretion of chloride ions and inhibition of sodium absorption from the intestines.

Mechanism of Toxicity

Although the *Salmonella* enterotoxin potentially can increase the secretion of an isotonic load into the intestines, there are few data to support this mechanism as the cause of diarrhea during human infections. Instead, these organisms produce an inflammatory diarrhea by invading the mucosa. *In vitro* studies suggest that an invasion-associated type III export system (encoded on *Salmonella* pathogenicity island 1) facilitates tissue invasion and triggers an inflammatory response that produces fluid accumulation and leukocyte influx.[55] The distal ileum probably is the portal of entry. In the mouse model, M cells located in the follicle-associated epithelium in the Peyer's patches are the primary intestinal epithelial cell type invaded by *Salmonella* in a process that morphologically resembles phagocytosis.[56] In most cases, the organisms remain in the intestine and the infection is localized and self-limited. Occasionally, an initial bacteremia causes the seeding of infectious foci distant to the gastrointestinal tract. The production of

TABLE 19.1. Recommendations for the Use of Eggs by Consumers

Storage:
- Use grade A or AA eggs that have been properly refrigerated (≤45°F/7°C), clean, and are not cracked during storage prior to purchase.
- Refrigerate eggs (<40°F/4°C) in original container as soon as possible after purchase.
- Washing eggs prior to storage is not necessary.
- Rotate eggs so that the oldest egg is used first.
- Use shell eggs within 2–3 weeks of purchase, yolks and egg whites within 4 days, and boiled eggs within one week.
- Refrigerate leftover cooked egg dishes and use within 3–4 days.
- Liquid, frozen, and dried eggs and egg products should be pasteurized.
- Freeze eggs without the shell and use frozen eggs within one year.
- Cooked eggs for a picnic should be packed in an insulated cooler with adequate amounts of ice or frozen gel packs.

Cooking:
- Cook (≥145°F/63°C for ≥15 seconds) eggs thoroughly until both yolk and white are firm. Casseroles and other egg-containing dishes should be cooked to at least 160°F/71°C, documented by a thermometer.
- Use pasteurized eggs rather than raw eggs in recipes that recommend raw eggs (Caesar salad, mayonnaise, ice cream, hollandaise sauce, and homemade eggnog).
- Wash hands, equipment, utensils, and work surfaces with hot, soapy water after contact with raw eggs.

Serving:
- Serve cooked eggs and egg-containing foods immediately after cooking or cooled to room temperature (70°F/21°C) within 2 hours and from room temperature to refrigeration (41°F/5°C) within an additional 4 hours.
- Refrigerate or reheat cooked eggs if the food sits at room temperature for <2 hours.
- Eggs and egg dishes (soufflés, quiches) should be reheated to 165°F (74°C) before serving if refrigerated.
- Susceptible individuals (infants, elderly, immunocompromised persons) should avoid eating raw or undercooked eggs (pasteurized egg products are appropriate substitutes).

Source: Adapted from Ref 5.

bacterial stress proteins can also produce postinfection inflammatory sequelae.

DOSE RESPONSE

In general, large numbers of *Salmonella* organisms are necessary to cause gastrointestinal illness because *Salmonella* is heat labile and acid labile in stomach fluid. A variety of factors affect the infectivity rate including the dose of the organism, the complex interaction of host resistance (e.g., fatty vehicles, achlorhydria, antacid use, state of health, age), and the virulence of the organism. *Salmonella* Typhi has a much lower minimum infective dose compared with other serovars (*Salmonella* Enteritidis, *Salmonella* Pullorum) that are not adapted to humans.[57] In an experimental study of volunteers from a penal institution using a variety of *Salmonella* serovars (*Salmonella* Anatum, *Salmonella* Bareilly, *Salmonella* Derby, *Salmonella* Meleagridis, *Salmonella* Newport, *Salmonella* Pullorum, *Salmonella* Typhi), the minimum infective dose exceeded 10^5–10^7 organisms.[58–60] Higher infective doses (10^8–10^9) increase the attack rate, shorten the incubation period, and may produce severe gastrointestinal disease depending on the susceptibility of the host.[61,62] The organism is ubiquitous in the environment and illness usually results from the ingestion of contaminated food. Small numbers (i.e., <1000) of bacteria do not usually survive the defenses of the gastrointestinal tract in healthy adults. However, case reports indicate that certain serovars (e.g., *Salmonella* Typhimurium, *Salmonella* Heidelberg, *Salmonella* Napoli, *Salmonella* Eastbourne) may produce gastrointestinal illness following inoculation with infective doses below 10^3.[63,64]

CLINICAL RESPONSE

The clinical manifestations of *Salmonella* infections include the following five main syndromes: enterocolitis, enteric fever, bacteremia, localized infection, and a chronic enteric or urinary carrier state. Complications of *Salmonella* infections include bacteruria, meningitis, septic arthritis, osteomyelitis, cholangitis, and pneumonia. These latter complications usually occur in immunocompromised patients. These patients typically develop these complications without obvious gastrointestinal symptoms, particularly after exposure to the serovar, *Salmonella* Choleraesuis.[13] Most healthy adults usually experience only self-limited diarrhea for 1–2 days.

Enterocolitis

The most common form of *Salmonella* infection is enterocolitis. *Salmonella* gastroenteritis is difficult to distinguish clinically from bacterial enterocolitis caused by other bacterial pathogens. The typical incubation period is about 6–72 hours with a median onset about 18–24 hours after eating contaminated food.[42,65,66] Patients present with the acute onset of fever, diarrhea, and abdominal cramps. Diarrhea is the cardinal manifestation of *Salmonella* gastroenteritis that ranges from a few loose stools to a severe watery diarrhea. In a majority of the cases, the stools are nonbloody and moderate in volume, although gross blood and tenesmus may occur in some cases. The presence of very bloody diarrhea suggests *Shigella* or enterohemorrhagic *E. coli* infection. Nausea and vomiting are brief (i.e., few hours) or absent, but myalgias and headache occur commonly. Chills, fever (38.0°C–39.0°C/100.4°F–102.2°F), and abdominal cramps develop in most patients with *Salmonella* enterocolitis. The fever usually resolves within 3 days and the diarrhea within 2–7 days.[67,68] Prolonged fever and diarrhea more than 7 days after the onset of illness suggest a complication or another diagnosis.

Complications of *Salmonella* enterocolitis in children and adults include reactive arthritis, Reiter's syndrome (arthritis, urethritis, conjunctivitis), and rash (erythema nodosum).[69–71] Reactive arthritis is an aseptic form of articular inflammation resulting from bacterial infections localized in the gastrointestinal or urogenital tracts. The typical manifestation of reactive arthritis is inflammation in a single joint of the lower extremity. A migratory, asymmetrical polyarthropathy typically may begin within 1–3 weeks of the onset of diarrhea.[72,73] The weight-bearing joints (knees, ankles) are the most frequently affected joints. Arthralgias typically resolve within 6 months, but symptoms may persist over one year.[74,75] Extraarticular manifestations (e.g., enthesitis, tenosynovitis, bursitis, dactylitis) can occur. Viable organisms are not detectable in synovial fluid, but immunoblotting and immunofluorescence testing of synovial fluid may demonstrate *Salmonella* lipopolysaccharide antibodies.[76] A substantial majority of these patients are HLA-B27 or HLA-B7 positive.[74,77] Other causes of reactive arthritis and Reiter's syndrome in susceptible individuals include yersiniosis, *Campylobacter* enteritis, shigellosis, and chlamydial urethritis.[78]

Bacteremia

Salmonella-induced bacteremia usually occurs in young children (<2 years),[79] sickle cell disease patients, the elderly, or immunosuppressed patients (e.g., HIV). The bacteremia is usually transient, but serious complications may develop including septicemia, septic arthritis, endarteritis, and meningitis, particularly in infants <3 months of age or in immunocompromised patients.[80] Watery diarrhea may precede the development of complications, such as septic arthritis up to 2 weeks. The *Salmonella* serovars Choleraesuis and Dublin demonstrate a higher predilection for producing bacteremia in humans, often without intestinal involvement.

Enteric Fever

The term enteric fever includes typhoid and paratyphoid fever with the former caused by *S. enterica* subsp. *enterica* serovar Typhi and the latter by *S. enterica* subsp. *enterica* serovars Paratyphi A, B (*Salmonella* Schottmuelleri), or C (*Salmonella* Hirschfeldii).[81] The clinical syndromes associated with these two types of infections are similar, but the paratyphoid fever is usually less severe than typhoid fever. Classically, the incubation period for the onset of enteric fever is about 7–21 days with active invasion of gut lymphoid tissue with the establishment of bacterial foci in the gut during the second and third week postexposure, respectively, to *Salmonella* Typhi and *Salmonella* Paratyphi. Symptoms of *Salmonella*-induced enteric fever include recurrent high fever, headache, meningismus, nausea, anorexia, myalgias, cough, and sore throat. Leukopenia, a relative bradycardia, and "rose spots" occur more often during typhoid fever than paratyphoid fever. The infection may be localized to the gastrointestinal tract, respiratory tract, urinary tract, or musculoskeletal system. Occasionally, overwhelming sepsis occurs. Intestinal bleeding during typhoid fever may occur in the ileum or proximal colon as a result of multiple ulcerations.[82] Other potential complications of enteric fever include coma, seizures, meningitis, intestinal perforation, septic arthritis, pneumonia, cholecystitis, hepatitis, osteomyelitis, parotitis, orchitis, and nephritis. The mortality rate of untreated patients is about 10%.

Local Infection

Soft tissue infections are relatively rare complications of exposure to *Salmonella* serovars with most cutaneous infections developing in immunocompromised patients. Case reports associated a hemorrhagic cellulitis with nontyphoid salmonellosis[83] and a cutaneous ulcer with *Salmonella* Typhi.[84]

Asymptomatic Carriage

Typically, clearance of Salmonella by asymptomatic carriers requires about one month depending on host factors, *Salmonella* serovar, and the extent of the *Salmonella* infection.[85] However, chronic or persistent carriers excrete *Salmonella* in the stool for more than one year. The incidence of chronic, asymptomatic carrier state following infection of healthy patients by nontyphoidal salmonella is relatively low (i.e., <1%).[86] Asymptomatic, infected food handlers, hospital staff, or patients may be sources of nosocomial *Salmonella* outbreaks.[87,88] Some survivors become asymptomatic carriers, and an Indian case-control study suggests that the asymptomatic carrier state increases the risk of carcinoma of the gallbladder.[89]

DIAGNOSTIC TESTING

Analytical Methods

There are approximately 40 AOAC (Association of Analytical Chemists International, Gaithersburg, MD; http://www.eoma.aoac.org/) official methods of analysis for *Salmonella,* primarily for the detection of these organisms in food.[90]

Biomarkers

Salmonella are facultative anaerobes that grow well in standard blood and tissue culture media. Although there are about 2,500 antigenically distinct *Salmonella* serovars, most salmonellosis results from about 10 *Salmonella* serovars (Typhimurium, Newport, Enteritidis, Infantis, Heidelberg, Agona, St. Paul, Derby, Oranienburg, Typhi). Up to two-thirds of subjects over 5 years of age continue to shed *Salmonella* in the stool 2–3 weeks after the onset of illness. In a review of 32 studies involving over 2,800 cases of salmonellosis, the median duration of salmonella excretion was 5 weeks.[86] In patients over the age of 5 years with symptomatic infections or patients infected with serovars other than *Salmonella* Typhimurium, persistent excretion of *Salmonella* in the stool is rare (<1%). Stool cultures establish the disease state, and sensitivities help select the appropriate antibiotic. Blood and urine cultures during the active phase of enteric fever frequently are positive for *Salmonella* serovars.

Abnormalities

The leucocyte count during *Salmonella* enterocolitis is usually normal. Febrile agglutinin tests to *Salmonella* group antigens are difficult to interpret and are less specific than clinical and epidemiological findings.[91]

TREATMENT

Cases of typhoid fever should be promptly treated with appropriate antibiotics (fluoroquinolone drugs with ceftriaxone and azithromycin as alternatives). Antibiotic therapy is *not* usually indicated in uncomplicated nontyphoidal *Salmonella* enterocolitis, and antibiotics should be administered selectively to bacteremic patients and high-risk individuals (sickle cell disease, diabetes, AIDS, cancer, immunosuppressive therapy, elderly, infants). Although treatment with quinolones may decrease the duration of symptoms of *Salmonella* enterocolitis by 1–3 days, the clinical significance of this improvement is uncertain.[94] The administration of quinolones in children or pregnant women should be done *only* after a careful evaluation of the risks and benefits of quinolone use. Consequently, most patients with mild *Salmonella* enterocolitis do not require antibiotic treatment.[92] Although quinolones do not unequivocally prolong the carrier state and there is limited evidence that antibiotics eradicate the carrier following *Salmonella* enterocolitis,[93] the use of antibiotics does not consistently shorten or eradicate the carriage of *Salmonella* organisms.[94] Antibiotics for the treatment of salmonellosis include ampicillin, trimethoprim-sulfamethoxazole, extended spectrum cephalosporins (cefotaxime, ceftriaxone), and fluoroquinolones (ciprofloxacin, ofloxacin). The choice of antibiotics depends on the susceptibilities or the presence of multidrug-resistant strains of *S. typhimurium*. Although fluoroquinolones and extended spectrum cephalosporins are alternatives for drug-resistant strains of *Salmonella*, low-level resistance is also emerging to these antibiotics.[95] Chloramphenicol resistance is a simple screening test for the 5-drug-resistant pattern (ampicillin, chloramphenicol, streptomycin, sulfonamides, tetracycline).[96] Either ampicillin or trimethoprim-sulfamethoxazole is the initial antibiotics of choice unless stool culture or local drug-resistance pattern indicate a lack of sensitivity to local organisms. Alternatives for patients with multiple allergies or unusually resistant organisms include azithromycin and aztreonam. Treatment with antibiotics probably does not induce a carrier state or alter the duration of *Salmonella* excretion in the stool.[97]

Treatment with an antimotility agent (e.g., loperamide) may decrease the number of diarrheal stools, and this treatment probably does not prolong the course or increase the incidence of the carrier state.[98,99] However, antimotility agents should be used with caution in patients with bloody diarrhea because the use of these agents are contraindicated in patients with infections resulting from Shiga toxins. Antimotility agents are not recommended in children under the age of 5 years. Because most bacterial gastrointestinal infections are self-limited, a positive stool culture does not mandate initiation of antibiotic therapy.[100] Such patients should be reevaluated and if symptoms persist or worsen, antibiotic therapy is appropriate.

Treatment of reactive arthritis involving *Salmonella* includes NSAIDs and sulfasalazine.[78] Steroids are administered when inflammatory symptoms are resistant to NSAIDs. Treatment for refractory cases includes the administration of azathioprine, methotrexate, cyclosporine, or tumor necrosis factor-alpha blockers (adalimumab, etanercept, infliximab).

References

1. Tauxe RV. *Salmonella*: a postmodern pathogen. J Food Protect 1991;54:563–568.
2. Rabsch W, Tschape H, Baumler AJ. Non-typhoidal salmonellosis: emerging problems. Microbes Infect 2001;3; 237–247.
3. Anderson ES. Drug resistance in *Salmonella typhimurium* and its implications. Br Med J 1968;3:333–339.
4. Vugia DJ, Mishu B, Smith M, Tavris DR, Hickman-Brenner FW, Tauxe RV. *Salmonella enteritidis* outbreak in a restaurant chain: the continuing challenges of prevention. Epidemiol Infect 1993;110:49–61.
5. Centers for Disease Control (CDC). Outbreaks of *Salmonella* serotype Enteritidis infection associated with eating raw or undercooked shell eggs—United States, 1996–1998. MMWR Morb Mortal Wkly Rep 2000;49: 73–79.
6. Angulo FJ, Johnson KR, Tauxe RV, Cohen ML. Origins and consequences of antimicrobial-resistant nontyphoid *Salmonella*: implications for the use of fluoroquinolones in food animals. Microb Drug Resist 2000;6:77–83.
7. Centers for Disease Control (CDC). *Salmonella* surveillance: annual tabulation summary, 1998. Atlanta: US Department of Health and Human Services, CDC; 1999.
8. Friedman CR, Malcolm G, Rigau-Perez JG, Armbulo PD, Tauxe RV. Public health risk from *Salmonella*-based rodenticides. Lancet 1996;347:1705–1706.
9. Painter JA, Molbak K, Sonne-Hanaen J, Barrett T, Wells JG, Tauxe RV. *Salmonella*–based rodenticides and public health. Emerging Infect Dis 2004;10:985–987.
10. Skerman VB, McGowan V, Sneath PH. Approved list of bacterial names (Amended) & Index of the Bacterial and Yeast Nomenclatural Changes. Herndon, VA; ASM Press; 1989:298 pp.
11. Tindall BJ, Grimont PA, Garrity GM, Euzeby JP. Nomenclature and taxonomy of the genus *Salmonella*. Int J Syst Evol Microbiol 2005;55:521–524.
12. Judicial Commission of the International Committee on Systematics of Prokaryotes. In: Proceedings of the Xth International (IUMS) Congress of Bacteriology and Applied Microbiology; Minutes of the meetings; 2002 July 28–August 1; Paris, France. Int J Syst Evol Microbiol 2005;55:525–532.
13. Chiu C-H, Su L-H, Chu C. *Salmonella enterica* serotype *choleraesuis*: epidemiology, pathogenesis, clinical disease, and treatment. Clin Microbiol Rev 2004;17:311–322.

14. Brenner FW, Villar RG, Angulo FJ, Tauxe RV, Swaminathan B. *Salmonella* nomenclature. J Clin Microbiol 2000;38:2465–2467.
15. Sneath PH: International Code of Nomenclature of Bacteria: Bacteriological Code, 1990 Revision. Herndon, VA; ASM Press; 1992;238 pp.
16. Centers for Disease Control (CDC). Preliminary FoodNet data on the incidence of infection with pathogens transmitted commonly through food—selected sites, United States, 2003. MMWR Morb Mortal Wkly Rep 2004;53: 338–343.
17. Duguid JP, North RAE. Eggs and salmonella food-poisoning: an evaluation. J Med Microbiol 1991;34:65–72.
18. Swanson SJ, Snider C, Braden CR, Boxrud D, Wunschmann A, Lockett J, Smith KE. Multidrug-resistant *Salmonella enterica* serotype Typhimurium associated with pet rodents. N Engl J Med 2007;356:21–28.
19. Smith K, Boxrud D, Leano F, Snider C, Braden C, Lockett J, et al. Outbreak of multidrug-resistant *Salmonella typhimurium* associated with rodents purchased at retail pet stores—United States, December 2003—October 2004. MMWR Morb Mortal Wkly Rep 2005;54:429–433.
20. Centers for Disease Control (CDC). Turtle-associated salmonellosis in humans—United States, 2006–2007. MMWR Morb Mortal Wkly Rep 2007;56:649–652.
21. Thomas AD, Forbes-Faulkner JC, Speare R, Murray C. Salmonelliasis in wildlife from Queensland. J Wildlife Dis 2001;37:229–238.
22. Centers for Disease Control (CDC). Outbreaks of multidrug-resistant *Salmonella typhimurium* associated with veterinary facilities—Idaho, Minnesota, and Washington, 1999. MMWR Morb Mortal Wkly Rep 2001;50:701–704.
23. Center for Disease Control (CDC). *Salmonella hadar* associated with pet ducklings—Connecticut, Maryland, and Pennsylvania, 1991. MMWR Morb Mortal Wkly Rep 1992;41:185–187.
24. Salna B, Monson T, Kurzynski T, Gundlach K, Fox PE, Kazmierczak J, et al. Salmonellosis associated with pet turtles—Wisconsin and Wyoming, 2004. MMWR Morb Mortal Wkly Rep 2005;54:223–226.
25. Centers for Disease Control (CDC). Reptile-associated salmonellosis—selected states, 1994–1995. MMWR Morb Mortal Wkly Rep 1995;44:347–350.
26. Mermin J, Hoar B, Angulo FJ. Iguanas and *Salmonella* Marina infection in children: a reflection of the increasing incidence of reptile-associated salmonellosis in the United States. Pediatrics 1997;99:399–402.
27. Ackman DM, Drabkin P, Birkhead G, Cielak P. Reptile-associated salmonellosis in New York State. Pediatr Infect Dis J 1995;14:955–959.
28. Olsen SJ, Bishop R, Brenner FW, Roels TH, Bean N, Tauxe RV, Slutsker L. The changing epidemiology of *Salmonella*: trends in serotypes isolated from humans in the Untied States, 1987–1997. J Infect Dis 2001;183:753-761.
29. Guard-Petter J. The chicken, the egg and *Salmonella enteritidis*. Environ Microbiol 2001;3:421–430.
30. Passaro DJ, Reporter R, Mascola L, Kilman L, Malcolm GB, Rolka H, Werner SB, Vugia DJ. Epidemic *Salmonella enteritidis* infection in Los Angeles County, California the predominance of phage type 4. West J Med 1996; 165:126–130.
31. St. Louis ME, Morse DL, Potter ME, DeMelfi TM, Guzewich JJ, Tauxe RV, Blake PA. The emergence of grade A eggs as a major source of *Salmonella enteritidis* infections new implications for the control of salmonellosis. JAMA 198;259:2103–2107.
32. Stewart DS, Reineke KF, Ulaszek JM, Tortorello ML. Growth of *Salmonella* during sprouting of alfalfa seed associated with salmonellosis outbreaks. J Food Protect 2001;64:618–622.
33. Van Beneden Ca, Keene WE, Strang RA, Werker DH, King AS, Mahon B, et al. Multinational outbreak of *Salmonella enterica* serotype Newport infection due to contaminated alfalfa sprouts. JAMA 1999;281:158–162.
34. Mohle-Boetani J, Werner B, Polumbo M, Farrar J, Vugia D, Anderson S, et al. Alfalfa sprouts–Arizona, California, Colorado, and New Mexico, February–April 2001. MMWR Morb Mortal Wkly Rep 2002;51:7–9.
35. Anderson SM, Verchick L, Sowadsky R, Sun B, Civen R, Mohle-Boetani JC, et al. Multistate outbreaks of Salmonella serotype Poona infections associated with eating cantaloupe from Mexico–United States and Canada, 2000–2002. MMWR Morb Mortal Wkly Rep 2002;51: 1044–1047.
36. Centers for Disease Control (CDC). Multistate outbreak of *Salmonella* serotype Tennessee infections associated with peanut butter—United States, 2006–2007. MMWR Morb Mortal Wkly Rep 2007;56:521–524.
37. Centers for Disease Control (CDC). Outbreak of salmonellosis associated with beef jerky—New Mexico, 1995. MMWR Morb Mortal Wkly Rep 1995;44:785–788.
38. Delarocque-Astagneau E, Bouillant C, Vaillant V, Bouvet P, Grimont PAD, Desenclos J-C. Risk factors for the occurrence of sporadic *Salmonella enterica* serotype Typhimurium infections in children in France: a national case-control study. Clin Infect Dis 2000;31:488–492.
39. Barrell RAE. Isolations of salmonellas from humans and foods in the Manchester area: 1981–1985. Epidemiol Infect 1987;998:277–284.
40. Centers for Disease Control (CDC). Salmonellosis from inadequately pasteurized milk—Kentucky. MMWR Morb Mortal Wkly Rep 1984;33:505–506.
41. Cody SH, Abbott SL, Marfin AA, Schulz B, Wagner P, Robbins K, et al. Two outbreaks of multidrug-resistant *Salmonella* serotype Typhimurium DT104 infections linked to raw-mild cheese in Northern California. JAMA 1999;281:1805–1810.
42. Gunn RA, Markakis G. Salmonellosis associated with homemade ice cream an outbreak report and summary of outbreaks in the United States in 1966 to 1976. JAMA 1978;240:1885–1886.

43. Sobel J, Hirshfeld AB, McTigue K, Burnett CL, Altekruse S, Brenner F, Malcolm G, Mottice SL, Nichols CR, Swerdlow DL. The pandemic of *Salmonella* Enteritidis phage type 4 reaches Utah: a complex investigation confirms the need for continuing rigorous control measures. Epidemiol Infect 2000;125:1–8.
44. Palmer S, Parry S, Perry D, Smith R, Evans M, Nehaul L, Roberts R, Walapu M, Wright D. The role of outbreaks in developing food safety policy: population based surveillance of salmonella outbreaks in Wales 1986–98. Epidemiol Infect 2000;125:467–472.
45. Fang FC, Fierer J. Human infection with *Salmonella* Dublin. Med 1991;70:198–207.
46. Small RG, Halliday GJ, Haxton AR, Hird MR,Sharp JCM, Wallace JM. *Salmonella* food–poisoning from turkeys. Health Bull (Edinb) 1976;34:206–211.
47. Opal SM, Mayer KH, Roland F, Brondum J, Heelan J, Lyhte L. Investigation of a food-borne outbreak of salmonellosis among hospital employees. Am J Infect Control 1989;17:141–147.
48. Hedberg CW, Angulo FJ, White KE, Langkop CW, Schell WL, Stobierski MG, et al: Outbreaks of salmonellosis associated with eating uncooked tomatoes: implications of public health. Epidemiol Infect 1999;122:385–393.
49. Mohle-Boetaini JC, Reporter R, Werner SB, Abbott S, Farrar J, Waterman SH, Vugia DJ. An outbreak of *Salmonella* serogroup Saphra due to cantaloupes from Mexico. J Infect Dis 1999;180:1361–1364.
50. Campbell JV, Mohle-Boetaini J, Reporter R, Abbott S, Farrar J, Brandl M, Mandrell R, Werner SB. An outbreak of *Salmonella* serotype Thompson associated with fresh cilantro. J Infect Dis 2001;183:984–987.
51. Humphrey TJ, Greenwood M, Gilbert RJ, Rowe B, Chapman PA. The survival of salmonellas in shell eggs cooked under simulated domestic conditions. Epidemiol Infect 1989;103:35–45.
52. Humphrey T. *Salmonella typhimurium* definitive type 104 a multi-resistant *Salmonella*. Int J Food Microbiol 2001;67:173–186.
53. Sharma M, Beuchat LR, Doyle MP, Chen J. Survival of Salmonellae in pasteurized, refrigerated calcium-fortified orange juice. J Food Protect 2001;64:1299–1304.
54. Yoshikawa TT, Herbert P, Oill PA. Clinical spectrum and management of salmonellosis. West J Med 1980;133: 408–417.
55. Wallis TS, Wood M, Watson P, Paulin S, Jones M, Galyov E. Sips, Sops, and SPIs but not stn influence *Salmonella* enteropathogenesis. Adv Exp Med Biol 1999;473: 275–280.
56. Santos RL, Tsolis RM, Baumler AJ, Adams LG. Pathogenesis of *Salmonella*-induced enteritis. Bras J Med Biol Res 2003;36:3–12.
57. McCullough NB, Eisele CW. Experimental human salmonellosis. IV. Pathogenicity of strains of *Salmonella* Pullorum obtained from spray-dried whole egg. J Infect Dis 1951;89:259–265.
58. McCullough NB, Eisele CW. Experimental human salmonellosis. III. Pathogenicity of strains of *Salmonella* Newport, *Salmonella* Derby, and *Salmonella* Bareilly obtained from spray-dried whole egg. J Infect Dis 1951;89:209–213.
59. McCullough NB, Eisele CW. Experimental human salmonellosis. I. Pathogenicity of strains of *Salmonella* Meleagridis and *Salmonella* Anatum obtained from spray-dried whole egg. J Infect Dis 1951;88:278–289.
60. Hook EW. Salmonellosis: certain factors influencing the interaction of *Salmonella* and the human host. Bull NY Acad Med 1961;37:499-512.
61. Taylor DN, Bopp C, Birkness K, Cohen ML. An outbreak of salmonellosis associated with a fatality in a healthy child: a large dose and severe illness. Am J Epidemiol 1984;119:907–912.
62. Blaser JM, Newman LS. A review of human salmonellosis: I. Infective Dose. Rev Infect Dis 1982;4:1096–1106.
63. Greenwood MH, Hooper WL. Chocolate bars contaminated with *Salmonella* Napoli: an infectivity study. Br Med J 1983;286;1394.
64. Fontaine RE, Cohen ML, Martin WT, Vernon TM. Epidemic salmonellosis from cheddar cheese: surveillance and prevention. Am J Epidemiol 1980;111:247–253.
65. Narain JP, Lofgren JP. Epidemic of restaurant-associated illness due to *Salmonella* Newport. Southern Med J 1989;82:837–840.
66. Tauxe RV, Tormey MP, Mascola L, Hargrett-Bean NT, Blake PA. Salmonellosis outbreak on transatlantic flights; foodborne illness on aircraft: 1947–1984. Am J Epidemiol 1987;125:150–157.
67. Spitalny KC, Okowitz EN, Vogt RL. Salmonellosis outbreak at a Vermont hospital. Southern Med J 1984;77: 168–172.
68. Centers for Disease Control (CDC). Outbreak of *Salmonella enteritidis* infection associated with consumption of raw shell eggs. MMWR Morb Mortal Wkly Rep 1992;41:369–372.
69. Steckelberg JM, Terrell CL, Edson RS. Laboratory-acquired *Salmonella typhimurium enteritis*: association with erythema nodosum and reactive arthritis. Am J Med 1988;85:705–707.
70. Saari KM, Vipppula A, Lassus A, Leirisalo M, Saari R. Ocular inflammation in Reiter's disease after *Salmonella enteritis*. Am J Ophthalmol 1980;90:63–68.
71. Inman RD, Johnston EA, Hodge M, Falk J, Helewa A. Postdysenteric reactive arthritis a clinical and immunogenetic study following an outbreak of salmonellosis. Arthritis Rheumatism 1988;31:1377–1383.
72. Maki-Ikola O, Granfors K. Salmonella-triggered reactive arthritis. Lancet 1992;339:1096–1098.
73. Stein HB, Abdullah A, Robinson HS, Ford DK. *Salmonella* reactive arthritis in British Columbia. Arthritis Rheumatism 1980;23:206–210.
74. Kanakoudi-Tsakalidou F, Pardalos G, Pratsidou-Gertsi P, Kansouzidou-Kanakoudi A, Tsangaropoulou-Stinga H.

Persistent or severe course of reactive arthritis following *Salmonella enteritidis* infection. Scan J Rheumatol 1998;27:431–434.

75. Hannu TJ, Leirisalo-Repo M. Clinical picture of reactive *Salmonella* arthritis. J Rheumatol 1988;15:1668–1671.
76. Isomaki O, Vuento R, Granfors K. Serological diagnosis of *Salmonella* infections by enzyme immunoassay. Lancet 1989;i:1411–1414.
77. Brewerton DA, James DCO. The histocompatibility antigen (HL-A27) and disease. Sem Arthritis Rheumatism 1975;4:191–207.
78. Palazzi C, Olivieri I, D'Amico E, Pennese E, Petricca A. Management of reactive arthritis. Expert Opin Pharmacother 2004;5:61–70.
79. Raucher HS, Eichenfield AH, Hodes HL. Treatment of *Salmonella* gastroenteritis in infants: The significance of bacteremia. Clin Pediatr 1983;22:601–604.
80. Jacobs JL, Gold JW, Murray HW, Roberts RB, Armstrong D. *Salmonella* infections in patients with acquired immunodeficiency syndrome. Ann Intern Med 1985;102: 186–188.
81. Crum NF. Current trends in typhoid fever. Curr Gastroenterol Rep 2003;5:279–286.
82. Lee JH, Kim JJ, Jung JH, Lee SY, Bae MH, Kim YH, et al. Colonoscopic manifestations of typhoid fever with lower gastrointestinal bleeding. Dig Liver Dis 2004;36:141–146.
83. Santos-Juanes J, Lopez-Escobar M, Galache C, Telenti M, Vidau P, Badillo A, Rio JS. Haemorrhagic cellulitis caused by *Salmonella enteritidis*. Scand J Infect Dis 2005;37: 309–310.
84. Gonzalez Quintela A, Sanchez Leira J, Torre JA, Barrio E. Isolation of *Salmonella typhi* in a cutaneous ulcer. J Infect 1995;31:81–82.
85. Buchwald DS, Blaser MJ. A review of human salmonellosis: II. Duration of excretion following infection with nontyphi *Salmonella*. Rev Infect Dis 1984;6:345–356.
86. Devi S, Murray CJ. *Salmonella* carriage rate amongst school children—a three year study. Southeast Asian J Trop Med Public Health 1991;22:357–361.
87. Dryden MS, Keyworth N, Gabb R, Stein K. Asymptomatic foodhandlers as the source of nosocomial salmonellosis. J Hosp Infect 1994;28:195–208.
88. Mehta G, Malik A, Singh S, Kumari S. Asymptomatic *Salmonella* Senftenberg carriage in a neonatal ward. J Hosp Infect 1992;22:317–322.
89. Shukla VK, Singh H, Pandey M, Upadhyay SK, Nath G. Carcinoma of the gallbladder—is it a sequel of typhoid? Dig Dis Sci 2000;45:900–903.
90. Horowitz W, editor. Official methods of analysis of AOAC International. 18th ed. Gaithersburg, MD: AOAC International; 2006.
91. Zuerlein TJ, Smith PW. The diagnostic utility of the febrile agglutinin tests. JAMA 1985;254:1211–1214.
92. Sirinavin S, Garner P. Antibiotics for treating salmonella gut infections. Cochrane Database Syst Rev 2000;(2): CD001167.
93. Gendrel D, Raymond J, Moulin F, Habib F, Iniguez JL, Chemillier-Truong M, Badoual J. [Eradication of asymptomatic carrier state of non-typhoid *Salmonella* with two doses of pefloxacin]. Arch Pediatr 1995;2:418–422. [French]
94. Hohmann EL. Nontyphoidal salmonellosis. Clin Infect Dis 2001;32:263–269.
95. Parry CM. Antimicrobial drug resistance in *Salmonella enterica*. Curr Opion Infect Dis 2003;16:467–472.
96. Glynn MK, Bopp C, Dewitt W, Dabney P, Mokhtar M, Angulo FJ. Emergence of multidrug-resistant *Salmonella enterica* serotype Typhimurium DT104 infections in the United States. N Engl J Med 1998;338:1333–1338.
97. Rossier P, Urfer E, Burnens A, Bille J, Francioli P, Mean F, Zwahlen A. clinical features and analysis of the duration of colonization during an outbreak of *Salmonella* Braenderup gastroenteritis. Schweiz Med Wochenschr 2000;130:1185–1191.
98. Murphy GS, Bodhidatta L, Echeverria P, Tansuphaswadikul S, Hoge CW, Imlarp S, Tamura K. Ciprofloxacin and loperamide in the treatment of bacillary dysentery. Ann Intern Med 1993;118:582–586.
99. Dryden MS, Gabb RJE, Wright SK. Empirical treatment of severe acute community-acquired gastroenteritis with ciprofloxacin. Clin Infect Dis 1996;2;1019–1025.
100. Karras DJ. Incidence of foodborne illnesses: preliminary data from the foodborne diseases active surveillance network (FoodNet), commentary. Ann Emerg Med 2000;35;93–95.

Chapter 20

SHIGELLA SPECIES (Shiga Enterotoxins)

CYRUS RANGAN, MD, FAAP

HISTORY

Shigella species are gram-negative, facultatively anaerobic, nonlactose fermenting, nonmotile bacilli. Japanese physician Kiyoshi Shiga isolated *S. dysenteriae* type 1 in 1896 during an outbreak of acute bacillary dysentery in Japan.[1] American bacteriologist Simon Flexner discovered *S. flexneri* in 1899 in Manila,[2] and subsequently the Danish bacteriologist Carl Olaf Sonne discovered *S. sonnei* in 1906 in Copenhagen.[3] American microbiologist Mark Frederick Boyd discovered *S. boydii* in 1921 in India.[3] All four of these *Shigella* species are pathogenic to humans. Infections by *Shigella* species differ from most other enteropathogenic infections by the abrupt onset of clinically severe symptoms and the high rates of bacterial invasion, especially in young children. *Shigella* species are genetically similar to both enteroinvasive (EIEC) and enterohemorrhagic (EHEC) *Escherichia coli.*

EXPOSURE

Shigellosis results from infection by the following four *Shigella* species: *S. dysenteriae* (Shiga, corrig.) Castellani and Chalmers (serogroup A), *S. flexneri* Castellani and Chalmers (serogroup B), *S. boydii* Ewing (serogroup C), and *S. sonnei* (Levine) Weldin (serogroup D).[4] Most of these serogroups are subdivided into multiple serotypes based on an *O*-antigen component of the lipopolysaccharide within the outer membrane of the cell wall. The exception is serogroup D (*S. sonnei*), which contains a single serotype. An estimated 500,000 cases of *Shigella* infections occur each year in the United States with over 150,000,000 annual cases worldwide.[5] In the United States and other developed countries, *S. sonnei* is the most common serogroup. The highest risk of shigellosis occurs in residents of developing countries and travelers to these nations, primarily as a result of *S. flexneri* infections with outbreaks occurring frequently secondary to *S. dysenteriae* infections. The majority of *Shigella* cases occur in children, primarily <5 years of age.[6] Consequently, family members are also at high risk of secondary transmission.[7] However, residents of high-risk regions acquire immunity to endemic serotypes with repeated infections.[8] *S. dysenteriae* is endemic to developing nations, whereas *S. boydii* occurs primarily in India. Although *S. sonnei* is the most commonly isolated *Shigella* species in the United States, *S. flexneri* is a common isolate in immunocompromised individuals and male homosexuals.[9] Other enteric illnesses with potential sexual transmission include *Salmonella, Campylobacter, Giardia lamblia, Entamoeba histolytica*, and hepatitis A.

Source

ENVIRONMENT

Humans are the only natural hosts for *Shigella*, and these organisms are endemic to developing nations with substandard sanitation and hygiene that facilitate fecal-oral transmission of bacteria.[10] Rarely, asymptomatic *Shigella* carrier states result in the excretion of *Shigella* organisms

Medical Toxicology of Natural Substances, by Donald G. Barceloux, MD

for up to one year after the initial infection.[11] *Shigella* outbreaks associated with drinking water typically result from the ingestion of well water contaminated with feces.[12] Shared water supplies in campgrounds, carnivals, and nonchlorinated swimming pools are also environmental sources of pathogenic *Shigella* species.[13–15] House flies may transport *Shigella* from feces to food on farms and in households.[16]

Animal

Humans represent the primary animal reservoir for *Shigella* transmission. Food animals do not typically carry pathogenic *Shigella* species.[17]

Food

Shigella sonnei infections from undercooked meat and contaminated water are responsible for the vast majority of US cases of shigellosis.[18] Any food, milk, or drinking water source with exposure to fecal contamination may harbor *Shigella*, including cold food on commercial airlines.[19,20] Contamination of food by *Shigella* can occur as a result of the handling of food by asymptomatic carriers of *Shigella*. Following inoculation of *S. sonnei* in foods during a laboratory study, the bacterial population declined rapidly to undetectable levels within 2 days when dried on smooth surfaces of tomatoes.[21] However, populations of *S. sonnei* populations did not decrease in potato salad and ground beef stored at 2.5°C (37°F) or 8°C (46°F) over the shelf life of the products.

Food Processing

Proper cooking and refrigeration of food are the most effective means to control *Shigella*. Because *Shigella* survives in highly acidic environments; acid additives in foods do not effectively eliminate *Shigella* growth in home-prepared or commercially prepared foods.[22,23]

TOXINS

Physiochemical Properties

Shigella grows optimally at pH 4.5, but survives in a pH range of 2 to 5, permitting easy transit of these bacteria through the stomach.[24] *Shigella* reaches the large intestine by forming the following plasmid-encoded proteins: 1) ompR-envZ (protects against bile salts and high osmolarity), 2) osmZ (protects against relatively high temperature of the human body), and 3) kcpA (maintains bacterial mobility through the small intestine).[25] *Shigella* strains produce at least two enterotoxins including a chromosome encoded *Shigella* enterotoxin 1 and a large plasmid-associated *Shigella* enterotoxin 2. These two plasmid-encoded Shiga enterotoxins act similarly to kill host cells. In patients with traveler's diarrhea, shiga enterotoxin 1 occurs almost exclusively in *S. flexneri*, whereas Shiga enterotoxin 2 is a plasmid-encoded toxin found in almost all *Shigella* species.[26] Shiga enterotoxins are very similar to the Shiga-like toxins of enteroinvasive (EIEC) and enterohemorrhagic (EHEC). Shiga enterotoxins are composed of an A subunit, which is responsible for the biochemical effect of shiga toxin, surrounded by a ring of five B subunits.[27] The B chains bind the toxin to cell surface receptors, allowing the A chains to enter the interior of the cell. Inside the cell, the A chains cleave the *N*-glycosidic bond of adenine at nucleotide position 4324 in the 28S rRNA of the 60S ribosomal subunit. The phage-borne Shiga toxin is a potent protein-synthesis inhibitor, primarily in the vascular endothelium that is produced mainly by *S. dysenteriae* serotype 1. This Shiga toxin and Shiga toxin produced by certain strains of *Escherichia coli* are genetically and antigenically identical, and both strains are capable of causing hemolytic-uremic syndrome (HUS).[28]

Mechanism of Toxicity

Typically, *Shigella* infections are limited to the intestinal mucosa. *Shigella* evades intestinal host defenses to reach basolateral surfaces intact.[29] These organisms invade columnar epithelial cells of the terminal ileum and colon, where they multiply intracellularly prior to the onset of symptoms. *Shigella* enters columnar epithelial cells through the basolateral surface rather than attaching to the mucosal surface of columnar epithelial cells. Lymphofollicular M cells transcytose *Shigella* to a subepithelial macrophage, which phagocytoses the organism. *Shigella* then degrades the phagosome via plasmid-encoded siderophores, superoxide dismutase, and somatic antigens. *Shigella* species deliver a plasmid-encoded invasion plasmid antigen complex (IpaA-IpaD) via a type-III secretion system similar to EHEC.[30] The insertion of IpaB and IpaC into the host membrane forms a 25A pore that allows the passage of other infective proteins and eventually causes cell lysis. IpaB is essential for the induction of apoptosis of the macrophage, release of interleukin-1β, and deposition of *Shigella* organisms along the basolateral surfaces of adjacent columnar epithelial cells.[31] Interleukin-1β recruits neutrophils into tight junctions, disrupts their integrity, and allows the direct passage of additional *Shigella* organisms from the intestinal lumen to the basolateral surfaces of columnar epithelial cells. Columnar epithelial cells promptly endocytose *Shigella* at the basolateral surface.[32] Lysis of epithelial cell vacuoles allows *Shigella* to multiply in the cytosol of epithelial cells. The type-III secretion system binds to the epithelial junctional complex of the adjacent columnar epithelial cell and subsequently rearranges the cytoskeleton by inducing actin filaments to form pedestal-like polymers.[33] The rearranged cytoskeleton buckles the epithelial cell membrane. A plasmid-encoded intercellular spread protein (IcsA) mediates transport of *Shigella* through the buckled membrane and into adjacent epithelial cells, where multiplication continues.[34] Shiga enterotoxins mediate clinical symptoms via two mechanisms. First, Shiga enterotoxins bind to membrane receptors to block the absorption of water and electrolytes, occasionally leading to brief distinct prodrome of osmotic diarrhea preceding dysenteric symptoms.[35] Second, Shiga enterotoxins exert a cytotoxic effect in which the B subunit binds to membrane glycolipid GB_3 and the A subunit enters the cytosol of the columnar epithelial cell. The A domain binds to the 60s ribosomal subunit and halts protein synthesis; these actions result in cell death.[35] Another plasmid-encoded intercellular spread protein (IcsB) lyses the epithelial cell membrane.[32] Cell death results in mucosal erythema, edema, and inflammation, leading to ulceration, microabscess, necrosis, hemorrhage, and purulent mucoid exudates with fecal leukocytes.[36] Histopathology of these lesions resembles those of ulcerative colitis.[37] Repeated infections in young children may result in scarring and chronic malabsorption.[38] Systemic dissemination of *Shigella* and Shiga enterotoxin results in systemic effects outside the gastrointestinal tract.[39–41] Binding of Shiga enterotoxin to renal endothelial cells may result in hemolytic-uremic syndrome in children.[42]

DOSE RESPONSE

The dose of *Shigella* organisms necessary to produce an infection depends on the virulence of the strain and the susceptibility of the host. The attack rate for 152 volunteers inoculated with one of four strains of *Shigella* ranged from about 25 to 50%.[43] In this case series, as few as 10 *Shigella* dysenteriae 1 organisms caused shigellosis in the human volunteers.

CLINICAL RESPONSE

The incubation period for *Shigella* bacteria is about 1–2 days. A brief nonspecific prodrome of osmotic, watery diarrhea may precede dysentery; in mild cases the watery diarrhea does not progress to dysentery.[44] Additional initial symptoms of shigellosis include anorexia, malaise, and weakness. Anorexia may persist into the convalescent period. Sudden onset of severe cramping, tenesmus, fever, myalgias, and frequent, small bloody, mucoid stools characterize bacillary dysentery.[45] Vomiting is rare, and large fluid volume losses with significant dehydration are not typical in most cases of shigellosis.[46] Seizures may occur during Shigella infections, particularly in children, and in this group of patients the presence of seizures does not necessarily indicate the presence of intracranial infection.[47] *S. dysenteriae* usually produces more severe disease, and patients with severe infections may pass >20 dysenteric stools in one day. Children may lose significant protein and immunoglobulins in their stools, leading to malnutrition, severe electrolyte disturbances, and increased susceptibility to further infections.[38] Intestinal perforation, rectal prolapse, or toxic megacolon may complicate ulcerations in the colonic mucosa.[48] Rare complications of disseminated *Shigella* include meningitis, osteomyelitis, and bacteremia as well as hemolytic-uremic syndrome in children and thrombotic thrombocytopenic purpura in adults.[41,49] *Shigella* septicemia occurs rarely in healthy individuals, but may occur more often in immunocompromised patients.[50,51] Complications of *S. flexneri* infections include postinfectious inflammatory arthritis or Reiter's syndrome, which is usually associated with human lymphocyte antigen B27 (HLA-B27).[52] Immunity to *Shigella* is serotype-specific. The differential diagnosis of patient with suspected shigellosis (i.e., bloody, mucoid stools) includes infections by enteroinvasive *E. coli*, *Campylobacter* species, *Salmonella enteritidis*, *Yersinia enterocolitica*, and *Entamoeba histolytica*.

DIAGNOSTIC TESTING

Analytical Methods

Shigella remain viable outside the human body for a limited time; therefore, stools samples should be processed for *Shigella* within a few hours of collection. Use of transport media (e.g., Cary-Blair, buffered glycerol saline) improves the yield of stool samples for *Shigella*. The typical culture media for *Shigella* is *Salmonella-Shigella* (SS) agar, but these organisms will also grow on MacConkey agar and other enteric media (e.g., Hektoen enteric, xylose lysine desoxycholate, desoxycholate citrate agar). Stool specimens may be preenriched in selenite gram-negative broth to improve the yield from culturing.[53] After incubation, surviving colonies are streaked onto triple sugar iron agar to isolate *Shigella* species. A characteristic feature of *Shigella* colonies is the production of acid, but not hydrogen sulfide, on triple sugar iron agar.[54] Polymerase chain reaction

(PCR) demonstrates superior sensitivity over stool culture (i.e., 96% vs. 54%).[55]

Biomarkers

Leukocytosis or leukopenia may be present; however, relative leukopenia with significant left shift should prompt the clinician to suspect shigellosis.[56] Standard blood cultures may be positive in rare cases of Shigella sepsis. Antibody titers are not useful for the diagnosis of shigellosis.

Abnormalities

Stool samples in shigellosis reveal blood, mucous, and sheets of fecal leukocytes.[57] Colonoscopy reveals multiple ulcerations with severe inflammation and erythema, similar to patients with ulcerative colitis.[36] Fluid and electrolyte loss can cause hypernatremia (serum Na > 150 mEq/L), hyponatremia (serum Na < 130 mEq/L), hypokalemia (serum K < 3.0 mEq/L), hypochloremia, elevated urine specific gravity (SpG > 1.020), and hyperphosphatemia. Low serum bicarbonate concentrations result from lactic acid production, whereas elevation of serum bicarbonate concentrations occur secondary to contraction alkalosis. Hypoglycemia may also occur, particularly in children. The presence of high numbers (i.e., >30%) of bands in the complete blood count in an adult with diarrhea is highly suggestive of shigellosis.[58] However, the band count in patients with shigellosis is not sensitive in children. ST-T flattening and arrhythmias may be observed on electrocardiogram.

TREATMENT

Shigellosis is usually a self-limited disease that responds to hydration. Treatment includes bowel rest and rehydration with fluid and electrolytes replacement. Oral replacement may be sufficient, but affected patients may experience significant anorexia, necessitating intravenous fluid replacement and hospitalization in cases of significant dehydration or severe electrolyte abnormalities, particularly in young children or elderly patients.[59,60] The clinical symptoms of shigellosis usually resolve spontaneously within 7 days; however, all cases of confirmed or suspected shigellosis should be treated with antimicrobials to avoid complications, sequelae, disease recurrence, and secondary transmission.[41] Antimicrobial therapy decreases illness duration from 7 days to 48–72 hours, significantly decreases bacterial shedding, and reduces the incidence of extraintestinal manifestations.[41] Intravenous or intramuscular ceftriaxone or cefixime are administered for 5 days in children.[61] Adults with shigellosis respond to ceftriaxone, azithromycin, or quinolones in susceptible strains; however, resistance to quinolones is prevalent.[62,63] Cefixime is not effective for the treatment of shigellosis in adults.[64] *Shigella* species commonly are resistant to sulfonamides, penicillins, and tetracyclines worldwide.[65,66] Many medical sources do not recommend opiates, loperamide, and other antimotility agents during shigellosis because of the concern about the potential for bacterial invasion following the use of the medications.[67,68] However, there are inadequate clinical data to determine the efficacy and complications associated with the use of these agents in shigellosis. In a study of 44 patients with shigellosis treated with ciprofloxacin 500 mg twice daily for 3 days, the use of loperamide decreased the number of diarrheal stools and shortened the duration of illness when compared with placebo. No prolonged fever or extended excretion of *Shigella* was noted in this study, but the sample size was relatively small. In endemic regions, maternal antibodies from breast milk protect young infants from Shigella infections.[69]

References

1. Shiga K. Observation on the epidemiology of dysentery in Japan. Philippine J Sci 1906;1:485–500.
2. Flexner S. On the etiology of tropical dysentery. Bull Johns Hopkins Hosp 1900;11:231–242.
3. Shiga K. The trend of prevention, therapy, and epidemiology of dysentery since the discovery of its causative organisms. N Engl J Med 1936;215:1205–1211.
4. Niyogi SK. Shigellosis. J Microbiol 2005;43:133–143.
5. Gerner-Smidt P, Hise K, Kincaid J, Hunter S, Rolando S, Hyytia-Trees E, et al. PulseNet Taskforce. PulseNet USA: a five-year update. Foodborne Pathog Dis 2006;3: 9–19.
6. Ashkenazi S. Shigella infections in children: new insights. Semin Pediatr Infect Dis 2004;15:246–252.
7. Khan AI, Talukder KA, Huq S, Mondal D, Malek MA, Dutta DK, et al. Detection of intra-familial transmission of shigella infection using conventional serotyping and pulsed-field gel electrophoresis. Epidemiol Infect 2006;134: 605–611.
8. Ferreeccio C, Prado V, Ojeda A, Cayyazo M, Abrego P, Guers L, et al. Epidemiologic pattern of acute diarrhoea and endemic *Shigella* infections in a poor periurban setting in Santiago, Chile. Am J Epidemiol 1991;134:614–627.
9. Centers for Disease Control (CDC). *Shigella flexneri* serotype 3 infections among men who have sex with men—Chicago, Illinois, 2003–2004. MMWR Morb Mortal Wkly Rep 2005;54:820–822.
10. Ashbolt NJ. Microbial contamination of drinking water and disease outcomes in developing regions. Toxicology 2004;198:229–238.

11. Levine MM, DuPont HL, Khodabandelou M, Hornick RB. Long-term *Shigella*-carrier state. N Engl J Med 1973;288: 1169–1171.
12. Centers for Disease Control (CDC). *Shigella sonnei* outbreak associated with contaminated drinking water—Island Park, Idaho, August 1995. MMWR Morb Mortal Wkly Rep 1996;45:229–231.
13. Coles FB, Kondracki SF, Gallo RJ, Chalker D, Morse DL. Shigellosis outbreaks at summer camps for the mentally retarded in New York State. Am J Epidemiol 1989;130: 966–975.
14. Centers for Disease Control (CDC). Shigellosis outbreak associated with an unchlorinated fill-and-drain wading pool—Iowa, 2001. MMWR Morb Mortal Wkly Rep 2001;50:797–800.
15. Lee LA, Ostroff SM, McGee HB, Johnson DR, Downes FP, Cameron DN, Bean NH, Griffin PM. An outbreak of shigellosis at an outdoor music festival. Am J Epidemiol 1991;133:608–615.
16. Levine OS, Levine MM. House flies (*Musca domestica*) as mechanical vectors of shigellosis. Rev Infect Dis 1991;13:688–696.
17. Gascon J. Epidemiology, etiology and pathophysiology of traveler's diarrhea. Digestion 2006;73:102–108.
18. Safdar A. Infectious diarrhea in the southeastern United States, 1998–2000. Clin Infect Dis 2003;36:533–534.
19. Hedberg CW, Levine WC, White KE, Carlson RH, Winsor DK, Cameron DN, et al. An international foodborne outbreak of shigellosis associated with a commercial airline. JAMA 1992;268:3208–3212.
20. Ollinger-Snyder P, Matthews ME. Food safety: review and implications for dietitians and dietetic technicians. J Am Diet Assoc 1996;96:163–168.
21. Warren BR, Yuk HG, Schneider KR. Survival of *Shigella sonnei* on smooth tomato surfaces, in potato salad and in raw ground beef. Int J Food Microbiol 2007;116: 400–404.
22. Kimura AC, Johnson K, Palumbo MS, Hopkins J, Boase JC, Reporter R. Multistate shigellosis outbreak and commercially prepared food, United States. Emerg Infect Dis 2004;10:1147–1149.
23. Ethelberg S, Olsen KE, Gerner-Smidt P, Molbak K. Household outbreaks among culture-confirmed cases of bacterial gastrointestinal disease. Am J Epidemiol 2004;159: 406–412.
24. Small P, Blankenhorn D, Welty D, Zinser E, Slonczewski JL. Acid and base resistance in *Escherichia coli* and *Shigella flexneri*: role of rpoS and growth pH. J Bacteriol 1994;176:1729–1737.
25. Sasakawa C, Buysse JM, Watanabe H. The large virulence plasmid of Shigella. Curr Top Microbiol Immunol 1992;180:21–44.
26. Vargas M, Gascon J, Jimenez De Anta MT, Vila J. Prevalence of *Shigella* enterotoxins 1 and 2 among *Shigella* strains isolated from patients with traveler's diarrhea. J Clin Microbiol 1999;37:3608–3611.
27. Donohue-Rolfe A, Acheson DW, Keusch GT. Shiga toxin: purification, structure, and function. Rev Infect Dis 1991;13:S293–S297.
28. Tesh VL, O'Brien AD. The pathogenic mechanisms of Shiga toxin and the Shiga-like toxins. Mol Microbiol 1991;5:1817–1822.
29. Adam T. Exploitation of host factors for efficient infection by Shigella. Int J Med Microbiol 2001;291:287–298.
30. Hueck CJ. Type III protein secretion systems in bacterial pathogens of animals and plants. Microbiol Mol Biol Rev 1998;62:379–433.
31. Hilbi H, Zychlinsky A, Sansonetti PJ. Macrophage apoptosis in microbial infections. Parasitology 1997;115: S79–S87.
32. Nhieu GT, Sansonetti PJ. Mechanism of *Shigella* entry into epithelial cells. Curr Opin Microbiol 1999;2:51–55.
33. Gruenheid S, Finlay BB. Microbial pathogenesis and cytoskeletal function. Nature 2003;422:775–781.
34. Goldberg MB, Sansonetti PJ. *Shigella* subversion of the cellular cytoskeleton: a strategy for epithelial colonization. Infect Immun 1993;61:4941–4946.
35. O'Brien AD, Tesh VL, Donohue-Rolfe A, Jackson MP, Olsnes S, Sandvig K. Shiga toxin: biochemistry, genetics, mode of action, and role in pathogenesis. Curr Top Microbiol Immunol 1992;180:65–94.
36. Phalipon A, Sansonetti PJ. Shigellosis: innate mechanisms of inflammatory destruction of the intestinal epithelium, adaptive immune response, and vaccine development. Crit Rev Immunol 2003;23:371–401.
37. Kobari K. Clinical aspects of shigellosis in reference of sigmoidoscopic findings and histopathological changes of the colon. Jpn J Med Sci Biol 1998;51:S23–S25.
38. Ashkenazi S. *Shigella* infections in children: new insights. Semin Pediatr Infect Dis 2004;15:246–252.
39. Yuhas Y, Shulman L, Weizman A, Kaminsky E, Vanichkin A, Ashkenazi S. Involvement of tumor necrosis factor alpha and interleukin-1beta in enhancement of pentylenetetrazole-induced seizures caused by *Shigella* dysenteriae. Infect Immun 1999;67:1455–1460.
40. Bhattacharya SK, Sinha AK, Sen D, Sengupta PG, Lall R, Pal SC. Extraintestinal manifestations of Shigellosis during an epidemic of bacillary dysentery in Port Blair, Andaman & Nicobar Island (India). J Assoc Physicians India 1988;36:319–320.
41. Bennish ML. Potentially lethal complications of shigellosis. Rev Infect Dis 1991;13:S319–S324.
42. Kaplan BS, Cleary TG, Obrig TG. Recent advances in understanding the pathogenesis of the hemolytic uremic syndromes. Pediatr Nephrol 1990;4:276–283.
43. DuPont HL, Levine MM, Hornick RB, Formal SB. Inoculum size in shigellosis and implications for expected mode of transmission. J Infect Dis 1989;159:1126–1128.
44. Kaur T, Ganguly NK. Modulation of gut physiology through enteric toxins. Mol Cell Biochem 2003;253: 15–19.

45. Mathan VI, Mathan MM. Intestinal manifestations of invasive diarrhoeas and their diagnosis. Rev Infect Dis 1991;13:S311–S313.
46. Prado D, Cleary TG, Pickering LK, Ericsson CD, Bartlett AV, DuPont HL. The relation between production of cytotoxin and clinical features in shigellosis. J Infect Dis 1986;154:149–155.
47. Ashkenazi S, Dinari G, Zevulunov A, Nitzan M. Convulsions in childhood shigellosis. Clinical and laboratory features in 153 children. Am J Dis Child 1987;141:208–210.
48. Upadhyay AK, Neely JA. Toxic megacolon and perforation caused by *Shigella*. Surg 1989;76:1217.
49. Keusch GT, Acheson DW. Thrombotic thrombocytopenic purpura associated with Shiga toxins. Semin Hematol 1997;34:106–116.
50. Morduchowicz G, Huminer D, Siegman-Igra Y, Drucker M, Block CS, Pitlik SD. *Shigella* bacteremia in adults a report of five cases and review of the literature. Arch Intern Med 1987;147:2034–2037.
51. Oh HM, Tan AL. *Shigella septicaemia* in adults: report of two cases and mini-review. Ann Acad Med Singapore 2001;30:668–670.
52. Hannu T, Mattila L, Siitonen A, Leirisalo-Repo M. Reactive arthritis attributable to *Shigella* infection: a clinical and epidemiological nationwide study. Ann Rheum Dis 2005;64:594–598.
53. Miyagi K, Takegaki Y, Nakano T, Nakano Y, Honda T, Sano K. Frequency of failure to isolate *Shigella* spp. by the direct plating technique and improvement of isolation by enrichment in selenite broth. Epidemiol Infect 2001;127:375–379.
54. Hoshiko M. Laboratory diagnosis of infectious diarrhea. Pediatr Ann 1994;23:570–574.
55. Dutta S, Chatterjee A, Dutta P, Rajendran K, Roy S, Pramanik KC, et al. Sensitivity and performance characteristics of a direct PCR with stool samples in comparison to conventional techniques for diagnosis of *Shigella* and enteroinvasive *Escherichia coli* infection in children with acute diarrhoea in Calcutta, India. J Med Microbiol 2001;50:667–674.
56. Fried D, Maytal J, Hanukoglu A. The differential leukocyte count in shigellosis. Infection 1982;10:13–14.
57. Echeverria P, Sethabutr O, Pitarangsi C. Microbiology and diagnosis of infections with *Shigella* and enteroinvasive *Escherichia coli*. Rev Infect Dis 1991;13:S220–S225.
58. Halpern Z, Averbuch M, Dan M, Giladi M, Levo Y. The differential leukocyte count in adults with acute gastroenteritis. Scand J Infect Dis 1992;24:205–207.
59. King CK, Glass R, Bresee JS, Duggan C. Centers for Disease Control (CDC). Managing acute gastroenteritis among children: oral rehydration, maintenance, and nutritional therapy. MMWR Morb Mortal Wkly Rep 2003;52:1–16.
60. Smith JL. Foodborne illness in the elderly. J Food Prot 1998;61:1229–1239.
61. Martin JM, Pitetti R, Maffei F, Tritt J, Smail K, Wald ER. Treatment of shigellosis with cefixime: two days vs. five days. Pediatr Infect Dis J 2000;19:522–526.
62. Khan WA, Seas C, Dhar U, Salam MA, Bennish ML. Treatment of shigellosis: V. Comparison of azithromycin and ciprofloxacin. A double-blind, randomized, controlled trial. Ann Intern Med 1997;126:697–703.
63. Sarkar K, Ghosh S, Niyogi SK, Bhattacharya SK. Shigella dysenteriae type 1 with reduced susceptibility to fluoroquinolones. Lancet 2003;361:785.
64. Salam MA, Seas C, Khan WA, Bennish ML. Treatment of shigellosis: IV. Cefixime is ineffective in shigellosis in adults. Ann Intern Med 1995;123:505–508.
65. Sack RB, Rahman M, Yunus M, Khan EH. Antimicrobial resistance in organisms causing diarrheal disease. Clin Infect Dis 1997;24:S102–S105.
66. Patwari AK. Multidrug resistant *Shigella* infections in children. J Diarrhoeal Dis Res 1994;12:182–186.
67. DuPont HL, Hornick RB. Adverse effect of Lomotil therapy in shigellosis. JAMA 1973;226:1525–1528.
68. Yates J. Traveler's diarrhea. Am Fam Physician 2005;71:2095–2100.
69. Passwell JH, Freier S, Shor R, Farzam N, Block C, Lison M, et al. *Shigella* lipopolysaccharide antibodies in pediatric populations. Pediatr Infect Dis J 1995;14:859–865.

Chapter 21

STAPHYLOCOCCUS AUREUS

CYRUS RANGAN, MD, FAAP

HISTORY

Staphylococcus aureus subsp. *aureus* Rosenbach (*S. aureus*) is a facultative anaerobic, nonspore-forming, gram-positive coccus that is one of the two most common causes of foodborne illness along with *Salmonella* serovars.[1] Although the bacterium was first isolated from furuncles by Louis Pasteur in 1878 and further classified by Rosenbach in 1884, *S. aureus* was not identified as a cause of foodborne toxicity until 1914.[2,3] In 1930, Dack et al demonstrated that a *Staphylococcus*-contaminated sponge cake with cream filling was the source of an outbreak of food poisoning.[4] Staphylococcal enterotoxins (SEs) were isolated after analyses of *S. aureus* strains in foodborne illness outbreaks by M.S. Bergdoll at the University of Wisconsin-Madison.[5] SEs comprise one group of toxins elaborated by *S. aureus*. Nonenterotoxigenic staphylococcal toxins include toxic shock syndrome toxin (TSST-1) and exfoliatins A and B with the latter being associated with scalded skin syndrome. A famous series of four US outbreaks of *S. aureus* food poisoning was traced to improperly canned mushrooms imported from China, prompting US trade restrictions on the import of Chinese mushrooms.[6] In 1992, a large outbreak of *S. aureus* food poisoning occurred in a Texas school system when 25% of nearly 6,000 children in 16 different schools developed vomiting after eating prepared lunches that were transported by trucks from a common source. Chicken salad was the food source for this outbreak. The chicken was boiled the day before and cooled slowly, while the salad ingredients were mixed the following day. After transport, the food was held at room temperature for at least an hour at the various schools. Enterotoxigenic *S. aureus* was subsequently identified in the chicken salad.[7] In another outbreak, over 50% of 8,000 individuals at a gathering in Brazil experienced acute gastroenteritis in 1998. Sixteen patients died from irreversible multisystem failure. Enterotoxigenic *S. aureus* was identified in food served at the gathering.[8] Although several staphylococcal species (*S. cohnii, S. epidermidis, S. haemolyticus, S. intermedius, S. xylosus*) other than *S. aureus* produce SEs, coagulase positive *S. intermedius* is the only other non-*S. aureus* species clearly associated with staphylococcal food poisoning.[9,10]

EXPOSURE

Source

ENVIRONMENT

Staphylococcus aureus is a ubiquitous pathogen found in air, dust, and sewage.[11] Populations of bacteria in these milieus generally originate from human and animals. Surfaces of equipment used in food gathering and harvesting may harbor *S. aureus* in fomites for several days.[12]

ANIMAL

Up to 50% of mammals carry *S. aureus* asymptomatically in the nares and throat, on skin, and in hair.[13] Mastitis in dairy cows is highly correlated to the appearance of *S. aureus* in milk.[14] Animals are the primary source of *S. aureus* contamination of raw foods, whereas food handlers are the primary source of contaminated prepared foods.[15]

Food

Staphylococcus aureus may be present in virtually any food because it can reproduce within wide ranges of temperature, acidity, and salt content.[16] Up to 50% of *S. aureus* isolates from various foods are enterotoxigenic strains, but enterotoxins are not usually produced until populations reach 10^6 CFU/gm.[17] The main sources of staphylococcal contamination are food handlers (i.e., asymptomatic carriers or persons with open lesions), and contamination typically occurs after heat treatment of the food.[18,19] However, some food sources (e.g., raw meat, cheese, sausage) of staphylococcal food poisoning result from animal contamination. Commonly identified sources of *S. aureus* foodborne illness include red meats and sausages; poultry; cheeses; cream-filled pastries; milk; dairy products; and salads made with chicken, mayonnaise, pasta with tomato sauce, egg, tuna, and macaroni.[20,21] Fermented dairy products (e.g., soft cheeses, creams) are well-recognized potential sources of *S. aureus* because of the prolific growth of these bacteria at temperatures required to make these foods.[22] Dressed animal carcasses are commonly contaminated with staphylococci.

Food Processing

Inadequate refrigeration and heating are the most common reasons for the proliferation of *S. aureus* in food. Refrigeration temperatures <45 °F (7 °C) are adequate to keep growth to a minimum, and bacteria are readily destroyed by routine cooking temperatures exceeding 140 °F (60 °C).[23] Existing commercial sanitization techniques are adequate to control growth of *S. aureus*. Therefore, growth of *S. aureus* during commercial food processing usually results from poor sanitary practices.[24] Because *S. aureus* does not form spores, regeneration of colonies does not occur during cooling and reheating of properly cooked food. However, food is easily recontaminated with *S. aureus* from improper preparation and processing, particularly from human carriers. Thus, foods at high risk for *S. aureus* contamination primarily involve food, which require frequent handling after initial cooking or baking (e.g., cream-filled pastries).[25] Kitchen utensils (e.g., knives, meat grinders and slicers, cutting boards) also harbor *S. aureus* in fomites.[26] SEs are destroyed by heat sterilization and canning, though they may escape destruction if present in high enough concentrations. Pasteurization kills *S. aureus*, but this process may not completely destroy preformed SEs.[27] Acid treatments in food processing (pH < 2) can also deactivate SEs, but many SEs regain their activity as soon as pH becomes more alkaline.[28] Because *S. aureus* is a facultative anaerobe, anaerobic processing conditions are not effective in inhibiting *S. aureus* growth in food.[29]

TOXINS

Physiochemical Properties

Staphylococcus aureus grows optimally at 86 °F–98 °F (30 °C–37 °C), but these bacteria may grow at temperatures from 45 °F–120 °F (7 °C–49 °C).[30] The pH range for staphylococcal proliferation is 4.5 to 9.3 with optimal growth between 7.0 and 7.5. Most strains of *S. aureus* are highly tolerant to sugars and salts, and these strains can grow over a water activity (a_w) range of 0.83 to >0.99. *S. aureus* exerts its effects by the production of staphylococcal enterotoxins. SEs are single polypeptides of 25–28 kDa that are highly heat resistant and remain active even when the staphylococci are destroyed. SE production is usually inhibited at pH < 5. Most SEs are resistant to proteolytic enzymes (e.g., pepsin, trypsin); consequently, these toxins remain intact during transit through the gastrointestinal tract.[31] Production of SEs is tightly governed by an accessory gene regulator (*agr*) system, which is dormant during the exponential growth phase of *S. aureus*. An autoinducible quorum-sensing system triggers the expression of *agr* when *S. aureus* food counts reach 10^6 CFU/gm.[32] SEs are heat-stable, but the high temperatures used in commercial food processing can inactivated these toxins. Fourteen different SEs have been identified: SEA, SEB, SEC1 and related isotypes (SEC2, SEC3), SED, SEE, SEG, SEH, SEI, SEJ, SEK, SEL, SEM, SEN, and SEO. In a study of commercial Italian foods containing SEs, the most common enterotoxins were SEA and SEC.[33] SEB (CAS RN: 11100-45-1) is a heat-stable enterotoxin, which is a potential terrorist weapon when released as an aerosol (SEB respiratory syndrome).[34] SEG, SEH, and SEI are relatively minor enterotoxins.[35] Frequently, food implicated in staphylococcal food poisoning contains more than one enterotoxin. Enterotoxin F was the original name given to the toxin that is now known as TSST-1. Although the SEs are antigenically distinct, they have a high degree of homology; 26–83% of their amino acid sequences are similar. The majority of SEs contains a distinctive cystine loop that may influence the severity of vomiting.[36]

Mechanism of Toxicity

Most SEs produce vomiting through stimulation of gastrointestinal motility via direct interaction of SEs with gut epithelial emetic receptors. SEs cause hyperemic, inflammatory reactions on the mucosal surfaces of the stomach, duodenum, and jejunum via generation of inflammatory mediators such as prostaglandin E2, leukotriene B4, and 5-hydroxyeicosatetraenoic acid. Neutrophils and macrophages subsequently infiltrate these lesions, but these cells do not penetrate the lamina propria. Mucopurulent fluid is then discharged into the duodenum while brush border integrity is disrupted in the jejunum.[37] SEA, SEB, SEC1, SED, SEE, and SEH are characterized as "superantigens," which induce symptoms via stimulating mitogenic activity and cytokine production. These superantigens activate the immune system, resulting in T cell proliferation by interaction with Class II MHC molecules on macrophages and specific V beta chains of the T cell receptor. The important feature of this interaction is the subsequent generation of interleukin-1 (IL-1), interleukin-2 (IL-2), tumor necrosis factor (TNF), and other lymphokines.[38] Although the mechanism of action is unknown, IL-2 likely plays a significant role in the pathogenesis of gastrointestinal symptoms. Evidence for the IL-2 mechanism is deduced from cancer patients, who experience significant vomiting when treated with IL-2 immunomodulation chemotherapy.[39] SEs occasionally cross mucosal surfaces when massive doses of SE are present, potentially leading to a clinical syndrome similar to toxic shock.[40]

DOSE RESPONSE

The minimum dose of SE required to induce symptoms depends on the potency of the particular SEs elaborated in food and individual sensitivities, and the dose ranges from 1 ng to 1.0 µg of total enterotoxins. Typically, 100–200 ng of staphylococcal enterotoxin in sensitive subjects causes symptoms of staphylococcal intoxication.[41,42] Sufficient SE production generally occurs when populations of enterotoxigenic *S. aureus* exceed 10^6 CFU/gm.[32] The refrigeration (<10 °C/50 °F) or the heating (>45 °C/113 °F) of food prevents the proliferation of staphylococcal organisms to the numbers necessary to form detectable concentrations of enterotoxins.

CLINICAL RESPONSE

The clinical response to SEs resembles the emetic syndrome induced by *Bacillus cereus*. *S. aureus* food poisoning is characterized by nausea, vomiting, retching, and abdominal cramping with diarrhea on rare occasions. Symptoms may arise as early as 30 minutes after exposure, but symptoms may be delayed up to 8 hours after exposure. In a large outbreak of staphylococcal food poisoning resulting from the ingestion of contaminated powdered milk, about 83% of the patients developed illness within 6 hours with a peak onset of 3–4 hours.[43] By 12 hours after ingestion, approximately 96% of these patients developed symptoms of food poisoning. Diarrhea virtually never occurs without vomiting; therefore, the occurrence of diarrhea without vomiting should prompt the clinician to seek other etiologies.[44] Associated symptoms of staphylococcal food poisoning include headache, dizziness, tachycardia, and other physical signs of dehydration.[45] The clinical response to SEs is self-limited, and resolves spontaneously in 24–48 hours in the majority of cases. Anorexia can persist for several days. Rare cases of systemic toxicity occur following massive SE doses with a clinical course similar to toxic shock syndrome manifested by fever, hypotension, and multiorgan failure. Death from systemic invasion of enterotoxins is extremely rare (i.e., typically < 0.03–0.4%).[46]

DIAGNOSTIC TESTING

Analytical Methods

Direct microscopic evaluation of food contaminated with *S. aureus* reveals spherical cocci in pairs, chains, or grape-like clusters (*staphyle* [Greek]: bunches of grapes). Direct plate count determines the size of the population. Surveillance of high-risk food products for *S. aureus* is also performed using direct plate counts.[47] There are at least 8 AOAC (Association of Analytical Chemists International, Gaithersburg, MD; http://www.eoma.aoac.org/) microbiological methods available for the detection and enumeration of *S. aureus* in various food matrixes.[48]

1. AOAC official method 966.23 determines the aerobic plate count, most probable number of coliform bacteria, and the number of *S. aureus* in food products.
2. AOAC official method 995.12 is a latex agglutination method for the identification of *S. aureus* isolates from foods.
3. AOAC official method 975.55 is a surface plating method for the isolation and enumeration of *S. aureus* in foods expected to contain ≥10 cells/g of food.
4. AOAC official method 987.09 is used for the detection and enumeration of small numbers of *S. aureus* in food ingredients and food expected to contain large populations of competing species.
5. AOAC official method 2001.05 is applicable to the rapid enumeration of confirmed *S. aureus* organisms in pasta filled with beef and cheese, frozen

hash browns, cooked chicken patty, egg custard, frozen ground raw pork, and instant nonfat dried milk.

6. AOAC official method 2003.11 is applicable to the enumeration of *S. aureus* in cooked and diced chicken, ham, salmon, and pepperoni.
7. AOAC official method 2003.08 is applicable to the enumeration of *S. aureus* in ice cream, raw milk, yogurt, whey powder, and cheese.
8. AOAC official method 2003.07 is used for the enumeration of *S. aureus* in frozen lasagna, custard, frozen mixed vegetables, frozen hash browns, and frozen batter-coated mushrooms.

The presence of large numbers of staphylococci in foods is not sufficient to incriminate a specific food as a source of staphylococcal food poisoning because not all staphylococcal strains are enterotoxigenic. Serologic testing can determine the enterotoxigenicity of specific isolates. In foodborne illness outbreaks, enterotoxins are separated from food by selective adsorption on ion exchange resins. Enterotoxins are then identified in food by tracer-labeled immunoassay. Enzyme-linked immunosorbent assay (ELISA), reverse passive latex agglutination, and radioimmunoassay (RIA) are the most commonly utilized methods for identification of enterotoxins.[49] These highly sensitive methods are able to detect enterotoxin concentrations as low as 1.0 ng/gm of food. AOAC official method 993.06 is a polyvalent enzyme immunoassay method for the detection of 1.3–3.3 ng/mL of staphylococcal enterotoxin in extracts prepared from selected foods containing 4–10 ng/mL of staphylococcal enterotoxin. Specific toxin serotypes A to E are not differentiated. Although commercial assays detect the more common staphylococcal enterotoxins (SEA–SEE), these assays do not necessarily detect less common staphylococcal enterotoxins (SEG–SEO).[50] Multiplex PCR with rapid DNA isolation quickly and comprehensively identifies the genes for SEs, TSST, and exfoliatins in *S. aureus* isolates from human specimens, but this method is not widely available.[24] Detection of enterotoxigenic strains of *S. aureus* $>10^6$ CFU/gm of food and enterotoxin $>1.0\,\mu$g/gm of food with adequate clinical correlation is sufficient to confirm the diagnosis of staphylococcal food poisoning.

Biomarkers

Standards for measurement of SEs concentrations in emesis or other body fluids are not established. Emesis, blood, urine, and stool are not routinely tested for SEs.

Abnormalities

Staphylococcus aureus food poisoning can produce severe serum electrolyte abnormalities, with a laboratory profile similar to dehydration from gastrointestinal fluid losses. Severe dehydration involves hypernatremia (serum Na > 150 mEq/L) or hyponatremia (serum Na < 130 mEq/L), hypokalemia (serum K < 3.0 mEq/L), hypochloremia, and elevated urine specific gravity (SpG > 1.020). Low serum bicarbonate results from lactic acid production or elevated bicarbonate secondary to contraction alkalosis. The short duration of illness induced by *S. aureus* limits the severity of fluid and electrolyte imbalances, which usually resolve spontaneously.[51]

TREATMENT

Treatment includes bowel rest and rehydration with spontaneous resolution of symptoms typically occurring in 24–48 hours. Oral rehydration is usually sufficient, but intravenous fluid replacement and hospitalization may be necessary in cases of severe dehydration or severe electrolyte abnormalities, particularly in young children or elderly patients.[52,53] There are no clinical trials that demonstrate an improvement in clinical outcome, avoidance of hospitalization, or shortened duration of symptoms following the use of antiemetic agents. The short duration of symptoms generally limits the need for these medications; however, some patients may experience symptomatic relief with these agents when symptoms are severe. Antibiotics are not recommended because the illness results from the ingestion of preformed toxins rather than from the *in vivo* release of toxins or true invasion of the gastrointestinal tract. Furthermore, the injudicious use of antibiotics potentially could contribute to the increased incidence of antimicrobial-resistant *S. aureus* as a result of the rapid appearance of resistant strains after the initiation of antibiotic therapy.

References

1. McCabe-Sellers BJ, Beattie SE. Food safety: emerging trends in foodborne illness surveillance and prevention. J Am Diet Assoc 2004;104:1708–1717.
2. Elbaze P, Ortonne JP. [Chronic furunculosis] Ann Dermatol Venereol 1988;115:859–864. [French]
3. Bergdoll MS, Surgalla MJ, Dack GM. Staphylococcal enterotoxin. Identification of a specific precipitating antibody with enterotoxin-neutralizing property. J Immunol 1959;83:334–338.
4. Dack GM, Cary WE, Woolpert O, Wiggers H. An outbreak of food poisoning proved to be due to a yellow hemolytic *Staphylococcus*. J Prev Med 1930;4:167–175.

5. Bergdoll MS, Huang IY, Schantz EJ. Chemistry of the staphylococcal enterotoxins. J Agric Food Chem 1974;22:9–13.
6. Levine WC, Bennett RW, Choi Y, Henning KJ, Rager JR, Hendricks KA, et al. Staphylococcal food poisoning caused by imported canned mushrooms. J Infect Dis 1996;173:1263–1267.
7. US Food and Drug Administration. (2007). Foodborne pathogenic microorganisms and natural toxins handbook. Available at http://www.cfsan.fda.gov/~mow/chap3.html. Accessed March 9, 2008.
8. Do Carmo LS, Cummings C, Linardi VR, Dias RS, De Souza JM, De Sena MJ, et al. A case study of a massive staphylococcal food poisoning incident. Foodborne Pathog Dis 2004;1:241–246.
9. Khambaty FM, Bennett RW, Shah DB. Application of pulsed-field gel electrophoresis to the epidemiological characterization of *Staphylococcus intermedius* implicated in a food-related outbreak. Epidemiol Infect 1994;113: 75–81.
10. Becker K, Keller B, von Eiff C, Bruck M, Lubritz G, Etienne J, et al. Enterotoxigenic potential of *Staphylococcus intermedius*. Appl Environ Microbiol 2001;67:5551 –5557.
11. Lewis DL, Gattie DK, Novak ME, Sanchez S, Pumphrey C. Interactions of pathogens and irritant chemicals in land-applied sewage sludges (biosolids). BMC Public Health 2002;2:11.
12. Zadoks RN, van Leeuwen WB, Kreft D, Fox LK, Barkema HW, Schukken YH, et al. Comparison of *Staphylococcus aureus* isolates from bovine and human skin, milking equipment, and bovine milk by phage typing, pulsed-field gel electrophoresis, and binary typing. J Clin Microbiol 2002;40:3894–3902.
13. Kluytmans J, van Belkum A, Verbrugh H. Nasal carriage of *Staphylococcus aureus*: epidemiology, underlying mechanisms, and associated risks. Clin Microbiol Rev 1997;10:505–520.
14. Cenci-Goga BT, Karama M, Rossitto PV, Morgante RA, Cullor JS. Enterotoxin production by *Staphylococcus aureus* isolated from mastitic cows. J Food Prot 2003;66:1693–1696.
15. Vandenbergh MF, Verbrugh HA. Carriage of *Staphylococcus aureus*: epidemiology and clinical relevance. J Lab Clin Med 1999;133:525–534.
16. Holeckova B, Holoda E, Fotta M, Kalinacova V, Gondol J, Grolmus J: Occurrence of enterotoxigenic *Staphylococcus aureus* in food. Ann Agric Environ Med 2002;9: 179–182.
17. Armand-Lefevre L, Ruimy R, Andremont A. Clonal comparison of *Staphylococcus aureus* isolates from healthy pig farmers, human controls, and pigs. Emerg Infect Dis 2005;11:711–714.
18. Colombari V, Mayer MD, Laicini ZM, Mamizuka E, Franco BD, Destro MT, Landgraf M. Foodborne outbreak caused by *Staphylococcus aureus*: phenotypic and genotypic characterization of strains of food and human sources. J Food Prot 2007;70:489–493.
19. Wei H-L, Chiou C-S. Molecular subtyping of *Staphylococcus aureus* from an outbreak associated with a food handler. Epidemiol Infect 2002;128:15–20.
20. Nychas GJ, Arkoudelos JS. Staphylococci: their role in fermented sausages. Soc Appl Bacteriol Symp Ser 1990;19:167S–188S.
21. Mossel DA, van Netten P. *Staphylococcus aureus* and related staphylococci in foods: ecology, proliferation, toxinogenesis, control and monitoring. Soc Appl Bacteriol Symp Ser 1990;19:123S–145S.
22. Gilmour A, Harvey J. Staphylococci in milk and milk products. Soc Appl Bacteriol Symp Ser 1990;19: 147S–166S.
23. Genigeorgis CA. Present state of knowledge on staphylococcal intoxication. Int J Food Microbiol 1989;9:327–360.
24. Letertre C, Perelle S, Dilasser F, Fach P. A strategy based on 5′ nuclease multiplex PCR to detect enterotoxin genes *sea* to *sej* of *Staphylococcus aureus*. Mol Cell Probes 2003;17:227–235.
25. Lindqvist R, Sylven S, Vagsholm I. Quantitative microbial risk assessment exemplified by *Staphylococcus aureus* in unripened cheese made from raw milk. Int J Food Microbiol 2002;78:155–170.
26. Exner M, Vacata V, Hornei B, Dietlein E, Gebel J. Household cleaning and surface disinfection: new insights and strategies. J Hosp Infect 2004;56(Suppl 2):S70–S75.
27. Glass KA, Kaufman KM, Johnson EA. Survival of bacterial pathogens in pasteurized process cheese slices stored at 30 degrees C. J Food Prot 1998;61:290–294.
28. Lopes JA. Susceptibility of antibiotic-resistant and antibiotic-sensitive foodborne pathogens to acid anionic sanitizers. J Food Prot 1998;61:1390–1395.
29. Belay N, Rasooly A. *Staphylococcus aureus* growth and enterotoxin A production in an anaerobic environment. J Food Prot 2002;65:199–204.
30. Schmitt M, Schuler-Schmid U, Schmidt-Lorenz W. Temperature limits of growth, TNase and enterotoxin production of *Staphylococcus aureus* strains isolated from foods. Int J Food Microbiol 1990;11:1–19.
31. Mauff G, Rohrig I, Ernzer U, Lenz W, Bergdoll M, Pulverer G. Enterotoxigenicity of *Staphylococcus aureus* strains from clinical isolates. Eur J Clin Microbiol 1983;2:321–326.
32. Gustafsson E, Nilsson P, Karlsson S, Arvidson S. Characterizing the dynamics of the quorum-sensing system in *Staphylococcus aureus*. J Mol Microbiol Biotechnol 2004;8:232–242.
33. Normanno G, Firinu A, Virgilio S, Mula G, Dambrosio A, Poggiu A, Decastelli L, et al. Coagulase-positive staphylococci and *Staphylococcus aureus* in food products marketed in Italy. Int J Food Microbiol 2005;98:73–79.
34. Salzman M, Madsen JM, Greenberg MI. Toxins: bacterial and marine toxins. Clin Lab Med 2006;26:397–419.

35. Chen T-R, Chiou C-S, Tsen H-Y. Use of novel PCR primers specific to the genes of staphylococcal enterotoxin G, H, I for the survey of *Staphylococcus aureus* strains isolated from food-poisoning cases and food samples in Taiwan. Int J Food Microbiol 2004;92:189–197.

36. Hovde CJ, Marr JC, Hoffmann ML, Hackett SP, Chi YI, Crum KK, et al. Investigation of the role of the disulphide bond in the activity and structure of staphylococcal enterotoxin C1. Mol Microbiol 1994;13:897–909.

37. Arbuthnott JP, Coleman DC, de Azavedo JS. Staphylococcal toxins in human disease. Soc Appl Bacteriol Symp Ser 1990;19:101S–107S.

38. Rich RR, Mollick JA, Cook RG. Superantigens: interaction of staphylococcal enterotoxins with MHC class II molecules. Trans Am Clin Climatol Assoc 1989;101: 195–204.

39. Johnson HM, Russell JK, Pontzer CH. Staphylococcal enterotoxin superantigens. Proc Soc Exp Biol Med 1991;198:765–771.

40. Hamad AR, Marrack P, Kappler JW. Transcytosis of staphylococcal superantigen toxins. J Exp Med 1997;185: 1447–1454.

41. Evenson ML, Hinds MW, Bernstein RS, Bergdoll MS. Estimation of human dose of staphylococcal enterotoxin A from a large outbreak of staphylococcal food poisoning involving chocolate milk. Int J Food Microbiol 1988;7: 311–316.

42. Bennett RW. Staphylococcal enterotoxin and its rapid identification in foods by enzyme-linked immunosorbent assay-based methodology. J Food Prot 2005;68:1264–1270.

43. Asao T, Kumeda Y, Kawai T, Shibata T, Oda H, Haruki K, et al. An extensive outbreak of staphylococcal food poisoning due to low-fat milk in Japan: estimation of enterotoxin A in the incriminated milk and powdered skim milk. Epidemiol Infect 2003;130:33–40.

44. Le Loir Y, Baron F, Gautier M. Staphylococcus aureus and food poisoning. Genetics Mol Res 2003;2:63–76.

45. Tranter HS. Foodborne staphylococcal illness. Lancet 1990;336:1044–1046.

46. Dinges MM, Orwin PM, Schlievert PM. Exotoxins of *Staphylococcus aureus*. Clin Microbiol Rev 2000;13: 16–34.

47. Stuhlmeier R, Stuhlmeier KM. Fast, simultaneous, and sensitive detection of staphylococci. J Clin Pathol 2003;56:782–785.

48. Horowitz W, editor. Official methods of analysis of AOAC International. 18th ed. Gaithersburg, MD: AOAC International; 2006.

49. Vernozy-Rozand C, Mazuy-Cruchaudet C, Bavai C, Richard Y. Comparison of three immunological methods for detecting staphylococcal enterotoxins from food. Lett Appl Microbiol 2004;39:490–494.

50. Ikeda T, Tamate N, Yamaguchi K, Makino S-I. Mass outbreak of food poisoning disease caused by small amounts of staphylococcal enterotoxins A and H. App Environ Microbiol 2005;71:2793–2795.

51. Guerrant RL, Van Gilder T, Steiner TS, Thielman NM, Slutsker L, Tauxe RV, et al. Infectious Diseases Society of America. Practice guidelines for the management of infectious diarrhea. Clin Infect Dis 2001;32:331–351.

52. King CK, Glass R, Bresee JS, Duggan C, Centers for Disease Control and Prevention (CDC). Managing acute gastroenteritis among children: oral rehydration, maintenance, and nutritional therapy. MMWR Recomm Rep 2003;52:1–16.

53. Smith JL. Foodborne illness in the elderly. J Food Prot 1998;61:1229–1239.

Chapter 22

STREPTOCOCCUS SPECIES

CYRUS RANGAN, MD, FAAP

HISTORY

Streptococci are gram-positive, facultatively anaerobic, nonmotile, nonspore forming, catalase-negative cocci, which were first discovered by Louis Pasteur in 1880. Although there are over 15 species of pathogenic streptococci, the Lancefield classification of *Streptococcus* defines the antigenic distinctions among bacteria.[1] Lancefield group A (*S. pyogenes* Rosenbach) is responsible for most foodborne streptococcal disease, whereas group C (*S. equi*), group D (*S. faecalis, S. bovis*), and group G (*S. intermedius, S. anginosus*) cause rare foodborne outbreaks.[2–4] Dairy products were the main source of foodborne streptococcal outbreaks before commercial pasteurization. Today, food handlers are the primary vectors of direct streptococcal contamination of prepared foods. Foodborne streptococcal infections typically involve the pharynx, and only rarely do these infections cause gastroenteritis. The first recorded US epidemic of foodborne pharyngitis occurred in 1968 at a US Air Force Academy, where 1,200 cadets were infected with Group A *Streptococcus* from tuna salad containing mayonnaise and hard-boiled eggs.[5] The salad was prepared by an asymptomatic food handler with a positive throat culture for the identical strain of Group A *Streptococcus* identified in throat swabs from these patients.

EXPOSURE

Source

ENVIRONMENT

Group A *Streptococcus* are distributed worldwide, and are more prevalent in the warmer months of spring and summer.[6]

ANIMAL

Dairy animals harbor group A and group C *streptococci*; however, commercial pasteurization has eliminated virtually all streptococcal contamination of dairy products.[7] Humans are the primary reservoir for modern streptococcal food contamination. The pharynx and skin lesions are commonly identified sources of streptococci in traceable outbreaks.[8,9]

FOOD

The typical sources of foodborne streptococcal pharyngitis are inadequately handled, cold salads contaminated by food handlers with *Streptococcus pyogenes*.[10] Eggs, egg salads, tuna, mayonnaise, and cheeses are the most common sources of foodborne illness associated with

Medical Toxicology of Natural Substances, by Donald G. Barceloux, MD

Streptococcus pyrogenes. The risk of group A *Streptococcus* infection dramatically increases in foods prepared the day before serving or food remaining at room temperature for several hours in salad bars, buffets, or picnics.[11] With the advent of pasteurization, milk is a rare cause of epidemic streptococcal pharyngitis.[12]

Food Processing

The most effective means to control group A streptococcal food contamination include the following: proper cooking and refrigeration, purification of water, pasteurization of dairy products, sanitation of food preparation surfaces with at least 1% bleach solution, and cooking foods to 120°C–170°C (248°F–338°F).[13] Food handlers or salad bar patrons carrying streptococci in the nasopharynx or on skin lesions may directly inoculate food before or after preparation via sneezing, coughing, or handling the food with contaminated utensils or hands.[8] Protective measures include strict hygiene and surveillance of food handlers; use of protective gloves and masks; and particular care in handling cold salads, especially those containing egg products.[14]

TOXINS

Physiochemical Properties

The genus *Streptococcus* contains over 15 species. Lancefield Group A class (*S. pyogenes*) is most important as a cause of foodborne illness. Group A *Streptococcus* possesses a hyaluronic acid containing an outer capsule that encases fimbrial M proteins, which are coiled dimers of alpha-helices on the bacterial surface.[15] M proteins are serotyped 1–61. Strains with M proteins 1, 3, 5, 6, 14, 18, 19, 24, and 29 are generally associated with throat infections and rheumatic fever, whereas M proteins 2, 49, 57, 59, 60, and 61 are usually associated with skin infections.[16,17] M proteins mediate most of the group A *Streptococcus* pathogenicity along with numerous virulence determinants, including lipoteichoic acid (LTA), protein F, fibronectin-binding proteins, serum opacity factor, glyceraldehyde-3-phosphate dehydrogenase, vitronectin-binding protein, galactose-binding protein, and collagen-binding protein.[18]

Mechanism of Toxicity

Lipoteichoic acid (LTA) on the outer bacterial membrane binds loosely with fatty acids residues on fibronectin and epithelial cells on the buccal and pharyngeal mucosae.[19] The hyaluronic acid capsule binds to CD44 on epithelial cells and preserves bacterial virulence determinants.[20] M proteins 1, 3, 5, 6, 14, 18, 19, and 24 anchor the organism to throat epithelial cells, promote colonization, and induce a local inflammatory response by stimulating the production of interleukin-6 (IL-6).[21,22] M proteins 2, 49, 57, 59, 60, and 61 adhere strongly to keratinocytes, promoting colonization and infection in skin lesions.[23] Protein F adheres to fibronectin and initiates invasion into epithelial cells.[24] M protein, protein F, fibronectin-binding proteins, and serum opacity factor interact with α5 and β1 integrins on epithelial cell surfaces and promote bacterial invasion. This internalization may account for treatment failures, asymptomatic carriage, and invasion of deeper tissues (i.e., tonsillopharyngitis) in human hosts.[25,26] Once adhesion occurs, the bacterium evades host defenses by attacking the complement system, and M proteins and the hyaluronic acid capsule inhibit activation of the complement system by binding regulatory factor H.[27] Streptococcal C5a peptidase blocks chemotaxis of polymorphonuclear cells by cleaving complement factor C5a;[28] and the binding of M proteins to fibrinogen shields the bacterium from opsonization by complement factor C3b, thus inhibiting phagocytosis.[29] Leukocidins, streptolysin S, and streptolysin O form pores in leukocyte cell membranes with subsequent cell lysis.[30] Streptococcal proteinase mediates tissue damage by activating interleukin-1ß (IL-1ß), degrading vitronectin and fibronectin, and activating endothelial metalloproteases. This cascade destroys the extracellular tissue matrix and promotes an acute inflammatory response.[31–33]

Bacterial superantigens are highly potent immunostimulatory molecules produced by *Streptococcus pyrogenes* as well as *Staphylococcus aureus*. These toxins interact with MHC class II molecules and CD4+ T cells to form trimolecular complex that cause intense T-cell proliferation. The subsequent production of cytokine results in epithelial damage, capillary leak, and hypotension. The streptococcal superantigens include the classical pyrogenic exotoxins A and C along with more recently identified toxins (G–J, SMEZ, SSA).[34]

DOSE RESPONSE

The inoculum of group A *Streptococcus* required to produce pharyngitis from foodborne infections is unknown. Colony counts of group A *Streptococcus* in implicated foods range from 10^5 to 10^7 CFU/gm; however, the minimum inoculum size may be much lower.[35] The estimated size of inoculum to cause streptococcal endocarditis based on rabbit models is 3.7×10^5 to 8.5×10^6 CFU/g.[36] Patients under the age of 5 years do not typically acquire streptococcal pharyngitis because they lack epithelial cell receptors for group A *Streptococcus* adhesins.[37]

CLINICAL RESPONSE

Foodborne group A *Streptococcus* incubates for 1–3 days before symptoms develop. A brief prodrome of muscle aches, nausea, and fatigue may precede throat symptoms.[38] In comparison to airborne streptococcal pharyngitis, the onset following foodborne streptococcal pharyngitis is more abrupt, the attack rate is higher (i.e., 51–90%), and complication rates are lower.[39,40] Additionally, symptoms in foodborne disease are confined primarily to sore throat, odynophagia, fever, pharyngeal erythema, tonsillar enlargement, and submandibular lymphadenopathy, whereas cough and other respiratory symptoms usually accompany sore throat secondary to airborne streptococcal disease. Although some cases of foodborne disease may begin with respiratory symptoms, the presence of these symptoms in an outbreak suggests secondary infection. Complications such as peritonsillar abscess may represent treatment failure.[41] Foodborne infections are not usually associated with sequelae such as acute glomerular nephritis, rheumatic fever, reactive arthritis, toxic shock syndrome, pediatric autoimmune neuropsychiatric disorders associated with streptococcus (PANDAS), or septicemia in healthy individuals. Isolated cases of toxic shock syndrome or other invasive, superantigen-mediated diseases may occur in patients with underlying chronic diseases.[42] Foodborne streptococcal gastroenteritis is relatively rare, and pharyngitis is the typical presentation of foodborne streptococcal disease.[43] The development of mild, self-limited nausea, vomiting, or diarrhea in patients with foodborne streptococcal pharyngitis is relatively uncommon compared with upper respiratory tract symptoms.[44,45]

DIAGNOSTIC TESTING

Analytical Methods

Throat culture is the most common method to confirm streptococcal pharyngitis. Rapid antigen screens are 95% specific, but only 70–90% sensitive. Negative antigen screens should be followed by routine culture in suspected cases.[46] Group A *Streptococcus* produces a zone of β-hemolysis on 5% sheep blood agar in the presence of trimethoprim-sulfamethoxazole, which inhibits the growth of normal flora. Anaerobic conditions enhance streptolysin O activity.[47] DNA-based methods allow the amplification of specific genetic signals from a few cells; therefore, these analytical methods are much more sensitive and specific than the use of selective microbiological media. Polymerase chain reaction (PCR) is the basis of many nucleic acid-based detection systems. This method detects both live and dead strains of *Streptococcus*, and PCR for this bacterium yields 99% specificity and 100% sensitivity.[48] Subtyping procedures are useful for tracking individual strains responsible for outbreaks of *Streptococcus*. Analytical methods for subtyping include phenotypic subtyping (biogrouping, serotyping, phage typing, esterase typing) and genotypic subtyping (pulsed-field gel electrophoresis, multilocus enzyme electrophoresis, ribotyping, PCR-based subtyping techniques, amplified fragment length polymorphism, PCR-restriction fragment length polymorphism, repetitive element PCR, DNA sequencing-based subtyping techniques such as multilocus sequence typing). These subtyping techniques are reserved for the investigation of epidemics. Rapid DNA hybridization is an alternative technique of correlating T-antigen agglutination factors and M protein factors in human specimens to those in food specimens.[49] Throat or skin lesion specimens obtained from asymptomatic human carriers (e.g., food handlers) are tested by similar methodologies to identify potential sources of an outbreak.[50]

Biomarkers

Antistreptolysin-O and anti-DNase B titers may remain elevated in postepidemic surveillance for several months.[30]

Abnormalities

Leukocytosis with or without left shift may be observed; however, leukocyte counts are generally unnecessary for diagnosis in uncomplicated cases.

TREATMENT

If present, mild gastrointestinal symptoms typically are self-limited. Treatment of group A streptococcal gastroenteritis includes bowel rest and rehydration with fluid and electrolytes replacement. Oral replacement is usually sufficient.[51,52] Group A streptococcal pharyngitis is uniformly treated with penicillins. Despite exquisite susceptibility to penicillins, treatment failure may result from poor antibiotic compliance, reinfection, penicillin tolerance, endothelial internalization of bacteria, or the coexistence of oropharyngeal beta-lactamase-producing bacteria that degrade penicillins during therapy.[41,53] Patients with penicillin allergies may be treated with erythromycin or azithromycin; however, up to 20% of patients treated with erythromycin have positive throat cultures posttherapy.[54] Patients under the age of 5 years should not be treated presumptively with antimicrobial agents.[37] Prophylaxis of close contacts with penicillin is not more effective at preventing secondary transmission than prompt treatment of affected victims, and prophylactic treatment may lead to unexpected penicillin allergies.[55]

References

1. Lancefield RC. Current problems in studies of streptococci. J Gen Microbiol 1969;55:161–163.
2. Stryker WS, Fraser DW, Facklam RR. Foodborne outbreak of group G streptococcal pharyngitis. Am J Epidemiol 1982;116:533–540.
3. Cohen D, Ferne M, Rouach T, Bergner-Rabinowitz S. Food-borne outbreak of group G streptococcal sore throat in an Israeli military base. Epidemiol Infect 1987;99:249–255.
4. Franz CM, Holzapfel WH, Stiles ME. *Enterococci* at the crossroads of food safety? Int J Food Microbiol 1999;47:1–24.
5. Hill HR, Zimmerman RA, Reid GV, Wilson E, Kilton RM. Food-borne epidemic of streptococcal pharyngitis at the United States Air Force Academy. N Engl J Med 1969;280:917–921.
6. Hofkosh D, Wald ER, Chiponis DM. Prevalence of non-group-A beta-hemolytic *streptococci* in childhood pharyngitis. South Med J 1988;81:329–331.
7. Tripathy SB, Fincher MG, Bruner DW. Viable *micrococci* and *streptococci* other than *Streptococcus agalactiae* that remain on the surface of cows' teats after the use of certain disinfectants as dipping agents. Cornell Vet 1963;53:434–444.
8. Farley TA, Wilson SA, Mahoney F, Kelso KY, Johnson DR, Kaplan EL. Direct inoculation of food as the cause of an outbreak of group A streptococcal pharyngitis. J Infect Dis 1993;167:1232–1235.
9. Ollinger-Snyder P, Matthews ME. Food safety: review and implications for dietitians and dietetic technicians. J Am Diet Assoc 1996;96:163–168.
10. Katzenell U, Shemer J, Bar-Dayan Y. Streptococcal contamination of food: an unusual cause of epidemic pharyngitis. Epidemiol Infect 2001;127:179–184.
11. McCormick JB, Hayes P, Feldman R. Epidemic streptococcal sore throat following a community picnic. JAMA 1976;236:1039–1041.
12. Centers for Disease Control (CDC). A foodborne outbreak of streptococcal pharyngitis—Portland, Oregon. MMWR Morb Mortal Wkly Rep 1982;31:3–5.
13. Christensen BB, Rosenquist H, Sommer HM, Nielsen NL, Fagt S, Andersen NL, et al. A model of hygiene practices and consumption patterns in the consumer phase. Risk Anal 2005;25:49–60.
14. Horan JM, Cournoyer JJ. Foodborne streptococcal pharyngitis. Am J Public Health 1986;76:296–297.
15. Flicker PF, Cohen C, Manjula BN, Fischetti VA. Phillips GN, Streptococcal M protein: alpha-helical coiled-coil structure and arrangement on the cell surface. Proc Natl Acad Sci U S A 1981;78:4689–4693.
16. Bessen D, Jones KF, Fischetti VA. Evidence for two distinct classes of streptococcal M protein and their relationship to rheumatic fever. J Exp Med 1989;169:269–283.
17. Gillis D, Cohen D, Beck A, Rouach T, Katzenelson E, Green M. A new *Streptococcus* group A M-29 variant isolated during a suspected common-source epidemic. Mil Med 1992;157:282–223.
18. Hasty DL, Courtney HS. Group A streptococcal adhesion: all of the theories are correct. In: Kahane I, Ofek I, editors. Toward anti-adhesion therapy for microbial diseases. New York: Plenum Press; 1996:81–94.
19. Ofek I, Simpson WA, Beachey EH. Formation of molecular complexes between a structurally defined M protein and acylated or deacylated lipoteichoic acid of *Streptococcus pyogenes*. J Bacteriol 1982;149:426–433.
20. Schrager HM, Alberti S, Cywes C, Dougherty GJ, Wessels MR. Hyaluronic acid capsule modulates M protein-mediated adherence and acts as a ligand for attachment of group A *Streptococcus* to CD44 on human keratinocytes. J Clin Investig 1998;101:1708–1716.
21. Hollingshead SK, Simecka JW, Michalek SM. Role of M protein in pharyngeal colonization by group A streptococci in rats. Infect Immunol 1993;61:2277–2283.
22. Courtney H, Ofek SI, Hasty DL. M protein mediated adhesion of M type 24 *Streptococcus pyogenes* stimulates release of interleukin-6 by HEp-2 tissue culture cells. FEMS Microbiol Lett 1997;151:65–70.
23. Perez-Casal J, Okada N, Caparon M, Scott JR. Role of the conserved C repeat region of the M protein of *Streptococcus pyogenes*. Mol Microbiol 1995;15:907–916.
24. Guzman CA, Talay SR, Molinari G, Medina E, Chhatwal GS. Protective immune response against *Streptococcus pyogenes* in mice after intranasal vaccination with the fibronectin-binding protein SfbI. J Infect Dis 1999;179: 901–906.
25. Neeman R, Keller N, Barzilai A, Korenman Z, Sela S. Prevalence of the internalization-associated gene, prtF1, among persisting group A *Streptococcus* strains isolated from asymptomatic carriers. Lancet 1998;352:1974–1977.
26. LaPenta D, Rubens C, Chi E, Cleary PP. Group A *streptococci* efficiently invade human respiratory epithelial cells. Proc Natl Acad Sci USA 1994;91:12115–12119.
27. Horstmann RD, Sievertsen HJ, Knobloch J, Fischetti VA. Antiphagocytic activity of streptococcal M protein: selective binding of complement control protein factor H. Proc Natl Acad Sci USA 1988;85:1657–1661.
28. Wexler D, Chenoweth E, Cleary P. Mechanism of action of the group A streptococcal C5a inactivator. Proc Natl Acad Sci USA 1985;82:8144–8148.
29. Horstmann RD, Sievertsen HJ, Leippe M, Fischetti VA. Role of fibrinogen in complement inhibition by streptococcal M protein. Infect Immun 1992;60:5036–5041.
30. Bhakdi S, Bayley H, Valeva A, Walev I, Walker B, Kehoe M, et al. Staphylococcal alpha-toxin, streptolysin-O, and

Escherichia coli hemolysin: prototypes of pore-forming bacterial cytolysins. Arch Microbiol 1996;165:73–79.

31. Kapur V, Majesky MW, Li LL, Black RA, Musser JM. Cleavage of interleukin 1b precursor to produce active IL-1b by a conserved extracellular cysteine protease from *Streptococcus pyogenes*. Proc Natl Acad Sci USA 1993;90:7676–7680.
32. Kapur V, Topouzis S, Majesky MW, Li LL, Hamrick MR, Hamill RJ, et al. A conserved *Streptococcus pyogenes* extracellular cysteine protease cleaves human fibronectin and degrades vitronectin. Microb Pathog 1993;15:327–346.
33. Burns EH, Marceil AM, Musser JM. Activation of a 66-kilodalton human endothelial cell matrix metalloprotease by *Streptococcus pyogenes* extracellular cysteine protease. Infect Immunol 1996;64:4744–4750.
34. Baker MD, Acharya KR. Superantigens: structure-function relationships. Int J Med Microbiol 2004;293:529–537.
35. Nakajima K, Okuyama M, Okuda K. Epidemiological study of hemolytic *streptococci* in pupils in seven prefectures in Japan. Osaka City Med J 1985;31:1–10.
36. van de Rijn I. Role of culture conditions and immunization in experimental nutritionally variant streptococcal endocarditis. Infect Immunol 1985;50:641–646.
37. Gerber MA. Diagnosis and treatment of pharyngitis in children. Pediatr Clin North Am 2005;52:729–747.
38. Claesson BE, Svensson NG, Gotthardsson L, Gotthardsson L, Garden B. A foodborne outbreak of group A streptococcal disease at a birthday party. Scand J Infect Dis 1992;24:577–586.
39. Center for Disease Control (CDC). Streptococcal foodborne outbreaks—Puerto Rico, Missouri. MMWR Morb Mortal Wkly Rep 1984;33:669–672.
40. Bar-Dayan Y, Bar-Dayan Y, Shemer J. Food-borne and air-borne streptococcal pharyngitis—a clinical comparison. Infection 1997;25:12–15.
41. Sela S, Barzilai A. Why do we fail with penicillin in the treatment of group A *Streptococcus* infections? Ann Med 1999;31:303–307.
42. Cockerill FR, MacDonald KL, Thompson RL, Roberson F, Kohner PC, Besser-Wiek J, et al. An outbreak of invasive group A streptococcal disease associated with high carriage rates of the invasive clone among school-aged children. JAMA 1997;277:38–43.
43. Dudding BA, Dillon HC, Wannamaker LW, Kilton RM, Chapman SS, Anthony BF. Postepidemic surveillance studies of a food-borne epidemic of streptococcal pharyngitis at the United States Air Force Academy. J Infect Dis 1969;120:225–236.
44. Elsea WR, Mosher WE, Lennon RG, Markellis V, Hoffman PF. An epidemic of food-associated pharyngitis and diarrhea. Arch Environ Health 1971;23:48–56.
45. Matsumoto M, Miwa Y, Matsui H, Saito M, Ohta M, Miyazaki Y. An outbreak of pharyngitis caused by foodborne group A *Streptococcus*. Jpn J Infect Dis 1999;52:127–128.
46. Gerber MA, Shulman ST. Rapid diagnosis of pharyngitis caused by group A *streptococci*. Clin Microbiol Rev 2004;17:571–580.
47. Tolliver PR, Roe MH, Todd JK. Detection of group A *Streptococcus*: comparison of solid and liquid culture media with and without selective antibiotics. Pediatr Infect Dis J 1987;6:515–519.
48. Beall B, Facklam R, Thompson T. Sequencing emm-specific PCR products for routine and accurate typing of group A *streptococci*. J Clin Microbiol 1996;34:953–958.
49. Kaufold A, Podbielski A, Blockpoel M, Schouls L. Rapid typing of group A streptococci by use of DNA amplification and non-radioactive allele-specific oligonucleotide probes. FEMS Microbiol Lett 1994;119:19–26.
50. Sarvghad MR, Naderi HR, Naderi-Nassab M, Majdzadeh R, Javanian M, Faramarzi H, et al. An outbreak of food-borne group A *Streptococcus* (GAS) tonsillopharyngitis among residents of a dormitory. Scand J Infect Dis 2005;37:647–650.
51. King CK, Glass R, Bresee JS, Duggan C, Centers for Disease Control (CDC). Managing acute gastroenteritis among children: oral rehydration, maintenance, and nutritional therapy. MMWR Morb Mortal Wkly Rep 2003;52:1–16.
52. Smith JL. Food-borne illness in the elderly. J Food Prot 1998;61:1229–1239.
53. Hayes CS, Williamson H. Management of group A beta-hemolytic streptococcal pharyngitis. Am Fam Physician 2001;63:1557–1564.
54. Gerber MA. Antibiotic resistance: relationship to persistence of group A *streptococci* in the upper respiratory tract. Pediatrics 1996;97:971–975.
55. Ryder RW, Lawrence DN, Nitzkin JL, Feeley JC, Merson MH. An evaluation of penicillin prophylaxis during an outbreak of foodborne streptococcal pharyngitis. Am J Epidemiol 1977;106:139–144.

Chapter 23

VIBRIO SPECIES

CYRUS RANGAN, MD, FAAP

HISTORY

Vibrio species (*Vibrios*) are gram-negative, facultatively anaerobic, oxidase-positive, motile, flagellated, curved rods.[1] During the Asiatic cholera epidemic of 1846–1863, Italian microbiologist Filippo Pacini identified *Vibrio cholerae* Pacini (*V. cholerae*) in 1854, after British physician John Snow described the epidemiologic basis for cholera transmission in 1849.[2] Robert Koch rediscovered the organism in 1884 after the Egyptian cholera outbreak of 1883.[3] *Vibrios* are ubiquitous in surface waters worldwide, and these organisms have caused recognized pandemics at least since 1817. Typically, these pandemics appeared in 7-year cycles, primarily affecting Asia and Africa. The last US epidemic occurred in 1911.[4] The classical biotype of *V. cholerae* O1 caused the sixth pandemic in Asia in 1950. Proliferation of *V. cholerae* O1 biotype El Tor initiated the seventh pandemic, which began in Asia during 1961 and spread to West Africa and southern Europe.[5] In 1991, the seventh pandemic spread to Peru, with sporadic cases occurring in the United States along the gulf coasts of Texas, Louisiana, Alabama, Mississippi, and Florida.[6] In 1993, *V. cholerae* serogroup O139 caused the eighth cholera pandemic in Bangladesh and India.[7] Subsequently, *V. cholerae* O1 El Tor emerged in the same area 3 years later.[8] *V. cholerae* is the most prevalent *Vibrio* species worldwide, whereas *Vibrio parahaemolyticus* (Fujino et al) Baumann et al is the most common *Vibrio* species responsible for gastrointestinal infections in the United States, Taiwan, and Japan. This illness is transmitted primarily by oysters and other shellfish, which regularly harbor *V. parahaemolyticus* and cause an estimated 8,000 annual cases with 60 deaths.[9] Other noncholera *Vibrio* species that cause rare cases of gastrointestinal illnesses include *V. alginolyticus* (Miyamoto et al) Sakazaki, *V. fluvialis* Lee et al., *V. furnissii* Brenner et al, *V hollisae* Hickman et al., *V. mimicus* Davis et al, and *V. vulnificus* (Reichelt et al.) Farmer.[10]

EXPOSURE

Source

ENVIRONMENT

Vibrios thrive globally in all salty surface waters, especially along coastlines and estuaries. The lifespan of these organisms in the water column is about 3 weeks. However, *Vibrios* are infrequently isolated from coastal waters between epidemics because *Vibrios* enter into a dormant, viable but nonculturable (VNC) state as water temperatures fall.[11] During warming trends, VNC *Vibrios* associate with plankton (copepods) that proliferate extensively during cholera epidemics. This phenomenon suggests that cholera epidemics are associated with both exposure to contaminated waters and global weather changes that support the growth of plankton.[12] *Vibrio vulnificus* adheres to a wide range of surfaces, including the exoskeleton of mollusks and human tissue.[13]

ANIMAL

Shellfish and plankton in warm marine and estuarine waters are the most prominent reservoirs of *Vibrio*

Medical Toxicology of Natural Substances, by Donald G. Barceloux, MD

species, whereas contaminated drinking water is the primary source of transmission to humans.[14] Humans act as short-term reservoirs for 3–4 weeks in endemic areas, but person-to-person transmission is unusual because of the high inoculum required to cause disease.[15]

Food

Drinking water is the primary source of transmission of *V. cholerae. Vibrio* species are common contaminants of shellfish. Improperly cooked oysters, crab, lobster, mussels, and clams are the most common foodborne sources of *V. cholerae*,[16] *V. parahaemolyticus*,[17,18] and *V. vulnificus*,[19] especially shellfish harvested from fecally contaminated coastal waters. In a study of 682 batches of cooked crustaceans and molluscan shellfish from retail stores and production facilities in the United Kingdom, 0.4% of these samples contained concentrations (10^2–10^6CFU/g) of *Vibrio parahaemolyticus* exceeding European Commission recommendations.[20] A study of blue mussels grown in German waters detected the presence of *Vibrio* species in about 74% of the samples.[21] The most common *Vibrio* species was *V. alginolyticus* (51.2% of *Vibrio* isolates) followed by *V. parahaemolyticus* (39.5%) and *V. vulnificus* (3.5%).

Food Processing

Cholera has been virtually eliminated from developed countries as a result of improvements in sanitation, sewage management, water delivery, and water treatment, which effectively control *V. cholerae*.[22] Proper cooking and refrigeration of seafood are the most effective means of controlling *V. cholerae* and *V. parahaemolyticus*.[15] Transmission of *V. parahaemolyticus* occurs in restaurants from the contamination of utensils and food preparations surfaces with organisms from raw food or the handling of cooked food with contaminated hands. To prevent the rapid increase in *V. parahaemolyticus* in shellfish, harvested clams and oysters should be rapidly refrigerated during transport, processing, and storage at temperatures below 50°F (10°C). Boiling water for 10 minutes inactivates *V. cholerae*.[23]

TOXINS

Physiochemical Properties

Vibrios tolerate a wide range of alkaline pH (i.e., up to 8.5–9.5) and salty habitats, allowing them to thrive in surface waters.[24] Gastric acid readily destroys *V. cholerae*, and therefore patients with decreased gastric acid production are more susceptible to cholera.[25] Alkaline pH and resistance to bile salts enable *Vibrios* to survive in the small intestine.[26] Most *Vibrios* possess a heat-labile, flagellar H antigen that governs bacterial motility. Over 140 serogroups of *V. cholerae* occur worldwide, but only three serogroups, *V. cholerae* O1, O139, and 0141, produce cholera toxin (CAS RN: 9012-63-9).[27] *V. cholerae* O1 variably expresses surface antigens that define two serotypes: Inaba (antigens A and B) and Ogawa (antigens A and C). A third serotype, Hikojima, is a transitional serotype of *V. cholerae* O1 (antigens A, B, and C) that allows the organism to shift between the Inaba and Ogawa serotypes during the life cycle.[28] *V. cholerae* O1 serotypes are further differentiated by morphologic properties into the classical biotype and the El Tor biotype. The former biotype is lysed by classical IV and FK bacteriophages and inhibited by polymyxin B, whereas the latter biotype causes *in vitro* hemolysis, erythrocyte agglutination, and a positive Voges-Proskaur test.[29] The latter test is positive as a result of the bacterial conversion of glucose to acetyl methylcarbinol. *V. cholerae* O1 does not possess a capsule; thus, this organism does not spread beyond the intestine. *V. vulnificus* possesses an acid polysaccharide capsule that promotes dissemination and systemic infection.[30] *Vibrio vulnificus* is a natural contaminant of warm seawater (>20°C/68°F) and low salinity (<1.6%).[31] In a study of 34 virulent stains of *V. parahaemolyticus* isolated from coastal waters off Oregon and Washington, there were six O serotypes (O1, O3, O4, O5, O10, and O11) as measured by pulsed-field gel electrophoresis.[32] The most common serotypes were O5 (19 isolates) and O1 (9 isolates).

Mechanism of Toxicity

Vibrio cholerae O1 causes diarrhea via the following five virulence factors: toxin coregulated pili, neuramidase, accessory cholera enterotoxin, zonula occludens toxin, and cholera toxin (CT). *Vibrio cholerae* adheres to microvilli of the brush border of gastrointestinal epithelial cells via toxin coregulated pili, which enable the organism to remain attached during extensive diarrhea (up to 20–30L/d).[33] Neuramidase enhances binding of cholera toxin to epithelial cells in the intestinal wall by degrading membrane polysialosyl gangliosides into the monosialosyl ganglioside (GM-1), which is a specific receptor for cholera toxin.[34,35] The *V. cholerae* genome contains multiple copies of genes coding for accessory cholera enterotoxin, which enhances chloride secretion and impairs sodium and chloride absorption,[36] whereas zonula occludens toxin relaxes epithelial cell tight junctions.[37] Cholera toxin is encoded on a bacteriophage (CTXΦ). CTXΦ first binds to toxin coregulated pili, which subsequently incorporates

CTXΦ into the *V. cholerae* genome for transcription.[38] Cholera toxin is structurally and functionally identical to the heat-labile toxin (LT) of enterotoxigenic *Escherichia coli* (ETEC).[39] Cholera toxin is composed of an A subunit (28 kDa) linked to a ring of five B subunits (12 kDa each) by a disulfide bridge, and both units are necessary for the production of gastrointestinal illness.[40] The A subunit consists of the virulent enzymatic piece (A1 peptide), and the bridge from A1 to the B ring (A2 peptide). The B subunit ring binds to the GM-1 ganglioside of the intestinal epithelial cell, and the epithelial cell internalizes the A subunit. Lipid rafts carry the intact toxin to the endoplasmic reticulum, where protein disulfide isomerase splits the A subunit into A1 and A2.[41] A1 then diffuses into the cytosol. Subunit A1 activates regulatory G_s proteins by blocking the hydrolysis of GTP to GDP, which permanently activates adenyl cyclase for the life of the epithelial cell.[42] Adenyl cyclase breaks down intracellular ATP into the intracellular messenger cyclic AMP (cAMP). Cyclic AMP accumulates along the cell membrane where it activates protein kinase A, resulting in phosphorylation of ion transport channels and secretion of sodium, chloride, potassium, bicarbonate, and water into the intestinal lumen.[43] Not all strains of *V. cholerae* O1 elaborate enterotoxin, resulting in asymptomatic colonization in human hosts.[44] The mechanism of toxicity of *Vibrio cholerae* O139 and *Vibrio cholerae* O1 is identical.[45] *Vibrio parahaemolyticus* produces a thermostable direct hemolysin (TDH), which increases intracellular calcium and chloride ion secretion by epithelial cells.[46] A polysaccharide capsule confers antiphagocytic state to *V. vulnificus*, permitting systemic invasion and proliferation mediated by siderophores, cytolysins, and collagenase.[47]

DOSE RESPONSE

The infective dose of *V. cholerae* or *V. parahaemolyticus* is relatively high (10^{11} organisms) in drinking water because of the destruction of most of these organisms by the acidic environment of the stomach.[48] The buffering capacity of food reduces the inoculum threshold to 10^4 to 10^6 organisms, whereas patients with achlorhydria require as few as 10^3 organisms to cause illness.[49,50] The symptoms associated with cholera are more severe in patients with blood group O than with other blood types.[51] The mortality rate of cholera is approximately 1% in adequately treated patients compared with about 60% in untreated patients.[52] As few as 100 organisms may cause septicemia following the ingestion of food contaminated by *V. vulnificus*.[53] The infective dose of *V. parahaemolyticus* probably is >100,000 CFU.[54]

CLINICAL RESPONSE

The incubation period for *V. cholerae* is about 1–4 days, and during this time the bacteria proliferate and elaborate enough enterotoxin to produce clinical effects.[48] The severity of the gastrointestinal symptoms depends on the size of the inoculum, the extent of cholera toxin production, and the classical biotype.[44] About 60% of exposures to *V. cholerae* are asymptomatic. In symptomatic cases, diarrhea begins abruptly with massive watery fluid interspersed with mucus and epithelial tissue ("rice-water" stools). Nausea, vomiting, bloating, cramping abdominal pain, and low-grade fever are commonly associated clinical findings. Patients may lose up to one liter of fluid per hour.[48] Profound dehydration may cause anuria, hypovolemia, shock, and cardiovascular collapse. Hypernatremic or hyponatremic dehydration and severe acidosis may cause mental status changes and seizures, while hypokalemia may cause cardiac arrhythmias. The illness resolves within 1–5 days when treated.[55] The clinical courses of infections with *V. cholerae* O139 and *V. cholerae* O1 are similar.[56] *Vibrio parahaemolyticus* causes milder, self-limited gastrointestinal symptoms lasting 1–3 days, with bloody diarrhea in 25% of reported cases.[57] The mean incubation period for *Vibrio parahaemolyticus*-induced illness is about 15 hours with a range of 4 to 96 hours.[58] *Vibrio vulnificus* causes mild gastroenteritis with or without blood, followed by systemic invasion and septicemia.[59]

DIAGNOSTIC TESTING

Analytical Methods

Methods to detect *V. cholerae* in stool include the oxidase test, Kligler iron agar, triple sugar agar, Gram stain, string test, and wet mount. After isolation of the bacteria, the use of antiserum allows the determination of serogroup and type. Phase-contrast and darkfield microscopy of *Vibrio* species demonstrate curved, motile, gram-negative rods.[60] The addition of anti-*V. cholerae* O1 antiserum (monoclonal or polyclonal antibodies) to the microscopy slide instantly halts motility in about 50% of examinations, and this technique quickly differentiates *V. cholerae* 0139 from other motile rods such as *Campylobacter*.[61] Vibrios grow poorly on MacConkey agar, but these bacteria grow readily in thiosulfate-citrate-bile salts-sucrose (TCBS) agar that is pre-enriched in alkaline (pH 8.5) peptone broth.[62] Stool samples typically carry 10^7 to 10^8 organisms, and thus stool cultures rarely require pre-enrichment. Food and water samples carry fewer organisms, often necessitating pre-enrichment for investigation of *Vibrio*-related foodborne or waterborne illnesses.[63] A positive oxidase test quickly differentiates *Vibrio* species from isolated colonies of enteric bacteria.[64] Serotyping via anti-O group antisera is reserved for epidemiologic investigations.[65] Polymerase chain reaction (PCR) provides rapid identification of as few as three organisms of cholera toxin-producing strains in broth cultures[66] and in food samples.[67] Because of the prevalence of numerous nonpathogenic *Vibrio* species, identification of these bacteria in stool or in food samples must be followed by confirmation of cholera toxin production.[68] *Vibrio parahaemolyticus* and *V. vulnificus* are identified by similar methods.[69–71] *Vibrio parahaemolyticus* thermostable direct hemolysin is detected by immunochromatographic assay.[72] AOAC (Association of Analytical Chemists International, Gaithersburg, MD; http://www.eoma.aoac.org/) official method of analysis 988.20 determines the presence of *Vibrio cholera* in oysters.[73] AOAC official method 994.06 identifies *Vibrio vulnificus* by gas chromatographic examination of microbial fatty acid profile.

Biomarkers

Vibrio cholerae does not enter the blood, but these bacteria produce a transient serum antibody response that peaks 8–10 days after infection. This response resolves within 2 months.[74]

Abnormalities

Severe fluid and electrolyte losses cause hypernatremia (serum Na > 150 mEq/L) or hyponatremia (serum Na < 130 mEq/L), hypokalemia (serum K < 3.0 mEq/L), hypochloremia, elevated urine specific gravity (SpG > 1.020), and hyperphosphatemia. Low serum bicarbonate results from lactic acid production or elevated bicarbonate secondary to contraction alkalosis. ST-T flattening and arrhythmias may be observed on electrocardiogram.

TREATMENT

Treatment includes rehydration and bowel rest. Extreme fluid losses from the gastrointestinal tract necessitate aggressive rehydration with fluid and electrolytes. Oral rehydration is usually sufficient in mild to moderate cases, but intravenous fluid replacement and hospitalization may be necessary in cases of severe dehydration or electrolyte abnormalities, particularly in young children or elderly patients.[75,76] In 1991, the World Health Organization achieved a 99% survival rate in 300,000 patients in Peru with a cholera treatment plan using powdered formulations reconstituted into oral rehydration solutions.[77] Antimicrobial therapy may curtail toxin production, reduce stool output, decrease bacterial shedding in epidemics, and shorten illness duration in severe cases; however, antimicrobials are minor adjuncts to fluid replacement therapy.[78] Children are treated with trimethoprim-sulfamethoxazole; adults are treated with tetracycline or doxycycline.[79] Prophylaxis with tetracycline may reduce infection rates in travelers to endemic areas,[80] but plasmid-mediated, tetracycline-resistant strains are prevalent in endemic areas.[81,82] Production of IgA antibodies to *V. cholerae* O1 follows infection with predictable protection from future exposures for up to 3 years; however, immunity to *V. cholerae* O1 does not confer immunity to *V. cholerae* O139 or other *Vibrio* species.[83] Maternal antibodies from breast milk protect young infants in endemic regions.[84]

References

1. Thompson FL, Klose KE; AVIB Group. *Vibrio* 2005: the First International Conference on the Biology of Vibrios. J Bacteriol 2006;188:4592–4596.
2. Bentivoglio M, Pacini P. Filippo Pacini: a determined observer. Brain Res Bull 1995;38:161–165.
3. Howard-Jones N. Robert Koch and the cholera vibrio: a centenary. Brit Med J 1984;288:379–381.
4. Blake PA. Epidemiology of cholera in the Americas. Gastroenterol Clin North Am 1993;22:639–660.
5. Greenough WB. The human, societal, and scientific legacy of cholera. J Clin Invest 2004;113:334–339.
6. Tickner J, Gouveia-Vigeant T. The 1991 cholera epidemic in Peru: not a case of precaution gone awry. Risk Anal 2005;25:495–502.
7. Swerdlow DL, Ries AA. *Vibrio cholerae* non-O1—the eighth pandemic? Lancet 1993;342:382–383.
8. Faruque SM, Ahmed KM, Abdul Alim AR, Qadri F, Siddique AK, Albert MJ. Emergence of a new clone of toxigenic *Vibrio cholerae* O1 biotype El Tor displacing *V. cholerae* O139 Bengal in Bangladesh. J Clin Microbiol 1997;35:624–630.
9. Daniels NA, MacKinnon L, Bishop R, Altekruse S, Ray B, Hammond RM, et al. *Vibrio parahaemolyticus* infections in the United States, 1973–1998. J Infect Dis 2000;181:1661–1666.
10. Morris JG, Black RE. Cholera and other vibrioses in the United States. N Engl J Med 1985;312:343–350.
11. Jiang X, Chai TJ. Survival of *Vibrio parahaemolyticus* at low temperatures under starvation conditions and subsequent resuscitation of viable, nonculturable cells. Appl Environ Microbiol 1996;62:1300–1305.
12. Colwell RR. Global climate and infectious disease: the cholera paradigm. Science 1996;274:2025–2031.
13. Plotkin BJ, Konaklieva MI. Surface properties of *Vibrio vulnificus*. Lett Appl Microbiol 2007;44:426–430.
14. Centers for Disease Control (CDC). *Vibrio parahaemolyticus* infections associated with consumption of raw

shellfish—three states, 2006. MMWR Morb Mortal Wkly Rep 2006;55:854–856.

15. Todd EC. Epidemiology of foodborne diseases: a worldwide review. World Health Stat Q 1997;50:30–50.
16. Rabbani GH, Greenough WB. Food as a vehicle of transmission of cholera. J Diarrhoeal Dis Res 1999;17:1–9.
17. Center for Disease Control (CDC). *Vibrio parahaemolyticus* infections associated with consumption of raw shellfish—three states, 2006. MMWR Morb Mortal Wkly Rep 2006;55:854–856.
18. Yeung PS, Boor KJ. Epidemiology, pathogenesis, and prevention of foodborne *Vibrio parahaemolyticus* infections. Foodborne Pathog Dis 2004;1:74–88.
19. Harwood VJ, Gandhi JP, Wright AC. Methods for isolation and confirmation of *Vibrio vulnificus* from oysters and environmental sources: a review. J Microbiol Methods 2004;59:301–316.
20. Sagoo SK, Little CL, Greenwood M. Microbiological study of cooked crustaceans and molluscan shellfish from UK production and retail establishments. Int J Environ Health Res 2007;17:219–230.
21. Lhafi SK, Kühne M. Occurrence of *Vibrio* spp. in blue mussels (*Mytilus edulis*) from the German Wadden Sea. Int J Food Microbiol 2007;116:297–300.
22. Nyamogoba HD, Obala AA, Kakai R. Combating cholera epidemics by targeting reservoirs of infection and transmission routes: a review. East Afr Med J 2002;79: 150–155.
23. Ashbolt NJ. Microbial contamination of drinking water and disease outcomes in developing regions. Toxicology 2004;198:229–238.
24. Singleton FL, Attwell RW, Jangi MS, Colwell RR. Influence of salinity and organic nutrient concentration on survival and growth of *Vibrio cholerae* in aquatic microcosms. Appl Environ Microbiol 1982;43:1080–1085.
25. Merrell DS, Camilli A. Acid tolerance of gastrointestinal pathogens. Curr Opin Microbiol 2002;5:51–55.
26. Hung DT, Zhu J, Sturtevant D, Mekalanos JJ. Bile acids stimulate biofilm formation in *Vibrio cholerae*. Mol Microbiol 2006;59:193–201.
27. Dalsgaard A, Serichantalergs O, Forslund A, Lin W, Mekalanos J, Mintz E, et al. Clinical and environmental isolates of *Vibrio cholerae* serogroup O141 carry the CTX phage and the genes encoding the toxin-coregulated pili. J Clin Microbiol 2001;39:4086–4092.
28. Sack RB, Miller CE. Progressive changes of *Vibrio* serotypes in germ-free mice infected with *Vibrio cholerae*. J Bacteriol 1969;99:688–695.
29. Schoolnik GK, Yildiz FH. The complete genome sequence of *Vibrio cholerae*: a tale of two chromosomes and of two lifestyles. Genome Biol 2000;1:1016.1–1016.3.
30. Bush CA, Patel P, Gunawardena S, Powell J, Joseph A, Johnson JA, et al. Classification of *Vibrio vulnificus* strains by the carbohydrate composition of their capsular polysaccharides. Anal Biochem 1997;250:186–195.
31. Mouzin E, Mascola L, Tormey MP, Dassey DE. Prevention of *Vibrio vulnificus* infections assessment of regulatory educational strategies. JAMA 1997;278:576–578.
32. Chiu TH, Duan J, Su YC. Characteristics of virulent *Vibrio parahaemolyticus* isolated from Oregon and Washington. J Food Prot 2007;70:1011–1016.
33. Herrington DA, Hall RH, Losonsky G, Mekalanos JJ, Taylor RK, Levine MM. Toxin, toxin-coregulated pili, and the toxR regulon are essential for *Vibrio cholerae* pathogenesis in humans. J Exp Med 1988;168:1487–1492.
34. Svennerholm L. Interaction of cholera toxin and ganglioside G(M1). Adv Exp Med Biol 1976;71:191–204.
35. Holmgren J, Lonnroth I, Mansson J, Svennerholm L. Interaction of cholera toxin and membrane GM1 ganglioside of small intestine. Proc Natl Acad Sci USA 1975;72:2520–2524.
36. Trucksis M, Galen JE, Michalski J, Fasano A, Kaper JB. Accessory cholera enterotoxin (Ace), the third toxin of a *Vibrio cholerae* virulence cassette. Proc Natl Acad Sci USA 1993;90:5267–5271.
37. Fasano A, Baudry B, Pumplin DW, Wasserman SS, Tall BD, Ketley JM, et al. *Vibrio cholerae* produces a second enterotoxin, which affects intestinal tight junctions. Proc Natl Acad Sci USA 1991;88:5242–5246.
38. Taylor RK, Miller VL, Furlong DB, Mekalanos JJ. Use of phoA gene fusions to identify a pilus colonization factor coordinately regulated with cholera toxin. Proc Natl Acad Sci USA 1987;84:2833–2837.
39. Sixma TK, Pronk SE, Kalk KH, Wartna ES, van Zanten BA, Witholt B, et al. Crystal structure of a cholera toxin-related heat-labile enterotoxin from *E. coli*. Nature 1991;351:371–377.
40. Blanco LP, DiRita VJ. Bacterial-associated cholera toxin and GM binding are required for transcytosis of classical biotype *Vibrio cholerae* through an *in vitro* M cell model system. Cell Microbiol 2006;8:982–998.
41. Fujinaga Y, Wolf AA, Rodighiero C, Wheeler H, Tsai B, Allen L, et al. Gangliosides that associate with lipid rafts mediate transport of cholera and related toxins from the plasma membrane to endoplasmic reticulum. Mol Biol Cell 2003;14:4783–4793.
42. Cassel D, Selinger Z. Mechanism of adenylate cyclase activation by cholera toxin: inhibition of GTP hydrolysis at the regulatory site. Proc Natl Acad Sci USA 1977;74: 3307–3311.
43. Field M. Intestinal secretion: effect of cyclic AMP and its role in cholera. N Engl J Med 1971;284:1137–1144.
44. Gangarosa EJ, Mosley WH. Epidemiology and surveillance of cholera. In: Barua D, Burrows W, editors. Cholera. Philadelphia: W.B Saunders Co.; 1974:381–403.
45. Ramamurthy T, Yamasaki S, Takeda Y, Nair GB. *Vibrio cholerae* O139 Bengal: odyssey of a fortuitous variant. Microbes Infect 2003;5.329–344.
46. Nishibuchi M, Fasano A, Russell RG, Kaper JB. Enterotoxigenicity of *Vibrio parahaemolyticus* with and without

genes encoding thermostable direct hemolysin. Infect Immun 1992;60:3539–3545.

47. Stelma GN, Reyes AL, Peeler JT, Johnson CH, Spaulding PL. Virulence characteristics of clinical and environmental isolates of *Vibrio vulnificus*. Appl Environ Microbiol 1992;58:2776–2782.
48. Cash RA, Music SI, Libonati JP, Snyder MJ, Wenzel RP, Hornick RB. Response of man to infection with *Vibrio cholerae*. I. Clinical, serologic, and bacteriologic responses to a known inoculum. J Infect Dis 1974;129:45–52.
49. Levine MM, Kaper JB, Herrington D, Losonsky G, Morris JG, Clements ML, et al. Volunteer studies of deletion mutants of *Vibrio cholerae* O1 prepared by recombinant techniques. Infect Immun 1988;56:161–167.
50. Levine MM, Black RE, Clements ML, Nalin DR, Cisneros L, Finkelstein RA. Volunteer studies in development of vaccines against cholera and enterotoxigenic *E. coli*: a review. In: Holme T, Holmgren J, Merson MH, Molby, editors. Acute enteric infections in children. New prospects for treatment and prevention. Amsterdam: Elsevier/North Holland Biomedical Press; 1981:443–459.
51. Glass RI, Holmgren J, Haley CE, Khan MR, Svennerholm AM, Stoll BJ, et al. Predisposition for cholera of individuals with O blood group. Possible evolutionary significance. Am J Epidemiol 1985;121:791–796.
52. Guerrant RL, Carneiro-Filho BA, Dillingham RA. Cholera, diarrhea, and oral rehydration therapy: triumph and indictment. Clin Infect Dis 2003;37:398–405.
53. Wong TW, Wang YY, Sheu HM, Chuang YC. Bactericidal effects of toluidine blue-mediated photodynamic action on *Vibrio vulnificus*. Antimicrob Agents Chemother 2005;49:895–902.
54. Sanyal SC, Sen PC. Human volunteer study on the pathogenicity of *Vibrio parahaemolyticus*. In: Fujino T, Sakaguchi G, Sakazaki R, Takeda Y, editors. International Symposium on *Vibrio parahaemolyticus*. Tokyo, Japan: Saikon Publishing Co., Ltd.; 1974:227–30.
55. Wang F, Butler T, Rabbani GH, Jones PK. The acidosis of cholera. Contributions of hyperproteinemia, lactic acidemia, and hyperphosphatemia to an increased serum anion gap. N Engl J Med 1986;315:1591–1595.
56. Bhattacharya SK, Bhattacharya MK, Nair GB, Dutta D, Deb A, Ramamurthy T, Garg S, et al. Clinical profile of acute diarrhoea cases infected with the new epidemic strain of *Vibrio cholerae* O139: designation of the disease as cholera. J Infect 1993;27:11–15.
57. Makino K, Oshima K, Kurokawa K, Yokoyama K, Uda T, Tagomori K, et al. Genome sequence of *Vibrio parahaemolyticus*: a pathogenic mechanism distinct from that of *V. cholerae*. Lancet 2003;361:743–749.
58. Fyfe M, Kelly MT, Yeung ST, Daly P, Schallie K, Waller P, et al. Outbreak of *Vibrio parahaemolyticus* infections associated with eating raw oysters–Pacific Northwest, 1997. MMWR Morb Mortal Wkly Rep 1998;47:457–462.
59. Chiang SR, Chuang YC. *Vibrio vulnificus* infection: clinical manifestations, pathogenesis, and antimicrobial therapy. J Microbiol Immunol Infect 2003;36:81–88.
60. Greenough W, Beneson AS, Islam MR. Experience of darkfield examination of stools from diarrheal patients. In: Proceedings of the Cholera Research Symposium. Washington, DC: US Government Printing Office; 1965:56–58.
61. Gustafsson B, Holme T. Rapid detection of *Vibrio cholerae* O1 by motility inhibition and immunofluorescence with monoclonal antibodies. Eur J Clin Microbiol 1985;4:291–294.
62. Lesmana M, Richie E, Subekti D, Simanjuntak C, Walz SE. Comparison of direct-plating and enrichment methods for isolation of *Vibrio cholerae* from diarrhea patients. J Clin Microbiol 1997;35:1856–1858.
63. Sukazaki R. Bacteriology of *Vibrio* and related organisms. In: Barua D, Greenough WB, editors, Cholera. New York: Plenum Medical Book Co.;1991:37–55.
64. Tamrakar AK, Goel AK, Kamboj DV, Singh L. Surveillance methodology for *Vibrio cholerae* in environmental samples. Int J Environ Health Res 2006;16:305–312.
65. Rahman M, Sack DA, Wadood A, Yasmin M, Latif A. Rapid identification of *Vibrio cholerae* serotype O1 from primary isolation plates by a coagglutination test. J Med Microbiol 1989;28:39–41.
66. Shirai H, Nishibuchi M, Ramamurthy T, Bhattachya SK, Pal SC, Takeda Y. Polymerase chain reaction for detection of the cholera enterotoxin operon of *Vibrio cholerae*. J Clin Microbiol 1991;29:2517–2521.
67. Koch WH, Payne WL, Wentz BA, Cebula TA. Rapid polymerase chain reaction method for detection of *Vibrio cholerae* in foods. Appl Environ Microbiol 1993;59:556–560.
68. Mukhopadhyay AK, Garg S, Saha PK, Takeda Y, Bhattacharya SK, Nair GB. Comparative analysis of factors promoting optimal production of cholera toxin by *Vibrio cholerae* O1 (classical & E1Tor biotypes) & O139. Indian J Med Res 1996;104:129–33.
69. Sanjuan E, Amaro C. Protocol for specific isolation of virulent strains of *Vibrio vulnificus* serovar E (biotype 2) from environmental samples. Appl Environ Microbiol 2004;70:7024–7032.
70. Kishishita M, Matsuoka N, Kumagai K, Yamasaki S, Takeda Y, Nishibuchi M. Sequence variation in the thermostable direct hemolysin-related hemolysin (trh) gene of *Vibrio parahaemolyticus*. Appl Environ Microbiol 1992;58:2449–2457.
71. Elliot EL, Kaysner CA, Jackson L, Tamplin ML. Vibrio cholerae, *V. parahaemolyticus, V. vulnificus*, and other *Vibrio* spp. In: U.S. Food and Drug Administration bacteriological analytical manual. Gaithersburg, MD: AOAC International; 1998:9.01–9.27.
72. Kawatsu K, Ishibashi M, Tsukamoto T. Development and evaluation of a rapid, simple, and sensitive immunochromatographic assay to detect thermostable direct hemolysin produced by *Vibrio parahaemolyticus* in enrichment cultures of stool specimens. J Clin Microbiol 2006;44:1821–1827.

73. Horowitz W, editor. Official methods of analysis of AOAC International. 18th ed., Gaithersburg, MD: AOAC International; 2006.
74. Fernandez-Miyakawa ME, Brero ML, Mateo NA. Cholera toxin modulates the systemic immune responses against *vibrio cholerae* surface antigens after repeated inoculations. Microbiol Immunol 2006;50:607–619.
75. King CK, Glass R, Bresee JS, Duggan C, Centers for Disease Control (CDC). Managing acute gastroenteritis among children: oral rehydration, maintenance, and nutritional therapy. MMWR Morb Mortal Wkly Rep 2003;52:1–16.
76. Smith JL. Foodborne illness in the elderly. J Food Prot 1998;61:1229–1239.
77. Carpenter CC. The treatment of cholera: clinical science at the bedside. J Infect Dis 1992;166:2–14.
78. Bhattacharya SK. An evaluation of current cholera treatment. Expert Opin Pharmacother 2003;4:141–146.
79. Swerdlow DL, Ries AA. Cholera in the Americas. Guidelines for the clinician. JAMA 1992;267:1495–1499.
80. Ghosh S, Sengupta PG, Gupta DN, Sirkar BK. Chemoprophylaxis studies in cholera: a review of selective works. J Commun Dis 1992;24:55–57.
81. Jesudason MV. Change in serotype and appearance of tetracycline resistance in *V. cholerae* O1 in Vellore, South India. Indian J Med Microbiol 2006;24:152–153.
82. Dromigny JA, Rakoto-Alson O, Rajaonatahina D, Migliani R, Ranjalahy J, Mauclere P. Emergence and rapid spread of tetracycline-resistant *Vibrio cholerae* strains, Madagascar. Emerg Infect Dis 2002;8:336–338.
83. Dhar U, Bennish ML, Khan WA, Seas C, Huq Khan E, Albert MJ, et al. Clinical features, antimicrobial susceptibility and toxin production in *Vibrio cholerae* O139 infection: comparison with *V. cholerae* O1 infection. Trans R Soc Trop Med Hyg 1996;90:402–405.
84. Holmgren J, Svennerholm AM, Lindblad M. Receptor-like glycocompounds in human milk that inhibit classical and El Tor *Vibrio cholerae* cell adherence (hemagglutination). Infect Immun 1983;39:147–54.

Chapter 24

YERSINIA ENTEROCOLITICA

CYRUS RANGAN, MD, FAAP

HISTORY

Yersinia enterocolitica subsp. *enterocolitica* (Schleifstein & Coleman) Frederiksen (*Y. enterocolitica*) is a nonlactose fermenting, gram-negative, facultatively anaerobic, urease-positive, oxidase-negative coccobacillus. The US microbiologists J.I. Schleifstein and M.B. Coleman first cultured foodborne *Y. enterocolitica* in 1939; they named this bacterium after the Swiss bacteriologist Alexandre J. Yersin, who discovered the cause of plague (*Yersinia pestis*).[1] The first major US outbreak of foodborne *Yersinia* gastroenteritis occurred in 1976 in Holland Patent, New York, where 222 children developed gastroenteritis and mesenteric lymphadenitis from chocolate milk contaminated with *Y. enterocolitica* serotypes O:5 and O:8.[2] *Yersinia pseudotuberculosis* subsp. *pseudotuberculosis* (Pfeiffer) Smith and Thal is a rare cause of bacterial food poisoning in humans.[3]

EXPOSURE

Source

Yersinia enterocolitica is an extremely heterogeneous species that is divided into six biogroups (1A, 1B, 2, 3, 4, and 5). Biogroups 2, 3, and 4 are more common in Europe, whereas biogroup 1B occurs primarily in North America and biogroup 1A has a separate, unknown origin.[4] All of these biogroups are pathogenic to human hosts with the possible exception of biogroup 1A, which lacks virulent plasmids (pYV), chromosomal invasion-associated genes (*ail, inv*), and the heat-stable enterotoxic *ystA* gene present in the other virulent biotypes. However, the production of enterotoxin *ystB* by 1A strains suggests that this biotype can cause gastroenteritis in humans.[5,6] Each biogroup contains several serotypes. Bioserotypes 1B/O:8, 2/O:5,27, 2/O:9, and 4/O:3 are responsible for the majority of human infections including all major worldwide outbreaks since 1976.[7] Serotypes are not necessarily unique to pathogenic species as the O:3 antigen exists in biovar 1A and nonpathogenic *Yersinia* species (e.g., *Y. frederiksenii* Ursing, *Y. kristensenii* Bercovier).[8] Most cases of yersiniosis in humans occur sporadically without an apparent source; foodborne outbreaks of gastroenteritis caused by *Yersinia* species are rare.[9] The prevalence of yersiniosis is particularly high in Europe, where the high prevalence of *Y. enterocolitica* O:3/O:9-specific antibodies in healthy blood donors suggest the possibility of a high rate of subclinical yersiniosis in healthy populations.[10] *Y. enterocolitica* proliferates at 4°C (39°F), and therefore high bacterial loads (10^6–10^9 CFU/mL) develop in blood products refrigerated over 21 days.[11] Additionally, the presence of anticoagulant-preservative and hemoglobin from lysed red blood cells promote the growth of this bacterium.

ENVIRONMENT

Animal feces contaminate soil, water, fertilizers, and effluents with *Yersinia* worldwide, especially in colder regions such as Scandinavia and Canada.[12]

Medical Toxicology of Natural Substances, by Donald G. Barceloux, MD

Animal

Yersinia enterocolitica causes minor, self-limited disease and lifelong asymptomatic carriage in pigs, cattle, deer, sheep, seafood, birds, and reptiles. Of these animals, pigs are a major reservoir of pathogenic *Y. enterocolitica* with prevalence of pathogenic *Y. enterocolitica* in stool samples being approximately 10–15%.[13] Most animals carry nonvirulent *Yersinia* species of biogroup 1A. However, the indistinguishable genotypes in human strains of *Y. enterocolitica* and strains in dogs, cats, sheep, and rodents suggest that these animals also may be sources of human yersiniosis.[9] In a study of 2,349 pigs from eight US swine operations, 120 (5.1%) of the pigs were positive for pathogenic *Y. enterocolitica*, as measured by the presence of attachment-invasion locus factor (Ail) with multiplex polymerase chain reaction (PCR).[14]

Food

Yersinia may contaminate drinking water, pork products, meat, poultry, and unpasteurized dairy products.[15,16] *Yersinia enterocolitica* serotypes O:5 (biogroup 1B), O:8 (biogroup 2), and O:9 (biogroup 4) are commonly involved with human foodborne infections, primarily from raw pork and dairy products.[9] The incidence of *Y. enterocolitica* contamination in ready-to-eat food products (e.g., pork) is relatively low.[17] Most human infections originate from undercooked pork, tofu, and unpasteurized dairy products; vegetable and soy products may harbor *Y. enterocolitica* after washing with contaminated water during food manufacturing and handling.[18]

Food Processing

The most effective controls of *Yersinia* contamination include the following methods: proper cooking and refrigeration, purification of water, pasteurization of dairy products, separation of cooked and uncooked foods, sanitation of food preparation surfaces with at least 1% bleach solution, washing of vegetables, and cooking foods to 120 °C–170 °C (248 °F –338 °F).[19] *Yersinia enterocolitica* proliferates readily at refrigeration temperatures. Hence, refrigeration does not effectively control the growth.[20]

TOXINS

Physiochemical Properties

Yersinia enterocolitica is one of five common, foodborne bacterial gastrointestinal pathogens, which include non-typhoid *Salmonella, Shigella* species, diarrhea-producing *E. coli*, and *Campylobacter* species.[21] The genus *Yersinia* contains 12 species (*Y. aldovae, Y. aleksiciae, Y. bercovieri, Y. enterocolitica, Y. frederiksenii, Y. intermedia, Y. kristensenii, Y. mollaretii, Y. pestis, Y. pseudotuberculosis, Y. rohdei*, and *Y. ruckeri*). These bacteria are gram-negative, nonspore-forming, rod-shaped, facultative anaerobes.[22] Only *Y. pestis, Y. pseudotuberculosis, and Y. enterocolitica* are pathogenic to healthy humans, whereas other species occur rarely as opportunistic infections.[23] Of these pathogenic bacteria, *Y. enterocolitica* is the most important cause of foodborne illness.[24] *Yersinia enterocolitica* grows over a wide temperature range (1 °C –45 °C/34 °F–113 °F), and thus this bacterium proliferates under refrigeration (0 °C–4 °C/32 °F–39 °F), room temperature (20 °C–25 °C/68 °F–77 °F), and human body temperature (37.0 °C/98.6 °F).[25,26] *Y. enterocolitica* grows optimally at 25 °C (77 °F), and under these conditions the organism expresses chromosomally derived virulence determinants. These determinants allow the bacterium to maintain a coccidial, hydrophilic morphology and high motility; thus, enhancing survival at room temperature. At human body temperature (37.0 °C/98.6 °F), *Y. enterocolitica* becomes a pleomorphic, nonmotile, hydrophobic organism, which synthesizes virulence determinants from a 70-kbp plasmid (pYV). These determinants promote intestinal proliferation and invasion of human cells by *Y. enterocolitica*.[27] Table 24.1 lists the virulence determinants expressed by *Y. enterocolitica* at these temperatures.

Mechanism of Toxicity

Yersinia enterocolitica synthesizes both chromosomal and plasmid-encoded virulence factors that mediate colonization, host cells invasion, and resistance to host defenses. Some synthesized factors have multiple functions.[28] At room temperature (25 °C/77 °F), *Y. enterocolitica* synthesizes chromosomally encoded factors, including membrane surface proteins [invasin (Inv), attachment-invasion locus factor (Ail)][29,30] and flagellae. Expression of the latter factor allows the bacterium to travel through the stomach and small intestine after ingestion.[31] As *Y. enterocolitica* adapts to human body

TABLE 24.1. Virulence Factors Expressed by *Yersinia Enterocolitica* at 25 °C (77 °F) and 37.0 °C (98.6 °F)

25 °C/77 °F	37.0 °C/98.6 °F
Flagella	Adhesion factor (YadA)
Invasin (Inv)	Secreted effectors (Yops proteins)
Attachment-invasion locus factor (Ail)	V-antigen

Source: From Ref 27.

temperature (37.0°C/98.6°F), the bacterium synthesizes temperature-inducible, plasmid-encoded membrane surface proteins. The first synthesized protein, YadA, adheres preferentially to the mucus layer overlying the ileocecal brush border.[32] Chromosomally encoded Inv and Ail bind to B1 integrins on the surfaces of lymphofollicular M cells in Peyer's patches, initiating invasion of M cells, and facilitating delivery of microcolonies of *Y. enterocolitica* into adjacent tissues.[33] Plasmid-encoded V-antigen induces interleukin-10 (IL-10) release, which suppresses macrophage-derived tumor necrosis factor-alpha (TNF-α), an important inflammatory mediator and chemoattractant.[34] V-antigen further mediates bacterial resistance to host defenses by translocating several plasmid-encoded virulence effectors (Yops) via a Type III secretion system.[35,36] These Yop effectors (YopE, YopT, YopO) disrupt the actin skeleton of macrophages and synergistically increase the resistance of the bacteria to phagocytosis by macrophages and polymorphonuclear leucocytes.[37] YopH (a tyrosine phosphokinase) inhibits phagocytosis by dephosphorylating $p130_{cas}$, a macrophage signal transduction protein.[38] YopJ prevents formation of activated nuclear factor-κB (NF-κB), by blocking the mitogen-activated protein kinase (MAPK) cascade; NF-κB regulates macrophage production of proinflammatory cytokines, intercellular adhesin molecule 1, and E selectin. Down-regulation of these inflammatory signal pathways activates macrophage apoptosis.[39] Proliferation in the ileocecal mucosa and evasion of phagocytosis enable *Y. enterocolitica* to form microabscesses and ulcers in the mucosa. These processes destroy the architecture of Peyer's patches and promote inflammation, fluid secretion, blood loss, and diarrhea.[40] *Yersinia enterocolitica* invades the lamina propria (i.e. enteritis, enterocolitis, terminal ileitis), and the infection extends to regional lymphoid tissue (i.e., mesenteric lymphadenitis) or systemically through the blood (i.e., septicemia).[41] Outer membrane proteins YadA and Ail confer resistance to serum bactericidal activity by binding complement proteins and preventing membrane attack complex (MAC) assembly.[42] *Yersinia enterocolitica* produces a plasmid-encoded heat-stable enterotoxin (Yst) that is similar to enterotoxigenic *E. coli* (ETEC) heat-stable toxin. However, Yst is only mildly toxic to the gastrointestinal tract. This enterotoxin enhances the persistence of the organism after systemic invasion, thus facilitating spread of the bacteria to other organs and long-term shedding after infection.[43] *Yersinia enterocolitica* secretes yersiniabactin, an iron-complexion siderophore, which promotes bacterial growth in iron-rich environments; patients with iron overload are at increased risk of developing *Y. enterocolitica* septicemia.[44] *Yersinia enterocolitica* expresses a natural receptor for desferrioxamine on the outer membrane called FoxA, which transports the iron-desferrioxamine complex into the cell; thus, patients undergoing therapy with desferrioxamine are also at increased risk of developing septicemia.[45,46] HLA-B27-positive patients may harbor persistent *Yersinia* antigens, which suppress T-cells and TNF-α, leading to autoimmune sequelae such as reactive arthritis.[47] Autoimmune sequelae are more prevalent with infection with *Y. enterocolitica* serotype O:3 phage type 8. Immune-mediated response to *Yersinia* antigens probably causes some of the cases of chronic undifferentiated arthritis due to the formation of *Yersinia* antigens in colonic and synovial tissues.[48]

DOSE RESPONSE

The amount of inoculum of *Y. enterocolitica* required to produce disease is unknown because host defenses destroy a significant portion of the inoculum before bacteria reach the distal ileum, adhere to the mucosa, and synthesize plasmid-encoded resistance factors. The estimated size of inoculum for clinical infection based on rabbit models is 1.6×10^{10} CFU.[49] *Yersinia enterocolitica* serotype O:8 produces more severe symptoms than other serotypes.[50] *Yersinia enterocolitica* biogroup 1A bacteria are generally nonpathogenic, but may become opportunistic nosocomial pathogens in neonates or immunocompromised hosts.[51]

CLINICAL RESPONSE

Although *Yersinia enterocolitica* gastroenteritis occurs predominately in children, this organism is an uncommon cause of gastroenteritis in this age group compared with viruses and other bacteria (e.g., *E. coli, Salmonella, Shigella, Campylobacter*). In a prospective cohort study of children presenting to a pediatric emergency department with diarrhea and providing a stool sample, *Y. enterocolitica* was detected in 2 of 372 specimens testing positive for a pathogen.[52] *Y. enterocolitica* incubates for 1–2 days before the development of gastrointestinal symptoms, primarily diarrhea in infants and young children. Initial gastrointestinal signs and symptoms include low-grade fever, chills, malaise, sharp or cramping abdominal pain, and watery or bloody diarrhea with vomiting occurring in 40% of cases.[53] Children under 5 years of age typically present with enterocolitis, sharing many clinical characteristics with inflammatory bowel disease, whereas children over 5 years and young adults present with acute mesenteric lymphadenitis or terminal ileitis.[54] Infrequently, patients with acute mesenteric lymphadenitis or terminal ileitis complain of pain and tenderness localized to the right lower abdominal quadrant, mimicking acute appendicitis. This pseudoappendicitis typically occurs in older children and adolescents.[55] Septicemia occurs primarily in immunocompromised patients or young infants, patients with iron overload or undergoing therapy with desferrioxamine, or in immunocompromised patients with transfusion-transmitted infections.[56,57] Mortality rate in cases of *Y. enterocolitica* septicemia is approximately 35–50%, depending on the underlying health of the patients.[58] Liver and spleen abscesses, meningitis, encephalitis, pneumonia, endocarditis, osteomyelitis, or septic arthritis may complicate septicemia in high-risk patient groups.[59–64] Sequelae include reactive arthritis, erythema nodosum, uveitis, and Reiter's syndrome (arthritis, conjunctivitis, urethritis) 1–2 weeks after recovery in 1–2% of patients; 80% of patients with sequelae are positive for HLA-B27. These postinfection complications occur primarily in young adults. Reactive arthritis is more common following infections by *Y. enterocolitica* and *Salmonella* species than *Campylobacter* species.[21] Case reports suggest that the arthritic symptoms last at least 2–4 months.[65] Circumstantial evidence suggests that *Yersinia* infections initiate or exacerbate chronic inflammatory bowel disease (ulcerative colitis, Crohn's disease).[66]

DIAGNOSTIC TESTING

Analytical Methods

Yersinia enterocolitica grows readily on most routine enteric media such as MacConkey agar; slowly on Hektoen-enteric and xylose-lysine-deoxycholate agars; and poorly on *Salmonella-Shigella* agar. However, conventional culture-dependent methods are slow and *Y. enterocolitica* colonies on these media are small and difficult to distinguish visually from normal flora and other pathogenic bacteria.[67] Laboratories should be specifically requested to evaluate the specimen for *Y. enterocolitica. Yersinia enterocolitica*-selective cefsulodin-Irgasan (triclosan, CAS No: 3380-34-5)-novobiocin agar (CIN-agar) inhibits the growth of normal flora and isolates *Y. enterocolitica* as "bull's-eye" colonies with dark pink-to-red centers and clear borders.[68] Cold preenrichment of specimens at 4 °C (39 °F) in phosphate-buffered saline may enhance *Y. enterocolitica* growth before culture, but this method requires extended incubation time up to 21 days. This method also enhances the growth of normal flora and other pathogenic organisms, whereas incubation of CIN-agar plates at 25 °C (77 °F) for 18–20 hours effectively and rapidly isolates *Y. enterocolitica* from other bacteria.[69]

Molecular (DNA-based) methods allow the amplification of specific genetic signals from a few cells, and therefore these analytical methods are much more sensitive and specific than the use of selective microbiological media. Polymerase chain reaction (PCR) is the most common nucleic acid amplification technique for the diagnosis of infectious diseases. This method detects pYV-positive organisms both in live and dead strains of *Y. enterocolitica*, and PCR for this bacterium yields 99% specificity and 100% sensitivity.[70,71] Because of the possible loss of plasmids, real-time PCR methods also target chromosomal virulence genes (*ail, inv, yst*) in the chromosome of pathogenic *Y. enterocolitica* strains. Subtyping procedures are useful for tracking individual strains responsible for outbreaks of *Y. enterocolitica*. Analytical methods for subtyping include the older phenotypic subtyping, such as biogrouping (biochemical properties), antibiotic susceptibility, serotyping (O and H antigens), phage typing (bacteriophage lysis patterns), and esterase typing. Newer genotypic subtyping includes pulsed-field gel electrophoresis, multilocus enzyme electrophoresis, ribotyping, PCR-based subtyping techniques, amplified fragment length polymorphism, PCR-restriction fragment length polymorphism, repetitive element PCR, and DNA sequencing-based subtyping techniques (multilocus sequence typing). Conventional PCR assays typically require lengthy enrichment and DNA-extraction protocols, whereas real-time PCR relies on cleavage of a fluorogenic probe for detection of amplicons. Real-time PCR is 100% specific to virulent *Y. enterocolitica* with a limit of sensitivity near 10^3 CFU/mL in stool samples and an assay time of 1.5 days.[72] These subtyping techniques are reserved for the investigation of epidemics.

Biomarkers

Analysis of serum antibodies to *Yersinia* antigens helps differentiate acute from chronic *Yersinia* infections. IgM-class antibodies are associated with acute infections, whereas IgA and IgG antibodies develop as the disease progresses. Serum IgA antibodies to lipopolysaccharide (LPS or "O" antigen) persist for several months after the onset of *Yersinia* infections, particularly in patients with reactive arthritis after exposure to serotype O:3.[73,74]

Abnormalities

Leukocytosis with or without left shift occurs during gastroenteritis, acute mesenteric lymphadenitis, and septicemia. Fecal leukocytes and red blood cells (with or without frank blood) are commonly observed on light microscopy. Severe dehydration involves hypernatremia (serum Na > 150 mEq/L) or hyponatremia (serum Na < 130 mEq/L), hypokalemia (serum K < 3.0 mEq/L), hypochloremia, elevated urine specific gravity (SpG > 1.020), and hyperphosphatemia. Low serum bicarbonate results from lactic acid production or elevated serum bicarbonate secondary to contraction alkalosis. Computed tomography helps distinguish true appendicitis from pseudoappendicitis caused by *Y. enterocolitica*.[75]

TREATMENT

Treatment of *Yersinia* infection includes bowel rest with fluid and electrolytes replacement. Gastrointestinal manifestations and associated flu-like symptoms of *Yersinia* gastroenteritis rarely require hospitalization, and these effects generally resolve spontaneously in 3–10 days.[53] Oral replacement may be sufficient, but affected patients may experience significant anorexia, necessitating intravenous fluid replacement and hospitalization in cases of significant dehydration or severe electrolyte abnormalities, particularly in young children or elderly patients.[76,77] Patients with severe cases of enterocolitis, septicemia, acute mesenteric lymphadenitis, or disseminated disease require prompt antimicrobial therapy.[78] *Yersinia enterocolitica* synthesizes β-lactamases, rendering most strains resistant to penicillins and first- and second-generation cephalosporins.[79] Combined therapy with a third-generation cephalosporin (e.g., cefotaxime, ceftriaxone) plus an aminoglycoside (e.g. gentamicin) is appropriate empiric treatment pending susceptibility studies, as rare cases of antimicrobial resistance have emerged with these medications.[80] Fluoroquinolones (e.g., ciprofloxacin) may be utilized in adult patients with confirmed susceptibilities.[81] Pseudoappendicitis may be difficult to distinguish from true appendicitis, resulting in surgical exploration.

References

1. Scleifstein JI, Coleman MB. An unidentified microorganism resembling *B. ligniere and Past. Pseudotuberculosis*, and pathogenic for man. NY State J Med 1939; 39:1749–1753.
2. Black RE, Jackson RJ, Tsai T, Medvesky M, Shayegani M, Feeley JC, et al. Epidemic *Yersinia enterocolitica* infection due to contaminated chocolate milk. N Engl J Med 1978;298:76–79.
3. Putzker M, Sauer H, Sobe D. Plague and other human infections caused by *Yersinia* species. Clin Lab 2001;47:453–466.
4. Virdi JS, Sachdeva P. Molecular heterogeneity in *Yersinia enterocolitica* and "*Y. enterocolitica*-like" species—implications for epidemiology, typing and taxonomy. FEMS Immunol Med Microbiol 2005;45:1–10.
5. Burnens AP, Frey A, Nicolet J. Association between clinical presentation, biogroups and virulence attributes of *Yersinia enterocolitica* strains in human diarrhoeal disease. Epidemiol Infect 1996;116:27–34.
6. Singh I, Virdi JS. Production of *Yersinia* stable toxin (YST) and distribution of *yst* genes in biotype 1A strains of *Yersinia enterocolitica*. J Med Microbiol 2004;53:1065–1068.
7. Wauters G, Kandolo K, Janssens M. Revised biogrouping scheme of *Yersinia enterocolitica*. Contrib Microbiol Immunol 1987;9:14–21.
8. Chiesa C, Pacifico L, Ravagnan G. Identification of pathogenic serotypes of *Yersinia enterocolitica*. J Clin Microbiol 1993;31:2248–2249.
9. Fredriksson-Ahomaa M, Stolle A, Korkeala H. Molecular epidemiology of *Yersinia enterocolitica* infections. FEMS Immunol Med Microbiol 2006;47:315–329.
10. Maki-Ikola O, Heesemann J, Toivanen A, Granfors K. High frequency of *Yersinia* antibodies in healthy populations in Finland and Germany. Rheumatol Int 1997;16: 227–229.
11. Stenhouse MA, Milner LV. *Yersinia enterocolitica*. A hazard in blood transfusion. Transfusion 1982;22:396–398.
12. Bottone EJ. *Yersinia enterocolitica*: overview and epidemiologic correlates. Microbes Infect 1999;1:323–333.
13. Bhaduri S, Wesley IV, Bush EJ. Prevalence of pathogenic *Yersinia enterocolitica* strains in pigs in the United States. Appl Environ Microbiol 2005;71:7117–7121.
14. Bowman AS, Glendening C, Wittum TE, LeJeune JT, Stich RW, Funk JA. Prevalence of *Yersinia enterocolitica* in different phases of production on swine farms. J Food Prot 2007;70:11–16.
15. Lal M, Kaur H, Gupta LK, Sood NK. Isolation of *Yersinia enterocolitica* from raw milk and pork in Ludhiana. Indian J Pathol Microbiol 2005;48:286–287.
16. Cocolin L, Comi G. Use of a culture-independent molecular method to study the ecology of *Yersinia* spp. in food. Int J Food Microbiol 2005;105:71–82.
17. Lambertz ST, Danielsson-tham M-L. Identification and characterization of pathogenic *Yersinia enterocolitica* isolates by PCR and pulsed-field gel electrophoresis. Appl Environ Microbiol 2005;71:3674–3681.
18. de Boer E. Isolation of *Yersinia enterocolitica* from foods. Int J Food Microbiol 1992;17:75–84.
19. Kapperud G. *Yersinia enterocolitica* in food hygiene. Int J Food Microbiol 1991;12:53–65.
20. Kennedy J, Jackson V, Blair IS, McDowell DA, Cowan C, Bolton DJ. Food safety knowledge of consumers and the microbiological and temperature status of their refrigerators. J Food Prot 2005;68:1421–1430.
21. Helms M, Simonsen J, Molbak K. Foodborne bacterial infection and hospitalization: a registry-based study. Clin Infect Dis 2006;42:498–506.
22. Gilmour A, Walker SJ. Isolation and identification of *Yersinia enterocolitica* and the *Yersinia enterocolitica*-like bacteria. Soc Appl Bacteriol Symp Ser 1988;17:213–236.
23. Sulakvelidze A. *Yersiniae* other than *Y. enterocolitica, Y. pseudotuberculosis*, and *Y. pestis*: the ignored species. Microbes Infect 2000;2:497–513.

24. Linde H-J, Neubauer H, Meyer H, Aleksic S, Lehn N. Identification of *Yersinia* species by the Vitek GNI Card. J Clin Microbiol 1999;37:211–214.
25. Logue CM, Sheridan JJ, McDowell DA, Blair IS, Hegarty T. The effect of temperature and selective agents on the growth of *Yersinia enterocolitica* serotype O:3 in pure culture. J Appl Microbiol 2000;88:1001–1008.
26. Bresolin G, Neuhaus K, Scherer S, Fuchs TM. Transcriptional analysis of long-term adaptation of *Yersinia enterocolitica* to low-temperature growth. J Bacteriol 2006;188: 2945–2958.
27. Cornelis GR. *Yersinia* pathogenicity factors. Curr Top Microbiol Immunol 1994;192:243–263.
28. Chatzipanagiotou S, Kyriazi Z, Ioannidis A, Trikka-Graphakos E, Nicolaou C, Legakis NJ. Detection of chromosomal- and plasmid-encoded virulence-associated epidemiological markers in *Yersinia enterocolitica* strains isolated from clinical cases: a comparative study. Mol Diagn 2004;8:131–132.
29. Grassl GA, Bohn E, Muller Y, Buhler OT, Autenrieth IB. Interaction of *Yersinia enterocolitica* with epithelial cells: invasin beyond invasion. Int J Med Microbiol 2003;293:41–54.
30. Paerregaard A. Interactions between *Yersinia enterocolitica* and the host with special reference to virulence plasmid encoded adhesion and humoral immunity. Dan Med Bull 1992;39:155–172.
31. Kapatral V, Olson JW, Pepe JC, Miller VL, Minnich SA. Temperature-dependent regulation of *Yersinia enterocolitica* Class III flagellar genes. Mol Microbiol 1996;19: 1061–1071.
32. Heise T, Dersch P. Identification of a domain in *Yersinia* virulence factor YadA that is crucial for extracellular matrix-specific cell adhesion and uptake. Proc Natl Acad Sci USA 2006;103:3375–3380.
33. Lian CJ, Pai CH. Inhibition of human neutrophil chemiluminescence by plasmid-mediated outer membrane proteins of *Yersinia enterocolitica*. Infect Immun 1985;49: 145–151.
34. Sing A, Roggenkamp A, Geiger AM, Heesemann J. *Yersinia enterocolitica* evasion of the host innate immune response by V antigen-induced IL-10 production of macrophages is abrogated in IL-10-deficient mice. J Immunol 2002;168:1315–1321.
35. Cornelis GR. *Yersinia* type III secretion: send in the effectors. J Cell Biol 2002;158:401–408.
36. Une T, Nakajima R, Brubaker RR. Roles of V antigen in promoting virulence in *Yersinia*. Contrib Microbiol Immunol 1986;9:179–185.
37. Grosdent N, Maridonneau-Parini I, Sory MP, Cornelis GR. Role of Yops and adhesins in resistance of *Yersinia enterocolitica* to phagocytosis. Infect Immun 2002;70: 4165–4167.
38. Black DS, Marie-Cardine A, Schraven B, Bliska JB. The *Yersinia* tyrosine phosphatase YopH targets a novel adhesion-regulated signalling complex in macrophages. Cell Microbiol 2000;2:401–414.
39. Zhang Y, Ting AT, Marcu KB, Bliska JB. Inhibition of MAPK and NF-kappa B pathways is necessary for rapid apoptosis in macrophages infected with *Yersinia*. J Immunol 2005;174:7939–7949.
40. O'Loughlin EV, Gall DG, Pai CH. *Yersinia enterocolitica*: mechanisms of microbial pathogenesis and pathophysiology of diarrhoea. J Gastroenterol Hepatol 1990;5:173–179.
41. Handley SA, Newberry RD, Miller VL. *Yersinia enterocolitica* invasin-dependent and invasin-independent mechanisms of systemic dissemination. Infect Immun 2005; 73:8453–8455.
42. Biedzka-Sarek M, Venho R, Skurnik M. Role of YadA, Ail, and lipopolysaccharide in serum resistance of *Yersinia enterocolitica* serotype O:3. Infect Immun 2005;73:2232–2244.
43. Gemmell CG. Comparative study of the nature and biological activities of bacterial enterotoxins. J Med Microbiol 1984;17:217–235.
44. Heesemann J, Hantke K, Vocke T, Saken E, Rakin A, Stojiljkovic I, et al. Virulence of *Yersinia enterocolitica* is closely associated with siderophore production, expression of an iron-repressible outer membrane polypeptide of 65,000 Da and pesticin sensitivity. Mol Microbiol 1993;8:397–408.
45. Deiss K, Hantke K, Winkelmann G. Molecular recognition of siderophores: a study with cloned ferrioxamine receptors (FoxA) from Erwinia herbicola and *Yersinia enterocolitica*. Biometals 1998;11:131–137.
46. Robins-Browne RM, Prpic JK. Effects of iron and desferrioxamine on infections with *Yersinia enterocolitica*. Infect Immun 1985;47:774–779.
47. Granfors K, Merilahti-Palo R, Luukkainen R, Mottonen T, Lahesmaa R, Probst P, et al. Persistence of *Yersinia* antigens in peripheral blood cells from patients with *Yersinia enterocolitica* O:3 infection with or without reactive arthritis. Arthritis Rheum 1998;41:855–862.
48. van der Heijden IM, Res PC, Wilbrink B, Leow A, Breedveld FC, Heesemann J, Tak PP. *Yersinia enterocolitica*: a cause of chronic polyarthritis. Clin Infect Dis 1997;25:831–837.
49. Pai CH, Mors V, Seemayer TA. Experimental *Yersinia enterocolitica* enteritis in rabbits. Infect Immun 1980; 28:238–244.
50. Zhang L, Radziejewska-Lebrecht J, Krajewska-Pietrasik D, Toivanen P, Skurnik M. Molecular and chemical characterization of the lipopolysaccharide O-antigen and its role in the virulence of *Yersinia enterocolitica* serotype O:8. Mol Microbiol 1997;23:63–76.
51. Burnens AP, Frey A, Nicolet J. Association between clinical presentation, biogroups and virulence attributes of *Yersinia enterocolitica* strains in human diarrhoeal disease. Epidemiol Infect 1996;116:27–34.
52. Klein EJ, Boster DR, Stapp JR, Wells JG, Qin X, Clausen CR, et al. Diarrhea etiology in a children's hospital emergency department: a prospective cohort study. Clin Infect Dis 2006;43:807–813.

53. Black RE, Slome S. *Yersinia enterocolitica*. Infect Dis Clin North Am 1988;2:625–641.

54. Abdel-Haq NM, Asmar BI, Abuhammour WM, Brown WJ. *Yersinia enterocolitica* infection in children. Pediatr Infect Dis J 2000;19:954–958.

55. Ehara A, Egawa K, Kuroki F, Itakura O, Okawa M. Age-dependent expression of abdominal symptoms in patients with *Yersinia enterocolitica* infection. Pediatr Int 2000;42:364–366.

56. Leclercq A, Martin L, Vergnes ML, Ounnoughene N, Laran J-F, Giraud P, Carniel E. Fatal *Yersinia enterocolitica* biotype 4 serovar O:3 sepsis after red blood cell transfusion. Transfusion 2005;45:814–818.

57. Foberg U, Fryden A, Kihlstrom E, Persson K, Weiland O. *Yersinia enterocolitica* septicemia: clinical and microbiological aspects. Scand J Infect Dis 1986;18:269–279.

58. Saebo A, Lassen J. A survey of acute and chronic disease associated with *Yersinia enterocolitica* infection. A Norwegian 10-year follow-up study on 458 hospitalized patients. Scand J Infect Dis 1991;23:517–527.

59. al-Mohsen I, Luedtke G, English BK. Invasive infections caused by *Yersinia enterocolitica* in infants. Pediatr Infect Dis J 1997;16:253–255.

60. Kane DR, Reuman DD. *Yersinia enterocolitica* causing pneumonia and empyema in a child and a review of the literature. Pediatr Infect Dis J 1992;11:591–593.

61. Nemoto H, Murabayashi K, Kawamura Y, Sasaki K, Wakata N, Kinoshita M. Multiple liver abscesses secondary to *Yersinia enterocolitica*. Intern Med 1992;31:1125–1127.

62. Giamarellou H, Antoniadou A, Kanavos K, Papaioannou C, Kanatakis S, Papadaki K. *Yersinia enterocolitica* endocarditis: case report and literature review. Eur J Clin Microbiol Infect Dis 1995;14:126–130.

63. Marasco WJ, Fishman EK, Kuhlman JE, Hruban RH. Splenic abscess as a complication of septic *Yersinia*: CT evaluation. Clin Imaging 1993;17:33–35.

64. Crowe M, Ashford K, Ispahani P. Clinical features and antibiotic treatment of septic arthritis and osteomyelitis due to *Yersinia enterocolitica*. J Med Microbiol 1996;45: 302–309.

65. Olson DN, Finch WR. Reactive arthritis associated with *Yersinia enterocolitica* gastroenteritis. Am J Gastroenterol 1981;76:524–526.

66. Saebo A, Vik E, Lange OJ, Matuszkiewicz L. Inflammatory bowel disease associated with *Yersinia enterocolitica* O:3 infection. Eur J Intern Med 2005;16:176–182.

67. Fredriksson-Ahomaa M, Korkeala H. Low occurrence of pathogenic *Yersinia enterocolitica* in clinical, food, and environmental samples: a methodological problem. Clin Microbiol Rev 2003;16:220–229.

68. Logue CM, Sherwood JS, Doetkott C. Growth studies of plasmid bearing and plasmid cured *Yersinia enterocolitica* GER O:3 in the presence of cefsulodin, irgasan and novobiocin at 25 and 37 degrees C. J Appl Microbiol 2006;100:1299–1306.

69. Weissfeld AS, Sonnenwirth AC. Rapid isolation of *Yersinia* spp. from feces. J Clin Microbiol 1982;15:508–510.

70. Ibrahim A, Liesack W, Griffiths MW, Robins-Browne RM. Development of a highly specific assay for rapid identification of pathogenic strains of *Yersinia enterocolitica* based on PCR amplification of the *Yersinia* heat-stable enterotoxin gene (*yst*). J Clin Microbiol 1997;35:1636–1638.

71. Greisen K, Loeffelholz M, Purohit A, Leong D. PCR primers and probes for the 16S rRNA gene of most species of pathogenic bacteria, including bacteria found in cerebrospinal fluid. J Clin Microbiol 1994;32:335–351.

72. Zheng HX, Zhang MJ, Sun Y, Jiang B. [Detection of *Yersinia enterocolitica* in diarrhea stool by real-time PCR]. Zhonghua Yi Xue Za Zhi. 2006;86:2281–2284. [Chinese]

73. Stahlberg TH, Heesemann J, Granfors K, Toivanen A. Immunoblot analysis of IgM, IgG, and IgA responses to plasmid encoded released proteins of *Yersinia enterocolitica* in patients with or without yersinia triggered reactive arthritis. Ann Rheum Dis 1989;48:577–581.

74. Chart H, Cheasty T. The serodiagnosis of human infections with *Yersinia enterocolitica* and *Yersinia pseudotuberculosis*. FEMS Immunol Med Microbiol 2006;47:391–397.

75. Perdikogianni C, Galanakis E, Michalakis M, Giannoussi E, Maraki S, Tselentis Y, et al. *Yersinia enterocolitica* infection mimicking surgical conditions. Pediatr Surg Int 2006;22:589–592.

76. King CK, Glass R, Bresee JS, Duggan C, Centers for Disease Control (CDC). Managing acute gastroenteritis among children: oral rehydration, maintenance, and nutritional therapy. MMWR Morb Mortal Wkly Rep 2003;52:1–16.

77. Smith JL. Foodborne illness in the elderly. J Food Prot 1998;61:1229–1239.

78. Tripoli LC, Brouillette DE, Nicholas JJ, Van Thiel DH. Disseminated *Yersinia enterocolitica*. Case report and review of the literature. J Clin Gastroenterol 1990;12:85–89.

79. Pham JN, Bell SM, Martin L, Carniel E. The beta-lactamases and beta-lactam antibiotic susceptibility of *Yersinia enterocolitica*. J Antimicrob Chemother 2000;46:951–957.

80. Abdel-Haq NM, Papadopol R, Asmar BI, Brown WJ. Antibiotic susceptibilities of *Yersinia enterocolitica* recovered from children over a 12-year period. Int J Antimicrob Agents 2006;27:449–452.

81. Capilla S, Ruiz J, Goni P, Castillo J, Rubio MC, Jimenez de Anta MT, et al. Characterization of the molecular mechanisms of quinolone resistance in *Yersinia enterocolitica* O:3 clinical isolates. J Antimicrob Chemother 2004;53:1068–1071.

B Other Microbes

Chapter 25

CYANOBACTERIA

HISTORY

Cyanobacteria are ancient inhabitants of brackish-, marine- and fresh-waters as well as terrestrial environments. They were the first organisms capable of converting carbon dioxide to oxygen via photosynthesis, and fossil records indicate these primitive organisms existed 3.3–3.5 billion years ago.[1] Consequently, cyanobacteria probably contributed to the formation of oxygen in the atmosphere.[2] Eutrophication of freshwater by industrial and agricultural effluent during warm periods in the late summer and fall promotes the growth of blooms or scum containing toxin-producing cyanobacteria. Accounts of these blooms date back to the 12th century AD when a scum occurred on Llangorse Lake in Wales.[3] In 1878, Francis documented the death of Australian livestock from the ingestion of water from a heavily contaminated lake that probably contained a bloom of cyanobacteria.[4] Cyanobacteria are a suspected cause of Barcoo fever, which is an acute gastrointestinal illness documented in the outback of northern Australia since the late 1800s.[5] In 1931, a massive bloom of *Microcystis* species in the Ohio and Potomac Rivers was associated with gastrointestinal illness in persons using drinking water from these sources.[6] In 1979, a gastrointestinal illness, subsequently called Palm Island mystery disease, resulted in the hospital admission of over 100 children from Aboriginal families. This mystery illness was later causally linked to cylindrospermopsin-producing stain of the cyanobacterium, *Cylindrospermopsis raciborskii* (Woloszynska) Seenayya et Subba Raju, which was present in the water supply at nearby Solomon Dam.[7] A chronic neurodegenerative disease of the indigenous Chamorro people of Guam has been associated with the production of the neurotoxic nonprotein amino acid, β-methylamino-L-alanine (BMAA), by cyanobacteria.[8] Biomagnification of this neurotoxin occurs when the roots of the cycad tree absorbs BMAA from the symbiotic cyanobacteria and concentrate the BMAA in the cycad seeds. The indigenous Chamorro people are exposed to BMAA when they eat flying foxes that feed on these seeds. An amyotrophic lateral sclerosis/Parkinsonism dementia complex develops in these people, sometimes decades after exposure to BMAA.

In 1996, 131 dialysis patients in Caruaru, Brazil, were exposed to microcystins in dialysis fluid as a result of contamination of the water used for their hemodialysis.[9] Most of these patients developed nausea, vomiting, and visual disturbances. Acute liver failure developed in 110 of these patients, and by the end of the year, 60 of these patients died.

EXPOSURE

Toxic cyanobacteria occur in numerous brackish, coastal, and marine environments as well as in freshwater of over 45 countries.[10] There are about 500–1,500 species and 40 genera of primitive photosynthetic microorganisms from the phylum Cyanobacteria.[11] Other common names of these simple microorganisms include blue-green algae or pond scum.[1] Although photosynthesis occurs both in cyanobacteria and in true algae, cyanobacteria are gram-negative, prokaryotic, single-cell organisms with no membrane enveloping their nuclear

Medical Toxicology of Natural Substances, by Donald G. Barceloux, MD

material or membrane-bound, interior bodies. These organisms probably do not invade or colonize animal or human tissues. Algae are more complex eukaryotic cells containing a definite nuclear membrane and internal organelles (e.g., mitochondria). True algae also form blooms, particularly in freshwater with abundant supplies of phosphorus and dissolved nitrogen in the form of nitrate or ammonia. As nitrogen supplies become limited (e.g., after an algal bloom) the growth of cyanobacteria is favored because these microorganisms fix nitrogen from the atmosphere.[11] Like certain other chemicals (alkaloids, phenols, sterols, tannins) in vascular plants, the toxins present in cyanobacteria probably provide some protection against feeding by other organisms (e.g., zooplankton).

Source

Environment

Cyanobacteria release large amounts of blue-pigmented proteins (phycobiliproteins) during lysis, and these organisms were originally called blue-green algae. Although the photosynthetic apparatus of cyanobacteria and algae is similar, these organisms are genetically distinct and cyanobacteria are classified as bacteria. Most toxin-producing cyanobacteria are free-floating (planktonic) columns or filaments present in freshwater lakes, stagnant water, and slow-moving streams. Many planktonic cyanobacteria possess specialized, intracellular gas vesicles that allow adjustment of their buoyancy and their depth in the water. Thus, these bacteria can overcome the spatial separation of deep nutrient layers in the water and light at the surface of the water. During the beginning of summer, a variety of cyanobacteria and eukaryotic algae coexist in lakes. However, this diversity typically decreases dramatically during the end of summer following the development of a bloom of one particular species from the phylum Cyanobacteria.[12] Cyanobacteria also grow on moist terrestrial surfaces and on the surface of water as dense, benthic mats.[13] Table 25.1 lists the types of cyanobacterial populations seen in fresh and marine waters as well as the cyanobacterial genera commonly observed in these masses. Some species from the phylum Cyanobacteria are adapted to particular environments including species of *Anabaena* (freshwater lakes with limited nitrogen supplies), *Nodularia* (brackish and estuarine environments), and *Planktothrix* (turbid shallow lakes, thermally stratified lakes).[12]

Conditions that favor the growth of cyanobacteria include low wind velocity, warm temperatures (15 °C–30 °C/59 °F–86 °F), pH 6–9, and abundant sources of nitrogen and phosphorus.[1] The concentration of cyanobacterial toxins varies with season, conditions, and dominant species. In a study of German lakes and reservoirs, the highest microcystin concentrations occurred in waters dominated by *Planktothrix rubescens* (D.C. *ex* Gomont) Anagnostidis and Komárek and *Planktothrix agardhii* (Gomont) Anagnostidis and Komárek.[14] Median total microcystin content per biomass (mg/g dry weight) of water samples dominated by these two species were 2.0 (range: 1.03–5.6) and 1.46 (range: 0.19–3.6), respectively.

The most common toxin-producing cyanobacteria belong to the following genera: *Anabaena* (anatoxin-a), *Cylindrospermopsis* (cylindrospermopsin, saxitoxins), *Lyngbya* (lyngbyatosin), *Microcystis* (microcystins), *Nodularia* (nodularins), and *Oscillatoria* (anatoxins, microcystins).[11]

Surface-dwelling, scum-forming genera include *Anabaena, Aphanizomenon, Nodularia, Microcystis,* whereas the genera *Cylindrospermopsis* and *Oscillatoria* are subsurface bloom-formers.[15] Table 25.2 lists some of the recognized toxin-producing cyanobacteria. Most exposures to toxin-producing cyanobacteria occur in lakes, rivers, and ponds of the great plains of North America and Australia. Genera from the phylum Cyanobacteria associated with toxic blooms in eutrophic Scandinavian lakes included *Anabaena, Aphanizomenon, Gomphosphaeria, Microcystis, Nodularia*, and *Oscillatoria.*[16] The disagreeable color and odor of scum in freshwater discourages human consumption, but symptoms may result from exposure during water-related activities

TABLE 25.1. Observable Types of Cyanobacterial Mass Populations

Type	Toxin-Producing Genera	Description
Blooms	Anabaena, Anabaenopsis, Aphanizomenon, Cylindrospermopsis, Microcystis, Nodularia, Planktothrix	Present throughout water column secondary to buoyancy-regulating properties
Scums	Anabaena, Anabaenopsis, Aphanizomenon, Microcystis, Planktothrix	Accumulate at surface in calm waters secondary to rapid increase in cyanobacterial buoyancy
Mats/Biofilms	Lyngbya, Oscillatoria, Phormidium	Grow on surface of sediment and rocks near water edge

Source: Adapted from Ref 10.

TABLE 25.2. Some Toxin-Containing Species of Cyanobacteria

Species
Anabaena flos-aquae (Lyngbye) Brébisson ex Bornet & Flauhault
Anabaena lemmermannii forma *laxa* (Skuja) G. Cronberg & J. KomÃirek
Anabaena torulosa Lagerheim ex Bornet & Flahault
Aphanizomenon ovalisporum Forti
Aphanocapsa incerta (Lemmermann) Cronberg & Komárek
Coelosphaerium kuetzingianum Naegeli
Cylindrospermopsis raciborskii (Woloszynska) Seenaya & Subba Raju
Gloeotrichia echinulata
Lyngbya birgei Smith
Lyngbya majuscula (Dillwyn) Harvey
Microcystis aeruginosa (Kützing) Kützing
Microcystis flos-aquae (Wittrock) Kirchner
Microcystis viridis (A. Braun) Lemmermann
Nodularia spumigena Mertens
Nostoc rivulare Kützing
Planktothrix agardhii (Gomont) Anagnostidis & KomÃirek
Schizothrix calcicola (C. Agardh) Gomont

Source: From Ref 17.

(e.g., boating, swimming) in contaminated freshwater bodies.[17] The odor of cyanobacteria blooms does not necessarily correlate to toxicity.[13] The production of odor-producing compounds (e.g., geosmin from blooms of *Anabaena circinalis* Rabenhorst) is variable;[18] therefore, odor from mass of cyanobacteria does not correlate to the formation of toxins.[10]

Hot weather encourages blooms and light winds concentrate the algae in slimy, blue-green masses near the shoreline. Blooms of cyanobacteria can form green, blue-green, milky-blue, red, or dark-brown scums. Treatment of water eliminates some of the toxins associated with cyanobacteria present in surface water. Although microcystin-LR and cylindrospermopsin in concentrations of 8 µg/L and 1.3 µg/L, respectively, occurred in Australian surface waters used for drinking water, analysis of the final drinking water detected only trace amounts (i.e., <1 µg/L) of these toxins.[19]

Food

Some nontoxic freshwater cyanobacteria (*Spirulina* spp., *Aphanizomenon flos-aquae* L. Ralfs ex Barnet et Flahault) are used as nutrient-rich food supplements (lipid, protein, chlorophyll, carotenoids, vitamins).[20] However, some health food products labeled as blue-green algae may also contain hepatotoxins (e.g., microcystin-LR) resulting from the contamination of these products by toxin-producing strains from the phylum Cyanobacteria. In a study of blue-green algae products harvested during a bloom of the microcystin-producing cyanobacteria, *Microcystis aeruginosa* forma *aeruginosa* Kützing microcystins were detected in 85 of 87 samples of blue-green algae products tested by enzyme-linked immunosorbent assay (ELISA) and high performance liquid chromatography (HPLC).[21] About 72% of the samples contained microcystins exceeding the provisional guideline of 1 µg/g. The most prominent microcystin was microcystin-LR. A Portuguese study of 12 samples of food supplements of *Aphanizomenon flos-aquae* from Internet distributors detected microcystins in all 12 samples as measured by molecular assays.[22] The concentration of microcystins ranged from 0.1 to 4.72 µg/g microcystin-LR equivalents. Decomposition of these toxins occurs with chlorination by sodium hypochlorite.[23] Analysis of blue-green algae and Spirulina food supplements from Canada, the United States, and Europe did not detect the presence of anatoxin-a or its transformation products as measured by HPLC/tandem mass spectometry.[24] The detection limits in this study ranged from about 0.5 to 1.5 pg.

TOXINS

Physiochemical Properties

Cyanobacteria produce a wide range of low and medium molecular weight metabolites that are primarily hepatotoxins (microcystins, nodularin, cylindrospermopsin), neurotoxins (anatoxin-a, anatoxin-a(s), homoanatoxin-a, saxitoxins), cytotoxins (cylindrospermopsin), or gastrointestinal irritants (aplysiatoxin, debromoaplysiatoxin, lyngbyatoxin).[10,11] These hepatotoxins and neurotoxins (i.e., low-molecular-weight anatoxins and saxitoxins) are secondary metabolites present only in certain species from the phylum Cyanobacteria. Cylindrospermopsin is a cyclic guanidine alkaloid that causes necrosis in various organs (e.g., liver) in animal models secondary to inhibition of protein synthesis.[25] Additionally, this toxin causes genotoxicity (DNA strand breakage, chromosome loss) in these animal models. Lipopolysaccharide endotoxins occur in the cell walls of all species from the phylum Cyanobacteria, and these compounds probably contribute to inflammation and gastrointestinal irritation associated with cyanobacteria. Other toxins isolated from freshwater cyanobacteria include aeruginopeptins, cyanopeptolins, and oscillamide.[26] Lyngbyatoxin A, aplysiatoxin, and debromoaplysiatoxin have been associated with the formation of local lesions. Figure 25.1 demonstrates the chemical structure of six important toxins isolated from

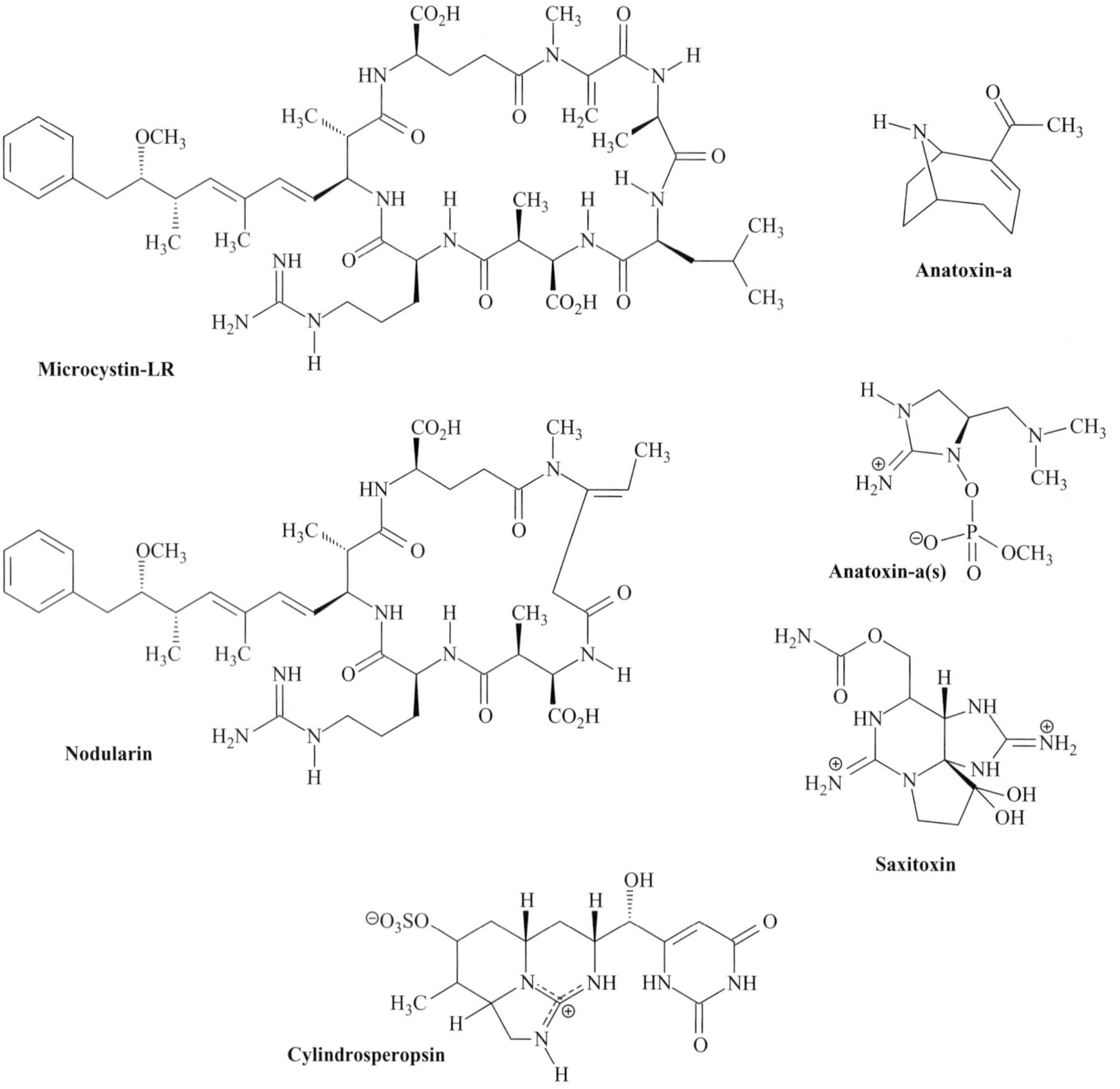

Microcystin-LR

Anatoxin-a

Anatoxin-a(s)

Nodularin

Saxitoxin

Cylindrosperopsin

FIGURE 25.1. Structures of cyanobacteria toxins.

cyanobacteria. The blue pigment of cyanobacteria becomes apparent as the cells deteriorate and the edges of infected ponds become blue as these microorganisms desiccate. The presence of toxin-producing cyanobacteria in water does not necessarily imply a health hazard because the toxins are rapidly dispersed in water as long as the organisms remain widely dispersed.[13] The formation of dense mats and surface scum suggests that a health hazard may exist if substantial numbers of toxin-producing cyanobacteria are present.

There are over 70 variants of microcystin toxins that share a general structure containing three D-amino acids (alanine, β-linked *erythro*-β-methylaspartic acid, γ-linked glutamic acid), *N*-methyldehydroalanine, two variable L-amino acids, and a rare C20 β-amino acid [(3-amino-9-methoxy-2,6,8-trimethyl-10-phenyldeca-4,6-dienoic acid).].[27] The nomenclature for microcystin variants involves the methylation pattern based on the two variable L-amino acids. For example, microcystin-LR contains a leucine (L) and arginine (R) substitution at these sites. The synthesis of microcystin toxins occurs via a mixed polyketide synthase/nonribosomal peptide synthetase system called microcystin synthetase. The synthesis of microcystin in cyanobacteria involves an integrated polyketide-peptide biosynthetic pathway with unusual structural and enzymatic features, such as integrated *C*- and *N*-methyltransferases, *O*-methyltransferase, and the formation of β- and γ-carboxyl-peptide bonds.[28] The ribosomes are not part of the synthesis of microcystin. Detoxification pathways of microcystin include conjugation by the glutathione S-transferase system.[29]

NEUROTOXINS

Cyanobacteria produce at least 19 structurally related neurotoxic compounds including anatoxins, saxitoxin, neosaxitoxin, C-toxins, G-toxins, and LW (*Lyngbya wollei*)-toxins. Marine bacteria and marine dinoflagellates also produce saxitoxins that are commonly associated with shellfish poisonings. Anatoxin-a is a low-molecular-weight, bicyclic secondary amine isolated most commonly from *Anabaena flos-aquae* and *Anabaena circinalis* Rabenhorst. Other species from the phylum Cyanobacteria also produce anatoxin-a including species in the genera *Aphanizomenon, Cylindrospermum, Microcystis*, and *Plankothrix*.[12] This neurotoxin is a highly water-soluble compound that binds to nicotinic acetylcholine receptors to produce flaccid paralysis. Anatoxin-a is a potent nicotinic agonist that is sensitive to heat, light, and alkali.[30] The methylene homologue of anatoxin-a is homoanatoxin-a, which was isolated from *Planktothrix formosa*.[31] Homoanatoxin-a enhances acetylcholine release from peripheral cholinergic nerves as a result of the opening of voltage-dependent L-type calcium channels in neurons.[32] Despite similar names, anatoxin-a and anatoxin-a(s) are not structurally similar. Anatoxin-a(s) is a phosphate ester of cyclic *N*-hydroxyguanine that is the only natural organophosphate compound known.[33]

HEPATOTOXINS

Microcystins and nodularins belong to the okadaic acid class of protein phosphatase inhibitors. There are at least 80 microcystin variants and at least 5 nodularin variants as a result of amino acid substitutions and modification of the structure of these cyclic peptides. Microcystin toxins are heat-stable, cyclic heptapeptides predominately from hepatotoxin-producing species of the genus *Microcystis* that demonstrate tumor-promoting activity by the inhibition of protein phosphatases 1 and 2A.[34] These cyclic peptides have molecular weights between 900 and 1,100 daltons, and these compounds are stable in a wide range of pH. Microcystins contain two variable L-amino acids and the two following novel D-amino acids: *N*-methyldehydroalanine (Mdha) and 3-amino-9-methoxy-2,6,8-trimethyl-10-phenyldeca-4,6-dienoic acid (Adda). These toxins are named according to the variable L-amino acids (e.g., microcystin-LR contains leucine and arginine at the variable sites).

Dose Response

There is no correlation between the composition or density of a particular bloom and the concentration of toxins within the cyanobacteria.[3] The World Health Organization (WHO) adopted provisional guideline values of 20 µg microcystin-LR/L in recreational water and 1 µg microcystin-LR/L drinking water.[35] The National Health and Medial Research Council of Australia established a drinking water guideline of 1.3 µg microcystin-LR/L drinking water. Estimated guideline values for anatoxin-a and cylindrospermopsin based on animal testing were 12.24 µg/L water and 0.81 µg/L water, respectively.[10]

The lethality of microcystins varies at least 10-fold between different variants.[36] In studies of mice, the oral LD_{50} (5,000 µg/kg body weight) of microcystin-LR is approximately 100-fold lower than the intraperitoneal LD_{50} of the same compound.[37] The lowest observable effect level (LOAEL) in a study of pigs administered seven different congeners of microcystins (excluding microcystin-LR and microcystin-RR) was 280 µg MC-LR equivalents/kg body weight daily for 44 days.[38] The effects observed were limited to histological changes in the liver of these animals. At least 52 deaths from acute liver failure were attributed to the use of microcystin-contaminated dialysis fluid by renal dialysis patients in Brazil.[39] Based on postmortem liver concentrations of microcystins and exposure volumes, the estimated concentration of microcystin-LR equivalents in dialysis fluid was approximately 20-fold higher (19.5 µg/L) than the current WHO provisional guideline (1 µg/L) for drinking water. In rodent studies, anatoxin-a(s) binds to acetylcholinesterases in the blood. Based on this study, this compound is 10 times more toxic than anatoxin-a.[40]

Mechanism of Toxicity

Some species from the phylum Cyanobacteria contain lipopolysaccharide endotoxins that are released during decomposition, and most, but not all, serious risks to human health from cyanobacteria result from the production of hepatotoxins or neurotoxins. Microcystin toxins are potent hepatotoxins that inhibit several serine/threonine protein phosphatases including phosphatase types 1 and 2A. This inhibition causes disruption of normal cell metabolism and function, leading to intrahepatic hemorrhage and cell death. In rodent models, chronic administration of microcystin causes hepatocyte degeneration with necrosis, mononuclear leukocyte infiltration, and progressive fibrosis.[41]

The filamentous blue-green cyanobacterium, *Lyngbya majuscula* Harvey ex Gomont, produces lyngbyatoxin A as well as over 70 other toxins.[42] These toxins have contaminated meat from marine turtles, and consumption of this meat was associated with acute inflammatory changes of the gastrointestinal tract, intraoral lesions, weakness, headache, lightheadedness, and salivation.[43] Contact with toxins from *Lyngbya majuscula* has also been associated with swimmer's dermatitis.[44,45]

Neurotoxins

A variety of freshwater species of *Anabaena, Aphanizomenon*, and *Oscillatoria* produce a nicotinic agonist (anatoxin-a) that binds irreversibly to the nicotinic acetylcholine receptor of the sodium channel of the postsynaptic neurotransmitter plate. In experimental animals, the action of this neurotoxin mimics acetylcholine and causes postsynaptic neuromuscular blockade and respiratory depression (fasciculations, muscle spasm, fatigue, and paralysis).[46,47] This potent acetylcholine receptor agonist essentially locks the sodium channel in the open position and produces overstimulation followed by paralysis. Strains of *Anabaena* species produce another neurotoxin unique to the phylum Cyanobacteria called anatoxin-a(s). This organophosphate compound causes a cholinergic-type syndrome by inhibiting acetylcholinesterase in a manner similar to synthetic organophosphate insecticides, but this unstable compound does not penetrate the blood–brain barrier.[1,48] Some species of *Anabaena* and *Aphanizomenon* also contain saxitoxin and neosaxitoxin. These two neurotoxins were first isolated from the marine algae (dinoflagellates) that produce "red tides" and cause outbreaks of paralytic shellfish poisoning. Saxitoxin blocks neuronal transmission and causes muscle paralysis by binding to the voltage-gated sodium channels in neurons.

Hepatotoxins

Hepatotoxins are the most common type of toxin formed by cyanobacteria. There are over 60 cyclic and ringed peptides produced by certain strains (e.g., *Microcystis, Nodularia*) from the phylum Cyanobacteria.[1] Microcystins are cyclic heptapeptide (seven amino acids), whereas nodularins contain five amino acids. Microcystins alter cellular structure by inhibiting protein phosphatases and disrupting the normal regulatory balance between phosphorylation and dephosphorylation. In experimental animals, these hepatotoxins are actively transported to the liver by the bile acid carrier transport system where these toxins produce liver damage and apoptosis. This damage includes lipid peroxidation, loss of membrane integrity, DNA fragmentation, hemorrhage, and massive centrilobular necrosis.[49,50]

CLINICAL RESPONSE

The symptoms associated with exposure to cyanobacteria are nonspecific (vomiting, diarrhea, malaise, rash, mucosal irritation).[51,52] Gastrointestinal illness associated with cyanobacteria in water supplies result either from the breakdown of a natural cyanobacterial bloom or the release of toxins from the lysis of cyanobacterial cells during the handling of bloom-associated water. Self-limited gastroenteritis (diarrhea, abdominal cramps, nausea, vomiting) occurs 3–5 hours after exposure to water containing certain species from the phylum Cyanobacteria.[17] Symptoms usually resolve within 1–2 days. Case reports associate local skin reactions (swelling, redness) and mucous membrane irritation (conjunctivitis, rhinitis) with exposure to cyanobacteria.[53,54] A few otherwise healthy individuals can develop a skin reaction to cyanobacteria during water recreational activities, but the rash is typically mild and resolves without treatment.[55] Contact with toxins from *L. majuscula* was associated with acute vesicular dermatitis similar to sunburn, but the rash appeared in the genital, perineum, and perianal areas.[44] Erythema and local pain occurred within several hours of contact followed by the formation of vesicles and desquamation. Despite the presence of neurotoxins (e.g., anatoxin-a) in some cyanobacteria, clinically significant neurological symptoms in humans to date have not been associated with exposure to cyanobacteria. A burning sensation in the mouth and throat, headache, myalgias, gastrointestinal symptoms developed 15–90 minutes after ingestion of seaweed contaminated with debromoaplysiatoxin and aplysiatoxin isolated from the red alga, *Gracilaria coronopifolia* J. Agardh.[56,57] Ingestion of seaweed contaminated by *L. majuscula* produced immediate burning sensation and an intraoral lesion that required up to 3 weeks to resolve.[58] Glycosidic macrolides (lyngbyaloside, lyngbyaloside B) isolated from *Lyngbya* species from the phylum Cyanobacteria are structurally similar to the polycavernosides associated with fatal seaweed poisoning from the ingestion of red alga.[59] Other fatal cases of seaweed poisoning have been associated with aplysiatoxin, which produces physiological effects similar to lyngbyatoxin A in animal models.[60] However, there are no well-documented human fatalities associated with the ingestion of cyanobacteria.

In a field study of Australian residents drinking water from a reservoir containing a heavy bloom of toxic *Microcystis aeruginosa*, consumption of contaminated water was associated with subclinical increase in the serum γ-glutamyl transpeptidase and alanine aminotransferase concentrations.[61] Increased fatalities from liver failure occurred in dialysis-dependent patients receiving treatment at a hemodialysis center where microcystins contaminated the water used for dialysis fluid.[62] Histological examination of liver tissue revealed disruption of liver plates, liver cell deformity, necrosis, cholestasis, cytoplasmic vacuolization, mixed leukocyte infiltration and multinucleated hepatocytes.[39]

The chronic effects of exposure to toxin-producing cyanobacteria are unclear. In animal studies, some of the hepatotoxins (e.g., microcystins) in certain cyanobacterial species are potent tumor promoters.[63] Although these toxins are potentially cofactors along with HIV and aflatoxins for the development of primary liver cancer, there is no direct evidence that these toxins produce tumors in humans.

DIAGNOSTIC TESTING

Analytical Methods

Light microscopy is the traditional method for monitoring water for cyanobacteria. Alternate methods for the detection of cyanobacteria in water columns include flow cytometry with autofluorescence/immunofluorescence and submersible spectrofluorometry with specificity for cyanobacterial pigments.[10] Limitations to the analysis of cyanobacterial toxins include the need for sample processing before analysis and the wide range of toxins (e.g., at least 80 variants of microcystins). Sample screening by mouse bioassay has been replaced by more sensitive and reliable assays (e.g., enzyme-linked immunosorbent assay, protein phosphatase assay).[64] Newer techniques for the detection, identification, and quantitation of cyanobacterial toxins include physicochemical methods (gas chromatography/mass spectrometry, high performance liquid chromatography [HPLC], HPLC/mass spectrometry with negative electrospray ionization),[65,66] biochemical techniques (protein phosphatase inhibition assay), and immunoassays (enzyme-linked immunosorbent assay [ELISA], fluorescent in situ hybridization, polymerase chain reaction [PCR]).[67,68] Covalent binding of microcystins to animal tissues limits the determination of microcystins to unbound microcystins unless special procedures (e.g., 2-methyl-3-methoxy-4-phenylbutyric acid oxidation) reduce tissue binding prior to analysis.[69] When stored at −70 °C (−94 °F), microcystins are relatively stable for at least 10 years.[65]

Currently, routine analysis of water samples for microcystin toxins involves sensitive biological screening procedures (e.g., antibody-based ELISA, protein phosphatase inhibition assays) followed by quantitation of the microcystin toxins by HPLC with photodiode array detection (PDA).[64] Solid-phase extraction with C_{18} silica or immunoaffinity columns enriches the concentration of microcystin toxins in environmental samples and helps eliminate contaminants. Most commercial tests analyze samples for the presence of cyanobacteria-related toxins (e.g., microcystins). By several orders of magnitude, polyclonal and monoclonal immunoassays are more sensitive for the detection of microcystin toxins compared with physicochemical methods, but these immunoassays did not detect the less toxic glucuronide conjugates or react with all the microcystins equally. Cross-reactivity of various microcystin and nodularin compounds as well as the occurrence of >70 variants of microcystin toxins limits the specificity of immunoassays for cyanobacterial toxins. Although constrained by the limited availability of purified standards for the hepatotoxins and neurotoxins, several physicochemical and biochemical methods complement the traditional mouse bioassay or immunoassays for the detection and confirmation of these toxins.[70] Liquid chromatography is the most widely available method for quantitation of microcystins and nodularins, primarily by reversed-phase C_{18} columns with UV detection. The use of gas chromatography with electron capture detection analyzes environment samples for anatoxin-a.[71] Analysis of samples by HPLC and ELISA are available to detect the presence of hepatotoxins (microcystin-LR, nodularins) in health food products. The use of photodiode-array HPLC with methanol extraction allows detection of microcystin concentrations as low as 250 ng/L.[72]

Biomarkers

Water

Current recommended guidelines for microcystin-LR equivalents and cylindrospermopsin in drinking water are 1 µg/L.[73] The amount of microcystin in algae blooms varies with water temperature, total phosphorus levels, and chlorophyll a concentrations. In a study of German recreational lakes, 26% of the lakes contained toxigenic cyanobacteria with the highest microcystin concentration in the waters outside scum areas ranging from 11–560 µg/L as measured by ELISA techniques.[74] The microcystin concentration in algae blooms changes during the seasons with water temperature. In a Chinese freshwater lake, the mean microcystin concentration varied between about 24 ± 4 µg/g dry algae and 97 ± 12 µg/g dry algae as measured by ELISA.[75] The highest microcystin concentrations occurred during high water temperatures, and the main microcystin was microcystin-LR. The concentration of cylindrospermopsin is relatively low compared with more dominant cyanobacterial toxins, such as microcystin-LR and anatoxin a.[76]

In the presence of species from the phylum Cyanobacteria, detection of chlorophyll-a concentrations exceeding 40 µg/L suggests that surface scums may form rapidly, particularly in large bodies of water subject to winds.[13] Severe scums are associated with chlorophyll-a concentration in open water samples exceeding 100–150 µg/mL. Toxic cyanobacteria blooms are unusual in water containing total phosphorus concentration below 0.01–0.02 mg/L. The presence of phosphorus concentrations exceeding this concentration and chlorophyll-a concentrations >100 μg/mL suggest the potential for cyanobacteria blooms. In these situations, warnings against the recreational use of these waters are appropriate.

HUMANS

Microcystins are detectable in serum from dialysis patients for up to approximately 60–90 days after cessation of exposure to contaminated water. In a Brazilian study of 44 hemodialysis patients exposed to sublethal doses of microcystins in contaminated dialysis fluid, 90% of the initial serum samples contained detectable concentrations of microcystins as measured by ELISA.[77] Twelve of these patients were monitored for 57 days after the discovery of the contaminated water, and the total serum microcystin concentrations ranged from <0.16 ng/mL to 0.96 ng/mL following MMPB (2-methyl-3-methoxy-4-phenylbutyric acid) oxidation to release bound microcystins. Peak microcystin concentrations occurred about 1 month after exposure ceased.

TREATMENT

Treatment is supportive.

References

1. Carmichael WW. The toxins of cyanobacteria. Sci Am 1994;270:78–86.
2. Echlin P. The blue-green algae. Sci Am 1966;214:75–81.
3. Elder GH, Hunter PR, Codd GA. Hazardous freshwater cyanobacteria (blue-green algae). Lancet 1993;341:1519–1520.
4. Francis G. Poisonous Australian lakes. Nature 1878;18: 11–12.
5. Hayman J. Beyond the Barcoo—probable human tropical cyanobacterial poisoning in outback Australia. Med J Austr 1992;157:794–796.
6. Tisdale ES. Epidemic of intestinal disorders in Charleston, W. Va., occurring simultaneously with unprecedented water supply conditions. Am J Publ Health 1931;21:198–200.
7. Bourke AT, Hawes RB, Neilson A, Stallman ND. An outbreak of hepato-enteritis (the Palm Island mystery disease) possibly caused by algal intoxication. Toxicon 1983;3: 45–48.
8. Murch SJ, Cox PA, Banack SA. A mechanism for slow release of biomagnified cyanobacterial neurotoxins and neurodegenerative disease in Guam. Proc Natl Acad Sci U S A 2004;101:12228–12231.
9. Azevedo SM, Carmichael WW, Jochimsen EM, Rinehart KL, Lau S, Shaw GR, Eaglesham GK. Human intoxication by microcystins during renal dialysis treatment in Caruaru-Brazil. Toxicology 2002;181–182:441–446.
10. Codd GA, Morrison LF, Metcalf JS. Cyanobacterial toxins: risk management for health protection. Toxicol Appl Pharmacol 2005;203:264–272.
11. Patocka J. The toxins of cyanobacteria. Acta Medic 2001;44:69–75.
12. Dittmann E, Wiegand C. Cyanobacterial toxins–occurrence, biosynthesis and impact on human affairs. Mol Nutr Food Res 2006;50:7–17.
13. Chorus I, Falconer IR, Salas HJ, Bartram J. Health risks caused by freshwater cyanobacteria in recreational waters. J Toxicol Environ Health 2000 (Part B);3:323–347.
14. Fastner J, Neumann U, Wirsing B, Weckesser J, Wiedner C, Nixdorf B, Chorus I. Microcystins (hepatotoxic hepatapeptides) in German fresh water bodies. Environ Toxicol 1999;14;13–22.
15. Paerl HW, Fulton RS 3rd , Moisander PH, Dyble J. Harmful freshwater algal blooms, with an emphasis on cyanobacteria. Sci World J 2001;1:76–113.
16. Berg K, Skulberg OM, Skulberg R, Underdal B, Willen T. Observations of toxic blue-green algae (cyanobacteria) in some Scandinavian lakes. Acta Vet Scand 1986;27; 440–452.
17. Spoerke DG, Rumack BH. Blue-green algae poisoning. J Emerg Med 1985;2:353–355.
18. Izaguirre G, Taylor WD. Geosmin and MIB events in a new reservoir in southern California. Water Sci Technol 2007;55:9–14.
19. Hoeger SJ, Shaw G, Hitzfeld B, Dietrich DR. Occurrence and elimination of cyanobacterial toxins in two Australian drinking water treatment plants. Toxicon 2004;43:639–649.
20. Kay RA. Microalgae as food and supplement. Crit Rev Food Sci Nutr 1991;30:555–573.
21. Gilroy DJ, Kauffman KW, Hall RA, Huang X, Chu FS. Assessing potential health risks from microcystin toxins in blue-green algae dietary supplements. Environ Health Perspect 2000;108:435–439.
22. Saker ML, Jungblut AD, Neilan BA, Rawn DF, Vasconcelos VM. Detection of microcystin synthetase genes in health food supplements containing the freshwater cyanobacterium *Aphanizomenon flos-aquae*. Toxicon 2005;46:555–562.
23. Tsuji K, Watanuki T, Kondo F, Watanabe MF, Nakazawa H, Suzuki M, Uchida H, Harada K. Stability of microcystins from Cyanobacteria—IV. Effect of chlorination on decomposition. Toxicon 1997;35:1033–1041.
24. Rawn DF, Niedzwiadek B, Lau BP, Saker M. Anatoxin-a and its metabolites in blue-green algae food supplements from Canada and Portugal. J Food Prot 2007;70:776–779.
25. Shen X, Lam PK, Shaw GR, Wickramasinghe W. Genotoxicity investigation of a cyanobacterial toxin, cylindrospermopsin. Toxicon 2002;40:1499–1501.

26. Harada K–I. Production of secondary metabolites by freshwater cyanobacteria. Chem Pharm Bull 2004;52: 889–899.
27. Botes DP, Kruger H, Viljoen CC. Isolation and characterization of four toxins from the blue-green alga, *Microcystis aeruginosa*. Toxicon 1982;20:945–954.
28. Tillett D, Dittmann E, Erhard M, von Dohren H, Borner T, Neilan BA. Structural organization of microcystin biosynthesis in Microcystis aeruginosa PCC7806: an integrated peptide-polyketide synthetase system. Chem Biol 2000;7:753–764.
29. Pflugmacher S, Wiegand C, Beattie KA, Krause E, Steinberg CE, Codd GA. Uptake, effects, and metabolism of cyanobacterial toxins in the emergent reed plant *Phragmites australis* (cav.) trin. ex steud. Environ Toxicol Chem 2001;20:846–852.
30. Molloy L, Wonnacott S, Gallagher T, Brough PA, Livett BGL. Anatoxin-a is a potent agonist of the nicotinic acetylcholine receptor of bovine adrenal chromaffin cells. Eur J Pharmacol 1995;289:447–453.
31. Furey A, Crowley J, Shuilleabhain AN, Skulberg OM, James KJ. The first identification of the rare cyanobacterial toxin, homoanatoxin-a, in Ireland. Toxicon 2003;41:297–303.
32. Lilleheil G, Andersen RA, Skulberg OM, Alexander J. Effects of a homoanatoxin-a-containing extract from *Oscillatoria formosa* (Cyanophyceae/cyanobacteria) on neuromuscular transmission. Toxicon 1997;35:1275–1289.
33. Dixit A, Dhaked RK, Alam SI, Singh L. Military potential of biological neurotoxins. Toxin Rev 2005;24:175–207.
34. Bates DP, Kruger H, Viljoen CC. Isolation and characterization of four toxins from blue-green algae, *Microcystis aeruginosa*. Toxicon 1982;20:945–954.
35. World Health Organization (WHO). Guidelines for drinking-water quality, first addendum to third edition, vol 1 recommendations. Geneva: WHO; 2006:492.
36. Dietrich D, Hoeger S. Guidance values for microcystins in water and cyanobacterial supplement products (blue-green algal supplements) a reasonable or misguided approach? Toxicol Appl Pharmacol 2005;203:273–289.
37. Fawell JK, Mitchell RE, Everett DJ, Hill RE. The toxicity of cyanobacterial toxins in the mouse: I microcystin-LR. Hum Exp Toxicol 1999;18:162–167.
38. Falconer IR, Burch MD, Steffensen DA, Choice M, Coverdale OR. Toxicity of the blue-green alga (cyanobacterium) *Microcystis aeruginosa* in drinking water to growing pigs, as an animal model for human injury and rick assessment. J Environ Toxicol Water Qual 1994;9:131–134.
39. Carmichael WW, Azevedo SMFO, An JS, Molica RJR, Jochimsen EM, Lau S, et al. Human fatalities from cyanobacteria: chemical and biological evidence for cyanotoxins. Environ Health Perspect 2001;109:663–668.
40. Mahmood NA, Carmichael WW. Anatoxin-a(s), an anticholinesterase from the cyanobacterium *Anabaena flos-aquae* NRC-525-17. Toxicon 1987;25:1221–1227.
41. Bhattacharya R, Sugendran K, Dangi RS, Rao PV. Toxicity evaluation of freshwater cyanobacterium *Microcystis aeruginosa* PCC 7806: II. Nephrotoxicity in rats. Biomed Environ Sci 1997;10:93–101.
42. Osborne NJT, Webb PM, Shaw GR. The toxins of *Lyngbya majuscula* and their human and ecological health effects. Environ Int 2001;27:381–392.
43. Yasumoto T. Fish poisoning due to toxins of microalgal origins in the Pacific. Toxicon 1998;36:1515–1518.
44. Grauer FH, Arnold HL. Seaweed dermatitis: first report of dermatitis-producing marine algae. Arch Dermatol 1961;84:720–732.
45. Serdula M, Bartilini G, Moore RE, Gooch J, Wiebenga N. Seaweed itch on windward Oahu. Hawaii Med J 1982;41:200–201.
46. Moore RE. Toxins from blue-green algae. Bioscience 1977;27:797–802.
47. Carmichael MW, Biggs DF, Gorham PR. Toxicology and pharmacological action of *Anabaena flos-aquae* toxin. Science 1975;187:542–544.
48. Beasley VR, Dahlem AM, Cook WO, Valentine WM, Lovell RA, Hooser SB, et al. Diagnostic and clinically important aspects of cyanobacterial (blue-green algae) toxicoses. J Vet Diagn Invest 1989;1:359–365.
49. Hawkins PR, Runnegar MT, Jackson AR, Falconer IR. Severe hepatotoxicity caused by the tropical cyanobacterium (blue-green alga) *Cylindrospermopsis raciborskii* (Woloszynska Seenaya and Subba Raju) isolated form a domestic water supply reservoir. Appl Environ Microbiol 1985;50:1292–1295.
50. DeVries SE, Galey FD, Namikoshi M, Woo JC. Clinical and pathological findings of blue-green algae (*Microcystis aeruginosa*) intoxication in a dog. J Vet Diagn Invest 1993;5:403–408.
51. Stewart I, Webb PM, Schluter PJ, Fleming LE, Burns JW Jr , Gantar M, et al. Epidemiology of recreational exposure to freshwater cyanobacteria–an international prospective cohort study. BMC Pub Health 2006;6:93.
52. Pilotto LS, Douglas RM, Burch MD, Cameron S, Beers M, Rouch GJ, et al. Health effects of exposure to cyanobacteria (blue-green algae) during recreational water-related activities. Aust N Z J Public Health 1997;21:562–566.
53. Hunter PR. Cyanobacteria and human health. J Med Microbiol 1992;36:301–302.
54. Bourke ATC, Hawes RB. Fresh water cyanobacteria (blue-green algae) and human health. Med J Aust 1983;1: 491–492.
55. Pilotto L, Hobson P, Burch MD, Ranmuthugala G, Attewell R, Weightman W. Acute skin irritant effects of cyanobacteria (blue-green algae) in healthy volunteers. Aust N Z J Public Health 2004;28:220–224.
56. Nagai H, Yasumoto T, Hokama Y. Aplysiatoxin and debromoaplysiatoxin as the causative agents of a red alga *Gracilaria coronopifolia* poisoning in Hawaii. Toxicon 1996;37:753–761.

57. Marshall KLE, Vogt RL. Illness associated with eating seaweed, Hawaii, 1994. West J Med 1998;169:293–295.
58. Sims JK, Zandee van Rilland RD. Escharotic stomatitis caused by "stinging seaweed" *Microcoleus lyngbyaceus* (formerly *Lyngbya majuscula*). Case report and literature review. Hawaii Med J 1981;40;243–248.
59. Yotsu-Yamashita M, Yasumoto T, Yamada S, Bajarias FF, Formeloza MA, Romero ML, Fukuyo Y. Identification of polycavernoside A as the causative agent of the fatal food poisoning resulting from ingestion of the red alga *Gracilaria edulis* in the Philippines. Chem Res Toxicol 2004;17:1265–1271.
60. Ito E, Satake M, Yasumoto T. Pathological effects of lyngbyatoxin A upon mice. Toxicon 2002;40:551–556.
61. Falconer IR, Bereford AM, Runnegar TO. Evidence of liver damage by toxin from a bloom of blue-green algae, *Microcystis aeruginosa*. Med J Aust 1983;1:511–514.
62. Jochimsen EM, Carmichael WW, Cardo DM, Cookson ST, Holmes CEM, et al. Liver failure and death after exposure to microcystins at a hemodialysis center in Brazil. N Engl J Med 1998;338:873–878.
63. Nishiwaki-Matsushima R, Ohta T, Hishiwaki S, Suganuma M, Kohyama K, Ishikawa T, et al. Liver tumor promotion by the cyanobacterial cyclic peptide toxin microcystin-LR. J Cancer Res Clin Oncol 1992;118(6):420–424.
64. McElhiney J, Lawton LA. Detection of the cyanobacterial hepatotoxins microcystins. Toxicol Appl Pharmacol 2005;203:219–230.
65. Maizels M, Budde WL. A LC/MS method for the determination of cyanobacteria toxins in water. Anal Chem 2004;76:1342–1351.
66. Yuan M, Carmichael WW, Hilborn ED. Microcystin analysis in human sera and liver from human fatalities in Caruaru, Brazil 1996. Toxicon 2006;48:627–640.
67. Hilborn ED, Carmichael WW, Yuan M, Azevedo SM. A simple colorimetric method to detect biological evidence of human exposure to microcystins. Toxicon 2005;46: 218–221.
68. Metcalf JS, Codd GA. Analysis of cyanobacterial toxins by immunological methods. Chem Res Toxicol 2003;16: 103–112.
69. Ott JL, Carmichael WW. LC/ESI/MS method development for the analysis of hepatotoxic cyclic peptide microcystins in animal tissues. Toxicon 2006;47:734–741.
70. Codd GA, Metcalf JS, Ward CJ, Beattie KA, Bell SG, Kaya K, Poon GK. Analysis of cyanobacterial toxins by physicochemical and biochemical methods. J AOAC Int 2001;84:1626–1635.
71. Stevens DK, Krieger RI. Analysis of anatoxin-a by GC/ECD. J Anal Toxicol 1988;12:126–131.
72. Lawton LA, Edwards C, Codd GA. Extraction and high-performance liquid chromatographic method for the determination of microcystins in raw and treated waters. Analyst 1994;119:1525–1530.
73. Falconer IR, Humpage AR. Health risk assessment of cyanobacterial (blue-green algal) toxins in drinking water. Int J Environ Res Public Health 2005;2:43–50.
74. Frank CA. Microcystin-producing cyanobacteria in recreational waters in Southwestern Germany. Environ Toxicol 2002;17:361–366.
75. Shen PP, Shi Q, Hua ZC, Kong FX, Wang ZG, Zhuang SX, Chen DC. Analysis of microcystins in cyanobacteria blooms and surface water samples from Meiliang Bay, Taihu Lake, China. Environ Int 2003;29:641–647.
76. Griffiths D, Saker ML. The Palm Island mystery disease 20 years on: a review of research on the cyanotoxin cylindrospermopsin. Environ Toxicol 2003;18:78–93.
77. Soares RM, Yuan M, Servaites JC, Delgado A, Magalhaes VF, Hilborn ED, et al. Sublethal exposure from microcystins to renal insufficiency patients in Rio de Janeiro, Brazil. Environ Toxicol 2006;21:95–103.

Chapter 26

PROTOZOA and INTESTINAL PARASITES

ENTAMOEBA HISTOLYTICA and AMEBIASIS

HISTORY

Molecular phylogeny places the genus *Entamoeba* on the lowermost branches of the eukaryotic tree near the genus *Dictyostelium*. Although this organism lacks mitochondria, *Entamoeba histolytica* possesses nuclear-encoded mitochondrial genes and remnant organelles.[1] Hippocrates first recognized amebiasis in a patient with fever and dysentery. In 1875, Fedor Losch isolated *E. histolytica* from a stool specimen of a patient with dysentery.[2] Leonard Rogers recommended emetine as the first effective treatment for amebiasis in 1912.[3]

EXPOSURE

Source

Entamoeba histolytica Schaudinn is the only pathogenic *Entamoeba* species, and humans are the primary reservoir of *E. histolytica*. Oral-fecal transmission is the main route of infection with the cyst spreading disease most commonly through contaminated water.[4] Some animal reservoirs of *E. histolytica* probably exist in dogs, monkeys, and pigs, but these sources are uncommon pathways for human infections.[5] Cockroaches and flies are potential but rare sources of the transmission of amebiasis to humans.[6] Male homosexuals and foreign travelers to developing countries are at the highest risk of acquiring amebiasis. In a study of recent symptomatic *E. histolytica* infections in large cities in Japan, amebiasis almost exclusively afflicted middle-aged men, probably as a result of exposure during homosexual activity and concomitant infection with HIV.[7]

PATHOGEN

Physiochemical Properties

Entamoeba histolytica cysts remain infective outside the host for weeks to months, particularly in damp conditions. However, low (<–5 °C/23 °F) and high (>40 °C/104 °F) temperatures destroy the cysts.

Mechanism of Toxicity

Within the intestinal lumen, ingested tetra-nucleate cysts of *E. histolytica* transform into trophozoites that disrupt the protective mucous layer overlying the mucosa of the colon, causing mucosal thickening and flask-shaped ulceration in the cecum and ascending colon.[8] The sigmoid colon and rectum rarely have lesions secondary to amebic infection.[6] These trophozoites excrete galactose and *N*-acetyl-D-galactosamine-specific lectin to adhere to colonic mucins. Subsequently, these trophozoites invade the large intestine, producing punc-

Medical Toxicology of Natural Substances, by Donald G. Barceloux, MD

tate, yellowish white ulcers despite the lack of symptoms in many cases of infection.[9] The presence of proteolytic enzymes (collagenase, neutral proteases) and cysteine proteases facilitate the penetration of the trophozoites into the tissue. A study of the transcription of nearly 40,000 human genes in sections of human colonic xenografts obtained 4 hours and 24 hours following infection with *E. histolytica* demonstrated a relatively stereotypical response. This response involved increased expression of genes activated in cells undergoing stress and/or hypoxic responses, genes encoding cytokines, chemokines, and mediators in immune and inflammatory responses, and genes encoding proteins in response to tissue injury and repair.[10] Aggregation of amoeba in the mucin layer of the colon triggers encystation and the subsequent excretion of cysts in the stool that completes the life cycle of the amoeba.

Colitis occurs when the trophozoite penetrates the intestinal mucous layer, which is the main barrier to the intestinal invasion of the trophozoites. The lectin engages *N*-acetyl-D-galactosamine on *O*-linked cell-surface oligosaccharides on epithelial cells, neutrophils, macrophages, and T-lymphocytes of the host. Activation of human caspase 3, a distal effector molecule in the apoptotic pathway, results in rapid cell death and progression of the invasion of host tissue.

DOSE RESPONSE

The infective dose of *E. histolytica* cysts is low (i.e., <100) compared with bacterial causes of gastroenteritis.[11]

CLINICAL RESPONSE

Amebiasis causes both intraluminal and disseminated disease. Risk factors include recent foreign travel to or immigration from developing countries and homosexual contact. Despite widespread infection with *E. histolytica*, the vast majority of individuals with amebiasis remain asymptomatic or mildly symptomatic.[12] Determination of the prevalence of *E. histolytica* infections is complicated by the presence of a morphologically identical species (*Entamoeba dispar* Brumpt) that is nonpathogenic.

Colitis

The incubation period of *E. histolytica* is typically 1–4 weeks, but symptoms may develop months after ingestion of the cysts. The onset of symptoms of amebic colitis is insidious and variable with the clinical diagnosis of amebic colitis complicated by the lack of specific clinical features. Typically, amebic colitis presents with several weeks of crampy abdominal pain, watery or bloody diarrhea, and weight loss. Fever and bloody stools are not usually prominent features of amebic colitis. The differential diagnosis of the bloody diarrhea associated with amebic colitis includes bacterial gastroenteritis (*Shigella, Salmonella, Campylobacter*, enteroinvasive and enterohemorrhagic *Escherichia coli*), diverticulitis, ischemic colitis, inflammatory bowel disease, and arteriovenous malformation.

Complications

Complications of invasive amebiasis include liver abscess, toxic megacolon, pneumatosis coli (intramural air), acute necrotizing colitis, perianal ulcerations with fistula formation, and peritonitis. Rarely, a reactive collection of granulation and fibrous tissue (i.e., ameboma) proliferates in the lumen, causing pain and obstruction.

Amebic liver abscess is the most common extraintestinal manifestation of infection with *Entamoeba histolytica*.[13] Symptoms of right upper quadrant pain, fever, and cough typically develop within 2–4 weeks after the development of the liver abscess. Many patients with amebic abscess lack a history of intestinal disease within the preceding year.[14] Jaundice is an uncommon complication of amebic abscess unless the liver abscess is very large or multiple abscesses exist. Right-sided shoulder and chest pain may develop when the abscess involves the diaphragm. The abscess may rupture into the pleural space. Most pleuropulmonary amebiasis occurs in patients with amebic liver abscesses that extend into the pleural space. A pulmonary amebic lesion without either a liver abscess or amebic colitis is extremely rare.[15]

DIAGNOSTIC TESTING

Analytical Methods

Methods for the diagnosis of amebiasis involve microscopic examination of fresh or frozen stool and serological tests including the detection of *E. histolytica*-specific antigens or DNA in stool samples or the presence of anti-amebic antibodies in the serum.[1] The traditional technique for identification of trophozoites and cysts involves the microscopic examination of stained (Lugol's, D'Antoni's iodine) or unstained stool samples, but this method is relatively nonspecific because of the difficulty distinguishing *E. histolytica* from other nonpathogenic *Entamoeba* species (e.g., *E. dispar* Brumpt, *E. moshkovskii* Tshalaia). The presence of Charcot-Leyden crystals during stool examination indicates the presence of eosinophils in the intestinal lumen. Occasionally, the presence of ingested erythrocytes in trophozoites helps distinguish *E. histolytica* from other *Entamoeba* species. Consequently, the diagnosis of *E. histolytica* requires confirmation with stool antigen or serum serology prior to initiation of treatment. The interpretation of serologic tests is complicated by the difficulty distinguishing current from past infections as a result of the persistence of positive serologic tests at least several years after infection. Serological tests are based on the presence of anti-galactose and *N*-acetyl-D-galactosamine-specific lectin IgG antibodies or the monoclonal anti-*E. histolytica* IgM antibodies. The former antibodies are detectable about 1 week after the onset of symptoms, whereas the latter usually develops during the first week of amebic colitis.[16] Antibody titers do not correlate well to the severity of the clinical manifestations of amebiasis or the response to therapy.[6] Recent DNA-based techniques include antigen detection methods and a single-step duplex polymerase chain reaction (PCR) procedure for rapid, specific, and sensitive identification of *Entamoeba histolytica*.[17] An Ethiopian study using PCR techniques suggests that some infections attributed to *E. histolytica* are mischaracterized because of the detection of non-pathogenic *E. dispar* using nonspecific techniques unable to distinguish between these two species.[18] The mischaracterization of *E. dispar* as *E. histolytica* results in unnecessary treatment with amoebicidal drugs, at least in some tropical and sub-Saharan countries.

Abnormalities

Colitis

During an acute infection with *E. histolytica*, the presence of Charcot-Leyden crystals, fecal erythrocytes, and the lack of fecal leukocytes are the most common laboratory findings. In a study of 289 Bangladeshi children (147 boys and 142 girls) between 2 and 5 years old followed for the development of diarrheal illness, visible blood or hemoccult-positive tests of stool samples occurred in 7% and 25%, respectively, of cases of *E. histolytica*-induced diarrhea.[19] Eosinophilia in the blood does not usually occur during amebic colitis.

Liver Abscess

Serum hepatic aminotransferase concentrations are usually normal or slightly elevated despite the presence of amebic liver abscess, but the serum alkaline phosphatase is usually normal. A mild to moderate leukocytosis along with anemia may occur in patients with amebic liver abscess. Ultrasonography and abdominal computed tomography are excellent imaging studies for the detection of liver abscesses, but the findings are nonspecific. A majority of patients with amebic liver abscess do not have detectable antigens in their stool.[1] Detection of serum amebic antigens and observation of trophozoites on liver biopsy/aspiration are the definitive diagnostic tests for amebic liver abscess.[20]

TREATMENT

Treatment of amebiasis includes the administration of both a luminal amebicide for cysts and a tissue amebicide for trophozoites. Common treatment of amebic liver abscess includes metronidazole (750 mg or 35–50 mg/kg every 8 hours for 7–10 days) or tinidazole (800 mg or 60 mg/kg 3 times daily for 5 days) and a luminal agent. Metronidazole is the drug of choice as a tissue amebicide. Administration of this antibiotic results in an 85–90% cure rate. Options for use of a luminal amebicide include iodoquinol [Yodoxin® (Glenwood, LLC, Englewood, NJ), 650 mg orally 3 times daily for 20 days], paromomycin (500 mg orally 3 times daily for 7 days), or diloxanide furoate (500 mg orally 3 times daily for 10 days). Dehydroemetine dihydrochloride is not currently used for amebiasis because of its cumulative skeletal and cardiac muscle toxicity.

Asymptomatic patients with confirmed *E. histolytica* infections should be treated with appropriate antimicrobial drugs. Treatment of *Entamoeba histolytica* infections depends on the presence of symptoms. For asymptomatic patients, appropriate antibiotic therapy includes paromomycin (25–35 mg/kg/day in 3 divided doses for 7 days) or diloxanide furoate (500 mg or 20 mg/kg orally 3 times daily for 10 days). Diarrhea is a common side effect of paromomycin; therefore, the treatment of symptomatic patients with paromomycin complicates the determination of whether persistence of diarrheal symptoms results from a continuing *E. histolytica* infection or from a side effect of paromomycin. The detection of *Entamoeba dispar* in stool samples requires no treatment.

CRYPTOSPORIDIUM PARVUM

HISTORY

Although Tyzzer identified *Cryptosporidium parvum* in the gastric glands of mice in 1905,[21] clinically significant infections in animals were not associated with this parasite until 1955.[22] By 1971, this organism was a recognized cause of chronic diarrhea and debility in calves and by 1976 *C. parvum* was a recognized cause of acute gastroenteritis in humans.[23] In the 1980s, cryptosporidiosis was associated with chronic gastroenteritis in HIV-infected patients. This parasite is now associated with disease in at least 79 animal species worldwide, particularly in calves.[24]

EXPOSURE

Cryptosporidium is prevalent both in developed and developing countries. Of at least 10 valid species of *Cryptosporidium, C. parvum* is the most common pathogenic species in humans and in animals.[36] Common sources of *C. parvum* infections include moist environments, such as contaminated drinking water and swimming pools.[25,26] In rural areas of England, fecal pollution of water sources by livestock feces is the leading cause of human sporadic cryptosporidiosis.[27] The oocysts of this organism survive in ice, and contaminated ice can transmit cryptosporidiosis.[28] Infectious *Cryptosporidium* oocysts occur in the stools of immunologically compromised persons as well as sporadically in healthy individuals, Russian travelers, animal handlers, and homosexual men without AIDS.[29] This disease is highly contagious; therefore, person-to-person transmission is common.[30] Animal-to-human transmission occurs rarely.[31] Food is an uncommon source of infection by *C. parvum*.[32] The oocysts can survive for extended periods. The largest outbreaks of *C. parvum* infections result from the ingestion of contaminated municipal water supplies.[33]

PATHOGEN

Physiochemical Properties

Cryptosporidium is an obligate, intracellular, protozoan parasite that replicates in the host and excretes oocysts into the stool. Transmission occurs via the fecal-oral route. Oocysts remain infectious up to several months in moist environments, and the prolonged survival of oocysts at 4 °C (39 °F) in experimental conditions suggest that these oocysts could survive for prolonged periods of time in groundwater of countries with cold climates.[34] These organisms are relatively resistant to chlorine and pool filtration systems. In an experimental study, exposure to 80 ppm chlorine or to 80 ppm monochloramine produced 90% inactivation of *C. parvum* oocysts after 90 minutes.[35] These oocysts were 14 times more resistant to these agents compared with *Giardia* cysts. Although oocysts of *C. parvum* can remain viable for many months under favorable conditions, freezing and desiccation kills the oocysts.[36] Microwaving at full power for at least 20 seconds completely inactivates *Cryptosporidium* oocysts.[37]

Mechanism of Toxicity

Cryptosporidium parvum invades the mucosa of the proximal small intestine causing abnormalities of crypts including villous atrophy and crypt hyperplasia. Oocysts are minimally invasive because this pathogen colonizes the surface epithelial cells rather than the deeper layers of the intestinal mucosa.[38] An inflammatory infiltrate develops in the lamina propria and epithelium. These changes cause a secretory diarrhea by reducing absorption (decreased surface area) and increasing intestinal secretions (impaired epithelial enzymes).[39]

DOSE RESPONSE

The infective dose of *Cryptosporidium parvum* is relatively small (i.e., 10–100 oocysts).[40] In an experimental study of 29 healthy adults, the median infective dose was 132 oocysts as calculated by linear regression.[41] All 7 subjects developed evidence of *Cryptosporidium* infection after ingesting 1,000 oocysts. Of the 18 subjects who excreted *Cryptosporidium* oocysts, 11 subjects had one enteric symptom and 7 subjects had clinical signs of cryptosporidiosis (i.e., diarrhea plus one enteric symptom). Based on a hierarchical predictive population dose-response Bayesian assessment of *C. parvum*, a dose of 6×10^{-6} oocysts per exposure corresponds to 10^{-4} infections per capita year.[42] This endpoint does not account for secondary transmission of the infection among hosts. Using Monte Carlo simulation with the field survey data from the Sagami River watershed in Japan, reduction of the 95% confidence interval for risk of *C. parvum* infection below 10^{-4} requires one oocyte per 80 liters of drinking water.[43]

CLINICAL RESPONSE

Infections by *Cryptosporidium parvum* cause watery diarrhea that is usually benign in healthy patients. However, a chronic debilitating gastroenteritis may develop in immunocompromised patients (e.g., AIDS).[44] In a study of 68 otherwise healthy patients with cryptosporidiosis, the mean incubation period was about 7 days (range: 1–12 days) and the duration of illness was approximately 12 days (range: 2–26 days).[45] During an outbreak of cryptosporidiosis from the consumption of contaminated apple cider, the median incubation period was about 6 days (range: 10 hours to 13 days) and the median duration of illness was about 6 days (range: 1–16 days).[32] In an outbreak of cryptosporidiosis involving

285 confirmed cases, the most common symptoms were watery diarrhea (93%), abdominal cramps (84%), and vomiting (48%).[28] The clinical course of cryptosporidiosis in immunocompromised patients depends on host resistance (e.g., CD4 count); thus, the course ranges from transient infection to asymptomatic infection, chronic diarrhea, and fulminant infections. Humans with persistent cryptosporidiosis may develop hepatobiliary or pancreatic duct infections.[40]

Watery diarrhea (four bowel movements per day), weight loss ranging from 1–5 kg, and anorexia characterize the illness. Fever, chills, and vomiting appear uncommonly. Acute pancreatitis has been associated with infections by the *Cryptosporidium* parasite.[46]

DIAGNOSTIC TESTING

Analytical Methods

During an acute illness, the laboratory diagnosis of cryptosporidiosis usually involves the detection of oocysts in stool with acid-fast staining or modification of this staining process.[47] Most clinical laboratories do not routinely screen for *C. parvum*, and the laboratory should be notified of the necessity to screen the stool specimen for *C. parvum*. Detection may require analysis of implicated water and food. In patients with chronic diarrhea, identification of *C. parvum* as well as other intestinal coccidian infections (e.g., *Cyclospora cayetanensis, Isospora belli*) requires the use of special methods using the Kinyoun modified acid-fast stain.[48] More sensitive and specific detection methods (e.g., immunofluorescence assays, enzyme-linked immunoassays, DNA-based techniques) have been developed as an alternative to the time-consuming method of stool analysis for the detection of common intestinal parasites, such as *Entamoeba histolytica, Giardia lamblia*, and *Cryptosporidium parvum*.[49] The DNA-based (e.g., polymerase chain reaction [PCR]) techniques are important for the detection of clinically inapparent infection involving the excretion of small numbers of oocysts.[50]

Abnormalities

Endoscopic examination of the proximal small intestine is usually normal.[39] Examination of mucosal biopsies by light microscopy detects forms of *C. parvum* at the epithelial cell surface.

TREATMENT

Treatment of *C. parvum* infection primarily is supportive (fluid/electrolyte replacement, loperamide). In healthy patients, this self-limited, enteric illness usually resolves without therapeutic intervention. There is no antimicrobial agent that effectively eradicates cryptosporidiosis. Macrolide antibiotics (azithromycin, clarithromycin) demonstrate variable effectiveness against *C. parvum*.[40] Nitazoxanide, a nitrothiazolyl-salicylamide derivative, is active against *C. parvum* in animal models, and a clinical trial suggests that this antimicrobial agent reduces diarrhea in most patients with cryptosporidiosis within 3–4 days.[51] Feces from patients with cryptosporidiosis should be considered infectious.

CYCLOSPORA CAYETANENSIS

HISTORY

Although Eimer first identified *Cyclospora* in the intestines of moles in 1870, recognition of *Cyclospora* oocysts in humans did not occur until the late 1970s.[52] During the 1980s, case reports of diarrheal illness documented *C. cayetanensis* as a human pathogen in travelers from tropical countries (Mexico, Haiti, South Africa, India), in residents of Nepal and Peru, and in AIDS patients.[53] In some of these outbreaks, infection with *C. cayetanensis* occurred after the ingestion of drinking water despite the chlorination and filtration of the water supply.[54] Before 1995, most cases of cyclosporiasis in the United States occurred in travelers. The transmission of *Cyclospora* is unusual compared with other enteric diseases because freshly excreted cysts require several days to weeks outside the host to sporulate and to become infective.

EXPOSURE

Cyclosporiasis is widespread throughout tropical and subtropical regions with endemic infections reported in Nepal, Guatemala, Peru, and Haiti.[55] Although cyclosporiasis has marked seasonality, seasons associated with increased infective rates vary from one endemic region to another.[52] Although cyclosporiasis usually occurs in immunoincompetent patients (e.g., HIV/AIDS patients), gastrointestinal infections occur occasionally in immunocompetent patients,[56] primarily in foreign travelers and children.[57,58] The major source of cyclosporiasis is fecally contaminated fruits (raspberries)[59] and vegetables (lettuce, basil, snow peas),[60] but occasionally waterborne infections of *C. cayetanensis* occur.[61] Analysis of samples of vegetables from foreign markets suggests that contamination by *Cyclospora* is substantially less than by *Cryptosporidium*. The contamination rate of *Cyclospora* and *Cryptosporidium* in market vegetables from an endemic area near Lima, Peru was <2% and 15–20%, respectively.[62] Simple washing of vegetables probably does not remove *Cyclospora*. Humans are the primary host for this species of *Cyclospora*, but other species (*C. talpae, C. caryolitica*) are present in the intestines of rodents and reptiles.[63] African primates are a reservoir for *C. cayetanensis*, but these primates probably are not a common source of human infections with this coccidian protozoan.[64] *Cyclospora* oocysts sporulate at temperatures of 25 °C–30 °C (77 °F–86 °F) in about 1–2 weeks, but temperatures <–20 °C (–4 °F) for 24 hours or 60 °C (140 °F) for 1 hour prevent sporulation.[65] Outbreaks of cyclosporiasis are not well documented in cooked or commercially frozen food.[52]

PATHOGEN

Physiochemical Properties

The genus Cyclospora is taxonomically similar to several other coccidian genera that cause human disease including Cryptosporidium, Isospora, Microspora, Nosema, Pleistophora, Microsporidium, Enterocytozon, Encephalitozoon, Trachypleistophora, Vittaforma, Sarcocystis, and Toxoplasma. The oocysts of *Cyclospora cayetanensis* are almost twice the size of *Cryptosporidium* oocysts. Extreme temperatures (e.g., pasteurization, commercial freezing) inactivate the former oocysts.[66]

Mechanism of Toxicity

Ingestion of sporulated oocysts of *C. cayetanensis* initiates an infection. In the small intestine, sporozoites invade the intestinal epithelial cells. This parasite is an intracellular human pathogen that causes inflammatory changes in the epithelium, mucosa, and lamina propria as well as villous atrophy and crypt hyperplasia. Sporulated oocysts are excreted in the stool; therefore, oral-fecal contact can transmit cyclosporiasis immediately.

DOSE RESPONSE

The attack rate for cyclosporiasis depends on the degree of immunity in the population as well as the number of oocysts. A pilot study of 7 healthy volunteers ingesting an inoculum of *Cyclospora* oocysts (approximately 200–49,000 oocysts) did not demonstrate gastrointestinal symptoms in the volunteers or oocysts in any stool samples during the 16-week observation period.[67] Person-to-person transmission is unlikely because the oocysts excreted in the stool are not sporulated.

CLINICAL RESPONSE

Characteristics of an infection by *Cyclospora* include a long incubation period and long duration of illness characterized by prolonged, relapsing diarrheal illness. The incubation period averages about 1 week with a range of 12 hours to 14 days, and disease patterns are typically seasonal depending on the locality.[52] The attack rates of cyclosporiasis are higher in adults than in children, and this disease is uncommon in children under 3 years of age.[53] Asymptomatic infections are common, particularly when *C. cayetanensis* is endemic. In healthy adults, cyclosporiasis is usually self-limited. Infection by *C. cayetanensis* usually produces prolonged, relapsing watery diarrhea, low-grade fever, nausea, and vomiting.[68] In a study of 180 Florida residents with laboratory-confirmed *Cyclospora* infections, the most common symptoms beside diarrhea were loss of appetite (92%), fatigue (91%), weight loss (90%), abdominal pain or cramps (70%), nausea (70%), fever (43%), and vomiting (30%).[69] Occasionally, a "flu-like" illness characterized by myalgias and arthralgias precedes the diarrhea. *Cyclospora* causes an opportunistic infection in HIV-seropositive patients associated with protracted, relapsing diarrhea and weight loss.[70] Complications of infections by *C. cayetanensis* include malabsorption syndrome, Guillain-Barré syndrome, acalculous cholecystitis in AIDS patients, and Reiter syndrome (ocular inflammation, sterile urethritis, and inflammatory arthritis).[71] Fatalities are rare.

DIAGNOSTIC TESTING

Analytical Methods

The laboratory diagnosis of cyclosporiasis depends primarily on the identification of oocysts in stool specimens, primarily by wet-mount phase-contrast microscopy of the 8- to 9-mm oocysts. Routine analysis of stool specimens for ova and parasites does not typically include the special techniques required to identify *Cyclospora cayetanensis* except the use of fluorescence microscopy to detect the natural autofluorescence of *Cyclospora* oocysts.[72] Consequently, the laboratory should be notified of the necessity to screen the stool specimen for *Cyclospora*. Identification of oocysts in the stool with wet preparations or modified acid-fast stain confirms the diagnosis of cyclosporiasis. Staining of *C. cayetanensis* with acid-fast techniques produces marked variability in color (light pink to deep red). The safranin procedure produces a uniform red to red-orange color when modified by heat during the staining. Other methods of detecting *Cyclospora* infections include flow cytometry and quantitative real-time polymerase chain reaction (PCR) analysis and sporulation of oocysts in 2.5% potassium dichromate solution.[73,74] Detection may require analysis of implicated water and food. Methods of recovering *Cyclospora* oocysts from water samples are similar to the methods used to detect *Cryptosporidium* oocysts and *Giardia* cysts.

Abnormalities

Endoscopic examination of infected patients usually demonstrates mild erythema and friability of the mucosa in the small intestine.[75] Ulcers and hemorrhagic areas are not usually present.

TREATMENT

In addition to supportive care, chronic diarrhea associated with cyclosporiasis usually responds to a one-week course of trimethoprim-sulfamethoxazole (160 mg/800 mg or one double-strength tablet) twice daily, even in patients with HIV infection. Although ciprofloxacin (500 mg b.i.d.) is not as effective as trimethoprim-sulfamethoxazole, ciprofloxacin is an alternative antibiotic when patients cannot tolerate trimethoprim-sulfamethoxazole.[76]

GIARDIA LAMBLIA

HISTORY

Van Leeuwenhoek discovered *Giardia* in 1681 during observation of his own stool.[77] This flagellated, binucleated protozoan is a parasite of mammals, birds, reptiles, and rodents. *Giardia* is an ancient eukaryote that does not possess mitochondria. Vilem Lambl published the first line drawings of both the trophozoite and the cyst forms of *G. lamblia* in 1859.

EXPOSURE

Giardia lamblia Kofoid and Christiansen is one of the most common parasitic infections worldwide. The spread of giardiasis occurs primarily by oral-fecal contact with the cysts becoming immediately infective when passed in the stool.[78] Giardiasis occurs in every region of the United States, particularly in mountain areas.[79] Waterborne sources are the usual method of transmission because *G. lamblia* is relatively resistant to chlorination and survives easily in cold mountain streams. Person-to-person transmission of *G. lamblia* can occur in day care centers and institutions.

Source

Animal

Giardiasis is a zoonosis that allows cross-infectivity between animal reservoirs (rodents, dogs, cattle, beaver, big horn sheep). However, most human infections do not result from exposure to *G. lamblia* in animal reservoirs.[78] Oral-fecal transmission occurs between humans, particularly in the setting of poor sanitary conditions. Individuals at risk of developing giardiasis include travelers, backpackers (e.g., Rocky Mountains, Sierra Nevada, California coastal ranges), institutionalized individuals, diaper-aged children in day care, immunocompromised patients (HIV), homosexuals, and the elderly with a history of gastric surgery or with hypoalbuminemia.[80]

Food

Foodborne transmission of *Giardia* is rare, but food may be a source of giardiasis when fecal-contaminated food is inadequately cooked. Flies are a potential source of contaminated food as a result of acquisition of *Giardia* cysts from unhygienic sources.[81]

PATHOGEN

Physiochemical Properties

Cysts remain viable for up to 3 months in moist environments including streams and lakes. Ozone and ultraviolet light potentially deactivate *Giardia* cysts.[82]

Mechanism of Toxicity

Trophozoites develop in the stomach when the acid environment of stomach causes excystation of the ingested cysts. The trophozoites migrate to the duodenum and proximal jejunum where they attach to the epithelium and overlying mucus layers via a ventral adhesive disk. This parasite does not usually invade the mucosa and *G. lamblia* causes little or no mucosal inflammation of the small bowel.[83] Trophozoites encyst in the jejunum as a result of exposure to biliary secretions, and the cysts pass into the stool for dissemination.

DOSE RESPONSE

Ingestion of 10–25 cysts may produce giardiasis depending on the susceptibility of the patient.

CLINICAL RESPONSE

The incubation of giardiasis is about 1–2 weeks with a range up to about 1 month. In comparison with bacterial gastroenteritides, the onset of diarrhea is usually gradual and relatively mild, and the onset of illness occurs frequently with a feeling of intestinal uneasiness followed by nausea and anorexia.[84] The clinical manifestations of *Giardia* infections range from none to a chronic malabsorption-like syndrome. Transient lactase deficiencies may develop in symptomatic patients that persist a few weeks after treatment.[77] Although many infections with *Giardia lamblia* are asymptomatic, symptoms and signs associated with giardiasis include nausea, diarrhea, epigastric pain, steatorrhea, flatulence, anorexia, and weight loss.[85] Vomiting is unusual. Gastrointestinal symptoms resolve gradually over 2–4 weeks as a result of host defense mechanisms including the production of secretory IgA antibodies. Chronic *G. lamblia* infection may develop after acute symptoms resolve. The clinical features of subacute giardiasis resemble abdominal malabsorption syndrome with malaise, fatigue, bloating, crampy abdominal pain, excessive flatus, steatorrhea, weight loss, loose stool, abdominal distension, and foul-smelling stools. The persistence of diarrhea beyond 2 weeks is consistent with giardiasis. Rarely, a reactive arthritis of the lower extremities, rash, and urticaria may complicate giardiasis.[86]

DIAGNOSTIC TESTING

Analytical Methods

Giardia lamblia is a pear-shaped, binucleated, flagellated protozoan. The diagnosis of giardiasis traditionally depended on the direct detection of trophozoites or cysts in stool samples. Trophozoites occur primarily in diarrheal stools whereas cysts appear in both formed and loose stools of infected cases. Examination of three stool specimens in infected patients increases the sensitivity of stool analysis for *G. lamblia* to 85–90%.[87] More sensitive, rapid, and specific detection methods (e.g., immunofluorescence assays, enzyme-linked immunoassays, DNA-based techniques) have been developed in recent years as an alternative to the time-consuming method of stool analysis for the detection of common intestinal parasites, such as *Entamoeba histolytica, Giardia lamblia*, and *Cryptosporidium parvum*.[49,88,89] Duodenal biopsy is highly sensitive option when the clinical presentation is highly suspicious of giardiasis and stool examinations are negative, but invasive procedures are not usually necessary for diagnosis. Duodenal aspiration is a less sensitive method for the detection of giardiasis than duodenal biopsy.[90]

Abnormalities

Eosinophilia, leukocytosis, and fecal leukocytes are not associated with giardiasis because the trophozoites do not invade the intestinal mucosa.

TREATMENT

Since the discontinuation of quinacrine production in 1992, metronidazole (250 mg every 8 hours for 5–7 days) has been the drug of choice for the treatment of giardiasis. Paromomycin (500 mg 3 times a day or 5–10 days) is an alternative for pregnant women afflicted with giardiasis, particularly during the first trimester. Tinidazole has a high cure rate (90–98%) in an easy to use dosage (2 g in adults), but this drug is not available in the United States. Furazolidone is an effective alternative to metronidazole, but this drug requires administration 4 times a day for 7–10 days.[91] Nitazoxanide (Romark Laboratories, Tampa, FL) has been approved by the US Food and Drug Administration (FDA) for the treatment of diarrhea caused by *Giardia* in children aged 1–11 years. The treatment of asymptomatic patients with giardiasis is controversial because of the high reinfection rate.

References

1. Haque R, Huston CD, Hughes M, Houpt E, Petri WA Jr. Amebiasis. N Engl J Med 2003;348:1565–1573.
2. Stilwell GG. Amebiasis: its early history. Gastroenterol 1955;28:606–622.
3. Rogers L. The rapid cure of amebic dysentery and hepatitis by hypodermic injection of soluble salts of emetine. Br Med J 1912;1:1424.
4. Ashbolt NJ. Microbial contamination of drinking water and disease outcomes in developing regions. Toxicology 2004;198:229–238.
5. Jackson TF, Sargeaunt PG, Visser PS, Gathiram V, Suparsad S, Anderson CB. *Entamoeba histolytica*: naturally occurring infections in baboons. Arch Invest Med (Mex) 1990;21(Suppl 1):153–156.
6. Tanyuksel M, Petri WA Jr. Laboratory diagnosis of amebiasis. Clin Microbiol Rev 2003;16:713–729.
7. Ohnishi K, Kato Y, Imamura A, Fukayama M, Tsunoda T, Sakaue Y, Sakamoto M, Sagara H: Present characteristics of symptomatic *Entamoeba histolytica* infection in the big cities of Japan. Epidemiol Infect 2004;132:57–60.

8. Huston CD. Parasite and host contributions to the pathogenesis of amebic colitis. Trends Parasitol 2004;20:23–26.
9. Blessmann J, Ali IK, Nu PA, Dinh BT, Viet TQ, Van AL, Clark CG, Tannich E. Longitudinal study of intestinal *Entamoeba histolytica* infections in asymptomatic adult carriers. J Clin Microbiol 2003;41:4745–4550.
10. Zhang Z, Stanley SL Jr. Stereotypic and specific elements of the human colonic response to *Entamoeba histolytica* and *Shigella flexneri*. Cell Microbiol 2004;6:535–554.
11. Cutting WA, Hawkins P. The role of water in relation to diarrhoeal disease. J Trop Med Hyg 1982;85:31–39.
12. Stauffer W, Ravdin JL. *Entamoeba histolytica*: an update. Cur Opin Inf Dis 2003;16:479–485.
13. Wells CD, Arguedas M. Amebic liver abscess. South Med J 2004;97:673–682.
14. Adams EB, MacLeod IN. Invasive amebiasis. I. Amebic dysentery and its complications. Medicine (Baltimore) 1977;56:315–323.
15. Hara A, Hirose Y, Mori H, Iwao H, Kato T, Kusuhara Y. Cytopathologic and genetic diagnosis of pulmonary amebiasis: a case report. Acta Cytol 2004;48:547–550.
16. Abd-Alla MD, Jackson TG, Ravdin JI. Serum IgM antibody response to the galactose-inhibitable adherence lectin of *Entamoeba histolytica*. Am J Trop Med Hyg 1998;59:431–434.
17. Freitas MA, Vianna EN, Martins AS, Silva EF, Pesquero JL, Gomes MA. A single step duplex PCR to distinguish *Entamoeba histolytica* from *Entamoeba dispar*. Parasitology 2004;128:625–628.
18. Kebede A, Verweij JJ, Endeshaw T, Messele T, Tasew G, Petros B, Polderman AM. The use of real-time PCR to identify *Entamoeba histolytica* and *E. dispar* infections in prisoners and primary-school children in Ethiopia. Ann Trop Med Parasitol 2004;98:43–48.
19. Haque R, Mondal D, Kirkpatrick BD, Akther S, Farr BM, Sack RB, Petri WA Jr. Epidemiologic and clinical characteristics of acute diarrhea with emphasis on *Entamoeba histolytica* infections in preschool children in an urban slum of Dhaka, Bangladesh. Am J Trop Med Hyg 2003;69:398–405.
20. Salles JM, Moraes LA, Salles MC. Hepatic amebiasis. Braz J Infect Dis 2003;7:96–110.
21. Tyzzer EE. A sporozoon found in the peptic glands of the common mouse. Proc Soc Exp Biol Med 1907;5;12–13.
22. Tzipori S. Cryptosporidiosis in animals and humans. Microbiol Rev 1983;47:84–96.
23. Nime FA, Burek JS, Page DL, Holscher MA, Yardley JH. Acute enterocolitis in a human being infected with the protozoan *Cryptosporidium*. Gastroenterology 1976; 70:592–598.
24. Mosier DA, Oberst RD. Cryptosporidiosis a global challenge. Ann NY Acad Sci 2000;916:102–111.
25. Puech MC, McAnulty JM, Lesjak M, Shaw N, Heron L, Watson JM. A statewide outbreak of cryptosporidiosis in New South Wales associated with swimming at public pools. Epidemiol Infect 2001;126:389–396.
26. McAnulty JM, Fleming SW, Gonzalez AH. A community-wide outbreak of cryptosporidiosis associated with swimming at a wave pool. JAMA 1994;272:1597–1600.
27. Goh S, Reacher M, Casemore DP, Verlander NQ, Chalmers R, Knowles M, Williams J, Osborn K, Richards S. Sporadic cryptosporidiosis, North Cumbria, England, 1996–2000. Emerg Infect Dis 2004;10:1007–1015.
28. MacKenzie WR, Hoxie NJ, Proctor ME, Gradus MS, Blair KA, Peterson DE, et al. A massive outbreak in Milwaukee of cryptosporidium infection transmitted through the public water supply. N Engl J Med 1994;331:161–167.
29. Soave R, Ma P. Cryptosporidiosis: travelers' diarrhea in two families. Arch Intern Med 1985;145:70–72.
30. Cordell RL, Addiss DG. Cryptosporidiosis in child care settings: a review of the literature and recommendations for prevention and control. Pediatr Infect Dis J 1994;13:310–317.
31. Miron D, Kenes J, Dagan R. Calves as a source of an outbreak of cryptosporidiosis among young children in an agricultural closed community. Pediatr Infect Dis J 1991;10:438–441.
32. Millard PS, Gensheimer KF, Addiss DG, Sosin DM, Beckett GA, Houck-Jankoski A, Hudson A. An outbreak of cryptosporidiosis from fresh-pressed apple cider. JAMA 1994;272:1592–1596.
33. Mac Kenzie WR, Hoxie NJ, Proctor ME, Gradus MS, Blair KA, Peterson DE, et al. A massive outbreak in Milwaukee of cryptosporidium infection transmitted through the public water supply. N Engl J Med 1994;331:161–167.
34. Nichols RA, Paton CA, Smith HV. Survival of *Cryptosporidium parvum* oocysts after prolonged exposure to still natural mineral waters. J Food Prot 2004;67:517–523.
35. Korich DG, Mead JR, Madore MS, Sinclair NA, Sterling CR. Effects of ozone, chlorine dioxide, chlorine, and monochloramine on *Cryptosporidium parvum* oocyst viability. Appl Environ Microbiol 1990;56:1423–1428.
36. Fayer R, Morgan U, Upton SJ. Epidemiology of *Cryptosporidium*: transmission, detection and identification. Int J Parasitol 2000;30:1305–1322.
37. Ortega YR, Liao J. Microwave inactivation of *Cyclospora cayetanensis* sporulation and viability of *Cryptosporidium parvum* oocysts. J Food Prot 2006;69:1957–1960.
38. Laurent F, McCole D, Eckmann L, Kagnoff MF. Pathogenesis of *Cryptosporidium parvum* infection. Microbes Infect 1999;2:141–148.
39. Farthing MJG. Clinical aspects of human cryptosporidiosis. Contrib Microbiol 2000;6:50–74.
40. Griffiths JK. Human cryptosporidiosis: epidemiology, transmission, clinical disease, treatment, and diagnosis. Adv Parasitol 1998;40:37–85.
41. DuPont HL, Chappell CL, Sterling CR, Okhuysen PC, Rose JB, Jakubowski W. The infectivity of *Cryptosporidium parvum* in healthy volunteers. N Engl J Med 1995;332:855–859.
42. Englehardt JD, Swartout J. Predictive population dose-response assessment for *Cryptosporidium parvum*: infec-

tion endpoint. J Toxicol Environ Health A 2004;67:651–666.

43. Masago Y, Katayama H, Hashimoto A, Hirata T, Ohgaki S. Assessment of risk of infection due to *Cryptosporidium parvum* in drinking water. Water Sci Technol 2002;46:319–324.
44. Petersen C. Cryptosporidiosis in patients infected with the human immunodeficiency virus. Clin Infect Dis 1992;15:903–909.
45. Jokipii L, Jokipii AM. Timing of symptoms and oocyst excretion in human cryptosporidiosis. N Engl J Med 1986;315:1643–1647.
46. Hawkins SP, Thomas RP, Teasdale C. Acute pancreatitis: A new finding in *Cryptosporidium* enteritis. Br Med J 1987;294:483–484.
47. Petry F. Laboratory diagnosis of *cryptosporidium parvum* infection. Contrib Microbiol 2000;6:33–49.
48. Ribes JA, Seabolt JP, Overman SB. Point prevalence of *Cryptosporidium, Cyclospora*, and *Isospora* infections in patients being evaluated for diarrhea. Am J Clin Pathol 2004;122:28–32.
49. Verweij JJ, Blange RA, Templeton K, Schinkel J, Brienen EA, van Rooyen MA, et al. Simultaneous detection of *Entamoeba histolytica, Giardia lamblia*, and *Cryptosporidium parvum* in fecal samples by using multiplex real-time PCR. J Clin Microbiol 2004;42:1220–1223.
50. Gomez-Couso H, Freire-Santos F, Amar CF, Grant KA, Williamson K, Ares-Mazas ME, McLauchlin J. Detection of *Cryptosporidium* and *Giardia* in molluscan shellfish by multiplexed nested-PCR. Int J Food Microbiol 2004;91:279–288.
51. Rossignol J-FA, Ayers MS. Treatment of diarrhea caused by *Cryptosporidium parvum*: A prospective randomized, double-blind, placebo-controlled study of nitazoxanide. J Infect Dis 2001;184:103–106.
52. Herwaldt BL. *Cyclospora cayetanensis*: a review, focusing on the outbreaks of cyclosporiasis in the 1990s. Clin Infect Dis 2000;31:1040–1057.
53. Mota P, Rauch CA, Edberg SC. Microsporidia and Cyclospora: epidemiology and assessment of risk from the environment. Crit Rev Microbiol 2000;26:69–90.
54. Rabold JG, Hoge CW, Shlim DR, Kefford C, Rajah R, Echeverria P. Cyclospora outbreak associated with chlorinated drinking water. Lancet 1994;344(8933):1360–1361.
55. Lopez AS, Bendik JM, Alliance JY, Roberts JM, da Silva AJ, Moura IN, Arrowood MJ, Eberhard ML, Herwaldt BL. Epidemiology of *Cyclospora cayetanensis* and other intestinal parasites in a community in Haiti. J Clin Microbiol 2003;41:2047–2054.
56. Turk M, Turker M, Ak M, Karaayak B, Kaya T. Cyclosporiasis associated with diarrhoea in an immunocompetent patient in Turkey. J Med Microbiol 2004;53:255–257.
57. Yu JR, Sohn WM. A case of human cyclosporiasis causing traveler's diarrhea after visiting Indonesia. J Korean Med Sci 2003;18:738–741.
58. Kansouzidou A, Charitidou C, Varnis T, Vavatsi N, Kamaria F. *Cyclospora cayetanensis* in a patient with travelers' diarrhea: case report and review. J Travel Med 2004;11:61–63.
59. Ho AY, Lopez AS, Eberhart MG, Levenson R, Finkel BS, Da Silva AJ, et al. Outbreak of cyclosporiasis associated with imported raspberries, Philadelphia, Pennsylvania, 2000. Emerg Infect Dis 2002;8:783–788.
60. Ness W, German D, Leyland S, Klein R, Bern C, Bartlett M. Outbreak of cyclosporiasis associated with snow peas–Pennsylvania, 2004. MMWR Morb Mortal Wkly Rep 2004;53:876–878.
61. Lopez AS, Dodson DR, Arrowood JM, Orlandi PA Jr, da Silva AJ, Bier JW, et al. Outbreak of cyclosporiasis associated with basil in Missouri in 1999. Clin Infect Dis 2001;32:1010–1017.
62. Ortega YR, Roxas CR, Gilman RH, Miller NJ, Cabrera L, Taquiri C, Sterling CR. Isolation of *Cryptosporidium parvum* and *Cyclospora cayetanensis* from vegetables collected in markers of an endemic region in Peru. Am J Trop Med Hyg 1997;57:683–686.
63. Taylor AP, Davis LJ, Soave R. Cyclospora. Curr Clin Top Infect Dis 1997;17:256–268.
64. Eberhard ML, Njenga MN, DaSilva AJ, Owino D, Nace EK, Won KY, Mwenda JM. A survey for *Cyclospora* spp. in Kenyan primates, with some notes on its biology. J Parasitol 2001;87:1394–1397.
65. Smith HV, Paton CA, Mitambo MM, Girdwood RW. Sporulation of *Cyclospora* sp. oocysts. Appl Environ Microbiol 1997;63:1631–1632.
66. Centers for Disease Control (CDC). Update: outbreaks of cyclosporiasis–United States and Canada, 1997. MMWR Morb Mortal Wkly Rep 1997;46:521–523.
67. Alfano-Sobsey EM, Eberhard ML, Seed JR, Weber DJ, Won KY, Nace EK, Moe CL. Human challenge pilot study with *Cyclospora cayetanensis*. Emerg Infect Dis 2004;10:726–728.
68. Kansouzidou A, Charitidou C, Varnis T, Vavatsi N, Kamaria F. *Cyclospora cayetanensis* in a patient with travelers' diarrhea: case report and review. J Travel Med 2004; 11:61–63.
69. Katz D, Kumar S, Malecki J, Lowdermilk M, Koumans EHA, Hopkins R. Cyclosporiasis associated with imported raspberries, Florida, 1996. Pub Health Rep 1999;114: 427–438.
70. Pape JW, Verdier RI, Boncy M, Boncy J, Johnson WD Jr. *Cyclospora* infection in adults infected with HIV. Clinical manifestations, treatment, and prophylaxis. Ann Intern Med 1994;121:654–657.
71. Connor BA, Johnson E, Soave R. Reiter syndrome following protracted symptoms of *Cyclospora* infection. Emerging Infect Dis 2001;7:453–454.
72. Eberhard ML, Pieniazek NJ, Arrowood MJ. Laboratory diagnosis of *Cyclospora* infections. Arch Pathol Lab Med 1997;121:792–797.

73. Hussein EM, El-Moamly AA, Dawoud HA, Fahmy H, El-Shal HE, Sabek NA. Real-time PCR and flow cytometry in detection of *Cyclospora* oocysts in fecal samples of symptomatic and asymptomatic pediatrics patients. J Egypt Soc Parasitol 2007;37:151–170.

74. Varma M, Hester JD, Schaefer FW 3rd, Ware MW, Lindquist HD. Detection of *Cyclospora cayetanensis* using a quantitative real-time PCR assay. J Microbiol Methods 2003;53:27–36.

75. Ortega YR, Nagle R, Gilman RH, Watanabe J, Miyagui J, Quispe H, et al. Pathologic and clinical findings in patients with cyclosporiasis and a description of intracellular parasite life-cycle stages. J Infect Dis 1997;176:1584–1589.

76. Verdier R-I, Fitzgerald DW, Johnson WD Jr, Pape JW. Trimethoprim-sulfamethoxazole compared with ciprofloxacin for treatment and prophylaxis of *Isospora belli* and *Cyclospora cayetanensis* infection in HIV-infected patients. A randomized controlled trial. Ann Intern Med 2000;132:882–888.

77. Lebwohl B, Deckelbaum RJ, Green PH. Giardiasis. Gastrointest Endosco 2003;57:906–913.

78. Ali SA, Hill DR. Giardia intestinalis. Curr Opin Infect Dis 2003;16:453–460.

79. Lee SH, Levy DA, Craun GF, Beach MJ, Calderon RL. Surveillance for waterborne-disease outbreaks–United States, 1999–2000. MMWR Surveill Summ 2002;51: 1–47.

80. Beaumont DM, James OF. Unsuspected giardiasis as a cause of malnutrition and diarrhoea in the elderly. Br Med J 1986;293:554–555.

81. Graczyk TK, Grimes BH, Knight R, Da Silva AJ, Pieniazek NJ, Veal DA. Detection of *Cryptosporidium parvum* and *Giardia lamblia* carried by synanthropic flies by combined fluorescent in situ hybridization and a monoclonal antibody. Am J Trop Med Hyg 2003;68:228–232.

82. Hayes SL, Rice EW, Ware MW, Schaefer FW 3rd. Low pressure ultraviolet studies for inactivation of *Giardia muris* cysts. J Appl Microbiol 2003;94:54–59.

83. Eckmann L. Mucosal defences against *Giardia*. Parasite Immunol 2003;25:259–270.

84. Mineno T, Avery A. Giardiasis: recent progress in chemotherapy and drug development. Curr Pharm Design 2003;9:841–855.

85. Handousa AE, El Shazly AM, Rizk H, Soliman M, Saker T, El-Alfy NM. The histopathology of human giardiasis. J Egyptian Soc Parasitol 2003;33:875–886.

86. Tupchong M, Simor A, Dewar C. Beaver fever–a rare cause of reactive arthritis. J Rheumatol 1999;26: 2701–2702.

87. Furness BW, Beach MJ, Roberts JM. Giardiasis surveillance–United States, 1992–1997. MMWR CDC Surveill Summ 2000;49:1–13.

88. Guy RA, Xiao C, Horgen PA. Real-time PCR assay for detection and genotype differentiation of *Giardia lamblia* in stool specimens. J Clin Microbiol 2004;42:3317–3320.

89. Garcia LS, Shimizu RY, Novak S, Carroll M, Chan F. Commercial assay for detection of *Giardia lamblia* and *Cryptosporidium parvum* antigens in human fecal specimens by rapid solid-phase qualitative immunochromatography. J Clin Microbiol 2003;41:209–212.

90. Oberhuber G, Kastner N, Stolte M. Giardiasis: a histologic analysis of 567 cases. Scand J Gastroenterol 1997;32: 48–51.

91. Gardner TB, Hill DR. Treatment of giardiasis. Clin Microbiol Rev 2001;14:114–128.

Chapter 27

GASTROINTESTINAL VIRUSES

HISTORY

Because of the difficulty identifying viruses, these organisms were not suspected causes of gastroenteritis until the 1940s.[1] In 1972, Kapikian et al isolated the first gastrointestinal virus (Norwalk virus) from a stool sample during an outbreak of infectious, nonbacterial diarrhea.[2] Soon afterward, several other viruses were identified in the stools of children with infectious diarrhea including rotavirus,[3] astrovirus,[4] and enteric adenovirus.[5] Subsequently, other viruses associated with infectious gastroenteritis included picobirnavirus, coronavirus, pestivirus, and torovirus. Rotavirus was identified as the leading cause of acute gastroenteritis in young children, and this illness is a major cause of morbidity and mortality in children under 3 years of age worldwide.[6] In 1998, a rhesus-based tetravalent rotavirus vaccine (Rotashield®, Wyeth Pharmaceuticals, Madison, NJ) was approved in the United States for the routine vaccination of infants with three doses at ages 2, 4, and 6 months. However, this vaccine was withdrawn from the US market within one year because of the association of the vaccine with an increased incidence of intussusception, particularly during the first 3–14 days after the first dose.[7] Subsequently, a monovalent vaccine based on an attenuated human rotavirus strain (P1A[8]G1) was developed by the Belgian firm, GSK Biologicals (in US, GlaxoSmithKline, Philadelphia, PA), and this vaccine is licensed in at least 30 countries for the prevention of severe rotavirus gastroenteritis.[8] In 2006, a live, oral, human-bovine reassortant rotavirus vaccine (Rotateq®, Merck & Co., Whitehouse Station, NJ) was approved for use in the United States. This pentavalent vaccine currently is recommended for initiation in infants between 6 and 12 weeks of age.[9] Preclinical trials on this pentavalent vaccine did not detect an increased incidence of intussusception in infants <8 months of age when compared with infants not receiving the vaccine.[10]

EXPOSURE

Taxonomy and Classification

Rotaviruses (Reoviridae)

Rotaviruses have a wheel-like shape characterized by a non-enveloped, 70-nm icosahedral structure. The double-layered capsid is composed of an outer capsid (VP7 and VP4 proteins) and an inner capsid (primarily V6) surrounding the genome-containing core (VP1-VP3).[11] The genome contains 11 segments of double-stranded RNA that encode these six structural VP proteins [VP1, VP2, VP3, VP4 (VP5 + VP8), VP6, VP7] and six nonstructural proteins (NSP1-NSP6).[12] Figure 27.1 displays the structure of a rotavirus. The virulence of rotaviruses is likely related to NSP4, which is probably an enterotoxin that causes diarrhea in experimental animals.[13] Rotaviruses are classified into at least seven groups (A–G) and two subgroups (I, II) based on the antigenic properties of the VP6 protein in the inner capsid. Human gastroenteritis is associated with groups A, B, and C.[14] Serotyping is based on the antigenic differences in the two following proteins in the outer capsid: VP7 glycoprotein (G-type) and VP4 protease

Medical Toxicology of Natural Substances, by Donald G. Barceloux, MD

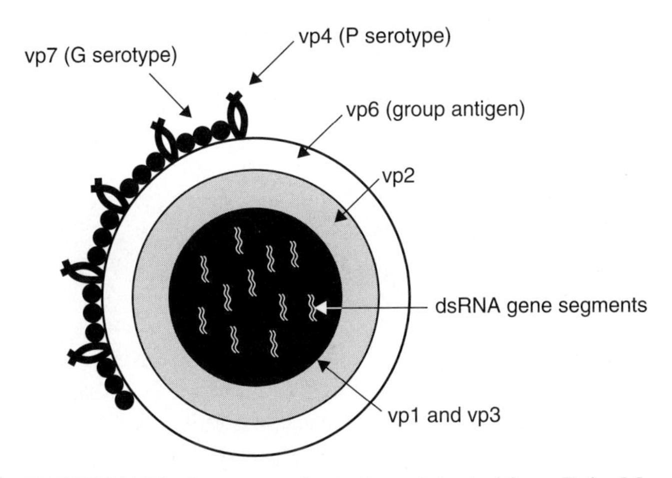

FIGURE 27.1. Structure of rotavirus. Adapted from Zahn M, et al. Clinical and epidemiological aspects of rotavirus infection. Pediatr Ann 2006;35:24.

(P-type). Because the two gene segments for VP7 and VP4 segregate independently, the typing system involves both G and P types. The characterization of P types by traditional methods is difficult; therefore, molecular methods are used to define genotype based on sequence analysis. These genotypes correlate well to serotypes, and these genotypes are placed in brackets. The most prevalent serotypes of rotavirus in the United States include P1A[8]G1, P1B[4]G2, P1A[8]G3, P1A[8]G4, P1A[8]G9, and P2A[6]G9.[15]

Enteric Adenoviruses (Adenoviridae)

Most enteric adenoviruses associated with gastroenteritis are 70-nm, nonenveloped viruses with an icosahedral structure. There are at least eight proteins in the core that maintain the integrity of the genome, which consists of a linear molecule of double-stranded DNA.[16] The most common serotypes involved with gastroenteritis belong to subgenus F (serotype 40, 41) and, rarely, to subgenus A (serotypes 12,18,32) and subgenus C (serotypes 1,2,5,6).[14]

Astroviruses (Astroviridae)

Under electron microscopy and physiological pH, these viruses have an icosahedral structure with obvious spikes and a diameter of 41 nm. In alkaline pH, astroviruses are small (28 nm) round organisms with the appearance of a five- or six-pointed star. The genome consists of single-stranded RNA with three open frames (ORFs). These ORFs encode a viral protease (ORF1a), viral polymerase (ORF1b), and a protein capsid precursor (ORF2). Serotyping is based on the immunoreactivity of capsid proteins (VP26) or the nucleotide sequence of a 348-bp region of the ORF2. There are at least eight serotypes (HAstV 1–8) with serotype 1 being the most common serotype associated with gastroenteritis.[17]

Caliciviruses (Caliciviridae)

The two genera of caliciviruses associated with gastroenteritis are *Norovirus* (Bristol virus, Desert Storm virus, Hawaii virus, Jena virus, Norwalk virus, Snow Mountain agent, Southampton virus, Toronto virus) and *Sapovirus* (London virus, Parkville virus, Sapporo virus). These viruses contain a single structural capsid protein with icosahedral symmetry that forms a continuous shell with arch-like protrusions. There are 32 cup-shaped depressions on the axes of the icosahedron that are characteristic of these viruses. The genome contains single-stranded RNA organized into three ORFs, which encode RNA-dependent RNA polymerase and helicase (ORF1) and a structural protein for the capsid (ORF2).[14] There are at least five norovirus genogroups (GI-GV), which contain at least 31 genetic clusters. GI, GII, and GIV strains infect humans, whereas GIII and GV strains infect cattle and mice, respectively.[18]

Torovirus and Coronavirus (Coronaviridae)

Toroviruses are enveloped, 100–140 nm viruses with a helicoidal symmetry to the capsid. The genome contains single-stranded RNA. Coronavirus are relatively large (60–220 nm) RNA viruses with helicoidal symmetry and a spiculated envelope. The role of coronaviruses in viral gastroenteritis is not well defined.[19]

Picobirnavirus

These small (30–40 nm), nonenveloped RNA viruses have a capsid with icosahedral symmetry. The genome is bi- or tri—segmented, double-stranded RNA.[20]

Source

The two classic types of viral gastroenteritis are epidemic and sporadic with the most common agents for these two types of gastroenteritides being 27-nm noroviruses and 70-nm group A rotaviruses, respectively.[21] Noroviruses are highly contagious organisms that cause epidemics of gastrointestinal illness in older children and adults, primarily genogroups GI and GII.[22] Rotavirus is the major cause of sporadic, viral gastroenteritis in young children primarily during the winter months, and there are five major serotypes (G1-G4, G9) with other serotypes (G12) occasionally involved.[23] The actual serotype involved with rotavirus-induced gastroenteritis varies with season and geography.[24] Group A rotaviruses sometimes cause epidemics of adult gastroenteritis in nursing homes, schools, and other institutions.[25] Other viral sources of sporadic gastroenteritis include enteric adenoviruses type 40 and 41, astrovirus, and sapovirus.[26] Adenovirus type 40 is prevalent year-round, whereas adenovirus 41 infections occur primarily in late autumn.[27]

Environment

The major route of transmission of the noroviruses is the fecal-oral route, primarily by direct person-to-person contact or the consumption of fecally contaminated food or water. Occasionally, both person-to-person and foodborne transmission of multiple strains of norovirus may cause an outbreak of gastroenteritis.[28] Although aerosolization of vomitus contaminated by noroviruses can cause infection, there is little evidence that transmission of noroviruses occurs by inhalation. Fomite contamination is also a source of norovirus infection. Viral shedding usually appears with the onset of symptoms and continues up to at least 5 days after illness begins. Occasionally, noroviruses can be recovered in stool samples collected 2 weeks after recovery. Community waterborne outbreaks of norovirus disease typically result from the contamination of private wells or recreational waters by sewage.[29] Transmission of rotaviruses also occurs by the fecal-oral route from close person-to-person contact or fomites.[30] Asymptomatic infants are a common source of rotavirus as a result of the high concentrations (i.e., 10^{12} viruses/g) of rotaviruses excreted in the stool.[31] By the age of 3 years, most (i.e., 90%) children in both developed and developing countries have antibodies against rotaviruses.[21] Hygienic measures have limited effect on the transmission of rotavirus infections, and this virus is an important cause of childhood gastroenteritis in both developed and developing countries.[32]

Animal

Although animal rotaviruses are probably not a significant reservoir for human disease, fecal samples from primates, cows, pigs, and rodents may contain Group B rotaviruses.[33] These animal rotavirus strains possess a high degree of genetic homology with human strains, but animal-to-human transmission is uncommon because *most* animal rotaviruses are species specific.[34] Human toroviruses associated with gastroenteritis in immunocompromised patients serologically resembles bovine (Breda virus)[35] and equine (Berne virus)[36] toroviruses. There are no known animal reservoirs of noroviruses, but waterborne sources of the disease occur frequently as a result of contamination by sewage. Picobirnavirus is widespread in both humans and animals (e.g., chickens).[20,37] Sapovirus does not have a known animal reservoir, and transmission presumably occurs by fecal-oral route, primarily in children.[38]

Food

The epidemic form of viral gastroenteritis affects institutions, communities, families, cruise ships, camps, and schools, primarily by noroviruses genogroups GII and, to a lesser extent, GI.[39,40] Noroviruses cause a majority of the cases of foodborne gastroenteritis in the United States.[41] Foodborne sources of gastrointestinal illness associated with noroviruses include cake frosting, drinking water, orange juice, salads, fresh produce, sandwiches, lettuce, and European oysters.[27,42,43] Waterborne outbreaks of norovirus-induced gastroenteritis usually occur as a result of the contamination of well water and, to a lesser extent, surface water by sewage.[44] Molluscan shellfish (oysters, mussels) are filter-feeders that accumulate environmentally stable, positive-stranded RNA viruses, such as norovirus.[45,46] Noroviruses are resistant to sewage treatment and high concentrations of these viruses occur in water contaminated by sewage, particularly during the epidemic season. The primary pathogens associated with shellfish-borne gastroenteritis are noroviruses, which bind specifically to digestive ducts by carbohydrate structures with a terminal *N*-acetylgalactosamine residue in an α-linkage.[47] Following exposure, oysters become rapidly contaminated, and natural depuration of the viruses from oysters is slow (i.e., several weeks). In an *in vivo* study of oysters contaminated with norovirus, the depuration of norovirus over 48 hours was minimal (7%) compared with the natural removal of bacteria (95%).[48] Most foodborne outbreaks of norovirus-induced illness result from direct contamination of food by a food handler immediately prior to consumption. Outbreaks of rotavirus gastroenteritis typically occur at day care centers, preschools, and other locations where there are large numbers of children under 5 years of age. Occasionally, outbreaks of rotavirus infections develop in adults (e.g., nursing homes), where fecal contamination of food (e.g., salad vegetables) occurs.[49]

Food Processing

Cooking does not necessarily destroy norovirus-contaminated food, including the cooking of raw oysters.[50] Temperatures above 60 °C (140 °F) are necessary to inactive noroviruses.

PATHOGEN

Physiochemical Properties

Both noroviruses and rotaviruses are stable in most typical environmental conditions. Noroviruses are relatively resistant to sodium hypochlorite (<300 ppm), 70% ethanol, and alteration in pH.[51] Freezing does not kill noroviruses, and gastrointestinal illness can occur after consumption of frozen oysters contaminated with noroviruses.[52] The ability of noroviruses to survive high concentrations of chlorine (<10 mg/L for <30 minutes) and temperatures from freezing to 60 °C (140 °F) facilitates the spread of these viruses in recreation and drinking water.[53] Rotavirus is relatively resistant to glutaraldehyde, iodine, and hexachlorophene, whereas calcium chelators and high concentrations of ethanol or chlorine are required to inactivate rotavirus.[54,55]

Mechanism of Toxicity

The mechanism of toxicity associated with viral gastroenteritis is not well defined because of the difficulty culturing these viruses *in vitro*. However, current evidence indicates that the etiology of the diarrhea associated with viral gastroenteritis is multifactorial. The enterocytes lining the small intestine are divided into crypt cells and mature enterocytes with the latter being differentiated by digestive and absorptive functions. Pathologically, rotaviruses infect mature enterocytes in the distal tips of the villi in the proximal two thirds of the ileum.[56] Trypsin-like proteases cleave VP4 and activate rotavirus. Animal studies indicate that multiplication of rotavirus particles in these mature enterocytes causes apoptosis, increased epithelial cell turnover, and a shutoff of gene expression in enterocytes showing viral replication.[57] The villous cells are primarily absorptive, whereas the crypt cells are secretory. Consequently, atrophy of the epithelium of these villi results in a net enhancement of secretion because of the imbalance between absorption of fluid by the enterocytes and the secretion by the crypt cells. Beside the malabsorption mechanism, other contributory mechanisms include reduced permeability to macromolecules (e.g., lactulose) and stimulation of the enteric nervous system with subsequent enhancement of intestinal fluid and electrolyte secretion by the rotavirus enterotoxin, NSP4.[58]

DOSE RESPONSE

There is substantial variation between different strains and between individuals receiving the same viral inoculum.[59] The infection rate and severity of illness depends on the type of virus (e.g., norovirus produce high frequency of symptoms), the virulence of the strain (e.g., rotavirus serotypes G2 and G3 produce more serious illness), the size of the inoculum, individual immunity, and the general health of the patient. The infectious dose of noroviruses is low with the ingestion of as few as 10–100 viral particles of norovirus causing gastrointestinal illness.[60] In a study of 59 healthy volunteers ingesting 30 mL of a 1:600 dilution of the 8F IIa strain of Norwalk agent, 34 (57%) subjects developed a gastrointestinal illness.[61] An additional 8 developed serum antibodies to noroviruses, but they remained asymptomatic. Seventeen (29%) of these volunteers did not develop any gastrointestinal symptoms and they remained seronegative to noroviruses.

CLINICAL RESPONSE

Adults

Epidemic viral gastroenteritis is typically a self-limited, mild gastrointestinal illness characterized by nausea, vomiting, non-bloody diarrhea, abdominal discomfort, malaise, anorexia, headache, and low-grade fever. Dehydration and electrolyte imbalance are the most serious complications of viral gastroenteritis. The mean incubation period for noroviruses involved in epidemic viral gastroenteritis is typically approximately 24–36 hours with a range of 10 to 61 hours.[21,28] The duration of illness typically ranges from 12–60 hours with vomiting occurring in over 50% of the cases, particularly in children.[62] Nonspecific symptoms (e.g., myalgias, fever, chills, headache) often develop during the illness. Immunity to noroviruses typically lasts 4–6 months, but re-infection after that period usually causes less severe illness. Long-term sequelae of norovirus-induced illnesses do not usually occur. In an outbreak of foodborne gastroenteritis caused by group A rotavirus, the median duration of illness was 4 days with a range of 1 to 8 days.[25] Rotavirus-induced gastroenteritis in adults typically occurs in travelers to developing countries, persons caring for children with rotavirus gastroenteritis, immunosuppressed patients, and the elderly.[63] Symptomatic astrovirus infections also occur in the elderly with infections rarely occurring in healthy adults.[64]

Children

The sporadic form of viral gastroenteritis typically involves the development of rotavirus-induced vomiting and diarrhea in infants and young children <2 years old.[65] In unvaccinated populations, group A rotavirus is the most common cause of severe gastroenteritis in child under 5 years of age.[66] In a Venezuelan study of mortality in children from diarrhea, rotavirus was the leading cause of death (21 of 100 cases).[67] Group B and group C rotavirus are less common causes of childhood gastroenteritis than group A rotavirus. Almost all children develop rotavirus infections by the age of 5 years, but the severe form of this illness usually occurs in children aged 3–36 months. Infection by rotavirus during the neonatal period is usually asymptomatic, probably as a result of the persistence of maternal antibodies. Additionally, up to about one-half of rotavirus infections during the first 2 years of life are asymptomatic.[68] Most cases of rotavirus-related gastroenteritis in temperate climates occur between October to April with US outbreaks beginning in the late autumn in the Southwest and ending early in the next calendar year in the Northeast.[69] The clinical course is highly variable ranging from subclinical infection to severe diarrhea and dehydration.[70] The clinical course of rotavirus-induced gastroenteritis is typically more severe than other viral gastroenteritides (adenovirus, astrovirus, calicivirus).[14] The incubation period for rotavirus-induced gastroenteritis is about 24–48 hours. The illness frequently begins abruptly with vomiting preceding diarrhea, which often has a characteristic smell.[70,71] The diarrhea may be explosive, numbering up to 10–20 stools during the severe phase of the illness. In some cases, diarrhea does not develop during rotavirus-induced gastroenteritis.[11] High temperatures (>102 °F/39 °C) can accompany the illness. Symptoms typically resolve in 3–7 days. Although vomiting and fever are more common in rotavirus-induced gastroenteritis than in norovirus-induced illness, there is no constellation of presenting symptoms that separates these two major causes of viral gastroenteritis.[72] Immunity after infection often is incomplete, but subsequent infections are usually less severe than the original infection.[73] Viremia can occur during viral gastroenteritis. Rarely, case reports associate extraintestinal infections (e.g., encephalitis, respiratory infections, poliomyelitis-like syndrome) with rotavirus or norovirus infections in children.[74–76] Although administration of the tetravalent rhesus-human rotavirus vaccine was associated with an increased incidence of intussusception, a retrospective study of 124 cases of intussusception and 450 cases of rotavirus found no clear evidence of an epidemiological association between rotavirus and intussusception.[77] Several case series associate the outbreak of hemorrhagic gastroenteritis and necrotizing enterocolitis with rotavirus infection both in neonates and in infants.[78,79] Astroviruses (i.e., predominately serotype 1) are the second most common cause of hospitalization for viral gastroenteritis in children during the first 2 years of life.[80] The incidence of astrovirus gastroenteritis is highest during the winter, similar to rotavirus.[81]

Immunosuppressed Patients

The main cause of severe viral gastroenteritis in immunosuppressed patients (AIDS, transplant recipients) are cytomegalovirus and Epstein-Barr virus.[14] When compared with rotavirus and astrovirus infections, torovirus is more often associated with persistent diarrhea and nosocomial infections in children, particularly in immunosuppressed children.[82] Additionally, bloody diarrhea is more common in torovirus-infected patients compared with children infected by rotaviruses or astroviruses. Picobirnavirus appears in fecal specimens collected from HIV-infected and noninfected patients with gastrointestinal symptoms in several countries, and some studies suggest that these viruses cause gastroenteritis in immunocompromised patients.[83]

DIAGNOSTIC TESTING

Analytical Methods

Clinically, viral gastroenteritis is a diagnosis of exclusion based on the clinical presentation and a history of an epidemic. Confirmation of viral gastroenteritis traditionally involved direct visualization of the virus by electron microscopy in a reference laboratory. The limit of detection of electron microscope for rotavirus or norovirus particles is about 10^6–10^7 viral particles/mL stool.[84] Detection of viral infections by electron microscopy requires refrigeration of the sample rather than freezing, which destroys the morphology of the viruses. Clinical specimens must be transported in sealed, insulated, waterproof containers on ice or frozen refrigerant packs. Virus particles are detectable in these refrigerated samples for about 2–3 weeks. Immune electron microscopy improves the sensitivity of electron microscopy by 10- to 100-fold.[60] Cell culture is a slow, difficult technique, particularly for human torovirus, coronavirus, and calicivirus.[85,86] The difficulty culturing noroviruses in cell cultures and laboratory animals limits the detection of these organisms in bioassays. The most important methods for the detection of viral gastroenteritis are molecular biology techniques and immunoassay using monoclonal or polyclonal antibodies. Currently, commercial tests for group A rotavirus, adenovirus, and astrovirus are available based on agglutination with latex particles, enzyme immunoassay (EIA) technology, and immunochromatography. In general, EIA techniques are highly specific and more sensitive (i.e., 10^4–10^6 viral particles/mL) for the detection of viruses than visualization by electron microscopy, particularly for the detection of group A rotavirus, group C rotavirus, adenovirus, and astrovirus using monoclonal antibodies.[87,88] The detection of recent infection involves seroconversion as defined by a ≥4-fold rise in IgG antibody titer between the acute phase serum sample (within 5 days of onset of illness) and the convalescent phase (3–6 weeks after illness begins). A simple rapid, enzyme-linked immunosorbent assay (ELISA) is available to detect rotavirus as a cause of diarrheal epidemics. Reverse transcriptase-polymerase chain reaction (RT-PCR) techniques are more sensitive (10^2–10^4 viral particles/mL) than immunoassays,[89] and PCR is useful to confirm positive results from immunoassay methods.[14] Rotaviruses are detectable by RT-PCR with type-specific primers for gentotypes G and P.[90]

Currently, the most sensitive analytical technique for the detection of noroviruses in stool, vomitus, and environmental swabs is the use of a RT-PCR-oligoprobe.[91] The detection limit of noroviruses using this method is >100-fold lower than radioimmunoassay. The optimum time for the detection of noroviruses in stool samples is 48–72 after the onset of symptoms. However, viral shedding continues beyond this period in most cases of rotavirus gastroenteritis, depending in part on the severity of the illness. In a study of 37 children hospitalized with rotavirus gastroenteritis, rotavirus was detectable by immunoassay and RT-PCR for 4–57 days with viral shedding ceasing in 43% (16) of the cases within 10 days of the onset of diarrhea.[92] A real-time RT-PCR method is available for the detection of astroviruses with relatively high sensitivity (0.0026 IU/μL) compared with conventional RT-PCR.[93]

Biomarkers

Rotavirus antigens are detectable in the serum of patients for about 3–7 days after the onset of illness as measured by enzyme immunoabsorbent assay, but most confirmatory studies for viral gastroenteritis usually involve the identification of viruses in fecal samples.[94]

Abnormalities

Fecal leukocytes and erythrocytes rarely occur in typical cases of viral gastroenteritis with the exception of human torovirus infections in immunosuppressed patients. Otherwise, the most common laboratory abnormalities are the electrolyte changes associated with dehydration.

TREATMENT

Treatment is supportive. Every child should be evaluated for signs of dehydration based on an assessment of urine output, clinical history of vomiting and diarrhea (frequency quantity, content), weight loss, physical findings (heart rate, blood pressure, respiratory rate, skin turgor, capillary refill, mucous membrane moisture, sunken eyes, mentation), and laboratory analysis (urine specific gravity, serum electrolytes). Table 27.1 displays the assessment of the degree of dehydration based on clinical and laboratory examinations. Oral administration of glucose-electrolyte solutions helps replace fluid and electrolyte losses. The current recommendation by the World Health Organization (WHO, Geneva, Switzerland) for oral rehydration solution is 75 mmol sodium/L, 20 mmol potassium/L, 65 mmol chloride/L, 30 mmol base/L, and 75 mmol glucose/L. Small, frequent feedings may be necessary when vomiting persists. The early introduction of age-appropriate foods helps reduce stool volume. Foods containing complex carbohydrates (rice, potatoes, bread, wheat, cereals) yogurt, fruits, lean meats, and vegetables are preferable to fatty food or

TABLE 27.1. Assessment of Degree of Dehydration

Symptom	Minimal or No Dehydration (<3%)*	Mile to Moderate Dehydration (3–9%)*	Severe Dehydration (>9%)*
Mental status	Alert, active	Normal to fatigued, restless, or irritable	Lethargic, apathetic, obtunded
Thirst	Drinks normally, may refuse liquids	Thirsty	Drinks poorly
Heart rate	Normal	Normal to increased	Tachycardia unless pre-terminal
Pulse Amplitude	Normal	Normal to decreased	Weak to absent
Breathing	Normal	Normal to tachypneic	Variable
Eyes	Normal	Slightly sunken	Deeply sunken
Tears	Good tearing	Decreased	Absent
Mouth/tongue	Moist	Dry	Parched
Skin fold	Instant recoil	Recoil <2 seconds	Recoil >2 seconds
Capillary refill	Normal	Prolonged	Minimal response
Extremities	Warm	Cool	Cold, mottled, cyanotic
Urine output	Normal to decreased	decreased/dry diapers	Minimal

Source: Adapted from Ref 96.
*% Body weight.

foods high in simple sugars (juices, gelatin deserts, large amounts of carbonated soft drinks). Milk products are acceptable if tolerated, and breast feeding should be continued on demand. There is no clinical advantage to using lactose-free or diluted formulas compared with standard formulation.[95] The goal of oral rehydration therapy during the rehydration phase is replacement of the fluid deficit within 3–4 hours, whereas during the maintenance phase both age-appropriate solid foods and fluids are administered. Older children and adults may benefit from the BRAT diet (bananas, rice, apples, toast). However, there is no clinical evidence that the BRAT diet is superior to other recommended diets and the diet is unnecessarily restrictive.[96] The severity of symptoms and dehydration determine the need for hospitalization. Volunteer studies suggest that the administration of bismuth subsalicylate to adults exposed to Norwalk agent reduces the severity and duration of abdominal cramps and the median duration of gastrointestinal symptoms, when compared with the placebo treatment group.[61] The median durations of illness in the treatment and placebo groups were 20 hours and 27 hours, respectively ($p < 0.05$). However, the dose of bismuth subsalicylate was relatively large (30 mL of a 17.6 mg/mL suspension every half-hour after the onset of symptoms for 8 doses, repeated 24 hours later), and clinical relevance of this dose is unclear. The routine use of antidiarrheal agents and antibiotics are not recommended for infants and children with viral gastroenteritis.[97]

References

1. Kapikian AZ. Overview of viral gastroenteritis. Arch Virol Suppl 1996;12:7–19.
2. Kapikian AZ, Wyatt RG, Dolin R, Thornhill TS, Kalica AR, Chanock RM. Visualization by immune electron microscopy of a 27-nm particle associated with acute infectious nonbacterial gastroenteritis. J Virol 1972;10: 1075–1081.
3. Bishop RF, Davidson GP, Holmes IH, Ruck BJ. Virus particles in epithelial cells of duodenal mucosa from children with acute non-bacterial gastroenteritis. Lancet 1973;2(7841):1281–1283.
4. Madeley CR, Cosgrove BP. Letter: 28 nm particles in faeces in infantile gastroenteritis. Lancet 1975;2(7932): 451–452.
5. Morris CA, Flewett TH, Bryden AS, Davies H. Epidemic viral enteritis in a long-stay children's ward. Lancet 1975;1(7897):4–5.
6. Parashar UD, Hummelman EG, Bresee JS, Miller MA, Glass RI. Global illness and deaths caused by rotavirus disease in children. Emerg Infect Dis 2003;9:565–572.
7. Centers for Disease Control (CDC). Withdrawal of rotavirus vaccine recommendation. MMWR Morb Mortal Wkly Rep 1999;48:1007.
8. Ruiz-Palacios GM, Perez-Schael I, Velazquez FR, Abate H, Breuer T, Costa Clemens S, et al. Safety and efficacy of an attenuated vaccine against severe rotavirus gastroenteritis. N Engl J Med 2006;354:11–22.
9. Parashar UD, Alexander JP, Glass RI. Prevention of rotavirus gastroenteritis among infants and children recommendations of the Advisory Committee on Immunization Practices (ACIP). MMWR Morb Mortal Wkly Rep 2006;55:1–13.
10. Clark HF, Offit PA, Plotkin SA, Heaton PM. The new pentavalent rotavirus vaccine composed of bovine (strain WC-3)-human rotavirus reassortants. Pediatr Infect Dis J 2006;25:577–583.
11. Zahn M, Marshall GS. Clinical and epidemiological aspects of rotavirus infection. Pediatr Ann 2006;35:23–28.
12. Coluchi N, Munford V, Manzur J, Vazquez C, Escobar M, Weber E, et al. Detection, subgroup specificity, and genotype diversity of rotavirus strains in children with acute diarrhea in Paraguay. J Clin Microbiol 2002;40:1709–1714.
13. Kirkwood CD, Palombo EA. Genetic characterization of the rotavirus nonstructural protein, NSP4. Virology 1997;236:258–265.
14. Wilhelmi I, Roman E, Sanchez-Fauquier A. Viruses causing gastroenteritis. Clin Microbiol Infect 2003;9:247–262.
15. Gentsch JR, Laird AR, Bielfelt B, Griffin DD, Banyai K, Ramachandran M, et al. Serotype diversity and reassortment between human and animal rotavirus strains: implications for rotavirus vaccine programs. J Infect Dis 2005;192(Suppl 1):S146–S159.
16. Favier AL, Schoehn G, Jaquinod M, Harsi C, Chroboczek J. Structural studies of human enteric adenovirus type 41. Virology 2002;293:75–85.
17. Jakab F, Meleg E, Banyai K, Melegh B, Timar L, Peterfai J, Szucs G. One-year survey of astrovirus infection in children with gastroenteritis in a large hospital in Hungary: occurrence and genetic analysis of astroviruses. J Med Virol 2004;74:71–77.
18. Karst SM, Wobus CE, Lay M, Davidson J, Virgin HW 4th. STAT1-dependent innate immunity to a Norwalk-like virus. Science 2003;299:1575–1578.
19. Horzinek MC. Molecular evolution of corona- and toroviruses. Adv Exp Med Biol 1999;473:61–72.
20. Chandra R. Picobirnavirus, a novel group of undescribed viruses of mammals and birds: a mini review. Acta Virol 1997;41:59–62.
21. Kapikian AZ. Viral gastroenteritis. JAMA 1993;269: 627–630.
22. Vidal R, Roessler P, Solari V, Vollaire J, Jiang X, Matson DO, et al. Novel recombinant norovirus causing outbreaks of gastroenteritis in Santiago, Chile. J Clin Microbiol 2006;44:2271–2275.

23. Pun SB, Nakagomi T, Sherchand JB, Pandey BD, Cuevas LE, Cunliffe NA, et al. Detection of G12 human rotaviruses in Nepal. Emerg Infect Dis 2007;13:482–484.
24. DiStefano DJ, Kraiouchkine N, Mallette L, Maliga M, Kulnis G, Keller PM, et al. Novel rotavirus VP7 typing assay using a one-step reverse transcriptase PCR protocol and product sequencing and utility of the assay for epidemiological studies and strain characterization, including serotype subgroup analysis. J Clin Microbiol 2005;43:5876–5880.
25. Centers for Disease Control (CDC). Foodborne outbreak of group A rotavirus gastroenteritis among college students – District of Columbia, March–April 2000. MMWR Morb Mortal Wkly Rep 2000;49:1131–1133.
26. Blanton LH, Adams SM, Beard RS, Wei G, Bulens SN, Widdowson MA, et al. Molecular and epidemiologic trends of caliciviruses associated with outbreaks of acute gastroenteritis in the United States, 2000–2004. J Infect Dis 2006;193:413–421.
27. Fleet GH, Heiskanen P, Reid I, Buckle KA. Foodborne viral illness—status in Australia. Int J Food Microbiol 2000;59:127–136.
28. Centers for Disease Control (CDC). Multistate outbreak of norovirus gastroenteritis among attendees at a family reunion—Grant County, West Virginia, October 2006. MMWR Morb Mortal Wkly Rep 2007;56:673–678.
29. Beller M, Ellis A, Lee SH, Drebot A, Jenkerson SA, Funk E, et al. Outbreak of viral gastroenteritis due to a contaminated well. JAMA 1997;278:563–568.
30. Butz AM, Fosarelli P, Dick J, Cusack T, Yolken R. Prevalence of rotavirus on high-risk fomites in day-care facilities. Pediatrics 1993;92:202–205.
31. Centers for Disease Control (CDC). Rotavirus surveillance–United States, 1989–1990. MMWR Morb Mortal Wkly Rep 1991;40:80–87.
32. Leung AK, Kellner JD, Davies HD. Rotavirus gastroenteritis. Adv Ther 2005;22:476–484.
33. Vonderfecht SL, Lindsay DA, Eiden JJ. Detection of rat, porcine, and bovine group B rotavirus in fecal specimens by solid-phase enzyme immunoassay. J Clin Microbiol 1994;32:1107–1108.
34. Palombo EA. Genetic analysis of Group A rotaviruses: evidence for interspecies transmission of rotavirus genes. Virus Genes 2002;24:11–20.
35. Lamouliatte F, du Pasquier P, Rossi F, Laporte J, Loze JP. Studies on bovine Breda virus. Vet Microbiol 1987;15:261–278.
36. Beards GM, Brown DW, Green J, Flewett TH. Preliminary characterisation of torovirus-like particles of humans: comparison with Berne virus of horses and Breda virus of calves. J Med Virol 1986;20:67–78.
37. Gallimore CI, Appleton H, Lewis D, Green J, Brown DW. Detection and characterisation of bisegmented double-stranded RNA viruses (picobirnaviruses) in human faecal specimens. J Med Virol 1995;45:135–140.
38. Rosenfeldt V, Vesikari T, Pang X-L, Zeng S-Q, Tvede M, Paerregaard A. Viral etiology and incidence of acute gastroenteritis in young children attending day-care centers. Pediatr Infect Dis J 2005;24:962–965.
39. Green KY, Belliot G, Taylor JL, Valdesuso J, Lew JF, Kapikian AZ, Lin FY. A predominant role for Norwalk-like viruses as agents of epidemic gastroenteritis in Maryland nursing homes for the elderly. J Infect Dis 2002;185:133–146.
40. Fankhauser RL, Monroe SS, Noel JS, Humphrey CD, Bresee JS, Parashar UD, et al. Epidemiologic and molecular trends of "Norwalk-like viruses" associated with outbreaks of gastroenteritis in the United States. J Infect Dis 2002;186:1–7.
41. Centers for Disease Control (CDC). Multisite outbreak of norovirus associated with a franchise restaurant—Kent County, Michigan, May 2005. MMWR Morb Mortal Wkly Rep 2006;55:395–397.
42. Widdowson M-A, Sulka A, Bulens SN, Beard RS, Chaves SS, Hammond R, et al. Norovirus and foodborne disease, United States, 1991–2000. Emerg Infect Dis 2005;11: 95–102.
43. Morse DL, Guzewich JJ, Henrahan JP, Stricof R, Shayegani M, Deibel R, et al. Widespread outbreaks of clam and oyster associated gastroenteritis: Role of Norwalk virus. N Engl J Med 1986;314:678–681.
44. Maunula L, Miettinen IT, von Bonsdorff C-H. Norovirus outbreaks from drinking water. Emerg Infect Dis 2005;11:1716–1721.
45. Ng TL, Chan PP, Phua TH, Loh JP, Yip R, Wong C, et al. Oyster-associated outbreaks of norovirus gastroenteritis in Singapore. J Infect 2005;51:413–418.
46. Boxman IL, Tilburg JJ, Te Loeke NA, Vennema H, Jonker K, de Boer E, Koopmans M. Detection of noroviruses in shellfish in the Netherlands. Int J Food Microbiol 2006;108:91–396.
47. Le Guyader FS, Loisy F, Atmar RL, Hutson AM, Estes MK, Ruvoen-Clouet N, et al. Norwalk virus-specific binding to oyster digestive tissues. Emerg infect Dis 2005;12:931–936.
48. Schwab KJ, Neill FH, Estes MK, Metcalf TG, Atmar RL. Distribution of Norwalk virus within shellfish following bioaccumulation and subsequent depuration by detection using RT-PCR. J Food Prot 1998;61:1674–1680.
49. Crerar SK, Dalton CB, Longbottom HM, Kraa E. Foodborne disease: current trends and future surveillance needs in Australia. Med J Aust 1996;165:672–675.
50. Center for Disease Control (CDC). Multistate outbreak of viral gastroenteritis associated with consumption of oysters—Apalachicola Bay, Florida, December 1994–January 1995. MMWR Morb Mortal Wkly Rep 1995:44: 37–39.
51. Duizer E, Bijkerk P, Rockx B, De Groot A, Twisk F, Koopmans M. Inactivation of caliciviruses. Appl Environ Microbiol 2004;70:4538–4543.

52. Murphy AM, Grohmann GS, Christopher PJ, Lopez WA, Davey GR, Millsom RH. An Australia-wide outbreak of gastroenteritis from oysters caused by Norwalk virus. Med J Aust 1979;2:329–333.
53. Keswick BH, Satterwhite TK, Johnson PC, DuPont HL, Secor SL, Bitsura JA, et al. Inactivation of Norwalk virus in drinking water by chlorine. Appl Environ Microbiol 1985;50:261–264.
54. Lloyd-Evans N, Springthorpe VS, Sattar SA. Chemical disinfection of human rotavirus-contaminated inanimate surfaces. J Hyg (Lond) 1986;97:163–173.
55. Abad FX, Pinto RM, Bosch A. Survival of enteric viruses on environmental fomites. Appl Environ Microbiol 1994;60:3704–3710.
56. Ramig RF. Pathogenesis of intestinal and systemic rotavirus infection. J Virol 2004;78:10213–10220.
57. Boshuizen JA, Reimerink JH, Korteland-van Male AM, Van Ham VJ, Koopmans MP, Buller HA, et al. Changes in small intestinal homeostasis, morphology, and gene expression during rotavirus infection of infant mice. J Virol 2003;77:13005–13016.
58. Lundgren O, Peregrin AT, Persson K, Kordasti S, Uhnoo I, Svensson L. Role of the enteric nervous system in the fluid and electrolyte secretion of rotavirus diarrhea. Science 2000;287:491–495.
59. Dolin R, Blacklow NR, DuPont H, Formal S, Buscho RF, Kasel JA, et al. Transmission of acute infectious nonbacterial gastroenteritis to volunteers by oral administration of stool filtrates. 1971. J Infect Dis 2004;189:2142–2147.
60. Centers for Disease Control (CDC). "Norwalk-like viruses" public health consequences and outbreak management. MMWR Morb Mortal Wkly Rep 2001;50:1–13.
61. Steinhoff MC, Douglas RG Jr , Greenberg HB, Callahan DR. Bismuth subsalicylate therapy of viral gastroenteritis. Gastroenterology 1980;78:1495–1499.
62. Kaplan JE, Feldman R, Campbell DS, Lookabaugh C, Gary GW. The frequency of a Norwalk-like pattern of illness in outbreaks of acute gastroenteritis. Am J Public Health 1982;72:1329–1332.
63. Hrdy DB. Epidemiology of rotaviral infection in adults. Rev Infect Dis 1987;9:461–469.
64. Kurtz JB, Lee TW, Craig JW, Reed SE. Astrovirus infection in volunteers. J Med Virol 1979;3:221–230.
65. Charles MD, Holman RC, Curns AT, Parashar UD, Glass RI, Bresee JS. Hospitalizations associated with rotavirus gastroenteritis in the United States, 1993–2002. Pediatr Infect Dis J 2006;25:489–493.
66. Blacklow NR, Greenberg HB. Viral gastroenteritis. N Engl J Med 1991;325:252–264.
67. Pérez-Schael I, Salinas B, González R, Salas H, Ludert JE, Escalona M, et al. Rotavirus mortality confirmed by etiologic identification in Venezuelan children with diarrhea. Pediatr Infect Dis J 2007;26:393–397.
68. Espinoza F, Paniagua M, Hallander H, Svensson L, Strannegard O. Rotavirus infections in young Nicaraguan children. Pediatr Infect Dis J 1997;16:564–571.
69. LeBaron CW, Lew J, Glass RI, Weber JM, Ruiz-Palacios GM. Annual rotavirus epidemic patterns in North America. Results of a 5-year retrospective survey of 88 centers in Canada, Mexico, and the United States. Rotavirus Study Group. JAMA 1990;264:983–988.
70. Rodriguez WJ, Kim HW, Brandt CD, Schwartz RH, Gardner MK, Jeffries B, et al. Longitudinal study of rotavirus infection and gastroenteritis in families served by a pediatric medical practice: clinical and epidemiologic observations. Pediatr Infect Dis J 1987;6:170–176.
71. Poulton J, Tarlow MJ. Diagnosis of rotavirus gastroenteritis by smell. Arch Dis Child 1987;62:851–852.
72. Coffin SE, Elser J, Marchant C, Sawyer M, Pollara B, Fayorsey R, et al. Impact of acute rotavirus gastroenteritis on pediatric outpatient practices n the Unites States. Pediatr Infect Dis J 2006;25:584–589.
73. Velazquez FR, Matson DO, Calva JJ, Guerrero L, Morrow AL, Carter-Campbell S, et al. Rotavirus infections in infants as protection against subsequent infections. N Engl J Med 1996;335:1022–1028.
74. Furuya Y, Katayama T, Miyahara K, Kobayashi A, Funabiki T. Detection of the rotavirus a genome from the cerebrospinal fluid of a gastroenteritis patient: a case report. Jpn J Infect Dis 2007;60:148–149.
75. Lynch M, Shieh WJ, Tatti K, Gentsch JR, Ferebee-Harris T, Jiang B, et al. The pathology of rotavirus-associated deaths, using new molecular diagnostics. Clin Infect Dis 2003;37:1327–1333.
76. Ito S, Takeshita S, Nezu A, Aihara Y, Usuku S, Noguchi Y, Yokota S. Norovirus-associated encephalopathy. Pediatr Infect Dis J 2005;25:651–652.
77. Chang EJ, Zangwill KM, Lee H, Ward JI. Lack of association between rotavirus infection and intussusception: implications for use of attenuated rotavirus vaccines. Pediatr Infect Dis J 2002;21:97–102.
78. Rotbart HA, Nelson WL, Glode MP, Triffon TC, Kogut SJ, Yolken RH, et al. Neonatal rotavirus-associated necrotizing enterocolitis: case control study and prospective surveillance during an outbreak. J Pediatr 1988;112:87–93.
79. Rotbart HA, Levin MJ, Yolken RH, Manchester DK, Jantzen J: An outbreak of rotavirus-associated neonatal necrotizing enterocolitis. J Pediatr 1983;103:454–459.
80. Glass RI, Noel J, Mitchell D, Herrmann JE, Blacklow NR, Pickering LK, et al. The changing epidemiology of astrovirus-associated gastroenteritis: a review. Arch Virol Suppl 1996;12:287–300.
81. Palombo EA, Bishop RF. Annual incidence, serotype distribution, and genetic diversity of human astrovirus isolates from hospitalized children in Melbourne, Australia. J Clin Microbiol 1996;34:1750–1753.
82. Jamieson FB, Wang EE, Bain C, Good J, Duckmanton L, Petric M. Human torovirus: a new nosocomial gastrointestinal pathogen. J Infect Dis 1998;178:1263–1269.
83. Giordano MO, Martinez LC, Rinaldi D, Guinard S, Naretto E, Casero R, et al. Detection of picobirnavirus in HIV-

infected patients with diarrhea in Argentina. J Acquir Immune Defic Syndr Hum Retrovirol 1998;18:380–383.

84. Roberton DM, Harrison M, Hosking CS, Adams LC, Bishop RF. Rapid diagnosis of rotavirus infection: comparison of electron microscopy and enzyme linked immunosorbent assay (ELISA). Aust Paediatr J 1979;15:229–232.
85. Atmar RL, Estes MK. Diagnosis of noncultivatable gastroenteritis viruses, the human caliciviruses. Clin Microbiol Rev 2001;14:15–37.
86. Duizer E, Schwab KJ, Neill FH, Atmar RL, Koopmans MP, Estes MK. Laboratory efforts to cultivate noroviruses. J Gen Virol 2004;85(Pt 1):79–87.
87. Lew JF, Moe CL, Monroe SS, Allen JR, Harrison BM, Forrester BD, et al. Astrovirus and adenovirus associated with diarrhea in children in day care settings. J Infect Dis 1991;164:673–678.
88. Dennehy PH, Gauntlett DR, Spangenberger SE. Choice of reference assay for the detection of rotavirus in fecal specimens: electron microscopy versus enzyme immunoassay. J Clin Microbiol 1990;28:1280–1283.
89. Roman E, Martinez I. Detection of rotavirus in stool samples of gastroenteritis patients. PR Health Sci J 2005;24:179–184.
90. Leite JP, Alfieri AA, Woods PA, Glass RI, Gentsch JR. Rotavirus G and P types circulating in Brazil: characterization by RT-PCR, probe hybridization, and sequence analysis. Arch Virol 1996;141:2365–2374.
91. De Leon R, Matsui SM, Baric RS, Herrmann JE, Blacklow NR, Greenberg HB, Sobsey MD. Detection of Norwalk virus in stool specimens by reverse transcriptase-polymerase chain reaction and nonradioactive oligoprobes. J Clin Microbiol 1992;30:3151–3157.
92. Richardson S, Grimwood K, Gorrell R, Palombo E, Barnes G, Bishop R. Extended excretion of rotavirus after severe diarrhoea in young children. Lancet 1998;351(9119): 1844–1848.
93. Royuela E, Negredo A, Sanchez-Fauquier A. Development of a one step real-time RT-PCR method for sensitive detection of human astrovirus. J Virol Method 2006;133:14–19.
94. Fischer TK, Ashley D, Kerin T, Reynolds-Hedmann E, Gentsch J, Widdowson MA, et al. Rotavirus antigenemia in patients with acute gastroenteritis. J Infect Dis 2005;192:913–919.
95. Brown KH, Peerson JM, Fontaine O. Use of nonhuman milks in the dietary management of young children with acute diarrhea: a meta-analysis of clinical trials. Pediatrics 1994;93:17–27.
96. King CK, Glass R, Bresee JS, Duggan C. Managing acute gastroenteritis among children. MMWR Morb Mortal Wkly Rep 2003;52:1–16.
97. Armon K, Stephenson T, MacFaul R, Eccleston P, Werneke U. An evidence and consensus based guideline for acute diarrhoea management. Arch Dis Child 2001;85: 132–142.

IV Seafood

Chapter 28

AMNESIC SHELLFISH POISONING and DOMOIC ACID

HISTORY

In 1987, over 100 Canadians developed gastrointestinal and/or neurological symptoms within 48 hours of ingesting cultured blue mussels (*Mytilus edulis* L.) harvested in the river estuaries of Cardigan Bay, eastern Prince Edward Island. There were 19 hospitalizations among the affected patients and death occurred in 4 elderly patients. Because the neurological symptoms included memory loss, the term *amnesic shellfish poisoning* was used to describe this syndrome. Subsequently, the source of amnesic shellfish poisoning was identified as an amino acid, domoic acid, which originated in some species of phytoplankton. Domoic acid entered the food chain as a result of the bioconcentration of this amino acid in the shellfish. In 1991, an outbreak of gastrointestinal and/or mild neurological symptoms occurred in 21 individuals in Washington state after they consumed Pacific razor clams (*Siliqua patula* Dixon). Although these clams were contaminated with domoic acid, the epidemic was not confirmed as amnesic shellfish poisoning because of the virtual absence of domoic acid from most mussels collected on the coasts of Washington and Oregon.[1] Domoic acid was responsible for the deaths of a large number of brown pelicans and Brandt's cormorants along the Santa Cruz coast in California during 1991.

Domoic acid was initially isolated from the red algae [*Chondria armata* (Kützing) Okamura] in Southern Japan during the late 1950s. This compound was later tested in children as a pediatric antihelmintic for ascariasis.[2] Mass mortality of sea lions and birds along the California and Baja California coast has been associated with high concentrations of domoic acid in food sources (anchovies, sardines).[3,4] In contrast to the contaminated anchovies and sardines, analysis of blue mussels (*Mytilus edulis* L.) collected during this outbreak did not contain toxicologically significant amounts of domoic acid.

EXPOSURE

Although domoic acid originally was isolated from the red algae [*Chondria armata* (Kützing) Okamura], the outbreak of amnesic shellfish poisoning in eastern Prince Edward Island, Canada was associated with marine pennate diatoms from the genus *Pseudo-nitzschia (Pseudo-nitzschia pungens multiseries Hasle)*.[5] Filter-feeding shellfish accumulate domoic acid by direct filtration of affected plankton or by feeding on contaminated organisms. A variety of shellfish bioconcentrate domoic acid including scallops (*Pecten maximus* L.), Pacific razor clams (*Siliqua patula* Dixon), mussels (*Mytilus edulis* L.), Dungeness crabs (*Cancer magister* Dana), cockles (*Cerastoderma edule* L.), and furrow shell (*Scrobicularia plana* da Costa). Some varieties of fish (e.g., anchovy, mackerel, sardine) also accumulate domoic acid, but the concentrations of this toxin in these species is much lower than shellfish, and the domoic acid occurs in the gut of the fish rather than the muscle tissues in shellfish.[6] Additionally, domoic acid accumulates in seaweed harvested from areas with blooms of

Medical Toxicology of Natural Substances, by Donald G. Barceloux, MD

Pseudo-nitzschia species, but the concentrations of toxins are usually not clinically significant. The increased concentrations of domoic acid in shellfish correlate to seasonal variations in phytoplankton blooms. High concentrations of phytoplankton appear in the spring and fall, particularly after periods of heavy rainfall and increased concentrations of nutrients in the water column. Light winds and low water currents allow phytoplankton to accumulate in warmer waters where periods of sunshine promote rapid phytoplankton growth. Other marine organisms capable of producing domoic acid include other *Pseudo-nitzschia* species [*Pseudo-nitzschia australis* Frenguelli, *P. pseudodelicatissima* (Hasle) Hasle], *Alsidium corallinum* C. Agardh, and *Amphora coffeaeformis* (C. Agardh) Kützing.[7,8] Table 28.1 outlines the differences in sources, toxins, and clinical features of amnesic shellfish poisoning and other illnesses associated with seafood-related toxins.

TOXINS

Physiochemical Properties

Domoic acid ($C_{15}H_{21}NO_6$, CAS RN: 14277-97-5, MW 111 Da) is an excitatory neurotransmitter that belongs to the kainoid class of amino acids. Figure 28.1 displays the chemical structure of domoic acid and related isomers. This compound is the principal toxin responsible for amnesic shellfish poisoning.[10] Marine organisms contain at least 10 isomers of domoic acid (isodomoic acids A-H) and domoic acid 5′-diastereomer.[11] However, these isomers are relatively minor constituents (i.e., 10–20%) in marine organisms compared with domoic acid.[12] Domoic acid possesses three carboxyl groups and a secondary amino group with pK_a of 2.10, 3.72, 4.97, and 9.82, respectively. Consequently, domoic acid can exist in valences ranging from −3 to +1 depending on

TABLE 28.1. Toxic Illnesses Associated with Seafood

Illness	Toxin-Producing Organism	Toxins	Clinical Features
Amnesic shellfish poisoning (ASP)	*Pseudo-nitzschia pungens* (Grunow ex Cleve) Hasle	Domoic acid	Nausea, vomiting, disorientation, memory loss, death
Neurotoxic shellfish poisoning (NSP)	*Karenia brevis* (C. C. Davis) G. Hansen et Moestrup	Brevetoxins	Paresthesia, abdominal pain, nausea, diarrhea, dysesthesias, visual changes
Paralytic shellfish poisoning (PSP)	*Alexandrium tamarense* (Lebour) E. Balech, *Alexandrium catenella* (Whedon & Kofoid) E. Balech, *Pyrodinium bahamense* Plate	Saxitoxins	Tingling in the mouth and extremities, dizziness, floating sensation, paralysis, respiratory distress, death
Diarrhetic shellfish poisoning (DSP)	*Prorocentrum lima* (Ehrenberg) Dodge, *Dinophysis fortii* Pavillard, *Dinophysis acuminata* Claparède & Lachmann, *Dinophysis norvegica* Claparède & Lachmann	Okadaic acid, dinophysis toxins, yessotoxin, pectenotoxins	Nausea, vomiting, diarrhea, abdominal pain
Ciguatera	*Gambierdiscus toxicus* Adachi & Fukuyo, *Prorocentrum concavum* Fukuyo, *Prorocentrum mexicanum* Tafall	Ciguatoxins, maitotoxin	Nausea, vomiting, diarrhea, abdominal pain, dizziness, blurred vision, dysesthesias, visual changes, hypotension, respiratory insufficiency; fatalities are rare
Red whelk poisoning	Salivary gland of whelk	Tetramine	Blurred vision, muscular twitching, weakness, headache, ataxia, weakness, paresthesias
Puffer fish poisoning	*Unknown source*[a]	Tetrodotoxin	Paresthesia, floating sensation, dysphagia, hypotension, bradycardia, respiratory distress, ascending paralysis, death
Scombroid fish poisoning	*Morganella morganii* subsp. *morganii* (Winslow et al) Fulton, *Hafnia alvei* Møller, *Klebsiella pneumoniae* subsp. *pneumoniae* (Schroeter) Trevisan, *Proteus* spp., *Vibrio* spp.	Histamine	Nausea, vomiting, diarrhea, headache, flushing, perioral burning, red rash, hypotension

Source: Adapted from Ref 9.

[a]Possible sources include bacteria and actinomycetes in skin gland and ovaries of certain marine puffer fish (e.g., *Fugu rubripes*), such as *Microbacterium arabinogalactanolyticum* (Yokota et al) Takeuchi and Hatano, *Serratia marcescens* subsp. *marcescens* Bizio, *Beneckea alginolytica* (Miyamoto et al) Baumann et al, and *Nocardiopsis dassonvillei* subsp. *dassonvillei* (Brocq-Rousseau) Meyer.

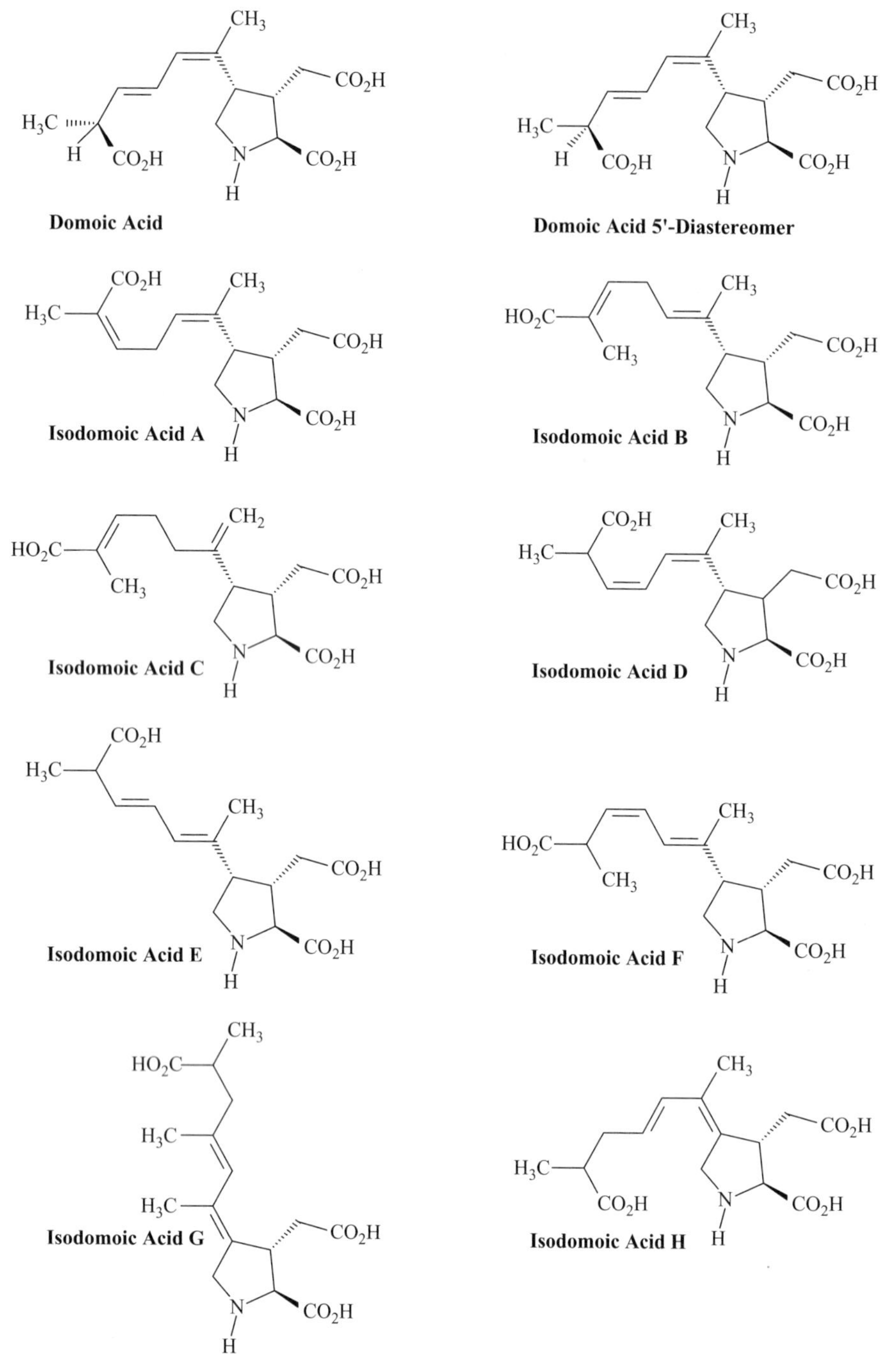

Domoic Acid

Domoic Acid 5'-Diastereomer

Isodomoic Acid A

Isodomoic Acid B

Isodomoic Acid C

Isodomoic Acid D

Isodomoic Acid E

Isodomoic Acid F

Isodomoic Acid G

Isodomoic Acid H

FIGURE 28.1. Chemical structure of domoic acid, isodomoic acids A–H, and 5′-diastereomer.

the pH. At neutral pH, all four groups are charged. Domoic acid and associated isomers are stable at ambient temperatures and in sunlight, but ultraviolet light and heat convert domoic acid to isodomoic acids. At pH 3, about 50% of domoic acid decomposes.[13] Cooking and frozen storage reduces, but does not eliminate domoic acid from food sources.[11]

Toxicokinetics

The elimination of domoic acid varies substantially between shellfish species, ranging from about 1 day in blue mussels (*Mytilus edulis*) to 3 months in razor clams (*Siliqua patula*).[11] Based on limited rodent and primate studies, the absorption of domoic acid is relatively poor

(<4–7%), but once absorbed the distribution and elimination is rapid with elimination half-lives of <2 hours.[14,15]

Mechanism of Toxicity

The two main families of inotropic excitatory amino acid receptors are the *N*-methyl-D-aspartate (NMDA) receptor family and the α-amino-3-hydroxy-5-phenyl-4-isoxazolepropionic acid (AMPA)/kainate receptor family. These ligand-gated cation channels allow the influx of sodium ions into the cell and the efflux of potassium ions from the cell. In the central nervous system, AMPA/kainate channels do not usually allow the entry of calcium into the cell. Receptor cloning studies indicate that there are a large number of potential subtypes of receptors in both of these families.[16] Domoic acid is a potent tricarboxylic agonist at the kainate and α-amino-5-methyl-3-hydroxyisoxazolone-4-propionate (AMPA) subclasses of glutamate receptors. These non-*N*-methyl-D-aspartic acid receptors agonists act as excitatory amino acids along with dicarboxylic amino acids (glutamic acid, aspartic acid).[17] Figure 28.2 demonstrates the structural similarity between domoic acid, kainic acid, and glutamate. Although AMPA receptors mediate fast excitatory synaptic transmission at most excitatory synapses in the central nervous system, the role of kainate receptors in synaptic transmission and neurotoxicity remains unclear. The affinities of domoic acids to the kainate receptors are substantially lower for isodomoic acids than for domoic acid. Although kainate receptors are widely distributed in mammalian brains, these receptors are highly concentrated in the CA3 area of the hippocampus and amygdala.[18] The suspected mechanism of toxicity is the binding of domoic acid to the glutamate receptor with subsequent stimulation causing massive intracellular calcium influx and neuronal death. In addition to neural lesions in the hippocampus, degenerating neuronal cell bodies and terminals also occur in other limbic structures following the administration of parenteral domoic acid including the entorhinal cortex, the subiculum, the piriform cortex, the lateral septum, and the dorsal lateral nucleus of the thalamus.[19] All postmortem examinations of the brains from the four deaths associated with amnesic shellfish poisoning demonstrated prominent necrosis and astrocytosis in the CA1 and CA3 regions of the hippocampus and amygdaloid nucleus. Neurological lesions were also present in the thalamus and subfrontal cortex in some of the postmortem examinations.

DOSE RESPONSE

In animal studies, the parenteral administration of 1–7 mg domoic acid/kg body weight produces characteristic pathological changes in the hippocampus. The toxicity of domoic acid is much lower following oral administration than parenteral use. The intravenous administration of 0.5–1 mg domoic acid/kg body weight to adult monkeys produces focal pathological changes in the CA2 region of the hippocampus, which contains the highest concentration of kainic acid receptor concentration in the brain. Higher doses cause widespread damage to pyramidal neurons and axon terminals of CA1-CA4 and subiculum subfields of the hippocampus.[20] In a study of children with ascariasis, domoic acid was administered at oral doses of 0.6 mg/kg body weight without evidence of adverse effects.[2]

The estimated amount of domoic acid ingested by confirmed cases of amnesic shellfish poisoning during the 1987 epidemic was 60–290 mg/patient. Based on estimations of the amount of mussel tissue ingested during this epidemic, the ingestion of domoic acid >1 mg/kg body weight produced gastrointestinal symptoms and >4.5 mg/kg body weight caused neurological symptoms. The presence of renal dysfunction increased the severity of the adverse effects. During this outbreak, the estimated ingestion of 15–20 mg domoic acid by adults did not produce symptoms.[21] The dose response for consumption of contaminated mussels was highly variable during this outbreak because most people eating these mussels did not develop symptoms, and the amount ingested by severely affected patients did not produce clinically significant symptoms in other individuals.

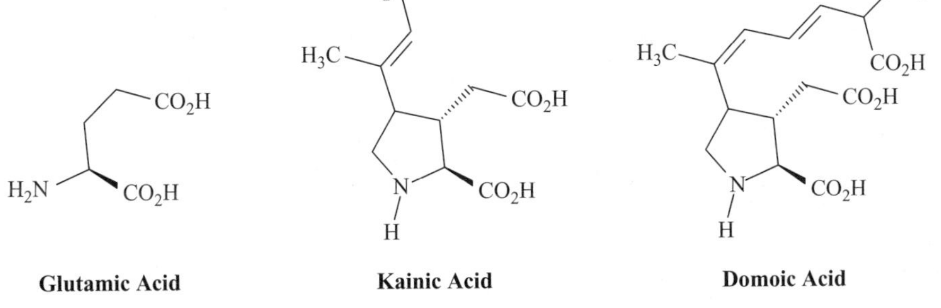

Glutamic Acid — Kainic Acid — Domoic Acid

FIGURE 28.2. Structural similarities of domoic acid, kainic acid, and glutamic acid.

CLINICAL RESPONSE

The single, well-documented case series of amnesic shellfish poisonings occurred in Canada during 1987. Clinical effects included gastrointestinal symptoms within 24 hours of mussel consumption and/or neurological symptoms within 48 hr of mussel consumption. The average onset of symptoms in this epidemic occurred about 5 hours after ingestion of the affected mussels. Of the 107 confirmed cases, the most common gastrointestinal symptoms were nausea (80%), vomiting (64%), abdominal cramping (41%), and diarrhea (41%), whereas the most common neurological effects were severe headache (34%), poor balance (33%), and short-term memory loss (26%).[22] The vast majority (i.e., 99%) of individuals consuming the contaminated mussels in this outbreak developed clinically significant illness.[23] Other neurological effects included unsteadiness, confusion, disorientation, agitation, piloerection, pupillary dilation/constriction, and generalized weakness. Most symptoms resolved within 72 hours with the exception of neurological symptoms, which improved up to 3 months after ingestion of affected mussels. Severe effects included focal motor seizures, coma, hypotension, and profuse bronchial secretions, primarily in elderly patients or patients with preexisting illnesses (hypertension, insulin-dependent diabetes, renal dysfunction). A few patients developed transient motor abnormalities (ophthalmoplegia, hemiparesis) during acute amnesic shellfish poisoning. Temporal lobe epilepsy is a potential complication of severe domoic acid intoxication.[24] The occurrence of seizures and memory loss are unique features of amnesic shellfish poisoning when compared with neurotoxic or paralytic shellfish poisoning.

DIAGNOSTIC TESTING

Analytical Methods

Bioassays with mice or rats are the most common methods of detecting algal toxins, but these tests have relatively low sensitivity.[25] Analytical methods for the detection of domoic acid include thin layer chromatography (semi-quantitative), gas chromatography/mass spectrometry after formation of the *N*-formyl-*O*-methyl derivative, and high performance liquid chromatography (HPLC) with UV, mass spectrometric, or fluorescence detection.[11] The limit of detection of the latter is in the range of 1 pg/g material.[26] The AOAC official method of analysis (Association of Analytical Chemists International, Gaithersburg, MD; http://www.eoma.aoac.org/) utilizes reverse phase liquid chromatography (RPLC) with UV detection at 242 nm with an analytical range of 20 pg/g of mussel tissue.[27] Competitive enzyme-linked immunoassays can detect domoic acid in plasma and urine at concentrations near 0.2 μg/mL.[28] The use of monoclonal antibodies against domoic acid reduces the limit of quantitation to <0.04 μg/g.[29]

Biomarkers

Domoic acid was not detected in 17 cases of human amnesic shellfish poisoning, when tested over 2 days after the ingestion of affected mussels.[30] The regulatory action limit for domoic acid in mussels developed following retrospective analysis of the Port Edwards outbreak of amnesic shellfish poisoning is 20 μg domoic acid/g mussel tissue, and this value includes a 10-fold safety factor.

Abnormalities

Electroencephalographic examination of severely affected individuals demonstrated moderate to severe generalized slowing of background activity that persisted at least 3 months in some affected patients.[31] Neuropsychological examination revealed anterograde memory disturbance in affected cases and retrograde amnesia in the more severely affected patients. Most patients did not develop global deterioration, and IQ scores remained at premorbid levels with intact language and frontal lobe function. Positron emission tomography (PET) scanning of 4 severely affect patients demonstrated severe reduction of glucose metabolism in the hippocampal and medial temporal lobes, consistent with the areas of the brain associated with short-term and long-term memory.[32]

TREATMENT

Treatment is supportive. Endotracheal intubation may be necessary to protect the airway in severely affected patients with respiratory insufficiency. Patients should be evaluated for oxygen saturation as well as dehydration as a result of the gastrointestinal symptoms associated with amnesic shellfish poisoning. Decontamination measures are not usually necessary because of the common occurrence of vomiting and the possibility of mental status changes. There is no known antidote for amnesic shellfish poisoning. Although there are few clinical data, theoretical considerations suggest that benzodiazepines (diazepam, lorazepam) may control seizures and excessive hippocampal activity.[7]

References

1. Todd EC. Seafood-associated diseases and control in Canada. Rev Sci Tech 1997;16:661–672.

2. Takemoto T, Daigo K. Constituents of *Chondria aramata*. Chem Pharmacol Bull 1958;6:578–580.
3. Sierra Beltran A, Palafox-Uribe M, Grajales-Montiel J, Cruz-Villacorta A, Ochoa JL. Sea bird mortality at Cabo San Lucas, Mexico: evidence that toxic diatom blooms are spreading. Toxicon 1997;35:447–453.
4. Scholin CA, Gulland F, Doucette GJ, Benson S, Busman M, Chavez FP, et al. Mortality of sea lions along the central California coast linked to a toxic diatom bloom. Nature 2000;403:80–84.
5. Bates SS, Bird CJ, de Freitas AS, Foxall R, Gilgan M, Hanic LA, et al. Pennate diatom *Nitzschia pungens* as the primary source of domoic acid, a toxin in shellfish from eastern Prince Edward Island, Canada. Can J Fish Aquat Sci 1989;46:1203–1215.
6. Vale P, Sampayo MA. Domoic acid in Portuguese shellfish and fish. Toxicon 2001;39:893–904.
7. Clark RF, Williams SR, Nordt SP, Manoguerra AS. A review of selected seafood poisonings. Undersea and Hyper Med 1999;26:175–185.
8. Walz PM, Garrison DL, Graham WM, Cattey MA, Tjeerdema RS, Silver MW. Domoic acid-producing diatom blooms in Monterey Bay, California: 1991–1993. Nat Toxins 1994;2:271–279.
9. Scoging AC. Illness associated with seafood. Communicable Dis Rep 1991;1:R117–R122.
10. Wright JL, Bird CJ, de Freitas AS, Hampson D, McDonald J, Quilliam MA. Chemistry, biology, and toxicology of domoic acid and its isomers. Can Dis Wkly Rep 1990;16(suppl 1):e21–e26.
11. Jeffery B, Barlow T, Moizer K, Paul S, Boyle C. Amnesic shellfish poisoning. Food Chem Toxicol 2004;42: 545–557.
12. Zhao JY, Thibault P, Quilliam MA. Analysis of domoic acid and isomers in seafood by capillary electrophoresis. Electrophoresis 1997;18:268–276.
13. Quilliam MA, Sim PG, McCulloch AW, McInnes AG. High-performance liquid chromatography of domoic acid, a marine neurotoxin, with application to shellfish and plankton. Int J Environ Anal Chem 1989;36:139–154.
14. Truelove J, Mueller R, Pulido O, Martin L, Fernie S, Iverson F. 30-day oral toxicity study of domoic acid In cynomolgus monkeys: lack of overt toxicity at doses approaching the acute toxic dose. Nat Toxins 1997;5: 111–114.
15. Truelove J, Iverson F. Serum domoic acid clearance and clinical observations in the cynomolgus monkey and Sprague-Dawley rat following a single I.V. dose. Bull Environ Contam Toxicol 1994;52:479–486.
16. Doble A. Excitatory amino acid receptors and neurodegeneration. Therapie 1995;50:319–337.
17. Hampson DR, Manalo JL. The activation of glutamate receptors by kainic acid and domoic acid. Nat Toxins 1998;6:153–158.
18. Porter RH, Eastwood SL, Harrison PJ. Distribution of kainate receptor subunit mRNAs in human hippocampus, neocortex and cerebellum, and bilateral reduction of hippocampal GluR6 and KA2 transcripts in schizophrenia. Brain Res 1997;751:217–231.
19. Schmued LC, Scallet AC, Slikker W Jr. Domoic acid-induced neuronal degeneration in the primate forebrain revealed by degeneration specific histochemistry. Brain Res 1995;695:64–70.
20. Scallet AC, Binienda Z, Caputo FA, Hall S, Paule MG, Rountree RL, et al. Domoic acid-treated cynomolgus monkeys (*M. fascicularis*): effects of dose on hippocampal neuronal and terminal degeneration. Brain Res 1993;627: 307–313.
21. Todd EC. Domoic acid and amnesic shellfish poisoning—a review. J Food Protect 1993;56:69–83.
22. Perl TM, Bedard L, Kosatsky T, Hockin JC, Todd EC, McNutt LA, Remis RS. Amnesic shellfish poisoning: a new clinical syndrome due to domoic acid. Can Dis Wkly Rep 1990;16(suppl 1)e7–e8.
23. Hynie I, Hockin J, Wright J, Iverson F. Panel discussion: evidence that domoic acid was the cause of the 1987 outbreak. Can Dis Wkly Rep 1990;16(suppl 1):e37–e40.
24. Cendes F, Andermann F, Carpenter S, Zatorre RJ, Cashman NR. Temporal lobe epilepsy caused by domoic acid intoxication: evidence for glutamate receptor-mediated excitotoxicity in humans. Ann Neurol 1995;37: 123–126.
25. Luckas B, Dahlmann J, Erler K, Gerdts G, Wasmund N, Hummert C, Hansen PD. Overview of key phytoplankton toxins and their recent occurrence in the North and Baltic Seas. Environ Toxicol 2005;20:1–17.
26. Sun T, Wong WH. Determination of domoic acid in phytoplankton by high performance liquid chromatography of the 6-aminoquinolyl-*N*-hydroxysuccinimidyl carbamate derivative. Agric Food Chem 1999;47:4678–4681.
27. Horwitz W, editor. Official methods of analysis of AOAC International. 17th ed. Gaithersburg, MD: AOAC International; 2000, Official method 991.26.
28. Smith DS, Kitts DD. A competitive enzyme-linked immunoassay for domoic acid determination in human body fluids. Food Chem Toxicol 1994;32:1147–1154.
29. Kawatsu K, Hamano Y, Noguchi T. Production and characterization of monoclonal antibody against domoic acid and its application to enzyme immunoassay. Toxicon 1999;37:1579–1589.
30. Perl TM, Bedard L, Kosatsky T, Hockin JC, Todd EC, Remis RS. An outbreak of toxic encephalopathy caused by eating mussels contaminated with domoic acid. N Engl J Med 1990;322:1775–1780.
31. Teitelbaum J, Zatorre RJ, Carpenter S, Gendron D, Cashman NR. Neurological sequelae of domoic acid intoxication. Can Dis Wkly Rep 1990;16(suppl 1):e9–e12.
32. Gjedde A, Evans AC. PET studies of domoic acid poisoning in humans: excitotoxic destruction of brain glutamatergic pathways, revealed in measurements of glucose metabolism by positron emission tomography. Can Dis Wkly Rep 1990;16(suppl 1): e105–e109.

Chapter 29

AZASPIRACID SHELLFISH POISONING and AZASPIRACID TOXINS

HISTORY

In 1995, 8 people in the Netherlands developed a diarrhetic shellfish-like poisoning after consuming mussels from Ireland. However, the concentrations of okadaic acid and dinophysistoxins were low. Subsequently, a lipophilic, highly oxygenated polyether was isolated from the contaminated mussels that contained a unique azaspiro ring (azaspiracid). A subsequent outbreak of azaspiracid poisoning occurred as a result of the consumption of contaminated mussels from Arranmore Island, Ireland during 1997. Other outbreaks of azaspiracid poisoning occurred in Italy, France, and England as a result of eating contaminated mussels from the Ireland coast.[1] In 2003, Nicolaou et al synthesized azaspiracid-1.[2]

EXPOSURE

Source

Although azaspiracid and azaspiracid analogs are the etiology of a diarrheal illness (i.e., azaspiracid poisoning) associated with eating shellfish, the source of this toxin remains unknown. The most likely source is a dinoflagellate because of the highly oxygenated polyether structure of the toxin and the seasonal occurrence of the diarrheal illness. The suspected producer of azaspiracid compounds is *Protoperidinium crassipes* (Kofoid) Balech, which belongs to the large, ubiquitous group of phytoplankton.[3] The outbreaks of azaspiracid poisoning occurred during the winter season with toxicity persisting from November through April in one outbreak and October through April in another outbreak. Despite widespread distribution of the suspected causative organism, most of the recognized cases of azaspiracid poisoning result from the ingestion of mussels from the northern coast of Europe (England, Ireland, Norway). A survey of blue Mediterranean mussels (*Mytilus galloprovincialis* Lamarck) and the common cockle (*Cerastoderma edule* L.) from the coast of Portugal detected no clinically significant amounts of azaspiracid compounds.[4] However, azaspiracid compounds were detected in mussels from the Spanish and French coasts with toxin profiles similar to contaminated Irish mussels.[5] Although maximum azaspiracid concentrations in shellfish occur in the late summer, azaspiracid poisoning can occur even when marine dinoflagellate populations are low.[6] Contamination of shellfish with azaspiracid compounds occurs in all seasons, but the heaviest contamination occurs in the summer.

Food Processing

Although contaminated mussels are the primary source of azaspiracid poisoning, other bivalve mollusks can accumulate azaspiracid compounds including oysters, scallops, cockles, and clams. Based on analysis of azaspiracid compounds in bivalves from the western coastal region of Ireland, only oysters accumulate similar amounts of these toxins as mussels.[6] However, the ability of oysters to accumulate these toxins is somewhat less than mussels because the latter are cultivated on ropes in relatively deep waters compared

Medical Toxicology of Natural Substances, by Donald G. Barceloux, MD

with the cultivation of oysters along the shoreline. Azaspiracid compounds (i.e., AZA1) occur primarily in the digestive tracts of bivalve mollusks, but some of these toxins (e.g., AZA3) can diffuse into surrounding tissues.[7] Consequently, the toxin profiles differ between different tissues as well as between individual specimens of bivalve mollusks.[8]

TOXINS

Physiochemical Properties

Azaspiracid (CAS RN: 214899-21-5, azaspiracid-1, AZA1) contains a unique spiro ring assembly, a cyclic amine and a carboxylic acid that is unique among nitrogen-containing toxins found in dinoflagellates and shellfish.[9] After extraction with 80% methanol, 40 kg of mussels (*Mytilus edulis* L.) involved with an outbreak of azaspiracid poisoning at Arranmore Island in November 1997 yielded 11 mg AZA1, 3 mg azaspiracid-2 (AZA2), and 0.6 mg azaspiracid-3 (AZA3). The latter two compounds are 8-methylazaspiracid and 22-demethylazaspiracid, respectively. Typically, AZA1 is the most prevalent azaspiracid compound detected in shellfish.[10] Mouse lethality studies indicate that AZA2 and AZA3 are slightly more potent than AZA1 (0.2 mg/kg).[11] Additional analogs of azaspiracid isolated from this outbreak include azaspiracid-4 (3-hydroxy-AZA3) and azaspiracid-5 (23-hydroxy-AZA3). Mouse lethality studies indicate that these two analogues are about 5 times less potent (i.e., 1 mg/kg) than azaspiracid via intraperitoneal administration in mice.[12] AZA6 is an isomer of azaspiracid, and AZA7–AZA10 are probably products of the biotransformation of azaspiracid compounds in shellfish. Typically, AZA4–AZA5 and AZA7–AZA10 account for <8% and <2%, respectively, of the total amount of azaspiracid compounds extracted from shellfish.[13] Figure 29.1 demonstrates the chemical structures of the five azaspiracid compounds.

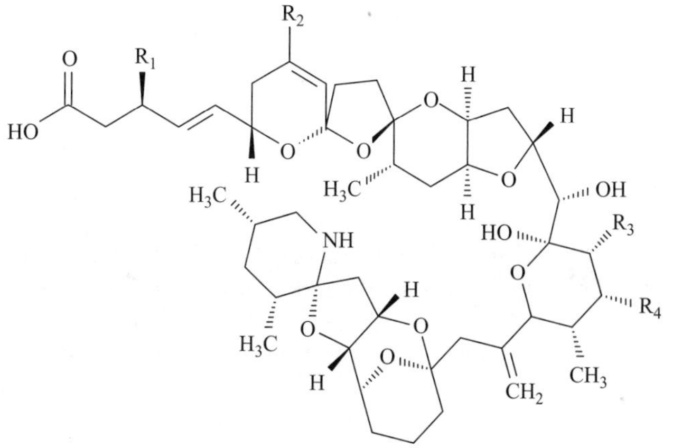

	R_1	R_2	R_3	R_4
Azaspiracid (AZA1)	H	H	CH_3	H
Azaspiracid-2 (AZA2)	H	CH_3	CH_3	H
Azaspiracid-3 (AZA3)	H	H	H	H
Azaspiracid-4 (AZA4)	OH	H	H	H
Azaspiracid-5 (AZA5)	H	H	H	OH

FIGURE 29.1. Chemical structures of azaspiracid and analogues.

Mechanism of Toxicity

The oral administration of azaspiracid to mice produces necrosis of the *lamina propria* of the small intestine as well as necrosis of B cell and T cell lymphocytes in lymphoid tissue of the thymus, Peyer's patches, and spleen. Additionally, fatty changes occur in the liver following the administration of azaspiracid compounds to experimental animals.[14] Mouse bioassays of hexane extracts of mussels associated with azaspiracid poisoning produce respiratory insufficiency, spasm, limb paralysis, and death at higher doses than the dose necessary to produce necrosis in the gastrointestinal tract. The neurological changes and necrosis in the gastrointestinal tract are distinct from histopathology of okadaic acid and diarrhetic shellfish poisoning, and neurological effects have not been reported in human cases of azaspiracid poisoning. The mechanism of these changes remains undefined, but preliminary data suggests that azaspiracid compounds modulate calcium and cyclic adenosine monophosphate (cAMP).[15] Mouse lethality studies indicate that oral and intraperitoneal administration of azaspiracid produce similar toxicity.[14]

DOSE RESPONSE

There are few human data on the dose response of azaspiracid compounds. The European Commission established a maximum limit of 0.16 μg AZA 1–3/g shellfish meat-based animal data.

CLINICAL RESPONSE

The clinical presentation of azaspiracid poisoning is similar to diarrhea shellfish poisoning including the occurrence of nausea, vomiting, severe diarrhea, and abdominal cramps.[1] Although neurological symptoms occur in animal studies of azaspiracid poisoning, human case reports to date do not associate neurologic symptoms with azaspiracid poisoning.

DIAGNOSTIC TESTING

Mouse bioassays are the traditional method for detecting the toxins associated with diarrhetic shellfish poisoning including azaspiracid compounds. However, the conventional mouse bioassay protocols are relatively insensitive to the total azaspiracid content of shellfish.[7] Based on directive 2002/225/EC, the European Commission established a maximum limit of 0.16 μg AZA 1–3/g shellfish meat.[16] Analytical determination of azaspiracid compounds usually involves the combination of liquid chromatography and mass spectrometry or tandem mass spectrometry.[17] Liquid chromatography/electrospray ionization mass spectrometry is a sensitive method of analyzing tissue samples for the major azaspiracid compounds and their hydroxyl analogues with a detection limit near 5 pg.[13] An alternate method for quantifying the amount of AZA-1 through AZA-3 in shellfish tissue involves the use of a single quadrupole mass spectrometer after solid phase extraction.[12] Stobo et al reported a rapid method for the detection of multiple algal toxins including AZA-1 through AZA3 based on liquid chromatography/mass spectrometry.[18] The best-validated method for the detection of AZA-1 is the multiresidue method for the determination of algal toxins in shellfish developed by McNabb et al based on liquid chromatography/tandem mass spectrometry.[19] The limit of detection with this method is <0.02 mg/kg.

TREATMENT

The treatment of azaspiracid poisoning is supportive (antiemetics, antidiarrheal agents). Patients with suspected azaspiracid poisoning should be evaluated for fluid and electrolyte imbalance depending on the severity of the presenting gastrointestinal symptoms. The illness is usually self-limited.

References

1. James KJ, Fidalgo Saez MJ, Furey A, Lehane M. Azaspiracid poisoning, the food-borne illness associated with shellfish consumption. Food Addit Contam 2004;21:879–892.
2. Nicolaou KC, Chen DY-K, Li Y, Qian W, Ling T, Vyskocil S, et al. Total synthesis of the proposed azaspiracid-1 structure, Part 2: coupling of the C1-C20, C21-C27, and C28-C40 fragments and completion of the synthesis. Angew Chem Int Ed Engl 2003;42:3649–3653.
3. James KJ, Moroney C, Roden C, Satake M, Yasumoto T, et al. Ubiquitous "benign" alga emerges as the cause of shellfish contamination responsible for the human toxic syndrome, azaspiracid poisoning. Toxicon 2003;41:145–151.
4. Vale P. Is there a risk of human poisoning by azaspiracids from shellfish harvested at the Portuguese coast? Toxicon 2004;44:943–947.
5. Magdalena AB, Lehane M, Krys S, Fernandez ML, Furey A, James KJ. The first identification of azaspiracids in shellfish from France and Spain. Toxicon 2003;42:105–108.
6. Furey A, Moroney C, Magdalena AB, Fidalgo Saez MJ, Lehane M, James KJ. Geographical, temporal, and species variation of the polyether toxins, azaspiracids, in shellfish. Environ Sci Technol 2003;37:3078–3084.
7. James KJ, Lehane M, Moroney C, Fernandez-Puente P, Satake M, Yasumoto T, Furey A. Azaspiracid shellfish poisoning: unusual toxin dynamics in shellfish and the increased risk of acute human intoxications. Food Addit Contam 2002;19:555–561.
8. Magdalena AB, Lehane M, Moroney C, Furey A, James KJ. Food safety implications of the distribution of azaspiracids in the tissue compartments of scallops (*Pecten maximus*). Food Addit Contam 2003;20:154–160.
9. Evans DA, Dunn TB, Kvaernø L, Beauchemin A, Raymer B, Olhava EJ, et al. Total synthesis of (+)-azaspiracid-1. Part II: synthesis of the EFGHI sulfone and completion of the synthesis. Angew Chem Int Ed Engl 2007;46:4698–4703.
10. James KJ, Furey A, Lehane M, Ramstad H, Aune T, Hovgaard P, et al. First evidence of an extensive northern European distribution of azaspiracid poisoning (AZ) toxins in shellfish. Toxicon 2002;40:909–915.
11. Ofuji K, Satake M, McMahon T, Silke J, James KJ, Naoki H, et al. Two analogs of azaspiracid isolated from mussels, *Mytilus edulis*, involved in human intoxication in Ireland. Nat Toxins 1999;7:99–102.
12. Ofuji K, Satake M, McMahon T, James KJ, Naoki H, Oshima Y, Yasumoto T. Structures of azaspiracid analogs, azaspiracid-4 and azaspiracid-5, causative toxins of azaspiracid poisoning in Europe. Biosci Biotechnol Biochem 2001;65:740–742.
13. Lehane M, Fidalgo Saez MJ, Magdalena AB, Ruppen Canas I, Diaz Sierra M, Hamilton B, et al. Liquid chromatography–multiple tandem mass spectrometry for the determination of ten azaspiracids, including hydroxyl analogues in shellfish. J Chromatogr A 2004;1024:63–70.
14. Ito E, Satake M, Ofuji K, Kurita N, McMahon T, James K, Yasumoto T. Multiple organ damage caused by a new toxin azaspiracid, isolated from mussels produced in Ireland. Toxicon 2000;38:917–930.
15. Roman Y, Alfonso A, Vieytes MR, Ofuji K, Satake M, Yasumoto T, Botana LM. Effects of azaspiracids 2 and 3 on intracellular cAMP, [Ca^{2+}], and pH. Chem Res Toxicol 2004;17:1338–1349.
16. Commission of the European Communities. Commission decision of 15 March 2002 laying down detailed rules for the implementation of Council Directive 91/492/EEC as regards the maximum levels and the methods of analysis of certain marine biotoxins in bivalve molluscs, echino-

derms, tunicates and marine gastropods. Off J Eur Commun L 2002;75:62–64.

17. Brombacher S, Edmonds S, Volmer DA. Studies on azaspiracid biotoxins. II. Mass spectral behavior and structural elucidation of azaspiracid analogs. Rapid Commun Mass Spectrom 2002;16:2306–2316.
18. Stobo LA, Lacaze JP, Scott AC, Gallacher S, Smith EA, Quilliam MA. Liquid chromatography with mass spectrometry—detection of lipophilic shellfish toxins. J AOAC Int 2005;88:1371–1382.
19. McNabb P, Selwood AI, Holland PT, Aasen J, Aune T, Eaglesham G, et al. Multiresidue method for determination of algal toxins in shellfish: single-laboratory validation and interlaboratory study. J AOAC Int 2005; 88:761–772.

Chapter 30

DIARRHETIC SHELLFISH POISONING and OKADAIC ACID

HISTORY

Although the first suspected cases of diarrhetic shellfish poisoning occurred in the Netherlands during 1961,[1] okadaic acid was first associated with diarrhetic shellfish poisoning in Japan during the late 1970s when Yasumoto and Oshima isolated this compound from the black sponge *Halichondria okadaic*.[2] The first documented outbreak of diarrhetic shellfish poisoning in Europe occurred in 1979.[3] During the 1980s, dinophysistoxins (DTX1, DTX3) were identified as additional causes of diarrhetic shellfish poisoning.[4] Subsequently, several other groups of toxins were associated with phytoplankton-induced diarrhetic shellfish poisoning including yessotoxins and pectenotoxins. Reports of diarrheic shellfish poisoning now occur worldwide, including North and South America, Australia, New Zealand, the former Soviet Union, and Europe (Italy, Sweden, Norway, Portugal, Belgium, Greece).[5]

EXPOSURE

Dinoflagellates, particularly species of the genera *Dinophysis* and *Prorocentrum,* are the primary sources of diarrhetic shellfish-associated toxins in contaminated bivalve mollusks, which accumulate these toxins. The three parent compounds in the group of diarrhetic shellfish poisoning toxins are okadaic acid, dinophysistoxin 1 (DTX1), and dinophysistoxin 2 (DTX2). The distribution of okadaic acid is worldwide (e.g., Japan, Europe, North America),[6] whereas the occurrence of DTX1 (Japan, Norway) and DTX2 (Ireland, Portugal) is more limited.[7,8] The concentrations of these toxins in the digestive glands of bivalve molluscs vary during different seasons depending on the occurrence of associated phytoplankton. In Portuguese coastal waters, *Dinophysis acuminata* Claparède & Lachmann and *Dinophysis acuta* Ehrenberg are the main phytoplankton associated with diarrhetic shellfish poisoning.[7] Okadaic acid predominates in contaminated bivalves during the spring coincident with the appearance of *D. acuminata*, whereas DXT2 appears in the late summer and early autumn when *D. acuta* occurs in coastal waters. During the occurrence of diarrhetic shellfish poisoning on the Spanish coast in 1993, okadaic acid and DTX2 were the primary toxins.[9] In mussels and scallops, other acetylated derivatives ("DTX3") of okadaic acid, DTX1, and DTX2 also occur.[10] In addition to okadaic acid and dinophysistoxins, additional toxins isolated from contaminated bivalves that cause adverse effects in animal studies include pectenotoxins and yessotoxins.[11] However, these latter two groups of toxins have not been associated with human illness, and these compounds do not produce diarrhea in animal studies.[12] Although the highest concentration of okadaic acid and related toxins occur in blue mussels (*Mytilus edulis* L.) and to a lesser extent in scallops, consumption of contaminated gastropods and crustaceans (e.g., brown crab, *Cancer pagurus*) also can cause diarrhetic shellfish poisoning.[13]

Okadaic acid was originally isolated from sponges on the Pacific coast of Japan (*Halichondria okadai* Kadota) and the coast of Florida (*Halichondria melanodocia* De Laubenfels). Sponges and bivalves accumulate okadaic

Medical Toxicology of Natural Substances, by Donald G. Barceloux, MD

acid from several species of dinoflagellates including *Dinophysis acuminata* Claparède & Lachmann, *Prorocentrum lima* (Ehrenberg) Dodge, and *Prorocentrum concavum* Fukuyo.[14] Other dinoflagellates (e.g., *Dinophysis fortii* Pavillard) in addition to *Prorocentrum lima* (Ehrenberg) Dodge contain a variety of dinophysistoxins including dinophysistoxin-1, a 35-(R)-methyl derivative of okadaic acid.[4,15] Dinophysistoxin-2 occurs in mussels (*Mytilus edulis* L.). The dinoflagellates, *Thalassiosira weissflogii* (Grunow) G. Fryxell & Hasle and *Prorocentrum maculosum* M.A. Faust, contain dinophysistoxin-4 and diophysistoxin-5a and b, respectively.[16,17] Pectenotoxins occur in scallops contaminated with the toxic dinoflagellate, *Dinophysis fortii* Pavillard.[4] Although shellfish poisoning develops from the ingestion of phytoplankton that bloom, the presence of a bloom is not necessary for the development of sufficient quantities of okadaic acid in shellfish to cause diarrhetic shellfish poisoning. A variety of factors affect the concentration of DSP toxins in contaminated bivalves including water temperature, rainfall, and salinity.[18] The marine dinoflagellate *Protoceratium reticulatum* (Claparède & Lachmann) Butschli is a source of yessotoxin.[19]

TOXINS

Physiochemical Properties

Okadaic acid is a lipophilic, polyether derivative of C38 fatty acids. Figure 30.1 displays the chemical structures of okadaic acid and dinophysistoxins 1–3 (DTX1–3). Okadaic acid, DTX1 (polyether carboxylic acid) and DTX 2A-C (isomers of okadaic acid) are the primary toxins involved in diarrhetic shellfish poisoning. Other congeners (DXT3–5, acyl derivatives) probably are precursors of okadaic acid or shellfish metabolites of the active toxins. In food sources (mussels, clams, cockles), these three toxins occur both in free and esterified forms. The esterification of these toxins is relatively less in mussels (family: Mytilidae) than in other commercial bivalves (scallops, oysters, clams, razor clams, cockles).[10] Outbreaks of diarrhetic shellfish poisoning from contaminated scallops and razor clams have been associated with esterified forms of okadaic acid and dinophysistoxins 1and 2.[20] Dinophysistoxin-3 is a mixture of 7-*O*-acyl derivatives of dinophysistoxin-1 also found in mussels.[21] Because of the absence of acylated forms in marine micro algae, the presumed origin of these acylated forms is biotransformation in the bivalves. Toxicokinetic data from intoxicated patients suggests that DTX-3 undergoes biotransformation to DTX-1 prior to the onset of gastrointestinal symptoms.[22] Dinophysistoxin 4, dinophysistoxin 5a, and dinophysistoxin 5b are water-soluble esters that probably result from the storage or biotransformation of primary parent toxins associated with diarrhetic shellfish poisoning.[23] At least 10 pectenotoxins including several metabolites of parent compounds are associated with extracts from outbreaks of diarrhetic shellfish poisoning.[24] These compounds are novel polyether lactones. Yessotoxins are novel polycyclic, disulfated ether toxins. Analogues of yessotoxin include homoyessotoxin and 45-hydroxy-homoyessotoxin. Yessotoxin is structurally related to ciguatoxin. Okadaic acid is a general tumor promotor in various organs via the inhibition of protein phosphatase-1 and protein phosphatase-2A.[25] The natural marine toxins, calyculin A and microcystin-LR, also inhibit these protein phosphatases, and these latter compounds are tumor promoter on mouse skin and rat glandular stomach.[26] The parent toxins (okadaic acid, DTX1, DTX2) associated with diarrhetic shellfish poisoning are lipophilic, and these compounds accumulate in the digestive glands of

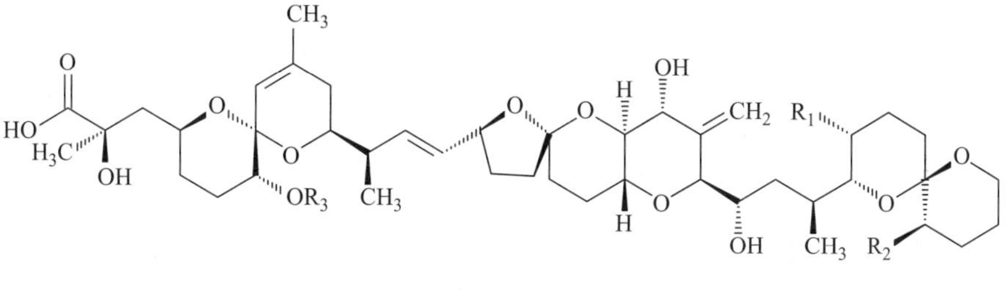

	R_1	R_2	R_3
Okadaic acid (OA)	CH_3	H	H
Dinophysistoxin-1 (DTX1)	CH_3	CH_3	H
Dinophysistoxin-2 (DTX2)	H	CH_3	H
Dinophysistoxin-3 (DTX3)	CH_3	CH_3	Acyl

FIGURE 30.1. Chemical structure of okadaic acid and dinophysistoxins.

bivalves filtering dinoflagellates contaminated with these toxins.

Mechanism of Toxicity

Okadaic acid and dinophysistoxins-1 inhibit serine and threonine protein phosphatases, which regulate metabolic processes in eukaryotic cells. Diarrhea develops from increased calcium influx, prostaglandin production, and the secretion of fluid into mucosal cells in the gastrointestinal tract as a result of the accumulation of phosphorylated proteins. Yessenotoxins do not inhibit protein phosphatases, and these compounds probably do not produce diarrhea based on animal studies.[27] Pectenotoxins are tumor promoters that cause apoptosis in rat hepatocytes, but their role in diarrhetic shellfish poisoning remains unclear.[24]

DOSE RESPONSE

The ingestion of 40–50 µg okadaic acid is the approximate threshold of toxicity for the development of diarrhea; this dose is equivalent to the ingestion of about 300 grams of mussel meat containing 160 µg okadaic acid/kg.[28] A maximum concentration of 80 µg okadaic acid/kg in whole shellfish meat is equivalent to a maximum concentration of 400 µg okadaic acid/kg hepatopancreas. During an outbreak of diarrhetic shellfish poisoning in Japan, the estimated ingestion of 12 mouse units (48 µg okadaic acid equivalents) was the lowest dose associated with mild gastrointestinal symptoms.[2] In animal studies, the oral administration of 50 µg okadaic acid/kg body weight did not produce observable adverse effects, whereas the oral administration of 90 µg/kg caused diarrhea within 8 hours of dosing.[29] Dose response depends on the toxin profile; therefore, there is substantial variation in the effect of a specific dose as defined by the mouse assay.

CLINICAL RESPONSE

Diarrhetic shellfish poisoning is a mild, self-limited gastrointestinal illness, manifest primarily by profuse diarrhea. The gastrointestinal symptoms of diarrhetic shellfish poisoning typically begin within a half hour to a few hours (i.e., up to 5–6 hours) after consumption of toxic shellfish. Adverse gastrointestinal effects include abdominal cramps, nausea, vomiting, and profuse diarrhea that can cause significant fluid and electrolyte loss.[20] Other symptoms include headache, fever, and chills.[30] Symptoms resolve within 3 days. Fatalities do not usually occur, and neurological symptoms are not associated with diarrhetic shellfish poisoning.

DIAGNOSTIC TESTING

Analytical Methods

Bioassay is the most common method of identifying toxins associated with diarrhetic shellfish poisoning. More-specific analytical methods include immunoassays, protein phosphatase assays, high performance liquid chromatography (HPLC) based on precolumn derivatization with fluorescence, and liquid chromatography with mass spectrometry (LC/MS).[3,31] The determination of individual toxins associated with diarrhetic shellfish poisoning requires chromatographic separation prior to analysis. Quantitative determination of toxins associated with diarrhetic shellfish poisoning by LC/MS allows determination of the concentration of these toxins in the range of 5–40 ng/g muscle tissue.[32] The use of combined LC/MS with electrospray ionization allows the separation of diarrhetic shellfish poisoning (DSP) toxins including okadaic acid, dinophysistoxins, pectenotoxins, and yessotoxins.[33] High performance capillary electrophoresis is an alternative to HPLC for the analysis of yessotoxins.[34] The inclusion of yessotoxins and pectenotoxins in extracts for testing the presence of diarrhetic shellfish poisoning toxins depends on the extraction process because acetone, but not ether, extracts these two toxins.

Biomarkers

Analysis of a marine phytoplankton biomass consisting primarily of *Dinophysis acuta* demonstrated average okadaic acid and the isomer (DTX-2) concentrations of approximately 60 pg/cell and 80 pg/cell, respectively, as measured by liquid chromatography coupled with electrospray mass spectrometry and tandem mass spectrometry.[35] Most countries use the mouse bioassay as a means to monitor the presence of okadaic acid in mussels and other commercial bivalves. One mouse unit corresponds to about 4–5 µg okadaic acid.[36] For most countries, the acceptance limit for okadaic acid is approximately 20–25 µg equivalents. Most countries set the acceptable limits for DSP toxins in the hepatopancreas of bivalve molluscs at 0.8–2 µg/g tissue. There is substantial variation in the DSP toxins in adjacent and geographically separated mussels.[8]

TREATMENT

Treatment is supportive. Patients should be evaluated for fluid and electrolyte imbalance as a result of profuse diarrhea and vomiting. Decontamination measures are not necessary.

References

1. Whittle K, Gallacher S. Marine toxins. Br Med Bull 2000;56:236–253.
2. Yasumoto T, Oshima Y. Occurrence of a new type of shellfish poisoning in the Tokoku district. Bull Jpn Soc Sci Fish 1978;44:1249–1255.
3. Luckas B, Dahlmann J, Erler K, Gerdts G, Wasmund N, Hummert C, Hansen PD. Overview of key phytoplankton toxins and their recent occurrence in the North and Baltic Seas. Environ Toxicol 2005;20:1–17.
4. Yasumoto T, Murata M. Diarrhetic shellfish toxins. Tetrahedron 1985;41:1019–1025.
5. Economou V, Papadopoulou C, Brett M, Kansouzidou A, Charalabopoulos K, Filioussis G, Seferiadis K. Diarrheic shellfish poisoning due to toxic mussel consumption: the first recorded outbreak in Greece. Food Addit Contam 2007;24:297–305.
6. Cembella AD. Occurrence of okadaic acid, a major diarrhetic shellfish toxin, in natural population of *Dinophysis* spp. from the eastern coast of North America. J Appl Phycol 1989;1:307–310.
7. Vale P, de M. Sampayo MA. Dinophysistoxin-2: a rare diarrheic toxin associated with *Dinophysis acuta*. Toxicon 2000;38:1599–1606.
8. Carmody EP, James KJ, Kelly SS. Dinophysistoxin-2: the predominant diarrhetic shellfish toxin in Ireland. Toxicon 1996;34:351–359.
9. Gago-Martinez A, Rodriguez-Vazquez JA, Thibault P, Quilliam MA. Simultaneous occurrence of diarrhetic and paralytic shellfish poisoning toxins in Spanish mussels in 1993. Nat Toxins 1996;4:72–79.
10. Vale P, Sampayo MA. Esterification of DSP toxins by Portuguese bivalves from the Northwest coast determined by LC-MS–a widespread phenomenon. Toxicon 2002;40:33–42.
11. Draisci R, Ferretti E, Palleschi L, Marchiafava C, Poletti R, Milandri A, et al. High levels of yessotoxin in mussels and presence of yessotoxin and homoyessotoxin in dinoflagellates of the Adriatic Sea. Toxicon 1999;37:1187–1193.
12. Aune T, Sorby R, Yasumoto T, Ramstad H, Landsverk T. Comparison of oral and intraperitoneal toxicity of yessotoxin towards mice. Toxicon 2002;40:77–82.
13. Torgersen T, Aasen J, Aune T. Diarrhetic shellfish poisoning by okadaic acid esters from Brown crabs (*Cancer pagurus*) in Norway. Toxicon 2005;46:572–578.
14. Dickey RW, Bobzin SC, Faulkner DJ, Bencsath FA, Andrzejewski D. Identification of okadaic acid from a Caribbean dinoflagellate, *Prorocentrum concavum*. Toxicon 1990;28:371–377.
15. Nascimento SM, Purdie DA, Morris S. Morphology, toxin composition and pigment content of *Prorocentrum lima* strains isolated from a coastal lagoon in southern UK. Toxicon 2005;45:633–649.
16. Macpherson GR, Burton IW, LeBlanc P, Walter JA, Wright JL. Studies of the biosynthesis of DTX-5a and DTX-5b by the dinoflagellate *Prorocentrum maculosum*: regiospecificity of the putative Baeyer-Villigerase and insertion of a single amino acid in a polyketide chain. J Org Chem 2003;68:1659–1664.
17. Windust AJ, Quilliam MA, Wright JL, McLachlan JL. Comparative toxicity of the diarrhetic shellfish poisons, okadaic acid, okadaic acid diol-ester and dinophysistoxin-4, to the diatom *Thalassiosira weissflogii*. Toxicon 1997;35:1591–1603.
18. Vale P, de M. Sampayo MA. Seasonality of diarrhetic shellfish poisoning at a coastal lagoon in Portugal: rainfall patterns and folk wisdom. Toxicon 2003;41:187–197.
19. Satake M, MacKenzie L, Yasumoto T. Identification of *Protoceratium reticulatum* as the biogenetic origin of yessotoxin. Nat Toxins 1997;5:164–167.
20. Vale P, de M. Sampayo MA. First confirmation of human diarrheic poisonings by okadaic acid esters after ingestion of razor clams (*Solen marginatus*) and green crabs (*Carcinus maenas*) in Aveiro lagoon, Portugal and detection of okadaic acid esters in phytoplankton. Toxicon 2002;40:989–996.
21. Marr JC, Hu T, Pleasance S, Quilliam MA, Wright JL. Detection of new 7-*O*-acyl derivatives of diarrhetic shellfish poisoning toxins by liquid chromatography-mass spectrometry. Toxicon 1992;30:1621–1630.
22. Garcia C, Truan D, Lagos M, Santelices JP, Diaz JC, Lagos N. Metabolic transformation of dinophysistoxin-3 into dinophysistoxin-1 causes human intoxication by consumption of *O*-acyl-derivatives dinophysistoxins contaminated shellfish. J Toxicol Sci 2005;30:287–296.
23. Hu T, Curtis JM, Walter JA, McLachlan JL, Wright JL. Two new water-soluble DSP toxin derivatives from the dinoflagellate *Prorocentrum maculosum*: possible storage and excretion products. Tetrahedron Lett 1995;36:9273–9276.
24. Burges V, Shaw G. Pectenotoxins—an issue for public health a review of their comparative toxicology and metabolism. Environ Int 2001;27:275–283.
25. Fujiki H, Suganuma M. Unique features of the okadaic acid activity class of tumor promoters. J Cancer Res Clin Oncol 1999;125:150–155.
26. Honkanen RE, Golden T. Regulators of serine/threonine protein phosphatases at the dawn of a clinical era? Curr Med Chem 2002;9:2055–2075.
27. Ogino H, Kumagai M, Yasumoto T. Toxicologic evaluation of yessotoxin. Nat Toxins 1997;5:255–259.
28. Jorgensen K, Jensen LB. Distribution of diarrhetic shellfish poisoning toxins in consignments of blue mussel. Food Addit Contam 2004;21:341–347.
29. Matias WG, Traore A, Crepy Eee. Variations in the distribution of okadaic acid in organs and biological fluids of mice related to diarrheic syndrome. Hum Exp Toxicol 1999;18:345–350.

30. De Schrijver K, Maes I, De Man L, Michelet J. An outbreak of diarrhoeic shellfish poisoning in Antwerp, Belgium. Euro Surveill 2002;7:138–141.
31. Paz B, Daranas AH, Cruz PG, Franco JM, Pizarro G, Souto ML, et al. Characterisation of okadaic acid related toxins by liquid chromatography coupled with mass spectrometry. Toxicon 2007;50:225–235.
32. Goto H, Igarashi T, Yamamoto M, Yasuda M, Sekiguchi R, Watai M, et al. Quantitative determination of marine toxins associated with diarrhetic shellfish poisoning by liquid chromatography coupled with mass spectrometry. J Chromatogr A 2001;907:181–189.
33. Draisci R, Palleschi L, Giannetti L, Lucentini L, James KJ, Bishop AG, et al. New approach to the direct detection of known and new diarrheic shellfish toxins in mussels and phytoplankton by liquid chromatography-mass spectrometry. J Chromatogr A 1999:847:213–221.
34. de la Iglesia P, Gago-Martinez A, Yasumoto T. Advanced studies for the application of high-performance capillary electrophoresis for the analysis of yessotoxin and 45-hydroxyyessotoxin. J Chromatogr A 2007;1156: 160–166.
35. James KJ, Bishop AG, Healy BM, Roden C, Sherlock IR, Twohig M, et al. Efficient isolation of the rare diarrheic shellfish toxin, dinophysistoxin-2, from marine phytoplankton. Toxicon 1999;37:343–357.
36. Aune T. Health effects associated with algal toxins from seafood. Arch Toxicol 1997;19(suppl):389–397.

Chapter 31

NEUROTOXIC SHELLFISH POISONING and BREVETOXINS

HISTORY

Neurotoxic shellfish poisoning was first recognized in bivalves from the Gulf of Mexico during the 1950s. Historically, neurotoxic shellfish poisoning results from the ingestion of bivalves from two areas: the Haruaki Gulf in New Zealand and the Gulf of Mexico near southwest Florida and Texas.[1,2]

EXPOSURE

Consumption of brevetoxin-contaminated bivalves (Pacific oysters, greenshell mussels, clams, cockles) produces neurotoxic shellfish poisoning. The source of neurotoxic shellfish poisoning is the dinoflagellate *Karenia brevis* (C.C. Davis) G. Hansen & Ø. Moestrup (*Gymnodinium breve, Ptychodiscus brevis*). Brevetoxins and some polar metabolites accumulate in filter-feeding bivalves. Blooms (red tides) of *Karenia brevis* occur annually in Florida coastal waters, and these events cause severe rhinitis, pulmonary hemorrhage and edema, hemosiderosis, nonsuppurative leptomeningitis, and death in marine animals including manatees.[3] These dinoflagellates are relatively fragile, and mild wave action along the beaches results in the release of toxins into the water and ocean spray. Marine organisms (e.g., whelks), which prey on contaminated filter-feeding bivalves (e.g., *Chione cancellata* L., *Mercenaria* spp.) can accumulate sufficient brevetoxins to produce neurotoxic shellfish poisoning.[4] Toxic red tides associated with *Karenia brevis* also occur in New Zealand.[5]

TOXINS

Physiochemical Properties

The ingestion of brevetoxins causes neurotoxic shellfish poisoning. These lipid soluble, cyclic polyether compounds are neuroexcitatory toxins similar to ciguatera toxin. Brevetoxins PbTx-1 (brevetoxin A, CAS RN: 98112-41-5), PbTx-2 (brevetoxin B, CAS RN: 79580-28-2), and PbTx-3 (brevetoxin T17, CAS RN: 85079-48-7) were isolated from cultures of *Karenia brevis* under controlled conditions in the ratio of 1:7:2; therefore, PbTx-2 and PbTx-3 probably are the major toxins produced by this dinoflagellate.[6,7] Brevetoxins are tasteless, odorless, heat stable, and acid stable. In a study of neurotoxic shellfish poisoning in the Gulf of Mexico, the predominant brevetoxin was PbTx-2 (brevetoxin B) with the concentrations of Pb-Tx-1 and Pb-Tx-3 being about 10 times less than the concentration of brevetoxin B in seawater extracts as measured by liquid chromatography/atmospheric pressure chemical ionization/mass spectrometry (HPLC/APCI/MS).[8] Metabolites (BTX) of these parent brevetoxin compounds occur as these toxins accumulate in the food chain, and bivalve mollusks biotransform parent brevetoxins via species-specific pathways.[9] Most bivalves have the ability to transform PbTx-2 to the reduction product, PbTx-3, and the oxidation product BTX-B5 with subsequent conversion of BTX-B5 to BTX-B1.[10] In a study of an outbreak of neurotoxic shellfish poisoning in New Zealand, brevetoxin metabolites present in New Zealand cockles

Medical Toxicology of Natural Substances, by Donald G. Barceloux, MD

(*Austrovenus stutchburyi* Wood) included BTX-B5 and BTX-B1 (N-taurine conjugate of BTX-B5). Analysis of the hepatopancreas of the greenshell mussel (*Perna canaliculus* Gmelin) indicated the presence of the following metabolites: BTX-2, BTX-3, and BTX-B4 (mixture of *N*-myristoyl-BTX-B2 and *N*-palmitoyl-BTX-B2).[11,12] Figure 31.1 demonstrates the chemical structure of brevetoxin metabolites and proposed metabolic pathway for the biotransformation of PbTx-2. In whelks preying on brevetoxin-contaminated bivalves, the predominant brevetoxin was PbTx-3, which is a metabolite of the parent toxin, PbTx-2.[4]

Mechanism of Toxicity

Brevetoxins are depolarizing compounds that bind to site 5 on the voltage-gated sodium channel, resulting in persistent activation and prolonged opening of the sodium channel of nerve and muscle cells. Some

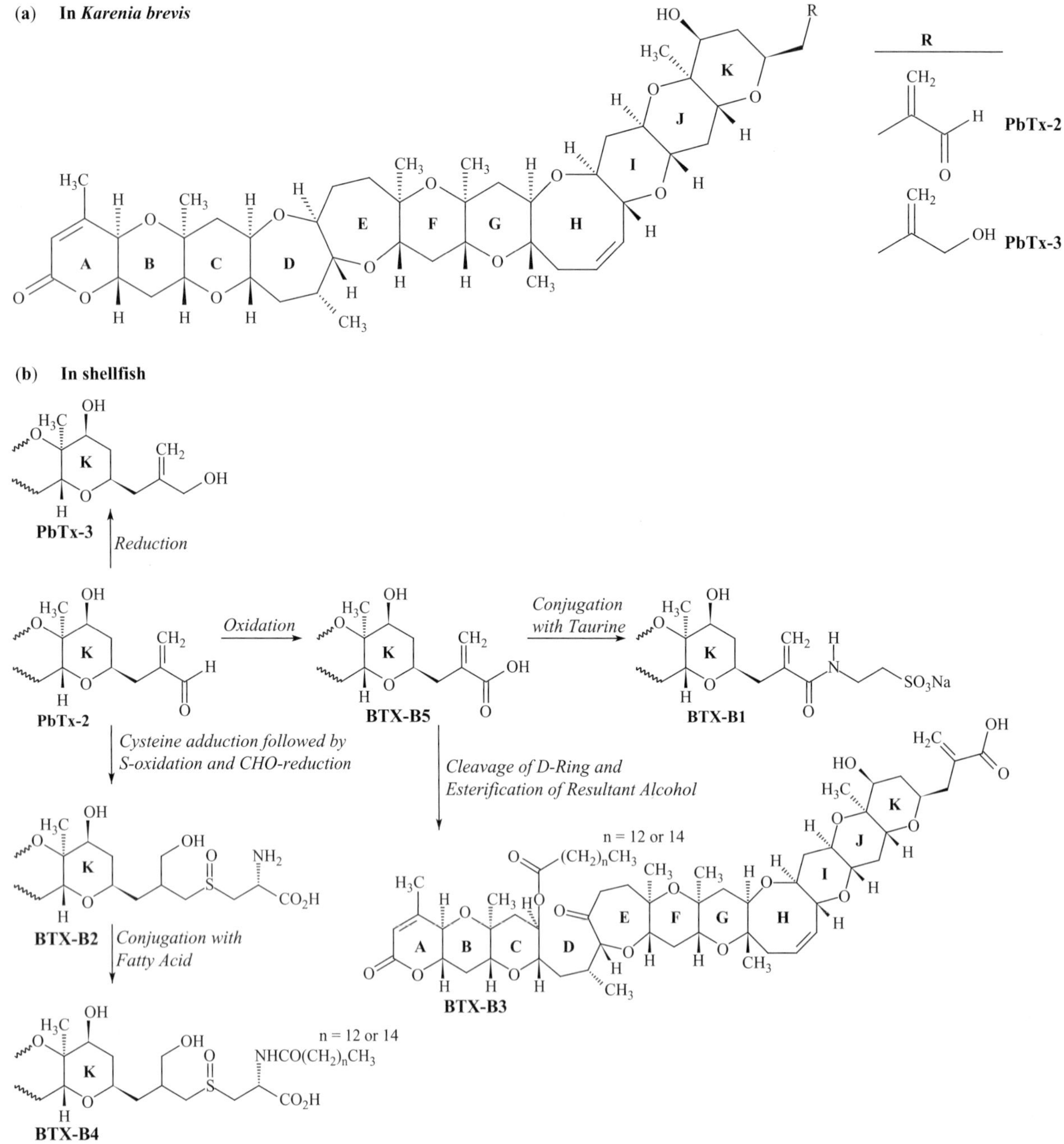

FIGURE 31.1. Chemical structures and suggested metabolic pathway of PbTx-2 biotransformation.

brevetoxin metabolites act as partial agonists at site 5.[13] Ciguatoxin also binds to the same site, but the affinity of ciguatoxin for site 5 is much greater than brevetoxins. Batrachotoxin, veratridine, and aconitine interact at a different site (i.e., site 2) on the voltage-gated sodium channel. Results of the effect of brevetoxin on the sodium channel include shift of activation potential to more negative membrane potentials, prolongation of open time, inhibition of channel inactivation, accumulation of intracellular sodium, and induction of subconductance states.[14]

DOSE RESPONSE

There are few data on the effects of specific doses of brevetoxins on humans. The current US guideline for prohibition of shellfish harvesting is >20 MU/100 g shellfish tissue based on animal studies.

CLINICAL RESPONSE

Neurological Effects

Neurotoxic shellfish poisoning is a relatively mild illness that causes gastrointestinal and excitatory neurological symptoms similar to mild ciguatera poisoning.[15] Both brevetoxins and ciguatoxins enhance sodium entry into cells by binding to site 5 on the voltage-sensitive sodium channel. Neurotoxic shellfish poisoning consists of gastrointestinal irritation and sensory disturbances of the face and extremities including painful sensations (cold allodynia) following exposure to cold objects similar to ciguatera fish poisoning. Symptoms of neurotoxic shellfish poisoning usually develop within a few minutes to 3 hours after ingestion of contaminated bivalves, manifest by perioral paresthesias, loss of coordination, myalgias, and gastrointestinal distress (nausea, vomiting, diarrhea, abdominal cramps). Case series suggest that some differences in clinical patterns between different outbreaks of neurotoxic shellfish poisoning can occur as a result of differences in the toxin profile. The median onset of illness in a series of 48 cases of neurotoxic shellfish poisoning from North Carolina was 3 hours with a range up to 18 hours.[15] The typical pattern of intoxication involves the onset of gastrointestinal distress and perioral numbness followed by progressive paresthesias of the trunk and extremities as well as mild muscle weakness.[16] Occasionally, ataxia, vertigo, or classical muscarinic symptoms (lacrimation, salivation, rhinorrhea, bronchorrhea, miosis, urination, defecation) can occur.[17] Symptoms usually resolve within 1–2 days. Muscle paralysis, cranial nerve dysfunction, respiratory insufficiency, and death from neurotoxic shellfish poisoning does not usually occur.

Respiratory Effects

Contact with aerosols generated by wave action during blooms of *Karenia brevis* can cause conjunctival and upper respiratory tract irritation, rhinitis, cough, and bronchoconstriction.[18] An observational study of 59 subjects with physician-diagnosed asthma demonstrated a small (i.e., 10%), acute reduction in forced expiratory volume in 1 second (FEV_1) during exposure to brevetoxins from a Florida red tide, when compared with baseline one hour prior to exposure.[19] These subjects reported mild respiratory irritation, but none of the subjects required medical treatment. A similar study of 28 healthy lifeguards also reported mildly increased respiratory symptoms (eye and throat irritation, cough, headache) during exposure to a Florida red tide, but there were no significant changes in pulmonary function during mild exercise.[20]

DIAGNOSTIC TESTING

Analytical Methods

Methods to detect brevetoxins include radioimmunoassay (RIA), enzyme-linked immunosorbent assay (ELISA), high performance liquid chromatography/mass spectrometry (HPLC/MS), and mouse bioassay.[4,21] The mouse bioassays of diethyl ether extracts from affected shellfish are the most common method of confirming the presence of brevetoxins in contaminated bivalves. One mouse unit is the amount of crude toxic residue that will kill 50% of the test animals (i.e., 20-gram mice) in about 15 hours. The mouse bioassay underestimates the composite toxicity of brevetoxin-contaminated shellfish compared with cytotoxicity assays, but the latter probably overestimates composite toxicity because of sensitivity to less toxic brevetoxin metabolites.[8] A competitive enzyme-linked immunosorbent assay using goat antibrevetoxin antibodies was developed for the monitoring of brevetoxins in bivalves.[22] The detection limit for brevetoxins in spiked oysters was 2.5 μg/100 g shellfish meat. More specific methods for the determination of brevetoxins involve liquid chromatography with mass spectrometry and tandem mass spectrometry. At a signal-to-noise ratio of 5, the limits of quantitation for BTX-B1 and PbTx-3 were 0.4 ng/g and 2 ng/g, respectively.[23]

Biomarkers

Brevetoxins are rapidly eliminated from the blood. Case reports indicate that brevetoxins remain in the urine <4 days with high concentrations (i.e., 100 ng/mL) present during the first 24 hours after onset of illness as measured by HPLC/MS.[4] The current US guidance level for prohibition of shellfish harvesting as a result of elevated brevetoxin concentrations is >20 MU/100 g shellfish tissue.

TREATMENT

Treatment is supportive. In contrast to ciguatera fish poisoning, complete recovery from neurotoxic shellfish poisoning usually occurs within a few days. Inhaled bronchodilators (e.g., albuterol) may be necessary for the treatment of bronchospasm.

References

1. Bates M, Baker M, Wilson N, Lane L, Handford S. Epidemiological overview of the New Zealand shellfish toxicity outbreak. Roy Soc N Zeal Misc Ser 1993;24:35–40.
2. Steindinger KA. Basic factors influencing red tide. In: LoCicero VR, editor. Proceedings of the First International Conference on Toxic Dinoflagellate Blooms. Wakefield, MA: Massachusetts Science and Technology Foundation; 1975:153–162.
3. Bossart GD, Baden DG, Ewing RY, Roberts B, Wright SD. Brevetoxicosis in manatees (*Trichechus manatus latirostris*) from the 1996 epizootic: gross, histologic, and immunohistochemical features. Toxicol Pathol 1998;26:276–282.
4. Poli MA, Musser SM, Dickey RW, Eilers PP, Hall S. Neurotoxic shellfish poisoning and brevetoxin metabolites: a case study from Florida. Toxicon 2000;38:981–993.
5. Ishida H, Nozawa A, Nukaya H, Tsuji K, Kaspar H, Berkett N, Kosuge T. Isolation and structure determination of a new marine neurotoxin from the New Zealand shellfish, *Austrovenus stutchburyi*. Toxicon 1994;32:1672–1674.
6. Lin YY, Risk M, Ray SM, Engen DV, Clardy J, Golik J, et al. Isolation and structure of brevetoxin B from the red tide dinoflagellate *Ptychodiscus brevis (Gymnodinium breve)*. J Am Chem Soc 1981;103:6773–6775.
7. Shimizu Y, Chou HN, Bando H, Duyne GV,Clardy JC. Structure of brevetoxin A (GB-1), the most potent toxin in Florida red time organism *Gymnodinium breve (Ptychodiscus brevis)*. J Am Chem Soc 1986;108:514–515.
8. Dickey R, Jester E, Granade R, Mowdy D, Moncreiff C, Rebarchik D, et al. Monitoring brevetoxins during a *Gymnodinium breve* red tide: comparisons of sodium channel specific cytotoxicity assay and mouse bioassay for determination of neurotoxic shellfish toxins in shellfish extracts. Nat Toxins 1999;7:157–165.
9. Ishida H, Nozawa A, Nukaya H, Rhodes L, McNabb P, Holland PT, Tsuji K. Confirmation of brevetoxin metabolism in cockle, *Austrovenus stutchburyi*, and greenshell mussel, *Perna canaliculus*, associated with New Zealand neurotoxic shellfish poisoning, by controlled exposure to *Karenia brevis* culture. Toxicon 2004;43:701–712.
10. Ishida H, Nozawa A, Nukaya H, Tsuji K. Comparative concentrations of brevetoxins PbTx-2, PbTx-3, BTX-B1 and BTX-B5 in cockle, *Austrovenus stutchburyi*, greenshell mussel, *Perna canaliculus*, and Pacific oyster, *Crassostrea gigas*, involved neurotoxic shellfish poisoning in New Zealand. Toxicon 2004;43:779–789.
11. Morohashi A, Satake M, Naoki H, Kaspar HF, Oshima Y, Yasumoto T. Brevetoxin B4 isolated from *Perna canaliculus*, the major toxin involved in neurotoxic shellfish poisoning in New Zealand. Nat Toxins 1999;7:45–48.
12. Murata Satake M, NaokiH, Kaspar HF, Yasumoto T. Isolation and structure of a new brevetoxin analog, brevetoxin B2, from greenshell mussels from New Zealand. Tetrahedron Lett 1998;54:735–742.
13. LePage KT, Baden DG, Murray TF. Bevetoxin derivatives act as a partial agonists at neurotoxin site 5 on the voltage-gated Na+ channel. Brain Res 2003;959:120–127.
14. Jeglitsch G, Rein K, Baden DG, Adams DJ. Brevetoxin-3 (PbTx-3) and its derivatives modulate single tetrodotoxin-sensitive sodium channels in rat sensory neurons. J Pharmacol Exp Ther 1998;284:516–525.
15. Morris PD, Campbell DS, Taylor TJ, Freeman JI. Clinical and epidemiological features of neurotoxic shellfish poisoning in North Carolina. Am J Public Health 1991;81:471–474.
16. McFarren EF, Silva FJ, Tanabe H, Wilson WB, Campbell JE, Lewis KH. The occurrence of a ciguatera-like poison in oysters, clams, and *Gymnodinium breve* cultures. Toxicon 1965;3:111–123.
17. Baden DG, Mende TJ. Toxicity of two toxins from the Florida red tide marine dinoflagellate, *Ptychodiscus brevis*. Toxicon 1982;20:457–461.
18. Baden DG, Mende TJ, Bikhazi G, Leung I. Bronchoconstriction caused by Florida red tide toxins. Toxicon 1982;20:929–932.
19. Fleming LE, Kirkpatrick B, Backer LC, Bean JA, Wanner A, Dalpra D, et al. Initial evaluation of the effects of aerosolized Florida red tide toxins (brevetoxins) in person with asthma. Environ Health Perspect 2005;113:650–657.
20. Backer LC, Kirkpatrick B, Fleming LE, Cheng YS, Pierce R, Bean JA, et al. Occupational exposure to aerosolized brevetoxins during Florida red tide events: effects on a healthy worker population. Environ Health Perspect 2005;113:644–649.
21. Baden DG, Melinek R, Sechet V, Trainer VL, Schultz DR, Rein KS, et al. Modified immunoassays for polyether toxins: implications of biological matrixes, metabolic states, and epitope recognition. J AOAC Int 1995;78:499–508.
22. Naar J, Bourdelais A, Tomas C, Kubanek J, Whitney PL, Flewelling L, et al. A competitive ELISA to detect brevetoxins from *Karenia brevis* (formerly *Gymnodinium breve*) in seawater, shellfish, and mammalian body fluid. Environ Health Perspect 2002;110:179–185
23. Nozawa A, Tsuji K, Ishida H. Implications of brevetoxin B1 and PbTx-3 in neurotoxic shellfish poisoning in New Zealand by isolation and quantitative determination with liquid chromatography-tandem mass spectrometry. Toxicon 2003;42:91–103.

Chapter 32

PARALYTIC SHELLFISH POISONING and SAXITOXINS

HISTORY

References to red tides have appeared since Biblical times as noted by the following passage from Exodus 7:20–21: "All the water of the river was changed into blood. The fish in the river died and the river itself became so polluted that the Egyptians could not drink the water." The Iliad describes red tides and ancient Greeks named the Red Sea for the occasional occurrence of reddish waters that probably occurred as a result of phytoplankton blooms.[1] Clinical syndromes consistent with paralytic shellfish poisoning occurred in Europe during the 17th century and during the exploration of North America in the 18th century. Scientific reports of paralytic shellfish poisoning appeared in the 19th century.[2] The red tides represent the proliferation or "bloom" of dinoflagellates that can produce neurotoxins including the saxitoxins associated with paralytic shellfish poisoning. Bivalve mollusks accumulate these neurotoxins during the filtering of water for food. The parent toxin, saxitoxin, was isolated from the giant Alaskan butter clam (*Saxidomus giganteus* Deshayes) and the California mussel (*Mytilus californianus* Conrad) in 1957.[3] Saxitoxin was first synthesized in 1977 by Tanino et al.[4]

EXPOSURE

Source

Filter-feeding bivalve mollusks (mussels, clams, oysters, scallops, cockles, warty venus) concentrate saxitoxins and gonyautoxins synthesized by species from several genera of dinoflagellates (plant kingdom division, Pyrrophycophyta) including *Alexandrium* [*A. tamarense* (Lebour) E. Balech, *A. catenella* (Whedon & Kofoid) E. Balech, *A. minutum* Halim], *Gymnodinium* (*G. catenatum* L.W. Graham),[5] *Gonyaulax*, and *Pyrodinium* (*P. bahamense* Plate).[6] The distribution of dinoflagellates from the genus *Alexandrium* is global compared with the more local distribution (e.g., Mexico, Argentina, Spain, Tasmania, Japan) of *Gymnodinium catenatum*. Mussels and clams are the most common source of paralytic shellfish poisoning,whereas the most prominent dinoflagellate in European waters is *Alexandrium tamarense*.[7] Dinoflagellates occur worldwide in temperate and tropical waters, and cases of paralytic shellfish poisoning occur throughout coast waters around the world (Europe, China, North America, South America, Australia, Mediterranean Sea). Other shellfish contaminated by these dinoflagellates include crustaceans, bivalve mollusks, gastropod mollusks, and cephalopods (octopus, cuttlefish, squid). Crustaceans (shrimp, lobster, crab), gastropod mollusks (whelk, abalone), cephalopods, and finfish do not feed on phytoplankton; therefore, paralytic shellfish poisoning is not usually associated with the consumption of these types of seafood.[8] Crabs and lobsters may transmit paralytic shellfish poisoning toxins when they prey on contaminated mussels and algae cysts. These dinoflagellates survive the winter as cysts, and then transform into motile vegetative forms that cause algal blooms during the summer when water temperatures, nutrients, and pH favor their growth. These blooms last approximately 2–3 weeks until

Medical Toxicology of Natural Substances, by Donald G. Barceloux, MD

nontoxic phytoplankton replace saxitoxin-forming dinoflagellates.[9] The digestive gland (hepatopancreas) of the lobster may contain paralytic shellfish poisoning toxins, but the amount of these toxins in the tail and claw is usually negligible compared with the hepatopancreas.[10] These toxins also have been reported in other marine animals that are not filter-feeders including the horseshoe crab, xanthid crabs,[11] marine snails,[12] starfish,[13] and puffer fish.[14] These animals also frequently contain tetrodotoxin, which produces pharmacologically similar effects as paralytic shellfish poisoning toxins and palytoxin.[15] In puffer fish, paralytic shellfish poisoning toxins (i.e., usually saxitoxin) always occur with tetrodotoxin, but the concentrations of paralytic shellfish poisoning toxins are substantially lower than tetrodotoxin depending on the puffer fish organ examined.[11,16] The bioconcentration of paralytic shellfish poisoning toxins in affected mollusks does not obviously affect the health of these animals.

About 300 of 5,000 marine species of phytoplankton can cause red tides, but most of these species do not produce clinically significant toxins.[17] Red tides become visible when the concentration of dinoflagellates exceeds 20,000/mL. Paralytic shellfish poisoning usually occurs during summer months along with relatively high water temperature from May to November. Some bivalve species contain paralytic shellfish poisoning toxins during the whole year.[5] Visible blooms are not necessary for bivalve mollusks to accumulate toxic concentrations of these toxins.[18] In addition to dinoflagellate sources, several species of cyanobacteria [e.g., *Aphanizomenon flos-aquae* (Linnaeus) Ralfs ex Bornet & Flahault, *Anabaena circinalis* Rabenhorst, *Lyngbya wollei* (Farlow) Farlow, *Cylindrospermopsis raciborskii* (Woloszynska) Seenaya & Subba Raju] can produce these toxins.[19,20]

Food Processing

Saxitoxin is a water-soluble and heat-stable guanidinium compound that decomposes in alkali media. Steaming and cooking do not significantly reduce the potency of saxitoxin, and commercial food processing does not eliminate the potential for developing paralytic shellfish poisoning from contaminated bivalves.[21] Although steaming and boiling can leach some of the water-soluble paralytic shellfish poisoning toxins from mollusks, heat does not destroy sufficient amounts of these compounds to prevent paralytic shellfish poisoning.[22] Boiling for 20 minutes reduced the saxitoxin concentration in the hepatopancreas of lobsters 60%, whereas the concentration of gonyautoxins 2 and 3 decreased by almost 100%.[17]

TOXINS

Physiochemical Properties

There are over 25 hydrophilic, nonprotein saxitoxin congeners associated with paralytic shellfish poisoning that include carbamates (saxitoxin, neosaxitoxin, gonyautoxins 1–4), sulfocarbamoyl compounds (gonyautoxins 5–6, fractions C1–4), and decarbamoyl derivatives. These compounds are relatively small, substituted dibasic tetrahydropurine compounds with molecular weights ranging from 280–450 Da. Saxitoxin (CAS RN: 35523-89-8) is a carbamate (tetrahydropurine derivative), which is colorless, odorless, water-soluble, and heat stable with a molecular weight of 299 Da. Strong alkali inactivates saxitoxin. The molecular formula of saxitoxin is $C_{10}H_{17}N_7O_4$, and saxitoxin has two pK_a values (8.2, 11.5). Neosaxitoxin is *N*-1-hydroxysaxitoxin, which can be converted to saxitoxin in bivalve mollusks.[23] This compound has local anesthetic properties based on preliminary human studies.[24] Gonyautoxins I–IV are sulfates of hydroxysaxitoxin (II, III) and hydroxyneosaxitoxin (I, IV).[25] The difference in toxin profiles between cultured extracts of dinoflagellates and bivalve mollusks suggests that some carbamate and decarbomyl toxins are converted by the mollusks from parent toxins in the filtered dinoflagellates.[26] Figure 32.1 displays the chemical structures of some paralytic shellfish poisoning toxins. In general, carbamate compounds (e.g., saxitoxin) are the most potent toxins with decarbamoyl and sulfocarbamoyl compounds being progressively less toxic, based on mouse bioassays.[27] For example, the toxin profile of the dinoflagellate *Gymnodinium catenatum* from a Spanish outbreak of paralytic shellfish poisoning contained about 90–95 mole % sulfocarbamoyl toxins.[26] However, these toxins accounted for 26% of the total toxicity as measured by saxitoxin equivalents compared with 73% for the much smaller amount of saxitoxin. Under acidic conditions, sulfocarbamoyl compounds convert to the corresponding carbamate. The profile of paralytic shellfish poisoning toxins varies between season, dinoflagellate, and bivalve species even in the same species harvested at the same location.[5,11]

Toxicokinetics

The biological half-life of these toxins is relatively short (90 minutes) with elimination primarily by glomerular filtration.[28,29] There is limited biotransformation of paralytic shellfish poisoning toxins. Postmortem analysis of these toxin profiles in samples from urine and stomach fluid suggest that biotransformation includes oxidation of the N_1 group on saxitoxin to form the hydroxy

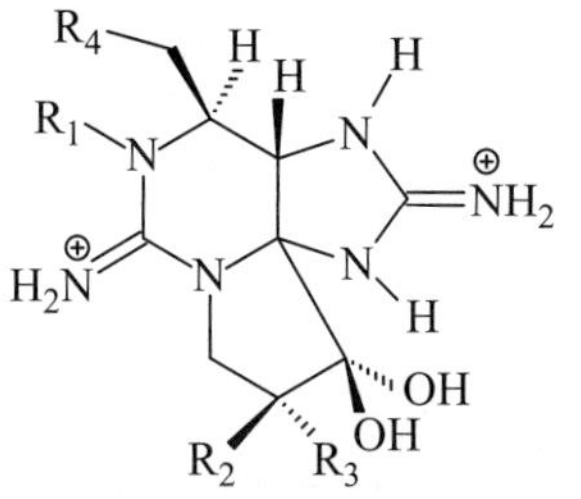

Toxin	R_1	R_2	R_3	R_4	
STX	H	H	H	H_2NCO_2—	Carbamoyl Toxins
NEO	OH	H	H		
GTX 1	OH	H	OSO_3^-		
GTX 2	H	H	OSO_3^-		
GTX 3	H	OSO_3^-	H		
GTX 4	OH	OSO_3^-	H		
B1	H	H	H	$^-O_3SONHCO_2$—	N-Sulfo-Carbamoyl Toxins
B2	OH	H	H		
C3	OH	H	OSO_3^-		
C1	H	H	OSO_3^-		
C2	H	OSO_3^-	H		
C4	OH	OSO_3^-	H		
dc-STX	H	H	H	HO—	Decarbamoyl Toxins
dc-NEO	OH	H	H		
dc-GTX 1	OH	H	OSO_3^-		
dc-GTX 2	H	H	OSO_3^-		
dc-GTX 3	H	OSO_3^-	H		
dc-GTX 4	OH	OSO_3^-	H		

FIGURE 32.1. Chemical structure of paralytic shellfish poisoning toxins. Adapted from Luckas B, et al. Overview of key phytoplankton toxins and their recent occurrence in the North and Baltic Seas. Environ Toxicol 2005;20:3.

analogue (neosaxitoxin) and hydrolysis of carbamoyl group to form decarbamoyl analogues (GTX3/DTX2 epimers).[30] Some of these metabolites are also excreted in the bile.[31]

Mechanism of Toxicity

Saxitoxin reversibly binds to site 1 on the outer ion conduction pore of the voltage-sensitive sodium channel, producing inhibition of conduction and disruption of the action potential.[32] Site 1 is located on the SS1–SS2 segments of all four domains of the channel protein. Gonyautoxins are sulfated saxitoxin analogs that also bind site 1 of the sodium channel.[33] Saxitoxin blocks the influx of sodium into the cell, but this toxin does not block potassium efflux. However, *in vitro* studies indicate that saxitoxin inhibits L-type Ca^{2+} currents,[34] and modifies the hERG K^+ channel, resulting in slower opening during depolarization and faster closure during repolarization.[35] The interruption of the excitation current in the presence of saxitoxin ultimately causes neuromuscular paralysis. Saxitoxin suppresses AV nodal conduction, reduces peripheral nerve conduction, and depresses the medullary respiratory center. Death usually results from respiratory depression and apnea. In animal studies, high doses of saxitoxin produce hypotension, cardiac failure, and cardiopulmonary arrest.[28]

DOSE RESPONSE

There is no direct correlation between the size of the mollusk and the presence of paralytic shellfish poisoning toxins.[36] During periods of red tides, mussels can accumulate up to 9 mg saxitoxin equivalents; however, the amount of saxitoxin equivalents in seafood does not correlate to toxicity. The amount of saxitoxin equivalents associated with illness and fatalities is highly variable as a result of individual susceptibility, specificity of analytical methods, composition of saxitoxin congeners, and recall bias. Reported ranges of saxitoxin equivalents associated with illness and fatalities from these toxins are 144–1660 µg/person and 300–12,400 µg/person, respectively.[37] A case report associated ingestion of 2,600 µg saxitoxin equivalents (36 µg/kg body weight) with complete paralysis and respiratory failure.[38] Severe illness (coma, respiratory failure, hypertension) was associated with the ingestion of an estimated 411 µg saxitoxin equivalents/kg composed primarily of C1, C2, and gonyautoxins.[39] The estimated ingestion by 4 patients with respiratory arrest from paralytic shellfish poisoning ranged from 230–411 μg saxitoxin equivalents/kg body weight (paralytic shellfish poisoning toxin profile not stated).[40] During an Alaskan epidemic of paralytic shellfish poisoning, all patients ingesting >37 mg saxitoxin equivalents became ill.[41] The estimated mean dose ingested by 33 ill and 10 asymptomatic individuals was 19.5 mg (range: 0.01–123 mg) and 13 mg (range: 0.02–37 mg), respectively.

CLINICAL RESPONSE

The clinical presentation of paralytic shellfish poisoning is similar to the effects of puffer fish poisoning, which results from the blockade of sodium channels in neurons by tetrodotoxin. The onset of symptoms following the ingestion of saxitoxin-contaminated bivalves is rapid with paresthesias of the face, tongue, and lips beginning within 1–2 hours after ingestion.[42] In a study of 224 confirmed cases in the Philippines, the median onset of symptoms was 5.6 hours with a range up to 34 hours.[36] Similar to tetrodotoxin intoxication, the rapid onset of symptoms from paralytic shellfish poisoning correlates to severity, and death usually occurs within 12 hours of ingestion. The paresthesia and tingling progressively involve the trunk and extremities over the next 1–2 hours, and cranial nerve dysfunction (diplopia, dysarthria, dysphagia, hoarseness) can develop. The onset of neurological symptoms >10–12 hours after ingestion suggests an etiology other than paralytic shellfish poisoning. Severe cases are associated with muscle weakness, ataxia, fixed and dilated pupils, absent or diminished deep tendon reflexes, and paralysis.[40,43] Case reports suggest that patients with paralytic shellfish poisoning retain consciousness at least during the early phase of complete paralysis.[38] In some case series, children under 6 years of age appear more susceptible than adults to these toxins.[44] Some autonomic dysfunction (diaphoresis, tachycardia, excessive salivation, hypertension) can occur. Nonspecific symptoms include headache, thirst, myalgias, low back pain, and vertigo. Peripheral vasodilation secondary to smooth muscle relaxation can cause hypotension. Table 32.1 lists the symptoms associated with paralytic shellfish poisoning in Alaska over an approximate 20-year period. Death is rare and results from respiratory insufficiency secondary to paralysis of muscles in the chest wall and diaphragm typically within 12 hours after ingestion. The patient remains alert during the course of intoxication while vital signs remain stable. Gastrointestinal complaints (nausea, vomiting, abdominal pain) are usually mild or absent. The ingestion of ethanol does not exacerbate the symptoms associated with paralytic shellfish poisoning. Neurological symptoms typically resolve in 12–48 hours.[42] The older medical literature suggests that

TABLE 32.1. Symptoms Associated with 117 Cases of Paralytic Shellfish Poisoning in Alaska 1973–1992

Symptom	Cases Present (*N* = 117)
Paresthesias	113
Perioral numbness	64
Perioral tingling	61
Nausea	45
Extremity numbness	43
Extremity tingling	39
Vomiting	34
Weakness	33
Ataxia	32
Shortness of breath	29
Dizziness	28
Floating sensation	24
Dry mouth	23
Diplopia	19
Dysarthria	16
Diarrhea	10
Dysphagia	6
Limb paralysis	4

Source: From Ref 41.

weakness and fatigue can persist for weeks after intoxication, but the extent of chronic sequelae are not well documented.[45] Chronic exposure to saxitoxin does not reduce toxicity. The differential diagnosis includes Guillain Barré syndrome, botulism, ciguatera, puffer fish poisoning (tetrodotoxin), and organophosphate poisoning.

DIAGNOSTIC TESTING

Analytical Methods

Contamination by saxitoxin is measured by a mouse bioassay in mouse units (MU/mg) as determined by the AOAC International official method of analysis for paralytic shellfish poisoning (Association of Analytical Chemists International, Gaithersburg, MD; http://www.eoma.aoac.org/).[46] Typically, the results of this bioassay are determined as mouse units and, less often as micrograms of poison per 100 g of shellfish tissue. The mouse unit is defined as the amount of toxin required to kill a 20-gram white mouse within 15 minutes following intraperitoneal administration. By definition, one mouse unit equals 0.18 μg saxitoxin equivalents.[21] In the European Union, the maximum acceptable content of paralytic shellfish poisoning toxins in bivalve mollusks is 80 μg/100 g tissue (EEC directive 91/492).[47] The current reference method in the European Union is the mouse bioassay. The presence of many congeners of saxitoxin complicates the analysis of paralytic shellfish poisoning toxin in biological tissues, and the biological assay does not determine the profile of the many paralytic shellfish poisoning toxins. More sensitive methods (e.g., chromatography) are preferable to biological assays. The most common analytical technique for the separation of PSP toxins is reversed phase high performance liquid chromatography (RP/HPLC) using ion-pairing reagents and postcolumn reaction permitting fluorescence detection. However, there is substantial interlaboratory variation in results from liquid chromatography methods with precolumn derivatization, followed by fluorescence detection.[48] Complete chromatographic separation (e.g., HPLC) of all paralytic shellfish poisoning toxins is necessary prior to quantitation with a fluorescence detector.[25] The use of standard methods of acid solvent extraction (0.1 N HCl) with heating to 100 °C (212 °F) for 5 minutes converts *N*-sulfocarbamoyl toxins into their respective carbamate toxins. Positive findings are usually confirmed with mass spectrometry. For diagnostic purposes, human specimens should be collected immediately because of the rapid elimination (<24 hours) of paralytic shellfish poisoning toxins from the blood. Both urine and blood samples should be collected and frozen for transport to specialty laboratories. The second AOAC official method of analysis for saxitoxin is a HPLC method with prechromatographic oxidation and fluorescence detection.[49] This method allows determination of individual toxins with results reported as μg/kg.

Abnormalities

Normal repetitive stimulation on electrophysiological testing distinguishes paralytic shellfish poisoning from botulism, and the F-wave remains normal during paralytic shellfish poisoning intoxication in contrast to Guillain-Barré syndrome. During an outbreak of paralytic shellfish poisoning that did not cause respiratory depression or hypotension, electrophysiological examination of 8 of 9 patients revealed normal motor and sensory conduction with normal amplitudes.[50] Proximal conduction times were delayed as assessed by decreased F wave latencies and frequencies. Peripheral and central sensory conduction was normal as assessed by somatosensory-evoked potentials. Routine biochemical and hematological tests are usually normal. A case report associated elevated creatine kinase MB fractions with paralytic shellfish poisoning, but there was no definite evidence of myocardial injury on electrocardiogram.[51]

TREATMENT

Treatment is supportive. The most serious complication is respiratory depression, and patients should be evaluated for the adequacy of respirations with pulse oximetry and arterial blood gases as needed. Emesis is not recommended because of the potential of respiratory depression and dysphagia. Although experimental studies suggest that activated charcoal (50 g adult, 1 g/kg children) can bind saxitoxin, there are no clinical data to indicate that the administration of activated charcoal improves clinical outcome. There are no antidotes. Metabolic acidosis should be corrected because of the potential of acidosis to exacerbate the toxicity of paralytic shellfish poisoning toxins in animal studies,[52] but there are inadequate clinical data to determine if serum alkalinization improves outcome. Patients with any signs of respiratory depression should be admitted for at least 24 hours of observation. Although limited animal data suggest that saxitoxins adhere to activated charcoal, there are no clinical data to indicate that charcoal hemoperfusion improves clinical outcome.[53]

References

1. Clark RF, Williams SR, Nordt SP, Manoguerra AS. A review of selected seafood poisonings. Undersea Hyperbar Med 1999;26:175–185.
2. Combe JS. On the poisonous effects of the mussel (*Mytilus edulis*). Edin Med Surg J 1828;29:86–96.
3. Schantz EJ, Mold JD, Stanger DW, Shavel J, Riel FJ, Bowden JP, et al. Paralytic shellfish poison. VI. A procedure for the isolation and purification of the poison from toxic clam and mussel tissues. J Am Chem Soc 1957;79:5230–5235.
4. Tanino H, Nakata T, Kaneko T, Kishi Y. A stereospecific total synthesis of *d,l*-saxitoxin. J Am Chem Soc 1977;99: 2818–2819.
5. Taleb H, Vale P, Jaime E, Blaghen M. Study of paralytic shellfish poisoning toxin profile in shellfish from the Mediterranean shore of Morocco. Toxicon 2001;39:1855–1861.
6. Usup G, Kulis DM, Anderson DM. Growth and toxin production of the toxic dinoflagellate *Pyrodinium bahamense* var. *compressum* in laboratory cultures. Nat Toxins 1994; 2:254–262.
7. Aune T. Health effects associated with algal toxins from seafood. Arch Toxicol 1997;19(suppl):389–397.
8. Hughes JM, Merson MH, Gangarosa EJ. The safety of eating shellfish. JAMA 1977;237:1980–1981.
9. Scoging AC. Illness associated with seafood. Communicable Dis Rep 1991;1:R117–R122.
10. Lawrence JF, Maher M, Watson-Wright W. Effect of cooking on the concentration of toxins associated with paralytic shellfish poison in lobster hepatopancreas. Toxicon 1994;32:57–64.
11. Llewellyn L, Negri A, Robertson A. Paralytic shellfish toxins in tropical oceans. Toxin Rev 2006;25:159–196.
12. Anderson DM, Kulis DM, Qi Y-Z, Zheng L, Lu S, Lin Y-T. Paralytic shellfish poisoning in southern China. Toxicon 1996;34:579–590.

13. Ito K, Asakawa M, Sida Y, Miyazawa K. Occurrence of paralytic shellfish poison (PSP) in the starfish *Asterina pectinifera* collected from the Kure Bay, Hiroshima Prefecture, Japan. Toxicon 2004;41:291–295.
14. Sato S, Ogata T, Borja V, Gonzales C, Fukuyo Y, Kodama M. Frequent occurrence of paralytic shellfish poisoning toxins as dominant toxins in marine puffer from tropical water. Toxicon 2000;38:1101–1109.
15. Alcala AC, Alcala LC, Garth JS, Yasumura D, Yasumoto T. Human fatality due to ingestion of the crab *Demania reynaudii* that contained a palytoxin-like toxin. Toxicon 1988;26:105–107.
16. Jen HC, Lin SJ, Lin SY, Huang YW, Liao IC, Arakawa O, Hwang DF. Occurrence of tetrodotoxin and paralytic shellfish poisons in a gastropod implicated in food poisoning in southern Taiwan. Food Addit Contam 2007;24:902–909.
17. Yan T, Zhou M-J. Environmental and health effects associated with harmful algal bloom and marine algal toxins in China. Biomed Environ Sci 2004;17:165–176.
18. Whittle K, Gallacher S. Marine toxins. Br Med Bull 2000;56:236–253.
19. Sivonen K. Cyanobacterial toxins and toxin production. Phycologia 1996:36:12–24.
20. Alam M, Shimizu Y, Ikawa M, Sasner JJ Jr. Reinvestigation of the toxins from the blue-green alga, *Aphanizomenon flos-aquae*, by a high performance chromatographic method. J Environ Sci Health A 1978;13:493–499.
21. Schantz EJ. Chemistry and biology of saxitoxin and related toxins. Ann N Y Acad Sci 1986;479:15–23.
22. Lehane L. Paralytic shellfish poisoning: a potential public health problem. Med J Aust 2001;175:29–31.
23. Okada K, Niwa M. Marine toxins implicated in food poisoning. J Toxicol Toxin Rev 1998;17:373–384.
24. Rodriguez-Navarro AJ, Lagos N, Lagos M, Braghetto I, Csendes A, Hamilton J, et al. Neosaxitoxin as a local anesthetic: preliminary observations from a first human trial. Anesthesiology 2007;106:339–345.
25. Luckas B, Dahlmann J, Erler K, Gerdts G, Wasmund N, Hummert C, Hansen PD. Overview of key phytoplankton toxins and their recent occurrence in the North and Baltic Seas. Environ Toxicol 2005;20:1–17.
26. Anderson DM, Sullivan JJ, Reguera B. Paralytic shellfish poisoning in northwest Spain: the toxicity of the dinoflagellate *Gymnodinium catenatum*. Toxicon 1989;27:665–674.
27. Shimizu Y. Microalgal metabolites: a new perspective. Annu Rev Microbiol 1996;50:431–465.
28. Andrinolo D, Iglesias V, Garcia C, Lagos N. Toxicokinetics and toxicodynamics of gonyautoxins after an oral toxin dose in cats. Toxicon 2003;40:699–709.
29. Andrinolo D, Michea LF, Lagos N. Toxic effects, pharmacokinetics and clearance of saxitoxin, a component of paralytic shellfish poison (PSP), in cats. Toxicon 1999;37: 447–464.
30. Llewellyn LE, Dodd MJ, Robertson A, Ericson G, de Koning C, Negri AP. Post-mortem analysis of samples from a human victim of a fatal poisoning caused by the xanthid crab, *Zosimus aeneus*. Toxicon 2003;40:1463–1469.
31. Garcia C, del Carmen Bravo M, Lagos M, Lagos N. Paralytic shellfish poisoning: post-mortem analysis of tissue and body fluid samples from human victims in the Patagonia fjords. Toxicon 2004;43:149–158.
32. Penzotti JL, Fozzard HA, Lipkind GM, Dudley SC Jr. Differences in saxitoxin and tetrodotoxin binding revealed by mutagenesis of the Na^+ channel outer vestibule. Biophys J 1998;75:2647–2657.
33. Choudhary G, Shang L, Li X, Dudley SC Jr. Energetic localization of saxitoxin in its channel binding site. Biophys J 2002;83:912–919.
34. Su Z, Sheets M, Ishida H, Li F, Barry WH. Saxitoxin blocks L-type ICa. J Pharmacol Exp Ther 2004;308:324–329.
35. Wang J, Salata JJ, Bennett PB. Saxitoxin is a gating modifier of hERG K^+ channels. J Gen Physiol 2003;121: 583–598.
36. Hartigan-Go K, Bateman DN. Redtide in the Philippines. Hum Exp Toxicol 1994;13:824–830.
37. van Egmond HP, Aune T, Lassus P, Speijers GJ, Waldock M. Paralytic and diarrheic shellfish poisons: occurrence in Europe, toxicity, and analysis and regulation. J Nat Toxins 1993;2:41–83.
38. Todd E, Avery G, Grant GA, Fenwick JC, Chiang R, Babiuk T. An outbreak of severe paralytic shellfish poisoning in British Columbia. Can Commun Dis Rep 1993;19: 99–102.
39. Gessner BD, Middaugh JP, Doucette GJ. Paralytic shellfish poisoning in Kodiak, Alaska. West J Med 1997;166:351–353.
40. Gessner BD, Bell P, Doucete GJ, Moczydlowski E, Poli MA, van Dolah F, Hall S. Hypertension and identification of toxin in human urine and serum following a cluster of mussel-associated paralytic shellfish poisoning outbreaks. Toxicon 1997;35:711–722.
41. Gessner BD, Middaugh JP. Paralytic shellfish poisoning in Alaska: a 20-year retrospective analysis. Am J Epidemiol 1995;141:766–770.
42. Sharifzadeh K, Ridley N, Waskiewicz R, Luongo P, Grady GF, De Maria A, et al. Paralytic shellfish poisoning—Massachusetts and Alaska. MMWR Morb Mortal Wkly Rep 1991;40:157–161.
43. Acres J, Gray J. Paralytic shellfish poisoning. Can Med Assoc J 1978;119:1195–1197.
44. Rodrigue DC, Etzel RA, Hall S, de Porras E, Velasquez OH, Tauxe RV, et al. Lethal paralytic shellfish poisoning in Guatemala. Am J Trop Med Hyg 1990;42:267–271.
45. Meyer KF. Food poisoning. N Engl J Med 1953;249:843–852.
46. Horwitz W, editor. Official methods of analysis of AOAC International. 17th ed. Gaithersburg, MD: AOAC International; 2000, Official method 959.08.

47. Poletti R, Milandri A, Pompei M. Algal Biotoxins of Marine Origin: New Indications from the European Union. Vet Res Commun 2003;27(suppl 1):S173-S182.
48. Van Egmond HP, Jonker KM, Poelman M, Scherpenisse P, Stern AG, Wezenbeek P, et al. Proficiency studies on the determination of paralytic shellfish poisoning toxins in shellfish. Food Addit Contam 2004;21:331–340.
49. Horwitz W, editor. Official methods of analysis of AOAC International. 17th ed. Gaithersburg, MD: AOAC International; 2000, Official method 2005.06.
50. de Carvalho M, Jacinto J, Ramos N, de Oliveira V, Pinho e Melo T, de Sa J. Paralytic shellfish poisoning: clinical and electrophysiological observations. J Neurol 1998;245:551–554.
51. Cheng H-S, Chua SO, Hung J-S, Yip K-K. Creatine kinase MB elevation in paralytic shellfish poisoning. Chest 1991; 99:1032–1033.
52. Franz DR, LeClaire RD. Respiratory effects of brevetoxin and saxitoxin in awake guinea pigs. Toxicon 1989;27:647–654.
53. Rand PW, Lawrence FH, Pirone LA Jr, Lavigne JR, Lacombe E. The application of charcoal hemoperfusion to paralytic shellfish poisoning. J Maine Med Assoc 1977;68: 147–155.

Chapter 33

CIGUATERA FISH POISONING and CIGUATOXINS

HISTORY

The name ciguatera derives from the Spanish name, "cigua." Spanish immigrants to the Caribbean used cigua to describe an illness associated with the poisonous sea snail *Turbo pica*, which produces a clinical syndrome similar to ciguatera in the Spanish Antilles.[1] Prohibitions on the eating of fish as a means of avoiding ciguatera reportedly date back to Alexander the Great and the T'ang Dynasty (618–907 AD) in China.[2] Although the journal of Captain William Bligh contains a description of symptoms consistent with ciguatera after eating dolphin (mahi-mahi), dolphin is an uncommon source of ciguatera poisoning. More accurate accounts of suspected ciguatera poisoning occurred in the log of the Spanish explorer Pedro de Quiros after eating snapper caught near the New Hebrides (i.e., Vanuatu Islands between Solomon Islands and New Caledonia) in 1606 and in the journal of Captain Cook's botanist after eating a similar fish in the same waters during 1774.[3,4]

EXPOSURE

Source

Ciguatera poisoning occurs following the ingestion of large carnivorous fish, which accumulate ciguatoxins through the food chain. This type of seafood-associated poisoning occurs worldwide as a result both of the migration of affected fish from endemic areas and the transport of affected fish harvested in endemic areas. Benthic dinoflagellates, primarily *Gambierdiscus toxicus* Adachi & Fukuyo, form ciguatoxin precursors (CTX4A) while living in association with bottom-dwelling macro algae attached to dead coral or in the detritus of coral reefs.[5,6] The dinoflagellate, *Ostreopsis lenticularis* Fukuyo, also produces ciguatoxins when grown in cultures with *Pseudomonas* species.[7] Biotransformation and bioaccumulation of these precursors occur as they accumulate in the food chain when carnivorous fish prey on the herbivorous fish that feed on the toxin-containing detritus, algae, or plants.[8] Fish do not appear affected by the presence of ciguatoxins. Weather patterns and natural disasters can destroy coral reefs and increase the concentration of benthic dinoflagellates, such as *Gambierdiscus toxicus* Adachi & Fukuyo, particularly on the windward side of tropical islands. Although scombroid poisoning is the most common seafood-associated illness, ciguatera poisoning is the most common cause of illness resulting from the ingestion of a marine toxin.[9] The most common areas where ciguatoxin is endemic are the western Indian Ocean, Caribbean, and the South Pacific (Solomon Islands, Marshall Islands, Kiribati, Tuvalu, Great Barrier Reef off the coast of Australia).[10] In a study of an outbreak of ciguatera fish poisoning in Hong Kong, fish implicated in the outbreak were coral fish from the South Pacific including brown-marbled grouper, camouflage grouper (*Epinephelus polyphekadion* Bleeker), humphead wrasse or tapiro (*Cheilinus undulatus* Rüppell), leopard coralgrouper (*Plectropomus leopardus*-Lacepède), Russell's snapper or Moses perch (*Lutjanus russelli* Bleeker), and two-spot red snapper (*Lutjanus*

Medical Toxicology of Natural Substances, by Donald G. Barceloux, MD

bohar Forsskål).[11] A relatively low number of strains of *Gambierdiscus toxicus* produce gambiertoxins, which are biotransformed by marine animals into more polar ciguatoxins; therefore, the distribution of ciguatera poisoning is uneven.[12] Substantial seasonal and yearly variation in ciguatoxin formation occurs as a result of changes in environmental and metabolic conditions, such as increased sea-surface temperature.[13] Consequently, the occurrence of ciguatera poisoning is sporadic and localized to specific areas regardless of the season. Visible blooms of *Gambierdiscus toxicus* are not necessary for the development of sufficient concentrations of ciguatoxins in carnivore fish to produce ciguatera poisoning. In addition, the varying toxin profile for ciguatoxins results in variation in the clinical presentation of ciguatera poisoning between different geographic locations. Although the distribution of ciguatera is worldwide, the areas of highest risk are the subtropical and tropical waters of the Pacific Ocean, Caribbean Sea, and the western Indian Ocean. Figure 33.1 demonstrates in dark shaded areas where there is a moderate to high risk of ciguatoxin in carnivorous fish.

Over 400 species of fish are vectors for ciguatera poisoning,[14] and this type of marine poisoning accounts for more than half of all foodborne outbreaks related to fish in the United States.[15] Table 33.1 lists the most common fish families associated with ciguatera. The fish most commonly involved in ciguatoxin poisoning are reef-dwelling benthic species (e.g., barracuda, grouper, red snapper, parrot fish, surgeon fish, sea bass, mackerel, emperor fish, moray eel, amber jack), which inhabit marine waters between the latitudes of 35°N and 35°S (North Carolina to Buenos Aires). The endemic areas of ciguatera in the United States are southern Florida and Hawaii. The shipping of flash frozen food allows the occurrence of ciguatera poisoning in nonendemic areas. Deep sea pelagic fish (e.g., tuna, dolphin, sailfish, marlin, flying fish) do not usually contain clinically significant amounts of ciguatoxins.[16] The most common season for ciguatera poisoning is April through August when the seawater is warm.

Food Processing

Because these toxins are tasteless and odorless, ciguatoxin-contaminated fish do not appear tainted. Consequently, contamination of fish by ciguatoxins is difficult to detect without diagnostic testing. Cooking, salting, drying, smoking, or freezing do not inactivate ciguatoxins; therefore, sufficient quantities of these toxins survive food preparation processes. Unlike marine toxins that are excreted from contaminated organisms (e.g., shellfish), ciguatoxin-contaminated fish remain toxic for long periods of time despite the absence of ciguatoxin in the diet of the affected fish. In a study of 66 specimens of the two-spot red snapper (*Lutjanus bohar* Forsskål) kept in holding ponds for 3–30 months, there was no significant decline in ciguatera toxins at the end of the study periods.[17]

TOXINS

Physiochemical Properties

Ciguatoxins are a family of highly oxygenated, cyclic polyether compounds. Ciguatoxins are acid stable, lipid

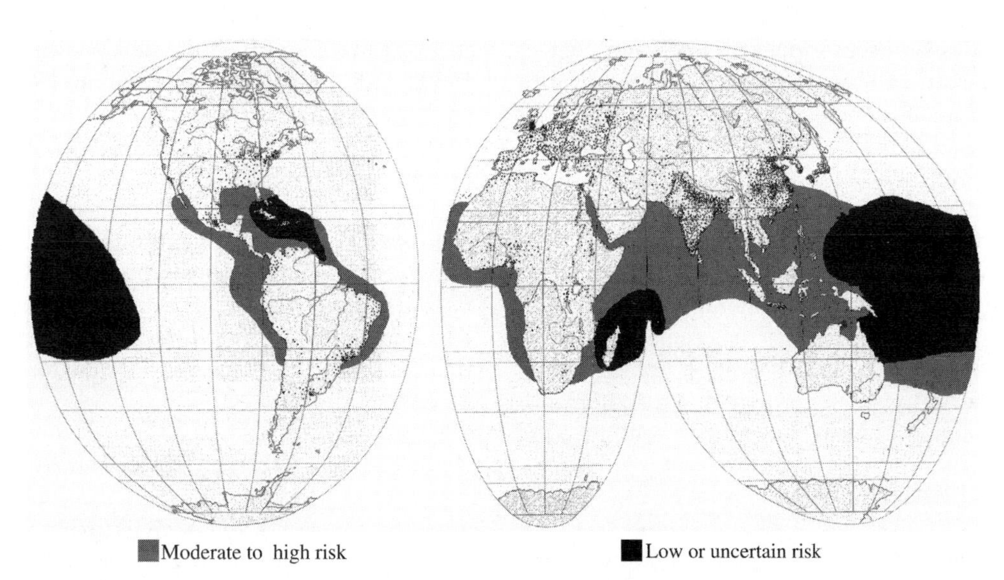

FIGURE 33.1. Global distribution of ciguatera with moderate to high risk in dark shaded areas and low to uncertain risk in light shaded areas. From Lewis RJ. The changing face of ciguatera. Toxicon 2001;39:98. Reprinted with permission from Elsevier.

TABLE 33.1. Fish Species Commonly Involved with Ciguatera Poisoning Worldwide

Family	Common Name	Americas	Western Pacific	Central Pacific	Eastern Pacific	Asia/India
Acanthuridae		+	+	+	+	
Balistidae		+	+	+	+	
Belonidae		+	+			
Carangidae		+	+	+	+	+
Carcharhinidae	Shark	+	+	+		
Holocentridae		+	+			
Labridae		+	+	+		
Lethrinidae		+	+	+		
Lutjanidae	Red snapper	+	+	+	+	+
Mugilidae		+				
Mullidae		+	+			
Murendiae		+	+	+		
Scaridae		+	+	+		
Scombridae	Mackerels	+	+	+	+	+
Serranidae	Groupers, coral trout	+	+	+	+	+
Sphyraenidae	Barracuda	+	+	+	+	+

Source: From Ref 8.

soluble, and heat-resistant up to 120°C (248°F).[18] Reef-dwelling fish may contain a variety of toxins beside ciguatoxins including maitotoxin, scaritoxin, okadaic acid, prorocentrolide, and palytoxin. In the Pacific Ocean, the main ciguatoxin is Pacific-ciguatoxin-1 (P-CTX-1), which has a molecular weight of 1112 Da and a molecular formula of $C_{60}H_{86}O_{19}$. Figure 33.2 displays the chemical structure of the main ciguatoxin (CTX-1), the precursor (CTX-4B), and five congeners. Maitotoxin-1 is a very large molecule (3422 Da, $C_{160}H_{225}S_2O_{74}$) isolated from cultures of *Gambierdiscus toxicus* Adachi & Fukuyo that is one of the most potent nonprotein toxins (intraperitoneal $LD_{50} \approx 50$ ng/ kg body weight).[19] This toxin is a water-soluble compound that activates voltage-sensitive and receptor-operated calcium channels in the plasma membrane of cells. Ciguatoxins are structurally similar to brevetoxins and yessotoxin.[20] Case reports suggest that ciguatoxins are distributed throughout various body fluids (e.g., breast milk, urine, sperm) as well as across the placenta.[21] Other case reports associated penile pain after ejaculation and the development of dysuria and vaginal discomfort in women following intercourse with men intoxicated with ciguatera.[22,23] Case reports suggest that ciguatoxins are excreted in breast milk, and breastfed infants can develop ciguatera poisoning when breastfed from mothers with ciguatera poisoning.[24,25]

Mechanism of Toxicity

Ciguatoxins prolong the opening of the sodium pores on excitable membranes by activating voltage-dependent sodium channels in muscle cells and nerve cells in small and large myelinated and unmyelinated fibers. Ciguatoxins and brevetoxins bind to site 5 on the neuronal voltage sensitive sodium channels that includes segments S5 of domain IV and S6 of domain I.[26] These sodium channels fail to close, resulting in spontaneous depolarizations, the influx of sodium and water into the cells, and swelling of the nodes of Ranvier. Spontaneous activation of the sodium channels at resting membrane potentials accounts for the neurological effects of ciguatoxin including an acute polyneuropathy, manifest by sensory, small-fiber dysfunction. Pretreatment with tetrodotoxin prevents the ciguatoxin-induced opening of the sodium channel. The role of maitotoxin, scaritoxin, and other polyether compounds in ciguatera intoxication is not well defined. Maitotoxin (CAS RN: 59392-53-9) is a large, complex molecule that affects calcium channels rather than sodium channels;[27] therefore, the pharmacological effects of these two toxins are probably different. However, the low oral toxicity and the limited bioaccumulation of maitotoxin in the food chain limit the contribution of maitotoxin to the etiology of ciguatera fish poisoning.[10] Spectroscopic data of compounds isolated from cultures of marine dinoflagellate *Gambierdiscus toxicus* indicate that scaritoxin is the ciguatoxin congener, ciguatoxin-4A (CTX4A, 52-epiciguatoxin-4B).[5]

DOSE RESPONSE

The clinical response to ciguatoxins depends on the dose, potency of the ciguatoxin, and individual sensitivity. The two main classes of ciguatoxins are the Pacific ciguatoxins and Caribbean ciguatoxins.[28] Differences in

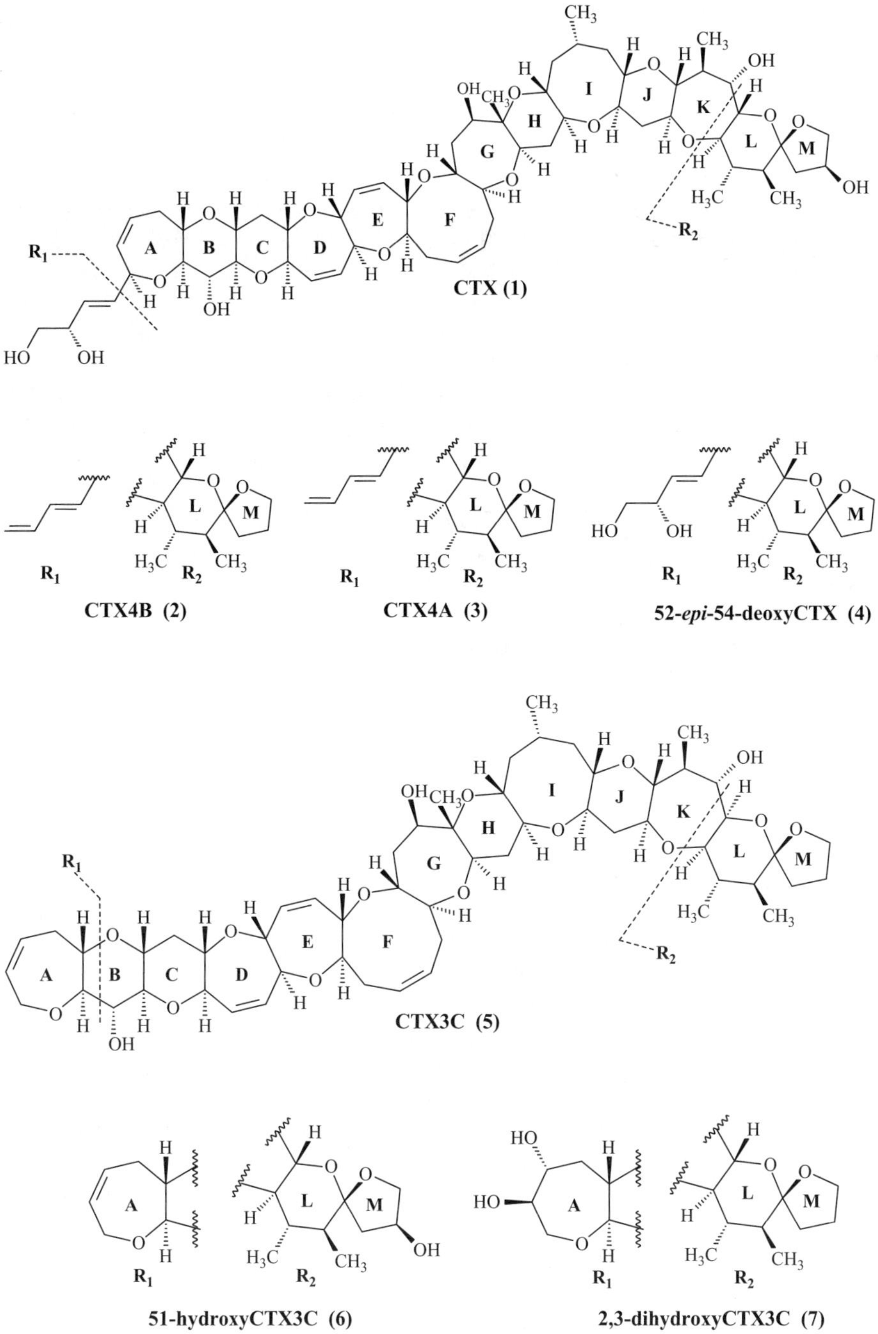

FIGURE 33.2. Ciguatoxin structures. (1) Ciguatoxin (CTX-1, P-CTX-1), (2) ciguatoxin-4B (CTX-4B), (3) ciguatoxin-4A (CTX-4A), (4) 52-epi-54-deoxyCTX, (5) ciguatoxin-3C (CTX-3C), (6) 51-hydroxyCTX3C, and (7) 2,3-dihydroxyCTX3. CTX-4B is the precursor of CTX-1.

the potencies of these two classes of ciguatoxins account for regional variation in the clinical presentation of ciguatera poisoning with Pacific ciguatoxin-1 being about 10 times more potent than Caribbean ciguatoxin-1.[29] Ciguatoxin-1 causes ciguatera poisoning at concentrations as low as 0.1 μg/kg (0.1 ng/g) flesh with a typical range up to approximately 5 μg/kg (5 ng/g) flesh.[30] The ingestion of an estimated 1 kg of coral trout (*Plectropomus maculatus* Bloch) by a 34-year-old man was associated with severe gastrointestinal symptoms, pruritus, anxiety, depression, burning, arthralgias, and lethargy, but his vital signs remained stable.[31] The estimated dose of ciguatoxin by bioassay was 0.25 MU/g flesh (1.3 ng ciguatoxin/g flesh). The estimated safe level for Pacific ciguatoxin-1 (P-CTX-1) in commercial fish is about 0.01 μg/kg (0.01 ng/g) tissue.[32] Mild toxicity probably develops following the ingestion of about 0.05 μg (P-CTX-1) or about 0.001 μg/kg body weight. The severity of ciguatera poisoning depends on several factors including cumulative dose, the profile of ciguatoxins and associated toxins (maitotoxin, scaritoxin, palytoxin, okadaic acid, brevetoxins), prior exposure, individual susceptibility, and age. For example, neurological symptoms predominate in cases of ciguatera from the Pacific Ocean, whereas gastrointestinal symptoms predominate in cases from the Caribbean Sea.[33] The highest risk of developing ciguatera poisoning is associated with the ingestion of large (>5 pounds) carnivorous fish, particularly the head, visceral organs, eggs, and ovaries of barracuda, grouper, red snapper, sharks, and amberjacks.

CLINICAL RESPONSE

The severity of symptoms and the clinical manifestations of ciguatera are highly variable.[34,35] The classic presentation of ciguatera poisoning is the combination of an early acute gastrointestinal phase (watery diarrhea, nausea, vomiting, abdominal cramps) within 3–6 hours after ingestion of contaminated fish followed by neurological symptoms.[36,37] Typical neurological symptoms include distal and perioral paresthesias, numbness, and cold allodynia (i.e., dysesthesias when touching cold objects). Neurological effects predominate in the Indo-Pacific region, whereas gastrointestinal effects are more prominent in the Caribbean.[28] Table 33.2 lists the most common effects of ciguatera poisoning in a case series of ciguatera poisoning from the Pacific Ocean. The onset of the first symptoms is highly variable with a range of 1 to 48 hours, but the onset of gastrointestinal symptoms more than 24 hours after ingestion of fish indicates that ciguatera poisoning is unlikely. Generalized symptoms include weakness, myalgias, pruritus, arthralgias, dizziness, and a desquamative or erythematous rash. Cardiorespiratory effects develop in a relatively small (<15%) subset of patients with ciguatera poisoning, primarily in patients ingesting large, ciguatoxin-contaminated fish[38] or fish eggs.[39] These effects include hypotension, respiratory insufficiency, and bradycardia in the more severe cases. Occasionally, intermittent hypertension develops. Fatalities occur rarely, primarily from respiratory depression and seizures. The differential diagnosis includes neurotoxic shellfish poisoning, paralytic shellfish poisoning, puffer fish poisoning (tetrodotoxin), botulism, bacteremia, and scombroid fish poisoning. The latter does not cause prominent neurotoxic effects, whereas ciguatera poisoning does not usually produce intense flushing. Although a case report associated transient facial palsy and possible weakness of the hands in a neonate born with ciguatoxin poisoning of the mother at term,[40] no detectable neonatal abnormalities occurred in children born to mothers experiencing ciguatoxin intoxication during their first and second trimesters.[21]

TABLE 33.2. Approximate Frequency of Adverse Effects Associated with Ciguatera Poisoning from Pacific Ocean Fish

Adverse Effect	Occurrence (%)
Extremity paresthesias	89
Circumoral paresthesias	88
Paradoxical dysesthesias	87
Arthralgia	86
Myalgia	85
Diarrhea	73
Asthenia	60
Chills and headache	60
Pruritus	44
Nausea	43
Abdominal pain	42
Vomiting	39
Perspiration	29
Tearing	25
Giddiness	23
Dental pain	21
Neck stiffness	19
Ataxia	15
Dysuria	13
Dyspnea	12
Paresis	10

Source: Adapted from Ref 58.

Gastrointestinal Effects

Gastrointestinal symptoms occur in a majority of patients with ciguatera poisoning. The onset of the gastrointestinal symptoms ranges from minutes to 24 hours after ingestion of ciguatoxin-containing fish with resolution of gastrointestinal symptoms occurring within approximately 12–48 hours. In some cases of ciguatera poisoning, gastrointestinal symptoms do not develop.[10] Animal studies indicate that the diarrhea accompanying ciguatera poisoning does not produce pathological changes in the mucosal lining of the gastrointestinal tract.[41]

Neurological Effects

Neurological symptoms can develop over the first few days after ingestion of ciguatoxin. Prominent neurological symptoms include paresthesias (i.e., abnormal sensations), dysesthesias (i.e., disagreeable sensation following ordinary stimulation), and headache. Paresthesias of the mouth and extremities are classic hallmarks of ciguatera poisoning, and dysesthesias are an early manifestations of the neurological effects of ciguatera. Several reviews indicate that reversal of hot and cold sensation is almost pathognomonic for this toxin, but this symptom (cold allodynia) results more from alteration of sensation (i.e., cold stimulus producing intense burning and sharp, lancinating pain) rather than cold objects feeling hot.[42] Case reports document painful sensations, such as electric shock, dry ice, burning, or tingling during ciguatera fish poisoning. A case series of 5 patients with ciguatera poisoning did not detect gross alteration of temperature perception, and reversal of hot and cold sensation did not occur during testing.[43] Similar dysesthesias occur during neurotoxic shellfish poisoning.[44] Additional adverse effects associated with ciguatera poisoning include hypersalivation, blurred vision, odontalgia, dysphagia, dysuria, tremor, athetosis, and ataxia.[38,45,46] Typically, ciguatera poisoning is not associated with cognitive changes.[47] Alteration of moods (e.g., depression, anxiety), nightmares, and hallucinations occur occasionally.[48] These latter symptoms should be differentiated from ichthyoallyeinotoxism, which is a rare type of seafood poisoning associated with hallucination and nightmares.[49] In this type of poisoning, symptoms of inebriation, poor coordination, loss of balance, and generalized malaise begin within a few hours of the ingestion of contaminated fish (sea bream, coral grouper, common mullet, convict surgeonfish, rabbitfish, goatfish) followed in a few hours by delirium hallucinations, depression, nightmares, and a sense of impending doom. Patients with ichthyoallyeinotoxism typically recover within 36 hours in contrast to the prolonged course of paresthesias and alteration of sensation associated with ciguatera poisoning.[48]

Sequelae

In a small subset of patients with ciguatera, adverse reactions persist for months to years.[50,51] These symptoms include protracted pruritus, fatigue, postural lightheadedness, arthralgias, and paresthesias. Certain foods (e.g., fish, chicken, pork, peanuts), ethanol consumption, heavy exercise, and periods of stress are associated with exacerbation of these chronic symptoms.[52] These sensitivities can persist long after the initial symptoms resolved. Ciguatera poisoning does not confer immunity to subsequent attacks, and subsequent attacks are often more severe than the original attack.[6] There are few data on the effect of ciguatoxin on the fetus. The occurrence of ciguatera poisoning in the first and second trimester of pregnancy was not associated with abnormal fetal development, and both infants were born full-term with APGAR scores of 9 (1 minute) and 9 (5 minutes).[31,53] Two days before expected birth, a term mother developed severe ciguatera poisoning.[54] She experienced tumultuous and bizarre fetal movements that resolved within 36 hours. Fetal heart monitoring did not reveal any abnormalities, but she subsequently delivered via Caesarean section a 3.8 kg male neonate with left-sided facial palsy. Examination the following day suggested some possible myotonia of the small muscles of the hands of the neonate. At 6 weeks of age, the child appeared normal with the exception that he had not smiled.

DIAGNOSTIC TESTING

Analytical Methods

Ciguatera poisoning is a clinical diagnosis based on the recent consumption of fish known to cause ciguatera along with the abrupt onset of gastrointestinal and neurological symptoms consistent with ciguatera poisoning. The most common method for the confirmation of ciguatoxins in fish is the mouse bioassay. *In vivo* testing with bioassays involves the oral dosing of flesh or the intraperitoneal injection of crude extracts of fish into mice. The detection of ciguatoxins is complicated because of a number of factors including multiple structural forms, low concentrations associated with ciguatera, and the difficulty synthesizing fragments of the basic structure of ciguatoxins.[55] There are no well-validated analytical methods for the rapid detection of ciguatoxins in human blood and urine samples. Immunoassays are complicated by the structural similarity of low-potency ciguatoxins, which cross-react with the antibodies to P-CTX-1; however, some immunoassays are available commercially (e.g., immunobead assay, Cigua-Check™, Oceanit Test Systems, Honolulu, HI).[56] Typically, these immunoassays do not detect ciguatoxin concentrations $<0.05\,\mu g/kg$ fish, and some false-positive reactions occur as a result of nonspecific binding.[29] Gradient reversed phase highperformance liquid chromatography/tandem mass spectrometry allows detection of Pacific ciguatoxin-1 (P-CTX-1) in the sub-ppb concentration of samples of fish flesh.[57]

Abnormalities

There are no characteristic laboratory abnormalities associated with ciguatera poisoning. The complete blood count, creatine kinase, and electrocardiogram are usually normal.[58] In cases with severe gastrointestinal symptoms, laboratory evidence of electrolyte imbalance and dehydration can occur. Reversible T-wave changes can develop during ciguatera poisoning, but myocardial injury does not usually occur. Severe cases of ciguatera poisoning are associated with slowing of sensory conduction velocity, prolongation of the absolute refractory period, and increased relative refractory period.[59] However, electrophysiological studies are usually normal in most cases of ciguatera poisoning with the exception of normocalcemic latent tetany manifest by doublets, triplets, and multiplets with relative electrical silence between groups of discharges.[60]

TREATMENT

Treatment for ciguatera poisoning primarily is supportive. Patients with severe gastrointestinal symptoms may require fluid and electrolyte replacement for dehydration. Bradycardia typically responds to atropine, and hypotension is treated with fluids and vasopressors as needed.

Decontamination

Decontamination measures are not usually necessary because of the delay between the ingestion of contaminated fish and the onset of symptoms. There are no clinical data to determine if the use of activated charcoal improves the outcome of ciguatera poisoning.

Antidotes

There are no antidotes for ciguatera, and most patients require only supportive treatment. Case reports suggest that the intravenous administration of 20% mannitol reduces some of the symptoms associated with ciguatera poisoning, particularly when administered within 24–48 hours of ingestion.[61] However, there is a lack of randomized clinical trials to support the efficacy of mannitol for the treatment of ciguatera poisoning. A double-blind randomized clinical trial comparing mannitol infusions with normal saline in 50 patients with sensory paresthesias from ciguatera poisoning did not demonstrate a difference in symptoms between the two groups at 24 hours after therapy.[62]

The usual intravenous mannitol dose for ciguatera poisoning is 10 mL 20% mannitol (1 g)/kg body weight over 30–45 minutes. Patients receiving mannitol should be well hydrated to correct any pre-existing fluid or electrolyte imbalances. A second dose of 5 mL 20% mannitol (0.5 g)/kg body weight can be repeated in 3–4 hours if symptoms do not improve. The lack of a response to two infusions indicates that further use of mannitol probably would not be efficacious. Mannitol infusions do not usually resolve all symptoms of ciguatera poisoning. Although serious complications of a mannitol infusion are unusual in well-hydrated patients, adverse effects include pain at the infusion site and thrombophlebitis.[63]

Supplemental Care

Symptomatic treatment includes the use of antiemetics (prochlorperazine, metoclopramide), antidiarrheal agents, and antipruritics (H_1 and H_2 histamine blockers). There are inadequate clinical data to determine if the use of antiemetics or antidiarrheal agents exacerbates the course of ciguatera poisoning. The bradycardia associated with ciguatera usually responds to atropine, if needed to support blood pressure.[64] Hospitalization is appropriate for those patients with significant volume depletion or those who are severely symptomatic after initial treatment. Unproven therapies for the treatment of persistent dysesthesias and pruritus include amitriptyline (25 mg twice daily),[65,66] tocainide, and gabapentin (400 mg orally t.i.d.).[67] There are inadequate data to support the use of calcium gluconate or corticosteroids during ciguatera intoxication.[42]

References

1. Gudger EW. Poisonous fishes and fish poisonings, with special reference to ciguatera in the West Indies. Am J Trop Med 1930;10:43–55.
2. Steinfeld AD, Steinfeld HJ. Ciguatera and the voyage of Captain Bligh. JAMA 1974;228:1270–1271.
3. Doherty MJ. Captain Cook on poison fish. Neurology 2005;65:1788–1791.
4. Withers NW. Ciguatera fish poisoning. Annu Rev Med 1982;33:97–111.
5. Satake M, Ishibashi Y, Legrand AM, Yasumoto T. Isolation and structure of ciguatoxin-4A, a new ciguatoxin precursor, from cultures of dinoflagellate *Gambierdiscus toxicus* and parrotfish *Scarus gibbus*. Biosci Biotechnol Biochem 1996;60:2103–2105.
6. Bagnis R, Chanteau S, Chungue E, Hurtel JM, Yasumoto T, Inoue A. Origins of ciguatera fish poisoning: A new dinoflagellate, *Gambierdiscus toxicus* Adachi and Fukuyo, definitely involved as the causal agent. Toxicon 1980;18:199–208.
7. Tosteson TR, Ballantine DL, Tosteson CG, Hensley V, Bardales AT. Associated bacterial flora, growth, and toxicity of cultured benthic dinoflagellates *Ostreopsis lenticularis* and *Gambierdiscus toxicus*. Appl Environ Microbiol 1989;55:137–141.
8. Glaziou P, Legrand A-M. The epidemiology of ciguatera fish poisoning. Toxicon 1994;32:863–873.
9. Clark RF, Williams SR, Nordt SP, Manoguerra AS. A review of selected seafood poisonings. Undersea Hyperbar Med 1999;26:175–185.
10. Lewis RJ. The changing face of ciguatera. Toxicon 2001;39:97–106.
11. Wong C-K, Hung P, Lee KL, Kam K-M. Study of an outbreak of ciguatera fish poisoning in Hong Kong. Toxicon 2005;46:563–571.
12. Holmes MJ, Lewis RJ, Poli MA, Gillespie NC. Strain dependent production of ciguatoxin precursors (gambier-

toxins) by *Gambierdiscus toxicus* (Dinophyceae) in culture. Toxicon 1991;29:761–775.

13. Tosteson TR, Ballantine DL, Durst HD. Seasonal frequency of ciguatoxic barracuda in southwest Puerto Rico. Toxicon 1988;26:795–801.
14. Halstead BW. Current status of marine biotoxicology—an overview. Clin Toxicol 1981;18:1–24.
15. Engleberg NC, Morris JG Jr , Lewis J, McMillan JP, Pollard RA, Blake PA. Ciguatera fish poisoning: a major common-source outbreak in the U.S. Virgin Islands. Ann Intern Med 1983;98:336–337.
16. Smith W, Lieber B, Perrotta DM. Ciguatera fish poisoning—Texas, 1997. MMWR Morb Mortal Wkly Rep 1998; 47:692–694.
17. Banner AH, Helfrich P, Piyakarnchana T. Retention of ciguatera toxin by the red snapper, *Lutjanus bohar*. Copeia 1966;2:297–301.
18. Pottier I, Vernoux JP, Jones A, Lewis RJ. Characterisation of multiple Caribbean ciguatoxins and congeners in individual specimens of horse-eye jack (*Caranx latus*) by high-performance liquid chromatography/mass spectrometry. Toxicon 2002;40:929–939.
19. Lewis RJ, Holmes MJ, Alewood PF, Jones A. Ionspray mass spectrometry of ciguatoxin-1, maitotoxin-2 and -3, and related marine polyether toxins. Nat Toxins 1994;2: 56–63.
20. Yasumoto T. The chemistry and biological function of natural marine toxins. Chem Rec 2001;1:228–242.
21. Senecal PE, Osterloh JD. Normal fetal outcome after maternal ciguateric toxin exposure in the second trimester. J Toxicol Clin Toxicol 1991;29:473–478.
22. Lange WR, Lipkin KM, Yang GC. Can ciguatera be a sexually transmitted disease? Clin Toxicol 1989;27:193–197.
23. Ting JY, Brown AF, Pearn JH. Ciguatera poisoning: an example of public health challenge. Austr NZ J Public Health 1998;2:140–142.
24. Swift AE, Swift TR. The transmission of ciguatera toxicity: another first isn't. JAMA 1991;265:2339.
25. Blythe DG, de Sylva DP. Mother's milk turns toxic following fish feast. JAMA 1990;264:2074.
26. Lewis RJ, Sellin M, Poli MA, Norton RS, MacLeod JK, Sheil MM. Purification and characterization of ciguatoxins from moray eel (*Lycodontis javanicus*, Muraenidae). Toxicon 1991;29:1115–1127.
27. Zheng W, DeMattei JA, Wu J-P, Duan JJ, Cook LR, Oinuma H, Kishi Y. Complete relative stereochemistry of maitotoxin. J Am Chem Soc 1996;118:7946–7968.
28. Isbister GK, Kiernan MC. Neurotoxic marine poisoning. Lancet Neurol 2005;4:219–228.
29. Lehane L, Lewis RJ. Ciguatera: recent advances but the risk remains. Int J Food Microbiol 2000;61:91–125.
30. Lewis RJ. Ciguatoxins are potent ichthyotoxins. Toxicon 1992;30:207–211.
31. Fenner PJ, Lewis RJ, Williamson JA, Williams ML. A Queensland family with ciguatera after eating coral trout. Med J Aust 1997;166:473–475.
32. Lehane L. Ciguatera update. Med J Aust 2000;172: 176–179.
33. Lewis RJ. The changing face of ciguatera. Toxicon 2001;39:97–106.
34. Pottier I, Vernoux JP, Lewis RJ. Ciguatera fish poisoning in the Caribbean islands and Western Atlantic. Rev Environ Contam Toxicol 2001;168:99–141.
35. Goodman A, Williams TN, Maitland K. Ciguatera poisoning in Vanuatu. Am J Trop Med Hyg 2003;68:263–266.
36. Chateau-Degat ML, Dewailly E, Cerf N, Nguyen NL, Huin-Blondey MO, Hubert B, et al. Temporal trends and epidemiological aspects of ciguatera in French Polynesia: a 10-year analysis. Trop Med Int Health 2007;12:485–492.
37. Barton ED, Tanner P, Turchen SG, Tunget CL, Manoguerra A, Clark RF. Ciguatera fish poisoning a Southern California epidemic. West J Med 1995;163:31–35.
38. Bagnis R, Kuberski T, Langier S. Clinical observations in 3009 cases of ciguatera (fish poisoning) in the South Pacific. Am J Trop Med Hyg 1979;28:1067–1073.
39. Hung Y-M, Hung S-Y, Chou K-J, Huang N-C, Tung C-N, Hwang D-F, et al. Short report: persistent bradycardias caused by ciguatoxin poisoning after barracuda fish eggs ingestion in southern Taiwan. Am J Trop Med Hyg 2005; 73:1026–1027.
40. Pearn J, Harvey P, De Ambrosis W, Lewis R, McKay R. Ciguatera and pregnancy. Med J Aust 1982;1:57–58.
41. Fasano A, Hokama Y, Russell R, Morris JG Jr, Diarrhea in ciguatera fish poisoning: preliminary evaluation of pathophysiological mechanisms. Gastroenterol 1991;100: 471–476.
42. Palafox NA, Buenconsejo-Lum LE. Ciguatera fish poisoning: review of clinical manifestations. J Toxicol Toxin Rev 2001;20:141–160.
43. Cameron J, Capra MF. The basis of the paradoxical disturbance of temperature perception in ciguatera poisoning. Clin Toxicol 1993;31:571–579.
44. Sakamoto Y, Lockey RF, Krzanowski JJ Jr. Shellfish and fish poisoning related to the toxic dinoflagellates. South Med J 1987;80:866–872.
45. Tatnall FM, Smith HG, Welsby PD, Turnbull PC. Ciguatera poisoning. Br Med J 1980;281:948–949.
46. Centers for Disease Control (CDC). Ciguatera fish poisoning—Texas, 1998, and South Carolina, 2004. MMWR Morb Mortal Wkly Rep 2006;55:935–937.
47. Friedman MA, Arena P, Levin B, Fleming L, Fernandez M, Weisman R, et al. Neuropsychological study of ciguatera fish poisoning: a longitudinal case-control study. Arch Clin Neuropsychol 2007;22:545–553.
48. Quod JP, Turquet J. Ciguatera in Reunion Island (SW Indian Ocean): epidemiology and clinical patterns. Toxicon 1996;34:779–785.
49. de Haro L, Pommier P. Hallucinatory fish poisoning (ichthyoallyeinotoxism): two case reports from the western

Mediterranean and literature review. Clin Toxicol 2006;44: 185–188.

50. Chan TY, Kwok TC. Chronicity of neurological features in ciguatera fish poisoning. Hum Exp Toxicol 2001;20: 426–428.
51. Ng S, Gregory J. An outbreak of ciguatera fish poisoning in Victoria. Commun Dis Intell 2000;24:344–346.
52. Gillespie NC, Lewis RJ, Pearn JH, Bourke AT, Holmes MJ, Bourke JB, Shields WJ. Ciguatera in Australia. Occurrence, clinical features, pathophysiology and management. Med J Aust 1986;145:584–590.
53. Senecal P-E, Osterloh JD. Normal fetal outcome after maternal ciguateric toxin exposure in the second trimester. Clin Toxicol 1991;29:473–478.
54. Pearn J, Harvey P, De Ambrosis W, Lewis R, McKay R. Ciguatera and pregnancy. Med J Aust 1982;1:57–58.
55. Hokama Y, Abad MA, Kimura LH. A rapid enzyme immunoassay for the detection of ciguatoxin in contaminated fish tissues. Toxicon 1983;21:817–824.
56. Hokama Y, Takenaka WE, Nishimura KL, Ebesu JS, Bourke R, Sullivan PK. A simple membrane immunobead assay for detecting ciguatoxin and related polyethers from human ciguatera intoxication and natural reef fishes. J AOAC Int 1998;81:727–735.
57. Pottier I, Hamilton B, Jones A, Lewis RJ, Vernoux JP. Identification of slow and fast-acting toxins in a highly ciguatoxic barracuda (*Sphyraena barracuda*) by HPLC/MS and radiolabeled ligand binding. Toxicon 2003;42: 663–672.
58. Swift AE, Swift RT. Ciguatera. Clin Toxicol 1993;31: 1–29.
59. Cameron J, Flowers AE, Capra MF. Electrophysiological studies on ciguatera poisoning in man (Part II). J Neurol Sci 1991;101:93–97.
60. Butera R, Prockop LC, Buonocore M, Locatelli C, Gandini C, Manzo L. Mild ciguatera poisoning: case reports with neurophysiological evaluations. Muscle Nerve 2000;23: 1598–1603.
61. Pearn JH, Lewis RJ, Ruff T, Tait M, Quinn J, Murtha W, et al. Ciguatera and mannitol: experience with a new treatment regimen. Med J Aust 1989;151:77–80.
62. Schnorf H, Taurarii M, Cundy T. Ciguatera fish poisoning a double-blind randomized trial of mannitol therapy. Neurol 2003;58:873–880.
63. Ting JY, Brown AF. Ciguatera poisoning: a global issue with common management problems. Eur J Emerg Med 2001;8:295–300.
64. Geller RJ, Olson KR, Senecal PE. Ciguatera fish poisoning in San Francisco, California, caused by imported barracuda. West J Med 1991;155:639–642.
65. Davis RT, Villar LA. Symptomatic improvement with amitriptyline in ciguatera fish poisoning. N Engl J Med 1986;315:65.
66. Bowman PB. Amitriptyline and ciguatera. Med J Aust 1984;140:802.
67. Perez CM, Vasquez PA, Perret CF. Treatment of ciguatera poisoning with gabapentin. N Engl J Med 2001;344: 692–693.

Chapter 34

PUFFER FISH POISONING and TETRODOTOXIN

HISTORY

The hieroglyphics of ancient Egypt contain references to puffer fish dating back to 2700 BC. Ancient Chinese literature also contains references to the dangers of eating puffer fish. The highest number of fatalities from the consumption of puffer fish occurs in Japan where puffer fish (fugu) is a traditional delicacy. The poems of the Edo era (17th–19th centuries) indicate that fugu was a popular food during this period. In Japan, the oldest record of puffer poisoning occurred in the Nara and Heian eras (200 AD).[1] The log of Captain Cook, written in September, 1774 off New Caledonia, contains a description of a puffer fish as well as weakness and paresthesias that resulted from the ingestion of the roe and liver of this fish.[2,3] Although a crude extract of puffer fish containing tetrodotoxin was isolated in 1909, the structure of tetrodotoxin was not elucidated until 1964.[4,5]

The administration of tetrodotoxin can cause the "locked-in" syndrome characterized by complete flaccidity with retention of consciousness. Investigation into zombies associated with Haitian voodoo culture suggests that the pharmacological agents used in these rituals contain ingredients from one or more species of tetrodotoxin-containing puffer fish (*Diodon hystrix* L., *Diodon holacanthus* L., *Sphoeroides testudineus* L.).[6] Gas chromatography/mass spectrometry and liquid chromatography/mass spectrometry analysis of a powder used by Haitian voodoo sorcerers demonstrated the presence of an alkaline degradation product of tetrodotoxin (2-amino-6-hydroxymethyl-8-hydroxyquinazoline) after base treatment.[7] Direct thermospray mass spectral activity detected tetrodotoxin and an isomer.

EXPOSURE

Source

Puffer fish belong to the Tetraodontidae family (blowfish, globefish, puffer, rabbitfish, swellfish, toadfish, sharp-nosed puffer) of the order Tetraodontiformes, which is distributed worldwide. Tetrodotoxin-containing puffer fish occur primarily in Japan and the Indo-Pacific Oceans. Although puffer fish near Japanese waters contain primarily tetrodotoxin, puffer fish from other locations contain saxitoxins. Puffer fish in southern Florida contain a sufficient amount of saxitoxin to produce symptoms consistent with paralytic shell fish poisoning.[8] Other families in the order Tetraodontiformes also contain tetrodotoxin including Balistidae (filefish, triggerfish), Molidae (ocean sunfish), Ostraciidae (boxfish), and Diodontidae (porcupine fish).[9] Puffer fish inhabit shallow waters of the temperate and tropical zones, and specimens range in size from 25 to 30 cm (10–12 inches). In addition to puffer fish, a variety of marine and terrestrial animals contain tetrodotoxin as listed in Table 34.1. The source of tetrodotoxin remains controversial. Certain bacteria (e.g., *Microbacterium arabinogalactanolyticum, Serratia marcescens, Vibrio alginolyticus, Vibrio fischeri*) and the actinomycete *Nocardiopsis dassonvillei* can produce tetrodotoxin.[10–12]

Medical Toxicology of Natural Substances, by Donald G. Barceloux, MD

TABLE 34.1. Presence of Tetrodotoxin in the Animal Kingdom

Phylum	Class	Family/Subfamily	Common Name
Chordata	Amphibia	Salamandridae	Newts
		Bufonidae	Atelopid Frogs
Chordata	Osteichthyes (Pisces)	Tetraodontidae	Blowfish, globefish, puffers, rabbitfish, swellfish, toadfish, sharp-backed puffers
		Diodontidae	burrfish, porcupine fish
		Gobiidae	yellowfin toxic goby
Mollusca	Gastropoda	Buccinidae, Bursidae, Terebridae, etc.	Marine snails
	Cephalopoda	Octopodidae	Blue-ringed octopus
Echinodermata	Asteroidea	Astropectinidae	Starfish
Arthropoda	Crustacea	Xanthidae	Crabs

Source: From Ref 9.

Salamanders, such as the California newt and the eastern salamander (*Ambystoma punctatum*), contain sufficient quantities of tetrodotoxin to cause death following ingestion.[13] Other species that contain tetrodotoxin include the yellowfin toxic goby [*Yongeichthys criniger* (Valenciennes in Cuvier and Valenciennes)], *Atelopus* frogs of Costa Rica, and the blue-ringed octopus (*Hapalochlaena maculosa*) as well as the ivory shell, the trumpet shell (*Charonia sauliae* Reeve), and gastropod mollusks (*Nassarius* spp., *Natica* spp.).[14,15] Seasonal and geographic variation occurs in the concentration of tetrodotoxin within puffer fish specimens. A study of two puffer fish species (*Takifugu niphobles* Jordan and Synder, *Takifugu alboplumbeus* Richard) in coastal waters near Hong Kong demonstrated moderate toxicity (100–1000 MU [mouse unit]) in the ovaries of both species and the liver of *T. niphobles* during non-spawning seasons.[16] The intestines and skin were weakly toxic (10–100 MU), whereas the flesh was not toxic (<10 MU) during the entire year. Contrary to popular belief, these two species were relatively less toxic during the spawning season. Accumulation of tetrodotoxin in the puffer fish does not alter the behavior of the puffer fish because of a mutation in the protein sequence of their sodium channels that results in insensitivity to the effects of tetrodotoxin.[17]

Food Processing

The Japanese puffer fish (fugu) is a delicacy in Japan, particularly tora-fugu (*Sphoeroides rubripes* Temminck & Schlegel) and ma-fugu (*Tetraodon porphyreus* Temminck & Schlegel). Specially licensed chefs in fugu restaurants prepare the puffer fish as thinly sliced "sashimi" fillets of muscle that contain only enough tetrodotoxin to cause mild perioral tingling and euphoria. Some gourmets prefer the liver ("kimo"), which can be highly toxic. Most cases of tetrodotoxin poisoning result from the ingestion of fish dishes prepared by inexperienced chefs. Tetrodotoxin is heat stable, and cooking does not inactivate this toxin. Although the importation of fugu into the United States is illegal, the smuggling of prepackaged fugu from Japan was associated with puffer fish poisoning that resolved without sequelae within one day.[18] US customs agents have also interdicted packages of fugu during attempted entry into the United States.[19]

TOXINS

Physiochemical Properties

Tetrodotoxin (CAS RN:4368-28-9) is the most potent marine nonprotein toxin with the exception of palytoxin and maitotoxin.[20] The molecular formula of tetrodotoxin is $C_{11}H_{17}N_3O_8$ and the molecular weight is 319.3 Da. Figure 34.1 demonstrates the chemical structure of tetrodotoxin. Analogues of tetrodotoxin present in puffer fish include 11-*nor*tetrodotoxin-6(R)-ol and other tetrodotoxin derivatives.[21] This highly polar, guanidine-substituted polysaccharide is soluble in acidic aqueous solutions, but is only slightly soluble in water and insoluble in most organic solvents. Strongly acidic or alkaline solutions deactivate tetrodotoxin. The distribution of this nonprotein, heat-stable neurotoxin within and between specimens of puffer fish is not uniform.[22] The highest concentrations of tetrodotoxin occur in the liver, roe, gonads, and skin. In a study of the Japanese marine puffer (*Chelonodon patoca* Hamilton, okinawafugu), all specimens contained tetrodotoxin.[23] For the specimens collected, the tetrodotoxin concentrations in various organs were as follows: skin, 60–6,700 MU/g; ovary, 25–670 MU/g; testes, 45–550 MU/g; muscle 2–390 MU/g; and liver, 5–380 MU/g. There is also substantial seasonal variability in tetrodotoxin concentrations in the same organs. Tetrodotoxin-producing strains of bacteria are

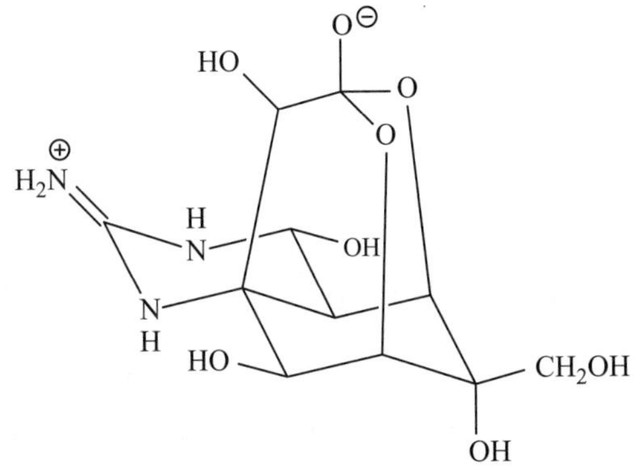

FIGURE 34.1. Chemical structure of tetrodotoxin.

present in the intestines of puffer fish, primarily from the family Vibrionaceae [e.g., *Beneckea alginolytica* (Miyamoto et al.) Baumann et al.].[24]

Mechanism of Toxicity

Tetrodotoxin blocks the voltage-gated sodium channel at the same site as saxitoxin. In contrast to most other fish toxins, which alter the sodium channel by binding to domains I–VI of the α-subunits of the sodium channel, tetrodotoxin occludes the outer pore of the sodium channel.[25] This marine toxin blocks tetrodotoxin-sensitive subtypes of the sodium channel in the single nanomolar range, particularly those neurons associated with peripheral nerve excitability.[26] Tetrodotoxin disrupts sodium conductance by stoichiometrically (i.e., one tetrodotoxin molecule to one sodium channel) blocking the fast sodium channel during cell depolarization, resulting in a competitively reversible block at the motor end plate. During severe intoxication impulse conduction ceases. Tetrodotoxin is slightly less neuroactive than saxitoxin, but its effects are more prolonged. Saxitoxin produces less hypotension and vomiting than tetrodotoxin.[22] Experimental reversal of the motor end plate blockade occurs when anticholinergic drugs increase the release of acetylcholine at the neuromuscular junction. This toxin affects the myelin sheaths, where selective blockade of voltage-gated sodium channels at the nodes of Ranvier inhibit motor and sensory conduction by reducing the generation of action potentials and impulse conduction. The axons are spared. Tetrodotoxin does not cross the blood–brain barrier; therefore, consciousness typically remains intact during the peripheral blockade of neurologic function.

DOSE RESPONSE

The effect of tetrodotoxin depends on the amount ingested, age, and pre-existing conditions. The estimated minimum lethal oral dose of tetrodotoxin is approximately 2 mg (10,000 MU equivalents). The threshold for toxicity is not well defined, but the threshold probably is near the minimum lethal dose.

CLINICAL RESPONSE

Toxicity from puffer fish results from ingestion of tetrodotoxin-containing tissues rather than envenomation by contact with dorsal spines. Most individuals ingesting tetrodotoxin-contaminated puffer fish develop no symptoms or mild perioral paresthesias, lightheadedness, and nausea that do not progress to weakness or paralysis. Numbness of the tongue and lips begin soon (i.e., 15 minutes to a few hours) after ingestion of the puffer fish. Occasionally, symptoms are delayed 3–6 hours. More rapid onset of symptoms, particularly weakness, suggests more severe poisonings, and symptoms usually begin within 1–2 hours during severe puffer fish poisoning. The paresthesias progress to involve the face and extremities along with sensations of floating, lightheadedness, and headache. Other neurological symptoms include blurred vision, vertigo, and diffuse paresthesias.[27] Gastrointestinal symptoms are variable with nausea, vomiting, and abdominal pain being the most common symptoms. Diarrhea is uncommon.

In more severe cases, an ascending paralysis progresses rapidly and cardiorespiratory dysfunction develops as outlined in Table 34.2.[28] Hypotension, bradycardia, depressed corneal reflexes, and fixed, dilated pupils indicate a serious poisoning. An ascending paralysis develops within 6–24 hours after ingestion, manifested by weakness, loss of coordination, cranial nerve dysfunction, and hyporeflexia. Despite the presence of complete paralysis, the patient usually remains lucid until cardiopulmonary instability develops. Death usually occurs during this period; however, the prognosis for recovery is good if the patient survives >24 hours after ingestion of the puffer fish. Cardiac abnormalities include dysrhythmias (bradycardia, tachycardia, depressed AV node conduction) and hypotension. Severe hypertension can complicate puffer fish poisoning, particularly in patients with pre-existing hypertensive disease.[29] Dermatological changes associated with puffer fish poisoning include petechiae, vesicles, and exfoliative dermatitis.[36] Tolerance does not develop following repeated exposure to tetrodotoxin.

DIAGNOSTIC TESTING

Analytical Methods

Bioassays determine the semi-quantitative concentration of tetrodotoxin based on the mouse units (MU).

TABLE 34.2. Progressive Stages of Tetrodotoxin Poisoning

Stage	Neurological	Gastrointestinal	Cardiorespiratory
First	Paresthesias (lips, tongue, throat), taste disturbances, dizziness, headache, diaphoresis, miosis	Nausea, vomiting, diarrhea, abdominal pain, drooling	
Second	Progressive neuromuscular symptoms (diffuse paresthesias, weakness/paresis of digits/extremities, hyporeflexia, mydriasis		
Third	Profound neuromuscular symptoms (dysarthria, dysphagia, cranial nerve palsies, ataxia, loss of coordination, ascending paralysis of extremities		Hypotension, hypertension, bradycardia, tachycardia, AV nodal block, dyspnea, cyanosis
Fourth	Seizure, altered mental status, complete paralysis, areflexia, fixed/dilated pupils		Respiratory failure, shock

Source: Adapted from Ref 28.

One MU is equivalent to the amount of toxin required to kill 20g-ICR strain male mice in 30 minutes after a single intraperitoneal injection. One MU corresponds to approximately 0.2 μg tetrodotoxin. Specific methods for the detection of tetrodotoxin include affinity chromatography, mass spectrometry of tetrodotoxin after alkaline degradation to the C9 derivative, and high performance liquid chromatography (HPLC) with post-column derivatization and fluorescence detection after clean up by solid-phase extraction cartridges.[30] The limit of quantitation for tetrodotoxin with the latter technique is approximately 5 ng/mL and 20 mg/mL for serum and urine samples, respectively.

Biomarkers

The serum concentrations of tetrodotoxin fall rapidly within 12–24 hours after ingestion of contaminated food. Case reports suggest that the elimination of tetrodotoxin from plasma is rapid (<24 hours), whereas tetrodotoxin remains detectable in urine samples for at least 4 days during severe tetrodotoxin poisoning as measured by gas chromatography.[31] A serum tetrodotoxin concentration of 36 ng/mL was associated with coma, bradycardia, hypotension, flaccid paralysis, and respiratory failure. The serum tetrodotoxin concentration was not detectable the following day despite profound weakness and stable cardiorespiratory status. Postmortem examination of fatal cases of puffer fish poisoning suggest that the concentrations of tetrodotoxin in the stomach and intestines are relatively low during postmortem examination of patients dying from puffer fish poisoning.[32]

Abnormalities

During the first 3 days of severe puffer fish poisoning manifest by coma, respiratory failure, and hypotension, nerve conduction studies demonstrated reduction or absence of conduction velocities and amplitudes of complex muscle (CMAPs) and sensory nerve action potentials (SNAPs).[31] Insertional activity is typically normal. The slowed conduction and reduced amplitudes are more profound in sensory nerves rather than motor nerves.[33] The F waves (proximal motor latencies) are also prolonged. These abnormalities resolved following clinical improvement within 5 days.[34] These changes in amplitude and conduction velocity are nonspecific indicators of nerve impairment. Laboratory abnormalities during severe tetrodotoxin poisoning include leucocytosis and respiratory acidosis.[35] Occasionally, serum creatine kinase concentrations are slightly elevated, but the MB fraction is usually normal in the absence of myocardial ischemia.[36]

TREATMENT

Treatment is primarily supportive. Patients should be evaluated for evidence of hypotension, shock, respiratory depression, and dysrhythmias with cardiac monitoring, electrocardiogram, and pulse oximetry. The presence of hypotension requires the use of intravenous fluids and inotropic agents (i.e., alpha agonists) as needed. The respiratory status should be carefully monitored for the development of progressive respiratory insufficiency, and assisted ventilation and supplemental oxygen should be initiated if respiratory failure develops. Fluid and electrolytes should be replaced as indicated by laboratory testing and physical examination. Bradycardia and hypotension in a patient suffering from mild intoxication usually respond to atropine, intravenous fluids, and oxygen. There are inadequate clinical data on the efficacy of activated charcoal. There are no antidotes. All patients with suspected puffer fish poisoning should be admitted for observation.

Based on case reports and theoretical considerations, the response of muscle weakness to the intravenous use of anticholinesterase drugs (e.g., 10 mg edrophonium) is equivocal.[37] A case report suggested some improvement in muscle strength after the intravenous administration of 10 mg edrophonium,[38] but the patient remained intubated for one more day. Another case report demonstrated no obvious improvement in neurological symptoms associated with tetrodotoxin poisoning after the intravenous administration of 0.05 mg neostigmine/kg.[39] There are no data to support the use of corticosteroids or antihistamines.

References

1. Miyazawa K, Noguchi T. Distribution and origin of tetrodotoxin. J Toxicol Toxin Rev 2001;20:11–33.
2. Doherty MJ. Captain Cook on poison fish. Neurology 2005;65:1788–1791.
3. Beaglehole JC, editor. The journals of Captain Cook. Vol II. Voyages of the resolution and adventure 1772–1775. London: Cambridge University Press, 1961:534–535.
4. Tsuda K, Ikuma S, Kawamura M, Tachikawa R, Sakai K, Tamura C, Amakasu O. Tetrodotoxin. VII. On the structures of tetrodotoxin and its derivatives. Chem Pharm Bull 1964;12:1357–1374.
5. Woodward RB. The structure of tetrodotoxin. Pure Appl Chem 1964;9:49–74.
6. Davis EW. The ethnobiology of the Haitian zombi. J Ethnopharmacol 1983;9:85–104.
7. Benedek C, Rivier L. Evidence for the presence of tetrodotoxin in a powder used in Haiti for zombification. Toxicon 1989;27:473–480.
8. Hammond R, Bodager D, Jackow MA, Minshew P, Siegenthaler C, Landsberg J, et al. Update: neurologic Illness associated with Eating Florida puffer fish, 2002. MMWR Morb Mortal Wkly Rep 2002;51:414–416.
9. Fuhrman FA. Tetrodotoxin, tarichatoxin, and chiriquitoxin: historical perspectives. Ann NY Acad Sci 1986;479: 1–14.
10. Wu Z, Xie L, Xia G, Zhang J, Nie Y, Hu J, et al. A new tetrodotoxin-producing actinomycete, *Nocardiopsis dassonvillei*, isolated from the ovaries of puffer fish *Fugu rubripes*. Toxicon 2005;45:851–859.
11. Simidu U, Noguchi T, Hwang DF, Shida Y, Hashimoto K. Marine bacteria which produce tetrodotoxin. Appl Environ Microbiol 1987;53:1714–1715.
12. Yu CF, Yu PH, Chan PL, Yan Q, Wong PK. Two novel species of tetrodotoxin-producing bacteria isolated from toxic marine puffer fishes. Toxicon 2004;44:641–647.
13. Bradley SG, Klika U. A fatal poisoning from the Oregon rough-skinned newt (*Taricha granulosa*). JAMA 1981;246: 247.
14. Lu F-M, Fu Y-M, Shih DY. Occurrence of tetrodotoxin poisoning in *Nassarius papillosus* Alectrion and *Nassarius gruneri* Niotha. J Food Drug Anal 2004;12:189–192.
15. Yang C-C, Han K-C, Lin T-J, Tsai W-J, Deng J-F. An outbreak of tetrodotoxin poisoning following gastropod mollusc consumption. Hum Exp Toxicol 1995;14:446–450.
16. Yu C-F, Yu PH. The annual toxicological profiles of two common puffer fish, *Takifugu niphobles* (Jordan and Snyder) and *Takifugu alboplumbeus* (Richardson), collected along Hong Kong coastal waters. Toxicon 2002;40: 313–316.
17. Yoshida S. Tetrodotoxin-resistant sodium channels. Cell Mol Neurobiol 1994;14:227–244.
18. Tanner P, Przekwas G, Clark R, Ginsberg M, Waterman S. Tetrodotoxin poisoning associated with eating puffer fish transported from Japan–California, 1996. MMWR Morb Mortal Wkly Rep 1996;45:389–391.
19. Alcaraz A, Whipple RE, Gregg HR, Andresen BD, Grant PM. Analysis of tetrodotoxin. Forensic Sci Int 1999; 99;35–45.
20. Stommel EW, Watters MR. Marine neurotoxins: ingestible toxins. Curr Treat Opt Neurol 2004;6:105–114.
21. Endo A, Khora SS, Murata M, Naoki H, Yasumoto T. Isolation of 11-*nor*tetrodotoxin-6(R)-ol and other tetrodotoxin derivatives from the puffer *Fugu niphobles*. Tetrahedron Lett 1988;29:4127–4128.
22. Mosher HS, Fuhrman FA, Buchwald HD, Fischer HG. Tarichatoxin-tetrodotoxin: a potent neurotoxin. Science 1964;144:1100–1110.
23. Mahmud Y, Tanu MB, Takatani T, Asayama E, Arakawa O, Noguchi T. *Chelonodon patoca*, a highly toxic marine puffer in Japan. J Nat Toxins 2001;10:69–74.
24. Noguci T, Hwang DF, Arakawa O, Sugita H, Deguchi Y, Shida Y, Hashimoto K. *Vibrio alginolyticus*, a tetrodotoxin-producing bacterium, in the intestines of the fish *Fugu vermicularis vermicularis*. Marine Biol 1987;94:625–630.
25. Ogata N, Ohishi Y. Molecular diversity of structure and function of the voltage-gated Na^+ channels. Jpn J Pharmacol 2002;88:365–377.
26. Kiernan MC, Isbister GK, Lin CS-Y, Burke D, Bostock H. Acute tetrodotoxin-induced neurotoxicity after ingestion of puffer fish. Ann Neurol 2005;57:339–348.
27. Ahasan HA, Al Mamun A, Rasul CH, Roy PK. Puffer fish poisoning (tetrodotoxin) in Bangladesh: clinical profile and role of anticholinesterase drugs. Trop Doct 2005; 35:235–236.
28. Noguchi T, Ebesu JS. Puffer poisoning: epidemiology and treatment. J Toxicol Toxin Rev 2001;20:1–10.
29. Deng J-F, Tominack RL, Chung H-M, Tsai W-J. Hypertension as an unusual feature in an outbreak of tetrodotoxin poisoning. Clin Toxicol 1991;29:71–79.
30. O'Leary MA, Schneider JJ, Isbister GK. Use of high performance liquid chromatography to measure tetrodotoxin in serum and urine of poisoned patients. Toxicon 2004;44:549–553.

31. Oda K, Araki K, Totoki T, Shibasaki H. Nerve conduction study of human tetrodotoxication. Neurol 1989;39: 743–745.
32. Tsunenari S, Uchimura Y, Kanda M. Puffer poisoning in Japan—a case report. J Forensic Sci 1980;25:240–245.
33. Isbister GK, Son J, Wang F, Maclean CJ, Lin CS, Ujma J, et al. Puffer fish poisoning: a potentially life-threatening condition. Med J Aust 2002;177:650–653.
34. Cong NH, Tuan LT. Electrodiagnosis in puffer fish poisoning—a case report. Electromyogr Clin Neurophysiol 2006;46:291–294.
35. Sun K, Wat J, So P. Puffer fish poisoning. Anaesth Intens Care 1994;22:307–308.
36. Sims JK, Ostman DC. Pufferfish poisoning: emergency diagnosis and management of mild human tetrodotoxication. Ann Emerg Med 1986;15:1094–1098.
37. Clark RF, Williams SR, Nordt SP, Manoguerra AS. A review of selected seafood poisonings. Undersea Hyperbar Med 1999;26:175–185.
38. Chew SK, Goh CH, Wang KW, Mah PK, Tan BY. Puffer fish (tetrodotoxin) poisoning: clinical report and role of anti-cholinesterase drugs in therapy. Singapore Med J 1983;24:168–171.
39. Tibballs J. Severe tetrodotoxic fish poisoning. Anaesth Intensive Care 1988;16:215–217.

Chapter 35

RED WHELK and TETRAMINE

HISTORY

Tetramine has been recognized as a cause of food poisoning following the ingestion of whelks since the 1950s. The appearance of an intense, self-limited illness manifest by severe headache, malaise, nausea, vomiting, lightheadedness, and urticaria was associated with the consumption of specimens of *Neptunea arthritica* Bernardi from the northern seas of Japan and *Neptunea intersculpta* Sowerby from the Japan Sea. Asano et al confirmed tetramine as the main toxic compound in the salivary glands of *Neptunea arthritica* Bernardi.[1] In the 1950s, Fange and colleagues confirmed the presence of tetramine in the salivary glands of *Neptunea antiqua* L. from Norwegian coastal waters.[2] Case reports of red whelk poisoning appeared in Scotland in the 1970s.[3]

EXPOSURE

Source

In the United Kingdom, food poisoning occurs when red whelks (*Neptunea antiqua* L.) are consumed instead of the edible whelk (*Buccinum undatum* L.). Red whelks have a large shell that is pale yellow to orange with no ridges compared with the smaller, smooth shell of the edible whelk. Several members of the Buccinidae family of gastropod mollusks contain the autonomic ganglionic agent, tetramine, in their salivary glands including *Buccinum leucostoma* Lischke, *Neptunea antiqua* L., *Neptunea arthritica* Bernardi, *Neptunea intersculpta* Sowerby, and *Neptunea lyrata* Martyn.[1,4] *Neptunea antiqua* is a necrophagous scavenger that inhabits the cold temperate waters of the Eastern Atlantic from the Bay of Biscay to the Arctic regions. The Oregon triton (*Fusitriton oregonensis* Redfield) from the Ranellidae family of gastropods also contains tetramine. Red whelks normally inhabit depths in the ocean up to 100–200 meters (325–650 feet), and persons collecting shellfish near the coastline do not usually encounter this marine gastropod. Although marine gastropods are associated with red whelk poisoning, tetramine occurs as metabolites in at least three marine phyla (Cnidaria, Mollusca, and Bryozoa), which include sea anemones (*Metridium dianthus* Ellis, *Actinia equina* L.), Portuguese man-of-war (*Physalia physalis* L.), hydroids (*Hydra littoralis* Hyman), giant Caribbean anemone (*Condylactis gigantea* Weinland), soft coral (*Plexaura flexosa* Lamouroux), and marine bryozoans (*Cellaria fistulosa* L., *Cellaria sinuosa* Hassall). Some plants (e.g., Sudanese *Courbonia virgata* Brongn.) also contain tetramine.

Food Processing

Although most of the tetramine occurs in the salivary glands of buccinid gastropods, other tissues can contain small amounts of tetramine as a result of the diffusion of tetramine from the salivary glands. Tetramine also diffuses into the broth during the boiling of tetramine-containing, marine gastropods. However, removing the contents of these marine gastropods and broiling does not prevent toxicity.[3] Broiling contaminated gastropods for 15 minutes reduces the mean tetramine content in the tissue by approximately 30%.[5] The heating of specimens of *Neptunea arthritica* and *N. intersculpta* to 100 °C (212 °F) for 5 minutes reduced the tetramine content by

Medical Toxicology of Natural Substances, by Donald G. Barceloux, MD

40 to 50% as measured by ion chromatography with a detection limit of 5 μg/mL.[6]

TOXINS

Physiochemical Properties

The red whelk contains the heat stable toxin, tetramine (CAS No. 3810-53-5), in their salivary glands. The molecular formula of this simple quaternary ammonium compound (tetramethylammonium ion) is $C_4H_{12}N^+$. This compound should not be confused with tetramethylenedisulfotetramine (CAS No. 80-12-6), which is a potent rodenticide that produces status epilepticus via non-competitive γ-aminobutyric acid (GABA) antagonism. Both these compounds have the common name, tetramine.[7] Although originally isolated as tetramethylammonium hydroxide from a sea anemone, the hydroxide salt is too basic to exist in nature.[8] In samples of salivary glands from four specimens of *Neptunea arthritica*, the tetramine concentration ranged from 8.4–12 mg/g tissue, whereas three samples of salivary glands from *Neptunea polycostata* Scarlato contained 0.85–4.9 mg/g tissue.[15] The tetramine concentration in muscle samples from these two species ranged from 0.016–0.18 mg/g tissue and 0.003–0.007 mg/g tissue, respectively. The tetramine concentration in salivary glands from *Fusitriton oregonensis* Redfield ranged from about 3–4 mg/g gland.[2] In addition to the significant variation in tetramine content between individual specimens, a study of salivary glands of *Neptunea antiqua* from the central western Irish Sea indicated that substantial seasonal variation in tetramine content occurred in the salivary glands.[9] The tetramine concentration ranged from undetectable to 6.53 mg/g gland during the season with the highest concentration after the breeding in July. Similar tetramine concentrations occurred in buccinid gastropods analyzed in specimens obtained from Japanese waters. The salivary glands from nine specimens of *Neptunea intersculpta* harvested during September contained 4.5–27 mg tetramine.[10] Two samples of salivary glands from *N. arthritica* obtained in November contained 1.8 mg and 12 mg tetramine. Tetramine was present in most samples of meat from *N. intersculpta*, but the total tetramine content typically was <1–2 mg. In experimental studies of rats, the elimination of tetramine is rapid ($t_{1/2} \approx 1$ hour) with most elimination occurring via the renal excretion of unchanged tetramine.[11]

Mechanism of Toxicity

The salivary glands of these marine gastropods empty into the mouth, and there is no venom apparatus for the rapid injection of tetramine. Tetramine blocks nicotinic acetylcholine receptors, which produces a complete blockade of neuromuscular transmission in experimental studies similar to conotoxins associated with cone shells. Other toxins (e.g., inhibitors of neuronal calcium channels) also may occur in the salivary glands of red whelk, but the effects of these toxins are unclear.[9]

DOSE RESPONSE

The consumption of red whelk containing 3.75 mg tetramine/100 g tissue (i.e., total estimated dose 1.75–2.5 mg) was associated with red whelk poisoning.[12] The estimated oral lethal dose of tetramine based on animal studies is 250–1000 mg.[13] However, there are no human data because of the lack of reported fatalities associated with red whelk poisoning.

CLINICAL RESPONSE

The clinical presentation of red whelk poisoning is similar to mild nicotine intoxication.[8] Symptoms of red whelk poisoning involve the gastrointestinal and neurological systems, usually within 30 minutes to 1 hour of the consumption of these gastropods. Case reports associated nausea, vomiting, diarrhea, paresthesias, diplopia, blurred vision, vertigo, headache, weakness, and ataxia with the consumption of red whelk.[12] Typically, symptoms resolve within 3–24 hours because of the rapid elimination of tetramine from the blood. Deaths are not usually associated with red whelk poisoning.

DIAGNOSTIC TESTING

Bioassays based on the dose-lethal time curves for mice are the most common method for determining the presence of toxins in tetramine-containing gastropods.[14] More specific and sensitive methods for the quantitation of tetramine include capillary zone electrophoresis/tandem mass spectrometry and the combination of liquid chromatography for separation and electrospray ionization-single quadrupole mass spectrometry for detection.[15]

TREATMENT

Treatment is supportive. There are no antidotes. There are inadequate human data on red whelk poisoning to recommend specific treatment.

References

1. Asano M, Itoh M. Salivary poison of a marine gastropod, *Neptunea arthritica* Bernardi, and the seasonal variation of its toxicity. Ann N Y Acad Sci 1960;90:674–688.

2. Fange R. The salivary gland of *Neptunea antiqua*. Ann N Y Acad Sci 1960;90:689–694.
3. Fleming C. Case of poisoning from red whelk. Br Med J 1971;3(773):520–521.
4. Shiomi K, Mizukami M, Shimakura K, Nagashima Y. Toxins in the salivary gland of some marine carnivorous gastropods. Comp Biochem Physiol 1994;107B:427–432.
5. Ayres PA, Wood PC. Toxins in the red whelk. Marine Pollut Bull 1973;4:157–159.
6. Saitoh H, Oikawa K, Takano T, Kamiura K. Determination of tetramethylammonium ion in shellfish by ion chromatography. J Chromatogr 1983;281:397–402.
7. Barrueto F Jr, Furdyna PM, Hoffman RS, Hoffman RJ, Nelson LS. Status epilepticus from an illegally imported Chinese rodenticide: "tetramine." J Toxicol Clin Toxicol 2003;41:991–994.
8. Anthoni U, Bohlin L, Larsen C, Nielsen P, Nielsen NH, Christophersen C. Tetramine: occurrence in marine organisms and pharmacology. Toxicon 1989;27:707–716.
9. Power AJ, Keegan BF, Nolan K. The seasonality and role of the neurotoxin tetramine in the salivary glands of the red whelk *Neptunea antiqua* (L.). Toxicon 2002;40:419–425.
10. Shindo T, Ushiyama H, Kan K, Saito H, Kuwahara Y, Uehara S-I, Yasuda K. Study on contents of tetramine in salivary gland, meat and internal organs of buccinid gastropods (Mollusca). J Food Hyg Soc Japan 2000;41:17–22.
11. Neef C, Oosting R, Meijer DK. Structure-pharmacokinetics relationship of quaternary ammonium compounds. Elimination and distribution characteristics. Naunyn Schmiedebergs Arch Pharmacol 1984;328:103–110.
12. Reid TM, Gould IM, Mackie IM, Ritchie AH, Hobbs G. Food poisoning due to the consumption of red whelks (*Neptunea antiqua*). Epidemiol Infect 1988;101:419–423.
13. Anthoni U, Bohlin L, Larsen C, Nielsen P, Nielsen NH, Christophersen C. The toxin tetramine from the "edible: whelk *Neptunea antiqua*. Toxicon 1989;27:717–723.
14. Saitoh H, Oikawa K, Takano T, Kamimura K. Determination of tetramethylammonium ion in shellfish by ion chromatography. J Chromatogr 1983;281:397–402.
15. Kawashima Y, Nagashima Y, Shiomi K. Determination of tetramine in marine gastropods by liquid chromatography/electrospray ionization-mass spectrometry. Toxicon 2004;44:185–191.

Chapter 36

SCOMBROID FISH, SCOMBROTOXIN, and HISTAMINE

HISTORY

The first recorded episode of scombroid fish poisoning involved the consumption of bonito by British sailors aboard the *Triton of Leith* during 1828.[1] In Japan, scombroid fish poisoning was a major cause of food-related illness during the 1950s.[2] The first recorded episode of scombroid fish poisoning in the United States occurred in the late 1960s. During the late 1970s, scombroid fish poisoning was recognized as a cause of seafood-related illness in the United Kingdom.[3]

EXPOSURE

Source

Worldwide, scombroid poisoning is the most common cause of illness related to the ingestion of seafood. The ingestion of freshwater fish does not usually cause scombroid fish poisoning. Deep-water fish from the Scombridae (tuna, bonito, mackerel, albacore) and Scomberesocidae (saury) families have high concentrations of the free amino acid, histidine, in their dark meat. The term scombroid fish poisoning is a misnomer because a variety of nonscombroid fish also cause scombroid fish poisoning including amberjack, mahi-mahi (bonito), bluefish, marlin, sockeye salmon, herring, pilchards, sardines, and anchovies. Inadequate refrigeration after these fish are caught results in the growth of bacteria, which synthesize histamine by the enzymatic action of histidine decarboxylase on histidine. A variety of bacteria produce histidine decarboxylase including *Proteus vulgaris, Escherichia coli, Morganella morganii, Clostridium* spp., *Salmonella* spp., *Enterobacter aerogenes, Klebsiella pneumoniae, Proteus mirabilis, Vibrio alginolyticus, Aeromonas spp., Serratia fonticola, Acinetobacter spp.,* and *Hafnia alvei*.[4] *Lactobacillus buchneri* is associated with histamine formation in cheese.[5] The rate of histidine decarboxylation varies between bacterial species.[6] Therefore, some species are more important histamine producers (e.g., *Hafnia alvei*) than others (e.g., *Morganella morganii*). Additionally, many bacteria produce enzymes that convert lysine to cadaverine, ornithine to putrescine, and arginine to spermidine and spermine. Consequently, the formation of histamine depends on several variables including the number and type of bacterial species, the fish species, the part sampled, time after catch, and temperature. Biogenic amines (i.e., histamine, cadaverine, tyramine, putrescine) also occur in meat and meat products, fruits, vegetables, cheese, sauerkraut, wine, beer, and soy sauce.[7,8]

Food Processing

Once formed, food processing (e.g., freezing, smoking, cooking, canning) does not inactivate the causative toxins associated with scombroid fish poisoning. Rapid cooling of fish after catching and the maintenance of adequate refrigeration during handling and storage substantially reduces the risk of scombroid fish poisoning. Histamine formation is minimal at temperatures below 0 °C (32 °F), but some bacteria can produce histamine in fish at temperatures between 0 °C (32 °F) and 10 °C

Medical Toxicology of Natural Substances, by Donald G. Barceloux, MD

(50 °F).[9] Refrigeration usually reduces the formation of histamine by enteric bacteria that produce histamine decarboxylase. An *in vitro* study indicated that the lower limit for production of toxicologically significant concentrations of histamine in tuna broth for *E. coli* and *K. pneumoniae* was 30 °C (86 °F) and 7 °C (45 °F), respectively.[10] Prolonged drying reduces histamine content, but freezing, smoking, and cooking do not destroy preformed histamine or prevent scombroid poisoning. Histamine production in smoked fish usually occurs prior to smoking. Although smoking probably reduces the histamine concentration in contaminated fish, storing the smoked fish at room temperature increases bacterial activity and histamine formation. In a study of 33 samples from retail smoked fish, storage for 2 days at room temperature resulted in histamine concentrations exceeding 200 ppm in four samples with a range up to about 1650 ppm.[11] An outbreak of scombroid fish poisoning occurred after the consumption of contaminated fish when allowed to stand for 3–4 hours at room temperature, but ingestion of the same fish allowed to stand 1 hour at room temperature did not cause illness.[12]

TOXINS

Physiochemical Properties

The toxin or toxins associated with scombroid fish poisoning are usually tasteless, odorless, and heat stable. Some heavily contaminated fish have a peppery or bitter taste. In humans, L-histidine ammonia lyase catalyzes the conversion of histamine to urocanic acid. Further biotransformation produces glutamate and α-ketoglutarate, which enters the citric acid cycle. The presence of diamine oxidase (histaminase) and histamine methyl transferase in the intestinal tract detoxify histamine prior to gastrointestinal absorption. Some medications, such as monoamine oxidase inhibitors, increase the severity or prolong the course of scombroid poisoning by inhibiting the metabolism of histamine.[13] Spoiling fish accumulate cadaverine and/or putrescine with the former being present in higher concentrations than histamine. However, fish that accumulate these biogenic amines are not known to be toxic when histamine concentrations are low.[16] In a study of 8 volunteers ingesting 90 mg histamine, histamine concentrations in plasma samples remained elevated above baseline concentrations for at least 24 hours after ingestion.[14]

Mechanism of Toxicity

The mechanism causing scombroid fish poisoning is complex and incompletely elucidated. Volunteer studies indicate that endogenous histamine, but not dietary (exogenous) histamine, is a major determinant of scombrotoxicosis.[15] Although histamine is an important part of scombroid fish poisoning, additional factors probably contribute to the development of this illness because histamine consumed in spoiled fish is more toxic than oral histamine.[16] The majority of 440 suspected incidents of scombroid fish poisoning in the United States between 1976 and 1990 were associated with the ingestion of fish that contained histamine concentration considered safe for consumption (i.e., <5 mg/100 g flesh).[17] The administration of oral histamine (e.g., 100 mg) far exceeding histamine doses associated with scombroid fish poisoning does not produce this illness. There are three types of histamine receptors (H_1, H_2, H_3), which produce a variety of effects including dilation of peripheral blood vessels resulting in hypotension, headaches, erythema, and urticaria. Although dysrhythmias are not a prominent feature of scombroid fish poisoning, potential arrhythmogenic actions of histamine receptors include H_1 receptor-mediated slowing of atrioventricular conduction and H_2-receptor-mediated increase in sinus and ventricular automaticity. Additional effects include nausea, vomiting, diarrhea, abdominal cramps, pruritus, and peripheral pain.

The exact chemical nature of scombrotoxin is not known. Potential mechanisms of toxicity include the facilitation of histamine absorption, the reduction of histamine absorption or decreased histamine metabolism, and the potentiation of histamine toxicity by other biogenic amines. However, in a study of mackerel fillets associated with an outbreak of scombrotoxicosis, the doses of amines (cadaverine, histamine, putrescine, spermidine, spermine, tyramine) did not correlate to the development of symptoms characteristic of scombroid fish poisoning.[18]

DOSE RESPONSE

The dose response associated with scombroid fish poisoning is not completely defined because histamine concentration do not account for all of the clinical features of scombroid fish poisoning. Furthermore, the response to histamine depends on a number of variables including individual susceptibility, route of administration, body weight, and the presence of certain drugs (e.g., isoniazid).[19] Therefore, the threshold of toxicity varies substantially between different individuals. The oral administration of 100–180 mg histamine in grapefruit juice to 4 volunteers produced headache and flushing in one volunteer while the other volunteers remained asymptomatic.[20] The addition of this dose of histamine to 100 grams of high-quality tuna resulted in headache and flushing in 4 of 8 volunteers. In a volunteer study, the addition of 300 mg histamine (about 5 mg/kg body weight) to 50g fresh mackerel produced symptoms of mild scombroid poisoning (flushing, headache, oral paresthesias) in some, but not all volunteers.[21]

CLINICAL RESPONSE

The clinical presentation of scombroid poisoning resembles an acute allergic response that is mild and self-limited. Symptoms begin rapidly (i.e., 10 minutes–2 hours) after consumption of contaminated fish.[22] Adverse effects include headache, dysphagia, diffuse erythema, perioral paresthesias, pruritus, diaphoresis, conjunctival erythema, and gastrointestinal upset (nausea, vomiting, diarrhea, abdominal cramps). The most common symptoms in a review of 94 confirmed incidents of scombroid fish poisoning in Britain were erythematous rash, flushing, diarrhea, and headache.[23] However, the type and severity of gastrointestinal symptoms varies substantially between different outbreaks.[24] Typically, flushing appears on the face and upper body. Urticaria, angioneurotic edema, hypotension, palpitations, tachycardia, and bronchospasm occur occasionally.[25,26] Symptoms usually resolve within 8–12 hours, and symptoms rarely persist more than 24 hours.[27] Rare case reports associate scombroid fish poisoning with hypotension,[28] acute pulmonary edema, myocardial ischemia, subendocardial myocardial infarction, and cardiogenic shock in patients with pre-existing cardiovascular disease.[29,30] The differential diagnosis includes allergic responses and salmonellosis. Food allergies typically involve only a very small portion of individuals who consume the fish. The diagnosis of scombroid fish poisoning involves the rapid onset of typical symptoms of histamine poisoning after eating fish known to cause scombroid fish poisoning. The detection of high concentrations of histamine in samples from implicated fish is confirmatory. Attack rates during outbreaks of scombroid fish poisoning vary from about 50 to 100%, depending primarily on the histamine concentrations in the spoiled fish.[31,32]

DIAGNOSTIC TESTING

Conversion Units (Histamine)

$$1\,\text{ng/mL} = 0.111\,\text{nmol/L}$$

$$1\,\text{nmol/L} = 8.997\,\text{ng/mL}$$

Analytical Tests

Analytical methods for the detection of histamine include gas chromatography with flame ionization (requires sample cleanup to avoid matrix interference),[33] high performance liquid chromatography (HPLC),[34] ion exchange chromatography with integrated pulsed amperometric detection,[35] fluorometric methods,[36] and capillary electrophoresis.[37] More rapid testing methods for the detection of histamine involve enzyme-linked immunosorbent assay (ELISA) kits with sensitivities near 1–5ppm.[16] There are three AOAC (Association of Analytical Chemists International, Gaithersburg, MD; http://www.eoma.aoac.org/) official methods of analysis for the quantitative determination of histamine in seafood. Official method 954.04 is a quantitative microbiological method, method 951.07 is a chromatographic method, and method 977.13 is a chromatographic method with fluorescence detection.[38] The latter method has been adopted by Codex Alimentarius as the internationally recognized method for the determination of this analyte in seafood. Food samples should be frozen if not analyzed promptly.[39] Histamine and *N*-methyl histamine concentrations in urine samples remain elevated at least 24 hours after the ingestion of fish contaminated with histamine concentrations up to about 20mg/100g tissue.[40]

Biomarkers

Although histamine concentrations in fish flesh is a good biomarker of spoilage secondary to inadequate storage, low histamine concentrations do not guarantee safety. Scombroid fish poisoning and histamine toxicity do not produce identical clinical syndromes; histamine concentrations in spoiled fish associated with scombroid fish poisoning usually exceed 50mg/100g tissue. Fresh fish typically contain negligible amounts of histamine (<0.1mg/100g).[41] Fish with histamine content below 5–10mg/100g tissue do not usually cause toxicity in most individuals, although symptoms of scombroid fish poisoning have been associated with histamine concentrations <5mg/100g tissue.[23] The ingestion of a 300g meal of fish containing 10mg/100g wet weight by a 60-kg (132lb) person results in a histamine dose of approximately 0.5mg/kg body weight. Only 2 of 8 adults ingesting 90mg (1.25–1.5mg/kg) histamine developed symptoms of histamine toxicity.[14] The plasma histamine concentrations in blood samples from 5 symptomatic patients with scombroid fish poisoning ranged between 21.5nmol/L and 47.8nmol/L as measured by radioimmunoassay.[42] The samples were drawn 3–4 hours after ingestion of contaminated tuna, and all the histamine concentrations were below the upper limit of normal (10.8nmol/L) by 24 hours after ingestion.

The US Food and Drug Administration (FDA) toxicity level for histamine in tuna is 50mg/100g (500ppm) tissue, while the action level is 5mg/100g (50ppm). European Union Directive 91/493 requires the following conditions: 1) the average histamine concentration is below 10 mg/100 g (100 ppm), 2) no more than two of nine samples contain histamine concentrations between 100–200 ppm, and 3) no sample contains histamine concentrations above 200 ppm.[43] Most spoiled fish from outbreaks of scombroid fish poisoning contain histamine concentrations above 200 ppm and most of these implicated samples contain histamine concentrations exceeding 1000 ppm.[23] In spoiled fish, the distribution of histamine is not uniform with the highest concentrations usually near the gut and gills.[44] There is marked seasonal and fish-to-fish variation in the content of biogenic amines as well as susceptibility to decomposition.

Abnormalities

Scombroid fish poisoning typically is a mild, self-limited illness. There are no characteristic laboratory changes associated with this seafood-related illness, and serum chemistry and blood counts are usually normal.

TREATMENT

Treatment is supportive, and most cases of scombroid fish poisoning are mild and self-limited. Symptomatic patients often respond to antihistamines. Case reports suggest that the intravenous administration of H_1 receptor (25–50 mg diphenhydramine)[45] and H_2 receptor (300 mg cimetidine, 20 mg famotidine)[46] antagonists are efficacious in symptomatic patients. There are no clinical data to indicate that decontamination with activated charcoal improves the clinical outcome of this usually mild illness. Fluids, antiemetics, and bronchodilators should be administered as indicated by the clinical presentation. There are inadequate clinical data to support the use of corticosteroids.

References

1. Henderson PB. Case of poisoning from the bonito (*Scomber pelamis*). Edinburgh Med J 1830;34:317–318.
2. Kawabata T, Ishizaka K, Miura T. Studies on the allergy-like food poisoning associated with putrefaction of marine products. III. Physiological and pharmacological action of saurine, a vagus stimulant of unknown structure recently isolated by the authors, and its characteristics in developing allergy-like symptoms. Jpn J Med Sci Biol 1955; 8:521–528.
3. Cruickshank JG, Williams HR. Scombrotoxic fish poisoning. Br Med J 1978;2:739–740.
4. Yoshinaga DH, Frank HA. Histamine-producing bacteria in decomposing skipjack tuna (*Katsuwonus pelamis*). Appl Environ Microbiol 1982;44:447–452.
5. Sumner SS, Speckhard MW, Somers EB, Taylor SL. Isolation of histamine-producing *Lactobacillus buchneri* from Swiss cheese implicated in a food poisoning outbreak. Appl Environ Microbiol 1985;50:1094–1096.
6. Tsai Y-H, Kung H-F, Lee T-M, Lin G-T, Hwang D-F. Histamine-related hygienic qualities and bacteria found in popular commercial scombroid fish fillets in Taiwan. J Food Prot 2004;67:407–412.
7. Chin KW, Garriga MM, Metcalfe DD. The histamine content of oriental foods. Food Chem Toxicol 1989; 27:283–287.
8. Stratton JE, Hutkins RW, Taylor SL. Biogenic amines in cheese and other fermented foods. A review. J Food Prot 1991;54:460–470.
9. Klausen NK, Lund E. Formation of biogenic amines in herring and mackerel. Z Lebensm Unters Forsch 1986; 182:459–463.
10. Behling AR, Taylor SL. Bacterial histamine production as a function of temperature and time of incubation. J Food Sci 1982;47:1311–1317.
11. Fletcher CG, Summers G, van Veghel PW. Levels of histamine and histamine-producing bacteria in smoked fish from New Zealand markets. J Food Prot 1998;61:1064–1070.
12. Kow-Tong C, Malison MD. Outbreak of scombroid fish poisoning, Taiwan. Am J Pub Health 1987;77:1335–1336.
13. Becker K, Southwick K, Reardon J, Berg R, MacCormack JN. Histamine poisoning associated with eating tuna burgers. JAMA 2001;285:1327–1330.
14. van Gelderen CE, Savelkoul TJ, Van Ginkel LA, Van Dokkum W. The effects of histamine, administered in fish samples to healthy volunteers. Clin Toxicol 1992;30:585–596.
15. Ijomah P, Clifford MN, Walker R, Wright J, Hardy R, Murray CK. The importance of endogenous histamine relative to dietary histamine in the aetiology of scombrotoxicosis. Food Addit Contam 1991;8:531–542.
16. Lehane L, Olley J. Histamine fish poisoning revisited. Int J Food Microbiol 2000;48:1–37.
17. Scoging AC. Illness associated with seafood. CDR (Lond Engl Rev) 1991;1:R117–R122.
18. Clifford MN, Walker R, Ijomah P, Wright J, Murray CK, Hardy R. Is there a role for amines other than histamines in the aetiology of scombrotoxicosis? Food Addit Contam 1991;8:641–651.
19. Taylor SL. Histamine food poisoning: toxicology and clinical aspects. Crit Rev Toxicol 1986;17:91–128.
20. Motil KJ, Scrimshaw NS. The role of exogenous histamine in scombroid poisoning. Toxicol Lett 1979;3:219–223.
21. Clifford MN, Walker R, Wright J, Hardy R, Murray CK. Studies with volunteers on the role of histamine in suspected scombrotoxicosis. J Sci Food Agric 1989;47:365–375.
22. Etkind P, Wilson ME, Gallagher K, Cournoyer J. Bluefish-associated scombroid poisoning an example of the expand-

ing spectrum of food poisoning from seafood. JAMA 1987;258:3409–3410.

23. Bartholomew BA, Berry PR, Rodhouse JC, Gilbert RJ. Scombrotoxic fish poisoning in Britain: features of over 250 suspected incidents from 1976 to 1986. Epidemiol Infect 1987;99:775-782.
24. Wu S-F, Chen W. An outbreak of scombroid fish poisoning in a kindergarten. Acta Paediatr Tw 2003;44:297–299.
25. Smart DR. Scombroid poisoning a report of seven cases involving the Western Australian salmon, *Arripis truttaceus*. Med J Aust 1992;157:748–751.
26. Eckstein M, Serna M, De la Cruz P, Mallon WK. Out-of-hospital and emergency department management of epidemic scombroid poisoning. Acad Emerg Med 1999;6:916–920.
27. Gilbert RJ, Hobbs G, Murray CK, Cruickshank JG, Young SE. Scombrotoxic fish poisoning: features of the first 50 incidents to be reported in Britain (1976–9). Br Med J 1980;281:71–72.
28. Borade PS, Ballary CC, Lee DK. A fishy cause of sudden near fatal hypotension. Resuscitation 2007;72:158–160.
29. Grinda J-M, Bellenfant F, Brivet FG, Carel Y, Deloche A. Biventricular assist device for scombroid poisoning with refractory myocardial dysfunction: a bridge to recovery. Crit Care Med 2004;32:1957–1959.
30. Ascione A, Barresi LS, Sarullo FM, De Silvestre G. [Two cases of "scombroid syndrome" with severe cardiovascular compromise.] Cardiologia 1997;42:1285–1288. [Italian]
31. Taylor SL, Stratton JE, Nordlee JA. Histamine poisoning (scombroid fish poisoning): an allergy-like intoxication. Clin Toxicol 1989;27:225–240.
32. Feldman KA, Werner SB, Cronan S, Hernandez M, Horvath AR, Lea CS, et al. A large outbreak of scombroid fish poisoning associated with eating escolar fish (*Lepidocybium flavobrunneum*). Epidemiol Infect 2005;133:29–33.
33. Antoine FR, Wei C-I, Otwell S, Sims CA, Littell RC, Hogle AD, Marshall MR. Gas chromatographic analysis of histamine in mahi-mahi (*Coryphaena hippurus*). J Agric Food Chem 2002;50:4754–4759.
34. Yen G, Hsieh C. Simultaneous analysis of biogenic amines in canned fish by HPLC. J Food Sci 1991;56:158–160.
35. Draisci R, Giannetti L, Boria P, Lucentini L, Palleschi L, Cavalli S. Improved ion chromatography-integrated pulsed amperometric detection method for the evaluation of biogenic amines in food of vegetable or animal origin and in fermented foods. J Chromatogr A 1998;798:109–116.
36. Rogers PL, Staruszkiewicz W. Gas chromatographic method for putrescine and cadaverine in canned tuna and mahimahi and fluorometric method for histamine (minor modification of AOAC Official Method 977.13): collaborative study. J AOAC Int 1997;80:591–602.
37. Trenerry VC, Marshall PA, Windahl K. Method optimization: determination of histamine in fish by capillary zone electrophoresis. J Capillary Electrophor 1998;5:27–32.
38. Horowitz W, editor. Official Methods of Analysis of AOAC International. 18th ed. Gaithersburg, MD: AOAC International; 2006.
39. Food and Agriculture Association of the United Nations/World Health Organization (FAO/WHO). Current food standards. Available at http://www.codexalimentarius.net/web/standard_list.do?lang=en. Accessed March 26, 2008.
40. Morrow JD, Margolies GR, Rowland J, Roberts LJ II. Evidence that histamine is the causative toxin of scombroid-fish poisoning. 1991;324:716–720.
41. Lopez-Sabater EI, Rodriguez-Jerez JJ, Hernandez-Herrero M, Mora-Ventura MT. Incidence of histamine-forming bacteria and histamine content in scombroid fish species from retail markets in the Barcelona area. Int J Food Microbiol 1996;28:411–418.
42. Bedry R, Gabinski C, Paty M-C. Diagnosis of scombroid poisoning by measurement of plasma histamine. N Engl J Med 2000;342:520–521.
43. The Council of European Communities. Directive on fishery products for human consumption. Available at http://www.cfsan.fda.gov/~acrobat/hp91493.pdf. Accessed March 26, 2008.
44. Lerke PA, Werner SB, Taylor SL, Guthertz LS. Scombroid poisoning. Report of an outbreak. West J Med 1978; 129:381–386.
45. Guss DA. Scombroid fish poisoning: successful treatment with cimetidine. Undersea Hyperb Med 1998;25:123–125.
46. Blakesley ML. Scombroid poisoning: prompt resolution of symptoms with cimetidine. Ann Emerg Med 1983; 12:104–106.

PART 2

FUNGAL TOXINS

PART 2 FUNGAL TOXINS

I Mushrooms

Chapter 37

AMATOXIN-CONTAINING MUSHROOMS

MUSHROOM IDENTIFICATION AND CLASSIFICATION

Fungi are neither plant nor animal. Because fungi lack chlorophyll and cannot manufacture their own food, these organisms must absorb nutrients from decaying organic matter (saprophyte) or from living organisms (parasite). Some species of fungi form symbiotic or mutually beneficial relationships with the roots of other plants in the form of a mycorrhiza. Mushrooms are the *fruiting bodies* of certain fungi. Most of the fungus consists of the *mycelium*, which is an extensive network of minute underground tubes (i.e., *hyphae*) that range in length from a few millimeters to hundreds of feet. Typically, the mycelium is hidden in a substrate (e.g., wood, soil, foods) where the hyphae secrete digestive enzymes and absorb nutrients through their cell walls. As the mycelium absorbs enough energy to reproduce, the mycelium generates new hyphae on the surface that develop into complex, spore-forming structures or fruiting bodies called *mushrooms*. The appearance of these fruiting bodies depends on heat, humidity, and the season. Variation in size, shape, color, and microscopic examination of spore morphology aids in the identification of mushroom species.

Poisonous mushrooms belong to the two higher fungus phyla: Basidiomycota (club fungi) and Ascomycota (sac fungi). The major difference between these two groups is the form of microscopic spores. The Basidiomycetes produce sexual spores (**basidiospores**) externally by budding on the ends of specialized cells called basidia. Basidiospores are forcibly discharged from basidia and then drop from the gills (or tubes for boletes from the order Boletales) located underneath the cap. Ascomycetes spores (**ascospores**) form by meiosis inside a tube-like structure called the ascus. The ascus expels the ascospores through a small lid or tip underneath the cap of the fungi. Although Ascomycetes contains over 50,000 diverse fungal species (e.g., *Penicillium* spp., *Gyromitra esculenta* (Pers.) Fr. or false morel), most poisonous mushrooms are the fruiting bodies of dikaryotic Basidiomycetes.

Identification

There is no single test that reliably distinguishes all poisonous mushrooms from nontoxic mushrooms ("no rule is the only rule"). Hence, identification of mushroom species requires careful observation of distinguishing characteristics species by species.[1] The use of rules (e.g., absence of tarnish on a silver spoon cooked in the same pot as the mushroom, presence of animal bite marks on the mushrooms) are unreliable methods to separate toxic from nontoxic mushrooms.[2] In fact, most poisonous mushroom species are morphologically similar to some edible species. Accurate identification of mushroom species requires careful observation of macroscopic characteristics including size, color, color changes, texture, odor, and anatomic characteristics of the mushroom. The size of mushroom spores is typically measured in microns. Microscopic analysis of the spores helps confirm the identification of the mushroom by evaluating the orientation of the hyphae in the gill tissue, the structure of the cap cuticle, and the shape and

Medical Toxicology of Natural Substances, by Donald G. Barceloux, MD

size of the spore-forming structures (asci, basidia). Guides are available to help identify poisonous mushrooms based on macroscopic and microscopic characteristics.[3–6]

MUSHROOM STRUCTURE

Figure 37.1 demonstrates the general features of a mushroom using an *Amanita* species as an example. Natural trauma or deterioration during storage may alter these characteristics. In addition to the color, texture, and thickness of the flesh, the following macroscopic characteristics help identify poisonous mushrooms:[7,8]

Cap (Pileus): The cap supports the spore-producing surface (gills, basidia) of the mushroom. The covering of the cap is called the cuticle or the pellicle if the cuticle is easily removed. Important features include size, shape (cylindrical, conical, bell-shaped, umbonate, convex), color, color changes, surface characteristics (wet, dry, smooth, scaly, irregular, warts), and margin (straight, striate, incurved). See Figure 37.2.

Gills (Lamellae): This structure appears on the undersurface of the cap as thin, radiating blades along with the spore-forming structures. Distinguishing features include the alignment of the blades in relation to the stalk (plane, uplifted, depressed, funnel-shaped, umbilicate), the spacing, thickness, depth, and color.

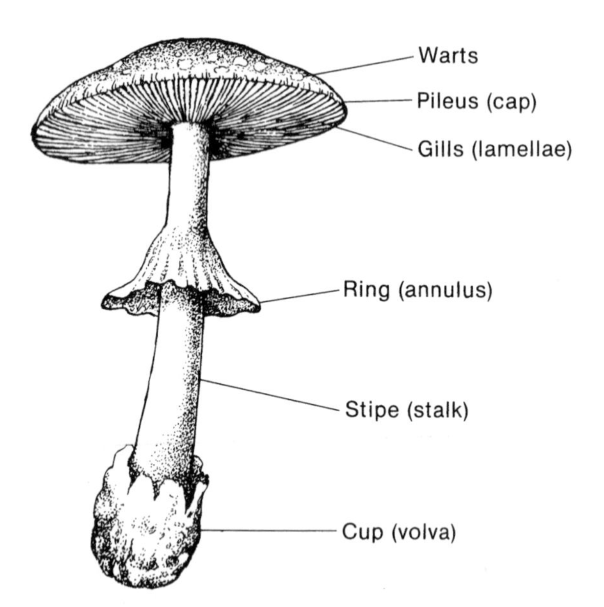

FIGURE 37.1. Morphologic structure of *Amanita phalloides*. Adapted from Ellenhorn MJ, Barceloux DG. Medical Toxicology: Diagnosis of Human Poisoning. 1st Edition, New York: Elsevier, 1988:1325.

Stalk (Stipe): This stem-like structure connects the cap to the ground and may contain a veil or annulus. The stalk pushes the spore-forming structures above the ground and allows the release of the spores into the air. The shape of the stalk varies from tapering downward to club-shaped and bulbous.

Ring (Annulus): Some mushrooms have a remnant of tissue, which remains on the stalk when the gills break through the veil. The shape may resemble a sheath, skirt, or collar. The position of the annulus may be oriented superior, inferior, medial, basal, or apical.

Veil: A specialized layer of tissue provides a protective cover for the developing mushroom. As the mushroom matures, a portion of the mushroom extrudes through this tissue to release the spores. The remnant of this tissue is the veil, which may or may not leave visible tissue on the stalk of the mature mushroom. There are two basic types of veils. A partial veil (inner veil) initially covers the gills, extending from the margin of the cap to the stalk. The remnant of the partial veil on the mature mushroom is the annulus. Occasionally, the partial veil adheres to the margin of the cap forming an appendiculate margin. The universal veil surrounds most of the immature mushroom. Remnants of the universal veil may form a volva at the base of the mature mushroom (e.g., most the *Amanita* species) or adhere to the underside of the partial veil. Pieces of the universal veil that adhere to the cap are called warts.

Cup (Volva): In some mushroom species (e.g., *Amanita phalloides*) visible remnants of the universal veil remain at the base of the stalk. These remains are called the volva. The shapes of these cups range from indistinct and scaly to sac-like and collar-like.

SPECIMEN COLLECTION

The mushroom specimen should be collected with as many features (cap, stem, cup) as intact as possible. The skirt and cup are important distinguishing features of *Amanita* species. The initial color of the flesh should be recorded along with any subsequent changes that result from trauma. Place the specimen in a paper bag and refrigerate. Plastic containers should be avoided. Samples of vomitus and lavage fluid should be filtered and refrigerated for later identification of spores.

SPORE IDENTIFICATION

Spores are the microscopic "seeds" of the fungi. Spore identification involves determination of the color of the

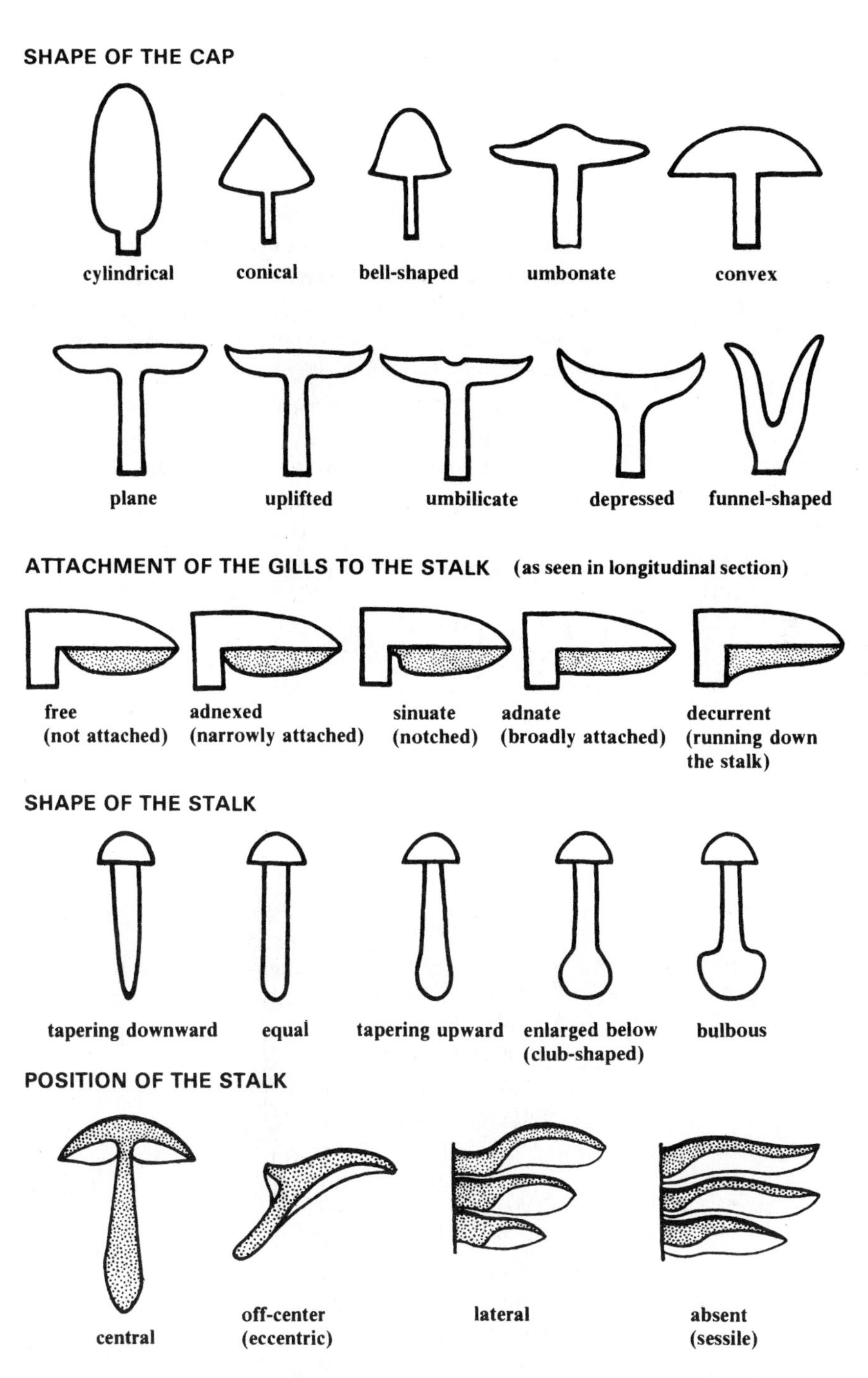

FIGURE 37.2. Nomenclature for the identification of mushrooms. From Arora D. Mushrooms demystified. A comprehensive guide to the fleshy fungi. 2nd ed. Berkeley, CA: Ten Speed Press; 1986:17.

spores through a spore print and microscopic examination of the shape and size of the spores. Because spores are not visible to the unaided eye, spores must be collected on a white paper for determination of color while using a standard color guide. The stalk of the mushroom is cut just below the cap, and the cap should be placed on the white paper with the gills facing the white paper. Covering the mushroom cap with a glass or bowl helps protect the spores from dispersing in air currents. After several hours, a characteristic color appears on the paper that is relatively resistant to environmental changes and constant within each mushroom species. The color of the mature gills does not always reflect the color of the spores. Spore prints may be difficult to obtain from young, soggy, or dried mushrooms. The application of various reagents helps to differentiate mushroom spores including the use of Melzer's reagent (20 g chloral hydrate, 0.5 g iodine, and 1.5 g of potassium iodide added to 20 mL of water) and less commonly Cotton Blue, aqueous potassium hydroxide, or sulphovanillin. Spores that turn grey, blue, or black after contact with Melzer's reagent are termed amyloid, whereas a more intense red-brown is termed pseudo-amyloid as occurs with *Lepiota* species. No color changes occur following application of Melzer's reagent to nonamyloid spores. Examination of gastric contents or stool for spores requires filtering, extraction (diethyl ether), washing, separation (glycerol), and centrifuging. The top layer is examined under the microscope for color, presence of spores, shape, size, and wall thickness of the spores. Colorless or white spores in this layer are evaluated with Melzer's reagent to determine if color changes occur.

Clinical Classification

Poisoning by fungi also is known as mycetism or mycetismus. This type of poisoning results from the consumption of unspoiled fruiting bodies (i.e., mushrooms) that contain toxic substances. Exposures to mushrooms and their spores also produce other adverse effects including gastrointestinal (GI) intolerance (i.e., trehalase deficiency), infectious diseases, bacterial-related toxicity (e.g., botulism), and allergic reactions (asthma, rhinitis).[9] Respiratory allergies result from the inhalation of fungal spores by sensitized individuals. These patients are usually sensitive to a variety of common allergens including grasses, dust, dust mites, and animal dander. Although mushrooms may contain environmental contaminants (e.g., lead, arsenic, mercury, pesticides), the amount of the contaminants are usually too small to cause toxicity. Some fungal species (e.g., *Agaricus*) accumulate heavy metals, but there are no well-documented cases of acute heavy metal poisoning following the ingestion of these mushrooms. The production of toxic secondary metabolites by fungal species causes mycotoxicosis (e.g., aflatoxicosis, ergotism) following ingestion of food stuffs contaminated by these toxins.

Frequently, samples of the mushroom involved in a case of poisoning are not available. For clinical evaluation, mushroom poisoning has been classified into eight categories based on clinical features and the onset of symptoms.[5,10] These clinical syndromes include the following: 1) amatoxin poisoning, 2) orellanine nephrotoxicity, 3) gyromitrin (hydrazine-like) poisoning, 4) pantherina syndrome associated with *Amanita muscaria* and *A. pantherina*, 5) muscarine toxicity, 6) coprine (disulfiram-like) toxicity, 7) other mushroom syndromes from exposure to *Paxillus* and other mushroom species, and 8) gastroenteritis-producing mushrooms. In the absence of adequate mushroom specimens, a clinical classification provides a satisfactory mechanism to guide medical treatment. However, individual variations in susceptibility and the possible ingestion of multiple mushroom species limit the usefulness of this classification as the sole basis for treatment. The most important aspect of treatment remains the accurate identification of the fungal species involved, particularly by an experienced mycologist. Several aspects of the history aid the clinical identification of the type of mushroom poisoning including onset of symptoms, alcohol consumption, thirst, diaphoresis, headache, and fatigue.[11] Of these features, the onset of symptoms provides the best basis to separate mushroom poisoning into specific categories. Table 37.1 and Table 37.2 list the syndromes associated with rapid (<6 hours) and delayed (>6 hours) onset of symptoms, respectively. As with any classification, the onset of symptoms is not a strict criterion for diagnosing mushroom poisonings because of the possible ingestion of multiple toxic mushrooms, the variability of toxins in mushroom species, and differences in individual responses.

AMATOXIN POISONING

HISTORY

One of the first historical accounts of mushroom poisoning was by the Greek poet Euripides, who documented the fatal ingestion of mushrooms by a woman and three children. Later, historians posit that Euripides had actually detailed the fate of his own wife and children as they traveled to Icarus in 430 BC.[12] Other historical accounts suggest that mushroom poisoning may have been involved in the deaths of Charles VI of France, Pope Clement II, and the widow of Tzar Alexis. The first mushroom-related deaths documented in the American medical literature occurred during 1871 near Chico, California.[13] California consistently leads the United States in annual mushroom fatalities. The estimated average annual death rate from mushroom ingestions in Europe and North America combined ranges from 100 to 200, with most fatalities occurring in Central Europe.[14] The ingestion of *Amanita phalloides* (Vaill. ex Fr.) Link accounts for most of the deaths related to mushroom consumption, particularly in Europe. In the United States, adult fatalities are limited to the consumption of several amatoxin-containing species of the genera *Amanita* [e.g., *A. ocreata* Peck, *A. phalloides*, *A. verna* (Bull.) Lam., *A. virosa* (Fr.) Bertill.] and *Galerina*.

Between 1957 and 1964, 30 deaths were associated with mushroom ingestions in the United States. In the

TABLE 37.1. Delayed Onset (>6 Hours) of Mushroom Poisoning

Latency Period[a]	Syndrome	Principal Mushrooms	Symptoms
6–12 Hours	Gyromitrin Poisoning	*Gyromitra esculenta*	Delayed onset of gastrointestinal symptoms, headache, hypoglycemia followed by hepatorenal dysfunction, seizures, coma[b]
6–24 Hours (Average 10–12 hours)	Hepatorenal (amatoxin)	*Amanita* spp. (*A. bisporigera, A. ocreata, A. phalloides, A. verna, A. virosa*) *Galerina* spp. (*G. fasciculata, G. marginata, G. venenata*) *Lepiota* spp. (*L. brunneoincarnata, L. brunneolilacea, L. castanea, L. helveola*)	Delayed onset of intense diarrhea and other gastrointestinal symptoms during the 1st day, followed by a quiescent period and then progressive hepatorenal dysfunction
>24 Hours	Acromelalgia	*Clitocybe amoenolens*	Distal paresthesias/dysesthesias
1–3 Days	Rhabdomyolysis	*Tricholoma equestre*	Fatigue, muscle weakness, myalgias, and rhabdomyolysis
1–4 Days	*Amanita* Nephropathy	*Amanita* spp. (*A. proxima, A. smithiana*)	Gastroenteritis followed by relatively rapid onset of renal dysfunction compared with *Cortinarius* spp.
36 hours–14 Days	Orellanus	*Cortinarius* spp. (*C. orellanus, C. rubellus*)	Delayed onset of gastrointestinal symptoms, headache, thirst followed by progressive renal failure

[a]Time between ingestion and onset of symptoms.
[b]Distinguished from amatoxin syndrome by less intense diarrhea and later by lack of hepatotoxicity.

next 10 years, there were at least 57 hospitalizations and eight deaths resulting from mushroom toxicity in California alone.[15] During the early 1980s, the San Francisco Bay Area Poison Control Center experienced a 5-fold increase in calls regarding wild mushroom ingestions compared with the previous year.[16,17] Analysis of poison control center data indicates that most of wild mushroom exposures are clinically insignificant ingestions by children less than 6 years of age. In a 5-year analysis of California poison control data, approximately one-half of the 6,317 reported mushroom exposures did not experience any symptoms.[18] Of all the reported cases, 61 patients (1%) were admitted to critical care units and one patient died.

MYCOLOGICAL DESCRIPTION

Of approximately 5,000 mushroom species, less than 100 species are poisonous and about 35 species are known to contain amatoxins. Table 37.3 lists the known amatoxin-containing species that belong to the following fungal families: Amanitaceae (*Amanita* spp.), Agaricaceae (*Lepiota* spp.), and Cortinariaceae (*Galerina* spp.). Mushroom species in other genera may contain amatoxins, but there are no clinical data that associate ingestion of these mushroom species with human toxicity. Analysis of specimens from The North American mushroom, *Conocybe filaris* (Fr.) Kühner (*Pholiotina filaris*) demonstrated the presence of amatoxins,[19] but the European *Conocybe crispa* (Longyear) Singer and the common North American species, *C. lactea* (J.E. Lange) Métrod did not contain detectable concentrations of amatoxins.[20] Hepatoxicity following the consumption of *Conocybe* species is not well documented in the medical literature.[5]

Distinguishing toxic and nontoxic mushrooms is difficult as suggested by the numerous hospitalizations that occur after the consumption of wild mushrooms.[21] No single test distinguishes poisonous and nonpoisonous mushrooms ("no rule is the only rule"). Amatoxin-containing mushrooms possess no characteristic smell or taste, and these poisonous mushrooms often resemble edible mushroom species.

AMANITA

Amanita species are large mushrooms that have an annulus (ring of death), volva (death cup), white gills unattached to the stalk, and colorless, broad ellipsoid spores that react to Melzer's reagent (i.e., amyloid). The color of the cap varies from white to olive, olive-gray, yellow-green, or yellow-brown. Although the mature *Amanita phalloides* mushrooms usually have a distinctive olive green cap, white gills, and veil at the base of

TABLE 37.2. Rapid Onset (<6 hr) of Mushroom Poisoning

Latency Period[a]	Syndrome	Principal Mushrooms	Symptoms
0.5–2 Hours	Gastrointestinal	*Chlorophyllum molybdates*, *Agaricus* spp. (*A. hondensis*, *A. xanthodermus*), *Dermocybe sanguinea*, *Entoloma* spp. (*E. lividum*, *E. sinuatum*), *Macrolepiota venenata*, *Hebeloma crustuliniforme*, *Tricholoma pardinum*, and many other mushroom species	Nausea, vomiting, diarrhea, and abdominal pain
0.5–2 Hours	Muscarine	*Inocybe* spp. (*I. dulcamara*, *I. fastigiata*, *I. geophylla*, *I. patouillardii*, *I. umbrina*, and other species) *Clitocybe* spp. (*C. candicans*, *C. dealbata*, *C. ericetorum*, *C. phyllophila*, and other species) *Mycena pura*	Muscarinic symptoms including nausea, vomiting, watery diarrhea, abdominal cramps, miosis, blurred vision, diaphoresis, salivation, and muscle fasciculations
0.5–3 Hours	Psilocybin	*Psilocybe* spp. (*P. baeocystis*, *P. cubensis*, *P. pelliculosa*, *P. semilanceata*, and other species) *Panaeolus* spp. (*P. bohemica*, *P. cyanescens*, and other species) *Inocybe aeruginascens* *Gymnopilus junonius* *Panaeolina foenisecii*	Euphoria, disorientation, vertigo, headache, weakness, hallucinations, altered mood, drowsiness, and occasionally seizures and coma
15–30 Minutes[b]	Coprinus	*Clitocybe clavipes*, *Coprinus atramentarius*	Antabuse-like reaction with nausea, vomiting, headache, flushing, vertigo, diaphoresis, tachycardia, and metallic taste
0.5–3 Hours	Pantherina	*Amanita* spp. (*A. aprica*, *A. gemmata*, *A. muscaria*, *A. pantherina*, *A. regalis*)	Euphoria, ataxia, agitation, disorientation, euphoria, anxiety tremor, ataxia followed by prolonged sleep
1–2 Hours	Paxillus	*Paxillus involutus*	Nausea, vomiting, diarrhea, weakness, abdominal pain along with hemolytic anemia, jaundice, and renal dysfunction[c]

[a]Time between ingestion and onset of symptoms.
[b]Minutes after ethanol consumption. Mushroom ingestion can occur up to 72 hours prior to ethanol ingestion.
[c]Minor reactions may occur during months to years prior to the development of the Paxillus syndrome.

the stem, several factors alter the appearance of this mushroom including weather and soil conditions, trauma, season, location, and age of the specimen.[16] The cup may not be present because of the loss of the lower portion of the stalk during collection of the specimen.

Common Name: Death Cap, Deadly Angel, Deadly Amanita, Destroying Angel, Deadly Agaric

Family: Amanitaceae

Scientific Name: *Amanita phalloides* (Vaill. ex Fr.) Link

Physical Description: The grayish olive to greenish yellow cap is 5–15 cm (~2–10 in.) with light greenish olive margin and a shiny, dry surface containing dense, dark grayish olive fibrils at the center. The gills are free with intercalated gills, which contain white spore masses that turn blue-black following application of Melzer's reagent. The stalk contains a membranous volva surrounding an enlarged bulb at the base. The stalk gradually tapers above the bulb where the remains of a partial veil (annulus) appear 1–1.5 cm (~0.5 in.) below the cap.

TABLE 37.3. Amatoxin-Containing Mushroom Species from the Genera *Amanita*, *Galerina*, and *Lepiota*. Other Amanita-Containing Mushrooms Include *Macrolepiota excoriata* (Schaeff.) M.M. Moser (*Lepiota heimii*)

Amanita	Galerina	Lepiota
bisporigera G. F. Atk.[a]	*badipes* (Fr.) Kühner	*brunneoincarnata* Chodat & C. Martín[a]
decipiens (Trimbach) Jacquet.	*beinrothii* Bresinsky	*brunneolilacea* Bon & Boiffard[a]
hygroscopica Coker	*fasciculata* Hongo	*boudieri* Guég[a]
ocreata Peck[a]	*helvoliceps* (Berk. & M.A. Curtis) Singer	*castanea* Quél.[a]
phalloides (Vaill. ex Fr.) Link[a]	*marginata* (Batsch) Kühner[b]	*clypeolaria* (Bull.) Quél.[a]
suballiacea (Murrill) Murrill	*sulciceps* (Berk.) Singer	*clypeolarioides* Rea
tenuifolia (Murrill) Murrill	*venenata* A.H. Sm.	*felina* (Pers.) P. Karst.
verna (Bull.) Lam.[a]		*fuscovinacea* F.H. Møller & J.E. Lange
virosa (Fr.) Bertill.[a]		*griseovirens* Maire
		helveola Bres.
		helveoloides Bon ex Bon & Andary
		jacobi Vellinga & Knudsen
		kuehneri Huijsman ex Hora
		lilacea Bres.
		locanensis Espinosa
		ochraceofulva P.D. Orton
		pseudohelveola Kühner
		rufescens Morgan
		subincarnata J.E. Lange
		xanthophylla P.D. Orton

Source: Adapted from Ref 14.
[a]Mushroom species most commonly associated with amatoxin poisoning.
[b]Synonyms: *G. autumnalis*, G. unicolor.

The white stalk usually contains a scattering of grey-olive scales. Spores are globose to ellipsoidal, ranging in size from 7–12.8 × 5.5–10.2 μm in diameter.

Distribution and Ecology: In the United States, *Amanita phalloides* occurs primarily in the cool coastal forests of the west coast (Northern California, Pacific Northwest) as documented by reports of fatal ingestions.[22] However, *A. phalloides* also inhabits the mid-Atlantic coast and oak woodlands of the northeastern United States. These mushrooms flourish during the fall, particularly after warm, heavy rains. *A. phalloides* is the predominant European poisonous mushroom, particularly in Central and Occidental Europe. This species occurs primarily in deciduous and mixed deciduous forest, especially near oak trees during August to October. *Amanita phalloides* is also widespread in southeastern Australia including the Melbourne suburbs, New South Wales, Tasmania, and Victoria.[23]

Common Name: Destroying Angel, Spring Amanita, White Death Cap, Fool's Mushroom

Family: Amanitaceae

Scientific Name: *Amanita verna* (Bull.) Lam.

Physical Description: This species is similar to *A. phalloides* with the exception of the overall white appearance. The cap is hemispherical in young specimens, and the 4–8 cm (2–3 in.) mature cap becomes more convex to flat than *A. phalloides*. The gills of *A. verna* are free and they do not contact the stem. The white stalk contains a ring in the upper portion and a cup at the base. Like spores from *A. phalloides*, the *A. verna* spores are colorless, broadly elliptical, and stain blue-black with Melzer's reagent.

Distribution and Ecology: *Amanita verna* occurs in the warmer regions of Europe from spring until autumn, whereas *Amanita virosa* inhabits the mountainous regions with conifer forests.

Common Name: Destroying Angel, White Amanita, Angel of Death, White Death Cap

Family: Amanitaceae

Scientific Name: *Amanita virosa* (Fr.) Bertill.

Physical Description: The appearance of this large, white mushroom is similar to *A. verna*. Like all *Amanita* species, the stalk contains a ring below the cap and a volva at the base. The gills are free with a white spore print. The 6–11 cm (2–4 in.) cap is smooth with a silky sheen in dry weather. The shape of the cap varies from cone- to egg-shaped in young specimens to cone-shaped in older specimens. The cap of mature specimens frequently has a central, navel-like depression.

Distribution and Ecology: The distribution of *A. virosa* is more localized in Europe (e.g., evergreen forests of Scotland and Upper Bavaria) compared with *A. phalloides*.[5]

Galerina

There are at least seven known species of *Galerina* that contain amatoxins including *G. badipes* (Fr.) Kühner, *G. beinrothii* Bresinsky, *G. fasciculata* Hongo, *G. helvoliceps* (Berk. & M.A. Curtis) Singer, *G. marginata* (Batsch) Kühner, *G. sulciceps* (Berk.) Singer, and *G. venenata* A.H. Sm.[5,24] These small brown mushrooms have a somewhat de-curved cap with brownish spore prints.

Common Name: Marginate Pholiota

Family: Cortinariaceae

Scientific Name: *Galerina marginata* (Batsch) Kühner (*Galerina autumnalis*, *Galerina unicolor*)

Physical Description: This small mushroom has a 1.4–5 cm (~0.5–2.0 in.) cap that contains ochre-colored gills. The spore print is dust-brown. The shape of the cap from immature specimens is almost hemispherical with a de-curved margin, but more mature specimens have a convex shape. The ochre-colored stalk contains a ring, but no volva. Below the ring, the flesh of the stalk is relatively smooth without scales.

Distribution and Ecology: This white-rot fungus is widely distributed, primarily on rotting wood or debris of conifers (hardwoods) in European and American forests.

Lepiota

Lepiota species produce small to medium-sized mushrooms with bare to conical, wart-containing caps typically <10 cm (4 in.) in size. The spores are pseudo-amyloid. Despite their relatively small size, case reports associate severe hepatorenal dysfunction with ingestion of these mushroom species including *Lepiota brunneoincarnata* Chodat & C. Martín, *Lepiota helveola* Bres., and *Lepiota subincarnata* J.E. Lange (*L. josserandii*).[25–27]

Common Name: Star dapperling

Family: Agaricaceae

Scientific Name: *Lepiota helveola* Bres.

Physical Description: This small to moderately large mushroom has a 1.5–6 cm (~0.5–2.5 in.) cap that is hemispherical to convex in young specimens and more umbonate with a slightly depressed center in older specimens. The pink to flesh-colored cap contains white to reddish tinged gills and elliptical spores that are pseudo-amyloid with Melzer's reagent. The free gills contain a white mass of spores. The stalk contains a ring, but no definite cup at the base.

Distribution and Ecology: This relatively rare fungus occurs in decaying matter at the edge of forests and nearby grassland in Europe.

EXPOSURE

The actual incidence of mushroom toxicity in the United States is difficult to estimate because poisoning often occurs in clusters, and the reporting of adverse reactions to mushroom ingestion is not required by US law. Amatoxin poisoning usually results from the misidentification of *A. phalloides* as an edible mushroom species, such as *Amanita fulva* (Schaeff.) Fr., *Agaricus* spp., *Macrolepiota procera* (Scop.) Singer (*Lepiota procera*), and *Tricholoma portentosum* (Fr.) Quél. Several other *Amanita* species (e.g., *A. virosa*, *A. verna*, *A. ocreata*, *A. bisporigera*) are more prevalent than *A. phalloides* in the United States. However, case reports of fatalities following consumption of other amatoxin-containing *Amanita* species (e.g., *A. verna*,[28] *A. virosa*[29]) are much less common than *A. phalloides*. Rare case reports associate fatalities with the ingestion of amatoxin-containing *Galerina* species (*G. marginata*, *G. venenata*).[30,31] In addition, several American *Lepiota* species (*L. helveola* Bres., *L. castanea* Quél., *L. felina* (Pers.) P. Karst.) contain amatoxin, but to date no cases of toxicity have been reported following ingestion of these mushrooms in the United States. Because of the wide variations in location and toxicity of *Amanita*, *Galerina*, and *Lepiota* species, the regional poison control center is an important source of information regarding poisonous mushrooms in the local area.

PRINCIPAL TOXINS

Physiochemical Properties

The three major types of toxic peptides in *Amanita* species are amatoxins, phallotoxins, and virotoxins.[32,33] These toxins contain a sulfur-linked tryptophan unit and unusual hydroxylated amino acids. Amatoxins, such as α-amanitin (CAS RN: 23109-05-9) or β-amanitin (CAS RN: 21150-22-1), are the bicyclic octapeptides responsible for the fatalities associated with the ingestion of amatoxin-containing mushrooms belonging to the genera *Amanita*, *Galerina*, and *Lepiota*.[34] Figure 37.3

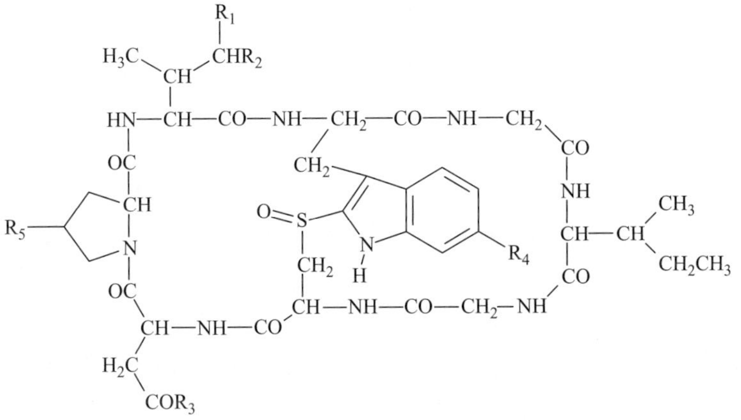

	R_1	R_2	R_3	R_4	R_5	LD_{50} (mg/kg)
α-Amanitin	CH_2OH	OH	NH_2	OH	OH	0.3
β-Amanitin	CH_2OH	OH	OH	OH	OH	0.5
γ-Amanitin	CH_3	OH	NH_2	OH	OH	0.2
ε-Amanitin	CH_3	OH	OH	OH	OH	0.3
Amanin	CH_2OH	OH	OH	H	OH	0.5
Amaninamide	CH_2OH	OH	NH_2	H	OH	0.3
Amanullin	CH_3	H	NH_2	OH	OH	>20
Amanullinic Acid	CH_3	H	OH	OH	OH	>20
Proamanullin	CH_3	H	NH_2	OH	H	>20

FIGURE 37.3. Chemical structure of amatoxins.

demonstrates the structure of eight amatoxins, which are composed of an amino acid ring bridged by a sulfur atom. The major amatoxins that cause hepatorenal toxicity are α-amanitin, β-amanitin, and γ-amanitin. These compounds are present in some *Galerina* and *Lepiota* species as well as *Amanita* species. The amatoxin-containing species of *Amanita* also contain several cyclic heptapeptides called phallotoxins (phalloidin, phallisin, phalloin, phallacidin, phallisacin).[35,36] Although phalloidin is a potent parenteral animal hepatotoxin, the poor GI absorption of phallotoxins probably limits any systemic toxicity following ingestion. Phallotoxins are bicyclic peptides containing cysteine, whereas virotoxins are monocyclic peptides containing serine. Virotoxins include ala–viroidin, viroisin, deoxoviroisin, viroidin, and deoxoviroidin. In experimental animals after intraperitoneal injection of virotoxins and phallotoxin, the histological changes of hepatocytic vacuolization and intracytoplasmic erythrocyte accumulation are similar;[37] however, the role of virotoxins in human toxicity is unclear. Amatoxins are stable compounds that do not degrade during cooking or drying at 250 °C–280 °C (480 °F–535 °F). These compounds resist freezing at −25 °C (13 °F), and ingestion of *Amanita phalloides* mushrooms stored in a freezer for 7–8 months produced fatal hepatoxicity.[38] However, amatoxins decompose slowly when stored in open, aqueous solutions or following prolonged exposure to sun or neon light.

TABLE 37.4. Approximate Content of Principal Amatoxins and Phallotoxins in *Amanita Phalloides*

Toxin	Weight (mg/g)*
α-Amanitin	1.0–2.5
β-Amanitin	1.2–3.25
γ-Amanitin	0.2–0.65
Phalloidin	0.9–2.2
Phallacidin	2.4–4.3
Phallisacin	1.7–6.0

Source: From Ref 5.
*Dry weight.

Poisonous Parts

There are approximately 35 amatoxin-containing mushroom species in three genera: *Amanita*, *Galerina*, and *Lepiota*. Amatoxins probably occur in most *Amanita* mushroom species, even in minute amounts in edible mushroom species. *Amanita phalloides* is the most toxic amatoxin-containing mushroom species with an amatoxin content of about 2–3 mg/g dry tissue. Table 37.4 demonstrates the approximate amatoxin and phallotoxin content of a several collections of samples from *A. phalloides* mushrooms. Consequently, 1–3 fresh mushrooms (i.e., about 50 g) probably contain lethal quantities of amatoxin depending on the amatoxin concentrations.

A variety of factors affects the amatoxin concentrations in *Amanita* species including plant part, environmental factors, location, type of toxin, and soil type as well as significant variation between *Amanita* species. A study of three *Amanita phalloides* specimens demonstrated the highest amatoxin content in the ring, gills, and cap, whereas the bulb and volva contained smaller amatoxin concentrations and larger concentrations of phallotoxins.[39] Similarly, examination of 25 fruiting bodies from *Amanita phalloides* specimens collected in France demonstrated relatively high concentrations of α-, β-, and γ-amatoxins in the cap, gills, ring, and stalk compared with the bulb and the volva (cup) as measured by high performance liquid chromatography (HPLC).[40] In contrast, the concentration of phallotoxins (phallacidin, phallisacin, phalloidin, phallisin, phalloin) was higher in the bulb and cup compared with the cap, gills, ring, and stalk. Specimens of amatoxin-containing species with the more toxic specimens typically contain about 0.1–0.4% amatoxins.[41] Generally, amatoxin-containing *Amanita* species other than *A. phalloides* contain less amatoxin (e.g., *A. virosa*: 1.2–2.6 mg/g dry tissue) compared with the amatoxin content of *A. phalloides*. Additionally, the ratio of α-amatoxin to β-amatoxin varies between the amatoxin-containing species of *Amanita* and individual specimens. For example, *A. virosa* contains primarily α-amatoxin, whereas *A. phalloides* contains relatively high concentrations of β-amatoxin.[42] Some American samples of *A. virosa* contain relatively small amounts of amatoxin, but relatively large concentrations of virotoxins, whereas some European specimens contain amatoxin concentrations similar to local specimens of *A. phalloides*.[5] *A. bisporigera* G.F. Atk. is one of the most hepatotoxic toxic American *Amanita* species, particularly in the Midwestern and Eastern United States.[43] *A. verna* contains variable amounts of amatoxins. Examination of a few American and European specimens of *Galerina marginata* (*G. autumnalis*) yielded small amounts of β-amanitin and 0.1–0.8 mg α-amanitin/g dry weight.[44] *Lepiota* species (e.g., *L. brunneoincarnata* Chodat & C. Martín, *L. subincarnata* J.E. Lange, *L. boudieri* Guég) can contain substantial amounts (1.75–3.5 mg/g dry weight) of amatoxins.[45] Typically, these species contain α-amanitin and lesser amounts of γ-amanitin, but β-amanitin and phallotoxins are usually not detectable.

Toxic *Galerina* species (e.g., *G. marginata*) contain 0.8–1.5 mg of α- and β-amanitin per gram dry weight. In a study of 27 specimens of *G. marginata* from Alsace-Lorraine and Franche-Comté, the mean concentrations (μg/g fresh weight) of total amatoxins in three groups of these specimens were as follows: 78.17 ± 10.08 ($n = 6$); 96.88 ± 12.82 ($n = 11$); and 243.61 ± 16.54 ($n = 10$).[46] The hepatorenal syndrome following the ingestion of amatoxin-containing *Galerina* mushrooms is similar to the toxic *Amanita* species despite the absence of phallotoxins in the former mushrooms. Consequently, amatoxins are the primary toxic constituents of these hepatotoxic mushrooms rather than phallotoxins or virotoxins.

Mechanism of Toxicity

Amatoxins are potent inhibitors of mammalian DNA-dependent RNA polymerase II at very low concentrations (5–10 ng/mL); therefore, amatoxins interfere with protein synthesis via inhibition of elongation essential to transcription.[47] Those cells (e.g., liver, kidney, and intestinal cells) with the highest replication rates and most direct contact with the amatoxins develop necrosis. There is substantial variation in the toxicity of amatoxins between animal species. Higher mammals including humans, dogs, cats, and guinea pigs are relatively sensitive to amatoxins, whereas lower mammals (e.g., rodents) are relatively resistant to the effects of amatoxins because nonmammalian polymerases show little sensitivity to amanitins.[41] Although parenteral administration of phallotoxins to animals stimulates the polymerization of G-actin and stabilizes the F-actin filaments, the role of phallotoxins in producing hepatorenal failure in humans is limited by the lack of GI absorption of these compounds.

Liver

Amatoxin poisoning produces massive centrilobular hepatic necrosis characterized by centrilobular hemorrhage with preservation of hepatocytes limited to the larger triads. *In vitro* studies suggest that the Na^+-taurocholate cotransporter polypeptide facilitates uptake of α-amanitin by liver cells;[48] recent studies indicate that the organic anion-transporting polypeptide 1B3 (OATP1B3) is the human hepatic uptake transporter for amatoxins.[49] Once inside the cell, α-amanitin binds DNA-dependent RNA polymerase II at each translocation step in the hepatocytes,[50] resulting in the inhibition of protein synthesis and cell necrosis.[51]

Postmortem examinations of patients dying from amanitoxin-induced hepatic failure demonstrate widespread vacuolar degeneration of the liver cells. Macroscopic changes include diffuse jaundice, cerebral edema, subserosal petechiae, pulmonary congestion, yellow atrophy of the liver with diffuse subcapsular hemorrhaging, and evidence of hemorrhage throughout the body.[52] Some periportal regenerative changes and lobular collapse may appear depending on the survival time since ingestion. The most striking anatomic changes occur in the nuclei, where chromatin clumps appear along the periphery leaving the central portion clear.[53] Typically, these changes are not permanent if the patient survives, and surviving patients do not usually develop cirrhosis.

Kidney

Characteristic renal lesions following ingestion of amatoxin-containing mushrooms involve acute tubular necrosis with deposition of massive quantities of hyaline casts in the tubules. Histological examination does not usually demonstrate glomerular lesions. Macroscopically, the kidney typically appears dark red with extravasation of blood, especially in the cortical region.[52] Dehydration and hypovolemia may contribute to the development of acute tubular necrosis.[54] A case report of a renal biopsy examined by electron microscopy suggests that diffuse interstitial renal fibrosis may develop 1–2 months after *Amanita phalloides* ingestion without evidence of inflammatory changes.[55] Both proximal and distal convoluted tubules contain large vesicles within their endoplasmic reticulum and exhibit a loss of brush borders. Basement membranes are thickened without disruption. Another case report associated the development of Fanconi-type renal tubular acidosis with *Amanita phalloides* ingestion.[56]

Central Nervous System

Central nervous system effects probably result from hepatorenal failure rather than direct toxic effects of amatoxins. Case reports of fatal amatoxin poisoning document cerebral edema and cerebellar tonsillar herniation with fatty infiltration and mild hemorrhage of the subendocardial region.[16] Both neuropathological and myocardial changes following fatal amatoxin poisoning are variable and nonspecific because of terminal events, such as hyperpyrexia, sepsis, and cardiovascular collapse.

DOSE RESPONSE

There is significant inter- and intraspecies variation in the concentration of amatoxins in mushrooms depending on a number of factors including growing conditions, time of year, and humidity. Therefore, accurate prediction of toxicity based on the quantity of mushrooms consumed is difficult. Although all animal species develop adverse effects following exposure to amatoxin, there is substantial variation in the response of various species to similar doses. The estimated lethal dose of amatoxin for humans is approximately 0.1–0.3 mg/kg.[57] This dose correlates to 50–300 g of amatoxin-containing mushrooms.

TOXICOKINETICS

Kinetics

The GI absorption of amatoxins is rapid based on animal data and the appearance of amatoxin in the urine of a suicidal patient within 1.5–2 hours after ingestion.[58] In animal studies, the volume of distribution of α- and γ-amanitin is similar to the volume of the extracellular space.[59] The plasma elimination half-lives of these amatoxins ranged from approximately 30–45 minutes. Amatoxins are not bound to plasma proteins.[60] Case series of patients with amatoxin poisoning support the rapid elimination of amatoxins from the blood.[61] These human studies indicate that most amatoxins are cleared from the plasma within 12–24 hours after ingestion.[62,63] In a study of amatoxin poisoning, analysis of plasma samples drawn from patients between 9–18 hours after amatoxin ingestion did not detect α- or β-amanitin in five of nine samples.[62] One sample contained 20 μg α-amanitin/L and three other samples contained β-amanitin concentrations ranging from 23.5–162 μg/L. Elimination of absorbed amatoxins occurs primarily by renal excretion with smaller amounts excreted in the bile.[64] Renal elimination of amatoxins continues for 2–3 days after ingestion as a result of the uptake of amatoxins by the kidney. Although the duodenal aspirate of some patients with amatoxin poisoning may contain high concentrations of α- or β-amanitin, the amount eliminated by the bile is limited by the relatively small amount of biliary fluid. In a study of the gastroduodenal fluid from 12 patients with amatoxin poisoning, the aspirate did not contain detectable amounts of α- or β-amanitin in 8 of the 12 patients, when the samples were obtained 18–48 hours after ingestion.[62] Only gastroduodenal fluid samples from 2 patients contained high concentrations of α- or β-amanitin, when obtained 30 hours and 84 hours after ingestion. Amatoxins in the feces probably result from amatoxins that are not absorbed by the GI tract.[62] Amatoxins probably do not cross the blood–brain barrier.

Pregnancy

Samples of blood and amniotic fluid withdrawn from a woman in the eighth month of pregnancy demonstrated that the α-amanitin concentration was detectable in maternal blood, but not in the amniotic fluid.[65] The child was born without complications or anomalies. A case report documented the delivery of a healthy infant despite the development of moderate amatoxin poisoning during the 1st trimester[66] and 2nd trimester[67] of pregnancy. Follow-up of the children indicated normal development. A small case-control study did not detect an increased prevalence of major or minor anomalies in women poisoned by amatoxin-containing mushrooms during pregnancy.[68] However, the statistical power of the study was limited by the inclusion of only 5 women poisoned during the 1st trimester of pregnancy.

CLINICAL RESPONSE

Onset of Symptoms

The clinical presentation of hepatorenal toxicity is similar between various amatoxin-containing species within the genera *Amanita*, *Galerina*, and *Lepiota*.[69] The onset of GI symptoms after the ingestion of amatoxin-containing mushrooms usually ranges from 6–24 hours with an average of approximately 10–12 hours. The delayed onset of GI symptoms is an important diagnostic tool, but the presence of a short onset of GI symptoms does not exclude the ingestion of amatoxin-containing mushrooms if other mushrooms or substances with GI effects were also ingested. The GI effects of most nonlethal mushrooms develop with 2 hours after ingestion. Characteristically, amatoxin poisoning has the following three stages after the onset of GI symptoms: 1) gastrointestinal phase, 2) latent period representing the time required for the amatoxins to noncompetitively bind to intranuclear RNA polymerase II and disrupt protein synthesis, and 3) the hepatorenal phase.

Gastroenteritis Phase

The initial features of toxicity involve a profuse, watery diarrhea manifest by the abrupt onset of severe abdominal pain, nausea, vomiting, and bloody diarrhea. This phase usually lasts about 12–36 hours. Fever, tachycardia, hyperglycemia, hypoglycemia, dehydration, and electrolyte imbalance may occur during this phase.

Latent Period

With adequate fluid and electrolyte replacement, a remission of symptoms occurs during a subclinical elevation of serum hepatic aminotransferase enzymes. If mushroom-related toxicity is not suspected during this second latent phase, the patient may be discharged with the diagnosis of gastroenteritis.

Hepatorenal Phase

Within 3 to 4 days after ingestion, hepatic dysfunction and fulminant hepatic failure becomes clinically obvious with jaundice, coagulopathy, hemorrhage, hypoglycemia, delirium, confusion, and coma.[70] Biochemical signs of liver damage may develop within 36–48 hours after ingestion of the amatoxins.[71] Coagulopathies, metabolic acidosis, conduction abnormalities, hemorrhage, and sepsis frequently complicate severe cases. Progression from stage I to stage III/IV hepatic encephalopathy may be rapid.[27] Renal failure accompanies liver failure in most fatal cases, but occasionally renal dysfunction rather than hepatic failure is the most prominent clinical feature of amanita toxicity.

Outcome

Current mortality rates vary between <10% and 30% because of the wide variation in the amount of toxin ingested, individual susceptibility, and the overall health of the patient. Recently, mortality rates have declined with improvements in supportive care for liver failure, particularly in the absence of cerebral edema.[72] In a series of 24 patients treated with supportive care and serial activated charcoal for the first 48–72 hours, the mortality rate was 9%.[56] Death usually results from hepatic encephalopathy, cerebral edema, and multiorgan failure 4–9 days after ingestion. Risk factors for fatal outcomes following ingestion of amatoxin-containing mushrooms include age <10 years, short latency between ingestion and onset of GI symptoms, and severe coagulopathy. In surviving patients, a slow resolution of the hepatic necrosis begins on day 4–day 5 with decreasing serum hepatic aminotransferases and improving prothrombin activity. Most surviving patients recover without sequelae.[73] However, follow-up studies of amatoxin poisoning suggest that some patients may develop chronic active hepatitis with immune complexes, antismooth muscle antibodies, and a persistent elevation of serum hepatic aminotransferases.[74] A case series of 64 Italian patients with amatoxin poisoning documented the persistence of chronic active hepatitis and a lobular hepatitis at least 6 months after moderate to severe *Amanita* mushroom poisoning.[75] Case reports associate *Lepiota* mushroom poisoning (*L. helveola*, *L. brunneoincarnata*) with the development of a mixed polyneuropathy beginning at least 8 days after mushroom ingestion.[76] Manifestations of the polyneuropathy include weakness in the lower extremities, loss of deep tendon reflexes, and alteration of sensation. Weakness in the lower extremities in several patients persisted at least one year.

Pregnancy

There are few data on the effect of amatoxins on the fetus. Case reports suggest that fetal damage may result from the ingestion of amatoxins by the mother, but the evidence for this damage is contradictory.[71] A 22-year old woman developed moderately severe amatoxin poisoning during her 11th week of pregnancy.[66] She was treated with silymarin and *N*-acetylcysteine, and she delivered a full-term, healthy infant without complications. Follow-up examination of the infant at 2 years detected no evidence of hepatic or neurological injury. Other case reports support the lack of liver injury in infants born to a mother poisoned by *A. phalloides* ingestion.[77]

DIAGNOSTIC TESTING

Analytical Methods

α-Amanitin can be identified by thin-layer chromatography, radioimmunoassay (RIA),[78] enzyme-linked immunosorbent assay (ELISA),[79] HPLC,[35,36] capillary zone electrophoresis with photodiode array detection,[80] gas chromatography (GC), gas chromatography/mass spectrometry (GC/MS), or multistage linear ion trap mass spectrometry.[81] The calculated limit of detection for the latter method in serum or liver samples is 0.50 ng/g. A rapid fluorescent method is available for the detection of α- and β-amanitin using ethidium bromide fluorophores at 560 nm and 525 nm, respectively.[82] Vomitus, blood, and urine should be collected from the patient and stored under refrigeration until analyzed. The Wieland-Meixner test (Table 37.5) is a crude field test for the presence of amatoxins that is relatively sensitive, but negative results do not guarantee the absence of other mushroom toxins.[83] The basis of this test is the color changes associated with the acid-catalyzed reaction of α-, β-, and γ-amanitin with lignin present in pulp paper. The detection limit is about 0.2 mg/mL, which is substantially below the concentration in most toxic species. Consequently, the Wieland-Meixner test is sensitive to the presence of amatoxins in mushrooms, but exposure to excessive heat and sunlight produces false-positive results. Additionally, *in vitro* studies suggest that the presence of psilocin (*Psilocybe* spp., *Panaeolus* spp., *Conocybe* spp.) and, to a lesser extent, 5-substituted tryptamines (*Panaeolus* spp.), cause false-positive reactions with the Wieland-Meixner test.[84] The presence of 5-substituted tryptamines (bufotenine) produces a reddish-brown color on the Wieland-Meixner test, whereas the presence of 4-substituted tryptamines (e.g., psilocin) produces a gray to pale blue color that can be confused with the blue color associated with amatoxins. Analysis of α-amanitin by HPLC has detection limits of approximately 10 ng/mL.[85,86] Radioimmunoassay techniques can detect α-amanitin concentrations in biological samples near 0.1 ng/mL in plasma, but some cross-reactive may occur with other compounds.[87]

TABLE 37.5. Wieland-Meixner Test

Squeeze the fresh mushroom, preferable in a garlic press, and place a drop of the juice onto a lignin-containing paper (e.g., print-free newspaper, cheap pulp paper).* If inadequate amounts of fresh tissue are present, thoroughly pulverize the remaining portion of the mushroom on the paper. A methanol-extract of a dry mushroom may be applied to the paper if a fresh mushroom is not available.

Circle the area of wetness with a marker, and also circle a blank control area on the paper outside the area where the tissue was pulverized.

Dry the wet spot with gentle, warm air (e.g., hair dryer).

Add a drop or two of concentrated (10–12 N) hydrochloric acid to the sample circle and to the control circle.

Observe the circles for the appearance of blue discoloration, usually within 1–2 minutes after application of the acid. The presence of trace amounts of α-amanitin requires longer (10–20 minutes) for the color change.

*Perform indoors away from sunlight and excessive heat. Low amatoxin concentrations can cause a delayed reaction and therefore a second reading 30 minutes after the first reading should be done to detect low amatoxin concentrations.

Biomarkers

α- and β-Amanitin concentrations in blood and urine samples confirm exposure to amatoxin-containing mushrooms, but there are few data to correlate blood amanitin concentrations to symptoms or outcome. Urine α-amanitin concentrations are usually higher than plasma α-amanitin concentrations, and urine samples are the preferred biological sample for confirmation of exposure to amatoxins because these compounds disappear rapidly from the blood.[88] These compounds are usually detectable in plasma and urine up to about 36 hours and 4 days, respectively, after ingestion.[62] Case reports indicate that α- and β-amanitin appear in the urine within 1½–2 hours after ingestion of amatoxin-containing mushrooms as measured by radioimmunoassay with limits of detection near 2.5 ng/mL.[58] However, even urine α-amanitin concentrations may be low (i.e., ≤6 ng/mL) 12 hours after ingestion despite a fatal ingestion of amatoxin-containing mushrooms.[34]

Abnormalities

Initial laboratory values may reflect the dehydration and electrolyte imbalance associated with the gastrointestinal phase of amatoxin poisoning. Amatoxin poisoning produces the onset of hepatorenal dysfunction 1–2 days after ingestion, manifest by increases in serum bilirubin, serum hepatic aminotransferases, serum creatinine, and a coagulopathy (increase prothrombin time, reduce fibrinogen and coagulation factors II, V,VII, and X). Serum hepatic aminotransferase values peak between 2–4 days after ingestion, whereas serum bilirubin concentrations peak in the blood from survivors 4–5 days after ingestion.[72] The coagulopathy and hyperbilirubinemia may continue to deteriorate in patients with severe amatoxin poisoning.[89] Hypoglycemia may complicate the onset of hepatic encephalopathy. Case reports indicate that clinical or biochemical evidence of pancreatitis occurs during the later stages of amatoxin poisoning.[90] In general, the serum hepatic aminotransferases are lower in survivors than in nonsurvivors, but the extent of the rise in serum hepatic aminotransferase values does not accurately predict survival.[16] Transient proteinuria may occur in the absence of renal insufficiency.[91] Serum ammonia concentrations reflect the severity of liver dysfunction, but this test alone does not necessarily predict the severity of amatoxin poisoning.

TREATMENT

Stabilization

Treatment of amatoxin poisoning is primarily supportive with initial attention directed toward the aggressive correction of the acid-base, fluid, and electrolyte imbalances caused by the gastrointestinal phase of the poisoning. Any patient with altered mental status should be immediately evaluated for hypoglycemia. Patients presenting late in the course of amatoxin poisoning should be evaluated for hepatorenal failure, hemorrhage, and coagulopathy. Fresh-frozen plasma, vitamin K (10mg daily for 3 days), and packed red blood cell transfusions may be necessary to correct the coagulopathy and associated bleeding. A careful history is necessary to evaluate the severity of the ingestion including time and location of mushroom ingestion, onset of symptoms in relation to ingestion, the number of mushroom species ingested, and the location of any remaining specimens. Any persons ingesting the same mushrooms should be evaluated immediately. Submitted mushroom remnants should be evaluated by an expert mycologist for identification.

Decontamination

Most patients with amatoxin poisoning develop severe gastrointestinal symptoms prior to presentation to a health care facility; therefore, decontamination measures are not necessary. Vomitus should be stored for evaluation for the presence of amatoxins or spores. Patients should receive a dose of activated charcoal. Because of the enterohepatic circulation of amatoxins, serial activated charcoal (20–40g every 3–4 hours) has theoretical value, but the efficacy of this therapy has not been tested in clinical settings. The toxicokinetics of amatoxins suggest that activated charcoal may be discontinued by 48 hours after ingestion. Gastroduodenal aspiration is a potential therapeutic modality to enhance the elimination of amatoxins, but there are limited data on the efficacy of this treatment, particularly when the amount of fluid removed is usually relatively small.

Elimination Enhancement

Amatoxins are dialyzable. However, the low concentration of amatoxins in plasma limits the usefulness of measures to enhance elimination, such as charcoal hemoperfusion and hemodialysis. Case reports associate a favorable outcome with the use of hemoperfusion within 36 hours of ingestion.[92] However, hemoperfusion performed 12–14 hours after ingestion of amatoxin-containing mushrooms eliminated <4% of the estimated absorbed dose of α-amanitin.[34] In 2 patients treated with hemodialysis 23 hours after ingestion of amatoxin and later with charcoal hemoperfusion, analysis of the blood before treatment, after treatment, and in the hemodialysis and hemoperfusion circuits did not detect any α- or β-amanitin.[93] Consequently, there are inadequate pharmacokinetic data to support the routine use of charcoal hemoperfusion or hemodialysis during amatoxin poisoning. The theoretical exception would be a patient with pre-existing renal dysfunction with confirmed amatoxin poisoning seen early in the course of the intoxication, when high concentrations of amatoxins are in the blood.

A review of detoxification techniques involving plasmapheresis suggested that mortality from amatoxin poisoning decreased substantially over the last 20 years following the use of plasmapheresis during the first 24–36 hours of intoxication in combination with supportive care.[94,95] However, there are few pharmacokinetic data or controlled clinical trials to support the use of plasmapheresis. Although glomerular filtration eliminates amatoxins, there is no definitive clinical evidence that forced diuresis enhances the elimination of amatoxins more than generous fluid replacement (i.e., urine flow = 100–200mL/h).[62] Case reports associated improvement in hepatic function after the use of hemodialysis and the Molecular Adsorbent Recycling System (MARS) during amatoxin-induced hepatic failure,[96] but the role of this technique during *Amanita*-induced liver failure remains to be defined.

Drug Therapy

There is no proven antidote for amatoxin poisoning. Case series document the use of a variety of drugs including the silymarin group, β-lactam antibiotics (penicillin G, ceftazidime), rifampicin, thioctic acid, cimetidine, insulin, human growth hormone, cytochrome c, aucubin (iridoid glucoside isolated from *Aucuba japonica* Thunb.), steroids, and vitamin C during treatment for amatoxin poisoning. However, these drugs are usually administered in combination with other drugs, and there are no controlled clinical trials to document the efficacy of any of these treatments. Thioctic acid (α-lipoic acid) is a coenzyme of the Krebs cycle that has few adverse effects except for hypoglycemia. Although thioctic acid has been used extensively in Eastern Europe since 1968,[97] there are few data to document the efficacy of this substance and studies in animals have not documented the beneficial effect of thioctic acid. Currently, thioctic acid is not recommended for the treatment of amatoxin poisoning, particularly because of associated hypoglycemia.[34]

Penicillin G

In the last 20 years, benzylpenicillin (penicillin G) was the most common drug used during amatoxin poisoning, in part because of the easy availability of benzylpenicillin compared with silibinin.[14] Adverse effects include allergic reactions, hyperkalemia, and electrolyte imbalance. There is very limited evidence to suggest that benzylpenicillin improves the outcome of patients with amatoxin poisoning. In dogs, the intravenous administration of benzylpenicillin 5 hours after the oral administration of sublethal dose of *Amanita phalloides* extract reduced liver inflammation and necrosis, as measured by serum hepatic aminotransferases.[98] The similar improvement of coagulation factors was noted, but the effect was less dramatic. An uncontrolled clinical case series suggested that the administration of intravenous benzylpenicillin within 24–48 hours of amatoxin ingestion may improve outcome.[97] Regression analysis of patients receiving multiple treatments for amatoxin poisoning suggested that the addition of benzylpenicillin treatment improves outcome as measured by decreased mortality, particularly in association with the use of silibinin.[99] However, there is no definitive evidence that the administration of high dose benzylpenicillin (300,000–1,000,000 U/kg/d) improves the outcome of patients with amatoxin poisoning. This drug is not approved by the US Food and Drug Administration (FDA) for the treatment of amatoxin poisonings. Penicillin is a treatment option primarily when silymarin is not available.

Silymarin Complex and Silibinin

Silymarin complex is a mixture of at least four closely related flavonolignans derived from the seeds and fruits of the milk thistle *Silybum marianum* (L.) Gaertn. This biennial herb is a member of the aster family (Compositae), which includes daisies and thistles. Historically, this herb was used for the treatment of liver, spleen, and gallbladder diseases. About 60–70% of this mixture is composed of two diastereomers of silibinin (silybin, CAS No. 22888-70-6) with the remainder consisting of isosilibinin, silydianin, and silychristin.[100] The oral absorption of silymarin is poor with estimated bioavailability of approximately 25–50% and an elimination half-life of about 6 hours.[101] Silibinin is the water-soluble preparation of the active agent in silymarin. Histochemical studies of isolated rat hepatocytes indicate that silymarin and silibinin block the uptake of α-amanitin as well as acting as a free-radical scavenger.[102] Some authors suggest that the use of both silibinin and benzylpenicillin improves the outcome of patients with amatoxin poisoning. In a retrospective, uncontrolled Austrian study, the administration of intravenous silibinin (20–50 mg/kg/d in four divided doses) and penicillin was associated with a favorable outcome in all but one of 18 cases of *Amanita phalloides* ingestion.[103] However, because of the study design and the administration of multiple drugs, the efficacy of silibinin was unclear. As a result of the relatively lack of serious adverse reactions following the administration of silibinin, the use of this compound is frequently recommended for the treatment of amatoxin poisoning despite the lack of controlled studies documenting the efficacy of this compound.[100] In the United States, silymarin is available only as a food supplement. In Europe, silibinin is available as the dihemisuccinate salt. The initial dose of silibinin dihemisuccinate is 5 mg/kg IV over 1 hour followed by a daily infusion of 20 mg/kg for 6 days or until the serum hepatic transaminase concentrations normalize.

Supplemental Care

Because the efficacy of drug therapy remains unproven, early diagnosis and good supportive care is the cornerstone of treatment. Attempts to enhance elimination (e.g., duodenal drainage, forced diuresis, serial activated charcoal) within the first 24 to 30 hours are theoretically beneficial. However, the amount of amatoxin removed is probably small and these procedures may worsen fluid and electrolyte imbalances. The patient should receive sufficient fluids to maintain a urine output of at least 100 mL/h for the first several days after amatoxin ingestion. Initial laboratory evaluation includes complete blood count, platelet count, prothrombin time, serum electrolytes, serum creatinine and BUN, serum calcium and phosphate, serum bilirubin, serum ammonia, serum alkaline phosphatase, and serum hepatic aminotransferases (ALT, AST). For potential serious ingestions, baseline laboratory tests should include arterial blood gases, serum lactate, fibrinogen, and factor analysis (Factor V, anti-thrombin III). Patients with hepatorenal failure should be admitted to an intensive care setting for careful monitoring of fluid and electrolyte balance, serum glucose, measures of hepatic encephalopathy (serum ammonia and bilirubin concentrations), coagulation parameters, and the development of sepsis. Treatment for hepatic encephalopathy includes oral lactulose (30–45 mL every 6–8 hours), protein restriction, and parenteral nutrition. Cerebral edema is the major cause of death following the development of hepatic encephalopathy, and treatment includes mannitol, elevation of the head to 10°–20°, adjustment of ventilator settings, and fluid restriction. Epidural intracranial pressure monitoring is usually necessary to guide treatment of patients with grade 3–4 hepatic encephalopathy. The efficacy of vitamin K_1 for elevation of the prothrombin time during fulminant hepatic failure is unclear. Fresh frozen plasma (rarely, recombinant human factor VIIa) should be used for evidence of current bleeding or prophylactically for invasive procedures. Prophylactic correction of coagulopathy in a nonbleeding patient is not usually recommended. Serial hemoglobin and stool guaiac tests should be followed for evidence of GI bleeding. The administration of proton pump inhibitors (e.g., omeprazole) should be administered to reduce the risk of GI hemorrhage.

Transplantation

The two surgical options for liver transplantation are orthotopic liver transplantation and auxiliary liver transplantation. Total orthotopic liver transplantation requires long-term immunosuppression to maintain the liver graft. Auxiliary partial liver transplantation is an alternative because some native livers may recover with partial hepatectomy and temporary support.[104] Consequently, the transplanted liver provides temporary assistance until the native liver regenerates. Evaluation of patients with amatoxin poisoning for liver transplantation should be considered whenever fulminant hepatic failure develops. Optimally, transport to a liver transport center should occur prior to progression to stage III–IV hepatic encephalopathy and the development cerebral edema. Therefore, patients with hepatic failure that progresses to stage II hepatic encephalopathy are candidates for liver transplant, particularly children. Factors that favor the need for transplantation include prolonged prothrombin time exceeding twice normal values despite vigorous replacement therapy, hypoglycemia requiring parenteral glucose infusions, and progression beyond stage II hepatic encephalopathy.[105] A retrospective study of 198 patients with amatoxin poisoning using receiver-operating-characteristic and sensitivity-specificity analysis indicated that the presence of a prothrombin index <25% and a serum creatinine >1.2 mg/dl (106 umol/L) between 3–10 days after amatoxin ingestion are highly sensitive and specific predictors of ensuing hepatic failure and the need for liver transplantation.[106] In a study of the criteria for transplantation in amatoxin poisoning, the presence of a prothrombin index <10% (INR ≥ 6) 4 days or more after ingestion was the most reliable criteria for the need for transplantation.[107] Neither the degree of encephalopathy nor elevation of the serum creatinine helped separate fatal from nonfatal *Amanita* intoxication in this study. The total serum bilirubin concentration is a less useful factor for the prediction of the need for transplantation because of the relatively late rise of serum bilirubin during the course of fatal amatoxin poisoning. The degree of serum hepatic aminotransferase elevation does not correlate to the need for liver transplant.

Factor V concentration <10% and inadequate response of the prothrombin time to the administration of clotting factors also indicates a poor prognosis without liver transplantation.[57,108] Case reports suggest that increased prothrombin time and decreased Factor V concentrations rather than coagulation inhibitors (antithrombin III, protein C, protein S) are better markers of the severity of amatoxin-induced liver damage.[109] In this report, Factor V, which began increasing 3–4 days after ingestion, was a better indicator of early hepatic recovery than other coagulation factors (fibrinogen, factors II, VII, X). The rapid clearance of amatoxins from the body within 4 days of ingestion limits any hepatotoxic effects of these compounds on liver transplants[62] or the transplantation of organs from donors dying of amatoxin poisoning.[110]

References

1. Lincoff G, Mitchel DH. Toxic and Hallucinogenic Mushroom Poisoning: A Handbook for Physicians and Mushroom Hunters. New York: Van Nostrand Reinhold Co; 1977.
2. Goos RD. Another case of mushroom poisoning involving *Tricholomopis platyphylla*. Mycologia 1984;76:350–351.
3. Lincoff GH. National Audubon Society Field Guide to Mushrooms North America. New York: Alfred A Knopf; 2000.

4. Bessette AL, Bessette AR, Fischer DW. Mushrooms of Northeastern North America. Syracuse, NY: Syracuse University Press; 1997.
5. Bresinsky A, Besl H. A colour atlas of poisonous fungi. A handbook for pharmacists, doctors, and biologists. London: Wolfe Publishing Ltd; 1990.
6. McNeil R. Le grand Livre des Champignons du Quebec et de l'est du Canada. Waterloo, Quebec, Canada: Editions Michel Quintin, Inc.; 2006.
7. Laessoe T, Del Conte A, Lincoff G. The Mushroom Book. How to Identify, Gather, and Cook Wild Mushrooms and Other Fungi. New York: DK Publishing, Inc.; 1996.
8. Arora D. Mushrooms demystified. A comprehensive guide to the fleshy fungi. 2nd ed. Berkeley, CA: Ten Speed Press; 1986.
9. Herxheimer H, Hyde HA, Williams DA. Allergic asthma caused by basidiospores. Lancet 1969;2(7612): 131–133.
10. Berger KJ, Guss DA. Mycotoxins revisited: Part I. J Emerg Med 2005;28:53–62.
11. Lampe KF. Mushroom poisoning in children updated. Paediatrician 1977;6:289–299.
12. Barbato MP. Poisoning from accidental ingestion of mushrooms. Med J Aust 1993;158:842–847.
13. Cheney WF. Cases of mushroom poisoning. Pacific Med J 1871;5:119–121.
14. Enjalbert F, Rapior S, Nouguier-Soule J, Guillon S, Amouroux N, Cabot C. Treatment of amatoxin poisoning: 20-year retrospective analysis. J Toxicol Clin Toxicol 2002;40:715–757.
15. Becker CE, Tong TG, Boerner U, Roe RL, Scott AT, MacQuarrie MB, Bartter F. Diagnosis and treatment of *Amanita phalloides*-type mushroom poisoning. West J Med 1976;125:100–109.
16. Cochran KW. Annual Report of the North American Mycological Association's Mushroom Poisoning Case Registry. Mcllvainea 2000;14:34–40.
17. Olson KR, Pond SB, Seward J, Healey K, Woo OF, Becker CE. *Amanita phalloides*-type mushroom poisoning. West J Med 1982;137:282–289.
18. Nordt SP, Manoguerra A, Clark RF. 5-Year analysis of mushroom exposures in California. West J Med 2000; 173:314–317.
19. Brady LR, Benedict RG, Tyler VE, Stuntz DE, Malone MH. Identification of *Conocybe filaris* as a toxic basidiomycete. Lloydia 1975;38:172–173.
20. Hallen HE, Watling R, Adams GC. Taxonomy and toxicity of *Conocybe lactea* and related species. Mycol Res 2003;107:969–979.
21. Jacobs J, von Behren J, Kreutzer R. Serious mushroom poisonings in California requiring hospital admission, 1990 through 1994. West J Med 1996;165:283–288.
22. Zevin S, Dempsey D, Olson K. *Amanita phalloides* mushroom poisoning—Northern California, January 1997. MMWR Morb Mortal Wkly Rep 1997;46:489–492.
23. Trim GM, Lepp H, Hall MJ, McKeown RV, McCaughan GW, Duggin GG, Le Couteur DG. Poisoning by *Amanita phalloides* ("deathcap") mushrooms in the Australian Capital Territory. Med J Aust 1999;171:247–249.
24. Muraoka S, Fukamachi N, Mizumoto K, Shinozawa T. Detection and identification of amanitins in the wood-rotting fungi *Galerina fasciculata* and *Galerina helvoliceps*. Appl Environ Microbiol 1999;65:4207–4210.
25. Feinfeld DA, Mofenson HC, Caraccio T, Kee M. Poisoning by amatoxin-containing mushrooms in suburban New York—report of four cases. J Toxicol Clin Toxicol 1994;32:715–721.
26. Meunier BC, Camus CM, Houssin DP, Messner MJ, Gerault AM, Launois BG. Liver transplantation after severe poisoning due to amatoxin-containing *Lepiota*—report of three cases. Clin Toxicol 1995;33:165–171.
27. Burton JR Jr, Ryan C, Shaw-Stifferl TA. Liver transplantation in mushroom poisoning. J Clin Gastroenterol 2002;35:276–280.
28. O'Brien BL, Khuu L. A fatal Sunday brunch: *Amanita* mushroom poisoning in a Gulf Coast family. Am J Gastroenterol 1996;91:581–583.
29. Broussard CN, Aggarwal A, Lacey SR, Post AB, Gramlich T, Henderson M, Younossi ZM. Mushroom poisoning—from diarrhea to liver transplantation. Am J Gastroenterol 2001;96:3195–3198.
30. Grossman CM, Malbin B. Mushroom poisoning: A review of the literature and report of two cases caused by a previously undescribed species. Ann Intern Med 1954;40: 249–259.
31. Tyler VE Jr, Smith AH. Chromatographic detection of Amanita toxins in *Galerina venenata*. Mycologia 1963;55: 358–359.
32. Wieland T. Peptides of poisonous amanita mushrooms. New York; Springer-Verlag; 1986.
33. Li C, Oberlies NH. The most widely recognized mushroom: chemistry of the genus *Amanita*. Life Sci 2005;78:532–538.
34. Koppel C. Clinical symptomatology and management of mushroom poisoning. Toxicon 1993;31:1513–1540.
35. Faulstich H, Georgopoulos D, Bloching M. Quantitative chromatographic analysis of toxins in single mushrooms of *Amanita phalloides*. J Chromatogr 1973;79:257–265.
36. Enjalbert F, Gallion C, Jehl F, Monteil H, Faulstich H. Simultaneous assay for amatoxins and phallotoxins in *Amanita phalloides* Fr. by high-performance liquid chromatography. J Chromatogr 1992;598:227–236.
37. Loranger A, Tuchweber B, Gicquaud C, St-Pierre S, Cote MG. Toxicity of peptides of *Amanita virosa* mushrooms in mice. Fundam Appl Toxicol 1985;5:1144–1152.
38. Himmelmann A, Mang G, Schnorf-Huber S. Lethal ingestion of stored *Amanita phalloides* mushrooms. Swiss Med Wkly 2001;131:616–617.
39. Enjalbert F, Gallion C, Jehl F, Monteil H. Toxin content, phallotoxin and amatoxin composition of *Amanita phalloides* tissues. Toxicon 1993;31:803–807.

40. Enjalbert F, Cassanas G, Salhi SL, Guinchard C, Chaumont J-P. Distribution of the amatoxins and phallotoxins in *Amanita phalloides*. Influence of the tissues and the collection site. Life Sci 1999;322:855–862.
41. Vetter J. Toxins of *Amanita phalloides*. Toxicon 1998;36: 13–24.
42. Tyler VE Jr, Benedict RG, Brady LR, Robbers JE. Occurrence of *Amanita* toxins in American collections of deadly amanitas. J Pharm Sci 1966;55:590–593.
43. Madhok M, Scalzo AJ, Blume CM, Neuschwander-Tetri BA, Weber JA, Thompson MW. *Amanita bisporigera* ingestion mistaken identity, dose-related toxicity, and improvement despite severe hepatotoxicity. Pediatr Emerg Care 2006;22:177–1180.
44. Faulstich H, Georgopoulos D, Bloching M, Wieland T. Analysis of the toxins of amanitin-containing mushrooms. Z Naturforsch 1974;29:86–88.
45. Haines JH, Lichstein E, Glickerman D. A fatal poisoning from an amatoxin containing *Lepiota*. Mycopathologia 1986;93:15–17.
46. Enjalbert F, Cassanas G, Rapior S, Renault C, Chaumont J-P. Amatoxins in wood-rotting *Galerina marginata*. Mycologia 2004;96:720–729.
47. Zanotti G, Petersen G, Wieland T. Structure-toxicity relationships in the amatoxin series. Structural variations of side chain 3 and inhibition of RNA polymerase II. Int J Pept Protein Res 1992;40:551–558.
48. Gundala S, Wells LD, Milliano MT, Talkad V, Luxon BA, Neuschwander-Tetri BA. The hepatocellular bile acid transporter Ntcp facilitates uptake of the lethal mushroom toxin alpha-amanitin. Arch Toxicol 2004;78:68–73.
49. Letschert K, Faulstich H, Keller D, Keppler D. Molecular characterization and inhibition of amanitin uptake into human hepatocytes. Toxicol Sci 2006;91:140–149.
50. Bushnell DA, Cramer P, Kornberg RD. Structural basis of transcription: alpha-amanitin-RNA polymerase II cocrystal at 2.8 A resolution. Proc Natl Acad Sci U S A 2002;99:1218–1222.
51. Gong XQ, Nedialkov YA, Burton ZF. Alpha-amanitin blocks translocation by human RNA polymerase II. J Biol Chem 2004;279:27422–27427.
52. Fineschi V, Di Paolo M, Centini F. Histological criteria for diagnosis of *Amanita phalloides* poisoning. J Forensic Sci 1996;41:429–432.
53. Kisilevsky R. Hepatic nuclear and nucleolar changes in *Amanita* poisoning. Arch Pathol 1974;97:253–258.
54. Costantino D, Falzi G, Langer M, Rivolta E. *Amanita-phalloides*-related nephropathy. Contrib Nephrol 1978;10: 84–97.
55. Myler RK, Lee JC, Hopper J. Renal tubular necrosis caused by mushroom poisoning. Renal biopsy findings by electron microscopy and use of peritoneal dialysis in treatment. Arch Intern Med 1964;114:196–204.
56. Pond SM, Olson KR, Woo OF, Osterloh JD, Ward RE, Kaufman DA, Moody RR. Amatoxin poisoning in Northern California 1982–1983. West J Med 1986;145: 204–209.
57. Scheurlen C, Spannbrucker N, Spengler U, Zachoval R, Schulte-Witte H, Brensing KA, Sauerbruch T. *Amanita phalloides* intoxications in a family of Russian immigrants. Case reports and review of the literature with a focus on orthotopic liver transplantation. Z Gastroenterol 1994;32:399–404.
58. Homann J, Rawer P, Bleyl H, Matthes KJ, Heinrich D. Early detection of amatoxins in human mushroom poisoning. Arch Toxicol 1986;59:190–191.
59. Faulstich H, Talas A, Wellhoner HH. Toxicokinetics of labeled amatoxins in the dog. Arch Toxicol 1985;56: 190–194.
60. Fiume L, Sperti S, Montanaro L, Busi C, Costantino D. Amanitins do not bind to serum albumin. Lancet 1977;1(8021):1111.
61. Piqueras J. Hepatotoxic mushroom poisoning: diagnosis and management. Mycopathologia 1989;105:99–110.
62. Jaeger A, Jehl F, Flesch F, Sauder P, Kopferschmitt J. Kinetics of amatoxin in human poisoning: therapeutic implications. Clin Toxicol 1993;31:63–80.
63. Vesconi S, Langer M, Iapichino G, Costantino D, Busi C, Fiume L. Therapy of cytotoxic mushroom intoxication. Crit Care Med 1985;13:402–406.
64. Busi C, Fiume L, Costantino D, Langer M, Vesconi F. Amanita toxins in gastroduodenal fluid of patients poisoned by the mushroom, *Amanita phalloides*. N Engl J Med 1979;300:800.
65. Belliardo F, Massano G, Accomo S. Amatoxins do not cross the placental barrier. Lancet 1983;1:1381.
66. Boyer J-C, Hernandez F, Estorc J, De la Coussaye J-E, Bali J-P. Management of maternal *Amanita phalloides* poisoning during the first trimester of pregnancy: a case report and review of the literature. Clin Chem 2001;47: 971–974.
67. Nagy I, Pogatsa-Murray G, Zalanyi S Jr, Komlosi P, Laszlo F, Ungi I. *Amanita* poisoning during the second trimester of pregnancy. A case report and a review of the literature. Clin Investig 1994;72:794–798.
68. Timar L, Czeizel AE. Birth weight and congenital anomalies following poisonous mushroom intoxication during pregnancy. Reprod Toxicol 1997;11:861–866.
69. Paydas S, Kocak R, Erturk F, Erken E, Zaksu HS, Gurcay A. Poisoning due to amatoxin-containing *Lepiota* species. Br J Clin Pract 1990;44:450–453.
70. Teutsch C, Brennan RW. Amanita mushroom poisoning with recovery from coma: a case report. Ann Neurol 1978;3:177–179.
71. Karlson-Stiber C, Persson H. Cytotoxic fungi—an overview. Toxicon 2003;42:339–349.
72. Rengstorff DS, Osorio R, Bonacini M. Recovery from severe hepatitis caused by mushroom poisoning without liver transplantation. Clin Gastroenterol Hepatol 2003;1: 392–396.

73. Giannini L, Vannacci A, Missanelli A, Mastroianni R, Mannaioni PF, Moroni F, Masini E. Amatoxin poisoning: a 15-year retrospective analysis and follow-up evaluation of 105 patients. Clin Toxicol 2007;45:539–542.
74. Fantozzi R, Ledda F, Caramelli L, Moroni F, Blandina P, Masini E, et al. Clinical findings and follow-up evaluation of an outbreak of mushroom poisoning—survey of *Amanita phalloides* poisoning. Klin Wochenschr 1986;64:38–43.
75. Bartoloni St Omer F, Giannini A, Botti P, Caramelli L, Ledda F, et al. *Amanita* poisoning: a clinical-histopathological study of 64 cases of intoxication. Hepatogastroenterology 1985;32:229–231.
76. Ramirez P, Parrilla P, Sanchez Bueno F, Robles R, Pons JA, Bizquert V, et al. Fulminant hepatic failure after *Lepiota* mushroom poisoning. J Hepatol 1993;19: 51–54.
77. Wu BF, Wang MM. Molecular adsorbent recirculating system in dealing with maternal Amanita poisoning during the second pregnancy trimester: a case report. Hepatobiliary Pancreat Dis Int 2004;3:152–154.
78. Andres RY, Frei W. ^{125}I-amatoxin and anti-amatoxin for radioimmunoassay prepared by a novel approach: chemical and structural considerations. Toxicon 1987;25:915–922.
79. Butera R, Locatelli C, Coccini T, Manzo L. Diagnostic accuracy of urinary amanitin is suspected mushroom poisoning: a pilot study. J Toxicol Clin Toxicol 2004;42:901–912.
80. Bruggemann O, Meder M, Freitag R. Analysis of amatoxins alpha-amanitin and beta-amanitin in toadstool extracts and body fluids by capillary zone electrophoresis with photodiode array detection. J Chromatogr A 1996; 744:167–176.
81. Filigenzi MS, Poppenga RH, Tiwary AK, Puschner B. Determination of alpha-amanitin in serum and liver by multistage linear ion trap mass spectrometry. J Agric Food Chem 2007;55:2784–2790.
82. Vlaskin DN, Gainullina ET, Klyuster OV, Rybal'chenko IV, Ryzhikov SB, Taranchenko VF. Express method for detection of *Amanita phalloides* amanitine toxins. Bull Exp Biol Med 2006;141:110–111.
83. Beutler JA, Vergeer PP. Amatoxins in American mushrooms. Evaluation of the Meixner test. Mycologia 1980; 72:1142–1149.
84. Beuhler M, Lee DC, Gerkin R. The Meixner test in the detection of α-amanitin and false-positive reactions caused by psilocin and 5-substituted tryptamines. Ann Emerg Med 2004;44:114–120.
85. Defendenti C, Bonacina E, Mauroni M, Gelosa L. Validation of a high performance liquid chromatographic method for alpha amanitin determination in urine. Forensic Sci Int 1998;92:59–68.
86. Rieck W, Platt D. High-performance liquid chromatographic method for the determination of alpha-amanitin and phalloidin in human plasma using the column-switching technique and its application in suspected cases of poisoning by the green species of amanita mushroom (*Amanita phalloides*). J Chromatogr 1988;425:121–134.
87. Andres RY, Frei W, Gautschi K, Vonderschmitt DJ. Radioimmunoassay for amatoxins by use of a rapid, 125I-tracer-based system. Clin Chem 1986;32:1751–1755.
88. Butera R, Locatelli C, Coccini T, Manzo L. Diagnostic accuracy of urinary amanitin in suspected mushroom poisoning: a pilot study. J Toxicol Clin Toxicol 2004;42:901–912.
89. Yamada EG, Mohle-Boetani J, Olson KR, Werner SB. Mushroom poisoning due to amatoxin—Northern California, winter 1996–1997. West J Med 1998;169:380–384.
90. Kaneko H, Tomomasa T, Inoue Y, Kunimoto F, Fukusato T, Muraoka S, et al. Amatoxin poisoning from ingestion of Japanese *Galerina* mushrooms. Clin Toxicol 2001;39: 413–416.
91. Harrison DC, Coggins CH, Welland FH, Nelson S. Mushroom poisoning in five patients. Am J Med 1965; 38:787–792.
92. Aji DY, Caliskan S, Nayir A, Mat A, Can B, Yasar Z, et al. Haemoperfusion in *Amanita phalloides* poisoning. J Trop Pediatr 1995;41:371–374.
93. Mullins ME, Horowitz BZ. The futility of hemoperfusion and hemodialysis in *Amanita phalloides* poisoning. Vet Hum Toxicol 2000;42:90–91.
94. Jander S, Bischoff J. Treatment of *Amanita phalloides* poisoning: I. Retrospective evaluation of plasmapheresis in 21 patients. Ther Apher 2000;4:303–307.
95. Jander S, Bischoff J, Woodcock BG. Plasmapheresis in the treatment of *Amanita phalloides* poisoning: II. A review and recommendations. Ther Apher 2000;4:308–312.
96. Rarah R, Farah R, Makhoul N, Kristal B. *Amanita phalloides* causing acute renal failure and liver disease resolved with intensive therapy. Clin Intensive Care 2003;14:105–107.
97. Moroni F, Fantozzi R, Masini E, Mannaioni P. A trend in the therapy of *Amanita phalloides* poisoning. Arch Toxicol 1976;36:111–115.
98. Floersheim GL, Eberhard M, Tschumi P, Duckert F. Effects of penicillin and silymarin on liver enzymes and blood clotting factors in dogs given a boiled preparation of *Amanita phalloides*. Toxicol Appl Pharmacol 1978; 46:455–462.
99. Floersheim GL, Weber O, Tschumi P, Ulbrich M. [Clinical death-cap (*Amanita phalloides*) poisoning: prognostic factors and therapeutic measures. Analysis of 205 cases]. Schweiz Med Wochenschr 1982;112:1164–1177. [German]
100. Wellington K, Jarvis B. Silymarin: a review of its clinical properties in the management of hepatic disorders. Bio Drugs 2001;15:465–489.
101. Pepping J. Milk thistle: *Silybum marianum*. Am J Health-Syst Pharm 1999;56:1195–1197.

102. Kroncke KD, Fricker G, Meier PJ, Gerok W, Wieland T, Kurz G. alpha-Amanitin uptake into hepatocytes. Identification of hepatic membrane transport systems used by amatoxins. J Biol Chem 1986;261:12562–12567.
103. Hruby K, Csomos G, Fuhrmann M, Thaler H. Chemotherapy of *Amanita phalloides* poisoning with intravenous silibinin. Hum Toxicol 1983;2:183–195.
104. Araz C, Karaaslan P, Esen A, Zeyneloglu P, Candan S, Torgay A, Haberal M. Successful treatment of a child with fulminant liver failure and coma due to *Amanita phalloides* poisoning using urgent liver transplantation. Transplant Proc 2006;38:596–597.
105. Klein AS, Hart J, Brems JJ, Goldstein L, Lewin K, Busuttil RW. *Amanita* poisoning: treatment and the role of liver transplantation. Am J Med 1989;86:187–193.
106. Ganzert M, Felgenhauer N, Zilker T. Indication of liver transplantation following amatoxin intoxication. J Hepatol 2005;42:202–209.
107. Escudien L, Francoz C, Vinel J-P, Moucari R, Cournot M, Paradis V, et al. *Amanita phalloides* poisoning: reassessment of prognostic factors and indication for emergency liver transplantation. J Hepatol 2007;46:466–473.
108. Bektas H, Schlitt HJ, Boker K, Brunkhorst R, Oldhafer KJ, Pichlmayr R. [Indications for liver transplantation in severe amanita phalloides mushroom poisoning]. Chirurg 1996;67:996–1001. [German]
109. Christen Y, Minazio P, de Moerloose P. Monitoring of haemostatic parameters in five cases of *Amanita phalloides* poisoning. Blood Coagulation Fibrinolysis 1993;4: 627–630.
110. Langer M, Gridelli B, Piccolo G, Markovic S, Quarenghi E, Gatti S, et al. A liver transplant candidate (fulminant hepatic failure from *Amanita phalloides* poisoning) as a multiorgan donor. Transplant Proc 1997;29: 3343–3344.

Chapter 38

FALSE MOREL and GYROMITRIN POISONING

HISTORY

Over 100 years ago Bohm and Kulz isolated an oily substance, called helvellic acid, from mushroom specimens belonging to *Gyromitra esculenta* (Pers.) Fr. (*Helvella esculenta*) that was associated with adverse effects following consumption of this mushroom. Later analysis of helvellic acid indicated that the composition of this substance included primarily nontoxic organic acids. In 1967, List and Luft isolated the main toxin, gyromitrin (*N*-methyl-*N*-formyl hydrazone).[1] Toxicity following ingestion of gyromitrin-containing mushrooms is highly variable, even among family members consuming the same meal.[2] Nevertheless, case reports have associated fatalities with the consumption of this mushroom as early as the 1920s.[3] *Gyromitra esculenta* was the most common cause of mushroom-associated toxicity in Poland after World War II based on epidemiologic data.[4] These data included some fatalities.

MYCOLOGICAL DESCRIPTION

The genus *Gyromitra* belongs to the division Ascomycota (sac fungi), which together share a common sac-like structure called the ascus. This structure contains four to eight ascospores in the sexual stage. Some species of *Gyromitra* are among the most valued mushrooms (i.e., the morels). There are at least eight species of *Gyromitra* in the United States that contain variable amounts of gyromitrin. *Gyromitra esculenta* (false morel, Discinaceae family) contains a relatively high concentration of gyromitrin; poisoning from the consumption of these mushrooms usually occurs as a result of confusion of the false morels with species from the true morel mushroom family (Morchellaceae). *Gyromitra esculenta* differs from true morels by the lack of ribs and cavities (pits) on the irregular, brain-shaped cap. Other *Gyromitra* species of toxicological concern include *G. fastigiate* (Krombh.) Rehm and *G. gigas* (Krombh.) Cooke.[5] Other mushroom species contain gyromitrin including *G. infula* (Schaeff.) Quél.), *Helvella crispa* (Scop.) Fr., *Helvella lacunosa* Afzel., *Leptopodia elastica* (Bull.) Boud., and *Otidea onotica* (Pers.) Fuckel,[6] but human toxicity from the consumption of these mushroom is not well documented in the medical literature.

Common Name: False morel, Turban fungus, Lorchel, Brain Mushroom, Beefsteak Morel, Elephant ears

Scientific Name: *Gyromitra esculenta* (Pers.) Fr.

Mycological Family: Discinaceae

Physical Description: *Gyromitra* species do not possess gills. The medium to large caps (3–12 cm/~1–5 in.) are red-brown to dark-brown with irregularly, deeply-wrinkled, convoluted shape. The whitish stem is relatively short compared with the cap. The genus *Gyromitra* belongs to the division Ascomycota, which contain internal ascospores rather than the external basidiospores that form distinguishing spore prints.

Distribution and Ecology: *Gyromitra esculenta* inhabits Europe and North America with specimens common in Germany and Poland. This mushroom prefers conifer forests with siliceous soils, but specimens of *G. esculenta* are found in the plains.

Medical Toxicology of Natural Substances, by Donald G. Barceloux, MD

EXPOSURE

In contrast to amatoxin poisoning, toxicity associated with the ingestion of mushrooms from *Gyromitra esculenta* usually occurs in the spring rather than the fall because the fruiting bodies of *G. esculenta* appear from March to May in the United States and in the beginning of summer in Europe. Reports of gyromitrin poisoning are more common in Central and Eastern Europe compared with Western and Southern Europe.[6] The industrial applications of the gyromitrin metabolite, *N*-methylhydrazine, include its use as a rocket propellant and corrosion inhibitor.

PRINCIPAL TOXINS

Physiochemical Properties

All *Gyromitra* species do not contain gyromitrin, and the gyromitrin concentration varies dramatically even within the same species as a result of variable climatic conditions and other undetermined variables. Consequently, adverse reactions can occur following previous, asymptomatic ingestions of this mushroom.[7] Gyromitrin (acetaldehyde-formyl-methyl-hydrazone, CAS RN: 16568-02-8) is the most toxic of the nine hydrazones isolated from *Gyromitra esculenta*.[8] Structurally, gyromitrin is an aldehyde of *N*-methyl-*N*-formyl hydrazine (MFH) as displayed in Figure 38.1. Following hydrolysis, gyromitrin yields aldehyde and *N*-methyl-*N*-formyl hydrazine (MFH), which further hydrolyzes to *N*-methylhydrazine (MMH, monomethylhydrazine).

Gyromitrin is a relatively unstable, colorless hydrazone that is heat labile, slightly volatile, and water soluble. This compound oxidizes at room temperature to acetaldehyde and *N*-methyl-*N*-formylhydrazine. Long-term dry storage and vigorous boiling removes the toxin from *Gyromitra* mushrooms. However, simple parboiling does not guarantee edibility. At least 10 minutes of boiling, preferably in at least 3L water/kg mushroom, is necessary to reduce the total *N*-methyl-*N*-formylhydrazone concentration below 1% of the value in fresh mushrooms.[9] The concentration of MMH also decreases during storage of these mushrooms at room temperature with reduction in the MMH concentration up to about 50% during 1–3 weeks of storage[10]

Poisonous Parts

Ingestion of gyromitrin-containing mushrooms produces variable toxicity that depends in part on individual metabolism. The ingestion of cooking juice or liquid that surrounds the mushrooms substantially increases toxicity. Although there are at least eight other hydrazone compounds in *Gyromitra esculenta,* gyromitrin is the most important toxic compound.[8] Thin-layer spectrophotometric studies suggest that the gyromitrin content of the false more (*G. esculenta*) varies substantially in different specimens with a typical range of about 40–730mg/kg wet weight.[11]

Mechanism of Toxicity

Hydrazine compounds inhibit the enzymatic activation of pyridoxine by reacting with pyridoxal-5-phosphate, resulting in decreased synthesis of GABA and enhanced susceptibility to seizures.[12] However, animal studies suggest that toxic doses of monomethylhydrazine (*N*-methyl hydrazine) cause minimal reduction in GABA formation in the brain;[13] consequently, monomethylhydrazine alone does not account for all of the toxic effects of gyromitrin. The specific metabolite associated with the neurological and gastrointestinal (GI) effects of gyromitrin remains undetermined. In animal studies, a metabolite of gyromitrin (*N*-methyl-*N*-formyl hydrazine) depleted cytochrome P-450 within 8–12 hours of the administration of gyromitrin.[14] Another metabolite (e.g., *N*-methyl hydrazine) of gyromitrin did not exhibit similar hepatotoxicity. Hydrazine compounds are irritating to mucous membranes and the GI tract. In animal studies, the inhalation of hydrazine causes anorexia, weight loss, liver and lung inflammation, and interstitial nephritis.[15] Additionally, hydrazine compounds deplete glutathione in erythrocytes and damage hepatic macromolecules, probably by the formation of free methyl radicals. Case reports associate chronic hydrazine exposure with severe tubular necrosis and mild hepatocellular damage.[16]

DOSE RESPONSE

The clinical response to the ingestion of gyromitrin-containing mushrooms is highly variable, depending on

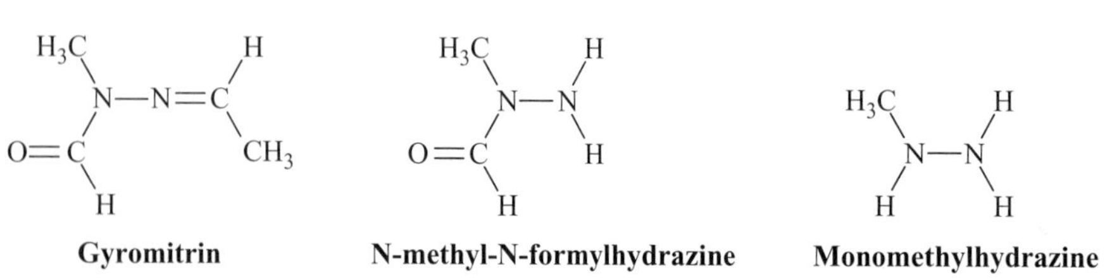

Gyromitrin | **N-methyl-N-formylhydrazine** | **Monomethylhydrazine**

FIGURE 38.1. Structure of gyromitrin and metabolites.

a number of factors including individual sensitivity, gyromitrin content of the mushrooms, and amount of gyromitrin removed during food preparation. There is also substantial variation in the response of different animal species to gyromitrin. In an animal study, the no-effect level of gyromitrin for rabbits and chickens varied 10-fold (i.e., 0.5 mg/kg/d and 0.05 mg/kg/d, respectively).[17] Based on these animal studies and the average gyromitrin content of mushrooms from *G. esculenta*, the estimated maximal tolerated dose in adults is 35 μg/kg and the estimated lethal dose in adults is approximately 20–50 mg/kg.[6]

TOXICOKINETICS

Hydrolysis of gyromitrin (acetaldehyde-*N*-formyl-*N*-methylhydrazone) in the acidic conditions of the stomach produces absorbable hydrazine compounds by conversion of gyromitrin to acetaldehyde and *N*-methyl-*N*-formylhydrazine (CAS RN: 758-17-8). The latter compound hydrolyzes to monomethylhydrazine (CAS RN: 60-34-4, methylhydrazine) and more slowly to formic acid.[18] There are few toxicokinetics data on the biotransformation of gyromitrin in humans. In a rodent study, about 25% of the gyromitrin was metabolized to monomethyl hydrazine.[19] Figure 38.1 displays the chemical structure of gyromitrin, *N*-methyl-*N*-formylhydrazine, and monomethylhydrazine.

CLINICAL RESPONSE

Clinical features of gyromitrin poisoning primarily involve gastrointestinal and neurological effects. Ingestions of gyromitrin-containing mushrooms usually produce mild GI symptoms, depending on a variety of variables including preparation techniques and gyromitrin content. The onset of GI symptoms is delayed 2–8 hours up to 2 days, and this delay may be similar to the onset of GI symptoms during amatoxin poisoning.[20] In mild cases, GI symptoms (nausea, vomiting, abdominal pain) may persist for several hours to several days including nonspecific symptoms (headache, fatigue, malaise). Symptoms usually resolve in 2–6 days without sequelae.[21] Some neurological symptoms (tremor, ataxia, nystagmus, dizziness, fatigue, slurred speech) also may occur. In severe cases, coma, seizures, and hepatic failure develop 1–3 days after ingestion complicated by hypoglycemia and hypovolemia.[22] Hemolysis with rhabdomyolysis and renal dysfunction are potential complications of gyromitrin poisoning, but these complications are not well documented in the recent medical literature. Fatalities from hepatorenal failure following the ingestion of gyromitrin-containing mushrooms are rare.[22]

The International Agency for Research on Cancer (IARC) lists gyromitrin as not classifiable with regard to its carcinogenicity in humans (group 3). Following the administration of gyromitrin by intragastric intubation to mice, increased incidences of tumors occurred in the forestomach, clitoral gland, and lung of females and in the preputial gland of males.[23] The lifetime oral administration of *Gyromitra esculenta* mushrooms to Swiss mice produced an increased incidence of tumors of the lung, blood vessels, and forestomach.[24] Although exposure potentially occurs during ingestion of gyromitrin-containing mushrooms, destruction of gyromitrin during cooking limits exposure in humans. Hydrolysis of gyromitrin in the stomach yields the methylating agents *N*-methyl-*N*-formyl hydrazine and *N*-methyl hydrazine. Based on studies in animals, the carcinogenic risk associated with the consumption of mushrooms from *G. esculenta* is small.[25] There are no case reports or epidemiological studies associating the consumption of gyromitrin-containing mushrooms with carcinogenicity.

DIAGNOSTIC TESTING

Chromatographic analytical methods are highly specific and sensitive for the detection of toxins in *Gyromitra esculenta*. Gas-liquid chromatography detects unbound gyromitrin and *N*-methyl-*N*-formyl hydrazine (MFH), but this method does not account for MFH bound to high-molecular-weight compounds. Thin-layer chromatography yields gyromitrin, MRH, and *N*-methyl hydrazine (monomethylhydrazine).[11] Extraction of dried gyromitrin-containing mushrooms with anhydrous solvents does not yield gyromitrin, whereas heating these mushrooms in a sealed tube at 120 °C (248 °F) liberates gyromitrin and *N*-methyl-*N*-formyl hydrazine.[26] Other methods for the quantitation of gyromitrin in mushroom specimens include gas chromatography/mass spectrometry (GC/MS). The limit of quantitation for gyromitrin using GC/MS and the pentafluorobenzoyl derivative of methylhydrazine is approximately 0.3 μg/g dry matter.[27] Potential laboratory abnormalities associated with gyromitrin poisoning include elevated serum hepatic aminotransferases, hypoglycemia, methemoglobinemia, and evidence of hemolysis. Typically, abnormalities of serum hepatic aminotransferases peak about 5 days after ingestion.[21,22]

TREATMENT

Fluid and electrolyte imbalance is the most common complication of serious gyromitrin poisoning. The occurrence of coma or convulsions is extremely rare, and these complications should be treated with the usual supportive care (e.g., respiratory support, lorazepam or diazepam) and rapid evaluation of blood glucose concentrations. Most patients do not require decontamination because presentation usually occurs >1–2 hours after ingestion. Although no human or animal data are available to guide the therapy of gyromitrin toxicity, the use of a single dose of activated charcoal is a reasonable treatment modality. Pyridoxine (25 mg/kg or 5 g intravenous dose up to a daily adult dose of 15–20 g) has been recommended for serious neurological symptoms (e.g., coma, seizures) of gyromitrin poisoning based on the similarity between gyromitrin and hydrazine toxicity.[28,29] Although some animal studies suggest that the use of pyridoxine may reduce the neurotoxicity of some hydrazine compounds (*N*-methyl hydrazine), these effects are minimal following the administration of other hydrazine compounds (e.g., *N*-methyl-*N*-formyl hydrazine).[30] There are inadequate clinical data to confirm the efficacy of pyridoxine for the treatment of neurotoxicity (seizures, coma) associated with the ingestion of gyromitrin-containing mushrooms.[31] Animal data do not support the efficacy of pyridoxine for the liver abnormalities caused by gyromitrin.[14] There are no clinical data to indicate that the use of methods to enhance elimination improve clinical outcome.

Gyromitrin-induced hepatic encephalopathy requires the same supportive treatment administered following amatoxin poisoning. In severe cases, baseline laboratory analysis of the blood should include a complete blood count, serum electrolytes, creatinine, ammonia, hepatic aminotransferases, and glucose as well as coagulation studies (e.g., prothrombin time). Animal studies suggest that hemoglobinuria may complicate serious gyromitrin toxicity, but there are no recent cases to document the occurrence of this complication in humans. Other suggested treatment regimens with few clinical data include the use of *N*-acetylcysteine and folic acid. Hydrazine compounds inhibit the chemical transformation of folic acid to the active form, folinic acid. Although some authors recommend the daily doses of 20–200 mg folinic acid, there are inadequate clinical data to confirm the efficacy of this drug for gyromitrin poisoning.

References

1. List PH, Luft P. Gyromitrin, das gift der fruhjahrslorchel *Gyromitra (Helvella) esculenta* Fr. Tetrahedron Lett 1967;20:1893–1894.
2. Hendricks HV. Poisoning by false morel (*Gyromitra esculenta*). JAMA 1940;114:1625.
3. Dearness J. Gyromitra poisoning. Mycologia 1924;16: 199.
4. Grzymala S. Les recherches sur la frequence des intoxications par les champignons. Bull Med Leg Toxicol Med 1965;2:200–210.
5. Michelot D. [Poisoning by *Gyromitra esculenta*]. J Toxicol Clin Exp 1989;9:83–99. [French]
6. Bresinsky A, Besl H. A Colour Atlas of Poisonous Fungi. A Handbook for Pharmacists, Doctors, and Biologists. London: Wolfe Publishing Ltd; 1990.
7. Buck RW. Mushroom toxins. A brief review of the literature. N Engl J Med 1961;265:681–686.
8. Michelot D, Toth B. Poisoning by Gyromitra esculenta–a review. J Appl Toxicol 1991;11:235–243.
9. Pyysalo H, Niskanen A. On the occurrence of *N*-methyl-*N*-formylhydrazones in fresh and processed false morel, *Gyromitra esculenta*. J Agric Food Chem 1977;25: 644–647.
10. Andary C, Privat G, Bourrier M-J. Variations of monomethylhydrazine content in *Gyromitra esculenta*. Mycologia 1985;77:259–264.
11. Andary C, Privat G, Bourrier MJ. [Thin-layer spectrofluorometric microanalysis of monomethylhydrazine in *Gyromitra esculenta*]. J Chromatogr 1984;287:419–424. [French]
12. Lampe KF. Toxic fungi. Annu Rev Pharmacol Toxicol 1979;19:85–104.
13. Maynert EW, Kaji HK. On the relationship of brain gamma-aminobutyric acid to convulsions. J Pharmacol Exp Ther 1962;137:114–121.
14. Braun R, Greeff V, Netter KJ. Liver injury by the false morel poison gyromitrin. Toxicology 1979;12:155–163.
15. Comstock CC, Lawson LH, Greene EA, Oberst FW. Inhalation toxicity of hydrazine vapor. AMA Arch Ind Health 1954;10:476–490.
16. Sotaniemi E, Hirvonen J, Isomaki H, Takkunen J, Kaila J. Hydrazine toxicity in the human. Ann Clin Res 1971; 3:30–33.
17. Niskanen A, Pyysalo H. Short-term peroral toxicity of ethylidene gyromitrin in rabbits and chickens. Food Cosmet Toxicol 1976;14:409–415.
18. Nagel D, Wallcave L, Toth B, Kupper R. Formation of methylhydrazine from acetaldehyde *N*-methyl-*N*-formylhydrazone, a component of *Gyromitra esculenta*. Cancer Res 1977;37:3458–3460.
19. von Wright A, Pyysalo H, Niskanen A. Quantitative evaluation of the metabolic formation of methylhydrazine from acetaldehyde-*N*-methyl-*N*-formylhydrazone, the main poisonous compound of *Gyromitra esculenta*. Toxicol Lett 1978;2:261–265.
20. Koppel C. Clinical symptomatology and management of mushroom poisoning. Toxicon 1993;31:1513–1540.
21. Leathem AM, Dorran TJ. Poisoning due to raw *Gyromitra esculenta* (false morels) west of the Rockies. Can J Emerg Med 2007:9:127–130.
22. Giusti GV, Carnevale A. A case of fatal poisoning by *Gyromitra esculenta*. Arch Toxicol 1974;33:49–54.

23. Toth B, Smith JW, Patil KD. Cancer induction in mice with acetaldehyde methylformylhydrazone of the false morel mushroom. J Natl Cancer Inst 1981;67:881–887.
24. Toth B, Patil K, Pyysalo H, Stessman C, Gannett. Cancer induction in mice by feeding the raw false morel mushroom *Gyromitra esculenta*. Cancer Res 1992;52:2279–2284.
25. Bergman K, Hellenas K–E. Methylation of rat and mouse DNA by the mushroom poison gyromitrin and its metabolite monomethylhydrazine. Cancer Lett 1992;61:165–170.
26. List PH, Luft P. [Detection and content determination of gyromitrin in fresh *Gyromitra esculenta*. 19. Mushroom contents]. Arch Pharm Ber Dtsch Pharm Ges 1969;302:143–146. [German]
27. Arshadi M, Nilsson C, Magnusson B. Gas chromatography-mass spectrometry determination of pentafluorobenzoyl derivative of methylhydrazine in false morel (*Gyromitra esculenta*) as a monitor for the content of the toxin gyromitrin. J Chromatogr A 2006;1125:229–233.
28. Kirklin JK, Watson M, Bondoc CC, Burke JF. Treatment of hydrazine-induced coma with pyridoxine. N Engl J Med 1976;294:938–939.
29. Hanrahan JP, Gordon MA. Mushroom poisoning: Case reports and a review of therapy. JAMA 1984;251:1057–1061.
30. Toth B, Erickson J. Reversal of the toxicity of hydrazine and analogues by pyridoxine hydrochloride. Toxicology 1977;7:31–36.
31. Zelnick SD, Mattie DR, Stepaniak PC. Occupational exposure to hydrazines: treatment of acute central nervous system toxicity. Aviat Space Environ Med 2003;74:1285–1291.

Chapter 39

GASTROENTERITIS-PRODUCING MUSHROOMS

EXPOSURE

The rapid onset of gastroenteritis is the most common adverse response to the ingestion of mushrooms. Most of the cases of mushroom-induced gastroenteritis are mild, and many of these cases are not reported in the medical literature. Despite food preparation techniques, a large group of mushrooms are capable of producing gastroenteritis as listed in Table 39.1.[1] The consumption of a variety of mushrooms can produce gastroenteritis when improperly prepared including *Amanita rubescens* var. *rubescens* Pers., *Amanita spissa* (Fr.) P. Kumm., *Amanita strobiliformis* (Paulet ex Vittad.) Bertill., *Amanita vaginata* (Bull.) Lam. var. *vaginata, Boletus queletii* Schulzer (*Boletus erythropus*), *Boletus luridus* Schaeff., *Choiromyces meandriformis* Vittad., and *Entoloma clypeatum* (L.) P. Kumm., pungent *Lactarius* and *Russula* species, *Clitocybe nebularis (Batsch)* Quél., *Lepista nuda* (Bull.) Cooke, *Paxillus involutus* (Batsch) Fr., *Russula olivacea* (Schaeff.) Fr. and *Sarcosphaera coronaria* (Jacq.) J. Schröt. (*S. crassa*).[1] These species are edible when properly prepared. In a small segment (<2%) of the population, individuals possess trehalase disaccharidase deficiency. Ingestion of trehalose-containing mushrooms (e.g., *Agaricus bisporus* (J.E. Lange) Pilát, *Boletus edulis* Bull.) by these susceptible patients causes the onset of diarrhea and abdominal cramps as a result of the failure to absorb trehalose through the intestinal brush-border.[2,3] Illudin S derivatives (i.e., irofulven and related compounds) are experimental chemotherapeutic agents for advanced solid tumors.[4,5]

PRINCIPAL TOXINS

Physiochemical Properties

There are few data on the chemical toxins associated with most cases of mushroom-induced gastroenteritis. Although many fungi contain terpenoid compounds capable of producing gastrointestinal (GI) effects, the chemical constituents of most GI toxins associated with the ingestion of mushrooms remains undetermined. The blood-red Cortinarius [*Cortinarius sanguineus* (Wulfen) Fr.] contains a variety of anthraquinone compounds including emodin, dermocybin, dermorubin, physcion, endocrocin, dermolutein, 5-chlorodermorubin, emodin-1-β-D-glucopyranoside, dermocybin-1-β-D-glucopyranoside, 7-chloroemodin, 5,7-dichloroemodin, 5,7-dichloroendocrocin, 4-hydroxyaustrocorticone and austrocorticone.[6] Anthraquinone compounds exhibit prominent cathartic properties. Several of the anthraquinone compounds in *Cortinarius sanguineus* are mutagenic in *in vitro* assays including the most prominent anthraquinone compound, emodin.[7] The sulfur tuft [*Hypholoma fasciculare* (Huds.) P. Kumm.] has been associated with mushroom poisoning,[8,9] but the toxic constituents have not been identified. Analytical studies indicate that this mushroom species contains sesquiterpenoid compounds (fascicularones A and B)[10,11] as well as potent calmodulin antagonists (fasciculic acids A, B, and C).[12,13] The clinical significance of these compounds remains undetermined.

The jack-o-lantern [*Omphalotus illudens* (Schwein.) Bresinsky & Besl] contains the sesquiterpene com-

Medical Toxicology of Natural Substances, by Donald G. Barceloux, MD

TABLE 39.1. Common Gastroenteritis-Producing Mushroom Species*

Genus	Species	Occurrence	Distinguishing Features
Agaricus	*hondensis* Murrill	Summer, fall	White to gray-brown cap with brown spore print and negative amyloid reaction
Agaricus	*xanthodermus* Genev.	Forests and occasionally meadows and gardens: summer to fall	Intense yellow coloration when flesh rubbed with phenol-like smell; forms large fairy rings in deciduous forests
Boletus	*satanas* Lenz	Warm climates underneath oak trees: July to September	Large pale cap (6–25 cm/~2–10 in.) and short stalk with swollen base; flesh has unpleasant smell and blue discoloration after trauma
Chlorophyllum	*molybdites* (G. Mey.) Massee *(Lepiota molybdites)*	Common mushroom found in the United States, Central and South America, Japan, Central Africa, Australia: Late summer, early fall	Large (20–25 cm), white, scaly cap with free, white-green gills, green spore print and negative amyloid reaction; No volva (cup)
Cortinarius	*sanguineus (Dermocybe sanguineus)*	Wet, boggy areas of conifer forests	Deep blood to wine-red flesh with rust-brown spore print
Entoloma	*sinuatum* (Bull.) P. Kumm. *(E. lividum)*	Deciduous forests in rows or rings: Summer to fall	Large, whitish to light ochre fungus with red-ochre gills on mature specimens; spore print salmon-pink
Hebeloma	*crustuliniforme* (Bull.) Quél.	Common in large groups or rings underneath trees	Medium-sized with pale, ochre curved cap with whitish border and brown spore mass; yellow to olive-brown spore print
Hypholoma	*fasciculare* (Huds.) P. Kumm. *(Naematoloma fasciculare)*	Appears as large tufts on stumps of conifer and deciduous trees	Sulfur yellow to green gills that become grey-lilac as the specimen matures; light yellowish cap with grey to brown-violet spore mass
Lactarius	*helvus* (Fr.) Fr.	Humus-rich soils beneath birch, pine, and spruce trees: Summer to fall	Large, dry, velvety cap with brown flesh to yellow-ochre color
Megacollybia	*platyphylla* (Pers.) Kotl. & Pouzar *(Tricholomopsis platyphylla)*	Large mushroom on wood of deciduous trees joined by mycelial strands	Grey to dark brown cap with white spore mass
Omphalotus	*illudens* (Schwein.) Bresinsky & Besl *(Clitocybe illudens)*	Wood-associated fungus near olive, chestnut, and oak trees, particularly in warm climates	Orange fungus with large red-brown cap and yellowish white spore mass
Ramaria[a]	*formosa* (Pers.) Quél.	On ground in deciduous forests near beech trees: August, September	Large, multicolored coral fungus with pink to salmon base and yellow tips. Spore mass is ochre to yellow with red-brown spore print.
Russula	*emetica* (Schaeff.) Pers.	Wet boggy forests and some drier deciduous and conifer forests	Pungent-tasting mushroom with large (5–10 cm/2–4 in.) blood to cherry-red cap and white spore print
Scleroderma	*citrinum* Pers.	Widespread in deciduous and conifer forests and the edges of woodlands: Summer to late fall	Thick, solid fruiting body that has an uneven, cream-yellow to ochre surface, known as poison puffball or common earth-ball; brown spore print
Tricholoma	*pardinum* Quél	Decomposes later in conifer and deciduous forests	Large mushroom with white spores and white, sinuate gills and grey, scaly cap

*This list is a partial one because of the large number of mushroom species capable to producing gastroenteritis and the lack of documentation in the medical literature.

[a]A similar species (e.g., *Ramaria pallida* Maire) is associated with gastroenteritis in Germany.

pound, iludin S, which is responsible for some of the GI effects from ingestion of this mushroom; this compound also has some antiviral properties.[14] In rodent studies, illudin S is a cytotoxic substance,[15] and the oral administration of illudin S causes inflammation and edema in the stomach and proximal small bowel as well as erosion and edema of the gastric mucosa.[16] *Boletus satanas* Lenz contains a nucleoside triphosphate phosphatase (bolesatine), which inhibits protein synthesis via the hydrolysis of guanosine-5′-triphosphate (GTP).[17] The clinical significance of this compound as a GI irritant requires more study. *Chlorophyllum molybdites* contains a partially heat labile, high-molecular-weight toxin based on animal studies.[18] However, the exact identity of this toxin is unknown.

Mechanism of Toxicity

Because most GI irritants remain unidentified, most mechanisms of action are also unknown. Although some mushrooms [e.g., *Amanita muscaria* (L.) Hook. or fly agaric] accumulate heavy metals (e.g., mercury) with bioconcentration factors up to 100–200, analysis of wild mushrooms for mercury indicate that mercury toxicity is unlikely following the normal consumption of mushrooms.[19]

DOSE RESPONSE

Predicting toxicity following the ingestion of gastroenteritis-producing mushrooms is complicated by variable responses. Some people consume varieties of gastroenteritis-producing mushrooms without effects, whereas other individuals develop gastroenteritis after previously eating similar mushrooms without adverse effects.[20] Similarly, some family members eating the same meal develop GI symptoms, while other family members do not. Factors that affect the severity of GI effects include geographic variations, individual sensitivity, food preparation techniques, and variability in toxin content.

CLINICAL RESPONSE

Because of the paucity of human data, precise descriptions of the gastrointestinal syndrome associated with each gastroenteritis-producing species are not well-documented. In general, the latent period following the ingestion of gastroenteritis-producing mushrooms is short with an average of 15 minutes to a few hours.[21] GI symptoms include nausea, vomiting, abdominal cramps, and profuse, watery diarrhea that may become blood-tinged.[22,23] Other non-specific symptoms include chills, myalgias, headache, and fatigue.[24] Occasionally, cholinergic-like symptoms develop including diaphoresis, lacrimation, blurred vision, and salivation, particularly following ingestion of European species of gastroenteritis-producing mushrooms.[25] The clinical features of this syndrome are self-limited with resolution within 24–48 hours without sequelae. Rarely, the GI symptoms are severe enough to produce clinically significant fluid and electrolyte imbalance, particularly in elderly patients with cardiovascular disease and young children.[26,27] The presence of fever suggests a secondary infectious process (e.g., pericolonic abscess).[20]

DIAGNOSTIC TESTING

In most cases, diagnostic testing is not available to identify the toxins associated with gastroenteritis-producing mushrooms. In serious cases, evaluation of serum electrolytes, serum creatinine, and serum BUN help determine the severity of the poisoning. Transient leukocytosis, proteinuria, microscopic hematuria, and fecal leukocytes may occur during the acute phase of gastroenteritis.[22] Rarely, minimal elevations of serum hepatic aminotransferases occur following the ingestion of *Omphalotus illudens* (Schwein.) Bresinsky & Besl (jack o' lantern), but these values quickly return to normal without liver dysfunction (i.e., prolonged prothrombin time).[28] A single case report associated profuse rectal bleeding with thrombocytopenia, mild hypofibrinogenemia, and slightly prolonged partial thromboplastin time after the ingestion of a portion of a raw specimen of *Chlorophyllum molybdites* (G. Mey.) Massee.[29] There was no evidence of hepatocellular damage (i.e., normal serum bilirubin and hepatic aminotransferases), and the patient recovered with supportive care. Methods are available to identify spores (e.g., *Chlorophyllum molybdites*) in stools and vomitus after sample filtration and staining with 1% acid fuchsin.[30]

TREATMENT

Treatment primarily involves the evaluation of fluid and electrolyte status. The most important part of patient care is the differentiation of gastroenteritis-producing mushrooms from amatoxin-containing mushrooms. Although both types can be ingested at the same time, the rapid onset of GI symptoms favors the ingestion of gastroenteritis-producing mushrooms. Severe cases should be monitored for electrolyte abnormalities (e.g., hypokalemia) and fluid depletion. Rarely, vasopressors and respiratory support are necessary. Most symptoms resolve within 24–48 hours. If the ingestion of amatoxin-containing mushrooms is excluded, patients with gastroenteritis may be discharged after control of the vomiting and diarrhea with antiemetics and antidiarrheal agents as needed.

References

1. Bresinsky A, Besl H. A Colour Atlas of Poisonous Fungi. A Handbook for Pharmacists, Doctors, and Biologists. London: Wolfe Publishing Ltd; 1990.
2. Madzarovova-Nohejlova J. Trehalase deficiency in a family. Gastroenterology 1973;65:130–133.
3. Bergoz R. Trehalose malabsorption causing intolerance to mushrooms. Report of a probable case. Gastroenterology 1971;60:909–912.
4. Seiden MV, Gordon AN, Bodurka DC, Matulonis UA, Penson RT, Reed E, et al. A phase II study of irofulven in women with recurrent and heavily pretreated ovarian cancer. Gynecol Oncol 2006;101:55–61.
5. Alexandre J, Raymond E, Kaci MO, Brain EC, Lokiec F, Kahatt C, et al. Phase I and pharmacokinetic study of irofulven administered weekly or biweekly in advanced solid tumor patients. Clin Cancer Res 2004;10:3377–3385.
6. Raisanen R, Bjork H, Hynninen PH. Two-dimensional TLC separation and mass spectrometric identification of anthraquinones isolated from the fungus *Dermocybe sanguinea*. Z Naturforsch 2000;55:195–202.
7. von Wright A, Raatikainen O, Taipale H, Karenlampi S, Maki-Paakkanen J. Directly acting geno- and cytotoxic agents from a wild mushroom *Dermocybe sanguinea*. Mutat Res 1992;269:27–33.
8. Herbich J, Lohwag K, Rotter R. [Fatal poisoning with the green-leaf sulfur cap (*Naematoloma fasciculars*).] Arch Toxikol 1966;21:310–320. [German]
9. Mortara M, Martinetti L. [Poisoning by *Hypholoma fasciculare* Fries (tufted agaric, bitter fungus)]. Arch Toxikol 1955;15:390–391. [German]
10. Ito Y, Kurita H, Yamaguchi T, Sato M, Okuda T. Naematolin, a new biologically active substance produced by *Naematoloma fasciculare* (Fr.) Karst. Chem Pharm Bull (Tokyo) 1967;15:2009–2010.
11. Shiono Y, Matsuzaka R, Wakamatsu H, Muneta K, Murayama T, Ikeda M. Fascicularones A and B from a mycelial culture of *Naematoloma fasciculare*. Phytochemistry 2004;65:491–496.
12. Kubo I, Matsumoto A, Kozuka M, Wood WF. Calmodulin inhibitors from the bitter mushroom *Naematoloma fasciculare* (Fr.) Karst. (Strophariaceae) and absolute configuration of fasciculols. Chem Pharm Bull (Tokyo) 1985;33:3821–3825.
13. Takahashi A, Kusano G, Ohta T, Ohizumi Y, Nozoe S. Fasciculic acids A, B and C as calmodulin antagonists from the mushroom *Naematoloma fasciculare*. Chem Pharm Bull (Tokyo) 1989;37:3247–3250.
14. Engler M, Anke T, Sterner O. Production of antibiotics by *Collybia nivalis, Omphalotus olearis*, a *Favolaschia* and a *Pterula* species on natural substrates. Z Naturforsch. 1998;53:318–324.
15. McMorris TC, Kelner MJ, Wang W, Moon S, Taetle R. On the mechanism of toxicity of illudins: the role of glutathione. Chem Res Toxicol 1990;3:574–579.
16. Tanaka K, Miyasaka S, Inoue T. Histopathological effects of illudin S, a toxic substance of poisonous mushroom, in rat. Hum Exp Toxicol 1996;15:289–293.
17. Ennamany R, Lavergne JP, Reboud JP, Dirheimer G, Creppy EE. Mode of action of bolesatine, a cytotoxic glycoprotein from *Boletus satanas* Lenz. Mechanistic approaches. Toxicology 1995;100:51–55.
18. Eilers FI, Nelson LR. Characterization and partial purification of the toxin of *Lepiota morganii*. Toxicon 1974;12:557–563.
19. Falandysz J, Lipka K, Kawano M, Brzostowski A, Dadej M, Jedrusiak A, Puzyn T. Mercury content and its bioconcentration factors in wild mushrooms at Lukta and Morag, northeastern Poland. J Agric Food Chem 2003;51:2832–2836.
20. Lehmann PF, Khazan U. Mushroom poisoning by *Chlorophyllum molybdites* in the Midwest United States. Mycopathologia 1992;118:3–13.
21. French AL, Garrettson LK. Poisoning with the North American jack o' lantern mushroom, *Omphalotus illudens*. Clin Toxicol 1988;26:81–88.
22. Goos RD, Shoop CR. A case of mushroom poisoning caused by *Tricholomopsis platyphylla*. Mycologia 1980;72:433–435.
23. Blayney D, Rosenkranz E, Zettner A. Mushroom poisoning from *Chlorophyllum molybdites*. West J Med 1980;132:74–77.
24. Goos RD. Another case of mushroom poisoning involving *Tricholomopsis platyphylla*. Mycologia 1984;76:350–351.
25. Maretic Z, Russell FE, Golobic V. Twenty-five cases of poisoning by the mushroom *Pleurotus olearius*. Toxicon 1975;13:379–381.
26. Vanden Hoek TL, Erickson T, Hryhorczuk D, Narasimhan K. Jack o'lantern mushroom poisoning. Ann Emerg Med 1991;20:559–561.
27. Stenklyft PH, Augenstein WL. *Chlorophyllum molybdites*—severe mushroom poisoning in a child. Clin Toxicol 1990;28:159–168.
28. Vanden TL, Ericksoln T, Hryhorczuk D, Narasimhan K. Jack o' lantern mushroom poisoning. Ann Emerg Med 1991;20:559–561.
29. Levitan D, Macy JI, Weissman J. Mechanism of gastrointestinal hemorrhage in a case of mushroom poisoning by *Chlorophyllum molybdites*. Toxicon 1981;19:149–180.
30. Eilers FI, Barnard BL. A rapid method for the diagnosis of poisoning caused by the mushroom *Lepiota morgani*. Am J Clin Pathol 1973;60:823–825.

Chapter 40

INKY CAP and COPRINE TOXICITY [*Coprinus atramentarius* (Bull.) Fr.]

HISTORY

The mushrooms of *Coprinus atramentarius* (Bull.) Fr. (ink cap) contain a protoxin, coprine, which following metabolism interacts with ethanol to produce a disulfiram-like reaction. In the early 1900s, amateur French mycologists discovered this mushroom–ethanol interaction after developing coprine toxicity following the consumption of ethanol and a meal containing *Coprinus atramentarius* mushrooms.[1] Later reports from France, Germany, and the United States confirmed the toxic effects of the combined consumption of *Coprinus atramentarius* mushrooms and ethanol.[2,3] Ingestion of ethanol before or concurrently with consumption of this otherwise edible species did not cause an adverse reaction.

MYCOLOGICAL DESCRIPTION

Although some related *Coprinus* species probably contain coprine including *C. acuminatus* (Romagn.) P.D. Orton, *C. romagnesianus* Singer, *C. quadrifidus* Peck, *C. variegatus* Peck, *C. comatus* (O.F. Müll.) Gray, *C. atramentarius* is the only species documented to contain sufficient amounts of coprine to produce human toxicity.[1] Other unrelated species are associated with a disulfiram-like reaction including *Boletus luridus* Schaeff., (club-footed Clitocybe), *Pholiota squarrosa* (Weigel) P. Kumm., *Tricholoma equestre* (L.) P. Kumm. (*T. auratum*), and *Ptychoverpa bohemica* (Krombh.) Boud., *Verpa bohemica,* and *Clitocybe clavipes* (Pers.) P. Kumm.,[4] but the relationship of coprine to these symptoms is not well defined.[5]

Common Name: Ink cap, Tipler's Bane

Scientific Name: *Coprinus atramentarius* (Bull.) Fr.

Mycological Family: Coprinaceae

Physical Description: This species has a moderately large (3–7 cm/~13 in.), oblong cap with a grey to grey-brown top covered with small brown scales. The margin is sulcate-folded over the stalk, and immature gills are white, but they rapidly turn black during maturity. The spore print is black-brown to black. The stem is lighter than the cap and young specimens occasionally have a ring in the lower portion. The common name "inky cap" derives from the propensity of the gills to dissolve into a black fluid during aging.

Distribution and Ecology: *C. atramentarius* is a common fungus that inhabits several continents (e.g., Europe, North America) in a variety of locations including open fields, waste areas, meadows, gardens, and thickets.

EXPOSURE

Coprinus mushrooms are common in autumn, but the fruiting body may also appear in spring and summer. Coprine was investigated as a natural substitute for disulfiram, but the mutagenic and gonadotoxic properties of this compound in studies of experimental animals prevented the commercial use of coprine.[6]

Medical Toxicology of Natural Substances, by Donald G. Barceloux, MD

PRINCIPAL TOXINS

Physiochemical Properties

Mushrooms from *C. atramentarius* do not contain disulfiram or calcium cyanamide.[7] The principal active constituent of *C. atramentarius* is coprine, but the lack of activity of this compound when administered to experimental animals indicates that the actual toxin is a metabolite of coprine. Coprine [N^5-(hydroxy-1 cyclopropyl) L-glutamine, CAS RN: 58919-61-2] is a derivative of glutamine with an empirical formula of $C_8H_{14}N_2O_4$. Figure 40.1 displays the chemical structure of coprine. This compound is water soluble and in neutral solutions heat-resistant. Coprine does not inhibit aldehyde dehydrogenase *in vitro*.

Poisonous Parts

The average coprine concentration in frozen specimens of *C. atramentarius* was approximately 160 mg/kg with young specimens containing about 50% of the coprine present in mature specimens. Frying or boiling does not significantly affect the concentration of coprine.[1]

Mechanism of Toxicity

The consumption of coprine-containing mushrooms produces an elevation of acetaldehyde concentrations as a result of the inhibition of aldehyde dehydrogenase. Coprine undergoes hydrolysis by an enzyme with glutaminase-like properties. The resulting hemiaminal compound (1-aminocyclopropanol) yields highly reactive species that form a saturable, reversible enzyme-inhibiting complex. This compound inactivates aldehyde dehydrogenase and blocks the oxidation of acetaldehyde.[8] *In vitro* studies indicate that 1-aminocyclopropanol produces much greater inhibition of aldehyde dehydrogenase activity than coprine.[9] As a result of the blockade of acetaldehyde oxidation, acetaldehyde concentrations increase substantially. However, the exact mechanism of toxicity remains unclear because during experimental studies elevated acetaldehyde concentrations alone do not reproduce the toxicity associated with the ingestion of coprine-containing mushrooms or the disulfiram-ethanol reaction.[10] In contrast to disulfiram, the administration of coprine to experimental animals does not inhibit dopamine decarboxylase.[11]

DOSE RESPONSE

In experimental animals administered coprine without ethanol, the oral toxicity of coprine is low with a LD_{50} of about 1,500 mg/kg.[6] The intensity of the disulfiram-like reaction depends on several factors including the dose of coprine, the amount of ethanol consumed, individual sensitivity, and the time between *Coprinus atramentarius* ingestion and ethanol consumption.

TOXICOKINETICS

Following absorption, coprine undergoes enzymatic hydrolysis of the amide bond to form the hemiaminal compound, 1-aminocyclopropanol (CAS RN: 54376-44-2). This reaction yields a highly reactive species, possibly a cyclopropyliminium intermediate, which interacts with aldehyde dehydrogenase to block the oxidation of acetaldehyde. Animal studies indicate that 1-aminocyclopropanol exerts potent inhibition of aldehyde dehydrogenase *in vitro* and *in vivo*, whereas the administration of coprine has no effect on rat liver aldehyde dehydrogenase *in vitro*.[12] Consequently, 1-aminocyclopropanol is probably the toxic metabolite of coprine. Figure 40.2 displays the chemical structure of 1-aminocyclopropanol.

CLINICAL RESPONSE

Sensitivity to ethanol begins as soon as 15 minutes–2 hours (average, 3–6 hours) after the consumption of *Coprinus atramentarius* mushrooms and may persist up to 48–72 hours. Symptoms of the disulfiram-like reaction in patients sensitized to ethanol by the consumption of *Coprinus atramentarius* mushrooms include headache, facial flushing, paresthesias, metallic taste in mouth, lightheadedness, vomiting, palpitations, tachycardia, diaphoresis, and chest pain. Diarrhea and abdominal pain are unusual. The intensity of symptoms depends

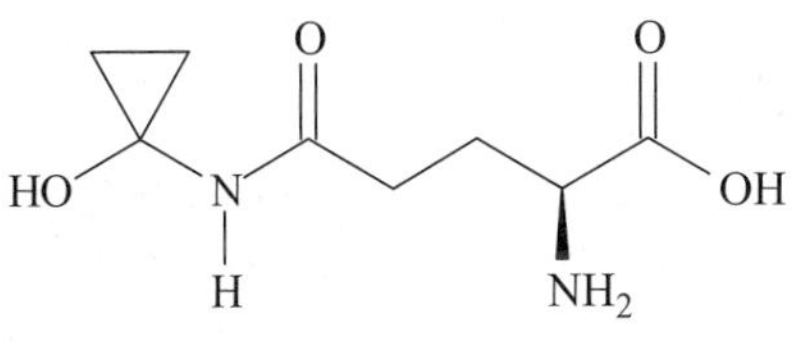

HO N H O O OH NH2

FIGURE 40.1. Chemical structure of coprine.

HO NH2

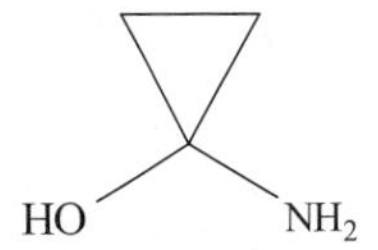

FIGURE 40.2. Chemical structure of the probable toxic metabolite, 1-aminocyclopropanol.

on the absorbed dose of coprine (e.g., food preparation techniques) and ethanol. Other factors include individual sensitivity and the length of time between ethanol consumption and the ingestion of *C. atramentarius* mushrooms. Symptoms of coprine toxicity are usually mild and brief with resolution occurring rapidly (minutes to several hours) without complications. Documented complications during this disulfiram-like reaction include an esophageal rupture confirmed by surgery[13] and the onset of atrial fibrillation in a previously healthy 37-year-old man that resolved spontaneously within 60 hours.[14] Shock, metabolic acidosis, life-threatening cardiac dysrhythmias, and myocardial infarction are complications of the disulfiram-ethanol reaction, but these serious adverse reactions are not well documented following the ingestion of coprine-containing mushrooms and ethanol. Fatalities after the ingestion of *Coprinus atramentarius* are also not well documented.

Coprine and the closely related compound, benzcoprine [CAS RN: 63805-80-1, N-(1-ethoxycyclopropyl)-benzamide], were tested in animals as a natural substitute for disulfiram. Animal studies indicated that chronic consumption of coprine produced bone marrow depression, lymphocytopenia, and reduced testicular weights due to the alkylating effects of coprine on germ cells.[6] However, there are no well documented reports of chronic toxicity resulting from the ingestion of *Coprinus atramentarius*.

DIAGNOSTIC TESTING

There are no biomarkers of coprine toxicity other than the determination of coprine by thin layer chromatography. There are no specific laboratory abnormalities associated with coprine poisoning. Most patients present with mild symptoms that are not associated with clinically significant laboratory abnormalities.

TREATMENT

The treatment of coprine toxicity is supportive, similar to the disulfiram-ethanol reaction. Although the clinical features of coprine toxicity are usually mild, potentially serious complications include hypotension and cardiac dysrhythmias based on the effects of disulfiram and ethanol. Patients with abnormal vital sign or pre-existing cardiovascular disease should receive cardiac monitoring, pulse oximetry, and an intravenous line along with fluids as necessary. Theoretically, hypotensive patients, who do not respond to fluid replacement, should receive a direct-acting vasopressor (e.g., norepinephrine) because norepinephrine stores may be depleted in the coprine-ethanol syndrome similar to the disulfiram-ethanol interaction. Decontamination measures are usually not necessary because of frequent vomiting and the rapid absorption of ethanol. There are no antidotes and no methods to enhance the elimination of coprine or the toxic metabolite. Correction of fluid and electrolyte imbalance and treatment of anxiety with benzodiazepines (lorazepam, diazepam) are the most important supportive measures. Patients may be discharged when comfortable. They should avoid all ethanol-containing products (e.g., ethanol-containing cold remedies and medical elixirs, alcoholic beverages) for at least 48 hours after the onset of symptoms. Theoretically, metoclopramide and odansetron are the antiemetics of choice because the α-adrenergic-blocking properties of phenothiazine antiemetics potentially exacerbate hypotension, but there are inadequate clinical data to confirm the clinical significance of this potential adverse effect.

References

1. Michelot D. Poisoning by *Coprinus atramentarius*. Nat Toxins 1992;1:73–80.
2. Josserand M. The ability of coprini to sensitize man to ethyl alcohol. Mycologia 1952:44:829–831.
3. Reynolds WA, Lowe FH. Mushrooms and a toxic reaction to alcohol: report of four cases. N Engl J Med 1965;272: 630–631.
4. Cochran KW, Cochran MW. *Clitocybe clavipes*: Antabuse-like reaction to alcohol. Mycologia 1978;70:1124–1126.
5. Kawagishi H, Miyazawa T, Kume H, Arimoto Y, Inakuma T. Aldehyde dehydrogenase inhibitors from the mushroom *Clitocybe clavipes*. J Nat Prod 2002;65:1712–1714.
6. Jonsson M, Lindquist NG, Ploen L, Ekvarn S, Kronevi T. Testicular lesions of coprine and benzcoprine. Toxicology 1979;12:89–100.
7. Hatfield GM, Schaumberg JP. Isolation and structural studies of coprine, the disulfiram-like constituent of *Coprinus atramentarius*. Lloydia 1975;38:489–496.
8. Wiseman JS, Abeles RH. Mechanism of inhibition of aldehyde dehydrogenase by cyclopropanone hydrate and the mushroom toxin coprine. Biochemistry 1979;18:427–435.
9. Helander A, Tottmar O. Effects of disulfiram, cyanamide and 1-aminocyclopropanol on the aldehyde dehydrogenase activity in human erythrocytes and leukocytes. Pharmacol Toxicol 1988;63:262–265.
10. Nakano J, Gin AC, Nakano SK. Effects of disulfiram on cardiovascular responses to acetaldehyde and ethanol in dogs. Q J Stud Alcohol 1974;35(Pt A):620–634.
11. Carlsson A, Henning M, Lindberg P, Martinson P, Trolin G, Waldeck B, Wickberg B. On the disulfiram-like effect of coprine, the pharmacologically active principle of *Coprinus atramentarius*. Acta Pharmacol Toxicol 1978;42:292–297.

12. Tottmar O, Lindberg P. Effects on rat liver acetaldehyde dehydrogenases *in vitro* and *in vivo* by coprine, the disulfiram-like constituent of *Coprinus atramentarius*. Acta Pharmacol Toxicol (Copenh) 1977;40:476–481.
13. Meyer JH, Herlocher JE, Parisian J. Esophageal rupture after mushroom alcohol ingestion. N Engl J Med 1971;285:1323.
14. Caley MJ, Clark RA. Cardiac arrhythmia after mushroom ingestion. Br Med J 1977;2:1633.

Chapter 41

ISOXAZOLE-CONTAINING MUSHROOMS and PANTHERINA SYNDROME (*Amanita muscaria, Amanita pantherina*)

HISTORY

Polychromatic rock paintings from the Paleolithic age (9000–7000 BC) contained images of *Amanita* mushrooms, which resembled *Amanita muscaria*.[1] DNA studies suggest that *A. muscaria* originated in Eurasia and migrated across land bridges to North America.[2] The use of psychoactive mushrooms from various genera (e.g., *Psilocybe, Conocybe, Gymnopilus, Panaeolus*) in Mexican and Central American religious rites continued into the middle of the 20th century. However, in contrast to the hallucinogenic mushrooms, fly agaric [*Amanita muscaria* (L.) Hook] has not been a common object of religious veneration.[3] In the Mayan civilization, legends linked *A. muscaria* with the thunderbolt.[4] *Amanita muscaria* and *Amanita pantherina* (DC.) Krombh. are the two major North American and European mushroom species that contain psychoactive isoxazole compounds.

MYCOLOGICAL DESCRIPTION

The Pantherina syndrome is associated with the ingestion of ibotenic- and/or muscimol-containing *Amanita* mushrooms, such as *A. aprica* J. Lindgr. and Tulloss, *A. gemmata* (Fr.) Bertill., *A. muscaria* (L.) Hook, *A. pantherina* (DC.) Krombh. and *A. regalis* (Fr.) Michael.[5,6] Like amatoxin-containing *Amanita* species, *Amanita muscaria* and *Amanita pantherina* possess an annulus (ring), white spore print, and free lamellae (gills). In contrast to the hepatotoxic *Amanita* species, *A. muscaria* and *A. pantherina* have a universal veil that remains on the cap in the form of scales. Rather than a cup, one or more verrucose zones encircle the bottom of the stalk. The spores of *Amanita muscaria* and *Amanita pantherina* are nonamyloid.

Common Name: Fly Agaric, Fly-Poison Agaric, False Orange, Scarlet Fly Cup, and Soma

Scientific Name: *Amanita muscaria* (L.) Hook

Mycological Family: Amanitaceae

Physical Description: This large mushroom has a bright red cap (5–15 cm/~2–6 in.) with large white conical scales. The white gills are free from the white stalk, which contains a hanging ring below the cap. Several warty rings encircle the base of the stalk, and warts can extend over the top of the cap. Figure 41.1 displays specimens of fly agaric.

Distribution and Ecology: *A. muscaria* grows in deciduous and conifer forests throughout several continents, particularly under spruce, birch, and pine trees during summer and autumn.

Common Name: The Panther, Panther Cap, Panther Agaric, Panther Amanita, False Blusher

Scientific Name: *Amanita pantherina* (DC.) Krombh.

Mycological Family: Amanitaceae

Physical Description: This mushroom has a distinctive cup-shaped bulb at the base of the stalk along with a brown cap (5–10 cm/~2–4 in.) that contains white scales. The margin of the cap is striped, and

Medical Toxicology of Natural Substances, by Donald G. Barceloux, MD

FIGURE 41.1. Fly agaric [*Amanita muscaria* (L.) Hook]. © 2006 Rune Midtgaard. See color insert.

the stalk has a fine ring just below the cap. The gills and stalk are white and the spore print is white.

Distribution and Ecology: *Amanita pantherina* inhabits a smaller geographic area compared with *A. muscaria*, but occasionally *A. pantherina* is abundant in local deciduous or conifer forests. *Amanita pantherina* appears most commonly during the rainy seasons of spring, summer, and fall.

EXPOSURE

Accidental ingestion of *Amanita muscaria* is unusual because of its distinctive features, and most ingestions of this mushroom occur when patients intentionally seek these mushrooms for their psychoactive properties.[7] Misidentification of *Amanita muscaria* for *Amanita caesarea* (Scop.) Pers. and of *Amanita pantherina* for *Amanita rubescens* var. *rubescens* Pers. also occurs. *Amanita pantherina* is more easily confused with edible *Amanita* species compared with *A. muscaria*.

PRINCIPAL TOXINS

Physiochemical Properties

The principal biologically active compounds in *Amanita muscaria* and *Amanita pantherina* probably are the

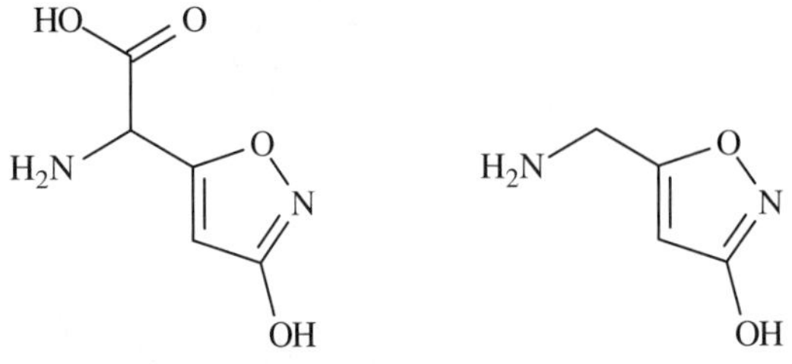

Ibotenic Acid **Muscimol**

FIGURE 41.2. Chemical structures of muscimol and ibotenic acid.

water soluble isoxazole compounds, ibotenic acid (α-amino-3-hydroxy-5-isoxazoloacetic acid, CAS RN: 2552-55-8) and muscimol (5—aminomethyl-3-hydroxy-isoxazole, CAS RN: 2763-96-4). Figure 41.2 displays the chemical structure of these isoxazole compounds. Other active compounds (e.g., muscazone) may also contribute to the effects of these mushrooms.[1] The fruiting bodies of these *Amanita* species contain substantially more ibotenic acid than muscimol. However, psychotropic effects probably result from muscimol, which is formed from the decarboxylation of ibotenic acid during food preparation and digestion in the gastrointestinal tract. Muscarine does not contribute to the toxic effects of these mushrooms.[8] Muscazone (α-amino-2,3-dihydro-2-oxooxazole-5-oacetic acid, CAS RN: 2255-39-2) is the lactame isomer of muscimol that probably exhibits minor pharmacological activity compared with muscimol.[1] Muscarufin is a terphenylquinone compound that is probably responsible for the bright color of *Amanita muscaria*. Specimens of *A. muscaria* accumulate relatively high concentrations of heavy metals including vanadium, cadmium, and selenium. Amavadin is a pale blue vanadium complex formed in *A. muscaria*, which may concentrate vanadium in concentrations up to 200 ppm.[9] The extracts from *A. muscaria* have weak insecticidal properties. Specimens of *A. pantherina* also contain two amino acids (stizolobic acid, stizolobinic acid), which probably are not toxicologically significant.[10]

Poisonous Parts

Chromatographic studies indicate that ibotenic acid and muscimol occur in *A. muscaria, A. pantherina*, A. *cothurnata* G.F. Atk., and to a lesser extent in *A. gemmata*.[6] The concentrations of isoxazole compounds vary both between these *Amanita* species and within individual specimens. The concentrations of isoxazole compounds are relatively high in the yellow tissue beneath the red cuticle and in the white flesh and gills compared with the stalk.[11,12] In a study of fresh specimens of *Amanita muscaria*, the concentrations of ibotenic acid and

TABLE 41.1. Ibotenic Acid and Muscimol Concentrations in Commercial Samples of Caps from Amanita pantherina and Amanita muscaria (ppm Dry Weight)*

Sample No.	Ibotenic Acid *A. muscaria*	Ibotenic Acid *A. pantherina*	Muscimol *A. muscaria*	Muscimol *A. pantherina*
1	612	188	286	1880
2	97	269	472	1554
3	342	—	254	—
4	<10	—	46	—
5	2845	—	1052	—

Source: From Ref 14.
*–, no sample.

muscimol were higher in caps (990 µg/g fresh weight and 380 µg/g fresh weight, respectively) than in the stalks (230 µg/g fresh weight and 80 µg/g fresh weight, respectively).[13] Table 41.1 lists the content of ibotenic acid and muscimol in samples of mushrooms purchased at Japanese smoke shops through the Internet.[14] The concentrations of these compounds in the stalks were dramatically lower than the caps with these compounds being undetectable in some stalks. This study also suggested that some purported extracts of *A. muscaria* actually contain psychoactive tryptamines (harmaline, 5-methoxy-*N,N*-diisopropyltryptamine) rather than ibotenic acid or muscimol. The concentration of muscarine in specimens of *A. muscaria* is several orders of magnitude below the muscarine concentrations found in certain *Inocybe* and *Clitocybe* species.

Mechanism of Toxicity

Ibotenic acid and muscimol are part of a naturally occurring group of excitatory amino acids that include domoic acid, tricholomic acid, kainic acid, and quisqualic acid. Ibotenic acid and muscimol are conformationally restricted derivatives of glutamic acid and γ-aminobutyric acid (GABA), respectively. In animal models, muscimol binds to GABA-receptors, whereas ibotenic acid affects glutamic acid receptors by changing membrane permeability and resting potential.[15,16] Specifically, ibotenic acid acts on inhibitory glutamate receptors, which are closely related to ionotropic glycine and gamma-aminobutyric acid (GABA) receptors.[17] Glutamic acid is a major excitatory neurotransmitter in mammalian central nervous systems. Muscimol is an inhibitor of neuronal and glial GABA uptake and a substrate for the GABA-metabolizing enzyme, GABA transaminase. In rodent models, muscimol is about five times more toxic than ibotenic acid.[18] *In vitro* studies of mouse brain homogenates indicate that ibotenic acid can undergo decarboxylation to muscimol in the brain.[19] Muscimol probably accounts for most of the central nervous system effects of these mushrooms due to the decarboxylation of ibotenic acid to muscimol.[1]

DOSE RESPONSE

Predicting toxicity based on a history of the amount ingested is difficult because of variation in the ibotenic acid and muscimol concentrations in these species as well as the preparation method and individual sensitivities. The concentration of toxins in these species is highly variable and food preparation may eliminate some, but not all of the toxins. Some people eat portions of *A. muscaria* after a variety of preparation techniques including consuming the entire fruiting body of *A. muscaria* after peeling the cuticle (outer membrane) from the cap and discarding the cooking water, sautéing the mushroom in brine after boiling and removing the cooking water, and drying and smoking the cap after peeling off the cuticle.[1] Based on an estimated average isoxazole content of 0.2%, 100 g of fresh psychoactive *Amanita muscaria* mushrooms contain sufficient muscimol equivalents (i.e., 12 mg) to cause mild intoxication (lightheadedness, ataxia, mood elevation, poor coordination).[20] Other authors estimated that the average size fruiting body of *A. muscaria* (60–70 g) contains up to 70 mg ibotenic acid (i.e., muscimol equivalents).[1] Based on the concentrations of ibotenic acid and muscimol in *Amanita* caps listed in Table 41.1, central nervous system effects develop after an estimated dose of about 7–30 g of *A. muscaria* caps or approximately 4–5 g of *A. pantherina* caps.[14]

TOXICOKINETICS

Based on the rapid onset of central nervous system symptoms after ingestion of these mushrooms, gastrointestinal absorption is rapid (i.e., 30–60 minutes). Both ibotenic acid and muscimol cross the blood–brain barrier. Decarboxylation of ibotenic acid (pantherin, agarin) produces muscimol (pyroibotenic acid). Renal excretion of muscimol is rapid with detectable quantities of unchanged muscimol appearing in the urine within 1 hour after ingestion.

CLINICAL RESPONSE

The clinical features of the pantherina syndrome are similar to ethanol intoxication. Initial symptoms begin about 30 minutes to a few hours after ingestion.[21] Toxic effects primarily involve the central nervous system manifest by euphoria, spacial distortion, increased sensitivity to visual and auditory stimuli, mydriasis, anxiety, depression, disorientation, ataxia, agitation, blurred vision, fasciculations, tremor, and convulsions.[22,23] Isoxazole compounds produce alterations in visual perception rather than well-formed hallucinations associated with the consumption of indole derivatives (psilocybin, psilocin). Consciousness may alternate between periods of severe lethargy and periods of agitation (screaming, confusion, incoherence, irritability). A deep sleep with vivid dreams often occurs before complete recovery. Vomiting is unusual, and classic features of anticholinergic or cholinergic intoxication do not occur.[24]

Symptoms are generally mild and self-limited, particularly for *A. pantherina* and *A. regalis*, with resolution occurring in 4–24 hours without sequelae.[25,26] Residual headaches may persist in some patients for several days. Respiratory depression does not usually occur unless the patient is heavily sedated with anticonvulsants for seizure activity. A 48-year-old man developed vomiting, seizures, and coma after ingesting a plate of mushrooms that included *A. muscaria*.[27] He awoke 10 hours after ingestion with a paranoid psychosis manifest by visual and auditory alterations. His mental status returned to baseline after 5 days.

DIAGNOSTIC TESTING

Analytical procedures for the detection of isoxazole compounds include paper chromatography,[28] gas chromatography/mass spectrometry (GC/MS),[14,29] high performance liquid chromatography (HPLC), high performance liquid chromatography/tandem mass spectrometry (HPLC/MS/MS),[30] and high performance thin layer chromatography (HPTLC). The latter method quantitates muscimol equivalents because the extraction procedure with formic acid facilitates the decarboxylation of ibotenic acid to muscimol. Ion-interaction HPLC allows the separation of ibotenic acid and muscimol with detection limits of 18 µg/L (ibotenic acid) and 30 µg/L (muscimol).[13] Muscimol and ibotenic acid are stable in dry mushroom specimens stored at room temperature for at least 90 days. Clinically significant laboratory changes are not usually associated with ingestion of these mushrooms.

TREATMENT

Treatment is supportive. Respiratory depression is not usually a feature of the pantherina syndrome, although intubation may be necessary following the treatment of seizures. Convulsions usually respond to benzodiazepines (lorazepam, diazepam). Decontamination measures are usually unnecessary. Most patients require only observation; drug therapy (atropine, physostigmine) is not indicated. Agitated patients should be placed in a quiet environment with monitoring of behavior by relatives or close friends. The administration of benzodiazepines (lorazepam, diazepam) may be necessary for severely agitated patients.

References

1. Michelot D, Melendez-Howell LM. *Amanita muscaria*: chemistry, biology, toxicology, and ethnomycology. Mycol Res 2003;107:131–146.
2. Oda T, Tanaka C, Tsuda M. Molecular phylogeny and biogeography of the widely distributed *Amanita* species, *A. muscaria* and *A. pantherina*. Mycol Res 2004;108:885–896.
3. Nyberg H. Religious use of hallucinogenic fungi: a comparison between Siberian and Mesoamerican cultures. Karstenia 1992;32:71–80.
4. Lowy B. *Amanita muscaria* and the thunderbolt legend in Guatemala and Mexico. Mycologia 1974;66:188–191.
5. Tulloss RE, Lindgren JE. *Amanita aprica*—a new toxic species from western North America. Mycotaxon 2005; 91:193–205.
6. Chilton WS, Ott J. Toxic metabolites of *Amanita pantherina, A. cothurnata, A. muscaria* and other *Amanita* species. Lloydia 1976;39:150–157.
7. Ott J. Psycho-mycological studies of *Amanita*—from ancient sacrament to modern phobia. J Psychedelic Drugs 1976;8:27–35.
8. Waser PG. The pharmacology of *Amanita muscaria*. Psychopharmacol Bull 1967:4:19–20.
9. Kneifel H, Bayer E. Stereochemistry and total synthesis of amavadin, the naturally occurring vanadium compound of *Amanita muscaria*. J Am Chem Soc 1986;108:3075–3077.
10. Chilton WS, Hsu CP, Zdybak WT. Stizolobic and stizolobinic acids in *Amanita pantherina*. Phytochemistry 1974;13: 1179–1181.
11. Gore MG, Jordan PM. Microbore single-column analysis of pharmacologically active alkaloids from the fly agaric mushroom *Amanita muscaria*. J Chromatogr 1982;243: 323–328.
12. Eugster CH. Isolation, structure, and syntheses of central-active compounds from *Amanita muscaria* (L. ex Fr.) Hooker. Psychopharmacol Bull 1967;4:18–19.
13. Gennaro MC, Giacosa D, Gioannini E, Angelino S. Hallucinogenic species in *Amanita muscaria*. Determination of muscimol and ibotenic acid by ion-interaction HPLC. J Liquid Chromatogr Related Technol 1997;20:413–424.
14. Tsujikawa K, Mohri H, Kuwayama K, Miyaguchi H, Kwata Y, Gohda A, et al. Analysis of hallucinogenic constituents in *Amanita* mushrooms circulated in Japan. Forensic Sci Int 2006;164:172–178.
15. Walker RJ, Woodruff GN, Kerkut GA. Effect of ibotenic acid and muscimol on single neurons of the snail *Helix aspersa*. Comp Gen Pharmacol 1971;2:168–174.

16. DeFeudis FV. Muscimol and GABA-receptors: basic studies and therapeutic implications. Rev Pure Appl Pharmacol Sci 1982;3:319–379.
17. Cleland TA. Inhibitory glutamate receptor channels. Mol Neurobiol 1996;13:97–136.
18. Scotti de CarolisA, LippariniF, LongoVG. Neuropharmacological investigations on muscimol, a psychotropic drug extracted from *Amanita muscaria*. Psychopharmacologia 1969;15:186–195.
19. Nielsen EO, Schousboe A, Hansen SH, Krogsgaard-Larsen P. Excitatory amino acids: studies on the biochemical and chemical stability of ibotenic acid and related compounds. J Neurochem 1985;45:725–731.
20. Hatfield GM, Brady LR. Toxins of higher fungi. Lloydia 1975:38:36–55.
21. Satora L, Pach D, Butryn B, Hydzik P, Balicka-Slusarczyk B. Fly agaric (*Amanita muscaria*) poisoning, case report and review. Toxicon 2005;45:941–943.
22. Satora L, Pach D, Ciszowski K, Winnik L. Panther cap *Amanita pantherina* poisoning case report and review. Toxicon 2006;47:605–607.
23. Bosman CK, Berman L, Isaacson M, Wolfowitz B, Parkes J. Mushroom poisoning caused by *Amanita pantherina*. S Afr Med J 1965;39:983–986.
24. Benjamin DR. Mushroom poisoning in infants and children: the *Amanita pantherina/muscaria* group. Clin Toxicol 1992;30:13–22.
25. Gelfand M, Harris C. Poisoning by *Amanita pantherina* a description of two cases. Central Afr J Med 1982;28: 159–163.
26. Elonen E, Tarssanen L, Harkonen M. Poisoning with brown fly agaric, *Amanita regalis*. Acta Med Scand 1979;205:121–123.
27. Brvar M, Mozina M, Bunc M. Prolonged psychosis after Amanita muscaria ingestion. Wien Klin Wochenschr 2006; 118:294–297.
28. Brady LR. Toxins of higher fungi. Lloydia 1975;38: 36–56.
29. Repke DB, Leslie DT, Kish NG. GLC–mass spectral analysis of fungal metabolites. J Pharm Sci 1978;67:485–487.
30. Tsujikawa K, Kuwayama K, Miyaguchi H, Kanamori T, Iwata Y, Inoue H, et al. Determination of muscimol and ibotenic acid in *Amanita* mushrooms by high-performance liquid chromatography and liquid chromatography-tandem mass spectrometry. J Chromatogr B 2007;852: 430–435.

Chapter 42

MUSCARINE-CONTAINING MUSHROOMS and MUSCARINE TOXICITY (*Clitocybe* and *Inocybe* Species)

HISTORY

In 1869, Schmiedeberg and Koppe isolated an impure compound from the Fly Agaric [*Amanita muscaria* (L.) Hook] that they believed was the principal toxin in this mushroom.[1] Almost a century later in 1954, this alkaloid was identified as muscarine by Eugster and Waser.[2] However, muscarine was a minor constituent of the Fly Agaric; muscarine was not the cause of the pantherina syndrome associated with this mushroom. Ingestions of muscarine-containing mushroom species (*Clitocybe* spp., *Inocybe* spp.) produce a syndrome characterized by mild cholinergic excess.[3] Serious toxicity is rare, in part, because of limited accessibility to many toxic species, the poor bioavailability of ingested muscarine, and undesirable physical characteristics (small, malodorous) of muscarine-containing mushrooms.

MYCOLOGICAL DESCRIPTION

Table 42.1 lists muscarine-containing mushrooms, primarily in the genera *Clitocybe* and *Inocybe*. The muscarine syndrome is not associated with the ingestion of mushrooms from the Fly Agaric [*Amanita muscaria* (L.) Hook]. The identification of muscarine-containing species frequently requires an expert mycologist because they often occur close to other similar-appearing fungi. *Inocybe* species are small to large fungi with glabrous, brown spores and radially striate caps. *Inocybe erubescens* A. Blytt (*Inocybe patouillardii* Bres., red-staining Inocybe) is the most common *Inocybe* species associated with toxicity in Europe.[1] They have white-tan to gray caps and fleshy central stalks. Most *Clitocybe* species are small, brown mushrooms with conical, fibrillose caps and adnate, whitish to tobacco brown gills. The toxic species, *Clitocybe dealbata* (Sowerby) Gillet, is whitish.[4] *Clitocybe* is a widely distributed genus that occasionally inhabits lawns and parks. *Inocybe* species occur in oak and pine woods.

Common Name: Ivory Clitocybe, White Clitocybe, Sweat-producing Clitocybe

Scientific Name: *Clitocybe dealbata* (Sowerby) Gillet

Mycological Family: Tricholomataceae

Physical Description: A small, white fungus with funnel-shaped caps (2.5–5 cm/~1–2 in.) and narrow, horizontal gills. The spore print is white. The stalk is cylindrical or slightly thickened at the base with whitish to pink-ochre flesh. Figure 42.1 displays specimens of this mushroom.

Distribution and Ecology: *Clitocybe dealbata* inhabits grasslands in meadow, lawns, pastures, and occasionally near trees, particularly in eastern Canada and the United States.

Common Name: Red-staining Inocybe, Patouillard's Inocybe

Scientific Name: *Inocybe erubescens* A. Blytt (*Inocybe patouillardii* Bres.)

Mycological Family: Cortinariaceae

Physical Description: This moderately large mushroom has a 3–9 cm (~1–4 in.) cap that initially is conical, but as the mushroom ages the cap flattens with involute and then de-curved, irregular, and finally fissured appearance. The cap of young

Medical Toxicology of Natural Substances, by Donald G. Barceloux, MD

TABLE 42.1. Muscarine-containing Mushrooms

Genus	Species	Synonym
Inocybe	*I. acuta* Boud.	
	I. agglutinata Peck	
	I. assimilata Britzelm.	*I. umbrina*
	I. bongardii (Weinm.) Quél. var. *bongardii*	
	I. calida Velen.	
	I. cervicolor (Pers.) Quél.	
	I. cincinnata (Fr.) Quél. var. cincinnata	*I. obscuroides*
	I. cinnamomea A.H. Sm.	
	I. dulcamara (Alb. & Schwein.) P. Kumm.	
	I. erubescens A. Blytt	*I. patouillardii*
	I. flocculosa (Berk.) Sacc.	*I. gausapata, I. lucifuga, I. pallidipes*
	I. geophylla (Pers.) P. Kumm. var. *geophylla*	
	I. geophylla var. lilacina Gillet	
	I. godeyi Gillet	
	I. griseolilacina J. E. Lange	
	I. heimii Bon	
	I. kaufmanii A. H. Sm.	
	I. lacera subsp. *lacera* (Fr.) P. Kumm.	
	I. langei R. Heim	
	I. lanuginosa var. *lanuginosa* (Bull.) P. Kumm.	
	I. lanuginosa var. *ovatocystis* (Boursier & Kühner) Stangl	*I. ovatocystis*
	I. maculata Boud.	
	I. muricellata Bres.	*I. hirtelloides*
	I. mixtilis (Britzelm.) Sacc.	
	I. napipes J.E. Lange	
	I. olympiana A.H. Sm.	
	I. praetervisa Quél	
	I. pusio P. Karst	
	I. queletii Konrad	
	I. rimosa (Bull.) P. Kumm.	
	I. sindonia (Fr.) P. Karst.	*I. kuehneri*
	I. soluta Velen.	
	I. sororia Kauffman	
	I. terrifera J.G. Kühn	
	I. terrigena (Fr.) Kühner	
	I. tristis	*I. georgii*
	I. whitei (Berk. & Broome) Sacc.	*I. pudica*
Clitocybe	*C. angustissima* (Lasch) P. Kumm.	
	C. candicans (Pers.) P. Kumm.	
	C. dealbata (Sowerby) Gillet	
	C. diatreta (Fr.) P. Kumm.	
	C. dilatate (Pers.) P. Karst.	
	C. dryadicola (J. Favre) Harmaja	
	C. festiva J. Favre	
	C. gibba (Pers.) P. Kumm.	
	C. gracilipes Lamoure	
	C. hydrogramma (Bull. & A. Venturi) P. Kumm.	
	C. marginella Harmaja	
	C. nebularis (Batsch) Quél.	
	C. nuoljea Lamoure	
	C. phyllophila (Pers.) P. Kumm.	*C. cerussata*
	C. rivulosa (Pers.) P. Kumm.	
	C. serotina Lamoure	
	C. suaveolens (Schumach.) P. Kumm.	
	C. vermicularis (Fr.) Quél.	
Mycena	*M. pura* (Pers.) P. Kumm.	

Source: Adapted from Ref 1.

FIGURE 42.1. Ivory Clitocybe [*Clitocybe dealbata* (Sowerby) Gillet], Photography courtesy of Daniel Sudakin, MD. See color insert.

specimens is white to light ochre, but older specimens have reddish brown to cinnamon-brown caps that have a fibrillose surface. The stem does not contain a bulb and the gills become red with age. The spore print is ochre-brown. Damage and age cause the flesh of the mushroom to turn red, and cutting the mushroom produces a fruity odor.

Distribution and Ecology: *I. erubescens* is found in parks and deciduous forests from late spring to autumn, primarily in May and June.

Common Name: Leaf-loving Clitocybe

Scientific Name: *Clitocybe phyllophila* (Pers.) P. Kumm.

Mycological Family: Tricholomataceae

Physical Description: This moderately large, white mushroom has a convex to flattened cap (5–11 cm/~2–4 in.). The flesh-colored to brown cap has a satin surface with slightly decurrent gills. The spore print is dirty flesh-colored to slightly pink.

Distribution and Ecology: *Clitocybe phyllophila* appears in deciduous forests among leaf litter during the autumn.

Common Name: Lilac or clean Mycena, Amethyst Agaric

Scientific Name: *Mycena pura* (Pers.) P. Kumm.

Mycological Family: Tricholomataceae

Physical Description: This small to moderate-sized mushroom with pink to pink-purple cap (2–6 cm/~1–2 in.) is initially conical and flattens as the mushroom ages. The gills are adenexed with a distinct radish-like smell. The spore print is white.

Distribution and Ecology: *Mycena pura* is a common mushroom found among leaf litter in a variety of diverse habitants including alpine meadows.

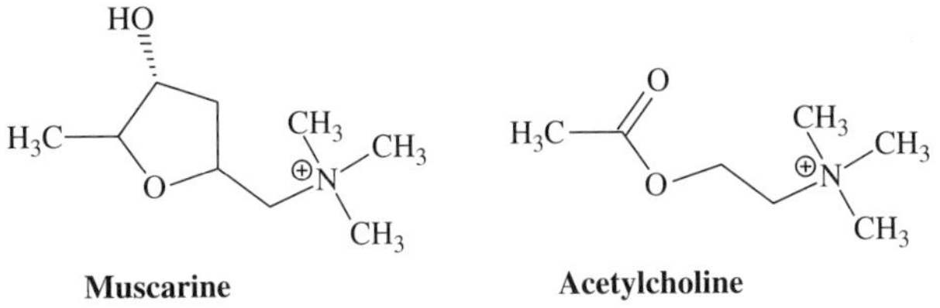

Muscarine **Acetylcholine**

FIGURE 42.2. Chemical structure of muscarine and acetylcholine.

EXPOSURE

In Europe, ingestion of the small white mushrooms of the genera *Inocybe* and *Clitocybe* typically results from the misidentification of small, white mushrooms of famous edible species, such as *Tricholoma terreum* (Schaeff.) Quél., *Clitopilus prunulus* (Scop.) P. Kumm., or *Marasmius oreade* (Bolton) Fr.

PRINCIPAL TOXINS

Physiochemical Properties

L-(+)-Muscarine, which is the quaternary trimethylammonium salt of 2-methyl-3-oxy-5-(aminomethyl) tetrahydrofuran, is the major constituent of toxic mushrooms from the genera *Clitocybe* and *Inocybe*.[5] This compound binds to the acetylcholine muscarinic receptor in the autonomic and postganglionic nerve fibers in the exocrine glands, eyes, bladder, gastrointestinal (GI) tract, lungs, and cardiovascular system. Depolarization results from the binding of muscarine to these receptors. Figure 42.2 demonstrates the structural similarities between muscarine and acetylcholine. Muscarine-containing mushrooms also contain other enantiomers of muscarine (*allo*-muscarine, *epi*-muscarine, *epi-allo*-muscarine). However, these enantiomers are relatively inactive compared with L-(+)-muscarine, and these enantiomers do not contribute to the cholinergic effects of muscarine-containing mushrooms. L-(+)-Muscarine is heat stable, colorless, and water soluble. Prolonged cooking does not reduce the toxicity of muscarine-containing mushrooms.[1]

Poisonous Parts

The muscarine content of *Inocybe* species is highly variable with alkaloid content ranging up to about 0.8% dry weight.[1] *Inocybe nappies* J.E. Lange contains a relatively high concentration (i.e., 0.71% dry weight) of muscarine compared with other North American *Inocybe* and *Clitocybe* species.[6,7]

Mechanism of Toxicity

The activity of L-(+)-muscarine results from the structural similarity of muscarine and acetylcholine. Muscarine stimulates muscarinic autonomic and postganglionic receptors in the central and peripheral nervous systems. The physiological consequence of this action is a variety of GI, cardiovascular, respiratory, genitourinary, and visual effects. Because muscarine does not contain an ester, acetylcholinesterase does not easily degrade muscarine; consequently, the cholinergic activity of muscarine is prolonged compared with acetylcholine.

TOXICOKINETICS

Gastrointestinal absorption of muscarine probably is low based on animal studies.[8]

CLINICAL RESPONSE

Characteristic features of muscarine poisoning include cholinergic effects, such as diaphoresis, salivation, urgency, diarrhea, and blurred vision.[9] Within minutes to 2 hours after ingestion of muscarine-containing mushrooms, muscarinic symptoms begin with GI (nausea, vomiting, diarrhea, abdominal pain) and autonomic (headache, salivation, lacrimation, fasciculations, diaphoresis, miosis, blurred vision) effects. Bronchoconstriction may cause wheezing and dyspnea. Cardiovascular effects primarily involve bradycardia, while the onset of hypotension is rare following the ingestion of muscarine-containing mushrooms. Symptoms are usually mild and resolve within 6 hours. Serious intoxication following ingestion of muscarine-containing mushrooms is rare. The ingestion of a mushroom *Inocybe tristis* Malencon and Bertault by a $2^1/_2$-year-old girl was associated with the rapid onset of diarrhea, vomiting, altered consciousness, and miosis.[10] Subsequently, hypertension and a purpuric rash developed along with transient hepatic, muscular, renal and cutaneous damage. She recovered with supportive care.

DIAGNOSTIC TESTING

Laboratory abnormalities are uncommon following the ingestion of muscarine-containing mushrooms. Case reports suggest that mildly increased serum hepatic aminotransferases may develop after ingestion of some muscarine-containing *Clitocybe* mushrooms.[11]

TREATMENT

Stabilization involves the evaluation of muscarinic effects associated with ingestion of these mushrooms. Serious poisoning is rare because of the low GI bioavailability of muscarine from the mushrooms and the small concentrations of muscarine. Potentially, life-threatening muscarinic symptoms include excessive respiratory secretion, bronchospasm, and cardiovascular collapse. Such patients require monitoring of oxygen saturations, intravenous fluids, cardiac monitoring, and, if indicated, intubation. Atropine is a specific antidote for the drying of bronchial secretions. Decontamination is usually not necessary because of the late presentation (i.e., >1–2 hours).

There are no effective methods to enhance the elimination of muscarine. Atropine antagonizes the effect of muscarine. Starting doses are 0.2 mg for children under 2 years of age, 0.3 mg for children 2–4 years, 0.4 mg for children 4–10 years, 0.5 mg for children 10–14 years, and 0.5 to 1.0 mg for adults depending on the severity of the intoxication. Additional smaller doses may be administered as dictated by the clinical response. Drying of secretions rather than mydriasis is an appropriate endpoint of atropine therapy to avoid atropinization. Patients may be discharged following resolution of muscarinic symptoms, typically in <24 hours.

References

1. Bresinsky A, Besl H. A Colour Atlas of Poisonous Fungi. A Handbook for Pharmacists, Doctors, and Biologists. London: Wolfe Publishing Ltd; 1990.
2. Eugster CH, Waser PG. [Properties of muscarine]. Experientia 1954;10:298–300. [German]
3. Bédry R, Saviuc P. Intoxications graves par les champignons à l'exception du syndrome phalloïdien. Réanimation 2002;11:524–532.
4. Bigelow HE. North American species of *Clitocybe*. Part 1. Beihefte zur Nova Hedwigia 1982;72:58–59.
5. Genest K, Hughes DW, Rice WB. Muscarine in *Clitocybe* species. J Pharm Sci 1968;57:331–333.
6. Hughes DW, Genest K, Rice WB. The occurrence of muscarine in *Clitocybe dealbata*. Lloydia 1966;29:328–332.
7. Brown JK, Malone MH, Stuntz DE, Tyler VE Jr. Paper chromatographic determination of muscarine in *Inocybe* species. J Pharm Sci 1962;51:853–856.
8. Fraser PJ. Pharmacological actions of pure muscarine chloride. Br J Pharmacol 1957;12:47–52.
9. Stallard D, Edes TE. Muscarinic poisoning from medications and mushrooms. A puzzling symptom complex. Postgrad Med 1989;85:341–345.
10. Amitai I, Peleg O, Ariel I, Binyamini N. Severe poisoning in a child by the mushroom *Inocybe tristis*, Malencon and Bertault. Isr J Med Sci 1982;18:798–801.
11. McCormick DJ, Avbel AJ, Gibbons RB. Nonlethal mushroom poisoning. Ann Intern Med 1979;90:332–335.

Chapter 43

ORELLANINE-CONTAINING MUSHROOMS and NEPHROTOXICITY
(*Cortinarius* Species)

HISTORY

The Orellanus syndrome involves the delayed development of renal failure following the ingestion of mushrooms from a group of orellanine-containing fungi, particularly from the genus *Cortinarius* [*C. orellanus* Fr., *C. rubellus* Cooke (*C. orellanoides*)]. Typically, these patients developed gastrointestinal symptoms (nausea, vomiting, diarrhea, abdominal cramps, anorexia) within several hours of ingesting these mushrooms followed by renal insufficiency within a few days to weeks after ingestion.[1] Some European and Japanese patients required renal transplantation after developing this syndrome.[2] Orellanus syndrome was first reported in Poland during 1952 following the onset of an epidemic of renal failure in the Konin province.[2] Subsequently, reports from Scotland, France, Switzerland, Finland, and Sweden associated the development of renal dysfunction with the ingestion of *Cortinarius* mushrooms.[4] Although *Cortinarius* species occur in North America, the development of delayed-onset of renal toxicity following the ingestion of mushrooms in North American is rare. However, case reports have associated a similar syndrome of acute gastroenteritis followed by the delayed-onset of nephrotoxicity a few days after the ingestion of mushrooms suspected to be *Amanita smithiana* in the Pacific Northwest of the United States.[5] Other mushrooms implicated in delayed nephrotoxicity include mushrooms from widely separated locations outside the United States, such as unidentified species of *Cortinarius* and *Amanita* in Canada[6] and Australia,[7] *Amanita pseudoporphyria* Hongo in Japan,[8] and *Amanita proxima* Dumée in France.[9] These *Amanita* mushrooms do not contain detectable amounts of orellanine.

MYCOLOGICAL DESCRIPTION

Cortinarius species comprise a large, diverse genus of over 250 species that usually form mycorrhizae with trees in forests. A small number of lemon-yellow *Cortinarius* species are nephrotoxic.[10] The mushrooms from these species have a partial veil composed of fine threads (cortina) connecting the margin of the cap to the stalk, particularly in young specimens. As the mushroom ages, the cortina disappears, resulting in almost indistinguishable remains of the cortina.

Common Name: Poznan Cort Mushroom, Fool's Webcap

Scientific Name: *Cortinarius orellanus* Fr.

Mycological Family: Cortinariaceae

Physical Description: This moderately large mushroom has a yellowish-brown cap (2–8 cm/~1–3 in.) that is conical and covered with fibrils on young specimens. As the mushroom ages, the cap becomes smooth. The spores are elliptical with a rust-brown spore print. The stem is lighter than the cap, manifest by a golden-yellow color.

Distribution and Ecology: Specimens of *C. orellanus* occur in warm deciduous forests near oak and pine trees in continental Europe.

Medical Toxicology of Natural Substances, by Donald G. Barceloux, MD

Common Name: Deadly Webcap

Scientific Name: *Cortinarius rubellus* Cooke (*Cortinarius speciosissimus* Kühner & Romagn., *Cortinarius orellanoides* Rob. Henry)

Mycological Family: Cortinariaceae

Physical Description: This moderately large mushroom has a pointed umbo on the rust- to yellow-brown cap (2–12 cm/~1–5 in.) that is conical in shape on young species and more flattened on older specimens. The broad, elliptical spores yield a rust-brown spore print. The cylindrical stalk is similar in color to the cap and the stalk sometimes contains lemon-yellow bands, which represents the remains of the cortina.

Distribution and Ecology: In northern Europe and mountainous areas of central Europe, *C. rubellus* species inhabit moist or wet conifer forests, particularly in moist acidic soil.

EXPOSURE

Nephrotoxic *Cortinarius* species are somewhat morphologically similar to mushrooms from the *Psilocybe* genus, and inexperienced mushroom hunters searching for hallucinogenic mushrooms may misidentify *Cortinarius orellanus* or *Cortinarius rubellus* as *Psilocybe semilanceata* (Fr.) P. Kumm.[11] Misidentification of *C. orellanus* as *Cantharellus cibarius* var. *cibarius* Fr. also occurs.

PRINCIPAL TOXINS

Physiochemical Properties

The nephrotoxic mushrooms of the genus *Cortinarius* contain the polar, bipyridyl *N*-oxide compound orellanine [2,2′-bipyridine-3,3′, 4,4′-tetrol (tetrahydroxy)-1,1′ dioxide, CAS RN: 37338-80-0], which is structurally related to paraquat (4,4′-bipyridine) and diquat (2,2-bipyridine).[12] Orellanine is a colorless, blue-fluorescing compound, which decomposes slowly in high heat (150 °C–180 °C/302 °F–356 °F) and UV light to the yellow, nontoxic bipyridyl compound, orelline.[13,14] Decomposition of orellanine is rapid at temperatures over 267 °C (513 °F). However, orellanine is stable for many years in the intact mushroom.[15] Figure 43.1 demonstrates the reduction of orellanine to the nontoxic compound, orelline via the intermediate orellinine. Cooking, freezing, and drying do not reduce the nephrotoxicity of *Cortinarius* species associated with renal damage.[16] *Cortinarius rubellus* contains other cyclic decapeptides (cortinarin A, B) that produce renal damage in animal species,[17] but the role of these decapeptides in causing the Orellanus syndrome remains unclear.[18] Cortinarin C is not associated with nephrotoxicity in animal models. Both orellanine and orelline are practically insoluble in organic solvents and water, but somewhat soluble in methanol and dimethyl sulfoxide (DMSO).[15]

Poisonous Parts

Species associated with the presence of orellanine include *C. orellanus, C. fluorescens* E. Horak, *C. henricii* Reumaux, *C. rainierensis* A.H. Sm. & D.E. Stuntz, and *C. rubellus*.[19] Orellanine content ranged from 0.45% (*C. henricii*) to 1.1–1.4% (*C. orellanus*) as measured by direct electron spin resonance (see Table 43.1).[20] The orellanine content in this study was approximately 2.5–3 times greater in the caps than in the stalks.[21] In dried specimens of *C. orellanus* and *C. rubellus* (*C. orellanoides, C. speciosissimus*), the orellanine content was 14 mg/g and 9 mg/g, respectively.[22]

Orellanine

Semiquinone Anion Radical

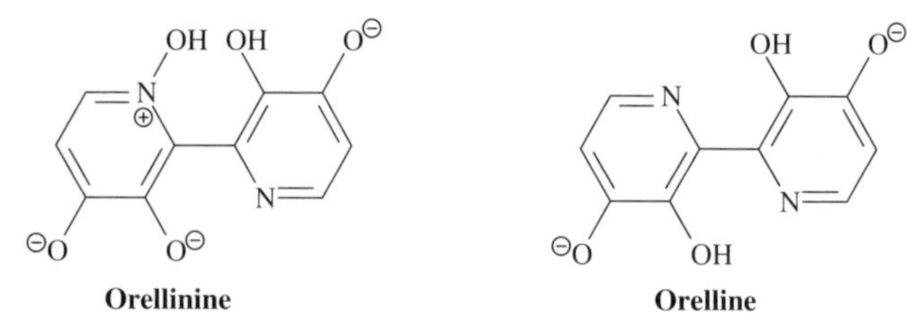

FIGURE 43.1. Chemical structures of orellanine, potential toxic intermediate, orthosemiquinone anion radical, and the reduction products, orelline and orellanine.

TABLE 43.1. Orellanine Content of Selected Cortinarius Species*

Species	Orellanine Content (% Dry Weight)
Cortinarius henricii Reumaux	0.45
Cortinarius orellanus Fr.	1.10
Cortinarius rainierensis A.H. Sm. & D.E. Stuntz	0.85
Cortinarius rubellus Cooke	0.20–0.50

Source: From Ref 20.

*Analytical method direct electron spin resonance with limit of detection = 5 µg.

Mechanism of Toxicity

The mechanism of action of orellanine is not well defined, but the target of orellanine or a toxic metabolite (e.g., orellinine) probably is the renal tubular epithelium.[23,24] Orellanine undergoes photochemical degradation to orellinine, which is reduced to the nontoxic metabolite, orelline. Consumption of nephrotoxic *Cortinarius* species produces a severe interstitial nephritis, acute focal tubular damage, and interstitial fibrosis. Glomeruli are minimally involved. The administration of orellanine produces orthosemiquinone anion and oxygen free radicals that increase oxidative stress by depletion of glutathione and ascorbate as demonstrated in Figure 43.1,[25] but the exact mechanism of toxicity remains unclear.[23] Renal biopsies of patients with renal failure after the ingestion of *Cortinarius* mushrooms demonstrate nonspecific tubulointerstitial changes with marked edema, patchy tubular necrosis, and mild mononuclear inflammatory infiltrates.[26] Immunofluorescence does not demonstrate immune complex deposition in the kidneys. The glomeruli are usually preserved with little if any mesangial reaction, and the renal vessels are patent. Some renal biopsies during the acute phase detected unusual multinucleated tubular giant cells and marked medial muscular hyperplasia of interlobular arteries with no evidence of injury to the intima.[27] Healing occurs via fibrosis with variable amounts of tubular atrophy, patchy tubular loss, and renal dysfunction. During the acute phase, electron microscopy demonstrates mitochondrial swelling, apical cytoplasmic vacuolization, and distorted brush borders in the proximal tubular epithelium.[28] Although a few case reports associate mild hepatocellular damage with *C. orellanus* ingestion,[29] hepatic damage is usually absent in contrast to amatoxin poisoning.[30]

DOSE RESPONSE

Based on case reports, there is considerable variation in the response to orellanine. The degree of renal dysfunction depends on a variety of factors including delay between ingestion and treatment, pre-existing renal disease, internal dose of orellanine, and the general health of the patient.[31] A group of 3 patients sharing a meal of mushrooms from *Cortinarius rubellus* (*C. orellanoides, C. speciosissimus*); 1 patient developed mild gastroenteritis and transitory renal dysfunction, whereas the other 2 patients developed end-stage renal failure that necessitated renal transplantation.[2] The ingestion of 15 *Cortinarius* mushrooms was associated with the development of acute renal failure with slow, partial resolution.[4] A 32-year-old man developed renal failure requiring dialysis after the ingestion of two fruiting bodies from a specimen of *Cortinarius orellanus*.[26] Although his renal function recovered, his serum creatinine remained elevated a month after ingestion of these mushrooms.

TOXICOKINETICS

Orellanine undergoes photoreduction to orellinine, which is subsequently reduced to the nontoxic metabolite, orelline.[32] In a series of 26 patients poisoned by *Cortinarius* mushrooms, the severity of the renal failure did not correlate to acetylation or hydroxylation status of hepatic isoenzyme function.[33] Based on limited testing, the clearance of orellanine from the blood is probably rapid (i.e., <2–3 days).[16]

CLINICAL RESPONSE

The ingestion of nephrotoxic *Cortinarius* mushrooms causes the delayed onset of an acute tubulointerstitial nephropathy that can progress to chronic renal insufficiency. During the initial latent period, some patients develop gastrointestinal symptoms that include nausea, vomiting, and less often abdominal pain, anorexia, and diarrhea. These symptoms are typically mild and begin within the first few days (median: 3 days, range: 12 hours–14 days) after ingestion of *Cortinarius* mushrooms.[34] The gastrointestinal phase of the Orellanus syndrome rarely causes dehydration or electrolyte imbalance. Other symptoms during the prerenal phase include anorexia, headache, chills, burning sensation in the mouth, intense thirst, polydipsia, and polyuria.[29] Fever is not usually part of the clinical presentation. In mild cases, these symptoms resolve over several days without the development of renal dysfunction. In more severe cases, myalgias, intense lumbar pain, flank pain, and oliguria occur over the next 1–2 weeks. Renal failure occurs 4–15 days (median: ~8 days) after ingestion depending on the dose ingested.[34] Diuresis typically begins about 7–10 days after the development of acute renal failure. Mortality is low following the use of hemodialysis. A substantial number of cases develop permanent renal dysfunction, but the actual number of patients with chronic renal failure after orellanine poisoning varies between different series of orellanine intoxication. In a study of 26 patients with nephrotoxicity secondary to *Cortinarius* mushroom ingestion, the incidence of end-stage renal failure requiring dialysis and renal transplantation was approximately 8%,[33] whereas 9 (41%) of 22 Swedish patients developed end-stage renal disease after ingesting mushrooms from *Cortinarius rubellus* (*C. orellanoides, C. speciosissimus*).[35] Anorexia and malaise may persist for months.

DIAGNOSTIC TESTING

Rapid analytical procedures are available to detect orellanine via thin layer chromatography (TLC) on silicagel or cellulose plates after extraction of the mushroom or tissue up to at least 60 days after ingestion.[36,37] Other analytical methods include electrophoresis on agarose gel (limit of detection [LOD] = 25 ng), direct electron spin resonance (LOD = 5 μg), and high performance liquid chromatography (HPLC; LOD = 50 pg).[20,38] In a woman who developed interstitial nephritis after the ingestion of mushrooms from *Cortinarius orellanus*, the orellanine concentration was 6.12 mg/L in a plasma sample drawn 10 days after ingestion as measured by direct spectrofluorometry on two-dimensional thin-layer chromatograms after specific photodecomposition to orelline.[39] The orellanine concentration in the first renal biopsy obtained 13 days after ingestion was 7 μg per 25 mm^3. The second renal biopsy taken 6 months later demonstrated 24 μg per 8 mm^3.

Laboratory abnormalities include leukocyturia, hematuria, and proteinuria along with elevation of the serum creatinine and serum potassium. Depending on the severity of the renal failure, hyperphosphatemia, hypocalcemia, and metabolic acidosis may also occur.

TREATMENT

The mainstay of treatment is supportive care with careful management of fluid and electrolyte balance as well as the use of hemodialysis as needed. Prolonged monitoring of renal function is necessary because of the slow resolution of kidney dysfunction. *In vitro* studies suggest that hemoperfusion materials adsorb orellanine, particularly activated charcoal.[40] However, there are inadequate clinical and toxicokinetic data, especially during the first 24 hours after ingestion, to confirm or exclude the efficacy of hemoperfusion or hemodialysis. Experimental studies in rats suggest that nephrotoxicity begins within 40 hours of ingestion,[41] and in animal studies the kidneys excrete detectable quantities of orellanine only for the first 24 hours after exposure.[42] Hemodialysis and hemoperfusion probably do not remove clinically significant amounts of orellanine more than 1–2 weeks after ingestion, and during the subsequent period the use of hemodialysis depends only on the need to support renal function. There are no antidotes.

Outcome depends on a variety of factors including delay until supportive therapy, pre-existing renal disease, internal dose of orellanine, and the general health of the patient.[31] Recovery of renal function is slow, and therefore renal transplantation usually occurs over 10 months after ingestion.[34] Case reports associate the use of corticosteroids and *N*-acetylcysteine with clinical improvement,[43] but the variable course of renal dysfunction after orellanine poisoning and the lack of clinical data limits conclusions about the efficacy of these treatment modalities.

References

1. Andary C, Rapior S, Delpech N, Huchard G. Laboratory confirmation of *Cortinarius* poisoning. Lancet 1989;1(8631):213.
2. Short AI, Watling R, MacDonald MK, Robson JS. Poisoning by *Cortinarius speciosissimus*. Lancet 1980;2(8201):942–944.
3. Grzymala S. Massenvergiftungen durch den orangefushsigen Hautkopf. Z Pilzkd 1957;23:139–142.
4. Holmdahl J, Mulec H, Ahlmen J. Acute renal failure after intoxication with *Cortinarius* mushrooms. Hum Toxicol 1984;3:309–313.
5. Warden CR, Benjamin DR. Acute renal failure associated with suspected *Amanita smithiana* mushroom ingestions: a case series. Acad Emerg Med 1998;5:808–812.
6. Raff E, Halloran PF, Kjellstrand CM. Renal failure after eating "magic" mushrooms. Can Med Assoc J 1992;147:1339–1341.
7. Mount P, Harris G, Sinclair R, Finlay M, Becker GJ. Acute renal failure following ingestion of wild mushrooms. Intern Med J 2002;32:187–190.
8. Iwafuchi Y, Morita T, Kobayashi H, Kasuga K, Ito K, Nakagawa O, et al. Delayed onset acute renal failure associated with *Amanita pseudoporphyria* Hongo ingestion. Intern Med 2003;42:78–81.
9. de Haro L, Jouglard J, Arditti J, David JM. [Acute renal insufficiency caused by *Amanita proxima* poisoning: experience of the Poison Center of Marseille]. Nephrologie 1998;19:21–24. [French]
10. Bresinsky A, Besl H. A Colour Atlas of Poisonous Fungi. A handbook for Pharmacists, Doctors, and Biologists. London: Wolfe Publishing Ltd; 1990.
11. Franz M, Regele H, Kirchmair M, Kletzmayr J, Sunder-Plassmann G, Horl WH, Pohanka E. Magic mushrooms: hope for a "cheap high" resulting in end-stage renal failure. Nephrol Dial Transplant 1996;11:2324–2327.
12. Kurnsteiner H, Moser M. Isolation of a lethal toxin from *Cortinarius orellanus* Fr. Mycopathologia 1981;74:65–72.
13. Andary C, Rapior S, Fruchier A, Privat G. Cortinaires de la section Orellani: Photodécomposition et hypothèse de la phototoxicité de l'orellanine. Cryptogamie Mycol 1986;7:189–200.
14. Antkowiak WZ, Gessner WP. Photodecomposition of orellanine and orellinine, the fungal toxins of *Cortinarius orellanus* Fries and *Cortinarius speciosissimus*. Experientia 1985;41:769–771.
15. Antkowiak WZ, Gessner WP. The structure of orellanine and orelline. Tetrahedron Lett 1979;21:1931–1934.

16. Karlson-Stiber C, Persson H. Cytotoxic fungi—an overview. Toxicon 2003;42:339–349.
17. Tebbett IR, Caddy B. Mushroom toxins of the genus *Cortinarius*. Experientia 1984;40:441–446.
18. Matthies L, Laatsch H. Cortinarins in *Cortinarius speciosissimus*? A critical revision. Experientia 1991;47:634–640.
19. Rapior S, Andary C, Privat G. Chemotaxonomic study of orellanine in species of *Cortinarius* and *Dermocybe*. Mycologia 1988;80:741–747.
20. Oubrahim H, Richard JM, Cantin-Esnault D, Seigle-Murandi F, Trecourt F. Novel methods for identification and quantification of the mushroom nephrotoxin orellanine. Thin-layer chromatography and electrophoresis screening of mushrooms with electron spin resonance determination of the toxin. J Chromatogr A 1997;758:145–157.
21. Rapior S, Andary C, Privat G. [Chemotaxonomic study of orellanine in species of *Cortinarius* and *Dermocybe*]. Mycologia 1988;80:741–747. [French]
22. Prast H, Werner ER, Pfaller W, Moser M. Toxic properties of the mushroom *Cortinarius orellanus*. I. Chemical characterization of the main toxin of *Cortinarius orellanus* (Fries) and *Cortinarius speciosissimus* (Kuhn & Romagn) and acute toxicity in mice. Arch Toxicol 1988;62:81–88.
23. Delpech N, Rapior S, Cozette AP, Ortiz JP, Donnadieu P, Andary C, Huchard G. [Outcome of acute renal failure caused by voluntary ingestion of Cortinarius orellanus]. Presse Med 1990;19:122–124. [French]
24. Oubrahim H, Richard JM, Cantin-Esnault D. Peroxidase–mediated oxidation, a possible pathway for activation of the fungal nephrotoxin orellanine and related compounds. ESR and spin-trapping studies. Free Radic Res 1998;28:497–505.
25. Richard JM, Cantin-Esnault D, Jeunet A. First electron spin resonance evidence for the production of semiquinone and oxygen free radicals from orellanine, a mushroom nephrotoxin. Free Radic Biol Med 1995;19:417–429.
26. Calvino J, Romero R, Pintos E, Novoa D, Guimil D, Cordal T, et al. Voluntary ingestion of *Cortinarius* mushrooms leading to chronic interstitial nephritis. Am J Nephrol 1998;18:565–569.
27. O'Donnell M, Fleming S. The renal pathology of mushroom poisoning. Histopathology 1997;30:280–282.
28. Holzl B, Regele H, Kirchmair M, Sandhofer F. Acute renal failure after ingestion of *Cortinarius speciosissimus*. Clin Nephrol 1997;48:260–262.
29. Schumacher T, Hoiland K. Mushroom poisoning caused by species of the genus *Cortinarius* Fries. Arch Toxicol 1983;53:87–106.
30. Richard JM, Taillandier G, Benoit-Guyod JL. A quantitative structure-activity relationship study on substituted pyridines as a contribution to the knowledge of the toxic effects of orellanine, a toxin from the mushroom *Cortinarius orellanus*. Toxicon 1985;23:815–824.
31. Horn S, Horina JG, Krejs GJ, Holzer H, Ratschek M. End-stage renal failure from mushroom poisoning with *Cortinarius orellanus*: report of four cases and review of the literature. Am J Kid Dis 1997;30:282–286.
32. Rapior S, Fruchier A. [Carbon-13 NMR parameters of orellanine]. An Quim 1989; 85C:69–71. [French]
33. Bouget J, Bousser J, Pats B, Ramee MP, Chevet D, Rifle G, Giudicelli CP, Thomas R. Acute renal failure following collective intoxication by *Cortinarius orellanus*. Intensive Care Med 1990;16:506–510.
34. Danel VC, Saviuc PF, Garon D. Main features of *Cortinarius* spp. poisoning: a literature review. Toxicon 2001; 39.1053–1060.
35. Holmdahl J, Blohme I. Renal transplantation after *Cortinarius speciosissimus* poisoning. Nephrol Dial Transplant 1995;10:1920–1922.
36. Delpech N, Rapior S, Donnadieu P, Cozette AP, Ortiz JP, Huchard G. [Voluntary poisoning by *Cortinarius orellanus*: usefulness of an original early treatment after determination of orellanine in the biological fluids and tissues]. Nephrologie 1991;12:63–66. [French]
37. Rohrmoser M, Kirchmair M, Feifel E, Valli A, Corrandini R, Pohanka E, et al. Orellanine poisoning: rapid detection of fungal toxin in renal biopsy material. Clin Toxicol 1997;35:63–66.
38. Cantin D, Richard JM, Alary J. Chromatographic behaviour and determination of orellanine, a toxin from the mushroom *Cortinarius orellanus*. J Chromatogr 1989;478: 231–237.
39. Rapior S, Delpech N, Andary C, Huchard G. Intoxication by *Cortinarius orellanus*: detection and assay of orellanine in biological fluids and renal biopsies. Mycopathologia 1989;108:155–161.
40. Fulde K, Bohler J, Keller E, Frahm AW. Efficiency of haemoperfusion materials at removing the fungal toxin orellanine from human plasma. Pharmazie 1998;53:58–59.
41. Nieminen L, Pyy K. Sex differences in renal damage induced in the rat by the Finnish mushroom, *Cortinarius speciosissimus*. Acta Pathol Microbiol Scand [A] 1976; 84:222–224.
42. Prast H, Pfaller W. Toxic properties of the mushroom *Cortinarius orellanus* (Fries). II. Impairment of renal function in rats. Arch Toxicol 1988;62:89–96.
43. Kilner RG, D'Souza RJ, Oliveira DB, MacPhee IA, Turner DR, Eastwood JB. Acute renal failure from intoxication by *Cortinarius orellanus*: recovery using anti-oxidant therapy and steroids. Nephrol Dial Transplant 1999;14: 2779–2780.

Chapter 44

PAXILLUS and OTHER MUSHROOM SYNDROMES

HISTORY

The Paxillus syndrome is a food allergy manifest by hemolytic anemia and renal failure. This idiosyncratic reaction was reported in Germany during the early 1960s after the development of this syndrome following ingestion of mushrooms previously considered edible.[1] Studies during the early 1970s indicated that the hemolysis associated with this syndrome resulted from an immune-mediated rather than a toxic reaction.[2]

MYCOLOGICAL DESCRIPTION

Common Name: Brown Roll-Rim, Common or Poisonous Paxillus, Brown Chanterelle

Scientific Name: *Paxillus involutus* (Batsch) Fr.

Mycological Family: Paxillaceae

Physical Description: This moderately large, ochre to red-brown ochre mushroom has a large cap (5–25 cm/~2–10 in.) with a margin that curves inward. The decurrent gills have a whitish to light ochre color that becomes olive-yellow to ochre-brown with aging. The spore mass is ochre-brown to rust. The sour-smelling, white to pale ochre flesh easily turns reddish-ochre after trauma.[3]

Distribution and Ecology: *Paxillus involutus* is a common mushroom in conifer and deciduous forests, particularly in damp areas including roadsides, parks, and gardens during late summer to late fall.

Common Name: Yellow Tricholoma, Man-on-Horseback, Chevalier

Scientific Name: *Tricholoma flavovirens* (Fries) Lundell (*Tricholoma equestre*)

Mycological Family: Tricholomataceae

Physical Description: *Tricholoma flavovirens* has a yellow to yellowish-brown, viscid cap, yellow notched gills, a robust stature, and pale yellow stipe. Fleshy stalk with sinuate to notched gills, and a white spore deposit.

Distribution and Ecology: This mushroom appears from late fall through mid-winter in sandy soils in conifer forests.

Common Name: Smith's Amanita, Smith's Lepidella

Scientific Name: *Amanita smithiana* Bas

Mycological Family: Amanitaceae

Physical Description: An unpleasant smelling white mushroom that stains light cinnamon to yellow-brown with age or bruising. The gills and spores are white, and the stem contains a ring. The cap has warts that may disappear in rain or hot, dry weather.

Distribution and Ecology: *Amanita smithiana* inhabits the Pacific northwestern United States and southwestern British Columbia. This fungal species prefers areas of moss under conifers in these regions.

Medical Toxicology of Natural Substances, by Donald G. Barceloux, MD

EXPOSURE

Matsutake (pine mushroom) is a common name applied to edible species from the genus *Tricholoma*. The American matsutake [*Tricholoma magnivelare* (Peck) Redhead], which grows primarily in the Pacific Northwest, is a relative of the Japanese matsutake [*Tricholoma matsutake* (S. Ito & S. Imai) Singer]. The mistaken identification of *Amanita smithiana* as a pine mushroom can result in poisoning, although the odor of this latter mushroom is not as spicy as matsutake mushrooms.

PRINCIPAL TOXINS

Paxillus Syndrome

The active constituent or antigen for the Paxillus syndrome has not been identified. *Paxillus involutus* contains a phenol, involutin, which oxidizes to a brown compound that accounts for the brown discoloration following damage of fresh specimens.[4] Figure 44.1 displays the chemical structure of involutin and the closely related compound, chamonixin. During an animal study, test animals received 50 or 250 mg/kg body weight of an aqueous suspension composed of dried homogenized mushrooms from *Paxillus involutus* collected near the Arctic Circle in Finland.[5] There were no observable differences between control and test rats. Experimental observations included general behavior, body weight, gross and microscopic changes in major organs, and the relative weights of the liver, kidney, and heart. Examinations of Polish mushroom specimens indicate that *Paxillus involutus* does not accumulate heavy metals, such as mercury.[6] *In vitro* studies indicate that the active fraction is a nonpolar, acid- and thermo-stable cytotoxic principle.[7] Some specimens of *Paxillus involutus* produce gastroenteritis when eaten in one geographical area, but in other locations this mushroom is edible when properly prepared. The onset of the Paxillus syndrome results from the ingestion of the *Paxillus involutus* mushroom by sensitized patients. The presence of an immune-complex directed at erythrocytes causes hemolytic anemia with subsequent renal damage. Attachment of immune complexes to the erythrocytes results in activation of complement, lysis of the red blood cells, and hypotension. Inadequate renal perfusion increases renal damage.

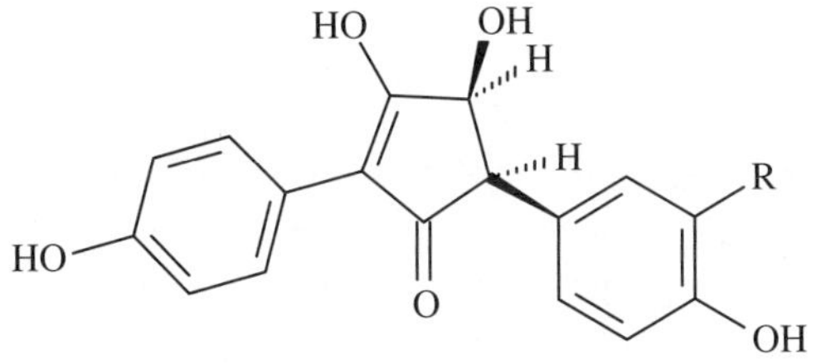

Chamonixin R = H
Involutin R = OH

FIGURE 44.1. Chemical structures of involutin and chamonixin.

Myotoxins

Animal studies suggest that the myotoxic effects of *Tricholoma flavovirens* (Pers.) S. Lundell results from a nonspecific response to large ingestions by sensitive individuals. The administration of dried extract from these mushrooms produces diarrhea, tachypnea, reduced motor activity, and histological changes in the muscle fibers of mice.[8] Similar reactions occur in rodent studies with other presumably edible species including *Russula subnigricans* Hongo, *Cantharellus cibarius* Fr. var. *cibarious* (chanterelle), *Leccinum versipelle* (Fr. & Hök) Snell (brown birch boletus), and *Albatrellus ovinus* (Schaeff.) Kotl. & Pouzar (sheep polypore).[9] The toxins in *T. flavovirens* and *Russula subnigricans* have not been identified. The latter mushroom species contains cytotoxic triphenyl ethers, russuphelins A–F.[10,11]

Renal Toxins

The renal toxins present in *Amanita smithiana* are not well defined. Mushrooms from *Amanita smithiana* and *Amanita pseudoporphyria* Hongo species contain a suspected nephrotoxin, but no detectable amounts of amatoxins are present in these mushrooms. Partial isolation of the renal toxin from *A. smithiana* indicates that this renal toxin has different physical properties (e.g., UV light stability, shorter latency period) than orellanine.[12] In contrast to most other *Amanita* species, specimens from *A. smithiana* contain 2(S)-amino-4,5-hexadienoic acid and trans-2-amino-5-chloro-4-hexenoic acid, but these compounds are not definitely nephrotoxic in animal models.[13] The role of allenic norleucine as a nephrotoxin also is unclear.

Acromelalgia

Analysis of specimens from *Clitocybe amoenolens* demonstrate neuro-excitatory amino acids (acromelic acids), primarily acromelic acid A–E. These compounds are structurally similar to nonprotein pyrrolidine dicarboxylic acids including kainic acid and domoic acid as demonstrated in Figure 44.2. In rodent studies, acromelic acid A is the most potent glutamate agonist, but a direct causal link between acromelic acids and acromelalgia has not been established. Analysis of dried specimens

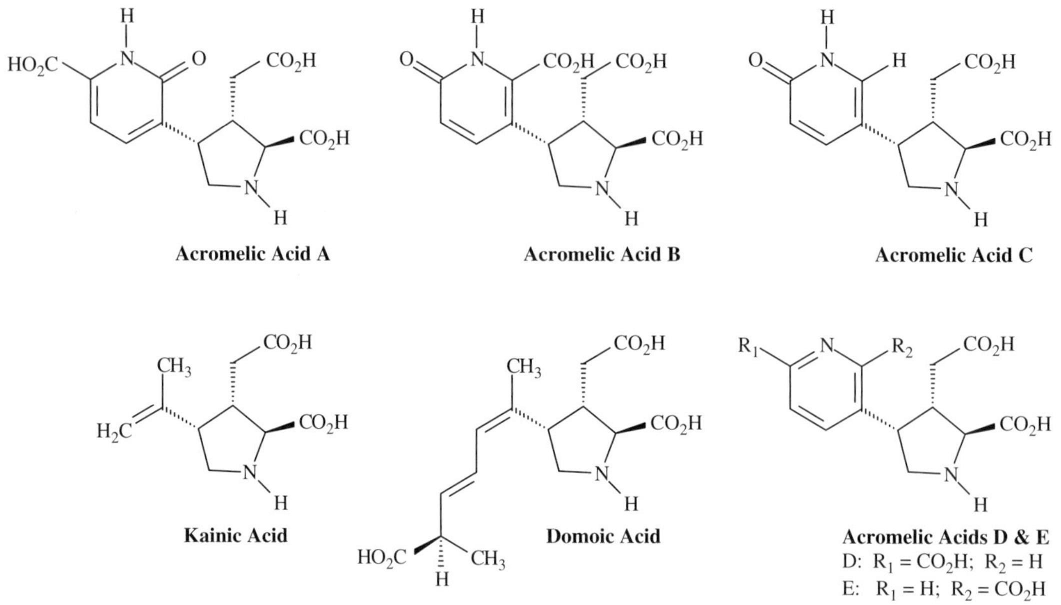

FIGURE 44.2. Chemical structures of acromelic acids A-E, kainic acid, and domoic acid.

of *C. amoenolens* and *C. acromelalga* detected acromelic acid A concentrations of 325 ng/mg and 283 ng/mg, respectively, as measured by liquid chromatography/mass spectrometry (LC/MS).[14] The limit of quantitation for this method was 10 ng/mg.

CLINICAL RESPONSE

Rare case reports associate anaphylaxis (angioedema, dyspnea) with the ingestion of the Japanese matsutake mushrooms [e.g., *Tricholoma matsutake* (S. Ito & S. Imai) Singer] by atopic individuals.[15]

Gastroenteritis

The ingestion of inadequately prepared *Paxillus involutus* mushrooms can produce the acute onset of gastroenteritis that is self-limited.[3] This toxic reaction should not be confused with the Paxillus syndrome, which occurs *only* in sensitized patients. This immune reaction commonly occurs in patients with a history of adverse reactions following the consumption of *Paxillus involutus* mushrooms. Initial symptoms include nausea, vomiting, diarrhea, and abdominal cramps followed by the development of the clinical features of hemolytic anemia and renal failure.

Rhabdomyolysis

Case reports have associated the ingestion of *Tricholoma equestre* (*T. flavovirens*) mushrooms with the delayed onset of muscle weakness and rhabdomyolysis.[16] These patients developed fatigue, muscle weakness, and myalgias 1–3 days after eating at least three consecutive meals of this mushroom.[8] The symptoms worsened over the next several days with the onset of facial erythema, nausea, diaphoresis, and dark urine. Three patients developed hyperthermia and died from multiple organ failure. Severe rhabdomyolysis (serum creatine phosphokinase >100,000 IU/L) and refractory heart failure preceded death in these patients. Some of these patients developed substantial elevation of serum hepatic aminotransferases, but there was no hepatorenal failure except in the patients with multiple organ failure *and* hyperthermia. The symptoms in survivors resolved without complications over several weeks.[17] A case report associated the development of coma, acute respiratory failure, convulsions, and generalized muscle weakness without significant elevation of serum creatine kinase in a pediatric patient following the ingestion of 300–400 g *Tricholoma equestre* (*T. flavovirens*) daily for 4 consecutive days.[16] Symptoms resolved over 12 days with supportive care including respiratory support that required intubation. These myotoxic reactions may be idiopathic because a case series did not detect rhabdomyolysis in adults ingesting similar doses of *Tricholoma equestre*. A study of 13 healthy volunteers did not demonstrate any significant elevations of serum creatine phosphokinase or serum hepatic aminotransferases following the ingestion of up to 1200 g *Tricholoma equestre* daily for 4 days.[18] Severe rhabdomyolysis with hyperkalemia, respiratory insufficiency, acute renal failure, and shock has been associated with the ingestion of *Russula subnigricans* Hongo in Japan.[19]

Renal Toxicity

Several case series associate the ingestion of mushrooms with the development of acute tubulointerstitial nephropathy and renal dysfunction 1–4 days after ingestion of mushrooms suspected to be *A. smithiana* Bas from the Pacific Northwest of the United States[20–22] and *Amanita proxima* Dumée from southern France.[23] The onset of gastrointestinal (GI) symptoms and renal dysfunction in these patients was more rapid than the renal failure associated with *Cortinarius orellanus*. Following consumption of these *Amanita* species, GI distress (nausea, vomiting, diarrhea, abdominal cramps, anorexia), malaise, diaphoresis, lightheadedness, and renal dysfunction developed within 12 hours of ingestion, in contrast to the delayed onset renal insufficiency associated with orellanine nephrotoxicity. Although the mushrooms reportedly were *Amanita smithiana* in the Pacific Northwest cases, no definitive identification of the actual mushrooms involved with these cases was made. Gastrointestinal symptoms typically began about 5–6 hours after ingestion and persisted for several days before renal failure developed. In a few case reports, the GI symptoms developed within 1–2 hours of the ingestion of the mushrooms, primarily following the ingestion of raw mushrooms.[20] There was moderate elevation of serum hepatic aminotransferases in some patients, but no significant hepatic dysfunction occurred and all patients recovered after at least several weeks with supportive care and hemodialysis as required.

Acromelalgia (Erythromelalgia)

A French case series of 7 patients associated the consumption of *Clitocybe amoenolens* Malencon with the development of distal paresthesias and intense, paroxysmal dysesthesias.[24] Symptoms developed >24 hours after ingestion of the mushrooms, particularly in the feet at night. Although there were no prominent skin lesions, a sensation of local heat, local erythema and edema occasionally occurred. Hepatorenal dysfunction and GI distress did not develop. Heat and pressure exacerbated the paresthesias, while immersion of the affected extremities in cold water provided some relief. Symptoms resolved in 1–4 months. Similar cases of erythromelalgia following the ingestion of *Clitocybe acromelalga* Ichimura have been reported in Japan since the end of the 19th century.

DIAGNOSTIC TESTING

Laboratory findings following the development of the Paxillus syndrome include evidence of hemolysis (increased serum haptoglobin, hemoglobinuria, anemia) and renal failure (elevated serum creatinine, electrolyte imbalance). Case reports indicate that specific IgG antibodies against membrane particles of *Paxillus involutus* are present in the serum of patients with the Paxillus syndrome as measured by serum hemagglutination tests.[25] Blood type potentially can alter the hemaglutination of human red blood cells by these IgG antibodies.[26] Patients, who ingest several meals of *Tricholoma equestre* (*T. falvovirens*) mushrooms, may develop rhabdomyolysis, but electrolytes and serum creatinine usually remain normal.[27] Electromyograms in these patients demonstrated isolated muscle damage with preservation of sensorimotor function. Some elevation of serum hepatic transaminases may occur in these patients, but hepatorenal failure usually does not occur unless the patient develops multiple organ failure.

TREATMENT

Treatment is supportive. Patients should be evaluated for hypotension secondary to anemia and renal dysfunction. Blood transfusions and hemodialysis may be necessary depending on the severity of the hemolytic anemia and renal dysfunction associated with the Paxillus syndrome. There are few data to guide management. A case report documented the reduction of free hemoglobin and immune complex levels by 60–75% following the use of plasma exchange with 3,000 mL albumin.[25] However, the patient required hemodialysis for the treatment of acute renal failure and the effectiveness of plasmapheresis in reducing renal damage or hemolysis remains undetermined. Patients ingesting multiple meals of *Tricholoma equestre* should be evaluated for rhabdomyolysis (serum creatine phosphokinase, urine myoglobin) as well as the development of multiple organ failure in symptomatic patients. Patients with rhabdomyolysis should receive generous fluid replacement to ensure adequate urine output (1–2 cc/kg/h) and alkalinization for the treatment of metabolic acidosis to prevent the excretion of highly acidic urine as needed.

Treatment of ingestions of mushrooms from *Amanita smithiana* is supportive including correction of fluid and electrolyte imbalance. These patients should be monitored for the development of renal failure within the first 2–4 days after ingestion. There are no antidotes. Renal function usually returns to baseline values after supportive care and hemodialysis as needed.

References

1. Bschor F, Mallach HJ. [Poisonings caused by Paxillus involutus, an edible mushroom]. Arch Toxikol 1963;20:82–95. [German]

2. Schmidt J, Hartmann W, Wurstlin A, Deicher H. [Acute kidney failure due to immunohemolytic anemia following consumption of the mushroom *Paxillus involutus*]. Dtsch Med Wochenschr 1971;96:1188–1191. [German]
3. Bresinsky A, Besl H. A Colour Atlas of Poisonous Fungi. A Handbook for Pharmacists, Doctors, and Biologists. London: Wolfe Publishing Ltd; 1990.
4. Edwards RL, Elsworthy GC, Kale N. Involutin, a diphenylcyclopenteneone from *Paxillus involutus* (Oeder ex Fries). J Chem Soc (C) 1967;405–407.
5. Nieminen L, Bjondahl K, Ojanen H, Ohenoja E. Short-term toxicity study of *Paxillus involutus* in the rat. Food Cosmet Toxicol 1977;15:445–446.
6. Falandysz J, Jedrusiak A, Lipka K, Kannan K, Kawano M, Gucia M, et al Mercury in wild mushrooms and underlying soil substrate from Koszalin, North-central Poland. Chemosphere 2004;54:461–466.
7. Habtemariam S. Cytotoxicity of extracts from the mushroom *Paxillus involutus*. Toxicon 1996;34:711–713.
8. Bedry R, Baudrimont I, Deffieux G, Creppy EE, Pomies JP, Ragnaud JM, et al. Wild-mushroom intoxication as a cause of rhabdomyolysis. N Engl J Med 2001;234:798–802.
9. Nieminen P, Kirsi M, Mustonen A-M. Suspected myotoxicity of edible wild mushrooms. Exp Biol Med 2006;231:221–228.
10. Takahashi A, Agatsuma T, Matsuda M, Ohta T, Nunozawa T, Endo T, Nozoe S. Russuphelin A, a new cytotoxic substance from the mushroom *Russula subnigricans* Hongo. Chem Pharm Bull 1992;40:3185–3188.
11. Takahashi A, Agatsuma T, Ohta T, Nunozawa T, Endo T, Nozoe S. Russuphelins B, C, D, E and F, new cytoxic substances from the mushroom *Russula subnigricans* Hongo. Chem Pharm Bull 1993;41:1726–1729.
12. Pelizzari V, Feifel E, Rohrmoser MM, Gstraunthaler G, Moser M. Partial purification and characterization of atoxic component of *Amanita smithiana*. Mycologia 1994;86:555–560.
13. Chilton WS, Ott J. Toxic metabolites of *Amanita pantherina, A. cothurnata, A. muscaria* and other *Amanita* species. Lloydia 1976;39:150–157.
14. Bessard J, Saviuc P, Chane-Yene Y, Monnet S, Bessard G. Mass spectrometric determination of acromelic acid A from a new poisonous mushroom: *Clitocybe amoenolens*. J Chromatogr A 2004;1055:99–107.
15. Ichikawa K, Ito R, Kobayashi Y, Aihara M, Osuna H, Aihara Y. A pediatric case of anaphylaxis caused by matsutake mushroom (*Tricholoma matsutake*) ingestion. Allergol Int 2006;55:85–88.
16. Chodorowski Z, Sein Anand J, Grass M. Acute poisoning with *Tricholoma equestre* of five-year old child. Przeglad Lekarski 2003;60:309–310.
17. Chodorowski Z, Waldman W, Sein Anand J. Acute poisoning with *Tricholoma equestre*. Przegl Lek 2002;59:386–387.
18. Chodorowski Z, Sein Anand J, Madalinski M, Rutkowski B, Cylkowska B, Rutkowski P, et al. Enzymatic examination of potential interaction between statins or fibrates and consumed *Tricholoma equestre*. Przegl Lek 2005;62:468–470.
19. Lee PT, Wu ML, Tsai WJ, Ger J, Deng JF, Chung HM. Rhabdomyolysis: an unusual feature with mushroom poisoning. Am J Kidney Dis 2001;38:E17.
20. Warden CR, Benjamin DR. Acute renal failure associated with suspected *Amanita smithiana* mushroom ingestions: a case series. Acad Emerg Med 1998;5:808–812.
21. Leathem AM, Purssell RA, Chan VR, Kroeger PD. Renal failure caused by mushroom poisoning. Clin Toxicol 1997;35:67–75.
22. Tulloss RE, Lindgren JE. *Amanita smithiana*—taxonomy, distribution, and poisonings. Mycotaxon 1992;45:373–387.
23. Leray H, Canaud B, Andary C, Klouche K, Beraud JJ, Mion C. [*Amanita proxima* poisoning: a new cause of acute renal insufficiency]. Nephrologie 1994;15:197–199. [French]
24. Saviuc PF, Danel VC, Moreau PA, Guez DR, Claustre AM, Carpentier PH, et al. Erythromelalgia and mushroom poisoning. J Toxicol Clin Toxicol 2001;39:403–407.
25. Winkelmann M, Stangel W, Schedel I, Grabensee B. Severe hemolysis caused by antibodies against the mushroom *Paxillus involutus* and its therapy by plasma exchange. Klin Wochenschr 1986;64:935–938.
26. Furukawa K, Ying R, Nakajima T, Matsuki T. Hemagglutinins in fungus extracts and their blood group specificity. Exp Clin Immunogenet 1995;12:223–231.
27. Saviuc P, Danel V. New syndromes in mushroom poisoning. Toxicol Rev 2006;25:199–209.

II Mycotoxins

Chapter 45

MYCOTOXINS

HISTORY

Historical literature documents the toxic effects of fungi since ancient times. In the 7th and 8th centuries BC, the festival of Robigalia honored the god Robigus, who protected the grain and trees during the spring when the crops were susceptible to attack by rust and mildew.[1] The Old Testament (Leviticus 14:37–45) contains passages concerning the presence of greenish or reddish streaks of mildew on the walls of houses that required removal. The first documented mycotoxicosis occurred in the Middle Ages during the epidemics of hallucinations, delirium, convulsions, and gangrene known as St. Anthony's Fire. By the 1850s, this disease was called ergotism, and the fungus, *Claviceps purpurea* var. purpurea (Fr.) Tul. was the recognized etiological agent.[2] In 1948, Freeman et al isolated the first trichothecene mycotoxin from the mold *Trichothecium roseum* (Pers.) Link.[3] The trichothecene mycotoxin, T-2 toxin, was a *suspected* etiological agent for numerous alimentary toxic aleukia (ATA)- related fatalities that occurred in Russia during the latter part of World War II.[4] Interest in mycotoxicoses increased after the isolation of a mycotoxin called aflatoxin from the mold *Aspergillus flavus* Link in Ugandan peanuts during the late 1950s.[5] Consumption of these contaminated peanuts by turkeys and ducks (Turkey X disease) caused hundreds of thousands of deaths in animals during the early 1960s. Subsequently, aflatoxin B_1 was identified as a potent natural carcinogen.[6,7] The World Health Organization cited three other classes of mycotoxins (ochratoxin, zearalenone, trichothecenes) as potential animal and human health hazards. In 1965, ochratoxin A was the first toxin isolated from *Aspergillus alutaceus* Berk. & M.A. Curtis.[8] In the last several decades, major outbreaks of human mycotoxicoses included hepatitis from the consumption of aflatoxin-contaminated maize in Kenya and India,[9] gastroenteritides from the consumption of ergot-contaminated pearl millet in India, a self-limited gastroenteritis from the ingestion of deoxynivalenol in bread made from moldy Indian wheat,[10] and a diarrheal disease associated with the consumption of fumonisin B_1-contaminated sorghum and maize.[11]

CLASSIFICATION AND IDENTIFICATION

The kingdom of fungi includes mushrooms, toadstools, puffballs, truffles, molds, mildew, smut, rusts, and yeast. Because fungi lack chlorophyll, these organisms must absorb food from outside their bodies by releasing digestive enzymes. Fungi account for about 25% of the biomass of the earth, and the decomposition of organic material by these ubiquitous organisms is an essential part of the energy cycle.[12] The taxonomical classification of fungi divides fungi into five main phyla: Ascomycota (e.g., yeasts, powdery mildews, the black and blue-green molds), Basidiomycota (e.g., most mushrooms, smut/rust in plants, bracket fungi, and puffballs), Chytridomycota (water molds), Glomeromycota (blastospore-producing fungi), and Zygomycota (e.g., black bread mold and other molds including *Mucor, Rhizopus*). Fungi may be unicellular or multicellular. Multicellular fungi are composed of microscopic, branched, interwoven filaments

Medical Toxicology of Natural Substances, by Donald G. Barceloux, MD

called hyphae that form a visible mass called the mycelium. Yeasts are unicellular fungi that do not produce a mycelium, and these organisms reproduce by budding rather than spores. Most yeast form round, pasty, or mucoid colonies rather than the fluffy colonies formed by the mycelium of multicellular fungi. Mold and mildew are not scientific terms because these terms are applied to visible fungal colonies that may belong to any one of the five fungal phyla. Mold commonly refers to visible fungal growth, and mildew frequently refers to fungal growth on household materials or plants. Fungi are ubiquitous both in indoor and outdoor air.

Properties

These filamentous organisms absorb nutrients by the digestion of food sources in the environment. Fungi are eukaryotic organisms that lack chlorophyll. The cell walls of these organisms primarily contain chitin (acetylglucosamine polymers) along with smaller concentrations of β-(1,3)-D-glucans, mucopolysaccharides, and polysaccharides. Fungi produce a variety of volatile organic compounds including alkanes, aldehydes, esters, ethers, alcohols, ketones, terpenes, terpene derivatives, and sulfur compounds (e.g., dimethyl disulfide).[13] These volatile compounds account for the musty smell associated with mold growth.

Mycotoxins are relatively nonvolatile, low-molecular-weight, secondary toxic metabolites of certain microscopic fungi (molds) that occur primarily on food as a result of fungal growth. At normal ambient temperatures, these complex molecules are usually nonvolatile.[14] These mycotoxins have no apparent physiologic function, but these compounds potentially increase food sources for the mycotoxin-producing fungi by decreasing competition for these food sources. The ingestion of these toxins by higher vertebrates produces toxic reactions known as mycotoxicoses. Mass outbreaks of mycotoxicoses are much more common in domestic animals than in humans, particularly during periods of food scarcity. Although allergic reactions to airborne molds are common, occurrences of mycotoxicoses following inhalation are not well documented. The appearance of mycotoxins result primarily from the growth of three genera (*Aspergillus, Fusarium, Penicillium*), and these toxic metabolites usually appear in the human food chain after replication of these fungi ceases.[15] Although species of *Cladosporium* contain biologically active metabolites (cladosporin, isocladosporin, cladospolides A and B, calphostins),[16] these compounds have not been associated with human or animal toxicity.[17] In contrast to mycotoxicoses, mycoses (e.g., blastomycosis, histoplasmosis, coccidioidomycosis) are infectious diseases, which require fungal replication in the host prior to the development of the disease. Mycotoxicoses are more common in developing countries than in developed countries because of the occurrence of food shortages, improper cultivation and storage methods, and poor cooking techniques. Well-known mycotoxins with medicinal benefits include the antimicrobial agents, penicillin (*P. chrysogenum* var. *chrysogenum* Thom), cephalosporins, and griseofulvin as well as immunosuppressive drugs (cyclosporin) and cholesterol-lowering medications (lovastatin).

Occurrence

Fungi are ubiquitous because of their ability to grow on almost any substrate. There are an estimated 100,000 species of fungi, which represents about one-half of the total species on the earth. Mycotoxins are nonvolatile, secondary metabolites of fungi that are not involved in any physiologic process within the fungus, and the mycotoxins occur in all parts of the fungal colony. Mycotoxins do not penetrate intact human skin, and inhalation of these molecules requires the generation of an aerosol of substrate, fungal elements, or spores. Mycotoxins do not represent a single class of chemical compounds because of their diverse chemical structures (e.g., polyketides, terpenes, indoles). These low-molecular-weight compounds range from single heterocyclic ring with a molecular weight of 50 Da to irregularly arranged 6–8 member rings with molecular weights exceeding 500 Da. Most mycotoxin-producing fungi belong to the class Ascomycetes in the phylum Ascomycota.

Food

Due to the widespread occurrence of fungi, the potential for ingesting mycotoxin-contaminated food is considerable when growing conditions are favorable. Most clinical syndromes associated with the mycotoxins involve the ingestion of contaminated food.[18] The most common fungal genera associated with mycotoxin production in food are *Aspergillus, Fusarium*, and *Penicillium*.[19] In a study of 84 cured fish samples collected in Nigeria, the most common fungi species were *Aspergillus niger* var. *niger* Tiegh. and *Aspergillus flavus* followed by *A. chevalieri* Thom & Church, *A. fumigatus, Penicillium* spp. including *P. citrinum* Thom, *Cladosporium herbarum* (pers.) Link, and *Basipetospora* species.[20,21] *Fusarium* species are destructive pathogens in cereal crops that produce mycotoxins before or shortly after harvest. Mycotoxin-producing species of *Aspergillus* and *Penicillium* may be pathogens or commensals, but mycotoxin production usually occurs during storage rather than crop production. Table 45.1 lists known mycotoxins potentially present

TABLE 45.1. Potential Mycotoxins in Human Foods and Animal Feed

Commodity	Situation	Mycotoxins
Cereals	Preharvest fungal infection	Deoxynivalenol, T-2 toxin, nivalenol, zearalenone, alternariol, alternariol monomethyl ether, tenuazonic acid fumonisins
Maize, peanuts	Preharvest fungal infection	Aflatoxins
Maize, sorghum	Preharvest fungal infection	Fumonisins
Stored cereals, nuts, spices	Damp, inadequate storage conditions	Aflatoxins, ochratoxin
Fruit juice	Fruit mold	Patulin
Dairy products	Animal consumption of mold-contaminated feeds	Aflatoxin M_1, cyclopiazonic acid, ochratoxin, compactin, cyclopaldic acid
Meat and eggs	Animal consumption of mold-contaminated feeds	Patulin, citrinin, ochratoxin, cyclopiazonic acid, cyclopaldic acid, citromycetin, roquefortine, fumonisins
Oilseeds	Preharvest fungal infection	Tenuazonic acid, alternariol

Source: Adapted from Ref 15.

TABLE 45.2. Mycotoxins Associated with Mold Isolated from Indoor Air Samples

Species	Toxins
Aspergillus flavus Link	Aflatoxins
Aspergillus fumigatus Fresen.	Fumiclavines, fumigatoxin, fumigillin, fumitrems, gliotoxin
Aspergillus versicolor (Vuill.) Tirab.	Sterigmatocystin, versicolorin, aspercolorin, averufin, cyclopiazonic acid
Penicillium aurantiogriseum Dierckx	Penicillic acid
Penicillium aurantiogriseum var. *viridicatum* (Westling) Frisvad & Filt.	Brevianamide A, citrinin, mycophenolic, acid, penicillic acid, viomellein, xanthomegnin
Penicillium brevicompactum Dierckx	Brevianamide *A,* mycophenolic acid
Penicillium chrysogenum var. chrysogenum Thom	Citrinin, penicillic acid, PR-toxin, roquefortine C
Penicillium citrinum Thom	Kojic acid, citrinin, citreoviridin
Penicillium corylophilum Dierckx	Gliotoxin
Penicillium dierckxii Biourge	Citreoviridin, citrinin
Penicillium expansum Link	Patulin, citrinin
Penicillium spinulosum Thom	Spinulosin
Stachybotrys chartarum (Ehrenb.) S. Hughes	Roridin E, satratoxin H, sporidesmin G, trichoverrins, verrucarol

Source: Adapted from Ref 24.

in foods and animal feed, including organically grown commodities.[22]

Table 45.2 lists the toxins associated with mycotoxin-producing fungi. *Trichoderma* species are infrequent producers of mycotoxins.[23]

Cultivation techniques, storage practices, and favorable weather conditions (e.g., >15% moisture content, 90–95% relative humidity, temperature range −8 °C–35 °C/18 °F–95 °F) at both harvest and storage time strongly influence the production of mycotoxins. Other factors affecting mycotoxin concentration include microbial genetics, minimum climatic conditions (moisture content, temperature), substrate, variations within species, the presence of other fungal species,[25] and the geographic distribution of mycotoxin-producing fungi.[5,26] The presence of mycoviruses also may affect the formation of mycotoxin by either inducing or reducing mycotoxin production through mechanical damage and alteration of mycotoxin metabolism.[27] Mycotoxins contaminate commercial and homemade food sources, particularly in Africa. In a survey of Tunisian patients who displayed symptoms compatible with mycotoxicoses, aflatoxins were detected in homemade and commercial food products; ochratoxin A and citrinin were found preferentially in homemade food sources.[28] Mycotoxin-induced diseases usually result from the ingestion of contaminated food sources.

Ambient Air

In well-constructed homes without water contamination, the indoor concentration of fungi generally reflects the outdoor concentration of fungi, typically in the range of 40–80% of the outdoor fungal concentration.[29] Although the presence of mold, moldy odors, and water damage usually indicates the presence of high concentrations of indoor fungi, epidemiological studies have not demonstrated a direct correlation between the subjective evaluation of respiratory complaints (e.g., questionnaires) and objective measures of fungal growth, in part because of systemic reporting bias.[30,31] In a prospective study of respiratory allergens and neonatal asthma, the characteristics of the home were not statistically correlated to the viable fungal count (propagules).[32] The questionnaire included design and age of house, ventilation (heating, air conditioning, humidifiers), flooring, observation of moisture problems (house dampness, flooding, mold growth), household occupants, pets, pests, cleaning techniques, and household income/educational level. There are limited data on the clinical relevance of the results from these sampling techniques to long-term health effects. Molds are indoor allergens, but other indoor and outdoor allergens are more important causes of hypersensitivity reactions than molds. In a study of the prevalence of allergic rhinitis and asthma in 395 members of a rural Israeli community, the most common genera of mold were *Aspergillus*, *Penicillium*, *Alternaria*, and *Cladosporium*.[33] Most of the subjects with allergies to molds in this study did not have significant respiratory complaints despite being allergic to molds present in the ambient air of their homes, and the abundance of molds did not correlate to the severity of respiratory symptoms. In general, the production of significant amounts of mycotoxins requires water activity (a_w) levels at the surface of the construction material exceeding 0.95. The mycotoxin content of spores of common indoor fungi (e.g., *Aspergillus, Penicillium*) is usually relatively low.[34] The exception is *Aspergillus versicolor* (Vuill.) Tirab., which can contain up to 1% sterigmatocystins wet weight, when the water activity is close to one. Bulk sampling helps confirm the potential severity of mold exposure, but chemical analyses are necessary to verify the presence of mycotoxins in mycotoxin-forming species.[35] Furthermore, exposure to indoor mold occurs primarily via inhalation of spores, and the presence of fungal species in bulk samples does not correlate to airborne concentrations of fungal spores.

Identification of Fungal Genera and Species

Molds are natural constituents of indoor and outdoor air, and most commercial and residential structures contain low concentrations of fungi on household surfaces and in ambient air. Mold-free environments do not occur naturally without interventions to produce a sterile environment. For example, there are about 20,000–25,000 species in the class Basidiomycetes (mushroom, puffballs, bracket fungi), and separation of the spores from individual species within the class Basidiomycetes is difficult because of the structural similarity of the spores. Species from the classes Ascomycetes and Basidiomycetes produce prolific numbers of spores that account for a majority of spores in the air.[36] Ascospores and basidiospores are associated with very humid conditions (e.g., fog, warm rain). After rain, the ascospore count rapidly decreases, but the basidiospore count usually remains elevated for 2–5 days.[37] During the 3–4 days after a rain, *Cladosporium* and, to a lesser extent, *Alternaria* spore counts also increase. In addition, *Alternaria* counts increase during hot, dry winds from the desert (e.g., Santa Ana winds of Southern California).

The concentrations of spores in ambient air depend primarily on outdoor mold concentrations and indoor humidity as well as time of day, season, climate, building materials, indoor temperatures, ventilation, indoor materials (furniture, carpeting), and the presence of plants. Indoor spore counts in commercial, urban buildings are usually less than residential buildings as a result of differences in cleaning schedules, natural ventilation, and the presence of pets.[38] Air sampling of home environments demonstrates the presence of a wide variety of filamentous fungal genera and species as well as gram-positive bacteria (*Micrococcus* spp., *Staphylococcus* spp., *Streptococcus* spp.), gram-negative bacteria (*Pseudomonas* spp., *Aeromonas* spp.), endospore-forming bacilli (*Bacillus* spp.), and yeasts.[39] In a prospective study of over 1000 houses in the Northeastern United States, air samples of fungal airborne propagules demonstrated 44 fungal genera and 116 fungal species.[32]

The specific concentration and type of fungi present in air samples depends on a variety of factors including the season, time of day, weather conditions (humidity, temperature, wind, snow), location, local flora and fauna, adjacent landscape near buildings, type of ventilation, wall-to-wall carpeting, the presence of pets or water leaks, and the analytical method.[40,41] The most common fungal genera in indoor air samples include *Alternaria, Aspergillus, Aureobasidium, Cladosporium*, and *Penicillium* depending primarily on humidity, season, and temperature.[42–44] Other factors that promote the growth of indoor fungi include wet plant containers (e.g., wicker), pets, absence of cleaning activities (wiping, dusting, vacuuming) or ventilation, and wet organic material (vegetable peels, jute-backed carpeting). The drying of wet organic materials within 3 days usually prevents the development of excessive mold growth, but the persistence of wet materials beyond this period encourages mold growth, particularly when the moist conditions exist longer than 2 weeks.[37] Some species of *Penicillium* and *Aspergillus* grow and reproduce effectively indoors without an outdoor source or an indoor water source. Most strains of *Penicillium* and *Aspergillus* species present in indoor air samples produce either no or low concentrations (i.e., few ng/g material) of mycotoxins.[34,45] Individual species of a specific fungal genus may vary substantially in samples from indoor and outdoor air.[46] *Cladosporium* was the most frequent genus of fungi in indoor ambient air samples from coastal sections of California, whereas *Alternaria* was more common in samples from the dry sections of interior California.[47] In another study of spore counts in coastal Southern California homes, the common spores in descending order of frequency were *Cladosporium*, basidiospores, *Alternaria*, smut, and ascospores.[37] The most common outdoor fungal species are *Cladosporium* and *Alternaria* as well as *Aspergillus*, and *Penicillium*. *Cladosporium* and *Alternaria* species grow on the surfaces of plant leaves, whereas *Penicillium* and *Aspergillus* species easily grow on topsoil and decaying litter. The concentration of total fungi colonies in indoor and outdoor air samples is usually lowest in winter, increases in spring, maximizes in summer, and declines in the fall.[42] In general, total fungal counts are highest indoors during the fall and winter, whereas the highest numbers of spores outdoors typically occur in the summer. The concentrations of *Cladosporium* and *Alternaria* are usually higher in samples from outdoor air than indoor air, whereas the concentrations of *Aspergillus* and *Penicillium* are usually higher in indoor air samples than in outdoor air samples.[30] The concentration of an individual fungal genus can vary substantially with seasons. In one location, the outdoor concentration of *Cladosporium* ranged from 26 CFU/m^3 in the winter to >11,000 CFU/m^3 in the summer.[48] Within a given locality the variation in the indoor fungi concentrations is greater in samples from the same house during different time periods than in air samples from nearby houses.[49,50]

Fungal Spore Concentrations

Current cross-sectional air sampling techniques from indoor and outdoor ambient air samples include culturable fungal spores (i.e., colony-forming units/m^3 or CFUs) and total (nonculturable) fungal propagule counts (nonviable and viable spores). Because outdoor air is the dominate source of mold, indoor air samples of fungi are qualitatively similar to outdoor air samples with the ratio of indoor to outdoor concentrations of specific fungal genera typically being <0.5–1 in air samples from buildings without heavy fungal growth.[51] Outdoor spores infiltrate buildings through openings (e.g., air intakes, windows, cracks). The concentrations of fungi in indoor and outdoor air samples typically are similar, except when outdoor air samples contain few fungi during periods of snow. Comparing indoor and outdoor air samples of fungi requires consideration of many factors including sampling techniques, weather, season, geographic region, and timing of the sample. Although molds grow during the whole year, cool temperatures during winter decrease total spore counts substantially. The highest mold counts in outdoor air samples usually occur during the summer and autumn when substantial accumulation of dead organic debris occurs.[30] Most indoor environments contain some fungi, and the assessment of indoor fungal concentrations should include comparable numbers of indoor and proximate outdoor air samples when comparing indoor and outdoor fungal concentrations.[29] Ideally, air sampling should include one or more control houses without complaints in the nearby community. Accurate comparisons of indoor and outdoor fungal samples require identification of fungal species rather than fungal genera. The presence of high concentration of fungi species in indoor air samples indicates the presence of excessive indoor moisture, but the lack of well-documented dose-response data for the development of clinical symptoms limits conclusions regarding the concentrations of fungi that cause adverse health effects. Spores remain airborne for various lengths of time, depending on the settling velocity, but these spores eventually settle on interior surfaces. Reservoirs for spores include carpeting, ducts, and filters for heating and air conditioning systems. Mechanical agitation of local surfaces (e.g., scraping, touching, vibration, air flow) and human activity (walking, sitting, moving household items) generate significant airborne release of spores to the ambient air.[52] In particular, the drying of spores (e.g., *Stachybotrys*) that normally reside in a sticky medium substantially increases the airborne count of these spores.

Colony-Forming Units. The culturable spore sampling technique (e.g., Anderson sampler) involves impaction of culturable organisms onto nutrient agar or impingement of the spores on transport media for later transfer to culture media as measured in CFU/m^3. The primary advantage of the culture techniques is the ability to identify *potentially* toxic fungal species. However, the use of different culture media and variation in growth rates among fungal species may result in the underestimation of the fungi based on viable counts alone, particularly when the fungal concentrations are low and the sample size is small. In a study of >12,000 fungal air samples collected during indoor air quality investigations in the United States between 1995 and 1998, the median indoor and outdoor fungal counts were 82 CFU/m^3 (95th percentile = 1,200 CFU/m^3) and 540 CFU/m^3 (95th percentile = 3,200 CFU/m^3), respectively.[53] A review of air sampling data from 149 commercial, urban buildings, where there were no health complaints, demonstrated that viable fungal counts in indoor and outdoor ambient air averaged 233 CFU/m^3 (range: 17–1,212 CFU/m^3) and 983 CFU/m^3 (range: 92–2,061 CFU/m^3), respectively.[38] For 820 residential structures the indoor and outdoor ambient air average was 1,252 CFU/m^3 (range: 17–9,100 CFU/m^3) and 1,524 CFU/m^3 (range: 20–11,883 CFU/m^3), respectively. A yearlong study of ambient air in two homes in Yokohama, Japan demonstrated substantial variation in the number of CFUs depending on the media used and the time of year.[54] The arithmetic mean of airborne fungi in indoor air samples on dichloran 18% glycerol agar and potato dextrose agar was 252 ± 383 CFU/m^3 (range: up to 3,750 CFU/m^3) and 57 ± 136 CFU/m^3 (range: up to 1,875 CFU/m^3), respectively. The total culturable fungal count in outdoor air samples often exceeds 1,000 CFU/m^3, and the total fungal count may reach 10,000 CFU/m^3, particularly during summer months.[55] The clinical significance of average viable fungal CFU concentrations in indoor air samples above 1,000/m^3 should be interpreted cautiously because there are limited data correlating these measurements to specific diseases, and there is no definite dose-response relationship between specific ambient mold concentrations and human health effects.[29] Viable fungal spores in ambient air samples from a composting plant ranged from 10^3–10^4 CFU/m^3 in the storage area to 10^4–10^7 CFU/m^3 in the loading area.[56]

Generally, outdoor air concentrations of fungi are very low when snow is present on the ground.[57] Table 45.3 lists the colony counts of the common fungal general in indoor and outdoor air samples from a large US study.

The concentration of viable fungal genera in homes without excessive fungal growth is usually <1,000–2,000 CFU/m^3.[58,59]

TABLE 45.3. Total Colony-Forming Units for Common Fungal Genera in Positive Air Samples from a Large, Nonrandom US Study

Fungal Genera	Positive Samples (%)	Median Count (CFU/m^3)	95% Confidence Interval (CFU/m^3)
Cladosporium			
Indoor	86	40	12–480
Outdoor	92	200	18–1,849
Penicillium			
Indoor	80	30	12–570
Outdoor	77	50	12–377
Aspergillus			
Indoor	62	20	12–373
Outdoor	49	20	12–170

Source: Adapted from Ref 32.

A cross-sectional study of indoor air samples from buildings with occupant complaints suggested that total fungal counts exceeding 2,000 CFU/m^3 are associated with respiratory irritation (cough, rhinitis, sore throat), but health complaints did not correlate to air concentrations of fungi.[60] Some studies of workers associated the presence of *Aspergillus*[61] and *Penicillium*[62] with respiratory complaints, but a direct causal relationship was not established. In these two studies, the suspected agent was different despite the presence of both fungal genera in ambient air samples from the affected buildings. The quantitative value of CFUs must be interpreted cautiously because of the high variability of colony forming units over very short sampling periods and the small percentage of culturable spores compared with the total spore count.[63]

TOTAL SPORE CONCENTRATIONS. The collection of noncultured, fungal propagules on a filter or sticky surface (e.g., Air-O-Cell cassette; Zefron, Inc., St. Petersburg, FL) allows the estimation of the total fungal mass (i.e., culturable and nonculturable) as measured in spores/m^3. Some fungi demonstrate distinctive spores, but the spores of many indoor fungal species (e.g., *Penicillium, Aspergillus*) are difficult to distinguish and characterizing fungi genera via gross microscopic features is unreliable. Consequently, total spore counts do not separate *potentially* toxic fungal species from nonmycotoxin-forming species because of the difficulty identifying the species of some spores. Disturbances of surface growth (i.e., wiping area of mold), dust control measures, pets, humidification, carpet type, and vacuum cleaning of carpets can cause large temporal increase in ambient air counts of spores. Comparison of indoor and outdoor spore samples must consider the number of samples and the season because of the variation of outdoor spore counts with vegetation, season, and weather. In particular, *Cladosporium* spore concentrations in outdoor air samples vary as much as 100-fold between summer and winter.[64]

A review of air sampling data from 47 commercial, urban buildings, where there were no health complaints, demonstrated total spore counts in indoor samples averaged 768 spore/m^3 (range: 610–1,040 spores/m^3), while total spore counts in outdoor samples ranges from 400–80,000 spores/m^3.[38] For 85 residential structures the indoor total spore count averaged 913 spores/m^3 (range: 68–2,307 spores/m^3), while the total spore count in outdoor samples ranges from 400–80,000 spores/m^3. The total fungal spore concentration in ambient air is substantially lower in moldy homes compared with some occupational environments (e.g., sawmills, composting facilities). The clinical significance of total spore concentrations in indoor air samples above 1,000/m^3 should be interpreted cautiously because there are limited data correlating these measurements to specific diseases, and there is no definite dose-response relationship between specific ambient mold concentrations and human health effects.[29,65] In a study of 10 healthy sawmill workers, personal sampling methods demonstrated a mean total fungal spore count of 668,400 spores/m^3 (range: 215,000–1,460,000 spores/m^3).[66] Although some of these workers reported nasal and lower airway symptoms, there was no clear dose-response relationship between spore counts and symptoms. Comparison of proinflammatory mediators in nasal lavage fluid of these workers during exposure (work) and nonexposed (vacation) conditions demonstrated no significant difference. In a study of air concentrations of farms with and without adverse health effects (i.e., allergic alveolitis, organic dust toxic syndrome) the average total spore count on farms without complaints was 1.2×10^8 spores/m^3 (120 million/m^3).[67] The presence of numerous spores with mycelia indicates active mold growth.[68]

WATER DAMAGE AND MOLD GROWTH

Although species of *Penicillium* and *Aspergillus* are common molds present in most homes, most of these species do not produce mycotoxins. Consequently, the presence of fungal species from these genera does not necessarily imply the presence of mycotoxins, and the proper interpretation of air samples requires the identification of fungal species.[57] Furthermore, the presence of a mycotoxin-producing species does not necessarily imply the presence of mycotoxins because the conditions for growth of the mold are often different from the conditions necessary for the formation of mycotoxins. Water damage increases the concentration and growth of fungi depending on the individual growth requirements of the particular fungal species. The growth of fungi on building materials always occurs as a result of high water activity (a_w) at the surface of the material with the minimal a_w being approximately 0.67–0.75 depending on temperature, time, and material.[69] Fungal genera are divided into the following classification based on water activity: <0.8 (primary colonizers-*Wallemia, Penicillium, Aspergillus*, and *Eurotium*; 0.8–0.9 (secondary colonizers—*Cladosporium*), and >0.9 (tertiary colonizers—*Stachybotrys, Trichoderma, Fusarium, Chaetomium, Ulocladium, Alternaria*).[70] *Ulocladium* species usually does not produce mycotoxins.[34]

Table 45.4 lists mycotoxins produced from fungal species present in water-damaged buildings. Although these fungal species are capable of producing mycotoxins, the amount of mycotoxins present in these fungal

TABLE 45.4. Potential Mycotoxin-producing Fungi in Water-Damaged Buildings

Genus	Species	Mycotoxin
Alternaria	spp.	Altenuene, alternariol, iso-altenuene, alternariol mono ethyl ether, tenuazonic acid, altertoxins I & II
Alternaria	*tenuissima*	Alternariol,[a] alternariol mono ethyl ether,[a] tenuazonic acid, altertoxins I
Alternaria	*infectoria*	No known toxic metabolites
Alternaria	*alternata*	Alternariol, alternariol mono ethyl ether, tenuazonic acid, altertoxins I
Aspergillus	*fumigatus*	Fumitoxin, gliotoxin,[a] tryptoquivalins, verrucologen
Aspergillus	*niger*	Naphthopyrones, malformins A, B, C, nigragillin,[a] orlandin,[a] ochratoxin A
Aspergillus	*ochraceus*	Ochratoxin A, penicillic acid, xanthomegnin, viomellein, vioxanthin
Aspergillus	*ustus*	Austamide, austdiol, austocystins A & B, kotanin[a]
Aspergillus	*versicolor*	Sterigmatocystin,[a] 5-methoxysterigmatocystin,[a] versicolorin
Chaetomium	*globosum*	Chaetoglobosins A, B and C,[a] chetomin
Paecilomyces	*variotii*	Patulin, viriditoxin
Penicillium	*aurantiogriseum*	Auranthine, nephrotoxic glycopeptides, penicillic acid, verrucosidin
Penicillium	*brevicompactum*	Mycophenolic acid[a]
Penicillium	*chrysogenum*	Roquefortine C, meleagrin,[a] chrysogin,[a] ochratoxin
Penicillium	*expansum*	Chaetoglobosins A and C, communesins A and B, roquefortine C, citrinin, patulin[a]
Penicillium	*polonicum*	3-methoxyviridicatin,[a] nephrotoxic glycopeptides, verrucosidin,[a] normethylverrucosidin, penicillic acid, verrucofortine[a]
Stachybotrys	*chartarum*	Macrocyclic trichothecenes (satratoxins H,[a] G, F, iso-F, roridin L-2,[a] roridin E,[a] verrucarins J and B), esters of verrucarol and trichodermol, phenylspirodrimanes
Trichoderma	*harzianum*	Alamethicin, emodin, suzukacillin, trichodermin
Wallemia	*sebi*	Walleminol A and B

Source: Adapted from Ref 69.
[a]Mycotoxins produced by fungal species grown on construction material. Although variable quantities of mycotoxins are produced by the fungal species above grown in culture, there are a limited number of mycotoxins produced by the growth of these fungal species on building materials.

species is highly variable depending on environmental conditions and the strain involved. Many strains of mycotoxin-producing species do not form mycotoxins. Most studies report the presence of the mycotoxin-producing fungi in samples from building material rather than from ambient air. The actual health risk to occupants from these water-damaged buildings is unclear because most known mycotoxins are nonvolatile,[71] and the development of mycotoxin-related disease would require the inhalation of an extremely large numbers of *airborne* mycotoxin-containing spores. The potential of mycotoxins to produce clinically significant diseases following inhalation depends on a variety of poorly defined factors including the volatility of the individual mycotoxins, the type of building materials,[72] the quantity of spores released into the air, and dose-response relationships for clinical effects. There are limited human data on the chronic health effects of exposure to low concentrations of mycotoxins in ambient air, and there is no definite clinical evidence that common concentrations of fungi in indoor air produce toxic effects.[73,74] Determination of a causal link between mycotoxins and respiratory disease requires the development of a sensitive assay for the specific mycotoxin; consequently, the evaluation of measurements of viable and nonviable spores in air samples must be interpreted cautiously. Studies of workers exposed to moldy, agricultural environments indicate that the inhalation of mycotoxins (e.g., ochratoxins) from ambient air is relatively small compared with dietary exposure to mycotoxins.[75]

Most fungal cell walls contain glucans, which are glucose polymers that are chemically bound to chitin. Common fungal genera that contain glucans [e.g., (1→3)-β-D-glucans] in the cell walls include *Alternaria*, *Aspergillus*, *Cladosporium*, and *Penicillium*. Although the exact role of these chemicals in human disease is unclear, the potent inflammatory properties of (1→3)-β-D-glucan suggest that the presence of this chemical in ambient indoor air may contribute to respiratory irritation and increased bronchial reactivity in sensitized individuals.[76,77] Normal background concentrations of (1→3)-β-D-glucan in ambient indoor air samples typically are below 5 ng/m^3 with values above this concentration associated with previous mold growth.[78] Houses with excessive mold growth may contain (1 → 3)-β-D-glucan concentration up to 100 ng/m^3.[79] However, the relationship between ambient concentrations of (1 → 3)-β-D-glucan and airway inflammation is complex. In a study of Swedish row houses, the baseline FEV_1 and the prevalence of respiratory symptoms did not correlate to the airborne (1 → 3)-β-D-glucan concentration either in atopic or normal individuals.[80]

TARGET ORGANS

Exposures to fungi are ubiquitous, and most healthy humans do not develop illness as a result of these exposures. The mechanisms of disease following these exposures include allergy, infection, and toxicity. Diseases associated with exposure to fungi include the following: 1) IgE-mediated hypersensitivity reactions (e.g., allergic rhinitis, allergic fungal rhinosinusitis, conjunctivitis, asthma, allergic bronchopulmonary aspergillosis), 2) hypersensitivity pneumonitis (non-IgE activation of Th1 lymphocytes), 3) infectious diseases (dermatophytes, *Blastomyces*, *Cryptococcus*, *Histoplasma*, *Coccidioides*, *Candida*), and 4) mycotoxicoses (organic dust toxic syndromes, aflatoxicosis, ochratoxicosis, ergotism).[81] Of the four types of reactions, Ig-E mediated hypersensitivity reactions are the most common. Although allergy to molds is not uncommon, mold exposure is a relatively minor cause of allergic reactions because most of these individuals demonstrate allergic responses to multiple allergens (e.g., dust mites, animal dander, house dust, pollens).[82] Hypersensitivity pneumonitis (IgG-mediated immune-complex response to pet birds, contaminated humidifiers, compost, moldy tobacco, cheese, moldy hay, moldy grapes, moldy dust) and IgE-mediated response to fungal growth in the host (allergic bronchopulmonary aspergillosis, allergic fungal sinusitis[83]) are rare forms of mold-related allergic diseases. A few outdoor fungi (*Blastomyces, Coccidioides, Cryptococcus, Histoplasma*) potentially cause serious systemic diseases as a result of the invasion of immunocompetent hosts. Other fungi or dermatophytes (*Tinea* spp., *Candida albicans*) may produce superficial infections, but most fungi in general do not produce serious systemic illness except in seriously immunocompromised patients. Organic dust toxic syndromes are a heterogeneous group of lower respiratory illnesses manifest by fever, chills, malaise, headache, myalgias, nonproductive cough, and dyspnea. Pulmonary function tests and chest radiographs are usually normal, and symptoms typically resolve within 48 hours of exposure. The high airborne concentrations of spores causing the organic dust toxic syndrome are unlikely to occur in home, school, or office environments.[67]

Most data linking mycotoxins with human disease result from extrapolation of animal data and limited human epidemiological data. Mycotoxins potentially may produce toxicity in numerous organ systems, particularly after the ingestion of mycotoxin-contaminated food stuffs. However, causal links between mycotoxins and human illness are limited because of the lack of direct methods to assess mycotoxins in indoor air, and the lack of validated clinical biomarkers for mycotoxin-induced illness. Additionally, there is no unequivocal evidence that the inhalation of mycotoxins produces illness in human inhabitants of dwellings that contain common molds. Modeling of airborne exposure to mycotoxins indicates that the delivery of mycotoxins via inhalation of mold spores is very inefficient, and unrealistic high airborne spore levels over extended periods of time are necessary to elicit a response.[84] Consequently, the associations between nonspecific human complaints (headache, nausea, vomiting, diarrhea, memory loss, weakness, moodiness, cough, rhinorrhea, cough, epistaxis) and mycotoxins are speculative.[81]

Liver

The ingestion of aflatoxins is a potential cause of acute hepatocellular disease and hepatocarcinoma in humans in developing countries. There are inadequate human data to link aflatoxins with illness in developed countries. Other animal hepatotoxins associated with mycotoxins include sporidesmin,[85] rubratoxin,[86] and ochratoxin.[87] Sporidesmin is a secondary metabolite of the saprophytic fungus, *Pithomyces chartarum* (Berk. & M.A. Curtis) M.B. Ellis. This fungus grows on dead pasture litter, particularly perennial rye grass.

Kidney

Both citrinin and ochratoxin A are nephrotoxic in pigs. High dietary intake of ochratoxin A occurs in geographic areas where Balkan nephropathy is endemic.[88] However, the contribution of ochratoxin A to the development of this disease in humans remains undefined.

Hematological and Immune System

Pancytopenia occurs in alimentary toxic ataxia together with hemorrhage and ulceration of the alimentary tract mucosa. A similar disease known as stachybotryotoxicosis occurs in horses, which consume moldy hay. Aflatoxin B_1 is a recognized human carcinogen, although the exact role of this mycotoxin in the development of cancer in humans remains undefined. The IARC lists fumonisin and ochratoxin A as possible carcinogens, but there are inadequate data to link these mycotoxins with cancer in humans.

Onyalai is a common cause of thrombocytopenic purpura in Africa. The consumption of mold-contaminated millet may contribute to the development of this disease. Experimental studies suggest that the ingestion of a mycotoxin from *Leptosphaeria sacchari* Breda de Haan [*Phoma sorghina* (Sacc.) Boerema, Dorenb. & Kesteren] may predispose these patients to onyalai, but there are inadequate epidemiological data for a causal connection.[89] Mycophenolic acid is a toxic metabolite formed by *Penicillium brevicompactum* Dierckx and several other species of the genus *Penicillium*. This mycotoxin causes a significant reduction of the intracellular guanine nucleotide pools by reversibility inhibiting inosine monophosphate dehydrogenase, which is the enzyme that catalyzes the first step in the synthesis of guanosine monophosphate.[90] This action causes immunosuppression by inhibiting T and B cell lymphocyte proliferation and function. The prodrug mycophenolate mofetil, which is the morpholinoethyl ester of mycophenolic acid, is approved by the US Food and Drug Administration (FDA) for the prevention of rejection in patients with allogeneic renal transplants.[91] An intentional overdose of 10 g mycophenolate produced only mild gastrointestinal distress with no hemodynamic, hepatic, renal, or hematologic toxicity.[92]

Cardiac and Nervous Systems

Citreoviridin is a mycotoxin isolated from a penicillin mold (*Penicillium citreonigrum* Dierckx) that grows on yellow rice during storage after harvest, particularly in the colder regions of Japan.[93] The administration of this mycotoxin to animals causes paralysis, convulsions, and cardiac failure. A similar human disease known as cardiac beriberi occurred during the 17th and 18th centuries in Japan and other Asian countries. Disappearance of this entity coincided with improved farming practices and the implementation of government programs to inspect rice. Sporadic outbreaks of cardiac beriberi (yellow rice disease, Shoshin-kakke disease) occurred after the consumption of decomposing rice contaminated with toxigenic *Fusarium* species (yellow rice toxins).[94] However, there are inadequate epidemiological data to causally link this disease with citreoviridin. There have been no outbreaks of yellow rice disease in the last several decades.[1]

Moldy sugarcane poisoning results from the ingestion 3-nitropropionic acid (3-NPA), which is a metabolite of the *Arthrinium* species present in moldy sugarcane during prolonged storage.[95] The ingestion of 3-NPA causes gastrointestinal symptoms in adults, but a severe encephalopathy was associated with 3-NPA ingestion by children.[96] The incubation period is about 2–3 hours with symptoms including nausea and vomiting that progress to dystonia, dysconjugate gaze, carpopedal spasm, and coma.

Skin

The trichothecene mycotoxins produce photosensitivity dermatitis along with hemorrhage and vesiculation followed by desquamation when applied to the skin of experimental animals. Similarly, photosensitizing furocoumarins (xanthotoxin, bergapten) are produced by the pink rot fungus, *Sclerotinia sclerotiorum* (Lib.) de Bary, which grows in celery.[97,98]

Reproductive System

Zearalenone (F-2 toxin) is a naturally occurring, nonsteroidal, estrogenic product of *Fusarium* species that causes vulvar and uterine enlargement in animals. This toxin has not produced documented adverse effects in humans.

Respiratory Tract

Stachybotrys chartarum (Ehrenb.) S. Hughes (*S. atra*) was isolated from homes of infants, who developed idiopathic pulmonary hemorrhage, and from buildings where occupants complained of building-related upper respiratory tract symptoms. The presence of mycotoxins in these cases was not confirmed; therefore, the association of these illnesses with mycotoxins remains speculative. Although the intratracheal instillation of *S. chartarum* spores produces a dose-dependent pulmonary inflammation,[99] no significant pulmonary damage occurs when mice are exposed to *S. chartarum* on moldy surfaces despite airflow at least 4 times greater than the airflow in a typical household.[100] Presently, there are inadequate data to support a causal connection between exposure to *Stachybotrys* mycotoxins and serious or permanent lung injury.[101]

Most mold-related respiratory disease is reversible.[66,102] Atopic individuals may develop allergic rhinitis, hypersensitivity pneumonitis, and asthma following exposure to mold. The prevalence for positive skin tests to mold in the general asthmatic population is approximately 5% with sensitization to fungi high in childhood and declining rapidly with age.[103] Relating mold exposure and mold sensitization to the risk of developing asthma is difficult because most individuals with specific sensitization to mold also are sensitized to other inhaled allergens (mites, house dust, pet dander). In addition, exposure of sensitized individuals to dust mites and to mold produces similar upper respiratory tract symptoms, and these two allergens frequently occur together.[104] Cross-sectional and prevalent case-control studies suggest that mold exposure based on questionnaires (visible mold, moldy odor) is associated with a slightly increased risk (mean OR <1.5) of asthma.[105,106]

Musculoskeletal System

Kashin-Beck Disease is a chronic, bilateral osteoarthritis believed to result from inadequate perfusion of the epiphyses.[107,108] The main feature of this disease is shortened stature, which results from multiple focal necroses in the growth plate of tubular bones. The disease occurred in Eastern Russia among the Cossacks during the 19th and early 20th centuries, but reports of the disease now originate exclusively from China. Mycotoxins from *Fusarium sporotrichiella* Bilai are suspected cofactors along with selenium deficiency, but there is insufficient epidemiological evidence to confirm this association.

EVALUATION MOLD-RELATED COMPLAINTS

Table 45.5 outlines the approach to a patient with complaints related to indoor mold exposure.[109] The medical

TABLE 45.5. Approach to the Evaluation of Mold-Related Complaints

Method	Components
Comprehensive Medical History	Symptoms onset, progression, persistence, exacerbation, improvement, resolution; medical and drug (licit, illicit) history, family history (allergies, related diseases)
Environmental History	Onset/duration of odors, water leaks (outside, roof, plumbing), ventilation (air/heat, temperature settings, humidifiers, vaporizers, purifiers), visible mold growth, pets (cats, birds), home use (use patterns, temperature settings, plants), hobbies and outdoor activities (camping, gardening, composting)
Occupational History	Symptoms at work, exposure to dusts, fumes, chemicals, particulate matter
Physical Examination	Complete examination to detect objective signs of disease
Diagnostic Testing	Chest x-ray, serologic testing (serum immunoglobulins, total IgE, rheumatoid factor, complete blood count, sedimentation rate, C-reactive protein, serological testing for specific IgG and IgE to fungi (radioallergosorbent test [RAST], skin testing, hypersensitivity panels), spirometry (pre- and postbronchodilator measurements), serum chemistry panel
Environmental Survey	Inspection of building (HVAC system, filters, ceiling tiles gypsum wallboard, paper/cellulose-containing surfaces), collection and analysis of airborne samples from appropriate inside and proximal outside locations

history is an important component of the evaluation, particularly relating the temporal onset and resolution to exposure in the building. The physical examination and diagnostic tests provide potential alternate explanations for the symptoms, correlate symptoms to objective evidence of disease, and determine exposure to specific molds. The extent of diagnostic testing depends on the mold and the organ systems involved with the complaints. Additional testing (e.g., CT sinuses, fungal IgE titers, methacholine challenge, neuropsychological testing, nerve conduction studies) may be important depending on the type and severity of complaints. Environmental surveys help determine the source of exposure based on inspection and collection of air samples from locations inside the building. An equal number of samples should be collected simultaneously from an area immediately outside the indoor location of sampling.

AFLATOXINS

HISTORY

Isolation of the first aflatoxins occurred in 1959. In the early 1960s, aflatoxin was a recognized animal carcinogen, when Turkey X disease was associated with the consumption of rosetti, an aflatoxin-contaminated groundnut meal from Brazil, by turkeys.[110]

PRINCIPAL TOXINS

Structure

Aflatoxins are highly substituted difuranocoumarin derivatives that contain an unusual fused dihydrofurofuran configuration. These mycotoxins are divided into two major groups based on their blue or green fluorescence under UV light (see Figure 45.1). The four major aflatoxins are the difurocoumarocyclopentenones B_1 [CAS No. 1162-65-8] and B_2 [CAS No. 7220-81-7], and the difurocoumarolactones G_1 [CAS No. 1165-39-5] and G_2 [CAS No. 7241-98-7]. Aflatoxins are densely fluorescent with the B referring to blue fluorescence and G signifying green fluorescence. The subscript numbers 1 and 2 indicate major and minor compounds, respectively. B_1 is the most potent and most common aflatoxin.[6] The other major aflatoxins are usually not present in the absence of B_1. M_1 (CAS No. 6795-23-9) and M_2 aflatoxins are active 4-hydroxylated metabolites present in milk and urine following the consumption of aflatoxins B_1 and B_2. Sterigmatocystin is a 7,8-dihydrofuro[2,3-β]furan, which is a precursor of aflatoxins found primarily in *A. versicolor* (Vuill.) Tirab. and *Emericella nidulans* (Eidam) Vuill. (*Aspergillus nidulans*). Cyclopiazonic acid is an indole-tetrameric acid mycotoxin found most commonly in *A. flavus*.

Physiochemical Properties

Aflatoxins are slightly soluble in water. The lactone ring in aflatoxins is susceptible to alkaline hydrolysis.

Aflatoxin B_1 and its derivatives

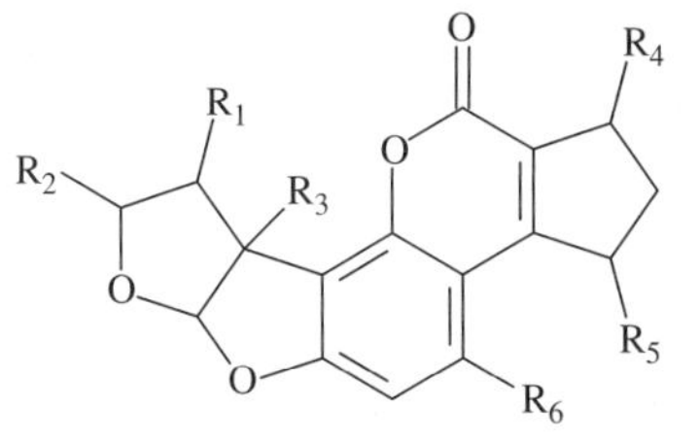

Aflatoxin	R_1	R_2	R_3	R_4	R_5	R_6
B_1	H	H	H	=O	H	OCH_3
B_2	H_2	H_2	H	=O	H	OCH_3
B_{2a}	HOH	H_2	H	=O	H	OCH_3
M_1	H	H	OH	=O	H	OCH_3
M_2	H_2	H_2	OH	=O	H	OCH_3
P_1	H	H	H	=O	H	OH
Q_1	H	H	H	=O	OH	OCH_3
R_0	H	H	H	OH	H	OCH_3

Aflatoxin G_1 and its derivatives

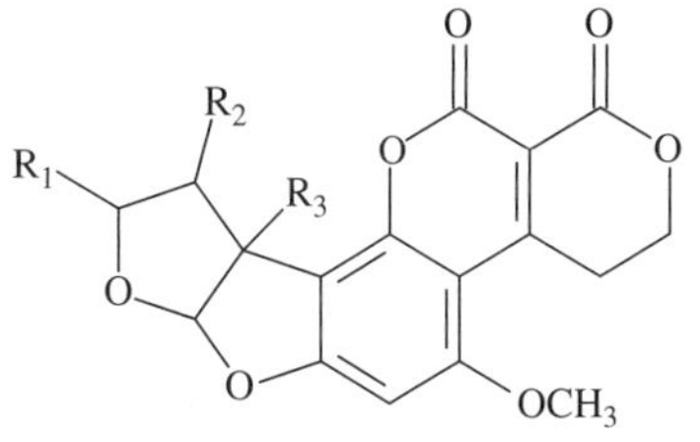

Aflatoxin	R_1	R_2	R_3
G_1	H	H	H
G_2	H_2	H_2	H
G_{2a}	OH	H_2	H
GM_1	H	H	OH

FIGURE 45.1. Structure of aflatoxin compounds and metabolites.

EXPOSURE

Source

About one-half of the strains of *Aspergillus flavus* Link produce aflatoxins B_1 and B_2, while *A. parasiticus* Speare and *A. nomius* Kurtzman, B.W. Horn & Hesselt. produce aflatoxins G_1 and G_2 in addition to B_1 and B_2.[111,112] Other less common aflatoxin-producing *Aspergillus* species include *A. bombycis* S.W. Peterson, Yoko Ito, B.W. Horn & T. Goto, *A. ochraceoroseus* Bartoli & Maggi, and *A. pseudotamari* Yoko Ito, S.W. Peterson, Wicklow & T. Goto. Aflatoxins M_1 and M_2 are hydroxylated metabolites of aflatoxins B_1 and B_2 that occur in milk or milk products after livestock ingest contaminated feed (groundnut meal, maize, cottonseed meal). *A. flavus* Link is a ubiquitous fungus that contaminates plants and plant products, whereas *A. parasiticus* Speare is less widely distributed. *A. nomius* Kurtzman, B.W. Horn & Hesselt. probably does not contribute significantly to the health impact of aflatoxins because this species is relatively rare.[19] Aflatoxin-producing strains of *Aspergillus* generally produce 2–3 aflatoxins, one of which is always aflatoxin B_1. Usually, these toxic strains produce substantially smaller amounts of aflatoxin $G_1$and G_2.[113] The presence of aflatoxin-producing *Aspergillus* species does not imply the presence of mycotoxins because not all of these strains produce aflatoxins. Approximately 30% (range: 20–98%) of *Aspergillus flavus* Link strains produce aflatoxins.[114] The use of heroin may increase the risk of exposure to aflatoxins. In a study of urine samples from 121 European heroin addicts, aflatoxins contaminated 20% of the urine samples of the heroin addicts compared with 2% of the urine samples from healthy volunteers.[115]

Environmental Conditions

Aflatoxin-producing *Aspergillus* species have an affinity for nuts, maize, and oilseeds. Various factors affect the rate of aflatoxin production. Ideal conditions for aflatoxin formation include moisture content between 18–20%, adequate amounts of zinc, and ambient temperature between 25°C–40°C (77°F–104°F).[110] Minimum environmental conditions for the production of aflatoxins include water activity >0.82, pH 3.0–8.5, and temperature 12°C–40°C (~54°F–104°F).[15] Other nutrition factors affecting the production of aflatoxins include carbohydrate and nitrogen sources, phosphates, lipoperoxide, and other trace metals.[116]

Inadequate drying or improper storage promotes the development of aflatoxins in humid climates. Significant concentrations of aflatoxins develop as a result of post-harvest spoilage of aflatoxin-contaminated commodities stored under warm conditions. In temperate climates, the invasion of susceptible crops (maize, nuts) by *A. flavus* before harvest contributes to the production of aflatoxins, particularly during periods of drought stress and insect invasions.[117] Small grain cereal spoilage by *A. flavus* is uncommon in the absence of poor handling.[19] Effective measures to limit the formation of aflatoxins include storage of crops in cool, dry, insect-free environment, harvesting of crops during the dry season with adequate drying, and prevention of chemical and insect damage to crops.[110] Soybeans rarely contain aflatoxin because of the presence of a strong zinc-chelating agent, phytic acid.[118]

Foodstuffs

Maize, cotton seed, and groundnuts (e.g., peanuts) are the major source of aflatoxins, particularly in India, South America, and the Orient.[119] Specific foodstuffs commonly contaminated by aflatoxins include topical and subtropical cereals, cassava, dried fish, rice, oilseeds, cotton-seed meal,[120] peanuts, Brazil nuts, and pistachios. Of the major grains used for human consumption in the United States (i.e., wheat, rye, oats, corn), significant aflatoxin contamination occurs only in corn. Contamination of peanuts occurs periodically in the Southeast; however, most of these products do not reach interstate commerce. Herbs and spices may contain relatively low aflatoxin concentrations (<10 μg/kg) compared with maize and groundnuts.[121] A study of dried-salted fish samples from markets in Sri Lanka demonstrated the presence of *Aspergillus flavus*, but these strains did not produce aflatoxins.[122]

Although wet and dry milling processes fractionate the aflatoxin between the germ, steep water, and the hull/fiber, these processes do not destroy significant quantities of the aflatoxin.[123] Aflatoxins are relatively heat-stable during the preparation of foods (e.g., boiling, autoclaving, microwave). Domestic cooking with microwave and conventional gas ovens does not reduce the aflatoxin concentration in contaminated food products. In an experimental study, roasting eliminated about 50% of the aflatoxin B_1 present in peanuts.[124] In general, aflatoxin ingestion is very low in the United States and Europe compared with developing countries. Although local beers brewed in Africa contain substantial amounts of aflatoxins, these mycotoxins are rare contaminates of European and North American beers.[125] In a study of 24 beer samples from the United States and Mexico, only one sample of Mexican beer contained aflatoxins at a concentration of 49 ng aflatoxin B_1/L.[126]

DOSE RESPONSE

A deliberate ingestion of 1.5 mg/kg of aflatoxin by a 25-year-old woman over approximately 2 weeks resulted in a transient nonpruritic macular rash, headache, and nausea, but no sequelae were detected during a follow-up examination 14 years later.[127] Maize grain samples from two Kenyan homes where 8 people died of hepatic failure revealed aflatoxin B_1 levels of 12 ppm and 3.2 ppm, while samples from unaffected homes had maximum aflatoxin B_1 concentrations of 0.5 ppm.[128] The estimated daily consumption of aflatoxins during an epidemic of acute hepatitis associated with aflatoxin consumption was 2–6 mg.[9]

TOXICOKINETICS

Absorption

Although few quantitative data on aflatoxin absorption exist, animal and human cases associated with aflatoxin to date involve ingestion rather than other routes of exposure. Experimental studies indicate that the dermal absorption of aflatoxin B_1 is slow and toxicologically insignificant.[129]

Biotransformation

The cytochrome P450 (CYP450) isoenzyme system in the liver metabolizes aflatoxin B_1 primarily via hydroxylation. The human CYP450 isoforms, CYP3A4 and CYP1A2, catalyze the epoxidation reaction of aflatoxin B_1 to aflatoxin B_1 exo-8,9-epoxide.[130] Oltipraz is a potential inhibitor of this epoxidation reaction.[131] This epoxide is highly unstable with a high affinity for the formation of adducts to guanine bases in DNA (i.e., aflatoxin-N^7-guanine). CYP1A2 also catalyzes the formation of the metabolite aflatoxin M_1. The most common metabolites of this biotransformation are aflatoxin M_1, which retains some acute oral toxicity in contrast to most other aflatoxin metabolites, and aflatoxin Q_1 (see Figure 45.2) Other metabolites of aflatoxin B_1 include aflatoxin P_1, aflatoxin B_2 , aflatoxicol, aflatoxicol H_1, aflatoxin B_1-2,3-dihydriol, and the unstable, reactive metabolite, aflatoxin B_1-2,3-epoxide. In general, <6% of the aflatoxin B_1 present in animal feed is converted to aflatoxin M_1 in milk.[111]

There is substantial variation in aflatoxin metabolism between species, and this difference probably accounts for the variable response of different species to aflatoxin-induced carcinogenesis. The formation of aflatoxicol from aflatoxin differs from other pathways of aflatoxin metabolism because of the requirement for cytoplasmic NADH-dependent dehydrogenase, which is sensitive to sex hormones. Such metabolic differences may account for some of the variation in response to aflatoxins between sexes (children and men are more susceptible) and species (rats highly sensitive, humans much less sensitive, mice least sensitive).[132]

Elimination

Renal excretion of aflatoxin M_1 (AFM_1) occurs after the ingestion of aflatoxin B_1,[133] but there are few human data on the toxicokinetics of aflatoxins or the relative amount of M_1 eliminated in human urine. In a study of

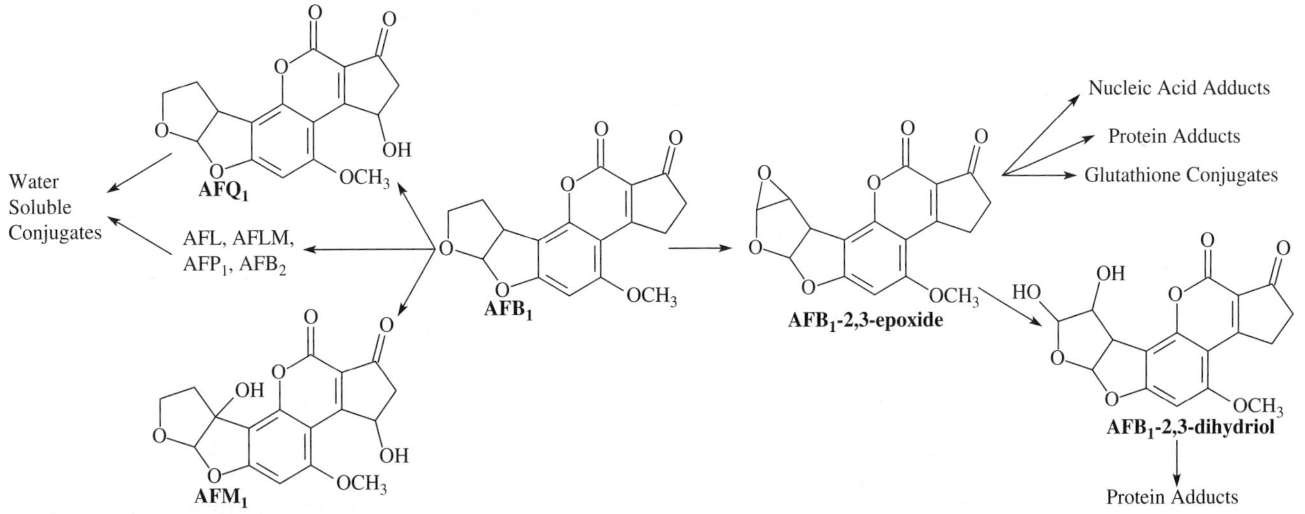

Water Soluble Conjugates
AFQ_1
AFL, AFLM, AFP_1, AFB_2
AFB_1
AFM_1
AFB_1-2,3-epoxide
Nucleic Acid Adducts
Protein Adducts
Glutathione Conjugates
AFB_1-2,3-dihydriol
Protein Adducts

FIGURE 45.2. Aflatoxin B_1 (AFB_1) Metabolism. Reduction of AFB_1 by cytoplasmic reductase produces aflatoxicol. Hydroxylation of AFB_1 forms AFQ_1 and AFM_1. *O*-demethylation of AFB_1 produces AFP_1. AFB_2 results from the hydration of AFB_1. Adapted from Denning DW. Aflatoxin and human disease. Adv Drug React Action Poison Rev 1987;4:178.

Chinese residents ingesting aflatoxins in corn and peanut oil, the renal excretion of AFM_1 accounted for about 1–2% of the dose of aflatoxin B_1 (AFB_1).[134] There is substantial variation in the urinary excretion of M_1 between different species. Aflatoxin M_1 is a minor urinary metabolite in monkeys,[135] whereas biliary excretion accounts for the majority of the absorbed dose of AFB_1 in rats. Animal studies indicate that the elimination of AFB_1 from the blood is rapid. Following the intravenous administration of 0.3 mg AFB_1/kg to monkeys, the plasma half-life of AFB_1 was approximately 36 minutes.[136] Fecal and renal excretion each accounted for the elimination of about one-third of the administered dose of AFB_1 in monkeys. Biliary excretion accounts for a substantially larger part of the elimination of aflatoxins in rodents than in monkeys.[136] In monkeys, about 80% of the administered dose of AFB_1 was eliminated within 1 week, but some radiolabeled AFB_1 remained in the blood and urine after 5 weeks.[137] A study of cord blood in Thailand infants suggests that transplacental transfer of aflatoxins occurs, and that aflatoxin concentrations may be higher in the placenta than in the maternal serum.[138]

HISTOLOGY AND PATHOPHYSIOLOGY

The most biologically active aflatoxins are aflatoxins B_1 (AFB_1) and aflatoxin G_1 (AFG_1). The liver is the main target organ of aflatoxins in animals, and animals fed high doses of aflatoxins usually develop hepatic necrosis. Characteristically, aflatoxins produce proliferation of the bile duct epithelium, fatty infiltration, and centrilobular necrosis.[139] The carcinogenic potential of aflatoxin probably depends on the metabolic activation of AFB_1 or other aflatoxins to DNA alkylating agents, such as the AFB_1-2,3-epoxide.[28] Aflatoxin B_1 reduces hepatic glycogen stores and causes the accumulation of hepatic lipids in experimental animals.[140] Aflatoxin B_1 strongly inhibits the salt-induced conversion of the right-and left-handed DNA forms. These DNA configurations may be associated with regulation of gene function; therefore, aflatoxin-induced changes in DNA equilibrium may cause permanent transcriptional changes as well as interfering with RNA synthesis via inhibition of RNA polymerase.[139] The main adduct formed by AFB_1 occurs at the 2-position of AFB_1 and the N-7 position of guanine in the DNA.

In experimental animal models, AFB_1 interferes with acquired immunity to *Pasteurella multocida* (Lehmann & Neumann 1899) Rosenbusch and Merchant, and this aflatoxin suppresses delayed cutaneous hypersensitivity, lymphokine activity, graft-versus-host reaction, phagocytic response to macrophages, and complement activity. Rather than act as a primary carcinogen, aflatoxin may suppress cell-mediated immunity and therefore increase the number of hepatitis B virus carriers. Other potential contributing factors to the carcinogenic potential of aflatoxins include chronic protein-deficiency and essential malnutrition associated with acquired immunocompetence syndrome.[141–143]

CLINICAL RESPONSE

Acute

Determining the exact cause of human illnesses associated with mycotoxins is difficult because of the lack of a suitable biomarker of exposure. Generally, reports use indirect measures of exposure (e.g., analysis of common foods) rather than direct measures of aflatoxin exposure. Reports of acute toxicity secondary to aflatoxins (aflatoxicosis) are rare, and these reports usually involve acute outbreaks of hepatitis. Because of the nonspecific symptoms associated with aflatoxicosis, outbreaks generally involve communities rather than individuals. Outbreaks of aflatoxicosis often occur in tropical countries among undernourished adults in rural populations that use maize as the staple food. In India during 1974, the consumption of maize containing aflatoxin concentrations of 6–15 μg/kg was associated with 397 cases of acute hepatitis.[9] Clinical features included jaundice, severe ascites, portal hypertension and death secondary to massive gastrointestinal bleeding. There were 106 deaths. A similar epidemic of acute hepatitis was reported in several Kenyan homes after the consumption of corn heavily contaminated with 3.5–12 ppm aflatoxin.[128] Although substantial mortality may occur during the acute phase of aflatoxicosis, follow-up studies indicate that the hepatic inflammation usually resolves without persistent jaundice.[144] An estimated ingestion of 1.5 mg aflatoxin/kg over 2 weeks was associated with headache, nausea, and a nonpruritic rash, but there was no documented liver disease at follow-up 14 years after the incident.[127] This case report suggests that the hepatotoxicity of aflatoxins may be less in well-nourished patients than in malnourished patients.

Chronic

The causal link between aflatoxin exposure and Reye's syndrome (acute encephalopathy with fatty degeneration of the viscera and particularly the liver) is speculative. There are pathological similarities between Reye's syndrome and experimental aflatoxin poisoning. However, most cases of aflatoxin-associated Reye's disease appear in Thailand and not in countries (e.g.,

Africa) where aflatoxins commonly contaminate food. Elevated aflatoxin levels in kwashiorkor probably result from aflatoxin exposure and the inability of the malnourished liver to metabolize the aflatoxins rather than from a direct toxic effect.[145] Few human data causally link liver cirrhosis with aflatoxin exposure.

Carcinogenesis

Both the US National Toxicology Program[146] and the International Agency for Research on Cancer (IARC)[147] recognize naturally occurring aflatoxins as human carcinogens. Animal studies indicate that the carcinogenic activity of aflatoxin B_1 requires activation of this aflatoxin by the cytochrome P450 enzymes to aflatoxin B_1-8,9-epoxide.[111] Aflatoxin B_1 is a potent carcinogen in animal species, but there is a wide variation in the carcinogenic response between different animal species. The carcinogenic potency of aflatoxin B_1 is three orders of magnitude greater in rats than in mice.[148] In animal models, exposure to aflatoxin B_1 produces tumors primarily in the liver, colon, and kidneys, and there is sufficient evidence for carcinogenicity for aflatoxins B_1, G_1, and M_1. Aflatoxin B_2 is substantially less potent in animal studies than these later hepatocarcinogens. The IARC lists the metabolite, aflatoxin M_1, as a possible human carcinogen (Group 2B).

Aflatoxin and hepatitis B are co-carcinogens and the presence of both factors in developing countries is associated with a high incidence of hepatocellular cancer compared with developed countries where these factors are uncommon. A cohort study in China detected statistically significant excess mortality from liver cancer in villages with heavy aflatoxin contamination. However, primary hepatocellular carcinoma is one of the most common cancers of developing countries where mycotoxins, hepatitis B virus, malnutrition, and parasitic infections are frequent confounding variables for epidemiological studies of liver cancer.[141] Several correlation studies from Swaziland and China indicated that the incidence of liver cancer correlated more to aflatoxin exposure than to hepatitis B viral infection. However, there was no correlation between the presence of urinary aflatoxin metabolites and the incidence of liver cancer. Aflatoxins probably do not contribute significantly to the development of liver cancer in developed countries because of the absence of cofactors, such as endemic hepatitis B infections.[149] A retrospective study of mortality data comparing US regions with differing aflatoxin exposure demonstrated a slight excess of age-adjusted mortality from primarily liver cancer despite dramatic differences in aflatoxin consumption.[150]

DIAGNOSTIC TESTING

Analytical Methods

Thin-layer chromatography is the most economical method for the detection of aflatoxins with limits of quantitation in the range 0.2–10 ng/g depending on the extraction solvent and clean-up procedures.[151] Aflatoxins are highly fluorescent and bidirectional high performance thin-layer chromatography achieves detection limits below 1 µg/kg.[152] High performance liquid chromatography (HPLC) coupled with postcolumn iodine derivatization and fluorescence detection is more sensitive and accurate than thin-layer chromatographic techniques with limits of detection near 1 ppb.[153,154] Commercial immunoassays for aflatoxins are available for determining the presence of aflatoxins,[155] but the limits of detection are above current regulatory concentrations.[156]

Biomarkers

The short half-life of aflatoxin B_1 limits the usefulness of this substance as a biomarker. Excretion of the metabolite, aflatoxin M_1, in milk may result from the presence of aflatoxin B_1 in animal feed. Suggested conversion ratios for aflatoxin B_1 in feed to aflatoxin M_1 in milk ranged from 66:1 to about 300:1.[158] In a study of children from Sierra Leone, the urinary concentrations of aflatoxin B_1 and aflatoxin M_1 ranged from 0.04–319 ng/mL and 0.1–374 ng/mL, respectively, with the urine concentration of aflatoxin B_1 being significantly greater during the dry season.[157]

Health agencies in most countries set permissible aflatoxin B_1 exposure limits at 5 µg/kg. In the United Kingdom, the limit for aflatoxin B_1 in nuts, dried figs, and products used for direct human consumption is 4 µg/kg; the limit is 10 µg/kg if further processing reduces aflatoxin concentrations.[158] Foodstuffs contain aflatoxins because of the ubiquitous nature of aflatoxin-producing fungal species. Current FDA guidelines limit aflatoxin levels to 20 ppb in all animal feed crops except cottonseed meal for which the limit was recently raised to 300 ppb. Aflatoxin M_1 levels in milk are restricted to levels below 0.5 ppb. Serum and urine aflatoxin B_1 concentrations are not reliable biomarkers of chronic aflatoxin exposure because of the rapid elimination of aflatoxin B_1 from the blood and urine.

TREATMENT

Management for exposure to aflatoxins is supportive along with reduction of aflatoxin exposure. Good cultivation and storage techniques represent the best method of reducing aflatoxin levels in food. In the United States, the FDA monitors interstate food shipments (e.g., peanuts, corn, pecans, walnuts) for aflatoxin concentrations. Under experimental conditions, activated charcoal avidly binds aflatoxins.[159] Alkaline conditions also destroy mycotoxins.

CITRININ

Citrinin (CAS No. 518-75-2) is a water-insoluble mycotoxin produced by a variety of genera including *Penicillium* (*P. canescens* Sopp), *P. chrzaszczii* K.M. Zalessky, *P. citreosulfuratum* Biourge, *P. citrinum* Thom, *P. claviforme* Bainier, *P. expansum* Link, *P. implicatum* Biourge, *P. lividum* Westling, *P. phaeojanthinellum* Biourge, *P. purpurascens* (Sopp) Biourge, *P. roqueforti* Thom, *P. velutinum* J.F.H. Beyma, *P. verrucosum* Dierckx, *P. viridicatum* Westling) and *Aspergillus* (*A. candidus* Link), *A. flavipes* (Bainier & R. Sartory) Thom & Church, *A. terreus* Thom) species.[160] This toxin was first isolated from *Penicillium citrinum* Thom in 1931. Although *P. citrinum* is the main source of citrinin, several other *Penicillium* species (*P. expansum* Link, *P. verrucosum* Dierckx) also produce relatively high concentrations of this mycotoxin. *P. citrinum* grows in a wide temperature range from 5°C–40°C (41°F–104°F) with an optimum growth rate between 26°C–30°C (~79°F–86°F) and a minimum water activity 0.80–0.83.[15] The optimum production of citrinin occurs at 30°C (86°F; range: 15°C–37°C/~59°F–99°F). Citrinin-producing fungi grow on cereals, peanuts, and fruits stored in high humidity, but the presence of these fungal species does not necessarily indicate the presence of citrinin. This mycotoxin occurs naturally in barley, oats, rye, and wheat grown in North American and Europe. Food processing destroys most citrinin and this mycotoxin does not survive the mashing stem in the beer-brewing process.[125] Renal damage in experimental rats is the most common pathological lesion reported after exposure to citrinin.[161] Despite more than 30 years of clinical research, the role of citrinin as a human toxin or food contaminant remains unclear. There are no well-documented cases of citrinin-induced illness in humans.

ERGOT and RELATED COMPOUNDS

HISTORY

Historically, the first recognized mycotoxicoses were human diseases produced by the consumption of rye products infected by the mold *Claviceps purpurea* var. *purpurea* (Fr.) Tul. and related fungi. As early as 600BC, Assyrian tablets referred to the "noxious pustules in the ear of grain," and one of the sacred books of the Pharisees (400–300BC) mentioned "grasses that cause pregnant women to drop the womb and die in childbirth."[162] At that time, the consumption of contaminated rye was discouraged as "the black malodorous product of Thrace and Macedonia." However, these taboos were lost during Roman times because rye products were not part of the Roman diet. The first written accounts of ergotism occurred during the Middle Age in Europe, when thousands of fatalities resulted from epidemics of ergotism.

From the 9th–19th centuries, recurrent epidemics of ergotism plagued Germany, France, Scandinavia, and other parts of Europe both as a gangrenous (chronic) disease and as a convulsive (acute) disorder.[163] The more widely recognized affliction was the gangrenous type characterized by severe hypesthesia, paresthesias, and hemorrhagic bullae that progressed to dry gangrene and black mummification of the involved extremity. The descriptive term, holy fire *(ignis sacer*, St. Anthony's Fire) was given to this malady. St. Anthony was the patron saint for an Order of Hospitallers, founded in Vienne, France in 1095. Although St. Anthony was a 3rd century (251–356AD) Egyptian ascetic with no direct connection to ergotism, these hospitals became pilgrimage centers for victims of ergotism. The diets at these hospitals probably did not contain rye infected with the mold *Claviceps*.

The less common type of ergotism was described as a convulsive syndrome and included symptoms such as vertigo, headache, tinnitus, sensual disturbances, hallucinations, muscle spasm, gastrointestinal distress, and seizures. One of the ergot alkaloids, isoergine (lysergic acid amide), has approximately one-tenth of the hallucinogenic properties of the structurally related compound LSD (lysergic acid diethylamide). Morning glory seeds contain the same alkaloid, and these seeds were used by the Aztec Indians as a ceremonial hallucinogen. Geographical differences in alkaloid production and host susceptibility probably accounted for the variation in expression of these two syndromes.

Growing conditions and agricultural practices in late 17th century New England favored the growth of *Claviceps* mold. Caporael suggested that ergotism may explain the physical afflictions and vivid imagery offered at the Salem Witchcraft Trials from 1691 to 1692.[164] The last European outbreak of ergotism occurred in the United Kingdom and in Russia during the late 1920s.[337] Improved agricultural and storage practices have virtually eliminated epidemic ergotism, although sporadic epidemics occur occasionally in less-developed countries. The last sporadic episode of ergotism involved a five-case outbreak in France in 1951.[165] Currently, ergotamine toxicity usually results in vascular insufficiency caused by excessive, chronic consumption of ergot alkaloids by patients with a history of migraine headaches. Acute massive overdose is rare.

Although ergot was a suspected cause of epidemics as early as 1677, the presence of ergot alkaloids in parasitic fungi was not recognized until the middle 19th century. In 1918, Stoll isolated the first active principle, ergotamine, from ergot.[166] Although ergotamine has some stimulant effects on the uterus, the alkaloid mixture from the sclerotium possesses greater stimulant properties. In 1935, Dudley and Moir isolated the major uterine stimulant, ergometrine (ergonovine) from other ergot alkaloids in the sclerotium.[167]

Ergometrine has a long history of obstetrical use during labor, and the use of ergot alkaloids to reduce postpartum hemorrhage has substantially decreased maternal deaths during labor.[163] Midwives used ergot as an herbal medicine during delivery long before Adam Lonicer recommended the use of three sclerotia (0.5 mg) of ergot for uterine atony in 1582.[168] In 1841, Francis Ramsbotham, founder of the London Obstetrical Society, noted that female midwives used ergot alkaloids as a promoter of labor pains for at least the preceding 150 years despite the recognized toxicity of ergot alkaloids.[169] In the early 19th century, the use of ergot to induce labor was discouraged in Europe because of the increased incidence of uterine rupture, stillbirths, and maternal death.[170] However, the use of ergot to accelerate labor was more widespread in the United States during the early 19th century. By the end of the 19th century, the detrimental effects (e.g., stillbirth) of prolonged, ergot-induced uterine contraction before delivery were widely recognized. In the early 20th century, the prophylactic use of ergotamine for the prevention of postpartum hemorrhage in the United States began. The treatment of migraine headaches with ergot began in the late 19th century in Germany, the United States, and England.[171] Although early medicinal use included the treatment of pruritus, diabetes, hyperthyroidism, and psychosis, as well as a uterine stimulant during labor, the administration of ergot was restricted to the treatment of migraine headaches and postpartum hemorrhage by the 1950s.[172]

PRINCIPAL TOXINS

Structure

The term *ergot* applies both to the sclerotium produced by the *Claviceps purpurea* mold and to the ergot alkaloids within the sclerotium. All natural ergot alkaloids (ergolines) are amide derivatives of the tetracyclic compound 6-methylergoline, which is a lysergic acid base chemical structurally similar to a number of biogenic amines (e.g., dopamine, serotonin, norepinephrine). The ergolines in *C. purpurea* primarily are lysergic and isolysergic acid derivatives, whereas ergolines from *C. fusiformis* primarily are derivatives of dimethylergolines (clavines). A previously extracted pharmacologically active compound, ergotoxine, contains four alkaloids (ergocristine, ergocornine, α- and β-ergocryptine). Ergoloid mesylates (dihydroergotoxine) is a mixture of four ergot alkaloids (dihydroergocornine, dihydroergocristine, α- and β-dihydroergocryptine). Figure 45.3 and Figure 45.4 display the chemical structures of ergot amine alkaloids and ergot amino acid alkaloids, respectively.

The majority of the alkaloids can be divided into three groups based on side-chain substitution (see Table 45.6).[173] Ergotamine is the prototype for ergot alkaloids derived from lysergic acid. Reduction of the double bond at the 9-10 position yields dihydroergotamine. In addition, the sclerotium contains several other pharmacologically active substances (e.g., tyrosine, tryptophan, tyramine, histamine, histidine, choline, acetylcholine), but their clinical significance is unclear. Bromocriptine is a synthetic brominated ergopeptide derivative with powerful dopaminergic agonist properties used to suppress postpartum lactation.

Physiochemical Properties

The shape of the sclerotia resembles a rooster's spur, and the term ergot is derived from the old French term for the cock's spur, *argot*. This structure remains the primary commercial source of ergot alkaloids, and these structures typically contain up to 1% ergot alkaloids along with other biologically active alkaloids, amino acids, carbohydrates, lipids, and pigments. The contents of the sclerotia are unique because of the presence of a large triglyceride fraction containing primarily ricinoleic acid. In the absence of other sources of ricinoleic acid (castor oil), the presence of ricinoleate in a foodstuff indicates the presence of *C. purpurea* sclerotia.[337] Hydrolysis of *C. purpurea* alkaloids produces lysergic or isolysergic acid, ammonia, a keto acid, and two amino

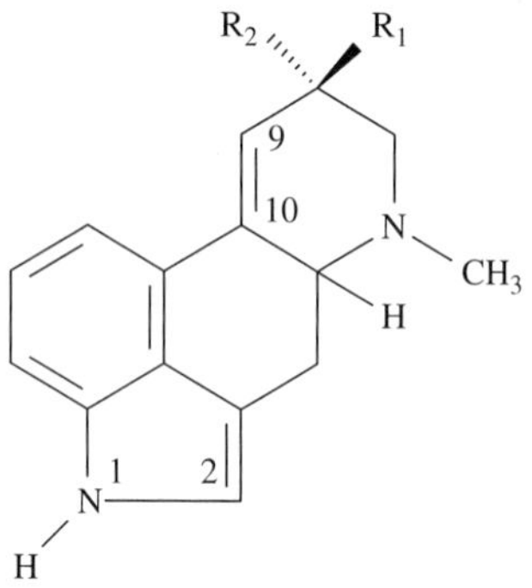

Alkaloid	R_1	R_2
d-Lysergic Acid	$—CO_2H$	$—H$
d-Isolysergic Acid	$—H$	$—CO_2H$
d-Lysergic Acid Diethylamide (LSD)	$—C(=O)—N(CH_2CH_3)_2$	$—H$
Ergometrine	$—C(=O)—NH—CH(CH_3)—CH_2OH$	$—H$
Methylergometrine	$—C(=O)—NH—CH(CH_2CH_3)(CH_2OH)$	$—H$
Methysergide[a]	$—C(=O)—NH—CH(CH_2CH_3)(CH_2OH)$	$—H$
Lisuride	$—H$	$—NH—C(=O)—N(CH_2CH_3)_2$
Lysergol	$—CH_2OH$	$—H$
Lergotrile[b,c]	$—CH_2CN$	$—H$
Metergoline[a,b]	$—CH_2—NH—C(=O)—O—CH_2—$Phenyl	$—H$

[a]Contains methyl substitution at N1; [b]Contains hydrogen atoms at C9 and C10; [c]Contains chlorine atom at C2

FIGURE 45.3. Chemical structures of ergot amine alkaloids. Adapted from de Groot AN, et al. Ergot alkaloids current status and review of clinical pharmacology and therapeutic use compared with other oxytocics in Obstetrics and Gynaecology. Drugs 1998;56:525. Reprinted with permission from Adis International.

TABLE 45.6. Classification of Ergot Alkaloids

Category	Ergot Derivative
Amine	Ergonovine (ergometrine), ergobasine, methylergonovine, methysergide (1-methyl methylergonovine)
Amino acid (ergopeptide)	Ergotamine, ergostine, ergosine, ergotoxine group (ergocornine, ergocristine, α-/β-ergocryptine), bromocriptine (2-bromo-α-ergocryptine)
Dihydrogenated[a]	Dihydroergotamine, dihydroergocristine, dihydroergosine, dihydroergocornine, α- and β-dihydroergocryptine

[a]Mostly semisynthetic.

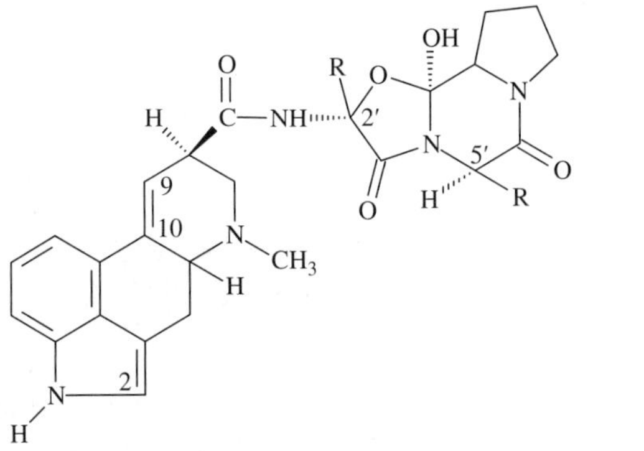

Alkaloid[a]	R(2′)	R(5′)
Ergotamine	$—CH_3$	$—CH_2—$Phenyl
Ergosine	$—CH_3$	$—CH_2CH(CH_3)_2$
Ergostine	$—CH_2CH_3$	$—CH_2—$Phenyl
Ergotoxine Group		
ergocornine	$—CH(CH_3)_2$	$—CH(CH_3)_2$
ergocristine	$—CH(CH_3)_2$	$—CH_2—$Phenyl
α-ergocryptine	$—CH(CH_3)_2$	$—CH_2CH(CH_3)_2$
β-ergocryptine	$—CH(CH_3)_2$	$—CH(CH_3)CH_2CH_3$
Bromocryptine[b]	$—CH(CH_3)_2$	$—CH_2CH(CH_3)_2$

[a]Dihydro derivatives contain hydrogen atoms at C9 and C10; [b]Contains bromine atom at C2

FIGURE 45.4. Chemical structures of ergot amino acid alkaloids. Adapted from de Groot AN, et al. Ergot alkaloids current status and review of clinical pharmacology and therapeutic use compared with other oxytocics in Obstetrics and Gynaecology. Drugs 1998;56:525. Reprinted with permission from Adis International.

acids. Lysergic acid derivatives are unstable in light or oxidizing agents, and the decomposition products are inactive. Oxidation products of ergot alkaloids frequently are colored. In solution at equilibrium, about 40% of the ergotamine undergoes isomerization to the relative inactive epimer, ergotaminine.[174]

EXPOSURE

Source

Wind carries fungal spores to the ovaries of young rye. These spores germinate into hyphal filaments that penetrate deeply into rye to form the sclerotia, which are purple-black or purple-red, curvilinear structures that replace the florets of grasses and cereals. The dark, purplish black mass called a *sclerotium* replaces the kernels. The densely packed mycelia represent the body of the mold. The superficial resemblance of sclerotia to kernels permits contaminated rye to enter the human food chain in bread. Exposure to ergot compounds is widespread. All of the leading cereals (barley, oats, rye, sorghum, wheat, corn, millet) are hosts for *C. purpurea* var. *purpurea* (Fr.) Tul., but the first five cereals are the most important hosts. The two important *Claviceps* species involved in human illness are *C. purpurea* and *C. fusiformis* Loveless. *C. fusiformis* is a parasite of pearl millet in Africa and East Asia. Other fungal genera containing ergolines include *Aspergillus* (*A. clavatus* Desm., *A. fumigatus* Fresen.), *Emericella nidulans* (Eidam) Vuill. (*A. nidulans*), *Rhizopus nigricans* Ehrenh., and *Penicillium* (*P. chermesinum* Biourge, *P. concavorugulosum* S. Abe, *P. dierckxii* Biourge). Some higher plants (e.g., *Ipomoea* species) also contain ergot alkaloids.

Environmental Conditions

Warm, rainy spring and summer months favor the sporadic production of ergot alkaloids. The bread baking process reduces, but does not destroy all the ergot alkaloids produced by sclerotia.

Foodstuffs

In general, the main ergolines present in ergot-contaminated cereals, such as wheat and rice are ergocristine, ergocristinine (isomer of ergocristine), and ergotamine.[175,176] Analysis of 14 samples of commercial rye and wheat flour demonstrated the presence of individual ergolines (ergometrine, ergosine, ergotamine, ergocornine, ergocryptine, ergocristine) ranging from 0.3–62 μg/kg.[177] Less common ergolines include ergocryptine, ergocornine, ergometrine, and ergosine. Total ergot alkaloid content for wheat and rye ranged from 0.013–0.307% and 0.011–0.452%, respectively. The total ergot alkaloid content is highly variable from one head to another. The average estimated daily intake of ergot alkaloids from Swiss diets was about 5 μg.[337] Food processing and baking significantly reduces the content of ergot alkaloids.[1] The baking of bread and pancakes using grain contaminated with ergot alkaloids resulted in a 59–100% reduction in the concentration of ergolines in the finished product.[178] The loses of ergot alkaloids in rye and whole wheat bread ranged up to 85% and 100%, respectively.

Medical Uses

Current therapeutic uses of ergot alkaloids include vascular headache prophylaxis and treatment, uterine atonia, senile dementia, orthostatic circulatory disturbances, and infertility secondary to hyperprolactinemia. Ergotamine tartrate is administered for the relief of migraine and related headaches usually in combination with caffeine. Dihydroergotamine (D.H.E.) is an intravenous preparation for the treatment of migraine headaches, but the efficacy of this treatment probably is lower than the use of sumatriptan.[179] The development of adverse effects following the administration of methysergide limits the usefulness of this drug for the treatment of migraine headaches. Methylergonovine (e.g., methylergometrine) is an effective treatment for postpartum hemorrhage. Dihydroergotoxine mesylate (CAS No. 8067-24-1) is a fixed combination of the following four dihydrogenated ergot alkaloids: dihydroergocornine mesylate (CAS No. 14271-04-6), dihydroergocristine mesylate (CAS No. 24730-10-7), α-dihydroergocryptine mesylate (CAS No. 14271-05-7), and β-dihydroergocryptine mesylate (ergoloid mesylate). Bromocriptine is used to manage amenorrhea, to suppress postpartum lactation, and for the treatment of neuroleptic malignant syndrome.[180] Previous use for hydrogenated ergot alkaloids (ergoloid mesylates) included the treatment of symptoms of idiopathic decline in mental activity of the elderly. Ergonovine was used for the assessment of coronary artery disease.

DOSE RESPONSE

The acute toxic dose of ergot alkaloids varies because of the marked individual susceptibility of patients to the vasoconstrictive properties of ergotamine. Factors that alter the dose response of ergot alkaloids include the presence of vascular disease, hypersensitivity of individual vascular beds, sepsis, smoking, malnutrition, pregnancy, hepatorenal disease, thyrotoxicosis, coronary artery disease, and the concurrent ingestion of certain medications (propranolol, oral contraceptives, xanthine derivatives, erythromycin, ritonavir).[181] Although case reports associated infant fatalities with the accidental ingestion of 12 mg ergotamine[182] and the intramuscular administration of 0.2 mg ergonovine to a newborn infant, the lack of biomarkers and limited laboratory data prevents conclusions regarding causation.[183]

Ergotamine

Substantial variation occurs in the toxicity of ergotamine following a specific dose. The administration of a single rectal suppository of ergotamine by a 32-year-old woman was associated with the development of paresthesias, weakness, difficulty walking, and painful muscle spasms,[184] whereas the chronic ingestion of approximately 4–7 g ergotamine tartrate over 15 years was not associated with any clinical features of ergotism.[185] Atrial fibrillation and precordial pain occurred in a 36-year-old woman after the intramuscular injection of 0.5 mg ergotamine.[186] The abnormalities resolved after 3–4 hours. A case report documented the development of an acute myocardial infarction followed by ventricular fibrillation in a 29-year-old man after the ingestion of 120 mg ergotamine *and* 6 g caffeine.[187] On the second day after admission, peripheral ischemia developed that eventually resolved.

Methylergonovine

The administration of 0.18 mg methylergonovine to a fullterm infant at birth resulted in the development of lethargy, hypercarbia, poor peripheral perfusion, and cyanosis, requiring intubation about 3.5 hours after birth.[188] There was no urine output for the first 16 hours. The infant was treated with nitroprusside infusion (0.3 μg/kg/min beginning 6 hours after birth increasing to 1.2 μg/kg/min) and furosemide. He was extubated at 51 hours after birth and discharged from the hospital after 10 days without obvious sequelae.

Methysergide

This drug is a potent central serotonin antagonist with limited emetic, vasoconstrictive, and uterine constrictive properties. The usual adult dose is 4–8 mg daily. There are few dose-response data on methysergide. A 30-mg adult ingestion resulted in no detrimental effects.[189]

Bromocriptine

Acute toxicity from bromocriptine overdose is relatively mild. An estimated bromocriptine ingestion of 1.5 mg/kg as part of a multiple overdose [Actifed® (Ortho-McNeil, Inc., Raritan, NJ), cyclobenzaprine, cortisone] resulted in a brief period of hallucinations, but there were no sequelae.[190] In an adult, the ingestion of 50 mg bromocriptine produced vomiting, mild hypotension, and sweating, whereas the ingestion of 75 mg caused vomiting in another adult.[191]

TOXICOKINETICS

There are limited data on the toxicokinetics of individual ergolines and ergot alkaloids.

Absorption

The gastrointestinal absorption of most ergot alkaloids with the exception of bromocriptine is relatively low. In addition, the large (>90%), first-pass effect of ergot alkaloids further limits bioavailability following the ingestion of therapeutic doses of ergot alkaloids.[192] Following the intramuscular administration of therapeutic doses of ergotamine, the mean bioavailability was about 47% with a range of 28–61%.[193] Peak plasma concentrations of ergot alkaloids following therapeutic doses occur between 0.5–2 hours after ingestion.[194–196] In a study of volunteers ingesting dihydroergotoxine mesylate tablets, the mean peak concentration of total dihydrogenated ergot compounds was 124 ± 16 pg/mL with a mean time to peak concentration of 1.15 ± 0.21 hours.[197] Caffeine increases the rate of absorption and the bioavailability of ergot alkaloids.[198]

Distribution

In a study of 10 healthy volunteers, the distribution half-life of ergotamine after intravenous administration was 3 minutes with an approximate volume of distribution of 2 L/kg.[193]

Biotransformation

Biotransformation of ergoloid alkaloids (dihydroergocornine, dihydroergocristine) occurs via oxidation and cleavage of proline in the peptide portion of the molecule as well as by the splitting of the amide bond yielding dihydrolysergic acid amide. A variety of active metabolites that retain the essential ring structures of ergot alkaloids (e.g., 8'-OH-dihydroergotamine) prolong the duration of action of ergot compounds.[192]

Elimination

Elimination of ergot alkaloids occurs almost exclusively by hepatic metabolism and biliary excretion. Mean initial elimination half-lives of most natural ergot alkaloids range from 1.5–6 hours with bromocriptine exhibiting an elimination half-life at the upper end of this range.[194] Following the ingestion of 2 mg ergotamine by health volunteers, the mean elimination half-life was 3.4 hours.[199] Excretion of ergot alkaloids in the bile accounts for approximately 80–90% of the intravenous dose of ergot alkaloid, primarily as metabolites.[194] Urinary elimination of unchanged ergot alkaloids is small (approximately 2%) compared with biliary excretion with most of the ergot metabolites appearing in the bile as conjugated compounds.

Drug Interactions

Propranolol blocks the vasodilating effects of epinephrine and may exacerbate the peripheral vasoconstricting properties of ergot alkaloids.[200] Macrolide antibiotics inhibit the metabolism of ergot alkaloids, potentially causing vasospasm and ischemia.[201] In addition, oral contraceptives and smoking may contribute to vasospasm. Case reports associated coronary vasospasm and cardiac arrest with the administration of ergonovine during coronary catheterization for the evaluation of variant angina.[202] Bromocriptine is a potent dopamine agonist. Case reports associate the therapeutic administration of bromocriptine with aggravation of puerperal mania[203] and affective disorders.[204]

HISTOLOGY AND PATHOPHYSIOLOGY

Mechanism of Action

The three classes of ergot alkaloids produce different pharmacological effects at therapeutic doses in experimental animals because of their action on a variety of receptors.[173] (See Table 45.7) Both amine alkaloids and amino-acid alkaloids cause uterine contraction, vasoconstriction, and stimulation of the central emetic center. In addition, the latter group (e.g., ergotamine) produces α-adrenergic blockade and vasodilation by inhibition of norepinephrine-induced vasoconstriction.[192] The differences in pharmacological effects between these three groups reflect varying agonist and antagonist actions on several different neuronal receptors including noradrenaline, dopamine, and serotonin (5-HT_{1A}, 5-HT_{1B}, 5-HT_{1D}, 5-HT_{1F}). Bromocriptine has moderate emetic activity and mild α-adrenergic-blocking properties in experimental animal models.[205]

Mechanism of Toxicity

Ergotamine is the prototype ergot alkaloid compounds, producing peripheral vasoconstriction. Vasospasm results both from central sympatholytic effects and from peripheral adrenergic effects mediated by α-adrenergic and serotonin agonist activity. Although vasospasm potentially occurs in all vessels, symmetrical involvement of the distal arteries is the most common form of

TABLE 45.7. Pharmacological Effects of Selected Ergot Alkaloids

Pharmacological Effect	Amines[a]	Amino Acid (Ergotamine)	Dihydrogenated (Dihydroergotamine)	Bromocriptine
Vasoconstriction (systemic)	+	+ + +	+ +	0
Myometrial stimulation	+++	+ + +	+	0
α-Adrenergic blockade	0	+ +	+ + +	+
Emesis	+	+ + +	+ +	+ +

Source: From Ref 173.
[a]Ergonovine, Methergine, methyl ergonovine.

ergotamine-induced vasospasm. Histological changes of the distal arteries following chronic ingestion of ergot alkaloids include a thickened elastica interna, subintimal and medial fibrosis, intimal lesions, and thrombosis. Occasionally, symptoms may persist despite the absence of pathological changes in the arteries.[206] Decreased sympathetic tone also may produce coronary or mesenteric spasm and bradycardia. Ergotamine directly stimulates the chemoreceptor trigger zone in the medulla oblongata, and nausea and vomiting are the most common adverse effects following the ingestion of therapeutic doses of ergotamine. [195]

Neurological effects of sporadic ergotism are not well documented. Theoretically, focal neuropathies may result from constriction of the nutrient vessels to peripheral nerves or from local arterial vasoconstriction (e.g. monocular blindness resulting from decreased retinal artery perfusion). However, the mechanism of toxicity remains unclear. Bromocriptine is a long-acting dopamine agonist that has central stimulating effects. Potentially, bromocriptine may exacerbate bipolar affective disorders (manic-depressive) or cause hypotension.

CLINICAL RESPONSE

Cases of ergotism result from an acute excessive ingestion, chronic therapeutic use of ergot alkaloids, or the recent ingestion of small amounts of ergot drugs in individuals hypersensitive to them. C. *purpurea* var. *purpurea* (Fr.) Tul. produces ergotamine and ergocristine alkaloids that cause the gangrenous form of ergotism, whereas C. *fusiformis* Loveless produces clavine alkaloids that cause gastroenteritis and a convulsive form of ergotism.[1] Toxicity secondary to ergot alkaloid ingestion is highly variable. Risk factors for the development of ergot toxicity include malnutrition, febrile illnesses including sepsis, coronary artery disease, hepatorenal insufficiency, peripheral vascular disease, smoking, hypertension, pregnancy, and thyrotoxicosis.

Adverse Drug Reactions

Adverse reactions to therapeutic doses of ergot alkaloids include nausea, vomiting, headache, lightheadedness, peripheral vasoconstriction, malaise, muscle spasms, tightness of the throat, and paresthesias. The incidence of vascular spasm after therapeutic doses of ergotamine is small (<0.01%).[207] The administration of therapeutic doses of ergotamine has been associated with vascular insufficiency in the cerebral, mesenteric,[208] renal, [209] ophthalmic,[210] and coronary arteries.[211] Side effects of therapeutic doses of methysergide involve the gastrointestinal (nausea, vomiting, diarrhea) and central nervous (lethargy, anxiety, insomnia, euphoria, confusion) systems. Similar to other ergot alkaloids, the chronic therapeutic use of methysergide has been associated with vascular occlusion at a variety of sites, including the lower and upper extremities, coronary arteries, superior mesenteric arteries, and the aorta.[212,213] Other adverse effects of chronic therapeutic use of methysergide include fibrosis of the heart valves, endocardium, pleura, lungs, great vessels, retroperitoneum, and pericardium.[214]

Acute Ergot Alkaloid Toxicity

Parenteral

Ergot toxicity in the newborn usually results from the inadvertent administration of methylergonovine instead of other medication (naloxone, vitamin K). Accidental parenteral administration of 0.2 mg ergonovine to newborn infants was associated with cutaneous erythema and hemorrhagic blebs on the extremities, groin, and genitalia.[183] The skin lesions appeared on the second day and subsided about the fifth day. The case report did not include any laboratory confirmation of the administration of ergonovine. Following the intramuscular or oral administration of ergometrine, neonates may develop respiratory depression, cyanosis, oliguria, coma, apnea, and seizures.[215,216] Symptoms of intoxication usually develop within 1 hour, but occasionally symptoms are delayed several hours.[188] Death usually results from respiratory failure.

Ingestion

Acute toxicity from the ingestion of ergot alkaloids is rare. Historically, acute (gangrenous) ergotism is associated with a convulsive syndrome that includes hallucinations and delusions as well as vertigo, headache, tinnitus, paresthesias, hyperesthesias, hemiplegia, seizures, muscle spasm, fatigue, and gastrointestinal distress.

Typically, cardiac and gastrointestinal effects begin within 4 hours of ingestion of ergotamine including restlessness, agitation or lethargy, coma, nausea/vomiting, tachycardia, and hypotension;[173] however, manifestations of peripheral ischemia often are delayed until the following day (e.g., cyanosis, cold extremities, paresthesias, pain). A case report associated a large dose of Cafergot® (Novartis, East Hanover, NJ; ergotamine, caffeine) with hemorrhagic gastritis, peripheral cyanosis, cerebral edema, and the death of an infant.[182] Bromocriptine intoxication usually is mild, and clinical features include vomiting, mild hypotension, and diaphoresis.[190] There are few data on the acute toxicity of methysergide. Based on the adverse effects of methysergide, the most likely effects associated with methysergide intoxication include the gastrointestinal and central nervous systems.

Chronic Ergotism

Chronic ergotism typically develops in young, female patients consuming excessive doses of ergotamine for migraine headaches.[217] This chronic, gangrenous form of ergotism usually presents as a symmetrical, bilateral peripheral vascular insufficiency affecting primarily the lower extremities. Common clinical features of chronic ergotism include intermittent claudication, cold extremities, pain, numbness, claudication, cyanosis, paresthesias, pallor, malaise, nausea, vomiting, diarrhea, abdominal pain, and edema. In women, amenorrhea, miscarriages, and poor lactation may occur. In a study of an epidemic of ergotism from the ingestion of food heavily contaminated with *C. purpurea* sclerotia, dry gangrene, reduced peripheral pulses, and desquamation of the skin occurred, particularly in the lower extremities.[218] The most common symptoms were weakness, formication, hyperesthesias, nausea, vomiting, and diarrhea. Less common symptoms included seizure, coma, confusion, diarrhea, syncope, diplopia, blurred vision, psychosis, and bradycardia.[172] Usually, improvement begins within 24 hours of the cessation of ergot alkaloid use, but ischemic symptoms may persist several days.[219] Although peripheral pulses may reappear early, abnormal distal blood flow as documented by angiography may remain several weeks.

Ischemia may occur in the upper extremities,[220] but ischemia is rare in the carotid, coronary, mesenteric, renal, or retinal arteries. Chronic, heavy use of ergotamine (i.e., 6 mg/d for 3 months) was associated with portal vein fibrosis and portal hypertension.[221] Cerebral manifestations are uncommon in sporadic ergotism because of the absence of the LSD-like ergot compound, isoergine, in therapeutic preparations. Neurological presentation may be local (e.g., facial palsy) or diffuse (e.g., confusion, hypoalgesias) depending on the affected vascular bed. The differential diagnosis of the peripheral ischemia associated with ergotism includes Buerger's disease, atherosclerosis, vasculitis (collagen vascular diseases), Raynaud's disease, trauma, aneurysms, drug-induced vasospasm (epinephrine, dopamine), arterial dissection, and thromboembolic events.

Symptoms of ergotism from the chronic ingestion of food contaminated with *C. fusiformis* may resemble an acute gastrointestinal illness. Clinical features included nausea, vomiting, and drowsiness within 1–2 hours after ingestion followed by resolution of symptoms within 48 hours.[222] There were no symptoms or signs of vasoocclusive disease. The pearl millet from affected villages contained 15–174 g ergot alkaloids/kg with 15–199 mg total ergolines/kg in the contaminated grain. The main ergolines were agroclavine, elymoclavine, chanoclavine, penniclavine, and setoclavine. After discontinuation of ergotamine, the course of peripheral ischemia is variable.

Reproduction

The teratogenic potential of ergotamine probably is relatively low.[223] Although some case reports associated high doses of ergotamine with malformation consistent with vascular injury,[224] epidemiological studies have not demonstrated a clear increase in the incidence of malformation among infants exposed to ergotamine *in utero*.[225]

DIAGNOSTIC TESTING

Biomarkers

The effects of ergotism (i.e., ischemia) persist after the concentrations of ergotamines are no longer detectable in the plasma, in part because of tight tissue binding and the production of active metabolites.[192] Consequently, there is no direct correlation between plasma ergotamine concentrations and the clinical features of ergotism.[226] In a study of 25 patients receiving 7–60 mg ergotamine tartrate weekly over 1.5–30 years for migraine headache, adverse reactions (cramps, paresthesias, tremor, limb pain) occurred in some of these patients despite the absence of detectable serum concentrations of ergotamine.[227] Although adverse reactions are more frequent when ergotamine concentrations exceed 2 ng/mL, these side effects also occur at therapeutic concentrations. During a pharmacokinetic study, adverse reactions developed in 39% of subjects with plasma ergotamine concentrations exceeding 1.8 ng/mL, whereas these reactions occurred in about 25% of subjects with plasma ergotamine concentrations below 1.8 ng/mL.[195]

Ancillary Tests

Peripheral angiography together with the clinical history and laboratory tests (electrocardiogram, serum cardiac enzymes) help exclude other causes of ischemia (e.g., arterial embolism, arteriosclerosis obliterans, thromboangiitis obliterans, vasculitis, reflex sympathetic dystrophy). Typical angiographic findings in chronic ergotism include generalized, intense arterial spasm, formation of collateral vessels (chronic toxicity), and thromboses.[228] The main feature of the vascular spasm is the smooth, tapered narrowing of the arteries with minimal irregularity of the lumen.[220] The vascular spasm usually involves the medium-sized muscular arteries and small arteries/arterioles rather than large arteries (e.g., aorta, iliac artery).[229] Vascular spasm commonly occurs in the superficial femoral artery as well as in the external carotid artery with arterial occlusion becoming progressively more severe distally. The absence of atherosclerotic plaques in association with ultrasonographic evidence of occlusive disease suggests the presence of vasculitis or ergotism. The relief of the vasospasm by intraarterial vasodilators in a patient with a history of chronic ergot alkaloid use indicates ergotism.

Abnormalities

Case reports associate mildly elevated serum creatine kinase and alkaline phosphatase concentrations with the ingestion of excessive doses of bromocriptine, but the clinical significance of these changes remains unclear.[165] Ischemic complications of acute ergotamine intoxication include elevated serum concentrations of amylase, lipase, and hepatic aminotransferases.[230]

TREATMENT

Stabilization

Case reports associate cardiac arrest secondary to ventricular dysrhythmias (ventricular fibrillation, asystole) with the therapeutic use of ergotamine during coronary catheterization for vascular disease[202] and following massive acute overdose of ergotamine *and* caffeine.[187] Patients with abnormal vital signs or significant cardiovascular disease should be monitored for cardiac arrhythmias, hypoxemia, and cardiac ischemia. Respiratory depression is uncommon except after the parenteral administration of ergometrine in neonates.[216]

Decontamination

Although there are limited data on the efficacy of decontamination measures for ergot alkaloids, the administration of activated charcoal within 1–2 hours after the ingestion of >0.5 mg ergotamine/kg or >1–1.5 mg methysergide or bromocriptine/kg is a reasonable therapeutic option. Because sensitive individuals may develop vasospasm at therapeutic levels, each patient or parent of the patient should be warned about seeking medical attention if signs or symptoms of vascular insufficiency (cyanosis, pain, paresthesias, mottled coloring, cool extremities) develop, particularly within 24 hours after ingestion.

Elimination Enhancement

There are inadequate pharmacokinetic data to evaluate the effectiveness of measures to enhance the elimination of ergot alkaloids. However, the large apparent volume of distribution and small concentration of ergotamine in the blood indicate that hemodialysis, peritoneal dialysis, diuresis, and hemoperfusion would probably not be effective.

Antidotes

There are no antidotes.

Supplemental Care

Peripheral Ischemia

In cases of limb-threatening vascular insufficiency, angiography is an important diagnostic test to determine the etiology of poor peripheral perfusion. Withdrawal of all ergot alkaloids is mandatory. Reduction of precipitating factors (hepatic and renal insufficiency, smoking, febrile illness, thyrotoxicosis, drugs such as oral contraceptives and β-blocking agents) may improve peripheral perfusion. There are few clinical data on the treatment of peripheral ischemia associated with ergotism. Case reports associated the use of intravenous or intraarterial nitroprusside therapy with reversal of severe peripheral ischemia over several days at doses of 2–5 μg/kg/min (0.3–1.2 μg/kg/min in a neonate).[188,231,232] Adjunctive treatment includes the use of oral vasodilators [prazosin (1 mg t.i.d.), captopril (50 mg t.i.d.), nifedipine] for at least several days after the peripheral ischemia resolves. In less severe cases, the use of oral agents may obviate the need for intravenous therapy. Case reports document the use of a variety of therapeutic modalities for refractory cases of ergot-induced vasospasm including sympathetic blockade,[233] surgical sympathectomy, low-molecular-weight dextrans, hyperbaric oxygen, mechanical intraarterial vasodilation,[234] and epidural spinal cord stimulation.[235] However, the use of these modalities remains controversial because of the lack of clinical trials. Heparin and antiplatelet drugs are recommended as prophylactic agents to prevent secondary thrombosis related to stasis.[231] Hospitalization is usually required to monitor the effectiveness of drug therapy.

Coronary Ischemia

Case reports indicate that ergonovine-induced coronary vasospasm during assessment of variant angina pectoris generally responds to sublingual or intravenous nitroglycerin.[202,236] Another case report suggested that the administration of intravenous nitroglycerin at maximum doses of 3.2 μg/kg can reverse peripheral ischemia associated with the use of ergot compounds.[237]

FUMONISIN

HISTORY

Gibberella moniliformis Wineland (*Fusarium verticillioide*) was the predominate fungus isolated from moldy corn involved with an epidemic of equine leukoencephalomalacia in South Africa during 1970. Subsequent studies of this fungus confirmed the pathological changes (i.e., liquefactive necrosis of cerebral white matter) associated with exposure to this fungal species.[238] As a result of investigations into the etiology of equine leukoencephalomalacia, fumonisin B_1 was isolated from *Gibberella moniliformis* Wineland in 1988.[239] During 1989, epidemics of equine leukoencephalomalacia and porcine pulmonary edema in the United States were associated with consumption of feed contaminated with fumonisins.[240]

PRINCIPAL TOXINS

Structure

The basic chemical structure is a C-20 diester of propane-1,2,3-tricarboxylic acid and a pentahydroxyicosane containing a primary amino group. These mycotoxins are long-chain polyhydroxyl alkylamines with two propane tricarboxylic acid moieties esterified to hydroxyls on adjacent carbons. Fumonisin B_1 is the most prevalent of the fumonisin family of mycotoxins that include at least 15 other compounds (e.g., FB_2, FB_3, FB_4, FA_1, and FA_2; see Figure 45.5).[241] Fumonisin B_2 is the C-10 deoxy analogue of fumonisin B_1. FA_1 and FA_2 are the *N*-acetyl derivatives of FB_1 and FB_2, respectively. These fumonisin mycotoxins form through the condensation of the amino acid alanine and an acetate-derived precursor.[242]

Physiochemical Properties

Fumonisin mycotoxins are highly polar, water-soluble compounds with no obvious chromophore in ultraviolet or visible spectrum. Fumonisin B_1 is relatively heat-stable with half-lives 10 minutes at 75 °C (167 °F) and 480 minutes at 150 °C (302 °F).[243] Ethanol fermentation of fumonisin-contaminated corn does not destroy these mycotoxins,[244] and food processing does not significantly alter the concentration of deoxynivalenol in food, even at high temperatures.[245] However, dry milling redistributes the fumonisin compounds into different constituents of maize. Wet milling extracts some of the mycotoxins into the steeping water.[111]

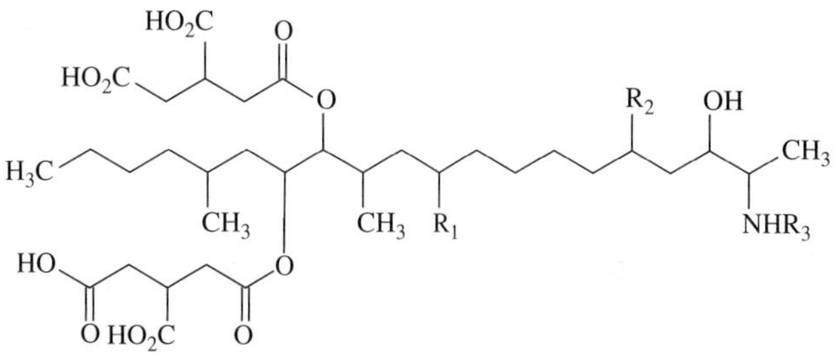

HO$_2$C, HO$_2$C, O, O, R$_2$, OH, CH$_3$, H$_3$C, CH$_3$, CH$_3$, R$_1$, NHR$_3$, HO, O, O, HO$_2$C, O

	R_1	R_2	R_3
Fumonisin B_1	OH	OH	H
Fumonisin B_2	H	OH	H
Fumonisin B_3	OH	H	H
Fumonisin A_1	OH	OH	$COCH_3$

FIGURE 45.5. Structure of common fumonisin mycotoxins.

EXPOSURE

Sources

Penicillium and *Aspergillus* species usually contaminate foods during drying and storage, whereas *Fusarium* species are widespread plant pathogens that invade the plant and produce mycotoxins before or immediately after harvest, primarily in maize.[15] Except under extreme conditions, the concentrations of fumonisin do not increase during storage. The main sources of fumonisin mycotoxins are *Gibberella moniliformis* Wineland (*Fusarium verticillioides, Fusarium moniliforme*), and, to a lesser extent, *F. proliferatum* (Matsush.) Nirenberg ex Gerlach & Nirenberg. Other species that produce substantially smaller amounts of mycotoxins include *F. napiforme* Marasas, P.E. Nelson & Rabie, *F. nygamai* L.W. Burgess & Trimboli, and *Alternaria alternata* (Fr.) Keissl.[246] Fumonisin B_1, B_2, and B_3 are the only fumonisin compounds produced in substantial quantities by *Gibberella moniliformis* (*F. verticillioides*) with B_1 being the most significant toxin.[247] The production of FB_1 in strains of *Gibberella moniliformis* is highly variable, and significant accumulation of FB_1 in maize occurs during adverse weather conditions (drought) and insect stress.[248] Fumonisin compounds are uncommon in food products other than maize and maize products; these mycotoxins are infrequent contaminates of rice, beer, sorghum, asparagus, and mung beans.[111,249] Clinically significant concentrations of FB_1 are not present in meat or dairy products from animals fed grain contaminated by fumonisin compounds.[250] These mycotoxins can be produced in liquid cultures.[251]

Environmental Conditions

Fumonisin compounds occur primarily in maize grown under warm, dry conditions, and heat stress is a determinant of the risk of fumonisin contamination of corn.[252] Many *Fusarium* species grow optimally at 24°C–26°C (~75°F–79°F) with a minimum water activity (a_w) of 0.90,[253] and *Gibberella zeae* (Schwein.) Petch (*Fusarium graminearum*, Gibberella ear rot) proliferates optimally under rainy conditions at 26°C–28°C (~79°F–82°F). Minimum growth requirements for *Gibberella moniliformis* include minimum temperature 2°C–5°C (~36°F–41°F, maximum temperature of 32°C–37°C (~90°F–99°F), and minimum water activity of 0.87.[15] Mycotoxin production occurs at water activity ranging from 0.92–0.97.[254]

Foodstuffs

Gibberella moniliformis is a pathogen of cereals (e.g., maize, rice), and fumonisin compounds are common contaminates of corn-based foods including beer.[255,256] Fumonisin B_1 is the most common fumonisin mycotoxin with concentrations up to 250mg/kg in maize.[158] The estimated daily intake of fumonisin B_1 ranges from about 0.02–1.0µg/kg body weight.[111] In a study of Brazilian maize, over 95% of the samples contained fumonisin B_1 at concentrations ranging from 0.6–18.5mg/kg.[257] Mean fumonisin B_1 concentrations were substantially higher in commercial Argentinean samples of corn meal (mean = 0.56mg/kg) than corn flakes (range: 0.02–0.038mg/kg).[258] The median and mean concentrations of fumonisin B_1 in samples of maize from the Netherlands were 420ng/g and 1,359ng/g, respectively.[259] In a study of corn-based human foods in Kansas, about 20% of the samples contained detectable (i.e., >40ppb) concentrations of fumonisins as measured by HPLC with a fluorescent detector.[260] About 50% of the corn flour samples contained detectable amounts of fumonisins at concentrations ranging from 0.765–9.995ppm. Corn syrup and grits are brewing adjuncts in some beer-production processes. In a study of *Gibberella moniliformis* strains isolated from European (Italy, Spain, Poland, France) corn, sorghum, wheat, barley, and mixed feed, the production of fumonisin B_1 in corn cultures ranged from 0.7–4,100µg/g.[261] Fumonisin B_1 and B_2 occur to a limited extent in beer. In a study of 26US commercial beers, 86% of the beer samples contained detectable concentrations of fumonisin B_1 with total fumonisin (B_1, B_2) concentrations in positive samples ranging from 0.3–12.7ng/mL.[262] Treatment of corn-based food products with calcium hydroxide solution substantially reduces the concentrations of fumonisins. Typically, alkali-processed corn foods (tortilla chips, nacho chips, taco shells, corn tortillas) contain <100ng total fumonisins/g.[263] Food processing methods currently used in North America and western Europe do not usually destroy fumonisin B_1.[241] Stability studies indicate that fumonisins are stable in γ-irradiated maize for at least 6 months at 25°C (77°F) and for at least 4 weeks at 40°C (104°F).[264]

DOSE RESPONSE

In rodent studies on male and female Sprague-Dawley rats, the administration of 0, 15, 50, and 150ppm fumonisin B_1 in the feed for 4 weeks was not associated with obvious signs of toxicity (e.g., weight loss, reduced feed consumption).[265] However, microscopic examination revealed renal toxicity (cortical nephrosis) in males fed ≥15ppm and females fed ≥50ppm along with hepatoxicity in males and females fed 150ppm. Liver and kidney samples contained elevated concentrations of sphinganine and the sphinganine/sphingosine ratio.

TOXICOKINETICS

The absorption of fumonisins from the gastrointestinal tract is relatively poor, and the distribution and elimination of these compounds is relatively rapid.[111] These mycotoxins are not metabolized by the cytochrome P450 enzyme system.

HISTOLOGY AND PATHOPHYSIOLOGY

Fumonisin compounds are the first mycotoxins known to inhibit sphingolipid biosynthesis and alter lipid metabolism. In experimental studies, fumonisins competitively inhibit substrates (sphinganine, fatty acyl coenzyme A) of sphinganine *N*-acetyltransferase. The inhibition of ceramide synthase and the disruption of *de novo* sphingolipids synthesis by fumonisin B_1 results in an increase of sphinganine levels and the sphinganine/sphingosine ratio.[266] This enzyme is an important part of the pathway leading to the formation of sphingomyelin and complex sphingolipids. Consequently, fumonisin mycotoxins inhibit cell growth and sphinganine accumulates as a result of the blocking of the synthesis of sphingolipids.[158]

CLINICAL RESPONSE

Animals

The administration of fumonisin to horses produces equine leukoencephalomalacia characterized by necrotic lesions in the brain.[267] Fumonisin B_1 acts as a competitive inhibitor of sphingosine *N*-acetyltransferase, and the resulting blockade of this enzyme causes inhibition of complex sphingolipid biosynthesis and accumulation of sphingosine.[268] Porcine pulmonary edema is another unique mycotoxicoses caused only by the ingestion of fumonisin compounds by swine.[269]

Humans

Despite cases of exposure to maize contaminated with FB_1 concentrations exceeding 100 mg/kg, there are no *confirmed* cases of acute fumonisin toxicity in humans.[241] The consumption of unleavened bread made from fumonisin-contaminated maize and sorghum produced a self-limited gastrointestinal illness characterized by diarrhea and abdominal pain. None of the subjects complained of vomiting. Fumonisin B_1 concentrations in samples of sorghum and maize ranged from 0.14–7.8 mg/kg and 0.25–64.7 mg/kg, respectively.[270] Although guidance levels are available for fumonisins,[271] there is no direct evidence that these mycotoxins produce epidemics of human illness.

Carcinogenicity

There is inadequate evidence of the carcinogenicity of fumonisin B_1 in humans and limited evidence of carcinogenicity in experimental animals. A 2-year study of rodents demonstrated an increased incidence of renal tubule neoplasms in male F344/N rats fed daily fumonisin doses 7.5 mg/kg and of hepatocellular carcinomas in female B6C3F1 mice fed 9.7 mg/kg.[266] The incidence of neoplasms was not increased in female F344/N rats and male B6C3F1 mice fed similar doses. Consequently, the IARC classifies this mycotoxin as possibly carcinogenic in humans (Group 2B). In Chinese provinces with high endemic rates of primary liver cancer, fumonisins frequently contaminate corn-based food along with other aflatoxins and trichothecene compounds.[272] However, the contribution of fumonisins to the increased rates of liver cancer present in these areas is unknown.

DIAGNOSTIC TESTING

Analytical Methods

Chromatographic methods for the detection of fumonisins include economical thin-layer chromatography, liquid chromatographic methods based on solid-phase extraction of solvent extracts, and HPLC using fluorescence detection.[273] Monoclonal and polyclonal antibodies are available to detect fumonisin $B_{1\text{-}3}$.[274,275] Detection limits for fumonisins using HPLC with detection by evaporative light scattering and electrospray detectors were 10–50 ng and <1 ng, respectively.[276] Fluorometric, thin-layer chromatographic and ELISA techniques have limits of quantitation in the range of 0.5–1 μg/g.[151]

Biomarkers

There are no validated biomarkers for human exposure to fumonisins.

TREATMENT

Treatment is supportive.

OCHRATOXINS

HISTORY

In 1965, ochratoxin A was the first mycotoxin isolated from *Aspergillus alutaceus* Berk. & M.A. Curtis.[8] Subsequently, ochratoxins were discovered over a dozen other *Aspergillus* and *Penicillium* species.

PRINCIPAL TOXINS

Structure

Ochratoxins are secondary metabolites produced by certain *Penicillium* and *Aspergillus* species that are substituted isocoumarin compounds linked to L-phenylalanine and classified as pentaketides (see Figure 45.6). Ochratoxin α is the only ochratoxin that lacks the phenylalanine moiety. Unlike most other mycotoxins, ochratoxins contain chlorine atoms. Ochratoxin A (CAS No. 303-47-9) is the most abundant and toxic of the five metabolites in the ochratoxin group.[277]

Physiochemical Properties

Ochratoxin A (7-L-β-phenylalanylcarbonyl-5-chloro-8-hydroxy-3,4-dihydro-3-*R*-methylisocoumarin, MW 403.82) is a colorless, crystalline compound with a melting point of about 90 °C (194 °F). This compound is slightly soluble in water, but ochratoxin A is soluble in dilute aqueous sodium bicarbonate and polar compounds. Ochratoxin A is weakly acidic with pK_a values of 4.2–4.4 and 7.0–7.3 for the carboxyl group of the phenylalanine moiety and the phenolic hydroxyl group of the isocoumarin portion, respectively.[277] Under experimental conditions acid hydrolysis of ochratoxin A

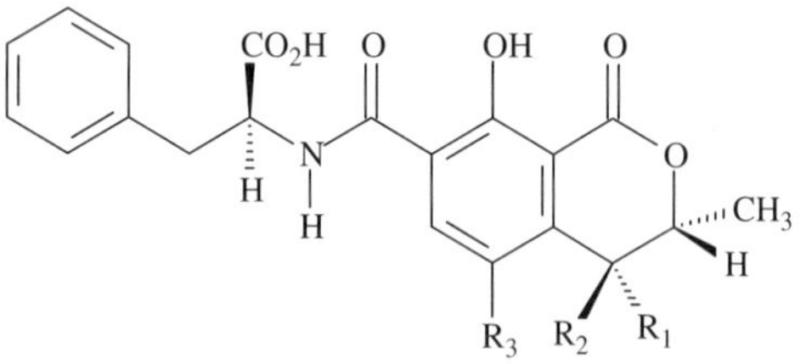

	R_1	R_2	R_3
Ochratoxin A	H	H	Cl
Ochratoxin B	H	H	H
(4*R*)-4-Hydroxyochratoxin A	OH	H	Cl
(4*S*)-4-Hydroxyochratoxin A	H	OH	Cl

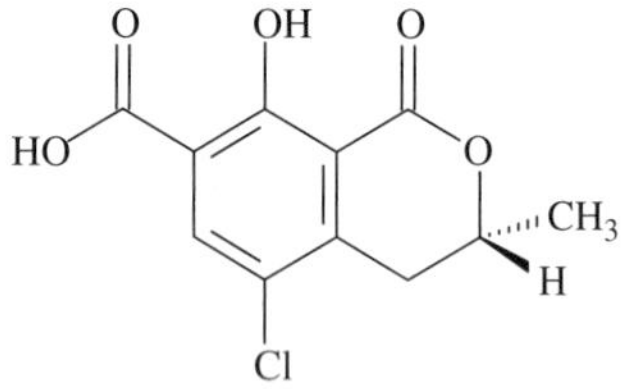

Ochratoxin α

FIGURE 45.6. Chemical structures of ochratoxins.

yields phenylalanine and the isocoumarin structure called ochratoxin α. The later product also occurs in the intestines of experimental animals fed ochratoxin A-containing feed as a result of microbial degradation.[278] Because ochratoxin A is relatively resistant to food processing techniques, this compound often occurs in cereal and meat or dairy products from animals reared on contaminated feed.[279] However, meats from ruminants do not contain ochratoxin A because the microflora in the gastrointestinal tract metabolizes this compound to ochratoxin α.[280]

EXPOSURE

Sources

Ochratoxin A is a contaminant in cereals, predominately in Europe, that results from the contamination of cereal during storage rather than during plant growth. Over a dozen *Aspergillus* species and *Penicillium verrucosum* Dierckx can produce ochratoxins. The main ochratoxin A-producing species are *Penicillium verrucosum* Dierckx, *Aspergillus ochraceus* G. Wilh., and several related *Aspergillus* species including a few isolates of *Aspergillus niger* var. *niger* Tiegh.[111] Ochratoxin contamination occurs in a wide range of agricultural products (wheat, barley, oats, maize, coffee beans, mixed feed) from Canada, Denmark, Poland, Sweden, Yugoslavia, and the United States.[337] Ochratoxin A and, to a much lesser extent, ochratoxin B are natural contaminants of plant material. The natural habitat of these ochratoxin-producing species is drying or decaying vegetation, seeds, nuts and fruits.

The most common fungi involved in the production of ochratoxin A in cool and temperate regions are *Penicillium verrucosum* Dierckx and *Penicillium nordicum* Dragoni & Marino, whereas *Aspergillus alutaceus* Berk. & M.A. Curtis is the primary producer of ochratoxins in topical and semitropical regions. The fungal species involved with ochratoxin production varies between locations because the growth requirements vary significantly between different ochratoxin-producing species. In a study of western Canadian grains (wheat, barley, oats, rye) using combined morphological and chemotaxonomic methods, the main ochratoxin-producing species were *P. aurantiogriseum* Dierckx and *P. freii* Frisvad & Samson, whereas the most common species in Denmark were *P. verrucosum* Dierckx, *P. aurantiogriseum* Dierckx (*P. cyclopium*), and *P. freii* Frisvad & Samson.[281] Approximately 72% of the isolates were mycotoxin producers in contrast to the typical range of 2–15% of the grain samples.[282] *P. verrucosum* occurs almost exclusively in grain from temperate zones.[19] Cereals from the tropics are unlikely to contain ochratoxin A because *P. verrucosum* Dierckx does not grow well in the tropics.[111] *Penicillium nordicum* Dragoni & Marino contaminates meat products and cheese, whereas *Aspergillus alutaceus* Berk. & M.A. Curtis contaminates a wide variety of foods (e.g., processed meat, smoked or salted fish, dried fruits, spices, beans, dried peanuts).[283] Other ochratoxin producing species include *A. sulphureus* (Fresen.) Thom & Church, *A. sclerotiorum* G.A. Huber, *A. melleus* Yukawa, *A. fonsecaeus* Thom & Raper, *A. niger* var. *niger* Tiegh., and *A. carbonarius* (Bainier) Thom.[158] *A. carbonarius* grows in high temperatures, and this species is a source of ochratoxin A in grapes, dried vine fruits, wine, and coffee.[111] *Penicillium viridicatum* Westling does not produce ochratoxins.[158] In a study of blood samples from the Lebanese Red Cross, about one-third of the samples demonstrated detectable amounts of ochratoxin A with a mean concentration of 0.17 ± 0.01 ng/mL and a range of 0.1–0.87 ng/mL.[284]

Environmental Conditions

Species, humidity, and temperature are important determinants of mycotoxin formation. The two ochratoxin-producing fungal genera differ in the conditions required to grow and produce mycotoxins.[285] The genus *Penicillium* grows well in cold environments, and species from this genus are the major ochratoxin producers in crops grown in colder climates (Scandinavia, Canada).[286] These species grow more slowly than the aflatoxin-producing *Aspergillus* species.[15] Ochratoxin-producing strains accounted for approximately 28–50% of the *A. alutaceus* Berk. & M.A. Curtis strains isolated from crops grown in warmer climates (e.g., Australia).[287] Under experimental conditions of optimum temperature range of ochratoxin production by *A. alutaceus* was 12°C–37°C (~54°F–99°F) compared with 4°C–31°C (39°F–88°F) for *P. aurantiogriseum* Dierckx (*P. cyclopium*).[288] The minimum water activity values for the formation of ochratoxin in *A. alutaceus and P. aurantiogriseum* were 0.83–0.87, and 0.87–0.90, respectively. The substrate also determines the amount of ochratoxin formed. *P. verrucosum* Dierckx strains produce more ochratoxin on carbohydrate-rich cereals (maize, wheat), whereas *A. alutaceus* produces higher ochratoxin concentrations on oilseeds and protein-rich seeds (groundnuts, soybeans).[289] Although ochratoxin A is stable during refrigeration when stored in ethanol, decomposition of ochratoxin occurs after a few days following exposure to fluorescent light.[290]

Foodstuffs

The main sources of ochratoxin A are cereal and cereal products with the level of contamination typically <200μg/kg.[291] Ochratoxin A is a contaminant of cereals and some beans (cocoa, coffee, soya) in many parts of the world as well as dried fruits, wheat, beer, pork and poultry meats, and nuts.[19] Coffee contains low concentrations of ochratoxin A. In a UK retail survey of 20 samples of roast and ground coffee, the ochratoxin concentration ranged from the limit of detection (0.1μg/kg) to 8μg/kg.[292] Most foodstuffs contain ochratoxin A concentrations below 200μg/kg.[337] A broad survey of over 500 samples of American wheat, barley, green coffee, and roasted coffee indicated that almost all samples contained ochratoxin A concentrations <5ng/g with only four wheat samples and one barley sample above 5ppb.[293] The maximum ochratoxin A concentration (31.1ng/g) occurred in a wheat sample. In a survey of Danish foods, pork contained the highest concentration of ochratoxin A with 81.5% of the samples containing concentrations above the limit of detection (0.02μg/kg).[291] Ochratoxin A rarely occurs in beers due to losses during the mashing step of the brewing process.[125] Ochratoxin B occurs very rarely in foodstuffs, whereas other ochratoxins generally do not occur in surveys of foodstuffs. Protozoan and bacterial enzymes in the forestomachs of ruminants cleave ochratoxin A to ochratoxin α, and therefore ochratoxin is not usually a contaminant of food sources from ruminants.[294] Following the ingestion of ochratoxin-containing feed, ochratoxin A is present in single-stomach animals (e.g., pigs), particularly in the kidney.[295] However, the ochratoxin concentrations are much lower in the liver and muscles.

DOSE RESPONSE

Ochratoxin-induced nephrotoxicity occurs in farm animals at feed concentrations of 200μg/kg. Ochratoxin intake in humans has not been causally related to specific diseases, and the effect of specific ochratoxin concentrations is unclear. The World Health Organization proposed a maximum limit for ochratoxin of 5μg/kg in cereals and 20μg/kg in cereal products.[111]

TOXICOKINETICS

There are few toxicokinetic data on humans. Animal studies suggest that the primary location of ochratoxin A absorption is the duodenum and jejunum, and the kidney is the tissue with the highest concentration of ochratoxin A.[296] The bioavailability of ochratoxin A in animal species ranges from about 40–65%.[297] Ochratoxin A is highly bound (i.e., 99%) to serum proteins, primarily albumin, with little ochratoxin A diffusing into erythrocytes. Biotransformation of ochratoxin A includes cleavage to ochratoxin α and phenylalanine as well as microsomal hydroxylation to nontoxic metabolites, such as 4-hydroxyochratoxin A.[298] Animal studies suggest that renal and, to a much lesser extent, fecal excretion account for most of the elimination of ochratoxin A. Excretion of ochratoxin in breast milk is small.[337] In a study of 111 samples of breast milk from donor mothers in Italy, about 20% (i.e., 22) were positive for ochratoxin A ranging from 0.1–12μg/kg.[299] Limited data on human toxicokinetics indicate that the elimination half-life of ochratoxin A is relative long (i.e., about one month) compared with rodents (1–5 days).[300,301]

HISTOLOGY AND PATHOPHYSIOLOGY

Ochratoxin A competitively inhibits phenylalanine-tRNA synthetase and protein synthesis in bacteria and hepatic cell cultures. The main target of toxicity is the kidney and the proximal renal tubule. In animal studies, ochratoxin A produces kidney damage when fed to ducklings, dogs, chickens, rats, mice, and trout. This mycotoxin is a suspected cause of porcine nephropathy in Denmark based on the replication of similar renal lesions in experimental animals fed subchronic doses (1mg/kg over 3 months) of ochratoxin A.[302]

Mycotoxic porcine nephropathy is a disease of swine kidneys that resembles Balkan endemic nephropathy. Because of the association of porcine nephropathy with ochratoxin A, this mycotoxin is a suspected cause of Balkan endemic nephropathy. However, there are inadequate epidemiological data to confirm a causal relationship between ochratoxin A with Balkan endemic nephropathy.

CLINICAL RESPONSE

Balkan Endemic Nephropathy

Although the incidence of Balkan endemic nephropathy is declining, this chronic tubulointerstitial kidney disease remains prevalent in rural areas of Serbia, Bosnia, Croatia, Bulgaria, and Romania.[303] This disease affects primarily adults with an age of onset at 30–50 years and a slow progression to end-stage renal failure.[304] There are no unique clinical features of this disease that separates Balkan endemic nephropathy from other causes of renal failure.[305] Clinical features reflect chronic, progressive renal failure with fatigue, anemia, lassitude, headache, weight loss, anorexia, and evidence of renal tubular damage (low-molecular-weight proteins). Table 45.8 lists the clinical characteristics of Balkan endemic nephropathy. Salt retention and hypertension are uncommon features of this disease.

Although sporadic cases of nephropathy occur in areas (e.g., Tunisia) with high ochratoxin content in foods outside the Balkans,[306,307] the disease appears primarily in geographic clusters of rural farming communities with high humidity and high rainfall.[308] Within the geographic distribution of the disease, cases continue to occur in afflicted villages, whereas adjacent villages without a history of Balkan endemic nephropathy remain disease-free. There is a slight predominance of female subjects (1.5:1).[305] Tumors of the urinary tract system frequently occur in patients with Balkan endemic nephropathy, particularly in the renal pelvis and ureters.

TABLE 45.8. Characteristics of Balkan Nephropathy

Characteristic	Features
Epidemiological	Residence in endemic village
	Family history renal disease or urothelial tumors
	Farming
Clinical	Progressive renal insufficiency
	Normochromic anemia
	Polyuria, polydipsia, nocturia, salt loss
	Urothelial tumors
	Abnormalities of urinalysis
	Hypertension (rare)
Physiological	Low-molecular-weight proteinuria
	Impaired concentrating ability and urinary acidification
	Increased uric acid excretion
	Decreased glomerular filtration rate
	Renal insufficiency
Pathophysiological	Focal interstitial edema and sclerosis
	Focal and segmental tubular atrophy
	Hyperplastic intima changes with arteriolar narrowing
	Dilation of renal lymphatics
	Glomerular sclerosis

Source: Adapted from Ref 308.

The exact etiology of this disease is unclear.[309] Because of the striking pathological similarities between ochratoxin A-induced porcine nephropathy and Balkan endemic nephropathy, ochratoxin A is a suspected cause of this disease; however, high concentrations of ochratoxin A occur both in healthy and diseased populations. Circumstantial evidence for a causal link between ochratoxin A and Balkan endemic nephropathy include a higher incidence of deaths from nephropathy during periods that favor fungal growth and a higher frequency of ochratoxin A in foodstuffs from villages with the disease compared with villages without the disease. Although there are inadequate scientific data to establish a causal link between Balkan endemic nephropathy and ochratoxin A, epidemiological studies suggest that the cause is multifactorial [e.g., genetic predisposition, contamination of well water (polycyclic aromatic hydrocarbon), coronavirus, citrinin].[310,311] Alternate etiologies for Balkan endemic nephropathy include the contamination of food sources by seeds of *Aristolochia clematitis*.[312] Seeds of *A. clematitis* (birthwort) contain about 0.65% aristolochic acid, which is associated with the onset of proximal renal tubular necrosis, renal failure, and urothelial cancer following the chronic use of aristolochic acid-containing herbs.[313]

Carcinogenicity

The IARC lists ochratoxin A as a possible human carcinogen (Group 2B) with inadequate evidence in humans and sufficient evidence in experimental animals. Ochratoxin A is carcinogenic to the renal tubular epithelium (renal cell adenomas and carcinomas) in male mice and in rats as well as hepatocellular tumors in mice.[282] *In vitro* studies suggest the ochratoxin A induces renal tumors in animals by a mechanism other than covalent interactions of a reactive metabolite with DNA.[314]

DIAGNOSTIC TESTING

Analytical Procedures

Extraction procedures for ochratoxins typically involve the use of polar solvents (e.g., chloroform) after acidification of the specimen, and identification procedures involve the use of thin-layer chromatography with UV irradiation for the visualization of the fluorescent ochratoxin compounds (detection limit near 12 µg/kg).[87] Other detection methods include HPLC (detection limit about 1 µg/kg) and reversed phase HPLC with fluorescence detection,[315] as well as enzyme-linked immunosorbent assay (limit of quantitation about 2.5 µg/kg) and radioimmunoassay for ochratoxin A (detection limit about 0.2 µg/kg).[316] The mean serum to whole blood ratio of ochratoxin A is approximately 1.9–2.0.[317]

Biomarkers

Blood

Ochratoxin A is a ubiquitous mycotoxin occurring in blood samples from residents of countries without a history of Balkan endemic nephropathy. A relatively high percentage of serum samples from these residents contain low concentrations of ochratoxin A. In most European countries, the mean concentration of ochratoxin A in blood samples from the general population does not usually exceed 1 ng/mL.[323] Ochratoxin A was detectable in all plasma samples from 406 Scandinavian blood donors with a mean plasma concentration of approximately 0.2 ng/mL and a 95th percentile of 0.48 ng/mL.[318] The plasma ochratoxin A did not correlate significantly with specific food items (e.g., cereal products, wine, beer pork, coffee, fruit). Using a sensitive HPLC procedure, about 57% of the serum samples from German residents contained detectable concentrations of ochratoxin A with a mean concentration of 0.6 ng/g (range: 0.1–14.4 ng/g).[319] In a study of serum from 368 Swiss subjects, the ochratoxin A concentration ranged from 0.06–6.02 ng/g with medians of 0.24 ng/g and 0.30 ng/g for women and men, respectively, as measured by HPLC with enhanced fluorescence detection.[317] Plasma samples from 144 Canadian residents of 16 different locations contained a mean ochratoxin A concentration of approximately 0.9 ng/mL with a range of 0.29–2.37 ng/mL.[320]

In general, the serum concentrations of ochratoxin A are higher in residents of eastern European countries where endemic nephropathy occurs than in other countries. In a study of Yugoslavia residents, approximately 7% of 639 serum samples from local residents contained ochratoxin A at concentrations ranging from 1–57 ng/g as measured by enzymatic spectrofluorometric procedures.[321] In a study of serum from Bulgarian residents, about 14% of the samples contained detectable concentration of ochratoxin A, and the mean serum concentration was approximately 14 ng/g.[322] However, the mean concentration of ochratoxin A in plasma samples from residents of five Croatian cities was 0.39 ng/mL with substantial variation between different cities.[323] In a nonrandom sample of workers exposed to ochratoxin A in processing of foods (coffee, cocoa beans, spices), the serum ochratoxin A concentrations ranged from 0.94–3.28 ng/mL, whereas ochratoxin A concentrations in air sampled from the breathing zones of these workers ranged from 0.006–0.087 ng/m^3.[324]

Urine

In a study of children from Sierra Leone the urinary concentrations of ochratoxin A ranged from 0.06–148 ng/mL.[157]

Abnormalities

Abnormalities in urine sediment from patients with Balkan endemic nephropathy usually are limited. Proteinuria is mild and intermittent. The proteinuria increases with progressive renal failure. Macroscopic hematuria is uncommon and usually suggests the presence of urothelial tumors. Tubular proteinuria and defects in urinary acidification precede reduction in glomerular filtration rates. Impaired concentration ability is a late manifestation of Balkan endemic nephropathy.

Health Surveillance

The tolerable daily intake (TDI) of ochratoxin A proposed by the Joint FAO/WHO Expert Committee on Food Additives is 16 ng/kg body weight.[325] The basis of this TDI is the lowest adverse effect level in pigs (kidney damage) and a safety factor of 500.

TREATMENT

Currently, federal regulations do not control ochratoxin levels because of the lack of proof that human disease results from ochratoxin consumption. The treatment of Balkan endemic nephropathy is supportive.

PATULIN

Patulin initially was evaluated as an antibiotic (claviformin, clavatin, clavacin, expansine, penicidin) in the early 1940s.[158] *Aspergillus clavatus* Desm. and *Penicillium expansum* Link are the most important patulin-producing species, but several other species of *Penicillium* [*P. griseofulvum* Dierckx, *P. roqueforti* Thom, *P. vulpinum* (Cooke & Massee) Seifert & Samson], *Aspergillus* and *Byssochlamys* also form patulin.[326] *A. clavatus* Desm.

contaminates cereal stubble and malted barley residues, whereas *P. expansum* Link grows on fruits, such as apples and pears. Patulin is relatively unstable, except in aqueous acidic conditions at warm temperatures. This mycotoxin is a lactone. Although the optimum growth temperature for *P. expansum* Link is 25°C (77°F), this fungal species can grow at low temperatures (−2°C/~28°F).[15] Thermal processing of foods contaminated with patulin causes only a modest reduction in the concentration of patulin.[327] The major source of exposure to patulin for the general population is the consumption of apple juice or other apple products made from contaminated fruits.[328] Occasionally, fruit products with spontaneous brown rot (e.g., cherry, tomatoes, apricots, peaches, pineapples, bananas, grapes) contain low concentrations of patulin. Obvious decay usually occurs in apples contaminated with patulin.[329] Samples of apple juice typically contain patulin concentrations below 50µg/L (ppb).[330,331] In a study of four Italian apple products, approximately one-third of the samples had detectable concentrations of patulin with the positive samples demonstrating a mean patulin concentration of 26.7µg/L (range: 1.4-74.2µg/L).[327] There was no difference in the patulin concentration between products grown under organic and nonorganic conditions. There are few data on the effects of patulin in humans.

Limited human data suggest that patulin rapidly degrades in the human body. In a subject ingesting apple juice containing 50µg patulin, no patulin was detected in a blood sample drawn 1 hour after ingestion as measured by GC/MS (limit of detection: 200ng/L).[332] Clinical trials involving the nasal insufflation of patulin did not detect any significant adverse effects.[333] The estimated maximum tolerable daily intake of patulin based on animal studies was 0.4µg/kg as determined by the Joint Expert Committee on Food Additives of the World Health Organization.[334]

RUBRATOXINS

Penicillium rubrum Sopp is the main fungi associated with the production of rubratoxins, primarily rubratoxin A (CAS RN:22467-31-8, $C_{26}H_{32}O_{11}$) and rubratoxin B (CAS RN:21794-01-4, $C_{26}H_{30}O_{11}$).[166] Moss et al determined the structure of rubratoxin A in 1970.[335] Figure 45.7 and Figure 45.8 display the chemical structures of rubratoxin A and B, respectively. These mycotoxins are present in a variety of animal feed, often in association with *Aspergillus flavus* Link. Rubratoxins cause liver, renal, and central nervous system damage in experimental animals. In a study of rats, the intraperitoneal injec-

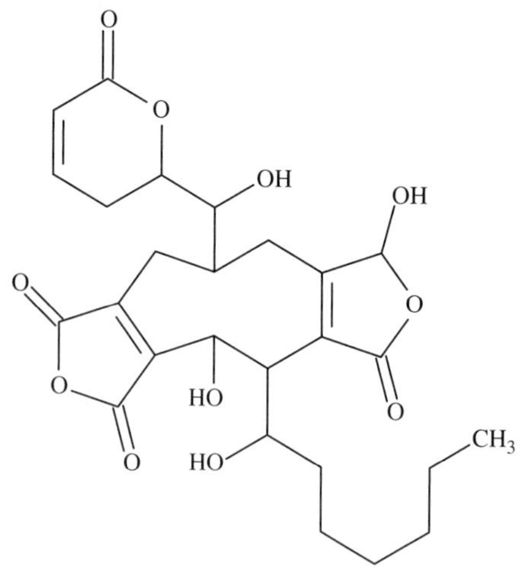

FIGURE 45.7. Chemical structure of rubratoxin A.

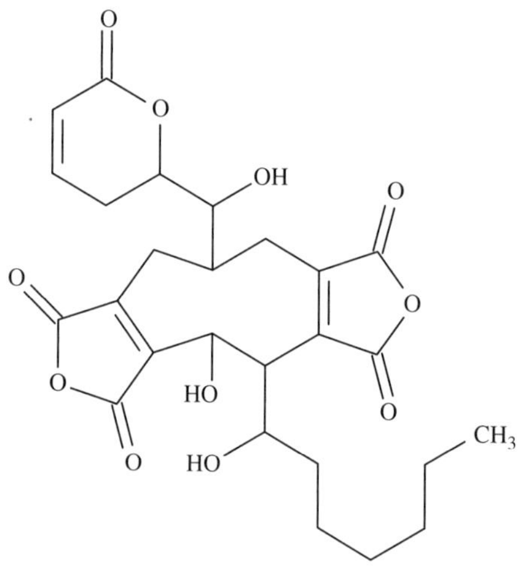

FIGURE 45.8. Chemical structure of rubratoxin B.

tion of 0.36mg rubratoxin B/kg produced hepatoxicity manifest by necrosis of the hepatocytes, disruption of cell membranes, irreversible degenerative changes in cytoplasmic organelles, and intra-sinusoidal hemorrhage.[336] The role of these mycotoxins in human disease is not well defined. Rubratoxins may act synergistically with aflatoxins to produce animal diseases including moldy corn toxicosis.[339]

TRICHOTHECENES

HISTORY

Trichothecene mycotoxins are a group of over 40 sesquiterpene compounds produced by a variety of widely distributed fungi.[337] In 1948, Freeman et al isolated the first trichothecene mycotoxin from the mold *Trichothecium roseum* (Pers.) Link.[3] In contrast to the more limited taxonomic distribution of aflatoxins, at least seven genera of fungi contain trichothecene mycotoxins. The trichothecene mycotoxin, T-2 toxin, was a suspected etiological agent for numerous ATA-related fatalities that occurred in Russia during the latter part of World War II.[4]

Stachybotryotoxicosis, a disease in Russian horses that caused pancytopenia and mucosal ulcerations, was the first well-described veterinary mycotoxicosis.[338] The recognition of the importance of clean, uncontaminated straw as a factor in preventing this disease in transport horses by Nikita Khrushchev was an important reason for his early rise to power, in part, because of the importance of these horses to the pre-World War II Russian economy.[339] Trichothecene compounds (i.e., T-2 toxin) were alleged chemical warfare agents in the Vietnam War. The American government described the release of small yellow particles (yellow rain) from low-flying planes on rural civilian populations in Cambodia, Laos, and Afghanistan. Victims reported painful skin lesions, lightheadedness, dyspnea, gastrointestinal irritation, and petechiae followed by leucopenia, anemia, sepsis, and death in susceptible individuals.[340] Survivors allegedly developed a radiation-type sickness that included resolution of bone marrow depression over 3 weeks to 2 months. An autopsy of a suspected victim, who died 31 days after the alleged exposure, revealed interstitial myocardial hemorrhage, acute myocarditis, early micronodular cirrhosis secondary to diffuse hepatic necrosis, and hepatorenal damage.[341] Typical symptoms associated with a yellow rain attack included chills, eye pain, blurred vision, coughing, respiratory distress, vomiting, and blood diarrhea.[342] Although these clinical features are consistent with mycotoxicosis, a causal link to trichothecene mycotoxins could not be established despite the presence of some trichothecene toxins (T-2, DAS, HT-2) in tissue samples from the kidney, liver, and intestine of suspected victims. Human and plant samples of yellow rain taken from contaminated areas by news media reportedly contained three trichothecene mycotoxins (T-2, diacetoxyscirpenol, 4-deoxynivalenol), zearalenone, and the emulsifier polyethylene glycol in a combination that was inconsistent with a natural occurrence.[343,344] In addition, a captured Soviet gas mask allegedly found in Afghanistan was tainted with T-2 mycotoxin.[345] Such findings remain controversial because mycotoxins are ubiquitous, the half-life of T-2 toxin is short, and mycotoxins occur naturally (e.g., honeybee excretions). Scientists from other countries also failed to confirm the presence of mycotoxins in battlefield samples.[346,347]

PRINCIPAL TOXINS

Structure

About 180 trichothecene compounds have been isolated from fungal cultures and plants.[337] Because all these compounds possess a double bond at C9-C10 and an epoxide at C12-C13, they are structurally designated as 12,13-epoxy trichothecenes; all these tetracyclic sesquiterpene compounds contain a 12,13-epoxy-Δ^9-trichothecene ring system. These naturally occurring toxins are all esters of parent alcohols.[348] There are four main categories based on the presence of specific functional groups (see Figure 45.9 and Figure 45.10). The largest category (type A) lacks a ketone group at C-8 and contains T-2 toxin (CAS No. 21259-20-1) and diacetoxyscirpenol (DAS). Type B trichothecene compounds, an epoxy-sesquiterpenoid, have a carbonyl group at C-8, such as 4-deoxynivalenol (DON, vomitoxin, RD-toxin, deoxynivalenol, CAS No. 51481-10-8) and nivalenol (NIV, CAS No. 23282-20-4). Satratoxin, which contains

Type-A Trichothecenes

Name	R_1	R_2	R_3	R_4	R_5
T-2 Toxin	OH	OAc	OAc	H	$OCOCH_2CH(CH_3)_2$
T-2 Tetraol	OH	OH	OH	H	OH
HT-2 Toxin	OH	OH	OAc	H	$OCOCH_2CH(CH_3)_2$
Diacetoxyscirpenol	OH	OAc	OAc	H	H
Neosolaniol	OH	OAc	OAc	H	OH

Type-B Trichothecenes

Name	R_1	R_2	R_3	R_4
Deoxynivalenol	OH	H	OH	OH
Nivalenol	OH	OH	OH	OH
Trichothecin	H	$OCOCH{=}CHCH_3$	H	H
Fusarenon-X	OH	OAc	OH	OH

FIGURE 45.9. Chemical structures of type A & B trichothecenes.

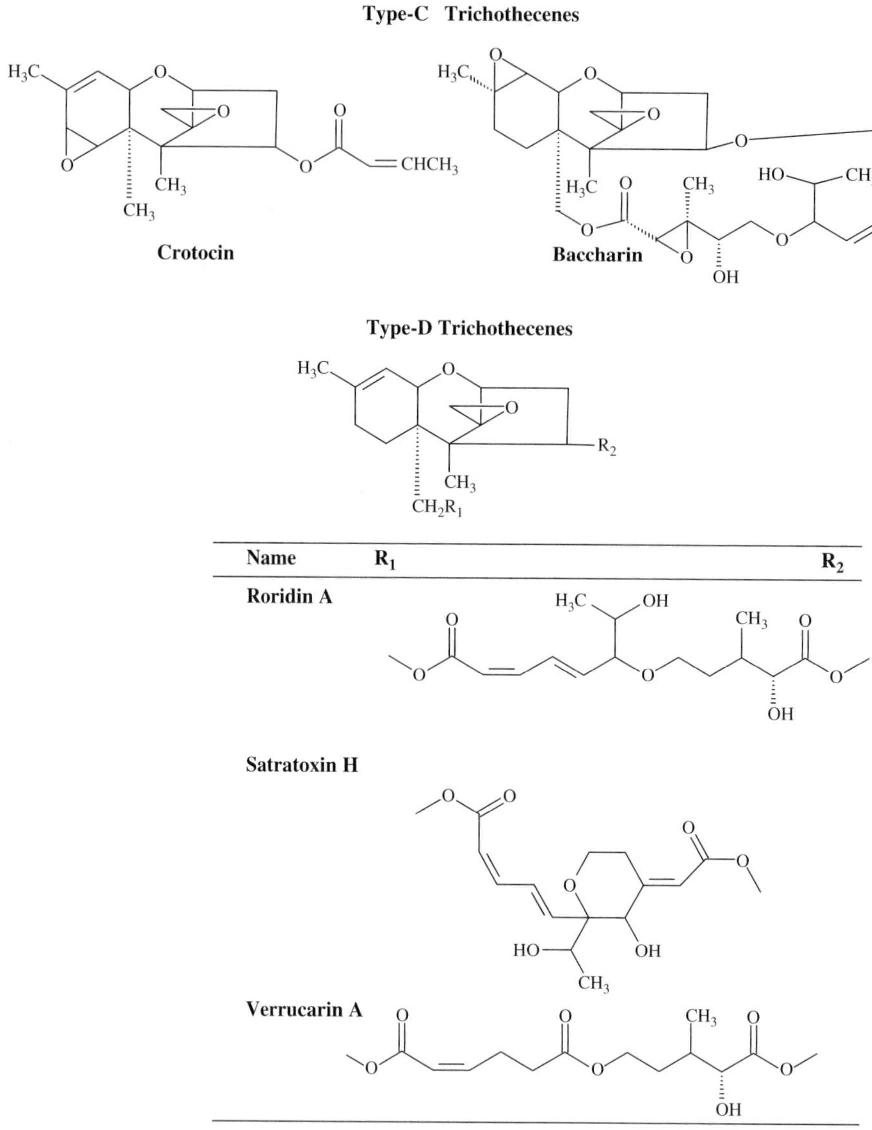

FIGURE 45.10. Chemical structures of type C & D trichothecenes.

a macrocyclic ring system between C-4 and C-5 with two ester bridges, is a type D trichothecene compounds. Only a few of the trichothecenes contaminate food or animal feed. Common trichothecene compounds present in agricultural products include T-2 toxin, 4-deoxynivalenol (DON, vomitoxin), diacetoxyscirpenol (DAS, anguidine), and nivalenol. Less frequent contaminants are 3-Ac-DON, 15-Ac-DON, fusarenon-X and HT-2 toxin. Some type D trichothecenes (satratoxin G and H, verrucarins) occasionally contaminate animal feed (hay, straw),[349] but these macrocyclic trichothecenes are not usually present in food.[337] Stachybocin A, B, and C are spirobenzofuran compounds isolated from *Stachybotrys chartarum* (Ehrenb.) S. Hughes that demonstrate endothelin receptor antagonism.[350]

Physiochemical Properties

Trichothecene mycotoxins are colorless, mostly crystalline compounds that vary in water solubility. These compounds lack native fluorescence and UV absorbance that occurs commonly in aflatoxins. Trichothecene compounds containing an ester group hydrolyze to the corresponding parent alcohol base following treatment with a base. The 12,13-epoxy group is resistant to nucleophilic attack, and trichothecene compounds are generally stable in organic solvents. Mycotoxins produced by *Stachybotrys* are water soluble. Deoxynivalenol and nivalenol are relatively stable, and decomposition of these compounds requires harsh conditions (pH 12, high salt concentration prolonged exposure to 80 °C/176 °F).[351]

Although trichothecene compounds are relatively resistant to acids, sunlight, UV light, x-rays, and temperatures up to 120 °C(248 °F), alkali rapidly degrades these mycotoxins.[101] *In vitro* studies indicate that the actions of coexisting fungi and bacteria in the environment can degrade and detoxify trichothecene compounds.[352]

EXPOSURE

Sources

STACHYBOTRYS

Stachybotrys is a common soil fungus, which is a minor component of indoor and outdoor air throughout the world from temperate climates to deserts.[53] This fungus is not pathogenic, and *Stachybotrys* species prefer stems of woody plants with soil acting as a reservoir.[353] This greenish-black saprophytic mold grows on nonliving, nitrogen-poor material (e.g., wall board) under conditions of high humidity. The fungus can survive over winter and the spores may be viable for many years. Several fungal genera (*Stachybotrys*, *Myrothecium*) produce macrocyclic trichothecene compounds, such as satratoxins, verrucarins, and roridins. *Stachybotrys* species produce >40 different toxins including trichothecenes, spirocyclic drimanes and closely related triprenyl phenolic compounds (e.g., stachybotrylactones, stachybotrylactams).[81] *Stachybotrys chartarum (S. atra)* frequently is associated with the trichothecenes compounds, satratoxin H, and to a lesser extent with satratoxin G. About one-third of *S. chartarum* strains are capable of producing mycotoxins, whereas most strains of *Stachybotrys chartarum* produce spirocyclic compounds.[354] The concentrations of spirocyclic compounds are usually substantially higher than the concentration of macrocyclic trichothecene compounds. Other mycotoxins isolated from *S. chartarum* include satratoxin F, isosatratoxin G and H, roridin E, and verrucarins J and B.[355,356] Roridin E is a precursor of satratoxins. The specific species of *Stachybotrys* are not well defined, and frequently *S. chartarum* (Ehrenb.) S. Hughes is used interchangeably with *S. atra*. Mycotoxin formation rarely is associated with *Stachybotrys* species other than *S. chartarum*.[355] In general, *Stachybotrys* spores are relatively large, and ambient air usually contains few spores even in areas with extensive growth of *Stachybotrys*.[73] In most studies of the general population, *Stachybotrys* has a low prevalence in air samples compared with other fungi (*Penicillium, Aspergillus, Cladosporium, Ulocladium*); and *Stachybotrys* is rarely, if every, found as the predominant species.[101] The prevalence of *Stachybotrys* increases in water-damaged buildings, but most air samples do not contain this mold. In a large US study of air samples collected during indoor air quality investigations, *Stachybotrys chartarum* (Ehrenb.) S. Hughes was identified in indoor air samples from 6% of the buildings and 1% of outdoor air samples surrounding these buildings.[32] The presence of *Stachybotrys chartarum* was not associated health complaints. Of 45 buildings investigated for health complaints, air samples from two buildings contained *Stachybotrys chartarum*. Consequently, *S. chartarum* occurs in air samples from buildings without reports of adverse health effects; therefore, the presence of *S. chartarum* in air samples does not necessarily imply the presence of fungal-related adverse health effects.[357] Animal studies indicate that trichothecene mycotoxins are not volatile because respiratory illness does not occur without the inhalation of spores.[100]

OTHER TRICHOTHECENE MYCOTOXINS

At least 24 fungal species produce over 40 trichothecene mycotoxins.[15,358] The fungal genera *Gibberella* [e.g., *G. intricans* Wollenw. (*Fusarium equiseti*), *G. zeae* (Schwein.) Petch (*F. graminearum*)] and *Fusarium* (e.g., *F. sporotrichioides* Sherb.) are the most important sources of these tricyclic sesquiterpene compounds because of their prolific mycotoxin production and diverse worldwide habitats. Type A trichothecene compounds (T-2, HT-2) are usually associated with *Gibberella tricincta* El-Gholl, McRitchie, Schoult. & Ridings (*F. tricinctum*), *Fusarium sporotrichioides* Sherb., *Fusarium poae* (Peck) Wollenw., and *Gibberella intricans* Wollenw. (*Fusarium equiseti*), whereas Type B trichothecenes are usually associated with *Gibberella zeae* (Schwein.) Petch (*F. graminearum*) and *F. culmorum* (W.G. Sm.) Sacc. *F. sporotrichioides* Sherb. is the most important source of type A trichothecene compounds. A few plant species (e.g., *Baccharis megapotamica* Sprengel) produce type C trichothecenes. *Gibberella zeae* and *Fusarium culmorum* (W.G. Sm.) Sacc. are plant pathogens that invade certain crops (e.g., *Fusarium* head blight in wheat, *Gibberella* ear rot in maize), and these pathogens produce DON and NIV. Of the numerous trichothecene mycotoxins, four economically important contaminants of agricultural products (wheat, corn, barley) include nivalenol, deoxynivalenol, diacetoxyscirpenol, and T-2 toxin.[345] Diacetoxyscirpenol (anguidine) is the principal mycotoxin in *Gibberella intricans*, and this mycotoxin was used unsuccessfully in preliminary clinical trials as a chemotherapeutic agent.[359]

Environmental Conditions

STACHYBOTRYS

Stachybotrys is a strongly hydrophilic, saprophytic fungi found in moist soil and cellulose-rich substrates (plant debris, wood pulp, paper, wallboard, cardboard, cotton, straw, hay, grains, cereals) that usually is a minor constituent of indoor air.[360] This fungus produces spores over a wide range of temperatures (<60 °C/140 °F), but the spores seldom appear in outdoor air samples because these spores usually occur in a cluster covered by dried slime. These spores require mechanical stimulation (brushing, washing, renovation, vacuuming) to enter the ambient air, particularly when wet. Although active fungal growth typically requires relative humidity >90%, the water activity of the substrate is a more important determinate of fungal growth.[361,362] The presence of pets, infrequent vacuuming, poor ventilation, and visible mold suggests the presence of high indoor fungal concentrations.[363] However, these factors do not necessarily correlate to high mycotoxin concentrations.

Temperature, nutritional status, and moisture content are the most important determinants of *Stachybotrys* growth. At 25 °C (77 °F), the moisture requirement for growth of *Stachybotrys* is about 93%.[364] Building materials of high cellulose and low nitrogen content (e.g., dust, fiberboard, ceiling tiles, lint, rice paddy grains, sorghum dust, straw) in moist conditions (>55% moisture content) and substantial temperature fluctuations (0–40 °C/32 °F–104 °F) favors the growth of mycotoxin-producing *S. chartarum*.[365] *Stachybotrys* does not compete well with *Aspergillus* and *Penicillium* species, and *S. chartarum* is an uncommon contaminate of indoor air. In a study of Southern California homes, *S. chartarum* was present in <3% of the indoor air samples.[366] Analysis of indoor dust samples for *S. chartarum* confirms the presence of these fungi in the environment, but this type of analysis does not prove that a toxic exposure occurred. Not all strains of *Stachybotrys* produce mycotoxins, and under some environmental conditions some species of *Stachybotrys* lose the ability to produce mycotoxins. Measurement of culturable airborne *Stachybotrys* spores does not necessarily correlate to trichothecene exposure because some nonviable spores may contain high concentrations of mycotoxins. Mycotoxin exposure from *S. chartarum* depends both on environmental conditions and the specific species involved.[355] Furthermore the presence of humid conditions does not imply that mycotoxin-producing *S. chartarum* (*S. atra*) are present. During a study of building-related illness in a water-damaged building, *S. chartarum* was present in only 1 of 19 bulk samples from areas of high moisture content.[367] Furthermore, the presence of *S. chartarum* in bulk samples does not necessarily indicate that satratoxins or verrucarins are present in ambient air at similar concentrations.[35] In a 22-month study of 48 schools with building-related complaints, *S. chartarum* was isolated from swab samples of 11 wet bulk samples.[62] However, no ambient air samples contained detectable concentrations of *S. chartarum*. In a rodent study, few biological effects occurred when the rodents were exposed to vapors from walls completely covered with *S. chartarum*.[100] Measurement of ambient air detected very low concentrations of liberated *S. chartarum* spores. This study suggests that there is no direct correlation between the amount of surface area covered with mold and biological effects.

Other Trichothecene Mycotoxins

Factors that affect trichothecene formation include fungal genetics, substrate, temperature, pH, crop practices, harvesting techniques, and humidity. Acidity apparently does not affect T-2 toxin production in contrast to aflatoxin production. *Gibberella zeae* (*F. graminearum*) is a plant pathogen that produces deoxynivalenol, nivalenol, and zearalenone in an optimum range of 24 °C–26 °C (~75 °F–79 °F) and a minimum water activity of 0.90. When grains are dried to ≤13%, moisture content fungal contamination primarily is a preharvest problem. Storage of grains under moist conditions, as occurs in underdeveloped countries, promotes mycotoxin formation. *F. sporotrichioides* Sherb., which produces T-2 toxin in addition to other trichothecene mycotoxins, has an optimum growth temperature ranging from 22.5–27.5 °C (72.5 °F–81.5 °F) with a minimum temperature of −2 °C (~28 °F).[15] This fungi is the most important source of type A trichothecene compounds, but these mycotoxins do not normally occur in grains during the growing period, harvest, or storage unless water damage develops.

Foodstuffs

DON, NIV, and DAS are more frequent contaminants of food and animal feed than T-2 toxin.[337] DON and NIV frequently contaminate grains (e.g., wheat, rye, barley, oats) at concentrations <1 mg/kg (ppm) and sporadically up to 5–20 ppm.[368] Less often, these mycotoxins occur in rice, sorghum, and triticale. In a survey of 500 cereal grain samples from 19 countries, about 45–50% contained both DON and NIV with barley being the most frequently contaminated grain.[369] Corn and wheat samples contained the highest concentrations of NIV and DON, respectively. There was substantial variation in the presence of these mycotoxins between different countries. Wheat is the major source of DON in most diets with average dietary intake ranging up to about 2.4 μg/kg/d.[111] Although reports document the presence of T-2 toxin in agricultural products at concentrations near 40 mg/kg,[370] analyses of most agricultural products indicate that T-2 toxin is present in detectable concentration in <10% of samples with concentrations usually <0.1 mg/kg.

Most confirmed cases of trichothecene contamination of human food involves DON in wheat or wheat products at concentrations <1 mg/kg.[337] DON can survive the brewing process, and beer occasionally contains DON concentrations exceeding 200 ng/mL.[125] Processing, milling, and baking do not destroy all trichothecene mycotoxins present in foods, although cleaning and milling remove some water-soluble trichothecene compounds. DON fractionates during the dry milling process with the highest concentrations in the bran and the lowest concentrations in the flour.[371] Dry milling reduces the concentrations of DON and NIV up to approximately 40% and 70%, respectively.[372] Processing of flour into baked or cooked products produces variable trichothecene losses. The overall reduction of DON concentrations during the processing of bread from uncleaned wheat ranged from 24–71%.[373] Losses of NIV and T-2 toxin during these processes probably are similar. DON is probably not a significant contaminant of commercial milk or milk products.[374] However, DON survives the beer brewing process and occasionally beer samples contain DON concentrations >200 ng/mL.[125]

DOSE RESPONSE

There are few data on the toxic concentrations of trichothecene compounds, in part, because of the lack of definite evidence to link human toxicity to trichothecene mycotoxins. No cases of alimentary toxic aleukia have been reported since the end of World War II probably because of improved agricultural and storage practices. Acute LD_{50} animal data provide the best available comparison of trichothecene mycotoxin toxicity, but significant variation between species limits extrapolation to humans.[87] Consumption of grains containing 0.02–3.5 mg DON/kg probably does not cause adverse effects.[111] The established limits for DON range between 0.5 ppm and 2 ppm.[375] In an early-phase clinical trial involving the administration of intravenous diacetoxyscirpenol (Anguidine) to patients with advanced malignancies, no biological effects were observed at daily doses of <2.4 mg/m^2 for 5 consecutive days.[376]

The toxicity of *S. chartarum* spores is highly variable (i.e., >10,000-fold) depending on the amount of satratoxins G and H, verrucarin J, roridin E, stachybotrylactones, and stachybotrylactams contained within the spores.[377] Following the intranasal installation of 10^5 toxic *S. chartarum* spores to rodents, there was histological evidence of definite pulmonary inflammation, but there was no evidence of necrosis. The intranasal instillation of 10^5 relatively nontoxic *S. chartarum* spores caused mild pulmonary inflammation.[378] The intranasal instillation of 10^3 of the less toxic spores produced no pathological evidence of pulmonary damage. Although pathological changes occurred in the lungs of the mice exposed to the toxic spores, the exposed mice did not demonstrate clinical changes or significant differences in weight gain when compared with nonexposed controls. During an experimental study, mice were exposed to a chamber that contained a *S. chartarum* load (i.e., surface area covered with mold per m^3 of air) several times greater than a normal room containing mold on all walls, floor, and ceiling.[100] Sensory irritation, bronchoconstriction and pulmonary irritation did not develop despite the presence of airflow 4 times greater than normal ventilation rates. The lack of clinical effects was consistent with the low concentration of volatile organic compounds emitted from the *S. chartarum*. These studies suggest that the concentrations of airborne spores of *S. chartarum* realistically attainable in indoor air are too low to produce clinical effects.

TOXICOKINETICS

There are limited data on the toxicokinetics of trichothecene compounds with most animal studies involving T-2 toxin and, to a lesser extent, DON. Extrapolation of animal data to humans is limited by substantial intraspecies variation in the response to trichothecene compounds. In general, these studies indicate that the absorption of these compounds from the gastrointestinal tract is rapid.[337] However, the bioavailability of trichothecene compounds is relatively low because of the physiological instability of these compounds and large first-pass metabolism.[379] *In vitro* studies suggest that absorption of trichothecene mycotoxins may occur through the skin with some variability between individual mycotoxins (e.g., DAS and T-2 toxin > verrucarin A).[380] However, the risk of systemic toxicity following dermal application of these mycotoxins is unclear because several of these studies utilized dimethylsulfoxide to enhance dermal absorption. Some metabolism of these mycotoxins occurs in the skin prior to systemic distribution. The distribution of trichothecene mycotoxins within the body is relatively even without accumulation in specific organs. Biotransformation of trichothecene mycotoxins to nontoxic metabolites involves hydrolysis, hydroxylation, de-epoxidation, and conjugation with glucuronides. Animal studies indicate that there is substantial variation in metabolic pathways between species. In a study of dogs, the elimination half-life T-2 toxin and its metabolite, HT-2 toxin, administered intravenously was about 5 minutes and 20 minutes, respectively.[381] Although there is substantial variation in the metabolism of DON, all animal species metabolize DON and bioaccumulation of DON does not occur.[375] In sheep,

DON was eliminated within 20–30 hours.[382] Animal studies indicate that <1% of T-2 toxin and DON administered to chickens and cows appears in eggs and in milk.[374,383]

HISTOLOGY AND PATHOPHYSIOLOGY

When administered chronically to animals, trichothecene compounds produce a wide variety of pathological responses including cutaneous inflammation and necrosis, mucous membrane lesions, gastrointestinal inflammation and hemorrhages, petechiae, ecchymosis, and hematological and immunological suppression (e.g., granulocytopenia, anemia). The relative systemic toxicity of trichothecene compounds in order of most potent to least potent are T-2 toxin, diacetoxyscirpenol (DAS or anguidine), NIV, and deoxynivalenol (DON or vomitoxin). Trichothecene compounds bind to the 60S ribosomal subunit and inhibits the enzyme peptidyltransferase. Deoxynivalenol (DON) is an exception because exposure to DON at low doses causes upregulation of protein synthesis as well as polyclonal immunoglobulin production.[384] The resulting interactions cause various degrees of disruption of the formation of peptide bonds. Sites of action of these compounds include the initiation of protein synthesis (T-2 toxin, DAS, scirpentriol, 15-acetoxyscripendiol, verrucarin A) and elongation or termination of protein synthesis (trichothecin, verrucarol).[101] In animal studies, T-2 toxin and DAS produce necrosis of epithelial tissue, cytotoxicity of rapidly proliferating tissue (e.g., intestinal crypt epithelium, hematological tissue), and coagulation defects.[385] In rodents, skin irritation occurs about 6 hours after the dermal application of <1 μg/cm^2 T-2 toxin and verrucarin A.[386] Diacetoxyscirpenol (DAS) and nivalenol (NIV) are less potent skin irritants compared with T-2 toxin and verrucarin A. *Stachybotrys* species also produce spirolactones and spirolactams that demonstrate endothelin receptor antagonism, immunosuppression, and anticomplement activity.[101,356]

Severe toxicity of T-2 toxin and DAS cause suppression of cell-mediated and humoral immunity, reduced immunoglobulin formation, depressed phagocytic activity, and pancytopenia.[387] Alteration of immune parameters decreases resistance to secondary infection of bacteria, yeast, and viruses. In studies of cats, the administration of T-2 toxin in feed caused a clinical syndrome similar to ATA.[388] Removal of this toxin from the *Fusarium* culture fed to the cats resulted in healthy animals.

Trichothecenes are potent inhibitors of protein synthesis and these compounds are particularly active at the peptidyl transferase center on the 60S ribosomal subunit. Experimental work suggests that some trichothecene mycotoxins interfere with chain termination by blocking the interaction between peptidyltransferase and releasing factor on the ribosome.[389] Experimental animal studies also suggest that satratoxins are more potent pulmonary toxins than spirolactones and spirolactams based on the amount of pulmonary inflammation following intranasal instillation.[390]

CLINICAL RESPONSE

Food-Related Illnesses

The main route of exposure to trichothecene mycotoxins is the ingestion of plant material contaminated with mycotoxins. Other theoretical sources of exposure include accidental dermal contact during laboratory procedures and inhalation of dust containing trichothecene mycotoxins. Although several case reports associate the outbreak of food-related illness with trichothecene exposure, there are no epidemiological studies documenting illness in patients with elevated trichothecene concentrations. An outbreak of an acute gastrointestinal illness in China was associated with the consumption of moldy cereals. The latency period was short (5–30 minutes) and symptoms included nausea, vomiting, diarrhea, abdominal pain, lightheadedness, and headache. The most prominent symptom was vomiting. There were no fatalities. Analysis of the moldy cereal from samples stored 6–8 years revealed high concentrations (up to 50 mg/kg) of DON as well as lower concentrations of NIV and other mycotoxins (zearalenone, fumonisin).[391] A similar outbreak of food poisoning occurred in India after the consumption of bread produced from flour contaminated with mycotoxins (DON, NIV Ac-DON, T-2 toxin).[10] The reported symptoms, which began within 1 hour after ingestion of the bread, in order of most frequent to least frequent, were abdominal pain, throat irritation, diarrhea, bloody stools, and vomiting. There were no fatalities. Trichothecene compounds are not recognized carcinogens based on limited data. Although T-2 toxin and DON cause immunosuppression in animal models, there are no clinical data that document the immunosuppression, opportunistic infections, or cancer in subjects exposed to trichothecenes.[337] The International Agency for Research on Cancer lists DON, NIV, and T-2 toxin as agents not classifiable as to their carcinogenicity (Group 3).[392]

Alimentary Toxic Aleukia

Although there is a lack of confirmatory scientific evidence to confirm Koch's postulates for establishing causation, strong circumstantial evidence associated the ingestion of moldy grain containing *Fusarium poae* (Peck) Wollenw. and *Fusarium sporotrichioides* Sherb. with fatal epidemics of alimentary toxic aleukia (ATA) occurring in Russia between 1931 and 1947. These grains remained in the fields during severe winter conditions as a result of war. The associated *Fusarium* species are capable of producing mycotoxins, such as T-2 toxin, HT-2, and neosolaniol. Multifactorial causes probably are involved in the occurrence of ATA because no similar outbreaks have occurred since the late 1940s. Clinical features of this illness were similar to radiation toxicity and to equine stachybotryotoxicosis with prominent features including necrotic lesions of the upper gastrointestinal tract (oral cavity, esophagus, and stomach), bone marrow hypoplasia, and leukopenia. Symptoms included fever, nausea, vomiting, diarrhea, leukopenia, bleeding, and sepsis. *Stachybotrys* species have not been implicated as a cause of ATA, and humans exposed to aerosolized mold during outbreaks of equine stachybotryotoxicosis developed dermal rather than systemic complaints.[101]

Clinically, ATA is divided into four stages (see Table 45.9).[393] Typically, stage 1 involves the delayed onset of gastrointestinal symptoms along with fatigue, weakness, and sore throat. During the latent period, fatigue and weakness may persist as increasing suppression of the bone marrow develops. About 3–4 weeks after exposure, clinical evidence of coagulopathy and granulocytopenia appear. Increased susceptibility to infection complicates the clinical picture. Death typically occurs during stage 3. Improving bone marrow function characterized by increasing white blood cell counts is a sign of recovery (stage 4). Permanent sequelae did not usually occur in surviving patients. Presently, the exact etiology of this disease remains unknown including the contribution of starvation, poor living conditions, and inadequate medical care.

Scaby Grain Toxicosis

Scaby grain toxicosis is a gastrointestinal disease associated with the consumption of *Fusarium*-contaminated food in Japan and Korea between 1946 and 1963.[394] Clinical features include nausea, vomiting, diarrhea, and abdominal pain that resolved over several days. There were no fatalities. The exact etiology of this disease has not been identified, but a trichothecene mycotoxin is a possible etiological agent.

Diacetoxyscirpenol

Diacetoxyscirpenol (anguidine) underwent unsuccessful clinical trials as a chemotherapeutic agent at 5-day/week intravenous doses of 5 mg/m^2 for 3 weeks.[359] Side effects recorded during these clinical trials included nausea, vomiting, diarrhea, hypotension, fever, chills, mucosal ulcerations, cutaneous erythema, hair loss, and bone marrow suppression. There were inadequate data to separate adverse reactions from treatment with anguidine and symptoms related to the progression of metastatic cancer. Some episodes of hypotension were severe enough to require termination of chemotherapy.[395] Dose-related myelosuppression was the limiting adverse effect following prolonged infusion.[396]

Stachybotrys Mycotoxicoses

Mycotoxicoses involving trichothecene mycotoxins usually result from the ingestion of mold-contaminated food or animal feed. The studies associating airborne exposure to *Stachybotrys* with primarily dermal toxicity are mainly descriptive, and there are no well-designed epidemiological data to confirm the association of airborne exposure to *Stachybotrys*-produced mycotoxin with human toxicity.[397,398] The lack of a simple, reliable biomarker of *Stachybotrys* exposure and of objective medical data on abnormalities in patients with *Stachybotrys* exposure limits conclusions regarding the causal link between these mycotoxins and human toxicity. Typically, sites contaminated with *Stachybotrys* also contain many other fungal species (e.g., *Penicillium, Alternaria, Aspergillus, Cladosporium*). The potential presence of irritancy, allergic reactions (dust mites), and infections (gram-negative bacteria, endotoxins, mycobacteria) in damp environments complicate the interpretation of *Stachybotrys* studies that do not exclude health effects associated with damp environments. For example,

TABLE 45.9. Clinical Stages of Alimentary Toxic Aleukia

Stage	Onset[a]	Phase	Symptoms
1	3–9	Gastrointestinal	Salivation, sore throat, headache, lightheadedness, weakness, fatigue, abdominal pain
2	10–14	Latent	Improved symptoms, increasing clotting abnormalities, bone marrow hypoplasia (thrombocytopenia, anemia, granulocytopenia)
3	21–28	Hematological	Bleeding gums, epistaxis, GI hemorrhage, secondary infection, sepsis, death
4		Recovery	Improved bone marrow function

[a]Days after initial exposure.

illnesses associated with damp environments that contain fungi include allergic rhinitis, asthma, hypersensitivity pneumonitis, Legionnaire's disease, and humidifier fever. The proliferation of mold and the chemical degradation of building materials (e.g., emission of 2-ethyl-1-hexanol from alkaline degradation of diethylhexyl phthalate) cause nasal inflammation in occupants of damp buildings.[399] The presence of excessive fungal growth also suggests the presence of inadequate building design and maintenance. Furthermore, the presence of mycotoxin-producing fungi does not necessarily imply exposure to a mycotoxin. Although there are case reports that include nonspecific, subjective complaints involving the neurological and immune systems, there is no conclusive data that links airborne exposure to *Stachybotrys* with neurological or immune damage.[101,109]

Stachybotryotoxicosis

Stachybotryotoxicosis is a disease described in Russian horses, which ate moldy hay contaminated with *Stachybotrys*. To date, this disease has not been reported in animals from North America.[400] The presentation of stachybotryotoxicosis is similar to alimentary toxic aleukia, but ATA is associated with *Fusarium* rather than *Stachybotrys* species.[166,401] Although some investigators associated human illnesses with exposure to *Stachybotrys*, reports from Russian investigators lacked reliable epidemiological data confirming the presence of this disease in humans.

Pulmonary Hemorrhage

A clinical series of 30 infants in Cleveland suggested the possibility of an association between acute pulmonary hemorrhage and water-damaged homes that contained *S. chartarum* among other potential causes (environmental tobacco smoke, sepsis with acute respiratory distress syndrome, general anesthesia).[402] A case-control study of 10 cases of pulmonary hemorrhage and hemosiderosis among infants demonstrated the presence of a high incidence of water damage, family members with hemoptysis, and exposure to environmental smoke in the cases compared with controls.[403] The statistically significant association between pulmonary hemorrhage and the presence of *S. chartarum* was relatively weak (OR = 1.6) compared with other environmental factors (e.g., smoking, OR = 7.9). There was no objective evaluation of the water damage or measurements of the actual exposure to airborne mycotoxins. Aggressive air sampling after the return of the infants demonstrated the presence of *Stachybotrys* in the homes of 5 of 9 patients at a mean concentration of 43 colony-forming units (CFU)/m^3 and in 4 or 27 control homes at a mean concentration of 4 CFU/m^3.[404] However, the use of aggressive sampling techniques that agitate ambient air prevents the quantitative evaluation of the actual exposure to trichothecene compounds in these homes. *S. chartarum* was detected in surface samples from 4 of 7 case homes and 10 of 19 control homes; therefore, the presence of the *S. chartarum* spores in surface samples does not imply the development of pulmonary hemorrhage. In addition, air sampling from some Cleveland and most Chicago homes of infants with pulmonary hemorrhage did not detect spores of *S. chartarum*.

Analysis of isolates from these *S. chartarum* spores revealed the presence of satratoxin H in 3 of 16 isolates along with trichoverroid trichothecene mycotoxins (roridin L-2, trichoverrol B).[405] Some of the isolates from control homes also contained satratoxins and trichoverroid trichothecenes including strains that produced the highest concentrations of trichothecene compounds. Comparison of *S. chartarum* strains from case homes and control homes did not demonstrate significant differences in the potency of the mycotoxins formed.[406] The frequency of highly toxic strains of *S. chartarum* from Cleveland was about the same as the frequency of similar strains from samples outside Cleveland, although three of the *S. chartarum* strains isolated from homes with pulmonary hemorrhage were both very highly toxic and hemolytic.

A subsequent review by a panel of outside experts and a panel of senior scientists organized by the Centers for Disease Control and Prevention (CDC) concluded that the association between acute pulmonary hemorrhage/hemosiderosis (AIPH) in infants and exposure to molds, specifically *Stachybotrys chartarum* was not proven.[407] The panels noted that serious shortcomings in the collection, analysis, and reporting of data resulted in inflated measures or association between *S. chartarum* and AIPH, resulting in a reduction in the OR from 9.8 to 1.5. Furthermore, the panels indicated that AIPH is not consistent with previous clinical effects associated with *S. chartarum* and that similar illnesses have not been reported in other flood-prone areas, which favor the growth of *S. chartarum*. The low spore counts and low trichothecene toxicity of the air samples in homes of infants developing pulmonary hemorrhage suggests the presence of multiple etiological factors (e.g., tobacco smoke, lack of breast feeding) in addition to the presence of *S. chartarum*.[408] The original cluster of cases occurred in a relatively small geographic area near Cleveland, Ohio. However, idiopathic pulmonary hemorrhage occurs in infants sporadically across the United States and the association of *S. chartarum* with these cases is unclear. Investigation of the home of a 40-day-old infant who developed idiopathic pulmonary hemorrhage in Florida, did not demonstrate the presence of *S. chartarum*.[409] A case report from Houston suggested that idiopathic pulmonary hemorrhage in infants might be associated with an invasive, hemolytic strain of *S. chartarum* rather than a mycotoxin.[410]

Building-Related Illness

The presence of excessive fungal growth in a building indicates inadequate building design, construction, and/or maintenance. Although some observational studies implicate mycotoxin-producing fungi (*Aspergillus, Penicillium, Stachybotrys*) as possible etiological factors of building-related illnesses, there is no clear evidence at the present time that mycotoxins contribute significantly to the symptoms associated with mold-contaminated buildings or with building-related illnesses.[73,101,411,412] Most homes of occupants with complaints related to building-related illness do not contain detectable concentrations of *Stachybotrys* in air samples.[367,413] A case report documented the presence of nonspecific symptoms (upper respiratory tract symptoms, sore throat, diarrhea, headache, dermatitis, focal alopecia, malaise) in occupants of a residence containing insulation and heating ducts heavily contaminated with *Stachybotrys chartarum*. Analysis of a sample of ceiling fiberboard contaminated with *S. chartarum* spores demonstrated the presence of macrocyclic trichothecenes (verrucarin B and J, satratoxin H, trichoverrins A and B). There was no sampling of ambient air to confirm the presence of airborne *S. chartarum* spores. Laboratory and physical examination of these patients did not detect any abnormalities. There is no substantive evidence that neurological injury including encephalopathy results from the inhalation of mycotoxins from the indoor environment.[14]

A case-control of 53 employees (female to male ratio = 2.8:1.0) and 21 controls (female to male ratio = 1:1) examined differences in 187 self-reported symptoms and a test battery of nonspecific serum tests (complete blood count, flow cytometry, lymphocyte function tests, natural killer cell activity).[414] There was no environmental data on the presence of *S. chartarum* in the work environment of controls. Reporting and selection bias are potential complications of the interpretation of this study because the cases presented to an occupational clinic for evaluation of potential workmen's compensation claims as a result of employment in building contaminated with a variety of fungi including *Stachybotrys*. There were statistically significant differences in the presence of symptoms related to the following areas: lower respiratory tract, skin, eye, and constitution. The study did not report any medical evaluations of the subjects enrolled in the study. However, there was no correlation between objective tests of respiratory and immune function and quantitative measures of mycotoxin exposure. There were some statistically significant differences in the white blood cell count, proportion of mature T-cell lymphocytes (CD3) and natural killer cell counts between workers in different locations. However, the differences were small and not dose-related. The subjects working in the basements, which contained the highest fungal concentrations, had elevated white blood cell counts and percentages compared with workers in other parts of the building. Although *Stachybotrys* was present in the buildings, no *in vitro* or *in vivo* assays confirmed that mycotoxins were present in ambient air. There is no convincing evidence that significant immunological suppression or modulation results from the inhalation of mycotoxins in the indoor setting.[14]

Several studies report an increased prevalence of chronic fatigue and respiratory symptoms in occupants of buildings containing fungi including *Stachybotrys*. These studies have not identified clinically significant immune dysfunction, antibodies directed against *S. chartarum*, or respiratory dysfunction. In general, these observational studies do not control important host and environmental confounding factors. A cross-sectional study of 197 occupants of a building with moisture-damaged interior demonstrated an excess number of respiratory symptoms (cough, dyspnea, feverishness, wheezing, and chest tightness when compared with controls from a building selected for convenience rather than confounding variables (e.g., smoking, asthma, atopy, mold allergies, job dissatisfaction, environmental contaminates).[415] Medical examinations of selected occupants detected asthma and reduced carbon monoxide diffusing capacity in a few cases, but there was no direct correlation between respiratory symptoms and decrements in pulmonary function (e.g., FEV_1, FVC). Ceiling tiles from one of three rooms tested in the contaminated building contained *S. chartarum* spores that produced satratoxins, but there were no direct measures of exposure to these mycotoxins. In this study, exposed occupants and matched controls did not have statistically significant differences in serum antibodies to fungi; consequently, there was no objective evidence that the occupants were exposed to clinical significant quantities of mycotoxins. Fungal cultures revealed other mycotoxin-producing organisms (*A. versicolor*) in addition to *S. chartarum*.

Carcinogenesis

The evidence for carcinogenicity following exposure to T-2 toxin is weak.[111] This mycotoxin is classified as IARC Group 3 (not classifiable with regard to carcinogenicity to humans). In general, studies of genotoxicity following

T-2 exposure are negative. Currently, there is no evidence that exposure to trichothecene mycotoxins (e.g., satratoxins) produces cancer in humans.[337]

DIAGNOSTIC TESTING

The complex physical and chemical properties of trichothecene compounds preclude the use of simple extraction and identification methods. Analytical methods include thin-layer chromatography, gas chromatography (GC), gas chromatography/mass spectrometry (GC/MS) with derivatization, high performance liquid chromatography (HPLC), and immunological techniques. Detection limits usually are <1 μg/g. Gas-liquid chromatography sensitivity for T-2 toxin is 0.03–0.05 ng/L (ppb); GC/MS is slightly more sensitive (0.02 ng/L). Sensitivity for trichothecenes analyzed by gas chromatography-negative chemical ionization tandem mass spectrometry with derivatization by heptafluorobutyryl esters was 1 ng/g.[416] The T-2 metabolite, T-2 tetraol, is stable in urine stored at −70 °C (−94 °F) for 6 months.[417] In blood samples stored at 4 °C (~39 °F), T-2 and HT-2 degrade substantially over 1 month. The limit of quantitation of deoxynivalenol using gas chromatographic methods and Florisil SPE cleanup is approximately 10 ng/g compared with 200 ng/g for thin-layer chromatographic techniques.[151] The mean deoxynivalenol concentrations in urine samples from a small, nonrandom sample of Chinese residents from suspected areas of high and low exposure to deoxynivalenol were 37 ng/mL (range: 14–94 ng/mL) and 12 ng/mL (range: 4–18 ng/mL), respectively.[418]

Classic bioassays for trichothecene compounds involve the skin necrosis test in rodents, rabbits, or guinea pigs based on the development of erythema, edema, intradermal hemorrhage, and necrosis. This test detects T-2 mycotoxins in the range of 0.2 μg.[419] Radioimmunoassay can detect T-2 in the range of 1–20 ng with minimal cross-reaction to other mycotoxins.[420]

Presently, there is no reliable commercial biomarker of acute or chronic exposure to trichothecene mycotoxins including satratoxins. Enzyme immunoassays demonstrate the presence of IgG and IgE antibodies against *S. chartarum*. In a study of 132 samples of sera from blood donors in Missouri, immunoblot analysis detected IgG and IgE antibodies in about 49% and 9% of the samples, respectively.[421] Because of the potential of cross-reaction with other fungal proteins, the presence of these antibodies does not confirm exposure to *S. chartarum*. In general, antigen-specific IgA concentrations are not useful adjuncts for the evaluation of allergic reactions. The results of commercial fungal immunological assays must be interpreted cautiously and correlated to the clinical presentation because current fungal allergen extracts lack standardization and specificity.[422] For example, some *S. chartarum* antigens cross-react with common fungi (e.g., *Aspergillus fumigatus, Alternaria alternata*).[423] Therefore, the presence of antibodies cross-reacting with *S. chartarum* antigen does not necessarily imply exposure to this fungus or to mycotoxins. Case reports of respiratory illness have not correlated *S. chartarum* antigens to clinical symptoms.[415] Alternatively, enzyme-linked immunosorbent assay methods have been developed for the detection of stachylysin in serum as a biomarker of exposure to *S. chartarum*.[424] However, there are inadequate data on the specificity and sensitivity of this potential biomarker to quantify the dose-exposure relationship between this assay and the amount of exposure to *S. chartarum*.

TREATMENT

There are little data on the treatment of trichothecene toxicity. Generally, the treatment of trichothecene exposure is supportive. Activated charcoal binds T-2 toxin in experimental models. The simultaneous administration of super-activated charcoal and T-2 toxin prevents the development of T-2 toxicosis in these animals.[425] Consequently, the use of activated charcoal as a decontamination measure for the ingestion of trichothecene compounds is a reasonable treatment when administered within 1–2 hours of ingestion. In animal models, polyethylene glycol 300 is a more effective dermal decontamination agent than aqueous soap solution.[386] Simple cleaning methods involving removal of damaged material and spraying affected areas with bleach usually control fungal contamination. The most important actions involved with the remediation of homes containing mold include the following: maintenance of heating, ventilation, and air conditioning; inspection and repair of water damage; regular cleaning of the home with dust removal and carpet cleaning; and the removal of visible mold. Sodium hypochlorite (i.e., one cup common bleach per gallon water applied for 15 minutes) degrades trichothecene compounds, but the toxicity of the breakdown products is unknown.[426] The presence of *Stachybotrys* requires no special precautions other than the usual methods for the elimination of mold problems.

ZEARALENONE

Zearalenone (F-2 toxin) is a phenolic resorcylic acid lactone derived from several *Fusarium* and *Gibberella* species including *F. crookwellense* L.W. Burgess, P.E.

Nelson & Toussoun, *F. culmorum* (W.G. Sm.) Sacc., *F. oxysporum* Schltdl., *F. sporotrichioides* Sherb., *Gibberella intricans* Wollenw. (*F. equiseti*), *Gibberella zeae* (Schwein.) Petch (*Fusarium graminearum*), *Gibberella tricincta* El-Gholl, McRitchie, Schoult. & Ridings (*Fusarium tricinctum*), and *Gibberella moniliformis* Wineland (*Fusarium verticillioides*). These lactones possess estrogenic and anabolic activity as a result of the chemical similarity of this mycotoxin to 17β-estradiol, the principal hormone of the human ovary. The reduced form of zearalenone, zearalanol, is an anabolic agent for beef cattle.[427] In several assays, the binding of zearalenone to estrogen receptors was approximately 20-fold lower than 17 β-estradiol.[427] Zearalenone is a natural contaminant of corn and occasionally appears in cereals and feeds depending on climatic conditions. Low temperatures (12 °C–14 °C/~54 °F–57 °F) enhance production of this mycotoxin. Zearalenone is a relative stable compound that requires harsh conditions (>12 days at 110 °C/230 °F in alkaline solution) to decompose.[351] In a sample of cornflakes, the zearalenone concentration was about 13 µg/kg.[428] Zearalenone concentrations ranging from 8–53 mg/kg contaminated South African beer produced from corn and sorghum. Because of the large doses of zearalenone required to excrete this mycotoxin in the milk of lactating cows, contamination of commercial milk with zearalenone is unlikely.[429] The estimated mean daily consumption of zearalenone by Canadians is about 0.05 µg/kg, which risk assessment data suggests is a safe dose.[427]

In several *Fusarium* species, zearalenone appears to control sexual development. Mammals and, in particular, swine are particularly sensitive to its estrogenic activity, displaying vulval, uterine, and breast enlargement as well as ovarian atrophy, vaginal prolapse, and infertility at feed levels of 1 mg/kg. There are no reports of hyperestrogenism in humans.

References

1. Peraica M, Radic B, Lucic A, Pavlovic M. Toxic effects of mycotoxins in humans. Bull World Health Org 1999;77:754–765.
2. Gloag O. Contamination of food: Mycotoxins and metals. Br Med J 1981;1:879–882.
3. Freeman GG, Morrison RI. Trichothecin: an antifungal metabolic product of *Trichothecium roseum* Link. Nature 1948;162:30.
4. Mirocha CJ, Pathre S. Identification of the toxic principle in a sample of poaefusarin. Appl Microbiol 1973;26:719–724.
5. Arstwick PKC. Mycotoxins. Br Med Bull 1975;31:222–229.
6. Nesbitt BF, O'Kelly J, Sargeant K, Sheridan A. Aspergillus flavus and turkey X disease. Toxic metabolites of *Aspergillus flavus*. Nature 1962;195:1062–1063.
7. Wogan GN. Aflatoxin carcinogenesis. Methods Cancer Res 1973;7:309–343.
8. van der Merwe KJ, Steyn PS, Fourie L. Mycotoxins. Part II. The constitution of ochratoxins A, B, and C, metabolites of *Aspergillus ochraceus* Wilh. J Chem Soc [Perkin 1] 1965; (Dec):7083–7088.
9. Krishnamachari KA, Bhat RV, Nagarajan V, Tilak TB. Hepatitis due to aflatoxicosis. An outbreak in Western India. Lancet 1975;1(7915):1061–1063.
10. Bhat RV, Beedu SR, Ramakrishna Y, Munshi KL. Outbreak of trichothecene mycotoxins associated with consumption of mould-damaged wheat products in Kashmir valley, India. Lancet 1989;1(8628):35–37
11. Bhat RV, Shetty PH, Amruth RP, Sudershan RV. A foodborne disease outbreak due to the consumption of moldy sorghum and maize containing fumonisin mycotoxins. Clin Toxicol 1997;35:249–255.
12. Miller JD. Fungi as contaminants in indoor air. Atm Environ 1992;26A:1263–2172.
13. McAfee BJ, Taylor A. A review of the volatile metabolites of fungi found on wood substrates. Nat Toxins 1999;7:283–303.
14. Fung F, Clark RF. Health effects of mycotoxins: a toxicological overview. J Toxicol Clin Toxicol 2004;42:217–234.
15. Sweeney MJ, Dobson ADW. Mycotoxin production by *Aspergillus, Fusarium* and *Penicillium* species. Int J Food Microbiol 1998;43:141–158.
16. Jacyno JM, Harwood JS, Cuytler HG, Lee M-K. Isocladosporin, a biologically active isomer of cladosporin from *Cladosporium cladosporioides*. J Nat Prod 1993;56:1397–1401.
17. Sharda DP, Wilson RF, Williams LE, Swiger LA. Effect of feeding corn inoculated with *Nigrospora oryzae* and *Cladosporium* on the performance of growing swine and rats. J Animal Sci 1971;33:1259–1262.
18. Revankar SG. Clinical implications of mycotoxins and *Stachybotrys*. Am J Med Sci 2003;325:262–274.
19. Pitt JI. Toxigenic fungi and mycotoxins. Br Med Bull 2000;56:184–192.
20. Joint Food and Agriculture Organization/United Nations Expert Committee on Food Additives. Safety evaluation of certain mycotoxins in food. Geneva: World Health Organization;2001:1–701.
21. Diyaolu SA, Adebajo LO. Effects of sodium chloride and relative humidity on growth and sporulation of moulds isolated from cured fish. Die Nahrung 1994;38:311–317.
22. Malmauret L, Parent-Massin D, Hardy J-L, Verger P. Contaminants in organic and conventional foodstuffs in France. Food Additive Contam 2002;19:524–532.
23. Niels KF, Thran U. Fast methods for screening of trichothecenes in fungal cultures using gas chromatography-tandem mass spectrometry. J Chromatogr A 2001;929:75–87.

24. Flannigan B, McCabe EM, McGarry F. Allergenic and toxigenic micro-organisms in houses. Soc Appl Bacteriol Symp Ser 1991;20:61S–73S.
25. Shantha T. Fungal degradation of aflatoxin B1. Nat Toxins 1999;7:175–178.
26. Nielsen KF, Thrane U, Larsen TO, Nielsen PA, Gravesen S. Production of mycotoxins on artificially inoculated building materials. Int Biodeter Biodegrad 1998;42:8–17.
27. Karlovsky P. Biological detoxification of fungal toxins and its use in plant breeding, feed and food production. Nat Toxins 1999;7:1–23.
28. Hadidane R, Roger-Regnault C, Bouattour H, Ellouze F, Bacha H, Creppy EE, Dirheimer G. Correlation between alimentary mycotoxin contamination and specific diseases. Hum Toxicol 1985;4:491–501.
29. Bardana EJ Jr. Indoor air quality and health: does fungal contamination play a significant role? Immunol Allergy Clin N Am 2003;2:291–309.
30. Verhoeff AP van Wijnen JH, Brunekreef B, Fischer P, Van Reenen-Hoekstra ES, Samson RA. Presence of viable mould propagules in indoor air in relation to house damp and outdoor air. Allergy 1992;47:83–91.
31. Dales RE, Miller D, McMullen E. Indoor air quality and health: validity and determinants of reported home dampness and moulds. Int J Epidemiol 1997;26:120–125.
32. Ren P, Jankun TM, Belanger K, Bracken MB, Leaderer BP. The relation between fungal propagules in indoor air and home characteristics. Allergy 2001;56:419–424.
33. Datz Y, Verleger H, Barr J, Rachmiel M, Kiviti S, Kuttin ES. Indoor survey of moulds and prevalence of mould atopy in Israel. Clin Exp Allergy 1999;29:186–192.
34. Nielsen KF. Mycotoxin production by indoor molds. Fungal Gen Biol 2003;39:103–117.
35. Tuomi T, Reijula K, Johnsson T, Hemminki K, Hintikka E-L, Lindroos O, et al. Mycotoxins in crude building materials from water-damaged buildings. Appl Environ Microbiol 2000:1899–1904.
36. Horner WE, O'Neil CE, Lehrer SB. Basidospore aeroallergens. Clin Rev Allergy 1992;10:191–211.
37. Ellis MH, Gallup J. Aeroallergens of Southern California. Immunol Allergy Clin N Am 1989;9:365–380.
38. Gots RE, Layton NJ, Pirages SW. Indoor health: background levels of fungi. Am Ind Hyg Assoc J 2003;64: 427–438.
39. Gorny RL, Dutkiewicz J. Bacterial and fungal aerosols in indoor environment in Central and Eastern European countries. Ann Agric Environ Med 2002;9:17–23.
40. Wu P-C, Su H-J, Lin C-Y. Characteristics of indoor and outdoor airborne fungi at suburban and urban homes in two seasons. Sci Total Environ 2000;253:111–118.
41. Dharmage S, Bailey M, Raven J, Mitakakis T, Thien F, Forbes A, et al. Prevalence and residential determinants of fungi within homes in Melbourne, Australia. Clin Exp Allergy 1999;29:1481–1489.
42. Vujanovic V, Smoragiewicz W, Krzysztyniak K. Airborne fungal ecological niche determination as one of the possibilities for indirect mycotoxin risk assessment in indoor air. Environ Toxicol 2001;16:1–8.
43. Ren P, Jankun TM, Leaderer BP. Comparisons of seasonal fungal prevalence in indoor and outdoor air and in house dust of dwellings in one Northeast American county. J Exp Anal Environ Epidemiol 1999;9:560–568.
44. Al-Doory Y. The indoor airborne fungi. NER Allergy Proc 1985;6:140–149.
45. Engelhart S, Loock A, Skutlarek D, Sagunski H, Lommel A, Farber H, Exner M. Occurrence of toxigenic *Aspergillus versicolor* isolates and sterigmatocystin in carpet dust from damp indoor environments. Appl Environ Microbiol 2002;68:3886–3890.
46. Fradkin A, Tobin RS, Tarlo SM, Tucic-Porretta M, Malloch D. Species identification of airborne molds and its significance for the detection of indoor pollution. JAPCA 1987;37:51–53.
47. Macher JM, Huang F-Y, Flores M. A two-year study of microbiological indoor air quality in a new apartment. Arch Environ Med 1991;46:25–29.
48. Flannigan B, McCabe EM, McGarry F. Allergenic and toxigenic micro-organisms in houses. Soc Appl Bacteriol Symp Ser 1991;20:61S–73S.
49. McGrath JJ, Wong WC, Cooley JD, Straus DC. Continually measured fungal profiles in sick building syndrome. Curr Microbiol 1999;38:33–36.
50. Verhoeff AP van Wijnen JH, Fischer P, Brunkreef B, Boleij JSM, Van Reenen ES, Samson RA. Presence of viable mould propagules in the indoor air of houses. Toxicol Ind Health 1990;6:133–145.
51. Burge HA, Pierson DL, Groves TO, Strawn KF, Mishra SK. Dynamics of airborne fungal populations in a large office building. Curr Microbiol 2000;40:10–16.
52. Rogers CA. Indoor fungal exposure. Immunol Allergy Clin North Am 2003;23:501–518.
53. Shelton BG, Kirkland KH, Flanders WD, Morris GK. Profiles of airborne fungi in buildings and outdoor environments in the United States. Appl Environ Microbiol 2002;68:1743–1753.
54. Takahashi T. Airborne fungal colony-forming units in outdoor and indoor environments in Yokohama, Japan. Mycopathology 1997;139:23–33.
55. Burge HA, Otten JA. Fungi. In: Macher J, editor. Bioaerosols: Assessment and Control. Cincinnati, OH: American Conference of Governmental and Industrial Hygienists; 1999:19.1–19.13.
56. Fischer G, Muller T, Schwalbe R, Ostrowski R, Dott W. Exposure to airborne fungi, MVOC and mycotoxins in biowaste-handling facilities. Int J Hyg Environ Health 2000;203:97–104.
57. Rao CY, Burge HA, Chang JCS. Review of quantitative standards and guidelines for fungi in indoor air. J Air Waste Manage Assoc 1996;36:899–908.

58. American Conference of Governmental Industrial Hygienists (ACGIH). Bioaerosol Committee Guidelines for the Assessment of Bioaerosols in the Indoor Environment. Cincinnati, OH: ACGIH;1989.
59. Kozak PP Jr, Gallup J, Cummins LH, Gillman SA. Currently available methods for home mold surveys. II. Examples of problem homes surveyed. Ann Allergy 1980;45:167–176.
60. Klanova K. The concentrations of mixed populations of fungi in indoor air: rooms with and without mould problems; rooms with and without health complaints. Cent Eur J Publ Health 2000;8:59–61.
61. Li C-S, Hsu C-W, Tai M-L. Indoor pollution and sick building syndrome symptoms among workers in day-care centers. Arch Environ Health 1997;52:200–207.
62. Cooley JD, Wong WC, Jumper CA, Straus DC. Correlation between the prevalence of certain fungi and sick building syndrome. Occup Environ Med 1998;55:579–584.
63. Dillon HK, Miller JD, Sorenson WG, Douwes J, Jacobs RR. Review of methods applicable to the assessment of mold exposure to children. Environ Health Perspect 1999;107(suppl 3):473–480.
64. Herbarth O, Schlink U, Muller A, Richter M. Spatiotemporal distribution of airborne mould spores in apartments. Mycol Res 2003;107:1361–1371.
65. Garrett MH, Rayment Pr, Hooper MA, Abramson MJ, Hooper BM. Indoor airborne fungal spores, house dampness and associations with environmental factors and respiratory health in children. Clin Exp Allergy 1998;28:459–467.
66. Roponen M, Seuri M, Nevalainen A, Hirvonen M-R. Fungal spores as such do not cause nasal inflammation in mold exposure. Inhalation Toxicol 2002;14:541–549.
67. Malmberg P, Rask-Andersen A, Rosenhall L. Exposure to microorganisms associated with allergic alveolitis and febrile reactions to mold dust in farmers. Chest 1993;103:1202–1209.
68. Kozak PP, Gallup J, Cummins LH, Gillman SA. Currently available methods for home mold surveys. I. Description of techniques. Ann Allergy 1980;45:85–89.
69. Nielsen KF, Gravesen S, Nielsen PA, Andersen B, Thrane U, Frisvad JC. Production of mycotoxins on artificially and naturally infested building materials. Mycopathologia 1999;145:43–56.
70. Grant C, Hunter CA, Flannigan B, Bravery AF. The moisture requirements of moulds isolated from domestic dwelling. Int Biodeter 1989;25:259–284.
71. Sorenson WG. Mycotoxins as potential occupational hazards. J Ind Microbiol 1990;31(suppl 5):205–211.
72. Meklin T, Husman T, Vepsalainen A, Vahteristo M, Koivisto J, Halla-Aho J et al. Indoor air microbes and respiratory symptoms of children in moisture damaged and reference schools. Indoor Air 2002;12:175–183.
73. Page EH, Trout DB. The role of *Stachybotrys* mycotoxins in building-related illness. Am Ind Hyg Assoc J 2001;62:644–648.
74. Burge HA. Fungi: toxic killers or unavoidable nuisances? Ann Allergy Asthma Immunol 2001;87(suppl):52–56.
75. Skaug MA. Levels of ochratoxin A and IgG against conidia of *Penicillium verrucosum* in blood samples from healthy farm workers. Ann Agric Environ Med 2003;10:73–77.
76. Beijer L, Thorn J, Rylander R. Effects after inhalation of (1 → 3)-beta-D-glucan and relation to mould exposure in the home. Mediators Inflamm 2002;11:149–153.
77. Fogelmark B, Goto H, Yuasa K, Marchat B, Rylander R. Acute pulmonary toxicity of inhaled beta-1,3-glucan and endotoxin. Agents Actions 1992;35:50–56.
78. Rylander R. Airborne (1 → 3)-β-D-glucan and airway disease in a day-care center before and after renovation. Arch Environ Health 1997;52:281–285.
79. Rylander R, Hsieh V, Courteheuse C. The first case of sick building syndrome in Switzerland. Indoor Environ 1994;3:159–162.
80. Thorn J, Rylander R. Airways inflammation and glucan in a rowhouse area. Am J Respir Crit Care Med 1998;157:1798–1803.
81. Nordness ME, Zacharisen MC, Fink JN. Toxic and other non-IgE-mediated effects of fungal exposures. Curr Allergy Asthma Rep 2003;3:438–446.
82. Assouline-Dayan Y, Leong A, Shoenfeld Y, Gershwin ME. Studies of sick building syndrome. IV. Mycotoxicosis. J Asthma 2002;39:191–201.
83. Stewart AE, Hunsaker DH. Fungus-specific IgG and IgE in allergic fungal rhinosinusitis. Otolaryngol Head Neck Surg 2002;127:324–332.
84. Kelman BJ, Robbins CA, Swenson LJ, Hardin BD. Risk from inhaled mycotoxins in indoor office and residential environments. Int J Toxicol 2004;23:3–10.
85. Hum S. Putative sporidesmin toxicity in an Eastern Grey kangaroo (*Macropus giganteus*). Aust Vet J 2005;83:678–679.
86. Lockard VG, Watson SA, Siraj MY, Hayes AW, O'Neal RM. Rubratoxin B hepatotoxicity: an electron microscopic study. Exp Mol Pathol 1981;34:94–109.
87. World Health Organization. Environmental Health Criteria. 11. Mycotoxins. Geneva: WHO; 1979.
88. Elling F. Ochratoxin A induced mycotoxic porcine nephropathy: alterations in enzyme activity in tubular cells. Acta Pathol Microbiol Scand (Sect A) 1979;87:237–243.
89. Rabie CJ, Van Rensburg SJ, Van Der Watt JJ, Lubben A. Onyalai—The possible involvement of a mycotoxin produced by *Phoma sorghina* in the aetiology. South Afr Med J 1975;49:1647–1650.
90. Shaw LM, Nowak I. Mycophenolic acid: measurement and relationship to pharmacologic effects. Ther Drug Monit 1995;17:685–689.

91. Hood KA, Zarembski DG. Mycophenolate mofetil: a unique immunosuppressive agent. Am J Health-Syst Pharm 1997;54:285–294.
92. Bebarta VS, Heard K, Nadelson C. Lack of toxic effects following acute overdose of CellCept (mycophenolate mofetil). J Toxicol Clin Toxicol 2004;42:917–919.
93. Ueno Y. The toxicology of mycotoxins. Crit Rev Toxicol 1985;14:99–132.
94. Uraguchi K, Tatsuno T, Tsukioka M, Sakai Y, Sakai F, Kobayashi Y, et al. Toxicological approach to the metabolites of *Penicillium islandicum* spp. growing on the yellowed rice. Jpn J Exp Med 1961;31:18.
95. Liu X, Luo X, Hu W. Studies on epidemiology and etiology of moldy sugarcane poisoning in China. Biomed Environ Sci 1992;5:161–177.
96. Ming L. Moldy sugarcane poisoning—a case report with a brief review. Clin Toxicol 1995;33:363–367.
97. Finkelstein E, Afek U, Gross E, Aharoni N, Rosenberg L, Halevy S. An outbreak of phytophotodermatitis due to celery. Int J Dermatol 1994;33:116–118.
98. Austad J, Kavli G. Phototoxic dermatitis caused by celery infected by *Sclerotinia sclerotiorum*. Contact Dermatitis 1983;9:448–451.
99. Rao CY, Brain JD, Burge HA. Reduction of pulmonary toxicity of *Stachybotrys chartarum* spores by methanol extraction of mycotoxins. Appl Environ Microbiol 2000;66:2817–2821.
100. Wilkins CK, Larsen ST, Hammer M, Poulsen OM, Wolkoff P, Nielsen GD. Respiratory effects in mice exposed to airborne emissions from *Stachybotrys chartarum* and implications for risk assessment. Pharmacol Toxicol 1998;83:112–119.
101. Kuhn DM, Ghannoum MA. Indoor mold, toxigenic fungi, and *Stachybotrys chartarum*: infectious disease perspective. Clin Microbiol Rev 2003;16:144–172.
102. Patovirta RL, Husman T, Haverinen U, Vahteristo M, Uitti JA, Tukiainen H, Nevalainen A. The remediation of mold damaged school—a three-year follow-up study on teachers' health. Cent Eur J Public Health 2004;12:36–42.
103. Kauffman HF, Tomee JFC, van der Werf TS, de Monchy JGR, Koeter GK. Review of fungus-induced asthmatic reactions. Am J Respir Crit Care Med 1995;151:2109–2116.
104. Menzies D, Comtois P, Pasztor J, Nunes F, Hanley JA. Aeroallergens and work-related respiratory symptoms among office workers. J Allergy Clin Immunol 1998;101:38–44.
105. Jaakkola MS, Nordman H, Piipari R, Uitti J, Laitinen J, Karjalainen A, et al. Indoor dampness and molds and development of adult-onset asthma: a population-based incident case-control study. Environ Health Perspect 2002;110:543–547.
106. Zock J-P, Jarvis D, Luczynska C, Sunyer J, Burney P, and the European Community Respiratory Health Survey. Housing characteristics, reported mold exposure, and asthma in the European Community Respiratory Health Survey. J Allergy Clin Immunol 2002;110:285–292.
107. Peng A, Yang C, Rui H, Li H. Study on the pathogenic factors of Kashin-Beck disease. J Toxicol Environ Health 1992;35:79–90.
108. Nesterov AI. The clinical course of Kashin—Beck's disease. Arthritis Rheum 1964;7:29–40.
109. Chapman JA, Terr AI, Jacobs RL, Charlesworth EN, Bardana EJ Jr. Toxic mold: phantom risk vs science. Ann Allergy Asthma Immunol 2003;91:222–232.
110. Denning DW. Aflatoxin and human disease. Adv Drug React Action Poison Rev 1987;4:175–209.
111. Klich MA, Pitt JI. Differentiation of *Aspergillus flavus* from *Aspergillus parasiticus* and other closely related species. Trans Br Mycol Soc 1988;91:99–108.
112. Joint FAO/WHO Expert Committee on Food Additives. Evaluation of certain mycotoxins in food. WHO Tech Rep Ser 2002;906:1–62.
113. Hesseltine CW, Rogers RF, Shokwell OL. Aflatoxin and mold flora in North Carolina in 1977 corn crop. Mycologia 1981;73:216–218.
114. Parrish FW, Wiley BJ, Simmons EG, Long L Jr. Production of aflatoxins and kojic acid by species of *Aspergillus* and *Pencillium*. Appl Microbiol 1966;14;139.
115. Hendrickse RG, Maxwell SM, Young R. Aflatoxins and heroin. Toxicol 1989;8:89–94.
116. Luchese RH, Harrigan WF. Biosynthesis of aflatoxin-the role of nutritional factors. J Appl Bacteriol 1993;74:5–14.
117. Cole RJ, Hill RA, Blankenship PD, Sanders TH, Garren H. Influence of irrigation and drought stress on invasion of *Aspergillus flavus* in corn kernels and peanut pods. Dev Ind Microbiol 1982;23:299–326.
118. Lillehoj EB, Garcia WJ, Lambron M. *Aspergillus flavus* infection and aflatoxin production in corn: influence of trace elements. Appl Microbiol 1974;28:763–767.
119. Yoshizawa T, Yamashita A, Chokethaworn N. Occurrence of fumonisins and aflatoxins in corn from Thailand. Food Addit Contam 1996;13:163–168.
120. Dandoy S. Aflatoxin contamination of cottonseed. JAMA 1980;243:731–732.
121. MacDonald S, Castle L. A UK retail survey of aflatoxins in herbs and spices and their fate during cooking. Food Addit Contam 1996;13:121–128.
122. Atapatu R, Samarajeewa UP Fungi associated with dried fish in Sri Lanka. Mycopathologia 1990;111:55–59.
123. Park DL. Effect of processing on aflatoxin. Adv Exp Med Biol 2002;504:173–179.
124. Waltking AE. Fate of aflatoxin during roasting and storage of contaminated peanut products. J Assoc Off Anal Chem 1976;54:533–539.
125. Scott PM. Mycotoxins transmitted into beer from contaminated grains during brewing. J AOAC Int 1996;79:875–882.
126. Scott PM, Lawrence GA. Determination of aflatoxins in beer. J AOAC Int 1997;80:1229–1234.

127. Willis RM, Mulvihill JJ, Hoofnagle JH. Attempted suicide with purified aflatoxin. Lancet 1980;1(8179):1198–1199.
128. Ngindu A, Johnson BK, Kenya PR, Ngira JA, Ocheng DM, Nandwa H, et al. Outbreak of acute hepatitis caused by aflatoxin poisoning in Kenya. Lancet 1982;1(8285): 1346–1348.
129. Riley RT, Kemppainen BW, Norred WP. Penetration of aflatoxins through isolated human epidermis. J Toxicol Environ Health 1985;15:769–777.
130. Guengerich FP, Johnson WW. Kinetics of hydrolysis and reaction of aflatoxin B_1 exo-8,9-epoxide and relevance to toxicity and detoxication. Drug Metab Rev 1999;31:141–158.
131. Sudakin DL. Dietary aflatoxin exposure and chemoprevention of cancer: a clinical review. J Toxicol Clin Toxicol 2003;41:195–204.
132. Wong ZA, Hsieh DPH. Aflatoxicol: Major aflatoxin B_1 metabolite in rat plasma. Science 1978;200:325–327.
133. Campbell TC, Caedo JP Jr, Bulatao-Jayme J, Salamat L, Engel RW. Aflatoxin M_1 in human urine. Nature 1970; 227:403–404.
134. Zhu J-Q, Zhang L-S, Hu X, Xiao Y, Chen J-S, Xu Y-C, Fremy J, Chu FS. Correlation of dietary aflatoxin B_1 levels with excretion of aflatoxin M1 in human urine 1987;47:1848–1852.
135. Dalezios JI, Hsieh PPH, Wogan GN. Excretion and metabolism of orally administered aflatoxin B_1 by rhesus monkeys. Food Cosmet Toxicol 1973;11:605–616.
136. Wong ZA, Hsieh DP. The comparative metabolism and toxicokinetics of aflatoxin B_1 in the monkey, rat, and mouse. Toxicol Appl Pharmacol 1980;55:115–125.
137. Dalezios JI, Hsieh DP, Wogan GN. Excretion and metabolism of orally administered aflatoxin B_1 by rhesus monkeys. Food Cosmet Toxicol 1973;11:605–616.
138. Denning DW, Allen R, Wilkinson AP, Morgan MR. Transplacental transfer of aflatoxin in humans. Carcinogenesis 1990;11:1033–1035.
139. Nordheim A, Hao WM, Wogan GN, Rich A. Salt induced conversion of B-DNA to Z-DNA inhibited by aflatoxin B_1. Science 1983;219:1434–1436.
140. Shank AC, Wogan GN. Acute effects of aflatoxin B_1 on liver composition and metabolism in the rat and duckling. Toxicol Appl Pharmacol 1966;9:468–476.
141. Enwonwu CO. The role of dietary aflatoxin in the genesis of hepatocellular cancer in developing countries. Lancet 1984;2(8409):956–958.
142. Chandra RK, Scrimshaw NS. Immunocompetence in nutritional assessment. Am J Clin Nutr 1980;33:2694–2697.
143. Ray PK, Singh KP, Prasad R, Prasad AK. Immunological responses to aflatoxins and other chemical carcinogens. J Toxicol Toxin Rev 1991;10:63–85.
144. Bhat RV, Krishnamachari KAVR. Follow-up study of aflatoxic hepatitis in parts of western India. Indian J Med Res 1977;66:55–58.
145. Editorial: Aflatoxins and kwashiorkor. Lancet 1984; 2(8412):1133–1134.
146. US National Toxicology Program. Report on carcinogens, ninth edition carcinogen profiles 2000. Research Triangle Park, NC: US Department of health and Human Service, Public Health Service; 2000.
147. International Agency for Research on Cancer. Aflatoxins: B_1, B_2, G_1, G_2, M_1. IARC Monogr 1993;56:245–396.
148. Hengstler JG, van der Burg B, Steinberg P, Oesch F. Interspecies differences in cancer susceptibility and toxicity. Drug Metab Rev 1999;31:917–970.
149. Stoloff L. Aflatoxin is not a probable human carcinogen: the published evidence is sufficient. Reg Toxicol Pharmacol 1989;10:272–283.
150. Stoloff L. Aflatoxin as a cause of primary liver-cell cancer in the Untied States: a probability study. Nutr Cancer 1983;5:165–186.
151. Trucksess MW, Pohland AE. Methods and method evaluation for mycotoxins. Mol Biotechnol 2002;22:287–292.
152. Tomlins KI, Jewers K, Coker RD, Nagler MJ. A bidirectional HPTLC development method for the detection of low levels of aflatoxin in maize extracts. Chromatographia 1989;27:49–52.
153. van Egmond HP, Heisterkamp SH, Paulsch WE. EC-collaborative study on the determination of aflatoxin B_1 in animal feeding stuffs. Food Addit Contam 1991;8:17–29.
154. Orti DL, Hill RH Jr, Liddle JA, Needham LL, Vickers L. High performance liquid chromatography of mycotoxin metabolites in human urine. J Anal Toxicol 1986;10: 41–45.
155. Degan P, Montagnoli G, Wild CP. Time-resolved fluoroimmunoassay of aflatoxins. Clin Chem 1989;35:2308–2310.
156. Ward CM, Morgan MRA. Reproducibility of a commercially available kit utilizing enzyme-linked immunosorbent assay for determination of aflatoxin in peanut butter. Food Addit Contam 1991;8:9–15.
157. Jonsyn-Ellis FE. Seasonal variation in exposure frequency and concentration levels of aflatoxins and ochratoxins in urine samples of boys and girls. Mycopathologia 2000;152:35–40.
158. Moss MO. Recent studies of mycotoxins. J Appl Microbiol 1998;84(Symp suppl):62S–76S.
159. Decker WJ. Activated charcoal absorbs aflatoxin B_1. Vet Hum Toxicol 1980;22:388–389.
160. International Agency for Research on Cancer. Citrinin. IARC Monogr Eval Carcinog Risk Chem Human 1986; 40:67–82.
161. Berndt WO, Hayes AW. Effects of citrinin on renal tubular transport function in the rat. J Environ Pathol Toxicol 1978; 1:93–103.
162. Thoms H. John Stearns and pulvis parturiens. Am J Obstet Gynecol 1931;22:418–423.

163. De Costa C. St Anthony's fire and living ligatures: a short history of ergometrine. Lancet 2002;359(9319): 1768–1770.
164. Caporael LR. Ergotism: the satan loosed in Salem? Science 1976;192:21–26.
165. Fuller JG. The Day of St. Anthony's Fire. New York: MacMillan Co; 1968.
166. Newberne PM. Mycotoxins: toxicity, carcinogenicity, and the influence of various nutritional conditions. Environ Health Perspect 1974;9:1–32.
167. Dudley HW, Moir C. The substance responsible for the traditional clinical effect of ergot. Br Med J 1935;1: 520–523.
168. van Dongen PW de Groot AN. History of ergot alkaloids from ergotism to ergometrine. Eur J Obstet Gynecol Reprod Biol 1995;60:109–116.
169. Ramsbotham F. The principles and practice of obstetric medicine and surgery. London: Churchill; 1841:702–706.
170. de Groot ANJA, van Dongen PWJ, Vree TB, Hekster YA, Van Roosmalen J. Ergot alkaloids current status and review of clinical pharmacology and therapeutic use compared with other oxytocics in Obstetrics and Gynaecology. Drugs 1998;56:523–535.
171. Rothlin E. Historical development of the ergot therapy of migraine. Int Arch Allergy 1955;7:205–209.
172. Harrison TE. Ergotaminism. JACEP 1978;7:162–169.
173. McGuigan MA. Ergot alkaloids. Clin Toxicol Rev 1984; 6(6):1–2.
174. Eadie MT. Ergotamine pharmacokinetics in man: an editorial. Cephalalgia 1983;3:135–138.
175. Young JC. Variability in the content and composition of alkaloids found in Canadian ergot. I. Rye. J Environ Sci Health B 1981;16:83–111.
176. Young JC. Variability in the content and composition of alkaloids found in Canadian ergot. II. Wheat. J Environ Sci Health B 1981;16:381–393.
177. Scott PM, Lawrence GA. Analysis of ergot alkaloids in flour. J Agric Food Chem 1980;28:1258–1261.
178. Scott PM, Lawrence GA. Losses of ergot alkaloids during making of bread and pancakes. J Agric Food Chem 1982;30:445–450.
179. Colman I, Brown MD, Innes GD, Grafstein E, Roberts TE, Rowe BH. Parenteral dihydroergotamine for acute migraine headache: a systematic review of the literature. Ann Emerg Med 2005;45:393–401.
180. Mueller PS, Vester JW, Fermaglich J. Neuroleptic malignant syndrome: Successful treatment with bromocriptine. JAMA 1983;249:386–388.
181. Bagby AJ, Cooper AD. Angiography in ergotism: report of two cases and review of the literature. Am J Roentgenol Rad Then Nucl Med 1972;116:179–186.
182. Jones EM, Williams B. Two cases of ergotamine poisoning in infants. Br Med J 1966;1:466.
183. Edwards WM. Accidental poisoning of newborn infants with ergonovine maleate. Clin Pediatr 1971;10:257–260.
184. Harrison TE. Ergotism after a single dose of ergotamine tartrate. J Emerg Med 1984;2:23–25.
185. Friedman AP, Brazil P, von Storch TJ. Ergotamine tolerance in patients with migraine. JAMA 1955;157:881–884.
186. Carter BJ. Cardiac manifestations following ergotamine tartrate therapy for migraine. JAMA 1940;114:2298–2299.
187. Carr P. Self induced myocardial infarction. Postgrad Med J 1981;57:654–655.
188. Bangh SA, Hughes KA, Roberts DJ, Kovarik SM. Neonatal ergot poisoning: a persistent iatrogenic illness. Am J Perinatol 2005;22:239–243.
189. Graham JR. Methysergide for prevention of headache. N Engl J Med 1964;270:67–72.
190. Warren DE, Nakfoor E. Acute overdose of bromocriptine. Drug Intell Clin Pharm 1983;17:394.
191. Descotes J, Frantz P, Bourrat C. Intoxication aigue par ingestion volontaire de bromocriptine. A propos de 2 observations personnelles. Bull Med Leg Toxicol 1979;22:487–490.
192. Silberstein SD. The pharmacology of ergotamine and dihydroergotamine. Headache 1997;37(suppl 1):S15–S25.
193. lbraheem JJ, Paalzow L, Tfelt-Hansen P. Kinetics of ergotamine after intravenous and intramuscular administration of migraine sufferers. Eur J Clin Pharmacol 1982;23:235–240.
194. Aellig WH, Nuesch E. Comparative pharmacokinetic investigations with tritium labeled ergot alkaloids after oral and intravenous administration in man. Int J Clin Pharmacol 1977;15:106–112.
195. Schran HF, McDonald S, Lehr R. Pharmacokinetics and bioavailability of ergoloid mesylates. Biopharmaceut Drug Dispos 1988;9:349–361.
196. Orton DA, Richardson RJ. Ergotamine absorption and toxicity. Postgrad Med J 1982;58:6–11.
197. Setnikar I, Schmid K, Rovati LC, Vens-Cappell B, Mazur D, Kozak I. Bioavailability and pharmacokinetic profile of dihydroergotoxine from a table and from an oral solution formulation. Arzeim-forsch Drug Res 2001;51:2–6.
198. Schmidt R, Fanchamps A. Effect of caffeine on intestinal absorption of ergotamine in man. Eur J Clin Pharmacol 1974; 7:213–216.
199. Sanders SW, Haering N, Mosberg H, Jaeger H. Pharmacokinetics of ergotamine in healthy volunteers following oral and rectal dosing. Eur J Clin Pharmacol 1986;30: 331–334.
200. Baumrucker JF. Drug interaction—propranolol and Cafergot. N Engl J Med 1973;288:915–916.
201. Horowitz RS, Dart RC, Gomez HF. Clinical ergotism with lingual ischemia induced by clarithromycin-ergotamine interaction. Arch Intern Med 1996;156: 456–458.
202. Buxton A, Goldberg S, Hirshfeld JW, Wilson J, Mann T, Williams DO, et al. Refractory ergonovine induced coro-

nary vasospasm: Importance of intracoronary nitroglycerin. Am J Cardiol 1980;46:329–334.

203. Brook NM, Cookson IB. Bromocriptine-induced mania? Br Med J 1980;1:790.
204. Johnson JM. Treated mania exacerbated by bromocriptine. Am J Psychol 1981;138:980–982.
205. Berde B. Pharmacology of ergot alkaloids in clinical use. Med J Aust 1978;2(suppl 3):3–13.
206. Glazer G, Myers KA, Davids ER. Ergot poisoning. Postgrad Med J 1966;42:562–568.
207. Merhoff GC, Porter JM. Ergot intoxication: historical review and description of unusual clinical manifestations. Ann Surg 1974;180:773–779.
208. Green FL, Aniyan 5, Stansel HC. Mesenteric and peripheral vascular ischemia secondary to ergotism. Surgery 1977;81:176–179.
209. Fedotin MS, Hartman C. Ergotamine poisoning producing renal artery spasm. N Engl J Med 1970;283:518–520.
210. Gupta DR, Strobos RJ. Bilateral papillitis associated with cat ergot therapy. Neurology 1972;22:793–797.
211. Goldfisher JD. Acute myocardial infarction secondary to ergot therapy. N Engl J Med 1960;202:860–863.
212. Ameli FM, Nathanson M, Elkan I. Methysergide therapy causing vascular insufficiency of the upper limb. Can J Surg 1977;20:158–160.
213. Vaughan-Lane T. Gangrene induced by methysergide and ergotamine: a case report. J Bone Joint Surg 1979; 61:213–214.
214. Orlando AC, Moyer P, Bannett TB. Methysergide therapy and constrictive pericarditis. Ann Intern Med 1978;88: 213–214.
215. Aeby A, Johansson A-B, De Schuiteneer B, Blum D. Methylergometrine poisoning in children: review of 34 cases. J Toxicol Clin Toxicol 2003;41:249–253.
216. Kenna AP. Accidental administration of syntometrine to a newborn infant. J Obstet Gynecol Br Common 1972;79:764–766.
217. Wells KE, Steed DL, Zajko AB, Webster MW. Recognition and treatment of arterial insufficiency from Cafergot. J Vasc Surg 1986;4:8–15.
218. Demeke T, Kidane Y, Wuhib E. Ergotism: a report on an epidemic. Ethiop Med J 1979;17:107–113.
219. Yao ST, Goodwin DP, Kenyon JR. Case of ergot poisoning: a case report. Br Med J 1960;3:86–87.
220. Safar HA, Alanezi KH, Cina CS. Successful treatment of threatening limb loss ischemia of the upper limb caused by ergotamine: a case report and review of the literature. J Cardiovasc Surg 2002;43:245–249.
221. Fisher PE, Silk DB, Menzies-Gow N, Dingle M. Ergotamine abuse and extra-hepatic portal hypertension. Postgrad Med J 1985;61:461–463.
222. Krishnamachari DAVR, Bhat RV. Poisoning of ergoty bajra (pearl millet) in man. Indian J Med Res 1976;4: 1624–1628.
223. Raymond GV. Teratogen update: ergot and ergotamine. Teratol 1995;51:344–347.
224. Peeden JN, Wilroy RS, Soper RG. Prune perineum. Teratol 1979;20:233–236.
225. Wainscott G, Sullivan FM, Volans GN, Wilkinson M. The outcome of pregnancy in women suffering from migraine. Postgrad Med J 1978;54:98–102.
226. Tfelt-Hansen P, Paalzow L. Intramuscular ergotamine. Plasma levels and dynamic activity. Clin Pharmacol Ther 1985;37:29–35.
227. Graham AN, Johnson ES, Persaud NP, Turner P, Wilkinson M. Ergotamine toxicity and serum concentrations of ergotamine in migraine patients. Hum Toxicol 1984;3:193–199.
228. Garcia GD, Goff JM Jr, Hadro NC, O'Donnell SD, Greatorex PS. Chronic ergot toxicity: a rare cause of lower extremity ischemia. J Vasc Surg 2000;31:1245–1247.
229. Tay JC, Chee YC. Ergotism and vascular insufficiency: a case report and review of literature. Ann Acad Med Singapore 1998;27:285–288.
230. Deviere J, Reuse C, Askenasi R. Ischemic pancreatitis and hepatitis secondary to ergotamine poisoning. J Clin Gastroenterol 1987;9:350–352.
231. Zavaleta EG, Fernandez BB, Grove MK, Kaye MD. St. Anthony's fire (ergotamine induced leg ischemia) a case report and review of the literature. Angiology 2001;52:349–356.
232. O'Dell CW, Davis GB, Johnson AD, Safdi MA, Brant-Zawadzki M, Bookstein JJ. Sodium nitroprusside in the treatment of ergotism. Radiology 1977;124:73–74.
233. Enge I, Sivertssen E. Ergotism due to therapeutic doses of ergotamine tartrate. Am Heart J 1965;70:665–670.
234. Shifrin E, Perel A, Olschwang D, Diamant Y, Cotev S. Reversal of ergotamine-induced arteriospasm by mechanical intra-arterial dilatation. Lancet 1980;2(8207): 1278–1279.
235. Lepantalo M, Rosenberg P, Pohjola J, Augustinsson LE, Holm J. Epidural spinal cord stimulation in the treatment of limb threatening vasospasm—report of a case with a five-year follow-up. Eur J Vasc Endovasc Surg 1996; 11:368–370.
236. Heupler FA Jr, Proudfit WL, Razavi M, Shirey EK, Greenstreet R, Sheldon WC. Ergonovine maleate: provocative test for coronary artery spasm. Am J Cardiol 1978;41:631–640.
237. Husum B, Metz P, Rasmussen JP. Nitroglycerin infusion for ergotism. Lancet 1979;2(8146):794–795.
238. Kellerman TS, Marasas WF, Pienaar JG, Naude TW. A mycotoxicosis of equidae caused by *Fusarium moniliforme* Sheldon. A preliminary communication. Onderstepoort J Vet Res 1972;39:205–208.
239. Marasas WF, Kellerman TS, Gelderblom WC, Coetzer JA, Thiel PG, van Der Lugt JJ. Leukoencephalomalacia in a horse induced by fumonisin B_1 isolated from *Fusarium moniliforme*. Onderstepoort J Vet Res 1988;55:197–203.

240. Marasas WF. Discovery and occurrence of the fumonisin: a historical perspective. Environ Health Perspect 2001; 109:239–243.
241. International Programme on Chemical Safety. Environmental health criteria 219 fumonisin B_1. Geneva: World Health Organization; 2000.
242. Branham BE, Plattner RD. Alanine is a precursor in the biosynthesis of fumonisin B_1 by *Fusarium moniliforme*. Mycopathologia 1993;124:99–104.
243. Dupuy J, Le Bars P, Boudra H, Le Bars J. Thermostability of fumonisin B_1, a mycotoxin form *Fusarium moniliforme*, in corn. Appl Exp Microbiol 1993;59:2864–2867.
244. Bothast RJ, Bennett GA, Vancauwenberge JE, Richard JL. Fate of fumonisin B_1 in naturally contaminated corn during ethanol fermentation. Appl Environ Microbiol 1992;58:233–236.
245. Rotter BA, Prelusky DB, Pestka JJ. Toxicology of deoxynivalenol (vomitoxin). J Toxicol Environ Health 1996;48:1–34.
246. Nelson PE, Dignani MC, Anaissie EJ. Taxonomy, biology, and clinical aspects of *Fusarium* species. Clin Microbiol Rev 1994;7:479–504.
247. Melcion D, Cahagnier B, Richard-Molard D. Study of the biosynthesis of fumonisins B_1, B_2, and B_3 by different strains of *Fusarium moniliforme*. Lett Appl Microbiol 1997;24:301–305.
248. Miller JD. Factors that affect the occurrence of fumonisin. Environ Health Perspect 2001;109(Suppl 2):321–324.
249. Abbas HK, Cartwright RD, Shier WT, Abouzied MM, Bird CB, Rice LG, et al. Natural occurrence of fumonisins in rice with *Fusarium* sheat rot disease. Plant Dis 1998;82:22–25.
250. Prelusky DB, Trenholm HL, Rotter BA, Miller JD, Savard ME, Yeung JM, Scott PM. Biological fate of fumonisin B_1 in food-producing animals. Adv Exp Med Biol 1996;392:265–278.
251. Miller JD, Savard ME, Rapior S. Production and purification of fumonisins from a stirred jar fermenter. Nat Toxins 1994;2:354–359.
252. Abbas HK, Williams WP, Windham GL, Pringle HC3rd , Xie W, Shier WT. Aflatoxin and fumonisin contamination of commercial corn (*Zea mays*) hybrids in Mississippi. J Agric Food Chem 2002;50:5246–5254.
253. Pitt JI, Basilico JC, Abarca ML, Lopez C. Mycotoxins and toxigenic fungi. Medical Mycol 2000;38(suppl 1):41–46.
254. Marin S, Sanchis V, Vinas I, Canela R, Magan N. Effect of water activity and temperature on growth and fumonisin B_1 and B_2 production by *Fusarium proliferatum* and *F. moniliforme* on maize grain. Lett Appl Microbiol 1995;21:298–301.
255. Marasas WFO. Fumonisins: their implications for human and animal health. Nat Toxins 1995;3:193–198.
256. Thiel PG, Marasas WF, Syndenham EW, Shephard GS, Gelderblom WC. The implications of naturally occurring levels of fumonisins in corn for human and animal health. Mycopathologia 1992;117:3–9.
257. Hirooka EY, Yamaguchi MM, Aoyama S, Sugiura Y, Ueno Y. The natural occurrence of fumonisins in Brazilian corn kernels. Food Addit Contam 1996;13:173–183.
258. Solovey MMS, Somoza C, Cano G, Pacin A, Resnick S. A survey of fumonisins, deoxynivalenol, zearalenone and aflatoxins contamination in corn-based food products in Argentina. Food Addit Contam 1999;16:325–329.
259. de Nijs M, van Egmons HP, Nauta M, Rombouts FM, Notermans SH. Assessment of human exposure to fumonisin B_1. J Food Protect 1998;61:879–884
260. Rubeiha WK, Oehme FW. Fumonisin exposure to Kansans through consumption of corn-based market foods. Vet Hum Toxicol 1997;39:220–225.
261. Visconti A, Doko MB. Survey of fumonisin production by *Fusarium* isolated from cereals in Europe. J AOAC Int 1994;77:546–550.
262. Hlywka JJ, Bullerman LB. Occurrence of fumonisin B_1 and B_2 in beer. Food Addit Contam 1999;16:319–324.
263. Scott PM, Lawrence GA. Determination of hydrolyzed fumonisin B1 in alkali-processed corn foods. Food Addit Contam 1996;13:823–832.
264. Visconti A, Solfrizzo M, Doko MB, Boenke A, Pascale M. Stability of fumonisins at different storage periods and temperatures in gamma-irradiated maize. Food Addit Contam 1996;13:929–938.
265. Norred WP, Voss KA, Riley RT, Plattner RD. Fumonisin toxicity and metabolism studies at the USDA. Fumonisin toxicity and metabolism. Adv Exp Med Biol 1996;392: 225–236.
266. National Toxicology Program. Toxicology and carcinogenesis studies of fumonisin B1 (CAS No. 116355-83-0) in F344/N rats and B6C3F1 mice (feed studies). Natl Toxicol Program Tech Rep Ser 2001;496:1–352.
267. Wilson TM, Ross PF, Owens DL, Rice LG, Green SA, Jenkins SJ, Nelson HA. Experimental reproduction of ELEM–a study to determine the minimum toxic dose in ponies. Mycopathologia 1992;117:115–120.
268. Sweeney MJ, Dobson AD. Molecular biology of mycotoxin biosynthesis. FEMS Microbiol Lett 1999;175: 149–163.
269. Harrison LR, Colvin BM, Greene JT, Newman LE, Cole JR. Pulmonary edema and hydrothorax in swine produced by fumonisin B_1, a toxic metabolite of *Fusarium moniliforme*. J Vet Diag Invest 1990;2:217–221.
270. Bhat RV, Shetty PH, Amruth RP, Sudershan RV. A foodborne disease outbreak due to the consumption of moldy sorghum and maize containing fumonisin mycotoxins. Clin Toxicol 1997;35:249–255.
271. Trucksess MW. Mycotoxins. J AOAC Int 2001;84: 202–211.
272. Ueno Y, Iijima K, Wang S-D, Sugiura Y, Sekijima M, Tanaka T, Chen C, Yu S-Z. Fumonisins as a possible

contributory risk factor for primary liver cancer: a 3-year study of corn harvested in Haimen, China, by HPLC and ELISA. Food Chem Toxicol 1997;35:1143–1150.

273. Shephard GS. Chromatographic determination of the fumonisin mycotoxins. J Chromatogr A 1998;815:31–39.

274. Azcona-Olivera JI, Abouzied MM, Plattner RD, Norred WP, Pestka JJ. Generation of antibodies reactive with fumonisins B_1, B_2, and B_3 by using cholera toxin as the carrier-adjuvant. Appl Environ Microbiol 1992;58: 169–173.

275. Elissalde MH, Kamps-Holtzapple C, Beier RC, Plattne RD, Rowe LD, Stanker LH. Development of an improved monoclonal antibody-based ELISA for fumonisins $B_{1\text{-}3}$ and the use of molecular modelling to explain observed detection limits. Food Agric Immunol 1995;7:109–122.

276. Plattner RD, Weisleder D, Poling SM. Analytical determination of fumonisins and other metabolites produced by *Fusarium moniliforme* and related species on corn. Adv Exp Med Biol 1996;392:47–64.

277. Ringot D, Chango A, Schneider Y-J, Larondelle Y. Toxicokinetics and toxicodynamics of ochratoxin A, an update. Chem Biol Interact 2006;159:18–46.

278. Galtier P, Alvinerie M. *In vitro* transformation of ochratoxin A by animal microbial floras. Ann Rech Vet 1976;7:91–98.

279. Beretta B, de Domenico R, Gaiaschi A, Ballabio C, Galli CL, Gigliotti C, Restani P. Ochratoxin A in cereal-based baby foods: occurrence and safety evaluation. Food Addit Contam 2002;19:70–75.

280. Kiessling K, Pettersson H, Sandholm K, Olsen M. Metabolism of aflatoxin, ochratoxin, zearalenone and three trichothecenes by intact rumen fluid, rumen protozoa and rumen bacteria. Appl Environ Microbiol 1984; 47:1070–1073.

281. Mills JT, Seifert KA, Frisvad JC, Abramson D. Nephrotoxigenic *Penicillium* species occurring on farm-stored cereal grains in western Canada. Mycopathologia 1995;130:23–28.

282. International Agency for Research on Cancer. Some naturally occurring substances: food items and constituents, heterocyclic aromatic amines and mycotoxins. IARC Monogr Eval Carcinog Risk Hum 1993;56:489–521.

283. Larsen TO, Svendsen A, Smedsgaard J. Biochemical characterization of ochratoxin A-producing strains of the genus *Penicillium*. Appl Environ Microbiol 2001;67:3630–3635.

284. Assaf H, Betbeder A-M, Creppy EE, Pallardy M, Azouri H. Ochratoxin A levels in human plasma and foods in Lebanon. Hum Exp Toxicol 2004;23:495–501.

285. Moss MO. Mode of formation of ochratoxin A. Food Addit Contam 1996;13(suppl):5–9.

286. Haggblom P. Production of ochratoxin A, in barley by *Aspergillus ochraceus* and *Penicillium viridicatum*; effect of fungal growth, time, temperature, and inoculum size. Appl Environ Microbiol 1982;43:1205–1207.

287. Connole MD, Blaney BJ, McEwan T. Mycotoxins in animal feeds and toxic fungi in Queensland 1971–1980. Aust Vet J 1981;57:314–318.

288. Northolt MD, van Egmond HP, Paulsch WE. Ochratoxin A production by some fungal species in relation to water activity and temperature. J Food Protect 1979;42:485–490.

289. Madhyastha SM, Marquardt RR, Frolich AA, Platford G, Abramson D. Effects of different cereal and oilseed substrates on the growth and production of toxins by *Aspergillus alutaceus* and *Penicillium verrucosum*. J Agric Food Chem 1990;38:1506–1510.

290. Neely WC, West AD. Spectroanalytical parameters of fungal metabolites. III. Ochratoxin A. J Assoc Off Anal Chem 1972;55:1305–1309.

291. Jorgensen K. Survey of pork, poultry, coffee, beer and pulses for ochratoxin A. Food Addit Contam 1998;15:550–554.

292. Patel S, Hazel CM, Winterton AGM, Gleadle AE. Survey of ochratoxin A in UK retain coffees. Food Addit Contam 1997;14:217–222.

293. Trucksess NW, Giler J, Young K, White KD, Page SW. Determination and survey of ochratoxin A in wheat, barley, and coffee–1997. J AOAC Int 1999;82:85–89.

294. Patterson DS, Shreeve BJ, Roberts BA, Berrett S, Brush PJ, Glancy EM, Krogh P. Effect on calves of barley naturally contaminated with ochratoxin A and groundnut meal contaminated with low concentrations of aflatoxin B_1. Res Vet Sci 1981;31:213–218.

295. Mortensen HP, Hald B, Madsen A. Feeding experiments with ochratoxin A contaminated barley for bacon pigs. V. Ochratoxin A in blood. Acta Agric Scand 1983;33:235–239.

296. Lee SC, Beery JT, Chu FS. Immunohistochemical fate of ochratoxin A in mice. Toxicol Appl Microbiol 1984;72:218–227.

297. Stoev SD. The role of ochratoxin A as a possible cause of Balkan endemic nephropathy and its risk evaluation. Vet Hum Toxicol 1998;40:352–360.

298. Stormer FC, Hansen CE, Pedersen JI, Hvistendahl G, Aasen AJ. Formation of (4R)- and (AS)-4-hydroxyochratoxin A from ochratoxin A by liver microsomes from various species. Appl Environ Microbiol 1981;42:1051–1056.

299. Micco C, Miraglia M, Brera C, Corneli S, Ambruzzi A. Evaluation of ochratoxin A level in human milk in Italy. Food Addit Contam 1995;12:351–354.

300. Schlatter C, Studer-Rohr J, Rasonyi T. Carcinogenicity and kinetic aspects of ochratoxin A. Food Addit Contam 1996;13(suppl):43–44.

301. Studer-Rohr I, Schlatter J, Dietrich DR. Kinetic parameters and intraindividual fluctuations of ochratoxin A plasma levels in humans. Arch Toxicol 2000;74:499–510.

302. Krogh P, Hold B, Pederson EJ. Occurrence of ochratoxin A and citrinin in cereals associated with mycotoxic

porcine nephropathy. Acta Pathol Microbiol Scand (Sect 8) 1973;81:689–695.

303. Cukuranovic R, Petrovic B, Cukuranovic Z, Stefanovic V. Balkan endemic nephropathy: a decreasing incidence of the disease. Pathol Biol 2000;48:558–561.
304. Bukvic D, Jankovic S, Dukanovic L, Marinkovic J. Survival of Balkan endemic nephropathy patients. Nephron 2000;86:463–466.
305. Tatu CA, Orem WH, Findelman RB, Feder GL. The etiology of Balkan endemic nephropathy: still more questions than answers. Environ Health Perspect 1998;106:689–700.
306. Abid S, Hassen W, Achour A, Skhiri H, Maaroufi K, Ellouz F, et al. Ochratoxin A and human chronic nephropathy in Tunisia: is the situation endemic? Hum Exp Toxicol 2003;22:77–84.
307. Maaroufi K, Achour A, Betbeder AM, Hammami M, Ellouz F, Creppy EE, Bacha H. Foodstuffs and human blood contamination by the mycotoxin ochratoxin A: correlation with chronic interstitial nephropathy in Tunisia. Arch Toxicol 1995;69:552–558.
308. Stefanovic V, Polenakovic MH. Balkan nephropathy kidney disease beyond the Balkans? Am J Nephrol 1991;11:1–11.
309. Pfohl-Leszkowicz A, Petkova-Bocharova T, Chernozemsky IN, Castegnaro M. Balkan endemic nephropathy and associated urinary tract tumours: a review on aetiological causes and the potential role of mycotoxins. Food Addit Contam 2002;19:282–302.
310. Vrabcheva T, Usleber E, Dietrich R, Martbauer E. Co-occurrence of ochratoxin A and citrinin in cereals from Bulgarian villages with a history of Balkan endemic nephropathy. J Agric Food Chem 2000;48:2483–2488.
311. Uzelac-Keserovic B, Spasic P, Bojanic N, Dimitrijevic J, Lako B, Lepsanovic Z, et al. Isolation of a coronavirus from kidney biopsies of endemic Balkan nephropathy patients. Nephron 1999;81:141–145.
312. Stefanovic V, Toncheva D, Atanasova S, Polenakovic M. Etiology of Balkan endemic nephropathy and associated urothelial cancer. Am J Nephrol 2006;26:1–11.
313. Hranjec T, Kovfac A, Kos J, Mao W, Chen JJ, Grollman AP, Jelakovic B. Endemic nephropathy: the case for chronic poisoning by *Aristolochia*. Croat Med J 2005; 46:116–125.
314. Zepnik H, Pahler A, Schauer U, Dekant W. Ochratoxin A-induced tumor formation: is there a role of reactive ochratoxin A metabolites? Toxicol Sci 2001;59:59–67.
315. Orti DL, Hill RH, Liddle JA, Neeham LL. High-performance liquid chromatography of mycotoxin metabolites in human urine. J Anal Toxicol 1986;10:41–45.
316. Rousseau DM, Candlish AAG, Slegers GA, van Peteghem CH, Stinson WH, Smith JE. Detection of ochratoxin A in porcine kidneys by a monoclonal antibody-based radioimmunoassay. Appl Environ Microbiol 1987;3:514–518.
317. Zimmerli B, Dick R. Determination of ochratoxin A at the ppt level in human blood, serum, milk and some foodstuffs by high-performance liquid chromatography with enhanced fluorescence detection and immunoaffinity column cleanup: methodology and Swiss data. J Chromatogr B 1995;666:85–99.
318. Thuvander A, Paulsen JE, Axberg K, Johansson N, Vidnes A, Enghardt-Barbieri H, Trygg K, et al. Levels of ochratoxin A, in blood from Norwegian and Swedish blood donors and their possible correlation with food consumption. Food Chem Toxicol 2001;39:1145–1151.
319. Hald B. Ochratoxin A in human blood in European countries. IARC Sci Publ 1991;115:159–164.
320. Scott PM, Kanhere SR, Lau BP-Y, Lewis DA, Hayward S, Ryan JJ, Kuiper-Goodman T. Survey of Canadian human blood plasma for ochratoxin A. Food Addit Contam 1998;15:555–562.
321. Hult K, Plestina R, Habazin-Novak V, Radic B, Ceovic S. Ochratoxin A in human blood and Balkan endemic nephropathy. Arch Toxicol 1982;51:313–321.
322. Petkova-Bocharova T, Chernozemsky IN, Castegnaro M. Ochratoxin A in human blood in relation to Balkan endemic nephropathy and urinary system tumours in Bulgaria. Food Addit Contam 1988;5:299–301.
323. Peraica M, Domijan A-M, Fuchs R, Lucic A, Radic B. The occurrence of ochratoxin A in blood in general population of Croatia. Toxicol Lett 1999;110:105–112.
324. Iavicoli I, Brera C, Carelli G, Caputi R, Marinaccio A, Miraglia M. External and internal dose in subjects occupationally exposed to ochratoxin A. Int Arch Occup Environ Health 2002;75:381–386.
325. Joint FAO/WHO Expert Committee on Food Additives (JECFA). Ochratoxin A. Safety evaluation of certain food additives and contaminants. WHO Food Addit 1995;35:363–376.
326. Pitt JI. Biology and ecology of toxigenic Penicillium species. Adv Exp Med Biol 2002;504:29–41.
327. Ritieni A. Patulin in Italian commercial apple products. J Agric Food Chem 2003;51:6086–6090.
328. Beretta B, Gaiaschi A, Galli CL, Restani P. Patulin in apple-based foods: occurrence and safety evaluation. Food Addit Contam 2000;17:399–406.
329. Lovett J, Thompson RG, Boutin BK. Trimming as a means of removing patulin from fungus-rotted apples. J Assoc Off Anal Chemists 1975;58:909–911.
330. Tangni EK, Theys R, Mignolet E, Maudoux M, Michelet JY, Larondelle Y. Patulin in domestic and imported apple-based drinks in Belgium: occurrence and exposure assessment. Food Addit Contam 2003;20:482–489.
331. Roach JAG, White KD, Trucksess NW, Thomas FS. Capillary gas chromatography/mass spectrometry with chemical ionization and negative ion detection for confirmation of identity of patulin in apple juice. J AOAC Int 2000;83:104–112.
332. Rychlik M. Rapid degradation of the mycotoxin patulin in man quantified by stable isotope dilution assays. Food Addit Contam 2003;20:829–837.

333. Gye WE. Preliminary trial in the common cold. Lancet 1943;2(6487):630–631.

334. van Egmond HP, Dekker HW. Worldwide regulations for mycotoxins in 1994. Nat Toxins 1995;3:332–336.

335. Moss MO, Wood AB, Robinson FV. The structure of rubratoxin A, a toxic metabolite of *Penicillium rebrum*. Tetrahedron Lett 1969;5:367–370.

336. Lockard VG, Watson SA, Siraj MY, Hayes AW, O'Neal RM. Rubratoxin B hepatotoxicity: an electron microscopic study. Exp Mol Pathol 1981;34:94–109.

337. International Programme on Chemical Safety. Environmental health criteria 105 selected mycotoxins: ochratoxins, trichothecenes, ergot. Geneva: World Health Organization; 1990.

338. Forgacs J. Mycotoxicoses: The neglected diseases. Feedstuffs 1962;34:124–134.

339. Ciegler A, Bennett JW. Mycotoxins and mycotoxicoses. BioScience 1980;30:512–515.

340. Harruff AC. Chemical biological warfare in Asia. JAMA 1983;250:497–498.

341. Stahl CJ, Green CC, Farnum JB. The incident at Tuol Chrey: pathologic and toxicologic examinations of a casualty after chemical attack. J Forensic Sci 1985;30: 317–337.

342. Watson SA, Mirocha CJ, Hayes AW. Analysis for trichothecenes in samples from Southeast Asia associated with "yellow rain." Fundam Appl Toxicol 1984;4:700–717.

343. Rosen RT, Rosen JD. Presence of four Fusarium mycotoxins and synthetic material in yellow rain. Evidence for the use of chemical warfare in Laos. Biomed Mass Spectrom 1982; 9:443–450.

344. Wade N. Yellow rain and the cloud of chemical war. Science 1981;214:1008–1009.

345. Walsh J. A cloud burst of yellow rain reports. Science 1982;21:1202–1204.

346. Marshall E. Yellow rain evidence slowly whittled away. Science 1986;233:18–19.

347. Nowicke JW, Meselson M. Yellow rain—a palynological analysis. Nature 1984;309:205–206.

348. Wogan GN. Mycotoxins. Annu Rev Pharmacol 1975;15: 437–451.

349. Harrach B, Mirocha CJ, Pathre SV, Palyusik M. Macrocyclic trichothecene toxins produced by a strain of *Stachybotrys atra* from Hungary. Appl Environ Microbiol 1981;41:1428–1432.

350. Ogawa K, Nakamura M, Hayashi M, Yaginuma S, Yamamoto S, Furihata K, et al. Stachybocins, novel endothelin receptor antagonists, produce by *Stachybotrys* sp. M6222. II. Structure determination of stachybocins A, B and C. J Antibiot (Tokyo) 1995;48:1396–1400.

351. Lauren DR, Smith WA. Stability of the *Fusarium* mycotoxins nivalenol, deoxynivalenol and zearalenone in ground maize under typical cooking environments. Food Addit Contam 2001;18:1011–1006.

352. Shima J, Takase S, Takahashi Y, Iwai Y, Fujimoto H, Yamazaki M, Ochi K. Novel detoxification of the trichothecene mycotoxin deoxynivalenol by a soil bacterium isolated by enrichment culture. Appl Environ Microbiol 1997;63:3825–3830.

353. Jong SC, Davis EE. Contribution to the knowledge of *Stachybotrys* and *Memnoniella* in culture. Mycotaxon 1976;3:409–485.

354. Jarvis BB. Chemistry and toxicology of molds isolated from water-damaged buildings. Adv Exp Med Biol 2002;504:43–52.

355. Nikulin M, Pasanen A-L, Berg S, Hintikka E-L. *Stachybotrys atra* growth and toxin production in some building materials and fodder under different relative humidities. Appl Environ Microbiol 1994;60:3421–3424.

356. Jarvis BB, Salemme J, Morais A. *Stachybotrys* toxins. 1. Nat Toxins 1995;3:10–16.

357. Harrison J, Pickering CAC, Faragher EB, Austwick PKC, Little SA, Lawton L. An investigation of the relationship between microbial and particulate indoor air pollution and the sick building syndrome. Respir Med 1992;86:225–235.

358. Grove JF. Non-macrocyclic trichothecenes. Nat Prod Rep 1988;5:187–209.

359. DeSimone PA, Greco FA, Lessner HF, Bartolucci A. Phase II evaluation of anguidine (NSC 141537) in 5-day courses in colorectal adenocarcinoma. A Southeastern Cancer Study Group Trial. Am J Clin Oncol 1986;9: 187–188.

360. Bata A, Harrach B, Ujszaszi K, Kis-Tamas A, Lasztity R. Macrocyclic trichothecene toxins produce by *Stachybotrys atra* strains isolated in Middle Europe. Appl Environ Microbiol 1985;49:678–681.

361. Austwick PK. Mycotoxins. Br Med Bull 1975;31:222–229.

362. Gravensen S. Fungi as a cause of allergic disease. Allergy 1979;32:135–154.

363. Dharmage S, Bailey M, Raven J, Mitakekis T, Thien F, Forbes A, et al. Prevalence and residential determinants of fungi within homes in Melbourne, Australia. Clin Exp Allergy 1999;29:1481–1489.

364. Brant C, Hunter CA, Flannigan B, Bravery AF. The moisture requirements of moulds isolated from domestic buildings. Int Biodeterior 1989;25:259–284.

365. Harrach B, Bata A, Bajmocy E, Benko M. Isolation of satratoxins from the bedding straw of a sheep flock with fatal stachybotryotoxicosis. Appl Environ Microbiol 1983;45:1419–1422.

366. Gallup J, Kozak P, Cummins L, Gillman S. Indoor mold spore exposure: characteristics of 127 homes in southern California with endogenous mold problems. Experientia Suppl. 1987;51:139–142.

367. Sudakin DL. Toxigenic fungi in a water-damaged building: an intervention study. Am J Ind Med 1998;34:183–190.

368. Jelinek CF, Pohland AE, Wood GE. Occurrence of mycotoxins in foods and feeds an update. J Assoc Off Anal Chem 1989;72:223–230.

369. Tanaka T, Hasegawa A, Yamamoto S, Lee U, Sugiura Y, Ueno Y. Worldwide contamination of cereals by the *Fusarium* mycotoxins nivalenol, deoxynivalenol and zearalenone. 1. Survey of 19 countries. J Agric Food Chem 1988;36:979–983.
370. Bhavanishankar TR, Shantha T. Natural occurrence of Fusarium toxins in peanuts, sorghum and maize from Mysore (India). J Sci Food Agric 1987;40:327–332.
371. Trigo-Stockli D. Effect of processing on deoxynivalenol and other trichothecenes. Adv Exp Med Biol 2002;504: 181–188.
372. Lee U-S, Lee M-Y, Park W-Y, Ueno Y. Decontamination of *Fusarium* mycotoxins, nivalenol, deoxynivalenol, and zearalenone in barley by the polishing process. Mycotoxin Res 1992;8:31–36.
373. Abbas HK, Mirocha CJ, Pawlosky RJ, Pucsh D. Effect of cleaning, milling, and baking on deoxynivalenol in wheat. Appl Environ Microbiol 1985;50:482–486.
374. Prelusky DB,Trenholm HL, Lawrence GA, Scott PM. Non-transmission of deoxynivalenol (vomitoxin) to milk following oral administration to dairy cows. J Environ Sci Health 1984;B19:593–609.
375. Pestka JJ, Smolinski AT. Deoxynivalenol: toxicology and potential effects on humans. J Toxicol Environ Health 2005;8(Part B):39–69.
376. Murphy WK, Burgess MA, Valdivieso M, Livingston RB, Bodey GP, Freireich EJ. Phase I clinical evaluation of anguidine. Cancer Treat Rep 1978;62:1497–1502.
377. el-Maghraby OM, Bean GA, Jarvis BB, Aboul-Nasr MB. Macrocyclic trichothecenes produced by *Stachybotrys* isolated from Egypt and eastern Europe. Mycopathologia 1991;113:109–115.
378. Nikulin M, Reijula K, Jarvis BB, Veijalainen P, Hintikka EL. Effects of intranasal exposure to spores of *Stachybotrys atra* in mice. Fundam Appl Toxicol 1997;35:182–188.
379. Yagen B, Bialer M. Metabolism and pharmacokinetics of T-2 toxin and related trichothecenes. Drug Metab Rev 1993;25:281–323.
380. Kemppainen BW, Riley RT, Biles-Thurlow S. Comparison of penetration and metabolism of [^{3}H]diacetoxyscirpenol, [^{3}H]verrucarin A, and [^{3}H]T-2 toxin in skin. Food Chem Toxicol 1987;25:379–386.
381. Sintov A, Bialer M, Yagen B. Pharmacokinetics of T-2 toxin and its metabolite HT-2 toxin, after intravenous administration in dogs. Drug Metab Disp 1986;14:250–254.
382. Prelusky DB, Veira DM, Trenholm HL. Plasma pharmacokinetics of the mycotoxin deoxynivalenol following oral and intravenous administration to sheep. J Environ Sci Health 1985;B20:603–624.
383. Prelusky DB, Trenholm HL, Hamilton RMG, Miller JD. Transmission of [^{14}C]deoxynivalenol to eggs following oral administration to laying hens. J Agric Food Chem 1987;35:182–186.
384. Pestka JJ, Smolinski AT. Deoxynivalenol: toxicology and potential effects on humans. J Toxicol Environ Health B Crit Rev 2005;8:39–69.
385. Johnsen H, Odden E, Johnsen BA, Boyum A, Amundsen E. Cytotoxicity and effects of T2-toxin on plasma proteins involved in coagulation, fibrinolysis and kallikrein-kinin system. Arch Toxicol 1988;61:237–240.
386. Fairhurst S, Maxwell SA, Scawin JW, Swanston DW. Skin effects of trichothecenes and their amelioration by decontamination. Toxicology 1987;46:307–319.
387. Pang VF, Lambert RJ, Felsburg PJ, Beasley VR, Buck WB, Haschek WM. Experimental T-2 toxicosis in swine following inhalation exposure: clinical signs and effects on hematology, serum biochemistry, and immune response. Fundam Appl Toxicol 1988;11:100–109.
388. Lutsky I, Mon N, Yagen B, Joffee AZ. The role of T-2 toxin in experimental alimentary toxic aleukia. A toxicity study in cats. Toxicol Appl Pharmacol 1978;43:111–124.
389. Wei C, McLaughlin CS. Structure function relationships in the 12,13-epoxytrichothecene. Novel inhibitors of protein synthesis. Biochem Biophys Res Commun 1974;57:838–844.
390. Nikulin M, Reijula K, Jarvis BB, Hintikka E-L. Experimental lung mycotoxicosis in mice induced by *Stachybotrys atra*. Int J Exp Pathol 1996;77:213–218.
391. Li F-Q, Luo X-Y, Yoshizawa T. Mycotoxins (trichothecenes, zearalenone and fumonisins) in cereals associated with human re-mold intoxication stored since 1989 and 1991 in China. Nat Toxins 1999;7:93–97.
392. International Agency for Research on Cancer. Toxins derived from *Fusarium sporotrichioides*: T-2 toxin. IARC Monogr 1993;56:467–488.
393. Spyker MS, Spyker DA. Yellow rain: Chemical warfare in Southeast Asia and Afghanistan. Vet Hum Toxicol 1983;25:335–340.
394. Chung HS. Cereal scab causing mycotoxicoses in Korea and present status of mycotoxin research. Korean J Mycol 1975;3:31–36.
395. Thomas AD. Protection of civilians and military personnel against trichothecene mycotoxins. Vet Hum Toxicol 1984;26:78–79
396. Thigpen JT, Vaughn C, Stuckey WJ. Phase II trial of anguidine in patients with sarcomas unresponsive to prior chemotherapy: a southwest oncology group study. Cancer Treat Rep 1981;65:881–882.
397. Hossain MA, Ahmed MS, Ghannoum MA. Attributes of *Stachybotrys chartarum* and its association with human disease. J Allergy Clin Immunol 2004;113:200–208.
398. Fung F, Clark R, Williams S. Stachybotrys, a mycotoxin-producing fungus of increasing toxicologic importance. Clin Toxicol 1998;36:79–86.
399. Walinder R, Wieslander G, Norback D, Wessen B, Venge P. Nasal lavage biomarkers: effects of water damage and microbial growth in an office building. Arch Environ Health 2001;56:30–36.

400. Croft WA, Jaris BB, Yatawara CS. Airborne outbreak of trichothecene toxicosis. Atmos Environ 1986;20:549–552.
401. Drobotko VG. Stachbotryotoxicosis. A new disease of horses and humans. Am Rev Soviet Med 1945;2:238–242.
402. Dearborn DG, Smith PG, Dahms BB, Allan TM, Sorenson WG, Montana E, Etzel RA. Clinical profile of 30 infants with acute pulmonary hemorrhage in Cleveland. Pediatrics 2002;110:627–637.
403. Centers for Disease Control (CDC). Update: pulmonary hemorrhage/hemosiderosis among infants—Cleveland, Ohio, 1993–1996. MMWR Morb Mortal Wkly Rep 1997;46:33–35.
404. Etzel RA, Montana E, Sorenson WG, Kullman GJ, Allan TM, Dearborn DG. Acute pulmonary hemorrhage in infants associated with exposure to *Stachybotrys atra* and other fungi. Arch Pediatr Adolesc Med 1998;152:757–762.
405. Jarvis BB, Sorenson WG, Hintikka E-L, Nikulin M, Zhou T, Jiang J, et al. Study of toxin production by isolates of *Stachybotrys chartarum* and *Memnoniella echinata* isolated during a study of pulmonary hemosiderosis in infants. Appl Environ Microbiol 1998;64:3620–3625.
406. Vesper SJ, Dearborn DG, Yike I, Sorenson WG, Haugland RA. Hemolysis, toxicity, and randomly amplified polymorphic DNA analysis of *Stachybotrys chartarum* strains. Appl Environ Microbiol 1999;65:3175–3181.
407. Center for Disease Control (CDC). Update: Pulmonary hemorrhage/hemosiderosis among infants–Cleveland, Ohio, 1993–1996. MMWR Morb Mortal Wkly Rep 2000;49:180–184.
408. Vesper S, Derarborn DG, Yike I, Allan T, Sobolewski J, Hinkley SF, et al. Evaluation of *Stachybotrys chartarum* in the house of an infant with pulmonary hemorrhage: quantitative assessment before, during, and after remediation. J Urban Health Bull NY Acad Med 2000;77:68–85.
409. Novotny WE, Dixit A. Pulmonary hemorrhage in an infant following 2 weeks of fungal exposure. Arch Pediatr Adolesc Med 2000;154:271–275.
410. Vesper SJ, Dearborn DG, Elidemir O, Haugland RA. Quantification of siderophore and hemolysin from *Stachybotrys chartarum* strains, including a strain isolated form the lung of a child with pulmonary hemorrhage and hemosiderosis. Appl Environ Microbiol 2000;66:2678–2681.
411. Terr AI. Are indoor molds causing a new disease? J Allergy Clin Immunol 2004;113:221–226.
412. Sudakin DL. Trichothecenes in the environment: relevance to human health. Toxicol Lett 2003;143:97–107.
413. Miller JD, Laflamme AM, Sobol Y, Lafontaine P, Greenalgh R. Fungi and fungal products in some Canadian houses. Int Biodeterior 1988;24:103–120.
414. Johanning E, Biagini R, Hull D, Morey P, Jarvis B, Landsbergis P. Health and immunology study following exposure to toxigenic fungi (*Stachybotrys chartarum*) in a water-damaged office environment. Int Arch Occup Environ Health 1996;68:207–218.
415. Hodgson MJ, Morey P, Leung W-Y, Morrow L, Miller D, Jaris BB, et al. Building-associated pulmonary disease from exposure to *Stachybotrys chartarum* and *Aspergillus versicolor*. J Occup Environ Med 1998;40:241–249.
416. Kostiainen R, Rizzo A, Hesso A. The analysis of trichothecenes in wheat and human plasma samples by chemical ionization tandem mass spectrometry. Arch Environ Contam Toxicol 1989;18:356–364.
417. Pace JG, Matson CF. Stability of T-2, HT-2, and T-2 tetraol in biological fluids. J Anal Toxicol 1988;12:48–50.
418. Meky FA, Turner PC, Ashcroft AE, Miller JD, Qiao Y-L, Roth MJ, Wild CP. Development of a urinary biomarker of human exposure to deoxynivalenol. Food Chem Toxicol 2003;41:265–273.
419. Balzer I, Bogdanic C, Muzic S. Natural contamination of corn (*Zea mays*) with mycotoxins in Yugoslavia. Ann Nutr Aliment 1977;31:425–430.
420. Gabal MA, Stahr M, Pfeifer R, Domoto M. Successful production and radioactive labeling of 2^{14}C acetate of "T-2" toxin on a liquid medium. Vet Hum Toxicol 1983;25:161–163.
421. Barnes C, Buckley S, Pacheco F, Portnoy J. IgE-reactive proteins from *Stachybotrys chartarum*. Ann Allergy Asthma Immunol 2003;89:29–33.
422. Trout DB, Seltzer JM, Page EH, Biagini RE, Schmechel D, Lewis DM, Boudreau AY. Clinical use of immunoassays in assessing exposure to fungi and potential health effects related to fungal exposure. Ann Allergy Asthma Immunol 2004;92:483–491.
423. Halsey J. Performance of a *Stachybotrys chartarum* serology panel. Allerg Asthma Proc 2000;21:174–175.
424. Van Emon JM, Reed AW, Yike I, Vesper SJ. ELISA measurement of stachylysin™ in serum to quantify human exposures to the indoor mold *Stachybotrys chartarum*. J Occup Environ Med 2003;45:582–591.
425. Bratich PM, Buck WB, Haschek WM. Prevention of T-2 toxin-induced morphologic effects in the rat by highly activated charcoal. Arch Toxicol 1990;64:251–253.
426. Price DL, Ahearn DG. Sanitation of wallboard colonized with *Stachybotrys chartarum*. Curr Microbiol 1999;38:21–26.
427. Kuiper-Goodman T, Scott PM, Watanabe H. Risk assessment of the mycotoxin zearalenone. Regul Toxicol Pharmacol 1987;7:253–306.
428. Scott PM, Panalaks T, Kanhere S, Miles WF. Determination of zearalenone in cornflakes and other corn-based foods by thin layer chromatography, high pressure liquid chromatography/ high resolution mass spectrometry. J Assoc Off Anal Chem 1978;61:593–600.
429. Prelusky DB, Scott PM, Trenholm HL, Lawrence GA. Minimal transmission of zearalenone to milk of dairy cows. J Environ Sci Health B25:87–103.

PART 3

MEDICINAL HERBS and ESSENTIAL OILS

PART 3 MEDICINAL HERBS and ESSENTIAL OILS

I Medicinal Herbs

Chapter 46

ALOE VERA
[*Aloe vera* (L.) Burm. f.]

HISTORY

Clay tablets discovered in Mesopotamia suggest the aloe vera was a pharmaceutical agent for early civilizations dating back to 1750 BC.[1] Aloe vera is a native plant of the Mediterranean region that extrudes a juicy clear mucilaginous gel (aloe vera gel) from the leaves and a yellow exudate containing high concentrations of anthraquinone compounds. Since Roman times, the latter agent has been ingested as a strong cathartic and medicinal purge; the gel has been applied to reduce infection and improve wound healing. Other uses of aloe vera in folk medicine have included the treatment of skin wounds, burns, hyperlipidemia, diabetes, and obesity.[2]

BOTANICAL DESCRIPTION

There are over 360 species in the genus *Aloe*. Of these species, plants used for medicinal purposes besides *A. vera* are relatively few including Cape aloe from South Africa (*A. ferox* Miller), Curacao aloe (*A. vera, Aloe barbadensis* Miller), candelabra aloe (*Aloe arborescens* Miller), and Zanzibar or Perry's aloe (*Aloe perryi* Baker).

Common Name: Aloe Vera

Scientific Name: *Aloe vera* (L.) Burm. f. *(A. barbadensis* Miller*)*

Botanical Family: Aloeaceae (formerly Liliaceae, lily family)

Physical Description: Aloe vera is an evergreen, perennial, succulent (cactus-like) plant that has gray-green, simple, lance-shaped leaves with a pointed apex (lanceolate). The leaves do not have stalks (petioles). The margins have small grayish teeth. The leaves are arranged alternately to form a rosette above a very short stem. The main rosette reaches approximately 2 feet (61 cm) in height, and the plant continually produces little offset rosettes. The flowers are regular, bisexual, yellow to orange in color, and appear in terminal series of flowers on lateral stalks (racemes). These flowers produce capsules with many flattened seeds. Figure 46.1 displays a specimen of the aloe vera plant.

Distribution and Ecology: Aloe adapts well to desert climates. This plant is native to Arabia and Madagascar, but the largest distribution occurs in South Africa. In the United States, aloe vera grows in the drier parts of Texas, Florida, Puerto Rico, and Hawaii.

EXPOSURE

Sources

This perennial succulent is a popular houseplant. The dried juice of aloe vera is a solidified substance originating in the cells of the pericycle and adjacent leaf parenchyma. After flowing spontaneously from the cut leaf, the fresh juice is dried with or without the aid of heat. Aloe vera gel is a clear, mucilaginous gel extracted from

Medical Toxicology of Natural Substances, by Donald G. Barceloux, MD

FIGURE 46.1. *Aloe vera* (L.). Photograph Courtesy of Paul Auerbach, MD. See color insert.

the parenchymatous cells in the leaves of *Aloe vera* (L.) Burm. f. Curacao; Barbados aloe is a dark chocolate-brown, dried juice extracted from *Aloe vera* (L.) Burm. f.; Cape aloe is a dark brown or greenish brown, dried juice derived from *A. ferox* Mill. or from hybrids of *A. ferox* and other aloe plants (e.g., *A. africana* Mill., *A. spicata* Baker).

Medicinal Uses

Traditional

Traditional uses of aloe vera include the treatment of arthritis, asthma, fungal infections, digestive and bowel disorders (e.g., constipation), lupus erythematosus, skin disorders (e.g., seborrheic dermatitis), musculoskeletal injuries, and gastroduodenal ulcers. Other uses include application as an insect repellant and the treatment of diabetes mellitus, antihemorrhoidal agent, gout, tuberculosis, gonorrhea, headaches, and eczema.[3]

Current

Aloe vera gel is a constituent of food and drink, an antiinflammatory in cosmetics and topical agents, and a moisturizer in lotions.[4] Although experimental studies[5,6] and anecdotal case reports suggest a wide range of medical applications for aloe vera gel,[7,8] the clinical effectiveness of the oral or topical application of this substance remains unsubstantiated.[9,10] In an experimental study, aloe vera did not alter the erythema or vasodilation produced by UVB radiation up to 24 hours after exposure.[11] Clinical trials did not demonstrate the efficacy of aloe vera gel for the treatment of radiation burns[12] or aphthous stomatitis.[13] The dried latex form of aloe is an approved oral agent based on the cathartic actions of aloe-emodins and other anthraquinone compounds. The plant portion most commonly used by the pharmaceutical industry is the latex, which stains clothes and fixtures purple.

Regulatory Status

The German Commission E approves the use of aloe for the limited (1–2 weeks) treatment of constipation. The US Food and Drug Administration (FDA) does not permit the use of aloe products as nonprescription drugs for treatment of constipation because of the lack of documented efficacy and safety. Aloe products are controlled as dietary supplements in the United States.

PRINCIPAL INGREDIENTS

Chemical Composition

The aloe vera gel is derived from the thin-walled mucilaginous cells of the leaves that contain a variety of carbohydrate polymers (glucomannans) along with organic and inorganic compounds.[14] Peripheral bundle sheath cells of aloe vera produce a bitter, yellow latex called aloe juice, aloe sap, or aloes. The aloe vera plant contains at least 75 potentially pharmacologically active compounds including vitamins, minerals, sugars, amino acids, enzymes, lignans, saponins, and salicylic acids as listed in Table 46.1.[15] The major constituent (i.e., about 15–40%) of dried aloe is a mixture of two diastereomeric C-glucosides [aloin A (10S) and aloin B (10R)], formerly known as barbaloin. Hydroxyaloin is a minor constituent (i.e., about 3%).

In a study of Cape aloes in South Africa, the major constituents were aloeresin A, aloesin, and aloin A and B.[16] These compounds accounted for 70–97% of the total dry weight of *Aloe capensis* (*ferox*) leaf exudate with aloin content of the juice samples averaging about 21%. The polysaccharides in the gel consist of glucomannans, mannans, and pectins with a wide range of molecular weights. The major carbohydrate fraction of aloe gel contains acemannan substances, which are β-1,4-linked acetylated mannan compounds interspersed with *O*-acetyl groups. Animal studies indicate that these compounds demonstrate relatively low toxicity.[17] Aloe vera gel does not contain the strong cathartics (anthraquinone compounds) present in aloe saps. Toxic low-molecular-weight compounds are present in aloe vera gel extracts,[18] but the concentration and pharmacological significance these substances remain unclear.

Physiochemical Properties

Figure 46.2 displays the structure of the common constituents of the dried latex extracted from the leaves of *Aloe* species. This extract has an aromatic aroma and

TABLE 46.1. Classification of Constituents of Aloe Vera

Class	Compounds
Anthraquinone	Aloe-emodin, aloetic acid, aloin A and B, hydroxyaloin, anthranol, chrysophanic acid, cinnamic acid ester, emodin, resistannol
Enzymes	Amylase, catalase, cyclooxygenase, lipase, alkaline, phosphatase, carboxypeptidase, oxidase
Essential amino acids	Isoleucine, leucine, lysine, methionine, phenylalanine, threonine, valine
Inorganic compounds	Calcium, chlorine, chromium, copper, iron, magnesium, manganese, potassium sorbate, sodium, zinc
Nonessential amino acids	Alanine, arginine, aspartic acid, glutamic acid, glycine, histidine, hydroxyproline, proline, tyrosine
Saccharides	Aldopentose, cellulose, glucose, mannose, L-rhamnose
Vitamins	B_1, B_2, B_6, C, choline, folic acid, α-tocopherol, β-carotene
Miscellaneous	Arachidonic acid, cholesterol, gibberellin, lectin-like substance, lignans, salicylic acid, β-sitosterol, steroids, triglycerides, uric acid

Source: Adapted from Ref 1.

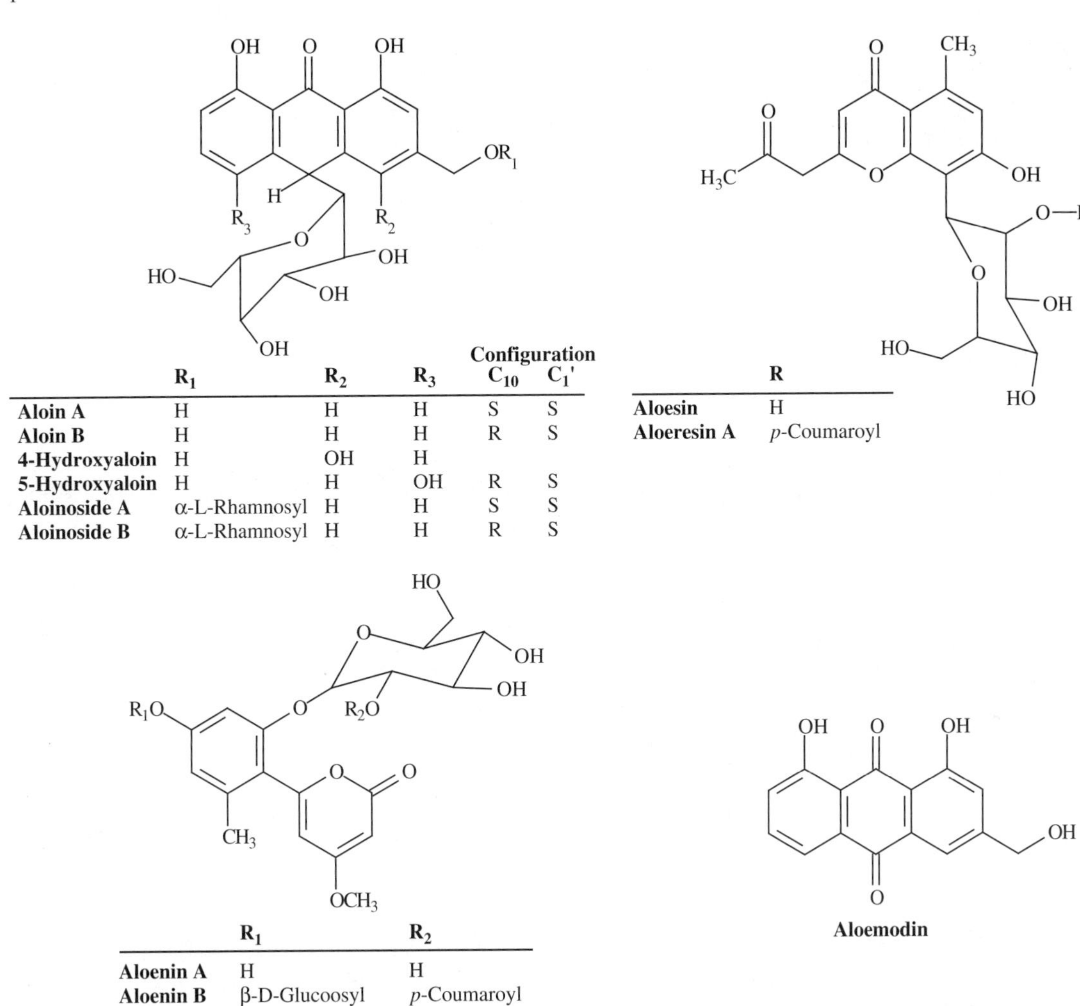

FIGURE 46.2. Structures of some compounds from *Aloe* species. Adapted from Zonta F, et al. High-performance liquid chromatographic profiles of aloe constituents and determination of aloin in beverages, with reference to the EEC regulation for flavouring substances. J Chromatogr A 1995;718:101.

bitter taste. Aloin A and B undergo rapid decomposition in solution, particularly in basic solutions. Aloe extracts have good emollient and moisturizing properties; however, there are limited data on the efficacy of these preparations. The compounds in aloe responsible for the physiological properties used for medicinal purposes remain uncertain.[19] Table 46.2 lists some physiochemical properties of aloin (CAS RN: 1415-73-2).

TABLE 46.2. Some Physiochemical Properties of Aloin

Physical Property	Value
log P (octanol-water)	0.550
Water Solubility	8,300 mg/L (20 °C/68 °F)
Vapor Pressure	5.28E-21 mm Hg (25 °C/77 °F)
Henry's Law Constant	2.61E-21 atm-m^3/mole (25 °C/77 °F)
Atmospheric OH Rate Constant	2.77E-10 cm^3/molecule-second (25 °C/77 °F)

Mechanism of Toxicity

The laxative effects of aloe result primarily from the effects of two main diastereomeric 1, 8-dihydroxyanthracene glycosides (aloin A and B), which are metabolized into the main active metabolite, aloe-emodin-9-anthrone, by colonic bacteria. This compound is a gastrointestinal stimulant and irritant, similar to senna.[20] Aloe-emodin-9-anthrone reduces fluid absorption from the fecal mass by stimulating colonic motility in addition to increasing water content in the large intestine, probably by inhibition of Na^+/K^+-adenosine triphosphatase or chloride channels.

DOSE RESPONSE

The daily laxative dose for adults and children over 10 years old is 0.04–0.11 g (Curacao or Barbados aloe) or 0.06–0.17 g (Cape aloe) of the dried juice, corresponding to 10–30 mg hydroxyanthraquinones per day, or 0.1 g as a single dose in the evening.[21] The recommended use of this laxative dose is limited to 1–2 weeks.

TOXICOKINETICS

The cathartic action of aloe usually begins at least 6–24 hours after ingestion. The absorption of aloin A and aloin B in the upper gastrointestinal tract is minimal, but animal studies indicate that bacteria metabolize these compounds to an active metabolite, aloe-emodin-9-anthrone, in the colon.[22]

CLINICAL RESPONSE

In the dried latex form, the aloe-emodins and other anthraquinone compounds are strong cathartics. Mucous membrane irritation may occur, but serious systemic symptoms following ingestion are unlikely. Rare case reports document the occurrence of relatively minor local reactions (eczematous and papular dermatitis) following the chronic oral or topical use of aloe preparations by hypersensitive patients.[23,24] The application of aloe vera preparations to skin during periods of enhanced susceptibility to dermal injury (e.g., dermabrasion, chemical peel) was associated with immediate local pain, erythema, and edema.[25] These adverse effects generally resolve promptly following cessation of use. There are limited data on the oral toxicity of *Aloe vera*, but systemic toxicity has not been associated with the oral administration of extracts of *A. vera* with the exception of the strong cathartic actions from the ingestion of the dried latex (aloe sap) containing anthraquinone compounds. Based on animal studies, the genotoxic risk of aloe-emodin is small.[26]

DIAGNOSTIC TESTING

A reversed-phase gradient elution system with high-performance liquid chromatography (HPLC) is available to simultaneously separate a variety of components in *Aloe* species including aloesin and aloeresin A.[27,28] Methods for the detection of anthraquinone compounds [aloe-emodin, aloin-A (barbaloin)] in aloe-based products include liquid chromatography/mass spectrometry and liquid chromatography with diode array detection in the UV range.[29] False-positive results for urinary urobilinogen can occur during aloe use because some anthranoid metabolites discolor the urine.

TREATMENT

Treatment is supportive. Based on limited data, the primary toxic reaction to aloe vera is catharsis. Therefore, patients with suspected toxicity should be evaluated for fluid and electrolyte balance if clinically significant diarrhea occurs. Decontamination measures are unnecessary.

References

1. Shelton RM. *Aloe vera* its chemical and therapeutic properties. Int J Dermatol 1991;30:679–682.
2. Morton FJ. Folk uses and commercial exploitation of *Aloe* leaf pulp. Econ Bot 1961;15:311–319.
3. Cole HN, Chen KK. *Aloe vera* in oriental dermatology. Arch Dermatol Syphilol 1943;47:250.
4. Briggs C. Herbal medicine. Aloe. Can Pharm J 1995;128:48–50.
5. Pugh N, Ross SA, ElSohly MA, Pasco DS. Characterization of aloeride, a new high-molecular-weight polysaccharide from *Aloe vera* with potent immunostimulatory activity. J Agric Food Chem 2001;49:1030–1034.
6. Choi S-W, Son B-W, Son Y-S, Park Y-I, Lee S-K, Chung M-H. The wound-healing effect of a glycoprotein fraction isolated from *Aloe vera*. Br J Dermatol 2001;145:535–545.

7. El Zawahry M, Rashad Hegazy M, Helal M. Use of aloe in treating leg ulcers and dermatoses. Int J Dermatol 1973;12:68–73.
8. Fulton JE Jr. The stimulation of postdermabrasion wound healing with stabilized aloe vera gel-polyethylene oxide dressing. J Dermatol Surg Oncol 1990;16:460–467.
9. Vermeulen H, Ubbink D, Goossens A, de Vos R, Legemate D. Dressings and topical agents for surgical wounds healing by secondary intention. Cochrane Database Syst Rev 2004(2):CD003554.
10. Vogler BK, Ernest E. Aloe vera: a systematic review of its clinical effectiveness. Br J Gen Pract 1999;42:823–828
11. Crowell J, Hilsenbeck S, Pennneys N. *Aloe vera* does not affect cutaneous erythema and blood flow following ultraviolet B exposure. Photodermatology 1989;6:237–239.
12. Williams MS, Burk M, Loprinzi CL, Hill M, Schomberg PJ, et al. Phase III double-blind evaluation of an aloe vera gel as a prophylactic agent for radiation-induced skin toxicity. Int J Radiation Oncology Biol Phys 1996;36:345–349.
13. Garnick JJ, Singh B, Winkley G. Effectiveness of a medicament containing silicon dioxide, aloe, and allantoin on aphthous stomatitis. Oral Surg Oral Med Oral Pathol Oral Radiol Endod 1998;86;550–556.
14. Spoerke DG, Ekins BR. Aloe vera—Fact or quackery? Vet Hum Toxicol 1980;20:418–424.
15. Atherton P. *Aloe vera* revisited. Br J Phytother 1998;4: 176–183.
16. van Wyk B-E, Van Rheede Van Oudtshoorn MCB, Smith GF. Geographical variation in the major compounds of *Aloe ferox* leaf exudate. Planta Med 1995;61:250–253.
17. Fogleman RW, Shellenberger TE, Balmer MF. Subchronic oral administration of acemannan in the rat and dog. Vet Hum Toxicol 1992;34:144–147.
18. Avila H, Rivero J, Herrera F, Fraile G. Cytotoxicity of a low molecular weight fraction from *Aloe vera (Aloe barbadensis* Miller) gel. Toxin 1997;35:1423–1430.
19. Eshun K, He Q. Aloe vera: a valuable ingredient for the food, pharmaceutical and cosmetic industries–a review. Crit Rev Food Sci Nutr 2004;44:91–96.
20. de Witte P. Metabolism and pharmacokinetics of anthranoids. Pharmacology 1993;47(Suppl 1):86–97.
21. World Health Organization. WHO monographs on selected medicinal plants. Vol. 1. Geneva: World Health Organization; 1999.
22. Ishii Y, Tanizawa H, Takino Y. Studies of aloe. IV. Mechanism of cathartic effect. (3). Biol Pharm Bull 1944;17: 495–497.
23. Shoji A. Contact dermatitis to *Aloe arborescens*. Contact Dermatitis 1982; 8:164–167.
24. Morrow DM, Rapaport JM, Strick RA. Hypersensitive to aloe. Arch Dermatol 1980;116:1064–1065.
25. Hunter D, Frumkin A. Adverse reactions to vitamin E and aloe vera preparations after dermabrasion and chemical peel. Cutis 1991;47:193–196.
26. Heidemann A, Völkner W, Mengs U. Genotoxicity of aloe emodin *in vitro* and *in vivo*. Mutat Res 1996;367:123–133.
27. Zonta F, Bogoni P, Masotti P, Micali G. High-performance liquid chromatographic profiles of aloe constituents and determination of aloin in beverages, with reference to the EEC regulation for flavouring substances. J Chromatogr A 1995;718:99–106.
28. Dell'Agli M, Giavarini F, Ferraboschi P, Galli G, Bosisio E. Determination of aloesin and aloeresin A for the detection of aloe in beverages. J Agric Food Chem 2007;55:3363–3367.
29. Elsohly MA, Gul W, Avula B, Khan IA. Determination of the anthraquinones aloe-emodin and aloin-A by liquid chromatography with mass spectrometric and diode array detection. J AOAC Int 2007;90:28–42.

Chapter 47

ARISTOLOCHIC ACID and CHINESE HERB NEPHROPATHY

HISTORY

Mu Tong is an ancient Chinese herb that was derived from *Akebia* species until the early 20th century.[1] After that time, *Clematis* species became the main source of Mu Tong until the 1950s. Then, *Aristolochia manshuriensis* became the main component of this herb. Since the substitution of *Aristolochia manshuriensis* for *Clematis* species in preparations of Mu Tong, case reports in China associated the development of renal failure with the use of Mu Tong.[2] Prior to the use of *Aristolochia manshuriensis* in preparations of Mu Tong, renal dysfunction was not reported in Chinese herbal formularies as a complication of the ingestion of Mu Tong.[1]

In 1993, Vanherweghem et al reported a rapidly progressive, fibrosing interstitial nephritis in two young women, who had been treated with an herbal slimming medication supposedly containing *Stephania tetrandra* and *Magnolia officinalis*.[3] An investigation revealed that seven other women, who also followed a slimming regimen at the same Brussels medical clinic, were admitted to a renal clinic for renal dialysis between 1991 and 1992. Renal biopsies from eight of the nine cases demonstrated extensive interstitial fibrosis without glomerular lesions. Phytochemical analysis of the pills from this weight-reduction clinic detected the presence of aristolochic acids instead of tetrandrine, and this substitution suggested the contamination of these herbal preparations by *Aristolochia fangchi*. Case reports described the development of interstitial nephritis following the use of other Chinese herbal preparations that contained aristolochic acid.[4] In 1999, a case report described two cases of end-stage renal failure in the United Kingdom following the use of Chinese herbal teas containing aristolochic acid.[5] Additional cases of Chinese herb (aristolochic acid) nephropathy occurred in Asia, Spain, and France.[6,7]

BOTANICAL DESCRIPTION

The genus *Aristolochia* contains over 400 species of perennial herbs and shrubs. Misidentification of *Aristolochia* species as a nonaristolochic acid–containing plant species occurs with a variety of Chinese herbs from *Akebia, Asarum, Clematis*, and *Cocculus* genera. Aristolochic acids also occur in several species of butterflies (e.g., *Atrophaneura, Battus, Pachliopta, Troides*), which feed on *Aristolochia* species.

Common Name: Kou-boui, Guang Fang Ji

Scientific Name: *Aristolochia fangchi* Y. C. Wu ex L. D. Chow & S. M. Hwang

Botanical Family: Aristolochiaceae (birthworts)

Physical Description: This herb is a perennial climbing vine with a slender, gray-brown stem that contains brown villi. The oblong or ovate-shaped leaves are petioled and relatively large, 3–11 cm (~1–4 in.) long and 2–6 cm (0.8–2.4 in.) wide. The tubular, purple flower is 2–3 cm (~.8–1.2 in.) long with yellow spots. The flowers bloom in May and June. The capsules contain many seeds.

Distribution and Ecology: The Aristolochiaceae family comprises seven genera of herbs and shrubs

Medical Toxicology of Natural Substances, by Donald G. Barceloux, MD

distributed throughout tropical, subtropical, and some northern temperate zones of Asia, Africa, South America, and North America. In Europe, *Aristolochia clematitis* is indigenous to the Mediterranean region. *A. fangchi* inhabits valleys near streams.

EXPOSURE

Sources

Aristolochic acid occurs primarily in herbs from the genus *Aristolochia* including *A. fangchi* (root), *A. manshuriensis* (stem, Guan Mu Tong), *A. contorta* Bge. (fruit or herb, Ma Dou Ling), and *A. debilis* Sieb et Zucc (radix, fruit, or herb). Table 47.1 lists some Chinese herbs, which contain ingredients from *Aristolochia* species. Other aristolochic acid-containing species include some *Asarum* species including *A. himalaicum* Hook. f. & Thomson ex Klotzsch (Xi Xin), *A. crispulatum* C. Y. Cheng & C. S. Yang, and *A. splendens* (F. Maekawa) C. Y. Cheng & C. S. Yang.[8] Chinese herb nephropathy develops after the chronic use of herbal preparations (e.g., Kampo prescriptions, *Stephania tetrandra, Magnolia officinalis*) contaminated with aristolochic acid-containing herbs. The *Aristolochia fangchi* root or the *Aristolochia manshuriensis* stem may be substituted for the Japanese herbal medicine *Akebia* stem or *Sinomenium* stem in Chinese-produced pharmaceutical preparations and health foods. Aristolochic acid is present in a variety of Chinese herbal preparations for weight reduction. Some raw Chinese herbs (Guan Mu Tong, Ma Dou Ling, Guang Fang Ji) still contain aristolochic acid I despite the ban on commercial products containing aristolochic acids.[9] A Swiss study of 42 Chinese herbal preparations for slimming demonstrated the presence of aristolochic acid I in four preparations in concentrations up to about 0.04%.[10]

TABLE 47.1. Chinese Herbs from Aristolochia Species

Aristolochia Species	Source	Name
A. contorta	Fruit	Ma Dou Ling
A. contorta	Herb	Tian Xian Teng
A. debilis	Fruit	Ma Dou Ling
A. debilis	Herb	Tian Xian Teng
A. debilis	Root	Qing Mu Xiang
A. fangchi	Root	Guang Fang Ji
A. manshuriensis	Stem	Guan Mu Tong

Source: From Ref 37.

Medicinal Uses

Traditional

The use of *Aristolochia* species is an ancient treatment for a wide variety of diseases including dysentery, skin infections, phlebitis, cancer, and arthritis.[11,12] Stems of *A. manshuriensis* are used in traditional Chinese medicine as a cardiotonic herb. Medicinal uses for the mature fruits of *A. debilis* (madouling) include the treatment of hypertension, tuberculosis, and snake bites.[13] In West Africa, *A. albida* is a traditional treatment for dysentery, colic, skin lesions, and snake bites.[14] The dry root of *A. indica* is an Indian folk medicine for inducing abortions. Charrua is a hydroalcoholic extract of *Aristolochia argentina* that is a traditional treatment for hemorrhoids and diarrhea as well as being a folk treatment for "purifying the blood."

Regulatory Status

Many countries now ban the use of herbal products containing aristolochic acid. After reports on the development of renal failure in Belgian patients given weight loss products containing aristolochic acid, the US Food & Drug Administration (FDA) issued a consumer advisory that recommended the cessation of use of any product containing aristolochic acid. Later in 2001, the FDA detained products suspected of containing aristolochic acid. Similar bans are now in effect in Canada, Australia, and parts of Europe including the United Kingdom.

PRINCIPAL INGREDIENTS

Chemical Composition

Aristolochia species contain a variety of aristolochic acids, aristololactams, aristoquinolines, aristogins, and sesquiterpenes in varying amounts depending on the species, plant part, and the geographic location.[15,16] In three commercial samples of *A. fangchi* (kou-boui) and *A. manshuriensis* (kan-mokutsu), the aristolochic acid I (CAS RN: 313-67-7) content ranged from 0.01–0.22% and 0.17–0.88%, respectively, compared with about 0.08–0.11% in samples of *A. debilis*.[17] The amount of aristolochic acid II (CAS RN: 475-80-9) was approximately 10% of the aristolochic acid I concentration as measured by high performance liquid chromatography (HPLC). As measured by capillary zone electrophoresis, the content of aristolochic acid I and II in samples of radix aristolochiae fangchi was approximately 0.06% and 0.03%, respectively.[18] Figure 47.1 displays the chemical structures of aristolochic acids I and II. Species of

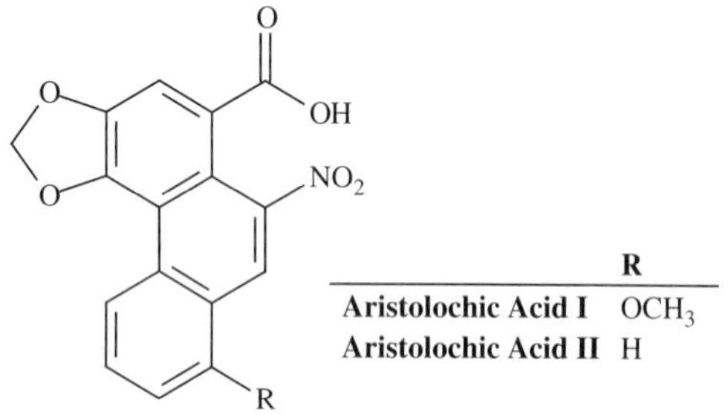

	R
Aristolochic Acid I	OCH_3
Aristolochic Acid II	H

FIGURE 47.1. Chemical structures of aristolochic acid I and II.

Aristolochia (e.g., *A. macrophylla, A. serpentaria*) in North America also contain detectable concentrations of aristolochic acids I and II. Samples from *A. macrophylla* Lam. and *A. serpentaria* L. contained 0.39% and 0.13%, respectively, aristolochic acid I by dry weight, while samples of *Asarum canadense* L. contained aristolochic acid I concentrations ranging from non-detectable to 0.037% dry weight.[19] However, these concentrations of aristolochic acids are low compared with *Aristolochia* species. Aristolochic acid II was not detected in samples of *Asarum canadense*. Seeds of *A. clematitis* (birthwort) contain about 0.65% aristolochic acid.[20] Essential oils from *Aristolochia* species contain a variety of monoterpene (camphanes, methanes, pinanes), diterpene (geranylgeranyl pyrophosphate derivatives), and sesquiterpene (aristolanes, bicyclogermacranes, cadinanes, germacranes) compounds.[11]

Physiochemical Properties

Aristolochic acids are slightly soluble in water with melting points in the range of 281 °C–286 °C (538 °F–547 °F). Aristolochic acids are nitrophenanthrene carboxylic acid derivatives that cause extensive necrosis of the squamous epithelium in the forestomach of rats, followed by hyperplasia, papilloma formation, and invasive squamous cell carcinoma.[21] Aristolochic acid I is a direct-acting mutagen in *Salmonella typhimurium* strains TA100, TA102, TA1537, and TM677, but this compound was not active in the nitroreductase-deficient strains TA98NR and TA100NR.[22]

Mechanism of Toxicity

Aristolochic acids are the probable proximal renal tubular toxin and carcinogen in Chinese herb associated nephropathy.[23] The activation of the metabolite of aristolochic acid I, aristolactam I, by peroxidase causes the formation of the major adducts associated with the development of urothelial carcinoma following the administration of aristolochic acid-containing herbs.[24] The cytochrome P450 isoenzymes, CYP1A1, CYP1A2, and possibly other CYP450 isoenzymes activate aristolochic acids to cyclic reactive nitrenium ions that form covalent DNA adducts on guanine and adenosine.[25] The oral administration of aristolochic acid to female rats causes renal lesions similar to the cases of aristolochic acid nephropathy, manifest by renal tubular necrosis, proteinuria, glucosuria, and azotemia.[26] The administration of decoctions from Chinese herbs [e.g., *Akebia quinata* (Houtt.) Dcne.] without detectable quantities of aristolochic acid did not cause renal lesions. There was evidence of renal toxicity, but not liver or bladder lesions in NIH mice administered the equivalent dose of up to 4 g raw herb (*Aristolochia manshuriensis*, 0.5–1% aristolochic acid) daily for 8 weeks.[27] This dose was equivalent to approximately 40 times the daily dose used in traditional Chinese medicine. The initial damage in the kidney occurs in the brush borders of the S3 segment of the proximal renal tubules, leading to the release of the lysosomal enzyme, *N*-acetyl-β-D-glucosaminidase, and increased urinary excretion of filtered microproteins (e.g., retinol-binding protein, β_2-microglobulin, α_1-microglobulin, Clara cell protein).[28] Animal studies suggest that there are substantial species differences in susceptibility to the renal damage caused by aristolochic acids.[29]

DOSE RESPONSE

The risk of developing renal failure and urothelial carcinoma after ingesting aristolochic acid is dose-related.[30] Cumulative doses of aristolochic acid exceeding 200 g were associated with increased risk of developing urothelial cancer.[31] Retrospective evaluation of patients with Chinese herb nephropathy indicated that the amount of aristolochic acid ingested by some of these patients was approximately 10 mg/kg body weight.[3] In female Wister rats, the administration of a single dose of 10 mg aristolochic acid/kg body weight caused renal dysfunction within 3 days manifest by necrosis of the renal tubular epithelium, elevation of plasma creatinine, and increased markers of renal tubular function (protein, *N*-acetyl-β-glucosaminidase, γ-glutamyl transferase, malate dehydrogenase).[32] Single oral doses of 120–300 mg/kg caused severe acute tubular necrosis within a few days. The intraperitoneal administration of 0.1 mg aristolochic acid/kg body weight to female New Zealand white rabbits 5 days/week for 17–21 months was associated with the development of renal hypocellular interstitial fibrosis.[33] Three of 12 rabbits in the treatment group developed tumors of the urinary tract. No significant pathologic changes were detected in 10 control rabbits.

TOXICOKINETICS

The principal metabolites of aristolochic acids are aristololactams, which are excreted in the urine and feces. The cytochrome P450 isoenzymes (CYP1A1, CYP1A2) and NADPH:CYP reductase metabolize aristolochic acids to the putative reactive cyclic nitrenium ion responsible for formation of the adducts, 7-(deoxyadenosin-N6-yl) aristolactam I, 7-(deoxyguanosin-N2-yl) aristolactam I, and 7-(deoxyadenosin-N6-yl) aristolactam II.[34] Other enzymes involved in the biotransformation of aristolochic acid to reactive nitrenium ions include NADPH:P450 reductase, NAD(P)H:quinone oxidoreductase, xanthine oxidase, and peroxidases.[35]

CLINICAL RESPONSE

Chinese herb nephropathy is a chronic renal disease characterized by progressive tubulointerstitial fibrosis that results in moderate to end-stage renal disease. Animal studies indicate that the S3 segment of the proximal tubule is the primary target with initial loss of the proximal tubule brush border.[36] Initially, the interstitial fibrosis begins in the proximal tubules of the superficial cortex of the kidney and spreads to the deep cortex. Dysplastic and neoplastic changes of the urothelial epithelium are also associated with this syndrome. In a study of 39 patients with aristolochic acid-induced nephropathy, only 2 patients had normal urothelium in their surgically removed kidneys.[31] There were 18 cases of urothelial carcinoma, which included 17 cases of carcinoma of the ureter and/or the renal pelvis and one case of papillary bladder tumor. Nineteen of the remaining patients had mild-to-moderate urothelial dysplasia. Tissue samples from the kidneys and ureters of these patients all had detectable concentrations of aristolochic acid-DNA adducts. The International Agency for Research on Cancer (IARC) lists herbal remedies containing plant species of the genus *Aristolochia* as carcinogenic to humans (Group 1) and naturally occurring mixtures of aristolochic acids as probably carcinogenic to humans (Group 2A).[37]

The contamination of food by seeds of *Aristolochia clematitis* is a possible cause (i.e., along with the mycotoxin, ochratoxin A) of Balkan endemic nephropathy, which is a familial chronic tubulointerstitial renal disease with slow progression to end-stage renal failure.[38] The presence of DNA adducts derived from aristolochic acid in renal tissues of patients with Balkan endemic nephropathy and in transitional cell cancers from this population suggest that chronic dietary exposure to aristolochic acids is a significant risk factor for this disease and the associated urothelial cancers.[39]

DIAGNOSTIC TESTING

Analytical Methods

Aristolochic acid is detectable in the 1 ppm range by HPLC with UV detection (390 nm) or photodiode array detector (DAD), liquid chromatography (LC) with electric ionization mass spectrometry,[40] and reversed-phase LC coupled with atmospheric pressure chemical ionization tandem mass spectrometry (MS) using positive ion detection.[41,42] The limits of detection for aristolochic acids and aristololactams using HPLC with DAD-UV (260 nm) were approximately 10–15 ng/mL.[43] Targeted analysis by LC/serial MS using a quadrupole ion-trap mass spectrometer can detect aristolochic acid in herbal samples contaminated with 0.1% dry weight *Aristolochia manshuriensis*.[44] Capillary zone and micellar electrokinetic electrophoresis are rapid methods for the determination of aristolochic acid content with limits of detection in the range of 10 μg/g material and 1 μg/mL, respectively.[15,45] HPLC and liquid chromatography/mass spectrometric methods are available for distinguishing aristolochy acid-containing species from the nontoxic species, e.g., *Stephania tetrandra*, based on the detection of aristolochic acid I (*Aristolochia* spp., *Asarum* species), tetrandrine, and/or fangchinoline (*S. tetrandra*).[46]

Biomarkers

Aristolochic acid-DNA adducts are specific markers of exposure to aristolochic acid in the kidneys and in the ureters after removal of the native organs for renal transplantation. The major DNA adduct of aristolochic acid, 7-(deoxyadenosin-N6-yl)-aristolactam I, is detectable in kidneys and ureters from patients with Chinese herb nephropathy using the nuclease P1 version of the ^{32}P-postlabeling method.[47] Other detectable minor adducts include 7-(deoxyguanosin-N2-yl)-aristolactam I and 7-(deoxyadenosin-N6-yl)-aristolactam II. The major DNA adduct is detectable at least 5–6 years after discontinuation of aristolochic acid-contaminated herbs.[48]

TREATMENT

Treatment is supportive. Patients should be evaluated for evidence of renal dysfunction and electrolyte imbalance. Dialysis and/or renal transplantation are therapeutic options for the treatment of end-stage renal disease associated with the chronic use of aristolochic acid-containing herbs.

References

1. Zhu Y-P. Toxicity of the Chinese herb Mu Tong (*Aristolochia manshuriensis*) what history tells us. Adv Drug React Toxicol Rev 2002;21:171–177.
2. Li X, Yang L, Yu Y. [An analysis of the clinical and pathological characteristics of mu-tong (a Chinese herb) induced tubulointerstitial nephropathy]. Zhonghua Nei Ke Za Zhi 2001;40:68168–68177. [Chinese]
3. Vanherweghem JL, Depierreux M, Tielemans C, Abramowicz D, Dratwa M, Jadoul M, et al. Rapidly progressive interstitial renal fibrosis in young women: association with slimming regimen including Chinese herbs. Lancet 1993;341(8842):387–391.
4. Gillerot G, Jadoul M, Arlt VM, van Ypersele De Strihou C, Schmeiser HH, But PP, et al. Aristolochic acid nephropathy in a Chinese patient: time to abandon the term "Chinese herbs nephropathy"? Am J Kidney Dis 2001;38: E26.
5. Lord GM, Tagore R, Cook T, Gower P, Pusey CD. Nephropathy caused by Chinese herbs in the UK. Lancet 1999;354(9177):481–482.
6. Yang CS, Lin CH, Chang SH, Hsu HC. Rapidly progressive fibrosing interstitial nephritis associated with Chinese herbal drugs. Am J Kidney Dis 2000;35:313–318.
7. Stengel B, Jones E. [End-stage renal insufficiency associated with Chinese herbal consumption in France]. Nephrologie 1998;19:15–20. [French]
8. Schaneberg BT, Khan IA. Analysis of products suspected of containing *Aristolochia* or *Asarum* species. J Ethnopharmacol 2004;94:245–249.
9. Cheung TP, Xue C, Leung K, Chan K, Li CG. Aristolochic acids detected in some raw Chinese medicinal herbs and manufactured herbal products—a consequence of inappropriate nomenclature and imprecise labelling? Clin Toxicol 2006;44:371–378.
10. Ioset J-R, Raoelison GE, Hostettmann K. Detection of aristolochic acid in Chinese phytomedicines and dietary supplements used as slimming regiments. Food Chem Toxicol 2003;41:29–36.
11. Mezger J. [Aristolochia clematitis as a remedy in purulent infections, wounds, and phlebitis.] Hippokrates 1956;27:450–453. [German]
12. Wu T-S, Damu AG, Su C-R, Kuo P-C. Terpenoids of *Aristolochia* and their biological activities. Nat Prod Rep 2004;21:594–624.
13. Vishwanath BS, Kini RM, Gowda TV. Characterization of three edema-inducing phospholipase A2 enzymes from habu (*Trimeresurus flavoviridis*) venom and their interaction with the alkaloid aristolochic acid. Toxicon 1987;25:501–515.
14. Otero R, Nunez V, Barona J, Fonnegra R, Jimenez SL, Osorio RG, et al. Snakebites and ethnobotany in the northwest region of Colombia. Part III: neutralization of the haemorrhagic effect of *Bothrops atrox* venom. J Ethnopharmacol 2000;73:233–241.
15. Shi LS, Kuo PC, Tsai YL, Damu AG, Wu TS. The alkaloids and other constituents from the root and stem of *Aristolochia elegans*. Bioorg Med Chem 2004;12:439–446.
16. Li W, Gong S, Wen D, Che B, Liao Y, Liu H, et al. Rapid determination of aristolochic acid I and II in *Aristolochia* plants from different regions by β-cyclodextrin-modified capillary zone electrophoresis. J Chromatogr A 2004;1049: 211–217.
17. Hashimoto K, Higuchi M, Makino B, Sakakibara I, Kubo M, Komatsu Y, et al. Quantitative analysis of aristolochic acids, toxic compounds, contained in some medicinal plants. J Ethnopharmacol 1999;64:185–189.
18. Ong ES, Woo SO. Determination of aristolochic acids in medicinal plants (Chinese) prepared medicine using capillary zone electrophoresis. Electrophoresis 2001;22:2236–2241.
19. Schaneberg BT, Applequist WL, Khan IA. Determination of aristolochic acid I and II in North American species of *Asarum* and *Aristolochia*. Pharmazie 2002;57:686–689.
20. Hranjec T, Kovfac A, Kos J, Mao W, Chen JJ, Grollman AP, Jelakovic B. Endemic nephropathy: the case for chronic poisoning by *Aristolochia*. Croat Med J 2005;46:116–125.
21. Mengs U. On the histopathogenesis of rat forestomach carcinoma caused by aristolochic acid. Arch Toxicol 1983;52:209–220.
22. Pezzuto JM, Swanson SM, Mar W, Che CT, Cordell GA, Fong HH. Evaluation of the mutagenic and cytostatic potential of aristolochic acid (3,4-methylenedioxy-8-methoxy-10-nitrophenanthrene-1-carboxylic acid) and several of its derivatives. Mutat Res 1988;206:447–454.
23. Arlt VM, Stiborova M, Schmeiser HH. Aristolochic acid as a probable human cancer hazard in herbal remedies: a review. Mutagenesis 2002;17:265–277.
24. Stiborova M, Frei E, Breuer A, Bieler CA, Schmeiser HH. Aristolactam I a metabolite of aristolochic acid I upon activation forms an adduct found in DNA of patients with Chinese herbs nephropathy. Exp Toxicol Pathol 1999;51: 421–427.
25. Arlt VM, Stiborova M, Schmeiser HH. Aristolochic acid as a probable human cancer hazard in herbal remedies: a review. Mutagenesis 2002;17:265–277.
26. Liu M-C, Maruyama S, Mizuno M, Morita Y, Hanaki S, Yuzawa Y, Matsuo S. The nephrotoxicity of *Aristolochia manshuriensis* in rats is attributable to its aristolochic acids. Clin Exp Nephrol 2003;7:186–194.
27. Hu S-L, Zhang H-Q, Chan K, Mei Q-X. Studies on the toxicity of *Aristolochia manshuriensis* (Guanmuton). Toxicol 2004;198:195–201.
28. Lebeau C, Debelle FD, Arlt VM, Pozdzik A, De Prez EG, Phillips DH, et al. Early proximal tubule injury in experimental aristolochic acid nephropathy: functional and his-

tological studies. Nephrol Dial Transplant 2005;20;2321–2332.

29. Sato N, Takahashi D, Chen SM, Tsuchiya R, Mukoyama T, Yamagata S, et al. Acute nephrotoxicity of aristolochic acids in mice. J Pharm Pharmacol 2004;56:221–229.
30. Martinez M-C, Nortier J, Vereerstraeten P, Vanherweghem J-L. Progression rate of Chinese herb nephropathy: impact of *Aristolochia fangchi* ingested dose. Nephrol Dial Transplant 2002;17:408–412.
31. Nortier JL, Martinez MC, Schmeiser HH, Arlt VM, Bieler CA, Petein M, et al. Urothelial carcinoma associated with the use of a Chinese herb (*Aristolochia fangchi*). N Engl J Med 2000;342:1686–1692.
32. Mengs U, Stotzem CD. Renal toxicity of aristolochic acid in rats as an example of nephrotoxicity testing in routine toxicology. Arch Toxicol 1993;67:307–311.
33. Cosyns JP, dehoux JP, Guiot Y, Goebbels RM, Robert A, Bernard AM, van Ypersele De Strihou C. Chronic aristolochic acid toxicity in rabbits: a model of Chinese herbs nephropathy? Kidney Int 2001;59:2164–2173.
34. Stiborova M, Sopko B, Hodek P, Frei E, Schmeiser HH, Hudecek J. The binding of aristolochic acid I to the active site of human cytochromes P450 1A1 and 1A2 explains their potential to reductively activate this human carcinogen. Cancer Lett 2005;229:193–204.
35. Rietjens IM, Martena MJ, Boersma MG, Spiegelenberg W, Alink GM. Molecular mechanisms of toxicity of important food-borne phytotoxins. Mol Nutr Food Res 2005;49:131–158.
36. Lebeau C, Debelle FD, Arlt VM, Pozdzik A, De Prez EG, Phillips DH, et al. Early proximal tubule injury in experimental aristolochic acid nephropathy: functional and histological studies. Nephrol Dial Transplant 2005;20:2321–2332.
37. International Agency for Research on Cancer. Aristolochia species and aristolochic acids. IARC Mongr 2002;82: 69–128.
38. Stefanovic V, Toncheva D, Atanasova S, Polenakovic M. Etiology of Balkan endemic nephropathy and associated urothelial cancer. Am J Nephrol 2006;26:1–11.
39. Grollman AP, Shibutani S, Moriya M, Miller F, Wu L, Moll U, et al. Aristolochic acid and the etiology of endemic (Balkan) nephropathy. Proc Natl Acad Sci U S A 2007;104: 12129–12134.
40. Yuan J, Liu Q, Wei G, Tang F, Ding L, Yao S. Characterization and determination of six aristolochic acids and three aristololactams in medicinal plants and their preparations by high-performance liquid chromatography-photodiode array detection/electrospray ionization mass spectrometry. Rapid Commun Mass Spectrom 2007;21:2332–2342.
41. Seto T, Hamano T, Shioda H, Kamimura H. Analysis of aristolochic acid I and II in Kampo medicine preparations. J Health Sci 2002;48:412–417.
42. Jong T-T, Lee M-R, Hsiao S-S, Hsai J-L, Wu T-S, Chiang ST, Cai S-Q. Analysis of aristolochic acid in nine sources of Xixin, a traditional Chinese medicine, by liquid chromatography/atmospheric pressure chemical ionization/tandem mass spectrometry. J Pharm Biomed Anal 2003;33:831–837.
43. Zhang C, Wang X, Shang M, Yu J, Xu Y, Li Z, et al. Simultaneous determination of five aristolochic acids and two aristololactams in *Aristolochia* plants by high-performance liquid chromatography. Biomed Chromatogr 2006;20: 309–318.
44. Kite GC, Yule MA, Leon C, Simmonds MS. Detecting aristolochic acids in herbal remedies by liquid chromatography/serial mass spectrometry. Rapid Comm Mass Spectrometry 2002;16:585–590.
45. Priestap HA, Iglesias SL, Desimone MR, Diaz LE. Determination of aristolochic acids by capillary electrophoresis. J Capillary Electrophor 2003;8:39–43.
46. Koh HL, Wang H, Zhou S, Chan E, Woo SO. Detection of aristolochic acid I, tetrandrine and fangchinoline in medicinal plants by high performance liquid chromatography and liquid chromatography/mass spectrometry. J Pharm Biomed Anal 2006;40:653–661.
47. Bieler CA, Stiborova M, Wiessler M, Cosyns JP, van Ypersele de Strihou C, Schmeiser HH. 32P-post-labelling analysis of DNA adducts formed by aristolochic acid in tissues from patients with Chinese herbs nephropathy. Carcinogenesis 1997;18:1063–1067.
48. Nortier JL, Vanherweghem JL. Renal interstitial fibrosis and urothelial carcinoma associated with the use of a Chinese herb (*Aristolochia fangchi*). Toxicol 2002;181–182: 577–560.

Chapter 48

BLACK COHOSH
(*Actaea racemosa* L.)

HISTORY

This plant was first described by Plukenet in 1696, and Linnaeus later classified black cohosh as *Actaea racemosa*.[1] Although Pursh later reclassified this plant into the genus *Cimicifuga*, recent DNA analysis resulted in the classification of black cohosh as *Actaea racemosa*.[2] Black cohosh is a native herb of eastern North America with a long history of traditional use by Native Americans, primarily for amenorrhea and menopause. In the Eclectic Dispensary of 1852, black cohosh was a treatment for joint pains, headache, inflammation, neuralgia, labor pains, and dysmenorrhea. Black cohosh was an ingredient of Lydia Pinkham's Vegetable Compound, which was a popular remedy for "female complaints" during the early 20th century. A fluid extract of black cohosh was listed in the US National Formulary from 1840 until 1946.

BOTANICAL DESCRIPTION

About 10 species of *Actaea* exist in the northern temperate zones of the world. The principal species in China are *A. dahurica*, *A. foetida*, and *A. heracleifolia*, whereas *A. acerina*, *A. japonica*, and *A. simplex* are the major species in Japan. Black cohosh should be distinguished from closely related species including blue cohosh [*Caulophyllum thalictroides* (L.) Michx.], summer cohosh or mountain bugbane (*Cimicifuga americana* Michx.), white cohosh or white baneberry (*Actaea pachypoda* Ell.), and red cohosh or red baneberry [*Actaea rubra* (Ait.) Willd.].

Common Name: Black Cohosh, Cohosh, Black Snakeroot, Squawroot, Rattle Weed, Rattle Root, Bugbane

Scientific Name: *Actaea racemosa* L. (formerly *Cimicifuga racemosa*)

Botanical Family: Ranuculaceae (buttercup)

Physical Description: Black cohosh is an erect perennial wildflower that produces a clump of quadrangular stems up to 3 ft (~90 cm) tall. The large, alternate, three-pinnately compound leaves have saw-tooth edges. Greenish-white, petalless flowers emerge from June through September. The rhizome is black and the surface is knotted. Seed pods remain on the raceme during the winter, and a distinctive noise occurs when the wind blows on these seed pods. This noise accounts for the common name of rattle weed.

Distribution and Ecology: *A. racemosa* occurs in the deciduous forests of North America as well as Europe. This plant grows in rich, shady soil in eastern North America from southern Ontario to northern Georgia and west to Arkansas, Missouri, and Wisconsin. The eastern United States and Canada are the sources for commercial stocks of black cohosh. The distribution of black cohosh is similar to blue cohosh, which inhabits the deep rich loam of shady woods from New Brunswick to South Carolina and westward to Nebraska, particularly in the Allegheny Mountain region.

Medical Toxicology of Natural Substances, by Donald G. Barceloux, MD

EXPOSURE

Sources

A. dahunica is the source of the traditional chinese medicine, sheng ma. Native Americans used the dried roots and rhizomes of *A. racemosa* for a variety of conditions ranging from menstrual problems to rattlesnake bites. Some commercial products of black cohosh contain Asian species of *Actaea* rather than *Actaea racemosa*. In a study of 11 commercial black cohosh products analyzed by high performance liquid chromatograph (HPLC)/photodiode array detection and liquid chromatography with selected ion monitoring mass spectrometry, 3 of the 11 samples contained the marker (cimifugin) for Asian *Actaea* species rather than the marker (cimiracemoside C) for *Actaea racemosa*.[3]

Medicinal Uses

Traditional

Traditional uses for black cohosh include the treatment of menopause and related symptoms (anxiety, insomnia, amenorrhea, hot flashes, depression). Cherokee and Iroquois Indians used this herb for rheumatism, sore throat, malaise, kidney diseases, and diseases related to the female reproductive system.[1] In Chinese medicine, sheng ma refers to rhizomes of other *Actaea* species [e.g., *Actaea dahurica* (Turcz. ex Fisch. & C. A. Mey.) Franch., *A. foetida* L., *A. heracleifolia* (Kom.) J. Compton] that are used for infectious diseases, uterine and rectal prolapse, fever, inflammation, and pain.

Current

Black cohosh is the principal herb used to treat menstrual disorders. An ethanolic extract of the rhizome of black cohosh is marketed as a 20 mg-tablet of Remifemin® (Enzymatic Therapy, Inc., Green Bay, WI), which contains 1 mg of triterpenes calculated as 27-deoxyactein. This commercial product is a dried 40% isopropanol extract of the roots and rhizomes of black cohosh (*Actaea racemosa*). The efficacy of black cohosh extracts for the treatment of menstrual symptoms remains unclear.[4] A double-blind, randomized, placebo-controlled clinical trial did not demonstrate a statistically significant, subjective improvement in hot flashes in the treatment group (20 mg *Cimicifuga racemosa* twice daily), when compared with placebo during the 4-week cross-over study.[5]

Regulatory Status

The German Commission E approved the use of black cohosh root for premenstrual discomfort, dysmenorrhea, and menopausal symptoms (hot flashes, diaphoresis, sloughing of skin, atrophic vaginitis, irritability, depression).

PRINCIPAL INGREDIENTS

Chemical Composition

Black cohosh contains alkaloids, at least 20 phenolic compounds (cimicifugic acid A, B, E, and F, fukinolic acid, ferulic acid, isoferulic acid, fukiic acid, and caffeic acid caffeic acid),[6,7] organic acids, palmitic and oleic acids, tannins, resins, volatile oils, flavonoids, and over 43 terpenoid glycosides (e.g., actein, 23-epi-26-deoxyactein, 26-deoxyactein, 12-acetylactein, cimiracemoside A, cimiaceroside C, cimigenoside, cimicifugoside H-1, cimicifugoside M).[8–10] In particular, black cohosh contains substantial concentrations of triterpene glycosides and caffeic acid derivatives. The caffeic acid derivatives consist of esters between the carboxyl group of caffeic acid, ferulic acid or isoferulic acid, and the hydroxyl group of fukiic or piscidic acid. Analysis of black cohosh extracts by HPLC/electrospray ionization tandem mass spectrometry demonstrated a variety of caffeic acid derivatives including caffeic acid, cimicifugic acid A, B, E, and F, dehydrocimicifugic acid A and B, ferulic acid, isoferulic acid, and fukinolic acid.[11] Figure 48.1 displays the chemical structures of some triterpene glycosides and phenolic compounds in black cohosh. Nonrandom sampling of extracts of an 85-year old sample of black cohosh contained triterpene glycoside concentrations ranging from about 3.1–5.2%, depending on the extraction method, whereas the amount of phenolic compounds range from approximately 5.2–5.8%.[12] In contrast to data from early studies on black cohosh, this herb and the root extract do not contain significant quantities of the estrogenic isoflavone, formononetin, or ononin (formononetin-7-glucoside).[13] Total phenolic compounds (caffeic acid, ferulic acid, isoferulic acid fukinolic acid, cimicifugic acid A, B, E, and F) in recently dried roots and rhizomes of black cohosh accounted for approximately 0.62% of the plant as measured by HPLC with diode array detection.[14]

Physiochemical Properties

Triterpene compounds are relatively stable compounds with the exception of compounds containing a free carboxylic group.[15] Additionally, triterpene compounds are not stable at the extreme ranges of pH. The total amount

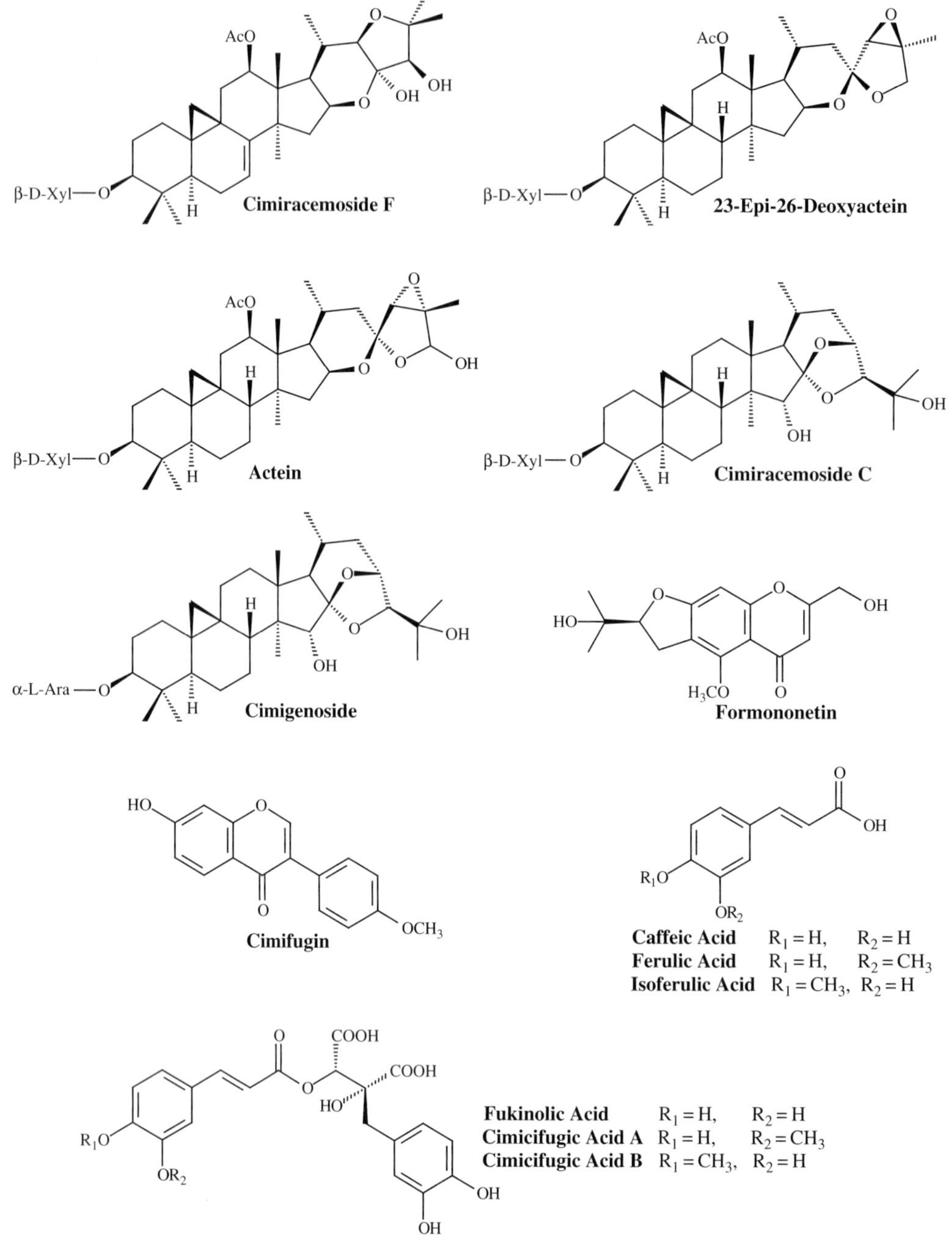

FIGURE 48.1. Chemical structures of triterpene glycosides and phenolic compounds in black cohosh. Adapted from Jiang B, et al. Evaluation of the botanical authenticity and phytochemical profile of black cohosh products by high-performance liquid chromatography with selected ion monitoring liquid chromatography-mass spectrometry. J Agric Food Chem 2006;54:3243.

of phenolic compounds in a 85-year-old sample of black cohosh stored under varying environmental conditions at a botanical garden was 4.25% compared with 5.68% in a modern sample of black cohosh.[12]

Mechanism of Toxicity

The mechanism of action (e.g., estrogenic activity, reduction of leuteinizing hormone) of black cohosh has not been defined, and evidence of the efficacy for the treatment of menstrual disorders is inconclusive.[16,17] However, the effects of black cohosh probably are not mediated through estrogen receptors.[18] Although earlier studies suggested that components in black cohosh have estrogenic activity,[19] more recent *in vitro* evidence indicates that extracts of black cohosh do not affect the estrogen receptors.[20,21] A study of ovariectomized Sprague-Dawley rats given a 40% isopropanol extract of black cohosh for 2 weeks with or without estradiol did not detect any differences in uterine weight or vaginal cellular cornification in the rats treated with black cohosh.[22] Subsequent testing of this extract against 10 subtypes of the serotonin receptor revealed strong binding to the 5-HT_{1A}, 5-HT_{1D}, and 5-HT_7 subtypes. Analysis of the binding data indicted that components of the black cohosh extract acted as a competitive ligand of the 5-HT_7 receptor, suggesting that the extract was a partial agonist at the receptor.

DOSE RESPONSE

Medicinal Use

The typical treatment for menopausal symptoms is two 20-mg standardized tablets of the root extract (Remifemin®) twice a day. Each tablet is standardized to contain 1 mg of triterpene glycosides calculated as 27-deoxyactein. There are no definite clinical data that two 20-mg standardized tablets of the root extract (Remifemin®) twice a day is more effective than one 20-mg standardized tablet of the root extract (Remifemin®) twice a day. Some herbalists combine black cohosh with blue cohosh for the induction of labor.[1] Many studies demonstrating the effectiveness of black cohosh for the treatment of menopausal symptoms are open clinical trials. Most randomized, placebo-controlled, double-blind studies of supplements containing soy isoflavones and *A. racemosa* do not detect statistically significant effects of these supplements on climacteric symptoms in perimenopausal women as measured by the modified Kupperman Index and the Greene Climacteric Scale.[23] In a multicenter, randomized, placebo-controlled double-blind study of 122 menopausal women, the primary efficacy analysis using the Kupperman Index and the Menopause Rating Scale did not demonstrate the superiority of black cohosh administration (mean dose 42 mg daily) over placebo.[24] However, in the subgroup of women with a Kupperman Index ≥20, black cohosh use decreased the index 47% compared with 21% for placebo ($P < 0.018$). A prospective, randomized, double-blind evaluation of black cohosh extract on hot flashes in women, who had completed primary breast cancer treatment, demonstrated a statistically significant reduction of sweating in the treatment group compared with the controls.[25] However, all other improvements in symptoms including declines in the number and intensity of hot flashes were similar in both groups. A randomized, multicenter, double-blind clinical trial compared the efficacy and tolerability of the isopropanol extract of black cohosh in the treatment of climacteric complaints with placebo.[26] The isopropanol extract of black cohosh extract was more effective than placebo ($P < 0.001$) depending on time from symptom onset ($P = 0.014$) and follicle-stimulating hormone level ($P = 0.011$) based on Menopause Rating Scale units. The effect size (0.03–0.05) on the Menopause Rating Scale was similar to recent hormone replacement therapy study results.

Toxicity

Rare case reports associate the use of black cohosh as well as other herbs with the development of severe hepatotoxicity.[27] These effects probably result from idiopathic reactions rather than dose-related hepatotoxicity.

TOXICOKINETICS

Preliminary evidence suggests that ingredients in commercial black cohosh can inhibit CYP3A4 enzymes, and therefore drug interactions potentially can occur with the concomitant administration of drugs metabolized by the same isoenzymes.[28,29] However, there is no human data to confirm the interaction of black cohosh with drugs requiring metabolism by CYP3A4 enzymes. An *in vivo* study detected a statistically significant reduction in CYP2D6 activity after 4 weeks of therapy with black cohosh, but the magnitude of the effect was small and probably clinically insignificant.[30] This *in vivo* study did not detect any inhibition of black cohosh on CYP3A4. Other *in vivo* studies of volunteers receiving 40 mg black cohosh daily for 14 days indicated that supplementation with this extract did not alter digoxin pharmacokinetics (P-glycoprotein transporter)[31] or midazolam (CYP3A4) pharmacokinetics.[32]

CLINICAL RESPONSE

Adverse reactions associated with the clinical use of pharmacological doses of black cohosh typically are rare, mild, and reversible with most symptoms involving mild gastrointestinal (GI) upset.[33] The most common adverse effects are GI disturbances, rashes, headaches, dizziness, mastalgia, and weight gain.[34] Rare case reports associate the use of black cohosh with the development of severe hepatitis requiring liver transplantation. A case report temporarily associated one week of back cohosh use with the development of severe hepatitis and fulminant hepatic failure requiring liver transplantation.[27] The liver biopsy demonstrated changes consistent with an idiosyncratic, immunological reaction rather than a toxic reaction manifest by eosinophilia, moderate portal and lobular inflammation, bridging necrosis, and early fibrosis. There was no bile duct damage and no other apparent cause (e.g., alcohol, drug use, blood transfusion, viral infection) of the liver failure.

A 52-year-old women developed severe hepatitis one week after stopping the ingestion of an herbal preparation that contained black cohosh.[35] Over the following week, she developed hepatic encephalopathy and hepatorenal failure that required liver transplantation. The herbal preparation also contained pulegone, a known hepatotoxin. Another 50-year-old woman developed hepatic failure after taking 500 mg black cohosh root daily for 5 months.[36] The history and laboratory data did not suggest an alternate explanation for the liver failure. She underwent successful orthotopic liver transplantation with the explanted liver demonstrating cell necrosis, cholestasis, and bridging fibrosis. Although there are few data on the effect of black cohosh on liver function or serum hepatic aminotransferases, clinical studies on the efficacy of black cohosh to date have not detected clinically significant adverse effects on the liver.[37]

DIAGNOSTIC TESTING

Triterpene glycosides are difficult to analyze by conventional HPLC UV methods because of weak UV absorbance. Methods to detect these compounds in biological samples include evaporative light scattering detection (ELSD)/HPLC, atmospheric performance chemical ionization (APCI)-LC/MS,[8] and LC/turbo ion spray (TIS)/MS.[38] High performance liquid chromatography/photodiode array/mass spectrometric/evaporative light scattering detection (HPLC/PDA/MS/ELSD) allows the identification of common, less-expensive Asian *Actaea* species and *A. racemosa* in herbal samples based on the HPLC/PDA/MS/ELSD fingerprints of triterpene glycosides (cimigenol-3-O-arabinoside, cimifugin, cimifugin-3-O-glucoside).[39] Using reversed-phase HPLC with diode array detection, the limits of detection and quantitation for phenolic compounds ranged from 24.8 ± 0.15 ng/mL (caffeic acid) to 78.3 ± 0.94 ng/mL (cimicifugic acid F) and 82.7 ± 0.51 ng/mL (caffeic acid) to 260.9 ± 3.13 ng/mL (cimicifugic acid F), respectively.[6]

TREATMENT

Treatment is supportive. Although the use of black cohosh has been associated with hepatitis only in case reports, patients presenting with possible adverse reactions to black cohosh should be evaluated for hepatic toxicity.

References

1. McKenna DJ, Jones K, Humphrey S, Hughes K. Back cohosh: efficacy, safety, and use in clinical and preclinical applications. Alter Ther 2001;7:93–100.
2. Compton JA, Culham A, Jury SLP. Reclassification of *Actaea* to include *Cimicifuga* and *Souliea* (Ranuculaceae): phylogeny inferred from morphology, nrDNA ITS, and cpDNA trnL-F sequence variation. Taxon 1998:47:593–634.
3. Jiang B, Kronenberg F, Nuntanakorn P, Qiu MH, Kennelly EJ. Evaluation of the botanical authenticity and phytochemical profile of black cohosh products by high-performance liquid chromatography with selected ion monitoring liquid chromatography-mass spectrometry. J Agric Food Chem 2006;54:3242–3253.
4. Borrelli F, Ernst E. *Cimicifuga racemosa*: a systematic review of its clinical efficacy. Eur J Clin Pharmacol 2002; 58:235–241.
5. Pockaj BA, Gallagher JG, Loprinzi CL, Stella PJ, Barton DL, Sloan JA, et al. Phase III double-blind, randomized, placebo-controlled crossover trial of black cohosh in the management of hot flashes: NCCTG Trial N01CC1. J Clin Oncol 2006;24:2836–2841.
6. Nuntanakorn P, Jiang B, Einbond LS, Yang H, Kronenberg F, Weinstein IB, Kennelly EJ. Polyphenolic constituents of *Actaea racemosa*. J Nat Prod 2006;69:314–318.
7. Stromeier S, Petereit F, Nahrstedt A. Phenolic esters from the rhizomes of *Cimicifuga racemosa* do not cause proliferation effects in MCF-7 cells. Planta Med 2005;71:495–500.
8. Chen SN, Li W, Fabricant DS, Santarsiero BD, Mesecar A, Fitzloff JF, et al. Isolation, structure elucidation, and absolute configuration of 26-deoxyactein from *Cimicifuga racemosa* and clarification of nomenclature associated with 27-deoxyactein. J Nat Prod 2002;65:601–605.
9. He K, Zheng B, Kim CH, Rogers L, Zheng Q. Direct analysis and identification of triterpene glycosides by LC/MS in black cohosh, Cimicifuga racemosa, and in several commercially available black cohosh products. Planta Med 2000;66:635–640.
10. Bedir E, Khan IA. Cimiracemoside a: a new cyclolanostanol xyloside from the rhizome of *Cimicifuga racemosa*. Chem Pharm Bull (Tokyo) 2000;48:425–427.
11. Li W, Sun Y, Liang W, Fitzloff JF, van Breemen RB. Identification of caffeic acid derivatives in *Actaea racemosa* (*Cimicifuga racemosa*, black cohosh) by liquid chromatography/tandem mass spectrometry. Rapid Comm Mass Spectrometry 2003;17:978–982.
12. Jiang B, Yang H, Nuntanakorn P, Balick MJ, Kronenberg F, Kennelly EJ. The value of plant collections in ethnopharmacology: a case of an 85-year-old black cohosh (*Actaea racemosa* L.) sample. J Ethnopharmacol 2005;96: 521–528.
13. Kennelly EJ, Baggett S, Nuntanakorn P, Ososki AL, Mori SA, Duke J, et al. Analysis of thirteen populations of black cohosh for formononetin. Phytomedicine 2002;9:461–467.
14. Nuntanakorn P, Jiang B, Yang H, Cervantes-Cervantes M, Kronenberg F, Kennelly EJ. Analysis of polyphenolic compounds and radical scavenging activity of four American *Actaea* species. Phytochem Anal 2007;18:219–228.

15. Tava A, Mella M, Bialy Z, Jurzysta M. Stability of saponins in alcoholic solutions: ester formation as artifacts. J Agric Food Chem 2003;51:1797–1800.
16. Huntley A. The safety of black cohosh (*Actaea racemosa, Cimicifuga racemosa*). Expert Opin Drug Saf 2004;3:615–623.
17. Zierau O, Bodinet C, Kolba S, Wulf M, Vollmer G. Antiestrogenic activities of *Cimicifuga racemosa* extracts. J Steroid Biochem Mol Biol 2002;80:125–130.
18. Borrelli F, Izzo AA, Ernst E. Pharmacological effects of *Cimicifuga racemosa*. Life Sci 2003;73:1215–1229.
19. Duker EM, Kopanski L, Jarry H, Wuttke W. Effects of extracts from *Cimicifuga racemosa* on gonadotropin release in menopausal women and ovariectomized rats. Planta Med 1991;57:420–424.
20. Lupu R, Mehmi I, Atlas E, Tsai MS, Pisha E, Oketch-Rabah HA, et al. Black cohosh, a menopausal remedy, does not have estrogenic activity and does not promote breast cancer cell growth. Int J Oncol 2003;23:1407–1412.
21. Liu J, Burdette JE, Xu H, Gu C, van Breemen RB, Bhat KP, et al. Evaluation of estrogenic activity of plant extracts for the potential treatment of menopausal symptoms. J Agric Food Chem 2001;49:2472–2479.
22. Burdette JE, Liu J, Chen S-N, Fabricant DS, Piersen CE, Barker EL, et al. Black cohosh acts as a mixed competitive ligand and partial agonist of the serotonin receptor. J Agric Food Chem 2003;51:5661–5670.
23. Verhoeven MO, van der Mooren MJ, Van de Weijer PH, Verdegem PJ, Van Der Burgt LM, Kenemans P. Effect of a combination of isoflavones and *Actaea racemosa Linnaeus* on climacteric symptoms in healthy symptomatic perimenopausal women: a 12-week randomized, placebo-controlled, double-blind study. Menopause 2005;12:412–420.
24. Frei-Kleiner S, Schaffner W, Rahlfs VW, Bodmer Ch, Birkhauser M. *Cimicifuga racemosa* dried ethanolic extract in menopausal disorders: a double-blind placebo-controlled clinical trial. Maturitas 2005;51:397–404.
25. Jacobson JS, Troxel AB, Evans J, Klaus L, Vahdat L, Kinne D, Lo KM, et al. Randomized trial of black cohosh for the treatment of hot flashes among women with a history of breast cancer. J Clin Oncol 2001;19:2739–2745.
26. Osmers R, Friede M, Liske E, Schnitker J, Freudenstein J, Henneicke-von Zepelin HH. Efficacy and safety of isopropanolic black cohosh extract for climacteric symptoms. Obstet Gynecol 2005;105:1074–1083.
27. Whiting PW, Clouston A, Kerlin P. Black cohosh and other herbal remedies associated with acute hepatitis. Med J Aust 2002;177:440–443.
28. Tsukamoto S, Aburatani M, Ohta T. Isolation of CYP3A4 Inhibitors from the Black Cohosh (*Cimicifuga racemosa*). Evid Based Complement Alternat Med 2005;2:223–226.
29. Tsukamoto S, Aburatani M, Ohta T. Isolation of CYP3A4 inhibitors from the black cohosh (*Cimicifuga racemosa*). eCam 2005;2:223–226.
30. Gurley BJ, Gardner SF, Hubbard MA, Williams DK, Gentry W, Khan IA, Shah A. *In vivo* effects of goldenseal, kava kava, black cohosh, and valerian on human cytochrome P450 1A2, 2D6, 2E1, and 3A4/5 phenotypes. Clin Pharmacol Ther 2005;77:415–426.
31. Gurley BJ, Barone GW, Williams DK, Carrier J, Breen P, Yates CR, et al. Effect of milk thistle (*Silybum marianum*) and black cohosh (*Cimicifuga racemosa*) supplementation on digoxin pharmacokinetics in humans. Drug Metabol Disp 2006;34:69–74.
32. Gurley B, Hubbard MA, Williams K, Thaden J, Tong Y, Gentry WB, et al. Assessing the clinical significance of botanical supplementation on human cytochrome P450 3A activity: comparison of a milk thistle and black cohosh product to rifampin and clarithromycin. J Clin Pharmacol 2006;46:201–213.
33. Dog TL, Powell KL, Weisman SM. Critical evaluation of the safety of *Cimicifuga racemosa* in menopause symptom relief. Menopause 2003;10:299–313.
34. Huntley A, Ernst E. A systematic review of the safety of black cohosh. Menopause 2003;10:58–64.
35. Lontos S, Jones RM, Angus PW, Gow PJ. Acute liver failure associated with the use of herbal preparations containing black cohosh. Med J Aust 2003;170:390–391.
36. Levitsky J, Alli TA, Wisecarver J, Sorrell MF. Fulminant liver failure associated with the use of black cohosh. Dig Dis Sci 2005;50:538–539.
37. Mahady GB. Black cohosh (*Actaea/Cimicifuga racemosa*). Review of the clinical data for safety and efficacy in menopausal symptoms. Treat Endocrinol 2005;4:177–184.
38. Wang H-K, Sakurai N, Shih CY, Lee K-H. LC/TIS-MS fingerprint profiling of *Cimicifuga* species and analysis of 23-epi-26-deoxyactein in *Cimicifuga racemosa* commercial products. J Agric Food Chem 2005;53:1379–1386.
39. He K, Pauli GF, Zheng B, Wang H, Bai N, Peng T, Roller M, Zheng Q. *Cimicifuga* species identification by high performance liquid chromatography-photodiode array/mass spectrometric/evaporative light scattering detection for quality control of black cohosh products. J Chromatogr A 2006;1112:241–254.

Chapter 49

BLUE COHOSH
[*Caulophyllum thalictroides* (L.) Michx.]

HISTORY

Linnaeus originally classified blue cohosh [*Caulophyllum thalictroides* (L.) Michx.] in 1753, and subsequently André Michaux placed this plant under the genus *Caulophyllum* in 1803. Blue cohosh was a traditional medicine of Native Americans that was used as an herb to facilitate childbirth. During the 1800s, the root of this herb was used as a uterine stimulant and antispasmodic for dysmenorrhea.[1]

BOTANICAL DESCRIPTION

Blue cohosh (*Caulophyllum thalictroides*) is a native species of eastern North America along with the other North American species, *C. giganteum* (Farw.) Loconte & Blackwell. The latter species differs from *C. thalictroides* by the presence of inflorescences with a few large flowers that have large reproductive structures. The Asian member of the genus is *C. robustum* Maxim.

Common Name: Blue Cohosh, Squaw Root, Papoose Root, Blue Ginseng, Leontice, Blueberry Root

Scientific Name: *Caulophyllum thalictroides* (L.) Michx.

Botanical Family: Berberidaceae (bayberries)

Physical Description: This herbaceous, perennial, rhizomatous plant grows up to 1–2 ft (~30–60 cm) in height. The leaves are ternately or biternately compound. The yellow-green to brown flowers appear in terminal panicles. The leaves (sepals) directly underneath the flowers are large and pedal-like. The six true pedals are smaller than the sepals, and the six stamens have short filaments. The ovary of the flower is green, unilocular, and contains three ovules. The mature fruit is a dark blue berry with one to two seeds.

Distribution and Ecology: This herb grows in rich, moist woodlands of eastern North America from east of Missouri, north to southeast Manitoba and New Brunswick, and south to Alabama. The other North American species is the giant blue cohosh [*Caulophyllum giganteum* (Farw.) Loconte & Blackwell]. A closely related species (*Caulophyllum robustum* Maxim.) of *Caulophyllum thalictroides* exists in eastern Asia and Japan.

EXPOSURE

Traditional uses of the root of blue cohosh include the treatment of menstrual cramps, dysmenorrhea, epilepsy, and rheumatism. Blue cohosh is used a homeopathic remedy to enhance the progression of labor and to relieve symptoms associated with menstruation.[2] However, there is inadequate scientific evidence to support the use of blue cohosh during labor.[3] Some practitioners administer both blue cohosh and black cohosh during labor, but these herbs are not usually used during pregnancy.[4]

Medical Toxicology of Natural Substances, by Donald G. Barceloux, MD

PRINCIPAL INGREDIENTS

Chemical Composition

The root of blue cohosh contains a variety of alkaloids in the range of 0.02–1.1% wet weight including quinolizidine alkaloids (*N*-methylcytisine, baptifoline, anagyrine) and aporphine alkaloids (magnoflorine, aporphine) along with thalictroidine, 5,6-dehydro-alpha-isolupanine, alpha-isolupanine, lupanine, taspine, and sparteine.[5,6] Based on three samples of roots/rhizomes of blue cohosh, magnoflorine is the major alkaloid (0.5–0.7 g/100 g dry weight) with the mean concentration of *N*-methylcytisine ranging from about 0.01–0.4 g/100 g dry weight.[7] The concentration of individual alkaloids in root extracts ranges from approximately 0.02–1.1% wet weight. Analysis of an ethanol extract of blue cohosh demonstrated relative high concentrations of magnoflorine (11 mg/g dry weight) followed by *N*-methylcytisine (0.9 mg/g dry weight), baptifoline (0.75 mg/dry weight), and anagyrine (0.18 mg/g dry weight).[8] Figure 49.1 displays the chemical structures of the major alkaloids and triterpene saponins in blue cohosh. There are at least seven triterpene saponins in the root of blue cohosh including caulosaponin [caulosapogenin, $C_{42}H_{62}O_2(OH)_4$, and dextrose] and caulophyllosaponin (hederagenin, $C_{30}H_{48}O_4$, and arabinose).[9]

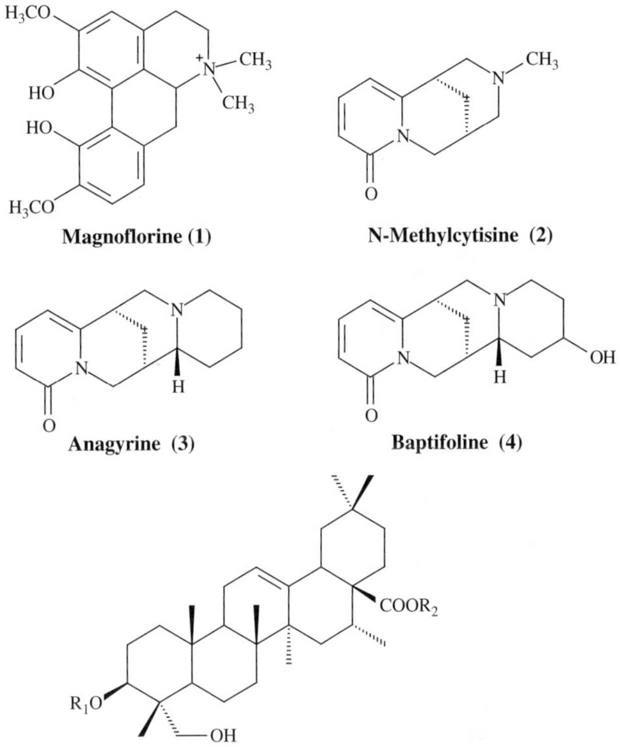

Magnoflorine (1) N-Methylcytisine (2) Anagyrine (3) Baptifoline (4)

	R_1	R_2	R_3
5	α-L-Ara-	α-L-Rha(1-->4)-β-D-Glc(1-->6)-β-D-Glc-	OH
6	β-D-Glc(1-->2)-α-L-Ara-	α-L-Rha(1-->4)-β-D-Glc(1-->6)-β-D-Glc-	H
7	α-L-Ara-	α-L-Rha(1-->4)-β-D-Glc(1-->6)-β-D-Glc-	H

FIGURE 49.1. Major alkaloids (1–4) and unnamed triterpene saponins (5–7) in blue cohosh. Adapted from Ganzera M, et al. Determination of saponins and alkaloids in *Caulophyllum thalictroides* (blue cohosh) by high-performance liquid chromatography and evaporative light scattering detection. Phytochem Anal 2003;14:2. Copyright John Wiley & Sons Limited.

Physiochemical Properties

In vitro studies in the early 1940s suggested that alcohol extracts of blue cohosh root contain glycosides with uterine stimulant and coronary vasoconstrictive properties.[10] In the rat embryo culture, the quinolizidine alkaloid, *N*-methylcytisine, exhibits teratogenic activity,[6] similar to other quinolizidine alkaloids.[11] However, thalictroidine, anagyrine, and alpha-isolupanine were not teratogenic at the tested concentrations, and taspine demonstrated high embryotoxicity, but no teratogenic activity. In animal models, *N*-methylcytisine exhibits mild nicotinic activity.

Mechanism of Toxicity

The toxic constituents in extracts of blue cohosh have not been determined; therefore, the mechanism of toxicity is unknown. To date, toxic effects of blue cohosh have been described in case reports and there is no consistent pattern of toxicity. The trauma associated with the use of this extract to induce illicit abortions could potentially cause sepsis.

DOSE RESPONSE

Typically, blue cohosh is consumed as a decoction (e.g., one teaspoon of dried root in cup of water brought to a boil and simmered for 10 minutes). However, because of safety issues and the lack of efficacy, this herb is not usually recommended during pregnancy.

TOXICOKINETICS

There are few human data on the toxicokinetics of components derived from blue cohosh.

CLINICAL RESPONSE

Acute Effects

A 21-year-old woman developed a nicotine-like intoxication (tachycardia, diaphoresis, abdominal pain, vomiting, muscle weakness, abdominal wall fasciculations) following the ingestion of blue cohosh (*Caulophyllum thalictroides*) for 4 consecutive days (10–20 doses/daily) as an abortifacient.[12] These symptoms were consistent with the nicotinic effects of *N*-methylcytisine present in blue cohosh. In addition to the blue cohosh, she ingested 15 cups of slippery elm (*Ulmus rubra* Muhl.) tea daily for 4 days and used slippery elm douches. Similar nicotinic effects have not been reported following the ingestion of slippery elm tea.

Chronic Effects

A case report associated the development of cardiomegaly, pulmonary edema, and profound hypokinesis of the left ventricular posterior wall of a newborn with the ingestion of three times the recommended dose of blue cohosh during the month before delivery.[13] An alternate explanation (e.g., congenital coronary artery anomaly, viral infection) could not be confirmed. At 2-year follow-up, the child was developing normally, but cardiomegaly and mild reduction in left ventricular function remained.

DIAGNOSTIC TESTING

Methods for the detection of alkaloids and triterpene saponins in blue cohosh include high performance liquid chromatography and evaporative light scattering detection.[7]

TREATMENT

Treatment is supportive.

References

1. Lloyd JU, Lloyd CG. Drugs and Medicines of North America. Cincinnati, OH: Robert Clarke & Co; 1884–1885.
2. Priestman KG. A few useful remedies in pregnancy, labour and the first few days of the babies' life. Br Homeopathy J 1988;77:172–173.
3. Smith CA. Homoeopathy for induction of labour (review). Cochrane Database Syst Rev 2003(4):CD003399.
4. Baillie N, Rasmussen P. Black and blue cohosh in labour. N Z Med J 1997;110:20–21.
5. Flom MS, Doskotch RW, Beal JL. Isolation and characterization of alkaloids from *Caulophyllum thalictroides*. J Pharm Sci 1967;56:1515–1517.
6. Kennelly EJ, Flynn TJ, Mazzola EP, Roach JA, McCloud TG, Danford DE, Betz JM. Detecting potential teratogenic alkaloids from blue cohosh rhizomes using an *in vitro* rat embryo culture. J Nat Prod 1999;62:1385–1389.
7. Ganzera M, Dharmaratne HR, Nanyakkara NP, Khan IA. Determination of saponins and alkaloids in *Caulophyllum thalictroides* (blue cohosh) by high-performance liquid chromatography and evaporative light scattering detection. Phytochem Anal 2003;14:1–7.
8. Woldemariam TZ, Betz JM, Houghton PJ. Analysis of aporphine and quinolizidine alkaloids from *Caulophyllum thalictroides* by densitometry and HPLC. J Pharmaceut Biomed Anal 1997;15:839–843.
9. Jhoo J-W, Sang S, He K, Cheng X, Zhu N, Stark RE, et al. Characterization of the triterpene saponins of the roots and rhizomes of blue cohosh (*Caulophyllum thalictroides*). J Agric Food Chem 2001;49:5969–5974.
10. Ferguson HC, Edwards LD. A pharmacological study of a crystalline glycoside of *Caulophyllum thalictroides*. J Am Pharm Assoc 1954;43:16–21.
11. Panter KE, Keeler RF. Quinolizidine and piperidine alkaloid teratogens from poisonous plants and their mechanism of action in animals. Vet Clin North Am Food Anim Pract 1993;9:33–40.
12. Rao RB, Hoffman RS. Nicotinic toxicity from tincture of blue cohosh (*Caulophyllum thalictroides*) used as an abortifacient. Vet Hum Toxicol 2002;44:221–222.
13. Jones TK, Lawson BM. Profound neonatal congestive heart failure caused by maternal consumption of blue cohosh herbal medication. J Pediatr 1998;132:550–552.

Chapter 50

BORAGE
(*Borago officinalis* L.)

BOTANICAL DESCRIPTION

The large, deep green, leaves of common borage are covered with white, stiff, prickly hairs, and occasionally the leaves of borage may be confused with the large, fussy, deep green leaves of foxglove. Misidentification of foxglove for borage resulted in digitoxin poisoning after the patient ingested a brewed tea from the misidentified foxglove leaves.[1]

Common Name: Common Borage, Bee Plant, Bee Bread, Ox's Tongue

Scientific Name: *Borago officinalis* L.

Botanical Family: Boraginaceae (borage, bourraches)

Physical Description: This hairy, annual herb grows up to 2–3 ft (~60–90 cm) in height. The leaves are alternate, wrinkled, oval, and pointed with dimensions of about 1.5 by 3 in (~4 cm by 8 cm). Bright blue to pink, star-shaped flowers appear in loose racemes during the summer. Borage has a salty flavor with an aroma of cucumbers.

Distribution and Ecology: This plant inhabits wide areas of the west coast and the northern and northeastern states of the United States, as well as southern Europe. Although this plant is native to Syria, common borage has naturalized in the warmer parts of central, eastern, and western Europe. The common name, bee plant, is derived from the use of this plant as an attractant for honeybees.

EXPOSURE

Sources

The leaves of borage are used raw in salads, steamed as a tea, or cooked in vinegar as a vegetable. Common borage is widely cultivated as an ornamental plant and as a pot herb

Medicinal Uses

Traditional

Borage oil (starflower oil) is an alternative to evening primrose as a source of γ-linolenic acid. Traditional herbal uses for borage include the treatment of arthritis, chest congestion with or without cough, inflammation, sore throat, depression, seborrheic dermatitis, and menopausal symptoms. Borage oil is a supplement increasing used by older patients for the treatment of premenstrual syndrome, diabetes, skin disorders, and rheumatoid arthritis.[2] Borage oil is a traditional treatment for a hangover.[3]

Current

The decreased production of anti-inflammatory dihomo-γ-linolenic acid metabolites (e.g., prostaglandin E_1, 15-hydroxyeicosatrienoic acid) in patients with atopic dermatitis suggests a role for γ-linolenic acid in the treatment of this disorder. The amount of γ-linolenic is

Medical Toxicology of Natural Substances, by Donald G. Barceloux, MD

substantially higher in borage oil than in evening primrose oil (i.e., about 23–25% vs. 8–10%, respectively). Studies on atopic dermatitis following the use of evening primrose yield conflicting results.[4] One hundred sixty patients with stable, moderate disease activity of atopic dermatitis were randomized to a treatment group (500 mg borage oil capsules) and to placebo (bland lipid miglyol) over 24 weeks.[5] Although several clinical symptoms (erythema, vesiculation, excoriation, lichenification, insomnia) improved in the treatment group compared with placebo, the overall response to borage oil did not reach statistical significance as measured by the amount of topical corticosteroid used and clinical symptoms (Costa score). However, there was some improvement in the subgroup of the treatment group that included patients with documented compliance and increased erythrocyte dihomo-γ-linolenic acid. A similar borage oil-induced reduction is postulated for the elevation of tumor necrosis factor-α associated with rheumatoid arthritis,[6,7] but there are inadequate clinical data to support the routine use of borage oil supplements for the treatment of rheumatoid arthritis. A randomized, double-blind, placebo-controlled, 24-week trial of 37 patients with rheumatoid arthritis and active synovitis suggested that daily treatment with 1.4 g γ-linolenic acid in borage seed oil improved the symptoms and signs of rheumatoid arthritis.[8]

REGULATORY STATUS

Borage oil is a dietary supplement in the United States. The German Commission E does not approve the use of borage oil because of concerns about potential hepatic and carcinogenic effects of the pyrrolizidine alkaloids in borage oil.

PRINCIPAL INGREDIENTS

Constituents of borage oil (CAS RN: 225234-12-8) include acids (acetic, lactic, malic silicic), alkaloids, fatty acids (γ-linolenic acid, linoleic acid, oleic acid, saturated fatty acids), mucilages, tannins, and saponins. Linoleic, linolenic, and oleic acids comprise about 75% of the total fatty acids in borage oil.[9] γ-Linolenic acid is an essential ω-6 polyunsaturated fatty acid that also occurs in evening primrose (*Oenothera biennis* L.) and black currant seed (*Ribes nigrum* L.). This compound is one of at least 79 triacylglycerols present in borage oil as measured by silver-ion high-performance liquid chromatography (HPLC)/atmospheric pressure chemical ionization/mass spectrometry.[10] In addition, borage oil contains small amounts (<10 ppm dry weight leaves) of pyrrolizidine alkaloids (e.g., amabiline, intermedine, lycopsamine, supinine), which are potentially hepatotoxic. As a result of the presence of pyrrolizidine alkaloids, the German Federal Health Agency limits the consumption of unsaturated pyrrolizidine alkaloids to ≤1 μg daily. Borage oils products are certified as being free of unsaturated pyrrolizidine alkaloids when the unsaturated pyrrolizidine alkaloid content is ≤0.5–1 μg/g.

DOSE RESPONSE

Clinical trials suggest that borage oil is a relatively safe dietary supplement in daily doses of 500–720 mg. In these studies, daily doses of borage oil administered as 500-mg borage capsules[5] or 720-mg γ-linoleic acid in gelatin capsules[11] were well tolerated.

TOXICOKINETICS

In vitro studies indicate that neutrophils rapidly elongate γ-linolenic acid by two carbons to dihomo-γ-linolenic acid (DGLA) following the use of borage oil as a dietary supplement.[12] Subsequently, biotransformation of DGLA by 15-lipoxygenase results in the formation of 15-hydroxy-eicosa-trienoic acid.[13]

Theoretically, the ingestion of borage oil may lower the seizure threshold based on the development of seizures with evening primrose oil, which contains 8–10% γ-linolenic acid.[14] However, there are no well-documented case reports associating the use of borage oil with seizures.

CLINICAL RESPONSE

In a study of 80 patients treated with 500 mg borage oil daily for 24 weeks, there were no definite herb-related adverse effects. One patient was withdrawn from the study for headache, diarrhea, and vomiting, but these effects could not be causally linked to borage oil. There are few data including case reports on the toxicity of borage or borage oil. A case report associated the development of atrial fibrillation, bradycardia, and gastrointestinal distress in a 72-year-old woman, who drank tea from leaves that she thought were borage.[1] Subsequent laboratory analysis indicated that she confused borage leaves for foxglove, and thus she probably had digitoxin poisoning.

DIAGNOSTIC TESTING

Methods for the detection of α- and γ-linolenic acid moieties include silver-ion HPLC followed by online atmospheric pressure chemical ionization/mass spectrometry (APCI/MS). HPLC/APCI/MS identified 79 triacylglycerols in borage oil derived from *Borago officinalis* seeds.[15]

TREATMENT

Treatment is supportive. Decontamination measures are unnecessary.

References

1. Brustbauer R, Wenisch C. [Bradycardiac atrial fibrillation after consuming herbal tea]. Dtsch Med Wochenschr 1997;122:930–932. [German]
2. Wold RS, Lopez ST, Yau CL, Butler LM, Pareo-Tubbeh SL, Waters DL, et al. Increasing trends in elderly persons' use of nonvitamin, nonmineral dietary supplements and concurrent use of medications. J Am Dietetic Assoc 2005;105:54–63.
3. Pittler MH, Verster JC, Ernst E. Interventions for preventing or treating alcohol hangover: systematic review of randomised controlled trials. BMJ 2005;331(7531):1515–1518.
4. Morse PF, Horrobin DF, Manku MS, Stewart JC, Allen R, Littlewood S, et al. Meta-analysis of placebo-controlled studies of the efficacy of Epogam in the treatment of atopic eczema. Relationship between plasma essential fatty acid changes and clinical response. Br J Dermatol 1989;121:75–90.
5. Henz BM, Jablonska S, van de Kerkhof PC, Stingl G, Blaszczyk M, Vandervalk PG, et al. Double-blind, multicentre analysis of the efficacy of borage oil in patients with atopic eczema. Br J Dermatol 1999;140:685–688.
6. Leventhal LJ, Boyce EG, Zurier RB. Treatment of rheumatoid arthritis with gammalinolenic acid. Ann Intern Med 1993;119:867–873.
7. Belch JJ, Hill A. Evening primrose oil and borage oil in rheumatologic conditions. Am J Clin Nutr 2000;71(suppl):352S–356S.
8. Leventhal LJ, Boyce EG, Zurier RB. Treatment of rheumatoid arthritis with gammalinolenic acid. Ann Intern Med 1993;119:867–873.
9. Barre DE. Potential of evening primrose, borage, black currant, and fungal oils in human health. Ann Nutr Metab 2001;45:47–57.
10. Laakso P, Voutilainen P. Analysis of triacylglycerols by silver-ion high-performance liquid chromatography-atmospheric pressure chemical ionization mass spectrometry. Lipids 1996;31:1311–1322.
11. Brosche T, Platt D. Effect of borage oil consumption on fatty acid metabolism, transepidermal water loss and skin parameters in elderly people. Arch Gerontol Geriatr 2000;30:139–150.
12. Chilton-Lopez T, Surette ME, Swan DD, Fonteh AN, Johnson MM, Chilton FH. Metabolism of gammalinolenic acid in human neutrophils. J Immunol 1996;156:2941–2947.
13. Miller CC, Ziboh VA, Wong T, Fletcher MP. Dietary supplementation with oils rich in (n-3) and (n-6) fatty acids influences *in vivo* levels of epidermal lipoxygenase products in guinea pigs. J Nutr 1990;120:36–44.
14. Miller LG. Herbal medicinals: Selected clinical considerations focusing on known or potential drug-herb interactions. Arch Intern Med 1998;158:2200–2211.
15. Laakso P, Voutilainen P. Analysis of triacylglycerols by silver-ion high-performance liquid chromatography-atmospheric pressure chemical ionization mass spectrometry. Lipids 1996;31:1311–1322.

Chapter 51

BURDOCK ROOT
(*Arctium lappa* L.)

HISTORY

The burdock root has been used since antiquity as a treatment for diabetes, venereal diseases, and skin lesions.

BOTANICAL DESCRIPTION

Common Name: Burdock, Gobo, Great Burdock, Greater Burdock

Scientific Name: *Arctium lappa* L.

Botanical Family: Compositae (Asteraceae)

Physical Description: Burdock is a tall biennial plant that grows in open areas as an introduced weed. Round pink-purple flowers appear during the second year. These flowers contain sharp spines (hooklets) that resemble thistle and develop into tenacious burrs. This plant has a large tap root.

Distribution and Ecology: This herb grows throughout the United States as an introduced annual weed. *Arctium* species are native plants in Europe and Asia.

EXPOSURE

Burdock is a popular edible vegetable in Japan called gobo, which contains a variety of fructans.[1] Traditional uses include the treatment of acne, psoriasis, venereal diseases, rheumatism, scurvy, whooping cough, conditions that require diuresis, and colds.[2] The tea brewed from burdock is a coffee and traditional tea substitute. Burdock is a constituent of some scalp and hair preparations as well as moisturizing creams. There are few data on the clinical efficacy of preparations containing burdock.

PRINCIPAL INGREDIENTS

Chemical Composition

Lignans are plant products of low molecular weight formed primarily from oxidative coupling of two *p*-propylphenol moieties (i.e., two C_6C_3 units) at their β-carbon atoms, whereas neolignans are plant products with units coupled in other ways. These lignans have a variety of biological activities. The seeds of *A. lappa* contain several dibenzylbutyrolactone lignans including arctigenin (MW 372, CAS RN: 7770-78-7) and the associated 4-glucoside, arctiin (MW 534, CAS RN: 20362-31-6) along with smaller concentrations of matairesinol, secoisolariciresinol, and neoarctin B.[3] The concentrations of these lignans varies between different plant parts. In a study of the stereochemistry of lignan biosynthesis in *Arctium lappa*, the seeds contained significant amounts of matairesinol and arctigenin compared with secoisolariciresinol, whereas the petioles contained only small amounts of secoisolariciresinol and no detectable concentrations of matairesinol and arctigenin.[4] The enantiomers of the secoisolariciresinol in the seeds and petioles were opposite to each other. Burdock root contains volatile oils, fatty oils, sucrose, resin, tannin, and inulin, but there are no significant amounts of atropine-like alkaloids. Figure 51.1 displays the chemical

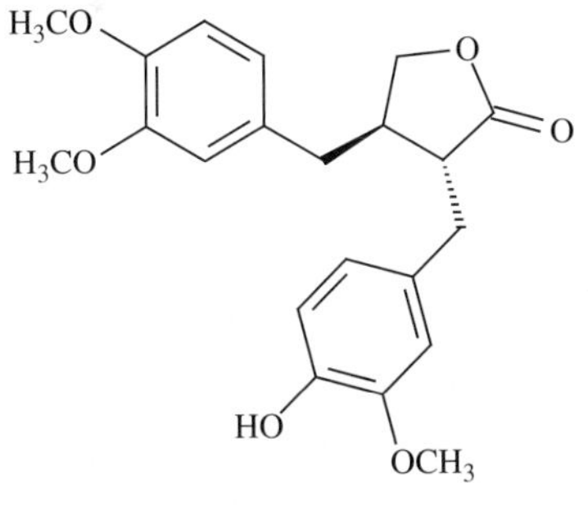

Arctigenin 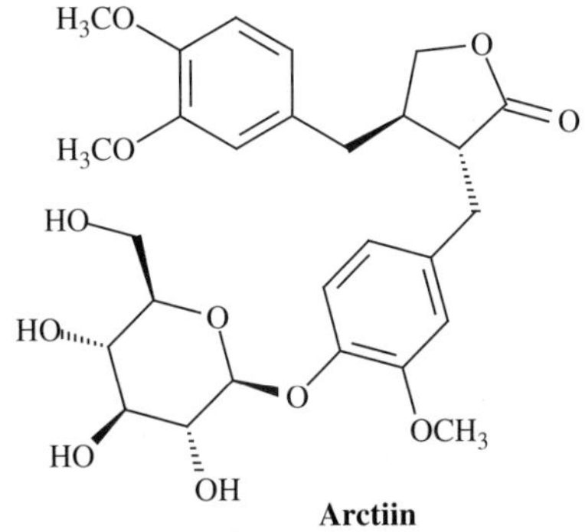 **Arctiin**

FIGURE 51.1. Chemical structures of arctigenin ($C_{21}H_{24}O_6$), and arctiin (arctigenin-4-glucoside, $C_{27}H_{34}O_{11}$).

structures of arctigenin and arctiin, the two major lignans in burdock seeds. Analysis of the commercial preparations of burdock root from case reports associated with toxicity demonstrated the presence of atropine at concentrations of 0.76 mg/g and 30 mg/g, probably as a contaminant.[5,6]

Physiochemical Properties

Lignans are present in almost all plants, and these compounds demonstrate a wide variety of biological responses. In many plants, arctigenin is a precursor in the synthesis of catechols.

DOSE RESPONSE

Use of burdock root involves drinking a decoction three times daily that is made by steeping one teaspoonful of the root into a cup of boiling water and simmering for 10–15 minutes. About 2–4 mL of the tincture is ingested three times daily. A toxic dose of burdock root extract has not been well defined because of a lack of human data.

TOXICOKINETICS

In vitro studies indicate that the lignan, arctiin, undergoes cleavage of the glycosidic bond to form the aglycone, arctigenin.[7] Further demethylation of arctigenin in the rat intestine at the 3″-position produces a secondary metabolite, 2-(3″, 4″-dihydroxybenzyl)-3-(3′, 4′-dimethoxyl)-butyrolactone.

CLINICAL RESPONSE

Several cases of mild anticholinergic poisoning (dizziness, dry mouth, blurred vision, mydriasis, weakness, urinary retention, confusion) developed after consumption of teas brewed from burdock root.[5] Analysis of the tea associated with these cases suggested that atropine contaminated the teas, but the source of this contamination was unclear. A case report associated the application of burdock root as a plaster for joint injuries with contact dermatitis (erythema, pruritus, exudation).[8] Patch tests for burdock were positive in these patients. Another case report associated the development of anaphylaxis (hypotension, diffuse erythema, stridor with the consumption of boiled burdock root.[9]

The burrs from the seed pod cause mechanical injuries. Case reports associate a conjunctival foreign body reaction (conjunctival erythema, tearing) and multiple, fine, linear corneal abrasions with embedded burdock spines in the palpebral conjunctiva.[10] The unbroken shaft from these spines is only 1 mm (~.04 in.) in length; therefore, slit lamp examination is required to identify the presence of these small foreign bodies in the palpebral conjunctiva. Serious secondary infections may develop characterized by chemosis and lid edema.[11] The ingestion of boiled burdock has been associated with the development of anaphylactic shock.[12]

DIAGNOSTIC TESTING

Analytical methods for determining the presence of lignans (arctiin, arctigenin) in commercial samples of burdock (*Arctium lappa*) include polyamide column chromatography with detection by high performance liquid chromatography/electrospray ionization mass spectrometry (HPLC/ESI/MS)[13] and reverse phase HPLC with photodiode array detection (280 nm).[14] The limits of detection and quantitation for the latter method were 1.22 ng and 3.69 ng, respectively.

TREATMENT

Treatment is supportive. Hypersensitivity reactions can develop following exposure to burdock root including anaphylactic reactions; therefore, patients should be evaluated for the presence of angioedema, wheezing, stridor, and hypotension associated with severe allergic reactions. Eye injuries should be evaluated by slit lamp examination for the presence of retained foreign body.

References

1. Kardosova A, Ebringerova A, Alfoldi J, Nosal'ova G, Franova S, Hribalova V. A biologically active fructan from the roots of *Arctium lappa* L., var. *Herkules*. Int J Biol Macromol 2003;33:135–140.
2. Sun WJ, Sha ZF, Gao H. Determination of arctiin and arctigenin in *Fructus arctii* by reverse-phase HPLC. Yao Xue Xue Bao 1992;27:549–551.
3. Wang HY, Yang JS. [Studies on the chemical constituents of *Arctium lappa* L.]. Yao Xue Xue Bao 1993;28:911–917. [Chinese]
4. Suzuki S, Umezawa T, Shimada M. Stereochemical diversity in lignan biosynthesis of *Arctium lappa* L. Biosci Biotechnol Biochem 2002;66:1262–1269.
5. Rhoads PM, Tong TG, Banner W Jr, Andersen R. Anticholinergic poisoning associated with commercial burdock root tea. Clin Toxicol 1984–1985;22:581–584.
6. Bryson PD, Watanabe AS, Rumack BH, et al. Burdock root tea poisoning. Case report involving a commercial preparation. JAMA 1978;239:2157.
7. Nose M, Fujimoto T, Takeda T, Nishibe S, Ogihara Y. Structural transformation of lignan compounds in rat gastrointestinal tract. Planta Med 1992;58:520–522.
8. Rodriquez P, Blanco J, Juste S, Garces M, Perez R, Alonso L, Marcos M. Allergic contact dermatitis due to burdock (*Arctium lappa*). Contact Dermatitis 1995;33:134–135.
9. Sasaki Y, Kimura Y, Tsunoda T, Tagami H. Anaphylaxis due to burdock. Int J Dermatol 2003;42:472–473.
10. Breed FB, Kuwabara T. Burdock ophthalmia. Arch Ophthalmol 1966;75:16–20.
11. Goodwin RA Jr. Burdock ophthalmia. J Maine Med Assoc 1968;59:53–54.
12. Sasaki Y, Kimura Y, Tsunoda T, Tagami H. Anaphylaxis due to burdock. Int Soc Dermatol 2003;42:472–473.
13. Liu S, Chen K, Schliemann W, Strack D. Isolation and identification of arctiin and arctigenin in leaves of burdock (*Arctium lappa* L.) by polyamide column chromatography in combination with HPLC-ESI/MS. Phytochem Anal 2005;16:86–89.
14. Yuan X, Koh HL, Chui WK. A high performance liquid chromatography method for the simultaneous determination of arctiin, chlorogenic acid and glycyrrhizin in a Chinese proprietary medicine. J Pharm Biomed Anal 2005;39:697–704.

Chapter 52

CALAMUS
(*Acorus calamus* L.)

HISTORY

Calamus has been used as a sedative, antipyretic, and antispasmodic since biblical times. This herb was introduced to Eastern Europe from Central Asia as a medicinal plant in the 16th century.[1] Historical uses of calamus include a spice for cooking, a flavoring for tooth powders and tonics, and an additive in pharmaceuticals and foods for taste and aroma. North American Indians used pulverized calamus root as snuff for the treatment of upper respiratory tract infections and as a stimulant for horses.[2] In colonial times, the fragrant oils were constituents of sacred oils and alcoholic beverages;[3] the aromatic leaves were popular room deodorizers and insect repellants. In 1974, the US Food & Drug Administration (FDA) banned the use of *Acorus calamus* as a result of animal studies demonstrating carcinogenic effects following chronic oral administration of this herb. These studies used the Jammu (Indian) variety of calamus oil, which contains about 80% β-asarone, rather than the European variety that contains much lower concentrations of β-asarone. However, the ban included all oils of calamus because of the lack of scientific data demonstrating the safety of the European or Kashmir varieties.[4]

BOTANICAL DESCRIPTION

In addition to *Acorus calamus* (calamus), plant species containing the flavoring agent, β-asarone, include *Acorus europaeum* L. (hazelwort).

Common Name: Calamus (Sweet Flag, Sweet Myrtle, Sweet Root, Sweet Sedge, Rat Root)

Scientific Name: *Acorus calamus* L.

Botanical Family: Acoraceae (formally under Araceae and genus *Arum*)

Physical Description: This perennial herbaceous monocot has branched, aromatic horizontal rhizomes that extend up to or near the soil surface. The leaves are erect with a cross-section in the shape of a rhomboid and veins along the whole length the leaf. These leaves are not differentiated into petiole and blade. The flowers are perfect and regular with six concave perianth segments. There are six stamens and 2–3 carpels, and a superior ovary with 2–3 locules. The mature fruit is hard and dry with a pyramid-shape (obpyramidal) that has an attachment near the narrow end.

Distribution and Ecology: Calamus is a semi-aquatic, aromatic herb, which inhabits moist areas near swamps, ponds, and streams in Asia, Europe, and North America at altitudes of 3,000–6,000 ft (~1,000–2,000 m).

EXPOSURE

Sources

Calamus oil is a flavoring additive to beer, distilled beverages (vermouth, liqueurs, bitters), and some food products (cakes, frozen desserts, yogurts) and drugs. Based on limited data, the maximum daily intake of

Medical Toxicology of Natural Substances, by Donald G. Barceloux, MD

β-asarone from the use of calamus oil in these products is approximately 2 μg/kg body weight.[5]

Medicinal Uses

Traditional

Calamus is a traditional Ayurvedic medicine in India, particularly for psychiatric disorders (depression, insomnia, anxiety, psychosis) and epilepsy.[6] Traditional Chinese medicine uses asarum oil (*Asarum sieboldii* Miquel, *Asarum heterotropoides* F. Schmidt), which contains substantial amounts of α-asarone and smaller amounts of the more toxic β-asarone, include the treatment of fever, pain, infection, and sedation.[7]

Current

Calamus is a common Ayurvedic medicine in India, but the distribution of this herb in the United States is limited. The fragrant oils from calamus rhizomes remain constituents of potpourri, perfume, and sacred oils.

Regulatory Status

The FDA banned the use of calamus as a flavoring agent in beverages, pharmaceuticals, and dental preparations because of the animal studies suggesting that β-asarone was carcinogenic following chronic oral administration. The presence of β-asarone in food is also not authorized in Canada. The European Council (Annex II of Directive 88/388/ECC) limits the β-asarone content in nonalcoholic and alcoholic beverages to 0.1 mg/kg and 1 mg/kg, respectively.[8]

PRINCIPAL INGREDIENTS

Chemical Composition

The essential oils of calamus (CAS RN: 8015-79-0) are mixtures of terpene and/or phenylpropanoic derivatives, and the composition of the essential oil from the rhizomes of *A. calamus* depends on the cytotypes. Karyological analysis of *Calamus* species indicates that these plants have four cytotypes (karyotypes): diploid (2x = 24), triploid (3x = 36), tetraploid (4x = 48), and hexaploid (6x = 72). The concentration of the carcinogen (i.e., β-asarone) depends on the cytotype; separation of the cytotypes of this species requires the chromosomal analysis rather than determination of anatomical characteristics. α- and β-Asarone are major components of triploid and tetraploid plants from calamus plants in the Old World, but the diploid calamus plants from the New World do not contain significant amounts of the carcinogenic compound, β-asarone. European and North American triploid cytotypes contain 3–19% β-asarone, whereas the Indian, Indonesian, and Taiwanese tetraploid cytotypes contain up to 96% β-asarone.[9] As measured by capillary electrophoresis, the β-asarone content of a diploid rhizome of *A. calamus* from the wetlands of the Midwestern United States was 0.2% wet weight compared with 4.4% wet weight for a commercial product derived from a triploid variety from India.[10] Analysis of alcoholic extracts of the triploid *A. calamus* by gas chromatography/mass spectrometry (GC/MS) demonstrated the following constituents: β-asarone (11%), α-selinene (5.02%), E-β-ocimene (3.28%), camphene (2.27%), τ-cadinol (2.00%), camphor (1.54%), and calarene (1.42%). The diploid specimen of *A. calamus* contained no detectable amounts of β-asarone as well as higher concentrations of acorone (26.33%), preiso calamendiol (22.81%), iso-shyobunone (8.62%), and β-sesquiphellandrene (3.28%), when compared with triploid specimens.[9] In an analytical study of Romanian oil of calamus and an alcoholic extract of Romanian *A. calamus*, the concentrations of α-asarone were 5.2–6.7 μg/mL and 2.7–5.7 μg/mL, respectively, and the concentration of β-asarone were 91–98 μg/mL and 88–97 μg/mL, respectively.[11] The essential oil of *Asarum europaeum* L. (European wild ginger) also contains substantial amounts of α-asarone and much smaller amounts of the more toxic β-asarone.[12] As measured by GC/MS, the content of α-asarone and β-asarone in essential oils from the steam distillation of *Asarum europaeum* rhizomes was 460–510 μg/mL and 24–29 μg/mL, respectively.[11]

Physiochemical Properties

The two geometrical isomers of the alkenylbenzene compound, asarone (CAS RN: 494-40-6), are α-asarone [*trans*-1,2,4-trimethoxy-5-(1-propenyl) benzene, CAS RN: 2883-98-9] and β-asarone [*cis*-1,2,4-trimethoxy-5-(1-propenyl) benzene, CAS RN: 5273-86-9]. Figure 52.1 demonstrates the chemical structures of α- and β-asarone. In animal studies, α-asarone and β-asarone potentiate the sedative properties of barbiturates and ethanol with the effects of β-asarone being much stronger than α-asarone.[13,14]

Mechanism of Toxicity

In a 2-year study, rats were fed 0, 500 ppm, 1000 ppm, 2,500 ppm and 5,000 ppm of oil of calamus (Jammu variety).[15] Dose-related toxic effects included moderate to marked growth retardation and degenerative changes in the liver and heart. Malignant tumors developed in the duodenum of rats at all oil of calamus dosages beginning in the 59th week of dosing.

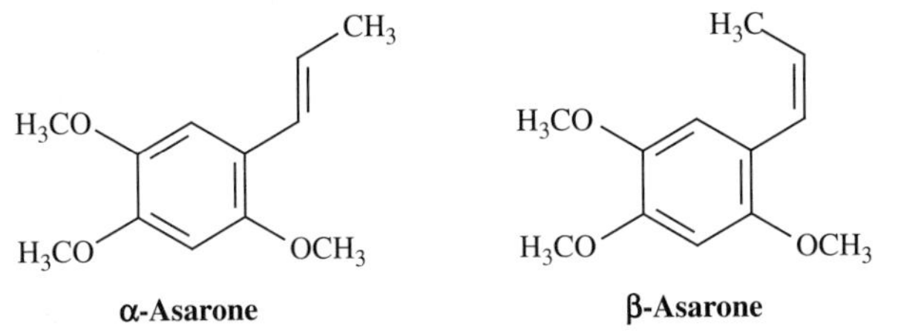

α-Asarone β-Asarone

FIGURE 52.1. Chemical structures of α-asarone and β-asarone.

DOSE RESPONSE

There are inadequate data to determine the toxic dose of calamus oil or calamus root. Large amounts of the root (45–350 mg) are ingested in the belief that hallucinations will occur as a result of the biotransformation of β-asarone to a derivative of the hallucinogen, trimethoxyamphetamine (TMA). However, there are inadequate data to associate specific doses with gastrointestinal or psychedelic effects, particularly because of the variable amount of asarone in calamus oil and in the root.

TOXICOKINETICS

There are few data on the human toxicokinetics of α- or β-asarone. In rodent studies, the intragastric administration of β-asarone results in the rapid absorption and elimination of this compound with a peak concentrations occurring within about 15 minutes.[16] The serum elimination half-life of β-asarone was approximately 1 hour.

CLINICAL RESPONSE

The ingestion of 8 in. (~20 cm) of *Acorus calamus* rhizome with water by a 19-year-old man produced gastrointestinal irritation, manifest by vomiting and diaphoresis.[17] The patient recovered within several hours without sequelae after supportive care. β-Asarone is an animal carcinogen based on limited animal studies, but there are no human data to confirm the carcinogenic potential of this constituent of oil of calamus. This compound has genotoxic and mutagenic properties in *in vitro* assays. In hepatocytes, β-asarone induces unscheduled DNA synthesis,[18] and commercial drugs containing β-asarone are mutagenic in a *Salmonella*/mammalian-microsome assay.[19]

DIAGNOSTIC TESTING

Analytical methods to determine the presence of α- and β-asarone include thin-layer chromatography (TLC) coupled with either UV-Vis absorbance or fluorescence detection,[11] high performance TLC,[20] high performance liquid chromatography (HPLC),[21,22] and gas chromatography/mass spectrometry.[23] TLC methods lack reproducibility and do not easily distinguish isomer content. Micellar electrokinetic capillary chromatography UV Vis absorbance is a rapid and inexpensive method for the determination of β-asarone content of natural products after simple alcohol extraction.[10]

TREATMENT

Treatment of acute exposure to *Acorus calamus* is supportive, primarily involving the replacement of fluid and correcting electrolyte imbalance as needed.

References

1. Raina VK, Srivastava SK, Syamasunder KV. Essential oil composition of *Acorus calamus* L. from the lower region of the Himalayas. Flavour Fragr J 2003;18:18–20.
2. de Smet PA. A multidisciplinary overview of intoxicating snuff rituals in the western hemisphere. J Ethnopharmacol 1985;13:3–49.
3. Dyer RH. Gas-liquid chromatographic determination of beta-asarone in wine: collaborative study. J Assoc Off Anal Chem 1977;60:1041–1043.
4. Goddard JL. Foods and drugs containing calamus, as the root, oil, or extract. Fed Regist 1968;33:6967.
5. Scientific Committee on Food. Opinion of the Scientific Committee on Food on the presence of β-asarone in flavourings and other food ingredients with flavouring properties. Brussels: European Commission, Health & Consumer Protection Directorate-General; 2002.
6. Dandiya PC, Sharma JD. Studies on *Acorus calamus*. Part V. Pharmacological actions of asarone and β-asarone on central nervous system. Indian J Med Res 1962;50:46–60.
7. Shuyan Q, Yingjie M, Yihua W. Inhibitory effect of asarum oil on the central nervous system. J Trad Chinese Med 1984;4:219–224.
8. Commission of the European Communities. Council directive of 22 June 1988 on the approximation of the laws of the Member States relating to flavourings for use in foodstuffs and to source materials for their production (88/388/EEC). Off J Eur Commun 1988;L184:61–67.
9. Bertea CM, Azzolin CM, Bossi S, Doglia G, Maffei ME. Identification of an EcoRI restriction site for a rapid and precise determination of beta-asarone-free *Acorus calamus* cytotypes. Phytochemistry 2005;66:507–514.
10. Hanson KM, Gayton-Ely M, Holland LA, Zehr PS, Soderberg BC. Rapid assessment of β-asarone content of *Acorus calamus* by micellar electrokinetic capillary chromatography. Electrophoresis 2005;26:943–946.
11. Oprean R, Tamas M, Roman L. Comparison of GC-MS and TLC techniques for asarone isomers determination. J Pharm Biomed Anal 1998;18:227–234.

12. Krogh A. The content of trans-aconitic acid in *Asarum europaeum* L. determined by means of a chromatogram spectrophotometer. Acta Chem Scand 1971;25:1495–1496.
13. Panchal GM, Venkatakrishna-Bhatt H, Doctor RB, Vajpayee S. Pharmacology of *Acorus calamus* L. Indian J Exp Biol 1989;27:561–567.
14. Menon MK, Dandiya PC. The mechanism of the tranquilizing action of asarone from *Acorus calamus* Linn. J Pharm Pharmacol 1967;19:170–175.
15. Taylor JM, Jones WI, Hagan EC, Gross MA, Davis DA, Cook EL. Toxicity of oil of calamus (Jammu variety). Toxicol Appl Pharmacol 1967;10:405.
16. Wu HB, Fang YQ. [Pharmacokinetics of beta-asarone in rats]. Yao Xue Xue Bao 2004;39:836–838. [Chinese]
17. Vargas CP, Wolf LR, Gamm SR, Koontz K. Getting to the root (*Acorus calamus*) of the problem. Clin Toxicol 1998;36:259–260.
18. Hasheminejad G, Caldwell J. Genotoxicity of the alkenylbenzenes a- and β-asarone, myristicin and elemicin as determined by the UDS assay in cultured rat hepatocytes. Food Chem Toxicol 1994;32:223–231.
19. Goggelmann W, Schimmer O. Mutagenicity testing of β-asarone and commercial calamus drugs with *Salmonella typhimurium*. Mutat Res 1983;121:191–194.
20. Widmer V, Schibli A, Reich E. Quantitative determination of beta-asarone in calamus by high-performance thin-layer chromatography. J AOAC Int 2005;88:1562–1567.
21. Tsai TH, Chen CM, Chen CF. Disposition of asarone after intravenous administration to rabbits assessed using HPLC. J Pharm Pharmacol 1992;44:620–622.
22. Curro P, Micali G, Lanuzza F. Determination of beta-asarone, safrole, isosafrole and anethole in alcoholic drinks by high-performance liquid chromatography. J Chromatogr 1987;404:273–278.
23. Oprean R, Tamas M, Sandulescu R, Roman L. Essential oils analysis. I. Evaluation of essential oils composition using both GC and MS fingerprints. J Pharm Biomed Anal 1998;18:651–657.

Chapter 53

CAMPHOR
(*Cinnamomum camphora* T. Nees & Eberm.)

HISTORY

In China, camphor has been used for centuries as an antiseptic, antipruritic, cold remedy, and abortifacient. *Linimentum camphoratum* was officially recognized as medicinal product in the first edition of the *US Pharmacopeia*, published in 1820. Up until the 1930s, camphor was used as a circulatory stimulant and analeptic. Reports on camphor toxicity were first reported in the English medical literature in the late 19th century.[1] Gastrointestinal irritation, seizures and altered consciousness were well-recognized complications of acute camphor poisoning in the early 20th century.[2] By 1954, over 130 cases of camphor poisoning had been reported in the medical literature, primarily involving the accidental ingestion of camphorated oil or camphor-containing cough and cold preparations.[3] These reports included approximately 20 deaths attributed to camphor intoxication.[4] In 1982, the US Food & Drug Administration (FDA) declared camphorated oil as "not generally recognized as safe," and the concentration of camphor in US over-the-counter products was limited to <11%. Camphor is no longer listed in the *US Pharmacopeia* or the National Formulary.

BOTANICAL DESCRIPTION

Common Name: Camphor Tree, Camphor Laurel

Scientific Name: *Cinnamomum camphora* T. Nees & Eberm. (*Camphora camphora* Karst., *Laurus camphora* L.)

Botanical Family: Lauraceae (laurel family)

Physical Description: The camphor tree is a broad-leafed evergreen in topical and subtropical regions of the world that grows to 50–100 ft (~15–30 m) in height. Young branches are green with tinges of red. The tops of the leaves are shiny with three distinct veins, whereas the underneath side of the leaves are somewhat pale. Crushing the leaves produces an aromatic odor. The flowers are yellow-white, borne in small inflorescences. The fruits are round, pea-sized berries attached to the branchlets by small, cup-like green cones. The immature fruits first turn red, and then turn black during maturation. The fruits contain a single seed.

Distribution and Ecology: This plant is indigenous to Taiwan, China, and Japan, but camphor is widely cultivated in subtropical regions of the world.

EXPOSURE

Sources

The traditional source of camphor is the wood of *Cinnamomum camphora*, an evergreen tree native to eastern Asia. Camphor from natural sources is the dextrorotatory isomer. Currently, the primary source of camphor is the steam distillation and crystallization of pinene, which is converted to camphene by treatment with acetic acid and nitrobenzene.

Medical Toxicology of Natural Substances, by Donald G. Barceloux, MD

Medicinal Uses

Traditional

Traditional uses of camphor include administration as an antiseptic, antipruritic, rubefacient, mild anesthetic, abortifacient, contraceptive, inhalant for cold preparations, and suppression of lactation. Rarely, this compound has been used as a homicidal or suicidal agent.

Current

The use of camphor is limited to inhalation of vaporized solutions and dermal application of topical compounds. Camphor is used as a rubefacient/liniment (0.1–3%), a constituent of embalming fluid, and a dental antiseptic (65% camphor, 35% parachlorophenol) for root canals. Frequently, dermal products containing camphor also contain other potentially toxic ingredients (phenol, eucalyptus oil, menthol). Topical products containing camphor include Campho-Phenique® antiseptic gel (10.8% camphor and 4.7% phenol; Bayer Consumer Products, Morristown, NJ) and Vicks Vaporub® (5% camphor, 5% turpentine oil, 2.75% levomenthol, 1.5% eucalyptus oil; Procter & Gamble Consumer Products, Cincinnati, OH). Commercial products that may contain camphor include lacquers, varnishes, plasticizers for cellulose esters and ethers, moth repellents, explosives and pyrotechnics, and preservatives for pharmaceuticals and cosmetics.

Regulatory Status

Camphorated oil typically contained 20% camphor weight/volume (w/v) in cottonseed oil. However, in 1982, the FDA banned the sale of over-the-counter products containing this concentration of camphor due to serious central nervous system (CNS) toxicity following accidental ingestion.[5] Camphor now appears in US over-the-counter topical preparations in concentrations ranging up to 11%.[6]

PRINCIPAL INGREDIENTS

Chemical Composition

Camphor (CAS RN: 76-22-2) is a cyclic ketone of the hydroaromatic terpene group. Figure 53.1 displays the chemical structure of this compound ($C_{10}H_{16}O$). Twenty grams of Vicks Vaporub® (i.e., about 5% camphor) contains about 1g of camphor, whereas 10mL of Campho-Phenique® (i.e., about 10.8% camphor) contains about 1g of camphor. Oxygenated monoterpene compounds were the major ingredients of the head-

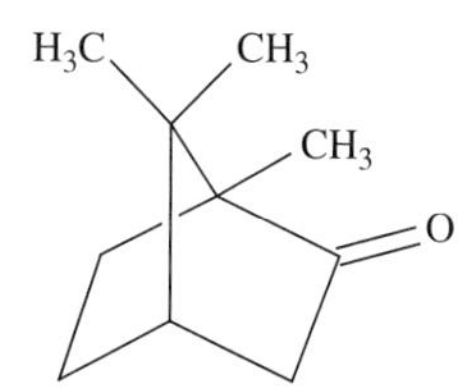

FIGURE 53.1. Chemical structure of camphor.

TABLE 53.1. Physiochemical Properties of Camphor

Physical Property	Value
Melting Point	180°C/356°F
Boiling Point	204°C/399°F
log P (Octanol-Water)	2.38
Water Solubility	1600mg/L (25°C/77°F)
Vapor Pressure	0.65mmHg (25°C/77°F)
Henry's Law Constant	8.10E-05 atm-m^3/mole (25°C/77°F)
Atmospheric OH Rate Constant	9.88E-12cm^3/molecule-second (25°C/77°F)

space constituents of a fresh sample of a Japanese specimen of *C. camphora* with camphor and 1,8 cineole accounting for 70% and 7.5% of the total, respectively.[7] The sample also contained about 8% monoterpenes (camphene, α-pinene, limonene) and approximately 9% sesquiterpene hydrocarbons. Cinnamomin is a ribosome-inactivating protein (RIP) isolated from mature seeds of the camphor tree that belongs to the type II class of RIPs.[8] These toxic proteins consist of an A-chain with RNA *N*-glycosidase activity and a galactose-specific lectin B-chain. The seeds of *Cinnamomum camphora* also contain a type I RIP, camphorin, with a molecular mass of 23kDa.[9] These ribosome-inactivating proteins consist of a single peptide chain with a molecular mass between 25–30kD.

Physiochemical Properties

Camphor is a translucent, crystalline solid at room temperature with a penetrating characteristic odor and a pungent, aromatic taste. Table 53.1 lists some of the physiochemical properties of camphor. Cinnamomin is a type II ribosome-inactivating protein, similar to other heterodimeric proteins (e.g., ricin) in this class. These 61kD-proteins inhibit protein synthesis by acting as a RNA *N*-glycosidase to remove an adenine at A4324 in the sarcin/ricin domain (S/R domain) of the largest RNA in the 28S ribosome.[10] This domain is responsible for the interaction of elongation factors with ribosomes. Although cinnamomin A-chain inhibits protein synthesis similar to ricin, *in vitro* studies in BA/F3β cells indicate that the cytotoxicity of cinnamomin is substantially lower than ricin.[8]

Mechanism of Toxicity

Following ingestion, camphor produces dose-dependent CNS excitation and gastrointestinal (GI) irritation. There are limited data on the toxic compounds in camphor, and the specific toxins (i.e., parent compound or toxic alcohol metabolite) in camphor that cause these CNS and GI effects remains unknown. Although cinnamomin is a type II ribosome-inactivating protein structurally and functionally similar to ricin, abrin, viscumin, ebulin 1, and nigrin b, the role of this protein in producing human toxicity remains undefined. Volunteer studies suggest that short-term exposure to vapors containing camphor stimulate cold receptors in the nose, but these vapors do not alter air flow through the nasal passages.[11]

DOSE RESPONSE

The ingestion of ≥1 g (5 mL of a 20% camphor solution) camphor by a child is a serious intoxication.[12] Camphor toxicity is unlikely to develop following the ingestion of ≤10 mg camphor/kg based a retrospective review of poison center data and the medical literature.[13] In a study from the 1920s of 80 postpartum women receiving camphor injections into engorged breasts, one women developed adverse reactions (nausea, vomiting) attributed to the treatment.[14] Adverse reactions were not reported for the rest of the subjects. The camphor regimen involved 195 mg camphor the first day, followed by three daily injections of 97 mg for a total camphor dose of 486 mg. Although minor GI symptoms may develop following the ingestion of 30–50 mg/kg, serious toxicity including convulsions is uncommon. The ingestion of 6–10 g camphor as a stimulant by two adults resulted in anxiety, agitation, and depersonalization.[15] There were no seizures. Major toxicity occurs following the consumption of camphor doses exceeding 150 mg/kg from preparations containing 20% camphor in cottonseed oil or camphor spirts (10% camphor in ethanol or isopropyl alcohol). The intentional ingestion of 68 mg camphor/kg (44 mL Campho-Phenique® – 10.8% camphor) by a 20-year-old man caused the onset of seizures within 10 minutes followed by vomiting and coma requiring intubation and respiratory support.[16] The patient recovered without sequelae within 24 hours. Another adult survived the ingestion of 42 g camphor with basic supportive care.[2] A survey of a camphor packaging facility indicated that exposure to ambient air concentrations exceeding 2 mg camphor/m^3 (2 ppm) causes mild to moderate eye, nose, and throat irritation.[17]

TOXICOKINETICS

Absorption

The absorption of camphor from the GI tract is relatively rapid based on limited data from volunteer studies and case reports. Seizures can develop within 5–10 minutes of the ingestion of camphor.[18] In fasted rats receiving one gram camphor as a 40% concentration of camphorated oil dissolved in cottonseed oil, the peak camphor concentration occurred about $1\frac{1}{2}$ hours after oral gavage.[19] The presence of food in the stomach delays absorption of camphor from the GI tract. Volunteer studies suggest that the systemic absorption of camphor from the skin is relatively low.[20]

Distribution

Camphor is highly tissue bound with a volume of distribution of approximately 2–4 L/kg.[21] This compound is widely distributed after absorption including diffusion into amniotic fluid, fetus, brain, liver, and kidney.

Biotransformation/Elimination

Based on animal studies, the biotransformation of camphor involves the oxidation of camphor to campherols (2-hydroxycamphor, 3-hydroxycamphor, borneol) followed by conjugation of these alcohols with glucuronic acid.[22] These inactive glucuronide conjugates are excreted in the urine. Analysis of urine following the ingestion of camphor as a stimulant by two adults demonstrated hydroxylated metabolites (3-OH-, 5-OH-, 8-OH-, 9-OH-camphor), which subsequently underwent oxidation to the corresponding ketone and carbonic acid with the latter compounds excreted as a glucuronides.[15] Very small amounts of unchanged camphor appear in the urine after absorption. The plasma elimination half-life of camphor following oral administration of 200 mg camphor to two volunteers was approximately 1.5 and 2.5 hours.[21] The mean elimination serum half-life of camphor was approximately 2 hours in a group of 50 rats receiving one gram camphor as a 40% concentration of camphorated oil dissolved in cottonseed oil.[19] The lungs excrete small amounts of camphor, resulting in the characteristic pungent odor of exhaled breath.

CLINICAL RESPONSE

Acute Effects

The primary effects of camphor intoxication are gastrointestinal irritation, seizures, and CNS depression. As a result of the rapid absorption of camphor, clinical effects begin soon (i.e., about 5–90 minutes) after ingestion.[23,24] The first symptoms of camphor intoxication occurred about 45 minutes after accidental ingestion of 1–1½ tablespoons of camphorated oil by a group of children (age 4–10 years).[25] Initially, local irritation of the GI tract occurred following ingestion, manifest by pain in the mouth and throat, nausea, vomiting, and a nonspecific feeling of warmth. Spontaneous nausea and vomiting occur in most patients ingesting more than a small amount of camphor. Other clinical features of camphor intoxication include headache, lightheadedness, anxiety, agitation, confusion, hallucinations, hyperreflexia, and myoclonus.[15,21,26] Toxicity following dermal application or inhalation is not usually associated with GI symptoms.[27] The odor of camphor may be present in the breath, vomitus, or urine, but the absence of this odor does not exclude camphor toxicity. Serious CNS effects include headache, confusion, vertigo, restlessness, delirium, hallucination, and muscle jerks that progress to tonic-clonic seizures.[28] Life-threatening effects include status epilepticus, coma, apnea, and pulmonary aspiration. Camphor is a local irritant of the eye and upper respiratory tract. Death from acute camphor ingestion is uncommon today,[29,30] but potentially fatal complications involve respiratory failure from status epilepticus or aspiration pneumonia. Symptoms usually resolve within 24 hours without sequelae.

Chronic Effects

Case reports have not associated camphor poisoning during pregnancy with teratogenesis or adverse neonatal effects. Serious camphor intoxication in a 32-year-old woman during the first trimester of pregnancy resulted in several seizures and altered consciousness.[31] Six months later she delivered a normal infant. Camphor is highly lipophilic, and this compound can cross the placenta. Although a neonatal death occurred shortly after the mother demonstrated signs of camphor poisoning, the role of camphor in the death of the child was unclear.[32] A 26-year-old mother with a history of preeclampsia delivered a normal infant one day after she demonstrated the effects of camphor intoxication (agitation, hyperreflexia, tremor, seizure, nausea, lethargy) following the accidental ingestion of 12 g camphorated oil.[33]

Authors have associated mild elevation of serum hepatic aminotransferases with the use of topical compounds containing camphor, but the contribution of camphor to these changes remains unclear.[34] A case report associated the development of acute hepatic encephalopathy (prodromal viral syndrome, diffuse interstitial infiltrates, rapid neurological deterioration, hypoglycemia, hepatomegaly) with the chronic administration of a home remedy containing camphor and whiskey to a 6-month child over 5 months.[35] Although certain features of this illness simulated Reye's syndrome, the liver histology did not demonstrate the pleomorphic, swollen mitochondria and loss of dense intramitochondrial bodies usually seen in fatal cases of Reye's syndrome.

DIAGNOSTIC TESTING

Analytical Methods

Near-infrared (NIR) technology is a sophisticated analytical method for the quality control of camphor-containing products that allows the rapid identification of individual components and the detection of adulteration.[36] Methods for quantitation of camphor in biological samples include reverse phase high performance liquid chromatography (RP/HPLC) with UV detection and gas chromatography with flame ionization detection (GC/FID).[37] The limits of detection and quantitation of camphor in plasma following analysis by GC/FID are about 1 ng/mL and 5 ng/mL, respectively.[38] The between-day coefficient of variation for camphor in this study was about 13.5%.

Biomarkers

There are few data correlating clinical effects with concentrations of camphor in blood. Consequently, the presence of camphor in blood samples confirms exposure, but these biomarkers are not usually available in clinical laboratories. Two adults developed agitation and anxiety after ingesting camphor about 2 hours prior to presentation to the emergency department. Their initial plasma samples contained 0.3 mg camphor/L and 0.4 mg camphor/L.[15] Approximately 3 hours after ingesting camphorated oil, the plasma camphor concentration in blood samples from a 60-year-old woman was 3.1 mg/L.[39] Prior to blood sampling, she had two grand mal seizures, and at the time the blood sample was drawn, she had stable vital signs and was comatose (i.e., responsive only to painful stimuli). Following the intentional ingestion about 44 mL Campho-Phenique® (68 mg camphor/kg) 5 hours prior to sampling, the urine sample from a 20-year-old man contained 1.5 μg/mL camphor as measured by GC.[16] The serum concentration of camphor 7 hours after the ingestion of Vicks Vaporub® by a 3-year-old girl was 1.95 mg/dL. Although the patient developed confusion, vomiting, and suffered a single seizure within 2 hours of the camphor ingestion, the patient had stable vital signs and regular respirations at the time the blood sample was drawn.[40] By 21 hours after ingestion, camphor was not detectable in the serum as measured by gas liquid chromatography.

Abnormalities

The chest x-ray is usually normal unless aspiration occurred during a seizure or altered consciousness. Most cases of camphor poisoning are not associated with laboratory changes with the exception of laboratory abnormalities related to seizures or respiratory depression during serious camphor intoxication.[41,42] Although some case reports associated changes in serum hepatic enzymes with the absorption of camphor, the role of camphor in these changes is unclear and any observed hepatic changes are usually transient.

TREATMENT

Stabilization

Because of the potential for rapid absorption of camphor and the quick onset of seizures, intravenous access should be established immediately in any patient with serious camphor intoxication (>30–50 mg/kg, the presence of alteration of consciousness, severe vomiting, ataxia, or seizures). Vital signs including pulse oximetry should be monitored closely. Sinus tachycardia is a common effect of camphor intoxication, but hypotension and shock are rare complications of camphor poisoning. Seizures should be treated with the usual doses of intravenous benzodiazepines. The use of intravenous phenobarbital is a second-line anticonvulsant based on older experimental studies and limited case reports.[3]

Decontamination

Referral to a health care facility is appropriate for camphor ingestions exceeding 30 mg/kg.[43] Because of the rapid absorption of camphor and the potentially rapid onset of seizures, the use of decontamination measures is not recommended. Although the low molecular weight of camphor suggests good adsorption to activated charcoal, there are limited data on the efficacy of activated charcoal. *In vivo* experiments in rats suggest that camphor is not well adsorbed to activated charcoal at a charcoal to camphor ratio of 2:1.[19] However, this study used a suboptimal dose of charcoal rather than the typical charcoal to toxin ratio of 10:1. Skin contaminated with camphor should be copiously washed with soap and water to prevent local irritant effects, although system toxicity is not expected to occur following dermal exposure to camphor.

Enhancement of Elimination

Camphor is highly lipophilic, and the large volume of distribution suggests that measures to enhance the elimination of camphor are not effective. Very small amounts of camphor appear in the urine unchanged, and diuresis does not significantly increase the renal elimination of camphor. Although a few case reports suggest that lipid hemodialysis[44] or resin hemoperfusion[45] improve the CNS effects (hyperexcitability, increased neuromuscular activity, seizures, coma) associated with serious camphor intoxication, the clinical data documenting the efficacy of these procedures are lacking. In a 54-year-old woman with coma, seizures, and respiratory failure after ingestion of 10% camphor spirits, 4 hours of hemoperfusion with Amberlite XAD4® (Sigma-Aldrich, St. Louis, MO) removed only 35 mg camphor.[21] Her seizures stopped during hemoperfusion, but she remained deeply comatose. Four hours of charcoal hemoperfusion in a comatose 60-year-old woman removed only about 1% of the estimated internal dose of camphor.

Supplemental Care

Although most laboratory values remain normal during camphor poisoning, patients with seizures or clinically significant CNS depression should be evaluated for any acid–base, electrolyte, glucose, or hepatorenal abnormalities. There are no specific antidotes for camphor poisoning. Patients with seizures during camphor intoxication should be monitored for at least 24 hours. Symptoms of camphor toxicity usually develop within 3 hours of ingestion; therefore, asymptomatic patients with a history of camphor ingestion may be discharged after 3–4 hours of observation.[3] In symptomatic patients, recurrence of toxic effects is not expected once the symptoms resolve. These patients may be discharged when symptoms of toxicity resolve.

References

1. Tidcombe FS. Severe symptoms following the administration of a small teaspoonful of camphorated oil. Lancet 1897;2:660.
2. Haft HH. Camphor liniment poisoning. JAMA 1925; 84:1571.
3. Smith AG, Margolis G. Camphor poisoning: Anatomical and pharmacologic study; report of a fatal case; experimental investigation of protective action of barbiturate. Am J Pathol 1954; 30:857–869.

4. Barker F. A case of poisoning by camphorated oil. Br Med J 1910;1:921.
5. New drugs. Camphorated oil drug products for human use. Fed Register 1982;47:11716–11720.
6. Committee on Drugs, American Academy of Pediatrics. Camphor revisited: focus on toxicity. Pediatrics 1994; 94:127–128.
7. Miyazawa M, Hashimoto Y, Taniguchi Y, Kubota K. Headspace constituents of the tree remain of *Cinnamomum camphora*. Nat Prod Lett 2001;15:63–69.
8. Xu H, Liu W-Y. Cinnamonin—a versatile type II ribosome-inactivating protein. Acta Biochimica Biophysica Sinica 2004;36:169–176.
9. Ling J, Liu W-Y, Wang TP. Simultaneous existence of two types of ribosome-inactivating proteins in the seeds of *Cinnamomum camphora*—characterization of the enzymatic activities of these cytotoxic proteins. Biochim Biophy Acta 1995;1252:15–22.
10. He W-J, Liu W-Y. Cinnamomin: a multifunctional type II ribosome-inactivating protein. Int J Biochem Cell Biol 2003;35:1021–1027.
11. Burrow A, Eccles R, Jones AS. The effects of camphor, eucalyptus and menthol vapour on nasal resistance to airflow and nasal sensation. Acta Otolaryngol 1983; 96:157–161.
12. Love JN, Sammon M, Smereck J. Are one or two dangerous? Camphor exposure in toddlers. J Emerg Med 2004;27:49–54.
13. Geller RJ, Spyker DA, Garrettson LK, Rogol AD. Camphor toxicity: Development of a triage strategy. Vet Hum Toxicol 1984;26(suppl 2):8–10.
14. Philpott NW. Intramuscular injections of camphor in the treatment of engorgement of the breasts. CMAJ 1929;20:494–495.
15. Koppel C, Tenczer J, Schirop TH, Ibe K. Camphor poisoning abuse of camphor as a stimulant. Arch Toxicol 1982;51:101–106.
16. Lahoud CA, March JA, Proctor DD. Campho-Phenique ingestion: an intentional overdose. Southern Med J 1997;90:647–648.
17. Gronka PA, Bobkoskie RL, Tomchick GJ, Rakow AB. Camphor exposures in a packaging plant. Am Ind Hyg Assoc J 1969;30:276–279.
18. Gibson DE, Moore GP, Pfaff JA. Camphor ingestion. Am J Emerg Med 1989;7:41–43.
19. Dean BS, Burdick JD, Geotz CM, Bricker JD, Krenzelok EP. *In vivo* evaluation of the adsorptive capacity of activated charcoal for camphor. Vet Hum Toxicol 1992;34:297–300.
20. Matin D, Valdez J, Boren J, Mayersohn M. Dermal absorption of camphor, menthol, and methyl salicylate in humans. J Clin Pharmacol 2004;44:1151–1157.
21. Koppel C, Martens F, Schirop TH, Ibe K. Hemoperfusion in acute camphor poisoning. Intensive Care Med 1988; 14:431–433.
22. Robertson JS, Hussain M. Metabolism of camphors and related compounds. Biochem J 1969;113:57–65.
23. Reid FM. Accidental camphor ingestion. JACEP 1979; 8:339–340.
24. Aronow R, Spigiel RW. Implications of camphor poisoning: therapeutic and administrative. Drug Intell Clin Pharm: 1976;10:631–634.
25. Benz RW. Camphorated oil poisoning with no mortality. Report of twenty cases. JAMA 1919;72:1217–1218.
26. Theis JG, Koren G. Camphorated oil: still endangering the lives of Canadian children. CMAJ 1995;152:1821–1824.
27. Skoglund RR, Ware LL Jr, Schanberger JE. Prolonged seizures due to contact and inhalation exposure to camphor. Clin Pediatr 1977;16:901–902.
28. Ginn HE, Anderson KE, Mercier RK, Stevens TW, Matter BJ. Camphor intoxication treated by lipid dialysis. JAMA 1968;203:230–231.
29. Davies R. A fatal case of camphor poisoning. Br Med J 1887;1:726.
30. Homan A. Fatal case of camphor poisoning. Aust Med J 1888;10:252–259.
31. Blackmon WP, Curry HB. Camphor poisoning report of case occurring during pregnancy. J Florida Med Assoc 1957;43:999–1000.
32. Riggs J, Hamilton R, Homel S, McCabe J. Camphorated oil intoxication in pregnancy; report of a case. Obstet Gynecol 1965;25:255–258.
33. Weiss J, Catalano P. Camphorated oil intoxication during pregnancy. Pediatr 1973;52:713–714.
34. Aliye UC, Bishop WP, Sanders KD. Camphor hepatotoxicity. Southern Med J 2000;93:596–598.
35. Jimenez JF, Brown AL, Arnold WC, Byrne WJ. Chronic camphor ingestion mimicking Reye's syndrome. Gastroenterology 1983;84:394–398.
36. Juliani HR, Kapteyn J, Jones D, Koroch AR, Wang M, Charles D, Simon JE. Application of near-infrared spectroscopy in quality control and determination of adulteration of African essential oils. Phytochem Anal 2006;17: 121–128.
37. Gallicano KD, Park HC, Young LM. A sensitive liquid chromatographic procedure for the analysis of camphor in equine urine and plasma. J Anal Toxicol 1985;9:24–30.
38. Valdez JS, Martin DK, Mayersohn M. Sensitive and selective gas chromatographic methods for the quantitation of camphor, menthol and methyl salicylate from human plasma. J Chromatogr B Biomed Sci Appl 1999;729:163–171.
39. Mascie-Taylor BH, Widdop B, Davison AM. Camphor intoxication treated by charcoal haemoperfusion. Postgrad Med J 1981;57:725–726.
40. Phelan WJ III. Camphor poisoning. Over-the-counter dangers. Pediatrics 1976;57:428–431.
41. Emery DP, Corban JG. Camphor toxicity. J Paediatr Child Health 1999;35:105–106.

42. Rabl W, Katzgraber F, Steinlechner M. Camphor ingestion for abortion (case report). Forensic Sci Int 1997;89:137–140.
43. Manoguerra AS, Erdman AR, Wax PM, Nelson LS, Caravati M, Cobaugh DJ, et al. Camphor poisoning: an evidence-based practice guideline for out-of-hospital management. Clin Toxicol 2006;44:357–370.
44. Auntman E, Jacob G, Volpe B, Finkel S, Savona M. Camphor overdosage. NY State J Med 1978;78:896–897.
45. Kopelman R, Miller S, Kelly R, Sunshine I. Camphor intoxication treated by resin hemoperfusion. JAMA 1979;241:727–728.

Chapter 54

CASCARA
(*Frangula purshiana* Cooper)

HISTORY

The dried bark from *Frangula purshiana* Cooper (cascara sagrada) has a long history of use by Native Americans in the Pacific Northwest as a laxative. Reputedly, Spanish priests gave the name, Cascara sagrada, to the herbal preparation. The eclectic physician, J. H. Bundy, from Colusa, California contacted Parke, Davis & Company about the medicinal properties of the bark from Cascara sagrada. In 1877, this company introduced the bark as a cathartic in their publication, *New Preparations*. The *US Pharmacopeia* officially listed Cascara sagrada as an herbal remedy for constipation in 1890.

BOTANICAL DESCRIPTION

Common Name: Cascara sagrada

Scientific Name: *Frangula purshiana* Cooper (*Rhamnus purshiana* DC.)

Botanical Family: Rhamnaceae (buckthorn)

Physical Description: Cascara sagrada is an erect shrub with multiple stems that reach a height up to about 35–40 ft (~9–12 m). Young branches are slender and covered with reddish brown hairs; older branches are smooth and grayish. The small flowers have five greenish-white petals arranged in loose clusters. Individual plants contain both unisexual and bisexual flowers. The fruits are small, round black drupe with a yellowish inner pulp and tree glossy black seeds.

Distribution and Ecology: *Frangula purshiana* (cascara sagrada) is a native plant of the US Pacific Northwest.

EXPOSURE

Sources

The dried bark of this tree is a source for several anthraquinone laxatives including cascara. Other related plant species historically used as anthraquinone laxatives include *Frangula alnus* P. Mill. (*Rhamnus frangula* L.) and *Rhamnus cathartica* L. (common buckthorn). *Frangula alnus* is a shrubby tree native to Europe, western Asia, and North Africa that also occurs in the eastern United States as a non-native plant species. *Rhamnus cathartica* is indigenous to Europe and Asia, but this species also occurs in the northern United States as an introduced species. In contrast to the bark of *Frangula purshiana* and *Frangula alnus* (glossy buckthorn), the fruit of *Rhamnus cathartica* is the source of the herbal laxative. In addition to the dried bark of cascara sagrada, other preparations derived from bark include casanthranol (purified mixture of Anthranol glycosides), cascara sagrada extract (boiled extract of macerated bark powder), cascara sagrada fluid extract (boiled bark powder preserved in 20% ethanol), and aromatic cascara fluid extract (aqueous extract of dried cascara sagrada bark powder and magnesium oxide).

Medicinal Uses

Traditional

Popular stimulant herbal laxatives include senna (*Cassia senna* L., *C. angustifolia* Vahl.) and cascara. Dried exudate from aloe vera (*Aloe barbadensis*) also contains anthraquinone compounds, which induce diarrhea. Less

Medical Toxicology of Natural Substances, by Donald G. Barceloux, MD

common herbal laxatives include Chinese rhubarb (*Rheum officinale* Baill.), frangula (*Frangula alnus*), and yellow dock (*Rumex crispus* L.).

Current

Oral laxatives include bulk-forming agents, stimulants, lubricants, and osmotic agents. Cascara contains anthracene glycosides, and this herbal laxative belongs to the stimulant class of laxatives along with senna. Diphenolic compounds (bisacodyl, phenolphthalein) are also stimulant laxatives. Osmotic laxatives include magnesium salts, sodium salts, and lactulose. Herbal laxatives (cascara, senna, frangula, aloe, rhubarb) are alternatives to other stimulant laxatives, such as synthetic diphenylmethane compounds (bisacodyl, phenolphthalein), docusate, dehydrocholic acid, and castor oil.

Regulatory Status

Cascara is an over-the-counter drug in the United States and Canada, and this substance is on the general sales list in the United Kingdom. Germany and France do not regulate the use of cascara.

PRINCIPAL INGREDIENTS

Chemical Composition

The basic structure for all anthranoid laxatives is an anthracene ring as demonstrated in Figure 54.1.[1] Anthranoid compounds refer to compounds with substitutions at C10 in the basic anthracene ring in Figure 54.1 as follows: anthrones, C10-H_2; anthraquinones, C10–O; dianthrones, C10—additional anthracene ring. Consequently, anthranoid compounds contain three aromatic rings joined in a linear arrangement that are termed anthraquinone compounds following the addition of two ketone groups.

Figure 54.2 demonstrates the general chemical structures of the three major groups (anthrones, anthraquinones, dianthrones) of anthranoid laxatives. The main

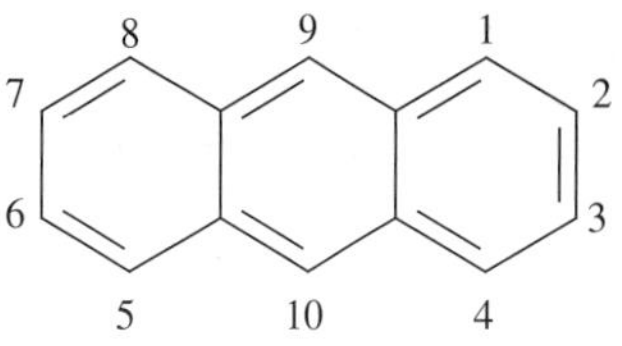

FIGURE 54.1. Basic anthracene ring and numbering system.

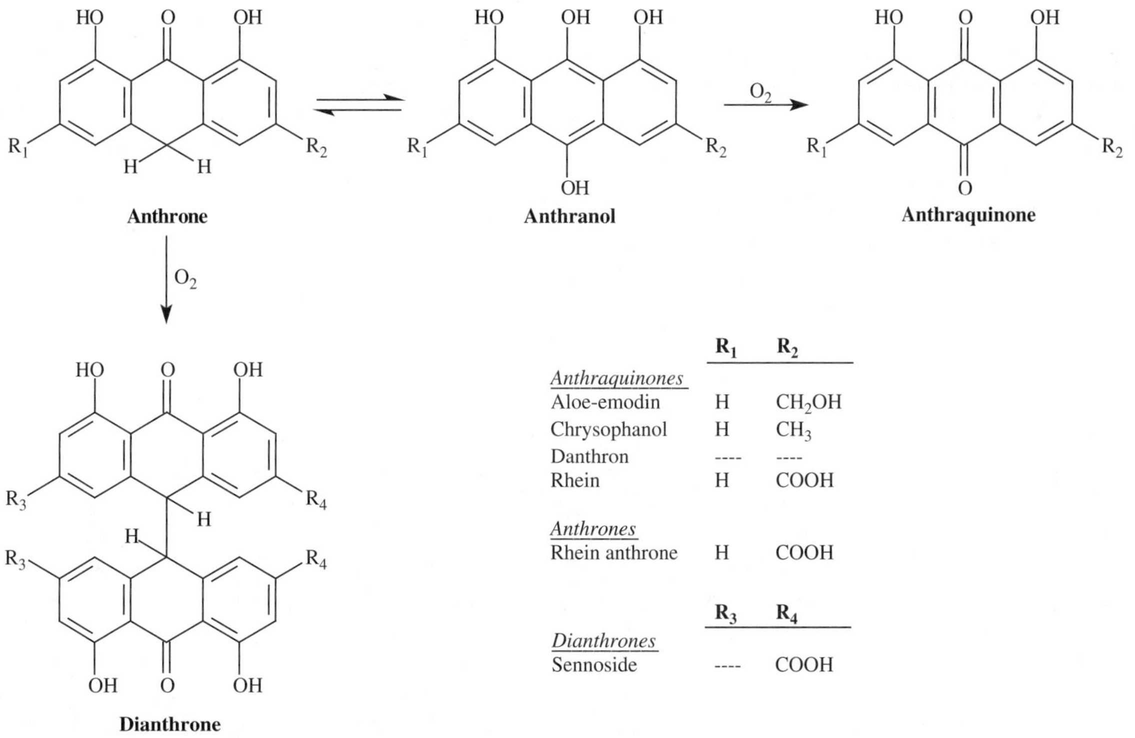

Anthrone

Anthranol

Anthraquinone

Dianthrone

	R_1	R_2
Anthraquinones		
Aloe-emodin	H	CH_2OH
Chrysophanol	H	CH_3
Danthron	----	----
Rhein	H	COOH
Anthrones		
Rhein anthrone	H	COOH
	R_3	R_4
Dianthrones		
Sennoside	----	COOH

FIGURE 54.2. General chemical structures of the three major groups (anthrones, anthraquinones, dianthrones) of anthranoid laxatives. Adapted from van Gorkom BA, et al. Review article: anthranoid laxatives and their potential carcinogenic effects. Aliment Pharmacol Ther 1999;13:445.

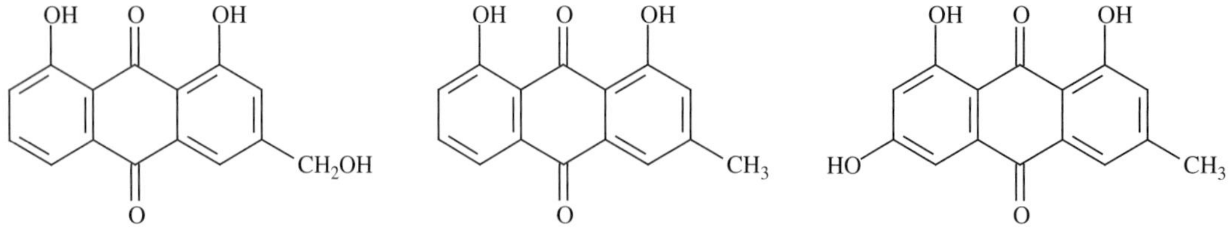

Aloe-emodin Chrysophanol Emodin

FIGURE 54.3. Chemical structures of three common anthraquinone compounds in cascara.

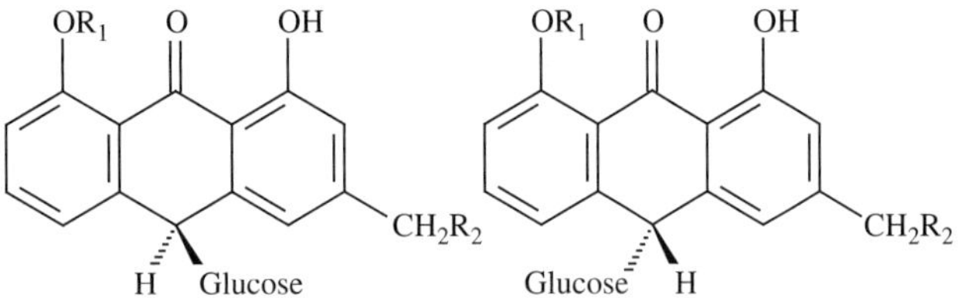

	R_1	R_2
Cascaroside A	Glucose	OH
Cascaroside C	Glucose	H
Barbaloin	H	OH
Chrysaloin	H	H

FIGURE 54.4. Chemical structures of cascaroside, barbaloin, and chrysaloin.

anthraquinone compounds in the bark extract from cascara sagrada are aloe-emodin, chrysophanol, and rhein, and the chemical structures of these anthraquinone compounds are displayed in Figure 54.3. Each of the major groups of anthranoid laxatives can be converted to another anthranoid compound by oxidative or reductive reactions as demonstrated in Figure 54.2. In plants, these anthranoid moieties (aglycon) are usually linked with a sugar (e.g., glucose, rhamnose) to a carbon at C10 (C-glycoside) or to a hydroxyl group at C1 or C8 (*O*-glycoside). The fresh bark of cascara sagrada contains free anthrones, anthrone-*O*-glycosides, anthrone-C-glycosides, aloe-emodin-*O*-glycoside, and dianthrones (i.e., two anthrones linked together).[2] During the drying process, the free anthrones and their corresponding *O*-glycosides are oxidized to dianthrone- and anthraquinone-*O*-glycosides to diminish the emetic properties of the bark extract. Bacterial β-glucosidase and reductase activity converts the inactive anthraquinone-*O*-glucosides into the pharmacologically active aglycon anthrones.

Dried bark from cascara sagrada contains about 8–10% hydroxyanthraquinone glycosides, which are hydroxyanthracene derivatives with hydroxyl groups at the C1 and C8 positions and sugar groups at the hydroxyl moiety (*O*-glycosides) or at the C10 position (C-glycosides).[3] Aloin compounds account for about 1–2% of the dried bark extract of cascara. A majority of the hydroxyanthracene compounds in dried bark extracts of cascara are cascarosides A–D (anthrone C-glycosides) as displayed in Figure 54.4. Cascarosides are primary glycosides of barbaloin. Cascaroside A is the C-10 isomer of 8-*O*-(β-D-glucopyranosyl)-(+) barbaloin, whereas cascaroside B is the corresponding diastereoisomer of 8-*O*-(β-D-glucopyranosyl)-(−) barbaloin.[4] Cascarosides C and D are diastereoisomers of primary C-glycosides of chrysaloin. Other ingredients of the dried bark extracts of cascara include barbaloin, chrysaloin aloe-emodin, deoxybarbaloin, chrysophanol, emodins, and oxanthrones as well as fatty acids (linoleic, myristic, syringic) and the volatile oil, rhamnol. Both barbaloin and chrysaloin are present in the dried bark extract of cascara as diastereoisomers. Figure 54.4 displays the chemical structure of the hydroxyanthracene compounds cascaroside A–D, barbaloin, and chrysaloin. During the extraction process, bianthraquinone glycosides are formed through the hydrolysis of bicascarosides in the extract.[5] These anthracene glycosides do not possess laxative activity.

Physiochemical Properties

Based on animal studies, cascarosides are more potent cathartics than the corresponding barbaloin.[6] Most primary glycosides in the bark extract of cascara are stable at pH 2; therefore, these compounds pass into the intestines with minimal degradation. Cascarosides have little taste in contrast to aloin compounds (e.g., barbaloin). The sugar moiety in anthraquinone glycosides confers hydrophilic properties to the molecule that limits absorption by intestinal epithelial cells.[7]

Mechanism of Toxicity

The exact mechanism of action by which anthranoid laxatives produce diarrhea is unclear. These compounds increase intestinal motility and intraluminal secretions by multiple mechanisms including stimulation of prostaglandin E2 formation,[8] platelet activating factor, and nitric oxide synthesis.[9] Rodent studies suggest the sennoside (dianthrone *O*-glycoside)-induced secretion of intestinal fluid is mediated, at least in part, by the release of serotonin.[10] Following oral administration of cascara, minimal absorption of anthranoid compounds occurs in the stomach and small intestine because of their large molecular size and hydrophilic properties. In the large intestine, bacteria metabolize these compounds to active aglycons, which directly affect intestinal absorption, secretion, and motility.[1] Chronic abuse of these compounds causes damage to the intestinal epithelium manifest by apoptotic bodies in the pigmented mucosa of the colon. These changes are characteristic of pseudomelanosis coli.

DOSE RESPONSE

Table 54.1 lists the standard dosage of cascara based on standard over-the-counter formulations defined by the *US Pharmacopeia*. Toxicity usually results from the chronic abuse of cascara, but the toxic dose is not well defined.

TOXICOKINETICS

Anthrone C-glycosides in cascara are prodrugs, which are not absorbed in the gastrointestinal tract until bacteria in the colon split off the sugar moiety.[11] Figure 54.5 displays the biotransformation of anthrone C-glycosides by bacteria in the gastrointestinal tract. In contrast to the dianthrone *O*-glycosides (e.g., sennosides A/B in senna), the anthrone C-glycosides in cascara are not easily metabolized by bacterial flora, and there is substantial variation in the response of different animal species to cascara.[6] For example, mice can break the *O*-glycoside link in cascarosides, but not the C-glycoside link. The released anthranoid aglycons (free anthrones) diffuse to the intestinal wall where they alter intestinal secretions and motility. The onset of action of the laxative effects is about 6–8 hours.[12] Animal studies suggest that some hydroxylation of these aglycons occurs after absorption in addition to the conjugation of the anthrone compounds to glucuronides and sulfates. At least some of these compounds are excreted in the bile and undergo enterohepatic recirculation, leading to prolonged action.

The maternal use of cascara is compatible with breastfeeding.[13] Stimulant laxatives may increase intestinal transit, theoretically reducing drug absorption. However, there are few data on the clinical significance of this potential drug interaction.

CLINICAL RESPONSE

Acute Effects

A single case report associated the use of cascara with the development of cholestatic hepatitis complicated by fibrosis, portal hypertension, and ascites.[14] Right upper quadrant pain, abdominal bloating, nausea, anorexia, and jaundice began 3 days after initiation of cascara (one 425 mg capsule of aged cascara sagrada bark with reported 5% cascaroside potency, 3 times daily). The case report did not include an analysis of the cascara capsules.

Chronic Effects

Chronic abuse of cascara is associated with melanosis coli, dehydration, and electrolyte imbalance (hypokalemia, mild systemic acidosis).[15] A case report associated the development of severe weakness and hypokalemia (serum potassium = 2.3 mEq/L) with the daily ingestion of 6 cascara tablets daily for many months.[16] Pseudomelanosis coli is a reliable marker of chronic laxative abuse (i.e., >9–12 months) that is specific for anthranoid drugs (aloe, cascara, frangula, and rheum); this marker occurs in about three-fourths of patients abusing anthraquinone laxatives.[17,18] The typical gross appearance of the mucosa of the colon is a reticular pattern of dark striations resembling alligator skin. The presence of dark brown-black granular pigment in macrophages residing in the lamina propria of the colon causes the characteristic histological appearance of pseudomelanosis coli.[19] Rarely, the abnormal pigmentation occurs in the gastric mucosa.[20] This pigment probably is lipofuscin rather than melanin; hence, the more appropriate name is pseudomelanosis coli rather than melanosis coli. No symptoms are typically associated with pseudomelanosis coli, except the adverse effects of chronic laxative abuse. The abnormal pigmentation associated

TABLE 54.1. Daily Dose of Cascara Based on Standard Preparations

Age	Daily Dose	Total Hydroxyanthracene Compounds	Daily Dose of Hydroxyanthracenes
Children (2–12 years)	150–500 mg	7–10%	10.5–50 mg
Adults	300–1000 mg	7–10%	21–100 mg

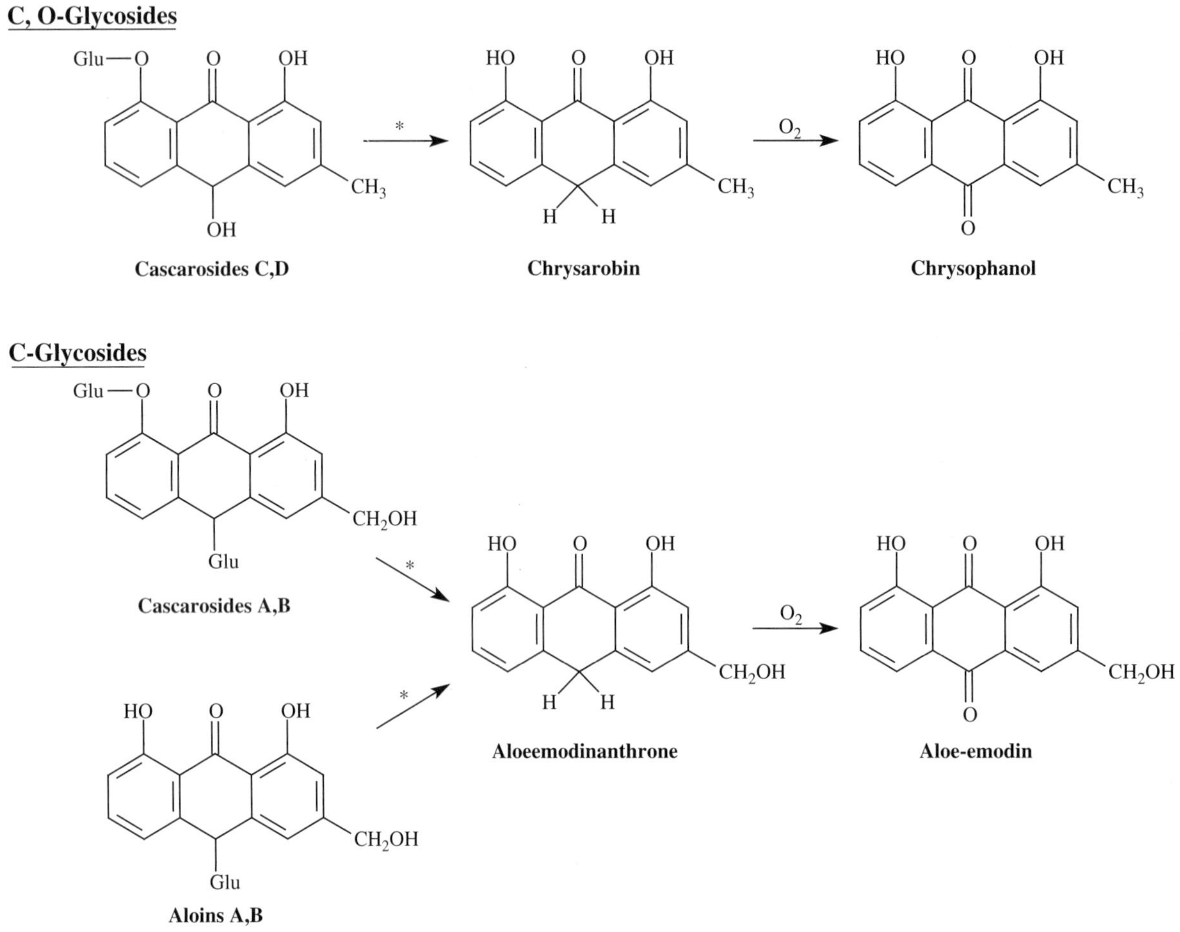

C, O-Glycosides

Cascarosides C,D → * → Chrysarobin → O_2 → Chrysophanol

C-Glycosides

Cascarosides A,B → * → Aloeemodinanthrone → O_2 → Aloe-emodin

Aloins A,B → * → Aloeemodinanthrone

FIGURE 54.5. Biotransformation of cascara anthrone C-glycosides in the gastrointestinal tract. *Bacterial degradation.

with pseudomelanosis coli usually resolves within 6–11 months after cessation of anthranoid laxative use.[21]

Analysis of some,[22] but not all,[23] epidemiological studies suggests that the presence of pseudomelanosis coli (and laxative abuse) increases the risk of colorectal cancer severalfold (i.e., about 1.5–3). Although some retrospective case-control studies indicate an increased risk of colorectal cancer in patients with constipation and commercial laxative use, other prospective case-control studies have not confirmed the suspected increased cancer risk.[24] A meta-analysis of 14 case-control studies demonstrated statistically significant risks for colorectal cancer associated with both constipation and the use of cathartics.[25] The pooled odds ratios were 1.48 (95% CI: 1.32–1.66) and 1.46 (95% CI: 1.33–1.61), respectively. These low odds ratios in comparison to various dietary components (fat, meat, alcohol, low-vegetable or low-residue diets) suggests that this increased risk reflects confounding factors associated with dietary habits rather than true causation. There is inadequate evidence for the carcinogenicity of cascara or other anthraquinone laxatives (with the exception of 1-hydroxyanthraquinone) in animals.[26] A 13-week rodent study using cascara (140 mg/kg/d and 420 mg/kg/d) on azoxymethane-induced aberrant crypt foci and tumors did not detect promoting or initiating activity for the laxative (140 mg) and diarrheal (420 mg) dose of cascara.[27] Similarly, there is no clear evidence that the anthranoid compounds in senna extracts (sennosides, rhein, aloe-emodin) are genotoxic or carcinogenic.[28]

DIAGNOSTIC TESTING

Colorimetric analysis of hydroxyanthraquinone compounds (rhein, aloe-emodin, emodin) after hydrolysis and oxidation determines these compounds as a whole group rather than individual hydroxyanthraquinone compounds. High performance liquid chromatography can separate individual hydroxyanthraquinone compounds.[29] A rapid, simple method for the detection of anthraquinone compounds in cascara along with other diphenolic laxatives (bisacodyl, phenolphthalein) is solid-phase extraction of urine samples coupled with high performance thin layer chromatography.[30] The detection limits are approximately 10–20 mg/L. Capillary gas chromatography/mass spectrometry allows the detection of anthraquinone metabolites in human urine samples along with other laxatives (e.g., bisacodyl, picosulfate, phenolphthalein).[31] The limit of detection in full scan mode ranges from 10–25 ng/mL with a coefficient of variation <15%.

TREATMENT

Treatment is supportive. Patients with chronic laxative abuse should be evaluated for evidence of fluid and electrolyte imbalance. Severe hypokalemia may develop following chronic abuse of cascara.

References

1. van Gorkom BA, de Vries EG, Karrenbeld A, Kleibeuker JH. Review article: anthranoid laxatives and their potential carcinogenic effects. Aliment Pharmacol Ther 1999; 13:443–452.
2. van Os FH. Anthraquinone derivatives in vegetable laxatives. Pharmacology 1976;14(Suppl 1):7–17.
3. Joint Committee of the Pharmaceutical Society and Society for Analytical Chemistry on Methods for the Evaluation of Drugs. Recommended methods for the evaluation of drugs. Analyst 1968;93:749–755.
4. Fairbairn JW, Evans FJ, Phillipson JD. Cascarosides A and B. J Pharm Sci 1977;66:1300–1303.
5. de Witte P, Cuveele J, Lemli J. Bicascarosides in fluid extracts of cascara. Planta Med 1991;57:440–443.
6. Fairbairn JW. Biological assay and its relation to chemical structure. Pharmacol 1976;14(Suppl 1):48–61.
7. Sendelbach LE. A review of the toxicity and carcinogenicity of anthraquinone derivatives. Toxicology 1989;57: 227–240.
8. Beubler E, Kollar G. Prostaglandin-mediated action of sennosides. Pharmacology 1988;36(Suppl 1):85–91.
9. Izzo AA, Gaginella TS, Mascolo N, Capasso F. Recent findings on the mode of action of laxatives: the role of platelet activating factor and nitric oxide. Trends Pharmacol Sci 1998;19:403–405.
10. Beubler E, Schirgi-Degen A. Serotonin antagonists inhibit sennoside-induced fluid secretion and diarrhea. Pharmacology 1993;47(Suppl 1):64–69.
11. de Witte P, Lemli L. The metabolism of anthranoid laxatives. Hepatogastroenterology 1990;37:601–605.
12. Rosengren JE, Aberg T. Cleansing of the colon without enemas. Radiologe 1975;15:421–426.
13. American Academy of Pediatrics Committee on Drugs. Transfer of drugs and other chemicals into human milk. Pediatrics 2001;108:776–789.
14. Nadir A, Reddy D, van Thiel DH. Cascara sagrada-induced intrahepatic cholestasis causing portal hypertension: case report and review of herbal hepatotoxicity. Am J Gastroenterol 2000;95:3634–3637.
15. Houghton BJ, Pears MA. Chronic potassium depletion due to purgation with cascara. Br Med J 1958; 14(5083): 1328–1330.
16. Aitchison JD. Hypokalaemia following chronic diarrhoea from overuse of cascara and a deficient diet. Lancet 1958;2:75–76.
17. Badiali D, Marcheggiano A, Pallone F, Paoluzi P, Bausano G, Iannoni C, et al. Melanosis of the rectum in patients with chronic constipation. Dis Colon Rectum 1985; 28:241–245.
18. Siegers CP. Anthranoid laxatives and colorectal cancer. Trends Pharmacol Sci 1992;13:229–231.
19. Wittoesch JH, Jackman RJ, Mcdonald JR. Melanosis coli: general review and a study of 887 cases. Dis Colon Rectum 1958;1:172–180.
20. Mitty RD, Wolfe GR, Cosman M. Initial description of gastric melanosis in a laxative-abusing patient. Am J Gastroenterol 1997;92:707–708.
21. Speare GS. Melanosis coli: experimental observations on its production and elimination in 23 cases. Am J Surg 1951;82:631–637.
22. Siegers CP, von Hertzberg-Lottin E, Otte M, Schneider B. Anthranoid laxative abuse—a risk for colorectal cancer? Gut 1993;34:1099–1101.
23. Nusko G, Schneider B, Ernst H, Wittekind C, Hahn EG. Melanosis coli—a harmless pigmentation or a precancerous condition? Z Gastroenterol 1997;35:313–318.
24. Nusko G, Schneider B, Schneider I, Wittekind C, Hahn EG. Anthranoid laxative use is not a risk factor for colorectal neoplasia: results of a prospective case control study. Gut 2000;46:651–655.
25. Sonnenberg A, Muller AD. Constipation and cathartics as risk factors of colorectal cancer: a meta-analysis. Pharmacology 1993;47(Suppl 1):224–233.
26. International Agency for Research on Cancer (IARC). Some traditional herbal medicines, some mycotoxins, naphthalene and styrene. IARC Monogr Eval Carcinog Risks Hum 2002;82:129–152.
27. Borrelli F, Mereto E, Capasso F, Orsi P, Sini D, Izzo AA, et al. Effect of bisacodyl and cascara on growth of aberrant crypt foci and malignant tumors in the rat colon. Life Sci 2001;69:1871–1877.
28. Brusick D, Mengs U. Assessment of the genotoxic risk from laxative senna products. Environ Mol Mutag 1997;29:1–9.
29. Grimminger W, Witthohn K. Analytics of senna drugs with regard to the toxicological discussion of anthranoids. Pharmacol 1993;47(Suppl 1):98–109.

30. Perkins SL, Livesey JF. A rapid high-performance thin-layer chromatographic urine screen for laxative abuse. Clin Biochem 1993;26:179–181.

31. Beyer J, Peters FT, Maurer HH. Screening procedure for detection of stimulant laxatives and/or their metabolites in human urine using gas chromatography-mass spectrometry after enzymatic cleavage of conjugates and extractive methylation. Ther Drug Monit 2005;27: 151–157.

Chapter 55

CAT'S CLAW [*Uncaria tomentosa* (Willd. ex Schult.) DC.]

HISTORY

Several Peruvian aborigine populations (e.g., Asháninka natives) have used extracts of cat's claw for centuries as a traditional treatment for infections and inflammatory reactions. The indigenous peoples in the Amazon rainforest and surrounding tropical areas (Colombia, Costa Rica, Ecuador, Guatemala, Guyana, Panama, Suriname, Venezuela) use two species of *Uncaria* [*U. tomentosa* (Willd. ex Schult.) DC., *U. guianensis* (Aubl.) J.F. Gmel.)] interchangeably for traditional medicinal applications.

BOTANICAL DESCRIPTION

The genus *Uncaria* contains at least 34 species widely distributed in tropical areas including Southeast Asia. The term cat's claw applies to two native Central and South American species, *Uncaria tomentosa* (yellowish-white flowers) and *Uncaria guianensis* (reddish-orange flowers). This genus is characterized by the appearance of short, hook-shaped peduncles on the side shoots that connect the flower calyxes to the main stems.

Common Name: Cat's Claw, Uña de Gato

Scientific Name: *Uncaria tomentosa* (Willd. ex Schult.) DC.

Botanical Family: Rubiaceae (madders family)

Physical Description: This tropical vine grows up to 100 ft (~30 m) in length with the main stem being up to 10 in. (~25 cm) in diameter. The fibrous bark is cinnamon in color with longitudinal striations. Young shoots have coarse hairs and curved hooks that resemble cat's claws. The yellowish-white flowers are tubular and the seed pods are long and sessile with oblong seeds.

Distribution and Ecology: *Uncaria tomentosa* inhabits areas at elevations of 1,000–2,500 ft (~300–800 m) in mature natural forests of Central (Panama, Costa Rica, Colombia, Guatemala, Honduras) and South (Peru, Ecuador, Venezuela) America, as well as Trinidad and Suriname.[2] *Uncaria guianensis* occurs at elevations of 300–1,500 ft (~100–450 m) in Peru, Venezuela, Bolivia, Paraguay, Guyana, Brazil, and Trinidad.

EXPOSURE

Medicinal Uses

TRADITIONAL

Decoctions of the bark of cat's claw are a traditional treatment for a variety of diseases including cancer, arthritis, gastritis, and infections.[1] Preparation of the decoction involves boiling the inner bark or the root bark, although traditional uses also include the addition of macerated bark to alcoholic beverages.[2] Asháninka priests in the Peruvian rainforest use extracts in rituals, and aqueous extracts of the bark are traditional treatments for a variety of illnesses including abscesses, allergies, arthritis, asthma, cancer, contraception, fever, gastric ulcers, menstrual irregularity, urinary tract inflammation, viral and bacterial infections, gastritis, and wounds.[3] Traditionally, indigenous groups produce an herbal decoction by boiling 20 g cat's claw root bark in

Medical Toxicology of Natural Substances, by Donald G. Barceloux, MD

a liter of water for 45 minutes, and then replacing the evaporated liquid with water to produce one liter of decoction.[4]

CURRENT

The main source of the active ingredients in herbal preparations of cat's claw is the root or stem bark. C-Med 100® (Human Genome Sciences, San Diego, CA) is a hot water extract of cat's claw (*U. tomentosa*) that contains compounds with molecular weights below 10kDa and excludes higher molecular weight compounds (e.g., flavonoids, tannins). Commercial preparations of bark extracts from cat's claw [C-Med-100®, Activar AC-11® (Optigenex, Inc., New York, NY)] are based on oxindole alkaloid content rather than bioavailability or clinical efficacy. Small clinical trials suggest a potential role for extracts of cat's claw in the treatment of osteoarthritis[5] and rheumatoid arthritis.[6] Although these controlled clinical studies demonstrated some reduction in pain associated with cat's claw intake, there is insufficient objective evidence to confirm the clinical efficacy of this anti-inflammatory activity.[7]

REGULATORY STATUS

In the United States, cat's claw is a dietary supplement as specified by the Definition of Certain Foods as Dietary Supplements, in Section 3 of the Dietary Supplement Health and Education Act, 1994. This substance has been classified as a dietary supplement since the late 1980s.

PRINCIPAL INGREDIENTS

Chemical Composition

Cat's claw contains at least 50 constituents including oxindole and indole alkaloids,[8] quinovic acid glycosides,[9,10] pyroquinovic acid glycosides (tomentosides A,B)[11], ursane-type pentacyclic triterpenes, procyanidins (cinchonain Ia and Ib), catechins, and sterols (β-sitosterol). The alkaloids of this plant have a spiro oxindole structure as demonstrated by Figure 55.1. The major alkaloids in crude extracts of bark and leaves of *Uncaria tomentosa* as determined by high performance liquid chromatography with photodiode array detection (HPLC/DAD) followed by HPLC coupled with electron spray ionization (HPLC/ESI) in positive ion mode include both pentacyclic (mitraphylline, isomitraphylline, pteropodine, isopteropodine, speciophylline, uncarine F) and tetracyclic alkaloids (ryncophylline, isoryncophylline).[12] Corynoxeine and isocorynoxeine are minor oxindole tetracyclic alkaloids, and uncarine C and E are the minor methyl esters of pteropodic and isopteropodic acids, respectively.[13] Oxindole alkaloids isomerize rapidly in aqueous solutions to form pH-dependent mixtures of isomers.[7] Variability in alkaloid content occurs with season, plant strains, and plant parts.[14] The major alkaloids in three samples of bark extract were pteropodine, isopteropodine, and isomitraphylline; mitraphylline and isomitraphylline were the major alkaloids in the leaves.[12] Total mean pentacyclic and tetracyclic alkaloid content in the bark extract were about 26.9mg/g dry weight and 2.8mg/g dry weight, respectively, compared with 14.2mg/ g dry weight and 3.2mg/g dry weight, respectively, in the leaves.

Physiochemical Properties

Several quinovic acid glycosides demonstrate anti-inflammatory activity in animal models. The bark is odorless with a slightly bitter, astringent flavor.[2] Animal studies suggest that components (e.g., quinic acid) in the hot water extract of the bark from *U. tomentosa* have some anti-inflammatory properties, such as inhibition of inflammatory cytokine tumor necrosis factor alpha and activation of the central transcription regulator, nuclear factor κB. The latter factor regulates the expression of proinflammatory cytokines and affects apoptosis. In a study of rodents, the administration of an industrial standardized *Uncaria tomentosa* fraction containing 95% oxindole alkaloids produced some systemic anti-nociceptive properties along with evidence of activation of 5-HT_2 receptors.[15]

Mechanism of Toxicity

Overall, animal studies using aqueous and hydroalcoholic extracts of cat's claw indicate a low degree of acute oral toxicity.[7]

DOSE RESPONSE

Establishing a dose-response curve for cat's claw is complicated by the variability in the concentration of active ingredients between different formulation, limited quantitative data, and inconsistent bioequivalence across different product formulas. The general range of the daily intake of cat's claw for inflammatory disorders is about 100–1000mg, which corresponds to a daily consumption of 10–30mg total alkaloids from *U. tomentosa*. The typical traditional daily dose for cat's claw decoction as described under Medicinal Uses is about 60mL. Daily oral doses of C-Med-100® (8–10% or 16–20% carboxy alkyl esters) in clinical studies range from 250–700mg.[16] There are inadequate data to determine the toxic dose of extracts from cat's claw.

Pentacyclic Alkaloids

Pteropodine	3S, 7R, 15S, 19S, 20S
Isopteropodine	3S, 7S, 15S, 19S, 20S
Speciophylline	3R, 7S, 15S, 19S, 20S
Uncarine F	3R, 7R, 15S, 19S, 20S
Mytraphylline	3R, 7R, 15S, 19S, 20R
Isomytraphylline	3S, 7S, 15S, 19S, 20R

Tetracyclic Alkaloids

Ryncophylline	3S, 7R, 15S, 20R	R = ethyl
Isoryncophylline	3S, 7S, 15S, 20R	R = ethyl
Corynoxeine	3R, 7R, 15S, 20R	R = vinyl
Isocorynoxeine	3S, 7S, 15S, 20R	R = vinyl

Quinovic Acid Glycosides

R_1	R_2	R_3
Glu-Fuc	Glu	H
Glu-Fuc	H	Glu
Fuc	H	Glu
Rha	H	Glu
Glu-Glu	H	H
H	H	Glu-Glu

Triterpenes

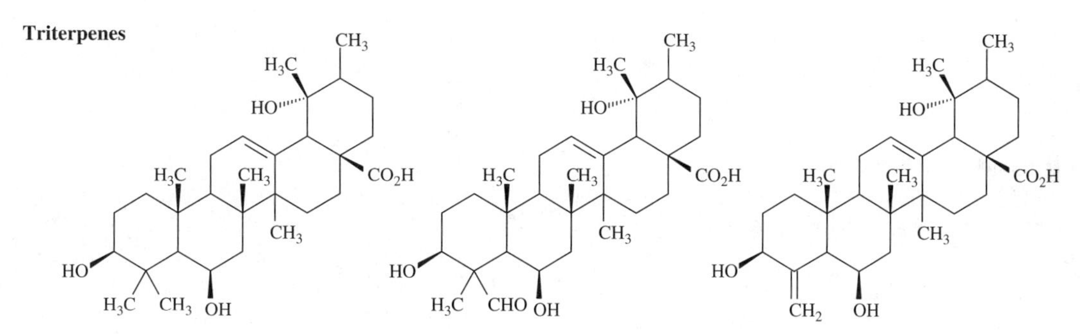

FIGURE 55.1. Chemical structures of pentacyclic and tetracyclic alkaloids, quinovic acid glycosides, and triterpenes in cat's claw. Adapted from Montoro P, et al. Identification and quantification of components in extracts of *Uncaria tomentosa* by HPLC-ES/MS. Phytochem Anal 2004;15;56. Copyright John Wiley & Sons Limited.

TOXICOKINETICS

There are few human data on the toxicokinetics of constituents in cat's claw. *In vitro* measurements of cytochrome P450 3A4 (CYP3A4) activity suggest some inhibitory activity through evaluation of extracts of cat's claw using a fluorometric microtiter plate assay.[17] However, there are inadequate clinical data to determine if metabolic inhibition of CYP3A4-dependent drugs (e.g., some benzodiazepines, cyclosporine, protease inhibitors, non-nucleoside reverse-transcriptase inhibitors) by cat's claw is clinically significant.

CLINICAL RESPONSE

In clinical trials, adverse effects associated with the administration of 20 mg of an aqueous acid-extracted dry extract of *Uncaria tomentosa* and 100 mg freeze-dried extract of *U. guianensis* were minimal, and there was no evidence of major side effects.[6] A case report temporally associated the development of an allergic interstitial nephritis (increased renal insufficiency, proteinuria, hematuria) with the daily use of cat's claw in a 35-year-old woman with systemic lupus erythematosus.[18] The medications for her condition included prednisone, atenolol, metolazone, furosemide, and nifedipine. The proteinuria and hematuria resolved after cessation of cat's claw use, and the renal insufficiency improved.

DIAGNOSTIC TESTING

Methods to detect alkaloid and flavonol compounds in extracts of cat's claw include reverse-phase HPLC[19] and high performance liquid chromatography with electrospray mass spectrometry (HPLC/EMS).[12]

TREATMENT

Treatment is supportive.

References

1. Keplinger K, Laus G, Wurm M, Dierich MP, Teppner H. *Uncaria tomentosa* (Willd.) DC.—ethnomedicinal use and new pharmacological, toxicological and botanical results. J Ethnopharmacol 1999;64:23–34.
2. Gattuso M, Di Sapio O, Gattuso S, Li Pereyra E. Morphoanatomical studies of *Uncaria tomentosa* and *Uncaria guyanensis* bark and leaves. Phytomedicine 2004;11:213–223.
3. Heitzman ME, Neto CC, Winiarz E, Vaisberg AJ, Hammond GB. Ethnobotany, phytochemistry and pharmacology of *Uncaria* (Rubiaceae). Phytochemistry 2005;66:5–29.
4. Reinhard KH. *Uncaria tomentosa* (Willd.) D.C.: cat's claw, una de gato, or saventaro. J Altern Complement Med 1999;5:143–151.
5. Piscoya J, Rodriguez Z, Bustamante SA, Okuhama NN, Miller MJ, Sandoval M. Efficacy and safety of freeze-dried cat's claw in osteoarthritis of the knee: mechanisms of action of the species *Uncaria guyanensis*. Inflamm Res 2001;50:442–448.
6. Mur E, Hartig F, Eibl G, Schirmer M. Randomized double blind trial of an extract from the pentacyclic alkaloid-chemotype of *Uncaria tomentosa* for the treatment of rheumatoid arthritis. J Rheumatol 2002;29:678–681.
7. Valerio LG Jr, Gonzales GF. Toxicological aspects of the South American herbs cat's claw (*Uncaria tomentosa*) and maca (*Lepidium meyenii*). Toxicol Rev 2005;24:11–35.
8. Wagner H, Kreutzkamp B, Jurcic K. [The alkaloids of *Uncaria tomentosa* and their phagocytosis-stimulating action]. Planta Med 1985;51:419–423. [German]
9. Sheng Y, Akesson C, Holmgren K, Bryngelsson C, Giamapa V, Pero RW. An active ingredient of cat's claw water extracts identification and efficacy of quinic acid. J Ethnopharmacol 2005;96:577–584.
10. Aquino R, De Feo V, De Simone F, Pizza C, Cirino G. Plant metabolites. New compounds and anti-inflammatory activity of *Uncaria tomentosa*. J Nat Prod 1991;54:453–459.
11. Kitajima M, Hashimoto K-I, Yokoya M, Takayama H, Sandoval M, Aimi N. Two new nor-triterpene glycosides from Peruvian "uña de gato" (*Uncaria tomentosa*). J Nat Prod 2003;66:320–323.
12. Montoro P, Carbone V, de Dioz Zuniga Quiroz J, De Simone F, Pizza C. Identification and quantification of components in extracts of *Uncaria tomentosa* by HPLC-ES/MS. Phytochem Anal 2004;15;55–64.
13. Muhammad I, Khan IA, Fischer NH, Fronczek FR. Two stereoisomeric pentacyclic oxindole alkaloids from *Uncaria tomentosa*: uncarine C and uncarine E. Acta Cryst 2001;C57:480–482.
14. Laus G. Advances in chemistry and bioactivity of the genus *Uncaria*. Phytother Res 2004;18:259–274.
15. Jurgensen S, DalBo S, Angers P, Soares Santos AR, Ribeiro-do-Valle RM. Involvement of 5-HT_2 receptors in the antinociceptive effect of *Uncaria tomentosa*. Pharmacol Biochem Behav 2005;81:466–477.
16. Lamm S, Sheng Y, Pero RW. Persistent response to pneumococcal vaccine in individuals supplemented with a novel water soluble extract of *Uncaria tomentosa*, C-Med-100. Phytomedicine 2001;8:267–274.
17. Budzinski JW, Foster BC, Vandenhoek S, Arnason JT. An *in vitro* evaluation of human cytochrome P450 3A4 inhibition by selected commercial herbal extracts and tinctures. Phytomedicine 2000;7:273–282.
18. Hilepo JN, Bellucci AG, Mossey RT. Acute renal failure caused by "cat's claw" herbal remedy in a patient with systemic lupus erythematosus. Nephron 1997;77:361.
19. Sandoval M, Okuhama NN, Zhang XJ, Condezo LA, Lao J, Angeles FM, et al. Anti-inflammatory and antioxidant activities of cat's claw (*Uncaria tomentosa* and *Uncaria guianensis*) are independent of their alkaloid content. Phytomedicine 2002;9:325–337.

Chapter 56

CHAMOMILE
[*Chamomilla recutita* (L.) Rauschert, *Chamaemelum nobile* L.]

HISTORY

The use of chamomile dates back at least to the ancient Greeks, who used this herb to treat fever. The word, chamomile, derives from the Greek word for the ground apple plant. The apple-like aroma of this plant was mentioned by Pliny the Elder in his book, *Naturalis Historia*, from the 1st century AD. Chamomile was part of monastery gardens during the Middle Ages, and European colonists brought chamomile to the United States in the 16th century.

BOTANICAL DESCRIPTION

Common Name: German Chamomile, Hungarian Chamomile, Wild Chamomile, Sweet False Chamomile, Mayweed

Scientific Name: *Chamomilla recutita* (L.) Rauschert (*Matricaria recutita*, *Matricaria chamomilla*)

Botanical Family: Asteraceae (Compositae, daisy)

Physical Description: German chamomile is an erect annual herb with branching stems that reaches 10–30 cm (~4–12 in.) in height. The leaves are alternate, tripinnately divided below and bipinnately divided leaves above. The small flower head (capitulum) grows up to 1.5 cm (0.6 in.) in diameter and is comprised of 12–20 white ligulate florets surrounding a conical hollow receptacle with numerous yellow tubular florets inserted. The fruit is small, yellow, and smooth.

Distribution and Ecology: German chamomile is a native herb of southeast Europe and west Asia that has spread throughout Europe, North America, and Australia.

Common Name: Roman Chamomile, English Chamomile, Common Chamomile

Scientific Name: *Chamaemelum nobile* L. (*Anthemis nobilis*)

Botanical Family: Asteraceae (Compositae, daisy)

Physical Description: This low-lying plant reaches up to about 1 ft (30 cm) in height. The gray-green leaves are thicker and closer to the ground than German chamomile. The flowers of Roman chamomile are similar to miniature daisies with yellow centers surrounded by white petals that smell like apples.

Distribution and Ecology: Roman chamomile is a perennial herb cultivated in western Europe and northern Africa.

EXPOSURE

Sources

Flos chamomillae consists of the dried flowering heads of *Chamomilla recutita* L. The yield of essential oil is typically at least 0.4% wet weight ranging up to about 1.5% wet weight. Powdered flos chamomillae is greenish yellow to yellowish brown.

Medical Toxicology of Natural Substances, by Donald G. Barceloux, MD

Medicinal Uses

Traditional

The water and ethanol extracts of pulverized flower heads from German and Roman chamomile are popular, old remedies used for spasmolytic, wound-healing, antibacterial, anti-inflammatory, and sedative effects.[1,2] Traditional indications for chamomile tea include insomnia, indigestion, anxiety, fever, upper respiratory tract infections, headaches, skin and eye irritation, and burns. Chamomile is used both as an herbal tea and as an essential oil in cosmetics.

Current

Chamomile teas and extracts are traditional treatment for a variety of illnesses related to inflammation of the skin, upper respiratory tract, lungs, and gastrointestinal tract. The essential oil of chamomile is popular in aromatherapy. Previously, chamomile was evaluated as a mouthwash for the treatment of 5-fluorouracil-induced stomatitis with equivocal results.[3] There is an absence of clinical data evaluating the sedative properties of chamomile. An uncontrolled clinical series of 10 cardiac patients associated a "deep sleep" for 90 minutes with the administration of chamomile tea.[4]

Regulatory Status

Chamomile is regulated as a dietary supplement in the United States. The German Commission E approved the use of camomile for the treatment of skin and mouth irritation, bacterial skin diseases, and irritation of the respiratory tract, as well as the internal treatment of gastrointestinal spasm.

PRINCIPAL INGREDIENTS

Chemical Composition

The major constituents of essential oils from German and Roman chamomile are the esters, isobutyl angelate and 2-methylbutyl angelate. Other compounds present in teas extracted from the flower heads of chamomile are coumarin compounds, caffeic acid, flavonoids (apigetrin, apiin, apigenin, apigenin-7-*O*-glucoside, luteolin, luteolin-7-*O*-glucoside, matricin, quercetin, rutin), volatile oils (α-bisabolol, azulene, chamazulene), amino acids, choline polysaccharide, plant and fatty acids, tannin, and triterpene hydrocarbons (α-pinene, camphene, myrcene).[5,6] Apigenin is the most abundant flavonoid in chamomile flowers. Apigenin accumulates in the white florets of the camomile anthodium bound as apigenin 7-*O*-glucoside and various acylated derivatives.[7] In a study of Roman chamomile tea produced by pouring boiling water over the flowers and steeping for 15 minutes, the decanted tea contained only about 7% of the original oil content of the flowers.[8] Most of the water-insoluble chemicals (esters of angelic and butyric acid, α-pinene, myrcene, pinocarvone) were not detectable, whereas the water soluble compound, *t*-pinocarveol, was a much larger part of the tea than the essential oil.

The essential oil content of plant parts from chamomile depends on the chemotype and the season with typical essential oil content of flowers ranging from about 0.4–1.5%.[9] Commercial chamomile cultivars are classified according to bisabolol oxide (CAS RN:11087-43-7) content. α-Bisabolol [CAS RN:515-69-5] and bisabolol oxides are potentially active ingredients of chamomile tea. Figure 56.1 displays the chemical structure of bisabolol. The maximum essential oil content in the flower occurs immediately prior to flowering, and the oil content decreases after the flowering process.[10] The root of chamomile contains minor amounts (i.e., <0.1%) of essential oils (e.g., *trans*-β-farnesene). There are few clinical data on the potential active ingredients (e.g., apigenin, apigenin glycosides, α-bisabolol, bisabolol oxides, chamazulene, *trans*-β-farnesene) of chamomile tea.

Species in the Compositae family including chamomile contain allergenic sesquiterpene lactones (e.g., desacetyl matricarin, anthecotulide) as well as other constituents (e.g., polyacetylenes, thiophenes, coumarins) that may contribute to an acquired hypersensitivity.[11] As a result of the poor water solubility of some allergenic sesquiterpene lactones, hot-water extracts (i.e., teas) of chamomile contain fewer allergens than essential oil of chamomile used for medicinal purposes.[12] Matricin (CAS RN: 29041-35-8) is 6,7-guaianolide ($C_{17}H_{22}O_5$). Figure 56.2 displays the chemical structure of this sesquiterpene lactone. Chamazulene carboxylic acid is a decomposition product of matricin that forms during the steam distillation of the essential oil from chamomile.

Physiochemical Properties

The essential oil of chamomile has an intense blue color secondary to the presence of chamazulene (1–15%).

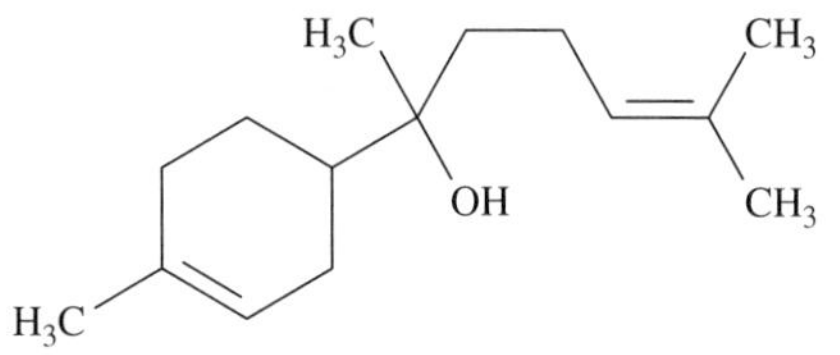

FIGURE 56.1. Chemical structure of bisabolol.

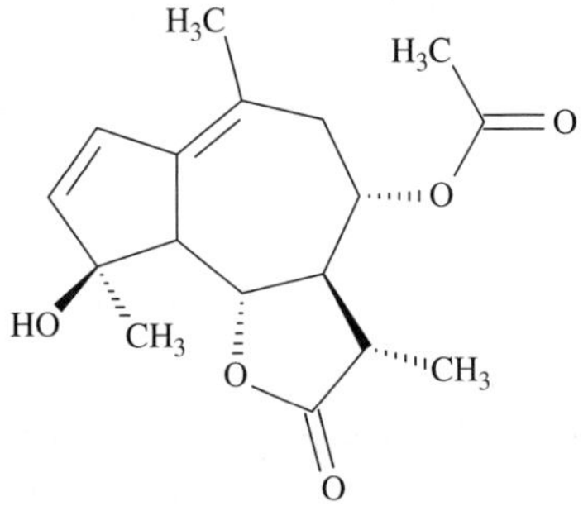

FIGURE 56.2. Chemical structure of matricin.

Mechanism of Toxicity

Radioreceptor binding assays for apigenin, a flavone constituent of chamomile, demonstrate some affinity for the central benzodiazepine binding site, but this affinity is relatively low. The $GABA_A$-benzodiazepine receptor antagonist, flumazenil, does not block this sedative action; consequently, the inhibitory activity of apigenin on locomotor behavior in rats probably does not result from an interaction with the $GABA_A$-benzodiazepine receptor.[13]

DOSE RESPONSE

The average adult dose is 2–8 g of the flower head (flos chamomillae) or 1–4 mL of the fluid extract (1:1 in 45% ethanol) 3 times daily.[9]

TOXICOKINETICS

There are few human data on the toxicokinetics of constituents in chamomile tea. *In vitro* data suggests that chamomile preparations contain constituents (e.g., chamazulene, *cis*-spiroether, *trans*-spiroether), which inhibit the activities of major human CYP450 isoenzymes, particularly CYP1A2.[14] Potential drug interactions resulting from the use of chamomile include increased anticoagulant effects due to coumarin content, inhibition of platelet aggregation, and the enhancement of sedative effects when used concomitantly with other central nervous system depressants. However, there are few data on the clinical significance of these potential drug interactions.[15] A case report associated the development of an elevated international normalized ratio (INR = 7.9) and internal hemorrhages with the use of chamomile tea by a 70-year-old woman, who was receiving coumarin for atrial fibrillation and a prosthetic heart valve.[16]

CLINICAL RESPONSE

Although the application of chamomile tea to the eye is a folk remedy for the treatment of conjunctivitis, allergic conjunctivitis can occur in sensitized patients following the application of these remedies.[17] Other hypersensitivity reactions to chamomile tea ranges from contact dermatitis, rhinitis, contact urticaria,[18] and facial edema[19] to acute anaphylaxis in individuals with allergies to ragweed or other members of the Compositae family, particularly following exposure to Roman chamomile (*Anthemis nobilis*).[20–22] A variety of allergens cross-react with chamomile including pollens from ragweed (*Ambrosia artemisiifolia* L.),[23] birch, and mugwort (*Artemisia vulgaris* L.).[25] In a series of 10 patients with documented allergies to chamomile, clinical effects resulting from the ingestion of chamomile tea included vomiting, diarrhea, abdominal cramps, rhinoconjunctivitis, dyspnea, and wheezing.[24] These patients also had variable cross-sensitivity to other allergens, such as celery roots, anise seeds, and pollen from mugwort, birch, and timothy grass. Life-threatening, immediate, IgE-mediated anaphylactic reactions can occur in hypersensitive patients. A case report associated the development of anaphylaxis (urticaria, laryngeal edema, hypotension, tachycardia) with the administration of an enema containing an oily extract of camomile flowers during labor.[25] An emergency cesarian section revealed an atonic, ischemic uterus, and the neonate died the following day. IgE-specific antibodies (RAST class = 2) were detected in blood samples from the mother.

DIAGNOSTIC TESTING

Methods for the analysis of the constituents of chamomile include high performance liquid chromatography (HPLC) with a diode-array detector or a fluorescence detector[26] and simultaneous isocratic HPLC.[27] Quantitation of apigenin in chamomile includes the use of gas chromatography/tandem mass spectrometry.[28] Additionally capillary electrophoresis can detect flavonoids, such as apigenin, with limits of detection and quantitation of approximately 4 μg/mL and 12 μg/mL, respectively.[29]

TREATMENT

Treatment is supportive and includes the use of standard treatment methods for contact dermatitis (topical steroids) or anaphylaxis (epinephrine, antihistamines).

References

1. Rossi T, Melegari M, Bianchi A, Albasini A, Vampa G. Sedative, anti-inflammatory and anti-diuretic effects induced in rats by essential oils of varieties of *Anthemis nobilis*: a comparative study. Pharmacol Res Commun 1988;20(Suppl 5):71–74.
2. Barene I, Daberte I, Zvirgzdina L, Iriste V. The complex technology on products of German chamomile. Medicina 2003;39:127–131.
3. Fidler P, Loprinzi CL, O'Fallon JR, Leitch JM, Lee JK, Hayes DL, et al. Prospective evaluation of a chamomile mouthwash for prevention of 5-FU-induced oral mucositis. Cancer 1996;77:522–525.
4. Gould L, Reddy CV, Gomprecht RF. Cardiac effects of chamomile tea. J Clin Pharmacol 1973;13:475–479.
5. Kunde R, Isaac O. On the flavones of chamomile (*Matricaria chamomilla* L.) and a new acetylated apigenin-7-glucoside. Planta Med 1979;37:124–130.
6. Redaelli C, Formentini F, Santaniello E. Reversed phase high performance liquid chromatography analysis of apigenin and its glucosides in flowers of *Matricaria chamomilla* and chamomile extracts. Planta Med 1981;56: 288–292.
7. Svehlikova V, Bennett RN, Mellon FA, Needs PW, Piacente S, Kroon PA, Bao Y. Isolation, identification and stability of acylated derivatives of apigenin 7-*O*-glucoside from chamomile (*Chamomilla recutita* [L.] Rauschert). Phytochemistry 2004;65:2323–2332.
8. Carnat A, Carnat AP, Fraisse D, Ricoux L, Lamaison JL. The aromatic and polyphenolic composition of Roman camomile tea. Fitoterapia 2004;75:32–38.
9. World Health Organization. Monographs on selected medicinal plants. Vol. 1. Geneva: WHO;1999:86–94.
10. Szoke E, Maday E, Tyihak E, Kuzovkina IN, Lemberkovics E. New terpenoids in cultivated and wild chamomile (*in vivo* and *in vitro*). J Chromatogr B 2004:800: 231–238.
11. Hausen BM. A 6-year experience with compositae mix. Am J Contact Dermatol 1996;7:94–99.
12. Paulsen E. Contact sensitization from Compositae-containing herbal remedies and cosmetics. Contact Dermatitis 2002;47:189–198.
13. Avallone R, Zanoli P, Puia G, Kleinschnitz M, Schreier P, Baraldi M. Pharmacological profile of apigenin, a flavonoid isolated from *Matricaria chamomilla*. Biochem Pharmacol 2000;59:1387–1394.
14. Ganzera M, Schneider P, Stuppner H. Inhibitory effects of the essential oil of chamomile (*Matricaria recutita* L.) and its major constituents on human cytochrome P450 enzymes. Life Sci 2006;78:856–861.
15. Miller LG. Herbal medicinals selected clinical considerations focusing on known or potential drug-herb interactions. Arch Intern Med 1998;158:2200–2211.
16. Segal R, Pilote L. Warfarin interaction with *Matricaria chamomilla*. CMAJ 2006;174:1281–1282.
17. Subiza J, Subiza JL, Alonso M, Hinojosa M, Garcia R, Jerez M, Subiza E. Allergic conjunctivitis to chamomile tea. Ann Allergy 1990;65:127–132.
18. Foti C, Nettis E, Panebianco R, Cassano N, Diaferio A, Pia DP. Contact urticaria from *Matricaria chamomilla*. Contact Dermatitis 2000;42:360–361.
19. Rudzki E, Rapiejko P, Rebandel P. Occupational contact dermatitis with asthma and rhinitis, from camomile in a cosmetician also with contact urticaria from both camomile and lime flowers. Contact Dermatitis 2003;49:162.
20. Sanchez-Perez J, Sanz T, Garcia-Diez A. Allergic contact dermatitis from hydrolyzed wheat protein in cosmetic cream. Contact Dermatitis 2000;42:360–361.
21. Rodriguez-Serna M, Sanchez-Motilla JM, Ramon R, Aliaga A. Allergic and systemic contact dermatitis from *Matricaria chamomilla* tea. Contact Dermatitis 1998; 39:192–209.
22. Giordano-Labadie F, Schwarze HP, Bazex J. Allergic contact dermatitis from chamomile used in phytotherapy. Contact Dermatitis 2000;42:247.
23. de la Torre Morin F, Sanchez Machin I, Garcia Robaina JC, Fernandez-Caldas E, Sanchez Trivino M. Clinical cross-reactivity between *Artemisia vulgaris* and *Matricaria chamomilla* (chamomile). J Invest Allergol Clin Immunol 2001;11:118–122.
24. Reider N, Sepp N, Fritsch P, Weinlich G, Jensen-Jarolim E. Anaphylaxis to camomile: clinical features and allergen cross-reactivity. Clin Exp Allergy 2000;30:1436–1443.
25. Jensen-Jarolim E, Reider N, Fritsch R, Breiteneder H. Fatal outcome of anaphylaxis to camomile-containing enema during labor: a case study. J Allergy Clin Immunol 1998;102:1041–1042.
26. Schulz H, Albroscheit G. High-performance liquid chromatographic characterization of some medical plant extracts used in cosmetic formulas. J Chromatogr 1988;442:353–361.
27. Pietta P, Manera E, Ceva P. Simultaneous isocratic high-performance liquid chromatographic determination of flavones and coumarins in *Matricaria chamomilla* extracts. J Chromatogr 1987;404:279–281.
28. Avallone R, Zanoli P, Puia G, Kleinschnitz M, Schreier P, Baraldi M. Pharmacological profile of apigenin, a flavonoid isolated from *Matricaria chamomilla*. Biochem Pharmacol 2000;59:1387–1394.
29. Fonseca FN, Tavares MF. Validation of a capillary electrophoresis method for the quantitative determination of free and total apigenin in extracts of *Chamomilla recutita*. Phytochem Anal 2004;15:65–70.

Chapter 57

CHAPARRAL

[*Larrea tridentata* (Sesse & Moc. ex DC.) Vail]

HISTORY

The general term, chaparral, usually describes a type of thick, dense, drought-tolerant vegetation composed of shrubs and small trees. However, the herb chaparral is a preparation of *Larrea tridentata*. Aqueous extracts of chaparral are folk treatments primarily for renal and gynecological disorders by traditional healers in the southwestern United States and northern Mexico.[1] The Pima Indians in the Southwest used oral decoctions and aqueous extracts of the leaves and stems of chaparral for the treatment of diabetes.[2] In 1945, Waller and Gisvold first isolated the active ingredient (nordihydroguaiaretic acid) from chaparral.[3] Until 1967, nordihydroguaiaretic acid was added to a variety of products including frozen foods, oils, baking goods, and pharmaceutical preparations as an antioxidant that inhibited aerobic glycolysis. The use of nordihydroguaiaretic acid was discontinued in the 1960s when long-term rodent studies indicated that the chronic administration of nordihydroguaiaretic acid caused cystic nephropathy.[4] During the late 1960s, a case series of 59 patients with advanced cancer demonstrated that this compound was not an effective chemotherapeutic agent for these advanced tumors.[5] Animal studies during this period indicated that nordihydroguaiaretic acid produced renal toxicity; subsequently, this compound was removed from the list of substances "generally recognized as safe."

BOTANICAL DESCRIPTION

Chaparral is a shrub (i.e., contains woody stems) rather than an herb (nonwoody stems), but the leaves of chaparral are ground for use in traditional herbal medications (teas, capsules). The term chaparral frequently refers to species other than *Larrea tridentate*, including the South American varieties of creosote bush (*Larrea nitida* Cav., *Larrea cuneifolia* Cav.).

Common Name: Creosote Bush, Greaseweed, El Gobernadora, Hideonodo, Palo Ondo, Tasajo, Gumis

Scientific Name: *Larrea tridentata* (Sesse & Moc. ex DC.) [Synonyms: *Covillea glutinosa* (Engelm.) Rydb, *Larrea divaricata Cav., Larrea glutinosa* Engelm., *Larrea mexicana* Moric.]

Botanical Family: Zygophyllaceae (caltrop)

Physical Description: This plant is a well-branched, evergreen shrub that reaches heights up to 1–3 m (~3–10 ft). The stems are gray with swollen, black bands at the nodes, and the young branches are slender. The leaves are opposite, compound with two olive-green leaflets, which are glossy with resinous coating and an oblong to obovate shape. These leaflets are about 1 cm (~0.4 in.) in length and 3–5 mm (~0.1–0.2 in.) in width. The small, solitary flowers have five bright-yellow sepals (not petals), which are twisted like the blades of a fan. The fruit is a pea-sized globose capsule, which is covered by dense, white hairs. These fruits separate into five indehiscent, one-seeded carpels at maturity.

Distribution and Ecology: Chaparral inhabits the arid lands in northern Mexico and the southwestern United States from southeastern California to

Medical Toxicology of Natural Substances, by Donald G. Barceloux, MD

southwestern Texas including New Mexico, Utah, Nevada, and Arizona. This ubiquitous shrub competes well with most other desert plants for environmental resources.

EXPOSURE

Sources

Chaparral is a Native American herb derived from the leaves and flowers of the desert shrub, *Larrea tridentata*. Ground leaves, stems, and/or flowers are usually ingested either as a brewed tea or a capsule. Although used primarily as a tea, other available forms of chaparral preparations include alcoholic extracts of the leaves and twigs, capsules, tablets, and poultices or fomentations of leaves and branches. The volatile oil represents about 0.1% of dried leaves and stems.

Medicinal Uses

Traditional

Traditional uses of this herb involve the purported anti-inflammatory, antimicrobial, antioxidant, and antineoplastic properties. Traditional healers in the southwestern United States and northern Mexico primarily used chaparral for the treatment of kidney stones, diabetes, and infertility.[1] Other herbal uses included the treatment of arthritis, rheumatism, inflammation, venereal diseases, abdominal pain, and infectious agents (e.g., varicella, skin infections).

Current

The use of chaparral as an herbal remedy is not recommended because of the lack of documented efficacy and the risk of hepatorenal damage. Clinical data does not support the use of this herb in the treatment of cancer, and animal studies suggest that chronic administration of the principal ingredient of chaparral, nordihydroguaiaretic acid (NDGA), may cause renal damage.[5] The chronic administration of feed containing 2% NDGA to rats produced renal cysts, vacuolation of the renal tubular epithelium, partial obstruction of nephrons, and hyperplasia of the collecting ducts.[6,7] Other herbal uses included the treatment of dandruff, bronchitis, arthritis, upper respiratory tract infections, tuberculosis, toothaches, venereal disease, rheumatism, menstrual cramps, and muscle spasms. Additionally, chaparral is ingested as a nutritional supplement.

Regulatory Status

Prior to 1967, nordihydroguaiaretic acid was a food additive that was used to prevent fermentation as a result of its antioxidant properties. The US Food & Drug Administration (FDA) removed chaparral from the list of substances "Generally Recognized as Safe" (GRAS) in 1968 when animal studies indicated that the chronic administration of NDGA in feed caused renal cysts and obstruction of nephrons. As a result of case reports associating serious hepatic damage with the use of chaparral teas, the FDA recommended the withdrawal of chaparral products in 1992.

PRINCIPAL INGREDIENTS

Chemical Composition

Nordihydroguaiaretic acid (NDGA, CAS RN: 500-38-9) is the principal ingredient in chaparral. Figure 57.1 displays the chemical structure of NDGA ($C_{18}H_{22}O_4$). Although the leaves, green stems, small woody stems (<5 mm/~0.2 in.), and flowers all contain NDGA, the highest concentrations of NDGA are found in the leaves and green stems. In a study of NDGA in chaparral, the leaves and green stems contained NDGA concentrations of 38.3 mg/g and 32.5 mg/g, respectively.[8] This plant contains a variety of compounds including volatile oils (calamine, eudesmol, limonene), lignans (nordihydroguaiaretic acid, norisoguaiacin, dihydroguaiaretic acid), flavonoids (isorhamnetin, kaempferol, quercetin, gossypetin, herbacetin), and amino acids (arginine, cystine, glutamic acid, glycine, isoleucine, leucine, phenylalanine, tryptophan, tyrosine, valine).

Physiochemical Properties

NDGA is a white crystalline solid with antioxidant properties. This compound also inhibits the lipoxygenase and cyclo-oxygenase pathways. Methyl and vinyl ketones contribute significantly to the characteristic, creosote-like odor of chaparral.[1]

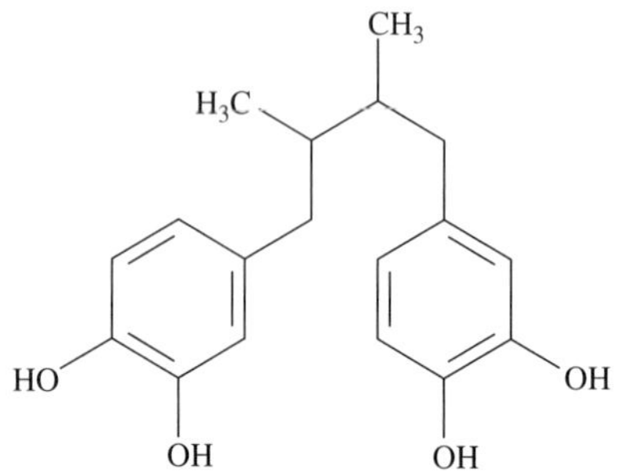

FIGURE 57.1. Chemical structure of nordihydroguaiaretic acid.

Mechanism of Toxicity

The mechanism of hepatotoxicity following chaparral use is probably idiosyncratic as the hepatotoxicity of NDGA has not been reproduced in animal models. Most biopsies from patients with suspected chaparral-induced liver damage are suggestive of cholestatic hepatitis.[9] Liver biopsies of patients with hepatitis after the use of chaparral demonstrate cholestatic injury manifest by acute inflammatory infiltrates, focal acute pericholangitis, mild ductal dilation, and proliferation of bile ductules in periportal areas as well as necrosis of hepatocytes. In severe cases, liver biopsies reveal severe acute hepatic necrosis with areas of lobular collapse, nodular regeneration, marked bile duct proliferation, mixed portal inflammation, and occasional fibrosis.[10]

DOSE RESPONSE

Medicinal Use

Although the FDA does not recommend the use of chaparral, some naturopathic practitioners use small doses of chaparral in complex herbal preparations.[11] In the United States between 1945 and 1967, nordihydroguaiaretic acid was an antioxidant added to food in concentrations of 0.01–0.02%. In preliminary, uncontrolled cancer chemotherapy studies, the administration of 250–3,000 g NDGA daily as a tea to humans did not produce any hematological or chemical abnormalities.[5]

Toxicity

The development of hepatitis following the use of chaparral tea is probably an idiosyncratic reaction; therefore, there is no clear dose response between the amount of chaparral tea and the severity of the hepatitis. Observational studies in humans did not detect hepatic damage in patients receiving daily doses of NDGA up to 400 mg /kg for the treatment of cancer.[5] In contrast, the development of cholestatic hepatitis was temporally associated with daily ingestion of 160 mg chaparral tablets for about 2 months.[12] Most case reports associating the use of chaparral with the development of hepatitis involve the use of chaparral within about 2–3 months of presentation to a health care facility for symptoms of hepatitis.[13]

TOXICOKINETICS

The major metabolite of NDGA is σ-quinone, which produces acute renal tubular dysfunction in rodents due to the accumulation of desquamated epithelial cells in the proximal tubules.[6,14] These studies indicate that the kidneys excrete little free NDGA in the urine, and the major detoxification pathway for this lignan is glucuronidation.[15] Excessive doses of chaparral may interfere with monoamine oxidase inhibitors as result of the presence of amino acid constituents, but there are few clinical data to confirm the clinical importance of this potential drug interaction.

CLINICAL RESPONSE

Adverse Reactions

Case reports associate dermal exposure to chaparral with the development of rash and stomatitis. Although the skin lesions occurred predominately in sun-exposed areas, the presence of lesions in other parts of the body and positive patch tests suggest an allergic contact dermatitis rather than a photodermatitis or irritant reaction.[16]

Toxic Effects

Sporadic case reports associate the ingestion of chaparral tea[17] and chaparral leaf[18] with the development of serious liver toxicity ranging from asymptomatic elevation of serum hepatic aminotransferases to fulminant hepatic failure.[19] Clinical manifestations of these hepatic effects include malaise, anorexia, nausea, retrosternal burning pain, pruritus, dark urine, jaundice, pedal edema, and distention of the abdomen. Resolution of these symptoms occurs over several months, and some focal scarring of the liver can persist.[18,20] In a case series of 18 patients with suspected chaparral-induced hepatitis, the range of chaparral use was 3–52 weeks, and resolution of symptoms ranged from 1–17 weeks after cessation of chaparral use in those patients without severe liver disease.[19] Four of these patients developed cirrhosis, and 2 patients required orthotopic liver transplantation due to fulminant hepatic failure. Renal damage is not usually associated with the ingestion of chaparral, but renal failure can occur as a complication of sepsis, hypotension, or the administration of nephrotoxic antibiotics associated with hepatic failure. Other herbal medicines associated with hepatitis include Chinese herbs (Jin Bu Huan, Ma-Huan, Sho-saiko-to), pyrrolizidine alkaloid-containing plants (*Crotalaria*, *Heliotropium*, *Senecio*, *Symphytum* spp.), skullcap (*Scutellaria laterifolia*), atractyloside-containing plants (*Atractylis gummifera* L., *Callilepis laureola* DC.), and germander (*Teucrium chamaedrys* L., *Teucrium polium* L.). Drugs associated with cholestasis and hepatitis include amoxicillin-clavulanate, azathioprine, carbamazepine, chlorpromazine, macrolide antibiotics, nonsteroidal anti-inflammatory drugs, oxypenicillins, and tricyclic antidepressants.[21] A case report associated the chronic consumption of chaparral tea with the development of cystic renal disease and cystic adenocarcinoma of the kidney.[22] However, there are insufficient human data to determine a causal relationship between chaparral use and cancer.

DIAGNOSTIC TESTING

Analytical Methods

High performance liquid chromatography with UV detection and mass spectrometry can determine the phenolic constituents (e.g., lignans, flavonoids) found in chaparral.[23]

Biomarkers

There are no specific biomarkers of chaparral-induced hepatitis because the specific hepatotoxin has not been identified.

Abnormalities

Laboratory abnormalities associated with hepatitis include elevation of serum hepatic aminotransferases, hypoprothrombinemia, hyperbilirubinemia, and elevated serum alkaline phosphatase. Elevations in serum alkaline phosphatase concentrations are typically minimal to mild compared with the substantial increases in the serum alanine aminotransferase concentration. Fulminant hepatic failure can develop along with hepatic encephalopathy (obtundation, elevated serum ammonia concentration).

TREATMENT

Treatment is supportive and includes withdrawal of chaparral in cases of allergic or idiosyncratic reactions.

References

1. Arteaga S, Andrade-Cetto A, Cardenas R. *Larrea tridentata* (creosote bush), an abundant plant of Mexican and US-American deserts and its metabolitc nordihydroguaiaretic acid. J Ethnopharmacol 2005;98:231–239.
2. Winkelman M. Ethnobotanical treatments of diabetes in Baja California Norte. Med Anthropol 1989;11: 255–268.
3. Waller C, Gisvold O. A phytochemical investigation of *Larrea divaricata* Cav. J Am Pharm Assoc 1945;34:78–81.
4. Grice HC, Becking G, Goodman T. Toxic properties of nordihydroguaiaretic acid. Food Cosmet Toxicol 1968;6:155–161.
5. Smart CR, Hogle HH, Vogel H, Broom AD, Bartholomew D. Clinical experience with nordihydroguaiaretic acid—"chaparral tea" in the treatment of cancer. Rocky Mt Med J 1970;67:39–43.
6. Goodman T, Grice HC, Becking GC, Salem FA. A cystic nephropathy induced by nordihydroguaiaretic acid in the rat. Light and electron microscopic investigations. Lab Invest 1970;23:93–107.
7. Evan AP, Gardner KD Jr. Nephron obstruction in nordihydroguaiaretic acid-induced renal cystic disease. Kidney Int 1979;15:7–19.
8. Hyder P, Fredrickson EL, Rick E, Estell R, Tellez M, Gibbens R. Distribution and concentration of total phenolics, condensed tannins, and nordihydroguaiaretic acid (NDGA) in creosote bush (*Larrea tridentata*). Biochem Syst Ecology 2002;30;905–912.
9. Batchelor WB, Heathcote J, Wanless IR. Chaparral-induced hepatic injury. Am J Gastroenterol 1995;90: 831–833.
10. Gordon DW, Rosenthal G, Hart J, Sirota R, Baker AL. Chaparral ingestion the broadening spectrum of liver injury caused by herbal medications. JAMA 1995;273: 489–490.
11. Heron S, Yarnell E. The safety of low-dose *Larrea tridentata* (DC) Coville (creosote bush or chaparral): a retrospective clinical study. J Altern Comp Med 2001;7: 175–185.
12. Alderman S, Kailas S, Goldfarb S, Singaram C, Malone DG. Cholestatic hepatitis after ingestion of chaparral leaf: confirmation by endoscopic retrograde cholangiopancreatography and liver biopsy. J Clin Gastroenterol 1994;19: 242–247.
13. Clark F, Reed R. Chaparral-induced toxic hepatitis—California and Texas, 1992. MMWR Morb Mortal Wkly Rep 1992;41:812–814.
14. Grice HC, Becking G, Goodman T. Toxic properties of nordihydroguaiaretic acid. Food Cosmet Toxicol 1968;6: 155–161.
15. Lambert JD, Zhao D, Meyers RO, Kuester RK, Timmermann BN, Dorr RT. Nordihydroguaiaretic acid: hepatotoxicity and detoxification in the mouse. Toxicon 2002; 40:1701–1708.
16. Leonforte JF. Contact dermatitis from *Larrea* (creosote bush). J Am Acad Dermatol 1986;14:202–207.
17. Gordon DW, Rosenthal G, Hart J, Sirota R, Baker AL. Chaparral ingestion. The broadening spectrum of liver injury caused by herbal medications. JAMA 1995;273:489–490.
18. Katz M, Saibil F. Herbal hepatitis: subacute hepatic necrosis secondary to chaparral leaf. J Clin Gastroenterol 1990;12:203–206.
19. Sheikh NM, Philen RM, Loved LA. Chaparral-associated hepatotoxicity. Arch Intern Med 1997;157:913–919.
20. Smith BC, Desmond PV. Acute hepatitis induced by ingestion of the herbal medication chaparral. Aust NZ J Med 1993;23:526.

21. Chitturi S, Farrell GC. Drug-induced cholestasis. Semin Gastrointestinal Dis 2001;12:113–124.
22. Smith AY, Feddersen RM, Gardner KD Jr, Davis CJ Jr. Cystic renal cell carcinoma and acquired renal cystic disease associated with consumption of chaparral tea: a case report. J Urol 1994;152:2089–2091.
23. Obermeyer WR, Musser SM, Betz JM, Casey RE, Pohland AE, Page SW. Chemical studies of phytoestrogens and related compounds in dietary supplements: flax and chaparral. Proc Soc Exp Biol Med 1995;208:6–12.

Chapter 58

CHASTE TREE
(*Vitex agnus-castus* L.)

HISTORY

Chaste tree is a native shrub of Mediterranean Europe that has been used since ancient Greek and Roman times as an herbal treatment for female reproductive problems. In the 4th century BC, Hippocrates recommended this herb for splenic injuries and inflammation.[1] The name of this herb derives from the Latin words for chastity (*castitas*) and lamb (*agnus*), and the name chaste tree refers to the belief that the use of this herb reduces sexual desire in women and encourages celibacy in members of religious orders. Pliny the Elder (AD 23–79) documented the use of this herb by Athenian women to preserve their chastity at the Thesmophoria by covering their beds with the foul-smelling leaves.[2] He also mentioned the use of this herb for the treatment of fever, flatulence, diarrhea, dropsy, and splenic diseases. In the *De Materia Medica* from about 55 AD, Dioscorides recommended a decoction of the seeds of the chaste tree for initiating lactation and menstruation. European herbalists used this herb to stimulate the uterus and promote menstruation, but the popularity of the herb waned until the middle 1900s.

BOTANICAL DESCRIPTION

Common Name: Chaste Tree, Chasteberry, Monk's Pepper

Scientific Name: *Vitex agnus-castus* L.

Botanical Family: Verbenaceae (verbena family)

Physical Description: This deciduous, aromatic plant ranges in size from a shrub to a small tree 3–9 ft (~1–3 m) tall. The leaves are opposite and palmately compounded with 5–7 leaflets and the underside is white-felted. Long spires of pale lilac or rose-colored flowers produce gray-brown fruits (drupes) with hard coverings. The dry fruits are the size of a black peppercorn, and these drupes contain four seeds.

Distribution and Ecology: This plant is indigenous to Mediterranean Europe and Central Asia.

EXPOSURE

Sources

Chaste tree extract is available as an herbal preparation, usually as a liquid (e.g., ethanol extract) or dried powder. The concentration of casticin in eight commercial samples of agni-casti fructus ranged from 0.025–0.212%, as measured by high performance liquid chromatography (HPLC).[3]

Medicinal Uses

TRADITIONAL

The fruit extract of chaste tree is a common folk medicine for female-related problems, including premenstrual syndrome, menstrual irregularity, infertility, acne, menopause, inadequate lactation, and hyperprolac-

Medical Toxicology of Natural Substances, by Donald G. Barceloux, MD

tinemia. Other uses of this herb include the treatment of fluid retention, indigestion, fungal infections, and anxiety. The chaste tree is not used in traditional Chinese or Indian (Ayurvedic) medicine.[2]

Current

This herb is typically administered as either an extract or as the dried ripe fruit of the chaste tree. The extract is usually standardized to contain a specified concentration of agnuside, aucubin, or casticin. Many commercial preparations contain 0.5% agnuside. Some studies support the use of chaste tree preparations for the treatment of menstrual disorders, but there is a lack of well-designed studies to document the efficacy of this herb for the treatment of female-related problems. In a prospective, open-label study of a 20 mg extract of the fruits of *Vitex agnus-castus* L., 20 of 43 women reported a reduction in premenstrual symptoms based on self-assessment using the Moos' menstrual distress questionnaire and a visual analogue scale.[4] The main response was related to the intensity rather than duration of symptoms. The extract was standardized to casticin. In a prospective, randomized, placebo-controlled study of 170 women with premenstrual syndrome, 52% of the treatment group responded to the treatment compared with 24% of the placebo group.[5] The treatment group received 20 mg daily of an ethanol extract of *Vitex agnus-castus* (Ze 440, Zeller AG, CH-Romanshorn). A responder was defined by a 50% reduction in premenstrual symptoms over three menstrual cycles based on self-assessment. A preliminary, single-blind, rater-blinded, prospective study compared the use of fluoxetine and *Vitex agnus-castus* extract to treat premenstrual dysphoric syndrome. The study did not demonstrate a statistically significant difference in response between the two treatment groups.[6] There was no placebo group. A double-blind, placebo-controlled study of the efficacy of a *Vitex agnus-castus* extract (Mastodynon®, Bionorica Arzneimittel GmbH, Neumarkt, Germany) for cyclical mastalgia detected an initial reduction in pain scores during the first menstrual cycle, but after three cycles the difference between treatment and placebo groups was not statistically significant ($p = 0.064$).[7] There are no clinical data to support the use of this herb to enhance breast size.[8]

Regulatory Status

The German Commission E approves the use of chaste tree extract for premenstrual tension, menstrual irregularities, and breast pain. Chaste tree extract is a dietary supplement in the United States and the United Kingdom.

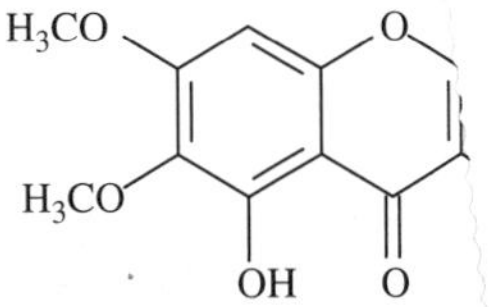

FIGURE 58.1. Chemical structur

PRINCIPAL INGREDIENTS

Chemical Composition

Extracts of chaste tree contain a variety of constituents including flavonoids (casticin, isovitexin, orientin), iridoid compounds (aucubin, agnuside, agnucastoside A, B, and C, mussaenosidic acid), monoterpene, diterpene (vitetrifolin B and C), and sesquiterpene compounds, castine (bitter principle), and linoleic acid.[9,10] Figure 58.1 displays the chemical structure of the flavonoid, casticin (CAS RN:479-91-4). The leaves, flowers, and fruits of chaste tree contain essential oils that vary in composition depending on distillation technique as well as geographical and environmental factors. The main constituents of a hydrodistillate of mature fruit from chaste tree were sabinene (16–44.1%), 1,8-cineole (8.4–15.2%), β-caryophyllene (2.1–5.0%), and *trans*-β-farnesene (5.0–11.7%).[11]

Mechanism of Toxicity

The active ingredient in chaste tree extracts has not been identified. *In vivo* human studies suggest that chaste tree preparations contain dopamine agonists that bind to D_2 receptors and inhibit the release of prolactin.[12] Extracts of *Vitex agnus-castus* contain diterpene compounds (clerodadienols) with potential dopaminergic activity.[13] *In vitro* studies also suggest that constituents in *Vitex agnus-castus* extract bind to mu and kappa opioid receptors.[14] Methanol extracts of the fruit from this herb probably do not have clinically significant estrogen agonist activity based on recombinant cell bioassay.[15] However, these extracts contain linoleic acid, which can bind to estrogen receptors (ERα, ERβ).[16]

DOSE RESPONSE

Recommended daily doses for relief of the symptoms of premenstrual tension include 175–225 mg/d of chaste tree or *Vitex agnus-castus* extract standardized to contain 0.5% agnuside and water/ethanol chaste tree extracts

…lent to 20–40 mg of fresh berries. The toxic dose …chaste tree extract is not well defined.

TOXICOKINETICS

The active ingredients in chaste tree extract are not well defined; therefore, there are few data on the toxicokinetics of these compounds. Although theoretically chaste tree extract can interact with dopamine antagonists, there are few data to indicate that the use of this herb causes clinically significant drug interactions.

CLINICAL RESPONSE

There are few data on the toxicity of chaste tree extracts. Adverse events following treatment with chaste tree extract are usually mild and reversible. The most common adverse reactions reported during the use of this herb are headache, nausea, abdominal pain, indigestion, acne, and allergic reactions (erythematous rash).[1]

DIAGNOSTIC TESTING

HPLC methods are available to determine the concentration of casticin in chaste tree extract.[3]

TREATMENT

Treatment is supportive with few toxic effects expected following exposure to chaste tree extracts.

References

1. Daniele C, Coon JT, Pittler MH, Ernst E. *Vitex agnus castus* a systematic review of adverse events. Drug Safety 2005;28:319–332.
2. Hobbs C. The chaste tree: *Vitex agnus castus*. Pharm History 1991;33:19–24.
3. Hoberg E, Meier B, Sticher O. Quantitative high performance liquid chromatographic analysis of casticin in the fruits of *Vitex agnus-castus*. Pharm Biol 2001;39:57–61.
4. Berger D, Schaffner W, Schrader E, Meier B, Brattstrom A. Efficacy of *Vitex agnus castus* L. extract Ze 440 in patients with pre-menstrual syndrome (PMS). Arch Gynecol Obstet 2000;264:150–154.
5. Schellenberg R. Treatment for the premenstrual syndrome with agnus castus fruit extract: prospective, randomized, placebo controlled study. BMJ 2001;322:134–137.
6. Atmaca M, Kumru S, Tezcan E. Fluoxetine versus *Vitex agnus castus* extract in the treatment of premenstrual dysphoric disorder. Hum Psychopharmacol Clin Exp 2003;18:191–195.
7. Halaska M, Beles P, Gorkow C, Sieder C. Treatment of cyclical mastalgia with a solution containing a *Vitex agnus castus* extract: results of a placebo-controlled double-blind study. Breast 1999;8:175–181.
8. Fugh-Berman A. "Bust enhancing" herbal products. Obstet Gynecol 2003;101:1345–1349.
9. Hajdú Z, Hohmann J, Forgo P, Martinek T, Dervarics M, Zupkó I, et al. Diterpenoids and flavonoids from the fruits of *Vitex agnus-castus* and antioxidant activity of the fruit extracts and their constituents. Phytother Res 2007;21:391–394.
10. Kuruuzum-Uz A, Stroch K, Demirezer LO, Zeeck A. Glucosides from *Vitex agnus-castus*. Phytochemistry 2003;63:959–964.
11. Sorensen JM, Katsiotis ST. Parameters influencing the yield and composition of the essential oil from Cretan *Vitex agnus-castus* fruits. Planta Med 2000;66:245–250.
12. Jarry H, Leonhardt S, Gorkow C, Wuttke W. *In vitro* prolactin but not LH and FSH release is inhibited by compounds in extracts of *Agnus castus*: direct evidence for a dopaminergic principle by the dopamine receptor assay. Exp Clin Endocrinol 1994;102:448–454.
13. Wuttke W, Jarry H, Christoffel V, Spengler B, Seidlova-Wuttke D. Chaste tree (*Vitex agnus-castus*)—pharmacology and clinical indications. Phytomedicine 2003;10;348–357.
14. Meier B, Berger D, Hoberg E, Sticher O, Schaffner W. Pharmacological activities of *Vitex agnus-castus* extracts *in vitro*. Phytomedicine 2000;7:373–381.
15. Klein KO, Janfaza M, Wong JA, Chang RJ. Estrogen bioactivity in fo-ti and other herbs used for their estrogen-like effects as determined by a recombinant cell bioassay. J Clin Endocrinol Metab 2003;88:4077–4079.
16. Liu J, Burdette JE, Sun Y, Deng S, Schlecht SM, Zheng W, et al. Isolation of linoleic acid as an estrogen compound from the fruits of *Vitex agnus-castus* L. (chaste berry). Phytomedicine 2004;11:18–23.

Chapter 59

CLOVE and EUGENOL
[*Syzygium aromaticum* (L.) Merr. & L. M. Perry]

HISTORY

Clove has been used as a spice and fragrance in China for over 2000 years. References to the medicinal use of clove oil for the treatment of toothache in Europe date back at least to the 17th century, when the use of clove oil was recommended in the *Practice of Physic*, published in France in the 1640s.[1] Clove oil and the main active ingredient, eugenol, have been popular ingredients of consumer products (i.e., soaps, detergents) since the 19th century in the United States. During the late 19th and early 20th centuries, clove oil was a popular source of eugenol for use in the synthesis of vanilla. Eugenol was isolated from clove oil in 1929, and commercial production of eugenol began in the United States during the early 1940s.[2] Clove cigarettes (kreteks) are an alternative form of tobacco use along with cigars and bidis (small hand-rolled cigarettes from India) that are particularly fashionable with youth. These cigarettes were popular in Australia during the 1970s and in the United States during the 1980s. In 1984 and 1985, the US Centers for Disease Control received 11 case reports associating the development of acute respiratory symptoms with the use of clove cigarettes.[3] These reports included two deaths, but the details of these cases were not well defined. Extensive publicity associated with these case reports and concerns about the health effects of kreteks resulted in a sharp decrease in the consumption of these cigarettes in the United States beginning in the late 1980s. As with tobacco cigarettes, the use of clove cigarettes (kreteks) among middle school and high school students currently is declining in the United States. In 2002, the current use (i.e., within the last 30 days) of clove cigarettes among middle school and high school students was 2.0% and 2.7%, respectively, as measured by National Youth Tobacco Survey.[4]

BOTANICAL DESCRIPTION

Common Name: Clove

Scientific Name: *Syzygium aromaticum* (L.) Merr. & L. M. Perry (*Eugenia caryophyllata*)

Botanical Family: Myrtaceae (myrtles)

Physical Description: The clove tree is a small, tropical evergreen that grows up to 20 ft (~6 m) tall. The large leaves are oblong up to 10 in. (~25 cm) long and 4 in. (~10 cm) wide.

Distribution and Ecology: The clove tree is indigenous to the Philippines and nearby islands, but this tree is widely cultivated in many tropical countries. Zanzibar and Mozambique are major producers of cloves and clove oil.

EXPOSURE

Sources

Commercial clove oil is a steam distillation product obtained from clove leaves or buds, which contain up to 84–88% eugenol. Eugenol is widely distributed in the plant kingdom, particular in cinnamon leaf and bark oil, basil oils (*Ocimum gratissimum* L.), and sweet basil essential oil (*Ocimum basilicum* L.). Further distillation

Medical Toxicology of Natural Substances, by Donald G. Barceloux, MD

of clove oil produces a refined product containing almost pure eugenol (i.e., >95%). Various whiskies aged in oak barrels contain eugenol (i.e., up to about 0.5 mg/L) as a result of the diffusion of the phenolic fraction of an ethanol extract of oak.[2] Clove cigarettes ("kreteks") are tobacco products exported from India that contain about 60% tobacco and 40% shredded clove buds. In Indonesia, clove cigarettes are smoked in a conventional manner. However, in the United States younger smokers typically inhale the smoke from kreteks deeply and then perform a Valsalva maneuver to maximize the delivery of the smoke to the alveoli in a manner similar to "toking" a marijuana cigarette.[5]

Uses

Traditional

Traditional uses for clove oil in Chinese medicine include use as an antimicrobial agent,[6] antispasmodic, antiparasitic agent, and a carminative for the treatment of excessive intestinal gas. Clove oil has insect-repellant properties. In a study of 38 essential oils, clove oil was the most effective mosquito repellant providing 100% repellency for 2–4 hours against three species of mosquitos. Clove oil was not directly compared with DEET (*N,N*-diethyl-meta-toluamide).[7]

Current

Clove oil is marketed over-the-counter in the United States as a topical dental analgesic and antiseptic, a flavoring agent in food, mouthwashes, and pharmaceutical products, and as an ingredient in aromatherapy. Eugenol is also a fragrance and flavoring agent, a zinc oxide-based dental analgesic, an insect attractant (Japanese beetles), and a chemical intermediate.

Regulatory Status

In the United States, eugenol and clove oil are generally recognized as safe food additives (GRAS), and the US Food and Drug Administration (FDA) has approved the use of eugenol and clove oil in foods.[8] Eugenol is also approved for use in the manufacture of textiles and textile fibers that contact food surfaces. Additionally, eugenol and clove oil are approved for use as over-the-counter dental analgesics.

PRINCIPAL INGREDIENTS

Chemical Composition

Clove oil represents slightly less than 25% of the weight of ground cloves as measured by ion trap gas chromatography/mass spectrometry.[9] Commercial clove oil contains eugenol in amounts ranging from approximately 50–88%, depending on the source. Other components of clove oil include eugenol acetate and β-caryophyllene. The content of eugenol in clove oil depends on the distillation process, geographic location, soil characteristics, and climate. Dried clove buds contain about 15% weight/volume clove oil (i.e., 120–130 mg/g dry clove buds). A commercial sample of clove oil contained about 70% eugenol and 15% β-caryophyllene, whereas a sample of clove bud extract from an Italian herb shop contained approximately 15% eugenol and 4% β-caryophyllene as measured by micellar electrokinetic capillary electrophoresis with diode array detection.[10]

Indonesian clove cigarettes contain approximately 60% tobacco and 40% shredded clove buds. In addition to the components from tobacco smoke, the smoke from clove cigarettes contains eugenol, β-caryophyllene, and minor amounts of eugenol acetate, α-humulene, and caryophyllene-epoxide.[24] The smoke from these cigarettes does not contain unique pyrolysis products when compared with tobacco smoke. Analysis of a popular clove cigarette (Djarum Special) indicated that the nicotine content of the rod filler of the cigarette was lower (7.4 mg) than conventional American cigarettes (13 mg).[11] However, the average smoking time and number of puffs are usually greater during clove cigarette smoking than conventional cigarette smoking. In a study of 10 smokers, the venous plasma nicotine concentrations and exhaled carbon monoxide concentrations were similar during clove cigarette and conventional cigarette smoking.[11]

Physiochemical Properties

Eugenol is a weakly acidic phenolic compound (CAS RN: 97-53-0, 4-allyl-2-methoxy-phenol), which is volatile and slightly water soluble. Figure 59.1 displays the structure of eugenol. This compound is colorless to pale-yellow liquid with an odor of cloves and a spicy, pungent taste. *In vitro* studies indicate that eugenol and acetyl eugenol inhibit cyclooxygenase and platelet aggrega-

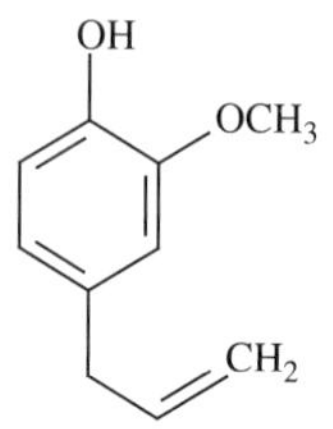

OH
OCH_3
CH_2

FIGURE 59.1. Chemical structure of eugenol ($C_{10}H_{12}O_2$, MW 164.2).

TABLE 59.1. Physical Properties of Eugenol

Physical Property	Value
Conversion Factor	1 ppm = 6.71 mg/m^3
Melting Point	−7.50 °C/18.5 °F
Boiling Point	253.2 °C/487.76 °F
Vapor Pressure	10 mmHg (123 °C/253 °F)
pK_a Dissociation Constant	10.19 (25 °C/77 °F)
log P (Octanol-Water)	2.27
Water Solubility	2,460 mg/L (25 °C/77 °F)
Henry's Law Constant	4.81E-08 atm-m^3/mole (25 °C/77 °F)
Atmospheric OH Rate Constant	6.50E-11 cm^3/molecule-second (25 °C/77 °F)

tion, reduce thromboxane synthesis in platelets,[12] and enhance lipoxygenase metabolism.[13,14] The smoke from clove cigarettes is sweetly aromatic with some numbing of the mouth occurring after inhalation of the smoke. Table 59.1 lists some of the physical properties of eugenol.

Mechanism of Toxicity

LUNGS

Animal studies to date have not demonstrated unique or highly toxic constituents in the smoke from clove cigarettes. A short-term study of rats exposed to smoke from tobacco and clove cigarettes for 30 minutes did not detect significant differences in body weights, food and water intake, respirations, or histological abnormalities between the tobacco-only group and the clove + tobacco group with the exception of higher carbon monoxide concentrations in the tobacco-only group.[15] A similar 14-day study of rats exposed to 15 minutes of smoke up to 6 times daily did not detect any unique histological or clinical effects in the kretek group.[16] There was a statistically significant increase in lung weights in the tobacco-only group, when compared with the kretek group. These studies were criticized because of the lack of statistical power, small group size (5 rats per group), and inability of the study to replicate actual human exposure to kretek smoke including the potential for aspiration.[17] The intratracheal instillation of 10% eugenol in doses up to 24 mg eugenol/kg produced interstitial hemorrhage, acute emphysema, and acute pulmonary edema in the lungs of the exposed rats.[18] However, exposure of male Syrian golden hamsters to kretek smoke from up to 30 cigarettes daily was not associated with any histological abnormalities in the lungs, trachea, larynx, nasal cavity, esophagus, kidney, or bladder. The estimated daily dose of eugenol was up to 2 mg/kg body weight.

LIVER

In vitro studies indicate that cytochrome P450 isoenzymes oxidize eugenol to a reactive quinone methide intermediate, which covalently binds to proteins or conjugates with glutathione.[19] These reactive intermediates can cause necrosis of hepatic cells. In animal models, depletion of glutathione prior to the administration of eugenol causes substantial, dose-dependent liver damage, whereas repletion of glutathione stores reduces the hepatotoxicity.[20]

DOSE RESPONSE

The estimated acceptable daily oral intake of eugenol is 2.5 mg/kg.[21] Case reports associate the accidental ingestion of 5–10 mL (5–10 g; density = 1.04 g/mL) of clove oil by infants and young children with the development of serious toxicity including seizures, coma, metabolic acidosis, hepatic failure, and disseminated intravascular coagulation (DIC).[22,23] The typical kretek smoker inhales about 7 mg eugenol per cigarette.

TOXICOKINETICS

There are few human data on the pharmacokinetics of eugenol, the active ingredient in clove oil. Animal studies suggest that the lungs absorb minor amounts of eugenol after inhalation of smoke from clove cigarettes.[24] *In vitro* studies of rat hepatocytes suggest that eugenol undergoes substantial biotransformation, primarily by the formation of sulfate, glucuronic acid, and glutathione conjugates.[25] The major metabolite formed in these studies was the glucuronic acid conjugate. Side-chain epoxidation is a minor metabolic pathway observed in these *in vitro* studies.[26,27]

CLINICAL RESPONSE

Acute Effects

Case reports associate the accidental ingestion of clove oil by infants with the development of vomiting, pulmonary aspiration, and central nervous system depression.[28] Patients usually recover without sequelae with timely supportive care. Additional case reports suggest that the ingestion of clove oil in young children can cause hepatorenal failure including hypoprothrombinemia, elevated serum hepatic aminotransferases, seizures, and metabolic acidosis.[23] The intravenous injection of an unknown amount of clove oil produced noncardiogenic pulmonary edema in a 32-year-old woman, manifest by bilateral interstitial and alveolar infiltrates without evidence of cardiac insufficiency.[29] She did not require intubation and she recovered without obvious sequelae.

Case reports associated the development of acute, severe respiratory symptoms in a few susceptible individuals with the recent use of clove cigarettes.[3] Frequently, these individuals had prodromal respiratory infections. These respiratory effects included hemoptysis, hemorrhagic and nonhemorrhagic pulmonary edema, pleural effusions, respiratory insufficiency, aspiration pneumonia, pulmonary infiltrates, and bronchospasm. However, the details of these case reports are not well documented, and there are limited new data on clove cigarette-associated respiratory distress since the middle 1980s.

Chronic Effects

Eugenol and clove leaf oil are weak sensitizers for the induction of allergic contact dermatitis.[30] Only one case of induced hypersensitivity and one case of preexisting hypersensitivity to eugenol were detected in patch test data from 11,632 subjects exposed to consumer products containing eugenol or clove leaf oil in concentrations up to 2.5%.[31] These cases of sensitization involved prolonged contact (i.e., 1–2 weeks) with dental products containing eugenol. Although rare, case reports associate the development of allergic reactions with the use of clove leaf oil. Stomatitis and intraoral ulceration can occur in sensitized individuals when exposed to clove oil in personal products, such as mouthwash.[32] Rarely, systemic reactions (e.g., urticaria) occur following the use of clove oil in dental products.[33,34]

The effects of long-term smoking of clove cigarettes are not well defined. There are limited data on the carcinogenicity of eugenol. The administration of eugenol to laboratory animals does not consistently produce tumors across animal species and gender. There is an absence of epidemiological data, and the IARC lists eugenol in Group 3 (not classifiable as to carcinogenicity).[2] *In vitro* test results for DNA damage following exposure to eugenol are inconsistent.[2] The US National Toxicology Program does not list eugenol as a compound reasonably anticipated to cause cancer.

DIAGNOSTIC TESTING

Analytical Methods

Eugenol is the principal active ingredient in clove oil, accounting for up to 80–85% of the product. Methods for the detection of eugenol in clove buds or clove oil include gas chromatography with flame ionization (GC/FID)[35] or gas chromatography/mass spectrometry (GC/MS). Micellar electrokinetic capillary electrophoresis with diode array detection is an alternative to gas chromatography that allows rapid separation and quantitation of eugenol and related compounds.[10] The limits of detection and quantitation for eugenol by this method are about 1 μg/mL and 3 μg/mL, respectively.

Biomarkers

The presence of eugenol helps confirm exposure, but there are few data correlating blood eugenol concentrations with clinical effects. Clove oil is not a unique source of eugenol; therefore, the presence of eugenol does not necessarily indicate exposure to clove oil or clove bud extracts.

Abnormalities

Laboratory abnormalities associated with case reports of acute clove oil poisoning include metabolic acidosis, elevated serum hepatic aminotransferases, hypoprothrombinemia, disseminated intravascular coagulation, and severe hypoglycemia.

TREATMENT

Stabilization

Patients with exposure to clove oil should be carefully evaluated for respiratory insufficiency and aspiration pneumonia with pulse oximetry and chest x-ray, particularly in the presence of altered consciousness. The analgesic properties of eugenol in clove oil may predispose these patients to aspiration pneumonia by decreasing the responsiveness of the gag reflex as a result of these local anesthetic effects. Severe hypoglycemia occasionally complicates the course of clove oil intoxication and all patients should be evaluated with a rapid blood glucose test. Severe hypoglycemia can occur during the first 8 hours after ingestion and before laboratory evidence of fulminant hepatic failure.[22] During severe intoxication, disseminated intravascular coagulation can develop despite the lack of clinical evidence of fulminant hepatic failure. These patients should be evaluated with a complete blood count (CBC), platelet count, prothrombin time (PT), partial thromboplastin time, D-dimer, and fibrin split products, as well as serum hepatic aminotransferases. The use of heparin for the treatment of DIC is controversial.

Antidote

There are few clinical data on the use of *N*-acetylcysteine for the treatment of hepatic failure associated with the ingestion of clove oil by children. Based on animal data, there is a theoretical basis for the use of *N*-acetylcysteine for the glutathione depletion and hepatic necrosis present in animals receiving eugenol.[20] Several case reports document improvement in hepatic function after the use of intravenous *N*-acetylcysteine,[23,36] but lack of data on the natural course of hepatic failure from clove oil intoxication limit determination of the efficacy of *N*-acetylcysteine in this setting. Treatment in these case reports involved the use of the standard doses of intravenous *N*-acetylcysteine, similar to the treatment of acetaminophen hepatotoxicity.

Supplemental Care

Patients hospitalized for serious clove oil poisoning should be monitored closely for hepatorenal function, hypoglycemia, hypophosphatemia, metabolic acidosis, and electrolyte imbalance. Baseline blood tests for hospitalized patients include CBC, PT, and serum chemistries including hepatic aminotransferase, bilirubin, glucose, electrolytes, and creatinine. Hepatorenal function should be monitored daily for 3 days until a clear pattern of improvement in these serum tests occurs. Typically, a rise in PT and unconjugated bilirubin concentrations are the earliest biochemical markers of hepatotoxicity; these abnormalities usually occur before an elevation of serum hepatic aminotransferases. Bedside glucose monitoring is necessary several times daily in patients with fulminant hepatic failure.

The efficacy of vitamin K_1 for elevation of the prothrombin time during fulminant hepatic failure is unclear. Fresh frozen plasma (rarely, recombinant human factor VIIa) should be used for evidence of current bleeding or prophylactically for invasive procedures.[37] Prophylactic correction of coagulopathy in a nonbleeding patient is not usually recommended. Serial hemoglobin and stool guaiac tests should be followed for evidence of gastrointestinal bleeding. The administration of proton pump inhibitors (e.g., omeprazole) should be administered to reduce the risk of gastrointestinal hemorrhage. Cerebral edema is the major cause of death following the development of hepatic encephalopathy, and treatment includes mannitol, elevation of the head to 10–20 degrees, adjustment of ventilator settings, and fluid restriction. Epidural intracranial pressure monitoring is usually necessary to guide treatment of patients with grade 3–4 hepatic encephalopathy, and early transfer to a liver transplant center should be considered for these patients.

References

1. Curtis EK. In pursuit of palliation: oil of cloves in the art of dentistry. Bull Hist Dent 1990;38:9–14.
2. International Agency for Research on Cancer (IARC). Eugenol. IARC Monogr Eval Risk Cancer 1985;36: 75–97.
3. US Centers for Disease Control (CDC). Illnesses possibly associated with smoking clove cigarettes. MMWR Morb Mortal Wkly Rep 1985;34:297–299.
4. Allen JA, Vallone D, Haviland ML, Healton C, Davis KC, Farrelly MC, et al. Tobacco use among middle and high school students—United States, 2002. MMWR Morb Mortal Wkly Rep 2003;52:1096–1098.
5. Guidotti TL, Laing L, Prakash UB. Clove cigarettes—the basis for concern regarding health effects. West J Med 1989;151:220–228.
6. Dorman HJ, Deans SG. Antimicrobial agents from plants: antibacterial activity of plant volatile oils. J Appl Microbiol 2000;88:308–316.
7. Trongtokit Y, Rongsriyam Y, Komalamisra N, Apiwathnasorn C. Comparative repellency of 38 essential oils against mosquito bites. Phytother Res 2005;19:303–309.
8. US Food and Drug Administration. Food and drugs. US Code Fed Reg, 1980, Title 21, Part 184.1257.
9. Huston CK, Ji H. Optimization of the analytical supercritical fluid extraction of cloves via an on-column interface to an ion trap GC/MS system. J Agric Food Chem 1991;39: 1229–1233.
10. Mandrioli R, Musenga A, Ferranti A, Lasaponara SS, Fanali S, Raggi MA. Separation and analysis of the major constituents of cloves by micellar electrokinetic chromatography. J Sep Sci 2005;28:966–972.
11. Malson JL, Lee EM, Murty R, Moolchan ET, Pickworth WB. Clove cigarette smoking: biochemical, physiological, and subjective effects. Pharmacol Biochem Behav 2003;74:739–745.
12. Saeed SA, Gilani AH. Antithrombotic activity of clove oil. J Pak Med Assoc 1994;44:112–115.
13. Srivastava KC, Justesen U. Inhibition of platelet aggregation and reduced formation of thromboxane and lipoxygenase products in platelets by oil of cloves. Prostaglandins Leukot Med 1987;29:11–18.
14. Srivastava KC, Malhotra N. Acetyl eugenol, a component of oil of cloves (*Syzygium aromaticum* L.) inhibits aggregation and alters arachidonic acid metabolism in human blood platelets. Prostaglandins Leukot Essent Fatty Acids 1991;42:73–81.
15. Clark GC. Comparison of the inhalation toxicity of kretek (clove cigarette) smoke with that of American cigarette smoke. I. One day exposure. Arch Toxicol 1989;63:1–6.
16. Clark GC. Comparison of the inhalation toxicity of kretek (clove cigarette) smoke with that of American cigarette smoke. II. Fourteen days, exposure. Arch Toxicol 1990;64: 515–521.
17. Guidotti TL. Critique of available studies on the toxicology of kretek smoke and its constituents by routes of entry involving the respiratory tract. Arch Toxicol 1989;63: 7–12.

18. LaVoie EJ, Adams JD, Reinhardt J, Rivenson A, Hoffmann D. Toxicity studies on clove cigarette smoke and constituents of clove: determination of the LD_{50} of eugenol by intratracheal instillation in rats and hamsters. Arch Toxicol 1986;59:78–81.
19. Thompson D, Constantin-Teodosiu D, Egestad B, Mickos H, Moldeus P. Formation of glutathione conjugates during oxidation of eugenol by microsomal fractions of rat liver and lung. Biochem Pharmacol 1990;39:1587–1595.
20. Mizutani T, Satoh K, Nomura H. Hepatotoxicity of eugenol and related compounds in mice depleted of glutathione: structural requirements for toxic potency. Res Commun Chem Pathol Pharmacol 1991;73:87–95.
21. World Health Organization. Evaluation of certain food additives. 23rd Report on the Joint FAO/WHO Expert Committee on Food Additives. Geneva: World Health Organization; 1980.
22. Hartnoll G, Moore D, Douek D. Near fatal ingestion of oil of cloves. Arch Dis Childhood 1993;69:392–393.
23. Janes SE, Price CS, Thomas D. Essential oil poisoning: *N*-acetylcysteine for eugenol-induced hepatic failure and analysis of a national database. Eur J Pediatr 2005;164: 520–522.
24. Council on Scientific Affairs. Evaluation of the health hazard of clove cigarettes. JAMA 1988;260:3641–3644.
25. Thompson DC, Constantin-Teodosiu D, Moldeus P. Metabolism and cytotoxicity of eugenol in isolated rat hepatocytes. Chem Biol Interact 1991;77:137–147.
26. Delaforge M, Janiaud P, Levi P, Morizot JP. Biotransformation of allylbenzene analogues *in vivo* and *in vitro* through the epoxide-diol pathway. Xenobiotica 1980;10:737–744.
27. Swanson AB, Miller EC, Miller JA. The side-chain epoxidation and hydroxylation of the hepatocarcinogens safrole and estragole and some related compounds by rat and mouse liver microsomes. Biochim Biophys Acta 1981;673: 504–516.
28. Lane BW, Ellenhorn MJ, Hulbert TV, McCarron M. Clove oil ingestion in an infant. Hum Exp Toxicol 1991;10: 291–204.
29. Kirsch CM, Yenokida GG, Jensen WA, Wendland R, Suh H, Bourgault M. Non-cardiogenic pulmonary oedema due to the intravenous administration of clove oil. Thorax 1990;45:235–236.
30. Kanerva L, Estlander T, Jolanki R. Occupational allergic contact dermatitis from spices. Contact Dermatitis 1996;35:157–162.
31. Rothenstein AS, Booman KA, Dorsky J, Kohrman KA, Schwoeppe EA, Sedlak RI, et al. Eugenol and clove leaf oil: a survey of consumer patch-test sensitization. Food Chem Toxicol 1983;21:727–733.
32. Vilaplana J, Grimalt F, Romaguera C, Conellana F. Contact dermatitis from eugenol in mouthwash. Contact Dermatitis 1991;24:223–224.
33. Grade AC, Martens BP. Chronic urticaria due to dental eugenol. Dermatologica 1989;178:217–220.
34. Barkin ME, Boyd JP, Cohen S. Acute allergic reaction to eugenol. Oral Surg Oral Med Oral Pathol 1984;57: 441–442.
35. Myint S, Daud WR, Mohamad AB, Kadhum AA. Gas chromatographic determination of eugenol in ethanol extract of cloves. J Chromatogr B Biomed Appl 1996;679:193–195.
36. Eisen JS, Koren G, Juurlink DN, Ng VL. N-acetylcysteine for the treatment of clove oil-induced fulminant hepatic failure. J Toxicol Clin Toxicol 2004;42:89–92.
37. Brown SA, Biggerstaff J, Savidge GF. Disseminated intravascular coagulation and hepatocellular necrosis due to clove oil. Blood Coag Fibrinol 1992;3:665–668.

Chapter 60

COLOCYNTH
[*Citrullus colocynthis* (L.) Schrad.]

HISTORY

Colocynth is an ancient purgative, which was mentioned in the Ebers papyrus from 1500 BC, in the Bible (II Kings 4:38–41), and in Dioscorides' writings. Egyptian physicians used this plant as a strong purgative. The Arab medical writer, Mesue, listed the use of two colocynth preparations, trochisi alhandal and electuarium majus hamech, in *The Pharmacopoea Augustana* of 1581.[1] In the 1800s, colocynth was listed in every issue of the *US Pharmacopeia*. The toxicity of colocynth was also well-known to medieval physicians, who recognized the occurrence of hemorrhagic ulcers in the intestines following the impaction of colocynth seeds.[2] In the middle 19th century, a fatal case of severe vomiting and diarrhea was associated with the use of colocynth.[3] Fatalities have been associated with the use of colocynth as an abortifacient.[4]

BOTANICAL DESCRIPTION

Common Name: Bitter Apple, Bitter Cucumber, Bitter Gourd, Egusi, Wild Watermelon

Scientific Name: *Citrullus colocynthis* (L.) Schrad.

Botanical Family: Cucurbitaceae (melon)

Physical Description: This annual vine has slender, hairy stems, which develop from a large, perennial fleshy root. The large (5–10 cm/~2–4 in.), obtuse, hairy, triangular leaves have long petioles and are similar in appearance to watermelon leaves. Long (4–10 cm/1.5-4 in.), pedunculated, solitary yellow flowers produce smooth, globular fruit with a thick yellow rind and a light, spongy pulp filled with numerous seeds.

Distribution and Ecology: This plant is indigenous to desert and semiarid areas of Africa, Western Asia, and the Mediterranean region. Arabs introduced this plant to Spain and Cyprus.

EXPOSURE

This extremely bitter purgative is not commonly available as an herbal preparation. The fruits and, in particular, the pulp of this plant are well-known natural cathartics since ancient times. Colocynth is one of several herbs used for the treatment of hypertension and diabetes mellitus in Morocco.[5,6] The leaves of this herb are used to treat asthma and jaundice, whereas the root is a traditional treatment for amenorrhea, breast inflammation, arthralgias, and ophthalmic diseases. Other medicinal uses include the treatment of seizures, tuberculosis, syphilis, and parasitic infections. The use of colocynth as a cathartic is not recommended by the German Commission E because of concern about adverse effects including severe gastrointestinal irritation, hemorrhagic diarrhea, renal dysfunction, and cystitis. Cucurbitacin I is a tetracyclic triterpenoid present in this plant that is a strong inhibitor of the janus kinase 2/signal transducer and an activator of transcription 3 (JAK2/STAT3) signaling pathway. STAT3 is a common oncogenic signaling pathway, and cucurbitacin compounds are potential chemotherapeutic agents.[7]

Medical Toxicology of Natural Substances, by Donald G. Barceloux, MD

PRINCIPAL INGREDIENTS

Ethanol extracts of colocynth and other members of the Cucurbitaceae family contain highly oxygenated tetracyclic triterpene compounds called cucurbitacins.[8] These bitter compounds possess cytotoxic properties.[9] Although originally isolated from species from this family, cucurbitacin compounds occur in a variety of plant families (e.g., Brassicaceae, Begoniaceae, Datiscaceae, Desfontainiaceae, Elaeocarpaceae, Rubiaceae, Rosaceae, Sterculiaceae, Thymelaeaceae), as well as several genera of mushroom (*Russula*, *Hebeloma*). Figure 60.1 displays the chemical structure of cucurbitacin I ($C_{30}H_{42}O_7$, CAS RN:2222-07-3). There are a variety of cucurbitacin compounds including cucurbitacin A, B, C, D, E, F, I, L and glucosides,[10] colocynthosides A and B, T, 23-dihydrocucurbitacin F, 24-dihydrocucurbitacin F, hexanorcucurbitacin F, and acetylated derivatives. However, not all of these compounds has been isolated from this plant species.[11,12] There are few data to determine which cucurbitacin compounds are the active or toxic ingredients present in colocynth.

DOSE RESPONSE

There are few data on the dose response of the toxin in colocynth, in part because of the lack of identification of the specific compound associated with the toxic colitis. The older medical literature suggests that 0.6–1 g of colocynth extract can produce bloody diarrhea.[13]

TOXICOKINETICS

There are few data on the toxicokinetics of cucurbitacin compounds. *In vitro* studies suggest that colocynth inhibits CYP1A-, CYP2B-, and CYP3A-dependent reactions, but there are few data to determine the clinical importance of these potential reactions.[14]

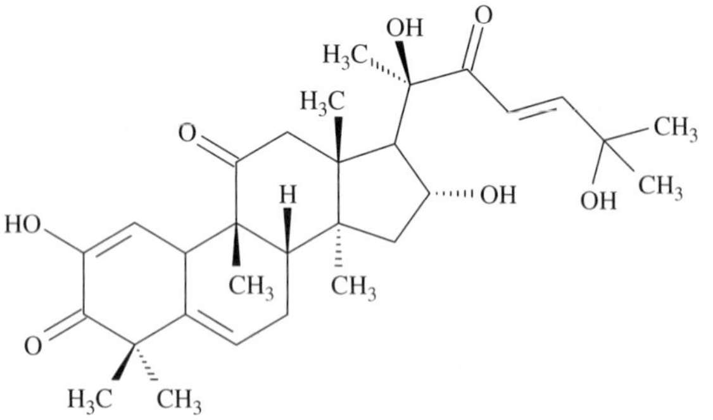

FIGURE 60.1. Chemical structure of cucurbitacin I.

CLINICAL RESPONSE

Case reports associate the use of extracts of *Citrullus colocynthis* with the development of bloody diarrhea, vomiting, colicky abdominal pain, and dehydration.[15] Pathological lesions primarily involve edema, erythema, superficial erosions and inflammatory exudates of the mucosa in the sigmoid and descending colon. Ulcerations and pseudopolyps are unusual features of this toxic colitis. Symptoms typically resolve within 3–6 days, and the pathological lesions resolve within 14 days without sequelae.[13]

DIAGNOSTIC TESTING

Biomarkers of the toxicity associated with colocynth currently are not available. Sensitive analytical methods are available for the detection of cucurbitacin compounds. The limit of quantitation of cucurbitacin I in plasma using liquid chromatography/mass spectrometry is about 10 ng/mL.[16]

TREATMENT

Treatment is supportive. Gut decontamination is usually unnecessary because of the diarrhea and vomiting associated with colocynth toxicity. Patients should be evaluated for fluid and electrolyte imbalance and any deficiencies should be corrected with fluid and/or electrolytes infusions.

References

1. Lloyd JU. *Citrullus colocynthis*. Chicago: The Western Druggist;1898.
2. Tidy CM. On poisoning by *Cueumis colocynthis*. Lancet 1868;i:158.
3. Roe RB. A case of colocynth poisoning. Lancet 1913;i:1527.
4. Patrick RL, Willey EN, Feter BF. Bitter apple (*Citrullus colocynthis*) poisoning. A discussion of its use as an abortifacient. N Carolina Med J 1960;21:23–27.
5. Nmila R, Gross R, Rchid H, Roye M, Manteghctti M, Petit P, et al. Insulinotropic effect of *Citrullus colocynthis* fruit extracts. Planta Med 2000;6:418–423.
6. Ziyyat A, Legssyer A, Mekhfi H, Dassouli A, Serhrouchni M, Benjelloun W. Phytotherapy of hypertension and diabetes in oriental Morocco. J Ethnopharmacol 1997;58: 45–54.
7. Blaskovich MA, Sun J, Cantor A, Turkson J, Jove R, Sebti SM. Discovery of JSI-124 (cucurbitacin I), a selective Janus kinase/signal transducer and activator of transcription 3 signaling pathway inhibitor with potent antitumor

activity against human and murine cancer cells in mice. Cancer Res 2003;63:1270–1279.

8. Seger C, Sturm S, Mair M-E, Ellmerer EP, Stuppner H. 1H and 13C NMR signal assignment of cucurbitacin derivatives from *Citrullus colocynthis* (L.) Schrader and *Ecballium elaterium* L. (Cucurbitaceae). Magn Reson Chem 2005;43:489–491.
9. Chen JC, Chiu MH, Nie RL, Cordell GA, Qiu SX. Cucurbitacins and curcurbitane glycosides: structures and biological activities. Nat Prod Rep 2005;22:386–399.
10. Nayab D, Ali D, Arshad N, Malik A, Choudhary I, Ahmed Z. Cucurbitacin glucosides from *Citrullus colocynthis*. Nat Prod Res 2006;20:409–413.
11. Yoshikawa M, Morikawa T, Kobayashi H, Nakamura A, Matsuhira K, Nakamura S, Matsuda H. Bioactive saponins and glycosides. XXVII. Structures of new cucurbitane-type triterpene glycosides and antiallergic constituents from *Citrullus colocynthis*. Chem Pharm Bull (Tokyo) 2007;55:428–434.
12. Che CT, Fang XD, Phoebe CH Jr, Kinghorn AD, Farnsworth NR, Yellin B, Hecht SM. High-field 1H-NMR spectral analysis of some cucurbitacins. J Nat Prod 1985;48: 429–434.
13. Goldfain D, Lavergne A, Galian A, Chauveinc L, Prudhomme F. Peculiar acute toxic colitis after ingestion of colocynth: a clinicopathological study of three cases. Gut 1989;30;1412–1418.
14. Barth A, Muller D, Durrling K. In vitro investigation of a standardized dried extract of *Citrullus colocynthis* on liver toxicity in adult rats. Exp Toxic Pathol 2002;54: 223–230.
15. Al Faraj S. Haemorrhagic colitis induced by *Citrullus colocynthis*. Ann Trop Med Parasitol 1995;89:695–696.
16. Molavi O, Shayeganpour A, Somayaji V, Hamdy S, Brocks DR, Lavasanifar A, et al. Development of a sensitive and specific liquid chromatography/mass spectrometry method for the quantification of cucurbitacin I (JSI-124) in rat plasma. J Pharm Pharmaceut Sci 2006;9:158–164.

Chapter 61

COLTSFOOT
(*Tussilago farfara* L.)

HISTORY

Herbal preparations of coltsfoot are traditional treatments for cough and bronchitis.

BOTANICAL DESCRIPTION

The development of veno-occlusive disease following the ingestion of teas brewed from pyrrolizidine-containing plants may occur as a result of the misidentification of coltsfoot for pyrrolizidine-containing plants (e.g., *Adenostyles alliariae* A. Kern.) or some *Petasites* species [*P. vitifolius* Greene (pro sp.), *P. frigidus* Fries].[1]

Common Name: Coltsfoot, Colts Foot

Scientific Name: *Tussilago farfara* L.

Botanical Family: Asteraceae (sunflowers or tournesol family)

Physical Description: This perennial herbaceous plant has a creeping rhizome and long runners. The basal leaves are hear-shaped and slightly toothed with width of up to 15 cm (~10 in.). The top of the mature leaves are glossy and dark green, whereas the underside is covered with gray, felt-like hairs. Young leaves have dense felt on both sides. Several, white woolly flowering stalks arise directly from the rootstock in the spring. Each stalk bears a single yellow flower head consisting of tubular disk flowers at the center surrounded by ray flowers. The appearance of the ripe seed head is similar to the dandelion. After the flowers appear, leaves develop on long erect stalks arising directly from the rootstock.

Distribution and Ecology: This plant is a common, invasive weed that grows in heavy clay soils throughout Britain, Europe, North Africa, Asia, and sporadically in the United States. This plant inhabits wet areas, such as creek beds and meadows.

EXPOSURE

The flower of coltsfoot (kuan donghua) is used as a traditional treatment for respiratory problems, such as bronchitis, asthma, and cough. Other traditional uses include the treatment of eye irritation, menstrual irregularities, varicose veins, wounds, and burns.[2] Coltsfoot contains pyrrolizidine alkaloids, which are potentially carcinogenic. The German Commission E limits the concentration of pyrrolizidine alkaloids in herbal products. However, the listing of pyrrolizidine alkaloids is not required in the United States.

PRINCIPAL INGREDIENTS

Chemical Composition

Extracts of coltsfoot contain pyrrolizidine alkaloids, primarily senkirkine (CAS RN: 2318-18-5) in an amount of approximately 0.15 g/kg.[3] Figure 61.1 displays the chemical structure of senkirkine ($C_{19}H_{27}NO_6$). The flowers and a decoction from the boiling of 5 g coltsfoot leaves in water contained about 0.45 ppm senkirkine as

Medical Toxicology of Natural Substances, by Donald G. Barceloux, MD

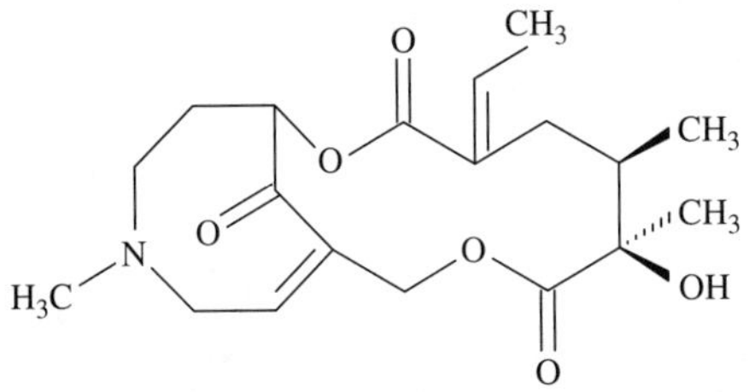

FIGURE 61.1. Chemical structure of the pyrrolizidine alkaloid, senkirkine.

measured by cation-exchange solid-phase extraction and ion-pair high performance liquid chromatography.[4] Senecionine (CAS RN: 130-01-8) was not detected in these samples by this method. Analysis of dried, young flowers of coltsfoot demonstrated a pyrrolizidine content (i.e., primarily senkirkine) of about 0.015% dry weight.[5] The methylene chloride extract of the flower buds from coltsfoot contains a vasoactive substance, tussilagone (L–652,469), which produces dose-dependent pressor effect in anesthetized animals.[6]

Physiochemical Properties

Tussilagone ($C_{23}H_{34}O_5$, CAS RN: 104012-37-5) is a sesquiterpene compound that inhibits the binding of platelet-activating factor to platelet membranes and the specific binding of experimental Ca^{2+} channel blockers in the cardiac sarcolemmal vesicles of rats.[7] This compound has low polarity and low water solubility.[6] Additionally, the ethyl acetate fraction of flower buds from *Tussilago farfara* inhibited lipid peroxidation and arachidonic acid metabolism, as well as demonstrating antioxidant properties.[8] Senkirkine is a hepatotoxic pyrrolizidine alkaloid present in variable amounts in traditional herbal medicines derived from coltsfoot.

DOSE RESPONSE

There are limited data on the dose response of the active ingredients in coltsfoot. For an ultrasonically prepared emulsion of soy bean phospholipid containing tussilagone, the intravenous LD_{50} and LD_{10} in mice was 28.9 ± 1.8 mg/kg and 25.3 ± 2.0 mg/kg, respectively.[6]

TOXICOKINETICS

There are inadequate data to determine the toxicokinetics of tussilagone in humans.

CLINICAL RESPONSE

There is limited evidence for the carcinogenicity of both senkirkine and coltsfoot (*Tussilago farfara*) in experimental animals. A small lifetime study of rats fed up to 32% coltsfoot in their diet detected a dose-related increase in hemangioendothelial sarcomas of the liver.[5] Histological examination of the livers demonstrated centrilobular necrosis, destruction of hepatocytes, intrahepatic bile proliferation, and cirrhosis even in the absence of liver tumors. Twenty rats received intraperitoneal injections of freshly prepared senkirkine at a dose of 10% of the median lethal dose twice weekly for 4 weeks and then once weekly for 52 weeks.[9] Over the lifetime of these rats, 9 of 20 rats developed liver adenomas, whereas no tumors developed in control rats receiving intraperitoneal injections of saline at the same time intervals. In the absence of epidemiological data, the International Agency for Research on Cancer listed senkirkine as an agent not classifiable as to its carcinogenicity to humans (Group 3).[10]

DIAGNOSTIC TESTING

Limits of detection and quantitation for the pyrrolizidine alkaloid, senkirkine, were 0.07 μg/mL and 0.20 μg/mL, respectively, when measured by cation-exchange solid-phase extraction and ion-pair high performance liquid chromatography.[4] Micellar electrokinetic chromatography is a simple, rapid method for the separation and determination of toxic pyrrolizidine alkaloids (senkirkine, senecionine, retrorsine, seneciphylline) potentially present in the traditional Chinese herbal medicines (kuan donghua) derived from coltsfoot.[11]

TREATMENT

Treatment is supportive. Gut decontamination measures are usually unnecessary.

References

1. Sperl W, Stuppner H, Gassner I, Judmaier W, Dietze O, Vogel W. Reversible hepatic veno-occlusive disease in an infant after consumption of pyrrolizidine-containing herbal tea. Eur J Pediatr 1995;154:112–116.
2. Uzun E, Sariyar G, Adsersen A, Karakoc B, Otuk G, Oktayoglu E, Pirildar S. Traditional medicine in Sakarya province (Turkey) and antimicrobial activities of selected species. J Ethnopharmacol 2004;95:287–296.
3. Culvenor CC, Edgar JA, Smith LW, Hirono I. The occurrence of senkirkine in *Tussilago farfara*. Aust J Chem 1976;29:229–230.
4. Mroczek T, Glowniak K, Wlaszczyk A. Simultaneous determination of N-oxides and free bases of pyrrolizidine alkaloids by cation-exchange solid-phase extraction and ion-pair high-performance liquid chromatography. J Chromatogr A 2002;949:249–262.

5. Hirono I, Mori H, Culvenor CC. Carcinogenic activity of coltsfoot, *Tussilago farfara* L. Gann 1976;67:125–129.
6. Li Y-P, Wang Y-M. Evaluation of tussilagone: a cardiovascular-respiratory stimulant isolated from Chinese herbal medicine. Gen Pharmacol 1988;19:261–263.
7. Hwang SB, Chang MN, Garcia ML, Han QQ, Huang L, King VF, Kaczorowski GJ, Winquist RJ. L-652,469—a dual receptor antagonist of platelet activating factor and dihydropyridines from *Tussilago farfara* L. Eur J Pharmacol 1987;141:269–281.
8. Cho J, Kim HM, Ryu J-H, Jeong YS, Lee YS, Jin C. Neuroprotective and antioxidant effects of the ethyl acetate fraction prepared from *Tussilago farfara* L. Biol Pharm Bull 2005;28:455–460.
9. Hirono I, Haga M, Fujii M, Matsuura S, Matsubara N, Nakayama M, et al. Induction of hepatic tumors in rats by senkirkine and symphytine. J Natl Cancer Inst 1979;63: 469–472.
10. International Agency for Research on Cancer. Senkirkine. IARC Monogr Eval Carcinog Risk Chem Hum 1983;31: 231–238.
11. Yu L, Xu Y, Feng H, Li SF. Separation and determination of toxic pyrrolizidine alkaloids in traditional Chinese herbal medicines by micellar electrokinetic chromatography with organic modifier. Electrophoresis 2005;26: 3397–3404.

Chapter 62

COMFREY and OTHER PYRROLIZIDINE-CONTAINING PLANTS

HISTORY

Hepatotoxic, pyrrolizidine-containing herbs are present in several plant families, and the consumption of these plants as teas or contaminates of cereal have caused epidemics of hepatic veno-occlusive disease in various countries throughout the world.[1] In 1920, Wilmot and Robertson described liver cirrhosis associated with the ingestion of *Senecio* species in South Africa.[2] Cases of pyrrolizidine alkaloid intoxication are sporadic in the developed Western countries, whereas epidemics of veno-occlusive disease occur in underdeveloped countries where pyrrolizidine alkaloids contaminate foodstuffs. During periods of famine in Afghanistan[3] and in southern Tajikistan,[4] the consumption of bread contaminated by seeds from *Heliotropium lasiocarpum* Fisch. & C. A. Mey. caused an outbreak of hepatic veno-occlusive disease. In India, the ingestion of bread contaminated by *Crotalaria* seeds produced an epidemic of veno-occlusive liver disease.[5] Case reports from India also associated the use of pyrrolizidine-containing herbs (*Heliotropium lasiocarpum*) with the development of veno-occlusive disease and hepatic failure.[6] The consumption of *Echium* and *Heliotropium* species by grazing animals also produces liver toxicity and death.[7] In Jamaica, the ingestion of bush teas brewed from local *Crotalaria* species caused veno-occlusive disease in the 1950s.[8]

BOTANICAL DESCRIPTION

There are over 350 pyrrolizidine alkaloid-containing plants in more than 6,000 plant species from the Boraginaceae, Asteraceae (Compositae), Convolvulaceae, Fabaceae, Hyacinthaceae, Lamiaceae, and Moraceae families.[9,10] Most of these plant species are invasive weeds that produce toxicity in animals consuming contaminated feeds and grains. Case reports of human toxicity from pyrrolizidine-containing plants are relatively uncommon compared with animal toxicity. Pyrrolizidine-containing plants involved in human toxicity include *Adenostyles alliariae* A. Kern., *Symphytum officinale* L., and *Tussilago farfara* L., as well as species from the genera *Senecio*, *Crotalaria*, and *Heliotropium*.[11]

Asteraceae (Sunflower)

There are approximately 3,000 species of *Senecio* plants distributed worldwide with over 30 species associated with pyrrolizidine toxicity in grazing animals.[12] One of the most common species is *S. jacobaea*.

Common Name: Tansy Ragwort, Stinking Willie

Scientific Name: *Senecio jacobaea* L.

Physical Description: This tall (up to 5 ft/~1.5 m), erect perennial weed has a tall, flowering stalk in the spring. Composite yellow heads produce thousands of seeds. The leaves are pinnate varying up to 9–10 in. (~23–25 cm) in length. Accidental exposure to tansy ragwort may result from the confusion of this plant with the herb tansy (*Tanacetum vulgare* L.).

Distribution and Ecology: *Senecio jacobaea* is a native plant in the British Isles, but this species has been introduced into Western Europe, North

Medical Toxicology of Natural Substances, by Donald G. Barceloux, MD

America (the West, the Northeast, Michigan, Illinois), New Zealand, Australia, and South Africa.

Common Name: Threadleaf Groundsel, Threadleaf Ragwort, Wooly Groundsel

Scientific Name: *Senecio flaccidus* Less. (*Senecio douglasii* species *longilobus*)

Physical Description: This low, perennial, woody shrub grows to 2 ft (0.6 m) in height. A white wooly pubescence covers the large, pinnate leaves. The composite heads contain numerous flat terminal clusters with yellow rays.

Distribution and Ecology: *Senecio flaccidus* is a native shrub of the southwestern United States.

Boraginaceae (Borage)

Echium species are hardy annual and biennial plants that possess erect, hairy stems and ovate to lanceolate, green leaves. *Heliotropium europaeum* L. (European heliotrope) is an annual plant that contains gray-green, oval leaves covered with short, white hairs and prominent veins. Heliotrope seeds are approximately 2 mm (~0.08 in.) long, occurring in groups of four single-seeded nutlets.

Common Name: Wild or Common Comfrey, Comfrey, Blackwort, Borraja, Bourrache, Buyuk Karakafesotu, Consoude, Consuelda, Hirehari-So, Liane Chique

Scientific Name: *Symphytum officinale* L.

Physical Description: This invasive perennial weed grows up to 5 feet (~1.5 m) in height and possesses long, hairy leaves with narrow ends, deep roots, and yellowish to red-violet flowers. Misidentification of this plant occurs with foxglove (*Digitalis* spp.), particularly when comfrey is not flowering.

Distribution and Ecology: *Symphytum* species inhabit areas in Europe, North America, Siberia, and Asia. Wild or common comfrey is native to England and this plant can be found throughout most of Europe, central Asia, western Siberia, and the United States. *Symphytum officinale* (common comfrey) is an invasive weed introduced into the western, midwestern, and northeastern United States.

Common Name: Russian Comfrey, Caucasian Comfrey, Blue Comfrey, Quaker Comfrey

Scientific Name: *Symphytum x uplandicum* Nyman (*S. asperum X officinale*)

Physical Description: This plant is a hybrid between the English species, *Symphytum officinale*, and the SW Asian species, *Symphytum asperum*. This hybrid is a tall, rank, herbaceous perennial with a poorly developed wing and blue, violet or purple tubular flowers. The poorly developed stem wing is the most characteristic feature distinguishing this hybrid from *S. officinale*.

Distribution and Ecology: This species is a natural hybrid of *S. officinale* L. (common comfrey) and *S. asperum* Lepechin (prickly comfrey) that was introduced into England from Russia in the 1800s and into Canada and the United States in the 1950s. Most comfrey plants in the United States derive from this hybrid. This plant easily naturalizes to waste areas and roadsides.

Fabaceae (Leguminoseae)

Most *Crotalaria* species are perennial branched legumes that reach up to 3 ft (~0.9 m) in height. The simple 1–2 in. (~2.5–5.0 cm) leaves contain short pubescence on both sides. The small yellow flowers produce 1–2 in. (~2.5–5.0 cm) seed pods that contain brown, kidney-shaped, 2–3 mm (~0.10 in.) seeds that rattle inside the pod when dried. The common name for these species is "rattle pod."

Common Name: Common Houndstongue, Gypsy-Flower, Gypsy Flower, Hound's Tongue, Houndstongue

Scientific Name: *Cynoglossum officinale* L.

Physical Description: An annual or biennial plant that forms rosettes the first year and regrows from a taproot the following year to produce a flowering stalk up to 3 ft (~1 m) in height and 2–3 in. (~5–8 cm) long, brown nutlets.

Distribution and Ecology: *Cynoglossum officinale* (hound's tongue) is an invasive weed in the United States that is native to Europe.

EXPOSURE

Most pyrrolizidine-containing plants are invasive weeds that may contaminate the feed of grazing animals. In Western Europe, comfrey is a traditional herb for the treatment of inflammatory conditions (gout, arthritis, thrombophlebitis), skin wounds (bruises, fungal infections, ulcers, fractures, strains), varicose veins, oral lesions (gargle), bronchitis, allergies, gastritis, gastroduodenal ulcers, and diarrhea.[13,14] Russian comfrey is consumed as salad and as a tea made from the powdered root.[15] The internal use of comfrey is not approved by US, Canadian, or South African health agencies. The

German Commission E limits the use of comfrey to external applications.

PRINCIPAL INGREDIENTS

Chemical Composition

Most pyrrolizidine-containing plants contain mixtures of alkaloids with varying potency and concentrations. In *Senecio* species, senecionine *N*-oxides form in the roots. These compounds are transported to the above-ground plant organs via the phloem where pyrrolizidine alkaloids form in a species-specific pattern.[16] Common patterns of pyrrolizidine alkaloids include jacobine, erucifoline, and mixed or senecionine types.[17] Typically, the pyrrolizidine content is highest in the roots with smaller concentrations in the leaves, particularly older leaves.[18] A study of *Senecio jacobaea* indicated that the pyrrolizidine content on the surface of the leaves was substantially higher than the alkaloids in the interior of the leaf.[19] In addition to pyrrolizidine alkaloids, comfrey contains allantoin, polyphenols, amino acids, phytosterols, saccharides, triterpenoids, and mucilage.[20] Commercial honey and honey samples containing two grains or less of Ragwort pollen per gram of honey do not usually contain detectable amounts of pyrrolizidine alkaloids.[21] Honey samples collected in late July and August contained Ragwort pollen at 15–21 grains/g, and these samples had total pyrrolizidine alkaloid concentrations of 0.011–0.056 mg/kg. The major alkaloids were jacobine, jacozine, seneciophylline, and senecionine with jacoline being a minor constituent.

Asteraceae

The pyrrolizine content in *Senecio jacobaea* is approximately 0.2% dry weight.[9] *S. flaccidus (S. douglasii* subsp. *longilobus)* contains four pyrrolizidine alkaloids comprising up to about 2% dry weight with retrorsine being the major alkaloid. Unlike most other *Senecio* species, Riddell's ragwort (*Senecio riddellii* Torr. & Gray) contains primarily one alkaloid (i.e., riddelliine) at concentrations ranging up to 18% dry weight.[9] Senkirkine is a constituent of numerous *Senecio* species including *Senecio kirkii* Hook. F. ex Kirk, *Senecio rodriguezii* Willk. ex Rodr., and *Senecio jacobaea* L. as well as *Tussilago farfara* L. (coltsfoot). The main pyrrolizidine alkaloids in *S. rodriguezii* are retronecine-based (seneciphylline, seneciphyllinine) and otonecine-based (senkirkine, acetylsenkirkine) alkaloids, whereas the main alkaloid in *S. leucanthemifolius* Poiret is integerrimine.[22] *Psacalium decompositum* (Gray) H.E. Robins. & Brett. (desert indianbush, matarique) is an indigenous herb of northern Mexico and southwestern United States that contains at least seven pyrrolizidine alkaloids.[23]

Boraginaceae

The pyrrolizidine content of *Symphytum* species varies with plant part, season, natural biological variation, and species. Pyrrolizidine alkaloids in common comfrey (*S. officinale*) include intermedine, acetylintermedine, lycopsamine, acetyllycopsamine, echimidine, symviridine, and symphytine (Figure 62.1). Small young leaves early in the season possess higher total alkaloid content than older leaves, and the roots contain greater concentrations of total pyrrolizidine alkaloids than above-ground plant parts.[24] In a study of commercial samples of common comfrey (*S. officinale*), analysis of the total pyrrolizidine alkaloid content demonstrated values ranging from 1,380–8,320 µg/g root compared with 15–55 µg/g leaf.[25] The major pyrrolizidine alkaloids were symphytine and symlandine along with lesser concentrations of echimidine, lycopsamine, and acetyl lycopsamine. Analysis of comfrey tablets indicated that total alkaloid concentrations are similar to the alkaloid content of comfrey roots. The total alkaloid content of fresh leaves from Russian comfrey (*Symphytum x uplandicum*) ranges from about 0.01–0.15% with lycopsamine and echimidine accounting for a majority of the eight pyrrolizidine alkaloids.[26] In general, the concentration of echimidine in *Symphytum x uplandicum* and *Symphytum asperum* Lepechin (prickly comfrey) are substantially higher than *Symphytum officinale*.[27]

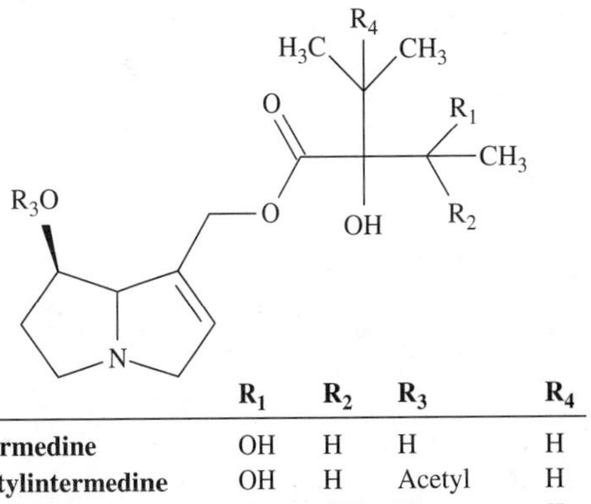

	R_1	R_2	R_3	R_4
Intermedine	OH	H	H	H
Acetylintermedine	OH	H	Acetyl	H
Lycopsamine	H	OH	H	H
Acetyllycopsamine	H	OH	Acetyl	H
Symphytine	H	OH	Tiglyol	H
Echimidine	H	OH	Angeloyl	OH
Symviridine	OH	H	Senecioyl	H

FIGURE 62.1. Chemical structures of pyrrolizidine alkaloids in common comfrey (*S. officinale*).

Fabaceae

Cynoglossum officinale (hound's tongue) contains four pyrrolizidine alkaloids at concentrations up to 2.2% dry weight with heliosupine being the most abundant.[9] The principal toxic alkaloids of *Heliotropium* species include heliotrine, lasiocarpine, and their *N*-oxides.

Physiochemical Properties

In general, naturally occurring pyrrolizidine alkaloids are esterified 8-membered necine compounds. These alkaloids are heterocyclic compounds composed of esters of basic alcohols (necine bases) connected to necic acid (C_5–C_{10} chain acids) via hydroxy groups at positions 7 and 9 by an ester function. Figure 62.2 demonstrates the basic structure of pyrrolizidine alkaloids. This diverse class of naturally occurring secondary metabolites include macrocyclic diesters (monocrotaline, retrorsine, senecionine, petasitenine), open diesters (lasiocarpine), monoesters (heliotrine), and necine bases.[28] Alkaline hydrolysis produces alkanolamines (necines) and necic acids. Total alkaloid content of these compounds includes both the free bases and the *N*-oxide forms. These compounds are colorless, optically active alkamine esters that are chemically stable and variably water soluble. Water-soluble salts of pyrrolizidine compounds include nitrates, sulfates, and chlorides. Drying and heating during food processing does not affect the toxicity of *Heliotropium* alkaloids.

Mechanism of Toxicity

Based on animal studies, the toxicity of pyrrolizidine alkaloids varies with individual compounds and animal species (e.g., pigs, chickens, and rats are much more sensitive than mice and sheep). In rodent studies, the most toxic pyrrolizidine compounds are macrocyclic diesters with heliotridine esters being 2–4 times more toxic than retronecine esters.[29] The primary toxicity of pyrrolizidine alkaloids is a sinusoidal-obstruction syndrome similar to the hepatotoxicity associated with bone marrow transplantation. During the acute phase of pyrrolizidine-induced hepatotoxicity, sinusoidal endothelial cells, central venular endothelial cells, and hepatic parenchymal cells undergo degeneration followed by a subacute phase consisting of fibrotic occlusion of central and sublobular veins along with sinusoidal fibrosis.[30] Most hepatotoxic pyrrolizidine are cyclic diesters (e.g., retrorsine, senecionine), and the structural requirement for pyrrolizidine toxicity is the presence of an unsaturated pyrrolizidine ring. Esterification of necine compounds containing a double bond in the 1,2 position causes the formation of toxic alkaloids.[31] All toxic pyrrolizidine alkaloids are ester derivatives of 1-hydroxymethyl-1,2-dehydropyrrolizidine with variable esterification at positions C1 and C7.[11] Reactive dehydroalkaloid (pyrrolic) intermediates resulting from the metabolism of pyrrolizidine compounds are potent electrophiles that can cross-link with cellular components (e.g., DNA, proteins, amino acids, glutathione).[32] These adducts are responsible for local cytotoxic and antimitotic effects.[33] These reactive pyrroles produce nonthrombolytic obliteration of small hepatic veins that may progress to cirrhosis and liver failure. Dehydroalkaloid intermediates with longer half-lives (e.g., retrosine, seneciphylline) may cause damage in the lungs (e.g., pulmonary endothelial cells, type II pneumocytes) similar to monocrotaline and fulvine.[34] Animal studies indicate that the hepatotoxic potential of rosmarinine, retronecine, and senaetnine is relatively low.[35]

Upon metabolic activation, pyrrolizidine alkaloids exhibit a variety of genotoxic properties including DNA binding, DNA-protein cross-linking, sister chromatid exchange, chromosomal aberrations, and carcinogenicity. The most potent genotoxic pyrrolizidine alkaloids are macrocyclic diesters (retronecine, heliotridine, otonecine).[36]

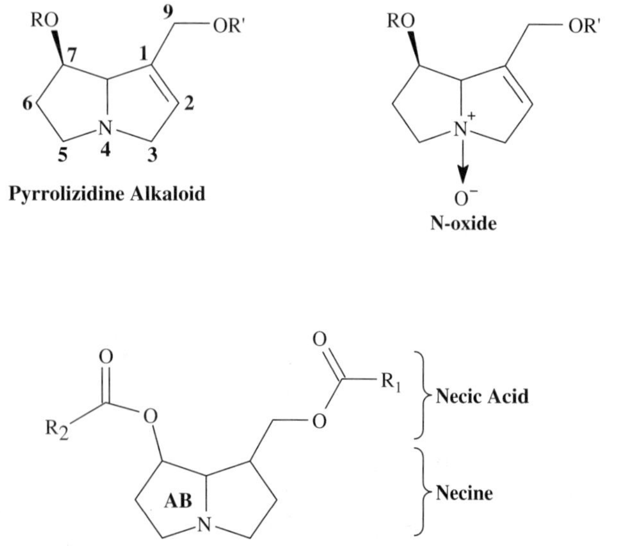

FIGURE 62.2. General structures of pyrrolizidine alkaloids.

DOSE RESPONSE

Medicinal Use

Typical daily herbal doses of comfrey leaf and root are about 5–30g and 0.5–10g, respectively.[37] There are no specific indications for the internal use of pyrrolizidine alkaloids. The German Federal Health Department limits the daily external application of total pyrrolizidine alkaloids to 0.1 μg for no more than 6 weeks per year. Because of potential toxicity, the internal use of pyrrolizidine-containing compounds is not recommended.

Toxicity

Due of the wide variability in total pyrrolizidine content in various plant species and interindividual variation in susceptibility, the dose response relationship for pyrrolizidine alkaloids is not well defined. Because the free base and the *N*-oxide forms of pyrrolizidine alkaloids demonstrate similar toxicity, the total alkaloid content of a plant or extract probably correlates reasonably well to severity of the hepatoxicity.[38] The daily consumption of comfrey containing an estimated 15 µg total pyrrolizidine alkaloids/kg body weight (total estimated minimal dose = 85 mg) over 6 months along with chamomile tea and vitamin supplements was associated with the development of veno-occlusive disease in a woman.[39] Three Chinese women developed fatal hepatic failure after ingesting an herbal tea containing pyrrolizidine alkaloids. The estimated mean cumulative dose of pyrrolizidine alkaloid base and *N*-oxide derivatives ingested prior to the onset of symptoms was 18 mg/kg.[40] A fourth woman, who ingested a total cumulative dose of 15 mg/kg, developed asymptomatic liver disease. During a famine, severe liver disease resulted from the daily ingestion of total pyrrolizidine alkaloids from *Heliotropium* alkaloids at an estimated dose of 30–40 µg/kg body weight.[3] An estimated dose of 70–147 mg of total pyrrolizidine alkaloids from *Senecio* tea (gordolobo yerba) was associated with liver cirrhosis in a 6-month-old infant.[41] The estimated daily ingestion of 60 µg total pyrrolizidine alkaloids (primarily seneciphylline)/kg over 15 months produced reversible veno-occlusive disease in an infant.[42] Children appear particularly susceptible to pyrrolizidine toxicity, but the specific dose response for the pediatric population has not been defined.

TOXICOKINETICS

Animal studies suggest that the percutaneous absorption of pyrrolizidine alkaloids is low (<0.5% of the applied dose).[43] Biotransformation of pyrrolizidine alkaloids involves *N*-oxidation to *N*-oxides, hydrolysis to free necic acid and necic bases, and dehydrogenation to toxic pyrroles (dehydroalkaloids).[44] Figure 62.3 displays the general biotransformation of pyrrolizidine alkaloids. The parent pyrrolizidine alkaloids are nontoxic, and the development of hepatic damage requires metabolism of the pyrrolizidine alkaloids by the cytochrome P450 oxidase system (e.g., CYP3A4, CYP2B6) to the hepatotoxic compounds, dehydropyrrolizidine alkaloids (pyrrolic esters).[45,46] These highly reactive compounds either form adducts with cell constituents or undergo further hydrolysis to dehydroamino alcohol compounds (pyrrolic alcohols). Detoxification involves *N*-oxidation of the pyrrolizidine alkaloid or conjugation of the pyrrole intermediate with glutathione.

The metabolic formation of reactive pyrrolic ester metabolites in humans occurs primarily through the actions of CYP3A and, to a lesser extent, CYP2B isoenzymes. Cleavage of pyrrolizidine alkaloids by nonspecific blood esterases produces necine compounds and

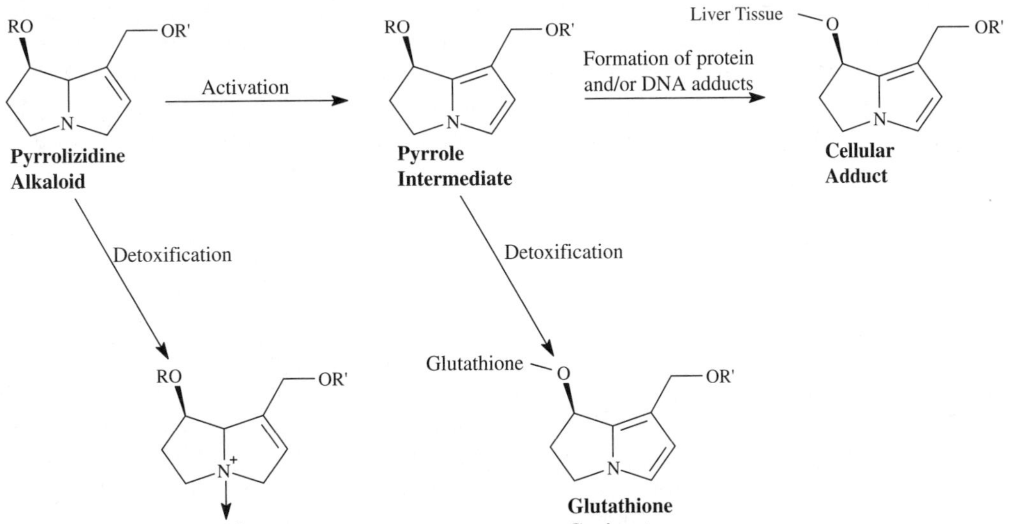

FIGURE 62.3. Activation and biotransformation of pyrrolizidine alkaloids (PA). Adapted from Rode D. Comfrey toxicity revisited. Trends Pharmacol Sci 2002;23:497.

necic acids, which ultimately appear in the urine as conjugates.[31] *In vitro* studies indicate that glutathione and *N*-acetylcysteine detoxify pyrrolizidine metabolites by forming water-soluble conjugates that are excreted by the kidney.[47] Animal studies indicate that the elimination of pyrrolizidine compounds is relatively rapid with most of the metabolites appearing in the urine within 24 hours after exposure.[48] Animal species vary substantially in their biotransformation of pyrrolizidine alkaloids to toxic metabolites.[15]

CLINICAL RESPONSE

Liver Toxicity

The clinical features of pyrrolizidine-induced veno-occlusive disease are similar to other diseases (Budd-Chiari syndrome, Reye's syndrome) that cause severe hepatic dysfunction with both an acute and chronic form of the disease.[49] Acute forms of veno-occlusive disease involve the rapid progression of hepatic necrosis, cirrhosis, portal hypertension, marked elevation of serum hepatic aminotransferases, and hepatic failure characterized clinically by vomiting, progressive hyperpnea, and encephalopathy.[50] The subacute form involves the gradual development of ascites, hepatomegaly, and gastrointestinal (GI) symptoms (abdominal pain, vomiting, diarrhea, pedal edema) that may resolve following the cessation of exposure to pyrrolizidine alkaloids.[42] Features of chronic forms of pyrrolizidine poisoning include the insidious onset of fatigue, right upper quadrant abdominal pain, anorexia, generalized weakness, weight loss, GI symptoms, ascites, hepatosplenomegaly, and progressive cirrhosis. Fever, jaundice, and bleeding are uncommon in this form of liver disease.[3] Prolonged pyrrolizidine alkaloid consumption can also cause pulmonary artery hypertension, leading to cor pulmonale and right ventricular hypertrophy. Other etiologies of veno-occlusive disease include methotrexate, azathioprine, radiotherapy, bone marrow transplantation, oral contraceptives, systemic lupus erythematosus, and alcoholic hepatitis.[9]

Carcinogenicity

The wide variety of pyrrolizidine compounds and the sparse carcinogenic testing of these compounds limit conclusions regarding the carcinogenic risk from ingesting comfrey and other plant materials containing pyrrolizidine compounds. The International Agency for Research on Cancer (IARC) lists lasiocarpine, riddelliine, and monocrotaline as possible human carcinogens (Group 2B), whereas retrorsine, hydroxysenkirkine, isatidine, jacobine, seneciphylline, and senkirkine are not classifiable as to carcinogenicity (Group 3).[51,52] Chronic rodent assays indicate that the consumption of dried leaves and roots of comfrey (*Symphytum officinale*) at doses of 1–16% of the total diet (dry weight/dry weight) produces an increased incidence of liver (adenomas, hemangioendothelial sarcomas) and urinary bladder tumors (papillomas, carcinomas).[53] The large doses of pyrrolizidine used in this study suggest that these compounds are weak carcinogens in rats. Although some pyrrolizidine alkaloids (e.g., symphytine, riddelliine) are positive in 2-year rodent studies,[54] there is no direct evidence that the ingestion of pyrrolizidine alkaloids causes cancer in humans.[55]

DIAGNOSTIC TESTING

Analytical Methods

Pyrrolizidine alkaloids can be extracted with classical methods (hot methanol, 95% ethanol).[11] *N*-oxides are more water soluble than the corresponding free bases; however, *N*-oxides are thermolabile. Colorimetry is a nonspecific, semiquantitative method for detecting pyrrolizidine alkaloids. The most sensitive and specific methods for the detection of pyrrolizidine compounds are liquid chromatography/mass spectrometry and gas chromatography/mass spectrometry (GC/MS). The coupling of liquid chromatography with electrospray ionization ion trap mass spectrometry allows the simultaneous analyses of these compounds and their corresponding *N*-oxides without the use of a reduction step required for GC/MS.[56,57] For pyrrolizidine compounds extracted from plant material, the use of gas chromatography-tandem mass spectrometry allows the analysis of toxic pyrrolizidine metabolites in blood samples at a detection limit of 2μg/mL.[11] The use of a C_{18} column with a water-acetonitrile gradient as the mobile phase allowed the identification of pyrrolizidine alkaloids and their *N*-oxides as measured by liquid chromatography-ion trap mass spectrometry.[58]

Abnormalities

Liver biopsies of patients with veno-occlusive disease demonstrate histological changes of a centrilobular hemorrhagic congestion and necrosis affecting primarily zones II and III along with reticulin fibers within the lumen of central and sublobular veins. Fibrosis occurs predominately in the perivenular areas rather than the portal tracts.[59] Thickening of the venules produces a functional outflow obstruction and congestion. Liver angiography demonstrates narrowing of the lumen of the small hepatic veins along with nonhomogeneous filling of the hepatic sinusoids.[60] The main hepatic veins and the portal veins typically remain patent. Electromicroscopy demonstrates the occurrence of hepatocyte blebs (clasmatosis) in sinusoidal borders, but these changes are not pathognomonic for pyrrolizidine-induced liver disease.[61] The hepatic lobular architecture usually remains intact until late in the course of the disease. Histological signs of inflammation are typically absent.[39]

Laboratory abnormalities associated with veno-occlusive disease include hypoglycemia, leukocytosis, mild hyperbilirubinemia (primarily indirect), variable increases of serum hepatic aminotransferase concentrations, thrombocytopenia, and hemolytic anemia. Ultrasound and imaging studies of the liver do not usually demonstrate significant abnormalities until portal hypertension becomes clinically significant.

TREATMENT

Management of veno-occlusive disease is supportive. Gut decontamination measures are usually unnecessary. There are no specific antidotes for pyrrolizidine-induced liver disease, and treatment is supportive. Patients hospitalized for serious pyrrolizidine-induced liver disease should be monitored closely for hepatorenal function, hypoglycemia, hypophosphatemia, metabolic acidosis, and electrolyte imbalance. Baseline blood tests for hospitalized patients include complete blood count, prothrombin time, and serum chemistries including hepatic aminotransferase, bilirubin, glucose, electrolytes, and creatinine. Hepatorenal function should be monitored daily for 3 days until a clear pattern of improvement in these serum tests occurs. Typically, a rise in prothrombin time and unconjugated bilirubin levels are the earliest biochemical markers of hepatotoxicity; these abnormalities usually occur before an elevation of serum hepatic aminotransferases. Bedside glucose monitoring is necessary several times daily in patients with fulminant hepatic failure.

The efficacy of vitamin K_1 for elevation of the prothrombin time during fulminant hepatic failure is unclear. Fresh frozen plasma (rarely, recombinant human factor VIIa) should be used for evidence of current bleeding or prophylactically for invasive procedures. Prophylactic correction of coagulopathy in a nonbleeding patient is not usually recommended. Serial hemoglobin and stool guaiac tests should be followed for evidence of GI bleeding. The administration of proton pump inhibitors (e.g., omeprazole) should be administered to reduce the risk of GI hemorrhage. Cerebral edema is the major cause of death following the development of hepatic encephalopathy, and treatment includes mannitol, elevation of the head to 10–20 degrees, adjustment of ventilator settings, and fluid restriction. Epidural intracranial pressure monitoring is usually necessary to guide treatment of patients with grade 3–4 hepatic encephalopathy. Hospitals without liver transplantation programs should transfer patients with fulminant hepatic failure as soon as possible to liver transplant centers because of the difficulty transferring patients with severe coagulopathy and cerebral edema. Orthotopic liver transplantation should be considered as a therapeutic option when severe liver failure occurs.

References

1. Stuart KL, Bras G. Veno-occlusive disease of the liver. Q J Med 1957;26:291–315.
2. Wilmot FC, Robertson GW. Senecio disease or cirrhosis of the liver due to senecio poisoning. Lancet 1920;ii: 828–829.
3. Mohabbat O, Youngos MS, Merzad AA, Srivastava RN, Sediz GG, Aram GN. An outbreak of hepatic veno-occlusive disease in north-western Afghanistan. Lancet 1976;2:269–271.
4. Chauvin P, Dillon JC, Moren A. [An outbreak of *Heliotrope* food poisoning, Tajikistan, November 1992–March 1993]. Sante 1994;4:263–268. [French].
5. Tandon RK, Tandon BN, Tandon HD, Bhatia ML, Bhargava S, Lal P, Arora RR. Study of an epidemic of venoocclusive disease in India. Gut 1976;17:849–855.
6. Datta DV, Khuroo MS, Mattocks AR, Aikat BK, Chhuttani PN. Herbal medicines and veno-occlusive disease in India. Postgrad Med J 1978;54:511–515.
7. Ketterer PJ, Glover PE, Smith LW. Blue heliotrope (*Heliotropium amplexicaule*) poisoning in cattle. Aust Vet J 1987;64:115–117.
8. Bras G, Jeliffe DB, Stuart KL. Veno-occlusive disease of the liver with non-portal type of cirrhosis, occurring in Jamaica. AMA Arch Pathol 1954;57:285–300.
9. Liddell JR. Pyrrolizidine alkaloids. Nat Prod Rep 2001;18;441–447.
10. Stegelmeier BL, Edgar JA, Colegate SM, Gardner DR, Schoch TK, Coulombe RA, Molyneux RJ. Pyrrolizidine alkaloid plants, metabolism and toxicity. J Nat Toxins 1999;8:95–116.
11. Stewart MJ, Steenkamp V. Pyrrolizidine poisoning: a neglected area in human toxicology. Ther Drug Monit 2001;23:698–708.
12. Johnson AE, Molyneux RJ, Ralphs MH. *Senecio*: A dangerous plant for man and beast. Rangelands 1989;11: 261–264.
13. Winship KA. Toxicity of comfrey. Adverse Drug React Toxicol Rev 1991;10:47–59.
14. Stickel F, Seitz HK. The efficacy and safety of comfrey. Publ Health Nutr 2000;3:501–508.

15. Abbott PJ. Comfrey: assessing the low-dose health risk. Med J Aust 1988;149:678–682.

16. Witte L, Ernst L, Adam H, Hartmann T. Chemotypes of two pyrrolizidine alkaloid-containing *Senecio* species. Phytochemistry 1992;31:559–565.

17. Macel M, Vrieling K, Klinkhamer GL. Variation in pyrrolizidine alkaloid patterns of *Senecio jacobaea*. Phytochemistry 2004;65:865–873.

18. Mattocks AR. Toxic pyrrolizidine alkaloids in comfrey. Lancet 1980;2:1136–1137.

19. Vrieling K, Derridj S. Pyrrolizidine alkaloids in and on the leaf surface of *Senecio jacobaea* L. Phytochemistry 2003;64: 1223–1228.

20. Furuya T, Hikichi M. Alkaloids and triterpenoids of *Symphytum officinale*. Phytochemistry 1971;10;2217–2220.

21. Crews C, Startin JR, Clarke PA. Determination of pyrrolizidine alkaloids in honey from selected sites by solid phase extraction and HPLC-MS. Food Addit Contam 1997;14:419–428.

22. Roeder E, Bourauel T. Pyrrolizidine alkaloids from *Senecio leucanthemifolius* and *Senecio rodriguezii*. Nat Toxins 1993;1:241–245.

23. Sullivan G. Detection of pyrrolizidine type alkaloids in matarique (*Psacalium decompositum*). Vet Hum Toxicol 1981;23:6–7.

24. Huxtable RJ, Luithy J, Zweifel U. Toxicity of comfrey-pepsin preparations. N Engl J Med 1986;315:1095.

25. Couet CE, Crews C, Hanley AB. Analysis, separation, and bioassay of pyrrolizidine alkaloids from comfrey (*Symphytum officinale*). Nat Toxins 1996;4;163–167.

26. Culvenor CC, Clarke M, Edgar JA, Frahn JL, Jago MV, Peterson JE, Smith LW. Structure and toxicity of the alkaloids of Russian comfrey (*Symphytum X uplandicum* Wyman), a medicinal herb and item of human diet. Experientia 1980;36:377–379.

27. Betz JM, Eppley RM, Taylor WC, Andrzejewski D. Determination of pyrrolizidine alkaloids in commercial comfrey products (*Symphytum* sp.). J Pharm Sci 1994;83:649–653.

28. Hirono I. Natural carcinogenic products of plant origin. Crit Rev Toxicol 1981;8:235–276.

29. Culvenor CC, Edgar JA, Jago MV, Qutteridge A, Peterson JE, Smith LW. Hepato- and pneumotoxicity of pyrrolizidine alkaloids and derivatives in relation to molecular structure. Chem Biol Interact 1976;12:299–324.

30. Copple BL, Ganey PE, Roth RA. Liver inflammation during monocrotaline hepatotoxicity. Toxicology 2003;190: 155–169.

31. Roeder E. Medicinal plans in Europe containing pyrrolizidine alkaloids. Pharmazie 1995;50:83–98.

32. Woo J, Sigurdsson ST, Hopkins PB. DNA interstrand cross-inking reactions of pyrrole-derived bifunctional electrophiles: evidence for a common target site in DNA. J Am Chem Soc 1993;115;3407–3415.

33. Cooper RA, Huxatable RJ. The relationship between reactivity of metabolites of pyrrolizidine alkaloids and extrahepatic toxicity. Proc West Pharmacol Soc 1999;42: 13–16.

34. Huxtable RJ. Activation and pulmonary toxicity of pyrrolizidine alkaloids. Pharmacol Ther 1990;47:371–389.

35. Mattocks AR, Driver E. Toxic actions of senaetnine, a new pyrrolizidine-type alkaloid, in rats. Toxicol Lett 1987;38: 315–319.

36. Fu PP, Xia Q, Lin G, Chou MW. Pyrrolizidine alkaloids–genotoxicity, metabolism enzymes, metabolic activation, and mechanism. Drug Metabol Rev 2004;36:1–55.

37. Rode D. Comfrey toxicity revisited. Trends Pharmacol Sci 2002;23:497–499.

38. Chojkier M. Hepatic sinusoidal-obstruction syndrome: toxicity of pyrrolizidine alkaloids. J Hepatol 2003;39: 437–446.

39. Ridker PM, Ohkuma S, McDermott WV, Trey C, Huxtable RJ. Hepatic venoocclusive disease associated with the consumption of pyrrolizidine-containing dietary supplements. Gastroenterology 1985;88:1050–1054.

40. Kumana CR, Ng M, Hsiang JL, Ko Wu, Wu P-C, Todd D. Herbal tea induced hepatic veno-occlusive disease: quantification of toxic alkaloid exposure in adults. Gut 1985;26:101–104.

41. Huxtable R, Stillman A, Ciramitaro D. Characterization of alkaloids involved in human senecio (pyrrolizidine) poisoning. Proc West Pharmacol Soc 1977;20:455–459.

42. Sperl W, Stuppner H, Gassner I, Judmaier W, Dietze O, Vogel W. Reversible hepatic veno-occlusive disease in an infant after consumption of pyrrolizidine-containing herbal tea. Eur J Pediatr 1995;154:112–116.

43. Brauchli J, Luthy J, Zweifel U, Schlatter C. Pyrrolizidine alkaloids from *Symphytum officinale* L. and their percutaneous absorption in rats. Experientia 1982;38:1085–1087.

44. JrCoulombe RA. Pyrrolizidine alkaloids in foods. Adv Food Nutr Res 2003;45:61–99.

45. Couet CE, Hopley J, Hanley AB. Metabolic activation of pyrrolizidine alkaloids by human, rat and avocado microsomes. Toxicon 1996;34:1058–1061.

46. Wiessler M. DNA adducts of pyrrolizidine alkaloids, nitroimidazoles and aristolochic acid. IARC Sci Publ 1994;125: 165–177.

47. Yan CC, Cooper RA, Huxtable RJ. The comparative metabolism of the four pyrrolizidine alkaloids, seneciphylline, retrorsine, monocrotaline, and trichodesmine in the isolated, perfused rat liver. Toxicol Appl Pharmacol 1995;133:277–284.

48. Mattocks AR. Toxicity of pyrrolizidine alkaloids. Nature 1968;217:723–728.

49. Fox DW, Hart MC, Bergeson PS, Jarrett PB, Stillman AE, Huxtable RJ. Pyrrolizidine (Senecio) intoxication mimicking Reye's syndrome. J Pediatr 1978;93:980–982.

50. Steenkamp V, Stewart M, Zuckerman M. Clinical and analytical aspects of pyrrolizidine poisoning caused by South African traditional medicines. Ther Drug Monit 2000;22: 302–306.

51. International Agency for Research on Cancer (IARC). *Senecio* species and riddelliine. IARC Mongr 2002;82: 153–170.
52. International Agency for Research on Cancer. Some naturally occurring substances. IARC Monogr Eval Carcinog Risk Chem Man 1976;10:1–342.
53. Hirono I, Mori H, Haga M. Carcinogenic activity of *Symphytum officinale*. J Nat Cancer Inst 1978;61:865–869.
54. National Toxicology Program. Toxicology and carcinogenesis studies of riddelliine (CAS No. 23246-96-0) in F344/N rats and B6C3F1 mice (gavage studies). Natl Toxicol Program Tech Rep Ser 2003;508:1–280.
55. Prakash AS, Pereira TN, Reilly PE, Seawright AA. Pyrrolizidine alkaloids in human diet. Mutat Res 1999;443: 53–67.
56. Altamirano JC, Gratz SR, Wolnik KA. Investigation of pyrrolizidine alkaloids and their N-oxides in commercial comfrey-containing products and botanical materials by liquid chromatography electrospray ionization mass spectrometry. J AOAC Int 2005;88:406–412.
57. Lin G, Zhou KY, Zhao XG, Wang ZT, But PP. Determination of hepatotoxic pyrrolizidine alkaloids by on-line high performance liquid chromatography mass spectrometry with an electrospray interface. Rapid Commun Mass Spectrom 1998;12:1445–1456.
58. Wuilloud JC, Gratz SR, Gamble BM, Wolnik KA. Simultaneous analysis of hepatotoxic pyrrolizidine alkaloids and N-oxides in comfrey root by LC-ion trap mass spectrometry. Analyst 2004;129:150–156.
59. Bach N, Thung SN, Schaffner F. Comfrey herb tea-induced hepatic veno-occlusive disease. Am J Med 1989;87:97–99.
60. Yeong ML, Swinburn B, Kennedy M, Gordon N. Hepatic veno-occlusive disease associated with comfrey ingestion. J Gastroenterol Hepatol 1990;5:211–214.
61. Yeong ML, Wakefield SJ, Ford HC. Hepatocyte membrane injury and bleb formation following low dose comfrey toxicity in rats. Int J Exp Path 1993;74:211–217.

Chapter 63

CRANBERRY
(*Vaccinium macrocarpon* Aiton)

HISTORY

Although Native Americans introduced Europeans to cranberries, canned cranberry sauce did not become commercially available until the 1920s and cranberry juice appeared in US markets during the 1940s. The use of cranberry extracts to treat urinary tract infections dates back at least to the 19th century.

BOTANICAL DESCRIPTION

Common Name: American Cranberry

Scientific Name: *Vaccinium macrocarpon* Aiton

Botanical Family: Ericaceae (éricacées, heaths)

Physical Description: A small, trailing evergreen shrub that reaches 1–4 in. (2.5–10 cm) in height. The simple, alternate leaves are small (0.25–0.75 in./0.6–1.9 cm). The young leaves are light green to maroon, whereas the mature leaves are deep green. Clusters of light pink flowers appear in early to mid-summer, and later small, bright red berries develop.

Common Name: European Cranberry, Guelder Rose, Crampbark, Cranberry Tree

Scientific Name: *Viburnum opulus* L.

Botanical Family: Caprifoliaceae (honeysuckle)

Physical Description: This tall, thick, deciduous shrub reaches up to 4 m (~13 ft) in height. The branches are smooth (glabrous) with maple-like leaves. Clusters of beautiful, white flowers produce small bright red berries.

EXPOSURE

Traditional

Cranberry juice is a common, traditional preventative, a treatment for the symptoms of urinary tract infection, and a deodorant for the odors associated with urinary incontinence.[1–3]

Current

Although there is limited evidence that the daily use of cranberry juice reduces the incidence of urinary tract infections, there are few data supporting the use of cranberry products to treat active urinary tract infections or to prevent the recurrence of urinary tract infections in patients with indwelling catheters or neurogenic bladders. In two acceptable, randomized, controlled clinical trials, cranberry products significantly reduced (RR = 0.61, 95% CI: 0.40–0.91) the incidence of urinary tract infections in women over a 12-month period compared with placebo control.[4] The dose of cranberry products in the two trials were 7.5 g cranberry in 50 mL cranberry-lingonberry juice in one trial[5] and 750 mL cranberry juice daily (or cranberry capsules twice daily) in the other trial.[6] There was no significant difference in the incidence of urinary tract infections between cranberry juice and cranberry capsule groups. The ingestion of cranberry juice may reduce the risk of urinary stone formation by increasing urinary citrate,[7] but there are inadequate clinical data supporting the use of cranberry juice to decrease the risk of calcium oxalate, urate, or brushite renal stones.

Regulatory

In the United States, cranberries and cranberry juice are a food and dietary supplement.

PRINCIPAL INGREDIENTS

Chemical Composition

Cranberries contain about 90% water along with flavonoids, anthocyanins, triterpenoids, catechin, phenolic compounds, glucose, fructose, and vitamin C. The carbohydrate load of cranberries is relatively large, particularly for diabetic patients. Cranberries contain large amounts of benzoic and phenolic acids (i.e., ~5–6 g/kg fresh weight) with the most abundant acids being benzoic acid (4.7 g/kg fresh weight), *p*-coumaric acid (0.25 g/kg fresh weight), and sinapic acid (0.2 g/kg fresh weight).[8] The benzoic and phenolic acids occur primarily in the bound forms, primarily to sugars with only about 10% of these compounds occurring as free acids. The major phenolic compound in freshly squeezed cranberry juice is benzoic acid, whereas quercetin and myricetin are the major flavonoids.[9]

Physiochemical Properties

The high citric and quinic acid content of cranberries imparts the characteristic tart taste. Blueberries are a more palatable source of proanthocyanidin compounds than cranberries. Although the presence of quinic acid in cranberries causes the urinary excretion of large amounts of hippuric acid,[10] this effect is relatively transient and probably does not produce significant antibacterial effect.[11] The ingestion of 360 mL cranberry juice daily can lower urinary pH, but the pH does not usually decrease below pH 5.5.[12]

The ingestion of cranberry juice produces mixed effects on the propensity to form urinary stones. In a study of 12 normal subjects and 12 calcium oxalate stone formers, the daily ingestion of 1 L cranberry juice reduced urinary pH (5.97 to 5.67), decreased urinary uric acid excretion, and increased calcium oxalate (i.e., ~18%) and undissociated uric acid in the urine.[13] Overall, this study suggested that the daily use of cranberry juice potentially increases the risk of calcium oxalate and uric acid stones, whereas the risk of brushite stones was potentially decreased. However, the clinical effect of these changes is probably small.

Mechanism of Action

The ingestion of cranberries results in the inhibition of bacterial adherence to mucosal surfaces. Most pathogenic *E. coli* express P fimbriae, which mediate the adherence of the bacteria to uroepithelial cells. Proanthocyanidin compounds in cranberries inhibit the ability of P-fimbriated *E. coli* to attach to isolated uroepithelial cells at concentrations of 10–50 μg/mL.[14] The inhibition of the papG fimbrial attachment of uropathogenic strains of *E. coli* to human cells resides in the A-type proanthocyanidin compounds rather than a B-type dimer or the (−)-epicatechin monomer.[15] The purified active fraction of the A-type proanthocyanidins contains over 200 compounds including 1-*O*-methylgalactose, prunin, and phlorizin.[16] Fructose also inhibits the adherence of type 1 (mannose-specific) fimbriated *E. coli*. However, the direct bacteriocidal activity of cranberry juice appears weak.[17]

DOSE RESPONSE

Treatment regimens for cranberry products include 480 mL cranberry juice daily for urine odors[1] and 300 mL cranberry juice cocktail (25–33% juice with sweetener) for the treatment of bacteriuria and pyuria.[18] The large number of withdrawals from clinical trials for the use of cranberry products to prevent urinary tract infections suggest that adverse effects occur with doses in the range of 750-mL cranberry juice over 12 months, but there are few data on the actual side effects. The daily ingestion of 300 mL cranberry juice was well tolerated by 376 older patients (>60 years old) in a randomized, placebo-controlled, double-blind clinical trial evaluating the use of cranberry juice for the reduction of the time to onset of first urinary tract infection.[19] The ingestion of large amounts (>3–4 L) of cranberry juice daily can produce gastrointestinal distress manifest by diarrhea and nausea.

TOXICOKINETICS

There are limited data on the toxicokinetics of the ingredients in cranberry products, in part because the active ingredients have not been identified. Flavonoids inhibit cytochrome P450 enzymes including CYP2C9; therefore, potential interactions occur between the use of cranberries and warfarin.[20] A case report associated a persistently elevated international normalized ratio (INR) with the use of cranberry juice in a patient on warfarin therapy for a prosthetic mitral valve.[21] The elevation in the INR was detected 2 weeks after the patient began drinking cranberry juice.

CLINICAL RESPONSE

The ingestion of large amounts of cranberry products may potentially cause gastrointestinal distress including nausea, vomiting, and diarrhea. There are no reports of serious toxic reactions associated with the use of cranberries except the development of dehydration and metabolic acidosis associated with the ingestion of large amounts of cranberry tea.[22]

DIAGNOSTIC TESTING

Methods for analysis of phenolic compounds in cranberries include gas-liquid chromatography, high performance liquid chromatography (HPLC), capillary electrophoresis (CE), and capillary gas chromatography/mass spectrometry (GC/MS). In contrast to gas-liquid chromatography, HPLC and CE do not require derivatization prior to quantitation, but these latter methods do not provide sufficient separating performance or identifying power. Consequently, capillary GC/MS remains the method of choice for determination of phenolic compounds in fruit juices, such as cranberry juice.

TREATMENT

Treatment is supportive. Serious adverse effects usually are not associated with the ingestion of cranberry products. Decontamination measures are not necessary.

References

1. Kraemer RJ. Cranberry juice and the reduction of ammoniacal odor of urine. Southwest Med 1964;45:211–212.
2. Super EA, Kemper KJ, Woods C, Nagaraj S. Cranberry use among pediatric nephrology patients. J Agric Food Chem 2005;53:8940–8947.
3. Bodel PT, Cotran R, Kass EH. Cranberry juice and the antibacterial action of hippuric acid. J Lab Clin Med 1959;54:881–888.
4. Jepson RG, Mihaljevic L, Craig J. Cranberries for preventing urinary tract infections. Cochrane Database Syst Rev 2004;(2):CD001321.
5. Kontiokari T, Sundqvist K, Nuutinen M, Pokka T, Koskela M, Uhari M. Randomised trial of cranberry-lingonberry juice and Lactobacillus GG drink for the prevention of urinary tract infections in women. BMJ 2001;322(7302): 1571–1573.
6. Stothers L. A randomized trial to evaluate effectiveness and cost effectiveness of naturopathic cranberry products as prophylaxis against urinary tract infection in women. Can J Urol 2002;9:1558–1562.
7. Pak CY. Citrate and renal calculi: an update. Miner Electrolyte Metab 1994;20:371–377.
8. Zuo Y, Wang C, Zhan J. Separation, characterization, and quantitation of benzoic and phenolic antioxidants in American cranberry fruit by GC-MS. J Agric Food Chem 2003;50:3789–3794.
9. Chen H, Zuo Y, Deng Y. Separation and determination of flavonoids and other phenolic compounds in cranberry juice by high-performance liquid chromatography. J Chromatogr A 2001;913:387–395.
10. Kinney AB, Blount M. Effect of cranberry juice on urinary pH. Nurs Res 1979;28:287–290.
11. McLeod DC, Nahata MC. Methenamine therapy and urine acidification with ascorbic acid and cranberry juice. Am J Hosp Pharm 1978;35:654.
12. Schultz A. Efficacy of cranberry juice and ascorbic acid in acidifying the urine in multiple sclerosis subjects. J Community Health Nurs 1984;1:159–169.
13. Gettman MT, Ogan K, Brinkley LJ, Adams-Huet B, Pak CY, Pearle MS. Effect of cranberry juice consumption on urinary stone risk factors. J Urol 2005;174:590–594.
14. Foo LY, Lu Y, Howell AB, Vorsa N. A-Type proanthocyanidin trimers from cranberry that inhibit adherence of uropathogenic P-fimbriated *Escherichia coli*. J Nat Prod 2000;63:1225–1228.
15. Foo LY, Lu Y, Howell AB, Vorsa N. The structure of cranberry proanthocyanidins which inhibit adherence of uropathogenic P-fimbriated *Escherichia coli in vitro*. Phytochemistry 2000;54:173–181.
16. Turner A, Chen S-N, Joike MK, Pendland SL, Pauli GF, Farnsworth NR. Inhibition of uropathogenic *Escherichia coli* by cranberry juice: a new antiadherence assay. J Agric Food Chem 2005;53:8940–8947.
17. Kahn HD, Panariello VA, Saeli J, Sampson JR, Schwartz E. Effect of cranberry juice on urine. J Am Diet Assoc 1967;51:251–254.
18. Avorn J, Monane M, Gurwitz JH, Glynn RJ, Choodnovskiy I, Lipsitz LA. Reduction of bacteriuria and pyuria after ingestion of cranberry juice. JAMA 1994;271: 751–754.
19. McMurdo ME, Bissett LY, Price RJ, Phillips G, Crombie IK. Does ingestion of cranberry juice reduce symptomatic urinary tract infections in older people in hospital? A double-blind, placebo-controlled trial. Age Ageing 2005; 34:256–261.
20. Suvarna R, Pirmohamed M, Henderson L. Possible interaction between warfarin and cranberry juice. BMJ 2003;327(7429):1454.
21. Grant P. Warfarin and cranberry juice: an interaction? J Heart Valve Dis 2004;13:25–26.
22. Garcia-Calatayud S, Larreina Cordoba JJ, Lozano De La Torre MJ. [Severe cranberry juice poisoning]. An Esp Pediatr 2002;56:72–73. [Spanish]

Chapter 64

DONG QUAI
[*Angelica sinensis* (Oliv.) Diels]

HISTORY

Dong quai is one of the most popular Chinese herbs with thousands of years of use in traditional Chinese medicine, primarily for gynecological complaints (e.g., "female ginseng"). In the 6th century AD, Lei Gong's *Treatise on Preparation of Materia Medica* described the herbal use of dong quai as a tonic for promoting the circulation, regulating the menstrual cycle, strengthening the heart, and lubricating the bowel.[1] In Europe during the late 1800s, an extract of dong quai (eumenol) was a popular herbal remedy for menstrual cramps and bleeding. There is a recent resurgence of interest in dong quai for the treatment of gynecological complaints based on proposed weak estrogen-like properties, but there are few clinical data to support the efficacy of dong quai for gynecological diseases.

BOTANICAL DESCRIPTION

Other similar plants used for the treatment of gynecological complaints include *Angelica acutiloba* (Siebold & Zucc.) Kitag. (dong dang gui, tang-kuei), *Angelica gigas* Nakai (Korean angelica), *Angelica dahurica* (Hoffm.) Franch. & Sav. (Bai zhi), and *Angelica pubescens* Maxim. (Du huo). However, *A. sinensis* remains the most popular type of *Angelica* in herbal therapy.

Common Name: Dong Quai (Chinese Angelica, Dang Gui, Tang Kuei, Tan Kue, Tanggwi, Toki)

Scientific Name: *Angelica sinensis* (Oliv.) Diels

Botanical Family: Apiaceae (Umbelliferae)

Physical Description: This fragrant, perennial herb grows to about 6ft (~2m) with purple, smooth stems, which contain linear striations. In June and July, white flowers appear in umbrella-like clusters.

Distribution and Ecology Dong quai (*Angelica sinensis*) is a fragrant, perennial herb, which grows in rich, well-drained soil on cold, damp mountain slopes in mainland China, Korea, and Japan. Commercial cultivation of *Angelica sinensis* occurs in the Gansu, Sichuan, Yunnan, Shaanxi, and Hubei provinces of China.

EXPOSURE

Sources

After 3 years of cultivation, the roots of *A. sinensis* are harvested and stored in sealed containers under cool, dry conditions to prevent mildew and evaporation of the volatile oils.

Medicinal Uses

TRADITIONAL

The herb, *Angelica sinensis*, is a traditional restorative tonic, blood purifier, and a treatment for premenstrual syndrome, menopause, and dysmenorrhea. Typically, dong quai is combined with other herbs (e.g., chamomile) for the treatment of these conditions.[2] The root is the most common portion of the plant used in Chinese medicine. The 10–25cm (~4–10in.), grayish-brown to

Medical Toxicology of Natural Substances, by Donald G. Barceloux, MD

dark brown root is divided into three parts (head, body, tail) with different medicinal properties ascribed to each part. Other traditional uses include the treatment of hypertension and other cardiovascular conditions (thromboangiitis, cor pulmonale, Berger's disease, Raynaud's disease), headache, neuralgias, chronic rhinitis, anxiety, insomnia, forgetfulness, chronic constipation, inflammation, and infectious diseases (herpes zoster, hepatitis, chronic glomerulonephritis).[3,4] The water soluble sodium salt of ferulic acid (sodium ferulate) is used in China as an antithrombolytic agent and as a treatment for cardiovascular and cerebrovascular diseases.[5] Native Americans used this herb for the treatment of headaches, renal dysfunction, cirrhosis, and bladder problems.

CURRENT

In the United States and Europe, dong quai is primarily used as a flavoring agent for a variety of commercial products including ice cream, candy, cream fillings, liqueurs, and vermouth. There is little clinical evidence to support the use of dong quai for the treatment of vasomotor symptoms associated with menopause.[6,7] A double-blind, placebo-controlled study of 71 postmenopausal women did not detect a significant difference between the dong quai groups (4.5 g dong quai root daily standardized to 0.5 mg ferulic acid/kg) and placebo.[8] The endpoints were the Kupperman Index scores, number of hot flashes, vaginal maturation index, and endometrial thickness.

REGULATORY STATUS

In the United States, dong quai is a dietary supplement. The German Commission E does not list dong quai as an approved herbal medication.

PRINCIPAL INGREDIENTS

Chemical Composition

Potentially bioactive components of dong quai include ferulic acid ($C_{10}H_{10}O_4$), ligustilide ($C_{12}H_{14}O_2$), and other alkylphthalide compounds (e.g., 6,7-dihydroxyligustilide, senkyunolide A).[9] The content of ferulic acid in extracts of dong quai depends on the extraction method as well as variations between individual plants with ranges of approximately 0.2–1.4 mg/g.[10] The ferulic acid concentration in an aqueous ethanol extract of two samples of raw root of dong quai was 0.31 mg/g and 0.48 mg/g as measured by capillary zone electrophoresis.[11] In 10 samples from the root of dong quai, the ferulic acid content ranged from about 0.17 mg/g to approximately 0.53 mg/g.[10] In addition to ligustilide and ferulic acid, the major components of the dried root of dong quai are *n*-butylidenephthalide, *n*-butylphthalide, vanillin, linoleic acid, nicotinic acid, and succinic acid.[1,12] The root of dong quai also contains the ester of ferulic acid (coniferyl ferulate). Figure 64.1 and Figure 64.2 display the chemical structures of ferulic acid and ligustilide, respectively. The essential oil represents about 0.4–0.7% of the root using standard Chinese aqueous-extracting methods, and this essential oil contains about 100 constituents.[1] Based on subwindow factor analysis and spectral correlative chromatography of an essential oil from the main root of dong quai, the major ingredient was ligustilide, which accounted for about 19% of the total.[13] Minor constituents of the essential oil included α-pinene, butylidenephthalide, carvacrol, alloocimene, 4-vinylguaiacol, and *d*-limonene. The aromatic properties of the volatile oil from dong quai result primarily from ligustilide compounds. In samples of three whole roots of dong quai (*Angelica sinensis*), the total content of ligustilide compounds ranged from about 10–16 mg/g dry weight as measured by liquid chromatography/mass spectrometry.[14] Potentially toxic compounds in dong quai include photosensitizing psoralens and the carcinogen, safrole. There are conflicting data on the amount of natural coumarin compounds (angelol I, angelol H, scopoletin, angelicone, bergapten, 7-desmethylsuberosin, oxypeucedanin, osthole, psoralen) present in dong quai with the amounts of these

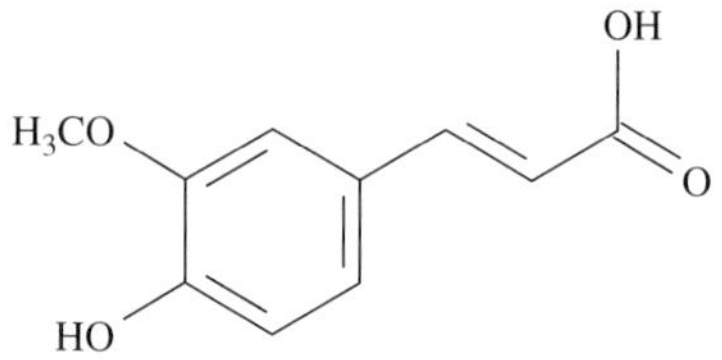

OH
H_3CO
O
HO

FIGURE 64.1. Chemical structure of ferulic acid (CAS RN: 1135-24-6).

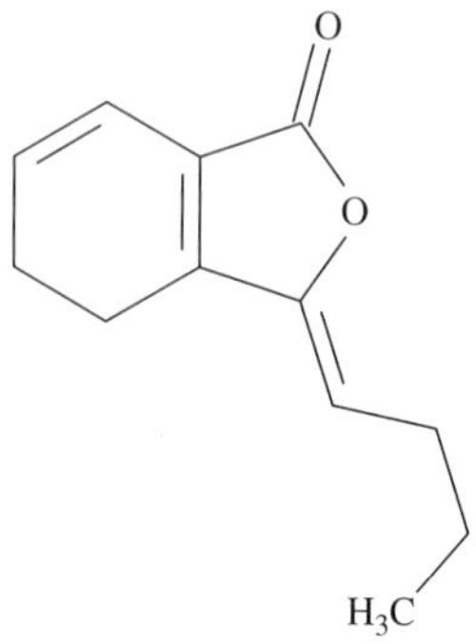

O
O
H_3C

FIGURE 64.2. Chemical structure of ligustilide (CAS RN:4431-01-0).

compounds in *Angelica sinensis* being relatively low compared with other related species, such as *A. dahurica*.[15,16]

Physiochemical Properties

In vitro studies indicate that dong quai has only weak estrogen receptor binding properties and little upregulation of progesterone receptor mRNA.[17] Table 64.1 lists some physiochemical properties of ferulic acid. This compound is a poorly water soluble, weak organic acid; therefore, ferulic acid is not a major component of aqueous extracts of dong quai.

DOSE RESPONSE

There is little standardization of the traditional dose of dong quai. Some commercial products use approximately 1% ligustilide or 0.5 ppm ferulic acid as their standard concentration for dong quai. Common traditional doses of dong quai in Chinese medicine range from about 1–5 g powdered or dried root 3 times daily.[16] Decoctions of dong quai typically involve the boiling or simmering of 1–3 teaspoons of the root in 1 cup of water for 2–5 minutes followed by 5–10 minutes of brewing. The daily dose of the decoction is about one to three cups. One gram of 100% aqueous extract of dong quai is equivalent to approximately 4 grams of the raw root.[1]

TOXICOKINETICS

There are few data on the pharmacokinetics of ferulic acid and ligustilide. Most herbal publications do not recommend the use of dong quai in patients on anticoagulants because of the presence of natural coumarin compounds in this herb and potential antiplatelet effects. A single case report temporally associated the use of dong quai with mild elevation of baseline prothrombin times in a patient on warfarin.[18] However, there are limited experimental data to confirm the clinical significance of any potential effect of dong quai on coagulation parameters in patients on anticoagulants. *In vitro* animal studies using washed platelets also suggest that some components (osthole, ferulic acid) of dong quai inhibit platelet aggregation by several mechanisms including inhibition of thromboxane A_2 formation and arachidonic acid metabolism.[19] Although the ingestion of dong quai could potentially enhance the effects of antiplatelet drugs, these activities are relatively weak, and there are insufficient data to conclude that such effects are clinically relevant.

TABLE 64.1. Some Physiochemical Properties of Ferulic Acid

Physical Property	Value
pKa Dissociation Constant	4.58 (25 °C/77 °F)
log P (Octanol-Water)	1.51
Water Solubility	5970 mg/L (25 °C/77 °F)
Vapor Pressure	2.69E-06 mmHg (25 °C/77 °F)
Henry's Law Constant	7.96E-14 atm-m^3/mole (25 °C/77 °F)
Atmospheric OH Rate Constant	4.83E-11 cm^3/molecule-second (25 °C/77 °F)

CLINICAL RESPONSE

There are few reports of toxicity associated with the use of dong quai. A case report temporally associated the daily use of dong quai pills for one month by a 35-year-old man with the development of gynecomastia that resolved after the cessation of the use of dong quai.[20] All laboratory tests were normal including plasma follicle stimulating hormone, luteinizing hormone, estradiol, testosterone, and prolactin concentrations.

DIAGNOSTIC TESTING

The root of dong quai contains at least 100 constituents, and separation of these components is a complex process that may require multidimensional separation techniques, such as two-dimensional liquid chromatography/mass spectrometry[12] and subwindow factor analysis/spectral correlative chromatography.[13] Methods for the detection of ferulic acid include high performance liquid chromatography and capillary zone electrophoresis.[11] Extracts of dong quai contain phenolic compounds (e.g., coniferyl ferulate), which can hydrolyze during the extraction process to form ferulic acid. In a study of 10 samples from the root of dong quai using methanol-formic acid extraction, the ratio of total measured ferulic acid to free (nonhydrolyzed) ferulic acid ranged from 1.53–9.73.[10]

TREATMENT

Treatment is supportive.

References

1. Zhu DP. Dong quai. Am J Chin Med 1987;15:117–125.
2. Kupfersztain C, Rotem C, Fagot R, Kaplan B. The immediate effect of natural plant extract, *Angelica sinensis* and *Matricaria chamomilla* (Climex) for the treatment of hot flushes during menopause. A preliminary report. Clin Exp Obstet Gynecol 2003;30:203–206.

3. Yim TK, Wu WK, Pak WF, Mak DH, Liang SM, Ko KM. Myocardial protection against ischaemia-reperfusion injury by a *Polygonum multiflorum* extract supplemented "Dang-Gui decoction for enriching blood", a compound formulation, *ex vivo*. Phytother Res 2000;14:195–199.
4. Qi-Bing M, Jing-Yi T, Bo C. Advances in the pharmacological studies of radix *Angelica sinensis* (Oliv) Diels (Chinese Danggui). Chinese Med J 1991;104: 776–781.
5. Wang B-H, Ou-Yang J-P. Pharmacological actions of sodium ferulate in cardiovascular system. Cardiovascular Drug Rev 2005;23:161–172.
6. Kronenberg F, Fugh-Berman A. Complementary and alternative medicine for menopausal symptoms: a review of randomized, controlled trials. Ann Intern Med 2002;137:805–813.
7. The North American Menopause Society. Position statement: treatment of menopause-associated vasomotor symptoms: position statement of The North American Menopause Society. Menopause 2004;11:11–33.
8. Hirata JD, Swiersz LM, Zell B, Small R, Ettinger B. Does dong quai have estrogenic effects in postmenopausal women? A double-blind, placebo-controlled trial. Fertil Steril 1997;68:981–986.
9. Dong ZB, Li SP, Hong M, Zhu Q. Hypothesis of potential active components in *Angelica sinensis* by using biomembrane extraction and high performance liquid chromatography. J Pharm Biomed Anal 2005;38:664–669.
10. Lu G-H, Chan K, Leung K, Chan C-L, Zhao Z-Z, Jiang Z-H. Assay of free ferulic acid and total ferulic acid for quality assessment of *Angelica sinensis*. J Chromatogr A 2005;1068:209–219.
11. Song-gang J, Yi-feng C, Yu-tian W, Xue-ping Y, Dong-sheng L, Zi-ming X, Xiao L. Determination of ferulic acid in *Angelica sinensis* and Chuanxiong by capillary zone electrophoresis. Biomed Chromatogr 1999;13:333–334.
12. Hu L, Chen X, Kong L, Su X, Ye M, Zou H. Improved performance of comprehensive two-dimensional HPLC separation of traditional Chinese medicines by using a silica monolithic column and normalization of peak heights. J Chromatogr A 2005;1092:191–198.
13. Huang L-F, Li B-Y, Liang Y-Z, Guo F-Q, Wang Y-L. Application of combined approach to analyze the constituents of essential oil from Dong quai. Anal Bioanal Chem 2004;378:510–517.
14. Lu G-H, Chan K, Chan C-L, Leung K, Jiang Z-H, Zhao Z-Z. Quantification of ligustilides in the roots of *Angelica sinensis* and related umbelliferous medicinal plants by high-performance liquid chromatography and liquid chromatography-mass spectrometry. J Chromatogr A 2004;1046:101–107.
15. Kwon YS, Shin SJ, Kim MJ, Kim CM. A new coumarin from the stem of *Angelica dahurica*. Arch Pharm Res 2002;25:53–56.
16. [No author listed]. Monograph. *Angelica sinensis*. Altern Med Rev 2004;9:429–433.
17. Liu J, Burdette JE, Xu H, Gu C, van Breemen RB, Bhat KP, et al. Evaluation of estrogenic activity of plant extracts for the potential treatment of menopausal symptoms. J Agric Food Chem 2001;49:2472–2479.
18. Page RLII , Lawrence JD. Potentiation of warfarin by dong quai. Pharmacotherapy 1999;19:870–876.
19. Ko FN, Wu TS, Liou MJ, Huang TF, Teng CM. Inhibition of platelet thromboxane formation and phosphoinositides breakdown by osthole from *Angelica pubescens*. Thromb Haemost 1989;62:996–999.
20. Goh SY, Loh KC. Gynaecomastia and the herbal tonic "dong quai." Singapore Med J 2001;42:115–116.

Chapter 65

FEVERFEW
(*Tanacetum parthenium* Schultz Bip.)

HISTORY

Feverfew is an ancient remedy for fever, pain, and menstrual problems that has been used since ancient Greek times. The 17th century herbalist Culpepper recommended the use of this herb for miscarriages, colic, and opium addiction.[1] Additionally, John Gerard's *Herball* recommended the use of feverfew for the treatment of inflammation and swelling including St. Anthony's Fire. Feverfew was considered the "medieval aspirin."

BOTANICAL DESCRIPTION

Common Name: Feverfew, Featherfew, Wild Chamomile, Wild Quinine, Bachelor's Button, Santa Maria

Scientific Name: *Tanacetum parthenium* (L.) Schultz Bip. [*Matricaria parthenium* (L.), *Leucanthemum parthenium* (L.) Gren. and Gordon, *Pyrethrum parthenium* (L.), *Chrysanthemum parthenium* (L.) Bernh.]

Botanical Family: Asteraceae (sunflowers, tournesols, Compositae)

Physical Description: This hardy, aromatic annual plant has bright green, chrysanthemum-like leaves. The height of feverfew reaches up to 2 ft (~60 cm). The deeply cut leaves have a sharp, unpleasantly bitter taste. Thick, daisy-like flowers with yellow centers appear from summer until mid-fall.

Distribution and Ecology: Feverfew is a native herb of Asia Minor that ranges from southeastern Europe to Asia. This plant is naturalized in well-drained soil in mountain scrub, rocky slopes, and disturbed areas of the United States and the United Kingdom.

EXPOSURE

Sources

Use of feverfew typically involves the use of fresh leaves or the ingestion of capsules or tablets composed of dried leaves. In a study of commercial Canadian and European herbal preparations of dried feverfew leaves, the parthenolide content ranged from 0 to >1%.[2] A study of 13 commercial feverfew capsules, tablets, and powders demonstrated parthenolide content ranging from 0–0.36%.[3] Only three of these products displayed accurate parthenolide concentrations on their labels. In a study of 21 commercial US capsules containing 25–500 mg dried feverfew leaves, the parthenolide content ranged from 0.14–0.74%.[4] Consequently, the daily dose of these products could vary by as much as 160-fold.

Medicinal Uses

Traditional

Traditional uses for feverfew include the treatment of migraine headache, fever, dysmenorrhea, psoriasis, and arthritis. Other uses include asthma, toothaches, insect bites, stomach cramps, and threatened abortions.

Medical Toxicology of Natural Substances, by Donald G. Barceloux, MD

CURRENT

Feverfew is a common herb administered as dried leaf capsules for prophylaxis of chronic migraine headaches, but there is limited evidence supporting efficacy.[5] In particular, there is insufficient evidence from randomized, double-blind clinical trials to conclude that the use of feverfew prevents migraine headaches or that the use of feverfew is more efficacious than other current prophylactic treatments (propanolol, topiramate). Several small, blinded, placebo-controlled studies suggest that the prophylactic use of dried feverfew capsules (i.e., approximately 50–80 mg daily) moderately reduces the frequency and severity of symptoms (nausea, vomiting, pain) of migraine headache, as measured by self-reported symptoms.[6,7] A randomized, double-blind, placebo-controlled, multicenter, parallel-group study followed 170 individuals who suffered from migraine headaches. They were treated with a CO_2 extract of feverfew (6.25 mg 3 times daily) for 16 weeks after a 4-week baseline period.[8] The primary effect was a reduction in the total number of migraine days and migraine duration rather than the intensity of the headache. When compared with baseline, a statistically significant reduction was barely achieved ($P = 0.0456$) in the frequency of migraine attacks (1.9 vs. 1.3, respectively). There was no significant increase in adverse effects during the treatment period. A double-blind, placebo-controlled multicenter randomized clinical trial compared three dosage regimens of CO_2 feverfew extract (2.08 mg, 6.25 mg, 18.75 mg) daily for 12 weeks with placebo.[9] When compared with the 4-week baseline, there was no statistically significant difference between the groups with respect to the total number of migraine attacks during the last 28 days of treatment. Additionally, there was no significant dose response relationship. There is also no clear scientific evidence that the use of feverfew reduces symptoms of rheumatoid arthritis.[10]

REGULATORY STATUS

Suggested standardization of feverfew preparations primarily involves the establishment of minimum parthenolide concentrations in the range of 0.1–0.2%. Currently, there is substantial variation in the concentration of parthenolide in various commercial preparations of feverfew-based high performance liquid chromatography (HPLC) and gas chromatography (GC) analyses.[11]

PRINCIPAL INGREDIENTS

Chemical Composition

Feverfew leaves contain substantial amounts of sesquiterpene lactones, principally the germacranolide, parthenolide, which occurs in concentrations in dried leaves of up to 1% by dry weight.[12] Figure 65.1 displays the sesquiterpene structure of parthenolide (CAS RN: 29552-41-8, CAS RN: 20554-84-1). The concentration of parthenolide in feverfew preparations varies depending on the source and a variety of genetic and environmental factors including strain, sunlight, water stress, location, season (preflowering vs. postflowering), and harvesting techniques.[13,14] The highest concentrations of parthenolide occur in the flowering tops and leaves, whereas the stalks and roots contain very small amounts of parthenolide. Analysis of plant parts from a UK specimen of feverfew revealed the following parthenolide concentrations: flowering tops, 1.38%; leaves, 0.95%; stalks, 0.08%, and roots, 0.01%.[2] Feverfew varieties with a light green-yellow colored variety have significantly higher mean parthenolide concentrations (1.61% ± 0.61%, dry weight) than darker-leafed varieties (<0.5%).[15] Other less common terpenoid compounds in feverfew include germacrene D, β-farnesene, chrysanthemonin, chrysanthemolide, magnoliolide, santamarine, and reynosin. The most common monoterpene compound in feverfew is camphor. Rare compounds in feverfew include the guaianolides, 8α-hydroxyestafiatin, endoperoxide tanaparthin-α-peroxide, canin, 10-epi-canin, and seco-tanapartholide A and B.[16] Lipophilic flavonols in the leaves and flowers of feverfew include tanetin (6-hydroxykaempferol 3,7,4′-trimethyl ether), 6-hydroxykaempferol 3,7-dimethyl ether, quercetagetin 3,7-dimethyl ether and quercetagetin 3,7,3′-trimethyl ether.[17] The essential oil of feverfew contains primarily camphor and *trans*-chrysanthenyl acetate along with smaller amounts of a variety of monoterpenes (camphene, α- and β-pinene, sabinene, myrcene, *p*-cymene, α- and γ-terpinene, terpinolene, terpine-4-ol, α-terpineol), oxidized monoterpenes (linalool, linalyl acetate, bornylacetate), sesquiterpene lactones (β-caryophyllene, *trans*-β-farnesene, germacrene), and eugenol.[18]

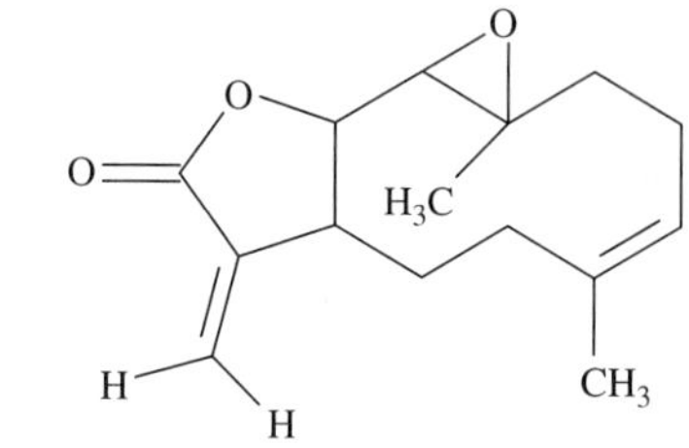

FIGURE 65.1. Chemical structure of parthenolide.

Physiochemical Properties

In vitro studies suggest that feverfew extracts inhibit platelet aggregation and the release of serotonin from platelets.[19] Thromboxane synthesis was unaffected by feverfew. Monoterpenes are the primary compounds released by aerial parts of feverfew including α-pinene, camphene, limonene, γ-terpinene, (*E*)-β-ocimene, linalool, *p*-cymene, (*E*)-chrysanthenol, camphor, and (*E*)-chrysanthenyl acetate.[20] The air around these plants did not contain detectable amounts of parthenolide or other sesquiterpene lactones as measured by dynamic headspace techniques along with GC and gas chromatography/mass spectrometry (GC/MS).

Mechanism of Toxicity

The identity of the active constituent in feverfew is unclear. A suggested mechanism of action is an anti-inflammatory effect from the major ingredient (parthenolide) in feverfew as a result of the attenuation of inflammatory processes including the degradation of the inhibitory protein-κBα, the activation of significant nuclear factor κB, and the expression of inducible nitric oxide synthase.[21] *In vitro* studies suggest that parthenolide blocks the activation of key central areas (i.e., nucleus trigeminalis caudalis) in the pathogenesis of migraine.[22]

DOSE RESPONSE

In England and Canada, standardization of feverfew products requires the presence of least 0.2% parthenolide. Traditional daily oral doses of feverfew include two to three dried leaves (approximately 60 mg) or 50–250 mg of dried leaf, standardized to 0.2% parthenolide.

TOXICOKINETICS

Kinetics

There are inadequate pharmacokinetic data to determine the bioavailability of parthenolide. In a phase I dose escalation trial of feverfew, parthenolide was not detectable in plasma following the administration of oral daily doses up to 4 mg feverfew capsules (Tanacet™, Ashbury Biologicals, Inc., Toronto, Ontario, Canada), as measured by solid phase extraction and MS.[23] The limit of detection was 0.5 ng/mL. The plasma concentration of parthenolide in mice 1 hour after the gavage of 4 mg/kg (i.e., about 10 times the dose administered to the patients) was 27.8 ng/mL. *In vitro* studies suggest that sesquiterpene lactones, such as parthenolide are about 30–50% bound in the blood.[24] Epoxidation and/or allylic oxidation with or without allylic rearrangement of parthenolide form a variety of metabolites in feverfew including β-hydroxyparthenolide, costunolide, β-hydroxycostunolide, artemorin, and 3β-hydroxyanhydroverlotorin.[12]

Drug Interactions

Based on laboratory research on the inhibition of platelet aggregation, feverfew theoretically can increase the risk of bleeding following the ingestion of other drugs affecting hemostasis (e.g., aspirin, warfarin). However, there are few clinical data to suggest that this potential effect is clinically significant.

CLINICAL RESPONSE

The adverse effects associated with the administration of feverfew during clinical trials are typically mild and transient. The most common adverse effect following the use of herbal preparations of feverfew leaves are oral ulceration and inflammation.[25] Feverfew is one of the major weed pollens along with ragweed (*Ambrosia* spp.), sunflower (*Helianthus* spp.), and mugwort (*Artemisia* spp.).[26] There is substantial cross-reactivity between members of the Asteraceae family. In a study of 190 Asteraceae-allergic patients determined by positive skin tests to an Asteraceae mix, feverfew elicited the most (81%) positive responses followed by tansy (77%), chamomile (64%), and yarrow (41%).[27] Most atopic patients are also hypersensitive to allergens other than members of the Asteraceae family, such as nickel.[28] The major allergen (Par h 1) in feverfew contains significant sequence identity with sunflower allergens and the major allergen (Art v 1) of mugwort.[29] The rash associated with allergic reactions to feverfew is a diffuse, erythematous, and occasionally lichenified lesion with a photodermatitis-like distribution.[30] However, this rash involves the upper eyelids, submental area, and retroauricular regions of the face in contrast to true photodermatitis. Sensitization to the sesquiterpene lactones of feverfew does not necessarily imply sensitization to the associated volatile monoterpene compounds in feverfew.[31] Patients with hypersensitivity to members of the Asteraceae family [e.g., aster (*Aster* spp.), chrysanthemums (*Chrysanthemum X morifolium* Ramat.), marguerite (*Leucanthemum vulgare* Lam.), marigold (*Calendula* ssp., *Tagetes* spp.) ragweed (*Ambrosia* spp.), sunflower (*Helianthus annuus* L.), tansy (*Tanacetum vulgare* L.), yarrow (*Achillea millefolium* L.)] may develop cross-sensitivity to feverfew, primarily as a result of the presence of sesquiterpene lactones.[32,33]

Case reports suggest that the abrupt cessation of feverfew use for migraine prophylaxis causes the exacerbation of migraine symptoms.[7] Other symptoms associated with withdrawal include muscle stiffness, insomnia, arthralgias, and anxiety.

DIAGNOSTIC TESTING

Parthenolide is the principal sesquiterpene lactone in feverfew, and standardization of feverfew preparations typically involves the analysis of parthenolide content. Minimum content of parthenolide is usually 0.1–0.2%. Methods to determine parthenolide content include liquid chromatography,[3] HPLC with UV detection at 210 nm,[34] GC with flame ionization,[11] and nuclear magnetic resonance.[2] Some deterioration of parthenolide occurs during storage. Following storage of three samples of feverfew at room temperature in a lighted room for 9 months, the parthenolide content decreased by 50–65%.

TREATMENT

Treatment is supportive. Allergic dermatitis is treated with topical steroids and antihistamines.

References

1. Groenewegen WA, Knight DW, Heptinstall S. Progress in the medicinal chemistry of the herb feverfew. Prog Med Chem 1992;29:217–238.
2. Heptinstall S, Awang DV, Dawson BA, Kindack D, Knight DW, May J. Parthenolide content and bioactivity of feverfew (*Tanacetum parthenium* (L.) Schultz-Bip.). Estimation of commercial and authenticated feverfew products. J Pharm Pharmacol 1992;44:391–395.
3. Abourashed EA, Khan IA. Determination of parthenolide in selected feverfew products by liquid chromatography. J AOAC Int 2000;83:789–792.
4. Nelson MH, Cobb SE, Shelton J. Variations in parthenolide content and daily dose of feverfew products. Am J Health-Syst Pharm 2002;59:1527–1531.
5. Pittler MH, Ernst E. Feverfew for preventing migraine. Cochrane Database Syst Rev 2004;(1):CD002286.
6. Murphy JJ, Heptinstall S, Mitchaell JR. Randomised double-blind placebo-controlled trial of feverfew in migraine prevention. Lancet 1988;2:189–192.
7. Johnson ES, Kadam NP, Hylands DM, Hylands PJ. Efficacy of feverfew as prophylactic treatment of migraine. Br J Med 1985;291:569–573.
8. Diener HC, Pfaffenrath V, Schnitker J, Friede M, Henneicke-von Zepelin HH. Efficacy and safety of 6.25 mg t.i.d. feverfew CO_2-extract (MIG-99) in migraine prevention—a randomized, double-blind, multicentre, placebo-controlled study. Cephalalgia 2005;25:1031–1041.
9. Pfaffenrath V, Diener HC, Fischer M, Friede M, Henneicke-von Zepelin HH. Investigators. The efficacy and safety of *Tanacetum parthenium* (feverfew) in migraine prophylaxis—a double-blind, multicentre, randomized placebo-controlled dose-response study. Cephalalgia 2002;22:523–532.
10. Pattrick M, Heptinstall S, Doherty M. Feverfew in rheumatoid arthritis: a double blind, placebo controlled study. Ann Rheum Dis 1989;48:547–549.
11. Abourashed EA, Khan IA. GC determination of parthenolide in feverfew products. Pharmazie 2001;56:971–972.
12. Knight DW. Feverfew: chemistry and biological activity. Nat Prod Rep 1995;12:271–276.
13. Awang DV, Dawson BA, Kindack DG. Parthenolide content of feverfew (*Tanacetum parthenium*) assessed by HPLC and ^{1}H-NMR spectroscopy. J Nat Prod 1991;54:1516–1521.
14. Fonseca JM, Rushing JW, Rajapakse NC, Thomas RL, Riley MB. Parthenolide and abscisic acid synthesis in feverfew are associated but environmental factors affect them dissimilarly. J Plant Physiol 2005;162:485–494.
15. Cutlan AR, Bonilla LE, Simon JE, Erwin JE. Intra-specific variability of feverfew: correlations between parthenolide, morphological traits and seed origin. Plant a Med 2000;66:612–617.
16. Todorova MN, Evstatieva LN. Comparative study of *Tanacetum* species growing in Bulgaria. Z Naturforsch [C] 2001;56:506–512.
17. Williams CA, Hoult JR, Harborne JB, Greenham J, Eagles J. A biologically active lipophilic flavonol from *Tanacetum parthenium*. Phytochemistry 1995;38:267–270.
18. Kalodera Z, Pepeljnjak S, Blazevic N, Petrak T. Chemical composition and antimicrobial activity of *Tanacetum parthenium* essential oil. Pharmazie 1997;52:885–886.
19. Heptinstall S, White A, Williamson L, Mitchell JR. Extracts of feverfew inhibit granule secretion in blood platelets and polymorphonuclear leucocytes. Lancet 1985;1(8437):1071–1074.
20. Christensen LP, Jakobsen HB, Paulsen E, Hodal L, Andersen KE. Airborne Compositae dermatitis: monoterpenes and no parthenolide are released from flowering *Tanacetum parthenium* (feverfew) plants. Arch Dermatol Res 1999;291:425–431.
21. Reuter U, Chiarugi A, Bolay H, Moskowitz MA. Nuclear factor-kappa B as a molecular target for migraine therapy. Ann Neurol 2002;51:507–516.
22. Tassorelli C, Greco R, Morazzoni P, Riva A, Sandrini G, Nappi G. Parthenolide is the component of *Tanacetum parthenium* that inhibits nitroglycerin-induced Fos activation: studies in an animal model of migraine. Cephalalgia 2005;25:612–621.
23. Curry EA III, Murry DJ, Yoder C, Fife K, Armstrong V, Nakshatri H, et al. Phase I dose escalation trial of feverfew with standardized doses of parthenolide in patients with cancer. Invest New Drugs 2004;22:299–305.
24. Wagner S, Kratz F, Merfort I. *In vitro* behaviour of sesquiterpene lactones and sesquiterpene lactone-containing plant preparations in human blood, plasma and human serum albumin solutions. Planta Med 2004;70:227–233.

25. Ernst E, Pittler MH. The efficacy and safety of feverfew (*Tanacetum parthenium* L.): an update of a systematic review. Public Health Nutr 2000;3:509–514.

26. Sriramarao P, Subba Rao PV. Allergenic cross-reactivity between Parthenium and ragweed pollen allergens. Int Arch Allergy Immunol 1993;100:79–85.

27. Paulsen E, Andersen KE, Hausen BM. Sensitization and cross-reaction patterns in Danish Compositae-allergic patients. Contact Dermatitis 2001;45:197–204.

28. Jovanovic M, Poljacki M, Duran V, Vujanovic L, Sente R, Stojanovic S. Contact allergy to Compositae plants in patients with atopic dermatitis. Med Pregl 2004;56: 209–218.

29. Gadermaier G, Dedic A, Obermeyer G, Frank S, Himly M, Ferreira F. Biology of weed pollen allergens. Curr Allergy Asthma Rep 2004;4:391–400.

30. Burry JN. Compositae dermatitis in South Australia: contact dermatitis from *Chrysanthemum parthenium*. Contact Dermatitis 1980;6:445.

31. Paulsen E, Christensen LP, Andersen KE. Do monoterpenes released from feverfew (*Tanacetum parthenium*) plants cause airborne Compositae dermatitis? Contact Dermatitis 2004;47:14–18.

32. Hausen BM, Osmundsen PE. Contact allergy to parthenolide in *Tanacetum parthenium* (L.) Schulz-Bip. (feverfew, Asteraceae) and cross-reactions to related sesquiterpene lactone containing Compositae species. Acta Derm Venereol 1983;63:308–314.

33. Schmidt RJ. Compositae. Clin Dermatol 1986;4:46–61.

34. Zhou JZ, Kou X, Stevenson D. Rapid extraction and high-performance liquid chromatographic determination of parthenolide in feverfew (*Tanacetum parthenium*). J Agric Food Chem 1999;47:1018–1022.

Chapter 66

GARLIC
(*Allium sativum* L.)

HISTORY

The use of garlic dates back at least 5,000 years to the Babylonians.[1] Many ancient civilizations used garlic including the Egyptians, Phoenicians, Greeks, Romans, Chinese, Hindus, and Vikings. Reportedly, the construction workers on the Egyptian pyramids ate garlic to protect themselves from diseases, and grave diggers drank wine containing crushed garlic to protect themselves from the plague during the Middle Ages.[2] When the Russians depleted their supplies of penicillin during World War II, they used garlic poultices to treat wounds.

BOTANICAL DESCRIPTION

The name garlic is derived from the Anglo-Saxon words *gar* (spear) and *lac* (plant) that refer to the shape of garlic leaves. The genus *Allium* contains about 450 species including a variety of chives, onions, and leeks.

Common Name: Cultivated garlic

Scientific Name: *Allium sativum* L.

Botanical Family: Liliaceae (lily)

Physical Description: This perennial herb has a bulb consisting of a cluster of smaller bulbs (bulblets or cloves). The leaves are long, linear, and sword-shaped. The flowers are small, flat umbels.

Distribution and Ecology: *Allium* species are widely distributed over temperate regions of the northern hemisphere. The cultivated garlic probably originated in Central Asia.

EXPOSURE

Sources

Commercial available forms of garlic include raw garlic homogenate, garlic oil, garlic oil macerate, dehydrated garlic powder, and aged garlic extract. Figure 66.1 lists the major ingredients in these garlic preparations. Raw garlic homogenate is the most common form of garlic, and this preparation is equivalent to the aqueous extract of garlic with the principal ingredients being allicin (diallyl thiosulfinate), allyl methyl thiosulfonate, 1-propenyl allyl thiosulfonate, γ-glutamyl-S-alkyl-L-cysteine.[3] The concentration of sulfur-containing compounds (e.g., S-allyl cysteine) varies with preparation processes. Garlic essential oil is a steam distillation product of garlic that contains primarily sulfides (diallyl trisulfide, diallyl disulfide, allyl methyl trisulfide, allyl methyl disulfide, diallyl tetrasulfide) and no allicin or water-soluble fraction. Garlic powder is a flavoring agent for condiments and processed foods that contains alliin (up to 1%) and small amounts of oil-soluble sulfur compounds. Oil macerates are encapsulated mixtures of whole garlic cloves ground in vegetable oil. These preparations contain decomposed products of allicin (ajoene, dithiins, sulfides, residual alliin). Typically, the storage of sliced raw garlic in 15–20% ethanol for 20 months produces aged garlic extract that is relatively low in allicin and high in S-allyl cysteine and S-allylmercaptocysteine. Consequently, the hydroalcoholic extract of aged garlic contains primarily water-soluble constituents (S-allyl cysteine, S-allylmercaptocysteine) and small amounts of oil-soluble sulfur compounds.

Medical Toxicology of Natural Substances, by Donald G. Barceloux, MD

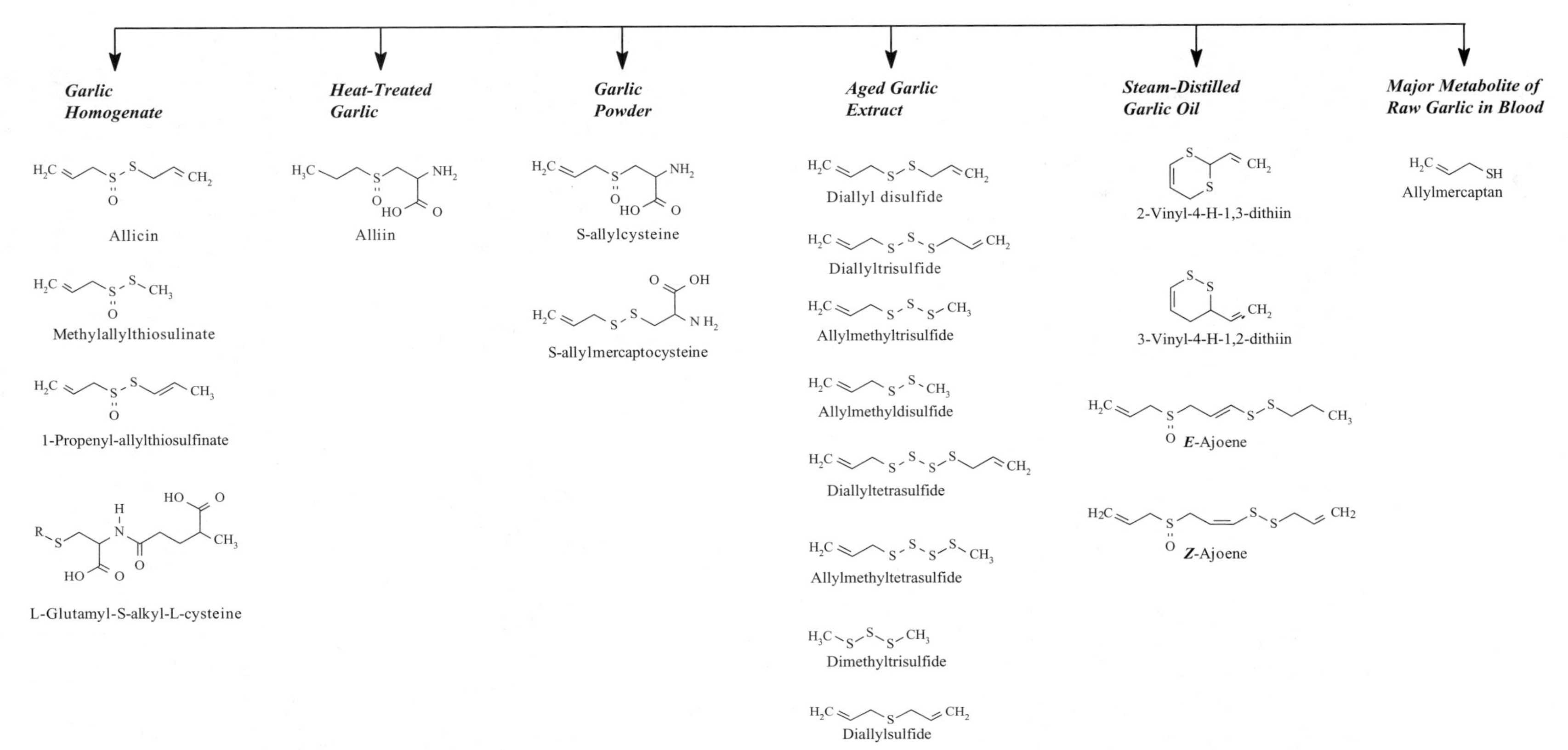

FIGURE 66.1. Major organosulfur compounds in various garlic preparations. Adapted from Banerjee SK, et al. Garlic as an antioxidant: the good, the bad and the ugly. Phytother Res 2003;17:98. Copyright John Wiley & Sons Limited.

Medicinal Uses

Traditional

Traditional uses of garlic include the treatment of parasites, skin diseases, wounds, gastrointestinal illnesses, flatulence, respiratory infections, intestinal parasites, leprosy, fever, and aging.

Current

The use of dried garlic powder causes some modest, short-term reduction (8–12 weeks) in total cholesterol concentrations, but these effects are not sustained over 6 months.[4] The consumption of garlic powder probably does not significantly lower plasma lipid concentrations when administered in conjunction with a low-fat, low-cholesterol diet.[5] Epidemiologic studies suggest some decreased risk of stomach and colon cancer following the high consumption of garlic and other allium vegetables (chives, leeks, onions, shallots),[6] but these studies are not well controlled for confounding variables. Although garlic has some modest antiplatelet aggregation activity *in vitro*,[7] there is no clear clinical evidence that the use of garlic reduces the risk of thrombosis. There is no reliable evidence that the use of garlic reduces blood pressure in hypertensive patients or serum glucose in diabetics. Additionally, there are insufficient data to support the use of garlic as an anti-infective agent.

Regulatory Status

Garlic products are marketed in the European Union (EU) as foodstuffs and as herbal medicinal products. The latter preparations are not available in all member nations. In the United States, garlic is generally recognized as safe (GRAS) as a dietary supplement and food additive. Garlic is also on the German Commission E list of approved herbs.

PRINCIPAL INGREDIENTS

Chemical Composition

The major flavoring constituents of garlic are sulfur compounds (diallyl disulfide, allyl sulfide, diallyl trisulfide), but preparation processes dramatically alter the type and concentrations of these compounds in garlic preparations.[8] These processes include steam distillation, simultaneous distillation and solvent extraction, solid-phase trapping solvent extraction, and headspace solid-phase microextraction. The primary sulfur-containing compounds in intact, whole garlic are γ-glutamyl-S-alk(en)yl-L-cysteines and S-alk(en)yl-L-cysteine sulfoxides (e.g., alliin, methiin). The odorless amino acid, alliin (S-allyl cysteine sulfoxide), accumulates during storage at cool temperatures, and garlic bulbs typically contain up to 1.8% alliin and 0.9% γ-glutamyl cysteine compounds along with small amounts of S-allyl cysteine.[9] Processing (i.e., crushing, cutting) whole, fresh garlic causes the release of the vacuolar enzyme, alliinase, which rapidly lyses cytosolic cysteine sulfoxides (e.g., alliin) into numerous alkyl alkane-thiosulfinate compounds including allicin (diallyl thiosulfinate, CAS RN:539-86-6). Allicin is usually absent from intact, fresh garlic bulbs. Figure 66.2 displays the chemical structure allicin. This odorous compound decomposes rapidly into various sulfides (e.g., ajoene, dithiins, diallyl sulfides), which contribute to the characteristic taste and flavor of garlic. Although freshly crushed garlic can contain small amounts of allicin, commercial products do not usually contain allicin, and allicin is not detectable in the blood after the ingestion of raw garlic or pure allicin.[9]

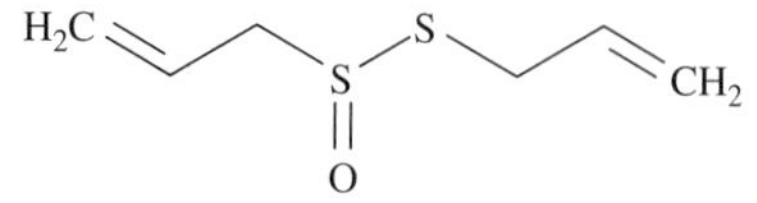

FIGURE 66.2. Chemical structure of allicin.

Additional constituents of garlic include steroidal glycosides, prostaglandins, lectins, fructan, essential oil, adenosine vitamins, essential amino acids, fatty acids, nicotinic acid, biotin, glycolipids, phospholipids, flavonoids, phenolics, and anthocyanins.[10] Garlic contains the allergen, alliin lyase, which cross-reacts with alliin lyases from other *Allium* species (e.g., leek, shallot, onion),[11] as well as the photocontact allergen, diallyl disulfide.[12] The main constituents of garlic oil are the hydrophobic compounds, allylpolysulfides (diallyl sulfide, diallyl disulfide, diallyl trisulfide). Flavonoids are polyphenols with diphenylpropane ($C_6C_3C_6$) skeletons. The four major classes of flavonoids are 4-oxo-flavonoids (flavones, flavonols), isoflavones, flavan-3-ol derivatives (catechin, tannins), and anthocyanins. *Allium* vegetables (onions, leek, chives, garlic) contain relatively high concentration of flavonoids. In a study of aqueous methanol extracts of edible plants, the total flavonoid content of garlic was 957 mg/kg dry weight consisting primarily of the flavonols, myricetin (639 mg/kg) and quercetin (47 mg/kg) along with the flavone, apigenin (217 mg/kg).[13]

Physiochemical Properties

The characteristic odor of garlic results from sulfur-containing compounds.[1] Aging of garlic reduces the content of sulfur compounds (e.g., alliin) and decreases the odor associated with garlic. The water-soluble precursor of the compound, alliin, is far more stable than its decomposition product, allicin. Both alliin (S-allyl cysteine sulfoxide) and alliinase are stable in dry environments. Allicin is a highly reactive, odorous liquid that decomposes rapidly into secondary products at temperatures above 60°C (140°F). At room temperature, the half-life of allicin is about $2^1/_2$ days in crushed garlic.[14] Consequently, heat-treated garlic contains mainly alliin. *In vitro* studies indicate that allicin easily diffuses across phospholipid membranes and into the cytoplasm of red blood cells.[15] Water soluble sulfur compounds (e.g., hydroalcoholic extracts) in garlic are probably less toxic than the oil-soluble organosulfur constituents of garlic.[16] S-allyl cysteine is an odorless, stable, water-soluble compound.

DOSE RESPONSE

Standard daily doses of garlic are as follows: raw garlic, 4g; dried garlic powder tablet, 600–1200mg (standardized to 1.3% alliin or 0.6% allicin); or aged garlic extract, 7.2g.[17] The use of aged garlic extract in doses up to 10g has not caused significant adverse effects in clinical trials.[9] In a clinical trial, the maximum tolerable dose of garlic extract was 25mL.[18]

TOXICOKINETICS

Kinetics

The absorption of the water-soluble compound, S-allyl cysteine, is relatively rapid and complete.[19] The oil-soluble organosulfur compounds (allicin, diallyl sulfide, ajoene, vinyldithiins) generally are not detectable in blood, even after ingestion of large amounts of garlic. Garlic preparations contain allicin, which decomposes in the stomach and intestines to release allyl sulfides, disulfides, and other volatile compounds. The main sulfur-containing volatile metabolite in the breath after garlic consumption is allyl methyl sulfide with allyl methyl disulfide being a minor volatile metabolite.[20] These studies also indicate that *d*-limonene and *p*-cymene are detectable in the breath of volunteers after consuming garlic. S-allyl cysteine appears in the blood after ingestion of garlic as a result of the hydrolysis of γ-glutamyl-S-allyl-L-cysteine by γ-glutamine-trans-peptidase. *N*-acetylation of S-allyl-cysteine by *N*-acetyltransferase produces the urinary biomarker, S-allyl-mercapturic acid. Other metabolites detectable after ingestion include methylates of *N*-acetyl-S-(2-carboxypropyl) cysteine and *N*-acetyl-S-allyl cysteine.[21]

Drug Interactions

In vitro studies suggest that sulfide constituents in garlic can potentially induce the cytochrome P450 isoenzymes, CYP3A4, CYP2C9, and CYP2C19.[22] The *in vitro* activity of CYP2D6 is not usually affected by constituents in garlic.[23] The clinical relevance of these potential interactions remains unclear. Human studies suggest that regular use of garlic can reduce the bioavailability of HIV-1 protease inhibitors (saquinavir and, to a lesser extent, ritonavir). In a study of 10 healthy volunteers, the use of garlic caplets (twice daily for 20 days) resulted in a significant decline in the plasma concentration of saquinavir, an HIV-1 protease inhibitor.[24] Although the exact mechanism of this effect was unclear, the similarity in the reduction of the AUC, the mean maximum concentration, and the 8-hour concentration of saquinavir suggested that garlic affected the bioavailability of saquinavir, rather than its systemic clearance. A study of 14 healthy volunteers receiving 3.6g garlic tablets (primarily alliin, alliinase) daily for 14 days did not detect significant differences in the pharmacokinetics of the probe substrates dextromethorphan (CYP2D6) or alprazolam (CYP3A4), when comparing baseline with garlic-treated phases.[25] A similar study of volunteers using garlic oil, which contains little alliin and alliinase, also did not demonstrate any significant effect of garlic oil on probe substrate, midazolam (CYP3A4),[26] but a similar study in elderly subjects detected a mean 22% decrease in CYP2E1 activity following the daily use of garlic oil for 28 days.[27]

The evidence for an interaction between garlic and coagulation remains conflicting. A case report suggested that the use of large amounts of garlic can cause platelet disorders and/or hemorrhage,[28] and another case report temporarily associated a reduction of the international normalized ratio (INR) below therapeutic range during the use of garlic tablets (600mg/daily) in addition to the ingestion of fluindione.[29] Normalization of the INR followed the cessation of garlic use. However, a double-blind, placebo-controlled study of 14 healthy men did not detect a clinical significant effect of typical garlic consumption on platelet aggregation.[30]

CLINICAL RESPONSE

Adverse Reactions

The most frequent side effect of garlic use is breath and body odor. Consumption of excessive amounts of raw garlic, particularly on an empty stomach, can cause flatulence, nausea, and abdominal discomfort.

Acute Effects

Fresh garlic is a potential source of *Salmonella virchow*;[31] therefore, the consumption of garlic contaminated with *Salmonella* species can cause acute gastroenteritis and salmonellosis. Additionally, case reports document chopped garlic in soybean oil as a potential source of botulism.[32] Case reports describe exposure to garlic with a variety of skin reactions including irritant contact dermatitis,[33] superficial and deep, partial-thickness chemical burns,[34,35] local urticaria,[36] photocontact allergies,[37] immediate hypersensitivity reactions,[38] and delayed hypersensitivity. Most topical reactions to garlic result from irritation rather than an allergic reaction, particularly following exposure to fresh crushed garlic or garlic oil.[39] An eczematous skin lesion can develop after chronic exposure to garlic in sensitized individuals.[40,41] The intentional application of crushed, raw garlic to the skin can produce second-degree chemical burns, particularly when applied as a poultice.[42,43] The degree of skin injury depends on a variety of factors including duration of exposure, pre-existing skin conditions, anatomic area of skin, and freshness of the garlic.[44] The irritant compound is heat sensitive; therefore, cooked garlic does not cause skin burns. Occasional case reports associate occupational contact dermatitis with exposure to garlic, particularly in workers who cut garlic. These case reports have described a low-grade, hyperkeratotic fingertip dermatitis with fissures on the thumb, index, and middle finger of the nondominant hand.[45,46] Other allergic reactions associated with garlic exposure include acute bronchospasm in atopic individuals,[47] rhinoconjunctivitis,[48] and anaphylaxis.[49] Potential allergens in garlic include diallyl disulfide and alliin lyase.[50]

DIAGNOSTIC TESTING

Analytical Methods

High performance liquid chromatography (HPLC) with UV detection and gas chromatography/mass spectrometry (GC/MS) are common methods for the determination of sulfur compounds (e.g., allicin) in garlic. Methods for the simultaneous detection of alliin, deoxyalliin, allicin and dipeptide precursors in garlic products are available using HPLC with a diode-array UV detector and ion trap mass spectrometry (electrospray ionization).[51] Limits of detection for these methods are in the range of 0.1–0.5 nmol.

Biomarkers

The urinary metabolite, S-allyl-mercapturic acid, is a biomarker of garlic use that results from the hydrolysis and *N*-acetylation of γ-glutamyl-S-allyl-L-cysteine.[52] In a study of 5 healthy volunteers administered garlic tablets, the mean elimination half-life of S-allyl-mercapturic acid in urine was 6.0 ± 1.3 hour.[53] The limit of detection for S-allyl-mercapturic acid in urine samples was 10 μg/mL, as measured by GC with sulphur-selective detection. This biomarker is stable in frozen urine samples at −20 °C (−4 °F) for at least 3 months. S-allyl cysteine is stable for at least 12 months at ambient temperature.[54] S-allyl-mercapturic acid is not a unique biomarker of garlic use because of the presence of similar sulfur compounds in other *Allium* species (e.g., onion, leek, chives). β-Chlorogenin is the sapogenin of eruboside B, which is the characteristic saponin of crushed raw garlic. Except for elephant garlic (*A. ampeloprasum*), this saponin is not usually present in other *Allium* vegetables.[55]

TREATMENT

Treatment of adverse reactions associated with garlic is supportive including the use of standard treatment methods for contact dermatitis (topical steroids) or anaphylaxis (epinephrine, antihistamines).

References

1. Block E. The chemistry of garlic and onions. Sci Am 1985;252:114–119.
2. Dubick MA. Historical perspectives on the use of herbal preparations to promote health. J Nutr 1986;116:1348–1354.
3. Banerjee SK, Mukee PK, Maulik SK. Garlic as an antioxidant: the good, the bad and the ugly. Phytother Res 2003;17:97–106.
4. Ackermann RT, Mulrow CD, Ramirez G, Gardner CD, Morbidoni L, Lawrence VA. Garlic shows promise for improving some cardiovascular risk factors. Arch Intern Med 2001;161:813–824.
5. Spigelski D, Jones PJ. Efficacy of garlic supplementation in lowering serum cholesterol levels. Nutr Rev 2001;59: 236–241.
6. Fleischauer AT, Arab L. Garlic and cancer: a critical review of the epidemiologic literature. J Nutr 2001(suppl);131: 1032S–1040S.
7. Ramsay NA, Kenny MW, Davies G, Patel JP. Complimentary and alternative medicine use among patients starting warfarin. Br J Haematol 2005;130:777–780.
8. Lee S-N, Kim N-S, Lee D-S. Comparative study of extraction techniques for determination of garlic flavor components by gas chromatography-mass spectrometry. Anal Bioanal Chem 2003;377:749–756.
9. Amagase H, Petesch BL, Matsuura H, Kasuga S, Itakura Y. Intake of garlic and its bioactive components. Recent advances on the nutritional effects associated with the use

of garlic as a supplement. J Nutr 2001;131(Suppl): 955S–962S.

10. Fenwick GR, Hanley AB. The genus *Allium*. Part 2. Crit Rev Food Sci Nutr 1985;22:273–377.
11. Kao SH, Hsu CH, Su SN, Hor WT, Chang TWH, Chow LP. Identification and immunologic characterization of an allergen, alliin lyase, from garlic (*Allium sativum*). J Allergy Clin Immunol 2004;113:161–168.
12. Alvarez MS, Jacobs S, Jiang SB, Brancaccio RR, Soter NA, Cohen DE. Photocontact allergy to diallyl disulfide. Am J Contact Dermatol 2003;14:161–165.
13. Miean KH, Mohamed S. Flavonoid (myricetin, quercetin, kaempferol, luteolin, and apigenin) content of edible tropical plants. J Agric Food Chem 2001;49;3106–3112.
14. Rybak ME, Calvey EM, Harnly JM. Quantitative determination of allicin in garlic: supercritical fluid extraction and standard addition of alliin. J Agric Food Chem 2004;52:682–687.
15. Miron T, Rabinkov A, Mirelman D, Wilchek M, Weiner L. The mode of action of allicin: its ready permeability through phospholipid membranes may contribute to its biological activity. Biochim Biophys Acta 2000;1463: 20–30.
16. Sumiyoshi H, Kanezawa A, Masamoto K, Harada H, Nakagami S, Yokota A, et al. [Chronic toxicity test of garlic extract in rats]. J Toxicol Sci. 1984;9:61–75. [Japanese]
17. Tattelman E. Health effects of garlic. Am Fam Physician 2005;72:103–106.
18. Caporaso N, Smith SM, Eng RH. Antifungal activity in human urine and serum after ingestion of garlic (*Allium sativum*). Antimicrob Agents Chemother 1983;23: 700–702.
19. Kodera Y, Suzuki A, Imada O, Kasuga S, Sumioka I, Kanezawa A, et al. Physical, chemical and biological properties of S-allyl cysteine, an amino acid derived from garlic. J Agric Food Chem 2002;50:622–632.
20. Rosen TR, Hiserodt RD, Fukuda E, Ruiz RJ, Zhou Z, Lech J, et al. The determination of metabolites of garlic preparations in breath and human plasma. Biofactors 2000;13:241–249.
21. Jandke J, Spiteller G. Unusual conjugates in biological profiles originating from consumption of onions and garlic. J Chromatogr 1987;421:1–8.
22. Delgoda R, Westlake AC. Herbal interactions involving cytochrome P450 enzymes a mini review. Toxicol Rev 2004;23:239–249.
23. Foster BC, Foster MS, Vandenhoek S, Krantis A, Budzinski JW, Arnason JT, et al. An *in vitro* evaluation of human cytochrome P450 3A4 and P-glycoprotein inhibition by garlic. J Pharm Pharmaceut Sci 2001;4:176–184.
24. Piscitelli SC, Burstein AH, Welden N, Gallicano KD, Falloon J. The effect of garlic supplements on the pharmacokinetics of saquinavir. Clin Infect Dis 2002;34:234–238.
25. Markowitz JS, DeVane CL, Chavin KD, Taylor RM, Ruan Y, Donovan JL. Effects of garlic (*Allium sativum* L.) supplementation on cytochrome P450 2D6 and 3A4 activity in healthy volunteers. Clin Pharmacol Ther 2003;74: 170–177.
26. Gurley BJ, Gardner SF, Hubbard MA, Williams DK, Gentry WB, Cui Y, Ang CY. Cytochrome P450 phenotypic ratios for predicting herb-drug interactions in humans. Clin Pharmacol Ther 2002;72:276–287.
27. Gurley BJ, Gardner SF, Hubbard MA, Williams DK, Gentry WB, Cui Y, Ang CY. Clinical assessment of effects of botanical supplementation on cytochrome P450 phenotypes in the elderly: St John's wort, garlic oil, *Panax ginseng* and *Ginkgo biloba*. Drugs Aging 2005;22:525–539.
28. Morris J, Burke V, Mori TA, Vandongen R, Beilin LJ. Effects of garlic extract on platelet aggregation: a randomized placebo-controlled double-blind study. Clin Exp Pharmacol Physiol 1995;22:414–417.
29. Pathak A, Leger P, Bagheri H, Senard J-M, Boccalon H, Montastruc J-L. Garlic interaction with fluindione: a case report. Therapie 2003;58:381–381.
30. Rose KD, Croissant PD, Parliament CF, Levin MB. Spontaneous spinal epidural hematoma with associated platelet dysfunction from excessive garlic ingestion: a case report. Neurosurgery 1990;26:880–882.
31. Bennett CM, Dalton C, Beers-Deeble M, Milazzo A, Kraa E, Davos D, et al. Fresh garlic: a possible vehicle for *Salmonella Virchow*. Epidemiol Infect 2003;131:1041–1048.
32. St Louis ME, Peck SH, Bowering D, Morgan GB, Blatherwick J, Banerjee S, et al. Botulism from chopped garlic: delayed recognition of a major outbreak. Ann Intern Med 1988;108:363–368.
33. Hughes TM, Varma S, Stone NM. Occupational contact dermatitis from a garlic and herb mixture. Contact Dermatitis 2002;47:48.
34. Lachter J, Babich JP, Brookman JC, Factor AY. Garlic: a way out of work. Mil Med 2003;168:499–500.
35. Baruchin AM, Sagi A, Yoffe B, Ronen M. Garlic burns. Burns 2001;27:781–782.
36. Pires G, Pargana E, Loureiro V, Almeida MM, Pinto JR. Allergy to garlic. Allergy 2002;57:957–958.
37. Alvarez MS, Jacobs S, Jiang SB, Brancaccio RR, Soter NA, Cohen DE. Photocontact allergy to diallyl disulfide. Am J Contact Dermatitis 2003;14:161–165.
38. Pires G, Pargana E, Loureiro V, Almeida MM, Pinto JR. Allergy to garlic. Allergy 2002;57:957–958.
39. Lee TY, Lam TH. Contact dermatitis due to topical treatment with garlic in Hong Kong. Contact Dermatitis 1991;24:193–196.
40. Bassioukas K, Orton D, Cerio R. Occupational airborne allergic contact dermatitis from garlic with concurrent type I allergy. Contact Dermatitis 2004;50:39–41.
41. Jappe U, Bonnekoh B, Hausen BM, Gollnick H. Garlic-related dermatoses: case report and review of the literature. Am J Contact Dermat 1999;10:37–39.
42. Rafaat M, Leung AK. Garlic burns. Pediatr Dermatol 2000;17:475–476.
43. Lachter J, Babich JP, Brookman JC, Factor AY. Garlic: a way out of work. Mil Med 2003;168:499–500.

44. Dietz DM, Varcelotti JR, Stahlfeld. Garlic burns: a not-so-rare complication of naturopathic remedy? Burns 2004;30; 612–613.
45. McGovern TW, LaWarre S. Botanical briefs: garlic—*Allium sativum*. Cutis 2001;67:193–194.
46. Kanerva L, Estlander T, Jolanki R. Occupational allergic contact dermatitis from spices. Contact Dermatitis 1996;35: 157–162.
47. Falleroni AE, Zeiss CR, Levitz D. Occupational asthma secondary to inhalation of garlic dust. J Allergy Clin Immunol 1981;68:156–160.
48. Seuri M, Taivanen A, Ruoppi P, Tukiainen H. Three cases of occupational asthma and rhinitis caused by garlic. Clin Exp Allergy 1993;23:1011–1014.
49. Perez-Pmiento AJ, Moneo I, Santaolalla M, de Paz S, Fernandez-Parra B, Dominguez-Lazaro AR. Anaphylactic reaction to young garlic. Allergy 1999;54:626–629.
50. Kao S-H, Hsu C-H, Su S-N, Hor W-T, Chang W-H, Chow L-P. Identification and immunologic characterization of an allergen, alliin lyase, from garlic (*Allium sativum*). J Allergy Clin Immunol 2004;113161–168.
51. Arnault I, Christides JP, Mandon N, Haffner T, Kahane R, Auger J. High-performance ion-pair chromatography method for simultaneous analysis of alliin, deoxyalliin, allicin and dipeptide precursors in garlic products using multiple mass spectrometry and UV detection. J Chromatogr A 2003;991:69–75.
52. Verhagen H, Hageman GJ, Rauma A-L, Versluis-de haan G, van Herwijnen MH, de Groot J, et al. Biomonitoring the intake of garlic via urinary excretion of allyl mercapturic acid. Br J Nutr 2001;86(Suppl 1):S111–S114.
53. de Rooij BM, Boogaard PJ, Rijksen DA, Commandeur JN, Vermeulen NP. Urinary excretion of *N*-acetyl-S-allyl-L-cysteine upon garlic consumption by human volunteers. Arch Toxicol 1996;70:635–639.
54. Lawson LD, Gardner CD. Composition, stability, and bioavailability of garlic products used in a clinical trial. J Agric Food Chem 2005;53:6254–6261.
55. Itakura Y, Ichikawa M, Mori Y, Okino R, Udayama M, Morita T. How to distinguish garlic from the other *Allium* vegetables. J Nutr 2001(Suppl);131:963S–967S.

Chapter 67

GERMANDER
(*Teucrium chamaedrys* L.)

HISTORY

Germander is an ancient herbal treatment for a variety of illnesses including stomach ailments, bronchitis, wounds, and diabetes mellitus. In 1992, the French Health Authority suspended the marketing of herbal preparations containing *Teucrium chamaedrys* after the publication of a report of 26 French cases of acute hepatitis associated with the consumption of herbal preparations containing wild germander.[1] Other countries in Europe and North America soon prohibited the use of herbal preparations containing *T. chamaedrys* L.

BOTANICAL DESCRIPTION

There are over 300 *Teucrium* species, which inhabit primarily the northern temperate and subtropical regions of the eastern hemisphere. Most *Teucrium* species are potentially hepatotoxic as a result of the presence of variable amounts of neoclerodane diterpenoid compounds including *T. canadense* L. (American or Canadian germander), *T. chamaedrys* L. (wild germander), *T. capitatum* L. (cat thyme), *T. polium* L. (golden germander), and *T. fruticans* L. (shrubby germander).[2,3] *Teucrium* species can taint herbal preparations of dog skullcap (*Scutellaria lateriflora* L.).[4] However, this contamination can be detected using digital photomicroscopy to identify abundant bristle-like trichomes (leaf of *T. canadense* L.) or a waxy cuticle with numerous glandular scales (leaf of *T. chamaedrys*).[5] *Scutellaria lateriflora* L. lacks both of these anatomical characteristics.

Common Name: Wild Germander, Wall Germander

Scientific Name: *Teucrium chamaedrys* L.

Botanical Family: Lamiaceae (Labiatae, Mint)

Physical Description: This somewhat hairy, evergreen shrub grows up to about 1 ft (~0.3 m) in height. The indented, opposite, dark green leaves are approximately 0.5 to 1.5 in. (~1.3–3.8 cm) long with a pattern similar to an oak leaf. The scented, rose-colored, labiate flowers appear from July to September in three to six flowered whorls on leafy, terminal spikes. The seeds ripen from August to September.

Distribution and Ecology: This plant prefers warm, dry areas on rocky soil or waste areas. The range of this plant is southern Europe and the Mediterranean with the plant naturalized in the United Kingdom and the United States.

Common Name: Golden Germander, Felty Germander

Scientific Name: *Teucrium polium* L.

Botanical Family: Lamiaceae (Labiatae, Mint)

Physical Description: This low-growing perennial plant grows up to about 1–1$\frac{1}{2}$ ft (~0.3–0.5 m) with erect, simple elongated ending in a paniculate or corymbose inflorescence. The sessile, oblong leaves are about 1 in. (2.5 cm) in length. The inconspicuous flowers are white to pale cream-colored and bloom from June to August.

Medical Toxicology of Natural Substances, by Donald G. Barceloux, MD

Distribution and Ecology: This small shrub inhabits dry, rocky grassland in southern Europe and southwestern Asia.

EXPOSURE

Sources

Teucrin A is not available commercially. Alcoholic extracts of *Teucrium chamaedrys* are bitter, aromatic flavoring ingredients. As a dietary aid, germander was available as 200mg or 270mg capsules. Germander is also available as a tea combined with other herbs or alone. Germander herbal teas typically contain about 75–150mg teucrin A/100mL.

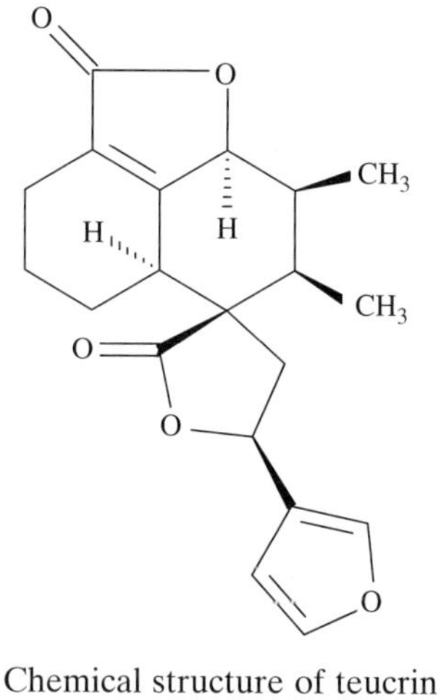

Medicinal Uses

TRADITIONAL

The use of *Teucrium polium* is a common herbal treatment for controlling diabetes mellitus.[6] In Qatar, a decoction of boiled leaves is a treatment for fever and stomach pain in children; crushed leaves are applied as compresses for wounds.[7] In Iran, *Teucrium polium* is a traditional medicine for the treatment of hypertension, hypercholesterolemia, inflammation, bacterial infections, seizures, pain, and diarrhea.[8] Other traditional uses of this herb include the treatment of anorexia, rhinitis, chronic bronchitis, gout, rheumatoid arthritis, fever, and dyspepsia. Animal studies indicate that anti-inflammatory activities of *T. polium* are low compared with other plant extracts.[9] The infusion of the leaves of *Teucrium* species also is considered a stimulant, antihelmintic, and a carminative (treatment of colic in infants).

REGULATORY STATUS

Europe and the United States banned the use of germander as a dietary supplement because of the association between the use of germander as a diet aid and hepatotoxicity. Teucrin A is approved by the US Food and Drug Administration only as a flavoring agent in alcoholic beverages to add a bitter aromatic taste, primarily to drinks such as vermouth, flavored wines, bitters, and less commonly soft drinks. Alcoholic extracts of germander are permitted as flavoring agents.

PRINCIPAL INGREDIENTS

Chemical Composition

The aerial parts of *Teucrium* species contain a variety of constituents including glycosides, saponins, flavonoids, and several furan-containing neoclerodane diterpenoid compounds. Teucrin A (CAS RN: 12798-51-5) and several other neoclerodane diterpenoid compounds present in the aerial parts of *Teucrium* species probably contribute to the hepatotoxicity of these herbs.[10,11] Figure 67.1 displays the chemical structure of the C19-compound, teucrin A ($C_{19}H_{20}O_6$). Other diterpenoid compounds include teucrin G, teucrin H1, dihydroteugin, isoteuflidin, teuflin, teucvidin, teucrolivin A, teucrolivin B, and 6-β-hydroxyteuscordin.[12,13] The major constituents in ethanol extracts of *T. canadense* are phenylpropanoid glycosides, primarily verbascoside, whereas teucrioside is the predominant compound in ethanol extracts of *T. chamaedrys*.[5] These compounds help detect adulteration of skullcap (*Scutellaria lateriflora* L.) by *Teucrium* species because skullcap does not contain these phenylpropanoid compounds. Analysis of bitter alcoholic beverages from the Italian market demonstrated teucrin A concentrations ranging from non-detectable to 10ppm as measured by high performance liquid chromatography (HPLC).[14] Essential oils from *Teucrium* species contain substantial amounts of monoterpene hydrocarbons (α-pinene, β-pinene, limonene, sabinene) and sesquiterpenes (elemol, germacrene D, T-cadinol, δ-cadinene, α-caryophyllene, isocaryophyllene, β-bisabolene).[15–17]

FIGURE 67.1. Chemical structure of teucrin A.

Physiochemical Properties

The molecular weight of teucrin A is 344Da.

Mechanism of Toxicity

Although the exact mechanism by which germander causes hepatic necrosis is unclear, animal studies suggest that cytochrome CYP3A-mediated metabolism produces electrophilic metabolites from furano neoclerodane diterpene compounds, such as teucrin A.[18] These electrophilic metabolites stimulate apoptosis by decreasing thiols (e.g., glutathione), increasing intracellular calcium, and activating Ca^{++}-dependent transglutaminase and endonucleases.[19] Animal studies indicate that the administration of teucrin A causes the same midzonal hepatic necrosis as powdered plant material from germander.[18] However, the exact role of teucrin A in producing germander-induced hepatotoxicity is undefined because germander contains a variety of furano neoclerodane diterpene compounds, and the amount (150 mg/kg) of teucrin A necessary to produce hepatoxicity in animal studies is much higher than the dose of germander in patients developing hepatotoxicity after ingesting germander. Consequently, hepatic injury may result from germander constituents other than teucrin A. In some case reports acute cholestatic hepatitis developed along with the transient appearance of serum autoantibodies (e.g., antimitochondrial M2, antimicrosomal epoxide hydrolase, antinuclear antibodies).[20,21] The presence of serum autoantibodies, eosinophilia, and eosinophilic infiltrates in liver samples from affected patients suggests the possibility of idiosyncratic reactions as well as indirect (i.e., through altered metabolism) hepatotoxic effects.[22]

DOSE RESPONSE

Medicinal Use

The recommended dose of germander (*T. chamaedrys*) for weight loss is 600–1600 mg daily, and case reports associate the use of germander in this dose range for 6 months with the development of jaundice and markedly elevated serum hepatic aminotransferases.[23,24] The Council of Europe established a NOAEL (no observed adverse effect level) of 0.4 mg/kg/day for teucrin A and a temporary tolerable daily intake of 0.12 mg/day based on a safety factor of 200.[25]

Toxicity

The dose and duration of germander required to produce hepatotoxicity is highly variable, and the appearance of liver toxicity is not clearly related to either the dose or the duration of use. The ingestion of germander tea up to 2 L daily for 6 months was associated with the development cholestatic hepatitis that resolved 1 month after stopping ingestion of the tea,[26] whereas two 10-day courses of germander tea were associated with the development of hepatic failure requiring liver transplantation.[27] A fatality resulting from massive hepatic necrosis and associated with the use of germander involved two courses of 450 mg daily for 2 weeks separated by 6 months.[28]

TOXICOKINETICS

The furano diterpenoid compounds in germander undergo metabolic activation to electrophilic metabolites capable of covalent binding to proteins in hepatocytes. The CYP3A isoenzymes catalyze the formation of reactive epoxides and/or unsaturated aldehydes from these aromatic compounds.[11]

CLINICAL RESPONSE

Occasional case reports temporally associate the herbal use of several *Teucrium* species with the development of hepatitis including *T. polium*,[27] *T. chamaedrys*,[23] and *T. capitatum*.[29] In some of these cases, readministration of germander was associated with the prompt recurrence of hepatitis.[23] These case reports involve both hepatocellular and cholestatic hepatitis, and the severity of the hepatotoxicity ranges from spontaneous recovery without sequelae to liver transplantation.[30] The clinical and histological changes associated with germander-induced hepatitis are nonspecific, and there are no unique features to distinguish this illness from viral or drug-induced hepatitis. Symptoms associated with germander-induced hepatitis include general malaise, anorexia, nausea, vomiting, painless jaundice, pruritus, dark urine, and light stools. Complete recovery typically occurs within 1–6 months after cessation of germander use.[23] A 37-year-old Greek woman was hospitalized 5 days after finishing the second 10-day course of *T. polium* tea following the development of jaundice, hypoprothrombinemia, and marked elevation of serum hepatic aminotransferases.[27] Histologic examination demonstrated massive centrilobular hepatic necrosis with portal tract fibrosis and bile duct proliferation. She developed severe liver failure (factor V = 12%) and she underwent successful liver transplantation. Rarely, germander use has been associated with fatal fulminant hepatitis.[28] Germander is one of a number of herbs that are associated with hepatotoxicity, including valerian, hops, skullcap, gentian, senna fruit extracts, mistletoe, chaparral leaf, and pyrrolizidine alkaloid-containing herbs.

DIAGNOSTIC TESTING

Neoclerodane diterpenoids in extracts of *Teucrium* species are detectable in concentrations of 0.24–0.90 μg/mL as measured by HPLC with UV-Vis detection at 220 nm.[12] The limits of detection and quantitation for teucrin A in alcoholic beverages were 0.1 ppm and

0.3 ppm, respectively, based on HPLC and liquid chromatography/electrospray ionization mass spectrometry.[14] Contamination of mad dog skullcap (*Scutellaria lateriflora* L.) by *Teucrium* species can be detected by analysis of ethanol extracts for phenylpropanoid compounds. This type of skullcap does not contain phenylpropanoid compounds, whereas the major phenylpropanoid compounds in ethanol extracts of *T. canadense* and *T. chamaedrys* are verbascoside and teucrioside, respectively.[5] The serum abnormalities associated with germander-induced hepatitis are nonspecific and include hyperbilirubinemia, elevated serum hepatic aminotransferases, and coagulation abnormalities (e.g., hypoprothrombinemia, factor V deficiency).

TREATMENT

Treatment of germander-associated hepatitis is supportive. The development of hepatic encephalopathy requiring liver transplantation may rarely occur. The role of *N*-acetylcysteine in replenishing depleted glutathione stores and limiting the progression of hepatic damage has not been investigated.

References

1. Castot A, Larrey D. [Hepatitis observed during a treatment with a drug or tea containing wild germander. Evaluation of 26 cases reported to the Regional Centers of Pharmacovigilance]. Gastroenterol Clin Biol 1992;16:916–922. [French]
2. Coll J, Tandron Y. Neo-clerodane diterpenes from *Teucrium fruticans*. Phytochemistry 2004;65:387–392.
3. Bruno M, Bondi ML, Rosselli S, Maggio A, Piozzi F, Arnold NA. Neoclerodane diterpenoids from *Teucrium montbretii* subsp. *libanoticum* and their absolute configuration. J Nat Prod 2002;65:142–146.
4. De Smet PA. Health risks of herbal remedies. Drug Saf 1995;13:81–93.
5. Gafner S, Bergeron C, Batcha LL, Angerhofer CK, Sudberg S, Sudberg EM, et al. Analysis of *Scutellaria lateriflora* and its adulterants *Teucrium canadense* and *Teucrium chamaedrys* by LC-UV/MS, TLC, and digital photomicroscopy. J AOAC Int 2003;86:453–460.
6. Gharaibeh MN, Elayan HH, Salhab AS. Hypoglycemic effects of *Teucrium polium*. J Ethnopharmacol 1988;24:93–99.
7. Rizk AM, Hammouda FM, Rimpler H, Kamel A. Iridoids and flavonoids of *Teucrium polium* herb. Planta Med 1986;52:87–88.
8. Suleiman MS, Abdul-Ghani AS, Al-Khalil S, Amin R. Effect of *Teucrium polium* boiled leaf extract on intestinal motility and blood pressure. J Ethnopharmacol 1988;22:111–116.
9. Capasso F, Cerri R, Morrica P, Senatore F. Chemical composition and anti-inflammatory activity of an alcoholic extract of *Teucrium polium* L. Boll Soc Ital Sper 1983;59:1639–1643.
10. Bedir E, Manyam R, Khan IA. *Neo*-clerodane diterpenoids and phenylethanoid glycosides from *Teucrium chamaedrys* L. Phytochemistry 2003:63:977–983.
11. Lekehal M, Pessayre D, Lereau JM, Moulis C, Fouraste I, Fau D. Hepatotoxicity of the herbal medicine germander: metabolic activation of its furano diterpenoids by cytochrome P450 3A depletes cytoskeleton-associated protein thiols and forms plasma membrane blebs in rat hepatocytes. Hepatology 1996;24:212–218.
12. Bruno M, Rosselli S, Maggio A, Piozzi F, Scaglioni L, Arnold NA, Simmonds MS. Neoclerodanes from *Teucrium orientale*. Chem Pharm Bull (Tokyo) 2004;52:1497–500.
13. Avula B, Manyam RB, Bedir E, Khan IA. HPLC analysis of *neo*-clerodane diterpenoids from *Teucrium chamaedrys*. Pharmazie 2003;58:494–496.
14. Bosisio E, Giavarini F, Dell'Agli M, Galli G, Galli CL. Analysis by high-performance liquid chromatography of teucrin A in beverages flavoured with an extract of *Teucrium chamaedrys* L. Food Addit Contam 2004;21:407–414.
15. Ricci D, Fraternale D, Giamperi L, Bucchini A, Epifano F, Burini G, Curini M. Chemical composition, antimicrobial and antioxidant activity of the essential oil of *Teucrium marum* (Lamiaceae). J Ethnopharmacol 2005;98:195–200.
16. Hassan MM, Muhtadi FJ, Al-Badr AA. GLC-mass spectrometry of *Teucrium polium* oil. J Pharm Sci 1979;68:800–801.
17. Cavaleiro C, Salgueiro LR, Miguel MG, Proenca da Cunha A. Analysis by gas chromatography-mass spectrometry of the volatile components of *Teucrium lusitanicum* and *Teucrium algarbiensis*. J Chromatogr A 2004;1033:187–190.
18. Kouzi SA, McMurtry RJ, Nelson SD. Hepatotoxicity of germander (*Teucrium chamaedrys* L.) and one of its constituent neoclerodane diterpenes teucrin A in the mouse. Chem Res Toxicol 1994;7:850–856.
19. Fau D, Lekehal M, Farrell G, Moreau A, Moulis C, Feldmann G, Haouzi D, Pessayre D. Diterpenoids from germander, an herbal medicine, induce apoptosis in isolated rat hepatocytes. Gastroenterology 1997;113:1334–1346.
20. De Berardinis V, Moulis C, Maurice M, Beaune P, Pessayre D, Pompon D, Loeper J. Human microsomal epoxide hydrolase is the target of germander-induced autoantibodies on the surface of human hepatocytes. Mol Pharmacol 2000;58:542–551.
21. Polymeros D, Kamberoglou D, Tzias V. Acute cholestatic hepatitis caused by *Teucrium polium* (golden germander) with transient appearance of antimitochondrial antibody. J Gastroenterol 2002;34:100–101.
22. Starakis I, Siagris D, Leonidou L, Mazokopalis E, Tsamandas A, Karatza C. Hepatitis caused by the herbal remedy

Teucrium polium L. Eur J Gastroenterol Hepatol 2006;18: 681–683.

23. Larrey D, Vial T, Pauwels A, Castot A, Biour M, David M, Michel H. Hepatitis after germander (*Teucrium chamaedrys*) administration: another instance of herbal medicine hepatotoxicity. Ann Intern Med 1992;117:129–132.
24. Laliberte L, Villeneuve J-P. Hepatitis after the use of germander, a herbal remedy. Can Med Assoc J 1996;154: 1689–1962.
25. De Vincenzi M, Maialetti F, Silano M. Constituents of aromatic plants: teucrin A. Fitoterapia 2003;74:746–749.
26. Mazokopakis E, Lazaridou S, Tzardi M, Mixaki J, Diamantis I, Ganotakis E. Acute cholestatic hepatitis caused by *Teucrium polium* L. Phytomedicine 2004;11:83–84.
27. Mattei A, Rucay P, Samuel D, Feray C, Reynes M, Bismuth H. Liver transplantation for severe acute liver failure after herbal medicine (*Teucrium polium*) administration. J Hepatol 1995;22:597.
28. Mostefa-Kara N, Pauwels A, Pines E, Biour M, Levy VG. Fatal hepatitis after herbal tea. Lancet 1992;340(8820): 674.
29. Dourakis SP, Papanikolaou IS, Tzemanakis EN, Hadziyannis SJ. Acute hepatitis associated with herb (*Teucrium capitatum* L.) administration. Eur J Gastroenterol Hepatol 2002;14:693–695.
30. Perez Alvarez J, Saez-Royuela F, Gento Pena E, Lopez Morante A, Velasco Oses A, Martin Lorente J. [Acute hepatitis due to ingestion of *Teucrium chamaedrys* infusions]. Gastroenterol Hepatol 2001;24:240–243. [Spanish]

Chapter 68

GINGER
(*Zingiber officinale* Roscoe)

HISTORY

Ginger has been a part of traditional Chinese medicine at least since the 4th century BC, particularly for hepatic and other gastrointestinal disorders.[1] Two millennia before the birth of Christ, the Vedic literature of India included information on the use of ginger as a spice and an herb; Marco Polo also documented the use of ginger in India during the 13th century. The ancient Greeks and Romans used ginger as a spice. Traditional Arabian medicine used ginger for treating constipation, colds, bronchitis, cataracts, and stomach acidity.

BOTANICAL DESCRIPTION

Common Name: Ginger, Jiang [Chinese], Gingembre [French], Ingwer [German], Zenzero [Italian], Shoga [Japanese], Gengibre [Portuguese], Jengibre [Spanish]

Scientific Name: *Zingiber officinale* Roscoe

Botanical Family: Zingiberaceae (ginger family)

Physical Description: This slender, perennial herb has upright stems and narrow medium green leaves arranged in two ranks on each stem. Ginger reaches up to 1.5 m (~5 ft) in height with leaves about 2 cm (~0.8 in.) wide and 20 cm (~8 in.) long. This plant grows from a thick, aromatic underground stem (rhizome), which has a branched, uneven (warty) surface. The inflorescence forms a dense spike up to 8 cm (~3 in.) tall on a different stem than the leaves. The bracts are green with translucent margins and the small flowers are yellow-green with purple lips and cream-colored blotches. The fruiting body is a triangular-oval capsule containing many irregular, black seeds.

Distribution and Ecology: Ginger is indigenous to the warm, humid, shady tropical parts of southern Asia, but now this plant is cultivated commercially in Jamaica, West Indies, Haiti, China, and Nigeria. Harvest occurs about 5–6 months after planting fresh ginger and 8–9 months after planting dry ginger. Major producers of dry ginger include India, China, Taiwan, Sierra Leone, Nigeria, Jamaica, Fiji, and Australia.[2]

EXPOSURE

Sources

Ginger is a spice, flavoring agent, and herbal medicine used as both fresh and dried preparations of the ginger rhizomes. The rhizome is the plant part used for culinary as well as medicinal purposes. Some reduction in the major constituents (i.e., gingerols) occurs during the commercial drying process as a result of the dehydration of gingerols to shogaols.[3] The main source of ginger is the root and rhizome of *Zingiber officinale*. Ginger and ginger extracts are common constituents of foods and beverages including chutney, pickles, liquors, and bakery products. Common medicinal forms of ginger include fresh root, dried root, tablets, capsules (50–550 mg), liquid extract, tincture, and teas. Ginger oil is a steam distillate of fresh ginger rhizome or powdered ginger composed primarily of sesquiterpene

Medical Toxicology of Natural Substances, by Donald G. Barceloux, MD

hydrocarbons (e.g., zingiberene).[2] Whole, intact rhizomes are the best source of the essential oil of ginger because of the presence of substantial amounts of the oil in the peel. Whole dried ginger has a loose corky layer; peeled ginger has a line of fiber bundles exposed by removal of the cork layer. Although there is no universal standard, ginger products are often standardized to gingerol content.

Medicinal Uses

Traditional

Traditional medicine in China, Japan, and India use the rhizomes and stems of ginger as a component of the herbal treatment of digestive disorders (indigestion, flatulence, constipation, nausea), headaches, rheumatism, colds, and cough. A large portion of traditional Chinese herbal remedies contain ginger. In the Ayurvedic medical practice in India, ginger is an herbal treatment for colds and other viral infections, poor appetite, digestive problems, arthritis, and headache.[4,5]

Current

Current medicinal uses of ginger include motion sickness, nausea and vomiting (postsurgical, pregnancy, chemotherapy-induced), and osteoarthritis. Ginger is one of the most common herbal treatments for nausea and vomiting in pregnancy along with chamomile, peppermint, and raspberry leaf. There is supportive evidence from a randomized controlled trial and an open-label study that ginger reduces the severity and duration of chemotherapy-induced nausea and vomiting.[6] The addition of ginger to a standard antiemetic regimen probably does not reduce the nausea or vomiting associated with chemotherapy.[7] Limited evidence supports the use of ginger for the reduction of nausea and vomiting associated with pregnancy, motion sickness, and surgery.[8] Most of the limited number of randomized clinical trials of ginger for the treatment of nausea and vomiting of early pregnancy demonstrate positive results, but these studies involve only several hundred patients and larger trials are necessary to determine the efficacy and safety of ginger relative to other treatments.[9] A randomized, controlled equivalence trial involving 291 pregnant women (<16 weeks gestation) administered daily doses of 1.05 g ginger capsules or 75 mg vitamin B6 did not detect a statistically significant difference between the two treatment groups as measured by the Rhodes index of nausea and vomiting scale.[10] There was no placebo group. No clinical trials of ginger for the treatment for hyperemesis gravidarum show consistent, long-term benefit.[11]

In a randomized, double-blinded trial involving 180 patients undergoing gynecologic laparoscopy, ginger did not reduce the incidence of postoperative nausea and vomiting after these procedures.[12] In a study of 28 volunteers, the administration of powdered ginger (whole root, 0.5 g or 1.0 g) or fresh ginger root (1 g) provided no protection against the motion sickness caused by a rotating chair.[13] Additionally, the administration of ginger did not significantly alter gastric function during motion sickness.

Regulatory Status

The German Commission E approves the use of ginger for dyspepsia, nausea and vomiting of pregnancy, and for prophylaxis of motion sickness. Ginger is listed on the General Sale List of the Medicines Control Agency of the United Kingdom. In the United States, ginger is a dietary supplement that is listed as generally recognized as safe (GRAS).

PRINCIPAL INGREDIENTS

Chemical Composition

The major pungent compounds in ginger are gingerols, particularly 6-gingerol. Other gingerols include 8-gingerol, 10-gingerol, methylgingerol, gingerdiol, dehydrogingerdione, gingerdiones, diarylheptanoids (curcuminoids), diterpene lactones, and galanolactone.[14] Gingerols are thermally unstable, and dehydrate (i.e., loss of hydroxyl group from C5 and a hydrogen from C4 with the formation of double bond between C4 and C5) to the corresponding shogaol compounds. Figure 68.1 displays the chemical structures of the major gingerol and shogaol compounds in ginger. Fresh ginger rarely contains zingerone and shogaols depending on the variety or cultivars, but the concentrations of these compounds increase significantly in dried preparations of ginger. Shogaol compounds also occur naturally in dried ginger along with vanillin, dihydroferulic acid, ferulic acid, and zingerone. Figure 68.2 displays the transformation of gingerols to zingerone, shogaols, and paradols.

In samples of fresh Australian ginger, 6-gingerol was the major pungent phenolic compound, while 8-gingerol and 10-gingerol were present in lower concentrations.[15] The mean concentration of these three gingerol compounds in fresh ginger rhizomes were $215 \pm 54\,\mu g/g$, $75 \pm 31\,\mu g/g$, and $73 \pm 26\,\mu g/g$, respectively. This study did not detect shogaol compounds in the ethanol extracts from fresh rhizomes, as measured by high performance liquid chromatography with UV detection (HPLC/UV). Analysis of fresh Hawaiian white and yellow ginger and a methylene chloride extract of commercially processed dry ginger by HPLC detected 115 compounds.[16] In addition to gingerol and shogaol compounds, other constituents included paradols (5-deoxygingerols), isogingerols, isoshogaols, gingerdiones, methoxy-gingerols, mono- and di-acetoxy-gingerdiols, dihydro-paradols, diarylheptanoids, 4-vinylguaiacol, acetovanillone, and some methyl ether derivatives.

Ginger contains about 1–3.5% essential oil following steam distillation.[2] This product contains relative high concentrations of sesquiterpene hydrocarbons (e.g., zingiberene), relatively small amounts of monoterpene hydrocarbons, and oxygenated compounds compared with dry ginger. There are a variety of compounds (e.g., camphene, β-phellandrene, β-bisabolene, α-farnesene) in the essential oil depending on the extraction process, strain, and the source (dried, fresh). The main sesquiterpene hydrocarbons in ginger oil prepared from fresh ginger rhizomes were α-zingiberene (27–30%), α-curcumene (8–9%), β-sesquiphellandrene (4.8%), and bisabolene (3.2%) as measured by gas chromatography/mass spectrometry (GC/MS).[17] Minor constituents of the essential oil of ginger include acyclic α-farnesene and a variety of monoterpene compounds (limonene, myrcene, α-pinene, borneol, citronellol, geraniol, linalool).[14] Deterioration of ginger root may cause the formation of mycotoxins (e.g., mycophenolic acid) from the degradation of the root by fungi (e.g., *Penicillium brevicompactum*).[18] Currently, there is no evidence that the formation of mycophenolic acid in ginger causes human disease.

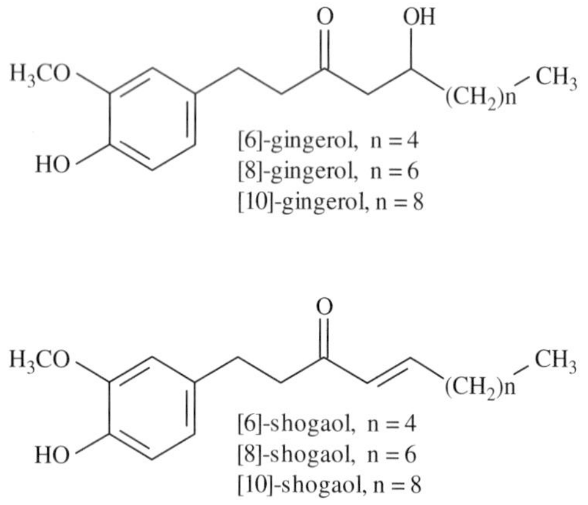

FIGURE 68.1. Chemical structures of major gingerol and shogaol compounds in ginger.

Physiochemical Properties

Gingerols exhibit novel reversible kinetics as a result of dehydration–hydration transformations with corresponding shogaols. The dehydration of gingerol is pH and temperature dependent with 6-gingerol being relatively stable at pH 4. The degradation of 6-gingerol is complete within 2 hours at pH 1 and 100 °C (212 °F).[19] 6-Gingerol is more pungent than other gingerol compounds (e.g., 8-gingerol, 10-gingerol). However, shogaol compounds are more pungent and lipophilic than corresponding gingerol compounds. *In vitro* studies suggest that gingerol inhibits thromboxane B_2 and prostaglandin D_2 formation by arachidonic acid.[20] However, the effect of *dried* ginger on thromboxane synthetase activity is dose-dependent, and up to 2 g dried ginger probably does not cause clinically significant platelet

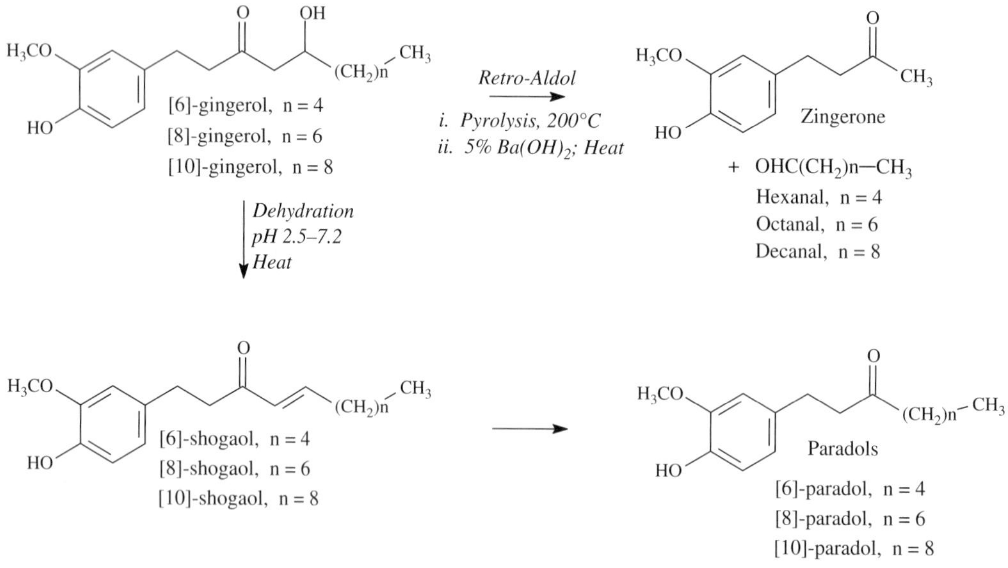

FIGURE 68.2. Transformation of gingerols. Adapted from Afzal M, et al. Ginger: an ethnomedical, chemical and pharmacological review. Drug Metab Drug Interact 2001;18:164.

dysfunction.[21] In a study of patients with coronary artery disease, the administration of 4 g powdered ginger daily for 3 months did not affect ADP- and epinephrine-induced platelet aggregation, fibrinolytic activity, or plasma fibrinogen concentration.[22] However, a single dose of 10 g powdered ginger administered produced a significant reduction in platelet aggregation induced by the ADP and epinephrine. *In vitro* studies suggest that ginger may produce anti-inflammatory effects by inhibiting arachidonic acid metabolism, cyclooxygenase, and lipoxygenase pathways.[23] *In vitro* studies also indicate that gingerol compounds are novel capsaicin-activated VR1 (vanilloid) receptor agonists.[24]

Mechanism of Toxicity

Potential active antiemetic agents in the rhizome of ginger include shogaol and gingerol. Some of the anti-emetic activity may result from the presence of gingerols, 6-shogaol, and galanolactone, which inhibit 5-HT_3 receptor function. *In vitro* studies indicate that these compounds are not competitive antagonists at the 5-HT_3 receptor, and the anti-emetic effect probably results from the binding of these compounds to a modulatory site distinct from serotonin.[25]

DOSE RESPONSE

The typical daily dose of fresh ginger is about 2–4 g (1.0 in./2.5 cm rhizome) or 0.5–1.0 g powdered dry rhizome in 3–4 divided doses. There are few data on the toxicity of ginger following the ingestion of doses of ginger exceeding typical therapeutic amounts. A clinical trial used a single dose of 10 g powdered ginger administered to patients with coronary artery disease, but the study did not report adverse effects.[22]

TOXICOKINETICS

Absorption

Based on *in vitro* studies, the dehydrated product (i.e., shogaol) undergoes almost complete reversion to gingerol in the acidic conditions of the stomach at body temperature (37.0 °C/98.6 °F).[19] Consequently, the dehydration of gingerol to shogaol may not reduce the bioavailability of gingerol.

Biotransformation

In rats, 6-gingerol undergoes conjugation, ω-1 oxidation and β-oxidation at the phenolic side chain.[26] The bile contains about 48% of an orally administered dose of 6-gingerol for over 60 hours, whereas about 12% appears in the urine as multiple metabolites (e.g., vanillic acid, ferulic acid, 9-OH-6-gingerol) and (S)-(+)-6-gingerol. There are few pharmacokinetic data on gingerol in humans.

Drug Interactions

Theoretically, ginger may increase the risk of bleeding when ingested with other drugs or herbs that affect coagulation. Because ginger can inhibit thromboxane formation and platelet aggregation, concomitant use with anticoagulants is not usually recommended. However, there is no clear evidence that these potential effects are clinically relevant. In an open-label, three-way crossover study of 12 healthy volunteers receiving pretreatment for 7 days with standard doses of ginger, there was no significant effect on clotting parameters or the pharmacokinetics of warfarin.[27] Outcomes included measurement of platelet aggregation, international normalized ratio (INR), warfarin protein binding, plasma warfarin concentrations, and S-7-hydroxywarfarin concentrations in the urine. A case report of a 76-year-old woman on long-term phenprocoumon therapy associated the development of an elevated INR (>10) and epistaxis with the use of ginger and tea from ginger powder.[28] Prior to the use of ginger, she had a stable INR, and her INR returned to the normal therapeutic range at the same phenprocoumon dose after cessation of ginger use. Other potential drug interactions include an increased effect of hypoglycemics and antihypertensive agents, and decreased efficacy of histamine$_2$-blockers (e.g., ranitidine) and proton-pump inhibitors (e.g., lansoprazole).

CLINICAL RESPONSE

There are few reports of toxicity associated with the use of ginger. Exposure to ginger is one of the most common contact allergens among spices based on patch testing.[29] Consumption of excessive amounts of ginger may cause sedation. Adverse effects of ginger during clinical trials include headache, diarrhea, abdominal discomfort, esophageal reflux, and drowsiness.[30] Although the number of subjects was limited, follow-up of randomized clinical trails on the use of ginger during pregnancy did not demonstrate an increased incidence of spontaneous abortions, stillbirths, neonatal deaths, low birth weight, or congenital abnormalities, when compared with the placebo group.[31,32] Ginger Jake paralysis was a peripheral neuropathy that was associated with the consumption of an alcoholic ginger extract, Jamaica ginger. This neurological condition resulted from contamination of this beverage with triorthocresyl phosphate rather than any constituent in ginger.

DIAGNOSTIC TESTING

The combination of high performance liquid chromatography (HPLC) with UV photodiode array detection and electrospray mass spectrometry allows the detection of constituents of ginger and ginger extracts. HPLC with mass spectrometry or nuclear magnetic resonance spectroscopy provides rapid determination of ginger constituents without sample purification or synthetic standards.[33] Thermal degradation and dehydration of gingerol compounds can occur during gas chromatographic analysis, resulting in the formation of aliphatic aldehydes, zingerones, and the corresponding shogaol compounds. Liquid chromatography/electrospray ionization/tandem mass spectrometry coupled with diode array detection identification is an alternative to GC/MS for the analysis of gingerol-related compounds, especially for thermally labile constituents.[34] At least during short-term storage, gingerol and shogaol are relatively stable at room temperature in the absence of strong acids or bases.[19] Aged essential oil of ginger contains relatively high amounts of α-curcumene and smaller amounts of zingiberene.[2]

TREATMENT

There are few data on the toxicity associated with excessive amounts of ginger. Treatment is supportive.

References

1. Afzal M, Al-Hadidi D, Menon M, Pesek J, Dhami MS. Ginger: an ethnomedical, chemical and pharmacological review. Drug Metab Drug Interact 18:159–190.
2. Govindarajan VS. Ginger—chemistry, technology, and quality evaluation: part 1. Crit Rev Food Sci Nutr 1982;17: 1–96.
3. Joland SD, Lantz RC, Solyom AM, Chen GJ, Bates RB, Timmermann BN. Fresh organically grown ginger (*Zingiber officinale*): composition and effects on LPS-induced PGE_2 production. Phytochemistry 2004;65: 1937–1954.
4. Mustafa T, Srivastava KC. Ginger (*Zingiber officinale*) in migraine headache. J Ethnopharmacol 1990;29:267–273.
5. Ghayur MN, Gilani AH. Pharmacological basis for the medicinal use of ginger in gastrointestinal disorders. Dig Dis Sci 2005;50:1889–1897.
6. Ernst E, Pittler MH. Efficacy of ginger for nausea and vomiting: a systematic review of randomized clinical trials. Br J Anaesth 2000;84:367–371.
7. Manusirivithaya S, Sripramote M, Tangjitgamol S, Sheanakul C, Leelahakorn S, Thavaramara T, Tangcharoenpanich K. Antiemetic effect of ginger in gynecologic oncology patients receiving cisplatin. Int J Gynecol Cancer 2004;14:1063–1069.
8. Chaiyakunapruk N, Kitikannakorn N, Nathisuwan S, Leeprakobboon K, Leelasettagool C. The efficacy of ginger for the prevention of postoperative nausea and vomiting: a meta-analysis. Am J Obstet Gynecol 2006;194: 95–99.
9. Borrelli F, Capasso R, Aviello G, Pittler MH, Izzo AA. Effectiveness and safety of ginger in the treatment of pregnancy-induced nausea and vomiting. Obstet Gynecol 2005;105:849–856.
10. Smith C, Crowther C, Willson K, Hotham N, McMillian V. A randomized controlled trial of ginger to treat nausea and vomiting in pregnancy. Obstet Gynecol 2004;103: 639–645.
11. Jewell D, Young G. Interventions for nausea and vomiting in early pregnancy. Cochrane Database Syst Rev 2003;(4):CD000145.
12. Eberhart LH, Mayer R, Betz O, Tsolakidis S, Hilpert W, Morin AM, et al. Ginger does not prevent postoperative nausea and vomiting after laparoscopic surgery. Anesth Analg 2003;96:995–998.
13. Stewart JJ, Wood MJ, Wood CD, Mims ME. Effects of ginger on motion sickness susceptibility and gastric function. Pharmacology 1991;42:111–120.
14. Chrubasik S, Pittler MH, Roufogalis BD. Zingiberis rhizoma: a comprehensive review on the ginger effect and efficacy profiles. Phytomedicine 2005;12:684–701.
15. Wohlmuth H, Leach DN, Smith MK, Myers SP. Gingerol content of diploid and tetraploid clones of ginger (*Zingiber officinale* Roscoe). J Agric Food Chem. 2005;53; 5772–5778.
16. Jolad SD, Lantz RC, Chen GJ, Bates RB, Timmermann BN. Commercially processed dry ginger (*Zingiber officinale*): composition and effects on LPS-stimulated PGE_2 production. Phytochemistry 2005;66:1614–1635.
17. Antonious GF, Kochhar TS. Zingiberene and curcumene in wild tomato. J Environ Sci Health B 2003;38:489–500.
18. Overy DP, Frisvad JC. Mycotoxin production and postharvest storage rot of ginger (*Zingiber officinale*) by *Penicillium brevicompactum*. J Food Protect 2005;68: 607–609.
19. Bhattarai S, Tran VH, Duke CC. The stability of gingerol and shogaol in aqueous solutions. J Pharm Sci 2001;90:1658–1664.
20. Guh J-H, Ko F-N, Jong T-T, Teng C-M. Antiplatelet effect of gingerol isolated from *Zingiber officinale*. J Pharm Pharmacol 1995;47:329–332.
21. Lumb AB. Effect of dried ginger on human platelet function. Thromb Haemost 1994;71:110–111.
22. Bordia A, Verma SK, Srivastava KC. Effect of ginger (*Zingiber officinale* Rosc.) and fenugreek (*Trigonella foenumgraecum* L.) on blood lipids, blood sugar and platelet aggregation in patients with coronary artery disease. Prostaglandins Leukot Essent Fatty Acids 1997;56:379–384.

23. Srivastava KC. Aqueous extracts of onion, garlic and ginger inhibit platelet aggregation and alter arachidonic acid metabolism. Biomed Biochim Acta 1984;43:S335–S346.
24. Dedov VN, Tran VH, Duke CC, Connor M, Christie MJ, Mandadi S, Roufogalis BD. Gingerols: a novel class of vanilloid receptor (VR1) agonists. Br J Pharmacol 2002; 137:793–798.
25. Abdel-Aziz H, Windeck T, Ploch M, Verspohl EJ. Mode of action of gingerols and s on 5-HT_3 receptors: binding studies, cation uptake by the receptor channel and contraction of isolated guinea-pig ileum. Eur J Pharmacol 2005;530:136–143.
26. Nakazawa T, Ohsawa K. Metabolism of [6]-gingerol in rats. Life Sci 2002;70:2165–2175.
27. Jiang X, Williams KM, Liauw WS, Ammit AJ, Roufogalis BD, Duke CC, et al. Effect of ginkgo and ginger on the pharmacokinetics and pharmacodynamics of warfarin in healthy subjects. Br J Clin Pharmacol 2005;59:425–432.
28. Druth P, Brosi E, Fux R, Monke K, Gleiter CH. Ginger-associated overanticoagulation by phenprocoumon. Ann Pharmacother 2004;38:257–260.
29. Futrell JM, Rietschel RL. Spice allergy evaluated by results of patch tests. Cutis 1993;52:288–290.
30. Boone SA, Shields KM. Treating pregnancy-related nausea and vomiting with ginger. Ann Pharmacother 2005;39: 1710–1713.
31. Vutyavanich T, Kraisarin T, Ruangsri R. Ginger for nausea and vomiting in pregnancy: randomized, double-masked, placebo-controlled trial. Obstet Gynecol 2001;97: 577–582.
32. Fischer-Rasmussen W, Kjaer SK, Dahl C, Asping U. Ginger treatment of hyperemesis gravidarum. Eur J Obstet Gynecol Reprod Biol 1991;38:19–24.
33. Saha S, Smith RM, Lenz E, Wilson ID. Analysis of a ginger extract by high-performance liquid chromatography coupled to nuclear magnetic resonance spectroscopy using superheated deuterium oxide as the mobile phase. J Chromatogr A 2003:991:143–150.
34. Jiang H, Solyom AM, Timmermann BN, Gang DR. Characterization of gingerol-related compounds in ginger rhizome (*Zingiber officinale* Rosc.) by high-performance liquid chromatography/electrospray ionization mass spectrometry. Rapid Commun Mass Spectrom 2005;19: 2957–2964.

Chapter 69

GINKGO TREE
(*Ginkgo biloba* L.)

HISTORY

The seeds and fruits of *Ginkgo biloba* have been part of traditional Chinese medicine for approximately three millennia before the birth of Christ, particularly for the treatment of asthma, indigestion, cough, and chilblains (i.e., pedal edema related to cold exposure).[1] In the 16th century, the Chinese *Materia Medica* described herbs from *Ginkgo biloba* as a treatment for senility in the royal families.[2] Gingko seed poisoning has been annotated in old Japanese scripture ("Yamato-honzo") at least since the early 18th century.[3] In 1881, the first case reports of gingko seed poisoning appeared in the Japanese medical literature.[4] During food shortages in Japan from 1930–1960, gingko seeds were an important source of food; there were numerous case reports of gingko seed poisoning (Gin-nan food poisoning).[5] In recent decades, medicinal uses for the extract of *Ginkgo biloba* include Alzheimer's disease, dementia, memory loss, cerebral ischemia, premenstrual syndrome, and altitude sickness.

BOTANICAL DESCRIPTION

The ginkgo tree is the last remaining species in the Ginkgoaceae family, which dates back at least 180–200 million years. Plants from the Ginkgoaceae family dominated the landscape 130 million years ago, but modern angiosperm seed plants gradually replaced all species in the Ginkgoaceae family except *Ginkgo biloba*. All other species in this family are now extinct. The ginkgo tree is a long-lived species (i.e., up to 1000 years) that is a popular ornamental tree worldwide. In Japanese, *ginkyo* describes the ginkgo nut as "silver apricot."

Common Name: Ginkgo, Ginkgo Tree, Maidenhair Tree

Scientific Name: *Ginkgo biloba* L.

Botanical Family: Ginkgoaceae

Physical Description: This well-branched, deciduous tree grows up to 40 m (~130 ft) in height after reaching maturity in 25 years. Male and female plants grow separately. The bark is usually light gray-brown with irregular ridges that eventually become deeply furrowed. Young twigs are light reddish-brown that become gray with numerous and obvious spur shoots. The reddish-brown buds are broadly conical to dome-shaped. Alternate, simple, fan-shaped leaves reach up to 5–8 cm (~2–3 in.) in length and width with parallel, fan-like venation. The light green leaves have long petioles with two or three irregular lobes and broad or undulating edges. Male flowers are very small and green, and these flowers appear on slim, cylindrical, leafless clusters (catkins) about 1 in. (2.5 cm) long. The female are "cones" on 4–5 cm (~1.5–2 in.) peduncles bearing one or two ovules. The fruit is a naked seed about 2.5 cm (~1 in.) long with a fleshy covering. After dropping off the tree, the fruit develops a strong, unpleasant odor. The inner hard seed of the mature fruit is edible after the first frost in the fall.

Medical Toxicology of Natural Substances, by Donald G. Barceloux, MD

Distribution and Ecology: Gingko is a native species in China. However, this ornamental shade tree has been introduced into Australia, Southeast Asia, Europe, Japan, and the United States. Commercial cultivation of the gingko tree occurs in both the United States and France.

EXPOSURE

Sources

The most commonly used form of *Ginkgo biloba* is the aqueous acetone extract of the dried and milled leaves. Ginkgo leaf extract is standardized to contain 24% ginkgo-flavone glycosides and 6% terpenoids. EGb761 is a standardized extract of the leaves of *Ginkgo biloba* that contains 5–7% terpene trilactones (ginkgolides, bilobalide), 22–24% flavonoids (isorhamnetin, kaempferol, quercetin), 0.5–1% organic acids (acetic acid, ascorbic acid, *p*-hydroxybenzoic acid, kynurenic acid, shikimic acid, vanillic acid), and <5ppm ginkgolic acids (anacardic acids).[6] Most commercial preparations contain about 40mg of *Ginkgo biloba* extract. The leaves of the *Gingko biloba* tree are harvested from July to September, dried, and then extracted with aqueous acetone. Drying of the extract results in the formation of one part extract from 50 parts raw leaves. Commercial trade names for this extract include BioGinkgo® (Pharmanex, Provo, UT), Flordis™ (Crow's Nest, New South Wales, Australia), Rökan® (Spitzner Arzneimittel, Ettlingen, Germany), Tavonin™ (Dr. Willmar Schwabe GmbH, Karlsruhe, Germany), Tanakan® (Beaufour Ipsen Pharma, Paris, France), and Tanamin® (YuYu, Inc., Seoul, Korea). Folium ginkgo consists of the dried whole leaf of *Ginkgo biloba* L.

Medicinal Uses

Traditional

The ginkgo nut is a culinary delicacy in China and Japan. Uses for ginkgo in traditional Chinese medicine include the treatment of asthma, cough, arthritis, edema, enuresis, and skin infections.

Current

The use of ginkgo primarily involves the use of this dietary supplement to improve cerebral and peripheral blood flow. The use of herbal extracts of ginkgo include the treatment of dementia, memory impairment, cerebral ischemia, peripheral vascular disorders, sexual dysfunction associated with selective serotonin reuptake inhibitors, asthma, bronchitis, stress, and hearing loss. Numerous studies suggest that the administration of ginkgo extract causes small improvements in claudication symptoms.[7,8] However, ginkgo may not be as efficacious as exercise therapy or prescription drugs. Overall, the scientific literature suggests that ginkgo (120–240mg daily) modestly improves memory function in the early stage of Alzheimer's disease and multi-infarct dementia,[9,10] but recent evidence on the efficacy of gingko extract remains inconsistent.[11,12] The effects on attention and memory in healthy and older adults appear short-lived as a result of tolerance developing with 6 weeks of treatment. In a placebo-controlled double-blind study of 52 students, a single dose of ginkgo (120mg standardized extract) modestly improved performance on a sustained-attention task and a pattern-recognition (episodic) memory task.[13] However, there were no statistically significant differences on any cognitive test, when baseline and posttreatment (6 weeks) scores were compared. Tests in this study measured sustained attention, episodic and working memory, mental flexibility, planning, and mood. There was no statistically significant difference in scores on eight tests of episodic and semantic memory between a group of 40 healthy subjects (age 35–80 years) using gingko supplements for 2 years and a control group.[14] In a group of 202 subjects over 60 years old without cognitive impairment, a 6-week randomized, double-blind, placebo-controlled, parallel-group trial did not demonstrate a statistically significant difference between the treatment (120mg gingko extract daily) and placebo groups on standard neuropsychological tests of learning, memory, attention, concentration, and verbal fluency.[15] Furthermore, there were no significant differences in self-reported memory function or global ratings by spouses or relatives.

There is little scientific evidence to support the use of ginkgo extracts for sexual dysfunction associated with selective serotonin reuptake inhibitors, stroke, postmenopausal syndrome, cocaine addiction, tinnitus, vertigo, depression, or the sustained enhancement of memory in healthy subjects. One week of gingko treatment in postmenopausal women improved attention, memory, and mental flexibility, but by 6 weeks of treatment these improvements disappeared except for mental flexibility in the older, poorer performing subjects.[16]

Regulatory Status

Gingko is a dietary supplement in the United States. Approved uses of gingko by the German Commission E include treatment of memory loss, poor concentration, dementia, depression, dizziness, vertigo, intermittent claudication, and headache.

PRINCIPAL INGREDIENTS

Chemical Composition

Constituents of the leaves of *Ginkgo biloba* include dimeric flavones (bilobetin, ginkgetin, isoginkgetin, sciadopitysin), flavonols (kaempferol, quercetin), proanthocyanidins, terpenoids (bilobalide, ginkgolides A, B, C, J, M), and amino acids. Ginkgolide J occurs only in the leaves of *Ginkgo biloba*, whereas the other ginkgolides are found primarily in the root bark.[2] The main active ingredients in ginkgo are probably bilobalide and ginkgolides.

Ginkgolides are diterpene compounds with a cage skeleton consisting of six 5-membered rings. Bilobalide (CAS RN: 33570-04-6) is a sesquiterpenoid compound with a molecular formula of $C_{15}H_{18}O_8$. Most of the flavonoid compounds (flavonols, flavones) are attached to sugars (glycosides), although occasionally they occur as aglycones. Analysis of 26 commercial ginkgo products indicates that there are severalfold variations in the composition and relative concentrations of ginkgolides and bilobalide in herbal products[17,18] as well as ginkgolic acids (i.e., <500 to 90,000 ppm).[19] Figure 69.1 and Figure 69.2 display the chemical structures of the main flavonoid and terpene trilactone compounds in the extract of *Gingko biloba*. The seeds of *Gingko biloba* contain the alkaloid, ginkgotoxin (4′-*O*-methylpyridoxine),[20] phenols, and cyanogenic glycosides as well as a 5′-glycoside of ginkgotoxin.[21] Although used as a medicinal herb, ginkgotoxin occurs in smaller concentrations in gingko leaves compared with gingko seeds. Figure 69.3 displays the chemical structure of ginkgotoxin. Gingko seeds contain variable amounts of the pyridoxine antagonist and convulsant, 4′-*O*-methylpyridoxine (ginkgotoxin). Analysis of the aqueous extract of ginkgo seeds by liquid chromatography with fluorescence detection demonstrated a mean free ginkgotoxin concentration of 179 μg/g and total ginkgotoxin concentration of 308 μg/g after β-glucosidase treatment.[21] Canned

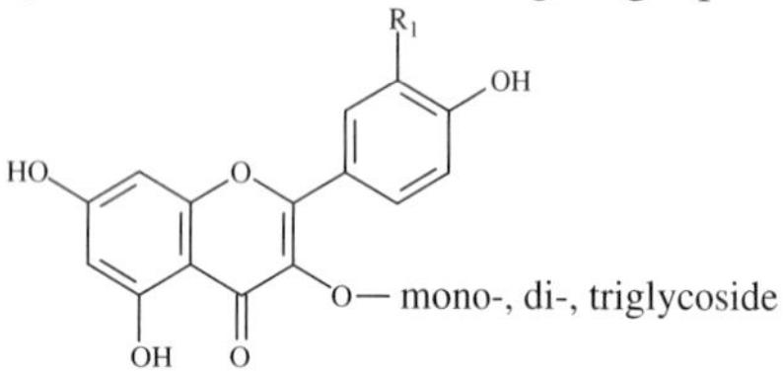

	R_1
Kaempferol derivatives	H
Quercetin derivatives	OH
Isorhamnetin derivatives	OCH_3

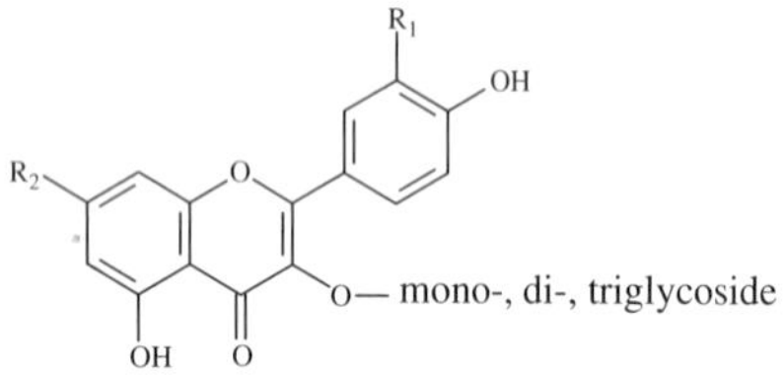

	R_1	R_2
Apigenin derivatives	H	O-glc
Luteolin derivatives	O-glc	OH

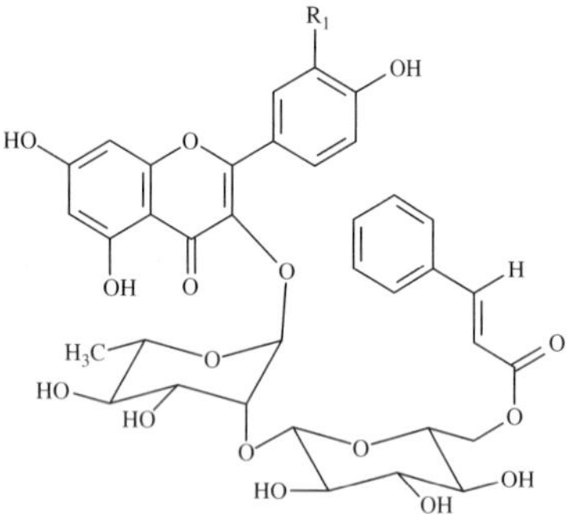

	R_1
Kaempferol derivatives	H
Quercetin derivatives	OH
R2	H or Glucose

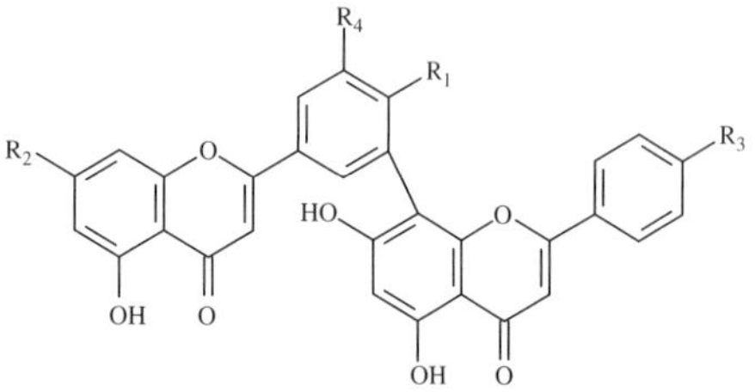

	R_1	R_2	R_3	R_4
Amentoflavon	H	H	H	H
Bilobetin	OCH_3	OH	OH	H
Ginkgetin	OCH_3	OCH_3	OH	H
Isoginkgetin	OCH_3	OH	OCH_3	H
Scaladopitysin	OCH_3	OCH_3	OCH_3	H
5'-Methoxybilobetin	OCH_3	OH	OH	OCH_3

FIGURE 69.1. Chemical structures of flavonoid glycosides in ginkgo biloba. Adapted from Smith JV, et al. Studies on molecular mechanisms of *Ginkgo biloba* extract. Appl Microbiol Biotechnol 2004;64:467.

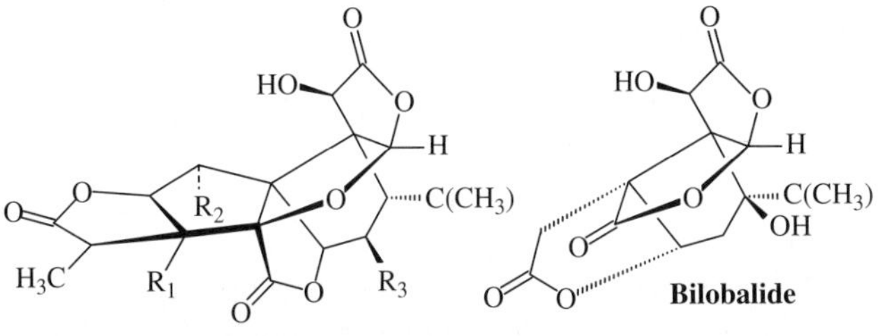

Ginkgolide	R_1	R_2	R_3
A	OH	H	H
B	OH	OH	H
C	OH	OH	OH
J	OH	H	OH
M	H	OH	OH

FIGURE 69.2. Chemical structures of terpene trilactone compounds (ginkgolides, bilobalide) in ginkgo biloba.

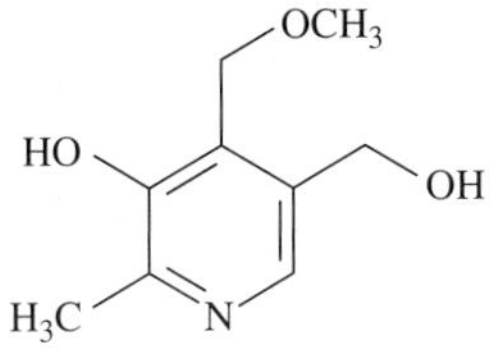

FIGURE 69.3. Chemical structure of ginkgotoxin.

ginkgo seeds contained no water-extractable ginkgotoxin, whereas water extracts of ginkgo capsules contained variable concentrations of ginkgotoxin ranging up to 457 g/g. In a study of 10 ginkgo natural health products, the ginkgotoxin concentration ranged between 3 µg/g (dry weight) in canned seeds and 181 µg/g (dry weight) in tablets, as measured by liquid chromatography with UV or fluorescence detection.[22] Although boiled Japanese ginkgo food may contain ginkgotoxin, the concentration of ginkgotoxin is generally below the concentration likely to cause toxicity. Concentrations of ginkgolides and ginkgotoxin in leaves and seeds are highest in late summer and early autumn. Analysis of fresh ginkgo seed albumin demonstrated approximately 105 µg ginkgotoxin (4′-*O*-methylpyridoxine)/g dry weight compared with 1.32 µg ginkgotoxin/dry weight in boiled seed albumin from canned Japanese food.[23] Ginkgolic acid and related alkylphenols are major components of the lipid fraction of the fruit pods of *Ginkgo biloba* L., but the aqueous extracts contain minimal amounts (<5 ppm) of these allergens.[24] Ginkgotoxin is not a unique compound in *Ginkgo biloba*, as this compound also occurs in the paperbark albizia (*Albizia tanganyicensis* Baker f.).[25]

Physiochemical Properties

All ginkgolides are acid-stable, bitter tasting solids without reported melting points. Glycine and $GABA_A$ receptors are members of the ligand-gated ion channel superfamily that mediate inhibitory synaptic transmission in the adult central nervous system. *In vitro* studies indicate that most of the ginkgolides are selective and potent antagonists of the glycine receptor.[26] Bilobalide is a less potent glycine receptor antagonist than ginkgolides, but bilobalide is a more potent inhibitor of the $GABA_A$ receptors than ginkgolides. Platelet-activating factor is a phospholipid synthesized in a variety of cells throughout the body for a variety of diverse physiological and pathological functions, particularly in the central nervous system. Ginkgolide B is a potent antagonist of the platelet-activating factor receptor, which is a 39 kDa G-protein coupled receptor.[27] Ginkgo extract also potentially inhibits platelet aggregation by increasing the concentrations of endothelium-derived thrombolytic agents (e.g., nitric oxide, prostacyclin) by inhibiting the synthesis of thromboxane B_2.[28,29] 4′-*O*-Methylpyridoxine is heat stable, and cooking does not detoxify gingko seeds.

Mechanism of Toxicity

The active ingredient of ginkgo extract has not been identified. However, the extracts are standardized to total flavonoid glycosides and terpenoid content, and the most clinically important compounds are these flavonoids (kaempferol, quercetin, isorhamnetin) and terpenoids (ginkgolides A/B, bilobalide).[30] Consumption of gingko seeds containing 4′-*O*-methylpyridoxine inhibits the formation of GABA and increases the formation of glutamate, resulting in the occurrence of seizures. Glutamate decarboxylase catalyzes the formation of γ-aminobutyric acid (GABA) from glutamate. Pyridoxal phosphate is a coenzyme of glutamate decarboxylase, and 4′-*O*-methylpyridoxine is a competitive antagonist of this coenzyme. Treatment of these seizures with pyridoxine (vitamin B6) terminates these seizures and prevents the development of status epilepticus.[31] Animal studies indicate that the main allergens in extracts of *Ginkgo biloba* are ginkgolic acids.[32]

DOSE RESPONSE

Medicinal Use

The usual daily dose of ginkgo extract is 80 to 240 milligrams in 2 to 3 divided doses of a 50:1 standardized leaf extract (standardized to 24–25% ginkgo flavone glycosides and 6% terpene lactones). Other dosage forms include the use of a tea (e.g., 30 mg extract/bag) or 3–6 mL of a 40 mg/mL extract daily in three divided doses.

Toxicity

Japanese adults typically ingest 5–10 ginkgo seeds during a meal, and boiled seed usually have small concentrations (i.e., ≤1 ppm) of ginkgotoxin (4′-*O*-methylpyridoxine). The ingestion of 50–60 raw ginkgo seeds by a 2-year-old was associated with the development of vomiting, diarrhea, irritability, and seizures.[3] The child recovered after receiving intravenous administration of vitamin B_6. A 36-year-old woman developed two generalized seizures and two episodes of vomiting within 24 hours after ingesting 70–80 gingko seeds cooked in a microwave.[33]

TOXICOKINETICS

Absorption

The bioavailability of bilobalide and ginkgolide A and B is relatively high (>75–80%), whereas the bioavailability of ginkgolide C is low.[34] Volunteer studies indicate that maximum plasma concentrations of free terpenoid compounds (ginkgolides A and B, bilobalide) occurs about 1–2 hours after ingestion by fasting volunteers.[35,36] The concomitant ingestion of the extract with food slightly delays (i.e., about 1 hour) the peak ginkgolide and bilobalide plasma concentration. In general, the bioavailability of flavonols (isorhamnetin, kaempferol, quercetin) is low.[37] In a study of 10 volunteers receiving a single dose of six gingko extract tablets, the bioavailability of kaempferol and quercetin was low (i.e., <1–2%).[38] P-glycoprotein is an ATP-driven efflux pump capable of transporting a wide variety of diverse compounds from the cell interior to the extracellular space. *In vitro* studies suggest that the P-glycoprotein type efflux pump limits the bioavailability of ginkgo flavonols.[39]

Distribution

The volumes of distribution of bilobalide and ginkgolide A and B are approximately 80–200 L, 40–60 L, and 60–100 L, respectively.[30,40]

Biotransformation

The biotransformation of flavonoids (flavonol glycosides, biflavones) in ginkgo extract is more extensive in humans than rats with urinary metabolites accounting for <30% of the flavonoid dose.[41] In contrast to rats, the major urinary metabolites are conjugated and free substituted benzoic acids (4-hydroxybenzoic acid, 3-methoxy-4-hydroxyhippuric acid, 3,4-dihydroxybenozoic acid, hippuric acid, vanillic acid) with little to no detectable phenylalkyl acids being present.

Elimination

The plasma elimination half-life of terpenoid compounds (ginkgolides A, B, bilobalide) is approximately 4–6 hours, depending on the specific terpenoid, the dose, and route of administration. In a study of healthy volunteers receiving a single dose of 240 mg EGb 761 extract, the mean elimination half-lives for ginkgolide A, ginkgolide B, and bilobalide were 5.1 ± 1.4 hours, 9.9 ± 2.0 hours, and 4.9 ± 2.5 hours, respectively.[40] The kidney excretes quercetin and kaempferol in the urine primarily as glucuronides, and there are relatively minor amounts of free quercetin and kaempferol in the urine after consumption of ginkgo extract.[42] In a study of 10 adult volunteers receiving a single oral dose of quercetin and kaempferol in ginkgo extract, approximately 0.17% and 0.22% of the oral dose was recovered in the urine.[38] All these compounds were present as glucuronides as reflected in the fact that neither of these compounds was detectable without hydrolysis. The plasma elimination half-life of these flavonoid compounds is about 1–4 hours.[38,40]

Drug Interactions

In vitro animal and *in vivo* clinical studies on the effect of ginkgo extracts on platelet aggregation, coagulation, and various cytochrome P450 isoenzymes are conflicting. Case reports suggest a *possible* adverse effect on bleeding after the use of gingko extract and aspirin, ibuprofen, or warfarin.[43,44] However, in an open-label, three-way crossover study of 12 healthy volunteers receiving pretreatment for 7 days with standard doses of ginkgo extract, there was no significant effect on clotting parameters or the pharmacokinetics of warfarin.[45] Outcomes assessed included measurement of platelet aggregation, international normalized ratio (INR) of prothrombin time, warfarin protein binding, plasma warfarin concentrations, and S-7-hydroxywarfarin concentrations in the urine.

In vitro studies indicate that ginkgolic acids I and II inhibit cytochrome P450 isoenzymes, CYP1A2, CYP2C9, CYP2C19, CYP2D6, and CYP3A4.[46] An *in vitro* study of human cytochrome P450 enzymes indicated the flavonoid fraction of ginkgo extract (EGb761) contains inhibitors of CYP2C9 and, to a lesser extent, CYP1A2 and CYP3A4.[47] However, a study of 12 volunteers receiving 120 mg gingko extract (EGb761) did not detect a statistically significant difference between baseline and posttreatment concentrations of dextromethorphan metabolite ratios (CYP2D6 activity) or alprazolam (CYP3A4 activity) metabolic ratios.[48] The plasma concentration *versus* time curve for alprazolam decreased 17%, but the unchanged elimination half-life of alprazolam indicated the there was no significant hepatic induction of CYP3A4. Using drug probes, a study of 12 healthy volunteers receiving a 28-day course of gingko extract did not detect any significant effect on CYP1A2, CYP2D6, CYP2E1, or CYP3A4 activity, when comparing baseline and posttreatment ratios of drug probes and their metabolites.[49] In a study of 18 healthy Chinese subjects, the administration of a 12-day course of 140 mg ginkgo extract twice daily induced CYP2C19 activity. This interaction could reduce the effect of drugs (e.g., omeprazole) metabolized by this enzyme.[50] However, the clinical significance of the interactions of ginkgo extract with cytochrome P450 isoenzymes remains undetermined.

CLINICAL RESPONSE

Adverse Reactions

The most common adverse effects associated with the use of ginkgo are diarrhea, flatulence, headache, dizziness, and palpitations. Cross-reactivity to urushiols (mango, sumac, poison ivy, poison oak, cashews) may occur in patients allergic to ginkgo. Rarely, case reports describe the use of ginkgo extract with isolated seizures, but a causal relationship between the use of ginkgo and the development of seizures has not been firmly established.[51] A case report associated the development of seizures in two epileptic patients stabilized on valproic acid. Seizures developed within 2 weeks of initiating treatment with ginkgo extract, and good control of the seizures returned following the cessation of ginkgo extract use.[52] Published case reports suggest a possible causal relationship between the use of gingko extract and the development of spontaneous bleeding including postoperative bleeding,[53] vitreous hemorrhage,[54] subarachnoid hemorrhage,[55] and intracranial hemorrhage.[56] Most of these case reports involve serious hemorrhage, but the timing of the bleeding event in relation to prolonged gingko exposure (i.e., months) do not suggest a strong causal association. However, bleeding did not recur in these case reports, which included evaluation following cessation of gingko use. The bleeding times were prolonged in the four case reports where bleeding times were measured, but other coagulation factors were normal.[57] Most large clinical trials do not report an increased incidence of spontaneous bleeding or other adverse effects in patients treated with gingko extract.[11]

Acute Effects

The ingestion of large amounts of gingko seeds is associated with the development of protracted nausea, vomiting, diarrhea, and seizures.[58] The gastrointestinal irritation begins with a few hours after the ingestion of ginkgotoxin, followed by the development of seizures. The convulsions are usually single, tonic-clonic seizures. Although status epilepticus does not typically develop, several seizures may occur during intoxication. Four hours after the ingestion of approximately 70–80 seeds, a 36-year-old woman developed a seizure.[33] She was transferred to the emergency department, where she had another episode of vomiting along with another seizure. Symptoms typically resolve within 24 hours without sequelae.

DIAGNOSTIC TESTING

Analytical Methods

Ginkgolides are difficult to separate because of similar solubilities and chromatographic behavior.[59] Analytical methods for the detection of ginkgotoxin include liquid chromatography with fluorescence detection, high performance liquid chromatography/electrospray ionization tandem mass spectrometry,[60] and gas chromatography/mass spectrometry (GC/MS) with solid-phase extraction (limit of detection = 50 pg).[61] Analysis of ginkgotoxin by ion-pair HPLC has greater speed and superior analytical sensitivity (i.e., limit of detection 5 pg) compared with conventional analytical methods using solid-phase extraction.[62]

Biomarkers

Serum concentrations of 4′-*O*-methylpyridoxine drawn near the time of seizure activity typically exceed 90 ng/mL. Approximately 9 hours after ingesting gingko seeds, a 2-year-old girl developed a tonic-clonic seizure. At that time, the serum 4′-*O*-methylpyridoxine concentration was 360 ng/mL.[3] Ten hours later, the serum 4′-*O*-methylpyridoxine concentration was below the limit of detection (15 ng/mL). However, the concentration of 4′-*O*-methylpyridoxine was 37 ng/mL in an admission serum sample from a 2-year-old child, who ingested 50 gingko nuts 4 hours prior to admission.[63] The child developed a second afebrile convulsion 30 minutes after admission, and 8 hours later the serum 4′-*O*-methylpyridoxine concentration was 157 ng/mL, as measured by HPLC. Eight hours after the ingestion of 10 pan-fried ginkgo seeds by a 17-month old girl, the serum 4′-*O*-methylpyridoxine concentration was 100 ng/mL, as measured by GC/MS.[61] She had a seizure 5 hours after ingestion that lasted about 10 minutes.

Abnormalities

Routine laboratory tests and electroencephalograms are typically normal after the ingestion of gingko seeds.

Transient hyperglycemia occurred in a 2-year old girl, who developed seizures following the ingestion of roasted ginkgo seeds.[3]

TREATMENT

Allergic reactions occasionally occur after the use of ginkgo, and the treatment is supportive. The most serious reactions occur with the ingestion of raw ginkgo seeds. Patients with suspected ginkgo seed poisoning and protracted vomiting should be evaluated for fluid and electrolyte imbalance as well as observed closely for the development of seizures and aspiration pneumonia. All of these patients should be placed under seizure precautions including intravenous access for the administration of anticonvulsants as needed. The seizures associated with gingko seed poisoning result from the competitive antagonist action of 4′-*O*-methylpyridoxine on GABA synthesis. Pyridoxine (vitamin B_6) is a suggested antidote for seizures caused by the ingestion of gingko seeds. There are few clinical data on the appropriate dose of intravenous pyridoxine following the ingestion of gingko seeds. A recommended dose of intravenous pyridoxal phosphate (i.e., nearly equivalent to pyridoxine hydrochloride) is one-fourth of the estimated ingestion of 4′-*O*-methylpyridoxine (i.e., based on 80 μg/seed).[3] When the amount of ingested seeds is unknown, the standard dose of pyridoxine hydrochloride for isoniazid (INH) intoxication in adults and children (70 mg/kg up to 2–5 g, intravenously) can be used. Patients with active seizures should always receive an intravenous dose of benzodiazepine (lorazepam, diazepam) in addition to the pyridoxine and supportive care (oxygen, observation for prevention of aspiration and trauma).

References

1. Bilia AR. *Ginkgo biloba* L. Fitoterapia 2002;73:276–279.
2. Nakanishi K. Terpene trilactones from *Gingko biloba*: from ancient times to the 21st century. Bioorg Med Chem 2005;13:4987–5000.
3. Kajiyma Y, Fujii K, Takeuchi H, Manabe Y. Ginkgo seed poisoning. Pediatr 2002;109:325–327.
4. Kudo K. Does the Ginkgo seed contain large amounts of cyanogenic glycosides? Tokyo Iji Shinshi 1881;149:19–21. [Japanese].
5. Sanada H, Kitatani H, Imaku M. A case report of ginkgo seed poisoning. Shoni Naika 1997;29:1187–1190. [Japanese].
6. Smith JV, Luo Y. Studies on molecular mechanisms of *Ginkgo biloba* extract. Appl Microbiol Biotechnol 2004; 64:465–472.
7. Pittler MH, Ernst E. Ginkgo biloba extract for the treatment of intermittent claudication: a meta-analysis of randomized trials. Am J Med 2000;108:276–281.
8. Horsch S, Walther C. *Ginkgo biloba* special extract EGb 761 in the treatment of peripheral arterial occlusive disease (PAOD)—a review based on randomized, controlled studies. Int J Clin Pharmacol Ther 2004;42:63–72.
9. Gertz H-J, Kiefer M. Review about *Ginkgo biloba* special extract EGb 761 (Ginkgo). Curr Pharm Design 2004;10: 261–264.
10. Le Bars PL, Velasco FM, Ferguson JM, Dessain EC, Kieser M, Hoerr R. Influence of the severity of cognitive impairment on the effect of the *Ginkgo biloba* extract EGb 761 in Alzheimer's disease. Neuropsychobiology 2002;45: 19–26.
11. Le Bars PL, Katz MM, Berman N, Itil TM, Freedman AM, Schatzberg AF. A placebo-controlled, double-blind, randomized trial of an extract of *Ginkgo biloba* for dementia. North American EGb Study Group. JAMA 1997;278: 1327–1332.
12. Birks J, Grimley EV, Van Dongen M. *Ginkgo biloba* for cognitive impairment and dementia. Cochrane Database Syst Rev 2002;(4):CD003120.
13. Elsabagh S, Hartley DE, Ali O, Williamson EM, File SE. Differential cognitive effects of *Gingko biloba* after acute and chronic treatment in healthy young volunteers. Psychopharmacol 2005;179:437–446.
14. Persson J, Bringlov E, Nilsson LG, Nyberg L. The memory-enhancing effects of Ginseng and *Ginkgo biloba* in healthy volunteers. Psychopharmacology (Berl) 2004;172: 430–434.
15. Solomon PR, Adams F, Silver A, Zimmer J, DeVeaux R. Ginkgo for memory enhancement a randomized controlled trial. JAMA 2002;288:835–840.
16. Elsabagh S, Hartley DE, File SE. Limited cognitive benefits in stage +2 postmenopausal women after 6 weeks of treatment with *Ginkgo biloba*. J Psychopharmacol 2005;19;173–181.
17. Ganzera M, Zhao J, Khan IA. Analysis of terpenelactones in *Ginkgo biloba* by high performance liquid chromatography and evaporative light scattering detection. Chem Pharm Bull (Tokyo) 2001;49:1170–1173.
18. Li XF, Ma M, Scherban K, Tam YK. Liquid chromatography-electrospray mass spectrometric studies of ginkgolides and bilobalide using simultaneous monitoring of proton, ammonium and sodium adducts. Analyst 2002;127: 641–646.
19. Kressmann S, Muller WE, Blume HH. Pharmaceutical quality of different *Ginkgo biloba* brands. J Pharm Pharmacol 2002;54:661–669.
20. Fiehe K, Arenz A, Drewke C, Hemscheidt T, Williamson RT, Leistner E. Biosynthesis of 4′-*O*-methylpyridoxine (Ginkgotoxin) from primary precursors. J Nat Prod 2000;63:185–189.
21. Scott PM, Lau BP-Y, Lawrence GA, Lewis DA. Analysis of *Ginkgo biloba* for the presence of ginkgotoxin and ginkgotoxin 5′-glycoside. J AOAC Int 2000;83:1313–1320.

22. Lawrence GA, Scott PM. Improved extraction of ginkgotoxin (4′-*O*-methylpyridoxine) from *Ginkgo biloba* products. J AOAC Int 2005;88:26–29.
23. Arenz A, Klein M, Fiehe K, Grob J, Drewke C, Hemscheidt T, Leistner E. Occurrence of neurotoxic 4′-*O*-methylpyridsxine in *Ginkgo biloba* leaves, ginkgo medications and Japanese ginkgo food. Planta Med 1996;62:548–551.
24. Jaggy H, Koch E. Chemistry and biology of alkylphenols from *Ginkgo biloba* L. Pharmazie 1997;52:735–738.
25. Steyn PS, Vleggaar R, Anderson LA. Structure elucidation of two neurotoxins from *Albizia tanganyicensis*. S Afr J Chem 1987;40: 191–192.
26. Ivic L, Sands TT, Fishkin N, Nakanishi K, Kriegstein AR, Stromgaard K. Terpene trilactones from Ginkgo biloba are antagonists of cortical glycine and GABA(A) receptors. J Biol Chem 2003;278:49279–49285.
27. Maclennan KM, Smith PF, Darlington CL. Platelet-activating factor in the CNS. Prog Neurobiol 1996;50: 585–596.
28. Diamond BJ, Shiflett SC, Feiwel N, Matheis RJ, Noskin O, Richards JA, Schoenberger NE. *Ginkgo biloba* extract: mechanisms and clinical indications. Arch Phys Med Rehabil 2000;81:668–678.
29. Kudolo GB, Wang W, Barrientos J, Elrod R, Blodgett J. The ingestion of *Ginkgo biloba* extract (EGb 761) inhibits arachidonic acid-mediated platelet aggregation and thromboxane B_2 production in healthy volunteers. J Herb Pharmacother 2004;4:13–26.
30. Kleijnen J, Knipschild P. Ginkgo biloba. Lancet 1992;340: 1136–1139.
31. Gammon GD, Gumnit R. Observations on the mechanism of seizures induced by a pyridoxine antagonist, methoxypyridoxine. Trans Am Neurol Assoc 1957–1958;82nd Meeting:57–59.
32. Lepoittevin JP, Benezra C, Asakawa Y. Allergic contact dermatitis to *Ginkgo biloba* L.: relationship with urushiol. Arch Dermatol Res 1989;281:227–230.
33. Miwa H, Iijima M, Tanaka S, Mizuno Y. Generalized convulsions after consuming a large amount of gingko nuts. Epilepsia 2001;42:280–281.
34. Fourtillan JB, Brisson AM, Girault J, Ingrand I, Decourt JP, Drieu K, Jouenne P, Biber A. [Pharmacokinetic properties of Bilobalide and Ginkgolides A and B in healthy subjects after intravenous and oral administration of *Ginkgo biloba* extract (EGb 761)]. Therapie 1995;50:137–144. [French]
35. Drago F, Floriddia ML, Cro M, Giuffrida S. Pharmacokinetics and bioavailability of a *Ginkgo biloba* extract. J Ocul Pharmacol Ther 2002;18:197–202.
36. Mauri P, Simonetti P, Gardana C, Minoggio M, Morazzoni P, Bombardelli E, Pietta P. Liquid chromatography/atmospheric pressure chemical ionization mass spectrometry of terpene lactones in plasma of volunteers dosed with *Ginkgo biloba* L. extracts. Rapid Commun Mass Spectrom 2001;15:929–934.
37. Ross JA, Kasum CM. Dietary flavonoids: bioavailability, metabolic effects, and safety. Annu Rev Nutr 2002;22: 19–34.
38. Wang FM, Yao TW, Zeng S. Disposition of quercetin and kaempferol in human following an oral administration of *Ginkgo biloba* extract tablets. Eur J Drug Metab Pharmacokinet 2003;28:173–177.
39. Wang Y, Cao J, Zeng S. Involvement of P-glycoprotein in regulating cellular levels of ginkgo flavonols: quercetin, kaempferol, and isorhamnetin. J Pharm Pharmacol 2005; 47:751–758.
40. Biber A. Pharmacokinetics of *Ginkgo biloba* extracts. Pharmacopsychiatry 2003;36(suppl 1):S32–S37.
41. Pietta PG, Gardana C, Mauri PL. Identification of *Ginkgo biloba* flavonol metabolites after oral administration to humans. J Chromatogr B 10997;693:249–255.
42. Watson DG, Oliveira EJ. Solid-phase extraction and gas chromatography-mass spectrometry determination of kaempferol and quercetin in human urine after consumption of *Ginkgo biloba* tablets. J Chromatogr B Biomed Sci Appl 1999;723:203–210.
43. Matthews MK Jr. Association of *Ginkgo biloba* with intracerebral hemorrhage. Neurology 1998;50:1933–1934.
44. Rosenblatt M, Mindel J. Spontaneous hyphema associated with ingestion of Ginkgo biloba extract. N Engl J Med 1997;336:1108.
45. Jiang X, Williams KM, Liauw WS, Ammit AJ, Roufogalis BD, Duke CC, et al. Effect of ginkgo and ginger on the pharmacokinetics and pharmacodynamics of warfarin in healthy subjects. Br J Clin Pharmacol 2005;59:425–432.
46. Zou L, Harkey MR, Henderson GL. Effects of herbal components on cDNA-expressed cytochrome P450 enzyme catalytic activity. Life Sci 2002;71:1579–1589.
47. Gaudineau C, Beckerman R, Welbourn S, Auclair K. Inhibition of human P450 enzymes by multiple constituents of the *Ginkgo biloba* extract. Biochem Biophys Res Comm 2004;318:1072–1078.
48. Markowitz JS, Donovan JL, DeVane CL, Sipkes L, Chavin KD. Multiple-dose administration of *Ginkgo biloba* did not affect cytochrome P-450 2D6 or 3A4 activity in normal volunteers. J Clin Psychopharmacol 2003;23:576–581.
49. Gurley BJ, Gardner SF, Hubbard MA, Williams DK, Gentry WB, Cui Y, Ang CY. Cytochrome P450 phenotypic ratios for predicting herb-drug interactions in humans. Clin Pharmacol Ther 2002;72:276–287.
50. Yin OQ, Tomlinson B, Waye MM, Chow AH, Chow MS. Pharmacogenetics and herb-drug interactions: experience with *Ginkgo biloba* and omeprazole. Pharmacogen 2004;14;841–850.
51. Gregory PJ. Seizure associated with *Ginkgo biloba*? Ann Intern Med 2001;134:344.
52. Granger AS. *Ginkgo biloba* precipitating epileptic seizures. Age Ageing 2001;30:523–525.
53. Bebbington A, Kulkarni R, Roberts P. *Ginkgo biloba* persistent bleeding after total hip arthroplasty caused by herbal self-medication. J Arthroplasty 2005;20;125–126.

54. MacVie OP, Harney BA. Vitreous haemorrhage associated with *Gingko biloba* use in a patient with age related macular disease. Br J Ophthalmol 2005;89:1378–1379.
55. Vale S. Subarachnoid haemorrhage associated with *Gingko biloba*. Lancet 1998;352:36.
56. Benjamin J, Muir T, Briggs K, Pentland B. A case of cerebral haemorrhage—can *Ginkgo biloba* be implicated? Postgrad Med J 2001;77:112–113.
57. Bent S, Goldberg H, Padula A, Avins AL. Spontaneous bleeding associated with *Ginkgo biloba* a case report and systematic review of the literature. J Gen Intern Med 2005;20:657–661.
58. Wada K, Ishigaki S, Ueda K, Sakata M, Haga M. An antivitamin B6, 4′-methoxypyridoxine, from the seed of *Ginkgo biloba* L. Chem Pharm Bull (Tokyo) 1985;33:3555–3557.
59. van Beek TA. Ginkgolides and bilobalide: their physical, chromatographic and spectroscopic properties. Bioorg Med Chem 2005;13:5001–5012.
60. Sun Y, Li W, Fitzloff JF, van Breemen RB. Liquid chromatography/electrospray tandem mass spectrometry of terpenoid lactones in *Ginkgo biloba*. J Mass Spectrom 2005;40:373–379.
61. Fujisawa M, Hori Y, Nakajima M, Shimada K, Yoshikawa H, Wada K. Gas chromatography-mass spectrometry analysis of 4′-*O*-methylpyridoxine (MPN) in the serum of patients with ginkgo seed poisoning. J Anal Toxicol 2002;26:138–143.
62. Hori Y, Fujisawa M, Shimada K, Oda A, Katsuyama S, Wada K. Rapid analysis of 4′-*O*-methylpyridoxine in the serum of patients with *Ginkgo biloba* seed poisoning by ion-pair high-performance liquid chromatography. Biol Pharm Bull 2004;27:486–491.
63. Hasegawa S, Oda Y, Ichiyama T, Hori Y, Furukawa S. Ginkgo nut intoxication in a 2-year-old male. Pediatr Neurol 2006;35:275–276.

Chapter 70

GINSENG

HISTORY

Ginseng is a popular Chinese herb that has been used for several thousand years as a panacea for promoting longevity.[1] Father Petrus Jartoux, a Jesuit missionary to northern China, was the first European to describe the herbal uses of ginseng, when he sent a letter to the Procurator General of the Missions of India and China from Peking in 1711.[2] American ginseng has been known in the United States since colonial times. George Washington mentioned the gathering of this plant in his diary; the American frontiersman and explorer, Daniel Boone, was active in the ginseng trade.[3]

BOTANICAL DESCRIPTION

Ginseng species from East Asia, Central Asia and North America include *Panax bipinnatifidus* Seem, *P. ginseng* C.A. May, *P. notoginseng* (Burkill) F.H. Chen, *P. omeiensis* J. Wen, *P. pseudoginseng* Wallich, *P. japonicus* (T. Nees) C.A. Meyer, *P. quinquefolius* L., *P. stipuleanatus* H.T. Tsai and K.M. Feng, *P. trifolius* L., *P. wangianus* S. C. Sun, *P. vietnamensis* Ha & Grushv., and *P. zingiberensis* C.Y. Wu and K.M. Feng. Most studies on the effects of ginseng involve American ginseng (*P. quinquefolius*), Asian or Chinese ginseng (*P. ginseng*), or Japanese ginseng (*P. japonicus*). The term ginseng often applies to herbal plants with similar appearance and uses as Asian or American ginseng, but some of these other types of "ginseng" belong to other plant families. Table 70.1 lists the names and origin of herbs commonly called ginseng.

Medical Toxicology of Natural Substances, by Donald G. Barceloux, MD

Common Name: American Ginseng

Scientific Name: *Panax quinquefolius* L.

Botanical Family: Araliaceae (ginseng)

Description: This plant is a smooth perennial herb, with a large, fleshy, very slow-growing, spindle-shaped root, which typically reaches 5–7 cm (~2–3 in.) in length and 1–2 cm (~0.4–0.8 in.) in diameter. The root has a round apex with a slight terminal, projecting point. Occasionally, a small branch appears in the fork formed near the curved end of the root. Some small rootlets exist upon the lower portion of the root. The color ranges from a pale yellow to a brownish colour. The stem is simple and erect with three leaves, which are divided into five finely-toothed leaflets. A single, terminal umbel has a few small, yellowish flowers that produce a cluster of bright red berries.

Distribution and Ecology: *P. quinquefolius* is a native perennial herb of shady hardwood forests of central and eastern United States and Canada, particularly in the mountains from Ontario and Quebec to Georgia. Commercial growth of this herb occurs in Canada (British Columbia, Ontario) and the United States (Wisconsin).

Common Name: Asian Ginseng, Chinese Ginseng, Korean Ginseng, Red or White Ginseng

Scientific Name: *Panax ginseng*

Botanical Family: Araliaceae (ginseng)

Description: This species is similar in appearance to American ginseng. Chinese ginseng has a thick, spindle-like brown-yellow root that is usually

TABLE 70.1. Herbs Commonly Called Ginseng

Common Name	Scientific Name	Family	Origin
Asian, Chinese, or Korean ginseng	*Panax ginseng* C.A. Meyer	Araliaceae	Korea, China, Japan, Russia
American ginseng	*Panax quinquefolius* L.	Araliaceae	United States, Southern Canada
Brazilian ginseng	*Pfaffia paniculata* (Mart.) Kuntze	Amaranthaceae	
Dwarf ginseng	*Panax trifolius* L.	Araliaceae	United States
Ginger ginseng	*Panax zingiberensis* C.Y. Wu and K.M. Feng	Araliaceae	China
Himalayan ginseng	*Panax pseudoginseng* Wallich		China, Himalayas
Indian ginseng	*Withania somnifera* (L.) Dunal	Solanaceae	Australia, East Asia, Africa
Japanese or Chikutsu ginseng	*Panax japonicus* C.A. Meyer		Japan, China
Notoginseng or Yunnan ginseng	*Panax notoginseng* (Burkill) F.H. Chen	Araliaceae	China
San-chi or Feather-leaf ginseng	*Panax bipinnatifidus* Seem	Araliaceae	Nepal, Eastern Himalayas
Siberian or Russian ginseng	*Eleutherococcus senticosus* (Rupr. & Maxim.) Maxim.	Araliaceae	Russia, China, Korea
Vietnamese or Bamboo ginseng	*Panax vietnamensis* Ha & Grushv.	Araliaceae	Vietnam

divided on the end. The simple, smooth (glabrous) stem forms a whorl of three or five palmately compound leaves consisting of five oblong-ovate, finely double-serrate leaflets. A single umbel of greenish-yellow flowers appears from June to August that produces small, pale red, fleshy berries (drupe) each with a solitary seed.

Distribution and Ecology: *P. ginseng* grows naturally in Korea, far-eastern Siberia, and northeastern China, but this species of Asian ginseng is rare now and the other Asian species of ginseng (e.g., *P. pseudoginseng* Wallich) are much more common.

EXPOSURE

Sources

Although ginsenosides are present in all parts of the plant, the fresh root is the main source of ginseng. Typically, the ginseng root is harvested after about 4–6 years of cultivation. Ginseng appears in the *Pharmacopoeias* of several countries including China, Japan, United Kingdom, Germany, Austria, and France as a treatment to restore well-being (adaptogenic). The two commercial forms of ginseng are called red and white ginseng. The latter preparation consists of dried, peeled roots of ginseng, whereas the former preparation is produced from the steaming of fresh, whole ginseng root prior to drying. Radix ginseng is the dried root of *Panax ginseng* C.A. Meyer. Forms of ginseng include crude plant material, capsules and tablets of [P1]powdered drugs, extracts, tonic drinks, teas, wines, and lozenges.

Medicinal Uses

In general, decoctions of the root of *Panax ginseng* and other *Panax* species are distributed as panaceas and as tonics for the restoration of strength. Ginseng is a traditional treatment for a variety of clinical conditions including cardiovascular collapse, edema, fatigue, sexual dysfunction, immune deficiencies, cardiac disease, Addison's disease, dyspepsia, neurosis, cancer, anemia, and leukopenia.[4] Animal studies suggest that ginseng contains a variety of active ingredients including anti-inflammatory compounds,[5] antioxidants,[1] antineoplastic agents,[6] memory aids,[7] and anxiolytics.[8] Although ginseng is a relatively safe herbal remedy,[9] there is little evidence from randomized clinical trials to document the efficacy of ginseng for any of these conditions,[10] including overall improvement in vitality[11] or the long-term enhancement of memory in healthy adult volunteers.[12] A randomized clinical trial of night-shift nurses did not demonstrate the improvement of mood or sense of well-being during the daily use of 1–2 g of Korean ginseng.[13] As measured by the Positive Affect-Negative Affect Scale and the Profile of Mood States Inventory, a double-blind, placebo-controlled study of 83 adults did not demonstrate a statistically significant difference in scores between controls and subjects receiving daily doses of 200 mg and 400 mg *Panax ginseng*.[14] There is no conclusive evidence in humans that the chronic administration of ginseng reduces cancer mortality.[15]

Regulatory Status

In the United States, ginseng is regulated as a dietary supplement. The German Commission E approves the use of *Panax ginseng* as an over-the-counter tonic for the treatment of fatigue, poor concentration, reduced work capacity, and general debility.

PRINCIPAL INGREDIENTS

Chemical Composition

The major active ingredients of ginseng are a diverse group of over 20 steroidal saponins, which primarily are tetracyclic triterpene dammarane compounds (ginsenosides) subdivided into (20S)-protopanaxadiol (Ra_{1-3}, Rb_{1-3}, Rc, Rc_2, Rd, Rd_2, Rh_2), (20S)-protopanaxatriol (Re_2, Re_a, Rf, Rg_1, Rg_2, Rh_1) along with an oleanolic acid ginsenoside (Ro) as demonstrated in Figure 70.1.[1,16] Ginsenoside Ro is a nonsteroidal saponin (i.e., oleanic acid). The ginsenoside content of ginseng varies substantially (i.e., 2–20%) depending on species, plant part, cultural conditions, age, and season with a typical range of 1–4%.[17,18] American ginseng contains at least 30 different ginsenoside compounds with the main functional group at the 3-OH and 17-OH positions. The major ginsenoside compounds include Rb_1 Rb_2, Rc, Rd, Rf, Rg_1, Rg_2, mRb_1, mRb_2, mRc, and the glycoside of Rc with Rb_1, Rb_2, and Rg_1 usually being the most abundant.[19] There are no estrogen compounds in ginseng. Additionally, commercial ginseng preparations contain variable amounts of ginsenosides.[20] In a study of 50 commercial ginseng preparations from 11 countries, the concentration of ginsenosides varied from 1.9–9% (weight/weight) as measured by gas chromatography/mass spectrometry (GC/MS).[21] Six samples sold in the United States, Sweden, and the United Kingdom did not contain detectable amounts of ginsenosides. Other compounds present in ginseng include fatty acids, peptides, polysaccharides, and polyacetylenic alcohols.

In a study of plant parts from *Panax ginseng* cultivated in China, the total saponin content of the main root, lateral root, rhizome, and fresh leaf averaged approximately 2%, 7%, 6%, and 5.5%, respectively.[22] Drying procedures during the processing of white ginseng do not alter the ratio of panaxadiol (Rb_1, Rb_2, Rc, Rd) and panaxatriol (Rg_1, Rg_2, Re, Rf) compounds. Typically, leaves contain Rd, Re, and Rg_1 ginsenosides. The ginsenoside Rg_1 content is relatively high in the root of Chinese ginseng (*Panax ginseng*) compared with American ginseng (*Panax quinquefolius*), which has a relatively low Rg to Rb ratio.[23] Although Siberian

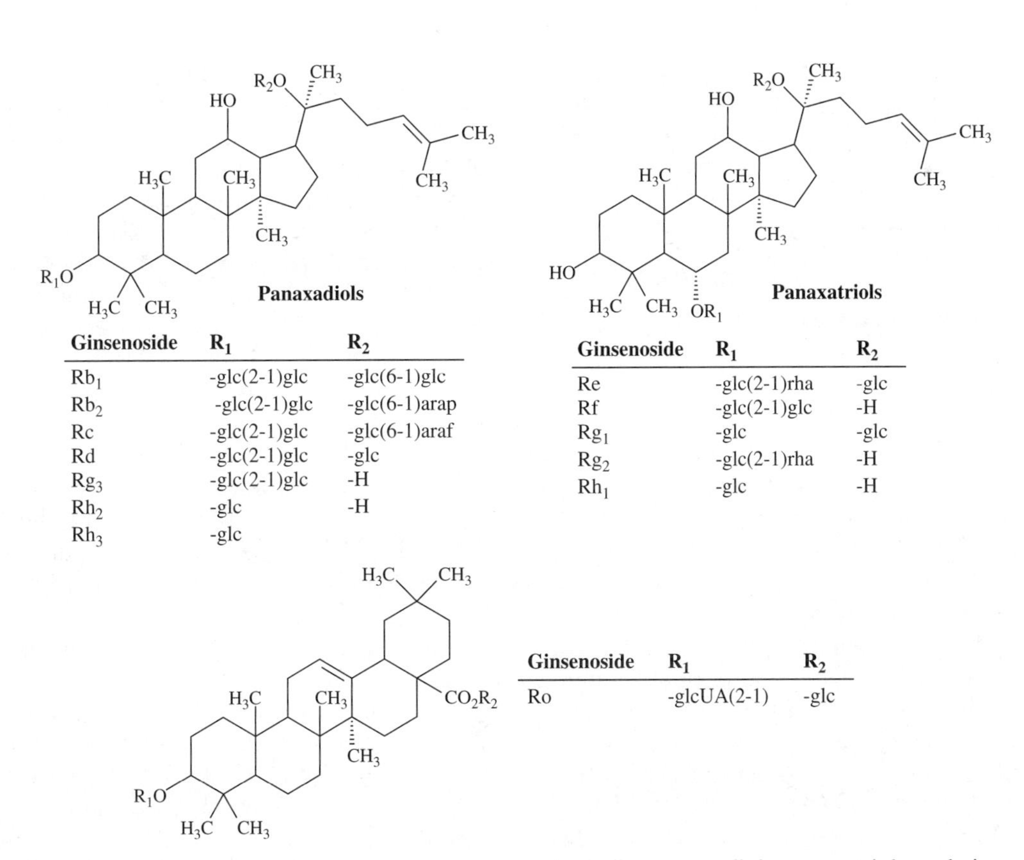

Ginsenoside	R_1	R_2
Rb_1	-glc(2-1)glc	-glc(6-1)glc
Rb_2	-glc(2-1)glc	-glc(6-1)arap
Rc	-glc(2-1)glc	-glc(6-1)araf
Rd	-glc(2-1)glc	-glc
Rg_3	-glc(2-1)glc	-H
Rh_2	-glc	-H
Rh_3	-glc	

Ginsenoside	R_1	R_2
Re	-glc(2-1)rha	-glc
Rf	-glc(2-1)glc	-H
Rg_1	-glc	-glc
Rg_2	-glc(2-1)rha	-H
Rh_1	-glc	-H

Ginsenoside	R_1	R_2
Ro	-glcUA(2-1)	-glc

FIGURE 70.1. The chemical structures of ginsenoside compounds including panaxadiols, panaxatriols, and ginsenoside R_0.

ginseng (*Eleutherococcus senticosus*) contains a variety of biologically active substances (e.g., sesamin, hederasaponin B, isofraxidin, β-sitosterol), most commercially available extracts of Siberian ginseng are standardized to the content of eleutheroside B (syringing 4-β-D-glucoside) and eleutheroside E (syringaresinol 4,4′-β-D-diglucoside).

Physiochemical Properties

The taste of the root is somewhat bitter with a hint of sweetness similar to liquorice, and there is little or no odor. Ginsenoside compounds are glycosylated steroids that contain four *trans*-ring rigid steroid skeletons with modification at the C20-side chain. These saponins are prone to hydrolysis under heat and acid conditions. Malonic esters of ginsenosides are present in substantial concentrations in ginseng root. The malonyl-ginsenosides are particularly unstable during heating in acidic conditions; consequently, the red form of Asian ginseng produced from the steaming and drying of the roots of *P. ginseng* contains substantially lower concentrations of malonyl-ginsenosides than does the white form.[24]

DOSE RESPONSE

The recommended, short-term dose for general stimulatory effects from Chinese ginseng is 2–3 g with doses up to 50 g producing insomnia, depression, and anxiety.[25] The estimated mean dose of ginseng (*P. ginseng, P. quinquefolius*) was 3 g with adverse effects (depression, depersonalization, confusion) more common with a 15 g dose. Headache, chest tightness, and gastrointestinal (GI) distress were associated with the ingestion of 25 g *Panax ginseng*.[26]

TOXICOKINETICS

Kinetics

Because ginseng contains over 20 structurally similar ginsenoside compounds, the toxicokinetic profile of ginseng is highly complex with limited data on individual ginsenosides. Animal studies suggest that the absorption of some ginsenoside compounds is relatively low (<1–2%) and some ginsenosides (e.g., Rg_1) undergo hydrolysis in the acidic milieu of the stomach.[27] During an experimental study, the absorption of orally administered ginsenosides Rb_1 and Rg_1 by rats was rapid with peak serum concentrations occurring about 30 minutes after oral administration.[28] The amount of ginsenoside Rg_1 appearing in the urine was about 1.9% with biliary excretion accounting for most of the elimination of this ginsenoside compound. Animal studies indicate that ginsenoside Rb_1 is poorly absorbed from the GI tract, and biliary excretion is not a significant route of elimination.[29] In a study of a single subject, the elimination half-lives of 20(S)-protopanaxadiol and 20(S)-protopanaxatriol extended over 13–17 hours.[30] Urinary excretion of these substances accounted for 0.24% and 1.2%, respectively, of the total dose of ginsenoside compounds. Metabolites of ginsenosides in humans include the protopanaxadiol-type ginsenoside G-Rh_1 and G-Rb_1, protopanaxatriol-type G-F_1, and compound K.[31]

Drug Interactions

Case reports suggest possible drug reactions between ginseng and ethanol, warfarin, and phenelzine,[32] but the clinical significance of these potential interactions is unclear. An open-label, nonrandomized clinical study of 14 healthy volunteers suggested that the administration of Chinese ginseng (*Panax ginseng*) reduces blood ethanol concentrations by increasing ethanol clearance.[33] The blood ethanol concentration at 40 minutes postingestion in most subjects decreased about 30–50% following the simultaneous administration of 3 g *P. ginseng* extract/65 kg body weight, when compared with controls (same subjects without ginseng). However, there was no statistical analysis of the data. Animal studies suggest that ginseng and/or ginsenosides reduce blood ethanol concentrations by delaying gastric emptying.[34] Case reports suggest that the simultaneous use of ginseng and warfarin reduces the anticoagulant effect of warfarin. Following the administration of 3 capsules Ginsana® (Pharmaton, Lugano, Switzerland; proprietary *P. ginseng* extract, G115) daily for 2 weeks, the international normalized ratio (INR) of a 47-year-old man decreased from 3.3 to 1.5; the INR normalized in therapeutic concentrations after the cessation of ginseng.[35] A randomized, double-blind, placebo-controlled trial of 20 healthy volunteers demonstrated that American ginseng (*Panax quinquefolius*) reduced the anticoagulant effect of warfarin (−0.19, 95% CI: −0.36 to −0.07, $P = 0.0012$).[36] However, the administration of Korean ginseng (3 g *P. ginseng* root daily) to 12 healthy subjects for 7 days did not alter coagulation parameters of a 25-mg dose of warfarin, when compared with controls (same volunteers without ginseng).[37] The clinical significance of a ginseng-warfarin drug interaction remains unclear.

Siberian ginseng probably does not significantly induce CYP2D6 and CYP3A4 isoenzymes.[38] Ginseng contains digoxin-like substances, which can interfere with digoxin immunoassays, but there is no clear pharmacokinetic or pharmacodynamic interaction between ginseng and digoxin.

CLINICAL RESPONSE

In general, adverse effects (e.g., indigestion, fatigue) during clinical trials involving ginseng are uncommon, mild, and transient.[39] The most common adverse effects are sleep disturbances, anxiety, fatigue, headache, and GI distress (diarrhea, nausea).[2] Although case reports associate the use of ginseng with psychological and ephedrine-like effects, the lack of biomarkers and analytical data limit conclusions regarding causality, particularly in patients taking multiple drugs and herbs. A case series associated the chronic use of ginseng with nervousness, generalized edema, hypertension, insomnia, skin eruptions, and morning diarrhea in 14 of the 133 subjects.[25] Chronic daily ingestion of high doses (≥15 g ginseng) was associated with increasing CNS effects (depression, confusion, depersonalization), but effect of ginseng was not well defined because of the variety of ingested preparations and the lack of analysis of herbal products. Several case reports document the development of manic symptoms (restlessness, insomnia, agitation irritability, pressured speech, and tangential thinking). These psychiatric abnormalities occurred in patients with[40–42] and without[43] prior psychiatric illness. A case report associated the onset of cerebral arteritis (severe headache, nausea, vomiting) with the ingestion of approximately 25 g ginseng root.[44] The clinical and angiographic features of this case report were similar to the cerebral arteritis associated with abuse of sympathomimetic drugs, but the patient denied drug use (the report did not document a urine drug screen). Other adverse effects associated with ginseng use include vaginal bleeding[45] and mastalgias.[46]

DIAGNOSTIC TESTING

Analytical Methods

The use of high performance thin-layer chromatography fingerprint analysis separates the various types of ginseng based on differences in the content of major ginsenosides (Rb_1, Re, Rg_1).[47] The most common analytical method for the quantitation of ginsenosides in herbal preparations is high performance liquid chromatography (HPLC) with UV absorbance detection at 205 nm. Other methods for the detection of ginsenosides include thin layer chromatography,[48] gas chromatography,[49] and HPLC with negative ion electrospray mass spectrometry.[50] The latter method allows the detection of minor ginsenosides as well as malonyl ginsenosides and hydrolysis products. Additionally, HPLC with positive mode electrospray mass spectrometry allows the separation of Asian ginseng (*P. ginseng*) from American ginseng (*P. quinquefolius*) by determining the presence of the acetylated form of ginsenoside Rb_1 (quinquenoside R_1) and the ratio of ginsenosides Rg_1 to Re.[51] In contrast to Asian ginseng, American ginseng contains the acetylated form of ginsenoside Rb_1 and a lower ratio of Rb_1 to Re. For quantitation of low concentrations of ginsenoside compounds, the limits of detection with GC/MS methods are a few ng/mL.[30] Mass spectral analyses of ginsenosides include electron impact MS, liquid secondary ion MS, LC/MS, and LC/MS/MS with nebulizer gas-assisted electrospray.[52] The latter technique has limits of detection for ginsenosides of about 2 pg. Electrospray ionization with HPLC/MS are most appropriate for polar thermo-labile ginseng compounds, whereas the use of atmospheric pressure chemical ionization is applicable with HPLC/MS to the analysis of weakly polar and stable ginsenosides.[53] Although there are no regulatory standards for ginseng, typically ginseng should contain at least 1.5% ginsenosides calculated as ginsenoside Rg_1 (D-glucopyranosyl-6β-glucopyranosyl-20S-protopanaxatriol, relative molecular mass 800).

Abnormalities

Ginseng contains digoxin-like immunoreactive substances, which can interfere with some digoxin immunoassays. Case reports suggest that the ingestion of ginseng by patients on digoxin may cause false elevation of digoxin concentrations depending on the reactivity of the individual immunoassay. An *in vitro* study examined the potential interference of Brazilian, Indian, Siberian, Asian, and North American ginseng with the following three digoxin immunoassays: fluorescence polarization (FPIA), microparticle enzyme (MEIA), and Tina-quant® (Roche Diagnostics, Indianapolis, IN).[54] Analysis of aliquots of serum pools prepared from patients receiving digoxin and supplemented with all ginsengs except Brazilian ginseng indicated that the addition of ginseng falsely elevated digoxin values with FPIA and falsely decreased digoxin values (negative interference) with MEIA. There was no interference of ginseng with digoxin as analyzed by Tina-quant®.

TREATMENT

Most adverse reactions to ginseng involve nonspecific complaints, such as headache, anxiety, and GI distress. Treatment is supportive.

References

1. Attele AS, Wu JA, Yuan C-S. Ginseng pharmacology multiple constituents and multiple actions. Biochem Pharmacol 1999;58:1685–1693.

2. Coon JT, Ernst E. *Panax ginseng* a systematic review of adverse effects and drug interactions. Drug Safety 2002;25:323–344.
3. Slazinski L. History of ginseng. JAMA 1979;242:616.
4. Keji C. The effect and abuse syndrome of ginseng. J Trad Chin Med 1981;1:69–72.
5. Hann BH, Han YN, Woo LK. Studies on the anti inflammatory glycosides of *Panax ginseng*. J Pharm Soc Korea 1972;16:129–136.
6. Xiaoguang C, Hongyan L, Xiaohong L, Zhaodi F, Yan L, Lihua T, Rui H. Cancer chemopreventive and therapeutic activities of red ginseng. J Ethnopharmacol 1998;60: 71–78.
7. Zhao R, McDaniel WF. Ginseng improves strategic learning by normal and brain-damaged rats. Neuroreport 1998;9:1619–1624.
8. Gillis CN. Panax ginseng pharmacology: a nitric oxide link? Biochem Pharmacol 1997;54:1–8.
9. Kitts DD, Hu C. Efficacy and safety of ginseng. Publ Health Nutr 2000;3:473–485.
10. Vogler BK, Pittler MH, Ernst E. The efficacy of ginseng. A systematic review of randomised clinical trials. Eur J Clin Pharmacol 1999;55:567–575.
11. Lieberman HR. The effects of ginseng, ephedrine, and caffeine on cognitive performance, mood and energy. Nutr Rev 2001;59:91–102.
12. Persson J, Bringlov E, Nilsson LG, Nyberg L. The memory-enhancing effects of Ginseng and *Ginkgo biloba* in healthy volunteers. Psychopharmacol (Berl) 2004;172:430–434.
13. Hallstrom C, Fulder S, Carruthers M. Effects of ginseng on the performance of nurses on night duty. Comp Med East West 1982;6:277–282.
14. Cardinal BJ, Engels H-J. Ginseng does not enhance psychological well-being in healthy, young adults: results of a double-blind, placebo-controlled, randomized clinical trial. J Am Dietetic Assoc 2001;101:655–660.
15. Shin HR, Kim JY, Yun TK, Morgan G, Vainio H. The cancer-preventive potential of *Panax ginseng*: a review of human and experimental evidence. Cancer Causes Control 2000;11:565–576.
16. Liu CX, Xiao PG. Recent advances on ginseng research in China. J Ethnopharmacol 1992;36:27–38.
17. Soldati F, Tanaka O. *Panax ginseng* C.A. Mayer: relation between age of plant and content of ginsenosides. Planta Med 1984;51:351–352.
18. Huang KC. The Pharmacology of Chinese Herbs. Boca Raton, FL: CRC Press; 1999.
19. World Health Organization. WHO monographs on selected medicinal plants. Geneva: World Health Organization; 1999.
20. Liberti LE, Der Marderosian A. Evaluation of commercial ginseng products. J Pharm Sci 1978;67:1487–1489.
21. Cui J, Garle M, Eneroth P, Bjorkhem I. What do commercial ginseng preparations contain? Lancet 1994;334: 134.
22. Yip TT, Lau CNB, But PPH, Kong YC. Quantitative analysis of ginsenosides in fresh *Panax Ginseng*. Am J Chin Med 1985;13:77–88.
23. Otsuka H, Morita Y, Ogihara Y, Shibata S. The evaluation of ginseng and its congeners by droplet counter-current chromatography (DCC). Planta Med 1977;32:9–17.
24. Kite GC, Howes M-J, Leon CJ, Simmonds MS. Liquid chromatography/mass spectrometry of malonyl-ginsenosides in the authentication of ginseng. Rapid Commun Mass Spectrom 2003;1:238–244.
25. Siegel RK. Ginseng abuse syndrome: Problems with the panacea. JAMA 1979;241:1614–1615.
26. Ryu S-J, Chien Y-Y. Ginseng-associated cerebral arteritis. Neurol 1995;45:829–830.
27. Odani T, Tanizawa H, Takino Y. Studies on the absorption, distribution, excretion and metabolism of ginseng saponins. IV. Decomposition of ginsenoside-Rg_1 and $-Rb_1$ in the digestive tract of rats. Chem Pharm Bull 1983;31: 3691–3697.
28. Odani T, Tanizawa H, Takino Y. Studies on the absorption, distribution, excretion and metabolism of ginseng saponins. II. The absorption, distribution and excretion of ginsenoside Rg_1 in the rat. Chem Pharm Bull 1983;31: 292–298.
29. Odani T, Tanizawa H, Takino Y. Studies on the absorption, distribution, excretion and metabolism of ginseng saponins. III. The absorption, distribution and excretion of ginsenoside Rb_1 in the rat. Chem Pharm Bull 1983;31: 1059–1066.
30. Cui JF, Bjorkhem I, Eneroth P. Gas chromatographic-mass spectrometric determination of 20(S)-protopanaxadiol and 20(S)-protopanaxatriol for study on human urinary excretion of ginsenosides after ingestion of ginseng preparations. J Chromatogr B Biomed Sci Appl 1997;689: 349–355.
31. Tawab MA, Bahr U, Karas M, Wurglics M, Schubert-Zsilavecz M. Degradation of ginsenosides in humans after oral administration. Drug Metab Disp 2003;31: 1065–1071.
32. Jones BD, Runikis AM. Interaction of ginseng with phenelzine. J Clin Psychopharmacol 1987;7:201–202.
33. Lee FC, Ko JH, Park JK, Lee JS. Effects of *Panax ginseng* on blood alcohol clearance in man. Clin Exp Pharmacol Physiol 1987;14:543–546.
34. Koo MW. Effects of ginseng on ethanol induced sedation in mice. Life Sci 1999;64:153–160.
35. Janetzky K, Morreale AP. Probable interaction between warfarin and ginseng. Am J Health Syst Pharm 1997;54: 692–693.
36. Yuan CS, Wei G, Dey L, Karrison T, Nahlik L, Maleckar S, Kasza K, Ang-Lee M, Moss J. Brief communication: American ginseng reduces warfarin's effect in healthy patients: a randomized, controlled Trial. Ann Intern Med 2004;141:23–27.
37. Jiang X, Williams KM, Liauw WS, Ammit AJ, Roufogalis BD, Duke CC, et al. Effect of St John's wort and ginseng

on the pharmacokinetics and pharmacodynamics of warfarin in healthy subjects. Br J Clin Pharmacol 2004;57: 592–599.

38. Donovan JL, deVane CL, Chavin KD, Taylor RM, Markowitz JS. Siberian ginseng (*Eleutherococcus senticosus*) effects on CYP2D6 and CYP3A4 activity in normal volunteers. Drug Metab Disp 2003;31:519–522.
39. Williams M. Immuno-protection against herpes simplex type II infection by *Eleutherococcus* root extract. Int J Altern Complement Med 1995;13:9–12.
40. Vazquez I, Aguera-Ortiz LF. Herbal products and serious side effects: a case of ginseng-induced manic episode. Acta Psychiatr Scand 2002;105:76–77.
41. Jones BD, Runikis AM. Interaction of ginseng with phenelzine. J Clin Psychopharmacol 1987;7:201–202.
42. Gonzalez-Seijo JC, Ramos YM, Lastra I. Manic episode and ginseng: report of a possible case. J Clin Psychopharmacol 1995;15:447–448.
43. Engelberg D, McCutcheon A, Wiseman S. A case of ginseng-induced mania. J Clin Psychopharmacol 2001;21: 535–537.
44. Ryu S-J, Chien Y-Y. Ginseng-associated cerebral arteritis. Neurol 1995;45:829–830.
45. Hopkins MP, Androff L, Benninghoff AS. Ginseng face cream and unexplained vaginal bleeding. Am J Obstet Gynecol 1988;159:1121–1122.
46. Palmer BV, Montgomery AC, Monteiro JC. Ginseng and mastalgia. Br J Med 1978;1:1284.
47. Xie P, Chen S, Liang Q-Z, Wang X, Tian R, Upton R. Chromatographic fingerprint analysis–a rational approach for quality assessment of traditional Chinese herbal medicine. J Chromatogr A 2006;1112:171–180.
48. Takino Y, Odani T, Tanizawa H, Hayashi T. Studies on the absorption, distribution, excretion and metabolism of ginseng saponins. I. Quantitative analysis of ginsenoside Rg_1 in rats. Chem Pharm Bull 1982;30:2196–2201.
49. Chen SE,Staba EJ. American ginseng. II. Analysis of ginsenosides and their sapogenins in biological fluids. J Nat Prod 1980;43:463–436.
50. Luchtefeld R, Kostoryz E, Smith RE. Determination of ginsenosides Rb_1, Rc, and Re in different dosage forms of ginseng by negative ion electrospray liquid chromatography-mass spectrometry. J Agric Food Chem 2004;52: 4953–4956.
51. Sloley BD, Lin YC, Ridgway D, Semple HA, Tam YK, Coutts RT, et al. A method for the analysis of ginsenosides, malonyl ginsenosides, and hydrolyzed ginsenosides using high-performance liquid chromatography with ultraviolet and positive mode electrospray ionization mass spectrometric detection. J AOAC Int 2006;89:16–21.
52. Wang X, Sakuma T, Asafu-Adjaye E, Shiu GK. Determination of ginsenosides in plant extracts from *Panax ginseng* and *Panax quinquefolius* L. by LC/MS/MS. Anal Chem 1999;71:1579–1584.
53. Ma X-Q, Liang X-M, Xu Q, Zhang X-Z, Xiao H-B. Identification of ginsenosides in roots of *Panax ginseng* by HPLC-APCI/MS. Phytochem Anal 2005;16:181–187.
54. Dasgupta A, Reyes MA. Effect of Brazilian, Indian, Siberian, Asian, and North American ginseng on serum digoxin measurement by immunoassays and binding of digoxin-like immunoreactive components of ginseng with Fab fragment of antidigoxin antibody (Digibind). Am J Clin Pathol 2005;124:229–236.

Chapter 71

GOLDENSEAL
(*Hydrastis canadensis* L.)

HISTORY

Goldenseal has a long history of use by Native Americans, especially by the Cherokee for the treatment of gastrointestinal (GI) and skin disorders. The Iroquois also used this herb for the treatment of diarrhea and whooping cough. Goldenseal use was popular among European settlers during the mid-19th century, and this herb was listed in the *United States Pharmacopeia*. By the beginning of the 20th century, supplies of goldenseal were dwindling as a result of excessive harvesting and loss of habitat. Wild goldenseal is now rare. This herb was officially recognized as an endangered species in 1991, and goldenseal now is listed in the Convention on International Trade in Endangered Species of Wild Fauna and Flora with associated trade restrictions.

BOTANICAL DESCRIPTION

As result of limitations in the supply of goldenseal, some herbal preparations of goldenseal may contain adulterants. These adulterants include Japanese goldthread [*Coptis japonica* (Thunb.)Makino.], Oregon grape [*Mahonia aquifolium* (Pursh) Nutt.], celandine (*Chelidonium majus* L.), barberry (*Berberis vulgaris* L.), and yellow root (*Xanthorhiza simplicissima* Marsh.). Berberine is a constituent of other herbs including *Berberis aristata* DC. (tree turmeric, Indian barberry), and *Coptis chinensis* Franch. (coptis root, Huang Lian).

Common Name: Goldenseal, Eye Root, Yellow Indian Plant, Turmeric Root, Yellow Paint Root, Orange Root, Goldenroot

Although synonyms for goldenseal include Indian turmeric and curcuma, this herb should not be confused with turmeric (*Curcuma longa* L.).

Scientific Name: *Hydrastis canadensis* L.

Botanical Family: Ranunculaceae (buttercups)

Physical Description: This plant is a small perennial herb with a horizontal, irregularly knotted, bright yellow root-stock, which produces a 6–12 in. (~15–30 cm) high, hairy flowering stem in early spring. Each stem has two dark green, hairy leaves with 5–7 lobes and irregular, finely toothed margins. The small, solitary flowers with three small greenish-white sepals appear in April. By July small, oblong, crimson berries develop, but these fruits are not edible.

Distribution and Ecology: Goldenseal is a native North American species, found primarily in the rich shady forests of the eastern United States and Canada. However, the range of goldenseal is shrinking due to deforestation and urbanization.

EXPOSURE

Sources

The root of goldenseal is the main portion of the plant used for herbal medicines, primarily as a tea. Goldenseal is a popular herbal component of dietary supplements, cold/cough/flu preparations, allergy remedies, eardrops, feminine hygiene products, laxatives, and digestive aids. The limited supply of goldenseal combined with high

Medical Toxicology of Natural Substances, by Donald G. Barceloux, MD

demand for this herb results in the adulteration of goldenseal preparations with other berberine-containing herbs [e.g., Chinese goldthread (*Coptis chinensis* Franch.)], and Oregon grape [*Mahonia aquifolium* (Pursh) Nutt.].[1] Berberine can be synthetically converted to hydrastine by a chemical process involving oxidation, hydration, *N*-alkylation, and reduction.[2]

Medicinal Uses

Traditional

Traditional uses for goldenseal rhizomes and roots (orange root, yellow root, tumeric root, goldenroot) include treatment of GI disturbances, anorexia, urinary symptoms, inflammation, and infections of the skin, mouth, respiratory tract, and eye. The root of goldenseal was also used as a clothing dye by Native Americans. In traditional Chinese medicine, goldenseal is an antimicrobial herb used for the treatment of dysentery and infectious diarrhea.

Current

Goldenseal tea is one of the most popular herbal preparations, particularly as an herbal supplement to improve immune function.[3] Purported uses for goldenseal include anorexia, cancer, cirrhosis, colitis, infectious diarrhea, diabetes, congestive heart failure, edema, fever, infections (colds, trachoma, conjunctivitis, malaria), and menorrhagia. Most clinical studies administer the alkaloids (e.g., berberine, hydrastine) rather than an extract of the goldenseal herb. Although goldenseal is one of the top-selling herbs in the United States, there is an absence of information from randomized, controlled clinical trials to support the medicinal use of this herb. Goldenseal is often formulated in herbal preparations with echinacea for the treatment of upper respiratory infections based in part on animal studies that suggested that the use of these two herbs increases immunoglobulin production in rats.[4] However, there is a lack of scientific evidence in humans that this combination prevents or improves upper respiratory tract infections.

Regulatory Status

In the United States, goldenseal is listed as a dietary supplement. Goldenseal is not generally recognized as safe or effective by the US Food and Drug Administration (FDA), and the use of this herb is not approved by the FDA in over-the-counter products. Specifically, the FDA does not approve the use of goldenseal as a blood purifier, herbal antibiotic, aphrodisiac, or decongestant.

PRINCIPAL INGREDIENTS

Chemical Composition

The principal compounds in goldenseal are the isoquinoline alkaloids, hydrastine (1.5–4%), berberine (0.5–6%), and canadine (tetrahydroberberine, CAS RN: 522-97-4).[1] The concentrations of these alkaloids vary by plant part with the roots and rhizomes of goldenseal containing the highest concentrations of hydrastine and berberine. The mean concentrations of these alkaloids in three samples of goldenseal from New Zealand are listed in Table 71.1.[5] Tetrahydroberberastine, berberastine (CAS RN: 2435-73-6), canadaline are minor alkaloids. Goldenseal does not contain detectable amounts of jatrorrhizine, coptisine, or palmatine. Figure 71.1 displays the chemical structure of hydrastine (CAS RN: 118-08-1, $C_{21}H_{21}NO_6$).

The typical ratio of berberine to hydrastine is about 3:2. However, the content of hydrastine and berberine

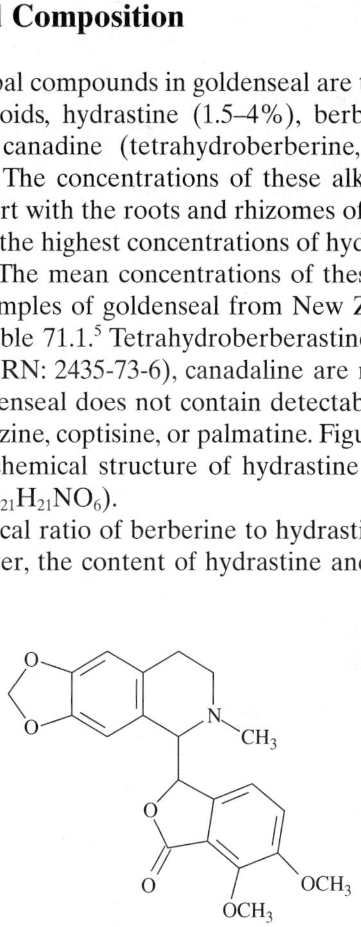

FIGURE 71.1. Chemical structure of hydrastine.

TABLE 71.1. Mean Concentration of Selected Alkaloid in New Zealand Goldenseal

Plant Part	Hydrastine (% w/w)	Berberine (% w/w)	Canadine (% w/w)	Palmatine (% w/w)
Rhizome	2.77	4.62	0.20	ND
Root	1.90	3.78	0.26	ND
Leaf	1.01	1.50	0.43	ND
Lower Stem	0.43	1.83	0.26	ND

Source: Adapted from Ref 5.
ND indicates not detected by high performance liquid chromatography (HPLC).

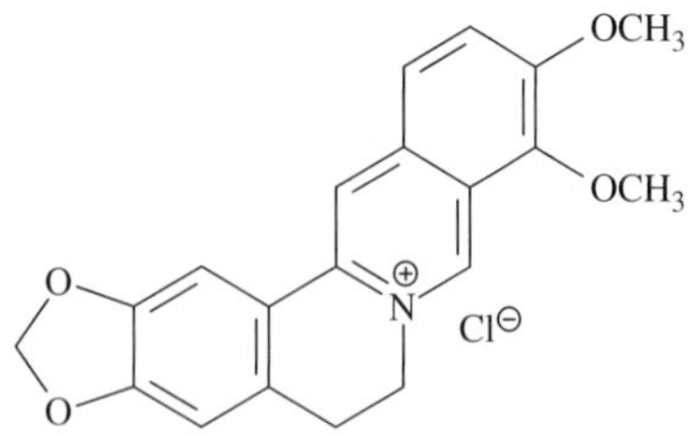

FIGURE 71.2. Chemical structure of berberine.

TABLE 71.2. Physiochemical Properties of Hydrastine

Physical Property	Value
Melting Point	132 °C/~270 °F
pKa	7.8
log P (Octanol-Water)	1.890
Water Solubility	30 mg/L (20 °C/68 °F)
Vapor Pressure	6.01E-11 mmHg (25 °C/77 °F)
Henry's Law Constant	1.34E-15 atm-m^3/mole (25 °C/77 °F)
Atmospheric OH Rate Constant	2.03E-10 cm^3/molecule-second (25 °C/77 °F)

TABLE 71.3. Physiochemical Properties of Berberine

Physical Property	Value
Melting Point	145 °C/~293 °F
pKa	2.47
log P (octanol-water)	2.080
Atmospheric OH Rate Constant	2.09E-10 cm^3/molecule-second (25 °C/77 °F)

in herbal products is highly variable. In a study of 20 samples of goldenseal herbal products purchased from Michigan pharmacies and health food stores, the mean hydrastine and berberine content were 1.57% (range: 0–2.93%) and 2.79% (range: 0.82–5.86%), respectively.[6] Figure 71.2 displays the chemical structure of berberine (CAS RN: 2086-83-1, $C_{20}H_{18}NO_4$). Berberine is found in several other plant species including *Mahonia aquifolium* (Oregon grape), *Berberis aristata* (tree turmeric), *Berberis vulgaris* (barberry), and *Coptis chinensis* (coptis or Chinese goldthread).

Physiochemical Properties

Goldenseal powder is dark yellow to moderate green-yellow in color. Both hydrastine and berberine are ionized, minimally water-soluble compounds. Table 71.2 and Table 71.3 list some physiochemical properties of hydrastine and berberine.

In vitro studies indicate that (+)-hydrastine is a $GABA_A$ receptor antagonist. This action is stereoselective with (+)-hydrastine being 180 times more potent than (−)-hydrastine. As measured by stimulation of [^{3}H]-diazepam binding to GABA receptors in rat brain membranes, the affinity of (+)-hydrastine for the $GABA_A$ receptor is five times greater than the competitive $GABA_A$ antagonist, bicuculline.[7]

Mechanism of Toxicity

In vitro studies suggest that berberine has positive inotropic effects as well as some antiarrhythmic actions by the blockade of delayed rectifier K^+ or K_{ATP} channels and increases in the effective refractory period.[8] Berberine-containing botanicals may cause QTc prolongation in patients with severe underlying heart disease, but there are few human data to confirm the clinical significance of this potential effect following the consumption of goldenseal.

DOSE RESPONSE

The typical oral doses of goldenseal are 0.5–1 gram tablets or capsules, 0.3–1 mL of liquid/fluid extract (1:1 in 60% ethanol), 0.5–1 gram as a decoction, or 2–4 mL as a tincture (1:10 in 60% ethanol) 3 times daily. Typical doses of berberine for the treatment of infectious diarrhea are 100–200 mg capsules 2–4 times or one 400 mg tablet daily. There few dose-response data for humans following consumption of goldenseal.

TOXICOKINETICS

Kinetics

The relatively high subcutaneous toxicity (LD_{50} 104 mg/kg) of berberine in animal studies compared with much lower oral toxicity suggests that the GI absorption of berberine is low.[9] Rodent studies also indicate that the oral bioavailabilities of the alkaloids, hydrastine and berberine, are relatively low.[10] Animal studies suggest that P-glycoprotein contributes to the poor intestinal absorption of berberine.[11] Based on animal studies, the kidneys excrete small amounts (i.e., 1%) of a berberine dose unchanged in the urine.[12]

Drug Interactions

Goldenseal strongly inhibits CYP2D6 and CYP3A4/5 activity *in vivo* and *in vitro*.[13] A study of 12 healthy volunteers used single-time point phenotypic metabolic ratios to determine the effect of 28 days of goldenseal supplementation on cytochrome P450 isoenzyme activities. The use of goldenseal resulted in an approximate 40% inhibition of CYP2D6 and CYP3A4/5 activity when compared with activities prior to supplementation.[14] There was no significant effect on CYP1A2 or CYP2E1 activity *in vivo*. The clinical significance of this potential drug interaction remains undetermined. There was no significance difference in the peak concentration or oral clearance of a single dose of 800 mg indinavir before and after 14 days of supplementation with 2,280 mg goldenseal daily.[15] The relatively high oral bioavailability of indinavir limits the effectiveness of this drug as a probe for testing CYP3A4/5 interactions. *In vitro* studies suggest that the major alkaloids responsible for inhibition of these two enzymes are hydrastine (CYP3A4) and berberine (CYP2D6).[16]

CLINICAL RESPONSE

Adverse effects of goldenseal are typically GI disturbances (e.g., nausea, vomiting). There are few human data on the acute toxicity of goldenseal. Based on animal studies with berberine, seizures and dysrhythmias are potential complications of excessive doses of goldenseal, but there are no human data to confirm these clinical effects. Although there are few human data, *in vitro* studies suggest that berberine and goldenseal extract can produce photosensitization following topical applications.[17]

In traditional Chinese medicine, the use of goldenseal is contraindicated during pregnancy. Some animal studies suggest adverse effects on maternal body weight and embryonic development at very high doses of goldenseal. In a study of CD-1® mice exposed to golden-seal root powder in feed (0, 3125, 12,500, or 50,000 ppm) from gestational day 6 to 17, a statistically significant (i.e., about 8%) reduction in average fetal body weight per litter occurred at the high dose only.[18] The calculated developmental toxicity NOAEL was 12,500 ppm and the LOAEL was 50,000 ppm. This dose represents about 300 times the estimated human intake from dietary supplements (26 mg/kg/d). Other animal studies using doses of goldenseal that did not produce maternal toxicity did not cause adverse fetal effects. A study of female Sprague-Dawley rats receiving 1.86 g goldenseal extract/kg (i.e., 65 times the daily human dose) during the pre-organogenic period (gestational day 1–8) and during the organogenic and early fetal period (gestational day 8–15) did not demonstrate reproductive abnormalities in the offspring.[19] There was no significant change in fetal body weights, increased implantation losses, or a statistically significant increase in fetal malformations in the treated group, when compared with the control group. The control group received the same dose of ethanol as the treatment group received in the goldenseal extract.

DIAGNOSTIC TESTING

Analytical Methods

Methods to analyze the alkaloid content (hydrastine, berberine) of goldenseal include colorimetry,[20] spectrophotometry,[21] thin layer chromatography with UV spectrophotometry,[22] gas chromatography/mass spectrometry (GS/MS),[23] capillary electrophoresis/mass spectrometry,[24] enzyme-linked immunosorbent assay (ELISA),[25] and high performance liquid chromatography (HPLC).[26] The presence of certain other alkaloids (e.g., jatrorrhizine, coptisine, palmatine) in goldenseal samples indicates adulteration of the sample by other plants.[23] Limits of detection and quantitation for berberine and hydrastine with the capillary zone electrophoresis method were 1 μg/mL and 2.5 μg/mL, respectively,[27] whereas these values for a HPLC with photodiode array detection method were 0.1 μg/mL and 0.4 μg/mL.[28] Isocratic liquid chromatography with UV detection is a rapid screening method for isoquinoline alkaloids in goldenseal that can be easily confirmed with LC/MS.[29] This procedure allows the quantitation of goldenseal alkaloids (berberine, berberastine, tetrahydroberberastine, canadine, canadaline, hydrastine) as well as non-goldenseal alkaloids (coptisine, jatrorrhizine, palmatine). The limits of quantitation were 1.65 μg/mL (berberine), 0.207 μg/mL (hydrastine), 0.122 μg/mL (canadine), and 0.27 (palmatine).

Biomarkers

Ingestion of large volumes of goldenseal tea is a popular method for concealing illicit drug use by causing darkening of the urine sample.[30] However, use of this tea probably does not alter the sensitivity or detection times of urine drug screens (e.g., marijuana, cocaine) other than the dilutional effect secondary to excessive fluids, when the tea is consumed with sufficient water to produce dilute urine specimens (creatinine $<$20 mg/mL, specific gravity $<$1.003).[31,32]

Abnormalities

In vitro studies indicate that berberine displaces bilirubin on a molar basis. The daily intraperitoneal administration of berberine (10 and 20 μg/g) for 1 week to adult rats produced a significant decrease in mean bilirubin–serum protein binding and an elevation in steady-state serum concentrations of unbound and total bilirubin.[33] Potentially, the ingestion of goldenseal can cause an exacerbation of hyperbilirubinemia in neonates or other diseases associated with hyperbilirubinemia depending on the concentration of berberine; however, there are inadequate data to determine the clinical importance of this potential interaction.

TREATMENT

The treatment of toxicity associated with goldenseal is supportive. There are few human data to guide management of potential complications of goldenseal use.

References

1. Govindan M, Govindan G. A convenient method for the determination of the quality of goldenseal. Fitoterapia 2000;71:232–235.
2. Moniot JL, Shamma M. The conversion of berberine into (+/–)-alpha- and (+/–)-beta-hydrastine. J Am Chem Soc 1976;98:6714–6715.
3. Hung OL, Shih RD, Chiang WK, Nelson LS, Hoffman RS, Goldfrank LR. Herbal preparation use among urban emergency department patients. Acad Emerg Med 1997;4: 209–213.
4. Rehman J, Dillow JM, Carter SM, Chou J, Le B, Maisel AS. Increased production of antigen-specific immunoglobulins G and M following *in vivo* treatment with the medicinal plants *Echinacea angustifolia* and *Hydrastis canadensis*. Immunol Lett 1999;68:391–395.
5. McNamara CE, Perry NB, Follett JM, Parmenter GA, Douglas JA. A new glucosyl feruloyl quinic acid as a potential marker for roots and rhizomes of goldenseal, *Hydrastis canadensis*. J Nat Prod 2004;67:1818–1822.
6. Edwards DJ, Draper EJ. Variations in alkaloid content of herbal products containing goldenseal. J Am Pharm Assoc 2003;43:419–423.
7. Huang JH, Johnston GA. (+)-Hydrastine, a potent competitive antagonist at mammalian $GABA_A$ receptors. Br J Pharmacol 1990;99:727–730.
8. Lau C-W, Yao X-Q, Chen Z-Y, Ko W-H, Huang Y. Cardiovascular actions of berberine. Cardiovascular Drug Rev 2001;19:234–244.
9. Poe CF, Johnson CC. Toxicity of hydrastine, hydrastinine, and sparteine. Acta Pharmacol Toxicol 1954;10:338–346.
10. Bhide MB, Chavan SR, Dutta NK. Absorption, distribution and excretion of berberine. Indian J Med Res 1969;57:2128–2131.
11. Pan GY, Wang GJ, Liu XD, Fawcett JP, Xie YY. The involvement of P-glycoprotein in berberine absorption. Pharmacol Toxicol 2002;91:193–197.
12. Schein FT, Hanna C. The absorption, distribution and excretion of berberine. Arch Int Pharmacodyn 1960;124:317–325.
13. Budzinski JW, Foster BC, Vandenhoek S, Arnason JT. An *in vitro* evaluation of human cytochrome P450 3A4 inhibition by selected commercial herbal extracts and tinctures. Phytomedicine 2000;7:273–282.
14. Gurley BJ, Gardner SF, Hubbard MA, Williams DK, Gentry WB, Khan IA, Shah A. *In vivo* effects of goldenseal, kava kava, black cohosh, and valerian on human cytochrome P450 1A2, 2D6, 2E1, and 3A4/5 phenotypes. Clin Pharmacol Ther 2005;77:415–426.
15. Sandhu RS, Prescilla RP, Simonelli TM, Edwards DJ. Influence of goldenseal root on the pharmacokinetics of indinavir. J Clin Pharmacol 2003;43:1283–1288.
16. Chatterjee P, Franklin MR. Human cytochrome P450 inhibition and metabolic-intermediate complex formation by goldenseal extract and its methylenedioxyphenyl components. Drug Metab Dispos 2003;31:1391–1397.
17. Inbaraj JJ, Kukielczak BM, Bilski P, Sandvik SL, Chignell CF. Photochemistry and photocytotoxicity of alkaloids from goldenseal (*Hydrastis canadensis* L.) 1. berberine. Chem Res Toxicol 2001;14:1529–1534.
18. National Toxicology Program. Developmental Toxicity Evaluation for Goldenseal (*Hydrastis Canadensis*) Root Powder Administered in the Feed to Swiss (CD-1®) Mice on Gestational Days 6–17. NTP Study: TER99004, Springfield, VA, National Technical Information Services, 2002.
19. Yao M, Ritchie HE, Brown-Woodman PD. A reproductive screening test of goldenseal. Birth Defects Res B 2005;74: 399–404.
20. El-Masry S, Korany MA, Abou-Donia AH. Colorimetric and spectrophotometric determinations of hydrastis alkaloids in pharmaceutical preparations. J Pharm Sci. 1980;69:597–598.
21. Stanislas E, Gleye J, Rouffiac R. [Spectrometric determination of the hydrastine content of the rhizome of *Hydrastis canadensis* L]. Ann Pharm Fr 1971;29:27–32. [French]
22. Datta DD, Bose PC, Ghosh D. Thin layer chromatography and U. V. spectrophotometry of alcoholic extracts of *Hydrastis canadensis*. Planta Med 1971;19:258–263.
23. Weber HA, Zart MK, Hodges AE, Molloy HM, O'Brien BM, Moody LA, et al. Chemical comparison of goldenseal (*Hydrastis canadensis* L.) root powder from three commercial suppliers. J Agric Food Chem 2003;51:7352–7358.
24. Sturm S, Stuppner H. Analysis of isoquinoline alkaloids in medicinal plants by capillary electrophoresis-mass spectrometry. Electrophoresis 1998;19:3026–3032.
25. Kim JS, Tanaka H, Shoyama Y. Immunoquantitative analysis for berberine and its related compounds using monoclonal antibodies in herbal medicines. Analyst 2004;129: 87–91.
26. Abourashed EA, Khan IA. High-performance liquid chromatography determination of hydrastine and berberine in dietary supplements containing goldenseal. J Pharm Sci 2001;90:817–822.
27. Unger M, Laug S, Holzgrae U. Capillary zone electrophoresis as a tool for the quality control of goldenseal extracts. Electrophoresis 2005;26:2430–2436.
28. Li W, Fitzloff JF. A validated high performance liquid chromatographic method for the analysis of goldenseal. J Pharm Pharmacol 2002;54:435–439.

29. Weber HA, Zart MK, Hodges AE, White KD, Barnes SM, Moody LA, et al. Method validation for determination of alkaloid content in goldenseal root powder. J AOAC Int 2003;86:476–483.
30. Mikkelsen SL, Ash KO. Adulterants causing false negatives in illicit drug testing. Clin Chem 1988;34:2333–2336.
31. Wu AH, Forte E, Casella G, Sun K, Hemphill G, Foery R, Schanzenbach H. CEDIA for screening drugs of abuse in urine and the effect of adulterants. J Forensic Sci 1995;40:614–618.
32. Cone EJ, Lange R, Darwin WD. In vivo adulteration: excess fluid ingestion causes false-negative marijuana and cocaine urine test results. J Anal Toxicol 1998;22:460–473.
33. Chan E. Displacement of bilirubin from albumin by berberine. Biol Neonate 1993;63:201–208.

Chapter 72

HAWTHORN
(*Crataegus* Species)

HISTORY

Chemical analysis of organic compounds absorbed into pottery jars from the early Neolithic village of Jiahu in Henan Province in China suggests that fermented beverages of rice, honey, and fruit were consumed by ancient Chinese as early as the seventh millennium before Christ (BC).[1] A potential source, other than grapes, for the tartaric acid in the Jiahu samples is the Chinese hawthorn (*Crataegus pinnatifida* Bunge, *Crataegus cuneata* Siebold & Zucc.). These plants contain four times the amount of tartaric acid in grapes. Dioscorides mentioned the use of hawthorn as a medicinal herb in the 1st century AD. Herbal use in traditional Chinese medicine dates back at least to 659 AD, when hawthorn was listed in the official Chinese pharmacopoeia, *Tang Ben Cao*.[2] Although treatment of heart disease with hawthorn began in Europe during the late 1800s, large clinical trials examining the efficacy of hawthorn extract for the treatment of congestive heart failure did not begin until the 1980s.

BOTANICAL DESCRIPTION

Hawthorn is the common name for plant species in the genus *Crataegus*. These plants are distributed throughout the world. Of the approximately 20 species of hawthorn used as herbs, the most common species used for medicinal purposes are *C. laevigata* (Poir.) DC. and *C. monogyna* Jacq.(*C. oxyacantha* auct. non L.). Chinese species include *Crataegus pinnatifida* and *Crataegus cuneata*, whereas European species of hawthorn include *Crataegus monogyna* Jacq., *C. pentagyna* Waldst. & Kit. ex Willd., *C. nigra* Waldst. & Kit., and *C. azarolus* L.

Common Name: Hawthorn, Aubépines

Scientific Name: *Crataegus* spp.

Botanical Family: Rosaceae (rose)

Physical Description: The dull gray stems of this thorny shrub have small, bright green, three- to five-lobed leaves with short, distinct stalks. Flat clusters of white flowers appear in spring with pink stamen, which turn brown and discharge pollen. The flower produces bright red berries.

Distribution and Ecology: There are over 280 species of hawthorn (*Crataegus* spp.) occurring primarily over northern temperate zones of East Asia, Europe, and eastern North America.

EXPOSURE

Sources

The medicinal parts of the hawthorn plant are the leaves, flowers, and fruits. The hawthorn fruit is also used as a foodstuff (e.g., jam, jelly, canned fruit, beverage, wine).

Medicinal Uses

Traditional

Hawthorn is a Chinese and European herb with a long history of use in traditional medicine, primarily as a digestive aid. In traditional Chinese medicine, hawthorn

Medical Toxicology of Natural Substances, by Donald G. Barceloux, MD

stimulates digestion, promotes the blood circulation, and reduces blood stasis. Herbal uses involve the treatment of diarrhea, indigestion, abdominal pain, hyperlipidemia, hypertension, and amenorrhea. In Europe, hawthorn extract is a treatment for fluid overload, hypertension, hyperlipidemia, congestive heart failure, and colic.

CURRENT

Experimental studies suggest that hawthorn extract has positive inotropic, negative chronotropic, and coronary artery vasodilatory properties similar to phosphodiesterase inhibitors (amrinone, milrinone), but there is no clear incremental benefit of adding hawthorn extract to standard regimens for congestive heart failure.[3,4] Typically, hawthorn is an adjunct to the conventional treatment (angiotensin-converting enzyme inhibitors, calcium antagonists, diuretics) for mild forms of chronic congestive heart failure. A meta-analysis of eight clinical trials (632 patients with chronic heart failure NYHA class I-III) demonstrated a statistically significant improvement in maximal workload along with reduced fatigue and dyspnea, when comparing hawthorn extract to placebo.[5] The weighted mean difference was 7 watts (95% CI: 3–11, $P < 0.01$). Seven of the eight trials used the hawthorn extract WS 1442 (standardized to 18.8% oligomeric). However, the number of patients in these trials is small, the follow-up period was short, and the improvement in maximal workload for the addition of hawthorn extract was small (9%) compared with angiotensin-converting enzymes alone (19%).[6]

An herbal preparation (Sympathyl, Laboratorier Innotech Internation, Arcueil, France) is an extract of hawthorn (*C. monogyna* Jacq. or *Crataegus oxyacantha*) and California poppy (*Eschscholzia californica* Cham.) that is used in France as a treatment for insomnia and anxiety.[7]

REGULATORY STATUS

In Germany, hawthorn extract is a prescription medication. The German Commission E approved the use of hawthorn extract (leaves, flowers) for the treatment of patients with New York Heart Association (NYHA) class II symptoms.

PRINCIPAL INGREDIENTS

Chemical Composition

The main compounds in hawthorn (leaves, flowers, berries) are flavonoids (catechin, oligomeric procyanidin, quercetin, quercitrin, rutin), triterpene saponins (crataegus acid, oleanolic acid, ursolic acid), organic acids (caffeic acid, chlorogenic acid), and amines (acetylcholine, β-phenethylamine, tyramine). Approximate amounts of these compounds in hawthorn include the following: oligomeric proanthocyanidins (1–3%, fruits or leaves with flowers), triterpene acids (0.5–1.4%, fruits), organic acids (2–6%), sterols, and trace amounts of cardioactive amines.[2] Ursolic and oleanolic acids account for about 90% of the pentacyclic triterpene compounds.[8] Flavonoids are naturally occurring antioxidants present in many fruits, vegetables, and beverages. This group of compounds is composed of five subclasses as follows: flavanols (catechin, epicatechin, procyanidins), flavanones (hesperidin, naringin), isoflavanones, flavones (apigenin, luteolin, vitexin), flavonols (quercetin, rutin, hyperoside), and anthocyanidins.[9] Figure 72.1 displays some representative flavones and flavonols in hawthorn.

The main flavonoid compounds in hawthorn are flavonols (i.e., oligomeric procyanidins).[10] The flavan-3-ols, (−)-epicatechin and (+)-catechin, occur extensively in foodstuffs (e.g., cocoa, peanuts) along with their related oligomers (dimers, trimers), the procyanidins. Figure 72.2 displays the chemical structure of procyanidin (CAS RN: 4852-22-6, $C_{30}H_{26}O_{13}$). In general, aqueous extracts of hawthorn contain lower and more variable concentrations of procyanidins, total flavonoids, and total phenols compared with alcohol (methanol, ethanol) extracts, but the concentrations of total vitexins are similar in these two types of extracts.[11]

Physiochemical Properties

Procyanidins are antioxidants that protect bilayer membranes by decreasing membrane fluidity.[12]

Mechanism of Toxicity

The mechanism of action of hawthorn is not well defined, but most authorities believe that the active ingredients in hawthorn extracts are flavanoid compounds. The trace amounts of amines in hawthorn extract are probably not absorbed by the gastrointestinal (GI) tract in clinically significant amounts.

DOSE RESPONSE

European clinical studies used daily doses of hawthorn extract from 160–900 mg that were standardized to 2.2% flavonoids or 18.75% oligomeric proanthocyanidins.[2] In China, typical doses of dried hawthorn fruit are 9–12 g 2 to 3 times daily.

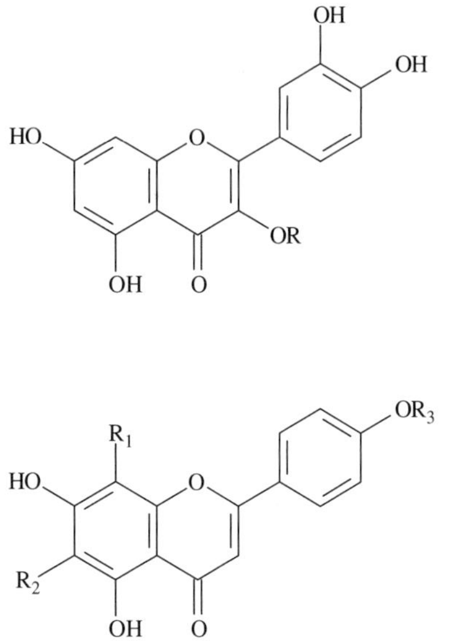

	R
Quercetin	H
Quercitrin	Rhamnose
Hyperoside	Galactose
Rutin	Glucose-Rhamnose

	R_1	R_2	R_3
Vitexin	Glucose	H	H
Vitexin-4'-*O*-rhamnose	Glucose	H	Glucose
Isovitexin	H	Glucose	H

FIGURE 72.1. Representative flavones and flavonols in hawthorn.

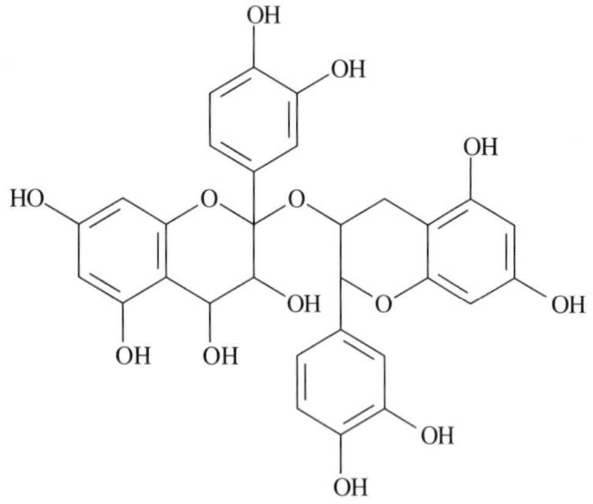

FIGURE 72.2. Chemical structure of procyanidin.

TOXICOKINETICS

Kinetics

There are few data on the pharmacokinetics of flavonoid compounds following ingestion by humans. Animal studies suggest that the oral bioavailability of epicatechin is low, while the GI absorption of other components of hawthorn extract (chlorogenic acid, hyperoside, isoquercitrin) is very low. In a study of rats, the bioavailability of epicatechin from hawthorn extract was about 10%, whereas the GI absorption of chlorogenic acid, hyperoside, and isoquercitrin was only about 1–2%.[13]

Drug Interactions

P-glycoprotein is a transmembrane protein that functions as an efflux transporter and is found in high concentrations in the gut and the kidney. The ingestion of hawthorn extract causes mild induction of P-glycoprotein activity, but this induction does not necessarily result in decreased absorption of P-glycoprotein dependent substrates, such as digoxin. A randomized, crossover trial with eight healthy volunteers did not detect a statistically significant difference in pharmacokinetic parameters (area under the concentration-time curve, maximum concentration, half-life, renal clearance) of digoxin, when comparing three weeks of daily hawthorn extract (WS 1442) to baseline.[14]

CLINICAL RESPONSE

Adverse effects associated with clinical trials using hawthorn extract at therapeutic doses are typically minor.[15] These effects include headache, dizziness, vertigo, rash, diaphoresis, palpitations, agitation, drowsiness, and GI disturbances. Larger doses of hawthorn extract in animals produce sedation, hypotension, and dysrhythmias.[16] There are few data on human toxicity associated with the ingestion of hawthorn extract. Based on animal data, potential toxic effects of hawthorn intoxication include cardiac abnormalities (hypotension, dysrhythmias). Rare case reports associate exposure to the fruits and leaves with contact dermatitis.[17] The concealed thorns on hawthorn stems can produce corneal abrasions and lacerations.

DIAGNOSTIC TESTING

Methods to determine the flavonoid content of hawthorn extract include high performance liquid chromatography and gas chromatography/mass spectrometry.[18,19] Separation of flavonoid compounds with capillary zone electrophoresis allowed detection limits of 0.35 μg/mL, 0.30 μg/mL, 0.40 μg/mL, and 0.29 μg/mL for vitexin-2′-rhamnoside, hyperoside, rutin, and vitexin, respectively.[20]

TREATMENT

Treatment is supportive. There are few clinical data on the toxicity of hawthorn extract to guide treatment.

References

1. McGovern PE, Zhang J, Tang J, Zhang Z, Hall GR, Moreau RA, et al. Fermented beverages of pre- and proto-historic China. PNAS 2004;101:17593–17598.
2. Chang Q, Zuo Z, Harrison F, Chow MS. Hawthorn. J Clin Pharmacol 2002;42:605–612.
3. Veveris M, Koch E, Chatterjee SS. *Crataegus* special extract WS 1442 improves cardiac function and reduces infarct size in a rat model of prolonged coronary ischemia and reperfusion. Life Sci 2004;74:1945–1955.
4. Schussler M, Holzl J, Fricke U. Myocardial effects of flavonoids from *Crataegus* species. Arzneimittelforschung 1995;45:842–845.
5. Pittler MH, Schmidt K, Ernst E. Hawthorn extract for treating chronic heart failure: meta-analysis of randomized trials. Am J Med 2003;114:665–674.
6. Narang R, Swedberg K, Cleland JG. What is the ideal study design for evaluation of treatment for heart failure? Insights from trials assessing the effect of ACE inhibitors on exercise capacity. Eur Heart J 1996;17:120–134.
7. Hanus M, Lafon J, Mathieu M. Double-blind, randomized, placebo-controlled study to evaluate the efficacy and safety of affixed combination containing two plant extracts (*Crataegus oxyacantha* and *Eschscholtzia californica*) and magnesium in mild-to-moderate anxiety disorders. Curr Med Res Opin 2004;20;63–71.
8. Griffiths DW, Robertson GW, Shepherd T, Birch AN, Gordon SC, Woodford JA. A comparison of the composition of epicuticular wax from red raspberry (*Rubus idaeus* L.) and hawthorn (*Crataegus monogyna* Jacq.) flowers. Phytochemistry 2000;55:111–116.
9. Rayan S, Fossen T, Nateland HS, Andersen OM. Isolation and identification of flavonoids, including flavone rotamers, from the herbal drug "crataegi folium cum flore" (hawthorn). Phytochem Anal 2005;16:334–341.
10. Zhang Z, Chang Q, Zhu M, Huang Y, Ho WK, Chen Z. Characterization of antioxidants present in hawthorn fruits. J Nutr Biochem 2001;12:144–152.
11. Vierling W, Brand N, Gaedcke F, Sensch KH, Schneider E, Scholz M. Investigation of the pharmaceutical and pharmacological equivalence of different hawthorn extracts. Phytomedicine 2003;10;8–16.
12. Verstraeten SV, Hammerstone JF, Keen CL, Fraga CG, Oteiza PI. Antioxidant and membrane effects of procyanidin dimers and trimers isolated from peanut and cocoa. J Agric Food Chem 2005;53:5041–5048.
13. Chang Q, Zuo Z, Ho WK, Chow MS. Comparison of the pharmacokinetics of hawthorn phenolics in extract versus individual pure compound. J Clin Pharmacol 2005;45:106–112.
14. Tankanow R, Tamer HR, Streetman DS, Smith SG, Welton JL, Annesley T, et al. Interaction study between digoxin and a preparation of hawthorn (*Crataegus oxyacantha*). J Clin Pharmacol 2003;43:637–642.
15. Tauchert M. Efficacy and safety of crataegus extract WS 1442 in comparison with placebo in patients with chronic stable New York Heart Association class-III heart failure. Am Heart J 2002;143:910–915.
16. Chang W-T, Dao J, Shao Z-H. Hawthorn: potential roles in cardiovascular disease. Am J Chin Med 2005:33:1–10.
17. Steinman HK, Lovell CR, Cronin E. Immediate-type hypersensitivity to *Crataegus monogyna* (hawthorn). Contact Dermatitis 1984;11:321.
18. Melikoglu G, Bitis L, Mericli AH. Flavonoids of *Crataegus microphylla*. Nat Prod Res 2004;18:211–213.
19. Hamburger M, Baumann D, Adler S. Supercritical carbon dioxide extraction of selected medicinal plants—effects of high pressure and added ethanol on yield of extracted substances. Phytochem Anal 2004;15:46–54.
20. Liu W, Chen G, Cui T. Determination of flavones in *Crataegus pinnatifida* by capillary zone electrophoresis. J Chromatogr Sci 2003;41:87–91.

Chapter 73

IMPILA, PINE THISTLE, and ATRACTYLOSIDE

HISTORY

The properties of pine thistle have been known at least since the first century AD, when the Greek pharmacologist and physician Pedanius Dioscorides of Anazarbos described the plant and its properties.[1] During the Renaissance, the plant was known as *chameleone*, and the use of this plant was suggested as an antidote for snake envenomation and other poisons, as well as urinary retention, somnolence, and dental pain. In 1867, Lefranc extracted a toxin from the rhizome of pine thistle that he named atractylic acid (atractyline), and later the compound was termed atractylate of potassium.[2] Stanislas and Vignais discovered a structurally related, more potent toxin called carboxy-atractyloside in 1956.[3] Brookes et al. isolated the main toxin, atractyloside, from the tubers of *C. laureola* in 1983.[4] *Callilepis laureola* (impila) is a popular traditional remedy in South Africa that was administered during the 1930s in low doses for whooping cough.[5]

BOTANICAL DESCRIPTION

Pine thistle (*Atractylis gummifera* L.) has been known since antiquity under a variety of names including ixia, chameleon leukos, chardon á glu (France), edded (North Africa), mascarida (Sardinia), and masticogna (Sicily).[6] This plant is often confused with the nontoxic plant, *Atractylis acaulis* Pers.

Common Name: Pine Thistle, Glue or Mediterranean Thistle, Blue Thistle, Birdlime, Chameleon

Scientific Name: *Atractylis gummifera* L.

Botanical Family: Asteraceae (sunflower)

Physical Description: This long-lived herbaceous plant has an aerial part (i.e., fern-like leaves, pink flowers, fruits) and a subterranean part (i.e., long, large rhizome of variable color depending on the season). The leaves are deeply divided into prickly lobes and grouped into rosettes.

Distribution and Ecology: *Atractylis gummifera* is a native of the Mediterranean region, particularly in dry areas of Tunisia, Morocco, Algeria, Spain, Italy, and Greece. Although this thistle is distributed worldwide, this plant is especially abundant in northern Africa and southern Europe.

Common Name: Ox-eye Daisy

Scientific Name: *Callilepis laureola* DC.

Botanical Family: Asteraceae (sunflower)

Physical Description: This woody tuber has a bulbous shape and a pungent odor.

Distribution and Ecology: *Callilepis* species are perennial, herbaceous plants that grow in the grassland of eastern South Africa.

EXPOSURE

Uses

In the Zulu culture, the tuber of *C. laureola* is a multipurpose remedy in the Kwa Zulu-Natal region of northeastern South Africa.[7] After boiling the powder of the

Medical Toxicology of Natural Substances, by Donald G. Barceloux, MD

root in water for 30 minutes, the decoction (i.e., impila) is administered orally or rectally. Traditional uses for this plant include stomach problems, constipation, tape worm infections, cough, induction of fertility in young women, and impotence.[8] Peoples of this region adhere to the magical belief that decoctions from this plant provide "protective powers" that ward off evil spirits. In Arab countries, the dried rhizome is burned as an incense to avoid bad fates. Tubers of *C. laureola* are harvested in the winter and processed into a dry powder. Alternatively, a piece of fresh material is cut and scraped from the tuber. Traditional uses of pine thistle include the treatment of syphilitic ulcers, intestinal parasites, syncope, drowsiness, abscesses, and discolored teeth, as well as an abortifacient, suicidal agent, and homicidal agent.[1,9] There are no approved medicinal uses of *C. laureola* (impila) or atractyloside.

PRINCIPAL INGREDIENTS

Chemical Composition

The two major toxic compounds in *C. laureola* are the diterpenoid glycosides, atractyloside and carboxy-atractyloside (gummiferin).[6] Atractyloside compounds also occur in other plant species including the pine thistle (*Atractylis gummifera*),[10] aerial parts of Asian *Atractylis* species (*A. carduus, A. lancea*), the South American herb *Pascalia glauca* Ortega (*Wedelia glauca*),[11] the Australia plant *Wedelia asperrima* (Decne.) Benth., and the common cocklebur (*Xanthium strumarium* L.).[12] Atractyloside and carboxy-atractyloside are the main toxic constituents of *Iphiona aucheri* (Boiss.) A. Anderberg, which has been used to poison racing camels in the Middle East.[13] *Atractylis acaulis* does not contain toxicologically significant amounts of atractyloside or carboxy-atractyloside.[9] Atractyloside may be a constituent of green and roasted coffee beans from *Coffea arabica* L.[14] Carboxy-atractyloside undergoes decomposition to atractyloside.[15] The concentration of atractyloside compounds varies with season, plant species, and environmental conditions; however all underground parts contain these two diterpenoid glucosides.[16] The green beans of *Coffea arabica* also contain substantial amounts of atractyloside, but the dry roasting process significantly decreases the content of atractyloside available in coffee extracts.[17]

Physiochemical Properties

Atractyloside consists of a relatively nontoxic aglycone with a perhydrophenanthrene structure and a glycoside moiety composed of glucose, sulfate, and isovalerate.[14] Figure 73.1 demonstrates the chemical structure of

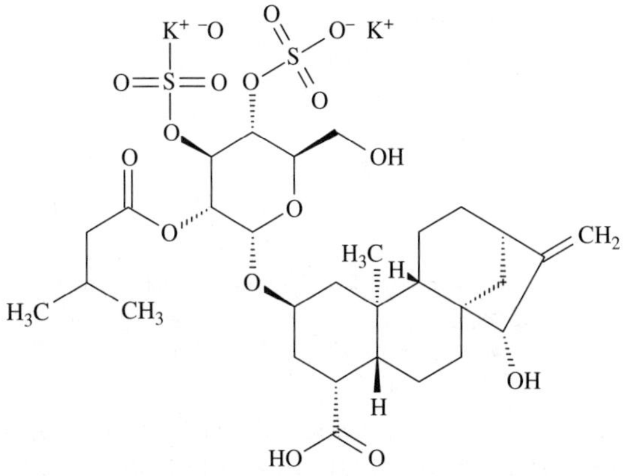

FIGURE 73.1. Chemical structure of atractyloside.

atractyloside. Carboxy-atractyloside occurs only in fresh roots because of the decarboxylation of this diterpenoid compound to atractyloside, particularly during heating and drying.[18] Other potential toxins in *Atractylis* species include carboxy-atractyloside, apoatractyloside, atractylgenine, and carboxy-atractylgenine.[6] Reduction of the carboxyl group on C4 of the diterpene ring to the corresponding alcohol produces a nontoxic compound (atractylitriol).[19]

Mechanism of Toxicity

Animal studies indicate that atractyloside and carboxy-atractyloside in *Atractylis* species inhibit mitochondrial oxidative phosphorylation and block the conversion of ADP to ATP by interacting with the adenine nucleotide translocator involved with the ATP/ADP antiport and mitochondrial membrane permeabilization.[20] Extracts of *C. laureola* probably contain both a liver toxin and a renal toxin (atractyloside), which cause hepatic necrosis and renal failure as a result of the inhibition of mitochondrial phosphorylation.[21] In experimental animals, atractyloside produces hypoglycemia and acute renal tubular necrosis, but the administration of this compound does not cause the centrilobular liver necrosis associated with impila.[17,22] Both atractyloside and carboxy-atractyloside bind selectively to phosphoryl transferase (nucleoside mono phosphokinase) in the outer mitochondria membranes resulting in the blockade of oxidative phosphorylation.[23] By binding to the adenine nucleotide translocator on the mitochondrial membrane, atractyloside and carboxy-atractyloside competitively inhibit the transport of ADP across the mitochondrial membrane. The result is a reduction in the synthesis of ATP and fatty acid oxidation, inhibition of gluconeogenesis, and acceleration of anaerobic glycolysis and glycogenolysis. The mechanism or mechanisms by which these compounds cause apoptosis of hepatocytes remain unclear, and the toxic effects of atractyloside involve mechanisms other than the blockade of oxidative phosphorylation.[14] *In vitro* studies indicate that aqueous extracts of *C. laureola* cause apoptosis via the caspase cascade and destruction of cytoplasmic tubulin architecture in human hepatocytes.[5,24] The histological features of hepatic damage are not consistent with lipid peroxidation and free radical formation.[9]

DOSE RESPONSE

The toxicity of decoctions from the tubers of *C. laureola* depends on variation of the toxic principals with season, climate, soil conditions, and plant part as well as on the sensitivity of individual patients (e.g., malnutrition).[25]

TOXICOKINETICS

There are few data on the toxicokinetics of atractyloside. Some unchanged atractyloside appears in the urine based on analyses of blood and urine samples from patients poisoned by atractyloside-containing herbs.[14]

CLINICAL RESPONSE

The classical presentation of toxicity following exposure to atractyloside-containing substances (impila, pine thistle) involves the sudden onset of abdominal pain, vomiting, diarrhea, altered consciousness, and fulminant hepatorenal failure.[26,27] In a retrospective review of admissions to a toxicology intensive care unit in Tunisia, ingestion of *Atractylis gummifera* accounted for about one-third of the cases.[28] Symptoms typically begin 6–36 hours after ingestion of the extract of *Atractylis gummifera* rhizome.[29] Convulsions occur in a substantial number of seriously poisoned patients. Although renal dysfunction usually occurs with substantial hepatic necrosis, case reports document the occurrence of early renal failure without prominent liver damage.[30] There are few data on the chronic toxicity of low doses of atractyloside. The differential diagnosis of atractyloside-induced hepatic dysfunction includes Jamaican vomiting sickness, Reye's syndrome, viral hepatitis, and sepsis.

DIAGNOSTIC TESTING

Analytical Methods

Thin layer chromatographic and enzyme-linked immunosorbent assay (ELISA) methods are screening procedures for the detection of amounts of atractyloside exceeding 0.5 µg.[31,32] These screening tests confirm exposure to atractyloside, but there are inadequate data to correlate atractyloside concentrations with clinical effects. Gas chromatography/mass spectrometry techniques are available to confirm the presence of atractyloside compounds.[33]

Abnormalities

The most common laboratory abnormalities in patients with severe atractyloside toxicity include hypoglycemia, metabolic acidosis, hypoprothrombinemia, and elevated serum hepatic aminotransferases. Other common laboratory abnormalities include leucocytosis, hyperkalemia, hyponatremia, and hypernatremia.[26] Jaundice is relatively mild compared with the severity of the liver disease, in part because of the rapid progression of liver failure.[31] Diagnosis is confirmed by liver biopsy demonstrating centrilobular necrosis of the liver.[34] Postmortem findings beside centrilobular liver necrosis include congestion of the renal cortex, acute tubular necrosis of the proximal convoluted tubules and the loops of Henle, and hemorrhage of the lungs and gastrointestinal tract.

TREATMENT

Treatment of atractyloside poisoning is supportive. Because of the frequency of hypoglycemia, all patients should be screened with rapid blood glucose tests and treated promptly if hypoglycemia is present. There are no clinical data to guide the treatment of hepatorenal failure other than the standard methods of treatment (i.e., fluid and electrolyte replacement, cardiorespiratory support, seizure control) and support of hepatorenal function.

References

1. Daniele C, Dahamna S, Firuzi O, Sekfali N, Saso L, Mazzanti G. *Atractylis gummifera* L. poisoning: an ethnopharmacological review. J Ethnopharmacol 2005;97:175–181.
2. Lefranc E. [De l'acide atractylique (2è etude)]. C R Hebd Seances Acad Sci 1873;76:438–440. [French]
3. Stanislas E, Vignais P. [On the principal toxins of Atractylis Gummifera L.] C R Hebd Seances Acad Sci 1964;259: 4872–4875. [French]
4. Brookes K, Candy H, Pegel K. Atractylosides in *Callilepis laureola* (Asteraceae). S Afr J Chem 1983;36:65–68.
5. Stewart MJ, Steenkamp V, van der Merwe S, Zuckerman M, Crowther NJ. The cytotoxic effects of a traditional Zulu remedy, impila (*Callilepis laureola*). Hum Exp Toxicol 2002;21:643–647.
6. Hamouda C, Hedhili A, Salah NB, Zhioua M, Amamou M. A review of acute poisoning from *Atractylis gummifera* L. Vet Hum Toxicol 2004;46:144–146.

7. Bye SN, Dutton MJ. The inappropriate use of traditional medicines in South Africa. J Ethnopharmacol 1991;34: 253–259.
8. Bryant AT. Zulu medicine and medicine-men. Ann Natal Museum 1909;2:1–103.
9. Stewart MJ, Steenkamp V. The biochemistry and toxicity of atractyloside: a review. Ther Drug Monit 2000;22: 641–649.
10. Georgiou M, Sianidou L, Hatzis T, Papadatos J, Koutselinis A. Hepatotoxicity due to *Atractylis gummifera*. J Toxicol Clin Toxicol 1988;26:487–493.
11. Schteingart C, Pomilio A. Atractyloside, toxic compound from *Wedelia glauca*. J Nat Prod 1984;47:1046–1047.
12. Witte ST, Osweiter GD, Staahr HM, Mobley G. Cocklebur toxicosis in cattle associated with consumption of mature *Xanthium strumarium*. J Vet Diag Invest 1990;2:263–267.
13. Roeder E, Bourauel T, Meier U, Wiedenfeld H. Diterpene glycosides from *Iphiona aucheri*. Phytochemistry 1994;37: 353–355.
14. Obatomi DK, Bach PH. Biochemistry and toxicology of the diterpenoid glycoside atractyloside. Food Chem Toxicol 1998;36:335–346.
15. Stedman C. Herbal hepatotoxicity. Sem Liver Dis 2002;22: 195–206.
16. Calmes M, Crespin F, Maillard C, Ollivier E, Balansard G. High performance liquid chromatographic determination of atractyloside and carboxy-atractyloside from *Atractylis gummifera*. L. J Chromatogr A 1994;663:119–122.
17. Obatomi DK, Bach PH. Biochemistry and toxicology of the diterpenoid glycoside atractyloside. Food Chem Toxicol 1998;36:335–346.
18. Bombardelli KB. Structure of the diterpenoid, carboxy-atractyloside. Phytochemistry 1972;11:3501–3504.
19. Vignais P, Duee E, Vignais P, Huet J. Effects of atractyligenin and its structural analogues on oxidative phosphorylation and on the translocation of adenine dinucleotides in mitochondria. Biochim Biophys Acta 1966;118:465–483.
20. Hedhili A, Warnet JM, Thevenin M, Martin C, Yacoub M, Claude JR. Biochemical investigation of *Atractylis gummifera* L. hepatotoxicity in the rat. Arch Toxicol 1989; 13(Suppl):312–315.
21. Wainwright J, Schonland MM, Candy HA. Toxicity of *Callilepis laureola*. S Afr Med J 1977;52:313–315.
22. Carpenedo F, Luciani S, Scaravilli F, Palatini P, Santi R. Nephrotoxic effect of atractyloside in rats. Arch Toxicol 1974;32:169–180.
23. Vancompernolle K, Van Herreweghe F, Pynaert G, Van de Craen M, De Vos K, Totty N, et al. Atractyloside-induced release of cathepsin B, a protease with caspase-processing activity. FEBS Lett 1998;438:150–158.
24. Popat A, Shear NH, Malkiewicz I, Stewart MJ, Steenkamp V, Thomson S, Neuman MG. The toxicity of *Callilepis laureola*, a South African traditional herbal medicine. Clin Biochem 2001;34:229–236.
25. Hutchings A, Terbalanche SE. Observations on the use of some known and suspected toxic Liliiflorae in Zulu and Xhosa medicine. S Afr Med J 1989;75:62–69.
26. Lemaigre G, Tebbi Z, Galinsky R, Michowitcz S, Abelanet R. [Fulminating hepatitis caused by glue thistle (*Atractylis glummifera* L.) poisoning. Anatomo-pathological study of 4 cases]. Nouv Presse Med 1975;4:2565–2568. [French]
27. Watson AR, Coovadia HM, Ghoola KD. The clinical syndrome of impila (*Callilepis laureola*) poisoning in children. South Afr Med J 1979;55:290–292.
28. Hamouda C, Amamou M, Thabet H, Yacoub M, Hedhili A, Bescharnia F, et al. Plant poisonings from herbal medication admitted to a Tunisian toxicologic intensive care unit, 1983–1998. Vet Hum Toxicol 2000;42:137–141.
29. Capdevielle P, Darracq R. [Poisoning by bird-line thistle (*Atractylis gummifera* L.)]. Med Trop (Mars) 1980;40:137–142. [French]
30. Seedat Y, Hitchcock P. Acute renal failure from *Callilepis laureola*. S Afr Med J 1971;45:832–833.
31. Bye SN, Coetzer HT, Dutton MF. An enzyme immunoassay for atractyloside, the nephrotoxin of *Callilepis laureola* (impila). Toxicon 1990;28:997–1000.
32. Steenkamp V, Stewart JM, Zuckerman M. Detection of poisoning by impila (*Callilepis laureola*) in a mother and child. Hum Exp Toxicol 1999;18;594–597.
33. Laurens JB, Bekker LC, Steenkamp V, Stewart MJ. Gas chromatographic-mass spectrometric confirmation of atractyloside in a patient poisoned with *Callilepis laureola*. J Chromatogr B 2001;765:127–133.
34. Wainwright J, Schonland MM. Toxic hepatitis in black patients in natal. S Afr Med J 1977;51:571–573.

Chapter 74

JIN BU HUAN and TETRAHYDROPALMATINE

HISTORY

Jin bu huan is a popular Chinese herb that has been used for over 1,000 years. In 1993, public health and healthcare providers in Colorado reported the development of central nervous system and respiratory depression in three children accidentally ingesting jin bu huan anodyne tablets.[1] All of these children recovered after receiving intensive supportive care. Subsequent analyses of these tablets indicated that they contained high concentrations of a compound, *l*-tetrahydropalmatine, which was not previously found in jin bu huan. Later that year, several cases of acute hepatitis were associated with the ingestion of these tablets over several months.[2] Subsequently, federal agencies in the United States and Canada banned the use jin bu huan in their respective countries.

BOTANICAL DESCRIPTION

Jin bu huan is an ancient Chinese herb with multiple herbal sources, including *Aristolochia yunnanensis* Franch, *Asystasiella neesiana* (Wallich) Lindau, *Huperzia serrata* var. *serrata* (Thunb. ex Murr.) Trevisan (*Lycopodium serratum*), *Panax pseudoginseng* Wallich, *Polygala chinensis* Hook L., *Rumex madaio* Makino, *Rumex patientia* L., *Selaginella involvens* (Swartz) Spring, *Stephania delavayi* Diels, and *Stephania sinica* Diels.[3] Of these plant species, only *Stephania* species contain the toxic ingredient (tetrahydropalmatine) in the jin bu huan anodyne tablets associated with central nervous system (CNS) and hepatic toxicity.

EXPOSURE

Sources

The jin bu huan anodyne tablets associated with hepatic and CNS toxicity were an herbal product purportedly containing *Polygala chinensis* L. (Chinese milkwort) imported from Kwangsi, China by the Kwangsi Pai Se Pharmaceutical Company. These tablets were sold as a sedative, analgesic, and sleeping tablet for individuals with nocturnal pain. Subsequent analysis of these tablets indicated that the sole active ingredient was *l*-tetrahydropalmatine (i.e., about 34–36% weight/weight), which is not present in detectable quantities in *Polygala chinensis*.[4] The suspected origin of tetrahydropalmatine in these tablets was probably a standard pharmaceutic source.[5]

Medicinal Uses

Jin bu huan is a traditional Chinese herbal product used as a sedative, antispasmodic, analgesic and decongestant. Tetrahydropalmatine (rotundine) is a pharmaceutical product in China.

PRINCIPAL INGREDIENTS

Chemical Composition

Analysis of tetrahydropalmatine-containing jin bu huan anodyne tablets indicated that the tetrahydropalmatine concentration in these tablets was approximately 34–

Medical Toxicology of Natural Substances, by Donald G. Barceloux, MD

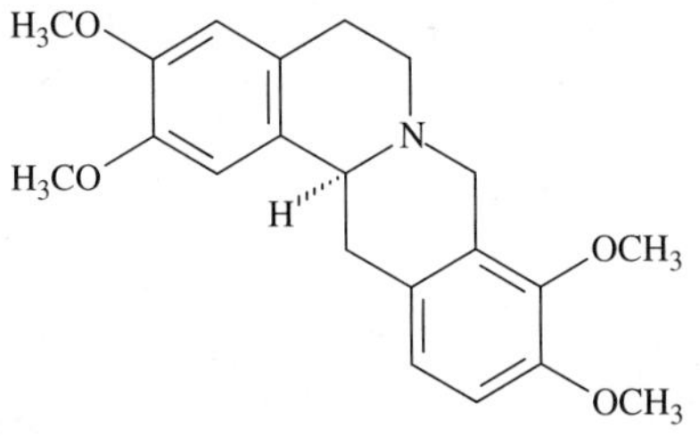

FIGURE 74.1. Chemical structure of tetrahydropalmatine.

36% weight/weight compared with 1.5% in herbs, which naturally contain this alkaloid.[6] The rest of the tablets contained starch, and there were no other plant alkaloids or pharmaceutical products detected. Figure 74.1 displays the chemical structure of tetrahydropalmatine (CAS RN: 10097-84-4, $C_{21}H_{25}NO_4$), which is similar to the structure of berberine. The latter compound has been associated with kernicterus and displacement of bilirubin in preterm neonates.[7]

Physiochemical Properties

Although tetrahydropalmatine in extractions of herbal plants (e.g., *Corydalis* spp., *Stephania* spp.) typically contain racemic mixtures (*d,l*-tetrahydropalmatine), rodent studies indicate that only the *l*-enantiomer is pharmacologically active.[8] In animal models, tetrahydropalmatine has reserpine-like properties resulting in the depletion of neurotransmitters (dopamine > norepinephrine > serotonin).[9] Rodent studies indicate that *l*-tetrahydropalmatine acts as a dopaminergic D_2 antagonist in nigro-striatal neurons, resulting in the blockage of both pre- and postsynaptic dopamine receptors.[10,11] This compound has little affinity for opioid receptors.[12]

Mechanism of Toxicity

The exact mechanism of hepatotoxicity associated with tetrahydropalmatine-adulterated jin bu huan remains unclear. Potential mechanisms include hypersensitivity reactions, idiosyncratic reactions, or direct toxicity from active metabolites. The abrupt recrudescence of hepatitis following reintroduction of jin bu huan anodyne tablets to 2 patients suggests a hypersensitivity reaction.[4] However, the significant decrease in serum hepatic aminotransferase following continuation of a reduced dosage of jin bu huan suggests a direct toxic effect. The clinical course of some of these patients is also consistent with an idiosyncratic reaction.

DOSE RESPONSE

The recommended adult dose of jin bu huan anodyne tablets was 2–4 tablets 1–3 times daily. The ingestion of 17 tetrahydropalmatine-adulterated jin bu huan anodyne tablets (28.8 mg *l*-tetrahydropalmatine/tablet) by a $2^1/_2$-year-old girl resulted in lethargy, respiratory depression requiring intubation, and bradycardia. The ingestion of 60 tablets by a 13-month-old boy caused lethargy, hypotonia, and transient bradycardia. Both children survived without sequelae after supportive care.[6]

TOXICOKINETICS

There are few data on the kinetics of tetrahydropalmatine in humans. The relatively rapid onset of symptoms in children ingesting tetrahydropalmatine-adulterated jin bu huan anodyne tablets suggests that the absorption of tetrahydropalmatine is rapid. The resolution of symptoms within 10 hours of a large overdose also suggests rapid elimination. Analysis of serum and urine from patients poisoned by tetrahydropalmatine suggests that biotransformation of this compound occurs by *O*-demethylation and elimination by excretion of glucuronide metabolites in the urine.[13]

CLINICAL RESPONSE

Acute Effects

Case reports associate the acute ingestion of tetrahydropalmatine in adulterated jin bu huan anodyne tablets by young children and the intentional overdose of tetrahydropalmatine-containing sleeping pills by adults with CNS depression. The clinical features of a case series of three young children ingesting 7–60 tablets included lethargy, hypotonia, bradycardia, dysconjugate gaze, and respiratory depression requiring intubation.[6] The children recovered within approximately 3–10 hours. A case series of 9 adult patients associated the intentional ingestion of up to 75 herbal sleeping pills containing 26 mg tetrahydropalmatine/tablet with disorientation, lethargy, and coma.[13] All patients were released within 1–3 days after supportive care.

Chronic Effects

Several case reports associate the development of acute hepatitis with the chronic ingestion of tetrahydropalmatine-adulterated jin bu huan anodyne tablets.[1,14] In a case series of 7 patients, the mean onset of acute hepatitis after beginning these tablets was 20 weeks (range: 7–52 weeks) and resolution occurred in 6 patients within 2–30 weeks (Mean: 8 weeks).[4] The seventh patient was recovering during the time the article was published. The clinical features of this illness included fever, fatigue, nausea, pruritus, abdominal pain, jaundice, and hepatomegaly. Liver biopsy in one patient suggested a drug reaction (i.e., eosinophilia), while other pathological abnormalities included focal hepatocellular necrosis, mixed periportal infiltrates, microvesicular steatosis, and moderate fibrosis. Although these biopsies occasionally demonstrate the presence of moderate fibrosis,[15] these patients usually recover without clinical or biochemical evidence of liver damage.

DIAGNOSTIC TESTING

Older methods for the detection of tetrahydropalmatine include thin layer chromatography (TLC) and reverse-phase high performance liquid chromatography (RP/HPLC). However, TLC has poor reproducibility and analysis of basic drugs by RP/HPLC is limited by excessive retention and poor column efficiency. The use of an unmodified silica column with reversed-phase eluents improves the speed and precision of HPLC.[16] Screening of urine or serum samples with uniform solid-phase extraction and HPLC with diode-array detection allows the detection of tetrahydropalmatine and metabolites.[13] The treatment of biological samples with hydrolysis or glucuronidase improves the yield, particularly in urine samples. Gas chromatography/mass spectrometry (GC/MS) confirms the presence of tetrahydropalmatine and specific metabolites.

Serum abnormalities associated with the consumption of tetrahydropalmatine-adulterated jin bu huan tablets include elevated serum hepatic aminotransferases and serum bilirubin.[17] The serum alkaline phosphatase concentration may be within the normal range, and the prothrombin time is usually normal unless the hepatitis is severe enough to alter synthesis of clotting factors by the liver. These abnormalities usually resolve within several months of cessation of the use of these tablets.

TREATMENT

The occurrence of respiratory depression requiring intubation and bradycardia indicates that symptomatic intoxications with tetrahydropalmatine-adulterated jin bu huan should be evaluated and monitored carefully for the adequacy of ventilation including pulse oximetry and cardiac monitoring.[6] The onset of symptoms is typically rapid (i.e., within 1 hour). Bradycardia in these cases was responsive to atropine. If the patient presents to the emergency department within 1 hour of ingestion, the use of activated charcoal is appropriate. The use of syrup of ipecac is contraindicated because of the rapid onset of CNS depression. Case reports indicate that the CNS toxicity associated with the acute ingestion of tetrahydropalmatine-adulterated jin bu huan anodyne tablets is not responsive to intravenous naloxone administration.[6] The treatment of acute hepatitis associated with tetrahydropalmatine-adulterated jin bu huan is supportive. Most of these cases respond to cessation of jin bu huan use.

References

1. Woolf GM, Rojter SE, Villamil FG, Vierling JM, Stermitz FR, Beck JJ, et al. Jin bu huan toxicity in adults–Los Angeles, 1993. MMWR Morb Mortal Wkly Rep 1993;42:920–922.
2. Horowitz RS, Feldhaus K, Dart RC, Stermitz FR, Beck JJ. The clinical spectrum of jin bu huan toxicity. Arch Intern Med 1995;156:899–903.
3. But PH, Choi KL. Jin bu huan anodyne tablets, a mislabelled and misclassified medicine. J Hong Kong Med Assoc 1994;46:302–305.
4. Woolf GM, Petrovic LM, Rojter SE, Wainwright S, Villamil FG, Katkov WN, et al. Acute hepatitis associated with the Chinese herbal product jin bu huan. Ann Intern Med 1994;121:729–735.
5. Brent J. Three new herbal hepatotoxic syndromes. Clin Toxicol 1999;37:715–719.
6. Horowitz RS, Dart RC, Gomez H, Moore LL, Fulton B, Feldhaus K, et al. Epidemiologic notes and reports jin bu huan toxicity in children—Colorado, 1993. MMWR Morb Mortal Wkly Rep 1993;42:633–636.
7. Chan E. Displacement of bilirubin from albumin by berberine. Biol Neonate 1993;63:201–208.
8. Hus B, Kin K-C. Further studies on the pharmacology of tetrahydropalmatine and its analogs. Int J Neuropharmacol 1964;2:283–290.
9. Liu GQ, Algeri S, Garattini S. D-L-tetrahydropalmatine as monoamine depletor. Arch Int Pharmacodyn Ther 1982;258:39–50.
10. Chueh F, Hsieh MT, Chen CF, Lin MT. *dl*-Tetrahydropalmatine-produced hypotension and bradycardia in rats through the inhibition of central nervous dopaminergic mechanisms. Pharmacol 1995;51:237–244.
11. Marcenac F, Jin GZ, Gonon F. Effect of *l*-tetrahydropalmatine on dopamine release and metabolism in the rat striatum. Psychopharmacol (Berl) 1986;89:89–93.
12. Zhang ZD, Jin GZ, Xu SX, Yu LP, Chen Y, Jiang FY, et al. [Effects of *l*-stepholidine on the central nervous and cardiovascular systems]. Zhongguo Yao Li Xue Bao 1986;7:522–526. [Chinese]
13. Lai C-K, Chan A Y-W. Tetrahydropalmatine poisoning: diagnoses of nine adult overdoses based on toxicology screens by HPLC with diode-array detection and gas chromatography-mass spectrometry. Clin Chem 1999;45:229–236.

14. McRae CA, Agarwal K, Mutimer D, Bassendine MF. Hepatitis associated with Chinese herbs. Eur J Gastroenterol Hepatol 2002;14:559–562.
15. Picciotto A, Campo N, Brizzolara R, Giusto R, Guido G, Sinelli N, et al. Chronic hepatitis induced by jin bu huan. J Hepatol 1998;28:165–167.
16. Yuan Y-F, Liu Z-L, Li X-L. Use of silica gel with reversed-phase eluents for the separation and determination of alkaloids in *Corydalis yanhusuo* W.T. Wang and its preparations. Biomed Chromatogr 1996;10:11–14.
17. Divinsky M. Case report: jin bu huan—not so benign herbal medicine. Can Fam Phys 2002;48:1640–1642.

Chapter 75

JUNIPER
(*Juniper communis* L.)

HISTORY

The use of juniper berries dates back at least to the ancient Greeks and Egyptians. The Greeks used these berries for medicinal purposes. Ancient Egyptian tombs, including the tomb of Tutankhamen, contained *Juniperus oxycedrus* berries even though this plant was not native to Egypt. The Romans used juniper berries as a domestic substitute for the expensive black pepper imported from India; Pliny the Elder mentioned the use of juniper berries in his book, *Historia Naturalis*. Extracts of juniper berries have a long history of use as flavoring agents, including as an additive to gin.

BOTANICAL DESCRIPTION

Common Name: Common Juniper, Dwarf Juniper, Genévrier Commun

Scientific Name: *Juniper communis* L.

Botanical Family: Cupressaceae (cypress, redwood)

Physical Description: This slow-growing coniferous evergreen shrub or a small columnar tree is usually decumbent or rarely upright. The height of juniper varies with the variety with shrub reaching 4m (~13ft) and trees reaching up to 10m (~33ft). The thin, fibrous, brown bark exfoliates in thin strips. The firm, green, simple leaves have a pungent odor and are arranged in whorls of three. The fruits are berrylike seed cones, and each cone contains 2–3 small (3–4mm/~0.1–0.2in.) seeds.

Distribution and Ecology: The genus *Juniperus* comprises about 70 species distributed throughout the North America, Europe, and Asia. These species include *Juniperus virginiana* L. (eastern red cedar from the eastern United States), *Juniperus ashei* Buchh. (Ashe juniper), *Juniperus phoenicea* L. (Phoenician cedar from the Mediterranean coast), and *Juniperus oxycedrus* L. (prickly or cade juniper from southwest Europe to northwest Africa).

EXPOSURE

Sources

Juniper oils are steam distillates of juniper berries or wood; these oils are fragrance ingredients in some cosmetics. The berries from the juniper tree contain about 0.5–1.5% essential oil.[1] The source of *Juniperus communis* extract (CAS RN: 84603-69-0) is the extract of the mature fruit of *J. communis*, whereas *Juniperus communis* oil (RN: 8002-68-4, juniper berry oil) is the volatile oil from the berries of *J. communis*. The destructive distillation of branches and wood from *Juniperus oxycedrus* L. is the source of cade oil (CAS RN: 8013-10-3, juniper tar). The concentration of juniper extract and juniper tar in cosmetics ranges from about 0.1–1% and 1–5%, respectively.[1] The chief flavoring agent in distilled liquor, gin, is the highly aromatic, blue-green berry of the juniper (*Juniperus* spp.), which is grown commercially in the United States, Canada, northern Italy, and Croatia.

Medical Toxicology of Natural Substances, by Donald G. Barceloux, MD

Medicinal Uses

TRADITIONAL

Cade oil is a traditional treatment for skin diseases (e.g., atopic dermatitis, chronic eczema) in humans and animals based on keratolytic, antipruritic, and antimicrobial properties. Other herbal uses of juniper include the treatment of diabetes mellitus, fluid overload (diuretic), and bronchopulmonary diseases.

CURRENT

Rectified cade oil is a fragrance in soaps, creams, lotions, detergents, and perfumes.

REGULATORY STATUS

Most countries do not regulate the use of juniper oil.

PRINCIPAL INGREDIENTS

Chemical Composition

Juniper oils are complex mixtures containing many monoterpene, sesquiterpene, oxygenated monoterpene, and oxygenated sesquiterpene hydrocarbons.[2] Essential oils from *Juniperus* species contain about 85–95% monoterpenes and approximately 0.1–12% alcohols. The major compounds in essential oils from *Juniperus* species are α-pinene, β-pinene, δ-3-carene, sabinene, myrcene, β-phellandrene, limonene, and D-germacrene.[3] Analysis of the essential oil from dried juniper berries (*J. communis*) by gas chromatography/mass spectrometry demonstrated the following approximate yield: α-pinene (29%), β-pinene (18%), sabinene (14%), limonene (6%), and myrcene (0.3%).[4] However, there are substantial quantitative and qualitative differences in the essential oils between different *Juniperus* species and between different plant parts. The main component in the leaf essential oil from *Juniperus communis* was sabinene (about 61%), whereas α-pinene (about 52%) was the main ingredient in the essential oil from the berries.[3] The yield of leaf essential oil from *Juniperus oxycedrus* is about 0.27%.[5] The major constituents are α-pinene (about 40%) and the diterpene, manoyl oxide (about 10%). The yield of berry essential oil is higher (i.e., 1.14%). α-Pinene is also the major constituent (about 65%) of berry essential oil with smaller amounts of myrcene (about 4%), limonene, β-pinene, manoyl oxide, and γ-muurolene.

The leaves of the eastern red cedar (*Juniperus virginiana* L.) contain podophyllotoxin. The leaves from the mature plants have about 1–2.5 mg podophyllotoxin/g material, but this concentration is much less than leaves from the American mayapple (*Podophyllum peltatum* L.).[6] The average podophyllotoxin content of juvenile plants was approximately 40% of the concentration in immature and mature specimens of *J. virginiana*.

Physiochemical Properties

Juniper berry oil is a yellowish, alcohol-soluble liquid with a density of 0.865 g/mL and a boiling point of 120 °C (248 °F).

Mechanism of Toxicity

Most adverse effects associated with exposure to juniper oil are hypersensitivity reactions.

DOSE RESPONSE

Toxic doses of juniper oil are not well defined. Many reactions to juniper oil involve hypersensitivity responses; therefore most of these reactions are not dose-related. A typical herbal dose of juniper oil is 100 mL of a 1:20 dilution in boiled water.

TOXICOKINETICS

There are few human or animal data on the toxicokinetics of constituents of juniper oil.

CLINICAL RESPONSE

In general, animal studies and clinical trails indicate that juniper oils produce little skin irritation or sensitization, although some sensitization can occur to juniper tar (cade oil).[1] Species in the Cupressaceae family, including the genus *Juniperus* (e.g., *Juniperus ashei* J. Buchholz, *J. virginiana*), are common causes of allergic rhinitis, particularly in the winter and early spring.[7] In a New York study of 102 patients with asthma and/or rhinitis as well as evidence of sensitivity to trees, grass, or ragweed pollens, 51% (52 patients) were sensitive to the eastern red cedar (*Juniperus virginiana*) or white cedar (*Thuja occidentalis* L.).[8] There is substantial cross-reactivity within juniper species and between juniper species and other cedar trees [*Libocedrus decurrens* Torr., *Chamaecyparis lawsoniana* (A. Murray bis) Parl., *Cryptomeria japonica* (Thunb. ex L. f.) D. Don or Japanese cedar] and cypress trees (*Cupressus arizonica* Greene).[9] However, there is little cross-reactivity with pines and angiosperms.

A case report associated the ingestion of a spoonful of homemade cade oil (juniper tar) by a 32-year old man with the development of fever, hypotension, gross hematuria, renal failure, white desquamated facial burns, thrombocytopenia, and elevated serum creatine phosphokinase concentrations.[10] His vital signs and hepatorenal function returned to normal after 8 days. These effects were attributed to phenol (carbolic acid) in the cade oil, but there was no chemical analysis of the cade oil. Although essential oils from *Juniperus* species contain phenolic compounds, phenol (carbolic acid) is not a major constituent of essential oils from *Juniperus* species, and the facial burns suggested exposure to phenol.

DIAGNOSTIC TESTING

Methods to analyze the components of juniper oil include high-performance liquid chromatography with UV detection,[11] capillary gas chromatography with flame ionization and mass spectrometry.[2]

TREATMENT

Treatment is supportive.

References

1. Cosmetic Ingredient Review Expert Panel. Final report on the safety assessment of *Juniperus communis* extract *Juniperus oxycedrus* extract, *Juniperus oxycedrus* tar, *Juniperus phoenicea* extract, and *Juniperus virginiana* extract. Int J Toxicol 2001;20(Suppl 2)41–56.
2. Barjaktarovic B, Sovilj M, Knez Z. Chemical composition of *Juniperus communis* L. fruits supercritical CO2 extracts: dependence on pressure and extraction time. J Agric Food Chem 2005;53:2630–2636.
3. Angioni A, Barra A, Russo Mt, Coroneo V, Dessi S, Cabras P. Chemical composition of the essential oils of *Juniperus* from ripe and unripe berries and leaves and their antimicrobial activity. J Agric Food Chem 2003;51:3073–3078.
4. Pepeljnjak S, Kosalec I, Kalodera Z, Blazevic N. Antimicrobial activity of juniper berry essential oil (*Juniperus communis* L., Cupressaceae). Acta Pharm 2005;55:417–422.
5. Salido S, Altarejos J, Nogueras M, Sanchez A, Pannecouque C, Witvrouw M, De Clercq E. Chemical studies of essential oils of *Juniperus oxycedrus* ssp. *badia*. J Ethnopharmacol 202;81:129–134.
6. Cushman KE, Maqbool M, Gerard PD, Bedir E, Lata H, Moraes RM. Variation of podophyllotoxin in leaves of eastern red cedar (*Juniperus virginiana*). Planta Med 2003;69:477–478.
7. Bousquet J, Knani J, Hejjaoui A, Ferrando R, Cour P, Dhivert H, Michel FB. Heterogeneity of atopy. I. Clinical and immunologic characteristics of patients allergic to cypress pollen. Allergy 1993;48:183–188.
8. Deane PM. Conifer pollen sensitivity in Western New York: cedar pollens. Allergy Asthma Proc 2005;26: 352–355.
9. Schwietz LA, Goetz DW, Whisman BA, Reid MJ. Cross-reactivity among conifer pollens. Ann Allergy Asthma Immunol 2000;84:87–93.
10. Koruk ST, Ozyilkan E, Kaya P, Colak D, Donderici O, Cesaretli Y. Juniper tar poisoning. Clin Toxicol 2005; 43:47–49.
11. Martin AM, Queiroz EF, Marston A, Hostettmann K. Labdane diterpenes from *Juniperus communis* L. berries. Phytochem Anal 2006;17:32–35.

Chapter 76

KAVA
(*Piper methysticum* Forster)

HISTORY

For several thousand years, South Pacific Islanders used the aqueous extracts of dried Kava roots for ceremonial and social purposes.[1] The use of this intoxicating beverage spread to Europe in the 1700s after two Swedish botanists described the plant on the first expedition (1768–1771) of Captain Cook. During his second voyage (1772–1775), botanist, Johann Georg Forster, named the plant *Piper methysticum* (intoxicating pepper). German colonists brought kava to Europe during the following century.[2] The kava rhizome appeared in the *British Pharmacopoeia* in 1914 and in the *US Dispensary* in 1950. Kava was listed as a treatment for gonorrhea in the *German Pharmacopoeia* prior to the discovery of penicillin. During the early 1980s, the recreational use of this beverage spread to Aboriginal communities in Australia, where heavy consumption of this beverage frequently occurred with ethanol. Since the 1990s, kava extracts prepared from nonaqueous solvents appeared in commercial markets of the United States and Europe as dietary supplements for the treatment of stress, insomnia, and anxiety. Case reports in the early 1990s associated the use of kava extracts with the development of hepatitis, cirrhosis, and fulminant liver failure. Consequently, several European countries and Canada banned the use of kava extracts.

BOTANICAL DESCRIPTION

Common Name: Kava, Kava Kava, Kawa-kawa, Yati, Yagona, Yangona

Scientific Name: *Piper methysticum* Forster

Botanical Family: Piperaceae (pepper)

Physical Description: This perennial shrub reaches 3 m (~10 ft) in height with long (15–20 cm/~6–8 in.), bright green, heart-shaped leaves about 15–20 cm (~6–8 in.) long. Small flowers appear in spikes, but these flowers do not produce fruit or seeds. The thick, succulent-like stems have prominent nodes.

Distribution and Ecology: Kava is a perennial shrub native to the South Pacific including Melanesia (Fiji, Papua New Guinea, the Solomon Islands and Vanuatu), Micronesia, and Polynesia. This shrub prefers well-drained, rich soil in the sun when mature.

EXPOSURE

Sources

Kava rhizomes undergo extraction using water, ethanol, or acetone. The aqueous extract of the kava plant (*Piper methysticum* Forster) is a traditional drink used by South Pacific Islanders for soporific and psychotropic effects.[3] This drink is an infusion of ground kava rhizomes (1–1.5 g) in cold water (100–150 mL) or in coconut milk. The kava root is pulverized by mastication for narcotic-like effects in some parts of Vanuatu and Papua New Guinea. Dried powdered kava imported from Fiji or Tonga is mixed with water by Aborigines in Arnhem Land, Australia. In Western cultures, commercial caplets of kava are used rather than the traditional kava rhizome infusion. The difference between the aqueous extraction

Medical Toxicology of Natural Substances, by Donald G. Barceloux, MD

method for the traditional drink and the acetone or alcohol extraction methods for commercial caplets results in substantial variation in the chemical composition of these extracts. The traditional aqueous beverage contains glutathione, which potentially provides some protection against hepatotoxicity in comparison to the more lipophilic acetone extracts of kava plant.[4] The commercial, acetone kava extract WS®1490 is a 90–110 mg extract of dry kava rhizome (11–20:1) that corresponds to about 70 mg kavalactones. The dried extracts are pressed into caplets or tablets for commercial use.

Medicinal Uses

Traditional

Traditional uses for kava include the treatment of asthma, rheumatism, urological problems, menopausal symptoms, gonorrhea, vaginitis, nocturnal incontinence, weight loss, and insomnia. As a traditional social beverage, ingestion of aqueous extracts of kava roots produces mild euphoria characterized by loquaciousness, increased sociability, and increased sensitivity to sounds. Kava is an important part of the Pacific Island societies as manifest by the use of kava in social processes, ceremonies, ritual practices, and the invocation of the supernatural.[5]

Current

Although the effect is small, several studies suggest that kava extract is an effective treatment for anxiety for 1–4 weeks.[6] A meta-analysis of seven double-blind, random clinical trials (380 patients) on kava extract suggested a significant reduction in total scores on the Hamilton Anxiety Scale in patients receiving kava extract, when compared with placebo (mean difference 3.9, 95% CI: 0.1–7.7, $P = 0.05$).[7] Limitations of these studies included the small size of the effect and the short duration of use. An Internet-based, double-blind, placebo-controlled study of 391 subjects did not detect a statistically significant difference in scores on the State-Trait Anxiety Inventory State subtest (anxiety) or the Insomnia Severity Index (insomnia), when comparing kava treatment and placebo groups.[8] The kava treatment group received 100 mg total kavalactones 3 times daily in kava extract for 28 days. Kava is present in some chocolates and beverages as a flavoring agent, but the kavalactone content in these products is low per serving (<10–12 mg) in most of these products compared with dietary supplements.[9]

Regulatory Status

The US Food and Drug Administration (FDA) issued a warning to consumers and physicians in 2002 regarding the hepatotoxicity of kava extracts, but this product is still available in the United States. However, Germany, the United Kingdom, France, Switzerland, and Australia banned the use of herbal products containing kava extracts during the early 2000s. These bans are being reevaluated in several countries (e.g., Germany).

PRINCIPAL INGREDIENTS

Chemical Composition

Commercial kava rhizomes typically contain about 3.5–8% kavalactones (kavapyrones) calculated as kavain (kawain), while kava extracts contain approximately 30–70% kavalactones. The actual content of kavalactones in kava rhizomes varies with the plant, cultivars, cultivation techniques, season, weather, and plant part sampled.[1] Kava lactone content decreases in higher plant parts being highest in the roots and smallest in the leaves. The major kavalactones in the leaves are dihydrokavain and dihydromethysticin, whereas kavain and methysticin are the major kavalactones in the roots and rhizomes. Although both aqueous and acetone/alcohol extracts contain similar kavalactones, the efficiency of the aqueous extracts using traditional methods is about 5–10% of direct extraction by organic solvents (acetone, ethanol, methanol).[10] Furthermore, the aqueous kava root extract contains the lower proportions of kavalactones (e.g., yangonin) compared with acetone/alcohol extracts because of the low water solubility of the kavalactones.[11] The latter extracts contain relatively small amounts of highly polar constituents present in the aqueous extracts. There are at least 18 kavalactones and three chalcones in alcohol/acetone extracts of kava rhizomes.[12] Six kavalactones account for approximately 96% of the total kavalactones in extracts of the kava plant. These kavalactones contain a 4-methoxy-2-pyrone ring system. Figure 76.1 displays the chemical structure of these six kavalactones. Water extraction methods (e.g., teas from kava) produce lower concentrations of kavalactones and higher variation in individual kavalactone content, when compared with methanol extracts of kava rhizomes.[13] Ethanol extracts of kava rhizomes also contain minor amounts (i.e., about 1% by weight) of the chalcones, flavokavins A–C.[14]

Physiochemical Properties

The water solubility of kavalactones is low. Animal studies indicate that kavalactones increase the number of $GABA_A$ receptor binding sites in the brains of rats, particularly in the hippocampus and the amygdala.[15] These findings suggest that $GABA_A$ receptor binding is a potential mechanism for the sedative effects of kava

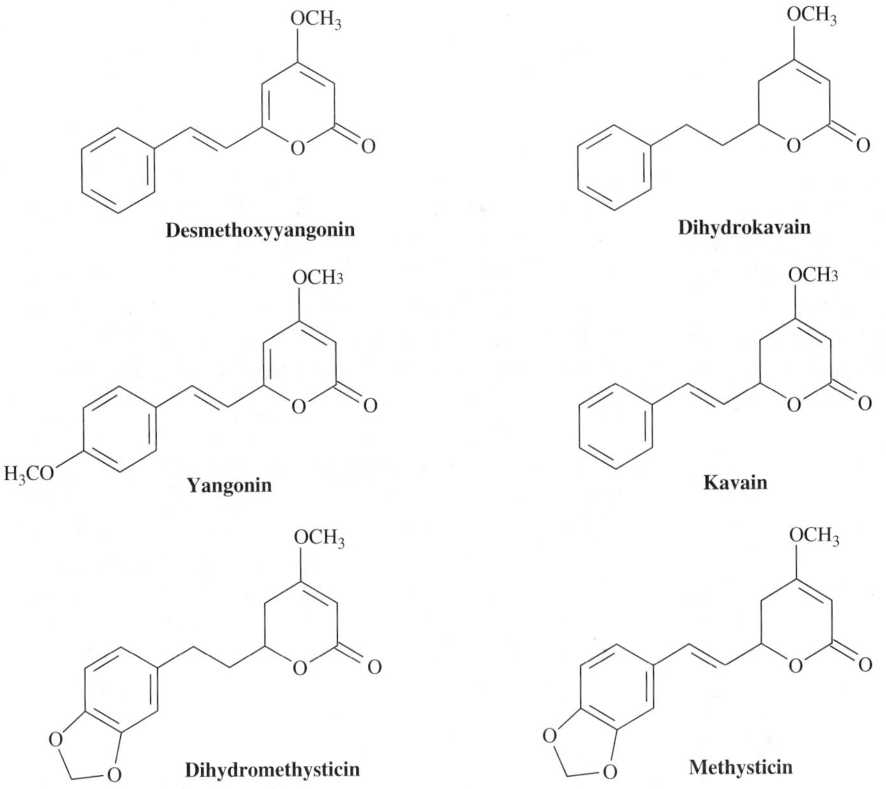

Desmethoxyyangonin

Dihydrokavain

Yangonin

Kavain

Dihydromethysticin

Methysticin

FIGURE 76.1. Chemical structures of the six common kavalactones in kava root.

extract. However, other rodent studies indicate that the binding of kavalactones to the $GABA_A$ receptor is weak; therefore, the mechanism of action of kava probably involves other paths.[16]

Mechanism of Toxicity

The exact mechanism of kava toxicity is unclear. Animal studies suggest that the anxiolytic action of kavalactones and dihydrokavain may result from $GABA_A$ receptor binding.[17] These studies also suggest that kavalactones directly alter neuronal excitability through voltage-dependent ion channels (e.g., sodium), resulting in a reduction of muscle tension.[18] The cause of the hepatotoxicity associated primarily with commercial kava extracts is unknown. Potential mechanisms of toxicity include immune-mediated idiosyncratic reaction, reduction of liver glutathione content, and drug interactions mediated by inhibition of cytochrome P450 enzymes.[19] Most case reports of hepatotoxicity involve the ingestion of multiple drugs rather than kava alone.

DOSE RESPONSE

Medicinal Use

Although kava extracts typically are standardized to 30% kavalactones, there is substantial variation between herbal preparations containing kava root extract. Many commercial herbal brands of kava use WS1490 as the ingredient of kava. The typical daily dose of kavalactones is up to about 250 mg. Doses up to 800 mg daily were tolerated over a short period without serious adverse effects.[20]

Toxicity

The mean weekly consumption of 11 very heavy users of kava beverages in Australian Aboriginal communities was about 440 g.[21] Observational studies indicate that the ingestion of large doses of aqueous kava extract (200 g powder over 14 hours) causes cerebellar dysfunction with significant reduction in cognitive abilities.[22] The rare association of kava use with hepatotoxicity suggests an idiopathic reaction that is not necessarily dose-related. However, most of the case reports of kava-associated hepatotoxicity occur in adults 20–60 years of age taking daily doses of 120–240 mg kavalactones from acetone or alcohol extracts over 4–16 weeks.[23] Liver failure requiring transplantation occurred in a 14-year old girl after the consumption of kava extract for 3 months,[24] and in another 14-year-old girl after the ingestion of an herbal preparation containing 200 mg commercial kava extract for 4 months.[25] Hepatotoxicity has been associated with the chronic consumption of kava beverages very rarely.[26]

TOXICOKINETICS

Absorption

In rodent studies, the gastrointestinal tract absorbs the kava lactone, kawain (kavain), quite well including rapid uptake of kawain and dihydrokawain by the brain.[27,28] Compared with the administration of kawain alone, the coadministration of kava extract increases the absorption of kawain two- to three-fold. In a human study (n = unreported), a single oral dose of 800 mg kavain produced peak serum free kavain concentration of 40 ng/mL within 30–60 minutes after ingestion.[29]

Biotransformation

In rats, the 5,6-dihydro-α-pyrones (dihydrokawain, kawain, methysticin) undergo variable amounts of hydroxylation, scission of the 5,6-dihydro-α-pyrone ring, and demethylenation, whereas biotransformation of the α-pyrones, 7,8-dihyroyangonin and yangonin, results from *o*-demethylation.[30] In a study of volunteers ingesting kava prepared by traditional aqueous methods, the urine contained all six major and some minor kavalactones.[31] Metabolites typically resulted from reduction of the 3,4-double bond and/or demethylation of the 4-methoxy group of the α-pyrone ring system. In contrast to rats, there were no dihydroxylated metabolites of the kavalactones in human urine as measured by methane chemical ionization gas chromatography/mass spectrometry. The main human metabolites of kavain are *p*-hydroxykavain, *p*-hydroxy-5,6-dehydrokavain and *p*-hydroxy-7,8-dihydrokavain.[29]

Elimination

Human studies suggest that the kidney excretes little unchanged kavain in the urine. *p*-Hydroxykavain is the main metabolite of kavain that is present in urine in both the free and conjugated (glucuronide, sulfate) forms.[29] 5,6-Dehydrokavain, *p*-hydroxy-5,6-dehydrokavain, and *o*-desmethyl-hydroxy-5,6-dehydrokavain are minor urinary metabolites.

Drug Interactions

In vitro studies indicate that the administration of kava extract inhibits CYP2C9, CYP2C19, CYP2D6, and CYP3A4 activity as well as the efflux transporter, P-glycoprotein, primarily as the result of the effects of kavalactones.[27,32] However, the results between *in vitro* and *in vivo* studies are somewhat inconsistent. Furthermore, the interaction varies between individual kavalactones with dihydromethysticin, methysticin, and desmethoxyyangonin demonstrating the greatest effect. Kawain and dihydrokawain had little effect on these CYP isoenzyme activities at concentrations up to 10 μM. IC_{50} is the concentration required to inhibit metabolism of surrogate substrates by 50%, and an IC_{50} <10 μM indicates potent inhibition of the particular CYP P450 isoenzyme. For *in vitro* studies the most potent kava lactone causing inhibition of CYP P450 isoenzymes was as follows: CYP1A2, desmethoxyyangonin (IC_{50} = 1.7 μM); CYP2C19, dihydromethysticin (IC_{50} = 0.43 μM); and CYP3A4, methysticin (IC_{50} = 1.49 μM).[33] In general, *in vitro* studies suggest that kavalactones do not affect the activities of CYP2E1, CYP2C8, or CYP2A6. However, an *in vivo* study of 12 healthy volunteers using single-time-point phenotypic metabolic ratios suggested that 28-day supplementation with kava root extract (1 g twice daily with no standardization claim) inhibited CYP2E1 activity approximately 40% when comparing ratios before and after supplementation.[34] This study did not detect changes in the activities of CYP1A2, CYP3A4, or CYP2D6. There are few data on the clinical significance of these potential drug interactions at usual doses of kava extracts. The action of kava may potentiate the effect of sedative-hypnotic drugs, alcohol, and barbiturates,[35] but there are no data to determine the clinical effect of standard herbal dose of kava with these drugs.

CLINICAL RESPONSE

Adverse Reactions

Adverse reactions associated with the use of commercial kava extracts include headache, malaise, rash, and stomach discomfort.[2] Case reports associate several dermatologic reactions with the use of kava extract including photosensitivity reactions, flushing, and urticaria. A 70-year-old man developed pruritic, erythematous papules and plaques on the face and trunk after sun exposure following the use of kava extracts for 3 weeks.[36] Biopsies of these plaques demonstrated lymphocytic infiltration and destruction of the associated sebaceous glands. Other reactions include delayed hypersensitivity reactions[37] and acute eczema.[38] Clinical features of acute kava intoxication include ataxia, euphoria, muscle weakness, hypoesthesias of the skin and mouth, sedation, and tremors. Although case reports associate the use of kava with sudden death in the Aboriginal community, a cross-sectional study did not demonstrate an increased risk of cardiovascular events when comparing kava and non-kava users in this population.[39] Some heavy kava drinkers develop a peculiar, dry, scaly skin eruption, which is reversible after cessation of kava use. This ichthyosiform eruption is called kava dermopathy.[40] In a nonrandom study of heavy kava drinkers, about 15% had the clinical features of kava dermopathy.[41]

Acute Effects

The ingestion of large doses of kava powder produces a state of intoxication in kava drinkers, characterized by ataxia, tremors, sedation, and blepharospasm.[22] Neuropsychological testing during this state of intoxication indicates that there is a reduction in coordination and visual attention, but performance of complex cognitive functions remains normal. A 37-year-old man developed ataxia, diaphoresis, mydriasis, severe vertigo, generalized muscle weakness, and vomiting after ingestion of a strong tea made from kava.[42] He recovered without sequelae within 4 hours after ingestion, and his serum hepatic aminotransferases remained normal.

Chronic Effects

Case reports associate the consumption of commercial kava caplets with hepatotoxicity. These reports include a few cases of acute fulminant hepatitis with extensive centrilobular necrosis and hepatic failure requiring liver transplant.[24,43] The development of hepatotoxicity has not been reported in clinical trials or following chronic use of the traditional infusion of kava roots. In a study of over 7,000 patients administered kava extract equivalent to 105–240mg kavalactones daily over 5–7 weeks, there were no cases of hepatotoxicity.[44]

DIAGNOSTIC TESTING

Analytical Methods

Thin-layer chromatographic methods with UV spectroscopy are available for the determination of kavalactones, but these methods lack precision with recoveries in the range of 80–95%.[45] Methods for the quantitation of kavalactones in kava extracts include gas chromatography (GC) and high performance liquid chromatography (HPLC), but each method has some limitations.[46] GC methods do not usually separate methysticin and yangonin, and the high injection port temperatures cause decomposition of methysticin. Normal-phase HPLC allows more rapid and better separation of constituents compared with reverse-phase HPLC, but the latter method is more sensitive. Limits of detection for kavalactones with ultrasonic extraction and isocratic reversed-phase liquid chromatography range from about 0.1–0.25 μg/g.[13] HPLC detection methods for kavalactones assays include UV (240nm and 254nm) with secondary detection by mass spectrometry or fluorescence spectroscopy. The UV detection of yangonin occurs at 355–360nm. Chiral active HPLC columns can separate enantiomers of kavalactones.[47] Reliable methods for the rapid detection of major kavalactones include near-infrared reflectance spectroscopy with partial least-squares regression.[48]

Abnormalities

Rarely, mild, subclinical elevation of serum hepatic aminotransferases occurs in some clinical studies of commercial kava extracts.[2] Elevated serum gamma glutamyl transferase, serum alkaline phosphatase, and mild elevations of serum hepatic aminotransferases occur in chronic consumers of aqueous kava drinks, in part as a result of the induction of cytochrome P450 isoenzymes.[26,49] These abnormalities resolve after 1–2 months of abstinence, and clinically significant liver damage does not usually occur.[50] Very rarely, case reports have associated the development of significant hyperbilirubinemia and markedly elevated serum hepatic aminotransferases with chronic use of aqueous kava extracts by kava drinkers.[26]

Neuropsychological testing of persons receiving kava suggest that consumption of usual doses (e.g., 1g/kg body weight) of kava powder does not produce significant impairment on test scores on short-term memory, reaction times (simple, choice), or attention (visual tracking, divided attention).[51] However, the use of kava with ethanol does modestly increase the deficits associated with blood ethanol concentrations near 100mg/dL, particularly in divided attention tasks.[52] These effects are not consistent among test subjects.

TREATMENT

The treatment of adverse reactions associated with the use of kava is supportive.

References

1. Singh YN. Kava: an overview. J Ethnopharmacol 1992;37:13–45.
2. Connor KM, Davidson JT, Churchill LE. Adverse-effect profile of kava. CNS Spectr 2001;6:848–853.
3. Bilia AR, Gallon S, Vincieri FF. Kava-kava and anxiety: growing knowledge about the efficacy and safety. Life Sci 2002;70:2581–2597.
4. Whitton PA, Lau A, Salisbury A, Whitehouse J, Evans CS. Kava lactones and the kava-kava controversy. Phytochemistry 2003;64:673–679.
5. Turner J. Substance, symbol and practice: the power of kava in Fijian society. Canberra Anthropol 1995;18: 97–118.

6. Witte S, Loew D, Gaus W. Meta-analysis of the efficacy of the acetonic kava-kava extract WS®1490 in patients with non-psychotic anxiety disorders. Phytother Res 2005;19; 183–188.
7. Pittler MH, Ernst E. Kava extract versus placebo for treating anxiety (review). The Cochrane base of Systematic Reviews 2003(1);CD003383.
8. Jacobs BP, Bent S, Tice JA, Blackwell T, Cummings SR. An internet-based randomized, placebo-controlled trial of kava and valerian for anxiety and insomnia. Medicine 2005;84:197–207.
9. de Jager LS, Perfetti GA, Diachenko GW. LC-UV and LC-MS analysis of food and drink products containing kava. Food Addit Contam 2004;21:921–934.
10. Cheng D, Lidgard RO, Duffield PH, Duffield AM, Brophy JJ. Identification by methane chemical ionization gas chromatography/mass spectrometry of the products obtained by steam distillation and aqueous acid extraction of commercial *Piper methysticum*. Biomed Environ Mass Spectrom 1988;17:371–376.
11. Cote CS, Kor C, Cohen J, Auclair K. Composition and biological activity of traditional and commercial kava extracts. Biochem Biophys Res Comm 2004;322:147–152.
12. Dharmaratne HR, Nanayakkara NP, Khan IA. Kavalactones from *Piper methysticum*, and their 13C NMR spectroscopic analyses. Phytochemistry 2002;59:429–433.
13. Hu L, Jhoo J-W, Ang CY-W. Determination of six kavalactones in dietary supplements and selected functional foods containing *Piper methysticum* by isocratic liquid chromatography with internal standard. J AOAC Int 2005;88:16–25.
14. Meissner O, Haberlein H. HPLC analysis of flavokavins and kavapyrones from *Piper methysticum* Forst. J Chromatogr B 2005;826:46–49.
15. Jussofie A, Schmiz A, Hiemke C. Kavapyrone enriched extract from *Piper methysticum* as modulator of the GABA binding site in different regions of rat brain. Psychopharmacology (Berl) 1994;116:469–474.
16. Davies LP, Drew CA, Duffield P, Johnston GA, Jamieson DD. Kava pyrones and resin: studies on $GABA_A$, $GABA_B$ and benzodiazepine binding sites in rodent brain. Pharmacol Toxicol 1992;71:120–126.
17. Yuan CS, Dey L, Wang A, Mehendale S, Xie JT, Aung HH, Ang-Lee MK. Kavalactones and dihydrokavain modulate GABAergic activity in a rat gastric-brainstem preparation. Planta Med 2002;68:1092–1096.
18. Cairney S, Maruff P, Clough AR. The neurobehavioral effects of kava. Austral N Z J Psychiatr 2002;36: 657–662.
19. Clouatre DL. Kava kava: examining new reports of toxicity. Toxicol Lett 2004;150:85–96.
20. Ulbricht C, Basch E, Boon H, Ernst E, Hammerness P, Sollars D, et al. Safety review of kava (*Piper methysticum*) by the Natural Standard Research Collaboration. Expert Opin Drug Saf 2005;4:779–794.
21. Mathews JD, Riley MD, Fejo L, Munoz E, Milns NR, Gardner ID, et al. Effects of the heavy usage of kava on physical health: summary of a pilot survey in an aboriginal community. Med J Aust 1988;148:548–555.
22. Cairney S, Maruff P, Clough AR, Collie A, Currie J, Currie BJ. Saccade and cognitive impairment associated with kava intoxication. Hum Psychopharmacol 2003;18: 525–533.
23. Gow PJ, Connelly NJ, Hill RL, Crowley P, Angus PW. Fatal fulminant hepatic failure induced by a natural therapy containing kava. Med J Aust 2003;178:442–443.
24. Campo JV, McNabb J, Perel JM, Mazariegos GV, Hasegawa SL, Reyes J. Kava-induced fulminant hepatic failure. J Am Acad Child Adolesc Psychiatry 2002;41:631–632.
25. Humberston CL, Akhtar J, Krenzelok EP. Acute hepatitis induced by kava kava. J Toxicol Clin Toxicol 2003;41: 109–113.
26. Russmann S, Barguil Y, Cabalion P, Kritsanida M, Duhet D, Lauterburg BH. Hepatic injury due to traditional aqueous extracts of kava root in New Caledonia. Eur J Gastroenterol Hepatol 2003;15:1033–1036.
27. Mathews JM, Etheridge AS, Valentine JL, Black SR, Coleman DP, Patel P, et al. Pharmacokinetics and disposition of the kavalactone kawain: interaction with kava extract and kavalactones *in vivo* and *in vitro*. Drug Metab Disp 2005;33:1555–1563.
28. Keledjian J, Duffield PH, Jamieson DD, Lidgard RO, Duffield AM. Uptake into mouse brain of four compounds present in the psychoactive beverage kava. J Pharm Sci 1988;77:1003–1006.
29. Tarbah F, Mahler H, Kardel B, Weinmann W, Hafner D, Daldrup T. Kinetics of kavain and its metabolites after oral application. J Chromatogr B Analyt Technol Biomed Life Sci 2003;789:115–130.
30. Rasmussen AK, Scheline RR, Solheim E, Hansel R. Metabolism of some kava pyrones in the rat. Xenobiotica 1979;9:1–16.
31. Duffield AM, Jamieson DD, Lidgard RO, Duffield PH, Bourne DJ. Identification of some human urinary metabolites of the intoxicating beverage kava. J Chromatogr 1989;475:273–281.
32. Weiss J, Sauer A, Frank A, Unger M. Extracts and kavalactones of *Piper methysticum* G. Forst (kava-kava) inhibit P-glycoprotein *in vitro*. Drug Metab Disp 2005;33: 1580–1583.
33. Zou L, Harkey MR, Henderson GL. Effects of herbal components on cDNA-expressed cytochrome P450 enzyme catalytic activity. Life Sci 2002;71:1579–1589.
34. Gurley BJ, Gardner SF, Hubbard MA, Williams DK, Gentry WB, Khan IA, Shah A. *In vivo* effects of goldenseal, kava kava, black cohosh, and valerian on human cytochrome P450 1A2, 2D6, 2E1, and 3A4/5 phenotypes. Clin Pharmacol Ther 2005;77:415–426.
35. Almeida JC, Grimsley EW. Coma from the health food store: interaction between kava and alprazolam. Ann Intern Med 1996;125:940–941.

36. Jappe U, Franke I, Reinhold D, Gollnick HP. Sebotropic drug reaction resulting from kava-kava extract therapy: a new entity? J Am Acad Dermatol 1998;38:104–106.
37. Schmidt P, Boehncke WH. Delayed-type hypersensitivity reaction to kava-kava extract. Contact Dermatitis 2000;42:363–364.
38. Suss R, Lehmann P. [Hematogenous contact eczema cause by phytogenic drugs exemplified by kava root extract]. Hautarzt 1996;47:459–461. [German]
39. Clough AR, Jacups SP, Wang Z, Burns CB, Bailie RS, Cairney SJ, et al. Health effects of kava use in an eastern Arnhem Land Aboriginal community. Intern Med J 2003;33:336–340.
40. Norton SA, Ruze P. Kava dermopathy. J Am Acad Dermatol 1994;31:89–97.
41. Ruze P. Kava-induced dermopathy: a niacin deficiency? Lancet 1990;335(8703):1442–1445.
42. Perez J, Holmes JF. Altered mental status and ataxia secondary to acute kava ingestion. J Emerg Med 2005;28: 49–51.
43. Centers for Disease Control and Prevention (CDC). Hepatic toxicity possibly associated with kava-containing products—United States, Germany, and Switzerland, 1999–2002. MMWR Morb Mortal Wkly Rep 2002;51: 1065–1067.
44. Connor KM, Davidson JR, Churchill LE. Adverse-effect profile of kava. CNS Spectr 2001;6:848, 850–853.
45. Csupor L. [Quantitative determination of kava lactones in *Piper methysticum* (Forster). 2. Determination of kawain, methysticin and yangonin]. Pharmazie 1970;25:197–198. [German]
46. Bilia AR, Scalise L, Bergonzi MC, Vincieri FF. Analysis of kavalactones from *Piper methysticum* (kava-kava). J Chromatogr B 2004;312:203–214.
47. Haberlein H, Boonen G, Beck MA. *Piper methysticum*: enantiomeric separation of kavapyrones by high performance liquid chromatography. Planta Med 1997;63: 63–65.
48. Gautz LD, Kaufusi P, Jackson MC, Bittenbender HC, Tang C-S. Determination of kavalactones in dried kava (*Piper methysticum*) powder using near-infrared reflectance spectroscopy and partial least-squares regression. J Agric Food Chem 2006;54:6147–6152.
49. Brown AC, Onopa J, Holck P, Kaufusi P, Kabasawa D, Craig WJ, et al. Traditional kava beverage consumption and liver function tests in a predominantly Tongan population in Hawaii. Clin Toxicol 2007:45:549–556.
50. Clough AR, Bailie RS, Currie B. Liver function test abnormalities in users of aqueous kava extracts. J Toxicol Clin Toxicol 2003;41:821–829.
51. Herberg KW. [Effect of kava-special extract WS 1490 combined with ethyl alcohol on safety-relevant performance parameters]. Blutalkohol 1993;30:96–105. [German]
52. Foo H, Lemon J. Acute effects of kava, alone or in combination with alcohol, on subjective measures of impairment and intoxication and on cognitive performance. Drug Alcohol Rev 1997;16:147–155.

Chapter 77

LAVENDER
(*Lavandula* Species)

HISTORY

In ancient Greece and Rome, lavender was an antiseptic and a fragrance for baths and cleansing agents. The ancient Egyptians used linen soaked in asphalt and oil of lavender to wrap mummies before drying the casts in the sun.[1]

BOTANICAL DESCRIPTION

For cosmetic and medicinal purposes, the most common variety is *L. angustifolia*. Other species of lavender include *Lavandula burmanni* Benth., *L. dentata* L., *L. dhofarensis* A.G. Miller, *L. lanata* Boiss., *L. latifolia* Medikus (broadleafed lavender), *Lavandula X intermedia* Emeric ex Lois. (lavandin), and *L. stoechas* L. (French lavender). Lavandin is a sterile hybrid of English lavender (*L. angustifolia*) and broadleafed lavender (*L. latifolia*). Currently, the two most common cultivars in the French lavender industry are Grosso and Super.

Common Name: English Lavender

Scientific Name: *Lavandula angustifolia* P. Miller (*L. officinalis, L. vera*)

Botanical Family: Lamiaceae (menthes, mints)

Physical Description: English lavender is an evergreen plant with a round, compact shape and opposite leaves that reach 2 in. (5 cm) in length. Small, fragrant, purple flowers appear in summer as interrupted spikes.

Distribution and Ecology: Although lavender (*Lavandula* spp.) is a native plant of the Mediterranean region, the Arabian Peninsula, Africa, and Russia, this plant is cultivated throughout the warm areas of the northern hemisphere.

EXPOSURE

Sources

Typically, lavender oil is a steam distillation product of the leaves and flowers of lavender species (*L. angustifolia, L. latifolia, L. stoechas, L. x intermedia*). For medicinal purposes, the flower is the most commonly used part of the lavender plant because the essential oil from this part of the plant is sweeter and more aromatic than the oil from other plant parts. Perfumes and cosmetic products with lavender usually contain essential oils of *Lavender* species (e.g., *L. angustifolia*) with relatively low camphor and high linalyl acetate content, whereas insect repellants with lavender typically contain essential oils from *Lavender* species (e.g., *L. stoechas*) with relative high camphor content. The constituents in Lavender products are not standardized in the United States.

Medicinal Uses

Traditional

Lavender has a long history of use as a fragrance and medicinal herb. Linen bags containing lavender were placed in pillows for soporific effects. Traditionally, this herb has been used as an expectorant, an antidepressant, carminative (smooth muscle relaxant), anti-

Medical Toxicology of Natural Substances, by Donald G. Barceloux, MD

spasmodic, sedative, antimicrobial agent, and healing agent for burns, insect bites, and other wounds.

CURRENT

The essential oils of lavender are common fragrances in a variety of products including soaps, candles, perfumes, shampoos, massage oils, colognes, and lotions. In food manufacturing, these essential oils are flavoring agents in ice cream, candy, beverages, chewing gum, and baked goods. Medicinal applications of lavender oil include use as anxiolytic agent (aromatherapy),[2] soporific agent (aromatherapy),[3,4] oral antispasmodic, topical antibiotic,[5] soothing agent for perineal discomfort after childbirth, adjunctive therapy for mild depression,[6] treatment for alopecia,[7] and chemotherapeutic agent.[8] There are few well-controlled clinical trials supporting the use of lavender for these indications. The weight of the evidence suggests a small positive effect of lavender aromatherapy for relaxation,[9] but the studies are not well designed and the strength of the evidence is weak.[1] A small, preliminary (phase I) trial suggested that a constituent of lavender (perillyl alcohol) is a safe chemotherapeutic drug in doses up to 1200 mg/m^2 4 times daily, but the efficacy of this compound as a chemotherapeutic agent remains unproven.[10]

REGULATORY STATUS

In the United States, lavender is regulated as a dietary supplement with GRAS (generally recognized as safe) status, when used in recommended amounts. Linalool, a constituent in lavender, is approved as a food additive (GRAS). This compound is used in fragrances, soaps, perfumes, beverages, ice cream, candy, gelatins, and chewing gum.

PRINCIPAL INGREDIENTS

Chemical Composition

Lavender contains over 100 compounds, including linalool, linalyl acetate, camphor limonene, tannins, triterpenes, cyclic monoterpenes (cineole, perillyl alcohol), hydroxycoumarins, and flavonoids. The main components of lavender oil are linalool (CAS RN: 78-70-6, 3,7-dimethyl-1,6-octadien-3-ol), linalyl acetate, 1,8-cineole (eucalyptol), β-ocimene, terpinen-4-ol, and camphor.[11] Figure 77.1 displays the chemical structure of linalyl acetate (CAS RN: 115-95-7, $C_{12}H_{20}O_2$). In a study of nine Australian samples of lavender essential oil derived from steam distillation, the approximate concentrations of the main components were as follows: linalool (23–57%), linalyl acetate (4–35%), 1,8-cineole (0.1–20%), β-ocimene (27%), terpinen-4-ol (0.1–3%), and camphor (0.1–7%).[12] Major *volatile* components of the leaves and flowers of *Lavandula* species include camphor, 1,8-cineole, and compounds with a necrodane structure, as measured by direct thermal desorption/gas chromatography/mass spectrometry.[13]

The chemical composition and fragrance of lavender essential oil depends on the extraction process, the cultivars, cultural practices, growing conditions, harvesting season, and species. Table 77.1 demonstrates the variability in linalyl acetate, linalool, camphor, and cineol content in four *Lavandula* species. Analysis of a sample of *L. angustifolia* essential oil by solid-phase trapping solvent extraction demonstrated linalyl acetate concentrations of 35.44% compared with 4.04% for reduced pressure steam distillation and 2.63% for simultaneous steam distillation-solvent extraction.[14] For linalool, the yields for these three extraction methods were 18.7%, 36.8%, and 43.5%, respectively. Analysis of a Turkish sample of lavender essential oil (*L. stoechas*) demonstrated the following major constituents: pulegone

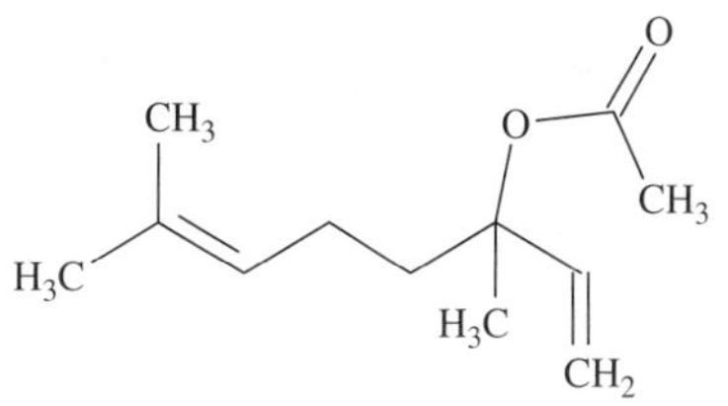

FIGURE 77.1. Chemical structure of linalyl acetate.

TABLE 77.1. Concentration of Major Components in Four Lavandula Species as Determined by Solid-Phase Trapping Solvent Extraction and Gas Chromatography/Mass Spectrometry

Chemical	*L. stoechas*[a]	*L. dentate*[a]	*L. angustifolia*[a]	*L. heterophylla*[a]
Linalyl acetate	0.5 ± 1.6	ND	35.4 ± 3.7	ND
Linalool	0.1 ± 1.8	4.4 ± 2.7	18.7 ± 2.1	1.4 ± 2.0
Camphor	53.4 ± 2.9	5.7 ± 4.1	0.5 ± 5.3	20.4 ± 9.5
Cineol	12.5 ± 1.9	47 ± 0.6	5.9 ± 6.3	50.1 ± 2.5

Source: From Ref 14.

[a]Normalized peak area (%) ± relative standard deviation (RSD) for three samples by solid-phase trapping solvent extraction for each *Lavandula* species; ND, not detected.

(40.4%), menthol (18.1%), and menthone (12.6%).[15] Typically, oils derived *L. lanata* and *L. stoechas* have higher camphor and lower terpene (e.g., β-phellandrene) and sesquiterpene (e.g., caryophyllene) concentrations than oils derived from *L. angustifolia* or *L. dentata*.[16] The concentrations of linalool and linalyl acetate are higher in the essential oil than in volatile fraction of lavender, whereas sesquiterpene compounds are higher in the volatile fraction than in the essential oils.

Physiochemical Properties

In studies on rats, constituents (e.g., cineole, perillyl alcohol) in lavender essential oil reduced cholesterol synthesis.[17] Linalool, linalyl acetate, and *L. angustifolia* essential oil (0.25% volume/volume) are cytotoxic to human skin cells (endothelial cells, fibroblasts) *in vitro*.[18] Table 77.2 lists some of the physical properties of linalyl acetate.

Mechanism of Toxicity

The active ingredient and mechanism of action of lavender oil is not well defined. Lavender essential oil has some spasmolytic activity on guinea pig ileum and rat uterus *in vitro* along with reduction in skeletal muscle tone in phrenic nerve-diaphragm preparations.[19] However, the clinical significance and mechanism of action remain poorly defined.

DOSE RESPONSE

Recommended adult doses of lavender include the following: tea, steeping 10 g (10 mL) of the leaves in 250 mL boiling water for 15 minutes; aromatherapy, inhalation of the vapors from 2–4 drops in 2–3 cups boiling water; message therapy, 1–4 drops per 15 mL of base oil; bath additive, 6 drops of lavender oil in bath.[1] In refractory cancer patients receiving perillyl alcohol as a chemotherapeutic agent, the maximum tolerated dose was 1200 mg/m^2 4 times daily.[8] Gastrointestinal toxicity (nausea, vomiting, satiety, eructation) was the dose-limiting factor.

TABLE 77.2. Physical Properties of Linalyl Acetate

Physical Property	Value
Melting Point	<25 °C/77 °F
Boiling Point	220 °C/428 °F
log P (Octanol-Water)	3.93
Water Solubility	8.200 mg/L (25 °C/77 °F)
Henry's Law Constant	1.74E-03 atm-m^3/mole (25 °C/77 °F)
Atmospheric OH Rate Constant	1.16E-10 cm^3/molecule-second (25 °C/77 °F)

TOXICOKINETICS

Lavender contains a variety of compounds, and the active ingredients have not been identified. Animal studies suggest that linalool is rapidly absorbed through the gastrointestinal tract and undergoes excretion in the bile and in the urine after metabolism.[20] The absorption of perillyl alcohol is relatively rapid with peak plasma concentrations occurring within 2 hours following ingestion.[21] Metabolites of limonene and perillyl alcohol include perillic acid, *cis*-dihydroperillic acid, and *trans*-dihydroperillic acid. In clinical trials using doses of perillyl alcohol up to 1600 mg/m^2 4 times daily, the major metabolite was perillic acid and <1% of the dose of perillyl alcohol appeared in the urine unchanged.[8] Lavender contains a variety of coumarin compounds, and theoretically these compounds could increase the effect of anticoagulants, but the clinical significance of this potential interaction remains unclear. Theoretically, lavender oil or aromatherapy could increase drowsiness caused by some drugs (e.g., benzodiazepines, narcotics, ethanol), but the clinical significance of this potential interaction is also undetermined.

CLINICAL RESPONSE

Side effects of the use of lavender oil include headache, nausea, vomiting, anorexia, constipation, chills, drowsiness, and confusion. Occasional case reports indicate that sensitization to lavender can occur after the use of lavender as a soporific agent on pillows (acute contact dermatitis, eczema),[22] as a shampoo by hairdressers,[23] or as aromatherapy (allergic airborne contact dermatitis).[24] Clinical features of these allergic reactions include erythema, lichenification, eczema, vesiculation, and local irritation.[25,26] Case reports also associate the chronic use of lavender oil with photosensitization[27] and alteration of skin pigments.[28] In general, local allergic reactions to lavender oil is relatively rare. However, a Japanese review of the positivity rate of patch tests to lavender oil indicated that this rate increased as lavender aromatherapy became more popular.[29]

DIAGNOSTIC TESTING

Extraction techniques for the determination of constituents in lavender essential oil include hydrodistillation, supercritical-fluid extraction, headspace solid-phase microextraction, solid-phase trapping solvent extraction, reduced pressure steam distillation, and simultaneous steam distillation-solvent extraction. Solid-phase trapping solvent extraction is a superior extraction technique because of effectiveness of this technique for the determination of linalyl acetate, which is an important measure of the quality of lavender essential oil.[14] Quantitation of constituents of lavender oil typically involves gas chromatography/mass spectrometry (GC/MS).[13,14] The use of multidimensional GC allows the qualitative and quantitative determination of enantiomers of various constituents in lavender.[12]

TREATMENT

Treatment of adverse reactions to lavender oil is supportive.

References

1. Basch E, Foppa I, Liebowitz R, Nelson J, Smith M, Sollars D, Ulbricht C. Lavender (*Lavandula angustifolia* Miller). J Herb Pharmacother 2004;4:63–78.
2. Field T, Diego M, Hernandez-Reif M, Cisneros W, Feijo L, Vera Y, Gil K. Lavender fragrance cleansing gel effects on relaxation. Int J Neurosci 2005;115:207–222.
3. Lehrner J, Marwinski G, Lehr S, Johren P, Deecke L. Ambient odors of orange and lavender reduce anxiety and improve mood in a dental office. Physiol Behav 2005;86: 92–95.
4. Lewith GT, Godfrey AD, Prescott P. A single-blinded, randomized pilot study evaluating the aroma of *Lavandula angustifolia* as a treatment for mild insomnia. J Altern Complement Med 2005;11:631–637.
5. Takarada K, Kimizuka R, Takahashi N, Honma K, Okuda K, Kato T. A comparison of the antibacterial efficacies of essential oils against oral pathogens. Oral Microbiol Immunol 2004;19:61–64.
6. Akhondadeh S, KIashani L, Fotouhi A, Jarvandi S, Mobaseri M, Moin M, et al. Comparison of *Lavandula angustifolia* Mill. tincture and imipramine in the treatment of mild to moderate depression: a double-blind, randomized trial. Prog Neuro-Psychopharmacol Psychiatr 2003;27: 123–127.
7. Hay IC, Jamieson M, Ormerod AD. Randomized trial of aromatherapy. Successful treatment for alopecia areata. Arch Dermatol 1998;134:1349–1352.
8. Ripple GH, Gould MN, Arzoomanian RZ, Alberti D, Feierabend C, Simon K, et al. Phase I clinical and pharmacokinetic study of perillyl alcohol administered four times a day. Clin Cancer Res 2000;6:390–396.
9. Buchbauer G, Jirovetz L, Jager W, Dietrich H, Plank C. Aromatherapy: evidence for sedative effects of the essential oil of lavender after inhalation. Z Naturforsch [C] 1991;46:1067–1072.
10. Ripple GH, Gould MN, Stewart JA, Tutsch KD, Arzoomanian RZ, Alberti D, et al. Phase I clinical trial of perillyl alcohol administered daily. Clin Cancer Res 1998;4: 1159–1164.
11. Flores G, Blanch GP, Ruiz del Castillo ML, Herraiz M. Enantiomeric composition of studies in *Lavandula* species using supercritical fluids. J Sep Sci 2005;28: 2333–2338.
12. Shellie R, Mondello L, Marriott P, Dugo G. Characterisation of lavender essential oils by using gas chromatography-mass spectrometry with correlation of linear retention indices and comparison with comprehensive two-dimensional gas chromatography. J Chromatogr A 2002;970:225–234.
13. Sanz J, Soria AC, Garcia-Vallejo MC. Analysis of volatile components of *Lavandula luisieri* L. by direct thermal desorption-gas chromatography-mass spectrometry. J Chromatogr A 2004;1024:139–146.
14. Kim N-S, Lee D-S. Comparison of different extraction methods for the analysis of fragrances from *Lavandula* species by gas chromatography-mass spectrometry. J Chromatogr A 2002;982:31–47.
15. Goren AC, Topcu G, Bilsel G, Bilsel M, Aydogmus Z, Pezzuto JM. The chemical constituents and biological activity of essential oil of *Lavandula stoechas* spp. *stoechas*. Z Naturforsch [C] 2002;57:797–800.
16. Cavanagh HM, Wilkinson JM. Biological activities of lavender essential oil. Phytother Res 2002;16:301–308.
17. Clegg RJ, Middleton B, Bell GD, White DA. The mechanism of cyclic monoterpene inhibition of hepatic 3-hydroxy-3-methylglutaryl coenzyme A reductase *in vivo* in the rat. J Biol Chem 1982;257:2294–2299.
18. Prashar A, Locke IC, Evans CS. Cytotoxicity of lavender oil and its major components to human skin cells. Cell Prolif 2004;37:221–229.
19. Lis-Balchin M, Hart S. Studies on the mode of action of the essential oil of lavender (*Lavandula angustifolia* P. Miller). Phytother Res 1999;13:540–542.
20. Parke DV, Qauddusur Rahman KM, Walker R. The absorption, distribution and excretion of linalool in the rat. Biochem Soc Trans 1974;2:612–615.
21. Zhang Z, Chen H, Chan KK, Budd T, Ganapathi R. Gas chromatographic-mass spectrometric analysis of perillyl alcohol and metabolites in plasma. J Chromatogr B 1999;728:85–95.
22. Coulson IH, Ali Khan AS. Facial "pillow" dermatitis due to lavender oil allergy. Contact Dermatitis 1999;41: 111.
23. Brandao FM. Occupational allergy to lavender oil. Contact Dermatitis 1986;15:249–250.
24. Schaller M, Korting HC. Allergic airborne contact dermatitis from essential oils used in aromatherapy. Clin Exp Dermatol 1995;20:143–145.
25. Rademaker M. Allergic contact dermatitis from lavender fragrance in Difflam® gel. Contact Dermatitis 1994;31: 58–59.

26. Varma S, Blackford S, Statham BN, Blackwell A. Combined contact allergy to tea tree oil and lavender Oil complicating chronic vulvovaginitis. Contact Dermatitis 2000;42:309–310.
27. Fisher AA. Patch testing with perfume ingredients. Contact Dermatitis 1975;1:166–168.
28. Nakayama H, Harada R, Toda M. Pigmented cosmetic dermatitis. Int J Dermatol 1976;15:673–675.
29. Sugiura M, Hayakawa R, Kato Y, Sugiura K, Hashimoto R. Results of patch testing with lavender oil in Japan. Contact Dermatitis 2000;43:157–160.

Chapter 78

LICORICE
(*Glycyrrhiza* Species)

HISTORY

The root of *Glycyrrhiza glabra* has been used at least for three millennia.[1] The Tomb of King Tut contained substantial amounts of licorice root. In ancient Egypt, a sweet licorice drink (mai sus) was prepared as part of a ritual ceremony to honor the spirits of the pharaohs. The army of Alexander the Great used licorice root as means to quench thirst, and Hippocrates used this root to heal wounds and sore throats. Theophrastus, the successor of Aristotle in the Peripatetic school, promoted the use of licorice root during the 4th–3rd century BC for the treatment of nonproductive cough, asthma, and other respiratory diseases, as well as to quench thirst.[2] His medical writing had a substantial impact on the practice of Medieval medicine. In the *Naturalis Historia*, Pliny the Elder mentioned the use of licorice in reducing hunger and thirst as well as treating asthma and sterility in women.[3] Celsus used licorice root to treat renal calculi, whereas the Latin medical writer, Marcellus Empiricus (5th century AD, Roman Gaul) used this herb for a variety of stomach, pulmonary, intestinal, renal, and back ailments. The first Chinese dispensary, Shen Nong Ben Cao Jing, contained references to the herbal use of licorice. During the Middle Ages, licorice root was a treatment for hypotension. In 1946, the Dutch physician F. E. Revers used the crude extract of licorice for the treatment of stomach ulcers. At that time, he noted that some of the treated patients developed pedal edema and weight gain.[4]

BOTANICAL DESCRIPTION

There are approximately 30 species of the genus *Glycyrrhiza* with many of these species inhabiting China. The *Glycyrrhiza* species commonly used for herbal therapy include cultivated licorice (*Glycyrrhiza glabra* L.) and Chinese licorice (*Glycyrrhiza uralensis* Fisch.). *G. lepidota* Pursh is another North American licorice species, which inhabits the Great Plains and western United States.

Common Name: Common Licorice, Cultivated Licorice, Sweetwood, Sweet Root

Scientific Name: *Glycyrrhiza glabra* L. (*Liquiritae officinalis*)

Botanical Family: Fabaceae (pea)

Physical Description: This shrub grows to 4–5 ft (~1.0–1.5 m) in height with oval leaflets, white-purplish flowers in clusters, and flat pods. The main taproot is yellow, soft and fibrous.

Distribution and Ecology: *G. glabra* is native species of southwest Asia and the Mediterranean region. In Europe (Spain, Italy, France), this species has been cultivated since at least the 16th century.

Common Name: Chinese Licorice, Gan Cao

Scientific Name: *Glycyrrhiza uralensis* Fisch.

Botanical Family: Fabaceae (Pea)

Medical Toxicology of Natural Substances, by Donald G. Barceloux, MD

Physical Description: This plant forms large clumps up to 4 ft (~1 m) in height with clusters of small lilac flowers that produce reddish pods. The flowers appear June to August, and the seeds ripen from July to October.

Distribution and Ecology: *G. uralensis* is a native of central Asia, China, and Japan.

EXPOSURE

Sources

Radix Glycyrrhizae consists of the dried roots and rhizomes of varieties of *Glycyrrhiza glabra* L. (common licorice) or of *Glycyrrhiza uralensis* Fisch. (Chinese licorice). The dried roots of *Glycyrrhiza* species are collected in autumn, and the fluid and dry extracts are prepared from powdered and finely cut licorice roots. Licorice contains glycyrrhizic acid, which is a sweet tasting glycoside used as a food additive and an herbal medicine. Various types of consumer products containing licorice include some sweets, chewing gum, cigarettes, chewing tobacco, and some alcoholic beverages, but the majority of the licorice in consumer items now occurs in tobacco products.[5] The flavoring agent in most licorice candies currently is anise oil. The licorice used in the manufacture of confectionery products is a semisolid extract obtained by treating dried roots of *Glycyrrhiza glabra* with boiling water and drying the extract. Herbal sources of licorice include laxatives, licorice tea, and Chinese medicines.

Medicinal Uses

Traditional

Chinese licorice (*G. uralensis*) is one of the most common herbs in traditional Chinese medicine that is frequently administered with other herbs.[6] Uses of Chinese licorice include anti-inflammatory agent, expectorant, carminative, and the treatment of skin eruptions (atopic dermatitis, pruritus, cysts, eczema) and allergies.[7] Other uses include the treatment of ulcers in the gastrointestinal tract, abdominal pain, Addison's disease, and a variety of infections (pharyngitis, malaria, tuberculosis, abscesses). In Ayurvedic medicine, licorice was a remedy for constipation, stomach ulcers, arthritis, and ocular diseases.

Current

Intravenous glycyrrhizic acid is an experimental treatment for reducing serum hepatic aminotransferases and the risk of progression to hepatocellular carcinoma in patient with chronic hepatitis C.[8] Previously, licorice had undergone clinical trials as a treatment for stomach ulcers. However, there is no clear evidence that licorice promotes the healing of stomach ulcers. In a randomized, placebo-controlled clinical trial of 96 patients with gastric ulcers, there were no differences in the healing rates between the treatment group (deglycyrrhizinated licorice) and the placebo group over 4 weeks.[9]

Regulatory Status

In the United States, licorice is used as flavoring and sweetening agents in food products, and the use of licorice is controlled as a food supplement. The US Food and Drug Administration (FDA) approved the use of licorice as a food additive (GRAS, generally recognized as safe).[10] The German Commission E approved the use of licorice for the treatment of peptic ulcers in daily doses of 200–600 mg glycyrrhizic acid. Treatment is limited to 5 weeks, and the use of licorice by patients with cardiovascular or renal disease is discouraged unless closely supervised by a physician.

PRINCIPAL INGREDIENTS

Chemical Composition

Licorice root contains triterpenes (glycyrrhizic acid), flavonoids (glabridin, glabrene, hispaglabridin A and B), polyamines, coumarins, polysaccharides, volatile oils, bitter principles, sterol, phytoestrogens, and vitamins.[11] The typical concentration of glycyrrhizic acid (glycyrrhizin) in licorice root is about 2–5%, but the amount of this compound can range up to approximately 15% depending on the plant species, climate, season, and location. In 1-year-old roots, the glycyrrhizic acid content increased substantially from October to November, but there was little change in this compound during May to August in 3-year-old plants.[12] The concentration of glycyrrhizic acid in licorice root and licorice extract is typically about 40 times higher than the pharmacologically active metabolite, glycyrrhetinic acid.[13] Polyphenol compounds account for about 1–5% of the dried licorice root with the main phenolic compounds being liquiritin (liquiritoside) and liquiritigenin and their chalcone-type derivatives, isoliquiritin and isoliquiritigenin.[14] In a study of nine samples of Italian licorice (*Glycyrrhiza glabra*), the yield of glycyrrhizic acid in methanol extracts of licorice ranged from 0.425–2.08%, whereas the yield of isoliquiritigenin ranged from 0.138–0.316%.[15] The active toxicological ingredient in licorice is the procompound, glycyrrhizic acid, which undergoes hydrolysis to the pharmacologically active compound, glycyrrhetinic acid (glycyrrhetic acid). The concentration of glycyrrhi-

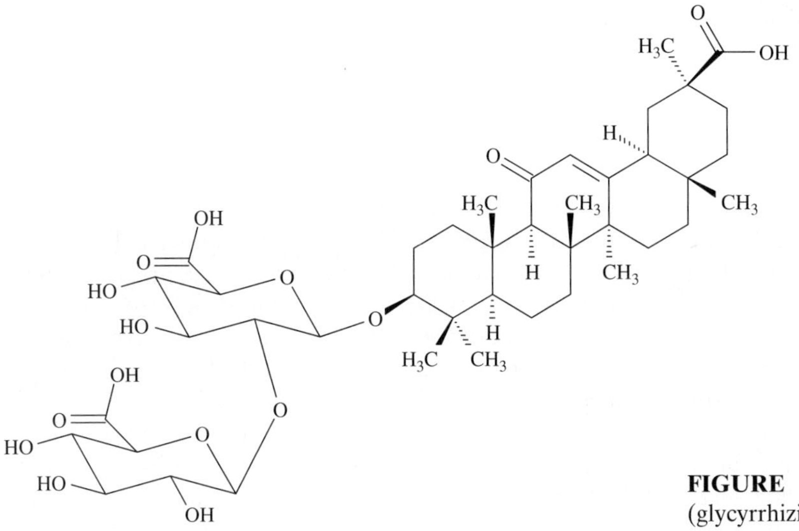

FIGURE 78.1. Chemical structure of glycyrrhizic acid (glycyrrhizin).

zic acid in the confectionary products ranged from 0.26 mg/g to 7.9 mg/g in a study of the glycyrrhizic acid (glycyrrhizin) content in 42 samples of English licorice-containing confectionery, health products, and raw materials, as measured by high performance liquid chromatography (HPLC).[16] The glycyrrhizic acid in health care products ranged from 0.3 mg/g to 47.1 mg/g in the throat pearls. The glycyrrhizic acid (glycyrrhizin) concentration in six samples of licorice root ranged from 22.2 mg/g to 32.3 mg/g. The highest glycyrrhizic acid concentrations occurred in samples of licorice extract powder (79 mg/g to 113 mg/g). Figure 78.1 displays the chemical structure of glycyrrhizic acid (CAS RN: 1405-86-3, $C_{42}H_{62}O_{16}$). Presystemic hydrolysis of glycyrrhizic acid by intestinal bacteria produces glycyrrhetinic acid, which has mineralocorticoid properties. Figure 78.2 displays the chemical structure of glycyrrhetinic acid (CAS RN: 471-53-4, $C_{30}H_{46}O_4$).

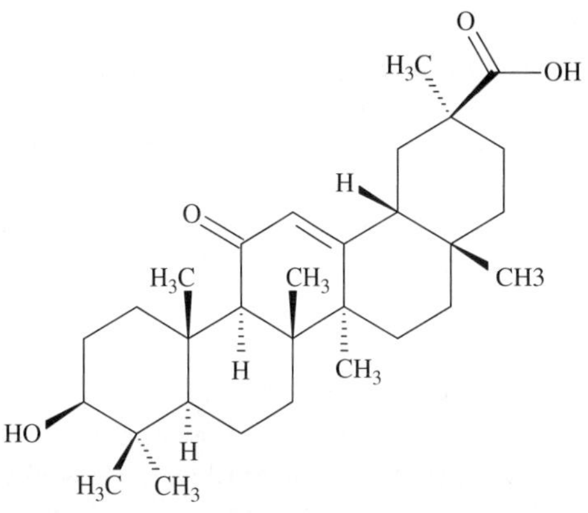

FIGURE 78.2. Chemical structure of glycyrrhetinic acid.

Physiochemical Properties

Glycyrrhizic acid accounts for the sweet taste of licorice, being about 50–170 times sweeter than sugar.[17] Table 78.1 lists some of the physical properties of glycyrrhizic acid.

TABLE 78.1. Physical Properties of Glycyrrhizic Acid

Physical Property	Value
Melting Point	220 °C/428 °F
log P (Octanol-Water)	2.8
Water Solubility	0.053 mg/L (25 °C/77 °F)
Vapor Pressure	5.22E-34 mmHg (25 °C/77 °F)
Henry's Law Constant	1.23E-35 atm-m³/mole (25 °C/77 °F)
Atmospheric OH Rate Constant	2.30E-10 cm³/molecule-second (25 °C/77 °F)

Mechanism of Toxicity

The chronic ingestion of large doses of licorice produces hypertension, hypokalemia, and pseudohyperaldosteronism in some patients. Cortisol is the active mineralocorticoid that binds to mineralocorticoid receptors in the renal cortex and salivary glands. The two 11-β-hydroxysteroid dehydrogenase isozymes catalyze the interconversion of cortisol and cortisone. Type I 11-β-hydroxysteroid dehydrogenase converts inactive cortisone to the active cortisol in the liver and adipose tissues. The other isoenzyme, type II 11-β-hydroxysteroid dehydrogenase, rapidly transforms the active mineralocorticoid, cortisol, to the inactive metabolite, cortisone. This biotransformation allows aldosterone, which is not metabolized by this enzyme, to bind to the mineralocorticoid receptors. The absorption of glycyrrhetinic acid after the ingestion of licorice inhibits the activity of type II 11-β-hydroxysteroid dehydrogenase. The aglycone, glycyrrhetinic acid, is approximately 200–1000 times a more potent inhibitor of 11-β-hydroxy-steroid dehydrogenase than the parent compound, glycyrrhizic acid.[18,19] This inhibition allows excess amounts of cortisol to bind to the renal mineralocorticoid receptors in the collecting ducts, resulting in excessive stimulation and a state of apparent mineralocorticoid excess. Although the secretion of aldosterone is suppressed, the excess mineralocorticoid activity causes a state of pseudohyperaldosteronism similar to primary hyperaldosteronism. Licorice ingestion increases the plasma atrial natriuretic peptide activity in healthy subjects, probably as a response to sodium retention and volume expansion. Following the ingestion of 100g licorice daily for 8 weeks by 12 healthy volunteers, the mean plasma atrial natriuretic peptide concentration increased about 80% compared with baseline.[20]

Glycyrrhetinic acid also inhibits the activity of 17-hydroxysteroid dehydrogenase and of 17–20 lyase. The former catalyzes the conversion of androstenedione to testosterone, whereas the latter converts 17-hydroxyprogesterone into androstenedione. After one week of receiving 7g licorice daily, the serum testosterone concentration in 17 healthy volunteers decreased about 25%.[21] However, the effect of licorice on 17-hydroxysteroid dehydrogenase and 17–20 lyase is relatively weak and transient, and the clinical significance of this reduction remains unclear because, even with the decrease, the serum testosterone concentrations generally remain within the normal range.[22] Glycyrrhetinic and glycyrrhizic acid do not bind to estrogen receptors,[23] but licorice ingestion can reduce plasma prolactin, and potentially cause gynecomastia.[24]

DOSE RESPONSE

Typical daily doses of licorice range from about 1–5g (2–10mg glycyrrhizinic acid based on 2% content). The acute ingestion of licorice containing an estimated 50mg glycyrrhizinic acid/kg body weight by a 35-year-old man was associated with hypertension (260/130mmHg) and vertigo that resolved within 24 hours.[25] Excessive intake of licorice can cause sodium retention and potassium loss, edema, increased blood pressure and depression of the renin–angiotensin–aldosterone system. There is substantial individual variation in the response to the ingestion of glycyrrhizic acid. The chronic daily dose of licorice in patients with signs of mineralocorticoid excess ranges from 1.5–250g licorice.[26] The typical daily dose of licorice ingestion in patients with severe hypokalemia is 100–200g over many months.[27] Predisposing factors associated with the development of licorice-induced hypokalemia at lower doses of chronic licorice ingestion include familial periodic paralysis, chronic diarrhea or laxative abuse, and chronic diuretic use without potassium replacement.[28]

The effect of licorice ingestion on blood pressure is dose-related. In a study of healthy volunteers, the ingestion of 50g licorice (75mg glycyrrhetinic acid) daily for 2 weeks increased the systolic blood pressure about 3mmHg compared with baseline measurements.[29] For subjects receiving 200g licorice (540mg glycyrrhetinic acid) daily for 2 weeks, the systolic blood pressure increased about 14mmHg at the end of 2 weeks of licorice use, when compared with baseline measurements. The maximum increase in systolic blood pressure occurred within about 2 weeks, and the systolic blood pressure in the low dose group (50g licorice) was not statistically different from baseline when licorice use was continued for 2 additional weeks. These studies suggest that the effect of licorice is greater on hypertensive subjects than normotensive subjects. Daily doses of 100g licorice (150mg glycyrrhetinic acid) produced a mean increase of 15mmHg in the systolic blood pressure of hypertensive subjects compared with a mean increase of approximately 3–4mmHg in normotensive subjects.[30] This difference in mean systolic blood pressure occurred despite similar cortisol/cortisone ratios between the two groups.

TOXICOKINETICS

Absorption

Following ingestion of licorice, the glycyrrhizic acid in the licorice undergoes presystemic hydrolysis primarily in the large intestine to glycyrrhetinic acid as a result of specialized glycyrrhizin β-glucuronidase activity by anaerobic intestinal bacteria (*Eubacterium* spp., *Ruminococcus* spp., *Clostridium innocuum*).[31] Common β-glucuronidases (e.g., *Escherichia coli*) are unable to hydrolyze glycyrrhizic acid.[32] After oral therapeutic doses of glycyrrhizic acid ranging from 100–800mg, this compound is not usually detectable in plasma.[33,34] Capacity-limited carriers transport glycyrrhetinic acid to the liver, where glycyrrhetinic acid undergoes conjugation to glucuronides and sulfates. The excretion of these conjugates in the bile and the subsequent hydrolysis of these conjugates by commensal bacteria results in the reabsorption and enterohepatic recirculation of glycyrrhetinic acid. Peak plasma concentrations of this compound occurs about 8–12 hours after the ingestion of glycyrrhizic acid.[35,36] In a study of six healthy adult volunteers ingesting approximately 200mg glycyrrhizic acid in licorice candy, the mean peak plasma glycyrrhetic acid concentration of 570 ng/mL occurred about 10 hours after ingestion.[37] For healthy volunteers receiving single doses of 500 mg glycyrrhetinic acid orally, the mean peak plasma glycyrrhetinic acid concentration of 4.5 ± 0.6 μg/mL occurred at 3.1 ± 0.3 hours.[38] In comparison, the mean peak plasma concentration of glycyrrhetinic acid was 9.0 ± 2.6 μg/mL at 4.3 ± 0.5 hours following the ingestion of 1500 mg glycyrrhetinic acid.

Biotransformation

The liver metabolizes glycyrrhetinic acid primarily to glucuronide and sulfate conjugates before excretion of these compounds into the bile.

Elimination

The enterohepatic recirculation of glycyrrhetinic acid causes a substantial delay in the terminal plasma elimination half-life of this compound. The plasma elimination rate of glycyrrhetinic acid is dose-dependent, and the accumulation of glycyrrhetinic acid may occur following multiple doses as a result of enterohepatic recirculation of this compound.[39] In a study of human volunteers receiving single doses of 1,000 mg and 1,500 mg glycyrrhetinic acid, the mean terminal plasma elimination half-life of glycyrrhetinic acid was 11.5 ± 1.2 hr and 38.7 ± 10.5 hours, respectively.[38]

Drug Interactions

In vitro studies indicate that flavonoids (e.g., glabridin) in licorice extract inhibit CYP3A4, CYP2C9, and CYP2B6 activities.[40] In an *in vitro* study of inhibitory activity of an extract of *Glycyrrhiza uralensis* on human CYP3A4, licopyranocoumarin, liquiritin apioside, and liquiritin demonstrated the highest inhibitory activity.[41] Glycyrrhetinic acid inhibits 5α-, 5β-reductase and 11-β-dehydrogenase, and these actions potentially decrease the inactivation of steroids. However, there are few clinical data to determine the clinical significance of these potential interactions. The use of licorice reduces the effectiveness of spironolactone, and patients with congestive heart failure (e.g., digoxin) or hypertension (e.g., diuretics) should be monitored closely for hypokalemia during the use of licorice.

CLINICAL RESPONSE

There are few data on the acute ingestion of licorice. A 35-year-old man developed hypertension and vertigo after the ingestion of licorice drops that resolved within 24 hours without sequelae.[25] Chronic, high-dose licorice consumption can cause a syndrome similar to primary hyperaldosteronism, characterized by hypertension, hypokalemia, metabolic alkalosis, low plasma renin activity, reduced plasma aldosterone concentrations, and normal urinary cortisol excretion.[42,43] Clinical features of this apparent mineralocorticoid excess include headache, fatigue, generalized muscle weakness, myalgias, edema, and myoclonus.[44] In cases associated with severe hypokalemia, rhabdomyolysis, renal dysfunction, and muscle paralysis can develop.[45,46] Rare case reports suggest that a substantial elevation of blood pressure can occur in sensitive individuals after the chronic ingestion of licorice.[47] A 38-yearold woman developed blood pressure up to 250/110 mmHg after the chronic ingestion of large doses of licorice candies.[48] After cessation of the ingestion of licorice, she became normotensive without the use of antihypertensive agents. Several case reports associated the chronic use of relatively low doses of licorice with the development of hypertension, headache, vomiting, focal weakness, and alteration of consciousness.[49,50] However, the changes in mentation did not immediately resolve after the normalization of the blood pressure, and complete resolution of some neurological abnormalities (slight hemiparesis, paresthesias) required 3–5 months.

Neither licorice nor glycyrrhizic acid are listed by the International Agency for Research on Cancer (IARC) or the US National Toxicology Program as suspected carcinogens. A cross-sectional study of Finnish women suggests the possibility of an association between heavy licorice consumption (>500 mg/week) and preterm delivery (OR = 3.07, 95% CI: 1.17–8.05), when compared with mothers reporting low licorice consumption (<250 mg/week).[51] There was no statistically significant association between licorice consumption and maternal blood pressure or birth weight.[52] The possible role of licorice, if any, in preterm deliveries requires further clarification.

DIAGNOSTIC TESTING

Analytical Methods

Methods for the detection of glycyrrhetinic and glycyrrhizinic acids include high performance liquid chromatography (HPLC) with UV detection,[53] capillary zone electrophoresis with diode array detection,[54] micellar electrokinetic chromatography,[55] and liquid chromatography/tandem mass spectrometry (LC/MS/MS).[56] The lower limit for quantitation of glycyrrhizinic and glycyrrhetinic acids by LC/MS/MS with solid phase extraction was 10 ng/mL in 0.2 mL plasma samples compared with 10 ng/mL and 20 ng/mL, respectively, for HPLC with isocratic elution.[13] The limits of quantitation are higher for capillary zone electrophoresis (5 μg/mL and 2.5 μg/mL, respectively),[54] whereas the limit of quantitation for glycyrrhetinic acid using solid-phase extraction with HPLC was 50 ng/ml.[57] The analytes in these samples were stable at room temperature for at least 24 hours.

Biomarkers

Measurements of plasma glycyrrhetinic acid concentrations confirm licorice ingestion. Case reports suggest that glycyrrhetinic acid persists in the urine at least one week after the cessation of chronic, high-dose licorice use.[58] Alternately, elevation of the ratio of cortisol to cortisone in saliva or urine also suggests the use of licorice and the subsequent inhibition of type II 11-β-hydroxysteroid dehydrogenase. In a placebo-controlled study of 20 normotensive subjects, the mean cortisol/cortisone ratios in arterial plasma at baseline levels 90–150 minutes after placebo was 4.9 ± 1.2 compared with 12.3 ± 3.4 during the same time period after the administration of 500 mg glycyrrhetinic acid.[59] This ratio is related to the dose of glycyrrhetinic acid ingested. In a study of healthy volunteers, the daily ingestion of 1,500 mg glycyrrhetinic acid produced consistent inhibition of type II 11-β-hydroxysteroid dehydrogenase, whereas the daily ingestion of 500 mg glycyrrhetinic acid caused transient inhibition.[38] Urine cortisol excretion and the cortisol/cortisone ratio gradually normalize over 3–4 weeks after cessation of licorice use. The chronic ingestion of high doses of licorice produces profound suppression of the renin–angiotensin–aldosterone system with low plasma renin activity and markedly reduced urinary excretion of aldosterone.[60] Plasma cortisol concentrations are usually normal or low normal. Suppression of renin and aldosterone concentrations in plasma begins with about one week after initiating licorice use, and complete resolution of the reduced renin and aldosterone concentrations may require 1–4 months.[61,62]

Abnormalities

Case reports indicate that profound hypokalemia and metabolic alkalosis can develop following the chronic ingestion of large doses of licorice. A 35-year-old man developed an acute myopathy with complete paralysis of the proximal muscles of his arms and shoulder girdles following the ingestion of 20–40 g licorice daily for 2 years.[63] His serum potassium concentration was 2.1 mEq/L, and the paralysis resolved within 3 day of the initiation of potassium replacement. In contrast to patients with licorice-induced excessive mineralocorticoid effect, patients with hypokalemic familial periodic paralysis do not usually have metabolic alkalosis. Elevation of serum creatine kinase can result from licorice-induced hypokalemia.[64]

TREATMENT

The most common serious complication of chronic licorice intoxication is hypokalemia. Rarer serious complications of chronic licorice ingestion include rhabdomyolysis, renal dysfunction, and hypertensive encephalopathy. Patients with chronic licorice intoxication should be evaluated with a complete blood count, serum electrolytes including serum magnesium and calcium, serum creatinine, and serum creatine kinase. Licorice-induced hypokalemia usually responds to spironolactone (25–100 mg PO daily) and cessation of licorice use. Potassium supplementation is necessary when the potassium concentration is critically low (<2.5–2.8 mEq/L). In cases of profound life-threatening hypokalemia (<2.0 mEq/L) associated with chronic licorice ingestion, patients have received 10–20 mEq KCL/h during initial potassium replacement.[65] Because of the long biological half-life of glycyrrhetinic acid, complete resolution of the electrolyte and acid–base disturbances associated with chronic licorice ingestion may require 2 weeks. Therefore, the serum potassium should be monitored during this period of time. Potential adjunctive therapy includes the use of dexamethasone to reduce the cortisol-induced stimulation of the mineralocorticoid receptors by suppressing endogenous cortisol production. However, there are few human data on the use of dexamethasone during treatment of chronic licorice intoxication.

References

1. Gibson MR. Glycyrrhiza in old and new perspectives. Lloydia 1978;41:348–354.
2. Fiore C, Calo LA, Ragazzi E, Bielenberg J, Armanini D. Licorice from antiquity to the end of the 19th century: applications in medical therapy. J Nephrol 2004;17:337–341.
3. Armanini D, Fiore C, Mattarello MJ, Bielenberg J, Palermo M. History of the endocrine effects of licorice. Exp Clin Endocrinol Diabetes 2002;110:257–261.
4. Revers FE. [Heeft succus liquiritae een genezende werking op de maagzweer?] Nederlands Tijdschr Geneeskunde 1946;90:135–137. [Dutch]
5. Blachley JD, Knochel JP. Tobacco chewer's hypokalemia: licorice revisited. N Engl J Med 1980;302:784–785.
6. Davis EA, Morris DJ. Medicinal uses of licorice through the millennia: the good and plenty of it. Mol Cell Endocrinol 1991;78:1–6.
7. Saeedi M, Morteza-Semnani K, Ghoreishi M-R. The treatment of atopic dermatitis with licorice gel. J Dermatol Treat 2003;14:153–157.

8. van Rossum TG, Vulto AG, Hop WC, Schalm SW. Glycyrrhizin-induced reduction of ALT in European patients with chronic hepatitis C. Am J Gastroenterol 2001;96:2432–2437.
9. Bardhan KD, Cumberland DC, Dixon RA, Holdsworth CD. Clinical trial of deglycyrrhizinised liquorice in gastric ulcer. Gut 1978;19:779–782.
10. US Department of Agriculture. GRAS status of licorice (*Glycyrrhiza*), ammoniated glycyrrhizin, and monoammonium glycyrrhizinate. Fed Regist 1985:21043–21045.
11. Kinoshita T, Tamura Y, Mizutani K. The isolation and structure elucidation of minor isoflavonoids from licorice of *Glycyrrhiza glabra* origin. Chem Pharm Bull 2005;53: 847–849.
12. Hayashi H, Hiraoka N, Ikeshiro Y, Yamamoto H, Yoshikawa T. Seasonal variation of glycyrrhizin and isoliquiritigenin glycosides in the root of *Glycyrrhiza glabra* L. Biol Pharm Bull 1998;21:987–989.
13. Sabbioni C, Ferranti A, Bugamelli F, Forti GC, Raggi MA. Simultaneous HPLC analysis, with isocratic elution, of Glycyrrhizin and glycyrrhetic acid in liquorice roots and confectionery products. Phytochem Anal 2006;17:25–31.
14. Wang ZY, Nixon DW. Licorice and cancer. Nutr Cancer 2001;39:1–11.
15. Statti GA, Tundis R, Sacchetti G, Muzzoli M, Bianchi A, Menichini F. Variability in the content of active constituents and biological activity of *Glycyrrhiza glabra*. Fitoterapia 2004;75:371–374.
16. Spinks EA, Fenwick GR. The determination of glycyrrhizin in selected UK liquorice products. Food Addit Contam 1990;7:769–778.
17. Mizutani K, Kuramoto T, Tamura Y, Ohtake N, Doi S, Nakaura M, Tanaka O. Sweetness of glycyrrhetic acid 3-*O*-beta-D-monoglucuronide and the related glycosides. Biosci Biotechnol Biochem 1994;58:554–555.
18. Buhler H, Perschel FH, Fitzner R, Hierholzer K. Endogenous inhibitors of 11 beta-OHSD: existence and possible significance. Steroids 1994;59:131–135.
19. Ploeger B, Mensinga T, Sips A, Seinen W, Meulenbelt J, DeJongh J. The pharmacokinetics of glycyrrhizic acid evaluated by physiologically based pharmacokinetic modeling. Drug Metab Rev 2001;33:125–147.
20. Forslund T, Fyhrquist F, Froseth B, Tikkanen I. Effects of licorice on plasma atrial natriuretic peptide in healthy volunteers. J Intern Med 1989;225:95–99.
21. Armanini D, Bonanni G, Mattarello JM, Fiore C, Sartorato P, Palermo M. Licorice consumption and serum testosterone in healthy man. Exp Clin Endocrinol Diabetes 2003;111:341–343.
22. Armanini D, Bonanni G, Palermo M. Reduction of serum testosterone in men by licorice. N Engl J Med 1999;341: 1158.
23. Armanini D, Karbowiak I, Funder JW. Affinity of liquorice derivatives for mineralocorticoid and glucocorticoid receptors. Clin Endocrinol (Oxf) 1983;19:609–612.
24. Werner S, Brismar K, Olsson S. Hyperprolactinaemia and liquorice. Lancet 1979;1(8111):319.
25. Belhadj-Tahar H, Nassar B, Coulais Y, Monsastruc J-L, Sadeg N. Acute pseudo-aldosteronism syndrome induced by liquorice. Therapie 2003;58:375–387.
26. Stormer FC, Reistad R, Alexander J. Glycyrrhizic acid in liquorice—evaluation of health hazard. Food Chem Toxicol 1993;31:303–312.
27. Nielsen I, Pedersen RS. Life-threatening hypokalaemia caused by liquorice ingestion. Lancet 1984;1(8389):1305.
28. Shintani S, Murase H, Tsukagoshi H, Shiigai T. Glycyrrhizin (licorice)-induced hypokalemic myopathy. Report of 2 cases and review of the literature. Eur Neurol 1992;32: 44–51.
29. Sigurjonsdottir HA, Franzson L, Manhem K, Ragnarsson J, Sigurdsson G, Wallerstedt S. Liquorice-induced rise in blood pressure: a linear dose-response relationship. J Hum Hypertens 2001;15:549–552.
30. Sigurjonsdottir HA, Manhem K, Axelson M, Wallerstedt S. Subjects with essential hypertension are more sensitive to the inhibition of 11 beta-HSD by liquorice. J Hum Hypertens 2003;17:125–131.
31. Akao T, Akao T, Hattori M, Kanaoka M, Yamamoto K, Namba T, Kobashi K. Hydrolysis of glycyrrhizin to 18 beta-glycyrrhetyl monoglucuronide by lysosomal beta-D-glucuronidase of animal livers. Biochem Pharmacol 1991;41:1025–1029.
32. Akao T, Akao T, Kobashi K. Glycyrrhizin beta-D-glucuronidase of *Eubacterium* sp. from human intestinal flora. Chem Pharm Bull (Tokyo) 1987;35:705–710.
33. de Groot G, Koops R, Hogendoorn EA, Goewie CE, Savelkoul TJ, van Vloten P. Improvement of selectivity and sensitivity by column switching in the determination of glycyrrhizin and glycyrrhetic acid in human plasma by high-performance liquid chromatography. J Chromatogr 1988;456:71–81.
34. Yamamura Y, Kawakami J, Santa T, Kotaki H, Uchino K, Sawada Y, et al. Pharmacokinetic profile of glycyrrhizin in healthy volunteers by a new high-performance liquid chromatographic method. J Pharm Sci 1992;81:1042–1046.
35. Cantelli-Forti G, Maffei F, Hrelia P, Bugamelli F, Bernardi M, D'Intino P, et al. Interaction of licorice on glycyrrhizin pharmacokinetics. Environ Health Perspect 1994;102 (Suppl 9):65–68.
36. Takeda S, Ono H, Wakui Y, Asami A, Matsuzaki Y, Sasaki H, et al. Determination of glycyrrhetic acid in human serum by high-performance liquid chromatography with ultraviolet detection. J Chromatogr 1990;530:447–451.
37. Gunnarsdottir S, Johannesson T. Glycyrrhetic acid in human blood after ingestion of glycyrrhizic acid in licorice. Pharmacol Toxicol 1997;81:300–302.
38. Krahenbuhl S, Hasler F, Frey BM, Frey FJ, Brenneisen R, Krapf R. Kinetics and dynamics of orally administered 18 beta-glycyrrhetinic acid in humans. J Clin Endocrinol Metab 1994;78:581–585.

39. Ploeger B, Mensinga T, Sips A, Meulenbelt J, DeJongh J. A human physiologically-based model for glycyrrhizic acid, a compound subject to presystemic metabolism and enterohepatic cycling. Pharm Res 2000;17:1516–1525.
40. Kent UM, Aviram M, Rosenblat M, Hollenberg PF. The licorice root derived isoflavan glabridin inhibits the activities of human cytochrome P450s 3A4, 2B6, and 2C9. Drug Metab Disp 2002;30:709–715.
41. Tsukamoto S, Aburatani M, Yoshida T, Yamashita Y, Elbeih AA, Ohta T. CYP3A4 inhibitors isolated from licorice. Biol Pharm Bull 2005;28:2000–2002.
42. Olukoga A, Donaldson D. Liquorice and its health implications. J R Soc Health 2000;120:83–89.
43. Janse A, van Iersel M, Hoefnagels WH, Olde Rikker MG. The old lady who liked liquorice: hypertension due to chronic intoxication in a memory-impaired patient. Neth J Med 2005;63:149–150.
44. van den Bosch AE, Van der Klooster JM, Zuidgeest DM, Ouwendijk RJ, Dees A. Severe hypokalaemic paralysis and rhabdomyolysis due to ingestion of liquorice. Neth J Med 2005;63:146–148.
45. Saito T, Tsuboi Y, Fujisawa G, Sakuma N, Honda K, Okada K, et al. An autopsy case of licorice-induced hypokalemic rhabdomyolysis associated with acute renal failure: special reference to profound calcium deposition in skeletal and cardiac muscle. Nippon Jinzo Gakkai Shi 1994;36: 1308–1314.
46. Ishiguchi T, Mikita N, Iwata T, Nakata H, Sato H, Higashimoto Y, et al. Myoclonus and metabolic alkalosis from licorice in antacid. Intern Med 2004;43:59–62.
47. Heikens J, Fliers E, Endert E, Ackermans M, van Montfrans G. Liquorice-induced hypertension—a new understanding of an old disease: case report and brief review. Neth J Med 1995;47:230–234.
48. Woywodt A, Herrmann A, Choi M, Goebel U, Luft FC. Turkish pepper (extra hot). Postgrad Med J 2000;76: 426–428.
49. van der Zwan A. Hypertension encephalopathy after liquorice ingestion. Clin Neuro Neurosurg 1993;95: 35–37.
50. Russo S, Mastropasqua M, Mosetti MA, Persegani C, Paggi A. Low doses of liquorice can induce hypertension encephalopathy. Am J Nephrol 2000;20:145–148.
51. Strandberg TE, Andersson S, Jarvenpaa A-L, McKeigue PM. Preterm birth and licorice consumption during pregnancy. Am J Epidemiol 2002;16:803–805.
52. Strandberg TE, Jarvenpaa A-L, Vanhanen H, McKeigue PM. Birth outcome in relation to licorice consumption during pregnancy. Am J Epidemiol 2001;153:1085–1088.
53. Raggi MA, Bugamelli F, Nobile L, Schiavone P, Cantelli-Forti G. HPLC determination of glycyrrhizin and glycyrrhetic acid in biological fluids, after licorice extract administration to humans and rats. Boll Chim Farm 1994;133:704–708.
54. Sabbioni C, Mandrioli R, Ferranti A, Bugamelli F, Saracino MA, Forti GC, et al. Separation and analysis of glycyrrhizin, 18 beta-glycyrrhetic acid and 18 alpha-glycyrrhetic acid in liquorice roots by means of capillary zone electrophoresis. J Chromatogr A 2005;1081:65–71.
55. Wang P, Li SF, Lee KH. Determination of glycyrrhizic acid and 18-beta-glycyrrhetinic acid in biological fluids by micellar electrokinetic chromatography. J Chromatogr A 1998;811:219–224.
56. Lin ZH, Qiu S-X, Wufuer A, Shum L. Simultaneous determination of glycyrrhizin, a marker component in Radix Glycyrrhizae, and its major metabolite glycyrrhetic acid in human plasma by LC-MS/MS. J Chromatogr B 2005;814: 201–207.
57. Russel FG, van Uum S, Tan Y, Smits P. Solid-phase extraction of 18 β-glycyrrhetinic acid from plasma and subsequent analysis by high-performance liquid chromatography. J Chromatogr B Biomed Sci Appl 1998;710:223–226.
58. Farese RV Jr, Biglieri EG, Shackleton CH, Irony I, Gomez-Fontes R. Licorice-induced hypermineralocorticoidism. N Engl J Med 1991;325:1223–1227.
59. van Uum SH, Walker BR, Hermus AR, Sweep CG, Smits P, de Leeuw PW, Lenders JW. Effect of glycyrrhetinic acid on 11 beta-hydroxysteroid dehydrogenase activity in normotensive and hypertensive subjects. Clin Sci (Lond) 2002;102:203–211.
60. Conn JW, Rovner DR, Cohen EL. Licorice-induced pseudoaldosteronism. Hypertension, hypokalemia, aldosteronopenia, and suppressed plasma renin activity. JAMA 1968;205:492–496.
61. Kageyama Y. A case of pseudoaldosteronism induced by glycyrrhizin. Nippon Jinzo Gakkai Shi 1992;34:99–102.
62. Epstein MT, Espiner EA, Donald RA, Hughes H. Liquorice toxicity and the renin-angiotensin-aldosterone axis in man. Br Med J 1977;1(6055):209–210.
63. Corsi FM, Galgani S, Gasparini C, Giacanelli M, Piazza G. Acute hypokalemic myopathy due to chronic licorice ingestion: report of a case. Ital J Neurol Sci 1983;4: 493–497.
64. Elinav E, Chajek-Shaul T. Licorice consumption causing severe hypokalemic paralysis. Mayo Clin Proc 2003;78: 767–768.
65. Lin S-H, Yang S-S, Chau T, Halperin ML. An unusual cause of hypokalemic paralysis: chronic licorice ingestion. Am J Med Sci 2003;325:153–156.

Chapter 79

MA HUANG
(*Ephedra* Alkaloids)

HISTORY

Ma huang has a long history of medicinal use in China dating back over 5,000 years. The use of ephedra juice as a drink of longevity was an ancient Indian custom mentioned in the sacred Sanskrit Vedas, Rigveda.[1] The Japanese chemist, Nagai, isolated ephedrine in 1887. A series of articles in the early 1930s popularized the use of ephedra alkaloids (ephedrine) for the treatment of asthma.[2] In 2004, the US Food and Drug Administration (FDA) issued a ban on the use of ma huang in herbal preparations.

BOTANICAL DESCRIPTION

The genus *Ephedra* contains over 50 species, but most of these species are not used for medicinal purposes because of the low concentration of ephedra alkaloids in their stems and leaves. The most commonly used *Ephedra* species is the Chinese species, *E. sinica* Stapf. Other less common medicinal herbal species include *E. equisetina* Bunge, *E. gerardiana* Wallich ex Stapf, and *E. intermedia* Schrenk. ex C.A. Mey. (zhong ma huang). Ephedra alkaloids are not usually detectable in *Ephedra* species (e.g., *E. nevadensis* S. Wats., Nevada Mormon tea) native to the New World.[3]

Common Name: Ma Huang, Chinese Ephedra, Herbal Ecstasy

Scientific Name: *Ephedra sinica* Stapf

Botanical Family: Ephedraceae

Physical Description: This small, almost leafless shrub has a thick woody base with many slender, jointed, yellow-green branches covered with tiny leaves. The average height of this plant is about 1 ft (~0.3 m), but some specimens may reach 4 ft (~1.2 m) in height. Flowers appear in the summer and produce fleshy, red, succulent, 2-seeded bracts.

Distribution and Ecology: This evergreen shrub is a native plant in central Asia, but this plant is cultivated throughout China, Japan, Southern Siberia, and India. These perennial shrubs grow primarily in arid regions.

EXPOSURE

Sources

The usual source of ma huang is the dried, green branches of Chinese ephedra (*Ephedra sinica*) because of the high ephedra alkaloid content of green branches compared with other plant parts. Most preparations containing ma huang also contain other herbal stimulants, such as caffeine in guarana (*Paullinia cupana* Kunth), kola nut [*Cola nitida* (Vent.) A. Chev.], or yerba mate (*Ilex paraguariensis* A. St. Hil.) and salicylates in white willow bark (*Salix alba* L.). Herba Ephedrae consists of the dried stem or aerial part of *Ephedra sinica* Stapf or other ephedrine-containing *Ephedra* species.

Medicinal Uses

TRADITIONAL

In traditional Chinese medicine, the main indication for *E. sinica* is the treatment of wheezing, cough, and

Medical Toxicology of Natural Substances, by Donald G. Barceloux, MD

congestions associated with bronchitis and asthma. Typically, ma huang is used in a decoction of several herbs (e.g., licorice, ginger, cinnamon, apricot seed, honey) for these purposes.[4] Other traditional uses include the treatment of urticaria, enuresis, narcolepsy, myasthenia gravis, and chronic postural hypotension.

Current

Ma huang and ephedrine-containing products have recently been promoted as an appetite suppressant, as a stimulant, and as an aphrodisiac. Most herbal products contain ephedra alkaloids combined with other stimulants (e.g., caffeine). Other herbal uses of ma huang include the treatment of low blood pressure, arthralgia, edema, enuresis, narcolepsy, cold and flu symptoms, asthma, and upper respiratory infections. Ephedrine-containing herbal preparations promote modest, short-term weight loss (<1 kg/month), but there are few data to support the use of these preparations for long-term weight loss or for the enhancement of athletic performance.[5] Additionally, the use of these products with caffeine is associated with a moderate (2- to 4-fold) increased risk of adverse effects (psychiatric, gastrointestinal, autonomic, palpitations, transient elevated heart rate/blood pressure).[6]

Regulatory Status

The German Commission E approved the use of ephedra only for the treatment for respiratory tract illness and mild bronchospasm for adults and children over 6 years of age. Health Canada banned the use of all combination products containing ephedra alkaloids and caffeine in 2002, including herbal supplements. In 2004, the FDA prohibited the sale of dietary supplements contain *Ephedra* species or ephedra alkaloids. However, the implementation of this ban was overturned by one of the lower courts in the United States.[7] The use of ma huang and ephedra alkaloids is banned as a performance enhancement supplement by the International Olympic Committee, The National Collegiate Athletic Association, and the National Football League.[8]

PRINCIPAL INGREDIENTS

Chemical Composition

The yield of ephedra alkaloids form commercially cultivated *Ephedra* species varies with species and ranges from about 0.5% to 3.5% with ephedrine accounting for approximately 30–90% of these alkaloids.[9] Concentrated extracts of *E. sinica* can contain up to 4–8% ephedra alkaloids.[10] The roots, fruits, and basal woody stems of *Ephedra* species generally contain none to minimal amounts of ephedra alkaloids. Figure 79.1 displays the chemical structures of ephedra alkaloids. Ephedra alkaloids include primarily (−)-ephedrine and (+)-pseudoephedrine with minor amounts of (−)-*N*-methyl ephedrine, (−)-*N*-pseudomethylephedrine, (+)-norpseudoephedrine (cathine), and (−)-norephedrine.[11] The latter compound is structurally identical to phenylpropanolamine, which typically is sold as the racemic mixture (*d,l*) in pharmaceutical preparations. The *l*-(levorotatory, −) isomer is the natural form of ephedrine in *Ephedra* species, whereas synthetic ephedrine is usually a racemic mixture of both the *l*- (−) and *d*- (+) isomers.

The ephedra alkaloid content of commercial preparations of *Ephedra* herbs depends on a variety of factors including species, lot, plant part, harvesting techniques, and extraction methods. (−)-Ephedrine and (+)-pseudoephedrine account for the vast majority of the alkaloid content of ephedra-containing commercial products.[12] In a study of 20 *Ephedra*-containing dietary supplements, the total ephedra alkaloid content ranged from 0–18.5 mg/dose with (−)-ephedrine and (+)-pseudoephedrine accounting for 1.1–15.3 mg/dose and 0.2–9.5 mg/dose, respectively.[13] A study of 25 lots of three ephedra-containing dietary supplements demonstrated an average amount of ephedrine was approximately 12–13 mg ± 1 mg per dose.[14] In general, the content of methyl ephedrine, norephedrine, and norpseudoephedrine is below 0.25–0.3 mg/dose in most dietary

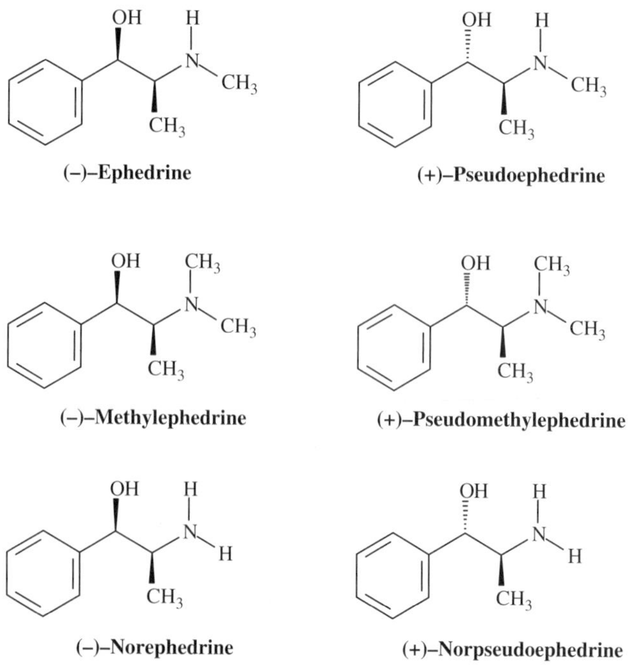

(−)-Ephedrine (+)-Pseudoephedrine

(−)-Methylephedrine (+)-Pseudomethylephedrine

(−)-Norephedrine (+)-Norpseudoephedrine

FIGURE 79.1. Chemical structures of ephedra alkaloids.

supplements containing ma huang as measured by high performance liquid chromatography (HPLC).[15] Occasionally, the amount of methyl ephedrine is higher (e.g., 3 mg).

Physiochemical Properties

The cut branches of ma huang (*E. sinica*) are yellow with a pine-like odor and a sharp bitter taste. Ephedrine is highly lipophilic, and this compound easily crosses the blood–brain barrier.

Mechanism of Toxicity

Ephedra alkaloids, which are phenethylamine derivatives, are the active ingredients in ma huang. Standard pharmacology texts typically emphasize that ephedrine is both a direct acting adrenergic agonist by the activation of adrenergic receptors and an indirect agonist by the release of norepinephrine via carrier-mediated exchange mechanism.[16] However, human *in vivo* data indicate that urinary catecholamine concentrations are not increased following the administration of typical doses of ephedrine in dietary supplements,[17] and this observation suggests that the physiologic effects of ephedrine are mediated by a direct adrenergic response. The central nervous system and cardiopulmonary stimulation are probably mediated predominantly by direct β-adrenergic agonist action with a much smaller degree of concomitant α-adrenergic involvement. There is little evidence of ephedra alkaloids having a significant indirect agonist action through the augmentation of norepinephrine release from the presynaptic neurons.

In vitro studies on cloned human receptors indicate that ephedrine acts as a substrate of the norepinephrine transporter with lesser substrate activity at the dopamine transporter.[18] These studies also indicate that ephedrine has weak affinity for α_2-adrenergic and 5-hydroxytryptamine receptors with little β-adrenergic or α_1-adrenergic receptor agonist activity. The reasons for this apparent mechanistic conflict with the human *in vivo* data are not clear. The stimulant and euphoric effects of ephedrine are less intense than amphetamines, but ephedra alkaloids have some abuse potential.[19,20] Pseudoephedrine has less potent cardiovascular effects than ephedrine. The relatively low rate of serious adverse effects compared with the high rate of use of these herbs suggests these adverse events may have a pharmacogenetic or idiosyncratic etiology.

DOSE RESPONSE

The FDA recommends limiting ephedrine doses to 8 mg as a single dose, 24 mg as a daily dose, and 7 days as the limit for the duration of treatment. Daily ephedrine doses in clinical trials on weight loss and physical endurance ranged from 90–150 mg without major cardiovascular side effects,[21,22] while a daily dose of pseudoephedrine ranged up to 240 mg.[23] Most of the case reports of serious cardiovascular or cerebrovascular complication involve the use of ma huang or ephedrine-containing herbal preparations in excessive doses for prolonged periods,[24,25] or the manufacturer's recommended doses for more than 1 month.[26]

TOXICOKINETICS

The pharmacokinetics of ephedra alkaloids in ma huang is similar to pharmaceutical preparations of these alkaloids with the possible exception of time to peak serum ephedrine concentrations.[27]

Absorption

The absorption of ephedrine from the gastrointestinal tract is relatively complete. Some studies suggest that the absorption of ephedrine in ma huang is substantially slower than the absorption of pharmaceutical preparations of ephedrine. In a study of 12 healthy, normotensive volunteers, the mean time from ingestion to peak plasma ephedrine concentrations was 3.9 hours for the ma huang preparation compared with 1.69 hours for the pharmaceutical ephedrine tablet.[28] Other studies of ephedrine-containing dietary supplements indicate that the time to peak plasma ephedrine concentrations for ephedrine-containing dietary supplements and pharmaceutical preparations of ephedrine are similar (i.e., about 3 hours).[14]

Distribution

The volume of distribution of ephedrine is large and protein binding in the blood is low. In a study of ephedrine-containing dietary supplements, the steady-state volume of distribution for ephedrine was about 3 ± 1 L/kg.[14]

Biotransformation

Pharmacokinetic studies indicate that substantial variation in the metabolism of major ephedra alkaloids occurs between individuals. In a study of eight healthy volunteers, the coefficient of variation (i.e., standard deviation divided by the mean) was 65% for the plasma AUC ratio of norephedrine to ephedrine and was 55% for the plasma AUC ratio of nor-pseudoephedrine to pseudoephedrine.[27] About 8–20% of the absorbed dose of ephedrine undergoes *N*-demethylation to norephedrine, whereas oxidative deamination of ephedrine to 1-phenylpropan-1,3-diol followed by side-chain oxidation to benzoic acid and hippuric acid accounts for approximately 4–13% of the absorbed dose.[29] The amount of aromatic hydroxylation, *N*-demethylation and oxidative deamination of ephedrine is species-specific.[30]

Elimination

The elimination of ephedrine occurs primarily via renal excretion of the unchanged compound. The elimination half-life of ephedra alkaloids increases as the urinary pH increases because of the ionizable amino group on the molecule and a pK_a of 9.6. Alkalinization of the urine increases the renal elimination half-life and prolongs the duration of action of these alkaloids as a result of the enhancement of renal tubular absorption of ephedra alkaloids in alkaline urine.[31] The plasma elimination half-life of ephedrine in ma huang is approximately 5–6 hours under normal conditions and urinary pH (6–7).[14]

Drug Interactions

The drug interactions between ephedrine and other drugs are primarily pharmacodynamic rather than pharmacokinetic. In a randomized, double-blind, placebo-controlled crossover study, 16 healthy adults ingested 25 mg ephedrine, 200 mg caffeine, or both drugs.[17] Pharmacodynamic effects (increased heart rate, increased blood pressure) were observed, but there were no clinically significant pharmacokinetic interactions between these two drugs. Similarly, the administration of 300 mg moclobemide, a reversible MAO-A inhibitor, twice daily and ephedrine (50 mg twice at a 4-hour interval) potentiates the pharmacodynamic effect of ephedrine on systolic blood pressure, but there were no clinically significant pharmacokinetic interactions.[32]

CLINICAL RESPONSE

Ma huang contains ephedra alkaloids and the clinical features associated with the use of ma huang are similar to the effects of ephedrine. The use of ephedra alkaloids produces dose-dependent sympathomimetic effects including headache, nausea, heartburn, anorexia, dry mouth, diaphoresis, mydriasis, tachycardia, palpitations, hypertension, bronchodilation, restlessness, insomnia, anxiety, and agitation.[22] Adverse reactions to the use of ephedrine-containing herbs exceed the rate of most other herbal preparations.[33,34] The sources of most of the data on the clinical effects of ma huang and ephedra alkaloids are case reports rather than clinical trails or epidemiological studies. Case reports associate a number of cardiovascular side effects with the use of herbal preparations containing ma huang including hypertension, ventricular dysrhythmias,[35] myocardial ischemia, myocardial infarction with or without underlying coronary artery disease,[36] cardiomyopathy,[37] hypersensitivity myocarditis,[38] and sudden death.[39] Neurological complication reportedly associated with the use of ma huang and ephedra alkaloids include seizures, cerebral vascular ischemic stroke,[40] intracranial hemorrhage, and vasculitis.[41] In some cases, the abuse of large doses of illicit ephedrine was associated with fatal intracranial hemorrhage.[42] A case-control study of ephedra use within 3 days of developing a hemorrhagic stroke did not detect a statistically significant increase in the risk of hemorrhagic stroke with ephedra use, although the odds ratio was the highest (OR = 3.59; 95% CI: 0.70–18.35) for the high-dose group (>32 mg daily).[43] Rare case reports associate the use of high doses of ephedra containing dietary supplements during risperidone and bupropion therapy with the development of mild delirium, autonomic hyperactivity, diaphoresis, and hyperreflexia.[44]

Reported psychiatric complications of excessive ma huang use include psychosis and affective disturbances, manifest by labile mode, agitation, sleep deprivation, suicidal ideations, and the development of acute paranoid psychosis with delusions and hallucinations.[45–47] These patients usually have clear consciousness, and the clinical presentation is similar to the psychiatric disturbances associated with the abuse of ephedrine.[48,49] In many of these case reports, there were pre-existing psychiatric illnesses or family histories of psychiatric disorders (schizoaffective, bipolar).[50] Many of the case reports associating cardiovascular or neurological complications with ma huang use indicate possible causal effects rather than probable causal effects based the Naranjo Adverse Drug Reaction Probability Scale. Clinical trials on the use of ephedra alkaloids for weight loss and enhancement of athletic performance have not reported serious adverse cardiovascular events (i.e., stroke, myocardial infarction, malignant dysrhythmias).[6] However, these clinical trials are typically relatively short-term (<6 months) with relatively high attrition rates. Rare case reports associate the use of ma huang with the development of acute hepatitis.[51] The presence of antinuclear and antismooth muscle antibodies along with eosinophils on the liver biopsy suggests an immune-mediated reaction. The role of ephedra alkaloids or other constituents of ma huang in these case reports remain unclear.

DIAGNOSTIC TESTING

Analytical Methods

Methods for the determination of ephedra alkaloids include liquid chromatography,[11] gas chromatography,[52] capillary electrophoresis,[53] HPLC,[54] liquid chromatography tandem mass spectrometry (LC/MS/MS),[55] and gas chromatography/mass spectrometry (GC/MS).[56] The limit of quantitation for ephedra alkaloids as analyzed by liquid chromatography/atmospheric pressure chemical ionization/tandem mass spectrometry is in the low nanogram-per-milliliter range.[57] Ephedra alkaloids in ma huang are a mixture of three diastereoisomeric pairs of ephedrine and pseudoephedrine along with their potential metabolites (norephedrine, norpseudoephedrine, methyl ephedrine, pseudomethylephedrine). Because these alkaloids have two chiral centers, each of the four diastereoisomers has a mirror image enantiomer for a total of eight isomers. Capillary electrophoresis with chiral discrimination allows for the separation and identification of ephedrine enantiomers, but LC/MS/MS does not separate these enantiomers.[58] The natural form of ephedrine in *Ephedra* species is the *l*-(levorotatory) isomer, whereas synthetic ephedrine is usually a racemic mixture of both the *l*- and *d*-isomers. Several analytical methods are available to identify and authenticate ma huang including HPLC chemical fingerprinting,[59] metabolic fingerprinting with ^{1}H-NMR spectroscopy,[60] chiral GC/MS with GC/matrix isolation/Fourier transform infrared spectroscopy,[61] and polymerase chain reaction analysis (PCR).[62] The molecular PCR method uses DNA as the template rather than ephedra alkaloids; therefore, chemical analyses (HPLC, GC/MS) are necessary to identify ephedra alkaloids. Alternatively, botanical identification of Chinese and North American species of *Ephedra* is possibly based on leaf and internode length using light microscopy.[63]

Biomarkers

Therapeutic

Preparations of ma huang contain several ephedra alkaloids, and the presence of only large amounts of ephedrine in an herbal preparation of ma huang without other ephedra alkaloids suggests adulteration with pharmaceutical ephedrine.[15,64] The ephedrine and methyl ephedrine in ma huang can produce positive results for amphetamines on urine drug screens using immunoassays, but these compounds can be separated from amphetamines using GC/MS.[65] The mean maximum serum ephedrine concentration in blood samples from 12 healthy volunteers receiving a single dose of ma huang was 81 ng/mL.[28] The mean dose of ephedrine in this study was 19.4 mg. In a study of volunteers receiving single doses of ephedrine-containing dietary supplements (i.e., about 24–27 mg ephedrine), the mean maximum serum ephedrine concentrations ranged from approximately 75–100 ng/mL.[14] After receiving 50 mg ephedrine orally, the mean maximum plasma concentration of ephedrine in blood samples from 16 volunteers was 138 ± 33 ng/mL (range: 80–207 ng/mL).[66]

Intoxication

A case report of a 22-year-old woman associated the ingestion of an herbal preparation containing 10 mg ephedrine and 100 mg caffeine with the development of vomiting, palpitations, tremulousness, and evidence of myocardial ischemia in her anterior leads.[67] The serum ephedrine concentration 8 hours after her last dose was 150 ng/mL. Methyl ephedrine is present in a variety of over-the-counter cough and cold preparations throughout the world, but this drug is not available in the United States.[68]

Postmortem

Ephedrine is a common constituent of postmortem samples from patients with chronic stimulant abuse, but there are few postmortem data on patients dying from ephedrine as the sole drug. In a case series of 127 decendents testing positive for ephedra alkaloids, the postmortem blood samples contained <0.49 mg/L in a majority of the cases.[69] There was no statistically significant difference in the postmortem ephedrine concentration in patients dying from traumatic (mean = 1.27 mg/L, range: 0.07–11.73 mg/L) or nontraumatic (mean = 1.61 mg/L, range: 0.02–12.35 mg/L) causes. Norephedrine (cathine) was detected in 33% of the samples tested with a mean concentration of 1.81 mg/L (range: 0.04–14.14 mg/L).

Abnormalities

Ephedra alkaloids in ma huang are stimulants with dose-dependent effects on the cardiovascular system (e.g., tachycardia, elevated blood pressure). The presence of electrolyte imbalance suggests a cause other than ma huang.

TREATMENT

The treatment of complications associated with the use of ephedrine-containing herbs is supportive. In general, decontamination measures are not recommended because of the rapid absorption of ephedrine. Rarely, cardiovascular or cerebrovascular complications develop during ephedrine intoxications. Agitation and seizures usually respond to diazepam or lorazepam. The decision to treat hypertension depends on clinical judgement regarding the severity of the hypertension, pre-existing cardiovascular disease, and the clinical effects observed.

There are few clinical data to guide the treatment of hypertension. Potentially useful agents include sodium nitroprusside and phentolamine. There are no definitive data on the use of beta-adrenergic blockers in the setting of an *Ephedra* overdose. Potentially, the use of nonselective beta-blockers can cause unopposed α_1-receptor stimulation on the vasculature and the blood pressure. Patients with potentially serious cardiovascular complication should be monitored for dysrhythmias. Laboratory testing includes complete blood count, serum electrolytes, serum glucose, serum creatinine, serum troponin, serum creatine phosphokinase, and electrocardiogram. There are no specific antidotes for the effects of ma huang. Methods to enhance the elimination of ephedrine are not usually efficacious including the use of dialysis, hemoperfusion, and acidification of the urine. Treatment of anxiety and psychotic manifestations of ephedrine toxicity includes the use of benzodiazepines and, for psychotic features unresponsive to benzodiazepines, antipsychotic drugs.

References

1. Mahdihassan S, Mehdi FS. Soma of the Rigveda and an attempt to identify it. Am J Chin Med 1989;17:1–8.
2. Chen K. Schmidt C. Ephedrine and related substances. Medicine 1930;9:117.
3. Caveney S, Charlet DA, Freitag H, Maier-Stolte M, Starratt AN. New observations on the secondary chemistry of world Ephedra (Ephedraceae). Am J Bot 2001;88: 1199–1208.
4. Mehendale SR, Bauer BA, Yuan C-S. Ephedra-containing dietary supplements in the US versus ephedra as a Chinese medicine. Am J Chin Med 2004;32:1–10.
5. Shekelle PG, Hardy ML, Morton SC, Maglione M, Mojica WA, Suttorp MJ, et al. Efficacy and safety of ephedra and ephedrine for weight loss and athletic performance: a meta-analysis. JAMA 2003;289:1537–1545.
6. Andraws R, Chawla P, Brown DL. Cardiovascular effects of ephedra alkaloids: a comprehensive review. Prog Cardiovasc Dis 2005;47:217–225.
7. Seamon MJ, Clauson KA. Ephedra: yesterday, DSHEA, and tomorrow—a ten year perspective on the Dietary Supplement Health and Education Act of 1994. J Herb Pharmacother 2005;5:67–86.
8. Bucci LR. Selected herbals and human exercise performance. Am J Clin Nutr 2000;72(Suppl 2): 624S–636S.
9. Soni MG, Carabin IG, Griffiths JC, Burdock GA. Safety of ephedra: lessons learned. Toxicol Lett 2004;150: 97–110.
10. Hurlbut JA, Carr JR, Singleton ER, Faul KC, Madson MR, Storey JM, Thomas TL. Solid-phase extraction cleanup and liquid chromatography with ultraviolet detection of ephedrine alkaloids in herbal products. J AOAC Int 1998;81:1121–1127.
11. Sagara K, Oshima T, Misaki T. A simultaneous determination of norephedrine, pseudoephedrine, ephedrine and methylephedrine in Ephedrae Herba and oriental pharmaceutical preparations by ion-pair high-performance liquid chromatography. Chem Pharm Bull (Tokyo) 1983;31:2359–2365.
12. Baker JI, Zhang X, Boucher TA, Keyler DE. Investigation of quality in ephedrine-containing dietary supplements. J Herb Pharmacother 2003;3:5–17.
13. Gurley BJ, Gardner SF, Hubbard MA. Content versus label claims in ephedra-containing dietary supplements. Am J Health Syst Pharm 2000;57:963–969.
14. Gurley BJ, Gardner SF, White LM, Wang P-L. Ephedrine pharmacokinetics after the ingestion of nutritional supplements containing *Ephedra sinica* (ma huang). Ther Drug Monit 1998;20:439–445.
15. Gurley BJ, Wang P, Gardner SF. Ephedrine-type alkaloid content of nutritional supplements containing *Ephedra sinica* (ma-huang) as determined by high performance liquid chromatography. J Pharm Sci 1998;87:1547–1553.
16. Hardman JG, Limbird LE, editors. Goodman & Gilman's The Pharmacological Basis of Therapeutics. 9th ed. New York: McGraw-Hill Co.;1996.
17. Haller CA, Jacob P, Benowitz NL. Enhanced stimulant and metabolic effects of combined ephedrine and caffeine. Clin Pharmacol Ther 2004;75:259–273.
18. Rothman RB, Vu N, Partilla JS, Roth BL, Hufeisen SJ, Compton-Toth BA, et al. *In vitro* characterization of ephedrine-related stereoisomers at biogenic amine transporters and the receptorome reveals selective actions as norepinephrine transporter substrates. J Pharmacol Exp Ther 2003;307:138–145.
19. Martin WR, Sloan JW, Sapira JD, Jasinski DR. Physiologic, subjective, and behavioral effects of amphetamine, methamphetamine, ephedrine, phenmetrazine, and methylphenidate in man. Clin Pharmacol Ther 1971;12: 245–258.
20. Gruber AJ, Pope HG Jr. Ephedrine abuse among 36 female weightlifters. Am J Addict 1998;7:256–261.
21. Pasquali R, Cesari MP, Melchionda N, Stefanini C, Raitano A, Labo G. Does ephedrine promote weight loss in low-energy-adapted obese women? Int J Obes 1987;11: 163–168.
22. Boozer CN, Daly PA, Homel P, Solomon JL, Blanchard D, Nasser JA, Strauss R, Meredith T. Herbal ephedra/caffeine for weight loss: a 6-month randomized safety and efficacy trial. Int J Obes Relat Metab Disord 2002;26:593–604.
23. Pleskow W, Grubbe R, Weiss S, Lutsky B. Efficacy and safety of an extended-release formulation of desloratadine and pseudoephedrine vs the individual components in the treatment of seasonal allergic rhinitis. Ann Allergy Asthma Immunol 2005;94:348–354.
24. To LB, Sangster JF, Rampling D, Cammens I. Ephedrine-induced cardiomyopathy. Med J Aust 1980;2:35–36.

25. Enders JM, Dobesh PP, Ellison JN. Acute myocardial infarction induced by ephedrine alkaloids. Pharmacotherapy 2003;23:1645–1651.
26. Samenuk D, Link MS, Homoud MK, Contreras R, Theohardes TC, Wang PJ, Mark Estes NA III. Adverse cardiovascular events temporally associated with ma huang, an herbal source of ephedrine. Mayo Clin Proc 2002;77: 12–16.
27. Haller CA, Jacob P III, Benowitz NL. Pharmacology of ephedra alkaloids and caffeine after single-dose dietary supplement use. Clin Pharmacol Ther 2002;71: 421–432.
28. White LM, Gardner SF, Gurley BJ, Marx MA, Wang PL, Estes M. Pharmacokinetics and cardiovascular effects of ma-huang (*Ephedra sinica*) in normotensive adults. J Clin Pharmacol 1997;37:116–122.
29. Sever PS, Dring LG, Williams RT. The metabolism of (−)-ephedrine in man. Eur J Clin Pharmacol 1975;9:193–198.
30. Sinsheimer JE, Dring LG, Williams RT. Species differences in the metabolism of norephedrine in man, rabbit and rat. Biochem J 1973;136:763–771.
31. Wilkinson GR, Beckett AH. Absorption metabolism and excretion of the ephedrines in man. I. The influence of urinary pH and urine volume output. J Pharmacol Exp Ther 1968;162:139–147.
32. Dingemanse J, Guentert T, Gieschke R, Stabl M. Modification of the cardiovascular effects of ephedrine by the reversible monoamine oxidase A-inhibitor moclobemide. J Cardiovasc Pharmacol 1996;28:856–861.
33. Woolf AD, Watson WA, Smolinske S, Litovitz T. The severity of toxic reactions to ephedra: comparisons to other botanical products and national trends from 1992–2002. Clin Toxicol 2005;43:347–355.
34. Bent S, Tiedt TN, Odden MC, Shlipak MG. The relative safety of ephedra compared with other herbal products. Ann Intern Med 2003;138:468–471.
35. Zahn KA, Li RL, Purssell RA. Cardiovascular toxicity after ingestion of "herbal ecstacy." J Emerg Med 1999;17: 289–291.
36. Rezkalla SH, Mesa J, Sharma P, Kloner RA. Myocardial infarction temporally related to ephedra—a possible role for the coronary microcirculation. WMJ 2002;101:64–66.
37. Peters CM, O'Neill JO, Young JB, Bott-Silverman C. Is there an association between ephedra and heart failure? A case series. J Cardiac Failure 2005;11:9–11.
38. Zaacks SM, Klein L, Tan CD, Rodriguez ER, Leikin JB. Hypersensitivity myocarditis associated with ephedra use. J Toxicol Clin Toxicol 1999;37:485–489.
39. Theoharides TC. Sudden death of a healthy college student related to ephedrine toxicity from a ma huang-containing drink. J Clin Psychopharmacol 1997;17:437–439.
40. Vahedi K, Domigo V, Amarenco P, Bousser MG. Ischaemic stroke in a sportsman who consumed ma huang extract and creatine monohydrate for body building. J Neurol Neurosurg Psychiatry 2000;68:112–113.
41. Yin PA. Ephedrine-induced intracerebral hemorrhage and central nervous system vasculitis. Stroke 1990;21: 1641.
42. Bruno A, Nolte KB, Chapin J. Stroke associated with ephedrine use. Neurology 1993;43:1313–1316.
43. Morgenstern LB, Viscoli CM, Kernan WN, Brass LM, Broderick JP, Feldmann E, et al. Use of ephedra-containing products and risk for hemorrhagic stroke. Neurology 2003;60:132–135.
44. Philibert R, Mac J. An association of ephedra use with psychosis and autonomic hyperactivity. Ann Clin Psychiatry 2004;16:167–169.
45. Walton R, Manos GH. Psychosis related to ephedra-containing herbal supplement use. South Med Assoc 2003;96:718–720.
46. Boerth JM, Caley CF. Possible case of mania associated with ma-huang. Pharmacotherapy 2003;23:380–383.
47. Jacobs KM, Hirsch KA. Psychiatric complications of ma-huang. Psychosomatics 2000;41:58–62.
48. Maglione M, Miotto K, Iguchi M, Hilton L, Shekelle P. Psychiatric symptoms associated with ephedra use. Expert Opin Drug Saf 20056;4:879–884.
49. Whitehouse AM, Duncan JM. Ephedrine psychosis rediscovered. Br J Psychiatry 1987;150:258–261.
50. Maglione M, Miotto K, Iguchi M, Jungvig L, Morton SC, Shekelle PG. Psychiatric effects of ephedra use: an analysis of Food and Drug Administration reports of adverse events. Am J Psychiatry 2005;162:189–191.
51. Nadir A, Agrawal S, King PD, Marshall JB. Acute hepatitis associated with the use of a Chinese herbal product, ma-huang. Am J Gastroenterol 1996;91:1436–1438.
52. Yamasaki K, Fujita K, Sakamoto M, Okada K, Yoshida M. Separation and quantitative analysis of ephedra alkaloids by gas chromatography and its application to evaluation of some ephedra species collected around Himalaya. Chem Pharm Bull (Tokyo) 1974;22:2898–2902.
53. Flurer CL, Lin LA, Satzger RD, Wolnik KA. Determination of ephedrine compounds in nutritional supplements by cyclodextrin-modified capillary electrophoresis. J Chromatogr B Biomed Appl 1995;669:133–139.
54. Imaz C, Carreras D, Navajas R, Rodriguez C, Rodriguez AF, Maynar J, Cortes R. Determination of ephedrines in urine by high-performance liquid chromatography. J Chromatogr 1993;631:201–205.
55. Lam JW, Gardner GJ, McCooeye M, Fraser CA, Sturgeon RE. A systematic approach to quantitation of ephedra alkaloids in natural health products. Anal Bioanal Chem 2005;383:268–281.
56. Marchei E, Pellegrini M, Pacifici R, Zuccaro P, Pichini S. A rapid and simple procedure for the determination of ephedrine alkaloids in dietary supplements by gas chromatography-mass spectrometry. J Pharm Biomed Anal 2006;41:1633–1641
57. Jacob P III, Haller CA, Duan M, Yu D, Peng M, Benowitz NL. Determination of ephedra alkaloid and caffeine

concentrations in dietary supplements and biological fluids. J Anal Toxicol 2004;28:152–159.

58. Iwata YT, Garcia A, Kanamori T, Inoue H, Kishi T, Lurie IS. The use of a highly sulfated cyclodextrin for the simultaneous chiral separation of amphetamine-type stimulants by capillary electrophoresis. Electrophoresis 2002;23: 1328–1334.
59. Schaneberg BT, Crockett S, Bedir E, Khan IA. The role of chemical fingerprinting: application to Ephedra. Phytochemistry 2003;62:911–918.
60. Kim HK, Choi YH, Erkelens C, Lefeber AW, Verpoorte R. Metabolic fingerprinting of *Ephedra* species using ^{1}H-NMR spectroscopy and principal component analysis. Chem Pharm Bull 2005;53:105–1091
61. Betz JM, Gay ML, Mossoba MM, Adams S, Portz BS. Chiral gas chromatographic determination of ephedrine-type alkaloids in dietary supplements containing Ma Huang. J AOAC Int 1997;80:303–315.
62. Techen N, Khan IA, Pan Z, Scheffler BE. The use of polymerase chain reaction (PCR) for the identification of *Ephedra* DNA in dietary supplements. Planta Med 2006;72:241–247.
63. Joshi VC, Khan I. Macroscopic and microscopic authentication of Chinese and North American species of *Ephedra*. J AOAC Int 2005;88:707–713.
64. Haller CA, Duan M, Benowitz NL, Jacob P III. Concentrations of ephedra alkalis and caffeine in commercial dietary supplements. J Anal Toxicol 2004;28;145–151.
65. Chen JS, Chang KJ, Charng RC, Lai SJ, Binder SR, Essien H. The development of a broad-spectrum toxicology screening program in Taiwan. J Toxicol Clin Toxicol 1995;33:581–589.
66. Berlin I, Warot D, Aymard G, Acquaviva E, Legrand M, Labarthe B, et al. Pharmacodynamics and pharmacokinetics of single nasal (5 mg and 10 mg) and oral (50 mg) doses of ephedrine in healthy subjects. Eur J Clin Pharmacol 2001;57:447–455.
67. Schier JG, Traub SJ, Hoffman RS, Nelson LS. Ephedrine-induced cardiac ischemia: exposure confirmed with a serum level. J Toxicol Clin Toxicol 2003;41:849–853.
68. Kunsman GW, Jones R, Levine B, Smith ML. Methylephedrine concentrations in blood and urine specimens. J Anal Toxicol 1998;22:310–313.
69. Blechman KM, Karch SB, Stephens BG. Demographic, pathologic, and toxicological profiles of 127 decedents testing positive for ephedrine alkaloids. Forensic Sci Int 2004;139;61–69.

Chapter 80

MILK THISTLE
[*Silybum marianum* (L.) Gaertner]

HISTORY

For over 2,000 years, milk thistle has been a treatment for hepatobiliary disease. The Roman naturalist, Pliny the Elder (AD 23–79), described this plant as "excellent for carrying off the bile." During the 16th century, milk thistle was a treatment for liver diseases. The English herbalist/physician, Nicolas Culpepper (1616–1654) recommended milk thistle for jaundice, and the Swiss physician, Albrecht von Haller (1708–1777) used milk thistle as a treatment for liver diseases. European colonists introduced milk thistle into the United States and South America. Medical herbalists (eclectics) used milk thistle for liver congestion, menstrual disorders, renal dysfunction, and varicose veins.[1]

BOTANICAL DESCRIPTION

Common Name: Milk Thistle, Blessed Milk-Thistle, Blessed Milkthistle, Spotted Thistle, Variegated Thistle

Scientific Name: *Silybum marianum* (L.) Gaertner

Botanical Family: Asteraceae (sunflowers, tournesols)

Physical Description: This annual or biennial plant reaches 3–10 ft (~1–3 m) in height with large, alternating leaves that contain spines on the edges. The veins on the leaves extrude a white, milky sap when ruptured. Single, large purple flowers with sharp spines appear at the end of each stem from June to August.

Distribution and Ecology: Milk thistle is indigenous to southern Europe, southern Russia, North Africa, and Asia Minor. This plant has naturalized to dry, rocky soils in North and South America, Australia, China, and Central Europe.

EXPOSURE

Sources

The seeds and dried fruits from milk thistle (*Silybum marianum*) are the source of the silymarin complex of flavonolignans, which are used to standardize milk thistle extract to contain 70–80% silymarin flavonolignans. The seeds and dried fruits usually contain about 1.5–3% silymarin. The typical silymarin extraction process involves the defatting of the seeds in a Soxhlet extraction with petroleum ether for 4 hours followed by a second Soxhlet extraction with methanol for 5 hours.[2] The maximum yields (mg/g seed) of flavonolignans in several samples of seeds from milk thistle using ethanol were as follows: silibinin A, 4.0; silibinin B, 7.0; silicristin (silychristin), 4.0; taxifolin, 0.6; and silidianin, 0.4.[3] The active flavonolignans in silymarin are poorly water soluble. Consequently, the extract is administered as a capsule rather than a tea, which typically contains <10% silymarin.

Medicinal Uses

Traditional

Milk thistle has been a traditional treatment of liver disease and dyspepsia since ancient times, including the

Medical Toxicology of Natural Substances, by Donald G. Barceloux, MD

treatment of gallstones and jaundice. Other traditional uses for milk thistle include amenorrhea, uterine hemorrhage, constipation, diabetes, hay fever, and varicose veins.

CURRENT

Interpretation of the clinical trials measuring the efficacy of milk thistle extract in hepatobiliary disease is difficult because of a number of flaws in these trials including small sample size, variability in the severity and type of hepatic disease, heterogeneous doses, frequent lack of control groups, and poorly defined end points. Based on well-designed clinical trials, the use of milk thistle extracts probably does not alter the clinical course of patients with alcoholic cirrhosis or hepatitis B/C liver disease.[4] Some studies suggest, but do not prove, that milk thistle extract improves liver injury. Pooled data from case reports involving 452 patients with *Amanita phalloides* poisoning suggest a significant difference in mortality favoring the use of silibinin (mortality 9.8% vs. 18.3%, $P < 0.01$).[5] However, there are no randomized clinical trials to document the efficacy of silibinin for amanita poisoning. Silymarin is possibly efficacious as an adjuvant in the therapy of alcoholic liver disease, but the clinical data are inconsistent. A randomized trial of 200 alcoholics with cirrhosis did not demonstrate a statistically significant reduction in mortality between the silymarin and control groups during the 2.5-year observation period.[6] Additionally, a randomized-controlled clinical trial failed to detect a significant difference in serum hepatic aminotransferases between the silymarin and control groups.[7] Despite some positive results in patients with acute viral hepatitis, the value of silymarin in the treatment of these infections is unproven. A systematic review of 148 studies involving the treatment of liver disease with silymarin noted that only seven studies focused on the outcome of patients with hepatitis B, hepatitis C, or unspecified viral hepatitis.[8] These studies suggested the possibility of silymarin treatment reducing serum hepatic aminotransferases in these patients, but there was no evidence of improvement in viral load or liver histology.

REGULATORY STATUS

The German Commission E approves the use of silymarin for the treatment of liver diseases, including hepatitis A, alcoholic cirrhosis, and chemically induced hepatitis.

PRINCIPAL INGREDIENTS

Chemical Composition

Silymarin is a mixture of flavonolignans present in milk thistle (*Silybum marianum*) seed and fruit extracts. The major constituents of the silymarin complex are the following four isomeric flavonolignans: silibinin (silybin, CAS RN: 22888-70-6), isosilibinin (isosilybin, CAS RN: 72581-71-6), silidianin (silydianin, CAS RN: 29782-68-1), and silicristin (silychristin, CAS RN: 33889-69-9) as demonstrated in Figure 80.1. There are at least two pairs of diastereoisomeric flavonolignans compounds (silibinin A and B, isosilibinin A and B).[9] These flavonolignans have phenylchromone skeletons (flavonoid moiety) with different oxidative links to coniferyl alcohols (lignan moiety).[10] Silibinin and isosilibinin occur as a pair of diastereomers, epimeric at C12 and C13. As measured by high performance liquid chromatography (HPLC) and capillary electrophoresis (CE), several samples of dried fruits from milk thistle contained about 2.5–3% weight/weight silymarin complex.[11] Silymarin accounts for about 40–80% of the bulk material for producing pharmaceutic preparations of the milk thistle, depending on the geographical source, season, and growing conditions.[12] Some, but not all, pharmaceutical preparations of milk thistle are standardized to 70% silymarin content. The principal flavonolignane in silymarin is silibinin, which comprises about 50–70% of the silymarin content. Minor flavonolignans in the silymarin complex include isosilicristin, silymonin, silandrin, 2,3-dehyrosilibinin, 2,3-dehydrosilicristin, and taxifolin.[13] The latter compound is a dihydroquercitin precursor to the flavonolignans. Remaining constituents of milk thistle extract are mostly polymeric and oxidized polyphenolic compounds that are not well delineated.[14]

Physiochemical Properties

Taxifolin (CAS RN: 480-18-2, $C_{15}H_{12}O_7$) and silicristin are more polar compounds than silibinin A and silibinin B, resulting in differences in the extraction rates for a given extraction process.[15] Degradation of flavonolignans (e.g., silibinin, silicristin) occurs at temperatures above 100 °C (212 °F).[16] Properties associated with the administration of silymarin include antioxidation, free radical scavenger, inhibition of lipid peroxidation, repletion of glutathione stores, and inhibition of phase I reactions.[17]

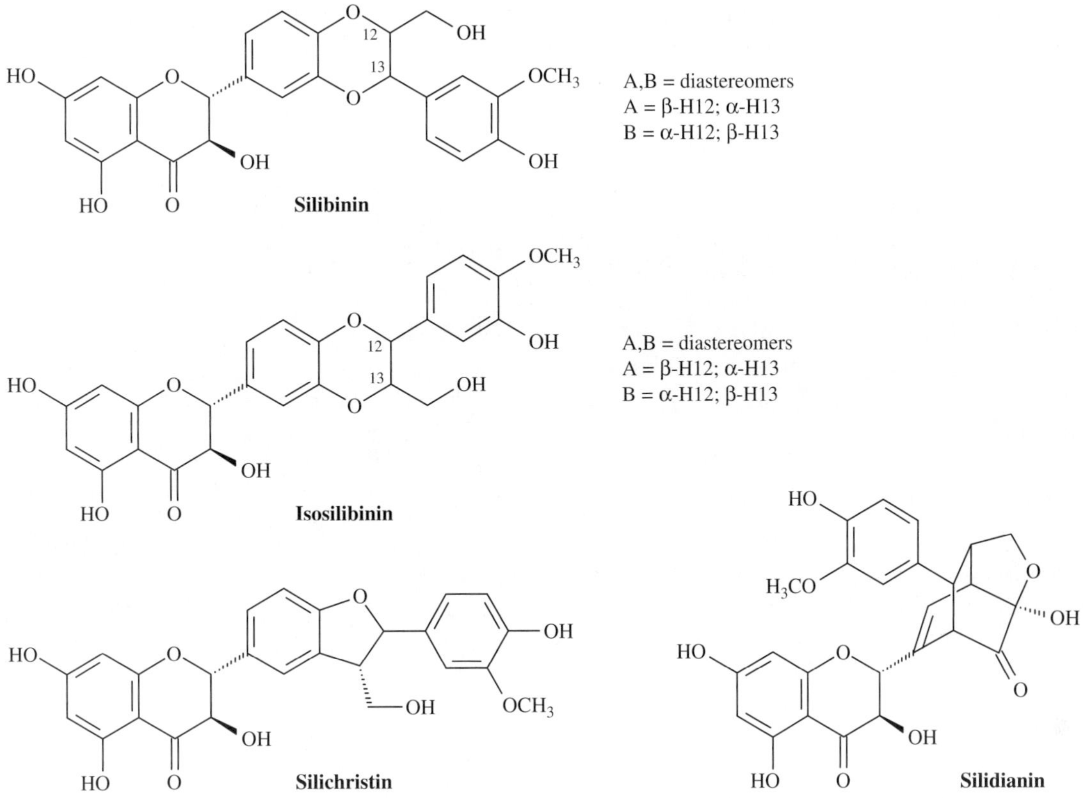

FIGURE 80.1. Chemical structures of major flavonolignans in milk thistle extract.

DOSE RESPONSE

The typical adult dose of the extract of milk thistle is 100–300 mg 3 times daily standardized to 70–80% silymarin.

TOXICOKINETICS

Kinetics

Absorption of silibinin from the gastrointestinal (GI) tract is relatively rapid with peak concentrations of free silibinin occurring about 1.5–2.5 hours after ingestion. The bioavailability of silibinin in milk thistle extract is relatively low (20–50%). The addition of phosphatidylcholine to silymarin improves the bioavailability of the flavonolignans.[18] Most (~90%) of the silibinin in plasma appears as conjugated silibinin as a result of the rapid conjugation of silibinin. The elimination half-life of total silibinin is relatively short. Following the administration of single doses ranging from 102–254 mg to six healthy volunteers, the elimination half-life of silibinin was approximately 6 hours[19] compared with about 4 hours for 12 healthy volunteers receiving a lipophilic silibinin-phosphatidylcholine complex.[20] The latter group received silipide® (IdB 1016, Inverni della Beffa Research and Development Laboratories, Milan, Italy) in a dose of 80 mg expressed as silibinin equivalents. Free silibinin is rapidly cleared from the plasma with a plasma elimination half-life of about 2 hours. The major metabolite of silibinin is a demethylated compound, while mono-hydroxy and di-hydroxy silibinin are minor metabolites.[21] The kidneys excrete <3–5% of the absorbed dose of silibinin as total silibinin (free + conjugated). Biliary excretion is a major route of silibinin elimination.[22]

Drug Interactions

In vitro studies suggest that silibinin inhibits both phase I and II reactions including CYP3A4, CYP2C9, and UGT1A1.[23] However, the clinical relevance of these effects is equivocal. A meta-analysis of three studies on the interaction of milk thistle extract on indinavir pharmacokinetics demonstrated a pooled mean difference of 1% in the AUC_{0-8} (95% CI: –53–55%) of indinavir before and after the administration of milk thistle extract.[24] This difference was not statistically significant

(P = 0.97). An *in vivo* study of 12 healthy volunteers receiving 700 mg milk thistle extract daily for 28 days did not detect a statistically significant difference between the presupplementation and postsupplementation phenotypic ratios for CYP1A2, CYP2D6, CYP2E1, or CYP3A4 activity.[25] Other *in vivo* studies of volunteers receiving 900 mg standardized milk thistle extract daily for 14 days indicated that supplementation with this extract did not alter digoxin pharmacokinetics (P-glycoprotein transporter)[26] or midazolam (CYP3A4) kinetics.[27] These studies suggest that any potential CYP-mediated herb–drug interaction between milk thistle extract and drugs metabolized by these isoenzymes would be minimal.

CLINICAL RESPONSE

With the exception of anaphylaxis, the adverse reactions associated with the use of milk thistle extract are usually mild. These effects typically involve headache, GI disturbances (nausea, vomiting, diarrhea, epigastric pain), and allergic reaction (pruritus, urticaria, arthralgias).[28] A case report temporally associated the use of milk thistle capsules with intermittent episodes of nausea, vomiting, diarrhea, colicky abdominal pain, sweating, and "collapse."[29] These symptoms resolved after cessation and recurred after re-challenge with the milk thistle extract. There was no analysis of the capsules for contaminants; therefore, a causal link could not be established between silymarin from milk thistle extract and the observed clinical effects.

DIAGNOSTIC TESTING

Methods for the quantitation of flavonolignans in the silymarin complex include capillary zone electrophoresis[30] and high performance liquid chromatography with UV,[31] and liquid chromatography with electrochemical or mass spectrometric detection.[32] Stereoselective assays can quantitate free plasma (unconjugated) silibinin using column-switching HPLC with electrochemical detection and total plasma (free and conjugated) silibinin using reverse-phase HPLC with UV detection.[33] The limit of quantitation for these methods is approximately 0.25 ng/mL per diastereomer. The limits of detection and quantitation for flavonolignans by HPLC with photodiode array detector and electron spray ionization-mass spectrometry ranged from 0.5–2 ng/mL and 2.5–10 ng/mL, respectively.[34] The thermal stability of the silymarin complex is relatively low, and long-term studies indicate that the shelf-life of milk thistle tinctures (hydro-alcohol extracts) is about 3 months at 25 °C (77 °F).[35]

TREATMENT

Adverse reactions to the use of milk thistle extracts are usually mild and treatment is supportive. Gut decontamination is unnecessary. Rarely, allergic reactions including anaphylaxis may occur that requires standard care (epinephrine, antihistamines, intravenous fluids, vasopressors).

References

1. Flora K, Hahn M, Rosen H, Benner K. Milk thistle (*Silybum marianum*) for the therapy of liver disease. Am J Gastroenterol 1998;93:139–143.
2. Wallace SN, Carrier DJ, Clausen EC. Batch solvent extraction of flavonolignans from milk thistle (*Silybum marianum* L. Gaertner). Phytochem Anal 2005;16:7–16.
3. Wallace SN, Carrier DJ, Clausen EC. Extraction of nutraceuticals from milk thistle Part II. Extraction with organic solvents. Appl Biochem Biotechnol 2003;105–108: 891–903.
4. Rambaldi A, Jacobs BP, Iaquinto G, Gluud C. Milk thistle for alcoholic and/or hepatitis B or C liver diseases—a systematic Cochrane hepato-biliary group review with meta-analyses of randomized clinical trials. Am J Gastroenterol 2005;100:2583–2591.
5. Saller R, Meier R, Brignoli R. The use of silymarin in the treatment of liver diseases. Drugs 2001;61:2035–2063.
6. Pares A, Planas R, Torres M, Caballeria J, Viver JM, Acero D, et al. Effects of silymarin in alcoholic patients with cirrhosis of the liver: results of a controlled, double-blind, randomized and multicenter trial. J Hepatol 1998;28: 615–621.
7. Lucena MI, Andrade RJ, De la Cruz JP, Rodriguez-Mendizabal M, Blanco E, Sanchez de la Cuesta F. Effects of silymarin MZ-80 on oxidative stress in patients with alcoholic cirrhosis. Results of a randomized, double-blind, placebo-controlled clinical study. Int J Clin Pharmacol Ther 2002;40:2–8.
8. Mayer KE, Myers RP, Lee SS. Silymarin treatment of viral hepatitis: a systematic review. J Viral Hepatitis 2005;12: 559–567.
9. Lee DY-W, Liu Y. Molecular structure and stereochemistry of silybin A, silybin B, isosilybin A, and isosilybin B, isolated from *Silybum marianum* (milk thistle). J Nat Prod 2003;66:1171–1174.
10. Crocenzi FA, Roma MG. Silimarin as a new hepatoprotective agent in experimental cholestasis: new possibilities for an ancient medication. Curr Medicinal Chem 2006;13: 1055–1074.
11. Quaglia MG, Bossu E, Donati E, Mazzanti G, Brandt A. Determination of silymarin in the extract from dried *Silybum marianum* fruits by high performance liquid chromatography and capillary electrophoresis. J Pharm Biomed Anal 1999;19:435–442.

12. Wagner H, Diesel P, Seitz M. [The chemistry and analysis of silymarin from *Silybum marianum* Gaertn]. Arzneimittelforschung 1974;24:466–471. [German]
13. Kim N-C, Graf TN, Sparacino CM, Wani MC, Wall ME. Complete isolation and characterization of silybins and isosilybins from milk thistle (*Silybum marianum*). Org Biomol Chem 2003;1:1684–1689.
14. Sonnenbichler J, Zetl I. Biochemical effects of the flavonolignane silibinin on RNA, protein and DNA synthesis in rat livers. Prog Clin Biol Res 1986;213:319–331.
15. Barreto JF, Wallace SN, Carrier DJ, Clausen EC. Extraction of nutraceuticals from milk thistle. Appl Biochem Biotechnol 2003;105–108:881–889.
16. Duan L, Carrier DJ, Clausen EC. Silymarin extraction from milk thistle using hot water. Appl Biochem Biotechnol 2004;113–116:559–568.
17. Carini R, Comoglio A, Albano E, Poli G. Lipid peroxidation and irreversible damage in the rat hepatocyte model. Protection by the silybin-phospholipid complex IdB 1016. Biochem Pharmacol 1992;43:2111–2115.
18. Barzaghi N, Crema F, Gatti G, Pifferi G, Perucca E. Pharmacokinetic studies on IdB 1016, a silybin-phosphatidylcholine complex, in healthy human subjects. Eur J Drug Metab Pharmacokinet 1990;15:333–338.
19. Weyhenmeyer R, Mascher H, Birkmayer J. Study on doselinearity of the pharmacokinetics of silibinin diastereomers using a new stereospecific assay. Int J Clin Pharmacol Ther Toxicol 1992;30:134–138.
20. Gatti G, Perucca E. Plasma concentrations of free and conjugated silybin after oral intake of a silybin-phosphatidylcholine complex (silipide) in healthy volunteers. Int J Clin Pharmacol Ther 1994;32:614–617.
21. Gunaratna C, Zhang T. Application of liquid chromatography-electrospray ionization-ion trap mass spectrometry to investigate the metabolism of silibinin in human liver microsomes. J Chromatogr B 2003;794:303–310.
22. Schandalik R, Gatti G, Perucca E. Pharmacokinetics of silybin in bile following administration of silipide and silymarin in cholecystectomy patients. Arzneimittelforschung 1992;42:964–968.
23. Sridar C, Gossen TC, Kent UM, Williams JA, Hollenberg PF. Silybin inactivates cytochromes P450 3A4 and 2C9 and inhibits major hepatic glucuronosyl transferases. Drug Metab Disp 2004;32:587–594.
24. Mills E, Wilson K, Clarke M, Foster B, Walker S, Rachlis B, et al. Milk thistle and indinavir: a randomized controlled pharmacokinetics study and meta-analysis. Eur J Clin Pharmacol 2005;61:1–7.
25. Gurley BJ, Gardner SF, Hubbard MA, Williams DK, Gentry WB, Carrier J, et al. *In vivo* assessment of botanical supplementation on human cytochrome P450 phenotypes: *Citrus aurantium, Echinacea purpurea*, milk thistle, and saw palmetto. Clin Pharmacol Ther 2004;76:428–440.
26. Gurley BJ, Barone GW, Williams DK, Carrier J, Breen P, Yates CR, et al. Effect of milk thistle (*Silybum marianum*) and black cohosh (*Cimicifuga racemosa*) supplementation on digoxin pharmacokinetics in humans. Drug Metab Disp 2006;34:69–74.
27. Gurley B, Hubbard MA, Williams K, Thaden J, Tong Y, Gentry WB, et al. Assessing the clinical significance of botanical supplementation on human cytochrome P450 3A activity: comparison of a milk thistle and black cohosh product to rifampin and clarithromycin. J Clin Pharmacol 2006;46:201–213.
28. Wellington K, Jarvis B. Silymarin: a review of its clinical properties in the management of hepatic disorders. BioDrugs 2001;15:465–489.
29. Adverse Drug Reactions Advisory Committee (Australia). An adverse reaction to the herbal medication milk thistle (*Silybum marianum*). Med J Aust 1999;170: 218–219.
30. Kvasnicka F, Biba B, Sevcik R, Voldrich M, Kratka J. Analysis of the active components of silymarin. J Chromatogr A 2003;990:239–245.
31. Rickling B, Hans B, Kramarczyk R, Krumbiegel G, Weyhenmeyer R. Two high-performance liquid chromatographic assays for the determination of free and total silibinin diastereomers in plasma using column switching with electrochemical detection and reversed-phase chromatography with ultraviolet detection. J Chromatogr B Biomed Appl 1995;670:267–277.
32. Ding T, Tian S, Zhang Z, Gu D, Chen Y, Shi Y, Sun Z. Determination of active component in silymarin by RP-LC and LC/MS. J Pharm Biomed Anal 2001;26: 155–161.
33. Rickling B, Hans B, Kramarczyk R, Krumbiegel G, Weyhenmeyer R. Two high-performance liquid chromatographic assays for the determination of free and total silibinin diastereomers in plasma using column switching with electrochemical detection and reversed-phase chromatography with ultraviolet detection. J Chromatogr B Biomed Appl 1995;670:267–277.
34. Zhao Y, Chen B, Yao S. Simultaneous determination of abietane-type diterpenes, flavonolignans and phenolic compounds in compound preparations of *Silybum marianum* and *Salvia miltiorrhiza* by HP-DAD-ESI MS. J Pharmaceut Biomed Anal 2005;38:564–570.
35. Bilia AR, Bergonzi MC, Gallori S, Mazzi G, Vincieri FF. Stability of the constituents of calendula, milk-thistle and passionflower tinctures by LC-DAD and LC-MS. J Pharmaceut Biomed Anal 2002;30:613–624.

Chapter 81

PASSIONFLOWER
(*Passiflora incarnata* L.)

HISTORY

Archeological excavations in North America suggest that the passionflower was used by prehistoric cultures in Virginia.[1] The Spanish explorer, Monardus, discovered *Passiflora incarnata* during an expedition to Peru in 1569. He associated the beautiful flower with the passion of Christ.[2] This plant became a popular traditional remedy and homeopathic medicine for insomnia and anxiety. During the late 18th century, the *Materia Medica Americana* mentioned the use of *Passiflora incarnata* for the treatment of sleeping problems in infants and adult epilepsy.

BOTANICAL DESCRIPTION

The genus, *Passiflora*, contains approximately 500 species. *Passiflora incarnata* L. and *Passiflora edulis* Sims are nearly identical *Passiflora* species that require sophisticated analysis for separation including microscopy (vein-inlet number, vein-termination number, stomatal number, stomatal index), physicochemical properties (ash values, extractive values), and thin layer chromatography profile.[3] The latter species is primarily an edible tropical fruit that does not possess significant central nervous system (CNS) depressant activity.[4] Some immature *Passiflora* fruits (e.g., *Passiflora adenopoda* DC, *P. edulis*) are toxic as a result of the presence of the cyanogen, prunasin.[5–6] However, most of the hydrogen cyanide is lost during maturation and the crushing of the fruit during processing. Although *Passiflora alata* Dryander has a lower total flavonoid concentration compared with *P. incarnata, P. alata* is an official herb and a substitute for *P. incarnata* in the *Pharmacopoeia* of Brazil.[7] *P. incarnata* does not grow well in the Brazilian climate.

Common Name: Purple Passionflower, Maypops, Maracuja, Apricot Vine, Passion Vine, Fleur de la Passion, Passiflore, Corona de Cristo

Scientific Name: *Passiflora incarnata* L.

Botanical Family: Passifloraceae (passionflower)

Physical Description: This woody, climbing vine has broadly cordate-ovate, serrated, palmately three-lobed leaves and glabrous stems with thin, green to purple bark. The flowers are solitary with a calyx containing five white petals and a purplish fringe called a corona. The fruit is ovoid and yellow.

Distribution and Ecology: This plant is indigenous to warm temperate and tropical regions in the southeastern United States.

EXPOSURE

Sources

This herbal remedy is administered as an infusion or an extract, often in combination with other herbs (valerian root, hops, *Melissa officinalis* L.).

Medical Toxicology of Natural Substances, by Donald G. Barceloux, MD

Medicinal Uses

Traditional

The dried aerial parts (leaves, flowers, fruit) from several *Passiflora* species are widely used in folk medicine as sedatives, anxiolytic agents, and antispasmodics.[8] Other traditional uses include the treatment of hysteria, neurasthenia, neuralgias, insomnia, epilepsy, dysmenorrhea, burns, hemorrhoids, diarrhea, cough, and asthma.[9]

Current

Preliminary clinical trials suggest that extract of passionflower is a possibly effective adjunctive agent for the treatment of opiate withdrawal and as a treatment for anxiety. However, there are too few randomized controlled clinical trials to determine the efficacy and safety of the use of passionflower for the treatment of anxiety.[10] A double-blind randomized controlled trial of 65 opiate addicts assigned to clonidine only and clonidine plus 60 drops passiflora extract daily detected no statistically significant difference between the two groups on the Short Opiate Withdrawal Scale.[11] There was a reduction in opiate withdrawal-associated psychological symptoms (nonstandardized observations by blinded physician) on the second day of treatment compared with baseline, and a higher psychological symptoms score than baseline (statistical significance not reported) for the clonidine only group on days 2,3,4,7, and 14 after withdrawal began. A pilot double-blind randomized controlled trial of 36 outpatients with generalized anxiety disorder assigned 18 patients to the oxazepam group (30 mg daily) and passiflora group (45 drops extract daily).[12] There was no statistically significant difference in the Hamilton Anxiety Rating Scale during the 28-day study. However, the reduction in these scores began earlier with the oxazepam group (day 4) compared with the passiflora group (day 7).

Regulatory Status

Passionflower extract is an approved dietary supplement and flavoring agent in the United States and Europe. The German Commission E approves the use of passionflower for nervous restlessness.

PRINCIPAL INGREDIENTS

Chemical Composition

Extracts of *Passiflora incarnata* contain flavonoids, β-carboline alkaloids, and terpene derivatives. The specific active ingredients are unknown. In rodent studies, both a lipophilic and hydrophilic fraction prolongs sleeping time and raises the nociceptive threshold in tail-flick and hot-plate tests.[13] Suspected bioactive compounds in passiflora extract include C-glycosyl flavonoid derivatives of apigenin and luteolin (vitexin, isovitexin, orientin, isoorientin, schaftoside). The most abundant flavonoids in passiflora extract include isovitexin, isovitexin-2″-*O*-β-D-glucopyarnoside, schaftoside, isoschaftoside, isoorientin-2″-*O*-β-D-glucopyranoside, and swertisin.[14] As measured by liquid chromatography with UV detection, the mean concentration of isovitexin in three samples of passiflora extract was 11.98 ± 0.09 mg/mL compared with vitexin (1.64 ± 0.0 mg/mL), hyperoside (0.32 ± 0.009 mg/mL), and apigenin (0.23 ± 0.03 mg/mL).[15] The concentrations of passiflora flavonoids vary substantially between various extracts, depending on the plant part, season, and *Passiflora* species. The flavonoid content in *Passiflora incarnata* is highest in the leaves, with peak isovitexin concentrations occurring just before the flowering stage.[16] In a study of 10 commercial *Passiflora incarnata* extracts analyzed by capillary electrophoresis, isovitexin-2″-*O*-β-D-glucopyarnoside was the major flavonoid in seven samples compared with isovitexin in two samples and swertisin in one sample.[17] The total mean flavonoid content in these extracts ranged from 0.37–3.75% with most flavonoid concentrations in the range of 1.18–2.48%. Figure 81.1 displays the chemical structures of some major and minor flavanoid compounds in passionflower extract.

	R_1	R_2	R_3	R_4
Swertisin	H	glc	CH_3	H
Isoscoparin-2"-O-β-D-glucopyranoside	OCH_3	saph	H	H
Vicenin-2	H	glc	H	glc
Isovitexin-2"-O-β-D-glucopyranoside	H	saph	H	H
Schaftoside	H	glc	H	ara
Leucenin-2	OH	glc	H	glc
Isoorientin-2"-O-β-D-glucopyranoside	OH	saph	H	H
Isoschaftoside	H	ara	H	glc
Isovitexin	H	glc	H	H
Isoorientin	OH	glc	H	H
Vitexin	H	H	H	glc
Orientin	OH	H	H	glc
Apigenin-6-C-glucosyl-8-β-D-ribofuranoside	H	glc	H	rib

FIGURE 81.1. Chemical structures of some major and minor flavonoids in passionflower extract.

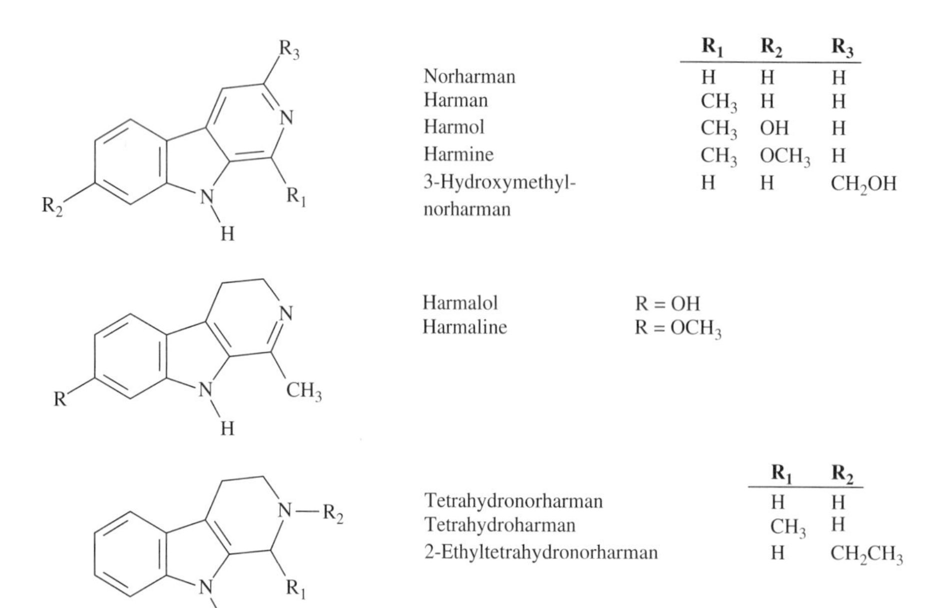

FIGURE 81.2. Chemical structures of β-carbolines in passionflower.

Indole harmala alkaloids in *Passiflora* species include β-carbolines (harman, norharman, harmol, harmine), and dihydro-β-carbolines (harmalol, harmaline).[18] Figure 81.2 displays the chemical structures of these β-carboline compounds. The concentration of these alkaloids in *P. incarnata* is relatively small as measured by reversed-phase high performance liquid chromatography (RP/HPLC), depending on growing conditions.[19] In a study of dry pulverized plant material using RP/HPLC with selective fluorometric detection, the range of harmala alkaloid concentrations (ng/g dried material) in specimens of passionflower were approximately as follows: harman, 48–224; norharman, 51–81; harmine, nondetectable to 79; harmol, 24–977; and harmalol, 745–776.[20] Harmaline was not detectable in these specimens. The leaves typically contain the highest concentration of harmala alkaloids.[21] *P. incarnata* contains several aromatic monoterpenes including limonene, cumene, and α-pinene.[2] The unripe fruits and stems of some species of *Passiflora* (*P. adenopoda*) contain cyanogenic glycosides, but there are inadequate data to determine if clinically significant amounts of cyanogenic glycosides are present in the fruits of *P. incarnata*. Ripe fruits of *Passiflora* species do not usually contain cyanogenic glycosides.

Physiochemical Properties

Infusions of *P. incarnata* have higher total flavonoid content per gram of dry plant material than hydroethanol extracts probably because of extraction requirements and the higher solubility of glycosylated flavonoids in water than in ethanol/water extracts.[7]

DOSE RESPONSE

Typically, daily oral doses of *Passiflora incarnata* include the following: dried herb, 0.5 to 2 g 3–4 times daily; tincture (1:8), 1–4 mL 3–4 times; tea, steeping 4–8 g dried herb; and infusion, 2.5 g 3–4 times. *Passiflora* extracts typically are not standardized, in part, because of the lack of clearly defined bioactive substances.

TOXICOKINETICS

Kinetics

There are few, if any, formal pharmacokinetic studies on the flavonoids or harmala alkaloids in *Passiflora incarnata*.

Drug Interactions

There are several potential drug interactions as a result of the ingredients in passiflora extracts including additive sedative effects (sedative-hypnotics, ethanol), enhancement of bleeding (aspirin, warfarin), and elevation of blood pressure (ephedra). However, there are few clinical data to suggest that these potential interactions are clinically relevant.

CLINICAL RESPONSE

Adverse effects associated with the use of passiflora extracts include sedation, dizziness, ataxia, and impaired cognitive function. A case report of a 34-year-old woman associated the development of severe nausea, vomiting, drowsiness, prolonged QTc interval (539 ms with nonspecific ST-T wave changes), and episodes of nonsustained ventricular tachycardia following the ingestion of 7 tablets of a commercial passiflora extract over 2 days.[22] The electrocardiogram normalized after several days of hospitalization. The tablets reportedly contained 500 mg of active ingredients, and the chromatographic profile of remaining tablets was similar to other tablets obtained from the manufacturer. Because of the lack of similar case reports, the authors suggested that this episode resulted from unique pharmacogenetics (e.g., CYP2D6) in this patient. However, no pharmacogenetic testing was done. Five patients required hospitalization for altered consciousness after ingesting an herbal product containing primarily passionflower fruit extract.[23] Several case reports associate the development of hypersensitivity skin and pulmonary reactions with exposure to *Passiflora* species. A case report documented the onset of cutaneous vasculitis and urticaria following the ingestion of passiflora extract.[24] A Brazilian pharmacy worker developed IgE-mediated occupational asthma and rhinitis following exposure to another species of *Passiflora* (*P. alata* Dryander).[25] Skin testing and Western blot confirmed the sensitization of the patient to this plant species and *Frangula purshiana* (cascara).

DIAGNOSTIC TESTING

Methods to determine flavonoid content in passiflora extracts include thin-layer chromatography, isocratic liquid chromatography,[26] RP/HPLC and capillary electrophoresis. The latter method has a limit of detection for passiflora flavonoids in the range of 4 μg/mL.[17] The limit of quantitation for flavonoids in passiflora extracts using liquid chromatography with UV detection was 1.29 μg/mL.[15] High performance thin layer chromatography is an alternative to HPLC for fingerprinting passiflora flavonoids.[7] The thermal stability of the passionflower tincture (60% volume/volume hydroalcoholic extract) is relatively low, and long-term studies indicate that the shelf-life of these tinctures is about 3 months at 25 °C (77 °F).[27]

TREATMENT

The treatment of adverse effects associated with passionflower is supportive, and there are few clinical data to guide management.

References

1. Gremillion KJ. The development of a mutualistic relationship between humans and maypops (*Passiflora incarnata* L.) in the southeastern United States. J Ethnobiol 1989; 9:135–158.
2. Dhawan K, Dhawan S, Sharma A. *Passiflora*: a review update. J Ethnopharmacol 2004;94:1–23.
3. Dhawan K, Kumar R, Kumar S, Sharma A. Correct identification of *Passiflora incarnata* Linn., a promising herbal anxiolytic and sedative. J Medicinal Food 2001;4:137–144.
4. Dhawan K, Kumar S, Sharma A. Comparative biological activity study on *Passiflora incarnata* and *P. edulis*. Fitoterapia 2001;72:698–702.
5. Saenz JA, Nassar M. Toxic effect of the fruit of *Passiflora adenopoda* D.C. on humans: phytochemical determination. Rev Biol Trop 1972;20:137–140.
6. Spencer KC, Seigler DS. Cyanogenesis of *Passiflora edulis*. J Agric Food Chem 1983;31:794–796.
7. Pereira CA, Yariwake JH, Lancas FM, Wauters J-N, Tits M, Angenot L. A HPTLC densitometric determination of flavonoids from *Passiflora alata, P. edulis, P. incarnata* and *P. caerulea* and comparison with HPLC method. Phytochem Anal 2004;15:241–248.
8. Dhawan K, Kumar S, Sharma A. Anxiolytic activity of aerial and underground parts of *Passiflora incarnata*. Fitoterapia 2001;72:922–926.
9. Dhawan K, Sharma A. Antitussive activity of the ethanol extract of *Passiflora incarnata* leaves. Fitoterapia 2002;73: 397–399.
10. Miyasaka LS, Atallah AN, Soares BG. Passiflora for anxiety disorder. Cochrane Database Syst Rev 2007;(1): CD004518.
11. Akhondzadeh S, Kashani L, Mobaseri M, Hosseini SH, Nikzad S, Khani M. Passionflower in the treatment of opiates withdrawal: a double-blind randomized controlled trial. J Clin Pharm Ther 2001;26:396–373.
12. Akhondzadeh S, Naghavi HR, Vazirian M, Shayeganpour A, Rashidi H, Khani M. Passionflower in the treatment of generalized anxiety: a pilot double-blind randomized controlled trial with oxazepam. J Clin Pharm Ther 2001;26: 363–367.
13. Speroni E, Minghetti A. Neuropharmacological activity of extracts from *Passiflora incarnata*. Planta Med 1988;54: 488–491.
14. Qimin L, van den Heuvel H, Delorenzo O, Corthout J, Pieters LA, Vlietinck AJ, Claeys M. Mass spectral characterization of C-glycosidic flavonoids isolated from a medicinal plant (*Passiflora incarnata*). J Chromatogr 1991;562:435–446.
15. Muller SD, Vasconcelos SB, Coelho M, Biavatti MW. LC and UV determination of flavonoids from *Passiflora alata* medicinal extracts and leaves. J Pharmaceut Biomed Anal 2005;37:399–403.
16. Menghini A, Mancini LA. TLC determination of flavonoid accumulation in clonal populations of *Passiflora incarnata* L. Pharmacol Res Commun 1988;20(Suppl 5):113–116.
17. Marchart E, Krenn L, Kopp B. Quantification of the flavonoid glycosides in *Passiflora incarnata* by capillary electrophoresis. Planta Med 2003;69:452–456.

18. Poethke W, Schwarz C, Gerlach H. [Contents of *Passiflora bryonioides*. 1. Alkaloids]. Planta Med 1970;18:303–314. [German]
19. Rehwald A, Sticher O, Meier B. Trace analysis of Harman alkaloids in *Passiflora incarnata* by reversed-phase high performance liquid chromatography. Phytochem Anal 1995;6:96–100.
20. Tsuchiya H, Hayashi H, Sato M, Shimizu H, Iinuma M. Quantitative analysis of all types of ß-carboline alkaloids in medicinal plants and dried edible plants by high performance liquid chromatography with selective fluorometric detection. Phytochem Anal 1999;10:247–253.
21. Lutomski J, Malek B. [Pharmacological investigations on raw materials of the genus passiflora. 4. The comparison of contents of alkaloids in some Harman raw materials]. Planta Med 1975;27:381–386. [German]
22. Fisher AA, Purcell P, Le Couteur DG. Toxicity of *Passiflora incarnata* L. Clin Toxicol 2000;38:63–66.
23. Solbakken AM, Rorbakken G, Gundersen T. [Nature medicine as intoxicant]. Tidsskr Nor Laegeforen 1997;117:1140–1141. [Norwegian]
24. Smith GW, Chalmers TM, Nuki G. Vasculitis associated with herbal preparation containing *Passiflora* extract. Br J Rheumatol 1993;32:87–88.
25. Giavina-Bianchi PF, Castro FF, Machado ML, Duarte AJ. Occupational respiratory allergic disease induced by *Passiflora alata* and *Rhamnus purshiana*. Ann Allergy Asthma Immunol 1997;79:449–454.
26. Pietta P, Manera E, Ceva P. Isocratic liquid chromatographic method for the simultaneous determination of *Passiflora incarnata* L. and *Crataegus monogyna* flavonoids in drugs. J Chromatogr 1986;357:233–237.
27. Bilia AR, Bergonzi MC, Gallori S, Mazzi G, Vincieri FF. Stability of the constituents of calendula, milk-thistle and passionflower tinctures by LC-DAD and LC-MS. J Pharmaceut Biomed Anal 2002;30:613–624.

Chapter 82

PENNYROYAL and PULEGONE (*Mentha pulegium* L.)

HISTORY

Dating back to Roman times, pennyroyal has a long history in folk medicine as an abortifacient and as a method to induce menses.[1] Toxic effects usually result from the misuse of the herb in folk medicine. Reports of pennyroyal toxicity date back to the late 1800s when the ingestion of pennyroyal oil was associated with syncope, seizures, diaphoresis, vomiting, and coma.[2,3] In 1897, a 23-year-old woman developed gastrointestinal (GI) distress 4 days after ingesting pennyroyal oil, and died 4 days later.[4] Pennyroyal exposure in these cases was not confirmed by any laboratory analysis.

BOTANICAL DESCRIPTION

Common Name: Pennyroyal, European Pennyroyal, Squaw Mint, Mosquito Plant

Scientific Name: *Mentha pulegium* L.

Botanical Family: Lamiaceae (mint)

Physical Description: This perennial mint has a variable habit, ranging from low-growing, spreading plants to lanky, upright shrubs. The stems are square in cross-section, ascending from rhizomes. Branches and simple leaves are opposite on the stems. The flowers are pale or deeper pink, blue, or violet, and these flowers cluster in dense whorls at the upper nodes. Unlike many mints, the flowers are not strongly bilateral, but only slightly two-lipped. The plant has a pungent minty odor.

Distribution and Ecology: Pennyroyal is a perennial weed that is native to Europe, and this plant has naturalized in the western and northeastern regions of the United States.

Common Name: American False Pennyroyal

Scientific Name: *Hedeoma pulegioides* (L.) Pers.

Botanical Family: Lamiaceae (mint)

Physical Description: This strongly aromatic herb is a low-growing plant from 1/2–1 ft (~0.2–0.3 m) in height with slender erect, branched, somewhat hairy stems. The leaves are small, thin, and rather narrow. Clusters of pale-bluish flowers appear from July to September.

Distribution and Ecology: The American false pennyroyal is a native annual present in the midwestern, northeastern, and southern parts of the United States with the exception of Louisiana and Florida.[5] Table 82.1 lists the pennyroyal species present in the United States.

EXPOSURE

Sources

Pennyroyal oil is a volatile oil derived by steam distillation from European (*M. pulegium*) and American (*H. pulegioides*) varieties of pennyroyal, primarily from the leaves and flowering tops. Although pennyroyal contains high concentrations of pulegone, the main commercial source of pulegone is peppermint (*Mentha* X *piperita*) and cornmint (*Mentha arvensis* L.).[6]

Medical Toxicology of Natural Substances, by Donald G. Barceloux, MD

TABLE 82.1. US Species of Pennyroyal

Species	Common Name
Hedeoma acinoides Scheele	Slender false pennyroyal, McKittrick's false pennyroyal
Hedeoma costata Gray	Ribbed false pennyroyal
Hedeoma dentata Torr.	Dentate false pennyroyal
Hedeoma diffusa Greene	Spreading false pennyroyal
Hedeoma drummondii Benth.	Drummond's false pennyroyal
Hedeoma hispida Pursh	Rough false pennyroyal
Hedeoma hyssopifolia Gray	Aromatic false pennyroyal
Hedeoma mollis Torr.	Softhair false pennyroyal
Hedeoma nana (Torr.) Briq.	Dwarf false pennyroyal, California false pennyroyal
Hedeoma oblongifolia (Gray) Heller	Oblongleaf false pennyroyal
Hedeoma pilosa Irving	Old Blue false pennyroyal, veiny false pennyroyal
Hedeoma pulcherrima Woot. & Standl.	White Mountain false pennyroyal
Hedeoma pulegioides (L.) Pers.	American false pennyroyal
Hedeoma reverchonii (Gray) Gray	Reverchon's false pennyroyal
Hedeoma todsenii Irving	Todsen's false pennyroyal

Source: Adapted from Ref 5.

Medicinal Uses

Traditional

Herbalists use pennyroyal oil as an abortifacient, an emmenagogue (i.e., induction of menses), a pesticide for pets, and a treatment for chronic inflammation (e.g., chronic bronchitis, influenza, upper respiratory tract infections).[7] Some cultures (e.g., Hispanic) use boiled leaves from mint plants to treat infant colic and other minor ailments in children, as well as a traditional medicine for abortions and the induction of menses.[8] These teas are called *yerba buena*, and these preparations are not usually toxic unless confused with pennyroyal oil.[9]

Current

Because of the mint-like odor, pennyroyal oil is a constituent of alcoholic beverages, foods (candy, baked goods, ice cream) and a fragrant for soaps and detergents. In an Italian study, the concentration of pulegone in mint syrup, mint-flavored candy, mint-flavored lozenges, and toothpaste ranged up to about 30 mg/kg, while the range for 5 samples of chewing gum was 1.63–138.15 mg/kg.[6] Pulegone was present almost exclusively as the (+)-isomer. There are inadequate data to support the use of pennyroyal oil for traditional treatments (i.e., abortifacient, induction of menses), and the toxicity associated with the administration of pennyroyal oil contraindicates these uses.

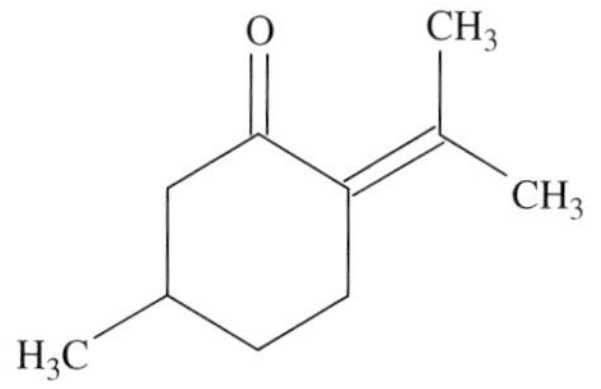

O CH$_3$ CH$_3$ H$_3$C

FIGURE 82.1. Chemical structure of pulegone.

Regulatory Status

There are no approved medicinal uses for pennyroyal oil. Pulegone is a component of peppermint oil, which is a fragrance component of cosmetic formulations. Recommended limits of pulegone in these formulations are ≤1%.[10] The European Commission (Directive 88/388/EEC) set maximum pulegone concentrations in the following foods and beverages: foodstuffs, 25 mg/kg; beverages, 100 mg/kg; peppermint or mint-flavored beverages, 250 mg/kg; and mint confectionary, 350 mg/kg.[11]

PRINCIPAL INGREDIENTS

Chemical Composition

The major constituent of pennyroyal oil is the cyclohexanone, (R)-(+)-pulegone (CAS RN: 89-82-7, $C_{10}H_{16}O$). This ketone constitutes about 80–90% and 16–30% of European and American pennyroyal oil, respectively. In addition, these oils contain several other monoterpenes including 3-methylcyclohexanone, α-pinene, β-pinene, p-cymene, limonene, menthone, isomenthone, isopulegone. Figure 82.1 displays the chemical structure of pulegone. Peppermint (*Mentha piperita* L.) also contains relatively high concentrations of pulegone, particular in young leaves. However, as the leaves mature, pulegone in peppermint undergoes metabolism to menthol and methone (dimedone).

Physiochemical Properties

Pennyroyal oil has a strong aromatic, mint-like odor and a pungent taste. Pulegone is a colorless, oily liquid that is almost insoluble in water. Table 82.2 demonstrates some physiochemical properties of pulegone. In animal studies, the pulegone metabolite, menthofuran, is more hepatotoxic than (R)-(+)-pulegone, whereas the (S)-(−) enantiomer of pulegone is about 30% as hepatotoxic as the (R)-(+) enantiomer of pulegone.[12]

TABLE 82.2. Physiochemical Properties of Pulegone

Physical Property	Value
Molecular Weight	152
Melting Point	<25 °C/77 °F
Boiling Point	224 °C/~435 °F
Specific Gravity	0.94
log P (Octanol-Water)	3.08
Atmospheric OH Rate Constant	1.12E-10 cm^3/molecule-second (25 °C/77 °F)

Mechanism of Toxicity

Experimental studies suggest that oxidative metabolites of pulegone cause hepatotoxicity because inhibitors of cytochrome P450 oxidases mitigate the hepatic damage caused by the administration of pennyroyal oil.[13] Reactive oxidative metabolites of pulegone and its oxidative metabolite, menthofuran bind to target cell proteins producing hepatocellular necrosis.[14] Pulegone forms reactive metabolites that deplete hepatic stores of glutathione to a substantially greater extent than menthofuran. Although menthofuran is a more potent hepatotoxin in animal studies, pulegone comprises 80–90% of pennyroyal oil compared with about 0.2% for menthofuran.

DOSE RESPONSE

There is no established safe dose of pennyroyal oil, and pennyroyal is not recommended for children as a result of potential toxicity and the lack of scientific data to support efficacy. In animal models, the intraperitoneal administration of pennyroyal oil produces a dose-dependent, centrilobular hepatic necrosis at doses exceeding 400 mg/kg body weight.[12] Depletion of glutathione exacerbates this hepatotoxicity. Case reports suggest that the ingestion of <10 mL of pennyroyal oil causes gastritis and mild central nervous system toxicity without hepatorenal damage, whereas fatalities result from the ingestion of 15–30 mL pennyroyal oil. The ingestion of ≤10 mL of pennyroyal oil by an adult was associated with GI distress, but no abnormalities of hepatorenal function or coagulation.[15] The ingestion of an estimated 30 mL of pennyroyal oil produced massive centrilobular hepatic necrosis, renal dysfunction, hepatic coma, and death in an 18-year-old woman.[16]

TOXICOKINETICS

Pulegone, the major monoterpene constituent of pennyroyal oil, undergoes extensive hepatic metabolism with the primary metabolite being menthofuran. Analysis of urine from rats administered pulegone demonstrated 11 metabolites including menthofuran, piperitone, piperitenone, p-cresol, 5-hydroxypulegone, 4-methylcyclohexenone, 3-methylcyclohexanone, isopulegone, pulegol, 7-hydroxypiperitone, and benzoic acid.[17] There are at least three distinct pathways involved in the metabolism of pulegone.[18] These three pathways include the following: 1) hydroxylation to monohydroxylated pulegone compounds followed by glucuronidation, 2) reduction of the carbon-carbon double bond to form diastereomeric menthone and isomenthone followed by hydroxylation and glucuronidation, and 3) Michael addition of glutathione to pulegone followed by further metabolism to produce diastereomeric 8-(*N*-acetylcystein-S-yl) menthone/isomenthone.[19]

Oxidation of the allylic methyl groups of pulegone produces the proximate hepatotoxic metabolite, menthofuran.[20] Subsequently, menthofuran undergoes oxidation by cytochrome P450 enzymes to form mintlactone and isomintlactone via several intermediaries (e.g., 2-hydroxymenthofuran). This metabolic pathway apparently does not involve the depletion of glutathione based on animal studies.[21] *In vitro* studies indicate that the oxidation of pulegone involves CYP1A2, CYP2E1, and CYP2C19, whereas the oxidation of menthofuran involves CYP2A6 in addition to these three isoenzymes.[22] A study using specific inhibitors of cytochrome P450 pathways suggested that CYP1A2 is more important than CYP2E1 in the development of a hepatotoxic metabolite responsible for pulegone-induced liver damage in the mouse model.[23] In mice administered 150 mg/kg, the plasma half-lives of pulegone and menthofuran were approximately 1 hour and 2 hours, respectively. There are few data on the toxicokinetics of pulegone during pennyroyal intoxication in humans other than documentation of the presence of pulegone and menthofuran.

CLINICAL RESPONSE

Toxicity results from the ingestion of pennyroyal oil as an abortifacient and as an infusion for the treatment of colic in children. The ingestion of pennyroyal oil produces direct toxic effects on the GI tract and the liver. Depending on the dose, the clinical presentation includes nausea, vomiting, abdominal pain, burning of the throat, and dizziness within 2 hours of ingestion followed by the delayed development of liver dysfunction in serious cases. In severe cases, lethargy, shock, consumptive coagulopathy (disseminated intravascular coagulation [DIC]), massive hepatic necrosis, and hepatorenal failure occur.[16,24] Multiple grand mal seizures and hypoglycemia occur rarely during pennyroyal intoxication.[25] Minor ingestions of pennyroyal oil are associated with GI distress and nonspecific symptoms (lightheadedness, generalized weakness). A rare case reports associated topical application of pennyroyal leaves with the development of contact dermatitis (erythema, vesiculation, edema, desquamation).[26] The patient had a positive direct patch test with pennyroyal leaves along with a history of atopic dermatitis (e.g., nickel allergy).

DIAGNOSTIC TESTING

Serum pulegone and menthofuran concentrations help confirm exposure to pennyroyal, but there are inadequate data to correlate these levels to clinical outcome. Pulegone is present in other members of the mint family, including peppermint oil. Therefore, the presence of pulegone and menthofuran alone does not necessarily confirm the ingestion of pennyroyal. Approximately $3^1/_2$ days after the ingestion of approximately 120mL of a tea brewed from a pennyroyal plant, the menthofuran concentration of a serum sample from an 8-week-old infant was 10ng/mL.[9] At that time, the infant had severe hepatic failure with cerebral edema. The following day the child died, and analysis of antemortem serum demonstrated a menthofuran concentration of 2ng/mL. Neither sample contained detectable concentrations of pulegone. The chronic ingestion of pennyroyal tea by a 6-month-old child produced serum pulegone and menthofuran concentrations of 25ng/mL and 41ng/mL, respectively, at the time of admission. On presentation to the emergency department the child was lethargic, post-ictal, hypoglycemic (serum glucose 7mg/dL), and hypoxic.[9] The child survived with residual muscle spastic rigidity and elevated serum hepatic aminotransferases 2 months after admission. Postmortem concentrations of pulegone and menthofuran in a sample of heart blood from a 24-year-old woman, who died after ingesting an unknown amount of pennyroyal herbal extract over 2 weeks, were 18ng/mL and 1ng/mL, respectively.[15] The last estimated ingestion of the pennyroyal herbal extract was approximately 2 days before death.

TREATMENT

Stabilization

The development of hepatorenal failure is delayed at least 24–48 hours after ingestion of pennyroyal. Seizures are usually self-limited and respond to diazepam or lorazepam. Severe hypoglycemia may occur due to hepatic dysfunction, particularly in children receiving toxic doses of teas brewed from pennyroyal.[9] All patients with altered mental status following pennyroyal ingestion should be promptly evaluated for blood glucose concentrations as well as respiratory function.

Decontamination

The administration of activated charcoal is a therapeutic option for alert patients who have ingested significant amounts of pennyroyal oil (i.e., more than a taste) within 1 hour of presentation; however, there are inadequate clinical data to determine if these decontamination measures improve outcome. Syrup of ipecac is not recommended because of the early onset of seizures and central nervous system (CNS) abnormalities.

Elimination Enhancement

No methods are available.

Antidotes

There are inadequate data to evaluate the efficacy of any antidotes for pennyroyal toxicity. The use of *N*-acetylcysteine is a potential antidote based on experimental studies suggesting that glutathione depletion enhances the hepatotoxicity of pennyroyal oil.[21] Case reports associate a benign clinical course with the administration of *N*-acetylcysteine (NAC) after the ingestion of pennyroyal oil.[15,27] However, efficacy of NAC is difficult to evaluate because of lack of knowledge about the clinical course of pennyroyal intoxication not treated with NAC. There are no clinical studies to guide treatment with NAC during pennyroyal intoxication. The case reports described the use of the same NAC doses as administered for acetaminophen toxicity.

Supplemental Care

Asymptomatic patients or those whose symptoms subside within 4 hours postingestion may be discharged. Symptomatic patients should be admitted. Patients admitted for observation should be monitored for mental status changes, respiratory depression, and hepatorenal failure. Laboratory tests include a complete blood count, electrolytes, creatinine, liver function tests, urinalysis, and coagulation profiles. Bedside glucose monitoring is necessary several times daily in patients with fulminant hepatic failure. The efficacy of vitamin K_1 for elevation of the prothrombin time during fulminant hepatic failure is unclear. Fresh frozen plasma (rarely, recombinant human factor VIIa) should be used for evidence of current bleeding or prophylactically for invasive procedures. Prophylactic correction of coagulopathy in a nonbleeding patient is not usually recommended. Serial hemoglobin and stool guaiac tests should be followed for evidence of GI bleeding. The administration of proton pump inhibitors (e.g., omeprazole) should be administered to reduce the risk of GI hemorrhage. Cerebral edema is the major cause of death following the development of hepatic encephalopathy, and treatment includes mannitol, elevation of the head to 10–20 degrees, adjustment of ventilator settings, and fluid restriction. Epidural intracranial pressure monitoring is usually necessary to guide treatment of patients with grade 3–4 hepatic encephalopathy. Patients with severe pennyroyal toxicity are candidates for liver transplantation, and these patients should be considered for liver transplantation early (i.e., before cerebral edema causes permanent sequelae).

References

1. Gunby P. Plant known for centuries still causes problems today. JAMA 1979;241:2246–2247.
2. Wingate UO. A case of poisoning by the oil of hedeoma (pennyroyal). Boston Med Surg J 1889;120:536.
3. Girling J. Poisoning by pennyroyal. Br Med J 1887;1: 1214.
4. Allen WT. Note on a case of supposed poisoning by pennyroyal. Lancet 1897;i:1022–1023.
5. US Department of Agriculture, NRCS. The PLANTS Database, Version 3.1; 2001. Available at http://plants.usda.gov. Accessed April 10, 2008.
6. Siano F, Catalfamo M, Cautela D, Servillo L, Castaldo D. Analysis of pulegone and its enanthiomeric distribution in mint-flavoured food products. Food Addit Contam 2005;22:197–203.
7. Ciganda C, Laborde A. Herbal infusions used for induced abortion. J Toxicol Clin Toxicol 2003;41:235–239.
8. Conway GA, Slocumb JC. Plants used as abortifacients and emmenagogues by Spanish New Mexicans. J Ethnopharmacol 1979;1:241–261.
9. Bakerink JA, Gospe SM Jr, Dimand RJ, Eldridge MW. Multiple organ failure after ingestion of pennyroyal oil from herbal tea in two infants. Pediatrics 1995;98: 944–947.
10. Nair B. Final report on the safety assessment of *Mentha piperita* (peppermint) oil, *Mentha piperita* (peppermint) leaf extract, *Mentha piperita* (peppermint) leaf, and *Mentha piperita* (peppermint) leaf water. Int J Toxicol 2001;20(Suppl 3):61–73.
11. Commission of the European Communities. Council directive of 22 June 1988 on the approximation of the laws of the Member States relating to flavourings for use in foodstuffs and to source materials for their production (88/388/EEC); 1991. Available at http://www.fsai.ie/legislation/food/eu_docs/Flavourings/Dir88.388.pdf. Accessed April 10, 2008.
12. Gordon WP, Forte AJ, McMurtry RJ, Gal J, Nelson SD. Hepatotoxicity and pulmonary toxicity of pennyroyal oil and its constituents terpenes in the mouse. Toxicol Appl Pharmacol 1982;65:413–424.
13. Mizutani T, Nomura H, Nakanishi K, Fujita S. Effects of drug metabolism modifiers on pulegone-induced hepatotoxicity in mice. Res Commun Chem Pathol Pharmacol 1987;58:75–83.
14. Gordon WP, Huitric AC, Seth CL, McClanahan RH, Nelson SD. The metabolism of the abortifacient terpene, (R)-(+)-pulegone, to a proximate toxin, menthofuran. Drug Metab Dispos 1987;15:589–594.
15. Anderson IB, Mullen WH, Meeker JE, Khojasteh-Bakht SC, Oishi S, Nelson SD, Blanc PD. Pennyroyal toxicity: Measurement of toxic metabolite levels in two cases and review of the literature. Ann Intern Med 1996;124: 726–734.
16. Sullivan JB, Rumack BH, Thomas H, Peterson RG, Bryson P. Pennyroyal oil poisoning and hepatotoxicity. JAMA 1979;242:2873–2874.
17. Madyastha KM, Gaikwad NW. Metabolic fate of S-(−)-pulegone in rat. Xenobiotica 1998;28;723–734.
18. Thomassen D, Pearson PG, Slattery JT, Nelson SD. Partial characterization of biliary metabolites of pulegone by tandem mass spectrometry. Drug Metab Dispos 1991;19:997–1003.
19. Chen LJ, Lebetkin EH, Burka LT. Metabolism of (R)-(+)-pulegone in F344 rats. Drug Metab Dispos 2001;29: 1567–1577.
20. Nelson SD, McClanahan RH, Thomassen D, Gordon WP, Knebel N. Investigations of mechanisms of reactive metabolite formation from (R)-(+)-pulegone. Xenobiotica 1992;22:1157–1164.
21. Thomassen D, Slattery JT, Nelson SD. Menthofuran dependent and independent aspects of pulegone hepatotoxicity: roles of glutathione. J Pharmacol Exp Ther 1990;253:567–572.
22. Khojasteh-Bakht SC, Chen W, Koenigs LL, Peter RM, Nelson SD. Metabolism of (R)-(+)-pulegone and (R)-(+)-menthofuran by human liver cytochrome P-450s: evidence for formation of a furan epoxide. Drug Metab Dispos 1999;27;574–580.
23. Sztajnkrycer MD, Otten EJ, Bond GR, Lindsell CJ, Goetz RJ. Mitigation of pennyroyal oil hepatotoxicity in the mouse. Acad Emerg Med 2003;10:1024–1028.
24. Vallance WB. Pennyroyal poisoning: A fatal case. Lancet 1955;2:850–851.
25. Early DF. Pennyroyal: a rare cause of epilepsy. Lancet 1961;2:580–581.
26. Roe E, Serra-Baldrich E, Dalmau J, Peramiquel L, Perez M, Granel C, Alomar A. *Mentha pulegium* contact dermatitis. Contact Dermatitis 2005;53:355.
27. Buechel DW, Haverlah VC, Gardner ME. Pennyroyal oil ingestion: report of a case. J Am Osteopath Assoc 1983;82:793–794.

Chapter 83

PURPLE CONEFLOWER and OTHER *ECHINACEA* SPECIES

HISTORY

Echinacea is one of the most popular herbs in the United States. Prior to the immigration of Europeans to North America, Native Americans (e.g., Blackfoot, Cheyenne, Choctaw, Comanche, Dakota, Sioux) used *E. angustifolia* to treat a variety of illnesses ranging from snakebite and toothaches to colds and arthritis.[1] The uses of *E. purpurea* by Native Americans varied by tribe with the herb being used as an antitussive (Choctaw), antisyphilitic agent (Delaware), and analgesic for sore throats and toothaches (Comanche). The Sioux used *E. pallida* for the treatment of pain and snake bites, whereas the Cheyenne used this herb for the treatment of colds, rheumatoid arthritis, and rashes. The use of these herbs spread to colonial pharmacopoeias (*Flora Virginica*, *Materia Medica Americana*) by the late 18th century.[1] Echinacea was a popular herb with eclectics, homeopaths, and naturopaths. During the 19th century, echinacea preparations were used as "blood purifiers." In the 1880s, the Lloyd Brothers from Cincinnati, Ohio manufactured an echinacea tincture and later developed an aqueous alcohol extract, called Specific Echinacea. The use of this herb for upper respiratory infections and other types of infection began in the early 20th century in the United States. During this period, the efficacy of echinacea was questioned,[2] and use of echinacea preparations in the United States waned until the late 20th century.

In the 1920s, the German pharmaceutical manufacturer, Gerhard Madaus, brought seeds of *E. purpurea* to Germany, believing them to be the more common *Echinacea* species, *E. angustifolia*.[1] *Echinacea purpurea* became a popular herb in Germany during the 20th century as a result of the ease and lower cost of cultivation compared with *E. angustifolia* and *E. pallida*.

BOTANICAL DESCRIPTION

Echinacea species are hardy, perennial herbs with either simple or branched stems. Linnaeus originally listed *Echinacea* as the genus, *Rudbeckia*. Of the nine American species of *Echinacea*, the main three medicinal species of *Echinacea* are *E. purpurea*, *E. angustifolia* (narrow-leaved echinacea), and *E. pallida* (pale-flowered echinacea) with the cultivated species, *E. purpurea*, being the most common constituent in herbal preparations containing *Echinacea*. Milled materials from different *Echinacea* species do not have distinctive morphological characteristics that separate these species.

Common Name: Purple Coneflower, Eastern Purple Coneflower

Scientific Name: *Echinacea purpurea* (L.) Moench

Botanical Family: Asteraceae (Compositae, daisy)

Physical Description: An erect herb reaching 3 ft (~1 m) in height with long, hairy stalks and alternate, oval leaves. The solitary reddish-orange flowers surrounded by purple bracts appear from June to September. The root is thick, black, and pungent. Figure 83.1 displays a specimen of the purple coneflower.

FIGURE 83.1. Purple coneflower [*Echinacea purpurea* (L.) Moench]. Photography Courtesy of Saralyn Williams, MD. See color insert.

Distribution and Ecology: Species from the genus *Echinacea* are native plants of midwestern North America. *Echinacea purpurea* is widely cultivated throughout the United States, Canada, and Europe, particularly in Germany. Common locations for this plant are open meadows or damp areas in woods, swamps, river banks, low-lying thickets, and ditches.

Common Name: Blacksamson, Blacksamson Echinacea, Kansas Snakeroot

Scientific Name: *Echinacea angustifolia* (DC.) Hell.

Botanical Family: Asteraceae (Compositae, daisy)

Physical Description: The smallest herb reaches about 1.5 ft (~0.5 m) in height.

Distribution and Ecology: This species is more predominant in the dry prairies and barren regions of Saskatchewan to Minnesota and south to Oklahoma and Texas.

Common Name: Pale Echinacea, Pale Purple Coneflower, Purple Coneflower

Scientific Name: *Echinacea pallida* (Nutt.) Nutt.

Botanical Family: Asteraceae (Compositae, daisy)

Physical Description: This perennial plant grows to 3 ft (~1 m) with a stout central gray to reddish green stem covered with coarse, white hairs. Most leaves originate at the base of the plant with a few along the lower one-third of the stem. These leaves are narrowly lanceolate, oblanceolate, or ovate with a length of up to 9 in. (~23 cm) and a diameter of 2 in. (5 cm). The margins are smooth, often curling upward. Fine white hairs cover the upper and lower surfaces of the leaves.

Distribution and Ecology: These plants inhabit sunny locations from Michigan to Nebraska and south to Texas.

EXPOSURE

Sources

Echinacea is available in a variety of commercial preparations including direct pressed juices, ethanolic or hydrophilic extracts, and freeze-dried whole or powdered leaves, flowers and roots that comprise a variety of tablets, capsules, tinctures, and teas. Herba Echinaceae Purpureae consists of the fresh or dried aerial parts of *Echinacea purpurea* (L.) Moench harvested in full bloom, whereas Radix Echinaceae consists of the fresh or dried roots of *Echinacea angustifolia* or *E. pallida* (Nutt.) Nutt. In Germany, intravenous preparations of echinacea extract are available. However, there is wide variation in the content, quality, and composition of echinacea preparations.[3] The fresh aerial parts of *Echinacea purpurea* are the most common source of herbal preparations.

Medicinal Uses

TRADITIONAL

Native Americans used echinacea for a variety of infections as well as snake bite, bee stings, wound infections, mumps, rabies, mouth sores, blood poisoning, and tumors. This herb was considered an immunostimulant, which promoted wound healing and increased resistance to colds. Additionally, this herb was a treatment for hemorrhoids, wound infections, syphilis, gangrene, malaria, typhoid, flu, and skin infections. Herbal preparations include ethanol extracts as well as fresh or dried *Echinacea* root or flowers. Frequently, echinacea is combined with other herbs (e.g., garlic, ginseng, goldenseal).

Current

Echinacea is a common herbal treatment for upper respiratory tract infections (URI) based on potential immunomodulatory properties, such as macrophage activation and enhanced neutrophil phagocytosis. Evaluation of studies of *Echinacea* herbs is complicated by the lack of standardized doses, the variety of preparations used in these studies, small study sizes, and other methodological flaws.[4] There is limited evidence for the efficacy of echinacea for the *treatment* of URI with most positive studies containing methodological flaws and nonstandardized dosage forms.[5] More recent studies have not confirmed the efficacy of echinacea herbs in treating URIs. A placebo-controlled, randomized, double-blind clinical trial of 80 employees of a German pharmaceutical firm detected a statistically significant reduction (i.e., 6 days vs. 9 days, $P = 0.01$) in the median duration of URI in the group treated with 5 mL EC31JO (pressed juice from fresh flowering purple coneflower) twice daily for 10 days, when compared with the placebo group.[6] However, a randomized, controlled clinical trial of 524 children aged 2–11 years did not detect a reduction in the duration or severity of URI in the children receiving 7.5–10 mL pressed *E. purpurea* juice daily, when compared with the placebo group.[7] A randomized, double-blind, placebo-controlled study examined the severity and duration of cold symptoms as recorded by 128 patients with the onset of URIs within the preceding 24 hours.[8] There was no statistically significant difference between treatment (100 mg freeze-dried pressed juice from the aerial portion of *E. purpurea* 3 times daily) and placebo groups for total symptoms scores, mean individual symptom scores, or time to resolution of symptoms. In a study of three *E. angustifolia* root extracts (supercritical carbon dioxide, 60% ethanol, 20% ethanol) involving subjects receiving both prophylactic and treatment for experimental infection with rhinovirus type 39, there was no statistically significant effects of any these extracts on the rates of infection, the severity of symptoms, or volume of nasal secretions when compared with placebo.[9] Analyses of the extracts demonstrated the following: supercritical carbon dioxide (73.8% alkamides, no polysaccharides), 60% ethanol extract (48.9% polysaccharides, 2.3% alkamides), and 20% ethanol extract (42.1% polysaccharides, 0.1% alkamides). There is little evidence that the *prophylactic* use of echinacea herbs reduces the frequency, duration, or severity of upper respiratory tract infections.[10]

Regulatory Status

Echinacea is approved as a dietary supplement in the United States, and the German Commission E list contains both *E. purpurea* herb (i.e., excluding the root) and *E. pallida* root. Uses in Germany include the treatment of colds, respiratory tract infections, poor-healing external wounds, urinary tract infections, and vaginal fungal infections. The fresh or dried root of *E. pallida* is approved by the German Commission E for the treatment of influenza-like illnesses.

PRINCIPAL INGREDIENTS

Chemical Composition

The chemical composition of extracts from medicinal *Echinacea* species is complex, consisting of alkamides (straight-chain fatty acids with olefinic and/or acetylenic bonds), polyacetylenes, phenolic caffeic acid conjugates (phenylpropanoids), polysaccharides, and glycoproteins. All plant tissues from these species contain acetaldehyde, camphene, dimethyl sulfide, hexanal, limonene, and β-pinene.[11] The chemical profile of these compounds in medicinal *Echinacea* species varies with the specific species, plant part, geographic location, season, processing, and type of extract. Terpenoids (camphene, limonene, β-myrcene, ocimene, α- and β-pinene, terpene) predominated (i.e., 82–91%) in the headspace analysis of volatile components from flowers and stems, whereas headspace analysis of volatile components from roots contain 41–57% aldehydes (2- and 3-methylbutanal, 2-methylpropanal, 2-propanal, acetaldehyde), alcohols, and 6–21% terpenoids.[11]

The active ingredients in Echinacea herbal preparations have not been defined. The major categories of chemicals in *Echinacea* species are alkamides and caffeic acid conjugates (caftaric acid, chlorogenic acid, cynarin, echinacoside, cichoric acid). In a study of roots from 3-year-old plants, the total phenolic compounds (i.e., caffeic acid derivatives) ranged from 13.92 ± 0.31 mg/g dry weight (*E. pallida*) to 22.79 ± 0.37 mg/g dry weight (*E. purpurea*). Caffeic acid was not detected in samples of roots from the three medicinal *Echinacea* species as measured by high performance liquid chromatography (HPLC). The composition of phenolic compounds in medical *Echinacea* species depends on growing season as well as plant part and species. The cichoric acid (chichoric acid) content in dried *E. purpurea* roots varied from 2.27% weight/weight in summer to 1.68% in autumn, whereas the cichoric acid content in dried tops from this species varied from 2.02% in summer to 0.52% in autumn.[12] Enzymatic browning during extraction substantially reduces the phenolic content of medicinal *Echinacea* species. Echinacea extracts also contain at least 20 alkamides, which are primarily undecenoic and dodecanoic acid derivatives differing in the degree of unsaturation and the configuration of the double bonds.[13] Fresh, dried *Echinacea purpurea* roots contain higher concentrations of alkamides (2.65–3.28 mg/g) than leaves (0.10–0.18 mg/g).[14] Alkamides occur in alcoholic extracts of the aerial parts and roots from *E. purpurea* and *E. angustifolia*. In a study of 11 healthy subjects administered a 60% ethanolic extract from the roots of *E. angustifolia*, the plasma contained at least six alkamides with dodeca-2E,4E,8Z,10Z/E-tetraenoic acid isobutylamides being the most prominent.[15] The major constituents of alcoholic extracts of the roots of *E. purpurea* were cichoric acid (2,3-dicaffeoyltartaric acid) and verbascoside, whereas echinacoside and 6-*O*-caffeolyechinacoside predominated in extracts of *E. pallida* roots.[16] Cynarine (1,3-dicaffeoylquinic acid) and dodeca-2E,4E,8Z,10Z/E-tetraenoic acid isobutylamides were the major chemicals in *E. angustifolia* root extracts. Volatile oils of *Echinacea* species contain a variety of alkenes, alkynes, and alkyl ketones with some of the former compounds subject to oxidation to 8-hydroxy derivatives. The volatile oils from the aerial parts of these species also contain borneol, bornyl acetate, caryophyllene, and germacrene D.[13] Saturated pyrrolizidine alkaloids (e.g., tussilagine, isotussilagine) occur in *Echinacea* species, but the toxicity of these compounds is low compared with the unsaturated pyrrolizidine compounds associated with venoocclusive disease and hepatotoxicity.[13]

Physiochemical Properties

A pilot study suggested that the alkamide, alkaloid, and polyacetylene fractions in echinacea preparations stimulate leukocytes and increase the release of cytokines (tumor necrosis factor-alpha, interleukin-1,-6, and -10).[17,18] Caffeic acid conjugates (echinacoside, chlorogenic acid, cichoric acid, cynarine) are free radical scavengers and antioxidants, particularly cichoric acid.[19] However, the bioavailability of these phenolic compounds is low.

Mechanism of Toxicity

Most serious reactions to echinacea preparations result from allergic reactions to echinacea preparations in atopic individuals. There are few data on dose-related toxicity to these preparations.

DOSE RESPONSE

Typical doses of liquid extract of *E. purpurea* root for the treatment of upper respiratory illness is 3 mL every 3–4 hours for the first 1–2 days, then 3 times daily for the following week.[20] The dose of echinacea tea from the roots of *E. angustifolia* or *E. purpurea* is 6–8 oz 4 times daily for the first 2 days, then once or twice daily for the rest of the week.

TOXICOKINETICS

Kinetics

Human studies indicate that the gastrointestinal tract rapidly absorbs alkamide compounds (e.g., dodeca-2E,4E,8Z,10E/Z-tetraenoic acid isobutylamides).[21] In volunteer studies, the time to maximum alkamide concentrations (T_{max}) ranged from about 30 minutes to $2\frac{1}{2}$ hours.[15,22] The mean plasma elimination half-life for total alkamides was about 3.5 ± 0.4 hour. In general, the bioavailability of caffeic acid conjugates is poor.

Drug Interactions

There is little clinical evidence of significant drug–herb interactions during the use of *Echinacea* herbs.[23] Short term *in vivo* studies in volunteers indicate that the administration of echinacea (400 mg *E. purpurea* root 4 times daily for 8 days) reduces the mean oral clearance of caffeine (CYP1A2) by approximately 27%.[24] The effect of *E. purpurea* root on CYP3A4 is complex because this study indicated that echinacea inhibited intestinal CYP3A activity, while inducing hepatic CYP3A activity. There was no alteration of CYP2C9 or CYP2D6 activity. However, the administration of botanical supplements (800 mg twice daily for 28 days, 2.7% cichoric acid) of *E. purpurea* to 12 healthy volunteers had no significant effect on CYP1A2, CYP2D6, CYP2E1, or CYP3A4 activity as measured by phenotypic ratios before and after supplementation.[25] Echinacea is not recommended for patients with autoimmune or collagen vascular disorders (e.g. rheumatoid arthritis, systemic lupus erythematosus), multiple sclerosis, tuberculosis, or HIV because of the potential alteration of immune response. However, there are few clinical data to evaluate the clinical relevance of this potential effect.

CLINICAL RESPONSE

Adverse Reactions

In general, echinacea herbal preparations are well tolerated with few adverse effects as evidenced by the similarity of adverse effects between treatment and placebo groups reported in most clinical studies.[26] Animal studies support the low toxicity of herbal preparations from *Echinacea* species at doses up to 60 times higher than typical doses administered to humans.[27] The most common serious adverse effects of treatment with these products are allergic reactions, particularly in individuals hypersensitive to plants from the Asteraceae family (chamomile, feverfew, ragweed, mugwort). In a randomized trial using pressed *E. purpurea* juice, the incidence of skin rash was significantly higher (7.1% vs. 2.7%, $P = 0.008$) in the group receiving echinacea than in the placebo group.[7] Case reports associate a variety of allergic reactions with ingestion of *Echinacea* herbal preparations including urticaria, acute bronchospasm, diffuse maculopapular rash, and angioedema.[28] These reactions occurred primarily in atopic individuals, some of whom had no prior exposure to *Echinacea* species. Other adverse effects associated with the use of echinacea preparations include unpleasant taste, headache, dizziness, nausea, vomiting, diarrhea, and difficulty swallowing.[29] Rare case reports temporally associate the use of echinacea herbal preparations with recurrent erythema nodosum.[30] There is little clinical information on dose-related toxicity of echinacea.

Chronic Effects

In a study of 206 women self-referred for evaluation of the use of echinacea preparations (i.e., up to 1g daily capsule/tablets or 60 drops tincture daily) during pregnancy, there were no statistical differences in the spontaneous abortion rate or major congenital malformations between the echinacea group and a control group matched for diseases, alcohol, and cigarette use.[31] The limitations of the study included small sample size (i.e., low statistical power) and potential misclassification (reporting bias as a result of self-reporting of exposure).

DIAGNOSTIC TESTING

Methods to distinguish the three main *Echinacea* species based on differences in the chemical profiles of alkamides and caffeoyl conjugates in extracts from these herbs include capillary electrophoresis with photodiode array UV detection and gradient elution reversed phase high performance liquid chromatography (HPLC) coupled to photodiode array UV or electrospray ionization mass spectrometric detection.[32,33] Cichoric acid is the predominant caffeoyl conjugate in extracts of *E. purpurea* root, whereas echinacoside and 6-caffeoylechinacoside are the main caffeoyl conjugates in extracts of *E. pallida* root.[16] The presence of cynarine and echinacoside and the absence of cichoric acid distinguish *E. angustifolia* roots from *E. purpurea* and *E. pallida* roots. The leaf extracts of these three *Echinacea* species also have distinctive chemical profiles. Cichoric acid is detectable in *E. purpurea* and *E. pallida*, but not in *E. angustifolia*, whereas echinacoside is detectable in *E. pallida*, but not in *E. angustifolia* or *E. purpurea* leaves.[34] Infrared reflectance spectroscopy is an alternative to HPLC techniques for the detection of adulteration of *Echinacea purpurea* roots with *Parthenium integrifolium* L. or other *Echinacea* species. This method can detect samples of *E. purpurea* root with >10% adulteration, and there are relative few false-positives with this method.[35] Although the impact of milling and drying on total alkamide content in *E. purpurea* roots is slight, storage at room temperature for prolonged periods can substantially reduce alkamide content. Over a 64-week period, the total alkamide content decreased about 80% in *E. purpurea* roots stored at 24°C (~75°F).[36] Freezing reduced, but did not eliminate, this loss. The content of the main phenolic compound, cichoric acid, is relatively stable at room temperature over at least 6 months in dry alcoholic extracts of *E. purpurea*.[37]

TREATMENT

Treatment is supportive. The most serious potential adverse effects associated with the use of echinacea preparations are allergic reactions that include acute bronchospasm and generalized urticaria.

References

1. Flannery MA. From rudbeckia to echinacea: the emergence of the purple cone flower in modern therapeutics. Pharm Hist 1999;41(2):52–59.
2. Puckner WA. Echinacea considered valueless: report of the council on pharmacy and chemistry. JAMA 1909;53: 1836.
3. Gilroy CM, Steiner JF, Byers T, Shapiro H, Georgian W. Echinacea and truth in labeling. Arch Intern Med 2003;163:699–704.
4. Barrett B. Medicinal properties of echinacea: a critical review. Phytomedicine 2003;10:66–86.
5. Melchart D, Linde K, Fischer P, Kaesmayr J. Echinacea for preventing and treating the common cold. Cochrane Database Syst Rev 2000;(2):CD000530.
6. Schulten B, Bulitta M, Ballering-Bruhl B, Koster U, Schafer M. Efficacy of *Echinacea purpurea* in patients with a common cold. Arzneim-Forsch Drug Res 2001;51: 563–568.
7. Taylor JA, Weber W, Standish L, Quinn H, Goesling J, McGann M, Calabrese C. Efficacy and safety of echinacea in treating upper respiratory tract infections in children. A randomized controlled trial. JAMA 2003;290: 2824–2830.
8. Yale SH, Liu K. *Echinacea purpurea* therapy for the treatment of the common cold. A randomized, double-blind, placebo-controlled clinical trial. Arch Intern Med 2004;164: 1237–1241.

9. Turner RB, Bauer R, Woelkart K, Hulsey TC, Gangemi JD. An evaluation of *Echinacea angustifolia* in experimental rhinovirus infections. N Engl J Med 2005;353: 341–348.
10. Grimm W, Muller H-H. A randomized controlled trial of the effect of fluid extract of *Echinacea purpurea* on the incidence and severity of colds and respiratory infections. Am J Med 1999;106:138–143.
11. Mazza G, Cotrell T. Volatile components of roots, stems, leaves, and flowers of *Echinacea* species. J Agric Food Chem 1999;3081–3085.
12. Perry NB, Burgess EJ, LeAnne Glennie V. Echinacea standardization: analytical methods for phenolic compounds and typical levels in medicinal species. J Agric Food Chem 2001;49:1702–1706.
13. Barnes J, Anderson LA, Gibbons S, Phillipson JD. Echinacea species (*Echinacea angustifolia* (DC.) Hell., *Echinacea pallida* (Nutt.) Nutt., *Echinacea purpurea* (L.) Moench): a review of their chemistry, pharmacology and clinical properties. J Pharm Pharmacol 2005;57:929–954.
14. Kim H-O, Durance TD, Scaman CH, Kitts DD. Retention of alkamides in dried *Echinacea purpurea*. J Agric Food Chem 2000;48:4187–4192.
15. Woelkart K, Koidl C, Grisold A, Gangemi JD, Turner RB, Marth E, Bauer R. Bioavailability and pharmacokinetics of alkamides from the roots of *Echinacea angustifolia* in humans. J Clin Pharmacol 2005;45:683–689.
16. Duff Sloley B, Urichuk LJ, Tywin C, Coutts RT, Pang PK, Shan JJ. Comparison of chemical components and antioxidant capacity of different *Echinacea* species. J Pharm Pharmacol 2001;53:849–857.
17. Burger RA, Torres AR, Warren RP, Caldwell VD, Hughes BG. Echinacea-induced cytokine production by human macrophages. Int J Immunopharmacol 1997;19:371–379.
18. Melchart D, Clemm C, Weber B, Draczynski T, Worku F, Linde K, et al. Polysaccharides isolated from *Echinacea purpurea* herbal cell cultures to counteract undesired effects of chemotherapy—a pilot study. Phytother Res 2002;16:138–142.
19. Facino RM, Carini M, Aldini G, Saibene L, Pietta P, Mauri P. Echinacoside and caffeoyl conjugates protect collagen from free radical-induced degradation: a potential use of Echinacea extracts in the prevention of skin photodamage. Planta Med 1995;61:510–514.
20. Klingler B. Echinacea. Am Fam Physician 2003;67:77–80.
21. Dietz B, Heilmann J, Bauer R. Absorption of dodeca-2E,4E,8Z,10E/Z-tetraenoic acid isobutylamides after oral application of Echinacea purpurea tincture. Planta Med 2001;67:863–864.
22. Matthias A, Addison RS, Penman KG, Dickinson RG, Bone KM, Lehmann RP. Echinacea alkamide disposition and pharmacokinetics in humans after tablet ingestion. Life Sci 2005;77:2018–2029.
23. Izzo AA, Ernst E. Interactions between herbal medicines and prescribed drugs: a systematic review. Drugs 2001;61: 2163–175.
24. Gorski JC, Huang S-M, Pinto A, Hamman MA, Hilligoss JK, Zaheer NA, et al. The effect of echinacea (*Echinacea purpurea* root) on cytochrome P450 activity *in vivo*. Clin Pharmacol Ther 2004;75:89–100.
25. Gurley BJ, Gardner SF, Hubbard MA, Williams DK, Gentry WB, Carrier J, et al. *In vivo* assessment of botanical supplementation on human cytochrome P450 phenotypes: *Citrus aurantium, Echinacea purpurea*, milk thistle, and saw palmetto. Clin Pharmacol Ther 2004;76:428–440.
26. Huntley AL, Coon JT, Ernst E. The safety of herbal medicinal products derived from *Echinacea* species. A systematic review. Drug Safety 2005;28:387–400.
27. Mengs U, Clare CB, Poiley JA. Toxicity of *Echinacea purpurea* acute, subacute and genotoxicity studies. Arzneim-Forsch Drug Res 1991;41:1076–1081.
28. Mullins RJ, Heddle R. Adverse reactions associated with echinacea: the Australian experience. Ann Allergy Asthma Immunol 2002;88:42–51.
29. Giles JT, Palat CT 3rd, Chien SH, Chang ZG, Kennedy DT. Evaluation of echinacea for treatment of the common cold. Pharmacotherapy 2000;20:690–697.
30. Soon SL, Crawford RI. Recurrent erythema nodosum associated with echinacea gerbil therapy. J Am Acad Dermatol 2001;44:298–299.
31. Gallo M, Sarkar M, Au W, Pietrzak K, Comas B, Smith M, et al. Pregnancy outcome following gestational exposure to Echinacea a prospective controlled study. Arch Intern Med 2000;160;3141–3143.
32. Molgaard P, Johnsen S, Christensen P, Cornett C. HPLC method validated for the simultaneous analysis of cichoric acid and alkamides in *Echinacea purpurea* plants and products. J Agric Food Chem 2003;51:6922–6933.
33. Laasonen M, Wennberg T, Harmia-Pulkkinen T, Vuorela H. Simultaneous analysis of alkamides and caffeic acid derivatives for the identification of *Echinacea purpurea, Echinacea angustifolia, Echinacea pallida* and *Parthenium integrifolium* roots. Planta Med 2002;68:572–574.
34. Luo W, Ang CY, Gehring TA, Heinze TM, Lin LJ, Mattia A. Determination of phenolic compounds in dietary supplements and tea blends containing echinacea by liquid chromatography with coulometric electrochemical detection. J AOAC Int 2003;86:202–208.
35. Laasonen M, Harmia-Pulkkinen T, Simard CL, Michiels E, Rasanen M, Vuorela H. Fast identification of *Echinacea purpurea* dried roots using near-infrared spectroscopy. Anal Chem 2002;74:2493–2499.
36. Perry NB, van Klink JW, Burgess EJ, Parmenter GA. Alkamide levels in *Echinacea purpurea*: effects of processing, drying and storage. Planta Med 2000;66: 54–56.
37. Livesey J, Awang DV, Arnason JT, Lechamo W, Barrett M, Pennyroyal G. Effect of temperature on stability of marker constituents in *Echinacea purpurea* root formulations. Phytomedicine 1999;5:347–349.

Chapter 84

ROSEMARY
(*Rosmarinus officinalis* L.)

HISTORY

Rosemary has been used as a medicinal and aromatic herb since ancient Greek and Roman times.[1] Although rosemary is a native species of the Mediterranean, this plant has been cultivated in Northern Europe and England at least since Medieval times where rosemary was a symbol of remembrance of events (birth, marriage, death).[2] Rosemary is mentioned in some classic books, such as *Hamlet* and *Don Quixote*.

BOTANICAL DESCRIPTION

Common Name: Rosemary, Polar Plant, Compass Weed, Compass Plant

Scientific Name: *Rosmarinus officinalis* L.

Botanical Family: Lamiaceae (menthes, mints)

Physical Description: This erect evergreen has pungent, pleasant-smelling, lavender-like, dark green leaves. The rigid branches bear fissured, ash-colored bark and long, linear, needle-like leaves. The small, pale blue flowers appear from April to August.

Distribution and Ecology: Rosemary is a native plant of the arid costal regions of the Mediterranean region. However, this plant is cultivated in warm, dry climates throughout the world.

EXPOSURE

Sources

The essential oil is a distillation product of rosemary leaves and flowers.

Medicinal Uses

Traditional

In folk medicine, rosemary extract is a treatment for urinary aliments, chronic weakness, nervous disorders, hair loss, and peripheral vascular diseases. In addition, rosemary is a traditional astringent, carminative, tonic, rubefacient, antispasmodic, anti-inflammatory, expectorant, emmenagogue, digestive, diaphoretic, and diuretic.[3,4]

Current

Rosemary extracts are natural sources of antioxidants (rosmarinic acid, carnosic acid, rosmanol, epirosmanol, carnosol), which are alternatives to the common synthetic antioxidants (butylated hydroxyanisole, hydroxytoluene, δ-tocopherol). Additionally, rosemary extract is a constituent of some perfumes, disinfectants, and insecticides. There is no conclusive evidence that rosemary is an effective insecticide.[5] Experimental uses for rosemary include the treatment of alopecia areata.[6] Rose-

Medical Toxicology of Natural Substances, by Donald G. Barceloux, MD

mary leaves are a seasoning for meat dishes, sauces, and salami.

Regulatory Status

The US Food and Drug Administration approved the use of rosemary in food [GRAS (generally recognized as safe)].

PRINCIPAL INGREDIENTS

Chemical Composition

The major polyphenolic constituents in rosemary are phenolic diterpenes (carnosic acid, carnosol, 12-*O*-methylcarnosic acid), caffeoyl derivatives (rosmarinic acid), and flavones (genkwanin, isoscutellarein 7-*O*-glucoside, eriocitrin, luteolin 3′-*O*-β-D-glucuronide, hesperidin, diosmin, hispidulin 7-*O*-glucoside).[7,8] Carnosic acid is a synonym for salvin (CAS RN: 3650-09-7). Figure 84.1 displays the chemical structures of the major constituents in rosemary. The flavonoid compounds primarily are derivatives of apigenin and luteolin. The concentration of polyphenolic compounds varies with season, age of the plant, solar radiation, growing conditions, and climate. These compounds are present in young leaves, but the concentrations of these polyphenolic compounds decrease with age.[9] Drought and high solar radiation increases the concentration of highly oxidized abietane diterpenes (e.g., isorosmanol, dimethyl isorosmanol) in rosemary leaves, while the concentration of carnosic acid decreases.[10]

The major constituents of hydro distillates of rosemary include α-pinene, β-pinene, *E*-caryophyllene, borneol, (−)-camphene, 1,8-cineole, camphor, verbenone, and bornyl-acetate.[11,12] The concentrations of these compounds also vary with distillation process, region, climate, environmental conditions (temperature, water stress), and strain.[13] In general, monoterpene compounds account for about 50% of the constituents in aqueous extracts of rosemary, whereas ketones and alcohols each represent <7–10% of the total concentration.[14] The main components in a sample of essential oil obtained from rosemary by supercritical CO_2 extraction were α-pinene, 1,8-cineole, camphor, verbenone, and borneol as measured by gas chromatography/mass spectrometry (GC/MS).[15] These constituents accounted for about 80% of the total oil.

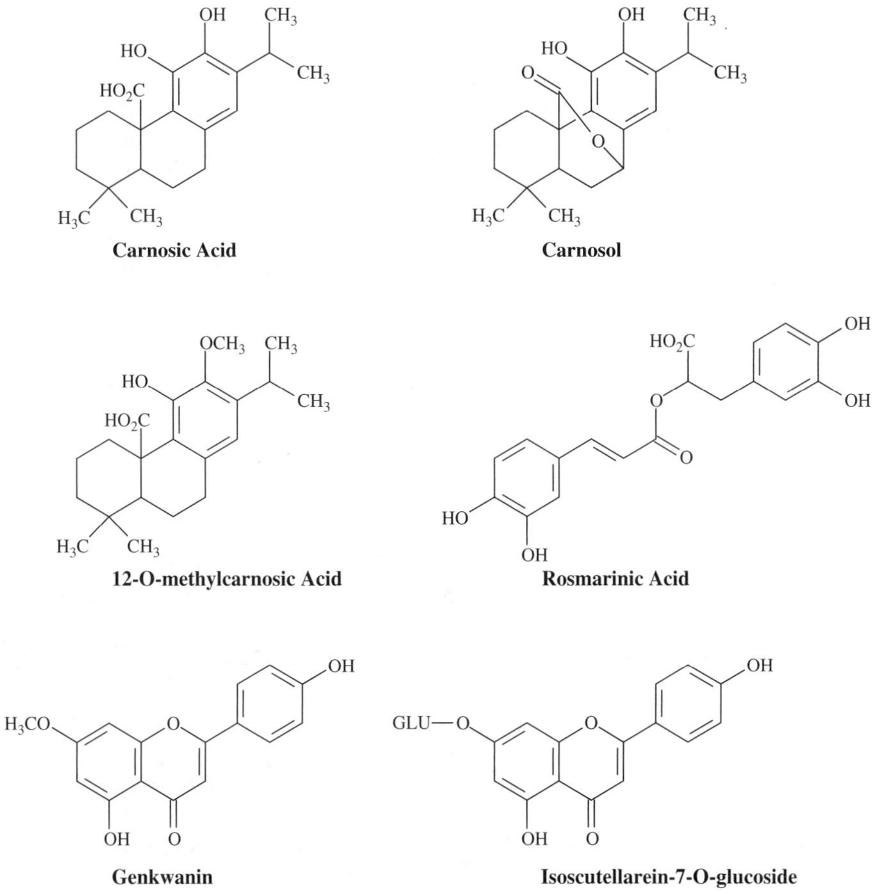

Carnosic Acid

Carnosol

12-O-methylcarnosic Acid

Rosmarinic Acid

Genkwanin

Isoscutellarein-7-O-glucoside

FIGURE 84.1. Chemical structures of major constituents of rosemary extracts.

Physiochemical Properties

Unsaturated fatty acids are sensitive to oxidation, and the resulting lipid oxidation reduces flavor and nutritive value of lipid-containing food products. Antioxidants prolong the shelf life and maintain the nutritional quality of lipid-containing foods.[16] The antioxidant properties of rosemary result primarily from the actions of phenolic diterpenes, such as carnosic acid, carnosol, methyl carnosate, rosmanol, epirosmanol, and 7-methyl-epirosmanol. *In vitro* studies indicate that the main diterpene antioxidant is carnosic acid.[17] Oxidation converts carnosic acid to carnosol, and further degradation produces other phenolic diterpene compounds with γ-lactone structures.[18] Consequently, the concentration of carnosol increases as the plant ages. The other active antioxidant in rosemary extract is the water-soluble phenolic diterpene, rosmarinic acid, which is also present in other members of the mint family (Lamiaceae) including oregano (*Origanum vulgare* L.), kitchen sage (*Salvia officinalis* L.), and common balm (*Melissa officinalis* L.). The concentration of carnosic and rosmarinic acids also depends on drying techniques and distillation processes (aqueous, organic solvent, microwave, supercritical fluid).[19]

Mechanism of Toxicity

Most of the adverse reactions reported from the use of rosemary are allergic reactions.

DOSE RESPONSE

Typical traditional doses of rosemary include 2–4 g dried leaves and twigs or 2–4 mL of 1:1 liquid extract in 45% ethanol.

TOXICOKINETICS

There are few data on the toxicokinetics of constituents in rosemary extracts. Rodent studies suggest the possibility of the induction of CYP1A, CYP2B, CYP2E1, and CYP3A along with some phase II enzymes (e.g., glutathione S-transferase, UDP-glucuronosyltransferase),[20] but the lack of human data limits conclusions regarding the clinical relevance of these potential interactions.

CLINICAL RESPONSE

Adverse reactions to rosemary typically involve hypersensitivity responses. Despite the frequent use of rosemary as an herb, case reports of allergic reactions are relatively rare. These case reports include the development of allergic contact dermatitis (edema, eczema, vesiculation) following direct contact with rosemary leaves,[21] the use of cosmetics and a cleansing gel containing rosemary leaf extract,[22] the application of rosemary alcohol directly to the skin,[23] and the use of rosemary leaf plasters.[24] Patch testing in these patients indicates that carnosol (CAS RN: 5957-80-2) is a major allergen in rosemary. The response of these patients to other members of the mint family (e.g., oregano, sage, mint, thyme, lavender) was variable. The chronic use of rosemary as a culinary spice in food was associated with the development of chronic contact cheilitis.[25] Exposure to rosemary extracts also has been associated with occupational contact dermatitis[26] and occupational asthma.[27] Although rosemary extracts contain variable quantities of camphor,[28] there are no human data to support the development of seizures as a complication of the ingestion of rosemary extract.

DIAGNOSTIC TESTING

Techniques for the separation of the major phenolic diterpene compounds responsible for the antioxidant properties of rosemary extract include capillary zone electrophoresis,[29] pressurized liquid extraction/capillary electrophoresis/electrospray-mass spectrometry,[30] high performance liquid chromatography with UV detection,[31] solid-phase extraction and capillary zone electrophoresis coupled to electrospray ionization/ion trap mass spectrometry,[32] and LC/MS.[33] Capillary electrophoresis with electrochemical detections allows the simultaneous determination of the active components (e.g., rosmarinic acid, caffeic acid, apigenin, luteolin) with detection limits in the range of 0.1–1 μg/mL.[34]

TREATMENT

Treatment is supportive.

References

1. Selmi G. [Therapeutic use of rosemary through the centuries]. Policlinico [Prat] 1967;74:439–441. [Italian]
2. Al-Sereiti MR, Abu-Amer KM, Sen P. Pharmacology of rosemary (*Rosmarinus officinalis* Linn.) and its therapeutic potentials. Indian J Exp Biol 1999;37:124–130.
3. Giron LM, Freire V, Alonzo A, Caceres A. Ethnobotanical survey of the medicinal flora used by the Caribs of Guatemala. J Ethnopharmacol 1991;34:173–187.
4. Haloui M, Louedec L, Michel JB, Lyoussi B. Experimental diuretic effects of *Rosmarinus officinalis* and *Centaurium erythraea*. J Ethnopharmacol 2000;71:465–472.
5. Waliwitiya R, Isman MB, Vernon RS, Riseman A. Insecticidal activity of selected monoterpenoids and rosemary oil

to *Agriotes obscurus* (Coleoptera: Elateridae). J Econ Entomol. 2005;98:1560–1565.

6. Hay IC, Jamieson M, Ormerod AD. Randomized trial of aromatherapy. Successful treatment for alopecia areata. Arch Dermatol 1998;134:1349–1352.
7. Mahmoud AA, AL-Shihry SS, Son BW. Diterpenoid quinones from rosemary (*Rosmarinus officinalis* L.). Phytochemistry 2005;66:1685–1690.
8. del Bano MJ, Lorente J, Castillo J, Benavente-Garcia O, Marin MP, Del Rio JA, et al. Flavonoid distribution during the development of leaves, flowers, stems, and roots of *Rosmarinus officinalis*. Postulation of a biosynthetic pathway. J Agric Food Chem 2004;52:4987–4992.
9. del Bano MJ, Lorente J, Castillo J, Benavente-Garcia O, Del Rio JA, Ortuno A, et al. Phenolic diterpenes, flavones, and rosmarinic acid distribution during the development of leaves, flowers, stems, and roots of *Rosmarinus officinalis*. Antioxidant activity. J Agric Food Chem 2003;51: 4247–4253.
10. Munne-Bosch S, Schwarz K, Alegre L. Enhanced formation of alpha-tocopherol and highly oxidized abietane diterpenes in water-stressed rosemary plants. Plant Physiol 1999;121:1047–1052.
11. Presti ML, Ragusa S, Trozzi A, Dugo P, Visinoni F, Fazio A, et al. A comparison between different techniques for the isolation of rosemary essential oil. J Sep Sci 2005; 28:273–280.
12. Angioni A, Barra A, Cereti E, Barile D, Coisson JD, Arlorio M, et al. Chemical composition, plant genetic differences, antimicrobial and antifungal activity investigation of the essential oil of *Rosmarinus officinalis* L. J Agric Food Chem 2004;52:3530–3535.
13. Wellwood CR, Cole RA. Relevance of carnosic acid concentrations to the selection of rosemary, *Rosmarinus officinalis* (L.), accessions for optimization of antioxidant yield. J Agric Food Chem 2004;52:6101–6107.
14. Flamini G, Cioni PL, Morelli I, Macchia M, Ceccarini L. Main agronomic-productive characteristics of two ecotypes of *Rosmarinus officinalis* L. and chemical composition of their essential oils. J Agric Food Chem 2002;50: 3512–3517.
15. Santoyo S, Cavero S, Jaime L, Ibanez E, Senorans FJ, Reglero G. Chemical composition and antimicrobial activity of *Rosmarinus officinalis* L. essential oil obtained via supercritical fluid extraction. J Food Prot 2005;68: 790–795.
16. Halliwell B, Aeschbach R, Loliger J, Aruoma OI. The characterization of antioxidants. Food Chem Toxicol 1995;33: 601–617.
17. Aruoma OI, Halliwell B, Aeschbach R, Loligers J. Antioxidant and pro-oxidant properties of active rosemary constituents: carnosol and carnosic acid. Xenobiotica 1992;22:257–268.
18. Schwarz K, Ternes W. Antioxidative constituents of *Rosmarinus officinalis* and *Salvia officinalis*. II. Isolation of carnosic acid and formation of other phenolic diterpenes. Z Lebensm Unters Forsch 1992;195:99–103.
19. Almela L, Sanchez-Munoz B, Fernandez-Lopez JA, Roca MJ, Rabe V. Liquid chromatographic-mass spectrometric analysis of phenolics and free radical scavenging activity of rosemary extract from different raw material. J Chromatogr A 2006;1120:221–229.
20. Debersac P, Vernevaut MF, Amiot MJ, Suschetet M, Siess MH. Effects of a water-soluble extract of rosemary and its purified component rosmarinic acid on xenobiotic-metabolizing enzymes in rat liver. Food Chem Toxicol 2001;39:109–117.
21. Sera E, Vila A, Peramiquel L, Dalmau J, Granel C, Alomar A. Allergic contact dermatitis due to rosemary. Contact Dermatitis 2005;53:179–180.
22. Inui S, Katayama I. Allergic contact dermatitis induced by rosemary leaf extract in a cleansing gel. J Dermatol 2005;32:667–669.
23. Gonzalez I, Lobesa T, del Pozo MD, Blasco A, Venturini M. Rosemary contact dermatitis and cross-reactivity with other labiate plants. Contact Dermatitis 2006;54: 210–212.
24. Fernandez L, Duque S, Sanchez I, Quinones D, Rodriguez F, Garcia-Abujeta JL. Allergic contact dermatitis from rosemary (*Rosmarinus officinalis* L.). Contact Dermatitis 1997;37:248–249.
25. Guin JD. Rosemary cheilitis: one to remember. Contact Dermatitis 2001;45:63.
26. Hjorther AB, Christophersen C, Hausen BM, Menne T. Occupational allergic contact dermatitis from carnosol, a naturally-occurring compound present in rosemary. Contact Dermatitis 1997;37;99–100.
27. Lemiere C, Cartier A, Lehrer SB, Malo J-L. Occupational caused by aromatic herbs. Allergy 1996;51:647–649.
28. Steinmetz MD, Vial M, Millet Y. [Actions of essential oils of rosemary and certain of its constituents (eucalyptol and camphor) on the cerebral cortex of the rat in vitro]. J Toxicol Clin Exp 1987;7:259–271. [French]
29. Sanez-Lopez R, Fernandez-Zurbano P, Tena MT. Capillary electrophoretic separation of phenolic diterpenes from rosemary. J Chromatogr 2002;953: 251–256.
30. Herrero M, Arraez-Roman D, Segura A, Kenndler E, Gius B, Raggi MA, et al. Pressurized liquid extraction-capillary electrophoresis-mass spectrometry for the analysis of polar antioxidants in rosemary extracts. J Chromatogr A 2005;1084:54–62.
31. Troncoso N, Sierra H, Carvajal L, Delpiano P, Gunther G. Fast high performance liquid chromatography and ultraviolet-visible quantification of principal phenolic antioxidants in fresh rosemary. J Chromatogr A 2005;1100: 20–25.
32. Arraez-Roman D, Gomez-Caravaca AM, Gomez-Romero M, Segura-Carretero A, Fernandez-Gutierrez A. Identification of phenolic compounds in rosemary honey using solid-phase extraction by capillary electrophoresis-electrospray ionization-mass spectrometry. J Pharmaceut Biomed Anal 2006; 41:1648–1656.

33. Senorans FJ, Ibanez E, Cavero S, Tabera J, Reglero G. Liquid chromatographic-mass spectrometric analysis of supercritical-fluid extracts of rosemary plants. J Chromatogr A 2000;870:491–499.

34. Peng Y, Yuan J, Liu F, Ye J. Determination of active components in rosemary by capillary electrophoresis with electrochemical detection. J Pharmaceut Biomed Anal 2005;39:431–437.

Chapter 85

RUE
(*Ruta graveolens* L.)

HISTORY

In addition to the medicinal use of rue for analgesia and inflammation, the use of this herb for defense against witches and evil spells dates back to ancient Greece and the Middle Ages.

BOTANICAL DESCRIPTION

Although the two most common *Ruta* species used for herbal medicine are *R. graveolens* L. (common rue) and *R. chalepensis* L. (fringed rue), other *Ruta* species (*R. corsica* DC., *R. montana* L.) are also traditional herbal treatments.

Common Name: Common Rue, Rue

Scientific Name: *Ruta graveolens L.*

Botanical Family: Rutaceae (rues, rutacées)

Physical Description: This evergreen shrub has bi- or tripinnate, bluish-green leaves, which emit a strong, disagreeable odor. The yellow flowers are about 1 cm (~0.4 in.) in diameter with toothed petals that appear from June to September.

Distribution and Ecology: Common rue is a native species of southern Europe and northern Africa that was introduced into North America after the Spanish Conquests. Common locations for this plant include moist fields, pastures, and roadsides.

EXPOSURE

Sources

The essential oil obtained by steam distillation of the dry aerial parts of *R. graveolens* typically yields <1% essential oil. Steam distillation of Italian rue (*R. graveolens*) produced 0.74% yellowish essential oil with an intense, penetrating odor.[1] This yield compares with about 4.5% following supercritical-fluid extraction of common rue.[2] Rue is also a seasoning for cooking and a source of methyl nonyl ketone, which is used in soaps, creams, and perfumes.

Medicinal Uses

Traditional

There are many traditional herbal uses for *Ruta* species throughout the world. Fresh aerial parts of rue are a traditional treatment for palpitations and circulatory disorders in Taiwan.[3] In Saudi Arabia, decoctions of the aerial parts of rue are a traditional treatment for fever, pain, rheumatism, and mental disorders.[4] This plant is a treatment for syncope, menstrual disorders, rheumatism, and neuralgias in Ayurvedic medicine. Decoctions of the roots of rue are traditional Chinese treatments for snake and scorpion envenomations. The aqueous extract of rue is an herbal treatment for fever in Africa. *Ruta* species are common herbs used as emmenagogues to initiate delayed menses, as abortifacients to terminate

Medical Toxicology of Natural Substances, by Donald G. Barceloux, MD

pregnancies,[5] and as a birth control method.[6] In a retrospective study on the use of herbal infusions to induce abortions, 86 cases were reported to the Montevideo Poison Center between 1986 and 1999.[7] *Ruta* species (*R. chalepensis* L., *R. graveolens* L.) were an ingredient in 26 of the cases with *Ruta* species being the single plant used in 12 cases. Rue also is used as an insect repellant and antihelminthic.[8]

Current

Extracts of *Ruta graveolens* are experimental chemotherapeutic agents, but the efficacy of these agents has not been established.[9] Other experimental uses include the treatment of parasitic infections.[10]

Regulatory Status

Most countries do not regulate the use of rue.

PRINCIPAL INGREDIENTS

Chemical Composition

Ruta species contain a variety of chemicals including quinoline alkaloids (graveoline), acridone alkaloids,[11] flavonoids, furanocoumarins (5-methoxypsoralen, 8-methoxypsoralen),[12] tannins, sterols, and terpene compounds. Volatile components of aqueous extracts of *Ruta graveolens* leaves, flowers, stems, foots, and fruits include primarily oxygenated compounds, such as ketones (2-undecanone), alcohols (2-undecanol), and aliphatic acids (pentanoic, hexanoic, octanoic, nonanoic).[13] The composition of the essential oil varies with plant part. Following subcritical (CO_2) extraction, the main constituents of the essential oil as measured by capillary chromatography with nitrogen-phosphorus, flame ionization, and mass selective detection were as follows: leaves, 2-nonanone (8.9%); flowers, 2-undecanone (13.4%); stems, chalepensin [3-(α-,α-dimethylallyl) psoralen,13%]; and roots, geijerene (19.3%).[2] Gas chromatography/mass spectrometry (GC/MS) detected 38 compounds in a steam distillate of dry aerial parts of *R. graveolens* with the primary constituents being 2-keto compounds.[1] Methyl nonyl ketone (undecan-2-one, RN: 112-12-9) and 2-nonanone (nonan-2-one, CAS RN: 821-55-6) represented 46.8% and 18.8% of the hydrodistillate, respectively. Triterpenoid compounds accounted for about 11% of the constituents with limonene, 1,8-cineole, and α-pinene being the most common monoterpene compounds. Minor constituents included octanoic acid, methyl salicylate, 2-decanone, and methyl undecyl ketone (2-tridecanone). The ethyl acetate extract of leaves from *Ruta graveolens* contains furanocoumarin compounds (5-methoxypsoralen, 8-methoxypsoralen) and quinolone alkaloids (graveoline) with 8-methoxypsoralen being the strongest photoirritant.[14] The biogenic precursor of these linear furanocoumarin compounds is umbelliferone.[15]

Physiochemical Properties

Animal studies suggest that extracts of rue have bacteriostatic, antihelminthic, anti-inflammatory, antipyretic, and analgesic properties.[4,16] However, there are few clinical data to support the use of this herb for any of these purposes.

Mechanism of Toxicity

Most adverse reactions to the use of rue result from photosensitization to the psoralen compounds, such as bergapten (5-methoxypsoralen) and xanthotoxin (8-methoxypsoralen) in the plant.

DOSE RESPONSE

Recommended doses for herbal preparations of *Ruta graveolens* include one cup of boiling water poured over 1–2 teaspoons of dried herb or the ingestion of 0.5–1 g dried herb 3 times daily. The development of allergic contact dermatitis is a hypersensitivity reaction that is not specifically dose-related. However, phytophototoxicity is an irritant type of photodermatitis that requires UV light and no prior sensitization. These dermal reactions increase with increasing doses of irritants.

TOXICOKINETICS

There are few data on the toxicokinetics of the components of rue extract.

CLINICAL RESPONSE

Rare case reports associate the use of rue with the development of phytophotodermatitis on sun-exposed portions of the face and extremities.[17] Skin lesions include the appearance of pruritic, erythematous, linear macules,[18] and acute vesiculobullous dermatitis[19] within 6–48 hours of exposure to rue.[20] These lesions are characterized by linear erythema with sharp demarcation between the lesion and unaffected skin. Potential complications include changes in pigmentation and infected bullae. Phytophotodermatitis occurs following exposure to several *Ruta* species in addition to *R. graveolens* including *R. montana* L. (mountain rue),[21] *R. corsica* (Corsican rue),[22] and *R. chalepensis* (fringed rue).[23,24] Postabortion sepsis with multiorgan failure is a potential complication of the use of *Ruta* infusions for the induction of abortions.[24] A case report associated the recent consumption of a decoction of *Ruta graveolens* with the development of dyspnea, bradycardia, renal dysfunction, and hyperkalemia in a 78-year-old woman. She began the decoction 1 day before development of dyspnea and 3 days before admission to an emergency department. A causal connection between rue and her symptoms was not established because of the presence of pre-existing disease and the lack of serum drug concentrations of her prescribed medications (i.e., bisoprolol, diltiazem).

DIAGNOSTIC TESTING

Methods for analyzing the constituents of extracts from *Ruta* species include GC with nitrogen-phosphorus or flame ionization detection,[1,2] gas/liquid chromatography,[25] and GC/MS.[13]

TREATMENT

Treatment is supportive. Therapy for skin lesions includes antihistamines, soothing baths to reduce pruritus, and topical corticosteroid creams. Oral steroids may be necessary for more severe cases of rue-induced phytophotodermatitis. The use of soap and water, sun screens, and avoidance of sun exposure for 48 hours reduces the severity of phytophotodermatitis if initiated before the appearance of skin lesions.

References

1. De Feo V, De Simone F, Senaore F. Potential allelochemicals from the essential oil of *Ruta graveolens*. Phytochemistry 2002;61:573–578.
2. Stashenko EE, Acosta R, Martinez JR. High-resolution gas-chromatographic analysis of the secondary metabolites obtained by subcritical-fluid extraction from Colombian rue (*Ruta graveolens* L.). J Biochem Biophys Methods 2000;43:379–390.
3. Chen C-C, Huang Y-L, Huang F-I, Wang C-W, Ou J-C. Water-soluble glycosides from *Ruta graveolens*. J Nat Prod 2001;64:990–992.
4. Al-Said MS, Tariq M, Al-Yahya MA, Rafatulla S, Ginnawi OT, Ageel AM. Studies on *Ruta chalepensis* an ancient medicinal herb still used in traditional medicine. J Ethnopharmacol 1990;28:305–312.
5. Conway GA, Slocumb JC. Plants used as abortifacients and emmenagogues by Spanish New Mexicans. J Ethnopharmacol 1979;1:241–261.
6. Kong YC, Lau CP, Wat KH, Ng KH, But PP, Cheng KF, Waterman PG. Antifertility principle of *Ruta graveolens*. Planta Med 1989;55:176–178.
7. Ciganda C, Laborde A. Herbal infusions used for induced abortion. J Toxicol Clin Toxicol 2003;41:235–239.
8. Guarrera PM. Traditional antihelmintic, antiparasitic and repellent uses of plants in Central Italy. J Ethnopharmacol 1999;68:183–192.
9. Pathak S, Multani AS, Banerji P, Banerji P. Ruta 6 selectively induces cell death in brain cancer cells but proliferation in normal peripheral blood lymphocytes: A novel treatment for human brain cancer. Int J Oncol 2003;23:975–982.
10. Banerji P, Banerji P. Intracranial cysticercosis: an effective treatment with alternative medicines. In Vivo 2001;15:181–184.
11. Touati D, Atta-ur-Rahman, Ulubelen A. Alkaloids from *Ruta montana*. Phytochemistry 2000;53:277–279.
12. Ulubelen A, Guner H, Cetindag M. Alkaloids and coumarins from the roots of *Ruta chalepensis* var. latifolia. Planta Med 1988;54:551–552.
13. Ivanova A, Kostova I, Rodriguez Navas H, Villegas J. Volatile components of some Rutaceae species. Z Naturforsch 2004;59:169–173.
14. Hale AL, Meepagala KM, Oliva A, Aliotta G, Duke SO. Phytotoxins from the leaves of *Ruta graveolens*. J Agric Food Chem 2004;52:3345–3349.
15. Ekiert H, Czygan F-C. Accumulation of biologically active furanocoumarins in agitated cultures of *Ruta graveolens* L. and *Ruta graveolens* ssp. divaricata (Tenore) Gams. Pharmazie 2005;60:623–626.
16. Mancebo F, Hilje L, Mora GA, Castro VH, Salazar R. Biological activity of *Ruta chalepensis* (Rutaceae) and *Sechium pittieri* (Cucurbitaceae) extracts on *Hypsipyla grandella* (Lepidoptera: Pyralidae) larvae. Rev Biol Trop 2001;49:501–508.
17. Eickhorst K, DeLeo V, Csaposs J. Rue the herb: *Ruta graveolens*-associated phytophototoxicity. Dermatitis 2007;18:52–55.
18. Heskel NS, Amon RB, Storrs FJ, White CR Jr. Phytophotodermatitis due to *Ruta graveolens*. Contact Dermatitis 1983;9:278–280.
19. Wessner D, Hofmann H, Ring J. Phytophotodermatitis due to *Ruta graveolens* applied as protection against evil spells. Contact Dermatitis 1999;41:232.
20. Gawkrodger DJ, Savin JA. Phytophotodermatitis due to common rue (*Ruta graveolens*). Contact Dermatitis 1983;9:224.
21. Ortiz-Frutos FJ, Sanchez B, Garcia B, Iglesias L, Sanchez-Mata D. Photocontact dermatitis from rue (*Ruta montana* L.). Contact Dermatitis 1995;33:284.
22. Ena P, Camarda I. Phytophotodermatitis from *Ruta corsica*. Contact Dermatitis 1990;22:63.
23. Brener S, Friedman J. Phytophotodermatitis induced by *Ruta chalepensis* L. Contact Dermatitis 1985;12:230–232.
24. Goncalo S, Correia C, Couto JS, Goncalo M. Contact and photocontact dermatitis from *Ruta chalepensis*. Contact Dermatitis 1989;21:200–201.
25. Gunaydin K, Savci S. Phytochemical studies on *Ruta chalepensis* (Lam.) lamarck. Nat Prod Res 2005;19:203–210.

Chapter 86

SASSAFRAS
[*Sassafras albidum* (Nutt.) Nees]

HISTORY

Sassafras was a common tonic and herb of many native North American Indian tribes for the treatment of upper respiratory tract infections, kidney dysfunction, rheumatism, and skin lesions.

BOTANICAL DESCRIPTION

The sassafras tree is native to the eastern United States, but other species also contain safrole. *Ocotea odorifera* (Vell.) Rohwer (Brazilian sassafras) is a medium-sized tree indigenous to southeastern Brazil, Columbia and Paraguay, whereas *Cinnamomum camphora* (L.) J. Presl (Chinese sassafras) occurs throughout Southeast Asia. The forest shrubs of the Piperaceae family (*Piper auritum* Kunth, *P. hispidinervum* C.DC., *P. callosum* Ruiz & Pav.) are indigenous to the humid forests of Central America and the Brazilian Amazon.

Common Name: Sassafras Tree, Ague Tree, Cinnamonwood, Saloop, Saxifrax, Smelling-Stick

Scientific Name: *Sassafras albidum* (Nutt.) Nees

Botanical Family: Lauraceae (laurel)

Physical Description: A large tree with alternate, deep green leaves up to 8–15 cm (~3–6 in.) long. This tree reaches 15 m (~49 ft) in height and 35 cm (~14 in.) in diameter. Older stems are deeply furrowed with reddish brown bark. The leaves, stems, and bark that contain pleasant-smelling oils. Small yellowish-green flowers appear in early spring in axillary racemes. The dark blue or black, oblong, fleshy fruit develops in August to September, and this drupe contains brown, oblong seeds.

Distribution and Ecology: This native eastern US species ranges from Florida to southern Maine and from western Iowa to Texas.

EXPOSURE

Sources

The roots of North American species, *Sassafras albidum*, are the source of true sassafras oil. This oil once was the main flavoring in root beer, but commercial production of this oil ceased as a result of concern about the health effects of safrole in this oil. The two important sources of commercial sassafras oil now are Brazilian sassafras oil and Chinese sassafras oil. The source of Brazilian sassafras oil is the woody trunk of *Ocotea odorifera* (*Ocotea cymbarum*) from southeastern Brazil, whereas the source of Chinese sassafras oil is the Southeast Asian species, *Cinnamomum camphora* (L.) J. Presl. Some forest shrubs of the Piperaceae family (*Piper auritum, P. hispidinervum, P. callosum*) contain high concentrations of safrole in their leaves, but there is limited commercial production of safrole from these sources.

Medicinal Uses

TRADITIONAL

Traditional uses of sassafras oil include topical anesthesia, muscle relaxation, blood purification, dental disin-

Medical Toxicology of Natural Substances, by Donald G. Barceloux, MD

fectant, and the treatment of skin disorders, hypertension, renal disorders, cancer, syphilis, and menstrual irregularity. Sassafras oil was a common flavoring agent for soft drinks and baked goods until the US Food and Drug Administration (FDA) banned the oral use of sassafras oil as a food additive because of the presence of an animal carcinogen in the oil. Some dietary supplements may contain detectable quantities of safrole.[6]

Current

Sassafras oil is a fragrance added to soap and toiletries. Brazilian and Chinese sassafras oils are raw materials for the isolation of safrole. Commercial safrole from these sources is a raw material for the production of the flavoring agent and fragrance, heliotropin, and the synergistic agent, piperonyl butoxide. The latter compound is an important ingredient of pyrethroid insecticides that enhances the effectiveness of pyrethroid insecticides. Safrole and isosafrole can be used as precursors in some synthetic processes that produce 3,4-methylenedioxymethylamphetamine (MDMA).[1]

Regulatory Status

Until 1960, safrole-containing sassafras extracts were flavoring agents in beverages, such as root beer in concentrations up to 20 ppm safrole. Because of the hepatotoxic effects of safrole in animal models and the classification of safrole as a weak hepatic carcinogen, the FDA prohibited the use of safrole, dihydrosafrole, oil of sassafras, and isosafrole as food additives in foods and beverages in the early 1960s.[2] This ban was extended to sassafras tea in 1977. Consequently safrole, isosafrole, and safrole-containing products are not recognized as safe for human consumption. However, the 1994 Dietary Supplement Health and Eduction Act allows the use of certain dietary supplements containing sassafras. Further, the FDA permits the use of nutmeg and mace, which contain small amounts of safrole. Uses of safrole as a pesticide also have been canceled by the EPA. In 1988, the Council of the European Communities also banned safrole as a flavoring agent in foodstuffs. Isosafrole is a minor constituent of some soaps, detergents, cream, and perfumes as a result of the addition of ilang-ilang (ylang-ylang) oil from the ilang-ilang tree [*Cananga odorata* (Lam.) Hook. F. & T. Thomson].[3] Brazilian and Chinese safrole oils contain about 80% safrole.

PRINCIPAL INGREDIENTS

Chemical Composition

The major constituent of sassafras oil is the alkenyl benzene compound called safrole (4-allyl-1,2-methylenedioxybenzene). The root bark from the sassafras tree contains about 5–9% volatile oil composed primarily of safrole (80–85%) along with camphor (3%), methyleugenol (1%), and minor amounts of monoterpenes (α-camphene, β-pinene, 1,8-cineole, *p*-cymene) and sesquiterpene compounds other than methyl eugenol.[4] Figure 86.1 displays the chemical structure of safrole. A commercial sample of essential oils from *Sassafras albidum* contained approximately 82% safrole, 3% methyleugenol, 1% α-cubebene, 1% α-calacorene, and <1% of a variety of terpenoid compounds (limonene, 1,8-cineole, carvone, eugenol) as measured by gas chromatography/mass spectrometry (GC/MS).[5] The safrole content in teas brewed from the root bark of the sassafras tree is highly variable with analysis of the safrole content in six samples of brewed tea ranging from 0.09–4.66 mg/cup.[6]

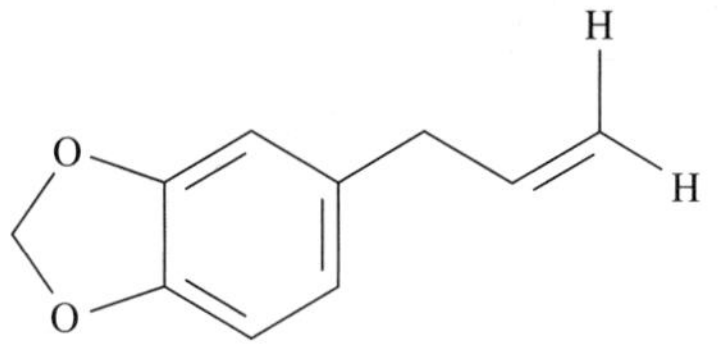

FIGURE 86.1. Chemical structure of safrole.

Concentrations of safrole in most leaf samples are not detectable (i.e., <1.5 ng/mL).[6] Safrole is also present in other spices (e.g., star anise, cumin, nutmeg, mace, black pepper, ginger) as well as betel quid (*Piper betle* inflorescence).[7,8] Methylsafrole (myristicin) is the major aromatic ingredient of nutmeg and mace. Isosafrole (4-propenyl-1,2-methylenedioxybenzene) is an isomer of safrole that occurs in the essential oil of ilang-ilang from the Philippines and in the fruit of Japanese star anise. However, isosafrole is not a constituent of sassafras oil.

Physiochemical Properties

A variety of food processing methods reduce >90% of the safrole content in food including washing and heating the seed at 70 °C (158 °F) for 30 minutes, boiling 5 minutes in distilled water, and sun drying for 48 hours.[8] Safrole is a clear, colorless to slightly yellow liquid with an aromatic odor of sassafras. This compound is water insoluble. Table 86.1 lists some physical and chemical properties of safrole.

Mechanism of Toxicity

Both safrole and its proximate metabolite, 1′-hydroxysafrole, are carcinogenic in animal models with the latter being more potent than the parent compound.[9] The

TABLE 86.1. Physical and Chemical Properties of Safrole

Property	Value
Molecular Formula	$C_{10}H_{10}O_2$
Formula Weight	162.18
CAS Registry Number	94-59-7
Density	1.09 g/cm^3
Melting Point	11.2 °C/~56.2
Boiling Point	232–234 °C/~450–453 °F
Flash Point	97 °C/~207 °F
Water Solubility	121 mg/L (25 °C/77 °F)
Vapor Pressure	0.062 mmHg (25 °C/77 °F)
log P (Octanol-Water)	3.450
Henry's Law Constant	9.07E-06 atm-m^3/mole (25 °C/77 °F)
Atmospheric OH Rate Constant	7.56E-11 cm^3/molecule-second (25 °C/77 °F)

carcinogenicity of safrole requires metabolic activation, probably to reactive quinone compounds via *O*-dealkylation to hydroxychavicol followed by oxidation to form the reactive quinone compounds.[10]

DOSE RESPONSE

Although the contents of a single sassafras tea bag may exceed the estimated toxic dose for humans as extrapolated from animal models, there are few clinical data to assess human toxicity. The inadvertent ingestion of 5 mL of sassafras oil was associated with tremulousness and vomiting, but the patient developed no evidence of hepatorenal dysfunction.[11] She recovered without sequelae. A 72-year-old woman developed diaphoresis and hot flashes after the daily ingestion of 10 cups of sassafras tea.[12] All symptoms resolved after she stopped drinking the tea.

TOXICOKINETICS

The absorption and elimination of low doses of safrole in humans is rapid. In a volunteer study, most of a 1.66-mg dose of safrole was eliminated in the urine within 24 hours with conjugated 1,2-dihyrodxy-4-allylbenzene being the main urinary metabolite.[13] This study suggested some differences in metabolism between species. The proximate carcinogen of safrole, 1′-hydroxysafrole, was present in the urine of rats, but not humans, following the oral administration of safrole. The major metabolic pathways for the formation of reactive metabolites are 1) allylic hydroxylation to 1′-hydroxysafrole followed by oxidation and *O*-dealkylation leading to the formation of 4-allyl-*O*-quinone, and 2) epoxidation of the allylic side chain or aromatic ring of safrole.[14] Several CYP450 isoenzymes are involved in the biotransformation of safrole to reactive intermediates. *In vitro* studies of human liver cells indicate that CYP2C9, CYP2A6, CYP2D6, and CYP2E1 are the main cytochrome P450 isoenzymes involved in the biotransformation of safrole to the carcinogen, 1′-hydroxysafrole.[15,16] The conversion of 1′-hydroxysafrole to 1′-sulfoxysafrole by sulfotransferase enzymes followed by the cleavage of the sulfate moiety results in the formation of a carbocation, which can produce DNA damage by covalently binding to nucleophilic sites on macromolecules.

CLINICAL RESPONSE

Acute Effects

There are limited data on the clinical effects of sassafras oil. The ingestion of 5 mL sassafras oil was associated with mild transient symptoms (tremulousness, vomiting),[11] whereas the daily ingestion of 10 cups of sassafras tea produced diaphoresis and hot flashes.[12]

Carcinogenicity

Sassafras oil contains 80% safrole, which is a weak liver carcinogen when fed to rodents in high doses.[17] The International Agency for Research on Cancer (IARC) lists safrole and dihydrosafrole as possible human carcinogens (Group 2B) based on the production of liver and lung tumors in rodents.[18] The US Toxicology Program lists safrole as reasonably anticipated to be a human carcinogen based on the increased incidences of liver hepatocellular carcinomas and cholangiocarcinomas in both sexes of rats and hepatocellular carcinomas in male mice following administration of safrole in the diet.[19] There are no case reports or epidemiological data on the carcinogenicity of safrole in humans.

DIAGNOSTIC TESTING

The use of reversed-phase liquid chromatography with UV detection at 235 nm allows the detection of safrole, isosafrole, and dihydrosafrole at concentrations of 1.5 ng/mL, 1.8 ng/mL, and 3.8 ng/mL, respectively.[6] Other methods to determine the content of safrole and isosafrole include GC and GC/MS.[4,20]

TREATMENT

Treatment is supportive. Gut decontamination measures are not usually necessary.

References

1. Renton RJ, Cowie JS, Oon MC. A study of the precursors, intermediates and reaction by-products in the synthesis of 3,4-methylenedioxymethylamphetamine and its application to forensic drug analysis. Forensic Sci Int 1993;60:189–202.
2. Food and Drug Administration. Refusal to extend effective date of statute for certain specified food additives. Fed Regist 1960;25:12412.
3. Opdyke DL. Isosafrole. Food Cosmet Toxicol 1976;14:329.
4. Kamdem DP, Gage DA. Chemical composition of essential oil from the root bark of *Sassafras albidum*. Planta Med 1995;61:574–575.
5. Simic A, Sokovic MD, Ristic M, Grujic-Jovanovic S, Vukojevic J, Marin PD. The chemical composition of some Lauraceae essential oils and their antifungal activities. Phytother Res 2004;18:713–717.
6. Carlson M, Thompson RD. Liquid chromatographic determination of safrole in sassafras-derived herbal products. J AOAC Int 1997;80:1023–1028.
7. Wang CK, Hwang LS. Phenolic compounds of betel quid chewing juice. Food Sci 1993;20:458–471.
8. Farag SEA, Abo-Zeid M. Degradation of the natural mutagenic compound safrole in spices by cooking and irradiation. Nahrung 1997;41:359–361.
9. Phillips DH. DNA adducts derived from safrole, estragole and related compounds, and from benzene and its metabolites. IARC Sci Publ 1994;125:131–140.
10. Johnson BM, Bolton JL, van Breemen RB. Screening botanical extracts for quinoid metabolites. Chem Res Toxicol 2001;14:1546–1551.
11. Grande GA, Dannewitz SR. Symptomatic sassafras oil ingestion. Vet Hum Toxicol 1987;29:447.
12. Haines JD Jr. Sassafras tea and diaphoresis. Postgrad Med 1991;90:75–76.
13. Benedetti MS, Malnoe A, Broill AL. Absorption, metabolism and excretion of safrole in the rat and man. Toxicology 1977;7:69–83.
14. Luo G, Guenthner TM. Covalent binding to DNA *in vitro* of 2′,3′-oxides derived from allylbenzene analogs. Drug Metab Dispos 1996;24:1020–1027.
15. , Jeurissen SM, Bogaards JJ, Awad HM, Boersma MG, Brand W, Fiamegos YC, et al. Human cytochrome P450 enzyme specificity for bioactivation of safrole to the proximate carcinogen 1′-hydroxysafrole. Chem Res Toxicol 2004;17:1245–1250.
16. Ueng YF, Hsieh CH, Don MJ, Chi CW, Ho LK. Identification of the main human cytochrome P450 enzymes involved in safrole 1′-hydroxylation. Chem Res Toxicol 2004;17:1151–1156.
17. , Borchert P, Wislocki PG, Miller JA, et al. The metabolism of naturally occurring hepatocarcinogen safrole to 1′-hydroxysafrole and the electrophilic reactivity of 1′-acetoxysafrole. Cancer Res 1972;33:573–589.
18. International Agency for Research on Cancer (IARC). Safrole, isosafrole and dihydrosafrole. IARC Monogr Eval Carcinog Risk Chem Man 1976;10:231–244.
19. National Toxicology Program (NTP). Report on carcinogens. 9th ed. Research Triangle Park, NC: US Department of Health and Human Services, Public Health Service; 2001.
20. Heikes DL. SFE with GC and MS determination of safrole and related allylbenzenes in sassafras teas. J Chromatogr Sci 1994;32:253–258.

Chapter 87

SAW PALMETTO
[*Serenoa repens* (Bartram) J.K. Small]

HISTORY

Native Americans used extracts of saw palmetto for the treatment of abdominal disorders and dysentery, whereas the fruit of the plant was a food and nutrient. The crude extracts have been used for centuries to improve breast size, sperm production, and sexual vigor. The American Indians used saw palmetto berries for genitourinary symptoms. In the late 19th century, saw palmetto was investigated for medicinal effects including stimulation of appetite and breast milk, amelioration of ovarian dysfunction, relief of dysmenorrhea and prostatitis, and the treatment of gonorrhea. Saw palmetto was listed as an official drug in the *US Pharmacopoeia* from 1906–1917 and as part of the *US National Formulary* from 1926–1950. Currently, saw palmetto is one of the top five most popular herbal products, primarily as a result of the use of saw palmetto to treat the symptoms related to benign prostatic hypertrophy.[1]

BOTANICAL DESCRIPTION

Common Name: Saw Palmetto, American Dwarf Palm Tree, Sabal

Scientific Name: *Serenoa repens* (Bartram) J.K. Small

Botanical Family: Arecaceae (Palmae, palm)

Physical Description: This palm plant grows up to 10 ft (3 m) tall with fan-shaped leaves being cleft at the apex. In spring, a seed stalk (spadix) sprouts and flowers. Subsequently, a hard green fruit, which is the size of an olive, develops during the summer and turns to varying shades of red and then to black. Harvesting of the mature fruit occurs in late summer and fall.

Distribution and Ecology: *Serenoa repens* grows in the southeastern United States, particularly in the swamps of eastern Florida and southern Georgia. This shrubby palm inhabits areas of the southern Atlantic coast (Carolina's region) and southern United States as far west as Texas. *Serenoa repens* is also indigenous to the West Indies. The berries are typically harvested in late fall and early winter.

EXPOSURE

Sources

Commercially, saw palmetto is available alone (160 mg–585 mg capsules, soft gelcaps, tablet) or in combination with other substances (zinc, pygeum, pumpkin seed oil extract, ginseng, bee pollen, pyridoxine, amino acids).

Medicinal Uses

Traditional

Traditional uses of *Serenoa repens* extract include breast enhancement, improvement of sexual desire, diuresis, and the treatment of cystitis. Alternative treatments of symptoms associated with prostatic hypertrophy include *Pygeum africanum* Hook.f. (African plum tree), *Hypoxis rooperi* T. Moore (African star grass), and rye grass pollens.

Medical Toxicology of Natural Substances, by Donald G. Barceloux, MD

Current

The primary use of saw palmetto is the short-term treatment of urinary symptoms associated with benign prostatic hypertrophy. Saw palmetto is a safe treatment that provides mild to moderate improvement in urinary symptoms and urinary flow measurements in some,[2,3] but not all randomized clinical studies.[4] A meta-analysis of 18 randomized, double-blind clinical trials (10 of them placebo controlled), which involved a 2,939 men with benign prostatic hypertrophy demonstrated superiority of saw palmetto over placebo in the treatment of nocturia. A 6-month double-blind randomized equivalence study compared the effects of saw palmetto extract (320mg Permixon®; Pierre Fabre Medicament, Boulogne, France) with 5mg finasteride in 1,098 men with moderate benign prostatic hypertrophy.[5] Both treatments were equally effective in increasing peak urinary flow rates and improving urinary tract symptoms as measured by the International Prostate Symptom Score. However, in contrast to finasteride, prostate size and prostatic specific antigen (PSA) concentrations in the saw palmetto group remained similar to baseline with fewer complaints of decreased libido and impotence in the saw palmetto group. The use of saw palmetto was less effective in clinical trials of more advanced cases of chronic prostatitis. In a prospective, 1-year study of 64 men with category III prostatitis and chronic pelvic pain syndrome, the daily administration of 325mg saw palmetto did not produce long-term improvement as measured by the National Institutes of Health Chronic Prostatitis Symptom Index and the American Urological Association Symptom Score.[6] In contrast, the use of 5mg finasteride improved the quality of life and pain scores, but did not produce a statistically significant improvement in symptoms related to urination. Similarly, a double-blind controlled clinical trial of the use of saw palmetto extract (160mg twice daily) for moderate-to-severe symptoms of benign prostatic hypertrophy did not detect a statistically significant difference between placebo and treatment groups in measures of the quality of life, prostate size, or residual volume.[4] There was no significant difference between the saw palmetto and placebo groups in the change in American Urological Association Symptom Index scores (mean difference = 0.04 point; 95% CI: —0.93–1.01), or in maximal urinary flow rate (mean difference = 0.43mL/min; 95% CI: –0.52–1.38).

Regulatory Status

Saw palmetto extracts are classified as dietary supplements by the US Food and Drug Administration (FDA). The German Commission E approved the use of saw palmetto for urinary problems associated with benign prostatic hyperplasia stages 1 (abnormal frequent urination, nocturia, delayed urination onset, weak urinary stream) and stage 2 (urge to urinate, residual urine). Although there are few data, saw palmetto is not recommended for use in children and pregnant women.[7] Permixon® is available in Europe, but not in the United States.

PRINCIPAL INGREDIENTS

Chemical Composition

Saw palmetto (Permixon®) is an *n*-hexane lipidosterolic extract of the berries from *Serenoa repens* (American dwarf palm tree). This drug is a complex mixture primarily of free fatty acids (90%) and related esters (7%) along with small amount of phytosterols (campesterol, β-sitosterol, stigmasterol),[8] aliphatic alcohols (C_{26}, C_{28}, C_{30}), and various other compounds (arabinose, flavonoids, galactose, glucose, uronic acid). The unsaturated and unsaturated fatty acids include capric, caproic, caprylic, lauric, linolenic, myristic, isomyristic, oleic, palmitic, and stearic acids, whereas the phytosterols include β-sitosterol, campesterol, stigmasterol, cycloartenol, lupeol, and lupenone. The major free fatty acids are oleic acid, lauric acid, myristic acid, and palmitic acid. The remaining components include carotene, lipase, and tannin. In contrast to most other herbs, there is no specific chemical constituent of saw palmetto that is a marker for product standardization. The composition of saw palmetto preparations in the United States is generally similar to marker compounds on the label. Analysis of eight US saw palmetto products indicated that the total free fatty acids content as measured by gas chromatography with flame ionization detection (GC/FID) ranged from 77–106% of the stated amounts.[9]

Physiochemical Properties

The lipidosterolic extract of saw palmetto berries has multiple effects including inhibition of 5α-reductase, which is the isoenzyme that converts testosterone to the biologically more potent dihydrotestosterone. The human prostate contains type I and type II isoenzymes of 5α-reductase with the latter being the most important isoform. Saw palmetto is a noncompetitive inhibitor of both type I and type II isoenzymes, whereas the synthetic 5α-reductase inhibitor, finasteride (Proscar®; Merck & Co., Whitehouse Station, NJ), is a selective competitive inhibitor of type II 5 α-reductase.[10] Saw palmetto extract also possesses anti-androgen effects by inhibiting the binding of dihydrotestosterone to cytosolic androgen receptors in prostatic cells as well as anti-inflammatory effects by reducing arachidonic acid metabolites by inhibiting aromatase cyclooxygenase and lipoxygenase.[11] Unlike other 5α-reductase inhibitors, lipophilic extracts of saw palmetto block the 5α-reductase activity of prostatic epithelial cells without interfering with the secretion of the prostate tumor marker, PSA.[12]

Mechanism of Action

Saw palmetto has several effects, but the exact mechanism responsible for improvement of urinary tract symptoms associated with benign prostatic hypertrophy is unclear.[13,14] Androgens have a significant effect on prostate growth, and dihydrotestosterone comprises about 95% of the total androgenic steroid content within the prostate cell nuclei.[15] Possible mechanisms of action of saw palmetto on the prostate include inhibition of 5α-reductase, which is the enzyme that converts testosterone to dihydrotestosterone and inhibition of the binding of dihydrotestosterone to androgen receptors in prostate cells. However, clinical trials comparing the 5α-reductase inhibitor, finasteride, with saw palmetto indicate that the saw palmetto extract, Permixon®, has little effect on androgen-dependent parameters compared with finasteride.[5] In a 1-week open, randomized, placebo-controlled study of 32 healthy male volunteers, 5 mg finasteride daily reduced the serum dihydrotesterone concentration an average of 65%, whereas there was no significant difference in serum dihydrotesterone concentrations between placebo and Permixon® 80 mg twice daily.[16] Other potential mechanisms based on animal studies involve inhibition of growth factors associated with proliferation of prostate cells, reduced prolactin binding, and decreased arachidonic acid metabolites from inhibition of 5-lipoxygenase.[17]

DOSE RESPONSE

Typical medicinal doses of saw palmetto are 160–320 mg daily of the lipophilic extract containing 80–90% volatile oils or 1–2 g of the dried saw palmetto fruit. Tea products are considered ineffective because of the lack of volatile oils.[18] Symptomatic relief may take up to 6 weeks to occur. The toxic dose of saw palmetto is not well defined.

TOXICOKINETICS

Kinetics

Because saw palmetto is a complex mixture of free fatty acids and other compounds, there are few data on the pharmacokinetics of this lipidosterolic extract. The dosage of saw palmetto refers to milligrams of the plant extract administered, and the portion of the extract studied is determined by the migration of the extract on the chromatograph. A study of 12 healthy volunteers ingesting 320 mg saw palmetto determined the pharmacokinetic parameters of the second component of saw palmetto based on high performance liquid chromatography (HPLC) with a retention time of 26.4 minutes.[19] Oral absorption was relatively low. The peak time to oral absorption is 1.5 hours with a mean peak plasma concentration of 2.58 μg/mL (range: 2.3–2.9 μg/mL). The elimination half-life was approximately 1.9 hours.

Drug Interactions

In vivo studies with single-time-point phenotypic ratios indicate that there is minimal risk of saw palmetto–drug interactions based on lack of effect on CYP1A2, CYP2D6, CYP2E1, and CYP3A4 isoenzymes.[20] The interaction of saw palmetto with other drugs during the treatment of urinary tract symptoms is rare.[21] In a study of 12 healthy volunteers receiving 320 mg saw palmetto daily for 14 days, the drug probe substrates dextromethorphan (CYP2D6 activity) and alprazolam (CYP3A4 activity) were administered to compare the baseline and posttreatment activity of these cytochrome P450 isoenzymes.[22] The lack of significant change in activity suggests that short-term saw palmetto therapy does not alter the metabolism of drugs dependent on CYP2D6 and CYP3A4 isoenzymes. Serious intraoperative bleeding developed during resection of a meningioma in a 53-year-old man, who used saw palmetto for benign prostatic hypertrophy.[23] The prothrombin (PT) and activated partial thromboplastin time (aPTT) were normal, but the bleeding time was prolonged to 21 minutes (normal: 2–10 minutes). After cessation of saw palmetto, the bleeding time returned to normal values within 5 days.

CLINICAL RESPONSE

Adverse effects associated with the use of saw palmetto are uncommon and typically involve mild gastrointestinal distress.[24] Reported adverse effects include gastrointestinal upset, abdominal pain, nausea (about 2%), vomiting, diarrhea, muscle pain, rhinitis (8.6%) headache (8%), dizziness (3%), and xerostomia (1%). Orthostatic hypertension was noted in 1.1 % of individuals treated with 320 mg for 12 months. Other cardiovascular effects include tachycardia and exacerbation of angina pectoris. Genitourinary effects include reduced libido, ejaculation disorders, and urinary retention. Case reports have associated the use of saw palmetto with cholestatic hepatitis, cholecystitis, and intraoperative hemorrhage.[23] None of the above adverse effects appear to result from a dose-effect relationship. Adverse effects occur in less than 5% of patients in clinical trials. Case reports indicate that rarely an allergic contact dermatitis can develop following the chronic topical application of saw palmetto extract.[25]

DIAGNOSTIC TESTING

There are no specific analytical methods for evaluating saw palmetto because there are no biomarkers or well-defined chemical constituents of saw palmetto. The use of saw palmetto does not affect determination of serum PSA concentrations.

TREATMENT

Gut decontamination is unnecessary. The treatment of adverse effects is symptomatic and supportive; there is no specific therapy for exposure to saw palmetto.

References

1. De Smet PA. Herbal remedies. N Engl J Med 2002;347: 2046–2056.
2. Wilt T, Ishani A, MacDonald R. *Serenoa repens* for benign prostatic hyperplasia. Cochrane Database Syst Rev 2002; (3):CD001423.
3. Boyle P, Robertson C, Lowe F, Roehrborn C. Meta-analysis of clinical trials of Permixon in the treatment of symptomatic benign prostatic hyperplasia. Urology 2000;55: 533–539.
4. Bent S, Kane C, Shinohara K, Neuhaus J, Hudes ES, Goldberg H, Avins AL. Saw palmetto for benign prostatic hyperplasia. N Engl J Med 2006;354:557–566.
5. Carraro JC, Raynaud JP, Koch G, Chisholm GD, Di Silverio F, Teillac P, et al. Comparison of phytotherapy (Permixon) with finasteride in the treatment of benign prostate hyperplasia: a randomized international study of 1,098 patients. Prostate 1996;29:231–240.
6. Kaplan SA, Volpe MA, Te AE. A prospective, 10-year trial using saw palmetto versus finasteride in the treatment of category III prostatitis/chronic pelvic pain syndrome. J Urol 2004;171:284–288.
7. Klepser TB, Klepser ME. Unsafe and potentially safe herbal therapies. Am J Health-Syst Pharm 1999;56: 125–138.
8. Sorenson WR, Sullivan D. Determination of campesterol, stigmasterol, and beta-sitosterol in saw palmetto raw materials and dietary supplements by gas chromatography: single-laboratory validation. J AOAC Int 2006;89: 22–34.
9. Krochmal R, Hardy M, Bowerman S, Lu Q-Y, Wang H-J, Elashoff RM, Heber D. Phytochemical assays of commercial botanical dietary supplements. eCAM 2004;1: 305–313.
10. Raynaud JP, Cousse H, Martin PM. Inhibition of type 1 and type 2 5-alpha-reductase activity by free fatty acids, active ingredients of Permixon. J Steroid Biochem Mol Biol 2002;82:233–239.
11. Paubert-Braquet M, Mencia Huerta JM, Cousse H, Braquet P. Effect of the lipidic lipidosterolic extract of *Serenoa repens* (Permixon) on the ionophore A23187-stimulated production of leukotriene B4 (LTB4) from human polymorphonuclear neutrophils. Prostaglandins Leukot Essent Fatty Acids 1997;57:299–304.
12. Habib FK, Ross M, Ho CK, Lyons V, Chapman K. *Serenoa repens* (Permixon®) inhibits the 5a-reductase activity of human prostate cancer cell lines without interfering with PSA expression. Int J Cancer 2004;114:190–194.
13. Buck AC. Is there a scientific basis for the therapeutic effects of *Serenoa repens* in benign prostatic hyperplasia? Mechanisms of action. J Urol 2004;172:1792–1799.
14. Polsker GL, Brogden RN. *Serenoa repens* (Permixon®) a review of its pharmacology and therapeutic efficacy in benign prostatic hyperplasia. Drugs Aging 1996:9: 379–395.
15. Siiteri PK, Wilson JD. Dihydrotestosterone in prostatic hypertrophy. I. The formation and content of dihydrotestosterone in the hypertrophic prostate of man. J Clin Invest 1970;49:1737–1745.
16. Strauch G, Perles P, Vergult G, Gabriel M, Gibelin B, Cummings S, et al. Comparison of finasteride (Proscar®) and *Serenoa repens* (Permixon®) in the inhibition of 5-alpha reductase in healthy male volunteers. Eur Urol 1994;26: 247–252.
17. Breu W, Hagenlocher M, Redl K, Tittel G, Stadler F, Wagner H. [Anti-inflammatory activity of sabal fruit extracts prepared with supercritical carbon dioxide. *In vitro* antagonists of cyclooxygenase and 5-lipoxygenase metabolism]. Arzneimittelforschung 1992;42:547–551. [German]
18. Gordon AE, Shaughnessy AF. Saw palmetto for prostate disorders. Am Fam Physician 2003;67:1281–1283.
19. De Bernardi di Valserra M, Tripodi AS, Contros S. *Serenoa repens* capsules: a bioequivalence study. Acta Toxicol Ther 1994;15:21–39.
20. Gurley BJ, Gardner SF, Hubbard MA, Williams DK, Gentry WB, Carrier J, et al. *In vivo* assessment of botanical supplementation on human cytochrome P450 phenotypes: *Citrus aurantium, Echinacea purpurea*, milk thistle, and saw palmetto. Clin Pharmacol Ther 2004;76: 428–440.
21. Izzo AA, Ernst E. Interactions between herbal medicines and prescribed drugs. A systematic review. Drugs 2001;61: 2163–2175.
22. Markowitz JS, Donovan JL, DeVane L, Taylor RM, Ruan Y, Wang J-S, Chavin KD. Multiple doses of saw palmetto

(*Serenoa repens*) did not alter cytochrome P450, 2D6, and 3A4 activity in normal volunteers. Clin Pharmacol Ther 2003;74:536–542.
23. Cheema P, El-Mefty O, Jazieh AR. Intraoperative haemorrhage associated with the use of extract of saw palmetto herb: a case report and review of literature. J Intern Med 2001;250:167–169.
24. Ernst E. The risk-benefit profile of commonly used herbal therapies: ginkgo, St. John's wort, ginseng, echinacea, saw palmetto, and kava. Ann Intern Med 2002;136:42–53.
25. Sinclair RD, Mallari RS, Tate B. Sensitization to saw palmetto and minoxidil in separate topical extemporaneous treatments of androgenetic alopecia. Australas J Dermatol 2002;45:311–312.

Chapter 88

SENNA
(*Senna alexandrina* P. Mill.)

HISTORY

The history of senna dates back at least to the 9th century AD, when Arabian physicians used both the pod and leaves for the treatment of constipation.

BOTANICAL DESCRIPTION

Common Name: Alexandrian Senna, Indian or Tinnevelly Senna, True Senna

Scientific Name: *Senna alexandrina* P. Mill. (*Cassia acutifolia* Delile, *Cassia angustifolia* Vahl, *Cassia lanceolata* Forssk., *Cassia senna* L.)

Cassia acutifolia Delile and *Cassia angustifolia* Vahl are no longer considered separate species in the genus *Senna*.

Botanical Family: Fabaceae

Physical Description: This annual herb reaches up to 1 m (~3 ft) in height. The leaves are alternate with 3–7 pairs of leaflets. Yellow flowers appear in erect terminal racemes that produce a broadly elliptical, flattened, parchment-like, dehiscent pod containing 6–10 seeds. The oblong pods are yellow-green to yellow-brown with a dark brown central area.

Distribution and Ecology: Senna is indigenous to tropical Africa, and this plant has naturalized near the Nile River from Aswan to Kordofan as well as the Arabian Peninsula, India, and Somalia. These small shrubs are cultivated in areas near Somalia, Sudan, the Arabian Peninsula, and the upper Nile River, as well as South India and Pakistan.

EXPOSURE

Sources

Senna consists of dried leaves or fruits of *Senna alexandrina* P. Mill. (formerly *Cassia senna* or *Cassia acutifolia*). Folium sennae consists of the dried leaflets of *Senna alexandrina*, whereas fructus sennae is derived from the dried ripe fruit. Branches with leaves and pods are dried in the sun before separating the leaves and pods for sale. Whole leaves are sold as herbs, while broken leaves are used as galenicals. Senna products include senna leaf powder (Herb-Lax®; Shaklee Corp., Pleasanton, CA), senna tea, senna pod concentrate, senna syrup, senna fluid extract, and senna fruit extract along with tablets containing sennosides A and B. A variety of commercial senna-containing products are available, such as Senokot® (Purdue Pharma, Stamford, CT), Dr. Caldwell's Senna Laxative® (Sterling Drug, New York, NY), and Fletcher's Castoria® (Rohto Pharmaceutical Co. Ltd, Osaka, Japan).

Medicinal Uses

TRADITIONAL

Senna is a well-known herbal drug in both Ayurvedic and Allopathic systems of medicine for the treatment of constipation. Other traditional uses of senna include the treatment of gonorrhea, bronchial congestion, wounds, diarrhea, intestinal gas, skin diseases, dyspepsia, fever, and hemorrhoids.

Medical Toxicology of Natural Substances, by Donald G. Barceloux, MD

CURRENT

Although senna has been used for centuries as a peristaltic stimulant laxative, there are few data on the efficacy of this herb for the treatment of chronic constipation.[1] In a study of 50 geriatric patients, senna (Senokot®) and sodium picosulfate (unavailable in the United States) were equally effective laxatives with few adverse effects.[2] A study of elderly psychiatric and mentally handicapped patients did not detect a significant difference in the efficacy of a bulk laxative (bran) and senna (Senokot® syrup).[3] Senna is also a constituent in some hair dyes.

REGULATORY STATUS

Most pharmacopoeias recommend evaluation of the total senna leaf glycoside content based on sennoside B (e.g., 2.5% in the German *Pharmacopoeia* for senna leaves). The German Commission E approves the use of senna pod and senna leaf as laxatives. In the United States, senna is available in a variety of over-the-counter products for the treatment of constipation.

PRINCIPAL INGREDIENTS

Chemical Composition

The main active ingredients in senna are dianthrone compounds from the anthraquinone family, called sennosides. These compounds are prodrugs for the peristalsis-stimulating dianthrone and anthraquinone compounds. The most common sennosides are sennoside A and B (i.e., about 1–2.5%), which hydrolyze to the corresponding aglycone (sennidin) and two molecules of glucose.[4,5] Smaller concentrations of sennoside C and D occur in the leaves and fruits along with naphthalene glycosides (6-hydroxymusicin glucoside, tinnevellin glucoside), monomeric glycosides, and free anthraquinone compounds (aloe-emodin, emodin, rhein).[6] Figure 88.1 displays the chemical structures of sennosides A–D. Minor constituents of senna include kaempferol (yellow flavonol) and related glucoside, isorhamnetin, calcium oxalate, and sterol glucosides. The concentration of sennosides depends on a variety of factors including the plant part, the growing season, and the extraction process.

Physiochemical Properties

Sennoside A and B are highly hydrophilic with a large molecular mass (863 Da). Neither sennosides nor rhein increase the permeability of the mucosa in the colon or enhance intestinal protein loss. Some degradation of sennosides occurs during the storage of crude senna plant materials as a result of the formation of sennidin glycosides from sennosides and the hydrolysis of rhein-8-*O*-glucoside to rhein by enzymatic processes.[7]

Mechanism of Toxicity

Sennosides and their degradation products affect both motility and secretory activity in the colon independent of any nonspecific irritation or increase in gut permeability. As a result of the decreased intestinal transit time, fluid absorption decreases. At lower doses, motility effects of senna predominate, whereas at higher doses the secretory actions of senna dominate. The effects of senna result from direct stimulation of neuronal

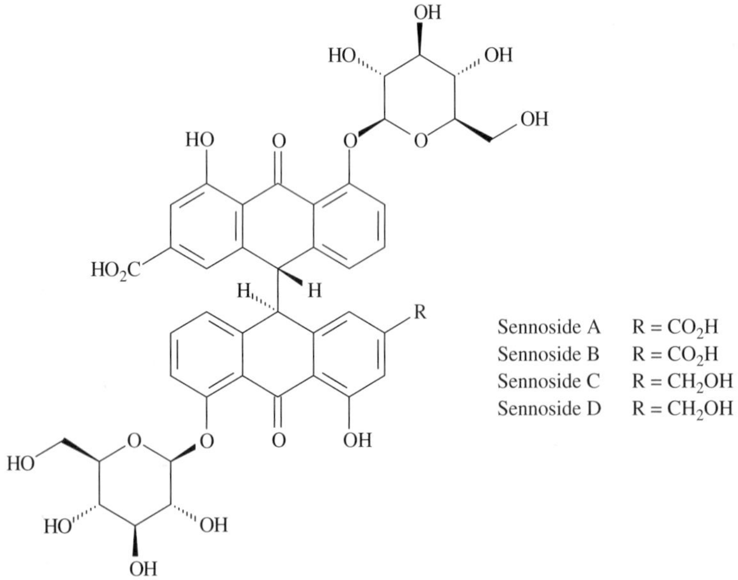

Sennoside A	$R = CO_2H$
Sennoside B	$R = CO_2H$
Sennoside C	$R = CH_2OH$
Sennoside D	$R = CH_2OH$

FIGURE 88.1. Chemical structures of sennosides A–D.

receptors, neurohumoral mechanisms, and alteration of prostaglandin E_2 concentrations.[8] In clinical studies, high doses of sennosides induce acute massive cell loss in colonic mucosa, probably by apoptosis.[9] This action reduces the length of crypts. To restore cellularity, a reduction in apoptosis and an increase in cell proliferation occur. Damaged epithelial cells appear as apoptotic bodies in the pigmented colonic mucosa that characterizes *pseudomelanosis coli*.[10]

DOSE RESPONSE

The typical dose for adults and children over 10 years of age is 1–2 g of the leaf powder at bedtime. The standardized daily dose is equivalent to 10–30 mg sennosides (calculated as sennoside B). The toxicity of senna is relatively low. The most serious adverse effects are fluid and electrolyte abnormalities from chronic abuse of senna products. In rats, gavage with purified sennosides (25, 40, 64 mg/kg) reversed the net absorption of fluid and produced diarrhea in a dose-dependent relationship.[11]

TOXICOKINETICS

Sennosides are natural prodrugs, which undergo degradation by intestinal bacteria to the active compound, rhein anthrone, in the lower gastrointestinal tract. The large molecular mass and poor lipid solubility of sennoside A and B limit the absorption of these dianthrone compounds. Human digestive enzymes do not split the β-glycoside bond in sennosides. Consequently, these compounds pass through the upper intestinal tract to lower gastrointestinal tract where intestinal bacteria hydrolyze sennosides A and B (dianthrone-*O*-glycosides) to the corresponding anthraquinone aglycone followed by the reduction of these compounds to the corresponding pharmacologically active anthrone (i.e., rhein anthrone). Figure 88.2 demonstrates the metabolism of sennosides by bacteria to rhein. *In vitro* experiments indicate that several bacteria in the human intestine can hydrolyze sennosides including *Streptococcus faecalis*, *Streptococcus faecium*, and *Bacteroides fragilis*.[12] Bacteria capable of reducing the anthraquinone aglycone to rhein include *Clostridium perfringens*, *Peptostreptococcus intermedius*, and *Eubacterium limosum*.[13]

Physiochemical factors and increased peristalsis limit the absorption of rhein anthrone by the large intestine; therefore, most of the rhein anthrone is excreted in the feces.[14] Rhein anthrone does not undergo first-pass metabolism. Peak concentrations of rhein (i.e., the oxidized product of rhein anthrone) occur about 3–5 hours and 10–11 hours after administration, probably as a result of the absorption of free rhein in the senna and the absorption of rhein after the degradation of sennosides.[15] The mean elimination half-life of rhein is about 7 hours.[15] Animal studies indicate that excretion of absorbed anthraquinone aglycone compounds occurs in the bile and urine after conjugation to the glucuronide or sulfate.[16] The excretion of rhein in breast milk after ingestion of senna is small. After the ingestion of a senna laxative containing 15 mg sennosides daily for 3 days, the concentration of rhein in breast milk was <10 ng/mL in 94 of 100 breast milk samples from 20 lactating women.[17] Aloe-emodin probably does not accumulate in the body after the administration of therapeutic doses of senna.

CLINICAL RESPONSE

Acute Effects

The onset of action of senna is about 8–10 hours after ingestion.[18] Potential adverse effects of chronic laxative abuse include fluid, electrolyte (hypokalemia), and acid-base imbalances (metabolic alkalosis). Several case reports associated the development of mild to moderate

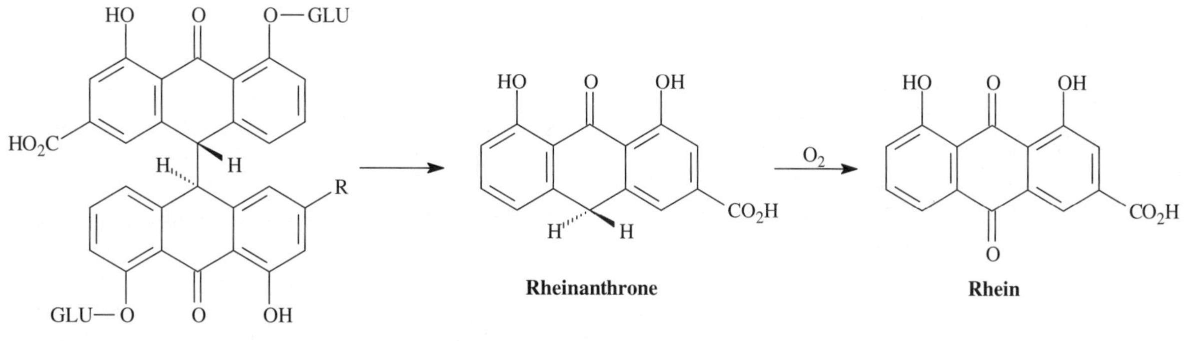

Sennosides A, B → Rheinanthrone → (O_2) Rhein

FIGURE 88.2. Metabolism of sennosides by bacteria to rhein.

hepatitis including cholestatic jaundice with chronic, large doses of senna.[19] In these case reports, reintroduction of senna laxatives was associated with mild elevation of serum hepatic aminotransferase concentrations.[20] Pharmacogenetic studies in one of these case reports suggested that inhibition of CYP2D6 activity (i.e., CYP2D6*4 variant) increased the risk of developing hepatitis following the use of senna laxatives.[21] Occasionally, sensitization occurs in workers exposed to dusts from senna during the manufacturing of senna laxatives. A cross-sectional study of 125 pharmaceutical workers involved in the manufacturing of bulk laxatives based on psyllium and senna pods demonstrated senna sensitization in about 15% of these workers.[22] The incidence of sensitization was particularly high in atopic workers and smokers. However, the risk of developing asthma was not increased in sensitized workers when compared with nonsensitized workers. Rare case reports associated occupational exposure to senna with asthma and allergic rhinitis, usually in atopic workers.[23]

Chronic Effects

Crude senna extracts and several minor anthranoid constituents (aloe-emodin, emodin) cause mutagenic effects in some, but not most *in vitro* studies.[24–25] Fluid and electrolyte imbalance are potential complications of chronic laxative abuse with senna extracts. A case report associated the chronic use of senna tea (daily ingestion of 1 L senna tea for over 3 years) with the development of hepatorenal failure and encephalopathy.[26] Analysis of concurrent tea samples failed to detect any adulterants including heavy metals. However, there was no rechallenge or analysis of serum samples for sennosides; therefore, the causal link remains weak and speculative.

Although some studies suggest that the chronic use of anthranoid laxatives increases the risk of colorectal cancer,[27] there is no direct evidence that the use of senna increases the risk of cancer. A 2-year carcinogenic study of Sprague-Dawley rats receiving 0, 5, 15, and 25 mg/kg purified senna extract via drinking water did not detect an increased incidence of neoplastic lesions in the exposed group.[28] Sennoside A and B account for about 25% of the purified extract. Although the high dose was near the maximum tolerated dose (i.e., 7% weight loss compared with controls), there were no ultrastructural changes in the myenteric nerve plexus of the colon or jejunum. No treatment-related neoplastic changes were detected in Sprague-Dawley rats gavaged with up to 300 mg/kg crude senna extract daily for 2 years.[29] All treatment groups had some minimal to slight hyperplasia of the mucosa in the colon and cecum, but these changes were reversible. The International Agency for Research on Cancer (IARC) does not list senna as a suspected carcinogen. There is no evidence that the ingestion of senna laxatives in therapeutic doses during pregnancy increases the risk of fetal abnormalities.

DIAGNOSTIC TESTING

High performance liquid chromatographic techniques with reverse-phase chromatography[30] or ion-pair liquid chromatography[31] can separate various sennosides with limits of detection of the former method in the range of 0.5 ng/mL.[4] Gas chromatography/mass spectrometry techniques can separate and quantitate various laxatives including senna with limits of detection in the range of 10–25 ng/mL in full-scan mode.[32] Using this procedure, rhein was detectable in the urine for 24 hours after the ingestion of a therapeutic dose of senna extract. Other procedures for the detection of sennosides include reversed-phase column liquid chromatography.[33] Metabolites of anthranoid compounds in senna produce discoloration of the urine that can cause false-positive tests for urinary urobilinogen.

TREATMENT

Treatment is supportive. Patients with chronic laxative abuse should be evaluated for evidence of fluid and electrolyte imbalance. Severe hypokalemia may develop following chronic abuse of senna.

References

1. Ramkumar D, Rao SS. Efficacy and safety of traditional medical therapies for chronic constipation: systematic review. Am J Gastroenterol 2005;100:936–971.
2. MacLennan WJ, Pooler AFWM. A comparison of sodium picosulphate ("Laxoberal") with standardised senna ("Senokot") in geriatric patients. Curr Med Res Opin 1974–75;2:641–647.
3. McCallum G, Ballinger BR, Presly AS. A trial of bran and bran biscuits for constipation in mentally handicapped and psychogeriatric patients. J Hum Nutr 1978;32:369–372.
4. Shah SA, Ravishankara MN, Nirmal A, Shishoo CJ, Rathod IS, Suhagia BN. Estimation of individual sennosides in plant materials and marketed formulations by an HPTLC method. J Pharm Pharmacol 2000;52:445–449.
5. Grimminger W, Witthohn K. Analytics of senna drugs with regard to the toxicological discussion of anthranoids. Pharmacol 1993;47(Suppl 1):98–109.
6. Franz G. The senna drug and its chemistry. Pharmacol 1993;47(Suppl 1):2–6.
7. Goppel M, Franz G. Stability control of senna leaves and senna extracts. Planta Med 2004;70:432–436.

8. Beubler E, Kollar G. Prostaglandin-mediated action of sennosides. Pharmacology 1988;36(Suppl 1):85–91.
9. van Gorkom BA, Karrenbeld A, Van der Sluis T, Koudstaal J, De Vries EG, Kleibeuker JH. Influence of a highly purified senna extract on colonic epithelium. Digestion 2000;61:113–120.
10. van Gorkom BA, de Vries EG, Karrenbeld A, Kleibeuker JH. Review article: anthranoid laxatives and their potential carcinogenic effects. Aliment Pharmacol Ther 1999;13: 443–452.
11. Beubler E, Schirigi-Degen A. Serotonin antagonists inhibit sennoside-induced fluid secretion and diarrhea. Pharmacology 1993;47(suppl 1)64–69.
12. Dreessen M, Lemli J. Qualitative and quantitative interactions between the sennosides and some human intestinal bacteria. Pharm Acta Helv 1982;57:350–352.
13. Hattori M, Namba T, Akao T, Kobashi K. Metabolism of sennosides by human intestinal bacteria. Pharmacology 1988;36(Suppl 1):172–179.
14. de Witte P. Metabolism and pharmacokinetics of anthranoids. Pharmacology 1993;47(Suppl 1):86–97.
15. Krumbiegel G, Schulz HU. Rhein and aloe-emodin kinetics from senna laxatives in man. Pharmacology 1993;47(Suppl 1):120–124.
16. de Witte P, Lemli L. The metabolism of anthranoid laxatives. Hepatogastroenterology 1990;37:601–605.
17. Faber P, Strenge-Hesse A. Relevance of rhein excretion in to breast milk. Pharmacology 1988;36(Suppl 1):212–220.
18. World Health Organization. WHO monographs on selected medicinal plants. Geneva:World Health Organization; 1999.
19. Yilmaz MI, Mas R, Ozcan A, Celasun B, Dogru T, Sonmez A, et al. Subacute cholestatic hepatitis likely related to the use of senna for chronic constipation. Acta Gastro-enterologica Belgica 2005;68:385–387.
20. Beuers U, Spengler U, Pape GR. Hepatitis after chronic abuse of senna. Lancet 1991;337:372–373.
21. Seybold U, Landauer N, Hillebrand S, Goebel F-D. Senna-induced hepatitis in a poor metabolizer. Ann Intern Med 2004;141:650–651.
22. Marks GB, Salome CM, Woolcock AJ. Asthma and allergy associated with occupational exposure to ispaghula and senna products in a pharmaceutical work force. Am Rev Respir Dis 1991;144:1055–1059.
23. Makinen-Kiljunen HT. Occupational asthma and rhinoconjunctivitis caused by senna. Allergy 1996;51:181–184.
24. Brusick D, Mengs U. Assessment of the genotoxic risk from laxative senna products. Environ Mol Mutag 1997;29:1–9.
25. Heidemann A, Miltenburger HG, Mengs U. The genotoxicity status of senna. Pharmacology 1993;47(Suppl 1):178–186.
26. Vanderperen B, Rizzo M, Angenot L, Haufroid V, Jadoul M, Hantson P. Acute liver failure with renal impairment related to the abuse of senna anthraquinone glycosides. Ann Pharmacother 2005;39:1353–1357.
27. Nusko G, Schneider B, Muller G, Kusche J, Hahn EG. Retrospective study on laxative use and melanosis coli as risk factors for colorectal neoplasms. Pharmacology 1993;47(Suppl 1):234–241.
28. Lyden-Sokolowski A, Nilsson A, Sjoberg P. Two-year carcinogenicity study with sennosides in the rat: emphasis on gastro-intestinal alterations. Pharmacology 1993;47(Suppl 1)209–215.
29. Mitchell JM, Mengs U, McPherson S, Zijlstra J, Dettmar P, Gregson R, Tigner JC. An oral carcinogenicity and toxicity study of senna (tinnevelly senna fruits) in the rat. Arch Toxicol 2005;80:34–44.
30. Hayashi S-I, Yoshida A, Tanaka H, Mitani Y, Yoshizawa K. Analytical studies on the active constituents in crude drugs. IV. Determination of sennosides in senna and formulations by high-performance liquid chromatography. Chem Pharm Bull 1980;28:406–412.
31. Sun S-W, Su H-T. Validated HPLC method for determination of sennosides A and B in senna tablets. J Pharmaceut Biomed Anal 2002;29:881–894.
32. Beyer J, Peters FT, Maurer HH. Screening procedure for detection of stimulant laxatives and/or their metabolites in human urine using gas chromatography-mass spectrometry after enzymatic cleavage of conjugates and extractive methylation. Ther Drug Monit 2005;27:151–157.
33. Bala S, Uniyal GC, Dubey T, Singh SP. An improved method for the analysis of sennosides in *Cassia angustifolia* by high-performance liquid chromatography. Phytochem Anal 2001;12:277–280.

Chapter 89

SKULLCAP (*Scutellaria lateriflora* L.)

HISTORY

American skullcap (*Scutellaria laterifolia*) has a long tradition of use by Native Americans in North America as a tonic and tranquilizer. The term mad-dog weed derives from the use of skullcap to treat rabies in the 18th century.

BOTANICAL DESCRIPTION

Scutellaria lateriflora L. is one American species of approximately 300 species of *Scutellaria* worldwide. Other *Scutellaria* species used as herbal medicines include *S. baicalensis* Georgi (eastern Asia, huang-qin), *S. barbata* D.Don (southern China, ban-zhi-lian), *S. pinnatifida* A. Hamilt. subsp. *alpina* (Iran), *S. planipes* Nakai & Kitagawa (Argentina, China), and *S. amabilis* Hara (Japan, yamaji-tatsunamisou). Contamination of skullcap with hepatotoxic species of germander (*Teucrium* spp.) can be recognized by digital photomicroscopy of the cell structures of these herbs.[1]

Common Name: Blue Skullcap, Mad Dog Skullcap, American Skullcap, Skullcap, Madweed

Scientific Name: *Scutellaria lateriflora* L.

Botanical Family: Lamiaceae (menthes, mints)

Physical Description: Skullcap is a slender, heavily branched plant that reaches 2–4 ft (~0.6–1.2 m) in height. In July, small, blue flowers appear that contain an outer whorl shaped like a cap. The name skullcap derives from the helmet-shaped upper lid of the seed pods from this perennial plant.

Distribution and Ecology: *Scutellaria lateriflora* is a perennial herb indigenous to North America that is cultivated in Europe. This plant grows in meadows and swampy woods from Canada to Florida and westward in North America to New Mexico, Oregon, and British Columbia.

EXPOSURE

Sources

Skullcap is available as a tea, liquid extract, tincture, capsule, and crude root extract. In health food stores, *S. lateriflora* typically is available as a tonic or as an herbal combination of valerian, passionflower, and skullcap for sleeping. Hydro-distillation of the dried aerial parts of *S. barbata* yields about 0.2–0.3% yellow essential oil.[2]

Medicinal Uses

TRADITIONAL

Extracts of *S. lateriflora* are traditionally used as a nerve tonic, antispasmodic, diuretic, anticonvulsant, sedative, and tranquilizer. In traditional Chinese medicine, huang-qin (*S. baicalensis* Georgi) is an herbal treatment for fever, inflammation, bacterial and viral infections, elevated cholesterol, hypertension, ulcers, neonatal jaundice, and cancer.

CURRENT

Although skullcap is promoted as a sedative and antispasmodic for the treatment of anxiety, menstrual

Medical Toxicology of Natural Substances, by Donald G. Barceloux, MD

tension, depression, and muscle spasms, there are few clinical data to support the use of this herb for these purposes.

Regulatory Status

The US Food and Drug Administration (FDA) regulates this herb as a dietary supplement, but the safety of skullcap has not been determined. Huang-qin (i.e., root of *Scutellaria baicalensis*) is officially listed in the Chinese *Pharmacopoeia*. Skullcap is approved as a nonprescription medication in Canada and is available in the United Kingdom. Skullcap in not listed in the German Commission E monographs.

PRINCIPAL INGREDIENTS

Chemical Composition

The dried roots from *Scutellaria* species contain over 30 flavone compounds with phenylethanoids, amino acids, sterols, and essential oils. Major flavonoids in *Scutellaria* species include baicalein, wogonin, baicalein glucuronide (baicalin), wogonin-6-*O*-glucuronide (wogonoside), chrysin, scutellarin, scutellarein (isocarthamidin), and oroxylin A.[3,4] Minor flavonoid compounds include dihydrobaicalin, ikonnikoside I, lateriflorin, oroxylin A-7-*O*-glucuronide, apigenin, wogonin-7-*O*-glucuronide, chrysin-7-*O*-glucuronide, and numerous other flavonoid compounds.[5] Figure 89.1 displays the chemical structures of several major flavonoid compounds in skullcap. The concentration of flavonoid compounds in roots from *Scutellaria* species depends on several variables including the extraction process, plant species, plant part, and growing conditions.[6] In a study of the methanol extract of roots from *Scutellaria baicalensis*, the most common flavanoid compounds were baicalin (8.12% of dry root mass) and wogonoside (2.52% of dry root mass). The concentration of baicalein was low (<0.01%). Typically, the concentration of baicalin is much greater in the root of *Scutellaria* species than baicalein, oroxylin A, or wogonin.[7] In a study of an aqueous ethanol extract of *S. baicalensis* root, the concentration of the major flavonoid compounds by dry weight were as follows: baicalin 144 μg/mg; baicalein, 29.6 μg/mg; and wogonin, 9.7 μg/mg.[8] The crude solvent extract of dried roots from *S. baicalensis* contained about 30% baicalein, 11% wogonin, and 2.5% oroxylin A, as measured by high-speed counter-current chromatography.[9]

Hydro-distillation of aerial parts from *Scutellaria* species produces essential oils consisting primarily of oxygenated monoterpene (menthol, linalool, α-terpineol, thymol) and sesquiterpene (α-*trans*-bergamotol, globulol) compounds.[2] Other compounds include methyl eugenol, hexahydrofarnesylacetone, 1-octen-3-ol, and 3,7,11,15-tetramethyl-2-hexadecen-1-ol. The predominate compounds in an Iranian sample of a hydro-distilled essential oil from dried aerial parts of Iranian skullcap (*S. pinnatifida* A. Hamilt. subsp. *alpina*) were germacrene-D (39.7%) and β-caryophyllene (15%).[10] All other compounds were present in concentrations ≤5%.

Physiochemical Properties

Baicalin and wogonoside are polar glycoside compounds with poor lipid solubility in contrast to their relatively nonpolar, lipophilic aglycones, baicalein and wogonin. Based on receptor binding studies, the affinity of flavonoid compounds in *Scutellaria* roots for the benzodiazepine site of the $GABA_A$ receptor is as follows: wogonin, K_i = 2.03 ± 0.24 μM; baicalein, K_i = 5.69 ± 0.95 μM; scutellarein, K_i = 12 ± 1.27 μM; and baicalin, K_i = 77.10 ± 4.79 μM.[11]

Mechanism of Toxicity

The toxicity of this herb has not been established.

DOSE RESPONSE

There is no standard dose of skullcap.

TOXICOKINETICS

Absorption

Bioactive flavonoid compounds in skullcap include baicalin and the aglycone, baicalein along with wogonoside and the associated aglycone, wogonin. Typically, absorption of the aglycone occurs after the hydrolysis

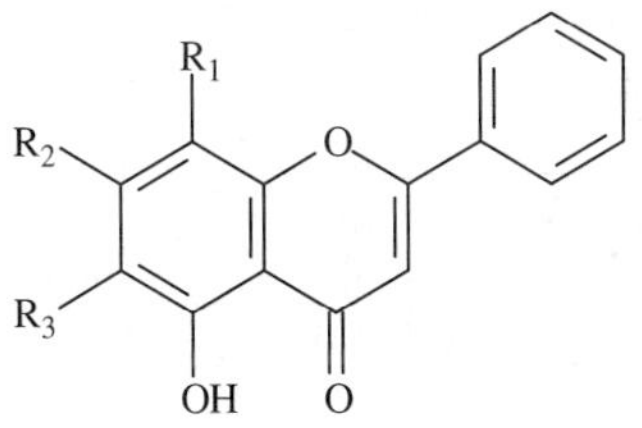

	R_1	R_2	R_3
Baicalein	H	OH	OH
Baicalin	H	O-glucuronyl	OH
Wogonin	OCH_3	OH	H
Wogonoside	OCH_3	O-glucuronyl	H
Oroxylin	H	OH	OCH_3

FIGURE 89.1. Chemical structures of baicalein, baicalin, wogonin, wogonoside, and oroxylin.

of the flavonoid glycoside by intestinal bacteria, but conjugation of the aglycone by hepatic microsomes during first-pass metabolism restores the original glycoside following absorption of the aglycone.[12] In human volunteers receiving baicalin, detectable (i.e., >5 ng/mL) and peak plasma baicalin concentrations occurred about 1 hour and 5 hours, respectively, after oral administration.[13]

Distribution

The low lipid solubility of baicalin probably limits the diffusion of this compound across the blood–brain barrier.

Biotransformation

The major metabolites of baicalin following oral administration are baicalein-6-*O*-sulfate and baicalein-7-*O*-glucuronide (baicalin).[13] Following the ingestion of herbs containing baicalin and wogonoside by 10 healthy male volunteers, conjugated metabolites of these two compounds accounted for 11.6% and 7.2%, respectively, of the total administered dose.[14]

Elimination

Based on animal studies, baicalin probably undergoes hepatobiliary excretion.[15] The plasma elimination half-lives of baicalein and wogonin conjugates were approximately 8 hours and 10 hours, respectively.[14] The kidneys excrete little free baicalin or wogonin in the urine.

Drug Interactions

In vitro studies suggest that extracts of *Scutellaria* species inhibit CYP1A2 and CYP3A4, but the clinical significance of these potential interactions remain undetermined.[16,17]

CLINICAL RESPONSE

Rare case reports associate the ingestion of skullcap with hepatic toxicity.[18] However, the role of skullcap in producing hepatotoxicity is unclear because of the potential contamination of this herb with germander (*Teucrium* spp.) and the lack of hepatotoxic diterpene compounds in skullcap. There are few experimental data to support the hepatotoxicity of skullcap.

DIAGNOSTIC TESTING

Methods for the detection of flavonoid compounds in *Scutellaria* species include thin-layer chromatography,[11] capillary electrophoresis, gas chromatography, high-speed counter-current chromatography,[19] solid-phase or reverse-phase high performance liquid chromatography (HPLC), HPLC with diode array and mass spectrometric detection,[20] and micellar electrokinetic capillary chromatography.[21] Extracts of skullcap contaminated with germander (*T. canadense, T. chamaedrys*) can be identified by the presence of phenylpropanoid compounds (verbascoside, teucrioside).[1] The major compounds in extracts from *Teucrium canadense* and *Teucrium chamaedrys* are verbascoside and teucrioside, respectively, whereas these compounds are not present in extracts of skullcap (*Scutellaria* spp.). The latter extracts contain predominantly flavonoid compounds (baicalin, wogonin, baicalein, wogonoside).

TREATMENT

Treatment is supportive.

References

1. Gergeron C, Batcha LL, Angerhofer CK, Sudberg S, Sudberg EM, Gafner S, et al. Analysis of *Scutellaria lateriflora* and its adulterants *Teucrium canadense* and *Teucrium chamaedrys* by LC-UV/MS, TLC, and digital photomicroscopy. J AOAC Int 2003;86: 453–460.
2. Yu J, Lei J, Yu H, Cai X, Zou G. Chemical composition and antimicrobial activity of the essential oil of *Scutellaria barbata*. Phytochemistry 2004;65:881–884.
3. Ishimaru K, Nishikawa K, Omoto T, Asai I, Yoshihira K, Shimomura K. Two flavone 2′-glucosides from *Scutellaria baicalensis*. Phytochemistry 1995;40:279–281.
4. Bochorakova H, Paulova H, Slanina J, Musil P, Taborska E. Main flavonoids in the root of *Scutellaria baicalensis* cultivated in Europe and their comparative antiradical properties. Phytother Res 2003;17:640–644.
5. Miyaichi Y, Hanamitsu E, Kizu H, Tomimori T. Studies on the constituents of *Scutellaria* species (XXII). Constituents of the roots of *Scutellaria amabilis* Hara. Chem Pharm Bull 2006;54:435–441.
6. Bergeron C, Gafner S, Clausen E, Carrier DJ. Comparison of the chemical composition of extracts from *Scutellaria lateriflora* using accelerated solvent extraction and supercritical fluid extraction versus standard hot water or 70% ethanol extraction. J Agric Food Chem 2005;53:3076–3080.
7. Lin SJ, Tseng HH, Wen KC, Suen TT. Determination of gentiopicroside, mangiferin, palmatine, berberine, baicalin, wogonin and glycyrrhizin in the traditional Chinese medicinal preparation sann-joong-kuey-jian-tang by high-performance liquid chromatography. J Chromatogr A 1996;730:17–23.

8. Don HY, Guo YZ, Ageta H, Harigaya Y, Onda M, Zhang YY, et al. Comparative study of *Scutellaria planipes* and *Scutellaria baicalensis*. Biomed Chromatogr 1998;12: 31–33.
9. Li H-B, Chen F. Isolation and purification of baicalein, wogonin and oroxylin A from the medicinal plant *Scutellaria baicalensis* by high-speed counter-current chromatography. J Chromatogr A 1074;2005:107–110.
10. Ghannadi A, Mehregan I. Essential oil of one of the Iranian skullcaps. Z Naturforsch 2003;58:316–318.
11. Hui KM, Wang XH, Xue H. Interaction of flavones from the roots of *Scutellaria baicalensis* with the benzodiazepine site. Planta Med 2000;66:91–93.
12. Kawabata K, Yanagisawa E, Ishihara K, Mizuhara Y, Wakui Y, Akao T, et al. Baicalin, the predominant flavone glucuronide of scutellariae radix, is absorbed from the rat gastrointestinal tract as the aglycone and restored to its original form. J Pharm Pharmacol 2000;52:1563–1568.
13. Muto R, Motozuka T, Nakano M, Tatsumi Y, Sakamoto F, Kosaka N. [The chemical structure of new substance as the metabolite of baicalin and time profiles for the plasma concentration after oral administration of sho-saiko-to in human]. Yakugaku Zasshi 1998;118:79–87. [Japanese]
14. Lai M-Y, Hsiu S-L, Chen C-C, Hou Y-C, Lee Chao P-D. Urinary pharmacokinetics of baicalein, wogonin and their glycosides after oral administration of Scutellariae Radix in humans. Bio Pharm Bull 2003;26:79–83.
15. Tsai P-L, Tsai T-H. Pharmacokinetics of baicalin in rats and its interactions with cyclosporin A, quinidine and SKF-525A: a microdialysis study. Planta Med 2004;70: 1069–1074.
16. Awad R, Arnason JT, Trudeau V, Bergeron C, Budzinski JW, Foster BC, Merali Z. Phytochemical and biological analysis of skullcap (*Scutellaria lateriflora* L.): a medicinal plant with anxiolytic properties. Phytomedicine 2003;10: 640–649.
17. Kim JY, Lee S, Kim DH, Kim BR, Park R, Lee BM. Effects of flavonoids isolated from Scutellariae radix on cytochrome P-450 activities in human liver microsomes. J Toxicol Environ Health A 2002;65:373–381.
18. MacGregor FB, Abernethy VE, Dahabra S, Cobden I, Hayes PC. Hepatotoxicity of herbal remedies. BMJ 1989;299(6708):1156–1157.
19. Wu S, Sun A, Liu R: Separation and purification of baicalin and wogonoside from the Chinese medicinal plant *Scutellaria baicalensis* Georgi by high-speed counter-current chromatography. J Chromatogr A 2005;1066: 243–247.
20. Horvath CR, Martos PA, Saxena PK. Identification and quantification of eight flavones in root and shoot tissues of the medicinal plant huang-qin (*Scutellaria baicalensis* Georgi) using high-performance liquid chromatography with diode array and mass spectrometric detection. J Chromatogr A 2005;1062:199–207.
21. Li H-B, Jiang Y, Chen F. Separation methods used for *Scutellaria baicalensis* active components. J Chromatogr B 2004;812:277–290.

Chapter 90

ST. JOHN'S WORT
(*Hypericum perforatum* L.)

HISTORY

Carolus Linnaeus first described *Hypericum perforatum* L. in 1753. This plant has been used for centuries as a medicinal plant to treat neuralgias and minor skin lesions.[1] In antiquity, Dioscorides cited St. John's wort as a remedy for sciatica and skin wounds during the Middle Ages.[2] In Europe, this herb was available commercially in the mid-20th century. In 1984, the German Commission E monograph recommended the use of *Hypericum* for "psychogenic disturbances, depressive states, anxiety, and/or nervous excitement." [2]

BOTANICAL DESCRIPTION

Common Name: St. John's Wort

This common name relates to the fact that the plant blooms around St. John the Baptist Day (June 24th).

Scientific Name: *Hypericum perforatum* L.

Botanical Family: Clusiaceae

Physical Description: St. John's wort is a perennial herbaceous plant with mature stems and branches that are smooth without hairs (glabrous). Young stems have two ridges on opposite faces. Leaves are simple, without stalks (apetiolate) and elliptic, oblong or ovate in shape; 10–25 mm (~0.4–1 in.) long and 2.5–11 mm (~0.1–0.4 in.) wide; usually with five lateral veins; apex obtuse and base clasping. Flowers are bisexual and radially symmetrical, borne in flat or round topped inflorescences; sepals 5, 2.5–4 mm (~0.1–0.2 in.) long; petals 5, yellow and usually with black-spots, 7–10 mm (~0.3–0.4 in.) long; the pollen receptive part (stigma) of the female organ of the flower is a head-shaped cluster (capitate) that connects to the ovary with three narrow, about 3–5 mm (~0.1–0.2 in.) long, separate tubes (styles), trisegmented ovary. Fruits are ovoid capsules, 3.5–5.5 mm (~0.1–0.2 in.) long, 2.5–4 mm (~0.1–0.2 in.) wide; the brown seeds are numerous, averaging about 1 mm (~0.04 in.) in length.

Distribution and Ecology: This plant is native to Europe, North Africa, and western Asia. In Africa, Asia, Australia, and the Americas, St. John's wort has become a naturalized plant. A common habitat for this plant includes neglected fields, dry pastures, rangelands, and the sides of country roads.

EXPOSURE

Sources

Hypericum extract (CAS RN 84082-80-4) is an extract of capsules, leaves, flowers, and stems of *Hypericum perforatum* (St. John's Wort) prepared by crushing these parts in stabilized olive oil over several weeks followed by dehydration and filtration.

Medicinal Uses

Traditional

St. John's wort consists of the leaves and flowering tops of *Hypericum perforatum* L. Traditional uses of *Hypericum* extract include the treatment of anxiety, depres-

Medical Toxicology of Natural Substances, by Donald G. Barceloux, MD

sion, insomnia, wounds, neuralgia, fibrositis, sciatica, gout, gastritis, rheumatism, bronchitis, hypothyroidism, and excitability.[3] Topically, this herb has been used as an astringent for burns, abrasions, hemorrhoids, vesicles, and bites.[4] Extracts and oils from *Hypericum perforatum* are biological additives in some cosmetic formulations at concentrations of 0.01–0.5%.[5]

Current

Botanical formulations are commercially available as herbal preparations in the forms of tea, tincture, tablet, and capsule usually containing about 300 mg hypericin. Hypericum extract is a popular alternative treatment for mild to moderate depression. Interpretation of clinical studies on the effectiveness of hypericum extract for the treatment of depression is complicated by numerous methodological flaws including lack of objective diagnostic criteria for depression, the use of subclinical doses of comparison antidepressants, high placebo responder rates of mild depression, limited analysis of adverse drug reactions, and lack of extended follow-up.[6] Several meta-analyses of the clinical effectiveness of hypericum extracts suggest that this herb is more efficacious for *mild* to *moderate* depression than placebo and as effective as low doses of some tricyclic antidepressants.[7,8] However, in a large-scale, multicenter, randomized, placebo-controlled trials of participants with *major* depressive illness, the use of St John's wort did not produce significant difference on any outcome measure when compared with placebo.[9,10]

Regulatory Status

Hypericum extracts are classified as dietary supplements by the US Food and Drug Administration (FDA). Some US manufacturers of herbal products obtain *US Pharmacopoeia* (*USP*) recognition for consistent product purity and potency. The German Commission E approved the use of hypericum as an herbal medicine for depression, insomnia, anxiety, and other psychogenic disturbances. Current hypericum preparations are standardized to a fixed hypericin content of 0.3%. Total hypericin content usually includes both hypericin and pseudohypericin.[3]

PRINCIPAL INGREDIENTS

Chemical Composition

Hypericum perforatum contains at least 10 classes of biologically active substances that vary in concentration among individual plants depending on a variety of factors including the extraction process, plant location, and environmental conditions (growing conditions, season, stage of plant development).[1] Biologically active constituents account for about 15–20% of the plant extract.[11,12] These compounds include phenylpropanes (chlorogenic acid, *p*-coumaric acid), flavonol glycosides (quercetin, hyperin, rutin, isoquercitrin, quercitrin, luteolin, campferol, myricetin), biflavones (biapigenin, amentoflavone), tannins-proanthrocyanidins (procyanidin B2, catechin, epicatechin), xanthones (kielcorin, 1,3,6,7-tetrahydroxyxanthone, mangiferin), phloroglucinols (hyperforin, adhyperforin, hydroperoxycadiforin), essential oils (α-pinene, β-pinene, myrcene, limonene), amino acids, γ-aminobutyric acid, and naphthodianthrones (hypericin, pseudohypericin, cyclopseudohypericin). Figure 90.1 displays the chemical structure of most important constituents, hypericin (CAS RN: 548-04-9) and hyperforin (CAS RN: 11079-53-1).

St. John's wort is an air-dried methanol or ethanol extract of the buds and flowers of *Hypericum perforatum*. These aerial parts are typically collected just before or during blossom. The composition of St. John's wort preparations in the United States is similar to marker compounds on the label. Analysis of 10 St. John's wort products from the US market indicated that the total hypericin content ranged from 88–110% of the stated amounts with one outlier at 139%.[13] There is substantial

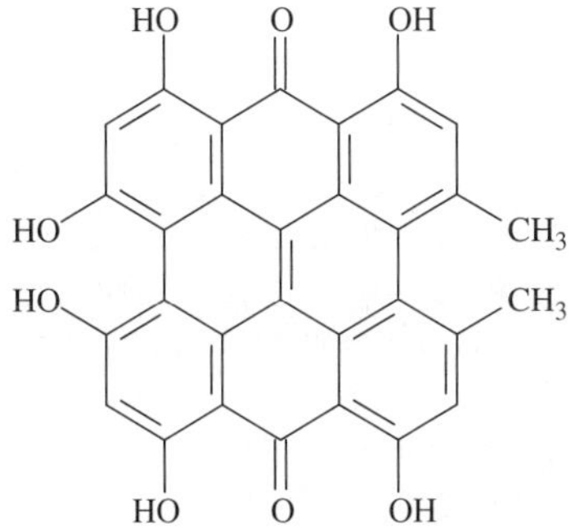

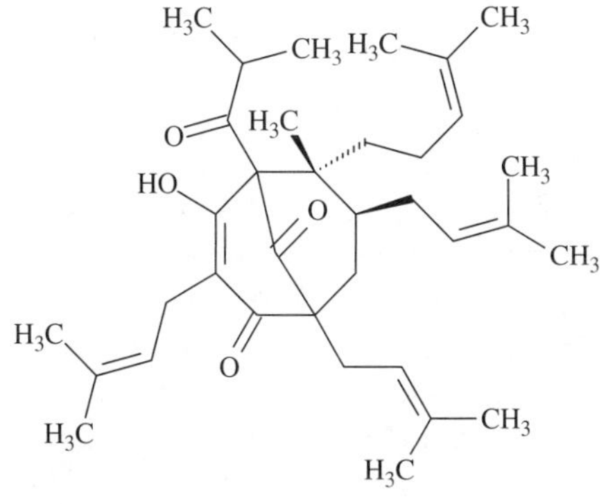

Hypericin

Hyperforin

FIGURE 90.1. Chemical structures of hypericin and hyperforin.

variation in the standardization of active ingredients (hypericin, pseudohypericin) in batches of St. John's Wort.[14] In a study of the major components of eight commercially available dry extracts of *Hypericum perforatum*, the approximate content of the active ingredients from the dry extract was as follows: hypericin, 0.19–0.3%; hyperforin, 1.3–3.9%; and flavonoids (calculated as rutin), 4.8–11.4%.[15] Other components included water-soluble sugars, citric acid, malic acid, tannins, and ash. The flowering portion of the plant contains naphthodianthrones (hypericin, pseudohypericin). Numerous flavonoid compounds (hyperoside, campferol, luteolin, isoquercitrin, quercitrin, myricetin, quercetin) are present in aboveground portions of the plant.[1] Hypericin and pseudohypericin account for almost all of naphthodianthrone compounds in the extract, whereas the raw plant contains very small amounts of these compounds. During the extraction process, a photodynamic reaction converts most of the protohypericin and protopseudohypericin to hypericin and pseudohypericin, respectively.[16] The flowers and buds also contain biologically active phloroglucinol compounds (hyperforin, adhyperforin).[17] Other compounds present in hypericum preparations that do not possess any antidepressant activity include amino acids, phenylpropanes, coumarin, tannins, and procyanidins.[1]

Physicochemical Properties

Some of the aromatic compounds in the leaves and flower buds of hypericum turn red when exposed to light. Hyperforin decomposes during storage at room temperature.[15] Because of the presence of hypericin, *H. perforatum* is a primary photosensitizer. A mixture of hypericum extract (10–25%) and propylene glycol (>75%) is a clear, slightly reddish, water soluble liquid with a faint herbal odor.

Mechanism of Action

The mechanism of action and active ingredients of *Hypericum perforatum* are not well defined. There is some experimental evidence that the administration of preparations of *Hypericum perforatum* affect the transmission of catecholamines (norepinephrine, dopamine) and serotonin. *In vitro* data suggest that the potential antidepressant action of hypericum extract results from alterations in central monoamine concentrations, similar to other antidepressants (e.g., selective serotonin reuptake inhibitors).[1] However, the mechanism of action probably is different from standard antidepressants because of the noncompetitive enhancement of intracellular sodium ion concentrations at neuronal synapses by hypericum extracts.[18] At the receptor level, chronic treatment with hypericum extracts may down-regulate β_1-adrenoceptors, and up-regulate post-synaptic 5-HT_{1A} and 5-HT_2 receptors. High concentrations of hypericin and flavonoid compounds inhibit MAO-A and MAO-B activity, but the concentrations of these compounds required to produce MAO inhibition are not obtainable in biological systems. Although hypericum extracts are frequently standardized to hypericin content, there is some clinical evidence that correlates antidepressant activity in a dose-dependent relationship with hyperforin content.[3,19,20] Hypericin is probably responsible for the photosensitivity reaction associated with the use of hypericum extract.[3]

DOSE RESPONSE

Medical Use

Most herbal preparations of *Hypericum perforatum* are standardized to hypericin content, but the hypericin content in commercial extracts varies substantially (e.g., 0.05–0.3%).[11] A 300mg-tablet of hypericum extract corresponds to about 15mg hyperforin.[21] Most clinical trials use daily doses of 900–1500mg of the standardized extract (i.e., 0.3% hypericin).

Toxicity

A review of photosensitivity reactions to hypericum extract indicated that the threshold for an increased risk of these adverse reactions was a daily dose of 2–4 grams of the commercial extract (i.e., about 5–10mg hypericin).[22] In a phase I clinical trial of hypericin for the treatment of HIV-infected adults, hypericin doses up to 0.5mg/kg twice weekly produced severe cutaneous photosensitivity in 11 of 23 patients.[23] The toxic reaction involved an erythematous rash with painful dysesthesias in the sun-exposed areas of the skin. There were few other adverse reactions.

TOXICOKINETICS

Absorption

The exact constituents responsible for the effects of *Hypericum* remain unclear. The absorption of hyperforin is relatively slow with mean peak plasma concentrations of about 150ng/mL occurring within $3^1/_2$ hours after ingestion.[21,24] In a study of 12 healthy male volunteers ingesting a single dose of 1,800mg dried hypericum extract containing 1.5mg hypericin, the median peak concentration of hypericin was approximately 14ng/mL and occurred about $2^1/_2$ hours after the ingestion.[25] The equivalent dose of pseudohypericin in the

1,800 mg dose of hypericum extract was about 3.2 mg, and the median peak pseudohypericin concentration of 30.6 ng/mL occurred approximately 1/2 hour after ingestion. The estimated bioavailability of hypericin and pseudohypericin in hypericum extract is approximately 15–20%.[26]

Biotransformation

There are few data on the metabolism of constituents in hypericum extracts. Being large molecules, the potential elimination pathways for hypericin and pseudohypericin include metabolism and hepatic glucuronidation followed by biliary excretion.[27]

Elimination

The elimination of both hypericin and hyperforin is relatively slow. In a study of 13 healthy volunteers administered 0.75 mg hypericin and about 1.6 mg pseudohypericin, the median elimination half-lives were approximately 43 hours and 25 hours, respectively.[27] After a single therapeutic dose of hypericin, pseudohypericin, and hyperforin, the mean elimination half-lives in 18 healthy volunteers were 18.71 hours, 17.19 hours, and 17.47 hours, respectively.[28]

Drug Interactions

Experimental and clinical studies suggest the possibility of drug interactions between constituents of hypericum extract and pharmaceutical drugs based on the induction of the cytochrome P450 isoenzymes, CYP3A4[29,30] and CYP2C19-dependent hydroxylation.[31] In a study of 12 healthy elderly volunteers, a 28-day course of St. John's wort induced the activity of CYP3A4 and CYP2E1 by a mean of 140% and 28%, respectively, when compared with phenotypic ratios prior to supplementation.[32] Additionally, there are potential pharmacodynamic interactions as well as alteration of the energy-dependent membrane-associated drug transporter, P-glycoprotein.[33] However, the clinical relevance of these interactions remains to be determined. Table 90.1 outlines potential drug interactions between St. John's Wort and pharmaceutical drugs. Hyperforin is a ligand for the orphan nuclear receptor SXR (steroid xenobiotic receptor), which is responsible for the induction of CYP3A4 and MDR1 gene expression. A 14-day course of St. John's wort significantly induced the activity of CYP3A4 as measured by changes in alprazolam pharmacokinetics.[34] CYP3A4 substrates include at least 50% of all marketed medications. Drugs metabolized by the CYP3A4 isoenzyme include calcium channel blockers, benzodiazepines, estrogens, macrolide antibiotics,

TABLE 90.1. Some Potential Clinically Important Drug Interactions between St. John's Wort and Other Drugs

Drug	Drug Interaction
HIV protease inhibitors: indinavir, nelfinavir, ritonavir, saquinavir	Reduced blood levels; possible loss of HIV suppression
HIV non-nucleoside reverse transcriptase inhibitors: efavirenz, nevirapine	Reduced blood levels; possible loss of HIV suppression
Warfarin	Reduced anticoagulant effect and need for increased warfarin dose
Cyclosporin	Reduced blood levels with risk of transplant rejection
Digoxin	Reduced blood levels and loss of efficacy
Theophylline	Reduced blood levels and loss of control of asthma
Anticonvulsants: carbamazepine, phenobarbitone, phenytoin	Reduced blood levels with increased risk of seizures
Oral contraceptives	Reduced blood levels with risk of unwanted pregnancy and breakthrough bleeding
Triptan compounds: sumatriptan, naratriptan, rizatriptan, zolmitriptan	Increased serotonergic effects with increased incidence of adverse reactions
Selective serotonin reuptake inhibitors: citalopram, fluoxetine, fluvoxamine, paroxetine, sertraline	Increased serotonergic effects with increased incidence of adverse reactions

Source: Adapted from Ref 29.

carbamazepine, and nonsedating antihistamines. Potential drug interactions include reductions of the plasma concentrations and clinical effects of cyclosporin, digoxin, HIV protease inhibitors, oral contraceptives, theophylline, and warfarin by the induction of cytochrome P450 isoenzymes CYP3A4, CYP2C9, CYP1A2, and the transport protein, P-glycoprotein.[35,36] Hypericum extract in daily doses of 900 mg significantly induced CYP3A4 isozyme activity in healthy volunteers following treatment for two weeks.[37]

Hyperforin, but not hypericin, is probably the chemical constituent of hypericum extract responsible for the activation of the CYP3A4 isozyme.[38] Theoretically, the use of hypericum extract reduces the efficacy of drugs by enhancing the clearance of drugs metabolized by the CYP3A4 isozyme.[39] Clinically important drugs affected by the CYP3A4 isozyme include indinavir sulfate, ethinyl estradiol, coumarin, and cyclosporin.[40] This interaction requires the administration of hypericum extract for more than a few days. The administration of hypericum extract concomitantly with alprazolam (CYP3A4) or dextromethorphan (CYP2D6) for 3 days did not produce clinically significant changes in the pharmacokinetics of these drugs.[41] However, a 14-day course of St. John's wort (900 mg daily) decreased the bioavailability of alprazolam by about 39%.[42] The induction of CYP3A4 by St. John's wort is greater after ingestion than after intravenous administration.[43]

In addition to the induction of CYP3A4, St. John's wort also induces CYP2C19, which catalyzes the metabolism of tricyclic antidepressants. The concomitant use of hypericum extract with digoxin potentially reduces the efficacy of digoxin, possibly by inducing the P-glycoprotein transporter and altering absorption rather than metabolism.[44] The interaction of St. John's wort and digoxin varies depending on the dose and preparation of St. John's wort as a result of differences in hyperforin concentrations.[45]

Pregnancy

A prospective, observational, cohort study of nursing mothers receiving St. John's Wort and their infants did not detect reduced milk production or decreased infant weight over the first year of life when compared with nursing mothers not using St. John's Wort. [46]

CLINICAL RESPONSE

Adverse Reactions

Adverse drug reactions associated with the use of hypericum extract are relatively mild and similar to placebo.[8] The commonly reported side effects of hypericum extract in a European drug-monitoring study of 3250 patients were as follows: gastrointestinal irritation, 0.6%; allergic reactions, 0.5%; fatigue, 0.4%; and restlessness, 0.3%.[47]

Acute Effects

Several case reports associate the use of hypericum extract with development of the serotonin syndrome in both young and elderly patients with (sertraline, nefazodone)[48] and without[49] the concomitant use of serotonin re-uptake inhibitor (SSRIs). Additionally, a case report associated the recent use of hypericum extract with the onset of serotonin syndrome in a patient on stable doses of fluoxetine and eletriptan.[50] However, no rechallenge with these medications confirmed a potential drug interaction. A case series of 17 patients described the onset of mania within 2 days to 6 months after initiating the use of St. John's wort.[51] The doses of St. John's wort were within typical pharmacological range in all but one case. Although the mania resolved in all cases following cessation of St. John's wort, none of the patients were rechallenged to determine a direct causal effect between St. John's wort and the mania. Seizures are a rarely reported adverse effect of the use of St. John's wort, and a case report described the onset of seizures and confusion in a 16-year old girl after the suicidal ingestion of a large dose of St. John's wort (up to 4,000 ug daily for 2 weeks plus 1,500 μg before admission).[52]

Chronic Effects

An increased incidence of photosensitivity occurs during the daily use of high doses (1,800–3,600 mg) of hypericum extract as manifest by a pruritic rash and epidermal erythema.[53] However, the incidence of photosensitization is relatively rare and resolves several days after discontinuation of therapy.[20] One case report associated painful paresthesias in areas of sun exposure (hands, face) during the use of *Hypericum* extract.[54] Another case report associated the chronic nasal insertion of St. John's wort with severe epistaxis requiring endoscopic sphenopalatine artery ligation and anterior ethmoidal artery ligation.[55]

DIAGNOSTIC TESTING

Methods for the determination of hyperforin, hypericin, and pseudohypericin in plasma samples include high performance liquid chromatographic (HPLC) with fluorescence and UV detection[56] and liquid chromatography/tandem mass spectrometry.[57] The limit of quantitation for the latter method was 0.05 ng/mL for hypericin and 0.035 ng/mL for hyperforin with precision ranging from 4.7–15.4% for both analytes. Following liquid-liquid extraction and separation by isocratic reversed-phase HPLC, the limits of quantitation for hypericin/pseudohypericin (fluorometric detection) and hyperforin (UV detection) were 0.25 ng/mL and 10 ng/mL, respectively.[58] Hyperforin decomposes during storage at room temperature. After 18 months, the hyperforin content of samples of St. John's Wort decreases from 17–77% compared with the initial values.[15]

TREATMENT

There are few clinical data on the treatment of toxic effects from the exposure to hypericum extract and treatment is primarily supportive. Decontamination measures are not necessary for casual exposure to hypericum extract.

References

1. Greeson JM, Sanford B, Monti DA. St. John's wort (*Hypericum perforatum*): a review of the current pharmacological, toxicological, and clinical literature. Psychopharmacology 2001;153:402–414.
2. Field HL, Monti DA, Greeson JM, Kunkel EJ. St. John's wort. Int J Psychiatry Med 2000;30:203–219.
3. Barnes J, Anderson LA, Phillipson JD. St John's wort (*Hypericum perforatum* L.). A review of its chemistry, pharmacology and clinical properties. J Pharm Pharmacol 2001;53:583–600.
4. Schwarz JT, Cupp MJ. St. John's Wort. In: Cupp JM, editor. Toxicology and clinical pharmacology of herbal products. Totowa, NJ: Humana Press; 2000:67–78.
5. Cosmetic Ingredient Review Expert Panel. Final report on the safety assessment of *Hypericum perforatum* extract and *Hypericum perforatum* oil. Int J Toxicol 2001;20(Suppl 2):31–39.
6. Deltito J, Beyer D. The scientific, quasi-scientific and popular literature on the use of St. John's wort in the treatment of depression. J Affect Disord 1998;51:345–351.
7. Kim HL, Streltzer J, Goebert D. St. John's wort for depression: a meta-analysis of well-defined clinical trials. J Nerv Ment Dis 1999;187:532–538.
8. Linde K, Ramirez G, Mulrow CD, Pauls A, Weidenhammer W, Melchart D. St. John's wort for depression—an overview and meta-analysis of randomized clinical trials. BMJ 1996;313:253–258.
9. Shelton RC, Keller MB, Gelengerg A, Dunner DL, Hirschfeld R, Thase ME, et al. Effectiveness of St John's wort in major depression a randomized controlled trial. JAMA 2001;285:1978–1986.
10. *Hypericum* Depression Trial Study Group. Effect of *Hypericum perforatum* (St John's Wort) in major depressive disorder a randomized controlled trial. JAMA 2002;287:1807–1814.
11. Nahrstedt A, Butterweck V. Biologically active and other chemical constituents of the herb *Hypericum perforatum* L. Pharmacopsychiatry 1997;30;129–134.
12. Vitiello B. *Hypericum perforatum* extracts as potential antidepressants. J Pharm Pharmacol 1999;51:513–517.
13. Krochmal R, Hardy M, Bowerman S, Lu Q-Y, Wang H-J, Elashoff RM, Heber D. Phytochemical assays of commercial botanical dietary supplements. eCAM 2004;1:305–313.
14. Draves AH, Walker SE. Analysis of the hypericin and pseudohypericin content of commercially available St John's Wort preparations. Can J Clin Pharmacol 2003;10: 114–118.
15. von Eggelkraut-Gottanka SG, Abu Abed S, Muller W, Schmidt PC. Quantitative analysis of the active components and the by-products of eight dry extracts of *Hypericum perforatum* L. (St John's Wort). Phytochem Anal 2002;13;170–176.
16. Wagner H, Bladt S. Pharmaceutical quality of *Hypericum* extracts. J Geriatr Psychiatry Neurol 1994;7(Suppl 1)S34–S38.
17. Maisenbacher P, Kovar KA. Adhyperforin: a homologue of hyperforin from *Hypericum perforatum*. Planta Med 1992;58:291–293.
18. Nathan PJ. *Hypericum perforatum* (St John's Wort): a non-selective reuptake inhibitor? A review of the recent advances in its pharmacology. J Psychopharmacol 2001;15: 47–54.
19. Laakmann G, Schule C, Baghai T, Kieser M. St. John's wort in mild to moderate depression: the relevance of hyperforin for the clinical efficacy. Pharmacopsychiatry 1998;31: 54–59.
20. Di Carlo G, Borrelli F, Ernst E, Izzo AA. St John's wort: Prozac from the plant kingdom. Trends Pharmacol Sci 2001;22:292–297.
21. Biber A, Fischer H, Romer A, Chatterjee SS. Oral bioavailability of hyperforin from *Hypericum* extracts in rats and human volunteers. Pharmacopsychiatry 1998;31(Suppl 1):36–43.
22. Schulz V. Incidence and clinical relevance of the interactions and side effects of hypericum preparations. Phytomedicine 2001;8:152–160.
23. Gulick RM, McAuliffe V, Holden-Wiltse J, Crumpacker C, Liebes L, Stein DS, Meehan P, Hussey S, Forcht J, Valentine FT. Phase I studies of hypericin, the active compound in St. John's Wort, as an antiretroviral agent in HIV-infected adults. AIDS Clinical Trials Group Protocols 150 and 258. Ann Intern Med 1999;130:510–514.
24. Agrosi M, Mischiatti S, Harrrasser PC, Savio D. Oral bioavailability of active principles from herbal products in humans. A study on *Hypericum perforatum* extracts using the soft gelatin capsule technology. Phytomedicine 2000;7: 455–462.
25. Staffeldt B, Kerb R, Brockmoller J, Ploch M, Roots I. Pharmacokinetics of hypericin and pseudohypericin after oral intake of the *Hypericum perforatum* extract LI 160 in healthy volunteers. J Geriatr Psychiatry Neurol 1994;(Suppl 7):S47–S53.
26. Bennett DA Jr, Phun L, Polk JF, Voglino SA, Zlotnick V, Raffa RB. Neuropharmacology of St. John's Wort (Hypericum). Ann Pharmacother 1998;32:1201–1208.
27. Kerb R, Brockmoller J, Staffeldt B, Ploch M, Roots I. Single-dose and steady-state pharmacokinetics of hypericin and pseudohypericin. Antimicrob Agents Chemother 1996;40:2087–2093.
28. Schulz HU, Schurer M, Bassler D, Weiser D. Investigation of pharmacokinetic data of hypericin, pseudohypericin, hyperforin and the flavonoids quercetin and isorhamnetin revealed from single and multiple oral dose studies with a hypericum extract containing tablet in healthy male volunteers. Arzneimittelforschung 2005;55:561–568.
29. Zhou S, Chan E, Pan S-Q, Huang M, Lee EJ. Pharmacokinetic interactions of drugs with St John's wort. J Psychopharmacol 2004;18:262–276.

30. Izzo AA. Drug interactions with St. John's Wort (*Hypericum perforatum*): a review of the clinical evidence. Int J Clin Pharmacol Ther 2004;42:139–148.

31. Wang LS, Zhou G, Zhu B, Wu J, Wang JG, Abd El-Aty AM, et al. St John's wort induces both cytochrome P450 3A4-catalyzed sulfoxidation and 2C19-dependent hydroxylation of omeprazole. Clin Pharmacol Ther 2004;75: 191–197.

32. Gurley BJ, Gardner SF, Hubbard MA, Williams DK, Gentry WB, Cui Y, Ang CY. Clinical assessment of effects of botanical supplementation on cytochrome P450 phenotypes in the elderly: St John's wort, garlic oil, *Panax ginseng* and *Ginkgo biloba*. Drugs Aging 2005;22:525–539.

33. Corns CM. Herbal remedies and clinical biochemistry. Ann Clin Biochem 2003;40:489–507.

34. Markowitz JS, Donovan JL, DeVane CL, Taylor RM, Ruan Y, Wang J-S, Chavin KD. Effect of St. John's wort on drug metabolism by induction of cytochrome P450 3A4 enzyme. JAMA 2003;290:1500–1504.

35. Wang EJ, Barecki-Roach M, Johnson WW. Quantitative characterization of direct P-glycoprotein inhibition by St. John's wort constituents hypericin and hyperforin. J Pharm Pharmacol 2004;56:123–128.

36. Henderson L, Yue QY, Bergquist C, Gerden B, Arlett P. St John's wort (*Hypericum perforatum*): drug interactions and clinical outcomes. Br J Clin Pharmacol 2002;54: 349–356.

37. Roby CA, Anderson GD, Kantor E, Dryer DA, Burstein AH. St. John's wort: effect on CYP3A4 activity. Clin Pharmacol Ther 2000;67:451–457.

38. Moore LB, Goodwin B, Jones SA, Wisely GB, Serabjit-Singh CJ, Willson TM, Collins JL, Kliewer SA. St John's wort induces hepatic drug metabolism through activation of the pregnane X receptor. Proc Natl Acad Sci USA 2000;97:7500–7502.

39. Durr D, Stieger B, Kullak-Ublick GA, Rentsch KM, Steinert HC, Meier PJ, Fattinger K. St John's Wort induces intestinal P-glycoprotein/MDR1 and intestinal and hepatic CYP3A4. Clin Pharmacol Ther 2000;68:598–604.

40. Obach RS. Inhibition of human cytochrome P450 enzymes by constituents of St. John's wort, an herbal preparation used in the treatment of depression. J Pharmacol Exp Ther 2000;294;88–95.

41. Markowitz JS, deVane CL, Boulton DW, Carson SW, Nahas Z, Risch SC. Effect of St. John's wort (*Hypericum perforatum*) on cytochrome P-450 2D6 and 3A4 activity in healthy volunteers. Life Sci 2000;66: PL133–PL139.

42. Wang Z, Gorski JC, Hamman MA, Huang SM, Lesko LJ, Hall SD. The effects of St John's wort (*Hypericum perforatum*) on human cytochrome P450 activity. Clin Pharmacol Ther 2001;70:317–326.

43. Dresser GK, Schwarz UI, Wilkinson GR, Kim RB. Coordinate induction of both cytochrome P4503A and MDR1 by St John's wort in healthy subjects. Clin Pharmacol Ther 2003;73:41–50.

44. Johne A, Brockmoller J, Bauer S, Maurer A, Langheinrich M, Roots I. Pharmacokinetic interaction of digoxin with an herbal extract from St. John's wort (*Hypericum perforatum*). Clin Pharmacol Ther 1999;66:338–345.

45. Mueller SC, Uehleke B, Woehling H, Petzsch M, Majcher-Peszynska J, Hehl EM, et al. Effect of St. John's wort dose and preparations on the pharmacokinetics of digoxin. Clin Pharmacol Ther 2004;75:546–557.

46. Lee A, Minhas R, Matsuda N, Lam M, Ito S. The safety of St. John's wort (*Hypericum perforatum*) during breast-feeding. J Clin Psychiatry 2003;64:966–968.

47. Woelk H, Burkard G, Grunwald J. Benefits and risks of the *Hypericum* extract LI 160: drug monitoring study with 3250 patients. J Geriatr psychiatry Neurol 1994;7: S34–S38.

48. Lantz MS, Buchalter E, Giambanco V. St. John's wort and antidepressant drug interactions in the elderly. J Geriatr Psychiatry Neurol 1999;12:7–10.

49. Brown TM. Acute St. John's wort toxicity. Am J Emerg Med 2000;18:231–232.

50. Bonetto N, Santelli L, Battistin L, Cagnin A. Serotonin syndrome and rhabdomyolysis induced by concomitant use of triptans, fluoxetine and hypericum. Cephalalgia 2007;27:1421–1423.

51. Stevinson C, Ernst E. Can St. John's wort trigger psychoses? Int J Clin Pharmacol Ther 2004;42:473–480.

52. Karalapillai DC, Bellomo R. Convulsions associated with an overdose of St John's wort. Med J Aust 2007;186: 213–214.

53. Schempp CM, Winghofer B, Langheinrich M, Schopf E, Simon JC. Hypericin levels inhuman serum and interstitial skin blister fluid after oral single-dose and steady-state administration of *Hypericum perforatum* extract (St. John's wort). Skin Pharmacol Appl Skin Physiol 1999;12; 299–304.

54. Bove GM. Acute neuropathy after exposure to sun in a patient treated with St John's wort. Lancet 1998;352: 1121–1122.

55. Crampsey DP, Douglas CM, Cooke LD. Nasal insertion of St John's wort: an unusual cause of epistaxis. J Laryngol Otol 2007;121:279–280.

56. Gioti EM, Skalkos DC, Fiamegos YC, Stalikas CD. Single-drop liquid-phase microextraction for the determination of hypericin, pseudohypericin and hyperforin in biological fluids by high performance liquid chromatography. J Chromatogr A 2005;1093:1–10.

57. Riedel KD, Rieger K, Martin-Facklam M, Mikus G, Haefeli WE, Burhenne J. Simultaneous determination of hypericin and hyperforin in human plasma with liquid chromatography-tandem mass spectrometry. J Chromatogr B Analyt Technol Biomed Life Sci 2004;813:27–33.

58. Bauer S, Stormer E, Graubaum HJ, Roots I. Determination of hyperforin, hypericin, and pseudohypericin in human plasma using high-performance liquid chromatography analysis with fluorescence and ultraviolet detection. J Chromatogr B Biomed Sci Appl 2001;76:29–35.

Chapter 91

STAR ANISE
(*Illicium verum* Hook. F. and *Illicium anisatum* L.)

HISTORY

Chinese star anise is a spice that closely resembles anise, and this herb has a long tradition of use in Chinese medicine for colic and indigestion. In contrast, the toxic herb, Japanese star anise is used as incense and as an ingredient for aromatherapy.

BOTANICAL DESCRIPTION

Separating the fruits of Chinese and Japanese star anise is difficult because these two species share common morphological characteristics including color, broad dehiscence, size, and beaked follicles. Chinese and Japanese star anise should not be confused with the European and North African herb, *Pimpinella anisum* L. (anise seed, aniseed).

Common Name: Chinese Star Anise, Staranise Tree

Scientific Name: *Illicium verum* Hook. f.

Botanical Family: Illiciaceae (illiciums)

Physical Description: This aromatic evergreen tree grows up 8–15 m (~26–49 ft) in height with alternate, simple leaves often clustered on the tip of the branch. The white-pink to red solitary flowers appear in June and in October. The star-shaped fruit matures in the following year.

Distribution and Ecology: Star anise (*I. verum*) inhabits subtropical and tropical areas of Asia. This tree prefers moist soil and sun or light shade.

Common Name: Japanese Star Anise, Shikimi (Japanese)

Scientific Name: *Illicium anisatum* L. (*Illicium religiosum*)

Botanical Family: Illiciaceae (illiciums)

Physical Description: This upright, rounded, bushy shrub grows up 10–15 ft (~3–5 m) in height. The evergreen leaves are 2–4 in. (~5–10 cm) long and medium to dark green in color with undulate margins. Yellowish-white or greenish-white flowers with up to 30 pedals appear in spring. These star-shaped flowers produce a star-shaped fruit that contains dark brown seeds.

Distribution and Ecology: Japanese star anise is an introduced plant species in the United States with a distribution in the western, southwestern, and southeastern United States. The native range of this shrub is eastern Asia, southern China, Japan, and Taiwan.

EXPOSURE

Sources

The fruit of Chinese star anise is a common commodity in Asian marketplaces.

Medicinal Uses

Traditional

Teas brewed from Chinese star anise (*I. verum*) are a traditional herbal medicine used to treat colic and exces-

Medical Toxicology of Natural Substances, by Donald G. Barceloux, MD

sive intestinal gas in infants as well as headaches, cough, insomnia, anxiety, rheumatism, back pain, hernia, abdominal pain and indigestion in adults. Hierba anís is a complex consisting of five herbs including star anise (*Illicium verum*) with the odor of anis. This herbal medicine is used to treat menstrual pain and irregularity, upper respiratory tract infections, intestinal colic, and liver-related illness.[1] Dried Chinese star anise (bajiao) is a spice with a licorice-like taste that is found in many Mandarin Chinese dishes. The fruit of star anise is chewed as a breath freshener. The oil is applied topically for the relief of rheumatism, otalgia, and skin infections.[2]

CURRENT

Chinese star anise is a flavoring agent in liqueurs, tobacco, confectionery, and pharmaceutical preparations. Fruits from the Chinese star anise contain substantial amounts of shikimic acid, which is a highly valued compound used as a starting material for the synthesis of the neuramidase inhibitor GS4104 (oseltamivir phosphate). This latter compound was developed under the name Tamiflu® (Roche Pharmaceuticals, Nutley, NJ) for the early treatment of viral infections.

REGULATORY STATUS

Chinese star anise is generally recognized as safe (GRAS) by the US Food and Drug Administration (FDA) when used as a spice or flavoring, whereas as Japanese star anise is considered toxic and unfit for human consumption. Japanese star anise potentially can contaminate Chinese star anise herbal preparations because of the difficulty distinguishing these two types of star anise. After reports of acute neurotoxicity, including seizures, in infants given star anise tea, the FDA issued a warning to avoid the use of these teas in children.[3]

PRINCIPAL INGREDIENTS

Chemical Composition

Illicium species contain prenylated C_6-C_3 compounds, neolignans, and *seco*-prezizaane-type sesquiterpene compounds. The stem bark, root bark and leaves accumulate the former two types of compounds, whereas the fruit contains relatively high concentrations of sesquiterpenes.[4] Star anise species (*I. anisatum, I. verum*) contain several neurotoxic *seco*-prezizaane terpenoids including veranisatin A (CAS RN: 153445-92-2, $C_{16}H_{22}O_8$), veranisatin B (CAS RN: 153445-93-3, $C_{16}H_{20}O_9$), anisatin (CAS RN: 5230-87-5, $C_{15}H_{20}O_8$), and neoanisatin (CAS RN: 15589-82-9, $C_{15}H_{20}O_7$).[5,6] Figure 91.1 and Figure 91.2 displays the chemical structures of veranisatin A and anisatin, respectively. The principal ingredient in Chinese star anise (*I. verum*) is anethole (CAS RN: 104-46-1, $C_{10}H_{12}O$), which accounts for about 72–92% of the herb. Other species of *Illicium* also contain anethole, but the concentration is usually <40%. Safrole and eugenol are not typically present in samples of *I. verum*. Figure 91.3 displays the chemical structure of anethole.

Physiochemical Properties

Anisatin and neoanisatin are picrotoxin-like, noncompetitive GABA-antagonists that produce similar toxicity in rodent studies.[7] In animal preparations (frog spinal cord, crude synaptic membrane from rat brain),[8] anisatin and neoanisatin produce seizure activity.[9] These *seco*-prezizaane terpenoids probably bind to the same

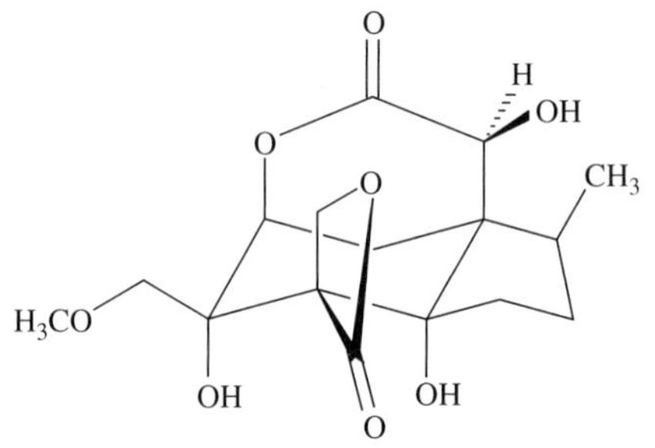

FIGURE 91.1. Chemical structure of veranisatin A.

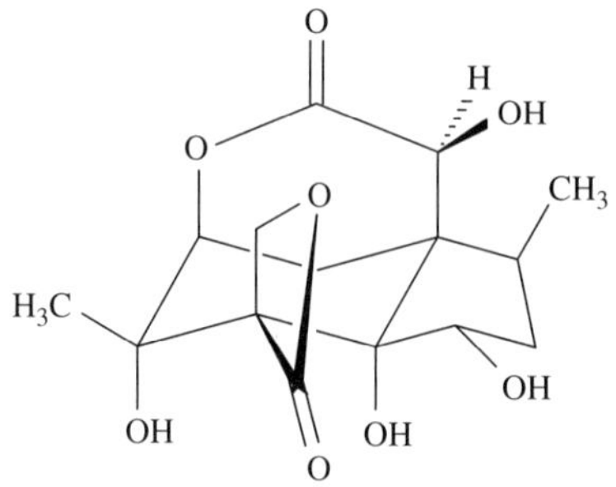

FIGURE 91.2. Chemical structure of anisatin.

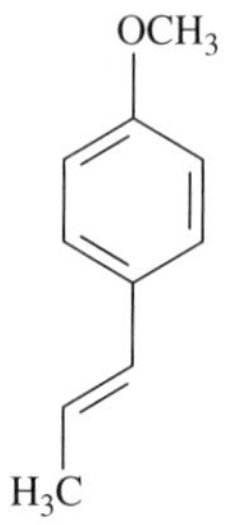

FIGURE 91.3. Chemical structure of anethole.

site as do picrotoxane terpenoids, such as picrotoxinin and picrodendrins. The neurotropic sesquiterpenoid compounds, veranisatins A, B, and C are also proconvulsant compounds in animal studies at high doses (3 mg/kg orally).[9]

Mechanism of Toxicity

Consumption of Japanese star anise is associated with seizures both in animals and in humans. Potential neurotoxins include anisatin and veranisatins. Chinese star anise is relatively safe for consumption, although this herb contains small amounts of potential neurotoxins, veranisatins A, B, and C.

DOSE RESPONSE

There is no recommended dose of Japanese star anise because of the presence of neurotoxins in this herb. In a case series of seven infants aged 2–12 weeks presenting with neurological symptoms ranging from jitteriness to seizures, the dose of Japanese star anise-contaminated herbs ranged from one to six stars boiled in water with the duration of administration varying from a single dose to a dose every 4 hours for 2 weeks.[10] The amount of anisatin present in these teas or in the serum of affected patients was not quantitated by analytical methods.

TOXICOKINETICS

There are few data on the human toxicokinetics of anisatin.

CLINICAL RESPONSE

Star anise oil in concentrations ≥1% is a strong irritant. Sensitization to the components (e.g., anethole, α-pinene, limonene, safrole) in star anise oil occurs in a few individuals, but cross-reactivity to other common essential oils is unusual.[11] A case series of seven infants fed tea made from Chinese star anise (*I. verum*) believed to be adulterated with the more toxic Japanese star anise (*I. anisatin*) resulted in the development of vomiting, jitteriness, myoclonic movements, poor sleep patterns, nystagmus, and seizures.[10] The presence of anisatin in the teas or in biological samples from these patients was not analytically confirmed. Similar symptoms (e.g., tremor, spasms, hypertonia, irritability, excessive crying, nystagmus, vomiting) developed in infants receiving star anise tea in Japan, China, Spain, Scotland, France, and the Netherlands.[12,13] Symptoms usually resolve within 24–48 hours. Contamination or adulteration of Chinese star anise teas with Japanese star anise (*Illicium anisatum*) was analytically confirmed in some of these cases.[14] An epidemic of seizures also occurred in adults consuming Chinese star anise contaminated with Japanese star anise.[15] Over 60 individuals developed nausea, vomiting and general malaise within 2–4 hours of ingesting Chinese star anise tea. Generalized tonic-clonic seizures developed in 16 of these patients. Nuclear magnetic resonance (NMR) analysis of samples of the herbal tea confirmed the presence of the neurotoxin, anisatin, from Japanese star anise.

DIAGNOSTIC TESTING

Analytical Methods

Analytical techniques for detecting the contamination of Chinese star anise by Japanese star anise include macroscopic examination of the herb and analysis of the extracted essential oil for myristicin, methoxyeugenol, eugenol, and safrole.[16] Under fluorescent and scanning electron microscopy there are anatomical differences in the epicarp cells that distinguish the fruits of Chinese star anise (*Illicium verum*) and Japanese star anise (*Illicium anisatum*). The follicle of *I. verum* has bluish-brown polygonal cells of cuticular striation with an indistinct collar, whereas the follicle of *I. anisatum* has thick-walled, frequently branched, beaded cuticular cells forming numerous pits.[17] These latter blue cells appear wavy and beaded with yellow pits. Chromatographically, the presence of methoxyeugenol, eugenol, and 2,6-dimethoxy-4-allyphenol indicate contamination by Japanese star anise as Chinese star anise does not typically contain detectable concentrations of these substances.

Biomarkers

There are few data on the concentrations of anisatin or veranisatins during intoxication with herbs contaminated by Japanese star anise. Anisatin is a biomarker for adulteration of Chinese star anise by the more toxic Japanese star anise. Typically, anisatin concentrations in Chinese star anise do not exceed 0.3 mg/kg. Concentrations of anisatin exceeding 1 mg/kg dry weight, as measured by high performance liquid chromatography/tandem mass spectrometry with thin layer chromatography, indicate contamination with Japanese star anise.[18] The limits of detection and quantitation for this method are approximately 1.2 μg/kg and 4 μg/kg, respectively.

Abnormalities

Laboratory tests and electroencephalograms are usually normal during star anise intoxication, when seizures are brief.

TREATMENT

Treatment is supportive. The main concern is the development of seizures following the ingestion of tea contaminated with Japanese star anise. Intravenous access for the administration of any necessary anticonvulsants (e.g., lorazepam, diazepam) should be obtained, and these patients should be closely observed for 24 hours to prevent trauma associated with seizures. There are few data to guide management after ingestion of Japanese star anise. Currently, there are no antidotes or methods to enhance elimination. Decontamination measures are not recommended because of the possibility of seizures and subsequent aspiration.

References

1. Linares E, Bye RA Jr. A study of four medicinal plant complexes of Mexico and adjacent United States. J Ethnopharmacol 1987;19:153–183.
2. De M, De AK, Sen P, Banerjee AB. Antimicrobial properties of star anise (*Illicium verum* Hook f). Phytother Res 2002;16:94–95.
3. United States Food and Drug Administration. FDA warns consumers about star anise teas. FDA Consumer 2003;37(6):4.
4. Huang J-M, Yang C-S, Zhao R, Takahashi H, Fukuyama Y. Seven novel *seco*-prezizaane-type sesquiterpenes from the pericarps of *Illicium merrillianum*. Chem Pharm Bull 2004;52:104–107.
5. Okuyama E, Nakamura T, Yamazaki M. Convulsants from star anise (*Illicium verum* Hook.f.). Chem Pharm Bull (Tokyo) 1993;41:1670–1671.
6. Kuriyama T, Schmidt TJ, Okuyama E, Ozoe Y. Structure-activity relationships of *seco*-prezizaane terpenoids in gamma-aminobutyric acid receptors of houseflies and rats. Bioorg Med Chem 2002;10:1873–1881.
7. Niwa H, Yamada K. Synthesis of (−)-neoanisatin, a neurotoxic sesquiterpenoid having a novel spiro β-lactone. Chem Lett 1991;22:639–640.
8. Kudo Y, Oka JI, Yamada K. Anisatin, a potent GABA antagonist, isolated from *Illicium anisatum*. Neurosci Lett 1981;25:83–88.
9. Nakamura T, Okuyama E, Yamazaki M. Neurotropic components from star anise (*Illicium verum* Hook. fil.). Chem Pharm Bull (Tokyo) 1996;44:1908–1914.
10. Ize-Ludlow D, Ragone S, Bruck IS, Bernstein JN, Duchowny M, Garcia Pena BM. Neurotoxicities in infants seen with the consumption of star anise tea. Pediatrics 2004;114:e653–e656.
11. Rudzki E, Grzywa Z. Sensitizing and irritating properties of star anise oil. Contact Dermatitis 1976;2:305–308.
12. Vandenberghe N, Pittion-Vouyouvitch S, Flesch F, Wagner M, Godet E. [An inaugural generalized tonic-clonic convulsive crisis following ingestion of Japanese star anise]. Presse Med 2003;32:27–28. [French]
13. Biessels GJ, Vermeij FH, Leijten FS. [Epileptic seizure after a cup of tea: intoxication with Japanese star anise]. Ned Tijdschr Geneeskd 2002;146:808–811. [Dutch]
14. Minodier P, Pommier P, Moulene E, Retornaz K, Prost N, Deharo L. [Star anise poisoning in infants]. Arch Pediatr 2003;10:619–621. [French]
15. Johanns ES, van der Kolk LE, van Gemert HM, Sijben AE, Peters PW, de Vries I. [An epidemic of epileptic seizures after consumption of herbal tea]. Ned Tijdschr Geneeskd 2002;146:813–816. [Dutch]
16. Saltron F, Langella C, Guerere M. Mise en evidence de contamination de badiane de Chine par d'autres espèces d'illicium. Ann Fals Exp Chim 2001;94:397–402.
17. Joshi VC, Srinivas PV, Khan IA. Rapid and easy identification of *Illicium verum* Hook. F. and its adulterant *Illicium anisatum* Linn. by fluorescent microscopy and gas chromatography. J AOAC Int 2005;88:703–706.
18. Lederer I, Schulzki G, Gross J, Steffen J-P. Combination of TLC and HPLC-MS/MS methods. Approach to a rational quality control of Chinese star anise. J Agric Food Chem 2006;54:1970–1974.

FIGURE 3.1. Akee fruit. Photograph courtesy of Stephan Russ.

FIGURE 6.1. Female specimen of *Cycas rumphii.* Miq. © 2005 Rune Midtgaard.

FIGURE 41.1. Fly agaric [*Amanita muscaria* (L.) Hook]. © 2006 Rune Midtgaard.

FIGURE 42.1. Ivory Clitocybe [*Clitocybe dealbata* (Sowerby) Gillet], Photography courtesy of Daniel Sudakin, MD.

FIGURE 46.1. *Aloe vera* (L.). Photograph Courtesy of Paul Auerbach, MD.

FIGURE 83.1. Purple coneflower [*Echinacea purpurea* (L.) Moench]. Photography Courtesy of Saralyn Williams, MD.

FIGURE 113.1. Castor bean (*Ricinus communis* L.). Photograph courtesy of Kimberly Ann Barceloux, MFCC.

FIGURE 117.1. Venus' chariot (*Aconitum napellus* L.). Photograph courtesy of Professor Sylvie Rapior.

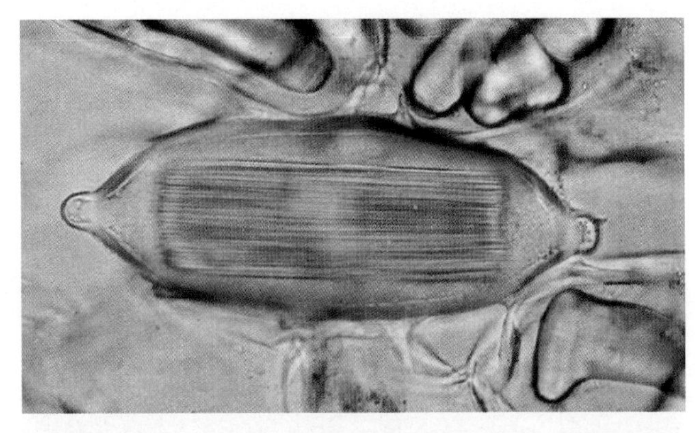

FIGURE 125.1. *Dieffenbachia* idioblast containing numerous rapides. Photograph courtesy of Professor Sylvie Rapior.

FIGURE 127.1. *Datura stramonium* L. (jimsonweed) with maturing seed pods. Photograph courtesy of Kimberly Ann Barceloux, MFCC.

FIGURE 132.1. Pokeweed (*Phytolacca americana* L). Photograph courtesy of Lewis Nelson, MD.

FIGURE 163.1. English yew (*Taxus baccata* L.). Photograph courtesy of Professor Sylvie Rapior.

FIGURE 164.1. European Castor-bean Tick (*Ixodes ricinus* L.). © 2006 Rune Midtgaard.

FIGURE 165.1. Tail (metasoma) and telson of black hairy scorpion (*Hadrurus spadix* Stahnke). © 2006 Rune Midtgaard.

FIGURE 166.1. The Chilean Rose tarantula (*Grammostola rosea* Walckenaer). © 2007 Rune Midtgaard.

FIGURE 169.2. White-eyed Assassin Bug (*Platymeris biguttata* L.). © 2007 Rune Midtgaard.

FIGURE 174.2. Golden Poison Frog (*Phyllobates terribilis* Myers, Daly, and Bibron). © 2005 Rune Midtgaard.

FIGURE 174.3. Red-spotted Newt (*Notophthalmus viridescens* Rafinesque). © 2007 Rune Midtgaard.

FIGURE 175.1. Beaded lizard (*Heloderma horridum* Wiegmann). © 2006 Rune Midtgaard.

FIGURE 176.2. Nose-horned Viper (*Vipera ammodytes* L.). © 2000 Rune Midtgaard.

FIGURE 176.5. Red Diamondback rattlesnake (*Crotalus ruber ruber* Cope). © 2005 Rune Midtgaard.

FIGURE 176.6. Mojave Rattlesnake (*Crotalus scutulatus scutulatus* Kennicott). In contrast to the Western Diamondback Rattlesnake, specimens of the Mojave Rattlesnake often have a greenish tinge, a diagonal facial stripe behind their eye that usually does not touch their mouth, two to three enlarged scales between their eyes on top of their head, and white bands on the tail that sometimes are wider than the black bands. Photograph Courtesy of Arizona Poison and Drug Information Center.

FIGURE 176.7. Western Diamondback rattlesnake (*Crotalus atrox* Baird and Girard). Photograph Courtesy of Dennis Caldwell.

FIGURE 176.10. Mangrove Pit Viper (*Trimeresurus purpureomaculatus* Gray). © 1997 Rune Midtgaard.

FIGURE 176.11. Siamese Spitting Cobra (*Naja siamensis* Laurenti). © 1996 Rune Midtgaard.

FIGURE 176.12. Common tiger snake (*Notechis scutatus* Peters). © 2002 Rune Midtgaard.

FIGURE 176.13. King brown snake (*Pseudechis australis* Gray). © 1997 Rune Midtgaard.

FIGURE 181.1. Crown-of-Thorns sea star (*Acanthaster planci* L.). Photograph courtesy of Paul Auerbach, MD.

FIGURE 181.2. Fire sea urchin (*Asthenosoma varium* Grube). Photograph courtesy of Paul Auerbach, MD.

FIGURE 183.1. Lionfish (*Pterois volitans* L.). Photograph courtesy of Paul Auerbach, MD.

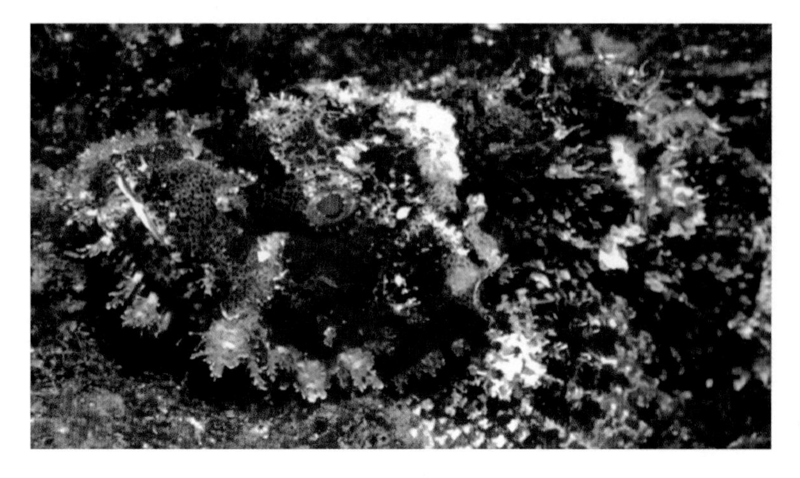

FIGURE 183.2. Spotted scorpionfish (*Scorpaena plumieri* Bloch). Photograph courtesy of Paul Auerbach, MD.

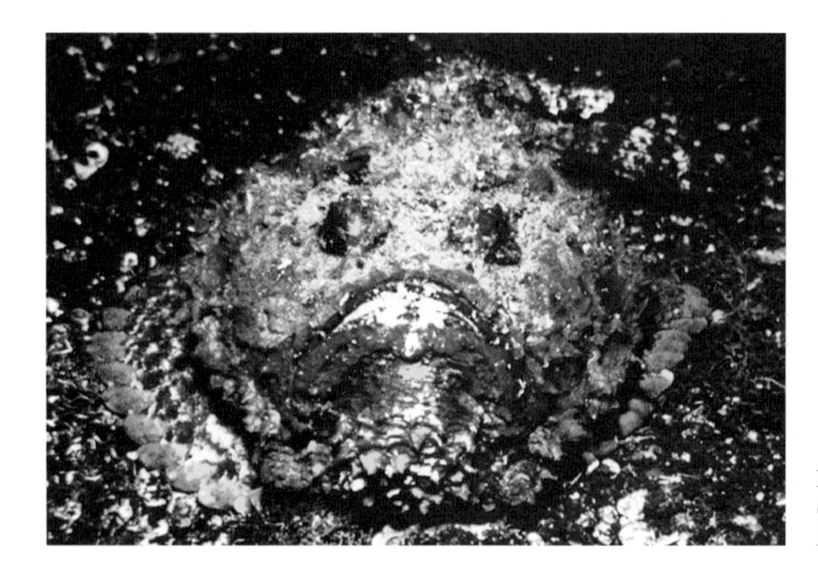

FIGURE 183.3. Australian or estuarine stonefish (*Synanceia horrida* L.). Photograph courtesy of Paul Auerbach, MD.

Chapter 92

STAR FRUIT
(*Averrhoa carambola* L.)

HISTORY

Star fruit originated in Southeast Asia, and this fruit has been cultivated in tropical areas of Southeast Asia and Malaysia for many centuries.

BOTANICAL DESCRIPTION

Common Name: Star fruit, carambola

Scientific Name: *Averrhoa carambola* L.

Botanical Family: Oxalidaceae (oxalis)

Physical Description: This slow-growing, short-trunked, subtropical evergreen tree has a branched, round, bushy canopy reaching up to 25–30 ft (~8–9 m) in height. The spirally arranged, alternate leaves are 6–10 in. (~15–25 cm) long with smaller, nearly opposite, ovate-oblong leaflets. The medium-green leaves are soft with a smooth upper surface and slightly hairy white underside. The yellow-green star fruit is ovate to ellipsoid in length with five to six prominent longitudinal ribs and star-shaped cross-section.

Distribution and Ecology: The star fruit probably originated in Sri Lanka and the Moluccas. This fruit tree is a commonly grown plant in southern China, Taiwan, India, and Brazil. In the United States, star fruit is cultivated in California, Hawaii, and Florida.

EXPOSURE

Star fruit is a popular tropical fruit available throughout the world as a fresh fruit, gourmet jam, fruit juice (sweet or sour), and constituent of fruit salad. Traditional uses for star fruit by Chinese herbalists and bomohs (Malay medicine men) include the treatment of headache, cough, sore throat, upper respiratory tract infections, vomiting, and restlessness. In India, star fruit is a traditional treatment for hemorrhage and hemorrhoids, while in Brazil this fruit is a diuretic. Star fruit has no approved medical uses in the United States or Europe.

PRINCIPAL INGREDIENTS

The neurotoxic principal in star fruit has not been identified.[1] Star fruit contains high concentrations of potassium and oxalates, but these substances have not been directly linked to the illness associated with the ingestion of star fruit by patients with renal dysfunction. The two types of star fruit are sour and (slightly) sweet. The potassium content in samples of homogenized sweet star fruit and star fruit juice were approximately 34 mEq/L and 3 mEq/L, respectively.[2] Sour star fruit juice contains substantially higher oxalate concentrations than sweet star fruit juice. The oxalate concentrations of sour star fruit juice consumed in two cases of acute renal failure were 308 mg/dL and 820 mg/dL.[3] Animal studies indicate that the neurotoxin in star fruit intoxication is a dialyzable, nonprotein with a molecular weight below

Medical Toxicology of Natural Substances, by Donald G. Barceloux, MD

500 Da. *In vitro* studies indicate that the neurotoxic fraction from star fruit inhibits GABA binding in a concentration-dependent manner (IC_{50} = 0.89 μM), but not glutamate binding. This fraction does not affect the uptake or release of either GABA or glutamate.[4]

DOSE RESPONSE

Star fruit intoxication occurs primarily in patients with chronic renal failure with the exception of individuals ingesting large amounts of sour star fruit juice.[3] There is no specific dose-response effect as patients with this syndrome have eaten a few pieces to several fruits without a direct correlation to the severity of illness. Case series of uremic patients developing star fruit intoxication report the ingestion of two to three fruits or 150–200 mL of the fruit juice with symptoms ranging from insomnia and hiccups to confusion and death.[5] The intraperitoneal injection of an aqueous methanol extract of star fruit to male albino mice produced convulsions and cardiopulmonary arrest at doses exceeding 8 g/kg body weight.[6] However, the study did not use an adequate control (normal saline) or analyses of serum chemistries (e.g., sodium, potassium) to determine the effect of such large volumes of fluid and electrolytes on the mice.

TOXICOKINETICS

There are no toxicokinetic data in humans on the specific neurotoxin associated with star fruit intoxication. Although star fruits contain high concentrations of potassium and oxalate, the effect of these substances do not account for all the clinical effects associated with the ingestion of star fruit by renal dialysis patients.

CLINICAL RESPONSE

The ingestion of star fruit or star fruit juice by patients with renal dysfunction can cause acute central nervous system (CNS) disturbances and deteriorating renal function, even in patients without advanced chronic renal failure.[7] Patients typically present within 6–12 hours after the ingestion of star fruit.[8] The rate of progression is highly variable, and death occurs in some patients with altered consciousness. Clinical features of this intoxication include hiccups, nausea, vomiting, fever, agitation, insomnia, lethargy, confusion, muscle weakness, seizures (grand mal, complex partial), and coma.[9] Intractable hiccups and agitation are the most common symptoms associated with this poisoning. The hiccups do not usually respond to haloperidol, metoclopramide, or diazepam. In a series of 32 uremic patients, the classification of the severity of the intoxication associated with star fruit ingestions was as follows: *mild* – hiccups, vomiting, insomnia; *moderate* – agitation, numbness and paraesthesias of the extremities, mild mental confusion; and *severe*– increasing mental confusion progressing to coma, seizures, hypotension, and shock.[8] Altered consciousness or seizures are poor prognostic signs. Recovery may require 7–14 days. A case report associated the development of increasing renal dysfunction and oxalate nephropathy with the chronic ingestion of an undetermined quantity of sour star fruit (i.e., relatively high oxalate content) over 3 years.[10] This 85-year old man had mild pre-existing renal dysfunction; he did not develop any neurological symptoms.

The differential diagnosis of star fruit intoxication in these patients includes hypertensive encephalopathy, dialysis dementia, dysequilibrium syndrome, uremic encephalopathy, CNS infections, hypoglycemia, hypoxia, electrolyte imbalance, cerebral vascular diseases, and drug intoxications. Individuals with normal renal function do not usually develop star fruit intoxication with the exception of individuals ingesting large amounts of sour star fruit juice containing high concentrations of oxalates. These patients develop gastrointestinal distress, back pain, and acute renal failure. However, these individuals do not usually have the CNS disturbances associated with star fruit intoxication in patients with renal dysfunction.[3]

DIAGNOSTIC TESTING

Hyperkalemia occurs in only a few of the patients presenting with star fruit intoxication despite the presence of renal failure and the large amount of potassium in star fruit.[2] Electroencephalograms during the acute illness typically demonstrate diffuse cortical dysfunction consistent with metabolic encephalopathy along with occasional focal regional cortical epileptic activities.[11] Computed tomography (CT) scans of the brain are usually normal, whereas magnetic resonance imaging (MRI) brain scans demonstrate hyperintense T2-weighted images in the thalamus and cortex.[12]

TREATMENT

Patients presenting with toxicity after star fruit ingestion typically require intensive supportive care as a result of CNS disturbances and the metabolic abnormalities (metabolic acidosis, hyperkalemia) associated with worsening renal failure. Seizures may be unresponsive to conventional anticonvulsants (e.g., lorazepam, diazepam) and may require deep sedation (e.g., propofol) or neuromuscular blockade with electroencephalographic monitoring.[13] Patients with mild intoxication (insomnia, intractable hiccups, agitation, slight confusion) usually respond to supportive care. Patients with more severe intoxication require dialysis. Case reports document of the use of plasma exchange in patients with severe star fruit intoxication, but the efficacy of this treatment is unclear because of the lack of toxicokinetic data and the simultaneous use of hemodialysis.[11] Gut decontamination is not recommended. Prompt hemodialysis is the treatment of choice, and most, but not all, patients respond to intensive dialysis within 2–3 days. In severe cases, daily hemodialysis or continuous methods (continuous venovenous hemodialysis, continuous arteriovenous hemodialysis) are preferred because symptoms (hiccups, confusion) recur during severe intoxication. A case report associated improvement in two patients with severe star fruit intoxication with the use of hemoperfusion.[1] However, there are few clinical data to guide treatment.

References

1. Wu M-Y, Wu IW, Wu S-S, Lin J-L. Hemoperfusion as an effective alternative therapy for star fruit intoxication: a report to 2 cases. Am J Kid Dis 2007;49:E1–E5.
2. Chang J-M, Hwang S-J, Kuo H-T, Tsai J-C, Guh J-Y, Chen H-C, et al. Fatal outcome after ingestion of star fruit (*Averrhoa carambola*) in uremic patients. Am J Kid Dis 2000;35:189–193.
3. Chen C-L, Fang H-C, Chou K-J, Wang J-S, Chung H-M. Acute oxalate nephropathy after ingestion of star fruit. Am J Kid Dis 2001;37:418–422.
4. Carolino RO, Beleboni RO, Pizzo AB, del Vecchio F, Garcia-Cairasco N, Moyses-Neto M, et al. Convulsant activity and neurochemical alterations induced by a fraction obtained from fruit *Averrhoa carambola* (Oxalidaceae: Geraniales). Neurochem Int 2005;46:523–531.
5. Neto MM, Robl F, Netto JC. Intoxication by star fruit (*Averrhoa carambola*) in six dialysis patients? (Preliminary report). Nephrol Dial Transplant 1998;13:570–572.
6. Muir CK, Lam CK. Depressant action of *Averrhoa carambola*. Med J Malaysia 1980;24:279–280.
7. Chang C-T, Chen Y-C, Fang J-T, Huang C-C. Star fruit (*Averrhoa carambola*) intoxication: an important cause of consciousness disturbance in patients with renal failure. Renal Failure 2002;24:379–382.
8. Neto MM, Cardeal Da Costa JA, Garcia-Cairasco N, Netto JC, Nakagawa B, Dantas M. Intoxication by star fruit (*Averrhoa carambola*) in 32 uraemic patients: treatment and outcome. Nephrol Dial Transplant 2003;18:120–125.
9. Tse K-C, Yip P-S, Lam M, Choy B-Y, Li F-K, Lui S-L, et al. Star fruit intoxication in uraemic patients: case series and review of the literature. Intern Med J 2003;33: 314–316.
10. Niticharoenpong K, Chalermsanyakorn P, Panvichian R, Kitiyakara C. Acute deterioration of renal function induced by star fruit ingestion in a patient with chronic kidney disease. J Nephrol 2006;19:682–686.
11. Yap H-J, Chen Y-C, Fang J-T, Huang C-C. Star fruit: a neglected but serious fruit intoxicant in the chronic renal failure. Dial Transplant 2002;31:564–567.
12. Chang C-H, Yeh J-H. Non-convulsive status epilepticus and consciousness disturbance after star fruit (*Averrhoa carambola*) ingestion in a dialysis patient. Nephrol 2004;9: 362–365.
13. Wang YC, Liu BM, Supernaw RB, Lu YH, Lee PY. Management of star fruit-induced neurotoxicity and seizures in a patient with chronic renal failure. Pharmacotherapy 2006;26:143–146.

Chapter 93

TANSY
(*Tanacetum vulgare* L.)

HISTORY

Common tansy (*Tanacetum vulgare* L.) was known to the ancient Greeks. This herb was present in the garden of Charlemagne the Great during the 8th century and in the herb gardens of Swiss Benedictine monks. Traditional uses for common tansy included treatment of gastrointestinal distress, parasitic infections, fever, and rheumatism. In Medieval times, common tansy in large doses was a common abortifacient. During colonial times, common tansy was a food preservative and a component of funeral shrouds and wreaths. In the 19th century, the *United States Pharmacopeia* listed tansy as a stimulant, emmenagogue, and anthelminthic.[1]

BOTANICAL DESCRIPTION

Common tansy is easily confused with tansy ragwort (*Senecio jacobaea* L.), which contains pyrrolizidine alkaloids.

Common Name: Common Tansy, Garden Tansy, Tansy, Golden Buttons

Scientific Name: *Tanacetum vulgare* L.

Botanical Family: Asteraceae (sunflowers, tournesols)

Physical Description: This aromatic plant has fern-like leaves and small, round yellow flowers. Purplish-red stems are hairless or mostly hairless and extensively branched toward the top. Leaves are finely divided into leaflets, ranging from 4–8 in. (~10–20 cm) in length and 1.5–3 in. (~4–8 cm) in width. Characteristic flat-topped clusters of small, button-like, petalless, yellow flowers appear in the summer.

Distribution and Ecology: This perennial herb is a common weed, which grows throughout the northern climate in North America, Europe, and Asia. Common tansy invades disturbed sites on roadsides, fences, pastures, stream banks, and waste areas throughout North America.

EXPOSURE

Sources

The yield of essential oil from tansy is about 1% volume/weight. In a study of 40 plants from wild populations of Norwegian tansy (*Tanacetum vulgare* L.), the average yield of essential oil was 0.81% volume/weight with a range of 0.35–1.90% volume/weight.[2]

Medicinal Uses

TRADITIONAL

Traditional uses of tansy include an insect repellant, insecticide, and the treatment of parasitic infection, kidney disease, respiratory infections, migraine headaches, neuralgias, anorexia, headache, diarrhea, sore throat, and rheumatism. The Micmac Indians used tansy as an abortifacient and diuretic.[1]

CURRENT

Tansy is a spice for cakes, salads, and omelettes as well as an aromatic constituent of some cosmetics, dyes, and insect repellants.

REGULATORY STATUS

There are no standard formulations of essential oils from tansy.

PRINCIPAL INGREDIENTS

Chemical Composition

Tansy contains a variety of compounds including sesquiterpene lactones (parthenolide), flavonoids (methyl ethers of scutellarein and 6-hydroxyluteolin),[3] phenols, coumarins, and polysaccharides (tanacetan TVF).[4] The chemical composition of tansy essential oil is highly diverse with marked geographic variation in monoterpene and sesquiterpene content. There are at least 15 chemotypes with camphor and thujone being the most common chemotypes. A study of 40 plants from wild populations in Norway revealed the following seven chemotypes: β-thujone (55%), camphor (15%), (*E*)-chrysanthenyl acetate/chrysanthenol (7.5%), artemisia ketone/artemisia alcohol (7.5%), α-thujone (5%), chrysanthenone (5%), and 1,8-cineole (5%).[2] The thujone chemotype had high concentrations of thujone (CAS RN: 546-80-5, $C_{10}H_{16}O$) with an average concentration of about 81%. Figure 93.1 displays the chemical structure of thujone. The average camphor concentration in the camphor chemotype was about 31%. In a study of air-dried flower heads from 20 Finnish tansy genotypes, the most common monoterpene was camphor with smaller concentrations of camphene, 1,8-cineole (eucalyptol), pinocamphone, chrysanthenyl acetate, bornyl acetate, and isobornyl acetate.[5] In 13 genotypes, the camphor concentrations ranged from 18.5–69.4%, while the camphor concentration in the remaining 7 genotypes ranged from 0.06–7.15%. The major sterol and triterpene compounds in aerial samples of tansy from Nova Scotia, Canada were β-sitosterol and α-amyrin, respectively, with other sterols including stigmasterol, campesterol, and cholesterol.[1] These samples also contained the triterpene compounds, β-amyrin and taraxasterol.

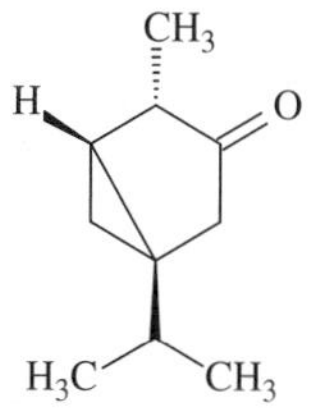

The composition of the essential oil from tansy depends on a number of factors including the plant part, ontogeny (i.e., chemotype) of the plant, seasonal changes, environmental conditions, and extraction method.[6] The essential oil contains over 100 volatile monoterpene and sesquiterpene compounds with camphor and thujone being the most common. However, the composition of the essential oil varies dramatically with the specific chemotype as indicated above. The essential oil of tansy contains optically pure (–)-camphor, which is a biosynthetic oxidation product of bicyclic monoterpene alcohol, (–)-borneol.[7]

FIGURE 93.1. Chemical structure of thujone.

Physiochemical Properties

Table 93.1 lists some physical properties of thujone.

TABLE 93.1. Some Physical Properties of Thujone

Physical Property	Value
Melting Point	<25 °C/77 °F
Boiling Point	203 °C/~397 °F
Log P (Octanol-Water)	2.650
Water Solubility	408 mg/L (25 °C/77 °F)
Henry's Law Constant	7.00E-05 atm-m³/mole (25 °C/77 °F)
Atmospheric OH Rate Constant	5.12E-12 cm³/molecule-second (25 °C/77 °F)

DOSE RESPONSE

Most of the clinical effects associated with tansy are hypersensitivity reactions. There are few data on the dose response of tansy essential oil.

TOXICOKINETICS

There are few data on the human toxicokinetics of camphor (See Chapter 53, Camphor), thujone, or eucalyptol (1,8-cineole) in tansy essential oil.

CLINICAL RESPONSE

Allergic contact dermatitis and allergic rhinitis are common complications of exposure to tansy in atopic patients with hypersensitivity to plants from the Asteraceae family as a result of substantial cross-reactivity between members of this plant family. In a study of 190 Asteraceae-allergic patients determined by positive skin tests to an Asteraceae mix, feverfew elicited the most (81%) positive responses followed by tansy (77%), chamomile (64%), and yarrow (41%).[8] Most atopic patients are hypersensitive to allergens other than members of the Asteraceae family, such as nickel.[9] Case reports associated airborne exposure to tansy pollens with allergic contact dermatitis manifest by pruritic, scaly lichenified plaques on the hands, scalp, posterior neck, arms, and legs.[10] Although the presence of oxygenated monoterpene compounds (camphor, thujone, eucalyptol) suggests the possibility of seizures following the ingestion of large doses of the essential oil from tansy,[11] there are no case reports that suggest that seizures are a clinically relevant complication of tansy ingestion.

DIAGNOSTIC TESTING

Gas chromatography[5] and gas chromatographic/mass spectrometry[2] can separate and quantitate the components of tansy essential oils including the main compounds, thujone and camphor.

TREATMENT

Treatment is supportive, including the use of antihistamines and steroids as indicated. Gut decontamination measures are unnecessary.

References

1. Chandler RF, Hooper SN, Hooper DL, Jamieson WD, Lewis E. Herbal remedies of the Maritime Indians: sterols and triterpenes of *Tanacetum vulgare* L. (tansy). Lipids 1982;17:102–106.
2. Rohloff J, Mordal R, Dragland S. Chemotypical variation of tansy (*Tanacetum vulgare* L.) from 40 different locations in Norway. J Agric Food Chem 2004;52:1742–1748.
3. Williams CA, Harborne JB, Geiger H, Hoult JR. The flavonoids of *Tanacetum parthenium* and *T. vulgare* and their anti-inflammatory properties. Phytochemistry 1999;51:417–423.
4. Polle AY, Ovodova RG, Chizhov AO, Shashkov AS, Ovodov YS. Structure of tanacetan, a pectic polysaccharide from tansy *Tanacetum vulgare* L. Biochemistry (Mosc) 2002;67:1371–1376.
5. Keskitalo M, Pehu E, Simon JE. Variation in volatile compounds from tansy (*Tanacetum vulgare* L.) related to genetic and morphological differences of genotypes. Biochem Syst Ecol 2001;29:267–285.
6. Dragland S, Rohloff J, Mordal R, Iversen T-H. Harvest regiment optimization and essential oil production in five tansy (*Tanacetum vulgare* L.) genotypes under a northern climate. J Agric Food Chem 2005;53:4946–4953.
7. Croteau R, Shaskus J. Biosynthesis of monoterpenes: demonstration of a geranyl pyrophosphate: (–)-bornyl pyrophosphate cyclase in soluble enzyme preparations from Tansy (*Tanacetum vulgare*). Arch Biochem Biophys 1985;236:535–543.
8. Paulsen E, Andersen KE, Hausen BM. Sensitization and cross-reaction patterns in Danish Compositae-allergic patients. Contact Dermatitis 2001;45:197–204.
9. Jovanovic M, Poljacki M, Duran V, Vujanovic L, Sente R, Stojanovic S. Contact allergy to Compositae plants in patients with atopic dermatitis. Med Pregl 2004;56:209–218.
10. Mark KA, Brancaccio RR, Soter NA, Cohen DE. Allergic contact and photoallergic contact dermatitis to plant and pesticide allergens. Arch Dermatol 1999;135:67–70.
11. Burkhardt PR, Burkhardt K, Haenggeli C-A, Landis T. Plant-induced seizures: reappearance of an old problem. J Neurol 1999;246:667–670.

Chapter 94

VALERIAN
(*Valeriana officinalis* L.)

HISTORY

Valerian was a recognized medicinal herb at least since the writing of *De Materia Medica* by the Greek physician and herbalist Pedanios Dioscorides of Anazarba.[1] In the late 16th century, Fabius Columna reported the effectiveness of valerian as an anticonvulsant, and subsequently this herb was used frequently as a treatment for a variety of nervous diseases. During the 19th century, the Eclectic physicians commonly recommended valerian for a variety of illnesses, including use as an antipyretic, aphrodisiac, sedative, and analgesic.

BOTANICAL DESCRIPTION

The Valerianaceae family contains approximately 250 species of the genus *Valeriana* with *V. jatamansi* Jones (Indian or Pakistan valerian), *V. edulis* Nutt. (Mexican valerian), and *V. officinalis* L. (European valerian) being the most common herbal sources. The latter species is the most common source of valerian extract in North America. Official medicinal sources of valerian include another European valerian species (*V. repens* Host). Adulteration of American valerian with species from other genera is uncommon.

Common Name: Valerian, Garden Heliotrope, Garden Valerian

Scientific Name: *Valeriana officinalis* L.

Botanical Family: Valerianaceae (valarians)

Physical Description: A tall, perennial herb that grows up to 2 m (~7 ft) in height with cylindrical, grooved, hollow stems. The oblong, yellow-brown to dark brown rhizome grows up to about 20 cm (~8 in.) in diameter and 50 cm (~20 in.) in length with a compressed base, one or more stolons, and many thick, brown rootlets. The large, dark green leaves are saw-toothed, and white to pale pink. Strongly scented flowers bloom from June to August; the fruits are oblong to ovate with four ridges and a single seed.

Distribution and Ecology: Valerian is a perennial native herb of western Asia and Europe that prefers streams, damp meadows, scrub, or woodlands. In the United States, valerian is an introduced species that has become naturalized. Valerian is cultivated in several countries including the United States, Eastern Europe, France, Germany, Britain, the Netherlands, Belgium, Russia, and Japan.

EXPOSURE

Sources

The yield of essential oil from the roots of valerian ranges from about 0.4–2.1% depending on seasonal variation, species and strain (genotype), drying and extraction technique, and growing conditions.[2] High-quality valerian contains about 1–1.5% essential oil and ≥0.5% valerenic acid. The crude drug Valerianae Radix consists of roots, stolons, and rhizomes of *Valeriana officinalis* dried at temperatures below 40 °C (104 °F).

Medical Toxicology of Natural Substances, by Donald G. Barceloux, MD

Medicinal Uses

Traditional

The most common herbal uses of valerian include use as a sedative, sleep aid, and antispasmodic. Other traditional uses include the treatment of muscle cramps, menstrual pain, tension headaches, seizures, yeast infections, sore throat, fluid retention, excessive perspiration, and excessive gastrointestinal gas.

Current

Valerian is a traditional herbal treatment for insomnia and anxiety. Although valerian is generally a safe herb, the efficacy of this herb in the treatment of insomnia and anxiety remains unproven.[3] Several clinical studies suggest that valerian is an effective treatment of insomnia, primarily by reducing sleep latency (i.e., time to fall asleep), night awakenings, and improving the quality of sleep,[4,5] whereas other clinical studies did not demonstrate improvement in these sleep parameters with the use of valerian.[6] The inconsistency of the results and design flaws (small sample sizes, poor assessment measures, inadequate control of confounding variables) limit conclusions about the efficacy of valerian for the treatment of insomnia. A double-blind, randomized crossover, placebo-controlled study compared the effects of single doses of valerian (400 and 800 mg), temazepam (15 and 30 mg), and diphenhydramine (50 and 75 mg) in 14 healthy elderly volunteers.[7] Assessment included validated measures of subjective sedation and mood along with psychomotor performance (manual tracking, digit symbol substitution). There was no statistically significant difference in any of these measures between valerian and placebo.

Frequently herbal sleep preparations combine the use of passionflower, balm, hops, and/or St. John's wort with valerian for the treatment of insomnia or anxiety.[8] There are limited scientific data to support the use of valerian for the alleviation of anxiety.[9,10] In a randomized, placebo-controlled, double-blind, cross-over trial of 10 healthy volunteers receiving single doses of valerian extract (600, 1,200, 1,800 mg), there was no significant difference between valerian treatment and placebo as assessed by questionnaire and psychomotor/cognitive testing (memory, eye–hand coordination, logical reasoning, Digit Symbol Substitution Test).[11] An Internet-based, double-blind, placebo-controlled study of 391 subjects did not detect a statistically significant difference in the scores on the State-Trait Anxiety Inventory State subtest (anxiety) or the Insomnia Severity Index (insomnia), when comparing valerian treatment and placebo groups.[12] The valerian treatment group received two valerian softgel capsules (3.2 mg valerenic acids, 1% valerenic acid in extract) 1 hour before bedtime for 28 days.

Regulatory Status

The US Food and Drug Administration (FDA) lists valerian as a food supplement, and regulatory control of valerian is limited to labeling requirements. The German Commission E approves the use of valerian for the treatment of insomnia and restlessness without any specific contraindications.

PRINCIPAL INGREDIENTS

Chemical Composition

The essential oil of valerian contains over 150 compounds including sesquiterpenes (valerenic acid, valeranone),[13] sesquiterpene derivatives (acetoxyvalerenic acid, 3β,4β-epoxyvalerenic acid, hydroxyvalerenic acid, valerane, valerenal, cryptofauronol),[14,15] bicyclic iridoid esters (valepotriates),[16] decomposition products of valepotriates (baldrinal, homobaldrinal, valtroxal), monoterpenes (bornyl acetate and isovalerate), flavonoids (6-methylapigenin, hesperidin),[17] furanofuran lignans, alkaloids (valerianine, actinidine), and free amino acids (γ-aminobutyric acid, arginine, glutamine, tyrosine). Figure 94.1 displays the chemical structures of the major types of compounds in valerian. Valepotriates are chemically unstable iridoid triesters (valtrate, isovaltrate, homovaltrate, acevaltrate, didrovaltrate). Valtrate and isovaltrate typically account for about 90% of the valepotriate compounds in valerian root.[18] Commercial root extracts of valerian usually contain small amounts of valepotriates (e.g., valtrate) because of their lipophilicity and instability in aqueous solutions. The composition of the essential oil of valerian depends on the chemovar, and the most common compounds include bornyl acetate, valerenal, valeranone, and cryptofauronol.[19] Minor constituents of the essential oil include α-pinene, β-pinene, camphene, β-caryophyllene, terpinolene, limonene, and carvyl acetate.

There is no official standard for herbal preparations of valerian and the content of main sesquiterpene compounds (valerenic acid, hydroxyvalerenic acid, acetoxyvalerenic acid) in these preparations is highly variable. Analysis of six commercial samples of valerian detected valerenic acid and hydroxyvalerenic acid in only one sample in concentrations of 0.54% and 0.13%, respectively.[20] In a study of 31 commercial Australian preparations (capsules, tablets, teas, liquids) of valerian, the total valerenic acid content ranged from <0.01 to 6.32 mg/g with 16% of these products containing

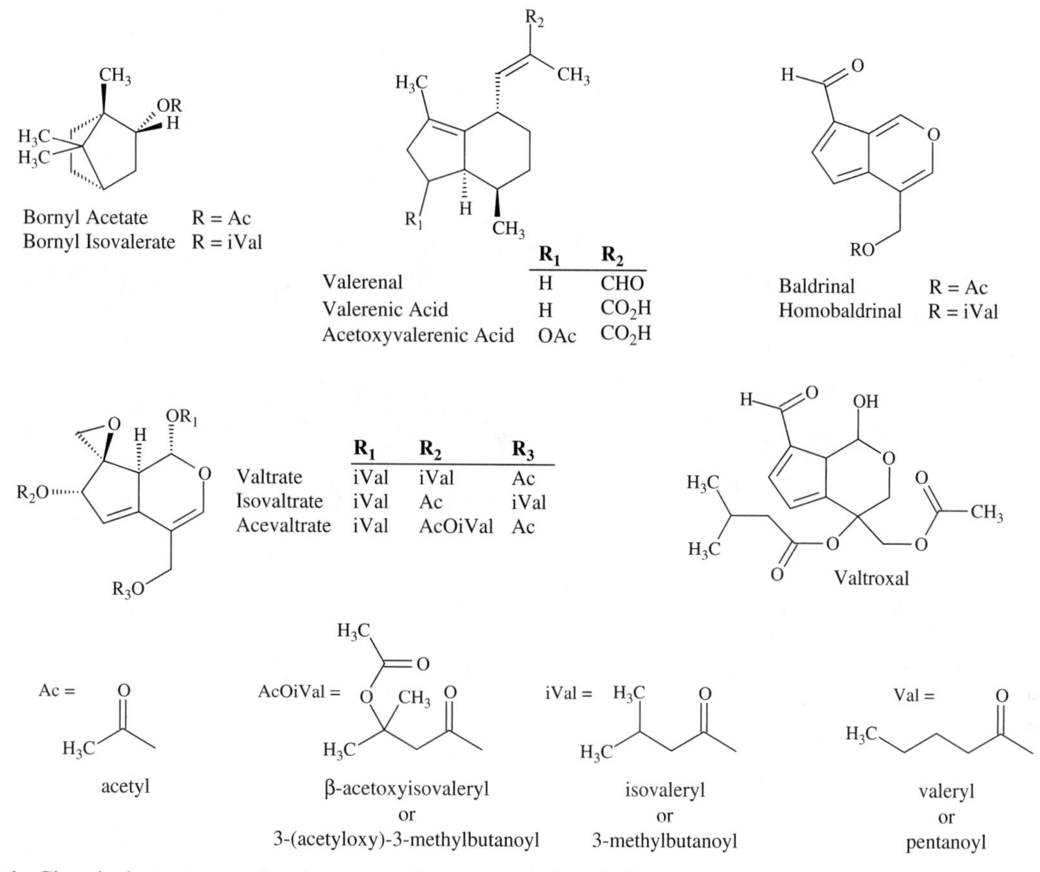

FIGURE 94.1. Chemical structures of major types of compounds in valerian.

<0.1 mg/g or mL.[21] Valepotriate compounds were present in low concentrations (<1 mg/g) in some teas, but most other valerian products did not contain detectable amounts of valepotriate compounds as measured by high performance liquid chromatography (HPLC). Figure 94.2 displays the chemical structure of valerenic acid (CAS RN: 3569-10-6, $C_{15}H_{22}O_2$).

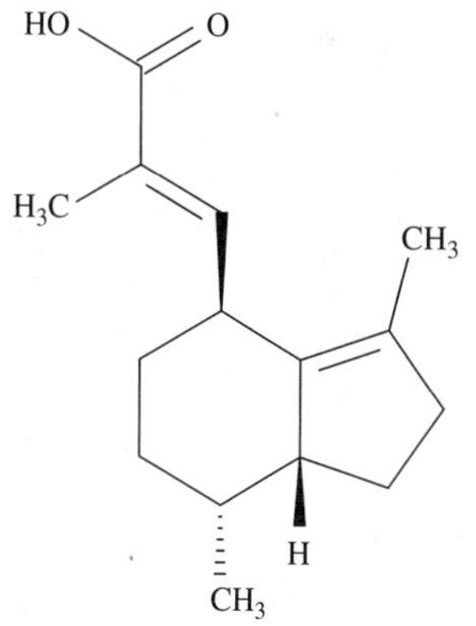

Physiochemical Properties

Valerenic acids are relatively stable compounds, whereas valepotriate compounds are thermolabile and unstable under acidic or alkaline conditions. Storage of valerian root in high humidity results in the degradation of acetoxyvalerenic acid to hydroxyvalerenic acid. The pungent smell associated with valerian root extract results from hydrolysis during storage and the subsequent formation of isovaleric acid. The fresh root is not typically aromatic. Preservation of the components in the essential oil requires storage in tight containers and a cool, dry, dark place. Exposure of valerian to humidity, high ambient temperatures (>40 °C/104 °F), alcoholic solutions, or acidic conditions (pH < 3) results in the decomposition of valepotriate compounds, and most commercial products do not contain these compounds more than 2 months after extraction.[22] High ambient temperatures also degrade valerenic acids. Exposure of an ethanol extract of dry ground valerian rhizomes to 70 °C/158 °F for 22 hours reduced the total valerenic acid content about 11–12%.[23]

FIGURE 94.2. Chemical structure of valerenic acid.

Mechanism of Toxicity

Although valepotriate compounds are frequently cited as the active components of valerian, currently there is no consensus for the mechanism of action for the possible sedative effects of valerian. *In vitro* studies indicate that the flavonoid, 6-methylapigenin in valerian root extract binds to the benzodiazepine binding site of the $GABA_A$ receptor,[17] whereas the lignan, 1-hydroxypinoresinol binds to the $5\text{-}HT_{1A}$ serotonin receptor.[24] These *in vitro* studies also indicate that valerian and valerenic acid are partial agonists of the $5\text{-}HT_{5a}$ receptor.

DOSE RESPONSE

Medicinal Use

The typical herbal dose of valerian is 300–600mg daily of the valerian root extract. An alternative is the steeping of 2–3 grams dried valerian root in one cup of hot water for 10–15 minutes. The preparation typically is ingested about 1/2–2 hours prior to bedtime.

Toxicity

The ingestion of an estimated 18.8–23.5g powdered valerian root by an 18-year-old woman caused crampy abdominal pain, chest tightness, lightheadedness, fatigue, fine hand tremor, and mydriasis.[25] All symptoms resolved within 24 hours, and the patient recovered without sequelae.

TOXICOKINETICS

Kinetics

In a study of six healthy adult volunteers receiving a single dose of 600mg valerian extract (Sedonium™, Lichtwer Pharma, Berlin, Germany), valerenic acid was detectable in the serum for at least 5 hours.[26] The peak serum concentration of valerenic acid occurred 1–2 hours after ingestion in five of the six subjects. The other subject had two peak valerenic acid concentrations occurring at 1 hour and 5 hours after ingestion. The serum elimination half-life for valerenic acid was 1.1 ± 0.6 hours for the five subjects with only a single peak serum valerenic acid concentration.

Drug Interactions

Valerian may potentiate the sedative effects of central nervous system depressants, sedative-hypnotic drugs, and anesthetics, but there are few clinical data to determine the clinical importance of these potential interactions. *In vitro* studies with the ATPase assay suggest that aqueous and ethanol extracts of valerian inhibit CYP3A4 activity,[13] but the clinical significance of this effect also remains unclear. In a study of 12 healthy volunteers using single-time-point phenotypic metabolic ratios to determine the effect of 28 days of valerian root extract (125mg 3 times daily, no standardization claims) on cytochrome P450 enzyme activities, no changes in phenotypic ratios of CYP2E1, CYP1A2, CYP2D6 or CYP3A4/5 activity was observed when compared with enzyme activities before supplementation.[27] Based on *in vivo* studies on the effect of valerian extract on the activities of CYP2D6 and CYP3A4, valerian is unlikely to produce clinically significant effects on the disposition of medications dependent on the CYP2D6 or CYP3A4 isoenzymes for biotransformation.[28]

CLINICAL RESPONSE

Adverse Reactions

Although most clinical trials to date are small, adverse reactions to valerian in these clinical studies are uncommon and mild, primarily involving headache and gastrointestinal distress.[29] A randomized, controlled, double-blind trial of 102 volunteers receiving 600mg valerian root extract (LI 156) nightly for 2 weeks did not detect any negative effects on reaction time, alertness, or concentration the morning after intake, when compared with baseline.[30] Adverse reactions associated with the use of valerian reported to the California Poison Control System included nausea, vomiting, drowsiness, agitation, and a probable anaphylactic reaction (urticaria, respiratory distress).[31] Occupational exposure to the dust from valerian causes relatively few hypersensitivity reactions with the exception of nonspecific reactions (conjunctivitis, nasal congestion).[32] Case reports associated liver toxicity with multi-herb preparations containing valerian.[33] However, the contribution of valerian to the development of hepatitis in these cases is not clear because of the potential liver toxicity of other ingredients (skullcap, chaparral) and potential contamination with hepatotoxic herbs.

Acute Effects

A case report about an 18-year-old woman associated the intentional ingestion of up to about 23g valerian root with mydriasis, fine hand tremor, lightheadedness, chest tightness, abdominal cramps, and fatigue.[25] Her vital signs remained normal (i.e., no tachycardia or tachypnea), and she recovered without sequelae within 24 hours. A case series documented the development of central nervous system depression and anticholinergic symptoms (vomiting, mydriasis, tachycardia, urinary retention, visual hallucination, dry mouth, flushing) after the intentional overdose with an herbal preparation containing 75 mg valerian dry extract, 0.25 mg hyoscine hydrobromide, and 2 mg cyproheptadine hydrochloride.[34] The toxic effects can be attributed to the two drugs other than valerian extract. Follow-up studies detected no evidence of hepatotoxicity.

Chronic Effects

Although some valepotriate compounds demonstrate alkylating activity *in vitro*,[35] these compounds degrade rapidly in stored valerian preparations and there is little known risk of genotoxicity associated with valerian use.[18]

DIAGNOSTIC TESTING

Analytical Methods

Methods to detect valerenic acid and derivatives (acetoxyvalerenic acid, hydroxyvalerenic acid) include thin layer chromatography, gas-liquid chromatography, and high performance liquid chromatography (HPLC) with diode array detection. Limits of detection for sesquiterpene compounds in valerian using capillary electrophoresis are in the range of 6 μg/mL.[20] Degradation products of valerian include hydroxyvalerenic acid, pinoresinol, and hydroxypinoresinol. Hydroxyvalerenic acid is a better marker of deterioration of valerian during storage than valerenic acid because acetoxyvalerenic acid hydrolyzes to hydroxyvalerenic acid during unfavorable storage conditions.[36]

Biomarkers

There are substantial differences in the chemical composition of various *Valeriana* species. The presence of valerenic acid, acetoxyvalerenic acid, and valerenal are biomarkers for valerian prepared from *Valeriana officinalis* rather than other *Valeriana* species (*V. edulis*, *V. jatamansi*). Although *V. officinalis* contains the highest concentration of these valerenic acids, some closely related species (e.g., *V. sitchensis* Bong.) contain small amounts of valerenic acids. Analysis of valerenic acids, flavonoids, and valepotriate compounds by liquid chromatography with UV detection provides unique chemical fingerprints that can separate various *Valeriana* species.[37]

Abnormalities

The complete blood count, serum electrolytes, renal function, and serum hepatic aminotransferases remained normal after the intentional ingestion of up to about 23 g valerian root.[25] In a case series of 23 patients with intentional overdoses of an over-the-counter preparations containing valerian root, there was no evidence of acute hepatitis.[34]

TREATMENT

Treatment is supportive. Adverse reactions to valerian are typically mild and respond to conventional treatment. Decontamination measures are unnecessary.

References

1. Upton R. *Valeriana officinalis*. J Altern Complement Med 2001;7:15–117.
2. Bos R, Woerdenbag HJ, van Putten FM, Hendriks H, Scheffer JJ. Seasonal variation of the essential oil, valerenic acid and derivatives, and valepotriates in *Valeriana officinalis* roots and rhizomes, and the selection of plants suitable for phytomedicines. Planta Med 1998;64:143–147.
3. Taibi DM, Landis CA, Petry H, Vitiello MV. A systematic review of valerian as a sleep aid: safe but not effective. Sleep Med Rev 2007;11:209–230.
4. Lindahl O, Lindwall L. Double blind study of a valerian preparation. Pharmacol Biochem Behav 1989;32: 1065–1066.
5. Leathwood PD, Chauffard F. Aqueous extract of valerian reduces latency to fall asleep in man. Planta Med 1985;51: 144–148.
6. Balderer G, Borbely AA. Effect of valerian on human sleep. Psychopharmacology (Berl) 1985;87:406–409.
7. Glass JR, Sproule BA, Herrmann N, Streiner D, Busto UE. Acute pharmacological effects of temazepam, diphenhydramine, and valerian in healthy elderly subjects. J Clin Psychopharmacol 2003;23:260–268.
8. Morin CM, Koetter U, Bastien C, Ware JC, Wooten V. Valerian-hops combination and diphenhydramine for treating insomnia: a randomized placebo-controlled clinical trial. Sleep 2005;28:1465–1471.
9. Ernst E. Herbal remedies for anxiety–a systematic review of controlled clinical trials. Phytomedicine 2006;13: 205–208.
10. Andreatini R, Sartori VA, Seabra ML, Leite JR. Effect of valepotriates (valerian extract) in generalized anxiety disorder: a randomized placebo-controlled pilot study. Phytother Res 2002;16:650–654.
11. Gutierrez S, Ang-Lee MK, Walker DJ, Zacny JP. Assessing subjective and psychomotor effects of the herbal medication valerian in healthy volunteers. Pharmacol Biochem Behavior 2004;78:57–64.
12. Jacobs BP, Bent S, Tice JA, Blackwell T, Cummings SR. An internet-based randomized, placebo-controlled trial of kava and valerian for anxiety and insomnia. Medicine 2005;84:197–207.

13. Lefebvre T, Foster BC, Drouin CE, Krantis A, Arnason JT, Livesey JF, Jordan SA. *In vitro* activity of commercial valerian root extracts against human cytochrome P450 3A4. J Pharm Pharmaceut Sci 2004;7:265–273.
14. Dharmaratne HR, Dhammika Nanayakkara NP, Khan IA. (–)-3β,4β-Epoxyvalerenic acid from *Valeriana officinalis*. Planta Med 2002;68:661–662.
15. Gao XQ, Bjork L. Valerenic acid derivatives and valepotriates among individuals, varieties and species of *Valeriana*. Fitoterapia 2000;71:19–24.
16. Hazelhoff B, Jellema R, Grobben H, Malingre TM. Separation of valtrate and isovaltrate by means of preparative liquid chromatography. Pharm Weekbl Sci 1982;4:21–24.
17. Marder M, Viola H, Wasowski C, Fernandez S, Medina JH, Paladini AC. 6-Methylapigenin and hesperidin: new valeriana flavonoids with activity on the CNS. Pharmacol Biochem Behav 2003;75:537–545.
18. World Health Organization. WHO monographs on selected medicinal plants. Volume I. Geneva: World Health Organization; 1999:267–276.
19. Violon C, Van Cauwenbergh N, Vercruysse A. Valepotriate content in different *in vitro* cultures of Valerianaceae and characterization of *Valeriana officinalis* L. callus during a growth period. Pharm Weekbl Sci 1983;5:205–209.
20. Mikell JR, Ganzera M, Khan IA. Analysis of sesquiterpenes in *Valeriana officinalis* by capillary electrophoresis. Pharmazie 2001;56:946–948.
21. Shohet D, Willis RB, Stuart DL. Valepotriates and valerenic acids in commercial preparations of valerian available in Australia. Pharmazie 2001;56:860–863.
22. Houghton PJ. The biological activity of Valerian and related plants. J Ethnopharmacol 1988;22:121–142.
23. Boyadzhiev L, Kancheva D, Gourdon C, Metcheva D. Extraction of valerenic acids from valerian (*Valeriana officinalis* L.) rhizomes. Pharmazie 2004;59:727–728.
24. Bodesheim U, Holzl J. [Isolation and receptor binding properties of alkaloids and lignans from *Valeriana officinalis* L.] Pharmazie 1997;52:386–391. [German]
25. Willey LB, Mady SP, Cobaugh DJ, Wax PM. Valerian overdose: a case report. Vet Hum Toxicol 1995;37:364–365.
26. Anderson GD, Elmer GW, Kantor ED, Templeton EI, Vitiello MV. Pharmacokinetics of valerenic acid after administration of valerian in healthy subjects. Phytother Res 2005;19:801–803.
27. Gurley BJ, Gardner SF, Hubbard MA, Williams DK, Gentry WB, Khan IA, Shah A. *In vivo* effects of goldenseal, kava kava, black cohosh, and valerian on human cytochrome P450 1A2, 2D6, 2E1, and 3A4/5 phenotypes. Clin Pharmacol Ther 2005;77:415–426.
28. Donovan JL, DeVane CL, Chavin KD, Wang JS, Gibson BB, Gefroh HA, Markowitz JS. Multiple night-time doses of valerian (*Valeriana officinalis*) had minimal effects on CYP3A4 activity and no effect on CYP2D6 activity in healthy volunteers. Drug Metab Dispos 2004;32: 1333–1336.
29. Donath F, Quispe S, Diefenbach K, Maurer A, Fietze I, Roots I. Critical evaluation of the effect of valerian extract on sleep structure and sleep quality. Pharmacopsychiatry 2000;33:47–53.
30. Kuhlmann J, Berger W, Podzuweit H, Schmidt U. The influence of valerian treatment on "reaction time, alertness and concentration" in volunteers. Pharmacopsychiatry 1999;32: 235–241.
31. Dennehy CE, Tsourounis C, Horn AJ. Dietary supplement-related adverse events reported to the California Poison Control System. Am J Health-Syst Pharm 2005;62: 1476–1482.
32. Skorska C, Golec M, Mackiewicz B, Gora A, Dutkiewicz J. Health effects of exposure to herb dust in valerian growing farmers. Ann Agric Environ Med 2005;12: 247–252.
33. MacGregor FB, Abernethy VE, Dahabra S, Cobden I, Hayes PC. Hepatotoxicity of herbal remedies. BMJ 1989;299:1156–1157.
34. Chan TY, Tang CH, Critchley JA: Poisoning due to an over-the-counter hypnotic, Sleep-Qik (hyoscine, cyproheptadine, valerian). Postgrad Med J 1995;71:227–228.
35. von der Hude W, Scheutwinkel-Reich M, Braun R, Dittmar W. *In vitro* mutagenicity of valepotriates. Arch Toxicol 1985;56:267–271.
36. Goppel M, Franz G. Stability control of valerian ground material and extracts: a new HPLC-method for the routine quantification of valerenic acids and lignans. Pharmazie 2004;59:446–452.
37. Navarrete A, Avula B, Choi Y-W, Khan IA. Chemical fingerprinting of *Valeriana* species: simultaneous determination of valerenic acids, flavonoids, and phenylpropanoids using liquid chromatography with ultraviolet detection. J AOAC Int 2006;89:8–15.

Chapter 95

YARROW
(*Achillea millefolium* L.)

HISTORY

Yarrow was one of the flowers discovered in an Iraqi tomb of a Middle Paleolithic Shanidar Neanderthal[1] Dioscorides first recorded the medicinal use of yarrow. Allegedly, the Greek hero, Achilles used yarrow leaves to stop the bleeding from the wounds of his soldiers, resulting in the name of the genus, *Achillea*.[2] The dried leaves and flowering tops were officially recommended in the *United States Pharmacopeia* from 1836–1882 as a tonic and emmenagogue. Yarrow leaves were reportedly applied to battle wounds during the US Civil War.

BOTANICAL DESCRIPTION

There are over 100 species in the genus, *Achillea*. Yarrow (*A. millefolium* L.) is a common species used in herbal therapy.

Common Name: Yarrow, Common Yarrow, Bloodwort, Carpenter's Weed, Hierba de las Cortaduras, Milfoil, Soldier's Woundwort, Herbe Militaris, Staunchweed, Plumajillo, Western Yarrow

Scientific Name: *Achillea millefolium* L.

Botanical Family: Asteraceae (sunflowers, tournesols)

Physical Description: This invasive, aromatic weed reaches 10–20 in. (~25–50 cm) in height with many finely divided, feathery dark-green leaves. The leaves are clustered at the base of angular, flowering stems with smaller leaves alternating upward. Flat-topped panicles consisting of numerous small, white to red flower heads appear from June to September.

Distribution and Ecology: Yarrow is an herbaceous perennial widely distributed throughout the temperate climates of the northern hemisphere and, to a lesser extent, in the southern hemisphere with substantial genetic diversity as a result of hybridization and polyploidization.[3] This plant is native to Europe and western Asia, but yarrow has naturalized to temperate areas worldwide including North America, Australia, and New Zealand. Yarrow is commonly found in sunny, disturbed areas along roadsides, old fields, pastures, and meadows in the eastern and central United States and Canada.

EXPOSURE

Sources

The source of yarrow essential oil is typically the aerial parts of *Achillea millefolium* L. The yield of this essential oil is about 0.1–0.3% with the yield increasing during full bloom.[4]

Medicinal Uses

Traditional

Yarrow has traditionally been ingested as a tea and applied as a lotion or poultice. In traditional European herbal medicine, yarrow is a treatment for hepatobiliary disorders, gastrointestinal illness, urinary tract infec-

Medical Toxicology of Natural Substances, by Donald G. Barceloux, MD

tions, fever, and wounds. Yarrow is also a diuretic, abortifacient, contraceptive, emmenagogue, and uterine stimulant in traditional herbal medicine. North American Indian tribes used yarrow for sprains, contusions, healing wounds, pruritic rashes, pain, and upper respiratory tract infections.[2]

CURRENT

Yarrow is a constituent of experimental herbal preparations for the treatment of cirrhosis[5] and atopic dermatitis,[6] but there is inadequate clinical evidence to conclude that yarrow is an effective treatment for any medical condition. Yarrow is a constituent of some beverages, beers, spices, shampoos, hair-grooming aids, skin fresheners, body lotions, and cosmetics. The essential oil of yarrow is probably not an effective insect repellant.[7]

REGULATORY STATUS

The Council of Europe Committee of Experts classifies yarrow in Group 3 ("safely used in cosmetic products for the purposes stated according to reported use levels").[8]

PRINCIPAL INGREDIENTS

Chemical Composition

The composition of the essential oil from yarrow depends on a variety of factors including season, growing conditions, geographical location or chemotype, plant part, and extraction technique.[9] As the plant matures, the concentrations of α-pinene, β-pinene, and α-thujone increase, while the concentrations of sabinene, borneol, and bornyl acetate decrease.[4] In contrast to steam distillation, solid-phase microextraction of aerial portions of yarrow produces substantial amounts of α-bisabolol, β-bisabolol, and δ-cadinene. The essential oil of yarrow contains over 100 compounds including monoterpenes (camphor, 4-terpineol, α- and β-pinene, α-thujone, sabinene), sesquiterpenes (β-caryophyllene, caryophyllene oxide, [*E*]-nerolidol, germacrene D, α-farnesene), sterols (β-sitosterol, stigmasterol, campesterol),[10] chlorogenic acid and related isomers, dicaffeoylquinic acids (e.g., cynarin),[11] and flavonoids (apigenin, leuteolin-7-glycoside, artemetin, casticin, centaureidin).[12,13] The major compounds in the essential oil of yarrow are camphor, sabinene, 1,8-cineole, α-pinene, and β-pinene.[14,15]

Physiochemical Properties

The estimated tropospheric lifetimes of monoterpene compounds (e.g., *d*-limonene, myrcene, α-pinene, sabinene, γ-terpinene, terpinolene) are relative short (<1–3.5 hours) with related compounds (e.g., 1,8-cinoeole, *p*-cymene) having somewhat longer half-lives.[16] Yarrow extract (CAS RN 84082-83-7) is a clear to brownish-green liquid with a faint herbal odor. This solution is soluble in water with a density ranging from 1.030–1.050 g/mL. The occurrence of bluish coloration to yarrow extract indicates the presence of azulene and related compounds (e.g., chamazulene).

Mechanism of Toxicity

Most adverse reactions to yarrow result from hypersensitivity responses.

DOSE RESPONSE

Medicinal Use

A typical herbal dose of yarrow is 2 to 4 grams dried herb in capsules or a tea (1–2 teaspoons dried herb in boiling water, steeped for 3–5 minutes) up to 3 times daily. However, there are no clinical studies to determine an efficacious dose.

Toxicity

A study of adult male Wistar rats receiving up to 1.2 g/kg body weight of an aqueous crude extract of yarrow leaves daily for 90 days did not detect any systemic or reproductive toxicity.[17]

TOXICOKINETICS

There are few data on the absorption or metabolism of constituents of yarrow following the ingestion of extracts from this herb.

CLINICAL RESPONSE

The most common complications associated with exposure to yarrow are hypersensitivity reactions. These adverse effects include allergic contact dermatitis, contact urticaria, rhinoconjunctivitis, and bronchial asthma.[18–20] There is substantial cross-reactivity between members of the Asteraceae family.[21] In a study of 190 Asteraceae-allergic patients determined by positive skin tests to an Asteraceae mix, feverfew elicited the most (81%) positive responses followed by tansy (77%), chamomile (64%), and yarrow (41%).[22] However, hypersensitivity reactions to yarrow are relatively uncommon, even in patients with multiple contact reactions, photosensitivity, or atopic dermatitis.[23] Rare case reports also associate sore throat, stomach aches, facial edema, and pruritus with the ingestion of yarrow tea.[24] Rodent studies suggest that high doses of yarrow (>50 times typical herbal doses) can produce fetal toxicity without maternal toxicity,[25] but the clinical significance of extrapolating this fetal toxicity in animals to humans remains unclear. This study detected no increase in fetal demise following the ingestion of yarrow before or after implantation suggesting that yarrow was neither an abortifacient nor a contraceptive; there was no increase in the rate of malformations when the treatment group was compared with controls.

DIAGNOSTIC TESTING

Analytical methods for the separation and identification of the constituents of the essential oil of yarrow include gas chromatography,[14] high performance liquid chromatography (HPLC) coupled with diode-array and fluorescence detectors,[26] and gas chromatography/mass spectrometry.[4,27] Methods for the detection and quantitation of flavonoids and phenolic compounds (e.g., dicaffeoylquinic acids) in yarrow include solid-phase extraction coupled with HPLC and UV detection.[28]

TREATMENT

Treatment is supportive. Most adverse reactions are hypersensitivity responses, which respond to cessation of use and antihistamines and/or steroids. Gut decontamination measures are unnecessary.

References

1. Lietava J. Medicinal plants in a Middle Paleolithic grave Shanidar IV? J Ethnopharmacol 1992;35:263–266.
2. Chandler RF, Hooper SN, Harvey MJ. Ethnobotany and phytochemistry of yarrow, *Achillea millefolium*, Compositae. Econ Bot 1982;36:203–223.
3. Guo Y-P, Saukel J, Mittermayr R, Ehrendorfer F. AFLP analyses demonstrate genetic divergence, hybridization, and multiple polyploidization in the evolution of *Achillea* (Asteraceae-Anthemideae). New Phytologist 2005;166: 273–290.
4. Rohloff J, Skagen EB, Steen AH, Iversen T-H. Production of yarrow (*Achillea millefolium* L.) in Norway: essential oil content and quality. J Agric Food Chem 2000;48: 6205–6209.
5. Huseini HF, Alavian SM, Heshmat R, Heydari MR, Abolmaali K. The efficacy of Liv-52 on liver cirrhotic patients: a randomized, double-blind, placebo-controlled first approach. Phytomedicine 2005;12:619–624.
6. Shapira MY, Raphaelovich Y, Gilad L, Or R, Dumb AJ, Ingber A. Treatment of atopic dermatitis with herbal combination of *Eleutherococcus*, *Achillea millefolium*, and *Lamium album* has no advantage over placebo: a double blind, placebo-controlled, randomized trial. J Am Acad Dermatol 2005;52:691–693.
7. Thorsell W, Mikiver A, Tunon H. Repelling properties of some plant materials on the tick *Ixodes ricinus* L. Phytomedicine 2006;13:132–134.
8. Cosmetic Ingredient Review Expert Panel. Final report on the safety assessment of yarrow (*Achillea millefolium*) extract. Int J Toxicol 2001;20(Suppl 2):79–84.
9. Agnihotri VK, Lattoo SK, Thappa RK, Kaul P, Qazi GN, Dhar AK, et al. Chemical variability in the essential oil components of *Achillea millefolium* Agg. from different Himalayan habitats (India). Planta Med 2005-71: 280–283.
10. Chandler RF, Hooper SN, Hooper DL, Jamiseson WD, Flinn CG, Safe LM. Herbal remedies of the Maritime Indians: sterols and triterpenes of *Achillea millefolium* L. (yarrow). J Pharmaceut Sci 1982;71:690–693.
11. Benedek B, Geisz N, Jager W, Thahammer T, Kopp B. Choleretic effects of yarrow (*Achillea millefolium* s.l.) in isolated perfused rat liver. Phytomedicine 2006;13:702–706.
12. Glasl S, Mucaji P, Werner I, Presser A, Jurenitsch J. Sesquiterpenes and flavonoid aglycones from a Hungarian taxon of the *Achillea millefolium* group. Z Naturforsch [C] 2002;57:976–982.
13. Falk AJ, Smolenski SJ, Bauer L, Bell CL. Isolation and identification of three new flavones from *Achillea millefolium* L. J Pharmaceut Sci 1975;64:1838–1842.
14. Bejnarowicz EA, Smolenski SJ. Gas chromatographic analysis of the essential oil from *Achillea millefolium* L. J Pharmaceut Sci 1968;57:2160–2161.
15. Falk AJ, Bauer L, Bell CL. The constituents of the essential oil from *Achillea millefolium* L. Lloydia 1974;37: 598–602.
16. Corchnoy SB, Atkinson R. Kinetics of the gas-phase reactions of OH and NO_3 radicals with 2-carene, 1,8-cineole, *p*-cymene, and terpinolene. Environ Sci Technol 1990;24:1497–1502.
17. Dalsenter PR, Cavalcanti AM, Andrade AJ, Araujo SL, Marques MC. Reproductive evaluation of aqueous crude extract of *Achillea millefolium* L. (Asteraceae) in Wistar rats. Reprod Toxicol 2004;18:819–823.
18. Guin JD, Skidmore G. Compositae dermatitis in childhood. Arch Dermatol 1987;123:500–502.
19. Uter W, Nohle M, Randerath B, Schwanitz HJ. Occupational contact urticaria and late-phase bronchial asthma caused by Compositae pollen in a florist. Am J Contact Dermatitis 2001;12:182–184.
20. Davies MG, Kersey PJ. Contact allergy to yarrow and dandelion. Contact Dermatitis 1986;14:256–257.
21. Hausen BM. A 6-year experience with Compositae mix. Am J Contact Dermat 1996;7:94–99.
22. Paulsen E, Andersen KE, Hausen BM. Sensitization and cross-reaction patterns in Danish Compositae-allergic patients. Contact Dermatitis 2001;45:197–204.

23. Wrangsjo K, Ros AM, Wahlberg JE. Contact allergy to Compositae plants in patients with summer-exacerbated dermatitis. Contact Dermatitis 1990;22: 148–154.
24. Hausen BM, Schulz KH. [Polyvalent contact allergy in a florist]. Derm Beruf Umwelt 1978;26:175–176. [German]
25. Boswell-Ruys CL, Ritchie HE, Brown-Woodman PD. Preliminary screening study of reproductive outcomes after exposure to yarrow in the pregnant rat. Birth Defects Res B Dev Reprod Toxicol 2003;68:416–420.
26. Schulz H, Albroscheit G. High-performance liquid chromatographic characterization of some medicinal plant extracts used in cosmetic formulation. J Chromatogr 1988;442:353–361.
27. Dokhani S, Cottrell T, Kjajeddin J, Mazza G. Analysis of aroma and phenolic components of selected *Achillea* species. Plant Foods Hum Nutr 2005;60:55–62.
28. Benedek B, Gjoncaj N, Saukel J, Kopp B. Distribution of phenolic compounds in Middle European taxa of the *Achillea millefolium* L. aggregate. Chem Biodivers 2007;4: 849–857.

Chapter 96

YOHIMBE BARK and YOHIMBINE (*Pausinystalia yohimbe* Pierre ex Beille)

HISTORY

Yohimbe is a native plant of western Africa, where Pygmies and Bushmen of the region traditionally used this plant as an aphrodisiac, antipyretic, and stimulant. German missionaries brought yohimbe bark from Africa to Europe during the late 19th century, where the plant became popular as the "love tree." Pharmaceutical manufacturers isolated yohimbine from the bark, and this drug has been available by prescription before the US Food and Drug Administration (FDA) began approving drugs in 1938. The popularity of this drug as an aphrodisiac has decreased since the introduction of Viagra® (Pfizer Pharmaceuticals, New York, NY) and similar drugs in the 1990s.

BOTANICAL DESCRIPTION

The main ingredient (yohimbine) of yohimbe bark is also present in other *Pausinystalia* species (e.g., *P. microceras* (K. Schum.) Pierre ex Beille, *P. trillesii* Beille). Although yohimbine and yohimbe are often used interchangeably, the term yohimbe refers to the yohimbe tree and yohimbine is a specific chemical compound.

Common Name: Yohimbe Tree, Corynine, Aphrodine, Quebrachine, Hydroaerogotocin

Scientific Plant Name: *Pausinystalia yohimbe* Pierre ex Beille (*Corynanthe yohimbi*)

Botanical Family: Rubiaceae (madders)

Physical Description: The yohimbe tree is a tall evergreen tree with gray-brown bark and oblong and elliptical leaves. This tree reaches up to 30m (~100ft) in height and 8m (~25ft) wide with erect, branching stems.

Distribution and Ecology: This tree (*Pausinystalia yohimbe*) is a native plant of western Africa including Nigeria, Cameroon, Congo Gaboon, and Nigeria.

EXPOSURE

Sources

Yohimbe bark (CAS RN: 85117-22-2, corynanthe yohimbe extract) consists of the dried bark of the trunk and/or branches of the yohimbe tree. Yohimbe bark extracts, which contain variable amounts of yohimbine, are available in health food stores and through direct-mail companies as capsules, tablets, and liquids. Yohimbine is also found in *Rauwolfia* root. Yohimbine hydrochloride is available as an FDA-approved prescription drug [Yocon® (Glenwood LLC, Englewood, NJ), Yohimex® (Kramer Laboratories, Miami, FL)] in 2mg and 4.5mg tablets.

Medicinal Uses

TRADITIONAL

Yohimbe bark is a traditional treatment for a variety of disorders in addition to the treatment of sexual dysfunction. These uses include the treatment of dementia, diabetic complications (e.g., neuropathy), exhaustion,

Medical Toxicology of Natural Substances, by Donald G. Barceloux, MD

fevers, insomnia, leprosy, low blood pressure, obesity, and syncope.

CURRENT

Health food stores have promoted the use of yohimbe bark extracts as an aphrodisiac for many years, but recently the use of this extract to enhance athletic performance and weight loss has become more popular. In veterinary medicine, yohimbe bark extract is a treatment for impotence in breeding stallions. There are inadequate data to support the use of yohimbe or yohimbine as a dieting agent,[1] and the results of the few clinical studies to date are conflicting.[2]

The main medical indication for yohimbine hydrochloride is the treatment of male impotence. Other suggested medical uses for yohimbine include the treatment of orthostatic hypotension secondary to autonomic failure,[3] dry mouth during tricyclic antidepressant therapy,[4] and the induction of lipid mobilization in obese subjects.[5] In a pilot study of 12 methadone-maintained, opiate-addicted patients, the daily administration of 30 mg yohimbine reduced the symptoms scores of naloxone-precipitated opiate withdrawal an average of 30%.[6] However, vital signs and objective measures of withdrawal were unaffected by yohimbine treatment. Yohimbine is a potential drug of abuse ("yo-yo") because of its alleged aphrodisiac properties.[7]

REGULATORY STATUS

The FDA classifies yohimbe as an unsafe herb, containing "the toxic alkaloid, yohimbine, and other alkaloids." In Germany, the expert panel that evaluates herbal medicines (Commission E) did not recommend the therapeutic use of yohimbe bark and its preparations because of insufficient proof of efficacy.

PRINCIPAL INGREDIENTS

Chemical Composition

The yohimbine concentration in commercial yohimbe products is highly variable. Chemical analysis of 26 commercial yohimbe products indicated that nine products contained no yohimbine and eight products contained only trace amounts (0.1–1 ppm) of yohimbine.[8] The concentrations of yohimbine ranged from <0.1 to 489 ppm compared with 7,089 ppm in the authentic yohimbe bark material. Analyses of authentic yohimbe bark indicate that the average total indole alkaloid content is about 3–6% with approximately 10–15% of the alkaloids being yohimbine. In addition to yohimbine and its isomers (α-yohimbine, β-yohimbine, allo-yohimbine), these alkaloids include ajmalicine, dihydroyohimbine, corynantheine, dihydrocorynantheine, and corynanthine (rauhimbin).

Physicochemical Properties

Yohimbine (CAS RN: 146-48-5, $C_{21}H_{26}N_2O_3$) is an indolealkylamine alkaloid extracted from the bark of the yohimbe tree. Figure 96.1 displays the chemical structure of yohimbine. Yohimbine is a selective α_2-adrenergic receptor antagonist with weak α_1-antagonist activity based on pharmacological studies and radioreceptor ligand binding assays.[9] Yohimbine potentiates the release of norepinephrine from noradrenergic nerve terminals. Additionally, yohimbine reduces the ability of norepinephrine to decrease cAMP and intracellular calcium at postsynaptic noradrenergic receptors, whereas the stimulation of postsynaptic noradrenergic receptors mediated by other adrenoceptor subtypes is unaffected.[10]

In animal studies, yohimbine is a pharmacological stressor, which is commonly used to produce anxiety-like effects.[11] The administration of yohimbine to humans causes elevated plasma catechol and cortisol concentrations, nausea, increased subjective feelings of anxiety, increased startle responses, and panic attacks.[12] Yohimbine is highly lipophilic and easily crosses the blood–brain barrier. Yohimbine is sparingly soluble in water, but soluble in alcohol and chloroform. Yohimbe bark extract is odorless with a bitter taste. Table 96.1 lists some identifying characteristics and physical properties of yohimbine.

Mechanism of Toxicity

Yohimbine is a selective α_2-adrenergic antagonist that produces tachycardia, mydriasis, elevated blood pressure, salivation, diaphoresis, nausea, vomiting, and facial flushing after intravenous administration.[13] The intravenous administration of yohimbine promotes unimpeded norepinephrine flow to α_1 and β receptors in the post-synaptic effector cells, resulting in a rise in blood

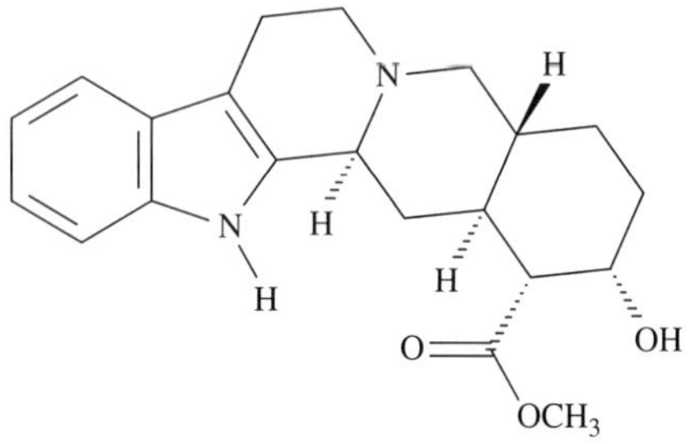

FIGURE 96.1. Chemical structure of yohimbine.

TABLE 96.1. Identifying Characteristics and Physical Properties of Yohimbine

Property	Value
CAS Registry Number	146-48-5; 65-19-0 (Hydrochloride salt)
CAS Names	Yohimban-16-carboxylic acid, 17-hydroxy-, methyl ester, (16α, 17α)
Synonyms	Aphrodyne; corynine; quebrachine
Melting Point	241 °C/~466 °F
log P (Octanol-Water)	2.73
Water Solubility	277 mg/L (25 °C/77 °F)
Vapor Pressure	1.61E-12 mmHg (25 °C/77 °F)
Henry's Law Constant	2.90E-17 atm-m^3/mole (25 °C/77 °F)
Atmospheric OH Rate Constant	2.74E-10 cm^3/molecule-second (25 °C/77 °F)

pressure and an increase in plasma norepinephrine. Yohimbine may also exert a partial block at α_1 receptor sites, dopamine receptor sites, and serotonin receptor sites.[9,14]

DOSE RESPONSE

In clinical studies on the treatment of sexual dysfunction, typical daily doses of yohimbine are 15–30 mg divided into three doses with patients tolerating daily doses up to 100 mg.[15] Side effects in these studies at doses of 20–45 mg include increased heart rate and blood pressure, anxiety and urinary frequency. A study of 20 commercial yohimbe bark preparations sold over the Internet as aphrodisiacs indicated that the maximum daily dose of yohimbine in these products ranged from about 1–23 mg.[16] Case reports indicate that the ingestion of approximately 200–250 mg yohimbine produces transient tachycardia, hypertension, tremulousness, diaphoresis, and anxiety, which resolve spontaneously without serious complications.[17,18]

There are few data on the dose of yohimbe bark when used as an herbal medicine. One preparation of a tea involves the simmering of 25–50 mL (5–10 teaspoons) of yohimbe bark in 360 mL (1 pint) water with the addition of up to 500–1,000 mg vitamin C to improve solubility as needed.

TOXICOKINETICS

The absorption and elimination of oral doses of yohimbine are rapid. Following the ingestion of therapeutic doses (e.g., 8–10 mg) of yohimbine, peak plasma concentrations occur within 45–60 minutes.[19] Oral bioavailability of pharmaceutical preparations of yohimbine is relatively low with substantial variability. In a study of seven healthy volunteers, the mean bioavailability of a 10-mg dose of yohimbine was 33% with a range of 7–87%.[20] The estimation of the volume of distribution (V_d) is complex as a result of the interaction of high protein binding, rapid metabolism, and high lipid solubility. Following therapeutic doses of yohimbine, the mean protein binding of yohimbine, 11-OH-yohimbine, and 10-OH-yohimbine was 82%, 43%, and 32%, respectively.[21] Elimination of yohimbine occurs primarily by hepatic metabolism with very small amounts (<1%) of yohimbine excreted unchanged by the kidneys.[15] The major hepatic metabolite in plasma is the active metabolite, 11-OH-yohimbine. Although the α_2-adrenoceptor binding of 11-OH-yohimbine is less than yohimbine, the α_2-adrenergic antagonist activities of these two compounds are similar because of the relatively lower protein binding of 11-OH-yohimbine.[21] The minor, inactive metabolite, 10-OH-yohimbine, is usually detectable only in urine samples. Following therapeutic doses of yohimbine, elimination of yohimbine is relatively rapid with a half-life ($t_{1/2}$) of about 1.5 hours compared with approximately 6 hours for the active 11-OH-yohimbine metabolite.[22] In a pharmacokinetic study of 16 healthy male participants receiving yohimbine tartrate and L-arginine glutamate in a single pharmaceutical preparation, the plasma elimination half-life of yohimbine was 1.0±0.34 hours with a range of 0.40–6.0 hours.[23] Consequently, the duration of the pharmacological effects of yohimbine may result from the prolonged half-life of 11-OH-yohimbine. The biotransformation of yohimbine in some subjects (i.e., slow metabolizers) with debrisoquine oxidative polymorphism is limited compared with other subjects.[22]

CLINICAL RESPONSE

Adverse effects associated with the use of yohimbe as a medicinal herb include anxiety, urinary frequency, agitation, headache, and elevated blood pressure.[15,24] A case report associated the development of severe hypertension (blood pressure 240/140 mmHg, headache, weakness) with the use of yohimbe as an herbal treatment for impotence.[25] The absorption of yohimbine is relatively rapid with symptoms of intoxication beginning within 1–2 hours. Clinical features of yohimbine intoxication include nausea, vomiting, anxiety, tremulousness, palpitations, diaphoresis, transient hypertension, and mild tachycardia. A 16-year-old girl developed a severe headache, weakness, paresthesias, poor coordination, tremors, and a dissociative state about 20–30 minutes after ingesting an estimated yohimbine dose of 250 mg.[18] Within 4 hours, she developed the sensation of severe pressure-like substernal chest pain that resolved after 2 hours. The electrocardiogram revealed a sinus tachycardia, but no evidence of ischemia or dysrhythmias. Other clinical features that resolved over the next 36 hours included palpitations, tachypnea, erythema, nausea, diaphoresis, tremulousness, anxiety, mild hypertension, and tachypnea. The ingestion of an estimated yohimbine dose of 200 mg with ethanol by a 62-year-old man produced vomiting, anxiety, lightheadedness, lower extremity paresthesias, mild hypertension, and tachycardia that resolved within 24 hours without complication.[26] Yohimbine intoxication may produce an acute dissociative reaction with anxiety, paresthesias, weakness, and incoordination.

DIAGNOSTIC TESTING

Methods for the quantitation of yohimbine include gas chromatography/nitrogen phosphorus detection[8] and high performance liquid chromatography (HPLC),[27] and HPLC with ultraviolet detection/mass spectrometry with atmospheric pressure chemical ionization and electrospray ionization.[16] During yohimbine intoxication, most laboratory tests (e.g., liver/kidney function, electrolytes) are normal, though serum catecholamine concentrations (norepinephrine, epinephrine) may be slightly elevated during the acute phase of intoxication.[18] The distribution of yohimbine between serum and red blood cells is uneven with only 20% of the yohimbine in whole blood residing in the erythrocytes.[19]

TREATMENT

Management of yohimbine intoxication is supportive. Decontamination measures (i.e., activated charcoal) are usually unnecessary, but the use of activated charcoal within one hour of ingestion of excessive doses of yohimbine is a therapeutic option. However, there are no clinical data to document the efficacy of decontamination measures in this setting. Benzodiazepines (diazepam, lorazepam) are the drugs of choice to alleviate anxiety and the initial sympathomimetic symptoms associated with yohimbine intoxication.[28]

There are few data on the efficacy of specific drugs to treat yohimbine intoxication. Clonidine is a *potential* antagonist of the effects of yohimbine. Clonidine acts centrally by stimulating alpha adrenergic receptors, causing a reduction in sympathetic outflow, and peripherally by agonist effects on α-adrenergic receptors, causing a reduction in vascular activity. However, clonidine possesses weak α_1-agonist activity (i.e., peripheral vasoconstriction) that may exacerbate hypertension following the administration of low clonidine doses during yohimbine intoxication. Consequently, sodium nitroprusside is safer than clonidine for the treatment of hypertensive emergencies associated with yohimbine toxicity.

References

1. Pittler MH, Ernst E. Complementary therapies for reducing body weight: a systematic review. Int J Obesity 2005;29:1030–1038.
2. Kucio C, Jonderko K, Piskorska D. Does yohimbine act as a slimming drug? Isr J Med Sci 1991;27:550–556.
3. Onrot J, Goldberg MR, Biaggioni I, Wiley RG, Hollister AS, Robertson D. Oral yohimbine in human autonomic failure. Neurol 1987;37:215–220.
4. Bagheri H, Picault P, Schmitt L, Houin G, Berlan M, Montastruc JL. Pharmacokinetic study of yohimbine and its pharmacodynamic effects on salivary secretion in patients treated with tricyclic antidepressants. Br J Clin Pharmacol 1994;37:93–96.
5. Berlan M, Galitzky J, Riiere D, Foureau M, Tran MA, Flores R, et al. Plasma catecholamine levels and lipid mobilization induced by yohimbine in obese and non-obese women. Int J Obesity 1991;15:305–315.
6. Hameedi FA, Woods SW, Rosen MI, Pearsall HR, Kosten TR. Dose dependent effects of yohimbine on methadone maintained patients. Am J Drug Alcohol Abuse 1997;23: 37–333.
7. Siegel RK. New trends in drug use among youth in California. Bull Narc 1985;37:7–17.
8. Betz JM, White KD, der Marderosian AH. Gas chromatographic determination of yohimbine in commercial yohimbe products. J AOAC Int 1995;78:1189–1194.
9. Goldberg MR, Robertson D. Yohimbe: A pharmacological probe for study of the alpha-$_2$-adrenoreceptor. Pharmacol Rev 1983;35:143–180.
10. Myers EA, Banihashemi L, Rinaman L. The anxiogenic drug yohimbine activates central viscerosensory circuits in rats. J Compar Neurol 2005;492:426–441.
11. Ghitza UE, Gray SM, Epstein DH, Rice KC, Shaham Y. The anxiogenic drug yohimbine reinstates palatable food seeking in a rat relapse model: a role of CRF1 receptors. Neuropsychopharmacology 2006;31:2188–2196.
12. Stine SM, Southwick SM, Petrakis IL, Kosten TR, Charney DS, Krystal JH. Yohimbine-induced withdrawal and anxiety symptoms in opioid-dependent patients. Biol Psychiatry 2002;51:642–651.
13. Holmberg G, Gershon S. Autonomic and psychic effects of yohimbine hydrochloride. Psychopharmacologia 1961;2: 93–106.
14. Goldberg MR, Hollister AS, Robertson D. Influence of yohimbe on blood pressure, autonomic reflexes and plasma catecholamines in human hypertension. Hypertension 1983;5:772–778.
15. Tam SW, Worcel M, Wyllie M. Yohimbine: a clinical review. Pharmacol Ther 2001;92:215–243.

16. Zanolari B, Ndjoko K, Ioset J-R, Marston A, Hostettmann KL. Qualitative and quantitative determination of yohimbine in authentic yohimbe bark and in commercial aphrodisiacs by HPLC-UV-API/MS methods. Phytochem Anal 2003;14:193–201.
17. Friesen K, Palatnick W, Tenebein M. Benign course after massive ingestion of yohimbine. J Emerg Med 1993;11: 287–288.
18. Linden CH, Vellman WP, Rumack B. Yohimbine: A new street drug. Ann Emerg Med 1985;14:1002–1004.
19. Owen JA, Nakatsu SL, Fenemore J, Condra M, Surridge DH, Morales A. The pharmacokinetics of yohimbine in man. Eur J Clin Pharmacol 1987;32:577–582.
20. Guthrie SK, Hariharan M, Grunhaus LJ. Yohimbine bioavailability in humans. Eur J Clin Pharmacol 1990;39: 409–411.
21. Berlan M, Le Verge R, Galitzky J, Le Corre P. Alpha$_2$-adrenoceptor antagonist potencies of two hydroxylated metabolites of yohimbine. Br J Pharmacol 1993;108: 927–932.
22. Le Corre P, Dollo G, Chevanne F, Le Verge R. Biopharmaceutics and metabolism of yohimbine in humans. Eur J Pharm Sci 1999;9:79–84.
23. Kernohan AF, McIntyre M, Hughes DM, Tam SW, Worcel M, Reid JL. An oral yohimbine/L-arginine combination (NMI 861) for the treatment of male erectile dysfunction: a pharmacokinetic, pharmacodynamic and interaction study with intravenous nitroglycerine in healthy male subjects. Br J Clin Pharmacol 2004;59: 85–93.
24. Pittler MH, Schmidt K, Ernst E. Adverse events of herbal food supplements for body weight reduction: systematic review. Obesity Rev 2005;6:93–111.
25. Ruck B, Shih RD, Marcus SM. Hypertensive crisis from herbal treatment of impotence. Am J Emerg Med 1999;17: 317–318.
26. Friesen K, Palatnick W, Tenebein M. Benign course after massive ingestion of yohimbine. J Emerg Med 1993;11: 287–288.
27. Mittal S, Alexander KS, Dollimore D. A high-performance liquid chromatography assay for yohimbine HCl analysis. Drug Dev Ind Pharm 2000;26:1059–1065.
28. Ingram CG. Some pharmacologic actions of yohimbine and chlorpromazine in man. Clin Pharmacol Ther 1962;3: 345–352.

II Essential Oils

Chapter 97

CITRONELLA OIL
[*Cymbopogon nardus* (L.) Rendle]

HISTORY

At the end of the 19th century, citronella was first used as an insecticide. At the beginning of the 20th century, the isolation of the aldehyde, hydroxycitronellal from citronella oil helped to launch the perfume industry.

BOTANICAL DESCRIPTION

Common Name: Citronella Grass, Ceylon Citronella, Nardus, Nard Grass, Nardus Grass, Mana Grass, Geranium Grass

Scientific Name: *Cymbopogon nardus* (L.) Rendle

Botanical Family: Poaceae (grass family)

Physical Description: Citronella grass is a perennial, coarse, clump-forming tropical grass that can reach 5–6 ft (~1–2 m) in height. The surface of the leafblade is smooth or scaberulous (small raised areas of roughness) and gray-green. The leaves droop for one-third of the length and are aromatic when crushed. Flowers are borne in terminal and axillary racemes.

Distribution and Ecology: Citronella grass is native to southeastern Asia, and this grass is grown commercially in Sri Lanka, India, Burma, Indonesia, and Java. This plant is an ornamental grass in southern Florida and California as well as a naturalized grass in tropical Asia.

EXPOSURE

Sources

Citronella oil is a steam distillation product of the fresh or partially dried leaves from citronella grass (*Cymbopogon nardus*) or Java citronella grass (*Cymbopogon winterianus* Jowitt). The latter is called Java-type citronella oil, whereas the former is called Ceylon-type citronella oil. The yield of essential oil from the leaves is approximately 0.4–1.4% volume/weight.

Medicinal Uses

TRADITIONAL

Traditional herbal treatments with citronella oil include use as a rubefacient (redness of skin), antispasmodic, and diaphoretic (increase sweat). Animal studies suggest that citronella oil produces some sedation and analgesic effects.[1]

CURRENT

Citronella oil and citronella candles are traditional mosquito repellants and household fumigants.[2,3] Citronella oil is also a strong food flavoring agent, which occurs in low concentrations in alcoholic and nonalcoholic beverages, baked goods, breakfast cereals, soft and hard

Medical Toxicology of Natural Substances, by Donald G. Barceloux, MD

candies, frozen dairy products, gelatins, and puddings. Additionally, this oil is a component of deterrents for dogs seeking to mate.

Regulatory Status

In the United States, citronella oil is listed as generally recognized as safe (GRAS), and the Flavor and Extract Manufacturers' Association (FEMA) approved GRAS status for citronella oil. The Council of Europe placed the Java type citronella oil in Category 1 (items are not considered a safety concern at levels consumed) and the Ceylon-type citronella oil in Category 3 (items with a safety concern associated with the active ingredient) based on the presence of isomethyleugenol.

PRINCIPAL INGREDIENTS

Chemical Composition

Both Ceylon- and Java-type citronella oils contain over 80 constituents, primarily monoterpene compounds and minor amounts of sesquiterpenes [(E)-nerolidol, β-caryophyllene, germacren-4-ol].[4] The major constituents vary in these two types of citronella oil depending on geographic location, climate, and strain. Citronellal (CAS RN: 106-23-0, $C_{10}H_{18}O$) and citronellol (RN: 106-22-9, $C_{10}H_{20}O$) are the main compounds in Ceylon-type citronella oil, accounting for about 30–45% of the total constituents depending on the source.[5] Figure 97.1 shows the chemical structure of citronellal. Geraniol is the other major constituent (i.e., about 25–35%) with minor amounts of geranyl acetate, citronellyl acetate, and limonene. Ceylon-type of citronella oil contains lower concentrations of citronellal and citronellol and higher concentrations of isomethyleugenol compared with Java-type citronella oil.

Physiochemical Properties

Ceylon-type of citronella oil is a pale-yellow to yellowish-brown liquid that has a characteristic citronellal-like smell, whereas Java-type is a clear, light-yellow to brownish liquid with a strong citrus smell. Table 97.1 lists some physical properties of citronellal.

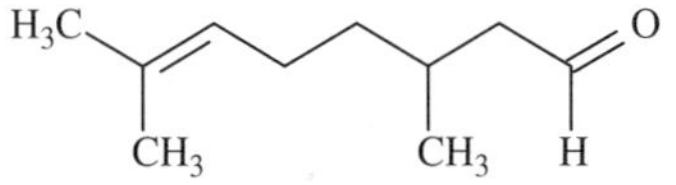

FIGURE 97.1. Chemical structure of citronellal.

Mechanism of Toxicity

The main constituents of citronella oil are citronellol, citronellal, and geraniol, but the specific toxic constituents have not been identified; therefore, the mechanism of action is unknown.

DOSE RESPONSE

The ingestion of citronella causes relatively low toxicity in the absence of pulmonary aspiration. The ingestion of an estimated 25 mL of citronella oil by a 16-month-old child was not associated with any systemic toxicity.[6]

TOXICOKINETICS

There are few data on the toxicokinetics of the major components of citronella oil.

CLINICAL RESPONSE

Most accidental ingestions of citronella oil do not produce symptoms, provided pulmonary aspiration does not occur.[6] The older medical literature associates the ingestion of citronella oil with respiratory distress, hypotension, and death.[7] However, the use of unconventional treatment, the possible role of pulmonary aspiration, and the lack of adequate documentation of medical events prevents conclusions about the causal role of citronella oil in the development of these serious effects. Contact dermatitis occurs following exposure to cosmetics containing the closely related compound, hydroxycitronellal (CAS RN: 107-75-5).[8,9] Long-term carcinogenesis studies in rodents with doses up to 2,000 mg/kg food-grade geranyl acetate containing approximately 29% citronellyl acetate did not detect evidence of carcinogenicity.[10]

DIAGNOSTIC TESTING

Methods to analyze the content of monoterpene compounds in citronella oil include gas chromatography and gas chromatography/mass spectrometry.[5]

TABLE 97.1. Some Physical Properties of Citronellal

Physical Property	Value
log P (Octanol-Water)	3.530
Water Solubility	70.2 mg/L (25 °C/77 °F)
Vapor Pressure	0.280 mmHg (25 °C/77 °F)
Henry's Law Constant	2.62E-04 atm-m³/mole (25 °C/77 °F)
Atmospheric OH Rate Constant	1.21E-10 cm³/molecule-second (25 °C/77 °F)

TREATMENT

Treatment is supportive. Accidental ingestions of citronella oil do not require decontamination. Patients should be evaluated carefully for the presence of pulmonary aspiration (cough, dyspnea, pulmonary infiltrates, reduced oxygen saturation).

References

1. Kokate CK, Rao RE, Varma KC. Pharmacological investigations of essential oil of *Cymbopogon nardus* (L.) Rendle: studies on central nervous system. Indian J Exp Biol 1971;9:515–516.
2. Lindsay LR, Surgeoner GA, Heal JD, Gallivan GJ. Evaluation of the efficacy of 3% citronella candles and 5% citronella incense for protection against field populations of *Aedes* mosquitoes. J Am Mosq Control Assoc 1996;12: 293–294.
3. Jantan I, Zaki AM, Ahmad AR, Ahmad R. Evaluation of smoke from mosquito coils containing Malaysian plants against *Aedes aegypti*. Fitoterapia 1999;70:237–243.
4. Mahalwal VS, Ali M. Volatile constituents of *Cymbopogon nardus* (Linn.) Rendle. Flavour Fragrance J 2003;18: 73–76.
5. Nakahara K, Alzoreky NS, Yoshihashi T, Nguyen HT, Trakoontivakorn G. Chemical composition and antifungal activity of essential oil from *Cymbopogon nardus* (citronella grass). Jpn Agric Res Q 2003;37: 249–252.
6. Temple WA, Smith NA, Beasley M. Management of oil of citronella poisoning. Clin Toxicol 1991;29:257–262.
7. Mant AK. Association proceeding VI. A case of poisoning by oil of citronella. Med Sci Law 1961;1:170–171.
8. Calnan CD. Unusual hydroxycitronellal perfume dermatitis. Contact Dermatitis 1979;5:123.
9. Heydorn S, Andersen KE, Johansen JD, Menne T. A stronger patch test elicitation reaction to the allergen hydroxycitronellal plus the irritant sodium lauryl sulfate. Contact Dermatitis 2003;49:133–139.
10. National Toxicology Program. NTP carcinogenesis studies of food grade geranyl acetate (71% geranyl acetate, 29% citronellyl acetate) (CAS No. 105-87-3) in F344/N rats and B6C3F1 mice (gavage study). Natl Toxicol Prog Tech Rep Ser 1987;252:1–162.

Chapter 98

CITRUS OIL and LIMONENE

HISTORY

Both the *d*- and *l*-isomers of limonene and the racemic mixture (dipentene) occur naturally. In Florida in the early 1940s, *d*-limonene was discovered as a by-product of citrus molasses production.[1] Since the 1980s, *d*-limonene has been an alternative to chlorinated hydrocarbon solvents and degreasing agents. In 2004, the US Food and Drug Administration (FDA) banned the use of ephedra products, and synephrine-containing herbal products of bitter orange have been promoted as alternative dietary supplements for energy enhancement and weight loss.

BOTANICAL DESCRIPTION

Bergamot is a variety of bitter orange cultivated on a variety of rootstocks.

Common Name: Bergamot Orange

Scientific Name: *Citrus* X *aurantium* L. subsp. *bergamia* (Risso & Poit.) Wight & Arn. ex Engler

Botanical Family: Rutaceae (rues, rutacées)

Physical Description: The bergamot orange is a small pear-shaped fruit that is a cross-product of the sour orange and the pear lemon.

Distribution and Ecology: This cultivar was originally developed in Italy. This acidic orange is now grown in the Ivory Coast, Brazil, and Argentina as a source of bergamot oil rather than a food.

Common Name: Sour Orange (Seville Orange, Bitter Orange, Green Orange, Zhi Shi)

Scientific Name: *Citrus* X *aurantium* L. ssp. *aurantium* (*Fructus aurantii*)

Botanical Family: Rutaceae (rues, rutacées)

Physical Description: This evergreen tree reaches up to 3–9 m (~10–30 ft) in height with a more erect and compact crown than the sweet orange. The trunk has smooth, brown bark, and the young, green twigs have relatively long thorns. The compound, ovate, pointed green leaves alternate on broad-winged, relatively long petioles compared with the sweet orange. Highly fragrant flowers appear singly or in small clusters in the leaf axils with widely separated petals surrounding a tuft containing many yellow stamens. The round, oblate or oblong-oval fruit has a rough surface and a thick, aromatic, bitter peel that becomes bright red-orange on maturity.

Distribution and Ecology: The sour orange is native to southeastern Asia. This orange was distributed widely to warm areas of the Pacific Islands (Guam, Samoa, Fiji), North Africa, Europe, and North America.

Common Name: Orange, Sweet Orange

Scientific Name: *Citrus* X *sinensis* (L.) Osbeck (pro sp.) [maxima X reticulata]

Botanical Family: Rutaceae (rues, rutacées)

Physical Description: The sweet orange is an old hybrid between the tangerine and similar fruit.

Medical Toxicology of Natural Substances, by Donald G. Barceloux, MD

This evergreen reaches about 10m (~33ft) in height with alternate, ovate, waxy green leaves that are pointed. Fragrant, white flowers appear in May, usually singly. There are several cultivars of the sweet orange including the Washington navel, Valencia (juice orange), and blood orange.

Distribution and Ecology: Sweet oranges are cultivated widely in warm climates throughout the world, where temperatures during winter do not persist below freezing.

Common Name: Tangerine

Scientific Name: *Citrus reticulata* Blanco

Botanical Family: Rutaceae (rues, rutacées)

Physical Description: The tangerine is a small (<6–7cm/2–3in.), oblate, loose-skinned, deep orange fruit.

Distribution and Ecology: Some tangerine trees (e.g., Satsuma tangerine) can tolerate cool subtropical environments incompatible with sweet oranges, where temperatures reach 15°F–18°F (~−9°C to ~−8°C).

Common Name: Key Lime, Mexican Lime

Scientific Name: *Citrus* X *aurantiifolia* (Christm.) Swingle (pro sp.) [medica X sp.]

Botanical Family: Rutaceae (rues, rutacées)

Physical Description: Limes grow on relatively small, much branched citrus trees. The Key lime is a round, green fruit about 3–5cm (~1–2in.) in diameter with a thin peel and an acid pulp.

Distribution and Ecology: The lime grows in most citrus-producing areas.

Common Name: Lemon

Scientific Name: *Citrus* X *limon* (L.) Burm. f. (pro sp.) [medica X aurantifolia]

Botanical Family: Rutaceae (rues, rutacées)

Physical Description: Lemon trees grow somewhat more rapidly than orange trees with more widespread branches than grapefruit or sweet oranges. The oval fruit is slightly oblong and turns yellow during maturity. The diameter of the lemon is about 6cm (~2in.).

Distribution and Ecology: The lemon grows in most citrus-producing areas, although this tree is slightly less resistant to cold than sweet oranges.

Common Name: Grapefruit, Paradise Citrus

Scientific Name: *Citrus* X *paradisi* Macfad. (pro sp.) [maxima X sinensis]

Botanical Family: Rutaceae (rues, rutacées)

Physical Description: The grapefruit tree has larger, thicker leaves than the sweet orange tree, and the grapefruit grows more vigorously than orange, particularly during the early years of growth. The relatively large (8–10cm/~3–4in.) fruit is round to oblate with a thick rind that turns yellow on maturity. Some varieties are seedless, and the pulp is white or pink.

Distribution and Ecology: The grapefruit grows in most citrus-producing areas.

EXPOSURE

Sources

There are a variety of sources for citrus oil including sweet orange (*Citrus sinensis*), mandarin (*Citrus deliciosa* Tenore), grapefruit (*Citrus paradisi*), lemon (*Citrus limon*), and sour or bitter orange (*Citrus aurantium*). *d*-Limonene is a non-nutritive dietary component of the essential oils of citrus fruits that occurs in smaller concentrations in turpentine, tea tree oil, and other essential oils including eucalyptus, lavender, rosemary, caraway, and peppermint. Commercial use of *d*-limonene includes the synthesis of spearmint oil and use as a lemon fragrance in soaps, detergents, creams, lotions, and perfumes as well as a lemon flavoring agent in foods, beverages and chewing gum. Commercial production processes for *d*-limonene involve alkalinization and steam distillation of citrus peel and pulp remaining from the production of citrus juice or cold-pressed citrus peel oils. Commercial production of *l*-limonene involves acid treatment of pinene or the purification of monoterpene compounds in pine oils. Synthetic processes for the racemic mixture include thermal isomerization of α-pinene or the removal of dipentene as a by-product of synthetic pine oils. As a result of the ubiquitous presence of *d*-limonene in commercial products, this compound is a common constituent of indoor air. In a study of indoor air samples from commercial office buildings in the midwestern United States, *d*-limonene was one of the five most prevalent volatile organic compounds present in concentration up to about 3μg/m^3.[2] Outdoor air samples from Houston, Texas contained limonene concentrations varying from 0–32μg/m^3 (0–5.7ppb).[3]

Medicinal Uses

Traditional

Common herbal uses for extracts of sour or bitter orange (zhi shi) include sedation and the treatment of gastrointestinal disorders, such as anorexia, abdominal pain, constipation, and nausea.[4]

Current

d-Limonene is an experimental chemotherapeutic compound undergoing testing as chemopreventative agent to enhance the detoxification of carcinogens by the induction of Phase II carcinogen-metabolizing enzymes.[5,6] The role of *d*-limonene in preventing cancer in humans is unproven. *Citrus aurantium* contains beta agonists, and herbal extracts of *C. aurantium* are used as weight-reduction agents.[7] Although extracts of Seville orange (*Citrus aurantium*) are marketed as a safe alternative to ephedra in herbal weight-loss products, there are few clinical data to support the efficacy of this product as a means to lose weight.[8,9] A study of 13 elderly patients with dementia did not support the use of sweet orange (*Citrus aurantium*) to reduce combative, resistive behaviors in individuals with dementia.[10] Of about 30 terpene compounds studied, *d*-limonene had the most desirable properties for the dissolution of gallstones in a Japanese study,[11] but there has been little use of this compound to solubilize gallstones in clinical practice. *d*-Limonene is a common constituent of home cleaning products, industrial hand cleaners, deodorants, disinfectants, wax and paint strippers, and lubricating oils, as well as some animal shampoos and topical insecticides.[12,13] In many histological laboratories, *d*-limonene is a substitute for xylene.

Regulatory Status

d-Limonene is on the US Food and Drug Administration's (FDA) list of compounds generally recognized as safe (GRAS), and the FDA approves the use of *d*-limonene as a food additive. *d*-Limonene is a common flavoring agent and fragrance in perfumes, soaps, beverages, and foods.

PRINCIPAL INGREDIENTS

Chemical Composition

The composition of essential citrus oils depends on the process (steam distillation, vacuum distillation, solvent extraction, cold pressing) as well as the place of origin, species, season, and weather. The most common methods for obtaining essential oils from citrus (bitter orange, sweet orange, mandrin, grapefruit, lime, lemon, bergamot) are cold mechanical pressing and steam distillation. The composition of various citrus oils is similar, consisting of 85–95% of the monoterpene *d*-limonene along with small quantities of terpene hydrocarbons (myrcene, α-pinene, sabinene), alcohols, aldehydes, phenolic acids, esters, and ketones. *d*-Limonene occurs naturally in both the dextrorotatory and levorotatory forms with the racemic mixture of the two isomers being known as dipentene. Figure 98.1 displays the chemical structure of limonene. About 90% of orange peel oil is *d*-limonene (1-methyl-4-isopropenyl-1-cyclohexene, CAS RN: 138-86-3). Synonyms for limonene include 4-isopropenyl-1-methyl cyclohexene, 1-methyl-4(1-methy lethenyl)cyclohexene, p-mentha-l,8-diene, carvene, cinene, and cajeputene. Citronella and lemon grass oils contain primarily *l*-limonene.

The presence of valencene as the principal sesquiterpene compound distinguishes sweet orange peel oil from other types of citrus oils. Compared with other citrus oils, the concentration of limonene in lemon peel oil is relatively low (70–80%) with higher concentrations of α-pinene, β-pinene, and γ-terpinene. The oxygenated fraction of lemon oil contains relatively high concentrations of the aldehyde, citral, which imparts the aroma of lemon oil. The *d*-limonene concentration in bergamot oil is substantially lower (25–32%) compared with other citrus oils, while the content of linalool and linalyl acetate are very high. A sample of lemonade juice contained about 0.5 mg *d*-limonene/mL.[14] Typical commercial products containing *d*-limonene include nonalcoholic beverages (31 ppm), ice cream (68 ppm), candy (49 ppm), baked goods (120 ppm), gelatins and puddings (48–400 ppm), and chewing gum (2,300 ppm).[15]

Extracts of *Citrus aurantium* (Seville orange, bitter orange) contain synephrine (*p*-synephrine, oxedrine), which is an indirect-acting sympathomimetic amine structurally similar to ephedrine and phenylpropanolamine.[8] Figure 98.2 displays the similarity between ephedrine, the neurotransmitter, epinephrine, and synephrine. Dry extracts of bitter orange contain about 3% synephrine as measured by high performance liquid chromatography (HPLC).[16] There are six possible isomers of synephrine including the *ortho*-, *meta*-, and

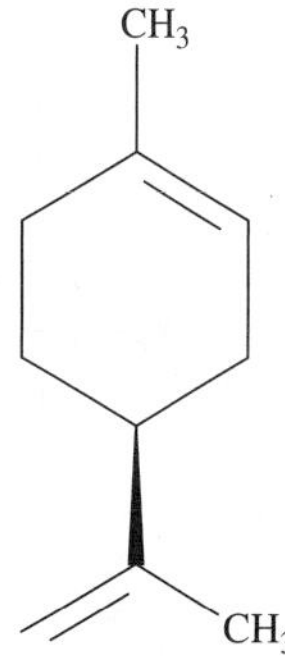

CH_3

CH_3

FIGURE 98.1. Chemical Structure of Limonene.

Ephedrine Epinephrine (adrenalin) Synephrine

FIGURE 98.2. Similarity of synephrine, ephedrine, and epinephrine.

para- isomers along with two optical isomers or chiral forms for each of these three positional isomers. At least some dietary supplements of bitter orange also contain *m*-synephrine (phenylephrine), which is a common over-the-counter nasal decongestant.[17] The exact composition of each isomer in citrus oils is not well defined. Bitter orange also contains other stimulant amines including *d,l*-octopamine and *N*-methyltyramine.[18] The low-volatile fraction of *Citrus* essential oils contains flavonoid (rutinosides, neohesperidosides) and coumarin compounds including neoeriocitrin, narirutin, naringin, hesperidin, neohesperidin, naringenin, aurapten, bergamottin, bergapten, nobiletin, tangeritin, and hesperetin.[19] In general, lime oil contains the highest concentration of furocoumarin compounds compared with other citrus oils.[20] High concentrations of bergapten occur in bergamot oil and lime oil; the concentration of oxypeucedanin is high in lime oil and undetectable in bergamot oil. Bergapten and oxypeucedanin are the main phototoxic furocoumarin compounds in lemon oil with the latter being about one-fourth as potent as bergapten in animal models of phototoxicity.[21] The concentration of these furocoumarin compounds varies substantially in lemon oil samples, ranging from 4–87 ppm and 26–728 ppm, respectively, in 13 samples from various parts of the world.[21] Most flavonoids in citrus are flavanones (rutinosides, neohesperidosides) bound as glycosides attached at the C7 position. The predominant flavanone in sweet orange, mandarin, lemon, and tangerine is the rutinoside compound, hesperetin, which is a relatively tasteless flavanone. Lime and lemon oils also contain substantial amounts of eriocitrin.[22] Sour or bitter orange and grapefruit contain different flavanones compared with sweet orange and mandarins. The neohesperidoside glycosides predominate in grapefruit (naringin, narirutin) and bitter orange (naringin, neohesperidin).[23] In one study, the concentration of naringin and neohesperidin in samples of bitter orange ranged from approximately 2–26 mg/g and 4–15 mg/g, respectively.[16] These neohesperidoside glycosides are intensely bitter and contribute to the characteristic taste of bitter orange and grapefruit. Linalool is a major terpene alcohol that contributes to the floral aroma of citrus peel oils.[24]

TABLE 98.1. Physical Properties of *d*-Limonene

Physical Property	Value
Conversion Factors	
1 ppm	5.56 mg/m^3
1 mg/m^3	0.18 ppm
Molecular Weight	136.26
Molecular Formula	$C_{10}H_{16}$
Density	0.8411 g/cm^3 (20 °C/68 °F)
Melting Point	−74.3 °C/−101.7 °F
Boiling Point	176 °C/~349 °F
Density (4 °C–20 °C)	0.8402
log P (octanol-water)	4.57
Water Solubility	13.8 mg/L (25 °C/77 °F)
Vapor Pressure	1.55 mmHg (25 °C/77 °F)
Henry's Law Constant	0.032 atm-m^3/mole (25 °C/77 °F)
Atmospheric OH Rate Constant	1.49E-10 cm^3/molecule-second (35 °C/95 °F)

Physiochemical Properties

Limonene is a colorless to slightly yellow liquid with a pleasant, lemon-like odor and a citrus taste. *d*-Limonene is the pharmacologically active isomer. Table 98.1 lists some of the physical properties of *d*-limonene. The high vapor pressure and low water solubility of limonene indicates a relatively high rate of evaporation. This compound is flammable with a flash point of 48 °C (~118 °F). In the atmosphere, ozone, nitrate radicals, and hydroxy radicals rapidly (i.e., few hours) oxidize *d*-limonene to monocyclic terpene compounds (e.g., *cis* and *trans*-limonene-2-hydroperoxide, carvone).[25] Although *d*-limonene is an irritant, the allergenic properties associated with *d*-limonene probably result from oxidation products (e.g., limonene oxide, limonene hydroperoxides, carvone) rather than the parent compound.[26,27] The pH of orange peel oil is about 5.6 compared with about 3.8 for orange juice concentrate.[28]

Mechanism of Toxicity

The contribution of synephrine to the toxicity associated with some case reports of bitter orange supplements is unclear. These supplements contain synephrine, which is an α_1-adrenergic receptor agonist that produces vasoconstriction and elevated blood pressure in *in vitro* animal preparations.[29] However, herbal preparations contain multiple ingredients including caffeine, and the role of synephrine in these cases remains undefined. A randomized, double-blind, placebo-controlled crossover study did not detect elevated blood pressure in 10 healthy subjects following single doses of synephrine, whereas the ingestion of weight loss supplements containing multiple ingredients including synephrine did produce significant cardiovascular stimulation.[30]

DOSE RESPONSE

Medicinal Use

In clinical trials on the use of *d*-limonene to dissolve gallstones, patients tolerated oral doses of 20 g *d*-limonene without obvious toxicity.[11] The ingestion of 100 mg *d*-limonene/kg did not produce clinically significant toxicity in seven healthy volunteers.[31] In a phase I clinical trial of *d*-limonene in patients with advanced cancer, the maximum tolerated dose was 8 g/m^2 with nausea, vomiting, and diarrhea being the dose-limiting effects.[32]

Toxicity

Based on the significant decreases in body weight gain associated with administration of *d*-limonene to male and female mice and rats and female rabbits, the World Health Organization (WHO) established an acceptable daily intake (ADI) for *d*-limonene of 1.5 mg/kg body weight. The daily intake of *d*-limonene varies considerably depending on the amount and type of citrus products consumed along with the processing procedures. The mean daily intake of *d*-limonene from citrus juices among US consumers in an Arizona study was about 13 mg based on questionnaires.[33]

TOXICOKINETICS

Absorption

Rodent studies indicate that the absorption of *d*-limonene is rapid with peak serum concentrations of this compound occurring within 2 hours of oral administration.[34] However, in a study of 32 patients with locally advanced metastatic solid tumors receiving oral *d*-limonene, the absorption of *d*-limonene was relatively slow with the time to peak concentration (T_{max}) ranging from 1–6 hours.[32] In eight healthy volunteers exposed to 450 mg/m^3 *d*-limonene for 2 hours, the relative pulmonary uptake of *d*-limonene was high (i.e., about 70%).[35]

Biotransformation

d-Limonene undergoes extensive species-dependent biotransformation (Figure 98.3).[34] The major metabolite in humans and guinea pigs is 8-hydroxy-*p*-meth-1-en-9-yl-β-D-glucopyranosiduronic acid (M-VI), whereas perillic acid-8,9-diol (M-IV) is the major metabolite in rats and rabbits.[36,37] In a phase I clinical trial of patients with advanced cancer, the predominant circulating metabolites of *d*-limonene were perillic acid, dihydroperillic acid, uroterpenol, limonene-1,2-diol, and an isomer of perillic acid.[32] *In vitro* studies indicate that CYP2C9 and, to a lesser extent, CYP2C19 catalyze the 6-hydroxylation of *d*-limonene to carveol and the 7-hydroxylation of *d*-limonene to perillyl alcohol.[38]

Elimination

The metabolism of *d*-limonene is extensive with relatively small amounts (<0.01%) of *d*-limonene excreted unchanged in the urine.[35] Animal studies indicate that renal excretion is the main route of elimination of *d*-limonene with about 75–90% of the absorbed dose eliminated within 2–3 days. Fecal excretion accounts for about 10% of the absorbed dose of *d*-limonene during this period. In a study of nine healthy volunteers ingesting lemonade rich in *d*-limonene, plasma perillic acid concentrations declined rapidly with a mean plasma half-life of 1.38 ± 0.40 hours.[14] Secretion of *d*-limonene occurs in breast milk as a result of nonoccupational exposure to this compound.[39]

Drug Interactions

Botanical supplements containing *C. aurantium* extracts probably do not produce clinical significant interactions with cytochrome P450 enzymes (CYP3A4, CYP1A2, CYP2E1, CYP2D6) in humans.[40]

CLINICAL RESPONSE

Acute Effects

d-Limonene is a skin irritant and skin sensitizer that produces various types of contact dermatitis including irritant, delayed allergic eczematous, immediate vesicular eruption, urticaria, and phototoxicity.[41,42] Case reports associate the topical exposure to *d*-limonene with the development of acute contact dermatitis.[43,44] Although these oxidized *d*-limonene compounds are the most common skin sensitizers in citrus oil, rare case reports associate contact dermatitis with minor components (e.g., geraniol, citral) of citrus peel oil.[45] Berloque dermatitis is a variant of phytophotodermatitis that

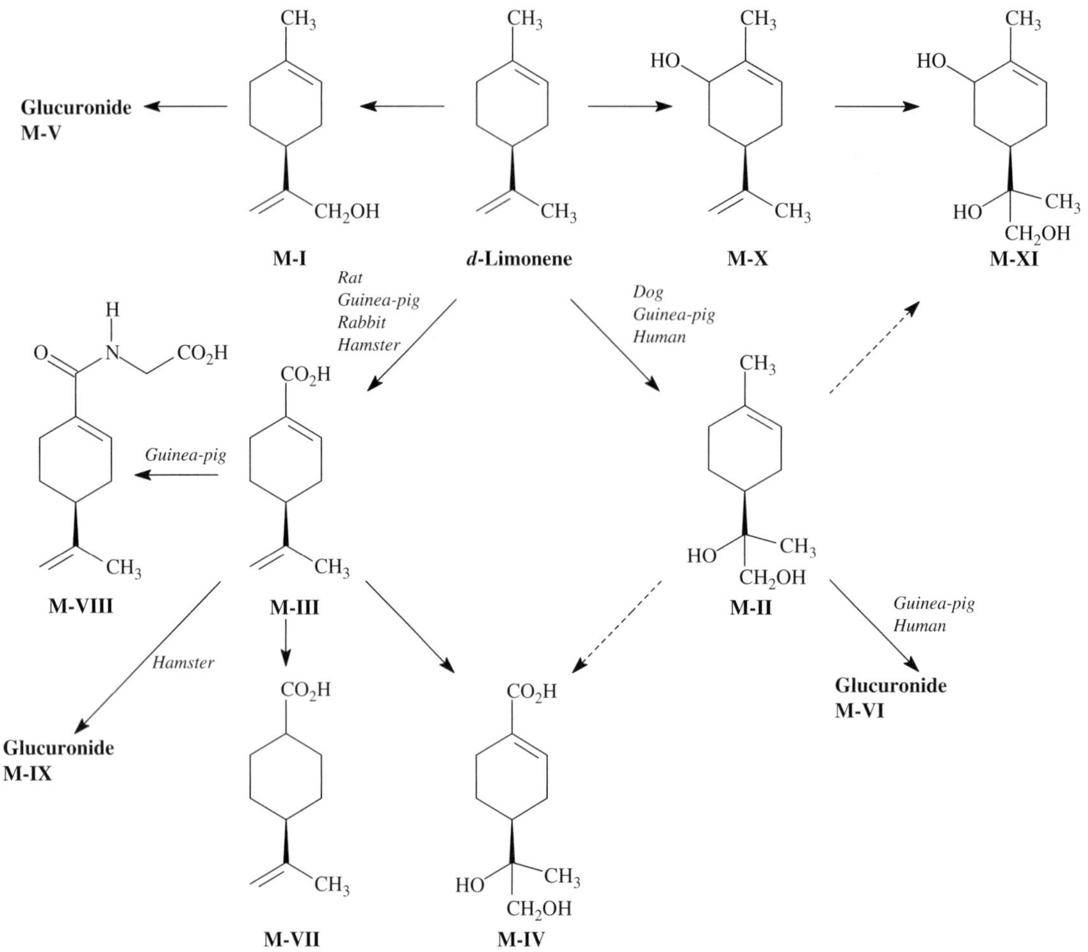

FIGURE 98.3. Major pathways of limonene biotransformation in various animal species. **Metabolites (M): M-I** = *p*-mentha-1,8-dien-10-ol, **M-II** = *p*-menth-1-ene-8,9-diol, **M-III** = perillic acid, **M-IV** = perillic acid 8,9-diol, **M-V** = *p*-mentha-1,8-dien-10-yl-beta-D-glucopyranosiduronic acid, **M-VI** = 8-hydroxy-p-menth-1-en-9-yl-beta-D-glucopyranosiduronic acid, **M-VII** = 2-hydroxy-p-menth-8-en-7-oic acid, **M-VIII** = perillylglycine, **M-IX** = perillyl-beta-D-glucopyrano-siduronic acid, **M-X** = *p*-mentha-1,8-dien-6-ol, **M-XI** = *p*-menth-1-ene-6,8,9-triol. Adapted from International Agency for Research on Cancer. *d*-Limonene. IARC Mongr Eval Carcinog Risk Hum 1999;73:313.

produces painful, red, edematous vesicles and bullae after contact with high concentrations of psoralen-containing fragrances, most commonly oil of bergamot.[46] Sequelae include hyperpigmentation of the affected skin. Although Berloque dermatitis is rarely seen today because of the removal of these fragrances from most cosmetic products, case reports associate these types of skin lesions with the use of bergamot oil during aromatherapy.[47] Gastrointestinal distress (nausea, vomiting, diarrhea) was the main adverse effect in a clinical trial of advanced cancer patients receiving daily *d*-limonene doses of 0.5–12 g/m^2 in 21-day cycles.[32]

Although the FDA banned the use of ephedra-containing products because of safety considerations in 2004, the ephedra ban did not preclude the use of other structurally related sympathomimetic amines. Bitter orange (*Citrus aurantium*) contains the indirect-acting sympathomimetic amine, synephrine, which is a common replacement for ephedra in herbal preparations for weight loss and energy enhancement. Several case reports associate the use of synephrine-containing supplements with adverse cardiovascular and cerebrovascular effects. A 38-year-old man developed an ischemic stroke (memory loss, unsteady gait) 1 week after initiating a weight-loss dietary supplement containing 6–12 mg synephrine.[48] The multivessel distribution of the stroke and normal magnetic resonance angiography suggested a vasospastic origin. A 55-year-old smoker developed an acute lateral myocardial infarction one year after beginning the daily use of 300 mg bitter orange (*C. aurantium*). Although she was hypertensive on presentation, she did not have a history of hypertension or other cardiovascular risk factors. A previously healthy 22-year-old woman developed exercise-induced syncope 1 hour after ingesting a second dose of an herbal weight loss supplement, which contained 3 mg synephrine.[49] The electrocardiogram demonstrated a prolonged QTc interval (516 milliseconds), which normalized within 24 hours. The exact causal role of synephrine in these case reports is unclear because of the ingestion of multiple ingredients and the lack of pharmacokinetic data on synephrine.

Chronic Effects

In a 2-year gavage study, there was evidence of an increased incidence of tubular cell hyperplasia, adenomas, and adenocarcinomas of the kidney in male F344/N rats receiving 150 mg *d*-limonene (99% pure)/kg 5 days/week.[50] This nephropathy was the only microscopic evidence of *d*-limonene-related toxicity in this study. There was no evidence of carcinogenic activity of *d*-limonene for female F344/N rats receiving 300 or 600 mg/kg, $B_6C_3F_1$ mice receiving 250 or 500 mg/kg, or female $B_6C_3F_1$ mice receiving 500 or 1,000 mg/kg on a similar schedule. *d*-Limonene is one of a diverse group of hydrocarbons, which produce a unique nephropathy in male rats secondary to the accumulation of α_{2u}-globulin following subchronic or chronic exposure. The International Agency for Research on Cancer (IARC) lists *d*-limonene as not classifiable as to carcinogenicity to humans (Group 3) based on the lack of evidence that any human protein can contribute to a renal syndrome similar to the α_{2u}-globulin nephropathy occurring exclusively in male rats.[51]

DIAGNOSTIC TESTING

Methods of analyzing the composition of citrus oils include liquid chromatography, high-speed gas chromatographic separation and time-of-flight mass spectrometry,[52] capillary electrophoresis,[53] and HPLC.[54] In the absence of antioxidants, the concentrations of monoterpene hydrocarbons (e.g., *d*-limonene) decrease during storage at room temperature, but these losses are small during refrigeration or frozen storage. Over one year, the concentration of monoterpene compounds in cold-pressed daidai peel oil decreased from 98% to approximately 66%, while the concentration of sesquiterpene hydrocarbons and alcohols increased.[55]

TREATMENT

Treatment is supportive. Patients with contact dermatitis from citrus oil benefit from topical steroids as a means of reducing the incidence of hyperpigmentation.

References

1. International Agency for Research on Cancer (IARC). *d*-Limonene. IARC Monogr Eval Carcinog Risk Hum 1993;56:135–162.
2. Reynolds SJ, Black DW, Borin SS, Breuer G, Burmeister LF, Fuortes LJ, et al. Indoor environmental quality in six commercial office buildings in the Midwest United States. Appl Occup Environ Hyg 2001;16:1065–1077.
3. Bertsch W, Chang RC, Zlatkis A. The determination of organic volatiles in air pollution studies: characterization of profiles. J Chromatogr Sci. 1974;12:175–182.
4. Hernandez L, Munoz RA, Miro G, Martinez M, Silva-Parra J, Chavez PI. Use of medicinal plants by ambulatory patients in Puerto Rico. Am J Hosp Pharm 1984;41:2060–2064.
5. Russin WA, Hoesly JD, Elson CE, Tanner MA, Gould MN. Inhibition of rat mammary carcinogenesis by monterpenoids. Carcinogenesis 1989;10:2161–2164.
6. Crowell PL. Prevention and therapy of cancer by dietary monoterpenes. J Nutr 1999;129:775S–778S.
7. Preuss HG, DiFerdinando D, Bagchi M, Bagchi D. *Citrus aurantium* as a thermogenic, weight-reduction replacement for ephedra: an overview. J Med 2002;33:247–264.
8. Bent S, Padula A, Neuhaus J. Safety and efficacy of *Citrus aurantium* for weight loss. Am J Cardiol 2004;94:1359–1361.
9. Fugh-Berman A, Myers A. *Citrus aurantium*, an ingredient of dietary supplements marketed for weight loss: current status of clinical and basic research. Exp Biol Med (Maywood) 2004;229:698–704.
10. Gray SG, Clair AA. Influence of aromatherapy on medication administration to residential-care residents with dementia and behavioral challenges. Am J Alzheimers Dis Other Demen 2002;17:169–174.
11. Igimi H, Hisatsugu T, Nishimura M. The use of *d*-limonene preparation as a dissolving agent of gallstones. Dig Dis 1976;21:926–939.
12. Powers KA, Hooser SB, Sundberg JP, Beasley VR. An evaluation of the acute toxicity of an insecticidal spray containing linalool, *d*-limonene, and piperonyl butoxide applied topically to domestic cats. Vet Hum Toxicol 1988;30:206–210.
13. Rastogi SC, Heydorn S, Johansen JD, Basketter DA. Fragrance chemicals in domestic and occupational products. Contact Dermatitis 2001;45:221–225.
14. Sherry Chow H-H, Salazar D, Hakim IA. Pharmacokinetics of perillic acid in humans after a single dose administration of a citrus preparation rich in *d*-limonene content. Cancer Epidemiol Biomark Prev 2002;11:1472–1476.
15. National Toxicology Program. National Toxicology Program technical report on the toxicology and carcinogenesis studies of *d*-limonene (CAS No. 5989-27-5) in F344/N rats and B6C3F1 mice (Gavage Studies), NTP Technical Report 347, NIH Publication No. 90-2802. Bethesda, MD: National Institutes of Health; 1990.

16. Pellati F, Benvenuti S, Melegari M. High-performance liquid chromatography methods for the analysis of adrenergic amines and flavanones in *Citrus aurantium* L. var. *amara*. Phytochem Anal 2004;15:220–225.
17. Allison DB, Ctter G, Poehlman ET, Moore DR, Barnes S. Exactly which synephrine alkaloids does *Citrus aurantium* (bitter orange) contain? Int J Obesity 2005;29:443–446.
18. Pellati F, Benvenuti S, Melegari M, Firenzuoli F. Determination of adrenergic agonists from extracts and herbal products of *Citrus aurantium* L. var. *amara* by LC. J Pharm Biomed Anal 2002;29:1113–1119.
19. Buiarelli F, Cartoni GP, Coccioli F, Leone T. Analysis of bitter essential oils from orange and grapefruit by high-performance liquid chromatography with microbore columns. J Chromatogr A 1996;730:9–16.
20. Ranganna S, Govindarajan VS, Ramana KV. Citrus fruits–varieties, chemistry, technology, and quality evaluation. Part II. Chemistry, technology, and quality evaluation. A. Chemistry. Crit Rev Food Sci Nutr 183;18:313–386.
21. Naganuma M, Hirose S, Nakayama Y, Nakajima K, Someya T. A study of the phototoxicity of lemon oil. Arch Dermatol Res 1985;278:31–36.
22. Montanari A, Chen J, Widmer W. Citrus flavonoids: a review of past biological activity against disease. Discovery of new flavonoids from Dancy tangerine cold pressed peel oil solids and leaves. Adv Exp Med Biol 1998;439: 103–116.
23. Gel-Moreto N, Streich R, Galensa R. Chiral separation of diastereomeric flavanone-7-*O*-glycosides in citrus by capillary electrophoresis. Electrophoresis 2003;24:2716–2722.
24. Bazemore R, Rouseff R, Naim M. Linalool in orange juice: origin and thermal stability. J Agric Food Chem 2003;51: 196–199.
25. Corchnoy SB, Atkinson R. Kinetics of the gas-phase reactions of OH and NO_3 radicals with 2-carene, 1,8-cineole, *p*-cymene, and terpinolene. Environ Sci Technol 1990;24: 1497–1502.
26. Karlberg A-T, Magnusson K, Nilsson U. Air oxidation of *d*-limonene (the citrus solvent) creates potent allergens. Contact Dermatitis 1992;26:332–340.
27. Matura M, Goossens A, Bordalo O, Garcia-Bravo B, Magnusson K, Wrangsjo K, Karlberg A-T. Oxidized citrus oil (*R*-limonene): a frequent skin sensitizer in Europe. J Am Acad Dermatol 2002;47:709–714.
28. Joslin CL, Bradley JE. Study on orange juice, orange juice concentrate, and orange peel oil in infants and children. J Pediatr 1951:39:325–329.
29. Brown CM, McGrath JC, Midgley JM, Muir AG, O'Brien JW, Thonoor CM, et al. Activities of octopamine and synephrine stereoisomers on alpha-adrenoceptors. Br J Pharmacol 1988;93:417–429.
30. Haller CA, Benowitz NL, Jacob P, III. Hemodynamic effects of ephedra-free weight-loss supplements in humans. Am J Med 2005;118:998–1003.
31. Crowell PL, Elson CE, Bailey HH, Elegbede A, Haag JD, Gould MN. Human metabolism of the experimental cancer therapeutic agent *d*-limonene. Cancer Chemother Pharmacol 1994;35:31–37.
32. Vigushin DM, Poon GK, Boddy A, English J, Halbert GW, Pagonis C, et al. Phase I and pharmacokinetic study of *d*-limonene in patients with advanced cancer. Cancer Research Campaign Phase I/II Clinical Trials Committee. Cancer Chemother Pharmacol 1998;42:111–117.
33. Hakim IA, Hartz V, Graver E, Whitacre R, Alberts D. Development of a questionnaire and a database for assessing dietary *d*-limonene intake. Pub Health Nutr 2002;5: 939–945.
34. Igimi H, Nishimura M, Kodama R, Ide H. Studies on the metabolism of *d*-limonene (*p*-mentha-1,8-diene). I. The absorption, distribution and excretion of *d*-limonene in rats. Xenobiotica 1974;4:77–84.
35. Falk-Filipsson A, Lof A, Hagberg M, Hjelm EW, Wang Z. *d*-Limonene exposure to humans by inhalation: uptake, distribution, elimination, and effects on the pulmonary function. J Toxicol Environ Health 1993;38:77–88.
36. Hardcastle IR, Rowlands MG, Barber AM, Grimshaw RM, Mohan MK, Nutley BP, Jarman M. Inhibition of protein prenylation by metabolites of limonene. Biochem Pharmacol 1999;57:801–809.
37. Kodama R, Yano T, Furukawa K, Noda K, Ide H. Studies on the metabolism of *d*-limonene (p-mentha-1,8-diene). IV. Isolation and characterization of new metabolites and species differences in metabolism. Xenobiotica 1976;6: 377–389.
38. Miyazawa M, Shindo M, Shimada T. Metabolism of (+)- and (−)-limonenes to respective carveols and perillyl alcohols by CYP2C9 and CYP2C19 in human liver microsomes. Drug Metab Disp 2002;30:602–607.
39. Pellizzari ED, Hartwell TD, Harris BS 3rd, Waddell RD, Whitaker DA, Erickson MD. Purgeable organic compounds in mother's milk. Bull Environ Contam Toxicol 1982;28:322–328.
40. Gurley BJ, Gardner SF, Hubbard MA, Williams DK, Gentry WB, Carrier J, et al. *In vivo* assessment of botanical supplementation on human cytochrome P450 phenotypes: *Citrus aurantium, Echinacea purpurea*, milk thistle, and saw palmetto. Clin Pharmacol Ther 2004;76:428–440.
41. Chang Y-C, Karlberg A-T, Maibach HI. Allergic contact dermatitis from oxidized *d*-limonene. Contact Dermatitis 1997;37:308–309.
42. Karlberg A-T, Dooms-Goossens A. Contact allergy to oxidized *d*-limonene among dermatitis patients. Contact Dermatitis 1997;36:201–206.
43. Falk A, Fischer T, Hagberg M. Purpuric rash caused by dermal exposure to *d*-limonene. Contact Dermatitis 1991;25:198–199.
44. Topham EJ, Wakelin SH. *d*-Limonene contact dermatitis from hand cleansers. Contact Dermatitis 2003;49:108–109.
45. Cardullo AC, Ruszkowski AM, DeLeo VA. Allergic contact dermatitis resulting from sensitivity to citrus peel, geraniol, and citral. J Am Acad Dermatol 1989;21: 395–397.

46. Wang L, Sterling B, Don P. Berloque dermatitis induced by "Florida water". Cutis 2002;70:29–30.
47. Kaddu S, Kerl H, Wolf P. Accidental bullous phototoxic reactions to bergamot aromatherapy oil. J Am Acad Dermatol 2001;45:458–461.
48. Bouchard NC, Howland MA, Greller HA, Hoffman RS, Nelson LS. Ischemic stroke associated with use of an ephedra-free dietary supplement containing synephrine. Mayo Clin Proc 2005;80:541–545.
49. Nasir JM, Durning SJ, Ferguson M, Barold HS, Haigney MC. Exercise-induced syncope associated with QT prolongation and ephedra-free Xenadrine. Mayo Clin Proc 2004;79:1059–1062.
50. Opdyke DL. Monographs on fragrance raw materials. Food Cosmet Toxicol 1975;13(Suppl):683–923.
51. IARC. *d*-Limonene. IARC Mongr Eval Carcinog Risk Hum 1999;73:307–327.
52. Veriotti T, Sacks R. High-speed GC and CG/time-of-flight MS of lemon and lime oil samples. Anal Chem 2001;73:4395–4402.
53. Cancalon PF. Analytical monitoring of citrus juices by using capillary electrophoresis. J AOAC Int 1999;82:95–106.
54. Buiarelli F, Cartoni G, Coccioli F, Jasionowska R, Mazzarino M. Analysis of limette and bergamot distilled essential oils by HPLC. Annali Chim 2002;92:363–372.
55. Njoroge SM, Ukeda H, Sawamura M. Changes of the volatile profile and artifact formation in daidai (*Citrus aurantium*) cold-pressed peel oil on storage. J Agric Food Chem 2003;51:4029–4035.

Chapter 99

EUCALYPTUS OIL
(*Eucalyptus* Species)

HISTORY

In Western civilization, eucalyptus oil has been an expectorant for cough and colds as well as a skin disinfectant, liniment for myalgias, and a counterirritant. A letter from First Fleet Surgeon Considen to Sir Joseph Banks in 1788 was the first recorded European reference to the medicinal use of eucalyptus oil.[1] *Eucalyptus* species dominate the flora of Australian forests. However, few mammalian herbivores use *Eucalyptus* leaves as a food source, probably as a result of the poor nutritional qualities and high concentration of phenols and terpenes in these leaves. There are four Australian marsupials that feed on *Eucalyptus* leaves including the koala (*Phascolarctos cinereus* Goldfuss), greater glider (*Petauroides volans* Kerr), ringtail possum (*Pseudocheirus peregrinus* Boddaert), and the brushtail possum (*Trichosurus vulpecula* Kerr).[2] The koala bear feeds almost exclusively on eucalyptus leaves from infancy, whereas the other marsupials also feed on leaves from other species.

BOTANICAL DESCRIPTION

There are over 50 species in the genus *Eucalyptus* including *Eucalyptus globules* (Tasmanian bluegum), *Eucalyptus fruticetorum* F. Muell. ex Miq., *Eucalyptus pulverulenta* Sims (silverleaf mountain gum), and *Eucalyptus robusta* Sm. (robust eucalyptus, swamp mahogany).

Common Name: Eucalyptus
Scientific Name: *Eucalyptus* species
Botanical Family: Myrtaceae (myrtle)
Physical Description: These mostly evergreen, hardwood trees produce copious amounts of oil in the leaves. Older leaves are alternate and slender with long petioles, whereas the younger leaves are round and opposite, occasionally without petioles. The distinctive flowers have many showy stamens rather than petals. The woody, cone-shaped fruits release seeds from valves at the end of the fruits. Some species have smooth bark, whereas other species have rough bark ranging from short fibers to deep furrows and long, stringy bark. The Australian Mountain-ash, (*Eucalyptus regnans* F. Muell.) is one of the tallest trees (i.e., up to 92 m/300 ft).
Distribution and Ecology: Eucalyptus species are dominant trees in temperate forests and woodlands in Australia. These trees are cultivated in temperate climates, such as California.

EXPOSURE

Sources

Distillation of *Eucalyptus* leaves is the source of eucalyptus oil. The three types of commercial eucalyptus oil are defined by the following uses: medicinal, perfumery, and industrial. The industrial type is derived from *E. dives* Schauer, but the production of this type of oil is very small. The most important type of eucalyptus oil is the medicinal type derived primarily from *Eucalyptus globulus* Labill. (Tasmanian blue gum) and *E. exserta*

Medical Toxicology of Natural Substances, by Donald G. Barceloux, MD

F. Muell. (Queensland peppermint). This type of oil has relatively high concentrations of cineole. The sources of perfumery eucalyptus oil are *E. citriodora* Hook. (lemon-scented gum) and, to a lesser extent, *E. staigeriana* F. Muell. ex Bailey (lemon-scented ironbark). Other plant species contain substantial amounts of the major ingredient, eucalyptol, in eucalyptus oil including *Artemisia pontica* L. (Roma mugwort), *Rosmarinus officinalis* L. (rosemary), *Salvia officinalis* L. (sage), *Hedychium flavum* Robx. (garland flower), and *Mentha piperita* L. (peppermint).[3]

Medicinal Uses

In Australia, eucalyptus oil is a discrete product and a component of vaporized liquids, candy, cough drops, ointments, salves, and cleaning products.[4] Australian Aborigines[5] and indigenous Brazilian people[6] traditionally used eucalyptus oil to treat upper respiratory tract infections, asthma, viral illness, fever, and bronchial infections as well as diarrhea, toothaches, snake bites, and skin lesions. Experimental studies suggest that eucalyptus-containing solutions have mosquito repellant properties.[7] Eucalyptol is approved by the US Food and Drug Administration (FDA) for use as a food additive and is a component of some mouthwashes (Listerine®, Johnson & Johnson, New Brunswick, NJ) and dental preparations. In Germany, eucalyptol is a licensed medicinal product (Soledum™ capsules–100 mg; Cassella-med, Cologne, Germany) for the treatment of bronchitis, sinusitis, and respiratory infections. A small clinical trial suggested that eucalyptol has mucolytic and anti-inflammatory properties.[8]

PRINCIPAL INGREDIENTS

Chemical Composition

About 10% of the dry leaf mass of eucalyptus leaves contain essential oils consisting primarily of terpene and phenolic compounds. Eucalyptus oil is a steam distillation product of eucalyptus leaves and terminal branches from varied species of *Eucalyptus*. The major constituent of eucalyptus oil is the 10-carbon bicyclic ether, 1,8-cineole (eucalyptol), and the concentration of eucalyptol varies with *Eucalyptus* species. In general, the essential oils from the leaves of *Eucalyptus globulus* contain high concentrations (70–80%) of eucalyptol. Figure 99.1 displays the chemical structure of saturated monoterpene compound, eucalyptol, which is an isoprenoid (C_5) compound consisting of two isoprene subunits (C_{10}). This essential oil also contains small amounts of hydrocyanic (prussic) acid as well as phellandrene, α-pinene, linalool, α-terpineol, eudesmol, and other terpene compounds.[9]

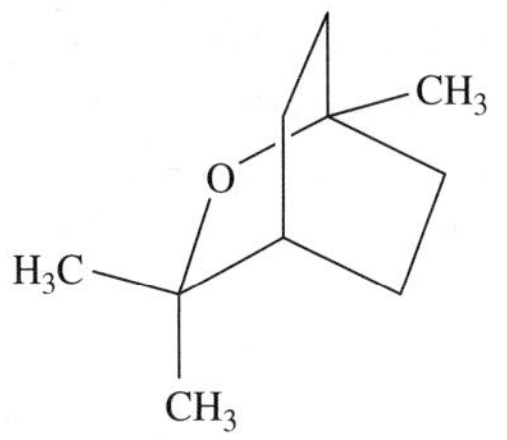

FIGURE 99.1. Chemical structure of eucalyptol.

TABLE 99.1. Some Physical Properties of Eucalyptol

Physical Property	Value
CAS RN	470-82-6
Molecular Weight	154.24
Molecular Formula	$C_{10}H_{18}O$
Melting Point	1.5 °C/~35 °F
Boiling Point	176.4 °C/~350 °F
log P (Octanol-Water)	2.74
Water Solubility	3,500 mg/L (21 °C/~70 °F)
Vapor Pressure	1.9 mmHg (25 °C/77 °F)
Henry's Law Constant	1.10E-04 atm-m³/mole (25 °C/77 °F)
Atmospheric OH Rate Constant	1.11E-11 cm³/molecule-second (25 °C/77 °F)

Analysis of leaves from *Eucalyptus ovata* demonstrated four flavonol glycosides, such as quercetin 3-*O*-β-arabinsoside, and nine tannins including catechin and gallocatechin.[10] The stem bark of *E. globulus* contains a variety of compounds including catechin, ellagic acid rhamnosides, engelitin, eriodictyol, methyl gallate, pinoresinol, quercetin, rhamnazin, rhamnetin, taxifolin, vomifoliol, and 3,4,5-trimethoxyphenol 1-*O*-beta-D-(6′-*O*-galloyl)glucopyranoside.[11]

Physiochemical Properties

Eucalyptus oil is a clear to pale-yellow liquid with a distinctive camphor-like odor and pungent taste. This steam-distillation product is water-insoluble and contains at least 70% 1,8-cineole. Synonyms for eucalyptol include cineole, 1,8-cineolcajeputol, 1,3,3-trimethyl-2-oxabicyclo(2.2.2.)octane, and 1,8-epoxy-*p*-menthane. Table 99.1 lists some physical properties of eucalyptol.

DOSE RESPONSE

Medicinal Use

The combination of 2 g eucalyptus oil, 3 g chloroform, and 40 g castor oil was a treatment for parasites in the early 20th century.[12] The recommended therapeutic dose of eucalyptus oil is up to 0.2 mL (184 mg) or 5 mL in a hot steam vaporizer. Topically, eucalyptus oil is applied in a 0.5–3% ointment. The daily oral dose of eucalyptol for the treatment of upper respiratory tract infections is 600 mg.

Toxicity

There are limited data on the correlation between the dose of eucalyptus oil and toxicity. The estimated tolerable daily intake (TDI) for eucalyptol is 0.1 mg/kg.[3] Because these data are from case reports, there is a wide range of reported toxic and fatal doses of eucalyptus oil. Case reports and a retrospective review of case histories of children admitted to an Australian hospital with the diagnosis of eucalyptus oil poisoning suggested that significant central nervous system depression may occur after the ingestion of 5–10 mL of eucalyptus oil.[13,14] This retrospective study used no biomarkers to document exposure to eucalyptus oil. In a retrospective review of 41 pediatric visits to emergency departments for eucalyptus oil poisoning, there was no correlation between estimated dose and clinical symptoms.[15] Approximately 80% of these cases were asymptomatic including four children with estimated doses of eucalyptus oil exceeding 30 mL. The suicidal ingestion of 200–250 mL eucalyptus oil by a 73-year-old woman was associated with coma, hypotension, seizures, and aspiration pneumonia.[16] The patient developed persistent pneumonia and died 3 months after ingestion.

TOXICOKINETICS

Eucalyptus oil usually contains at least 70% 1,8-cineole (eucalyptol), which is absorbed rapidly from the gastrointestinal tract. The dermal application of eucalyptus oil does not usually cause systemic symptoms.[17] A case report suggests that the dermal application of large amounts (i.e., compress applied to whole body continually for 2 days) may cause ataxia, obtundation, muscle weakness, and slurred speech.[18] However, the solution also contained ethanol, and no serum ethanol concentration was reported to document the role of ethanol in the development of signs of intoxication. There are few data on the toxicokinetics of eucalyptol in humans. Following the oral administration of 800 mg eucalyptol to rats, the main metabolites were 1,8-dihydroxy-10-carboxy-p-methane, 2-hydroxy-cineole, and 3-hydroxy-cineole.[19] Other potential metabolites include 3-hydroxycineole. In studies of rats, eucalyptol crosses the placenta, but not the blood–milk barrier.[20] *In vitro* testing suggests that the inhibitory activity of eucalyptus oil on several cytochrome P450 enzymes including CYP1A2 CYP2C8 CYP2C9 CYP2C19 CYP2D6 and CYP3A4 is relatively weak,[21] but there are limited data to confirm any drug interactions. The main potential drug interaction is the enhancement of the sedation associated with the use of eucalyptus oil and other sedatives.

CLINICAL RESPONSE

Most dermal and pulmonary exposures to eucalyptus oil do not produce systemic complaints.[17] The clinical features of poisoning from the ingestion of eucalyptus oil vary depending on the dose and the presence of complications (e.g., aspiration pneumonia). Mild poisoning typically involves burning of the mouth and throat, nausea, vomiting, and abdominal pain. Central nervous system symptoms include lightheadedness, slurred speech, ataxia, headache, and drowsiness.[22] More severe ingestions involve obtundation, coma, convulsions, hypotension, and respiratory distress.[16] Cutaneous reactions following the ingestion of eucalyptus oil include generalized erythema, pruritus, urticaria, eczema, and vesicular lesions.[18] In sensitized individuals, inhalation of eucalyptus pollen or the ingestion of eucalyptus infusions can exacerbate underlying asthma and/or rhinoconjunctivitis.[23] A case report suggests that vocal cord dysfunction mimicking asthma can occur in sensitive individuals following airborne exposure to dried eucalyptus leaves.[24] Testing indicated that *this* reaction results from irritation rather than an immune response.

DIAGNOSTIC TESTING

Methods to detect 1,8 cineole include gas chromatography with flame ionization detection[25] and isocratic high performance liquid chromatography.[26] The limit of detection for the latter method in solid material is 100 µg/g material.

TREATMENT

Treatment primarily is supportive. There are few data to guide the use of decontamination measures. Patients with eucalyptus oil ingestion should not receive syrup of ipecac because of the risk of seizures and aspiration pneumonia. The efficacy of activated charcoal is also equivocal because of the rapid onset of seizures and coma as well as the lack of data on the adsorption of the major ingredient in eucalyptus oil (1,8-cineole) to activated charcoal. Patients with central nervous system depression should be observed for the development of aspiration pneumonia and respiratory insufficiency.[12] The treatment of seizures includes airway control as needed, supplemental oxygen and anticonvulsants (diazepam, lorazepam).

References

1. Pearn J. The enchanted herb: the work of early medical botanists in Australia. Med J Aust 1987;147:568–572.
2. Pass GJ, McLean S, Stupans I, Davies N. Microsomal metabolism of the terpene 1,8-cineole in the common brushtail possum (*Trichosurus vulpecula*), koala (*Phascolarctos cinereus*), rat and human. Xenobiotica 2001;31: 205–221.
3. De Vincenzi M, Silano M, De Vincenzi A, Maialetti F, Scazzocchio B. Constituents of aromatic plants: eucalyptol. Fitoterapia 2002;73:269–275.
4. Day LM, Ozanne-Smith J, Parsons BJ, Dobbin M, Tibballs J. Eucalyptus oil poisoning among young children: mechanisms of access and the potential for prevention. Aust NZ J Publ Health 1997;21:297–302.
5. Ghisalberti EL. Bioactive acylphoroglucinol derivatives from *Eucalyptus* species. Phytochemistry 1996;41: 7–22.
6. Silva J, Abebe W, Sousa SM, Duarte VG, Machado MI, Matos FJ. Analgesic and anti-inflammatory effects of essential oils of Eucalyptus. J Ethnopharmacol 2003;89: 277–283.
7. Trigg JK. Evaluation of a eucalyptus-based repellent against *Culicoides impunctatus* (Diptera:Ceratopogonidae) in Scotland. J Am Mosq Control Assoc 1996;12: 329–330.
8. Juergens UR, Dethlefsen U, Seinkamp G, Gillissen A, Repges R, Vetter H. Anti-inflammatory activity of 1.8-cineol (eucalyptol) in bronchial asthma: a double-blind placebo-controlled trial. Respir Med 2003;97:250–256.
9. Dayal R, Ayyar KS. Analysis of medicinal oil from *Eucalyptus globulus* ssp. *biocostanta* leaves. Planta Med 1986;52:162.
10. Santos SC, Waterman PG. Polyphenols from Eucalyptus ovata. Fitoterapia 2001;72:316–318.
11. Yun BS, Lee IK, Kim JP, Chung SH, Shim GS, Yoo ID. Lipid peroxidation inhibitory activity of some constituents isolated from the stem bark of Eucalyptus globulus. Arch Pharm Res 2000;23:147–150.
12. Phillips LP. On eucalyptus oil as a vermifuge in ankylostomiasis. Lancet 1906;2:285–286.
13. Patel S, Wiggins J. Eucalyptus oil poisoning. Arch Dis Child 1980;55:405–406.
14. Tibballs J. Clinical effects and management of eucalyptus oil ingestion in infants and young children. Med J Aust 1995;163;177–180.
15. Webb JNA, Pitt WR. Eucalyptus oil poisoning in childhood: 41 cases in south-east Queensland. J Paediatr Child Health 1993;29:368–371.
16. Anpalahan M, Le Couteur DG. Deliberate self-poisoning with eucalyptus oil in an elderly woman. Aust NZ J Med 1998;28:58.
17. Spoerke DG, Vandenberg SA, Smolinske SC, Kulig K, Rumack BH. Eucalyptus oil: 14 cases of exposure. Vet Hum Toxicol 1989;31:166–168.
18. Darben T, Cominos B, Lee CT. Topical eucalyptus oil poisoning. Aust J Dermatol 1998;39:265–267.
19. Madyastha KM, Chadha A. Metabolism of 1,8-cineole in rat: its effects on liver and lung microsomal cytochrome P-450 systems. Bull Environ Contam Toxicol 1986;37: 759–766.
20. Jori A, Briatico G. Effect of eucalyptol on microsomal enzyme activity of foetal and newborn rats. Biochem Pharmacol 1973;22:543–544.
21. Unger M, Frank A. Simultaneous determination of the inhibitory potency of herbal extracts on the activity of six major cytochrome P450 enzymes using liquid chromatography/mass spectrometry and automated online extraction. Rapid Comm Mass Spectrometry 2004;18: 2273–2281.
22. Gurr FW. Eucalyptus oil poisoning treated by dialysis and mannitol infusion. Aust Ann Med 1965;14:238–249.
23. Galdi E, Perfetti L, Calcagno G, Marcotulli MC, Moscato G. Exacerbation of asthma related to eucalyptus pollens and to herb infusion containing eucalyptus. Monaldi Arch Chest Dis 2003;59:220–221.
24. Huggins JT, Kaplan A, Martin-Harris B, Sahn SA. Eucalyptus as a specific irritant causing vocal cord dysfunction. Ann Allergy Asthma Immunol 2004;93:299–303.
25. Nozal MJ, Bernal JL, Jimenez JJ, Gonzalez MJ, Higes M. Extraction of thymol, eucalyptol, menthol, and camphor residues from honey and beeswax. Determination by gas chromatography with flame ionization detection. J Chromatogr A 2002;954:207–215.
26. Tao L, Pereira MA. Quantification of carvone, cineole, perillaldehyde, perillyl alcohol and sobrerol by isocratic high-performance liquid chromatography. J Chromatogr A 1998;793:71–76.

Chapter 100

NEEM OIL (MARGOSA OIL)
(*Azadirachta indica* Adr. Juss.)

HISTORY

Neem oil (margosa oil) has a long history of use in Indian folk medicine as a treatment for upper respiratory tract infections, leprosy, and intestinal parasites. Since ancient times, Hindus have used an extract of the neem tree as a panacea and as a bitter additive to seeds to reduce insect infestations in the seeds.

BOTANICAL DESCRIPTION

The neem tree (*Azadirachta indica*) is sometimes confused with the chinaberry (*Melia azedarach* L.) However, these two species from the Mahogany (Meliaceae) family are easily distinguished by their leaves. *Azadirachta* species have simple pinnate leaves, in contrast to the 2- to 3-pinnate leaves from *Melia* species.

Common Name: Indian Neem Tree, Indian Lilac, Neem Tree, Margosa Tree

Scientific Name: *Azadirachta indica* Adr. Juss. (Melia Azadirachta L., Melia indica Margosa)

Botanical Family: Meliaceae (mahogany)

Physical Description: This fast-growing, evergreen tree reaches 15–20m (~49–66ft) in height with long spreading branches. The leaves are alternate, spiral, compound and pinnate with the leaf blade being glabrous. The white flowers are arranged in inflorescences with all sepals joined. The yellow, fleshy fruit grows to about 1–2cm (~0.4–0.8in.) and contains one (rarely, two) seeds.

Distribution and Ecology: The neem tree is native to India and the Indo-Pakistan subcontinent, and is currently distributed throughout southern Malaysia, Thailand, and southeastern Asia. This tree is a cultivated plant species in many semiarid areas of Asia, Africa, Australia, Philippines, Central America (Nicaragua), and southern United States below 700m (~2,300ft) elevation.

EXPOSURE

Neem (margosa) oil is an extract from the seeds of the neem tree that is used as a traditional medicine by Indians in India, Burma, Thailand, Malaysia, Indonesia, and Sri Lanka, primarily for external applications (scabies, sores, leucoderma, pruritus, bacterial/fungal infections, local contraceptive). There are over 700 herbal preparations in Ayurveda, Amchi, Siddha, and Unani health traditions.[1] Other uses include the treatment of diabetes, fever, respiratory disorders, constipation, rheumatism, intestinal parasites, and decreased libido.[2] Neem oil also is a folk remedy in Africa and other parts of Southeast Asia for the treatment of fever and parasitic infections as well as the prevention of insect bites.[3,4] Neem oil extracts (e.g., Neemix®, W.R. Grace and Co., Baltimore, MD) are used for the insecticidal properties of the bioactive tetranortriterpenoid compounds (e.g., azadirachtin A, B, H) that include growth regulation, fecundity suppression and sterilization, oviposition, repellence, changes in biological fitness, and blocking development of vector-borne pathogens.[5,6] The neem oil for these natural alternatives to synthetic pesticides and mosquito repellants is a methanolic

Medical Toxicology of Natural Substances, by Donald G. Barceloux, MD

extract of the defatted neem (*Azadirachta indica* Adr. Juss.) seed kernels.[7–9]

PRINCIPAL INGREDIENTS

Chemical Composition

Neem seeds contain about 45% oil. Margosa oil is synonym for neem oil (CAS RN: 8002-65-1). Neem oil is a yellowish, fatty acid-rich extract of neem tree seeds (*Azadirachta indica*) that contains at least 35 pharmacologically active compounds (triterpenoid, tetranortriterpenoid) as well as glycerides, flavonoids (kaempferol, myricetin, quercetin), fatty acids, and sulfur-containing compounds.[10] Bitter components of neem oil include the major sulfur-containing compound, nimbin, which accounts for about 1.5% weight/weight.[11] In addition to nimbin, azadirachtin-A, azadirachtin-B, and azadirachtin-H, other compounds in neem oil include azadirachtin-D, azadirachtin-I, deacetylnimbin, deacetylsalannin, and salannin.[12] Minor bitter constituents include nimbinin, nimbidin, gedunin, nimolicinol, epoxyazadiradione, and nimbidol. Commercial insecticidal formulations of neem oil typically list the concentration of azadirachtin A ($C_{35}H_{44}O_{16}$, CAS RN: 11141-17-6). The concentration of the other major tetranortriterpenoid compounds as listed in Figure 100.1 vary in these formulations depending on the source of the neem oil. The azadirachtin content in neem seeds depend primarily on genetic differences among neem trees rather than climatic factors. In a study of 28 neem trees in India, the azadirachtin A and B concentrations in kernels ranged from approximately 550–3,000 mg/kg and 40–590 mg/kg, respectively.[13] The major long-chained saturated and unsaturated fatty acids in neem oil are stearic, oleic, linoleic, and palmitic with smaller concentrations of arachidic, myristic, and behenic fatty acids.[14]

Physicochemical Properties

Neem oil is a yellow, malodorous extract with an unpleasant taste as a result of the presence of sulfur compounds. Tetranortriterpenoid compounds (e.g., azadirachtin, salannin, and deacetylgedunin) extracted from neem oil disrupt the metamorphosis of insects.[15] Although azadirachtin A decomposes in sunlight, acid/alkaline media, heat, and moisture, this compound is the currently accepted reference standard for neem-based products.[16]

Mechanism of Toxicity

The administration of neem oil from local commercial sources to mice produces fatty infiltration of the liver, cerebral edema, mitochondrial changes in the hepatocytes, and proximal renal tubule necrosis similar to Reye's syndrome. In animal models, the administration of neem oil (margosa oil) to mice produces the rapid onset of microvesicular liver steatosis and glycogen depletion as a result of mitochondrial injury manifest histologically by swelling, pleomorphism, loss of dense bodies, and loss of ribosomes.[17] Postmortem examination of children dying after the ingestion of margosa oil demonstrates swelling of hepatocytes, fatty metamorphosis of the liver, glycogen depletion, mitochondrial pyknosis, increased numbers of peroxisomes, and proliferation of smooth endoplasmic reticulum.[18] The ingredient(s) or contaminant(s) responsible for the hepatic effects of the neem oil have not been defined.

DOSE RESPONSE

Medicinal Use

There are no uses of neem oil approved by US health agencies. The estimated safe daily dose of azadirachtin is 15 mg/kg body weight.[19]

Toxicity

Typically, infants who developed encephalopathy received doses of neem oil (margosa oil) in the range of 5–30 mL.[20] The role of individual hypersensitivity, contaminants, and concomitant illness in the development of the encephalopathy remains undefined. Few of these case reports analyzed the ingredients in the neem oil involved with these poisonings. Common adverse effects of the medicinal use of neem oil include nausea and diarrhea. A dose of 10 mg/kg in rats produced stupor tremulousness, seizures, and death in 4 of 5 rats.[21] In subchronic rodent studies, the no-observed-effect level (NOEL) of azadirachtin technical 12% was 1,500 mg/kg per day for 90 days.[22] The intentional ingestion of one liter of neem leaf extract by a 24-year-old woman was associated with ventricular fibrillation and cardiac arrest in the emergency department that responded to standard measures of resuscitation.[23] She developed pulmonary edema in the hospital, but she responded to supportive care and she was discharged on the seventh hospital day.

TOXICOKINETICS

There are few data on the toxicokinetics of the toxic ingredients of neem oil including azadirachtin compounds.

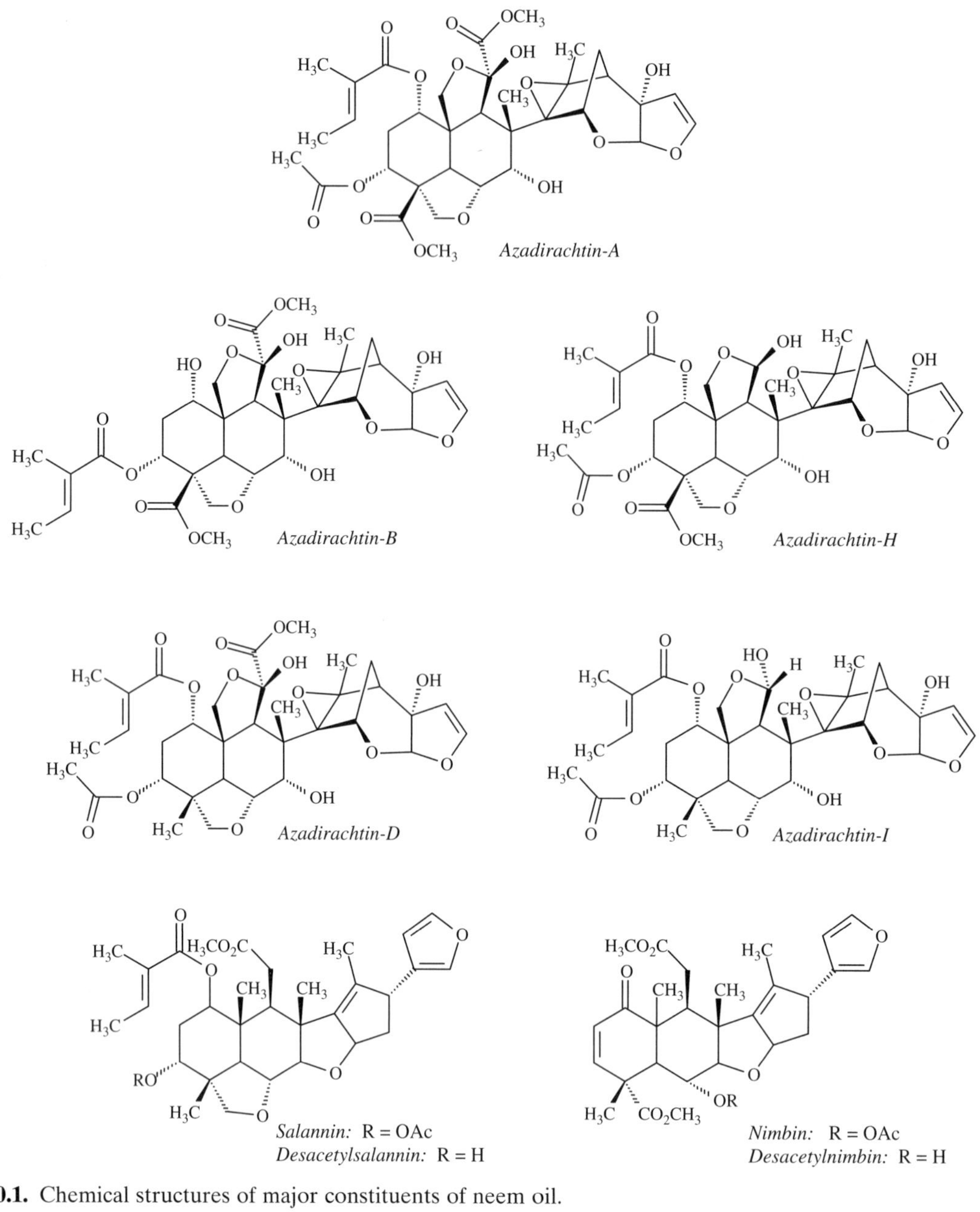

FIGURE 100.1. Chemical structures of major constituents of neem oil.

CLINICAL RESPONSE

Acute Effects

Case reports associate the onset of an encephalopathy similar to Reye's syndrome with the ingestion of neem oil (margosa oil) by infants and young children.[24,25] Clinical features include vomiting, drowsiness, tachypnea, and recurrent seizures that progress to coma or cardiopulmonary arrest.[26,27] These episodes typically involved the ingestion of unrefined neem oil that, at least in some cases, was contaminated with camphor.[24] Vomiting usually starts 15 minutes to 4 hours after ingestion. Although mild elevation of serum hepatic transaminases may occur, liver failure is unusual.[20] Common adverse effects from the use of neem oil as an herbal remedy include nausea and diarrhea.[26]

Chronic Effects

Animal studies suggest that the oral toxicity of azadirachtin compounds, the main biopesticides in neem oil, is relatively low.[22] In rodent studies, the daily administration of 1,500 mg azadirachtin technical 12%/kg during gestation days 6–15 did not produce evidence of embryo/fetotoxicity or teratogenic effects.[28] A three-generation rat study indicated that debitterized neem oil does not produce adverse reproductive or systemic effects after the ingestion of doses similar to the doses ingested by humans.[29]

DIAGNOSTIC TESTING

Analytical methods for separating and identifying compounds in neem extracts include enzyme-linked immunosorbent assay (ELISA),[30] reversed-phase liquid chromatography,[31] liquid chromatography/mass spectrometry with UV detection,[12] and high performance liquid chromatography/atmospheric pressure chemical ionization/mass spectrometry.[32] The limit of detection of the latter method for azadirachtin compounds is about 0.2 μg/mL. The most common laboratory abnormalities during serious cases of neem oil toxicity include leukocytosis, hypoglycemia, electrolyte imbalance, and metabolic acidosis.[24] Postmortem findings resemble Reye's syndrome with fatty infiltration of liver and proximal renal tubules, mitochondrial damage, and cerebral edema.[18] Fatty infiltration does not occur in Indian childhood cirrhosis, and the histological features of this disease differ markedly from neem oil poisoning.[33]

TREATMENT

Treatment is supportive. Decontamination measures are unnecessary because of the presence of vomiting and the rapid absorption of the liquid. Immediate management usually involves the control of seizures and associated hypoxemia with traditional anticonvulsants (diazepam, lorazepam) and respiratory support. Other serious abnormalities that require evaluation and treatment include hypoglycemia, metabolic acidosis, and dehydration. Cerebral edema may complicate the clinical course. The effect of neem oil on fatty acid metabolism suggests the L-carnitine as a potential antidote,[34] but there are inadequate clinical data to evaluate the efficacy of this substance during neem oil poisoning. The presence of a concomitant infection (sepsis, meningitis) should be excluded with appropriate laboratory tests.

References

1. Brahmachari G. Neem—an omnipotent plant: a retrospection. Food Chem Toxicol 2005;43:707–712.
2. van der Nat JM, van der Sluis WG, De Silva KT, Labadie RP. Ethnopharmacognostical survey of *Azadirachta indica* A. Juss (Meliaceae). J Ethnopharmacol 1991;3:1–24.
3. Ajaiyeoba EO, Oladepo O, Fawole OI, Bolaji OM, Akinboye DO, Ogundahunsi OA, et al. Cultural categorization of febrile illnesses in correlation with herbal remedies used for treatment in Southwestern Nigeria. J Ethnopharmacol 2003;85:179–185.
4. Seyoum A, Palsson K, Kung'a S, Kabiru EW, Lwande W, Killeen GF, et al. Traditional use of mosquito-repellent plants in western Kenya and their evaluation in semi-field experimental huts against *Anopheles gambiae*: ethnobotanical studies and application by thermal expulsion and direct burning. Trans R Soc Trop Med Hyg 2002;96:225–231.
5. Batra CP, Mittal PK, Adak T, Sharma VP. Efficacy of neem oil-water emulsion against mosquito immatures. Indian J Malariol 1998;35:15–21.
6. Mulla MS, Su T. Activity and biological effects of neem products against arthropods of medical and veterinary importance. J Am Mosq Control Assoc 1999;15:133–152.
7. Sharma V, Walia S, Kumar J, Nair MG, Parmar BS. An efficient method for the purification and characterization of nematicidal azadirachtins A, B, and H, using MPLC and ESIMS. J Agric Food Chem 2003;51:3966–3972.
8. Sharma SK, Dua VK, Sharma VP. Field studies on the mosquito repellent action of neem oil. Southeast Asian J Trop Med Public Health 1995;26:180–182.
9. Brahmachari G. Neem—an omnipotent plant: a retrospection. Chembiochem 2004;5:408–421.
10. Hallur G, Sivramakrishnan A, Bhat SV. Three new tetranortriterpenoids from neem seed oil. J Nat Prod 2002;65:1177–1179.
11. Prasad RBN, Venkob Rao S. Phospholipid composition of neem seeds. J Oil Technol Assoc India 1981;13:101–104.
12. Barrek S, Paisse O, Grenier-Loustalot MF. Analysis of neem oils by LC-MS and degradation kinetics of azadirachtin-A in a controlled environment. Characterization of degradation products by HPLC-MS-MS. Anal Bioanal Chem 2004;378:753–763.
13. Sidhu OP, Kumar V, Behl HM. Variability in neem (*Azadirachta indica*) with respect to azadirachtin content. J Agric Food Chem 2003;51:910–915.
14. Skellon JH, Thornburn S, Spence J, Chatterjee SN. The fatty acids of neem oils and their reduction products. J Sci Food Agric 1962;13:639–643.
15. Senthil Nathan S, Kalaivani K, Sehoon K, Murugan K. The toxicity and behavioural effects of neem limonoids on *Cnaphalocrocis medinalis* (Guenee), the rice leaffolder. Chemosphere 2006;62:1381–1387.
16. Kumar J, Parmar BS. Stabilization of azadirachtin A in neem formulations: effect of some solid carriers, neem oil, and stabilizers. J Agric Food Chem 1999;47:1735–1739.
17. Sinniah R, Sinniah D, Chia L-S, Baskaran G. Animal model of margosa oil ingestion with Reye-like syndrome. Pathogenesis of microvesicular fatty liver. J Pathol 1989;159:255–264.
18. Sinniah D, Baskaran G, Looi LM, Leong KL. Reye-like syndrome due to margosa oil poisoning: report of a case

with postmortem findings. Am J Gastroenterol 1982;77: 158–161.

19. Boeke SJ, Boersman MG, Alink GM, van Lon JJ, van Huis A, Dicke M, Rietjens IM. Safety evaluation of neem (*Azadirachta indica*) derived pesticides. J Ethnopharmacol 2004;94:25–41.
20. Sinniah D, Baskaran G. Margosa oil poisoning as a cause of Reye's syndrome. Lancet 1985;1:487–488.
21. Gandhi M, Lal R, Sankaranarayanan A, Banerjee CK, Sharma PL. Acute toxicity study of the oil from *Azadirachta indica* seed (neem oil). J Ethnopharmacol 1988;23: 39–51.
22. Raizada RB, Srivastava MK, Kaushal RA, Singh RP. Azadirachtin, a neem biopesticide: subchronic toxicity assessment in rats. Food Chem Toxicol 2001;39;477–483.
23. Sivashanmugham R, Bhaskar N, Banumathi N. Ventricular fibrillation and cardiac arrest due to neem leaf poisoning. J Assoc Physicians India 1984;32:610–611.
24. Sunderavalli N, Bhaskar Raju B, Krishnamoorthy KA. Neem oil poisoning. Indian J Pediatr 1982;49:357–359.
25. Lai SM, Lim KW, Cheng HK. Margosa oil poisoning as a cause of toxic encephalopathy. Singapore Med J 1990;31: 463–465.
26. Dhongade RK, Kavade SG, Damle RS: Neem oil poisoning. Indian Pediatr 2008;45:56–57.
27. Paranjothy M. Suspected contaminated margosa oil poisoning. Med J Malaysia 1978;33:17–19.
28. Srivastava MK, Raizada RB. Assessment of embryo/fetotoxicity and teratogenicity of azadirachtin in rats. Food Chem Toxicol 2001;39:1023–1027.
29. Chinnasamy N, Harishankar N, Kumar PU, Rukmini C. Toxicological studies on debitterized Neem oil (*Azadirachta indica*). Food Chem Toxicol 1993;31:297–301.
30. Hemalatha K, Venugopal NB, Rao BS. Determination of azadirachtin in agricultural matrixes and commercial formulations by enzyme-linked immunosorbent assay. J AOAC Int 2001;84:1001–1010.
31. Rao AD, Devi KN, Thyagaraju K. Isolation of antioxidant principle from *Azadirachta* seed kernels: determination of its role on plant lipoxygenases. J Enzyme Inhib 1998;14: 85–96.
32. Schaaf O, Jarvis AP, van der Esch SA, Giagnacovo G, Oldham NJ. Rapid and sensitive analysis of azadirachtin and related triterpenoids from Neem (*Azadirachta indica*) by high-performance liquid chromatography-atmospheric pressure chemical ionization mass spectrometry. J Chromatogr A 2000;886:89–97.
33. Sinniah D, Baskaran G, Vijayalakshmi B, Sundaravelli N. Margosa oil poisoning in India and Malaysia. Trans Roy Soc Trop Med Hyg 1981;75:903–904.
34. Koga Y, Yoshida I, Kimura A, Yoshino M, Yamashita F, Sinniah D. Inhibition of mitochondrial functions by margosa oil: possible implication in the pathogenesis of Reye's syndrome. Pediatr Res 1987;22:184–187.

Chapter 101

PEPPERMINT OIL
(*Mentha* x *piperita* L.)

HISTORY

Peppermint has a long history of medicinal use dating back to ancient Egypt, Greece, and Rome. The scientific name for peppermint (*Mentha* x *piperita*) derives from a mythological Greek nymph, Mintha, who transformed herself into the plant, and from the Latin word for pepper (*piper*). Peppermint is a hybrid plant; the first hybrids were propagated in Mitcham, England in 1696.[1]

BOTANICAL DESCRIPTION

Commercial peppermint is a sterile hybrid between spearmint (*Mentha spicata* L.) and water mint (*Mentha aquatica* L.). The sterility of this plant requires the spread of this clonal plant by vegetive propagation.

Common Name: Peppermint

Scientific Name: *Mentha* x *piperita* L. (pro sp.)

Botanical Family: Lamiaceae (menthes, mints)

Physical Description: This low-spreading plant grows up to 2–3 ft (~0.6–0.9 m) in height during the bloom of purple flowers. The lance-shaped, aromatic leaves are deeply notched when mature.

Distribution and Ecology: Peppermint is a perennial plant widely cultivated in Europe and North America. The northwestern United States is a major commercial area for growing peppermint as a result of favorable climatic conditions for this hybrid.

EXPOSURE

Sources

Products of *Mentha piperita* (peppermint) include *Mentha piperita* oil (peppermint oil, CAS RN: 8006-90-4), *Mentha piperita* leaf extract (peppermint leaf extract, CAS RN: 84082-70-2), *Mentha piperita* (peppermint) leaf, and *Mentha piperita* (peppermint) leaf water. American and European peppermint oils are volatile oils, steam distilled from the aerial parts of the flowering peppermint plant (*Mentha* x *piperita* L.). These oils are primarily fragrance components (i.e., 0.02–3%) of cosmetic formulations, such as dentifrices, mouthwash, breath fresheners, and skin-cleansing agents. The leaf water previously was used as a flavoring agent and fragrance component, but currently this formulation is not used. (–)-Menthol is the principal component of peppermint oil because this oil is not dementholized. Peppermint oil is available as enteric-coated capsules, soft gelatin capsules, a liquid, and bulk herb oil. Japanese peppermint oil is a steam distillation product of *Mentha arvensis* L. var. *piperascens*. This oil is primarily a source of menthol because of the poor flavor of the oil compared with American and European peppermint oils. (–)-Menthol is also prepared synthetically and crystallized from the essential oil of cornmint (*Mentha arvensis* var. *piperascens*).

The biosynthesis of (–)-menthol is a complex chemical process that involves eight enzymatic steps, beginning with the formation and cyclization of the monoterpene precursor, geranyl diphosphate, to the parent olefin,

Medical Toxicology of Natural Substances, by Donald G. Barceloux, MD

(–)-(4S)-limonene.[2] Monoterpene synthesis occurs in highly specialized secretory cells of the epidermal oil glands during two temporally distinct periods. The early biosynthetic process is the filling of the epidermal oil glands with (–)-menthone. During the late period of leaf development coincident with flower initiation, reduction of the C3-carbonyl of menthone by NADPH-dependent ketoreductases yields (–)-menthol and (+)-neomenthol.[3] Formation and storage of the essential oil of peppermint occurs in the oil glands (peltate glandular trichomes) on the aerial surfaces of *Mentha* species.

Medicinal Uses

TRADITIONAL

Traditional herbal uses of peppermint oil include the treatment of dyspepsia, bile duct disorders, bronchial spasm, cholelithiasis, intestinal colic, common cold, cough, cramps, dysmenorrhea, flatulence, inflammation of oral mucosa, influenza, and urinary tract infections. Caraway oil [(–) carvone] is frequently added to peppermint oil as a flavoring agent, particularly in preparations for the treatment of indigestion.

CURRENT

The major monoterpene, menthol, is a constituent of many pharmaceutical and oral health care products, teas, tobacco products, cosmetics and confectionary goods (chewing gum, candies). The use of peppermint oil is possibly efficacious for the treatment of irritable bowel syndrome.[4] A meta-analysis of five placebo-controlled, double-blind, clinical trails demonstrated a statistically significant ($P < 0.001$) global improvement in the symptoms associated with irritable bowel syndrome in patients treated with peppermint oil when compared with placebo groups.[5] However, the small number of patients ($n = 175$), short-term follow-up (<1 month), and methodological design flaws limit definitive conclusions regarding the efficacy of peppermint oil for the treatment of irritable bowel syndrome.

REGULATORY STATUS

The US Food and Drug Administration (FDA) approves the use of peppermint oil as a food additive and dietary supplement with GRAS (generally recognized as safe) status.

PRINCIPAL INGREDIENTS

Chemical Composition

(–)-Menthol is the major component (30–55%) of peppermint oil. Other constituents include menthyl acetate (2.8–10%), (–)-menthone (14–32%), (+)-isomenthone (1.5–10%), (+)-menthofuran (1–9%), 1,8-cineole (3.5–14%), along with minor amounts of *l*-limonene, *p*-menthane, α-pinene, phellandrene, menthyl isovalerate, amyl alcohol, acetaldehyde, dimethyl sulfide, α- and β-thujone, citronellol, γ-terpinene, and valeric acid.[6,7] Menthol usually refers to (–)- or *l*-menthol, which is one of eight possible stereoisomers of 1-methyl-4-(1-methylethyl)cyclohexan-3-ol. Figure 101.1 displays the chemical structure of (–)-menthol (CAS RN: 1490-04-6, $C_{10}H_{20}O$). Menthol accumulation in peppermint leaves occurs in leaves younger than 3 weeks with low catabolic losses after maximal leaf expansion. In general, high-quality peppermint oil (i.e., high menthol/menthone ratio) develops during full bloom.[8] This ratio increases somewhat during storage at room temperatures (i.e., 22 °C–24 °C/~72–75 °F).[9] The less desirable ketones (pulegone, carvone) are present in trace amounts during this period, and the concentration of pulegone in most commercial products is usually below 1% as a result of safety considerations.

Peppermint also contains (+)-pulegone, particularly during early plant development. In several samples of dried peppermint leaves from Italy, the (+)-pulegone concentration was approximately 7 mg/kg, whereas the concentration in several samples of peppermint oil was about 1–2.5%.[10] In the samples isolated from natural sources, (+)-pulegone represented about 99% of the isomers compared to about 1% for (–)-pulegone. The pulegone content in peppermint oil depends on season, growing conditions, extraction processes, and harvesting practices. In a 3-year study of pulegone content in American peppermint oil, the (+)-pulegone concentration ranged from 1.2–5.7% with a mean of about 3.6% ($n = 17$).[11] Isopulegone is usually present in smaller concentrations.

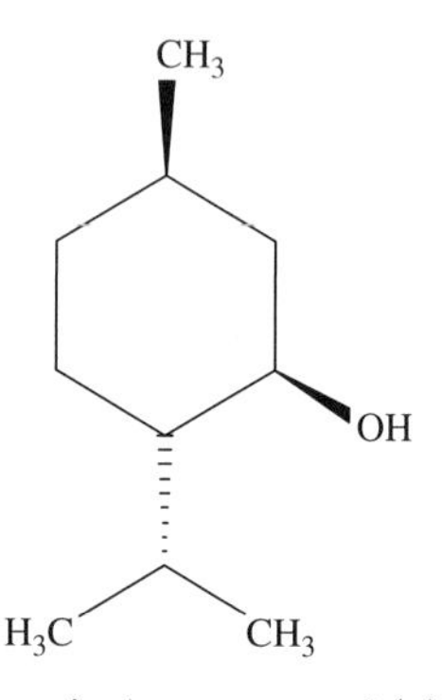

FIGURE 101.1. Chemical structure of (–)-menthol.

TABLE 101.1. Some Physiochemical Properties of Menthol

Physical Property	Value
Molecular Weight	156.26
log P (Octanol-Water)	3.380
Water Solubility	420 mg/L (25 °C/77 °F)
Atmospheric OH Rate Constant	2.41E-11 cm^3/molecule-second (25 °C/77 °F)

Physiochemical Properties

Peppermint oil is a colorless to pale yellow liquid with a strong, penetrating peppermint odor and a pungent taste. The enantiomer, (+)- or *d*-menthol has a musty, unpleasant odor; therefore, this enantiomer has little commercial value. Table 101.1 lists some physiochemical properties of (−)-menthol.

Mechanism of Action

In vitro studies in animal preparations suggest that menthol reduces intraluminal pressures in the colon by relaxing intestinal smooth muscle as a result of the calcium channel blocking properties of this compound.[12] However, the actual mechanism of action is unclear because some observational studies indicate that intraluminal administration of peppermint oil does not reduce directly measured intraluminal pressures in the colon of volunteers.[13]

DOSE RESPONSE

Typical oral doses of peppermint oil for indigestion are 0.2–0.4 mL in enteric-coated capsules or suspensions three times daily. Topical use of 10% peppermint oil (in methanol) involves the application of this solution to the forehead several times daily for relief of headache. The traditional use of peppermint leaves as an infusion entails the use of 3–6 grams of peppermint leaf daily. Doses of peppermint oil exceeding 1 g/kg body weight are considered potentially toxic, but there are few clinical data of the dose-response of peppermint oil. Concentrated peppermint oil is irritating to the oral mucosa, and high concentrations of peppermint oil can cause burns.

TOXICOKINETICS

Kinetics

The monoterpenes in peppermint oil are rapidly absorbed orally (i.e., maximum concentrations about 2 hours after ingestion) as a result of their high lipid solubility, and pharmaceutical preparations typically contain a sustained-release formulation to deliver these compounds to the lower gastrointestinal tract.[14] Following absorption, menthol undergoes conjugation with glucuronic acid to form menthol glucuronide, which is excreted by the kidney.[15] In a study of four fasted volunteers receiving 180 mg enteric-coated peppermint oil, about 40% of the menthol dose was recovered as total (free + conjugated) menthol within the first 14 hours after ingestion.[16] Menthol glucuronide also undergoes enterohepatic recirculation.[6]

Drug Interactions

Peppermint oil (K_i = 36 ± 3 µmol/L), menthol (K_i = 87 ± 7 µmol/L), and menthyl acetate (Ki = 124 ± 7 µmol/L) are moderately potent reversible inhibitors of *in vitro* CYP3A4 activity. In a study of 12 healthy volunteers, the use of peppermint oil (600 mg daily) increased the bioavailability of felodipine, CYP3A4-dependent drug by a mean of 140% (range: 77–262%) when compared with placebo (water).[17] However, the use of peppermint oil did not alter the AUC_{0-8}; therefore, the clinical relevance and underlying mechanism (e.g., inhibition of presystemic metabolism) remain undetermined.

CLINICAL RESPONSE

Adverse Reactions

Adverse effects associated with the use of peppermint oil include heartburn, headache, dizziness, and anal burning. Peppermint is a rare cause of hypersensitivity reactions following the use of personal care products, such as toothpaste.[18,19] The allergic reactions associated with the use of products containing peppermint oil include eczematous contact dermatitis, urticaria, recurrent intra-oral ulcerations, and lichenoid reaction.[20,21] Constituents (e.g., α-pinene, dipentene, phellandrene) other than menthol are usually the sensitizers, despite the high concentration of menthol in peppermint oil.[22] Case reports also associate chronic, excessive use of peppermint-containing compounds with stomatitis and hypertrophy of papillae on the tongue.[23]

Acute Effects

The intravenous injection of peppermint oil can produce adult respiratory distress syndrome (ARDS) and hypoxia as a result of increased pulmonary vascular permeability. An 18-year-old woman injected 5 mL of peppermint oil intravenously during a suicide attempt.[24] Within 4 hours she developed fulminant pulmonary edema and hypoxemia requiring ventilatory support for

9 days. After discharge from the hospital, she was lost to follow-up. The ingestion of pure peppermint oil can cause mucosal burns. A 49-year-old woman developed multiple burns in the entire oropharynx and edema of the tongue, soft palate, and epiglottis after ingesting 40 drops of pure peppermint oil.[25] She required endotracheal intubation, but she was extubated 24 hours later. The chemical burns resolved within 2 weeks.

DIAGNOSTIC TESTING

Methods for the detection and quantitation of menthol include gas chromatography with flame ionization detection,[16] reverse-phase high-performance liquid chromatography, and capillary gas chromatography/mass spectrometry.[10] The use of supercritical fluid chromatography and electrochromatography allows the enantioselectivity required to individually quantitate isomers of the chiral compounds in peppermint oil.[26]

TREATMENT

Treatment is supportive. The most serious adverse effects are anaphylactic reactions and chemical burns involving the oropharynx and epiglottis. Patients should be evaluated carefully for the adequacy of the airway and ventilation, and treated with standard measures (epinephrine, antihistamines, intubation) as needed.

References

1. Murray MJ, Lincoln DE, Marble PM. Oil composition of *Mentha aquatica* x *M. spicata* F1 hybrids in relation to the origin of *M.* x *piperita*. Can J Genet Cytol 1972;14:13–29.
2. Croteau RB, Davis EM, Ringer KL, Wildung MR. (−)-Menthol biosynthesis and molecular genetics. Naturwissenschaften 2005;92:562–577.
3. Davis EM, Ringer KL, McConkey ME, Croteau R. Monoterpene metabolism. Cloning, expression, and characterization of menthone reductases from peppermint. Plant Physiol 2005;137:873–881.
4. Grigoleit H-G, Grigoleit P. Peppermint oil in irritable bowel syndrome. Phytomedicine 2005;12:601–606.
5. Pittler MH, Ernst E. Peppermint oil for irritable bowel syndrome: a critical review and metaanalysis. Am J Gastroenterol 1998;93:1131–1135.
6. Analytical Methods Committee, Essential Oils Sub-Committee. Application of gas-liquid chromatography to the analysis of essential oils. Part VI. Determination of limonene and 1,8-cineole in oils of peppermint (varieties *Mentha*). Analyst 1978;103:375–381.
7. Grigoleit H-G, Grigoleit P. Pharmacology and preclinical pharmacokinetic of peppermint oil. Phytomedicine 2005;12:612–616.
8. Rohloff J, Dragland S, Mordal R, Iversen T-H. Effect of harvest time and drying method on biomass production, essential oil yield, and quality of peppermint (*Mentha* x *piperita* L.). J Agric Food Chem 2005:53:4143–4148.
9. Nair B. Final report on the safety assessment of *Mentha piperita* (Peppermint) Oil, *Mentha piperita* (Peppermint) leaf extract, *Mentha piperita* (Peppermint) leaf, and *Mentha piperita* (Peppermint) leaf water. Int J Toxicol 2001;20(Suppl 3):61–73.
10. Siano F, Catalfamo M, Cautela D, Servillo L, Castaldo D. Analysis of pulegone and its enantiomeric distribution in mint-flavoured food products. Food Addit Contam 2005;22:197–203.
11. Farley DR, Howland V. The natural variation of the pulegone content in various oils of peppermint. J Sci Food Agric 1980;31:1143–1151.
12. Hawthorn M, Ferrante J, Luchowski E, Rutledge A, Wei XY, Triggle DJ. The actions of peppermint oil and menthol on calcium channel dependent processes in intestinal, neuronal and cardiac preparations. Aliment Pharmacol Ther 1988;2:101–118.
13. Rogers J, Tay HH, Misiewicz JJ. Peppermint oil. Lancet 1988;ii:98–99.
14. Mascher H, Kikuta C, Schiel H. Pharmacokinetics of menthol and carvone after administration of an enteric coated formulation containing peppermint oil and caraway oil. Arzneimittelforschung 2001;51:465–469.
15. Somerville KW, Richmond CR, Bell GD. Delayed release peppermint oil capsules (Colpermin) for the spastic colon syndrome: a pharmacokinetic study. Br J Clin Pharmacol 1984;18:638–640.
16. Kaffenberger RM, Doyle MJ. Determination of menthol and menthol glucuronide in human urine by gas chromatography using an enzyme-sensitive internal standard and flame ionization detection. J Chromatogr 1990;527:59–66.
17. Dresser GK, Wacher V, Wong S, Wong HT, Bailey DG. Evaluation of peppermint oil and ascorbyl palmitate as inhibitors of cytochrome P4503A4 activity *in vitro* and *in vivo*. Clin Pharmacol Ther 2002;72:247–255.
18. Smith IL. Acute allergic reaction following the use of toothpaste. Br Dent J 1968;125:304–305.
19. Papa CM, Shelley WB. Menthol hypersensitivity; diagnostic basophil response in a patient with chronic urticaria, flushing, and headaches. JAMA 1964;189:546–548.
20. Foti C. Contact dermatitis from peppermint and menthol in a local action transcutaneous patch. Contact Dermatitis 2003;49:312–313.
21. Morton CA, Garioch J, Todd P, Lamey PJ, Forsyth A. Contact sensitivity to menthol and peppermint in patients with intra-oral symptoms. Contact Dermatitis 1995;32:281–284.
22. Dooms-Goossens A, Degreef H, Holvoet C, Maertens M. Turpentine-induced hypersensitivity to peppermint oil. Contact Dermatitis 1977;3:304–308.

23. Rogers SN, Pahor AL. A form of stomatitis induced by excessive peppermint consumption. Dental Update 1995; 22:36–37.
24. Behrends M, Beiderlinden M, Peters J. Acute lung injury after peppermint oil injection. Anesth Analg 2005;101: 1160–1162.
25. Tamir S, Davidovich Z, Attal P, Eliashar R. Peppermint oil chemical burn. Otolaryngol Head Neck Surg 2005;133: 801–802.
26. Gubitz G, Schmid MG. Chiral separation principles in chromatographic and electromigration techniques. Mol Biotechnol 2006;32:159–180.

Chapter 102

TEA TREE OIL
(*Melaleuca* Species)

HISTORY

Indigenous populations in Australia used extracts of the Australian tea tree (*Melaleuca alternifolia*) for the treatment of skin infections many centuries before the arrival of the Europeans. During the late 1800s, Captain James Cook documented the use of the leaves from this tree to brew tea. Commercial production of tea tree oil began in Australia during the 1920s after the publication of reports by Penfold and Grant on the antimicrobial properties of Australian essential oils.[1] These studies suggested that tea tree oil was a more efficacious antibacterial agent than phenol, which was the most common antimicrobial agent during this period. For several decades, tea tree oil was produced by the hand cutting the plant material followed by distillation in mobile, wood-fired bush stills. Production of tea tree oil declined during the 1940s as effective antibiotics became available.[2] As natural products became more popular during the 1970s and 1980s, high-density, commercial plantations of tea trees were established in Western Australia, Queensland, and New South Wales that produced the essential oil after extraction from cut and chipped plant material.

BOTANICAL DESCRIPTION

There are at least 100 species in the genus *Melaleuca*, and most of these species are indigenous to Australia. Although *Melaleuca alternifolia* (Maiden and Betche) Cheel is the common commercial source of tea tree oil, the hydro-distillation of other *Melaleuca* species can produce essential oils similar to tea tree oil including *M. linariifolia* Sm. (cajeput tree), *M. dissitiflora* F. Muell. and *M. uncinata* R.Br (broom honeymyrtle). Other *Melaleuca* species producing essential oils include *Melaleuca cajuputi* Powell (cajuput oil) and *Melaleuca quinquenervia* (Cav.) Blake (niaouli oil).

Common Name: Australian Tea Tree, Narrow-Leaved Paperbark

Scientific Name: *Melaleuca alternifolia* (Maiden and Betche) Cheel

Botanical Family: Myrtaceae (myrtles, myrtacées)

Physical Description: This hardwood tree grows to 5–8 m (~16–26 ft) in height. The bark of the tree is flaky (paperbark), and the lower bark becomes charred as a result of wildfires. Loose, white to cream-colored terminal spikes of small flowers appear from October to November. The stamens of these flowers are prominent, and the petals are small and inconspicuous.

Distribution and Ecology: The Australian tea tree is a native plant of eastern Australia. Other species of *Melaleuca* inhabit forests, woodlands, and swamps in New Zealand, New Guinea, New Caledonia, Indonesia, and Malaysia. The typical habitat of this tree is swampy, low-lying subtropical coastal areas of northeastern New South Wales and southern Queensland.

Medical Toxicology of Natural Substances, by Donald G. Barceloux, MD

EXPOSURE

Sources

Steam distillation or hydro-distillation (i.e., immersion in boiling water) of leaves and terminal branches of *M. alternifolia* produce tea tree oil (melaleuca oil) with a typical yield of about 1–2% weight of wet plant material. This essential oil should not be confused with essential oils from other *Melaleuca* species, such as cajuput oil, manuka oil, kanuka oil, or niaouli oil. These essential oils differ significantly in the composition of cyclic hydrocarbons when compared with tea tree oil.

Medicinal Uses

Traditional

Bundjalung Aborigines of northern New South Wales inhaled vapors from crushed tea tree leaves; they applied a poultice made of tea tree leaves to treat coughs, colds, insect bites, abrasions, and skin infections. An infusion of tea tree leaves was a traditional treatment for skin diseases and sore throats. The essential oil was not used until the publication of the reports by Penfold and Grant as mentioned earlier. Tea tree oil was added to cutting oils as a method to reduce infections associated with skin trauma during the use of these oils.

Current

Tea tree oil is an active ingredient of many topical preparations for the treatment of cutaneous infections including wound infections,[3] fungal dermatoses,[4] otitis media,[5] and acne.[6] However, there are presently inadequate clinical data to support the use of tea tree oil for any of these conditions.[7] Although several clinical studies suggest that tea tree oil possesses antimicrobial activity,[8] the poor quality of the studies (low numbers, lack of blinding due to the characteristic odor of tea tree oil, lack of statistical significance testing, absence of comparison to current therapy) prevent conclusions regarding the efficacy of tea tree oil as an antimicrobial agent.[2] Tea tree oil is present in a variety of household products including moisturizing cream, laundry detergents, fabric softeners, mouthwashes, massage oils, and candles, often as a substitute for nutmeg oil. Tea tree oil is an investigational drug as an antimicrobial, anti-inflammatory, and chemotherapeutic agent.

Regulatory Status

Current standardization recommendations suggest that tea tree oil preparations contain >30% terpinen-4-ol (terpinenol-4) and <15% 1,8-cineole (see Table 102.1).

TABLE 102.1. Typical Composition and International Organization for Standardization (ISO) Standard Composition of Tea Tree Oil

Compound	ISO Standard (%)[a]	Composition (%)[b]
Terpinen-4-ol	≥30	40.1
γ-Terpinene	10–28	23.0
α-Terpinene	5–13	10.4
1,8-Cineole	≤15	5.1
Terpinolene	1.5–5	3.1
ρ-Cymene	0.5–12	2.9
α-Pinene	1–6	2.6
α-Terpineol	1.5–8	2.4
Aromadendrene	Trace-7	1.5
δ-Cadinene	Trace-8	1.3
Limonene	0.5–4	1.0
Sabinene	Trace-3.5	0.2
Globulol	Trace-3	0.2
Viridiflorol	Trace-1.5	0.1

[a]ISO standard no. 4730.
[b]Mean composition of about 800 tea tree oil samples from Russell, 2003.[9]

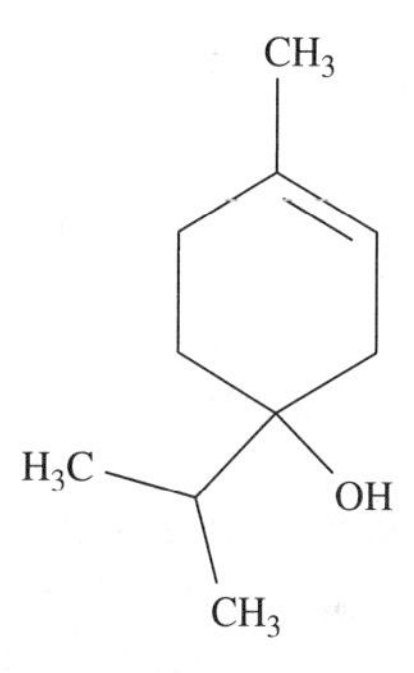

FIGURE 102.1. Chemical structure of terpinen-4-ol.

PRINCIPAL INGREDIENTS

Chemical Composition

Tea tree oil (melaleuca oil) contains over 100 compounds, consisting primarily of monoterpene and sesquiterpene hydrocarbons with their associated tertiary alcohols. A majority of the oil consists of terpinen-4-ol (CAS RN: 562-74-3, $C_{10}H_{18}O$) and γ-terpinene (CAS RN:8013-00-1, $C_{10}H_{16}$). Figure 102.1 displays the chemical structure of terpinen-4-ol. The composition and toxicity of tea tree oil varies with chemotype, season, growing conditions, and extraction techniques.[9,10] The leaves of seedlings contain higher concentrations of terpinolene, α-pinene, and β-pinene than mature leaves, while the concentrations of terpinen-4-ol, sabinene, and *cis*-sabinene hydrate are lower in the leaves of seedlings than in mature leaves.[11] The concentration of 1,8-cineole (eucalyptol) remained constant during the studied growing period. The chemical composition of mature and seedling leaves were similar when the seedling reached 4 months. Table 102.1 lists the average composition of terpinen-4-ol (terpinenol-4) chemotype of tea tree oil based on over 800 samples analyzed by gas chromatography/mass spectrometry (GC/MS).[12] This table also lists the standards set by the International Organization for Standardization (ISO) for the composition of 14 components of "oil of Melaleuca–terpinen-4-ol type."[13] The species of *Melaleuca* is not stipulated by the ISO standard, and potential sources of tea tree oil other than *M. alternifolia* include *M. dissitiflora, M. linariifolia*, and *M. uncinata*. However, essential oils from some *Melaleuca* species differ dramatically from the essential oils derived from *M. alternifolia*. For example, the essential oil from *Melaleuca ericifolia* Sm. (swamp paperbark) contains about 97% methyl eugenol,[14] and the leaves have unique phenolic compounds (ericifolin, 2-*O*-*p*-hydroxybenzoyl-6-*O*-galloyl-(α,β)-4C1-glucopyranose, 3-methoxyellagic acid 4-*O*-rhamnopyranoside).[15] There are at least six varieties (chemotypes) of *M. alternifolia* that produce unique essential oils including the terpinen-4-ol, terpinolene, and four types of 1,8-cineole (eucalyptol).[16] Terpinen-4-ol is the most common chemotype of tea tree oil in commercial production.

Physiochemical Properties

Tea tree oil is a transparent, colorless to pale yellow oil with a characteristic odor. This oil is soluble in nonpolar solvents and has a specific gravity ranging from about 0.885–0.906 g/mL.[13] Although terpinen-4-ol is soluble in water, α- and γ-terpinene are poorly water soluble. Table 102.2 lists some physical properties of the major component of tea tree oil, terpinen-4-ol. Tea tree oil undergoes photo-oxidation when exposed to light, moisture, heat, or air during storage. Under these conditions, the concentrations of α- and γ-terpinene decrease over a few days to several months, while the concentration of ρ-cymene increases.[12] *In vitro* studies suggest that tea tree oil has antimicrobial properties (terpinen-4-ol, α-terpineol) as a result of the disruption of bacterial cell wall integrity and respiration.[17] The estimated tropospheric lifetimes of monoterpene compounds (e.g., *d*-limonene, myrcene, α-pinene, γ-terpinene, terpinolene) is relatively short (<1–3.5 hours) with the lifetimes of related compounds (e.g., 1,8-cinoeole, *p*-cymene) being longer.[18]

TABLE 102.2. Some Physical Properties of Terpinen-4-ol

Physical Property	Value
Boiling Point	209 °C/~408 °F
Water Solubility	1491 ppm
log P (Octanol-Water)	3.26
Atmospheric OH Rate Constant	1.04E-10 cm^3/molecule-second (25 °C/77 °F)

Mechanism of Toxicity

The Melaleuca tree is probably not a significant source of aeroallergens.[19] The specific sensitizing agent or agents in tea tree oil are not well defined, but several compounds both in the monoterpene and sesquiterpene fractions probably are involved.[20] Potential sensitizing compounds in tea tree oil include *d*-limonene, α-pinene, α-terpinene, terpinen-4-ol, α-phellandrene, terpinolene, aromadendrene, and ρ-cymene.[21] Based on animal studies, fresh tea tree oil is a very weak sensitizing agent compared with degradation products, such as peroxides, epoxides and endoperoxides including ascaridol and 1,2,4-trihydroxy menthane.[22,23] 1,8-Cineole (eucalyptol) is probably not a major irritant or sensitizing agent in tea tree oil.[24]

DOSE RESPONSE

Medicinal Use

The use of tea tree oil typically involves the topical application of 5–10% solutions to the affected area with the exception of burns. The use of oral preparations is not recommended.

Toxicity

There are few data on the dose-related toxicity of tree tea oil in humans. The ingestion of an estimated volume of <10 mL of pure tea tree oil was associated with disorientation and ataxia that resolved spontaneously 5 hours after ingestion.[25] The ingestion of about 10 mL of pure tea tree oil by a 4-year-old child produced ataxia and respiratory depression that required intubation.[26] The patient was discharged 24 hours after ingestion with no sequelae. The ingestion of an estimated 0.5–1 mL/kg body weight by an adult was associated with coma for 12 hours followed by hallucinations and persistent, colicky diarrhea.[27]

TOXICOKINETICS

There are few human data on the toxicokinetics of components of tea tree oil.

CLINICAL RESPONSE

Cutaneous Reactions

Allergic reactions to tea tree oil include irritant and allergic contact dermatitis, systemic hypersensitivity reactions, and erythema multiforme-like eruptions, particular after the use of improperly stored oil on damaged skin. Case reports also associate the ingestion of tea tree oil with systemic contact dermatitis[28] and the inhalation of essential oil vapors with airborne contact dermatitis.[29] The typical skin lesion of the contact dermatitis associated with tea tree oil is an eczematous plaque in the area of topical application, but bullous lesions and an erythema multiforme-like reaction (target lesions distal to the initial dermatitis) can also occur after dermal use of tea tree oil.[30,31] Skin biopsy from these target lesions distal to the contact dermatitis demonstrated spongiotic dermatitis rather than the histological features of erythema multiforme.

Systemic Effects

Most accidental ingestions of tea tree oil cause no or minor symptoms as reported to US Poison Control Centers.[32] Case reports of accidental and intentional ingestions of tea tree oil indicate that the main complications of exposure result from central nervous system depression, loss of coordination, and respiratory insufficiency. The most common symptom following accidental ingestion of tea tree oil is central nervous system depression manifest by ataxia, drowsiness, and unsteady gait. Symptoms usually resolve within 6–12 hours depending on the dose of tree oil ingested.[33] Anaphylactic reactions may occur in sensitized individuals. Case reports associate the topical application of tea tree oil with the development of anaphylactic reactions manifest by facial erythema, throat constriction, lightheadedness, and pruritus.[34]

Chronic Effects

There is little evidence to suggest that tea tree oil is genotoxic or carcinogenic.[35]

DIAGNOSTIC TESTING

GC/MS methods are available to separate and quantitate the monoterpene hydrocarbons, sesquiterpene hydrocarbons, and related alcohols in tea tree oil.[12,36]

TREATMENT

Treatment is supportive. Because of the potential for central nervous system depression associated with the ingestion of tea tree oil, patients should be observed closely for evidence of respiratory insufficiency during the first few hours after ingestion. Decontamination measures are not recommended because of the lack of documented efficacy and the risk of pulmonary aspiration. Symptoms usually resolve within 6–12 hours with supportive care.

References

1. Penfold AR, Grant R. The germicidal values of the pure constituents of Australian essential oils, together with those for some essential oil isolates and synthetics. Part II. J Roy Soc New South Wales 1924;58:117–123.
2. Carson CF, Hammer KA, Riley TV. *Melaleuca alternifolia* (tea tree) oil: a review of antimicrobial and other medicinal properties. Clin Microbiol Rev 2005;19:50–62.
3. Halcon L, Milkus K. *Staphylococcus aureus* and wounds: a review of tea tree oil as a promising antimicrobial. Am J Infect Control 2004;32:402–408.
4. Buck DS, Nidorf DM, Addino JG. Comparison of two topical preparations for the treatment of onychomycosis: *Melaleuca alternifolia* (tea tree) oil and clotrimazole. J Fam Pract 1994;38:601–605.
5. Farnan TB, McCallum J, Awa A, Khan AD, Hall SJ. Tea tree oil: *in vitro* efficacy in otitis externa. J Laryngol Otol 2005;119:198–201.
6. Bassett IB, Pannowitz DL, Barnetson RS. A comparative study of tea-tree oil versus benzoyl peroxide in the treatment of acne. Med J Aust 1990;153:455–458.
7. Ernst E, Huntley A. Tea tree oil: a systematic review of randomized clinical trials. Forsch Komplementarmed Klass Naturheilkd 2000;7:17–20.
8. Tong MM, Altman PM, Barnetson RS. Tea tree oil in the treatment of *tinea pedis*. Australas J Dermatol 1992;33: 145–149.
9. Russell MF, Southwell IA. Monoterpenoid accumulation in 1,8-cineole, terpinolene and terpinen-4-ol chemotypes of *Melaleuca alternifolia* seedlings. Phytochemistry 2003; 62:683–689.
10. Baker GR, Lowe RF, Southwell IA. Comparison of oil recovered from tea tree leaf by ethanol extraction and steam distillation. J Agric Food Chem 2000;48:4041–4043.
11. Russell M, Southwell I. Monoterpenoid accumulation in *Melaleuca alternifolia* seedlings. Phytochemistry 2002;59: 709–716.
12. Brophy JJ, Davies NW, Southwell IA, Stiff IA, Williams LR. Gas chromatographic quality control for oil of *Melaleuca* terpinen-4-ol type (Australian tea tree). J Agric Food Chem 1989;37:1330–1335.

13. International Organisation for Standardisation. ISO4730:2004. Oil of Melaleuca, terpinen-4-ol type (tea tree oil). Geneva: International Organisation for Standardisation; 2004.
14. Farag RS, Shalaby AS, El-Baroty GA, Ibrahim NA, Ali MA, Hassan EM. Chemical and biological evaluation of the essential oils of different *Melaleuca* species. Phytother Res 2004;18:30–35.
15. Hussein SA, Hashim AN, El-Sharawy RT, Seliem MA, Linscheid M, Lindequist U, Nawwar MA. Ericifolin: an eugenol 5-*O*-galloylglucoside and other phenolics from *Melaleuca ericifolia*. Phytochemistry 2007;68:1464–1470.
16. Homer LE, Leach DN, Lea D, Slade Lee L, Henry RJ, Baverstock PR. Natural variation in the essential oil content of *Melaleuca alternifolia* Cheel (Myrtaceae). Biochem Syst Ecol 2000;28:367–382.
17. Hammer KA, Dry L, Johnson M, Michalak EM, Carson CF, Riley TV. Susceptibility of oral bacteria to *Melaleuca alternifolia* (tea tree) oil *in vitro*. Oral Microbiol Immunol 2003;18:389–392.
18. Corchnoy SB, Atkinson R. Kinetics of the gas-phase reactions of OH and NO_3 radicals with 2-carene, 1,8-cineole, *p*-cymene, and terpinolene. Environ Sci Technol 1990;24: 1497–1502.
19. Stablein JJ, Bucholtz GA, Lockey RF. Melaleuca tree and respiratory disease. Ann Allergy Asthma Immunol 2002; 89:523–530.
20. Rubel DM, Freeman S, Southwell IA. Tea tree oil allergy: what is the offending agent? Report of three cases of tea tree oil allergy and review of the literature. Austral J Dermatol 1998;59:244–247.
21. Crawford GH, Sciacca JR, James WD. Tea tree oil: cutaneous effects of the extracted oil of *Melaleuca alternifolia*. Dermatitis 2004;15:59–66.
22. Harkenthal M, Hausen BM, Reichling J. 1,2,4-Trihydroxy menthane, a contact allergen from oxidized Australian tea tree oil. Pharmazie 2000;55:153–154.
23. Hausen BM, Reichling J, Harkenthal M. Degradation products of monoterpenes are the sensitizing agents in tea tree oil. Am J Contact Dermat 1999;10:68–77.
24. Knight TE, Hausen BM. *Melaleuca* oil (tea tree oil) dermatitis. J Am Acad Dermatol 1994;30;423–427.
25. Jacobs MR, Hornfeldt CS. Melaleuca oil poisoning. J Toxicol Clin Toxicol 1994;32:461–464.
26. Morris MC, Donoghue A, Markowitz JA, Osterhoudt KC. Ingestion of tea tree oil (*Melaleuca* oil) by a 4-year-old boy. Pediatr Emerg Care 2003;19;169–171.
27. Seawright A. Tea tree oil poisoning—comment. Med J Aust 1993;159:831.
28. Elliott C. Tea tree oil poisoning. Med J Aust 1993;159: 830–831.
29. de Groot AC. Airborne allergic contact dermatitis from tea tree oil. Contact Dermatitis 1996;35:304–305.
30. Khanna M, Qasem K, Sasseville D. Allergic contact dermatitis to tea tree oil with erythema multiforme-like id reaction. Am J Contact Dermatol 2000;11: 238–242.
31. Bhushan M, Beck MH. Allergic contact dermatitis from tea tree oil in a war paint. Contact Dermatitis 1997;36: 117–118.
32. Watson WA, Litovitz TL, Rodgers GC Jr, Klein-Schwartz W, Reid N, Youniss J, et al. 2004 Annual report of the American Association of Poison Control Centers Toxic Exposure Surveillance System. Am J Emerg Med 2005;23: 589–666.
33. Del Beccaro MA. Melaleuca oil poisoning in a 17-month-old. Vet Hum Toxicol 1995;37:557–558.
34. Mozelsio NB, Harris KE, McGrath KG, Grammer LC. Immediate systemic hypersensitivity reaction associated with topical application of Australian tea tree oil. Allergy Asthma Proc 2003;24:73–75.
35. Hammer KA, Carson CF, Riley TV, Nielsen JB. A review of the toxicity of *Melaleuca alternifolia* (tea tree) oil. Food Chem Toxicol 2005;44:616–625.
36. Kim H-J, Chen F, Wu C, Wang X, Chung HY, Jin Z. Evaluation of antioxidant activity of Australian tea tree (*Melaleuca alternifolia*) oil and its components. J Agric Food Chem 2004;52:2849–2854.

Chapter 103

TUNG OIL
(*Aleurites fordii* Hemsl.)

HISTORY

Tung oil is an important raw material in the dye and oil-based paint industries. Prior to World War II, the major producers of tung oil were China, Japan, the United States, Mexico, and Indonesia. After the War, production of the oil increased in Africa (Madagascar, Malawi) and South America (Argentina, Brazil, Paraguay).

BOTANICAL DESCRIPTION

Misidentification of tung nuts as chestnuts has been associated with acute tung nut poisoning.[1]

Common Name: Tung Nut, Tung Oil Tree, China Wood Oil Tree

Scientific Name: *Aleurites fordii* Hemsl.

Botanical Family: Euphorbiaceae (spurge)

Physical Description: The tung oil tree reaches 12m (~39ft) in height with a wide trunk and smooth bark. The dark green leaves are up to 15cm (~6in.) wide with heart-shaped or occasionally lobed-shaped. The foliage turns yellow in the fall. White flowers appear in clusters during early spring with male and female flowers in same inflorescence. The green to purple fruits are pear-shaped at maturity with four to five individual seeds consisting of a hard outer shell and a kernel. The latter is the source of tung oil.

Distribution and Ecology: This tree is indigenous to central and western China. In 1932, tung oil trees were introduced into the southeastern United States and these trees are now present from eastern Texas to Florida. The current distribution of the tung oil tree also includes Japan, Taiwan, and India. Cultivation of tung oil trees requires long, hot summers with abundant moisture.

EXPOSURE

Sources

The thick, woody endocarp layer of the tung nut is the source of tung oil. The oil-rich seeds are produced inside a thick, woody endocarp layer of the tung nut, and these seeds are the source of tung oil used on fine furniture.

Medicinal Uses

TRADITIONAL

The tung oil tree and the oil extract are traditional herbal treatments for parasitic infections, scabies, burns, skin infections (erysipelas), congestion, constipation, mental illness, and metal poisoning.

CURRENT

Tung oil is a component of lacquers, varnishes, paints, inks, dyes, linoleum, resins, artificial leather, felt-base floor coverings, polishing compounds, and greases. The commercial use of tung oil primarily results from the unique ability of this oil to dry to a clear, hard finish following oxidation of α-eleostearic acid in the oil.[2]

Medical Toxicology of Natural Substances, by Donald G. Barceloux, MD

Additionally, tung oil is a constituent of insulating wires, metallic surfaces, and coatings for beverage, food, and medicine containers.

PRINCIPAL INGREDIENTS

Chemical Composition

Tung oil contains 75–80% α-eleostearic acid (CAS RN:13296-76-9; 9,11,13-*cis, trans, trans*-octadecatrienoic acid), 10–15% oleic acid, and 4% palmitic acid as well as tannins, phytosterols, and a saponin.[3] Figure 103.1 displays the chemical structure of eleostearic acid. Tung Δ^{12} oleate desaturase catalyzes the formation of α-eleostearic acid from linoleic acid.[4]

Physiochemical Properties

Tung oil is a viscous, dark yellow, opalescent oil with a specific gravity of 0.9341 at 15 °C–25 °C (59 °F–77 °F). This oil is relatively insoluble in alcohol and chloroform. The development of gastroenteritis after the ingestion of food cooked in tung oil suggests that the toxic constituent of tung oil is heat stable.[5] The conjugated triene in α-eleostearic acid is an intrinsic chromophore, which causes strong UV absorbance (269 nm) and mild fluorescent properties.[6]

Mechanism of Toxicity

The toxic components of tung oil have not been identified. Animal studies suggest that the tung nut contains the gastrointestinal toxin, but tung oil does not contain this toxin.[7] However, cooking food in tung oil has been associated with gastrointestinal toxicity.

DOSE RESPONSE

The ingestion of 6–15 tung nuts produces a severe gastroenteritis including colicky abdominal pain, protracted vomiting, and diarrhea.[1,8] Patients usually recover uneventfully with supportive care.

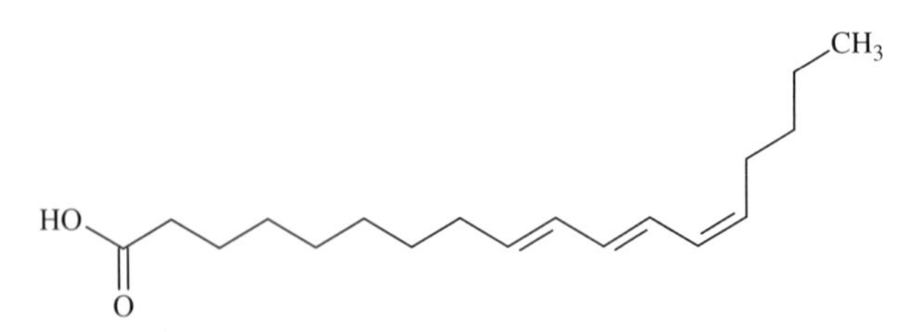

FIGURE 103.1. Chemical structure of eleostearic acid.

TOXICOKINETICS

The toxic component in tung oil has not been identified, and there are no data on the toxicokinetics of components in tung oil.

CLINICAL RESPONSE

Rarely, acute hypersensitivity reactions occur after exposure to dust from the tung nut including conjunctivitis, urticaria, rhinitis, and acute bronchospasm.[9] The ingestion of tung nuts causes severe gastrointestinal irritation.[10] In addition, the ingestion of tung oil or food cooked in tung oil produces a severe gastroenteritis. Profuse vomiting, diaphoresis, and palpitations developed within 15 minutes after the ingestion of food cooked in tung oil followed by abdominal pain and diarrhea.[5] Nonspecific symptoms include malaise, fatigue, lightheadedness, somnolence, sore throat, and chills.[1] Older medical reports suggest that central nervous system depression, mydriasis, respiratory insufficiency, and renal dysfunction can occur after the ingestion of tung oil,[7] but the etiology of these symptoms remains unclear because of poor documentation and the use of unconventional treatment methods by current medical standards.

DIAGNOSTIC TESTING

The primary laboratory abnormality associated with acute tung nut poisoning is electrolyte imbalance due to gastrointestinal losses. Occasionally, mild hyperglycemia and glucosuria occurs.[8] Typically, serum creatinine and hepatic aminotransferases are normal.

TREATMENT

Treatment is supportive, primarily involving the evaluation of the patient for fluid and electrolyte imbalances and the treatment of vomiting, diarrhea, and fluid/electrolyte disturbances caused by the gastroenteritis. Gastrointestinal decontamination measures are unnecessary.

References

1. Lin T-J, Hsu C-I, Lee K-H, Shiu L-L, Deng J-F. Two outbreaks of acute tung nut (*Aleurites fordii*) poisoning. Clin Toxicol 1996;34:87–92.
2. Li F, Larock RC. Synthesis, structure and properties of new tung oil-styrene-divinylbenzene copolymers prepared by thermal polymerization. Biomacromol 2003;4:1018–1025.

3. Lee YC, Nobles WL. A chemical study of American tung oil. J Am Pharm Assoc (Baltimore) 1959;48:162–165.
4. Dyer JM, Chapital DC, Kuan J-C, Mullen RT, Turner C, McKeon TA, Pepperman AB. Molecular analysis of a bifunctional fatty acid conjugase/desaturase from tung. Implications for the evolution of plant fatty acid diversity. Plant Physiol 2002;130:2027–2038.
5. Huisman J, Vasbinder H. A case of food poisoning caused by Chinese wood oil (tung oil). Trop Geogr Med 1961;13:183–185.
6. Pencreac'h G, Graille J, Pina M, Verger R. An ultraviolet spectrophotometric assay for measuring lipase activity using long-chain triacylglycerols from *Aleurites fordii* seeds. Anal Biochem 2002;303:17–24.
7. Balthrop E, Gallagher WB, McDonald TF, Camariotes S. Tung nut poisoning. J Florida Med Assoc 1954;11:813–820.
8. Balthrop E. Tung nut poisoning. South Med J 1952;45:864–865.
9. Macaulay DB. A case of tung nut sensitivity. Acta Allergologica 1952;5:304–307.
10. Erickson JL, Brown JH Jr. A study of the toxic properties of tung nuts. J Pharmacol Exp Ther 1942;74:114–117.

Chapter 104

TURPENTINE and PINE OIL (*Pinus* Species)

HISTORY

Oil of turpentine (terebinthinates) has been an herbal treatment for a variety of illnesses since ancient times. Hippocrates (460–370 BC) mentioned the value of turpentine oil as an emmenagogue (stimulant of menstrual flow) and inhibitor of nasal discharge.[1] The Greek physician, Dioscorides used turpentine oil as an aphrodisiac, diuretic, decongestant, and antiarthritic agent. The Romans used this oil for a wide variety of internal and external diseases including stroke, lethargy, depression, and pleurisy. From the late 17th century through the US Civil War, surgeons used turpentine in the treatment of wounds and amputations. In the Confederate Army, the external application of turpentine was a substitute for quinine in the treatment of malaria.[2] During the 18th and 19th centuries, turpentine oil was administered for the treatment of a variety of infections including croup, yellow fever, typhus, dysentery, and various parasites. Turpentine enemas were a treatment for abdominal distension, hysteria, and amenorrhea during the late 19th century.[3]

BOTANICAL DESCRIPTION

Common Name: Pines

Scientific Name: *Pinus* species

Botanical Family: Pinaceae

Physical Description: Pine trees have distinctive bundles of long, narrow needles and large, woody cones with tough scales. Branches commonly grow in rings around the trunk.

Distribution and Ecology: Pines are coniferous species, which are native to Central and North America, Europe, Asia, and small areas of North Africa. Pines are widely cultivated around the world including in Africa and South America. There are a variety of *Pinus* hybrids, and *Pinus* species are the most common conifers worldwide.

EXPOSURE

Sources

Oil of turpentine (gum turpentine) is the volatile, oily, liquid fraction derived from the steam distillation of pine resin. The latter material is obtained by tapping living species in the genus *Pinus* including *P. brutia* Tenore (calabrian pine), *Pinus elliottii* Engelm. (slash pine), *P. halepensis* P. Miller (Aleppo pine), *P. massoniana* D. Don (Masson pine, Chinese red pine), *P. merkusii* Junghuhn & Vriese ex Vriese (Merkus pine), *P. pinaster* Aiton (maritime pine), *P. radiata* D. Don (Monterey pine), and *P. sylvestris* L. (Scots pine). Other sources of turpentine include the extraction or destructive distillation of wood (wood turpentine) and by-products of chemical pulping of pine trees. Although technically not a petroleum distillate, most reviews include turpentine with petroleum distillates because turpentine is an aromatic hydrocarbon with properties, toxic effects, and uses similar to petroleum distillates. The skin sensitizer, colophony, is the nonvolatile, solid residue (rosin, gum rosin) that remains after the evaporation of volatiles from the distillation of crude turpentine.[4] Both colophony (rosin) and turpentine oil are distillation products of *Pinus* species, and the rosin used on violins and equestrian equipment includes both products. The high-temperature distillation of turpentine or other pine oil resins produces pine oil. Tall oil is dark-brown, liquid rosin, which is a by-product of pinewood in the wood pulp industry.

Uses

Traditional

Traditional uses of turpentine include use as an abortifacient,[5] antihelminthic agent,[6] rubefacient, decongestant, antipyretic, insect repellant,[7] and a treatment for human myasis (i.e., infection with dipterous larvae, such as maggots).[8] A tea brewed from the tips of pine branches was a substitute for coffee.

Current

Turpentine is a constituent of paint and other coatings, paint thinners, and solvents. The use of turpentine for these applications has decreased as other, less-expensive solvents have become available. However, the use of turpentine as a solvent remains common in developing countries. Other products containing turpentine include perfumes, liniments, and cleaning products. Turpentine is a raw product for the isolation of chemicals (e.g., camphor, citral, citronellal, isobornyl acetate, linalool, menthol, pine oil) in a wide range of commercial products. Pine oil is a solvent for varnish and polish, and synthetic pine oil is a constituent of fragrances and flavorings. Purification of monoterpene compounds in pine oil or the acid catalysis of pinene produces *l*-limonene, whereas citrus pulp or peels are the sources for the *d*-isomer of limonene. More recent formulations of pine oil-cleaning products contain smaller concentrations of pine oil and higher concentrations of constituents that reduce the viscosity and aspiration potential of the product.

PRINCIPAL INGREDIENTS

Chemical Composition

Turpentine is the volatile, predominantly terpenic, fraction or distillate resulting from the solvent extraction of softwoods (e.g., pine). The exact composition of turpentine oil depends on a variety of factors including extraction and refining methods, plant part, species, geographical location, and season. The major constituents of turpentine (CAS RN:8006-64-2) are terpene compounds ($C_{10}H_{16}$), such as α-pinene, β-pinene, camphene, 3-carene, and dipentene (*d,l*-limonene).[9,10] Other constituents include other acyclic, monocyclic, or bicyclic terpenes, oxygenated terpenes, and anethole. Monoterpenes in turpentine are volatile substances that occur naturally in air as a result of the emission of these substances from vegetation.[11] *d*-Limonene is a minor constituent of turpentine, but this compound is a major constituent of caraway, dill, celery, and oil from several fruits of the genus *Citrus* (lemon, orange, grapefruit).

Pine oil contains highly lipophilic secondary and tertiary cyclic terpene alcohols. Depending on the source of turpentine and the production methods, pine oil can contain terpene ethers and other hydrocarbons. Analysis of a sample of pine oil demonstrated 57% α-pinene, 26% carene, 8% β-pinene, 6% limonene, and 3% other hydrocarbons as measured by gas chromatography/mass spectrometry.[12] Individual constituents of the essential oil vary with *Pinus* species, extraction methods and plant parts. Table 104.1 lists the common constituents of the essential oil from the Balkan pine (*Pinus peuce* Gris.), North American pine (*Pinus ponderosa*, subspecies not reported), red pine (*Pinus resinosa* Aiton), and the eastern white pine (*Pinus strobus* L.).[13,14] Pine oil is usually present in household products along with other constituents (e.g., isopropyl alcohol, chloroxylenol), which may increase CNS and pulmonary toxicity.[15] Pine bark extract contains a variety of phenolic acid glucosides with anti-inflammatory properties, catechin compounds, taxifolin and taxifolin derivatives, procyanidins, and lignan glucosides.[16]

Physiochemical Properties

d-Limonene is an excellent solvent, but air oxidation of this substance produces potent sensitizing agents.[17] Turpentine also contains the strong sensitizer, Δ-3-carene.[18] These unsaturated cyclic hydrocarbons are very lipophilic. Turpentine is soluble in ether and alcohol, but insoluble in water. The water solubility of turpentine at 20°C–25°C (68°F–77°F) is 0.023% by weight and the vapor pressure at 25°C (77°F) is 5mmHg.[19] The ingestion of turpentine produces a violet-like odor of the breath and urine. The viscosity of pine oil is greater than turpentine, and therefore the aspiration hazard of pine oil is less than the aspiration hazard of the turpentine. The terpene compounds present in turpentine are released into the air both from natural (bark, cortex of conifers) and anthropogenic (wood processing) sources. Monoterpenes (α-pinene, β-pinene, 3-carene) are highly photoreactive in the atmosphere with relatively short atmospheric half-lives.[20,21] Some evaporation of monoterpene compounds occurs during the use of cleaning products containing pine oil, particularly α-phellandrene, terpinolene, γ-terpinene, *d*-limonene, camphene, and α-pinene.[22]

TABLE 104.1. Most Common Constituents as a Percentage of the Total Essential Oil from Several *Pinus* Species

Compound	*Pinus peuce*[a]	*Pinus peuce*[b]	*Pinus strobes*[b]	*Pinus ponderosa*[b]	*Pinus resinosa*[b]
β-Phellandrene	26.93	6.78	3.0[+]	2.4[+]	2.5[+]
β-Pinene	12.46	22.00	7.9	45.7	42.4
Citronellol	12.48	13.42	NR	NR	NR
α-Pinene	7.38	23.07	17.7	10.2	23.3
β-Caryophyllene	4.48	3.05	3.8	0.2	2.2
Myrcene	3.41	2.04	3.6	1.4	14.5
3-Carene	2.58	0.46	Trace	8.4	0.5
δ-Cadinene	1.60	0.65	7.5	3.1	0.4
Bornyl acetate	0.57	9.76	NR	NR	NR
Camphene	0.24	5.52	3.2	0.5	1.6
Terpinyl acetate	0.56	2.02	NR	NR	NR
Germacrene D	NR	NR	12.2	0.3	4.9
α-Cadinol	NR	NR	5.7	2.7	0.3
γ-Cadinene	1.16	0.36	2.8	0.9	0.1
α-Humulene	0.97	0.53	0.9	0.3	0.4

Source: Data from Refs 13 and 14.
[a]Twig oil.
[b]Needle oil.
[+]Includes limonene.
NR, not reported.

Mechanism of Toxicity

In vivo animal experiments indicate that the central nervous system is the primary organ of toxicity, whereas alveolar damage results from the aspiration of less volatile gums and resins.[23] Hence, lower volatility means aspiration risk is much lower for turpentine than with more volatile petroleum distillates. Colophony is also a potent skin sensitizer that previously contaminated turpentine. Older preparations of turpentine contained irritants (formic acid, aldehydes) and relatively high concentrations of skin sensitizers (δ-3-carene, α-pinene).[24,25]

DOSE RESPONSE

Ingestion

The toxic dose of turpentine or pine oil is not well defined because of the effects of aspiration, the variable composition of essential oils, and the presence of other toxic compounds in most turpentine and pine oil products. Most accidental exposures to turpentine or pine oil produce mild symptoms, if any, provided pulmonary aspiration does not occur. Ingestions exceeding 2 mL/kg probably are potentially serious.[26] The ingestion of the equivalent of about 200 mL pure pine oil by an adult was associated with coma, whereas the ingestion of the equivalent of approximately 30 mL by an 18-month-old child caused lethargy and ataxia.[27] Both patients recovered without sequelae. A 14-month-old child developed tachypnea, lethargy, and seizures after reportedly ingesting 120 mL turpentine.[28] Within 24 hours, the child was asymptomatic.

Inhalation

Volunteer studies indicate that eye, nose, and throat irritation occurs following exposure to turpentine concentrations in the range of 125–175 ppm.[29] Following exposure of volunteers to turpentine vapors at 75 ppm, several subjects developed mild nasal and throat irritation. Although 175 ppm was intolerable, most subjects could tolerate daily exposure to 100 ppm. The American Conference on Governmental and Industrial Hygienists (ACGIH) recommends a threshold limit value–time weighted average (TLV-TWA) of 100 ppm to minimize the potential for upper respiratory tract irritation. There is no short-term exposure limit (STEL) because of limited data. Both the US Occupational Safety and Health Administration permissible exposure limit (OSHA PEL) and the National Institute for Occupational Safety and Health recommended exposure limits (NIOSH REL) are consistent with the TLV-TWA. The NIOSH immediately dangerous to life or health (IDLH) value is 1500 ppm. There are no well-recognized chronic effects from inhalational exposure to turpentine other than contact dermatitis.

TOXICOKINETICS

The main components of turpentine are monoterpenes (e.g., α-pinene, β-pinene, 3-carene, *d*-limonene). The high solubility of monoterpenes in blood indicates that these compounds are rapidly absorbed from the lungs and stored in adipose tissues.[30] Based on volunteer studies, the respiratory uptake of these monoterpenes ranges from about 60–70%. In a study of eight volunteers exposed to turpentine at 450 mg/m^3 for 2 hours, the mean relative uptakes of α-pinene, β-pinene, and 3-carene were 62%, 66%, and 68%, respectively.[31] Metabolism accounts for most of the elimination of the ingested monoterpenes with relatively small amounts of unchanged monoterpenes appearing in the urine and in exhaled air. During experimental studies, the amount of unchanged α-pinene,[32] 3-carene,[33] and *d*-limonene[36] excreted in expired air were 8%, 3%, and 1%, respectively, whereas the amount of unchanged monoterpene excreted in the urine was <0.001%. Following the ingestion of pine oil, the metabolism and renal elimination of monoterpenes is relatively slow. The peak urinary excretion of the main metabolite, bornyl acetate, occurred approximately 5–6 days after the ingestion of pine oil.[34] Following exposure of volunteers to α-pinene, the kidneys eliminated about 4% of the dose as *cis*- and *trans*-verbenol.[35] Although a slow terminal elimination phase exists, volunteer studies indicate that most of a single, ingested dose of these monoterpenes is eliminated within 3–4 days.[36] α-Terpineol glucuronide disappeared rapidly from the urine of an 18-month-old child who was surreptitiously administered pine oil, and this compound was not detectable in urine samples obtained 12 days after admission.[37]

CLINICAL RESPONSE

Adverse Effects

Turpentine oil is both an irritant and a sensitizer. As reported to the German Information Network of Departments of Dermatology Multicentre Project, the rate of sensitization to oil of turpentine was approximately 2% in tested patients with a slight decline in recent years.[38] Although relatively rare, allergic contact dermatitis results from exposure to turpentine oil, particularly in atopic individuals working as painters, potters, perfumery workers, or violinists (fiddler's neck).[39,40] A painter with prior sensitization to colophony developed hand eczema after handling lottery tickets that contained colophony.[41] Cross-reaction occurs between turpentine and a variety of other essential oils including tea tree oil, peppermint, balsam of Peru, pyrethrum, chrysanthemum, and ragweed.[42,43]

Acute Toxicity

Most internal human exposures to turpentine or pine oil involve the accidental ingestion of turpentine-containing products by children. Turpentine oil is a marked gastrointestinal irritant that produces more gastrointestinal and central nervous system symptoms than a similar ingestion of a petroleum distillate.[37] However, the ingestion of turpentine products produces a significantly lower incidence of pneumonitis compared with petroleum distillates.[44] Clinical features following the ingestion of mixtures containing *Pinus* distillates (turpentine oil, pine oil) include mucosal irritation (nausea, vomiting, abdominal pain, sore throat, chest pain), mild respiratory tract irritation (cough, transient shortness of breath) and central nervous symptoms (headache, elation, confusion, ataxia, lethargy).[27] A violet-like odor to the breath and urine suggests ingestion of pine oil, although this odor is not always present. Coma, seizures, and death can result from large, intentional ingestions of turpentine or pine oil.[45] Symptoms usually develop within 2–3 hours following ingestion, and the toxic effects typically resolve within 24 hours.[26] Rare complications of turpentine or pine oil aspiration include acute respiratory distress syndrome (ARDS), bronchopleural fistula, pneumatoceles, lung necrosis, empyema,[46] pneumothorax, sepsis, and multiorgan failure.[47,48] Urinary tract irritation (dysuria, hematuria, frequency) may also occur.[49,50] The parenteral injection of approximately 5 mL of turpentine was associated with the development of hypoxemia and noncardiogenic pulmonary edema, followed by cellulitis and a sterile abscess at the injection site.[51] Similar respiratory abnormalities occurred in horses injected with pine oil for malicious purposes.[52]

Chronic Toxicity

Preliminary animal studies indicate that turpentine oil has weak tumor promoting properties.[53] Neither the International Agency for research on Cancer (IARC) nor the US Toxicology Program list turpentine oil or major ingredients in turpentine oil as suspected carcinogens.

DIAGNOSTIC TESTING

Analytical Methods

Turpentine oil contains many compounds. Terpenoid compounds increased the flexibility of the oil binders in oil paintings.[54] Analysis of these compounds in pine resins (e.g., Venice turpentine from *Larix decidua* Mill.) by gas chromatography (GC) and mass spectrometry

(MS) helps determine the best compounds for restorations of old paintings.[55]

Biomarkers

The blood turpentine concentration in an 85-year-old woman as measured by GC with flame ionization detector was 28 μg/mL, when compared with a standard consisting of the material involved with the poisoning.[56] At the time the sample was drawn, she was comatose, hypotensive, and apneic. 1-α-Terpineol is a major constituent of pine oil and a biomarker for ingestion of substances containing pine oil.[52] Postmortem blood from a 89-year-old Alzheimer's patient, who was found at home in cardiac arrest after drinking about 100 mL Pine-Sol® (Uline, Waukegan, IL), contained 25 mg/dL 1-α-terpineol as measured by gas chromatography/mass spectrometry.[57] Pine oil composed about 17–22% of this product, and about 50% of the pine oil consisted of 1-α-terpineol.

Abnormalities

Complications of the ingestion of turpentine include hematuria, thrombocytopenia,[58] and pneumatoceles.[59] The typical course of the pneumatocele is slow, progressive resolution of the radiographic changes over a period of weeks. Laboratory abnormalities following the ingestion of mixtures of monoterpene compounds can also include leukocytosis and mild elevation of hepatic aminotransferases.[12]

TREATMENT

The treatment of accidental exposures to turpentine is similar to the treatment of petroleum distillates. Patients should be observed for the development of signs of aspiration because the ingestion of turpentine frequently causes vomiting. Decontamination measures are not recommended because of the lack of documented efficacy and the risk of pulmonary aspiration. These patients should be admitted for observation for the development of respiratory depression and seizure activity. Asymptomatic children may be discharged after 3–4 hours of observation if the 2-hour postingestion chest x-ray is normal. Because monoterpenes are lipophilic compounds with relatively low blood concentration, the use of hemodialysis and hemoperfusion probably does not alter clinical outcome.[34]

References

1. Haller JS Jr. Sampson of the terebinthinates: medical history of turpentine. South Med J 1984;77:750–754.
2. Hasegawa GR. Quinine substitutes in the confederate army. Mil Med 2007;172:650–655.
3. Doyle D. Per rectum: a history enemata. J R Coll Physicians Edinb 2005;35:367–370.
4. Cronin E. Oil of turpentine—a disappearing allergen. Contact Dermatitis 1979;5:308–311.
5. Quander MF, Moseley JE. Abortion, chemical peritonitis and pulmonary edema following intrauterine injection of turpentine. Obstet Gynecol 1964;24:572–574.
6. Lanski SL, Greenwald M, Perkins A, Simon HK. Herbal therapy use in a pediatric emergency department population: expect the unexpected. Pediatrics 2003;111:981–985.
7. Ansari MA, Mittal PK, Razdan RK, Sreehari U. Larvicidal and mosquito repellent activities of pine (*Pinus longifolia*, Family: Pinaceae) oil. J Vect Borne Dis 2005;42:95–99.
8. Shaunik A. Pelvic organ myiasis. Obstet Gynecol 2006;107:501–503.
9. Mirov NT, Zavarin E, Bicho JG. Composition of gum turpentine of pines *Pinus nelsonii* and *Pinus occidentalis*. J Pharm Sci 1962;51:1131–1135.
10. Williams AL, Bannister MH. Composition of gum turpentines from twenty-two species of pines grown in New Zealand. J Pharm Sci 1962;51;970–975.
11. Fehsenfeld F, Calvert J, Fall R, Goldan P, Guenther AB, Hewitt CN, et al. Emissions of volatile organic compounds from vegetation and the implications for atmospheric chemistry. Global Biogeochem Cycles 1992;6:389–430.
12. Koppel C, Tenczer J, Tonnesmann U, Schirop TH, Ibe K. Acute poisoning with pine oil– metabolism of monoterpenes. Arch Toxicol 1981;49:73–78.
13. Koukos PK, Papadopoulou KI, Patiaka DT, Papagiannopoulos AD. Chemical composition of essential oils from needles and twigs of Balkan pine (*Pinus peuce* Grisebach) grown in Northern Greece. J Agric Food Chem 2000;48:1266–1268.
14. Krauze-Baranowska M, Maradarowicz M, Wiwart M, Poblocka L, Dynowska M. Antifungal activity of the essential oils from some species of the genus *Pinus*. Z Naturforsch [C] 2002;57:478–482.
15. Chan TY, Critchley AJ, Lau JT. The risk of aspiration in Dettol poisoning: a retrospective cohort study. Hum Exp Toxicol 1995;14:190–191.
16. Karonen M, Hamalainen M, Nieminen R, Klika KD, Loponen J, Ovcharenko VV, et al. Phenolic extractives from the bark of *Pinus sylvestris* L. and their effects on inflammatory mediators nitric oxide and prostaglandin E2. J Agric Food Chem 2004;52:7532–7540.
17. Karlberg A-T, Magnusson K, Nilsson U. Air oxidation of *d*-limonene (the citrus solvent) creates potent allergens. Contact Dermatitis 1992;26:332–340.
18. Dooms-Goossens A, Degreef H, Holvoet C, Maertens M. Turpentine-induced hypersensitivity to peppermint oil. Contact Dermatitis 1977;3:304–308.
19. Anonymous. Turpentine. Am Ind Hyg Assoc J 1967;28:297–300.

20. Stromvall A-M, Petersson G. Monoterpenes emitted to air from industrial barking of Scandinavian conifers. Environ Poll 1993;79;215–218.
21. Stromvall A-M, Petersson G. Photooxidant-forming monoterpenes in air plumes from Kraft pulp industries. Environ Poll 1993;79:219–223.
22. Singer BC, Destaillats H, Hodgson AT, Nazaroff WW. Cleaning products and air fresheners: emissions and resulting concentrations of glycol ethers and terpenoids. Indoor Air 2006;16:179–191.
23. Sperlin F. *In vivo* and *in vitro* toxicology of turpentine. Clin Toxicol 1969;2:21–35.
24. Lear JT, Heagerty AH, Tan BB, Smith AG, English JS. Transient re-emergence of oil of turpentine allergy in the pottery industry. Contact Dermatitis 1996;35:169–172.
25. Pirila V, Kilpio O, Olkkonen A, Pirila L, Siltanen E. On the chemical nature of the eczematogens in oil of turpentine. Dermatologica 1969;139:183–194.
26. Jacobziner H, Raybin HW. Turpentine poisoning. Arch Pediatr 1961;78:357–364.
27. Brook MP, McCarron MM, Mueller JA. Pine oil cleaner ingestion. Ann Emerg Med 1989;18:391–395.
28. Craig JO. Poisoning by the volatile oils in childhood. Arch Dis Child 1953;28:475–483.
29. Nelson KW, Ege JF Jr , Ross M, Woodman LE, Silverman L. Sensory response to certain industrial solvent vapors. J Ind Hyg Toxicol 1943;25:282–285.
30. Falk A, Gullstrand E, Lof A, Wigaeus-Hjelm E. Liquid/air partition coefficients of four terpenes. Br J Ind Med 1990;47:62–64.
31. Filipsson AJ. Short term inhalation exposure to turpentine: toxicokinetics and acute effects in men. Occup Environ Med 1996;53:100–105.
32. Falk AA, Hagberg MT, Lof AE, Wigaeus-Hjelm EM, Wang ZP. Uptake, distribution and elimination of alpha-pinene in man after exposure by inhalation. Scand J Work Environ Health 1990;16:372–378.
33. Falk AA, Lof AE, Hagberg MT, Wigaeus-Hjelm EM, Wang ZP. Human exposure to 3-carene by inhalation: toxicokinetics, effects on pulmonary function and occurrence of irritative and CNS symptoms. Toxicol Appl Pharmacol 1991;110:198–205.
34. Koppel C, Tenczer J, Tonnesmann U, Schirop T, Ibe K. Acute poisoning with pine oil: metabolism of monoterpenes. Arch Toxicol 1981;49:73–78.
35. Levin J-O, Eriksson K, Falk A, Lof A. Renal elimination of verbenols in man following experimental α-pinene inhalation exposure. Int Arch Occup Environ Health 1992;63:571–573.
36. Falk-Filipsson A, Lof A, Hagberg M, Wigaeus-Hjelm E, Wang Z. *d*-Limonene exposure to humans by inhalation: uptake, distribution, elimination, and effects on the pulmonary function. J Toxicol Environ Health 1993;38: 77–88.
37. Hill RM, Barer J, Hill LL, Butler CM, Harvey DJ, Horning MG. An investigation of recurrent pine oil poisoning in an infant by the use of gas chromatographic-mass spectrometric methods. J Pediatrics 1975;87:115–118.
38. Schnuch A, Lessmann H, Geier J, Frosch PJ, Uter W. Contact allergy to fragrances: frequencies of sensitization from 1996 to 2002. Results of the IVDK. Contact Dermatitis 2004;50:65–76.
39. Kuner N, Jappe U. Allergic contact dermatitis from colophonium, turpentine and ebony in a violinist presenting as fiddler's neck. Contact Dermatitis 2004;50:258–259.
40. Vente C, Fuchs T. Contact dermatitis due to oil of turpentine in a porcelain painter. Contact Dermatitis 1997;37: 187.
41. Pereira F, Manuel R, Gafvert E, Lacerda MH. Relapse of colophony dermatitis from lottery tickers. Contact Dermatitis 1997;37:43.
42. Heine G, Schnuch A, Uter W, Worm M. Frequency of contact allergy in German children and adolescents path tested between 1995 and 2002: results from the Information Network of Department of Dermatology and the German Contact Dermatitis Research Group. Contact Dermatitis 2004;51:111–117.
43. Laube S, Tan BB. Contact dermatitis from turpentine in a painter. Contact Dermatitis 2004;51:41–42.
44. Beamon RF, Siegel CJ, Landers G, Green V. Hydrocarbon ingestion in children: A six year retrospective study. JACEP 1976;5:771–775.
45. Pande TK, Pani S, Hiran S, Rao VV, Shah H, Vishwanathan KA. Turpentine poisoning: a case report. Forensic Sci Int 1994;65:47–49.
46. Khan AJ, Akhtar RP, Faruqui ZS. Turpentine oil inhalation leading to lung necrosis and empyema in a toddler. Pediatr Emerg Care 2006;22:355–357.
47. Welker JA, Zaloga GP. Pine oil ingestion a common cause of poisoning. Chest 1999;116:1822–1826.
48. Rodricks A, Satyanarayana M, D'Souza GA, Ramachandran P. Turpentine-induced chemical pneumonitis with broncho-pleural fistula. J Assoc Physicians India 2003;51: 729–730.
49. Chapman EM. Observations of the effect of paint on the kidneys with particular reference to the role of turpentine. J Ind Hyg Toxicol 1941;23:277–289.
50. Klein FA, Hackler RH. Hemorrhagic cystitis associated with turpentine ingestion. Urology 1980;16:187.
51. Wason S, Greiner PT. Intravenous hydrocarbon abuse. Am J Emerg Med 1986;4:543–544.
52. Tobin T, Swerczek TW, Blake JW. Pine oil toxicity in the horse: drug detection, residues and pathological changes. Res Commun Chem Pathol Pharmacol 1976;15:291–301.
53. Roe FJ, Field WE. Chronic toxicity of essential oils and certain other products of natural origin. Food Cosmet Toxicol 1965;3:311–324.
54. Cartoni G, Russo MV, Spinelli F, Talarico F. GC-MS characterisation and identification of natural terpenic resins employed in works of art. Ann Chim 2004;94:767–782
55. Osete-Cortina L, Domenech-Carbo MT, Mateo-Castro R, Gimeno-Adelantado JV, Bosch-Reig F. Identification

of diterpenes in canvas painting varnishes by gas chromatography-mas spectrometry with combined derivatisation. J Chromatogr A 2004;1024:187–194.

56. Troulakis G, Tsatsakis AM, Tzatzarakis M, Astrakianakis A, Dolapsakis G, Kotas R. Acute intoxication and recovery following massive turpentine ingestion: clinical and toxicological data. Vet Hum Toxicol 1997;28:155–157.

57. Cording DJ, Vallaro GM, DeLuca R, Camporese, Spratt E. A fatality due to accidental PineSol™ ingestion. J Anal Toxicol 2000;24:664–667.

58. Wahlberg P, Nyman D. Turpentine and thrombocytopenic purpura. Lancet 1969;2:215–216.

59. Bray A, Pirronti T, Marano P. Pneumatoceles following hydrocarbon aspiration. Eur Radiol 1998;8:262–263.

PART 4

TOXIC PLANTS

PART 4 TOXIC PLANTS

I Plant Dermatitis

Chapter 105

PLANT DERMATITIS

The plant kingdom represents a highly diversified biological system containing over 300,000 plant species. Based on the plant taxonomy developed by the Swedish botanist, Linné, Latin names describe families (capitalized, plural, no italic), genus (capitalized, italic, singular), and species (italic, singular). Exposure of humans to chemicals in plants causes several types of dermal lesions, collectively called phytodermatitis. Based on the mechanism of action, phytodermatitis is divided in the following classifications: (1) irritant contact dermatitis, (2) allergic contact and airborne allergic dermatitis, (3) phytophotodermatitis, (4) pharmacological injury, and (5) mechanical injury.[1] The stinging nettle (*Urtica dioica* L.) causes a mechanical injury secondary to contact with tiny, type-like bristles followed by a transitory (few minutes to hours) pruritic, urticarial rash. These lesions result from the injection of chemicals (e.g., histamine, serotonin, acetylcholine, leukotriene B_4, leukotriene C_4) following penetration of the skin by these bristles.[2,3] Isolated mechanical injuries result from contact with nontoxic bristles and spines (e.g., cactus) that may cause the formation of slow healing, foreign body granulomas.[4] Other sources of skin lesions following exposure to plants include pseudophytodermatitis (e.g., contact with arthropods or insecticides on the plant) and pseudophytophotodermatitis (e.g., contact with phototoxic chemicals from fungal infections in the plant).

Skin eruptions occur during all seasons, particularly during the warmer months when exposure to plants and pollens are high. Weeds, ornamental flowers, native vegetation, vegetables, sawdust, and plant products contain thousands of chemicals that can produce a contact dermatitis either by direct irritation or by an allergic reaction.[5] Host susceptibility and variation in the concentration of sensitizing chemicals caused by genetic factors, plant parts, environmental conditions, and season result in different clinical responses to dermatitis-inducing plants. Potent plant sensitizers (e.g., castor bean, western red cedar) can produce both contact dermatitis and asthma in the occupational setting.[6,7]

IRRITANT CONTACT DERMATITIS

Plant dermatitis (phytodermatitis) results both from allergic reactions and from direct irritation (i.e., nonallergic mechanisms). Irritant dermatitis is the most frequent dermatitis associated with plant exposure, and this type of dermatitis involves chemical and physical injury rather than an immunological response.[8,9] Table 105.1 lists the common plants that cause primarily irritation following direct exposure. The most common plant families involved with irritant contact dermatitis are Agavaceae (*Agave*) Brassicaceae (*Brassica*, *Eruca*, *Raphanus*, *Sinapis*), Ranunculaceae (*Aconitum*, *Anemone*, *Ranunculus*), Euphorbiaceae (*Croton*, *Euphorbia*, *Hippomane*), and Rutaceae. Skin lesions

Medical Toxicology of Natural Substances, by Donald G. Barceloux, MD

TABLE 105.1. Common Plant Irritants

Family	Genus and Species	Common Name
Agavaceae	*Agave americana* L.	Century plant, American aloe, maguey
Araceae	*Dieffenbachia seguine* (Jacq.) Schott	Dumb cane
	Philodendron cordatum Kunth ex Schott	Philodendron
Amaryllidaceae	*Narcissus jonquilla* L.	Jonquil
	Narcissus pseudonarcissus L.	Daffodil
	Narcissus poeticus L.	Pheasant's eye
Brassicaceae (Cruciferae)	*Armoracia rusticana* P.G. Gaertn., B. Mey. & Scherb.	Horseradish
	Brassica nigra (L.) W.D.J. Koch	Black mustard
	Sinapis alba L.	White mustard
	Rorippa nasturtium-aquaticum (L.) Hayek	Watercress
	Raphanus sativus L.	Radish
Euphorbiaceae	*Codiaeum variegatum* (L.) Blume	Croton
	Euphorbia helioscopia L.	Sun spurge
	Euphorbia pulcherrima Willd. ex Klotzsch	Poinsettia
	Euphorbia milii var. splendens (Bojer ex Hook.) Ursch & Leandri	Crown-of-thorns
	Hippomane mancinella L.	Manchineel tree
Liliaceae	*Hyacinthus orientalis* L.	Hyacinth
Ranunculaceae	*Anemone hupehensis* (hort. ex Lem.) Lem. ex Boynton	Japanese anemone
	Clematis vitalba L.	Clematis, Ol' man's beard
	Pulsatilla vulgaris Miller	Pasqueflower
	Ranunculus acris L.	Buttercup

Source: Adapted from Ref 1.

result from the combined effects of mechanical trauma by sharp plant parts and chemical reactions. These reactions involve low molecular weight acids (acetic, citric, formic, malic), glycosides, proteolytic enzymes, or calcium oxalate crystals that may become embedded in the skin as a result of trauma (e.g., use of chain saws).[8] In contrast to allergic contact dermatitis, almost all exposed individuals develop skin lesions after exposure to direct irritants. Reaction begins soon after exposure, and the clinical response depends both on the concentration of the irritant and on the condition of the skin. Clinical manifestations of irritant contact dermatitis resemble a chemical burn and include erythema with or without edema, hyperkeratosis, linear crusty abrasions, pruritic erythematous papules,[10] vesicles, urticaria, pruritic subepidermal bullae, purpura,[11] necrotic ulcers, and granulomas. Typically, these cutaneous lesions develop within minutes to several hours after exposure. Common histological findings include spongiosis, eosinophils, superficial perivascular lymphocytic infiltrates, and subcorneal pustules. Irritant and allergic types of reactions are not mutually exclusive, and mixed reactions may occur because some chemicals are both sensitizers and irritants. Although systemic symptoms are unusual, severe cases may be associated with headache, diarrhea, myalgias, fever, and leucocytosis.[12] Treatment is supportive.

ALLERGIC PHYTODERMATITIS

Oleoresins (essential oils) occur in a variety of plants and plant parts depending on genetic background, growing conditions, and season. These compounds are complex mixtures of aldehydes, aromatic alcohols, aliphatic and aromatic esters, phenols (catechol, hydroquinone, resorcinol), and terpenes.[1] These low-molecular-weight compounds are haptens that induce cell-mediated inflammatory reactions involving prior sensitization and the priming of memory and effector T lymphocytes.

The distribution and shape of eczematous skin lesions suggests the possibility of allergic contact dermatitis. Plant sources of allergic contact dermatitis include species in the Anacardiaceae (poison oak, ivy, sumac), Asteraceae (ragweed, feverfew, wormwood, burdock, chamomile, artichoke, chrysanthemum, yarrow, tansy) and Primulaceae (primrose) families as well as genera of liverwort (*Frullania*) and lichens (*Cladonia, Evernia, Parmelia, Usnea*).[13,14] Occupational exposure to the latter two group produces a chronic, lichenified, dermatitis of the hands, forearms, face, and neck in forest workers and wood cutters (e.g., wood cutter's dermatitis).[15] These lesions often are photosensitive. Common plants associated with skin reactions in the occupational settings include Christmas cactus (*Schlumbergera* hybrids), barberton daisy (*Gerbera* hybrids), poinsettia (*Euphorbia pulcherrima* Willd. ex Klotzsch), Chinese hibiscus (*Hibiscus rosa-sinensis* L.), *Campanula* spp., marguerite daisy (*Argyranthemum frutescens (L.)* Sch. Bip.), geranium (*Pelargonium* hybrids), tomato plant (*Solanum lycopersicum* var. *lycopersicum* L.), spathe flower (*Spathiphyllum* spp.), weeping fig (*Ficus benjamina* L.), primrose (*Primula* spp.), and *Begonia* hybrids.[16–18] Although allergic rhinitis and bronchial hyperreactivity as a result of exposure to *Asteraceae* pollens are common, immediate-type hypersensitivity responses to contact with these plants are relatively rare.[19]

Table 105.2 and Table 105.3 list common sensitizing agents in plants and trees. Typically, linear streaks of erythema, papules, and vesicles appear on the exposed skin of sensitized patients after a latent period up to 5–6 days. The severity of the dermatitis depends on the degree of exposure to the sensitizing agent as well as individual sensitivity. Cross-reactivity to antigens within the same plant genera (e.g., *Toxicodendron*) often occurs, but the presence of contact dermatitis does not correlate to symptoms of allergic rhinitis. In Oriental medicine, ingestion of lacquer from *Toxicodendron* species is a general health supplement and a treatment for gastrointestinal symptoms.[20]

TOXICODENDRON DERMATITIS (POISON OAK, SUMAC, AND IVY)

BOTANICAL DESCRIPTION

Poison ivy, poison sumac, and poison oak are indigenous to the United States, Canada, and northern Mexico. Although poison oak and poison ivy can grow in most

TABLE 105.2. Common Plant Sensitizers

Family	Plant	Common Name	Sensitizing Agent
Alliaceae	*Allium sativum* L. *Allium triquetrum* L.	Garlic onion	Diallyl disulfide, allylpropyldisulfide, allicin
Alstroemeriaceae	*Alstroemeria* spp.	Peruvian lily	Tulipalin A
Anacardiaceae		Poison ivy, poison oak, poison sumac, cashew nut (un-roasted), oriental lacquers, ginkgo tree, mango (stem, skin of fruit only), Brazilian pepper	Catechols (e.g., urushiol)
Araliaceae	*Dendropanax trifidus* (Thunb.) Makino ex Hara		Falcarinol
Brassicaceae		Wild garlic, horseradish, black mustard, winter cress, candy tuft, wild mustard	
Caryophyllaceae		Carnation, cauzeflower	
Asteraceae[a]	*Tanacetum parthenium* (L.) Schultz-Bip.	Feverfew	Parthenolide
		Elecampane	Alantolactone
	Inula helenium L.	Chrysanthemum, gerbera daisy	Sesquiterpene lactone
	Chrysanthemum spp.		
Euphorbiaceae		Spurge, castor bean	
Liliaceae	*Tulip* spp.	Tulip	Tulipalin A
	Hyacinthus spp.	Hyacinth	
	Lilium longiflorum Thunb.	Easter lily	
Primulaceae	*Primula* spp.	Primula	Primin (2-methoxy-6-pentylbenzoquinone)
Proteaceae	*Grevillea robusta* A. Cunningham ex R. Br.	Silky oak	Urishiol
Rutaceae		Citrus fruit (lime, orange, grapefruit)	

[a]The main sensitizers in the Asteraceae family are sesquiterpene lactones with the α-methylene-γ-butyrolactone moiety.[21] These compounds also occur in other families including magnolias, laurel bay (*Laurus nobilis* L.),[22] and liverworts.

TABLE 105.3. Common Sensitizing Agents in Trees

Sensitizing Agent	Sources	Common Name
Deoxylapachol, lapachol	*Tectona grandis* L. f.	Teak
(R)-3,4-Dimethoxy-dalbergione	*Machaerium scleroxylon* Tul.	Caviuna vermelha
2,6-Dimethoxy-1,4-benzoquinone	>50 Plant sources	Mahogany, macoré
Chlorophorin	*Chlorophora excelsa* Benth.	Kambala
Mansonone A	*Mansonia altissima A.* Chev.	Mansonia bété
Obtusaquinone	*Dalbergia retusa* Hemsl.	Cocobolo
Thymoquinone	*Cedar* species	Cedar
Cardol, anacardic acid	*Anacardium occidentale* L.	Cashew tree
Urushiol	*Toxicodendron verniciflurum* (Stokes) F. Barkley	Japanese lacquer trees
Ginkgolic acid	*Semecarpus anacardium* L. f.	Marking nut tree
	Mangifera indica L.	Mango tree
	Schinus terebinthifolius Raddi	Florida holly
	Ginkgo biloba L.	Ginkgo tree, Maidenhair tree
	Grevillea robusta A. Cunningham ex R. Br.	Silky oak

temperate zones throughout North America under the proper climatic conditions, these plants are not usually found in the deserts of Texas, Arizona, and central Mexico, the rain forests of Washington, Hawaii, Alaska, elevations above 4,000–5,000 feet (~1,200–1,500 m), or northern Canada.[23] These plants do not usually exist in the Rocky or Sierra Nevada Mountains or in the high plateaus of Nevada and Idaho.[24] Identification of poison oak and poison ivy is complicated by the regional variation of these plants, such as deviations from the clusters of three leaves. All plants in the genus *Toxicodendron* have pinnate leaves (i.e., multiple leaflets arranged along a midrib), but the number of leaves varies with species. Plants with an appearance similar to *Toxicodendron* species include squawbash (*Rhus trilobata* Nutt.), fragrant sumac (*Rhus aromatica* Ait.), Virginia creeper (*Parthenocissus quinquefolia* (L.) Planch.), and Boston ivy (*Parthenocissus tricuspidata* (Sieb. & Zucc.) Planch.).[25]

Poison Ivy

Common Name: Poison Ivy, Poison Vine, Mark Weed, Three-Leaved Ivy

Scientific Name: *Toxicodendron radicans* (L.) Kuntze

Physical Description: Although occasionally an erect shrub reaching 15–20 ft (~5–6 m) in height, poison ivy commonly grows as a woody vine. Hair-like, aerial rootlets anchor the vines of eastern poison ivy (*T. radicans*) to supporting structures. Northern or western poison ivy [*T. rydbergii* (Small ex Rydb.) Greene] grows as small shrub without aerial rootlets. The bright green, hairless, serrated or smooth leaves are pointed and grow in three leaflet clusters on a short stalk. In the fall, the leaves become various shades of red, yellow, and brown. Small greenish white flowers produce clusters of grayish brown fruit. Cultivated varieties of garden ivies belong to different botanical families than poison ivy.

Distribution and Ecology: Poison ivy is found in North and South America as well as eastern Asia. This species is found in every state except California and Nevada. Poison ivy is most abundant along the rocky coasts of New England, sandbars off New Jersey and Long Island, and throughout the Carolinas.

Eastern Poison Oak

Common Name: Eastern Poison Oak, Oakleaf Ivy

Scientific Name: *Toxicodendron quercifolium* (Michx.) Greene

Physical Description: This erect, deciduous, woody shrub has triple leaflets similar to those of poison ivy, but the leaf edges are sinuate like oak leaves.

Distribution and Ecology: This plant is present less commonly than poison ivy, and eastern poison oak inhabits an area bounded by Texas, northern Florida, and New Jersey.

Western Poison Oak

Common Name: Western Poison Oak, Pacific Poison Oak

Scientific Name: *Toxicodendron diversilobum* (Torr. & Gray) Greene

Physical Description: This woody, deciduous shrub grows to about 3–6 ft (~1–2 m) in wooded areas below 4,000–5,000 ft (~1,200–1,500 m). Many thornless woody stems arise from the underground roots. The lobed leaves usually appear in clusters of three, similar to oak leaves. Small white flowers appear from August to November followed by relatively large (7.5 mm) fruit that resemble cherries. In contrast to eastern poison oak (*T. pubescens* P. Mill.), western poison oak has aerial rootlets to anchor the vines to supporting structures (e.g., trees).

Distribution and Ecology: The habitat of western poison oak is the Pacific Coast from Canada to Mexico, particularly in canyons with abundant water.

Poison Sumac

Common Name: Poison Sumac, Poison Elder, Poison Ash, Swamp Sumac, Thunderwood

Scientific Name: *Toxicodendron vernix* (L.) Kuntze

Physical Description: Grows as a wood shrub (5–8 ft/~1–2 m) or as a small tree (up to 25 ft/~8 m) with light gray bark. Leaves appear as 7- to 13-leaflet clusters arranged in pairs, with a terminal leaflet at the end of the stem. Leaflet margins are smooth and pointed. The fruit is red.

Distribution and Ecology: Poison sumac occurs primarily in the southeastern and northeastern United States and in some midwestern areas, particularly in swamps and bogs east of the Mississippi River. This plant grows primarily in undisturbed areas where human activity is minimal.

ALLERGEN

Physiochemical Properties

Urushiol compounds are the oleoresin components of *Toxicodendron* species that cause contact dermatitis. Inside the undamaged plant, urushiol is a transparent oil that contains a mixture of diphenols (catechols) with long, hydrophobic alkyl side chains at C3 on the catechol ring. Upon exposure to oxygen, urushiol polymerizes to become a shiny, black lacquer. The volatility of urushiol is low, and this compound remains antigenic after drying on fomites.[26] The length of the side chain varies with species. Although the genetic and environment factors alter the concentration of catechols between different species and individual plants, all catechols in the family *Anacardiaceae* are structurally similar.[27] The antigenic structure in this family is the side chain at the C3 position of the orthocatechol, which is a pentadecyl (15-carbon) chain in poison ivy, a heptadecyl (17-carbon) side chain in poison oak, and a 13-carbon side chain in poison sumac. Although the chemical structure of urushiols differs between various species, sensitization to one species of *Toxicodendron* usually results in cross-sensitivity to other *Toxicodendron* species. Reducing the saturation and increasing the length of the side chain causes greater antigenicity. On the skin, the reduction of urushiol at the C1 and C2 positions produces highly reactive quinones, which are transformed into potent contact allergens.

In moist aerobic environments, urushiol oxidizes and polymerizes to form a brown-black lacquer. The sap from poison ivy, poison oak, and poison sumac turns dark brown within about 10 minutes and black by about 24 hours.[28] Urushiol adheres to many surfaces, including gloves, skin, clothes, equipment, and animal fur. Because of the solubility of catechols in rubber, urushiol penetrates rubber gloves, but not heavy-duty vinyl gloves.[29] Soap and water degrade urushiol, and a thorough machine washing removes dried urushiol from clothes. Once the urushiol-protein complex forms in the skin, washing will not reduce the antigenicity of the complex. Urushiol is nonvolatile; therefore, airborne contact dermatitis occurs only when smoke carries the urushiol on dust particles. Many members of the Anacardiaceae family of tropical and subtropical plants contain urushiol or similar substances. Cardol in mango (*Mangifera indica* L.) and urushiol demonstrate cross-reactivity with contact allergens of a variety of species from this family including poison oak and ivy (*Toxicodendron* species), rengas or black varnish tree (*Gluta renghas* L.), cashew nut tree (*Anacardium occidentale* L.), and the Indian marking nut tree (*Semecarpus anacardium* L. f.), and the Chinese lacquer tree [*Toxicodendron vernicifluum* (Stokes) F. Barkley].[26]

Sensitizers

Urushiol occurs throughout poison ivy and poison oak plants including the stems, roots, leaves, and vines. During the fall, the red, brittle leaves contain high concentrations of urushiol. In the winter season, the bare twigs of poison oak and poison ivy retain significant concentrations of urushiol in the sap. In a study of seasonal urushiol content of poison oak leaves, the urushiol content remained relatively stable (0.02–0.4%) from Spring to early November, when the urushiol content increased to 0.9–2.6%.[30] By December when the leaves fell, the urushiol content was low; therefore, fallen leaves are not allergenic. Urushiol flows through the resin channels under the exocarp of the fruit and diffuses into the leaves, stems, and roots. At high concentrations, urushiol is a contact irritant.

Mechanism of Action

Hypersensitivity to urushiol compounds in *Toxicodendron* species is a delayed cell-mediated immunological reaction that involves T lymphocytes. Urushiol, the immunogenic hapten of *Toxicodendron* species, elicits a predominantly $CD8^+$ T-cell response in the epidermis. Catechols from *Toxicodendron* plants are haptens that penetrate the skin to form reactive quinones. These compounds bind to a ligand on Langerhans cells in the epidermis and to macrophages and dendrocytes in the dermis.[31] $CD4^+$ T lymphocytes in the regional lymph nodes process the antigen presenting cells and begin to divide into effector and memory T cells that circulate throughout the body. Subsequent exposure to urushiol produces a cytotoxic response from the release of a variety of cytokines by the memory T-cells. The amount of redness, swelling, vesiculation and pruritus depends on the balance between the effector T cells and the T suppressor lymphocytes. Subsequently, a delayed hypersensitivity reaction occurs.

Susceptible Populations

Of all members of the animal kingdom, only humans and some other primates are sensitive to *Toxicodendron* species. The severity of the reaction to these plants depends on the degree of sensitization, the site of exposure (i.e., skin thickness, sun-exposed skin), and the quantity of antigen in contact with the skin. Children under the age of 5 years do not usually develop sensitivity to *Toxicodendron* species. Beyond 6–8 years of age, experimentally induced hypersensitivity is possible in up to 85% of that population.[32] Based on experimental studies in healthy volunteers, about 50% of the adult population develops hypersensitivity dermatitis following a transient contact (i.e., approximately 2.5 μg purified urushiol) to *Toxicodendron* species.[33] Increasing the dose of urushiol to 50 μg produces contact dermatitis in an additional 30–40% of the population. Consequently, about 15% of the population does not develop reactions to urushiol despite exposure to doses far exceeding the amount of urushiol normally present following exposure to *Toxicodendron* plants. Occupational groups with a high incidence of sensitivity to *Toxicodendron* species include farm workers, forestry workers, and fire fighters.[24] The high incidence of urushiol sensitivity in children of urushiol-sensitive parents suggests a significant genetic predisposition.[34]

CLINICAL RESPONSE

Typically, contact dermatitis produces a burning pruritic erythema over the area of contact in the configuration of linear scratch marks about 2–4 days after exposure to *Toxicodendron* species.[35] Highly sensitized patients may react to contact with *Toxicodendron* species within 8 hours, and patients without previous skin lesions may require 5–6 days to react. New lesions can appear up to 14 days after exposure.[36] The erythematous lesions progress to papules and vesicles. The skin lesions usually appear first on exposed areas in the configuration of contact (e.g., linear scratch marks). Secondary lesions may develop on areas distal from the initial contact as a result of repeated contact with oleoresins on animal fur or clothing. These lesions may develop in clumps because of variation in skin sensitivity (e.g., the eyelids and dorsum of the hand are more sensitive than the palm and hairy areas) and in the amount of secondary exposure. The fluid in the skin vesicles does not contain sensitizing agents. Rarely, black spots or linear lesions appear on the skin as a result of the deposition of urushiol on the skin that oxidizes and polymerizes to form a black lacquer.[37] Nausea, vomiting, diarrhea, and abdominal pain may accompany the skin lesions in highly sensitive individuals. Typically, the skin lesions resolve within approximately 10–14 days. The development of new skin lesions 2–3 weeks after initial contact suggests re-exposure to antigen reservoirs (e.g., pets, equipment, tools, clothing, fingernails). Sensitivity to urushiol decreases with age. In general, the onset of the hypersensitivity reaction is longer and the severity of the reaction is milder in older patients compared with a similar exposure in younger individuals.[38]

The most common complication following the development of contact dermatitis is secondary infection. Rarely, the systemic absorption of catechols causes urticaria, scarlatiniform eruptions, and erythema multiforme. A rare case report associated membranous nephropathy, proliferative glomerulonephritis, and arteritis with *Toxicodendron* exposures, particularly following the administration of commercial poison oak extract.[39] Hyperpigmentation and vitiligo may occur, particularly in dark-skinned races.

Ingestion of leaves may produce a severe gastroenteritis and pruritus ani. The development of immunity to *Toxicodendron* species following the chewing of its leaves by American Indians has not been documented.[40] Rarely, *Toxicodendron* species may produce urticaria, allergic rhinitis, and asthma. Differential diagnosis includes phytophotodermatitis, nummular eczema, chemical-induced irritant dermatitis, bedbugs (*Cimex lectularius* L.), Portuguese jellyfish envenomation, and caterpillar dermatitis. Skin lesions associated with the ingestion of *Toxicodendron* plants as herbal remedies include generalized maculopapular rash, erythroderma, vesiculobullous lesions, and erythema multiforme.[20]

TREATMENT

Decontamination

Urushiol binds quickly to the skin (i.e., <30–60 minutes), and the skin should be rinsed immediately with cool water beginning with the hands or other contaminated areas first. Urushiol is neutralized by alkaline soap, such as ordinary face soap or dishwashing liquid (pH 8.8–10.5); the affected skin should be washed with soap and water after the initial rinse, particularly the palms of the hands.[36] Vigorous scrubbing of the affected area should be avoided. Volunteer studies suggest that the removal of urushiol with soap or solvents reduces the severity of urushiol-induced lesions when initiated within 2 hours of exposure, but decontamination does not eliminate reactions in sensitized individuals.[41,42] The use of an organic solvent [acetone, isopropanol, Tecnu® outdoor skin cleanser (Tec Laboratories, Inc., Albany, OR)] may leach some of the urushiol from the skin up to 4–6 hours after an exposure; however, further exposure to urushiol must be avoided immediately after the use of these solvents because these solvents reduce the protective skin lipids and the natural barrier to urushiol. Following the use of these solvents, the affected skin should be washed with soap and water to remove any urushiol remaining on the surface of the skin. The addition of hydrogen peroxide or sodium hypochlorite to the water increases the oxidation of urushiol. All clothing should be changed and the contaminated clothing washed with soap and water.

Topical Steroids

High-potency fluorinated steroid gels 2–3 times per day are appropriate for localized skin lesions, but medium-potency steroid gels are more appropriate for large lesions. Topical steroids should be started before the appearance of the vesicles and continued for 2–3 weeks.[36] Moderate-strength topical steroids should be avoided on the face and genitalia.[26] Systemic steroids (adult dose 1–2 mg/kg daily for 7–10 days) may be required for more severe lesions.

Soaks

Cold water compresses or soaks 3–4 times daily for 15 minutes are recommended for weeping or pruritic lesions during the acute phase. The addition of aluminum subacetate [Burow's solution, Domeboro™ solution (Bayer HealthCare, Morristown, NJ)] may provide additional relief when applied as wet compresses of a 21/2% solution. Tepid (32 °C/90 °F) showers are preferred to hot showers because heat may cause vasodilatation, erythema, and edema.

Shake Lotions

Shake lotions (e.g., calamine) are topical preparations, which are combinations of powders and liquids that require physical agitation (i.e., shaking) prior to application. Calamine lotion helps dry oozing, wet skin lesions after the application of soaks. Shake lotions used to sooth the dermal lesions associated with poison ivy or poison oak should not contain antihistamines because of potential sensitization. Additionally, these topical products should not contain zirconium because of the potential to cause granulomas. Over-the-counter (OTC) soap mixtures (e.g., Zanfel™; Zanfel Laboratories, Inc., Morton, IL) are available to reduce the pruritus associated with *Toxicodendron* hypersensitivity reactions.

Systemic Steroids

Adequate doses of systemic corticosteroids given early in severe, edematous cases provide quick relief and may blunt the natural course of the skin lesions. At least 40 mg (0.5–0.75 mg/kg) of prednisone daily is often necessary. The dose is tapered during the second and third weeks. The dose is tapered by 10 mg every 4 days until the daily dose is 20 mg. If no rebound occurs after the administration of 20 mg every other day for one week, the prednisone is discontinued. At times, prolonged therapy (i.e., 5–6 weeks) is required.

Oral Antihistamines

Although the use of oral antihistamines may reduce the sensation of itch and sedate the patient, these drugs do not alter the course of the skin lesions.

Supportive Care

All contaminated clothing and domestic animals should be washed with soap and water. The lesions of poison oak are not usually covered with an occlusive dressing (i.e., left open), but large bullous lesions may require protection by a bulky dressings [e.g., Kerlix® (Kendall Company, Mansfield, MA) and nonadherent dressing]. These bullous lesions should not be lanced. The dressing may be soaked off with cool water to prevent disruption of the bullae. The skin lesions should be observed carefully for signs of secondary infection (i.e., increasing redness, warmth, pain).

Hyposensitization

No well-controlled clinical trials have demonstrated the efficacy of commercial extracts, and potential adverse reactions (gastrointestinal distress, pruritus ani, local inflammation) limit the usefulness of hyposensitization methods.[43] US Food and Drug Administration-approved barrier cream (IvyBlock®; Enviroderm Pharmaceuticals, Louisville, KY) is available when exposure of sensitized individuals to *Toxicodendron* plants is anticipated. These organoclay barrier compounds reduce, but do not eliminate hypersensitivity reactions.[44]

BULBS

There are approximately 250 genera of bulbous plants that are divided into true bulbs (daffodils, onions), tubers (dahlia), rhizomes (*Iris*), and corms (*Crocus, Gladiolus*). Bulbs commonly producing dermatitis include tulip, narcissus, hyacinth, and members of the Alliaceae (Allium) family (garlic, onion, chives).[45] Rare causes of contact dermatitis include contact with species from the family Ranunculaceae[46] and the genera *Iris, Ornithogalum*,[45] and *Scilla*.[47] These skin lesions usually result from irritation rather than an allergic reaction.

TULIP

The genus *Tulipa* is member of the Liliaceae family that is native to southwestern central Asia (Turkey). In 1593, the French botanist and physician, Carolus Clusius started the bulb trade in Holland by introducing *Tulipa gesneriana* L. into the medical garden of the University of Leiden. During severe food shortages in Holland at the end of World War II, tulips were a food source. Now there are over 2500 cultivars of *Tulip*. Contact dermatitis occurs following exposure to the plant sap, pollen, and the hard outer skin (tecta) of the tulip bulb, particularly in professional growers (digging, peeling, packaging, sorting) and to a lesser extent in florists.[48]

Typical features of contact dermatitis produced by tulips range from paronychia and dry erythematous, hyperkeratotic plaques on the finger tips to more extensive eczema.[49] The characteristic fissured, hyperkeratotic finger tips and pustular lesions under the nail are termed "tulip finger."[50] Paresthesias and discomfort ("tulip fire") may develop in the fingertips of exposed worker within several hours of handling tulip bulbs.[51] Occasionally, contact urticaria and bronchial hyperreactivity develop following exposure to tulips.[52,53] Hydrolysis of tuliposides in the tulip bulbs forms tulipalin A (α-methylene-γ-butyrolactone) and tulipalin B with the former being the primary sensitizer.[54] Although tulipalin A occurs throughout the tulip plant, the bulb and the pistils contain the highest concentrations.[55] Patch testing suggests that tulipalin A penetrates vinyl and latex gloves, but not nitrile gloves.[56] Tulipalin A also occurs in other members of the Liliaceae and Alstroemeriaceae families including *Alstroemeria* species (Peruvian lily),[57] glacier lily (*Erythronium grandiflorum* Pursh),[58] and *Lilium longiflorum* Thunb. (Easter lily).[59] Contact with the sap from these plants produces a contact dermatitis similar to tulip finger and, rarely, bronchial hyperreactivity.[60,61] The tulip does not contain clinically significant quantities of calcium oxalate crystals. Patch tests are available to determine individual sensitivities to allergens in bulbs.[62]

NARCISSUS

Narcissus is a native plant of the Mediterranean, North Africa, and southwest Asia. The source of most cultivars is *Narcissus pseudonarcissus* L. (wild daffodil), *N. jonquilla* L. (joquil), *N. poeticus* L. (pheasant's eye), *N. tazetta* L. (tazetta), and *N. triandrus* L. (angel's tears).[45] Narcissus bulbs contain needle-shaped calcium oxalate crystals as well as several weak sensitizers (homolycorine, masonin).[63] These sensitizers are not present in the stems and leaves of narcissus. True contact allergen to narcissus is relatively rare, and almost all skin lesions following exposure to narcissus results from an irritant dermatitis.[63] Consequently, most severe reactions result from the contact of the sap with damaged skin.

HYACINTH

The native habitat of hyacinth is the eastern Mediterranean. There are many cultivars, probably derived from *Hyacinthus orientalis* L. The brown outer covering (tecta) of the hyacinth bulb has several layers that contain high concentrations (i.e., up to 6%) calcium oxalate crystals. Mechanical microtrauma caused by these crystals produces skin irritation and "hyacinth itch." Although there is substantial individual variation, most individuals develop an irritant dermatitis following exposure. Pruritus and urticaria are the most common features of hyacinth itch, primarily in the volar surface of the forearms, the face, thighs, or inguinal region.[64] The itching usually resolves following removal of the irritant from the skin. Eczema, hyperkeratosis, and fissuring of the finger tips occur occasionally. True allergic dermatitis or occupational asthma as a result of exposure to hyacinth is rare.[65]

ONION AND GARLIC (*Allium Species*)

Important species in the genus *Allium* include chives (*A. schoenoprasum* L.), garlic (*A. sativum* L.), and onion (*A. cepa* L.). Bulbs from many *Allium* species contain irritant sulfur compounds (isothiocyanates). Potential allergens in garlic include diallyldisulfide, allylpropyldisulfide, and allicin.[45] Most skin lesions from exposure to onions cause an irritant dermatitis rather than allergic contact dermatitis.

TREES

Table 105.3 lists common sensitizing compounds in trees. The oil between the inner kernel and the outer layer of the cashew nut from the cashew tree (*Anacardium occidentale* L.) contains a variety of antigenic substances, including cardol and anacardic acid.[66] Roasting destroys these sensitizing agents. Contact dermatitis, stomatitis, or pruritus ani may result from the consumption of inadequately processed cashew nuts[67] or from exposure of sensitized individuals to swizzle sticks or voodoo dolls made from the cashew nut tree.[68] Cross-reactivity to *Toxicodendron* species occurs in patients sensitized to cashew nut oils because of the structural similarity between cardol and catechols in *Toxicodendron* species.[39] The antigenic oleoresins (urushiol, cardol) in mango demonstrate cross-reactivity with similar compounds in *Toxicodendron* species.[69]

Oriental lacquers derived from lacquer trees contain the same sensitizers (urushiols) that are present in *Toxicodendron* species.[70,71] These lacquers retain sensitizing properties indefinitely, but the sensitizing agent is sensitive to heat.[72] The fruit from the marking nut tree (*Semecarpus anacardium* L. f.) also contains an urushiol that produces contact dermatitis. The oil hardens into a black substance used by the laundrymen (dhobis) of India to mark clothes. The dhobi mark remains indefinitely despite repeated washings. A case report associated the development of acute renal failure and acute cortical necrosis with exposure to sap of the marking-nut tree (*Semecarpus anacardium* L. f.).[73] The fruit pulp of the maidenhair tree and ginkgo tree (*Ginkgo biloba* L.) contains catechols that produce cheilitis, stomatitis, pruritus ani, and proctitis in sensitized individuals. The sensitizing agents are not present in the seeds.

AIRBORNE CONTACT PHYTODERMATITIS

During dry windy conditions, sensitizing chemicals (e.g., monoterpenes, sesquiterpene lactones) concentrate in dried portions of the plant and subsequently these chemicals become airborne.[74,75] Occupational exposure accounts for most cases of airborne allergic phytodermatitis. Sources of airborne contact dermatitis include plants from the Asteraceae (Aster) family including ragweed (*Ambrosia* species), feverfew (*Tanacetum parthenium* L.),[76] and *Parthenium hysterophorus* L. (wild feverfew), garlic, sunflower, citrus fruits, thyme, soybean, rosewood, tea tree oil, essential oils, colophonium, yarrow, and lichens.[77] Seasonal peaks of allergic rhinitis and airborne allergic dermatitis are similar. Airborne allergic dermatitis may develop following fires as a result of the adsorption of sensitizing oleoresins to airborne ash. The deposition of these allergenic particles on exposed skin surfaces (upper eye lid, neck, upper chest, hands, wrists) produces pruritic macules and papules on exposed areas. Vesicles and bullae are uncommon. Lesions may appear on the trunk and lower extremities as a result of contact with contaminated clothing. In contrast to the linear lesions of direct contact dermatitis, airborne allergens produce a more uniform involvement of the hands, face, and other exposed areas. Although airborne contact phytodermatitis commonly occurs with a prominent demarcation at the midbiceps line and the upper sternal V line similar to phytophotodermatitis, airborne allergic dermatitis often involves the submental and postauricular regions, whereas phytophotodermatitis does not typically produce lesions in these areas. Airborne allergic dermatitis also involves other shaded portions of the body (e.g., upper eyelids, scalp covered by hair).[78] The treatment of airborne contact dermatitis and allergic contact dermatitis are similar depending primarily on the severity of the skin lesions.

PHYTOPHOTODERMATITIS

The ability of plants to cause hyperpigmentation was recognized by the ancient Egyptians, who used juice from *Ammi majus* L. (false Bishop's weed) and sunlight to treat vitiligo.[79] Phytophotodermatitis is a skin eruption that results from vigorous contact with crushed plants containing heterocyclic furan compounds including linear furocoumarin (psoralen, 5-methoxypsoralen, 8-methoxypsoralen) and angular furocoumarin (angelicin, pimpinellin). In the presence of sunlight (UVA), these compounds form covalent bonds with pyrimidine bases that result in single DNA adducts and cross-linked DNA strands. The cross-linking of DNA in the epidermis causes a rapid, temporary decline in RNA and DNA synthesis.

Clinically, the combination of sunlight, exposure to furocoumarin compounds, and moist skin produces a burning erythema in irregular patches on sun-exposed skin followed by edema and vesicles. This reaction involves a toxic rather than an immunological response, and most patients respond in a similar fashion to the same dose of psoralens.[80] A variety of plant families contain furocoumarin compounds (psoralens) including Brassicaceae (Cruciferae or mustard), Moraceae (figs), Ranunculaceae (buttercup), Rutaceae (bergamot, common rue, gas plant, lemon, lime), and Apiaceae or Umbelliferae (angelica, carrots, celery, dill, fennel, giant hogweed, wild parsnip).[8] Some of these plant species contain photosensitizing coumarin compounds other than psoralens, such as bergapten and xanthotoxin present in limes.[81] Table 105.4 lists the common causes of phytophotodermatitis. This disease occurs in occupational settings, such as food processing (celery, parsley)[82] and gardening (wild parsnip, gas plant).[83] Fresh, healthy celery plants do not contain clinically significant amounts of psoralen.[84]

The appearance of phytophotodermatitis is similar to severe sunburn[85] or jellyfish envenomation.[86] Typically, linear, burning macules develop soon after exposure and these lesions progress to papules and, in severe cases, bullae over 12–36 hours. The occurrence of maximal photosensitivity depends on the toxicokinetics of the involved furocoumarin range from 2–3 hours (5-methoxypsoralen) to approximately 32–36 hours for 8-methoxypsoralen. These lesions persist several days and often hyperpigmentation begins about 1–2 weeks after exposure. The major differential diagnosis involves *Toxicodendron* species, which produce more pruritus and a delayed response. The treatment for poison ivy and phytophotodermatitis are similar with immediate decontamination with soap and water being the first priority.

TABLE 105.4. Common Sources of Phytophotodermatitis

Family	Species	Vernacular Name
Apiaceae		
	Ammi majus L.	Bishop's weed, Artrillal
	Anethum graveolens L.	Dill
	Angelica archangelica L.	Angelica
	Apium graveolens L.	Celery
	Daucus carota L.	Carrot
	Foeniculum vulgare P. Mill.	Fennel
	Heracleum spp.	Giant hogweed, cow parsnip
	Pastinaca sativa L.	Parsnip
	Peucedanum spp.	Masterwort
	Pimpinella anisum L.	Anise
Clusiaceae		
	Hypericum spp.	St. John's wort
Brassicaceae		
	Brassica spp.	Mustard
Fabaceae		
	Psoralea corylifolia L.	Babchi
Chenopodiaceae		
	Chenopodium spp.	Goosefoot
Moraceae		
	Ficus carica L.	Fig tree
Ranunculaceae		
	Ranunculus spp.	Buttercup
Rosaceae		
	Agrimonia eupatoria L.	Agrimony
Rutaceae		
	Citrus aurantium subsp. *bergamia* (Risso & Poit.) Wight & Arn. ex Engler	Bergamot orange
	Citrus aurantifolia (Christm.) Swingle	Lime
	Cneoridium dumosum (Nutt.) Hook. f. ex Baill.	Coast spice bush, berryrue
	Dictamnus albus L.	Gas plant
	Pelea anisata Mann	Hawaiian lei
	Phebalium argenteum Smith	Blister bush
	Ruta graveolens L.	Common rue

Topical steroids are effective only for minor cases, whereas more severe cases require oral steroids. Wet compresses with cool water or Burrow's solution for 15–20 minutes several times daily may improve patient comfort. The use of hydroquinone compounds for persistent hyperpigmentation is an option for hyperpigmented lesions that do not fade with time. Avoidance of exposure to the sensitizing agent is the most important reason to use patch testing to differentiate phytophotodermatitis from allergic contact dermatitis.

References

1. Sasseville D. Phytodermatitis. J Cut Med Surg 1999;3: 263–279.
2. Anderson BE, Miller CJ, Adams DR. Stinging nettle dermatitis. Am J Contact Dermatitis 2003;14:44–46.
3. Emmelin N, Feldberg W. The mechanism of the sting of the common nettle (*Urtica urens*). J Physiol 1947;106: 440–455.
4. Winer LH, Zeilenga RH. Cactus granulomas of the skin: report of a case. Arch Dermatol 1955;72:566–569.
5. Mitchell JC, Rosk AJ. Diagnosis of contact dermatitis from plants. Int J Dermatol 1977;16:257–266.
6. Topping MD, Tyrer FII, Lowing RK. Castor bean allergy in the upholstery department of a furniture factory. Br J Ind Med 1981;38:293–296.
7. Chan-Yeung M, Lam S, Koener S. Clinical features and natural history of occupational asthma due to western red cedar (*Thuja plicata*). Am J Med 1982;72:411–415.
8. Brazzelli V, Romano E, Balduzzi A, Borroni G. Acute irritant contact dermatitis from *Agave americana* L. Contact Dermatitis 1995;33:60–61.
9. Santucci A, Picardo M. Occupational contact dermatitis to plants. Clin Dermatol 1992;10:157–165.
10. High WA. Agave contact dermatitis. Am J Contact Dermatitis 2003;14:213–214.
11. Cherpelis BS, Fenske NA. Purpuric irritant contact dermatitis induced by *Agave americana*. Cutis 2000;66: 287–288.
12. Brenner S, Landau M, Goldberg I. Contact dermatitis with systemic symptoms from *Agave americana*. Dermatology 1998;196:408–411.
13. Lamminpaa A, Estlander T, Jolanki R, Kanerva L. Occupational allergic contact dermatitis caused by decorative plants. Contact Dermatitis 1996;34:330–335.
14. Logan PC, Parker GF. Poison ivy (*Rhus*) dermatitis. Am Fam Physician 1972;6:62–66.
15. Mitchell JC. Industrial aspects of 112 cases of allergic contact dermatitis from *Frullania* in British Columbia during a 10-year period. Contact Dermatitis 1981;7: 268–269.
16. Kanerva L, Estlander T, Petman L, Makinen-Kiljunen S. Occupational allergic contact urticaria to yucca (*Yucca aloifolia*), weeping fig (*Ficus benjamina*), and spathe flower (*Spathiphyllum wallisii*). Allergy 2001;56:1008–1011.
17. Kanerva L, Estlander T, Aalto-Korte K. Occupational protein contact dermatitis and rhinoconjunctivitis caused by spathe (*Spathiphyllum*) flowers. Contact Dermatitis 2000;42:369–370.
18. Paulsen E, Skov PS, Andersen KE. Immediate skin and mucosal symptoms from pot plants and vegetables in gardeners and greenhouse workers. Contact Dermatitis 1998; 39:166–170.
19. Uter W, Nohle M, Randerath B, Schwanitz HJ. Occupational contact urticaria and late-phase bronchial asthma caused by *Compositae* pollen in a florist. Am J Contact Dermatitis 2001;12:182–184.
20. Oh S-H, Haw C-R, Lee M-H. Clinical and immunologic features of systemic contact from ingestion of *Rhus (Toxicodendron)*. Contact Dermatitis 2003;48:251–254.
21. Stampf J-L, Schlewer G, Ducombs G, Foussereau J, Benezra C. Allergic contact dermatitis due to sesquiterpene lactones. Br J Dermatol 1978;99:163–169.
22. Cheminat A, Stampf JL, Benezra C. Allergic contact dermatitis to laurel (*Laurus nobilis* L.): isolation and identification of haptens. Arch Dermatol Res 1984;276:178–81.
23. Guin JD, Beaman JH. Toxicodendrons of the United States. Clin Dermatol 1986;4:137–148.
24. Epstein WL. Occupational poison ivy and oak dermatitis. Dermatol Clin 1994;12:511–516.
25. McGovern TW, LaWarre SR, Brunette C. Is it, or Isn't it? Poison ivy look-a-likes. Am J Contact Dermatitis 2000;11: 104–110.
26. Gladman AC. *Toxicodendron* dermatitis: poison ivy, oak, and sumac. Wilderness Environ Med 2006;17:120–128.
27. Corbett M, Billets S. Characterization of poison oak urushiol. J Pharm Sci 1975;64:1715–1718.
28. Guin JD. The black spot test for recognizing poison ivy and related species. J Am Acad Dermatol 1980;2:332–333.
29. Fisher AA. Poison ivy/oak dermatitis. Part: prevention–soap and water, topical barriers, hyposensitization. Cutis 1996;57:384–386.
30. Gartner BL, Wasser C, Rodriguez E, Epstein WL. Season variation of urushiol content in poison oak leaves. Am J Contact Dermatitis 1993;4:33–36.
31. Kalish RS. Recent developments in the pathogenesis of allergic contact dermatitis. Arch Dermatol 1991;127: 1558–1563.
32. Epstein WL. Contact-type delayed hypersensitivity in infants and children: induction of *Rhus* sensitivity. Pediatrics 1961;27:51–53.
33. Epstein WL, Baer H, Dawson CR, Khurana RG. Poison oak hyposensitization: evaluation of purified urushiol. Arch Dermatol 1976;109:356–360.
34. Walker FB, Smith PD, Maibach HI. Genetic factors in human allergic contact dermatitis. Int Arch Allergy Appl Immunol 1967;32:453–462.

35. Resnick SD. Poison-ivy and poison-oak dermatitis. Clin Dermatol 1986;4:208–212.
36. Williford PM, Sheretz EF. Poison ivy dermatitis. Nuances in treatment. Arch Fam Med 1994;3:184–188.
37. Kurlan JG, Lucky AW. Black spot poison ivy: A report of 5 cases and a review of the literature. J Am Acad Dermatol 2001;45:246–249.
38. Lejman E, Stoudemayer T, Grove G, Kligman AM. Age differences in poison ivy dermatitis. Contact Dermatitis 1984;11:163–167.
39. Devich KB, Lee JC, Epstein WL, Spitler LE, Hopper J Jr. Renal lesions accompanying poison oak dermatitis. Clin Nephrol 1975;3:106–113.
40. Fisher AA. The notorious poison ivy family of Anacardiaceae plants. Cutis 1977;20:570–595.
41. Kligman AM. Poison ivy (*Rhus*) dermatitis; an experimental study. AMA Arch Derm 1958;77:149–180.
42. Stibich AS, Yagan M, Sharma V, Herndon B, Montgomery C. Cost-effective post-exposure prevention of poison ivy dermatitis. Int J Dermatol 2000;39:515–518.
43. Marks JG, Trautlein MD, Epstein WL, Laws DM, Sicard GR. Oral hyposensitization to poison ivy and poison oak. Arch Dermatol 1987;123:476–478.
44. Marks JG Jr, Fowler JF Jr, Sheretz EF, Rietschel RL. Prevention of poison ivy and poison oak allergic contact dermatitis by quaternium-18 bentonite. J Am Acad Dermatol 1995;33:212–216.
45. Bruynzeel DP. Bulb dermatitis. Dermatological problems in the flower bulb industries. Contact Dermatitis 1997;37: 70–77.
46. Rudzki E, Dajek Z. Dermatitis caused by buttercups (*Ranunculus*). Contact Dermatitis 1975;1:322.
47. Pohl RW. Contact dermatitis from the juice of *Ornithogalum caudatum*. Toxicon 1965;3:167–168.
48. Gette MT, Marks JEJr. Tulip fingers. Arch Dermatol 1990;126:203–205.
49. Guin JD, Franks H. Fingertip dermatitis in a retail florist. Cutis 2001;67:328–330.
50. Hjorth N, Wilkinson DS. Contact dermatitis IV tulip fingers, hyacinth itch and lily rash. Br J Dermatol 1968; 80:696–698.
51. Crawford GH, McGovern TW. Botanical briefs: tulips—*Tulipa* species L. Cutis 2003;71:347–348.
52. Piirila P, Keskinen H, Leino T, Tupasela O, Tuppurainen M. Occupational asthma caused by decorative flowers: review and case reports. Int Arch Occup Environ Health 1994;66:131–136.
53. Lahti A. Contact urticaria and respiratory symptoms from tulips and lilies. Contact Dermatitis 1986;14:317–319.
54. Verspyck Mijnssen GAW. Pathogenesis and causative agent of "tulip finger." Br J Dermatol 1969;81:737–745.
55. Slob A, Jekel B, De Jong B. On the occurrence of tuliposides in the Liliflorae. Phytochemistry 1975;14:1997–2005.
56. Marks JG. Allergic contact dermatitis to *Alstroemeria*. Arch Dermatol 1988;124:914–916.
57. Hausen BM, Prater E, Schubert H. The sensitizing capacity of *Alstroemeria* cultivars in man and guinea pig. Contact Dermatitis 1983;9:46–54.
58. Diamond KB, Warren GR, Cardellina JH II. Native American food and medicinal plants. 3. a-methylene butyrolactone from *Erythronium grandiflorum* Pursh. J Ethnopharmacol 1985;14:99–101.
59. Adams RM, Daily AD, Brancaccio RR, Dhillon I, Gendler EC. *Alstroemeria* a new and potent allergen for florists. Contact Dermatitis 1990;8:73–76.
60. Thiboutot DM, Hamory BH, Marks JG Jr. Dermatoses among floral shop workers. J Am Acad Dermatol 1990;22:54–58.
61. Piirila P, Kanerva L, Alanko K, Estlander T, Keskinen H, Pajari-Backas M, Tuppurainen M. Occupational IgE-mediated asthma, rhinoconjunctivitis, and contact urticaria caused by Easter lily (*Lilium longiflorum*) and tulip. Allergy 1999;54:273–277.
62. Jerschow E, Hostynek JJ, Maibach HI. Allergic contact dermatitis elicitation thresholds of potent allergens in humans. Food Chem Toxicol 2001;39:1095–1108.
63. Gude M, Hausen BM, Heitsch H, Konig WA. An investigation of the irritant and allergenic properties of daffodils (*Narcissus pseudonarcissus* L., *Amaryllidaceae*). Contact Dermatitis 1988;19:1–10.
64. Van der Werff PJ. Occupational diseases among workers in the bulb industries. Acta Allergol 1959;14:338–355.
65. Piirila P, Hannu T, Keskinen H, Tuppurainen M. Occupational asthma to hyacinth. Allergy 1998;53:328–329.
66. Rosen T, Fordice DB. Cashew nut dermatitis. South Med J 1994;87:543–546.
67. Maje HA, Freedman DO. Cashew nut dermatitis in a returned traveler. J Travel Med 2001;8:213–215.
68. Centers for Disease Control (CDC). Dermatitis associated with cashew nut consumption—Pennsylvania. Morbid Mortal Week Rep 1983;32:129–130.
69. Weinstein S, Bassiri-Tehrani S, Cohen DE. Allergic contact dermatitis to mango flesh. Int Soc Dermatol 2004;43: 195–196.
70. Powell SM, Barrett DK. An outbreak of contact dermatitis from *Rhus verniciflua (Toxicodendron vernicifluum)*. Contact Dermatitis 1986;14:288–289.
71. Warshaw EM, Zug KA. Sesquiterpene lactone allergy. Am J Contact Dermatitis 1996;7:1–23.
72. Kawai K, Nakagawa M, Kawai K, Miyakoshi T, Miyashita K, Asami T. Heat treatment of Japanese lacquerware renders it hypoallergenic. Contact Dermatitis 1992;27: 244–249.
73. Matthai TP, Date A. Renal cortical necrosis following exposure to sap of the marking-nut tree (*Semecarpus anacardium*). Am J Trop Med Hyg 1979;28:773–774.
74. Hausen BM. Airborne contact dermatitis caused by tulip bulbs. J Am Acad Dermatol 1982;7:500–503.
75. Christensen LP, Jakobsen HB, Paulsen E, Hodal L, Andersen KE. Airborne Compositae dermatitis: monoterpenes

and no parthenolide are released from flowering *Tanacetum parthenium* (feverfew) plants. Arch Dermatol Res 1999;291:425–431.

76. Hjorth N, Roed-Petersen J, Thomsen K. Airborne contact dermatitis from Compositae oleoresins simulating photodermatitis. Br J Dermatol 1976;96:613–620.
77. Huygens S, Goossens A. An update on airborne contact dermatitis. Contact Dermatitis 2001;44:1–6.
78. Lotti T, Menchini G, Teofoli P. The challenge of airborne dermatitis. Clin Dermatol 1998;16:27–31.
79. Bowers AG. Phytophotodermatitis. Am J Contact Dermatitis 1999;10:89–93.
80. Takashima A, Yamamoto K, Kimura S, Takakuwa Y, Mizuno N. Allergic contact and photocontact dermatitis due to psoralens in patients with psoriasis treated with topical PUVA. Br J Dermatol 1991;124:37–42.
81. Nigg HN, Nordby HE, Beier RC, Dillman A, Macias C, Hansen RC. Phototoxic coumarins in limes. Food Chem Toxicol 1993;31:331–335.
82. Finkelstein E, Afek U, Gross Eaharoni N, Rosenberg L, Halevy S. An outbreak of phytophotodermatitis due to celery. Int J Dermatol 1994;3:116–118.
83. Sommer AG, Jilison OF. Phytophotodermatitis (solar dermatitis from plants). Gas plant and the wild parsnip. N Engl J Med 1967;276:1484–1486.
84. Ashwood-Smith MJ, Ceska O, Chaudhary SK. Mechanism of photosensitivity reaction to diseased celery. Br Med J 1985;290:1249.
85. Weber IC, Davis CP, Greeson DM. Phytophotodermatitis: the other "lime" disease. J Emerg Med 1999;17:235–237.
86. Burnett JW, Horn TD, Mercado F, Niebyl PH. Phytophotodermatitis mimicking jellyfish envenomation. Acta Derm Venereol (Stockh) 1988;68:168–171.

Chapter 106

BUTTERCUP FAMILY

BOTANICAL DESCRIPTION

The family Ranunculaceae (buttercup) has various species in a number of genera that contain the skin irritant, protoanemonin, including species of the genera *Anemone*, *Caltha* (marsh marigold), *Clematis* (clematis, leather flower), *Pulsatilla*, *Helleborus* (hellebore), and *Ranunculus* (buttercup). There are at least 30 species of *Anemone* that are distributed in the United States, Europe, and Asia. *Anemone* species are distributed primarily in northern temperate and Arctic climates of the Old and New World with about 15–20 *Anemone* species in the United States.

Common Name: Cutleaf Anemone, Pacific Anemone, Red Windflower

Scientific Name: *Anemone multifida* Poir.

Botanical Family: Ranunculaceae (buttercup)

Physical Description: Basal leaves with petioles reach up to 15 cm (~10 in.) in length and 6.5 cm (~3 in.) in width. Typically, there are several flowering stems reaching up to 45 cm (~18 in.) in height that produce a single, white to reddish-pink flower about 2.5 cm (~1 in.) in diameter. Three small leaf lobes (2–3 mm/~0.1 in. in width) develop at the end of the erect, silky-haired stems. Seed heads are globose.

Distribution and Ecology: These plants inhabit prairie hillsides and moist meadows in North America.

Common Name: American pasqueflower, pasqueflower

Scientific Name: *Pulsatilla patens* (L.) P. Mill. (*Anemone patens*)

Botanical Family: Ranunculaceae (buttercup)

Physical Description: This perennial plant has a single flowering stalk with a whorl of three leaves that is covered with long, white silky hairs. The basal leaves are palmately divided with a few, coarse teeth along the margins. Each basal leaf develops from a long, hairy petiole. When fully open, the flower is 7–8 cm (~3 in.) in diameter with five to eight petal-like, purple sepals and a ring of yellow stamens. The root system consists of a woody taproot, which swells into a caudex on older plants that can produce flowering stalks. The flower appears for a few weeks during the early spring.

Distribution and Ecology: This plant is a native species to both North America and Eurasia that inhabits dry hill or gravel prairies and barren areas with scant, woody vegetation.

EXPOSURE

There are a variety of herbal uses in traditional medicine for protoanemonin-containing species including *Anemone* and *Pulsatilla* species. The most common use by Native American peoples for these plants was as a counter-irritant in the form of an external poultice for

Medical Toxicology of Natural Substances, by Donald G. Barceloux, MD

abrasions, boils, cuts, and skin sores.[1] Other traditional uses for extracts from these plants included rheumatism, stomach troubles, and promoting childbirth. The Teton Sioux Indians also used extracts from *Clematis hirsutissima* Pursh (sugarbowls, hairy clematis) to stimulate horses by inserting the scraped root from *C. hirsutissima* into the nostril of the horse.[2]

PRINCIPAL TOXINS

Physiochemical Properties

Protoanemonin (CAS RN: 108-28-1, $C_5H_4O_2$) is an unstable, pale yellow oil with a water solubility of approximately 1 g/dL.[3] This simple lactone compound is volatile. Protoanemonin possess some antibacterial[4] and antifungal[5,6] properties *in vitro*. Anemonin (CAS RN:508-44-1, $C_{10}H_8O_4$) is an inactive crystalline solid with limited water solubility. Drying or boiling of protoanemonin-containing herbs destroys the irritant properties in *Anemone* and *Pulsatilla* species by accelerating the polymerization of protoanemonin.

Poisonous Parts

Protoanemonin occurs in varying concentrations in different species with the highest concentrations (0.5–0.8%) in bulbs from *Ranunculus bulbosus* L. and *Helleborus niger* L. as displayed in Table 106.1.[3,7] Species of *Anemone* and *Pulsatilla* species do not contain clinically significant concentration of photosensitizing furocoumarin compounds. The roots of *Pulsatilla* species contain a variety of oleanane-type glycosides and other triterpene saponins.[8]

Mechanism of Toxicity

Toxicity to this plant usually results from hypersensitivity responses.

DOSE RESPONSE

Allergic responses to this plant depend on individual sensitivity rather than a direct toxic effect.

TOXICOKINETICS

Ranunculin (CAS RN:644-69-9, $C_{11}H_{16}O_8$) is a nonirritating glycoside, which forms protoanemonin by enzymatic hydrolysis after maceration of plant tissue or alkaline hydrolysis. Spontaneous polymerization results in the formation of a nonirritating substance, anemonin. Figure 106.1 displays the structures of protoanemonin, the precursor (ranunculin), and metabolite (anemonin).

CLINICAL RESPONSE

Contact with plants containing protoanemonin causes an irritant contact dermatitis characterized by erythema, edema, vesicles, and postinflammatory hyperpigmentation.[9] A case report associated the development of local erythema and bullae with the application of a compress of pasque flower (*Pulsatilla patens*).[10] The skin eventually healed after sloughing. Ingestion of these plants can cause stomatitis and mucosal ulcerations.

DIAGNOSTIC TESTING

Analytical methods for the separation of protoanemonin include reversed-phase high pressure liquid chromatography.[7,11]

TABLE 106.1. Protoanemonin Content of Ten Species of the Ranuculaceae Family

Genus	Species	Protoanemonin Content (mg/g Wet Weight)
Anemone	*nemorosa* L.	0.33
	trifolia L.	0.17
Clematis	*flammula* L.	0.50
	recta L.	0.01
	vitalba L.	0.15
Helleborus	*foetidus* L.	0.67
	odorus Waldtst et Kit	0.004
	niger L.	5.82
Ranunculus	*bulbosus* L.	7.77
	repens L.	0.13

Source: Adapted from Ref 7.

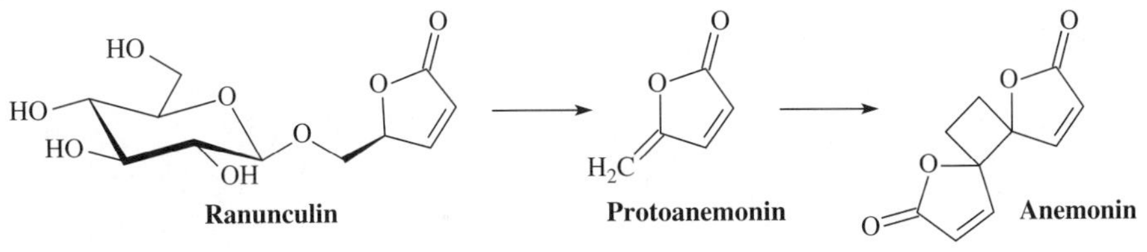

FIGURE 106.1. Protoanemonin formation and degradation.

TREATMENT

Treatment is supportive. The most common toxic effect associated with exposure to protoanemonin-containing plants is an irritant contact dermatitis that responds to wound cleansing and topical steroids (0.05% fluocinonide). Gut decontamination measures are unnecessary.

References

1. Turner NJ. Counter-irritant and other medicinal uses of plants in Ranunculaceae by native peoples in British Columbia and neighbouring areas. J Ethnopharmacol 1984;11:181–201.
2. Kern JR, Cardellina JH II. Native American medicinal plants. Anemonin from the horse stimulant *Clematis hirsutissima*. J Ethnopharmacol 1983;8:121–123.
3. Smolinske SC, Spoerke DG Jr. Toxicity of Houseplants. Boca Raton, FL: CRC Press; 2000.
4. Bruckmann M, Blasco R, Timmis KN, Pieper DH. Detoxification of protoanemonin by dienelactone hydrolase. J Bacteriol 1998;180:400–402.
5. Martin ML, Roman LS, Dominguez A. *In vitro* activity of protoanemonin, an antifungal agent. Plant Med 1990;56: 66–69.
6. Mares D. Antimicrobial activity of protoanemonin, a lactone from ranunculaeous plants. Mycopathologia 1987; 98:133–140.
7. Bonora A, Dall'Olio G, Bruni A. Separation and quantitation of protoanemonin in Ranunculaceae by normal- and reversed phase HPLC. Planta Med 1985;51:364–367.
8. Ye W, Pan G, Zhang Q, Che CT, Wu H, Zhao S. Five new triterpene saponins from *Pulsatilla patens* var. *multifida*. J Nat Prod 1999;62:233–237.
9. Vance JC. Toxic plants of Minnesota. Skin toxicity of the prairie crocus (*Anemone patens* L.). Minnesota Med 1982; 65:149–151.
10. Aaron TH, Muttitt EL. Vesicant dermatitis due to prairie crocus (*Anemone patens* L.). Arch Dermatol 1964;90: 168–171.
11. Minakata H, Komura H, Nakanishi K, Kada T. Protoanemonin, an antimutagen isolated from plants. Mutation Res 1983;116:317–322.

Chapter 107

COLCHICINE-CONTAINING PLANTS

HISTORY

The toxicity of *Colchicum autumnale* has been recognized for over 2,000 years. Theophrastus of Eresus, a Greek botanist, provided the first written account of the toxicity of extracts from this plant in 370 BC.[1] He described the toxic ingredient as Ephemeron. In 150 BC, Nicander named this substance after Colchis, a district near the Black Sea known for sorcery, and he called this extract "the destructive fire of Colchican Medea." Alexander of Trallis first recommended the use of extracts of *Colchicum autumnale* in 550 AD.[2] Known as Surugen to Arabian physicians and Herbstzeit-lose to European physicians during the Middle Ages, this extract was a common treatment for gout in many parts of the world. However, purified colchicine was not available until John Dorsey isolated colchicine from *Colchicum autumnale* tubers in 1820.

BOTANICAL DESCRIPTION

Misidentification of *colchicum autumnale* as alpine wild garlic (*Allium ursinum* L.) can result in colchicine poisoning. The ingestion of tubers from the glory lily is a common cause of admission to local hospitals for plant poisonings in Sri Lanka.[3]

Common Name: Autumn Crocus, Meadow Saffron, Wild Saffron, Naked Lady, Son-before-the-Father

Scientific Name: *Colchicum autumnale* L.

Family: Liliaceae (lily)

Physical Description: This bulb is cultivated in the United States for its clusters of white or light violet crocus-like flowers that appear in the autumn after the long, lanceolate leaves detach. They are often sold as a curiosity because they bloom without soil or water.

Distribution and Ecology: The autumn crocus is a native plant of Europe that was introduced into the United States. The habitat of this plant includes Oregon, Utah, Kentucky, North Carolina, and parts of the northeastern United States.

Common Name: Glory Lily, Gloriosa, Climbing Lily, Superb Lily, Flame Lily

Scientific Name: *Gloriosa superba* L.

Family: Liliaceae (lily)

Physical Description: This slender plant or vine has a large tuberous root, alternate leaves, and colorful flowers on a long stalk. Occasionally, the tubers of the glory lily have been mistakenly consumed because of their similarity to sweet potatoes (*Ipomoea batatas* (L.) Lam.) and wild garlic (*Allium ursinum* L.).

Distribution and Ecology: The flame or glory lily is a tropical plant native to Asia and Africa that grows outdoors in the southern United States including Florida. In colder regions, cultivation of the plant requires a greenhouse.

Medical Toxicology of Natural Substances, by Donald G. Barceloux, MD

EXPOSURE

Colchicine is one of the drugs used for the relief of acute gouty arthritis and currently colchicine is administered for the prophylactic treatment of familial Mediterranean fever. The effectiveness of colchicine for amyloidosis, sarcoid arthritis, primary biliary cirrhosis, Behçet's disease, Paget's disease, pulmonary fibrosis, systemic scleroderma, vasculitis, and liver/biliary cirrhosis remains unproven. In experimental studies, the antimitotic properties of colchicine are useful to evaluate cell division and function. Product formulations of colchicine include 0.5- and 0.6-mg tablets and an injectable solution of 0.5 mg/mL.

PRINCIPAL TOXINS

Physiochemical Properties

Colchicine is a white, crystalline alkaloid extracted from the following two members of the Liliaceae family: autumn crocus (*Colchicum autumnale*) and glory lily (*Gloriosa superba*). Table 107.1 lists some physical properties of colchicine. This substance is soluble in water and alcohol. Figure 107.1 displays the chemical structure of colchicine (CAS RN: 64-86-8, $C_{22}H_{25}NO_6$).

TABLE 107.1. Some Physical Properties of Colchicine

Physical Property	Value
Melting Point	156 °C/~313 °F
pKa Dissociation Constant	1.85
log P (Octanol-Water)	1.3
Water Solubility	4.50E+04 mg/L (25 °C/77 °F)
Vapor Pressure	3.13E–12 mmHg (25 °C/77 °F)
Henry's Law Constant	1.79E–17 atm-m³/mole (25 °C/77 °F)
Atmospheric OH Rate Constant	3.32E–10 cm³/molecule-second (25 °C/77 °F)

FIGURE 107.1. Chemical structure of colchicine.

Poisonous Parts

AUTUMN CROCUS

Colchicine and related compounds are found throughout the plant, but the highest concentrations occur in the bulb (about 0.8% by weight). The pale purple flowers contain about 0.1–0.6% colchicine by weight.[4]

GLORY LILY

All parts of the plant are poisonous. The tubers of the glory lily contain about 0.02–0.3% colchicine by dry weight, depending on season and location.[5] About 10 g fresh tubers contain approximately 6 mg colchicine depending on the variety. In addition to colchicine, the glory lily contains a structurally related toxin known as gloriosine. This compound possesses antimitotic properties similar to those of colchicine.

DOSE RESPONSE

Medicinal Dose

Recommended oral doses of colchicine begin in the range of 0.5–1 mg every 2–3 hours until the symptoms of gout resolve or gastrointestinal symptoms develop. Typically, the intravenous dose of colchicine is about half the oral dose. Single intravenous doses of colchicine should not exceed 2–3 mg with the cumulative intravenous dose over 7 days limited to 4–5 mg.[6] Gastrointestinal symptoms are much less common after intravenous than oral doses.

Toxic Dose

COLCHICINE

The ingestion of equivalent amounts of autumn crocus (*C. autumnale*) and colchicine tablets produces similar toxicity.[7] In clinical series on colchicine toxicity, survival with supportive care was associated with the ingestion of <0.5 mg colchicine/kg, whereas fatalities occurred when the colchicine dose exceeded 0.8 mg/kg.[8] This series did not include patients who vomited or received decontamination measures within the first 4 hours after ingestion. Case reports of colchicine intoxication associated the ingestion of 0.37 mg/kg and 1.72 mg/kg with survival and death, respectively, in pediatric patients presenting after decontamination measures would be expected to significantly reduce colchicine absorption.[9] Fatal oral doses in healthy adults typically exceed 30–40 mg.[10,11] The ingestion of 20 mg colchicine by a 47-year-old man with gout caused gastroenteritis, dehydration, leukopenia, thrombopenia, mild hepatocellular injury, and renal dysfunction.[12] He survived without sequelae and all laboratory abnormalities resolved. The intravenous administration of 8–12 mg colchicine may cause death, particularly in patients with hepatorenal dysfunction.[13] A 70-year-old man developed a fatal marrow aplasia and pancytopenia after he received 10 mg of intravenous colchicine over 5 days.[14]

Colchicine-Containing Plants

Clinical features of colchicine toxicity following the ingestion of colchicine-containing plants depend on a variety of factors other than dose including season, plant part, and species. The ingestion of an estimated 40 flowers of *C. autumnale* by a 44-year-old man produced gastrointestinal symptoms without any significant systemic abnormalities.[4] However, the intentional ingestion of approximately 12 *Colchicum autumnale* flowers resulted in hypovolemia, renal failure, adult respiratory distress syndrome, and death in a 16-year-old girl.[15]

TOXICOKINETICS

Analysis of the toxicokinetics of colchicine following massive overdose indicates that the pharmacokinetic parameters of therapeutic and toxic doses are similar.[16]

Absorption

Colchicine undergoes extensive presystemic metabolism, resulting in a bioavailability below 50% of the administered oral dose. The absorption of colchicine from the gastrointestinal tract is rapid with peak plasma concentrations occurring about 1–1.5 hours after administration.[17] Analysis of toxicokinetics data during colchicine poisoning indicates that the ingestion of large amounts of colchicine does not delay gastrointestinal absorption.[16] Case reports of colchicine intoxication suggest that absorption of colchicine through nasal membranes is sufficient to produce serious toxicity.[18]

Distribution

The clearance of colchicine from the blood is rapid and extensive with complete distribution into the body occurring in 3–6 hours.[22] In a study of volunteers administered 0.5–1.5 g colchicine orally, the volume of distribution (V_d) ranged from about 11–15 L/kg.[19] Following the intravenous administration of 0.5 mg colchicine, the average volume of distribution in six healthy males and four elderly female volunteers was 6.7 L/kg and 6.3 L/kg, respectively.[20] The V_d is slightly larger during colchicine intoxication. In a study of 2 patients with colchicine poisoning, the V_d averaged approximately 21 L/kg.[16] Plasma protein binding of colchicine is about 50%.

Biotransformation

Metabolism of colchicine occurs primarily by deacetylation in the liver mediated by the cytochrome P450 isoenzyme, CYP3A4.

Elimination

Clearance of colchicine occurs primarily by hepatic metabolism with renal clearance accounting for about 10–20% of absorbed dose.[17] Renal clearance increases moderately in patients ingesting large doses of colchicine.[16] The enterohepatic circulation is a significant route of excretion with relatively large amounts of both parent drug and metabolites excreted in the bile and intestines. The presence of liver or renal failure reduces the clearance of colchicine. The terminal plasma elimination half-life of colchicine in a series of 5 patients ingesting 8–20 g colchicine ranged from approximately 20–32 hours.[16] The average plasma elimination half-life of colchicine following intravenous administration to six healthy adult males and four elderly females was 30 hours and 34 hours, respectively.[20]

Drug Interactions

Grapefruit juice, erythromycin, cimetidine, and ketoconazole inhibit the cytochrome P450 isoenzyme (CYP3A4) involved with the metabolism of colchicine.[21] The ingestion of large amounts of grapefruit juice by an 8-year-old child on chronic colchicine therapy for familial Mediterranean fever was associated with severe colchicine toxicity.[22]

HISTOPATHOLOGY AND PATHOPHYSIOLOGY

Mechanism of Toxicity

The disruption of microtubule function following the ingestion of colchicine results from the binding of colchicine to the protein subunit of microtubules. Colchicine prevents the polymerization of tubulin into microtubules by the reversible, noncovalent binding to tubulin.[23] This antimitotic property disrupts the spindle apparatus that separates chromosomes during metaphase. The most seriously affected organs involve tissues containing cells (e.g., intestinal epithelium, hair follicles, bone marrow) with high metabolic rates. However, the inhibition of cellular division does not account for the appearance of multiorgan failure during severe colchicine overdoses.[13]

Colchicine produces gastrointestinal dysfunction leading to significant water loss and sequestration of fluid. The probable mechanism of toxicity is neurogenic rather than direct irritation. In experimental studies, colchicine has no direct stimulant action on isolated segments of the small bowel, and atropine abolishes the diarrhea associated with experimentally induced colchicine toxicity.[24] Although depressed myocardial function occurs in patients with normal right-sided filling pressures, hypotension probably results from hypovolemia.[25] Experimental studies indicate that colchicine produces muscle weakness and direct toxic effects on skeletal muscle.[26] Direct myocardial toxic effects are less well documented.

Postmortem Examination

Postmortem findings after fatal colchicine poisoning typically demonstrate multiorgan failure characterized by hemorrhagic pulmonary edema, cerebral edema, centrilobular fatty necrosis of the liver, blunting and shortening of the intestinal villi, petechial bleeding in the fatty tissues, hypocellular bone marrow, and necrosis of the proximal convoluted tubules of the kidney.[27–29] In a patient who died 33 hours after ingesting colchicine, the pathological examination demonstrated a bilateral bronchopneumonia with marked hypocellularity of the bone marrow and decreased megakaryocytes and myeloid cells.[30] Histological examination confirmed the presence of numerous mitotic figures in the metaphase of cells from the gastrointestinal epithelia. Typically, tissues with rapidly regenerating epithelium (e.g., gastrointestinal mucosa, testes, urinary tract) contain numerous blocked mitoses.[31] Other histological findings include diffuse sloughing of the surface epithelium of the colon, collapse of the crypt cells of Lieberkühn, extensive autolysis of the stomach and small intestine, and disruption of the sinusoidal architecture of the red pulp in the spleen. Tissue from the heart, skeletal muscle, and nervous system frequently are normal.[32] In some cases of acute colchicine intoxication involving multisystemic failure, focal areas of contraction band necrosis and pale, anuclear myocytes appear consistent with ischemia. Cardiac changes may occur in the absence of coronary artery disease.[33] Clumps of chromatin (i.e., "colchicine bodies") in nuclei of hepatocytes may occur in pathological specimens,[29] but these changes are not invariably present.

CLINICAL RESPONSE

Adverse effects associated with the therapeutic use of colchicine include gastrointestinal irritation, bone marrow depression with aplastic anemia, thrombocytopenia, alopecia, peripheral neuropathy, myopathy, and rhabdomyolysis, particularly in patients with renal dysfunction.[34] Colchicine is a potent gastrointestinal toxin that causes intractable multiorgan failure following the absorption of large doses.[35] The clinical features of colchicine toxicity appear in three stages depending on the severity of the ingestion.[13] Similar toxic effects also result from the consumption of parts from the autumn crocus and tubers from glory lily (*Gloriosa superba*); however, the severity of these reactions are usually less prominent than toxicity secondary to colchicine tablets because of the large volume of plant material needed to cause a serious toxic response. The ingestion of crude extract of tubers from *Gloriosa superba* with suicidal intent produced burning sensation in the GI tract, nausea, vomiting, diarrhea, abdominal pain, and sinus bradycardia.[36]

Gastrointestinal Phase (Phase I)

The presence of gastrointestinal symptoms is a sensitive, but not specific, sign of serious colchicine poisoning.[10] Almost all seriously poisoned patients develop signs of gastrointestinal distress (nausea, vomiting, anorexia, diarrhea, cramping abdominal pain), usually within 2–6 hours after ingestion. Occasionally, the emesis or diarrhea may be bloody. Volume depletion, electrolyte imbalance, and hypotension may occur during this period. During this stage, a peripheral leukocytosis frequently occurs characterized by numerous immature myeloid cells, damaged cells with karyorrhexis, and unsegmented granulocytes (Pelger-Huet cells).[8]

Multisystem Failure (Phase II)

Life-threatening complications often develop 24–72 hours after ingestion. Respiratory insufficiency may develop from multiorgan failure, liver dysfunction, pancreatitis, sepsis, hypovolemia, muscle weakness or adult respiratory distress syndrome (ARDS).[37,38] Muscle weakness may last several weeks and contribute to ventilatory depression. Sepsis is a well-recognized and common fatal complication of severe colchicine overdose. Intestinal ileus commonly develops during phase II of severe colchicine intoxication, and gastrointestinal dysfunction may persist for one week. Figure 107.2 demonstrates the time course in days of multiorgan failure developing after a patient ingested *Autumn crocus*.

CARDIOVASCULAR SYSTEM

Sudden cardiac arrest resulting from asystole or ventricular dysrhythmias has occurred between 36 and 54

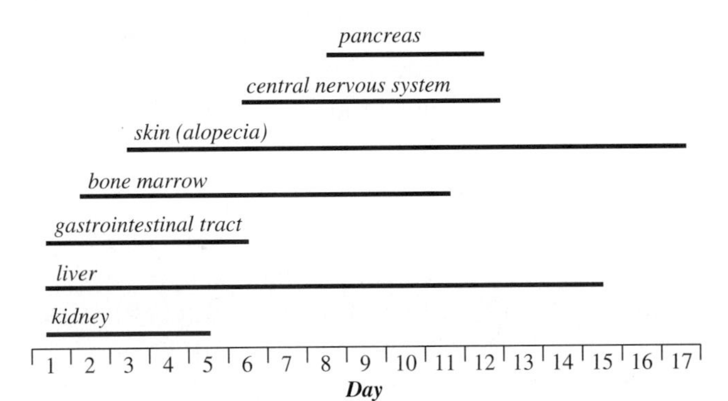

FIGURE 107.2. Time course of colchicine poisoning in a patient, who developed multiorgan failure after autumn crocus ingestion. From Gabrscek L, et al. Accidental poisoning with *Autumn crocus*. J Toxicol Clin Toxicol 2004;42:86. Copyright Taylor and Francis Ltd. Reproduced with permission.

hours after ingestion despite adequate oxygenation and fluid/electrolyte replacement.[13,39] Most deaths result from cardiogenic shock within the first 72 hours. Patients with hemodynamic features of cardiogenic shock (decreased cardiac index, increased systemic vascular resistance despite adequate filling pressures) have a poor prognosis.[25]

CENTRAL NERVOUS SYSTEM

Confusion and delirium usually results from poor cerebral perfusion rather than direct central nervous system toxicity.[40] Papilledema, coma, seizures, and cerebral edema may occur during severe colchicine intoxication.[41] Case reports associate the development of peripheral neuropathy (i.e., loss of deep tendon reflexes and ascending paralysis) with severe colchicine toxicity.[42] The ingestion of *Gloriosa* tubers was associated with an ascending polyneuropathy characterized by stocking-glove anesthesia, weakness, and loss of deep tendon reflexes that resolved after 4 weeks.[43]

KIDNEY

Oliguric renal failure is a common complication of severe colchicine intoxication.[44] Dysuria, hematuria, and proteinuria are common; however, renal dysfunction is probably secondary to hypoxia, hypotension, and myoglobinuria. The contribution of direct colchicine toxicity to renal failure is unclear, but severe renal dysfunction is usually a terminal event rather than a prominent part of most fatal overdose cases.

BLOOD

Severe leukopenia and thrombocytopenia are common during the second phase of severe colchicine intoxication. The lack of bone marrow suppression suggests mild toxicity. Peak bone marrow suppression occurs between the fourth and seventh day after ingestion. Case reports document the development of a consumptive coagulopathy with elevated fibrin degradation products, decreased fibrinogen, and prolonged coagulation profiles.[45,46] Heinz body hemolytic anemias develop rarely after acute ingestions of colchicine.[47]

Recovery Phase (Phase III)

If the patient survives, most organ systems show definite signs of improvement by the 10th day. A rebound leukocytosis often occurs during this phase. Alopecia begins about the 12th day, reaching completion by 3 weeks after ingestion.[7,43] Replacement hair growth begins after the first month. Case reports suggest that complete recovery from acute complications of colchicine poisoning often occurs after a convalescent period of several months to a year, but a neuropathy, muscle weakness, and intermittent hair loss can persist a few years after ingestion.[37,48] Fever may be a persistent part of the clinical picture for the first several weeks, but the body temperature rarely exceeds 39°C/~102°F. Myopathy can occur about 3 months after ingestion along with elevation of serum creatine kinase concentrations.

DIAGNOSTIC TESTING

Analytical Methods

Sensitive and specific methods for the detection of colchicine in biological fluids include high performance liquid chromatography with diode array detector (HPLC/DAD),[49] HPLC coupled to ion spray mass spectrometry,[50] and HPLC combined with electrospray ionization/tandem mass spectrometry,[51,52] The limits of detection and quantitation for the latter method are approximately 0.1 ng/mL and 0.5 ng/mL, respectively, whereas the limit of quantitation for HPLC/DAD is about 4 ng/mL.

Biomarkers

THERAPEUTIC

Following therapeutic doses of colchicine, the plasma colchicine concentrations are relatively low (<10 ng/mL), depending on the duration and time after ingestion.[20] Typical therapeutic concentrations of *plasma* colchicine are <2.4 ng/L and <13 ng/mL following the daily administration of 1 mg and 3 mg colchicine, respectively.[53] Fifteen minutes after the intravenous administration of 2 mg colchicine to 16 volunteers, the *plasma* colchicine concentration averaged 11.4µg/L with a range of 4.5–33µg/L.[54]

Overdose

Serum colchicine concentrations document exposure to *Colchicum autumnale* or colchicine medication, but there are few clinical data on the prognostic value of these biomarkers. Concentrations of colchicine in the blood decline rapidly, and multiorgan failure during severe colchicine intoxication occurs after peak colchicine concentrations. In a *serum* sample drawn from a hypotensive, comatose patient 24 hours after the ingestion of a large amount of colchicine, the colchicine concentration was 4.5ng/mL.[55] The patient died on the second day of hospitalization. Another patient died 2 days after mistakenly ingesting *Colchicum autumnale* following the development of hypotension, disseminated intravascular coagulation, and asystolic cardiac arrest. The serum colchicine concentration of the second day after ingestion was 9ng/mL.[56] In a patient, who survived after developing multiorgan failure, the serum colchicine concentration 48 hours following the ingestion of *Autumn crocus* was 70ng/mL.[48] Two days later colchicine was not detectable in the serum.

In a case report on the ingestion of flowers from *C. autumnale* that produced only gastrointestinal symptoms, the *plasma* colchicine concentration 6 hours after ingestion was 3.8ng/mL as measured by HPLC coupled to ion spray mass spectrometry.[4] The peak *plasma* colchicine concentration was 4.34ng/mL at 13 hours after ingestion, whereas the maximum colchicine concentration in erythrocytes was 5.43ng/mL at 16 hours postingestion. After an intentional multiple-drug overdose that included 7.5g of colchicine, the blood colchicine level 6 hours and 24 hours after ingestion was 21ng/mL and <5ng/mL, respectively.[57]

Postmortem

Postmortem colchicine blood concentrations do not correlate to the clinical severity or the amount ingested because of the rapid distribution of colchicine from the blood and the extended length of time between ingestion and death.[58] Frequently, the colchicine concentrations are nondetectable in postmortem samples from fatal colchicine intoxications depending on the time between absorption, the time of death, and medical interventions.[32] Two hours after ingesting an estimated 18–21g colchicine, the colchicine concentration in blood samples from a 39-year-old woman was 250µg/L. She died 40 hours after ingestion and her postmortem blood samples did not contain detectable concentrations of colchicine.[59] A 45-year-old man died in the hospital 3–4 days after ingesting colchicine.[30] Colchicine was detectable in blood as measured by HPLC with photodiode array detection, but not in postmortem samples from liver, stomach contents, or vitreous humor (limit of detection <5µg/L). A postmortem bile sample contained 4,200µg colchicine/L. An antemortem plasma sample collected 30 hours before death contained 60µg colchicine/L, but subsequent antemortem samples did not contain detectable concentrations of colchicine. The postmortem peripheral and heart blood from a 57-year-old man found dead contained 21.9ng colchicine/mL and 22.8ng colchicine/mL, respectively, whereas similar samples from a 33-year-old nurse contained 17.4ng colchicine/mL and 5.2ng colchicine/mL, respectively.[52] The latter patient died after 61 hours of intensive care following the ingestion of 80mg colchicine. Postmortem vitreous humor samples contained much lower (e.g., 0.5ng/mL and 3ng/mL, respectively) colchicine concentrations than postmortem blood.

The extensive distribution of colchicine into cells suggests that colchicine concentration may be higher in whole blood than in plasma. Compared with other tissue samples, bile may contain high concentrations of colchicine because of the extensive hepatobiliary recirculation of colchicine.[60] Analysis of postmortem blood samples from a 44-year-old man, who died of multiorgan failure 2 days after ingesting a meal containing leaves from *Colchicum autumnale*, did not detect colchicine as measured by HPLC with UV-detection.[27] Postmortem samples from the bile contained 7.5µg colchicine/mL.

Abnormalities

Transient elevation of serum hepatic aminotransferase concentrations may occur, but fulminant hepatic failure is not a common occurrence.[36] Electrolyte abnormalities during the various stages of colchicine poisoning include hypocalcemia, hypophosphatemia, hyponatremia, hypomagnesemia, mild hyperamylasemia, hyperbilirubinemia, elevated hepatic aminotransferases, hypokalemia thrombocytopenia, and leukopenia.[42,45] Metabolic acidosis usually occurs as a result of lactate accumulation secondary to hypovolemia or shock. Rhabdomyolysis and myoglobinuria occur during acute colchicine poisoning as a result of direct muscle damage associated with dramatic increases in muscle enzymes and myoglobinuria.[42] The presence of rhabdomyolysis may contribute to the metabolic acidosis and electrolyte abnormalities. Rising serum cardiac troponin I concentrations indicate the myocardial injury and the potential to develop cardiogenic shock.[33] Urinary abnormalities include hematuria, proteinuria, myoglobinuria, and oliguria during severe toxicity.[61] The ECG may demonstrate signs of myocardial injury (e.g., ST elevation in limb and precordial leads) with both creatine kinase and initial MB isoenzyme fractions increased.[45,61]

TREATMENT

Stabilization

Initially, hypotension represents the most serious risk. An intravenous line with fluid replacement (normal saline or Ringer's lactate) should be established, and then vasopressors given if the hypotension does not respond to fluid challenges. All patients should receive cardiac and pulse oximeter monitoring in an intensive care unit. Hemodynamic monitoring is often required in seriously ill patients. Respiratory dysfunction is usually secondary to the adult respiratory distress syndrome, which commonly presents between 24 and 36 hours after ingestion.

Decontamination

All symptomatic patients should be medically evaluated. The usual measures of decontamination (lavage, charcoal) are often unnecessary because of the severe gastrointestinal symptoms associated with colchicine intoxication and the delay in seeking medical care. Decontamination measures (e.g., activate charcoal) are a therapeutic option even beyond the usual 1–2 hours postingestion in colchicine plant ingestions, but there are no clinical data to confirm the efficacy of these measures. A postmortem examination 30 hours after a glory lily flower ingestion revealed 196 mg of colchicine in the victim's stomach.[15] Because colchicine is a potent poison, children who ingest more than one tablet (0.5 mg) and present < 2 hours often ingestion should receive decontamination measures. Plant toxicity is more difficult to predict, but decontamination measures should also be used in all patients except those who ingest small amounts during exploratory activity. Intravenous fluids and electrolytes must be given to maintain adequate peripheral perfusion. The use of cathartics is not necessary because of spontaneous diarrhea.

Elimination Enhancement

Serial activated charcoal (0.5 g/kg every 3–4 hours) is a therapeutic option because of the enterohepatic circulation of colchicine, but there are inadequate human data to evaluate the efficacy of serial-activated charcoal for colchicine intoxication. Paralytic ileus may complicate the use of serial-activated charcoal therapy. Duodenal drainage is also an alternative method of reducing the enterohepatic circulation of colchicine, but there are no data to document the efficacy of this method. There is no evidence that hemoperfusion or hemodialysis increases elimination of colchicine, especially in view of its large volume of distribution and limited renal excretion.

Antidotes

No effective antidotes are available. A case report suggests that the administration of experimental goat-derived, colchicine-specific Fab fragments may reverse serious colchicine poisoning,[62] but these antibodies are not commercially available. In this case report, the use of Fab fragments did not prevent transient bone marrow suppression.

Supplemental Care

Hemodynamic Status

Symptomatic patients (e.g., those with nausea, vomiting, diarrhea, dyspnea) require hospitalization for fluid replacement and medical observation in a monitored setting. Fluid administration often requires hemodynamic monitoring to replace losses accurately in the setting of cardiogenic shock. The hemodynamic response to fluid administration has prognostic value because intoxicated patients with cardiogenic shock (low cardiac index, high systemic vascular resistance) despite adequate filling pressures have a poor prognosis.[24] Close monitoring of oxygen saturation is important because the presence of increasing hypoxemia suggests the onset of adult respiratory distress syndrome. Most fatalities occur within the first 24–72 hours.

Laboratory Monitoring

Daily complete blood counts are important to detect either hemolytic anemia or upper gastrointestinal hemorrhages. Transfusions of both platelet packs and packed red blood cells may be needed. A consumptive coagulopathy may appear within the first several days. Treatment with fresh-frozen plasma is usually reserved for patients with active bleeding rather than prophylactic use because of the interference of activated factors with the prothrombin time and the lack of data on the effect of this treatment on clinical outcome. The use of vitamin K_1 is a therapeutic option, but the effect of prophylactic vitamin K_1 on outcome has not been defined. The prothrombin time, partial thromboplastin time, and platelet count should be monitored daily for the first several days and then as indicated. Fibrinogen and fibrin split product levels should also be obtained if a consumptive coagulopathy is suspected.

Frequent (twice daily) monitoring of electrolytes (e.g., sodium, potassium, calcium, phosphorus) is necessary during the first several days. Electrolyte abnormalities may persist for one week. Hypoglycemia, myoglobinuria, and creatine kinase levels also should be monitored. The presence of myoglobinuria indicates the need for generous fluid replacement to maintain good urine output (1–2 mL/kg/h) along with furosemide as need to maintain adequate urine flow. The use of urinary alkalinization is a therapeutic option, but the efficacy of this treatment is not well documented.

SEPSIS

Sepsis may complicate the clinical course, especially during the nadir of bone marrow suppression between the fourth and seventh days. Patients who have fever and/or hemodynamic instability should have blood cultures and broad-spectrum antibiotic coverage if sepsis is suspected. Granulocyte transfusions are usually unnecessary because of the short period of leukocytopenia. Granulocyte colony-stimulating factor is an option to granulocyte transfusions during severe colchicine intoxications.[63,64] Fever is not a specific sign of infection because colchicine intoxication produces elevated temperatures. Hypothermia is a poor prognostic sign and should be treated with the usual passive rewarming measures.

References

1. Hartung EF. History of the use of colchicum and related medicaments in gout. Ann Rheum Dis 1954;13:190–201.
2. Rodnan GP, Benedek TG. The early history of antirheumatic drugs. Arthritis Rheum 1970:13;145–152.
3. Fernando R, Fernando DN. Poisoning with plants and mushrooms in Sri Lanka: a retrospective hospital based study. Vet Hum Toxicol 1990;31:579–581.
4. Danel VC, Wiart JF, Hardy GA, Vincent FH, Houdret NM. Self-poisoning with *Colchicum autumnale* L. flowers. Clin Toxicol 2001;39:409–411.
5. Thakur RS, Santavy F. Substances from plants of the subfamily Wurmbaeoideae and their derivatives. Part LXXIX. Alkaloids of the plant *Gloriosa superba* L. Planta Med 1975;28:201–209.
6. Wallace SL, Singer JZ. Systemic toxicity associated with intravenous administration of colchicine—guidelines for use. J Rheum 1988;15:495–499.
7. Gooneratne BW. Massive generalized alopecia after poisoning by *Gloriosa superba*. Br Med J 1966;2:1023–1024.
8. Bismuth C, Gaultier M, Conso F. Aplasie medullaire apres intoxication aigue a la colchicine. Nouv Presse Med 1977; 6:1625–1629.
9. Atas B, Caksen H, Tuncer O, Kirimi E, Akgun C, Odabas D. Four children with colchicine poisoning. Hum Exp Toxicol 2004;23:353–356.
10. Milne ST, Meek PD. Fatal colchicine overdose: report of a case and review of the literature. Am J Emerg Med 1998;16:603–608.
11. Mullins ME, Carrico EA, Horowitz Z. Fatal cardiovascular collapse following acute colchicine ingestion. Clin Toxicol 2000;38:51–54.
12. Borras-Blasco J, Enriquez R, Sirvent AE, Amoros F, Navarro-Ruiz A, Reyes A. Acute renal failure associated with an accidental overdose of colchicine. Int J Clin Pharmacol Ther 2005;43:480–484.
13. Stapczynski JS, Rothstein RJ, Gaye WA, Niemann JT. Colchicine overdose: report of two cases and review of the literature. Ann Emerg Med 1981;10:364–368.
14. Liu YK, Hymowitz R, Carroll MG. Marrow aplasia induced by colchicine. A case report. Arthritis Rheum 1978;21: 731–735.
15. Ellwood MG, Robb GH. Self poisoning with colchicine. Postgrad Med J 1971;47:129–131.
16. Rochdi M, Sabouraud A, Baud FJ, Bismuth C, Scherrmann JM. Toxicokinetics of colchicine in humans: analysis of tissue, plasma and urine data in ten cases. Hum Exp Toxicol 1992;11:510–516.
17. Achtert G, Scherrmann JM, Christen MO. Pharmacokinetics/bioavailability of olchicine in healthy male volunteers. Eur J Drug Metab Pharmacokinet 1989;14: 317–322.
18. Baldwin LR, Talbert RL, Samples R. Accidental overdose of insufflated colchicine. Drug Safety 1990:5;305–312.
19. Girre C, Thomas G, Scherrmann JM, Crouzette J, Fournier PE. Model-independent pharmacokinetics of colchicine after oral administration to healthy volunteers. Fund Clin Pharmacol 1989;3:537–543.
20. Rochdi M, Sabouraud A, Girre C, Venet R, Scherrmann JM. Pharmacokinetics and absolute bioavailability of colchicine after i.v. and oral administration in healthy human volunteers and elderly subjects. Eur J Clin Pharmacol 1994;46:351–354.
21. Ben Chetrit E, Levy M. Colchicine 1998 update. Semin Arthritis Rheum 1998;28:48–59.
22. Goldbart A, Press J, Sofer S, Kapelushnik J. Near fatal acute colchicine intoxication in a child. A case report. Eur J Pediatr 2000;159:895–897.
23. Borisy GG, Taylor EW. The mechanism of action of colchicine. J Cell Biol 1967;34:525–533.
24. Ferguson FC. Colchicine: I. General pharmacology. J Pharmacol Exp Ther 1952;106:261–270.
25. Sauder P, Kopferschmitt J, Jaeger A, Mantz JM. Haemodynamic studies in eight cases of acute colchicine poisoning. Hum Toxicol 1983;2:169–173.
26. Markand ON, D'Agostino AN. Ultrastructural changes in skeletal muscle induced by colchicine. Arch Neurol 1971;24:72–82.

27. Sundov Z, Nincevic Z, Definis-Gojanovic M, Glavina-Durdov M, Jukic I, Hulina N, Tonkic A. Fatal colchicine poisoning by accidental ingestion of meadow saffron-case report. Forensic Sci Int 2005;149:253–256.

28. Brnčić Višković I, Perić R, Dirlić A, Vitezić D, Cuculić D. Accidental plant poisoning with *Colchicum autumnale*: report of two cases. Croat Med J 2001;42:673–675.

29. Klintschar M, Beham-Schmidt C, Radner H, Henning G, Roll P. Colchicine poisoning by accidental ingestion of meadow saffron (*Colchicum autumnale*): pathological and medicolegal aspects. Forensic Sci Int 1999;106:191–200.

30. McIntyre IM, Ruszkiewicz AR, Crump K, Drummer OH. Death following colchicine poisoning. J Forensic Sci 1994; 39:280–286.

31. Davies HO, Hyland RH, Morgan CD, Laroye GJ. Massive overdose of colchicine. CMAJ 1988;138:335–336.

32. Clevenger CV, August TF, Shaw LM. Colchicine poisoning: report of a fatal case with body fluid analysis by GC/MS and histopathologic examination of postmortem tissues. J Anal Toxicol 1991;15:151–154.

33. van Heyningen C, Watson ID. Troponin for prediction of cardiovascular collapse in acute colchicine overdose. Emerg Med J 2005;22:599–600.

34. Altman A, Szyper-Kravitz M, Shoenfeld Y. Colchicine-induced rhabdomyolysis. Clin Rheumatol 2007;26:2197–2199.

35. Maxwell MJ, Muthu P, Pritty PE. Accidental colchicine overdose. A case report and literature review. Emerg Med J 2002;19:265–267.

36. Aleem HMA. *Gloriosa superba* poisoning. J Assoc Physician India 1992;40:541–542.

37. Brvar M, Kozelj G, Mozina M, Bunc M. Acute poisoning with autumn crocus (*Colchicum autumnale* L.). Wien Klin Wochenschr 2004;116:205–208.

38. Hill RN, Spragg RG, Wedel MK, Moser KM. Adult respiratory distress syndrome associated with colchicine toxicity. Ann Intern Med 1975;83;523–524.

39. Allender WJ. Colchicine poisoning as a mode of suicide. J Forensic Sci 1982;27:944–947.

40. Nagaratnam N, De Silva DPKM, De Silva N. Colchicine poisoning following ingestion of *Gloriosa superba* tubers. Trop Geo Med 1973;25:15–17.

41. Heaney D, Derghazarian CB, Pineo GF, Ali MA. Massive colchicine overdose: A report on the toxicity. Am J Med Sci 1976;271:233–238.

42. Naidus RM, Roduien R, Mielke H. Colchicine toxicity. A multisystem disease. Arch Intern Med 1977;137:394–396.

43. Angunawela RM, Fernando HA. Acute ascending polyneuropathy and dermatitis following poisoning by tubers of *Gloriosa superba*. Ceylon Med J 1971;16:233–235.

44. DeVillota ED, Galdos P, Mosquera JM, Tomas MI. Colchicine overdose: an unusual origin of multi-organ failure. Crit Care Med 1979;7:278–279.

45. Murray SS, Kramlinger KG, McMichan JC, Mohr DN. Acute toxicity after excessive ingestion of colchicine. Mayo Clin Proc 1983;58:528–532.

46. Stanley MW, Taurog JD, Snover DC. Fatal colchicine toxicity: report of a case. Clin Exp Rheumatol 1984;2: 167–171.

47. Stahl N, Weinberg A, Benjamin D, Pinkhas J. Fatal colchicine poisoning in a boy with familial Mediterranean fever. Am J Med Sci 1979;278:77–81.

48. Gabrscek L, Lesnicar G, Krivec B, Voga G, Sibanc B, Blatnik J, Jagodic B. Accidental poisoning with *Autumn crocus*. J Toxicol Clin Toxicol 2004;42:85–88.

49. Deveaux M, Hubert N, Demarly C. Colchicine poisoning: case report of two suicides. Forensic Sci Int 2004;143: 219–222.

50. Tracqui A, Kintz P, Ludes B, Rouge C, Douibi H, Mangin P. High-performance liquid chromatography coupled to ion spray mass spectrometry for the determination of colchicine at ppb levels in human biofluids. J Chromatogr B Biomed Appl 1996;675:235–242.

51. Jiang Y, Wang J, Wang Y, Li H, Fawcett JP, Gu J. Rapid and sensitive liquid chromatography-tandem mass spectrometry method for the quantitation of colchicine in human plasma. J Chromatogr B Analyt Technol Biomed Life Sci 2007;850:564–568.

52. Cheze M, Deveaux M, Pepin G. Liquid chromatography-tandem mass spectrometry for the determination of colchicine in postmortem body fluids. Case report of two fatalities and review of the literature. J Anal Toxicol 2006;30:593–598.

53. Halkin H, Dany S, Greenwald M, Shnaps Y, Tirosh M. Colchicine kinetics in patients with familial Mediterranean fever. Clin Pharmacol Ther 1980;28:82–87.

54. Wallace SL, Omokoku B, Ertel NH. Colchicine plasma levels. Implications as to pharmacology and mechanism of action. Am J Med 1970;48:443–448.

55. Dehon B, Chagnon J-L, Vinner E, Pommery J, Mathieu D, Lhermitte M. Colchicine poisoning: report of a fatal case with body fluid and post-mortem tissue analysis by high-performance liquid chromatography. Biomed Chromatog 1999;13:235–238.

56. Brvar M, Ploj T, Kozelj G, Mozina M, Noc M, Bunc M. Case report: fatal poisoning with *Colchicum autumnale*. Crit Care 2004;8:R56–R59.

57. Jarvie P, Park J, Stewart MJ. Estimation of colchicine in a poisoned patient using high performance liquid chromatography. Clin Toxicol 1979;14:375–381.

58. Devaux M, Hubert N, Demarly C. Colchicine poisoning: case report of two suicides. Forensic Sci Int 2004;143: 219–222.

59. Caplan YH, Orloff KG, Thompson BC. A fatal overdose with colchicine. J Anal Toxicol 1980;4:153–155.

60. Kintz P, Jamey C, Tracqui A, Mangin P. Colchicine poisoning: report of a fatal case and presentation of an HPLC procedure for body fluid and tissue analyses. J Anal Toxicol 1997;21:70–72.

61. Mendis S. Colchicine cardiotoxicity following the ingestion of *Gloriosa superba* tubers. Postgrad Med J 1989;65: 752–755.
62. Baud FJ, Sabouraud A, Vigaut E, Taboulet P, Lang J, Bismuth C, et al. Brief report: treatment of severe colchicine overdose with colchicine-specific Fab fragments. N Engl J Med 1995;332:642–645.
63. Biçer S, Soysal DD, Ctak A, Uçsel R, Karaböcüoğlu M, Uzel N. Acute colchicine intoxication in a child: a case report. Pediatr Emerg Care 2007;23:314–317.
64. Folpini A, Furfori P. Colchicine toxicity–clinical features and treatment. Massive overdose case report. Clin Toxicol 1995;33:71–77.

Chapter 108

DAFFODILS and OTHER EMETIC BULBS

HISTORY

These bulbs are native species of Western Europe. The Romans brought daffodils to Britain for medicinal purposes. However, now the bulb is widely cultivated as an ornamental plant.

BOTANICAL DESCRIPTION

The Amaryllidaceae family contains approximately 725 species of perennial bulbs that are native to tropical and subtropical regions in Europe, South American Andes, and southern Africa.

Common Name: Daffodil, Wild Daffodil

Scientific Name: *Narcissus pseudonarcissus* L.

Botanical Family: Amaryllidaceae (amaryllis)

Physical Description: Narcissus plants are usually similar to lilies and onions with numerous narrow leaves and a single flowering stalk arising from a true bulb. The flowers appear in early spring with petals that form a prominent tubular "corona" in the center.

Distribution and Ecology: Daffodils are native plants of Western Europe, United Kingdom, Iraq, and Turkey. They grow in moist areas of lightly shaded woodlands.

EXPOSURE

Galanthamine occurs in several members of the Amaryllidaceae (*Leucojum* spp., *Narcissus* species, *Galanthus* spp.). In Eastern European countries (e.g., Bulgaria), this compound was used in the treatment of poliomyelitis; recently, galanthamine was introduced into Western markets as a treatment for Alzheimer's disease.[1]

PRINCIPAL TOXINS

Physiochemical Properties

The daffodil bulb contains the heat-stable, gastrointestinal irritant alkaloid lycoramine as well as lesser amounts of narcissamine and galanthamine.[2] Galanthamine (galantamine, CAS RN: 357-70-0) is a specific, competitive, and reversible acetylcholinesterase inhibitor that is approved for the treatment of Alzheimer's disease along with the synthetic drug, galantamine hydrobromide (CAS RN: 69353-21-5). Figure 108.1 displays the chemical structure of galanthamine. Lycorine-type alkaloids (e.g., 1-*O*-acetyllycorine) from the Amaryllidaceae family exhibit greater acetylcholinesterase inhibitor than galanthamine.[3] There are few data on the contribution of this alkaloid to the gastrointestinal toxicity associated with the ingestion of daffodil

Medical Toxicology of Natural Substances, by Donald G. Barceloux, MD

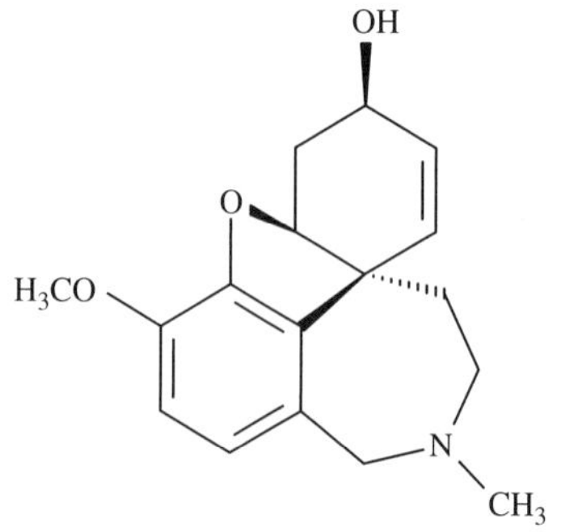

FIGURE 108.1. Chemical structure of galanthamine (galantamine).

bulbs. The sap from *Narcissus* bulbs and hollow stems contain large amounts of calcium oxalate crystals.[4] These crystals are water insoluble. California species of buttercup (*Ranunculus*) contain irritant oils (e.g., protoanemonine), which are mild gastrointestinal irritants. Hydrolysis of the glycoside ranunculin produces protoanemonin.

Poisonous Parts

Narcissus pseudonarcissus contains at least seven major alkaloids including hemanthamine, galanthamine, galanthine, narcissamine, masonin, homolycorine, lycoramine, pseudolycorine, and lycorine. Lesser amounts of lycoramine occur in the stems, leaves, and flowers of *Narcissus* cultivars. Galanthamine is widespread within species of the Amaryllidaceae family with galanthamine content typically averaging from trace to 0.5% dry weight. Some *Narcissus* species (e.g., *N. confusus* Pugsley) contain high concentrations of galanthamine exceeding 1.2% (dry weight) with the highest concentrations occurring in the emerging bulb.[5] The concentration of galanthamine varies with different *Narcissus* cultivars with the concentration typically <1–2.5%.[6] In an Egyptian species of *Ranunculus* (*R. sceleratus* L.), the protoanemonin and ranunculin content was approximately 0.25% and 0.67%, respectively, after steam distillation.[7] There are few data on the alkaloids responsible for the irritant and toxic effects of daffodil bulbs.[8]

DOSE RESPONSE

Galanthamine and the synthetic drug, galantamine hydrobromide are used to treat Alzheimer's disease with maximum daily doses of galantamine ranging up to 24–32 mg.[9] Galanthamine was first isolated from snow drop (*Galanthus* spp.), but today other sources for the production of galanthamine include *Narcissus* species, *Leucojum* species, and chemical synthesis.[1]

MECHANISM OF TOXICITY

In animal models, galanthamine is a powerful gastrointestinal irritant and central nervous system (CNS) stimulant that can cause death after large doses.[2] Galanthamine is a reversible, competitive inhibitor of acetylcholinesterase and an allosteric modulator of nicotinic receptors.[10] This compound increases the sensitivity of the nicotinic receptors, facilitating both excitatory and inhibitory transmissions in human brain tissue.[11]

TOXICOKINETICS

The gastrointestinal absorption of galanthamine is high with a bioavailability of about 90%. Galanthamine has a relatively large (i.e., about 2.5 L/kg) volume of distribution and low protein binding (i.e., 15–20%).[12] Biotransformation of galantamine occurs primarily via the cytochrome P450 isoenzymes (i.e., CYP2D6, CYP3A4) to inactive *O*-desmethyl galantamine and galantamine *N*-oxide metabolites.[13] Inhibitors of CYP2D6 (e.g., paroxetine) and CYP3A4 (ketoconazole, erythromycin) increase the bioavailability of galantamine.

CLINICAL RESPONSE

Daffodils

Ingestion of daffodil bulbs produces the rapid onset of gastroenteritis (nausea, vomiting, abdominal pain, diarrhea, lightheadedness) that usually resolves within several hours depending on the amount ingested.[14] Profuse vomiting and diarrhea developed in a family following the consumption of a soup prepared from jonquil (*Narcissus jonquilla* L.) leaves.[2] The gastrointestinal symptoms resolved after approximately 24 hours. In contrast to the cardiovascular and neurological effects present in experimental animals administered large quantities of *Narcissus* bulbs, cardiovascular and neurological toxicity are uncommon in these human case reports.

Occupational exposure of bulb handlers to daffodil bulbs is a common cause of a rash called "daffodil itch."[15] This rash is primarily an irritant contact dermatitis that affects areas of the skin (hands, wrists) in direct contact with the sap during picking.[16] Allergic responses to daffodils occur rarely. Typically, pruritus is localized to the tips of the fingers, first dorsal web space, and the periungual region (i.e., paronychia). Lesions also occur on the sides of the fingers, the hands, and the forearms. Occasionally, lesions involve the face, neck or genitals as a result of contact with contaminated gloves or hands. Erythema, edema, painful fissures, and a dry, hyperkeratotic dermatitis may develop. Skin lesions sometimes appear on distant sites, such as the neck, abdominal wall, face, and genital region. Rarely, delayed asthmatic symptoms (cough, dyspnea) may occur within 48 hours after exposure to daffodil bulbs.[17]

Other Bulbs

The ingestion of other bulbs (amaryllis, hyacinth, buttercup) may produce a self-limited gastroenteritis, but there are limited data on the toxic effects of these bulbs in humans. Ingestion of plants from *Iris* species may cause mucosal irritation as well as gastrointestinal symptoms; the clinical course of these symptoms is also mild and self-limited. Illness resulted from the prolonged use of bulbs as food sources during World War II in Holland.[18] Native *Anemone* species also produce mild gastrointestinal irritation similar to the effects following consumption of bulbs from *Ranunculus* species. The South African bulb, *Boophane disticha* (L.f.) Herb. contains buphanine, which produces an anticholinergic syndrome similar to scopolamine characterized by hallucination, agitation, and altered mental status.[19]

DIAGNOSTIC TESTING

Methods to quantitate the concentration of galanthamine in plasma include high performance liquid chromatography with UV detection[20] and liquid chromatography/tandem mass spectrometry.[21] The blood to plasma ratio of galantamine averages about 1.2. Following administration of daily doses of 32 mg galantamine to healthy volunteers, the mean steady state galantamine concentration in plasma samples drawn with 10 hours of administration was approximately 115–125 µg/L.[13]

TREATMENT

Treatment of the gastrointestinal complications following the ingestion of daffodil bulbs is supportive. Fluid and electrolyte imbalance represents the most serious toxic effect in those patients experiencing protracted vomiting or diarrhea. Decontamination measures are not usually necessary because of the protracted vomiting associated with substantial ingestions.

References

1. Heinrich M, Lee Teoh H. Galanthamine from snowdrop—the development of a modern drug against Alzheimer's disease from local Caucasian knowledge. J Ethnopharmacol 2004;92:147–162.
2. Vigneau CH, Tsao J, Chamaillard C, Galzot J. Accidental absorption of daffodils (*Narcissus jonquilla*): two common intoxications. Vet Hum Toxicol 1982;24(suppl):133–135.
3. Elgorashi EE, Stafford GI, Van Staden J. Acetylcholinesterase enzyme inhibitory effects of Amaryllidaceae alkaloids. Planta Med 2004;70:260–262.
4. Stoner JG. Miscellaneous dermatitis-inducing plants. Clin Dermatol 1986;4:94–101.
5. Lopez S, Bastida J, Viladomat F, Codina C. Galanthamine pattern in *Narcissus confusus* plants. Planta Med 2003;69:1166–1168.
6. Moraes-Cerdeira RM, Burandt CL Jr, Bastos JK, Nanayakkara NP, Mikell J, Thurn J, McChesney JD. Evaluation of four *Narcissus* cultivars as potential sources for galanthamine production. Planta Med 1997;63:472–474.
7. Mahran GH, Saber AH, El-Alfy T. Spectrophotometric determination of protoanemonin, anemonin and ranunculin in *Ranunculus sceleratus* L. Planta Med 1968;16:323–328.
8. Gude M, Hausen BM, Heitsch H, Konig WA. An investigation of the irritant and allergenic properties of daffodils (*Narcissus pseudonarcissus* L., Amaryllidaceae) a review of daffodil dermatitis. Contact Dermatitis 1988;19:1–10.
9. Raskind MA, Peskind ER, Truyen L, Kershaw P, Damaraju CV. The cognitive benefits of galantamine are sustained for at least 36 months: a long-term extension trial. Arch Neurol 2004;61:252–256.
10. Dajas-Bailador FA, Heimala K, Wonnacott S. The allosteric potentiation of nicotinic acetylcholine receptors by galantamine is transduced into cellular responses in neurons: Ca^{2+} signals and neurotransmitter release. Mol Pharmacol 2003;64:1217–1226.
11. Santos MD, Alkondon M, Pereira EF, Aracava Y, Eisenberg HM, Maelicke A, et al. The nicotinic allosteric potentiating ligand galantamine facilitates synaptic transmission in the mammalian central nervous system. Mol Pharmacol 2002;61:1222–1234.
12. Piotrovsky V, van Peer A, van Osselaer N, Armstrong M, Aerssens J. Galantamine population pharmacokinetics in patients with Alzheimer's disease: modeling and simulations. J Clin Pharmacol 2003;43:514–523.
13. Farlow MR. Clinical pharmacokinetics of galantamine. Clin Pharmacokinet 2003;42:1383–1392.
14. Litovitz TL, Fahey BA. Please don't eat the daffodils. N Engl J Med 1982;306:547.
15. McGovern TW, Barkley TM. Botanical briefs: daffodils–*Narcissus* L. Cutis 2000;65:130–132.
16. Julian CG, Bowers PW. The nature and distribution of daffodil pickers' rash. Contact Dermatitis 1997;37:259–262.
17. Goncalo S, Freitas JD, Sousa I. Contact dermatitis and respiratory symptoms from *Narcissus pseudonarcissus*. Contact Dermatitis 1987;16:115–116.
18. O'Leary SB. Poisoning in man from eating poisonous plants. Arch Environ Health 1964;9:216–242.

19. du Plooy WJ, Swart L, van Huysteen GW. Poisoning with *Boophane disticha*: a forensic case. Hum Exp Toxicol 2001;20:277–278.
20. Maláková J, Nobilis M, Svoboda Z, Lísa M, Holcapek M, Kvetina J, et al. High-performance liquid chromatographic method with UV photodiode-array, fluorescence and mass spectrometric detection for simultaneous determination of galantamine and its phase I metabolites in biological samples. J Chromatogr B Analyt Technol Biomed Life Sci 2007;853:265–274.
21. Nirogi RV, Kandikere VN, Mudigonda K, Maurya S. Quantitative determination of galantamine in human plasma by sensitive liquid chromatography-tandem mass spectrometry using loratadine as an internal standard. J Chromatogr Sci 2007;45:97–103.

Chapter 109

DEATH CAMAS

HISTORY

Early settlers, Native Americans, and explorers of the western United States and Canada were familiar with the toxicity of death camas. Diaries from the Lewis and Clark Expedition documented the presence of a poisonous bulb growing on hillsides of the upper Missouri River plains that probably was from the Melanthiaceae family.[1] Consequently, members of the Expedition were required to buy roots from local Indians rather than dig roots by themselves when adequate food was unavailable. Indians of Northern California recognized the toxicity of death camas bulbs, and several fatalities were associated with the use of this bulb by these Indians during the late 19th century.[1] Similar edible species (e.g., common camas, sego lily) were important staples for Pacific Northwest Indians and early settlers.

BOTANICAL DESCRIPTION

There are approximately 15 species of death camas in the United States; however, the taxonomy of death camas remains unsettled. These bulbs are listed in several genera including *Zigadenus, Toxicoscordion, Stenanthium*, and *Anticlea*. Most authorities list these genera under the Melanthiaceae family and sometimes under the Liliaceae family. Table 109.1 lists common species of death camas and associated synonyms.

These perennial herbs have tall, narrow leaves that are clustered mostly at the base of a thick horizontal rootstock. Yellow or whitish green flowers form during the spring along the top of a central stalk depending on the species. These bulbs are distinguished from wild onion by the lack of the characteristic onion odor. Most species are found in moist grassy meadows in the western United States, but these species also occur in foothills, mountains, and deserts. Toxicity usually occurs when these bulbs are misidentified as edible wild onion or Sego lily.

Common Name: Foothill death camas, Sand-corn, Deadly Zigadenus, Meadow death camas

Scientific Name: *Toxicoscordion paniculatum* (Nutt.) Rydb.

Botanical Family: Melanthiaceae

Physical Description: This glabrous perennial herb arises from an onion-like bulb with a simple stem reaching 30–50 cm (~12–20 in.) in height. The mostly basal, linear leaves are tapered at the ends. Clusters of white to cream-colored flowers develop on a panicle up to 10–30 cm (~4–12 in.) long. The fruits are capsules about 1.4–1.8 cm (~0.5–0.8 in.) long.

Distribution and Ecology: This species inhabits moist grassy meadows in the western United States including Utah, New Mexico, Colorado, Nevada, and California.

EXPOSURE

There are no commercial uses for death camas because of the toxicity associated with ingestion of plant parts.

Medical Toxicology of Natural Substances, by Donald G. Barceloux, MD

TABLE 109.1. Scientific and Common Names of Death Camas

Species	Synonym	Common Name
Stenanthium densum (Desr.) Zomlefer & Judd	*Zigadenus densus* (Desr.) Fern., *Zigadenus leimanthoides* A. Gray	Osceola's plume, black snakeroot, crowpoison, pinebarren death camas
Anticlea elegans (Pursh) Rydb.	*Zigadenus elegans* Pursh	mountain death camas, white camas
Toxicoscordion brevibracteatum (M.E. Jones) R.R. Gates	*Zigadenus brevibracteatus* (M.E. Jones) H.M. Hall	Desert death camas,
Toxicoscordion exaltatum (Eastw.) A. Heller	*Zigadenus exaltatus* Eastw.	Giant death camas
Toxicoscordion fontanum (Eastw.) Zomlefer & Judd	*Zigadenus fontanus* Eastw.	
Toxicoscordion fremontii (Torr.) Rydb.	*Zigadenus fremontii* (Torr.) Torr. ex S. Watson	Fremont's death camas, star-lily, star zygadene
Zigadenus glaberrimus Michx.		Sandbog death camas,
Toxicoscordion gramineum (Rydb.) Rydb.	*Zigadenus acutus* Rydb., *Zigadenus falcatus* Rydb., *Zigadenus gramineus* Rydb, *Zigadenus intermedius* Rydb.	Grassy death camas, death camas
Toxicoscordion micranthum (Eastw.) A. Heller	*Zigadenus micranthus* Eastw.	Smallflower death camas
Zigadenus mogollonensis W.J. Hess & R.C. Sivinski		Mogoll death camas
Toxicoscordion nuttallii (A. Gray) Rydb.	*Zigadenus nuttallii* (A. Gray) S. Watson	Nuttall death camas, poison camas, poison sego, death camas
Toxicoscordion paniculatum (Nutt.) Rydb.	*Zigadenus paniculatus* (Nutt.) S. Watson	Foothill death camas, sand-corn, deadly zigadenus, meadow death camas
Zigadenus vaginatus (Rydb.) J.F. Macbr.		Sheathed death camas
Toxicoscordion venenosum (S. Watson)	*Zigadenus venenosus* S. Watson	Elegant death camas, grassy death camas, meadow death camas, Watson's death camas
Zigadenus virescens (Kunth) J.F. Macbr.	*Anticlea virescens* (Kunth) Rydb.	Green death camas
Zigadenus volcanicus Benth.	*Anticlea virescens* (Kunth) Rydb.	Lava death camas

Intoxication usually results from misidentification of these plants as similar edible plants, such as common camas [*Camassia quamash* (Pursh) Greene] or the sego lily (*Calochortus nuttallii* Torr. & Gray).

PRINCIPAL TOXINS

Physiochemical Properties

Death camas contain veratrum-like steroidal alkaloids that produce hypotension, bradycardia, and respiratory depression in experimental animals.[2] Major ester alkaloids include veratroyl zygadenine, and vanilloyl zygadenine, and neogermitrine.[3] The bitter taste of these bulbs discourages consumption.

Poisonous Parts

Although all parts of the plant contain veratrum-like alkaloids, toxicity usually results from ingestion of the bulb and, less often the flowers.[4]

Mechanism of Toxicity

The primary effects of these veratrum-like alkaloids are hypotension, bradycardia, gastrointestinal distress, and central nervous system (CNS) depression. Hypotension results from peripheral vasodilation from reduced α-adrenergic activity rather than bradycardia, and changes in blood pressure are not related to the heart rate.[2,5]

DOSE RESPONSE

The typical bulb weighs approximately 10–15 g.[1] The ingestion of 1–2 bulbs by adults produces gastrointestinal distress and bradycardia.[4] Although the ingestion of death camas bulbs is potentially lethal, there are no well-documented fatalities in the modern medical literature. The ingestion of large numbers of bulbs (e.g., four to five) by a young child and of much larger doses by adults probably is necessary to cause serious intoxication.[1]

TOXICOKINETICS

There are few data on the toxicokinetics of the veratrum-like alkaloids present in the these species. The rapid onset of symptoms suggests that these alkaloids are easily absorbed from the gastrointestinal tract.[4]

CLINICAL RESPONSE

The classical presentation following the ingestion of bulbs or other parts of these plants includes profound gastrointestinal distress (nausea, vomiting, diarrhea, abdominal pain), bradycardia, and hypotension.[6,7] Other symptoms include paresthesias, ataxia, gait disturbances, muscle fasciculations, mydriasis, and obtundation.[8] Gastrointestinal symptoms typically begin within 1 hour of ingestion, and resolve within 4–24 hours. Although sinus bradycardia is common, serious dysrhythmias do not usually occur. Fatalities following the ingestion of death camas are not well documented in the medical literature.

DIAGNOSTIC TESTING

Laboratory abnormalities are usually absent during intoxication from these plants.

TREATMENT

Treatment is supportive. Decontamination measures are usually unnecessary because of the early onset of severe gastrointestinal symptoms. There are no clinical data to document the efficacy of the administration of activated charcoal within 1 hour after ingestion. Hypotension usually responds to intravenous fluids. If hypotension persists despite fluid therapy and signs of inadequate perfusion appear, vasoconstrictive agents (dopamine, norepinephrine) may be necessary. The administration of atropine usually improves the heart rate, but often the increased heart rate is not associated with increases in blood pressure.[6] Patients with vital sign abnormalities or CNS depression should be observed for at least 12 to 24 hours.

References

1. Wagstaff DJ, Case AA. Human poisoning by *Zigadenus*. Clin Toxicol 1987;25:361–367.
2. Yaffe S, Kupchan SM. Veratrine-like properties of the alkaloidal fraction from *Zigadenus venenosus*. Fed Proc 1950;9:326.
3. Krayer O, Kupchan SM, Deliwala CV, Rogers BH. [Studies on *Veratrum* alkaloids. XVIII. Chemical and pharmacological relations between *Zygadenus* and *Veratrum* alkaloids.] Naunyn Schmiedebergs Arch Exp Pathol Pharmakol 1953;219:371–385. [German]
4. Spoerke DG, Spoerke SE. Three cases of *Zigadenus* (death camas) poisoning. Vet Hum Toxicol 1979;21:346–347.
5. Oda T. Effects of zygadenine derivatives upon the circulatory system. Osaka City Med J 1963;17:31–52.
6. Heilpern KL. *Zigadenus* poisoning. Ann Emerg Med 1995;25:259–262.
7. Peterson MC, Rasmussen GJ. Intoxication with foothill camas (*Zigadenus paniculatus*). J Toxicol Clin Toxicol 2003;41:63–65.
8. Cameron K. Death camas poisoning. Northwest Med 1952;51:682–683.

Chapter 110

KAFFIR LILY
[*Clivia miniata* (Lindley) Bosse]

HISTORY

The Kaffir lily is a native species of South Africa that was imported into England from Kwazulu-Natal during the 1800s. During Victorian times, this beautiful plant was a popular indoor plant in England and Europe.

BOTANTICAL DESCRIPTION

Common Name: Kaffir lily, Bush Lily, Umayime (Zulu), Boslelie (Afrikaans)

Scientific Name: *Clivia miniata* (Lindley) Bosse

Botanical Family: Amaryllidaceae

Physical Description: This perennial plant forms clumps of dark green, shiny elongated leaves that originate from a fleshy underground stem. Brilliant orange, trumpet-shaped flowers appear in spring and occasionally at other times of the year with a sweet smell. The fruit is a red berry. The plant reaches up to 45cm (~18in.) in height.

Distribution and Ecology: The Kaffir lily is an indigenous plant of South Africa. This plant prefers partial shade or deep shade in rich soil of the forest floor or between boulders on slopes.

EXPOSURE

In South Africa, traditional herbal uses of this plant include initiation of parturition, fever, and treatment of snake bites. Additionally, extracts of the Kaffir lily are used as oxytocic agents in South African traditional herbal medicine.

PRINCIPAL TOXINS

Physiochemical Properties

5-Hydroxymethyl-2-furancarboxaldehyde is a component of this extract that has cholinergic agonist properties.[1] *In vitro* studies indicate that the extract of the Kaffir lily has uterotonic activity.[2] Table 110.1 lists some physical properties of lycorine, which is the major alkaloid in this plant.

Poisonous Parts

Alkaloids isolated from the Kaffir lily are primarily derivatives of the 2-benzopyrano(3,4-g) indole nucleus. The major alkaloid in the Kaffir lily is lycorine (CAS RN: 476-28-8, $C_{16}H_{17}NO_4$). Figure 110.1 displays the chemical structure of lycorine. Following ethanol extraction and ion exchange chromatography, leave samples of *Clivia miniata* contained lycorine concentrations of approximately 0.043% by wet weight.[3] Other components in extracts of the Kaffir lily include clivonine (CAS RN: 477-16-7, C_{17}-H_{19}-N-O_5,), clivatine (CAS RN: 1355-66-4, C_{21}-H_{25}-N-O_7), linoleic acid and 5-hydroxymethyl-2-furancarboxaldehyde.

Mechanism of Toxicity

The mechanism of action of lycorine has not been determined.

Medical Toxicology of Natural Substances, by Donald G. Barceloux, MD

TABLE 110.1. Some Physical Properties of Lycorine

Physical Property	Value
Melting Point	280 °C/535 °F
log P (Octanol-Water)	0.930
Atmospheric OH Rate Constant	2.03E-10 cm^3/molecule-second (25 °C/77 °F)

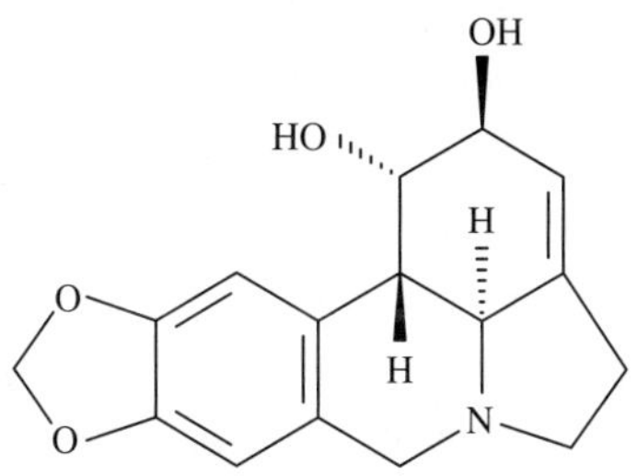

FIGURE 110.1. Chemical structure of lycorine.

DOSE RESPONSE

There are few data on the mechanism of toxicity associated with the gastrointestinal distress that occurs following the ingestion of this plant.

TOXICOKINETICS

There are inadequate data on the toxicokinetics of lycorine to determine the absorption, distribution, or elimination of this compound in humans.

CLINICAL RESPONSE

There are few clinical data on the effects of ingesting parts of the Kaffir lily. Based on the presence of lycorine in the bulbs, the most likely clinical effects following ingestion are minor gastrointestinal symptoms (e.g., nausea, vomiting, diarrhea).

DIAGNOSTIC TESTING

Previously published methods to identify active components of extracts of the Kaffir lily include spectrophotometry and ion exchange chromatography.[1]

TREATMENT

Treatment is supportive. Because gastrointestinal irritation is the major effect of ingestion of these bulbs, the patient should be evaluated for fluid and electrolyte balance as indicated by the clinical presentation. However, the clinical effects of exposure to the Kaffir lily are usually minimal, and only observation frequently is required. Decontamination measures are unnecessary because of the vomiting associated with more severe cases of poisoning by these bulbs.

References

1. Sewram V, Raynor MW, Mulholland DA, Raidoo DM. Supercritical fluid extraction and analysis of compounds from *Clivia miniata* for uterotonic activity. Planta Med 2001;67:451–455.
2. Veale DJ, Oliver DW, Havlik I. The effects of herbal oxytocics on the isolated "stripped" myometrium model. Life Sci 2000;67:1381–1388.
3. Ieven M, Vlietinck AJ, Banden Berghe DA, Totte J. Plant antiviral agents. III. Isolation of alkaloids from *Clivia miniata* regel (Amaryllidaceae). J Nat Prod 1982;45: 564–573.

Chapter 111

SNOWDROP
(*Galanthus nivalis* L.)

HISTORY

Snowdrop (*Galanthus nivalis* L.) has been used as a treatment for pain and neurological disorders since ancient Greek times.[1] In the epic poem, the Odyssey, Homer describes the plant, "moly," which was given to Odysseus by the god Hermes as an antidote to the potions of the enchantress, Circe. Plaitakis and Duvoisin hypothesized that Circe's potion contained a belladonna alkaloid-containing Solanaceae plant (*Atropa belladonna* L. or *Datura stramonium* L.) and the antidote was the acetylcholinesterase-containing snowdrop plant.[2] Hence, Homer's Odyssey may be the first recorded use of an anticholinesterase to reverse the muscarinic symptoms associated with belladonna alkaloids. This small perennial bulbous herb most likely originated in Turkey, Iran, and the Caucasus, and the plant was probably introduced into Great Britain during the early medieval period by religious communities. Saint Francis of Assisi called the snowdrop an emblem of hope as a result of the appearance of the flower at the end of winter. Because the snowdrop flowers near the feast of the Purification of the Blessed Virgin Mary, the flower was used in churches on Candlemas as a symbol of virginity and purity. The flower was also considered a symbol of death because a single bloom resembled a shrouded corpse. In the early 1950s, Russian pharmacologists isolated galantamine from a species of *Galanthus* after investigating the use of snowdrop in local Caucasian villagers for the treatment of poliomyelitis in children.[3] By the late 1950s, preclinical studies documented the antagonist effects of galantamine on nondepolarizing neuromuscular blocking agents, and galantamine hydrobromide was registered in Bulgaria under the trade name Nivalin® (Sopharma Pharmaceuticals, Sofia, Bulgaria). By the 1990s, clinical investigations indicated that galantamine was an effective treatment for early Alzheimer's disease.

BOTANICAL DESCRIPTION

Common Name: Snowdrop, Candlemas Bell, Mary's Taper, Snow-Piercer, Dingle-Dangle, February Fairmaids, Purification Flower

Scientific Name: *Galanthus nivalis* L.

Botanical Family: Liliaceae (lily)

Physical Description: Narrow linear leaves protrude from the bulb with three long, outer tepals surrounding three green inner tepals. The snowdrop flowers in early February near Candlemas (February 2nd), which is feast of the Purification of the Blessed Virgin Mary and the source of several common names for this plant. Many cultivars exist.

Distribution and Ecology: The common snowdrop probably originated in the Caucasus region of eastern Europe or western Asia. This plant now inhabits the United Kingdom, the northeastern United States, and Washington. Snowdrops grow in damp, deciduous woodlands, particularly in shady areas near rivers, pastures, woods, and streams.

Medical Toxicology of Natural Substances, by Donald G. Barceloux, MD

Common Name: Giant Snowdrop

Scientific Name: *Galanthus elwesii* Hook.f.

Botanical Family: Liliaceae (lily)

Physical Description: Slender, oblong leaves about 6in. (~15cm) long arise from a small bulb. The flower stalks produce solitary, white flowers with three spreading outer petals surrounding three smaller, less convex inner pedals that contain distinctive green markings. The flower is milky-white (in Greek, *galanthus* means milk flower—*gala* is milk, *anthos* is flower).

Distribution and Ecology: The plant inhabits areas of Bulgaria, Greece, Turkey, and Romania, and the giant snowdrop is an introduced species in Ohio, Pennsylvania, and New York.

EXPOSURE

The snowdrop has a long tradition of use as a folk medicine for pain and neurological conditions. Galantamine is a tertiary alkaloid present in snowdrop and other members of the Liliaceae family. This acetylcholinesterase inhibitor has been used during anesthesia to reverse the neuromuscular paralysis induced by tubocurarine and other paralytic agents.[4] The use of galantamine (Reminyl®, Shire Pharmaceuticals, UK) is approved in many European countries and the United States for the treatment of Alzheimer's disease, but improvement in measures of cognition and global assessment of dementia is usually clinically marginal.[5] A meta-analysis of studies on the use of drugs including galantamine on the outcome of Alzheimer's disease indicated that the standardized estimates of effect size across diverse functional outcome measures for drug treatment in patients with Alzheimer's disease were small.[6] The current lack of alternative treatments for Alzheimer's disease and the good safety profile of galantamine suggest a role for this drug in the treatment of Alzheimer's disease and possible other dementias.[7,8] Although galantamine was first isolated from *Galanthus* species, the principal sources are now synthetic production as well as species of *Narcissus* and *Leucojum*. Galantamine is an experimental treatment for alcoholism.[9]

PRINCIPAL TOXINS

Physiochemical Properties

Galantamine (galanthamine, galanthine, lycoremine) is one of several Liliaceae alkaloids with anticholinesterase inhibitor activity that reverses the central actions of tropane alkaloids (hyoscine, atropine). This compound

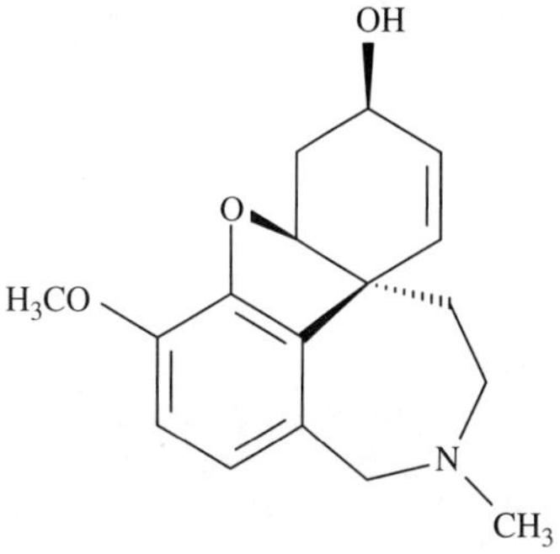

OH
O
H_3CO
N
CH_3

FIGURE 111.1. Chemical structure of galantamine.

(CAS RN: 357-70-0, $C_{17}H_{21}NO_3$) is a phenanthrene alkaloid that has a chemical structure similar to codeine as displayed in Figure 111.1. The pK_a of galantamine is 8.32. This compound has substantially greater affinity for human erythrocyte acetyl cholinesterase than plasma acetyl cholinesterase, and the affinity of galantamine is about 10 times greater for human brain acetyl cholinesterase than human erythrocyte acetyl cholinesterase.[10] Lycorine-type Liliaceae alkaloids also exhibit acetylcholinesterase activity with 1-*O*-acetyllycorine being about 2-fold more potent than galantamine.[11,12]

Poisonous Parts

Galantamine occurs in a variety of plant genera besides *Galanthus* including *Narcissus*, *Crinum*, *Hippeastrum*, *Leucojum*, *Lycoris*, and *Hemerocallis*.[1] This compound is one of several Liliaceae alkaloids present in *Galanthus* species that include hamayne, ismine, lycorenine, lycorine, tazettine, and ungeremine as well as crinane and phenanthridine type alkaloids.[13] The concentrations of these alkaloids vary substantially between *Galanthus* species with haemanthamine, galantamine, and crinine being the principal alkaloids in the giant snowdrop (*Galanthus elwesii*).[14,15] The common snowdrop (*G. nivalis*) contains a unique lectin that possesses a carbohydrate-binding specificity for nonreducing terminal α-D-mannosyl groups.[16]

Mechanism of Toxicity

Galantamine is a long-acting, selective, reversible, and competitive inhibitor of acetylcholinesterase. This compound potentiates nicotinic neurotransmission by allosteric modulation on nicotinic acetyl cholinergic receptors as well as inhibiting acetylcholinesterase.[17] Galantamine can penetrate the blood—brain barrier and augment cholinergic transmission.

DOSE RESPONSE

Intravenous doses of 5 mg to 20 mg galantamine reverse therapeutic doses of pancuronium, gallamine, and tubocurarine, although the recovery of muscle is slower compared with similar doses of neostigmine. In a study of 10 medical students administered 2 mg scopolamine IV, the subsequent IV administration of galantamine hydrobromide 0.5 mg/kg reversed the central effects (drowsiness, disorientation, visual hallucination, delirium) of scopolamine within 5–10 minutes.[18] In placebo-controlled trials of Alzheimer's patients, the dose of galantamine hydrobromide ranged up to 100 mg daily with adverse effects including gastrointestinal distress, agitation, and insomia.[4]

TOXICOKINETICS

The gastrointestinal absorption of therapeutic doses of galantamine is excellent, and the volume of distribution is relatively large with little protein binding based on volunteer studies.[19] Renal clearance accounted for about 25% of the elimination of galantamine. The mean terminal elimination half-life was approximately 5–6 hours. Biotransformation of galantamine involves chiral conversion to epigalantamine and oxidative metabolism with cytochrome P450 isoenzymes (CYP2D6, CYP3A4) to *O*-desmethyl-galantamine, *N*-desmethyl-galantamine, and galantamine-*N*-oxide.[20]

CLINICAL RESPONSE

Side effects of the therapeutic use of galantamine include parasympathetic effects, such as nausea, vomiting, salivation, and blurred vision.[21] The ingestion of lycorine is associated with profuse vomiting.

DIAGNOSTIC TESTING

Analytical methods for the detection and quantitation of galantamine in plasma and tissues include liquid chromatography/tandem mass spectrometry[22] and high performance liquid chromatographic method with UV photodiode array, fluorescence, and mass spectrometric detection.[23] The lower limit of quantitation of the former method was 0.5 ng/mL. There are inadequate data in the medical literature to associate the ingestion of snowdrop bulbs with laboratory abnormalities, such as abnormalities in plasma or red blood cell (RBC) cholinesterases.

TREATMENT

Treatment is supportive. Because gastrointestinal irritation is the major effect of ingestion of these bulbs, the patient should be evaluated for fluid and electrolyte balance as indicated by the clinical presentation. Decontamination measures are unnecessary because of the vomiting associated with poisoning by these bulbs. Although snowdrop bulbs contain anticholinesterase inhibitors, there are few data in the medical literature to suggest that cholinergic crisis results from the ingestion of snowdrop bulbs.

References

1. Lee MR. The snowdrop (*Galanthus nivalis*): from Odysseus to Alzheimer. Proc R Coll Physicians Edinb 1999;29:349–352.
2. Plaitakis A, Duvoisin RC. Homer's moly identified as *Galanthus nivalis* L.: physiologic antidote to stramonium poisoning. Clin Neuropharmacol 1983;6:1–5.
3. Heinrich M, Teoh HL. Galanthamine from snowdrop–the development of a modern drug against Alzheimer's disease from local Caucasian knowledge. J Ethnopharmacol 2004;92:147–162.
4. Sramek JJ, Frackiewicz EJ, Cutler RN. Review of the acetylcholinesterase inhibitor galanthamine. Exp Opin Invest Drugs 2000;9:2393–2402.
5. Raina P, Santaguida P, Ismaila A, Patterson C, Cowan D, Levine M, et al. Effectiveness of cholinesterase inhibitors and memantine for treating dementia: evidence review for a clinical practice guideline. Ann Intern Med 2008;148:379–397.
6. Hansen RA, Gartlehner G, Lohr KN, Kaufer DI. Functional outcomes of drug treatment in Alzheimer's disease: a systematic review and meta-analysis. Drugs Aging 2007; 24:155–167.
7. Razay G, Wilcock GK. Galantamine in Alzheimer's disease. Expert Rev Neurother 2008;8:9–17.
8. Auchus AP, Brashear HR, Salloway S, Korczyn AD, De Deyn PP, Gassmann-Mayer C, GAL-INT-26 Study Group. Galantamine treatment of vascular dementia: a randomized trial. Neurology 2007;69:448–458.
9. Mann K, Ackermann K, Diehl A, Ebert D, Mundle G, Nakovics H, et al. Galantamine: a cholinergic patch in the treatment of alcoholism: a randomized, placebo-controlled trial. Psychopharmacol 2006;184:115–121.
10. Harvey AL. The pharmacology of galanthamine and its analogues. Pharmacol Ther 1995;68:113–128.
11. Lopez S, Bastida J, Viladomat F, Codina C. Acetylcholinesterase inhibitory activity of some Amaryllidaceae alkaloids and Narcissus extracts. Life Sci 2002;71:2521–2529.
12. Elgorashi EE, Stafford GI, Van Staden J. Acetylcholinesterase enzyme inhibitory effects of Amaryllidaceae alkaloids. Planta Med 2004;70:260–262.
13. Berkov S, Codina C, Viladomat F, Bastida J. Alkaloids from *Galanthus nivalis*. Phytochemistry 2007;68:1791–1798.

14. Sidjimova B, Berkov S, Popov S, Evstatieva L. Galanthamine distribution in Bulgarian *Galanthus* spp. Pharmazie 2003;58:935–936.
15. Berkov S, Sidjimova B, Evstatieva L, Popov S. Intraspecific variability in the alkaloid metabolism of *Galanthus elwesii*. Phytochemistry 2004;65:579–586.
16. Kaku H, Goldstein IJ. Snowdrop lectin. Methods Enzymol 1989;179:327–331.
17. Santos MD, Alkondon M, Pereira EF, Aracava Y, Eisenberg HM, Maelicke A. Albuquerque EX. The nicotinic allosteric potentiating ligand galantamine facilitates synaptic transmission in the mammalian central nervous system. Mol Pharmacol 2002;61:1222–1234.
18. Baraka A, Harik S. Reversal of central anticholinergic syndrome by galanthamine. JAMA 1977;238:2293–2994.
19. Bickel U, Thomsen T, Weber W, Fischer JP, Bachus R, Nitz M, Kewitz H. Pharmacokinetics of galanthamine in humans and corresponding cholinesterase inhibition. Clin Pharmacol Ther 1991;50:420–428.
20. Bachus R, Bickel U, Thomsen T, Roots I, Kewitz H. The *O*-demethylation of the antidementia drug galanthamine is catalysed by cytochrome P450 2D6. Pharmacogenetics 1999;9:661–668.
21. Rockwood K, Mintzer J, Truyen L, Wessel T, Wilkinson D. Effects of a flexible galantamine dose in Alzheimer's disease: a randomised, controlled trial. J Neurol Neurosurg Psychiatry 2001;71:589–595.
22. Nirogi RV, Kandikere VN, Mudigonda K, Maurya S. Quantitative determination of galantamine in human plasma by sensitive liquid chromatography-tandem mass spectrometry using loratadine as an internal standard. J Chromatogr Sci 2007;45:97–103.
23. Maláková J, Nobilis M, Svoboda Z, Lísa M, Holcapek M, Kvetina J, et al. High-performance liquid chromatographic method with UV photodiode-array, fluorescence and mass spectrometric detection for simultaneous determination of galantamine and its phase I metabolites in biological samples. J Chromatogr B Analyt Technol Biomed Life Sci 2007;853:265–274.

Chapter 112

ZEPHYR LILY (*Zephyranthes* Species)

BOTANICAL DESCRIPTION

There are at least 11 species of the genus *Zephyranthes* in the United States. Zephyr lily refers generally to most species in this genus including *Zephyranthes grandiflora* Lindl. (rosepink zephyr lily) and *Zephyranthes simpsonii* Chapman (redmargin zephyr lily); whereas atamasco lily refers to *Zephyranthes atamasco* (L.) Herbert.

Common Name: Rosepink Zephyr Lily, Rain Lily

Scientific Name: *Zephyranthes grandiflora* Lindl.

Botanical Family: Liliaceae (lily)

Physical Description: Narrow, pointed green leaves arise from bulb along with a length of about 25–30 cm (10–12 in.). Funnel-shaped pink flower appear on the end of the stem that reach up to about 10–18 cm (~2–7 in.) in length. This perennial plant blooms several times during the summer and autumn, usually after a heavy rain; hence, the common name rain lily.

Distribution and Ecology: This plant is a native species of Mexico and Central America that inhabits moist, open areas near woodlands.

Common Name: Atamasco Lily, Fairy Lily

Scientific Name: *Zephyranthes atamasco* (L.) Herbert

Botanical Family: Liliaceae (lily)

Physical Description: This plant arises from an ovoid to oblong bulb with a short neck. The light green, glossy leaves are shaped like a blade. Funnel-shaped white or purple-tinged flowers appear on a long stem. In contrast to the rain lily, the atamasco lily blooms once in the early spring.

Distribution and Ecology: This bulb is a native species of North America ranging from Missouri and Virginia to Florida. These plants prefer moist meadows and woods.

EXPOSURE

Zephyr lilies are popular cultivated plants. Traditional uses of *Zephyranthes* extract in herbal medicine included the treatment of cancer and diabetes mellitus. Pancratistatin (CAS RN: 96281-31-1) is a biosynthetic product of *Zephyranthes grandiflora* and some other Liliaceae plants.[1] Derivatives of this compound have been investigated for antineoplastic activity.[2]

PRINCIPAL TOXINS

Lycorine is a glucosidic alkaloid common to many species in the Liliaceae family including the zephyr lily (*Zephyranthes* species). Other alkaloids include tazettine (CAS RN: 507-79-9, $C_{18}H_{21}NO_5$) as displayed in Figure 112.1.[3]

DOSE RESPONSE

Most human exposures to the zephyr lily produce no symptoms or minor gastrointestinal distress.

Medical Toxicology of Natural Substances, by Donald G. Barceloux, MD

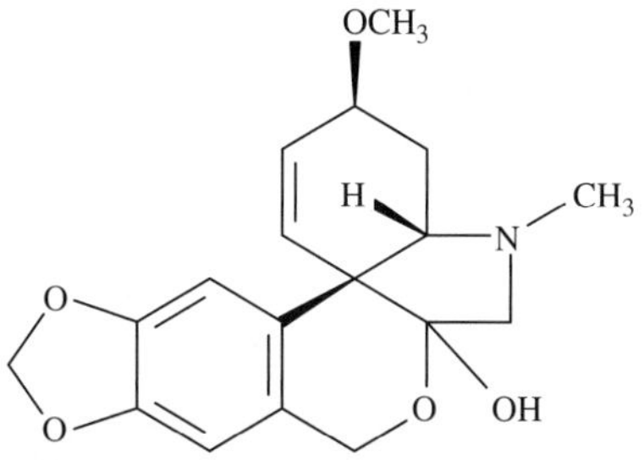

FIGURE 112.1. Chemical structure of tazettine.

TOXICOKINETICS

There are few data on the toxicokinetics of lycorine or tazettine.

CLINICAL RESPONSE

There are few clinical data on the clinical effects of the ingestion of parts from *Zephyranthes* species. Based on the presence of lycorine and similar alkaloids, the predominant effect probably would be gastrointestinal irritation including nausea, vomiting, and diarrhea. The veterinary literature suggests the grazing animals can develop bloody diarrhea, ataxia, and death following the ingestion of *Zephyranthes* bulbs.[4]

DIAGNOSTIC TESTING

Analytical methods for the identification of alkaloids in the *Zephyranthes* species include solid-phase extraction and reversed-phase high performance liquid chromatography.[5]

TREATMENT

Treatment is supportive. Because gastrointestinal irritation is the major effect of the ingestion of these bulbs, the patient should be evaluated for fluid and electrolyte balance as indicated by the clinical presentation. Decontamination measures are unnecessary because of the vomiting associated with poisoning by these bulbs. Although some *Zephyranthes* bulbs (*Z. simpsonii*) contain anticholinesterase inhibitors (galanthamine), there are few data in the medical literature to suggest that cholinergic crisis results from the ingestion of snowdrop bulbs.

References

1. Pettit GR, Gaddamidi V, Cragg GM. Antineoplastic agents, 105, *Zephyranthes grandiflora*. J Nat Prod 1984;47:1018–1020.
2. Rinner U, Hillebrenner HL, Adams DR, Hudlicky T, Pettit GR. Synthesis and biological activity of some structural modifications of pancratistatin. Bioorg Med Chem Lett 2004;14:2911–2915.
3. Wenkert E. The structure of tazettine. Experientia 1954;10:476–480.
4. Burrows GE, Tyrl RJ. Toxic Plants of North America. Ames, IA: Iowa State University Press; 2001.
5. Lopez S, Bastida J, Viladomat F, Codina C. Solid-phase extraction and reversed-phase high-performance liquid chromatography of the five major alkaloids in *Narcissus confusus*. Phytochem Anal 2002;13:311–315.

III Beans

Chapter 113

CASTOR BEAN and RICIN (*Ricinus communis* L.)

HISTORY

Castor bean extract was a constituent of classical Greek medicine, and the early Egyptians used castor bean extract as a salve, purgative, and oil.[1] The medicinal use of castor bean was described in the Susruta Ayurveda, which is a Sanskrit work on medicine dating back to the 6th century BC.[2] Raw or roasted castor beans were chewed by indigenous peoples of Africa, China, India, Jamaica, and Mexico for medical purposes (leprosy, syphilis, laxative). In 1978, a 49-year-old Bulgarian journalist, Georgi Markov died 3 days after the injection of a 1.52-mm platinum sphere into his thigh at a London bus stop. The West London coroner ruled that his death was a homicide, most likely the result of ricin poisoning based on circumstantial evidence (i.e., no tissue confirmation of ricin).[3] Within 1 month of the death of Georgi Markov, another Bulgarian exile, Vladimir Kostov survived a febrile illness that began after sustaining a small wound at a Paris metro station.[2] Examination of the wound revealed a platinum pellet similar to the one associated with the death of Georgi Markov.

The US Chemical Warfare Service studied ricin as a potential chemical weapon during World War I, and the British military developed a ricin bomb called the W-bomb.[2] In 1952, the US Army filed a patent on the preparation of ricin for use as an inhalational weapon (i.e., purified, respiratory particles stored in artillery shells for aerosolization during a mass attack). Ricin was also developed as a biological weapon by the former Soviet Union, and ricin was found in Al-Qaeda caves in Afghanistan. During the late 1980s, Iraq manufactured and tested weapons-grade ricin.[4] The US Centers for Disease Control (CDC) lists ricin as a category B agent (i.e., modestly easy to disseminate, causes moderate morbidity and low mortality, requires specific surveillance, and special diagnostic techniques for detection) along with Q fever, brucellosis, epsilon toxin of *Clostridium perfringens, Staphylococcus* enterotoxin B, and waterborne pathogens (*Salmonella* spp. *Shigella dysenteriae, Escherichia coli* 0157:H7, *Vibrio cholerae, Cryptosporidium parvum*). Ricin is the only protein listed under the Chemical Weapons Convention (CWC). Under the CWC, ricin is listed as a schedule I agent (toxic chemical developed primarily for military purposes with few or no legitimate uses). There are no documented uses of ricin as a biological warfare agent against large populations, although there have been instances where ricin was developed for use against specific individuals. On October 15, 2003, a mail processing and distribution facility in Greenville, South Carolina received a sealed container together with a letter threatening to poison water supplies. Although laboratory testing confirmed the presence of ricin in the container, investigation of the postal facility did not detect any ricin in the building or any ricin-related illness among the postal workers.[5]

BOTANICAL DESCRIPTION

Common Name: Castor Bean, Palma Christi, Mole Bean

Scientific Name: *Ricinus communis* L.

Botanical Family: Euphorbiaceae (spurge)

Medical Toxicology of Natural Substances, by Donald G. Barceloux, MD

FIGURE 113.1. Castor bean (*Ricinus communis* L.). Photograph courtesy of Kimberly Ann Barceloux, MFCC. See color insert.

Physical Description: A large shrub that grows on woody stalks to 4–5 ft (~1.2–1.5 m) in height with wide palmate green-red leaves. Three of the mottled brown and gray-white, oblong seeds are found in a soft-spine brown capsule clustered together on a central stalk. Figure 113.1 displays a specimen of the castor bean plant with seed pods, which contain three seeds.

Distribution and Ecology: In California, castor bean plants grow abundantly in disturbed soils (e.g., along roadsides). This species is native to tropical Africa, but castor bean has been naturalized in most subtropical and temperate climates.

EXPOSURE

An estimated one million tons of *Ricinus* seeds are processed annually in the world with India being a major exporter of castor oil. This country accounts for about one-quarter of the world production of *Ricinus* seeds.[6] Other major exporters include Brazil and China. Cold mechanical pressing or solvent extraction of the castor beans produces components used in a variety of products including high-temperature lubricants [Castrol-R (BP Lubricants, Wayne, NJ), racing motor oil], coatings, antifungal additives, paints, and plastics. The chemical composition of castor oil including the presence of glycerine results in castor oil being an excellent emollient and pigment carrier. Oxidation of the extracted oil from the castor bean plant is an important commercial process for the production of sebacic acid, which is an important component of a variety of products including corrosion inhibitors, plasticizers, lubricants, polymers (nylon), and personal care products.[7] The oil extraction process destroys the ricin in the oil portion. Castor oil is a purgative that increases peristalsis of the small intestine by the action of ricinoleic acid on the small intestine. Following oil extraction, the resultant meal or pomace is used as an additive to fertilizer.[8]

PRINCIPAL TOXINS

Physiochemical Properties

Purified ricin is a water soluble, white powder that is stable over a wide range of pH. In aqueous solution, ricin degrades after heating to 80 °C (176 °F) for 1 hour, whereas higher temperatures are necessary to destroy ricin in powder form.[9] Although ricin is relatively stable after extraction, heat or any method that breaks the disulfide bond inactivates ricin. Because of the low volatility of ricin, delivery of ricin via inhalation requires aerosolization of the ricin.[10] The technical difficulty aerosolizing of ricin limits the use of ricin as a large-scale terrorist agent. Ricin is unlikely to persist in the environment after aerosolization, but low micron-sized particles of ricin can remain in undisturbed areas or become resuspended after many hours. Although dermal absorption occurs, massive quantities of ricin necessary to produce toxicity via the dermal route limits the use of this route of exposure for terrorist acts.

After the separation of oil from the castor bean, ricin concentrates in the fibrous residue (bean pulp, mash) rather than the oil because ricin is water soluble. Consequently, the oil extract of castor bean (castor seed oil or castor oil, CAS RN: 8001-79-4) does not contain significant amounts of ricin. Ricinoleic acid composes about 85–90% of castor oil resulting from the organic solvent extraction of the mash from the castor bean plant.[11] In addition to ricinoleic acid, castor oil contains smaller amounts of glycerides of ricinoleic acid, isoricinoleic acid, and other fatty acids. Estimated concentrations of the fatty acids include oleic acid (2–7%), linoleic acid (3–5%), palmitic acid (1–2%), stearic acid (1%), and dihydroxystearic acid (1%).[12]

Ricin is a highly toxic, heterodimeric type II ribosome-inactivating protein.[13,14] This toxin (CAS RN: 9009-86-3) is a 66 kDa glycoprotein composed of a cell-binding, galactose-specific lectin subunit (Chain B) linked by a disulfide bridge to a rRNA N-glycosidase-subunit (Chain A) that inhibits protein synthesis.[15] The A-chain contains 267 amino acid residues with eight alpha helices and eight beta sheets, whereas the B-chain consists of 262 amino acid residues with a barbell-like tertiary structure.[16]

The pulp of the castor bean contains a variety of allergenic glycoproteins, which account for the allergic responses that occur in workers exposed to the pulp.[17] The major allergen of castor bean is probably a 2S albumin storage protein consisting of two heterodimeric proteins, Ric c 1 (11,212 Da) and Ric c 3 (12,032 Da).[18,19] The castor bean plant contains a glycoprotein lectin, ricin communis agglutinin; this lectin causes hemolysis following intravenous administration, but not following ingestion or inhalation.[20]

Poisonous Parts

Although ricin is present throughout the plant, seeds contain the highest concentration of ricin. In addition to ricin, the leaves, stems, and seeds of the castor bean plant contain potassium nitrate and hydrocyanic acid. Toxicity occurs after the ingestion of well-masticated seeds, whereas whole seeds are relatively nontoxic. The castor seed contains about 40% oil, 1–5% ricin, and 0.3–0.8% ricinine. Analysis of freshly dried leaves of *Ricinus communis* demonstrated the presence of ricinine (0.7%), *N*-demethylricinine (0.008%), and at least six flavonol glycosides (kaempferol-3-*O*-β-D-xylopyranoside, quercetin-3-*O*-β-D-xylopyranoside, kaempferol-3-*O*-β-D-glucopyranoside, quercetin-3-*O*-β-D-glucopyranoside, kaempferol-3-*O*-β-rutinoside, quercetin-3-*O*-β-rutinoside).[21]

Mechanism of Toxicity

The ingestion of castor oil produces diarrhea. Animal studies suggest that the major component of castor oil, ricinoleic acid, increases both intraluminal secretions and peristalsis.[22] Parenteral administration of ricin causes widespread disruption of cell processes as a result of the action of ricin, a type 2 ribosome-inactivating protein.[23] The target of ricin is the eukaryotic ribosome; hence, ricin is called a ribosome inactivating protein (RIP) or ribotoxin. Ricin is a large (66 kDa) glycoprotein composed of an enzymatically active A-protein chain and a receptor-binding B-protein chain bound together by a single disulfide bridge.[24] Chain B binds reversibly to cell surface galactose-terminated oligosaccharides (i.e., β-D-galactopyranoside) via lectin-binding sites, and the bound glycoprotein facilitates the entry of Chain A into the cell and the endoplasmic reticulum through endocytosis.[25] Chain A is an *N*-glycosidase, which catalytically inactivates protein synthesis by hydrolysis of a specific adenosine residue at position 4324 in the 28 s ribosomal RNA in the cytosol.[26] This action halts the binding of elongation factors and ultimately protein synthesis. The larger rRNA 60s subunit is responsible for protein synthesis, and this larger subunit contains the 28s rRNA. Inactivation of the smaller 28s rRNA inhibits protein synthesis by the larger 60s rRNA subunit, and the inhibition of protein synthesis ultimately causes cell death by necrosis or apoptosis.[27] Isotoxins of ricin (e.g., ricin D, ricin E) occur in cultivars of the castor bean plant.[28]

Entry of ricin into the cell involves the following steps: 1) binding to a range of cell surface glycolipids or glycoproteins, which contain beta-1,4-linked galactose residues, via the ricin B chain; 2) uptake of ricin into the cell by endocytosis; 3) entry of the toxin into early endosomes; 4) transfer of ricin from early endosomes to the trans-Golgi network by vesicular transport; 5) retrograde vesicular transport through the Golgi complex to reach the endoplasmic reticulum; 6) reduction of the disulfide bond connecting the ricin A and B chains; 7) partial unfolding of ricin A chain to cross the endoplasmic reticulum membrane via the Sec61p translocon; 8) avoidance of rapid degradation by cytosolic proteasomes after membrane translocation; 9) refolding into its protease-resistant, biologically active conformation; and 10) interaction with the ribosome to catalyze the depurination reaction.[29] Brefeldin A protects cells from the toxic effects of ricin by disrupting the structure of the Golgi apparatus, but research indicates that brefeldin A is not an effective treatment for ricin intoxication.[16]

The red cell agglutinating properties are independent of the diffuse, systemic effects of ricin, and these properties probably do not contribute significantly to the toxic effects of ricin.[30] The pulp of the seed contains allergenic glycoproteins, which cause allergic dermatitis, rhinitis, and asthma in sensitized industrial workers. The 2-S albumin storage protein of *Ricinus communis* is the major allergen of castor bean, and this protein resides in the CB-1A fraction of the castor bean extract.[18]

The pathology associated with inhalation of ricin is limited to the lungs with little evidence of significant systemic toxicity. The target cell for inhaled ricin is the type I pneumocyte. Animal studies indicate that *Ricinus communis* agglutinin I binds to cell surface receptors on type I pneumocytes and initiates acute alveolitis and necrosis of the lower respiratory tract epithelium.[31] A rapidly progressive pulmonary edema develops that causes severe hypoxia and death. Proliferation of type II pneumocytes occurs in response to the injury during the resolution phase of the pulmonary damage.

DOSE RESPONSE

Predicting toxicity based on the number of seeds ingested is difficult because a variety of factors affect the toxicity of castor beans including growing conditions, season, species, host susceptibility (i.e., immunity), cooking techniques, degree of mastication, bioavailability of ricin, and route of exposure (e.g., parenteral >> oral).[32] Anaphylaxis may occur following exposure to a single bean.

Parenteral

Ricin is several hundred times more toxic when administered parenterally than by ingestion. In animal models, ricin is one of the three most toxic *parenteral* substances by weight with an LD_{50} in the range of 0.1–1 µg/kg. Botulinum neurotoxin is about 1,000 times more toxic than ricin based on the LD_{50} in animals.[13] The estimated lethal dose of ricin in humans is about 1–10 mg/kg.[32] The injection of an estimated 2 mg ricin/kg body weight caused fever, nausea, anorexia, mild elevation of hepatic aminotransferases, and leucocytosis that resolved over 10 days.[33] The sphere found during the autopsy of the Bulgarian journalist, who died 3 days after the injection of the sphere, had a 0.28-mm^3 hole that contained an estimated 500 µg ricin.[34] During phase I cancer clinical trials, the intravenous injection of 23 µg ricin/m^2 every 2 weeks was tolerated by the participants with mild complaints that included fatigue, myalgias, nausea, and vomiting.[35,36] The dose-limiting side effect of ricin A-chain-containing immunotoxins in these clinical trials was a vascular leak syndrome, characterized by hypoalbuminemia and edema.[37]

Oral

Historical accounts of the number of beans ingested do not necessarily estimate the actual dose of ricin ingested because of a number of factors including the maturity of the bean, the degree of mastication, seasonal variation in moisture and ricin, age of patient, and associated diseases. However, most case reports denote the number of beans ingested rather than ricin content. These case reports suggest that the ingestion of 2–30 seeds produces gastrointestinal symptoms in a dose-dependent manner with the ingestion of 10–15 seeds by an adult causing a moderate gastroenteritis that resolves in 2–3 days.[38] Clinical features of the sepsis-like illness associated with parenteral injection of ricin are unlikely following ingestion of castor beans. In an Italian report, ingestion of two seeds by a 12-year-old boy caused protracted nausea and vomiting, but the child recovered without complications.[39] His father ingested three seeds, and he displayed similar symptoms as well as subclinical hyperbilirubinemia. Ingestion of eight seeds by the mother caused a hemorrhagic gastritis, and ingestion of 10 seeds by the 8-year-old sister caused dehydration. An Indian series of 57 children revealed that 22 children displayed evidence of hypovolemic shock, but there were no fatalities.[40] The average ingestion was four to five seeds up to a maximum of 15. A 38-year-old woman intentionally ingested 24 chopped seeds, but she remained asymptomatic after gut decontamination.[32] Ingestion of 30 partially masticated seeds by a 21-year-old resulted in severe hemorrhagic gastritis and dehydration, which responded to supportive care.[41]

Inhalation

In animal studies, large doses (20.95–41.8 µg/kg) of aerosolized ricin produce necrotizing tracheitis, bronchitis, bronchiolitis, interstitial pneumonia, and alveolar edema.[42] The toxicity of aerosolized ricin is highly dependent on particle-size. In a study of mice exposed to aerosolized ricin, all mice died within 72 hours of exposure to 1-µm aerosols (mass median aerodynamic diameter) at 55 µg ricin/kg (i.e., approximately 4.5 times LD_{50}) with about 60% of the total ricin dose being deposited in the lungs.[43] In the comparison group, none of the mice died after exposure to aerosols containing three times the LD_{50} when delivered by a device that produced particles in the range of 1–50 µm. With this device, about 20% of the total ricin dose was deposited in the lungs of the experimental animals. In contrast to humans, mice are obligate nose-breathers and the turbinates of rodents trap more large particulates than humans. However, the inhalation LD_{50} of rodents and humans are the same order of magnitude.[44]

TOXICOKINETICS

There are few clinical data on the toxicokinetics of ricin in humans. The usual route of exposure is gastrointestinal, and, rarely, parenteral. Animal studies indicate that the oral absorption of ricin is low, and the gastrointestinal tract absorbs only small amounts of ricin. Absorption of ricin through the skin is negligible. Inhalation is a potential route of a toxic exposure during a bioterrorism attack, but clinical effects from the absorption of ricin following inhalation are not well documented.[5] The effects of ricin following inhalation depend primarily on particle size with low micron-sized particles producing the most serious effects. Although larger-sized particles can be transported to the gastrointestinal tract via the mucociliary system, clinically significant systemic absorption of ricin does not usually occur in animal models following pulmonary exposures.[31]

A relatively large portion (i.e., approximately 20–45%) of the oral dose of ricin is excreted in the feces over 72 hours.[45] A case report suggested that urinary excretion of ricin was slow with a calculated urinary elimination half-life of 8 days.[41] However, experimental data in rats indicate that the elimination of ^{125}I-labelled ricin following intravenous administration is rapid with only 11% remaining after 24 hours.[46] Renal elimination accounted for 70% of the parenteral dose during the first 24 hours. Following nose-only aerosol inhalation exposure of ^{125}I-ricin in mice, the lungs sequester most of the ricin until degradation.[47] In the small intestine of humans following ingestion of castor oil, pancreatic enzymes hydrolyze components in castor oil primarily to glycerol and ricinoleic acid.[48]

CLINICAL RESPONSE

Oral

The latent period for toxic effects following ingestion of castor beans varies between 1–24 hours depending on the absorbed dose of ricin. In the large series from India, the mean onset of gastrointestinal symptoms was 1–3 hours with 10 patients admitted within 1 hour.[40] The most common initial symptoms result from a severe gastroenteritis that includes burning in the alimentary tract, nausea, vomiting, diarrhea, and colicky abdominal pain.[38,49] In severe oral poisoning, these symptoms progress to hemorrhagic gastritis, hypovolemia, and hypotension. The primary target organs are the kidney, liver, and pancreas, although the inhibition of protein synthesis by ricin may affect any organ. Death usually results from prolonged dehydration, hypotension, and electrolyte imbalance.[50] Hematuria and elevated free serum hemoglobin level were reported in a 4-year-old girl who ingested four seeds.[51] However, hemolysis is a rare complication of castor bean ingestion.[50] Ricin poisoning is not contagious and person-to-person transmission does not occur.

Parenteral

Compared with oral ricin, parenteral administration of ricin causes more serious toxicity as a result of clinical effects similar to septic shock (fever, multisystem failure, cardiovascular collapse). In the famous case report of the Bulgarian journalist, he developed fever, leukocytosis, tachycardia, and lymphadenopathy, followed by progressive hypotension and a septic shock-like condition.[3] All these effects developed within 36 hours of the injection of a sphere reputedly containing ricin. A necrotizing fascitis containing *Enterococcus faecalis* developed after a 53-year-old man, who masticated 13 castor bean seeds and injected the pulverized material into his left thigh.[52] No evidence of multisystem failure developed, and there were no specific symptoms that could be directly attributed to the injected material other than an infectious process. The differential diagnosis of ricin poisoning includes biological toxins (*Staphylococcus aureus, Salmonella* spp., *Shigella* spp., Norwalk virus, adenovirus, influenza), heavy metal poisoning (arsenic, inorganic mercury, thallium, iron), acute radiation sickness, sepsis, capillary leak syndromes, overdose with chemotherapeutic agents, paraquat poisoning, and phosgene intoxication.

Inhalation

There are few human data on the effects of pulmonary exposure to ricin, but experimental studies indicate that large doses of inhaled ricin can cause irreversible pathological pulmonary changes that result in epithelial necrosis of type I and II pneumocytes, pulmonary edema, and death.[53] In laboratory animals, anorexia and progressive weakness develop within 8–24 hours after inhalation of ricin followed by death about 36–48 hours after exposure. Mortality following pulmonary exposure to ricin usually results from respiratory failure rather than the multisystem failure associated with parenteral ricin. In a study of five monkeys administered approximately 21–42 μg/kg of aerosolized ricin, all monkeys developed fibrinopurulent tracheitis and pneumonia, diffuse necrosis, and acute inflammation of airways, diffuse peribronchiolar edema, alveolar exudates, and purulent mediastinal lymphadenitis.[42] Sensitized individuals may develop allergic responses to inhaled ricin ranging from urticaria and wheezing to anaphylaxis.

Hypersensitivity

Castor bean dust is highly allergenic as a result of potent allergens in the seed, seed coat, and pollen.[54] Inhalation of this dust in facilities using castor beans can cause rhinitis, conjunctival irritation, nasal and throat congestion, urticaria, cough and wheezing.[55] Although gastrointestinal toxicity from ingesting masticated ricin occurs within several hours, allergic responses in sensitized individuals may occur immediately after exposure. Allergic reactions in workers handling these products include rhinitis, urticaria,[56] conjunctivitis, dermatitis,[57] facial edema, wheezing, and anaphylaxis.[58] Case reports document the development of severe orbital edema and urticaria after contact with a castor bean necklace.[59]

DIAGNOSTIC TESTING

Analytical Methods

Methods to detect ricin include radioimmunoassay, ELISA, polymerase chain reaction (PCR),[60] immunoaffinity and liquid chromatography/tandem mass spectrometry (IAC/LC/MS/MS),[61] time-resolved fluorescence immunoassay (TRF), and liquid chromatography/mass spectrometry (LC/MS) for the analysis of ricinine. The lower limit of detection was 0.1 ng ricin/mL using IAC/LC/MS/MS. The analysis of ricin in biological samples is complicated because ricin is a large heterogeneous protein with glycosylation. The most common screening methods for detection of ricin exposure include radioimmunoassay and ELISA. The limits of detection for radioimmunoassays of ricin in blood samples range from about 0.05–0.1 ng/mL with little cross-reactivity between ricin and abrin.[62] Colorimetric and chemiluminescence ELISA assays for ricin are available from specialized laboratories with limits of detection in the range of 1 and 0.1 ng/mL, respectively, along with limits of quantitation of approximately 10 ng/mL.[63] A colloidal gold-based immunochromatography assay uses a pair of high-affinity anti-ricin monoclonal antibodies to detect ricin at concentrations of 50 ng/mL.[64] Presumptive identification of ricin involves determination of the molecular weight and trypsin-digested peptide mapping from LC/MS with matrix assisted laser desorption/ionization time-of-flight analysis.[65] The detection of ricinine by LC/MS provides additional evidence of exposure to ricin.[66] However, interference with similar compounds (e.g., *R. communis* agglutinin, biological variants of ricin–ricin D, ricin E) during the purification process limits the identification of ricin during these procedures. Protein sequencing with high performance liquid chromatography/electrospray/mass spectrometry (HPLC/ES/MS) provides core structural information on the peptide backbone of ricin, and this method provides definitive determination of ricin.[67] Confirmatory tests for detection of suspected cases of ricin poisoning as a result of bioterrorist acts include nasal or throat swabs within 24 hours postexposure (aerosol exposure) and serum tests for toxin assay and antibody response within 36–48 hours. Other proposed rapid methods for the detection of exposure to ricin include immunochromatography with a reported limit of detection in the range of 1 ng/mL buffer.[68] Methods for the detection of fatty acids in castor oil include thin-layer chromatography,[69] reverse-phase high performance liquid chromatography/mass spectrometry,[70] and matrix-assisted laser desorption/ionization time-of-flight mass spectrometry.[71]

Biomarkers

There are no commercial biomarkers for exposure to ricin, although ricinine is a potential confirmatory biomarker as determined by isocratic HPLC/MS/MS.[72] The advantage of ricinine is that it is a low-molecular-weight component (i.e., $C_8H_8N_2O_2$, MW 164.1) of the castor bean that requires only simple extraction compared with the complicated extraction of the large ricin molecule. In a blood sample from a patient who survived ricin poisoning, the ricin concentration was approximately 0.3 ng/mL 14 hours after the ingestion of ricin seeds as measured by radioimmunoassay.[62]

Abnormalities

Mild, transient elevation of serum hepatic transaminases may occur 1–3 days after ricin ingestion, but these changes are not usually associated with liver dysfunction.[73] Rhabdomyolysis with elevation of serum creatine kinase may also occur.

TREATMENT

Decontamination

Exposure to ricin does not require any unique decontamination techniques.[74] There are limited data on the use of decontamination measures following castor bean ingestion. Vomiting and diarrhea are early symptoms of castor bean ingestion, and these patients usually require no decontamination measures. If the patient presents less than 1 hour after the ingestion of several masticated beans and the patient has not vomited, the use of one dose of activated charcoal is a therapeutic option, but there are inadequate data to confirm the efficacy of this decontamination measure. For aerosol exposure, treatment involves standard measure for the detection and the treatment of noncardiogenic pulmonary edema with respiratory support (oxygen, bronchodilators, endotracheal intubation, positive end-expiratory pressure, continuous positive expiratory pressure) as needed. Patients with substantial airborne exposure to ricin should have a chest x-ray and pulse oximetry to detect the early development of alveolar edema and hypoxemia. Gut decontamination is unnecessary for exposure to airborne toxins. Standard methods for the decontamination of the skin of patients directly contaminated with ricin include removal of contaminated clothes, washing exposed skin with soap and copious amounts of water, and double-bagging contaminated clothes. The risk to health care workers is minimal.[9]

Antidotes

There are no antidotes to ricin. Rabbit antibodies directed against ricin have been described by Houston[75] and tested parenterally in mice models. The antiserum provided protection at five times the LD_{50} dose of ricin only if given within 1 hour; however, at this dose, only 30% of the antiserum-treated group survived. Although there are no specific ricin antidotes at the present time, research continues into the use of pterins (e.g., pteroic acid), which bind in the specificity pocket of ricin and inhibit the action of ricin.[76] Several antiricin vaccines currently are being investigated. Potential vaccine candidates include the vascular leak syndrome site (V76M) and the recombinant ribotoxic A chain with amino acid substitutions that disrupt the ribotoxic site (Y80A).[77,78]

Supplemental Care

The treatment of ricin poisoning is supportive, primarily directed at the presenting symptoms. Following ingestion or inhalation, the clinical course typically progresses over 4–36 hours. The maintenance of fluid and electrolyte balance is the most important aspect of supportive care. Useful laboratory tests for the treatment of ricin toxicity include a complete blood count, blood urea nitrogen, prothrombin time, urinalysis, and measurement of serum electrolytes, creatinine, glucose, amylase, creatine phosphokinase, and hepatic amino transaminases. Symptomatic patients should be hospitalized and monitored for dehydration, liver dysfunction, pancreatic inflammation, and rhabdomyolysis. Asymptomatic patients who have ingested more than several beans should be observed for at least 4–6 hours after ingestion. If a patient remains asymptomatic during this period, discharge is appropriate with instructions to return immediately should symptoms develop. Initial daily outpatient follow-up for several days is important to identify target organ toxicity (e.g., liver, kidney). Ricin is not dialyzable as a result of its large molecular weight; therefore, measures to enhance elimination of ricin are probably not effective.[17] Patients with aerosol exposure to ricin should be observed for 24 hours if abnormalities appear on the pulse oximetry or chest x-ray. Otherwise, asymptomatic patients can be discharged after 6–12 hours with the provision to return if pulmonary symptoms develop. Although there are few human data, animal experiments suggest that the onset of respiratory symptoms can occur up to 24 hours after exposure.[42]

References

1. Scarpa A, Guerci A. Various uses of the castor oil plant (*Ricinus communis* L.). A review. J Ethnopharmacol 1982;5:117–137.
2. Olsnes S. The history of ricin, abrin and related toxins. Toxicon 2004;44:361–370.
3. Knight B. Ricin: a potent homicidal poison. Br Med J 1979:1:350–351.
4. Zilinskas RA. Iraq's biological weapons. The past as future? JAMA 1997;278:418–424.
5. Gibson J, Drociuk D, Fabian T, Brundage S, Ard L, Fitzpatrick N, et al. Investigation of a ricin-containing envelope at postal facility—South Carolina, 2003. MMWR Morb Mortal Wkly Rep 2003;52:1129–1131.
6. Marsden CJ, Smith DC, Roberts LM, Lord JM. Ricin: current understanding and prospects for an antiricin vaccine. Expert Rev Vaccines 2005;4:229–237.
7. Doan LG. Ricin: mechanism of toxicity, clinical manifestations, and vaccine development. A review. J Toxicol Clin Toxicol 2004;42:201–208.
8. Snell MA. Castor bean pomace exposure. Arch Ind Hyg Occup Med 1952;6:113–115.
9. Audi J, Belson M, Patel M, Schier J, Osterloh J. Ricin poisoning a comprehensive review. JAMA 2005;294:2342–2351.
10. Winder C. Toxicity of ricin. J Toxicol Toxin Rev 2004;23: 97–103.
11. McKeon TAQ, Chen GQ, Lin J-T. Biochemical aspects of castor oil biosynthesis. Biochem Soc Trans 2000;28: 972–974.
12. Johnson W Jr and Cosmetic Ingredient Review Expert Panel. Final report on the safety assessment of Ricinus Communis (Castor) Seed Oil, Hydrogenated Castor Oil, Glyceryl Ricinoleate, Glyceryl Ricinoleate SE, Ricinoleic Acid, Potassium Ricinoleate, Sodium Ricinoleate, Zinc Ricinoleate, Cetyl Ricinoleate, Ethyl Ricinoleate, Glycol Ricinoleate, Isopropyl Ricinoleate, Methyl Ricinoleate, and Octyldodecyl Ricinoleate. Int J Toxicol 2007;26(Suppl 3):31–77.
13. Mosher HS, Furman FA, Buckwald HD, Fishcher HG. Tarichatoxin-tetrodotoxin: a potent neurotoxin. Science 1964;144:1100–1110.
14. Edwards RO. Poisoning from plant ingestions. J Fla Med Assoc 1965;52:875–881.
15. Frankel AE, Burbage C, Fu T, Tagge E, Chandler J, Willingham MC. Ricin toxin contains at least three galactose-binding sites located in B chain subdomains 1a, 1β, 1? Biochem 1996;35:14749–14756.
16. Hicks RP, Hartell MG, Nichols DA, Bhattacharjee AK, van Hamont JE, Skillman DR. The medicinal chemistry of botulinum, ricin and anthrax toxins. Curr Medicinal Chem 2005;2:667–690.
17. Balint GA. Ricin: the toxic protein of castor oil seeds. Toxicology 1974;2:77–102.
18. Bashir ME, Hubatsch I, Leinenbach HP, Zeppezauer M, Panzani RC, Hussein IH. Ric c 1 and Ric c 3, the allergenic 2S albumin storage proteins of *Ricinus communis*: complete primary structures and phylogenetic relationships. Int Arch Allergy Immunol 1998;115:73–82.
19. Thorpe SC, Kemeny DM, Panzani RC, McGurl B, Lord M. Allergy to castor bean. II. Identification of the major allergens in castor bean seeds. J Allergy Clin Immunol 1988;82:67–72.
20. Olsnes S, Refsnes K, Christensen TB, Pihl A. Studies on the structure and properties of the lectins from *Abrus precatorius* and *Ricinus communis*. Biochim Biophys Acta 1975;405:1–10.
21. Kang SS, Cordell GA, Soejarto DD, Fong HH. Alkaloids and flavonoids from *Ricinus communis*. J Nat Prod 1985; 48:155–156.
22. Mathias J R, Martin J L, Burns T W, Carlson G M, Shields, R P. Ricinoleic acid effect on the electrical activity of the small intestine in rabbits. J Clin Invest 1978; 61:640–644.

23. Barbieri L, Battelli MG, Stirpe F. Ribosome-inactivating proteins from plants. Biochim Biophys Acta 1993;1154: 237–282.
24. Olsnes S, Kozlov JV. Ricin. Toxicon 2001;39:1723–1728.
25. van Deurs B, Sandvig K, Petersen OW, Olsnes S, Simons K, Griffiths G. Estimation of the amount of internalized ricin that reaches the trans-Golgi network. J Cell Biol 1988;106:253–267.
26. Endo Y, Tsurugi K. The RNA *N*-glycosidase activity of ricin A-chain. The characteristics of the enzymatic activity of ricin A-chain with ribosomes and with rRNA. J Biol Chem 1988;263:8735–8739.
27. Nilsson L, Nygard O. The mechanism of the protein-synthesis elongation cycle in eukaryotes. Effect of ricin on the ribosomal interaction with elongation factors. Eur J Biochem 1986;161:111–117.
28. Mice T, Funatsu G, Ishiguro M, Fanatsu M. Isolation and characterization of ricin E from castor beans. Agric Biol Chem 1977;41:2041–2046.
29. Lord MJ, Jolliffe NA, Marsden CJ, Pateman CS, Smith DC, Spooner RA, et al. Ricin. Mechanisms of cytotoxicity. Toxicol Rev 2003;22:53–64.
30. Cawley DB, Hedblom ML, Houston LL. Homology between ricin arid *Ricinus communis* agglutinin: amino terminal sequence analysis and protein synthesis inhibition studies. Arch Biochem Biophys 1978;190:744–755.
31. Griffiths GD, Rice P, Allenby AC, Upshall DG. Inhalation toxicology and histopathology of ricin and abrin toxins. Inhal Toxicol 1995;7:269–288.
32. Rauber A, Heard J. Castor bean toxicity re-examined: a new perspective. Vet Hum Toxicol 1985;27:498–502.
33. Fine DR, Shepherd HA, Griffiths GD, Green M. Sublethal poisoning by self-injection with ricin. Med Sci Law 1992;32:70–72.
34. Crompton R, Gall D. Georgi Markov—death in a pellet. Med Leg J 1980;48:51–62.
35. Lambert JM, Goldmacher VS, Collinson AR, Nadler LM, Blattler WA. An immunotoxin prepared with blocked ricin: a natural plant toxin adapted for therapeutic use. Cancer Res 1991;51:6236–6242.
36. Fodstad O, Kvalheim G, Godal A, Lotsberg J, Aamdal S, Host H, Pihl A. Phase I study of the plant protein ricin. Cancer Res 1984;44:862–865.
37. Soler-Rodriguez AM, Ghetie MA, Oppenheimer-Marks N, Uhr JW, Vitetta ES. Ricin A-chain and ricin A-chain immunotoxins rapidly damage human endothelial cells: implications for vascular leak syndrome. Exp Cell Res 1993;206:227–234.
38. Aplin PJ, Eliseo T. Ingestion of castor oil plant seeds. Med J Aust 1997;167:260–261.
39. Malizia E, Sarcinelli L, Adrencci G. Ricin poisoning: a familiar epidemy. Acta Pharmacol Toxicol 1977;41(suppl): 351–361.
40. Ingle VN, Kale VG, Taiwalkar YB. Accidental poisoning in children with particular reference to castor beans. Indian J Pediatr 1966;33:237–240.
41. Kopferschmitt J, Flesch F, Lugnier A, Sauder P, Jaeger A, Mantz JM. Acute voluntary intoxication by ricin. Hum Toxicol 1983;2:239–242.
42. Wilhelmsen CL, Pitt ML. Lesions of acute inhaled lethal ricin intoxication in rhesus monkeys. Vet Pathol 1996; 33:296–302.
43. Roy CJ, Hale M, Hartings JM, Pitt L, Duniho S. Impact of inhalation exposure modality and particle size on the respiratory deposition of ricin in BALB/c mice. Inhal Toxicol 2003;15:619–638.
44. Griffiths GD, Phillips GJ, Holley J. Inhalation toxicology of ricin preparations: animal models, prophylactic and therapeutic approaches to protection. Inhal Toxicol 2007; 19:873–887.
45. Ishiguro M, Tanabe S, Matori Y, Sakakibara R. Biochemical studies on oral toxicity of ricin. IV. A fate of orally administered ricin in rats. J Pharmacobiodyn 1992;15: 147–156.
46. Ramsden CS, Drayson MT, Bell EB. The toxicity, distribution and excretion of ricin holotoxin in rats. Toxicol 1989;55:161–171.
47. Doebler JA, Wiltshire ND, Mayer TW, Estep JE, Moeller RB, Traub RK, et al. The distribution of [^{125}I]ricin in mice following aerosol inhalation exposure. Toxicology 1995;98: 137–149.
48. Thompson WG. Laxatives: clinical pharmacology and rational use. Drugs 1980;19:49–58.
49. Wedin GP, Neal JS, Everson EP, Krenzelok EP. Castor bean poisoning. Am J Emerg Med 1986;4:259–261.
50. Challoner KR, McCarron MM. Castor bean intoxication. Ann Emerg Med 1990;19:1177–1183.
51. Henry GW, Schwenk GR, Bohnert PG. Umbrellas and mole beans: A warning about acute ricin poisoning. J Indiana Med Assoc 1981;43:572–573.
52. Passeron T, Mantoux F, Lacour J-P, Roger P-M, Fosse T, Iannelli A, Ortonne J-P. Infectious and toxic cellulitis due to suicide attempt by subcutaneous injection of ricin. Br J Dermatol 2004;150:154.
53. DaSilva L, Cote D, Roy C, Marinez M, Duniho S, Pitt ML, Downey T, Dertzbaugh M. Pulmonary gene expression profiling of inhaled ricin. Toxicon 2003;41:813–822.
54. Singh A, Panzani RC, Singh AB. Specific IgE to castor bean (*Ricinus communis*) pollen in the sera of clinically sensitive patients to seeds. J Invest Allergol Clin Immunol 1997;7:169–174.
55. Topping MD, Henderson RT, Luczynska CM, Woodmass A. Castor bean allergy among workers in the felt industry. Allergy 1982;37:603–608.
56. Metz G, Bocher D, Metz J. IgE-medicated allergy to castor bean dust in a landscape gardener. Contact Dermatitis 2001;44:367.
57. Tan BB, Noble AL, Roberts ME, Lear JT, English JS. Allergic contact dermatitis from oleyl alcohol in lipstick cross-reacting with ricinoleic acid in castor oil and lanolin. Contact Dermatitis 1997;37:41–42.

58. Kanerva L, Estlander T, Jolanki R. Long-lasting contact urticaria from castor bean. J Am Acad Dermatol 1990;23: 351–355.
59. Lockey SD. Anaphylaxis from an Indian necklace. JAMA 1968;206:2900–2901.
60. He X, Brandon DL, Chen GQ, McKeon TA, Carter JM. Detection of castor contamination by real-time polymerase chain reaction. J Agric Food Chem 2007;55: 545–550.
61. Becher F, Duriez E, Volland H, Tabet JC, Ezan E. Detection of functional ricin by immunoaffinity and liquid chromatography-tandem mass spectrometry. Anal Chem 2007;79:659–665.
62. Godal A, Olsnes S, Pihl A. Radioimmunoassays of abrin and ricin in blood. J Toxicol Environ Health 1981;8: 409–417.
63. Poli MA, Rivera VR, Hewetson JF, Merrill GA. Detection of ricin by colorimetric and chemiluminescence ELISA. Toxicon 1994;32:1371–1377.
64. Shyu R-H, Shyu H-F, Liu H-W, Tang S-S. Colloidal gold-based immunochromatographic assay for detection of ricin. Toxicon 2002;40:255–258.
65. Darby SM, Miller ML, Allen RO. Forensic determination of ricin and the alkaloid marker ricinine from castor bean extracts. J Forensic Sci 2001;46:1033–1042.
66. Mouser P, Filigenzi MS, Puschner B, Johnson V, Miller MA, Hooser SB. Fatal ricin toxicosis in a puppy confirmed by liquid chromatography/mass spectrometry when using ricinine as a marker. J Vet Diagn Invest 2007;19:216–220.
67. Fredriksson S-A, Hulst AG, Artursson E, de Jong AL, Nilsson C, van Baar BL. Forensic identification of neat ricin and of ricin from crude castor bean extracts by mass spectrometry. Anal Chem 2005;77:1545–1555.
68. Guglielmo-Viret V, Splettstoesser W, Thullier P. An immunochromatographic test for the diagnosis of ricin inhalational poisoning. Clin Toxicol 2007;45:505–511.
69. Srinivasulu C, Mahapatra SN. A rapid method for detecting groundnut oil in castor oil. J Chromatogr 1973;86: 261–262.
70. Stübiger G, Pittenauer E, Allmaier G. Characterisation of castor oil by on-line and off-line non-aqueous reverse-phase high-performance liquid chromatography-mass spectrometry (APCI and UV/MALDI). Phytochem Anal 2003;14:337–346.
71. Ayorinde FO, Garvin K, Saeed K. Determination of the fatty acid composition of saponified vegetable oils using matrix-assisted laser desorption/ionization time-of-flight mass spectrometry. Rapid Commun Mass Spectrom 2000; 14:608–615.
72. Johnson RC, Lemire SW, Woolfitt AR, Ospina M, Preston KP, Olson CT, Barr JR. Quantification of ricinine in rat and human urine: a biomarker for ricin exposure. J Anal Toxicol 2005;29:149–155.
73. Palatnick W, Tenenbein M. Hepatotoxicity from castor bean ingestion in a child. Clin Toxicol 2000;38:67–69.
74. Spivak L, Hendrickson RG. Ricin. Crit Care Clin 2005; 21:815–824.
75. Houston LL. Protection of mice from ricin poisoning by treatment with antibodies directed against ricin. Clin Toxicol 1982;19:385–389.
76. Miller DJ, Ravikumar K, Shen H, Suh JK, Kerwin SM, Robertus JD. Structure-based design and characterization of novel platforms for ricin and shiga toxin inhibition. J Med Chem 2002;45:90–98.
77. Vitetta ES, Smallshaw JE, Coleman E, Jafri H, Foster C, Munford R, Schindler J. A pilot clinical trial of a recombinant ricin vaccine in normal humans. Proc Natl Acad Sci 2006;103:2268–2273.
78. Smallshaw JE, Richardson JA, Pincus S, Schindler J, Vitetta ES. Preclinical toxicity and efficacy testing of RiVax, a recombinant protein vaccine against ricin. Vaccine 2005;23: 4775–4784.

Chapter 114

COWITCH and HORSE EYE BEAN (*Mucuna* Species)

HISTORY

The ancient Ayurveda Indian and Chinese systems of traditional medicine used *Mucuna pruriens* for the treatment of stress, male sexual disorders, aging, and general susceptibility to infections.[1] In 1937, Damodaran and Ramaswamy isolated L-dopa from the seeds of *Mucuna pruriens*.[2] In the late 1970s, preliminary clinical trials evaluated the administration of seeds from *Mucuna pruriens* as an inexpensive substitute for medicinal doses of L-dopa for the treatment of Parkinson's disease. However, no subsequent clinical trials were conducted.[3]

BOTANICAL DESCRIPTION

Common Name: Cowitch, Cowhage, Kewanch, Pic-Pica, Feijão Macaco, Velvet Bean

Scientific Name: *Mucuna pruriens* (L.) DC.

Botanical Family: Fabaceae (pea)

Physical Description: This hardy annual vine has trifoliate leaves and purple flowers that produce 10–13 cm (~4–5 in.) fruiting pods with many barbed spicules.

Distribution and Ecology: *M. pruriens* is an annual vine indigenous to North and South Carolina, Florida, Puerto Rice, and the Virgin Islands. This plant grows well in tropical areas, such as India and the Caribbean Islands.

Common Name: Horse Eye Bean, Ojo de Venado (eye of the deer)

Scientific Name: *Mucuna sloanei* Fawcett & Rendle

Botanical Family: Fabaceae (pea)

Physical Description: This perennial climbing vine produces yellow flowers and a nondescript brown bean about 2 × 1 cm (~0.8 × 0.4 in.).

Distribution and Ecology: *M. sloanei* is a native plant in Florida, Hawaii, and Puerto Rico. This plant is now distributed worldwide in temperate and tropical areas of South American, Central America, the Caribbean, and Africa.

EXPOSURE

Previously, components of the seeds from *M. pruriens* were constituents of itching powder and anthelmintic agents. Other traditional herbal uses of *M. pruriens* include uterine stimulation, treatment of dysentery, diuretic, hypoglycemic agent, aphrodisiac, and nerve tonic.[4,5] The seeds from *M. pruriens* are also traditional remedies for developing resistance to snake venom (e.g., *Echis carinatus*)[6,7] and the treatment of Parkinson's disease.[3] During famines in India, the pods of *M. pruriens* are a food source after repeated boiling and thorough rinsing with water.[8]

PRINCIPAL TOXINS

Physiochemical Properties

The pods, seed, leaves, and roots of *M. pruriens* contain indole alkylamine compounds (*N,N*-

Medical Toxicology of Natural Substances, by Donald G. Barceloux, MD

dimethyltryptamine, bufotenine, 5-methoxy-*N,N*-dimethyltryptamine, β-carboline compounds)[4] and 1,2,3,4-tetrahydroisoquinoline alkaloids.[8] Seeds of *M. pruriens* are well-know sources of L-dopa (L-3,4-dihydroxyphenylalanine).[3,9]

Poisonous Parts

Each fruiting pod from *M. pruriens* contains approximately 5,000 barbed spicules that easily detach and produce an inflammatory response. These spicules measure about 20 µm × 2 mm, and they contain a protease, mucunain, that is the suspected cause of pruritus.[10] The average L-dopa content is somewhat greater in samples of unripe seeds than mature seeds, and the average L-dopa content ranges up to approximately 3–4% dry weight depending on the extraction technique.[11] The L-dopa yield of *Mucuna pruriens* seeds using an ethanol extraction with ascorbic acid was 1.78% compared with 0.98% for water extraction with sulfur dioxide.[12] Seed powder from *M. pruriens* contains up to approximately 350 mg L-dopa.

DOSE RESPONSE

Other than local skin reactions, the ingestion of *M. pruriens* seeds produces clinical effects similar to the ingestion of L-dopa. The administration of doses of *M. pruriens* seed powder up to 60 g (3 g L-dopa) produced minimal side effects that included giddiness, diaphoresis, dry mouth, and diarrhea.[3]

CLINICAL RESPONSE

Dermal contact with hairs from the dried seed pods of *M. pruriens* causes an immediate inflammatory response characterized by erythematous, pruritic macular lesions.[13] Symptoms usually resolve within 1–2 hours of decontamination. There are few data on the human toxicity of *M. sloanei*. Because of the high concentration of L-dopa in the seeds from *Mucuna* species, the clinical features of intoxication from *Mucuna* seeds may resemble L- dopa toxicity. However, there are few data in the medical literature regarding *Mucuna* intoxication in humans.

DIAGNOSTIC TESTING

There are few data on the analytical methods or biomarkers for toxicity associated with the cowitch bean.

TREATMENT

All skin exposed to the pods of *M. pruriens* should be washed with soap and water; all pods and contaminated material should be handled with caution. Treatment is supportive. Decontamination measures are unnecessary.

References

1. Kumar KV, Srinivasan KK, Shanbhag T, Rao SG. Aphrodisiac activity of the seeds of *Mucuna pruriens*. Indian Drug 1994;31:321–327.
2. Damodaran M, Ramaswamy R. Isolation of L-dopa from the seeds of *Mucuna pruriens*. Biochemistry 1937;31: 2149–2152.
3. Vaidya AB, Rajagopalan TG, Mankodi NA, Antarkar DS, Tathed PS, Purohit AV, Wadia NH. Treatment of Parkinson's disease with the cowhage plant–*Macuna pruriens* Bak. Neurol India 1978;26:171–176.
4. Donati D, Lampariello LR, Pagani R, Guerranti R, Cinci G, Marinello E. Antidiabetic oligocyclitols in seeds of *Mucuna pruriens*. Phytother Res 2005;19:1057–1060.
5. Ghosal S, Singh S, Bhattacharya SK. Alkaloids of *Mucuna pruriens* chemistry and pharmacology. Planta Med 1971;19: 280–284.
6. Guerranti R, Aguiyi JC, Ogueli IG, Onorati G, Neri S, Rosati F, et al. Protection of *Mucuna pruriens* seeds against *Echis carinatus* venom is exerted through a multiform glycoprotein whose oligosaccharide chains are functional in this role. Biochem Biophys Res Commun 2004;323: 484–490.
7. Guerranti R, Aguiyi JC, Neri S, Leoncini R, Pagani R, Marinello E. Proteins from *Mucuna pruriens* and enzymes from *Echis carinatus* venom. J Biol Chem 2002;277: 17072–17078.
8. Misra L, Wagner H. Alkaloidal constituents of *Mucuna pruriens* seeds. Phytochemistry 2004;65:2565–2567.
9. Daxenbichler ME, van Etten CH, Earle FR, Tallent WH. L-dopa recovery from *Mucuna* seed. J Agric Food Chem 1972;20:1046–1047.
10. Shelly WB, Arthur RP. Studies on cowhage (*Mucuna pruriens*) and its pruritogenic protease mucunain. Arch Dermatol 1955;72:399–406.
11. Prakash D, Niranjan A, Tewari SK. Some nutritional properties of the seeds of three *Mucuna* species. Int J Food Sci Nutr 2001;52:79–82.
12. Misra L, Wagner H. Extraction of bioactive principles from *Mucuna pruriens* seeds. Indian J Biochem Biophys 2007;44:56–60.
13. Fairbrothers D, Kirby E, Lester RM, Wegmann PC. *Mucuna pruriens*-associated pruritus—New Jersey. MMWR Morb Mortal Wkly Rep 1985;48:732.

Chapter 115

JEQUIRITY BEAN and ABRIN (*Abrus precatorius* L.)

HISTORY

In India, there is a long history of the use of abrin from the jequirity bean plant to poison animals and humans.[1] Administration of small doses of seeds (beans) from *Abrus precatorius* to protect animals from abrin poisoning is one of the earliest examples of active immunization.[2] The seeds from *Abrus precatorius* are called gumchi in Hindustani and gunja in Sanskrit. The physical properties (e.g., uniformity, durability) of these seeds were used in Southeast Asia for weighing jewels and gold. An old system of weights called the Ganda system was based on multiples of the weight of four *A. precatorius* seeds.[2] In addition, jequirity beans were used by some African and Madagascar tribes as an ordeal poison, but the delayed toxicity and the development of immunity limited the effectiveness of the these seeds as an ordeal poison. In Arab countries, jequirity beans (coq's eye) were an aphrodisiac. Extracts of these seeds were used to treat chronic eye diseases during the 19th century, particularly trachoma.[3] In India and Sri Lanka during the early 20th century, abrin was used as a homicidal agent by stabbing the victim with a hardened, needle-shaped abrin paste.[21]

BOTANICAL DESCRIPTION

Common Name: Jequirity Bean, Rosary Pea, Buddhist Rosary Bead, Indian Bead, Corral Peas, Paternoster Beans, Lucky Beans, Minnie-Minnies, Seminole Bead, Prayer Bead, Crab's Eye, Weather Plant, Ojo de Pajaro, Pois Rouge, Tento Muido

Scientific Name: *Abrus precatorius* L.

Botanical Family: Fabaceae (pea)

Physical Description: Small (1 cm/~0.4 in.) alternate compound leaflets develop on a twisting vine that grows up to 20 ft (6 m) in length. The bright scarlet seeds are 3–8 mm (0.1–0.3 in.) long and ovoid with a jet black end. The three color variations of these seeds include a red seed with a black eye (Indian native), a black seed with a white eye, and a white seed with a black eye.

Distribution and Ecology: *Abrus precatorius* is a vine used as an ornamental plant in tropical regions, such as southern Florida and the Florida Keys. *Abrus precatorius* also inhabits tropical and subtropical climates in Southeast Asia, India, Hawaii, Virgin Islands, and Puerto Rico.

EXPOSURE

Traditional herbal uses of *Abrus precatorius* include an anticonvulsant, insecticide, and the treatment of corneal opacities and trachoma by oculists.[4,5] In the Andaman Islands of India, aborigines eat boiled jequirity beans during extreme famines.[6] The jequirity bean is also a decorative bead in jewelry (e.g., necklace), shoes, and rosaries. Potential uses of abrin as a chemical weapon include aerosolization as a dry powder or liquid droplets and/or the contamination of food and water sources.[9] However, to date there is no documentation of the use of abrin as an aerosolized weapon.

Medical Toxicology of Natural Substances, by Donald G. Barceloux, MD

PRINCIPAL TOXINS

Physiochemical Properties

Constituents of the jequirity bean include *N*-methyl tryptophan, abric acid, glycyrrhizin (the active principle of licorice), a lipolytic enzyme, a heterodimeric glycoprotein (abrin), and a heterotetrameric glycoprotein (*Abrus* agglutinin-I). The latter two structurally similar substances are class II ribosome-inactivating proteins (ribotoxin). However, *Abrus* agglutinin-I is significantly less toxic than abrin, probably as a result of a difference in the secondary structural elements in chain A.[7] Like ricin, abrin has two dissimilar, disulfide-linked polypeptide chains composed of a lectin with two D-galactose moiety-binding sites (B chain) and an RNA-specific *N*-glycosidase that inhibits protein synthesis (A chain).[8] The A chain contains 251 residues divided into three folding domains with a molecular weight of approximately 30 kDa.[9] The B chains share the same 256 residues in addition to 12 other amino acids. There are at least three variants of abrin within *Abrus precatorius* species that differ in toxicity, binding ability, and lag period.[10] The toxic constituents of the jequirity bean are heat-labile and water soluble.[11] A thorough boiling removes these toxic ingredients.[6] The abrin compounds consist of four isolectins (A to D), which are monovalent compounds with molecular weights ranging from 63,000 to 67,000 Da. However, ricin and abrin are not identical disulfide-linked polypeptide chains. Based on minimal lethal intravenous doses to mice, abrin (0.7 µg/kg) is approximately four times more potent than ricin (2.7 µg/kg).[12]

Poisonous Parts

The seeds of the jequirity bean plant (*Abrus precatorius*) contain four toxic lectins (i.e., abrin compounds) that bind to carbohydrates containing terminal nonreducing galactose residues.[13]

DOSE RESPONSE

There is a marked variation between animal species and abrin toxicity with an intravenous minimal lethal dose (MLD) in mice and rabbits of 0.7 µg/kg and 0.06 µg/kg, respectively.[14] Predicting the fatal oral dose of jequirity beans is difficult because of the lack of data on the bioavailability of abrin in jequirity beans. A 20-year-old male died about 4 days after the ingestion of part of a mixture containing 20 pulverized jequirity beans.[15] The patient did not seek medical treatment until one day before his death, and the case report did not include a detailed case history or results of any laboratory analyses (e.g., drug screens) except the hematocrit.

Mechanism of Toxicity

Most cases of human poisoning after the ingestion of jequirity beans involve gastrointestinal toxicity. These clinical features are consistent with abrin-induced damage to vascular endothelial cells, interstitial edema, and extravasation of fluids and proteins similar to the vascular leak syndrome associated with ricin toxicity.[16] The general structure of abrin compounds is similar to ricin with two polypeptide chains (A chain and B chain). At the cellular level, abrin inhibits protein synthesis and causes cell death. Like ricin, the A chain inactivates the 60 S ribosomal subunits enzymatically after the B chain attaches the A chain to cell surface receptors. *In vitro* studies indicate that the B chain of each abrin molecule rapidly binds to nonreducing β-galactosyl residues of cell surface glycoproteins, particularly to mannose receptors on cells of the reticuloendothelial system.[17] Following the entry of the A chain into the cell, this *N*-glycosidase cleaves adenine from positions 4 and 324 from a loop on the 28S rRNA.[18] The result is the inhibition of protein synthesis after a 30-minute delay and subsequent cell death.[19] The roots of *Abrus precatorius* also contain glycyrrhizin, the active ingredient in licorice; therefore, the clinical syndrome of hyperaldosteronism characterized by sodium retention and hypertension potentially may develop following large doses of roots. However, there are no data to confirm this adverse affect in humans.

Like ricin, the pathology associated with inhalation of abrin is limited to the lungs with little evidence of significant systemic toxicity. The target cell for inhaled abrin and ricin is the type I pneumocyte. Animal studies indicate that abrin binds to cell surface receptors on type I pneumocytes and initiates acute alveolitis and necrosis of the lower respiratory tract epithelium.[8] A rapidly progressive pulmonary edema develops that produces severe hypoxia and death in exposed animals. Proliferation of type II pneumocytes occurs in response to the injury during the resolution phase of the pulmonary damage.

TOXICOKINETICS

There are few data on the toxicokinetics of abrin. The hard coat surrounding the jequirity bean limits the gastrointestinal absorption of abrin. Release of abrin from jequirity beans requires chewing or grinding of the bean prior to ingestion, and nonmasticated seeds pass through the gastrointestinal tract without causing toxicity. The high molecular weight (i.e., about 65 kDa) of abrin also limits the gastrointestinal absorption of this toxin. Based on animal studies, elimination of abrin probably occurs by the renal excretion of metabolites.[20]

CLINICAL RESPONSE

The ingestion of jequirity beans frequently involves children, who are attracted to the bright colors of the seeds. These ingestions are usually asymptomatic because the whole bean passes through the gastrointestinal tract without absorption of abrin from the bean.[21] Serious abrin ingestion produces a severe gastroenteritis several hours after consumption, followed by the development of bloody diarrhea. Delayed symptoms do not usually occur during abrin intoxication; however, rare case reports associated the development of cerebral edema, altered sensorium, and seizures 4–6 days after ingesting 7–10 crushed jequirity beans.[22] One of these patients died, but the cause of death was unclear because of the lack of imaging studies or an autopsy. The latter case is typical of fatalities associated with exposure to jequirity beans. These case reports lack sufficient clinical details to determine the cause and mechanism of death. Jequirity beans are highly antigenic and exposure to material in these beans may produce allergic or anaphylactic responses, particularly in atopic patients.

DIAGNOSTIC TESTING

Older analytical techniques include the detection of aqueous extracts of *A. precatorius* by characteristic ultraviolet absorption on spectrophotometry.[23] Radioimmunoassays have limits of detection near 50–100 pg abrin/mL, and there is little cross-reactivity of abrin with ricin on radioimmunoassay.[24] In animal experiments, sublethal parenteral doses of abrin cause a leukocytosis and mild to moderate elevation of serum hepatic aminotransferases.[25] There are few data on the laboratory abnormalities in humans following the ingestion of jequirity beans.

TREATMENT

The treatment of jequirity bean poisoning is supportive.[26] Decontamination measures are usually unnecessary after the ingestion of whole seeds. The administration of activated charcoal is a therapeutic option for patients, who present to a health care facility within one hour of the ingestion of well-masticated seed, but there are no clinical data to determine the efficacy of decontamination measures including activated charcoal in this clinical setting. There are no commercial antidotes or efficacious methods to enhance the elimination of abrin. Merck and Company (Whitehouse Station, NJ) developed an antiserum in large animals (Antiabrin® or Jequiritol®) to control abrin-induced ocular inflammations (i.e., secondary to trachoma treatment), but no clinical trials have ever been conducted in humans. Consequently, there is no commercial source of the antiserum. Careful fluid and electrolyte replacement is the most important aspect of management. In general, supportive care is similar to treatment for ricin (castor bean) poisoning. Asymptomatic children may be observed at home; symptomatic patients should be hospitalized. Gram-negative sepsis from the passage of intestinal bacteria across damaged intestinal mucosa should be considered when a patient with serious abrin ingestion develops fever and hypotension.

References

1. Stirling RF. Poisoning by gumchi or suis (needles). Vet J 1924;80:473–484.
2. Olsnes S. The history of ricin, abrin and related toxins. Toxicon 2004;44:361–370.
3. Martin S. The proteins of the seeds of *Abrus precatorius*. Proc Roy Soc 1887;42:3331–3333.
4. Moshi MJ, Kagashe GA, Mbwambo ZH. Plants used to treat epilepsy by Tanzanian traditional healers. J Ethnopharmacol 2005;97:327–336.
5. Molgaard P, Nielsen SB, Rasmussen DE, Drummond RB, Makaza N, Andreassen J. Anthelmintic screening of Zimbabwean plants traditionally used against schistosomiasis. J Ethnopharmacol 2001;74:257–264.
6. Rajaram N, Janardhanan K. The chemical composition and nutritional potential of the tribal pulse, *Abrus precatorius* L. Plant Foods Human Nutr 1992;42:285–290.
7. Bagaria A, Surendranath K, Ramagopal UA, Ramakumar S, Karande AA. Structure-function analysis and insights into the reduced toxicity of *Abrus precatorius* agglutinin I in relation to abrin. J Biol Chem 2006;281:34465–34474.
8. Griffiths GD, Rice P, Allenby AC, Bailey SC, Upshall DG. Inhalation toxicology and histopathology of ricin and abrin toxins. Inhal Toxicol 1995;7:269–288.
9. Dickers KJ, Bradberry SM, Rice P, Griffiths GD, Vale JA. Abrin poisoning. Toxicol Rev 2003;22:137–142.
10. Hedge R, Podder SK. Studies on the variants of the protein toxins ricin and abrin. Eur J Biochem 1992;204:155–164.
11. Niyogi SK, Rieders F. Toxicity studies with fractions from *Abrus precatorius* seed kernels. Toxicon 1969;7:211–216.
12. Gill DM. Bacterial toxins: a table of lethal amounts. Microbiol Rev 1982;46:86–94.
13. Wu AM, Wu JH, Herp A, Chow L-P, Lin J-Y. Carbohydrate specificity of a toxic lectin, abrin A, from the seeds of *Abrus precatorius* (jequirity bean). Life Sci 2001;69: 2027–2038.
14. Fodstad O, Johannessen JV, Schjerven L, Pihl A. Toxicity of abrin and ricin in mice and dogs. J Toxicol Environ Health 1979;5:1073–1084.
15. Davis JH. *Abrus precatorius* (rosary pea) the most common lethal plant poison. J Florida Med Assoc 1978;65: 188–191.

16. Baluna R, Rizo J, Gordon BE, Ghetie V, Vitetta ES. Evidence for a structural motif in toxins and interleukin-2 that may be responsible for binding to endothelial cells and initiating vascular leak syndrome. Proc Natl Acad Sci U S A 1999;96:3957–3962.

17. Tsuzuki J, Wu HC. Temporal behavior of abrin in the intoxication of Chinese hamster cells (line CHO). J Cell Physiol 1982;113:94–98.

18. Liu C-L, Tsai C-C, Lin S-C, Wang L-I, Hsu C-I, Hwang M-J, Lin J-Y. Primary structure and function analysis of the *Abrus precatorius* agglutinin A chain by site-directed mutagenesis. J Biol Chem 2000;275:1897–1901.

19. Benson S, Olsnes S, Pihl A, Skorve J, Abraham AK. On the mechanism of protein synthesis inhibition of abrin and ricin. Inhibition of the GTP hydrolysis site on the 60 S ribosomal subunit. Eur J Biochem 1975;59:573–580.

20. Fodstad O, Olsnes S, Pihl A. Toxicity, distribution and elimination of the cancerostatic lectins abrin and ricin after parenteral injection into mice. Br J Cancer 1976;34:418–425.

21. Hart M. Hazards to health jequirity-bean poisoning. N Engl J Med 1963;268:885–886.

22. Subrahmanyan D, Mathew J, Raj M. An unusual manifestation of *Abrus precatorius* poisoning: a report of two cases. Clin Toxicol (Phila) 2008;46:173–175.

23. Niyogi SK. The toxicology of the *Abrus precatorius* Linnaeus. J Forensic Sci 1970;15:529–536.

24. Godal A, Olsnes S, Pihi A. Radioimmunoassays of abrin and ricin in blood. J Toxicol Environ Health 1981;8:409–417.

25. Niyogi SK. Elevation of enzyme levels in serum due to *Abrus precatorius* (jequirity bean) poisoning. Toxicon 1977;15:577–580.

26. Kinamore PA, Jaeger RW, deCastro FJ. Abrus and ricinus ingestion: management of three cases. Clin Toxicol 1980;17:401–405.

Chapter 116

MESCAL BEAN
[*Sophora secundiflora* (Ortega) Lagasca ex DC.]

HISTORY

Anthropomorphs during prehistoric times depict shamans of the Great Plains Indians engaged in mescal bean cult activities.[1] Consistent with the same archaic tradition that produced these rock paintings, caves in Texas contained the same materials (mescal beans, red pigments).[2] Prior to the discovery of peyote, many southwestern American Indian tribes used mescal beans to induce visions during ceremonies. Native Americans continue to use the mescal bean as part of the ceremonial dress for the leaders of the peyote ceremony.[3] There is no modern experimental evidence that mescal beans contain hallucinogens.

BOTANICAL DESCRIPTION

Common names for the seeds include mescal bean, red bean, coral bean, red-hot, frijolillo, burn bean, and dry whiskey.

Common Name: Texas mountain laurel, Mescal Bean Plant

Scientific Name: *Sophora secundiflora* (Ortega) Lagasca ex DC.

Botanical Family: Fabaceae (pea)

Physical Description: This large woody, evergreen shrub or small tree grows up to 15–25 ft (~5–8 m) in height with a 10 ft (3 m) spread. The thick, dark green leaves are alternate and pinnate. Pendulous clusters of blue to purple flowers form hairy seedpods that contain yellow to scarlet seeds about 1 cm (~0.4 in.) long.

Distribution and Ecology: This shrub is indigenous to the southwestern United States and northern Mexico along the lower Pecos River region.

EXPOSURE

American and Mexican Indians used the seeds from this plant in religious ceremonies for many centuries, reportedly for hallucinogenic purposes. However, there is no clear clinical evidence to support the association of mescal bean ingestion with hallucinations.[4] Today, the seeds are most commonly used as a good luck charm.[5]

PRINCIPAL TOXINS

Physiochemical Properties

Of the seven quinolizidine alkaloids found in mescal beans, the major alkaloids are cytisine (0.25%), methylcytisine, and sparteine.[4] Figure 116.1 displays the chemical structure of cytisine (CAS RN: 485-35-8, $C_{11}H_{14}N_2O$). By contrast, golden chain seeds (*Laburnum anagyroides* Medikus) contain cytisine concentrations of approximately 1%. Minor alkaloids found in *Sophora secundiflora* include lupinine and Δ_5-dehydrolupanine. Pharmacologically, cytisine resembles nicotine. Analysis of the alkaloids in *S. secundiflora* did not reveal any

Medical Toxicology of Natural Substances, by Donald G. Barceloux, MD

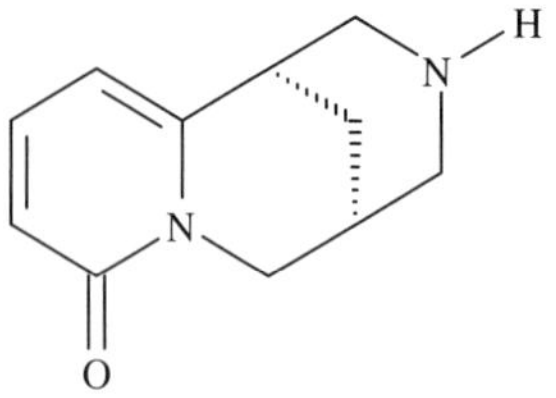

FIGURE 116.1. Chemical structure of cytisine.

TABLE 116.1. Some Physical Properties of Cytisine

Physical Property	Value
Melting Point	152.5 °C/306.5 °F
log P (Octanol-Water)	0.600
Water Solubility	4.39E + 05 mg/L (16 °C/60.8 °F)
Vapor Pressure	1.57E-05 mmHg (25 °C/77 °F)
Henry's Law Constant	4.01E-12 atm-m^3/mole (25 °C/77 °F)
Atmospheric OH Rate Constant	8.59E-11 m^3/molecule-second (25 °C/77 °F)

known hallucinogenic compounds.[6] Table 116.1 lists some physical properties of cytisine.

Poisonous Parts

The mescal bean contains about 0.25% cytisine, whereas seeds from the golden chain tree contain cytisine concentrations about 4 times higher.[4]

Mechanism of Toxicity

In animal studies, the administration of alkaloids (cytisine, sparteine, methylcytisine) produces neurological effects similar to mescaline, *N,N*-dimethyltryptamine, and psilocybin.[7] However, there are no well-documented human studies that demonstrate hallucinogenic effects following the ingestion of these alkaloids at concentrations found in the mescal bean.

DOSE RESPONSE

Cytisine has an oral LD_{50} dose in mice of 50–100 mg/kg.[4] The subcutaneous injection of 0.5 mL of a 50% ethanolic extract of 250 mg of *S. secundiflora* seeds was rapidly fatal to 150- to 200-g rats.[8] The lethal human dose is not known. Most children who ingest the more toxic laburnum seeds do not develop symptoms. In those children who are symptomatic, the most common presentation is vomiting and abdominal pain with delirium, drowsiness, diarrhea, and sore mouth occurring occasionally.

TOXICOKINETICS

There are few toxicokinetic data on cytisine in humans. Based on animal studies, the peak concentration of cytisine occurs about 2 hours after ingestion with bioavailability being about 40%.[9] The plasma elimination half-life of cytisine in these animal studies is about 3–4 hours. Following intravenous administration, the concentration of cytisine in the bile is high compared with the blood.

CLINICAL RESPONSE

Mescal bean intoxication is primarily an affliction of range sheep, which is manifest by stiff gait, trembling with exercise, and somnolence. Mescal bean poisoning may occur in domestic animals.[10] The clinical course of this illness is rarely fatal. Human poisoning is not well documented in the modern medical literature. A case report from the late 19th century associated the ingestion of powder from the crushing of one-quarter of a seed from *Sophora speciosa* Benth. with headache, paresthesias, lightheadedness, and difficulty in walking within 2 hours of ingestion.[11] Subsequently diarrhea developed. A "peculiar numbness in the spinal column" lasted over 24 hours. Because of the issue of plant identification during the 19th century, the relevance of this case report to mescal bean toxicity remains unclear.

DIAGNOSTIC TESTING

Cytisine is the major toxic constituent of the mescal bean, and confirmation of mescal bean poisoning requires analysis of this compound in blood samples. Methods for the quantitation of cytisine in plasma samples include liquid chromatography/atmospheric pressure chemical ionization/mass spectrometry and liquid chromatography/electrospray ionization/tandem mass spectrometry.[12] There are few data on the laboratory abnormalities following mescal bean intoxication because of the lack of clinical case reports in the modern medical literature associating the ingestion of mescal beans with toxicity.

TREATMENT

Management is primarily supportive. Decontamination measures are usually unnecessary, and there are no clinical data to support the efficacy of decontamination measures in the treatment of mescal bean ingestions. Serious poisoning is unlikely unless mescal teas or large quantities (probably over 20 beans) are ingested.

References

1. Wellmann KF. North American Indian rock art and hallucinogenic drugs. JAMA 1978;239:1524–1527.
2. Campbell TN. Origin of the mescal bean cult. Am Anthropologist 1958;60:156–160.
3. Schultes RE. Hallucinogens of plant origin. Science 1969; 163:245–254.
4. Hatfield GM, Valdes LJ, Keller WJ, Merrill WL, Jones VH. An investigation of *Sophora secundiflora* seeds (mescalbeans). Lloydia 1977;40:374–483.
5. Sullivan G, Chavez P1. Mexican good-luck charm potentially dangerous. Vet Hum Toxicol 1981;23: 259–261.
6. Bourn WM, Keller WJ, Bonfiglio JF. Comparisons of mescal bean alkaloids with mescaline, delta 9-THC and other psychotogens. Life Sci 1979;25:1043–1054.
7. Bourn WM, Keller WJ, Bonfiglio JF. Comparisons of mescal bean alkaloids with mescaline, Δ^9-THC and other psychotogens. Life Sci 1979;25:1043–1054.
8. Izaddoost M, Harris BG, Gracy RW. Structure and toxicity of alkaloid and amino acids of *Sophora secundiflora*. J Pharm Sci 1976;65:352–354.
9. Klöcking HP, Richter M, Damm G. Pharmacokinetic studies with 3H-cytisine. Arch Toxicol 1980;4(suppl): 312–314.
10. Knauer KW, Reagor JC, Bailey EM Jr, Carriker L. Mescalbean (*Sophora secundiflora*) toxicity in a dog. Vet Hum Toxicol 1995;37:237–239.
11. Weigand TS. Minutes of the pharmaceutical meeting. Am J Pharm 1878;50:39.
12. Beyer J, Peters FT, Kraemer T, Maurer HH. Detection and validated quantification of toxic alkaloids in human blood plasma—comparison of LC-APCI-MS with LC-ESI-MS/MS. J Mass Spectrom 2007;42:621–633.

IV Nonwoody Plants (Herbs)

Chapter 117

ACONITE POISONING and MONKSHOOD

HISTORY

Aconite is a crude extract of dried leaves and roots from various species of *Aconitum* (e.g., *A. japonicum* Thunb., *A. carmichaelii* Debeaux) that contain aconitine and other diterpenoid ester alkaloids. During ancient times, aconite was a common poison in Europe and in Asia. Aconite was a medicinal drug as well as a homicidal agent and arrow poison in Asia. The use of aconite may account for some deaths in antiquity attributed to stroke, even though substantial physical pain accompanies fatal aconite poisoning. The death of the Jewish High Priest Alkimos in 159 BC was regarded as stroke-related (collapse and loss of speech followed shortly by death). However, the description of severe pain suggests another cause, such as aconite poisoning.[1] The manner of death of the Roman Emperor Claudius in 54 AD was consistent with some of the clinical effects of acute aconite poisoning; his spouse, Agrippina, was the suspected assassin. The Ainu inhabitants from the Japanese island of Hokkaido concocted an arrow poison called *surku* from *Aconitum japonicum* and other local plant species, such as the marsh marigold (*Caltha palustris* L.) and mugwort (*Artemisia vulgaris* L.).[2] The crude extract of the aconite root (*A. Japonicum*), called *bushi*, has long been a traditional treatment for debilitated patients to improve metabolism and to cure dysuria, cardiac weakness, gout, neuralgias, and rheumatism.

BOTANICAL DESCRIPTION

The genus *Aconitum* contains over 350 species. North American species of *Aconitum* include *A. columbianum* Nutt. (Columbia monkshood), *A. reclinatum* Gray (trailing white monkshood), and *A. uncinatum* L. (southern blue monkshood). *Aconitum napellus* L. (Venus' chariot) is a naturalized species from Europe. Figure 117.1 displays the flowers and branches of *A. napellus*. The Asian varieties of *Aconitum* include *A. carmichaeli* Debeaux, *A. japonicum* Thunb., *A. kusnezoffii* Reichb. (bei wu tou), and *A. vulparia* Reichb. ex Sprengel (yellow-flowered wolfsbane). The misidentification of *Aconitum* root as horseradish (*Armoracia rusticana* P.G. Gaertn., B. Mey. & Scherb.) or the leaves as parsley [*Petroselinum crispum* (Mill.) Nym.] can cause poisonings.

At least half of *Aconitum* species grow in northern latitudes including North America, Europe, and China.[3] *Aconitum columbianum* (Western or Columbia monkshood) is a native of moist woodlands in the Rocky Mountains and Sierra Nevada Mountains from southwestern Canada to New Mexico. *Aconitum reclinatum* (blue flowers) and *Aconitum uncinatum* (white flowers) are natives of mountains in the eastern United States.

Common Name: Cultivated Aconite, Friar's Cap, Monkshood, Wolfsbane, Venus' Chariot

Scientific Name: *Aconitum napellus* L.

Botanical Family: Ranunculaceae (buttercup)

Physical Description: This perennial herb grows 2–6 ft (0.6–1.8 m) in height with palmate leaves and a tuberous root. The blue or white trumpet-shaped flowers are bilaterally symmetrical with a prominent upper hood and grow in a raceme on a long stem.

Medical Toxicology of Natural Substances, by Donald G. Barceloux, MD

FIGURE 117.1. Venus' chariot (*Aconitum napellus* L.). Photograph courtesy of Professor Sylvie Rapior. See color insert.

Distribution and Ecology: *A. napellus* is a cultivated European species present in many US and Canadian gardens.

EXPOSURE

Aconitum tubers are a common component of traditional Chinese medicines including fu zi or fuzi (lateral root or tuber of *A. carmichaelii* Debeaux), caowu (*A. kusnezoffii* Reichb.), chuanwu (*A. carmichaelii* Debeaux.), futxu, wu tao, and zue shang yi zhi gao.[4] In Hong Kong, chuanwu (*A. carmichaelii*), fuzi (*A. carmichaelii*), and caowu (the root of *A. kusnezoffii*) are prescribed by traditional health care providers for musculoskeletal disorders (rheumatism, arthritis, fractures, bruises, trauma) based on potential anti-inflammatory, analgesic, and cardiotonic effects. A retrospective review of admissions to a Hong Kong hospital suggested that these two traditional Chinese remedies accounted for about 70% of the admissions for severe poisonings from Chinese herbal remedies.[4] In China and Hong Kong, traditional uses for herbal medicines derived from *Aconitum* species beside musculoskeletal disorders (e.g., gout) include hypertension, malaise, generalized weakness, and epigastric pain.[5] Aconite is also an abortifacient and suicidal agent. Experimental studies in rats suggest that a hetisane-type diterpenoid alkaloid in *Aconitum* root, called ignavine (CAS RN: 1357-76-2), has anti-inflammatory properties.[6]

PRINCIPAL TOXINS

Physiochemical Properties

There are over 170 *Aconitum* alkaloids that have been isolated from 20 *Aconitum* species in China alone.[7] Figure 117.2 and Figure 117.3 displays the structures of

Alkaloid	R_1	R_2	R_3	MW
Aconitine	Ethyl	Hydroxyl	Hydrogen	645
Mesaconitine	Methyl	Hydroxyl	Hydrogen	631
Jesaconitine	Ethyl	Hydroxyl	*O*-Methyl	675
Hypaconitine	Methyl	Hydrogen	Hydrogen	615
Deoxyaconitine	Ethyl	Hydrogen	Hydrogen	629

FIGURE 117.2. Structures of aconitum alkaloids.

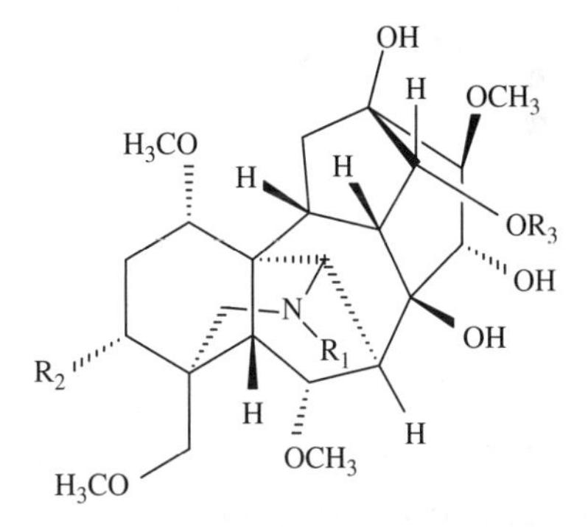

Alkaloid	R_1	R_2	R_3	MW
Benzoylaconine	Ethyl	Hydroxyl	Benzoyl	603
Benzoyl mesaconine	Methyl	Hydroxyl	Benzoyl	589
Anisoylaconine	Ethyl	Hydroxyl	Anisoyl	633
Benzoylhypaconine	Methyl	Hydrogen	Benzoyl	573
Benzoyldeoxyaconine	Ethyl	Hydrogen	Benzoyl	587
Aconine	Ethyl	Hydroxyl	Hydrogen	499
Mesaconine	Methyl	Hydroxyl	Hydrogen	485
Hypaconine	Methyl	Hydrogen	Hydrogen	469
Deoxyaconine	Ethyl	Hydrogen	Hydrogen	483

FIGURE 117.3. Structures of hydrolysis products of aconitum alkaloids.

common *Aconitum* alkaloids and hydrolysis productions of *Aconitum* alkaloids, respectively. Aconite is a crude extract of dried leaves and roots from various *Aconitum* species that contains aconitine and other diterpenoid ester alkaloids (aconitine, mesaconitine, jesaconitine, hypaconitine).[8] Aconitine (CAS RN: 302-27-2) is a methylated and benzoylated ester of the hexacyclic norditerpenic amino alcohol, aconine (CAS RN: 509-20-6). Hydrolysis of these alkaloids forms less toxic substances.[3] Soaking and boiling during the preparation of herbal decoctions of dried aconite roots reduces the toxicity of the compound by hydrolyzing aconitine and related alkaloids to less toxic benzylaconine and aconine derivatives. However, these preparations may still cause toxicity.[4] In animal studies, the direct application of aconitine to the myocardium produced atrial and ventricular dysrhythmias.[9]

In addition to the cardioactive alkaloids, *Aconitum* roots contain physiologically active catecholamine analogs.[10] *A. carmichaelii* contains small amounts of the quaternary ammonium compound coryneine chloride (dopamine methochloride); *A. japonicum* possesses the triphenolic 1-benzyl-tetrahydroisoquinoline base, higenanine.[3] In animal studies, ignavine produces substantially less sedation and loss of motor coordination than aconitine, mesaconitine, or jesaconitine.[6] Other aconite alkaloids with low toxicity include ignavine, hypognavine, isohypognavine, kobusine, and pseudokobusine.[6] *Aconitum* alkaloids, which lack a benzoylester side chain (e.g., heteratisine, lappaconitine, napelline), produce lower toxicity compared with other *Aconitum* alkaloids.[11]

Poisonous Parts

Aconitine (CAS RN: 8063-12-5), yunaconitine (guayewuanine B, CAS RN: 70578-24-4), and jesaconitine (CAS RN: 16298-90-1) are the main toxic alkaloids in *Aconitum* species.[12] Other less cardiotoxic alkaloids include *N*-desacetyllappaconitine, 6-benzoylheteratisine, heteratisine, lappaconitine, and napelline. All parts of the plant contain toxic alkaloids including seeds,[13] but the alkaloid content and composition depends on the species, plant part, environmental conditions, and season. Generally, the relative alkaloid content is distributed as follows: roots > flowers > leaves > stems. Fresh *Aconitum* plants typically contain 0.4–2% alkaloids, depending on season, species, and environmental conditions.

Mechanism of Toxicity

The voltage-gated sodium channels generate the action potentials in nerve and in muscle tissue. Aconitine binds with high affinity to site 2 of the α-subunit of the Na^+ channel protein. This binding of aconitine to the open Na^+ channels shifts the voltage dependence of activation to more hyperpolarized potentials and reduces channel inactivation from the open state. As a consequence of prolonged sodium channel activation, sustained Na^+ influx causes a lack of excitability. Thus, aconitine initially stimulates and then inactivates the voltage-gated sodium channels in the heart and nervous system, resulting in interference with the propagation of the action potential.[14] Veratridine, grayanotoxin, and batrachotoxin also cause inactivation of the sodium channel by sustained activation at this transmembrane receptor.[15] The arrhythmogenic effects of these alkaloids result from the initiation of premature or triggered excitations secondary to a delay in the final repolarization phase of the action potential.

Histopathology

Postmortem examinations of fatalities following *Aconitum* toxicity demonstrate nonspecific finding (e.g., congestion of main organs) often without evidence of coronary artery disease.[16,17]

DOSE RESPONSE

The recommended dose of aconite decoctions of processed Chinese herbs is 1–3 grams. The exact composition and type of *Aconitum* alkaloids necessary to produce toxicity are not well defined. Inadequate curing of these decoctions increases toxicity. Most of the toxic effects of aconite result from the accidental, suicidal, or homicidal ingestion of aconite powder, especially in India and Asia where the paste is used to relieve neuralgias and dental pain. The estimated lethal dose of aconite is about 1–2 mg, but there are limited clinical data on the dose response of crude aconite and *Aconitum* alkaloids. Serious toxicity may result from the ingestion of a decoction from *Aconitum* species containing more than 6 g cured *Aconitum* alkaloids or consumption of more than 0.2–0.6 mg aconitine.[18] The estimated ingestion of 3–10 mg of aconite was associated with paresthesias, nausea, vomiting, weakness, and cardiac dysrhythmias.[19,20] The ingestion of an estimated dose of 11 mg *Aconitum* alkaloids caused ventricular tachycardia and cardiac arrest, but the patient survived with intensive supportive care.[21] A case report associated the reported ingestion of 3 capsules of dried *Aconitum napellus* root (i.e., about 57 μg aconite) with the development of paresthesias, nausea, diarrhea, vertigo, bradycardia, and ventricular ectopy.[22] Symptoms resolved within 12 hours. The concentrations of other terpenoid compounds (e.g., mesaconitine, jesaconitine) were not reported.

TOXICOKINETICS

Metabolic pathways for *Aconitum* alkaloids include de-esterification at the C8 and C14 groups by esterase and the *N*- and/or *O*-dealkylation by cytochrome P450 oxidases. Urinary metabolites of aconitine in humans include 16-*O*-demethylaconitine and benzoylaconine, whereas benzoylhypaconine and 16-*O*-demethylhypaconitine are metabolites of hypaconitine. Benzoylmesaconitine is a metabolite of mesaconitine.[23] Elimination of *Aconitum* alkaloids from the blood is rapid with most of the absorbed dose of aconitine, mesaconitine, and hypaconitine appearing unchanged or hydrolyzed in the urine within 24 hours after ingestion. In a 21-year-old man with moderate aconitine poisoning (i.e., bradycardia without hypotension), the plasma elimination half-life of aconitine was about 3 hours.[22] In a case series of 5 patients admitted to an intensive care unit with aconite poisoning, the serum half-life of aconitine ranged from 3.7 hours to 17.8 hours.[24] The serum half-life of jesaconitine in these patients ranged from 5.8 hours to 15.4 hours.

Peak concentrations of these *Aconitum* alkaloids and their metabolites in the urine occurred about 24 hours after the ingestion of an estimated 11 mg of *Aconitum* alkaloids with small concentrations (<0.03 ng/mL) present 6 days after ingestion.[21] Analysis of urine samples 6 days after ingestion demonstrated trace amounts of the *Aconitum* alkaloids in the urine with aconine having the highest concentration (0.5 ng/mL). There are inadequate data to characterize the percentage of these *Aconitum* alkaloids excreted unchanged in the urine.

CLINICAL RESPONSE

The clinical presentation of aconite poisoning involves the gastrointestinal, cardiac, and nervous systems. Ingestion of aconite produces intoxication similar to veratrum alkaloids (veratrine), except that paresthesias are more prominent and persistent during aconite intoxication. Numbness and paresthesias of tongue, throat, and mouth begin soon after ingestion followed within several hours by nausea, salivation, and generalized weakness.[25] During serious intoxication, convulsions, quadriplegia, loss of deep tendon reflexes, and coma may develop within 6 hours after ingestion.[26,27] Changes in sensorium result from secondary causes rather than direct central nervous system (CNS) depression. Other neurological symptoms include blurred vision, mydriasis, and yellow-green vision. Most symptoms usually resolve within 24–30 hours.[28]

Cardiotoxicity usually occurs during serious aconite poisoning, characterized by hypotension, shock, conduction delays, and dysrhythmias beginning within 6 hours of ingestion.[29,30] Reported cardiac dysrhythmias include sinus bradycardia with first-degree atrioventricular block,[31] supraventricular tachycardia and hypotension,[32] multiple premature ventricular contractions,[20] bundle branch block, junctional escape rhythms,[26] ventricular tachycardia, and *torsade de pointes*.[33] Serious ventricular dysrhythmias typically resolve within 12–24 hours of ingestion,[34] but ST-T wave abnormalities and conduction delays may persist for several days. Death usually results from cardiovascular collapse and persistent ventricular fibrillation unresponsive to both defibrillation and the usual pharmacological agents.[18] The differential diagnosis of aconite poisoning includes poisoning by cardiac glycosides, andromedotoxin, or veratrine, ciguatera intoxication, and paralytic shellfish poisoning.[35]

DIAGNOSTIC TESTING

Analytical Methods

High performance liquid chromatography (HPLC) with UV detection and mass spectrometry detects aconitine, hypaconitine, jesaconitine, and mesaconitine in blood and urine samples at limits of detection approaching 50 ng/mL.[36] The limit of detection and limit of quantitation for aconitine in blood samples after solid-phase extraction using mixed-mode C8 cation exchange columns followed by liquid chromatography/tandem mass spectrometry (LC/MS/MS) were 0.1 ng/g and 0.5 ng/g, respectively.[37] The use of capillary liquid chromatographic-frit fast atom bombardment mass spectrometric analysis allows the detection of 1 pg aconitine.[38] Urinary metabolites of aconitine and hypaconitine are detectable by liquid chromatography/electrospray ionization/multistage tandem mass spectrometry.[23] After solid-phase extraction and analysis by LC/MS/MS, the limits of quantitation for urinary metabolites were as follows: yunaconitine, 0.15 ng/mL; crassicauline A, 0.20 ng/mL; and foresaconitine, 0.20 ng/mL.[39] The limits of detection for these urinary metabolites were 0.03 ng/mL, 0.05 ng/mL, and 0.05 ng/mL, respectively.

Biomarkers

There are little clinical data on the interpretation of serum *Aconitum* alkaloid concentrations. The clearance of aconitine, mesaconitine, and hypaconitine from the blood is relatively rapid. A 45-year-old man developed ventricular tachycardia and cardiovascular collapse after ingesting approximately 11 mg *Aconitum* alkaloids.[21] Analyses of blood samples obtained 2 hours after the ingestion demonstrated the following concentrations: aconitine, 1.72 ng/mL; mesaconitine, 4.04 ng/mL; and hypaconitine, 0.60 ng/mL. Analysis of blood samples from samples drawn the following day did not detect these alkaloids, although aconite alkaloids and metabolites remained in the urine for several more days. A 61-year-old man ingested boiled and seasoned *Aconitum* stems and leaves.[40] Within 6 hours, he developed refractory ventricular fibrillation and anoxic brain damage leading to death on the sixth day as a result of cerebral edema and probable brain herniation. Approximately 24 hours after ingestion, the serum jesaconitine concentration was 0.6 ng/mL, whereas the aconitine, mesaconitine, and hypaconitine concentrations were not detectable. Mesaconine remained detectable at least 4 days after ingestion. Postmortem blood samples drawn from a 40-year-old woman, who died 4 hours after ingesting *Aconitum* tubers, demonstrated serum jesaconitine and aconitine concentrations of 69.1 ng/mL and 1.1 ng/mL, respectively.[41] Hypaconitine and mesaconitine were not detectable. A 60-year-old man developed a fatal cardiac arrest 2 hours after ingesting liquified tissue from *Aconitum napellus*.[42] The aconitine concentration in a sample of postmortem femoral blood was 10.8 ng/mL, as measured by HPLC. Aconitine was detectable in urine and liver samples 5 years after initial analysis of postmortem samples as measured by LC/MS/MS.[43]

Abnormalities

Occasional hypokalemia, hyperglycemia, and metabolic acidosis occur during aconite toxicity.

TREATMENT

Stabilization

Symptomatic patients (e.g., those with protracted vomiting, weakness) should receive an intravenous line, oxygen, and monitoring of the cardiac rhythm and oxygen saturation. Cardiovascular collapse and ventricular dysrhythmias represent the greatest risk to life, and prolonged cardiopulmonary resuscitation may be necessary for survival of severely intoxicated patients.[44] There are few clinical data on the most effective antiarrhythmic agent for ventricular dysrhythmias during aconite poisoning. Consequently, there is no universally accepted, first-line antiarrhythmic agent for the treatment of dysrhythmias associated with aconite toxicity.[29] Standard treatment includes amiodarone, magnesium, and lidocaine for ventricular dysrhythmias and atropine for sinus bradycardia.[45] Typically, agents that worsen QT prolongation (e.g., procainamide, disopyramide, quinidine, flecainide, encainide, sotalol) are avoided during intoxication by aconite because of concern that these drugs may prolong the QT interval and cause *torsade de pointes*. However, a case report of aconite poisoning documented the resolution of bidirectional tachycardia after the intravenous administration of 40 mg flecainide.[46] In this patient, analysis of serum, urine, and gastric samples did not detect digoxin, salicylates, or tricyclic antidepressants. A case series of 17 patients with aconitine poisoning suggested that amiodarone and flecainide were first line antiarrhythmic drugs, whereas lidocaine was a second-line antiarrhythmic drug.[35] Procainamide and phenytoin have been clinically and experimentally successful in suppression of ventricular dysrhythmias resulting from some cases of aconite poisoning,[9] whereas other cases of aconite-induce ventricular dysrhythmias have not responded to intravenous lidocaine.[47] A 32-year-old man survived aconite intoxication associated with bidirectional ventricular tachycardia and persistent ventricular fibrillation following cardiopulmonary bypass.[48] The presence of respiratory or CNS depression is unusual during aconite poisoning unless there is poor cerebral perfusion, and the presence of these clinical features suggests the ingestion of a CNS depressant.

Decontamination

Because all parts of the *Aconitum* plant contain aconitine, decontamination with activated charcoal is a therapeutic option for alert patients ingesting more than a taste of the plant and presenting for treatment within one hour of ingestion. However, there are no clinical data to confirm the efficacy of decontamination measures to improve outcome during aconite poisoning, and decontamination measures are usually unnecessary.

Supplemental Care

Laboratory assessment includes electrolytes, creatine phosphokinase, arterial blood gases, and an electrocardiogram. The severity of toxicity following aconite ingestion is difficult to predict based on history, and asymptomatic patients should be observed for at least 4–6 hours. Symptomatic patients should be hospitalized for 24 hours with cardiac monitoring.

References

1. Moog FP, Karenberg A. Toxicology in the Old Testament. Did the High Priest Alcimus die of acute aconite poisoning? Adv Drug React Toxicol Rev 2002;21:151–156.
2. Bisset NG. Hunting poisons of the North pacific region. Lloydia 1976;39:87–124.
3. Bisset NG. Arrow poisons in China: Part II. *Aconitum*—botany, chemistry and pharmacology. J Ethnopharmacol 1981;4:247–336.
4. Chan TY, Tomlison B, Tse LK, Chan JC, Chan WW, Critchley JA. Aconitine poisoning due to Chinese herbal medicines: a review. Vet Hum Toxicol 1994;36: 452–455.

5. Chan TY. Aconitine poisoning: a global perspective. Vet Hum Toxicol 1994;36:326–328.
6. Saito H, Ueyama T, Naka N, Yagi J, Okamoto T. Pharmacological studies of ignavine, an *Aconitum* alkaloid. Chem Pharm Bull 1982;30:1844–1850.
7. Zhu D-Y, Bai D-L, Tang X-C. Recent studies on traditional Chinese medicinal plants. Drug Dev Res 1996;39: 147–157.
8. Hikino H, Takata H, Konno C. Anabolic principles of *Aconitum* roots. J Ethnopharmacol 1983;7:277–286.
9. Scherf D, Blumenfeld S, Taner D, Yildiz M. The effect of diphenylthydantoin (Dilantin®) sodium on atrial flutter and fibrillation provoked by focal application of aconite and delphinine. Am Heart J 1960;60:936–947.
10. Konno C, Shirasaka M, Hikino H. Cardioactive principle of *Aconitum carmichaelii* roots. Planta Med 1979;35: 150–155.
11. Ameri A. The effects of *Aconitum* alkaloids on the central nervous system. Prog Neurobiol 1998;65:211–235.
12. Poon WT, Lai CK, Ching CK, Tse KY, So YC, Chan YC, et al. Aconite poisoning in camouflage. Hong Kong Med J 2006;12:456–459.
13. Imazio M, Belli R, Pomari F, Cecchi E, Chinaglia A, Gashino G, et al. Malignant ventricular arrhythmias due to *Aconitum napellus* seeds. Circulation 2000;102: 2907–2908.
14. Friese J, Gleitz J, Gutser UT, Heubach JF, Matthiesen T, Wilffert B, Selve N. *Aconitum* sp. alkaloids: the modulation of voltage-dependent Na+ channels, toxicity and antinociceptive properties. Eur J Pharmacol 1997;337:165–174.
15. Adams ME, Olivera BM. Neurotoxins: overview of an emerging research technology. Trends Neurosci 1994;17: 151–155.
16. Yi-gu Z, Guang-zhao H. Poisoning by toxic plants in China report of 19 autopsy cases. Am J Forensic Med Pathol 1988;9:313–319.
17. Dickens P, Tai YT, But PPH, Tomlison B, Ng HK, Yan KW. Fatal accidental aconitine poisoning following ingestion of Chinese herbal medicine: a report of two cases. Forensic Sci Int 1994;67:55–58.
18. But PP-H, Tai Y-T, Young K. Three fatal cases of herbal aconite poisoning. Vet Hum Toxicol 1994;36:212–215.
19. Fiddes FS. Poisoning by aconitine. Report of two cases. Br Med J 1956;2:779–780.
20. Nicolas G, Desjars P, Godon JF, Rozo L. Intoxication par l'anacotine. Toxicol Eur Res 1978;1:45–49.
21. Mizugaki M, Ito K, Ohyama Y, Konishi Y, Tanaka S, Kurasawa K. Quantitative analysis of *Aconitum* alkaloids in the urine and serum of a male attempting suicide by oral intake of aconite extract. J Anal Toxicol 1998;22: 336–340.
22. Moritz F, Cmpagnon P, Kaliszczak IG, Kaliszczak Y, Caliskan V, Girault C. Severe acute poisoning with homemade *Aconitum napellus* capsules: toxicokinetic and clinical data. Clin Toxicol 2005;43:873–876.
23. Zhang H-G, Sun Y, Duan M-Y, Chen Y-J, Zhong D-F, Zhang H-Q. Separation and identification of *Aconitum* alkaloids and their metabolites in human urine. Toxicon 2005;46:500–506.
24. Fujita Y, Terui K, Fujita M, Kakizaki A, Sato N, Oikawa K, et al. Five cases of aconite poisoning: toxicokinetics of aconitines. J Anal Toxicol 2007;31:132–137.
25. Chan TY, Tomlinson B, Chan WW, Yeung VT, Tse LK. A case of acute aconitine poisoning caused by chuanwu and caowu. J Trop Med Hyg 1993;96:62–63.
26. Chan TY, Tomlinson B, Critchley JA, Cockram CS. Herb-induced aconitine poisoning presenting as tetraplegia. Vet Hum Toxicol 1994;36:133–134.
27. Merchant HC, Choksi ND, Ramamoorthy K, Parihar LM, Shikaripurkar NK. Aconite poisoning and cardiac arrhythmias: report of 3 cases. Indian J Med Sci 1963;17: 857–865.
28. Chan TY, Tomlinson B, Critchley JA. Aconitine poisoning following the ingestion of Chinese herbal medicines: a report of eight cases. Aust NZ J Med 1993;23:268–271.
29. Tai Y-T, But PP-H, Young K, Lau C-P. Cardiotoxicity after accidental herb-induced aconite poisoning. Lancet 1992; 340:1254–1256.
30. Fatovich DM. Aconite: a lethal Chinese herb. Ann Emerg Med 1992;21:309–311.
31. Hartung EF. A case of aconite poisoning causing cardiac collapse. JAMA 1930;95:1265.
32. Ffrench G. Case reports: aconite induced cardiac arrhythmia. Br Heart J 1958;20:140–142.
33. Kapoor SC, Sen AK. Cardiovascular aspects of aconite poisoning in human beings. Indian Heart J 1969;21: 329–338.
34. Guha S, Dawn B, Dutta G, Chakraborty T, Pain S. Bradycardia, reversible panconduction defect and syncope following self-medication with a homeopathic medicine. Cardiol 1999;91:268–271.
35. Lin CC, Chan TY, Deng JF. Clinical features and management of herb-induced aconitine poisoning. Ann Emerg Med 2004;43:574–579.
36. Ohta H, Seto Y, Tsunoda N. Determination of *Aconitum* alkaloids in blood and urine samples. I. High-performance liquid chromatographic separation, solid-phase extraction and mass spectrometric confirmation. J Chromatogr B Biomed Sci Appl 1997;691:351–356.
37. Beike J, Frommherz L, Wood M, Brinkmann B, Kohler H. Determination of aconitine in body fluids by LC-MS-MS. Int J Legal Med 2004;118:289–293.
38. Ohta H, Seto Y, Tsunoda N, Takahashi Y, Matsuura K, Ogasawara K. Determination of *Aconitum* alkaloids in blood and urine samples. II. Capillary liquid chromatographic-frit fast atom bombardment mass spectrometric analysis. J Chromatogr B Biomed Sci Appl 1998;714: 215–221.
39. Wang Z, Wang Z, Wen J, He Y. Simultaneous determination of three *Aconitum* alkaloids in urine by LC-MS-MS. J Pharm Biomed Anal 2007;45:145–148.

40. Yoshioka N, Gonmori K, Tagashira A, Boonhooi O, Hayashi M, Saito Y, Mizugaki M. A case of aconitine poisoning with analysis of aconitine alkaloids by GC/SIM. Forensic Sci Int 1996;81:117–123.
41. Ito K, Tanaka S, Funayama M, Mizugaki M. Distribution of *Aconitum* alkaloids in body fluids and tissues in a suicidal case of aconite ingestion. J Anal Toxicol 2000;24: 348–353.
42. Elliot SP. A case of fatal poisoning with the aconite plant: quantitative analysis in biological fluid. Sci Just 2002;42; 111–115.
43. Van Landeghem AA, De Letter EA, Lambert WE, Van Peteghem CH, Piette MH. Aconitine involvement in an unusual homicide case. Int J Legal Med 2007;121: 214–219.
44. Weijters BJ, Verbunt RJ, Hoogsteen J, Visser RF. Salade malade: malignant ventricular arrhythmias due to an accidental intoxication with *Aconitum napellus*. Neth Heart J 2008;16:96–99.
45. Kolev ST, Leman P, Kite GC, Stevenson PC, Shaw D, Murray VSG. Toxicity following accidental ingestion of *Aconitum* containing Chinese remedy. Hum Exp Toxicol 1996;15:839–842.
46. Tai Y-T, Lau C-P, But PP-H, Fong P-C, Li JP-S. Bidirectional tachycardia induced by herbal aconite poisoning. Pacing Clin Electrophysiol 1992;15:831–839.
47. Lin C-C, Chou H-L, Lin J-L. Acute aconitine poisoned patients with ventricular arrhythmias successfully reversed by charcoal hemoperfusion. Am J Emerg Med 2002;20: 66–67.
48. Fitzpatrick AJ, Crawford M, Allan RM, Wolfenden H. Aconite poisoning managed with a ventricular assist device. Anaesth Intens Care 1994;22:714–717.

Chapter 118

AFRICAN BLUE LILY (*Agapanthus* Species)

BOTANICAL DESCRIPTION

Common Name: African Blue Lily

Scientific Name: *Agapanthus orientalis* F.M. Leight

Botanical Family: Liliaceae (lily)

Physical Description: This erect perennial herb forms large clumps of leaves and smooth, flowering stems up to about 1 m (~3 ft) arise from underground rhizomes. The leaves are long, elongated, glossy, and dark green. The tubular flowers are blue or white in large spherical clusters at the top of the flowering stems. Leathery, green capsules form from the flowers that contain black, winged seeds. As the capsule dries, the color changes to pale brown.

Distribution and Ecology: This plant was introduced into the United States, and now is cultivated in Florida, Arizona, and coastal California.

Common Name: Lily of the Nile

Scientific Name: *Agapanthus africanus* (L.) Hoffmanns.

Botanical Family: Liliaceae (lily)

Physical Description: This evergreen plant reaches 50 cm (~20 in.) in height with clumps of 8–18 linear-lanceolate leaves about 1–1.5 cm (0.4–0.6 in.) wide. The flowering stalks are umbel (nearly equal in length, arising from a common point) with funnelform, deep violet-blue flowers up to 4 cm (~2 in.) in length.

Distribution and Ecology: The plant is native to South Africa.

EXPOSURE

Agapanthus africanus is a traditional medicine (Isihlambezo) for Zulu women in South Africa as a health tonic during pregnancy and as a means to induce or augment contractions during labor.[1] The herbal remedy is prepared by boiling the herb in water (decoction). In the Transkei region of South Africa, women consume a decoction of the root of *Agapanthus africanus* beginning in the sixth month of pregnancy as means of ensuring an easy childbirth.[2]

PRINCIPAL TOXINS

The sap of these plants contains a direct mucosal and skin irritant rather than a potent allergen. The chemical structure of the irritant is not known. *In vitro* studies of isolated rat uterus preparations indicate that aqueous extracts of *A. africanus* produce some agonist activity on uterine muscarinic receptors and increase synthesis of prostaglandins in estrogenized rat uterus.[3]

DOSE RESPONSE

There are insufficient data on the toxic constituents of *Agapanthus* species to determine dose-response data.

TOXICOKINETICS

The toxic constituents of *Agapanthus* species have not been identified.

Medical Toxicology of Natural Substances, by Donald G. Barceloux, MD

CLINICAL RESPONSE

Contact of the oral mucosa or skin with parts of the plant can cause local irritation, but systemic symptoms are unlikely. Potentially irritation of the eyes, mouth, throat, or skin can develop following exposure to sap from the leaves or roots of *Agapanthus* species.

DIAGNOSTIC TESTING

There are few data on the use of diagnostic methods to detect biomarkers in *Agapanthus* species.

TREATMENT

Treatment is supportive. All plant material should be removed from the mouth and demulcents (milk, antacids, cool water) used as needed. Symptomatic irritation of the eye requires saline irrigation and examination for corneal abrasions. Contact with plant material does not usually cause substantial edema (i.e., intraoral edema, drooling), and treatment with topical analgesics [e.g., Cepacol® (Bayer Consumer Care, Morristown, NJ), Orabase Plain® (Colgate-Palmolive Company, New York, NY)] is usually sufficient.

References

1. Varga CA, Veale DJ. Isihlambezo: utilization patterns and potential health effects of pregnancy-related traditional herbal medicine. Soc Sci Med 1997;44:911–924.
2. Kaido TL, Veale DJ, Havlik I, Rama DB. Preliminary screening of plants used in South Africa as traditional herbal remedies during pregnancy and labour. J Ethnopharmacol 1997;55:185–191.
3. Veale DJ, Havlik I, Oliver DW, Dekker TG. Pharmacological effects of *Agapanthus africanus* on the isolated rat uterus. J Ethnopharmacol 1999;66:257–262.

Chapter 119

RED BANEBERRY
[*Actaea rubra* (Aiton) Willd.]

BOTANICAL DESCRIPTION

The European variety (Eurasian baneberry, Herb Christopher) is a separate species (*Actaea spicata* L.). Black cohosh (*Actaea racemosa* L.) and *Actaea rubra* are closely related species.

Common Name: Red Baneberry, Snakeberry, White Cohosh, Doll's Eyes, Coral Berry

Scientific Name: *Actaea rubra* (Aiton) Willd.

Botanical Family: Ranunculaceae (buttercup)

Physical Description: This tall perennial herb reaches up to about 50 cm (~20 in.). The roots are thick and the large leaves are compound with toothed edges. The white flowers appear in June to July and produce attractive red berries, which ripen from July to August.

Distribution and Ecology: The western baneberry (red fruit) inhabits areas from Alaska to central California and Arizona and east to Montana. The eastern baneberry (*Actaea pachypoda* Ell.) produces white fruit in rich woodlands from Canada to Georgia and west to the northern Rocky Mountains.

EXPOSURE

This North American native plant was used by Native Americans as an analgesic and rubefacient, particularly the root. An infusion of the root was a treatment for upper respiratory tract infections, intestinal pain, postpartum pain, and constipation.

PRINCIPAL TOXINS

The rootstock, sap, and berries of baneberry contain glycosides or essential oils that produce a severe gastroenteritis. White baneberry contains a variety of unique 9,10-seco-9,19-cyclolanostane arabinosides compounds (podocarpasides A-J).[1,2] The contribution of these novel triterpenoid compounds to the gastrointestinal toxicity of the baneberry plant has not been determined.

DOSE RESPONSE

There are insufficient data on the toxic constituents of the baneberry plant to determine dose-response data.

TOXICOKINETICS

The toxic constituents of the baneberry plant have not been identified.

CLINICAL RESPONSE

A case report from the older medical literature associated the development of gastroenteritis with the ingestion of six baneberries.[3] Gastrointestinal symptoms began within 1/2 hour after ingestion. Confusion, headache, and distorted visual perception developed within 1 hour followed by tachycardia, abdominal pain, lethargy, dry mouth, and indigestion. Symptoms resolved within 3 hours after ingestion. The toxicity of baneberry in not well documented in the modern medical literature.

Medical Toxicology of Natural Substances, by Donald G. Barceloux, MD

DIAGNOSTIC TESTING

Analytical methods for the detection of polyphenol compounds in the roots and rhizomes of baneberry include reversed-phase high performance liquid chromatography with diode array detection.[4]

TREATMENT

Treatment is supportive. Decontamination measures are usually unnecessary.

References

1. Ali Z, Khan SI, Khan IA. New cycloartane-type triterpene arabinosides from the roots of *Actaea podocarpa* and their biological study. Planta Med 2007;73:699–703.
2. Ali Z, Khan SI, Fronczek FR, Khan IA. 9,10-seco-9,19-Cyclolanostane arabinosides from the roots of *Actaea podocarpa*. Phytochemistry 2007;68:373–382.
3. Bacon AE. An experiment with the fruit of red baneberry. Rhodora 1903;5:77–79.
4. Nuntanakorn P, Jiang B, Yang H, Cervantes-Cervantes M, Kronenberg F, Kennelly EJ. Analysis of polyphenolic compounds and radical scavenging activity of four American *Actaea* species. Phytochem Anal 2007;18:219–228.

Chapter 120

BEGONIAS

HISTORY

The begonias are a large group of cultivated ornamental plants with many beautiful hybrids and variations. These plants are indigenous to Mexico, Central and South America, Asia, and South Africa; begonias were introduced into England during the late 18th century as an ornamental plant.

BOTANICAL DESCRIPTION

Common Name: Wax Begonia

Scientific Name: *Begonia semperflorens* Link & Otto

Botanical Family: Begoniaceae (begonia)

Physical Description: Begonias have fibrous or tuberous tubers and simple, alternate, broad leaves with lobed margins. The color and texture of the leaves are highly variable. The stems are thick and the leaves shiny. Small white, pink, or red flowers appear as two small and two large petal-like parts in axillary clusters. There are many stamens. The inconspicuous fruit is a winged capsule.

Distribution and Ecology: This native of tropical parts of Central and South America is widely cultivated throughout temperate zones.

Common Name: Tuberous or Non-stop Begonia

Scientific Name: *Begonia* x. *tuberhybrida*

Botanical Family: Begoniaceae (begonia)

Physical Description: This hybrid appears similar to *B. semperflorens* with flowers displaying a variety of colors (white, pink, salmon, yellow, orange)

Distribution and Ecology: This tuber is cultivated as an ornamental plant in a wide variety of temperate climates.

EXPOSURE

This plant is a common ornamental plant in temperate climates. Cucurbitacin compounds from begonias have been tested in *in vitro* antitumor assays because of the ability of these compounds to depolymerize F-actin in proliferating endothelial cells and thereby disrupt the cytoskeleton.[1] However, the relatively low therapeutic index of these compounds limits the medical use of cucurbitacins.

PRINCIPAL TOXINS

The tuber of the begonia contains calcium oxalate crystals, which are potential mucosal irritants when ingested. The alcoholic extract of the *Begonia* x *tuberhybrida* contains cucurbitacin B along with minor amounts of cucurbitacin D and dihydrocucurbitacin.[2] Rhizomes of the starleaf begonia (*Begonia heracleifolia* Cham. & Schlecht.) from Southern Mexico contain at least six cucurbitacin compounds including cucurbitacin B, cucurbitacin D, 23,24-dihydrocucurbitacin D, 23,24-dihydrocucurbitacin F, 2-*O*-β-glucopyranosylcucurbitacin B, and 2-*O*-β-glucopyranosylcucurbitacin F.[3]

Medical Toxicology of Natural Substances, by Donald G. Barceloux, MD

Cucurbitacin compounds are cytotoxic sesquiterpenoid sterols that inhibit the steroid hormones (ecdysteroids) in insects responsible for the regulation of many developmental, biochemical, and physiological processes.[4,5] These compounds occur primarily in the families Cucurbitaceae (gourd) and Brassicaceae (mustard). Cucurbitacin compounds are very bitter.

DOSE RESPONSE

Most adverse reactions to begonias result from individual hypersensitivity responses rather than direct toxicity with the exception of irritant contact dermatitis.

TOXICOKINETICS

There are few data on the toxicokinetics of cucurbitacin compounds in begonias.

CLINICAL RESPONSE

Begonias are relatively safe houseplants with little serious toxicity following exposure. Most oral exposures to begonia plants result in no symptoms. The tuber contains calcium oxalate crystals that potentially cause mucosal erythema, pain, and edema when ingested. Systemic symptoms do not usually develop after exposure to begonias. Contact with begonia plants is a rare cause of allergic dermatitis, and most contact dermatitis from exposure to begonia plants results from an irritant contact dermatitis.[6] Rarely, exposures to begonias are a source of occupational dermatitis.[7]

DIAGNOSTIC TESTING

Liquid chromatographic/mass spectrometric methods are available for the quantitation of the chemotherapeutic drug, cucurbitacin I, in plasma with a limit of quantitation of about 10 ng/mL.[8]

TREATMENT

Treatment is supportive. Most exposures to this plant cause no symptoms. Irritation of the mucosa can be treated with topical anesthetics [Chloraseptic® (Proctor & Gamble Pharmaceuticals, Cincinnati, OH), Cepacol lozenges® (Bayer Consumer Care, Morristown, NJ), Orabase® (Colgate-Palmolive Company, New York, NY)]. Decontamination measures are unnecessary.

References

1. Duncan KL, Duncan MD, Alley MC, Sausville EA. Cucurbitacin E-induced disruption of the actin and vimentin cytoskeleton in prostate carcinoma cells. Biochem Pharmacol 1996;52:1553–1560.
2. Doskotch RW, Malik MY, Beal JL. Cucurbitacin B, the cytotoxic principle of *Begonia tuberhybrida* var. alba. Lloydia 1969;32:115–122.
3. Frei B, Heinrich M, Herrmann D, Orjala JE, Schmitt J, Sticher O. Phytochemical and biological investigation of *Begonia heracleifolia*. Planta Med 1998;64:385–386.
4. Chen JC, Chiu MH, Nie RL, Cordell GA, Qiu SX. Cucurbitacins and cucurbitane glycosides: structures and biological activities. Nat Prod Rep 2005;22:386–399.
5. Dinan L, Whiting P, Girault J-P, Lafont R, Dhadialla TS, Cress DE, et al. Cucurbitacins are insect steroids hormone antagonists acting at the ecdysteroid receptor. Biochem J 1997;327:643–650.
6. Paulsen E. Occupational dermatitis in Danish gardeners and greenhouse workers (II). Etiological factors. Contact Dermatitis 1998;38:14–19.
7. Schubert H, Prater E. [Pollen allergy as an occupational disease in gardeners]. Dermatol Monatsschr 1990;176:97–104. [German]
8. Molavi O, Shayeganpour A, Somayaji V, Hamdy S, Brocks DR, Lavasanifar A, et al. Development of a sensitive and specific liquid chromatography/mass spectrometry method for the quantification of cucurbitacin I (JSI-124) in rat plasma. J Pharm Pharmaceut Sci 2006;9:158–164.

Chapter 121

YELLOW BIRD-OF-PARADISE [*Caesalpinia gilliesii* (Hook.) Wallich ex D. Dietr.]

BOTANICAL DESCRIPTION

Worldwide, there are over 150 species in the genus *Caesalpinia*. *Caesalpinia pulcherrima* L. Sw. (pride-of-Barbados, dwarf poinciana) is a smaller *Caesalpinia* species. *Caesalpinia* is primarily a tropical genus found in lowland rain forests or open, dry scrub areas.

Common Name: Bird-of-Paradise, Paradise Poinciana, Yellow Bird-of-Paradise, Desert Bird-of-Paradise

Scientific Name: *Caesalpinia gilliesii* (Hook.) Wallich ex D. Dietr.

Botanical Family: Fabaceae (pea)

Physical Description: This large, irregular shrub reaches 5–10 ft (~2–3 m) in height and spreads to 4–6 ft (~1– 2 m). The sparsely branched, alternate leaves are pinnate with numerous small leaflets resembling a fern. The long, showy, glandular flowers have bright red stamens that produce a tan, fuzzy pod.

Distribution and Ecology: *C. gilliesii* is a native of South America introduced into the warmer areas of the southern portions of the United States as an ornamental plant.

EXPOSURE

C. gilliesii and *C. pulcherrima* are cultivated as ornamental plants for their showy yellow, orange or red blossoms with brilliant red stamens. Aqueous extracts of *C. gilliesii* have been investigated for antitumor activity as part of routine screening programs,[1] but there are no current clinical applications for extracts of this plant. In India and the Philippines, *C. pulcherrima* is a common ornamental shrub. This plant has a variety of medicinal uses in India including anthelminthic, antispasmodic, antipyretic, hemostatic agent, abortifacient, and treatment for liver disease and mental disorders.

PRINCIPAL TOXINS

The main toxic constituents in *Caesalpinia* species are probably tannins. Analyses of *C. pulcherrima* indicate that this plant contains a variety of flavanoid and diterpene compounds including pulcherralpin, bonducellin, isobonducellin, caesaldekarin A, spathulenol, caryophyllene oxide, phytol, and sitosterol.[2–4] The pulcherralpin, tannins, and irritant diterpenoids in this plant cause dermatitis and gastrointestinal (GI) irritation.

DOSE RESPONSE

There are insufficient data on the toxic constituents of *Caesalpinia* species to determine a dose response between parts of this plant and clinical effects.

TOXICOKINETICS

There are few toxicokinetic data on the tannins, flavanoid, or diterpenoid compounds in these plants.

Medical Toxicology of Natural Substances, by Donald G. Barceloux, MD

CLINICAL RESPONSE

There limited human data on the ingestion of seeds or parts of *Caesalpinia* species. Based on the presence of tannins, the primary adverse effects of exposure to these plants are probably GI irritation including nausea, vomiting, and diarrhea. Severe systemic symptoms are unusual, and the persistence of symptoms beyond 24 hours suggests another etiology.

DIAGNOSTIC TESTING

Biomarkers of GI toxicity associated with the ingestion of parts of these plants have not been identified.

TREATMENT

Treatment is supportive with efforts directed toward the assessment and correction of fluid and electrolyte balance. Gastrointestinal decontamination is unnecessary because of the irritant effects of this plant on the GI tract.

References

1. Ulubelen A, McCaughey WF, Cole JR. Proteinaceous antitumor substances from plants III. *Caesalpinia gilliesii* (Leguminosae). J Pharm Sci 1967;56:914–916.
2. Srinivas KV, Koteswara Rao Y, Mahender I, Das B, Rama Krishna KV, Hara Kishore K, Murty US. Flavanoids from *Caesalpinia pulcherrima*. Phytochemistry 2003;63:789–793.
3. Ragasa CY, Ganzon J, Hofilena J, Tamboong B, Rideout JA. A new furanoid diterpene from *Caesalpinia pulcherrima*. Chem Pharm Bull (Tokyo) 2003;51:1208–1210.
4. Roach JS, McLean S, Reynolds WF, Tinto WF. Cassane diterpenoids of *Caesalpinia pulcherrima*. J Nat Prod 2003;66:1378–1381.

Chapter 122

CENTURY PLANT
(*Agave americana* L.)

HISTORY

The fermented beverage, pulque, was depicted in Native American stone carvings from 200 AD.[1] This native plant of Mexico and the southwestern United States was introduced into Europe during the 16th century as an ornamental plant, and now this plant occurs throughout the Mediterranean region. *Agave* species were an important food source and commodity for both prehistoric and historic Native American peoples across the Southwest and Mexico. *Agave tequilana* A. Weber (i.e., an *Agave* species other than the century plant) is an important distillation stock for tequila liquors.

BOTANICAL DESCRIPTION

The Agavaceae family contains about 20 genera and 300 species with about 200 species inhabiting Mexico.

Common Name: Century Plant, American Agave, American Century Plant, Maguey Americano

Scientific Name: *Agave americana* L.

Botanical Family: Agavaceae

Physical Description: This large plant has a low basal rosette composed of thick succulent leaves up to 6 ft (~2 m) in length that contain hooked spines along the margins. The flowering stalk appears after 10 or more years, and this stalk may reach 15–20 ft (~5–6 m) in height. Some varieties have yellow or white stripes. The plant dies shortly after flowering. The plant is nearly stemless.

Distribution and Ecology: *Agave americana* is a native of Mexico that has been naturalized into open, arid areas in a variety of soils of the southwestern United States and California.

EXPOSURE

This plant is a popular ornamental in warm areas. The stems of the century plant are also consumed by native peoples after cooking. The stems are then chewed for the sweet juice before expectorating the residue. *Agave* species are sources of popular food (quiote), fermented beverages (pulque), and distilled products (tequila). Tequila is produced by the distillation of fermented juice from the heads (i.e., leafless stem) of *Agave tequilana* A. Weber and up to 49% additional sugars (sugar cane, molasses, hydrolyzed maize syrup). Tequilas differ as a result of variation in the proportion of agave material (i.e., 51–100%), strain of fermenting bacteria, production processes, distillation equipment, aging times, and growing conditions. The ethanol content ranges from about 40% to 55%. The production process involves cooking to hydrolyze inulin to fructose in stone ovens or autoclaves, extraction of sugars from the cooked agave by milling, bacterial fermenting to convert the sugars to ethanol and organoleptic compounds, and a two-step distillation of the fermented wort using pot stills or rectification columns.[2] The head of an 8- to 9-year-old *Agave tequilana* contains high concentrations of the linear fructose polymer, inulin, compared with other *Agave* species. Mezcal is a rudimentary

Medical Toxicology of Natural Substances, by Donald G. Barceloux, MD

fermentation and distillation product of *Agave potatorum* Zucc. or *Agave angustifolia* Haw. from the Mexican state of Oaxaca. During the production process some brands of mezcal add a worm to the bottle, primarily as a commercial ploy.

Pulque is a viscous fermentation product of several *Agave* species (*A. atrovirens* Karw. ex Salm-Dyck, *A. salmiana* Salm-Dyck) that results from the actions of a complex series of yeasts and bacteria. Because the product is not distilled or pasteurized, pulque deteriorates relatively quickly. The ethanol content of this fermentation product ranges from approximately 4–6%.[3] Fibers from *Agave fourcroydes* Lemaire and *Agave sisalana* Perrine are used to form rope and carpets. Folk uses for *Agave* species include diuresis, purgative, treatment of rheumatism, and birth control.[1]

PRINCIPAL TOXINS

The agave plant contains calcium oxalate crystals that produce local irritation upon contact with mucosal or dermal surfaces along with oxalic acid, saponins, and acrid volatile oil. *Agave* species including *A. americana* contain hecogenin and tigogenin, which previously was used as a natural precursor for the synthesis of corticosteroids.[4] Other compounds include steroidal saponins (agamenosides) and cholestane steroids (agavegenin).[5] The sap contains saponins and needle-like raphides composed of calcium oxalate crystals that probably contribute to the development of the irritant contact dermatitis (agave worker's disease).[6] Histological examination of skin lesions demonstrate spongiotic dermatitis with superficial and deep perivascular lymphohistiocytic infiltrates with occasional eosinophils.[7]

CLINICAL RESPONSE

Dermal contact with the sap of the century plant produces an acute irritant contact dermatitis.[8] Case reports indicate that the direct application of the sap from the century plant produces the immediate onset of a burning sensation followed by the development of a confluent vesiculopapular rash and an exudative dermatitis of the area in contact with the sap.[9] Rarely, the vesiculopapular rash progresses to purpura suggestive of a leukocytoclastic or purpuric vasculitis.[10,11] Biopsies of the affected skin demonstrate superficial perivascular infiltrates, superficial crusts, spongiosis, extravasated erythrocytes, and mild suppurative folliculitis.[12]

Systemic symptoms following contact with the sap are unusual, but mild fever and nonspecific constitutional symptoms can occur.[7,13] Direct contact with the oral mucosa causes inflammatory lesions. Symptoms following the chewing of unprocessed leaves from the century plant include intraoral erythema and edema. Mechanical trauma following contact with agave spines usually heals without infections, although rarely infections develop involving common plant organisms (e.g., *Enterobacter agglomerans*). Because of the high fiber content of *Agave* species, phytobezoars may develop following the chronic ingestion of material from *Agave* species.[14] Typically, the cooked stems are masticated to extract the sweet juice and the remainder is expectorated.

DIAGNOSTIC TESTING

Agave spines appear as elongated low-signal-intensity lesions during magnetic resonance imaging (MRI).[15]

TREATMENT

Treatment is supportive. The irritant contact dermatitis usually responds to topical steroids (0.05% fluocinonide) and oral antibiotics are not usually necessary. Gastrointestinal decontamination measures are unnecessary.

References

1. Hackman DA, Giese N, Markowitz JS, McLean A, Ottariano SG, Tonelli C, et al. Agave (*Agave americana*): an evidence-based systematic review by the Natural Standard Research Collaboration. J Herbal Pharmacother 2006;6:101–122.
2. Cedeno M. Tequila production. Crit Rev Biotechnol 1995;15:1–11.
3. Kuhnlein HV. Karat, pulque, and gac: three shining stars in the traditional food galaxy. Nutr Rev 2004;62: 439–442.
4. Applezweig N. Steroids. Chem Week 1969;104:57–72.
5. Jin J-M, Zhang Y-J, Yang C-R. Four new steroid constituents from the waste residues of fibre separation from *Agave americana* leaves. Chem Pharm Bull 2004;52: 654–658.
6. Salinas ML, Ogura T, Soffchi L. Irritant contact dermatitis caused by needle-like calcium oxalate crystals, raphides, in *Agave tequilana* among workers in tequila distilleries and agave plantations. Contact Dermatitis 2001;44:94–96.
7. Brenner S, Landau M, Goldberg I. Contact dermatitis with systemic symptoms from *Agave americana*. Dermatol 1998;196:408–411.
8. Brazzelli V, Romano E, Balduzzi A, Borroni G. Acute irritant contact dermatitis from *Agave americana* L. Contact Dermatitis 1995;33:60–61.
9. Kerner J, Mitchell J, Maibach HI. Irritant contact dermatitis from *Agave americana* L. Arch Dermatol 1973;108: 102–103.

10. Genillier-Foin N, Avenel-Audran M. [Purpuric contact dermatitis from Agave Americana]. Ann Dermatol Venereol 2007;134:477–478. [French]
11. Cherpelis BS, Fenske NA. Purpuric irritant contact dermatitis induced by *Agave americana*. Cutis 2000;66: 287–288.
12. High WA. Agave contact dermatitis. Am J Contact Dermatitis 2003;14:213–214.
13. Ricks MR, Vogel PS, Elston DM, Hivnor C. Purpuric agave dermatitis. J Am Acad Dermatol 1999;40:356–358.
14. Villarreal R, Martinez O, Berumen U Jr. Phytobezoar from the stem ("quiote") of the cactus *Agave americana*: report of case. Am J Gastroenterol 1985;80:838–840.
15. Borup LH, Meehan JJ, Severson JM, Kaufman K. Terminal spine of *Agave* plant extracted from patient's spinal cord. AJR 2003;181:1155–1156.

Chapter 123

CLEISTANTHIN, DITERPENE ESTERS, and the SPURGE FAMILY (Euphorbiaceae)

HISTORY

Many species in the spurge family (Euphorbiaceae) contain a milky sap that is mildly toxic and irritating to the skin. The name derives from Euphorbus, the physician to King Juba II of Mauritania (50 BC–23 AD), who discovered the therapeutic properties of Euphorbiaceae plants grown in nearby mountains. Passages from the Greek herbalist, Dioscorides, and by the Roman writer, Pliny the Elder, suggest that the sap from *Euphorbia* species has long been used to treat skin diseases.[1] During the Medieval Ages, the acrid yellow-brown resin from several *Euphorbia* species (euphorbium) was an irritant remedy brought to Europe by Arab traders from Morocco. *Euphorbium* was used as a vesicant plaster and a sneezing powder during the Renaissance. Some species of the spurge family contain potent systemic toxins including *Ricinus communis* L. (ricin) and *Cleistanthus collinus* (Roxb.) Hook. f. (cleistanthin A and B, diphyllin, collinusin). The extracts of these two species are ancient homicidal agents in India. The poinsettia is a member of the *Euphorbia* genus that was introduced into the United States in the 1800s by the American Ambassador to Mexico, Joel R. Poinsett, and this plant has been a popular ornamental since the early 1900s.

BOTANICAL DESCRIPTION

Euphorbiaceae is a highly diversified family with substantial differences in morphologically, chemical, and ecological characteristics. This family is widely distributed throughout both hemispheres with approximately 8,000 species ranging from large desert succulents to small herbaceous plants and trees in grasslands and forests. The presence of latex (milky sap) is characteristic of plants from the spurge family. This milky sap is mildly irritating to the skin. In general, the leaves are small and the flowers are inconspicuous. African milkbush (*Synadenium grantii* Hook. f.) plants are closely related to *Euphorbia* species.

Common Name: Poinsettia

Scientific Name: *Euphorbia pulcherrima* Willd. ex Klotzsch

Botanical Family: Euphorbiaceae (spurge)

Physical Description: This popular indoor Christmas ornamental plant has large (3–6in./~8–15cm) alternating leaves and prominent red, pink, yellow, or cream bracts clustered at the tops of the stems.

Distribution and Ecology: Poinsettia is native to the tropical areas of Central America and Mexico, but now this plant is cultivated worldwide in temperate climates or indoors.

Common Name: Candelabra Cactus, Mottled Spurge

Scientific Name: *Euphorbia lactea* Haw.

Botanical Family: Euphorbiaceae (spurge)

Physical Description: This plant is a multibranched succulent with serrated edges.

Distribution and Ecology: This ornamental plant from tropical areas is often grown as a houseplant.

Medical Toxicology of Natural Substances, by Donald G. Barceloux, MD

Common Name: Pencil Tree, Milk Bush, Malabar Tree, Indiantree Spurge

Scientific Name: *Euphorbia tirucalli* L.

Botanical Family: Euphorbiaceae (spurge)

Physical Description: This spineless, succulent shrub has green branching stems and a few, small, linear leaves confined to the branch tips. The finely striated branches form dense, tangled masses.

Distribution and Ecology: This ornamental plant is found in tropical habitats and is often grown as a houseplant.

Common Name: Crown of Thorns

Scientific Name: *Euphorbia milii* Des Moulins

Botanical Family: Euphorbiaceae (spurge)

Physical Description: This spiny, woody plant grows as a shrub up to 4ft (~1m) high with attractive small floral bracts one centimeter across.

Distribution and Ecology: This ornamental plant is found in tropical areas, and is often grown as a houseplant.

Common Name: African Milkbush

Scientific Name: *Synadenium grantii* Hook. f.

Botanical Family: Euphorbiaceae (spurge)

Physical Description: This succulent shrub has obovate, red-tinged leaves, short stalks, and dark red clusters of flowers. The shrub may reach 12ft (~4m) in height.

Distribution and Ecology: The African milkbush is a succulent shrub that is indigenous to East Africa, Madagascar, and the Mascarene Islands.

Common Name: Oduvan, Garari, Karada, Pasu, Karlajuri

Scientific Name: *Cleistanthus collinus* (Roxb.) Hook. f.

Botanical Family: Euphorbiaceae (spurge)

Physical Description: This bushy, thorny shrub has dark brown bark and long spines.

Distribution and Ecology: *Cleistanthus collinus* is a native plant of India, Malaysia, and Africa.

Common Name: Dog's Mercury

Scientific Name: *Mercurialis perennis* L.

Botanical Family: Euphorbiaceae (spurge)

Physical Description: This plant has serrated, opposite leaves and clusters of flowers on a central stalk.

Distribution and Ecology: Dog's mercury is a woodland plant indigenous to Britain.

Common Name: Annual Mercury

Scientific Name: *Mercurialis annua* L.

Botanical Family: Euphorbiaceae (spurge)

Physical Description: This slender, erect annual reaches up to 60cm (~2ft) in height. The leaves are ovate to oblong leaves within serrated margins. Flowers appear in the spring. The male flowers are long, terminal inflorescences, whereas female flowers occur in the upper leaves.

Distribution and Ecology: Annual mercury is a common weed found in uncultivated areas of California, northeastern United States, Alabama, and South Carolina. This annual herb was introduced into the United States.

EXPOSURE

Many species of the spurge family are houseplants and ornamental plants in temperate climates including *Synadenium grantii, Euphorbia lactea, and Euphorbia tirucalli.* In India, the extract of crushed oduvan leaves (*Cleistanthus collinus*) has been used as an abortifacient, animal poison, suicidal agent, and homicidal agent. Plants from the Euphorbiaceae family have been used for many centuries in Indian and Brazilian traditional medicine for the treatment of tumors, metastatic cancer, and a depilatory for the removal of skin blemishes and warts.[2] Resiniferatoxin is a potent capsaicin analog isolated from the sap of *Euphorbia resinifera* O. Berg that interacts with vanilloid receptors and reduces neurogenic inflammation. Recently, the administration of resiniferatoxin has been investigated as an alternative therapy for neurogenic and nonneurogenic forms of detrusor overactivity in the bladder.[3]

PRINCIPAL TOXINS

Physiochemical Properties

Toxins from the Euphorbiaceae family belong to a group of diterpene hydrocarbons that include tigliane, daphnane, and ingenane groups that demonstrate irritant and tumor-promoting activities.[4] Oxygenated derivatives of macrocyclic diterpenes with effects similar to toxins in these groups include the following diterpene types: casbane, jatrophane, lathyrane, jatropholane, and crotofolane.[5] Potential irritants in the sap of plants from the spurge family include a variety of diterpene esters (e.g., phorbols, ingenols, daphnanes). The poinsettia plant (*E. pulcherrima*) contains small amounts of triterpenes compounds (e.g., 9,19-cycloart-25-ene-3β, 24-diol; 9,19-cycloart-23-ene-3β, 25-diol) that possess cytotoxic properties as demonstrated in tumor cell cultures.[6] The concentration of skin and gastrointestinal irritants in the milky sap varies between *Euphorbia* species.[7] The latex of *Euphorbia tirucalli* originating from Madagascar contains irritant constituents (e.g., ingenane- and tigliane-type diterpene esters) derived from the parent alcohols, ingenol and phorbol.[8] The main irritant constituents are isomeric 12,13-acetates, acylates of phorbol and 3-acylates of ingenol, whereas other *Euphorbia* species contain higher concentrations of 4-deoxyphorbol esters in their latex.[9] The latex of *Synadenium grantii* contains an unusual irritant ester of 4-deoxyphorbol, 12-*O*-tigloyl-4-deoxyphorbol-13-isobutyrate.[10] The active principle in Dog's mercury is thought to be the volatile basic oil mercurialine. Latex from several species in the spurge family including the sandbox tree (*Hura Crepitans* L.)[11] and the Mediterranean spurge (*Euphorbia characias* L.)[12] contains ribosome-inactivating proteins (ribotoxins), but the role of these chemicals in causing toxicity is unclear. These lectins share a similar mechanism of action to ricin, but the lectins from the *Euphorbia* species are less potent. *In vitro* studies suggest that some members of the spurge family contain variable amounts of tumor-promoting (nongenotoxic) phorbol esters, but the clinical relevance of these substances is unclear.[13,14]

The major toxic compounds in *Cleistanthus collinus* are the lignan lactone glycosides, cleistanthin A and B. The genin, diphyllin, is the major metabolite of these lignan lactone glycosides. The leaves of *C. collinus* also contain another lignan lactone glycoside (collinusin). Cleistanthin A (CAS RN: 25047-48-7) is a $C_{28}H_{28}O_{11}$ compound that is also known as sheng bai xin in traditional Chinese medicine. Cleistanthin B (CAS RN: 30021-77-3) is a $C_{27}H_{26}O_{12}$ compound that is structurally similar to Cleistanthin A. Figure 123.1 demonstrates the chemical structure of the major lignan lactone glycosides (Cleistanthin A and B), the major metabolite (diphyllin), and another lignan lactone glycoside (collinusin) present in the leaves of *C. collinus*.

Poisonous Parts

Unicellular or multicellular tubes (laticifers) in the stems, leaves, and fruits of species from the spurge family produce the milky latex seen following damage to these plant parts. Within the sap are esters of polycyclic diterpene alcohols, but the identity of the exact irritant compound remains unclear. Ingenol and ingenol esters are common compounds in members of the spurge family, and ability of these diterpenes to activate protein kinase C may contribute to the toxicity of these plants.[15] Although most parts of the oduvan plant (*Cleistanthus collinus*) contain toxic arylnaphthalene lignan lactones (cleistanthin A and B, collinusin), the most serious poisonings occur following ingestion of the extract of the leaves.[16]

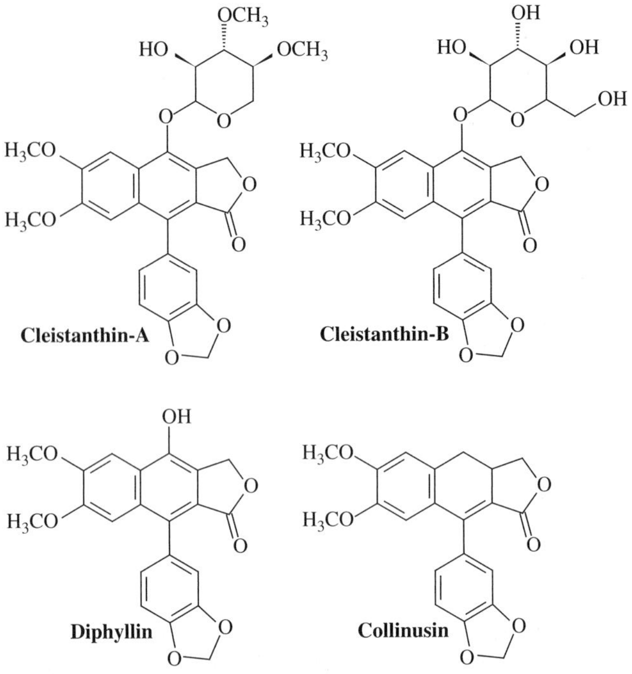

FIGURE 123.1. Chemical structures of *Cleistanthus collinus* toxins.

Mechanism of Toxicity

The primary mechanism for the development of vesiculopapular lesions following contact with the sap from the spurge family commonly involves an irritant contact dermatitis rather than an allergic response.[17] Mechanical irritation may occur following contact with stiff hairs, but the clinical consequences of this exposure is usually minimal.

The toxicity of *Cleistanthus collinus* results from the cytotoxic effects of lactones on a number of cellular, tissue, and organ systems. In experimental studies, cleistanthin A inhibits DNA synthesis, induces DNA damage, and causes apoptosis.[18] These genotoxic properties have been evaluated as potential chemotherapeutic agents.[19] The severe hypokalemia results from the renal tubular loss of potassium rather than intracellular translocation of the potassium.[20]

DOSE RESPONSE

Parts of the poinsettia plant are relatively nontoxic based on animal studies and human case series. The feeding of fresh leaves and bracts of poinsettia plants in doses up to 125 mg/kg over 5 days did not produce any alterations of behavior, pathological changes in the gastrointestinal tract, or decrease in weight when compared with controls.[21,22] Repeated exposure of albino rats to poinsettia plant materials resulted in mild skin irritation and skin photosensitivity, but there was minimal mucosal irritation.[23]

CLINICAL RESPONSE

Local

Exposure to the sap from the *Euphorbia* plants (e.g., poinsettia) produce primarily an irritant contact dermatitis manifest by erythema, edema, and urticaria.[24,25] The ability of species from the spurge family to cause erythematous papulovesicular skin lesions varies with individual species. Other spurge species capable of causing an irritant contact dermatitis following exposure to the sap include the croton [*Codiaeum variegatum* (L.) Blume],[26] African milk bush (*Synadenium grantii*),[10,27] *Euphorbia milii* (crown-of-thorns), *E. trigona* Mill. (African Milk Tree),[28] *E. lactea* (candelabra cactus), *E. tirucalli* (pencil tree), *E. marginata* Pursh (snow on the mountain),[29] and *Euphorbia myrsinites* L. (cushion spurge).[30,31] During the chronic phase of healing, hyperpigmentation can develop in the areas of vesiculobullous eruptions.[32] Rarely, mucosal ulceration may occur after prolonged contact with the sap of *Euphorbia* species including the poinsettia plant.[33] Contact of the eye with sap from *E. lactea* (candelabra cactus),[34] *E. cyparissias* L. (cypress spurge), *E. helioscopia* L. (madwoman's milk), *E. lathyris* L. (moleplant), *E. peplus* L. (petty spurge), or *E. tirucalli* (pencil tree) can produce mild to moderately severe keratoconjunctivitis and uveitis.[35,36] Symptoms of irritation include immediate conjunctival burning, tearing, and photophobia. Chemosis of the conjunctiva may develop within 8–12 hours after exposure along with increased pain and blurred vision as epithelial sloughing occurs. The healing of corneal ulcerations may require up to 9 days.[37] Rarely, uveitis occurs that usually responds to steroid therapy without the development of permanent corneal damage.[38]

Systemic

Most exposures to plants from the spurge family do not produce systemic symptoms other than mild gastrointestinal irritation. Self-limited, spontaneous vomiting may occur following ingestion of the poinsettia plant, but systemic symptoms of toxicity do not usually occur[39] contrary to the older medical literature.[40,41] Exposure to plant parts from a few species of the Euphorbiaceae family causes systemic toxicity. In addition to the castor bean plant (*Ricinus communis* L.), ingestion of extracts of *Cleistanthus collinus* produces serious cardiotoxicity.[42] Clinical manifestations of severe intoxication from *C. collinus* include gastrointestinal distress, neuromuscular weakness with respiratory failure, renal dysfunction, metabolic acidosis, severe hypokalemia, serious ventricular dysrhythmias, and sudden death.[16] The onset of clinically significant organ dysfunction usually occurs several days after ingestion, and serious cardiac abnormalities are usually associated with severe hypokalemia (i.e., <2 meq/L).[43]

Ingestion of boiled stems and leaves from *Mercurialis perennis* (Dog's mercury) produced severe gastroenteritis, bilateral flank pain, anisocytosis, and trace hemoglobinuria without anemia or hepatorenal dysfunction.[44] In grazing animals, (e.g., cattle) excessive consumption of *M. perennis* causes methemoglobinemia and hemolytic anemia,[45] but these effects have not been reported in humans. Allergic responses may occur following exposure to pollens from *Mercurialis* species, including the development of respiratory allergies.[46] A case series of 18 Chinese patients associated the ingestion of a soup containing parts of *Breynia officinalis* Hemsley (Chi R Yun) from the Euphorbiaceae family with gastroenteritis (nausea, vomiting, diarrhea), mild hyperbilirubinemia, and elevated serum hepatic aminotransferases.[47] Progression to marked jaundice, ascites, and hepatic failure did not develop, and these patients recovered within 6 months. The latex from some Euphorbiaceae species are strongly sensitizing including the castor bean (*Ricinus communis* L.), the annual mercury (*Mercurialis annua*), and the rubber tree [*Hevea brasiliensis* (Willd. ex A.Juss.) Müll.Arg.]. There is substantial cross-reactivity with these allergenic species.[48]

DIAGNOSTIC TESTING

In experimental studies, an enzyme-linked immunosorbent assay (ELISA) has been developed in rabbits for the detection of the toxic lignan glycosides, cleistanthin A and cleistanthin B, present in serum after the ingestion of plant parts from *Cleistanthus collinus*.[49,50] This assay cross-reacts with other phytotoxins (e.g., diphyllin, collinusin) in this plant, but not with other plant toxins (e.g., digitoxin, digoxin, gitoxin, neriifolin, oleandrigenin, oleandrin, thevetin). The limit of detection is approximately 2 ng/mL.

Electrocardiographic abnormalities during severe intoxication by *C. collinus* reflect severe hypokalemia, including QT_c prolongation, ST wave depression, inverted T waves, and malignant cardiac rhythms (pulseless electrical activity, ventricular fibrillation, asystole). AV conduction block does not usually occur during *Cleistanthus collinus* leaf poisoning.[20] Case reports of *C. collinus* intoxication document clinically significant laboratory abnormalities in addition to severe hypokalemia including coagulopathy, hyperchloremic anion gap metabolic acidosis, and alkaline urinary pH.[16]

TREATMENT

Treatment of local irritation secondary to the sap from species of the spurge family is supportive (e.g., removal of all plant material as soon as possible), particularly from the mouth. Skin irrigation and cleansing reduces the risk of minor irritation associated with exposure to irritant sap from species in Euphorbiaceae family. The use of demulcents (milk, antacids, cool water) may reduce the effects of the sap on the oral mucosa. Although healing may be delayed up to approximately 1 week, corneal erosions following contact with the sap of Euphorbiaceae species usually respond to topical antibiotics and pressure patching.[37]

Exposure to *Cleistanthus collinus* extract can produce severe cardiotoxicity and hypokalemia. These patients require intravenous access and cardiorespiratory monitoring for the development of malignant ventricular dysrhythmias, respiratory failure, and sudden cardiac arrest. Additionally, these patients should be monitored frequently for the development of fluid and electrolyte imbalance including hypokalemia. Decontamination measures are usually unnecessary because of the typical delay between ingestion and the onset of symptoms. The administration of activated charcoal to alert patients, who present to the emergency department less than 1 hour prior after ingestion, is a therapeutic option, but there are no clinical data to confirm the clinical efficacy of activated charcoal on outcome.

Substantial amounts of potassium may be excreted during intoxication, and these patients often require large doses of intravenous potassium as well as continuous monitoring of serum potassium and creatinine concentrations. Potassium replacement may be complicated by the development of renal failure. There are no clinical data on antidotes for *C. collinus* poisoning, although changes in glutathione stores during intoxication suggest a potential role for *N*-acetylcysteine.[51]

References

1. Appendino G, Szallasi A. Euphorbium: modern research on its active principle, resiniferatoxin, revives an ancient medicine. Life Sci 1997;60:681–696.
2. Hartwell JL. Plants used against cancer. A survey. Lloydia 1969;32:153–205.
3. el-Mahrouky AS, Elashry OM, Emran MA. The effect of intravesical capsaicin and resiniferatoxin in neurogenic bladder dysfunction. Adv Exp Med Biol 2003;539(Pt A):359–379.
4. Evans FJ, Soper CJ. The tigliane, daphnane and ingenane diterpenes, their chemistry, distribution and biological activities. A review. Lloydia 1978;41:193–233.
5. Cateni F, Falsone G, Zilic J. Terpenoids and glycolipids from Euphorbiaceae. Mini Rev Med Chem 2003;3: 425–437.
6. Smith-Kielland I, Bornish JM, Malterud KE, Hvistendahl G, Romming C, Bockman OC, et al. Cytotoxic triterpenoids from the leaves of *Euphorbia pulcherrima*. Planta Med 1996;62:322–325.
7. Marston A, Hecker E. Active principles of the Euphorbiaceae. VII. Milliamines H and I, peptide esters of 20-deoxy-5 epsilon-hydroxyphorbol from *Euphorbia milii*. Planta Med 1984;50:319–322.
8. Furstenberger G, Hecker E. On the active principles of the Euphorbiaceae, XII. Highly unsaturated irritant diterpene esters from *Euphorbia tirucalli* originating from Madagascar. J Nat Prod 1986;49:386–397.
9. Kinghorn AD. Characterization of an irritant 4-deoxyphorbol diester from *Euphorbia tirucalli*. J Nat Prod 1986;42:112–115.
10. Kinghorn AD. Major skin-irritant principle from *Synadenium grantii*. J Pharm Sci 1980;69:1446–1447.
11. Stirpe F, Gasperi-Campani A, Barbieri L, Falasca A, Abbondanza A, Stevens WA. Ribosome-inactivating proteins from the seeds of *Saponaria officinalis* L. (soapwort), of *Agrostemma githago* L. (corn cockle) and of *Asparagus officinalis* L. (asparagus), and from the latex of *Hura crepitans* L. (sandbox tree). Biochem J 1983;216:617–625.
12. Barbieri L, Falasca A, Franceschi C, Licastro F, Rossi CA, Stirpe F. Purification and properties of two lectins from the latex of the euphorbiaceous plants *Hura crepitans* L. (sand-box tree) and *Euphorbia characias* L. (Mediterranean spurge). Biochem J 1983;215:433–439.
13. Vogg G,Mattes E, Polack A, Sandermann H Jr. Tumor promoters in commercial indoor-plant cultivars of the Euphorbiaceae. Environ Health Perspect 1999;107: 753–756.
14. Cruz CM, Kasper P, Cataldo A, Zamith HP, Paumgartten FJ. Tumor promoter-like activity of the molluscicidal latex of "Crown-of-Thorns" (*Euphorbia milii* var. *hislopii*) in the V79 metabolic cooperation assay. Brazil J Med Biol Res 1996;29:1519–1523.
15. Halaweish FT, Kronberg S, Hubert MB, Rice JA. Toxic and aversive diterpenes of *Euphorbia esula*. J Chem Ecol 2002;28:1599–1611.
16. Eswarappa S, Chakraborty AR, Palatty BU, Vasnaik M. *Cleistanthus collinus* poisoning: case reports and review of the literature. J Toxicol Clin Toxicol 2003;41: 369–372.
17. Webster GL. Irritant plants in the spurge family (Euphorbiaceae). Clin Dermatol 1986;4:36–45.
18. Pradheepkumar CP, Panneerselvam N, Shanmugam G. Cleistanthin A causes DNA strand breaks and induces

apoptosis in cultured cells. Mutat Res 2000;464:185–193.
19. Prabhakaran C, Kumar P, Panneerselvam N, Rajesh S, Shanmugam G. Cytotoxic and genotoxic effects of cleistanthin B in normal and tumour cells. Mutagenesis 1996;11:553–557.
20. Kurien T, Dayal AK, Ganesh A, Sheshadri MS, Cherian AM, Bhanu M. Metabolic and cardiac effects of *Cleistanthus collinus* poisoning. J Assoc Physicians India 1991;39:312–314.
21. Runyon R. Toxicity of fresh poinsettia (*Euphorbia pulcherrima*) to Sprague-Dawley rats. Clin Toxicol 1980;16:167–173.
22. Stone RP, Collins WJ. *Euphorbia pulcherrima*: Toxicity in rats. Toxicol 1971;9:301–302.
23. Winek CL, Butala J, Shanor SP, Fochtman FW. Toxicology of poinsettia. Clin Toxicol 1978;13:27–45.
24. Paulsen E, Skov PS, Andersen KE. Immediate skin and mucosal symptoms from pot plants and vegetables in gardeners and greenhouse workers. Contact Dermatitis 1998;39:166–170.
25. Santucci B, Picardo M, Cristaudo A. Contact dermatitis from *Euphorbia pulcherrima*. Contact Dermatitis 1985;12:285–286.
26. Hausen BM, Schulz KH. Occupational contact dermatitis due to Croton (*Codiaeum variegatum* (L.) A. Juss var pictum (Lodd.) Muell. Arg.): Sensitization by plants of the Euphorbiaceae. Contact Dermatitis 1977;3:289–292.
27. Spoerke DG, Montanio CD, Rumack BH. Pediatric exposure to the houseplant *Synadenium grantii*. Vet Hum Toxicol 1985;28:283–284.
28. Worobec SM, Hickey TA, Kinghorn AD, Soejarto DD, West D. Irritant contact dermatitis from an ornamental *Euphorbia*. Contact Dermatitis 1981;7:19–22.
29. Urushibaata O, Kase K. Irritant contact dermatitis from *Euphorbia marginata*. Contact Dermatitis 1991;24:155–156.
30. Cammu G, de Boulle K, De Cleene M, De Doncker R, van Landeghem L, Demeyer I. Toxic effects of cushion spurge after exogenous contact. Eur J Emerg Med 2000;7:155–157.
31. Spoerke DG, Temple AR. Dermatitis after exposure to a garden plant (*Euphorbia myrsinites*). Am J Dis Child 1979;133:28–29.
32. Asilian A, Faghihi G. Severe irritant contact dermatitis from Cypress spurge. Contact Dermatitis 2004;51:37–39.
33. Edwards N. Local toxicity from a poinsettia plant: A case report. J Pediatr 1983;102:404–405.
34. Merani R, Sa-Ngiampornpanit T, Kerdraon Y, Billson F, McClellan KA. *Euphorbia lactea* sap keratouveitis: case report and review of the literature. Cornea 2007;26:749–752.
35. Hsueh K-F, Lin P-Y, Lee S-M, Hsieh C-F. Ocular injuries from plant sap of genera *Euphorbia* and *Dieffenbachia*. J Chin Med Assoc 2004;67:93–98.
36. Desatnik H, Ashkenazi I, Avni I, Abraham F, Blumenthal M. Acute conjunctivokeratouveitis caused by latex from the pencil tree. Am J Ophthalmol 1991;112:464–466.
37. Scott IU, Karp CL. Euphorbia sap keratopathy: four cases and a possible pathogenic mechanism. Br J Opthalmol 1996;80:823–826.
38. Crowder JI, Sexton RR. Keratoconjunctivitis resulting from the sap of candelabra cactus and the pencil tree. Arch Ophthalmol 1964;72:476–484.
39. Krenzelok EP, Jacobsen TD, Aronis JM. Poinsettia exposures have good outcomes just as we thought. Am J Emerg Med 1996;14:671–674.
40. Rock JF. The poisonous plants of Hawaii. Hawaiian Forest Agric 1920;17:61.
41. Arnold HL. Poisonous plants of Hawaii. Honolulu, HI: Tong Publishing Company;1944.
42. Benjamin SP, Fernando ME, Jayanth JJ, Preetha B. *Cleistanthus collinus* poisoning. J Assoc Physicians India 2006;54:742–744.
43. Kurien T, Dayal AK, Alan G, Seshadri MS, Cherian AM. Oduvanthalai leaf poisoning. J Assoc Physicians India 1987;35:769–771.
44. Rugman F, Melchan J, Edmondson J. *Mercurialis perennis* (dog's mercury) poisoning. A case of mistaken identity. Br Med J 1983;287:1924.
45. Watson PJ. Suspected dog's mercury (*Mercurialis perennis*) poisoning in cattle. Vet Rec 1998;142:116–117.
46. Ariano R, Panzani RC, Falagiani P, Chiapella M, Augeri G. Respiratory allergy to the pollen of *Mercurialis annua* (Euphorbiaceae). Ann Allergy 1993;70:249–254.
47. Lin T-J, Su C-C, Lan C-K, Jiang D-D, Tsai J-L, Tsai M-SP Acute poisonings with *Breynia officinalis*—an outbreak of hepatotoxicity. J Toxicol Clin Toxicol 2003;41:591–594.
48. Palosuo T, Panzani RC, Singh AB, Ariano R, Alenius H, Turjanmaa K. Allergen cross-reactivity between proteins of the latex from *Hevea brasiliensis*, seeds and pollen of *Ricinus communis*, and pollen of *Mercurialis annua*, members of the Euphorbiaceae family. Allergy Asthma Proc 2002;23:141–147.
49. Ragupathi G, Prabhasankar P, Sekharan PC, Annapoorani KS, Damodaran C. Enzyme-linked immunosorbent assay (ELISA) for the determination of the toxic glycoside cleistanthin B. Forensic Sci Int 1992;56:127–136.
50. Ragupathi G, Prabhasankar P, Sekharan PC, Annapoorani KS, Damodaran C. Enzyme-linked immunosorbent assay for the phytotoxin cleistanthin A. J Immunoassay 1992;13:321–338.
51. Sarathchandra G, Balakrishnamurthy P. Perturbations in glutathione and adenosine triphosphatase in acute oral toxicosis of *Cleistanthus collinus*: an indigenous toxic plant. Ind J Pharmacol 1997;29:82–85.

Chapter 124

CYANOGENIC PLANTS and LAETRILE®

HISTORY

From a toxicological perspective, amygdalin and the commercial extract, Laetrile®, are some of the most important natural cyanogenic substances. The ancient Egyptians probably used amygdalin extracted from peach pits as a means of capital punishment known as "penalty of the peach." In 1830, French chemists isolated the diglucoside of mandelonitrile, called amygdalin, from bitter almonds. Emulsin was the crude extract of bitter almond that contained the plant enzymes necessary to hydrolyze amygdalin and release hydrogen cyanide.

Laetrile® is a cyanide-containing substance allegedly synthesized by Ernst T. Krebs, Jr. that was registered with the US Patent Office for the treatment of certain intestinal disorders. Subsequently, this compound was marketed illicitly as an intravenous cancer chemotherapeutic agent.[1] Although Laetrile® was patented as mandelonitrile-β-glucuronide, analysis of samples of Laetrile® demonstrated mostly *d*-amygdalin, similar to the compound isolated by French chemists.[2] Laetrile® itself does not contain the hydrolytic enzymes present in emulsin necessary to release cyanide from amygdalin; therefore, the intravenous administration of laetrile does not usually produce hydrogen cyanide poisoning. The US Food and Drug Administration banned the sale of Laetrile® to consumers because this compound lacks efficacy as a chemotherapeutic agent. Additionally, the ingestion of laetrile is associated with a risk of cyanide toxicity in patients receiving this therapy.[3] Despite the legal ban on the use of Laetrile®, proponents of this compound continued to advocate the use of Laetrile® for cancer including labeling this compound, "vitamin B-17," and promoting of the use of Laetrile®, vitamins, and special diets as a natural cure for cancer. Interest in the "Laetrile Movement" peaked during the late 1970s following an adverse Supreme Court decision, the lack of efficacy in the National Cancer Institute (NCI) study on Laetrile®, and the cancer-related death of Steve McQueen despite his use of Laetrile® in a Mexican facility. However, illicit marketing of this compound continues.

BOTANICAL DESCRIPTION

The distribution of cyanogenic glycosides is widespread throughout the plant kingdom, occurring in at least 2,500 plants species. These plants reside primarily in the families Compositae, Fabaceae, Linaceae, and Rosaceae.[4] Amygdalin occurs primarily in Rosaceae species (peach, apricot, apple seeds), whereas the closely related compound, prunasin occurs in at least six plant families (Myoporaceae, Myrtaceae, Polypodiaceae, Rosaceae, Saxifragaceae, Scrophulariaceae).

Prunus Species

Common Name: Black Cherry

Scientific Name: *Prunus serotina* Ehrh.

Botanical Family: Rosaceae (rose)

Physical Description: Black cherry is an oval-shaped tree that reaches 60–90 ft (~18–27 m) in height. The drooping branches reach the ground. The fine-toothed, waxy, dark green leaves turn yellow, orange, and red before dropping in the fall. The

Medical Toxicology of Natural Substances, by Donald G. Barceloux, MD

glabrous leaves are alternate, simple, and oval to oblong-lanceolate, being 5–15 cm (~2–6 in.) long and 2.5–5 cm (~1–2 in.) wide. Small, white blossoms appear in March to June, and produce a fleshy, round, purple fruit that attracts many birds from June to October.

Distribution and Ecology: This native North American species inhabits valleys and floodplains of the southwestern, midwestern, and eastern United States. This ornamental tree is an invasive species in European forests.

Common Name: Chokecherry, Common Chokecherry, Virginia Chokecherry

Scientific Name: *Prunus virginiana* L.

Botanical Family: Rosaceae (rose)

Physical Description: Mature plants reach up to 1–6 m (~3–18 ft) in height with smooth, reddish brown bark that turns gray with age. The bright green, hairless leaves are alternate and oblong with a pointed tip. Small white flowers appear in drooping clusters from May to June, and each flower produces a single-seeded red cherry that becomes red-purple upon maturity in August or September.

Distribution and Ecology: This native species occurs over most of Canada and the United States with the exception of the southern United States. This large shrub inhabits streams, river banks, wood edges, and open woods, particularly in the central and northern United States.

Common Name: Bitter Almond

Scientific Name: *Prunus amygdalus* Batch var. amara (DC.) Focke

Botanical Family: Rosaceae (rose)

Physical Description: The physical characteristics of the bitter almond and sweet almond are similar. The presence of substantial concentrations of amygdalin distinguishes the bitter almond from the edible sweet almond [*Prunus dulcis* (P. Mill.) D.A. Webb], which is grown commercially in California.

Distribution and Ecology: The bitter almond tree is a native of the temperate desert parts of Western Asia that spread westward to warm, dry areas in the Mediterranean region.

Elderberry

Common Name: European Black Elderberry

Scientific Name: *Sambucus nigra* L.

There are several subspecies of the European black elderberry including the blue elder (*Sambucus nigra* L. ssp. *caerulea* (Raf.) R. Bolli), the common elderberry (*Sambucus nigra* L. ssp. *canadensis* (L.) R. *Bolli*), and the European black elderberry (*Sambucus nigra* L. ssp. *nigra*).

Botanical Family: Caprifoliaceae (honeysuckle)

Physical Description: The common elderberry is a perennial shrub that grows up to about 4 m (~13 ft) in height. The large, compound leaves have 5–11, toothed, elliptical leaflets. Small white flowers appear in flat-topped clusters that produce purple-black berries. The twigs are hollow. The European elderberry is a small tree that produces larger fruit than the common elderberry.

Distribution and Ecology: The common elderberry is a native species of Canada and the eastern United States.

Hydrangea

This family contains 17 genera, and the genus, *Hydrangea*, contains about 23 species worldwide.

Common Name: Hydrangea

Scientific Name: *Hydrangea macrophylla* Ser. (cultivar)

Botanical Family: Hydrangeaceae (hydrangea)

Physical Description: This cultivated species is a deciduous shrub that reaches up to 2 m in height. The color of the flowers range from white, pink, mauve, and blue depending on soil conditions (e.g., pH). These flowers form beautiful, round clusters. The broad, deeply lobed leaves are alternate and relatively large (i.e., 10–25 cm/~4–10 in. in length). The underside of the leaf is pale green and lighter than the top. The capsular fruit contains many small, brown seeds.

Distribution and Ecology: *H. macrophylla* is a popular garden and ornamental plant in temperate climates. The wild varieties of hydrangea grow in shady, moist habitat in the eastern United States.

EXPOSURE

The almond is a drupe that is closely related to the peach, apricot, and cherry. Unlike these other fruits, the edible portion of the sweet almond is the seed rather than the outer layer. The bitter almond is not edible, but this tree is cultivated for the extraction of bitter almond oil, which is used in cosmetics and flavorings. The bitter almond seed is banned for consumption in the United States. A flavoring extract of the fragrant, bitter bark of the black cherry is a constituent of wild cherry cough syrup. Black cherry wood has been used since colonial times in the United States for the construction of quality furniture. *Hydrangea chinensis* Maxim. is a Chinese medicinal plant used for the treatment of malaria and cardiovascular disease.

Laetrile® is a semisynthetic derivative of amygdalin. The US patent for Laetrile® defines this compound as D-mandelonitrile-β-glucuronide. The Mexican version of this product is D-mandelonitrile-β-gentiobioside, which is derived from crushed apricot pits.[5] Both parenteral and oral products may contain a variety of contaminates. Although there are inadequate data to support the use of Laetrile® as a chemotherapeutic agent,[6] illicit use of this substance in the United States continues from illicit sources.[7]

PRINCIPAL TOXINS

Physiochemical Properties

The cyanogenic glycosides, sambunigrin (*Sambucus* spp.), vicianin (*Vicia* spp.), and amygdalin (*Prunus* spp.) are structurally similar to prunasin (*Prunus* spp.) with the exception of the sugar moiety.

Bitter Almond

Amygdalin is the cyanogenic diglucoside, D-mandelonitrile-β-D-gentiobioside. The active form of amygdalin is the dextrorotatory (*R*) configuration, whereas the *S*-isomer (neo-amygdalin) is inactive. Only the former isomer occurs naturally. Amygdalin hydrolase and prunasin hydrolase are two glycosidases, which catalyze the hydrolysis of amygdalin in a stepwise reaction by removing two glucose residues and releasing benzaldehyde and hydrogen cyanide (HCN).[8] Amygdalin hydrolase removes the terminal glucose of amygdalin and forms the monoglucoside prunasin, which is subsequently hydrolyzed by prunasin hydrolase. These enzymes are released following the crushing of moistened kernels containing amygdalin hydrolase. Certain human intestinal bacteria may also possess glycosidase activity, so that ingestion of amygdalin in the absence of this plant enzyme can cause cyanide toxicity. *In vitro* studies indicate that the addition of β-glucuronidase to amygdalin does not liberate hydrogen cyanide.[9]

Black Cherry

In black cherry (*Prunus serotina* Ehrh.) seed homogenates, (*R*)-amygdalin is degraded to HCN, benzaldehyde, and glucose by the sequential action of amygdalin hydrolase, prunasin hydrolase, and mandelonitrile lyase during a two-step process.[10] Upon seed disruption, amygdalin hydrolase catalyzes the hydrolysis of the β-1,6-glycosidic bond of amygdalin, yielding prunasin and D-glucose. Similarly, the action of prunasin hydrolyase on prunasin forms mandelonitrile and D-glucose. The complete hydrolysis of amygdalin requires both enzymes from the maceration of the cherry seeds. Subsequently, mandelonitrile dissociates either spontaneously or enzymatically in the presence of the α-hydroxynitrile lyase, mandelonitrile lyase, generating HCN and benzaldehyde.[11] In a similar way in disrupted black cherry leaves, prunasin hydrolase and mandelonitrile lyase degrade prunasin to HCN.

Elderberry

Sambunigrin is the main cyanogenic glycoside present in *Sambucus* species, but many species and strains of this genus do not contain cyanogenic glycosides.[12] The galactose-specific lectin, nigrin b, is a constituent of elderberry bark that contains a two-chain structure similar to ricin. However, the inhibitory enzymatic activity of nigrin b on ribosomes and protein synthesis is substantially less than ricin.[13] Elderberries contain relatively high concentrations of anthocyanin compounds (e.g., cyanidin-3-glucoside, cyanidin-3-sambubioside), which are food colorants with antioxidant properties.[14]

Hydrangea

Hydrangea species produce alkaloids, steroids, triterpenes, isocoumarins, secoiridoid glycosides, and flavanoids.[15] The isocoumarin fraction contains hydrangenol, a strong contact sensitizer.[16] Two secoiridoid glucoside complexes, hydramacrosides A and B possess some antihistamine properties.[17]

Poisonous Parts

Prunus Species

Bitter Almond. In a review of several studies on the HCN content of bitter almond, the estimated amount of HCN released from bitter almond was 0.8–4.7 mg/g.[9] Analysis of a homemade spirit produced from soaking chokecherries demonstrated cyanide concentrations of about 0.045 mg/g.[18]

Black Cherry. Like other rosaceous stone fruits, the leaves, bark, seeds, and immature fruits in black cherry accumulate the cyanogenic monoglucoside, (*R*)-prunasin.[10] The leaves are particularly cyanogenic because they possess prunasin hydrolase and mandelonitrile lyase in addition to (*R*)-prunasin. The related diglucoside, (*R*)-amygdalin, also occurs in high concentrations in the cotyledons of the mature seeds.[19] A study of seed development indicated that amygdalin concentrations in the cotyledon declined by about 80% during the first 3 weeks following imbibition.[20] Although prunasin was not present in mature, nongerminated seeds, the seedling epicotyls, hypocotyls, and cotyledons did contain this monoglucoside compound. The mature fruit is edible when the seeds are discarded prior to consumption.

CHOKECHERRY. Although a case report in the older medical literature associated the ingestion of a large quantity of chokecherry seeds with a fatal intoxication similar to cyanide poisoning,[21] there are few data in the modern literature to indicate that chokecherries contain sufficient cyanide to produce serious toxicity. In Italy, chokecherries are soaked in alcohol to produce homemade spirits. A case report associated the onset of obtundation, respiratory insufficiency, and metabolic acidosis with the ingestion of 300 g chokecherries steeped in alcohol the previous evening.[22] Laboratory analysis of the chokecherries and the associated liqueur demonstrated cyanide concentrations of approximately 0.005–0.015 mg/g cherries and about 0.045 mg/g liqueur.

ELDERBERRY

The content of cyanogenic glycosides in American species of *Sambucus* varies much more than European species.[23] Although the mature elderberries are edible, leaves, stems, and especially roots contain cyanogenic glycosides. The consumption of a drink distilled from the crushed leaves, stems, and berries of an elderberry bush produced nausea, vomiting, abdominal pain, weakness, and numbness, but cyanide concentrations and metabolic acidosis were not demonstrated in these patients.[24]

Mechanism of Toxicity

Cyanogenic glycosides of these plants consist of a α-hydroxynitrile aglycone and a sugar moiety (e.g., D-glucose). Hydroxylation of nitriles to α-hydroxynitrile followed by glycosylation forms the cyanogenic glycosides. The initial metabolism of cyanogenic glycosides involves the cleavage of the carbohydrate moiety by one or more β-glycosidases, which yield the corresponding α-hydroxynitrile.[25] Hydroxynitrile compounds may decompose either spontaneously or enzymically in the presence of a hydroxynitrile lyase to produce hydrogen cyanide and an aldehyde or ketone. The separation of cyanogenic glycosides and the degrading enzymes in different compartments of the plant prevents the formation of significant amounts of cyanide until the physical disruption of the plant compartments. The production of cyanide occurs rapidly only when maceration of cyanogenic glycoside-containing plant tissues allows contact between these glycosides and the corresponding enzymes.

DOSE RESPONSE

Predicting toxicity of cyanogenic glycoside-containing products is difficult because of the variation in the content both of cyanogenic glycosides and degrading enzymes. The intravenous (IV) administration of Laetrile® does not usually produce cyanide poisoning because of the lack of hydrolytic enzymes to cleave hydrogen cyanide from the cyanogenic glycosides. Case reports suggest that the concomitant ingestion of cyanogenic glycosides and amygdalin hydrolase from foods (e.g., crushed bitter almond kernels) can cause fatal cyanide toxicity.[26] Frequently, the amygdalin content in illicit pharmaceutical formulations of Laetrile® varies significantly.[2,9] Laetrile® is much more toxic via the oral than the intravenous (IV) route of administration. The oral administration of 500 mg amygdalin to six volunteers 3 times daily produced blood cyanide concentrations up to 2.1 μg/mL, but there was no clinical or laboratory evidence of a toxic reaction.[27] A case report associated a fatal ingestion of an estimated dose of 45 mL or 10.5 g of an intravenous solution of Laetrile® with the immediate onset of severe headache, dizziness, coma, and convulsions.[28] Another case report indicated that an estimated ingestion of 0.5–2.5 g of Laetrile® (amygdalin) tablets by an 11-month-old child caused vomiting and listlessness within $\frac{1}{2}$ hour followed quickly by obtundation, depressed respirations, shock, and death 72 hours after ingestion.[29] An 18-year-old man developed coma, respiratory insufficiency, and hypotension following the ingestion of 3 grams (30 tablets) of Laetrile®.[30] He survived with supportive care.

The ingestion of vitamin C with amygdalin potentially increases the conversion of amygdalin to cyanide and reduces the detoxifying effects of cysteine. The ingestion of 3 grams amygdalin along with the daily use of 4.8 g vitamin C resulted in the development of severe cyanide poisoning (severe lactic acidosis, coma, respiratory failure).[31]

Ingestion of 12 bitter almonds produced vomiting, abdominal pain, coma, lactic acidosis, and transient pulmonary edema within 15 minutes after ingestion.[32]

TOXICOKINETICS

Absorption/Distribution

There are few human data on the toxicokinetics of cyanogenic glycosides. Limited animal studies suggest that the gastrointestinal tract poorly absorbs (i.e., <3%) oral doses of amygdalin, whereas the oral bioavailability of prunasin is about 50%.[33] In animal models, the occurrence of peak whole blood cyanide concentrations after oral administration of amygdalin and linamarin depends on the rate of enzymatic hydrolysis. A study of female golden Syrian hamsters indicates that peak whole blood cyanide concentrations following the oral administration of amygdalin and linamarin occur 1 hour and 3 hours after administration, respectively.[34] The protein binding of amygdalin is minimal.

Biotransformation/Elimination

The biotransformation of cyanogenic glycosides begins with the cleavage of the carbohydrate moiety by one or more β-glycosidases that form α-hydroxynitrile. Depending on the plant species, two distinct pathways involve 1) the hydrolysis of cyanogenic disaccharides by the stepwise removal of the sugar residues, and 2) hydrolysis of the aglycone-disaccharide bond producing mandelonitrile and the corresponding disaccharides.[25] Figure 124.1 demonstrates these two pathways. Hydrolysis of amygdalin produces the active metabolite, prunasin (D-mandelonitrile-β-D-glucoside). Animal studies indicate that the elimination of amygdalin is rapid with most of absorbed dose (62–96%) excreted in the urine during the first 24 hours after ingestion as unchanged drug.[27,35] The hepatic excretion of amygdalin is relatively small compared with renal excretion.[36] Following intravenous administration of 500 mg amygdalin to 6 terminal cancer patients, the mean elimination half-life of amygdalin was approximately 1–2 hours with a mean plasma clearance of about 100 mL/min.[37]

CLINICAL RESPONSE

The clinical features associated with the acute ingestion of cyanogenic glycosides mimics cyanide poisoning, and the severity of the intoxication correlates to the dose of active cyanogenic compounds. The onset of vomiting, abdominal pain, weakness, dyspnea, diaphoresis, and lightheadedness followed by convulsions, stupor, disorientation, hypotension, metabolic acidosis, coma, respiratory failure, and cardiovascular collapse develops after exposure to high doses of cyanogenic glycoside-containing foods[38,39] and from the ingestion of Laetrile®.[26,40] Symptoms typically start within $^1/_2$ hour of consumption and progress quickly. A case series of wild apricot seed ingestions by Turkish children associated the sudden onset of vomiting and crying followed by fainting, lethargy, and coma with the ingestions.[41]

Rare case reports associate contact with hydrangeas with the development of contact dermatitis. Hydrangeas contain a strong contact sensitizer, hydrangenol, which causes chronic contact dermatitis (scaly erythema, painful fissures), particularly in the hands of workers chronically exposed to hydrangeas.[42–44]

DIAGNOSTIC TESTING

Analytical Methods

Methods to detect amygdalin include thin layer chromatography, high performance liquid chromatography (HPLC), and gas chromatography/mass spectrometry (GC/MS).[45] Amygdalin and prunasin can be quantitated in biological samples (urine, plasma) by HPLC on a reversed phase column with detection at 215 nm and

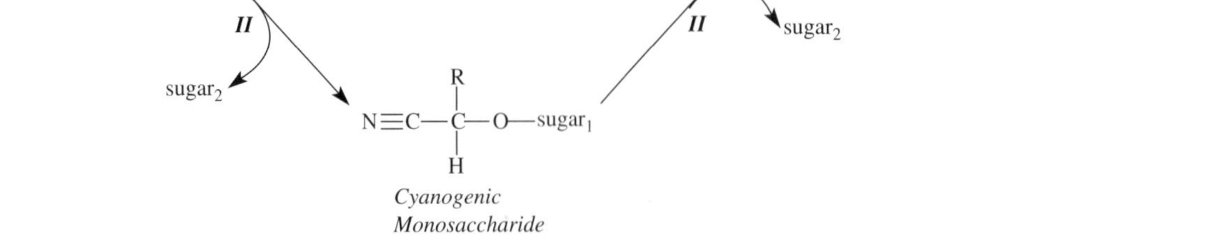

FIGURE 124.1. Biotransformation of cyanogenic glycosides.

confirmation by MS.[46] Methods for detecting hydrogen cyanide in biological samples include spectrometry, electrochemical analysis with cyanide ion-selective electrode, GC, GC/MS, and HPLC/MS.[47,48] The latter two methods are widely used methods for quantitating cyanide after headspace extraction or derivatization. More recent and rapid methods of detecting cyanide in biological specimens include capillary gas chromatography with cryogenic oven trapping [limit of detection (LOD): 2 ng/mL],[49] capillary electrophoresis with fluorescence detection (LOD: 0.1 ng/mL),[50] automated headspace GC (LOD: 13.8 μg/L),[51] and isotope-dilution GC/MS (LOD: 0.003–0.007 μg/g).[52]

Biomarkers

There are few data correlating blood cyanide concentrations to the ingestion of specific amounts of cyanogenic plants. In the presence of a compatible history, the presence of elevated blood cyanide concentrations confirms exposure to cyanogenic plants, but the results are not usually available to guide clinical management.

Abnormalities

Cyanide poisoning inevitably produces a metabolic acidosis that results in increased serum lactate concentrations from anaerobic metabolism.[53] The lack of metabolic acidosis in a symptomatic patient suggests a diagnosis other than cyanide poisoning.

TREATMENT

Treatment of poisoning from exposure to cyanogenic plants is similar to the treatment of cyanide poisoning. Decontamination measures are usually unnecessary because of the limited absorptive capacity of activated charcoal for cyanide and the risk of aspiration from the rapid onset of mental status changes. Mild to moderate ingestions usually require only supportive care. Indications for the use of an antidote include alteration in mental status, severe acidosis, continuous seizures, and refractory hypotension. The administration of the cyanide antidote kit to children requires the appropriate adjustments for weight and hemoglobin content to avoid methemoglobinemia. Clinical features that suggest an appropriate response to nitrites and thiosulfate include the following: normalization of vital signs (blood pressure, pulse), normal mentation, cessation of seizure activity, spontaneous respirations, and resolution of metabolic acidosis.[54] The presence of persistent metabolic acidosis may require a continuous infusion of sodium thiosulfate (2 g/h) for 24 hours.[55] The use of hydroxocobalamin is a safe and probably efficacious alternative to the cyanide antidote kit,[56] but there are few data on the efficacy of either antidote in the clinical setting of cyanide poisoning, particularly associated with cyanogenic plants or Laetrile®. The intravenous dose of hydroxocobalamin in adults without an allergy to vitamin B_{12} is 5–15 g over 30 minutes (more rapidly if clinically indicated), whereas the pediatric dose is 70 mg/kg. Additionally, adults should receive 12.5 g (50 mL 25% solution) sodium thiosulfate over 10–20 minutes. The pediatric dose of thiosulfate is 400 mg/kg up to the adult dose of 12.5 g. Survival has occurred in patients with Laetrile®-induced, severe cyanide intoxication (coma, metabolic acidosis, profound hypotension requiring aggressive support with vasopressors) following only supportive care.[57]

References

1. Davignon JP, Trissel LA, Kleinman LM. Pharmaceutical assessment of amygdalin (Laetrile) products. Cancer Treat Rep 1978;62:99–104.
2. Fenselau C, Pallante S, Batzinger RP, Benson WR, Barron RP, Sheinin EB, Maienthal M. Mandelonitrile beta-glucuronide: synthesis and characterization. Science 1977;198:625–627.
3. Moertel CG, Fleming TR, Rubin J, Kvols LK, Sarna G, Koch R, et al. A clinical trial of amygdalin (laetrile) in the treatment of human cancer. N Engl J Med 1982;306: 201–206.
4. Vetter J. Plant cyanogenic glycosides. Toxicon 2000;38: 11–36.
5. Dorr RT, Paxinos J. The current status of laetrile. Ann Intern Med 1978;89:389–397.
6. Milazzo S, Lejeune S, Ernst E. Laetrile for cancer: a systematic review of the clinical evidence. Support Care Cancer 2007;15:583–595.
7. Milazzo S, Ernst E, Lejeune S, Schmidt K. Laetrile treatment for cancer. Cochrane Database Syst Rev 2006;(2): CD005476.
8. Haisman DR, Knight DJ. The enzymic hydrolysis of amygdalin. Biochem J 1967;103:528–534.
9. Holzbecher MD, Moss MA, Ellenberger HA. The cyanide content of laetrile preparations, apricot, peach and apple seeds. Clin Toxicol 1984;22:341–347.
10. Zhou J, Harmann S, Shepherd BK, Poulton JE. Investigation of the microheterogeneity and aglycone specificity-conferring residues of black cherry prunasin hydrolases. Plant Physiol 2002;129:1252–1264.
11. Hu Z, Poulton JE. Molecular analysis of (R)-(+)-mandelonitrile lyase microheterogeneity in black cherry. Plant Physiol 1999;119:1535–1546.
12. Buhrmester RA, Ebinger JE, Seigler DS. Sambunigrin and cyanogenic variability in populations of *Sambucus*

canadensis L. (Caprifoliaceae). Biochem Systematic Ecol 2000;28:689–695.

13. Battelli MG, Citores L, Buonamici L, Fereras JM, de Benito FM, Stirpe F, Girbes T. Toxicity and cytotoxicity of nigrin b, a two-chain ribosome-inactivating protein from *Sambucus nigra*: comparison with ricin. Arch Toxicol 1997;71:360–364.
14. Murkovic M, Adam U, Pfannhauser W. Analysis of anthocyane glucosides in human serum. Fresenius J Anal Chem 2000;366:379–381.
15. Patnam R, Chang F-R, Chen C-Y, Kuo R-Y, Lee Y-H, Wu Y-C. Hydrachine A, a novel alkaloid from the roots of *Hydrangea chinensis*. J Nat Prod 2001;64:948-949.
16. Hausen BM. Hydrangenol, a strong contact sensitizer found in hydrangea (*Hydrangea* sp.; Hydrangeaceae). Contact Dermatitis 1991;23:233–235.
17. Matsuda H, Shimoda H, Uemura T, Ueda T, Yamahara J, Yoshikawa M. Chemical constituents from the leaves of *Hydrangea macrophylla* var. *thunbergii* (III): Absolute stereostructures of hydramacrosides A and B, secoiridoid glucoside complexes with inhibitory activity on histamine release. Chem Pharm Bull 1999;47:1753–1758.
18. Pentore R, Venneri A, Nichelli P. Accidental chokecherry poisoning: early symptoms and neurological sequelae of an unusual case of cyanide intoxication. Int J Neurol Sci 1996;17:233–235.
19. Zheng L, Poulton JE. Temporal and spatial expression of amygdalin hydrolase and (R)-(+)-mandelonitrile lyase in black cherry seeds. Plant Physiol 1995;109:31–39.
20. Swain E, Poulton JE. Utilization of amygdalin during seedling development of *Prunus serotina*. Plant Physiol 1994;106:437–445.
21. Pijoan M. Cyanide poisoning from chokecherry seeds. Am J Med Sci 1942;204:550–553.
22. Pentore R, Venneri A, Nichelli P. Accidental chokecherry poisoning: early symptoms and neurological sequelae of an unusual case of cyanide intoxication. Int J Neurol Sci 1996;17:233–235.
23. Aikman K, Bergman D, Ebinger J, Seigler D. Variation of cyanogenesis in some plant species of the midwestern United States. Biochem Systematic Ecol 1996;24: 637–645.
24. Kunitz S, Melton RJ, Updyke T, Breedlove D, Werner SB. Poisoning from elderberry juice—California. MMWR Morb Mortal Wkly Rep 1984;33:173–174.
25. Poulton JE. Localization and catabolism of cyanogenic glycosides. Ciba Found Symp 1988;140:67–91.
26. Braico KT, Humbert JR, Terplan KL, Lehotay JM. Laetrile intoxication report of a fatal case. N Engl J Med 1979;300:238–240.
27. Moertel CG, Ames MM, Kovach JS, Moyer TP, Rubin JR, Tinker JH. A pharmacologic and toxicological study of amygdalin. JAMA 1981;245:591–594.
28. Sadoff L, Fuchs K, Hollander J. Rapid death associated with laetrile ingestion. JAMA 1978;239:1532.
29. Humbert JR, Tress JH, Braico KT. Fatal cyanide poisoning: accidental ingestion of amygdalin. JAMA 1977;238: 482.
30. Lee M, Berger HW, Givre HL, Jayamanne DS. Near fatal laetrile intoxication: complete recovery with supportive treatment. Mt Sinai J Med 1982;49:305–307.
31. Bromley J, Hughes BG, Leong DC, Buckley NA. Life-threatening interaction between complementary medicines: cyanide toxicity following ingestion of amygdalin and vitamin C. Ann Pharmacother 2005;39:1566–1569.
32. Shragg TA, Albertson TE, Fisher CJ Jr. Cyanide poisoning after bitter almond ingestion. West J Med 1982;136: 65–69.
33. Rauws AG, Olling M, Timmerman A. The pharmacokinetics of prunasin, a metabolite of amygdalin. J Toxicol Clin Toxicol 1982–1983;19:851–856.
34. Frakes RA, Sharma RP, Willhite CC. Comparative metabolism of linamarin and amygdalin in hamsters. Food Chem Toxicol 1986;24:417–420.
35. Ames MM, Kovach JS, Flora KP. Initial pharmacologic studies of amygdalin (laetrile) in man. Res Commun Chem Pathol Pharmacol 1978;22:175–185.
36. Rauws AG, Olling M, Timmerman A. The pharmacokinetics of amygdalin. Arch Toxicol 1982;49:311–319.
37. Ames MM, Moyer TP, Kovach JS, Moertel CG, Rubin J. Pharmacology of amygdalin (laetrile) in cancer patients. Cancer Chemother Pharmacol 1981;6:51–57.
38. Akintonwa A, Tunwashe OL. Fatal cyanide poisoning from cassava-based meal. Hum Exp Toxicol 1992;11: 47–49.
39. Espinoza OB, Perez M, Ramirez MS. Bitter cassava poisoning in eight children: a case report. Vet Hum Toxicol 1992;34:65.
40. Beamer WC, Shealy RM, Prough DS. Acute cyanide poisoning from laetrile ingestion. Ann Emerg Med 1983;12; 449–451.
41. Sayre JW, Kaymakcalan S. Cyanide poisoning of apricot seeds among children in central Turkey. N Engl J Med 1964;270:1113–1115.
42. Kuligowski ME, Chang A, Leemreize JH. Allergic contact hand dermatitis from hydrangea: report of a 10th case. Contact Dermatitis 1992;26:269–270.
43. Avenel-Audran M, Hausen BM, Le Sellin J, Ledieu G, Verret JL. Allergic contact dermatitis from hydrangea—is it so rare? Contact Dermatitis 2000;43:189–191.
44. Rademaker M. Occupational contact dermatitis to hydrangea. Australas J Dermatol 2003;44:220–221.
45. Balkon J. Methodology for the detection and measurement of amygdalin in tissues and fluids. J Anal Toxicol 1982;6:244–246.
46. Rauws AG, Gramberg LG, Olling M. Determination of amygdalin and its major metabolite prunasin in plasma and urine by high pressure liquid chromatography. Pharm Weekbl Sci 1982;4:172–175.

47. Tracqui A, Raul JS, Geraut A, Berthelon L, Ludes B. Determination of blood cyanide by HPLC-MS. J Anal Toxicol 2002;26:144–148.
48. Kage S, Nagata T, Kudo K. Determination of cyanide and thiocyanate in blood by gas chromatography and gas chromatography-mass spectrometry. J Chromatogr B 1996;675: 27–32.
49. Ishii A, Seno H, Watanabe-Suzuki K, Suzuki O. Determination of cyanide in whole blood by capillary gas chromatography with cryogenic oven trapping. Anal Chem 1998;70:4873–4876.
50. Chinaka S, Tanaka S, Takayama N, Tsuji N, Takou S, Ueda K. High-sensitivity analysis of cyanide by capillary electrophoresis with fluorescence detection. Anal Sci 2001;17:649–652.
51. Calafat AM, Stanfill SB. Rapid quantitation of cyanide in whole blood by automated headspace gas chromatography. J Chromatogr 2002;772:131–137.
52. Murphy KE, Schantz MM, Butler TA, Genner BA Jr, Wood LJ, Turk GC. Determination of cyanide in blood by isotope-dilution gas chromatography-mass spectrometry. Clin Chem 2006;52:458–467.
53. Ruangkanchanasetr S, Wananukul V, Suwanjutha S. Cyanide poisoning, 2 cases report and treatment review. J Med Assoc Thai 1999;82(Suppl 1):S162–S166.
54. Hall AH, Linden CH, Kulig KW, Rumack BH. Cyanide poisoning from laetrile ingestion: role of nitrite therapy. Pediatrics 1986;78:269–272.
55. Suchard JR, Wallace KL, Gerkin RD. Acute cyanide toxicity caused by apricot kernel ingestion. Ann Emerg Med 1998;32:742–744.
56. Shepherd G, Velez LI. Role of hydroxocobalamin in acute cyanide poisoning. Ann Pharmacother 2008;42:661-669.
57. O'Brien B, Quigg C, Leong T. Severe cyanide toxicity from "vitamin supplements." Eur J Emerg Med 2005;12: 257–258.

Chapter 125

DIEFFENBACHIA and OTHER OXALATE-CONTAINING PLANTS

HISTORY

The irritant properties of plants from the Araceae family (arum) have been known since the medical writing of the Greek physician and botanist, Dioscorides, during the 1st century. The 17th century physician, Bernardino Ramazzini, reported that an apothecary "has touched his genitals with an arum root and they became so acutely inflamed that gangrene resulted and a hemorrhage so acute that he nearly died."[1] The roots were boiled in wine as a concoction for the treatment of gout in the 18th century, and a tincture of the *Dieffenbachia* plant was a homeopathic treatment for sexual impotence in the 19th century.[2] Amazon Indians applied *Dieffenbachia* juice to their curare arrows, and rubbing the juice of this plant on the mouths of slaves was a method of punishment practiced by Jamaican slave traders.[3] In the Bahamas, anecdotal reports suggest that criminals rubbed *Dieffenbachia* stem on the lips of witnesses to prevent adverse testimony in court.[4] Medieval uses for extracts from the arum family included the treatment of gout, dropsy, and sexual impotence, and contraception. The Nazis conducted experiments on animal sterility induced by juice extracts, but they were unable to produce enough plants to expand their experimentation to concentration camp victims.

BOTANICAL DESCRIPTION

The Araceae family contains over 1,800 known species including a number of plants that produce mucosal irritation. *Dieffenbachia* species contain the most potent mucosal irritant within the Araceae family. Scientific (common) names for some common species in this family include the following: *Alocasia* spp. (elephant ears); *Arisaema* spp. (jack-in-the-pulpit, green dragon); *Caladium* spp. (caladium); *Calla palustris* L. (wild calla); *Colocasia esculenta* (L.) Schott (elephant ears, taro); and *Symplocarpus foetidus* (L.) Salisb. ex Nutt. (skunk cabbage).

Common Name: Dumbcane, Dumbplant, Mother-in-Law Plant

Scientific Name: *Dieffenbachia seguine* (Jacq.) Schott.

Botanical Family: Araceae (arum)

Physical Description: *Dieffenbachia* species grows up to 6 ft (~2 m) and has broad, shiny leaves that are variegated or spotted. The arrangement and number of ovules in the ovaries separate the genera *Colocasia* and *Alocasia* from the genus *Dieffenbachia*.

Distribution and Ecology: The natural habitat of the *Dieffenbachia* species is the Antilles archipelago and the tropical forests of South America.

Common Name: Schefflera, Brassaia, Ivy Palm, Octopus Tree, Umbrella Tree

Scientific (Common) Name: *Schefflera actinophylla* (Endl.) H.A.T. Harms

Botanical Family: Araliaceae

Physical Description: This fast-growing tree reaches 6–9 m (~20–30 ft) in height with ascending thick

Medical Toxicology of Natural Substances, by Donald G. Barceloux, MD

branches marked by conspicuous leaf-scars. The leaves are palmate, containing an average of nine leaflets (range 7–15) with main petioles about 15–45 cm (~6–18 in.) long. The waxy, dark green leaflets are approximately 10–20 cm (~4–8 in.) long, elliptic-obovate, and all glabrous. Red flowers appear in small heads that produce fruit containing 10–12 seeds.

Distribution and Ecology: Most species of *Schefflera* are native inhabitants of the tropics and subtropics, but these species are houseplants in temperate climates.

EXPOSURE

The arum family contains popular ornamental plants with large leaves that are suitable for indoor locations. *Dieffenbachia, Caladium*, and *Philodendron* species are popular indoor plants that together are the most common plant exposures reported to US poison control centers.[5] Native Hawaiians prepared *Colocasia* root as a staple in their diet after detoxifying the root by boiling for 5 minutes. In Papua, New Guinea, sapal is a traditional food produced by fermenting a mixture of cooked, grated taro (*Colocasia esculenta*), corn, and coconut cream.[6] Reported traditional uses for *Dieffenbachia* species include the use of the extracts of this plant to treat dropsy, sexual frigidity, impotence, gout, and general inflammation.[4] In Thailand and China, extracts of the leaves from *Schefflera leucantha* R. vig. are used for the prevention and treatment of asthma attacks.[7] The ethanolic leaf extract of *Schefflera arboricola* (Hayata) Merr. is a traditional herbal treatment for analgesia, sprains, fractures, back pain, hypertension, and anxiety.[8]

PRINCIPAL TOXINS

Physiochemical Properties

Three major protein components of the juice from *Dieffenbachia* stalk have been separated by paper chromatography on the basis of molecular weight (above 150,000; 4,400 ± 300; 1,000–4,400).[9] These compounds are relatively unstable when exposed to air or to heat. The exact structure of these toxic components remains unknown. *Schefflera* species contain a variety of triterpenoid compounds including oleanane and ursane glycosides, benzyl glycosides, asiatic acid, and 3-epi-betulinic acid.[10–12]

Poisonous Parts

Exposure to fresh juice from the stalk of *Dieffenbachia* produces the most severe local reaction compared with other parts of the plant.[13] The leaves and stems are less irritating. The toxicity of roots of other members of the Araceae family is not well documented. The approximate concentrations of oxalic acid in samples of the whole plant and juice from *Dieffenbachia seguine* (Jacq.) Schott (*D. picta*) were 0.21% and 0.10%, respectively.[2] Table 125.1 lists some genera that contain calcium oxalate crystals.[14]

Mechanism of Toxicity

The most common toxic effects of the ingestion of plants from the arum family involve mucosal irritation. The exact cause of mucosal irritation is controversial, and several mechanisms of toxicity may be involved. *Dieffenbachia* species contain many insoluble calcium oxalate crystals. However, a number of edible plants contain similar concentrations of calcium oxalate crystals (e.g., beets, rhubarb), and the presence of oxalate crystals alone does not account for the toxicity of *Dieffenbachia* species.[2] *Dieffenbachia* leaves contain special cells known as idioblasts composed of needle-shaped calcium oxalate crystals (raphides) arranged in compact bundles.[15] Figure 125.1 displays a *Dieffenbachia* idioblast containing numerous raphides. When stimulated, idioblasts forcefully extrude the raphides for a distance of about two to three cell lengths.[16] In animal studies, an insoluble fraction of crude *Dieffenbachia* juice produced substantially more edema of the mouse paw than the insoluble fraction.[17] Consequently, the inflammation associated with plants from the arum family may involve the interaction of chemical (i.e., trypsin-like enzymes, proteases) and mechanical mechanisms.[3,18] In contrast to soluble calcium oxalate crystals, the insoluble crystals in these plants do not produce systemic toxicity because of poor bioavailability.

Schefflera species contain variable amounts of the sensitizer, falcarinol. Although there are three structurally related polyacetylene compounds (falcarindiol, falcarinone, dehydrofalcarinone), falcarinol probably is the primary sensitizer.[19] Falcarinone is an oxidation product of falcarinol.

DOSE RESPONSE

Exposure to plants from the arum family usually produces local irritation rather than systemic symptoms because the poor bioavailability of insoluble oxalate crystals. Variation in the severity of local reaction to oxalate crystals is expected between various genera based on differences in oxalate content and other factors, but there are few human data.

TABLE 125.1. Some Genera that Contain Oxalate Crystals Along with Oxalate Concentrations When Available

Family	Genus	Form	Amount
Agavaceae	*Agave*	Calcium oxalate	
Amaranthaceae	*Amaranthus*	Potassium acid oxalate	0.74–1%
Araceae	*Aglaonema*	Calcium oxalate	
	Alocasia	Calcium oxalate	
	Amorphophallus	Calcium oxalate	
	Anthurium	Calcium oxalate	
	Arisaema	Calcium oxalate	
	Arum	Calcium oxalate	
	Caladium	Calcium oxalate	
	Colocasia	Calcium oxalate	
	Dieffenbachia	Calcium oxalate	0.4% (stem)
	Epipremnum	Calcium oxalate	
	Monstera	Calcium oxalate	
	Philodendron	Calcium oxalate	0.7%
	Spathiphyllum	Calcium oxalate	
	Syngonium	Calcium oxalate	
	Xanthosoma	Calcium oxalate	
	Zantedeschia	Calcium oxalate	
Araliaceae	*Schefflera*		
Begoniaceae	*Begonia*	Calcium oxalate	
Bromeliaceae	*Ananas*	Calcium oxalate	
Caprifoliaceae	*Symphoricarpos*	Calcium oxalate	
Chenopodiaceae	*Beta vulgaris*	Soluble oxalate	>10%
	Halogeton	Sodium oxalate	>10–34.5%
	Sarcobatus	Sodium oxalate	>10%
Liliaceae	*Ornithogalum*	Calcium oxalate	
Oxalidaceae	*Oxalis*	Potassium acid oxalate	7–10%
Palmae	*Caryota*	Calcium oxalate	0.05% (berry)
Poaceae	*Setaria*	Ammonium oxalate	
Polygonaceae	*Rheum*	Potassium oxalate	0.28%
	Rumex	Potassium acid	>10%
Rubiaceae	*Psychotria*	Calcium oxalate	
Vitaceae	*Parthenocissus*	Soluble oxalate	

Source: Adapted from Ref 14.

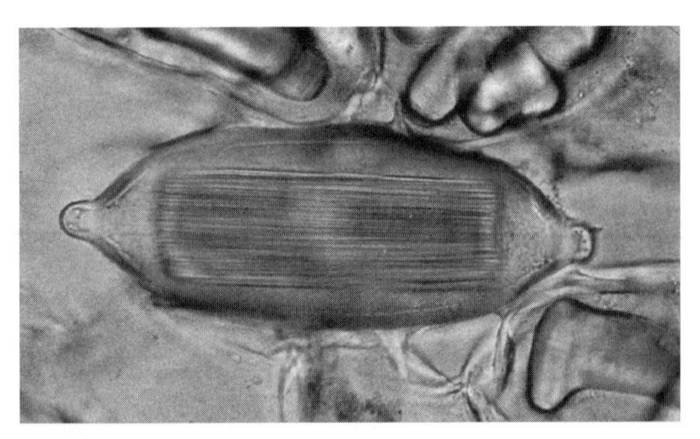

FIGURE 125.1. *Dieffenbachia* idioblast containing numerous rapides. Photograph courtesy of Professor Sylvie Rapior. See color insert.

CLINICAL RESPONSE

Local reaction to contact with oxalate-containing plants varies between genera based on differences in oxalate content with exposures to arum plants among the best described cases. Most exposures to plants from the arum family are asymptomatic as reported to poison control centers.[20] Contact with sap from these plants can produce irritation of the oral mucous membranes, the conjunctiva, and the cornea.[21,22] Ocular toxicity following accidental contact with the sap from *Dieffenbachia* species includes severe pain, lacrimation, photophobia, blepharospasm, corneal abrasion, chemosis, and keratoconjunctivitis.[23] The chemical chemosis and keratoconjunctivitis resulting from ocular contact with

Dieffenbachia species usually resolves within 1 week after exposure.[24] The *potentially* most serious toxic effects following the ingestion of plants from the arum family involve salivation and local edema that may interfere with swallowing and/or breathing. Symptoms of potential airway obstruction include drooling, dysphagia, stridor, and facial edema. A case report documented the development of severe laryngeal and epiglottic edema after a 69-year-old man bit into the stem from a species of *Dieffenbachia*.[25] He developed stridor and respiratory distress unresponsive to intravenous steroids, inhaled racemic epinephrine, and nebulized albuterol; subsequently, an emergency tracheostomy was necessary. An 11-month-old boy died 17 days after ingesting leaves of a philodendron plant, and the autopsy demonstrated esophageal erosions and hyperemia.[26] Although the authors speculated that vagotonia secondary to the esophageal lesions was the cause of the sudden cardiac arrest, such a cause of death is unlikely, particularly in a witnessed cardiac arrest. After a 40-year-old woman bit into a *Dieffenbachia* stalk, she experienced sialorrhea, dysphagia, and intense burning of her mouth.[27] Physical examination revealed bullae of the mouth and tongue, severe edema of the face and tongue, and thick, almost unintelligible speech. However, no systemic or respiratory symptoms developed, but granulating oral lesions remained after 11 days. Case reports indicate that these lesions resolve without scar formation or permanent dysfunction.[28] European and African case studies document similar local toxic effects in children without systemic complaints except for mild transient central nervous system symptoms.[29,30] Occasionally, severe paresthesias and numbness are associated with local inflammation caused by ingesting plants from the arum family.[31]

Rarely, an allergic contact dermatitis or asthma develops in gardeners exposed to philodendron plants[32] or to *Schefflera* species.[33] In sensitized individuals, contact with these plants produces an acute vesiculobullous skin eruption in the areas of contact.[34] These individuals are usually allergic to a variety of antigens from fungi, dust mites, animal dander, trees, weeds, and pollens.

DIAGNOSTIC TESTING

There are no specific biomarkers for toxicity related to these plant species.

TREATMENT

Most exposures to these plants involve transient contact with the plant material, and treatment is supportive.[35] Therapeutic measures include the removal of all plant material from the mouth and the use of demulcents (milk, antacids, cool water).[29] Eye involvement requires irrigation and examination for corneal abrasions. The most serious sequela is the development of respiratory obstruction, which progresses over the first 6 hours. However, only those patients with substantial edema (i.e., intraoral edema, drooling) require direct medical observation. Patients without evidence of significant intraoral edema require only topical analgesics [e.g., Cepacol® (Bayer Consumer Care, Morristown, NJ), Orabase Plain® (Colgate-Palmolive Company, New York, NY)]. Gut decontamination measures are unnecessary.

Criteria for admission include the presence of respiratory obstruction and the inability to maintain fluid/electrolyte balance. Antihistamines are not indicated because of the lack of data supporting the role of histamine-induced edema.[27] The development of hypocalcemia is unlikely except in very large ingestions because the calcium oxalate crystals in the arum family are relatively insoluble. For patients ingesting large amounts of calcium oxalate-containing plants, generous fluid replacement (>2–3 $L/m^2/d$) is the most effective prophylactic treatment to prevent the deposition of calcium oxalate crystals in the kidney, bones, and muscles.[36] Sodium citrate (0.15 g/kg/d) is used as a treatment for primary and secondary oxalosis to inhibit the crystallization of calcium oxalate.

References

1. Mitchell JC. Ramazzini and skin injury from the Arum lily. Br J Dermatol 1974;91:235–236.
2. Fochtman FW, Manno JE, Winek CL, Cooper JA. Toxicity of the genus *Dieffenbachia*. Toxicol Appl Pharmacol 1969;15:38–45.
3. Arditti J, Rodriguez E. Dieffenbachia's uses, abuses and toxic constituents. J Ethnopharmacol 1982;5:293–302.
4. Barnes BA, Fox LE. Poisoning with Dieffenbachia. J Hist Med Allied Sci 1955;10:173–181.
5. Lai MW, Klein-Schwartz W, Rodgers GC, Abrams JY, Haber DA, Bronstein AC, Wruk KM. 2005 Annual Report of the American Association of Poison Control Centers' national poisoning and exposure database. Clin Toxicol (Phila) 2006;44:803–932.
6. Gubag R, Omoloso DA, Owens JD. Sapal: a traditional fermented taro [*Colocasia esculenta* (L.) schott] corm and coconut cream mixture from Papua New Guinea. Int J Food Microbiol 1996;28:361–367.
7. Witthawaskul P, Panthong A, Kanjanapothi D, Taesothikul T, Lertprasertsuke N. Acute and subacute toxicities of the saponin mixture isolated from *Schefflera leucantha* Viguier. J Ethnopharmacol 2003;89:115–121.

8. Melek FR, Miyase T, Abdel Khalik SM, El-Gindi MR. Triterpenoid saponins from *Schefflera arboricola*. Phytochemistry 2003;63:401–407.
9. Cherian S, Smith WT, Stolte LP. Soluble proteins of Dieffenbachia. Trans Ky Acad Sci 1976;37:16–19.
10. Braca A, Autore G, De Simone F, Marzocco S, Morelli I, Venturella F, De Tommasi N. Cytotoxic saponins from *Schefflera rotundifolia*. Planta Med 2004;70:960–966.
11. Maeda C, Ohtani K, Kasai R, Yamasaki K, Duc NM, Nham NT, Cu NK. Oleanane and ursane glycosides from *Schefflera octophylla*. Phytochemistry 1994;37:1131–1137.
12. Sung TV, Steglich W, Adam G. Triterpene glycosides from *Schefflera octophylla*. Phytochemistry 1991;30:2349–2356.
13. Draize JH, Woodard G, Calvary HO. Methods for the study of irritation and toxicity of substances applied topically to the skin and mucous membranes. J Pharmacol Exp Ther 1944;82:377–390
14. Spoerke DG Jr., Smolinske SC. Toxicity of Houseplants. Boca Raton, FL: CRC Press;1990.
15. Dore WG. Crystalline raphides in the toxic houseplant *Dieffenbachia*. JAMA 1963;185:1045.
16. Gardner DG. Injury to the oral mucous membranes caused by the common houseplant, *Dieffenbachia*. A review. Oral Surg Oral Med Oral Pathol 1994;78:631–633.
17. Kubafla B, Lugnier AJ, Anton R. Study of *Dieffenbachia*-induced edema in mouse and rat hindpaw. Respective role of oxalate needles and trypsin-like protease. Toxicol Appl Pharmacol 1983;55:444–451.
18. Rauber A. Observations on the idioblasts of *Dieffenbachia*. Clin Toxicol 1985;23:79–90.
19. Hansen L, Hammershoy O, Boll PM. Allergic contact dermatitis from falcarinol isolated from *Schefflera arboricola*. Contact Dermatitis 1986;14:91–93.
20. Mrvos R, Dean BS, Krenzelok EP. Philodendron/dieffenbachia ingestions: are they a problem? Clin Toxicol 1991;29:485–491.
21. Gardner DG. Injury to the oral mucous membranes caused by the common houseplant, dieffenbachia. A review. Oral Surg Oral Med Oral Pathol 1994;78:631–633.
22. Evans CR. Oral ulceration after contact with the houseplant *Dieffenbachia*. Br Dent J 1987;162:467–468.
23. Seet B, Chan WK, Ang CL. Crystalline keratopathy from *Dieffenbachia* plant sap. Br J Ophthalmol 1995;79: 98–99.
24. Hsueh KF, Lin PY, Lee SM, Hsieh CF. Ocular injuries from plant sap of genera *Euphorbia* and *Dieffenbachia*. J Chin Med Assoc 2004;67:93–98.
25. Cumpston KL, Vogel SN, Leikin JB, Erickson TB. Acute airway compromise after brief exposure to a *Dieffenbachia* plant. J Emerg Med 2003;25:391–397.
26. McIntire MS, Guest JR, Porterfield JF. Philodendron—an infant death. Clin Toxicol 1990;28;177–183.
27. Drach G, Maloney WA. Toxicity of the common houseplant *Dieffenbachia*: Report of a case. JAMA 1963;184: 1047–1048.
28. Wiese M, Kruszewska S, Kolacinski Z. Acute poisoning with *Dieffenbachia picta*. Vet Hum Toxicol 1996;38: 356–357.
29. Tagwireyi D, Ball DE. The management of Elephant's Ear poisoning. Hum Exp Toxicol 2001;20:189–192.
30. Brzeski Z, Budzynska H. Clinical picture of poisoning with Arum. Pol Tzg Lek 1968;23:1442–1443.
31. Chan TYK, Chan LY, Tam LS, Critchley JAJH. Neurotoxicity following the ingestion of a Chinese medicinal plant, *Alocasia macrorrhiza*. Hum Exp Toxicol 1995;14: 727–728.
32. Hammershoy O, Verdich J. Allergic contact dermatitis from *Philodendron scandens*. Contact Dermatitis 1980;6: 95–99.
33. Hammershoy O. Allergic contact dermatitis from *Schefflera*. Contact Dermatitis 1981;7:57–58.
34. Calnan CD. Dermatitis from *Schefflera*. Contact Dermatitis 1981;7:341.
35. Tagwireyi D, Ball DE. The management of Elephant's Ear poisoning. Hum Exp Toxicol 2001;20:189–192.
36. Broyer M, Jouvet P, Niaudet P, Daudon M, Revillon Y. Management of oxalosis. Kidney Int 1995;49(Suppl 53): S93–S98.

Chapter 126

DIGITALIS-CONTAINING FLOWERS
(Foxglove, Lily of the Valley)

HISTORY

Digitalis purpurea was well known to medieval herbalists of the 16th and 17th centuries.[1] As a result of his extensive botanical knowledge, William Withering, the son of an English apothecary, was able to identify *Digitalis purpurea* as the principal ingredient in the herbal preparation of a local herbalist that was used to treat congestive heart failure (dropsy).[2] In 1785, he published the first scientific treatise on the use of foxglove called *An Account of the Foxglove, and Some of its Medical Uses: With Practical Remarks on Dropsy, and Other Diseases.*[3] He attributed the herb's effectiveness in the treatment of congestive heart failure to diuresis. Hall Jackson, a physician in Portsmouth, New Hampshire read this treatise, and he received *D. purpurea* seeds from William Withering for use in North America. By 1799, John Ferriar recognized that the diuresis associated with the ingestion of foxglove resulted from the effect of the plant constituents on the heart.[4]

BOTANICAL DESCRIPTION

Common Name: Foxglove, Purple Foxglove

Scientific Name: *Digitalis purpurea* L.

Botanical Family: Scrophulariaceae (snapdragon)

Physical Description: This beautiful biennial herb has simple toothed leaves and a central stalk of purple, pink, yellow, or white tubular flowers. The small fruit are ovoid capsules that contain many minute seeds.

Distribution and Ecology: Foxglove is an ornamental plant in Michigan, Illinois, the US Pacific Coast, and northern United States. In many of these areas, this plant has become naturalized. Foxglove is native to Western Europe, the Mediterranean region, and northwest Africa. In addition to the United States and Canada, this plant has become naturalized in Asia, Africa, South America, and New Zealand.

Common Name: Lily of the Valley

Scientific Name: *Convallaria majalis* L.

Botanical Family: Liliaceae (lily)

Physical Description: This perennial plant has slender rootstock and fragrant white bell-shaped flowers originating from a central stalk. The leaves are long and broad, typical of the lily family.

Distribution and Ecology: Lily of the valley is a native European plant that is grown as a garden flower in the United States. Related species include *Convallaria majuscula* Greene (American lily of the valley) from the Appalachian Mountains and *Speirantha convallarioides* Baker. from eastern China.[5]

Common Name: Star-of-Bethlehem, Pyrenees Star-of-Bethlehem, Sleepydick

Scientific Name: *Ornithogalum umbellatum* L.

Medical Toxicology of Natural Substances, by Donald G. Barceloux, MD

Botanical Family: Liliaceae (lily)

Physical Description: This perennial plant arises from a bulb basal, narrow, strap-like leaves that contain a light green midrib. The white to pale yellow flowers appear in a cluster at the top of a naked stem.

Distribution and Ecology: This bulb grows in gardens and moist meadows in the northeastern United States and eastern England.

EXPOSURE

Digitalis leaf is obtained for medicinal use by the grinding and milling of *D. purpurea* leaves. The European leaf preparations are derived from *Digitalis lanata* Ehrh., which contains primarily digoxin rather than digitoxin. Foxglove is a popular ornamental flower in the United States.

PRINCIPAL TOXINS

Physiochemical Properties

Cardioactive steroids are classified as cardenolide (5-membered lactone ring) or bufadienolide (6-membered lactone ring). The substitution of a sugar residue for a 3β-OH group on the cardenolide produces a cardiac glycoside (digitoxin, digoxin). Foxglove contains a number of cardiac glycosides, the most prominent of which is digitoxin. Each cardiac glycoside is a combination of an aglycone or genin (e.g., digoxigenin, digitoxigenin) and one to four molecules of a sugar (e.g., digitoxose or 2,6-dideoxyhexose). The lily of the valley plant contains cardiac glycosides, called convallatoxins, but apparently these glycosides are less potent than digoxin or digitoxin.

Poisonous Parts

D. purpurea contains primary glycosides including purpurea glycoside A, purpurea glycoside B, glucogitaloxin, and digitalinum verum. The plant enzyme, digipurpidase, transforms these primary glycosides into the corresponding secondary glycosides (digitoxin, gitoxin, gitaloxin, and strospeside, respectively). The mean concentration of purpurea glycoside A, purpurea glycoside B, and glucogitaloxin in dry leaf powder from *D. purpurea* was about 0.74 μg/mg, 0.39 μg/mg, and 1.91 μg/mg, respectively, as measured by high performance liquid chromatography (HPLC).[6] In five samples of dry *D. purpurea* leaf, the mean amounts of digitoxin, gitoxin, and gitaloxin in the dry leaf powder was approximately 0.22 μg/mg, 0.14 μg/mg, 0.55 μg/mg, and 0.02 μg/mg, respectively, as measured by HPLC.[7] The cardenolide content of foxglove plants varies with a variety of conditions including location and strain.[8]

DOSE RESPONSE

Most serious intoxications from digitalis-containing plants involve the ingestion of teas brewed from these plants. There are few data on dose response of these teas or plant parts.

TOXICOKINETICS

The cardiac glycosides of *D. purpurea* are hydrolyzed to digitoxin, gitoxin, and gitalin with only small amounts of digoxin formed from hepatic metabolism. Hence, *D. purpurea* toxicity may be prolonged because of the long serum half-life of digitoxin (4–5 days). Digitoxin undergoes metabolism to digitoxigenin, digitoxigenin monodigitoxoside, and gitoxin. Additionally, the kidneys excrete some digitoxin unchanged. Liquid chromatography/electrospray/mass spectrometry (LC/ES/MS) analysis of samples from a patient during the first 5 days after the ingestion of foxglove tea demonstrated digitoxin, digitoxigenin, gitoxin, and digitoxigenin monodigitoxoside in the serum and gitaloxin, digitoxin, digitoxigenin, gitoxin, and digitoxigenin monodigitoxoside in the urine.[9] Digitoxin was the digitalis glycoside present in the highest concentration (112.6 ng/mL) in the serum 8 hours after the ingestion. Relatively high serum concentrations of gitaloxin appeared only on the fifth day, when the serum gitaloxin concentration (112.6 ng/mL) exceeded the serum concentrations of all other cardiac glycosides.

CLINICAL RESPONSE

Foxglove toxicity resembles oleander and digitoxin poisoning. Gastrointestinal symptoms develop within several hours and are followed by changes in sensorium (e.g., confusion), cardiac conduction defects, bradycardia, hyperkalemia, and ventricular dysrhythmias.[10] Visual disturbances (e.g., yellow haloes) are classic symptoms of *Digitalis* leaf poisoning, and these visual changes do not usually occur following intoxication with pharmaceutical preparations of digoxin. Ventricular tachycardia, junctional rhythms, and atrial fibrillation with high-grade atrioventricular block requiring 6 days to revert to normal were reported in a foxglove tea poisoning.[11] After the ingestion of a foxglove tea, confusion and visual disturbances lasted 5 days and electrocardiogram (ECG) changes resolved in 10 days.[12] Two fatalities resulted from the ingestion of foxglove tea—an elderly woman, who was dead on arrival, and her husband, who died of refractory ventricular fibrillation 17 hours after admission.[13] There are limited clinical data on intoxications associated with lily of the valley.

DIAGNOSTIC TESTING

Methods to quantitate cardiac glycosides in this herb include reverse-phase thin layer chromatography[14] and HPLC.[15] Near-infrared spectroscopy is a rapid technique for confirmation of the presence of cardiac glycosides.[16] Elevated serum digitoxin concentrations confirm the ingestion of digitalis-like compounds (in the absence of simultaneous consumption of *Digitalis* preparations), but digitoxin or digoxin concentrations do not necessarily guide management because of variable cross-reactivity with cardenolides in these plants. The presence of hyperkalemia indicates serious digitalis poisoning.

TREATMENT

Management of intoxication with digitalis-containing plants is similar to the management for oleander and digoxin poisoning. Gut decontamination measures are usually unnecessary. Because of the low potency of *C. majalis*, small, accidental ingestions probably do not require decontamination. The administration of activated charcoal to alert patients is a therapeutic option, when the patient ingests large amounts of foxglove less than 1 hour prior to presentation; however, there are no clinical data to indicate that activated charcoal improves the outcome of patients with foxglove intoxication. The efficacy of digoxin Fab fragments in foxglove or lily of the valley intoxication has not been established. Although the use of digoxin Fab fragments is a therapeutic option, these plants contain a variety of cardiac glycosides other than digitalis, and poisoning with extracts from foxglove have not been consistently reversed by the use of digoxin-specific Fab fragments.[10]

References

1. Cook GC. Erasmus Darwin FRS (1731–1802) and the foxglove controversy. J Med Biogr 1999;7:86–92.
2. Krikler DM. The foxglove, "The old woman from Shropshire" and William Withering. J Am Coll Cardiol 1985;5: 3A–9A.
3. Goldman P. Herbal medicines today and the roots of modern pharmacology. Ann Intern Med 2001;135:594–600.
4. Eichna LW. The foxglove. Pharos 1994:57:14–16.
5. Pauli GF. The cardenolides of *Speirantha convallarioides*. Planta Med 1995;61:162–166.
6. Ikeda Y, Fujii Y, Nakaya I, Yamazaki M. Quantitative HPLC analysis of cardiac glycosides in *Digitalis purpurea* leaves. J Nat Prod 1995;58:897–901.
7. Fujii Y, Ikeda Y, Yamazaki M. High-performance liquid chromatographic determination of secondary cardiac glycosides in *Digitalis purpurea* leaves. J Chromatogr 1989; 479:319–325.
8. Usai M, Atzei AD, Marchetti M. Cardenolides content in wild Sardinian *Digitalis purpurea* L. populations. Nat Prod Res 2007;21:298–304.
9. Lacassie E, Marquet P, Martin-Dupont S, Gaulier J-M, Lachatre G. A non-fatal case of intoxication with foxglove, documented by means of liquid chromatography-electrospray-mass spectrometry. J Forensic Sci 2000;45: 1154–1158.
10. Thierry S, Blot F, Lacherade J-C, Lefort Y, Franzon P, Brun-Buisson C. Poisoning with foxglove extract: favorable evolution without Fab fragments. Intensive Care Med 2000;26:1586.
11. Dickstein ES, Kunkel FW. Foxglove tea poisoning. Am J Med 1980;69:167–169.
12. Bain RJ. Accidental digitalis poisoning due to drinking herbal tea. Br Med J 1985;290:1624.
13. Grunenfelder G, Blackman J, Fretwell M. Poisoning associated with herbal teas—Arizona, Washington. MMWR Morb Mortal Wkly Rep 1977;26:257–258.
14. Ikeda Y, Fujii Y, Umemura M, Hatakeyama T, Morita M, Yamazaki M. Quantitative determination of cardiac glycosides in *Digitalis lanata* leaves by reversed-phase thin-layer chromatography. J Chromatogr A 1996;746: 255–260.
15. Barrueto F, Jortani SA, Valdes R, Hoffman RS, Nelson LS. Cardioactive steroid poisoning from an herbal cleansing preparation. Ann Emerg Med 2003;41:396–399.
16. Kudo M, Watt RA, Moffat AC. Rapid identification of *Digitalis purpurea* using near-infrared reflectance spectroscopy. J Pharm Pharmacol 2000;52:1271–1277.

Chapter 127

JIMSON WEED and OTHER BELLADONNA ALKALOIDS

HISTORY

The toxic, mind-altering properties of belladonna alkaloids have been known since ancient times. The remains of mandrake were found in the casket of the Egyptian king, Tutankhamen. References to *Datura* and mandrake appeared in Homer's Odyssey,[1] and the Romans used mandrake as an anesthetic.[2] The Romans also used a sponge dipped in mandrake wine called *spongia somnifera* for administration to victims of crucifixion, and this "death wine" caused a stuporous state in the victims.[3] The Greek herbalist and surgeon to Nero's armies, Dioscorides, recommended the use of this plant with wine also as an anesthetic during the 1st century AD.[4,5] In the 11th century AD, Avicenna recommended the use of mandrake as an anesthetic, but he warned that overuse of this plant caused hallucinations, madness, and difficulty in breathing.[6] The use of mandrake (mandragora) as an aphrodisiac, fertility drug, hallucinogen, and herbal cure continued through medieval times and the Renaissance. The "cursed hebenon" of Shakespeare's Hamlet probably was the anticholinergic poison henbane.[7] During the Renaissance, Italian women used belladonna extracts to dilate their eyes for cosmetic purposes. References to mandrake appeared in Shakespeare's Romeo and Juliet as well as Anthony and Cleopatra.[8] Linnaeus named the deadly nightshade plant, "Atropo belladonna" after the eldest of the three fates, Atropos, whose duty was to cut the thread of life.[9,10]

Historical accounts of *Datura* poisoning in the United States date back to 1676 when British soldiers sent to Jamestown to extinguish the Bacon rebellion mistakenly ate *Datura*, "the effect of which was a very pleasant comedy, for they turned natural fools upon it for several days."[11] The intentions of the soldiers were suspect because the toxic effects lasted much longer (i.e., 11 days) than expected for a single, accidental ingestion. Indians from the southwestern United States used teas brewed from jimsonweed during religious rites and tribal ceremonies.[12] The aboriginal rock paintings by shamans from the California Chumash Indian tribe demonstrate designs of plants that suggest the shamans were under the influence of jimsonweed during the painting of these forms.[13] The use of *Datura* and *Brugmansia* species was probably imported to South America by Indian immigrants from the US Southwest and Mexico. Mayan uses for *Datura* included the treatment of hemorrhoids, skin ulcers, wounds, muscle spasms, gout, and night terrors.[14] In western South America and the upper Amazon, *Brugmansia* species were valued by local Indians for the medicinal and psychotropic properties, especially in their shamanistic religion that emphasized malevolent magic and supernatural causes of illness.[15] The preconquest Chibchas of Colombia concocted a mixture of *Brugmansia* species mixed with *chicha* (fermented beer) and tobacco that was administered to the slaves and wives of dead kings to induce a stupor prior to being buried alive with their dead masters and husbands.[15]

Modern medical reports of *Datura* intoxication date back over 200 years.[16] During the late 18th century in India, the British Physician-General of the East Indian Company, James Anderson, discovered the traditional

Medical Toxicology of Natural Substances, by Donald G. Barceloux, MD

use of *Datura* in Ayurvedic medicine for asthma, and *Datura* preparations became popular anti-asthma treatments in apothecary shops during the 19th century in England.[17] Today, the most common cause of *Datura* poisoning involves the abuse of the mind-altering, anticholinergic properties of *Datura* species by ingesting *Datura* seeds or tea.[18] "Autumnal high" is a common name for this intoxication because the seeds appear in the fall.[19]

BOTANICAL DESCRIPTION

Anticholinergic plants include *Datura* species, *Atropa belladonna* L. (deadly nightshade), *Cestrum nocturnum* L. (night jasmine), *Mandragora officinarum* L. (mandrake), *Hyoscyamine niger* L. (henbane), and *Lycium halimifolium Mill.* (boxthorn, matrimony vine). Over 12 species of *Datura* grow as weeds throughout the United States, particularly near roadsides, waste areas, cornfields, and pastures. Common US *Datura* species other than jimsonweed include *D. inoxia* P. Mill. (angel's trumpet, moonflower, desert thornapple, sacred datura, prickly burr), and *D. suaveolens* (Humb. & Bonpl. ex Willd.) Bercht. & K. Presl (angel's tears). European species include *Datura sanguinea, Brugmansia arborea* (L.) Lagerh., and *Datura aurea* (Lagerh.) Saff. These European species cause anticholinergic symptoms similar to those produced by *D. stramoni*um.[20] In addition to *M. officinarum* and *M. autumnalis*, there are four other species of the genus *Mandragora* in the Eurasian continent including *M. acaulis* Gaertn., *M. caulescens* (C. B. Clarke) Diels, *M. turcomanica* Mizgir., and *M. vernalis* Bertol.

Common Name: Jimsonweed, Thornapple, Jamestown Weed, Mad Apple, Stinkwort, Moonflower

Scientific Name: *Datura stramonium* L.

Botanical Family: Solanaceae (potato, nightshade)

Physical Description: Jimsonweed is a common annual weed that grows 3–6 ft (~1–2 m) in height and emits a rank odor. It has a simple, large, white taproot and a strong, dichotomous stem. The leaves are large, waxy and dark green with pointed margins and scalloped borders. Attractive, white tubular flowers bloom in late spring and grow 3–4 in. (~8–10 cm) long. The green fruit grows to 5 cm (~2 in.) with spicules and the fruit matures in early fall. The dried pod may contain up to five hundred 2- to 3-mm kidney-shaped, brown-black seeds. Figure 127.1 displays a specimen of jimsonweed with maturing seed pods.

Distribution and Ecology: *Datura stramonium* is native to Asia and naturalized to Canada, West Indies, and the United States. This plant is the most common *Datura* species in the eastern United States.

FIGURE 127.1. *Datura stramonium* L. (jimsonweed) with maturing seed pods. Photograph courtesy of Kimberly Ann Barceloux, MFCC. See color insert.

Common Name: Deadly Nightshade

Scientific Name: *Atropa belladonna* L.

Botanical Family: Solanaceae

Physical Description: *A. belladonna* is a freely branching bush with oval leaves and purple-black fruit that reaches 2–4 ft (~1 m) in height.

Distribution and Ecology: Deadly nightshade is a native European plant that is cultivated in the eastern United States as an ornamental plant. This species does not grow well in wild areas.

Common Name: Black Henbane

Scientific Name: *Hyoscyamus niger* L.

Botanical Family: Solanaceae

Physical Description: The black henbane plant is a biennial herb averaging about 8–32 in. (~20–81 cm) in height. The lower leaves are oval to lance-shaped with winged stems, and the flowers are dark yellow with violet veins. The thimble-shaped fruit is cone-shaped with about 200 grayish-brown seeds. The flowers appear from June to August and the seeds mature in August and September. The plant has an unpleasant odor that discourages use.

Distribution and Ecology: Henbane grows in sandy loam soils in waste areas and near roadsides.

Common Name: European Mandrake, Love Apple, Devil's Apple

Scientific Name: *Mandragora officinarum* L.

Botanical Family: Solanaceae

Physical Description: The roots of mandrake are similar to *Atropa belladonna*, except mandrake contains irregular crystals of calcium oxalate and sclereids.[21] Misidentification of this species for *Podophyllum peltatum* L. (American mandragora American mandrake) or *Borago officinalis* L. can cause toxicity.[22] The European mandrake has light green leaves that have irregular surfaces. Yellowish flowers appear in the spring in contrast to the fall for the autumn mandrake (*M. autumnalis*).

Distribution and Ecology: Mandrake is a native plant of the Mediterranean region and southern Europe.

EXPOSURE

Prior over-the-counter asthma preparations that contained extracts of *Datura* species include Asthmador (0.23–0.32% belladonna alkaloids),[23] Barter's Powder, Kinsman's Asthmatic Powder, Green Mountain Asthmatic Compound, and Haywood's Powder. Medicinal uses of these plants as Chinese herbs include the treatment of bronchitis, asthma, and pain.[24] Both *Datura* species and henbane are abused as aphrodisiacs and hallucinogens.[25]

PRINCIPAL TOXINS

Physiochemical Properties

The principal toxins in anticholinergic plant species are tropane belladonna alkaloids that primarily include hyoscyamine and scopolamine (*l*-hyoscine).[26] The ratio of hyoscyamine to scopolamine varies between anticholinergic plant species. The average ratios of hyoscyamine/scopolamine are relatively high for *A. belladonna* compared with *H. niger* and *D. stramonium*.[27] The distribution of tropane alkaloids in the plant kingdom is widespread outside the genera (*Datura*, *Brugmansia*, *Duboisia*) in the Solanaceae family including the following families: Brassicaceae, Convolvulaceae, Erythroxylaceae, Proteaceae, and Rhizophoraceae.[28] Analysis of antemortem and postmortem urine samples from patients ingesting *Datura* tea indicates that hyoscyamine and scopolamine (MW 303.36) are the main toxic compounds in *D. stramonium*.[29,30] Minor constituents include hyoscine and atropine (MW 289.37). Atropine is a racemic mixture of *d* (+) and *l* (−) hyoscyamine, but only the *l*-hyoscyamine is pharmacologically active. Figure 127.2 demonstrates the chemical structures of atropine and scopolamine.

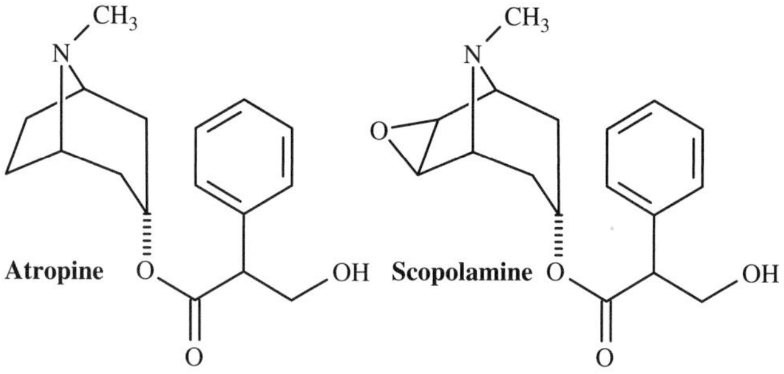

FIGURE 127.2. Chemical structures of scopolamine and atropine (*d,l* hyoscyamine).

Poisonous Parts

All parts of the *Datura* plant contain belladonna alkaloids, and case reports associate anticholinergic poisoning with teas brewed from plant parts,[29] ingestion of leaves and flowers,[31] and smoking of stramonium cigarettes.[32] However, seeds contain the highest atropine concentration (~0.1 mg/seed). A variety of factors affect the alkaloid concentrations including cultivation practices, environmental factors (temperature, moisture), storage techniques, age of the plant, part of the plant, and interspecies variation.

Datura Species

In a study of tropane alkaloid content in *D. stramonium*, tender stems from young plants and mature seeds contained mean atropine concentrations of 0.915 µg/mg and 0.384 µg/mg, respectively.[27] During a subchronic feeding study using seeds from *D. stramonium*, the mean atropine and scopolamine content of the seeds were 2.71 µg/mg and 0.66 µg/mg, respectively.[33] The ratio of atropine to scopolamine varies between *Datura* species. In a sample of seeds from *Datura stramonium* and *Datura ferox*, this ratio was approximately 5:1–7:1 in *D. stramonium* seeds, whereas the seeds from *D. ferox* had an atropine to scopolamine ratio of approximately 1:6–1:8.[34] The formation of atropine may occur during the extractive process by racemization of hyoscyamine.[27]

Atropa belladonna (Deadly Nightshade)

The alkaloid *l*-hyoscyamine represents almost all (90–95%) of the total alkaloid content, with most of the remainder being scopolamine.[35] Minor alkaloids include apoatropine, aposcopolamine, hydroxy-hyoscyamine, and tigloyloxytropane. The roots and leaves have the highest total alkaloid content (0.6% and 0.4%, respectively), followed by seeds and berries. Stems have an approximate total alkaloid concentration of 0.05% by weight. *A. belladonna* does not contain clinically significant amounts of solanine. Cooking berries from deadly nightshade does not prevent atropine toxicity.[36]

Hyoscyamus niger (Henbane)

All plant parts contain tropane alkaloids with the leaves containing from 0.05–0.15% of the total alkaloid content. Although most of the tropane alkaloids in henbane are *l*-hyoscyamine similar to *A. belladonna*,[37] the scopolamine (*l*-hyoscine) content of *H. niger* is substantially higher than *A. belladonna*. The alkaloid content averages about 0.08% in roots, 0.17% in leaves, and 0.3% in seeds.[38]

Mandragora officinarum (Mandrake)

All plant parts contain *l*-hyoscyamine and lesser amounts of scopolamine in addition to small amounts of the alkaloid, mandragorine (cuscohygrine). The estimated alkaloid content of the underground parts (rhizome, root) from mandrake varies from approximately 0.2–0.6%.[2]

Mechanism of Toxicity

Tropane belladonna alkaloids are strong anticholinergic compounds, which act as competitive antagonists to acetylcholine at central and peripheral muscarinic receptors. Peripheral muscarinic receptors innervate exocrine glands (sweating, salivation), smooth muscle, and cardiac muscle. These tertiary amines penetrate the blood–brain barrier and bind to acetylcholine receptors in the cortex and subcortical regions of the brain, producing acute psychosis. At low doses, tropane alkaloids cause peripheral anticholinergic signs including decreased salivation and sweating, mydriasis, loss of accommodation, increased pulse rate, and sinus tachycardia. As the dose increases, ileus, urinary retention, and elevated temperature occurs followed by the beginning of central nervous system (CNS) effects (confusion, agitation). Volunteer studies indicate that the peripheral effects on heart rate and blood pressure begin before CNS effects (impairment of vision, cognitive performance, and psychomotor skills), and the CNS effects persist longer than the cardiovascular effects.[39] High doses of belladonna alkaloids produce delirium, hallucination, and coma with the CNS effects being somewhat greater for scopolamine than for atropine.

DOSE RESPONSE

Because of the variable tropane alkaloid content of these plants, predicting toxicity based on the ingestion of plant parts is difficult. Based on calculations of hyoscyamine content, 50–100 *Datura* seeds contain enough atropine (3–6 mg) to produce serious anticholinergic poisoning.[11] Hall et al analyzed *Brugmansia suaveolens* (formerly *Datura suaveolens*; angel's trumpet) flowers and stated that paralysis and convulsions were likely after ingestion of six flowers or the tea from nine flowers.[31] Consumption of one half to two seed pods and alcohol produced symptoms of atropine poisoning that lasted 18 hours to 9 days. Blurred vision is the most persistent symptom. Consumption of 10 mandrake fruits (*Mandragora officinarum*) caused confusion, tachycardia, mydriasis, blurred vision, and facial flushing, which resolved in 20 hours.[40]

TOXICOKINETICS

The absorption of belladonna alkaloids is rapid through the gastrointestinal tract and the skin based on experimental studies and case reports of toxicity by these routes of exposure.[41,42] Case reports also document the onset of hallucinations and confusion after the transdermal application of scopolamine.[43] Volunteer studies indicate that the kidneys excrete about one-half of a therapeutic dose of atropine unchanged; hepatic metabolism accounts for the remainder of the dose.[44] Major metabolites include noratropine and atropine-*N*-oxide, whereas tropine and tropic acid are minor metabolites. The elimination half-life of atropine following therapeutic doses is about 2–4 hours.[45] During acute intoxication with cooked berries from *Atropa belladonna*, the plasma elimination half-life of atropine was about 2–2½ hours with most of the belladonna alkaloids eliminated from the blood within 24 hours.[44] However, the clinical effects of belladonna poisoning may last up to 48 hours, and tropane alkaloids are present in urine samples during this period.[36]

CLINICAL RESPONSE

Exposure to plants containing tropane belladonna alkaloids classically produce an anticholinergic poisoning syndrome described by the phrase "blind as a bat, hot as a hare, dry as a bone, red as a beet and mad as a wet hen."[46,47] Clinical features range from asymptomatic mydriasis to severe alteration of mental status (agitation, hallucination, coma).[48] Symptoms typically begin within 1–6 hours after ingestion of plant parts, but the onset of symptoms is much more rapid (<1 hour) after the ingestion of teas brewed from plant parts containing belladonna alkaloids.[49–51] The classical presentation of a severe intoxication includes fever, erythema, dilated pupils, blurred vision, myoclonic jerks, delirium, and hallucinations. In a case series of 31 Turkish children hospitalized after ingesting black henbane (*Hyoscyamus niger*), most patients demonstrated flushing, pupillary dilation, dizziness, ataxia, agitation, and tachycardia.[52] The differential diagnoses of anticholinergic poisoning secondary to belladonna alkaloids include intoxication with antihistamines (cold preparations, over-the-counter sleep aids), antispasmodic preparations [Donnatal® (A.H. Robins, CO., Richmond, VA), dicyclomine], benactyzine, muscle relaxants, tricyclic antidepressants (doxepin, amitriptyline, desipramine, nortriptyline, protriptyline), and antipsychotics.

Jimsonweed

Case series indicate that agitation, confusion, hallucinations, and combative behavior are common clinical manifestations of jimsonweed poisoning in patients presenting to emergency departments for treatment, particularly in adolescents who drank tea brewed from these plants.[53,54] In a review of case studies of jimsonweed intoxication from the smoking of *D. stramonium* cigarettes, dryness of the mouth and pupillary dilation were the earliest signs of intoxication.[32] Other clinical signs of toxicity in these cases included hallucinations (83%), disorientation (54%), dilated pupils (54%), dry mucous membranes (46%), fever (31%), ataxia (26%), rapid pulse (20%), amnesia (19%), and flushing (19%). Hallucinations usually involve simple images in natural colors, in contrast with the brilliant visual displays seen with lysergic acid diethylamide (LSD). The presence of bowel sounds and diaphoresis suggests an adrenergic (cocaine, amphetamine) excess rather than an anticholinergic (jimsonweed) poisoning. Tactile hallucinations are uncommon. Amnesia may last up to 24 hours during anticholinergic intoxication.[55] Seizures occur rarely during jimsonweed intoxication,[31,56,57] but patients may present in coma, particularly following the ingestion of teas.[58] Fatalities usually result from trauma or drowning rather than medical complications (respiratory failure, arrhythmias).[32,49,59] Symptoms typically resolve in 24 to 48 hours, although pupillary dilation may continue up to 1 week.[60,61] Proliferation of jimsonweed in cornfields has led to an occupational disease known as cornpicker's pupil in which the pulverized jimsonweed dust caused mydriasis for several days after mechanical harvesting.[62,63] Asymmetrical anisocoria can occur in patients following ocular contact with parts of *Datura* species, and this effect usually resolves within 3 days.[64]

Deadly Nightshade

The ingestion of berries from deadly nightshade (*Atropa belladonna*) produces the clinical features of atropine poisoning. Following the ingestion of deadly nightshade berries by an elderly man was associated with delirium, hallucinations, tachycardia, fever, diffuse hypertonia, and amnesia that resembled delirium tremens.[46] His symptoms resolved the day after ingestion. Respiratory failure associated with the ingestion of deadly nightshade berries may require ventilatory support for 24–30 hours. The delirium and hallucination usually resolve within 48 hours, but mydriasis may persist for several more days.[36]

Mandrake

Ingestion of mandrake causes gastrointestinal distress and anticholinergic symptoms similar to other plant species containing tropane alkaloids. The ingestion of leaves from *Mandragora autumnalis* by a 72-year-old woman was associated with nausea, hallucination, confusion, agitation, tachypnea, tachycardia, and mydriasis.[65] She recovered with supportive care. Other clinical effects of the ingestion of mandrake include hypertension, dry mouth, urinary retention, and flushed skin.[66]

DIAGNOSTIC TESTING

Analytical methods to quantitate tropane alkaloids (hyoscyamine, scopolamine) include thermospray high performance liquid chromatography/mass spectrometry (HPLC/MS),[67] HPLC/tandem mass spectrometry (MS/MS),[68] and gas chromatography/mass spectrometry techniques using special columns and dichloromethane extract with a limit of detection ranging down to 5 ng/mL.[69] There are limited forensic data on scopolamine and atropine concentrations. In a serum sample from a mild case of *Datura inoxia* poisoning, the serum hyoscyamine concentration was 12 ng/mL, while the scopolamine concentration was below the limit of quantitation (<5 ng/mL).[69] A 19-year-old man, who was found dead, intentionally ingested an unknown quantity of *Datura stramonium* seeds.[70] Postmortem blood samples contained hyoscyamine and scopolamine concentrations of 1.1 μg/mL, and 0.2 μg/mL, respectively. Two case reports detailing the postmortem blood samples of two men following the fatal ingestion of atropine and scopolamine demonstrated atropine and scopolamine concentrations of 0.4 mg/kg and 0.13 mg/kg, respectively.[71] Both men were found dead and their stomach contents contained detectable amounts of these substances. Laboratory abnormalities include elevated concentrations of serum hepatic aminotransferases, lactate dehydrogenase (LDH), and bilirubin;[56] however, the role of liver necrosis in these cases is unclear because postmortem examinations of the liver demonstrate only subcapsular petechiae.[29] Some of these changes may result from rhabdomyolysis secondary to agitation, decerebrate posturing, and myoclonic jerking during anticholinergic intoxication. Electroencephalographic changes during periods of altered mental status include prominent lambda activity, high-amplitude slow wave activity, and occasionally periodic short bursts of high-voltage sharp waves.[72] No marked suppression of background activity usually appears between bursts of sharp wave activity.

TREATMENT

Stabilization

Most patients respond well to supportive and protective measures. The patient should be placed in a reassuring, low-stimulus environment with a familiar person. Pharmacological intervention should be limited to those patients who remain severely agitated after supportive measures. Hypertension is usually transient and antihypertensive therapy is typically unnecessary.

Decontamination

There are few clinical data to guide management of patients with anticholinergic intoxication after exposure to belladonna alkaloids in plants from the Solanaceae family. Although tropane belladonna alkaloids are well absorbed from solution, the anticholinergic properties of these compounds may delay the absorption of these compounds from plant parts and seeds. Theoretically, patients may benefit from decontamination measures (i.e., activated charcoal) beyond the usual 1 hour after ingestion. However, the occurrence of altered mental status (delirium, confusion, agitation) during anticholinergic poisoning increases the risk of aspiration. Consequently, decontamination measures are usually unnecessary.

Measures to Enhance Elimination

There is no clinical evidence to support the use of hemodialysis or forced diuresis to increase the elimination of belladonna alkaloids during anticholinergic poisoning.

Antidotes

Physostigmine is an antagonist to serious atropine poisoning (severe CNS or peripheral anticholinergic symptoms, dysrhythmias, seizures, hypertension or hallucinations). The initial adult dose is 0.5–2 mg (0.01–0.03 mg/kg) intramuscularly or intravenously over 2–5 minutes. Improvement usually occurs within 15 minutes and a second dose may be repeated in 20–30 minutes. The onset of action for intravenous physostigmine is 3–8 minutes and the duration of action is 30–60 minutes. Signs of cholinergic excess (bradycardia, heart block, excessive secretions) indicate excessive doses of physostigmine. Repeated doses should be given only if symptoms reappear. Relative contraindications to the administration of physostigmine include a history of cardiovascular disease, gangrene, asthma, glaucoma, and gastrointestinal or genitourinary obstruction. Because symptoms usually resolve within 24 hours, only three to four doses are usually necessary.[56] Physostigmine is not routinely administered for mild cases of anticholinergic poisoning. Patients with persistent signs of anticholinergic poisoning (agitation, hallucination, seizures) should be admitted for at least 24 hours of observation.

Supplemental Care

Benzodiazepines (diazepam, lorazepam) are the initial drugs of choice for agitation during anticholinergic intoxication. Phenobarbital, phenothiazine compounds, and haloperidol should be administered cautiously, if at all, because of potential potentiation of anticholinergic effects (phenobarbital, phenothiazines) or lowering of seizure thresholds (haloperidol).[73] Vital signs should be monitored every 15–20 minutes until improvement occurs. Hypertension is usually transient and requires no therapy. Hyperthermia may be treated with a cooling blanket or a sponge bath. Catheterization may be necessary for urinary retention. Baseline laboratory tests for patients with severe anticholinergic poisoning (seizures, muscle rigidity, hyperthermia) include hepatic transaminases, bilirubin, creatine kinase, prothrombin time, and urinalysis for myoglobin. Patients who exhibit CNS symptoms (e.g., disorientation, hallucinations, delusions) should be observed until those symptoms resolve. Patients with mild symptoms may be discharged from the emergency department after 4–6 hours of observation, if symptoms resolve.

References

1. Weintraub S. Stramonium poisoning. Postgrad Med 1960; 28:364–367.

2. Hanus LO, Rezanka T, Spizek J, Dembitsky VM. Substances isolated from *Mandragora* species. Phytochemistry 2005;66:2408–2417.
3. Carter AJ. Narcosis and nightshade. Br Med J 1996;313:1630–1632.
4. Peduto VA. The mandrake root and the Viennese Dioscorides. Minerva Anestesiol 2001;67:751–766.
5. Cule J. The devil's apples. Vesallus 1997;3:95–105.
6. Aziz E, Nathan B, McKeever J. Anesthetic and analgesic practices in Avicenna's Canon of Medicine. Am J Chin Med 2000;28:147–151.
7. Kotsias BA. Scopolamine and the murder of King Hamlet. Arch Otolaryngol Head Nec Surg 2002;128:847–849.
8. Lazenby E. Mandrake. Rep Proc Scott Hist Med 1994–96:39–52.
9. Lee MR. Solanaceae IV: *Atropa belladonna*, deadly nightshade. J R Coll Physicians Edinb 2007;37:77–84.
10. Holzman RS. The legacy of Atropos, the fate who cut the thread of life. Anesthesiology 1998;89:241–249.
11. Brown JK, Malone MH. "Legal highs"—Constituents, activity, toxicology, and herbal folk lore. Clin Toxicol 1978;12:1–31.
12. Shultes RE. Hallucinogens of plant origin. Science 1969;163:245–254.
13. Wellman KE. North American Indian rock art and hallucinogenic drugs. JAMA 1978;239:1524–1527.
14. Litzinger WJ. Yucateco and Lacandon Maya knowledge of *Datura* (Solanaceae). J Ethnopharmacol 1994;42: 133–134.
15. Lockwood TE. The ethnobotany of *Brugmansia*. J Ethnopharmacol 1979;1:147–164.
16. Rush B. An account of the effects of the stramonium or thorn apple. Clin Pediatr 1973; 12:50–53.
17. Cohen SG. Asthma among the famous. James Anderson (1738–1809). Allergy Asthma Proc 1996;17:165–167.
18. Klein-Schwartz W, Oderda GM. Jimsonweed intoxication in adolescents and young adults. Am J Dis Child 1984;138: 737–739.
19. Moore DW. The autumnal high: Jimson weed in North Carolina. NC Med J 1976;37:492–494.
20. Belton PA, Gibbons DO. Datura intoxication in West Cornwall. Br Med J 1979;1:585–586.
21. Berry MI, Jackson BP. European mandrake (*Mandragora officinarum* and *M. autumnalis*) the structure of the rhizome and root. Planta Med 1976;30:281–290.
22. Frasca T, Brett AS, Yoo SD. Mandrake toxicity a case of mistaken identity. Arch Intern Med 1997;157:2007–2009.
23. Gabel MC. Purposeful ingestion of belladonna for hallucinatory effects. J Pediatr 1968;72:865–866.
24. Chan TY. Anticholinergic poisoning due to Chinese herbal medicines. Vet Hum Toxicol 1995;37;156–157.
25. Tugrul L. Abuse of henbane by children in Turkey. Bull Narcot 1985;37:75–78.
26. Hudson MJ. Acute atropine poisoning from ingestion of *Datura rosei*. NZ Med J 1973;77:245–248.
27. Miraldi E, Masti A, Ferri S, Camparini IB. Distribution of hyoscyamine and scopolamine in *Datura stramonium*. Fitoterapia 2001;72;644–648.
28. Griffin WJ, Lin GD. Chemotaxonomy and geographical distribution of tropane alkaloids. Phytochemistry 2000; 53:623–637.
29. Nogue S, Pujol L, Sanz P, de la Torre R. *Datura stramonium* poisoning. Identification of tropane alkaloids in urine by gas chromatography-mass spectrometry. J Int Med Res 1995;23:132–137.
30. Urich RW, Bowerman DL, Levisky JA, Pflug JL. *Datura stramonium*: A fatal poisoning. J Forensic Sci 1982;27:948–954.
31. Hall RW, Popkin MK, McHenry LE. Angel's trumpet psychosis: A central nervous system anticholinergic syndrome. Am J Psychiatry 1977;134:312–314.
32. Gowdy JM. Stramonium intoxication: A review of symptomatology in 212 cases. JAMA 1972;221:585–587.
33. Dugan GM, Gumbmann MR, Friedman M. Toxicological evaluation of jimson weed (*Datura stramonium*) seed. Food Chem Toxicol 1989;27:501–510.
34. Steenkamp PA, Harding NM, van Heerden FR, Van Wyk B-E. Fatal *Datura* poisoning: identification of atropine and scopolamine by high performance liquid chromatography/ photodiode array/mass spectrometry. Forensic Sci Int 2004;145:31–39.
35. Pestalozzi BC, Caduff F. Group poisoning by belladonna. Schweiz Med Wochenschr 1986;116:924–926.
36. Schneider F, Lutun P, Kintz P, Astruc D, Flesch F, Tempe J-D. Plasma and urine concentrations of atropine after the ingestion of cooked deadly nightshade berries. Clin Toxicol 1996;34:113–117.
37. Eeva M, Salo J-P, Oksman-Caldentey K-M. Determination of the main tropane alkaloids from transformed *Hyoscyamus muticus* plants by capillary zone electrophoresis. J Pharmaceutical Biomed Anal 1998;16:717–722.
38. Spoerke DG, Hall AH, Dodson CD, Stermitz FR, Swanson CH Jr, Rumack BH. Mystery root ingestion. J Emerg Med 1987;5:385–388.
39. Ellinwood EH Jr, Nikaido AM, Gupta SK, Heatherly DG, Nishita JK. Comparison of central nervous system and peripheral pharmacodynamics to atropine pharmacokinetics. J Pharmacol Exp Ther 1990;255:1133–1139.
40. Vlachos P, Poulos L. A case of mandrake poisoning. J Toxicol Clin Toxicol 1982;19:521–522.
41. Pereira CA, Nishioka Sde A. Poisoning by the use of *Datura* leaves in a homemade toothpaste. Clin Toxicol 1994;32:329–331.
42. Janes RG, Stiles JF. The penetration of C^{14} labeled atropine into the eye. Arch Ophthalmol 1959;67:97–102.
43. Wilkinson JA. Side effects of transdermal scopolamine. J Emerg Med 1987;5:389–392.
44. Van der Meer MJ, Hundt HK, Muller FO. The metabolism of atropine in man. J Pharm Pharmacol 1986;38:781–784.

45. Hayden PW, Larson SM, Lakshminarayan S. Atropine clearance from human plasma. J Nucl Med 1979;20; 366–367.
46. Trabattoni G, Visintini D, Terzano GM, Lechi A. Accidental poisoning with deadly nightshade berries: a case report. Hum Toxicol 1984;3:513–516.
47. Sands JM, Sands R. Henbane chewing. Med J Aust 1976;2:55–58.
48. Tiongson J, Salen P. Mass ingestion of jimson weed by eleven teenagers. Del Med J 1998;70;471–476.
49. Coremans P, Lambrecht G, Schepens P, Vanwelden J, Verhaegen H. Anticholinergic intoxication with commercially available thorn apple tea. Clin Toxicol 1994;32:589–592.
50. Chang S-S, Wu M-L, Lee C-C, Chin T-F, Liao S-J. Poisoning by *Datura* leaves used as edible wild vegetables. Vet Hum Toxicol 1999;41:242–245.
51. Hayman HJ. *Datura* poisoning. The angel's trumpet. Pathology 1985;17:465–466.
52. Kurkcuoglu M. Henbane (*Hyoscyamus niger*) poisonings in the vicinity of Erzurum. Turkish J Pediatr 1970;12: 48–56.
53. Gopel C, Laufer C, Marcus A. Three cases of angel's trumpet tea-induced psychosis in adolescent substance abusers. Nord J Psychiatry 2002;56:49–52.
54. Birmes P, Chounet V, Marzeolles M, Cathala B, Schmitt L, Lauque D. Self-poisoning with *Datura stramonium*: 3 case reports. Presse Med 2002;31:69–72 [French].
55. Johnson RT. Jimson weed toxicity. Clin Med 1977;84: 14–15.
56. Mikolich JR, Paulson GW, Cross CJ. Acute anticholinergic syndrome due to jimson weed ingestion. Clinical and laboratory observation in six cases. Ann Intern Med 1975;83: 321–325.
57. Rosen CS, Lechner M. Jimson weed intoxication. N Engl J Med 1967;267:448–450.
58. Diker D, Markovitz D, Rothman M, Sendovski U. Coma as a presenting sign of *Datura stramonium* seed tea poisoning. Eur J Intern Med 2007;18:336–338.
59. Blattner RJ. Jimson weed poisoning. Stramonium intoxication. J Pediatr 1962;61:941–943.
60. Al-Shaikh AM, Sablay ZM. Hallucinogenic plant poisoning in children. Saudi Med J 2005;26:118–121.
61. Francis PD, Clarke DF. Angel trumpet lily poisoning in five adolescents: clinical findings and management. J Paediatr Child Health 1999;35:93–95.
62. Thompson HS. Cornpicker's pupil: Jimson weed mydriasis. J Iowa Med Soc 1971;61:475–478.
63. Savitt DL, Roberts JR, Siegel EG. Anisocoria from jimson weed. JAMA 1986;255:1439–1440.
64. Firestone D, Sloane C. Not your everyday anisocoria: angel's trumpet ocular toxicity. J Emerg Med 2007;33: 21–24.
65. Piccillo GA, Miele L, Mondati E, Moro PA, Musco A, Forgione A, et al. Anticholinergic syndrome due to "devil's herb": when risk come from the ancient time. Int J Clin Pract 2006;60:492–494.
66. Piccillo GA, Mondati EG, Moro PA. Six clinical cases of *Mandragora autumnalis* poisoning: diagnosis and treatment. Eur J Emerg Med 2002;9:342–347.
67. Auriola S, Martinsen A, Oksman-Caldentey KM, Naaranlahti T. Analysis of tropane alkaloids with thermospray high-performance liquid chromatography-mass spectrometry. J Chromatogr 1991;562:737–744.
68. Xu A, Havel J, Linderholm K, Hulse J. Development and validation of an LC/MS/MS method for the determination of L-hyoscyamine in human plasma. J Pharm Biomed Anal 1995;14:33–42.
69. Namera A, Yashiki M, Hirose Y, Yamaji S, Tani T, Kojima T. Quantitative analysis of tropane alkaloids in biological materials by gas chromatography-mass spectrometry. Forensic Sci Int 2002;130:34–43.
70. Boumba VA, Mitselou A, Vougiouklakis T. Fatal poisoning from ingestion of *Datura stramonium* seeds. Vet Hum Toxicol 2004;46:81–82.
71. Sticht G, Kaferstein H, Staak M. Results of toxicological investigations of poisonings with atropine and scopolamine. Acta Med Leg Soc 1989;39:441–447.
72. Hanna JP,Schmidley JW, Braselton WE Jr. *Datura* delirium. Clin Neuropharmacol 1992;15:109–113.
73. Shenoy RS. Pitfalls in the treatment of jimsonweed intoxication. Am J Psychiatry 1994;151:1396–1397.

Chapter 128

LUPINES and FALSE LUPINE

HISTORY

Many species of lupines contain teratogenic quinolizidine and piperidine alkaloids. Although lupine seeds contain these toxic alkaloids, a "debittering" process used by natives in the Andean Highlands allowed the use of these seeds as a food crop for many years.[1] Since the late 1800s, western US stockmen recognized the toxicity associated with the ingestion of lupine hay by grazing animals. Large losses of livestock occurred in Montana and other western states as a result of the heavy grazing of lupine pods and seeds by cattle and sheep during the late summer and early fall.[2] In the 1950s, the eponym "crooked calf disease" was given to the lupine-induced skeletal malformations associated with the ingestion of lupines by pregnant cows.[3]

BOTANICAL DESCRIPTION

Common Name: Lupines, Lupins

Scientific Name: *Lupinus* (genus)

Botanical Family: Fabaceae (Leguminosae)

Physical Description: Lupines have alternate, palmately compound leaves (5–17 oblong to lanceolate leaflets) and a variety of colored flowers (red, yellow, blue, white, purple). The plant flowers in early to mid-summer with pods appearing in late summer and early fall.

Distribution and Ecology: There are more than 500 species of annual, perennial, and soft woody lupines worldwide with about 200–300 species in North and South America and approximately 12 species in Europe and Africa.[4] In the United States, lupines are most common in the West, with 150 species growing west of the Rocky Mountains.

Common Name: False Lupine, False Lupin, Prairie Buckbean, Golden Banner, Mountain Thermopsis, Mountain Pea, Yellow Pea, Buffalo Pea, Golden Pea

Scientific Name: *Thermopsis rhombifolia* (Pursh) Richardson (*T. montana*)

Botanical Family: Fabaceae (Leguminosae)

Physical Description: The foliage and inflorescence of this species resembles lupines. This erect perennial herb has palmately compound leaves with three leaflets. The pea-like, showy flowers are yellow or purple with lupine-like racemes. Flowers bloom in May to July and subsequently produce long pods reaching 5–8 cm (~2–3 in.) in length.

Distribution and Ecology: *Thermopsis rhombifolia* is adapted to a variety of habitats including grasslands, desert shrub areas, and mountain meadows. This plant inhabits areas from the Pacific Coast to the eastern Rocky Mountains. Several other similar species inhabit meadows of Southeast woodlands including Carolina lupine [*T. villosa* (Walt.) Fern. & Schub.] and Allegheny Mountain goldenbanner [*T. mollis* (Michx.) M.A. Curtis ex Gray]. Central and eastern Asia has other varieties of *Thermopsis* species.

Medical Toxicology of Natural Substances, by Donald G. Barceloux, MD

EXPOSURE

Lupines are cultivated in Australia, South America, Eastern Europe, and the Mediterranean region as supplemental feed for livestock with regulatory limitations of 100 μg alkaloids/g material. Humans also consume the "sweetened" seeds in dishes containing lupine flour. Increasing the protein and total amino acid content of wheat flour by the addition of legume seed flour improves the nutritional content of the bread. The high-lysine, low-methionine content of lupine seed flour complements the low-lysine, high-sulfur-containing amino acids in wheat flour.[5] Sweetened lupine seeds are an appetizer in southern Spain. In southern Europe and the Middle East, boiled lupines are home remedies for diabetes. In the Russian *Pharmacopoeia*, extracts of *Thermopsis lanceolata* were substitutes for ipecac.[6]

PRINCIPAL TOXINS

Physiochemical Properties

The lupine (*Leguminosae*) family contains more than 150 quinolizidine alkaloids including lupanine, 13-hydroxylupanine, sparteine, and the teratogen, anagyrine. Lupanine (CAS RN: 550-90-3, $C_{15}H_{24}N_2O$) and sparteine (CAS RN: 90-39-1, $C_{15}H_{26}N_2$) are structurally similar compounds as demonstrated in Figure 128.1 and Figure 128.2. Lupanine is a ketonic derivative of sparteine, and animal studies suggest that lupinine is substantially less toxic than sparteine.[7] Analysis of lupine seeds indicates that lupanine is the most prevalent alkaloid in the seeds.[1] Other major alkaloids include sparteine, 4-hydroxylupanine, and 13-hydroxylupanine. Alkaloid content varies with environmental conditions, season, stage of growth, and species with high alkaloid content during the early growth stages, decreasing during the flowering stage, and increasing during the seeding.[4,8] Prolonged soaking in water and cooking are necessary to remove toxic amounts of alkaloids in lupine seeds.[9] Lupine extracts contain low-molecular-weight compounds that are waste products from the removal process for bitter (high-alkaloid) lupine seeds.[10] Table 128.1 lists some of the physical properties of sparteine. Alkaloids in *T. rhombifolia* include cystisine, *N*-methylcystisine, hydroxylupanine, thermopsine, and anagyrine.[6] Trace amounts of lupinine (CAS RN: 486-70-4, $C_{10}H_{19}NO$) and 17-oxosparteine also occur in this plant species.

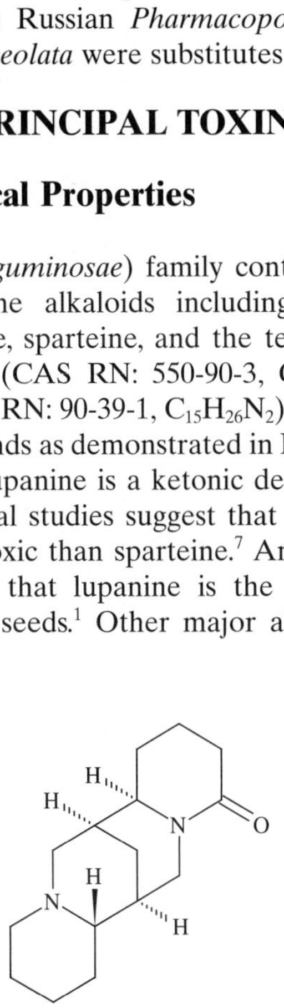

FIGURE 128.1. Chemical structure of lupanine.

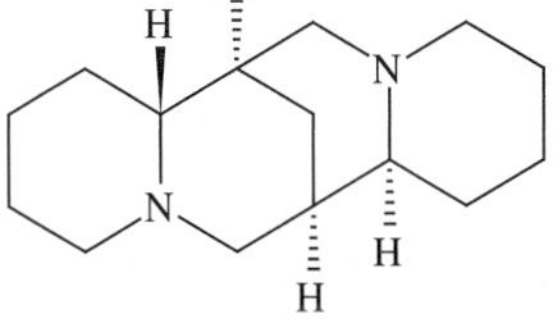

FIGURE 128.2. Chemical structure of sparteine.

Poisonous Parts

The concentrations of major alkaloids are highest in the stems and leaves early in the growth phase, and these alkaloids decrease as the plant matures.[11] The exception is the intact mature seed, which retains the highest alkaloid concentrations of all plant parts. Sweet lupine refers to seeds from cultivars of the narrow-leafed lupine (*Lupinus angustifolius* L.) with low alkaloid content compared with the bitter, wild-type lupine seeds. The "sweeting" (i.e., "debittering") process that converts bitter lupine seeds to edible food sources reduces the alkaloid content from about 3% to approximately 0.02%.[1] The seeds of modern cultivars of *L. angustifolius* contain <200 mg alkaloids/kg compared with bitter varies that contain 5–40 g alkaloid/kg.[12] The typical alkaloid content of sweet lupines is approximately 42–59% lupanine, 24–45% 13-hydroxylupanine, 7–15% angustifoline, and 1–1.5% α-isolupanine.[12] Tracer studies indicate that shoot tissues synthesize the majority of quinolizidine alkaloids (e.g., lupanine, 13-hydroxylupanine), and redistribution of these alkaloids occurs by translocation in the phloem.[13]

TABLE 128.1. Some Physical Properties of Sparteine

Property	Value
Melting Point	30.5 °C/86.9 °F
Boiling Point	325 °C/617 °F
log P (Octanol-Water)	2.730
Water Solubility	3040 mg/L (22 °C/71.6 °F)
Vapor Pressure	1.79E-04 mmHg (25 °C/77 °F)
Henry's Law Constant	1.16E-08 atm-m^3/mole (25 °C/77 °F)
Atmospheric OH Rate Constant	1.26E-10 cm^3/molecule-second (25 °C/77 °F)

The highest concentrations of alkaloids in *Thermopsis rhombifolia* occur in the sprouts with decreasing amounts in the flowering and fruiting tops. Analysis of alkaloid content indicate that *N*-methylcytisine is abundant in early collections of leaves and stems, but the flowers and fruiting tops contain minimal amounts of this substance.[6] This study indicated that maximum hydroxylupanine concentrations occurred late in the growing season. The ingestion of the seeds or flowers is the most common cause of poisoning.[14]

Mechanism of Toxicity

Some of the quinolizidine alkaloids in *Lupinus* species possess varying degrees of muscarinic and nicotinic activity. Lupanine, 13-hydroxylupanine, and sparteine block nicotinic cholinergic receptors and possess weak antagonism of muscarinic cholinergic receptors.[7] In animal studies, both lupanine and sparteine are substantially (i.e., 6- to 10-fold) more toxic following intravenous than oral administration.[15]

Lupinus species contain quinolizidine (anagyrine) and piperidine (ammodendrine) alkaloids that produce congential birth defects (cleft palate, multiple congential contractures, bowed limbs, scoliosis, kyphosis, torticollis) in pregnant goats gavaged with plant material during gestation days 30–60.[16,17] Lupinosis is a veterinary disease characterized by hepatic steatosis, anorexia, weight loss, jaundice, and stupor. The disease occurs in animals grazing on *Lupinus* species contaminated by a mycotoxin-producing species of the genus *Phomopsis*. There are no case reports of a similar disease in humans.

In animals, ingestion of dry foliage of *T. rhombifolia* produces a severe muscle degeneration manifest by persistent recumbency.[18] The presence of the teratogen, anagyrine, suggests that the grazing of animals on this plant may produce defects of limb rotation, arthrogryposis, or other teratogenic effects similar to those that occur from the grazing of lupines.

DOSE RESPONSE

The ingestion of an estimated 24 mg lupanine/kg body weight produced acute toxicity in an adult ingesting bitter lupine beans.[19] Pediatric case reports associate the ingestion of six peas and two to three flowers from *T. rhombifolia* with nausea, vomiting, lethargy, and headache.[14] The ingestion of an estimated 20 flowers produced similar symptoms.

TOXICOKINETICS

Volunteer studies indicate that the kidneys excrete most of a 10 mg dose of lupanine and 13-hydroxylupanine unchanged in the urine.[12] During this study, the mean elimination half-lives of these compounds were approximately 6 hours.

CLINICAL RESPONSE

The ingestion of "debittering" water used to remove toxic alkaloids from lupine seeds produced anticholinergic poisoning manifest by generalized weakness, palpitations, sinus tachycardia, premature ventricular contractions, dry mouth, and urinary retention within 1 hour of the ingestion.[20] Most of the symptoms resolved within several hours. Case reports associate an anticholinergic syndrome (tachycardia, blurred vision, anxiety, generalized weakness, tachypnea, mydriasis) with the ingestion of seeds[21] or the bitter water from boiling lupine seeds in water.[22] Analysis of the bitter water demonstrated the dibasic quinolizidine alkaloid, sparteine. Impaired hydroxylation of sparteine may predispose some patients (i.e., slow metabolizers) to the toxic effects of sparteine. Ingestion of lupine beans can cause pupillary dilation, blurred vision, orthostatic hypotension, and an elevated heart rate.[23] Allergic responses to lupine seeds may occur, particularly in patients with allergies to peanuts.[24] Rarely, anaphylaxis (e.g., angioedema, dyspnea) develops following the ingestion of lupine flour.[25]

The peas from *Thermopsis rhombifolia* are strong emetics. Case reports associated the ingestion of peas and flowers from this plant with nausea, vomiting, and headache that usually resolves within 12–24 hours. Almost all ingestions result in nausea and vomiting. Symptoms of dehydration may occur, but systemic effects usually are minimal after the incidental ingestion of parts of this plant.[26]

DIAGNOSTIC TESTING

Analytical methods for the detection and quantitation of lupine alkaloids (lupanine, anagyrine, sparteine) include gas chromatography with flame ionization or nitrogen phosphorus detection and gas chromatography/mass spectrometry (GC/MS).[27] Detection limits for the latter method of detection ranged from 0.025–0.2 μg/g sample.[28] A high performance liquid chromatographic (HPLC) method can detect sparteine and the 2-dehydro and 5-dehydro metabolites in biological samples with a limit of detection in the range of 2 ng/mL.[29]

TREATMENT

Treatment is supportive. Pediatric patients should be evaluated for dehydration and electrolyte imbalance. The incidental ingestion of plant parts by children is not usually associated with systemic complications, and decontamination measures are unnecessary. Most causal ingestions may be managed as outpatients unless dehydration occurs.

References

1. Hatzold T, Elmadfa I, Gross R, Wink M, Hartmann T, Witte L. Quinolizidine alkaloids in seeds of *Lupinus mutabilis*. J Agric Food Chem 1983;31:934–938.
2. Chestnut VK, Wilcox EV. The stock-poisoning plants of Montana: A preliminary report. USDA Bull 1901;26:100–110.
3. Binns W, James LF. A congenital deformity in calves, similar to "crooked calf disease," has been experimentally produced by feeding heifers lupine and lead. Proc Western Sec Am Soc Animal Prod 1961;66(12):1–3.
4. Panter KE, James LF, Gardner DR. Lupines, poison-hemlock and *Nicotiana* spp: Toxicity and teratogenicity in livestock. J Nat Toxins 1999;8:117–134.
5. Mubarak AE. Chemical, nutritional and sensory properties of bread supplemented with lupin seed (*Lupinus albus*) products. Nahrung 2001;45:241–245.
6. Keller WJ, Cole FR. The alkaloids of *Thermopsis montana*. Lloydia 1969;32:498–502.
7. Yovo K, Huguet F, Pothier J, Durand M, Breteau M, Narcisse G. Comparative pharmacological study of sparteine and its ketonic derivative lupinine from seeds of *Lupinus albus*. Planta Med 1984;50:420–424.
8. Wink M, Meibner C, Witte L. Patterns of quinolizidine alkaloids in 56 species of the genus *Lupinus*. Phytochemistry 1995;38:139–153.
9. Smith RA. Potential edible lupine poisonings in humans. Vet Hum Toxicol 1987;29:444–445.
10. Stobiecki M, Blaszczyk B, Kowalczyk-Bronisz SH, Gulewicz K. The toxicity of seed extracts and their fractions from *Lupinus angustifolius* L. and *Lupinus albus* L. J Appl Toxicol 1993;13:347–352.
11. Keeler RF, Cronin EH, Shupe JL. Lupin alkaloids from teratogenic and nonteratogenic lupins. IV. Concentration of total alkaloids, individual major alkaloids, and the teratogen anagyrine as a function of plant part and stage of growth and their relationship to crooked calf disease. J Toxicol Environ Health 1976;1:899–908.
12. Petterson DS, Greirson BN, Allen DG, Harris DJ, Power BM, Dusci LJ, Ilett KF. Disposition of lupanine and 13-hydroxy lupanine in Man. Xenobiotica 1994;24:933–941.
13. Lee MJ, Pate JS, Harris DJ, Atkins CA. Synthesis, transport and accumulation of quinolizidine alkaloids in *Lupinus albus* L. and *L. angustifolius* L. J Exp Bot 2007;58:935–946.
14. Spoerke DG, Murphy MM, Wruk KM, Rumack BH. Five cases of *Thermopsis* poisoning. Clin Toxicol 1988;26:397–406.
15. Petterson DS, Ellis ZL, Harris DJ, Spadek ZE. Acute toxicity of major alkaloids of *Lupinus angustifolius* seed to rats. J Appl Toxicol 1987;7:51–53.
16. Panter KE, Keeler RF, Bunch TD, Callan RJ. Congenital skeletal malformations and cleft palate induced in goats by ingestion of *Lupinus*, *Conium* and *Nicotiana* species. Toxicon 1990;28:1377–1385.
17. Panter KE, Keeler RF. *Conium*, *Lupinus*, and *Nicotiana* alkaloids: fetal effects and the potential for residues in milk. Vet Hum Toxicol 1990;32(suppl):89–94.
18. Keeler RF, Johnson AE, Chase RL. Toxicity of *Thermopsis montana* in cattle. Cornell Vet 1986;76:115–127.
19. Hadorn H. Ueber die zusammensetzung von frischei und gefriervollei. Mitt Lebensmitt Hyg 1973;64:187–205.
20. Marquez LR, Guitierrez-Rave M, Miranda FI. Acute poisoning by lupine seed debittering water. Vet Hum Toxicol 1991;33:265–267.
21. Di Grande A, Paradiso R, Amico S, Fulco G, Fantauzza B, Noto P. Anticholinergic toxicity associated with lupin seed ingestion: case report. Eur J Emerg Med. 2004;11:119–120.
22. Tsiodras S, Shin RK, Christian M, Shaw LM, Sass DA. Anticholinergic toxicity associated with lupine seeds as a home remedy for diabetes mellitus. Ann Emerg Med 1999;33:715–717.
23. Lowen RJ, Alam FKA, Edgar JA. Lupin bean toxicity. Med J Aust 1995;162:256–257.
24. Moneret-Vautrin D-A, Guerin L, Kanny G, Flabbee J, Fremont S, Morisset M. Cross-allergenicity of peanut and lupine: the risk of lupine allergy in patients allergic to peanuts. J Allergy Clin Immunol 1999;104:883–888.
25. Wassenberg J, Hofer M. Lupine-induced anaphylaxis in a child without known food allergy. Ann Allergy Asthma Immunol 2007;98:589–590.
26. McGrath-Hill CA, Vicas IM. Case series of *Thermopsis* exposures. Clin Toxicol 1997;35:659–665.
27. Reinhard H, Rupp H, Sager F, Streule M, Zoller O. Quinolizidine alkaloids and phomopsins in lupin seeds and lupin containing food. J Chromatogr A 2006;1112:353–360.
28. Holstege DM, Seiber JN, Galey FD. Rapid multiresidue screen for alkaloids in plant material and biological samples. J Agric Food Chem 1995;43:691–699.
29. Moncrieff J. Simultaneous determination of sparteine and its 2-dehydro and 5-dehydro metabolites in urine by high-performance liquid chromatography with electrochemical detection. J Chromatogr 1990;529:194–200.

Chapter 129

MAYAPPLE
(*Podophyllum peltatum* L.)

HISTORY

The resin extract of *Podophyllum peltatum* was used as a cathartic during the 19th century. The crystalline substance, podophyllotoxin was isolated in 1880, but the chemical structure of podophyllotoxin was not determined as a diastereoisomer until 1951.[1] In 1942, Kaplan introduced the use of 25% podophyllum as a single, topical treatment for venereal warts (condyloma acuminatum).[2]

BOTANICAL DESCRIPTION

The US variety of mayapple is *Podophyllum peltatum*; the Indian variety is *Podophyllum hexandrum* Royle (Himalayan mayapple or *Podophyllum emodi*).

Common Name: Mayapple

Scientific Name: *Podophyllum peltatum* L.

Botanical Family: Berberidaceae (bayberry)

Physical Description: Mayapple is a perennial plant with a horizontal fleshy rootstock, thick fibrous roots, and large, circular, multilobed leaves. Solitary flowers produce a fruit that becomes yellow upon ripening.

Distribution and Ecology: Mayapple grows in wet meadows, open woodland, and disturbed areas along the roadside, often as a group of plants in the eastern half of the United States. *P. hexandrum* is indigenous to Asia, and this plant does not grow naturally in the United States.

EXPOSURE

Native Americans used the root from *P. peltatum* as a cathartic. USP Podophyllum resin is an alcohol extract of the dried rhizomes and roots of *Podophyllum peltatum* used to treat condyloma acuminatum in the United States since 1942.[3] The self-administration of 0.15–0.5% purified podophyllotoxin (podofilox) preparations applied twice daily for 3 days is a safe alternative for the treatment of condyloma acuminatum compared with the use of nonstandardized USP podophyllum resin.[4] Etoposide and teniposide are chemotherapeutic agents that are semisynthetic derivatives of podophyllotoxin derived from the Himalayan mayapple (*P. hexandrum*) and the US species (*P. peltatum*).[5,6] These semisynthetic podophyllotoxins (e.g., etoposide 7a, teniposide 7c) are used to treat a variety of tumors including germ cell tumors, Hodgkin's disease, small cell lung cancer, acute myelogenous leukemia (AML), acute lymphocytic leukemia (ALL), Kaposi's sarcoma, ovarian and testicular cancers, and retinoblastoma.[7] Bajiaolian is a traditional Chinese remedy derived from a species of the Mayapple family that has been used for the treatment of condyloma acuminata, snake bite, generalized weakness, lymphadenopathy, and tumors.

PRINCIPAL TOXINS

Physiochemical Properties

Traditional podophyllum resin solutions contain 25% podophyllum extract as well as vehicles such as tincture

Medical Toxicology of Natural Substances, by Donald G. Barceloux, MD

of benzoin, salicylic acid, cantharidin, penederm, castor oil, ethanol, and acetone. The podophyllum extract is a complex, nonstandardized mixture of organic substances that contains at least 16 active physiological compounds including podophyllotoxin (podofilox), picropodophyllin (cis isomer of podophyllotoxin), deoxypodophyllotoxin, 4-demethylpodophyllotoxin, lactones (α- and β-peltatin), rutin, and quercetin.[8] The active toxic ingredient in podophyllum extract is podophyllotoxin (CAS RN: 518-28-5, $C_{22}H_{22}O_8$,), and the amount of podophyllotoxin varies from about 25–100 mg/mL depending on the source and storage conditions (light, heat).[9,10] Figure 129.1 displays the chemical structure of podophyllotoxin. Officially, podophyllum is a resin extract of the American source (*Podophyllum peltatum*) rather than the Indian source (*Podophyllum hexandrum*). Podophyllotoxin (podofilox) is available as a stable, low-concentration (0.5% solution, 5 mg/mL) commercial gel. Table 129.1 lists some physical properties of podophyllotoxin.

Poisonous Parts

Most human poisonings result either from ingestion of the commercial extract or excessive topical application rather than from the ingestion of plant parts. The green fruit, stem, leaves, and roots contain podophyllotoxin. Analysis of a sample of pulverized plant parts that produced severe nausea and vomiting after ingestion demonstrated a podophyllotoxin content of approximately 0.4%.[11] In samples of *P. hexandrum* from the Kumaun region of the Indian Central Himalaya, the podophyllotoxin content of rhizomes ranged between 0.36–1.08% (dry weight).[12] The ripe fruit is edible and ingestion of the ripe fruit does not produce toxicity. Podophyllum poisoning may result from the ingestion of herbs contaminated by *Podophyllum* species.[13]

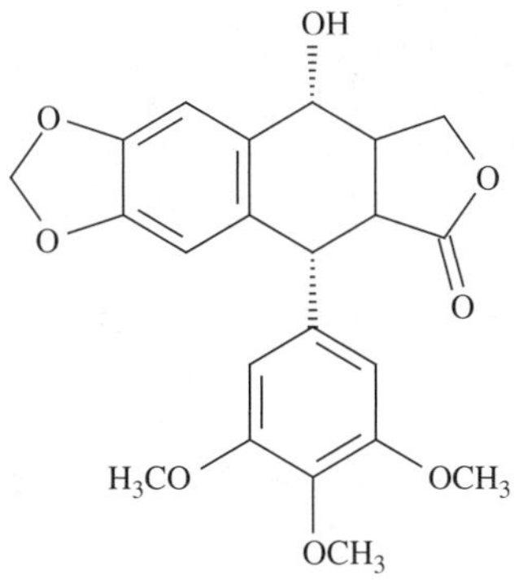

OH
O
O
O
O
H_3CO
OCH_3
OCH_3

FIGURE 129.1. Chemical Structure of Podophyllotoxin.

TABLE 129.1. Some Physical Properties of Podophyllotoxin

Physical Property	Value
Melting Point	183 °C/361.4 °F
log P (Octanol-Water)	2.01
Water Solubility	155 mg/L (25 °C/77 °F)
Vapor Pressure	8.34E-15 mmHg (25 °C/77 °F)
Henry's Law Constant	3.80E-18 atm-m³/mole (25 °C/77 °F)
Atmospheric OH Rate Constant	2.20E-10 cm³/molecule-second (25 °C/77 °F)

Mechanism of Toxicity

Podophyllotoxin is a potent spindle poison that blocks mitosis in metaphase. Both colchicine and podophyllotoxin disrupt the assembly and function of microtubules and other microtubule-dependent processes.[14] Additionally, podophyllotoxin demonstrates a wide variety of actions including inhibition of enzymes involved in the tricarboxylic acid cycle, inhibition of purine synthesis, blockade of RNA synthesis, damage to the endothelial lining of small vessels, induction of interleukin-1 and interleukin-2, suppression of immune responses, and inhibition of lymphocyte response to mitogens.[15–17]

DOSE RESPONSE

The recommended maximum therapeutic dose of podophyllum extract is 2 mL of 20% podophyllum resin (100 mg/m²). However, case reports associate bone marrow suppression and a severe sensorimotor neuropathy in a 20-year-old woman with topical application of 3 mL of a 20% podophyllum resin solution that remained on the skin after use.[18] The calculated dose based on patient recall was 150 mg (100 mg/m²). The ingestion of 10 mL of an 8% podophyllum resin solution by a child produced fever, hypotension, and coma.[19] Based on animal studies, the estimated toxic dose of podophyllum extract is above 300 mg/m² body surface area.[20] Case reports from the modern medical literature associate fatalities with the estimated ingestion of 10–11 g podophyllum extract,[21] and survival with the estimated ingestion of 2.8 g podophyllum extract.[22] During early clinical studies of podophyllotoxin as a chemotherapeutic agent at daily doses of 0.5–1 mg/kg, an average cumulative dose of 6 mg/kg produced bone marrow depression and gastrointestinal toxicity.[23]

TOXICOKINETICS

The dermal absorption of podophyllotoxin is limited. In a study of adults with anogenital warts, the topical application of 0.05 mL of 0.5% podophyllotoxin (0.25 mg) was not associated with detectable serum concentrations of podophyllotoxin as measured by high performance liquid chromatography (HPLC).[24] About 1–2 hours after topical application of podophyllotoxin doses of 7.5 mg, the maximum serum concentration of podophyllotoxin ranged up to 17 ng/mL. The serum half-life of podophyllotoxin following dermal application was approximately 1–4.5 hours.

CLINICAL RESPONSE

The ingestion of podophyllum produces neurologic, gastrointestinal, and hematological toxicity. Features of systemic toxicity include nausea, vomiting, ileus, fever, peripheral neuropathy, and obtundation. Severe intoxication produces coma, hypotension, oliguria, and death. Tachypnea usually results from the presence of a metabolic acidosis rather than direct damage to the lungs.

Clinical Pattern

Classically gastrointestinal symptoms begin within 2 hours of ingestion, and the persistence of gastrointestinal symptoms after several days suggests the presence of a paralytic ileus.[25] Central nervous system (CNS) symptoms develop about 5–20 hours after exposure.[19] Signs of peripheral neuropathy (ataxia, paraesthesias, hyporeflexia) occur after symptoms of CNS involvement resolve (i.e., about 24–48 hours after ingestion). Hematological changes (leukopenia, thrombocytopenia) develop about 5–7 days after exposure and resolve within 3 weeks. Although mild hepatorenal dysfunction commonly develops, liver or kidney failure is uncommon unless multiple organ failure occurs.

Central and Peripheral Nervous System

During podophyllum intoxication, patients develop changes primarily in sensorium and in peripheral nervous system function. In serious cases, lethargy and obtundation progress to convulsions, confusion, delirium, and prolonged coma.[20] Over the course of the first several days, a symmetrical, distal sensorimotor peripheral neuropathy develops manifest by profound bilateral weakness in upper and lower extremities. In addition, loss of deep tendon reflexes, vibration, position sense, and distal light touch occurs.[19] Changes in sensorium may persist 1–2 weeks.[26] Peripheral neuropathies usually resolve, but residual effect may continue for several years after the incident.[18,27] Postural hypotension may complicate recovery.

Pregnancy

Podophyllum is contraindicated during pregnancy because of the potential effects from the diffusion of a spindle poison (i.e., podophyllotoxin) from the maternal to fetal circulation. Case reports associate fetal demise and premature labor with the use of podophyllum resin extract.[28]

DIAGNOSTIC TESTING

Analytical methods to quantitate podophyllotoxin and derivatives in plasma include HPLC[24] and HPLC with UV detection.[29] Bone marrow aspiration during podophyllum intoxication demonstrates sudden maturational arrest of both erythroid and myeloid precursors.[20,25] Initially, a profound leukocytosis occurs along with the presence of metamyelocytes and immature band forms in the blood. The leukocyte and platelet counts reach a nadir within the first week after exposure. Severe poisonings produce mild to moderate elevation of serum creatine kinase and hepatic aminotransferase concentrations, but clinically significant hepatic or renal failure is rare. Other laboratory abnormalities include glycosuria, pyuria, and hyperamylasemia.

TREATMENT

Decontamination

Any topical podophyllum resin solution should be removed with soap and water. There are limited data on the efficacy of decontamination measures following the ingestion of podophyllum extract. Because of the potential of serious and delayed toxicity, all alert patients who ingest podophyllum resin within 1–2 hours of presentation to the emergency department should receive activated charcoal. Symptomatic patients should be hospitalized. Children who ingest potentially toxic quantities (i.e., one swallow or 5 mL of the 20% podophyllum resin solution) should be directly observed for at least 8 hours.[19]

Elimination Enhancement

There are inadequate data documenting that hemodialysis or hemoperfusion improve clinical outcome. The high lipid solubility of podophyllotoxin theoretically limits the effectiveness of hemodialysis. A case report on podophyllum toxicity associated clinical improvement with the use of hemoperfusion.[30] However, the use of hemoperfusion removed only 29 mg of podophyllotoxin compared with an estimated ingestion of 400 mg podophyllotoxin.

Supplemental Care

Replacement of fluid and correction of electrolyte imbalance (Na^{++}, K^{+}, Ca^{++}) is the most important aspect of supportive care. The acid-base balance should be monitored, particularly in dyspneic patients. Symptomatic patients should be observed for at least 24 hours. Although fever may occur during podophyllum intoxication, patients should be evaluated for a potential source of infection (e.g., sepsis, pneumonia), particularly when the fever occurs during the periods of leukopenia (i.e., fourth through seventh day after exposure). Clinical judgment dictates the necessity of broad-spectrum antibiotics. During severe podophyllum poisonings, complete blood counts, platelet counts, and coagulation profiles should be determined daily until evidence of hematological improvement occurs typically at the end of the first week after exposure.

References

1. Hartwell JL, Schrecker AW. Components of podophyllin. V. The constitution of podophyllotoxin. 1951;73:2909–2916.
2. Kaplan IW. Condyloma acuminata. N Orleans Med Surg J 1942;94:388–390.
3. Sullivan M. Podophyllotoxin. Arch Dermatol Syphil 1949;60:1–13.
4. Longstaff E, von Krogh G. Condyloma eradication: self-therapy with 0.15–0.5% podophyllotoxin versus 20–25% podophyllin preparations—an integrated safety assessment. Regul Toxicol Pharmacol 2001;33:117–137.
5. Sinkule JA. Etoposide: A semisynthetic epipodophyllotoxin. Chemistry, pharmacology, pharmacokinetics, adverse effects and use as an antineoplastic agent. Pharmacotherapy 1984;4:61–73.
6. Moraes RM, Bedir E, Barrett H, Byurandt C Jr, Canel C, Khan IA. Evaluation of *Podophyllum peltatum* accessions for podophyllotoxin production. Planta Med 2002;68:341–344.
7. Botta B, Delle Monache G, Misiti D, Vitali A, Zapia G. Aryltetralin lignins: chemistry, pharmacology and biotransformations. Curr Medicinal Chem 2001;8:1363–1381.
8. Rosenstein G, Rosenstein H, Freeman M, Weston N. Podophyllum—a dangerous laxative. Pediatrics 1976;57:419–421.
9. Beutner KR, von Krough G. Current status of podophyllotoxin for the treatment of genital warts. Sem Dermatol 1990;9:148–151.
10. Sullivan M, Follis RH, Hilgartner M. Toxicology of podophyllin. Proc Soc Exp Biol Med 1951;77:269–272.
11. Frasca T, Brett AS, Yoo SD. Mandrake toxicity. A case of mistaken identity. Arch Intern Med 1997;157:2007–2009.
12. Nadeem M, Palni LM, Kumar A, Nandi SK. Podophyllotoxin content, above- and below-ground biomass in relation to altitude in *Podophyllum hexandrum* populations from Kumaun region of the Indian Central Himalaya. Planta Med 2007;73:388–391.
13. But PP. Herbal poisoning caused by adulterants or erroneous substitutes. J Trop Med Hyg 1994;97:371–374.
14. McIntosh JR, Cande Z, Snyder J, Vanderslice K. Studies on the mechanism of mitosis. Ann N Y Acad Sci 1975;253:407–427.
15. Brigati C, Sander B. CPH-86, a highly purified podophyllotoxin, efficiently suppresses *in vivo* and *in vitro* immune responses. J Immunopharmacol 1985;7:285–302.
16. Zheng QY, Wiranowska M, Sadlik JR, Hadden JW. Purified podophyllotoxin (CPH-86) inhibits lymphocyte proliferation but augments macrophage proliferation. Int J Immunopharmacol 1987;9:539–549.
17. Kao W-F, Hung D-Z, Tsai W-J, Lin K-P, Deng J-F. Podophyllotoxin intoxication: toxic effect of Bajiaolian in herbal therapeutics. Hum Exp Toxicol 1992;11:480–487.
18. Rate RG, Leche J, Chervenak C. Podophyllin toxicity. Ann Intern Med 1979;90:723.
19. Campbell AN. Accidental poisoning with podophyllin. Lancet 1985;ii:206–207.
20. Moher LM, Mauer SA. Podophyllum toxicity. Case report and literature review. J Fam Pract 1979;9:237–240.
21. Cassidy DE, Drewry J, Fanning JP. Podophyllum toxicity: A report of a fatal case and a review of the literature. J Toxicol Clin Toxicol 1982;19:35–44.
22. Clark AN, Parsonabe MJ. A case of Podophyllum poisoning with involvement of the nervous system. Br Med J 1957;2:1155–1157.
23. Savel H. The metaphase-arresting plant alkaloids and cancer chemotherapy. Prog Exp Tumor Res 1966;8:189–224.
24. von Krogh G. Podophyllotoxin in serum: absorption subsequent to three-day repeated applications of a 0.5% ethanolic preparation on condylomata acuminata. Sex Transm Dis 1982;9:26–33.
25. Stoehr GP, Peterson AL, Taylor WJ. Systemic complications of local podophyllin therapy. Ann Intern Med 1978;89:362–363.
26. Dobb GJ, Edis RH. Coma and neuropathy after ingestion of herbal laxative containing podophyllin. Med J Aust 1984;140:495–496.
27. McFarland MF, MacFarland J. Accidental ingestion of Podophyllum. Clin Toxicol 1981;18:973–977.
28. Powell LC. Condyloma acuminatum. Clin Obstet Gynecol 1972;15:948.
29. Hüttmann A, Mross K, Hossfeld DK. Determination of the new podophyllotoxin derivative NK 611 in plasma by high-performance liquid chromatography with ultraviolet detection. J Chromatogr 1993;620:233-238.
30. Heath A, Mellstrand T, Ahlmin J. Treatment of podophyllin poisoning with resin hemoperfusion. Hum Toxicol 1982;1:373–378.

Chapter 130

MISTLETOE

HISTORY

Mistletoe has a long medicinal history including use by herbalists for infertility, epilepsy, asthma, and nervous tension. Until 1978, a mistletoe-containing prescription drug, Felsol, was available in Britain as a treatment for asthma.[1]

BOTANICAL DESCRIPTION

Mistletoe is a hemiparasite (facultative parasite) on the branches of trees or shrubs that can live either as a parasite or as an evergreen plant capable of photosynthesis. These plants rely on their host primarily for water and minerals, and these parasitic plants are adapted to specific hosts. There are over 20 species of the genus *Phoradendron* (mistletoe) in the United States beside the oak mistletoe, including the Christmas mistletoe [*Phoradendron tomentosum* (DC.) Engelm. ex Gray], Pacific mistletoe [*Phoradendron villosum* (Nutt.) Nutt.], mesquite mistletoe (*Phoradendron californicum* Nutt.), and the Colorado Desert mistletoe [*Phoradendron macrophyllum* (Engelm.) Cockerell]. The Christmas mistletoe is a native species of scrub areas of Oklahoma, Texas, and Louisiana.

Common Name: Oak Mistletoe

Scientific Name: *Phoradendron leucarpum* (Raf.) Reveal & M.C. Johnston (*Phoradendron flavescens, Phoradendron serotinum*)

Botanical Family: Viscaceae (mistletoe)

Physical Description: The oak mistletoe is a perennial, facultative parasite that grows primarily on oak trees. The light green leaves and hairy stems grow as a bush 1–4 ft (~0.3–1 m) in height among oak branches. Small white berries appear in grape-like clusters.

Distribution and Ecology: The oak mistletoe is distributed across oak forests in the southwestern, southern, and eastern regions of the United States.

Common Name: European Mistletoe

Scientific Name: *Viscum album* L.

Botanical Family: Viscaceae (mistletoe)

Physical Description: The European mistletoe has oval leaves with smooth edges. The leaves appear as pairs on a woody stem. In contrast to the long cluster (>10) of berries on the oak mistletoe, the European species has shorter, dense clusters (2–6) of waxy, white berries.

Distribution and Ecology: The European mistletoe is the only native mistletoe species is England and most of Europe.

EXPOSURE

Mistletoe is a common, herbal remedy traditionally used for treating hypertension and cancer, particularly breast cancer.[2] Commercial European mistletoe preparations are aqueous extracts of the whole plant, capsules containing dried portions of the whole plant, and teas

Medical Toxicology of Natural Substances, by Donald G. Barceloux, MD

brewed from leaves. Iscador® (Weleda AG, Arlesheim, Switzerland) is a European extract of *Viscum album* L. distributed by anthroposophic clinics for the treatment of metastatic cancer and HIV as part of a therapeutic program that includes homeopathic extracts, dietary manipulation, and movement therapies.[3] Other European mistletoe extracts include Isorel, Vysorel, or Helixor (sterile extract for subcutaneous injection). Although mistletoe extracts are widely used in the alternative treatment of cancer, rigorous trials of mistletoe extracts fail to demonstrate efficacy of these therapies as an adjuvant to cancer chemotherapy.[4,5] In the semi-arid areas of northern Mexico, aqueous extracts of *Phoradendron tomentosum* (Raf.) Engelm. ex Gray are used to treat hyperglycemia associated with diabetes mellitus.[6] *Phoradendron liga* (Gill. Ex H. et A.) Eichl. is an herbal substitute for European mistletoe (*Viscum album* L.) that is used in Argentina for the treatment of hypertension.[7] In the United States, the Food and Drug Administration (FDA) lists mistletoe as a food additive that should not be used until safety has been established. The berries and leaves are commonly used as Christmas decorations.

PRINCIPAL TOXINS

Physiochemical Properties

The major constituents of mistletoe are carbohydrate-binding proteins or lectins (e.g., viscumin), polypeptides (i.e., viscotoxins with a basic chemical structure of thionins), and a number of phenolic compounds (digallic acid, *o*-coumaric acid). The three classes of potential toxins in mistletoe include gastric irritating alkaloids, cardiac toxins (viscotoxins, phoratoxins), and lectins. Viscotoxins are toxic peptides with 46 amino acids and a molecular weight of about 5kDa. These compounds are present in European mistletoe and are physiologically similar to cobra venom.[8] However, the amino acid sequences and acetylcholine agonist/antagonist properties of mistletoe and cobra toxins are quite different.[9] Viscotoxins contain three disulfide bridges (six cysteine residues) with a chemical structure related to α- and β-thionins. There are at least six isoforms (A1, A2, A3, B, 1-PS, U-PS) of these low-molecular-weight, basic polypeptides.[10] These compounds cause vasoconstriction, bradycardia, and negative inotropic effects when administered intravenously to cats. Substantial variation in physiological effects occurs between the cardiotoxins in different mistletoe species. Phoratoxins (A-F) are small basic proteins extracted from the American oak mistletoe (*Phoradendron leucarpum*).[11] These proteins cause effects quantitatively different from those of viscotoxins from the European mistletoe (*Viscum album*).[12] To cause bradycardia, hypotension, and decreased inotropy, the required parenteral dose of phoratoxin is 10 times higher than the dose of viscotoxins.

Poisonous Parts

All parts of the plants are poisonous. The concentration of viscotoxins in European mistletoe varies with individual strains, but the concentration of these compounds is relatively stable during the harvesting season.[13] Dried stems and leaves contain viscotoxins[12] and lectins.[14] The concentration of viscotoxins varies between plant parts. For example, the concentration of viscotoxin A1 is substantially higher in seeds than in young shoots.[15] The berries of mistletoe contain a gastrointestinal irritant.

Mechanism of Toxicity

Viscotoxins apparently alter cell membrane permeability and polarity by displacing calcium.[8] In high doses, viscotoxins are cardiotoxic as a result of the inhibition of papillary muscle function. A toxic lectin, viscumin, occurs in the European mistletoe. Like ricin, viscumin is composed of an A chain (type II ribosome-inactivating protein, ribotoxin) and a D-galactose binding B chain that allows entry into the cell rather than inhibit protein synthesis.[16,17] Unlike those of ricin, the toxic and hemagglutinating properties cannot be separated. The contribution of viscumin to human toxicity is unclear.

DOSE RESPONSE

In a retrospective case series of 92 US patients reporting exposure to mistletoe, there were six cases of gastrointestinal upset, two cases of mild drowsiness, and one case of ataxia.[18] The reported amount ingested ranged up to 20 berries and five leaves. A review of 1,754 exposures to mistletoe reported to the American Association of Poison Control Centers national data collection system from 1985–1992 did not demonstrate any major effects attributed to exposure to mistletoe.[19]

TOXICOKINETICS

Although mistletoe contains a number of potential oral toxins (e.g., phenylethylamine, tyramine, viscotoxins, phoratoxins), the degree of gastrointestinal absorption remains poorly documented.

CLINICAL RESPONSE

Although ingestion of large quantities of mistletoe berries potentially causes gastroenteritis, the toxic effects resulting from ingestions of mistletoe are not well documented in the literature.[1] In 1874, a 14-year-old boy reportedly ate some mistletoe berries and developed a "state of intoxication" characterized by conjunctival erythema, amnesia, dilated pupils, and slow, bounding pulse that resolved by the following day.[20] The National Clearinghouse for Poison Control Centers reported a death from tea brewed from berries 12 hours after the onset of acute gastroenteritis. This case was the first mistletoe poisoning case reported in the Quarterly Cumulative Medical Index since 1930. The family reported that a 28-year-old woman suddenly developed copious vomiting and diarrhea several hours after drinking an unknown quantity of mistletoe tea. The patient presented to the emergency department in shock with confusion, diaphoresis, and no detectable blood pressure. She died within 10–15 minutes after arrival, and no dysrhythmias were recorded. Significant autopsy findings included a slightly enlarged uterus containing placental tissue that suggested the possibility of sepsis or the use of abortifacients. Rare adverse reactions associated with the use of European mistletoe extract include eosinophilia[21] and severe anaphylaxis (generalized urticaria, angioedema, dyspnea, vomiting).[22] Immunoblotting of serum from the patient with severe anaphylaxis demonstrated IgE binding to 5-kDa proteins in mistletoe extract that corresponds to the molecular weight of viscotoxins.

DIAGNOSTIC TESTING

High performance liquid chromatographic methods are available for the quantitation of viscotoxins in mistletoe extracts, such as Iscador®.[23]

TREATMENT

Because mistletoe causes significant gastrointestinal irritation following ingestion, decontamination measures are usually unnecessary. The ingestion of one to three berries or one to two leaves probably would not cause toxic effects in a child.[24] Symptomatic patients should be monitored for vital sign and cardiac rhythm abnormalities. There are no specific antidotes to mistletoe toxicity. The potential for adverse cardiac effects theoretically exists based on the presence of parenteral cardiotoxins in mistletoe, but there are few human data to indicate that these cardiotoxins cause clinically significant cardiac abnormalities following causal ingestions. Patients with dysrhythmias or electrolyte imbalance should be hospitalized for observation overnight. Asymptomatic patients may be discharged after several hours of observation.

References

1. Harvey J, Colin-Jones DG. Mistletoe hepatitis. Br Med J 1981;282:186–187.
2. Fritz P, Dippon J, Kierschke T, Siegle I, Mohring A, Moisa A, Murdter TE. Impact of mistletoe lectin binding in breast cancer. Anticancer Res 2004;24:1187–1192.
3. Gorter RW, van Wely M, Reif M, Stoss M. Tolerability of an extract of European mistletoe among immunocompromised and healthy individuals. Alter Ther 1999;5:37–48.
4. Ernst E, Schmidt K, Steuer-Vogt MK. Mistletoe for cancer? A systematic review of randomised clinical trials. Int J Cancer 2003;107:262–267.
5. Steuer-Vogt MK, Bonkowsky V, Ambrosch P, Scholz M, Neiss A, Strutz J, et al. The effect of an adjuvant mistletoe treatment programme in resected head and neck cancer patients: a randomized controlled clinical trial. Eur J Cancer 2001;37;23–31.
6. Calzado-Flores C, Hurtado-Ramirez MB, Flores-Villanueva Z, Verde-Star MJ, Segura-Luna JJ, et al. Preliminary chronic toxicological study of aqueous extract of *Phoradendron tomentosum*. Proc West Pharmacol Soc 2002;45:162–163.
7. Varela BG, Fernandez T, Ricco RA, Zolezzi PC, Hajos SE, Gurni AA, et al. *Phoradendron liga* (Gill. Ex H. et A.) Eichl. (Viscaceae) used in folk medicine: anatomical, phytochemical, and immunochemical studies. J Ethnopharmacol 2004;94:109–116.
8. Andersson KE, Johannsson M. Effects of viscotoxin on rabbit and aorta, and on frog skeleton muscle. Eur J Pharmacol 1973;23:223–231.
9. Smythies JR, Robinson CR, Al-Zahid SA. On the tertiary structure and mechanism of action of the viscotoxins. Ala J Med Sci 1976;13:240–246.
10. Romagnoli S, Ugolini R, Fogolari F, Schaller G, Urech K, Giannattasio M, et al. NMR structural determination of viscotoxin A3 from *Viscum album* L. Biochem J 2000;350:569–577.
11. Mellstrand ST, Samuelsson G. Phoratoxin, a toxic protein from the mistletoe *Phoradendron tomentosum* subsp. *macrophyllum* (Loranthaceae). Acta Pharm Suecica 1974;11:347–360.
12. Rosell S, Samuelsson G. Effect of mistletoe viscotoxin and phoratoxin on blood circulation. Toxicon 1966;4:107–110.
13. Schaller G, Urech K, Grazi G, Giannattasio M. Viscotoxin composition of the three European subspecies of *Viscum album*. Planta Med 1998;64:677–678.
14. Franz H, Ziska P, Kindt A. Isolation and properties of three lectins from mistletoe (*Viscum album* L.) Biochem J 1981;195:481–484.
15. Orru S, Scaloni A, Giannattasio M, Urech K, Pucci P, Schaller G. Amino acid sequence, S-S bridge arrangement and distribution in plant tissues of thionins from *Viscum album*. Biol Chem 1997;378:989–996.
16. Wacker R, Stoeva S, Pfuller K, Pfuller U, Voelter W. Complete structure determination of the A chain of mistletoe

lectin III from *Viscum album* L. ssp. *album*. J Pept Sci 2004;10:138–148.

17. Olsnes S, Stirpe F, Sandvig K, Pihl A. Isolation and characterization of viscumin, a toxic lectin from *Viscum album* L. (mistletoe). J Biol Chem 1982;257:13263–13270.
18. Spiller HA, Willias DB, Gorman SE, Sanftleban J. Retrospective study of mistletoe ingestion. Clin Toxicol 1996;34:405–408.
19. Krenzelok EP, Jacobsen TD, Aronis J. American mistletoe exposures. Am J Emerg Med 1997;15:516–520.
20. Dixon J. Case of poisoning by berries of the mistletoe: Recovery. Br Med J 1874;1:224.
21. Huber R, Klein R, Berg PA, Ludtke R, Werner M. Effects of a lectin- and a viscotoxin-rich mistletoe preparation on clinical and hematologic parameters: a placebo-controlled evaluation in healthy subjects. J Altern Complement Med 2002;8:857–866.
22. Bauer C, Oppel T, Rueff F, Przybilla B. Anaphylaxis to viscotoxins of mistletoe (*Viscum album*) extracts. Ann Allergy Asthma Immunol 2005;94:86–89.
23. Jordan E, Wagner H. Detection and quantitative determination of lectins and viscotoxins in mistletoe preparations. Arzneimittelforschung 1986;36:428–433. [German]
24. Hall AH, Spoerke DG, Rumack BH. Assessing mistletoe toxicity. Ann Emerg Med 1986;15:1320–1323.

Chapter 131

POISON HEMLOCK
(*Conium maculatum* L.)

HISTORY

The early Greeks used poison hemlock as a means of suicide. In the Socratic dialogue *Phaedo*, Plato implied that Socrates died in 399 BC from *Conium* poisoning as was the custom in Greek executions.[1] The passage described a slowly ascending paralysis beginning in the feet and progressing to the chest with preservation of mental function until close to death. However, Plato's *Phaedo* did not specifically mention hemlock as the poison, and his description of Socrates' death was more metaphorical than historical.[2] The calm death characterized by ascending paralysis and terminal asphyxia suggests that the death cup drunk by Socrates may have contained other compounds, such as opium.[3] Gieseche isolated coniine in 1827 and Landenburg synthesized this alkaloid in 1886.[4] The London and Edinburgh *Pharmacopoeias* contained medical references to the dried leaf and juice of *Conium maculatum* from 1864 to 1898, and the last officially recognized passage to the medicinal use of hemlock appeared in the British *Pharmacopoeia* of 1934.[5]

BOTANICAL DESCRIPTION

The term hemlock has been loosely applied to several plants and trees that vary in toxicity. Other toxic varieties of hemlock include the water hemlock plant (*Cicuta* spp.) and the ground hemlock (*Taxus canadensis* Marshall). The hemlock tree is a conifer (*Tsuga* spp.) and is nontoxic.

Common Name: Poison Hemlock, Poison Fool's Parsley, California or Nebraska Fern

Scientific Name: *Conium maculatum* L.

Botanical Family: Apiaceae (carrot, formerly Umbelliferae)

Physical Description: This large biennial weed produces a large, forked, white to yellow taproot. Lacy, fern-like leaves reach 8–10ft (~2–3m) in height. The whole plant exudes a fetid odor. The stems are erect, smooth, stout, and hollow with characteristic brownish-red to purple spots on the lower portions. The numerous triangular leaves are multiply pinnate and deeply divided into minute leaflets. The flowers are small, white, and clustered into multiple umbrella-shaped groups. The plant is mistaken for wild parsnip, the seeds for anise, and the leaves for parsley. The water hemlock (*Cicuta virosa* L.) lacks the purplish spots and mousy odor present on *Conium maculatum*, and the water hemlock has a branched root system with a lateral tuber rather than a single taproot present on the poison hemlock.

Distribution and Ecology: *Conium* species grow as prolific weeds along roadsides, ditches, and open areas throughout the United States, especially in the north. This plant prefers damp ground, river banks, roadsides, pastures, and meadows including waste land. *C. maculatum* was introduced into the United States from Europe as an ornamental plant.

Medical Toxicology of Natural Substances, by Donald G. Barceloux, MD

EXPOSURE

C. maculatum was listed in the *British Pharmaceutical Codex* until 1934 and in the *US National Formulary* until 1936 as an antispasmodic, sedative, and analgesic.[6] Prior herbal uses of *C. maculatum* include the treatment of gout, herpes, erysipelas, breast tumors, edema, and stomach pains. Poisonings usually result from mistaken identification during foraging for natural food; however, the bitter taste of the plant usually discourages consumption.

PRINCIPAL TOXINS

Physiochemical Properties

Conium toxicity results from piperidine alkaloids, which are structurally related to the pyridine nicotine. These propyl piperidine alkaloids are among the simplest alkaloids and appear in other plant families (Crassulaceae, Punicaceae). At least eight piperidine alkaloids are present in *C. maculatum* with the major alkaloids being coniine (2-propylpiperidine), *N*-methylconiine, and γ-coniceine.[6] γ-Coniceine is the precursor of conhydrinone, *N*-methylconiine, conhydrine, *N*-methylconhydrine, pseudoconhydrine, and 2-methylpiperidine.[7] Investigations with $^{14}CO_2$ revealed that the biosynthetic sequence is CO_2 → γ-coniceine → coniine → *N*-methylconiine.[8] The interconversion of γ-coniceine and coniine suggests that these compounds are members of an oxidation-reduction system.[9] Figure 131.1 displays the chemical structures of various *Conium* alkaloids. Coniine is heat stable.

Poisonous Parts

Although all vegetative organs, flowers, and fruits contain alkaloids, the alkaloid content varies significantly between species, plant parts, season, time of day, moisture, temperature, and geographic locations. The alkaloid content of plant parts during the flowering stage is relatively low. As the plant matures, the alkaloid concentration increases respectively in the stems, leaves, and fruit, but the roots consistently contain the largest alkaloid concentrations. Analysis of *C. maculatum* during the growing season indicates that γ-coniceine (an unsaturated alkaloid) predominates in the wet season during active growth, whereas saturated alkaloids (e.g., coniine, conhydrine) predominate during the dry season.[10] Thus, γ-coniceine predominates in the early (vegetative and flowering) stages of plant growth, whereas coniine is the major alkaloid during the development of fruits. During the latter stage, the coniine content of the fruit is about 10–20 times greater than the γ-coniceine content (e.g., 87 μg vs. 5 μg, respectively).[7] In the ripe fruits, *N*-methylconiine is the predominant alkaloid.[9] The concentration of leaves from a *C. maculatum* plant associated with *Conium* intoxication averaged about 850 μg γ-coniceine/g plant, wet weight.[11] The estimated average alkaloid content of various parts of *Conium maculatum* is as follows: roots, 0–0.5%; stems,

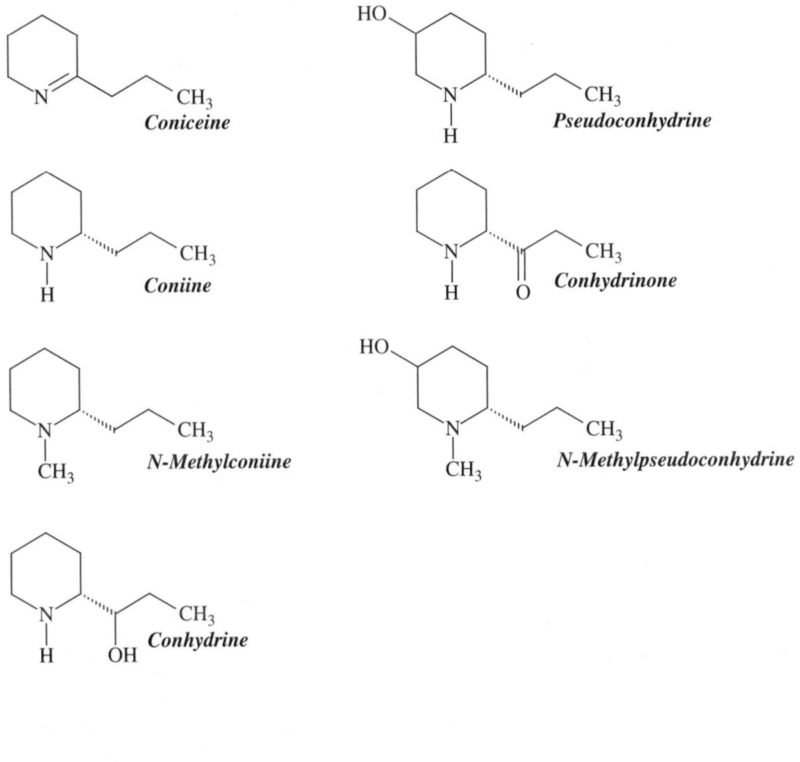

Coniceine

Pseudoconhydrine

Coniine

Conhydrinone

N-Methylconiine

N-Methylpseudoconhydrine

Conhydrine

FIGURE 131.1. Conium alkaloids.

0.02–0.7%; leaves, 0.3–1.5%; flowers, 1%; unripe fruit 1.6–3%; ripe fruit 0.2–1%; and seeds, 0.02–0.9%.[12]

Mechanism of Toxicity

The pharmacological actions of *Conium* alkaloids (coniine, *N*-methylconiine, γ-coniceine) are similar to nicotine.[5] These alkaloids are nondepolarizing blocking agents in high doses that produce respiratory failure as well as peripheral nicotinic effects (mydriasis, salivation, tachycardia). Postmortem examinations of fatalities associated with *Conium* poisoning demonstrate nonspecific findings (e.g., pulmonary congestion).[13]

DOSE RESPONSE

There are inadequate human data to determine the dose response for the ingestion of poison hemlock.

TOXICOKINETICS

There are no data on the toxicokinetics of coniine, a known animal teratogen, in humans. The variable response of different animal species (e.g., rats, hamsters, and chickens are not susceptible) to *Conium* alkaloids suggests that there are species-related differences in the metabolism of these compounds.[14]

CLINICAL RESPONSE

Conium intoxication in animals is commonly manifest by ataxia, salivation, convulsions, and coma. In animal studies, congenital joint and spinal abnormalities similar to those of crooked calf disease associated with consumption of lupine developed after the feeding of fresh green *Conium* plants to pregnant cows.[15] Based on animal data and the structural similarity of *Conium* alkaloids, the expected clinical features of *Conium* poisoning would include initial stimulation (tremor, ataxia, mydriasis), nausea, vomiting, and sore throat followed by cardiorespiratory depression (bradycardia, paralysis, coma) and ascending paralysis. Respiratory failure and death potentially result from the ingestion of high doses of *Conium* alkaloids.[16] Case reports associate exposure to *Conium maculatum* with respiratory failure,[17] coma, and death.[13] Although laboratory analysis confirmed exposure to *Conium maculatum*, there were limited clinical data and the exact contribution of *Conium maculatum* to the clinical course of these individuals is unclear. In contrast to *Cicuta* (water hemlock) poisoning, convulsions are not usually associated with *Conium* intoxication.

DIAGNOSTIC TESTING

Gas chromatographic methods with nitrogen phosphorus detectors are available to confirm exposure to *Conium* alkaloids (e.g., coniine) at detection limits ranging between 0.025–0.2 μg/g sample.[18] Intratubular deposition of myoglobin and acute tubular necrosis may complicate *Conium* poisoning as a result of rhabdomyolysis.[19] The presence of γ-coniceine in blood and urine samples confirms exposure to *Conium* species.

TREATMENT

Treatment is supportive. Decontamination measures are usually unnecessary.

References

1. Ober WB. Did Socrates die of hemlock poisoning? NY State J Med 1977;77:254–258.
2. Gill C. The death of Socrates. Classical Q 1973;23:25–28.
3. Daugherty CG. The death of Socrates and the toxicology of hemlock. J Med Biog 1995;3:178–182.
4. Landenburg H. [Research with the synthesis of coniine]. Chem Ber 1886;19:439–441. [German]
5. Bowman WC, Sanghvi IS. Pharmacological actions of hemlock (*Conium maculatum*) alkaloids. J Pharm Pharmacol 1963;15:1–25.
6. Panter KE, James LF, Gardner DR. Lupines, poison-hemlock and *Nicotiana* spp: Toxicity and teratogenicity in livestock. J Nat Toxins 1999;8:117–134.
7. Fairbairn JW, Suwal PN. The alkaloids of hemlock (*Conium maculatum* L.) II. Evidence for a rapid turnover of the major alkaloids. Phytochemistry 1961;1:38–46.
8. Dietrich SMC, Martin RO. The biosynthesis of *Conium* alkaloids using carbon 14 dioxide. The kinetics of 14C incorporation into the known alkaloids and some new alkaloids. Biochemistry 1969;8:4163–4172.
9. Vetter J. Poison hemlock (*Conium maculatum* L.) Food Chem Toxicol 2004;42:1373–1382.
10. Fairbairn JW, Challen SB. The alkaloids of hemlock (*Conium maculatum* L.) distribution in relation to the development of the fruit. Biochem J 1959;72:556–561.
11. Frank BS, Michelson WB, Panter KE, Gardner DR. Ingestion of poison hemlock (*Conium maculatum*). West J Med 1995;163:573–574.
12. Lopez TA, Cid MS, Bianchini ML. Biochemistry of hemlock (*Conium maculatum* L.) alkaloids and their acute and chronic toxicity in livestock. A review. Toxicon 1999;37:841–865.
13. Drummer OH, Roberts AN, Bedford PJ, Drump KL, Phelan MH. Three deaths from hemlock poisoning. Med J Aust 1995;162:592–593.

14. Forsyth CS, Frank AA. Evaluation of developmental toxicity of coniine to rats and rabbits. Teratology 1993;48: 59–64.
15. Keeler RF, Balls LD. Teratogenic effects in cattle of *Conium maculatum* and *Conium* alkaloids and analogs. Clin Toxicol 1978;12:49–64.
16. Fraser NC. Accidental poisoning deaths in British children 1958–77. Br Med J 1980;290:1595–1598.
17. Biberci E, Altuntas Y, Cobanoglu A, Alpinar A. Acute respiratory arrest following hemlock (*Conium maculatum*) intoxication. Clin Toxicol 2002;40:517–518.
18. Holstege DM, Seiber JN, Galey FD. Rapid multiresidue screen for alkaloids in plant material and biological samples. J Agric Food Chem 1995;43:691–699.
19. Rizzi D, Basile C, Di Maggio A, Sebastio A, Introna F Jr, Rizzi R, et al. Clinical spectrum of accidental hemlock poisoning: neurotoxic manifestations, rhabdomyolysis, and acute tubular necrosis. Nephrol Dial Transplant 1991;6:939–943.

Chapter 132

POKEWEED
(*Phytolacca americana* L.)

HISTORY

This native North American plant was used by Native Americans in the eastern United States for the treatment of a variety of illnesses including dysentery, cancer, and rheumatism. The powdered root was an emetic and cathartic. During the 19th century, this herb was one of the treatments for syphilis. The name pokeweed is derived from the Virginian Indian word "pocon," which means plant with red dye.

BOTANICAL DESCRIPTION

Common Name: Pokeweed, Pokeberry, Poke, Pokeroot, Inkberry, Crowberry, Pigeonberry, American Cancer, American Nightshade, Scoke, Jalap, Garget

Scientific Name: *Phytolacca americana* L.

Botanical Family: Phytolaccaceae (pokeweed)

Physical Description: Pokeweed is a large perennial herb that grows from a long fleshy taproot each spring to reach a height of 3–10ft (~1–3m). The smooth, purple-green leaves have smooth margins and sharp tips. In late summer and early fall, white to purplish flowers produce round clusters of dark purple berries with red juice. The roots may be mistaken for parsnips or horseradish, and the shoots unwittingly substituted for asparagus. Figure 132.1 displays the branches and berries of the pokeweed plant.

Distribution and Ecology: Pokeweed is a native weed of the eastern United States that grows abundantly in rich, moist areas in open fields and roadsides. Occasionally, pokeweed grows in the western United States.

EXPOSURE

Folk remedies containing pokeweed are used as a salve for pruritus and as a bronchodilator. The fresh root was applied as a poultice on bruises and painful joints. Poke salad is a regional delicacy of the rural southern United States that requires special preparation. In the spring, young, tender leaves are boiled twice, the washing water is discarded, and the leaves are rinsed carefully before eating. Powdered poke root is available as an herbal tea.

PRINCIPAL TOXINS

Physiochemical Properties

Pokeweed contains the glycoside, phytolaccine, which produces gastrointestinal irritation. The concentration of phytolaccine increases as the plant matures. Salt extraction of pokeweed produces five nonspecific mitogens that possess hemagglutinating and mitotic activity.[1] Noncardioactive steroids and triterpenoid glycosides (saponins) also occur in pokeweed,[2] but the contribution of these substances to pokeweed poisoning remains unclear. In general, saponins are irritants and produce foaming in solutions. Pokeweed lectin is a lectin specific for *N*-acetylglucosamine-containing saccharides that stimulates peripheral lymphocytes to undergo mitosis by binding to their cell surfaces.[3] The name inkberry was

Medical Toxicology of Natural Substances, by Donald G. Barceloux, MD

FIGURE 132.1. Pokeweed (*Phytolacca americana* L). Photograph courtesy of Lewis Nelson, MD. See color insert.

given to pokeweed because of the purplish stains that result from contact with the mature berries. Pokeweed contains a ribosome-inactivating protein (ribotoxin) that has a mechanism of action similar to ricin; however, the effects are much less severe.[4] The role of this lectin in producing toxicity after ingestion of pokeweed is unclear.

Poisonous Parts

All parts of the pokeweed plant are potentially toxic depending on preparation, individual susceptibility, plant part, and growing season. Toxicity generally increases with maturity with the exception that green berries are more toxic than ripe berries. The root is particularly toxic with the leaves, the stems, and the berries demonstrating progressively less toxicity. Double washing of pokeweed leaves does not guarantee edibility.

Mechanism of Toxicity

Although phytolaccine is a recognized gastrointestinal (GI) irritant, the important toxins in pokeweed are not well delineated. Saponins may potentiate GI toxicity. Pokeweed mitogen is a glycoprotein that promotes leukocytosis and stimulates the production of interferon.[5]

DOSE RESPONSE

The severity of clinical symptoms depends on individual sensitivity as well as the dose of pokeweed. Of 46 campers who ate "properly prepared" pokeweed on a New Jersey nature expedition, 20 developed gastroenteritis.[6] In a case series of boy scouts eating pokeberry pancakes, only a few individuals developed mild diarrhea.[7] Anecdotal case reports associate the ingestion of a large amount of juice from pokeberries with a fatal outcome for a 5-year-old girl, but this case report is not well documented.[8]

TOXICOKINETICS

The toxins in pokeweed have not been well defined, and there are few toxicokinetic data on suspected toxins.

CLINICAL RESPONSE

The GI effects of pokeweed ingestion are highly variable depending on a variety of variables including the preparation method and season. Although properly prepared pokeweed is a delicacy, the ingestion of pokeweed can produce symptoms ranging from a burning sensation of the GI tract to severe hemorrhagic gastritis. Early gastrointestinal symptoms predominate in pokeweed poisoning depending on the amount of pokeweed ingested. Reported clinical features of an outbreak of pokeweed poisoning from the ingestion of salad containing pokeweed included nausea (86%), stomach cramps (86%), vomiting (81%), headache (52%), dizziness (48%), burning of mouth or stomach (38%), and diarrhea (29%).[6] The mean onset of illness was 3 hours with a range of $^1/_2$ to $5^1/_2$ hours. Symptoms resolved within 1–48 hours. The diarrhea may contain foam as a result of the action of saponins on the contents of the gastrointestinal tract. A case report documented the onset of lethargy, weakness, paresthesias, blurred vision, vertigo, sweating, salivation, urgency, syncope, respiratory depression, and convulsions 30 minutes after the ingestion of a brewed tea of pokeweed leaves and roots.[9]

Peripheral and central cholinergic signs may occur during severe pokeweed intoxications. Hypotension and tachycardia commonly occur after serious pokeweed poisoning.[10] However, serious cardiac dysrhythmias are uncommon.[11] Rate-related ST depression in a 65-year-old woman was associated with the ingestion of a folk remedy containing pokeweed.[12] She recovered without sequelae. Most case reports of fatalities associated with pokeweed poisoning appeared in the older medical literature before modern medical diagnostic testing and therapy. There are no reports of long-term sequelae from pokeweed intoxication.

DIAGNOSTIC TESTING

A peripheral plasmacytosis may occur after pokeweed ingestion as a result of the effects of pokeweed mitogen.

In four of five children exposed to pokeweed, a significant increase in the number of both immature and mature plasma cells occurred after the ingestion of pokeweed.[13]

TREATMENT

The treatment of pokeweed poisoning is supportive. Dehydration and hypotension are the most common serious effects of pokeweed intoxication that require immediate medical treatment. Patients symptomatic during pokeweed intoxication usually respond to antiemetic drugs and restoration of fluid and electrolyte balance. Occasionally, the administration of vasopressors may be necessary. Because pokeweed contains a powerful gastrointestinal irritant, decontamination measures are unnecessary. Serious pokeweed intoxications typically involve ingestion of pokeweed teas and these patients may require hospitalization for the monitoring of hypotension and electrolyte imbalance. Symptoms usually resolve within 24–36 hours. There are no specific antidotes to pokeweed toxins and atropine does not effectively reverse the effects of pokeweed intoxication.

References

1. Waxdal MJ. Isolation, characterization and biological activities of five mitogens from pokeweed. Biochemistry 1974;18:3671–3676.
2. Kang SS, Woo WS. Triterpenes from the berries of *Phytolacca americana*. J Nat Prod 1980;43;510–513.
3. Fujii T, Hayashida M, Hamasu M, Ishiguro M, Hata Y. Structures of two lectins from the roots of pokeweed (*Phytolacca americana*). Acta Crystallogr D Biol Crystallogr 2004;60:665–673.
4. Stirpe F, Barbieri L, Gorini P, Valbonesi P, Bolognesi A, Polito L. Activities associated with the presence of ribosome-inactivating proteins increase in senescent and stressed leaves. FEFS Lett 1996;382:309–312.
5. Friedman RM, Cooper HL. Stimulation of interferon production in human lymphocytes by mitogens. Proc Soc Exp Biol Med 1967;125:901–905.
6. Callahan R. Plant poisonings—New Jersey. MMWR Morb Mortal Wkly Rep 1981;30:65–67.
7. Edwards N, Rogers GC. Pokeberry pancake breakfast or it's gonna be a great day. Vet Hum Toxicol 1982;24(Suppl): 135–137.
8. Hardin JW, Arena JIM. Human Poisoning from Native and Cultivated Plants, 2nd ed. Durham, NC: Duke University Press; 1974:73.
9. Jaeckle KA, Freeman FR. Pokeweed poisoning. South Med J 1981;5:639–640.
10. Lewis WH, Smith PR. Pokeroot herbal tea poisoning. JAMA 1979;242:2759–2760.
11. Hamilton RJ, Shih RD, Hoffman RS. Mobitz type I heart block after pokeweed ingestion. Vet Hum Toxicol 1995;37: 66–67.
12. Roberge R, Brader E, Martin ML, Jehle D, Evans T, Harchelroad F Jr, et al. The root of evil—pokeweed intoxication. Ann Emerg Med 1986;15:470–473.
13. Barker BE, Parnes P, La Marche. Peripheral blood plasmacytosis following systemic exposure to *Phytolacca americana* (pokeweed). Pediatrics 1966;38:490–493.

Chapter 133

EUROPEAN BITTERSWEET and OTHER *SOLANUM* SPECIES

BOTANICAL DESCRIPTION

The Solanaceae (potato) family contains annual and perennial herbs, shrubs, and trees that grow worldwide. The leaves are simple and alternate. The flowers have five widely spreading petals; the stamens are grouped conically in the center. The ovary of the flower matures into a fruit. In the United States, *Solanum* (nightshade) contains over 100 native and introduced species. This genus has a number of species that are toxic to grazing animals including *Solanum carolinense* L. (Carolina horsenettle), *Solanum elaeagnifolium* Cav. (silverleaf nightshade, tropillo), *Solanum nigrum* L. (black or common nightshade), and *Solanum triflorum* Nutt. (cutleaf nightshade).

Common Name: European Bittersweet, Woody Nightshade, Climbing Nightshade, Blue Nightshade

Scientific Name: *Solanum dulcamara* L.

Botanical Family: Solanaceae (potato)

Physical Description: This slender, woody, climbing perennial vine grows from rhizomes. The leaves are ovate, alternate, simple or deeply lobed with pointed tips. The bluish to deep purple flowers arise from stalks. The mature fruit is orange-red and almost round.

Distribution and Ecology: This native species of Eurasia inhabits thick woods and moist environments of low woods, thickets, and roadsides from Nova Scotia and Minnesota south to Texas and Florida.

Common Name: Jerusalem Cherry

Scientific Name: *Solanum pseudocapsicum* L.

Botanical Family: Solanaceae (potato)

Physical Description: This perennial plant produces red berries during Christmas that are similar in appearance to cherry tomatoes.

Distribution and Ecology: This native, frost-tolerant Peruvian plant is an ornamental house and garden plant that has been naturalized in areas of Australia, southeastern United States, Texas, and New England.

Common Name: Bhankatiya, Devil's Fig, Fausse Aubergine, Kausoni, Prickly Solanum, Terongan, Turkey Berry, Susumber Berry

Scientific Name: *Solanum torvum* Sw.

Botanical Family: Solanaceae (potato)

Physical Description: This broad-leaved, evergreen, shrub or small tree grows up to 4 m (~13 ft) in height. The stems contain straight or lightly curved thorns. The alternate leaves have a prominent midvein that also has thorns. Leaf margins are shallowly and irregularly lobed. Small, white flowers appear in large clusters that produce small yellow berries about 10–15 mm in diameter. The glabrous, globose fruit contains many seeds.

Distribution and Ecology: This plant is a major weed in pastures, roadsides, and wasteland that prefers moist, fertile soil. In tropical areas of the United States (e.g., Hawaii, Florida), this weed forms

Medical Toxicology of Natural Substances, by Donald G. Barceloux, MD

impenetrable thickets. Susumber berry is native to tropical areas from Mexico to Peru and Venezuela as well as the Caribbean Islands and Bermuda. *S. torvum* is widely cultivated throughout the world for its edible fruit.

EXPOSURE

Leaves and fruit from *Solanum nigrum* have been used in the Middle East, North Africa, and India as herbal treatments for toothache, skin lesions, burns, scabies, and anxiety.[1] The susumber berry is used in traditional Jamaican cooking after washing and boiling in water to a tender, crisp state. The berries are picked in various stages of development, and the cooked berries are added to sautéed salted codfish, peppers, onions, and tomatoes.

PRINCIPAL TOXINS

There are few data on the toxic constituents of *Solanum* species, including potentially toxic glycoalkaloids (α-chaconine, α-solamargine, α-solanine, solasonine, sycophantine, α- and β-solamarine, soladulcidine, (tomatine), and aglycones (e.g., solanidine, solanocapsine, solasodine, tomatidine). Analysis of unripe berries from *S. dulcamara* by gas chromatography/mass spectrometry demonstrated the presence of the glycoalkaloid aglycones, solasodine (CAS RN: 126-17-0) and diosgenin (CAS RN: 512-04-9).[2] The latter compound is the steroidal aglycone of the saponin, dioscin. Tropane alkaloids (e.g., atropine, hyoscyamine) were not detected in the berries. Other glycoalkaloids include α-, β-, γ-, and δ-solamarine, soladulcidine, tomatine, and tomatidenol. The specific compounds responsible for the toxic effects associated with exposure to *Solanum* species are not well defined, in part, because of the variable concentration of these compounds in different strains grown under diverse conditions. Animal studies and epidemiological data indicate that ripe berries of the Jerusalem cherry do not usually contain substantial amounts of solanine or other toxic glycoalkaloids.[3] The ingestion of unripe berries typically causes a mild gastroenteritis. The toxin in susumber berries has not been identified, but analysis of toxic berries by high performance liquid chromatography (HPLC) with UV detection suggests that these berries contain toxins in addition to α-chaconine and solanine.[4]

DOSE RESPONSE

There are inadequate data to determine a dose-response relationship between exposure to *Solanum* species and toxic effects.

TOXICOKINETICS

The toxic principals in *Solanum* species have not been identified; therefore, there are few toxicokinetic data.

CLINICAL RESPONSE

There are few reported cases of poisoning associated with woody nightshade or the Jerusalem cherry in the medical literature. The existing case reports do not confirm the presence of biomarkers (e.g., glycoalkaloids) of *Solanum* species in biological samples from these patients. A case report of a 7-year-old girl associated the ingestion of berries from *S. dulcamara* with a self-limited gastroenteritis and mydriasis,[5] whereas a 4-year-old girl developed sinus tachycardia, confusion, disorientation, mydriasis, elevated temperature, and tremor after ingesting woody nightshade berries.[2] More serious clinical effects associated with peripheral and central anticholinergic syndrome include tachycardia, weakness, headache, diaphoresis, mydriasis, urinary retention, fever, and central nervous system (CNS) effects (disorientation, confusion, ataxia, agitation, seizures, coma, respiratory failure). However, there are few well-documented reports of serious central anticholinergic symptoms temporally related to exposure to woody nightshade or Jerusalem cherry plants. A four-year-old girl developed memory disturbances, hallucinations, ataxia, seizures, fever (39.4 °C/102.9 °F), tachycardia, facial flushing, mydriasis, and dry mucous membranes.[6] The parents reported gastrointestinal symptoms in the girl several days *prior* to the onset of CNS effects. The patient stated that she ate seeds and other parts of *Solanum pseudocapsicum*, and her urinalysis contained solanocapsin-related compounds consistent with ingestion of this plant. Fatalities from solanine poisoning are unusual with modern supportive care, but case reports before 1950 associate fatalities with woody nightshade (*Solanum dulcamara*). An older case report associated the ingestion of berries from woody nightshade by a 9-year-old girl with a hemorrhagic gastritis, dehydration, and dyspnea.[7] She died after 36 hours of treatment by methods currently not used, including stomach lavage without respiratory protection, soap-and-water enemas, nikethamide, and rectal fluids. Rare case reports associate the consumption of susumber berries with the development of gastrointestinal and neurological symptoms. Three adult members of a family ate a traditional evening meal containing codfish, ackee, and susumber berries.[4] All three family members became ill the following morning with symptoms ranging from mild diarrhea to facial droop, blurred vision, dry mouth, dysarthria, and facial paresthesias. The most seriously ill member of the family developed hypertension, confusion, proximal upper extremity weakness, and respiratory failure that required intubation. She recovered after a prolonged hospitalization.

DIAGNOSTIC TESTING

Methods to identify and quantify the content of glycoalkaloid compounds in solid samples include HPLC/electrospray ionization/mass spectrometry[8] and HPLC.[9] The limits of detection for α-solanine and α-chaconine using the latter method were 1.2 and 1.3 ng/mL, respectively.

TREATMENT

Treatment is supportive. Decontamination measures are usually unnecessary. The administration of activated charcoal to alert patients is a therapeutic option, when the patient ingests large amounts of *Solanum* species less than 1 hour prior to presentation; however, there are no clinical data to indicate that activated charcoal improves the outcome of patients with *Solanum* intoxication. The use of physostigmine is not usually indicated because of the lack of a clear anticholinergic syndrome following exposure to these plants.

References

1. Dafni A, Yaniv Z. Solanaceae as medicinal plants in Israel. J Ethnopharmacol 1994;44:11–18.
2. Ceha LJ, Presperin C, Young E, Allswede M, Erickson T. Anticholinergic toxicity from nightshade berry poisoning responsive to physostigmine. J Emerg Med 1997;15:65–69.
3. Hornfeldt CS, Collins JE. Toxicity of nightshade berries (*Solanum dulcamara*) in mice. Clin Toxicol 1990;28:185–192.
4. Smith SW, Giesbrecht E, Thompson M, Nelson LS, Hoffman RS. Poisoning by *Solanum torvum*, the normally edible susumber berry. Clin Toxicol 2007;45:344–355.
5. Rubinfeld RS, Currie JN. Accidental mydriasis a from blue nightshade "lipstick". J Clin Neuroophthalmol 1987;7:34–37.
6. Parisi P, Francia A. A female with central anticholinergic syndrome responsive to neostigmine. Pediatr Neurol 2000;23:185–187.
7. Alexander RF, Forbes GB, Hawkins ES. A fatal case of solanine poisoning. Br Med J 1948;2:518–519.
8. Matsuda F, Morino K, Miyazawa H, Miyashita M, Miyagawa H. Determination of potato glycoalkaloids using high-pressure liquid chromatography-electrospray ionisation/mass spectrometry. Phytochem Anal 2004;15:121–124.
9. Kodamatani H, Saito K, Niina N, Yamazaki S, Tanaka Y. Simple and sensitive method for determination of glycoalkaloids in potato tubers by high-performance liquid chromatography with chemiluminescence detection. J Chromatogr A 2005;1100:26–31.

Chapter 134

SWEET PEA and OSTEOLATHYRISM

HISTORY

Neurolathyrism has been known to occur after the ingestion of the grass pea or chickling pea (*Lathyrus sativus*) at least since the time of Hippocrates, but the occurrence of skeletal abnormalities from the ingestion of sweet peas (*Lathyrus odoratus*) is less well documented. In animal experiments during the early 1930s, the administration of peas from *Lathyrus odoratus* (sweet pea) produced growth retardation, skeletal deformities, and aortic dissections when the sweet peas accounted for more than 25% of the diet.[1]

BOTANICAL DESCRIPTIONS

There are more than 100 *Lathyrus* species in the United States and Europe, and a number of these species (*Lathyrus hirsutus* L., *Lathyrus latifolius* L., *Lathyrus odoratus* L., *Lathyrus pusillus* Elliot, *Lathyrus sphaericus* Retz., *Lathyrus splendens* Kellogg, *Lathyrus alefeldii* (T.G.White) Bradshaw, *Lathyrus sylvestris* L.) have been associated with toxic effects in animals. There are several species of the genus *Lathyrus* associated with osteolathyrism (odoratism) including *Lathyrus odoratus* L., *Lathyrus hirsutus* L. (singletary pea, caley pea), and *Lathyrus pusillus* Elliot (low pea vine, tiny pea).[2]

Common Name: Sweet Pea

Scientific Name: *Lathyrus odoratus* L.

Botanical Family: *Fabaceae* (pea)

Physical Description: This commonly cultivated vine has showy flowers in erect racemes that mature into legume pods containing many broad flat seeds.

Ecology and Distribution: This plant is an introduced species in North America including northeastern United States, California, Ohio, and Mexico. This climbing annual plant inhabits sunny areas and light woodlands in southern Europe (Italy, Sicily, Crete), Asia, northern Africa, and Australia.

Common Name: Caley Pea, Singletary Pea, Hairy Bitter Vetch

Scientific Name: *Lathyrus hirsutus* L.

Botanical Family: *Fabaceae* (pea)

Physical Description: This legume has winged stems that reach up to about 100 cm (~3 ft) in length. The compound leaves have two long, narrow leaflets that end in a branched tendril up to approximately 10 cm (~4 in.) in length. The small, reddish-blue, pea-like flowers produce distinctive pods, covered with hairs and small, raised areas. Each pod is about 3–4 cm (~1–2 in.) and contains 4–10 mottled, round seeds.

Ecology and Distribution: This plant is an introduced species in the United States, inhabiting areas in California, the Southwest, the Southeast, and north central United States. The singletary pea is widely distributed in Europe, North Africa, temperate areas of Asia, and the Indian subcontinent.

Common Name: Low Peavine, Tiny Pea

Scientific Name: *Lathyrus pusillus* Elliot

Botanical Family: *Fabaceae* (pea)

Physical Description: This slender, annual climbing plant grows to 60 cm (~2 ft) in length three-branched tendrils. The outer ring of flower parts (calyx) is green and there are one to two blue flowers. The brown pods are 3–6 cm (~1–2 in.) long, and they contain many blue seeds with a broad hilum.

Ecology and Distribution: This native US species inhabits Oregon, Texas, the southern midwestern United States, Virginia, North Carolina, and Florida as well as areas of Central and South America (Argentina, Peru).

EXPOSURE

L. odoratus is widely cultivated as a sweet-smelling ornamental plant. An essential oil from L. *odoratus* is an ingredient in perfumery. The seeds of this plant are not edible. There are no definite medicinal uses of this plant. The singletary pea is a traditional cover crop and winter forage that becomes naturalized in open areas.

PRINCIPAL TOXINS

Physiochemical Properties

Osteolathyrism is a disease of collagen tissue that affects animals grazing on certain *Lathyrus* species (e.g., *L. odoratus*). Compounds associated with osteolathyrism in experimental animals include β-aminopropionitrile (*L. odoratus*), γ-glutamyl-β-aminopropionitrile (*L. pusillus*), and L-2,4-diaminobutanoic acid (*L. latifolius* L.).[3,4] Table 134.1 lists some physical properties of β-aminopropionitrile (CAS RN: 151-18-8, $C_3H_6N_2$). Neurolathyrism in humans results from the ingestion of chickling peas (*Lathyrus sativus* L.), which do not contain β-aminopropionitrile.[5] However, the shoots from *L. sativus* do contain an unstable compound [2-(2-cyanoethyl)isoxazolin-5-one] that potentially may be metabolized to β-aminopropionitrile.[6] Cooking sweet peas for several hours in water does not prevent the development of osteolathyrism in experimental animals when sweet peas represent >25% of the diet.[1]

TABLE 134.1. Some Physical Properties of β-Aminopropionitrile

Physical Property	Value
Melting Point	<25 °C/77 °F
Boiling Point	185 °C/365 °F
log P (Octanol-Water)	−1.130
Atmospheric OH Rate Constant	1.26E-12 cm³/molecule-second (25 °C/77 °F)

Poisonous Parts

The ingestion of both seedlings and seeds from sweet pea can produce osteolathyrism.[7]

Mechanism of Toxicity

In experimental animals, β-aminopropionitrile inhibits lysyl oxidase, which causes a decrease in the formation of cross-links in collagen and elastin.[8] The development of skeletal abnormalities in animals is called osteolathyrism. The destruction of cross-linking produces abnormal collagen and elastin with poor tensile strength as a result of the replacement of normal amino acids with β-aminopropionitrile and the inhibition of lysyl oxidase. The human variety of lathyrism is a chronic neurologic disease (neurolathyrism) that results from the excessive consumption of the chick pea (*L. sativus*) during periods of food shortages.[9] Skeletal lesions in patients with neurolathyrism are minor because the most common species responsible for neurolathyrism (*L. sativus*) does not contain significant amounts of β-aminopropionitrile. Furthermore, β-aminopropionitrile is not neurotoxic, and osteolathyrism does not occur in humans consuming sweet peas.[8] Other compounds also produce osteolathyrism in experimental animals including thiosemicarbazide derivatives.[10]

DOSE RESPONSE

The dose response for acute or chronic toxicity from the ingestion of sweat peas is not well defined. In rodent studies, rats feed diets containing 0.2% β-aminopropionitrile or 50% crude sweet pea meal develop signs of osteolathyrism.[11]

TOXICOKINETICS

There are few data on the toxicokinetics of β-aminopropionitrile in humans. In a study of one volunteer receiving 250 mg β-aminopropionitrile, unchanged β-aminopropionitrile appeared in the urine within 1 hour of ingestion, and renal excretion of β-aminopropionitrile was completed within 7 hours.[12] Following the administration of 250 mg β-aminopropionitrile 4 times daily for 21 days, the renal excretion of the deaminated, inactive metabolite, cyanoacetic acid, continued at least 2 days after the cessation of the 21-day treatment. During treatment, unchanged β-aminopropionitrile in the urine accounted for approximately 16% of the dose.

CLINICAL RESPONSE

In osteolathyrism, connective tissues fibers do not polymerize properly and animals grazing on sweet peas may develop osteolathyrism. However, the occurrence of osteolathyrism or neurolathyrism is not well documented in humans chronically ingesting sweet peas (*Lathyrus odoratus*). The toxic compound in sweet pea (β-aminopropionitrile) is not neurotoxic; therefore, neurolathyrism is unlikely to occur in humans ingesting sweet peas. Skeletal abnormalities resembling osteolathyrism in animals occurs rarely in humans ingesting chick peas (*Lathyrus sativa*).[13] In a follow-up study of 45 former German concentration camp victims with neurolathyrism from the ingestion of chick peas, five former prisoners had skeletal abnormalities including nonfusion of the iliac crest, ischial apophysis, and vertebral epiphysis, as well as bowing and structural changes of the femur.[14] At the time of exposure to chick peas, the secondary centers of ossification probably were not fused. Rare case reports associate occupational exposure to *Lathyrus odoratus* with rhinitis and asthma in sensitized workers.[15]

DIAGNOSTIC TESTING

The toxic principals in *Lathyrus* species are not well defined. Isoxazolinone compounds are *N,O*-heterocyclic compounds derived from this genus that are precursors to potentially toxic metabolites. The presence of β-(isoxazolin-5-on-2-yl)propionitrile and 2-(3-amino-3-carboxypropyl)isoxazolin-5-one are highly suggestive of *L. sativus* or *L. odoratus*. The isoxazolinone compounds are sensitive to alkaline conditions and UV radiation. Identification of these compounds involves high performance liquid chromatography with pre-column phenyl isothiocyanate derivatization or capillary zone electrophoresis.[16] Using the latter method, limits of detection for these compounds range between 3 μg/mL and 7.7 μg/mL.

TREATMENT

Management involves primarily supportive care. Decontamination measures are unnecessary.

References

1. Geiger BJ, Steenbock H, Parson HT. Lathyrism in the rat. J Nutr 1935;6:427–442.
2. Selye H. Lathyrism. Rev Can Biol 1957;16:1–82.
3. McKay GF, Lalich JJ, Schilling ED, Strong FM. A crystalline from *Lathyrus odoratus*. Arch Biochem 1954;52:313–322.
4. Dupuy HP, Lee JG. Isolation of a material capable of producing experimental lathyrism. J Am Pharm Assoc Am Pharm Assoc (Baltim). 1954;43:61–62.
5. Bell EA. Relevance of biochemical taxonomy to the problem of lathyrism. Nature 1964;203:378–380.
6. Lambein F, Khan JK, Kuo YH. Free amino acids and toxins in *Lathyrus sativus* seedlings. Planta Med 1992;58: 380–381.
7. van Rompuy L, Lambein F, Van Parijs R, Ressler C. Lathyrogenic activity of 2 (2-cyanoethyl)-3-isoxazolin-5-one from *Lathyrus odoratus* seedlings. Experientia 1974; 30:1379–1380.
8. Wilmarth KR, Froines JR. *In vitro* and *in vivo* inhibition of lysyl oxidase by aminopropionitriles. J Toxicol Environ Health 1992;37:411–423.
9. Weaver AL, Spittell JA. Lathyrism. Mayo Clin Proc 1964;39:485–489.
10. Dawson DA, Schultz TW, Baker LL, Mannar A. Structure-activity relationships for osteolathyrism. III. Substituted thiosemicarbazides. J Appl Toxicol 1990;10:59–64.
11. Bachhuber TE, Lalich JJ, Angevine DM, Schilling ED, Strong FM. Lathyrus factor activity of beta-aminopropionitrile and related compounds. Proc Soc Exp Biol Med 1955;89:294–297.
12. Fleisher JH, Peacock EE, Chvapil M. Urinary excretion of beta-aminopropionitrile and cyanoacetic acid. Clin Pharmacol Ther 1978;23:520–524.
13. Haque A, Hossain M, Wouters G, Lambein F. Epidemiological study of lathyrism in northwestern districts of Bangladesh. Neuroepidemiology 1996;15:83–91.
14. Streifler M, Cohn PF. Chronic central nervous system toxicity of the chickling pea (*Lathyrus sativus*). Clin Toxicol 1981;18:1513–1517.
15. Jansen A, Vermeulen A, van Toorenenbergen AW, Dieges PH. Occupational asthma in horticulture caused by *Lathyrus odoratus*. Allergy Proc 1995;16:135–139.
16. Chowdhury B, Rozan P, Kuo Y-H, SuM, Lambein F. Identification and quantification of natural isoxazolinone compounds by capillary zone electrophoresis. J Chromatogr A 2001;933:129–136.

Chapter 135

TREE TOBACCO and OTHER PIPERIDINE-CONTAINING PLANTS

BOTANICAL DESCRIPTION

Nicotiana genus consists of about 60 plant species in North and South America, Australia, and the South Pacific. *N. rustica* L. (Aztec tobacco) is native to Virginia, but the commercial value of this plant is substantially less than the South American native, *N. tabacum* L. (cultivated tobacco). Cultivated tobacco does not persist well in the wild in contrast to native species (e.g., *Nicotiana obtusifolia* Mertens & Galeotti or desert tobacco from the southwestern United States).

Common Name: Tree Tobacco
Scientific Name: *Nicotiana glauca* Graham
Botanical Family: Solanaceae
Physical Description: This plant is a large, evergreen, loose branching tree that grows up to 6–18 ft (~2–6 m) in height. The pithy, hairless stems have ovate leaves and a waxy, blue-green cuticle. The distinctive flowers are tubular and yellow.
Distribution and Ecology: *N. glauca* is a native plant of South American (Paraguay, Argentina) that is naturalized in temperate climates in Central America, Mexico, California, Texas, New Mexico, and Arizona. This drought-resistant plant tolerates a wide variety of habitats, particularly in waste areas, dry river beds, and roadsides.

Common Name: Coyote Tobacco
Scientific Name: *Nicotiana attenuata* Torr. ex S. Wats. (*Nicotiana trigonophylla* Dunal)
Botanical Family: Solanaceae
Physical Description: This branching, annual plant has long, tubular white flowers that open at night.
Distribution and Ecology: This native plant of the United States inhabits dry areas and streambeds of the West and Southwest.

Common Name: Indian Tobacco
Scientific Name: *Lobelia inflata* L.
Botanical Family: Lobeliaceae (lobelia)
Physical Description: An annual weed that has simple alternate leaves with inconspicuous blue flowers in a terminal raceme.
Distribution and Ecology: This plant is a native species of the midwestern and eastern United States that inhabits dry areas of waste areas, open woods, meadows, fields, and roadsides. Several other *Lobelia* species (*L. berlandieri* A. DC., *L. cardinalis L. syphilitica* L.) are native species in moist woodlands.

EXPOSURE

Previously, nicotine was used as a pesticide and veterinarian anesthetic, but less toxic products are now available for these applications. As a substitute for smoking, nicotine transdermal patches deliver about 7–21 mg nicotine per day. Anabasine was extracted from *Anabasis aphylla* L. in Russia for use as a pesticide.[1] Tree tobacco (*N. glauca*) was one of several *Nicotiana* species

Medical Toxicology of Natural Substances, by Donald G. Barceloux, MD

used as smoking tobacco by early California Native Americans, particularly northern California tribes (e.g., Karok). Northeastern Native Americans used the dried herb, *Lobelia inflata*, as a purgative and emetic. These Native Americans also incorporated the herb into tobacco leaves for smoking. Lobeline sulfite has been advocated as a tobacco substitute to aid people in breaking the smoking habit, but there are inadequate data to support the use of lobeline as a tobacco substitute.

PRINCIPAL TOXINS

Physiochemical Properties

Nicotiana Glauca

This plant contains primarily piperidine alkaloids including an isomer of nicotine, anabasine [1-(3-pyridyl) piperidine]. Anabasine (CAS RN: 494-52-0, $C_{10}H_{14}N_2$) is a minor alkaloid in the common tobacco plant leaf, but this alkaloid is a major constituent of the *N. glauca* leaf as well as the stalks of *N. tabacum*.[2] Figure 135.1 displays the chemical structure of anabasine. Native Australian species of *Nicotiana* that contain anabasine include *N. debneyi* Domin. and *N. rotundifolia* Lindl. Anabasine is a diprotic base with a pK_{a2} of 3.21. Table 135.1 lists some physical properties of anabasine and Table 135.2 lists some physical properties of nicotine.

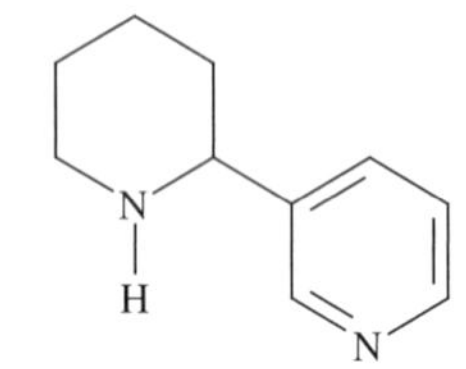

FIGURE 135.1. Chemical structure of anabasine.

TABLE 135.1. Some Physical Properties of Anabasine

Physical Property	Value
Melting Point	90 °C/194 °F
Boiling Point	276 °C/528.8 °F
Molecular Mass	162.23 g/mol
pKa Dissociation Constant	11 (30 °C/86 °F)
log P (Octanol-Water)	0.97
Water Solubility	1.00E + 06 mg/L (25 °C/77 °F)
Vapor Pressure	3.01E-03 mm Hg (25 °C/77 °F)
Henry's Law Constant	1.81E-09 atm-m^3/mole (25 °C/77 °F)
Atmospheric OH Rate Constant	1.01E-10 cm^3/molecule-second (25 °C/77 °F)

Lobelia Inflata

Lobelia inflata contains at least 14 different piperidine alkaloids including lobeline and the emetic alkaloids, lobelanine and lobelanidine. Lobeline is structurally and pharmacologically similar to nicotine.

Poisonous Parts

Nicotine (CAS RN: 54-11-5, $C_{10}H_{14}N_2$) is the major toxic alkaloid in *N. tabacum* (tobacco) leaves averaging about 2–4% with a range of approximately 1.5–5.4% dry weight.[3] Figure 135.2 displays the chemical structure of nicotine. The nicotine isomer, anabasine, is the primary alkaloid in *Nicotiana glauca* with nicotine occurring in minor amounts. The approximate anabasine content in the dry plant, root, and leaves is about 1.3%, 1%, and 0.2%, respectively.[4,5] Analysis of three South African specimens of *N. glauca* demonstrated anabasine concentrations in flowers and leaves of approximately 1.4–1.5 mg/g wet weight and 1.8–2.0 mg/g wet weight, respectively.[6] Lobeline is the principal alkaloid in *Lobelia inflata* with a typical concentration of approximately 0.3%. Although cytisine and lobeline produce nicotine-like peripheral effects, animal studies indicate that the potency of these two alkaloids is much lower than nicotine.[7]

Mechanism of Toxicity

Nicotinic cholinergic receptors are composed of a ligand-gated ion channel consisting of five subunits;

FIGURE 135.2. Chemical structure of nicotine.

TABLE 135.2. Some Physical Properties of Nicotine

Physical Property	Value
Melting Point	−7.90E + 01 °C/33.8 °F
Boiling Point	247 °C/476.6 °F
Molecular Mass	162.23 g/mol
pKa Dissociation Constant	3.1
log P (Octanol-Water)	1.17
Water Solubility	1.00E + 06 mg/L
Henry's Law Constant	3.00E-09 atm-m^3/mole (25 °C/77 °F)
Atmospheric OH Rate Constant	9.10E-11 cm^3/molecule-second (25 °C/77 °F)

these receptors occur in the brain, autonomic ganglia, and the neuromuscular junction.[8] Nicotinic receptors appear throughout the brain including the amygdala, septum, locus ceruleus, and brainstem motor nuclei; however, the highest concentrations of binding sites occur in the cortex, interpeduncular nucleus, and the thalamus.[9] Initially, nicotine stimulates the nicotinic cholinergic receptors stereoselectively in the sympathetic and parasympathetic ganglia of autonomic nervous system as well as in the neuromuscular junction.[10] *l*-Nicotine is the active stereoisomer, whereas *d*-nicotine is a weak agonist at the cholinergic receptors. Depending on the dose of nicotine, the stimulatory phase may be transient with prolonged ganglionic blockade resulting from persistent membrane depolarization.

DOSE RESPONSE

Although the commonly quoted estimated lethal adult dose of nicotine is approximately 60 mg,[11] there are few data correlating nicotine toxicity with dose. The suicidal ingestion of a nicotine dose exceeding 5 grams (71 mg/kg) by a 17-year-old male smoker resulted in hypotension, intractable seizures, ventricular fibrillation, and death.[12] A case report associated the ingestion of 10–12 fried, tree tobacco leaves with the development of severe muscle weakness, bulbar palsy, hypertension, vomiting, and respiratory insufficiency.[13] The patient recovered with supportive care.

TOXICOKINETICS

Absorption

Case reports document the absorption of clinically significant amounts of nicotine from the skin,[14] lungs, and gastrointestinal tract.[15] The absorption of nicotine is highly pH dependent and toxicity depends on pH, formulation, mastication, and nicotine content. Experimental studies demonstrated that absorption of nicotine from a solution buffered to a pH of 1, 7.4 and 9.8 was 3.3%, 8.2%, and 18.6%, respectively.[16] Absorption of nicotine from the stomach is limited because of acid pH, whereas nicotine absorption is much higher in the alkaline environment of the duodenum. Symptoms of nicotine toxicity develop rapidly after large doses. Case reports from the older medical literature indicates that death can occur within 5–10 minutes of the ingestion of liquid pesticide containing 40% nicotine.[17,18] Dermal absorption of a 95% nicotine solution caused coma and respiratory failure.[19]

Distribution

Nicotine distributes extensively and rapidly into tissues with a volume of distribution (V_d) ranging from about 1–3 L/kg.[20] The average V_d following the intravenous administration of a nicotine dose equivalent similar to the dose of nicotine present in cigarettes was approximately 1.7 times body weight.[21] Although there is substantial interindividual variation, nicotine concentrations in plasma samples are generally higher in arterial blood than venous blood, with the latter double the former shortly after the beginning of smoking.[22]

Biotransformation/Elimination

Hepatic cytochrome P450 enzymes catalyze the metabolism of about 70–80% of administered dose of nicotine to inactive metabolites (e.g., cotinine, nicotine-1′-*N*-oxide) via C-oxidation. Experimental studies of poor and extensive metabolizers of dextromethorphan, a probe for CYP2D6 activity, suggests that CYP2D6 is not a major cytochrome P450 isoenzyme for the metabolism of nicotine or cotinine.[23] The kidneys excrete nicotine metabolites as 3′-hydroxycotinine, unchanged compounds or glucuronide conjugates.[9] About 5–10% of the absorbed dose of nicotine from cigarette smoking appears in the urine unchanged.[24] The average plasma elimination half-life of nicotine following intravenous administration is about 2–3 hours, whereas the half-life of the inactive metabolite cotinine is about 16 hours. Elimination of nicotine is similar in smokers and nonsmokers.[25]

CLINICAL RESPONSE

Nicotine

The onset of symptoms is usually rapid (15–30 minutes) beginning with gastrointestinal distress (vomiting, nausea, salivation, abdominal pain, diarrhea). Vomiting is almost always present following the ingestion of tobacco products, but the absence of vomiting does not necessary exclude the development of potentially serious nicotine toxicity.[26] Symptoms in mild nicotine poisoning resolve in several hours. Severe nicotine poisoning involves a stimulatory phase followed by an inhibitory phase that occurs over 24 hours. Transiently, hypertension and tachycardia may develop followed by hypotension, bradycardia, decreased deep tendon reflexes, fasciculations, paralysis and cardiovascular collapse. Neurologic symptoms include headache, diaphoresis, ataxia, lightheadedness, weakness, confusion, and, in severe poisonings, convulsions, coma, and respiratory failure.[15] Nicotine acts unpredictably as an agonist of the muscarinic receptors; therefore, the occurrence of cholinergic symptoms (miosis, salivation, lacrimation, bronchorrhea, diarrhea, bradycardia) is variable.[27]

Tree Tobacco

The ingestion of tree tobacco leaves (*N. glauca*) produces clinical features similar to nicotine poisoning with initial hypertension, tachycardia, vomiting, headache, salivation, and altered mentation followed by neuromuscular blockade and respiratory failure sometimes requiring intubation.[13,28] The onset of symptoms following the ingestion of tree tobacco leaves are slightly delayed (i.e., up to several hours) compared with commercial tobacco because the absorption of anabasine is slower than nicotine. Sudden cardiac arrest may occur during severe poisoning from the ingestion of anabasine in tree tobacco.[29]

In experimental piglets, consumption of *N. glauca* produces congenital defects, including cleft palate and limb defects. The principal alkaloid of the common tobacco plant, nicotine, apparently is not the teratogenic agent in this plant. Anabasine is a potential animal teratogen based on its structural similarity to other known piperidine teratogens (e.g., coniine and γ-coniceine in *Conium maculatum* L.).[30]

DIAGNOSTIC TESTING

Analytical Methods

Methods to detect toxic alkaloids (e.g., anabasine, nicotine) following ingestion of tree tobacco include gas chromatography/mass spectrometry,[4,5] capillary electrophoresis with sequential light-emitting diode-induced fluorescence and electrochemiluminescence detection,[31] and high performance liquid chromatography (HPLC)/photodiode array/mass spectrometry.[6] The limits of detection and quantitation with the latter method were 250 ng/mL and 500 ng/mL, respectively. A reverse-phase HPLC method is available for the determination of lobeline and cytisine in plasma.[7]

Biomarkers

The mean plasma nicotine concentration following the smoking of one cigarette ranges from about 5–30 ng/mL.[32] In a study of volunteers undergoing treatment for nicotine addiction, the steady state serum nicotine and cotinine concentrations two weeks after the daily administration of 22 mg nicotine via transdermal patches ranged from approximately 10–50 μg/L and 35–249 μg/L, respectively.[33] Serum nicotine concentrations exceeding 2,000 μg/L indicate serious toxicity. A case report of a woman, who died of asphyxia from a plastic bag following the placement of multiple nicotine patches on her body, suggests that some postmortem distribution of nicotine may occur; however, the distribution of the nicotine patches on the body may also alter the heart to femoral blood ratio.[34] Case reports of victims found dead after ingesting *Nicotiana glauca* document the presence of anabasine concentrations in postmortem blood of 1.15 mg/L[4] and 2.2 mg/L.[5]

Abnormalities

Leukocytosis and glycosuria may develop following nicotine intoxication, but most laboratory values are normal.

TREATMENT

Stabilization

Respiratory failure represents the greatest risk, and patients with suspected *Nicotiana* intoxication should be evaluated carefully for signs of respiratory distress (e.g., ventilatory rate, oxygen saturation, dyspnea). Patients with altered sensorium should receive intravenous lines, cardiac monitoring, oxygen, pulse oximetry, and arterial blood gases in addition to rapid evaluation of serum glucose concentrations and intravenous naloxone, if indicated. The initial hypertension and tachycardia should not be aggressively treated unless signs of cardiovascular compromise are present because a depressive phase of bradycardia and hypotension often follows the initial stimulatory phase.

Decontamination

Because nicotine stimulates the emetic center in the medulla oblongata, spontaneous vomiting occurs in most nicotine ingestions and decontamination measures are usually unnecessary. Most children only taste parts of *Nicotiana* or *Lobelia* plants, and these children do not usually develop symptoms. Additionally, the potential for rapid onset of changes in consciousness and seizures during serious *Nicotiana* or *Lobelia* intoxications increases the risk of using decontamination measures. Consequently, most of these ingestions require no decontamination measures. All skin contaminated with nicotine solutions should be washed thoroughly with a nonalkaline soap and cool water.

Elimination Enhancement

There are inadequate clinical data to recommend any methods to enhance the elimination of toxins associated with ingestions of *Nicotiana* or *Lobelia* species. Continuous nasogastric suction and serial activated charcoal are theoretically advantageous, but there are few clinical data to guide management. Acidification of the urine also has theoretical advantages, but the lack of data to support this method, the short duration of nicotine action, and complications associated with acid diuresis limit the usefulness of this method.

Antidotes

Atropine may improve bradycardia and hypotension, but this drug does not reverse neuromuscular weakness during nicotine intoxication.

Supplemental Care

The treatment of nicotine poisoning is supportive. Serious poisoning must be followed closely in an intensive care setting for the development of respiratory failure within the first 12 hours. Seizures are treated with benzodiazepines (diazepam, lorazepam).

References

1. Jacobson M. Plants, insects, and man—their interrelationships. Econ Bot 1982;36:345–354.
2. Keeler RF, Balls LD, Panter K. Teratogenic effects of *Nicotiana glauca* and concentration of anabasine, the suspect teratogen in plant parts. Cornell Vet 1981;71: 47–53.
3. Tso TC. Effect of farm production practices on nicotine and total particulate matter in cigarette smoke. Natl Cancer Inst Monogr 1968;28:97–111.
4. Castorena JL, Garriott JC, Barnhardt FE. A fatal poisoning from *Nicotiana glauca*. Clin Toxicol 1987;25:429–435.
5. Sims DN, James R, Christensen T. Another death due to ingestion of *Nicotiana glauca*. J Forensic Sci 1999;44: 447–449.
6. Steenkamp PA, van Heerden FR, Van Wyk B-E. Accidental fatal poisoning by *Nicotiana glauca*: identification of anabasine by high performance liquid chromatography/ photodiode array/mass spectrometry. Forensic Sci Int 2002;127:208–217.
7. Reavill C, Walther B, Stolerman IP, Testa B. Behavioural and pharmacokinetic studies on nicotine, cytisine and lobeline. Neuropharmacology 1990;29:619–624.
8. Changeux JP, Galzi JL, Devillers-Thiery A, Betrand D. The functional architecture of the acetylcholine nicotinic receptor explored by affinity labeling and site-directed mutagenesis. Q Rev Biophys 1992;25:395–432.
9. Benowitz NL. Pharmacology of nicotine: addiction and therapeutics. Annu Rev Pharmacol Toxicol 1996;36: 597–613.
10. Grenhoff J, Svensson TH. Pharmacology of nicotine. Br J Addict 1989;84:477–492.
11. Larson PS, Haag HB, Silvette H. Tobacco—experimental and clinical studies. Baltimore: Williams & Wilkins; 1961.
12. Lavoie FW, Harris TM. Fatal nicotine ingestion. J Emerg Med 1991;9:133–136.
13. Mellick LB, Makowski T, Mellick GA, Borger R. Neuromuscular blockade after ingestion of tree tobacco (*Nicotiana glauca*). Ann Emerg Med 1999;34:101–104.
14. Woolf A, Burkhart K, Caraccio T, Litovitz T. Self-poisoning among adults using multiple transdermal nicotine patches. Clin Toxicol 1996;34:691–698.
15. Oberst BB, McIntyre RA. Acute nicotine poisoning. Pediatrics 1953;11:338–340.
16. Ivy KJ, Triggs EJ. Absorption of nicotine by the human stomach and its effect on gastric ion fluxes and potential difference. Dig Dis 1978;23:809–814.
17. Moore HW. Poison case report of the month. Acute nicotine poisoning. SC Med Assoc J 1962;58:445.
18. McNally WD. A report of seven cases of nicotine poisoning. J Lab Clin Med 1922;8:83–85.
19. Lockhart LP. Nicotine poisoning. Br Med J 1933;1: 246–247.
20. Svensson CK. Clinical pharmacokinetics of nicotine. Clin Pharmacokinet 1987;12:30–40.
21. Rosenberg J, Benowitz NL, Jacob P, Wilson KM. Disposition kinetics and effects of intravenous nicotine. Clin Pharmacol Ther 1980;28:517–522.
22. Henningfield JE, Stapleton JM, Benowitz NL, Grayson RF, London ED. Higher levels of nicotine in arterial than in venous blood after cigarette smoking. Drug Alcohol Depend 1993;33:23–29.
23. Benowitz NL, Jacob P III, Perez-Stable E 3rd. CYP2D6 phenotype and the metabolism of nicotine and cotinine. Pharmacogenetics 1996;6:239–242.
24. Benowitz NL, Jacob P III, Fong I, Gupta S. Nicotine metabolic profile in man: comparison of cigarette smoking and transdermal nicotine. J Pharmacol Exp Ther 1994;268: 296–303.
25. Benowitz NL, Jacob P III. Nicotine and cotinine elimination pharmacokinetics in smokers and nonsmokers. Clin Pharmacol Ther 1993:53;316–323.
26. Smolinske SC, Spoerke DG, Spiller SK, Wruk KM, Kulig K, Rumack BH. Cigarette and nicotine chewing gum toxicity in children. Hum Toxicol 1988;7:27–31.
27. Davies P, Levy S, Pahari A, Martinez D. Acute nicotine poisoning associated with a traditional remedy for eczema. Arch Dis Child 2001;85:500–502.
28. Manoguerra AS, Freeman D. Acute poisoning from the ingestion of *Nicotiana glauca*. J Toxicol Clin Toxicol 1982–1983;19: 861–864.
29. Mizrachi N, Levy S, Goren Z. Fatal poisoning from *Nicotiana glauca* leaves: identification of anabasine by gas-chromatography/mass spectrometry. J Forensic Sci 2000;45:736–741.

30. Keller RF, Crowe MW. Congenital deformities in swine induced by wild tree tobacco, *Nicotiana glauca*. Clin Toxicol 1983;20: 47–58.
31. Chang PL, Lee KH, Hu CC, Chang HT. CE with sequential light-emitting diode-induced fluorescence and electro-chemiluminescence detections for the determination of amino acids and alkaloids. Electrophoresis 2007;28: 1092–1099.
32. Benowitz NL. Drug therapy. Pharmacologic aspects of cigarette smoking and nicotine addition. N Engl J Med 1988;319:1318–1330.
33. Hurt RD, Dale LC, Offord KP, Lauger GG, Basking LB, Lawson GM, et al. Serum nicotine and cotinine levels during nicotine-patch therapy. Clin Pharmacol Ther 1993;54:98–106.
34. Kemp PM, Sneed GS, George CE, Distefano RF. Postmortem distribution of nicotine and cotinine from a case involving the simultaneous administration of multiple nicotine transdermal systems. J Anal Toxicol 1997;21:310–313.

Chapter 136

VERATRUM ALKALOIDS

HISTORY

The use of *Veratrum* alkaloids in sorcery dates back hundreds of years in Europe. During the Middle Ages, *Veratrum* alkaloids were folk treatments for cardiac aliments and counterirritants for neuralgias. *Veratrum viride* Aiton was part of the potions used by American Indians to test the endurance of young braves. During the 1950s, *Veratrum* alkaloids were tested as antihypertensive agents, but the administration of these agents was discontinued because of adverse effects.

BOTANICAL DESCRIPTION

The genus *Veratrum* includes the following species: *Veratrum album* L. (white false hellebore), *Veratrum californicum* Dur. (California false hellebore), *Veratrum fimbriatum* Gray (fringed false hellebore), *Veratrum insolitum* Jepson (Siskiyou false hellebore), *Veratrum nigrum* L. (false hellebore), *Veratrum tenuipetalum* Heller (Colorado false hellebore), and *Veratrum viride* Aiton (green false hellebore). The most common *Veratrum* species in the United States is *V. viride*. Intoxication with *Veratrum* alkaloids occurs when *Veratrum* species are misidentified as skunk cabbage [*Symplocarpus foetidus* (L.) Salisb. ex Nutt.], pokeweed (*Phytolacca americana* L.), or ramps (*Allium tricoccum* Ait.). *Veratrum californicum* Dur. is the related *Veratrum* species that grows in moist woodlands of the coniferous forest in the Pacific Coast and northern Rockies up to an elevation of 11,000 ft (3,353 m). The European and Asian species are *Veratrum album* L. and *Veratrum japonicum* (Baker) Loesener, respectively. *Veratrum album* also inhabits Alaska. In western and central Europe, misidentification of *Veratrum* species as *Gentiana lutea* L. (yellow gentian) results in *Veratrum* toxicity when *Veratrum* species are used to make homemade gentian wine.

Common Name: Green False Hellebore, Hellebore, White Hellebore, Indian Poke

Scientific Name: *Veratrum viride* Aiton

Botanical Family: Liliaceae (lily)

Physical Description: This tall perennial herb grows to 4–6 ft (~1–2 m) in height with alternate, pleated leaves. Small flowers emerge from a large terminal panicle.

Distribution and Ecology: *V. viride* is an indigenous plant present in moist woodlands of Canada as well as the Pacific Coast and eastern United States. This perennial herb is found from Canada and northern United States to North Carolina, Georgia, and Tennessee.

EXPOSURE

Past medicinal uses of *Veratrum* alkaloids include the treatment of hypertension, cardiac failure, and toxemia of pregnancy. Li-lu is a crude Chinese drug prepared from the dried roots and rhizomes of several *Veratrum* species and used to treat vascular disease. In agriculture, these alkaloids have been used as pesticides. Cyclopa-

Medical Toxicology of Natural Substances, by Donald G. Barceloux, MD

mine (11-deoxojervine, CAS RN: 4449-51-8) is an experimental chemotherapeutic agent for several types of cancer including medulloblastoma as well as a potential dermatological agent (psoriasis, basal cell carcinoma) that inhibits Sonic hedgehog (Shh) signal transduction.[1,2]

PRINCIPAL TOXINS

Physiochemical Properties

Veratrum species contain over 200 alkaloids.[3] The most important constituents of these species include the following classes of alkaloids: veratranine (veratridine, veratramine, veratrosine), jervanine (jervine, pseudojervine, isorubijervine), and cevanine [cevadine, cevine, zygadenine, veracevine, germine, germerine, germidine, 15-*O*-(2-methylbutyroyl)-germine, protoverine].[4] Figure 136.1 displays the chemical structure of veratridine (CAS RN: 71-62-5, $C_{36}H_{51}NO_{11}$). Table 136.1 lists some chemical properties of veratridine. Jervine and 11-deoxojervine (cyclopamine) are suspected teratogens in grazing animals feeding on *V. californicum* during Day 14 of gestation.[5]

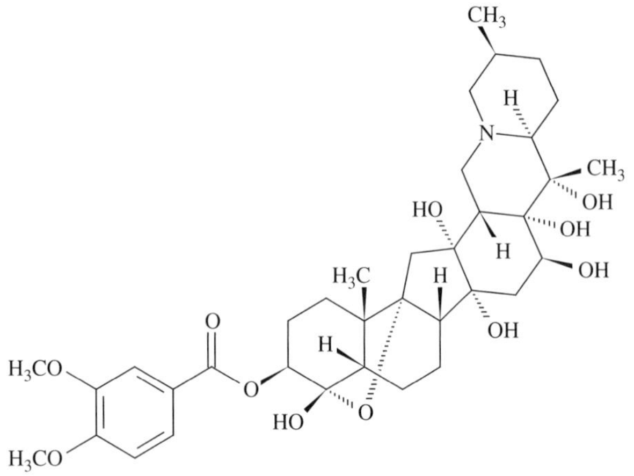

FIGURE 136.1. Chemical structure of veratridine.

TABLE 136.1. Some Physical Properties of Veratridine

Physical Property	Value
Melting Point	180 °C/356 °F
pKa Dissociation Constant	9.54
log P (Octanol-Water)	3.760
Atmospheric OH Rate Constant	1.60E-10 cm^3/molecule-second (25 °C/77 °F)

Poisonous Parts

All parts of *Veratrum* species contain a mixture of *Veratrum* alkaloids, but the concentration of specific *Veratrum* alkaloids varies substantially between *Veratrum* species and plant parts. For example, *V. viride* contains significant concentration of the hypotensive alkaloids, whereas *V. californicum* does not. In general, higher concentrations of *Veratrum* alkaloids occur in the root (i.e. ~2.0%) than in the leaves (i.e., ~0.5%).[6]

Mechanism of Toxicity

The steroidal alkaloid veratridine is a depolarizing agent that activates voltage-sensitive sodium channels resulting in hyperexcitability and depolarization. Voltage-gated Na^+ channels are heteromeric membrane glycoproteins responsible for the generation of action potentials. Veratridine is a lipid-soluble toxin that binds to receptor sites on the S6 segments of the Na^+ channel alpha-subunit on voltage-gated Na^+ channels. This binding affects gating function of Na^+ channels, and increases permeability to calcium and sodium.[7] Veratridine also binds to μ-opioid ligands on receptor sites in the rat cerebral cortical and hypothalamic slices, but the clinical significance of this property is not clear.[8] This compound causes the release of GABA, perhaps by increasing the permeability of the nerve terminals to sodium. The afferent vagal fibers of the coronary sinus and left ventricle appear to be most sensitive to the depolarizing effects of *Veratrum* alkaloids. Cardiac parasympathetic stimulation results in a decrease in peripheral α-adrenergic tone, bradycardia, and hypotension (Bezold-Jarisch reflex). The hypotensive effect of *Veratrum* alkaloids is independent of heart rate, and the use of atropine in *Veratrum*-intoxicated experimental animals only partially reverses the hypotension associated with the use of *Veratrum* alkaloids.[9] Direct stimulation of the peripheral receptors by *Veratrum* alkaloids produces sneezing.[10] Postmortem examination of fatal *Veratrum* intoxications usually reveals nonspecific findings.[11]

DOSE RESPONSE

The clinical response to the ingestion of *Veratrum* alkaloids depends on species, form, and extraction methods (e.g., soup, powder), plant part, and individual hypersensitivity.[12,13] Ingestion of 0.75–1 g of *Veratrum* alkaloids by adults produces gastrointestinal distress, bradycardia, lightheadedness, and salivation.[10]

TOXICOKINETICS

There are few data on the toxicokinetics of toxic *Veratrum* alkaloids.

CLINICAL RESPONSE

Veratrum toxicity classically causes vomiting, bradycardia, and hypotension.[11,14] An initial burning sensation of the oropharynx begins about $^1/_2$–4 hours after ingestion followed by pain in the upper abdomen, nausea, vomiting, diaphoresis, blurred vision, lightheadedness, headache, and confusion.[15] Massive *Veratrum* intoxication causes atrioventricular block, junctional and idioventricular escape rhythms, atrial dysrhythmias, ST abnormalities, T wave inversions, and QT prolongation.[16] Neurological signs are not usually a prominent part of *Veratrum* intoxication. Other clinical features of *Veratrum* intoxication include dilated pupils, paresthesias, muscle cramps, weakness, and syncope. Myoclonus and seizures are uncommon except during severe *Veratrum* intoxication. Dermal contact with powder containing *Veratrum* alkaloids may cause severe conjunctival irritation, lacrimation, and erythema. The clinical effects of *Veratrum* intoxication usually resolve within 24–48 hours. Although there are no data associating *Veratrum* alkaloids with reproductive abnormalities in humans, congenital limb malformations occurred in sheep grazing on *V. californicum*.[17,18]

DIAGNOSTIC TESTING

Analytical Methods

Concentrations of veratridine and cevadine below 1 ng/mL are detectable in biological samples by liquid chromatography/electrospray ionization/mass spectrometry.[6]

Biomarkers

During the postmortem examination of two fatalities from the ingestion of *Veratrum* seeds, the heart blood contained veratridine concentrations of 0.17 ng/mL and 0.40 ng/mL and cevadine concentrations of 0.32 ng/mL and 0.48 ng/mL.[6] The bodies were submerged in water for 1 month before discovery. The postmortem findings were nonspecific except for the presence of a large amount of *Veratrum* seeds in the stomachs.

Abnormalities

Electrocardiographic changes are the most common abnormalities associated with intoxication by *Veratrum* alkaloids including bradycardia and variable degrees of conduction block.

TREATMENT

Treatment is supportive.[19] Decontamination measures are usually unnecessary because of the rapid onset of profuse vomiting. The administration of activated charcoal to alert patients is a therapeutic option, when the patient ingests large amounts of *Veratrum* alkaloids less than 1 hour prior to presentation; however, there are no clinical data to indicate that activated charcoal improves the outcome of patients with *Veratrum* intoxication. Atropine usually reverses the bradycardia, but the effect of atropine on hypotension is usually limited. Fluids and vasopressors may be required. Cardiac monitoring and electrocardiograms should be instituted in severely poisoned patients because of the potential of digitalis-like dysrhythmias. Symptoms and signs (with the exception of eye irritation) usually resolve within 12–15 hours.[10]

References

1. McFerren MA. Useful plants of dermatology. VIII. The false hellebore (*Veratrum californicum*). J Acad Dermatol 2006;54:718–720.
2. Romer J, Curran T. Targeting medulloblastoma: small-molecule inhibitors of the Sonic Hedgehog pathway as potential cancer therapeutics. Cancer Res 2005;65: 4975–4978.
3. Schep LJ, Schmierer DM, Fountain JS. Veratrum poisoning. Toxicol Rev 2006;25:73–78.
4. Kimura I, Islam A, Honda R, Nojima H, Tezuka Y, Zhao W. Blood-pressure lowering, positive chronotropy and inotropy by the *Veratrum* alkaloids germidine and germerine but negative chronotropy by veratridine in mice. J Asian Nat Prod Res 2000;2:133–144.
5. James LF, Panter KE, Gaffield W, Molyneux RJ. Biomedical applications of poisonous plant research. J Agric Food Chem 2004;52:3211–3230.
6. Gaillard Y, Pepin G. LC-EI-MS determination of veratridine and cevadine in two fatal cases of *Veratrum album* poisoning. J Anal Toxicol 2001;25:481–485.
7. Wang SY, Wang GK. Voltage-gated sodium channels as primary targets of diverse lipid-soluble neurotoxins. Cell Signal 2003;15:151–159.
8. Van Huizen F, Wilkinson M, Cynader M, Shaw C. Sodium channel toxins veratrine and veratridine modify opioid and muscarinic but not β-adrenergic binding sites in brain slices. Brain Res Bull 1988;21:129–132.
9. Freis ED, Stanton JR, Culbertson JW, Litter J, Halperin MH, Burnett CH, Wilkins RW. The hemodynamic effects of hypotensive drugs in man: I. *Veratrum viride*. J Clin Invest 1949;28:353–368.

10. Fogh A, Kulling P, Wickstrom E. *Veratrum* alkaloids in sneezing powder. A potential danger. J Toxicol Clin Toxicol 1983;20:175–179.
11. Quatrehomme G, Bertrand F, Chauvet C, Ollier A. Intoxication from *Veratrum album*. Hum Exp Toxicol 1993;12:111–115.
12. Carlier P, Efthymiou ML, Garnier R, Hoffelt J, Fournier E. Poisoning with *Veratrum* containing sneezing powder. Hum Toxicol 1983;2:321–325.
13. Nelson DA. Accidental poisoning by *Veratrum japonicum*. JAMA 1954;156:33–34.
14. Crummett D, Bronstein D, Weaver Z III. Accidental *Veratrum viride* poisoning in three "ramp" foragers. North Carolina Med J 1985;46:469–471.
15. Jaffe AM, Gephardt D, Courtemanche L. Poisoning due to ingestion of *Veratrum viride* (false hellebore). J Emerg Med 1990;8:161–167.
16. Garnier R, Carlier P, Hoffelt J, Savidan A. [Acute dietary poisoning by white hellebore (*Veratrum album* L.). Clinical and analytical data. A series of 5 cases]. Ann Med Interne (Paris) 1985;136:125–128 [French].
17. Keeler RF. Teratogenic compounds of *Veratrum californicum* (Durand). XIV. Limb deformities produced by cyclopamine. Proc Soc Exp Bio Med 1973;142:1287–1291.
18. Binns W, Keller RF, Balls LD. Congenital deformities in lambs, calves and goats resulting from maternal ingestion of *Veratrum californicum*: hare lip, cleft palate, ataxia and hypoplasia of metacarpal and metatarsal bones. Clin Toxicol 1972;5:245–261.
19. Zagler B, Zelger Z, Salvatore C, Pechlaner C, de Giorgi F, Wiedermann CJ. Dietary poisoning with *Veratrum album*—a report of two cases. Wien Klin Wochenschr 2005;117:106–108.

Chapter 137

WANDERING JEW
(*Tradescantia fluminensis* Vellozo)

BOTANICAL DESCRIPTION

Species in the genus *Tradescantia* include two common houseplants, the Mexican native, *Tradescantia pallida* (Rose) D. Hunt and the silver-lined, variegated *Tradescantia zebrina* Bosse.

Common Name: Small-Leaf Spiderwort, Small Leaf Spiderwort, White-Flowered Wandering Jew

Scientific Name: *Tradescantia fluminensis* Vellozo [*T. albifora* (Kunth)]

Botanical Family: Commelinaceae (spiderwort)

Physical Description: This common houseplant is a creeping, perennial, branching plant with erect tips that forms a dense ground cover when naturalized. The leaves are simple, alternate, and glossy green with purple-tinged tips. The oblong to ovate leaf blades arise from a short sheath. Small clusters of white flowers appear at the tip of the stems. The flowers produce small, three-part capsules that contain pitted, black seeds.

Distribution and Ecology: *Tradescantia fluminensis* is a native species of the tropical rainforests of southeastern Brazil that has been introduced into New Zealand, southeastern Australia, Portugal, Italy, Russia, Japan, and the southeastern United States. In these areas, this plant is a persistent invasive weed that disrupts native vegetation.

EXPOSURE

The wandering Jew is a popular indoor ornamental plant, which has become a serious threat to natural vegetation. In subtropical regions, this plant forms dense covers in forested areas that smother low-growing plants and disrupts the regeneration of taller native species.

PRINCIPAL TOXINS

The toxic principal in *Tradescantia* species has not been identified. Case reports suggest that these plants contain allergens.

DOSE RESPONSE

There are inadequate data to determine a dose-response relationship between *Tradescantia* species and clinical effects. Some responses to exposures to these plants result in allergic responses rather than toxic effects.

TOXICOKINETICS

The toxic constituents of *Tradescantia* species have not been identified, and there are no toxicokinetic data.

CLINICAL RESPONSE

Casual exposure to this plant does not usually cause clinical effects. Rarely, vomiting occurs within 2 hours

Medical Toxicology of Natural Substances, by Donald G. Barceloux, MD

of ingestion. Atopic individuals can develop an immediate hypersensitivity reaction manifest by pruritus, facial swelling, wheezing, conjunctival irritation, dyspnea, and wheezing following contact with this plant.[1]

DIAGNOSTIC TESTING

There are no specific biomarkers for exposure to *Tradescantia* species.

TREATMENT

Treatment is supportive. Decontamination measures are unnecessary. The contact dermatitis associated with *Tradescantia* species usually responds to topical steroids (0.05% fluocinonide) and antihistamines as necessary.

Reference

1. Wüthrich B, Johansson SG. Allergy to the ornamental indoor green plant *Tradescantia* (*albifloxia*). Allergy 1997; 52:556–559.

Chapter 138

WATER HEMLOCK and WATER DROPWORT

HISTORY

The 17th century Swiss pathologist, Johann Jakob Wepfer, reported the first cases of *Cicuta* toxicity in his book *Cicutae Aquaticae Historia Et Noxae Commentario Illustrata*.[1] In 1814, Stockbridge reported the first three cases in the United States. In 1876, Boehm isolated a toxic substance from the European species of hemlock water dropwort, *Oenanthe crocata*, which he named oenanthotoxin. In 1949, Clark et al purified oenanthotoxin,[2] and Anet et al identified oenanthotoxin as the highly unsaturated higher alcohol, *trans*-heptadeca-2:8:10-triene-4:6-diene-1:14-diol in 1953.[3] The hemlock water dropwort (*O. crocata*) probably is the most poisonous indigenous plant in the United Kingdom, and ingestion of the roots from this plant caused death in 9 of 13 reported cases since 1900.[4] Beside amatoxins, cicutoxin is the most lethal plant toxin in North America. *Cicuta* toxicity is one of the best described serious plant poisonings in the American literature. Between 1900 and 1975, Starrevekl and Hope identified 83 cases of *Cicuta* poisoning with an overall mortality rate of 30%.[5]

BOTANICAL DESCRIPTION

Common Name: Water Hemlock, Cowbane, Snakeweed, Wild Carrot, Poison Parsnip, Spotted Hemlock, Masquash Root, Beaver Poison, False Parsley

Scientific Name: *Cicuta* spp.

Botanical Family: Apiaceae (carrot)

Physical Description: These perennial weeds have multiple, thick tuberous roots and a hollow stalk reaching 8 ft (~2 m) in height. The leaves are large and pinnate with narrow, saw-toothed leaflets several inches long. Small white clustered flowers appear in an umbrella shape by June or July. The combination of fasciculated, tuberous roots, multicolored swollen root-stock, yellowish oily exudate, and pungent odor of raw parsnips distinguishes *Cicuta* species from other members of the carrot family.[6]

Distribution and Ecology: At least nine distinct species of *Cicuta* occur throughout the United States, and all these species grow along moist areas such as marshes or streams. Common US species include *Cicuta bulbifera* L. (bulb water hemlock), *Cicuta douglasii* (DC.) Coult. & Rose (Douglas water hemlock), *Cicuta maculata* L. (common water hemlock), and *Cicuta virosa* L. (Mackenzie's water hemlock).

Common Name: Hemlock Water Dropwort, Dead Men's Fingers, Five Finger Death

Scientific Name: *Oenanthe crocata* L.

Botanical Family: Umbelliferae

Physical Description: This tall plant grows up to 4–6 ft (~1–2 m) in height. The hollow, grooved stems contain leaves that are similar to celery. White flowers occur in terminal compound umbels with 12–40 rays and a wine-like scent. The root (tuber)

Medical Toxicology of Natural Substances, by Donald G. Barceloux, MD

resembles small clusters of parsnips. The cut surface of the tuber extrudes a yellow oil liquid from oil ducts, and exposure to air turns this liquid brown. Microscopic examination of the numerous oil ducts after staining with iodine helps identify this plant.

Distribution and Ecology: Common European species of water hemlock are *Oenanthe crocata* and *Cicuta virosa*. *O. crocata* grows in moist areas (e.g., canal banks, drainage ditches, wet meadows) of western and southern Britain.

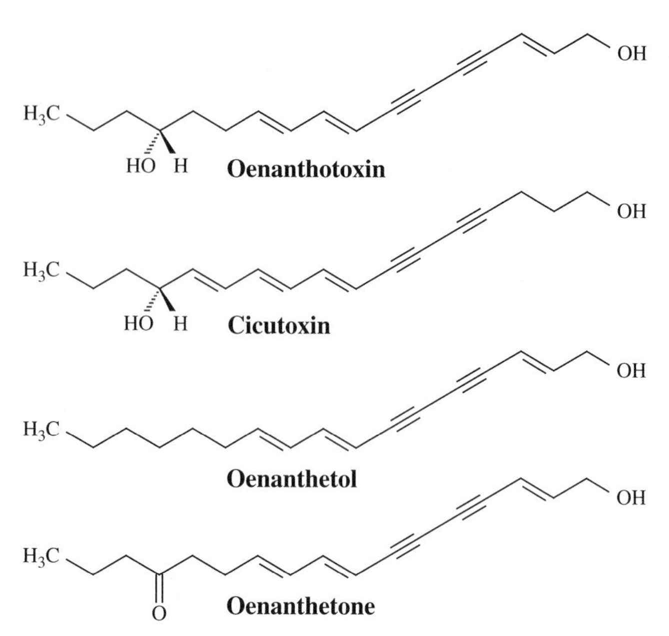

FIGURE 138.1. Comparative chemical structures of cicutoxin and major components of *Oenanthe crocata*.

EXPOSURE

Most human toxic reactions result from causal contact by children or the misidentification of *Cicuta* species or *O. crocata* for wild edible species (e.g., wild carrot, wild parsnip, Jerusalem artichoke).[7] There are no medicinal uses for water hemlock.

PRINCIPAL TOXINS

Physiochemical Properties

Cicutoxin, as isolated from *C. virosa*, is structurally similar to oenanthotoxin with exception that cicutoxin has the triene substitutions in the C8, C10, and C12 positions and an opposite configuration at C14.[3] Figure 138.1 displays the chemical structure of cicutoxin and oenanthotoxin. Oenanthetol (C_{14}-deoxyoenanthotoxin) and oenanthetone (C_{14}-ketone of C_{14}-deoxyoenanthotoxin) are also major constituents of *O. crocata*, but these compounds possess relatively mild toxicity.[8] In mice, oenanthotoxin is more toxic than cicutoxin, but the clinical syndrome is similar. Seasonal variation in toxin content is a more important variable of toxicity than relative potency.

Case reports suggest that cooking does not destroy these toxins.[9]

Poisonous Parts

Although toxins occur throughout the plant, the root is the most toxic portion of these plants. Case reports indicate that ingestions of the root produce convulsions, whereas ingestions of flowers and leaves produce nonspecific symptoms and gastrointestinal complaints.[10,11] Most poisonings occur in the spring when roots contain the highest concentration of oenanthotoxin.[3]

Histopathology and Pathophysiology

Mechanism of Toxicity

Ingestion of cicutoxin and oenanthotoxin produce similar clinical syndromes. The major toxicity results from the central nervous system (CNS) stimulant properties of these toxins; the proconvulsant properties of these toxins distinguish cicutoxin and oenanthotoxin from coniine (poison hemlock), which causes CNS depression rather than seizures. Death following ingestion of water hemlock results from status epilepticus, perhaps caused by excessive stimulation of cholinergic receptors in the basal ganglia and brainstem. Respiratory failure then ensues. Hypotension usually results from fluid depletion secondary to gastrointestinal distress. Intravenous alcohol extracts of hemlock water dropwort (*O. crocata*) cause hypotension within 1 minute of intravenous injection into animals, but the applicability of this model to clinical situations is unclear.[12]

Postmortem Changes

Pathological abnormalities in a 14-year-old boy, who died 20 hours after ingesting root material from water hemlock, included bilateral, blood-tinged pleural fluid and congestion of the lungs, liver, and brain.[13] Neuropathological changes included cerebellar and cerebral edema with cortical gyral flattening and obliteration of the ventricular system, cerebellar tonsillar swelling with medullary compression, diffuse ischemic injury, and eosinophilia.

DOSE RESPONSE

The ingestion of the tubers of *Cicuta* species or *O. crocata* produces serious toxicity. A case report associated the death of an 8-year-old boy with the ingestion of a piece of one *Cicuta* root.[14]

TOXICOKINETICS

Although most cases involve ingestion, a case report suggests the possibility that dermal absorption of the toxin occurs following direct contact with the juice from plant parts. In 1911, Egdahl reported two deaths from the use of *Cicuta* extracts as a topical antipruritic agent.[9] However, the report lacked descriptive and laboratory data, and there are no other clinical or experimental data to determine the extent of dermal absorption of toxins in water hemlock.

CLINICAL RESPONSE

Typically, severe gastrointestinal symptoms (i.e., nausea, vomiting, abdominal pain) begin within 15–90 minutes of ingestion of the root. Depending on the severity of the ingestion, convulsions and coma may occur.[15–17] Occasionally, seizures develop in these patients up to 10 hours after ingestion.[7] Profuse salivation, diaphoresis, low-grade fever, mydriasis, flushing, muscle spasm, trismus, and dizziness commonly appear.[4] The patient may progress to status epilepticus, respiratory arrest, and death. Several recent case studies[1,5] document the presence of severe opisthotonus and hemiballismus. These movements appear centrally mediated because human nerve conduction studies were normal. Furthermore, spinal studies in animals demonstrate abolishment of spastic movement after CNS destruction.[18]

Cardiovascular abnormalities are uncommon with the exception of orthostatic hypotension and volume depletion.[1,19] Significant dysrhythmias or conduction disturbances do not usually occur during intoxication by these plants except during the terminal phase characterized by status epilepticus, apnea, bradycardia, and hypotension.[20] Aspiration pneumonia may complicate the clinical course.

DIAGNOSTIC TESTING

Analytical Methods

UV absorption spectrometry is a sensitive method for analysis of extracts of stomach contents for cicutoxin or oenanthotoxin with confirmation by high performance liquid chromatography with photodiode array/tandem mass spectrometry,[21] or gas chromatography/mass spectrometry.[8,22] Oenanthotoxin has an intense, highly characteristic UV absorption spectrum, particularly at 313 nm.

Abnormalities

A severe metabolic acidosis occurs in patients with seizure activity after the ingestion of toxins from these plants.[23] Evidence of rhabdomyolysis and acute oliguric renal failure may complicate the course of *Cicuta* toxicity.[24] Other laboratory abnormalities include transient hematuria, hyperbilirubinemia, mildly elevated hepatic transaminases, glycosuria, and proteinuria.[1,4]

TREATMENT

Stabilization

Patients with *Cicuta* intoxication may develop seizures suddenly; therefore, intravenous access should be established and all patients should be monitored for cardiac dysrhythmias, hypotension, and hypoxia. There are limited data on the efficacy of any specific anticonvulsant during intoxication by these plants.[29] In the emergency department, actively seizing adult patients should receive lorazepam or diazepam. Both these compounds have similar time to cessation of seizures, but the duration of action of lorazepam is longer than diazepam.[25] The recommended adult dose of lorazepam is 0.05–0.2 mg/kg at the rate of 2 mg/min up to 8 mg, whereas the adult dose for diazepam is 0.15–0.25 mg/kg. Phenobarbital is probably a second-line anticonvulsant because of the synergistic action of phenobarbital with GABA agonists and the limited effect of phenytoin on the lowered seizure threshold commonly associated with toxin-induced seizures.[26] The recommended dose of phenobarbital is 20 mg/kg intravenously infused at 50–75 mg/min.[27] Adverse effects of phenobarbital use include respiratory depression and hypotension; therefore, these patients must be monitored carefully for the need for intubation and measures to treat hypotension (fluids, vasopressors). Serum electrolyte and acid–base status should also be monitored closely. Therapeutic options for refractory seizures include a midazolam infusion (0.2 mg/kg slow intravenous bolus followed by 0.75–10 µg/kg/min), an additional intravenous phenobarbital dose (5–10 mg/kg), or a propofol infusion (1–2 mg/kg loading dose, followed by 2–10 mg/kg/h). Experimental therapy for refractory seizures includes intravenous valproate and intragastric topiramate.[25]

Decontamination

The presence of vomiting limits the efficacy of decontamination measures. Additionally, the potential for rapid onset of changes in consciousness and seizures during serious water hemlock intoxication increases the risk of using decontamination measures. Consequently, most of these ingestions require no decontamination measures.

Elimination Enhancement

There are no data on the elimination of cicutoxin or oenanthotoxin. Hemodialysis improves the hyperosmolality and metabolic acidosis associated with cicutoxin-induced status epilepticus, but efficacy of hemodialysis to enhance the elimination of these toxins and to improve the clinical course is unproven.[28]

Antidotes

No specific antidotes are available. In animal studies, anticholinergic agents do not alleviate seizures or the cholinergic effects of *Cicuta* toxicity.[29]

Supplemental Care

Muscle spasticity usually responds to diazepam. Supportive care for these cases involves monitoring ventilation, acid–base status, electrolytes, and volume status. Appropriate laboratory studies include a complete blood count, electrolytes, creatine phosphokinase (CK), creatinine, hepatic transaminases, arterial blood gases, and urinalysis. Grossly elevated serum CK concentrations (>1000 Units/L) suggest the need for generous fluid replacement and possibly alkalinization of the urine. All patients with CNS symptoms (including spasticity) or cardiac instability should be admitted. All other patients should be observed at least 4–6 hours before discharge.

References

1. Costanza DJ, Hoversten VW. Accidental ingestion of water hemlock. Report of two patients with acute and chronic effects. Calif Med 1973;119:78–82.
2. Clarke EG, Kidder DE, Robertson WD. The isolation of the toxic principle of *Oenanthe crocata*. J Pharm Pharmacol 1949;1:377–381.
3. Anet EFLJ, Lythgoe B, Silk MH, Trippett S. Oenanthotoxin and cicutoxin. Isolation and structure. J Chem Soc 1953;62:309–322.
4. Mitchell MI, Rutledge PA. Hemlock water dropwort poisoning. A review. Clin Toxicol 1978;12:417–426.
5. Starreveld E, Hope CE. Cicutoxin poisoning (water hemlock). Neurology 1975;25:730–734.
6. Kingsbury JM. Poisonous plants of the United States and Canada. Englewood Cliffs, NJ: Prentice-Hall Inc;1964.
7. Downs C, Phillips J, Ranger A, Farrell L. A hemlock water dropwort curry: a case of multiple poisoning. Emerg Med J 2002;19:472–473.
8. King LA, Lewis MJ, Parry D, Twitchett PJ, Kilner EA. Identification of oenanthotoxin and related compounds in hemlock water dropwort poisoning. Hum Toxicol 1985;4:355–364.
9. Egdahl A. A case of poisoning due to eating poison hemlock (*Cicuta maculata*) with a review of reported cases. Arch Intern Med 1911;7:348–366.
10. Undine CA. Poisoning with *Cicuta maculata* or water hemlock. Minn Med 1938;21:262–263.
11. Gompertz LM. Poisoning with water hemlock (*Cicuta maculata*). A report of 17 cases. JAMA 1926;87:1277–1278.
12. Roboson P. Water hemlock poisoning. Lancet 1965;2:1274–1275.
13. Heath KB. A fatal case of apparent water hemlock poisoning. Vet Hum Toxicol 2001;43:35–36.
14. Withers LM, Cole FR, Nelson RB. Water hemlock poisoning. N Engl J Med 1969;281:566–567.
15. Haggerty DR, Conway JA. Report of poisoning by *Cicuta maculata*. NY State J Med 1936;36:1511–1514.
16. Miller MM. Water hemlock poisoning. JAMA 1933;101:852–853.
17. Landers D, Seppi K, Blauer W. Seizures and death on a White River float trip. Report of water hemlock poisoning. West J Med 1985;142:637–640.
18. Grundy HF, Howarth F. Pharmacologic studies on hemlock dropwort. Br J Pharmacol 1956;11:225–230.
19. Applefield JJ, Caplan ES. A case of water hemlock poisoning. JACEP 1979;8:401–403.
20. Sweeney K, Gensheimer KF, Knowlton-Field J, Smith RA. Water hemlock poisoning—Maine, 1992. MMWR Morb Mortal Wkly Rep 1994;43:229–231.
21. Kite GC, Stoneham Ca, Veitch NC, Stein BK, Whitwell KE. Application of liquid chromatography-mass spectrometry to the investigation of poisoning by *Oenanthe crocata*. J Chromatogr B 2006;838:63–70.
22. del Castillo B, Garcia De Marina A, Martinez-Honduvilla MP. Fluorimetric determination of oenanthotoxin. Ital J Biochem 1980;29:233–237.

23. Pallares JM, Saban J, Bouza C, Diaz JM, Rodriguesz, de La Morena JCL, et al. Reversible autonomic dysfunction in *Oenanthe crocata* poisoning evaluated by simple bedside tests. Hum Toxicol 1985;4:521–526.
24. Carlton BE, Tufts E, Girard DE. Water hemlock poisoning complicated by rhabdomyolysis and renal failure. Clin Toxicol 1979;14:87–93.
25. Lowenstein DH. The management of refractory status epilepticus: an update. Epilepsia 2006;47(Suppl 1): 35–40.
26. Tunnicliff G. Basis of the antiseizure action of phenytoin. Gen Pharmacol 1996;27:1091–1097.
27. Treiman DM. Treatment of convulsive status epilepticus. Int Rev Neurobiol 2007;81:273–285.
28. Knutsen OH, Paszkowski P. New aspects in the treatment of water hemlock poisoning. Clin Toxicol 1984;22:157–166.
29. Nelson RB, North DS, Kaneriya M, Fletcher CV. The influence of biperiden, benztropine, physostigmine and diazepam on the convulsive effects of *Cicuta douglasii*. Proc West Pharmacol Soc 1978;21:137–139.

Chapter 139

WHITE SNAKEROOT [*Ageratina altissima* (L.) King & H.E. Robins]

HISTORY

During colonial times and the first half of the 19th century, milk sickness caused numerous deaths in humans and livestock in the eastern and midwestern United States.[1] Dr. Thomas Barbee first described the symptoms associated with milk sickness in 1809.[2] In 1818, Abraham Lincoln's mother, Nancy Hanks Lincoln, died of what is now recognized as milk sickness.[3] Thomas Lincoln then left Indiana for Illinois with his new wife, Sarah, and Abraham Lincoln. Despite the seriousness of poisoning produced by ingesting milk from cow that graze on white snakeroot, the causal relationship between milk sickness and white snakeroot was not recognized until Dr. Anna Pierce Hobbs advocated the eradication of white snakeroot in the 1840s. Although settlers continued to suffer from milk sickness during the 19th century, the incidence of milk sickness decreased as cultivated land replaced the forests of the Midwest. The etiology of milk sickness remained unclear until the early 20th century, when Couch isolated a crude extract (tremetol) that reproduced a disease similar to trembles in livestock and milk sickness in humans.[4] The specific toxin in this crude extract remains unknown, and toxicity in horses and goats grazing on white snakeroot remains a problem in the United States.[5] The appearance of milk sickness was rare during the 20th century.[6]

BOTANICAL DESCRIPTION

Common Name: White Snakeroot, Fall Poison, Richweed, Pool Root, White Sanicle, Indian Sanicle, Deer Wort, Pool Wort, Squaw Weed

Scientific Name: *Ageratina altissima* (L.) King & H.E. Robins (*Eupatorium rugosum* Houtt., *E. urticaefolium*)

Botanical Family: Asteraceae (sunflower)

Physical Description: This slender, erect, branching plant grows up to 1–5 ft (~0.3–2 m) in height. The large ovate leaves have serrated edges and a purplish color with large veins on the undersurface. Clusters of white flowers at the ends of numerous branches appear in late summer and produce small fruit.

Distribution and Ecology: These native US plants are shade-loving weeds found along steep canyons, damp open areas, and rivers in the East and Midwest from North Dakota to Texas, Georgia, and Maine.

EXPOSURE

Native Americans used the root as a diaphoretic, diuretic, antipyretic, stimulant, and an herbal remedy for diarrhea, toothache, kidney stones, snake bite, and urinary diseases. The plant is not edible.

PRINCIPAL TOXINS

Physiochemical Properties

Tremetol is the crude extract of the snakeroot plant that causes illness both in humans and animals. This substance is not a pure compound and contains several ketones, such as tremetone (CAS RN:4976-25-4,

Medical Toxicology of Natural Substances, by Donald G. Barceloux, MD

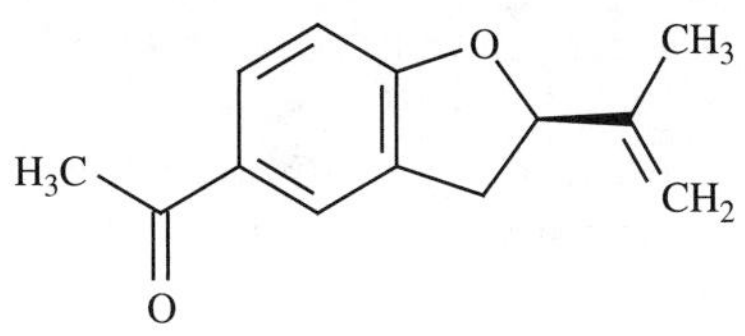

FIGURE 139.1. Chemical structure of tremetone.

$C_{13}H_{14}O_2$), dehydrotremetone, and hydroxytremetone.[7] Figure 139.1 displays the chemical structure of tremetone. The identity of the specific toxin responsible for the illness associated with the ingestion of snakeroot by livestock remains unknown.[8] The southern goldenbush [*Isocoma pluriflora* (Torr. & Gray) Greene] is a perennial shrub indigenous to the southwestern United States that also contains tremetone and causes milk sickness. Synonyms for this plant include rayless goldenrod [*Isocoma wrightii* (Gray) Rydb., *Haplopappus heterophyllus* (Gray) Blake]. The odor and consistency of tremetol resembles turpentine.[1] The 2-isopropenyl-2,3-dihydrobenzofuran skeleton of tremetone has been synthesized.[9]

Poisonous Parts

This disease (milk sickness) results from the ingestion of milk from cows grazing on the white snakeroot plant.

Mechanism of Toxicity

The entire plant contains ketones and glycosides that accumulate in the milk of cows grazing on this plant. The afebrile epidemic illness (milk sickness) results from ingestion of contaminated milk. Modern milk-processing procedures have eliminated milk sickness, but consumption of unprocessed milk represents a potential danger. Postmortem findings are nonspecific with fatty degeneration of the liver and kidney.[10]

DOSE RESPONSE

Milk sickness is a chronic illness. The toxin in tremetol has not been identified and consequently there are inadequate dose-response data.

TOXICOKINETICS

In vitro studies indicate that tremetone is biotransformed to a toxic metabolite.[11] In plants, tremetone is converted to dehydrotremetone, but these microsomal activation studies suggest that dehydrotremetone is not physiologically active.

CLINICAL RESPONSE

Milk sickness is an insidious disease causing weakness, anorexia, nausea, vomiting, tremors, headache, constipation, dizziness, delirium, and, in serious cases, coma, convulsions, and death. Older synonyms for milk sickness include puking fever, sick stomach, tires, slows, and trembles (grazing animals). Symptoms are typically prolonged.

DIAGNOSTIC TESTING

Milk sickness is associated with ketosis, mild hypoglycemia, and lipemia without evidence of diabetes mellitus, salicylate intoxication or infectious processes. The acidosis is sometimes refractory to therapy.[12]

TREATMENT

Treatment is supportive. Decontamination measures are unnecessary because milk sickness is a chronic illness.

References

1. Snively WD Jr, Furbee L. Discoverer of the cause of milk sickness (Anna Pierce Hobbs Bixby). JAMA 1966;196: 1055–1060.
2. Niederhofer RE. The milk sickness. JAMA 1985;254: 2123–2125.
3. Christensen WI. Milk sickness; a review of the literature. Econ Bot 1965;19:293–300.
4. Couch JF. The toxic constituent of richweed or white snakeroot (*Eupatorium urticaefolium*). J Agric Res 1927; 35:547–576.
5. Sharma OP, Dawra RK, Kurade NP, Sharma PD. A review of the toxicosis and biological properties of the genus *Eupatorium*. Nat Toxins 1998;6:1–14.
6. Bryan FL. Epidemiology of milk-borne diseases. J Food Protect 1983;46:637–649.
7. Bowne DM, Degraw JI Jr, Shah VR, Bonner WA. The synthesis and pharmacological action of tremetone. J Med Chem 1963;6:315–319.
8. Beier RC, Norman JO. The toxic factor in white snakeroot: identity, analysis and prevention. Vet Hum Toxicol 1990;32(suppl):81–88.
9. Pelly SC, Govender S, Fernandes MA, Schmalz HG, De Koning CB. Stereoselective syntheses of the 2-Isopropenyl-2,3-dihydrobenzofuran nucleus: potential chiral building blocks for the syntheses of tremetone, hydroxytremetone, and rotenone. J Org Chem 2007;72: 2857–2864.

10. Jordan EO, Harris MN. Milk sickness. J Infect Dis 1909;6: 401–505.
11. Beier RC, Norman JO, Reagor JC, Rees MS, Mundy BP. Isolation of the major component in white snakeroot that is toxic after microsomal activation: possible explantation of sporadic toxicity of white snakeroot plants and extracts. Nat Toxins 1993;1:286–293.
12. Hartmann AF Sr, Hartmann AF Jr, Purkerson ML, Wesley ME. Tremetol poisoning—not yet extinct. JAMA 1963;185: 78–81.

V Shrubs and Vines

Chapter 140

BARBADOS NUT
(*Jatropha curcas* L.)

BOTANICAL DESCRIPTION

The genus *Jatropha* (nettle spurge) contains over 20 species. Several species have been associated with human toxicity including *J. curcas* L. (Barbados nut), *J. gossypiifolia* L. (bellyache bush), *J. macrorhiza* Benth. (ragged jatropha, ragged nettle spurge), and *J. multifida* L. (coral bush).[1]

Common Name: Barbados Nut, Purging Nut Tree, Black Vomit Nut, Physic Nut, Purgeerboonijie (South Africa)

Scientific Name: *Jatropha curcas* L.

Botanical Family: Euphorbiaceae (spurge)

Physical Description: This large shrub reaches 6m (~20ft) in height, and has undulating, five-lobed leaves. The thick stem is green with a woody base, and the yellow-green flowers are inconspicuous. The elliptical fruit is about 2.5cm (~1 in.) in length and contains black seeds when ripe.

Distribution and Ecology: This succulent probably originated in Mexico and Central America. This evergreen plant now inhabits wastelands in arid and semiarid areas (grassland-savannah, thorn forest scrub) throughout tropical Africa, the Indian peninsula, and Asia.

EXPOSURE

Jatropha curcas is a perennial shrub commonly used as a hedge in tropical and subtropical climates. Traditional medicinal uses for extracts (yellow oil or hell oil) of *Jatropha* seeds include a purgative, abortifacient, antipyretic, anthelminthic, treatment of gout and gonorrhea, a styptic for the control of local hemorrhage, a chemotherapeutic agent for cancerous growths (root), and a diuretic. Animal studies indicate that the administration of extracts of the fruit of *Jatropha curcas* increases fetal resorptions when administered soon after implantation.[2] In some areas of Africa and Mexico, properly prepared young leaves and nuts are food sources, and the ashes from the woody parts of the plant are a cooking salt. Jatropha oil is a potential fuel and burning oil. Physic nut oil was used in engines in Mali during World War II. Leaf juice stains red and constituents of this plant are useful as marking ink.

PRINCIPAL TOXINS

Viscous, astringent latex exudes from damaged portions of *Jatropha* species that dries to a bright red-brown film resembling shellac. The toxins in *Jatropha* species are not well defined. This latex contains a proteolytic enzyme, curcin, with a molecular weight of about 22,000Da,[3] and an octapeptide (curcacycline A) with *in vitro* anticomplement activity.[4] The effect of the latex on coagulation is complex. *In vitro* testing indicates that the whole latex has procoagulant activity, whereas dilute latex demonstrates anticoagulant activities.[5] The oil (oleum ricini majoris, hell oil, oleum infernale, ratanjyot) extract from the seeds of *Jatropha curcas* contains curcanoleic acid, which is structurally similar to ricinoleic acid in the castor bean (*Ricinus comminus* L.) and polyunsaturated esters of the tigliane-type diterpenoids (e.g., 12-deoxy-16-hydroxyphorbol).[6] Animal models indicate that this oil produces severe diarrhea and gastrointestinal inflammation.[7] The purgative oil represents about 40% of the seed by weight, and one of the constituents of this oil is the purgative and skin irritant, 12-deoxy-10-hydroxyphorbol.[8] Other constituents include lathyrane and podocarpane type diterpenoid compounds, triterpenes, and oleanolic acid.[9] The material remaining after the extraction of the oil contains a toxalbumin, curcin, which demonstrates hemostatic properties.[10] Experimental studies indicate that curcin inhibits protein synthesis similar to ricin, but curcin is approximately 1,000 times less potent than ricin.[11] This compound has antitumor activity in *in vitro* studies.[12]

DOSE RESPONSE

In a series of 20 children admitted to an emergency department with vomiting and loose stools, the median number of Barbados nuts ingested was two with a range of one to four.[13] There was no evidence of dehydration and all children recovered without sequelae within 5–6 hours.

TOXICOKINETICS

There are few human data on curcin or other potential toxins in the Barbados nut.

CLINICAL RESPONSE

Ingestion of seeds from *Jatropha curcas* causes severe gastroenteritis manifest by profuse vomiting, diarrhea, and abdominal cramps along with burning of the throat.[14] The gastrointestinal symptoms begin within 30 minutes to several hours of ingestion, and the severity of the vomiting and diarrhea can cause significant dehydration in children.[15] Recovery usually occurs within 6–24 hours without sequelae. Other *Jatropha* species (e.g., *Jatropha multifida* L., coralbush) also produce severe gastroenteritis, myalgias, dehydration, lethargy, and hypotension.[1,16]

DIAGNOSTIC TESTING

There are no commercial biomarkers of exposure to *Jatropha* species. Evidence of subclinical hepatic inflammation (i.e., elevated serum hepatic aminotransferases) may develop during acute intoxication, but liver dysfunction does not usually occur.[16]

TREATMENT

Treatment of acute intoxication from *Jatropha* species involves supportive care for the fluid and electrolyte imbalance associated with severe gastroenteritis. Decontamination is unnecessary because of the presence of profuse vomiting and diarrhea. There are no specific antidotes. Intravenous fluids may be necessary to correct the fluid and electrolyte deficits associated with gastrointestinal inflammation. Diagnostic tests include monitoring of electrolytes, acid–base status, complete blood count, and hepatorenal function. Patients can be safely discharged after 6–8 hours if the gastroenteritis subsides and fluid and electrolyte balance is restored.

References

1. Koltin D, Uziel Y, Schneidermann D, Kotzki S, Wolach B, Fainmesser P. A case of *Jatropha multifida* poisoning resembling organophosphate intoxication. Clin Toxicol 2006;44:337–338.
2. Goonasekera MM, Gunawardana VK, Jayasena K, Mohammed SG, Balasubramaniam S. Pregnancy terminating effect of *Jatropha curcas* in rats. J Ethnopharmacol 1995;47:117–123.
3. Nath LK, Dutta SK. Extraction and purification of curcin, a protease from the latex of *Jatropha curcas* Linn. J Pharm Pharmacol 1991;43:111–114.
4. van den Berg AJ, Horsten SF, Kettenes-van den Bosch JJ, Kroes BH, Beukelman CJ, Leeflang BR, Labadie RP. Curcacycline A—a novel cyclic octapeptide isolated from the latex of *Jatropha curcas* L. FEBS Lett 1995;358:215–218.
5. Osoniyi O, Onajobi F. Coagulant and anticoagulant activities in *Jatropha curcas* latex. J Ethnopharmacol 2003;89:101–105.
6. Adolf W, Opferkuch HJ, Hecker E. Irritant phorbol derivatives from four *Jatropha* species. Phytochemistry 1984;23:129–132.
7. Gandhi VM, Cherian KM, Mulky MJ. Toxicological studies on ratanjyot oil. Food Chem Toxicol 1995;33:39–42.
8. Adolf W, Opferkuch HJ, Hecker E. Irritant phorbol derivatives from four *Jatropha* species. Phytochemistry 1984;23:129–132.
9. Ravindranath N, Reddy MR, Ramesh C, Ramu R, Prabhakar A, Jagadeesh B, Das B. New lathyrane and podocarpane diterpenoids from *Jatropha curcas*. Chem Pharm Bull 2004;52:608–611.
10. Mourgue M, Delphaut J, Baret R, Kassab R. [Study of the toxicity and localization of toxalbumin (curcin) in the seeds of *Jatropha curcas* Linn]. Bull Soc Chim Biol (Paris). 1961;43:517–531. [French]
11. Stirpe F, Pession-Brizzi A, Lorenzoni E, Strocchi P, Montanaro L, Sperti S. Studies on the proteins from the seeds

of *Croton tiglium* and of *Jatropha curcas*. Toxic properties and inhibition of protein synthesis *in vitro*. Biochem J 1976;156:1–6.

12. Luo MJ, Yang XY, Liu WX, Xu Y, Huang P, Yan F, Chen F. Expression, purification and anti-tumor activity of curcin. Acta Biochim Biophys Sin (Shanghai) 2006;38: 663–668.
13. Kulkarni ML. *Jatropha curcas*—poisoning. Indian J Pediatr 2004;72:75–76.
14. Joubert PH, Brown JM, Hay IT, Sebata PD. Acute poisoning with *Jatropha curcas* (purging nut tree) in children. S Afr Med J 1984;65:729–730.
15. Abdu-Aguye AS, Alafiya-Tayo RA, Bhusnurmath SR. Acute toxicity studies with *Jatropha curcas* L. Hum Toxicol 1986;5:269–274.
16. Levin Y, Shrer Y, Bibi H, Schlesinger M, Hay E. Rare *Jatropha multifida* intoxication in two children. J Emerg Med 2000;19:173–175.

Chapter 141

BOXWOOD
(*Buxus sempervirens* L.)

BOTANICAL DESCRIPTION

Common Name: Boxwood, Boxtree, Box, Common Box

Scientific Name: *Buxus sempervirens* L.

Botanical Family: Buxaceae

Physical Description: This broad-leaf, evergreen shrub is commonly cultivated as a hedge that possesses thick, leathery leaves and small white flowers.

Distribution and Ecology: *Buxus sempervirens* is a perennial shrub that occurs naturally throughout Eurasia as well as sections of the eastern United States (Virginia, Tennessee, Ohio, North Carolina, New York). The habitat of this plant includes dry scrub and forested areas.

EXPOSURE

Herbal uses of *Buxus sempervirens* include the treatment of rheumatism, fever, and malaria. This plant is a popular ornamental for hedges and sculptured plants.

PRINCIPAL TOXINS

Buxus species contain over 135 steroidal alkaloids containing 9,19-cyclo-9,10 secobuxus, buxane (9,19-cyclobuxus), and 9,10-seco-buxa-9(11),10(19)diene skeletons.[1,2] In particular, *B. sempervirens* contains numerous steroidal alkaloids including spirobuxus alkaloids, buxenone, buxene, sempervirone, and benzamide alkaloids as well as volatile oils in its leaves and stems.[3–5]

DOSE RESPONSE

The principal toxin in boxwood has not been identified.

TOXICOKINETICS

There are few data on the toxicokinetics of toxic constituents of boxwood.

CLINICAL RESPONSE

There are few data on the clinical features of intoxication from *Buxus sempervirens*. Potentially, this perennial shrub is a gastrointestinal irritant.

DIAGNOSTIC TESTING

There are few data on the diagnostic testing of samples from patients with exposure to boxwood. Clinically significant laboratory abnormalities after exposure to boxwood are unlikely with the possible exception of abnormalities related to fluid and electrolyte imbalance.

TREATMENT

Treatment is supportive. Decontamination measures are unnecessary.

Medical Toxicology of Natural Substances, by Donald G. Barceloux, MD

References

1. Ata A, Naz S, Iqbal Choudhary M, Atta-ur-Rahman, Sener B, Turkoz S. New triterpenoidal alkaloids from *Buxus sempervirens*. Z Naturforsch 2002;57:21–28.

2. Loru F, Duval D, Aumelas A, Akeb F, Guedon D, Guedj R. Four steroidal alkaloids from the leaves of *Buxus sempervirens*. Phytochemistry 2000;54:951–957.

3. Rahman AU, Ahmed D, Choudhary I. Alkaloids from the leaves of *Buxus sempervirens*. J Nat Prod 1988;51: 783–786.

4. Rahaman AR, Ahmed D, Choudhary I, Turkoz S, Sener B. Chemical constituents of *Buxus sempervirens*. Planta Med 1988;54:173–174.

5. Kupchan SM, Kennedy RM, Schleigh WR, Ohta G. *Buxus* alkaloids—XII benzamide alkaloids from *Buxus sempervirens* L. Tetrahedron 1967;23:4563–4586.

Chapter 142

BUCKTHORN
[*Karwinskia humboldtiana* (J.A. Schultes) Zucc.]

BOTANICAL DESCRIPTION

The buckthorn family contains several genera of buckthorn including *Frangula, Karwinskia, Rhamnus*, and *Sageretia*.

Common Name: Coyotillo, Wild Cherry, Tullidora, Buckthorn

Scientific Name: *Karwinskia humboldtiana* (J.A. Schultes) Zucc.

Botanical Family: Rhamnaceae (buckthorn)

Physical Description: This spineless, woody shrub or small tree reaches up to 1–5 ft (~30–150 cm) in height and has opposite, 1–3 in. (~3–8 cm) long leaves on a stalk. Clusters of greenish flowers produce brownish black berries.

Distribution and Ecology: This shrub grows in semi-arid environments along canyons and gullies of the southwestern United States, northern and central Mexico, Central America, and northern Colombia. The species of common buckthorn (*Rhamnus* species) are ornamental hedges that have been naturalized in the eastern United States. Cascara buckthorn [*Frangula purshiana* (DC.) Cooper] is a small deciduous tree indigenous to the Pacific Northwest.

EXPOSURE

There are no specific medical indications for the use of *K. humboldtiana. In vitro* studies have investigated the potential antineoplastic effects of a constituent (i.e., T-514 toxin) of this plant.[1]

PRINCIPAL TOXINS

Physicochemical Properties

The fruit from *K. humboldtiana* contains at least four dimeric anthracenone neurotoxins identified by their molecular weight (T-496, T-514, T-516, T-544).[2] The exact role of each compound in the production of neurotoxicity is unclear. In animal models, the administration of the green fruit produces liver, lung, and renal damage with T-514 being more cytotoxic than T-544.[3] Experimental animals given T-544 develop classic ascending paralysis as well as somewhat less hepatic and pulmonary damage than T-514. Species of common buckthorn (*Rhamnus* spp.) also contain glycosides that are strong cathartics.

Poisonous Parts

All the plant parts are toxic, but the green fruit contains the highest concentration of neurotoxins. The amount of T-544 neurotoxin varies with season and location. In a series of seven fruit samples from different locations in Mexico, the T-544 content of the fruits from *K. humboldtiana* ranged from 0.69–1.6% (green fruit) and 0.4–1.26% (ripe fruit).[4]

Medical Toxicology of Natural Substances, by Donald G. Barceloux, MD

Mechanism of Toxicity

In experimental animals, the administration of anthracenone neurotoxins from the fruit of *K. humboldtiana* for 2–3 weeks causes a progressive, ascending paralysis similar to Guillain-Barré syndrome.[5] However, the specific toxin has not been identified. Observable effects during the first week after administration of fruit from the buckthorn included lethargy, hypersensitivity to stimuli, and weight loss followed by a quiescent period.[6,7] An ascending paralysis developed by the second to third week characterized by segmental demyelination of large peripheral nerve fibers, probably as a result of the disruption of Schwann cell metabolism and axon transport.[8] Animal studies also indicate that this fruit produces acute nephrotoxicity characterized by marked reduction in renal perfusion pressure, hydropic degeneration of the proximal convoluted tubules, and increased sodium excretion in the urine.[9] *In vitro* studies suggest that oxidative stress mediated by reactive intermediates may cause the cytotoxicity produced by T-514.[10]

DOSE RESPONSE

The specific neurotoxin has not been identified. Potentially toxic anthracenone compounds include T-514 (peroxisomicine A1), T-516, T-544, and T-496.[11] The amount of buckthorn fruit necessary to produce neurotoxicity is not well defined.

TOXICOKINETICS

There are few data on the toxicokinetics of suspected toxins (T-496, T-514, T-516, T-544) in buckthorn.

CLINICAL RESPONSE

Case reports indicate that chronic consumption of *K. humboldtiana* causes a progressive, symmetrical polyneuropathy characterized by a flaccid quadriplegia without sensory changes.[12] Deep tendon reflexes are usually absent, and cranial nerve signs are normal. Occasionally, respiratory insufficiency and pneumonia complicate the ascending neuropathy.[13] The onset of paralysis ranges between about 1–6 weeks after ingestion of the fruit.[14] A slow, gradual recovery usually occurs, although some reflex deficits may remain. Exposure to other *Rhamnus* species may cause diarrhea, but neurological abnormalities occur only following exposure to *K. humboldtiana*. The differential diagnosis of chronic buckthorn neuropathy includes Guillain-Barré syndrome and poliomyelitis.

DIAGNOSTIC TESTING

In case reports of chronic buckthorn intoxication, sural nerve biopsy demonstrated distal segmental demyelination and axon degeneration with swelling of the Schwann cells due to phagocytosis of the myelin.[12,14] Lymphocytic infiltration does not usually occur. Animal studies suggest the changes in the Schwann cells may be secondary to direct axon toxicity.[15] Motor conduction velocities are reduced with an absence of conduction in the extremities in severe cases. Marked reduction in compound muscle action and fibrillation potentials consistent with axonal damage occurs rarely.[12] Analysis of cerebrospinal fluid (CSF) does not demonstrate any abnormalities, and CSF protein and glucose concentrations are typically normal.

TREATMENT

Treatment is supportive. Decontamination is unnecessary.

References

1. Peneyro-Lopez A, Martinez de Villarreal L, Gonzalez-Alanis R. *In vitro* selective toxicity of toxin T-514 from *Karwinskia humboldtiana* (buckthorn) plant on various human tumor cell lines. Toxicology 1994;92:217–227.
2. Dreyer D, Arai I, Bashuman C, Anderson W, Smith R, Daves D. Toxins causing non-inflammatory paralytic neuropathy. Isolation and structure elucidation. J Am Chem Soc 1975;97:4986–4990.
3. Bermudez MV, Gonzalez-Spencer D, Guerrero M, Waksman N, Pineyro A. Experimental intoxication with fruit and purified toxins of buckthorn (*Karwinskia humboldtiana*). Toxicon 1986;24:1091–1097.
4. Guerrero M, Pineyro A, Waksman N. Extraction and quantification of toxins from *Karwinskia humboldtiana* (tullidora). Toxicon 1987;25:565–568.
5. Bermudez MV, Martínez FJ, Salazar ME, Waksman N, Pieyro A. Experimental acute intoxication with ripe fruit of *Karwinskia humboldtiana* in rat, guinea-pig, hamster and dog. Toxicon 1992;30:1493–1496.
6. Muñoz-Martínez EJ, Cueva J, Joseph-Nathan P. Denervation caused by tullidora (*Karwinskia humboldtiana*). Neuropathol Appl Neurobiol 1983;9:121–134.
7. Mitchell J, Weller RO, Evans H, Arai I, Daves GD Jr. Buckthorn neuropathy. Effects of intraneural injection of *Karwinskia humboldtiana* toxins. Neuropathol Appl Neurobiol 1978;4:85–97.
8. Muñoz-Martínez EJ, Cuéllar-Pedroza LH, Rubio-Franchini C, Jáuregui-Ricón J, Joseph-Nathan P. Depression of fast axonal transport in axons demyelinated by intraneural

injection of a neurotoxin from *K. humboldtiana*. Neurochem Res 1994;19:1341–1348.

9. Jaramillo-Juarez F, Ortiz GG, Rodriguez-Vazquez ML, Falcon-Franco MA, Feria-Velasco A. Renal failure during acute toxicity produced by tullidora ingestion (*Karwinskia humboldtiana*). Gen Pharmacol 1995;26:649–653.
10. Garza-Ocanas L, Zanatta-Calderon MT, Acosta D, Torres-Alanis O, Pineyro-Lopez A. Production of reactive oxygen species by toxin T-514 of genus *Karwinskia in vitro*. Toxicol In Vitro 2003;17:19–25.
11. Salazar-Leal ME, Flores MS, Sepulveda-Saavedra J, Romero-Diaz VJ, Becerra-Verdin EM, Tamez-Rodriguez VA, et al. An experimental model of peripheral neuropathy induced in rats by *Karwinskia humboldtiana* (buckthorn) fruit. J Peripher Nerv Syst 2006;11:253–261.
12. Martínez HR, Bermudez MV, Rangel-Guerra RA, De Leon Flores L. Clinical diagnosis in *Karwinskia humboldtiana* polyneuropathy. J Neurol Sci 1998;154: 49–54.
13. Calderon-Gonzalez R, Rizzi-Hernandez H. Buckthorn polyneuropathy. N Engl J Med 1967;277:69–71.
14. Ocampo-Roosens LV, Ontiveros-Nevares PG, Fernandez-Lucio O. Intoxication with buckthorn (*Karwinskia humboldtiana*): report of three siblings. Pediatr Dev Pathol 2007;10:66–68.
15. Heath JW, Ueda S, Bornstein MB, Daves GD, Raine CS. Buckthorn neuropathy *in vitro*: evidence for a primary neuronal effect. J Neuropathol Exp Neurol 1982;41: 204–220.

Chapter 143

CACTUS

BOTANICAL DESCRIPTION

The Cactaceae family (cactus) contains about 40 genera and about 2,000 species of plants. *Opuntia* is the largest genus of the cactus family. Members of the genus *Euphorbia* (Euphorbiaceae) differ from cacti by the absence of areoles at the base of the spines and by the presence of an irritant, white, milky sap. The cactus family is indigenous to the arid and semiarid regions of the western hemisphere. Cacti adapt easily to areas with low rainfall in the Mediterranean region, Africa, Europe, western and midwestern United States, and northern Mexico.

Common Name: Prickly Pear, Cholla

Scientific Name: *Opuntia* spp.

Botanical Family: Cactaceae (cactus)

Physical Description: The surfaces of leaves from cactus specimens have cushion-like structures (areoles, nipple) that contain medium-sized spines along with smaller spines (glochidia) at the base of the larger spines (See Figure 143.1). Prickly pears have flat joints, whereas chollas have cylindrical joints. The glochidia (glochids) arise from the base of the larger spines in a pin-cushion-like structure that contains hundreds of short (4–8 mm), barbed hairs. These glochidia appear as a soft, fluffy dot on the pad of the cactus. These tiny, profusely barbed hairs are translucent to yellow-white.

Distribution and Ecology: *Opuntia polyacantha* Haw. (prickly pear) is a perennial native of the western and midwestern United States.

Common Name: Barrel Cactus

Scientific Name: *Echinocactus* spp., *Ferocactus* spp., *Sclerocactus* spp.

Botanical Family: Cactaceae (cactus)

Physical Description: Long, sturdy spines up to 8–10 cm (~3–4 in.) arise from the central portion of the cactus. Although the end of the spine can be curved, the ends are not barbed. These spines do not easily break; therefore, foreign bodies are less common complications of contact with these cacti compared with other types of cactus spines. Radial spines of the barrel cactus are smaller than long spines in the central portion. These spines easily penetrate soft tissue and occasionally foreign bodies result from tangential pressure on the spine during contact.

Distribution and Ecology: The barrel cacti (*Ferocactus* spp.) are perennial native plants of the southwestern United States.

Common Name: Saguaro

Scientific Name: *Carnegiea gigantea* (Engelm.) Britt. & Rose

Botanical Family: Cactaceae (cactus)

Physical Description: Both central and radial spines of this cactus are long and sufficiently sturdy to limit dislodgement during contact.

Medical Toxicology of Natural Substances, by Donald G. Barceloux, MD

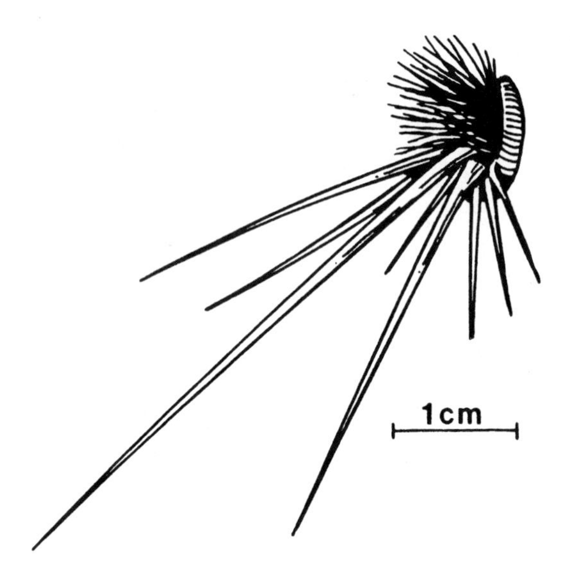

FIGURE 143.1. Glochidia. From Lindsey D, et al. Cactus spine injuries. Am J Emerg Med 1988;6:363. Reprinted from Elsevier with permission.

Distribution and Ecology: The saguaro cactus is a perennial native species of deserts in California and Arizona.

Common Name: Hedgehog

Scientific Name: *Echinocereus* spp., *Pediocactus* spp.

Botanical Family: Cactaceae (cactus)

Physical Description: These small to medium sized cacti have prominent ribs. The flowers are relatively large and produce edible fruit.

Distribution and Ecology: There are over 25 species of the hedgehog cactus in the western and southwestern United States, and Mexico.

EXPOSURE

The prickly pear fruit is a food source containing fructose and glucose.[1] In Mexico, the prickly pear cactus (nopal) has been an important food source since pre-Hispanic times. The leaves of prickly pear cacti are used by the Mexican population for medicinal purposes (e.g., diabetes, scurvy) and food.[2] The young stems (napalitos) are a vegetable in a variety of recipes. When ingested 5 hours prior to ethanol consumption, a double-blind, placebo-controlled, cross-over trial of 64 volunteers suggested that an extract of *Opuntia ficus-indica* (L.) P. Mill. (prickly pear) reduces the severity of a hangover (i.e., dry mouth, nausea, anorexia).[3] A concomitant reduction in C-reactive protein concentrations suggested that the mechanism of action was a decrease in the inflammatory state induced by alcohol impurities and metabolic byproducts of alcohol metabolism.

PRINCIPAL TOXINS

Physiochemical Properties

Calcium oxalate is the most common mineral in higher plants, occurring in two hydration states as monohydrate (whewellite) or as dihydrate (weddellite).[4] Crystalline forms of calcium oxalate in plants include druses, crystal sand, prisms, raphides, and styloids. Specimens from the cactus family contain highly pure and well-crystallized calcium oxalate crystals that form druses in both hydration states.[5] These druses appear as aggregates of thousands of individual crystallites in the shape of tetragons.

Mechanism of Toxicity

The adverse effect associated with exposure to cactus primarily involves mechanical trauma. Although infections can develop as a result of multiple puncture wounds, bacteria and acid-fast organisms do not usually colonize most cactus spines; therefore, infections are an uncommon complication of contact with cactus spines.[6] However, these spines often contain fungi (e.g., *Alternaria, Aspergillus*). The transfer of glochidia to human skin results from direct contact with surfaces (e.g., clothing or pets) containing glochidia, and contact of unprotected skin with glochidia produces an irritant contact dermatitis.[7] Wetting the prickly pear fruit and the use of protective clothing during windy days limits contact with the glochidia. Retention of *Opuntia* bristle fragments in the dermis can produce granulomas consisting of epithelioid cells, Langhans type and foreign body type giant cells, and retained bristles that react strongly with periodic acid-Schiff's (PAS) stain.[8] The fragments have angulated microscopic barbs, which remain in the skin after removal of the bristles.

CLINICAL RESPONSE

The initial sign of embedded cactus spines is the development of small (2–5 mm), asymptomatic, glistening papules within 24–72 hours after penetration of the spines.[8,9] Occasionally, the development of the lesions is delayed 7–10 days. Typically, the granulomas persist 2–8 months before resolving spontaneously.[10] Secondary infections following contact with cactus spines are relatively rare. Complications associated with foreign bodies secondary to cactus spines include chronic synovitis (afebrile, negative joint aspirate),[11] acute aseptic monoarticular arthritis,[12] chronic granulomas,[13] and postinflammatory hyperpigmentation. An occupational dermatitis (sabra dermatitis) results from contact with glochidia during the picking of prickly pear fruit characterized by a pruritic papular eruption primarily in the intertriginous web space of the fingers.[14,15] Retention of these spines can produce a delayed hypersensitivity reaction that persists several months and simulates scabies.[16] A pruritic rash can also develop on unprotected areas of the skin when the glochidia become airborne. Occasionally, portions of retained glochidia produce a keratotic plug that requires scraping or a miniature cutaneous horn that requires excision. Bezoars in the stomach and rectum can develop following the chronic consumption of seeds from the prickly pear fruit.[17] Symptoms include dyspepsia, malaise, bloating, anorexia, early satiety, weight loss, and fullness in the epigastrium.

DIAGNOSTIC TESTING

Although xerography occasionally identifies cactus spines, these foreign bodies are not usually seen with conventional radiography. A case report suggests that magnetic resonance imaging (MRI) can detect subcutaneous cactus spines embedded in an inflammatory nodule.[18]

TREATMENT

Foreign Bodies

Successful removal of cactus spines depends on identification of the location of the spine, particularly in relation to important subcutaneous structures.[19] Axial traction should be applied to the portion of the spine protruding above the surface of the skin. The use of tangential lighting and splinter forceps held flat along the surface of the skin aids removal of the spines. The initial approach for spines, which cannot be seen on the surface of the skin, is an incision that intercepts the spine similar to the approach for the removal of subcutaneous needles. The spines of a few cacti (e.g., *Opuntia fulgida* Engelm., *Opuntia bigelovii* Engelm.) are sufficiently tenacious that a portion of the stem breaks off the plant with the spines. Because the spines are directed in multiple planes, direct pressure on the stem embeds more spines. The spiney stem should be inverted over a bucket and spines cut next to the surface of the stem. Then the embedded spines are removed with axial pressure. Removal of spines from *Opuntia* species involves initial removal of the larger spines by forceps under direct visualization, followed by the techniques to peel off the numerous minute spines. These techniques include warm wax, glue, cellophane tape and water soluble, commercial facial gels. Rubber cement is applied with an applicator, covered with gauze, and allowed to dry.[20] Although somewhat difficult, the edge is peeled upward by a fingernail to remove the cement and small spines. Hospital cement (e.g., ostomy cement) or Super Glue® (Super Glue Corp., Rancho Cucamonga, CA) should be avoided because of the adherence of these compounds to the skin. The removal of cactus spines in the cornea may require the use of a fiberoptic illuminator in the operating room.[21]

Chronic Granulomata

Unroofing a granulomatous papule with the aid of a dissecting microscope helps identify and remove the embedded cactus spine. The glochidia of *Opuntia* species contain microscopic, posterior-angulated barbs that break off and form granulomatous inflammation in the skin. Treatment of chronic granulomatous lesions involves the use of topical corticosteroids (0.05% fluocinonide) several times a day for up to 1 month.

References

1. Kossori RL, Villaume C, El Boustani E, Sauvaire Y, Mejean L. Composition of pulp, skin and seeds of prickly pears fruit (*Opuntia ficus-indica* sp.). Plant Foods Hum Nutr 1998;52:263–270.
2. Silos-Espino H, Fabian-Morales L, Osuna-Castro JA, Valverde ME, Guevara-Lara F, Paredes-Lopez O. Chemical and biochemical changes in prickly pears with different ripening behaviour. Nahrung 2003;47:334–338.
3. Wiese J, McPherson S, Odden MC, Shlipak MG. Effect of *Opuntia ficus-indica* on symptoms of the alcohol hangover. Arch Intern Med 2004;164:1334–1340.
4. Frey-Wyssling A. Crystallography of the two hydrates of crystalline calcium oxalate in plants. Am J Bot 1981;68:130–141.
5. Monje PV, Baran EJ. Characterization of calcium oxalates generated as biominerals in cacti. Plant Physiol 2002;128:707–713.
6. Karpman RR, Spark RP, Fried M. Cactus thorn injuries to the extremities: their management and etiology. Arizona Med 1980;37:849–851.
7. Goodheart HP, Huntley AC. Cactus dermatitis. Dermatol Online J 2001;7(2):10.
8. Suzuki H, Baba S. Cactus granuloma of the skin. J Dermatol 1993;20:424–427.
9. McGovern TW, Barkley TM. Botanical briefs: prickly pear cactus—*Opuntia* Miller. Cutis 2001;68:188–190.
10. Doctoroff A, Vidimos AT, Taylor JS. Cactus skin injuries. Cutis 2000;65:290–292.
11. Zoltan JD. Cactus thorn synovitis. J Arthroscop Rel Surg 1991;7:244–245.
12. Miller EB, Gilad A, Schattner A. Cactus thorn arthritis: case report and review of the literature. Clin Rheumatol 2000;19:490–491.

13. Spoerke DG, Spoerke SE. Granuloma formation induced by spines of the cactus, *Opuntia acanthocarpa*. Vet Hum Toxicol 1991;33:242–344.
14. Shanon J, Sagher F. Sabra dermatitis. An occupational dermatitis due to prickly pear handling simulating scabies. Arch Dermatol 1956;74:269–274.
15. Banerjee K. A case report of Sabra dermatitis. Indian J Dermatol 1977;22:159–162.
16. Sagher F, Shanon J. Sabra dermatitis; an occupational dermatitis due to prickly pear handling stimulating scabies. AMA Arch Derm 1956;74:269–275.
17. Steinberg JM, Eitan A. Prickly pear fruit bezoar presenting as rectal perforation in an elderly patient. Int J Colorectal Dis 2003;18:365–367.
18. Stevens MA, De Coster TA, Renwick SE. Cactus thorn embedded in the cartilaginous proximal tibia. West J Med 1995;162:57–59
19. Lindsey D, Lindsey WE. Cactus spine injuries. Am J Emerg Med 1988;6:362–369.
20. Marinez TT, Jerome M, Barry RC, Jaeger R, Xander G. Removal of cactus spines from the skin. A comparative evaluation of several methods. Am J Dis Child 1987;141: 1291–1292.
21. Chen W-L, Tseng C-H, Hu F-R. Removal of semi-translucent cactus spines embedded in deep cornea with the aid of a fiberoptic illuminator. Am J Ophthalmol 2002;134:769–771.

Chapter 144

CAROLINA JESSAMINE
[*Gelsemium sempervirens* (L.) St. Hil.]

HISTORY

Extracts of *Gelsemium sempervirens* were used as herbal medications during the 19th century for neuralgias, fever, and nervousness. In 1862, a Detroit physician was indicted for administering this extract to a servant, who died about 2 hours after ingestion the extract. In 1885, a historical case series associated gelsemium poisoning with muscular weakness, staggering gait, mydriasis, respiratory depression, seizures, and death.[1] Animal experimentation at that time indicated that the extract of *Gelsemium sempervirens* was toxic, but there were no laboratory analyses of the toxic ingredients.

BOTANICAL DESCRIPTION

Common Name: Carolina Jessamine, Yellow Jessamine, Evening Trumpetflower

Scientific Name: *Gelsemium sempervirens* (L.) St. Hill.

Botanical Family: Loganiaceae (logania)

Physical Description: This woody, perennial, evergreen climbing vine has paired lance-shaped leaves and clusters of bright yellow, funnel-shaped flowers. Yellow jessamine should not be confused with day jessamine (*Cestrum diurnum* L.) or night jasmine (*Cestrum nocturnum* L.) from the Solanaceae family.

Distribution and Ecology: Carolina jessamine is found throughout the southeastern coastal plain from Virginia to Texas as a native vine. *G. sempervirens* is the state flower of South Carolina. This plant is cultivated as a climbing vine in California.

EXPOSURE

The Carolina jessamine is not commonly used as a medicinal herb. This plant is an ornamental plant in temperate climates.

PRINCIPAL TOXINS

Physiochemical Properties

All parts of *G. sempervirens* contain a mixture of 17-indole alkaloids including gelsemine, gelsevirine, gelsedine, 21-oxoglesemine, 14-β-hydroxygelsedine, gelsenicine, humantenidine, humantenirine, koumine, koumidine, kumantenine, and rankinidine.[2,3] Gelsemine (CAS RN:509-15-9, $C_{20}H_{22}N_2O_2$) is complex hexacyclic compound with an azatricyclodecane ring system that can be synthesized from 1-triisopropylsiloxy-3-methyl-1,3-cyclohexadiene via 26 intermediates.[4] Figure 144.1 displays the chemical structure of gelsemine.

Poisonous Parts

All parts of the plant contain alkaloids including gelsemine, gelseminine, and gelsemoidin. However, the blossoms and roots contain primarily high concentrations of gelsemine.[5] Honey made from the flowers is potentially poisonous, but there are no well-documented cases of

Medical Toxicology of Natural Substances, by Donald G. Barceloux, MD

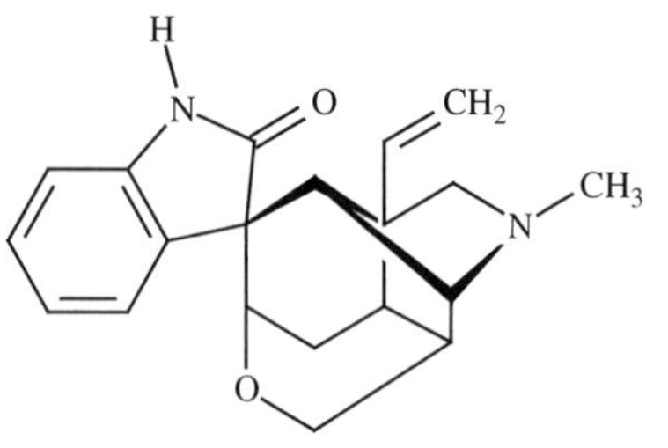

FIGURE 144.1. Chemical structure of gelsemine.

human intoxication from the ingestion of gelsemine-contaminated honey.

Mechanism of Toxicity

Gelsemine is a potent antagonist of central and peripheral muscarinic and nicotinic acetylcholine receptors.[5] This toxin interferes with cholinergic transmission at the neuromuscular junction and central nervous system by binding to the acetylcholine receptor.[5]

DOSE RESPONSE

There are inadequate data to determine the dose response of toxic constituents in this plant.

TOXICOKINETICS

The toxic constituents in this plant have not been identified.

CLINICAL RESPONSE

Most clinical reports on the toxicity of Carolina jessamine were published before 1900 when *G. sempervirens* extracts were used by herbalists to induce abortions. Patients developed profound muscle weakness, nausea, vomiting, diaphoresis, and respiratory depression leading to asphyxia.[6] A case report on a 3½-year-old girl associated the development of wide-based gait, bilateral ptosis, mydriasis, brisk deep tendon reflexes, generalized weakness and obtundation with the ingestion of five blossoms from *G. sempervirens*.[5] Additional symptoms included headache, dry mouth, dysphagia, blurred vision, and lightheadedness. Symptoms resolved within 10 hours after ingestion. Increased muscle tone and hyperactive deep tendon reflexes were reported in cases of *G. sempervirens* intoxication prior to 1900, but the etiology of these clinical findings is unclear.[1]

DIAGNOSTIC TESTING

Experimental studies use nuclear magnetic resonance (NMR) spectrometry to identify hexacyclic compounds in this plant.[7]

TREATMENT

Management is supportive. Decontamination measures are not usually recommended. The administration of activated charcoal to alert children who ingest more than one blossom or several leaves less than 1 hour prior to presentation is a therapeutic option, but there are no clinical data to confirm the clinical efficacy of decontamination measures following the ingestion of Carolina jessamine.

References

1. Rehfuss EG. Gelsemium and its reputed antidotes, with experiments and collection of cases of poisoning. Ther Gazette 1885;9:655–666.
2. Schun Y, Cordell GA. Cytotoxic steroids of *Gelsemium sempervirens*. J Nat Prod 1987;50:195–198.
3. Lin LZ, Cordell GA, Ni CZ, Clardy J. New humantenine-type alkaloids from *Gelsemium elegans*. J Nat Prod 1989;52:588–594.
4. Madin A, O'Donnell CJ, Oh T, Old DW, Overman LE, Sharp MJ. Total synthesis of (±)-gelsemine. Angew Chem Int Ed 1999;38:19–21.
5. Blaw ME, Adkinson MA, Levin D, Garriott JC, Tindall RS. Poisoning with Carolina jessamine (*Gelsemium sempervirens* (L.) Act.). J Pediatr 1979;94:998–1001.
6. Ringer S. Murrel W. On *Gelsemium sempervirens*. Lancet 1878;1:858–860.
7. Schun Y, Cordell GA. Studies on the NMR spectroscopic properties of gelsemine—revisions and refinements. J Nat Prod 1985;48:969–971.

Chapter 145

DAPHNE
(*Daphne mezereum* L.)

BOTANICAL DESCRIPTION

Common Name: Daphne, February Daphne, Paradise Plant, Spurge Laurel, Mezereon, Dwarf Bay

Scientific Name: *Daphne mezereum* L.

Botanical Family: Thymelaeaceae (mezereum)

Physical Description: This small deciduous shrub grows 1–4 ft (~0.3–1 m) in height. Small lilac-purple, fragrant flowers develop before the simple alternate leaves appear.

Distribution and Ecology: *D. mezereum* is an ornamental shrub throughout the United States, and this perennial shrub has naturalized in the northeastern United States from Europe and Western Asia.

EXPOSURE

The compound, mezerein, isolated from seed extracts of *D. mezereum* demonstrates some activity against lymphocytic leukemia in mice, but there are limited clinical data on effects in humans.[1,2] Traditional herbal uses of this plant include the treatment of cancer since the time of Aphrodisias around 200 AD.[3]

PRINCIPAL TOXIN

Active constituents of *Daphne mezereum* include mezerein (CAS RN: 34807-41-5, $C_{38}H_{38}O_{10}$), the diterpenoid compound daphnetoxin (CAS RN: 28164-88-7) and 12-hydroxydaphnetoxin.[4] Figure 145.1 displays the chemical structure of mezerein. All plant parts have glycosides (e.g., daphnin, daphnetin, esculin) that contain dihydroxycoumarin as the aglycone.[5] Mezerein is a second-stage tumor promoter and an experimental antileukemic agent that activates protein kinase C (PKC) both *in vitro* and *in vivo*.[6] *In vitro* studies indicate that daphnetoxin is also a PKC activator, but this compound is a more selective and less potent PKC activator than mezerein.[2]

DOSE RESPONSE

There inadequate data to determine a dose response for ingestion of *D. mezereum.*

TOXICOKINETICS

There are few data on the toxicokinetics of active constituents in *D. mezereum.*

CLINICAL RESPONSE

There are few data in the modern medical literature on the toxicity of *Daphne mezereum.* Most reports of exposure to *D. mezereum* in children are not associated with symptoms. Frequently mentioned symptoms following ingestion of parts of this plant from textbooks on plant poisonings include local irritation, nausea, vomiting, abdominal pain, weakness, diarrhea, and occasionally bloody diarrhea. A case report associated the exposure of a 1-year-old girl to *D. mezereum* with the

Medical Toxicology of Natural Substances, by Donald G. Barceloux, MD

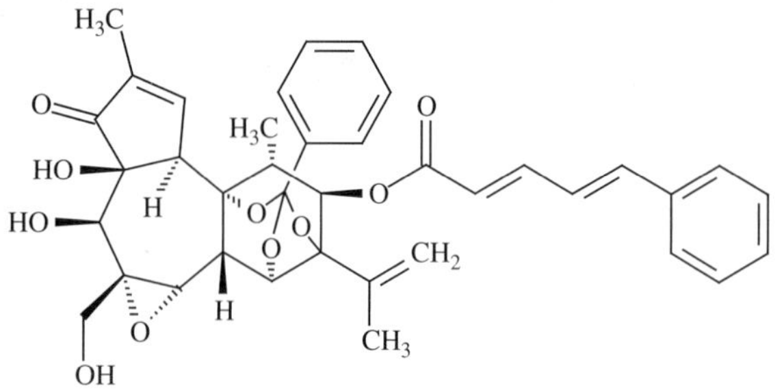

FIGURE 145.1. Chemical structure of mezerein.

development of increased salivation, hematuria, and diarrhea that resolved with conservative treatment.[7]

DIAGNOSTIC TESTING

There are few data on the detection of toxic constituents of laboratory abnormalities following ingestion of *Daphne mezereum*.

TREATMENT

Treatment is entirely supportive. Gastrointestinal decontamination measures are unnecessary. There are no clinical data to determine the efficacy of decontamination measures.

References

1. Kupchan SM, Baster RL. Mezerein: Antileukemic principles isolated from *Daphne mezereum*. Science 1975;187:652–653.
2. Saraiva L, Fresco P, Pinto E, Portugal H, Goncalves J. Differential activation by daphnetoxin and mezerein of PKC-isotypes α, β_1, δ, and γ. Planta Med 2001;67:787–790.
3. Hartwell JL. Plants used against cancer. A survey. Lloydia 1971;34:204–255.
4. Ronlan A, Wickberg B. The structure of mezerein, a major toxic principle of *Daphne mezereum* L. Tetrahedron Lett 1970;49:4261–4264.
5. Ulubelen A, Terem B, Tuzlaci E. Coumarins and flavonoids from *Daphne gnidioides*. J Nat Prod 1986;49:692–694.
6. Miyake R, Tanaka Y, Tsuda T, Kaibuchi K, Kikkawa U, Nishizuka Y. Activation of protein kinase C by non-phorbol tumor promoter, mezerein. Biochem Biophys Res Commun 1984;121:649–656.
7. Lamminpää A, Kinos M. Plant poisoning in children. Hum Exp Toxicol 1996;15:245–249.

Chapter 146

DOG LAUREL
(*Leucothoe* Species)

BOTANICAL DESCRIPTION

There are about 40 species in the genus *Leucothoe.*

Common Name: Dog Laurel, Doghobble

Scientific Name: *Leucothoe* spp.

Botanical Family: Ericaceae

Physical Description: These shrubs or small trees have alternated leaves and white, fused petals. This evergreen shrub has green, slightly arching stems and pointed leaves. The fruit has a globular, five-lobed capsule with a slightly depressed apex.

Distribution and Ecology: *L. axillaris* (Lam.) D. Don forms dense thickets in moist woodlands of the coastal plain of the southeastern United States and in California.

EXPOSURE

Most serious exposures to this plant occur in grazing animals. There are few human data.

PRINCIPAL TOXINS

Older Japanese literature indicates that the main toxic ingredient in dog laurel is grayanotoxin I, which was structurally similar to andromedotoxin isolated from the *Rhododendron* species.[1] Leaves of *Leucothoe grayana* Maxim. contain isograyanotoxin II,[2] but the clinical significance of this tetracyclic diterpenoid is unclear.

DOSE RESPONSE

The toxic constituents in this plant have not been isolated.

TOXICOKINETICS

There are inadequate data on the principal toxins in dog laurel.

CLINICAL RESPONSE

Most casual exposures to *Leucothoe* plants do not produce toxicity in humans, and summaries of plant exposures reported to US poison centers do not indicate that this plant causes clinically significant toxicity.[3] Based on the presence of grayanotoxins in these plants, ingestions of these plants may cause gastrointestinal distress (nausea, vomiting, diarrhea, abdominal pain), lethargy, neuromuscular weakness, and cardiac depression. However, there are inadequate data to indicate that these toxins are present in sufficient quantities to cause serious toxicity. Grazing animals consuming large quantities of *Leucothoe* species can develop serious toxicity including vomiting, anemia, staggers, and focal hepatic necrosis.[4]

DIAGNOSTIC TESTING

The toxic constituents have not been well defined.

Medical Toxicology of Natural Substances, by Donald G. Barceloux, MD

TREATMENT

Most patients ingesting parts of dog laurel remain asymptomatic, and treatment is supportive. The most likely adverse effect is gastrointestinal irritation, and symptomatic patients should be evaluated and treated for any fluid or electrolyte imbalance. Decontamination measures are unnecessary.

References

1. Miyajima S, Takei S. [The active constituents of *Lecothoe grayana* Max]. J Agr Chem Soc Jap 1934;10:1093–1103. [Japanese]
2. Terai T, Araho D, Osakabe K, Katai M, Narama I, Matsuura T, et al. Isolation of iso-grayanotoxin II from leaves of *Leucothoe grayana* Max. Its x-ray crystallographic analysis and acute toxicity in mice. Chem Pharm Bull 2000;48:142–144.
3. Krenzelok EP, Jacobsen TD, Aronis JM. Plant exposure: a state profile of the most common species. Vet Hum Toxicol 1996;38:289–298.
4. Kohanawa M, Shoya S, Ikeda K, Sugiura K, Okuyama T. Experimental *Leucothoe grayana* poisoning in goats. Nat Inst Anim Health Q 1970;10:160–170.

Chapter 147

DOGBANE FAMILY and CARDENOLIDES

COMMON OLEANDER

HISTORY

Since antiquity, the Greeks, Romans, and Hindus recognized the medicinal properties of oleander as a cardiotonic and suicidal agent.[1] The early Greeks considered the oleander plant poisonous to "all four-footed beasts."[2] Preparations of oleander have been used as rodenticides and insecticides as well as remedies for indigestion, malaria, and venereal disease. The most commonly mentioned folk use was as an abortifacient. The Chinese used oleander extract for the treatment of mental disorders. Oleander extracts (Oleander Perpurat, Knoll) were treatments for heart failure and atrial fibrillation. African arrow poisons often contained cardenolides (ouabain, acolongifloroside K, acovenoside A) obtained from the wood and roots of other species of the dogbane (Apocynaceae) family including the genus *Carissa* (*Acokanthera*).[3,4]

BOTANICAL DESCRIPTION

Common Name: Common Oleander, Rose Laurel, Soland, Lorier Bol, Rosebay

Species Name: *Nerium oleander* L. [*Nerium indicum* (Mill.), *Nerium odorum* Aiton]

Botanical Family: Apocynaceae (dogbane)

Physical Description: The common oleander is a large ornamental evergreen shrub that may reach 20–25 ft (~6–8 m) in height with oblong, pointed leaves that contain a prominent midrib. During the summer, large clusters of white, pink, or red flowers appear at the ends of the branches. The fruit consists of two long, narrow pods.

Distribution and Ecology: Oleander plants occur throughout the temperate regions of the United States from California to Florida. The oleander plant is native to Asia and the Mediterranean regions, including India and China. In temperate climates of Africa, South America, and the southern United States, these plants are cultivated as flowering shrubs. A similar plant, called the yellow oleander [*Thevetia peruviana* (Pers.) K. Schum., *Thevetia thevetioides* (Kunth) Schumann], grows in Australia, Hawaii, and southern Florida.

EXPOSURE

Common oleander is an ornamental shrub that adapts well to dry areas along roadsides and in gardens. These shrubs are popular parts of gardens because of their prolific growth and decorative flowers. The root and the bark have been used in India for skin diseases and leprosy. Extracts of oleander have been used as insecticides, rodenticides, and as an abortifacient, as well as a cure for leprosy, malaria, ringworm, venereal disease, indigestion, fever, and alcoholism.[5] African arrow poisons typically contain toxic cardenolides (e.g., ouabain, acolongifloroside K) from *Carissa* (*Acokanthera*) species.[3]

In European folk medicine, herbal preparations from *Adonis* species are used as diuretics and cardiac stimulants.[6] Experimental studies in the 1950s suggested that

Medical Toxicology of Natural Substances, by Donald G. Barceloux, MD

alcohol extracts of star of Bethlehem (*Ornithogalum umbellatum* L.) caused positive inotropic actions similar to digitoxin.[7] Preliminary studies investigated the efficacy of these extracts for the treatment of congestive heart failure, but the use of these preparations did not demonstrate a clear superiority over the administration of digitalis preparations (digitoxin).[8,9]

PRINCIPAL TOXINS

Physiochemical Properties

The two major classes of cardiac glycosides are 1) the bufadienolides from the poisonous glands in the skin of the cane toad (*Bufo marinus* L.), and 2) the cardenolides in plant tissues.[10] The major toxins in oleander are cardiac glycosides that contain a five-membered steroid nucleus with a lactone ring in contrast to the six-membered lactone ring in bufadienolides. Digitalis is the prototypical cardenolide, and oleander-derived cardenolides share similar physiological and structural characteristics with digitalis.

The three structural subunits of cardenolides are 1) steroid backbone, 2) lactone ring, and 3) sugar or carbohydrate moiety in glycosidic linkage.[1] The steroid component and the lactone ring represent the "aglycone" or "genin" component of the cardenolide compound. These plant cardenolides have the carbohydrate or sugar moiety attached to the three-carbon of the "A" ring of the five-membered steroid nucleus. The sugar moieties of cardenolides are specific, and this portion of the compound defines the name of the cardenolide (e.g., digitoxose, digitoxin, oleandrose, oleandrin). Figure 147.1 compares the chemical structures of common cardiac glycosides present in oleander.

Common oleander contains several cardiac glycosides (oleandrin, digitoxigenin, neriantin, rosagenin) that produce a digitoxin-like effect by inhibiting the sodium-potassium adenosine triphosphatase enzyme system. These cardiac glycosides bind to an extracellular portion of the Na^+/K^+ ATPases and cause increased intracellular calcium concentrations. Minor glycosides in oleander with limited cardiac activity include adynerigenin, adigoside, digitalin, and uzarigenin.[11] Because of the lack of toxicokinetics data, the identities of the toxic glycosides or toxic metabolites remain unclear.

Poisonous Parts

Of over 400 cardiac glycosides isolated from the plant kingdom, most of the cardiac glycosides occur in extracts from the milkweed (Asclepiadaceae), figwort (Scrophulariaceae), lily (Liliaceae), and dogbane (Apocynaceae) families. The highest concentrations of cardiac

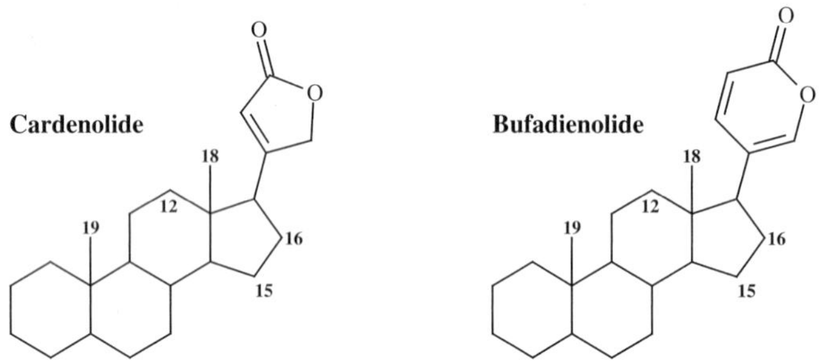

Cardenolide

Bufadienolide

Cardiac Glycoside			Aglycone Substituent		
	C-12	C-15	C-16	C-18	C-19
Cardenolide					
Digoxin	OH			CH_3	CH_3
Digitoxin				CH_3	CH_3
Oleandrin			OAc	CH_3	CH_3
Neriantin		OH		CH_3	CH_3
Thevetin A				CH_3	HC=O
Thevetin B				CH_3	CH_3
Adonitoxin			OH	CH_3	HC=O
Bufadienolide					
Bufalin				CH_3	CH_3
Bufogenin B			OH	CH_3	CH_3
Bufotalin			OAc	CH_3	CH_3

FIGURE 147.1. Structural similarities between the prominent cardiac glycosides present in oleander and secretions from the cane toad (*Bufo marinus*). Adapted from Cheung K, et al. Detection of poisoning by plant-origin cardiac glycoside with the Abbott TDx Analyzer. Clin Chem 1989;35:297.

glycosides are found in seed, stems, and roots, followed by the fruit and leaves. However, the leaves contain the greatest oleandrin concentrations. The concentration of oleandrin varies widely between various strains and species of oleander. The estimated oleandrin content of leaves from common oleander ranges from approximately 0.02–0.4%.[11] In a study of 20 *Nerium oleander* L. from various countries, the oleandrin content of fresh, dried leaves ranged from 0.007–0.42%.[12] Using a radioimmunoassay that measured apparent digoxin concentrations as a surrogate for cardiac glycoside content, the value for stems from *Nerium oleander* was 620 μg/g compared with 52 μg/g for leaves from *Thevetia peruviana*.[10] Differences in the apparent digoxin concentration exist between plant species as measured by different immunoassays. The apparent digoxin concentration of leaves from the *Nerium oleander* was 2.21 μg/L as measured by Abbott TDx Analyzer (Abbott Diagnostics, Abbott Park, IL) and 2.86 μg/L as measured by in-house radioimmunoassay, whereas the apparent digoxin concentrations of the leaves from *Thevetia peruviana* were similar (3.51 μg/L and 3.56 μg/L, respectively).[13]

The highest concentrations of total cardiac glycosides occur during the flowering stage in April to May with red flowers possessing the highest glycoside content.[14] The roots contain larger concentrations of total cardenolides and smaller concentrations of oleandrin compared with the leaves, flowers, and stems. By unconfirmed historical accounts, 7 of 12 French soldiers died in 1809 when they roasted meat on oleander stems from which the bark had been stripped.[15] However, there are no well documented reports to indicate that sufficient amounts of oleandrin are transferred from oleander stems to food during the cooking process. *Thevetia* fruit and *Nerium* leaves were the most common parts of the oleander ingested in an Australian study of children with oleander exposure.[16] The children, who ate parts of whole plants, remained relatively asymptomatic compared with adults, who develop serious cardiac toxicity after intentional ingestions of teas made from oleander.

DOSE RESPONSE

Animals

There is substantial variation in the sensitivity to oleander poisoning between animal species. Dogs, cats, goats, monkeys, and humans are relatively sensitive to the effects of oleander-derived cardenolides, whereas rodents and avian species are relatively insensitive. The intravenous administration of oleander tincture in dogs produces a digitalis-like poisoning during experimental studies manifest by a combination of delayed conduction and increased automaticity terminating in intractable ventricular fibrillation.[17] In a study of monkeys receiving 30 mg dried oleander leaves per kg body weight, vomiting developed within 2 hours and bradycardia within 1–6 hours.[18] All the monkeys died, and postmortem examination revealed necrosis in the heart, kidneys, adrenals, liver, and pancreas.

Humans

Although one leaf has been considered potentially lethal, ingestion of larger amounts is probably necessary to produce serious toxicity. Using calculations on the digoxin pharmacokinetic in the antemortem serum from a 96-year-old woman who died after ingesting oleander tea, Osterloh et al estimated that the apparent digoxin concentration of 5.8 ng/mL was equivalent to the absorption of cardiac glycosides from 5–15 *Nerium* leaves.[19] The postmortem examination did not demonstrate any evidence of coronary artery disease. Using a crude oleander extract prepared by ethanol extraction, lead acetate purification, and solvent partition, the group estimated that 100 ng/mL of crude extract produced a 1 ng/mL false elevation of the digoxin concentration. Assuming the V_d of digitoxin was 1 L/kg, the estimated absorbed dose of cardiac glycosides was approximately 4 grams, which was equivalent to 5–15 oleander leaves.

In a study of children hospitalized for oleander ingestions in Australia from 1972–1978, the reported ingestion of three leaves by a 7-year-old resulted in a partial right bundle branch block.[16] The patient remained asymptomatic. Three other toddlers remained asymptomatic after ingesting leaves, flowers, or seed husks.

TOXICOKINETICS

There are few data on the kinetics of cardenolides during oleander intoxication. The apparent half-life of cardiac glycosides as calculated from the apparent digoxin concentration was 44 hours in the serum from a 17-year-old girl with oleander poisoning. She developed mild conduction abnormalities after ingesting five oleander leaves that were crushed and boiled.[20]

CLINICAL RESPONSE

Oleander poisoning resembles digitoxin poisoning with the predominant symptoms involving the gastrointestinal and cardiac systems. Serious oleander intoxication usually occurs following the ingestion of teas prepared from oleander leaves rather than from the ingestion of plant parts.[21–23] The bitter taste of the oleander leaves limits the ingestion of oleander leaves in children. Characteristic features of serious oleander poisoning include confusion, marked hyperkalemia, conduction abnormalities, and ventricular arrhythmias. Nausea and vomiting usually occur within several hours, and gastrointestinal symptoms are more prominent following oleander poisoning than following digoxin toxicity. Increased salivation, dry mouth, burning of the mouth, and paraesthesias of the tongue may develop as a result of the local effects of oleander extracts.[22] Case reports indicate that the astringent properties of oleander sap produce mucosal irritation.[24]

The most common serious complication of oleander poisoning is disruption of cardiac conduction.[25] Cardiogenic shock, ventricular fibrillation, and cardiovascular collapse may occur following severe oleander toxicity. Conduction delays may persist 3–6 days, and these cardiac effects resemble classical digitalis toxicity (e.g., conduction abnormalities with ectopy, such as supraventricular tachycardia with atrioventricular block).[23,26] Following the intentional ingestion of oleander tea, a 96-year-old woman developed cardiovascular collapse 15 minutes after arrival in the emergency department.[19] Cardiac monitoring demonstrated a variety of rhythms including ventricular tachycardia, ventricular fibrillation, and asystole that failed to respond to the standard protocols outlined in the American Heart Association's Advanced Cardiac Life Support guidelines. Her admission potassium level was 8.6 mEq/L. A 30-year-old woman presented in cardiogenic shock with an idioventricular rhythm 10 hours after ingesting an oleander tea.[21] She died despite 1 hour of resuscitation that included fluids, endotracheal intubation, sodium bicarbonate, atropine, and a transthoracic pacemaker. A 33-year-old woman developed complete heart block and hyperkalemia (6.7 mEq K^+/L) after the ingestion of oleander tea as an abortifacient.[27] Within 24 hours, the complete heart block reverted to bradycardia that persisted for 6 days along with diarrhea and dizziness. She recovered completely with only supportive care.

Ornithogalum species contain calcium oxalate crystals arranged in bundles. An irritant dermatitis was associated with the direct contact of mucilaginous juice from *Ornithogalum caudatum* Aiton.[28] Erythema developed within 5 minutes at the site of contact along with pruritus and a sensation of burning that persisted 2 days. These symptoms were related to raphides in the mucilage rather than the presence of cardenolide compounds.

DIAGNOSTIC TESTING

Analytical Methods

Chromatography

The only cardiac glycoside in *Nerium oleander* available as a pure standard is oleandrin. The use of thin layer chromatography (TLC) and fluorescence spectrophotometry for the identification of oleandrin provides verification of oleander poisoning for forensic purposes. Oleandrin is the cardiac glycoside identified as the *o*-arabinoside of the 16-β-(acetyloxy)-14-hydroxy-card-20-(22)-enolide. Similar cardenolides include the glycoside, neriantin, and several odoroside glycoside derivatives of the aglycone oleandrigenin.[29] Because of the structural similarities of these cardenolides, the fluorometric assays for oleandrin also measure other cardiac glycosides. The thermolability and poor volatility of the cardiac glycosides present in oleander complicate the quantitation of these substances by gas chromatography/mass spectrometry. A more sensitive method for the detection of oleander poisoning is the use of high pressure liquid chromatography/mass spectrometry (HPLC/MS) to identify oleandrin and the aglycone metabolite, oleandrigenin.[30,31] The limit of quantitation for oleandrin in whole blood was 0.4 ng/mL.

Immunoassays

The presence of a positive digoxin immunoassay is consistent with the ingestion of oleander, but the apparent digoxin concentrations vary between assays depending on the assay. The relationship between the measured apparent digoxin concentration and the actual concentration of cardiac glycosides in serum depends on the specificity and heterogeneity of the antibody in each immunoassay as well as the type of cardenolide present in the ingested material. Consequently, the apparent digoxin concentration does not correlate to the usual therapeutic and toxic concentration of digoxin in serum. In a variety of plant genera, cardiac glycosides cross-react with digoxin immunoassays to a variable degree.[32] Table 147.1 lists plants and animals that contain cardiac glycosides capable of cross-reacting with digoxin assays.

The cardiac glycosides present in oleander cross-react with digoxin radioimmunoassays and commercial immunoassays. However, the identity of these cardiac glycosides and the extent of these cross-reactions are not well defined. Furthermore, these tests are not sensitive for all cardioactive glycosides, and sensitivity to a specific cardiac glycoside varies with the specific antigen used. For example, antidigoxin antiserum cross-reacts well with glycosides of digoxigenin that share the aglycone portion of digoxin, but this antiserum is less sensitive to digitoxin-like cardiac glycosides.[33,34] Radford et al demonstrated the variable sensitivity of immunoassays to cardiac glycoside concentrations present in samples of various cardenolide-containing plants including oleander by using four different antibodies (12-acetyl-3-succinyl digoxigenin, 3,12-disuccinyl

TABLE 147.1. Plants and Animals Containing Cardiac Glycosides that Potentially Interact with Digoxin Immunoassays

Scientific Name	Common Name	Principal Cardenolides
Plants		
Acokanthera oppositifolia Lam. Codd	Common poison bush	Acovenoside A, ouabain, acolongifloroside K
Acokanthera oblongifolia (Hochst.) Codd (A. spectabilis), *Carissa acokanthera* Pichon	Bushman's poison bush, dune poison bush, wintersweet	Ouabain (G- strophanthin, acocantherin)
Adenium obesum (Forssk.) Roem. & Schult.	Desert rose	Digitalin
Adonis spp.	Pheasant's eye	Adonin, vernadigin, adonitoxin, cymarin, strophanthidin
Antiaris toxicaria Lesch.	Upas tree	Convallatoxin, α-antiarin,
Apocynum cannabinum L.	Indian hemp (US), dogbane	Strophanthidin
Asclepias curassavica L.	Blood flower, swallow wort, Indian root	Calactin
Asclepias eriocarpa Benth., *A. labriformis* M. E. Jones	Wollypod milkweed, Labriform milkweed	Eriocarpin, labriformidin, labriformin
Asclepias spp. [*Asclepias physocarpa* (E. Mey.) Schlechter]	Balloon cotton bush	Asclepin
Bersama abyssinica Fresen.	Winged bersama	Hellebrigenin-3-acetate
Bryophyllum pinnatum (Lam.) Oken	Air plant	Bryophyllin A
Calotropis procera (Ait.) Ait. f. (*Asclepias procera*)	King's crown, rubber bush	Calatropin, usharidin, calactin
Castilla elastica Sessé	Black rubber tree	Cymarin
Cerbera floribunda K. Schum.		Cerbertin
Cerbera manghas L.	Sea mango	Cerberin (monoacetylneriifolin)
Cerbera odollam Gaertn.	Odollam tree	Cerberoside
Convallaria majalis L.	Lily of the valley	Convallerin, convallamarin, convallatoxin, (strophanthidin α-L-rhamnoside)
Cryptostegia grandiflora (Roxb. ex R. Br.) R. Br.	Rubber vine, India rubber vine, Palay rubbervine, purple allamanda	
Digitalis lanata Ehrh.	Grecian foxglove	Digoxin, digitoxin, digitonin, gitoxin, digitalin, lanatoside C
Digitalis grandiflora P. Mill.	Yellow foxglove	Digoxin
Digitalis purpurea L.	Purple foxglove	Digitoxin, digitonin, gitoxin, digitalin
Erysimum spp. (E. cheiranthoides L.)	Sibertian wallflower, treacle mustard	Helveticoside
Helleborus niger L.	Christmas rose	Helleborin, hellebrin and helleborein
Helleborus foetidus L.	Stinking hellebore	Helleborin, hellebrin and helleborein
Helleborus viridis L.	Green hellebore	Helleborin, hellebrin and helleborein
Kalanchoe delagoensis Ecklon & Zeyh. (*Bryophyllum tubiflorum* Harv.)	Mother of millions	Bryotoxins B and C
Kalanchoe lanceolata (Forssk.) Pers.		Lanceotoxin A, lanceotoxin B
Nerium oleander L.	Oleander	Oleandrin, oleandrigenin, neriine, diginoside, digitoxigenin, adynerin
Ornithogalum umbellatum L.	Star of Bethlehem	Convallatoxin, convalloside, rhodexin A
Rhodea japonica Roth	Nippon lily	Rhodexin A
Strophanthus divaricatus		Musaroside, decoside, divaricoside, divostroside, sarmentoloside
Strophanthus gratus (Wallich & Hook. ex Benth.) Baillon	Dogbane	Ouabain
Strophanthus kombe Oliver	Kombe, kombe arrow-poison	Cymarin, K-strophanthoside
Strophanthus sarmentosus DC.		Sarmentoloside, bipindoside
Sirophanthus tholloni Franchet		Bipindoside
Strophanthus spp.	Arrow poisons	Lokundjoside
Thevetia peruviana (Pers.) K. Schum., *Thevetia thevetioides* (Kunth) Schumann	Yellow oleander, be-still tree, tiger apple, luckynut	Thevetin A and B, thevetoxin, digitoxigenin, neriifolin
Urginea maritima (L.) Baker (U. scilla)	Red squill, sea onion	Scillaren A and B, scilliroside
Animals		
Bufo spp. (*B. alvarius* Girard in Baird, *B. gargarizans* Cantor, *B. marinus* L.)	Chinese toad, marine toad, Colorado river toad	Bufalin, cinobufotalin, cinobufagin

Source: Adapted from Ref 32.

digoxigenin, digoxin, and an unknown commercial assay; see Table 147.2).[10] An *in vitro* study on the cross-reactivity of five common digoxin immunoassays for oleandrin indicated that fluorescence polarization immunoassay (FPIA) was the most sensitive for oleandrin, whereas the chemiluminescent assay (Bayer Diagnostics, Tarrytown, NY) was insensitive.[35] The Beckman digoxin assay on the Synchron LX and the turbidimetric assay on the ADVIA analyzer (Bayer) demonstrated intermediate sensitivity for oleandrin. Although the measurement of serum oleandrin concentrations with the digitoxin fluorescence polarization immunoassay is linear, the result does not correlate to toxicity.[11] Consequently, immunoassays for digoxin and digitoxin confirm exposure to cardiac glycosides present in oleander, but the results do not correlate linearly to the actual concentration of cardiac glycosides in these plants.

Biomarkers

Chromatography

A 45-year-old woman presented to the emergency department with nausea, vomiting, abdominal pain, sinus bradycardia, and cardiovascular shock.[31] The oleandrin concentration in a plasma sample drawn at the time of admission was 1.1 ng/mL, as measured by HPLC/MS. The patient survived and the following day the concentration of oleandrin in a blood sample was not detectable. Eighteen hours after ingesting a bowl of *Nerium oleander* blooms, a 47-year-old woman presented to the emergency department with bradycardia, intermittent third-degree AV block, and hyperkalemia.[36] The serum oleandrin concentration on admission was 1.6 ng/mL, as measured by HPLC/MS/MS. Sixteen hours later the serum oleandrin concentration was below the limit of quantitation (<1.0 ng/mL). The postmortem oleandrin concentration in heart blood from a 49-year-old woman, who died 1 day after ingesting an aqueous extract of oleander, was 9.8 ng/mL, as measured by liquid chromatography/three-dimensional quadrupole mass spectrometry with sonic spray ionization.[37] The limit of detection was 3 ng/mL.

Immunoassays

The apparent digoxin concentration in a serum sample from an asymptomatic 17-year-old boy with suspected ingestion of oleander leaves was 0.45 µg/L as measured by radioimmunoassay and 0.19 µg/L, as measured by the Abbott TDx Analyzer. The apparent digoxin concentration in the serum of a patient with asymptomatic bradycardia and a normal potassium level following the ingestion of several oleander leaves was 1.2 ng/mL, as measured by radioimmunoassay.[38] A 37-year-old woman ingested seven *Nerium oleander* leaves 8 hours prior to arrival at the hospital.[39] A blood sample drawn at admission contained an apparent digoxin concentration of 4.4 ng/mL, as measured by radioimmunoassay. The clinical course of this patient was unremarkable. Case reports have documented similar apparent digoxin concentrations following the ingestion of oleander tea, which

TABLE 147.2. Cardiac Glycoside Concentrations in Various Plant Species as Measured by Three Different Immunoassays

Source	Common Name	Plant Part	Antibody A[a] (µg/g)	Antibody B[a] (µg/g)	Antibody C[a] (µg/g)
Asclepias curassavica L.	Redheaded cotton-bush	Leaves/stems	1400	460	89
Asclepias fruticosa (L.) W. T. Aiton	Balloon cotton	Fruit			20
Bufo marinus L.	Cane toad	Blood	15	9	0.3
		Venom	9200	2300	320
Calotropis procera (Ait.) Ait. f.	King's crown	Stem	160	150	16
Carissa acokanthera Pichon	Bushman's poison	Seed			400
Cerbera manghas L.	Sea-mango	Stem	150	660	11
		Fruit	70	200	10
Nerium oleander L.	Common oleander	Stem leaves	430	620	33
Thevetia peruviana (Pers.) K. Schum.	Yellow oleander	Leaves		52	36
		Stems		24	

Source: Adapted from Ref 33.

[a]Antibody A (goat) 12-acetyl-3-succinyl digoxigenin; antibody B (goat) 3,12-disuccinyl digoxigenin; antibody C (cow) digoxin. All three cardiac glycoside assays demonstrate 100% cross-reactivity to digoxin, whereas the cross-reactivity to digitoxin was 350%, 100%, and 23%, respectively.

caused severe toxicity (third degree heart block, serum potassium 8.8 mEq/L).[40] Following the ingestion of oleander extract the night before, the serum from an 83-year-old lady contained an apparent digoxin concentration of 7.1 ng/mL, as measured by radioimmunoassay.[22] The serum potassium concentration was 6.1 mEq/L. Using a digoxin radioimmunoassay from Abbott Diagnostics, Osterloh et al detected an apparent digoxin level of 5.8 ng/mL in the serum sample from a 96-year-old woman, who developed cardiac arrest within 15 minutes of arrival at the emergency department.[19] The serum potassium concentration was 8.6 mEq/L. The digoxin concentration was 6.4 ng/mL in postmortem blood from a 30-year-old woman, who died about 10–11 hours after ingesting oleander tea, as measured by radioimmunoassay (RIA).[21] There was some evidence of postmortem redistribution because the apparent digoxin concentration (method not specified) in serum drawn during resuscitation and analyzed 6 days later was 1.2 ng/mL.

Abnormalities

Classic electrolyte disturbances include hyponatremia and hyperkalemia as a result of the blockade of the Na^+-K^+ ATPase pump. Severe oleander toxicity is associated with hyperkalemia, but not with hypermagnesemia. Electrocardiographic changes include a wide range of conduction abnormalities and dysrhythmias. These changes range from bundle branch block to complete heart block and from ventricular premature beats and junctional escape rhythms to atrial and ventricular fibrillation.

TREATMENT

Stabilization

Serious oleander toxicity usually results from the ingestion of oleander teas rather than the ingestion of plant parts. These patients require careful evaluation for adequacy of ventilation, conduction abnormalities, ventricular ectopy, and hyperkalemia with cardiac monitoring and pulse oximetry. Lidocaine is the antiarrhythmic of choice for the treatment of digitalis-induced tachyarrhythmias. Quinidine, procainamide, and beta-blockers should be avoided because of adverse effects on cardiac conduction and contractility. There are few data on the use of amiodarone during digitalis intoxication. Symptomatic bradycardias should be treated first with atropine, and then a cardiac pacemaker. Patients who do not respond to supportive care are candidates for the administration of digoxin-specific Fab fragments, although the efficacy for this treatment has not been established.

Decontamination

Casual contact with common oleander leaves and flowers by children does not usually result in serious morbidity. If the child is alert and the parents are absolutely convinced that the child ingested less than a leaf or flower, then the child may be observed at home. The administration of activated charcoal to alert children who ingest more than one leaf within 1 hour prior to presentation is a therapeutic option, but there are no clinical data to confirm the clinical efficacy of decontamination measures on outcome following the ingestion of common oleander leaves. These patients should be monitored for evidence of conduction abnormalities and dysrhythmias.

Antidote

Although potentially useful, the role of digoxin-specific Fab fragments in the treatment of serious oleander poisoning remains undefined as a result of the pharmacokinetic differences between cardenolides and the lack of good quality controlled clinical trials.[41] Several case reports document the use of Fab fragments following oleander toxicity, but the efficacy of the Fab fragments in these cases is unclear. Currently, the use of digoxin-specific fragments is a reasonable therapeutic option in oleander-intoxicated patients with serious cardiac toxicity (e.g., complete heart block, hypotension, ventricular tachyarrhythmias) who do not respond to supportive care. The use of these fragments in dogs administered fatal doses of cardiac glycosides from *Nerium oleander* successfully reversed the digoxin toxicity, but the dose administered was equivalent to the use of 105 vials in a 70-kg (~154-lb) person.[42] Consequently, large quantities of Fab fragments may be necessary to reverse the digoxin-like effects of oleander poisoning.

A Haitian woman died at a hospital after failing to respond to supportive care including the use of 20 vials of Digibind® (digoxin-specific Fab fragments; GlaxoSmithKline Pharmaceuticals, Mississauga, Ontario, Canada). A 37-year-old man developed asymptomatic sinoatrial nodal arrest, a junctional escape rhythm, and a serum potassium concentration of 4.3 mEq/L after ingesting a few oleander leaves.[38] Thirty minutes after the intravenous administration of 5 vials (200 mg) of Digibind®, the heart rate and conduction improved, but a mild sinus bradycardia persisted for 4 days. No atropine was given. A 24-year-old man presented to the emergency department 10 hours after ingesting a mixture of orange juice and six ground *Nerium oleander* leaves.[43] He had marked miosis, bilateral ptosis, diaphoresis, and glossal fasciculations. The apparent digoxin concentration was 0.8 ng/mL, as measured by digoxin radioimmunoassay. Hypotension, coma, and hyperkalemia (potassium 6.8 meq/L) developed despite the use of supportive care (atropine, pacemaker, insulin-glucose therapy, bicarbonate). The patient received 480 mg digoxin-specific Fab antibody fragments intravenously 18 hours after admission, and reversal of the coma, hyperkalemia, diplopia, and complete A-V block occurred within 1 hour of administration. The first-degree A-V block and the diffuse ST depression persisted for 6 days. Recovery from serious oleander poisoning has occurred without the use of digoxin-specific Fab fragments. Following the ingestion of oleander tea the previous night, an 83-year-old woman presented to the emergency department with a bradycardia, conduction delay (QRS = 138 milliseconds), blood pressure of 102/50 mmHg, and an apparent digoxin concentration of 7.1 ng/mL.[22] The administration of atropine converted her bradycardia to atrial fibrillation with a normal ventricular response. The hyponatremia (121 meq/L) and hyperkalemia (6.1 meq/L) resolved without the use of Digibind®.

Supplemental Care

Symptomatic patients should be evaluated and treated for fluid and electrolyte imbalance. Serial serum potassium and creatinine concentrations should be monitored until a downward trend develops. Hyperkalemia may not respond to sodium bicarbonate or insulin-glucose therapy because the inhibition of the Na^+/K^+ ATPase pump by cardiac glycosides in oleander prevents the movement of potassium into the cell.[21] Hemodialysis is an option if the hyperkalemia contributes to cardiac instability. However, the use of hemoperfusion and hemodialysis in a 17-year-old girl, who developed mild conduction abnormalities after ingesting five oleander leaves, did not alter the apparent serum digoxin concentration.[20] Cardiac monitoring may be discontinued after normal sinus rhythm returns.

YELLOW OLEANDER

HISTORY

Ingestion of yellow oleander seeds is a relatively common cause of accidental poisoning among children worldwide.[16] Currently, the ingestion of yellow oleander seeds is a common method of attempted suicide in northern Sri Lanka where the case fatality rate is about 10% for the thousands of cases reported every year there.[44] Thevetin was a cardiac glycoside tested in the 1930s for cardiotonic effects.[45] Although this extract of yellow oleander was efficacious in the treatment of heart failure and atrial fibrillation, the frequency of gastrointestinal side effects limited the use of this agent.

BOTANICAL DESCRIPTION

Common Name: Yellow Oleander, Be-Still Tree

Scientific Name: *Thevetia peruviana* (Pers.) K. Schum., *Thevetia thevetioides* (Kunth) K. Schum.

Botanical Family: Apocynaceae (dogbane)

Physical Description: This large ornamental evergreen shrub grows up to 20 ft (~6 m) in height and has narrow, oblong leaves with a prominent midrib similar to common oleander. Clusters of dull orange to bright yellow tubular flowers appear in the summer followed by the appearance of fleshy, triangular, globular black fruit. The fruit contains an outer flesh and a hard stone center.

Distribution and Ecology: The yellow oleander grows in Australia, South America, Hawaii, and southern Florida, whereas the common oleander (*Nerium oleander*) appears more frequently in the dry, temperate regions of the United States.

EXPOSURE

At one time, peruvoside, an extract of the seed, underwent clinical trials as a substitute for ouabain in India.[46] Past uses for extracts of yellow oleander include suicidal and homicidal agents, abortifacients and herbal remedies, as well as an experimental drug for heart failure and atrial fibrillation.[1]

PRINCIPAL TOXINS

Physiochemical Properties

The yellow oleander contains at least eight cardiac glycosides that are cardenolides with a chemical structure similar to digitoxin.[16] Cardenolides contain a five-membered, steroid nucleus with a lactone ring in contrast to the six-membered steroid nucleus present in bufadienolides (see Figure 147.1). These cardiac glycosides have a carbohydrate or sugar moiety attached to the three-carbon on the "A" ring of the steroid by an oxygen bridge. The cardiac glycosides in extracts from yellow oleander include thevetin A, thevetin B (cerberoside), thevetoxin, peruvoside, ruvoside, and neriifolin. Figure 147.2 displays the chemical structures of some *Thevetia* cardenolides, digoxin, digitoxin, and ouabain. The

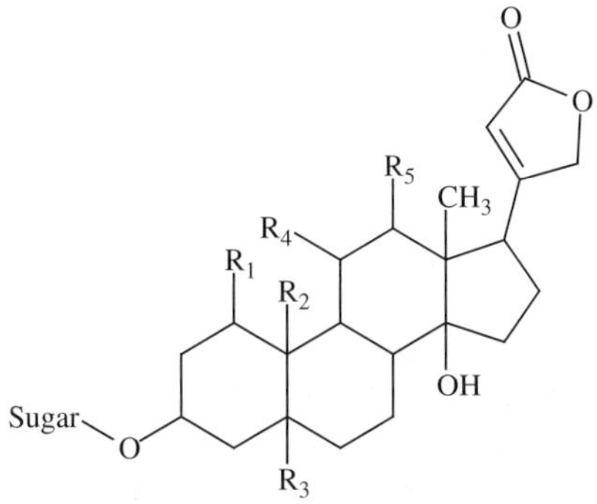

Cardenolide	Sugar	R_1	R_2	R_3	R_4	R_5	MW	Log P	Bio-availability (%)	PO Onset	$t_{1/2}$	V_D (L/kg)	Protein Binding (%)
Digoxin	Di-Di-Di	H	CH_3	H	H	OH	780.95	0.85	60–70	1–2 hrs	20–50 hrs	6	20
Digitoxin	Di-Di-Di	H	CH_3	H	H	H	764.95	2.44	>90	2 hrs	3–16 days	0.5	>90
Thevetin A	Th-Gl-Gl	H	CHO	H	H	H	872.96	–0.08	NK	2–7 hrs	NK	NK	NK
Thevetin B	Th-Gl-Gl	H	CH_3	H	H	H	858.95	1.39	NK	2–7 hrs	NK	NK	NK
Peruvoside	Th	H	CHO	H	H	H	548.67	2.25	50	1 hr	40–90 hrs	NK	40–50
Neriifolin	Th	H	CH_3	H	H	H	534.70	3.716	NK	NK	NK	NK	NK
Ouabain	Rh	OH	CH_2OH	OH	OH	H	584.66	–1.64	0–5	N/A	5 hrs	NK	0–5

Di = Digitoxose; Th = Thevetose; Gl = Glucose; Rh = Rhamnose; NK = Not Known; N/A = Not Applicable

FIGURE 147.2. Chemical structures and pharmacokinetic parameters of digitalis and thevetia cardenolides. Adapted from Roberts DM, et al. Pharmacokinetics of digoxin cross-reacting substances in patients with acute yellow oleander (*Thevetia peruviana*) poisoning, including the effect of activated charcoal. Ther Drug Monit 2006;28:785.

estimated potency of thevetin is about one-eighth the potency of ouabain on a weight for weight basis.[47]

Poisonous Parts

There are few data regarding the concentration of cardiac glycosides in yellow oleander. In a study of Australian children sent to medical care following exposure to oleander, *Thevetia* fruit and *Nerium* leaves were the most common plant parts associated with oleander exposure.[34] These children remained relatively asymptomatic and none of these children developed serious toxicity. Based on radioimmunoassay of digoxin-equivalents, Ansford and Morris demonstrated higher concentrations of digoxin-equivalents (i.e., cardiac glycosides) in *Thevetia* leaves than in mature fruit.[54] Although one leaf is considered potentially lethal, the ingestion of larger amounts is probably necessary to cause serious toxic effects. The kernel of the yellow oleander seed reportedly contains the highest concentration of cardiac glycosides in the seed.[57] The estimated percentage by weight of cardiac glycosides in different parts of *Thevetia peruviana* was as follows: sap, 0.036%; fruit, 0.045%; leaf, 0.07%; and kernel of seed, 4.8%.[47]

Mechanism of Toxicity

Similar to common oleander, the cardiac glycosides in yellow oleander inhibit the transmembrane Na^+/K^+ ATPase pump, and this action produces increased intracellular concentrations of Ca^{++} and Na^+. Subendocardial and perivascular hemorrhage along with focal myocardial edema are the common pathological findings during postmortem examination of fatal cases of yellow oleander poisoning.[48]

DOSE RESPONSE

Predicting outcome based on the number of seeds ingested is problematic because of the lack of reliable historical data. Based on radioimmunoassay of a broad cardiac glycoside-specific antibody, and assuming that *Thevetia* glycosides are concentrated 20 times more in heart muscle than in blood, Ansford and Morris estimated that the absorption of cardiac glycosides from two *Thevetia* leaves resulted in the death of a 12.5-kg (~28-lb) child.[54] However, there are inadequate human data to confirm this estimation. Based on historical estimates of the amount of yellow oleander seed ingested,

Saraswat et al estimated that the ingestion of one to two yellow oleander seeds caused mild gastrointestinal symptoms and the ingestion of three to four seeds produced moderate gastrointestinal and cardiac toxicity.[49] They estimated that the lethal dose of yellow oleander seeds was 8 to 10 seeds. However, the number of seeds reportedly ingested by 6 patients who died ranged from 1 to 10, and only 2 of 7 patients ingesting >10 seeds died.[44] In a 65-year-old patient with hypertension treated with metoprolol, the ingestion of two *Thevetia* seeds produced mild bradycardia, peak serum potassium concentration of 5.6 meq/L, non-specific ST-T wave changes, and atrioventricular conduction block, which resolved over 2 days.[50]

TOXICOKINETICS

There are limited data on the toxicokinetics of *Thevetia* cardenolides in intoxicated humans. Toxicokinetic data from a clinical study of 104 patients with *Thevetia* intoxication indicated that the absorption of *Thevetia* cardenolides is highly variable and prolonged with the time to maximum concentration exceeding 24 hours in some patients.[51] The apparent terminal half-life of *Thevetia* cardenolides as estimated by a digoxin immunoassay in patients receiving supportive care and no activated charcoal was also highly variable with a median time of 42.9 hours (range: 13.27–94.24 hours). Figure 147.2 displays some pharmacokinetic parameters for individual *Thevetia* and *Digitalis* cardenolides.

CLINICAL RESPONSE

Although the bitter taste of the leaves usually limits the quantity of oleander *leaves* consumed by children, the ingestion of yellow oleander *seeds* may produce serious toxicity in children. Like common oleander poisoning, yellow oleander intoxication closely resembles digitoxin poisoning as manifest by gastrointestinal and cardiac effects. Nausea, vomiting, abdominal pain, burning and dryness of the throat, and mucosal erythema may develop within several hours of the ingestion of yellow oleander. The most common symptoms in a series of 79 patients with yellow oleander poisoning were vomiting and abdominal pain.[44] Other symptoms (e.g., paresthesias, weakness, hypotension, hypertonia, excessive salivation) may occur depending on the severity of the poisoning.[46,52,53] Serious toxicity usually develops within 12 hours of ingestion, and fatalities usually occur within about 24 hours after ingestion.

Most abnormal electrocardiograms demonstrate bradycardia, and serious toxic effects usually result from cardiac complications, such as delayed conduction in sinoatrial and AV nodes.[54,55] Although serious ventricular dysthymias (ventricular tachycardia, ventricular fibrillation) occur in severe yellow oleander toxicity, ventricular ectopy is more common during digoxin poisoning than during oleander poisoning.[54] In a series of 168 patients admitted to a secondary care hospital in Sri Lanka with yellow oleander poisoning, about 15% of the patients developed serious conduction disturbances (e.g., nodal bradycardia, type II second degree AV block, third degree AV block).[56] Conduction block and sinus bradycardia may persist as long as 4–5 days after ingestion in seriously toxic patients.[46,57] These disturbances usually resolve without sequelae unless complete heart block or cardiogenic shock develops.[50]

DIAGNOSTIC TESTING

As with common oleander, digoxin radioimmunoassays in yellow oleander poisoning confirm the presence of cardiac glycosides, which cross-react with digoxin.[54] The apparent concentration of digoxin does not correlate to the usual toxic and therapeutic concentrations of digoxin. Although apparent digoxin concentrations exceeding 1.5 ng/mL suggests serious intoxication, variability in the cross-reactivity between these assays and the cardenolide compounds in yellow oleander limit the correlations to clinical outcome.[35] The extent of the rise in serum potassium is probably the most practical biochemical measure of the severity of yellow oleander poisoning. A $2^1/_2$-year-old child died of complete atrioventricular block and cardiogenic shock approximately 18 hours after the ingestion of a large quantity of yellow oleander seeds.[58] The apparent digoxin and digitoxin concentrations in a sample of blood drawn in the emergency department were 11 μg/L and 170 μg/L, respectively. The serum potassium concentration was 9.0 meq/L. The cardiac (conduction blocks, dysrhythmias) and electrolyte (hyperkalemia) abnormalities following yellow oleander poisoning are similar to the abnormalities that follow common oleander poisoning. Magnesium concentrations are usually within the normal range during yellow oleander poisoning.

TREATMENT

The management of yellow oleander poisoning is similar to the medical treatment for common oleander (*Nerium oleander*) ingestions. The efficacy of digoxin-specific Fab fragments for yellow oleander poisoning is variable, with improvement of the cardiac abnormalities in some, but not all patients with serious yellow oleander poisoning. The administration of 190 mg digoxin-specific Fab antibody fragments (5 ampules Digibind®) to a 7-year-old girl with sinus bradycardia and variable second- and third-degree AV block after the ingestion of yellow oleander seeds resulted in conversion of the rhythm to sinus tachycardia with first-degree AV block.[59] Her rhythm did not revert to normal sinus rhythm until 16 hours after ingestion despite the administration of an additional 190mg digoxin-specific Fab antibody fragments. The administration of 1,200mg digoxin-specific Fab fragments to 34 patients with serious yellow oleander toxicity resulted in the resolution of the dysrhythmias in 15 of the treated patients.[60] About 23% of the patients in the trial developed adverse reactions primarily involving urticarial rashes and a few cases of bronchospasm and angioedema. The administration of serial-activated charcoal is an alternative treatment of oleander intoxication,[61] when digoxin-specific Fab fragments are not available. Although the administration of multiple dose-activated charcoal (50g every 6 hours for 3 days) usually causes minimal adverse reactions, there are no unequivocal clinical data to confirm the efficacy of this treatment. Given the high tissue binding of digoxin-like compounds, more clinical data are necessary to prove the efficacy of serial-activated charcoal.[51]

ODOLLAM TREE

The odollam tree (*Cerbera odollam* Gaertn., Apocynaceae) is a native tree in coastal salt swamps and creeks in south India and along river banks in Vietnam, Cambodia, Myanmar, and Sri Lanka.[62] *Cerbera manghas* L. (sea mango) is a closely related species that is very similar in appearance to the odollam tree. The odollam tree grows to 6–15m (~20–49ft) in height and possesses dark green, lanceolate leaves that produce large white flowers with a jasmine-like scent. The fruit has a green fibrous shell surrounding an ovoid kernel. On exposure to air, the color of the white kernel changes to violet and then to brown-black. *C. odollam* and *C. manghas* are differentiated by the color of the corolla, the shape of the fruit, and the color of the ripe fruit. *C. odollam* possesses a yellow-eyed, white corolla with a spherical to ovoid-shaped fruit that is green when ripe. *C. manghas* (sea mango) has a red-eyed, white corolla with oblong to ellipsoid-shaped fruit that is red when ripe. Traditional uses for extracts from oil extracts of the seed from the odollam tree include lighting oil, cosmetic, insect repellant, insecticide, rubefacient, antipruritic agent, and ordeal poison. The seeds of odollam tree (yellow-eyed cerbera, pink-eyed cerbera) and *Cerbera manghas* (sea mango) from the Apocynaceae family contain cardenolide compounds, such as cerberin (monoacetylneriifolin), 17β-neriifolin, deacetyltanghinin, tanghinin, 2′-*O*-acetyl-cerleaside A, cerberoside, and odollin.[63,64] The use of *Cerbera odollam* seeds is a common means of suicidal poisoning in India; The ingestion of these seeds produces cardiotoxicity similar to oleander poisoning manifest by hyperkalemia, bradycardia, and AV nodal conduction block.[65,66] The leaves from this tree are substantially less toxic than the seeds. Treatment is supportive and similar to common oleander poisoning.

URGINEA SANGUINEA SCHINZ

This Southern African plant from the Liliaceae family is a notorious livestock poison and a well-known herbal medicine. Common names include Burke's slangkop, red slangkop, and Transvaal slangkop. Traditional uses for this herb include the treatment of venereal disease, hypertension, abdominal pain, backache, impotence, menstrual disorders, asthma, bronchitis, and blood poisoning.[67] Extracts from this plant are also used as an arrow poison and an abortifacient. Poisoning of grazing animals (e.g., sheep, cattle, goats) by *Urginea sanguinea* are characteristic of cardiac glycoside intoxication.[68] *Urginea* species are bulbous plants with a large flower stalk, which emerges before the leaves in spring. The large, reddish brown pear-shape bulb is about 15cm (~6in.) in diameter. This onion-like bulb is perennial with the leaves dying back in the winter. The red bulb extrudes a reddish material when cut, and hence the name sanguinea. *U. sanguinea* is a highly invasive pant that inhabits southern Africa, particularly in hilly areas. All parts of the plant are poisonous, particularly the flowers and bulb. The main toxin in *U. sanguinea* is the bufadienolide, scillaren A (transvaalin), which occurs in concentration of approximately 0.01–0.05%.[69] Scillaren A (CAS RN: 124-99-2) has a molecular formula of $C_{36}H_{52}O_{13}$. Figure 147.3 displays the chemical structure of scillaren A. Spectrographic analysis of fresh bulbs of *Urginea sanguinea* yielded stigmasterol, phloroglucinol, phloroglucinol 1-β-D-glucopyranoside (phlorin), scillaren A, 5-α-4,5-dihydroscillaren A, salicylic acid, and 3-hydroxy-4-methylbenzoic acid with scillaren A and 5-α-4,5-dihydroscillaren A being the major compounds.[70]

Scillaren A is Na^+-K^+ pump poison that alters membrane depolarization and elevates extracellular potassium concentrations. There are few clinical data on *U. sanguinea* poisoning in humans. Ingestion of herbal remedies containing *U. sanguinea* extracts was associated with gastrointestinal distress, dehydration, weakness, abdominal pain, hematuria, tachycardia, and renal failure, but the classic signs and symptoms (hyperkale-

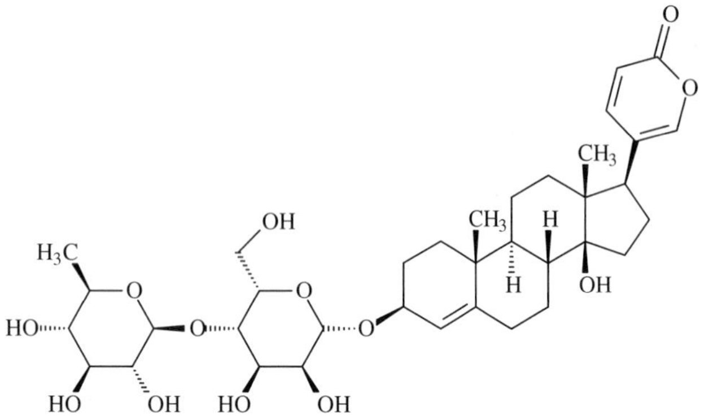

FIGURE 147.3. Chemical structure of scillaren A.

mia, bradycardia, conduction block) of digoxin poisoning were not documented.[67] Treatment is supportive and similar to the treatment of oleander poisoning if signs of digoxin poisoning are present. There are no clinical data on the affinity of digoxin-specific Fab fragments for scillaren A.

References

1. Langford SD, Boor PJ. Oleander toxicity: an examination of human and animal toxic exposures. Toxicol 1996;109: 1–13.
2. Kirtikar KR. The poisonous plants of Bombay. J Bombay Natural Hist Soc 1897;11:251–261.
3. Cassels BK. Analysis of a Maasai arrow poison. J Ethnopharmacol 1985;14:273–281.
4. Omino EA, Kokwaro JO. Ethnobotany of *Apocynaceae* species in Kenya. J Ethnopharmacol 1993;40:167–180.
5. Kaojarern S, Sukhuunyarak S, Mokkhavesa C. Oleander Yee Tho poisoning. J Med Assoc Thai 1986;69:108–112.
6. Buchner S, Jackowska H, Kikuchi K, Chen KK. A new glycoside of *Adonis vernalis*. Life Sci 1965;4:37–40.
7. Waud RA. The action of *Ornithogalum umbellatum* on the heart. J Pharmacol Exp Ther 1954;111:147–151.
8. Volelsang A. Clinical trial of *Ornithogalum umbellatum* on the human heart (preliminary report). Can Med Assoc J 1955;73:295–296.
9. Volelsang A. *Ornithogalum umbellatum* in the treatment of congestive heart failure: progress report. J Am Geriatr Soc 1961;9:1096–1099.
10. Radford DJ, Gilles AD, Hinds JA, Duffy P. Naturally occurring cardiac glycosides. Med J Aust 1986;144: 540–544.
11. Gaillard Y, Pepin G. Poisoning by plant material: review of human cases and analytical determination of main toxins by high-performance liquid chromatography-(tandem) mass spectrometry. J Chromatogr B Biomed Sci Appl 1999;733:181–229.
12. Yamauchi T, Abe F, Tachibana Y, Atal CK, Sharma BM, Imre Z. Quantitative variations in the cardiac glycosides of oleander. Phytochemistry 1983;22:2211–2214.
13. Blum LM, Rieders F. Oleandrin distribution in a fatality from rectal and oral Nerium oleander extract administration. J Anal Toxicol 1987;11:219–221.
14. Karawya MS, Balbaa SI, Khayyal SE. Estimation of cardenolides in *Nerium oleander*. Planta Med 1973;23:70–73.
15. Kingsbury JM. Poisonous plants of the United States and Canada. Englewood Cliffs, NJ: Prentice Hall Inc;1964.
16. Shaw D, Pearn J. Oleander poisoning. Med J Aust 1979;2: 267–269.
17. Szabuniewicz M, McCrady JD, Camp BJ. Treatment of experimentally induced oleander poisoning. Arch Int Pharmacodyn 1971;189:12–21.
18. Schwartz WL, Bay WW, Dollahite JW, Storts RW, Russell LH. Toxicity of *Nerium oleander* in the monkey. Vet Pathol 1974;11:259–277.
19. Osterloh J, Herold S, Pond S. Oleander interference in the digoxin radioimmunoassay in a fatal ingestion. JAMA 1982;247:1596–1597.
20. Durakovic Z, Durakovic A, Durakovic S. Oleander poisoning treated by resin haemoperfusion. J Indian Med Assoc 1996;94:149–150.
21. Haynes BE, Bessen HA, Wightman WD. Oleander tea: herbal draught of death. Ann Emerg Med 1985;14: 350–353.
22. Driggers DA, Solbrig R, Steiner JF, Swedberg J, Jewell GS. Acute oleander poisoning. A suicide attempt in a geriatric patient. West J Med 1989;151:660–662.
23. Spevak L, Soc M. [2 cases of poisoning by tea from Oleander leaves]. Arh Hig Rada Toksikol 1975;26:147–150. [Croatian]
24. Dorsey CS. Plant dermatitis in California. California Med 1962;96:412–413.
25. Ts'ang K. Heart block in oleander poisoning. Chin Med J 1953;71:287–292.
26. Chin D, Wei-Lang C. Auricular tachycardia with auriculoventricular block in oleander leaf poisoning. Chin Med J 1957;75:74–77.
27. de A Nishioka S, Resende ES. Transitory complete atrioventricular block associated to ingestion of *Nerium oleander*. Rev Ass Med Brasil 1995;41:60–62.
28. Pohl RW. Contact dermatitis from the juice of *Ornithogalum caudatum*. Toxicon 1965;3:167–168.
29. Chen KK, Henderson FG. Pharmacology of sixty-four cardiac glycosides and aglycones. J Pharmacol Exp Ther 1954;111:365–383.
30. Rule G, McLaughlin LG, Henion J. A quest for oleandrin in decayed human tissue. Anal Chem 1993;65:857–863.
31. Tracqui A, Kintz P, Branche F, Ludes B. Confirmation of oleander poisoning by HPLC/MS. Int J Legal Med 1998;111:32–34.
32. Flanagan RJ, Jones AL. Fab antibody fragments some applications in clinical toxicology. Drug Safety 2004;27: 1115–1133.

33. Cheung K, Hinds JA, Duffy P. Detection of poisoning by plant-origin cardiac glycoside with the Abbott TDx Analyzer. Clin Chem 1989;35:295–297.
34. Nickel SL, Staba EJ. Suitability of antidigoxin antiserum for digoxin in plant extracts. Lloydia 1977;40:230–235.
35. Dasgupta A, Datta P. Rapid detection of oleander poisoning using digoxin immunoassays. Comparison of five assays. Ther Drug Monit 2004;26:658–663.
36. Pietsch J, Oertel R, Trautmann S, Schulz K, Kopp B, Dressler J. A non-fatal oleander poisoning. Int J Legal Med 2005;1119:236–240.
37. Arao T, Fuke C, Takaesu H, Nakamoto M, Morinaga Y, Miyazaki T. Simultaneous determination of cardenolides by sonic spray ionization liquid chromatography-ion trap mass spectrometry—a fatal case of oleander poisoning. J Anal Toxicol 2002;26:222–227.
38. Shumaik GM, Wu AW, Ping AC. Oleander poisoning: treatment with digoxin-specific Fab antibody fragments. Ann Emerg Med 1988;17:732–735.
39. Romano GA, Momelli G. Intoxikation mit oleanderblattern. Schweiz Med Wschr 1990;120:596–597.
40. Monzani V, Rovellini A, Schinco G, Rampoldi E. Acute oleander poisoning after a self-prepared tisane (Letter). Clin Toxicol 1997;35:667–668.
41. Roberts DM, Buckley NA. Antidotes for acute cardenolide (cardiac glycoside) poisoning (review). Cochrane Database Sys Rev 2006;(4):CD005490.
42. Clark RF, Selden BS, Curry SC. Digoxin-specific Fab fragments in the treatment of oleander toxicity in a canine model. Ann Emerg Med 1991;20:1073–1077.
43. Safadi R, Levy I, Amitai Y, Caraco Y. Beneficial effect of digoxin-specific Fab antibody fragments in oleander intoxication. Arch Intern Med 1995;155:2121–2125.
44. Eddleston M, Ariaratnam CA, Meyer WP, Kularatne AM, Attapattu S, Sheriff MHR, Warrell DA. Epidemic of self-poisoning with seeds of the yellow oleander tree (*Thevetia peruviana*) in northern Sri Lanka. Trop Med Int Health 1999;4:266–273.
45. Middleton WS, Chen KK. Clinical results from oral administration of thevetin, a cardiac glycoside. Am Heart J 1936;11:75–88.
46. Bhattacharya SK, Somani PN, Srivastava PK. Cardiac changes in *Thevetia nerifolia* poisoning. Acta Cardiol 1976;31:169–174.
47. Saravanapavananthan N, Ganeshamoorthy J. Yellow oleander poisoning—a study of 170 cases. Forensic Sci Int 1988;36:247–250.
48. Bose TK, Basu RK, Biswas B, De JN, Majumdar BC, Datta S. Cardiovascular effects of yellow oleander ingestion. J Indian Med Assoc 1999;97:407–410.
49. Saraswat DK,Garg PK, Saraswat M. Rare poisoning with *Cerebra thevetia* (yellow oleander) review of 13 cases of suicidal attempt. J Assoc Physicians India 1992;40: 628–629.
50. Maringhini G, Notaro L, BArberi O, Giubilato A, Butera R, Di Pasquale P. Cardiovascular glycoside-like intoxication following ingestion of *Thevetia nereifolia/peruviana* seeds: a case report. Ital Heart J 2002;8:137–140.
51. Roberts DM, Southcott E, Potter JM, Roberts MS, Eddleston M, Buckley NA. Pharmacokinetics of digoxin cross-reacting substances in patients with acute yellow oleander (*Thevetia peruviana*) poisoning, including the effect of activated charcoal. Ther Drug Monit 2006;28: 784–792.
52. Sarma AVS. Oleander poisoning. Antiseptic 1953;48: 55–57.
53. Kakrani AL, Rajput CS, Khandare SK, Redkar VE. Yellow oleander seed poisoning with cardiotoxicity. A case report. Indian Heart J 1981;33:31–33.
54. Eddleston M, Ariaratnam CA, Sjostrom L, Joyalath S, Rajakanthan K, Rajapakse S, et al. Acute yellow oleander (*Thevetia peruviana*) poisoning: cardiac arrhythmias, electrolyte disturbances, an serum cardiac glycoside concentrations on presentation to hospital. Heart 2000;83: 301–306.
55. Ansford AJ, Morris H. Fatal oleander poisoning. Med J Aust 1981;1:360–361.
56. Fonseka MMD, Seneviratne SL, de Silva CE, Gunatilake SB, de Silva HJ. Yellow oleander poisoning in Sri Lanka: outcome in a secondary care hospital. Hum Exp Toxicol 2002;21:293–295.
57. Ahlawat SK, Agarwal AK, Wadhwa S. Rare poisoning with *Cerebra thevetia* (yellow oleander): a report of three cases. Trop Doc 1994;24:37–38.
58. Brewster D. Herbal poisoning: a case report of a fatal yellow oleander poisoning from the Solomon Islands. Ann Trop Paediat 1986;6:289–291.
59. Camphausen C, Haas NA, Mattke AC. Successful treatment of oleander intoxication (cardiac glycoside) with digoxin-specific Fab antibody fragments in a 7-year-old child. Case report and review of literature. Z Kardiol 2005;94:817–823.
60. Eddleston M, Rajapakse S, Rajakanthan K, Jayalath S, Shostrom L, Santharaj W, et al. Anti-digoxin Fab fragments in cardiotoxicity induced by ingestion of yellow oleander: a randomised controlled trial. Lancet 2000;355: 967–972.
61. de Silva HA, Fonseka MM, Pathmeswaran A, Alahakone DG, Ratnatilake GA, Gunatilake SB, et al. Multiple-dose activated charcoal for treatment of yellow oleander poisoning: a single-blind, randomized, placebo-controlled trial. Lancet 2003;361:1935–1938.
62. Gaillard Y, Krishnamoorthy A, Bevalot F. *Cerbera odollam*: a "suicide tree" and cause of death in the state of Kerala, India. J Ethnopharmacol 2004;95:123–126.
63. Cheenpracha S, Karalai C, Rat-a-pa Y, Ponglimanont C, Chantrapromma K. New cytotoxic cardenolide glycoside from the seeds of *Cerbera manghas*. Chem Pharm Bull 2004;52:1023–1025.
64. Narendranathan M, Krishna KV, Vijayaraghavan G. Electrocardiographic changes in *Cerbera odollum* poisoning. J Assoc Phys India 1975;23:757–762.

65. Narendranathan M, Krishna Das KV, Vijayaraghavan G. Prognostic factors in *Cerbera odollum* poisoning. Indian Heart J 1975;27:283–286.
66. Guruswami MN, Ganapathy MN, Thampai CK. A preliminary study of the pharmacological actions and toxicity of *Cerbera odollam*. Indian J Med Sci 1970;24:82–87.
67. Foukaridis GN, Osuch E, Mathibe L, Tsipa P. The ethnopharmacology and toxicology of *Urginea sanguinea* in the Pretoria area. J Ethnopharmacol 1995;49:77–79.
68. Joubert JP, Schultz RA. The treatment of *Urginea sanguinea* Schinz poisoning in sheep with activated charcoal and potassium chloride. J S Afr Vet Assoc 1982;53:25–28.
69. Marx J, Pretorius E, Espag WJ, Bester MJ. *Urginea sanguinea*: medicinal wonder or death in disguise? Environ Toxicol Pharmacol 2005;20:26–34.
70. Majinda RR, Waigh RD, Waterman PG. Bufadienolides and other constituents of *Urginea sanguinea*. Planta Med 1997;63:188–190.

Chapter 148

HOLLY
(*Ilex* Species)

BOTANICAL DESCRIPTION

Common Name: English Holly (*Ilex aquifolium* L.), American Holly (*Ilex opaca* Ait.)

Scientific Name: *Ilex* species

Botanical Family: Aquifoliaceae (holly)

Physical Description: These plants grow as deciduous and evergreen shrubs and trees. They have bright green leaves, often with hard teeth on the leaf margins, which discourage leaf consumption.

Distribution and Ecology: The genus *Ilex* contains 300 to 350 species found throughout regions with temperate and tropical climates.

EXPOSURE

These plants typically are grown as ornamental plants. The red or black berries (Christmas holly) are attractive decorations, especially to curious children.

PRINCIPAL TOXINS

Polyphenols, saponins, triterpenes, steroids, and alkaloids have been extracted from *I. opaca*.[1] In animal experiments, ethanol leaf extracts of holly were lethal when injected into mice, whereas *in vitro* studies indicated that these extracts were hemolytic to human erythrocytes. Fruit extracts did not produce toxic reactions in mice. Although saponins are present in both fruit and leaf extracts, the limited oral bioavailability of saponins makes systemic toxicity by the oral route unlikely.

DOSE RESPONSE

There are inadequate data to determine the dose response between the ingestion of holly berries or active constituents and clinical effects.

TOXICOKINETICS

The gastrointestinal toxin in holly berries has not been identified.

CLINICAL RESPONSE

Ilex species produce gastrointestinal distress in some, but not all ingestions. In fact, early settlers of the Southeast enjoyed a very mild brew from *Ilex vomitoria* Ait. leaves called "yaupon tea."[2] Case studies of toxic effects from the ingestion of *Ilex* species in humans are rare. A set of identical twins ingested a "handful of holly berries" after which 15 mL of ipecac was given. Over the next 6 hours, the twins vomited 40 and 50 times. The following day, several loose stools were passed and mild vomiting continued, but symptoms resolved completely by 30 hours after ingestion.[3] Although one twin was lethargic, his neurological symptoms could not be differentiated from a side effect of ipecac administration.

DIAGNOSTIC TESTING

There are few data on the detection of toxic constituents of *Ilex* species.

Medical Toxicology of Natural Substances, by Donald G. Barceloux, MD

TREATMENT

Treatment is supportive. Decontamination measures are unnecessary. Casual ingestions (e.g., a few berries) require no decontamination measures.

References

1. West LG, McLaughlin JL, Eisenbeiss GK. Saponins and triterpenes from *Ilex opaca*. Phytochemistry 1977;16:1846–1847.
2. Arena JA. Are holly berries toxic? JAMA 1979;242:2341.
3. Rodrigues TD, Johnson PN, Jeffrey LP. Holly berry ingestion: Case report. Vet Hum Toxicol 1984;26:157–158.

Chapter 149

HONEYSUCKLE
(*Lonicera* Species)

BOTANICAL DESCRIPTION

There are over 150 species of the genus *Lonicera* distributed over the northern hemisphere with approximately 35 native and introduced species in the United States.[1] *Lonicera sempervirens* L. is a native species common throughout the southeastern United States, whereas the native species, *L. involucrata* Banks ex Spreng. inhabits moist areas of Canada and the western United States.

Common Name: Chinese Honeysuckle, Japanese Honeysuckle

Scientific Name: *Lonicera japonica* Thunb.

Botanical Family: Caprifoliaceae (honeysuckle)

Physical Description: These shrubs or vines have showy, pleasantly scented flowers that form red to blue-black berries. The flowers are red to yellow or white, and the leaves are simple.

Distribution and Ecology: Japanese honeysuckle is a native species of eastern Asia. This plant is an introduced species in the United States that has naturalized to the roadsides, abandoned fields, and margins of mature woodlands in the eastern, southern, and western United States.

EXPOSURE

Jinyinhua is the Chinese herbal medicine containing extracts of the honeysuckle plant. Medicinal uses for this herb include the treatment of fever, pneumonia, diuresis, dysentery, furuncles, and other skin infections.

PRINCIPAL TOXINS

Extracts of the buds and flowers from *Lonicera japonicum* contains saponin (lupine-triterpene glycosides, iridoid glycosides, loniceroside C), flavones, and phenolic compounds.[2–4] The major compounds detected in a traditional Chinese medicinal herb (Caulis Lonicerae Japonicae) derived from the caulis of *Lonicera japonica* were chlorogenic acid, caffeic acid, loganin, sweroside, secoxyloganin, rutin, and luteolin 7-*O*-glucoside.[5] Some phenolic compounds (methyl caffeate, 3,4-di-*O*-caffeolquinic acid, methyl 3,4-di-*O*-caffeolyquinate) inhibit ADP-induced platelet aggregation in *in vitro* studies.[6]

DOSE RESPONSE

There are inadequate data to determine the dose response associated with exposure to this plant or its extracts.

TOXICOKINETICS

A case report of a 61-year-old man receiving warfarin associated the development of easy bruising, epistaxis, bleeding gums, and elevated international normalized ratio (INR) with the ingestion of an herbal product containing extracts of *Lonicera japonica*.[7] However, the herbal product contained many herbs and the contribution of jinyinhua to the apparent drug interaction between warfarin and the herbal product remains unclear.

CLINICAL RESPONSE

There are few case reports of toxicity following exposure to the honeysuckle despite the reporting of over 1,000 annual exposures to US national data collection system for poison control centers.[8] Several case reports document the development of a pruritic, maculopapular rash following exposure to the sap of *Lonicera japonica*.[9] Although casual contact with *Lonicera* berries frequently does not produce adverse effects, ingestion of the berries can cause mydriasis, myalgias, and mild gastrointestinal irritation manifest by nausea, vomiting, and diarrhea.[10]

DIAGNOSTIC TESTING

There are few data on the detection of toxic constituents in honeysuckle or its extracts. Methods for detecting bioactive constituents in *Lonicera* species include high performance liquid chromatography (HPLC).[5]

TREATMENT

Treatment is supportive. The contact dermatitis associated with honeysuckle usually responds to topical steroids (0.05% fluocinonide) and antihistamines as necessary. Decontamination measures are unnecessary.

References

1. Burrows GE, Ryrl RJ. Toxic plants of North America. Ames, IA: Iowa State University Press; 2001.
2. Kwak WJ, Han CK, Chang HW, Kim HP, Kang SS, Son KH. Loniceroside C, an antiinflammatory saponin from *Lonicera japonica*. Chem Pharm Bull 2003;51:333–335.
3. Machida K, Sasaki H, Iijima T, Kikuchi M. Studies on the constituents to *Lonicera* species. XVII. New Iridoid glycosides of the stems and leaves of *Lonicera japonica* Thunb. Chem Pharm Bull 2002;50:1041–1044.
4. Peng L-Y, Mei S-X, Jiang B, Zhou H, Sun H-D. Constituents from *Lonicera japonica*. Fitoterapia 2000;71:713–715.
5. Qian ZM, Li HJ, Li P, Chen J, Tang D. Simultaneous quantification of seven bioactive components in Caulis Lonicerae Japonicae by high performance liquid chromatography. Biomed Chromatogr 2007;21:649–654.
6. Chang W-C, Hsu F-L. Inhibition of platelet activation and endothelial cell injury by polyphenolic compounds isolated from *Lonicera japonica* Thunb. Prostaglandins Leukot Essent Fatty Acids 1992;45:307–312.
7. Wong AL, Chan TY. Interaction between warfarin and the herbal product quilinggao. Ann Pharmacother 2003;37: 836–838.
8. Krenzelok EP, Jacobsen TD, Aronis JM. Plant exposure: a state profile of the most common species. Vet Hum Toxicol 1996;38:289–298.
9. Webster RM. Honeysuckle contact dermatitis. Cutis 1993;51:424.
10. Lamminpaa A, Kinos M. Plant poisonings in children. Hum Exp Toxicol 1996;15:245–249.

Chapter 150

IVY and FALCARINOL

English or common ivy (*Hedera helix* L.) and Algerian ivy (*H. canariensis* Willd., canary ivy) are popular ground covers and indoor plants that have been recognized causes of contact dermatitis since approximately 1900.[1] *H. helix* contains sterols, saponosides, bidesmosides (bayogenin, hederagenin, oleanolic acid), and flavonoids, as well as emetine alkaloids.[2] The major saponins in English ivy include kalopanax saponin B (hederacoside C), hederagenin, and α-hederin. Other saponins include helixoside A and B.[3] Saponins from the leaves of *Hedera helix* are used by herbalists for anti-infective and antitussive properties. The role of these saponins in causing human toxicity is unsubstantiated. *H. helix* contains falcarinol (heptadeca-1,9-diene-4,6-diyne-3-ol, CAS RN: 21852-80-2), didehydrofalcarinol (heptadeca-1,9,11,16-tetraene-4,6-diyne-3-ol, CAS RN: 110927-49-6), and the ketone falcarinone. Extraction of powdered, dry leaves from a specimen of *H. helix* yielded 21.83 mg hederacoside C/g, 0.41 mg α-hederin/g, and 0.02 mg hederagenin/g as measured by liquid chromatography coupled to electrospray ionization/tandem mass spectrometry.[4] Falcarinol and didehydrofalcarinol are direct irritants and moderate sensitizers, particularly falcarinol. In adult volunteers, the application of a nonirritating 5% solution of falcarinol produced sensitization in 10 of the 20 subjects.[5] Falcarinol also occurs in members of the carrot family (Apiaceae) including carrots, celery, and parsnips, as well as ginseng (*Panax ginseng* C.A.Mey.) and *Schefflera arboricola* (Hayata) Merr. Some shampoos and therapeutic creams also contain this sensitizer.[1] *Hedera* dermatitis is a contact dermatitis characterized by linear and vesicular eczema on the forearms and hands, usually in the spring time following the pruning of actively growing ivy.[6,7] The stems and roots also contain high concentrations of the main allergen, falcarinol. Contact dermatitis following exposure to *Hedera helix* is uncommon in children.[8]

Older European medical literature suggests that children develop significant toxicity including vomiting, diarrhea, respiratory depression, and coma after ingestion of berries or leaves from *H. helix*. However, most exposures reported to poison control centers are not associated with clinically significant toxicity. The most common symptoms reported to the American Association of Poison Control Centers Toxic Exposure Surveillance System after undocumented exposure to *Hedera helix* included vomiting, cough, and oral irritation.[9] There were no reports of serious systemic complaints. Children exposed to *Hedera* species usually require only observation, and decontamination measures are unnecessary.

Reference

1. Hausen BM, Brohan J, Konig WA, Faasch H, Hahn H, Bruhn G. Allergic and irritant contact dermatitis from falcarinol and didehydrofalcarinol in common ivy (*Hedera helix* L.). Contact Dermatitis 1987;17:1–9.
2. Mahran HG, Hilal SH, El-Alfy TS. The isolation and characterization of emetine alkaloid from *Hedera helix*. Planta Med 1975;27:127–132.
3. Bedir E, Kirmizipekmez H, Sticher O, Calis I. Triterpene saponins from the fruits of *Hedera helix*. Phytochemistry 2000;53:905–909.

Medical Toxicology of Natural Substances, by Donald G. Barceloux, MD

4. Gaillard Y, Blaise P, Darre A, Barbier T, Pepin G. An unusual case of death: suffocation caused by leaves of common ivy (*Hedera helix*). Detection of hederacoside C, α-hederin, and hederagenin by LC-EI/MS-MS. J Anal Toxicol 2003;27:257–262.
5. Gafner F, Epstein W, Reynolds G, Rodriguez E. Human maximization test of falcarinol, the principal contact allergen of English ivy and Algerian ivy (*Hedera helix, H. canariensis*). Contact Dermatitis 1988;19:125–128.
6. Garcia M, Fernandez E, Navarro JA, del Pozo MD, Fernandez de Corres L. Allergic contact dermatitis from *Hedera helix* L. Contact Dermatitis 1995;33:133–134.
7. Yesudian PD, Franks A. Contact dermatitis from *Hedera helix* in a husband and wife. Contact Dermatitis 2002;46: 125–126.
8. Massmanian A, Valcuende Cavero F, Ramirez Bosca A, Castells Rodellas A. Contact dermatitis from variegated ivy (*Hedera helix* subsp. *canariensis* Willd.). Contact Dermatitis 1988;18:247–148.
9. Lai MW, Klein-Schwartz W, Rodgers GC, Abrams JY, Haber DA, Bronstein AC, Wruk KM. 2005 Annual Report of the American Association of Poison Control Centers' national poisoning and exposure database. Clin Toxicol (Phila) 2006;44:803–932.

Chapter 151

LANTANA
(*Lantana camara* L.)

BOTANICAL DESCRIPTION

Varieties of *Lantana* species (e.g., *L. canescens* Kunth, *L. involucrata* L., *L. montevidensis* (Spreng.) Briq., *L. trifolia* L.) inhabit areas throughout the world, but *L. camara* is probably the most widespread and toxic to grazing animals (cattle, sheep, goats, buffalo).[1]

Common Name: Lantana, Red or Yellow Sage, Wild Saga, Bunchberry, West Indian Lantana

Scientific Name: *Lantana camara* L. (*Lantana aculeata* L.)

Botanical Family: Verbenaceae (vervain)

Physical Description: This sprawling, perennial shrub has simple, serrated leaves and heads of multicolored red, white, yellow, pink, and violet flowers. Flowering starts in April/May, and the fruiting continues until November with the development of green-blue to dark blue berries.[2]

Distribution and Ecology: Lantana is a hardy shrub in subtropical, tropical, and temperate climates at elevations up to about 1,500 ft (~450 m) elevation. This plant occurs in parts of California, southern United States, tropical Central and South America, West Indies, Australia, New Zealand, South Africa, East Indies, and India. In many warm climates, this plant became naturalized weed, particularly near fences and water.

EXPOSURE

Hepatotoxicity and photosensitization following the ingestion of lantana by grazing animals is common, but human poisoning is rare.[3] Medicinal uses of lantana include the treatment of gastrointestinal diseases (antispasmodic, carminative), fever, malaria, and infection.[4,5]

PRINCIPAL TOXINS

Physiochemical Properties

Lantana contains a number of pentacyclic triterpenoid compounds that probably produce hepatotoxicity in animal studies, particularly lantadene A and lantadene C (dihydrolantadene A).[6,7] The major triterpenoid compounds in *L. camara* are lantadene A (22-β-angeloxy-3-oxoolean-12-en-28-oic acid), lantadene B (22-β-dimethylacryloyloxy-3-oxoolean-12-en-28-oic acid), lantadene C (dihydrolantadene A) lantadene D, reduced lantadene A and B, and 22-β-hydroxyoleanonic acid.[8] Lantadene A (rehmannic acid, CAS RN:467-81-2) is the most abundant triterpenoid compound.[9] Figure 151.1 displays the chemical structure of lantadene A. Lantadene B is structurally similar to lantadene A, but lantadene B is not apparently pharmacologically active.[10]

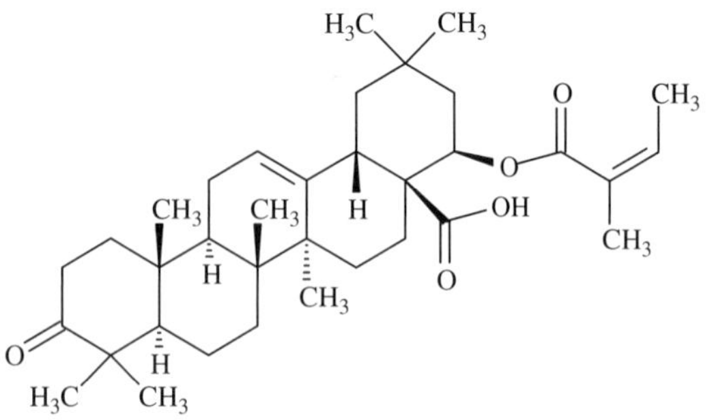

FIGURE 151.1. Chemical structure of lantadene A.

Poisonous Parts

Lantadene A content varies between plant parts and season with the lantadene content increasing with maturity of the leaves.[11] Total lantadene content varied from 0.2% in spring to 1.4%–1.7% in late summer and early fall. In a study of *L. camara* in July and August, the flowers, fruits, and shoots contained no detectable amounts of lantadene A, whereas the plant leaves contained 0.6% by dry weight lantadene A.[12]

Mechanism of Toxicity

The exact mechanism of toxicity is unknown. Lantana poisoning in animals causes widespread changes in mitochondrial, microsomal, and membrane enzymes resulting in cholestasis, hepatotoxicity, and photosensitization.[13]

DOSE RESPONSE

Lantana consumption by grazing animals causes widespread livestock loss in South Africa, Australia, India, Mexico, and the United States. In children, the ingestion of green, but not ripe, berries has been associated with serious toxicity including a fatality.[14]

TOXICOKINETICS

There are few data on the toxicokinetics of lantadene compounds in humans.

CLINICAL RESPONSE

Animals

In animals grazing on lantana, an acute cholestatic jaundice develops characterized by weakness, vomiting, anorexia, jaundice, conjunctivitis, blindness, mucosal ulcerations, and photosensitization. Death occurs within 7–10 days. The chronic form is associated with dehydration and renal failure.[15,16]

Humans

There are few data on the toxicity of lantana in humans despite the abundance of this plant. In a series of 17 children presenting to a Florida poison control center after ingesting an unknown quantity of lantana berries, four children became symptomatic with nausea, vomiting, lethargy, depressed respiration, decreased deep tendon reflexes, and dilated pupils.[14] Symptoms begin within 2–3 hours after ingesting *green* berries, and one patient died. The postmortem examination demonstrated marked congestion of the lungs and green *Lantana* berries in the small intestine. In this series, 13 of 17 patients ingesting *Lantana* berry ingestions remained asymptomatic.

DIAGNOTIC TESTING

Methods to quantitate lantadene A and related compounds include high performance liquid chromatography (HPLC).[8]

TREATMENT

The management of patients with exposure to *Lantana* species is supportive. Decontamination measures are usually unnecessary. The administration of activated charcoal to alert children, who ingest more than several *Lantana* berries less than 1 hour prior to presentation, is a therapeutic option, but there are no clinical data to confirm the clinical efficacy of these measures on outcome of *Lantana* ingestions. Children should be observed closely for 4–6 hours after ingestion, and symptomatic children should be evaluated by a physician. Most symptoms resolve within 6–8 hours.

References

1. Seawright AA. Toxicity of *Lantana* spp. in Queensland. Aust Vet J 1965;41:235–238.
2. Sharma OP, Makkar HPS, Dawra RK. A review of the noxious plant *Lantana camara*. Toxicon 1988;26:975–987.
3. Sharma OP, Sharma S, Pattabhi V, Mahato SB, Sharma PD. A review of the hepatotoxic plant *Lantana camara*. Crit Rev Toxicol 2007;37:313–352.
4. Hernandez T, Canales M, Avila JG, Duran A, Caballero J, Romo de Vivar A, Lira R. Ethnobotany and antibacterial activity of some plants used in traditional medicine of Zapotitlan de las Salinas, Puebla (Mexico). J Ethnopharmacol 2003;88:181–188.
5. Tabuti JR. Herbal medicines used in the treatment of malaria in Budiope county, Uganda. J Ethnopharmacol 2008;116:33-42.

6. Sharma OP, Sharma S, Pattabhi V, Mahato SB, Sharma PD. A review of the hepatotoxic plant *Lantana camara*. Crit Rev Toxicol 2007;37:313–352.
7. Sharma OP, Vaid J, Pattabhi V, Bhutani KK. Biological action of lantadene C, a new hepatotoxicant from *Lantana camara* var. *aculeate*. J Biochem Toxicol 1992;7:73–79.
8. Sharma S, Sharma OP, Singh B, Bhat TK. Biotransformation of lantadenes, the pentacyclic triterpenoid hepatotoxins of lantana plant, in guinea pig. Toxicon 2000;38: 1191–1202.
9. Barton DH, DeMayo P, Orr J. Triterpenoids. Part XXIII. The nature of lantadene A. J Chem Soc 1956;(pt 4): 4160–4162.
10. Sharma OP, Dawra RK, Makkar HP. Isolation and partial purification of lantana (*Lantana camara* L.) toxins. Toxicol Lett 1987;37:165–172.
11. Sharma OP, Singh A, Sharma S. Levels of lantadenes, bioactive pentacyclic triterpenoids, in young and mature leaves of *Lantana camara* var. *aculeate*. Fitoterapia 2000;71:487–491.
12. Sharma OP, Makkar HP, Pal RN, Negi SS. Lantadene A content and toxicity of the lantana plant (*Lantana camara* Linn.) to guinea pigs. Toxicon 1980;18:485–488.
13. Sharma OP. Review of the biochemical effects of *Lantana camara* toxicity. Vet Hum Toxicol 1984;26:488–493.
14. Wolfson SL, Solomons TW. Poisoning by fruit of *Lantana camara*. Am J Dis Child 1964;107:109–112.
15. McSweeney CS, Pass MA. Effect of lantana on the composition of the extracellular fluid of sheep. Vet Hum Toxicol 1983;25:330–334.
16. Sharma OP. A review of the toxicity of *Lantana camara* (Linn) in animals. Clin Toxicol 1981;18:1077–1094.

Chapter 152

RHODODENDRONS and GRAYANOTOXINS

HISTORY

The toxicity of honey collected from bees frequenting rhododendron leaves has been recognized since ancient times. The ancient Greek soldier, Xenophon, reported in the *Anabasis* the poisoning of his troops by the honey of *Rhododendron ponticum* flowers. While disabled after honey ingestion, Pompey's army was massacred by King Pontius's army in northern Asia Minor.[1] The disease was characterized by nausea, vomiting, lethargy, weakness, and paresthesias that lasted several days.

BOTANICAL DESCRIPTION

Azaleas and rhododendrons belong to the same genus, *Rhododendron*, which contains over 800 species. Toxic *Rhododendron* species in grazing animals include the following plants: *Kalmia angustifolia* L. (lambkill, sheepkill, dwarf laurel), *Kalmia latifolia* L. (sheepkill laurel, mountain laurel, ivy bush, mountain ivy),[2,3] *Kalmia microphylla* (Hook.) Heller (pale laurel, alpine bog laurel), *Ledum glandulosum* Nutt. (Western labrador tea), *Ledum columbianum* Piper (Pacific or coast Labrador tea), *Leucothoe davisiae* Torr. ex Gray (Sierra laurel, black laurel), *Menziesia ferruginea* Sm. (mock azalea, rustyleaf), *Pieris japonica* (Thunb.) D. Don ex G. Don (Japanese pieris), *Rhododendron albiflorum* Hook. (white-flowered rhododendron), *Rhododendron occidentale* (Torr. & Gray ex Torr.) Gray (Western azalea), *Rhododendron macrophyllum* D. Don ex G. Don (California rosebay), and *Rhododendron maximum* L. (rhododendron, great laurel, rose bay).[4] The latter two *Rhododendron* species are the state flowers of Washington and West Virginia, respectively. There are few data to indicate which of these *Rhododendron* species causes human toxicity.

Common Name: Rhododendron, Azalea, Rosebay

Scientific Name: *Rhododendron* spp.

Botanical Family: Ericaceae (Heath)

Physical Description: These evergreen, semi-evergreen, and deciduous shrubs have flowers of various colors. Commercial hybridization produces new rhododendron plants that are difficult to identify.

Distribution and Ecology: *Rhododendron* species grow in the mountainous areas of Turkey near the eastern Black Sea, Japan, Brazil, Nepal, and parts of North America and Europe.

EXPOSURE

Rhododendron dauricum L. and *Rhododendron latoucheae* Franchet are medicinal herbs used in China as expectorants for chronic tracheitis.[5,6] In Turkey, honey from rhododendron plants is a traditional treatment for diabetes mellitus, bowel disorders, hypertension, abdominal pain, and sexual dysfunction.[7]

PRINCIPAL TOXINS

Physiochemical Properties

Grayanotoxins are unique tetracyclic diterpenes that occur only in the Ericaceae family. There are over 30

Medical Toxicology of Natural Substances, by Donald G. Barceloux, MD

diterpenic polyalcohols in this family that possess an tetracyclic A-*nor*-B-*homo*-*ent*-kaurane skeleton including the most prominent toxins, grayanotoxin I (andromedotoxin, acetylandromedol, rhodotoxin) and grayanotoxin III (andromedol). Other structurally similar Ericaceae compounds include asebotoxins, rhodojaponins, and lyoniatoxins (lyoniols).[8] Volatile constituents of solvent extracts of 15 different organs (leaves, flowers, fruits) of five *Rhododendron* species grown in Turkey demonstrated 200 volatile compounds as analyzed by gas chromatography/mass spectrometry (GC/MS).[9] About 34 volatile compounds were identified in an organic solvent extract of *R. luteum* (L.) Sweet. (Honeysuckle azalea) flowers with the major compounds being benzyl alcohol (16.6%), limonene (14.6%), and *p*-cymene (8.4%).

Poisonous Parts

Grayanotoxins are present in nectar, flowers, stems, and the leaves of plants in the family Ericaceae. The grayanotoxin I (andromedotoxin) content of a fresh *Rhododendron ponticum* L. leaf is about 0.024%.[10] Grayanotoxins in honey produced by bees that feed on *Rhododendron* species are responsible for toxicity associated with mad honey poisoning, particularly in Turkey.[11]

Mechanism of Toxicity

Grayanotoxin, batrachotoxin, veratridine, and aconitine have similar structural groups that modulate the sodium gate.[12] These toxins bind the sodium channel in the open state, modify the sodium channel to prevent activation, and shift the sodium channel in the direction of hyperpolarization.[13] In animal models, the intravenous administration of grayanotoxins produces dose-dependent bradycardia, hypotension, respiratory depression, conduction delay, ventricular tachycardia and fibrillation, emesis, and central nervous system depression similar to the effects of veratrum poisoning.[14] These animal studies suggest that grayanotoxin may bind to the M_2-muscarinic receptors.[15] Grayanotoxin I is a potent parenteral hypotensive agent in experimental animal studies.[16] Animal studies suggest that the bradycardia association with grayanotoxin results from peripheral vagal stimulation.[17]

DOSE RESPONSE

Animals

In animal studies, grayanotoxins I, II, and VI are the most toxic grayanotoxin compounds following parenteral administration.[18] There is substantial variation in toxicity depending on the route of exposure. The minimum *parenteral* dose necessary to produce hypotension and bradycardia was 1–2 μg/kg, whereas more than 35 μg/kg was required to induce ventricular dysrhythmias. In contrast, the LD_{50} for grayanotoxin I in mice *per os* was 5.1 mg/kg.[18] Prediction of toxicity to humans based on a history of ingestion is limited by the lack of knowledge of diterpenoid pharmacokinetics and human toxicity data.

Humans

Although rhododendron plants cause significant animal toxicity, exposure of humans to these plants do not usually result in serious toxicity except under unusual circumstances (e.g., ingestion of teas and extracts).[19] The ingestion of 75 mL of Turkish honey containing grayanotoxin produced coma, convulsions, bradycardia, and atrioventricular block in a 17-year-old woman.[20] Not all *Rhododendron* species produce toxic nectar, and predicting which hybrids will secrete toxic nectar is difficult.[21] Hypotension, bradycardia and syncope were associated with the ingestion 10–15 g contaminated honey.[22] There are few data on the dose-response relationship between parts of the rhododendron and toxicity.

TOXICOKINETICS

There are few data on the toxicokinetics of grayanotoxins in humans. The elimination of grayanotoxin is relatively rapid with symptoms usually resolving within 24 hours.[8,15]

CLINICAL RESPONSE

Most casual exposures to rhododendron plants do not produce toxicity in humans. In a series of 152 exposures to azalea plants reported to a regional poison control center, only one patient became symptomatic (protracted vomiting, transient hypertension).[23] Consumption of an unknown amount of rhododendron leaves and flowers by a 4-year-old child resulted in vomiting and abdominal pain about 2½ hours after ingestion, followed by apnea and perioral cyanosis for 30 seconds.[24] Subsequent cardiac monitoring revealed a sinus bradycardia. No neuromuscular weakness, ventricular dysrhythmias, or hypotension occurred. The child recovered without sequelae. Toxic effects associated with the ingestion of mad honey derived from *Rhododendron* species include nausea, vomiting, salivation, blurred vision, dyspnea, generalized weakness, circumoral paresthesias, syncope, bradycardia, hypotension, seizures, supraventricular tachycardia, and conduction disturbances including complete heart block responsive to atropine.[11,25,26] Cardiac abnormalities in serious poisonings generally resolve within 1 to 2 days.[15,27] However, symptoms from minor intoxications usually resolve within 2 to 6 hours.[28] Fatalities are rare following the ingestion of grayanotoxin-contaminated honey.

DIAGNOSTIC TESTING

Analytical methods for the detection of grayanotoxin I include thin-layer chromatography,[10] GS/MS,[29] and liquid chromatography/mass spectrometry.[30] The low vapor pressure and heat instability of grayanotoxins necessitates derivatization prior to analysis when using gas chromatography.

TREATMENT

Most children who sample rhododendron plants remain asymptomatic. Decontamination measures are usually unnecessary. The administration of activated charcoal to alert children, who ingest more than a whole leaf or flower less than 1 hour prior to presentation, is a therapeutic option; however, there are no clinical data to confirm the clinical efficacy of activated charcoal on outcome of *Rhododendron* ingestions. Symptomatic patients should be observed by a physician for 4–6 hours with monitoring of cardiac rhythm and vital signs. Hypotension usually responds to fluid administration.[22] The administration of atropine for bradycardia is appropriate if the dysrhythmia is associated with a reduction in cardiac output.[15,26]

References

1. Leach DG. The history of rhododendron poisoning. Garden J 1967;17:15–18.
2. Mancini SD, Edwards JM. Cytotoxic principles from the sap of *Kalmia latifolia*. J Nat Prod 1979;42:483–488.
3. Hadley WJ, Haden HH. Alabama's "sheep kill" laurel. *Kalmia latifolia* L. A preliminary study. J Am Pharm Assoc 1946;35:340–342.
4. Kingsbury JM. Poisonous Plants of the United States and Canada. Englewood Cliffs, NJ: Prentice Hall; 1964.
5. Cao Y, Lou C, Fang Y, Ye J. Determination of active ingredients of *Rhododendron dauricum* L. by capillary electrophoresis with electrochemical detection. J Chromatogr A 2001;943:153–157.
6. Fan CQ, Zhao WM, Ding BY, Qin GW. Constituents from the leaves of *Rhododendron latoucheae*. Fitoterapia 2001;7:449–452.
7. Koca I, Koca AF. Poisoning by mad honey: a brief review. Food Chem Toxicol 2007:45:1315–1318.
8. Hikino H, Ohizumi Y, Konno C, Hashimoto K, Wakasa H. Subchronic toxicity of ericaceous toxins and rhododendron leaves. Chem Pharm Bull 1979;27:874–879.
9. Tasdemir D, Demirci B, Demirci F, Donmez AA, Baser KH, Ruedi P. Analysis of the volatile components of five Turkish *Rhododendron* species by headspace solid-phase microextraction and GC-MS (HS-SPME-GC-MS). Z Naturforsch [C] 2003;58:797–803.
10. Humphreys DJ, Stodulski JB. Detection of andromedotoxins for the diagnosis of *Rhododendron* poisoning in animals. J Appl Toxicol 1986;6:121–122.
11. Lampe KF. Rhododendrons, mountain laurel, and mad honey. JAMA 1988;259:2009.
12. Masutani T, Seyama I, Narahashi T, Iwasa J. Structure-activity relationship for grayanotoxin derivatives in frog skeletal muscle. J Pharmacol Exp Ther 1981;217:812–819.
13. Kimura T, Kinoshita E, Yamaoka K, Yuki T, Yakehiro M, Seyama I. On site of action of grayanotoxin in domain 4 segment 6 of rat skeletal muscle sodium channel. FEBS Lett 2000;465:18–22.
14. Moran NC, Dresel PE, Perkins ME, Richardson AP. The pharmacological actions of andromedotoxin, an active principle from *Rhododendron maximum*. J Pharmacol Exp Ther 1954;110:415–432.
15. Onat FY, Yegen BC, Lawrence R, Oktay A, Oktay S. Mad honey poisoning in man and rat. Rev Environ Health 1991;9:3–9.
16. Scott PM, Coldwell BB, Wiberg GS. Grayanotoxins. Occurrence and analysis in honey and a comparison of toxicities in mice. Food Cosmet Toxicol 1975;9:179–184.
17. Onat F, Yegen BC, Lawrence R, Oktay A, Oktay S. Site of action of grayanotoxins in mad honey in rats. J Appl Toxicol 1991;1:199–201.
18. Hikino H, Ohta T, Ogura M, Ohizumi Y, Konno C, Takemoto T. Structure activity relationship of ericaceous toxins on acute toxicity in mice. Toxicol Appl Pharmacol 1976;35:303–310.
19. O'Leary SB. Poisoning in man from eating poisonous plants. Arch Environ Health 1964;9:216–242.
20. Gossinger H, Hruby K, Haubenstock A. Cardiac arrhythmias in a patient with grayanotoxin-honey poisoning. Vet Hum Toxicol 1983;25:328–329.
21. Carey FM, Lewis JJ, MacGregor JL, Martin-Smith M. Pharmacological and chemical observations on some toxic nectars. J Pharm Pharmacol 1959;11:269T–274T.
22. Yilmaz O, Eser M, Sahiner A, Altintop L, Yesildag O. Hypotension, bradycardia and syncope caused by honey poisoning. Resuscitation 2006;68:405–408.
23. Klein-Schwartz W, Litovitz T. Azalea toxicity. An overrated problem. Clin Toxicol 1985;23:91–101.
24. McGee MD. Rhododendron ingestion. Natl Clearinghouse Poison Control Center Bull 1973;Sept–Oct:1–2.

25. Gossinger H, Hruby K, Haubenstock A. Cardiac arrhythmias in a patient with grayanotoxin-honey poisoning. Vet Hum Toxicol 1983;25:328–329.

26. Ergun K, Tufekcioglu O, Aras D, Korkmaz S, Pehlivan S. A rare cause of atrioventricular block: mad honey intoxication. Int J Cardiol 2005;99;347–348.

27. Sutlupinar N, Mat A, Satganoglu Y. Poisoning by toxic honey in Turkey. Arch Toxicol 1993;67:148–150.

28. Gunduz A, Turedi S, Uzun H, Topbas M. Mad honey poisoning. Am J Emerg Med 2006;24:595–598.

29. Tasdemir D, Demirci B, Demirci F, Dönmez AA, Baser KH, Rüedi P. Analysis of the volatile components of five Turkish *Rhododendron* species by headspace solid-phase microextraction and GC-MS (HS-SPME-GC-MS). Z Naturforsch [C] 2003;58:797–803.

30. Holstege DM, Puschner B, Le T. Determination of grayanotoxins in biological samples by LC-MS/MS. J Agric Food Chem 2001;49;1678–1651.

Chapter 153

SNOWBERRY [*Symphoricarpos albus* (L.) Blake]

BOTANICAL DESCRIPTION

Common Name: Snowberry, Common Snowberry

Scientific Name: *Symphoricarpos albus* (L.) Blake (*Symphoricarpos racemosus*)

Botanical Family: Caprifoliaceae (honeysuckle)

Physical Description: Snowberry is a low deciduous shrub that reaches about 3–6 ft (~1–2 m) in height and produces waxy, white berries in late summer and fall. These berries usually remain on the shrub until the leaves drop in the winter.

Distribution and Ecology: This deciduous shrub grows in woodlands and open areas of the northern United States and Canada. In Europe, this plant species is cultivated as an ornamental hedge.

EXPOSURE

The bark of *S. albus* is used by Pacific Northwest Indians as a compress for pruritus.[1]

PRINCIPAL TOXINS

The alkaloid chelidonine occurs in extracts from *S. albus*, but this alkaloid does not cause the toxicity associated with snowberry ingestion.[2] Additional compounds contained in *S. albus* include secologanin, aesculin, sitosterol saponins, tannins, terpenes, triglycerides, sugars, coumarins, and flavonoids. However, the compounds associated with toxicity remain unknown.

DOSE RESPONSE

There are inadequate data to determine the dose response of snowberry and related parts following consumption by humans.

TOXICOKINETICS

The toxic constituents in snowberry have not been identified.

CLINICAL RESPONSE

The ingestion of snowberries by children does not usually produce symptoms. Occasionally, nausea and vomiting occur after the ingestion of snowberries.[3] Although the older medical literature associates gastrointestinal distress, delirium, and altered mental status with the consumption of large amounts of snowberries,[4] there are no well-documented serious complications or fatalities in the modern medical literature. The ingestion of snowberries by a 4-year-old girl was associated with somnolence, mydriasis, difficulty urinating, and coma.[5] The patient was treated supportively with an intravenous infusion, and she recovered without sequelae within 24 hours.

DIAGNOSTIC TESTING

There are few data on the detection of toxic constituents in snowberry.

Medical Toxicology of Natural Substances, by Donald G. Barceloux, MD

TREATMENT

Treatment is supportive. Decontamination measures are unnecessary.

References

1. Turner NJ, Hebda RJ. Contemporary use of bark for medicine by two Salishan native elders of Southeast Vancouver Island, Anada. J Ethnopharmacol 1990;29:59–72.
2. Szaufer M, Kowalewski Z. Chelidonine from *Symphoricarpos albus*. Phytochemistry 1978;17:1446–1447.
3. Lewis WH. Snowberry (*Symphoricarpos*) poisoning in children. JAMA 1979;242:2663.
4. Amyot TE. Poisoning by snowberries. Br Med J 1885;1: 986.
5. Lamminpää A, Kinos M. Plant poisonings in children. Hum Exp Toxicol 1996;15:245–249.

Chapter 154

SQUIRTING CUCUMBER
[*Ecballium elaterium* (L.) A. Richard]

HISTORY

The use of the juice from the squirting cucumber for the treatment of chronic sinusitis dates back to the writings of Dioscorides (20–79 AD) in the *Materia Medica*. The Greek philosopher Theophrastus mentioned this plant in his history of plants.

BOTANICAL DESCRIPTION

Common Name: Squirting Cucumber, Jumping Cucumber, Donkey's Green

Scientific Name: *Ecballium elaterium* (L.) A. Richard

Botanical Family: Cucurbitaceae (citrouilles, gourdes, gourds, squashes)

Physical Description: This hairy vine has palmately lobed, bristly leaves. The funnel-shaped, yellow flowers produce hairy, blue-green seed pods, which eject dark seeds and juice after maturity in response to light pressure. Both male and female flowers appear on the same plant.

Distribution and Ecology: This perennial vine grows in hot, dry areas on disturbed ground and roadsides in the Mediterranean region and Micronesia. It is cultivated in central Europe and England.

EXPOSURE

The diluted (i.e., up to 30:1) aqueous extract from the fruit of *Ecballium elaterium* is a traditional anti-inflammatory and analgesic for chronic sinusitis by nasal aspiration.[1,2] Other traditional uses include the treatment of fever, cancer, liver cirrhosis, constipation, hypertension, dropsy, and rheumatic diseases.[3]

PRINCIPAL TOXINS

Physiochemical Properties

Cucurbitacin compounds (CAS RN: 60137-06-6, $C_{30}H_{42}O_6$) are a group of colorless, bitter, highly oxygenated triterpenoid substances characterized by the tetracyclic cucurbitane nucleus [19-(10→9β)-abeo-10α-lanost-5-ene].[4,5] These compounds are present in several other plant species including *Citrullus colocynthis* (L.) Schrad. (colocynth, bitter apple). Figure 154.1 displays the chemical structure of cucurbitacin B (RN: 6199-67-3, $C_{32}H_{46}O_8$), which is the principal anti-inflammatory compound in the squirting cucumber.[6,7]

Poisonous Parts

All parts of the squirting cucumber are toxic, particularly the ovoid green fruits. The juice of the fruit produces dermal erythema, inflammation, and edema.

Mechanism of Toxicity

Although cucurbitacin B has been identified as the active anti-inflammatory compound in the squirting cucumber, the ingredient or ingredients causing severe irritation to mucus membranes are unknown. The mechanism of toxicity (i.e., direct toxic effect, hypersensitivity response) is also not well defined, but the clinical

Medical Toxicology of Natural Substances, by Donald G. Barceloux, MD

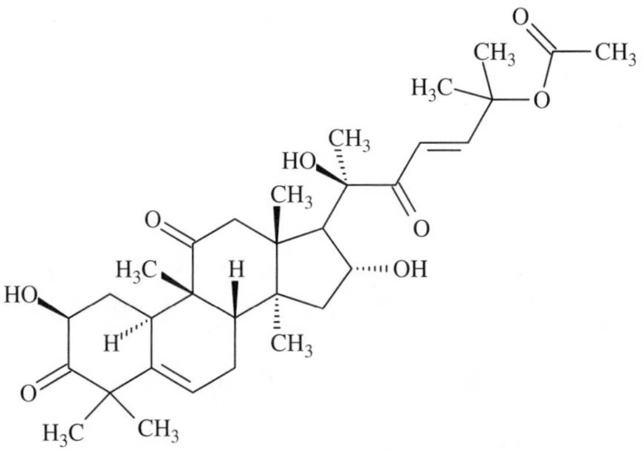

FIGURE 154.1. Chemical structure of cucurbitacin B.

effects often do not respond to antihistamines or epinephrine.[8]

DOSE RESPONSE

Cases of severe mucus membrane swelling typically involve the use of undiluted juice from the squirting cucumber. Traditionally, the juice is diluted 3–30 times before nasal aspiration for the treatment of chronic sinusitis.

TOXICOKINETICS

The toxic ingredient in squirting cucumber has not been identified, and there are no human data on the toxicokinetics of constituents (e.g., cucurbitacins) in the juice of squirting cucumber.

CLINICAL RESPONSE

The undiluted juice from the squirting cucumber causes marked inflammation of mucous membranes and localized edema following ingestion, instillation in the nasal cavities, or ocular exposure. Most patients with adverse responses to this juice have multiple allergies, and previously may have used the juice without complications.[9] Symptoms begin within 10–20 minutes of exposure and persist for 24–48 hours.[10] Progression of the inflammation can cause conjunctival irritation, corneal edema and erosions, sore throat, dysphagia, drooling, dyspnea, or respiratory distress secondary to upper airway edema (soft palate, uvula).[11] Obstruction of the upper airway is a potentially fatal complication of the nasal installation of undiluted juice from the squirting cucumber.[12] A case report associated the development of progressive facial and pharyngeal edema, fever, respiratory distress, renal insufficiency, and cardiac failure over 5 days with the use of four to five nasal aspirations of diluted juice from the squirting cucumber.[13] The patient died on the sixth day after exposure to the juice, but the role of the exposure to this juice is unclear because of the lack of medical details (blood cultures, oxygen saturations, imaging studies) and the failure to exclude other causes (sepsis, cardiovascular/renal disease).

DIAGNOSTIC TESTING

Laboratory measurements of hepatorenal function and serum electrolytes typically are normal after exposure to the undiluted juice of the squirting cucumber with the exception of laboratory evidence of airway obstruction.[14]

TREATMENT

Patients should be evaluated carefully for the presence of upper airway obstruction (stridor, drooling, dyspnea, tachypnea, reduced oxygen saturation). The respiratory effects associated with exposure to the squirting cucumber usually occur within minutes to a few hours after exposure. Although the exact mechanism of toxicity is unclear, treatment of the severe upper airway edema following exposure to undiluted juice from the squirting cucumber typically involves medicines to treat anaphylactic responses including the use of epinephrine (0.3–0.5 mL 1:1000 solution SQ), intravenous H_1- (diphenhydramine) and H_2- (ranitidine or cimetidine) histamine blockers, and intravenous methylprednisolone. However, the swelling may not respond to these agents, and prophylactic intubation may be necessary in severe cases. Ocular exposure to the undiluted juice should be treated with copious irrigation followed by slit lamp examination to detect corneal erosions.

References

1. Sezik E, Yesilada E. Clinical effects of the fruit juice of *Ecballium elaterium* in the treatment of sinusitis. J Toxicol Clin Toxicol 1995;33:381–383.
2. Uslu C, Karasen RM, Sahin F, Taysi S, Akcay F. Effect of aqueous extracts of *Ecballium elaterium* rich, in the rabbit model of rhinosinusitis. Int J Pediatr Otorhinolaryngol 2006;70:515–518.
3. Sezik E, Yesilada E, Honda G, Takaishi Y, Takeda Y, Tanaka T. Traditional medicine in Turkey X. Folk medicine in Central Anatolia. J Ethnopharmacol 2001;75:95–115.
4. Chen JC, Chiu MH, Nie RL, Cordell GA, Qiu SX. Cucurbitacins and cucurbitane glycosides: structures and biological activities. Nat Prod Rep 2005;22:386–399.
5. Seger C, Sturm S, Haslinger E, Stuppner H. A new cucurbitacin D related 16,23-epoxy derivative and its isomerization products. Organic Lett 2004;6:633–636.

6. Yesilada E, Ustun O, Sezik E, Takaishi Y, Ono Y, Honda G. Inhibitory effects of Turkish folk remedies on inflammatory cytokines: interleukin-1alpha, interleukin-1beta and tumor necrosis factor alpha. J Ethnopharmacol 1997; 58:59–73.
7. Yesilada E, Tanaka S, Sezik E, Tabata M. Isolation of an anti-inflammatory principle from the fruit juice of *Ecballium elaterium*. J Nat Prod 1988;51:504–508.
8. Eken C, Ozbek K, Yildirim CK, Eray O. Severe uvular edema and nasal mucosal necrosis due to *Ecbalium elaterium* (squirting cucumber): an allergic reaction or direct toxic effect? Clin Toxicol 2008;46:257–258.
9. Kloutsos G, Balatsouras DG, Kaeros AC, Kandiloros D, Ferekidis E, Economou C. Upper airway edema resulting from use of *Ecballium elaterium*. Laryngoscope 2001;111: 1652–1655.
10. Raikhlin-Eisenkraft B, Bentur Y. *Ecbalium elaterium* (squirting cucumber)—remedy or poison? Clin Toxicol 2000;38:305–308.
11. Koussidis GA, Mountantonakis S, Petrichou CC. Folk remedies still in use: a case of soft palate and uvular oedema due to *Ecballium elaterium*. Int J Clin Pract 2002;56:817.
12. Satar S, Gokel Y, Toprak N, Sebe A. Life-threatening uvular angioedema caused by *Ecballium elaterium*. Eur J Emerg Med 2001;8:337–339.
13. Vlachos P, Kanitsakis NN. Fatal cardiac and renal failure due to *Ecballium elaterium* (squirting cucumber). Clin Toxicol 1994;32:737–738.
14. Eray O, Tuncok Y, Eray E, Gunerli A, Guven H. Severe uvular angioedema caused by intranasal administration of *Ecballium elaterium*. Vet Hum Toxicol 1999;41:376–378.

Chapter 155

WISTERIA

BOTANICAL DESCRIPTION

Japanese wisteria [*Wisteria floribunda* (Willd.) DC.] is similar to Chinese wisteria with the exception that the white-barked Japanese wisteria vines twine clockwise around the host plant and Chinese wisteria twines counter-clockwise. Additionally, the Chinese species has 7–13 compound leaves compared with 13–19 for the Japanese species. The American wisteria [*Wisteria frutescens* (L.) Poir.], which is native to the southeastern United States, flowers June through August and produces a nonhairy seed pod 2–4 in. (~5–10 cm) long.

Common Name: Wisteria, Chinese Wisteria, Kidney Bean Tree

Scientific Name: *Wisteria sinensis* (Sims) DC.

Botanical Family: Fabaceae (pea)

Physical Description: This climbing vine has sweet-scented, lavender, pea-like flowers that produce pods 12.5–15 cm (~5–6 in.) in length. The flat, brown seeds are slightly oblong, and velvety seed pods forcefully eject the seeds up to 2–3 m (~7–10 ft) when ripe in late summer and early autumn.

EXPOSURE

Wisteria is a native plant of China, Japan, and the southeastern United States. This plant is grown as an ornamental vine covering balconies, walls, and trellises in temperate climates of Europe and the United States. There are no well-recognized medicinal uses for wisteria. In Japan, the gall from *Wisteria brachybotrys* Siebold & Zucc. is a folk remedy for gastric cancer.[1]

PRINCIPAL TOXINS

Tendrils are slender, coiled extension of a modified leaf or stem that functions as a means of support for climbing vines. Some species of wisteria (*Wisteria sinensis*, Chinese wisteria) have pointed, transparent hairs on their tendrils that produce an irritant contact dermatitis following direct exposure of the skin to these tendrils.[2] *Wisteria* species contain a variety of triterpenoid saponins (wisteria saponins) and isoflavonoids.[3] There are few data on toxic constituents in wisteria seeds or other parts of the plant.

DOSE RESPONSE

The ingestion of 10 raw seeds by a 50-year-old woman produced headache, vomiting, vertigo, and syncope.[4] The patient was released from the emergency department after symptomatic treatment and administration of activated charcoal. Similar gastrointestinal symptoms and drowsiness developed in two young teenagers after the ingestion of at least 5–6 seeds from a wisteria plant.[5]

TOXICOKINETICS

The toxic constituents in *Wisteria* species have not been identified.

Medical Toxicology of Natural Substances, by Donald G. Barceloux, MD

CLINICAL RESPONSE

The primary target organ after ingestion of *Wisteria* seeds is the gastrointestinal tract manifest by nausea, vomiting, and abdominal pain.[5] The vomitus can become bloody or bile-stained. Diarrhea is less common than other gastrointestinal symptoms. Symptoms develop with several hours of ingestion, and these symptoms usually resolve within 6–24 hours.[6]

DIAGNOSTIC TESTING

The ingestion of wisteria seeds has been associated with leucocytosis and some abnormalities of the prothrombin times, but no clinically significant abnormalities of serum hepatic aminotransferases or serum creatinine.[5]

TREATMENT

Treatment is supportive, primarily involving correcting any fluid or electrolyte imbalance resulting from the gastrointestinal irritant properties of *Wisteria* seeds. Decontamination measures are unnecessary because of the presence of profuse vomiting. Patients can be discharged from the emergency department after control of the vomiting and replacement of fluid and/or electrolyte deficits.

References

1. Konoshima T, Takasaki M, Kozuka M, Tokuda H, Nishino H, Matsuda E, Nagai M. Anti-tumor promoting activities of isoflavonoids from *Wisteria brachybotrys*. Biol Pharm Bull 1997;20:865–868.
2. Southcott RV, Haegi LA. Plant hair dermatitis. Med J Aust 1992;156:623–632.
3. Kinjo J, Fujishima Y, Saino K, Tian R-H, Nohara T. Five new triterpene glycosides from *Wisteria brachybotrys* (Leguminoseae). Chem Pharm Bull 1995;43:636–640.
4. Rondeau ES. Wisteria toxicity. Clin Toxicol 1993;31: 107–112.
5. Piola C, Ravaglia M, Zoli MP. Intossicazione da semi di *Wisteria siniensis*. Boll Soc Intal Farm Osp 1983;29:333–337.
6. Jacobziner H, Raybin HW. Briefs on accidental chemical poisonings in New York City. New York State J Med 1961;61: 2463–2466.

VI Trees

Chapter 156

BLACK LOCUST
(*Robinia pseudoacacia* L.)

BOTANICAL DESCRIPTION

Common Name: Black Locust, False Acacia

Scientific Name: *Robinia pseudoacacia* L.

Botanical Family: Fabaceae (pea)

Physical Description: This large, fast-growing tree that grows to 75 ft (~23 m) in height with alternate, compound leaves and unbranched spines resembling rose thorns. Showy white flowers mature into straight, flat pods. The fragrant, drooping, pea-like flowers occur in clusters that appear in May and June. The smooth fruit pods (legumes) contain four to eight seeds and reach about 10 cm (~4 in.) in length.

Distribution and Ecology: This native tree of the southeastern United States prefers dry woodlands and occurs throughout the United States, particularly in the eastern and central United States and on the Pacific Coast from Washington to central California. The black locust is widely cultivated as a source of nectar, and this tree has become naturalized in forests, prairies, and savannas in temperate climates of Europe and Asia, as well as North America.

EXPOSURE

This large tree is a popular ornamental plant along streets and in parks. The black locust is also a source of honey, such as the French acacia monofloral honey. This durable wood is a source of firewood, fence posts, and wooden crafts.

PRINCIPAL TOXINS

The toxic constituents in *R. pseudoacacia* are not well defined, but at least some of the adverse effects probably result from toxic glycoproteins (lectins). Compounds isolated from this species include lectins (robin, robitin, phasin), robinlin (C_{11}-homomonoterpene)[1] and triterpene glycosides (robiniosides A-J, sophoraflavoside II).[2,3] Analysis of an ethanolic extract of *R. pseudoacacia* indicated that the major constituent was robinetin (CAS RN:490-31-3, $C_{15}H_{10}O_7$) with smaller amounts of myricetin and quercetin as measured by nuclear magnetic resonance and mass spectroscopy.[4] The toxic constituents probably contain heat-labile proteins. In animal models, extracts of bark from *R. pseudoacacia* produce gastrointestinal (GI) distress and lethargy.[5] *In vitro* studies in rats indicate that water-soluble extracts of *R. pseudoacacia* increase the frequency of smooth muscle contraction in uterine tissue, but not in intestinal or vascular tissues.[6] Several high-molecular-weight lectins are nonspecific agglutinators of the human erythrocyte.[7] Lectins from the black locust stimulate *in vitro* blast transformation by binding to specific receptors on the cell surface and activating T-cells.[8] Studies of patients with seasonal allergies suggest that substantial cross-reactivity occurs between *R. pseudoacacia* pollen and common testing pan allergens; consequently, *R. pseudoacacia* pollen may not produce potent allergic responses.[9]

Medical Toxicology of Natural Substances, by Donald G. Barceloux, MD

DOSE RESPONSE

There are inadequate data to determine the dose-response relationship between ingestion of plant parts of black locust and clinical effects. These effects are primarily gastrointestinal.

TOXICOKINETICS

The GI toxin in black locust has not been well defined.

CLINICAL RESPONSE

Reports of toxicity from exposure to the black locust are rare in the modern medical literature. GI symptoms and lethargy are the most common clinical features of *Robinia pseudoacacia* intoxication.[10] In an orphanage where 32 children ate bark from the black locust, the clinical presentation included vomiting, mild hematemesis, pallor, peripheral cyanosis, dilated pupils, and stupor.[11] This case series did not report any laboratory values or vital signs. There were no fatalities. In another case series of seven boys, the chewing of bark from the black locust tree was associated with the development of nausea, vomiting, diarrhea, abdominal pain, dry throat, mydriasis, headache, dizziness, and muscle weakness in a group of children.[12] There were no sequelae. A case report during modern times associated the development of vomiting with the chewing of bark from the black locust.[13] The vomiting began about $2^1/_2$ hours after chewing the bark, and the vomiting responded to anti-emetic medication. There were no other symptoms.

DIAGNOSTIC TESTING

Abnormalities associated with ingestion of plant parts from *R. pseudoacacia* include hematuria, mild elevation of serum hepatic aminotransferases, and electrolyte imbalance.

TREATMENT

There are limited data on the treatment of intoxication from *R. pseudoacacia,* and management is supportive. Substantial ingestions of plant parts from the black locust are usually associated with vomiting and therefore GI decontamination is unnecessary. Casual contacts require no treatment. The main aspect of supportive care is the maintenance of fluid and electrolyte balance. Asymptomatic patients may be discharged after 3–4 hours after ingestion.

References

1. Tian F, Chang C-J, Grutzner JB, Nichols DE, McLaughlin JL. Robinlin: a novel bioactive homo-monoterpene from *Robinia pseudoacacia* L. (Favaceae). Bioorg Med Chem Lett 2001;11:2603–2606.
2. Cui B, Kinjo H, Nohara T. Triterpene glycosides from the bark of *Robinia pseudo-acacia* L. I. Chem Pharm Bull 1992;40:2995–2999.
3. Cui B, Kinjo J, Nohara T. Triperpene glycosides from the bark of *Robinia pseudo-acacia* L. II. Chem Pharm Bull 1993;41:553–556.
4. Nasir H, Iqbal Z, Hiradate S, Fujii Y. Allelopathic potential of *Robinia pseudo-acacia* L. J Chem Ecol 2005;31:2179–2192.
5. Power FB. The poisonous constituent of the bark of *Robinia pseudoacacia*. Am J Pharm 1913;85:339–344.
6. Pharris BB, Russell RL. Effects of a water soluble extract of *Robinia pseudoacacia* leaves on uterine smooth muscle. Experientia 1967;23:637–638.
7. Horejsi V, Haskovec C, Kocourek S. Studies of lectins. XXXVIII. Isolation and characterization of the lectin from black locust bark (*Robinia pseudoacacia* L). Biochim Biophys Acta 1978;532:98–104.
8. Sharif A, Brochier J, Bourrillon R. Specific activation of human T lymphocytes by *Robinia pseudoacacia* seeds' lectin. Cell Immunol 1977;31:302–310.
9. Compés E, Hernández E, Quirce S, Palomares O, Rodríguez R, Cuesta J, et al. Hypersensitivity to black locust (Robinia pseudoacacia) pollen: "allergy mirages." Ann Allergy Asthma Immunol 2006;96:586–592.
10. Costa Bou X, Soler I Ros JM, Seculi Palacios JL. Poisoning *by Robinia pseudoacacia*. An Esp Pediatr 1990;32:68–69. [Spanish]
11. Emery AT. Report of thirty-two cases of poisoning by locust bark. NY Med J 1887;45:92.
12. Sivera AA, Pena AA, Cubo AP. Estudio clinicoepidemioló de un envenenamiento accidental por *Robinia pseudoacacia* L. en escolares. An Esp Pediatr 1989;30:191–194.
13. Hui A, Marraffa JM, Stork CM. A rare ingestion of the black locust tree. J Toxicol Clin Toxicol 2004;42:93–95.

Chapter 157

BUCKEYE
(*Aesculus* Species)

BOTANICAL DESCRIPTION

The genus *Aesculus* includes *Aesculus hippocastanum* L. (horse chestnut), *Aesculus glabra* Willd. (Ohio buckeye), and *Aesculus flava* Ait. (yellow buckeye, *Aesculus octandra* Marsh.).

Common Name: Buckeye, Horse Chestnut

Scientific Name: *Aesculus* spp.

Botanical Family: Hippocastanaceae (buckeye)

Physical Description: These deciduous trees and shrubs contain opposite leaves palmately divided into five to seven leaflets (toothed margins) on a long stalk. Yellow, red, and white flowers produce a characteristic three-compartment fruit surrounded by a leathery husk that encloses brown, round, glossy seeds.

Distribution and Ecology: Seven native species appear throughout the United States. *Aesculus hippocastanum* L. (horse chestnut) is widely planted as an ornamental shade tree in the United States, Europe, and the United Kingdom. Other species include *Aesculus californica* (Spach) Nutt. (California buckeye), *Aesculus sylvatica* Bartr. (painted buckeye), *Aesculus parviflora* Walt. (bottlebrush buckeye), *Aesculus pavia* L. (red buckeye), *Aesculus glabra* Willd. (Ohio buckeye), and *Aesculus flava* Ait. (yellow buckeye). The horse chestnut (*A. hippocastanum*) is indigenous to southeastern Europe, and the horse chestnut is also present in Iran, Northern India, and Asia Minor.

EXPOSURE

Traditional herbal uses for horse chestnut are as an antipyretic and a hemorrhoidal cream. Randomized clinical trials suggest that the use of horse chestnut seed extract is an efficacious, short-term treatment for chronic venous insufficiency based on reduction in leg size and leg pain, when compared with placebo.[1] However, the US Food and Drug Administration (FDA) has not approved the use of horse chestnut seed extract containing aescin for the treatment of chronic venous insufficiency; compression therapy (e.g., pumps, bandaging, and/or graded compression stockings) remains the primary treatment for this disease.[2] Horse chestnut contains saponins that have been tested in experimental animals for cholesterol lowering effects.[3]

PRINCIPAL TOXINS

This genus contains the bioactive saponins, such as escins (Ia, Ib, IIa, IIb, IIIa, IIIb, IV, V, VI) and isoescins (I, II, V), aesculin, bioflavonoids (quercetin, kaempferol), coumarin compounds (esculin), and antioxidants (proanthocyanidin A_2).[4,5] The bioactive saponins (escin) are derived from the seed extracts of the horse chestnut. Escin (aescin, CAS RN:6805-41-0) is a complex mixture of chemically related compounds called triterpenoid saponins. β-escin (β-aescin, CAS RN:11072-93-8) is a fraction of escin that contains about 30 constituents, and this relatively nontoxic extract is administered as a treatment for chronic venous insufficiency.[6] In experimental animals, the administration of extracts from

Medical Toxicology of Natural Substances, by Donald G. Barceloux, MD

Aesculus produces muscle weakness, twitching, lethargy, ataxia, paralysis, respiratory depression, coma, and death.[7]

DOSE RESPONSE

The recommended dose of β-escin for the treatment of varicose veins is 100 mg as a sustain-release preparation, which corresponds to about 500–600 mg of the extract.[8]

TOXICOKINETICS

The bioavailability of β-escin is relatively high with peak plasma concentrations occurring about 2–3 hours after dosing.[9] However, there are large inter- and intra-individual variations in the absorption of β-escin following the administration of single doses to volunteers.[10] Animal studies suggest that the elimination of escin is relatively slow with only 1–2.5% of the dose excreted in the urine during the first 24 hours after ingestion.[11] Unchanged escin accounts for about 50% of the renal elimination with various metabolites accounting for the remainder including the main metabolite, escinol.

CLINICAL RESPONSE

Most ingestions of horse chestnut by children produce no significant symptoms other than mild gastrointestinal distress. In grazing animals, the ingestion of parts of plants from the *Aesculus* genus produces ataxia, gait disturbances, muscle stiffness, and seizures.[12] The pollen of *Aesculus hippocastanum* (horse chestnut) is highly allergenic and sensitization to the pollen may cause allergic rhinitis as well as contact dermatitis and anaphylactic shock.[13,14] Rare case reports associate occupational exposure to escin with the development of occupational asthma.[15] Clinical trials using horse chestnut seed extract suggest that the doses associated with these trials produce minimal side effects.[16]

DIAGNOSTIC TESTING

A thin layer chromatographic (TLC) method is available to determine the total escin (saponin) content of herbal products containing horse chestnut that lacks the interference associated with colorimetric methods.[17] The use of high pressure liquid chromatography (HPLC) method with UV detection is limited by the poor UV absorbance of triterpene glycosides.[18] Other analytical methods for the quantitation of saponins (e.g., escins) include accelerated solvent extraction followed by HPLC/diode array detector and positive ion electrospray/time-of-flight/mass spectrometry.[19] Using this method, the limits of detection of escin Ia, escin Ib, isoescin Ia and isoescin Ib ranged from 0.40–0.75 μg/mL. Radioimmunoassay (RIA) methods can detect aescin concentrations in extracts of *Aesculus hippocastanum* in the range of 100 pg.[20] This method does not separate escin from the main metabolite, escinol.

TREATMENT

Treatment is supportive. Decontamination measures are unnecessary.

References

1. Pittler M, Ernst E. Horse chestnut seed extract for chronic venous insufficiency. Cochrane Database Syst Rev 2006;(1): CD003230.
2. Rathbun SW, Kirkpatrick AC. Treatment of chronic venous insufficiency. Curr Treat Options Cardiovasc Med 2007;9: 115–126.
3. Dworschak E, Antal M, Biro L, Regoly-Merei A, Nagy K, Szepvolgyi J, et al. Medical activities of *Aesculus hippocastaneum* (horse-chestnut) saponins. Adv Exp Med Biol 1996;404:471–474.
4. Yoshikawa M, Murakami T, Yamahara J, Matsuda H. Bioactive saponins and glycosides. XII. Horse chestnut. (2): Structures of escins IIIb, IV, V, and VI and isoescins Ia, Ib, and V, acylated polyhydroxyoleanene triterpene oligoglycosides, from the seeds of horse chestnut tree (*Aesculus hippocastanum* L., Hippocastanaceae). Chem Pharm Bull (Tokyo) 1998;46:1764–1769.
5. Yoshikawa M, Murakami T, Matsuda H, Yamahara J, Murakami N, Kitagawa I. Bioactive saponins and glycosides. III. Horse chestnut. (1): The structures, inhibitory effects on ethanol absorption, and hypoglycemic activity of escins Ia, Ib, IIa, IIb, and IIIa from the seeds of *Aesculus hippocastanum* L. Chem Pharm Bull (Tokyo) 1996;44: 1454–1464.
6. Sirtori CR. Aescin: Pharmacology, pharmacokinetics and therapeutic profile. Pharmacol Res 2001;44:183–193.
7. Williams MC, Olsen JD. Toxicity of seeds of three *Aesculus* spp to chicks and hamsters. Am J Vet Res 1984;45: 539–542.
8. Bassler D, Okpanyi S, Schrodter A, Loew D, Schurer M, Schulz H-U. Bioavailability of β-aescin from horse chestnut seed extract: comparative clinical studies of two galenic formulations. Adv Ther 2003;20:295–304.
9. Kunz K, Lorkowski G, Petersen G, Samcova E, Schaffler K, Wauschduhn CH. Bioavailability of escin after administration of two oral formulations containing *Aesculus* extract. Arzneimittalforschung 1998;48:822–825.
10. Bassler D, Okpanyi S, Schrodter A, Loew D, Schurer M, Schlz H-U. Bioavailability of β-aescin from horse chestnut seed extract: comparative clinical studies of two galenic formulations. Adv Nat Ther 2003;20:295–304.

11. Lang W. Studies on the percutaneous absorption of ^{3}H-aescin in pigs. Res Exp Med (Berl) 1977;169:175–187.
12. Casteel SW, Wagstaff DJ. *Aesculus glabra* intoxication in cattle. Vet Hum Toxicol 1992;34:55.
13. Jaspersen-Schib R, Theus L, Guirguis-Oeschger M, Gossweiler B, Meier-Abt PJ. Serious plant poisonings in Switzerland 1966–1994. Case analysis from the Swiss Toxicology Information Center. Schweiz Med Wochenschr 1996;126: 1085–1098. [German]
14. Popp W, Horak F, Jager S, Reiser K, Wagner C, Zwick H. Horse chestnut (*Aesculus hippocastanum*) pollen: a frequent cause of allergic sensitization in urban children. Allergy 1992;47:380–383.
15. Munoz X, Culebras M, Cruz MJ, Morell F. Occupational asthma related to aescin inhalation. Ann Allergy Asthma Immunol 2006;96:494–496.
16. Siebert U, Brach M, Sroczynski G, Berla K. Efficacy, routine effectiveness, and safety of horse chestnut seed extract in the treatment of chronic venous insufficiency. A meta-analysis of randomized controlled trials and large observational studies. Int Angiol 2002;21:305–315.
17. Apers S, Naessens T, Pieters L, Vlietinck A. Densitometric thin-layer chromatographic determination of aescin in a herbal medicinal product containing *Aesculus* and *Vitis* dry extracts. J Chromatogr A 2006;1112:165–170.
18. Profumo P, Caviglia AM, Gastaldo P, Dameri RM. Aescin content in embryogenic callus and in embryoids from leaf explants of *Aesculus hippocastanum*. Planta Med 1991;57: 50–52.
19. Chen J, Li W, Yang B, Guo X, Lee FS, Wang X. Determination of four major saponins in the seeds of *Aesculus chinensis* Bunge using accelerated solvent extraction followed by high-performance liquid chromatography and electrospray-time of flight mass spectrometry. Anal Chim Acta 2007;596:273–280.
20. Lehtola T, Huhtikangas A. Radioimmunoassay of aescine, a mixture of triterpene glycosides. J Immunoassay 1990;11: 17–30.

Chapter 158

CHINABERRY
(*Melia azedarach* L.)

BOTANICAL DESCRIPTION

There are about 5–10 species of the genus *Melia* with the common US species being *Melia azedarach*. This deciduous tree should not be confused with the margosa or neem tree (*Azadirachta indica* Adr. Juss.). Previously this latter tree was named *Melia azedarach*, but now the evergreen neem tree belongs to its own genus.

Common Name: Chinaberry, China Tree, White Cedar, Cape Lilac, Pride-of-India

Scientific Name: *Melia azedarach* L.

Botanical Family: Meliaceae (mahogany)

Physical Description: A rapidly growing, deciduous tree that grows to 50 ft (~15 m) with a stout trunk and finely furrowed, brown-gray bark. The leaves are large, pinnate, and compound. The small, numerous purple flowers produce a round, yellow berry about 1 cm (~0.4 in.) in diameter inside drupes (berry-like seedpods).

Distribution and Ecology: This plant is native to northern Australia, eastern Africa, and southwestern Asia. In warm regions, chinaberry is also a popular ornamental plant. The chinaberry is naturalized in the southeastern United States, California, Hawaii, and other warm climates around the world.

EXPOSURE

The medicinal use of *Melia indica* (neem or margosa oil) is much more common than the use of *Melia azedarach*. Uses of the latter include the treatment of parasitic and bacterial infections, leprosy, skin lesions (eczema), asthma, and fever. Meliatoxins A and B demonstrate some insect repellant properties *in vitro* and *in vivo*.[1,2] The chinaberry tree is an ornamental shade tree.

PRINCIPAL TOXINS

The toxins in *Melia azedarach* are not well defined. The chinaberry fruit contains several tetranortriterpene limonoid compounds, such as meliatoxin A_1 (Figure 158.1), A_2 (Figure 158.2), B_1, and B_2. The latter two compounds are probably isomers of the two former compounds, respectively, that lack an epoxide group. Experimental studies suggest that meliatoxins A_1 and A_2 are the primary toxins, whereas meliatoxins B_1 and B_2 are probably artifacts of the analytical process. Other limonoid compounds present in the seeds, fruit, and bark of this plant include salannin, meldenin, meliacarpinins, and meliantriol.[3,4] In experimental animal studies, the administration of these compounds produced muscle contractions, weakness, ataxia, paresis, abdominal distress, seizures, emaciation, and death.[5,6] The leaves of the chinaberry also contain the tetracyclic diterpenoid compound, grayanotoxin III, which produces salivation, dyspnea, ataxia, hind leg paralysis, and death following intraperitoneal doses to animals.[7] This compound increases sodium permeability in various excitable tissues.[8] The oil extracted from the fruit of *Melia azedarach* contains linoleic and oleic fatty acids along with lesser amounts of palmitic, stearic, and palmitoleic fatty acids. In an analysis of a chloroform extract of green berries from chinaberry, the concentration of mystic

Medical Toxicology of Natural Substances, by Donald G. Barceloux, MD

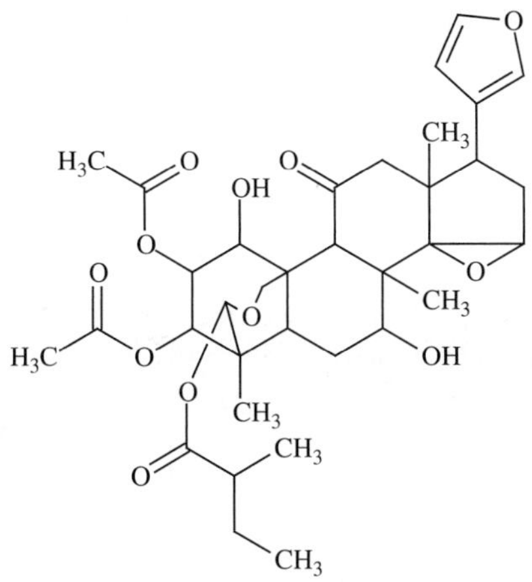

FIGURE 158.1. Meliatoxin A_1 (CAS RN: 87725-70-0).

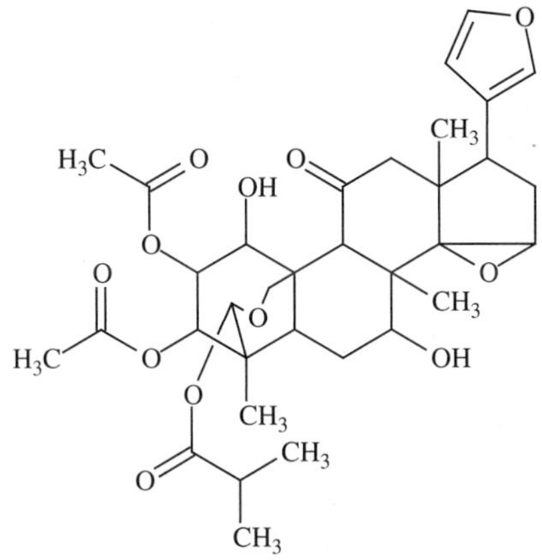

FIGURE 158.2. Meliatoxin A_2 (CAS RN: 87617-82-1).

acid was <0.1%.[9] Although most cases of exposure involve ingestion of the fruits, the bark and flowers also contain meliatoxins. There is considerable variability in toxicity following the consumption of the fruits and foliage of the chinaberry plant, probably as a result of differences in toxin content between location and stage of growth. The yield of meliatoxins in one sample of fruit ranged up to 0.5% by wet weight.[6]

DOSE RESPONSE

The toxicity of the chinaberry varies substantially as a result of several factors including location and the part of the plant. The LD_{50} of pigs administered an ethanol extract of the meliatoxins was 6.4 mg/kg, whereas the LD_{50} following intraperitoneal injection of mice was 16 mg/kg.[6] Animal feeding studies suggest that the lethal dose of chinaberry fruit is approximately 0.5% of body weight.[10]

TOXICOKINETICS

There are inadequate data to determine the toxicokinetics of meliatoxins.

CLINICAL RESPONSE

There are few case reports of toxicity following exposure to the chinaberry despite the reporting of over 1,400 exposures to the data collection system for state poison control centers in the United States.[11] Case reports of the older medical literature associate the development of respiratory insufficiency, cardiac depression, coma, and death with ingestion of fruit[12] and aqueous extracts of the bark[13] from chinaberry. Case reports in the veterinary literature indicate that the ingestion of fruit from the chinaberry causes ataxia, diarrhea, trembling, respiratory distress, bradycardia, and death in grazing animals.[14,15] Pathological examination demonstrated hemorrhagic enteritis and coagulation necrosis in the liver, kidney, and myocardium.

DIAGNOSTIC TESTING

There are few data on the use of analytical methods to detect the presence of meliatoxins in the clinical setting.

TREATMENT

Treatment is supportive.

References

1. Carpinella MC, Miranda M, Almiron WR, Ferrayoli CG, Almeida FL, Palacios SM. *In vitro* pediculicidal and ovicidal activity of an extract and oil from fruits of *Melia azedarach* L. J Am Acad Dermatol 2007;56:250–256.
2. Macleod JK, Moeller PD Molinski TF, Koul O. Antifeedant activity against *Spodoptera litura* larvae and [^{13}C]NMR spectral assignments of the meliatoxins. J Chem Ecol 1990;16:2511–2518.
3. Fukuyama Y, Ogawa M, Takahashi H, Minami H. Two new meliacarpinins from the roots of *Melia azedarach*. Chem Pharm Bull 2000;48:301–303.
4. Srivastava SD. Limonoids from the seeds of *Melia azedarach*. J Natural Prod 1986;49:56–61.
5. Bahri S, Sani Y, Hooper PT. Myodegeneration in rats fed *Melia azedarach*. Aust Vet J 1992;69:33.
6. Oelrichs PB, Hill MW, Vallely PJ, MacLeod JK, Molinski TF. Toxic tetranortriterpenes of the fruit of *Melia azedarach*. Phytochemistry 1983;22:631–534.

7. Terai T, Sato M, Narama I, Matuura T, Katakawa J, Tetsumi M. Transformation of grayanotoxin III to 10-*epi*-grayanotoxin III. Its x-ray crystallographic analysis and acute toxicity in mice. Chem Pharm Bull 1996;44:1245–1247.
8. Narahashi T, Seyama I. Mechanism of nerve membrane depolarization caused by grayanotoxin I. J Physiol 1974; 242:471–487.
9. Kaplan ER, Sapeika N. Chemical composition of the fruit of *Melia azedarach* L. S Afr J Med Sci 1971;36:83–84.
10. Kingsbury JM. Poisonous Plants of the United States and Canada. Englewood Cliffs, NJ: Prentice-Hall;1964.
11. Krenzelok EP, Jacobsen TD, Aronis JM. Plant exposure: a state profile of the most common species. Vet Hum Toxicol 1996;38:289–298.
12. Caratala RE. Fatal intoxication by fruit from *Melia azedarach* Linn. Chem Abstr 1939;33:6951.
13. Kiat TK. *Melia azedarach* poisoning. Singapore Med J 1969;10:24–28.
14. Hare WR, Schutzman H, Lee BR, Knight MW. Chinaberry poisoning in two dogs. JAVMA 1997;210:1638–1640.
15. Kwatra MS, Singh B, Hothi DS, Dhingra PN. Poisoning by *Melia azedarach* in pigs. Vet Record 1974;95:421.

Chapter 159

GOLDEN CHAIN TREE
(*Laburnum anagyroides* Medikus)

HISTORY

The toxicity of *Laburnum* seeds was recognized during the 19th century.[1] During this period, case reports associated convulsions and death with the ingestion of *Laburnum* seeds, but the validity of this association is unclear because of the lack of adequate data (e.g., postmortem examination) to exclude other causes of death.[2] Studies during the early 1900s indicated that cytisine was the principal toxin in *Laburnum* species.[3] Ing identified the chemical structure of cytisine in 1931.[4] *Laburnum* ingestion was a common occurrence in England during the late 1970s, primarily due to the attraction of children to the beautiful seeds and pods. During this period, the ingestion of parts from the golden chain tree accounted for an estimated 3,000 admissions to hospitals in England and Wales, but serious toxicity was not well documented.[5]

BOTANICAL DESCRIPTION

British species of *Laburnum* include *L. alpinum* (Mill.)Bercht. & J.Presl (Scottish laburnum) and *L. anagyroides.*

Common Name: Golden Chain Tree

Scientific Name: *Laburnum anagyroides* Medikus

Botanical Family: Fabaceae (pea)

Physical Description: This deciduous tree reaches up to 12–30 ft (~4–9 m) and has alternate leaves on a long stalk with three leaflets. The golden, pea-like flowers appear in long drooping clusters during May and June. The green fruit is a long, flattened pod that contains two to seven kidney-shaped seeds.

Distribution and Ecology: This beautiful flowering tree is a native species in southern and central Europe that was introduced in the northern United States and southern Canada as an ornamental tree.

EXPOSURE

In the United States, Native Americans used *Laburnum* seeds as emetics and purgatives during rituals. Uses for alcoholic extracts of *Laburnum* species in traditional European medicine included the treatment of cough, constipation, respiratory depression, neuralgias, headache, asthma, and insomnia. In Europe during World War II, the leaves of *Laburnum anagyroides* were a substitute for tobacco.[6] In Eastern European countries, cytisine is currently an aid for smoking cessation.[7] However, the evidence for the efficacy of cytisine for smoking cessation is inconclusive.[8] Most poisonings occur in children between July and September when the attractive pods and seeds appear.[9]

PRINCIPAL TOXINS

Laburnum contains the toxic quinolizidine alkaloid, cytisine, which is structurally and pharmacologically similar to nicotine and lobeline.[3] The pK_a of cytisine is 7.92. *Laburnum* seeds contain approximately 1%

Medical Toxicology of Natural Substances, by Donald G. Barceloux, MD

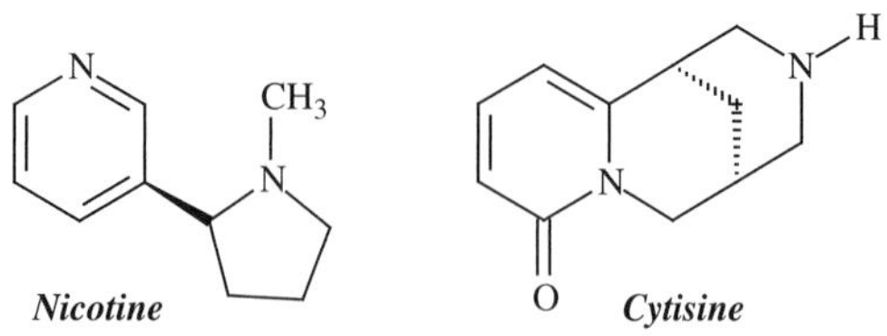

Nicotine Cytisine

FIGURE 159.1. Chemical structures of cytisine and nicotine.

cytisine by weight; the estimated lethal dose is approximately 0.5 mg/kg.[10] Animal studies indicate that cytisine produces toxic effects similar to nicotine.[11] Figure 159.1 compares the chemical structures of cytisine and nicotine. Cytisine is a nicotine agonist that binds to nicotinic receptors (e.g., $\alpha_4\beta_2$),[12] but the physiological and behavioral effects are complex, and cytisine is probably not active at all nicotine receptor subtypes.[13]

DOSE RESPONSE

In a retrospective review of 49 children exposed to *Laburnum* seeds, the estimated ingestion of 1–10 seeds from *L. anagyroides* was not associated with serious toxicity.[14] Of the 49 children, 34 children remained asymptomatic. There are inadequate data to correlate the ingestion of a specific number of *Laburnum* seeds to toxic effects.

TOXICOKINETICS

In rodent studies, the average bioavailability of cytisine following a dose of 2 mg/kg body weight was 42% with peak blood concentrations occurring about 2 hours after ingestion.[15] About 18% of an oral dose of cytisine was excreted in the urine during the first 24 hours after oral administration. The bile contained approximately 11% of the oral dose during the first 6 hours. In animal studies, the plasma elimination of cytisine is relatively rapid (i.e., 1–3 hours), and limited biotransformation of cytisine occurs.[6]

CLINICAL RESPONSE

Despite frequent exposures to *Laburnum*, particularly in children, there are few data on the clinical features of *Laburnum* (cytisine) intoxication.[16] Symptoms classically associated with cytisine intoxication are similar to mild nicotine toxicity and the onset of symptoms usually begins about $^1/_2$–4 hours after ingestion with nausea, vomiting, abdominal pain, pallor, lightheadedness, drowsiness, and ataxia.[17] Other clinical features include dilated pupils, tachycardia, delirium, paralysis, and respiratory depression. Symptoms resolve within 12–24 hours. An adult paranoid schizophrenic died suddenly at an inpatient facility without any obvious signs or symptoms of toxicity.[10] The death was associated with *Laburnum* toxicity because the autopsy demonstrated 23 *Laburnum* seedpods in the stomach. A comprehensive analysis of postmortem blood for drugs was not done. He had appeared in his usual state of health 10 minutes prior to death, and he did not display symptoms of cytisine toxicity immediately prior to death.

DIAGNOSTIC TESTING

Analytical methods for determination of cytisine in *Laburnum* extracts include reversed-phase high performance liquid chromatography,[18] gas chromatography/mass spectrometry,[19] liquid chromatography/atmospheric pressure chemical ionization/mass spectrometry, and liquid chromatography/electrospray ionization/tandem mass spectrometry.[20]

TREATMENT

Management is supportive. Gut decontamination measures are usually unnecessary, and there are no clinical data to determine the efficacy of activated charcoal following the *Laburnum* ingestion. Most children can be safely discharged after 4–6 hours of observation, providing the child is asymptomatic and the vital signs are normal. There are no antidotes to cytisine toxicity.

References

1. Joll BB. On poisoning by Laburnum seeds. Br Med J 1879; 1:737.
2. Usher Somers NL. Death from *Laburnum* poisoning. Lancet 1883;ii:1114.
3. Dale HH, Laidlaw PP. The physiological action of cytisine, the active alkaloid of laburnum (*Cytisus laburnum*). J Pharmacol Exp Ther 1912;3:205–221.
4. Ing HR. Cytisine. Part I. J Chem Soc 1931:2195–2203.
5. Forrester RM. Have you eaten laburnum? Lancet 1979;i: 1073.
6. Tutka P, Zatonski W. Cytisine for the treatment of nicotine addiction: from a molecule to therapeutic efficacy. Pharmacol Rep 2005;58:777–798.
7. Etter J-F. Cytisine for smoking cessation a literature review and a meta-analysis. Arch Intern Med 2006;166: 1553–1559.
8. Cahill K, Stead L, Lancaster T. Nicotine receptor partial agonists for smoking cessation. Cochrane Database Syst Rev 2007;(1):CD006103.
9. Mitchell RG. *Laburnum* poisoning in children. Report of 10 cases. Lancet 1951;2:57–58.

10. Richards HG, Stephens A. A fatal case of laburnum seed poisoning. Med Sci Law 1970;10:260–266.
11. Barlow RB, McLeod LJ. Some studies on cytisine and its methylated derivatives. Br J Pharmacol 1969;35:161–174.
12. Zhang J, Steinbach JH. Cytisine binds with similar affinity to nicotinic $alpha_4beta_2$ receptors on the cell surface and in homogenates. Brain Res 2003;959:98–102.
13. Chandler CJ, Stolerman IP. Discriminative stimulus properties of the nicotinic agonist cytisine. Psychopharmacology 1997;129;257–264.
14. Bramley A, Goulding R. Laburnum "poisoning." Br Med J 1981;283:1220–1221.
15. Klöcking H-P, Richter M, Damm G. Pharmacokinetic studies with ^{3}H-cytisine. Arch Toxicol 1980;4(Suppl): 312–314.
16. Chin KC, Beattie TJ. *Laburnum* poisoning. Lancet 1979;i:1299.
17. Mitchell RG. *Laburnum* poisoning in children report of ten cases. Lancet 1951;ii:57–58.
18. Reavill C, Walther B, Stolerman IP, Testa B. Behavioural and pharmacokinetic studies on nicotine, cytisine and lobeline. Neuropharmacology 1990;29:619–624.
19. Nihei K, Shibata K, Kubo I. (+)-2,3-dehydro-10-oxo-alpha-isosparteine in *Uresiphita reversalis* larvae fed on *Cytisus monspessulanus* leaves. Phytochemistry 2002;61: 987–990.
20. Beyer J, Peters FT, Kraemer T, Maurer HH. Detection and validated quantification of toxic alkaloids in human blood plasma—comparison of LC-APCI-MS with LC-ESI-MS/MS. J Mass Spectrom 2007;42:621–633.

Chapter 160

KARAKA NUT
(*Corynocarpus laevigatus* J.R. & G. Forst.)

BOTANICAL DESCRIPTION

Common Name: New Zealand Laurel, Karaka Nut

Scientific Name: *Corynocarpus laevigatus* J.R. & G. Forst.

Botanical Family: Corynocarpaceae (karaka)

Physical Description: This evergreen tree reaches up to 12m (~40ft) in height and 8m (~25ft) in width. The alternate, obovate leaves have a smooth edge over the entire leaf margin and they reach 12–20cm (~5–8in.) in length. The bark is brownish to grayish fibrous tissues with patches of reddish-brown cork cells. The pistil develops into a club-shaped drupe containing one seed.

Distribution and Ecology: This tree is a native species of New Zealand that inhabits moist woodlands and is an invasive species in Hawaii.

EXPOSURE

Karaka nuts are one of the staple vegetables of the Maori people of New Zealand. They prepare this food by baking the fruit in earthen ovens for several hours and then washing the baked fruit for 1–2 days to remove the covering of the fruits.[1] The kernels are then dried in the sun.

PRINCIPAL TOXINS

Karakin is the bitter principle isolated from raw portions of the karaka berry.[2] The primary compound in karakin probably is a 1,4,6-triester compound (1,4,6-tris-(β-nitropropionyl)-D-glucopyranose) structurally similar to endecaphyllin B [1,4,6-tri-*O*-(3-nitropropanoyl)-β-D-glucopyranose] isolated from *Indigofera spicata* Forssk. (trailing indigo, Fabaceae).[3,4] Karakin (CAS RN:1400-11-9) has a molecular formula of $C_{15}H_{21}N_3O_{15}$, and Figure 160.1 displays the chemical structure of this compound. Hydrolysis of the glycoside karakin produces three molecules of β-nitropropionic acid (CAS RN:504-88-1, hiptagenic acid). When administered to laboratory animals, both karakin and β-nitropropionic acid cause convulsions and death.[5] Other nitropropanoyl glucopyranose compounds are present in the seeds of *C. laevigatus* including cibarian [1,6-di-(3-nitropropanoyl)-β-D-glucopyranose], coronarian [2,6-di-(3-nitropropanoyl)-α-D-glucopyranose, CAS RN: 63505-68-0], and 2,3,6-tri-(3-nitropropanoyl)-α-D-glucopyranose.[6]

DOSE RESPONSE

In a study involving experimental animals, the lethal dose of β-nitropropionic acid was about 100mg/kg with a range of 65-120mg/kg.[5]

TOXICOKINETICS

There are inadequate data to determine the toxicokinetics of karakin or other potentially toxic principals in the karaka nut.

Medical Toxicology of Natural Substances, by Donald G. Barceloux, MD

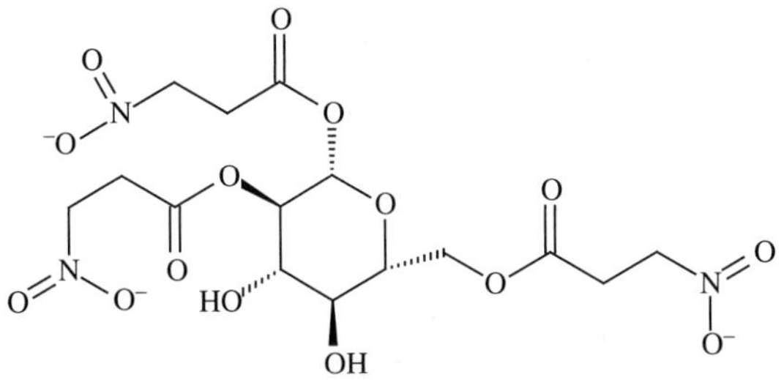

FIGURE 160.1. Chemical structure of karakin.

CLINICAL RESPONSE

Consumption of raw karaka nuts has been associated with the development of convulsions.[1] However, there are few data in humans to determine the toxicity of karaka nuts.

DIAGNOSTIC TESTING

There are few data on analytical data for the determination of karakin concentrations in biological samples.

TREATMENT

Treatment is supportive. Although the older medical literature suggests the potential of seizures following ingestion of karaka nuts, there are inadequate clinical data to guide management. Decontamination measures are unnecessary.

References

1. Skey W. Preliminary notes on the isolation of bitter substances of the nut of the karaka tree (*Corynocarpus laevigata*). Trans Proc. NZ Inst 1871;4:316–321.
2. Finnegan RA, Stephani RA. The structure of karakin. Lloydia 1970;33:491–492.
3. Cabalion P, Poisson J. [*Corynocarpus similis* Hemsley, dietary and toxic plant of Vanuatu (formerly New Hebrides)]. J Ethnopharmacol 1987;21:189–191. [French].
4. Finnegan RA, Stephani RA. Structure of hiptagin as 1,2,4,6-Tetra-o-(3-nitropropanoyl)-beta-D-glucopyranoside, its identity with endecaphyllin X, and the synthesis of its methyl ether. J Pharm Sci 1968;57:353–354.
5. Bell ME. Toxicology of karaka kernel, karkin, and β-nitropropionic acid. N Zeal J Sci 1974;17:327–334.
6. Moyer BG, Pfeffer PE, Valentine KM, Gustine DL. 3-Nitropropanoyl-D-glucopyranoses of *Corynocarpus laevigatus*. Phytochemistry 1979;18:111–113.

Chapter 161

OAKS
(*Quercus* Species)

BOTANICAL DESCRIPTION

Oak species involved in poisonings of grazing animals include the following: United States, *Q. havardii* Rydb. (Harvard oak), *Q. gambelii* Nutt. (Gambel oak); Europe, *Q. robur* L. (English oak), *Q. sinuata* Walter (bastard oak); Israel, *Q. coccifera* L. (Palestine oak); and Himalayas, *Q. incana* Bartram (bluejack oak).[1]

Common Name: Oaks, Live Oaks (evergreens)

Scientific Name: *Quercus* speciesThere are over 200 species of oaks with about 60 species of evergreen oaks in the United States and Canada.

Botanical Family: Fagaceae (beech)

Physical Description: The *Quercus* species range from bushes to trees and are characterized by their fruit, called the acorn. Small, yellowish-green flowers (catkins) typically produce a single seed that matures slowly. This fruit has a detachable cap covering a hard oval body.

Distribution and Ecology: This genus contains both evergreen and deciduous species native to the northern hemisphere. Oaks are widely distributed from cold latitudes to tropical areas of North America, South America, Europe, and Asia.

EXPOSURE

Ethanolic extracts of *Quercus* species have been used in Jordanian folk medicine for the treatment of digestive disorders, including ulcers.[2] Tannic acid was introduced as a cleansing agent in barium enemas for the delineation of mucosal lesions.[3] The origin of commercial tannic acid used in barium enemas in the middle 1900s was Turkish nutgalls gathered from young twigs of *Quercus infectoria* Oliver (aleppo oak).[4] The use of tannic acid continued into the 1960s, when case reports and animal studies indicated that high doses of tannic acid produce coagulative necrosis of the colon and hepatotoxicity.

PRINCIPAL TOXINS

Physiochemical Properties

The primary toxic agents in acorns are polyhydroxyphenolic compounds, which are commonly called tannins. This complex mixture consists of polyhydroxyphenolic compounds that vary in molecular weight between 500 and 3,000 Da. Tannins include hydrolyzable tannins (e.g., gallotannins, ellagitannins) and condensed or nonhydrolyzable tannins (proanthocyanidins).[5] The most important toxins in tannins include tannic acid, pyrogallol, and gallic acid.[6] The extraction of powdered galls with ether, alcohol, and water produces a complex mixture of substances including about 50–70% tannic acid. Hydrolysis of commercial tannic acid forms primarily glucose and gallic acid. Tannic acids oxidize easily in air or light to create dark substances.

Poisonous Parts

The concentration of total phenolics in acorns varies between different species of oaks. The concentrations of these chemicals are higher in green acorns than in

Medical Toxicology of Natural Substances, by Donald G. Barceloux, MD

mature ones.[7] Mature acorns are edible after processing removes the tannins. For example, the mean concentration of total phenolics by weight in acorns of *Q. velutina* Lam. (black oak) was 4.51% (range: 3.29–6.13%) compared with 1.09% (range: 0.41–2.54%) for acorns from *Q. alba* L. (white oak).[8] Mature oak leaves contain a small amount of tannins, and the consumption of mature oak leaves by grazing animals does not usually produce toxicity.[9]

DOSE RESPONSE

Animals

Rodent studies indicate that enemas containing concentrations of tannic acid >2% cause edema and inflammation of the colon as well as coagulative necrosis and hemorrhage at concentrations of 16%.[10]

Humans

Several case reports associated the use of tannic acid with fatal hepatotoxicity. A case report associated the deaths of three children with the administration of two or three enemas containing 0.75% tannic acid during the same morning.[11] Five fatalities also occurred in a group of patients receiving barium enemas containing 2% tannic acid.[12] The use of enemas containing <0.25–0.5% were not associated with hepatotoxicity in a study of human volunteers.[13]

TOXICOKINETICS

There are few data on the pharmacokinetics of tannins in humans. Volunteer studies suggest that ellagitannins (e.g., ellagic acid, castalagin, vescalagin) are metabolized (e.g., urolithin B) and conjugated with glucuronic acid before excretion in the urine.[14] The kidneys excrete little unchanged ellagic acid in the urine.

CLINICAL RESPONSE

Chronic consumption of *Quercus* parts (immature leaves, buds, acorns) by grazing animals (e.g., cattle, sheep) produces gastrointestinal toxicity and renal damage, leading to numerous deaths.[1,15,16] In most human ingestions, toxic effects are not reported.[17] Loose stools were the most common adverse effect in calls to a poison control center concerning acorn consumption. The oak processionary moth (*Thaumetopoea processionea* L.) inhabit *Quercus* (oak) species, and the third to sixth larval instars possesses poisonous hairs (setae) that contain an urticating toxin (thaumetopoein, CAS RN: 99332-51-1) potentially harmful to humans.[18] An anaphylactic response (generalized urticaria, angioedema, vomiting, diarrhea, hypotension) occurred in a 45-year-old man about 15 minutes after ingesting three acorn nuts.[19]

DIAGNOSTIC TESTING

Analytical methods to analyze extractable tannins (e.g., gallic and ellagic acid esters) include liquid chromatography/electrospray ionization/mass spectrometry (LC/ESI/MS),[20] nuclear magnetic resonance,[21] and reversed-phase liquid chromatography/electrospray ionization/tandem mass spectrometry (LC/ESI/MS/MS).[22] Radioallergosorbent (RAST) testing revealed IgE-specific antibodies to acorn, peanut, grass pollen, and *Quercus ilex* pollen extracts.[19]

TREATMENT

Casual ingestions require only symptomatic care for gastrointestinal distress (e.g., clear liquid diet). Decontamination measures are unnecessary.

References

1. Garg SK, Makkar HP, Nagal KB, Sharma SK, Wadhwa DR, Singh B. Oak (*Quercus incana*) leaf poisoning in cattle. Vet Hum Toxicol 1992;34:161–164.
2. Alkofahi A, Atta AH. Pharmacological screening of the anti-ulcerogenic effects of some Jordanian medicinal plants in rats. J Ethnopharmacol 1999;67:341–345.
3. Hamilton JB. Use of tannic acid in barium enemas. Am J Roentgenol Radium Ther 1946;56:101–103.
4. Krezanoski JZ. Tannic acid: chemistry, analysis, and toxicology. Radiology 1966;87:655–657.
5. Niemetz R, Gross GG. Enzymology of gallotannin and ellagitannin biosynthesis. Phytochemistry 2005;66: 2001–2011.
6. Sandusky GE, Forsnaugh C, Smith JB, Mohan R. Oak poisoning of cattle in Ohio. J Am Vet Med Assoc 1977;171: 627–629.
7. Wiseman A, Thompson H. Acorn poisoning. Vet Rec 1984;115:605.
8. Basden KW, Dalvi RR. Determination of total phenolics in acorns from different species of oak trees in conjunction with acorn poisoning in cattle. Vet Hum Toxicol 1987;29:305–306.
9. Flaoyen A, Handeland K, Arnemo JM, Vikoren T. Toxicity testing of leaves from oak (*Quercus robur*) harvested in Aust-Agder county, Norway. Vet Res Comm 1999;23: 317–321.
10. Zboralske FF, Harris PA, Riegelman S, Rambo ON, Margulis AR. Toxicity studies on tannic acid administered by enema. III. Studies on the retention of enemas in

humans. IV. Review and conclusions. Am J Roentgeol Radium Ther Nucl Med 1966;96:505–509.

11. McAlister WH, Anderson MS, Bloomberg GR, Margulis AR. Lethal effects of tannic acid in barium enema: report of three fatalities and experimental studies. Radiology 1963;80:765–773.
12. Lucke HH, Hodge KE, Patt NL. Fatal liver damage after barium enemas containing tannic acid. Can Med Assoc J 1963;89:1111–1114.
13. Burhenne J, Vogelaar P, Arkoff RS. Liver function studies in patients receiving enemas containing tannic acid. Am J Roentgeol Radium Ther Nucl Med 1966;96: 510–518.
14. Cerda B, Tomas-Barberan FA, Espin JC. Metabolism of antioxidant and chemopreventive ellagitannins from strawberries, raspberries, walnuts, and oak-aged wine in humans: identification of biomarkers and individual variability. J Agric Food Chem 2005;53:227–235.
15. Yeruham I, Avidar Y, Perl S, Yakobson B, Shlosberg, Hanji V, Bogin E. Probable toxicosis in cattle in Israel caused by the oak *Quercus calliprinos*. Vet Hum Toxicol 1998;40: 336–340.
16. Spier SJ, Smith BP, Seawright AA, Norman BB, Ostrowski SR, Oliver MN. Oak toxicosis in cattle in northern California: clinical and pathological findings. J Am Vet Med Assoc 1987;191:958–964.
17. Bronstein AC, Spyker DA, Cantilena LRJr , Green J, Rumack BH, Heard SE. 2006 Annual Report of the American Association of Poison Control Centers' National Poison Data System (NPDS). Clin Toxicol (Phila) 2007;45:815–917.
18. Maier H, Spiegel W, Kinaciyan T, Krehan H, Cabaj A, Schopf A, Honigsmann H. The oak processionary caterpillar as the cause of an epidemic airborne disease: survey and analysis. Br J Dermatol 2003;149:990–997.
19. Vega A, Dominguez C, Cosmes P, Martinez A, Bartolome B, Martinez J, Palacios R. Anaphylactic reaction to ingestion of *Quercus ilex* acorn nut. Clin Exp Allergy 1998; 28:739–742.
20. Mammela P, Savolainen H, Lindroos L, Kangas J, Vartiainen T. Analysis of oak tannins by liquid chromatography-electrospray ionisation mass spectrometry. J Chromatogr A 2000;891:75–83.
21. Glabasnia A, Hofmann T. Identification and sensory evaluation of dehydro- and deoxy-ellagitannins formed upon toasting of oak wood (*Quercus alba* L.). J Agric Food Chem 2007;55:4109–4118.
22. Zywicki B, Reemtsma T, Jekel M. Analysis of commercial vegetable tanning agents by reversed-phase liquid chromatography-electrospray ionization-tandem mass spectrometry and its application to wastewater. J Chromatogr A 2002;970:191–200.

Chapter 162

PEPPER TREE
(*Schinus* Species)

BOTANICAL DESCRIPTION

Common Name: Peruvian Peppertree, California Peppertree, Molle

Scientific Name: *Schinus molle* L.

Botanical Family: Anacardiaceae (cashew)

Physical Description: This evergreen shrub or tree reaches 3–15 m (~10–50 ft) in height and has pinnately compound leaves that emit a malodorous smell when crushed similar to black pepper. The trunk is relatively short with deeply fissured, dark brown bark. Viscous, white latex extrudes from damaged bark. The branches have a graceful, weeping habit similar to a willow, and the small, white to yellow flowers have four to five petals. The flowers produce small, red berries. The appearance of this tree is similar to the mastic tree (*Pistacia lentiscus* L.)

Distribution and Ecology: A native tree of South America, the *Schinus* species is naturalized in Mexico, Florida, southern Arizona, and southern California.

Common Name: Brazilian Peppertree, Florida Holly, Christmas Berry

Scientific Name: *Schinus terebinthifolius* Raddi

Botanical Family: Anacardiaceae (cashew)

Physical Description: This tree is smaller (up to 30 ft/~9 m) and has a more erect habit than *S. molle* with fewer, but larger, leaflets. White flowers produce clusters of glossy, green berries, which turn red during maturity. The crushed leaves and fruits emit a less intense pepper-like odor compared with *S. molle*.

Distribution and Ecology: The Brazilian peppertree is a native species of South America that is cultivated in the temperate climates of California, Texas, Florida, New Zealand, Australia, Caribbean, and the South Pacific Islands.

EXPOSURE

In Brazil, medicinal uses for crushed, dried leaves of *S. terebinthifolius* include antiseptic poultices on ulcers, gouty joints, painful muscles, and strained tendons.[1] Leaf infusions are given to treat bronchitis and other respiratory ailments.

PRINCIPAL TOXINS

Extracts of crushed leaves of *S. terebinthifolius* contain triterpenes, such as ursolic, masticadienonic, and hydroxymasticadienoic acids.[2] The neutral, oily portion of crushed berries contains monoterpene hydrocarbons, such as α-pinene, β-pinene, δ3-carene, α-phellandrene, β-phellandrene, limonene, and *p*-cymene.[3] The triterpene compounds [e.g., masticadienonic acid, masticadienolic acid (schinol)] are specific inhibitors of phospholipase A_2. This latter substance affects a variety of inflammatory processes as a result of the mobilization of eicosanoids and the metabolism of phospholipids.[4]

Medical Toxicology of Natural Substances, by Donald G. Barceloux, MD

DOSE RESPONSE

There are inadequate data to determine the dose response following exposure to *Schinus* species.

TOXICOKINETICS

There are few data on the pharmacokinetics of polyphenol compounds in *Schinus* species.

CLINICAL RESPONSE

The pepper tree (*Schinus* spp.) belongs to the cashew (Anacardiaceae) family that contains potent sensitizers. Contact with plant parts from the pepper tree may cause a mild allergenic dermatitis in sensitized individuals characterized by pruritus and a maculopapular rash, particularly in woodcutters involved with cutting branches of the Brazilian pepper tree. Ingestion of large amounts of berries produces gastrointestinal irritation characterized by nausea, vomiting, and diarrhea.

DIAGNOSTIC TESTING

Methods to identify polyphenol compounds in *Schinus* species include high performance liquid chromatography (HPLC).[5]

TREATMENT

Treatment is supportive. Skin lesions following contact with parts of the pepper tree should respond to topical steroids (0.05% fluocinonide) and antihistamines after the skin is cleaned. Decontamination measures are unnecessary.

References

1. Morton JF. Brazilian pepper—its impact on people, animals and the environment. Econ Bot 1978;32:353–359.
2. de Paivo Campello J, Marsaioli AJ. Triterpenes of *Schinus terebinthifolius*. Phytochemistry 1974;13:659–660.
3. Lloyd HA, Jaquni TM, Evans SL, Morton JF. Terpenes of *Schinus terebinthifolius*. Phytochemistry 1977;16:1301–1302.
4. Jain MK, Yu B-Z, Rogers JM, Smith AE, Boger ET, Ostrander RL, Rheingold AL. Specific competitive inhibitor of secreted phospholipase A_2 from berries of *Schinus terebinthifolius*. Phytochemistry 1995;39:537–547.
5. Queires LC, Fauvel-lafetve F, Terry S, De la Taille A, Kouyoumdjian JC, Chopin DK, et al. Polyphenols purified from the Brazilian aroeira plant (*Schinus terebinthifolius* Raddi) induce apoptotic and autophagic cell death of DU145 cells. Anticancer Res 2006;26:379–387.

Chapter 163

YEW
(*Taxus* Species)

HISTORY

The use of the yew predates Christian churches, and knowledge of the toxicity of this plant dates back to classical times. The Celts used yew as agent for ritual suicides and a poison on the tips of arrows during the Gaelic Wars.[1] The yew was a sacred tree and symbol of immortality for Druid priests. Julius Caesar documented the poisoning of the King of Eburones, Catibulus, by yew "juice in the 2nd century BC".[2] Yew trees were often planted on sacred sites, and Christians adopted this custom. The Third Witch in Macbeth mentions yew as a constituent of the caldron's brew.[3] Decoctions of yew leaf were abortifacients in Europe and India during the 18th and 19th centuries.[4] Lucas was the first scientist to recover a white, noncrystalline powder from European yew (*Taxus baccata*) in 1856; he called the residual powder taxine.[5] Twenty years later, the French scientist, Marmé, isolated a crystalline form of taxine using a similar extraction method as Lucas. By 1890, Hilger and Brande determined the basic molecular structure of taxine.[6]

The first description of yew toxicity appeared in the English medical literature in 1836 when Hurt reported the death of a 3-year-old child following the ingestion of yew berries.[7] In the 1800s, the mastication of yew seeds was associated with convulsions.[8] Taxanes (paclitaxel, docetaxel) are novel antitumor agents isolated from the Pacific yew tree and the European yew. During the US National Cancer Institute drug screening program of the 1960s, Wani et al discovered the chemical structure and cytotoxic properties of paclitaxel,[9] and by 1979 paclitaxel was a recognized spindle poison. Preclinical studies on the antitumor activity of paclitaxel began in the 1980s. Docetaxel, a semisynthetic derivative from the needles of the European yew, entered clinical trials as a chemotherapeutic agent for metastatic cancer in the early 1990s.[10] Currently, taxanes are widely administered for the medical treatment of cancer.[11]

BOTANICAL DESCRIPTION

Common (Scientific) Name: Western or Pacific Yew (*Taxus brevifolia* Nutt.), Canadian Yew (*Taxus canadensis* Marsh.), English Yew (*Taxus baccata* L.), Hicks Yew (*Taxus* x *media* Rehder), Japanese Yew (*Taxus cuspidata* Sieb. & Zucc.)

Botanical Family: Taxaceae (yew)

Physical Description: The Pacific yew is a conifer that grows up to 5–15 m (~15–50 ft) with drooping branches and alternate branchlets. The characteristic flattened, stiff, green, needlelike leaves are about 2–3 cm (~0.8–1 in.) long. The upper surface is dark green with a prominent midrib, and the undersurface is more pale yellow. The large, hard seeds are partially enclosed by a fleshy red cup called an aril. English yew (*Taxus baccata*) is an evergreen shrub several feet tall with linear spirally arranged dark green leaves. The bark is thin and red or red-brown with a scaly appearance. Figure 163.1 displays a specimen of the English yew (*T. baccata* L.).

Distribution and Ecology: Western or Pacific yew is a tall tree native to mountain gorges below 7,000 ft (~2,000 m) in mixed evergreen forests from

FIGURE 163.1. English yew (*Taxus baccata* L.). Photograph courtesy of Professor Sylvie Rapior. See color insert.

northern California to Alaska and Montana. *Taxus cuspidata* Sieb. & Zucc. (Japanese yew) and *Taxus baccata* L. (English yew) are cultivated as ornamental shrubs. *Taxus canadensis* Marsh. and *Taxus brevifolia* Nutt. are native species in the eastern and western United States, respectively. Most species of yew are dark green evergreen shrubs that flourish in many soil types.

EXPOSURE

Aqueous extracts of yew are an Indian folk medicine (i.e., Zarnab) used for their cardiotonic, expectorant, antispasmodic, diuretic, and antiseptic properties.[12] Extracts of *Taxus celebica* (Warb.) H. L. Li contain a flavonoid (sciadopitysin) used in traditional Chinese medicine for the treatment of diabetes mellitus. Paclitaxel (Taxol®, Bristol-Myers Squibb Company, Princeton, NJ) and docetaxel (Taxotere®, Sanofi-Aventis, Bridgewater, NJ) are semisynthetic derivatives of the bark of *T. brevifolia* (Pacific yew) and *T. baccata* (European yew), respectively. These compounds are spindle poisons that possess antitumor activities. Paclitaxel is also used as a starting skeleton for other chemotherapeutic agents and in paclitaxel-eluting stents in cardiovascular therapies.[13] Japanese and English yew plants are ornamental shrubs with red berries.

PRINCIPAL TOXINS

Physiochemical Properties

Taxus species contain more than 350 naturally occurring taxoids (taxane diterpenoids).[14] These diterpenoid compounds are a mixture of alkaloids formed from nitrogen-free polyhydroxyl diterpenes (taxicins) esterified with β-dimethylamino-β-phenylpropionic acid and acetic acid.[15] Most taxoids possess a pentamethyl tricyclopentadecane (taxane) skeleton, although some taxoids have bicyclic structures or rearranged skeletons (abeotaxoids). Isotaxine B, a structural isomer of taxine B, is the major constituent of the alkaloid fraction of *Taxus* species with minor constituents including taxine A, 2-deacetyltaxine A, deoxytaxine B, 1-deoxyisotaxine B, and taxine B pseudoalkaloids.[16] Taxanes are novel compounds that share a complex taxane ring (baccatin III). This class of compounds includes the chemotherapeutic agent, paclitaxel from *T. brevifolia* and docetaxel from *T. baccata*. The concentration of taxanes in the low trailing shrub, Canadian yew (*Taxus canadensis*) differs for other yews. This plant contains a variety of uncommon taxanes including 15-benzoyl-10-deacetyl-2-debenzoyl-10-dehydro-abeo-baccatin III; 15-benzoyl-2-debenzoyl-7, 9-dideacetyl-abeo-baccatinVI; *N*-acetyl-*N*-debenzoyltaxol; 7,9,13-trideacetylbaccatin VI; 10-deacetyl-10-glycolylbaccatin IV; 1-β-hydroxy-10-deacetyl-10-glycolylbaccatin I; and 7-deacetyltaxuspine L in addition to taxanes more commonly present in other yew species (e.g., 9-dihydro-13-acetylbaccatin III, taxacustin, taxagifine, 2-deacetyl-7,10-diacetyl-5-deaminoacyl taxine A).[17,18]

Poisonous Parts

Taxine alkaloids are present in all parts of the yew plant except the edible scarlet aril (berry).[4] The taxine content of yew plants varies with season and species. Maximal taxine concentrations occur in winter, and taxines are relatively abundant in English yew (*T. baccata*) and Japanese yew (*T. cuspidata*) compared with Pacific yews (*T. brevifolia*).[2]

Mechanism of Toxicity

Taxine B is the principal cardiotoxin in *Taxus* species. In isolated papillary muscle preparations, taxine B significantly reduces the maximum rate of depolarization of the action potential similar to class I antiarrhythmic drugs (flecainide, quinidine, procainamide).[19] The mechanism of toxicity is an alteration of calcium and sodium channel conductance that increases cytosolic calcium concentrations. The clinical consequence of these alterations is hypotension, bradycardia, and depressed myocardial contractility and conduction delay, similar to digitalis poisoning.[20] Taxanes (paclitaxel, docetaxel) are spindle poisons that possess antitumor activity as a result of the binding of these compounds to β-subunit of tubulin causing microtubule polymers, mitotic arrest, and cell death. Postmortem examination of patients dying of yew poisoning is usually nonspecific with generalized congestion of internal organs, dilation of the ventricles, and unremarkable microscopic findings. Histological evidence of taxine poisoning on postmortem examinations included alveolar hemorrhagic edema, focal pulmonary hemorrhages, and pronounced interstitial myocardial edema with early signs of ischemia (swollen nuclei, homogenization and eosinophilia of the sarcoplasm, perinuclear vacuolization).[21] Immunohistochemical staining for troponin C was positive with depletion of troponin C in many areas of the left ventricles.

DOSE RESPONSE

Most casual ingestions of berries result either in no symptoms or in mild gastrointestinal distress. After the ingestion of at least 5 whole leaves and 10 seeds, a 19-month-old toddler remained asymptomatic following the administration of syrup of ipecac.[22] The fatal dose of yew leaves is unknown, but the estimated lethal dose of chopped yew leaves is about 50–100 g. The ingestion of 150 yew leaves (*Taxus baccata*) by a 40-year-old woman[23] and four to five handfuls of yew leaves by a psychiatric patient produced ventricular tachycardia, refractory cardiogenic shock, and death.[20]

Paclitaxel (Taxol®, Bristol-Myers Squibb Company, Princeton, NJ) and the synthetic analogue docetaxel (Taxotere®, Sanofi-Aventis, Bridgewater, NJ) are novel chemotherapeutic drugs used in the treatment of advanced breast cancer. The dosages in phase II/III trials were an infusion of 135–200 mg/m^2 and 100 mg/m^2, respectively, once every 3 weeks.[24] During clinical trials, paclitaxel doses in the range of 200–250 mg/m^2 produced neutropenia and myelosuppression, whereas doses of 390 mg/m^2 caused severe mucositis.[25]

TOXICOKINETICS

Elimination of taxanes (paclitaxel, docetaxel) occurs primarily by hepatic metabolism and biliary excretion. Paclitaxel undergoes stereospecific hydroxylation by CYP2C8 to form the major metabolite, 6-α-OH-paclitaxel and by CYP3A4 to the minor metabolite, 3′-*p*-hydroxyphenyl-paclitaxel.[26]

CLINICAL RESPONSE

Most cases of yew berry ingestion result in no symptoms. Symptomatic patients typically develop gastrointestinal distress about 1–3 hours after ingestion. Serious cardiac toxicity may occur after large ingestions including ventricular dysrhythmias, intraventricular conduction delay, cardiogenic shock, and death.[27] Lightheadedness, nausea, and abdominal pain developed within 1 hour after the ingestion of four to five handfuls of yew leaves ingestion, followed by coma, wide complex ventricular tachycardia, QRS prolongation, and cardiopulmonary arrest.[20] Several case reports associate the administration of the flavonoid, sciadopitysin, from traditional Chinese extracts of *Taxus celebica* (Warburg) H. L. Li with acute interstitial nephritis, hemolysis, and acute tubular necrosis.[28] Allergic responses may develop following exposure to yew. Case reports associate chewing yew needles with the immediate onset of anaphylactic shock,[29] and cutting yew wood with a severe contact dermatitis.[30]

DIAGNOSTIC TESTING

Analytical methods to quantitate ingredients in *Taxus* species include high performance liquid chromatography with photodiode detection (HPLC/PDA),[31] immunoaffinity chromatography/HPLC,[32] liquid chromatography/mass spectrometry (LC/MS),[33] gas chromatography/mass spectrometry (GC/MS),[34] and liquid chromatography/tandem mass spectrometry (LC/MS/MS).[35] Other methods include competitive-inhibition enzyme immunoassay[36] and an enzyme-linked immunosorbent assay (ELISA) method. The latter method detects free, biologically active taxine compounds in patients receiving paclitaxel or docetaxel for advanced cancer.[37] The toxic alkaloid, taxine B, is unstable and difficult to detect in postmortem blood.

Gas chromatography/mass spectrometry analysis of postmortem blood samples from a 19-year-old man, who was found dead after ingesting both leaves and tea brewed from the yew leaves, did not detect taxine compounds.[38] However, the detection of 3,5-dimethoxyphenol in heart blood at a concentration of 320 μg/kg provided *indirect* evidence of yew ingestion. 3,5-Dimethoxyphenol is the aglycon of the *Taxus* ingredient, taxicatine, which results from the glucosidic cleavage of taxicatine. In a case series of five individuals found dead with evidence of yew poisoning, cardiac blood samples contained the following concentrations of 3,5-dimethoxyphenol: 31 ng/mL, 47 ng/mL, 97 ng/mL, 110 ng/mL, and 528 ng/mL.[39] Although analyses of samples from femoral blood suggested postmortem redistribution, there were insufficient samples to determine the extent of postmortem redistribution. Analysis of heart blood from a 43-year-old man found dead after ingesting tea brewed from *Taxus baccata* demonstrated the presence of total taxines (taxine B, isotaxine B) equivalent to 11 ng/g tissue as measure by high performance liquid chromatography/mass spectrometry.[21] Microscopic examination of contents from the gastrointestinal tract may reveal characteristic stomata on undigested yew leaves.[40]

TREATMENT

Most exposures to *Taxus* species require no decontamination because whole seeds are relatively nontoxic.[41] The administration of activated charcoal to alert children, who ingest more than several *Taxus* seeds less than 1 hour prior to presentation, is a therapeutic option, but there are no clinical data to confirm the clinical efficacy of activated charcoal following the ingestion of yew. Asymptomatic patients or those with only mild gastrointestinal distress may be released after 4 hours of observation. There are limited clinical data to guide management in serious ingestions involving the ingestion of large amounts of yew leaves or extracts of the yew plant. Serious ingestions require monitoring of oxygen saturation, an intravenous line, and cardiac monitoring in an intensive care setting. Cardiac pacing may be required for severe conduction disturbances. There is no clinical evidence to support the use of antidotes including digoxin-specific Fab antibody fragments.[42]

References

1. Hartzell H. Yew and us: a brief history of the yew tree. In: Suffness M, editor. Taxol: science and applications. Boca Raton, FL: CRC Press; 1995;27–34.
2. Wilson CR, Sauer J-M, Hooser SB. Taxines: a review of the mechanism and toxicity of yew (*Taxus* spp.) alkaloids. Toxicon 2001;39:175–185.
3. Lee MR. The yew tree (*Taxus baccata*) in mythology and medicine. Proc R Coll Physicians Edinb 1998;28;569–575.
4. Bryan-Brown T. The pharmacological actions of taxine. Quar J Pharm Pharmacol 1932;5:205–219.
5. Lucas H. Ueber ein in den blätter von *Taxus baccata* L. enhaltenes alkaloid (das taxin). Archiv Pharmaz 1856;135: 145–149.
6. Hilger A, Brande F. Ueber taxin, das alkaloid des eibenbaumes (*Taxus baccata*). Berichte Deutsch Chem Ges 1890;23:464–468.
7. Hurt S. Poisonous effects of the berries or the seeds of the yew. Lancet 1836;i:394–395.
8. Thompson J. Poisoning by yew berry. Lancet 1868;95:530.
9. Wani MC, Taylor HL, Wall ME, Coggon P, McPhail AT. Plant antitumor agents. VI. The isolation and structure of taxol, a novel antileukemic and antitumor agent from *Taxus brevifolia*. J Am Chem Soc 1971;93:2325–2327.
10. Ringel I, Horwitz SB. Studies with RP56976 (Taxotere): a semisynthetic analogue of taxol. J Natl Cancer Inst 1991;83:288–291.
11. Ferlini C, Gallo D, Scambia G. New taxanes in development. Expert Opin Investig Drugs 2008;17:335–347.
12. Vohora SB, Kumar I. Studies on *Taxus baccata*. I. Preliminary phytochemical and behavioral investigations. Planta Med 1971;2:100–107.
13. Kingston DG, Newman DJ. Taxoids: cancer-fighting compounds from nature. Curr Opin Drug Discov Devel 2007;10:130–144.
14. Arslanian RL, Bailey DT, Kent MC, Richheimer SL, Thornburg KR, Timmons DW, Zheng QY. Brevitaxin, a new diterpenolignan from the bark of *Taxus brevifolia*. J Nat Prod 1995;58:583–585.
15. Baxter JN, Lythgoe B, Scales B, Scrowston RM, Trippett S. Taxine. Part I: Isolation studies and the functional groups of *o*-cinnamoyltaxicin-I. J Chem. Soc 1962;2964–2972.
16. Adeline MT, Wang XP, Poupat C, Ahond A, Potier P. Evaluation of toxoids from *Taxus* spp. Crude extracts by high performance liquid chromatography. J Liq Chromatogr Rel Technol 1997;20:3135–3145.
17. Zhang J, Sauriol F, Mamer O, Zamir LO. Taxoids from the needles of the Canadian yew. Phytochemistry 2000;54: 221–230.
18. Shi Q-W, Sauriol F, Mamer O, Zamir LO. New taxanes from the needles of *Taxus canadensis*. J Nat Prod 2003;66:470–476.
19. Alloatti G, Penna C, Levi RC, Gallo MP, Appendino G, Fenoglio I. Effects of yew alkaloids and related compounds on guinea-pig isolated perfused heart and papillary muscle. Life Sci 1996;58:845–854.
20. Shulte T. Lethal intoxication with leaves of the yew tree (*Taxus baccata*). Arch Toxicol 1975;34:153–158.
21. Beike J, Karger B, Meiners T, Brinkmann B, Kohler H. LC-MS determination of *Taxus* alkaloids in biological specimens. Int J Legal Med 2003;117:335–339.
22. Porter KA, Kroll S. Yewberry ingestion. Br Med J 1982;284:116.
23. Yersin B, Frey J-G, Schaller M-D, Nicod P, Perret C. Fatal cardiac arrhythmias and shock following yew leaves ingestion. Ann Emerg Med 1987;1396–1397.
24. Mantle D, Lennard TWJ, Pickering AT. Therapeutic applications of medicinal plants in the treatment of breast cancer: a review of their pharmacology, efficacy, and tolerability. Adv Drug React Toxicol Rev 2000;19:223–240.
25. Guchelaar H-J, Ten Napel CHH, de Vries EGE, Mulder NH. Clinical, toxicological and pharmaceutical aspects of the antineoplastic drug Taxol: a review. Clin Oncol 1994;6:40–48.
26. Walle T, Walle UK, Kumar GN, Bhalla KN. Taxol metabolism and disposition in cancer patients. Drug Metab Dispos 1995;23:506–512.
27. Janssen J, Peltenburg H. [A classical way of committing suicide with *Taxus baccata*.] Ned Tijdschr Geneeskd 1985;129:603–605. [Dutch]
28. Lin J-L, Ho Y-S. Flavonoid-induced acute nephropathy. Am J Kid Dis 1994;23:433–440.
29. Burke MJ, Siegel D, Davidow B. Anaphylaxis. Consequence of yew (*Taxus*) needle ingestion. NY J Med 1979;79:1576–1577.

30. Woods B, Calnan CD. Toxic woods. Br J Dermatol 1976;95(Suppl 13):1–12.
31. Mroczek T, Glowniak K. Solid-phase extraction and simplified high-performance liquid chromatographic determination of 10-deacetylbaccatin III and related taxoids in yew species. J Pharm Biomed Anal 2001;26: 89–102.
32. Theodoridis G, Haasnoot W, Cazemier G, Schilt R, Jaziri M, Diallo B, et al. Immunoaffinity chromatography for the sample pretreatment of Taxus plant and cell extracts prior to analysis of taxanes by high-performance liquid chromatography. J Chromatogr A 2002;948:177–185.
33. Kite GC, Lawrence TJ, Dauncey EA. Detecting *Taxus* poisoning in horses using liquid chromatography/mass spectrometry. Vet Hum Toxicol 2000;42:151–154.
34. Tiwary AK, Puschner B, Kinde H, Tor ER. Diagnosis of *Taxus* (yew) poisoning in a horse. J Vet Diagn Invest 2005;17:252–255.
35. Frommherz L, Kintz P, Kijewski H, Köhler H, Lehr M, Brinkmann B, Beike J. Quantitative determination of taxine B in body fluids by LC-MS-MS. Int J Legal Med 2006;120:346–351.
36. Grothaus PG, Bignami GS, O'Malley S, Harada KE, Byrnes JB, Waller DF, et al. Taxane-specific monoclonal antibodies: measurement of taxol, baccatin III, and "total taxanes" in *Taxus brevifolia* extracts by enzyme immunoassay. J Nat Prod 1995;58:1003–1014.
37. Svojanovsky SR, Egodage KL, Wu J, Slavik M, Wilson GS. High sensitivity ELISA determination of taxol in various human biological fluids. J Pharm Biomed Anal 1999;20: 549–555.
38. Musshoff F, Jacob B, Fowinkel C, Daldrup T. Suicidal yew leave ingestion—phloroglucindimethylether (3,5-dimethoxyphenol) as a marker for poisoning from *Taxus baccata*. Int J Leg Med 1993;106:45–50.
39. Pietsch J, Schulz K, Schmidt U, Andresen H, Schwarze B, Dreßler J. A comparative study of five fatal cases of *Taxus* poisoning. Int J Legal Med 2007;121:417–422.
40. van Ingen G, Visser R, Peltenburg H, van der Ark AM, Voortman M. Sudden unexpected death due to *Taxus* poisoning. A report of five cases, with review of the literature. Forensic Sci Int 1992;56:81–87.
41. Krenzelok EP, Jacobsen TD, Aronis J. Is the yew really poisonous to you? Clin Toxicol 1998;36:219–223.
42. Cummins RO, Haulman J, Quan L, Graves JR, Peterson D, Horan S. Near-fatal yew berry intoxication treated with external cardiac pacing and digoxin-specific Fab antibody fragments. Ann Emerg Med 1990;19:38–43.

Part 5

VENOMOUS ANIMALS

PART 5 VENOMOUS ANIMALS

I Arthropods

A Arachnids

Chapter 164

MITES and TICKS (Order: Acari)

Mites and ticks belong to the phylum Arthropoda, class Arachnida, and order Acari (Acarina). This taxon contains the following two separate lineages: Acariformes (mites) and the Parasitiformes (bird mites, soft-bodied and hard-bodied ticks).

MITES

Mites are ubiquitous creatures that are barely visible to the naked eye and cause significant pruritic dermatoses. These animals are easily distinguished from other arachnids by a distinctive gnathosoma (capitulum with the mouth parts) and the absence of a division between the abdomen and the cephalothorax.[1] Mites can infest cheese, flour, sugar, cereals, grains (*Pyemotes ventricosus* Newport), dried meats, vegetables, feathers, birds (*Dermanyssus gallinae* De Geer, *Dermanyssus avium* De Geer), rats, cats, dogs, and rabbits. With the exception of the human follicle mites (*Demodex folliculorum* Simon) and the scabies mite [*Sarcoptes scabiei* (L.)], most mites leave the human host soon after feeding rather than burrow into human skin. Mites do not possess any specialized venom apparatus, and they usually return to their natural hosts after feeding. The *Demodex* mites feed on sebum in hair follicles and sebaceous glands primarily on the nose, forehead, and cheeks. There is no definite association between the presence of these nonblood sucking mites and human disease.[2] Bites from a variety of these arachnids produce pruritic, excoriated lesions with and without central puncta as a result of the injection of reactive salivary proteins deposited during feeding on human blood.[3] Although only a small portion of people infested with these mites develop symptoms, extensive crusted lesions of the scalp may appear during an infestation.[4] Clinical characteristics of mite bites include unexplained pruritus, closely associated persons with similar lesions, absence of other insects (e.g., bedbugs, lice, mosquitoes, fleas, spiders), and the lack of burrows.[5] Human eruptions range from mild, pruritic, papular lesions to severe vesicles and pustules. Infestations of the eyelids may cause conjunctival inflammation, superficial corneal vascularization, marginal corneal infiltration, superficial corneal opacity, and nodular corneal scarring.[6] Lesions resulting from animal mites are typically symmetric in the area of greatest contact with the pet (i.e., inner forearms, chest, abdomen). Identification of mites requires a careful search of the environment because these mites do not remain on human skin for identification.[7] Infected cats and dogs may be asymptomatic.[8] Beside elimination of the mites, treatment of mite-induced lesions includes the use of topical antipruritic agents (e.g., menthol, camphor, pramoxine), and oral antihistamines.

Trombiculidae (Chiggers/Harvest Mites)

This family includes chiggers (harvest mites, red bugs). Chiggers are the parasitic larval form of harvest mites (e.g., *Trombicula alfreddugesi* Oudemans) in the

Medical Toxicology of Natural Substances, by Donald G. Barceloux, MD

Trombiculidae family. The larval stages produce crusted lesions similar to chicken pox. They possess a bright red, hairy body 0.5 mm × 0.25 mm. Contact usually occurs during periods of peak outdoor activity during late summer and fall when the larvae crawl along the ground and grass after emerging from eggs laid in the soil.[9] Small (1–2 mm) macules usually appear near belt lines, ankles, or moist intertriginous areas where the larva pierces the skin with its stylostome and releases proteolytic enzymes to liquefy epidermal cells. These mites feed on lymph and tissue dissolved by their proteolytic enzymes rather than blood. The initial bite is painless, but an intensely pruritic papule surrounded by a clear vesicle develops within 3–24 hours. The lesion may become ecchymotic. The mite typically leaves the human host after feeding and before the pruritus develops. A 0.5- to 2.0-cm (~0.2–0.8 in.) lesion persists for several weeks and eventually crusts, similar to chicken pox. Treatment is supportive with prevention of secondary infection. Anecdotal evidence suggests that application of an ointment containing 5–10% sulfur and 1% phenol may relieve pruritus and remove any remaining mites. Insect repellents (e.g., diethyltoluamide or *N,N*-diethyl-meta-toluamide [DEET]) effectively repel biting insects, such as biting flies, gnats, chiggers, and ticks.

Scabies

IDENTIFYING CHARACTERISTICS

Common Name: Scabies

Scientific Name: *Sarcoptes scabiei* L.

Physical Description: The scabies mite is barely visible to the unaided eye.

EXPOSURE

These mites remain on humans during their whole life cycle unless interrupted by an effective miticide. After mating on the skin, the female scabies mite burrows into the epidermis at a rate of several millimeters per day. The female mite lays up to 40 eggs over 1–2 months while leaving a trail of debris and feces in the burrow. The female then dies. Although the male scabies mite remains on the skin, the life span is shorter for the male than the female. Additionally, the male produces only shallow burrows. About 3 weeks after the eggs are laid, larvae emerge from tunnels, where impregnation occurs; the females burrow again, repeating the cycle. Infestation requires close personal contact because the mite lives only several days when not in direct human contact. These mites can neither jump nor fly, but these insects can crawl as fast as 2.5 cm (1 in.)/minute on warm human skin.[2] Therefore, the transmission of scabies between humans requires direct contact with infected persons. Norwegian scabies commonly occurs in immunocompromised or debilitated patients in institutional settings including patients with human immunodeficiency virus (HIV).[10,11]

PRINCIPAL TOXINS

The female mite has specialized jaws and cutting claws on the forelegs that allow easy entrance into the epidermis. Scabies mites subsist on dissolved human tissue rather than blood.

HISTOPATHOLOGY AND PATHOPHYSIOLOGY

The lesions of scabies result from both mechanical irritation and hypersensitivity.

CLINICAL RESPONSE

These burrowing mites produce intensely pruritic lesions in sensitized individuals, primarily on the hands and feet. Often the pruritus is worse at night. These nodular lesions classically appear in skin folds, such as the fingers, wrists, elbows, umbilicus, lower abdomen, genitalia, gluteus, and breast. In infants, the plantar region of the foot is frequently affected with vesicles. The burrows resemble a wavy, gray-white papule several millimeters in length with a small vesicle at one end. The vesicle at the end of the gray threadlike trail covers the mite resting at the end of the channel. These lesions typically occur in the interdigital spaces of the hand, flexor surfaces of the wrists and elbows, areola of infected women, umbilicus, waist, and male genitalia.[12] Secondary lesions (papules, pustules, vesicles, excoriations) develop from hypersensitivity or bacterial infection; these secondary lesions do not correlate to the distribution of the adult mites. These secondary lesions are more numerous and prominent than the burrows when the infestation persists. The intensely pruritic lesions usually begin about 10–30 days after the onset of the infestation. Pruritic lesions on the head or neck are uncommon, except in immunocompromised patients. Norwegian or crusted scabies describes a heavy infestation of hundreds to thousands of the scabies mite with severe skin lesions, primarily in homeless AIDS patients.[2] Health workers are at risk for contracting scabies from direct contact with patients infested with Norwegian (crusted) scabies.[13]

TABLE 164.1. Method of Skin Scraping for Scabies

1. Obtain mineral oil or microscope immersion oil, microscope slide, and #15 scalpel blade.
2. Apply mineral oil or microscope immersion oil to suspicious skin lesions and to the scalpel blade.
3. After holding the skin taut around the lesion, scrape the noninflamed, nonexcoriated portions of the skin lesions vigorously until the stratum corneum is removed.
4. Apply skin scrapings from at least six different areas to a single slide.
5. After placing a slipcover over the scrapings on the slide, examine the slide under low (2×–4×) and high magnification (25×–50×).

DIAGNOSTIC TESTING

For laboratory confirmation of a scabies infestation, the burrows, vesicles, and nonexcoriated papules are scraped with a mineral-oil-coated Joseph knife or small curved blade from a scalpel. Scraping of linear burrows from nonexcoriated areas with a sterile scalpel may contain mites, mite parts, or black fecal material. Table 164.1 lists the steps required to scrape the skin for suspected scabies. Areas of highest yield include the nonexcoriated lesions in fingers, web space, wrists, or the ulnar aspect of the hand. Inspection of the scraping in oil under low power on the microscope reveals actively moving mites, eggs, and/or feces in a majority of patients with lesions highly suspicious of scabies.[14] Failure to identify mites in skin scraping samples does not eliminate the diagnosis of scabies.

TREATMENT

Options for the topical treatment of scabies infestations include 5% permethrin cream, 1% lindane (gamma benzene hexachloride) lotion, 6% precipitated sulfur in petrolatum, crotamiton, malathion, allethrin spray, and benzyl benzoate, as well as the only oral treatment, ivermectin.[15] The treatment of scabies varies between locations depending on the presence of resistant strains of *Sarcoptes scabiei*.[16] The usual treatment of choice for scabies in adults and children over the age of 2 months is 5% permethrin cream applied at bedtime. This cream contains 0.1% formaldehyde, which is a potential sensitizing agent. The patient should take a warm shower or tub bath using soap and then apply the lotion over the entire body including the scalp, face (except eyes, nose, mouth), postauricular area, nails, fingers, and toes. The lotion should remain on the skin overnight until washed off in the morning. The process is usually repeated in 1 week, but repeat applications are not always necessary. All family members should be treated for scabies regardless of the absence of symptoms. The clothing, bed linen, and towels in intimate contact with the patient should be washed and dried simultaneously with treatment, preferably on high heat settings. Papular and pustular skin lesions usually resolve within 1 week following treatment, although nodular lesions may persist for months. Pruritus may continue up to 6 weeks after treatment because of hypersensitivity to dead mite parts. The presence of new skin lesions 2 weeks after initial therapy suggests reinfestation or inadequate treatment.

Precipitated sulfur (6%) in petrolatum applied for three consecutive nights is the treatment of choice for patients under 2 months of age; this treatment is an alternative to permethrin cream in adults when permethrin is not available. Lindane is also a treatment option, but the potential of toxicity in pediatric patients limits the use of this compound as a scabicide.[17] The most frequent adverse effects associated with the use of lindane for *pediculosis* as reported to a poison control center were relative minor adverse effects including nausea, vomiting, and ocular irritation or pain.[18] Less common complications included lethargy, conjunctivitis, and seizures. Symptomatic treatment for pruritus includes pramoxine-containing lotions and oral antihistamines at night.

Treatment of Norwegian scabies is complicated by the presence of hyperkeratosis that prevents the adequate penetration of topical scabicides. Options to traditional treatment include the use of adjunctive keratolytic agents, repeated daily application of the scabicide, and the sequential use of several scabicides. Ivermectin in a single oral dose of 150–200 μg/kg may be useful for resistant cases of scabies in HIV patients, who do not respond to topical scabicides (e.g., 25% benzyl benzoate).[19,20]

TICKS

Ticks are specialized groups of mites that are classified into the following two families: *Ixodidae* (hard-bodied ticks) and *Argasidae* (soft-bodied ticks). Of the approximately 800 species of ticks, almost 100 tick species can transmit infectious diseases to humans.[2] These bloodsucking parasites are common inhabitants of rural areas worldwide. They possess an ovoid body (fused thorax and abdomen) and have prominent legs; they reach 1–1.5 cm (~0.4–0.6 in.) in size when engorged with blood. The adult tick has a mite-like body with tough skin and four pairs of clawed legs. After contact with animal hosts (cats, dogs) or with tall grass or brush containing hungry ticks, these insects (i.e., larva, nymph, or adult) attack humans by painlessly attaching to the host with strong barbed mouth parts (chelicerae) reinforced by cement-like secretions. Attractants for ticks include the odor of sweat, body heat, and the color white.[21] The tick may spend up to 24 hours on the host searching for a protected site to feed (e.g., skinfold, hairline). The mouth parts consist of a paired anchoring organ (rostrum) covered with posterior-directed, curved hooks and a pair of sharp mandibles. Soft-bodied ticks primarily infest birds, and the mouth parts of these ticks reside underneath the body. These insects sense the presence of a host primarily by detecting increased carbon dioxide concentrations and movements. Tick bites commonly occur in the spring and early summer.[22] The tick typically feeds up to 1 week until engorged with blood, and then the tick leaves the human host.

Hard-bodied ticks are parasites primarily of mammals, and these tick species are disease vectors for typhus, viral encephalitis, ehrlichiosis, babesiosis, tularemia, Q fever, hemorrhagic fevers (e.g., Rocky Mountain spotted fever, Colorado tick fever), Lyme disease (deer tick—*Ixodes dammini*),[23] and a symmetrical ascending paralysis known as tick paralysis. The latter disease results from the secretion of a neurotoxin by certain species including *Dermacentor andersoni* Stiles and *Dermacentor variabilis* Say in North America, *Ixodes holocyclus* Neumann in Australia, and *Ixodes rubicundus* Neumann, *Rhipicephalus evertsi* Neumann *evertsi*, and *Argas walkerae* Kaiser & Hoogstraal in Africa. Ataxia and weakness in the cranial nerves occurs before the flaccid paralysis develops. Tick paralysis usually resolves within 24 hours of tick removal. Relapsing-fever results from the bite of a soft-bodied tick (*Ornithodoros turicata* Dugès). The European castor bean tick (*Ixodes ricinus* L.) is endemic to Europe, parts of Asia, and North Africa. This hard-bodied tick infests a variety of mammals (e.g., livestock, dogs, deer, humans), and this tick transmits a variety of diseases including tick-borne encephalitis, Lyme disease, babesiosis, and rickettsial illnesses. Figure 164.1 displays a specimen of the European castor bean tick.

Tick bites induce foreign body and hypersensitivity reactions, as well as delayed hypersensitivity reaction with fever, urticaria, and pruritus. The most common reaction to a tick bite is a red papule that resolves over 2–3 weeks. Local swelling and pruritus may occur, but necrosis and ulcerations are rare. Occasionally, a persistent nodule or granuloma may develop that requires months to resolve. Rarely, tick bites may produce severe allergic reactions manifest by hypotension and dyspnea.[24] Removal of the tick requires gentle traction on the mouth parts with care to avoid excretion of fluid from the body of the tick. Dislodgement of the chelicerae in the skin may produce a granuloma that necessitates surgical removal. Permethrin-based repellents kill ticks on contact, and these products can be safely applied to clothes, tents, and sleeping bags. The combination of permethrin-based repellents on clothes and DEET on the skin provides the best protection against ticks.[25]

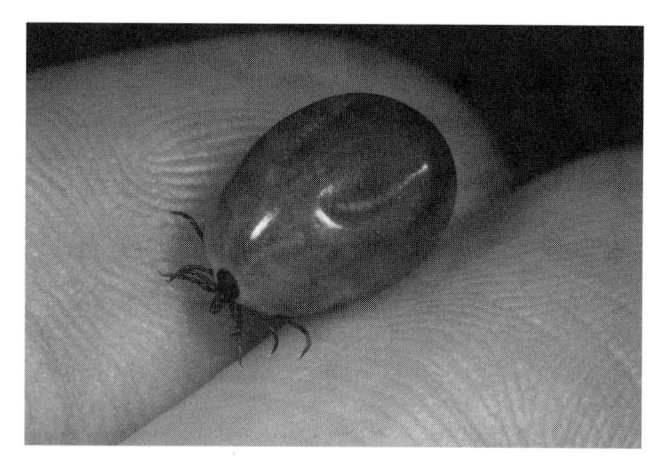

FIGURE 164.1. European Castor-bean Tick (*Ixodes ricinus* L.). © 2006 Rune Midtgaard. See color insert.

References

1. Blankenship ML. Mite dermatitis other than scabies. Dermatol Clin 1990;8:265–275.
2. Steen CJ, Carbonaro PA, Schwartz RA. Arthropods in dermatology. J Am Acad Dermatol 2004;50:819–842.
3. Krinsky WL. Dermatoses associated with the bites of mites and ticks (Arthropoda: Acari). Int J Dermatol 1983;22:75–91.
4. García-Vargas A, Mayorga-Rodríguez JA, Sandoval-Tress C. Scalp demodicidosis mimicking favus in a 6-year-old boy. J Am Acad Dermatol 2007;57(2 Suppl):S19–S21.
5. Shelley ED, Shelley WB, Pula JF, McDonald SG. The diagnostic challenge of nonburrowing mite bites: *Cheyletiella yasguri*. JAMA 1984;251:2690–2691.
6. Kheirkhah A, Casas V, Li W, Raju VK, Tseng SC. Corneal manifestations of ocular demodex infestation. Am J Ophthalmol 2007;143:743–749.
7. Shelley WB, Shelly ED, Welbourn WC. Polypodium fern wreaths (*Hagnaya*): A new source of occupational mite dermatitis. JAMA 1985;253:3137–3138.
8. Keh B, Lane RS, Shachter SP. *Cheyletiella blakei* an ectoparasite of cats, as cause of cryptic arthropod infestations affecting humans. West J Med 1987;146:192–194.
9. Parkhurst HJ. Trombidiosis (infestation with chiggers). Arch Dermatol Syphilol 1937;35:1011–1036.
10. Perna AG, Bell K, Rosen T. Localised genital Norwegian scabies in an AIDS patient. Sex Transm Infect 2004;80: 72–73.

11. Sterling GB. Janniger CK, Kihiczak G, Schwartz RA, Fox MD. Scabies. Am Fam Physician 1992;46:1237–1241.
12. Molinaro MJ, Schwartz RA, Janniger CK. Scabies. Cutis 1995;56:317–321.
13. Obasanjo OO, Wu P, Conlon M, Karanfil LV, Pryor P, Moler G, et al. An outbreak of scabies in a teaching hospital: lessons learned. Infect Control Hosp Epidemiol 2001;22:13–18.
14. Metry DW, Hebert AA. Insect and arachnid stings, bites, infestations, and repellents. Pediatr Ann 2000;29:39–48.
15. Scheinfeld N. Controlling scabies in institutional settings: a review of medications, treatment models, and implementation. Am J Clin Dermatol 2004;5:31–37.
16. Heukelbach J, Feldmeier H. Ectoparasites—the underestimated realm. Lancet 2004;363:889–891.
17. Sievert J, Lackovic M, Becker A, Lew DH, Blondell J, Kim-Jung LY, et al. Unintentional topical lindane ingestions—United States, 1998–3003. MMWR Morb Mortal Wkly Rep 2005;54:533–535.
18. Forrester MB, Sievert JS, Stanley SK. Epidemiology of lindane exposures for pediculosis reported to poison centers in Texas, 1998–2002. J Toxicol Clin Toxicol 2004;42: 55–60.
19. Sule HM, Thacher TD. Comparison of ivermectin and benzyl benzoate lotion for scabies in Nigerian patients. Am J Trop Med Hyg 2007;76:392–395.
20. Dourmishev AL, Serafimova DK, Dourmishev LA, Mualla MA, Papaharalambous V, Malchevsky T. Crusted scabies of the scalp in dermatomyositis patients: three cases treated with oral ivermectin. Int J Dermatol 1998;37: 231–234.
21. Middleton DB. Tick-borne infections. What starts as a tiny bite may have a serious outcome. Postgrad Med 1994;95: 131–139.
22. Eisen L. Seasonal pattern of host-seeking activity by the human-biting adult life stage of *Dermacentor andersoni* (Acari: Ixodidae). J Med Entomol 2007;44:359–366.
23. Steere AC, Coburn J, Glickstein L. The emergence of Lyme disease. J Clin Invest 2004;113:1093–1101.
24. Solley GO. Allergy to stinging and biting insects in Queensland. Med J Aust 1990;153:650–654.
25. Brown M, Hebert AA. Insect repellents: an overview. J Am Acad Dermatol 1997;36:243–249.

Chapter 165

SCORPIONS
(Order: Scorpiones)

HISTORY

Scorpions are one of the oldest terrestrial animals in the order Arachnida based on fossil records dating back almost 400 million years.[1] Written medical accounts of painful scorpion stings date back to the 3rd century BC.[2] Although all species of scorpions cause painful stings, only a few are very toxic (e.g., Mexican *Centruroides suffusus* Pocock, Brazilian *Tityus serrulatus* Lutz & Mello, North African *Leiurus quinquestriatus* Ehrenberg, Indian *Hottentotta tamulus* Fabricius). The only dangerous species in the United States is *Centruroides exilicauda* Wood (*C. sculpturatus*, bark scorpion). Although numerous fatalities were associated with *C. exilicauda* Wood envenomations of young children during the 1950s, fatalities from scorpion envenomations in the United States have not been well documented since the late 1960s.[3]

IDENTIFYING CHARACTERISTICS

There are approximately 650–800 species of scorpions distributed throughout the tropical and subtropical regions of the world. Scorpions are arachnids that form one of the largest orders of Arthropoda. Medically important arachnids include species from the order Araneida (spiders) and order Acari (ticks, mites) as well as the order Scorpiones (scorpions). Arachnids are terrestrial, ambulatory creatures with two body segments, whereas insects have wings, antennae, and three body segments. Scorpions possess a hard exoskeleton, two anterior pinching claws, and a tail with a terminal bulbous enlargement. Most medically important scorpions belong to the family Buthidae that is characterized by a triangular sternal plate. The bark scorpion (*Centruroides exilicauda* Wood) is the only scorpion native to the United States that produces life-threatening envenomation, particularly in children.

Common Name: Bark Scorpion

Scientific Name: *Centruroides exilicauda* Wood (*C. sculpturatus*)

Physical Description: The body of the scorpion consists of a cephalothorax, an abdomen, and a six-segmented tail that terminates in a bulbous enlargement called the telson and contains the stinger. This scorpion is a small (13–75 mm/0.5–3 in. long), hairless creature that passes through four color phases ranging from the characteristic yellow-brown to striped.[4] Features distinguishing the bark scorpion from other scorpions include a rectangular proximal tail segment, a small spine or tubercle at the base of the stinger that requires magnification for identification, and long, slender pedipalps (pincers).

EXPOSURE

Geographic Distribution

The effect of scorpion envenomation varies among scorpion species and geographic location with the most severe clinical effects resulting from certain buthid species (Buthidae family). Highly toxic scorpions include

Centruroides suffussus Pocock (Mexico), *Androctonus crassicauda* Oliver (Middle East, North Africa), *Androctonus australis* Hector *(Israel), Hottentotta tamulus* Fabricius (*Mesobuthus tamulus*, *Buthus tamulus*, India), *Tityus serrulatus* Lutz & Mello (Brazil), *Tityus trinitatis* Pocock (Trinidad), *Parabuthus granulatus* Ehrenberg (South Africa), and *Leiurus quinquestriatus* Ehrenberg (Middle East, North Africa). The geographic distribution of scorpion species is between the 45° latitudes. Occasionally, scorpion envenomations occur at locations distant to their native habitat as a result of inadvertent transport of the scorpion.[5,6]

Australia

In Australia, scorpion stings are not usually associated with severe clinical effects.[7] Most scorpion envenomations in Australia result from stings of species in the family Buthidae (e.g., *Lychas* spp., mostly mainland Australia), and to a lesser extent the family Bothriuridae (*Cercophonius* spp.) and the family Scorpionidae (*Urodacus* spp.). Usually stings from Australian buthid scorpions cause more severe effects than from the larger species in the families Scorpionidae (*Urodacus* spp.) and Hemiscorpiidae (*Liocheles* spp.).[8] Most of the scorpions that cause serious toxicity reside in North Africa and the tropics.

United States and Mexico

Medically important *Centruroides* species in the western United States and Mexico include *C. exilicauda* Wood (bark scorpion, *C. sculpturatus*) and *C. suffusus* Pocock (Durango scorpion). The bark scorpion inhabits areas throughout the dry, arid regions of Arizona and parts of Texas, New Mexico, Northern Mexico, small desert sections of southeastern California, and Lake Mead in Nevada. These scorpions tolerate some subfreezing temperatures by seeking warm shelter during the cold temperatures. Other US species of scorpions associated with less severe envenomations compared with *C. exilicauda* include *Isometrus maculatus* de Geer (California, Florida), *Diplocentrus* species (California, Texas, Florida), *Hadrurus hirsutus* Wood (southwestern United States), and *Vejovis* species (the Southwest). The most important venomous species of *Centruroides* in Mexico beside *C. suffusus* and *C. exilicauda* include *C. elegans* Thorell, *C. infamatus* C. L. Koch, *C. limpidus* Karsch, *C. noxius* Hoffman, and *C. pallidiceps* Pocock.

South America

Medically important species of scorpions in South America include species from the genus *Tityus* including *T. discrepans* Karsch, *T. serrulatus* Lutz & Mello, *T. trinitatis* Pocock, and *T. bahiensis* Perty.

Middle East and North Africa

The most dangerous scorpions in this region include *Androctonus crassicauda* Olivier (Black scorpion), *Androctonus amoreuxi* Audouin, *Androctonus bicolor* Ehrenberg, *Androctonus australis* L., *Leiurus quinquestriatus* Ehrenberg (Yellow scorpion), and *Buthus occitanus* Amoreux.

South Africa

Parabuthus granulatus Ehrenberg is one of the primary scorpions involved in serious scorpion envenomation in South America.

India

Serious scorpion envenomations in India typically involve *Hottentotta tamulus* Fabricius and *Palamneus swammerdami* Krishnan.[9] Evenomation by these scorpions may cause death despite the use of local antivenom, particularly in children and older adults.[10]

Behavior

Scorpions are nocturnal hunters that hide in rocks, leaf litter, and burrows during the day. These creatures grasp their prey (i.e., insects, spiders, small animals) with their pedipalpus, and they paralyze prey with the venom injected from their caudal stinger. Stings frequently occur indoors during the night; a scorpion will typically hide in bedding or clothing. Scorpions detect prey by air movements and vibrations rather than by their limited vision.[4]

Bark Scorpion

This scorpion is a climbing rather than burrowing scorpion species that prefers areas around trees. Climbing characteristics cause it to sting unsuspecting humans, who fail to look beneath objects hiding the scorpion. The bark scorpion often rests the telson on the ground rather than carry it arched overhead like most scorpions. This scorpion is a nocturnal feeder that eats small insects and spiders by clasping them in its claws and stinging them to death by inserting its stinger from the arched tail. Scorpions hibernate during winter and seek shelter during the day, preferring moist, cool areas under rocks, debris, houses, or household items (e.g., shoes, clothing, beds).

PRINCIPAL TOXINS

Venom Composition

Scorpion venom is a complex, species-specific mixture of low-molecular-weight neurotoxic proteins, phospholipase A_2, acetylcholinesterase, hyaluronidase, amino acids, protease inhibitors, histamine releasers, and serotonin.[11] In contrast to spider and snake venoms, most scorpion venoms lack clinically significant amounts of tissue-destructive enzymes with few exceptions (e.g., South Indian scorpion, *Heterometrus scaber* Kraepelin).[12] The most toxic portion of scorpion venom consists of small, neurotoxic peptides that recognize and bind to integral membrane proteins involved with the control of ion transport. To date, experimental studies suggest that most of these peptides bind to activated Na^+ and voltage-dependent K^+-channels[13] and, to a lesser extent, to Cl^-- and Ca^{++}-channels.[14] One class of scorpion neurotoxins induces slow, inward sodium currents that prolong the action potential so that a given stimulus causes greater sodium channel activation. Other neurotoxins are stabilizers that reduce sodium channel flow during depolarization. These effects on sodium gating occur at preterminal sites, resulting in the release of neurotransmitters at the neuromuscular junction and repetitive, enhanced axonal discharges.[3]

Venom Apparatus

Two venom glands are located within the telson with independent ducts leading to an opening near the stinger. The metasoma (tail) of the scorpion contains five segments. Distal to the metasoma is the round, terminal segment (telson) that contains the venom glands and the stinger as demonstrated in Figure 165.1. Scorpions catch prey in their two anterior pincers, and they inject their venom by raising their caudal stinger over their head and inserting the stinger into the prey.

FIGURE 165.1. Tail (metasoma) and telson of black hairy scorpion (*Hadrurus spadix* Stahnke). © 2006 Rune Midtgaard. See color insert.

Dose Response

The severity of scorpion envenomation depends on the potency of the venom (species-specific) and volume of venom injected (i.e., amount of venom and number of stings). Most scorpion envenomations produce mild effects.

TOXICOKINETICS

The absorption, peak concentration, and elimination half-life of scorpion venom is variable and species-specific.[15] Animal studies suggest that substantial amounts of venom are absorbed within 15 minutes of the sting, and distribution is complete within 1–4 hours.[16,17] These animal studies indicate that the absorption and distribution phases are relatively rapid while the elimination half-life is relative long (i.e., about 4–24 hours).[17,18]

HISTOPATHOLOGY AND PATHOPHYSIOLOGY

Mechanism of Toxicity

The principal targets of scorpion venom are the voltage-dependent ion channels involved with nerve impulse conduction. The clinical features of scorpion envenomation result from a complex interaction of parasympathetic and sympathetic stimulation along with the release of a variety of endogenous compounds including catecholamines, angiotensin II, glucagon, corticosteroids, bradykinins, prostaglandins, and cytokines (e.g., interleukin-8). Scorpion venom increases sodium influx into the cell, resulting in prolonged duration and increased amplitude of the action potential. The subsequent increase in calcium permeability at presynaptic and postsynaptic nerve fibers increases the release of acetylcholine. Scorpion venom also causes excess catecholamine release resulting in autonomic stimulation, the accumulation of endothelin, and generalized vasoconstriction. Symptoms generally reflect sympathetic, parasympathomimetic, and neuromuscular excitation followed by exhaustion. Scorpion venom contains a mixture of short, neurotoxic proteins that open sodium channels at presynaptic nerve terminals and inhibit Ca^{++}-dependent potassium channels leading to excess autonomic stimulation. During severe envenomation, these actions cause transitory myocardial ischemia that manifests as depression of the left ventricular ejection

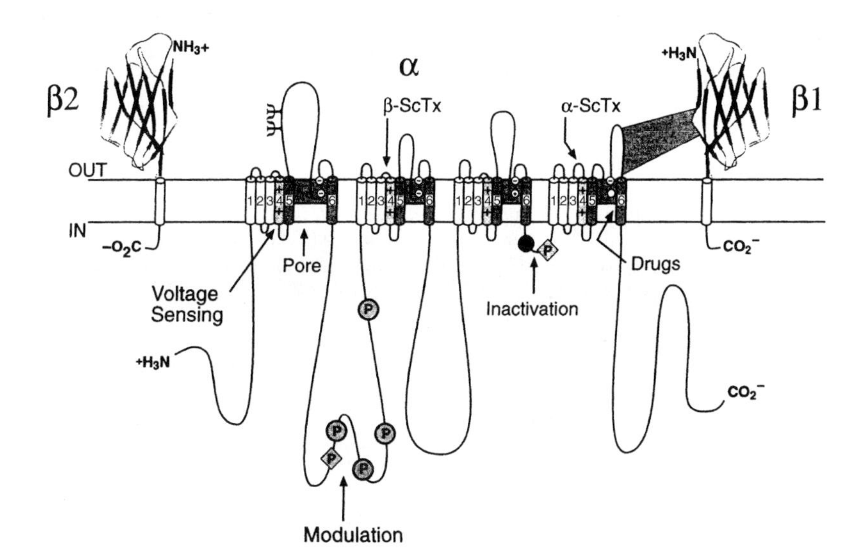

FIGURE 165.2. Structure of voltage-gated Na^+-channels in mammalian brains. The extracellular domains, β1 and β2, are folds of α-helical segments. Bold lines represent polypeptide chains. ψ is the probable site of *N*-linked glycosylation and P represents areas of protein phosphorylation. The lightly shaded S4 segment is the voltage sensor and the dark circle represents the inactivation particle in the inactivation gate loop. The sites of binding of β- and α-scorpion toxins are domains II and IV, respectively. From Catterall WA. Molecular mechanisms of gating and drug block of sodium channels. Novartis Found Symp 2002;241:207. Reproduced with permission from John Wiley & Sons Limited.

fraction and acute pulmonary edema.[19] These perfusion defects normalize as the effects of the venom cease.

Sodium ion channels from mammalian brain are complexes of α-, β_1-, and β_2-subunits as demonstrated in Figure 165.2.[20] The molecular weights of these subunits are 260kDa, 36kDa, and 33kDa, respectively. Each of the four homologous domains (I–IV) of the sodium channels consist of six α helical transmembrane segments (S1–S6) as well as a re-entrant loop that dips into the transmembrane region of the protein between transmembrane segments S5 and S6. Domain II is involved with activation of the membrane, whereas domain IV is involved with the coupling of activation to inactivation. The β_1- and β_2-subunits modulate channel gating and cell-cell interaction. Membrane depolarization causes a voltage-dependent conformational change of the channel that induces an increase in Na^+ permeability. Voltage sensors (gating charges) in the S4 transmembrane segments initiate the voltage-dependent activation of the sodium channel by a conformational change that moves the S4 segments outward along a spiral path causing the opening of the pore.[21] Voltage-dependent coupling of channel activation to inactivation results from voltage-dependent multiple transitions during the activation phase including the translocation of the S4 segment outward. This conformational change initiates the closure of inactivation gates formed by the short intracellular loop connecting domains III and IV of the α-subunit of the sodium channel as demonstrated in Figure 165.2. Inactivation produces closure of the pore and the restoration of pre-excitation sodium ion concentrations.

Scorpion venoms contain two different toxins (α- and β-toxins) that produce voltage sensor-trapping. β-Scorpion toxins bind to a receptor site (neurotoxin receptor site 4) in the S3–S4 extracellular loop in domain II of the α-subunit of the sodium channel as demonstrated in Figure 165.2.[22] These β-scorpion toxins modify the activation process of voltage-gated sodium channels by trapping the IIS4 segment in an activated position, and thereby enhance activation by shifting the voltage dependence to more negative membrane potentials.[23] Normally, Na^+-channels inactivate within a few milliseconds of opening. Once the membrane is activated, the presence of β-scorpion toxins slows deactivation of the channel and creates large tail currents.

α-Scorpion toxins bind to a receptor site (neurotoxin receptor site 3) in the S3–S4 extracellular loop in domain IV of the α-subunit of the sodium channel. The α-scorpion toxin-binding site is the same site bound by sea anemone[24] and funnel web spider toxins.[25] These α-scorpion toxins (e.g., makatoxin I, bukatoxin) slow inactivation of the Na^+ channel and prolong the action potentials. Voltage-sensor trapping results from the effects of α-scorpion toxins on Na^+-channel inactivation. These toxins prevent the outward translocation or movement of the S4 segment in domain II during repolarization and trap the IVS4 segment in an inward, partially activated position. The physiological effect is the slow or delayed closure of the inactivation gate.[20]

Potassium channels are divided into voltage-gated K^+-channels and ligand-gated-K^+-channels. These channels are composed of two main subunits: α and β. The α-subunit is involved with the conduction of the potassium ion across the membrane, whereas the β-subunit modulates the properties of the channel.[23] Scorpion venom contains toxins that bind to the outer vestibule of potassium channels and block ion conduction by physical occlusion of the K^+ channel pore without altering the gating kinetics of the K^+ channel.[22] Voltage-gated and calcium-dependent K^+-channels are tetrameric molecules that contain four α-subunits. These potassium channels control a variety of physiological processes involved with cell excitation including processes that regulate neuronal and cardiac electrical patterns, muscle contraction, neurotransmitter release, hormone secretion, and cell secretions.[26] A variety of scorpion toxins [e.g., noxiustoxin (CAS RN: 85205-49-8), maurotoxin, tamapin] alter conductance through potassium channels[27–29] as well as through T-type Ca^{2+} channels (e.g., kurtoxin).[30,31] The venom from the Israeli scorpion, *Leiurus quinquestriatus hebraeus* Ehrenberg, contains the 37-amino acid neurotoxin, charybdotoxin (CAS RN: 115422-61-2). This polypeptide inhibits potassium channels on nerve cell membranes by a pore-blocking mechanism.[32]

Target Organs

Local histological changes at the site of scorpion stings include superficial and deep mixed inflammatory cell infiltrate with a few eosinophils dispersed among collagen bundles in an edematous dermis.[33] Most scorpion envenomations do not cause clinically significant changes in the tissue surrounding the envenomation. In rats, the injection of purified venom from *Tityus serrulatus* causes acute and chronic pancreatitis characterized by necrosis, interstitial fibrosis, lymphocyte infiltration, ductal and ductular dilation, acinar cell atrophy, periductal ductular hyperplasia, and hyperplasia of the islets of Langerhans.[34]

Cardiovascular

Stimulation of the autonomic nervous system with the release of catecholamines causes a predominance of sympathetic stimulation and cardiorespiratory dysfunction. In experimental animals and in children, scorpion venom causes marked, prolonged hypertension and vasoconstriction.[35] Severe scorpion envenomation produces predominant left ventricular dysfunction with normal systemic vascular resistance that manifests as pulmonary edema or severe hypotension depending on the fluid balance.[36] These cardiovascular effects may progress to myocardial damage and multisystem organ failure secondary to biventricular dysfunction and vasodilation.[37,38] Pathological changes associated with severe scorpion envenomation include degeneration of the muscle fibers, local necrosis, hemorrhage, coagulative myocytolysis (contraction bands), and interstitial edema.[39] Electromicroscopy of the heart demonstrates intracellular edema, scattered destruction of the I-band, contraction band necrosis with ruptured and hypercontracted sarcomeres, dilation of the tubular system, and lipid deposition along with preservation of the mitochondria, nucleus and sarcolemma.[40] These abnormalities are qualitatively different from the histopathological changes associated with massive catecholamine release.[35]

Central Nervous System

Neurological complications associated with scorpion envenomation include dysarthria, dysphagia, tongue fasciculations, paresthesias, hemiplegia, muscular weakness, abnormal eye movements, blurred vision, and seizures.[41] Case reports document central nervous system (CNS) changes in fatal cases of scorpion envenomation, including cerebral edema, diffuse punctate hemorrhages, cerebral infarction, and cerebrovascular thrombosis.[42,43] Because scorpion venom diffuses poorly across the blood–brain barrier, these CNS changes probably are secondary to the cardiorespiratory effects (e.g., hypertension, pulmonary edema, shock) associated with severe scorpion envenomation.

CLINICAL RESPONSE

Stings from most scorpion species produce only mild pain, even following envenomation by dangerous species (e.g., *Centruroides suffussus, Androctonus crassicauda, Androctonus amoreuxi, Parabuthus liosoma* Ehrenberg, *Nebo hierichonticus* Simon, *Leiurus quinquestriatus, Hottentotta tamulus, Tityus serrulatus*).[44–46] The severity of a scorpion envenomation depends on the presence of neurological and cardiorespiratory dysfunction, and the development of this type of dysfunction varies with the species, age, and size of the scorpion as well as the age of the victim and the amount of injected venom.[47] Severity of envenomation by dangerous scorpion species ranges from local pain and paresthesias to encephalopathy and fatal cardiotoxicity depending on the above factors. Local pain develops almost immediately (i.e., seconds to minutes) after scorpion envenomation. Although many envenomations in healthy adults and older children result only in local pain, significant respiratory distress can develop, particularly in young children (i.e., <6 years of age).[48,49] Scorpion envenomations are not usually associated with clinically significant edema or erythema with the exception of stings by *Androctonus crassicauda*.[18] Systemic symptoms of envenomation typically begin within about 45–60 minutes after envenomation with common signs including diaphoresis, hypertension, tachycardia, and salivation.[50]

Local

Most scorpion stings produce local pain and paraesthesias, but swelling, erythema, and ecchymosis do not usually develop.[51] Patients with *Centruroides exilicauda* envenomations almost always demonstrate moderately severe local pain following mild palpation of the site of envenomation (i.e., positive "tap" test). Local symptoms typically resolve within 10 hours after *C. exilicauda* envenomation. Even envenomations by other dangerous scorpion species (e.g., *Leiurus quinquestriatus*) frequently produce only local pain and paresthesias.[52]

Systemic

Nervous System

Mild envenomations produce agitation, diaphoresis, and anxiety. Neurotoxic effects following serious scorpion envenomation include excitability, exaggerated startle response, roving eye movements, ptosis, hyperreflexia, ataxia, hemiplegia, seizures, and visual disturbances. The most common initial symptoms are irritability, motor restlessness and anxiety.[44] Envenomations by some scorpion species (e.g., *Parabuthus transvaalicus* Purcell, *Centruroides exilicauda*) cause more prominent neuromuscular than cardiac symptoms.[53] Children, especially infants, typically develop more severe CNS symptoms compared with adults, and these symptoms are usually occur soon (i.e., 15 minutes) after envenomation.[54,55] The typical sequence of severe envenomation by neurotoxic scorpion species is initial intense pain followed within a few minutes to several hours by hyperexcitability, increased salivation, tachypnea, muscle fasciculations, muscle spasm, opisthotonus, erratic/wandering eye movements, disconjugate gaze, nystagmus, convulsions, paralysis, and respiratory failure. Sympathetic effects include tachycardia, diaphoresis, mydriasis, piloerection, priapism, hyperthermia, and hyperactivity.[56] Parasympathomimetic signs are salivation, lacrimation, urination, defecation, vomiting, and gastric distension.

Cardiorespiratory System

Tachycardia and hypertension are common toxic effects of scorpion envenomation, but these effects usually respond to conservative therapy.[57] Patients with moderate to severe cardiotoxicity following scorpion envenomations usually present to medical facilities within 3 hours, and almost always within 6 hours of the sting.[58] During severe scorpion envenomations by cardiotoxic scorpion species (e.g., *L. quinquestriatus*), severe hypertension, pulmonary edema, myocardial infarction, and cardiogenic shock may develop along with serious dysrhythmias (AV dissociation, ventricular tachycardia) and sudden cardiac arrest (i.e., ventricular fibrillation).[59] Toxic myocardial effects (i.e., acute pulmonary edema secondary to congestive heart failure) are more common than neurotoxic effects following envenomation by some Indian (*Hottentotta tamulus*)[60,61] and Middle East/North African (*Androctonus crassicauda, Leiurus quinquestriatus*)[56] scorpion species. Respiratory effects in patients with severe envenomation, particularly in young children, include stridor, cyanosis, dyspnea, Cheyne-Stokes breathing, respiratory depression, and the development of pulmonary edema.[62] The presence of tachypnea (respiratory rate >30), diaphoresis, agitation, and dyspnea suggests the development of pulmonary edema.[63]

Clinical Severity

Curry et al. graded the severity of envenomation by the native US species, *Centruroides exilicauda*, based on the presence of pain, paresthesias, and cranial nerve or somatic/neuromuscular dysfunction.[49] Other species of scorpions (e.g., *Centruroides suffussus, Androctonus crassicauda, Androctonus australis, Hottentotta tamulus, Tityus serrulatus, Tityus trinitatis, Leiurus quinquestriatus*) produce more cardiovascular dysfunction compared to *C. exilicauda* and *Parabuthus transvaalicus* Purcell. Table 165.1 displays the four grades of severity modified to include cardiovascular dysfunction. Cranial nerve dysfunction may include blurred vision, disconjugate gaze, hypersalivation, dysphagia, slurred speech, tongue fasciculations, or upper airway obstruction, whereas somatic/neuromuscular dysfunction includes increased involuntary activity (e.g., myoclonic jerks, tremors), restlessness, and hemiballismus.

TABLE 165.1. Clinical Grades of Scorpion Envenomation

Grade	Clinical Features
I	Local pain and/or paresthesias limited to site of envenomation.
II	Local signs plus pain and/or paresthesias remote from site of envenomation.
III	Presence of cranial nerve dysfunction (blurred vision, dysconjugate gaze, hypersalivation, dysphagia, slurred speech, fasciculations) **or** somatic signs (muscle jerking, restlessness, involuntary shaking) **or** cardiovascular dysfunction (moderate to severe hypertension, cardiac dysrhythmias, myocardial ischemia, pulmonary edema).
IV	Presence of any combination of cranial nerve, cardiac, or somatic/neuromuscular dysfunction.

Source: Adapted from Ref 49.

DIAGNOSTIC TESTING

Analytical Methods

Enzyme-linked immunosorbent assays (ELISA) have been developed from polyclonal Fab fragments for the detection of envenomation by North African (*Androctonus australis garzonii* Goyffon & Lamy, *Buthus occitanus* Amoreux, *Androctonus mauritanicus* Pocock),[64,65] and Brazilian (*Tityus serrulatus*) scorpions.[66] In general, the degree of envenomation is proportional to the amount of venom present in the serum of the victim. These tests can detect moderate to severe envenomations, but the sensitivity of some of these assays for mild envenomation is relatively low.[66]

Abnormalities

A variety of cardiac arrhythmias are associated with scorpion envenomation ranging from premature atrial and ventricular contractions to AV nodal block and ventricular tachycardia. In a case series of 17 patients admitted to a Turkish emergency department for scorpion envenomation, the most common electrocardiographic abnormalities in order of frequency were as follows: sinus tachycardia, premature ventricular contractions, premature atrial contractions, and paroxysmal supraventricular tachycardia.[67] Less common electrocardiographic abnormalities were sinus bradycardia, first-degree and second degree atrioventricular block, and atrial fibrillation. Occasionally, during severe envenomations by cardiotoxic scorpion species, electrocardiographic abnormalities may resemble a myocardial infarction with inverted or peaked T-waves, ST elevation or depression, and Q waves, particularly following envenomation by dangerous Middle Eastern and Indian scorpion species.[68] A case series of severe envenomation by *Buthus occitanus* and *Leiurus quinquestriatus* scorpions indicated that serum concentrations of cardiac troponin I are a specific marker of myocardial injury during envenomation by these species.[69] Echocardiographic abnormalities involve segmental hypokinesis and left ventricular dysfunction (ejection fraction <0.55, increased left ventricular end-systolic dimensions, interventricular septal thickening).[70] Myocardial perfusion scans by thallium-201 scintigraphy during severe scorpion envenomation by cardiotoxic scorpion species demonstrates evidence of myocardial hypoperfusion consistent with echocardiographic abnormalities that may persist for at least 2 weeks.[71]

Common laboratory abnormalities include leukocytosis and hyperglycemia.[37] Hyponatremia, hypocalcemia, and a relative hyperkalemia occasionally develop following scorpion envenomation. Rarely, complications of severe scorpion envenomation include acute pancreatitis (hyperamylasemia) after *Tityus trinitatis* evenomation and rhabdomyolysis (elevated creatine kinase).[72,73]

TREATMENT

With the exception of species-specific antivenom, the treatment of scorpion envenomations is supportive including the maintenance of fluid and electrolyte balance, analgesics, and sedatives. During the hypertensive phase of severe scorpion envenomation, the use of cardiovascular agents (vasodilators, adrenergic antagonists, calcium channel blockers) may be necessary depending on the severity of the clinical response. The administration of corticosteroids probably does not alter the course of scorpion envenomation.[74]

Stabilization

Most US envenomations (grade I or II) produce no serious side effects and may be treated with local ice application, local anesthetics or regional nerve block, analgesics (e.g., codeine), antiemetics (metoclopramide, chlorpromazine), and occasionally sedation (benzodiazepines). The immediate, life-threatening complications associated with scorpion envenomation include upper airway obstruction, respiratory depression, pulmonary edema, and cardiogenic shock. Patients should be evaluated carefully for the adequacy of the airway, ventilation, and perfusion. Those patients with signs of systemic toxicity require cardiorespiratory monitoring, intravenous access, and supplemental oxygen, continuous positive airway pressure (CPAP), or intubation as indicated by pulse oximetry and the clinical presentation. These patients should be observed for deterioration of respiratory function secondary to respiratory paralysis, upper airway obstruction, and pulmonary edema.

There are limited data on the efficacy of specific cardiac drugs during scorpion envenomation. In the absence of hypovolemia, intravenous diuretics are appropriate for the treatment of pulmonary edema. The use of a Swan-Ganz catheter may be necessary to optimize fluid replacement and hemodynamic parameters. Vasodilators for the treatment of hypertension include nitroprusside, angiotensin-converting enzyme (ACE) inhibitors (e.g., captopril), and calcium channel blockers (nifedipine).[37] Prazosin is a α_1-blocker that reduces preload and left ventricular impedance without raising heart rate or renin secretion. Case series suggest that prazosin (250 mg, children; 500 mg, adults) orally every 6 hours reduces morbidity and shortens recovery time following scorpion envenomations by the Indian red scorpion (*Hottentotta tamulus*), which is associated with severe cardiotoxicity.[75,76] Afterload reduction with vasodilators (e.g., nitroglycerin, nitroprusside) is the treatment of choice for scorpion envenomations that develop hypertensive encephalopathy or pulmonary edema.[77] A case series suggests that dobutamine in dosages of 7–20 μg/kg/min titrated to achieve the best compromise between hemodynamic and tissue oxygenation improves left and right ventricular dysfunction following scorpion envenomation associated with cardiogenic shock and pulmonary edema.[78]

Both benzodiazepines (diazepam, lorazepam) and phenobarbital enhance GABA-induced increases in chloride conductance and augment the anticonvulsant effects of GABA neurons in the CNS. Benzodiazepines are the drug of choice to treat seizures associated with scorpion envenomation because of relatively less respiratory depression compared with phenobarbital. The administration of phenobarbital (5–10 mg/kg intravenously) is an option when seizures do not respond to benzodiazepines. Large phenobarbital doses may depress respirations and patients receiving phenobarbital should be monitored closed for respiratory failure, particularly when the patient also receives benzodiazepines.[48]

First Aid

First aid primarily involves rest, reassurance, and transport to a health care facility, particularly if any systemic symptoms develop. There are few data on the efficacy of restrictive bands to retard the distribution of scorpion toxins. Young children should be transported immediately for a medical evaluation when envenomated by dangerous species from scorpion genera (e.g., *Centruroides, Leiurus, Androctonus, Tityus, Hottentotta*) because of the vulnerability of young children to scorpion toxins.

Antidotes

The use of antivenom following scorpion envenomation is controversial, and there are limited data to support the use of scorpion antivenom in addition to supportive care (systemic and local analgesics, ice, cardiorespiratory support), particularly in adults.[50,64] Case series from various parts of the world suggest the scorpion-specific antivenom from animal species (goat, horse, donkey) improves clinical outcome following *severe* scorpion envenomation (e.g., grades III and IV for US bark scorpion) including studies from Arizona (*Centruroides exilicauda*),[79] Brazil (*Tityus serrulatus*),[80] Mexico (*Centruroides* spp. including *C. suffusus suffusus*),[81] Morocco (*Androctonus mauritanicus, Buthus occitanus*),[64] and Saudi Arabia (*L. quinquestriatus, A. crassicauda*).[82,83] However, other studies did not confirm the efficacy of scorpion-specific antivenom for the treatment of cardiovascular and neurological complications of scorpion envenomations, particularly in adults.[84] Variables that affect the efficacy of antivenom include specificity of the antivenom, the potency of the scorpion venom, dose of antivenom administered, the severity of the envenomation, and the delay between envenomation and antivenom administration. Some of these antivenoms are genus-specific. The polyclonal equine antivenom raised against *Tityus discrepans* Karsch is highly cross-reactive with venoms from other *Tityus* species, but not other genera of scorpions (e.g., *Centruroides, Chactas, Brotheas, Rhopalurus, Diplocentrus*).[85] Adverse effects associated with scorpion antivenom include urticaria, diffuse maculopapular rash, and bronchial hyperreactivity.[82] Anaphylaxis is relatively rare following the administration of scorpion antivenom.

In the United States, *Centruroides exilicauda*-specific antivenom has been developed in goats by the Antivenom Product Laboratory, Department of Botany and Microbiology, Arizona State University, Tempe, Arizona. However, this product was never approved by the US Food and Drug Administration (FDA) for transportation across state lines. Current supplies of this whole immunoglobulin G scorpion-specific goat antivenom are limited because of curtailment of production. A case series of 151 bark scorpion envenomations indicated that the administration of one to two vials of antivenom relieved symptoms in 71% of the patients within 30 minutes.[79,86] Adverse effects included immediate hypersensitivity reactions occurring in 8% of the patients receiving antivenom, but these acute reactions were relatively mild. Typically, one vial of antivenom is diluted in 50 mL normal saline, and the infusion is titrated slowly for the first 10 minutes to observe for adverse reaction. If no reaction occurs the remainder is infused over 30–45 minutes. Delayed reactions (serum sickness, rash) are relatively common following the administration of this antivenom. In a retrospective chart review with follow-up contact, 7 of 12 (58%) children receiving *Centruroides* scorpion antivenom had rash, hives, or serum sickness within 2 weeks of the administration of the antivenom.[87] These symptoms usually resolved with a few days after the initiation of oral steroid and antihistamine therapy. Anascorp® (Rare Disease Therapeutics, Inc, Nashville, TN) is an equine $F(ab)_2$ antivenom currently undergoing clinical trials for the treatment of *Centruroides* envenomation in the United States and Mexico.

Supplemental Care

Atropine is not recommended for the treatment of most scorpion envenomations (i.e., scorpions associated with cardiovascular toxicity) because of the potential to exacerbate adrenergic toxicity. There are inadequate data to support the use of antihistamines, corticosteroids, or calcium during scorpion envenomation without evidence of an allergic response. Hyperthermia is a frequent complication of patients with scorpion envenomation, and patients should be monitored for the development of hyperthermia. Although there are limited data on the effectiveness of specific measures, commonly used measures for the treatment of scorpion-induced hyperthermia include cooling measures (sponge bath, evaporative cooling) and possibly antipyretics (acetaminophen). Routine laboratory tests in severe envenomations include complete blood count, electrolytes, creatinine, amylase, creatine phosphokinase, troponin, and urinalysis.

Children under 3 years of age are at the greatest risk of developing respiratory failure following envenomation by *C. exilicauda*; these children frequently require hospital admission for monitoring, particularly following envenomation by the bark scorpion (*C. exilicauda*).[44] All patients below one year of age and older children with systemic signs of envenomation (i.e., hypersalivation, hypertonus, priapism, fever, dyspnea) require monitoring in an intensive care unit (ICU).[88] For patients with severe, involuntary motor activity and agitation that does not respond to an initial dose of benzodiazepines (e.g., midazolam 0.05–0.1 mg/kg up to 1–2 mg), a midazolam infusion (0.1 mg/kg/h titrated to light sleep) is a treatment option.[89] These patients require careful monitoring in an ICU for respiratory depression, and they are candidates for antivenom therapy. Patients without signs of clinically significant cardiotoxicity can be discharged after a 6-hour observation period.[58]

References

1. Minton SA. Venom diseases. Springfield, IL: Charles C. Thomas; 1974:24–37.
2. Yarom R. Scorpion venom: a tutorial review of its effects in men and experimental animals. Clin Toxicol 1970;3:561–569.
3. Rachesky IJ, Banner W, Dansky J, Tong T. Treatments for *Centruroides exilicauda* envenomation. Am J Dis Child 1984;138:1136–1139.
4. Stahnke HL. A key to the genera of Buthidae (Scorpionida). Entomol News 1972;83:121–133.
5. Bush SP. Envenomation by the scorpion (*Centruroides limbatus*) outside its natural range and recognition of medically important scorpions. Wilderness Environ Med 1999;10:161–164.
6. Trestrail JH. Scorpion envenomation in Michigan: three cases of toxic encounters with poisonous stowaways. Vet Hum Toxicol 1981;23:8–11.
7. Isbister GK, Volschenk ES, Balit CR, Harvey MS. Australian scorpion stings: a prospective study of definite stings. Toxicon 2003;41:877–883.
8. Isbister GK, Volschenk ES, Seymour JE. Scorpion stings in Australia: five definite stings and a review. Intern Med J 2004;34:427–430.
9. Das S, Nalini P, Ananthakrishnan S, Ananthanarayanan PH, Balachander J, Sethuraman KR, Srinivasan S. Scorpion envenomation in children in southern India. J Trop Med Hyg 1995;98:306–308.
10. Bawaskar HS, Bawaskar PH. Clinical profile of severe scorpion envenomation in children at rural setting. Indian Pediatr 2003;40;1072–1081.
11. Wang GK, Strichartz GA. Purification and physiological characterization of neurotoxins from venom of the scorpions *Centruroides sculpturatus* and *Leiurus quinquestriatus*. Mol Pharmacol 1983;23:519–533.
12. Gwee MC, Gopalakrishnakone P, Cheah LS, Wong PT, Gong JP, Kini RM. Studies on venom from the Black scorpion *Heterometrus longimanus* and some other scorpion species. J Toxicol Toxin Rev 1996;15:37–57.
13. Rodriguez de la Vega RC, Possani LD. Current views on scorpion toxins specific for K^+-channels. Toxicon 2004;43:865–875.
14. Possani LD, Merino E, Corona M, Bolivar F, Becerril B. Peptides and genes coding for scorpion toxins that affect ion-channels. Biochimie 2000;82:861–868.
15. de Rezende NA, Chavez-Olortegui C, Amaral CF. Is the severity of *Tityus serrulatus* scorpion envenoming related to plasma venom concentrations? Toxicon 1996;34:820–823.
16. Ismail M, Shibl AM, Morad AM, Abdullah ME. Pharmacokinetics of ^{125}I-labelled antivenin to the venom from the scorpion *Androctonus amoreuxi*. Toxicon 1983;21:47–56.
17. Ismail M, Abd-Elsalam MA. Are the toxicological effects of scorpion envenomation related to tissue venom concentration? Toxicon 1988;26:233–256.
18. Ismail M, Abd-Elsalam MA, Al-Ahaidib MS. *Androctonus crassicauda* (Olivier), a dangerous and unduly neglected scorpion—I. Pharmacological and clinical studies. Toxicon 1994;32:1599–1618.
19. Cupo P, Figueiredo AB, Filho AP, Pintya AO, Tavares GA Jr , Caligaris F, et al. Acute left ventricular dysfunction of severe scorpion envenomation is related to myocardial perfusion disturbance. Int J Cardiol 2007;116:98–106.
20. Catterall WA. Molecular mechanisms of gating and drug block of sodium channels. Novartis Found Symp 2002;241:206–225.
21. Catterall WA, Cestèle S, Yarov-Yarovoy V, Yu FH, Konoki K, Scheuer T. Voltage-gated ion channels and gating modifier toxins. Toxicon 2007;49:124–141.
22. Gwee MC, Nirthanan S, Khoo H-E, Gopalakrishnakone P, Kini RM, Chaeh L-S. Autonomic effects of some scorpion venoms and toxins. Clin Exp Pharmacol Physiol 2002;29:795–801.

23. Goudet C, Chi C-W, Tytgat J. An overview of toxins and genes from the venom of the Asian scorpion *Buthus martensi* Karsch. Toxicon 2002;40:1239–1258.
24. Norton RS. Structure and function of peptide and protein toxins from marine organisms. J Toxicol Toxin Rev 1998;17:99–130.
25. Rash LD, Hodgson WC. Pharmacology and biochemistry of spider venoms. Toxicon 2002;40:225–254.
26. Biggin PC, Roosild T, Choe S. Potassium channel structure: domain by domain. Curr Opin Struct Biol 2000;10: 456–461.
27. Dhawan R, Varshney A, Mathew MK, Lala AK. BTK-2, a new inhibitor of the Kv1.1 potassium channel purified from Indian scorpion *Buthus tamulus*. FEBS Lett 2003;539: 7–13.
28. Castle NA, London DO, Creech C, Fajloun Z, Stocker JW, Sabatier J-M. Maurotoxin: a potent inhibitor of intermediate conductance Ca^{2+}-activated potassium channels. Mol Pharmacol 2003;63:409–418.
29. Pedarzani P, D'hoedt D, Doorty KB, Wadsworth JD, Joseph JS, Jeyaseelan K, et al. Tamapin, a venom peptide from the Indian red scorpion (*Mesobuthus tamulus*) that targets small conductance Ca^{2+}-activated K^+ channels and afterhyperpolarization currents in central neurons. J Biol Chem 2002;277:46101–46109.
30. Sidach SS, Mintz IM. Kurtoxin, a gating modifier of neuronal high- and low-threshold Ca channels. J Neurosci 2002;22:2023–2034.
31. Olamendi-Portugal T, Garcia BI, Lopez-Gonzalez I, van der Walt J, Dyason K, Ulens C, et al. Two new scorpion toxins that target voltage-gated Ca^{2+} and Na^+ channels. Biochem Biophys Res Comm 2002;299:562–568.
32. Visan V, Sabatier JM, Grissmer S. Block of maurotoxin and charybdotoxin on human intermediate-conductance calcium-activated potassium channels (hIKCa1). Toxicon 2004;43:973–980.
33. Carbonaro PA, Janniger CK, Schwartz RA. Scorpion sting reactions. Cutis 1996;57:139–141.
34. Novaes G, de Queiroz AC, das Neves MM, Cardozo C, Ribeiro-Filho L, De Carvalho MH, et al. Induction of acute and chronic pancreatitis with the use of the toxin of the scorpion *Tityus serrulatus*: experimental model in rats. Arq Gastroenterol 1998;35:216–222.
35. Ismail M. The scorpion envenoming syndrome. Toxicon 1995;33:215–858.
36. Karnad DR. Haemodynamic patterns in patients with scorpion envenomation. Heart 1998;79:485–489.
37. Gueron M, Ilia R, Sofer S. The cardiovascular system after scorpion envenomation. A review. J Toxicol Clin Toxicol 1992;30:245–258.
38. Gueron M, Yarom R. Cardiovascular manifestations of severe scorpion stings. Chest 1970;57:156–162.
39. Benvenuti LA, Douetts KV, Cardoso JL. Myocardial necrosis after envenomation by the scorpion *Tityus serrulatus*. Trans Roy Soc Trop Med Hyg 2002;96:275–276.
40. Yarom R, Braun K. Cardiovascular effects of scorpion venom, morphological changes in the myocardium. Toxicon 1970;8:41–46.
41. Sadeghian H. Transient ophthalmoplegia following envenomation by the scorpion *Mesobuthus eupeus*. Neurol 2003;60:346–347.
42. Tiwari SK, Gupta GB, Gupta SR, Mishra SN, Pradhan PK. Fatal stroke following scorpion bite. J Assoc Physicians India 1988;36:225–226.
43. Sofer S, Gueron M. Respiratory failure in children following envenomation by the scorpion *Leiurus quinquestriatus*: hemodynamic and neurological aspects. Toxicon 1988;26:931–939.
44. Forrester MB, Stanley SK. Epidemiology of scorpion envenomations in Texas. Vet Hum Toxicol 2004;46: 219–221.
45. Likes K, Banner W, Chavez M. *Centruroides exilicauda* envenomation in Arizona. West J Med 1984:141:634–637.
46. Otero R, Navio E, Cespedes FA, Nunez MJ, Lozano L, Moscoso ER, et al. Scorpion envenoming in two regions of Colombia: clinical, epidemiological and therapeutic aspects. Trans Roy Soc Trop Med Hyg 2004;98:742–750.
47. Amitai Y. Clinical manifestations and management of scorpion envenomation. Public Health Rev 1998;26: 257–263.
48. Rimsza ME, Zimmerman OR, Bergeson PS. Scorpion envenomation. Pediatrics 1980:66:298–301.
49. Curry SC, Vance MV, Ryan PJ, Kunkel DB, Northey WT. Envenomation by the scorpion *Centruroides sculpturatus*. J Toxicol Clin Toxicol 1983–1984:21:417–449.
50. Al-Asmari AK, Al-Saif AA. Scorpion sting syndrome in a general hospital in Saudi Arabia. Saudi Med J 2004;25: 64–70.
51. Diaz P, Chowell G, Ceja G, D'Auria TC, Lloyd RC, Castillo-Chavez C. Pediatric electrocardiograph abnormalities following *Centruroides limpidus tecomanus* scorpion envenomation. Toxicon 2005;45:27–31.
52. Bogomolski-Yahalom V, Amitai Y, Stalnikowicz R. Paresthesia in envenomation by the scorpion *Leiurus quinquestriatus*. Clin Toxicol 1995;33:79–82.
53. Bergman NJ. Clinical description of *Parabuthus transvaalicus* scorpionism in Zimbabwe. Toxicon 1997;35:759–771.
54. Adiguzel S, Ozkan O, Inceoglu B. Epidemiological and clinical characteristics of scorpionism in children in Sanliurfa, Turkey. Toxicon 2007;49:875–880.
55. LoVecchio F, McBride C. Scorpion envenomations in young children in Central Arizona. J Toxicol Clin Toxicol 2003;41:937–940.
56. Bawaskar HS, Bawaskar PH. Scorpion sting: a review of 121 cases. J Wilderness Med 1991;2:164–174.
57. Groshong TD. Scorpion envenomation in Eastern Saudi Arabia. Ann Emerg Med 1993;22:1431–1437.
58. Bentur Y, Taitelman U, Aloufy A. Evaluation of scorpion stings: the poison center perspective. Vet Hum Toxicol 2003;45:108–111.

59. Dittrich K, Ahmed R, Ahmed QA. Cardiac arrest following scorpion envenomation. Ann Saudi Med 2002;22: 87–90.

60. Kumar BU, Mukesh T, Sanjib S, Rajesh P. Scorpion sting envenomation presenting with pulmonary edema in adults: a report of seven cases from Nepal. Indian J Med Sci 2006;60:19–23.

61. Bawaskar HS. Diagnostic cardiac premonitory signs and symptoms of red scorpion sting. Lancet 1982;1:552–554.

62. El-Amin EO. Issues in management of scorpion sting in children. Toxicon 1992;30:111–115.

63. Bouaziz M, Bahloul M, Hergafi L, Kallel H, Chaari L, Hamida CB, et al. Factors associated with pulmonary edema in severe scorpion sting patients—a multivariate analysis of 428 cases. Clin Toxicol 2006;44:293–300.

64. Ghalim N, El-Hafny B, Sebti F, Heikel J, Lazar N, Moustanir R, Bensilmane A. Scorpion envenomation and serotherapy in Morocco. Am J Trop Med Hyg 2000;62:277–283.

65. Krifi MN, Kharrat H, Zghal K, Abdouli M, Abroug F, Bouchoucha S, et al. Development of an ELISA for the detection of scorpion venoms in sera of humans envenomed by *Androctonus australis garzonii* (Aag) and *Buthus occitanus tunetanus* (Bot): correlation with clinical severity of envenoming in Tunisia. Toxicon 1998;36:887–900.

66. de Rezende NA, Dias MB, Campolina D, Chavez-Olortegui C, Amaral CF. Standardization of an enzyme linked immunosorbent assay (ELISA) for detecting circulating toxic venom antigens in patients stung by the scorpion *Tityus serrulatus*. Rev Inst Med Trop Sao Paulo 1995;37:71–74.

67. Alan S, Ulgen MS, Soker M, Geyik F, Karabulut A, Toprak N. Electrocardiologic and echocardiographic features of patients exposed to scorpion bite. Angiology 2004;55: 79–84.

68. Bawaskar HS, Bawaskar PH. Indian red scorpion envenoming. Indian J Pediatr 1998;654:383–391.

69. Meki AR, Mohamed ZM, Mohey El-deen HM. Significance of assessment of serum cardiac troponin I and interleukin-8 in scorpion envenomed children. Toxicon 2003;41: 129–137.

70. Kumar EB, Soomro RS, Al Hamdani A, El Shimy N. Scorpion venom cardiomyopathy. Am Heart J 1992; 123:725–729.

71. Bahloul M, Hamida CB, Chtourou K, Ksibi H, Dammak H, Kallel H, Chaari A, et al. Evidence of myocardial ischaemia in severe scorpion envenomation myocardial perfusion scintigraphy study. Intensive Care Med 2004; 30:461–467.

72. Sofer S, Shalev H, Weizman Z, Shahak E, Gueron M. Acute pancreatitis in children following envenomation by the yellow scorpion *Leiurus quinquestriatus*. Toxicon 1991;29:125–128.

73. Berg RA, Taratino MD. Envenomation by the scorpion *Centruroides exilicauda* (*C sculpturatus*): severe and unusual manifestations. Pediatrics 1991;87:930–933.

74. Abroug F, Nouira S, Haguiga H, Elatrous S, Belghith M, Boujdaria R, et al. High-dose hydrocortisone hemisuccinate in scorpion envenomation. Ann Emerg Med 1997;30: 23–27.

75. Bawaskar HS, Bawaskar PH. Utility of scorpion antivenin vs prazosin in the management of severe *Mesobuthus tamulus* (Indian red scorpion) envenoming at rural setting. J Assoc Physicians India 2007;55:14–21.

76. Bawaskar HS, Bawaskar PH. Severe envenoming by the Indian red scorpion *Mesobuthus tamulus*: the use of prazosin therapy. Q J Med 1996;89:701–704.

77. Sofer S, Gueron M. Vasodilators and hypertensive encephalopathy following scorpion envenomation in children. Chest 1990;97:118–120.

78. Elatrous S, Nouira S, Besbes-Ouanes L, Boussarsar M, Boukef R, Marghli S, Abroug F. Dobutamine in severe scorpion envenomation effects on standard hemodynamics, right ventricular performance, and tissue oxygenation. Chest 1999;116;748–753.

79. Gateau T, Bloom M, Clark R. Response to specific *Centruroides sculpturatus* antivenom in 151 cases of scorpion stings. Clin Toxicol 1994;32:165–171.

80. De Rezende NA, Dias MB, Campolina D, Chavez-Olortegui C, Diniz CR, Amaral CF. Efficacy of antivenom therapy for neutralizing circulating venom antigens in patients stung by *Tityus serrulatus* scorpions. Am J Trop Med Hyg 1995;52:277–280.

81. Dehesa-Davila M, Possani LD. Scorpionism and serotherapy in Mexico 1994;32:1015–1018.

82. Hamed MI. Treatment of scorpion envenoming syndrome: 12-years experience with serotherapy. Int J Antimicrob Agents 2003;21:170–174.

83. Ismail M. The treatment of the scorpion envenoming syndrome: the Saudi experience with serotherapy. Toxicon 1994;32:1019–1026.

84. Abroug F, El Atrous S, Nouira S, Haguiga H, Touzi N, Bouchoucha S. Serotherapy in scorpion envenomation: a randomized controlled trial. Lancet 1999;354:906–909.

85. D'Suze G, Moncada S, Gonzalez C, Sevcik C, Alagon A. Antigenic cross-reactivity between sixteen venoms from scorpions belonging to six genera. Clin Toxicol 2007;45: 158–163.

86. LoVecchio F, Welch S, Klemens J, Curry SC, Thomas R. Incidence of immediate and delayed hypersensitivity to *Centruroides* antivenom. Ann Emerg Med 1999;34: 615–619.

87. Bond GR. Antivenin administration for *Centruroides* scorpion sting: risk and benefits. Ann Emerg Med 1992;21: 788–791.

88. Ben-Abraham R, Eschel G, Winkler E, Weinbroum AA, Barzilay Z, Paret G. Triage for *Leiurus quinquestriatus* scorpion envenomation in children—is routine ICU hospitalization necessary. Hum Exp Toxicol 2000;19; 663–666.

89. Gibly R, Williams M, Walter FG, McNally J, Conroy C, Berg RA. Continuous intravenous midazolam infusion for *Centruroides exilicauda* scorpion envenomation. Ann Emerg Med 1999;34:620–625.

Chapter 166

SPIDERS
(Order: Araneae)

Spiders are a successful, ancient group of arachnids from the phylum Arthropoda with fossil records dating back 300 million years (Carboniferous period). Although only a few spider species possess fangs large enough to envenomate humans, there are almost 40,000 species of spiders in the order Araneae. Only about 200 species from 20 genera produce clinically significant envenomations in humans even though all spiders are venomous with the exception of spiders from the family Uloboridae. Arthropods (e.g., centipedes, crabs, insects, lobsters, mites, ticks) are bilaterally symmetrical invertebrates that have true body segments, a chitinous exoskeleton, and jointed appendages (antennae, mouth parts, spinnerets, legs). The class Arachnida also includes ticks, mites, vinegaroons, scorpions, and solpugids (Camel spiders). Solpugids are large, quick arachnids that inhabit hot arid regions, such as the Arabian Peninsula. These nocturnal creatures subdue their prey with large chelicerae (jaws) that do not inject venom. The order Araneae is divided into two suborders: Orthognatha (mygalomorph) and Labidognatha (araneomorphs). Species (e.g., Australian funnel-web spiders, tarantulas) from the former are primitive or trapdoor spiders with chelicera projecting forward from the cephalothorax, while the fangs move downward. The chelicera on the araneomorphs (true spiders) project vertically and move together with the fangs like pincers. Figure 166.1 displays a specimen of the Chilean Rose Tarantula (*Grammostola rosea* Walckenaer).

Most bites by spiders cause minimal, if any, toxic effects.[1] Of the approximately 20,000 species of spiders in the United States, at least 50–60 species have fangs capable of penetrating the human skin.[2] Spiders have a cephalothorax and abdomen along with four legs in contrast to most insects (e.g., bees, wasps, mosquitoes, flies), which possess a head, thorax, abdomen, three pairs of legs, one or two pairs of wings attached to the thorax, and one pair of antennae. A hard convex coat (carapace) covers the dorsal aspect of the body, whereas a ventral plate (sternum) envelopes the underside of the cephalothorax. Eight legs and a pair of pedipalps (tactile sensing organ) emerge from the sternum. The anterior end of the cephalothorax contains a pair of jaws (chelicerae) composed of a basal segment and distal fangs. These fangs are mobile and heavily chitinized. A small opening at the tip of the fangs is connected to a venom gland located in the basal segment of the chelicerae in tarantulas. This poison gland extends to the cephalothorax in most spider species. A narrow pedicle connects the cephalothorax and the abdomen, which usually contains spinning organs (spinnerets) on the ventral surface.

Most spiders depend primarily on a sense of touch mediated by mechanoreceptive hairs covering the appendages and other parts of the body because they are extremely shortsighted. However, some hunting spiders have relatively good vision. Spiders inhabit almost every niche in terrestrial habitats including seashores, alpine meadows, deserts, and tropical rain forests. They generally live for 1–2 years. Spiders molt 5 to 10 times before maturity. During this period, they shed their hard shells and may change coloration or markings.

Most spiders are not aggressive or harmful to humans. They feed predominately on insects or other animals including frogs, lizards, and rodents. The major

Medical Toxicology of Natural Substances, by Donald G. Barceloux, MD

FIGURE 166.1. The Chilean Rose tarantula (*Grammostola rosea* Walckenaer). © 2007 Rune Midtgaard. See color insert.

constituents of spider venom are proteins, enzymes, polypeptide and polyamine neurotoxins, nucleic acids, monoamines, free amino acids, and inorganic salts. These components can cause cytotoxic, hemolytic, or neurotoxic effects. The targets for the neurotoxic components of spider venom are the neuronal receptors, neuronal ion channels, or the presynaptic membrane proteins involved in neurotransmitter release.[3] Components in these venoms can interact with glutamate receptors, ion channels (potassium, sodium, chloride, calcium), presynaptic neurotransmitter release, or cholinergic transmission. However, the weak venom apparatus and the small amount of injected venom limit toxicity in humans. Systemic toxicity of American spiders is limited to two genera (*Latrodectus* and *Loxosceles*), despite the fact that spider venom is more potent than pit viper venom on a weight for weight basis. Spiders are not known to transmit communicable diseases.

WIDOW SPIDERS and OTHER THERIDIID SPIDERS

HISTORY

Superstitions regarding spiders appear in Greek mythology, where the jealous goddess Aphrodite transformed the beautiful young weaver, Arachne, into a spider. *Latrodectus tredecimguttatus* Rossi is native to the northern Mediterranean region, where envenomation by this species probably accounted for the hysteria surrounding the tarantula in the 17th century.[4] Beginning in the 17th century in the southern Italian city of Taranto and then throughout Europe, panic spread concerning spider bites that caused muscle contractions and pain. These envenomations were attributed to the large European wolf spider (*Lycosa tarantula* L.), commonly called a tarantula. The tarantella was a dance devised to protect inhabitants from what was erroneously believed to be bites by the wolf spider. The mild-mannered North and South American tarantula species produce a bite similar to that of a hymenoptera sting without systemic toxicity.[5] Many species of tarantulas possess urticaria-producing hairs on the dorsal surface of the abdomen. A case series of 515 individuals bitten by the South American wolf spider indicates that local necrosis or severe systemic symptoms are not usually associated with bites by these species.[6] In the 1970s, *Latrodectus* venom was an experimental treatment to accelerate recovery from botulism.[7] In mice paralyzed by botulinum toxin, the injection of black widow spider venom was associated with enhanced sprouting of the axons and earlier return of muscle function compared with the injection of botulinum toxin alone.[8]

IDENTIFYING CHARACTERISTICS

Widow spiders (*Latrodectus* spp.) include about 40 species belonging to the family Theridiidae (comb-footed spiders). Other theridiid spiders capable of producing less severe human envenomation compared with widow spiders include spiders from the genera Steatoda and Achaearanea. The female widow spider is one of the largest female spiders with a leg span up to 5 cm (~2 in.) and a body length of 1–1.5 cm (~0.4–0.6 in.). *L. mactans* Fabricius and *L. geometricus* C. L. Koch females have the characteristic red or orange-red hourglass spot on the ventral surface of their black, shiny abdomen, whereas the *L. tredecimguttatus* Rossi (European widow spider) has two red marks that do not coalesce. Three US species are predominantly black (i.e., *L. mactans* Fabricius, *L. hesperus* Chamberlin & Ivie, *L. variolus* Walckenaer), whereas other *Latrodectus* species are brown (*L. geometricus* C. L. Koch, *L. indistinctus* O. P.-Cambridge) or orange-red (*L. bishopi* Kaston). *Latrodectus geometricus* C. L. Koch has well-demarcated geometric markings, ranging in color from cream to brown-orange. Marked intraspecies variation may occur as a result of molting. These black spiders become darker with less prominent red spots with each molt. In contrast to most other black spiders, the legs of

Latrodectus species are smooth. The substantially smaller male spider retains the more colorful white markings and geometric patterns during the immature stages of development.

Widely distributed cosmopolitan spiders from the genus *Achaearanea* are common inhabitants of houses and bushland in Australia and the United States. *Steatoda* species (cupboard spider, false widow spider, brown house spider) are small brown or black spiders that usually have white markings without red coloration. These spiders are typically smaller than black widow spiders, but larger specimens (e.g., *Steatoda grossa* C. L. Koch) may resemble the black widow spider.

EXPOSURE

Geographic Distribution

Latrodectus species are cosmopolitan spiders that inhabit all parts of the world except ecological extremes (e.g., high mountains above 8,000 feet, hot deserts, polar ice caps), in part, as a result of human transport.[9] Members of *Latrodectus* species are present in North and South America, Southern Europe, Southern Africa, the Middle-East, Asia, New Zealand, and Australia. There are about 30 species of *Latrodectus* distributed throughout the world as demonstrated by Table 166.1. Widow spiders include the red-back spider in Australia, the brown and black button spider of South Africa, the New Zealand katipo, and the widow spiders of Europe and South America.

TABLE 166.1. Distribution of Some Relatively Well-Known Species of *Latrodectus*

***Latrodectus* Species/ Authority**	**Distribution**
antheratus Badcock	Paraguay, Argentina
apicalis Butler	Galapagos Islands
atritus Erquhart	New Zealand
bishopi Kaston	United States, Florida primarily
cinctus Blackwell	Cape Verde Islands, Africa, Kuwait
corallinus Abalos	Argentina
curacaviensis Muller	Lesser Antilles, South America
dahli Levi	Middle East to Central Asia
erythromelas Schmidt & Klaas	Sri Lanka
geometricus C. L. Koch	Worldwide
hasselti Thorell	Southeast Asia, Australia, New Zealand
hesperus Chamberlin & Ivie	Western United States
hystrix Simon	Yemen, Socotra Island
indistinctus O.P.-Cambridge	Namibia, South Africa
karrooensis Smithers	South Africa
katipo Powell	New Zealand
lilianae Melic	Spain
mactans Fabricius	North America
menavodi Vinson	Madagascar, Comoro Islands
mirabilis Holmberg	Argentina
obscurior Dahl	Cape Verde Islands, Madagascar
pallidus O.P.-Cambridge	Libya to Russia and Iran
quartus Ablos	Argentina
renivulvatus Dahl	Africa, Saudi Arabia, Yemen
revivensis Shulov	Israel
rhodesiensis MacKay	Southern Africa
tredecimguttatus Rossi	Mediterranean to China
variegatus Nicolet	Chile, Argentina
variolus Walckenaer	United States, Canada

Source: Adapted from Ref 9.

NORTH AND SOUTH AMERICA

One of the five US species of *Latrodectus* is present in every state except Alaska. These species include *Latrodectus mactans* Fabricius (black widow from the South and East), *Latrodectus hesperus* Chamberlin & Ivie (western black widow), *Latrodectus variolus* Walckenaer (eastern United States, Canada), *Latrodectus bishopi* Kaston (red or red-legged widow spider from limited area of scrub pines in central Florida), and *Latrodectus geometricus* C. L. Koch (brown widow). The latter species is found throughout the world. The prominent *Latrodectus* species in South America is *L. curacaviensis* Muller.

ASIA–PACIFIC

Medically important widow spiders (family Theridiidae) from the Asia–Pacific region include the red-back widow spider (*L. hasselti* Thorell), the New Zealand katipo (*L. katipo* Powell), and the cupboard or false black widow spiders (*Steatoda* spp.). Spiders in this region other than Theridiidae that cause potentially serious neurotoxicity include the Australian funnel-web spiders (Atracinae subfamily) and Australian mouse spiders (Actinopodidae family).[10]

SOUTH AFRICA

The three species of *Latrodectus* found in South Africa are *L. indistinctus* O. P.-Cambridge (black widow or black button spider), *L. geometricus* (brown widow or brown button spider), and *L. rhodesiensis* MacKay (Rhodesian widow spider). Macroscopically, the latter two species are indistinguishable, but the egg sac of *L. geometricus* is several times larger and wooly. *L. indistinctus* O. P.-Cambridge is an endemic species to South Africa found throughout rural areas (bushes, shrubs, grass tufts, loose debris), whereas *L. geometricus* C. L. Koch is primarily an urban species.

Behavior

Latrodectus species are general predators that feed on a variety of prey including other arachnids, insects, crustaceans, and small vertebrate (i.e., mice, geckos, lizards).[11] The male is smaller, shorter lived, and less aggressive than the female; consequently, the male does not usually produce latrodectism. Female widow spiders are venomous from birth, but their venom toxicity is seasonal, peaking in the warmer months. These shy, sedentary, trapping spiders use their venom to paralyze and extract the hemolymph from prey, primarily at night. A period of postcopulation lassitude allows the male spider to escape unless the female has a voracious appetite. Female widow spiders form irregular, untidy three-dimensional webs in dark, hidden protected places (e.g., crevices, ground cover, stone walls, woodpiles, barns, stables, outdoor toilets). *Latrodectus* bites usually occur when the widow spider is disturbed while residing in man-made objects (clothes, shoes, furniture, closets, sheds, attics, woodpiles), primarily on an extremity. Contrary to popular belief, both females and males often live together.

PRINCIPAL TOXINS

Venom Composition

Medically important *Latrodectus* species produce potent neurotoxins. Although *Latrodectus* venom contains low-molecular-weight purine derivatives and at least seven protein fractions (e.g., hyaluronidase, phosphodiesterase, GABA, and 5-hydroxytryptamine),[12] examination of the venom of European widow spider (*L. tredecimguttatus*) indicates that the major toxin is a high-molecular-weight (i.e., 120,000 Da), presynaptic neurotoxin called α-latrotoxin.[13] Other high-molecular-weight neurotoxins (e.g., α-latroinsectotoxin, δ-latroinsectotoxin) also are present, but the role of these neurotoxins following envenomation requires further study.[14,15] Although the presence of α-latrotoxin in all *Latrodectus* species has not been confirmed, *Latrodectus* venoms are probably structurally similar because of the clinical similarity of toxic effects produced by most *Latrodectus* venoms.[1] The low-molecular-weight compounds probably do not contribute to the neurotoxicity associated with latrodectism.[16] Variation in venom potency occurs between species and during different seasons based on animal studies.[17] Animal susceptibility is variable with sensitivity relatively high in camels and horses compared with rats.

Venom Apparatus

The small size of the male fangs prevents envenomation under most circumstances. The female widow spider has a pair of venom glands in the cephalothorax with the walls of these glands containing striated muscles that control the delivery of the venom.[18] Figure 166.2 demonstrates the venom apparatus of *Latrodectus* species.

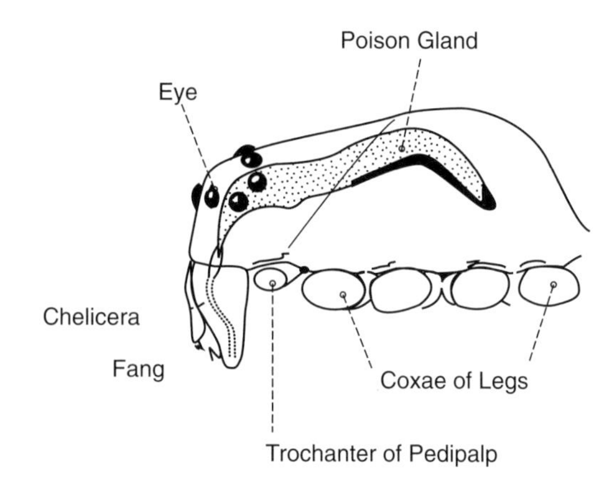

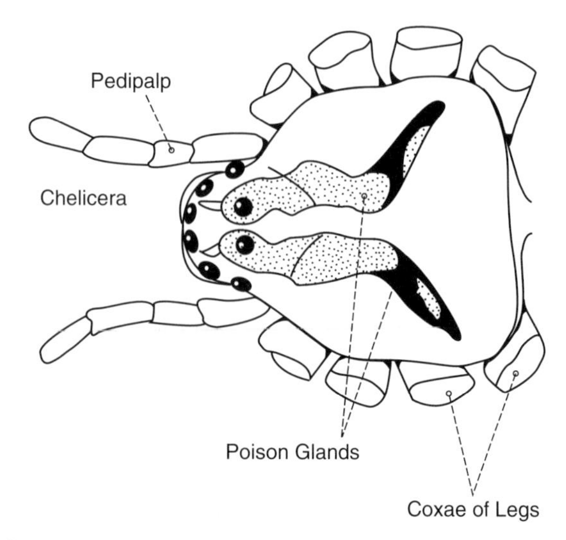

FIGURE 166.2. The venom apparatus of *Latrodectus* species. Adapted from Wong RC, et al. Spider bites. Arch Dermatol 1987;123:99.

Dose Response

The severity of an envenomation depends on a number of factors including the size of the spider, the physical condition of the spider, season, locations of the bite, amount of venom injected, susceptibility of the victim, and, to a lesser extent, on the species. In animals, the LD_{50} of *L. geometricus* venom is lower than *L. mactans* (i.e., 0.43 mg/kg vs. 1.39 mg/kg, respectively). Black button spiders (*L. indistinctus*) characteristically produce more severe envenomation compared with brown button spiders (*L. geometricus*).[19] Based on LD_{50} tests in mice, the venom of *L. indistinctus* was 3–4 times more potent than the venom of *L. geometricus*.[20]

HISTOPATHOLOGY AND PATHOPHYSIOLOGY

Black widow spider venom contains several large proteins (i.e., latrotoxins), which selectively target different classes of animals including vertebrates, crustaceans, and insects.[21] Latrotoxins act by several Ca^{++}-dependent and Ca^{++}-independent mechanisms based on pore formation and receptor activation.[22] All latrotoxins including α-latrotoxin produce massive, spontaneous release of several types of neurotransmitters (GABA, norepinephrine, acetylcholine) by facilitating synaptic vesicle exocytosis from presynaptic nerve terminals.[23] These neurotoxins cause the clinical features (e.g., diaphoresis, muscle fasciculation) associated with latrodectism. The postsynaptic membranes remain unaffected. α-Latrotoxin is a 120 kDa protein that depletes presynaptic vesicles by binding to two distinct families of neuronal cell-surface receptors to trigger exocytosis.[10,24] Major high-affinity receptors for α-latrotoxin include a G protein-coupled receptor (latrophilin) and a single-transmembrane receptor (neurexin). Exocytosis of classical neurotransmitters (glutamate, GABA, acetylcholine) is calcium-independent as a result of the direct intracellular action of α-latrotoxin, whereas exocytosis of catecholamines requires extracellular calcium. High concentrations of calcium theoretically facilitate the release of neurotransmitters by endogenous electrical stimulation while inhibiting venom-induced release.[4] However, the clinical efficacy of calcium for relief of pain associated with widow spider envenomation is questionable. Both botulinum toxin and black widow spider venom cause presynaptic blockade of neuromuscular transmission. However, botulinum toxin impairs the release of acetylcholine and persistent paralysis of skeletal muscle that requires sprouting of axons and formation of new motor end-plates prior to recovery.

CLINICAL RESPONSE

Latrodectus Species

Envenomations by all species of *Latrodectus* produce similar clinical features depending on the degree of envenomation and slight differences in the potency of the venom.[25] The predominant feature of *Latrodectus* envenomation is pain. Although *Latrodectus* bites may be painless, most bites cause a mild pinprick sensation that produces local pain, sweating, erythema, and piloerection in the affected extremity within 1 hour of envenomation. The characteristic skin reaction is a halo or target lesion characterized by two central punctuate fang marks 1 mm apart surrounded by an outer erythematous ring. Slight swelling may develop along with urticaria and pruritus, particularly following bites by *L. hesperus*. Necrosis does not occur following envenomation. Local symptoms often subside within a few minutes, and cutaneous signs of envenomation may be absent when the patient presents for medical care. Approximately 75% of *Latrodectus* bites involve only local symptoms that do not progress to systemic effects.[26] The presence of localized piloerection and sweating is highly suggestive of latrodectism, particularly when a target lesion is present.

In about 25% of *Latrodectus* bites, systemic evidence of envenomation develops with severe pain being the most consistent feature of *Latrodectus* envenomation.[27] Within the first few hours after envenomation, the gradual onset of progressive, local pain, sweating, piloerection, and muscle fasciculations occurs.[28] Diaphoresis may be generalized or limited to the local area of the envenomated extremity. Common symptoms of severe envenomations involve intense, generalized abdominal, back, and leg pains.[29] Muscle spasms begin on the upper body following upper extremity bites or in the abdominal wall following bites in lower extremities. Chest or abdominal pains may become severe with the intensity of the abdominal pain simulating acute peritonitis (i.e., without direct or indirect abdominal tenderness to palpation). The sudden onset of irritability, hypertension and sweating in a child is suggestive of *Latrodectus* envenomation.[30]

Associated clinical features of latrodectism include nausea, vomiting, headache, tremulousness, anxiety, tachycardia, chest tightness, sweating, mild pyrexia, lacrimation, salivation, and increased bronchorrhea.[31] Hyperesthesias (especially on the feet), paresthesias, fasciculations, ptosis, hyperreflexia, urinary retention, uterine contractions, and convulsions also occur. The course of symptoms may follow a cyclical pattern with spontaneous improvement and then deterioration.[28] Most envenomations by the *Latrodectus* species produce symptoms that peak in 3–4 hours, and resolve within 2–3 days.[32,33] However, subjective symptoms of weakness, headache, insomnia, myalgias, paresthesias, and malaise may persist for weeks, particularly in patients not receiving antivenom.[34,35] Death is rare, even in young children.[36] Susceptible patients include the elderly, infants, patients with preexisting cardiovascular disease, and pregnant women. Hypertension occurs frequently, but few case reports associate complications secondary to hypertension during latrodectism. Rare complications include renal dysfunction,[37] priapism, herpes zoster,[38] and pulmonary edema.[39] Several case reports of pregnant women envenomated during the second and third trimesters of pregnancy did not describe adverse fetal effects.[40]

Other Theridiidae Species

Some theridiid spiders other than *Latrodectus* species (e.g., *Steatoda* spp.) produce a neurotoxin that causes nausea, vomiting, feverishness, intense radiating pain, and diffuse myalgias. The envenomation by spiders from the genus *Steatoda* may cause prolonged pain and the systemic effects similar, but usually less severe, than envenomations by *Latrodectus* species.[41] However, diaphoresis is not usually part of the clinical course following envenomation by *Steatoda* species.[42] These spiders inhabit many parts of the world including the Americas, Australia, Europe, and South Africa. Most bites by the theridiid spiders (false button spiders, false widow spiders, cob-web spiders) produce no or mild local symptoms without systemic complaints. Bites by spiders from the genera *Achaearanea* may cause moderate to severe, prolonged pain along with some mechanical trauma (puncture marks, local bleeding).

DIAGNOSTIC TESTING

Analytical Methods

There are no clinical tests to detect latrodectism. The diagnosis of *Latrodectus* envenomation is based on clinical features and the identification of the spider. The presence of localized piloerection and sweating is almost pathognomonic of latrodectism, particularly when a target lesion is present. However, the appearance of these clinical features is not required for the diagnosis of latrodectism.

Abnormalities

Most laboratory tests remain normal following *Latrodectus* envenomation. Leukocytosis and mildly elevation of the serum creatine kinase (CK) may occur in moderate to severe cases. Albuminuria may develop, but renal function usually remains normal unless prerenal azotemia occurs.

TREATMENT

Prehospital treatment of latrodectism involves limitation of activity, comfort measures, and transport. There are inadequate clinical data to support the use of any specific first-aid measures (tourniquet, local incision, suction). Wounds should be cleaned and tetanus prophylaxis given as indicated. Antibiotics are not usually indicated. The treatment of envenomation by other theridiid spiders (e.g., *Steatoda* species) usually involves only supportive care, except following severe envenomation that may require antivenom.

Antivenom

Latrodectus antivenom is available in the United States (Merck & Co., West Point, PA), Australia (Commonwealth Serum Laboratories, Melbourne, Australia), South Africa (South African Institute for Medical Research, South Africa), and Mexico (Instituto Bioclon, Calzada de Telapan, Mexico; Aracmyn®).[43] The latter antivenom is a polyvalent antivenom for latrodectism and loxoscelism. There is substantial cross-reactivity between *Latrodectus* venom within US species and throughout the world. *In vitro* data suggests that antivenom for numerous *Latrodectus* species (e.g., *L. mactans, L. hesperus, L. hasselti, L. katipo, L. tredecimguttatus*) produce antibodies against α-latrotoxin and provide at least partial protection from envenomation by these species based on animal studies[44,45] and case reports.[46] The antivenom from the United States is a hyperimmune, equine whole IgG (150 kDa) serum product derived from *L. mactans* venom, and this antivenom is effective for all US species. Skin testing prior to the use of the antivenom is recommended by the US manufacturer, but not in Australia where the antivenom is a trypsin-digested, equine $F(ab)_2$ IgG fragment (100 kDa). Although skin testing may detect a highly allergic individual, a negative skin test does not exclude the possibility of an allergic response following the administration of antivenom. Prior or current evidence of horse serum sensitivity is a contraindication to the use of *Latrodectus* antivenom or skin testing because most envenomated patients respond to supportive care. Most patients with mild to moderate envenomations require only supportive care.

Anecdotal evidence suggests that the use of *Latrodectus* antivenom reduces the duration of symptoms associated with severe envenomations by *Latrodectus* species.[29] Patients with signs of systemic envenomation or severe local symptoms unresponsive to parenteral analgesics are candidates for the administration of antivenom. Therefore, indications for antivenom use include pain unresponsive to parenteral analgesics, particularly in pediatric or geriatric patients, pregnant women,[47] patients with cardiovascular disease and uncontrolled hypertension, and patients with respiratory depression. However, few controlled studies of humans document the efficacy and safety of *Latrodectus* antivenom.[32] The usual dosage is 1 or 2 vials diluted in 50–100 mL of saline and infused over 30 minutes to 1 hour. A preliminary study of Red-back spider antivenom (Commonwealth Serum Laboratories) suggested the average amount of antivenom used for patients presenting to Western Australian hospitals with Red-back spider envenomation was 2–3 vials with only 21% receiving one vial to control symptoms.[48] This study did not detect a significant difference between the intramuscular and intravenous administration of antivenom, but the small number of cases and the large number of patients lost to follow-up limits conclusions regarding the most efficacious route of antivenom administration. Rapid infusion of antivenom may cause histamine release and an anaphylactoid reaction. Symptoms usually resolve within an hour after the infusion ends, and the presence of systemic signs of envenomation over 2 hours after the infusion ends suggests the need for an additional vial of antivenom.[26,29] Pregnancy is not a contraindication to the use of *Latrodectus* antivenom.[49] Case reports suggest that patients with signs of systemic envenomation may respond to antivenom therapy even though envenomation occurred several days earlier.[50] Patients, who respond to the administration of *Latrodectus* antivenom without complications, do not require admission to the hospital because relapse of systemic symptoms is unusual.

Serum sickness and anaphylactoid reactions are rare (i.e., 1–2%) following the intramuscular use of Australian purified *Latrodectus* antivenom, but patients should be re-evaluated within 1 week for the development of serum sickness.[51,52] Case reports suggest that some envenomated patients, who fail to respond to intramuscular injections of the antivenom, will respond to *Latrodectus* antivenom administered by intravenous infusion.[53] Fatalities are rare during the use the antivenom, although a case report documented the occurrence of an asthmatic dying of severe bronchospasm during treatment with US *Latrodectus* antivenom. Severe envenomation by the female Australian Cupboard spider (*Steatoda grossa* C. L. Koch) produces clinical features similar to latrodectism, and the administration of red-backed widow spider venom from Australia (Commonwealth Serum Laboratories) to these patients resulted in rapid clinical improvement.[54]

Supplemental Care

Generalized spasms may respond to the judicious use of intravenous opiates (morphine 0.1–0.2 mg/kg every 2–4 hours, hydromorphone 1–2 mg for an adult every 3–4 hours) and/or muscle relaxants (diazepam 0.04–0.2 mg/kg, lorazepam 0.01–0.015 mg/kg) with careful observation of the adequacy of respirations. Warm moist compresses are often soothing. Some authors recommend 10% intravenous calcium gluconate (0.2 mL/kg or 0.1 mEq Ca^{++}/kg up to 10 mL) given slowly with continuous cardiac monitoring. Calcium gluconate is the preferred calcium solution because calcium chloride is more irritating to tissues. The analgesia provided by intravenous calcium is often inconsistent and short-acting. Case series of latrodectism suggest that calcium administration in this setting frequently does not relieve the pain associated with *Latrodectus* envenomation.[29] Intravenous methocarbamol (Robaxin®, Whitehall-Robins, Richmond, VA) is probably inferior to intravenous calcium for the pain associated with latrodectism based on case reports.[55] There are inadequate clinical studies to evaluate the efficacy of benzodiazepine compounds (diazepam, lorazepam) or dantrolene (1 mg/kg over 15 minutes every 2 hours).[56]

ARMED (BANANA) SPIDER

These large (30 mm/~1 in.) South American banana (wandering) spiders belong to the genus *Phoneutria* (Ctenidae family) that inhabit eastern areas of Central and South America. In Brazil, the clinically most important species is *P. nigriventer* Keyserling (west central, southeast, south).[57] Other species of Brazilian wandering spiders include *P. fera* Perty (Amazon, Ecuador, Peru, Guiana) and *P. reidyi* F.O.P.-Cambridge (Amazon). The genus is characterized by the location of three rows of eyes distributed as follows: the first row of two eyes near the edge of the chelicerae, the second row containing four eyes, and the third row composed of two eyes. These large spiders (body length: 3 cm/~1 in., leg span: 15 cm/~6 in.) are covered with short, gray to gray brown hairs on the body and red hairs on the basal segment of the chelicerae. These large neurotoxic spiders can be confused with large, harmless *Cupiennius* spiders, which also inhabit banana clusters and possess similar distinctive brushes of red hairs surrounding the chelicerae. These aggressive, hunting *Phoneutria* spiders actively defend themselves by menacingly holding the front pair of legs toward the offending stimulus. These venomous spiders are found primarily along the coast of Brazil. They wander into rural dwellings and attack inhabitants during sleep or work. They do not build webs.[58]

The venom of *Phoneutria* spiders interferes with transmission of impulses in calcium and sodium channels found in excitable cells. The neurotoxic venom of *P. nigriventer* Keyserling is a complex mixture of polypeptides (T×1, T×2, T×3, T×4) that are potent activators and blockers of ionic channels.[59,60] The main neurotoxic effect is the action of the venom on voltage-gated Na^+ channels that causes repetitive-action potential discharge and depolarization of the muscle membrane. These neurotoxins produce a variety of effects in animal models including activation of voltage-dependent sodium channels, neuromuscular blockade, release of acetylcholine and norepinephrine, contraction of smooth muscle, and local edema.[61,62] In experimental animals, polypeptides of molecular mass ranging from 4–9 kDa also block calcium channels, resulting in flaccid paralysis.[63] In these animal models, T×3–3 inhibits ω-agatoxin IVA-sensitive calcium channels, which controls exocytosis by decreasing calcium influx and glutamate release.[64] Animal studies indicate that *P. nigriventer* venom temporarily disrupts the blood–brain barrier and allows the passage of molecules into the brain.[65]

Most envenomations by these spiders produce mild pain, edema, local sweating, elevated blood pressure, nausea, and piloerection that resolve within 24–48 hours.[57] Rarely, severe envenomations may occur, particularly in young children. Clinical features of severe envenomations include parasympathetic stimulation, profuse sweating, muscle spasms, priapism, bradycardia, hypotension, and pulmonary edema.[57] Antivenom is available in Brazil (Butantan Soro Antiarachnidico Butantan Institute, Sao Paulo, Brazil) for intravenous administration in doses of 2–5 ampules intravenously or intramuscularly. Supportive care includes wound cleansing, tetanus prophylaxis, elevation, and immobilization of the affected extremity, and warm compresses.

AUSTRALIAN FUNNEL-WEB SPIDERS

The Australian funnel-web spiders (*Atrax* spp., *Hadronyche* spp.) belong to the family: Hexathelidae: subfamily: Atracinae. These large (body length: 24–32 mm/~1-1¼ in.), highly venomous spiders inhabit the moist, eastern seaboard of Australia and Tasmania, as well as limited areas in southern Australia near Adelaide. Most reports of funnel-web spider bites involve the Sydney funnel-web spider (*Atrax robustus*) found in densely populated areas of the Sydney basin. A review of documented funnel-web spider envenomation indicated that the funnel-web spider species most commonly associated with severe envenomations included *Atrax robustus* O.P.-Cambridge (Sydney), *Hadronyche formidabilis* Rainbow (Northern tree), *H. cerberea* L. Koch (Southern tree), *H. infensa* Hickman (Toowoomba), and *H. versuta* Rainbow (Blue mountains).[66] Funnel-web spiders build long silk retreats in the ground with two large spinnerets, frequently near a tree. These nocturnal spiders adopt an upright stance when provoked. Mygalomorph spiders other than species in the Hexathelidae family include the mouse spiders (*Missulena* species) and other funnel-web spiders of the Hadronyche species. In general, these spiders are medium to large in size with characteristic prominent forward facing jaws (chelicerae) and large parallel fangs. These mygalomorphs (i.e., mouse, diplurid, idiopid, and nemesiid spiders) are difficult to distinguish from funnel-web spiders.[67]

PRINCIPAL TOXINS

The venom from the Australian funnel-web spider (*Hadronyche versuta* Rainbow) contains at least 100 components.[68] Adverse effects of funnel-web spider envenomations on mammals result primarily from the actions of δ-atracotoxins on the peripheral nervous system. In a manner similar to scorpion α-toxins, these excitatory neurotoxins cause spontaneous repetitive firing and prolongation of action potentials in excitable cells via slowing of voltage-gated Na^+ channel inactivation and the hyperpolarizing shift of the voltage-dependence of activation.[69] The spontaneous repetitive firing of action potentials and the release of excessive amounts of neurotransmitters (e.g., epinephrine, acetylcholine, norepinephrine) cause muscle fasciculations and autonomic instability.[70] The voltage-dependent binding of δ-atracotoxins to neurotoxin receptor site-3 is similar, but not identical, to the binding of α-scorpion toxins and sea anemone toxins.[71]

Primates are much more susceptible to the effects of δ-atracotoxins than other vertebrates.[72] Although the female funnel-web spider secretes much more venom than a male spider, the venom of the female spider demonstrates much lower neurotoxicity compared with male spider venom.[73] Janus-faced atracotoxins are a unique family of invertebrate-specific potassium channel blockers that are being investigated as insecticides.[74] The venom of the Eastern mouse spider (*Missulena bradleyi* Rainbow) contains a neurotoxin that facilitates neurotransmitter release by modifying tetrodotoxin-sensitive sodium channel gating, similar to δ-atracotoxin in funnel-web spider venom.[75] Consequently, bites from this spider potentially produce similar clinical effects as funnel-web spider envenomation.

CLINICAL RESPONSE

The clinical features of funnel-web spider bites are difficult to predict because many bites do not cause systemic envenomation. In a case series of 4 bites by *Atrax robustus* O.P.-Cambridge and 12 bites by *Hadronyche* species, 10 patients had only local effects that resolved within 3 hours, whereas 4 patients had moderate envenomation and 2 patients had severe envenomation requiring antivenom therapy.[67] Typically, the bite of a funnel-web spider is very painful as a result of the relatively large size of the fangs and the acidic pH of the venom. The initial pain over the puncture site may last for 30–60 minutes along with local bleeding from the puncture wound. Initial local features of envenomation include the rapid onset of local muscle fasciculation, sweating, and piloerection within minutes after envenomation. Local muscle fasciculation and paraesthesias are highly characteristic of envenomation by this spider. Early symptoms of systemic envenomation involve profuse sweating, salivation, lacrimation, dyspnea, and neuromuscular excitation including paresthesias and generalized muscle fasciculations. Rapid progression to confusion, agitation, coma, metabolic acidosis, cerebral edema, and noncardiogenic pulmonary edema may occur within 1–2 two hours after envenomation. Other features of funnel-web spider envenomation include muscle spasms, tachycardia, hypertension, cardiac arrhythmias, and gastrointestinal distress. Late effects include persistent hypotension, coma, rhabdomyolysis, coagulopathy, disseminated intravascular coagulation, and multiorgan failure. Death may result from noncardiogenic pulmonary edema or cerebral edema.[76] In a review of severe cases of documented funnel-web spider envenomations, clinical effects associated with a majority of severe envenomations included diaphoresis, hypertension, tachycardia, fasciculations, and pulmonary edema.[66]

Envenomations by some *Hadronyche* species can produce a life-threatening clinical syndrome indistinguishable from that of *Atrax robustus* including Port Macquarie funnel-web, Southern Tree funnel-web (*Hadronyche cerberea*), Northern Tree funnel-web (*Hadronyche formidabilis*), and the Darling Downs funnel-web (*Hadronyche infensa*).[77] Clinical features of envenomation by *H. infensa* (Toowoomba funnel-web) included generalized paresthesias, muscle spasm, blurred vision, fasciculations, local induration and pain, salivation, lacrimation, and altered consciousness.[78] Envenomation by mouse spiders (*Missulena* spp.) typically cause local effects, occasionally with paraesthesias and diaphoresis, but the envenomation is not usually serious.[67]

TREATMENT

Recommended first aid treatment in Australia includes the use of pressure-immobilization (i.e., wrapping of the extremity with a large pressure bandage similar to an ankle sprain along with immobilization).[79] Although there are few data on the efficacy of pressure-immobilization to limit the spread of atraxotoxin, case reports associate the removal of pressure-immobilization bandages with clinical deterioration.[80] Rabbit-derived IgG antivenom (Commonwealth Serum Laboratories) to the Sydney funnel-web spider (*Atrax robustus*) was introduced in the 1980s. *In vitro* studies suggest that the venom of most Australian funnel-web spiders share common antigens.[81] Symptomatic patients are candidates for the administration of antivenom, particularly if the symptoms progress rapidly. The initial adult and pediatric dose is 2–4 vials (100 mg IgG/vial) of antivenom intravenously, repeated in 15 minutes if symptoms do not significantly improve. Additional doses may be necessary for severe envenomations. Skin testing is not usually recommended. Although some physicians administer H_1 and H_2 receptor antagonists prior to antivenom administration, there are few clinical data to guide management. Patients should be carefully monitored in an intensive care setting for the development of urticaria, stridor, bronchospasm, and hypotension. The decision to continue antivenom therapy despite the appearance of clinical signs of anaphylaxis depends on the clinical evaluation of the risks and benefits of antivenom therapy. Pre-treatment of antihistamines, concurrent administration of epinephrine, and a slow rate of antivenom administration are therapeutic options for restarting antivenom therapy in these patients when anaphylactic responses occur.[82]

This antivenom demonstrates cross-reactivity with other funnel-web spiders (*Hadronyche* spp.), although antivenom requirements may be greater.[77] Envenomations by mouse spiders (*Missulena* spp.) rarely produce systemic symptoms as severe as funnel-web spiders,[83] but the administration of funnel-web spider antivenom has been associated with remission of a severe envenomation by the Eastern mouse spider (*Missulena bradleyi*).[25] All patients with signs of Australian funnel-web envenomation should be observed in an intensive care unit for the development of delayed noncardiogenic pulmonary edema.

BROWN RECLUSE SPIDERS

HISTORY

Caveness first described *Loxosceles*-induced, necrotic skin lesions in the US medical literature in 1872.[84] In 1929, Schmaus associated a case of necrotic arachnidism with the bite of *Loxosceles reclusa* Gertsch & Mulaik (i.e., initially identified as *L. rufescens*, but probably *L. reclusa*).[85] However, the bite of the *Loxosceles* spider was not confirmed as a cause of necrotic lesions until 1947 when Macchiavello confirmed necrosis in a guinea pig model following the bite of *Loxosceles laeta* Nicolet in Chile.[86,87] Although clinical cases of fever associated with brown spider bites were reported in the United States from the late 1800s[84] to the 1940s,[88] the causal relationship between cutaneous necrosis and *Loxosceles* bites was not confirmed in the United States until the late 1950s.[89]

IDENTIFYING CHARACTERISTICS

Loxosceles species belong to the Sicariidae family (formally in Loxoscelidae or Scytodidae), which includes primitive hunters in the genera *Loxosceles* (100 species) and *Sicarius* (21 species).[90] These medium-sized (6–10mm/~0.2–0.4in.), nondescript spiders possess a uniformly colored cephalothorax that is dull yellow to light brown to gray. The cephalothorax contains a dark-brown, violin-shaped marking on the dorsal surface that is somewhat variable. The presence of the fiddle-shaped mark on the anterior carapace varies depending on species, age, and condition of the specimen. Other brown spiders (*Kukulcania* species, Filistatidae family) may exhibit similar markings. The abdomen usually is olive-tan. The legs and abdomen are covered with fine hairs, and the legs lack the large leg spines commonly present on most brown spiders.[91] The presence of the violin-shaped marking and six eyes arranged in three diads (one anterior, two lateral) suggests that the spider belongs to the genus *Loxosceles*. Two closely related species (*Scytodes* spp., *Sicarius* spp.) have similar eye patterns; therefore, the presence of six eyes alone does not confirm the presence of *Loxosceles* spiders. Spiders of this genus seldom have a prominent colored pattern. There are 11 species of native *Loxosceles* spiders in North America.[92]

Common Name: Brown Recluse, Fiddleback, Violin or Brown Spider

Scientific Name: *Loxosceles reclusa* Gertsch & Mulaik

Physical Description: These medium-sized, brown spiders possess short, hairy bodies about 7–12mm (average 9mm/~0.35in.) long with long, thin, hairy legs up to a span of 1–5cm (~0.4–2in.). The hairs are fine and dark. Their color varies from fawn to dark brown depending on the stage of molting and on the species. For example, *L. laeta* Gertsch & Mulaik is pale brown, whereas specimens of *L. gaucho* Gertsch are dark chocolate. The anterodorsal portion of the cephalothorax has a violin-shaped dark brown to yellow marking with the larger end cephalad. Brown recluse spiders have three pairs of eyes in contrast to most spiders, which have four pairs of eyes. The size and venom apparatus of female and male brown recluse spiders are similar; therefore, both sexes of violin spiders are equally capable of producing loxoscelism.

EXPOSURE

Geographic Distribution

The distribution of the genus *Loxosceles* is worldwide in temperate and tropical regions with about 100 species throughout America, Europe, Africa, Asia, and Oceania. There are 11 native and 2 nonnative *Loxosceles* species in the United States that inhabit primarily the Southeast and warmer portions of the Midwest. Endemic *Loxos-*

celes species in North America include *L. deserta* Gertsch and *L. reclusa* Gertsch & Mulaik. The highest number of *Loxosceles* envenomations in the United States occurs in Arkansas, Tennessee, Missouri, and Kansas.[93] At least 17 *Loxosceles* species (e.g., *L. spinulosa* Purcell, *L. speluncarum* Simon, and *L. parrami* Newlands) inhabit Africa that can produce necrotic lesions.[94] *Loxosceles rufescens* Dufour occurs in countries in the Mediterranean region, particularly in Israel.[95] *Loxosceles* species also appear in Asian–Pacific countries (e.g., China, Japan, Australia),[96] South Africa (*L. spinulosa* Purcell, *L. speluncarum* Simon, *L. parrami* Newlands),[94,97] and South America (*L. gaucho* Gertsch, *L. laeta* Nicolet, *L. similis* Moenkhaus, *L. adelaida* Gertsch, *L. hirsuta* Mello-Leitao, *L. amazonica* Gertsch, *L. intermedia* Mello-Leitao).[98]

Loxosceles Reclusa Gertsch & Mulaik

This spider is endemic to the midwestern and southeastern portions of the United States except Florida. The range of this spider is primarily the Missouri–Ohio–Mississippi River basin from southeastern Nebraska to southern Ohio and south to Texas and across to Georgia. Other states containing *Loxosceles reclusa* include Alabama, Arkansas, Illinois, Indiana, Iowa, Kansas, Louisiana, Tennessee, and South Carolina.[99] The northern range of *Loxosceles* species in the United States is Illinois, Indiana, and Iowa.[100] The southern extent of their geographic distribution is the Gulf of Mexico. These spiders are not endemic to Canada. *Loxosceles reclusa* does not survive through the seasons in hot desert climates. The diagnosis of *Loxosceles reclusa* spider bites outside endemic areas is suspect.[92] Figure 166.3 demonstrates the distribution of *Loxosceles* species in the United States.

Loxosceles Laeta Nicolet

Although this brown spider is endemic to South America (Argentina, Brazil, Chile, Ecuador, Peru, Uruguay), modern transportation allowed the establishment of isolated colonies in the United States including California during the 1960s. However, the clinical significance of these colonies is unclear because of the lack of well-documented cases of envenomation by *L. laeta* in the United States. Other *Loxosceles* species in South American include *L. intermedia* and *L. gaucho*.

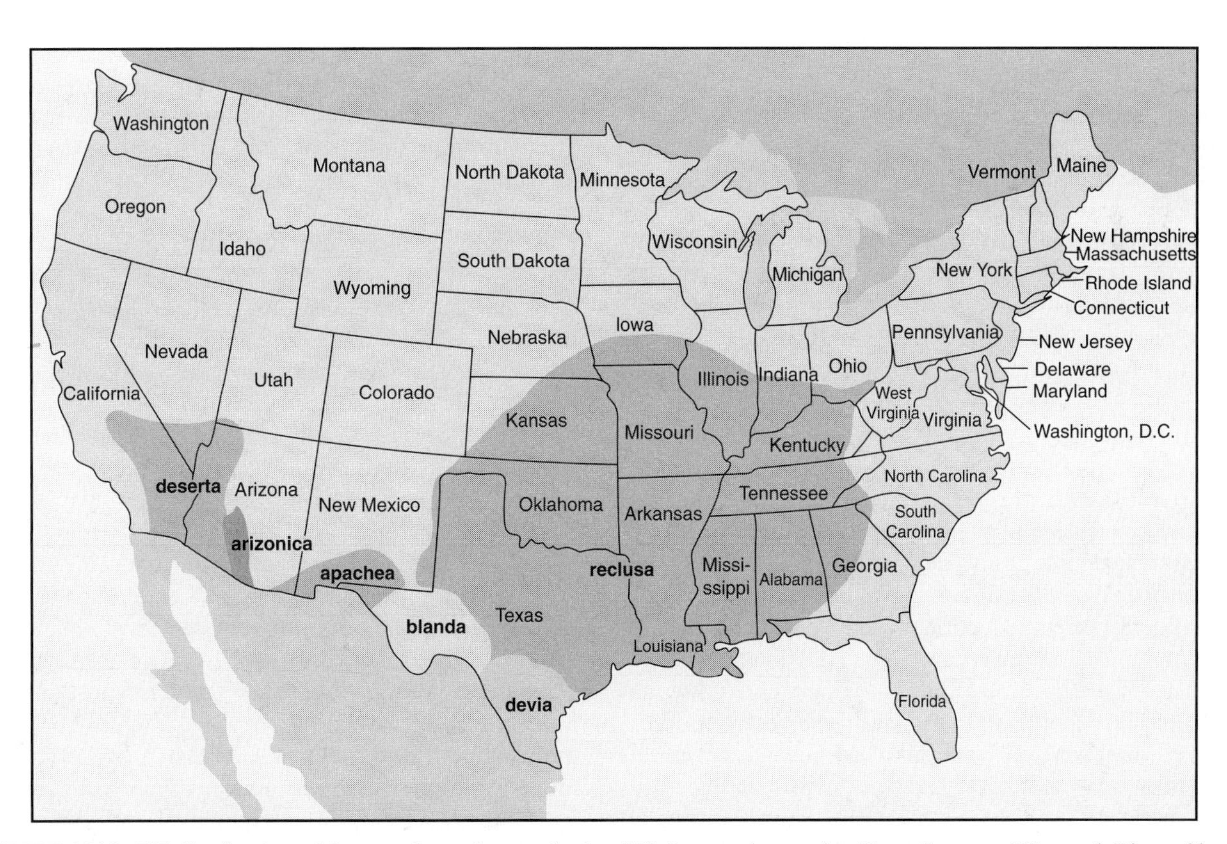

FIGURE 166.3. US distribution of *Loxosceles reclusa* and other US *Loxosceles* species. From Swanson DL, et al. Bites of brown recluse spiders and suspected necrotic arachnidism. N Engl J Med 2005;352:701.

Loxosceles Deserta Gertsch

This desert recluse inhabits rural areas of western deserts of Arizona, Nevada, New Mexico, Texas, Utah, and southern California including the Mohave Desert. These spiders rarely reside indoors except in homes surrounded by desert habitats. Consequently, human contact is rare.

Loxosceles Arizonica Gertsch & Mulaik

The distribution of *L. arizonica* is similar to the desert habitat of *L. deserta*, particularly in Arizona.

Loxosceles Rufescens Dufour

This species is native to the Mediterranean region. However, this species has been widely dispersed by modern transportation, and colonies are found in the eastern United States, California, Washington, Ohio, Georgia, and Texas, as well as areas throughout Europe, Africa, and Australia.

Behavior

Loxosceles spiders are sedentary and nocturnal, and these spiders prefer dark, quiet areas, such as wood piles, construction materials, paper, the underside of stones, and holes.[101] They adapt easily to dark, dry, indoor areas (clothes, stored personal items, stairwells, basements, boxes, closets). These spiders primarily are hunters, but occasionally they weave irregular, white, flocculent webs to sense prey. Brown recluse spiders (*L. reclusa*) can tolerate months without food and water depending on ambient temperature.[102] In an experimental study, food was withheld from brown recluse spiders. The survival of the spiders ranged between 2 weeks at 40 °C (104 °F) and 4–7 months at 5 °C (41 °F).[103] The average life span of *Loxosceles reclusa* Gertsch & Mulaik is about 1–3 years,[104] whereas specimens of *L. laeta* can live 3–7 years.[105] The long lifespan, resistance to starvation, and propensity to hide in dark places allows survival during accidental transport.[90] However, these spiders do not easily disperse from their new environment because they do not travel on air currents and they experience a reduction of egg production and fecundity unless mating recurs. These spiders are usually absent from areas surrounding buildings infested with brown recluse spiders. In northern climates, brown recluse spiders do not survive in low winter temperatures.

Most envenomations occur from April through October. The brown recluse spiders usually hibernate during the winter months. Envenomations also tend to occur in the morning when one is most likely to open the closet and choose the day's clothing. Multiple bites are not characteristic of *Loxosceles* species, and the presence of multiple lesions suggests an etiology other than spider bites. Large numbers of *L. reclusa* species may be present in a building without causing envenomations in occupants; consequently, the presence of *Loxosceles* species does not necessarily implicate these spiders as a cause of skin lesions.[106]

PRINCIPAL TOXINS

Venom Composition

The venoms of the native US *Loxosceles* species (*L. deserta, L. rufescens, L. arizonica*) are less potent compared with the venom of *L. reclusa*. The venom is primarily cytotoxic, particularly to endothelial tissue. At least three *Loxosceles* species have similar venom,[107] and the dermonecrotic and lethal factors in the venom reside in homologous proteins in the range of 32–35 kDa based on animal studies.[108] Almost all of the venom consists of spreading factors and enzymes (e.g., hyaluronidase, collagenase, esterase, protease, phospholipase, deoxyribonuclease, ribonuclease, dipeptides), but several of these enzymes (hyaluronidase, alkaline phosphatase, 5′ ribonucleotide phosphohydrolase) are apparently not responsible for the dermal necrosis.[109] The nonproteolytic 30–35 kDa protein, sphingomyelinase D, is probably the main toxin in *Loxosceles* venom.[101] Animal studies indicate that sphingomyelinase D causes activation of the classical pathway of complement resulting in platelet aggregation, dermal necrosis, and hemolysis.[110] This compound is rare in other species of spiders and venomous animals.[111] Sphingomyelinase D also activates the alternative pathway of complement on human erythrocytes by removing glycophorins secondary to the activation of endogenous metalloproteinase. This compound produces profound dermal inflammation as a result of the interaction with cell membrane sphingomyelin and the release of choline and *N*-acylsphingosine phosphate. These latter compounds cause serotonin release, platelet aggregation, and local ischemia. *Loxosceles* venom induces the release of interleukin 8, E-selectin, and granulocyte-macrophage colony-stimulating factor similar to the effects of tumor necrosis factor-α.[112] The dermonecrotic fraction of *L. gaucho* contains several proteins (e.g., loxnecrogin A and B) that produce necrosis in rabbit models, but the mechanism of necrosis probably involves more than one or two proteins.[113] Compounds involved with the hematological effects of *Loxosceles* venom include metalloproteinases (loxolysin A and B), serine protease, and other proteases.[114]

Venom Apparatus

The venom glands of *L. intermedia* include an external layer of striated muscle fibers and an internal muscle layer in contact with simple glandular epithelium. The epithelial cells project into the lumen of the glands and these projections contain venom-rich, secretory vesicles. A basal lamina separates the secretory epithelial cells from the multinucleated muscular cells, and this layer contains laminin, entactin, glycoproteins, and glycosaminoglycan sulfated residues.[115] The total volume of the venom is minute (i.e., about 4 μL), and the venom contains about 65–100 μg proteins.[114] The estimated amount of venom injected with each bite has not been well-studied, but the anecdotal data suggests that the *Loxosceles reclusa* injects about 2 μg venom/bite.[116]

Dose Response

As a result of variations in the biological activity of proteins, glycoproteins, and sphingomyelinase activity, the severity of cutaneous and system loxoscelism depends on the *Loxosceles* species as well as the sex (female > male).[117] Envenomations by *L. arizonica, L. deserta*, and *L. rufescens* cause relatively mild lesions.[93] However, animal models suggest that the size and depth of the eschar following subcutaneous injection of *Loxosceles* venom are dose-related.[118] The estimated LD_{50} of *L. intermedia* venom in four strains of mice ranged between 4.6–24.5 μg.[119] In a study of rabbits, large skin lesions developed following the intradermal injection of 3 μg venom from three species of *Loxosceles* (*L. gaucho, L. laeta, L. intermedia*).[108] The mean size of these lesions at 48 hours ranged between $1.8 \pm 0.8\,cm^2$ (*L. laeta*) to $5.4 \pm 3.6\,cm^2$ (*L. gaucho*). The intradermal LD_{50} for these species ranged between 0.48 mg/kg (*L. intermedia*) to 1.45 mg/kg (*L. laeta*), similar to the LD_{50} for the Brazilian species *L. similis* Moenkhaus.[120]

HISTOPATHOLOGY AND PATHOPHYSIOLOGY

The exact mechanism of necrosis produced by *Loxosceles* venom requires further delineation, but overall *Loxosceles* envenomation produces an inflammatory infiltrate, thrombosis, hemorrhage, acute inflammation and induration, erythema, and a liquefaction necrosis similar to pyoderma gangrenosum.[114] The dermal effect of *Loxosceles* venom is species-specific with mice and rats relatively resistant to the venom and humans, rabbits, and guinea pigs susceptible to the venom. In animal models, *Loxosceles* venom aggregates platelets, liberates thromboxane B2, causes intravascular fibrin deposition and endothelial thickening, and enhances leukocyte chemotaxis. Sphingomyelinase D is a phospholipase in *Loxosceles* venom that interacts with plasma membranes of various cells including erythrocytes, platelets, and endothelial cells. Experimental data indicate that the dermal necrosis following *Loxosceles* envenomation results from vascular thrombosis secondary to platelet thrombi.[121] The venom induces expression of E-selectin, and stimulates the release of pro-inflammatory mediators (chemokines) including interleukin-8, growth-related oncogene α, granulocyte-macrophage-colony-stimulating factor, and monocyte chemoattractant protein-1.[122] These markers amplify the inflammatory response to *Loxosceles* venom. Animal studies indicate that histological changes include interstitial edema, local hemorrhage, pronounced infiltration of polymorphonuclear leukocytes, degeneration of vascular walls, plasma exudation, thrombosis, coagulative necrosis, and vasculitis.[123,124] Light microscopy demonstrates early dermoepidermal separation, full-thickness dermal edema, thrombosis of small arterioles, infiltration of neutrophils and eosinophils, and extravasation of erythrocytes.[171] Electron microscopy of tissue biopsies from *Loxosceles*-envenomated animals demonstrates subendothelial blebs, vacuoles, and intravascular endothelial cell membrane degeneration along with thrombus formation within blood vessels.[125] A well-delineated zone of eosinophilic coagulative necrosis develops in the skin surrounded by a dense band of neutrophils. Necrosis is greatest in fatty areas of the buttocks, thighs, and abdomen. *In vitro* experiments indicate that *L. reclusa* venom attaches directly to red blood cell membranes and causes structural changes.[126] Venom attachment to erythrocyte membranes is apparently a prerequisite for hemolysis of human cells.

CLINICAL RESPONSE

The hallmark of *Loxosceles* envenomation is the development of a necrotic skin lesion with gravitational spreading. Clinical features of *Loxosceles* bites range from no or mild local necrosis to severe systemic effects including hemolysis and disseminated intravascular coagulopathy. The frequent lack of identification of *Loxosceles* specimens following envenomation complicates the determination of the incidence and clinical features of loxoscelism. A substantial number of *Loxosceles* bites are clinically insignificant.[127]

Cutaneous Loxoscelism

Typically, *Loxosceles* bites produce mild initial pain similar to a pinprick, but the bite may be painless.[127] The progression of cutaneous responses to *Loxosceles* envenomation is highly variable. Most *Loxosceles* envenomations in the United States produce only a mild inflammatory reaction including a mild urticarial reaction.[128] A rapidly expanding, single lesion may spontaneously regress before necrosis occurs. The presence of multiple necrotic lesions indicates an etiology other than *Loxosceles* envenomation, such as pyoderma gangrenosum (i.e., a rare skin disease with recurrent deep skin ulcers and impaired cell-mediated immunity). A mild skin response may occur characterized by the formation of a small, erythematous papule that becomes very firm before healing. Typically, severe lesions become obvious within 12–24 hours, and the lack of necrotic signs within 48–96 hours indicates that necrosis will not occur.[129] Localized urticaria may persist several days after the bite.[130] More severe cutaneous reactions to *Loxosceles* bites result from cytotoxicity. Variable degrees of pain and erythema begin within 2–6 hours of envenomation, followed by formation of an irregular bleb at the bite site surrounded by a white ring of ischemia 12–24 hours after envenomation. Although occasionally a transient, bluish discoloration develops within the first 24 hours, the skin reaction usually progresses from red to violet forming a dependent, eccentric-shaped lesion that displays a characteristic gravitational spread. The central bleb develops necrosis after 3–4 days, and a black eschar forms within one week that sloughs within several weeks. Eventually, the eschar sinks below the surface of the skin in a pattern called the "red, white, and blue sign." The indurated ulcer heals by secondary intention over 6–8 weeks depending on the size of the ulcer.[131] The lowering of the eschar below the surface of the skin differentiates necrotic from nonnecrotic lesions, which remain red and elevated above the surrounding skin. Large areas of necrosis may occur, particularly in areas surrounded by thin skin and abundant, subcutaneous fat. Skin grafting may be necessary to cover the lesion. Secondary infection of the eschar is uncommon. In patients with cutaneous loxoscelism, nonspecific symptoms may develop within 48 hours after envenomation characterized by mild fever and malaise, but laboratory evidence of hemolysis is usually absent.[132]

The appearance of diffuse macular erythema over the trunk is a good prognostic sign. Rarely, a generalized vasculitis without cutaneous necrosis occurs after *Loxosceles* envenomation.[133]

Color and configuration differentiate *Loxosceles* bites from nonnecrotic lesions because the latter bites remain raised and erythematous. Hence, the rapid expansion of the blue macular area along with severe pain signifies necrosis and probable *Loxosceles* envenomation in endemic areas. However, the diagnosis of Loxosceles envenomation by clinical examination alone is difficult, and frequently biopsies are necessary to exclude other causes of necrotic skin lesions.[134] *Loxosceles* lesions are often confused with other necrotic spider bites, other arthropod bites, and systemic disease. Each suspected *Loxosceles* bite should be carefully evaluated for other potential sources, in part, because the diagnosis of loxoscelism does not correlate to known populations of *Loxosceles* species.[135] Important differential diagnoses for the suspected *Loxosceles*-induced lesions include neoplasia, cutaneous anthrax, tularemia, necrotizing fasciitis, Rocky Mountain spotted fever, Lyme disease, pyoderma gangrenosum, and sporotrichosis.[136] Other differential diagnoses of suspected *Loxosceles*-induced cutaneous lesion include various arthropod bites (ticks, mites, bedbugs, fleas, ants, mosquitoes, biting flies, bees, caterpillars), erysipelas, vasculitis, necrotizing fasciitis, pyoderma gangrenosum, adverse drug reactions, herpes zoster, thermal or chemical burns, and urticaria. Table 166.2 lists conditions associated with *Loxosceles*-like dermal lesions described in the medical literature. There is no clinical evidence to indicate that the risk of adverse effects from *Loxosceles* envenomation increases during pregnancy.[137]

Viscerocutaneous Loxoscelism

Systemic reactions are uncommon, and the occurrence of systemic symptoms does not correlate to the severity

TABLE 166.2. Conditions Associated with Loxosceles-like Dermal Lesions

Disease Type	Cause
Bacterial	*Pseudomonas aeruginosa, Staphylococcus* (methicillin resistant), *Streptococcus*, cutaneous anthrax, gonococcemia, Lyme disease, Rocky Mountain spotted fever, tularemia
Fungal	Aspergillosis, cryptococcosis, sporotrichosis
Parasitic	Leishmaniasis
Viral	Herpes simplex, herpes zoster
Vascular disease	Erythema nodosum, small-vessel occlusive disease, vasculitis, statis ulcer, polyarteritis nodosa, Wegener's granulomatosis, pressure ulcers, Stevens-Johnson syndrome, erythema multiforme, polyarteritis nodosa
Neoplastic disease	Primary cutaneous neoplasms, lymphoma, lymphomatoid papulosis
Other diseases	Chemical or thermal burns, toxic epidermal necrolysis, phytodermatitis, radiation, cryoglobulinemia, diabetic ulcer, pemphigus vegetans, pyoderma gangrenosum, septic emboli, arthropod bites (flea, mite, biting fly, bees, ants)

Source: Adapted from Ref 90.

of cutaneous lesions.[138] Typically, systemic symptoms develop within 24–48 hours of envenomation. Children appear more susceptible than adults.[139] Clinical features of a systemic reaction to *Loxosceles* envenomation based on case reports include fever, chills, malaise, nausea, vomiting, myalgias, arthralgias, petechiae, hemolysis,[140] hypotension,[141] hemoglobinuria,[142] and consumptive coagulopathy (DIC).[143,144] Low-grade fever and arthralgias can occur during cutaneous *Loxosceles* envenomation, and these symptoms are not specific to viscerocutaneous loxoscelism. The most serious complications of *Loxosceles* envenomation are sepsis, hemolysis, and disseminated intravascular coagulation. Coma, convulsions, and renal dysfunction are probably secondary to venom-induced hemolysis.[145] Hemolytic anemia may be insidious, often requiring 2–3 days to develop. Deaths from *Loxosceles* envenomation in the United States have not been well-documented during modern times when intensive supportive care has been available, in part, because of the lack of confirmed *Loxosceles* envenomation.[146] However, the presence of significant hemolysis is potentially life-threatening.[147] Oliguria and dark urine suggest the intravascular hemolysis and rhabdomyolysis that can cause anemia and acute renal failure. Several case reports of pregnant women envenomated during the second and third trimesters of pregnancy did not report adverse fetal effects.[40]

DIAGNOSTIC TESTING

Analytical Methods

There are no routine laboratory tests available to diagnose necrotic arachnidism or *Loxosceles* envenomation. Research methods have been developed for the detection of *Loxosceles* venom including the passive hemagglutination inhibition test[148] and a *Loxosceles* species enzyme-linked immunosorbent assay (ELISA) based on animal models.[116,149] Complicated laboratory techniques and delayed results (6–24 hours) limit the clinical application of the passive hemagglutination inhibition tests for *L. reclusa* envenomations. The ELISA test demonstrates moderate cross-reactivity with related spider genera depending on the amount of venom present. In rabbit models, this assay can detect venom in hair, wound aspirate, and dermal biopsies up to 7 days after envenomation, but this method did not demonstrate evidence of envenomation in the serum.[150] There is substantial cross-reactivity on immunoassays between North American *Loxosceles* species.[151] The diagnosis of *Loxosceles* envenomation is based on a history of a relatively painless bite, characteristic lesion (increasingly painful solitary plaque with central, white ischemia surrounded by erythema), and the identification of the spider in an area containing endemic species of *Loxosceles*.

Abnormalities

Intravascular hemolysis may develop during the first 96 hours after envenomation and persist up to 1 week. Reduced hemoglobin concentrations and increased plasma-free haptoglobin concentrations develop during intravascular hemolysis, and normal serum haptoglobin concentrations exclude the presence of intravascular hemolysis.[152] Analysis of the urine during hemolysis may or may not demonstrate free haptoglobin with a positive urine dipstick for blood and minimal evidence of erythrocytes on microscopic examination because hemolysis may occur in extravascular areas (e.g., spleen). The anemia is usually Coombs-negative, but case reports of *Loxosceles* envenomation document the presence of Coombs-positive hemolytic anemia.[153,154] Characteristically, the serum hemoglobin decreases during hemolysis while the reticulocyte count increases. Substantial leukocytosis (i.e., up to 30,000 cells/cm^3) may also occur without evidence of a secondary infection. Rarely, case reports associated a severe envenomation with rhabdomyolysis,[155] myoglobinuria, hepatorenal dysfunction, and/or a consumptive coagulopathy (decreased fibrinogen, clotting factors, and platelets; increased fibrin degradation products). Antithrombin III is an important physiological inhibitor of coagulation, and antithrombin III decreases substantially during clinically significant coagulopathies.

TREATMENT

Systemic Reaction

All patients with systemic reactions should be followed in the hospital for evidence of increasing hemolysis, coagulopathy, and renal failure with serial hemoglobin, free plasma haptoglobin, serum creatinine, serum electrolytes, and serum creatine kinase. Treatment for hemolysis and coagulopathy is supportive (i.e., red blood cell transfusions, platelets, cryoprecipitate, fresh-frozen plasma, monitoring of fluid/electrolytes). Generous fluid replacement is important following the development of hemolysis and hemoglobinuria. Urine alkalinization is also a treatment option, but there is limited data on the efficacy of this treatment in preventing hemoglobin-induced renal damage. Anecdotal data suggests that corticosteroids (1–2 mg methylprednisolone/kg loading dose, 0.5–1 mg/kg maintenance dose every 6 hours for 5–10 days) may reduce hemolysis from *Loxosceles* envenomation, but clinical data are limited.

There are no clinically effective antidotes for *Loxosceles* envenomation. Evaluation of patients with systemic symptoms should include a complete blood count, including platelets, coagulation profile (prothrombin time, partial thromboplastin time, fibrinogen level, fibrin degradation products), urinalysis, blood electrolytes, liver function tests, serum creatinine and serum creatine phosphokinase. Children with moderate to a severe systemic reaction should be observed for the development of hemolysis, renal dysfunction, and coagulation abnormalities. The presence of venom-induced antibody interference may complicate cross-matching for blood products during *Loxosceles* envenomation. Transfusions with packed red cells, fresh frozen plasma, and cryoprecipitate should be used cautiously, and the administration of whole blood should be avoided because the complement in whole blood may increase hemolysis.[156] Antithrombin III concentrates may be needed to reverse severe coagulopathies, if available. Hyperkalemia and renal failure may require hemodialysis.[153]

Cutaneous Reaction

There are no proven treatments for brown recluse spider bites based on controlled, randomized studies. Most suspected brown recluse spider bites heal with supportive care alone; therefore, conservative, local wound care is the most important treatment of cytotoxic responses to *Loxosceles* venom.[157] Supportive measures include cold applications, elevation, and antibiotics when indicated by the presence of secondary infections. General wound care (cleansing with bactericidal agents, tetanus prophylaxis as indicated, limited motion, neutral or elevated position, cold compresses), serial observation, and antipruritic agents (e.g., hydroxyzine, diphenhydramine) are the main treatments.[158] Antibiotics are limited to documented evidence of secondary skin infections (*Staphylococcus, Streptococcus, Clostridium perfringens*, other anaerobes). Examination of the fangs and venom of some *Loxosceles* species (*L. intermedia*) demonstrated the presence of *Clostridium perfringens*,[159] but the role of this anaerobe in the clinical progression of the skin lesions associated with loxoscelism remains undetermined.

The progression of skin lesions to necrosis is difficult to predict based on the initial cutaneous lesion. Surgical incision is not usually necessary until 2–8 weeks after envenomation, when the acute-phase reactants are degraded and the wound is fully mature. Rarely, surgical intervention may be necessary 2–3 weeks after envenomation for cosmetic purposes, especially when the ulcer exceeds 1–2 cm (~0.4–0.8 in.). Although *Loxosceles* antivenom is available in Mexico and Brazil, there are limited experimental and clinical data to support the efficacy of the antivenom, particularly against local responses more than several hours after envenomation.[43,160]

There are few clinical data to support the use of adjunctive therapy (dapsone, steroids, hyperbaric oxygen, heparin, electric shock) for the treatment of dermal lesions associated with *Loxosceles* envenomation.[93] Animal experiments do not support the use of intra-lesional or systemic steroids for the treatment of cutaneous skin reactions when administered more than a few hours after envenomation. Experimental studies in guinea pigs suggested that treatment with dapsone, an inhibitor of the local infiltration of polymorphonuclear leucocytes, reduced the average size of *Loxosceles*-induced skin lesion.[161,162] A retrospective study suggested that a better clinical outcome results from the use of dapsone (4,4′-diaminodiphenyl-sulfone) therapy combined with delayed surgical excision rather than early surgical excision alone.[163] However, numerous design flaws including the use of different surgical techniques in the two groups limits extrapolation of this study to other groups of patients. Other experimental studies of *Loxosceles*-induced skin lesions in the rabbit[164] and guinea pig[165] model did not detect clinically significant differences between dapsone-treated group and controls, when administered 4 hours and 36 hours, respectively, after envenomation. Complications of dapsone therapy include hemolysis (glucose-6-phosphate dehydrogenase deficiency), agranulocytosis, aplastic anemia, hypersensitivity syndrome, toxic epidermal necrolysis, methemoglobinemia, and rarely a hypersensitivity reaction (fever, headache, nausea, vomiting, hepatitis, hemolysis, leukopenia).[166,167] Inadequately controlled and documented human cases, as well as the difficulty in predicting the course of the cutaneous lesions, indicates that incisional therapy prior to the development of lesions is contraindicated. The use of hyperbaric oxygen for the treatment of *Loxosceles*-induced skin lesions remains controversial because of the lack of supportive experimental[168] and clinical data.[164,169]

For patients with signs of necrotic lesions, baseline complete blood count and urinalysis should be followed daily for the first several days to exclude the development of hemolysis and hemoglobinuria. All patients receiving dapsone must be screened for glucose-6-phosphate dehydrogenase deficiency, anemia, and liver dysfunction. The appearance of fever or dark urine requires immediate follow-up by a physician.

NECROTIZING ARACHNIDISM

Necrotizing arachnidism applies to necrotic lesions produced by spiders other than *Loxosceles* species. However, the clinical disorder of necrotic arachnidism is poorly defined, and the sources of necrotic lesions after spider bites are not well documented.[170] These lesions typically are single, rarely clustered, and usually are located on an extremity. A central punctum or vesicle may be present along with vasodilation, ischemia, and thrombosis manifest as a classic "red, white, and blue sign." Dependent erythema with irregular borders develops as venom spreads in the direction of gravity.[171] At least four American species of spiders, excluding *Loxosceles* species, are capable of producing necrotic lesions.[172] Most spider bites do not cause necrotic lesions, and spider bites are frequently confused with other medical conditions (e.g., bacterial infections, insect bites). Minor features of bites from these spiders include immediate, transient pain occasionally associated with local effects (bleeding, puncture marks, pruritus, redness).[173,174] Rarely, infections may develop in the area of the bite. Table 166.3 lists spiders associated with painful, indurated lesions that do not typically produce necrosis. The differential diagnosis of necrotic spider bites includes infective ulcers (*Staphylococcus, Streptococcus, Mycobacterium ulcerans, Chromobacterium violaceum, Sporotrichosis, Cutaneous nocardiosis, Pseudomonas* species), pyoderma gangrenosum, vasculitis, allergic reactions, and arthropod envenomation (sandfly, blister beetle).

TABLE 166.3. Spider Genera Associated with Painful, Indurated, Nonnecrotic Lesions

Genus and Species	Common Name	Family
Agelenopsis aperta Gertsch	Grass spider	Agelenidae
Avicularia avicularia L.	Pink toe tarantula	Theraphosidae
Herpyllus ecclesiasticus Hentz	Parson spider	Gnaphosidae
Lampona	White-tailed spider	Lamponidae
Lycosa	Wolf spider	Lycosidae
Megaphobema	Tarantula, Giant Leg spider	Theraphosidae
Neosparassus	Sparassid spider, huntsmen spider	Sparassidae
Pamphobeteus	Tarantula	Theraphosidae
Phormictopus	Tarantula	Theraphosidae
Tegenaria agrestis Walckenaer	Hobo spider	Agelenidae
Thiodina sylvana Hentz	Jumping spider	Salticidae
Trachelas	Broad-faced sack spider	Corinnidae (running spiders)
Xenesthis	Tarantula	Theraphosidae

Black-and-Yellow Garden Spiders

These large, brightly colored spiders weave large, conspicuous webs that trap prey (insects).[175] Some orb weavers build relatively permanent webs and store prey in the webs, whereas other spiders from this family build new webs daily.[176] The golden orb weaver (black-and-yellow garden spider, *Argiope aurantii* Lucas) is a common inhabitant of gardens in California, Oregon, and the eastern United States. This black spider has bright yellow and orange spots on the dorsal surface of the abdomen. The females are about 4 times larger than the males, reaching 7.5 cm (~3 in.) in length with legs extended. Bites usually occur at night or during outdoor activities. Although the venom from some *Argiope* species contain low-molecular-weight compounds (argiotoxins) that block postsynaptic glutamate receptors,[177,178] envenomation by these spiders does not usually result in systemic effects. Envenomation may produce local pain, surrounding erythema, and vesicles, but necrosis does not usually occur.[179,180] Treatment involves conservative wound care (elevation, tetanus prophylaxis as indicated, antibiotics for secondary infection, analgesics).

Running or Sac Spiders

The genus *Cheiracanthium* contains about 200 species belonging to the family Miturgidae. These long legged, nondistinct spiders average about 7–10 mm in length with olive-green to pale yellow or brown coloration. There are eight eyes arranged in two parallel rows of four eyes. *C. inclusum* Hentz is a common garden spider that inhabits most of the United States with exception of the cold, northern latitudes. *C. mildei* L. Koch (sac spider, cream house spider) is a common house spider throughout the United States that moves indoors during autumn and winter months. These nocturnal spiders do not weave webs, and they usually rest in fine, white, silken sacs woven in curled leaves or crevices. Most bites occur while the victim is asleep or while dressing in clothes containing these spiders.

Although medical references frequently list spiders from the genus *Cheiracanthium* as a cause of necrotic arachnidism, necrotic lesions following envenomation by these spiders are not well documented. In a review of 20 verified bites by *Cheiracanthium* spiders from the United States and Australia, there were no necrotic lesions.[181] This group also reviewed 39 verified *Cheiracanthium* bites from the international literature with documentation of only one case of mild necrosis by a European species (*C. punctorium* Villers). In cases of suspected (i.e., not necessarily documented) cases of *Cheiracanthium* bites, case reports associated bites by *Cheiracanthium* species with sharp pain followed by a red wheal and the formation of a crusted, slightly necrotic lesion within several days.[182,183] Induration and erythema surrounds the lesion, and the lesion may be painful or pruritic.[184] The size of the lesion is variable, and pain may occur without clinical signs of inflammation.[185] Most *Cheiracanthium* bites do not ulcerate, and the lesion becomes asymptomatic after about 2 days.[186] Occasionally, mild systemic complaints can develop within 3 days following envenomation by *Cheiracanthium* spiders including low-grade fever, headache, abdominal cramps, vomiting, and rash.[97]

Wolf Spider

These medium to large spiders (Lycosidae family) have elongated, cylindrical bodies (1–30 mm/~0.4–1.2 in.) with colored bands on the carapace and small clumps of white hairs on the abdomen that form spots. They are superb hunters with good vision to distinguish movement (but not form) and a characteristic pattern of eyes. Four small, anterior eyes are set close together. Above these four eyes are two large eyes that are directed forward and two large eyes that are directed upward. These diurnal hunters use their strong front legs and powerful jaws to hold and crush prey. They do not hunt in packs, and they prefer open, disturbed habitats (e.g., urban parks, gardens). *Lycosa* venom contains peptides (lycotoxin I, lycotoxin II) that promote efflux of Ca^{2+} from synaptosomes, induce erythrocyte hemolysis, and dissipate voltage gradients across muscle membranes *in vitro*.[187] The venom of *Lycosa* species also contains substantial amounts of histamine.[188] In animal models, the venom of *Lycosa godeffroyi* L. Koch causes necrotic lesions.[189] However, necrosis following bites by *Lycosa* species is unusual, and most symptoms following envenomation involve varying degrees of pain rather than ulceration.[190,191] Case reports associate bites by *Lycosa* species with pain, swelling, and lymphangitis.[192] Treatment primarily involves conservative wound care including the use of antihistamines.

Green Lynx Spider

The lynx spider (*Peucetia Viridans* Hentz) belongs to the family Oxyopidae that inhabits large areas of the southern United States, particularly around plants. These diurnal hunters average about 2 cm (~1 in.) in length, and the bright, translucent green body usually contains rows of small, red dots. Aids to the identification of this spider include bright green coloration, spindly legs covered with spiny hairs, and large size. Their typical habitat includes grassy areas, cultivated fields (cotton, soybeans), and ornamental gardens in the lower latitudes of the United States to Central America. Typically, bites by this spider produce local pain, pruritus, erythema, and induration without necrosis.[193] Occasionally, a pustule may develop following envenomation, but systemic effects do not develop.[194] Treatment involves supportive wound care.

Black Jumping Spiders

The genus *Phidippus* contains the largest spiders in the family Salticidae with some specimens reaching 1 cm (~0.4 in.) in length. The identification of the jumping spider (*Phidippus audax* Hentz) is based on the characteristic short legs, white stripes, bands of hair on the carapace, and fluorescent green mouth parts (chelicerae). These spiders also have a characteristic pattern of eyes. The first row of four eyes are enlarged, particularly the middle two eyes, with a second row of two small eyes above and a third row of larger eyes posterior to the second row. These spiders are excellent hunters, and they can jump over 40 times their body length.[195] They do not build webs. *In vitro* studies suggest that *Phidippus* venom is cytotoxic and capable of producing necrotic lesions.[196] Envenomation produces a sharp, painful site that may develop into urticarial swelling and pruritus lasting several days to a week.[197] A small ulcer potentially may develop in the area of the bite, but the formation of necrotic lesions in humans is not well documented. Treatment is supportive.

Six-Eyed Crab Spiders

These flat, medium-sized (17 mm) spiders inhabit remote areas of Africa and South America. They usually avoid human contact by burying themselves in sand. Their venom is proteolytic.

Hobo Spiders

The hobo spider, *Tegenaria agrestis* Walckenaer, and the related species *Tegenaria duellica* Simon are very similar European spiders that inhabit the northwestern United States (Washington, Oregon, Idaho, northern Utah, the western Montana, western Wyoming) and southern British Columbia.[198] These members of the Agelenidae family have a distinctive, funnel-shaped web, but this spider should not be confused with the more dangerous, unrelated Australian funnel-web spider (*Atrax robustus* O.P.-Cambridge). These medium-sized (7–14 mm/~0.3–0.6 in.), brown spiders often have a dorsal herringbone pattern on their abdomen, but the lack of distinctive features requires identification by qualified experts. They wait for prey inside a horizontal, trampoline-like mat of silk.

Several case reports associate bites from the hobo spider (*Tegenaria agrestis* Walckenaer) with necrotic lesions found on patients from the northwestern United States, but these case reports did not include adequate data to exclude infectious causes of these lesions.[199,200] These case reports delineate a clinical course involving painless erythema and induration followed by vesicle formation that subsequently forms an eschar and subeschar necrosis; however, these reports lack expert identification of the involved spiders.[201] Furthermore, hobo spiders are common spiders that do not usually bite, and the presence of these spiders in the vicinity of the patient does not necessarily imply envenomation. Systemic complaints associated with hobo spider bites in these case reports include dry mouth, anorexia, disorientation, arthralgias, paresthesias, and headaches.[202] The ability of these spiders to cause necrotic lesions remains equivocal, in part, because this spider is a native European species, where necrotic arachnidism is not associated with this species. Examinations of the venom from European and American specimens of the hobo spider did not demonstrate substantial differences in their venoms.[203]

References

1. Isbister GK, White J. Clinical consequences of spider bites: recent advances in our understanding. Toxicon 2004;43:477–492.
2. Wong RC, Hughes SE, Voorhees JJ. Spider bites. Arch Dermatol 1987;123:98–104.
3. Rash LD, Hodgson WC. Pharmacology and biochemistry of spider venoms. Toxicon 2002;40:225–254.
4. Rauber A. Black widow spider bites. J Toxicol Clin Toxicol 1983–1984;21:473–485.
5. Lucas SM, Da Silva PI Jr, Bertani R, Costa Cardoso JL. Mygalomorph spider bites: a report on 91 cases in the state of Sao Paulo, Brazil. Toxicon 1994;32:1211–1215.
6. Ribeiro LA, Jorge MT, Piesco RV, Nishioka Sde A. Wolf spider bites in Sao Paulo, Brazil: a clinical and epidemiological study of 515 cases. Toxicon 1990;28:715–717.
7. Stern P, Valjevac K, Dursum K, Ducic V. Increased survival time in botulinum toxin poisoning by treatment with a venom gland extract from black widow spider. Toxicon 1975;13:197–198.
8. Gomez S, Queiroz LS. The effects of black widow spider venom on the innervation of muscles paralysed by botulinum toxin. Q J Exp Physiol 1982;67:495–506.
9. Garb JE, Gonzalez A, Gillespie RG. The black widow spider genus *Latrodectus* (Araneae: Theridiidae): phylogeny, biogeography, and invasion history. Mol Phylogenet Evolution 2004;31:1127–1142.
10. Nicholson GM, Graudins A. Spiders of medical importance in the Asia-Pacific: atracotoxin, latrotoxin and related spider neurotoxins. Clin Exp Pharmacol Physiol 2002;29:785–794.
11. Hodar JA, Sanchez-Pinero F. Feeding habits of the black widow spider *Latrodectus lilianae* (Araneae: Theridiidae) in an arid zone of south-east Spain. J Zool (London) 2002;257:101–109.
12. Horni A, Weickmann D, Hesse M. The main products of the low molecular mass fraction in the venom of the spider *Latrodectus menavodi*. Toxicon 2001;39:425–428.
13. Henkel AW, Sankaranarayanan S. Mechanisms of alpha-latrotoxin action. Cell Tissue Res 1999;296:229–233.
14. Grishin EV. Black widow spider toxins: the present and the future. Toxicon 1998;36:1693–1701.
15. Magazanik LG, Fedorova IM, Kovalevskaya GI, Pashkov VN, Bulgakov OV, Grishin EV. Selective presynaptic insectotoxin (alpha-latroinsectotoxin) isolated from black widow spider venom. Neuroscience 1992;46: 181–188.
16. Grishin EV. Neurotoxin from black widow spider venom structure and function. Adv Exp Biol 1996;391:231–236.
17. McCrone JD, Netzloff ML. An immunological and electrophoretical comparison of the venoms of the North American *Latrodectus* spiders. Toxicon 1965;3:107–110.
18. Binder LS. Acute arthropod envenomation incidence, clinical features and management. Med Toxicol Adverse Drug Exp 1989;4:163–173.
19. Muller GJ. Black and brown widow spider bites in South Africa. A series of 45 cases. S Afr Med J 1993;83: 399–405.
20. Muller GJ, Koch HM, Kriegler AB, Van der Walt BJ, Van Jaarsveld PP. The relative toxicity and polypeptide

composition of the venom of two southern African widow spider species; *Latrodectus indistinctus* and *Latrodectus geometricus*. S Afr J Sci 1989;85:44–46.

21. Rohou A, Nield J, Ushkaryov YA. Insecticidal toxins from black widow spider venom. Toxicon 2007;49: 531–549.
22. Ushkaryov YA, Volynski KE, Ashton AC. The multiple actions of black widow spider toxins and their selective use in neurosecretion studies. Toxicon 2004;43:527–542.
23. Sudhof TC. alpha-Latrotoxin and its receptors: neurexins and CIRL/latrophilins. Annu Rev Neurosci 2001;24:933–962.
24. Geppert M, Khvotchev M, Krasnoperov V, Goda Y, Missler M, Hammer RE, et al. Neurexin I alpha is a major alpha-latrotoxin receptor that cooperates in alpha-latrotoxin action J Biol Chem 1998;273:1705–1710.
25. Nicholson GM, Graudins A. Antivenoms for the treatment of spider envenomation. J Toxicol Toxin Rev 2003;22:35–59.
26. Jelinek GA. Widow spider envenomation (latrodectism): a worldwide problem. Wilderness Environ Med 1997;8: 226–231.
27. Isbister GK, Gray MR. A prospective study of 750 definite spider bites, with expert spider identification. QJM 2002;95:723–731.
28. Maretic Z. Latrodectism: Variation in clinical manifestations provoked by *Latrodectus* species of spiders. Toxicon 1983:21:457–466.
29. Clark RF, Wethern-Kestner S, Vance MV, Gerkin R. Clinical presentation and treatment of black widow spider envenomation: a review of 163 cases. Ann Emerg Med 1992;21:782–787.
30. Trethewy CE, Bolisetty S, Wheaton G. Red-back spider envenomation in children in Central Australia. Emerg Med 2003;15:170–175.
31. Moss HS, Binder LS. A retrospective review of black widow spider envenomations. Ann Emerg Med 1987;16: 188–191.
32. Isbister GK, Gray MR. Latrodectism: a prospective cohort study of bites by formally identified redback spiders. Med J Aust 2003;179:88–91.
33. Timms PK, Gibbons RB. Latrodectism–effects of the black widow spider bite. West J Med 1986:144:315–317.
34. Sutherland SK, Trinca JC. Survey of 2144 cases of red-back spider bites: Australia and New Zealand, 1963–1976. Med J Aust 1978;2:620–623.
35. Vutchev D. A case of intoxication after a bite by *Latrodectus tredecimguttatus*. Scand J Infect Dis 2001;33: 313–314.
36. Woestman R, Perkin R, van Stralen D. The black widow: is she deadly to children? Pediatr Emerg Care 1995;12: 360–364.
37. Karcioglu O, Gumustekin M, Tuncok Y, Celik A. Acute renal failure following latrodectism. Vet Hum Toxicol 2001;43:161–163.
38. Heller AW, Kelly AP. Herpes zoster developing after a spider bite. Cutis 1980:26:417–419.
39. Visser LH, Khusi SN. Pulmonary oedema from a widow spider bite. A case report. S Afr Med J 1989;75:338–339.
40. Langley RL. A review of venomous animal bites and stings in pregnant patients. Wilderness Environ Med 2004;15:207–215.
41. Isbister GK, Gray MR. Effects of envenoming by comb-footed spiders of the genera *Steatoda* and *Achaearanea* (family Theridiidae: Araneae) in Australia. J Toxicol Clin Toxicol 2003;41:809–819.
42. Warrell DA, Shaheen J, Hillyard PD, Jones D. Neurotoxic envenoming by an immigrant spider (*Steatoda nobilis*) in southern England. Toxicon 1991;29:1263–1265.
43. Isbister GK, Graudins A, White J, Warrell D. Antivenom treatment in arachnidism. J Toxicol Clin Toxicol 2003;41:291–300.
44. Keegan HL. Effectiveness of *Latrodectus tredecimguttatus* antivenin in protecting laboratory mice against effects of intraperitoneal injection of *Latrodectus mactans*. Am J Trop Med Hyg 1955;4:762–764.
45. Daly FF, Hill RE, Bogdan GM, Dart RC. Neutralization of *Latrodectus mactans* and *L. hesperus* venom by redback spider (*L. hasselti*) antivenom. Clin Toxicol 2001;39:119–123.
46. Graudins A, Padula M, Broady K, Nicholson GM. Red-back spider (*Latrodectus hasselti*) antivenom prevents the toxicity of widow spider venoms. Ann Emerg Med 2001;37:154–160.
47. Handel CC, Izquierdo LA, Curet LB. Black widow spider (*Latrodectus mactans*) bite during pregnancy. West J Med 1994;150:261–262.
48. Ellis RM, Sprivulis PC, Jelinek GA, Banham ND, Wood SV, Wilkes GJ, et al. A double-blind, randomized trial of intravenous versus intramuscular antivenom for Red-back spider envenoming. Emerg Med Australas 2005;17: 152–156.
49. Russell FE. Black widow spider envenomation during pregnancy: Report of a case. Toxicon 1979;17:188–189.
50. Banham ND, Jelinek GA, Finch PM. Late treatment with antivenom in prolonged red-back spider envenomation. Med J Aust 1994;161:379–381.
51. Sutherland SK. Antivenom use in Australia. Premedication, adverse reactions and the use of venom detection kits. Med J Aust 1992;157:734–739.
52. Sutherland SK, Trinca JC. Survey of 2144 cases of red back spider bites. Med J Aust 1978:2:620–623.
53. Isbister GK. Failure of intramuscular antivenom in Red-back spider envenoming. Emerg Med (Fremantle) 2002;14:436–439.
54. Graudins A, Gunja N, Broady KW, Nicholson GM. Clinical and *in vitro* evidence for the efficacy of Australian red-back spider (*Latrodectus hasselti*) antivenom in the treatment of envenomation by a Cupboard spider (*Steatoda grossa*). Toxicon 2002;40:767–775.

55. Key GF. A comparison of calcium gluconate and methocarbamol (Robaxin®) in the treatment of latrodectism (black widow spider envenomation). Am J Trop Med Hyg 1981;30:273–277.
56. Ryan PJ. Preliminary report: Experience with the use of dantrolene sodium in the treatment of bites by the black widow spider *Latrodectus hesperus*. J Toxicol Clin Toxicol 1983–1984;21:487–489.
57. Bucaretchi F, de Deus Reinaldo CR, Hyslop S, Madureira PR, de Capitani EM, Vieira RJ. A clinico-epidemiological study of bites by spiders of the genus *Phoneutria*. Rev Inst Med Trop S Paulo 2000;42:17–21.
58. Lucas S. Spiders in Brazil. Toxicon 1988;26:759–772.
59. Cardoso FC, Pacifico LG, Carvalho DC, Victoria JM, Neves AL, Chavez-Olortegui C, et al. Molecular cloning and characterization of *Phoneutria nigriventer* toxins active on calcium channels. Toxicon 2003;41:755–763.
60. Cordeiro Mdo N, de Figueiredo SG, Valentim Ado C, Diniz CR, von Eickstedt VR, Gilroy J, Richardson M. Purification and amino acid sequences of six T×3 type neurotoxins from the venom of the Brazilian "armed" spider *Phoneutria nigriventer* (Keys). Toxicon 1993;31: 35–42.
61. de Paula Le Sueur L, Kalapothakis E, da Cruz-Hofling MA. Breakdown of the blood–brain barrier and neuropathological changes induced by *Phoneutria nigriventer* spider venom. Acta Neuropathol 2003;105:125–134.
62. Cruz-Hofling MA, Love S, Brook G, Duchen LW. Effects of *Phoneutria nigriventer* spider venom on mouse peripheral nerve. Q J Exp Physiol 1985;70:623–640.
63. TroncOne LR, Lebrun I, Magnoli F, Yamane T. Biochemical and pharmacological studies On a lethal neurotoxic polypeptide from *Phoneutria nigriventer* spider venom. Neurochem Res 1995;20:879–883.
64. Guatimosim C, Romano-Silva MA, Cruz JS, Beirao PS, Kalapothakis E, Moraes-Santos T, et al. A toxin from the spider *Phoneutria nigriventer* that blocks calcium channels coupled to exocytosis. Br J Pharmacol 1997;122: 591–597.
65. Rapôso C, Zago GM, da Silva GH, da Cruz Höfling MA. Acute blood–brain barrier permeabilization in rats after systemic *Phoneutria nigriventer* venom. Brain Res 2007; 1149:18–29.
66. Isbister GK, Gray MR, Balit CR, Raven RJ, Stokes BJ, Porges K, et al. Funnel-web spider bite: a systematic review of recorded clinical cases. Med J Aust 2005;182: 407–411.
67. Isbister GK, Gray MR. Bites by Australian mygalomorph spiders (Araneae, Mygalomorphae), including funnel-web spiders (Atracinae) and mouse spiders (Actinopodidae: *Missulena* spp). Toxicon 2004;43:133–140.
68. Wang X, Smith R, Fletcher JI, Wilson H, Wood CJ, Howden ME, King GF. Structure-function studies of omega-atracotoxin, a potent antagonist of insect voltage-gated calcium channels. Eur J Biochem 1999;264: 488–494.
69. Szeto TH, Birinyi-Strachan LC, Smith R, Connor M, Christie MJ, King GF, Nicholson GM. Isolation and pharmacological characterisation of delta-atracotoxin-Hv1b, a vertebrate-selective sodium channel toxin. FEBS Lett 2000;470:293–299.
70. Sutherland SK. Venomous Australian creatures: the action of their toxins and the care of the envenomated patient. Anaesth Intensive Care 1974;2:316–328.
71. Nicholson GM, Little MJ, Birinyi-Strachan LC. Structure and function of delta-atracotoxins: lethal neurotoxins targeting the voltage-gated sodium channel. Toxicon 2004;43:587–599.
72. Atkinson RK. Naturally occurring inhibitors of the venom of funnel web spiders (*Atrax* species). Aust J Exp Biol Med Sci 1981;59:317–325.
73. Atkinson RK. Comparisons of the neurotoxic activity of the venom of several species of funnel web spiders (*Atrax*). Aust J Exp Biol 1981;59:307–316.
74. Tedford HW, Sollod BL, Maggio F, King GF. Australian funnel-web spiders: master insecticide chemists. Toxicon 2004;43:601–618.
75. Rash LD, Birinyi-Strachan LC, Nicholson GM, Hodgson WC. Neurotoxic activity of venom from the Australian eastern mouse spider (*Missulena bradleyi*) involves modulation of sodium channel gating. Br J Pharmacol 2000;130:1817–1824.
76. Fisher MM, Carr GA, McGuinness R, Warden JC. *Atrax robustus* envenomation. Anaesth Intensive Care 1980;8: 410–420.
77. Miller MK, Whyte IM, White J, Keir PM. Clinical features and management of *Hadronyche* envenomation in man. Toxicon 2000;38:409–427.
78. Harrington AP, Raven RJ, Bowe PC, Hawdon GM, Winkel KD. Funnel-web spider (*Hadronyche infensa*) envenomations in coastal south-east Queensland. Med J Aust 1999;171:651–653.
79. Dieckmann J, Prebble J, McDonogh A, Sara A, Fisher M. Efficacy of funnel-web spider antivenom in human envenomation by *Hadronyche* species. Med J Aust 1989;151:706–707.
80. Browne GJ. Near fatal envenomation from the funnel-web spider in an infant. Pediatr Emerg Care 1997;13: 271–273.
81. Graudins A, Wilson D, Alewood PF, Broady KW, Nicholson GM. Cross-reactivity of Sydney funnel-web spider antivenom: neutralization of the *in vitro* toxicity of other Australian funnel-web (*Atrax* and *Hadronyche*) spider venoms. Toxicon 2002;40:259–266.
82. Sutherland SK. Antivenom use in Australia. Premedication, adverse reactions and the use of venom detection kits. Med J Aust 1992;157:734–739.
83. Isbister GK. Mouse spider bites (*Missulena* spp.) and their medical importance. A systematic review. Med J Aust 2004;180:225–227.
84. Caveness W. Insect bite complicated by fever. Nashville J Med Surg 1972;10:333–337.

85. Schmaus LF. Case of arachnoidism (spider bite). JAMA 1929;92:1265–1266.
86. Macchiavello A. La *Loxosceles laeta*, causa del aracnoidismo cutaneo, o mancha gangrenosa, de Chile. Rev Chilena His Nat 1937;41:11–19.
87. Macchiavello A. cutaneous arachnidism or gangrenous spot in Chile. PRJ Public Health Trop Med 1947;22: 425–466.
88. Gotten H, McGowen JJ. Black water fever (hemoglobinuria) caused by spider bite. JAMA 1940;114:1547–1550.
89. Atkins JA, Wingo CW, Sodeman WA, Flynn JE. Necrotic arachnidism. Am J Trop Med Hyg 1958;7:165–184.
90. Vetter RS. Spiders of the genus *Loxosceles* (Araneae, Sicariidae): a review of biological, medical and psychological aspects regarding envenomation. J Arachnol 2008;36:150–163.
91. Wendell RP. Brown recluse spiders: a review to help guide physicians in nonendemic areas. South Med J 2003;96:786–490.
92. Swanson DL, Vetter RS. Bites of brown recluse spiders and suspected necrotic arachnidism. N Engl J Med 2005;352:700–707.
93. Hogan CJ, Barbaro KC, Winkel K. Loxoscelism: old obstacles, new directions. Ann Emerg Med 2004;44: 608–624.
94. Newlands G, Isaacson C, Martindale C. Loxoscelism in the Transvaal, South Africa. Trans R Soc Trop Med Hyg 1982;76:610–615.
95. Efrati P. Bites by *Loxosceles* spiders in Israel. Toxicon 1969;6:239–241.
96. Southcott RV. Spiders of the genus *Loxosceles* in Australia. Med J Aust 1976;1:406–408.
97. Newlands G, Atkinson P. Behavioural and epidemiological considerations pertaining to necrotic araneism in southern Africa. S Afr Med J 1990;77:92–95.
98. Malaque CM, Castro-Valencia JE, Cardoso JL, Franca FO, Barbaro KC, Fan HW. Clinical and epidemiological features of definitive and presumed loxoscelism in Sao Paulo, Brazil. Rev Inst Med Trop Sao Paulo 2002;44: 139–143.
99. Majeski JA, Durst GG Sr. Necrotic arachnidism. South Med J 1976;69:887–891.
100. Gertsch WJ, Ennik F. The spider genus *Loxosceles* in North America, Central America, and the West Indies (Araneae, Loxoscelidae). Bull Am Mus Nat His 1983;175: 265–359.
101. Swanson DL, Vetter RS. Loxoscelism. Clin Dermatol 2006;24:213–221.
102. Foil LD, Norment BR. Envenomation by *Loxosceles reclusa*. J Med Entomol 1979;16:18–25.
103. Eskafi FM, Frazier JL, Hocking RR, Norment BR. Influence of environmental factors on longevity of the brown recluse spider. J Med Entomol 1977;14:221–228.
104. Horner NV, Stewart KW. Life history of the brown spider, *Loxosceles reclusa* Gertsch and Mulaik. Texas J Sci 1967;19:333–347.
105. Andrade RM, Lourenco WR, Tambourgi DV. Comparison of the fertility between *Loxosceles intermedia* and *Loxosceles laeta* spiders. J Arachnol 2000;28:245–247.
106. Vetter RS, Barger DK. An infestation of 2,055 brown recluse spiders (Araneae: Sicariidae) and no envenomations in a Kansas home: implications for bite diagnoses in nonendemic areas. J Med Entomol 2002;39:948–951.
107. Stochosky B. Necrotic arachnidism. West J Med 1979;131: 143–148.
108. Barbaro KC, Sousa MV, Morhy L, Eickstedt VR, Mota I. Compared chemical properties of dermonecrotic and lethal toxins from spiders of the Genus *Loxosceles* (Araneae). J Protein Chem 1996;15:337–343.
109. Wright AP, Elgert KD, Campbell BJ, Barrett JT. Hyaluronidase and esterase activities of the venom of the poisonous brown recluse spider. Arch Biochem Biophys 1973;159:415–426.
110. Tambourgi DV, Magnoli FC, van den Berg CW, Morgan BP, de Araujo PS, Alves EW, Da Silva WD. Sphingomyelinases in the venom of the spider *Loxosceles intermedia* are responsible for both dermonecrosis and complement-dependent hemolysis. Biochem Biophys Res Commun 1998;251:366–373.
111. Binford GJ, Wells MA. The phylogenetic distribution of sphingomyelinase D activity in venoms of Haplogyne spiders. Comp Biochem Physiol B Biochem Mol Biol 2003;135:25–33.
112. Patel KD, Modur V, Zimmerman GA, Prescott SM, McIntyre TM. The necrotic venom of the brown recluse spider induces dysregulated endothelial cell-dependent neutrophil activation. Differential induction of GM-CSF, IL-8, and E-selectin expression. J Clin Invest 1994;94: 631–642.
113. Cunha RB, Barbaro KC, Muramatsu D, Portaro FC, Fontes W, de Sousa MV. Purification and characterization of loxnecrogin, a dermonecrotic toxin from *Loxosceles gaucho* brown spider venom. J Protein Chem 2003;22: 135–146.
114. da Silva PH, da Silveira RB, Appel MH, Mangili OC, Gremski W, Veiga SS. Brown spiders and loxoscelism. Toxicon 2004;44:693–709.
115. dos Santo VL, Franco CR, Viggiano RL, da Silveira RB, Cantao MP, Mangili OC, Veiga SS, Gremski W. Structural and ultrastructural description of the venom gland of *Loxosceles intermedia* (brown spider). Toxicon 2000;38: 265–285.
116. Gomez HF, Krywko DM, Stoecker WV. A new assay for the detection of *Loxosceles* species (brown recluse) spider venom. Ann Emerg Med 2002;39:469–474.
117. de Oliveira KC, Goncalves de Andrade RM, Piazza RM, Ferreira JM Jr, van den BergCW, Tambourgi DV. Variations in *Loxosceles* spider venom composition and toxicity contribute to the severity of envenomation. Toxicon 2005;45:421–429.
118. McGlasson DL, Harroff HH, Sutton J, Dick E, Elston DM. Cutaneous and systemic effects of varying doses of

brown recluse spider venom in a rabbit model. Clin Lab Sci 2007;20:99–105.

119. Tambourgi DV, Petricevich VL, Magnoli FC, Assaf SL, Jancar S, Dias Da Silva W. Endotoxemic-like shock induced by *Loxosceles* spider venoms: pathological changes and putative cytokine mediators. Toxicon 1998;36:391–403.
120. Silvestre FG, de Castro CS, de Moura JF, Giusta MS, De Maria M, Alvares ES, et al. Characterization of the venom from the Brazilian Brown Spider *Loxosceles similis* Moenkhaus, 1898 (Araneae, Sicariidae). Toxicon 2005;46: 927–936.
121. Berger RS, Adelstein EH, Anderson PC. Intravascular coagulation: The cause of necrotic arachnidism. J Invest Dermatol 1978;61:142–150.
122. Desai A, Miller MJ, Gomez HF, Warren JS. *Loxosceles deserta* spider venom induces NF-κB-dependent chemokine production by endothelial cells. Clin Toxicol 1999;37: 447–456.
123. Ospedal KZ, Appel MH, Neto JF, Mangili OC, Veiga SS, Gremski W. Histopathological findings in rabbits after experimental acute exposure to the *Loxosceles intermedia* (brown spider) venom. Int J Exp Pathol 2002;84: 287–294.
124. Elston DM, Eggers JS, Schmidt WE, Storrow AB, Doe RH, McGlasson D, Fischer JR. Histological findings after brown recluse spider envenomation. Am J Dermatopathol 2000;22:242–246.
125. Zanetti VC, da Silveira RB, Drefuss JL, Haoach J, Mangili OC, Veiga SS, Gremski W. Morphological and biochemical evidence of blood vessel damage and fibrinogenolysis triggered by brown spider venom. Blood Coagul Fibrinolysis 2002;13:135–148.
126. Futrell JM, Morgan PN, Su SP, Roth SI. Localization of brown recluse venom attachment sites on human erythrocytes by the ferritin-labeled antibody technique. Am J Pathol 1979;95:675–682.
127. Berger RS. The unremarkable brown recluse spider bite. JAMA 1973;225:1109–1111.
128. Cacy J, Mold JW. The clinical characteristics of brown recluse spider bites treated by family physicians: an OKPRN Study. Oklahoma Physicians Research Network. J Fam Pract 1999;48:536–542.
129. Wilson DC, King LE Jr. Spider and spider bites. Dermatol Clin 1990;8:277–286.
130. Wasserman GS, Siegel C. Loxoscelism (brown recluse spider bite): A review of the literature. Clin Toxicol 1979;14:353–358.
131. Sams HH, Hearth SB, Long LL, Wilson DC, Sanders DH, King LE Jr. Nineteen documented cases of *Loxosceles reclusa* envenomation. J Am Acad Dermatol 2001;44: 603–608.
132. Morena P, Nonoyama K, Cardoso JL, de O. Barretto OC. Search of intravascular hemolysis in patients with the cutaneous form of loxoscelism. Rev Inst Med Trop Sao Paulo 1994;36:149–151.
133. Robb CW, Hayes BB, Boyd AS. Generalized vasculitic exanthem following *Loxosceles reclusa* envenomation. J Cutan Pathol 2007;34:513–514.
134. Vetter RS, Swanson DL. Of spiders and zebras: publication of inadequately documented loxoscelism case reports. J Am Acad Dermatol 2007;56:1063–1064.
135. Vetter RS, Cushing PE, Crawford RL, Royce LA. Diagnoses of brown recluse spider bites (loxoscelism) greatly outnumber actual verifications of the spider in four western American states. Toxicon 2003;42:413–418.
136. Osterhoudt KC, Zaoutis T, Zorc JJ. Lyme disease masquerading as brown recluse spider bite. Ann Emerg Med 2002;39:558–561.
137. Anderson PC. Loxoscelism threatening pregnancy: five cases. Am J Obstet Gynecol 1991;165:1454–1456.
138. Gendron BP. *Loxosceles reclusa* envenomation. Am J Emerg Med 1990;8:51–54.
139. Madrigal GC, Ercolani RL, Wenzl JE. Toxicity from a bite of the brown spider (*Loxosceles reclusus*): skin necrosis, hemolytic anemia and hemoglobinuria in a nine year old child. Clin Pediatr 1972;11:641–644.
140. Chu JY, Rush CT, O'Connor DM. Hemolytic anemia following brown spider (*Loxosceles reclusa*) bite. Clin Toxicol 1978;12:531–534.
141. Bey TA, Walter FG, Lober W, Schmidt J, Spark R, Schlievert PM. *Loxosceles arizonica* bite associated with shock. Ann Emerg Med 1997;30:701–703.
142. Sezerino UM, Zannin M, Coelho LK, Gonclalves J Jr, Grando M, Mattosinho SG, et al. A clinical and epidemiological study of *Loxosceles* spider envenoming in Santa Catarina, Brazil. Trans Roy Soc Trop Med Hyg 1998;92:546–548.
143. Murray LM, Seger DL. Hemolytic anemia following a presumptive brown recluse spider bite. Clin Toxicol 1994;32:451–456.
144. Vorse H, Seccareccio P, Woodruff K, Humphrey GB. Disseminated intravascular coagulopathy following fatal brown spider bite (necrotic arachnidism). J Pediatr 1972;80:1035–1037.
145. de Souza AL, Malaque CM, Sztajnbok J, Romano CC, Duarte AJ, Seguro AC. Loxosceles venom-induced cytokine activation, hemolysis, and acute kidney injury. Toxicon 2008;51:151–156.
146. Wasserman GS, Anderson PC. Loxoscelism and necrotic arachnidism. J Toxicol Clin Toxicol 1983–1984;21:451–472.
147. Williams ST, Khare VK, Johnston GA, Blackall DP. Severe intravascular hemolysis associated with brown recluse spider envenomation a report of two cases and review of the literature. Am J Clin Pathol 1995;104: 463–467.
148. Barrett SM, Romine-Jenkins M, Blick KE. Passive hemagglutination inhibition test for diagnosis of brown recluse spider bite envenomation. Clin Chem 1993;39: 2104–2107.

149. Miller MJ, Gomez HF, Snider RJ, Stephens EL, Czop RM, Warren JS. Detection of *Loxosceles* venom in lesional hair shafts and skin: application of a specific immunoassay to identify dermonecrotic arachnidism. Am J Emerg Med 2000;18:626–628.

150. Krywko DM, Gomez HF. Detection of *Loxosceles* species venom in dermal lesions: a comparison of 4 venom recovery methods. Ann Emerg Med 2002;39:475–480.

151. Gomez HF, Miller MJ, Waggener MW, Lankford HA, Warren JS. Antigenic cross-reactivity of venoms from medically important North American *Loxosceles* spider species. Toxicon 2001;39:817–824.

152. Morena P, Nonoyama K, Cardoso JL, Barretto OC. Search of intravascular hemolysis in patients with the cutaneous form of loxoscelism. Rev Inst Med Trop Sao Paulo 1994;36:149–151.

153. Elbahlawan LM, Stidham GL, Bugnitz MC, Storgion SA, Quasney MW. Severe systemic reaction to *Loxosceles reclusa* spider bites in a pediatric population. Pediatric Emerg Care 2005;21:177–180.

154. Lane DR, Youse JS. Coombs-positive hemolytic anemia secondary to brown recluse spider bite: a review of the literature and discussion of treatment. Cutis 2004;74: 341–347.

155. Franca FO, Barbaro KC, Abdulkader RC. Rhabdomyolysis in presumed viscero-cutaneous loxoscelism: report of two cases. Trans R Soc Trop Med Hyg 2002;96: 287–290.

156. Hardman JT, Beck ML, Hardman PK, Stout LC. Incompatibility associated with the bite of a brown recluse spider (*Loxosceles reclusa*). Transfusion 1983;23:233–236.

157. Wright SW, Wrenn KD, Murray L, Seger D. Clinical presentation and outcome of brown recluse spider bite. Ann Emerg Med 1997;30:28–32.

158. Futrell JM. Loxoscelism. Am J Med Sci 1992;304: 261–267.

159. Monteiro CL, Rubel R, Cogo LL, Mangili OC, Gremski W, Veiga SS. Isolation and identification of *Clostridium perfringens* in the venom and fangs of *Loxosceles intermedia* (brown spider): enhancement of the dermonecrotic lesion in loxoscelism. Toxicon 2002;40: 409–418.

160. Gomez HF, Miller MJ, Trachy JW, Marks RM, Warren JS. Intradermal anti-Loxosceles Fab fragments attenuate dermonecrotic arachnidism. Acad Emerg Med 1999;6: 1195–1202.

161. Barrett SM, Romine-Jenkins M, Fisher DE. Dapsone or electric shock therapy of brown recluse spider envenomation? Ann Emerg Med 1994;24:21–25.

162. King LE, Rees PS. Dapsone treatment of a brown recluse spider. JAMA 1983;250:648.

163. Rees RS, Altenbem DP, Lynch JB, King LE Jr. Brown recluse spider bites: a comparison of early surgical excision versus dapsone and delayed surgical excision. Ann Surg 1985:202:659–663.

164. Phillips S, Kohn M, Baker D, Vander Leest R, Gomez H, McKinney P, et al. Therapy of brown spider envenomation: a controlled trial of hyperbaric oxygen, dapsone, and cyproheptadine. Ann Emerg Med 1995;25:363–368.

165. Beilman GJ, Winslow CL, Teslow TW. Experimental brown spider bite in the guinea pig: results of treatment with dapsone or hyperbaric oxygen. J Wilderness Med 1994;5:287–294.

166. Wille RC, Morrow JD. Case report: dapsone hypersensitivity syndrome associated with treatment of the bite of a brown recluse spider. Am J Med Sci 1988;296: 270–271.

167. Iserson KV. Methemoglobinemia from dapsone therapy for a suspected brown spider bite. J Emerg Med 1985;3: 285–288.

168. Hobbs GD, Anderson AR, Greene TJ, Yealy DM. Comparison of hyperbaric oxygen and dapsone therapy for *Loxosceles* envenomation. Acad Emerg Med 1996;3: 758–761.

169. Strain GM, Snider TG, Tedford BL, Cohn GH. Hyperbaric oxygen effects on brown recluse spider (*Loxosceles reclusa*) envenomation in rabbits. Toxicon 1991;29: 989–996.

170. Isbister GK. Necrotic arachnidism: the mythology of a modern plague. Lancet 2004;364:549–553.

171. Sams HH, Dunnick CA, Smith ML, King LE Jr. Necrotic arachnidism. J Am Acad Dermatol 2001;44:561–573.

172. Anderson PC. Necrotizing spider bites. Am J Fam Pract 1982;26:198–203.

173. Isbister GK, Gray MR. White-tail spider bite: a prospective study of 130 definite bites by *Lampona* species. Med J Aust 2003;179:199–202.

174. Isbister GK, Hirst D. A prospective study of definite bites by spiders of the family Sparassidae (huntsmen spiders) with identification to species level. Toxicon 2003;42: 163–171.

175. Reed CF, Witt PN, Scarboro MB. The orb web during the life of *Argiope aurantia* (Lucas). Dev Psychobiol 1969;2:120–129.

176. Champion de Crespigny FE, Herberstein ME, Elgar MA. Food caching in orb-web spiders (Araneae: Araneoidea). Naturwissenschaften 2001;88:42–45.

177. Budd T, Clinton P, Dell A, Duce IR, Johnson SJ, Quicke DL, et al. Isolation and characterisation of glutamate receptor antagonists from venoms of orb-web spiders. Brain Res 1988;448:30–39.

178. Grishin EV, Volkova TM, Arseniev AS. Isolation and structure analysis of components from venom of the spider *Argiope lobata*. Toxicon 1989;27:541–549.

179. Isbister GK. Prospective cohort study of definite spider bites in Australian children. J Paediatr Child Health 2004;40:360–364.

180. Newlands G, Atkinson P. Review of southern African spiders of medical importance, with notes on the signs and symptoms of envenomation. S Afr Med J 1988;73: 235–239.

181. Vetter RS, Isbister GK, Bush SP, Boutin LJ. Verified bites by yellow sac spiders (genus *Cheiracanthium*) in the United States and Australia: where is the necrosis? Am J Trop Med Hyg 2006;74:1043–1048.
182. Newlands G, Martindale C, Berson SD, Rippey JJ. Cutaneous necrosis caused by the bite of *Chiracanthium* spiders. S Afr Med J 1980;57:171–173.
183. Minton SA Jr. Poisonous spiders of Indiana and a report of a bite by *Chiracanthium mildei*. J Indiana State Med Assoc 1972;65:425–426.
184. Gorham JA, Rheney TB. Envenomation by the spiders *Chiracanthium inclusum* and *Argiope aurantia*. JAMA 1968;206:1958–1962.
185. Furman DP, Reeves WC. Toxic bite of a spider: *Chiracanthium inclusum* (Hentz). Calif Med 1957;87:114.
186. Krinsky WL. Envenomation by the sac spider *Chiracanthium mildei* Cutis 1987;40:127–129.
187. Yan L, Adams ME. Lycotoxins, antimicrobial peptides from venom of the wolf spider *Lycosa carolinensis*. J Biol Chem 1998;273:2059–2066.
188. Rash LD, King RG, Hodgson WC. Evidence that histamine is the principal pharmacological component of venom from an Australian wolf spider (*Lycosa godeffroyi*). Toxicon 1998;36:367–375.
189. Atkinson RK, Wright LG. A study of the necrotic actions of the venom of the wolf spider, *Lycosa godeffroyi*, on mouse skin. Comp Biochem Physiol C 1990;95:319–325.
190. Campbell DS, Rees RS, King LE. Wolf spider bites. Cutis 1987;39:113–114.
191. Isbister GK, Framenau VW. Australian wolf spider bites (Lycosidae): clinical effects and influence of species on bite circumstances. J Toxicol Clin Toxicol 2004;42: 153–161.
192. Redman JF. Human envenomation by a lycosid. Arch Dermatol 1974;110:111–112.
193. Bush SP, Giem P, Vetter RS. Green lynx spider (*Peucetia viridans*) envenomation. Am J Emerg Med 2000;18: 64–66.
194. Hall RE, Madon MB. Envenomation by the green lynx spider *Peucetia viridans* (Hentz 1832), in Orange County, California. Toxicon 1973;11:197–199.
195. Hunt GR. Bites and stings of uncommon arthropods. 1. Spiders. Postgrad Med 1981;70:91–102.
196. Cohen E, Quistad GB. Cytotoxic effects of arthropod venoms on various cultured cells. Toxicon 1998;36: 353–358.
197. Russell FE. Bite of the spider *Phidippus formosus*: Case history. Toxicon 1970;8:193–194.
198. Vetter RS, Roe AH, Bennett RG, Baird CR, Royce LA, Lanier WT, Antonelli AL, Cushing PE. Distribution of the medically-implicated hobo spider (Araneae: Agelenidae) and a benign congener, *Tegenaria duellica*, in the United States and Canada. J Med Entomol 2003;40: 159–164.
199. Sadler MA, Force RW, Solbrig RM, Sommer S. Suspected *Tegenaria agrestis* envenomation. Ann Pharmacother 2001;35:1490–1491.
200. Center for Disease Control (CDC). Necrotic arachnidism —Pacific Northwest, 1988–1996. MMWR Morb Mortal Wkly Rep 1996;45:433–436.
201. Fisher RG, Kelly P, Krober MS, Weir MR, Jones R. Necrotic arachnidism. West J Med 1994;160:570–572.
202. Vest DK. Necrotic arachnidism in the northwest United States and its probable relationship to *Tegenaria agrestis* (Walckenaer) spiders. Toxicon 1987;25:175–184.
203. Binford GJ. An analysis of geographic and intersexual chemical variation in venoms of the spider *Tegenaria agrestis* (Agelenidae). Toxicon 2001;39:955–968.

B Centipedes

Chapter 167

CENTIPEDES
(Subclass: Chilopoda)

IDENTIFYING CHARACTERISTICS

Centipedes belong to the phylum Arthropoda, class: Chilopoda. Fossil records dating back 420 million years document the worldwide distribution of these animals, and there are over 3,300 species of centipedes belonging to five extant orders.[1] Centipedes occur in all continents except Antarctica, primarily in warm temperate regions and the tropics. In some classifications, Chilopoda is part of the subphylum Myriapoda that includes Diplopoda (millipedes), Symphyla, and Pauropoda. Centipedes are slender, multisegmented arthropods that possess one pair of short legs per flattened segment in contrast to millipedes, which possess two sets of legs per segment and no venomous claws. Although some giant millipedes (*Polyconoceras* spp.) from Papua, New Guinea can spray noxious secretions that irritate mucous membranes and eyes,[2] millipedes are harmless vegetarians that do not bite. The odd number of segments on centipedes range between 15–191, and the first pair of legs is modified into a venom-delivering organ. There is one set of antennae. Their length ranges between 1–30cm (~0.4–12in.), and their coloration varies from bright yellow to brown black. The last pair of legs is elongated. All four orders of centipedes are venomous including Geophilomorpha (soil centipedes), Lithobiomorpha (rock or garden centipedes), Scolopendromorpha (tropical or giant centipedes), and Scutigeromorpha (house centipedes). The order Scolopendromorpha contains the most aggressive species and voracious predators.[1] Some classifications include a fifth order (Craterostigmomorpha) that contains only one genus (*Craterostigmus*). This genus represents a stage between scolopendrids and lithobids. They have only 15 pairs of legs and only seven sets of spiracles. The largest centipedes belong to the Scolopendridae family. These yellow-brown creatures with orange and blue cephalic parts reach 8–15cm (~3–6in.) in length.

Common Name: American House Centipede

Scientific Name: *Scutigera coleoptrata* L.

Family: Scutigeridae

Physical Description: These centipedes are elongated, dorsoventrally flattened arthropods with 15 segments. The hindmost segment is modified for sensory and defensive purposes. This arthropod matures in 3 years.

Common Name: Centipede

Scientific Name: *Scolopendra subspinipes* Leach

Family: Scolopendridae

Physical Description: This large, aggressive centipede ranges in length up to 10–20cm (~4–8in.). Their color varies, but most of specimens are red to brown with yellow to orange legs. These centipedes have 18 to 19 segments on their antennae with the first 6 segments being smooth. The terminal legs have three spines. There are at least eight subspecies of *S. subspinipes*.

EXPOSURE

Geographic Distribution

Centipedes inhabit warm temperate and tropical climates throughout the world. Centipedes occur commonly in the southern portions of the United States from California, Arizona, and Texas to Alabama, Louisiana, and Georgia as well as Hawaii. The distribution of *Scolopendra subspinipes* includes tropical and subtropical areas of eastern and southern Asia as well as Hawaii.

Behavior

Centipedes have no waxy waterproofing layer on their cuticle, and almost all centipedes are terrestrial. Hence, they prefer dark, damp environments (e.g., undersurface of stones, rotting logs, leaf litter, tree bark) and during rainy weather these creatures will invade indoor habitats. The females deposit eggs in the soil. They are nocturnal carnivores with a wide range of prey including insects, worms, slugs, and small snakes. Centipedes are typically solitary except during breeding and the care of hatchlings.[1] In contrast to centipedes, millipedes are herbivores. Centipedes paralyze their prey prior to capture and ingestion. These creatures do not typically hold onto the victim following the sting, except in rare circumstances.[3] Centipedes hide in bedding or clothing; therefore, humans are most likely exposed to these creatures in bed or as they dress. Exposure to centipedes in the United States typically occurs July through September.

Uses

Centipedes (*Scolopendra subspinipes mutilans* L. Koch) are a traditional part of Chinese and Korean medicine for the treatment of cardiovascular diseases.[4] Extracts of centipede are used in traditional Chinese medicine to treat cognitive decline, and chloroform extracts of *Scolopendra subspinipes mutilans* L. Koch demonstrate some anticholinesterase activity *in vitro*.[5]

PRINCIPAL TOXINS

Venom Composition

The composition of centipede venom is not well studied. These venoms are diverse and contain a variety of unique proteins.[6] The venom of these creatures contains a variety of compounds including pain mediators (serotonin, histamine), acetylcholine, a cardiodepressant lethal factor (Toxin-S), proteinases, lipoproteins, and cytolysin.[7,8] The tissue of the Korean centipede (*Scolopendra subspinipes mutilans* L. Koch) contains a novel serine protease that demonstrates potent fibrinolytic activity as well as the unique quinoline alkaloid, scolopendrine or 2-hydroxy-7-[(4-hydroxy-3-methoxyphenyl)methyl]-3-methoxy-8-quinolyl sulfate.[9,10]

Venom Apparatus

Centipedes are carnivores that prey on soil invertebrates (earthworms, insects) by capturing their prey with modified appendages. The venom apparatus of a centipede resides in the telopodites of the postcephalic segment, which is the first segment behind and ventral to the head. This pair of legs is modified into two sharp stinging structures connected to a muscular venom gland. The venom apparatus consists of 1) venom gland, 2) venom duct ending in the tip of the structure, 3) venom claw, and 4) muscles that operate the claw.[11] The venom duct carries the venom into the venom claw, which injects the venom through a subterminal pore in the outer curvature of the claw. The structure of the venom claw maximizes the size of the wound and the depth of venom injection. This venom apparatus typically produces two tiny puncture wounds on the victim several millimeters to 1.5 cm apart depending on the size of the centipede. These puncture wounds may bleed or develop subcutaneous hemorrhage. Double bite marks may appear after envenomation.[4]

Millipedes secrete odoriferous droplets from the glands on both lateral surfaces of the body segments. The extrusion of these malodorous secretions is a defensive maneuver that produces irritation. These secretions contain phenols, aldehydes, and nitroethylbenzene compounds,[12] but toxicity of these substances is not well documented.

Dose Response

The venom of the centipede does not usually produce serious systemic toxicity in humans. Based on extrapolation from rodent studies, over 1,000 centipede bites would be necessary to cause death in a human.[11]

TOXICOKINETICS

The main toxins in centipede venom have not been identified, and there are few data on the toxicokinetics of the toxic constituents.

HISTOPATHOLOGY AND PATHOPHYSIOLOGY

The venom from centipedes causes local pain and edema. There are no lethal factors identified in the venom to date. Components of some centipede venoms are structurally similar to allergens in bee venom. Potentially, these components can cause hypersensitivity responses as well as some cross-reactivity between these venoms.[13]

CLINICAL RESPONSE

Centipede bites typically cause moderate to severe local pain, localized swelling, and erythema, but systemic symptoms are usually absent or mild.[14,15] In contrast to viper envenomations, centipede bites do not typically produce hemorrhagic blebs or progressive swelling more than 10–20 cm (~4–8 in.) beyond the bite. Local effects include the rapid onset of erythema, edema, intense pain, paresthesias, lymphangitic steaks, and lymphadenitis.[16,17] The skin response usually remains localized to the area immediately (i.e., 10 cm/~4 in.) around the bite,[4] and evidence of envenomation does not usually progress beyond the affected extremity.[18,19] Skin necrosis may occur at the site of the bite, but these lesions typically heal spontaneously without the need for grafts. Nonspecific systemic complaints include mild constitutional symptoms (e.g., nausea, lightheadedness, weakness, anxiety, palpitations, fever). Symptoms usually resolve within a few hours to a week.[20] Fatalities following centipede envenomation are not well documented.[21] The ingestion of a centipede by a 6-month-old infant produced vomiting, hypotonia, and pallor that resolved within 48 hours.[22]

Patients usually recover from centipede envenomations without sequelae.[4] Complications include a type II hypersensitivity reaction (immune complex deposition) manifest by recurrent swelling and pruritus several weeks after envenomation as well as the development of nephrotic syndrome.[23] A case report documented the development of a compartment syndrome, rhabdomyolysis, and acute renal failure following massive swelling of the lower extremity of a 44-year-old woman following a centipede envenomation.[24] She was allergic to bee stings, and she developed wheezing after the envenomation. An additional case report associated the ingestion of an alcoholic beverage made from a Korean centipede soaked in 53% ethanol with rhabdomyolysis, bullous skin lesions, renal failure, and an upper extremity neuropathy.[25] However, the patient was lying comatose in a narrow space overnight, and a compartment syndrome with a compression neuropathy may account for the clinical features of this case report. A case report associated the development of eosinophilic cellulitis (Wells' syndrome) with a centipede bite.[26] Rare case reports from Turkey temporally associate centipede bites with the development of transient increases in cardiac enzymes and ST-T wave changes.[27,28] These patients did not have underlying heart disease as determined by coronary angiography and exercise stress testing. Symptoms and signs of myocardial ischemia resolved within 12–24 hours.

DIAGNOSTIC TESTING

The specific toxins associated with centipede envenomation have not been identified, and there are no commercial biomarkers for the confirmation of envenomations by centipedes.

TREATMENT

Most centipede envenomations are benign and self-limited. Treatment of centipede envenomation is supportive including local wound care, antihistamines for pruritus, local and systemic analgesics as necessary, and tetanus prophylaxis. Case reports suggest that heat may reduce the intensity of the sting similar to some marine envenomations, but there are inadequate clinical data to confirm this observation of centipede envenomations.[14,16]

References

1. Edgecombe GD, Giribet G. Evolutionary biology of centipedes (Myriapoda: Chilopoda). Annu Rev Entomol 2007;52:151–170.
2. Hudson BJ, Parsons GA. Giant millipede 'burns' and the eye. Trans R Soc Trop Med Hyg 1997;91:183–185.
3. Mumcuoglu KY, Leibovici V. Centipede (*Scolopendra*) bite: a case report. Isr J Med Sci 1989;25:47–49.
4. Lin T-J, Yang C-C, Yang G-Y, Ger J, Tsai W-J, Deng J-F. Features of centipede bites in Taiwan. Trop Geo Med 1995;47:300–302.
5. Ren Y, Houghton P, Hider RC. Relevant activities of extracts and constituents of animals used in traditional Chinese medicine for central nervous system effects associated with Alzheimer's disease. J Pharm Pharmacol 2006;58:989–996.
6. Rates B, Bemquerer MP, Richardson M, Borges MH, Morales RA, de Lima ME, Pimenta AM. Venomic analyses of *Scolopendra viridicornis nigra* and *Scolopendra angulata* (centipede, Scolopendromorpha): shedding light on venoms from a neglected group. Toxicon 2007;49: 810–826.
7. Mohamed AH, Abu-Sinna G, El-Shabaka HA, El-Aal AA. Proteins, lipids, lipoproteins and some enzyme characterizations of the venom extract from the centipede *Scolopendra morsitans*. Toxicon 1983;21:371–377.
8. Gomes A, Datta A, Sarangi B, Kar PK, Lahiri SC. Isolation, purification & pharmacodynamics of a toxin from the venom of the centipede *Scolopendra subspinipes dehaani* Brandt. Indian J Exp Biol 1983;21:203–207.
9. Noda N, Yashiki Y, Nakatani T, Miyahara K, Du X-M. A novel quinoline alkaloid possessing a 7-benzyl group from the centipede, *Scolopendra subspinipes*. Chem Pharm Bull 2001;49;930–931.

10. You W-K, Sohn Y-D, Kim K-Y, Park D-H, Jang Y, Chung K-H. Purification and molecular cloning of a novel serine protease from the centipede, *Scolopendra subspinipes mutilans*. Insect Biochem Mol Biol 2004;34:239–250.
11. Menez A, Zimmerman K, Zimmerman S, Heatwole H. Venom apparatus and toxicity of the centipede *Ethmostigmus rubripes* (Chilopoda, *Scolopendridae*). J Morphol 1990;206:303–312.
12. Kuwahara Y, Omura H, Tanabe T. 2-Nitroethenylbenzenes as natural products in millipede defense secretions. Naturwissenschaften 2002;89:308–310.
13. Harada S, Yoshizaki Y, Natsuaki M, Shimizu H, Fukuda H, Nagai H, Ikeda T. Three cases of centipede allergy—analysis of cross reactivity with bee allergy. Arerugi 2005;54:1279–1284. [Japanese]
14. Balit CR, Harvey MS, Waldock JM, Isbister GK. Prospective study of centipede bites in Australia. J Toxicol Clin Toxicol 2004;42:41–48.
15. Mohri S, Sugiyama A, Saito K, Nakajima H. Centipede bites in Japan. Cutis 1991;47:189–190.
16. Rodriguez-Acosta A, Gassette J, Gonzalez A, Ghisoli M. Centipede (Scolopedra gigantea Linneaus 1758) envenomation in a newborn. Rev Inst Med Trop S Paulo 2000;42:341–342.
17. Bush SP, King BO, Norris RL, Stockwell SA. Centipede envenomation. Wild Environ Med 2001;12:93–99.
18. Guerrero AP. Centipede bites in Hawai'i: a brief case report and review of the literature. Hawaii Med J 2007;66:125–127.
19. Bouchard NC, Chan GM, Hoffman RS. Vietnamese centipede envenomation. Vet Hum Toxicol 2004;46:312–313.
20. Burnett JW, Calton GJ, Morgan RJ. Centipedes. Cutis 1986;37:241.
21. Remington DL. The bite and habits of a giant centipede (*Scolopendra subspinipes*) in the Philippine Islands. Am J Trop Med 1950;30:453–455.
22. Barnett PL. Centipede ingestion by a six-month-old infant: toxic side effects. Pediatr Emerg Care 1991;7:229–230.
23. Hasan S, Hassan K. Proteinuria associated with centipede bite. Pediatr Nephrol 25;20:550–551.
24. Logan JL, Ogden DA. Rhabdomyolysis and acute renal failure following the bite of the giant desert centipede *Scolopendra heros*. West J Med 1985;142:549–550.
25. Wang I-K, Hsu S-P, Chi C-C, Lee K-F, Lin PY, Chang H-W, Chuang F-R. Rhabdomyolysis, acute renal failure, and multiple focal neuropathies after drinking alcohol soaked with centipede. Renal Failure 2004;26:93–97.
26. Friedman IS, Phelps RG, Baral J, Sapadin AN. Wells' syndrome triggered by centipede bite. Int J Dermatol 1998;37:602–605.
27. Yildiz A, Biceroglu S, Yakut N, Bilir C, Akdemir R, Akilli A. Acute myocardial infarction in a young man caused by centipede sting. Emerg Med J 2006;23:e30.
28. Ozsarac M, Karcioglu O, Ayrik C, Somuncu F, Gumrukcu S. Acute coronary ischemia following centipede envenomation: case report and review of the literature. Wilderness Environ Med 2004;15:109–112.

C Insects

Chapter 168

BEES, WASPS, and ANTS (Order: Hymenoptera)

IDENTIFYING CHARACTERISTICS

Insects have three major body parts including the head, thorax, and abdomen with three pairs of legs (subphylum: Hexapoda). Common characteristics of Hymenoptera include two pairs of wings (bees, vespids), one pair of antennae, three body segments (head, abdomen, thorax) and, in females, a tubular ovipositor that evolved into a stinging apparatus. The large, diverse order of Hymenoptera contains over 70 families with over 100,000 species. The medically most important families in the order Hymenoptera include bees (Apidae), vespids (Vespidae), and fire ants (Formicidae).

Bees

There are at least five species of honey bees ranging in size from the dwarf (1 cm/~0.4 in.) honey bee (*Apis florae* Fabricius) to the large (3 cm/~1 in.) Himalayan or giant tree honey bee (*Apis laboriosa* Smith). The Apidae Family includes the European or common honey bees (*Apis mellifera* L.) and the Asian honey bees (*Apis cerana* Fabricius, *Apis dorsata* Fabricius), as well as the North American bumble bees (*Bombus pennsylvanicus* De Geer). Commercial production of honey is limited to *A. mellifera* L. and *A. cerana* Fabricius because of the behavioral limitations of other *Apis* species, particularly open-air nesting. Africanized honey bees are hybrids of the African honey bee and the European honey bee, and the former subspecies is slightly smaller than European honey bees. Africanized honey bees are grossly indistinguishable from European honey bees. Identification of the former requires morphometry or analysis of mitochondrial DNA.

Vespids

The Vespidae Family includes the genera *Vespula* (yellow jackets), *Dolichovespula* (aerial yellow jackets, bald-faced or white-faced hornet, yellow hornet), *Vespa* (hornets), and *Polistes* (paper wasps). Wasps have an egg-shaped abdomen supported on the thorax by a slender stalk. Hornets possess a truncated abdomen and a shorter waist than wasps. Characteristic yellow and black stripes of the male yellow jacket distinguish the nonstinging male from the stinging, black-faced female.

Fire Ants

There are about 18 species of fire ants (genus *Solenopsis*) with workers ranging in size between 1.8–6 mm/~0.1–0.2 in. in size. The three species of native US fire ants (*S. geminata* Fabricius, *S. xyloni* McCook, *S. aurea* Wheeler) produce small colonies that do not cause economic or medical problems. The imported red fire ant (*Solenopsis wagneri* Santschi) now dominates large areas of the southeastern United States. The imported red fire ant is bright red in color, and this insect inhabits loose dirt mounds about 30 cm (~12 in.) in diameter and 5–10 cm (~2–4 in.) in height. Imported red fire ants are inconspicuous, reddish brown ants that appear indistinguishable from common native species of ants. Identifying morphological characteristics of *S. wagneri* Santschi (*S.*

Medical Toxicology of Natural Substances, by Donald G. Barceloux, MD

invicta) under magnification include two humps on the petiole that connects the thorax to the abdomen and clubbing of the last two segments of the antennae.[1]

EXPOSURE

Geographic Distribution

Bees

Honey bees are native to Europe, western Asia, and Africa, but with domestication the distribution of honey bees is now worldwide. Subspecies of the European honey bees (*Apis mellifera mellifera, Apis mellifera ligustica, Apis mellifera carnica, Apis mellifera iberica*) occur throughout North and South America with the exception of large tropical regions. The African honey bee (*Apis mellifera scutellata* Lepeletier) was accidentally introduced into Brazil during experimental studies in 1957,[2] and the habitat of this hybrid honey bee gradually expanded northward to include Texas, New Mexico, Arizona, Nevada, and California. Figure 168.1 displays the migration of the Africanized honey bee from Brazil to the southwestern United States. Theoretically, the habitat of these hybrid honey bees is limited by the requirement of a mean, high temperature exceeding 60 °F (~16 °C). This climatic condition includes areas from coastal California to Arizona, Texas, and portions of the southern United States. The African subspecies *Apis mellifera scutellata* Lepeletier (Africanized bee) is not present in Australia. The European hornet (*Vespa crabro* L.) occurs both in Europe and in North America whereas the Asian hornet (*Vespa mandarina* Smith) is limited to Asia. In the United States, the yellow jacket (*Vespula* species) is the most common cause of allergic reactions to insect stings.[3]

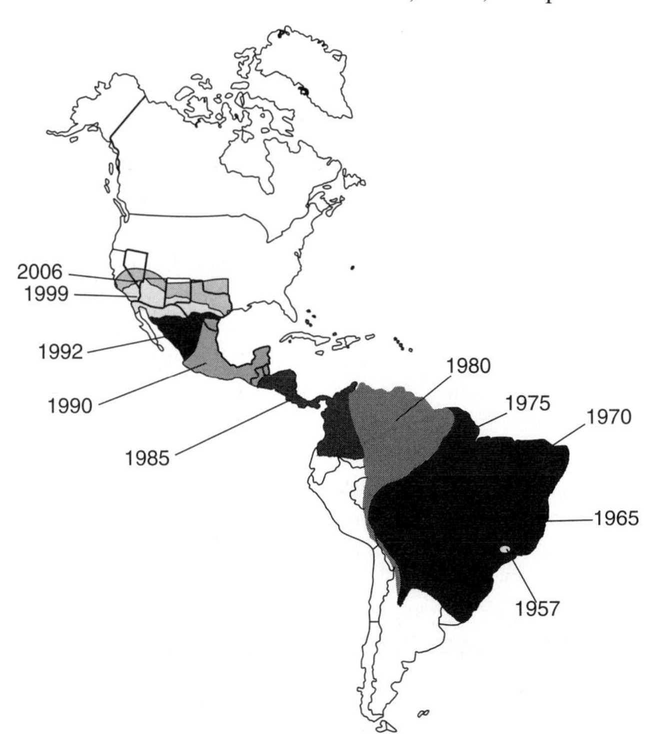

FIGURE 168.1. Migration of the Africanized honey bee. Adapted from Rodriguez-Lainz A, et al. Animal and human health risks associated with Africanized honeybees. JAVMA 1999;215:1800.

Wasps and Hornets

Hornets (*Vespa* spp.) are distributed mostly in Europe and Asia. The hornet species, *Vespa crabro*, is endemic in the northeastern United States. Yellow jackets are established in the United States as well as tropical and subtropical regions of Hawaii, Australia, New Zealand, South American, and South Africa.

Fire Ants

The black imported fire ant (*Solenopsis richteri* Forel) is a native of Argentina and Uruguay, whereas the red imported fire ant (*Solenopsis wagneri* Santschi, previously *Solenopsis invicta*) is a native species in Argentina, Paraguay, and Brazil.[4] The black fire ant was imported into the United States in the late 1910s or early 1920s, probably through the port of Mobile, Alabama.[5] In the late 1930s, importation of the red fire ant (*Solenopsis wagneri* Santschi) resulted in the gradual displacement of the imported black fire ant (*S. richteri*). The range of the imported black fire ant is now limited to northern Mississippi and Alabama. The red imported fire ant (*Solenopsis wagneri* Santschi) is the dominant species occupying millions of acres in Alabama, Arkansas, Florida, Georgia, Louisiana, Mississippi, North Carolina, Oklahoma, South Carolina, Tennessee, Texas, and Virginia. Colonies also occur in Arizona, California, New Mexico, and Puerto Rico. Figure 168.2 demonstrates the areas in the United States quarantined by the US Department of Agriculture because of infestation with imported red fire ants. The ultimate range of the imported red fire ant is unknown, depending on the climate (average minimum annual temperature >0 °F/~18 °C) and the development of cold tolerant red fire ant species.[6] Fire ants expand their habitat by mating flights, floating on flood waters, and the shipment of infested agricultural and nursery goods. Harvester ants (genus *Pogonomyrmex*) live in the arid sections of the western United States and Mexico. These ants produce painful stings, but are less aggressive than fire ants. Consequently, the medical significance of the presence of harvester ants is substantially less compared with fire ants. In addition to these ants, other causes of adverse effects of ants include the painful bites of the genus *Paraponera* from Central and South America, and the

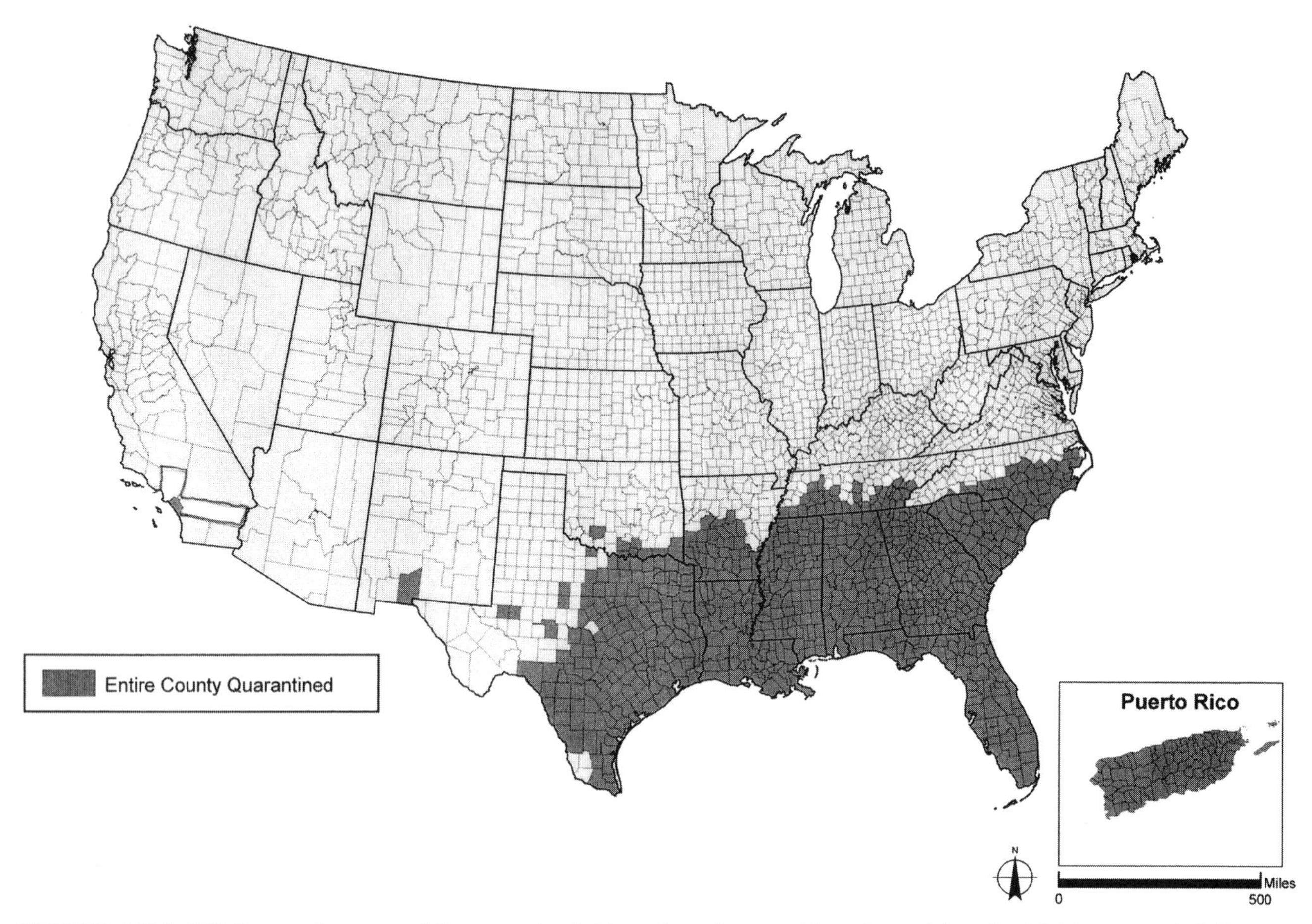

FIGURE 168.2. US Quarantine map of fire ants. Available at http://www.aphis.usda.gov/plant_health/plant_pest_info/fireants/downloads/fireant.pdf

severe allergic responses associated with the bites of the southeastern Australian bulldog ants (genus *Myrmecia*) including the jack jumper ant (*Myrmecia pilosula* Smith) and the bull ant (*Myrmecia pyriformis* Smith).[7] Other immunogenic ant species include the green-head ant (*Rhytidoponera metallica* Smith) from Australia and adjacent areas, as well as the Korean ant (*Pachycondyla chinensis* Emery).[8]

Behavior

Bees

Most bees are solitary, and their nomenclature reflects their particular habitat (e.g., miner, carpenter, mason, cuckoo). These bees do not usually defend their nests, but they will sting as a defensive maneuver when caught. Honey bees are social bees that form honey in a sophisticated colony (hive). Generally, honey bees are docile, and these insects usually sting only when disturbed or when their hive is threatened. In contrast to European honey bees, Africanized honey bees are easily agitated near their nest even with slight disturbances, and these honey bees remain agitated for hours. The Africanized honey bees reproduce and swarm at an accelerated rate compared with European honey bees, and the former frequently inflict hundreds of stings within several minutes.[9] Africanized honey bees are relatively indiscriminate about the selection of nesting habitat that includes open ground, tree branches, boxes, old tires, and buildings.[10] Although swarming honey bees are usually docile, these honey bees fiercely defend the hive once the colony settles on a location. Bees do not usually sting when foraging for pollen on flowers including Africanized honey bees. The bumble bee is a primitive bee that is relatively docile despite their large size and loud noises. Although they possess painful stingers and aggressively defend their nest, the relatively small nest population (i.e., about 200) limits their threat to humans.

Wasps and Yellow Jackets

Most wasps are solitary and harmless. Paper wasps (genus *Polistes*) may seek indoor shelter during winter, but these insects are not usually aggressive. These wasps build small umbrella-shaped hanging nests, commonly under the eaves of houses. These colonies are smaller than yellow jacket colonies. The nests of paper wasps typically contain only 100–200 insects; therefore, paper wasps are not usually involved in mass envenomations. Wasps consume only liquid food. Yellow jackets are aggressive scavengers that forage food around areas (e.g., picnics) containing garbage. Their nests typically contain a few thousand individuals. These insects are common causes of anaphylactic reactions because of their aggressive behavior and highly antigenic venom. Hornets are relatively docile insects that nest either in the ground or above ground. Stings from these insects are uncommon unless their hive is threatened.

Fire Ants

Although almost all members of the ant family bite, the fire ant is the most vicious species with a relatively high attack rate and a well-developed stinging apparatus.[11] Fire ants are extremely prolific, in part, because of the presence of polygyne colonies (i.e., multiple fertile queens) in large conspicuous, above-ground, dome-shaped mounds. After mating in the air, the male fire ants die and the fertilized female flies about 6–24 hours (1–10 miles) before landing and shedding their wings. The queens that survive burrow into the soil to lay eggs. Over the next 25 days, the queen survives on the energy from wing muscles until worker ants mature enough to forage. A well-nourished queen produces up to 800 or more workers daily during her lifetime. Communication between ants results from the release of species-specific chemical pheromones as well as visual and vibrational stimuli. Fire ants contain magnetite, which orients the insect to north-south direction and attracts these insects to magnetic and electric fields present in underground power line. Fire ants fiercely defend their nest, stinging both humans and animals near their nest. A fire ant swarm can inflict 3,000–5,000 stings within seconds. Stings occur most commonly in the summer, particularly on the lower extremities of children.[4] While foraging for food or seeking shelter, fire ants may invade dwellings and attack infants or frail residents of long-term health care facilities.[12–14]

PRINCIPAL TOXINS

Venom Composition

Insect venoms have unique allergens as well as homologous allergen with partial sequence identity.[15] Consequently, multiple sensitivities to different insects may occur in an individual patient as a result of antigenic cross-reactivity of venom allergens or multiple exposures.

Bees

Honey bee venom is a complex mixture of vasoactive amines (dopamine, histamine, norepinephrine), enzymes (phospholipase A_2, hyaluronidase, acid phosphatase, lysophospholipase, α-D-glucosidase), and toxic peptides (melittin, apamin, protease inhibitor, secapin, peptide 401).[16] Histamine occurs in concentrations up to 1.5% of the venom. Peptide 401 is a mast cell-degranulating peptide that releases histamine and other vasoactive peptides. Melittin is a membrane-active polypeptide containing 26 amino acid residues. This substance causes local pain via histamine release and enhances phospholipase A_2-induced intravascular hemolysis as well as catecholamine release. Melittin disrupts membranes and then phospholipase A_2 cleaves bonds in the fatty acid portion of the bilipid membrane layer. Melittin accounts for about 50% of the dry weight of honey bee venom. The concentration of phospholipase A_2 in honey bee venom averages approximately 10% (dry weight). However, there is substantial variation in phospholipase A_2 content between individual European honey bees and between different honey bee colonies. In a study of phospholipase A_2 concentrations of 103 bees from approximately 10 colonies, the concentration of this enzyme varied between 1.8% and 27.4% (wt/wt).[17]

Hyaluronidase allows the spread of the venom through the dermis by disrupting the hyaluronic acid connective-tissue matrix and altering cell membranes. Apamin is a unique 18-amino-acid, single-chain structure with neurotoxic properties in small animals,[18] but the role of this toxin in humans has not been determined. In animal models, apamin in a convulsant that induces hyperactivity by blocking Ca^{++}-dependent potassium channels in the spinal cord and brain.[19] The venom of Africanized bees is similar to the venom of European honey bees with the latter containing slightly more hyaluronidase and less volume.[20] Consequently, the greater toxicity associated with Africanized honey bees results from the greater number of stings compared with European honey bees rather than a difference in the potencies of the venoms.

Although the composition of venom varies substantially among families within the order Hymenoptera, there are some homologous venom allergens.[21] Honey bee venom contains at least six allergens that include acid phosphatase, the glycoprotein phospholipase A_2 (PLA_2), the peptide melittin, apamin, peptide 401, and hyaluronidase.[22] Phospholipase A_2 and hyaluronidase are the major immunogenic components of honey bee venom that induce IgE-mediated anaphylaxis. The honey bee queens use venom only against other queens, and their venom does not contain the major allergens present in the venom of the worker honey bees.[16]

Bumble bee venom also contains venom protease, acid phosphatase, hyaluronidase, and PLA_2.[23] The latter allergen is structurally distinct from vespid phospholipase A_1 and partially cross-reactive with honey bee PLA_2.[24]

Wasps and Hornets

Vespid venom has some structural similarities to bee venom, but the variability of components in wasp venoms is greater than in bee venoms. Phospholipase A_2 and melittin occur only in bee venom, whereas antigen 5 occurs only in vespid venom. Both wasp and hornet venoms also contain phospholipase B.[25] Studies of hornet venom (*Vespula arenaria* Fabricius, *Vespula maculata* L.) indicated that the antigenic and allergic properties of vespid venom did not change significantly during the peak season (summer) of allergic responses to Hymenoptera stings.[26] *In vitro* studies indicate that venom from the Oriental hornet (*Vespa orientalis* L.) contains anticoagulants that prolong the activated partial thromboplastin time (aPTT) and, to a lesser extent, the prothrombin time (PT). However, there was no evidence of fibrinogenolytic or antithrombin activity.[27] Case reports of serious toxic reactions following multiple oriental hornet stings have not documented bleeding or clinically significant prolongation of the aPTT or the PT.[28]

Vespid venoms include at least four allergens (antigen 5, hyaluronidase, PLA_1, a protease) and two low-molecular-weight proteins (kinin, mastoparan). The peptide, mastoparan (CAS RN: 72093-21-1), causes histamine release by degranulating mast cells. The biological significance of antigen 5 is unclear. Although honey bee and vespid venoms are relatively distinct antigenically, both venoms contain immunologically cross-reactive hyaluronidase. The sequence identity of vespid and bee venom hyaluronidases is approximately 55%.[24] Patients sensitized to wasp venom are not usually allergic to bee venom.[29]

Fire Ants

In contrast to winged Hymenoptera venoms, fire ant (genus *Solenopsis*) venoms contain primarily water-insoluble alkaloids (solenopsins) with distinctive piperidine bases. These venoms contain only small amounts (i.e., about 0.1%) of proteins. The alkaloid portion of the venom consists of 2,6-dialkylpiperidine compounds that produce mast cell degranulation and histamine release via a nonallergenic mechanism.[30] There are age- and size-related changes in the alkyl and alkenyl groups in the venom from different fire ant workers as well as differences in isomers between different species.[31] Imported red fire ant venom contains at least four antigens (Sol i 1-4) with two of these antigens being homologous to vespid antigen 5s protein (Sol i 3) and phospholipase A_1 (Sol i 1). The venoms of the imported red and black fire ants are highly cross-reactive, but the venom of the imported black fire ant lacks the Sol i 4 protein.[32,33] Yellow jacket (*Vespula*) venom cross-reacts the Sol i 1 component of fire ant (genus *Solenopsis*) venom, and systemic allergic responses may occur in individuals sensitized to *Vespula* venom following the first-documented fire ant sting.

Venom Apparatus

Venomous members of the order Hymenoptera (ants, bees, wasps) inject venom via a tapered, posterior structure that is attached to a venom gland. Scorpions have a similar venom apparatus, but insects inject venom through their mouth parts. The sting is designed primarily as a defensive mechanism to discourage contact with the organism.

Winged Hymenoptera

The stinger of the European honey bee contains two reverse-barbed lancets connected to a venom sac, two membranous pump diaphragms, nerve ganglions, and a muscle. The curved barbs provide one-way traction that causes the barb to penetrate deeper into tissue as a result of the action of the detached muscles. Figure 168.3 demonstrates the stinging mechanism of the honey bee. A valve and piston on the proximal ends of the lancets forces the venom (i.e., about 50 µg) from the venom sac into an opening near the tip of the lancets. Unlike the stinging apparatus of other members of the order Hymenoptera, the stinging apparatus separates from the abdomen during the sting. Withdrawal of the stinger from the victim causes the tearing of the distal segment of the abdominal wall of the bee. During this process, the release of alarm pheromones (e.g., isoamyl acetate) from the area of the sting attracts other honey bees to the site of the sting. These pheromones evaporate from the surface of the sting with a banana-like smell. The eviscerated bee soon dies, but the reflex action of the pulsating muscles continues to inject venom. The Africanized honey bee has a smaller venom gland than the European honey bee.[20] In contrast to honey bee venom apparatuses, vespid stingers do not usually detach from their body during envenomation.[16]

The most primitive members of the order Hymenoptera (i.e., wasps) possess ovipositors to insert eggs into plant tissue. In some parasitic wasps, this structure evolved into a stinging apparatus to inject venom to obtain prey for developing larvae. Specialization of the

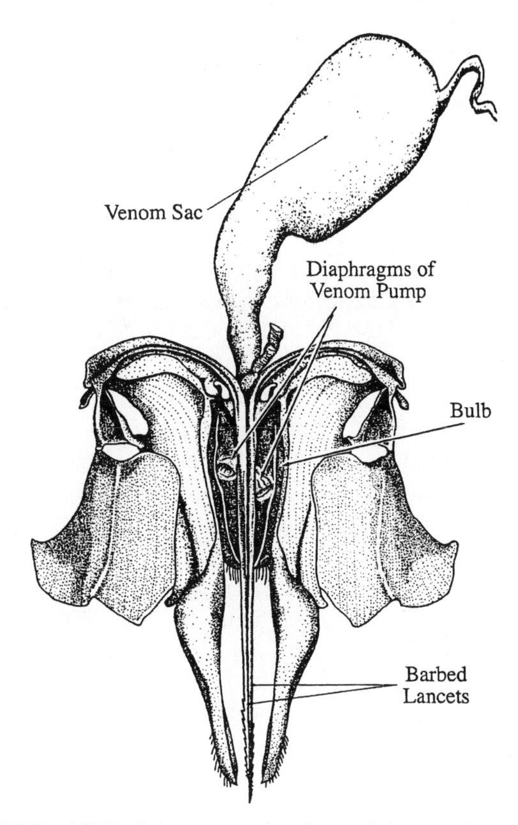

FIGURE 168.3. Stinger mechanism of honey bee. From Vetter RS, et al. Bites and stings of medically important venomous arthropods. Int J Dermatol 1998;37:484. Reproduced with permission from John Wiley & Sons Limited.

stinging apparatus by modern wasps resulted in an exclusively defensive weapon. Bees and ants evolved from the wasps, and the stinging apparatuses of these more modern organisms are primarily defensive.[16] The stinging apparatus of the wasp family lacks the barbs present on the stingers of honey bees. This feature allows the wasp to withdraw the stinger easily and to reinsert the stinger. Hence, multiple stings may occur from a single wasp, and retained wasp stingers are uncommon.

Fire Ants

Fire ant workers have well-developed stinging organs. The venom apparatus of the imported fire ant (*Solenopsis* spp.) is located on the distal abdomen as a modification of the ovipositor. Most ants (e.g., harvester or agricultural ants) grasp the skin with their jaws, and then these insects sting or spray locally irritating venom. During envenomation, the fire ant attaches to the skin with powerful mandibles, arches its body, and injects about 0.04–0.11 µL of venom containing approximately 10–100 ng proteins.[34] Compared with winged Hymenoptera, the sting is slower and lacks the immediate onset of pain. When undisturbed, the fire ant may rotate after

TABLE 168.1. Types of Allergic Reactions

Type of Reaction	Clinical Features
Limited, local reaction	Pain, erythema, and swelling in immediate area of sting
Large, local reaction	Spread of local reaction *contiguously* beyond the local site of sting
Anaphylaxis	Urticaria, angioedema, dyspnea, bronchospasm, lightheadedness, hypotension
Delayed hypersensitivity	Maculopapular rash with serum sickness (arthralgias, malaise)

20–30 seconds and inflict several more stings in a circular pattern.[35] This maneuver produces a characteristic arc-shaped lesion.

Dose Response

About 50–100 simultaneous Hymenoptera stings produce systemic symptoms in a nonsensitized adult, whereas the estimated fatal dose of bee venom involves 500–1500 stings. Most patients survive attacks from hundreds of wasps or approximately 1,000 bees,[36] and case reports document the survival of patients stung over 2,000 times.[37] A 52-year-old man developed transient hypotension and acute tubular necrosis after being stung by an estimated 1,200 Africanized bees.[38] He recovered with minimal renal dysfunction. A 30-year-old man survived over 2,000 honey bee stings following the development of shock and multiorgan failure.[39] Envenomation of a six-year-old boy by at least 60 Oriental hornets (*Vespa orientalis* L.) was associated with intravascular hemolysis, rhabdomyolysis, acute renal failure, and thrombocytopenia. The child survived without sequelae after peritoneal dialysis.[28] Fatal Hymenoptera evenomation in allergic individuals typically involves one to four stings.[40] The venoms of the Africanized honey bee and the European honey bee are similar; therefore, toxic responses from multiple envenomations by the Africanized honey bee are typically a dose-related phenomenon.[41] Fire ant stings do not usually produce toxic reactions despite massive numbers (>5,000–10,000) of stings.[42,43]

HISTOPATHOLOGY AND PATHOPHYSIOLOGY

Mechanism of Toxicity

Table 168.1 lists the four types of allergic reactions associated with Hymenoptera envenomation. Most allergic responses to Hymenoptera envenomation are either local toxic responses or systemic type I immediate hypersensitivity reactions mediated by IgE antibodies directed against the specific venom.[29] Limited local responses result from release of biological amines in the venom rather than an IgE-mediated response.[44] Although IgE is probably the primary immunoglobulin responsible for anaphylaxis, homocytotropic IgG or activated complement contribute to the degranulation of mast cells and basophils. Sensitized individuals synthesize excessive IgE antibodies when re-exposed to Hymenoptera venoms. Such antibodies cause the release of vasoactive substances that cause hypotension and smooth muscle contraction.

Type III (Arthus) hypersensitivity produces delayed reactions mediated by precipitating IgG and IgM antibodies. These proteins form immune complexes with the antigen that subsequently fix complement. Local tissue damage may result when platelet aggregates form on vascular endothelium and produce thrombi. Following massive envenomation by winged Hymenoptera, dose-related toxic reactions may result from the injection of large amounts of vasoactive substances. Toxic reactions do not usually develop following multiple stings by imported fire ants.

Pathological Changes

Massive Envenomation

Massive envenomations by species of winged Hymenoptera are associated with renal damage secondary to acute tubular necrosis, myoglobinuria, and hemoglobinuria. Histological changes following massive Hymenoptera envenomation include hyaline membrane disease, fatty degeneration of myocardium, liver, and kidneys, acute tubular necrosis, splenic hemorrhage, and cerebral vascular thromboembolism.

Fire Ant Lesions

The pustule formed by fire ant stings is pathognomonic for fire ant envenomation. Within the first hour, early spongiosis of the epidermis occurs with a surrounding red halo deepening in color following dilation of superficial dermal blood vessels.[45] The central papule evolves into a pustule as the overlying epidermis erodes during the first 24 hours. About one-half of these stings produce a biphasic reaction characterized by the resolution of the initial wheal-and-flare reaction followed within about 6 hours by a hypersensitivity reaction (edema, erythema, pruritus, pain) that peaks at 24 hours.[46] After 48–72 hours, healing begins from the base of the pustule. Microscopic examination of the lesions demonstrates a late-phase, mast-cell-dependent reaction characterized by infiltration of eosinophils and neutrophils along with the deposition of fibrin.[46] Superficial scars may form over the immediate area of atrophic skin.[47] A variety of ant species are capable of causing serious anaphylactic reactions in addition to fire ant species (*Solenopsis* spp.) including ant species belonging to at least six different subfamilies (Formicinae, Myrmeciinae, Ponerinae, Ectatomminae, Myrmicinae, Pseudomyrmecinae) and nine genera (*Formica, Hypoponera, Myrmecia, Odontomachus, Pachycondyla, Pogonomyrmex, Pseudomyrmex, Rhytidoponera, Tetramorium*).[48]

Postmortem

Most fatalities from Hymenoptera envenomation result from anaphylaxis, primarily massive edema, bronchoconstriction, and secretions causing bronchiolar obstruction.[49] Less often anaphylactic shock, airway obstruction, and vascular events (myocardial infarction, cerebral hemorrhage) cause death. There are no pathognomonic signs of anaphylaxis and postmortem findings in these cases of anaphylaxis are usually nonspecific.[50] Marked angioedema and laryngeal edema may or may not be present during autopsy.[40] The presence of stings or local reactions may be difficult to detect on postmortem examination despite a careful search, except following imported fire ant envenomation.[51,52] Common abnormalities present in postmortem examinations of victims of anaphylaxis include hyperemia, edema and hemorrhage of the larynx, pulmonary congestion and hemorrhage, and pleural petechiae.[40] Postmortem findings in fatal cases of anaphylaxis following fire ant envenomation involve characteristic skin lesions as well as increased blood concentrations of imported fire ant venom-specific IgE antibodies and tryptase.

CLINICAL RESPONSE

The most common response to winged Hymenoptera envenomations is a local reaction manifest by a sharp, stabbing pain followed within several minutes by intense, burning pain. The release of vasoactive amines causes a toxin-related wheal and flare reaction around the sting characterized by central erythema and a circumferential, white ring and edema. In most patients, the edema usually resolves within several hours. Pruritus and a warm sensation may persist several hours.

Fire ant envenomations produce a wheal-and-flare reaction within 20 minutes, followed in 24 hours by the formation of a characteristic, painful, sterile pustule. This lesion usually resolves within 3–10 days. Characteristically, multiple bites occur in a circular pattern; however, the absence of pustules does not rule out fire ant bites. Sensitized individuals may develop large local reactions (erythema, edema). The differential diagnosis of fire ant lesions includes blister beetle bites (bullous) and the nonpruritic, nonpustular papular eruptions of *Molluscum contagiosum* (poxvirus).

Allergic Reactions

Large Local Reactions

Hymenoptera envenomations can produce an exaggerated response secondary to an allergic response manifest by a lack of systemic complaints and contiguous swelling that extends from the site of envenomation. These large local reactions seldom cause serious sequelae unless obstruction of vital structures (upper airway, esophagus) occurs. Severe local symptoms (pruritus, pain) may develop during the first 48–72 hours after envenomation and resolution may require up to one week. Vesiculation may occur in dependent areas of the extremities. An allergic response is usually associated with less pain and more pruritus compared with an infectious process. Painful, red streaks radiating up the extremity indicate the occurrence of lymphangiitis, particularly following yellow jacket stings. Rarely, case reports associate bullae formation, rhabdomyolysis, and myoglobinuria with large local reactions following a single hornet envenomation.[53] Acute bullous lesions may develop after fire ant envenomation, which progress to scar formation.[54]

Anaphylaxis

Approximately 0.4–4% of a random adult population report systemic allergic responses after stings by winged Hymenoptera.[55] Predicting which adults will develop severe anaphylactic reactions is difficult, and children appear to develop less severe, recurrent systemic reactions than adults.[56] Less than one-half of adults with positive venom skin tests to Hymenoptera and previous severe allergic reactions (laryngeal edema, severe bronchospasm, hypotension) will develop the same severe reaction after a subsequent hymenoptera sting.[57] Systemic symptoms of an immediate hypersensitivity reaction range from mild (urticaria, diffuse erythema, angioedema, nausea, vomiting, pruritus, anxiety) to severe (hypotension, laryngeal edema, severe bronchospasm, dyspnea). Systemic symptoms usually begin within 10 minutes after envenomation.[29] The presence of a sense of impending doom, wheezing, hoarseness, profound weakness, or confusion indicate a severe reaction, particularly during the first hour after envenomation. Similar anaphylactic reactions also occur in individuals with sensitivity to bites of the imported red fire ant including the occurrence of anaphylaxis without a history of prior sensitization of Hymenoptera stings.[58] However, the incidence of anaphylactic reactions per sting is less compared with winged Hymenoptera.[59]

Delayed Hypersensitivity

Delayed symptoms of serum sickness may occur about 7–14 days after envenomation. The clinical features of delayed hypersensitivity reactions include fever, malaise, lymphadenopathy, headache, arthralgias, myalgias, and urticaria.[60] Local tissue necrosis may result from delayed hypersensitivity.

Fatalities

Annually, anaphylactic responses to Hymenoptera envenomations account for approximately 40–50 US deaths, but the actual number of fatalities may be higher as a result of unsuspected deaths related to anaphylaxis.[61] Frequently, these patients do not have a history of prior anaphylactic reactions.[49] Symptoms of a severe allergic reaction typically develop immediately, and death occurs within one hour of envenomation secondary to respiratory obstruction or anaphylactic shock. Vascular events including coronary artery occlusion or cerebral thromboembolism may contribute to fatal outcomes in susceptible patients. Fatalities from massive Hymenoptera envenomations usually involve multiorgan failure including intravascular hemolysis, respiratory distress syndrome (ARDS), hepatocellular necrosis, acute renal failure with myoglobinuria and hemoglobinuria, focal subendocardial necrosis, and disseminated intravascular coagulation.[62] Fatalities secondary to toxic reactions from multiple fire ant stings are not well documented, and fatalities from fire ant envenomations usually result from anaphylaxis after a few stings.[32,63]

Toxic Reactions

Severe, immediate toxic reactions may result from massive envenomation by bees, hornets or wasps, particularly Africanized honey bees. The clinical features of a massive envenomation are similar to anaphylaxis, and symptoms include fatigue, headache, weakness, fever, muscle spasm, diarrhea, nausea, vomiting, lightheadedness, and transient loss of consciousness. Rarely, convulsions may occur usually along with multiorgan failure.[64] Within 24 hours, hemolysis, hemoglobinuria, rhabdomyolysis, and elevation of serum hepatic aminotransferases can develop.[65] The presence of hypotension, rhabdomyolysis, and hemoglobinuria may cause acute tubular necrosis, hyperkalemia, and renal failure.[66,67] Subendocardial damage and elevation of serum cardiac enzymes can occur in the absence of hypotension or anaphylaxis.[68] Serious complications of massive envenomations from vespids and bees also include intravascular hemolysis, disseminated intravascular coagulation, hepatic necrosis, acute renal failure secondary to hemolysis, rhabdomyolysis, or direct nephrotoxicity,[69] intra-cranial hemorrhage, and myocardial infarction.[70,71] Case reports associate the development of acute interstitial nephritis after wasp stings secondary to hypersensitivity response in the absence of hemolysis or rhabdomyolysis.[72]

Signs of multiorgan failure after massive envenomation (>50–100 stings) may be delayed up to 18 hours, particularly in older patients with underlying systemic diseases.[73] Although case reports associate some neurologic disorders (multiple sclerosis, optic neuritis, myasthenia gravis, Guillain-Barré syndrome, Parkinson's disease) with recent hymenoptera envenomation, the causal role of Hymenoptera venom in the etiology of these neurological diseases is questionable. Fire ant stings do not usually produce toxic reactions despite massive numbers of stings.[42,74] Rarely, seizures are associated with fire ant stings, usually in the setting of anaphylaxis.[75] Complications of fire ant stings include secondary infection, primarily in immunocompromised hosts (e.g., diabetics, peripheral vascular disease), and rarely cutaneous sporotrichosis (dimorphic fungus).[76] Rare case reports associate massive fire ant envenomation with anaphylaxis, altered consciousness, rhabdomyolysis, and acute renal failure.[77]

Reproduction

Although there are few data on anaphylaxis during pregnancy, Hymenoptera-related anaphylaxis is a potential cause of fetal abnormalities and fetal loss. A case series of pregnant mothers with a history of anaphylaxis did not detect any significant abnormalities after anaphylaxis (one case) or the use of venom immunotherapy (26 women, two mild adverse reactions).[78]

DIAGNOSTIC TESTING

Analytical Methods

Hypersensitivity Testing

The gold standard for detecting hypersensitivity to Hymenoptera envenomations is intradermal testing. Venom-specific IgE antibody measurements with radioallergosorbent tests (RAST) or nonisotopic variants (e.g., the capsulated hydrolic carrier polymer radioallergosorbent test or CAP-RAST System) are alternative diagnostic methods to intradermal testing. Although 50–80% of patients with large local reactions to Hymenoptera envenomations will demonstrate positive skin tests or venom-specific antibody measurements, only about 1–2% of these patients will develop life-threatening systemic reactions to subsequent stings.[79] The presence of venom-specific IgE does not imply that an allergic response will occur following each sting. Venom-specific IgE eventually disappears from the blood, although the elimination of venom-specific IgE may require years following envenomation.

About 30% of adult patients with clear clinical histories of anaphylaxis following envenomation by Hymenoptera species do not demonstrate positive response to intradermal testing.[22] However, some of these patients do demonstrate positive venom-specific IgE-antibody serology, such as RAST or the CAP-RAST System (Pharmacia Diagnostics, Uppsala, Sweden).

Postmortem

Positive RAST tests and serum tryptase concentrations confirm the presence of hypersensitivity in postmortem serum samples.[80] The radioallergosorbent test (RAST) for venom-specific IgE is positive in most, but not all postmortem samples from patients dying from anaphylaxis; however, the concentration of venom-specific IgE antibodies does correlate to the severity of the anaphylactic reaction. In a study of postmortem samples from fatal anaphylactic reactions, the radioallergosorbent test did not detect specific IgE antibodies in about 10% of the victims.[81] Tryptase is mast-cell-specific enzyme released during the degranulation of the mast cell. Peak concentrations of tryptase occur within 1–2 hours after anaphylaxis begins, and the serum half-life of tryptase is about 2 hours.[82] Although this test is highly specific for anaphylaxis in antemortem specimens,[83] the lack of specificity in postmortem samples indicates that this test must be interpreted with other evidence of anaphylaxis (venom-specific IgE, clinical history, differential diagnosis, pathological findings).[84]

Abnormalities

Intravascular hemolysis may occur after massive Hymenoptera evenomation. Case reports associate transient, nonspecific ST-T wave changes, elevated CPK-MB fractions, and myocardial wall motion abnormalities with winged Hymenoptera envenomation despite the presence of normal coronary arteries.[85] These changes usually resolve within about one week without complications.[68]

TREATMENT

Stabilization

Anaphylaxis

Anaphylactic reactions are medical emergencies that require the prompt use of epinephrine. The most important therapeutic measures are the prompt use of epinephrine, the maintenance of blood pressure, and the support of oxygenation. Most deaths occur within the first hour, and these fatalities typically result from respiratory failure in patients who did not received parenteral epinephrine. Patients with systemic symptoms of anaphylaxis should immediately receive oxygen, intravenous access, cardiac monitoring, and ventilatory support as needed. Substantial fluid replacement may be necessary in addition to epinephrine to treat hypotension. Patients with massive envenomation (>50–100 stings) may develop signs of anaphylaxis, and the treatment is similar to the treatment for anaphylaxis.

Ventilatory Support Patients with signs of upper airway edema should immediately receive parenteral epinephrine and oxygen. Inhaled racemic epinephrine may transiently reduce upper airway edema, but the limited absorption of epinephrine by the respiratory route limits reversal of other systemic symptoms. Bronchospasm should be treated with inhaled β_2-agonists (salbutamol, albuterol).

Epinephrine Epinephrine is the treatment of choice for anaphylaxis, and all patients with signs of a systemic response (generalized urticaria, angioedema, respiratory distress, shock) should receive epinephrine immediately.[86] For patients without obvious signs of shock, 0.1 mL/kg up to 0.3–0.5 mL of 1:1,000 epinephrine (adult dose: 0.3–0.5 mg) should be administered subcutaneously or intramuscularly. Studies in adults without anaphylaxis indicate that intramuscular epinephrine into the deltoid muscle produces higher and earlier peak plasma epinephrine concentrations higher than subcutaneous epinephrine.[87] Repeat doses of epinephrine at 5–15 minute intervals may be necessary to reduce symptoms and maintain blood pressure because of the short elimination half-life of epinephrine.[88]

Patients with cardiovascular disease and those over 40 years of age require careful cardiac monitoring for evidence of ischemia or dysrhythmias. Patients receiving β-blocker drugs may demonstrate a blunted response to epinephrine. The use of the mixed α and β agonist, epinephrine, in the presence of β blockade, produces primarily α-adrenergic effects, resulting in release of vasoactive mediators. Hence, patients with β blockade may be refractory to epinephrine and these patients may require an intravenous isoproterenol drip infused at a higher dose than the usual adult dosage of 1–20 μg/min.[89] The administration of glucagon is an option for these patients. The recommended dose for glucagon is 1–5 mg (20–30 μg/kg up to 1 mg for children) intravenously over 5 minutes followed by an infusion (5–15 μg/min) titrated to clinical effect.[86] The patients should be observed closely for the onset of emesis during the administration of glucagon.

Refractory Hypotension Aggressive fluid resuscitation is necessary for severely hypotensive patients unresponsive to epinephrine. Large volumes of crystalloid or colloid may be necessary to correct fluid extravasation and vasodilatation resulting from anaphylaxis. Intravenous epinephrine is an option for severe hypotension unresponsive to fluid infusion or cardiac arrest associated with anaphylaxis. There are few data on the clinical efficacy of intravenous epinephrine during the treatment of anaphylaxis, and intravenous epinephrine must be administered judiciously because of the potential for lethal arrhythmias. Epinephrine mixtures in resuscitation carts are usually a 1:10,000 dilution. For cardiac arrest *only* secondary to anaphylaxis, the recommended dose of epinephrine is 0.1–0.5 mg intravenously as a bolus compared with the standard 1-mg intravenous dose of epinephrine (10 mL of the 1:10,000 typically on crash carts) recommended by the American Heart Association for cardiac arrest. For patients with refractory hypotension, an intravenous epinephrine infusion by an infusion pump is an appropriate option. An epinephrine solution of 10 μg/mL (1 mg in 100 mL saline) is infused at 5–15 μg/min in adults according to the severity of the reaction.[90] The rate is titrated up or down based on response and side effects (i.e., tachycardia, tremor, pallor in a setting of normal or increased blood pressure). The infusion is stopped 30 minutes after resolution of the symptoms and signs of anaphylaxis. Vasopressors with strong α-adrenergic agonist properties (e.g., norepinephrine) may be necessary to support blood pressure if the patient does not respond to fluids and parenteral epinephrine. Case reports suggest that vasopressin (10–40 IU, intravenously) may restore blood pressure in patients with severe anaphylactic shock and no palpable blood pressure,[91] but there are few clinical data to determine if vasopressin is superior to norepinephrine for severe hypotension.

Antihistamines Antihistamines are slower in onset than epinephrine, and these H_1 and H_2 antagonists improve the cutaneous manifestations of anaphylaxis rather than hypotension.[86] Diphenhydramine hydrochloride (Benadryl®; Johnson & Johnson Consumer Companies, Inc., Skillman, NJ) should be administered in a dosage of 1 mg/kg (50–100 mg in an adult) either intramuscularly or intravenously depending on the adequacy of circulation. Frequently, an H_2-antihistamine blocker (e.g., ranitidine, 1 mg/kg in adults; 12.5–50 mg infused over 10 minutes for children) is administered intravenously for the treatment of cutaneous manifestations of anaphylaxis,[92] but there are few clinical data on the efficacy of adding this drug to the outcome of anaphylaxis. Cyproheptadine is an alternative to diphenhydramine for the treatment of persistent local reactions.[53]

STEROIDS Corticosteroids are frequently administered to reduce the effects of delayed hypersensitivity reactions, despite the slow onset of action of these drugs. Typical steroid doses for serious systemic symptoms include are methylprednisolone 1–2 mg/kg intravenously.

Massive Envenomation

Toxic reactions resemble some of the manifestations of anaphylaxis, and treatment, in general, is similar to the treatment of anaphylaxis including antihistamines (H_1- and H_2-antagonists), epinephrine, steroids, and respiratory support as needed. The infliction of >50 stings in an adult or >1 sting/kg in a child indicated the need for at least 24 hours of monitoring of serum chemistries, serum creatine phosphokinase, and adequacy of urine output.[65] These patients should be evaluated for the development of renal failure, hepatic necrosis, intravascular hemolysis, and rhabdomyolysis. The presence of serum creatine phosphokinase concentrations exceeding 1000 μ/L indicates the need for generous fluid therapy to prevent renal damage.

Decontamination

Bees, but not wasps or hornets, may leave a stinger in the skin. Any stinger should be scraped off as soon as possible because the rhythmic action of the detached muscles continues for several minutes after envenomation. Experimental studies suggest that at least 90% of the venom sac contents are injected within 20 seconds and that venom delivery is almost complete within one minute.[93] Efforts to remove the stinger without touching the venom sac should *not* preclude the rapid removal of the stinger or interfere with treatment priorities.[94]

Supplemental Care

Patients with life-threatening allergic symptoms should be admitted for at least 24–48 hours of observation because of the potential for the onset of delayed biphasic reactions in up to 20% of patients with anaphylaxis.[95] Patients with massive envenomation (50–100 stings) by winged Hymenoptera may develop delayed signs of multiorgan failure, and these patients should be observed for 24 hours, particularly elderly patients or patients with underlying systemic diseases.[73] These patients should be evaluated for the presence of hemolysis, thrombocytopenia, hepatorenal dysfunction, rhabdomyolysis, and disseminated intravascular coagulation (DIC). Blood tests include complete blood count with differential, hepatic aminotransferase, electrolyte panel, DIC screen, creatinine, and creatine phosphokinase as well as a urinalysis. These tests should be repeated every 6 hours during the first day of admission.

Local Wound Care

The skin surrounding the sting may be cleaned with an antiseptic agent (e.g., povidone–iodine; Betadine™, Purdue Pharma, Stamford, CT) after inspection to ensure that the stinger has been removed. Remove constrictive jewelry immediately because of the potential for rapid swelling. Cold compresses, calamine lotion, and oral antihistamines (diphenhydramine, hydroxyzine) may help relieve pruritus and pain. A randomized, controlled trial of topical aspirin indicated that the application of a topical aspirin paste did not reduce the duration of swelling or pain after bee and wasp stings.[96] Systemic steroids are appropriate for patients with large local reactions (40 mg prednisone for 3 days), systemic symptoms, or delayed hypersensitivity reactions. Animal studies suggest that topical and systemic steroids, antibacterial medications, and antihistamines do not significantly alter the clinical course of fire ant stings,[97] but there are few clinical data in humans to confirm experimental data. Fingernails of children should be trimmed to reduce excoriation of the wound. The presence of pruritus and the absence of tenderness around the vesicles distinguish the sterile pustules from secondary infection.

Prevention

Patients with serious systemic reactions should receive a prescription of epinephrine self-injectors. EpiPen® and EpiPen JR® (Dey Inc., Napa, CA) are designed for self-administration of 0.3 and 0.15 mg epinephrine, respectively, subcutaneously at home, work, or recreational facilities.[98] These patients should wear a warning bracelet or tag at all times.

Sensitized individuals should wear muted, tight clothing that covers their extremities including their feet when outdoors. Bright, flowered clothing and attractive scents, soaps, and shampoos should be avoided. Long-sleeved clothing should be worn, especially for garden work. Materials or plants (e.g., clover, dandelions, uncovered sweet drinks) that attract stinging Hymenoptera species should be eliminated where possible, and Hymenoptera nests removed from close contact with living spaces of sensitized individuals. When confronted with winged Hymenoptera species, excessive noise and motion should be avoided. Gently brush off the insect from the skin. Slowly retreat or stand still unless attacked by a swarm. When attacked by a swarm, run as quickly as possible in a zig-zag pattern to an indoor area (i.e., car, house). The face should be protected to avoid stings to the nose, mouth, and throat. Insect repellents (e.g., *N,N*-diethyl-3-methylbenzamide or *N,N*-diethyl-*meta*-toluamide [DEET]) do not repel stinging insects, such as bees, wasps, or fire ants.

Immunotherapy

Most authorities recommend desensitization with venom immunotherapy for those patients with previous immediate serious systemic symptoms (respiratory distress, asthma, marked hypotension, sudden collapse), and occasionally for patients with moderate systemic reactions (asthma, diffuse angioedema).[29,99] Commercial preparations of venom from honey bees, yellow jackets, white-faced hornets, yellow hornets, and wasps are available for diagnosis and venom immunotherapy. In contrast to winged Hymenoptera venom therapy, commercial fire ant immunotherapy involves only whole body extracts because of the lack of commercial fire ant venom extract.[100,101]

Mechanism of Action The exact mechanism for the reduction of symptoms following immunotherapy with increasing doses of insect venom is not known. Immune responses to protein antigens depend on the specific recognition of antigenic determinants (epitopes) by T and B lymphocytes. For example, bee venom phospholipase A_2 is the major antigen (allergen) of honey bee venom with three dominant immunogenic peptide and one glycopeptide T cell recognition sites. Specific antigen recognition by B cells can trigger distinct cytokine profiles in T cells leading to increased interleukin 4 (IgE synthesis), interleukin 5 (activation/attraction of eosinophils), and interleukin 3 (activation/attraction mast cells). Following immunotherapy, reduction of cytokine release by T_2-helper cells and increased cytokine release by T_1-helper cells occurs.[102] The reduced concentration of T_2-helper cells reduces the release of interleukins 3, 4, and 5. Additionally, increased T_1-helper cells increase the concentration of interleukin 2 and interferon gamma. The latter opposes the effects of interleukin 4 and reduces the secretion of venom-specific IgE.

Adverse Reactions In general, the use of venom immunotherapy is efficacious and the administration of immunotherapy does not usually cause serious adverse effects.[103,104] Mild systemic reactions may occur in up to 15% of patients during the initial course, but these patients are treated effectively with epinephrine and antihistamines.[29] Local reactions are common, but these reactions diminish with continued immunotherapy. Rarely, anaphylactic reactions may occur.

Indications Venom immunotherapy is a safe, effective therapy for the prevention of recurrent systemic reactions from Hymenoptera envenomations in patients with positive skin tests to Hymenoptera venom[105] as well as some fire ant and native ant [e.g., jack jumper ant (*Myrmecia pilosula* Smith)].[106] Patients with large local reactions and positive skin tests for venom-specific IgE-specific antibodies are not candidates for immunotherapy because very few (i.e., 1–2%) of these patients will develop life-threatening reactions following subsequent stings.[22] The decision to use venom immunotherapy in patients with positive skin tests and no prior reaction or with negative skin tests and prior systemic reactions requires physician judgment and experience.

References

1. Ettershank G. A generic revision of the world Myrmicinae related to *Solenopsis* and *Pheidologeton* (Hymenoptera: Formicidae). Aust J Zool 1966;14:73–171.
2. Michener CD. The Brazilian bee problem. Annu Rev Entomol 1975;20:399–416.
3. Reisman RE. Insect stings. N Engl J Med 1994;331: 523–527.
4. Kemp SF, de Shazo RD, Moffitt JE, Williams DF, Buhner WA II. Expanding habitat of the imported fire ant (*Solenopsis invicta*): a public health concern. J Allergy Clin Immunol 2000;105:683–691.
5. Patterson R, Valentine M. Anaphylaxis and related allergic emergencies including reactions due to insect stings. JAMA 1982;248:2632–2636.
6. Vinson SB. Invasion of the red imported fire ant (Hymenoptera: Formicidae): spread, biology, and impact. Am Entomologist 1997;43;23–39.
7. Brown SG, Wu QX, Kelsall GR, Heddle RJ, Baldo BA. Fatal anaphylaxis following jack jumper ant sting in southern Tasmania. Med J Aust 2001;175:644–647.
8. Shek LP, Ngiam NS, Lee BW. Ant allergy in Asia and Australia. Curr Opin Allergy Clin Immunol 2004;4: 325–328.
9. Sherman RA. What physicians should know about Africanized honeybees. West J Med 1995;163:541–546.
10. Rodriguez-Lainz A, Fritz CL, McKenna WR. Animal and human health risks associated with Africanized honeybees. JAVMA 1999;215:1799–1804.
11. Tracy JM, Demain JG, Quinn JM, Hoffman DR, Goetz DW, Freeman TM. The natural history of exposure to the imported fire ant (*Solenopsis invicta*). J Allergy Clin Immunol 1995;95:824–828.

12. Goddard J, Jarratt J, deShazo RD. Recommendations for prevention and management of fire ant infestation of health care facilities. South Med J 2002;95:627–633.
13. deShazo RD, Williams DF, Moak ES. Fire ant attacks on residents in health care facilities: a report of two cases. Ann Intern Med 1999;131:424–429.
14. Rupp MR, deShazo RD. Indoor fire ant sting attacks: a risk for frail elders. Am J Med Sci 2006;331:134–138.
15. King TP, Guralnick M. Hymenoptera allergens. Clin Allergy Immunol 2004;18:339–353.
16. Vetter RS, Visscher PK. Bites and stings of medically important venomous arthropods. Int J Dermatol 1998;37:481–496.
17. Schumacher MJ, Schmidt JO, Egen NB, Dillon KA. Biochemical variability of venoms from individual European and Africanized honeybees (*Apis mellifera*). J Allergy Clin Immunol 1992;90:59–65.
18. Dotimas EM, Hamid KR, Hider RC, Ragnarsson U. Isolation and structure analysis of bee venom mast cell degranulating peptide. Biochim Biophys Acta 1987;911: 285–293.
19. Grissmer S, Tytgat J. Toxins and potassium channels. Toxicon 2004;43:863.
20. Owen MD. The venom system and venom hyaluronidase of the African honeybee (*Apis mellifera adansonii*). Toxicon 1983;21:171–174.
21. King TP, Spangfort MD. Structure and biology of stinging insect venom allergens. Int Arch Allergy Immunol 2000;123:99–106.
22. Hamilton RG. Diagnosis of Hymenoptera venom sensitivity. Curr Opin Allergy Clin Immunol 2002;2: 347–351.
23. Hoffman DR, El-Choufani SE, Smith MM, de Groot H. Occupational allergy to bumblebees: allergens of *Bombus terrestris*. J Allergy Clin Immunol 2001;108:855–860.
24. King TP, Lu G, Gonzalez M, Qian N, Soldatova L. Yellow jacket venom allergens, hyaluronidase and phospholipase: sequence similarity and antigenic cross-reactivity with their hornet and wasp homologs and possible implications for clinical allergy. J Allergy Clin Immunol 1996;98:588–600.
25. Watala C, Kowalczyk JK. Hemolytic potency and phospholipase activity of some bee and wasp venoms. Comp Biochem Physiol 1990;97C:187–194.
26. Reisman RE, Littler SJ, Wypych JI. Comparison of the biochemical, immunologic and allergenic properties of vespid venoms collected in early and late summer. Toxicon 1984;22:148–153.
27. Kornberg A, Kaufman S, Silber Lk, Ishay JS. Effect of venom sac extract of the oriental hornet (*Vespa orientalis*) on coagulation factors. Toxicon 1988;26:1169–1176.
28. Ghosh K, Singh S, Pereira DJG, Singhi SC. Acute systemic toxic reactions caused by hornet stings. Indian Pediatr 1988;256:796–798.
29. Ewan PW. Venom allergy. BMJ 1998;316:1365–1368.
30. Lind NK. Mechanism of action of fire ant (*Solenopsis*) venoms. I. Lytic release of histamine from mast cells. Toxicon 1982;20:831–840.
31. Deslippe RJ, Guo Y-J. Venom alkaloids of fire ants in relation to worker size and age. Toxicon 2000;38: 223–232.
32. Hoffman DR. Fire ant venom allergy. Allergy 1995;50: 535–544.
33. Hoffman DR. Reactions to less common species of fire ants. J Allergy Clin Immunol 1997;100:679–683.
34. deShazo RD, Butcher BT, Banks WA. Reactions to the stings of the imported fire ant. N Eng J Med 1990;323:462–466.
35. Caro MR, Derves VJ, Jung R. Skin responses to the sting of the imported fire ant (*Solenopsis invicta*). Arch Dermatol 1957:75:475–488.
36. Vetter RS, Visscher PK, Camazine S. Mass envenomations by honey bees and wasps. West J Med 1999; 170:223–227.
37. Murray JA. A case of multiple bee stings. Central Afr J Med 1964;10:249–251.
38. Beccari M, Castiglione A, Cavaliere G, d'Aloya G, Fabbri C, Losi B, et al. Unusual case of anuria due to African bee stings. Int J Artif Organs 1992;15:281–283.
39. Diaz-Sanchez CL, Lifshitz-Guinzberg A, Ignacio-Ibarra G, Halabe-Cherem J, Quinones-Galvan A. Survival after massive (>2000) Africanized honeybee stings. Arch Intern Med 1998;158:925–927.
40. Riches KJ, Gillis D, James RA. An autopsy approach to bee sting-related deaths. Pathology 2002;34:257–262.
41. Nelson DR, Collins AM, Hellmich RL, Jones RT, Helm RM, Squillace DL, Yunginger JW. Biochemical and immunochemical comparison of Africanized and European honeybee venoms. J Allergy Clin Immunol 1990;85: 80–85.
42. Diaz JD, Lockey RF, Stablein JJ, Mines HK. Multiple stings by imported fire ants (*Solenopsis invicta*), without systemic effects. South Med J 1989;82:775–777.
43. Smith JD, Smith EB. Multiple fire ant stings. A complication of alcoholism. Arch Dermatol 1971;103:438–441.
44. Riches HR. Insect bites and stings. Practitioner 1977;219: 199–203.
45. Goddard J, Jarratt J, de Castro FR. Evolution of the fire ant lesion. JAMA 2000;284:2162–2163.
46. deShazo RD, Griffing C, Kwan TH, Banks WA, Dvorak HF. Dermal hypersensitivity reactions to imported fire ants. J Allergy Clin Immunol 1984;74:841–847.
47. Prahlow JA, Barnard JJ. Fatal anaphylaxis due to fire ant stings. Am J Forensic Pathol 1998;19:137–142.
48. Klotz JH, deShazo RD, Pinnas JL, Frishman AM, Schmidt JO, Suiter DR, et al. Adverse reactions to ants other than imported fire ants. Ann Allergy Asthma Immunol 2005;95:418–425.
49. Barnard JH. Studies of 400 hymenoptera sting deaths in the United States. J Allergy Clin Immunol 1973;52: 259–264.

50. Delage C, Irey NS. Anaphylactic deaths: a clinicopathologic study of 43 cases. J Forensic Sci 1972;17:525–540.
51. Riches KJ, Gillis D, James RA. An autopsy approach to bee sting-related deaths. Pathology 2002;34:257–262.
52. Johansson B, Eriksson A, Ornehult L. Human fatalities caused by wasp and bee stings in Sweden. Int J Legal Med 1991;104:99–103.
53. Lin C-C, Chang M-Y, Lin J-L. Hornet sting induced systemic allergic reaction and large local reaction with bullae formation and rhabdomyolysis. J Toxicol Clin Toxicol 2003;41:1009–1011.
54. Albright DD, Napoli DC, Hagan LL. Acute bullous skin reactions after imported fire ant envenomation. Pediatr Asthma Allergy Immunol 2006;19:31–35.
55. Charpin D, Birnbaum J, Vervloet D. Epidemiology of hymenoptera allergy. Clin Exp Allergy 1994;24: 1010–1015.
56. Valentine MD, Schuberth KC, Kagey-Sobotka A, Graft DF, Kwiterovich KA, Szklo M, Lichtenstein LM. The value of immunotherapy with venom in children with allergy to insect stings. N Engl J Med 1990;323: 1601–1603.
57. van der Linden P-W, Struyvenberg A, Kraaijenhagen RJ, Hack CE, van der Zwan JK. Anaphylactic shock after inset–sting challenge in 138 persons with a previous insect-sting reaction. Ann Intern Med 1993;118:161–168.
58. Fernández-Meléndez S, Miranda A, García-González JJ, Barber D, Lombardero M. Anaphylaxis caused by imported red fire ant stings in Málaga, Spain. J Investig Allergol Clin Immunol 2007;17:48–49.
59. Solley GO, Vanderwoude C, Knight GK. Anaphylaxis due to red imported fire ant sting. Med J Aust 2002;176:521–523.
60. Reisman RE, Livingston A. Late-onset allergic reactions, including serum sickness, after insect stings. J Allergy Clin Immunol 1989;84:331–337.
61. Litovitz TL, Klein-Schwartz W, Rodgers GC Jr, Cobaugh DJ, Youniss J, Omslaer JC, et al. 2001 Annual report of the American Association of Poison Control Centers Toxic Exposure Surveillance System. Am J Emerg Med 2002;20:391–452.
62. Franca FO, Benvenuti LA, Fan HW, Dos Santos DR, Hain SH, Picchi-Martins FR, et al. Severe and fatal mass attacks by "killer" bees (Africanized honey bees—*Apis mellifera scutellata*) in Brazil: clinicopathological studies with measurement of serum venom concentrations. Q J Med 1994;87:269–282.
63. deShazo RD, Kemp SF, deShazo MD, Goddard J. Fire ant attacks on patients in nursing homes: an increasing problem. Am J Med 2004;116:843–846.
64. Watemberg N, Weizman Z, Shahak E, Aviram M, Maor E. Fatal multiple organ failure following massive hornet stings. Clin Toxicol 1995;33:471–474.
65. Betten DP, Richardson WH, Tong TC, Clark RF. Massive honey bee envenomation-induced rhabdomyolysis in an adolescent. Pediatrics 2006;117:231–235.
66. Mejia G, Arbelaez M, Henao JE, Sus AA, Arango JL. Acute renal failure due to multiple stings by Africanized bees. Ann Intern Med 1986;104:210–211.
67. Bourgain C, Pauti MD, Fillastre JP, Godin M, Francois A, Leroy JP, et al. Envenimation massive après piqûres d'abeilles africaines. Presse Med 1998;27: 1099–1101.
68. Ceyhan C, Ercan E, Tekten T, Kirilmaz B, Onder R. Myocardial infarction following a bee sting. Int J Cardiol 2001;80:251–253.
69. Vikrant S, Pandey D, Machhan P, Gupta D, Kaushal SS, Grover N. Wasp envenomation-induced acute renal failure: a report of three cases. Nephrol 2005;10: 548–552.
70. Chao S-C, Lee Y-Y. Acute rhabdomyolysis and intravascular hemolysis following extensive wasp stings. Int J Dermatol 1999;38:131–141.
71. Levine HD. Acute myocardial infarction following wasp sting. Report of two cases and critical survey of the literature. Am Heart J 1976;91:365–374.
72. Zhang R, Meleg-Smith S, Batuman V. Acute tubulointerstitial nephritis after wasp stings. Am J Kidney Dis 2001;38:E33.
73. Kolecki P. Delayed toxic reaction following massive bee envenomation. Ann Emerg Med 1999;33:114–116.
74. Reschly MJ, Ramos-Caro F, Mathes BM. Multiple fire ant stings: report of 3 cases and review of the literature. Cutis 2000;66:1779–182.
75. Candiotti KA, Lamas AM. Adverse neurologic reactions to the sting of the imported fire ant. Int Arch Allergy Immunol 1993;102:417–420.
76. Miller SD, Keeling JH. Ant sting sporotrichosis. Cutis 2002;69:439–442.
77. Koya S, Crenshaw D, Agarwal A. Rhabdomyolysis and acute renal failure after fire ant bites. J Gen Intern Med 2007;22:145–147.
78. Schwartz HJ, Golden DBK, Lockey RF. Venom immunotherapy in the Hymenoptera-allergic pregnant patient. J Allergy Clin Immunol 1990;85:709–712.
79. Portnoy JM, Moffitt JE, Golden DB, Bernstein IL, Berger WE, Dykewicz MS, et al. Stinging insect hypersensitivity: A practice parameter. J Allergy Clin Immunol 1999;103: 963–980.
80. Yunginger JW, Nelson DR, Squillace DL, Jones RT, Holley KE, Hyma BA, et al. Laboratory investigation of deaths due to anaphylaxis. J Forensic Sci 1991;36: 857–865.
81. Hoffman DR. Fatal reactions to hymenoptera stings. Allergy Asthma Proc 2003;24:123–127.
82. Schwartz LB, Yunginger JW, Miller J, Bokhari R, Dull D. Time course of appearance and disappearance of human mast cell tryptase in the circulation after anaphylaxis. J Clin Invest 1989;83:1551–1555.
83. Kucharewicz I, Bodzenta-Lukaszyk A, Szymanski W, Mroczko B, Szmitkowski M. Basal serum tryptase level

correlates with severity of hymenoptera sting and age. J Investig Allergol Clin Immunol 2007;17:65–69.

84. Edston E, van Hage-Hamsten M. beta-Tryptase measurements post-mortem in anaphylactic deaths and in controls. Forensic Sci Int 1998;93:135–142.
85. Brasher GW, Sanchez SA. Reversible electrocardiographic changes associated with wasp sting anaphylaxis. JAMA 1974;229:1210–1211.
86. Sampson HA, Munoz-Furlong A, Campbell RL, Adkinson NF Jr, Bock SA, Branum A, et al. Second symposium on the definition and management of anaphylaxis: summary report—Second National Institute of Allergy and Infectious Disease/Food Allergy and Anaphylaxis Network Symposium. Ann Emerg Med 2006;47: 373–380.
87. Simons FE, Gu X, Simons KJ. Epinephrine absorption in adults: intramuscular versus subcutaneous injection. J Allergy Clin Immunol 2001;108:871–873.
88. Korenblat P, Lundie MJ, Dankner RE, Day JH. A retrospective study of epinephrine administration for anaphylaxis: how many doses are needed? Allergy Asthma Proc 1999;20:383–386.
89. Awai LE, Mekori YA. Insect sting anaphylaxis and beta-adrenergic blockade: a relative contraindication. Ann Allergy 1984;53:48–49.
90. Brown SG, Blackman KE, Stenlake V, Heddle RJ. Insect sting anaphylaxis; prospective evaluation of treatment with intravenous adrenaline and volume resuscitation. Emerg Med J 2004;21:149–154.
91. Kill C, Wranze E, Wulf H. Successful treatment of severe anaphylactic shock with vasopressin. Two case reports. Int Arch Allergy Immunol 2004;134:260–261.
92. Simons FE. Advances in H_1-antihistamines. N Engl J Med 2004;351:2203–2217.
93. Schumacher MJ, Tveten MS, Egen NB. Rate and quantity of delivery of venom from honeybee stings. J Allergy Clin Immunol 1994;93:831–835.
94. Visscher PK, Vetter RS, Camazine S. Removing bee stings. Lancet 1996;348:301–302.
95. Lieberman P. Biphasic anaphylactic reactions. Ann Allergy Asthma Immunol 2005;95:217–226.
96. Balit CR, Isbister GK, Buckley NA. Randomized controlled trial of topical aspirin in the treatment of bee and wasp stings. J Toxicol Clin Toxicol 2003;41:801–808.
97. Parrino J, Kandawalla NM, Lockey RF. Treatment of local skin response to imported fire ant sting. South Med J 1981;74:1361–1364.
98. Goldberg A, Conino-Cohen R. Insect sting-inflicted systemic reactions: attitudes of patients with insect venom allergy regarding after-sting behavior and proper administration of epinephrine. J Allergy Clin Immunol 2000;106:1184–1189.
99. Muller UR. New developments in the diagnosis and treatment of Hymenoptera venom allergy. Int Arch Allergy Immunol 2001;124:447–453.
100. Moffitt JE, Barker JR, Stafford CT. Management of imported fire ant allergy: results of a survey. Ann Allergy Asthma Immunol 1997;79:125–130.
101. Tankersley MS, Walker RL, Butler WK, Hagan LL, Napoli DC, Freeman TM. Safety and efficacy of an imported fire ant rush immunotherapy protocol with and without prophylactic treatment. J Allergy Clin Immunol 2002;109:556–562.
102. Blaser K, Carballido J, Faith A, Crameri R, Akdis C. Determinants and mechanisms of human immune responses to bee venom phospholipase A_2. Int Arch Allergy Immunol 1998;117:1–10.
103. Golden DB, Kagey-Sobotka A, Norman PS, Hamilton RG, Lichtenstein LM. Outcomes of allergy to insect stings in children, with and without venom immunotherapy. N Engl J Med 2004;351:668–674.
104. Carballada F, Martin S, Boquete M. High efficacy and absence of severe systemic reactions after venom immunotherapy. J Invest Allergol Clin Immunol 2003;13: 43–49.
105. Graft DF. Maintenance venom immunotherapy. Curr Opin Allergy Clin Immunol 2002;2:359–362.
106. Brown SG, Wiese MD, Blackman KE, Heddle RJ. Ant venom immunotherapy: a double-blind, placebo–controlled, crossover trial. Lancet 2003;361:1001–1006.

Chapter 169

BUGS and BLISTER BEETLES

BUGS

IDENTIFYING CHARACTERISTICS

Most species in the order Hemiptera feed on plants. The only medically important families in the order Hemiptera (true bugs) are Reduviidae (assassin/kissing bugs, wheel bugs) and Cimicidae (bedbugs). The Reduviidae bugs vary substantially in body size and shape, ranging from a small, robust habitus to long, slender "walking sticks." The stout, three-segmented beak that fits into a grove on the prosternum distinguishes these insects from other members of the order Hemiptera. Of at least 10 species of kissing or cone-nosed bugs (*Triatoma* spp.), there are five common species in the continental United States including *T. gerstaeckeri* Stal, *T. protracta* Uhler, *T. rubida* Uhler, *T. rubrofasciata* De Geer, and *T. sanguisuga* LeConte.[1]

Common Name: Assassin Bug, Kissing Bug, Reduviid Bug, Western Cone-Nosed Bug

Scientific Name: *Triatoma protracta* Uhler

Family: Reduviidae

Physical Description: This black or brown bug is 2.5–4 cm (~1–1.5 in.) in length and has a long triply segmented proboscis (beak). Other species of *Triatoma* range in length from 1.2–3 cm (~0.5–1 in.). The elongated head has a transverse groove behind the eyes. This elongated head and short curved proboscis distinguishes the assassin bugs from plant-feeding bugs, which possess a proboscis that rests flat under their head. The meeting of the membranous wings just behind the head forms a distinguishing triangular shape.

Common Name: Wheel Bug

Scientific Name: *Arilus cristatus* L.

Family: Reduviidae

Physical Description: The wheel bug is a dark robust, bizarre insect with long legs and antennae, stout beak, and large eyes on a slim head. This 2–4 cm (~1–1.5 in.) long bug has a characteristic semicircular crest suggesting a cogwheel or chicken's comb on its prothorax along with a narrow head and stout proboscis. Females are longer and wider than males with the abdominal margins being more widely exposed in females. A fine yellowish pubescence occurs over most of the body with the exception of the bronze-colored elytral membrane over the wingcase. The background color is mostly dark brown. Variable amounts of tiny white patches or granules appear throughout the pubescence.

Common Name: Bedbug

Scientific Name: *Cimex lectularius* L.

Family: Cimicidae

Physical Description: The bedbug has a very flat dorsoventral shape with a broad, oval body of about 3–6 mm (~0.1–0.2 in.) in length.

Medical Toxicology of Natural Substances, by Donald G. Barceloux, MD

EXPOSURE

Geographic Distribution

Fifteen species of assassin bugs occur in the United States distributed in a crescent-shape from Florida to northern California. Kissing bugs (*Triatoma* spp.) belong to the subfamily Triatominae of the insect family Reduviidae, and Figure 169.1 demonstrates the approximate US distribution of the five common kissing bug species. Kissing bug species are found in Mexico, Central America, and South America where they can inhabit the cracks and cervices of human dwellings. The wheel bug occurs in areas from Rhode Island westward through Iowa and Nebraska to California, and southward to Florida, Texas, Mexico, and Guatemala. There are four genera of wheel bugs in the New World, but only *Arilus cristatus* L. occurs in the United States. The common bedbugs (*Cimex lectularius* L.) occur in both tropical and temperate regions. The Asian species is *Cimex hemipterus* Usinger. Dissemination of bedbugs results from contact with infested clothing, baggage, beds, and laundry.

Behavior

Assassin and Kissing Bugs

Assassin bugs are predators, which feed on small animal (snails, insects, spiders, other arthropods). The proboscis of the assassin bug is curved outwards from the head, and the proboscis swings forward to attack prey. All members of the genus *Triatoma* (kissing bugs) feed on blood from vertebrate hosts (small- to medium-sized mammals, birds, humans) including males, females, and all developmental stages. The host for *T. protracta* Uhler is the wood rat (*Neotoma* spp.). The usual habitat of *T. protracta* Uhler is the den of wild rodents, armadillos, or opossums, but this insect readily adapts to human habitats. During summer, this bug leaves the host, usually during warm summer nights. These bugs are attracted to lights in human habitats. Although these nocturnal insects can fly when provoked, they prefer to crawl. The primary stimulus for flights of *Triatoma* bugs is starvation. High after-dark temperatures and calm air promote these flights, which are usually less than one mile (~1.6 km).[2] *Triatoma* bugs spend daylight hours in dark protected areas and feed nocturnally, inflicting multiple, painless bites on open or loosely covered areas of skin. The multiple bite marks probably represent temporary disruptions of feeding. The victim usually remains asleep during the 10–25 minutes of feeding. An adult kissing bug can ingest 2–4 times their body weight in blood, and the kissing bug deposits feces and undigested portions of the meal on the skin and wound after completion of feeding.[1] The name *kissing bug* is derived from the predilection of these nocturnal, blood-sucking insects for skin near the lips.

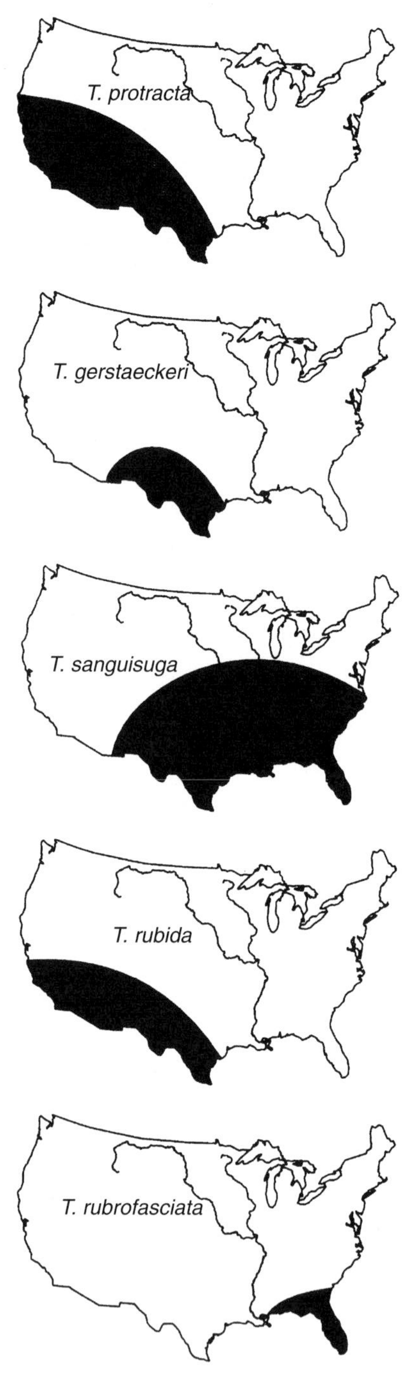

FIGURE 169.1. Distribution of five common kissing bug species in the United States. From Moffitt JE, et al. Allergic reactions to *Triatoma* bites. Ann Allergy Asthma Immunol 2003;91:123. Reproduced with permission from Adis International.

The wheel bugs produce one generation per year with the eggs laid during winter. This insect is diurnal, but lights attract prey and this bug. The wheel bug and most other reduviids produce "chirping" sounds by rubbing the tip of the rostrum back and forth over transverse ridges on a longitudinal groove on the prosternum.

Bedbugs

Bedbugs cannot jump or fly because of the lack of wings. However, they can run relatively rapidly on their six legs during warm weather.[3] These large, flat brown bugs hide in protected crevices, roosts, or other protected places during the day and feed at night on exposed skin. After completing a meal in 4–10 minutes, bedbugs return to their hiding places (mattress seams/tufts, bed frames, springs, floor cracks, furniture, wall surfaces).[4] Bites are usually multiple and painless in an irregular, linear fashion involving groups of three. They cluster on dry, rough surfaces (wood, paper) rather than stone, plaster, metal, or textiles. They avoid wet surfaces. Well adapted to human habitats, the female bugs lay eggs in cracks on floors or on furniture. The mature bug emerges after 1–2 months to feed nocturnally.

PRINCIPAL TOXINS

Venom Composition

The composition of the saliva from assassin bugs and bedbugs are not well-characterized. Apyrases are nucleoside triphosphate-diphosphohydrolases found in the saliva of hematophagous insects (e.g., kissing bugs) that probably prevent ADP-induced platelet aggregation of the host during blood sucking.[5]

Venom Apparatus

Assassin bugs paralyze their prey by puncturing them with sharp stylets in their proboscis and injecting saliva. Thereafter, the assassin bug extracts the body fluids. Figure 169.2 displays the thorax and head including proboscis of a white-eyed assassin bug (*Platymeris biguttata* L.). *Triatoma* bugs feed by capillary puncture after the injection of anticoagulants in their saliva. The secretions of the salivary glands contain antigens capable of initiating allergic reactions.[6] Wheel bug saliva contains a toxic, paralytic substance that immobilizes and kills the prey soon after injection.

HISTOPATHOLOGY AND PATHOPHYSIOLOGY

The lesions occurring after assassin bug bites probably results from an immune response rather than a toxic reaction.[7] A foreign protein in the saliva of kissing bugs

FIGURE 169.2. White-eyed Assassin Bug (*Platymeris biguttata* L.). © 2007 Rune Midtgaard. See color insert.

causes allergic response ranging from pruritus, urticaria, lightheadedness, and gastrointestinal distress to anaphylaxis.[7]

CLINICAL RESPONSE

Assassin and Kissing Bugs

Bites from members of the Reduviidae family produce distinctive, but not pathognomonic lesions. Commonly, the kissing bug inflicts multiple erythematous, pruritic nodules clustered on a single, uncovered area (arms, neck, shoulders, face).[7] Bug bites produce marked variability in skin reactions as a result of differences in individual hypersensitivity to these bites.[8] One or more puncta may be visible on the skin lesions. A delayed hypersensitivity reaction produces erythematous, pruritic papules 24–48 hours after the bite. Kissing bug bites (*T. protracta*) may be associated with substantial swelling (i.e., 2–6cm/~1–2.5in.), ecchymosis, and pain compared with most nonstinging insects.[9] These lesions usually resolve over 24–48 hours, but lesions may last up to one week in sensitized patients.[7] Victims may awake with an anaphylactic reaction (urticaria, angioedema, flushing, pruritus, respiratory distress, syncope, wheezing, hypotension, nausea, vomiting, abdominal cramps) following an undetected bite by the kissing bug.[10] This insect is a vector for American trypanosomiasis and Chagas' disease. The mild-mannered wheel bug uses its snout to produce a temporarily painful laceration that usually heals without complications. These bugs produce several linear, often hemorrhagic lesions that do not usually cause a hypersensitivity reaction.

Bedbugs

Bedbugs (*Cimex lectularius*) typically inflict linear bites along with a characteristic track of black or brown excrement on the bed linen. These bites produce a wheal and papule with a small, central hemorrhagic punctum.[11] In contrast to bedbugs, kissing bug bites are nonlinear and grouped together; these lesions may be misdiagnosed as spider bites. Bedbug bites often do not produce sensitization in adults, but a papular urticaria is associated with flea and bedbug bites in children aged 2–7 years.[12]

DIAGNOSTIC TESTING

There are no biomarkers for the detection of bites by these bugs. Clinical laboratory abnormalities are not usually associated with bites by these insects.

TREATMENT

Management of the assassin bug bite is supportive. Medications containing menthol, phenol, or camphor may relieve itching. The presence of systemic reactions necessitates a search for the offending organism and its eradication from the home including removal of piles of paper, clothing, blankets, and sheets. *Triatoma* bugs rarely bite covered skin; therefore, victims should wear long sleeve pajamas to cover all the skin. Lesions should be observed for secondary infections. Immunotherapy may be necessary for patients developing serious anaphylactic reactions.[13] Bedbug treatment is supportive, with cool, wet compresses and mild analgesics.

BLISTER BEETLES

HISTORY

Cantharidin has a history of use in Medieval medicine as a skin irritant, vesicant, abortifacient, and aphrodisiac. Hippocrates prescribed cantharidin for the treatment of dropsy and amenorrhea. Dioscorides, a Greek army physician in the 1st century AD, advocated the use of cantharidin as an aphrodisiac. In a treatise published in 1698, Groeneveldt recommended the use of cantharidin for bladder and kidney infections, dropsy, renals stones, and some venereal diseases.[14] Widespread use of cantharidin as an aphrodisiac occurred during the 18th century, probably because of the priapism and pelvic congestion produced by the marked genitourinary irritation resulting from toxic doses of cantharidin.[15] Roviquet first isolated cantharidin in 1810. The dermatitis associated with the blister beetle has been known in the United States since the middle 1800s, when rove beetle epidemics occurred during flooding along the Colorado River.[16] Classic descriptions of blister beetle dermatitis were reported in the 1940s and 1950s.[17,18]

IDENTIFYING CHARACTERISTICS

There are three major families of blister beetles in the order Coleoptera including Meloidae (e.g., Spanish Fly), Staphylinidae (rove beetles), and Oedemeridae (e.g., false blister beetle, Nairobi fly). The Meloidae family contains the Spanish fly (*Lytta vesicatoria* L.) from Southern Europe. Of the three families, Oedemeridae is the smallest with members including the genera *Ananca* (*Eobia* spp., coconut beetles), *Hypasclera* (*Alloxacis*), *Oxacis*, and *Thelyphassa*.[19] These light-colored beetles are 5–12 mm (~0.2–0.5 in.) in length with the prothorax widest in the front, narrowing to the wing-case (elytron). The abdomen is relatively large, and the antennae are usually thread-like. The family Staphylinidae contains the African blister beetle called the Nairobi fly (*Paederus sabaeus* Erichson). This beetle is about 5–7 mm (~0.2–0.3) long with striking red, dark green, and black colorations.

EXPOSURE

More than 200 species of blister beetle occur worldwide; these beetles are most abundant in southern Europe and western Asia. The distribution of members of the Oedemeridae family primarily is the Pacific Basin (Papua New Guinea, Solomon Islands, Pago Pago, Samoa, Fiji, New Caledonia) and the Caribbean (Florida keys, Bahamas, Cuba, West Indies). This insect lives in brush, and exposure is accidental. Commonly, contact occurs painlessly during handling of garden plants in the eastern United States. Blister beetles do not bite or sting; the blisters associated with these beetles results from contact with cantharidin in the crushed beetle. Rove (Staphylinid) beetles are nocturnal predators attracted to light with about 26,000 species distributed worldwide. They do not bite or sting.

Cantharidin has been used for centuries as a sexual stimulant to enhance sexual pleasure.[20] Vascular congestion and inflammation result from the inhibition of phosphodiesterase and protein phosphatase activity as well as the stimulation of beta-receptors. However, the abuse of cantharidin causes substantial morbidity. Other traditional methods to increase sexual pleasure include the ingestion of live beetles (*Palembus dermestoides*

Farmaire) in Southeast Asia and triatomids in Mexico. Cantharidin is the active ingredient in a flexible collodion for the removal of warts. Blister beetles do not bite or sting.

PRINCIPAL TOXINS

All parts of the blister beetle in the families Meloidae and Oedemeridae contain the powerful vesicant, cantharidin (CAS RN: 56-25-7).[21] This volatile compound is a simple double-ringed structure that is the crystalline, anhydrous form of cantharidic acid as demonstrated in Figure 169.3. This vesicant is a colorless, odorless, glistening, water-insoluble compound. Cantharidin is the active ingredient in the apocryphal aphrodisiac "Spanish fly." Although the highest concentration of cantharidin occurs in the Chinese blister beetle (*Mylabris cichorii* L.), the source of most cantharidin is the Spanish fly (*Cantharis vesicatoria* De Geer). Table 169.1 lists some physical properties of cantharidin.

The dermatitis associated with beetles from the family Staphylinidae (e.g., *Paederus* spp.) result from contact with pederin (CAS RN: 27973-72-4) in the hemolymph, and the skin reaction is delayed substantially longer (i.e., 12–36 hours) compared with contact with cantharidin. The cutaneous reaction to pederin produces a more prominent urticarial dermatitis than cantharidin. Figure 169.4 displays the chemical structure of pederin.

The site of cantharidin and pederin synthesis is not known. All 10 stages of the blister beetle contain cantharidin, and the larvae extrude a milky oral fluid that contains this chemical when disturbed.[22] The adult beetles reflexively discharge toxic chemicals in the hemolymph extruded near leg joints. Although female beetles do not produce cantharidin, female beetles are able to extrude cantharidin as a result of the deposition of cantharidin in seminal fluid during copulation.[23] The mature adult male blister beetle contains about 17 mg of cantharidin, which represents up to about 10% of the beetle's live weight.[22] The highest concentration of cantharidin in male specimens of *Epicauta conferta* and *Epicauta occidentalis* were 2.48%/dry weight and 3.31% dry weight, respectively.[24]

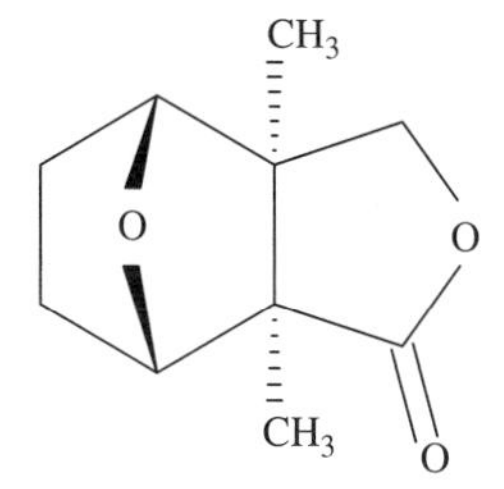

FIGURE 169.3. Chemical structure of cantharidin.

TABLE 169.1. Some Physical Properties of Cantharidin

Physical Property	Value
Melting Point	218 °C/424.4 °F
log P (Octanol-Water)	1.220
Water Solubility	30 mg/L (20 °C/68 °F)
Atmospheric OH Rate Constant	1.68E-11 cm^3/molecule-second (25 °C/77 °F)

HISTOPATHOLOGY AND PATHOPHYSIOLOGY

Contact with hemolymph from the blister beetles causes irritant contact dermatitis characterized by intraepidermal and subepidermal blistering, epidermal necrosis, and acantholysis.[25] Cantharidin causes the activation of neutral proteases that destroy dense desmosomal plaques, and this compound is a potent inhibitor of protein phosphatases 1 (PP1) and 2A (PP2A). The loss of these plaques results in the detachment of tonofilaments from the desmosomes and the appearance of intraepidermal blisters. Skin biopsies of these lesions demonstrate large intraepidermal vesicles with fibrin, polymorphonuclear cells, and acanthocytes.[21] A light, superficial, perivascular infiltrate may be present in the area underneath the intact stratum corneum along with necrotic keratinocytes, but eosinophils are generally few. Because the action of cantharidin is entirely epidermal, lesions heal without scarring. Large doses of cantharidin may produce acute renal tubular necrosis with loss of brush borders and varying degrees of cloudy swelling and hydropic degeneration.[26] Pederin in the

FIGURE 169.4. Chemical structure of pederin.

hemolymph from beetles from the family Staphylinidae causes a spectrum of histopathologic changes ranging from acute epidermal necrosis and vesiculation to marked acanthosis with mitotic figures in later stages.

CLINICAL RESPONSE

Cantharidin is a skin vesicant and a potent irritant of the gastrointestinal and genitourinary mucosa.[27] Following oral exposure, this irritation causes vesiculobullous lesions, mucosal edema, and hematuria.[28]

Cutaneous Effects

Blister beetle dermatosis is a seasonal vesiculobullous skin disorder resulting from contact with hemolymph from beetles in three beetle families (Meloidae, Staphylinidae, Oedemeridae). There is large individual variation in the response to cantharidin. Within several hours of contact with the hemolymph of the blister beetle, 5–50 mm (~0.2–2 in.) vesicles or bullae develop. The vesiculation associated with beetles of the *Paederus* species (whiplash rove beetles) is delayed 12–96 hours compared with the occurrence of vesicles within 2–3 hours after exposure to cantharidin from the Meloidae and Oedemeridae families.[29] The vesicles and bullae contain clear fluid, and the skin surrounding the lesion demonstrates no or minimal erythema. The initial contact is painless, and the victim is usually unaware of contact with the beetle. The flaccid, asymmetric blebs may become painful and hemorrhagic, but the lesions are usually only mildly or moderately pruritic. The geographic shape resembles the pattern caused by hot water splashed on the skin. The skin on the neck and exposed extremities are the most common site of injury. Secondary blisters may appear in opposing tissue (e.g., popliteal fossa). Skin lesions usually resolve over 5–7 days without scarring.[30]

A periorbital, vesicular dermatitis and keratoconjunctivitis may develop after transfer of the toxin from contaminated fingers to the facial skin.[31] These adverse effects usually resolve over several weeks. The differential diagnosis of this dermatitis includes irritant or allergic contact dermatitis, thermal or chemical burns, pemphigus, dermatitis herpetiformis, herpes simplex, herpes zoster, impetigo, and porphyria. Contact with rove beetles is associated with the development of vesicles with large areas of surrounding erythema in the area of contact. The lesions may become pustular.[32]

Systemic Effects

Systemic effects (fever, lymphangitis, lymphadenopathy, wheezing, dyspnea, lightheadedness, joint swelling) are usually absent following dermal exposure to blister beetles. A case report associated the presence of cantharidin-induced skin lesions with the development of nephritis (edema, renal failure, hematuria), but there is no clear evidence that transdermal absorption of cantharidin causes renal abnormalities.[33] Although experimental animal data suggest that cantharidin may be cardiotoxic,[34] there is no clinical data to confirm a causal link between cantharidin exposure in humans and cardiac damage.

Ingestion of blister beetle produces profound gastrointestinal irritation manifest by abdominal pain, nausea, and vomiting along with vesiculation, hemorrhage, and ulceration of the mucosa in the oropharynx and in the gastrointestinal tract.[18,35] Hematemesis may occur as a result of excoriation of the gastrointestinal tract. Blistering of the mouth and throat may cause dysphagia and dysphonia. A fatality has been associated with the use of cantharides from dried Chinese medicine (Ban Mao) as an abortifacient.[36] A 23-year-old female died of renal failure and disseminated intravascular coagulation 2 days after ingesting the boiled extract of 200 blister beetles (*Mylabris phalerata* Pallas). Postmortem examination of the kidneys indicated acute tubular necrosis, primarily involving the proximal renal tubules.

DIAGNOSTIC TESTING

Analytical methods for the quantitation of cantharidin in biological samples include gas chromatography (GC), gas chromatography/mass spectrometry (GC/MS), and high performance liquid chromatography (HPLC).[37] The postmortem cantharidin concentration in a patient, who died 2 days after the ingestion of a boiled extract containing cantharidin as an abortifacient, was 0.11 μg/mL as determined by GC using trichloroacetic acid as an extractant.[36] The antemortem blood and urine cantharidin concentrations were 0.27 μg/mL and 4.45 μg/mL, respectively. The cantharidin concentration in the postmortem blood from a patient who died shortly after admission to the hospital was 0.073 μg/mL as measured by GC/MS with selected ion monitoring technique.[38] Reportedly, the deceased ingested cantharides powder containing 0.87% cantharidin prior to the onset of illness. Laboratory abnormalities associated with oral exposure to blister beetles include hematuria and mild proteinuria with normal renal function.[39]

TREATMENT

Treatment is supportive. Application of steroid cream to skin blebs may be helpful. Lesions usually subside in 2–3 days.

References

1. Moffitt JE, Venarske D, Goddard J, Yates AB, de Shazo RD. Allergic reactions to *Triatoma* bites. Ann Allergy Asthma Immunol 2003;91:122–128.
2. Ekkens DB. Nocturnal flights of Triatoma (Hemiptera: Reduviidae) in Sabino Canyon, Arizona. I. Light collections. J Med Entomol 1981;18:211–227.
3. Elston DM, Stockwell S. What's eating you? Bedbugs. Cutis 2000;65:262–264.
4. Crissey JT. Bedbugs: An old problem with a new dimension. Int J Dermatol 1981;20:411–414.
5. Faudry E, Lozzi SP, Santana JM, D'Souza-Ault M, Kieffer S, Felix CR, et al. *Triatoma infestans* apyrases belong to the 5′-nucleotidase family. J Biol Chem 2004;279: 19607–19613.
6. Chapman MD, Marshall NA, Saxon A. Identification and partial purification of species-specific allergens from *Triatoma protracta* (Heteroptera: Reduviidae). J Allergy Clin Immunol 1986;78:436–442.
7. Lynch PJ, Pinnas JL. "Kissing bug" bites: *Triatoma* species as an important cause of insect bites in the Southwest. Cutis 1978;22:585–591.
8. Costa CH, Costa MT, Weber JN, Gilks GF, Castro C, Marsden PD. Skin reactions to bug bites as a result of xenodiagnosis. Trans Roy Soc Trop Med Hyg 1981;75: 405–408.
9. Moran ME, Ehreth JT, Drach GW. Venomous bites to the external genitalia: an unusual cause of acute scrotum. J Urol 1992;147:1085–1086.
10. Edwards L, Lynch PJ. Anaphylactic reaction to kissing bug bites. Ariz Med 1984;41:159–161.
11. Masetti M, Bruschi F. Bedbug infestations recorded in central Italy. Parasitol Int 2007;56:81–83.
12. Steen CJ, Carbonaro PA, Schwartz RA. Arthropods in dermatology. J Am Acad Dermatol 2004;50:819–842.
13. Rohr AS, Marshall NA, Saxon A. Successful immunotherapy for *Triatoma protracta*-induced anaphylaxis. J Allergy Clin Immunol 1984;73:369–375.
14. Groeneveldt J. De tuto cantharidum in medicina usu interno. London: J. Taylor; 1698.
15. Robertson J. A practical treatise on the powers of cantharides. Edinburgh: Mundell Doig & Stevenson; 1806:52–77.
16. Austin EP. On the species of *Sunius* and *Paederus* found in the United States. Proc Boston Soc Nat Hist 1876; 4–11.
17. Swarts WB, Wanamaker JF. Skin blisters caused by vesicant beetles. JAMA 1946;131:594–595.
18. Lehmann CF, Pipkin JL, Ressmann AC. Blister beetle dermatosis. Arch Dermatol 1955;71:36–38.
19. Fleisher TL, Fox I. Oedemerid beetle dermatitis. Arch Dermatol 1970;101:601–605.
20. Sandroni P. Aphrodisiacs past and present: a historical review. Clin Auton Res 2001;11:303–307.
21. Nicholls DS, Christmas TI, Greig DE. Oedemerid blister beetle dermatosis: a review. J Am Acad Dermatol 1990;22: 815–819.
22. Carrel JE, McCairel MH, Slagle AJ, Doom JP, Brill J, McCormick JP. Cantharidin production in a blister beetle. Experientia 1993;49:171–174.
23. Schlatter C, Waldner E, Schmid H. [On the biosynthesis of cantharidin. I.] Experientia 1968;24:994–995. [German]
24. Edwards WC, Edwards RM, Ogden L, Whaley M. Cantharidin content of two species of Oklahoma blister beetles associated with toxicosis in horses. Vet Hum Toxicol 1989;31:442–444.
25. Borroni G, Brazzelli V, Rosso R, Pavan M. *Paederus fuscipes* dermatitis. A histopathological study. Am J Dermatopathol 1991;13:467–474.
26. Oaks WW, DiTunno JF, Magnani T, Levy HA, Mills LC. Cantharidin poisoning. AMA Arch Intern Med 1960;105: 574–582.
27. Presto AJ III, Muecke EC. A dose of Spanish fly. JAMA 1970;214:591–592.
28. Melen DR. Haematuria due to cantharides poisoning. Cystoscopic findings. Urol Cutaneous Rev 1922;26:337–340.
29. Millard PT. Whiplash dermatitis produced by the common rove beetle. Med J Aust 1954;41:741–744.
30. Olson PE, Claborn DM, Polo JM, Earhart KC, Sherman SS. Staphylinid (Rove) beetle dermatitis outbreak in the American Southwest? Mil Med 1999;164:209–213.
31. Poole TR. Blister beetle periorbital dermatitis and keratoconjunctivitis in Tanzania. Eye 1998;12:883–885.
32. Dursteler BB, Nyquist RA. Outbreak of rove beetle (Staphylinid) pustular contact dermatitis in Pakistan among deployed U.S. personnel. Mil Med 2004;169:57–60.
33. Browne SG. Cantharidin poisoning due to a blister beetle. Br Med J 1960;2:1290–1291.
34. Rabkin SW, Friesen JM, Ferris JA, Fung HY. A model of cardiac arrhythmias and sudden death: cantharidin-induced toxic cardiomyopathy. J Pharmacol Exp Ther 1979;210:43–50.
35. Tagwireyi D, Ball DE, Loga PJ, Moyo S. Cantharidin poisoning due to "blister beetle" ingestion. Toxicon 2000;38: 1865–1869.
36. Cheng KC, Lee HM, Shum SF, Yip CP. A fatality due to the use of cantharides from *Mylabris phalerata* as an abortifacient. Med Sci Law 1990;30:336–340.
37. Ray AC, Tamulinas SH, Reagor JC. High pressure liquid chromatographic determination of cantharidin, using a derivatization method in specimens from animals acutely poisoned by ingestion of blister beetles, *Epicauta lemniscata*. Am J Vet Res 1979;40:498–504.
38. Hundt HK, Steyn JM, Wagner L. Post-mortem serum concentration of cantharidin in a fatal case of cantharides poisoning. Hum Exp Toxicol 1990;9:35–40.
39. Mallari RQ, Saif M, Elbualy MS, Sapru A. Ingestion of a blister beetle (Meloidae family). Pediatrics 1996;98: 458–459.

Chapter 170

FLEAS
(Order: Siphonaptera)

IDENTIFYING CHARACTERISTICS

Of the more than 2,400 species and subspecies of fleas, only a few species are medically important to humans including *Ctenocephalides felis* Bouche (cat flea), *Ctenocephalides canis* Curtis (dog flea), *Xenopsylla* spp. (oriental rat flea and vector of endemic typhus and bubonic plague), *Tunga penetrans* L. (sand flea), *Echidnophaga gallinacea* Westwood (stick-fast flea of poultry), *Ceratophyllus niger* C. Fox (chicken flea), and *Pulex irritans* L. (human flea and vector of plague).[1] The flea is an intermediate host for the tapeworm *Dipylidium caninum* L., which infects dogs, cats, and occasionally, humans. Most fleas prefer a particular host, but these ectoparasites will attack a range of hosts when hungry.

Fleas are small (2–3 mm/~0.1 in.), dark, shiny, wingless, blood-sucking insects with a highly compact, laterally compressed body and specialized head. The flattened lateral body and posteriorly directed spines allows easy traversal of skin containing hair or feathers.

EXPOSURE

Geographic Distribution

Fleas are distributed worldwide in association with their hosts.

Behavior

Fleas usually prefer specific hosts (e.g., dogs, cats, birds, humans), but they remain on their host only long enough to obtain a blood meal. These wingless ectoparasites attack humans in the absence of their normal host (e.g., during absence of a host pet in a home). They are intermittent, blood-sucking insects that prefer covered areas of skin to avoid light. Fleas can live for several months in rugs and furniture without a meal. Their long, strong hind legs allow vertical jumps of 7–8 in. (~18–20 cm) and horizontal jump of 14–16 in. (~36–41 cm).[2]

Cat and dog fleas undergo four separate stages (egg, larva, pupa, adult) in their life cycle as depicted in Figure 170.1. The time required for each cycle ranges from 3 months to several years depending on the species and on environmental factors (e.g., food sources, humidity, temperature). Adult cat and dog fleas mate on the host, and the female must ingest a blood meal to lay fertile eggs. During feeding, the female fleas excrete long tubular coils or fine pellets containing incompletely digested blood. Flea eggs are oval, white, nonadherent objects that typically hatch in 1–10 days. The eggs drop from the host and become widely distributed in the surrounding area. The larvae develop in the house, and they feed on organic debris in dark places (rugs, floor cracks, furniture, pet bedding). The larvae enter the pupal stage after forming silken cocoons, which can protect the flea for about 6 months in the absence of desiccation. The adult flea emerges from the cocoon after appropriate stimulation (e.g., vibrations).

PRINCIPAL TOXINS

Fleas penetrate skin with a needle-lick proboscis (labrum epipharynx), and they feed on blood extravasated from surrounding capillaries. The oral secretions injected from the salivary gland during the bite include

Medical Toxicology of Natural Substances, by Donald G. Barceloux, MD

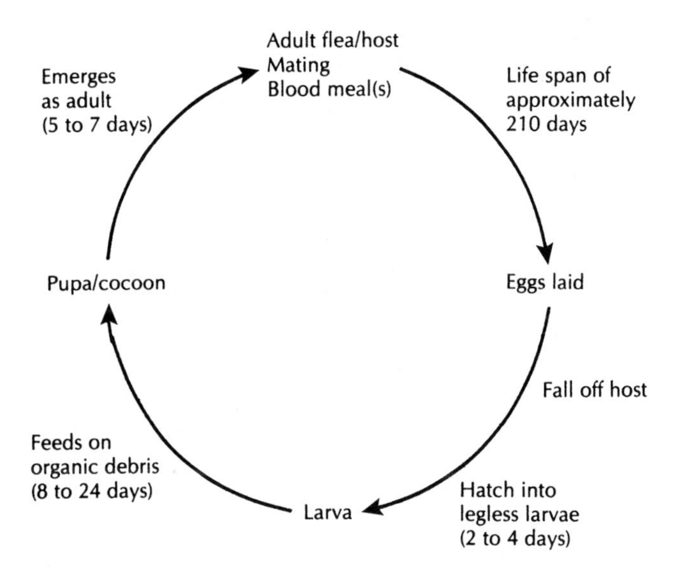

FIGURE 170.1. Life cycle of dog and cat fleas.

anticoagulants, local anesthetics, and antigens. Cat fleas and flea debris may contribute to the complex allergens associated with house dust.[3]

HISTOPATHOLOGY AND PATHOPHYSIOLOGY

The bite of the flea involves penetration of the mouth parts that usually causes a light-colored, punctate lesion surrounded by erythema.

CLINICAL RESPONSE

Although fleas are important disease vectors (e.g., murine typhus, bubonic plague), their primary clinical manifestation is the appearance of bite sensitivity. The inflammatory papules caused by flea bites characteristically are arranged in linear groups of three or four limited by tight clothing. A central punctum occurs where the mouth parts penetrate the skin. The roseate papule may become hemorrhagic or bullous in highly sensitized individuals. Cat and dog fleas typically produce lesions on the ankles in ambulatory individuals, whereas bird fleas usually produce more widespread lesions. Flea bites are common in children, particularly in children with papular urticarial lesions on the lower extremities. Papular urticaria is the term for chronic or recurrent eruption of irregular, pruritic papules affecting primarily children between 2–7 years of age.[4] These lesions typically persist for 2–10 days with temporary hyperpigmentation. Excoriations of these intensely pruritic lesions frequently cause secondary infections. The most common causes of these lesions are the cat flea, dog flea, or human flea.[5] Other causes include the bedbug and less commonly mosquitoes or various species of mites. Infants and adults in the same household may not develop papular urticaria despite the presence of this condition in young children from the affected household.

Flea bites on the genital, perianal, and axillary regions are unusual. Severe attacks generally occur in patients moving into a new residence that previously contained pets because fleas become aggressive when they lose their host. Tungiasis is a cutaneous ectoparasitic infestation by the female sand flea (*Tunga penetrans* L.) that primarily involves the feet where the flea burrows underneath the stratum corneum of the toes and causes paronychia.[6] This disease occurs mostly in impoverished areas of sub-Saharan Africa, the Caribbean, Latin America, and South America.[7]

DIAGNOSTIC TESTING

The diagnosis of flea bites is a clinical rather than laboratory diagnosis. The presence of a group of several hemorrhagic puncta surrounded by an erythematous urticarial papule is characteristic of flea bites, particularly when located on the ankles and legs.[8]

TREATMENT

Most flea bites resolve without treatment. Calamine lotion and oral antihistamines are effective antipruritic agents. Cool wet compresses (Epsom salts or Domeboro tablets) provide symptomatic relief. Application of a menthol ointment 3–4 times a day provides relief from the burning and pain associated with the bites. The treatment of papular urticaria involves the use of mild topical steroids and system antihistamines for control of pruritus. Secondary infections may occur that necessitate the use of oral antibiotics. Removal of fleas is the most important preventive measure.[9] Eradication of fleas requires a multistep approach including treatment of the infested animals with insecticide and the cleaning of the surrounding habitat. The pet bedding should be washed in hot water, and the area in contact with the infested pet should be vacuumed thoroughly. The vacuum cleaner bag should be sealed and discarded after cleaning. All pets in contact with the infested pet also should be treated. After the house has been cleaned, the house should be treated with an approved pesticide spray (resmethrin, malathion, carbaryl, diazinon). Newer pet products for the topical treatment include imidacloprid (Advantage™, Bayer HealthCare, LLC, Shawnee Mission, KS) and fipronil (Frontline®, Merial, Ltd., London, UK), whereas orally administered pet products include insect growth regulators (methoprene, fenoxycarb, pyriproxyfen) and insect development inhibitors (lufenuron). Repeat treatment may be necessary, especially during heavy infestations because of the relative resistance of the eggs and pupae to treatment.

References

1. Laudisoit A. Plague and the human flea, Tanzania. Emerg Infect Dis 2007;13:687–693.
2. Jones JE. Fleas. Am Fam Physician 1984;29:143–147.
3. Trudeau WL, Fernandez-Caldas E, Fox RW, Brenner R, Bucholtz GA, Lockey RF. Allergenicity of the cat flea (*Ctenocephalids felis felis*). Clin Exp Allergy 1993;23: 377–383.
4. Steen CJ, Carbonaro PA, Schwartz RA. Arthropods in dermatology. J Am Acad Dermatol 2004;50:819–842.
5. Stibich A, Schwartz RA. Papular urticaria. Cutis 2001;68: 89–91.
6. Kehr JD, Heukelbach J, Mehlhorn H, Feldmeier H. Morbidity assessment in sand flea disease (tungiasis). Parasitol Res 2007;100:413–421.
7. Sachse MM, Guldbakke KK, Khachemoune A. *Tunga penetrans*: a stowaway from around the world. J Eur Acad Dermatol Venereol 2007;21:11–16.
8. Hutchins ME, Burnett JW. Fleas. Cutis 1993;51:241–243.
9. Millikan LE. Papular urticaria. Semin Dermatol 1993;12: 53–56.

Chapter 171

FLIES and MOSQUITOES (Order: Diptera)

IDENTIFYING CHARACTERISTICS

A single pair of wings characterizes the members of this order of blood-sucking insects (biting flies and mosquitoes). A small knobbed structure that functions as a stabilizer appears behind the wings.

Common Name: Mosquito
Scientific Name: Family *Culicidae*
Physical Description: Mosquitoes have slender bodies about 3–6mm (~0.1–0.2in.) in length that is divided into a head, thorax, and abdomen. There are six long, thin legs with tiny claws at the ends and two wings covered with flat scales. Most other flies have four wings. The head has two antennae and a long proboscis, which allows the mosquito to suck liquid from prey. The abdomen is a long, slender tube with eight pairs of spiracles.

Common Name: Sandfly
Scientific Name: *Lutzomyia* spp. (*Phlebotomus* spp.)
Physical Description: Sand flies are tiny, hairy gnats (1–3mm/~0.1in.) that breed in dark, damp crevices during the day.

Common Name: Black Flies, Midges, Buffalo Flies
Scientific Name: *Simuliidae* spp.
Physical Description: Black flies, midges, and buffalo flies are small (1–5mm/~0.1–0.2in.) flies that have a hump-backed appearance.

Common Name: Stable Fly
Scientific Name: *Stomoxys calcitrans* L.
Physical Description: The stable fly is physically similar to the common housefly.

Common Name: Horsefly Family
Scientific Name: Family: *Tabanidae*
Physical Description: The horse fly is a large (20mm/~0.8in.), slow-moving, blood-sucking insect that is usually easily repelled.

Common Name: Deer Fly
Scientific Name: *Chrysops* spp.
Physical Description: These stout-bodied, fast-flying insects range in size from 6–30mm (~0.2–1.25in.).

EXPOSURE

Geographic Distribution

Mosquitoes are aquatic breeders found worldwide, even beyond the Arctic Circle.

Important mosquito species (Diptera: Culicidae) include *Aedes vexans* Meigan, *Aedes aegypti* NPV, and *Culex quinquefasciatus* Say. The latter two mosquito species are distributed worldwide throughout the tropical regions, whereas *Aedes vexans* Meigan inhabits North America, Asia, Africa, and Eurasia. The insect family Tabanidae comprises approximately 3,000 species of flies throughout the world except at the extreme

Medical Toxicology of Natural Substances, by Donald G. Barceloux, MD

northern and southern latitudes. Deer flies are found in wet, underdeveloped areas throughout the United States.

Behavior

Mosquitoes are most active during cooler, shady times of the day. These insects are attracted by smell as well as sight and temperature.[1] The female *Aedes* mosquito is responsible for most bites because the male *Aedes* mosquito lacks the penetrating mouth parts present on female mosquitoes. Attractants for female mosquitoes include moisture, carbon dioxide, warmth, odor, and estrogen.[2] Tabanids (horse flies, deer flies, clegs, gadflies) are wasp- or bee-like insects that often land unnoticed on humans while outdoors. Most females of the Tabanidae family require a blood meal to complete egg development. Female sand flies feed exclusively on blood at night. Female deer flies feed on animals and occasionally on humans. These flies are attracted by carbon dioxide and large, dark, moving objects. Bites from deer flies (*Chrysops* spp.) occur from late May until the middle of September.[3] These insects produce painful bites, and frequently bite unremittingly. The male tabanids feed on nectar; they do not possess the modified mouth parts capable of biting humans. Tabanids are holometabolous (i.e., undergo complete metamorphosis) insects that undergo four life stages (egg, larva or maggot, pupa, adult) similar to other true flies.

Black flies, midges, and buffalo flies breed in fast-flowing streams. Black flies swarm in late spring and early summer. Biting midges swarm at sunset, forming dense clouds. This behavior distinguishes biting midges from the night-biting sand flies and day-biting black flies. The stable fly is a blood-sucking fly that breeds around farm animals and feeds at one spot for several minutes before moving to a different location.

PRINCIPAL TOXINS

Venom Composition

The salivary ducts of mosquitoes contain local anesthetics as well as antigens that can cause immediate or delayed hypersensitivity reactions in sensitized individuals.[4] Salivary extracts of tabanids (horseflies, clegs, deerflies) include peptides and small proteins that inhibit thrombin activity in a dose-dependent manner.[5] The salivary secretions of *Tabanus bovinus* L. contain a 7-kD peptide (tabanin) that is a thrombin inhibitor. Other anticoagulants present in the salivary secretions from deer flies (*Chrysops* spp.) include platelet aggregation inhibitors and glycoprotein IIb/IIIa fibrinogen receptor antagonists (chrysoptin).[6] The salivary secretions of deer flies also contain a 69kD IgE-binding protein capable of producing an anaphylactic reactions.[7]

Venom Apparatus

Deer flies (*Chrysops* spp.) and other tabanids (horseflies, clegs) possess a scissor-like proboscis that incises the skin of the prey; they feed on the blood extruded from these wounds. Anticoagulants in the saliva help extract blood from the victim. Injected fluids produce local tissue damage as well as mild local hypersensitivity reactions.

Dose Response

The reactions to bites from mosquitoes and flies result from hypersensitivity reactions; therefore, the response to these bites is not dose-related.

HISTOPATHOLOGY AND PATHOPHYSIOLOGY

Clinical and experimental data suggests that various reactions to mosquito bites result from sensitization to the mosquito saliva injected into the skin during feeding.[8–10] These reactions include both IgE-mediated and T-lymphocyte-mediated hypersensitivity responses. The size of the immediate wheal and flare reaction correlates to the presence of IgE-specific antibodies, and there are strong cross-reactive skin and IgE-mediated responses among common mosquito species.[11] The immediate skin reaction involves an irregular, central punctate lesion in the epidermis surrounded by superficial edema.[12] Marked vasodilation develops in the dermis along with perivascular infiltration of leukocytes (polymorphonuclear, eosinophils). In addition to an Arthus-type mechanism, cutaneous late-phase reactivity and cell-mediated immunity may also be involved in the formation of delayed papules.[13] Frequent exposure to mosquito bites may produce blocking (IgG) antibodies and desensitization to the bites.[14] The saliva of deer flies contains immunogenic proteins that cause IgE-mediated anaphylactic reactions.[7]

CLINICAL RESPONSE

Cutaneous responses to mosquito bites range from immediate wheals and delayed papules to severe Arthus-type reactions with systemic complaints.[15] Individual responses to mosquito bites are highly variable; the type of response includes vesicular, urticarial, eczematoid, and granulomatous lesions. These lesions may become bullous and excoriated. The initial mosquito bite causes no response, but subsequent bites produce delayed papular lesions. The typical response to mosquito bites in sensitized individuals involves the formation of pruritic wheals 2–10 mm (~0.1–0.4 in.) in diameter. Pruritic papules commonly develop several hours after the wheal and flare reaction, and the symptoms peak over 24–48 hours. These delayed papules typically resolve over several days, but occasionally these papules may persist a few weeks. Repeat mosquito bites can produce cutaneous reactions at old bite sites. Large local reactions occur occasionally. The appearance and intensity of delayed hypersensitivity reactions decreases with age and the frequency of bites.[16]

Rarely in sensitized asthmatic patients, mosquito bites may produce Arthus-type reactions (joint swelling, fever, adenopathy, nausea, headache) or anaphylaxis (wheezing dizziness, lethargy, urticaria, angioedema, hypotension).[17,18] A seasonal bullous eruption of the lower legs has been associated with bites by the *Aedes detritus* Haliday mosquito.[19] Mosquito bites can transmit filariasis, dengue fever, malaria, yellow fever, chikungunya fever, and viral encephalitis to humans.[20,21] However, mosquitos probably do not transmit HIV disease because of the absence of T_4 antigen on cell surfaces and low titer of the virus in human body fluids.[22]

Rarely, severe anaphylactic responses (e.g., hypotension, dyspnea) may occur after bites from horse flies (*Tabanus* spp.)[23,24] or from deer flies (*Chrysops* spp.).[25] Common house flies (*Musca domestica* L.) are rarely associated with respiratory sensitization in asthmatic patients.[26] Black flies, midges, and buffalo flies produce an extremely painful, pruritic, erythematous lesion that usually begins one hour after the bite. Later, the lesion may develop into a nodular, eczema-like patch or a hard pigmented nodule.

DIAGNOSTIC TESTING

Clinical abnormalities are not usually associated with bites from these insects other than those associated with infectious or allergic processes. There are no commercial biomarkers for these insect bites.

TREATMENT

Supportive care for mosquito and fly bites is similar to the treatment of flea bites. Washing horse and stable fly bites with an antiseptic solution [e.g., povidone-iodine (Betadine®, Purdue Pharma, Stamford, CT)] or soap and water and the application of a topical antibiotic ointment may prevent infection. Topical corticosteroid ointment may reduce swelling, redness, and itching associated with the bites, but the use of topical antihistamine creams is probably not efficacious.[9] Oral antihistamines (e.g., cetirizine 10 mg) reduce the pruritus and wheal associated with mosquito bites, but these antihistamines do not alter the intensity or duration of the delayed symptoms.[27] A volunteer study suggests that dilute ammonia solution (3.6%) applied minutes after the mosquito bite may reduce the wheal and flare reaction.[28] Other topical agents include preparations with menthol and camphor.

Insect repellents containing DEET (*N,N*-diethyl-*m*-toluamide) effectively repel biting insects, such as mosquitos, biting flies, gnats, chiggers, and ticks. Currently available non-DEET repellents do not provide the duration of protection from mosquito or fly bites compared with DEET-based repellents.[29] The repellent should be applied to all exposed areas of the skin except the eyelids and any skin lesions. Field entomological studies indicate that electronic mosquito repellants have no effect on preventing mosquito bites.[30]

References

1. Metry DW, Hebert AA. Insect and arachnid stings, bites, infestations, and repellents. Pediatr Ann 2000;29:39–48.
2. Brown AW. The attraction of mosquitoes to hosts. JAMA 1966;196:249–252.
3. Pratt GK, Pratt HD. Notes on deer flies and horse flies (Diptera: Tabanidae) from southern Vermont. J Am Mosq Control Assoc 1986;2:365–367.
4. Newsome WH, Jones JK, French FE, West AS. The isolation and properties of the skin-reactive substance in *Aedes aegypti* oral secretion. Can J Biochem 1969;47:1129–1136.
5. Kazimirova M, Sulanova M, Kozanek M, Takac Pk, Labuda M, Nuttall PA. Identification of anticoagulant activities in salivary gland extracts of four horsefly species (Diptera, Tabanidae). Haemostasis 2001;31:294–305.
6. Reddy VB, Kounga K, Mariano F, Lerner EA. Chrysoptin is a potent glycoprotein IIb/IIIa fibrinogen receptor antagonist present in salivary gland extracts of the deerfly. J Biol Chem 2000;275:15861–15867.
7. Hemmer W, Focke M, Vieluf D, Berg-Drewniok B, Gotz M, Jarisch R. Anaphylaxis induced by horsefly bites: identification of a 69 kd IgE-binding salivary gland protein from *Chrysops* spp. (Diptera, Tabanidae) by Western blot analysis. J Allergy Clin Immunol 1998;101:134–136.
8. Hudson A, Bowman L, Orr CM. Effects of absence of saliva on blood feeding by mosquitoes. Science 1960;131:1730–1731.
9. Reunala T, Brummer-Korvenkontio H, Lappalainen P, Rasanen L, Palosuo T. Immunology and treatment of mosquito bites. Clin Exp Allergy 1990;20(suppl 4):19–24.
10. McKiel JA. Sensitization to mosquito bites. Can J Zool 1959;37:341–351.

11. Peng Z, Simons FE. Cross-reactivity of skin and serum specific IgE responses and allergen analysis for three mosquito species with worldwide distribution. J Allergy Clin Immunol 1997;100:192–198.
12. Goldman L, Rockwell E, Richfield DF. III. Histopathological studies on cutaneous reactions to the bites of various arthropods. J Invest Dermatol 1952;19:514–525.
13. Tokura Y, Tamura Y, Takigawa M, Koide M, Satoh T, Sakamoto T, Horiguchi D, Yamada M. Severe hypersensitivity to mosquito bites associated with natural killer cell lymphocytosis. Arch Dermatol 1990;126:362–368.
14. Ailus K, Palosuo T, Brummer-Korvenkontio M, Rantanen T, Reunala T. Demonstration of antibodies to mosquito antigens in man by immunodiffusion and ELISA. Int Arch Allergy Appl Immunol 1985;78:375–379.
15. Rockwell EM, Johnson P. The insect bite reaction. II. Evaluation of the allergic reactions. J Invest Dermatol 1952;19: 137–155.
16. Oka K, Ohtaki N. Clinical observations of mosquito bite reactions in man: a survey of the relationship between age and bite reaction. J Dermatol 1989;16:212–219.
17. Gaig P, Garcia-Ortega P, Enrique E, Benet A, Bartolome B, Palacios R. Serum sickness-like syndrome due to mosquito bites. Invest Allergol Clin Immunol 1999;9: 190–192.
18. Gluck JC, Pacin MP. Asthma from mosquito bites: a case report. Ann Allergy 1986;56:492–493.
19. Walker GB, Harrison PV. Seasonal bullous eruption due to mosquitoes. Clin Exp Dermatol 1985;10:127–132.
20. Thwing J, Skarbinski J, Newman RD, Barber AM, Mali S, Roberts JM, et al. Malaria surveillance—United States, 2005. MMWR Surveill Summ 2007;56:23–40.
21. Centers for Disease Control and Prevention (CDC). West Nile virus update—United States, January 1–July 24, 2007. MMWR Morb Mortal Wkly Rep 2007;56:740–741.
22. Iqbal MM. Can we get AIDS from mosquito bites? J La State Med Soc 1999;151:429–433.
23. Freye HB, Litwin C. Coexistent anaphylaxis to Diptera and Hymenoptera. Ann Allergy Asthma Immunol 1996;76: 270–272.
24. Solley GO. Allergy to stinging and biting insects in Queensland. Med J Aust 1990;153:650–654.
25. Hrabak TM, Dice JP. Use of immunotherapy in the management of presumed anaphylaxis to the deer fly. Ann Allergy Asthma Immunol 2003;90:351–354.
26. Focke M, Hemmer W, Wohrl S, Gotz M, Jarisch R, Kofler H. Specific sensitization to the common housefly (*Musca domestica*) not related to insect panallergy. Allergy 2003;58: 448–451.
27. Karppinen A, Kautiainen H, Petman L, Burri P, Reunala T. Comparison of cetirizine, ebastine and loratadine in the treatment of immediate mosquito-bite allergy. Allergy 2002;57:534–537.
28. Zhai H, Packman EW, Maibach HI. Effectiveness of ammonium solution in relieving type I mosquito bite symptoms: a double-blind, placebo-controlled study. Acta Derm Venereol 1998;78:297–298.
29. Fradin MS, Day JF. Comparative efficacy of insect repellents against mosquito bites. N Engl J Med 2002;347: 13–18.
30. Enayati AA, Hemingway J, Garner P. Electronic mosquito repellents for preventing mosquito bites and malaria infection. Cochrane Database Syst Rev 2007;(2): CD005434.

Chapter 172

LICE
(Order: Phthiraptera)

HISTORY

Lice have been present since the beginning of recorded history as documented in the Old Testament and ancient Egyptian writings.[1]

IDENTIFYING CHARACTERISTICS

There are two species of lice (Anoplura: Pediculidae) that infect humans: *Pediculus humanus* L. and *Pthirus pubis* L. Synonyms for the former species include *P. humanus capitis* De Geer (head lice) and *P. humanus humanus* L. or *P. humanus corporis* (body lice).

Common Name: Head Lice
Scientific Name: *Pediculus humanus* L. (*Pediculus humanus capitis* De Geer)
Physical Description: The head louse is a small (2–3 mm/~0.1 in.), pale gray insect that lives on the hair closest to the scalp. Both body lice and head lice have flattened, elongated, triple-segmented thoraces and long slender legs that allow greater mobility than the body of the pubic louse. Two short antennae with five segments protrude from the lateral aspect of the cylindrical head.

Common Name: Body Louse
Scientific Name: *Pediculus humanus* L. (*P. humanus corporis, P. humanus humanus*)
Physical Description: Body lice are morphologically similar to head lice, except for the location of this louse on the body. The tan to grayish-white, adult body louse has six legs and is about the size of a sesame seed.

Common Name: Pubic Louse, Crab Louse
Scientific Name: *Phthirus pubis* L.
Physical Description: The pubic louse has a shorter body (≤2 mm/<0.1 in.) compared with head lice. Three sets of legs are attached to the anterior portion of the abdomen with the middle and hind sets possessing specialized claws and opposing tibial thumb to grasp hair shafts. The body of the pubic louse is more appropriate for grasping pubic, axillary, and rarely eyelid hair shafts than crawling on the body.

EXPOSURE

Geographic Distribution

Infestations with head lice occur at any age and within all socioeconomic groups. Epidemics of pediculosis occur worldwide. The incidence of head lice is highest in school-age children, particularly Caucasians and women.[2] Head lice are less common in African Americans than in Caucasians probably because of differences in the shape of the hair shaft.[3] However, head lice in Africa have adapted claws for grasping the oval-shaped hair shafts of African Blacks. Body lice are more common in colder months, whereas head lice are more common in warmer parts of the year.[4]

Medical Toxicology of Natural Substances, by Donald G. Barceloux, MD

Behavior

Eyes are present on the lateral portion of the head near the base of each antenna, and these organs help the louse avoid light. The peg organ is located on the distal segment of the antenna. This organ detects odor and humidity. Lice crawl rather than jump or fly. Pets are not vectors for the transmission of lice, and these insects require human contact to complete their life cycle. Lice also require symbiotic bacteria to digest their blood meals, probably as a source of B vitamins.[5] The female louse lays up to 300 eggs or nits in her 30-day life cycle. Although lice can survive up to 2 days away from the human host, lice typically pierce the skin every few hours to obtain a blood meal.

Head Louse

The spread of head lice results primarily from direct head-to-head contact.[6] However, there are some data to support the spread of lice by contact with personal belongs (e.g., brushes, combs, hats) of an infested individual,[7] although lice found on these objects are often unhealthy.[8] Fertilized females deposit approximately one to six oval egg sacs (nits)/day on base of the hair shafts. These 1-mm-long egg sacs are initially translucent, but after hatching in 7–14 days, the empty egg cases become white. Viable nits are camouflaged with pigment similar to the hair of the host. Empty nits are more visible, particularly in the posterior hairline. Nits are firmly attached to the hair shafts, and the nits move away from the scalp as the hair grows. The distance of the nit from the scalp is a measure of the duration of the infestation based on hair growth of 1 cm/month.

Body heat incubates the eggs within 7–14 days. The nymphs grow during the next 9–12 days before mating. The louse lives on the head by injecting small amounts of saliva and extracting tiny amounts of blood from the scalp every few hours. The life cycle repeats every 3–4 weeks. Head lice usually survive only 1–2 days away from the scalp, and the eggs do not hatch at temperatures below the ambient temperature of the scalp.

Body Louse

Most body lice live on clothing and bedding with a few attached to body hairs. Body lice are found in clothing seams of infested patients. These ectoparasites live preferentially in cold climates and infest individuals in overcrowded, impoverished conditions.[9] Body lice do not survive over 10 days at 15 °C (59 °F), and survival is shortened during colder and warmer periods.[1] Body lice are not usually detected on body surfaces during feeding. These insects hide in the seams of clothing where they lay eggs and complete their life cycle.

Pubic Louse

Pediculosis pubis is a venereal disease acquired by direct physical contact with infested sexual partners or clothing. These lice usually remain localized in the pubic and inguinal region, but they may infest the axilla or rarely the eyelashes, beard, areolar hair, and scalp hair. In daylight, the louse remains motionless or moves into the relative darkness of skin fold. The bodies of pubic lice are modified to attach to hair shafts rather than move on the skin, and the pubic louse does not survive over 24 hours away from the human host.[10] Infestations with pubic lice are more common in low socioeconomic classes. Frequently, pubic lice are present along with other sexually transmitted infections.[11]

PRINCIPAL TOXINS

Skin lesions associated with lice bites result from hypersensitivity reactions rather than direct toxic effects.

HISTOPATHOLOGY AND PATHOPHYSIOLOGY

Lice bites cause excoriation and occasionally urticarial lesions secondary to a hypersensitivity response. Clinical manifestations of this hypersensitivity response may require up to 30 days from the first blood meal.

CLINICAL RESPONSE

Most lice bites do not cause immediate pain, except during a heavy infestation of lice when a stinging sensation may occur. The initial skin lesion is a minute red macule that does not cause inflammation. Repeated exposure to lice over a course of 7–10 days results in a hypersensitivity response and pruritus. Consistent with most hypersensitivity responses, the individual response to lice bites is highly variable.

Head Lice

The most common symptom of head lice infestation is variable numbers of pruritic papules on the scalp, particularly in the occipital and retroauricular regions. After an asymptomatic period of about 24–48 hours, sensitization develops and pruritic papules with central puncta appear. A wheal then occurs followed by delayed hypersensitivity response at the bite site.[12] Systemic reactions to lice bites do not usually develop. Excoriation may produce secondary infections, impetigo-like lesions, posterior cervical and postauricular adenopathy, and fever. The sequence of skin responses following body lice and head lice bites is similar.[13] Unlike body lice, head lice do not transmit any communicable diseases; therefore, head lice are not a health hazard.[14]

Body Lice

Typical lesions associated with the bites of body lice involve small, pruritic, red papules under the arm and on the back. Heavy infestations and frequent itching produce pyoderma in areas normally covered with loose clothes. Ulcerations, excoriations, and secondary infections occur, particularly in the trunk, axilla, and groin. Body lice can transmit such diseases as epidemic typhus (*Rickettsia prowazekii*), trench fever (*Bartonella quintana*), and relapsing fever (*Borrelia recurrentis*).[15]

Pubic Lice

Pubic lice (*Pthirus pubis* L.) produce an infestation limited to the short hairs, primarily in the pubic region. Nits may also be present on eyelashes, eyebrows, beards, axillary hair, body hair, and occasionally on the short hair of the occipital region of the scalp. Pubic lice are not disease vectors for humans, but the incidence of sexually transmitted diseases is higher in these patients compared with patients without pubic lice.[16] Occasionally, bluish brown, irregular macules (maculae caeruleae) about 5 mm/–0.2 in. in diameter occur in the lower abdomen or eyelid after pubic or body lice infestations, probably as a result of the deposition of hemosiderin at sites of intradermal hemorrhage.[17] Phthiriasis palpebrarum is an uncommon cause of blepharitis and conjunctivitis associated with infestation of the eye lashes with pubic lice.[18] Symptoms range from pruritic lid margins to blepharitis with marked conjunctival inflammation.

DIAGNOSTIC TESTING

Confirmation of head lice infestation involves the detection of multiple, oval, grayish-white egg capsules (nits) firmly attached to the hair shafts. A white towel is placed around the neck and the tangles are removed with a standard comb or brush. The hair should then be examined by through-combing for at least 5 minutes with a louse comb.[19] Nits are transparent, flask-shaped egg cases that are tightly cemented to hair shafts (head lice, pubic lice) or clothing (body lice). Nits are more visible at the nape of the neck and behind the ear within 1 cm (~0.4 in.) of the scalp. Unlike dandruff, the nits are firmly attached to the hair shafts. Hair casts (pseudonits) are opaque amorphous remnants of the inner root sheath that remain near the hair shaft and may be confused with nits. These casts are also mobile in contrast to nits. Viable nits develop an "eye spot" visible under a microscope within several days after deposition on the skin, and these nits seldom occur over 1 cm (~0.4 in.) from the scalp. Nits found more than several millimeters from the scalp are usually nonviable. The demonstration of nits within ¼ in. (~5 mm) of the scalp suggests an active infestation. However, the presence of nits without live lice does not necessarily indicate an active infestation because nits may remain firmly attached to the hair shafts for months.[20]

Under the low power of the microscope, the presence of adult lice, immature nymphs, and/or viable eggs indicates an active infestation of head lice. Live head lice may be difficult to find because they can crawl up to 6–30 cm (~2–12 in.) /minute.[3] Taping the affected area and trapping the lice with transparent adhesive tape may be helpful. The use of a fine-toothed or louse comb improves the sensitivity and facilitates the detection of live lice when compared with visual inspection of the scalp hair.[21] The separation between teeth on these combs is approximately 0.2–0.3 mm (~0.01 in.). After removal of tangles with a standard brush or comb, the fine-toothed comb is inserted near the hairline on the face next to the scalp, and the comb is drawn firmly over the head. The entire head should be combed at least twice. The nits of pubic lice and head lice are similar, and lice may be detected on hair shafts with microscopic examination of the hair shafts. Body lice are usually detected by examination of the seams of clothing from the infected individual.

TREATMENT

Head Lice/Pubic Lice

Successful treatment of lice infestations requires destruction of both the adult lice and the eggs. Permethrin, synergized pyrethrin, and malathion (Ovide®, Taro Pharmaceuticals, Hawthorne, NY) are effective agents for the treatment of head lice, and the choice of the pediculicide depends on local patterns of resistance.[22,23] There is substantial resistance of head lice to 1% lindane, and this product lacks significant ovicidal properties.[24,25] Consequently, 1% lindane is not currently indicated for head lice infestation. Case reports indicate that repeated, excessive use of topical lindane over one week causes neurotoxicity including seizures and altered mental status.[26] These neurological effects usually resolve after cessation of lindane use. There are inadequate data to support the use of physical methods ("bug-busting," wet-combing) or herbal preparations alone to eliminate lice.[22]

Preparation

The hair should be washed with a regular shampoo (no creme rinse or conditioner), rinsed with water, and dried thoroughly. The hair should not be rewashed for 1–2 days after treatment. These agents should not be applied to wet hair because dilution, decreased contact time, and poor penetration of the agent reduce the effectiveness of the pediculicide. Hair removal is not necessary to control lice infestations.

Permethrin

1% Permethrin (synthetic pyrethroid) cream rinse is applied generously, left on for 10 minutes, and rinsed with water. Pyrethroids are not consistently ovicidal, and successful eradication of lice requires two to three treatment cycles to eradicate lice.[27] Synergized natural pyrethrins (0.3% pyrethrin with 3% piperonyl butoxide) are an over-the-counter product available for the treatment of head lice using the same techniques as permethrin with the exception that the hair is shampooed *after* the 10-minute application. Pyrethrins are natural extracts of the chrysanthemum plant; patients with chrysanthemum allergies should not be treated with pyrethrins. Synergized pyrethrin is less ovicidal than 1% permethrin; therefore, a second treatment one week later is necessary. Itching or mild burning of the scalp may persist for days after the head lice are destroyed, and the persistence of these symptoms is not necessarily an indication for repeat treatment. Topical corticosteroids and oral antihistamines should relieve these symptoms.

Malathion

If treatment with natural pyrethrins or permethrin fails to eradicate the infestation, 0.5% malathion lotion is the treatment of choice.[28] This product is applied to hair and allowed to dry for 8–12 hours before thorough rinsing. The lotions should be reapplied in 7–10 days if live lice reappear. Malathion is ovicidal, and eradication of the lice requires only one or two treatments.

Nit Removal

After insecticidal treatment, nits can be removed with sturdy metal combs and acidic solutions (a 1:1 mixture of 5% acetic acid or vinegar, specialized over-the-counter 8% formic acid preparations). The evidence for the effectiveness of these adjunctive methods remains anecdotal.[29] Acid shampoos (pH 4.5–5.5) are applied along with water to the hair for 15 minutes; then gently combed with a fine-toothed comb (e.g., louse comb). Combing is primarily a cosmetic maneuver because this adjunctive therapy does not increase the efficacy of applied medication (e.g., permethrin cream).[30] Lubrication facilitates combing and removal of lice and eggs from the scalp, but suffocating agents (mayonnaise, oils, hair gels) do not usually kill substantial numbers of lice.

Mechanical removal of head lice by combing does not eradicate an infestation or eliminate the need for a pediculicide.[31] The hair should be checked every 2–3 days with a nit comb to remove nits and lice for the first 2–3 weeks after initial treatment. Mechanical removal of nits from the eyelashes requires the use of the slit lamp. Ophthalmic ointments (erythromycin, physostigmine) have been recommended for local application to the eyelids, but there are few clinical data to confirm the effectiveness of these agents.

Supportive Care

All household contacts and current sexual partners should be examined and treated if evidence of a head lice infestation exists. The US Center for Disease Control and Prevention, but not UK authorities, recommends cleaning of personal and household items touched by infected patients within 2 days prior to treatment.[32] Lice and eggs are destroyed by temperatures above 52 °C (~126 °F) for 5 minutes. Options for destroying lice include washing clothes in hot water (130 °F/54 °C), drying for 20 minutes, ironing, drycleaning, and storage for 2 weeks in plastic bags (life cycle is 30 days). Consequently, all bedding, clothing, pillows, hair items (brushes, combs), and head gear that had contact with an infested patient within 48 hours of treatment should be washed, drycleaned, or stored for 2 weeks in plastic. Floors, play areas, and furniture should be vacuumed thoroughly to remove any shed hairs. However, fumigation is not necessary.

Patients should be reexamined in one week for the continued presence of lice. The incubation period for pubic lice is 5 days to several weeks, and sexual contact should be avoided until the treatment is completed. The incidence of coexisting sexually transmitted diseases is relatively high in patients with pubic lice; therefore, these patients should be evaluated for the presence of sexually transmitted diseases.[33]

Body Lice

The treatment of body lice infestation primarily is supportive consisting of good personal hygiene, calamine lotion for itching, and the laundering of all clothing and bed linen in hot water (130 °F/54 °C). The patient should shower and receive a clean set of clothes. The use of a pediculicide (permethrin, synergized pyrethrin) is optional.

References

1. Elgart ML. Pediculosis. Dermatol Clin 1990;8:219–228.
2. Roberts RJ. Head lice. N Engl J Med 2002;346: 1645–1650.
3. Frankowski BL, Weiner LB. Committee on School Health the Committee on Infectious Diseases. American Academy of Pediatrics. Head lice. Pediatrics 2002;110:638–643.
4. Mimouni D, Ankol OE, Gdalevich M, Grotto I, Davidovitch N, Zangvil E. Seasonality trends of *Pediculosis capitis* and *Phthirus pubis* in a young adult population: follow-up of 20 years. J Eur Acad Dermatol Venereol. 2002;16: 257–259.
5. Eberle MW, McLean DL. Observation of symbiote migration in human body lice with scanning and transmission electron microscopy. Can J Microbiol 1983;29:755–762.
6. Chunge RN, Scott FE, Underwood JE, Zavarella KJ. A review of the epidemiology, public health importance, treatment and control of head lice. Can J Public Health 1991;82:196–200.
7. Burkhart CN, Burkhart CG. Fomite transmission in head lice. J Am Acad Dermatol 2007;56:1044–1047.
8. Chunge RN, Scott FE, Underwood JE, Zavarella KJ. A pilot study to investigate transmission of head lice. Can J Public Health 1991;82:207–208.
9. Nutanson I, Steen C, Schwartz RA. Pediculosis corporis: an ancient itch. Acta Dermatovenerol Croat 2007;15: 33–38.
10. Kraus S, Glassman LH. The crab louse—review of physiology and study of anatomy as seen by the scanning electron microscope. J Am Venereal Dis Assoc 1976;2:12–18.
11. Varela JA, Otero L, Espinosa E, Sanchez C, Junquera ML, Vazquez F. *Phthirus pubis* in a sexually transmitted diseases unit a study of 14 years. Sex Trans Dis 2003;30: 292–296.
12. Mumcuoglu KY, Klaus S, Kafka D, Teiler M, Miller J. Clinical observations related to head lice infestations. J Am Acad Dermatol 1991;25:248–251.
13. Peck SM, Wright WW, Gant JQ. Cutaneous reactions to the body louse. JAMA 1943;123:821–826.
14. Fournier PE, Ndihokubwayo JB, Guidran J, Kelly PJ, Raoult D. Human pathogens in body and head lice. Emerg Infect Dis 2002;8:1515–1518.
15. Roux V, Raoult D. Body lice as tools for diagnosis and surveillance of reemerging diseases. J Clin Microbiol 1999;37:596–599.
16. Varela JA, Otero L, Espinosa E, Sanchez C, Junquera ML, Vazquez F. *Phthirus pubis* in a sexually transmitted diseases unit: a study of 14 years. Sex Transm Dis 2003;30: 292–296.
17. Nuttall GH. The biology of *Phthirus pubis*. Parasitol 1919;10:383–405.
18. Couch JM, Green WR, Hirst LW, De La Cruz ZC. Diagnosing and treating *Phthirus pubis* palpebrarum. Surv Opthalolmol 1982;26:219–225.
19. Mumcuoglu KY. Prevention and treatment of head lice in children. Paediatr Drugs 1999;1:211–218.
20. Williams LK, Reichert A, MacKenzie WR, Hightower AW, Blake PA. Lice, nits, and school policy. Pediatrics 2001;107:1011–1015.
21. Mumcuoglu KY, Friger M, Ioffe-Uspensky I, Ben-Ishai F, Miller J. Louse comb versus direct visual examination for the diagnosis of head louse infestations. Pediatr Dermatol 2001;18:9–12.
22. Burkhart CG. Relationship of treatment-resistant head lice to the safety and efficacy of pediculicides. Mayo Clin Proc 2004;79:661–666.
23. Dodd CS. Interventions for treating head lice. Cochrane Database Syst Rev 2001;(2):CD001165.
24. Meinking TL, Entzel P, Villar ME, Vicaria M, Lemard GA, Porcelain SL. Comparative efficacy of treatments for *Pediculosis capitis* infestations: update 2000. Arch Dermatol 2001;137:287–2892.
25. Burgess IF. Human lice and their management. Adv Parasitol 1995;36:271–342.
26. Bhalla M, Thami GP. Reversible neurotoxicity after an overdose of topical lindane in an infant. Pediatr Dermatol 2004;21:597–599.
27. Lebwohl M, Clark L, Levitt J. Therapy for head lice based on life cycle, resistance, and safety considerations. Pediatrics 2007;119:965–974.
28. Jones KN, English JC 3rd. Review of common therapeutic options in the United States for the treatment of *Pediculosis capitis*. Clin Infect Dis 2003;36:1355–1361.
29. Elston DM. What's eating you? *Pediculus humanus* (head louse and body louse). Cutis 1999;63:259–264.
30. Meinking TL, Clineschmidt CM, Chen C, Kolber MA, Tipping RW, Furtek CI, et al. An observer-blinded study of 1% permethrin creme rinse with and without adjunctive combing in patients with head lice. J Pediatr 2002;141:665–670.
31. Roberts RJ, Casey D, Morgan DA, Petrovic M. Comparison of wet combing with malathion for treatment of head lice in the UK: a pragmatic randomized controlled trial. Lancet 2000;356:540–544.
32. Treating head lice. Atlanta: Centers for Disease Control and Prevention; 2001. Available at http://www.cdc.gov/ncidod/dpd/parasites/lice/factsht_head_lice_treating.htm. Accessed May 12, 2008.
33. Opaneye AA, Jayaweera DT, Walzman M, Wade AA. *Pediculosis pubis*: a surrogate marker for sexually transmitted diseases. J Roy Soc Health 1993;113: 6–7.

Chapter 173

MOTHS and BUTTERFLIES (Order: Lepidoptera)

HISTORY

The ancient Greeks attributed some sporadic dermatological disorders to species within the order Lepidoptera. During the Roman era, Galen and Pliny the Elder recorded the irritant potential of species within Lepidoptera. Biting caterpillars were applied as a medicinal treatment during the Middle Ages, and the irritant potential of these caterpillars was recognized again in the 18th century.[1] In Germany during the 1860s, conjunctivitis and iritis were recognized complications of exposure to caterpillar hairs, and the term *ophthalmia nodosa* was applied to the ocular complications of caterpillar hairs during the early 1900s.[2] From 1868 to 1869, the gypsy moth caterpillar (*Lymantria dispar* L.) was intentionally imported from Europe to the Boston area in an attempt to develop an American silk industry. The attempt failed because of the poor quality of silk in their cocoons, but the gypsy moth spread throughout the northeastern United States. White documented the first US case of caterpillar dermatitis from a brown-tail moth (*Euproctis chrysorrhoea* L.) in 1901.[3] In 1914, von Ihering reported the first case of a bleeding diathesis associated with exposure to the South American caterpillars.[4] Caffrey reported the occurrence of a case of occupational asthma (tearing, rhinitis, asthmatic bronchitis) in a biologist studying the New Mexico buck moth caterpillars (*Hemileuca nevadensis* Strech.).[5] During the 1910s and 1920s, outbreaks of caterpillar dermatitis in Texas corresponded to large increases in the caterpillar population, resulting in the closing of public schools for several days in San Antonio during the spring of 1923.[6] During the 1920s, Gilmer published his classic microscopic studies on the urticating hairs of several US caterpillars.[7] Epidemics of caterpillar dermatitis continued sporadically, causing the closing of public schools for several days in Galveston, Texas in 1951.[8] These epidemics occurred primarily in the springtime when aerosolization of the urticating hairs and hemolymph from the first larval instars of the caterpillars cause contact with human skin and mucous membranes. In 1969, Arocha-Pinango first associated a fibrinolysis with exposure to the Venezuelan caterpillars (*Lonomia achelous* Cramer).[9] Human contact with caterpillars from *Lonomia* species causes inflammation at the site of contact, followed by systemic symptoms (e.g., headache, malaise, fever, vomiting). A severe coagulopathy began within 24 hours in some of these patients. Other severe complications include intracranial hemorrhage, acute renal failure, and hemoptysis.

IDENTIFYING CHARACTERISTICS

The order Lepidoptera consists of scaly, winged insects that belong to the large phylum Arthropoda and the class Insecta. Of the estimated 140,000–165,000 butterflies and moths worldwide, most of these insects are harmless. There are at least 11 families [Aganaidae (Hypsidae), Arctiidae, Limacodidae, Lasiocampidae, Lymantriidae, Megalopygidae, Noctuidae, Notodontidae, Saturniidae, Thaumetopoeidae, Zygaenidae] of moths and one family (Nymphalidae) of butterflies that contain about 50–150 venomous species.[10] The four main species involved with human envenomations are

Automeris io Fabricius, *Hemileuca maia* Drury, *Megalopyge opercularis* Smith & Abbot, and *Sibine stimulea* Clemens.[11] Caterpillars are the larval stage of moths and butterflies from the order Lepidoptera.

Common Name: Flannel Moth, Puss Caterpillar (larval form of moth), Woolly Slug, Tree Asp, Italian Asp, El Perrito, Opossum Bug, Asp, Bicho Peludo Negro

Scientific Name: *Megalopyge opercularis* Smith & Abbot

Physical Description: The mature larva has a flat, ovoid, pale yellow to gray-brown body reaching up to 35mm (~1.4in.) in length and 20mm (~0.75in.) in width. The spines are yellow with black tips. This species has a distinctive posterior tuft of hair that does not occur in other *Megalopyge* species. The female moth is larger than the male moth with a wing span of about 4cm (~1.5in.). The yellow-brown adults emerge from cocoons April through June with a second generation appearing September to November.

Common Name: Io Moth, Spiny Caterpillar, Corn Emperor Moth

Scientific Name: *Automeris io* Fabricius

Physical Description: This pale yellow-green caterpillar is 5–6cm (~2in.) in length with red true legs and pro-legs. There is a broad white strip on each side surrounded by a maroon stripe above and a thinner red-purple strip below. Each dorsal segment has raised tubercles or scoli containing a whorl of black-tipped green spines.

Common Name: Gypsy Moth

Scientific Name: *Lymantria dispar* L.

Physical Description: This, hairy blue-gray caterpillar is about 20–30mm (~0.75–1.2in.) in length with lateral tufts of dull yellow hair and two dorsal white stripes. There are round, raised umps or tubercles on each segment with the first five dorsal tubercles being blue and the remaining seven tubercles being red.

Common Name: Browntail Moth

Scientific Name: *Euproctis chrysorrhoea* L.

Physical Description: This hairy browntail moth caterpillar is 40–50mm (~1.5–2in.) long with distinctive dorsal red spots and two dorsolateral white strips.

Common Name: Saddleback Caterpillar

Scientific Name: *Sibine stimulea* Clemens (*Acharia stimulea*)

Physical Description: This brown-red caterpillar reaches 20–30mm (~0.75–1.2in.) in length with a green dorsal midsection surrounded by white and dorsal brown-purple centration spot (saddle) also trimmed in white. Adult moths lay eggs during the summer. The cocoon is a brown, oval encasement that retains stinging hairs.[12]

Common Name: Saturniid Caterpillar, Giant Silkworm Moth

Scientific Name: *Lonomia achelous* Cramer, *Lonomia obliqua* Walker

Physical Description: These large, well-camouflaged caterpillars are about 4.5–5.5cm (~2in.) in length with background colors ranging from green to brown. They have rows of raised tubercles crowned with whorls of easily detachable spines of different sizes that are present on all larval stages (first–sixth larval stage or instar).

EXPOSURE

Geographic Distribution

Lepidopterism is a common problem in the southern United States, where moths cause symptoms similar to those associated with venomous caterpillars.[13] The habitat of the puss caterpillar (*Megalopyge opercularis* Smith & Abbot) includes areas of the southern coastal states from Texas to Florida as well as Maryland, New York, North Carolina, South Carolina, Virginia, Missouri, Arkansas, and Oklahoma. The io moth ranges from eastern Canada to the Atlantic seaboard and along the Gulf Coast to Texas and northern Mexico. The gypsy moth occurs in the northeastern United States.[14] The browntail moth ranges from southeastern Canada to the northeastern United States. Related species include the Oriental tussock moth (*Euproctis flava* Bremer) from Asia and Siberia, Douglas-fir tussock moth (*Orgyia pseudotsugata* McDunnough) in the Pacific northwest and western Canada, and the mulberry tussock moth (*Euproctis similis* Moore). The saddleback caterpillar lives in the southeastern United States. The processionary caterpillars (*Thaumetopoea pityocampa* Schiffermüller, *Thaumetopoea wilkinsoni* Tams, *Thaumetopoea pinivora* Treitschke) inhabit Europe, Israel, and the Mediterranean coast, particularly in pine forests.[15] The white-stemmed gum moth (*Chelepteryx collesi* Gray) of eastern Australia is a cause of minor skin irritation following contact with many spine-like hairs of the cocoon.

TABLE 173.1. Distribution of Common Envenomating Caterpillars of Moths and Butterflies

Moth/Butterfly Family	Species	Distribution
Arctiidae (moth)	*Arctia* spp.,	Worldwide
	Hyphantria spp.	United States, Europe, Japan
Lasiocampidae (moth)	*Malacosoma* spp.	United States, southern Canada
	Dendrolimus pini L.	Central Asia, north Africa
Limacodidae (moth)	*Euclea delphinii* Boisduval	North America
	Sibine stimulea Clemens	Eastern and southern United States
	Phobetron pithecium Abbot & Smith	North America
Lymantriidae (moth)	*Euproctis chrysorrhoea* L.	United States, United Kingdom, Europe, North Africa
	E. pseudoconspersa Strand	Japan
	E. edwardsii Newman	Australia
	E. flava Fabricius	Asia
	E. similis Moore	Worldwide
	Lymantria dispar L.	Eastern United States, Europe
	Orgyia spp.	North Africa, Europe, Asia, North America
	O. leukostigma J. E. Smith	North America
	O. pseudotsugata McDunnough	United States and Canadian Pacific Northwest
Megalopygidae (moth)	*Megalopyge* spp.	United States, Latin America
Notodontidae (moth)	*Lochmaeus* spp.	New World
(processionary	*Ochrogaster contraria* Walker	Australia
caterpillars)	*Thaumetopoea pityocampa* Schiffermüller	Asia, North Africa, Europe
	Thaumetopoea processionea L.	Europe
	Thaumetopoea wilkinsoni Tams	United States. Asia, Africa. Europe
Nymphalidae (butterfly)	*Nymphalis* spp.	North America, Europe, Asia
Saturniidae (moth)	*Automeris io*	New World (Southern Canada, United States, Mexico
	Hemileuca spp.	Venezuela, French Guyana, Northern Brazil
	Lonomia achelous Cramer	Southern Brazil)
	Lonomia obliqua Walker	
Zygaenidae (moth)	*Neoprocris* spp.	New World
	Zygaena spp.	Central and southern Europe

Table 173.1 lists the taxonomy and distribution of common envenomating species of caterpillars.

Behavior

After the fertilized egg is deposited onto appropriate foliage, the eggs hatch into larval forms (caterpillars) that undergo a series (i.e., usually five to six) of molts and transformations (instars). The last instar pupates into a cocoon. Typically, caterpillars forage in the spring and early summer after winter hibernation, and the adult emerges from the cocoon in late summer.[16] An exception is the Megalopygidae family (e.g., the puss caterpillar or *Megalopyge opercularis* Smith & Abbot), which molt in late summer to early fall and the adults emerge from the cocoons in spring. The habitat of the puss caterpillar includes citrus, sycamore, wild plum, hackberry, plum, elm, oak, and corn, as well as household flowers including rose bushes. The io moth caterpillar (*Automeris io* Fabricius) is a solitary feeder that does not cause significant economic damage. The gypsy moth (*Lymantria dispar* L.) feeds communally on deciduous trees, particularly oaks. The saddleback caterpillar (*Sibine stimulea* Clemens) is a solitary feeder that prefers bushes (oleander, croton) and palm trees, but this caterpillar will eat the leaves of many trees and shrubs including apple, basswood, cherry, chestnut, dogwood, elm, maple, oak, and plum. These caterpillars are most active during the late summer and early fall. Moths from the genus *Hylesia* are native to Central and South America, where these nocturnal creatures are attracted to the lights of ships anchored in ports. The dermatitis associated with this moth is called the *Caripito itch* because of the frequent occurrence of this dermatitis in sailors from ships anchored in Caripito, Venezuela. *Euproctis chrysorrhoea* L. inhabits communal silken nests on blackthorn and hawthorn ornamental shrubs in the eastern United States, but this caterpillar will feed on most fruit and ornamental trees. The mistletoe browntail moth (*Euproctis edwardsii* Newman) of Australia marches down the host tree at dawn to hide under rubbish, bark, or grass until nighttime when the caterpillar returns to the mistletoe to feed.[17] Severe urticarial lesions may result from the handling of this material. The processionary tree caterpillars have voracious appetites and cause deforestation. These caterpillars follow a queue, head to tail, often meandering single file following a trailing silk thread like a series of train cars.

PRINCIPAL TOXINS

Venom Composition

The mechanism of action and the composition of secretions from the venomous glands at the base of the specialized hairs are not well defined. The secretions of some caterpillars (e.g., gypsy moth, Oriental tussock moth, browntail) contain histamine, but the use of antihistamines does not usually alter the clinical course of caterpillar dermatitis.[18,19] Fibrinolytic proteases and clotting activators occur in the hemolymph and saliva of the saturniid moth caterpillar (*Lonomia achelous*) from South America.[20] Low molecular weight, basic substances (achelase I, achelase II) activate the fibrinolytic system via plasminogen. These plasmin-like substances are probably activator serine proteinases (Lopap).[21] The hemorrhagic activity of other *Lonomia* species probably differs slightly from the effects of venom from *L. achelous*. The cause of hemorrhage from *Lonomia obliqua* probably results from a consumptive coagulopathy with secondary fibrinolysis rather than primary fibrinolytic activity. An *in vitro* study of venom from *L. obliqua* demonstrated no tissue plasminogen activator-like or fibrinolytic activity as well as no degradation or inhibition of factor XIII.[22]

Venom Apparatus

The skin lesions associated with caterpillar dermatitis result from contact with specialized hollow hairs or spines that are connected to venom glands.[23,24] Hypodermal, glandular epithelial cells produce the venom, which circulates in the hemolymph and collects in external tegument, tubercles (scoli), urticating hairs and spines (setae).[25] Most venomous hairs are straight with sharp ends, although the io caterpillar has a thick, distal chitinous plug.[26] The hairs penetrate the skin and portions of the hairs remain embedded. The poisonous setae or spines may be solid, capped by a distal, detachable tip, or hollow with venom underneath a pigmented plug.[27] Release of the venom occurs when trauma ruptures the poisonous setae or spines, releasing venoms from subapical pores or proximal bases.

Microscopic examination of *Lonomia obliqua* caterpillars indicates that there is no single gland that produces the venom.[28] All spines have hollow internal canals with weak articulations at the tips that allow contact with the venom following rupture of the fragile tips. A secretory epithelium, which underlies the tegument and the spines, secretes the venom. These cells contain vesicles that increase in size and number as they reach the apical region. Deposition of the venom occurs in the subcuticular space and at the tips of the spines. All larval stages (i.e., third to sixth developmental stage or instar) can envenomate, but toxicity increases with each molt (instar) as the size and number of spines increase.

HISTOPATHOLOGY AND PATHOPHYSIOLOGY

The mechanism of dermal toxicity of caterpillars is not well defined. Potential mechanisms include mechanical irritation, hypersensitivity, and envenomation with toxic secretions. Biopsies of lesions associated with caterpillar dermatitis demonstrate nonspecific pathological changes including hydropic degeneration, spongiosis, edema, and a mixed lymphohistiocytic infiltrate.[29,30] A delayed contact dermatitis may result from exposure to large amounts of caterpillar hairs.[14]

Although the venoms from *Lonomia achelous* and *Lonomia obliqua* produce similar clinical effects, the mechanisms of action are somewhat different. The venom of *L. achelous* has both procoagulant and anticoagulant activity, whereas the venom of *L. obliqua* is primarily procoagulant. The main toxins in the former venom are lonomin II (direct fibrinolytic activity) and lonomin V (degrades coagulation factor XIII).[31] In *L. obliqua* venom, the principal procoagulant toxins are a prothrombin activator, Lopap (*L. obliqua* prothrombin activator protease) and a factor X activator, Losac (*L. obliqua* Stuart-factor activator).[32] Acute renal failure complicates envenomation by *L. obliqua*; potential pathogenic factors causing renal dysfunction include hypotension, renal ischemia, disseminated intravascular coagulation with deposition of fibrin in the glomeruli, and direct nephrotoxic effects.[33] The venom from *L. obliqua* also contains factors, which release kinins from low-molecular-weight kininogen. These substances contribute to the edema and hypotension associated with envenomation.[34]

CLINICAL RESPONSE

Approximately 50–150 species within the order Lepidoptera produce inflammatory lesions of the skin and mucous membranes following direct contact. Most adverse effects involve accidental contact of the skin (caterpillar dermatitis) with the larvae.[35] Rarely, caterpillar dermatitis (rhinitis, pruritic maculopapular rash) results from airborne exposure following the appearance of exceptionally large caterpillar populations and favorable climatic conditions.[14,36] Following ingestion of venous caterpillars, drooling and dysphagia may develop as a result of a local reaction to embedded spines, but systemic symptoms do not usually occur.[37] There are at least five syndromes associated with exposure to caterpillars including erucism, lepidopterism, dendrolimiasis, ophthalmia nodosa, and blood disorders characterized by consumptive coagulopathy with secondary fibrinolysis.[25]

Caterpillar Dermatitis (Erucism)

Erucism is the term applied to the classic lesion of caterpillar dermatitis. The skin lesions associated with exposure to venomous species of Lepidoptera typically develop immediately (i.e., within several hours) after contact with the hairs, but direct contact with the caterpillar is not required to develop the skin lesions. The skin lesions begin as localized or generalized erythematous, pruritic macules that rapidly evolve into urticarial lesions. Small papules or papulovesicles replace the wheals, and occasionally bullae develop.[38] Although most of the lesions are pruritic, some species produce painful lesions including the puss (*Megalopyge opercularis*), io (*Automeris io*),[26] saddleback,[29] and oak processionary caterpillars.[39,40] The puss caterpillar produces characteristic grid-like, hemorrhagic skin lesions within two to three hours following envenomation along with intense, radiating pain and generalized muscle cramps.[41,42] Contact with species of the family Megalopygidae (e.g., puss caterpillar) may produce serous or hemorrhagic bullae as well as the rapid onset of lymphangitis and regional lymphadenopathy, but the skin lesions usually resolve without sequelae.[43,44] The duration of the lesions depends on the species. Most symptoms of envenomation by saddleback and io caterpillars subside within two to eight hours, whereas resolution of the lesions associated with contact by gypsy moths requires up to about two days.[45] Other lesions may require seven to ten days to heal including the dermatitis associated with puss caterpillar, browntail moth,[35] processionary caterpillar,[46] and the yellowtail moth.[10]

Lepidopterism

Lepidopterism is the systemic illness associated with direct or airborne contact with the urticarial hairs or body fluids of caterpillars. In a case series of over 40 patients developing dermal lesions after exposure to the oak processionary caterpillar, only a few patients had direct contact with the caterpillars.[47] Systemic symptoms include paresthesias, vomiting, malaise, headache, generalized urticaria, conjunctivitis, pharyngitis, cough, muscle spasms, and radiating pains.[48,49] Although systemic symptoms are rare, angioedema of the upper respiratory tract, dyspnea, bronchoconstriction, and wheezing may develop in atopic patients, particularly in patients with documented allergies to bees.[18,50] The onset of the contact dermatitis days after exposure secondary to delayed sensitivity complicates the diagnosis of lepidopterism.[51] There are no well documented cases of deaths from lepidopterism other than deaths associated with blood disorders.[52] Complications include secondary bacterial infection, lichen simplex chronicus, and postinflammatory dyschromia.[10] IgE-medicated anaphylactic reactions may occur, particularly in workers chronically exposed to pine processionary caterpillar.[53] The onset of lepidopterism is more common following contact with the gypsy moth (*Lymantria dispar*) or the browntail moth (*Euproctis* spp.) caterpillar rather than the io moth (*Automeris io*) or the saddleback (*Sibine stimulea*) caterpillar.[25,54]

Dendrolimiasis

This syndrome is a chronic form of lepidopterism caused by direct contact with the hemolymph, cocoon, or urticating hairs of the central Asian pine-tree lappet moth or caterpillar (*Dendrolimus pini* L.).[55] Clinical effects include a maculopapular dermatitis, migratory polyarthritis, inflammatory polychondritis, osteoarthritis, and rarely, acute scleritis. The clinical course of this illness suggests an IgE-mediated acute phase secondary to hypersensitivity to foreign proteins followed by an autoimmune-mediated, chronic phase involving the bones and joints. The mean onset of the acute phase is about 1–3 days.

Ophthalmia Nodosa

This disease is a chronic ocular disorder characterized by conjunctivitis followed by panuveitis resulting from the intraocular penetration of urticating hairs from lymantriid caterpillars.[56] However, the most common cause of ophthalmia nodosa is contact with tarantula hairs.[57] Airborne exposure to caterpillar setae usually causes mild, superficial ocular injuries compared with the intraocular injuries associated with application of caterpillars directly to the eye. A 15-year-old male presented to an emergency department 2 days after a caterpillar was rubbed in his eye.[2] Under general anesthesia, 150–250 caterpillar setae were removed from his cornea. Despite repeated removal of caterpillar setae from his eye, he developed persistent iritis and subsequently chorioretinitis from the migration of setae into the interior of his eye. Ultimately, he underwent left pars plana vitrectomy, but his visual acuity remained below normal.

A variety of clinical effects can result from exposure to setae including the following: 1) acute chemosis and conjunctivitis, 2) chronic mechanical keratoconjunctivitis secondary to embedded setae in the conjunctiva, 3) subconjunctival or corneal granulomas, 4) iritis, and 5) chorioretinitis secondary to penetration of the setae into the posterior chamber.[58]

Blood Disorders

In addition to severe local pain, envenomations by *Lonomia* caterpillars of South American produce coagulopathy, bleeding, and renal dysfunction. The venom of the spiny-hair Venezuelan (*Lonomia achelous*) or Brazilian saturniid (*Lonomia obliqua*) caterpillars contains anticoagulant and procoagulant factors that potentially may cause serious bleeding abnormalities. The clinical features of envenomation by these two caterpillars are similar, although renal complications are less common following envenomation by *L. achelous* than by *L. obliquá*.[59] Although there are limited clinical data, case series associate envenomation by these caterpillars with internal bleeding, renal dysfunction, and intracranial hemorrhage,[60] as well as painful skin lesions, nausea, vomiting, lightheadedness, and headache.[61,62] The clinical effects of *L. achelous* envenomation resemble a combination of fibrinolytic activity (plasmin-like, plasminogen activation) and mild disseminated intravascular coagulation (procoagulation activity).[63] *L. obliqua* envenomation results in a consumptive coagulopathy and secondary activation of fibrinolysis.

DIAGNOSTIC TESTING

Radiographic features of dendrolimiasis include epiphyseal separation, initial swelling of small and intermediate joints and adjacent soft tissue, osteoporosis near insertion of tendons and cartilage, osteosclerosis, soft tissue ossifications, and fusion of ankle joints.

Laboratory abnormalities associated with envenomation by *Lonomia obliqua* typically include anemia, prolonged prothrombin time, decreased coagulation factors V, VIII, and XIII, decreased fibrinogen and plasminogen, and increased fibrin degradation products.[33] These abnormalities are consistent with a consumptive coagulopathy with secondary fibrinolysis. In contrast to disseminated intravascular coagulation associated with multisystem failure, the platelet count is relatively normal following envenomation by *L. obliqua*.[64]

TREATMENT

Management of caterpillar dermatitis is supportive (local wound care, elevation, ice). There are few clinical data to guide management. Gently wash the area of the sting with soap and water followed by "no touch" drying of the sting site with a hair dryer rather than a towel. Methods to remove embedded urticating spines include the application of adhesive duct tape, rapidly drying clear fingernail polish, or commercial facial peel solutions. These peeling solutions should dry thoroughly prior to removal. Application of ice packs may improve the pain associated with envenomation. Antihistamines (H_1-, H_2-blockers) are usually effective only in mild cases of lepidopterism. Case reports suggest that high-dose steroids may alleviate severe pain and pruritus associated with caterpillar dermatitis, particularly if a hypersensitivity reaction is suspected.[65] Envenomation by the puss caterpillar may produce intense pain that requires parenteral narcotic analgesics for adequate control.[41] General supportive measures include the use of calamine lotion and soothing soaks (e.g., Epsom salts or Domeboro solution) as needed. The clinical effects of the white-stemmed gum moth (*Chelepteryx collesi* Gray) are generally minor with mild local pain, and complete removal of the hairs is neither possible nor necessary.[66] Treatment of ophthalmia nodosa requires referral to an ophthalmologist for medical and surgical management.

A specific heterologous $F(ab')_2$ antivenom for envenomation by *Lonomia obliqua* is available in Brazil for the treatment of systemic effects associated with envenomation.[67] Animal and human studies suggest that the antivenom (antilonomic serum; Instituto Butantan, Sao Paulo, Brazil) reverses the hemorrhagic syndrome associated with contact by *Lonomia obliqua* caterpillars from Brazil.[67] A randomized, prospective study of 44 patients with grade I (external bleeding only) or II (external and internal bleeding) hemorrhagic syndrome indicated that the administration of 3 vials of this antivenom resolves the coagulation abnormalities within a mean of approximately 15 hours after administration.[67] Early treatment of the consumptive coagulopathy associated with Lonomia caterpillars is important to reduce complications. The administration of whole blood or fresh frozen plasma usually causes thrombocytopenia and renal dysfunction.

References

1. Fabre JH. The life of the caterpillar. New York: Dodd Mead & Co.; 1918.
2. Fraser SG, Dowd TC, Bosanquet RC. Intraocular caterpillar hairs (setae): clinical course and management. Eye 1994;8:596–598.
3. White J. Dermatitis induced by the caterpillar of the browntail moth. Boston Med Surg J 1901;144:599.

4. von Ihering R. Estudo biologica das largartas urticantes ou tatoranas. Ann Paulistas Med Cirugia 1914;3:129–139.
5. Caffrey DJ. Notes on the poisonous urticating spines of *Hemileuca oliviae* larvae. J Econ Entomol 1918;11: 363–367.
6. Foot NC. Pathology of the dermatitis caused by *Megalopyge opercularis*, a Texas caterpillar. J Exp Med 1922;35: 747–753.
7. Gilmer PM. A comparative study of the poison apparatus of certain lepidopterous larvae. Ann Entomol Soc Am 1925;18:203–239.
8. Micks DW. Clinical effects of the sting of the "puss caterpillar" (*Megalopyge opercularis* S & A) on man. Tex Rep Biol Med 1952;10:399–405.
9. Archoa-Pinango CL. Fibrinolysis produced by contact with a caterpillar. Lancet 1969;i:810–812.
10. Rosen T. Caterpillar dermatitis. Dermatol Clinics 1990;8: 245–252.
11. Everson GW, Chapin JB, Normann SA. Caterpillar envenomations: a prospective study of 112 cases. Vet Hum Toxicol 1990;32:114–119.
12. Elston DM. What's eating you? Saddleback caterpillar (*Acharia stimulea*). Cutis 80;110–112
13. Green VA, Siegal CJ. Bites and stings of Hymenoptera, caterpillar and beetles. J Toxicol Clin Toxicol 1983–1984;21: 491–502.
14. Beaucher WN, Farnham JE. Gypsy moth caterpillar dermatitis. N Engl J Med 1982;306:1301–1302.
15. Maier H, Spiegel W, Kinaciyan T, Honigsmann H. Caterpillar dermatitis in two siblings due to the larvae of *Thaumetopoea processionea* L., the oak processionary caterpillar. Dermatology 2004;208:70–73.
16. Smith WD. Contact urticaria due to the brown-tail moth. Practitioner 1966;196:690–694.
17. Dunlop K, Freeman S. Caterpillar dermatitis. Austral J Dermatol 1997;38:193–195.
18. Etkind PH, Odell TM, Canada AT, Shama SK, Finn AM, Tuthill R. The gypsy moth caterpillar: a significant new occupational and public health problem. J Occup Med 1982;24:659–662.
19. Kawamoto F, Suto C, Kumada N. Studies on the venomous spicules and spines of some moth caterpillars in Japan. Jpn J Med Sci Biol 1979;32:114–117.
20. Kelen EM, Picarelli ZP, Durarte AC. Hemorrhagic syndrome induced by contact with caterpillars of the genus *Lonomia* (Saturniidae, Hemileucinae). J Toxicol Toxin Rev 1995;14:283–308.
21. Amarant T, Burkhart W, LeVine H III , Arocha-Pinango CL, Parikh I. Isolation and complete amino acid sequence of two fibrinolytic proteinases from the toxic *Saturniid* caterpillar *Lonomia achelous*. Biochim Biophys Acta 1991;1079:214–221.
22. Fritzen M, Schattner M, Quintana Ribeiro AL, de Fatima Correia Batista I, Ventura J, Prezoto BC, Chudzinski-Tavassi AM. *Lonomia obliqua* venom action on fibrinolytic system. Thomb Res 2003;112:105–110.
23. Lamdin JM, Howell DE, Kocan KM, Murphey DR, Arnold DC, Fenton AW, Odell GV, Ownby CL. The venomous hair structure, venom and life cycle of *Lagoa crispata*, a puss caterpillar of Oklahoma. Toxicon 2000;38: 1163–1189.
24. Henwood BP, MacDonald DM. Caterpillar dermatitis. Clin Exp Dermatol 1983;8:77–93.
25. Diaz JH. The evolving global epidemiology, syndromic classification, management, and prevention of caterpillar envenoming. Am J Trop Med Hyg 2005;72:347–357.
26. Hughes G, Rosen T. *Automeris io* (caterpillar) dermatitis. Cutis 1980;26:71–73.
27. Wirtz RA. Allergic and toxic reactions to non-stinging arthropods. Ann Rev Entomol 1984;29:47–69.
28. Veiga AB, Blochtein B, Guimaraes JA. Structures involved in production, secretion and injection of the venom produced by the caterpillar *Lonomia obliqua* (Lepidoptera, Saturniidae). Toxicon 2001;39:1343–1351.
29. Edwards EK Jr, Edwards EK, Kowalczyk AP. `Contact urticaria and allergic contact dermatitis to the saddleback caterpillar with histologic correlation. Int J Dermatol 1986;25:467.
30. de Jong MC, Hoedemaeker J, Jongebloed WL, Nater JP. Investigative studies of the dermatitis caused by the larva of the brown-tail moth (*Euproctis chrysorrhoea* Linn.) II. Histopathology of skin lesions and scanning electron microscopy of their causative setae. Arch Dermatol Res 1976;255:177–191.
31. Guerrero BA, Arocha-Pinango CL, Gil San Juan A. Degradation of human factor XIII by lonomin V, a purified fraction of *Lonomia achelous* caterpillar venom. Thromb Res 1997;87:171–181.
32. Donato JL, Moreno RA, Hyslop S, Duarte A, Antunes E, Le Bonniec BF, et al. *Lonomia obliqua* caterpillar spicules trigger human blood coagulation via activation of factor X and prothrombin. Thromb Haemost 1998;79: 539–542.
33. Carrijo-Carvalho LC, Chudzinski-Tavassi AM. The venom of the *Lonomia* caterpillar: an overview. Toxicon 2007;49: 741–757.
34. Bohrer CB, Junior JR, Fernandes D, Dordi R, Guimaraes JA, Assreuy J, Termignoni C. Kallikrein-kinin system activation by *Lonomia obliqua* caterpillar bristles: involvement in edema and hypotension responses to envenomation. Toxicon 2007;49:663–669.
35. Blair CP. The browntail moth, its caterpillar and their rash. Clin Exp Dermatol 1979;4:215–222.
36. De-Long S. Mulberry tussock moth dermatitis. A study of an epidemic of unknown origin. J Epidemiol Community Health 1981;35:1–4.
37. Lee D, Pitetti RD, Casselbrant ML. Oropharyngeal manifestations of lepidopterism. Arch Otolaryngol Head Neck Surg 1999;125:50–52.

38. Redd JT, Voorhees RE, Török TJ. Outbreak of lepidopterism at a Boy Scout camp. J Am Acad Dermatol 2007;56: 952–955.
39. Maier H, Spiegel W, Kinaciyan T, Krehan H, Cabaj A, Schopf A, Honigsmann H. The oak processionary caterpillar as a cause of an epidemic airborne disease: survey and analysis. Br J Dermatol 2003;149:990–997.
40. Garty BZ, Danon YL. Processionary caterpillar dermatitis. Pediatr Dermatol 1985;2:194–196.
41. Holland DL, Adams DP. "Puss caterpillar" envenomation: a report from North Carolina. Wild Environ Med 1998;9: 213–216.
42. Stipetic ME, Rosen PB, Borys DJ. A retrospective analysis of 96 "asp" (*Megalopyge opercularis*) envenomations in Central Texas during 1996. Clin Toxicol 1999;37:457–462.
43. McGovern JP, Barkin GB, McElhenney TR, Wende R. *Megalopyge opercularis.* Observations of its life history, natural history of its sting to man, and reports of epidemic. JAMA 1961;175:1155–1158.
44. Gardner TL, Elston DM. Painful papulovesicles produced by the puss caterpillar. Cutis 1997;60:125–126.
45. Tuthill RW, Canada AT, Wilcock K, Etkind PH, O'Dell TM, Shama SK. An epidemiologic study of gypsy moth rash. Am J Public Health 1984;74:799–803.
46. Ducombs G, Lamy M, Mollard S, Guillard JM, Maleville J. Contact dermatitis from processional pine caterpillar (*Thaumetopoe pityocampa* Schiff Lepidoptera). Contact Dermatitis 1981;7:287–278.
47. Gottschling S, Meyer S. An epidemic airborne disease caused by the oak processionary caterpillar. Pediatr Dermatol 2006;23:64–66.
48. Gottschling S, Meyer S, Dill-Mueller D, Wurm D, Gortner L. Outbreak report of airborne caterpillar dermatitis in a kindergarten. Dermatology 2007;215:5–9.
49. Finkelstein Y, Raikhlin-Eisenkraft B, Taitelman U. Systemic manifestations of erucism: a case report. Vet Hum Toxicol 1988;30:573–574.
50. Mulvaney JK, Gatenby PA, Brookes JG. Lepidopterism: two cases of systemic reactions to the cocoon of a common moth, *Chelepteryx collesi*. Med J Aust 1998;168:610–611.
51. Hellier FF, Warin RP. Caterpillar dermatitis. Br Med J 1967;2:346–348.
52. Burnett JW, Calton GJ, Morgan RJ. Caterpillar and moth dermatitis. Cutis 1986;37:320.
53. Vega J, Vega JM, Moneo I, Armentia A, Caballero ML, Miranda A. Occupational immunologic contact urticaria from pine processionary caterpillar (*Thaumetopoea pityocampa*): experience in 30 cases. Contact Dermatitis 2004;50:60–64.
54. Lamy M, Werno J. [The brown-tail moth of Bombyx *Euproctis chrysorrhoea* L. (Lepidoptera) responsible for lepidopterism in France: biological interpretation]. C R Acad Sci III 1989;309:605–610. [French]
55. Huang DZ. Dendrolimiasis: an analysis of 58 cases. J Trop Med Hyg 1991;94:79–87.
56. Teske SA, Hirst LW, Gibson BH, O'Connor PA, Watts WH, Carey TM. Caterpillar-induced keratitis. Cornea 1991;10:317–321.
57. Belyea DA, Tuman DC, Ward TP, Babonis TR. The red eye revisited: ophthalmia nodosa due to tarantula hairs. South Med J 1998;91:565–567.
58. Cadera W, Pachtman MA, Fountain JA, Ellis FD, Wilson FM 2nd. Ocular lesions caused by caterpillar hairs (ophthalmia nodosa). Can J Ophthalmol 1984;19:40–44.
59. Gamborgi GP, Metcalf EB, Barros EJ. Acute renal failure provoked by toxin from caterpillars of the species *Lonomia obliqua*. Toxicon 2006;47:68–74.
60. Kowacs PA, Cardoso J, Entres M, Novak EM, Werneck LC. Fatal intracerebral hemorrhage secondary to *Lonomia obliqua* caterpillar envenoming. Case report. Arq Neuropsiquiatr 2006;64:1030–1032.
61. Zannin M, Lourenco DM, Motta G, Dalla Costa LR, Grando M, Gamborgi GP, et al. Blood coagulation and fibrinolytic factors in 105 patients with hemorrhagic syndrome caused by accidental contact with *Lonomia obliqua* caterpillar in Santa Catarina, southern Brazil. Thromb Haemost 2003;89:355–364.
62. Arocha-Pinango CL, de Bosch NB, Torres A, Goldstein C, Nouel A, Arguello A, et al. Six new cases of a caterpillar-induced bleeding syndrome. Thromb Haemost 1992;67: 402–407.
63. Arocha-Pinango CL, Guerrero B. *Lonomia* genus caterpillar envenomation: clinical and biological aspects. Haemost 2001;31:288–293.
64. Zannin M, Lourenco DM, Motta G, Dalla Costa LR, Grando M, Gamborgi GP, et al. Blood coagulation and fibrinolytic factors in 105 patients with hemorrhagic syndrome caused by accidental contact with *Lonomia obliqua* caterpillar in Santa Catarina, southern Brazil. Thromb Haemostasis 2003;89:355–364.
65. el-Mallakh RS, Baumgartner DL, Fares N. "Sting" of puss caterpillar, *Megalopyge opercularis* (Lepidoptera: Megalopygidae): first report of cases from Florida and review of literature. J Fla Med Assoc 1986;73:521–525.
66. Balit CR, Geary MJ, Russell RC, Isbister GK. Clinical effects of exposure to the white-stemmed gum moth (*Chelepteryx collesi*). Emerg Med Australas 2004;16:74–81.
67. Rocha-Campos AC, Goncalves LR, Higashi HG, Yamagushi IK, Fernandes I, Oliveira JE, et al. Specific heterologous F(ab′)2 antibodies revert blood incoagulability resulting from envenoming by *Lonomia obliqua* caterpillars. Am J Trop Med Hyg 2001;64:283–289.
68. Caovilla JJ, Guardao Barros EJ. Efficacy of two different doses of antilonomic serum in the resolution of hemorrhagic syndrome resulting from envenoming by *Lonomia obliqua* caterpillars: a randomized controlled trial. Toxicon 2004;43:811–818.

Chapter 174

AMPHIBIANS—TOADS, FROGS, SALAMANDERS, and NEWTS
(Class: Amphibia)

TOADS and FROGS

Colorado River Toad

HISTORY

Anthropologists suggest that pre-Columbian peoples used the common marine toad, *Bufo marinus* L., as an intoxicant during ritualistic ceremonies. Toads from genus *Bufo* contain the first hallucinogenic agent documented in the animal kingdom.[1]

IDENTIFYING CHARACTERISTICS

Over 200 species of toads throughout the world produce a wide variety of biologically active substances stored in parotid glands on their backs. The giant marine toad of South and Central America is *Bufo marinus* L., whereas the main European and Asian species are *B. bufo* L. and *B. gargarizans* Cantor, respectively.

Common Name: Colorado River Toad, Sonoran Desert Toad

Scientific Name: *Bufo alvarius* Girard in Baird

Family Name: Bufonidae

Physical Description: Compared with frogs, toads are stout, slower-moving amphibians with short limbs and small webs on their feet. The dorsal skin is rough with numerous dermal protuberances and cryptic coloration. The Colorado River toad is a large amphibian with prominent parotid glands that secrete viscous, milky-white venom. This toad and the cane toad (*Bufo marinus* L.) are morphologically similar except cane toads have a distinctive head and face manifest by a visor extending over each eye and a high, angular boney ridge extending from the eyes to the nose.

Common Name: Cane Toad

Scientific Name: *Bufo marinus* L.

Family Name: Bufonidae

Physical Description: Adult cane toads are relatively large toads with weights up to about 4 pounds (~2 kg) and length of approximately 4–9 in. (~10–23 cm) and their skin is warty. The female is usually bigger than the male. The ventral surfaces of their round flat bodies are yellow-white with darker mottling, whereas their backs and sides vary from olive-brown or reddish-brown to gray and yellow along with light middorsal stripes. The front feet do not have webs between the toes.

EXPOSURE

Geographic Distribution

The Colorado River Toad ranges from southeastern California to Arizona and Northwestern Mexico. The

Medical Toxicology of Natural Substances, by Donald G. Barceloux, MD

natural habitat of the cane toad (*B. marinus* L.) is the Amazon basin in South America through Central America to Mexico and southern Texas along the Rio Grande River. In 1935, this species was introduced into eastern Australia for pest control. Cane toads were also introduced in many other parts of the world (e.g., Philippines, Florida, Puerto Rico, Caribbean Islands, Venezuela). This tropical toad prefers forested areas with nearby water sources. *Bufo compactilis* Wiegmann ranges from Northern Arizona to Texas and Northeastern Mexico. The Chinese toad is *Bombina maxima* Boulenger (*Bufo bufo gargarizans*). The guttural toad (*Bufo gutturalis* Power) and the raucous toad (*Bufo rangeri* Hewitt) reside in South Africa.[2]

Behavior

The Colorado River toad is a semiaquatic amphibian found only in the Sonoran desert. This toad remains underground in a dormant state from September to April until the breeding season starts. As the summer rains begin, the toads emerge from hibernation, particularly at night. The cane toad (*B. marinus* L.) is primarily nocturnal and ground-dwelling. These toads are particularly attracted to lights at night and feed both on live and dead matter. They have a voracious appetite that includes eating almost any animal (e.g., insects, small lizards, frogs, birds, fish, mice, bees, worms, and beetles) they are capable of ingesting.

Uses

Chan Su is a traditional Chinese medicine that is prepared from the dried white secretions of the auricular and skin glands of Chinese toads including the Chinese cane toad. Other traditional Chinese medicines (e.g., Lin-Shen-Wan, Kyrushin) also contain substantial amounts of Chan Su. These preparations are constituents of traditional expectorants, diuretics, cardiotonic agents, antiarrhythmics, local analgesics (sore throat, tonsillitis, teeth), and detoxifying agents for the blood and skin.[3,4] The West Indian aphrodisiac "Love Stone" and Chan Su are derived from similar sources based on virtually identical gas chromatography/mass spectrometry (GC/MS) profiles.[5] Besides the presence of cardiotonic steroids (bufalin, cinobufagin, bufogenin), these products contain the hallucinogenic substance, bufotenine, which is an isomer of psilocin. Toad licking is a method of drug abuse popularized in the 1980s, particularly in areas where the toads are indigenous.[6]

PRINCIPAL TOXINS

Venom Composition

Toad venoms contain a complex variety of toxins including biogenic amines (epinephrine, norepinephrine), indole alkylamines [bufotenin (bufotenine, cinobufotenine), bufothionine, dehydrobufotenine, serotonin], C24 cardioactive steroids (bufadienolides), and conjugates of bufogenins (bufotoxins).[7] The major bufadienolides in these preparations are bufalin, cinobufagin, cinobufotalin, and bufogenin (resibufogenin).[8] The main toxins in toad venoms are probably these bufadienolides, which have cardiac effects similar to digitoxin; these compounds vary substantially between different species of toads.[9] Bufadienolides are structurally similar to cardiac glycosides (cardenolides), such as digitoxin, digoxin, and oleandrin present in the plant kingdom.[10] Digitoxin is an analog of digoxin with one less hydroxyl group on the steroid part of the molecule. Both bufadienolides and cardenolides have a steroid nucleus; however, bufadienolides have a six-member lactone ring compared with a five-member lactone ring in cardenolides.[11] The cardiac genins (aglycones) in toads do not have the attached sugars present in cardiac glycosides in the plant kingdom. Figure 174.1 demonstrates the structural similarities between bufadienolides (bufalin, cinobufotalin), plant cardenolides (oleandrin, oleandrigenin), and cardiac glycosides (digoxin, digitoxin).

Experimental studies in humans suggest that the psychedelic substance, bufotenin (5-hydroxy-*N,N*-dimethyltryptamine, CAS RN: 487-93-4), does not produce the severe cardiac toxicity associated with toad poisoning.[12] Bufotenin is also a constituent of Cohoba seeds (*Anadenanthera peregrina* (L.) Speg.) from Haiti and some fungi in concentrations up to 1.9% including the fungal species, *Amanita citrina* var. *citrina* (Schaeff.) Pers., *Amanita porphyria* Alb et Schw.:Fr., and *Amanita rubescens* (Pers.:Fr.) Gray.[13] *O*-methyl transferase in the Sonoran desert toad (*B. alvarius* Girard in Baird) catalyzes the conversion of bufotenin to the more potent psychedelic substance, 5-methoxy-*N,N*-dimethyltryptamine. Tropical toads from the genus *Atelopus* contain tetrodotoxin including *Atelopus chiriquiensis* Shreve, *Atelopus oxyrhynchus* Boulenger, *Atelopus peruensis* Gray and Cannatella, *Atelopus subornatus* Werner, and *Atelopus zeteki* Dunn.[14,15]

Venom Apparatus

When threatened, toads secrete venom from parotid-like glands in the skin near their eyes. The cane toad (*B. marinus* L.) has a large, protuberant parotid gland behind each ear. The parotid and tibial glands of the

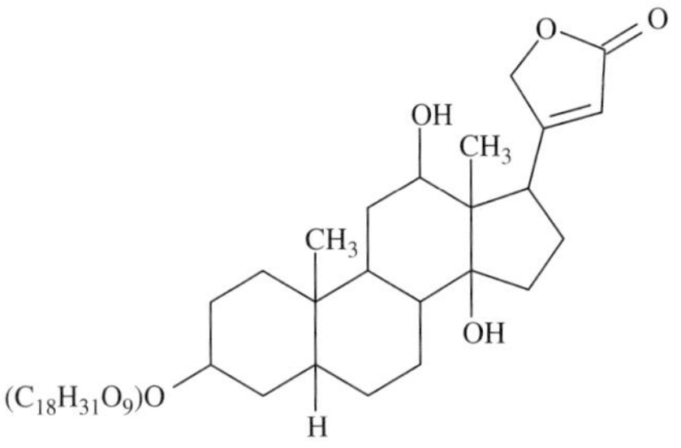

Digoxin 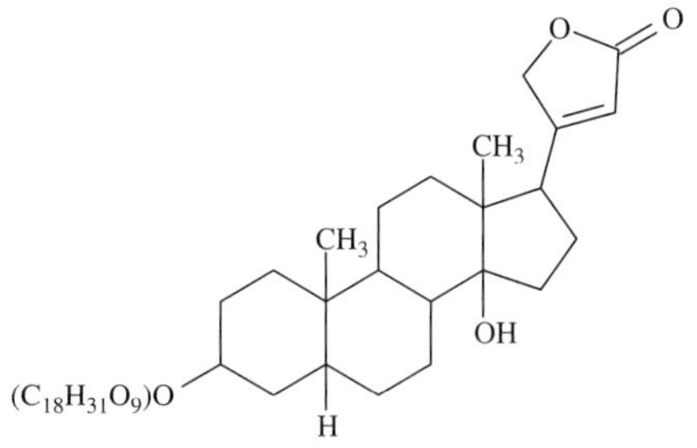 **Digitoxin**

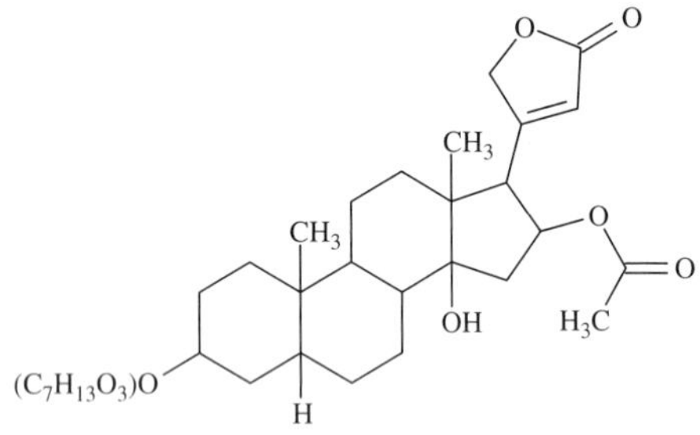

Oleandrin 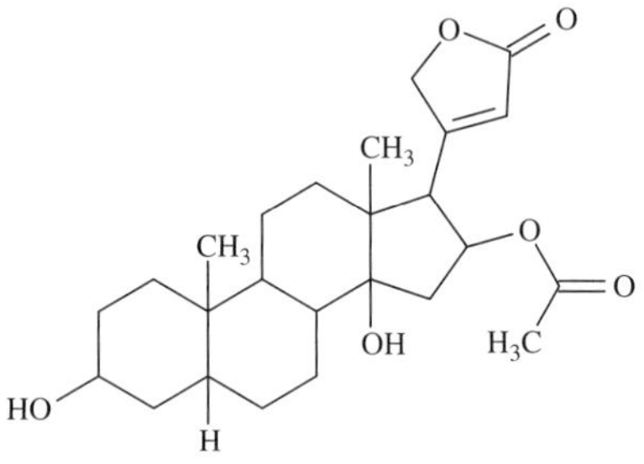 **Oleandrigenin**

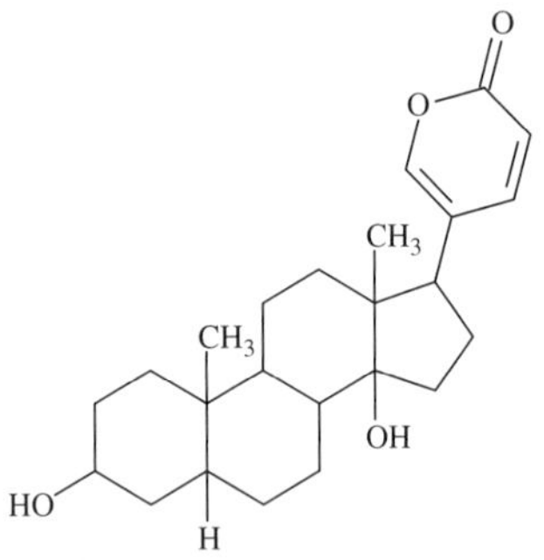

Bufalin 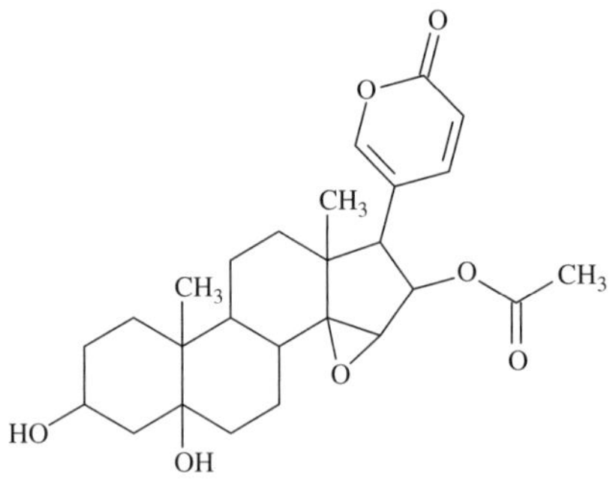 **Cinobufatalin**

FIGURE 174.1. Chemical structures of bufadienolides and similar cardiac glycosides.

Sonoran desert toad can contain relatively large amounts (i.e., up to 15% by dry weight) of the psychedelic substance, 5-methoxy-*N,N*-dimethyltryptamine, as a result of the *O*-methylation of bufotenin.[16]

Dose Response

The toxicity of the Colorado River toad (*Bufo alvarius* Girard in Baird), the cane toad (*Bufo marinus*) and the Chinese toad (*Bombina maxima* Boulenger) are similar. Substantial variation in sensitivity to digoxin-like compounds and presumably to bufadienolides occurs between animal species. Humans, cats, dogs, and rabbits are highly sensitive, whereas rats, mice, and toads are relatively insensitive to these compounds.[17] In clinical studies on healthy volunteers, intravenous doses of 4–16 mg bufotenin produced hallucination, distortion of time and space, depersonalization, mydriasis, and facial erythema.[12] The higher doses also caused shivering, ataxia, and purple-black visual illusions.

TOXICOKINETICS

Following ingestion of Chinese medications containing toad venom, peak concentrations of bufadienolides in healthy volunteers occur about 1–2 hours after ingestion.[18] The elimination half-lives of bufalin, cinobufagin, and bufogenin (resibufogenin) are approximately 16 hours, 9 hours, and 11½ hours, respectively. Bufalin is strongly bound to serum albumin. Animal studies indicate that cinobufagin is deacetylated to less active metabolites (desacetylcinobufagin, epidesacetylcinobufagin), but there are substantial interspecies variation in the metabolism of cinobufagin.[19]

HISTOPATHOLOGY AND PATHOPHYSIOLOGY

Although toad secretions contain a variety of toxins, the major toxins are probably the bufadienolides (bufalin, cinobufagin, bufogenin). These compounds are specific inhibitors of the plasma membrane bound Na^+/K^+ ATPase, and this action produces a digoxin-like toxicity. Toxins from the Colorado River toad (*B. alvarius*), African toad (*Bufo regularis* Reuss), common marine or cane toad (*Bufo marinus*), and American toad (*Bufo americanus* Holbrook) also enhance the adenosine 3′, 5′-monophosphate content in mononuclear leukocytes.[20]

CLINICAL RESPONSE

Acute Toxicity

The clinical course of toad poisoning is similar to digoxin toxicity,[21] and the ingestion of high doses of traditional Chinese medicines containing toad venom has been associated with cardiac dysrhythmias, dyspnea, convulsions, and death.[22] In a series of 6 patients ingesting a topical aphrodisiac containing toad venom, all patients developed vomiting and bradycardia.[11] Other clinical features included generalized weakness, salivation, abdominal pain, diarrhea, coldness and numbness of the lower extremities, mental status changes, hypotension, and noncardiogenic pulmonary edema. Four of the 6 patients died following intractable ventricular fibrillation. Within 15 minutes after placing a toad in his mouth, a five-year-old boy developed drooling and generalized tonic-clonic seizures.[23] Additional effects included total left-sided hemiparesis, Babinski sign, slurred speech, tachypnea, tachycardia, systolic hypertension, and hypokalemia. The neurologic deficits resolved within one week. Nausea, vomiting blurred vision, general weakness, perioral numbness, coma, high-grade A-V block, and sudden cardiopulmonary arrest occurred in a 31-year-old male who ingested a bowl of toad soup stewed with rice wine.[24] A 40-year-old man ingested an aphrodisiac containing toad venom, and he developed bradycardia, A-V block, hyperkalemia, respiratory distress, and terminal ventricular fibrillation.[25]

Abuse

"Toad licking" is a colloquial euphemism for the ingestion of the glandular secretions of the bufo toad. This activity was popularized by the lay press in the late 1980s after the publication of a book by Albert Most that explained the use of the bufo toad for ritual and pleasure.[26] One of the constituents of these glandular secretions is bufotenin, which has been classified as a Schedule I drug by the US Drug Enforcement Administration since 1967. There are limited clinical data on the prevalence and the adverse effects of this form of drug abuse.[27] Toad licking can transmit bacterial diseases, such as salmonellosis.

DIAGNOSTIC TESTING

Biomarkers

Postmortem analysis of liver samples from a 30-year-old woman, who died after ingesting a traditional Chinese medicine derived from the skin and parotid venom glands of toad (Chan Su), demonstrated detectable concentrations of several bufadienolides including bufalin, cinobufotalin, and cinobufagin as determined by reversed-phase high performance liquid chromatography and photodiode array detection(RP-HPLC/PDA).[28] Patients ingesting high doses of traditional Chinese medicine (Chan Su, Dan Shen) containing toad venom usually demonstrate positive apparent digoxin concentrations as detected by polyclonal digoxin immunoassays. The apparent digoxin concentration of a woman, who died after the ingestion of a tea brewed from Chan Su, was 4.9 ng/mL as measured by fluorescence polarization digoxin immunoassay.[22] Two hours after the ingestion of toad soap, a 54-year-old man developed vomiting, hypotension, bradycardia, AV block, and hyperkalemia (6.09 mmol/L).[29] His serum dioxin concentration was 1.62 ng/mL. He survived with supportive care despite developing nonsustained ventricular tachycardia 4 hours after ingestion of the soup. These traditional Chinese medications also cross-react with the digitoxin fluorescence polarization immunoassay,[30] but these medications do not cross-react with the digitoxin chemiluminescent immunoassay.[31] The apparent elevation in serum digitoxin immunoassays is much less for cinobufagin and cinobufotalin than for bufalin.

Because of the structural difference between bufadienolides in toad venom and digoxin, the apparent digoxin assays following toad poisoning is relative low even during serious toxicity. Following exposure to bufalin and cinobufotalin, the apparent serum concentrations are much higher for digitoxin than digoxin, as measured by fluorescence polarization immunoassay.[32] The administration of Chinese medications containing toad venom displaces digitoxin from serum protein binding sites causing an increase in free digitoxin concentrations.[33] Consequently, the presence of a positive apparent digoxin or digitoxin immunoassay confirms exposure to bufadienolides, but these concentrations do not correlate to toxicity.[11] A negative apparent digoxin assay does not necessarily exclude exposure to bufadienolides depending on the specificity of the assay; therefore, polyclonal digoxin assays are preferable to monoclonal digoxin assays.

Bufotenin (*N,N*-dimethyl-5-hydroxytryptamine) is a dimethylated indoleamine derivative of serotonin that occurs naturally in human urine samples as a result of the biotransformation of serotonin by indolethylamine *N*-methyltransferase.[34]

Abnormalities

The presence of hyperkalemia suggests serious toad poisoning.[35] Although hyperkalemia is a poor prognostic sign in digoxin toxicity and in toad poisoning,[36] ventricular fibrillation and asystole may occur in patients with normal serum potassium concentrations. A 23-year-old man developed noncardiogenic pulmonary edema, ventricular fibrillation, and asystole the day after ingesting a topical aphrodisiac containing dried toad venom.[11] The serum potassium in a blood sample drawn in the emergency department was 4.3 meq/L.

TREATMENT

Gastric emptying is not usually necessary because profuse vomiting often occurs during toad poisoning. The administration of activated charcoal to alert patients who present to the emergency department within 1 hour after ingestion is a therapeutic option, but there are no clinical data to confirm the clinical efficacy of decontamination measures on outcome following the ingestion of bufadienolides. The treatment of toad poisoning is similar to oleander toxicity. The sudden onset of ventricular fibrillation and cardiovascular collapse may occur during serious toad poisoning; therefore, all patients should be monitored for ventricular dysrhythmias at least 6 hours after ingestion. There are limited clinical data on the effectiveness of digoxin Fab fragments for toad poisoning. The structural differences between bufadienolide and digoxin suggest that cross-reactivity between these two compounds may be limited. *In vitro* binding studies indicate that Digibind© (GlaxoSmithKline Pharmaceuticals, Mississauga, Ontario, Canada) is less effective in neutralizing free bufalin than free cinobufotalin.[32] Case reports associate the administration of digoxin Fab fragments with clinical improvement of toad venom poisoning, but these patients did not demonstrate serious ventricular dysrhythmias or hyperkalemia prior to the use of digoxin Fab fragments.[11] The amount of digoxin Fab fragments administered should be based on empiric estimates (i.e., 10 vials for serious intoxication) rather than serum digoxin concentrations because the serum digoxin concentration underestimates the concentration of bufadienolide compounds. Some differences in efficacy may exist between different digoxin Fab preparations (e.g., Digibind®; DigiFab®, Protherics, Inc., Brentwood, TN) as a result of varying cross-reactivity with the cardioactive steroids present in toad venom. In a study of mice receiving digoxin Fab (Digibind®) 30 minutes after the administration of a lethal dose of bufadienolides, digoxin Fab fragments bound to some of the bufadienolides.[37] However, 7 of the 15 treated mice died compared with all 30 controls. All symptomatic patients or patients with laboratory abnormalities (dysrhythmias, conduction disturbances, hyperkalemia) should be admitted for observation in a monitored setting.

Colombian Poison-Dart Frog

HISTORY

Since pre-Colombian times, the Emerá Chocó and the Noanamá Chocó Indians of western Columbia poisoned blow darts with the secretions from three toxic species of the genus *Phyllobates*.[38] Batrachotoxin is the primary toxin in the skin secretions from the three species of Colombian poison-dart frogs (*Phyllobates terribilis* Myers, Daly and Malkin, *Phyllobates bicolor* Duméril and Bibron, *Phyllobates aurotaenia* Boulenger).[39] There is no evidence that other Central or South American Indian tribes used skin secretions from neotropical frogs on poison darts or arrows.

IDENTIFYING CHARACTERISTICS

Common Name: Colombian Poison-Dart Frog

Scientific Name: *Phyllobates bicolor* Duméril and Bibron, *Phyllobates aurotaenia* Boulenger, *Phyllobates terribilis* Myers, Daly and Malkin

Family: Dendrobatidae

Physical Description: The black-legged dart frog (*Phyllobates bicolor*) typically has golden-yellow skin with varying color patterns and black spots on the hind legs. This frog has terminal disks on the digits for climbing plants and trees. The golden poison dart frog (*P. terribilis*) is a small frog ranging in size up to 5 cm (~2 in.). The skin has a uniform metallic golden yellow color, and tiny disks on the end of the digits facilitate climbing as demonstrated in the specimen of *Phyllobates terribilis* in Figure 174.2. A bony plate in the upper jaw separates this frog from other poison dart frogs. The kokoe poison dart frog (*P. aurotaenia*) is the smallest poison arrow frog. They have a black body with

FIGURE 174.2. Golden Poison Frog (*Phyllobates terribilis* Myers, Daly, and Bibron). © 2005 Rune Midtgaard. See color insert.

a green, orange, or gold U-shaped pattern starting at the snout that persists into maturity along with small, light blue spots on a black background.

EXPOSURE

The genus, *Phyllobates*, contains five species of frogs, which live in tropical areas from Costa Rica (*P. vittatus*) through Panama (*P. lugubris* Schmidt) and western Colombia (*P. bicolor*, *P. aurotaenia*, *P. terribilis*). All three poison-dart frogs inhabit the rain forests of western Columbia. *Phyllobates bicolor* lives in tropical rain forests of Central and South America, where the conditions are moist and humid. They live in colonies on the forest floor, particularly near small streams. Like other poison dart frogs, *P. terribilis* inhabits tropical rain forests. The kokoe dart frog is a highly social frog that requires high humidity and cool temperatures in tropical rain forests.

PRINCIPAL TOXINS

Venom Composition

Cutaneous granular glands from the neotropical poison frogs (family: Dendrobatidae) contain biogenic amines, peptides, bufadienolides (bufogenins) and alkaloids.[40] The latter substances are cyclic nitrogen-containing compounds that include water-soluble alkaloids (tetrodotoxins), lipophilic alkaloids, and alkaloids derived from biogenic amines. The major classes of dendrobatid alkaloids are batrachotoxins, histrionicotoxins, indolizidines, pumiliotoxins, and decahydroquinolines.[40] Minor classes include gephyrotoxins, 2,6-disubstituted piperidines, pyrrolidine alkaloids, pyridylpiperidines (noranabasamine), indole alkaloids (frog calycanthine, frog chimonanthine), azatricyclododecenes, and amidine alkaloids. The skin secretions of all species of *Phyllobates* including the Colombian poison-dart frogs contain the highly toxic steroidal alkaloid, batrachotoxin, whereas species of the genus *Dendrobates* secrete pumiliotoxins A, B, and C, histrionicotoxins, and gephyrotoxins.[41] Some species of the family Dendrobatidae of neotropical frogs do not contain significant amounts of toxins in their glandular secretions.[42]

Tetrodotoxin is the primary toxin in the skin of colorful, Central and South American frogs from the genus *Atelopus* (*A. chiriquiensis* Shreve, *A. peruensis* Gray and Cannatella, *A. oxyrhynchus* Boulenger, *A. subornatus* Werner, *A. varius* Lichtenstein and Martens) of the Bufonidae family.[15,43] This toxins also occurs in the Panamanian frog *Colostethus inguinalis* Cope from the Dendrobatidae family as well as an analogue of tetrodotoxin (chiriquitoxin).[44] This latter substance blocks sodium and potassium channels. Batrachotoxins occur in the contour feathers of the belly, breast, and legs of New Guinean birds including the hooded pitohui (*Pitohui dichrous* Bonaparte), variable pitohui (*Pitohui kirhocephalus* Lesson & Garnot), and the blue-capped ifrita (*Ifrita kowaldi* De Vis).[45] Other feathers from these species contain lesser amounts of batrachotoxins. Batrachotoxin (CAS RN: 23509-16-2) and homobatrachotoxin (CAS RN: 23509-17-3) were not detectable in samples of skin from these birds.

Venom Apparatus

Because of a lack of speed and armament, amphibians rely predominately on chemicals to deter predators. The structure of the secretory glands in the skins of poison-dart frogs from the Dendrobatidae family is similar, even though different frogs secrete substances with varying toxicity. Skin secretions from the genus *Phyllobates* contain the highly toxic batrachotoxins, whereas the secretions from frogs from the genus *Dendrobates* are less toxic.

Dose Response

On the basis of LD_{50} experiments in mice, amphibians contain highly toxic substances.[46] The mouse LD_{50} for batrachotoxin is 2 pg/kg compared with 8 pg/kg for tetrodotoxin and 500 pg/kg for curare.[47] Although the concentration of these toxins in amphibians is highly variable, the golden poison frog (*Phyllobates terribilis*) generally contains the highest batrachotoxin concentrations. Analysis of one specimen from this species detected 1.9 mg of batrachotoxin in comparison to the estimated human lethal dose of 0.02 to 0.2 mg.[48]

HISTOPATHOLOGY AND PATHOPHYSIOLOGY

Mechanism of Toxicity

Batrachotoxin is a potent modulator of voltage-gated sodium channels that can produce dysrhythmias and the depolarization of muscles and nerves. This toxin depolarizes neurons by irreversibly binding to open voltage-dependent sodium channels; continual activation of the receptor results in spastic paralysis. Tetrodotoxin and saxitoxin are sodium channel blockers that bind to a separate and distinct site on the sodium channel protein.[41] The result of the binding of batrachotoxin is a stabilization of the sodium channel in an open conformation. Batrachotoxin also modulates the voltage-gated sodium channel Na(v)1.8, which is an important part of nociception. Alteration of these channels produces altered sensation (i.e., paresthesias).[49] Other natural toxins (e.g., aconitine, veratridine, grayanotoxin) interact at the same site as batrachotoxin, but these compounds are less potent than batrachotoxin.[50] Batrachotoxin does not bind to the calcium or potassium channels.[51]

Target Organs

In guinea pig atria, batrachotoxin demonstrates positive chronotropic and inotropic properties. Dysrhythmias and atrial arrest develop after the application of batrachotoxin to isolated cardiac muscle preparations. Morphologically, batrachotoxin produces swelling of the axon at the node of Ranvier and extracellular fluid accumulation along the axon.[52]

CLINICAL RESPONSE

Case reports of poisoning by the poison dart frogs and batrachotoxin are not well described in the medical literature. Based on similarity in the mechanism of toxicity between batrachotoxin and veratridine, intoxication by batrachotoxin and the poison dart frog would be expected to produce vomiting, bradycardia, and hypotension. Massive poisoning could produce serious cardiac abnormalities including atrioventricular block, junctional and idioventricular escape rhythms, atrial dysrhythmias, ST abnormalities, T wave inversions, QT prolongation, and ventricular fibrillation.

DIAGNOSTIC TESTING

There are no commercially available laboratory tests to detect poisoning by neotropical poison frogs.

TREATMENT

Treatment is supportive. Atropine is the initial drug of choice for symptomatic bradycardia, but the effect of atropine on hypotension is usually limited. Fluids and vasopressors may be necessary to correct hypotension. Cardiac monitoring and electrocardiograms should be instituted in severely poisoned patients because of the potential for the development of conduction blocks and ventricular dysrhythmias.

SALAMANDERS and NEWTS

IDENTIFYING CHARACTERISTICS

The order Caudata contains over 500 species of primitive salamanders, advanced salamanders, newts, and sirens as outlined in Table 174.1.[53] Salamanders have slender bodies, four short legs of equal size, and long tails that are typically as long as their bodies. Some salamanders do not have legs, but all members of this order have tails (tailed amphibians). The primitive salamanders include members of the Cryptobranchidae and Hynobiidae families. The former family contains the largest salamanders (*Andrias davidianus* Blanchard), which grow up to 1.6 m (~5.3 ft.) in length. Members of the other family of primitive salamanders (Hynobiidae) are small- to medium-sized salamanders. These two families of primitive salamanders contain only three species each, whereas the family of lungless salamanders (Plethodontidae) contains over one-half of the species in this order. Salamanders from the genus *Onychodactylus* in this family do not possess lungs. Advanced salamanders comprise the largest number of salamander species, and these salamanders have fused angular and prearticular bones in the lower jaw in contrast to primitive salamanders. Terrestrial mole salamanders (family: Ambystomatidae) have prominent, protruding eyes, thick arms, prominent costal grooves, thick arms, and round tails. Skin colorations range from dark blue patches to large yellow bars on dark backgrounds.

Newts are members of the true salamander family, Salamandridae. Newts are salamanders with a unique pattern of teeth in the roof of their mouths, whereas

TABLE 174.1. Members of the order Caudata

Family	Common Names of Members	Primary Habitat
Ambystomatidae	Mole salamander, axolotl	North America
Amphiumidae	Amphiuma	Southeastern US
Cryptobranchidae	Giant salamander	America, Asia
Dicamptodontidae	Pacific giant salamander	Western US, southwestern Canada
Hynobiidae	Asiatic salamander	Asia, Russia, Middle East
Plethodontidae	Lungless salamander	North, Central, South America
Proteidae	Mudpuppy, olm, waterdog	North America
Rhyacotritonidae	Torrent salamander, seep salamander	Northern California, Oregon, Washington
Salamandridae	True salamander, newt	North America, Europe, Russia, Middle East
Sirenidae	Siren	Midwestern and eastern US, northeastern Mexico

other salamanders have teeth in the upper and lower jaws. Figure 174.3 displays a specimen of the red-spotted newt (*Notophthalmus viridescens* Rafinesque). Unlike salamanders, newts have no costal grooves and their skin is rough in contrast to the moist, mucoid secretions on the skin of most salamanders. The absence of hind legs separates sirens from other species in the order Caudata. Although they have forelimbs, these appendages are reduced in size compared with other species in this order. They also retain their larval gills and one to three gill slits.

EXPOSURE

There are only six living species of primitive salamanders, and these amphibians live in Asia and America with isolated species in the Middle East and Russia. Most advanced salamanders and newts inhabit the cooler regions of the northern hemisphere with the exception of some species in the Plethodontidae family that inhabit tropical areas of Central and South America. Most members of the order Caudata begin as aquatic larvae, but adults may be aquatic, semi-aquatic, or terrestrial. The Plethodontidae family contains a variety of salamander species, which inhabit aquatic, terrestrial, and arboreal habitats. Salamander species in the Amphiumidae family are aquatic, and these species live in swamps and ponds with dense vegetation in the southeastern United States. Species in the Dicamptodontidae and Rhyacotritonidae families inhabit conifer forests, particularly along streams and creeks during the mating season. Many newts inhabit aquatic areas. In general, species in the order Caudata prefer to walk or run rather than leap like frogs or toads.

FIGURE 174.3. Red-spotted Newt (*Notophthalmus viridescens* Rafinesque). © 2007 Rune Midtgaard. See color insert.

PRINCIPAL TOXINS

Venom Composition

The dermal secretions of amphibians contain a variety of compounds including indolealkylamine compounds,

TABLE 174.2. Relative Toxicity of Newts of the Family Salamandridae

Species	Relative Toxicity	Habitat
Taricha torosa Rathke in Eschscholtz*	High	California
Taricha rivularis Twitty*	High	Northern California
Taricha granulosa Skilton*	High	Northwestern US
Notophthalmus viridescens Rafinesque*	Medium	Eastern United States
Cynops pyrrhogaster Boie	Low	Japan
Cynops ensicauda Hallowell	Low	Ryukyu Islands
Paramesotriton hongkongensis Myers and Leviton	Low	China
Triturus vulgaris L.	Slight	Europe and Western Asia
Triturus cristatus Laurenti	Slight	Europe
Triturus alpestris Laurenti	Slight	Europe
Triturus marmoratus Latreille	Slight	Europe

Source: Adapted from Ref 43.
*Primary toxin definitely identified as tetrodotoxin.

alkaloids, biogenic amines, vasoactive peptides, and cardenolide-like steroids. The skin, internal organs and embryos of newts contain tarichatoxin, which is structurally identical to tetrodotoxin. Tetrodotoxin is a colorless amino perhydroquinazoline compound with a pKa of 8.5. This substance rapidly degrades in alkaline media to several quinazoline compounds. The concentration of tetrodotoxin in newts varies between different newt species, sex (i.e., female > male), and specific organ.[43] Table 174.2 lists the relative toxicity of newt species. In general, the skin and muscle contain a higher concentration of tetrodotoxin than the liver, stomach, intestines, and gonads. In a study of the Japanese newt (*Cynops pyrrhogaster* Boie) from western Japan, the concentration of tetrodotoxin in tissues ranged between 5–370 mouse units (MU)/g.[54]

Certain genera of salamanders from the Salamandridae family (*Taricha*, *Notophthalmus*, *Cynops*, *Triturus*) contain tetrodotoxin and related toxins (6-epitetrodotoxin, 11-deoxytetrodotoxin, 11-oxotetrodotoxin) in their skin secretions, muscle, and eggs.[55,56] Tetrodotoxin is uniformly distributed in the dorsal skin of the newt.[57] The concentration of tetrodotoxin varies substantially between individual specimens from the same species.[58] In a study of 12 adult specimens of the red-spotted newt (*Notophthalmus viridescens* Rafinesque), the tetrodotoxin concentration ranged between 0.15 μg/g and 23.5 μg/g as measured by postcolumn fluorescent HPLC with mass spectrometry in selected ion monitoring mode.[59] The tetrodotoxin concentration of specimens of newts from *Taricha torosa* Rathke in Eschscholtz ranged between 2–36 μg/g body weight, whereas specimens of *Taricha granulosa* Skilton contained higher concentrations of tetrodotoxin, ranging between 8–328 μg/g body weight.[60] Although 6-epitetrodotoxin was detectable in all specimens, the concentration of this substance was minor. Young newts (efts) also contain substantial amounts of tetrodotoxin as well as some analogues (11-oxotetrodotoxin).[61] The analogue, 6-epitetrodotoxin, is usually a minor constituent of these newts.

Tetrodotoxin occurs in a wide range of marine and terrestrial animals, and this neurotoxin is the cause of puffer fish poisoning. Although the ingestion of livers from certain species of puffer fish is the most common cause of tetrodotoxin poisoning, a variety of animals other than puffer fish, salamanders, and newts contain tetrodotoxin including Goby fish, frogs (*Atelopus*, *Brachycephalus*, *Colostethus* species from South and Central America), horseshoe crab, xanthid crabs, blue-ringed octopus, gastropods trumpet shell, arrow worms, and annelids.[62] These studies indicate that these species contain tetrodotoxin as well as tetrodotoxin analogs including 6-epitetrodotoxin, 4-epitetrodotoxin, 4,9-anhydro-6-epitetrodotoxin, and 4,9-anhydrotetrodotoxin. In general, the concentration of tetrodotoxin is higher in the skin, muscles, and blood of newts (e.g., *Taricha torosa* Rathke in Eschscholtz) compared with tetrodotoxin-containing fish.[63] The origin of tetrodotoxin in salamanders and newts remains controversial. Experimental studies did not detect the *de novo* synthesis of tetrodotoxin after the feeding of radiolabeled metabolites to captive newts. These studies suggest that tetrodotoxin is synthesized following infection by bacteria similar to the presence of tetrodotoxin in other marine animals.[60] However, subsequent testing of newts with polymerase chain reaction (PCR) primers suggests that symbiotic bacteria in newts are not the source of the tetrodotoxin.[64]

Venom Apparatus

Newts do not possess a venom apparatus. Intoxication results from the ingestion of tetrodotoxin from the skin or muscles of these animals.

Dose Response

An average adult newt contains about 250 μg tetrodotoxin.[65] Ingestion of ethanol and a 20-cm (~8 in.) newt (*Taricha granulosa* Skilton) produced paresthesias within 15 minutes and a fatal cardiopulmonary arrest within 2 hours.[66] However, patients have survived the consumption of five newts. Variations in the presence of the neurotoxin probably account for differences in clinical presentation. The East Coast species of newt found in the Carolinas (*Notophthalmus* spp.) contains less tetrodotoxin compared with West Coast newts (*Taricha* spp.).[56,65]

HISTOPATHOLOGY AND PATHOPHYSIOLOGY

Tetrodotoxin blocks the sodium ion channels with subsequent inhibition of nerve and muscle activity. This toxin produces rapid, reversible flaccid paralysis of skeletal muscle, primarily by blockade of nerve transmission as a result of the inhibition of sodium channels.

CLINICAL RESPONSE

Certain species of newts contain tetrodotoxin and the ingestion of these reptiles produces classical tetrodotoxin poisoning similar to puffer fish. Clinical features include headache, perioral paresthesias, weakness, vomiting, diarrhea, excessive salivation, and lightheadedness. During severe tetrodotoxin poisoning, fixed and dilated pupils, ascending paralysis, bradycardia, hypotension, respiratory depression and cardiovascular collapse may occur.[66] Alteration of consciousness does not usually develop unless hypotension or respiratory failure occurs. Contact with the skin of a newt usually produces an immediate burning sensation.[67]

DIAGNOSTIC TESTING

Commercial laboratory tests for the detection of tetrodotoxin are not available. Methods to detect tetrodotoxin in biological specimens include postcolumn fluorescent HPLC and GC/MS.[59] In a study of four hospitalized victims of an unknown fish poisoning, blood samples were analyzed 10 hours after the ingestion of the fish by HPLC/MS with electrospray ionization interface.[68] The limit of detection was 15.6 nM. The blood tetrodotoxin concentrations in two patients requiring assisted ventilation were 344 ± 17 nM and 325 ± nM. Two other patients had neurological symptoms and respiratory distress, but they did not require intubation and assisted ventilation. Their blood tetrodotoxin concentrations were 47 ± 5 nM and 190 ± 16 nM.

TREATMENT

Treatment is supportive. Gut decontamination measures are usually unnecessary. The administration of activated charcoal to alert patients who present to the emergency department within 1 hour after ingestion is a therapeutic option, but there are no clinical data to confirm the clinical efficacy of decontamination measures on outcome following the ingestion of newts. Hypotension should be treated with intravenous saline and vasopressors if the patient fails to respond to fluid infusions. Symptomatic patients should be observed for 24 hours for signs of respiratory depression and hypotension. There are no antidotes.

References

1. Weil A, Davis W. *Bufo alvarius*: a potent hallucinogen of animal origin. J Ethnopharmacol 1994;41:1–8.
2. Pantanowitz L, Naude TW, Leisewitz A. Noxious toads and frogs of South Africa. S Afr Med J 1998;88: 1408–1414.
3. Jensen H, Evans EA Jr. Chemical studies on toad poisons. IV. Ch'an su, the dried venom of the Chinese toad and the secretion of the tropical toad, *B. marinus*. J Biol Chem 1934;104:307–316.
4. Morishita S, Shoji M, Oguni Y, Ito C, Higuchi M, Sakanashi M. Pharmacological actions of "kyushin," a drug containing toad venom: cardiotonic and arrhythmogenic effects, and excitatory effect on respiration. Am J Chin Med 1992;20:245–256.
5. Barry TL, Petzinger G, Zito SW. GC/MS comparison of the West Indian aphrodisiac "Love Stone" to the Chinese medication "Chan Su": bufotenine and related bufadienolides. J Forensic Sci 1996;41:1068–1073.
6. Lyttle T, Goldstein D, Gartz J. Bufo toads and bufotenine: fact and fiction surrounding an alleged psychedelic. J Psychoactive Drugs 1996;28:267–290.
7. Mebs D. Chemistry of animal venoms, poisons and toxins. Experientia 1973;29:1328–1334.
8. Hong Z, Chan K, Yeung HW. Simultaneous determination of bufadienolides in traditional Chinese medicines preparations, Lin-Shen-Wan by liquid chromatography. J Pharm Pharmacol 1992;44:1023–1026.
9. Mebs D, Wagner MG, Pogoda W, Maneyro R, Kwet A, Kauert G. Lack of bufadienolides in the skin secretion of red bellied toads, *Melanophryniscus* spp. (Anura, Bufoni-

dae), from Uruguay. Comp Biochem Physiol C Toxicol Pharmacol 2007;144:398–402.

10. Chen KK, Kovarikova A. Pharmacology and toxicology of toad venom. J Pharmaceut Sci 1967;56:1535–1541.
11. Brubacher JR, Ravikumar PR, Bania T, Heller MB, Hoffman RS. Treatment of toad venom poisoning with digoxin-specific Fab fragments. Chest 1996;110:1282–1288.
12. Fabing HD, Hawkins JR. Intravenous bufotenine injection in the human being. Science 1956;13:886–887.
13. Wurst M, Kysilka R, Koza T. Analysis and isolation of indole alkaloids of fungi by high-performance liquid chromatography. J Chromatogr 1992;593:201–208.
14. Yotsu-Yamashita M, Mebs D, Yasumoto T. Tetrodotxin and its analogues in extracts from the toad *Atelopus oxyrhynchus* (family: Bufonidae). Toxicon 1992;30:1489–1492.
15. Mebs D, Yotsu-Yamashita M, Yasumoto T, Lotters S, Schluter A. Further report of the occurrence of tetrodotoxin in *Atelopus* species (family: Bufonidae). Toxicon 1995;33:246–249.
16. Erspamer V, Vitali T, Roseghini M, Cei JM. 5-Methoxy- and 5-hydroxyindoles in the skin of *Bufo alvarius*. Biochem Pharmacol 1967;16:1149–1164.
17. Gupta RS, Chopra A, Stetsko DK. Cellular basis for the species differences in sensitivity to cardiac glycosides (digitalis). J Cell Physiol 1986;127:197–206.
18. Shimizu Y, Morishita S. Metabolism and disposition of kyushin, a drug containing senso (ch'an su). Am J Chin Med 1996;24:289–303.
19. Toma S, Morishita S, Kuronuma K, Mishima Y, Hirai Y, Kawakami M. Metabolism and pharmacokinetics of cinobufagin. Xenobiotica 1987;17:1195–1202.
20. Stoic V. Stimulation of adenosine 3′,5′-monophosphate formation in mononuclear leukocytes by food venoms. Biochim Biophys Acta 1982;721:236–239.
21. Yei CC, Deng JF. Toad or toad cake intoxication in Taiwan: report of four cases. J Formosa Med Assoc 1993;92(Suppl): S135–S139 (English abstract).
22. Ko RJ, Greenwald MS, Loscutoff SM, Au AM, Appel BR, Kreutzer RA, Haddon WF, Jackson TY, Boo FO, Presicek G. Lethal ingestion of Chinese herbal tea containing ch'an su. West J Med 1996;164:71–75.
23. Hitt M, Ettinger DD. Toad toxicity. N Engl J Med 1986;314:1517–1518.
24. Chern M-S, Ray C-Y, Wu D. Biologic intoxication due to digitalis-like substance after ingestion of cooked toad soup. Am J Cardiol 1991;67:443–444.
25. Gowda RM, Cohen RA, Khan IA. Toad venom poisoning: resemblance to digoxin toxicity and therapeutic implications. Heart 2003;89:e14.
26. Most A. *Bufo alvarius*: Psychedelic toad of the Sonoran Desert. New Mexico: Venom Press; 1984.
27. Lyttle T. Misuse and legend in the "toad licking" phenomenon. Int J Addict 1993;28;521–538.
28. Wang Z, Wen J, Zhang J, Ye M, Guo D. Simultaneous determination of four bufadienolides in human liver by high-performance liquid chromatography. Biomed Chromatogr 2004;18:318–322.
29. Hyun DW, Kwon TG, Kim KY, Bae JH. Toad venom poisoning resembling digitalis intoxication and hyperkalemia: a case report. Korean Circulation J 2007;37:283–286.
30. Fushimi R, Koh T, Iyama S, Yasuhara M, Tachi J, Kohda K, Amino N, Miyai K. Digoxin-like immunoreactivity in Chinese medicine. Ther Drug Monit 1990;12:242–245.
31. Biddle DA, Datta P, Wells A, Dasgupta A. Falsely elevated serum digitoxin concentrations due to cross-reactivity of water-extractable digitoxin-like immunoreactivity of Chinese medicine Chan SU: elimination of interference by use of a chemiluminescent assay. Clinica Chim Acta 2000;300:151–158.
32. Dasgupta A, Emerson L. Neutralization of cardiac toxins oleandrin, oleandrigenin, bufalin, and cinobufotalin by Digibind: monitoring the effect by measuring free digitoxin concentrations. Life Sci 1998;63:781–788.
33. Datta P, Dasgupta A. Interactions between drugs and Asian medicine: displacement of digitoxin from protein binding site by bufalin, the constituent of Chines medicines Chan Su and Lu-Shen-Wan. Ther Drug Monit 2000;22:155–159.
34. Raisanen M, Karkkainen J. Mass fragmentographic quantification of urinary N,N-dimethyltryptamine and bufotenine. J Chromatogr 1979;162:579–584.
35. Kuo H-Y, Hsu C-W, Chen J-H, Wu Y-L, Shen Y-S. Life-threatening episode after ingestion of toad eggs: a case report with literature review. Emerg Med J 2007;24: 215–216.
36. Chi H-T, Hung D-Z, Hu W-H, Yang D-Y. Prognostic implications of hyperkalemia in toad toxin intoxication. Hum Exp Toxicol 1998;17;343–346.
37. Brubacher JR, Lachmanen D, Ravikumar PR, Hoffman RS. Efficacy of digoxin specific Fab fragments (Digibind(r)) in the treatment of toad venom poisoning. Toxicon 1999;37:931–942.
38. Marki F, Witkop B. The venom of the Colombian arrow poison from *Phyllobates bicolor*. Experientia 1963;19: 329–338.
39. Myers CW, Daly JW, Malkin B. A dangerously toxic new frog *(Phyllobates)* used by Enbera Indians of western Colombia, with discussion of blowgun fabrication and dart poisoning. Bull Am Mus Nat Hist 1978;101:311–365.
40. Daly JW, Myers CW, Whittaker N. Further classification of skin alkaloids from neotropical poison frogs (Dendrobatidae), with a general survey of toxic/noxious substances in the Amphibia. Toxicon 1987;265:1023–1095.
41. Daly JW. Alkaloids of neotropical poison frogs (Dendrobatidae). Fortschr Chem Org Naturst 1982;41:205–340.
42. Neuwirth M, Daly JW, Myers CW, Tice LW. Morphology of the granular secretory glands in skin of poison-dart frogs (Dendrobatidae). Tissue Cell 1979;11:755–771.

43. Fuhrman FA. Tetrodotoxin, tarichatoxin, and chiriquitoxin: historical perspectives. Ann NY Acad Sci 1986;479: 1–14.
44. Daly JW, Gusovsky F, Myers CW, Yotsu-Yamashita M, Yasumoto T. First occurrence of tetrodotoxin in a Dendrobatid frog (*Colostethus inguinalis*), with further reports for the bufonid genus *Atelopus*. Toxicon 1994;32:279–285.
45. Dumbacher JP, Spande TF, Daly JW. Batrachotoxin alkaloids from passerine birds: a second toxic bird genus (*Ifrita kowaldi*) from New Guinea. Proc Natl Acad Sci U S A 2000;97:12970–12975.
46. Daly JW. Biologically active alkaloids from poison frogs (Dendrobatidae). J Toxicol Toxin Rev 1982;1:33–86.
47. Daly J, Witkop B. Batrachotoxin, an extremely active cardio- and neurotoxin from the Colombian arrow poison frog *(Phyllobates aurotaenia)*. Clin Toxicol 1971;4: 331–342.
48. Daly JW, Myers CW, Warnick JE, Albuquerque EX. Levels of batrachotoxin and lack of sensitivity to its action in poison-dart frogs (*Phyllobates*). Science 1980;208: 1383–1385.
49. Bosmans F, Maertens C, Verdonck F, Tytgat J. The poison Dart frog's batrachotoxin modulates Na(v)1.8. FEBS Lett 2004;577:245–248.
50. Brown GB. ^{3}H-Batrachotoxin—a benzoate binding To voltage-sensitive sodium channels: inhibition by the channel blockers tetrodotoxin and saxitoxin. J Neurosci 1986;6:2064–2070.
51. Narahashi T, Deguchi T, Albuquerque EX. Effects of batrachotoxin on nerve membrane potential and conductances of giant squid axons. J Gen Physiol 1971;58;54–70.
52. Moore GR, Boegman RJ, Robertson DM, Raine CS. Acute stages of batrachotoxin-induced neuropathy: a morphologic study of a sodium channel toxin. J Neurocytol 1986;15:573–583.
53. Order: Caudata. Amphibian Resource Center. Available at http://www.livingunderworld.org/caudata/. Accessed May 13, 2008.
54. Tsuruda K, Arakawa O, Noguchi T. Toxicity and toxin profiles of the newt, *Cynops pyrrhogaster* from western Japan. J Nat Toxins 2001;10:79–89.
55. Hanifin CT, Brodie ED 3rd, Brodie ED Jr. Tetrodotoxin levels of the rough-skin newt, *Taricha granulosa*, increase in long-term captivity. Toxicon 2002;40:1149–1153.
56. Brodie ED, Hensel JL Jr, Johnson JA. Toxicity of the Urodele amphibians *Taricha, Notophthalmus, Cynops* and *Paramesotriton* (Salamandridae). Copeia 1974;1974(2): 506–511.
57. Hanifin CT, Brodie ED 3rd, Brodie ED Jr. A predictive model to estimate total skin tetrodotoxin in the newt *Taricha granulosa*. Toxicon 2004;43:243–249.
58. Yotsu-Yamashita M, Mebs D, Kwet A, Schneider M. Tetrodotoxin and its analogue 6-epitetrodotoxin in newts (*Triturus* spp.; Urodela, Salamandridae) from southern Germany. Toxicon 2007;50:306–309.
59. Yotsu-Yamashita M, Mebs D. The levels of tetrodotoxin and its analogue 6-*epi*tetrodotoxin in the red-spotted newt, *Notophthalmus viridescens*. Toxicon 2001;39: 1261–1263.
60. Shimizu Y, Kobayashi M. Apparent lack of tetrodotoxin biosynthesis in captured *Taricha torosa* and *Taricha granulosa*. Chem Pharm Bull 1983;31:3625–3631.
61. Yotsu-Yamashita M, Mebs D. Occurrence of 11-oxotetrodotoxin in the red-spotted newt, *Notophthalmus viridescens*, and further studies on the levels of tetrodotoxin and its analogues in the newt's efts. Toxicon 2003;41:893–897.
62. Miyazawa K, Noguchi T. Distribution and origin of tetrodotoxin. J Toxicol Toxin Rev 2001;20:11–33.
63. Kao CY. Tetrodotoxin, saxitoxin and their significance in the study of excitation phenomena. Pharmacol Rev 1966;18:997–1049.
64. Lehman EM, Brodie ED Jr, Brodie ED 3rd. No evidence for an endosymbiotic bacterial origin of tetrodotoxin in the newt *Taricha granulosa*. Toxicon 2004;44:243–249.
65. Brodie ED Jr. Investigations of the skin toxins of the adult rough-skinned newt, *Taricha granulosa*. Copeia 1968;(1): 307–313.
66. Bradley SG, Kilka UJ. A fatal poisoning from the Oregon rough-skinned newt *(Taricha granulosa)*. JAMA 1981;246: 247.
67. King BR, Hamilton RJ, Kassutto Z. "Tail of newt". an unusual ingestion. Pediatr Emerg Care 2000;16:268–269.
68. Tsai YH, Hwang DF, Cheng CA, Hwang CC, Deng JF. Determination of tetrodotoxin in human urine and blood using C18 cartridge column, ultrafiltration and LC-MS. J Chromatogr B 2006;832:75–80.

Chapter 175

GILA MONSTER and BEADED LIZARD

IDENTIFYING CHARACTERISTICS

Of about 2,000 species of lizards, only two species of lizard previously were considered venomous. These lizards included the Gila monsters (*Heloderma suspectum* Cope) and the Mexican beaded lizards (*Heloderma horridum* Wiegmann). Recent evidence suggests that the lace monitor lizard (*Varanus varius* White) and the iguanian lizard (*Pogona barbata* Cuvier) also possess venom.[1] The taxonomic order Squamata contains both venomous snakes and the family of venomous lizards called Helodermatidae. Construction of gland complementary-DNA libraries and phylogenetic analysis of transcripts suggest that the venom from lizards and snakes contain nine similar toxin types. The shared lineages possessing toxin-secreting oral glands suggest a single early origin of the venom system in lizards and snakes.

Common Name: Gila Monster

Scientific Name: *Heloderma suspectum*

Infraspecies: *H. s. cinctum* Bogert and Martin del Campo (Banded Gila monster), *H. s. suspectum* Cope (Reticulate Gila monster)

Family: Helodermatidae

Physical Description: These large (25-50 cm/~10–12 in.), stout creatures possess sharp claws on short, stout legs, and flat, relatively large heads with blunt snouts and thick, black, forked tongues. The skin contains thick, beaded (bone-impregnated) scales on the dorsal surfaces that range in color from yellow and orange to pink and brown.

Common Name: Beaded Lizard

Scientific Name: *Heloderma horridum* Wiegmann

Infraspecies: H. *h. exasperatum* Bogert and Martin del Campo (Rio Fuerte beaded lizard), *H. h. alvarezi* Bogert and Martin del Campo (Black beaded lizard), *H. h. horridum* Wiegmann (Mexican beaded lizard)

Family: Helodermatidae

Physical Description: This large, heavy bodied lizard has scales resembling black and yellow beads. Features distinguishing the beaded lizard from the Gila monster include a larger size (70–100 cm/~2–3 ft.), longer tail, pink tongue, and finer speckles. The beaded lizard grows 1 m (~3 ft.) in length and weighs up to 2 kilograms (~2 lbs). Figure 175.1 displays a specimen of the beaded lizard (*Heloderma horridum* Wiegmann).

EXPOSURE

Geographic Distribution

The two species of Gila monsters inhabit arid regions of the southwestern United States and western Mexico from sea level to over 5,000 ft (1,500 m). The Gila monster appears primarily in the Great Sonoran Desert. Two infraspecies are the banded Gila monster (southeastern California, southern Nevada, Arizona, Mojave Desert) and the reticulate Gila monster (Sonoran and Chihuahua deserts in southern Arizona and northern Mexico). The habitat of the Mexican beaded lizard is Sonoran Mexico, the Pacific coast of Mexico from the

Medical Toxicology of Natural Substances, by Donald G. Barceloux, MD

FIGURE 175.1. Beaded lizard (*Heloderma horridum* Wiegmann). © 2006 Rune Midtgaard. See color insert.

Isthmus of Tehuantepec to southern Sonora, and Guatemala.

Behavior

These large, slow-moving lizards live 10–25 years. The skin of the Gila monster is more permeable to water than the skin of many other lizards; therefore, these shy creatures retreat into burrows and rocky shelters during the hot daytime sun. Although they are primarily nocturnal during the warm summer months, they are diurnal in the cooler temperatures during the spring and winter. When threatened, these lizards display an open mouth while moving venom in preparation of envenomation. Gila monsters are binge feeders, consuming very large meals (i.e., up to one-third body weight) at infrequent intervals. They prey upon the eggs, newborn, and the young of many species of small mammals, reptiles, and birds.

PRINCIPAL TOXINS

Venom Composition

Helodermatidae venom is a complex mixture of proteins and enzymes (phospholipase A_2, hyaluronidase, L-amino acid oxidase, proteolytic enzymes, serotonin) with diverse biological activities.[2] The venom of the Gila monster and Mexican beaded lizard are similar.[3] Unique substances in this venom include gilatoxin (CAS RN: 77641-30-6), *N*-benzoyl-L-arginine ethyl ester hydrolase, exendin-3 (*H. horridum*) exendin-4 (*H. suspectum*), and helothermine (CAS RN: 127670-24-0). Exendin-4 (exenatide) is 39-amino acid peptide isolated from the saliva of the Gila monster (*H. suspectum*) that is structurally similar to glucagon-like peptide-1.[4] This compound stimulates glucose-induced insulin secretion in animal models,[5] and exendin-4 is a potential treatment for type II diabetes mellitus.[6] Rapid increases in the plasma concentration of exendin-4 occur in the Gila monster after feeding, primarily as a result of mechanical stimulation.[7] Helothermine (CAS RN: 127670-24-0) is a 25.5 kDa serine protease that inhibits Ca^{2+} channels in cerebellar granule cells of newborn rats.[8] These experiments suggest that helothermine produces a high-potency blockage of at least three pharmacologically distinct Ca^{2+} current components (i.e., ω-conotoxin GVIA-sensitive, ω-Aga-IVA-sensitive, dihydropyridine-sensitive). Additionally, *in vitro* experiments suggest that helothermine inhibits the intracellular ryanodine receptor, which is associated with the release of intracellular calcium stores in response to voltage or inotropic agents.[9] Whole-cell voltage-clamp experiments on cerebellar granule neurons also suggest that helothermine inhibits at least two different potassium current components similar to an IA-type current and a classical delayed-rectifier current.[10] Helothermine has kallikrein-like properties.[11]

Gilatoxin is a serine protease with kallikrein-like and thrombin-activating activity.[12] There are a variety of phospholipase A_2 (PLA) enzymes in animal venoms including at least five anionic phospholipases in the venom of the Gila monster.[13] The acidic PLA in the venom of the beaded lizard (*Heloderma horridum*) differs from snake PLA by specifically inhibiting thromboxane-induced platelet aggregation *in vitro*.[14] *N*-benzoyl-L-arginine ethyl ester hydrolase stimulates the release of vasoactive kinins and causes hypotension in animal models.[15] Horridum toxin (i.e., a purified glycoprotein from the crude venom from the Mexican beaded lizard) is hemorrhagic toxin with 210 amino acid residues that produces exophthalmia, internal hemorrhage, and paralysis when injected into experimental animals.[16,17] In contrast, gilatoxin has 245 amino acid residues and does not cause exophthalmia in experimental animals.

Venom Apparatus

The primitive injection apparatus of the Gila monster and the Mexican beaded lizard involves, but is not necessarily limited to, grooved teeth that are connected by a duct to the lower mandibular venom glands. Located subcutaneously in the lower anterior jaw, these glands deliver venom via ducts to openings located at the base of the enlarged, central mandibular teeth. Glandular compression during chewing motions moves the venom along the ducts. Alongside each mandible, dental sacs serve as reservoirs for the secreted venom. The venom diffuses primarily by simple capillary action up the rudimentary grooves present on the sharp, recurved teeth.

In contrast to Gila monsters, most venomous snakes have compressive muscular tissue surrounding the venom glands. Envenomation by these venomous lizards requires prolonged contact and a grinding motion. The teeth of the Gila monster are poorly anchored and frequently are shed during envenomation. Bites may be tenacious, requiring minutes to several hours to disengage. Bites in humans are rare and usually result from the careless handling of captive lizards; lizards use their venom primarily for defensive purposes.

Dose Response

The rat LD_{50} of lizard venom (10.3 mg/kg) is comparable to that of rattlesnake venom (10.6 mg/kg), but lizard venom has much lower enzymatic activity. The severity of the envenomation depends on several factors including the size of the lizard, the duration of contact, the location and depth of the bite, and the aggressiveness of the lizard. The amount of venom in the gland of a Gila monster exceeds the estimated lethal dose for humans, but the above factors limit the amount of venom injected during a human evenomation.[18]

HISTOPATHOLOGY AND PATHOPHYSIOLOGY

In experimental animals, *Heloderma* venom produces lethal changes in cardiopulmonary parameters without obvious effects on coagulation. Although some local bleeding occurs from puncture sites following human envenomation, animal studies indicate that the injection of lethal doses of Gila monster venom does not alter clotting or prothrombin time (PT).[19]

CLINICAL RESPONSE

Medically significant envenomations by members of the Helodermatidae family are rare. Most envenomations by Gila monsters cause local effects manifest by erythema and pain along with minor systemic complaints (weakness, lightheadedness, nausea, diaphoresis, chills).[20] A case report associated envenomation by the Mexican beaded lizard with the immediate onset of excruciating pain, nausea, and vomiting that resolved within 24 hours.[3] Tissue edema begins about 15–45 minutes after envenomation and progresses slowly up the extremity over the next 4–8 hours. Local pain may be severe during the first few hours after envenomation, but the pain usually resolves within 6–10 hours. In contrast to rattlesnake envenomation, local edema is usually minimal and skin necrosis is absent.[18] Neurovascular compromise of the involved extremity does not usually develop. Paresthesias, muscle fasciculations, and diaphoresis may also occur after envenomation, and some local hyperesthesia typically persisting for several days after the swelling resolves. The presence of hypotension, decreased vascular resistance, and tachycardia indicate a serious envenomation.[21] A 23-year-old herpetologist with a prior history of cocaine use developed changes suggestive of an anterolateral myocardial infarction (increased serum creatine kinase with positive MB band and electrocardiographic changes of right bundle branch block and ST-segment elevation) after a Gila monster envenomation.[22] A stress test demonstrated decreased contractility in the apex of the heart, but the patient refused a cardiac catheterization. Case reports associate bites from the Gila monster[23] and Mexican beaded lizard[24] with the occurrence of an anaphylactic-like reaction involving hypotension, dyspnea, drooling, dysphagia, angioedema, diaphoresis, and lightheadedness. There were no laboratory abnormalities or sequelae. These injuries occurred in patients with prior exposure and previous envenomations by these lizards.

Sequelae are rare except mild local hyperesthesia.[25] Lymphangitis may complicate the envenomation.[21] The wound may remain tender for several weeks. Fatalities following envenomation by the Gila monster or by the Mexican beaded lizard are not well documented in the modern medical literature with the last death recorded in 1930.[26]

DIAGNOSTIC TESTING

Laboratory abnormalities are usually minimal with leucocytosis and hypokalemia occurring occasionally. Coagulation studies are typically normal, although a case report documented the presence of hypofibrinogenemia, thrombocytopenia, elevated fibrin split products, and increased PT and partial thromboplastin time (PTT) following an envenomation by a Gila monster.[27] Transient intraventricular conduction defects and T-wave inversions may occur.[27,28] Serum cardiac enzymes are usually normal. Teeth shed into the wound during envenomation are often radiopaque, and the absence of teeth on radiographs does not exclude the presence of foreign bodies in the wound.[21]

TREATMENT

The most important first-aid measure is detachment of the tenacious lizard from the victim with the best available method (e.g., pliers, crowbar, strong stick placed in near posterior teeth, submerging the lizard in cold water, hot water on the head of the lizard, placing an open flame under jaw). A cast spreader is appropriate if the lizard remains attached during arrival to the emergency department. Pulling the reptile by the tail should be avoided because of the potential for increased tissue destruction. After removal of the lizard, the affected extremity should be placed in a neutral position and the patient transported to a medical facility as soon as possible. The wound should be cleaned with 1% povidone-iodine, irrigated with saline, and débrided to remove all embedded teeth. If a foreign body is suspected, appropriate radiographs should be obtained. Prophylactic antibiotics are usually unnecessary unless there is evidence of lymphangitis. Other measures of supportive care include tetanus prophylaxis as indicated and analgesics (i.e., opiates, fentanyl, hydromorphone, nonsteroidal anti-inflammatory agents) for pain relief. There are theoretical concerns about the use of nonsteroidal anti-inflammatory drugs because of their antiplatelet properties, but there are no documented clinical complications associated with the use of these drugs following Gila monster envenomations. Peripheral nerve blocks are occasionally necessary to relieve pain.

Serious toxicity following Gila monster envenomations is rare. Hypotension usually responds to fluid replacement (10–20 mL saline/kg). Laboratory studies should include complete blood count, serum chemistry panel (electrolytes, blood urea nitrogen [BUN], creatinine, glucose), serum creatine kinase, PT, PTT, fibrinogen, fibrin degradation products, and D-dimer. Bleeding secondary to clotting abnormalities is rare, but outpatients should be instructed to return if abnormal bruising or bleeding develops. The platelet count rarely decreases below 100,000/mm^3; thrombocytopenia usually resolves within several days without complications.

References

1. Fry BG, Vidal N, Norman JA, Vonk FJ, Scheib H, Ramjan SF, et al. Early evolution of the venom system in lizards and snakes. Nature 2006;439:584–588.
2. Zarafonetis CJD, Kalas JP. Serotonin degradation by homogenates of tissues from *Heloderma horridum*, the Mexican beaded lizard. Nature 1962;195:707–708.
3. Albritton DC, Parrish HM, Allen ER. Venenation by the Mexican beaded lizard (*Heloderma horridum*): report of a case. South Dakota J Med 1970;23:9–11.
4. Bray GM. Exenatide. Am J Health-Syst Pharm 2006;63: 411–418.
5. Goke R, Fehmann HC, Linn T, Schmidt H, Krause M, Eng J, Goke B. Exendin-4 is a high potency agonist and truncated exendin-(9-39)-amide an antagonist at the glucagon-like peptide 1-(7-36)-amide receptor of insulin-secreting beta-cells. J Biol Chem 1993;268:19650–19655.
6. Gallwitz B. Glucagon-like Peptide-1-based therapies for the treatment of type 2 diabetes mellitus. Treat Endocrinol 2005;4:361–370.
7. Christel CM, DeNardo DF. Release of exendin-4 is controlled by mechanical action in Gila monsters, *Heloderma suspectum*. Comp Biochem Physiol A Mol Integr Physiol 2006;143:85–88.
8. Nobile M, Noceti F, Prestipino G, Possani LD. Helothermine, a lizard venom toxin, inhibits calcium current in cerebellar granules. Exp Brain Res 1996;110: 15–20.
9. Morrissette J, Kratzschmar J, Haendler B, El-Hayek R, Mochca-Morales J, Martin BM, et al. Primary structure and properties of helothermine, a peptide toxin that blocks ryanodine receptors. Biophys J 1995;68:2280–2288.
10. Nobile M, Magnelli V, Lagostena L, Mochca-Morales J, Possani LD, Prestipino G. The toxin helothermine affects potassium currents in newborn rat cerebellar granule cells. J Membr Biol 1994;139:49–55.
11. Alagon A, Possani LD, Smart J, Schleuning WD. Helodermatine, a kallikrein-like, hypotensive enzyme from the venom of *Heloderma horridum horridum* (Mexican beaded lizard). J Exp Med 1986;164:1835–1845.
12. Utaisincharoen P, Mackessy SP, Miller RA, Tu AT. Complete primary structure and biochemical properties of gilatoxin, a serine protease with kallikrein-like and angiotensin-degrading activities. J Biol Chem 1993;268: 21975–21983.
13. Vandermeers A, Vandermeers-Piret M-C, Vigneron L, Rathe J, Stievenart M, Christophe J. Differences in primary structure among five phospholipases A_2 from *Heloderma suspectum*. Eur J Biochem 1991;196:537–544.
14. Huang T-F, Chiang H-S. Effect on human platelet aggregation of phospholipase A_2 purified from *Heloderma horridum* (beaded lizard) venom. Biochim Biophys Acta 1994;1211:61–68.
15. Alagon AC, Maldonado ME, Julia JZ, Sanchez CR, Possani LD. Venom from two sub-species of *Heloderma horridum* (Mexican beaded lizard): general characterization and purification of *N*-benzoyl-L-arginine ethyl ester hydrolase. Toxicon 1982;20:463–475.
16. Patterson RA. Some physiological effects caused by venom from the Gila monster, *Heloderma suspectum*. Toxicon 1967;5:5–10.
17. Datta G, Tu AT. Structure and other chemical characterizations of Gila toxin, a lethal toxin from lizard venom. J Peptide Res 1997;50:443–450.
18. Russell FE, Bogert CN. Gila monster: its biology, venom and bite. A review. Toxicon 1981;19:341–359.
19. Patterson RA, Lee IS. Effects of *Heloderma suspectum* venom on blood coagulation. Toxicon 1969;7:321–324.
20. Strimple PD, Tomassoni AJ, Otten EJ, Bahner D. Report on envenomation by a Gila monster (*Heloderma suspectum*) with a discussion of venom apparatus, clinical findings, and treatment. Wilderness Environ Med 1997;8: 111–116.
21. Hooker KR, Caravati EM. Gila monster envenomation. Ann Emerg Med 1994;24:731–735.

22. Bou-Abboud CF, Kardassakis DG. Acute myocardial infarction following a Gila monster (*Heloderma suspectum cinctum*) bite. West J Med 1988;148:577–579.
23. Piacentine J, Curry SC, Ryan PJ. Life-threatening anaphylaxis following Gila monster bite. Ann Emerg Med 1986;15:959–961.
24. Cantrell FL. Envenomation by the Mexican beaded lizard: a case report. J Toxicol Clin Toxicol 2003;41: 241–244.
25. Stahnke HL, Heifron WA, Lewis DU. Bite of the Gila monster. Rocky Mt Med J 1970;67:25–30.
26. Storer TI. *Heloderma* poisoning in man. Bull Antivenin Inst Am 1931;5:12–15.
27. Preston CA. Hypotension, myocardial infarction, and coagulopathy following Gila monster bite. J Emerg Med 1989;7;37–40.
28. Roller JA, Davis DH. Gila monster bite. JAMA 1976;235: 249–250.

Chapter 176

TERRESTRIAL SNAKES
(Suborder: Serpentes)

All snakes belong to the order Squamata and the suborder Serpentes that is characterized by scaly skin, absence of movable eyes, and elongated bodies without appendages. Venomous snakes inhabit all continents *except* Antarctica and some islands (Ireland, New Zealand). Approximately 15% of over 3,000 snake species are venomous, and the two main families of venomous snakes are **Elapidae** (elapids) and **Viperidae** (vipers).[1] The family **Atractaspididae** contains side-fanged snakes (stiletto vipers, mole vipers) that live in subterranean areas of Africa and the Middle East. The **Colubridae** family (colubrids) contains only a few species that are venomous to humans.

Elapids are characterized by short-fixed fangs that inject venom during a series of chewing movements. The smooth scales, large head shields, and mostly terrestrial habitat are similar to species in the Colubridae family. The main subfamily of elapids is Elapinae, which contains cobras (*Naja* spp.), kraits (*Bungarus* spp.), mambas (*Dendroaspis* spp.), and coral snakes (*Micrurus* spp., *Micruroides* spp.). Marine snakes were formerly classified as a separate family (**Hydrophiidae**), but now these snakes belong to the other two subfamilies of **Elapidae**, Hydrophiinae (sea snakes), and Laticaudinae (sea kraits).[2] The two main subfamilies of **Viperidae** are Crotalinae (pit viper) and Viperinae (true viper). The pit vipers (e.g., rattlesnakes, water moccasins, copperheads) possess heat-sensing foramen or pits, whereas true vipers (e.g., small adder [*Vipera berus* L.], large puff adder [*Bitis arietans* Merrem]) do not possess this heat-sensing organ. The third subfamily of **Viperidae** consists of only one ancient viper, Fea's viper (*Azemiops feae* Boulenger). The large family of **Colubridae** (colubrids) contains a majority of snake species, but most of these species are not venomous to humans. Several species of this family including the African boomslang (*Dispholidus typus* Smith) and the twig snakes (*Thelotornis* spp.) have back-fangs that inject potent venom.[3]

ATRACTASPIDS

Species of Atractaspidal are usually small, innocuous snakes. Most venomous members of this family are side-fanged, mole or burrowing vipers including *Atractaspis bibronii* Smith (Bibron's burrowing adder), *Atractaspis congica* Peters (Congo burrowing adder), and *Atractaspis engaddensis* Haas (burrowing asp). *Atractaspis bibronii* inhabits most of sub-Saharan Africa and arid regions extending northeast into Saudi Arabia and Jordan, whereas *A. engaddensis* inhabits Jordan and Israel. These snakes burrow into the ground, expose their frontal fangs protruding from the side of their partially closed mouth, and hook nearby prey. Bites frequently involve only one fang, and the envenomation is usually mild compared with other venomous snakes.[4] The venom of *A. engaddensis* contains 21-amino acid residue peptides called sarafotoxins that are structurally similar to the vasoactive peptides, endothelins.[5] Endothelins modulate the cardiovascular system including the regulation of blood pressure through smooth muscle contraction.[6] Although these snakes possess unique

Medical Toxicology of Natural Substances, by Donald G. Barceloux, MD

venoms, envenomation by these snakes is an uncommon cause of mortality and morbidity in these regions.[7] Clinical features of these envenomations usually involve localized pain, hemorrhage and swelling, regional lymphadenopathy with occasional discoloration, and blistering or necrosis at the bite site.[8] Less common complaints include headache, myalgias, dry mouth, and hoarseness. Neurological symptoms are usually absent, and systemic symptoms are uncommon. Hematological abnormalities are minimal, and cardiovascular effects (i.e., P-R prolongation, nonspecific ST-T wave changes, T-wave inversions as noted on the electrocardiogram) do not usually produce clinically significant hemodynamic effects.[9] Treatment is supportive with analgesics, wound care, and fluids as necessary. There is no commercially available antivenom for envenomation by these snakes.

VIPERS

The family Viperidae consists of folding-fanged snakes that are the most important cause of serious snake envenomations globally.[10] Taxonomically, this family is divided into the following subfamilies: 1) Azemiopinae containing only the ancient Fea's viper (*Azemiops feae* Boulenger), 2) Viperinae or true vipers (e.g., European asp, common viper, puff adders, Russell's viper), 3) Crotalinae or pit vipers (e.g., rattlesnakes, fer-de-lance, Asian lance-headed vipers, tree vipers), and Causinae containing only species of the genus *Causus*. Species within the Viperidae family are wide-ranging snakes that live in many different habitats with the exception of New Guinea and Australia. Members of this snake family include the Malayan pit viper, Gaboon viper, Asian habu, Chinese mamushi, Brazilian pit viper, North American rattlesnakes, Central American bushmaster, European adder, and the African puff adder. In Europe, species in the genus *Vipera* account for most snake envenomations. These vipers include *V. aspis* L. (asp viper or European asp), *V. ammodytes* L. (sand viper), *V. berus* L. (common viper), and *V. latastei* Bosca (Lataste's viper).[11] Mortality from these envenomations is very low. In the Middle East, the most common species involved in snakebites belong to the genera *Vipera* (*V. palaestinae* or Palestine viper, *V. xanthina* Gray), *Macrovipera* (*M. lebetina* L. or Lavantine viper), and *Cerastes* (*Cerastes vipera* L. or Cleopatra's asp). The annual incidence of snakebite in the United States and Europe is similar, and most US snake envenomations result from bites by rattlesnakes (*Crotalus* spp.), copperheads (*Agkistrodon contortrix* L.), or cottonmouths (*Agkistrodon piscivorus* Lacépède). The prevalence of snakebites is significantly higher in Central and South America than in the United States. Pit vipers account for most of the serious bites in this region (e.g., *Bothrops* spp. in Brazil, *Crotalus durissus* subsp. *terrificus* Laurenti in the savanna areas of South America, and *Crotalus durissus* subsp. *durissus* L. in Central America). The prevalence of snakebites in Africa is probably underestimated because of inadequate reporting systems, but existing evidence suggests that envenomations by species from the Viperidae family account for most deaths from snakebites. The common species involved in snake envenomations depend on the following locations: banana plantations (*Causus maculatus* Hallowell or spotted night adder), palm tree and rubber plantations (black cobras or *Naja melanoleuca* Hallowell, green mambas from the *Dendroaspis* species), forests and rice fields (*Bitis* spp. such as Gaboon viper), and savanna (*Echis* spp.). There is wide variation in the prevalence of snakebites in Asia. The majority of snakebites in Southern Japan and south China involve *Trimeresurus* species or *Protobothrops flavoviridis* Hallowell (habu or *Trimeresurus flavoviridis*). Dangerous snake species in Southeast Asia include *Calloselasma rhodostoma* Boie (Malayan pit viper), *Daboia russelii* Shaw & Nodder (Russell's viper, *Vipera russelii*), and *Naja naja* L. (Asian cobra). Almost all deaths from snakebites in Australia occur from envenomations by Elapidae species (*Pseudonaja, Notechis, Oxyuranus*). Figure 176.1 compares the mortality from snakebites across the world.

Old World Vipers (Viperinae)

Table 176.1 includes the African, European, and Middle Eastern vipers (adders, asps, vipers) that are venomous species of the Viperinae subfamily of the family Viperidae (viper). Viper venom contains a wide variety of components including pre- and postsynaptic neurotoxins, myotoxins, and toxins directed at the coagulation system and blood vessels, depending on the specific viper species.

EUROPE

EXPOSURE

Table 176.2 lists the common European venomous vipers. Indigenous venomous snakes in Europe belong

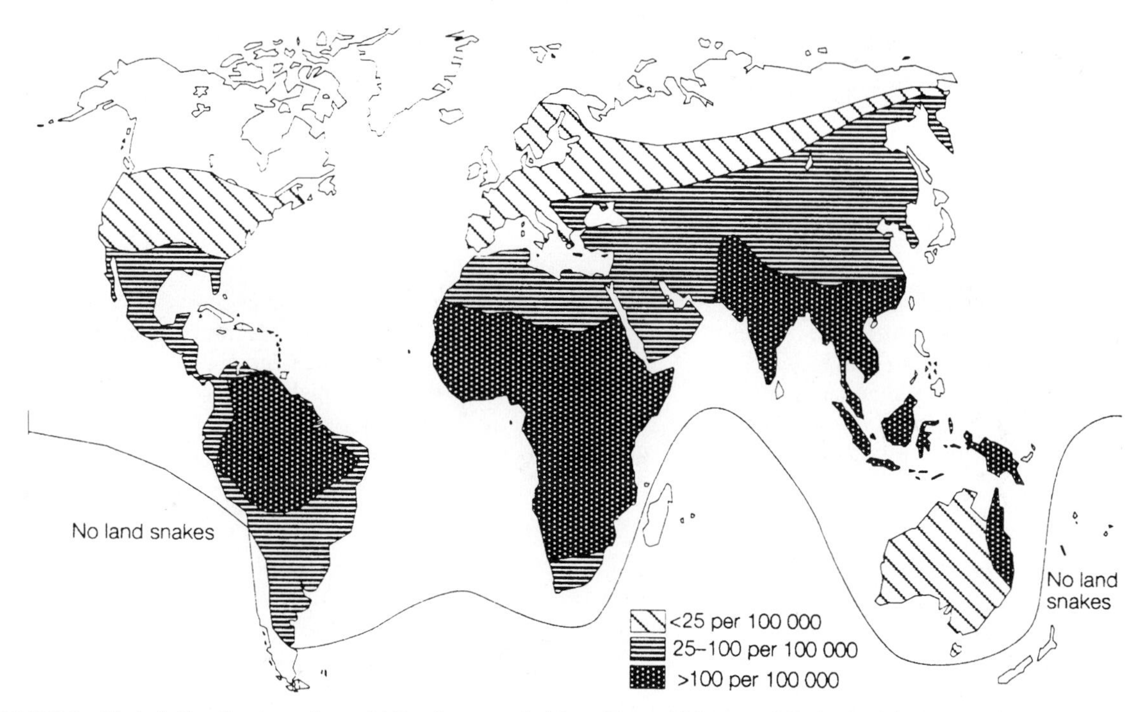

FIGURE 176.1. Global distribution of morbidity from snakebites. From Chippaux J-P. Snake-bites: appraisal of the global situation. Bull World Health Organ 1998;76:520. Reproduced with permission from the World Health Organization.

almost exclusively to the family Viperidae, subfamily Viperinae. *Vipera berus* L. (common viper) inhabits most parts of Europe, whereas *Vipera aspis* L. (asp viper) is found in France, Italy, and the Alpine region of Southern Germany, Austria, and Switzerland. Other important venomous viper species in Europe include the nose-horned or sand viper (see specimen in Figure 176.2) in the Balkans and the Orsini's viper (*Vipera ursinii* Bonaparte) in Eastern Europe (see Table 176.2). The adder is the only venomous snake indigenous to the United Kingdom.

VENOM

The venom from European vipers is similar to the composition of venom from tropical vipers. These venoms contain proteases, esterases, L-amino acid oxidases, and hyaluronidase, but the concentration of myotoxic phospholipases A_2 and hemorrhagic factors are usually clinically insignificant.[12] Venom from the sand viper (*Vipera ammodytes* L.) does contain neurotoxic phospholipase A_2, which occasionally causes presynaptic neuromuscular blockage (i.e., ptosis ophthalmoplegia, generalized weakness) following envenomation.[13,14]

CLINICAL RESPONSE

Table 176.3 lists the gradation of envenomations by European vipers as outlined by Audebert et al. In a series of 147 cases of envenomations by *V. ammodytes* presenting to a Greek hospital, the most common clinical features of envenomation were pain, swelling, and ecchymosis with only 3% of cases presenting with blurred vision.[15] There were only two cases of severe envenomation (grade 3). The clinical features of envenomation by European vipers (e.g., *V. berus*) primarily involve cytotoxic and slight hemorrhagic effects resulting in mild local edema and pain, gastrointestinal symptoms (nausea, vomiting, diarrhea, abdominal pain), and

FIGURE 176.2. Nose-horned Viper (*Vipera ammodytes* L.). © 2000 Rune Midtgaard. See color insert.

TABLE 176.1. Scientific and Common Names of Venomous Species of the Viperinae Subfamily[a]

Group	Scientific Name	Common Name
Adders		
	Bitis arietans Merrem	Puff adder
	Bitis atropos L.	Berg adder
	Bitis caudalis Smith	Horned puff adder
	Bitis cornuta Daudin	Horned adder
	Bitis gabonica Schmidt	Gaboon adder/viper
	Bitis inornata Smith	Plain mountain adder
	Bitis nasicornis Shaw	Rhinoceros or horn-nosed adder/viper
	Bitis peringueyi Boulenger	Peringuey's adder
	Causus defilippii Jan	Snouted night adder
	Causus lichtensteinii Jan	Lichtenstein night adder
	Causus rhombeatus Lichtenstein	Common night adder
	Vipera kaznakovi Nikolsky	Caucasus adder
	Vipera superciliaris Peters	African lowland adder
Asps		
	Cerastes vipera L.	Cleopatra's asp
	Vipera aspis L.	European asp, asp viper
Vipers/Other Adders		
	Atheris species	African bush vipers
	Atropoides nummifer Rüppell	Jumping viper
	Bothriechis schlegelii Berthold	Eyelash viper
	Cerastes cerastes L.	Horned viper
	Cerastes vipera L.	Sahara sand viper
	Daboia russelii Shaw & Nodder	Russell's viper
	Echis carinatus Schneider	Saw-scaled viper
	Echis coloratus Günther	Saw-scaled or Arabian viper
	Echis pyramidum Geoffroy Saint-Hilaire	Egyptian carpet viper
	Macrovipera lebetina L. (*Vipera lebetina*)	Desert adder, blunt-nosed viper, Levantine viper
	Porthidium nasutum Bocourt	Hognose viper
	Pseudocerastes persicus Duméril, Bibron & Duméril	Palestine or Persian horned viper
	Vipera ammodytes L.	Long-nosed or nose-horned viper
	Vipera aspis L.	Asp viper
	Vipera berus L.	European or common adder, common viper
	Vipera latastei Bosca	Lataste's or snub-nosed viper
	Vipera ursinii Bonaparte	Meadow adder or meadow viper, Orsini's viper
	Vipera xanthina Gray	Near East viper
	Vipera palaestinae Werner	Palestine viper

[a]Source: Adapted from Ref 1.

occasionally, hypovolemia.[16,17] In a case series of 231 *Vipera berus* bites, about 11% of the bites did not result in envenomation, while 13% of the bites produced severe envenomation (grade 3).[18] In a study of 99 bites by *Vipera aspis* L. presenting to a Swiss regional hospital, there were no cases of neurotoxicity or hemolysis.[19] The percentages of dry bites (grade 0) and severe envenomation (grade 3) were 8% and 10%, respectively. Rarely, case reports associate the development of severe local swelling and elevated compartment pressures with envenomation by *V. berus* L.[20] or *V. aspis* L.[21] The composition of the venom and the clinical presentation following envenomation are similar for both the asp and the adder with clinical effects ranging between local edema and rare cases of severe hypotension and bleeding.[22,23] Although ptosis occurs occasionally following envenomation by the European vipers,[24] flaccid paralysis and clinically significant coagulopathies are unusual. Severe envenomations are uncommon and these envenomations produce rapidly progressive edema, angioneurotic edema of face and pharynx similar to anaphylaxis, severe gastrointestinal symptoms, hypotension, shock, renal failure, coma, and death.[25,26] Rare case reports associate European viper bites with myocardial infarction,[27] cerebrovascular bleeding, and thrombosis of the iliac vessels with intestinal infarction.[28]

TABLE 176.2. Most Common European Venomous Vipers[a]

Species	Common Name	Description	Distribution	Habitat
Vipera ammodytes L.	Sand viper, nose-horned viper	Males; light-gray to gray with well developed black vertebral zig-zag strip; females: reddish brown to yellow with brown or gray bands. Distinct fleshy horn on snout and 60–80 cm length	Austria, Italy, Balkan States, Turkey, Transcaucasia	Dry, rocky slopes 200–600 m up to 2000 m (~6,500 ft)
Vipera aspis L.	Asp viper	60–75 cm (~25–30 in.) with variable coloration (light gray to gray-brown with black zig-zag strips). Snout slight upturned, but not erect	West, Central, Southern Europe	Sunny, dry, rocky hillsides up to 3,000 m (~10,000 ft)
Vipera berus L.	Adder, common viper	Thick-bodied 50–65 cm (~20–25 in.) in length with pale gray color and zig-zag stripes. Y- or X-pattern on distal head	England, France, Central and Northern Europe up to about 1,000 km.	Cool, moist habitats (dunes, marsh, meadows, forest edges up to polar circle)
Vipera latastei Bosca	Lataste's viper	50–60 cm (~20–25 in.) with coloration similar to asp viper, except darker edges, black tail tip and distinct nose-horn	Central and Eastern Spain, Southern Portugal	Rocky hillsides, open forest up to 1,500 m (~5,000 ft)
Macrovipera lebetina L.	Blunt-nosed viper	Largest (80–100 cm/~30–40 in.) European viper with yellow-, gray-, or red-brown coloration and vertebral zig-zag stripe	Cyprus, Greek Islands, Near East, North Africa	Rocky, dry hillsides with some vegetation up to 1,500 m
Vipera ursinii Boneparte	Meadow viper, Orsini's viper	Smallest (40–50 cm/~15–20 in.) European viper with yellow, brown, gray or dark green coloration and vertebral zig-zag stripes	Southern France, Italy, Austria, Balkan States	Dry meadows, hillsides up to 2,000 m (~6,500 ft)
Vipera xanthina Gray (*Montivipera xanthina*)	Ottoman viper	Large (80–100 cm/~30–40 in.) viper with light gray or brown coloration with dark brown or olive vertebral zig-zag stripes	Greece, West Turkey Islands, Turkey	Hillsides, rocky meadows up to 2,500 m (~8,000 ft)

[a]Source: Adapted from Ref 2.

TABLE 176.3. Grading of European Viper Envenomations[a]

Grade	Envenomation	Clinical Features
0	None	Fang marks without edema
1	Minimal	Local edema around bite with systemic symptoms
2	Moderate	Regional edema involving major portion of limb with moderate systemic symptoms (mild hypotension, vomiting diarrhea)
3	Severe	Extensive edema spreading to trunk with severe systemic symptoms (persistent hypotension, shock, bleeding)

[a]Source: Adapted from Ref 22.

DIAGNOSTIC TESTING

Laboratory abnormalities following envenomation by European vipers include leukocytosis and hemoconcentration, as well as anemia and thrombocytopenia in severe envenomations. Coagulopathy, internal bleeding, and rhabdomyolysis are rare.[29,30] Enzyme-linked immunosorbent assays (ELISA) can detect European viper venom antigens (*Vipera aspis*, *Vipera berus*, *Vipera ammodytes*) for at least 24 hours in serum samples from moderately envenomated patients.[31] This test does not cross-react with venom from *Vipera russelii, Bothrops jararaca* Wied, or *Crotalus durissus* subsp. *terrificus* Laurenti.

TREATMENT

Therapy for European viper envenomation is primarily supportive. Antivenom administration (e.g., ovine ViperaTAb® [Protherics PLC, Macclesfield, Cheshire, UK] administered intravenously) is often restricted to cases of moderate to severe envenomation with rapidly progressive edema and persistent hypotension depending on availability.[32] A few patients with the clinical appearance of minor or moderate viper envenomation during initial evaluation can develop severe envenomation. In general, early use of antivenom reduces both morbidity and mortality. Web databases (http://www.toxinfo.org/antivenoms/Index_Indication.html) are available for locating antivenom in Europe for vipers and other venomous snakes. Monitoring of laboratory data in these cases involves complete blood count, coagulation parameters (fibrinogen, fibrin split products, prothrombin time, activated partial thromboplastin time), serum electrolytes, and renal function. Supportive care includes tetanus immunization as needed, elevation, splinting in a position of rehabilitation, daily cleansing and wound dressings, and early, gradual mobilization with physical therapy.

MIDDLE AND NEAR EAST/AFRICA

EXPOSURE

Table 176.4 lists important venomous vipers in this area of the world. *Vipera palaestinae* Werner (Palestine viper) accounts for most of the serious venomous snakebites in Israel,[33] whereas the desert horned vipers (*Cerastes* species) are the most abundant venomous snakes of the great deserts of North Africa and the Middle East. The genus *Cerastes* comprises three species of desert vipers including the Saharan sand viper (*C. vipera* L.), Gasperetti's horned viper (*C. gasperettii* Leviton & Anderson), and the Saharan horned viper (*C. cerastes* L.). In contrast to the Saharan horned viper, the Saharan sand viper lacks the distinctive supraocular horns. In Israel and southwestern Asia, envenomations by venomous snakes from the subfamily Viperinae produce both hemorrhagic and neurotoxic effects depending on the particular species.[34] Most of these snakes exhibit warning sounds (e.g., puff adders hiss, saw-scaled vipers produce a rasping sound by rubbing their scales together). Nocturnally active species include the carpet vipers (*Echis* spp.) and night adders (*Causus* spp.). Important causes of morbidity and mortality from African vipers include carpet and saw-scaled vipers (*Echis* spp.), night adders (*Causus* spp.), and puff adders (*Bitis* spp.).[35] Vipers possess erectile fangs that may reach 3 cm (~1 in.) in length in large vipers (e.g., Gaboon viper, *Bitis gabonica* Duméril Bibron & Duméril). In contrast to crotaline snakes (i.e., pit vipers), non-crotaline vipers (e.g., *Bitis* spp.) do not possess the ability to strike prey with thermal sensors.[36]

VENOM

Viper venoms produce local tissue damage and coagulation abnormalities as a result of endothelial damage, platelet aggregation and dysfunction, increased capillary permeability, and inhibition of thrombin and thromboplastin along with some fibrinolysis and hypofibrinogenemia. The venom of snakes from the genus *Cerastes* contains a variety of toxins that affect coagulation including serine proteases (cerastotin),[37] hemorrhagic protease (cerastase F-4), fibrinogenases (afaâcytin),[38] protein C activator, and other thrombin-like enzymes (cerastocytin).[39] *Echis* venoms from the saw-scaled and carpet vipers contain an enzyme, ecarin, which directly activates prothrombin and initiates a consumption coagulopathy by forming fibrin after the splitting of fibrinopeptide A and B from fibrinogen. These venoms produce severe local signs of envenomation (edema, vesiculation, ecchymosis, necrosis) along with coagulopathy. *Bitis* (puff adder) venoms produce local hemorrhage and necrosis, but these venoms do not cause clinical significant coagulation disorders with the exception of venom from the Gaboon viper (*Bitis gabonica*)[40] and the puff adder (*Bitis arietans*).[41] The venom of tree vipers (*Atheris* species) can produce a consumptive coagulopathy because of platelet-aggregating and fibrinogen-converting activity.[42]

CLINICAL RESPONSE

Local Tissue Damage

Symptoms and signs of envenomation usually begin within a few minutes to approximately 1 hour after the bite.[43] With the exception of the Gaboon viper (*Bitis gabonica*), puff adder (*Bitis* spp.) envenomations produce local edema and hemorrhage with vesiculation around the bite site, but systemic and hematological complications are usually mild. Occasionally, severe envenomations may cause hypotension, respiratory failure, thrombocytopenia, severe coagulopathy, and necrosis. Envenomation by the puff adder (*Bitis arietans*) may cause extensive local tissue destruction.[44] Severe envenomation manifests by progressive swelling and local necrosis. Prolonged coagulation times may

TABLE 176.4. Important African, Middle Eastern, and Near Eastern Venomous Vipers[a]

Species	Common Name	Description	Distribution	Habitat
Echis Species	Carpet and saw-scaled vipers	Small (50–60 cm), slender snakes with variable coloration ranging from red, brown to olive gray with light dorsal spots	Near East: *E. carinatus*, (*E. multisquamatus, E. sochureki*); *E. pyramidum* Middle East: *E. coloratus* East/North Africa: *E. megalocephalus, E. jogeri, E. pyramidum* West Africa: *E. hughesi, E. leucogaster, E. ocellatus*	Dry areas (savannah, deserts), agricultural areas (West Africa)
Bitis Species	Puff adders	Largest African vipers up to 2 m length with triangular head and variable coloration from light brown and white/yellow strips to highly diverse coloration (green, purple, brown)	*B. arietans:* Africa to Arabian Peninsula *B. gabonica:* Forests of West/Central/East Africa *B. nasicornis:* West/Central Africa *B. caudalis:* Southern Africa	*B. arietans*—dry habitats and river beds, *B. gabonica* and *B. nasicornis*—forest floors
Cerastes and *Pseudocerastes* Species	Horned vipers, sand vipers	Thick-bodied vipers up to 70 cm in length with yellow or gray background, marked serrated scales, and irregular dorsal coloration; most have distinctive horn above eyes	North Africa, Sahara, Middle East to Pakistan	Sandy deserts, dry rock areas with sparse brush
Atheris, Montatheris, Proatheris Species	Bush or tree vipers	Small to medium vipers up to 80 cm with overlapping scales and variable coloration ranging from gray brown to yellow green	West/East Africa to Angola	*Montatheris, Proatheris* species—terrestrial; *Atheris* species—arboreal
Causus Species	Night adders	Medium sized vipers up to 1 m with gray brown or olive background and dark spots resembling a rhomboid	Southwest Africa below the Sahara	Forests and lowlands near water

[a]Source: Adapted from Ref 2.

occur after envenomation by the Gaboon viper; death may occur in envenomated patients unable to obtain medical care.[45] Envenomations by *Causus* species (night adders) usually cause mild local edema and vesiculation with occasional necrosis, but systemic symptoms are usually mild.[46] Envenomations by *Cerastes* species (horned or sand vipers) and European vipers are similar. Local edema, vesiculations, and hemorrhage occur, but impairment of coagulation and systemic symptoms are usually minimal to mild.[47] However, case reports suggest that severe *Cerastes* envenomations may produce clinically significant coagulopathy, increased fibrinolysis, thrombocytopenia, microangiopathic hemolytic anemia, and acute renal failure.[48,49] Envenomations by *Pseudocerastes* species (horned vipers) usually result in mild local effects without systemic complications.

Local Tissue Damage and Coagulopathy

Because carpet vipers (*Echis* species) may bite during the night, envenomation may not be recognized until the following day when slight local swelling and bleeding from the mouth or wounds occur. Complications include thrombocytopenia, severe consumption coagulopathy, massive hematomas, intracerebral hemorrhage, gastrointestinal bleeding, hematuria, renal failure, and death.[50,51] Envenomation by tree and bush vipers (*Atheris* species) can produce severe coagulopathy, systemic symptoms (drowsiness, respiratory impairment, vomiting, diarrhea), and hypotension. However, internal bleeding is not well documented following *Atheris* envenomations.

TREATMENT

Local treatment for viper envenomations is similar to the treatment for crotaline envenomations. Initially, the envenomated extremity should be immobilized at or below the level of the heart in a position of rehabilitation. The wound should be cleansed and dressed with sterile dressings, and tetanus immunization updated as necessary. Systemic antibiotics are usually necessary only for secondary infections. Victims should be evaluated for hypovolemia, respiratory depression, and coagulopathy. Antivenom is indicated in life-threatening conditions (e.g., coagulopathies, respiratory depression, refractory hypotension) or marked progressive local signs including envenomation by the common viper (*Vipera berus*)[52] or *Vipera palaestinae*.[53] The recommended dose of the equine *Vipera palaestinae* antivenom (Felsenstein Medical Research Institute, Rabin Medical Center, Petah Tikva, Israel) starts with a very slow test dose over 10 minutes followed by a fixed dose of 50mL infused over 2 hours intravenously Additional intravenous doses (30–50mL) may be administered as dictated by the progression of symptoms. Envenomations by carpet and saw-scaled vipers (*Echis* species),[54] Gaboon viper (*Bitis gabonica*), and tree and bush vipers (*Atheris* species) usually require antivenom (South African polyvalent antivenom). Specific monovalent antivenom is preferable to polyvalent antivenom because of the interspecies variation of venom composition, particularly for *Echis* species. However, there are limited supplies of antivenom for African vipers. For African viper species, only the South African polyvalent antivenom is available. This antivenom has limited efficacy for North African viper species. Antivenom for other vipers (*Bitis, Causus, Cerastes, Pseudocerastes*) is usually necessary only for severe envenomations. Supportive care in viper envenomation is similar to crotaline envenomation.

ASIA

EXPOSURE

Russell's viper (*Daboia russelii,* a.k.a. *Vipera russelii*) is the most dangerous viper in Southeast Asia[55] with the following two main subspecies: a western species (India, Pakistan, Sri Lanka, Bangladesh) named *Daboia russelii* subsp. *russelii* Shaw, and an eastern species (China, Taiwan, Thailand, Burma, Indonesia) named *Daboia russelii* subsp. *siamensis* Smith. *Echis carinatus* Schneider, *Echis carinatus* subsp. *multisquamatus* Cherlin, *Echis carinatus* subsp. *sochureki* Stemmler, and related species (*E. coloratus* Günther, *E. ocellatus* Stemmler, *E. leucogaster* Roman, *E. pyramidum* Geoffroy Saint-Hilaire) are distributed over Africa and Asia as listed in Table 176.5.[56]

VENOM

The venom from Russell's viper (*Daboia russelii*) contains protease inhibitors, hemorrhagins (i.e., factors that increase vascular permeability or damage vascular endothelium), phospholipase A_2, and procoagulant factors that activate factor X to factor Xa and initiate the coagulation cascade. There is substantial geographical variation in venom composition and clinical manifestations of envenomation by Russell's viper with neurotoxicity less common in the western (Indian) species of Russell's viper.[57] Saw-scaled vipers (*Echis carinatus*) range between the Middle East and India, and their venom

TABLE 176.5. Venomous Asian Vipers[a]

Species	Common Name	Description	Distribution	Habitat
Daboia russelii (subsp. *russelii* and *siamensis*)	Russell's viper	Large snake up to 2m (~7ft) with background coloration from gray to light yellow and brown and dark spots surrounded by white	6 species in Indian subcontinent, Burma, Thailand, South China, Taiwan	Open terrain with dense grass or low bushes up to 3,000m
Echis carinatus Complex	Saw-scaled viper	Small, thin snakes up to about 80cm (~300in.) in length with red, brown, grey or olive colorations and small light spots on the dorsum. The flanks contain V-shaped markings	*E. multisquamatus:* Uzbekistan, Turkmenistan, Iran, Afghanistan and Pakistan *E. sochureki:* northern India, Pakistan, Afghanistan, Arabian Peninsula	

[a]Source: Adapted from Ref 2.

contains an enzyme, ecarin, which directly activates prothrombin and initiates a consumption coagulopathy. The presence of hemorrhagins in addition to procoagulant factors in *Echis* species frequently produces clinically significant bleeding. Local tissue damage and coagulopathy (i.e., defibrination) are the most common complications of envenomations by Asian vipers with low fibrinogen concentrations and thrombocytopenia being the most common laboratory abnormalities.[58] The massive consumption of clotting factors following serious envenomations typically results in clinical anticoagulation rather than thrombosis and embolic disease.

CLINICAL RESPONSE

Clinical features of envenomation usually develop within a few hours, and delay of the onset of signs of envenomation beyond 6 hours postenvenomation is rare.[59] Acute tubular necrosis, renal failure, coagulopathy (prolonged prothrombin time with depletion of factors V and X, thrombocytopenia, hypofibrinogenemia, increased D-dimer), disseminated intravascular coagulation, hypotension, flaccid paralysis, and hemolysis are the major complications of severe envenomation by Russell's viper (*Daboia russelii* subsp. *russelii*).[60] Neurological effects are variable with the most common neurological signs of envenomation including ptosis, bulbar palsy, and weakness in neck flexors.[61] Rhabdomyolysis is rare. Sheehan's syndrome may result from pituitary infarction and panhypopituitarism following envenomation by Russell's viper. Some regional variation occurs following envenomation by Russell's viper with the Formosan Russell's viper (*Daboia russelii* subsp. *siamensis*) producing more systemic thrombosis and little neurological symptoms compared with the eastern variety (*D. russelii* subsp. *russelii*).[62]

TREATMENT

Treatment of Asian viper envenomation is similar to crotaline envenomation, particularly supportive care (cleansing wound, sterile dressing, immobilization, early physical therapy) for local tissue damage. Antibiotics are usually reserved for secondary infections. Studies of snake flora indicate that the mouth of Asian vipers contains enteric organisms (i.e., Gram-negative rods, *Enterobacter* and *Pseudomonas* species) as well as *Staphylococci* and *Clostridia* species.[63] Most envenomations require the use of antivenom. The specificity and efficacy of antivenom varies with individual localities and species. The administration of polyvalent antivenom produced for the Russell's viper in India causes mixed results following envenomations by Russell's vipers in Sri Lanka.[64] A new, monovalent ovine Fab antivenom (Polonga TAb Therapeutic Antibodies, Inc., London, UK) has been developed against the venom of Russell's viper from Sri Lanka. An open, randomized clinical trial comparing the two venoms suggests that the monovalent Fab antivenom is more effective against species of Russell's vipers from Sri Lanka.[65] However, this antivenom is available primarily for research purposes, and a polyvalent antivenom is imported from India for the treatment of envenomation by the Russell's viper in Sri Lanka. A polyvalent antivenom for the eastern species of Russell's viper (*Daboia russelii siamensis*, Formosan Russell's viper) is available in Taiwan, as well as monospecific antivenom raised against the Formosan Russell's viper.[66] The polyvalent antivenom effectively reverses the coagulation disorder (disseminated intravascular coagulation with secondary fibrinolysis and thrombocytopenia), but administration of the antivenom may not reverse the acute renal failure associated with severe envenomations.[67] Coagulation profiles and complete blood count should be followed serially until the parameters substantially improve. The presence of dark urine indicates the need to evaluate for the presence of myoglobinuria, hematuria, and renal dysfunction. Generous intravenous fluid replacement should be given to maintain adequate urine flow if myoglobinuria occurs. The efficacy of urine alkalinization for the prevention of myoglobin-induced renal disease remains unclear.

Pit Vipers (Crotalinae)

Table 176.6 lists some members of the *Crotalinae* (pit vipers) subfamily of the Viperidae family that includes the Asian (e.g., Okinawan and Chinese habu of the genus *Trimeresurus*) and the North American pit vipers (rattlesnake, copperhead, cottonmouth), as well as the Central and South American bushmaster (*Lachesis* spp.) and fer-de-lance (*Bothrops* spp.).[68,69] There are no native vipers in the New World. The hump-nose viper (*Hypnale hypnale* Merrem) is a pit viper native to Southeast Asia, and envenomation by this snake usually produces only local effects (pain, swelling, hemorrhagic vesicles, regional lymphadenopathy).[70] However, occasionally coagulopathy, spontaneous hemorrhage, acute renal failure, and respiratory failure can develop following envenomation by the hump-nose viper.[71]

NORTH AMERICA

Of about 120 species of snakes in the United States, approximately 20 are venomous.[72] At least one species

TABLE 176.6. Scientific and Common Names of Venomous Pit Vipers (Crotalinae) Subfamily[a]

Group	Scientific Name	Common Name
American		
	Agkistrodon bilineatus bilineatus Günther	Common cantil
	Agkistrodon bilineatus taylori Burger and Robertson	Taylor's cantil
	Agkistrodon contortrix contortrix L.	Southern copperhead
	Agkistrodon contortrix laticinctus Gloyd and Conant	Broad-banded copperhead
	Agkistrodon contortrix mokasen Palisot de Beauvois	Northern copperhead
	Agkistrodon contortrix phaeogaster Gloyd	Osage copperhead
	Agkistrodon contortrix pictigaster Gloyd and Conant	Trans-Pecos copperhead
	Agkistrodon piscivorus conanti Gloyd	Florida cottonmouth
	Agkistrodon piscivorus piscivorus Lacépède	Eastern cottonmouth
	Agkistrodon piscivorus leucostoma Troost	Western Cottonmouth
American Lanceheads		
	Atropoides nummifer Rüppell	Jumping viper
	Bothriechis lateralis Peters	Yellow-lined pit viper
	Bothriechis nigroviridis Peters	Black spotted pit viper
	Bothriechis schlegelii Berthold	Eyelash viper
	Bothriopsis bilineatus bilineatus Wied-Neuwied	Amazonian tree-viper
	Bothriopsis peruviana Boulenger	Peruvian pit viper
	Bothrops alternatus Duméril & Bibron	Urutu or Yarara
	Bothrops andianus Amaral	Andean pit viper, Terciopelo (Fer-de-lance)
	Bothrops asper Garman	Barba amarilla (Fer-de-lance)
	Bothrops atrox L.	Caissaca (Fer-de-lance)
	Bothrops barnetti Parker	Barnett's pit viper
	Bothriechis bicolor Bocourt	Bocourt's pit viper
	Bothrops brazili Hoge	Brazil's pit viper
	Bothrops caribbaeus Garman	St. Lucia pit viper
	Bothrops cotiara Gomes	Cotiara
	Bothrops fonsecai Hoge and Belluomini	Fonseca's *pit* viper
	Bothrops insularis Amaral	Island jararaca
	Bothrops jararaca Wied-Neuwied	Jararaca
	Bothrops jararacussu Lacerda	Jararacussu
	Bothrops lanceolatus Bonnaterre	Fer-de-lance
	Bothrops lojanus Parker	Lojan lancehead
	Bothrops neuwiedii Wagler	Jararaca pintada or Wied's lancehead
	Bothrops pirajai Amaral	Piraja's pit viper, Jararacucu
	Cerrophidion barbouri Dunn	Barbour's montane pit viper
	Cerrophidion godmani Günther	Godman's pit viper
	Ophryacus melanurus Müller	Black-tailed pit viper
	Porthidium dunni Hartweg and Oliver	Dunn's pit viper
	Porthidium lansbergii Schlegel	Lansberg's hog nose viper
	Porthidium nasutum Bocourt	Hog-nosed pit viper
	Porthidium ophryomegas Bocourt	Western hog-nosed pit viper
	Porthidium yucatanicum Smith	Yucatan pit viper
Rattlesnake		
	Crotalus adamanteus Palisot de Beauvois	Eastern Diamondback rattlesnake
	Crotalus atrox Baird and Girard	Western Diamondback rattlesnake
	Crotalus basiliscus Cope	Mexican west-coast rattlesnake, Oaxacan rattlesnake
	Crotalus catalinensis Cliff	Santa Catalina Island or rattleless
	Crotalus cerastes cerastes Hallowell	Mojave Desert sidewinder
	Crotalus cerastes cercobombus Savage and Cliff	Sonoran Desert sidewinder
	Crotalus cerastes laterorepens Klauber	Colorado Desert sidewinder
	Crotalus durissus durissus L.	Central American rattlesnake
	Crotalus durissus culminates Klauber	Northwestern neotropical rattlesnake

TABLE 176.6. *Continued*

Group	Scientific Name	Common Name
	Crotalus durissus terrificus Laurenti	South American rattlesnake
	Crotalus durissus totonacus Gloyd and Kauffeld	Totonacan rattlesnake
	Crotalus enyo enyo Cope	Baja California rattlesnake
	Crotalus enyo cerralvensis Cliff	Cerralvo Island rattlesnake
	Crotalus enyo furvus Lowe and Norris	Rosario rattlesnake
	Crotalus horridus atricaudatus Latreille in Sonnini and Latreille	Canebrake rattlesnake
	Crotalus horridus horridus L.	Timber rattlesnake
	Crotalus intermedius intermedius Troschel	Totalcan small-headed rattlesnake
	Crotalus lepidus klauberi Gloyd	Banded rock rattlesnake
	Crotalus lepidus lepidus Kennicott	Mottled rock rattlesnake
	Crotalus lepidus morulus Klauer	Tamaulipan rock rattlesnake
	Crotalus mitchellii mitchellii Cope	San Lucan speckled rattlesnake
	Crotalus mitchellii pyrrhus Cope	Southwestern speckled rattlesnake
	Crotalus mitchellii stephensi Cope	Panamint rattlesnake
	Crotalus molossus molossus Baird and Girard	Northern black-tailed rattlesnake
	Crotalus molossus nigrescens Gloyd	Mexican black-tailed rattlesnake
	Crotalus polystictus Cope	Mexican lance-headed rattlesnake
	Crotalus pricei pricei Van Denburgh	Twin-spotted rattlesnake
	Crotalus ruber lucansensis Van Denburgh	San Lucan diamond rattlesnake
	Crotalus ruber ruber Cope	Red diamond rattlesnake
	Crotalus scutulatus scutulatus Kennicott	Mojave rattlesnake
	Crotalus scutulatus salvini Günther	Huamantlan rattlesnake
	Crotalus stejnegeri Dunn	Long-tailed rattlesnake
	Crotalus tigris Kennicott in Baird	Tiger rattlesnake
	Crotalus tortugensis Van Denburgh and Slevin	Tortuga Island diamond rattlesnake
	Crotalus unicolor Lidth de Jeude	Aruba Island rattlesnake
	Crotalus vegrandis Klauber	Uracoan rattlesnake
	Crotalus viridis abyssus Klauber	Grand Canyon rattlesnake
	Crotalus viridis caliginis Klauber	Coronado Island rattlesnake
	Crotalus viridis cerberus Coues in Wheeler	Arizona black rattlesnake
	Crotalus viridis concolor Woodbury	Midget faded rattlesnake
	Crotalus viridis helleri Meek	Southern Pacific rattlesnake
	Crotalus viridis lutosus Klauber	Great Basin rattlesnake
	Crotalus viridis nuntius Klauber	Hopi rattlesnake
	Crotalus viridis oreganus Holbrook	Northern Pacific rattlesnake
	Crotalus viridis viridis Rafinesque	Prairie rattlesnake
	Crotalus willardi silus Klauber	Chihuahuan ridge-nosed rattlesnake
	Crotalus willardi willardi Meek	Arizona ridge-nosed rattlesnake
	Sistrurus catenatus catenatus Rafinesque	Eastern massasauga
	Sistrurus catenatus edwardsi Baird and Girard	Desert massasauga
	Sistrurus catenatus tergeminus Say in James	Western massasauga
	Sistrurus miliarius barbouri Gloyd	Southeastern pygmy rattlesnake
	Sistrurus miliarius miliarius L.	Carolina pygmy rattlesnake
	Sistrurus miliarius streckeri Gloyd	Western pygmy rattlesnake
	Sistrurus ravus Cope	Mexican pygmy rattlesnake
Bushmaster	*Lachesis muta* L.	Bushmaster
Asian		
	Calloselasma rhodostoma Boie	Malayan pit viper
	Deinagkistrodon acutus Günther	Hundred-pace snake
	Gloydius halys Pallas	Mamushi (Japan and Korea)
	Gloydius himalayanus Günther	Himalayan pit viper
	Gloydius intermedius Strauch	Central Asian pit viper
	Hypnale hypnale Merrem	Hump-nosed viper
	Ovophis okinavensis Boulenger	Himehabu

TABLE 176.6. *Continued*

Group	Scientific Name	Common Name
	Protobothrops elegans Gray	Japanese lace-headed viper, Sakishima-habu
	Protobothrops flavoviridis Hallowell	Habu
	Trimeresurus albolabris Gray	White-lipped tree viper
	Trimeresurus gramineus Shaw	Green tree viper
	Trimeresurus mucrosquamatus Cantor	Chinese or Taiwan habu
	Trimeresurus popeorum Smith	Pope's tree viper
	Trimeresurus purpureomaculatus Gray	Mangrove pit viper
	Trimeresurus stejnegeri Schmidt	Chinese green tree viper
	Tropidolaemus wagleri Wagler	Wagler's pit viper

[a]Source: Adapted from Ref 1.

of venomous snake resides in every state of the United States except Hawaii, Alaska, and Maine. The vast majority (95%) of poisonous snakebites involve members of the subfamily Crotalinae (pit viper). Eastern and Western Diamondback rattlesnakes are dangerous species because of their large size and potential for large envenomations.[73] The only venomous species of the Elapidae (cobra) family indigenous to the United States are the Northern coral snakes (*Micrurus fulvius* L.), Texas coral snakes (*Micrurus tener* Baird & Girard), and the Western or Sonoran coral snakes (*Micruroides euryxanthus* Kennicott). These elapids account for less than 3% of US snakebites.

American pit vipers belong to the subfamily Crotalinae of the Viperidae family, and this subfamily includes 25 species divided into the following three genera: *Crotalus* (true rattlesnakes), *Sistrurus* (pygmy rattlesnake, massasauga), and *Agkistrodon* (copperhead, cottonmouth). The rattle, which functions as a warning device to predators, is the main characteristic common to species in the genera *Crotalus* and *Sistrurus*; hence, these snakes are called rattlesnakes. The rattle consists of a series of interlocking horny rings that vibrate to produce a buzzing sound following stimulation of the rattlesnake. *Agkistrodon* species do not possess rattles. The following four characteristics separate poisonous American pit vipers (subfamily Crotalinae) from most of the other nonpoisonous American snake species (See Figure 176.3):

1. Triangular or arrow-shaped head due to the temporal location of the venom glands
2. Facial pits between the nostril and eye that contain heat-sensing organs
3. Vertical-oriented, elliptical pupils ("cat's" eyes)
4. Single row of subcaudal scales rather than a double row

Coloration and diamondback patterns do not reliably identify rattlesnakes because nonpoisonous snakes mimic these markings for defensive purposes. Furthermore, the characteristics of nonvenomous species in Figure 176.3 do *not* separate poisonous Elapidae species from nonpoisonous snakes. Coral snakes have round pupils rather than elliptical pupils and smooth tapered bodies with narrow heads. All pit vipers have movable fangs, whereas Elapidae species have fixed fangs.

Reactions to snakebite are not reportable in the United States; therefore, there are limited data on the actual incidence of snakebites in the United States. During the 1950s and 1960s, there were approximately 45,000 snakebites annually in the United States with about 7,000–8,000 venomous bites and an estimated 10–15 fatalities.[74,75] Currently, rattlesnake-induced fatalities, which usually result from prolonged hypotension and multiple organ system failure, are rare.[76] Although pediatric patients sustain more serious envenomations than adults because the relative higher amount of venom per kilogram, the case-fatality rates are similar between children and adults.[77] Improved supportive care and the aggressive use of antivenom significantly reduced mortality and morbidity (e.g., limb and skin necrosis) since the 1960s when the use of antivenom was less common.[78] In the United States, fatalities from allergic reactions to *Hymenoptera* (bees, wasps, ants) stings are at least 10 times greater than the fatalities resulting from snakebites.[79,80]

The majority of rattlesnake bites in California occur in the warm, highly populated regions of southern California, whereas the highest incidence of venomous snakebite occurs in the southern areas of the United States (i.e., North Carolina, Texas, Arkansas, Mississippi, Georgia, West Virginia, Louisiana, Oklahoma).[81] Not all rattlesnake bites result in envenomation because rattlesnakes fail to inject venom in approximately 20% of bites.[82]

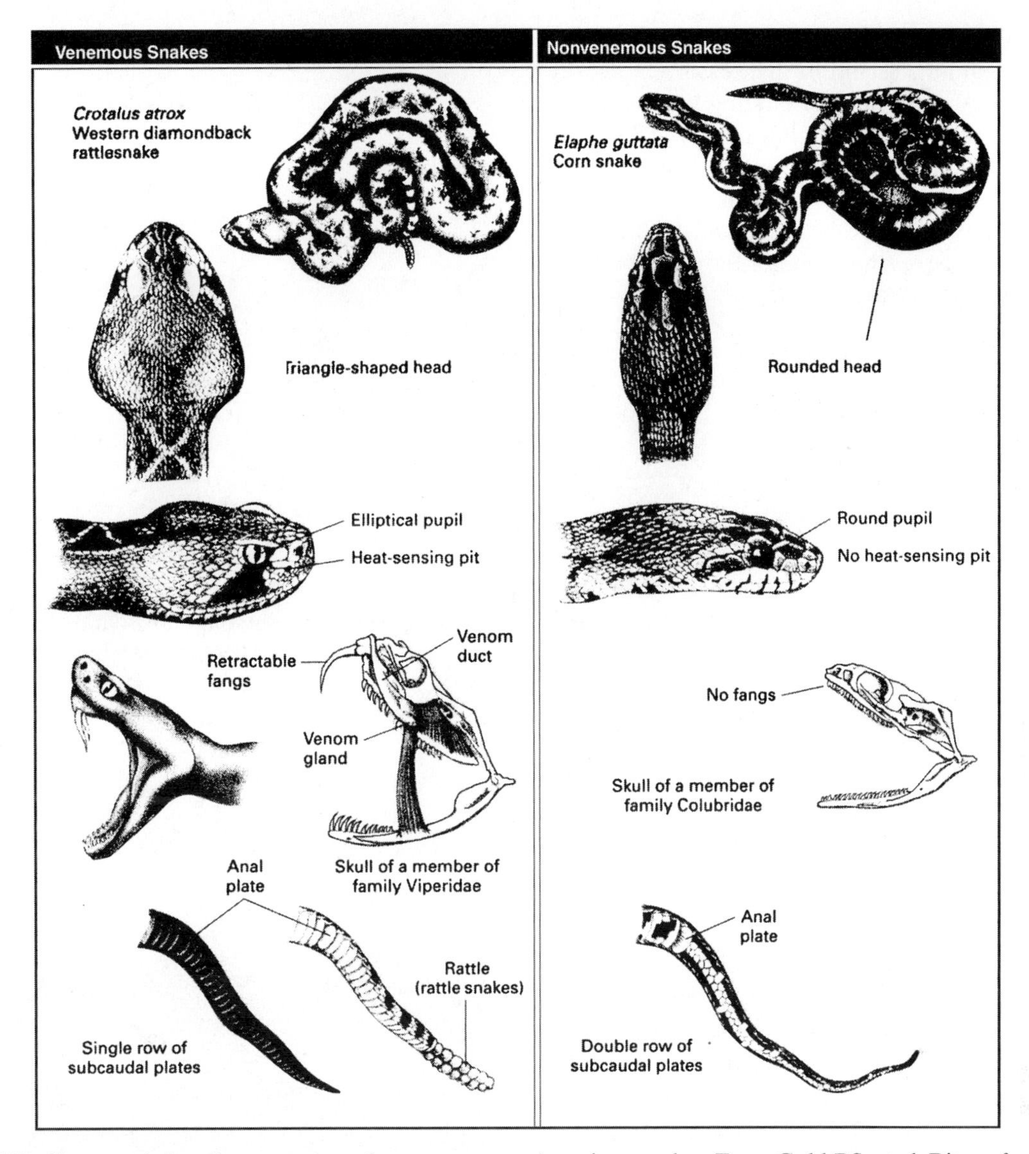

FIGURE 176.3. Characteristics of venomous and nonvenomous American snakes. From Gold BS, et al. Bites of venomous snakes. N Engl J Med 2002;347:350.

Because of the inactivity of snakes during the winter, most snakebites occur from April through October depending on the temperatures in specific localities. Nearly three-quarters of the rattlesnake bite victims are men. The typical victim of a pit viper envenomation is a young man bitten on the hand while handling the snake. Illegitimate bites (i.e., failure to move away from the snake, snake handling) represent a substantial portion of snakebites, particularly in association with the use of alcohol. In a study of 86 consecutive rattlesnake bites in patients presenting to a central Arizona medical center, 56% of the victims had illegitimate (i.e., provoked) bites.[83] In a study of 36 patients hospitalized in New Mexico for rattlesnake envenomation, about 28% of the patients were envenomated while consuming alcohol and handling the snake.[84] Most accidental (i.e., legitimate) snakebites occur in the afternoon or early evening of the summer months (July, August) during recreational activities. Although most bites occur during daytime hours, the intense heat of southwestern deserts and the nocturnal habits of some rattlesnakes result in nighttime bites.[85] Bites to the lower extremity are usually legitimate.

True Rattlesnake (*Crotalus spp.*)

Although some species differences occur in the coloration and body of rattlesnakes, all pit vipers share the four main characteristics demonstrated in Figure 176.3. Almost all rattlesnakes (genus *Crotalus*) have a set of interlocking, terminal horny segments known as rattles.

The exception is *Crotalus catalinensis*, which is found only on Santa Catalina Island in the Sea of Cortez off Baja, California. A characteristic buzzing similar to the sound of escaping steam usually, but not always, occurs before a rattlesnake strike. Each molting (one to four times per year) adds a rattle, but trauma or congenital abnormalities may reduce the expected number of rattles. This genus represents some of the most biologically advanced of all snakes.

EXPOSURE

Geographic Distribution

Of the 32 species of rattlesnakes comprising 65–70 subspecies, 21 species inhabit the United States. Rattlesnakes appear in every state on the US mainland except Maine and Delaware including areas above the timberline (i.e., 10,000 ft/~3000 m) during the summer. Most rattlesnake bites result from the Western Diamondback (*C. atrox*), the Southern Pacific (*Crotalus viridis helleri*), Northern Pacific (*Crotalus viridis oreganus*), the prairie (*C. viridis viridis*), the Eastern Diamondback (*C. adamanteus*), the timber (*C. horridus horridus*), and the pygmy (*Sistrurus miliarius*) rattlesnakes. In the eastern United States, the latter three species inflict most of the rattlesnake bites.[86] The Mojave rattlesnake (*Crotalus scutulatus scutulatus*) is a clinically important species because its venom typically produces less tissue damage and substantial neurotoxicity compared with most other rattlesnake bites. Proper treatment of neurological sequelae requires recognition of envenomations by this species. The range of this species includes southeastern California, southern Nevada, Arizona, New Mexico, and southwestern Texas (Figure 176.4). Although components of the venom from some specimens of *Crotalus lepidus lepidus* (Mottled rock rattlesnake) resemble the neurotoxin (Mojave toxin) present in the venom of the Mojave rattlesnake, rare case reports of envenomation by *C. lepidus lepidus* do not document neurological complications.[87] The distribution of *C. lepidus lepidus* includes southwest Texas, southeastern New Mexico, and the Mexican Plateau to San Louis Potosi. Rattlesnake bites may occur in areas distant (e.g., Europe) to the natural habitat because of the importation of species from the Crotalinae subfamily.[88]

FIGURE 176.4. Geographic distribution of Mojave rattlesnake. Adapted from Ellenhorn MJ, Barceloux DG. Medical Toxicology: Diagnosis of Human Poisoning. 1st Edition, New York: Elsevier, 1988:1117.

FIGURE 176.5. Red Diamondback rattlesnake (*Crotalus ruber ruber* Cope). © 2005 Rune Midtgaard. See color insert.

Species Differences

Variations between rattlesnake species occur in size, coloring, and distribution as well as in venom potency. Figure 176.5 displays a picture the Red Diamondback rattlesnake (*Crotalus ruber ruber Cope*). The Eastern Diamondback rattlesnake (*Crotalus adamanteus*) is the largest rattlesnake with a maximum length up to 7 ft (~2 m). The highly specialized sidewinder species uses a peculiar type of lateral motion (i.e., side winding). This motion is well designed for rapid movement across smooth, hot desert terrain because this movement requires less contact with the hot ground than crawling. Venom from sidewinders is substantially less potent than venom from the Eastern and Western Diamondback rattlesnakes.

Identification of the Mojave rattlesnake (Figure 176.6) is important because this venom from this species possesses more neurotoxicity and less tissue toxicity than the venom from the similar appearing Western

Diamondback species (Figure 176.7). The Mojave rattlesnake favors flat grassland or scrub-covered desert. These snakes rarely inhabit rocky, hilly terrain. Rhomboid markings are found on a brownish yellow to green background, with a tail characterized by white rings up to twice the size of the black rings. The basal rattle is yellow or white. The Western Diamondback has white and black tail rings of equal width and a black basal rattle. This snake is the largest rattlesnake in the western United States.

Table 176.7 lists the distinguishing characteristics of the Mojave and Western Diamondback rattlesnakes. Coloration alone does not distinguish these two rattlesnake species because at least some hybrids of the Western Diamondback and the Mojave rattlesnakes probably exist in the overlapping ranges of these species.[89]

FIGURE 176.6. Mojave Rattlesnake (*Crotalus scutulatus scutulatus* Kennicott). In contrast to the Western Diamondback Rattlesnake, specimens of the Mojave Rattlesnake often have a greenish tinge, a diagonal facial stripe behind their eye that usually does not touch their mouth, two to three enlarged scales between their eyes on top of their head, and white bands on the tail that sometimes are wider than the black bands. Photograph Courtesy of Arizona Poison and Drug Information Center. See color insert.

Behavior and Senses

Pit vipers are carnivores that use their venom primarily as an offensive weapon to obtain food. The pit vipers' keen sense of smell and their ability to sense the temperature of their prey and surrounding areas allows them to track their prey effectively. The potency of the venom allows the prey several minutes to withdraw from the scene before its death. The snake cannot accurately follow prey more than 20 ft (~6 m); consequently, the snake usually stays within that distance following envenomation. Rattlesnakes usually, but not always, strike from the coiled position. The strike averages 8 ft (~2.5 m)/s and is in a slightly downward position. A rattlesnake rarely strikes accurately over a distance exceeding one-half their body length (i.e., about 3–4 ft/~1 m maximum). Hence, a safe distance is 10–15 ft (~3–5 m) from the snake. Reflex actions produce envenomation up to 1 hour after the death of the snake despite decapitation.

The facial pit contains a unique thermal sensory organ that detects temperature changes as small as 0.003 °C.[90] These organs efficiently collect infrared photons to map the environment, similar to visual cues. Consequently, pit vipers simultaneously receive cues

FIGURE 176.7. Western Diamondback rattlesnake (*Crotalus atrox* Baird and Girard). Photograph Courtesy of Dennis Caldwell. See color insert.

TABLE 176.7. Distinguishing Characteristics of Mojave (*C. scutulatus scutulatus*) and Western Diamondback (*C. atrox*) Rattlesnakes

Characteristic	Mojave	Western Diamondback
Body size	Heavy-bodied	Heavy-bodied
Dorsal Markings	Single row of large dark diamond-shapes on lighter background with light margins	Single row of large dark diamond-shapes on lighter background with light margins
Facial strips	Bilateral with postocular light strip extending backwards above the angle of the mouth	Bilateral with postocular light strip intersecting mouth
Tail rings	Alternating circumferential black and white rings with white rings wider than black rings	Circumferential black rings encircling the tail about equal width as white rings
Intersupraocular scales	Less than 4 scales (usually 2–3)	4–7 scales (usually 4–5)

from both the visible and infrared parts of the electromagnetic spectrum. However, the maximum range of thermal detection using this organ is short (~1 ft). Sensitivity to changes in radiant temperature allows pit vipers to direct strikes at warm-blooded prey within a close range or to identify appropriate areas for thermoregulation. The presence of humans probably overwhelms the sensory mechanism because of our relatively large size. Consequently, the strike is defensive rather than offensive, and the rattlesnake does not usually inject the whole contents of the venom gland. The vision of rattlesnakes is poor because rattlesnakes lack a fovea and color receptors. Although pit vipers are deaf, they are quite sensitive to ground vibrations at distances up to 100 ft. Most rattlesnakes crawl at maximum speeds of about 3 miles/h, which is equivalent to a moderately paced walk by a man.

Venom Apparatus

Fangs and Bite Marks

Pit vipers have retractable, canalized fangs that retract into the roof of the mouth and rotate forward during a strike. In a study of 48 Western Diamondback snakes (*C. atrox*), the mean distance between fangs was about 11 mm (~0.4 in.) with a range between 4–24 mm (~0.15–1 in.).[91] At least three pairs of replacement fangs lie behind each functional fang. These replacement fangs are in various stages of development, and they move forward after the loss of the functional fang. Adult snakes shed their fangs every 6–10 weeks. Most envenomations involve the subcutaneous deposition of venom. Although subfascial injections are uncommon, intramuscular penetration is theoretically possible, especially in children. The pattern of fang marks depends on the strike. The most common pattern is two marks, 2 mm to 4 cm (~0.1–1.5 in.) apart, depending on the size of the snake. Because of the presence of reserve fangs, bite marks may contain one, two, three, or more fang marks in the case of multiple bites. Bite patterns alone do not adequately discriminate between venomous and nonvenomous snakes because of the variability in the presence of functional and replacement fangs. In general, nonvenomous snakes produce horseshoe-shaped bite marks characterized by four rows of small abrasions (upper jaw) separated from two rows of small abrasions (lower jaw). A definitive diagnosis of envenomation requires identification of the snake and/or classic clinical signs of envenomation.

Venom Gland and Envenomation

Venom glands are modified parotid glands that are surrounded by a muscle (compressor glandulae) in rattlesnakes. The amount of venom excreted from the venom gland varies depending on a number of circumstances, but the amount of venom extruded from the gland is not necessarily under neuronal control of the rattlesnake. In a study of the Western Diamondback rattlesnake (*C. atrox*), the amount of venom injected by the snake correlated to the pressure created by the compressor glandulae, the peripheral resistance of the target tissue, and compression of the fang sheath during the strike.[92] Contraction of the compressor glandulae accounted for approximately 30% of the variation in venom flow. The venom sac contains a lethal adult dose of venom, but the rattlesnake usually injects less than the full contents of the venom gland. There are variations in the amount of venom available for envenomation as a result of seasonal variation, previous strikes, and the size of the snake. Rattlesnakes are venomous from birth, but some changes occur in the composition of the venom during maturation.

PRINCIPAL TOXINS

Venom Composition

Venom is a complex mixture of proteins and polypeptides that is highly stable and resistant to drying and temperature fluctuations. The venom contains a lethal component (toxins) designed to immobilize the prey while digestive enzymes circulate through the prey to enhance subsequent digestion. Rattlesnake venom is a complex mixture of 5–15 enzymes, metal ions, biogenic amines, lipids, free amino acids, large and small proteins, and polypeptides. Lethal proteins and peptides consist of 20–80 amino acids that damage endothelial cells and plasma membranes. Polypeptides are low-molecular-weight proteins that lack enzymatic activity and are 5–20 times more lethal in animal models than crude venom. The weight of these polypeptides ranges from 6,000–30,000 daltons, and the concentration of these lethal proteins is higher in cobra (Elapidae) than in rattlesnake (Crotalinae) venom. Digestive enzymes include the following: 1) phospholipase A_2 that hydrolyzes the ester bond of lecithin and damages fatty molecules in cell membranes, 2) hyaluronidase that decreases the viscosity of connective tissue, 3) amino acid esterase that promotes fibrin formation, and 4) proteolytic enzymes and 5′ nucleotidase that damage proteins in muscle fibers.

Phospholipases A, B, C, and D hydrolyze lipids with subsequent disruption of neurotransmission at both the presynaptic and postsynaptic areas. Hyaluronidase allows the rapid spread of venom through tissue by hydrolyzing connective tissue hyaluronic acid. Proteases produce damage by dissolving proteins and peptides in the tissues. Hemolysis results from the destruction of lecithin in the cell membranes. L-Amino acid oxidase produces local tissue destruction by catalyzing the oxidation of amino acids. Other enzymes include transaminases, ribonuclease, L-arginine ester hydrolases, deoxyribonuclease, phosphomonoesterase, phosphodiesterase, DNAse, ATPase, alkaline phosphatase, acid phosphatase, and endonuclease. The venom from the Mojave rattlesnake (*C. scutulatus scutulatus*) contains few digestive enzymes compared with other rattlesnake species, and therefore envenomations by Mojave rattlesnakes produce less tissue destruction than envenomation by other true rattlesnakes. The venom from *Crotalus* species does not contain acetylcholinesterase enzymes.

Dose Response

There is substantial interspecies, intraspecies, and seasonal variation in rattlesnake venom composition and toxicity.[93] The intraperitoneal LD_{50} in mice exposed to the venom of *Crotalus* species ranges from 0.23 mg/kg (*C. scutulatus scutulatus*) to 6.69 mg/kg (*C. ruber ruber*).[94] Seasonal variation in venom potency results from changes in feeding patterns and physiological responses (e.g., hibernation).[95] Factors producing intraspecies variation include age and time from last feeding. Rattlesnakes are venomous from birth.

TOXICOKINETICS

Crotalus venom is a complex mixture of substances with varying molecular weights and activities. There are very limited data on the toxicokinetics of these components. Detection of whole venom depends on the time since envenomation, the administration of antivenom, and the sensitivity of the analytical method. Case reports and case series of rattlesnake envenomations suggest that coagulopathic components of rattlesnake venom may remain active up to 1–2 weeks.[96,97] In a study of the urine from a patient envenomated by the bite from the Western Diamondback rattlesnake (*C. atrox*), venom was detectable by ELISA from days 3–6 postenvenomation.[98] The duration of detectable venom was not determined. As measured by enzyme-linked immunosorbent assay, whole venom from the South American rattlesnake (*Crotalus durissus durissus*) was not detectable in plasma when the time elapsed between the bite and hospital admission was over 8 hours.[99] Crotoxin was not detectable in most of the patients, who were admitted to the hospital at times ranging between 4 hours and 8 hours after the snakebite. Rattlesnake venom does not penetrate the blood–brain barrier.

HISTOPATHOLOGY AND PATHOPHYSIOLOGY

Mechanism of Action

Venom affects almost every organ system because of the increased permeability of capillary membranes that results in the extravasation of albumin, electrolytes, and red blood cells. However, the brain is relatively resistant to the direct toxic effects of rattlesnake venom. The primary effects of venom include local tissue damage, edema, hypoalbuminemia, coagulopathy, thrombocytopenia, hypovolemia, and, in a few pit viper species, neurotoxicity. Snake venom metalloproteinases are a series of zinc-dependent proteins that disrupt the interaction between the basement membranes and endothelial cells leading to increased capillary membrane permeability, extravasation of fluid, electrolytes, proteins, and erythrocytes.[100] Rarely, pulmonary edema and renal failure result from multiorgan failure and the transudation of fluid into alveoli. The venom of the Mojave rattlesnake contains a fraction of fluid that produces flaccid paralysis by blocking presynaptic neuromuscular transmission.

Target Organs

Neurotransmission

Most rattlesnake envenomations cause relatively minor changes in neuromuscular transmission as compared with coral snake envenomations. The major exception is envenomation by the Mojave rattlesnake (*C. scutulatus scutulatus*), which can produce respiratory paralysis. Experimental studies suggest that the neurotoxic component results from calcium channel blockade of presynaptic neurons and the noncompetitive inhibition of the release of acetylcholine.[101] Neurotoxicity requires the synergistic action of phospholipase and an acidic, inert component.[102] The venom from some *Crotalus* species (e.g., *C. lepidus lepidus*) contains this neurotoxin, but neurological complications are rare because the amount of this component is substantially less than the Mojave rattlesnake.

Skin and Soft Tissue

Crotalus envenomation produces serious local damage due to the action of proteases and hyaluronidases. Venom metalloproteinase activates tumor necrosis factor alpha, which is a mediator of the inflammatory response that causes neutrophil degranulation, leukocyte migration, and the release of mediators of inflammation (e.g., interleukins, human metalloproteinase).[103] The venom from some *Crotalus* species (e.g., *C. viridis,*

C. atrox) contains phospholipases A_2-type enzymes, called myotoxins (e.g., myotoxin-a).[104] Myotoxin-a produces increased intracellular calcium concentrations that ultimately cause muscle necrosis.[105]

Blood

The most common coagulation abnormality following *Crotalus* envenomation is hypofibrinogenemia and thrombocytopenia due to venom-induced consumption of fibrinogen and platelets. The injection of thrombin-like enzymes in rattlesnake venom converts fibrinogen to an unstable fibrin clot that is subsequently lysed by the fibrinolytic system. A consumptive coagulopathy may develop following the excessive activation of the fibrinolytic system. Amino acid esterases and thrombin-like enzymes (e.g., crotalase) promote clotting of fibrinogen by the hydrolytic splitting of fibrinopeptide A or B from fibrinogen.[106,107] The incomplete splitting of fibrin inhibits the release of clot-stabilizing factor, and crotaline venom may enhance anticoagulation by reversibly binding to prothrombin. Factor XIII is not activated and the unstable clots trap platelets. Crotalase acts at a different fibrinogen recognition site than thrombin, and crotalase does not directly activate platelets. Plasmin and proteolytic enzymes in the venom readily lyse the clot, and a consumptive coagulopathy develops characterized by excessive consumption of fibrin, thrombocytopenia, and increased fibrin split products. The destabilizing action of phospholipase on erythrocyte membranes contributes to the hemolysis associated with venom-induced coagulopathies. However, intravascular coagulation (disseminated intravascular coagulation [DIC]) does not usually occur as reflected in normal concentrations of antithrombin III, factor XIII, and D-dimer. A venom fraction of *C. horridus horridus* (timber rattlesnake) causes platelet aggregation, and this effect contributes to the severe thrombocytopenia often seen after envenomations by this species.[108] Decreased platelet survival results from local consumption of the platelets at the site of injury.[109]

Cardiovascular System

Although some rattlesnake venoms (e.g., Mojave, Southern Pacific rattlesnakes) contain a direct cardiotoxin, hypotension usually results from fluid extravasation caused by intimal injury as well as changes in vascular resistance. Large volumes of fluid may diffuse across injured capillaries. Consequently, hypotension usually results from hypovolemia. In experimental animal models, rattlesnake venom produces myocardial necrosis and ischemia as demonstrated by elevation of cardiac enzymes and histological changes in the heart.[110] However, clinical evidence of myocardial injury in humans following envenomation by rattlesnakes is rare. The venom damages the endothelial cells of the vascular walls as manifest by endothelial blebs, dilation of the perinuclear space, and deterioration of the plasma membranes.

Pulmonary System

Pulmonary edema may result from extravasation of proteins through damaged alveolar membranes. Pulmonary hemorrhage and intraalveolar edema may cause hypoxia.

Kidney

Renal dysfunction may result from hypoperfusion, intravascular hemolysis with hemoglobinuria, myoglobinuria, disseminated intravascular coagulation, and/or a direct nephrotoxicity.[111]

CLINICAL RESPONSE

The clinical presentation of rattlesnake envenomation varies substantially depending on a variety of factors including the potency of the venom, the amount of venom injected, and patient-related factors. The latter factors involve age, health, and individual susceptibility of the patient as well as the use of protective clothing. Consequently, the clinical features of a rattlesnake bite range from an asymptomatic, "dry" bite to skin necrosis, weakness, lethargy, paresthesias, fasciculations, and life-threatening coagulopathy. About 20% of *Crotalus* bites do not result in envenomation, and therefore no toxic effects develop. The lack of burning pain and edema over 2 cm from a fang mark more than 1 hour after the snakebite suggests that either no or minimal envenomation occurred, whereas the absence of edema and erythema at the bite site more than 6–8 hours after the snakebite indicates a lack of evenomation for most rattlesnake bites.

The most common response to a snakebite is fear, which may cause cold, clammy skin, tachycardia, nausea, vomiting, or syncope. In general, children have more severe responses because of the increased amount of venom per kilogram of body weight compared with adults, but serious morbidity other than restricted range of motion and muscle weakness following rattlesnake bites is uncommon.[112] Topical exposure or ingestion of rattlesnake venom does not usually cause systemic envenomation. Contact of rattlesnake venom to the eye often produces transient irritation without systemic effects, and the local irritation usually responds to standard first-aid measures.[113] *Salmonella* species (e.g.,

Salmonella arizonae) are part of the normal intestinal flora of reptiles including rattlesnakes.[114] In parts of the southwestern United States, *Salmonella* gastroenteritis is a complication of the use of rattlesnake meat as a folk remedy by individuals with normal and compromised immune systems.[115,116] The ingestion of rattlesnake capsules produced *Salmonella* sepsis in two immunocompromised, elderly patients.[117]

Local Effects

Crotalus envenomation classically presents with visible fang marks (one or more), immediate burning without severe pain, and localized edema beginning about 5 minutes after the bite. The number of fang marks depends on a number of factors including the number of strikes, the accuracy of the strike, and the number of reserve fangs. *Crotalus* bites that produce superficial abrasions and lacerations do not usually cause envenomation because the orifice of the fang lies slightly proximal to the tip of the fang. Local findings after envenomation are usually obvious within 30–60 minutes, and the edema continues to worsen during the first several hours after envenomation. Swelling progresses over 6–8 hours depending on the severity of the envenomation, and pain correlates to the amount of edema. Mild cyanosis commonly occurs at the site of injury without the presence of ischemia. Ecchymosis typically appears within several hours because of the extravasation of blood into subcutaneous tissue from damaged vessels. Clear or hemorrhagic vesicles and petechiae may appear distant to the fang marks on the same extremity. However, generalized petechiae and purpura appear rarely, and these findings are usually associated with thrombocytopenia, especially in previously damaged skin (e.g., sunburn).[118] Necrosis is usually localized following the use of adequate doses of antivenom, but return to full function of the envenomated extremity typically requires approximately 1 month despite the use of antivenom. The amount of local tissue damage and edema is species-dependent ranging from minimal changes following some Mojave rattlesnake envenomations to the severe swelling produced by the Eastern Diamondback. Pressures in muscle compartment are usually normal in the absence of muscle necrosis from a subfascial envenomation. Despite significant discoloration and swelling of extremities, arterial perfusion generally increases after envenomation. Decreased skin temperature in the extremity suggests the possibility of ischemia. Tender regional lymph nodes may develop in the absence of infection and signs of lymphangitis usually reflect the proximal diffusion of venom rather than a local infection.

Systemic Effects

Minty, rubbery, or metallic taste is a frequent symptom following envenomation by prairie (*C. viridis viridis*), Northern Pacific (*C. viridis oreganus*), and Southern Pacific (*C. viridis helleri*) rattlesnakes. Weakness, diaphoresis, lightheadedness, and nausea occur commonly. However, the development of perioral paresthesias, muscle paralysis, or facial fasciculations suggests a moderate to severe envenomation. Muscle fasciculations may involve the head and neck in addition to the extremities, and these fasciculations can persist for a few days despite the use of antivenom.[119] Paralysis and respiratory insufficiency may develop following envenomation by the Eastern Diamondback, Mojave, or Southern Pacific rattlesnakes,[120] but these findings are relatively rare except following envenomation by the Mojave rattlesnake. Other neurological effects of envenomation by this rattlesnake include dysphagia, ptosis, diplopia, and respiratory failure. Myokymia is a fine, spontaneous, involuntary, undulating movement of muscle fibers that occurs in some demyelinating diseases (Guillain-Barré syndrome, multiple sclerosis). Myokymia can also result from envenomation by a few rattlesnake species (e.g., timber rattlesnake—*C. horridus horridus* and midget faded rattlesnake—*C. viridis concolor*).[121,122] These low-amplitude muscle movements occur in muscle groups close to the bite and the face, but muscle fibers in nonbitten extremities also can develop myokymia. Convulsions do not usually develop following rattlesnake envenomations. Severe systemic effects (hypotension, respiratory distress, angioedema of the oropharynx) can occur with minimal local effects following the intravenous injection of venom.[123]

During severe envenomation, pulmonary edema secondary to adult respiratory distress syndrome (ARDS) may develop from the extravasation of fluid across damaged alveolar membrane. Most fatalities occur within 48 hours of envenomation due to ARDS, persistent hypotension unresponsive to fluid administration, and disseminated intravascular coagulation. Renal dysfunction usually results from hypoperfusion, myoglobinuria, or multisystemic failure. Although rare except following severe envenomations from intravascular injection of rattlesnake venom, a consumptive coagulopathy may develop as characterized by petechiae, diffuse bleeding from puncture sites, hematomas, melena, epistaxis, and hematuria. Intracerebral hemorrhage, intracardiac bleeding, retinal hemorrhage, and bowel infarction are infrequent complications of rattlesnake envenomation.

Classical Grading of Envenomation

Classification of the severity of rattlesnake evenomation involves separating the clinical features of envenomation into four categories (none, minimal, moderate, severe). Typical snakebites produce one or more fang marks with puncture wounds circumscribed by ragged edges. Envenomation produces local subcutaneous edema and variable amounts of pain that correlate to the extent of the edema. Slight bleeding often occurs at the puncture site that later is associated with subcutaneous hemorrhage. Common systemic signs of envenomation include muscle fasciculations and paraesthesia. The presence of cranial nerve signs (drooling, diplopia, dysphagia) suggests envenomation by the Mojave rattlesnake. Without antivenom administration, minimal envenomations may progress to severe envenomations over several hours. Consequently, the severity of envenomation is frequently determined by the time to definitive intervention (i.e., antivenom).

- *None*: A dry bite produces a puncture wound without surrounding edema, significant local pain, systemic complaints, or laboratory abnormalities.
- *Minimal*: A minimal envenomation results in the presence of fang marks with local swelling confined to the bite area. Systemic symptoms and laboratory abnormalities are absent or clinically insignificant.
- *Moderate*: A moderate envenomation causes local swelling, which progresses beyond the immediate bite area, but the edema does not involve the entire extremity. Significant systemic symptoms (paresthesias, fasciculations) develop along with moderate laboratory changes including mild thrombocytopenia (<80,000/m^3), decreased fibrinogen (<80 mg/dL), and mildly increased fibrin split products. There is no bleeding from remote sites. Nausea and vomiting may be present.
- *Severe*: Severe envenomations involve fang marks with rapidly progressive edema, subcutaneous hemorrhage along with severe generalized symptoms and signs, particularly hypotension. Subcutaneous changes extend over the entire extremity. Marked laboratory abnormalities include decreases in all coagulation factors, proteinuria, hematuria, and increased serum creatine kinase. Envenomation by large, highly toxic rattlesnake species (e.g., Eastern or Western Diamondback) usually produces this type of clinical presentation.

Anaphylaxis

Anaphylactic responses to *Crotalus* envenomation are extremely rare. A 22-year-old man with a history of rattlesnake envenomation one year earlier developed urticaria, hypotension, and bronchospasm within 30 minutes of an unidentified rattlesnake bite.[124] He responded to subcutaneous (SQ) epinephrine, intravenous (IV) diphenhydramine, and fluid administration. The administration of antivenom was terminated after he developed wheezing and shortness of breath following the administration of 60 mL of a 100 mL solution containing 5 vials of Wyeth (Wyeth Pharmaceuticals, Collegeville, PA) antivenom. Hypersensitivity responses are common in snake handlers who sustain multiple bites.[125] Severe anaphylaxis developed in a professional snake handler who had a prior history of snake venom sensitivity.[126]

DIAGNOSTIC TESTING

Analytical Methods

Double diffusion technique is a simple, rapid method of identifying snake venom, but the test is qualitative rather than quantitative with detection limits of about 1 μg for *C. atrox*, whereas the ELISA is more sensitive (1-nanogram range).[127] However, an ELISA test for the detection of rattlesnake envenomation is not commercially available. Although these tests are available in some experimental laboratories, the low concentration of venom in urine and plasma samples frequently limits detection in clinical settings.[128] Potentially, such assays can differentiate snakebites from other injuries and envenomations, determine the genus and occasionally the species involved, and detect the presence of specific antibodies in human populations.

Abnormalities

Bleeding abnormalities following rattlesnake envenomation include defibrination, venom-induced thrombocytopenia, and disseminated intravascular coagulation (see Table 176.8). The most common coagulopathy following envenomation by North American rattlesnakes is defibrination without DIC, particularly following envenomation by Eastern and Western Diamondback, timber, blacktail, or Mojave rattlesnakes.[129] Early hematological changes include coagulopathy, hypofibrinogenemia, and venom-induced thrombocytopenia.[130] Case series indicated that hypofibrinogenemia and thrombocytopenia may develop without other evidence of a significant coagulopathy, particularly after timber rattlesnake envenomations.[131] Thrombocytopenia reaches a nadir 2–4 days after envenomation and may persist for 10–14 days. A rebound thrombocytosis typically occurs approximately 36 hours after envenomation.

A consumptive coagulopathy may occur during the first few days manifest by prolonged clotting times

TABLE 176.8. Abnormalities Separating Defibrination (DF), Venom-Induced Thrombocytopenia (VIT), and Disseminated Intravascular Coagulation (DIC) during Rattlesnake Envenomation[a]

Test	Defibrination (DF)	Venom-Induced Thrombocytopenia (VIT)	DF and VIT	Disseminated Intravascular Coagulation (DIC)
Fibrinogen	Low	Normal	Low	Low
Fibrin Split Products	Elevated	Normal	Elevated	Elevated
Prothrombin Time	Prolonged	Normal	Prolonged	Prolonged
Activated Partial Thromboplastin Time	Prolonged	Normal	Prolonged	Prolonged
Platelet Count	Normal	Low	Low	Low
Hemolysis[b]	Absent	Absent	Absent	Present
Diffuse Bleeding	Absent	Severe only	Severe only	Present
Organ Infarction	Absent	Absent	Absent	Present

[a]Source: Adapted from Ref 129.
[b]Microangiopathic

(partial thromboplastin time [PTT] > prothrombin time [PT]), hypofibrinogenemia, elevated fibrin split products, and thrombocytopenia, but hemorrhagic complications and DIC are uncommon except in severe cases (e.g., multiorgan failure, intravascular envenomation).[132] The presence of DIC is indicated by intravascular fibrin deposition with subsequent fibrinolysis, hemolysis, thrombocytopenia, reduced coagulation factors, and bleeding from diffuse sites. D-dimer is a direct measure of cross-linked fibrin derivatives from intravascular fibrin deposition and plasmin degradation. The thrombin-like enzymes in rattlesnake venom do not form cross-linked fibrin, and therefore D-dimer concentrations remain normal after envenomation unless DIC occurs.

Anemia occurs primarily during severe envenomations and results from internal bleeding or intravascular hemolysis secondary to DIC.[133] Laboratory manifestations of rhabdomyolysis include elevated serum creatine kinase (CK) and myoglobinuria. Case reports suggest that increased serum concentrations of the CK-MB fraction following an envenomation do not necessarily reflect myocardial damage, particularly following bites by the canebrake rattlesnake.[134] Urine abnormalities include hematuria, glycosuria, proteinuria, or myoglobulinuria. These abnormalities are transient and usually resolve within 2 weeks unless renal damage develops.[135]

TREATMENT

First-Aid

There are limited data regarding the efficacy of various first-aid measures following crotalid envenomation. Tourniquets, incision, electric shock therapy,[136,137] and cryotherapy are ineffective and potentially dangerous first-aid measures for rattlesnake bites.[138] Most authorities advise against the use of ice packs, cold packs, ice water soaks, ethanol, and stimulants.

General recommendations for first aid in the field are as follows:

1. Retreat from the snake at least 10–15 ft (~3–5 m) and remain calm. Arrange to move the victim to a medical facility as soon as possible to begin the administration of antivenom.
2. Remove all constricting items, such as rings, watches, and restrictive clothing immediately.
3. Put the victim at rest, give reassurance, and keep him or her warm. Excessive activity may increase the spread of venom. However, keeping the victim at rest should not delay the transport of the patient to the nearest medical facility.
4. Lightly immobilize the injured part below the heart in a position of rehabilitation (e.g., full extension of phalangeal joints, 90° flexion of metacarpal-phalangeal joints, wrist in neutral position). Typically, the extremity is elevated in the hospital when antivenom is available. Identification of the snake is not usually necessary for the proper treatment of *Crotalus* envenomation, unless more than one dangerous species of rattlesnake exists in the area (e.g., Mojave rattlesnake). The snake should not be handled directly because the snake's reflexes remain intact 30–60 minutes after death and bites may occur even after decapitation of the snake.[139]

Although the use of pressure dressings has been a common first-aid recommendation for rattlesnake snakebites, there are no clinical data validating the efficacy of this modality and there are theoretical disadvantages to restricting the venom to the area of the bite.[140]

Application of a lightly constricting band above the swelling that admits a finger beneath it is commonly recommended *only* when definitive treatment will be delayed 1–2 hours, such as occurs in remote locations. Available experimental data suggest that a constrictive band limits systemic spread of the venom, but there is no clinical evidence that the use of a constrictive band improves outcome.[141,142] Clinical experience suggests that previously applied tourniquets should *not* be removed in the emergency department until intravenous lines are established, a blood pressure cuff is inflated proximal to the tourniquet, and antivenom therapy is available. The blood pressure cuff should be deflated gradually as the patient receives antivenom and fluids.

The use of incision and drainage is discouraged because of potential contamination of the wound and the lack of clinical evidence regarding a clear benefit for this first-aid measure. Suction is potentially useful *only* in the unusual circumstance where 1) medical help is over 1–2 hours away, and 2) the procedure begins within 5 minutes of the bite. Use of a commercial suction device (e.g., Sawyer Extractor Vacuum Pump; Sawyer Products, Inc., Safety Harbor, FL) is preferable to suction by mouth based on theoretical concerns about contaminating the wound with human oral flora. However, there are inadequate experimental and clinical data to support the efficacy of this device.[143,144] If an extractor pump is present and contains fluid, the pump should be left in place until arrival in a health care facility. Otherwise, the extractor pump may be removed during transport.

Stabilization

Life-threatening local edema may result from airway obstruction following intraoral or facial snakebites.[145,146] Anaphylactic responses to snakebites are rare, except in those patients, such as snake handlers, who sustain snakebites on multiple occasions. A case report documented the development of life-threatening glossal and epiglottic edema in a 36-year old man envenomated on the thumb by the head of a decapitated timber rattlesnake.[147] Hypotension should be treated by starting two intravenous lines for the administration of crystalloid solution. Large volumes may be required. Vasopressors should be reserved as a short-term measure, and necessity of using vasopressors indicates the need to monitor pulmonary wedge or central venous pressures. Pulmonary edema usually results from adult respiratory distress syndrome. Respiratory depression is uncommon except as a complication of shock and Mojave rattlesnake envenomations. Vital signs including oxygen saturation initially must be monitored every 15 minutes in the acute care setting with direct, continuous observation.

Elimination Enhancement

There are no clinical data to support the use of methods to enhance the elimination of venom as a treatment for rattlesnake envenomation.

Antivenom

Two antivenoms approved for use following *Crotalus* envenomations include Crotalidae polyvalent antivenom produced in horses and Crotalidae polyvalent Fab fragments (CroFab™; Protherics, Inc., Brentwood, TN) produced in sheep. There are limited supplies of the former antivenom, and the use of the latter antivenom may completely replace the use of the former depending on the resolution of issues on the incidence of hypersensitivity responses and the efficacy of treatment of delayed coagulopathies. Although both antivenoms reduce systemic symptoms following North American Crotalinae envenomation, the efficacy of these antivenoms to reduce postenvenomation edema remains unclear, in part because of the lack of large, controlled clinical trials.[148] The antivenom, Antivipmyn, is a polyclonal antivenom $F(ab')_2$ fragment produced in horses and available from the Instituto Bioclon, Mexico City, Mexico.[149,150]

Crotalidae Polyvalent Immune (Ovine) Fab (CroFab™)

Formulation. Crotalidae Polyvalent Immune Fab (Ovine) is a lyophilized powder produced in sheep as a monospecific, polyvalent antivenom for the treatment of North American Crotalinae envenomations. Flocks of sheep are immunized with one of four crotaline venoms (Eastern Diamondback, Western Diamondback, Mojave, Cottonmouth). Ovine immune serum from each flock is digested by papain to separate the Fab and Fc fragments. After an affinity-purification process to remove the Fc fragments, the four individual monospecific Fab preparations are mixed to form the final product. The advantages of CroFab™ compared with the Wyeth antivenom include faster and easier administration and fewer allergic responses.[151,152] However, the shorter elimination half-life of CroFab™ requires more frequent administration compared with the equine antivenom.

Mechanism of Action. CroFab™ is a polyvalent antivenom that is a specific antidote to North American pit viper venom. This antivenom neutralizes the whole pit viper venom rather than specific components. Consequently, the antivenom may be more effective against some components. The administration of antivenom does not reverse local tissue injury caused by snake venom prior to the use of antivenom. Antivenom limits the progression of local swelling and reverses coagulopathy. The effect on thrombocytopenia is variable.

INDICATIONS. The indications for the administration of CroFab™ to victims of snakebites are similar to the indications for antivenin (Crotalidae) polyvalent. Antivenom is indicated in patients who demonstrate progressive signs of envenomation during the first hour after envenomation. Progression of swelling and pain in a digit suggests the need for antivenom, particularly in children. Although premarketing and postmarketing studies of the use of CroFab™ have excluded children, retrospective studies indicate that the use of this antivenom in pediatric patients with crotalid envenomations is safe.[153] The antivenom is most effective when administered within 4 hours of the bite. Although CroFab™ production uses the venoms of four snake species (Western and Eastern Diamondback rattlesnakes, Mojave rattlesnake, water moccasin), clinical experience suggests that CroFab™ is effective against envenomation by other rattlesnakes (e.g., Southern Pacific rattlesnake or *Crotalus viridis helleri*).[154]

CAUTIONS. Sensitization potentially may occur in patients administered a foreign protein, such as CroFab™. Therefore, patients receiving repeat courses of CroFab™ should be evaluated carefully for the development of allergic responses. Occasionally, fasciculations, muscle weakness, and respiratory insufficiency persist despite the administration of adequate doses of antivenom and cessation of the spread of extremity edema.

ADMINISTRATION. Each vial should be reconstituted with 10 mL sterile water and mixed by gentle, continuous swirling. The initial dose of antivenom is diluted in 250 mL normal saline, and the total dose is infused intravenously over 1 hour with the initial infusion beginning at 25–50 mL/h to observe for signs of any allergic response. The antivenom is relatively stable to heat and to motion.[155]

Dosage. The initial dose of CroFab™ is 4–6 vials and this dose should be repeated if signs of envenomation progress as defined by continued progression of edema or lack of improvement in coagulation abnormalities and systemic signs. An additional 2-vial dose is recommended 6 hours, 12 hours, and 18 hours after the initial control of the envenomation to prevent recurrence of local and systemic signs (e.g., coagulopathy) of envenomation. The typical *Crotalus* envenomation involves the administration of 14–18 vials of CroFab™. Although ideally CroFab™ should be administered as soon as possible after envenomation, case reports suggest that the administration of antivenom may be effective up to 2 days after envenomation, particularly for the venom-induced coagulopathy.[156] Recurrence of signs of envenomation following CroFab™ treatment results from the pharmacokinetic mismatch between the receptors for the venom components and the antivenom.[157] This situation occurs because the clearance is faster for unbound antivenom than for some venom components. In general, the distribution times and the elimination half-lives increase as the size of the molecule increases. The prolonged absorption of nonneutralized venom from deposition sites near the bite enhances the phenomena of recurrence. Consequently, recurrence occurs more frequently following the administration of CroFab™ than Wyeth antivenom because of the relatively smaller size of the Fab molecules compared with the IgG molecules present in the Wyeth antivenom. The clinical significance of recurrent coagulopathy is unclear. Table 176.9 lists suggested recommendations for the treatment of recurrent coagulopathy.[158] Standard treatment regimens for CroFab™ usually limit the recurrence of local symptoms and signs. However, fasciculations may persist despite the use of adequate doses of CroFab™.

Sensitivity Testing. The affinity-purified Crotaline antivenom ovine Fab does not usually contain the immunogenic Fc and IgG molecules present in the Wyeth antivenom (Crotalidae) polyvalent. Because of the lack of immunogenicity and allergic responses following the use of the ovine Fab antivenom, sensitivity testing is not recommended prior to the administration of this antivenom.[159]

Adverse Reactions. Both anaphylactoid reactions, immediate hypersensitivity, and serum sickness may occur after the administration of CroFab™, but the

TABLE 176.9. Recommendations for Repeat Treatment of Recurrent Coagulopathy with CroFab™[a,b]

Test	Lab Value
Fibrinogen	<50 µg/mL
Platelet count	<25,000/mm^3
INR[c]	>3.0
aPTT[d]	>50 s

[a]Source: Adapted from Ref 158.
[b]Additional indications for repeat treatment include high-risk for trauma, recurrence of coagulopathy in patient with history of severe coagulopathy, comorbid conditions that increase the risk of internal hemorrhage, and multicomponent coagulopathy.
[c]INR = international normalized ratio.
[d]aPTT = activated partial thromboplastin time.

immune responses are usually mild without sequelae.[160,161] Serious anaphylactic reactions may occur in patients allergic to sheep. In a case series of 31 patients given CroFab™, 6 patients developed acute responses ranging from transient urticaria to mild cough and wheezing.[162] Of the 4 patients requiring treatment, all responded promptly. Over a one-month follow-up period, 6 of 26 patients developed serum sickness ranging from mild (two cases) to severe (two cases). All patients were asymptomatic 28 days after treatment. Persistent, recurrent, or late coagulopathy (thrombocytopenia, hypofibrinogenemia) may occur 2–14 days after rattlesnake envenomation despite the administration of CroFab™.[163] Although these abnormalities are not usually associated with spontaneous bleeding, monitoring the coagulation should continue during the first week and elective surgical procedures should be avoided at least for several weeks. As a result of platelet-aggregating constituents of venom (e.g., crotalocytin) and platelet sequestration in areas of vascular damage, thrombocytopenia may persist 5–7 days after envenomation despite the use of antivenom. Following evenomation by *Crotalus ruber ruber* (Red Diamond rattlesnake), a 42-year-old man developed severe thrombocytopenia (platelet count 5,000–10,000/mm^3) within 4 days that did not respond to 46 vials of CroFab™ and 10 vials of antivenin (Crotalinae) polyvalent (Wyeth) despite the normalization of other coagulation factors (fibrinogen, prothrombin time).[164] Despite receiving large amounts of antivenom, the patient did not develop serum sickness.

Follow-up Care. Patients with thrombocytopenia or hypofibrinogenemia should be reevaluated for evidence of delayed coagulopathy within 4 days of initial therapy or whenever signs of coagulopathy (e.g., hematuria, petechiae, bleeding) develop. If the patient developed an abnormal coagulation profile during the first 36 h, the patient should be reexamined for recurrent coagulopathy 48 hours after antivenom administration. Redosing with CroFab™ is appropriate when the platelet count is <25,000/mm^3 and the fibrinogen level is below 50 mg/dL.

Antivenin (Crotalidae) Polyvalent (Equine by Wyeth)

Formulation. Antivenin (Crotalidae) polyvalent (antivenom) is a lyophilized powder produced by the immunization of horses with the venoms of *Crotalus atrox* (Western Diamondback), *Crotalus adamanteus* (Eastern Diamondback), *Crotalus durissus terrificus* (South American rattlesnake), and *Bothrops atrox* (fer-de-lance). Because of the common antigens shared by pit vipers, the antivenom is effective against poisonous American pit vipers, including rattlesnakes, pygmy rattlesnakes, cottonmouth, and copperhead as well as the fer-de-lance, cantil, and Central and South American bushmaster. This antivenom also provides protection against the habu (*Protobothrops flavoviridis* Hallowell) and *Agkistrodon* species in Japan and Korea. Commercial venom is freeze-dried and must be reconstituted in sterile water. One vial of antivenom yields 10 mL of concentrated serum.

Mechanism of Action. Antivenom neutralizes pit viper venom, but the administration of antivenom does not reverse local tissue injury caused by snake venom prior to the administration of antivenom. The use of antivenom limits the progression of local swelling, reverses coagulopathy and thrombocytopenia, and improves muscle strength in patients with generalized weakness. The persistence of DIC after the administration of antivenom suggests the presence of organ infarction.[129]

Indications. Antivenom is indicated in patients who demonstrate progressive signs of envenomation during the first hour after envenomation unless the risks outweigh the benefits (e.g., horse serum-sensitive patients). Progression of swelling and pain in a digit suggests the need for antivenom, particularly in children. The antivenom is most effective when administered within 4 hours of the bite. Early administration of antivenom is necessary to prevent local and systemic injury before vascular destruction restricts the delivery of antivenom to the tissues. Antivenom is less effective when administered over 12 hours after envenomation, but some reversal of coagulation defects occurs 24 hours after envenomation. Thrombocytopenia frequently improves after antivenom therapy, but the response is usually partial and platelet counts often remain below normal 7–10 days.[165]

Cautions. The only absolute contraindication to antivenin (Crotalidae) polyvalent use is the lack of signs of envenomation. Another contraindication is horse serum sensitivity and the option to use CroFab™. The use of antivenin (Crotalidae) polyvalent in sensitized patients requires careful consideration of the risks and potential benefits. Consultation, when time permits, is advisable to decide the need for antivenom and the best method of administration (e.g., desensitization or simultaneous administration of diphenhydramine and epinephrine with diluted antivenom). Pregnancy is not a contraindication to the administration of antivenom and the use of antivenom may prevent envenomation-induced abortions that are common in untreated, pregnant women.[166,167]

ADMINISTRATION. Antivenin (Crotalidae) polyvalent should be administered only by the intravenous route because of the superior delivery of antivenom to the tissues using this route, and antivenom should not be injected into the area of the bite. Ten mL of diluent or normal saline is added to each vial, and the solution is mixed by rolling the vial between the hands rather than shaking. For adults, the appropriate number of antivenom vials is added to 250–500 mL of normal saline (about 1:5 dilution) and then infused into the main intravenous line, ideally via a four-way stopcock located near the catheter site. For children, the volume is decreased proportionally so that the reconstituted vials may be diluted 1:1 in normal saline and added to a microdrop set. Because of the false-negative results (i.e., up to about 20%) of hypersensitivity testing, the initial infusion begins at a rate of 0.25–1 mL/min during the first 10–15 minutes. If allergic symptoms (pruritus, urticaria, flushing) develop, the infusion is stopped and the main line started to give fluid and diphenhydramine. If no reaction occurs within the first 10–15 minutes, the infusion rate should be increased to administer the entire dose of antivenom over the next 1 to 2 hours. The initial dose of antivenom should be repeated every 2 hours if local or systemic signs of envenomation continue to progress. Antivenin (Crotalidae) polyvalent is less effective 8–12 hours after envenomation and of questionable value after 24 hours, except in reversing coagulation defects (e.g., thrombocytopenia).

If a reaction occurs, stop the infusion and give 50 mg (1.25 mg/kg) of both diphenhydramine (50–100 mg, 1.25 mg/kg) and ranitidine (50 mg, 1–2 mg/kg) intravenously and, if needed, epinephrine subcutaneously (0.01 mL/kg of a 1:1,000 dilution to a maximum of 0.3–0.5 mL in adults). Epinephrine must be given in more severe anaphylactic responses because of the inability of antihistamines to inactivate leukotrienes during anaphylaxis.[168] For anaphylactic shock, an initial intravenous epinephrine dose of 0.1 mg (0.1 mL of a 1:1,000 solution diluted in 10 mL of normal saline) infused over 5–10 minutes is recommended. For poor adult responders, a constant infusion of a 4 µg/mL solution (1 mg of a 1:1,000 solution added to 250 mL of 5% dextrose in water [D_5W]) may be infused at a rate of 1–4 µg/min.[169] Subcutaneous epinephrine may not be adequately absorbed when the patient is severely hypotensive. Myocardial ischemia is a potential complication in patients with coronary artery disease. Vasopressors, oxygen, and intubation equipment should be available for immediate use. There are no well-documented anaphylaxis-related fatalities in the medical literature from the administration of Wyeth antivenom.

Dosage. The clinical grading of the severity of envenomation determines the initial amount of antivenom. Children typically have more severe reactions to a given envenomation because of the greater amount of venom per body weight; therefore, children usually require more antivenom (i.e., up to 50%) than adults on a weight per weight basis. However, the use of clinical grading accounts for the difference in size between children and adults. The total dose of antivenom depends on the initial clinical presentation, the progression of edema, and systemic symptoms after the administration of antivenom. The typical average *initial* dose of equine antivenin (Crotalidae) polyvalent is 5 vials for minimal, 10–12 vials for moderate, and 15–20 vials for severe rattlesnake bites. If local or systemic signs progress after 2 hours, infuse an additional 5-vial incremental dose diluted in 500 mL saline. The circumference of the bitten extremity should be measured at a distance 10 and 20 cm above the bite every 30–60 minutes to monitor of effectiveness of the antivenom. The neurovascular status of the affected extremity should be evaluated regularly until swelling ceases. If signs of a compartment syndrome develop, the compartment pressures should be measured directly.

Mojave rattlesnake envenomation can produce respiratory paralysis 8–12 hours after envenomation despite the initial absence of local pain or systemic symptoms. If an envenomation by the Mojave rattlesnake is strongly suspected, the administration of 10 vials of antivenom should be considered despite the initial lack of local or systemic symptoms. All patients with suspected Mojave rattlesnake bites should be observed in an intensive care unit and additional doses of antivenom administered if neurological symptoms develop.[170] In patients receiving antivenom for coagulopathy after a crotaline envenomation, a retrospective analysis of 354 consecutive cases of North American crotaline snake envenomation suggests that coagulopathy may recur within several days after the administration of antivenom.[171] Repeat administration of antivenom is usually unnecessary unless clinical evidence of internal hemorrhage develops.

Sensitivity Testing. The decision to use hypersensitivity testing prior to the administration of Antivenin (Crotalidae) polyvalent depends on physician judgment. The efficacy of skin testing prior to antivenom therapy is questionable because of the insensitivity of the test to predict which patients will develop anaphylactic reactions. In a study of 26 patients receiving skin testing prior to antivenom administration, 3 of 6 patients with positive skin tests and 2 of 20 patients with negative skin tests developed anaphylaxis.[172] Therefore, skin testing prior to the administration of Crotalidae antivenom (equine) is not necessary, particularly when life-threatening symptoms are present. Sensitivity testing should not be administered to patients unlikely to receive antivenom because hypersensitivity testing may complicate future antivenom administration by producing sensitivity to horse serum. Before testing, a history should be obtained about horse serum sensitivity because patients with a history of horse serum reactions are at higher risk of anaphylactic responses during antivenom administration.

The protocol for hypersensitivity testing for patients without horse serum sensitivity involves the injection of 0.02 mL of horse serum diluted 1:10 (Wyeth antivenom kits contain this 1:10 dilution of horse serum) intradermally at a site contralateral from the bite and the injection of 0.02 mL of normal saline in the opposite extremity as a control. A wheal and erythema indicate a positive reaction. Occasionally, patients may develop generalized urticaria, pruritus, or bronchospasm. Along with intravenous access, epinephrine and diphenhydramine should be immediately available prior to the skin testing. While awaiting skin test results, the initial doses of antivenom may be reconstituted with saline from the infusion bag. No reaction by 20–30 minutes indicates a negative reaction. For patients suspected of horse serum sensitivity, the horse serum should be diluted 1:100 (10 times weaker) and administered in the same acute care setting under direct observation. Not all patients with positive skin tests will have immediate hypersensitivity reactions when antivenom is given slowly.

Adverse Reactions. The most common acute reaction to antivenom administration is urticaria, which usually responds to slowing the infusion and intravenous antihistamines.[173] Although rare, the most serious response is a type I anaphylactic response to the horse serum, especially in patients with a history of horse serum sensitivity. During the production process, ammonium sulfate precipitation destroys up to one-half of the neutralizing IgG antibodies while most antigenic material remains. A majority of patients, who receive antivenin (Crotalidae) polyvalent, develop serum sickness (type III) within 3 weeks (on average within 7–14 days) of antivenom administration. In a retrospective chart review of 181 rattlesnake envenomations receiving antivenin (Crotalidae) polyvalent, 56% developed serum sickness.[174] For patients receiving <20 vials and ≥40 vials the incidence of serum sickness was 34% and 100%, respectively. Clinical manifestations of serum sickness include fever, headache, myalgias, arthralgias, lymphadenopathy, and urticaria. Patients who receive more than 7 vials invariably develop serum sickness.[175] These patients usually respond to a 7- to 10-day, tapering course of prednisone beginning at 60 mg/day.

Skin Test-Positive Patients. The use of antivenom in a patient with a positive skin test requires careful clinical evaluation because withholding antivenom in severe envenomations may result in serious morbidity unless CroFab™ is immediately available.[176] Pretreatment with diphenhydramine and careful, slow titration of antivenom, as well as the use of subcutaneous epinephrine for reactions, may avoid a serious response. The patient should be monitored in an intensive care setting. Highly allergic patients may respond to antivenom infusions and simultaneous epinephrine infusion in a separate line, but this method requires considerable clinical judgment. Patients have survived serious envenomation with only supportive care; permanent sequelae in these patients may be limited despite severe initial swelling of the involved extremity. The option of using CroFab™ should be seriously considered in these patients.

Supplemental Care

The management of rattlesnake envenomation depends, in part, on the patient population and the type of local venomous snakes. All asymptomatic patients suspected of being envenomated should be observed for at least 6–8 hours.[177] Before the release of asymptomatic patients, all laboratory studies should be repeated at least 4 hours after the first tests to confirm the absence of a coagulopathy. Severe thrombocytopenia and defibrination may occur in the absence of significant local swelling.[129] Suspected Mojave rattlesnake envenomations should be observed for 24 hours. All envenomated patients require observation in an intensive care setting. Smoking and drug use should be discouraged during the first few days following envenomation, and dental work, contact sports, and elective surgery should be avoided for the first few weeks. All patients should be reevaluated in 1–2 days. Repeat laboratory tests (complete blood count [CBC], PT/aPTT, fibrinogen) should be done within the first 2 to 3 days, and then repeated within the first 2 weeks if abnormalities develop. The patient should be instructed to return to the Emergency Department if recurrent swelling/pain, abnormal bleeding/bruising, wound infection, rhabdomyolysis (decreased urination, dark-colored urine), or serum sickness develops.

Ancillary Tests

Depending on severity, the following routine laboratory tests should be obtained on admission and monitored daily unless the clinical picture dictates more frequent intervals: CBC, urinalysis, serum electrolytes, creatinine, blood urea nitrogen, creatine kinase, coagulation profile (platelet count, PT, aPTT, fibrin split products), and chest x-ray for envenomations with pulmonary symptoms.

Most mild to moderate envenomations require only CBC and a coagulation profile unless the envenomating rattlesnake species (e.g., canebrake rattlesnake, *Crotalus horridus atricaudatus*) are associated with rhabdomyolysis., Electrolytes, kidney function, and serum creatine kinase should be monitored closely if there is evidence of rhabdomyolysis. Type and cross-match is necessary only for patients with severe envenomation and signs of active bleeding. For patients over 40 years of age or for severe envenomations, cardiac monitoring and an electrocardiogram are indicated. Transient hyperkalemia may develop due to cell lysis.

Wound Care

Circumferential measurements of the bitten extremity should be repeated every 30–60 minutes initially, and then as needed to monitor the effectiveness of antivenom therapy. There are few clinical data on the efficacy of varying methods of wound care during rattlesnake envenomation. The envenomated area should be cleaned by gentle irrigation with saline. The wound may be gently covered with nonadherent, hydrophilic dressings (e.g., Adaptic™, Johnson & Johnson, Inc., New Brunswick, NJ; Xeroform™, Kendall Company, Mansfield, MA). For the first 3 days, the extremity should be immobilized with a padded splint in a position that facilitates rehabilitation below the heart. Edema usually resolves about 7–10 days after a moderate envenomation. Local treatment includes sterile whirlpool baths and cleansing with 1:20 aluminum acetate soaks (Burow's solution) for 15 min. Hemorrhagic blebs and necrotic tissue should not be débrided until hemostasis returns to normal (i.e., after the third to fifth days postenvenomation).[178] Rehabilitation should start after the first surgical débridement (i.e., after the first 3 days) with aggressive passive and active range of motion exercises. Skin grafting may be necessary to cover large areas of skin necrosis.

In general, observational studies indicate that the incidence of wound infections is low following envenomation by both North American and South American rattlesnake species.[179,180] Prophylactic antibiotics may be considered for high-risk wounds (e.g., severe envenomations, first-aid measures involving incision and drainage, puncture wounds on the hand or foot). Abrasions and mild envenomations do not require antibiotic coverage. The presence of gram-negative organisms in the mouth of the North American pit viper requires broad-spectrum coverage (e.g., quinolones). Studies of the bacterial flora of North American pit vipers indicate that *Proteus vulgaris, Escherichia coli, Corynebacterium* spp. (diphtheroids), *Streptococcus*, and *enterobacteria* are the most common organisms.[181] Diphtheria-tetanus prophylaxis is similar to other puncture wounds (i.e., every 5 years).

Coagulation Disorders

Monitoring of coagulation defects requires the determination of the complete blood count with platelet count, PT, activated PTT [aPTT], fibrinogen concentration, and fibrin degradation products as well as serum electrolytes, serum creatinine, and blood typing as necessary. Laboratory coagulation parameters should be repeated every 4–6 hours until the patient stabilizes. All patients with serious clinical bleeding should receive antivenom as well as platelet packs, cryoprecipitate, and fresh frozen plasma as needed.[182] However, multiple coagulation defects (i.e., international normalized ratio [INR] > 3, aPTT > 50, fibrinogen <75 mg/dL, *and* platelet count <50,000) suggest the potential for serious clinical bleeding, and the administration of antivenom is an option when these multiple coagulation defects develop. If adequate amounts of antivenom have been administered to control local tissue abnormalities and systemic effects, close observation is an option to further administration of antivenom as long as there is *no* evidence of internal or external bleeding.[183] Consequently, the administration of more antivenom is not necessary to correct laboratory evidence of *isolated* coagulation defects.

The consumptive coagulopathy is not responsive to the replacement of coagulation factors (i.e., fresh-frozen plasma) until there is sufficient antivenom to neutralize circulating snake venom. Furthermore, isolated coagulation defects are not usually associated with clinically significant bleeding because thrombin generation usually remains intact.[184] The infusion of platelet packs is not usually necessary when the platelet count exceeds 20,000/m^3 and the rest of the coagulation profile is normal. Marked reduction in platelet count is common. An increase in the platelet count usually begins within a few hours after the administration of CroFab™, but the platelet count may not return to normal for 7 days. In the presence of clinically significant bleeding, cryoprecipitate is indicated when the fibrinogen concentration decreases below 100 mg/dL. Hypofibrinogenemia is relatively resistant to the infusion of CroFab™.[185] Heparin is not effective for venom-induced coagulopathy because the thrombin-like glycoproteins in crotaline venom are not inhibited by antithrombin III (heparin cofactor).

Analgesia

Mild sedatives and analgesics (e.g., acetaminophen, codeine) are appropriate for anxiety and pain. More potent analgesics may be necessary (e.g., fentanyl, hydromorphone). Aspirin is contraindicated because of the effect of aspirin on platelet aggregation.

Compartment Syndrome and Fasciotomy

Most envenomations result in superficial rather than subfascial swelling, and muscle necrosis usually results from the direct toxic effects of venom on muscle rather than elevated muscle compartment pressure.[186,187] The blood flow usually increases in most envenomated extremities, and there are inadequate data to determine if the use of fasciotomy improves clinical outcome. The clinical diagnosis of a compartment syndrome is difficult because rattlesnake envenomation produces tense facial compartments and loss of two-point discrimination without the presence of a compartment syndrome. The presence of painful passive stretching of the involved compartment, distal and localized paresthesias, or persistent pain despite the use of parental analgesics and antivenom suggest the possibility of a compartment syndrome, particularly following the cessation of antivenom therapy.[188] The presence of muscle compartment pressures >30 mm Hg indicates the need for continued limb elevation and additional therapy. The use of additional doses of antivenom (e.g., 3–5 vials of CroFab™) and 1–2 g mannitol/kg is an option to fasciotomy if the compartment pressures are slightly elevated (i.e., 40–50 mm Hg). Surgical incisions and fasciotomies are treatment options *only* when there is objective evidence (i.e., compartment pressures >30 mm Hg for over 1 h, reduced Doppler flow) of vascular compromise that does not respond to adequate doses of antivenom and mannitol.[189] The effect of mannitol infusion on the enhancement of CroFab™ clearance has not been evaluated. Commercial compartment pressure monitors (e.g., Stryker, Kalamazoo, MI) are available for measurement of elevated compartment pressures in the arm, but there is no practical method for measuring compartment pressures in the digits. Measurement techniques include the use of needle manometer, slit catheter, or wick catheter. Variation in oxygen saturation as measured by pulse oximetry is not a sensitive measure of elevated compartment pressures.[190] Long-term complications of rattlesnake envenomations are rare with the exception of joint stiffness and muscle weakness.[191]

Steroids

Steroids should be reserved for those patients who develop serum sickness. No clinical data document an improvement in the clinical course of rattlesnake envenomation after the administration of steroids. Clinical features of serum sickness include malaise, fever, headache, gastrointestinal upset, arthralgias, myalgias, lymphadenopathy, skin lesions, and, less commonly, peripheral neuritis, meningismus, and renal dysfunction. Symptoms typically begin 7–14 days after the administration of antivenom, beginning with urticaria. Prednisone, 40–60 mg daily (1–1.5 mg/kg in children) in four divided doses for 2 to 3 days and then tapered over 7–10 days. Antihistamines are not usually effective.

Pygmy Rattlesnake and Massasauga

The *Sistrurus* genus (pygmy rattlesnakes, a.k.a. pigmy rattlesnakes) contains more primitive snakes compared with true rattlesnakes (*Crotalus*). The pygmy rattlesnake (*Sistrurus miliarius*) has a small, slender body that seldom exceeds 75 cm (~30 in.) in length. Like the massasauga, the pygmy rattler displays the characteristic dorsal head scales. The massasauga is distinguished from true rattlesnakes (*Crotalus* spp.) by nine large shields arranged in rows across the crown of the head, whereas rattlesnakes from the genus *Crotalus* have small crown scales. These shy and highly secretive snakes (*Sistrurus catenatus*) are the most gentle of the rattlesnake species. Generally, the massasauga inhabits bogs or swamps except during the summer when these snakes move to drier habitat. The desert massasauga (*S. catenatus* subspecies *edwardsi*) lives in the Sonoran desert region of the southwestern United States.

EXPOSURE

Geographical Distribution

The pygmy rattlesnake (*Sistrurus miliarius*) comprises three subspecies distributed over the southeastern United States from Texas to Florida (See Figure 176.8). The western pygmy (*Sistrurus miliarius* subsp. *streckeri*) rattlesnake ranges over Texas, Louisiana, and parts of Missouri, Arkansas, Tennessee, and Alabama. The Carolina pygmy rattlesnake (*Sistrurus miliarius* subsp. *miliarius*) is found from the Carolinas to Alabama. The southeastern pygmy (*Sistrurus miliarius* subsp. *barbouri*) appears in the Deep South. The massasauga (*Sistrurus catenatus*) species comprises three subspecies, which range from the Great Lakes to desert regions of the southwestern United States (see Figure 176.9). These ranges of these three subspecies include the following: from the desert to the Southwest (*Sistrurus catenatus* subsp. *edwardsi*); the western seaboard to the middle Midwest (*Sistrurus catenatus* subsp. *tergeminus*); the eastern seaboard to the Great Lakes region in marshy habitats and flood plains (*Sistrurus catenatus* subsp. *catenatus*).

Habitat/Characteristics

Generally, the snakes of this genus avoid human contact and these snakes are much more reclusive than the true

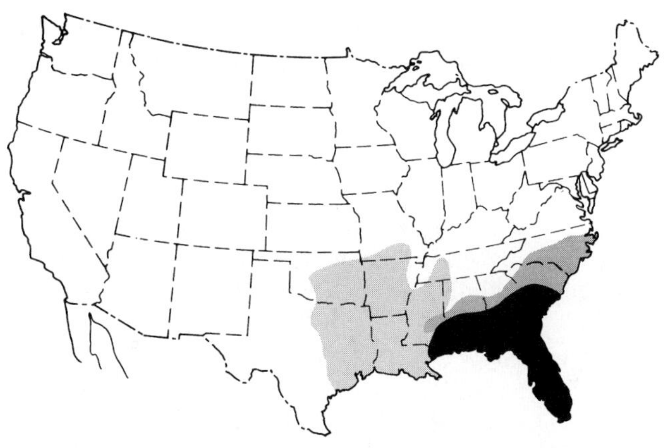

- Western pigmy rattlesnake, *Sistrurus miliarius streckeri*
- Carolina pigmy rattlesnake, *Sistrurus miliarius miliarius*
- Southeastern pigmy rattlesnake, *Sistrurus miliarius barbouri*

FIGURE 176.8. Distribution of pygmy rattlesnake. Adapted from Ellenhorn MJ, Barceloux DG. Medical Toxicology: Diagnosis of Human Poisoning. 1st Edition, New York: Elsevier, 1988:1124.

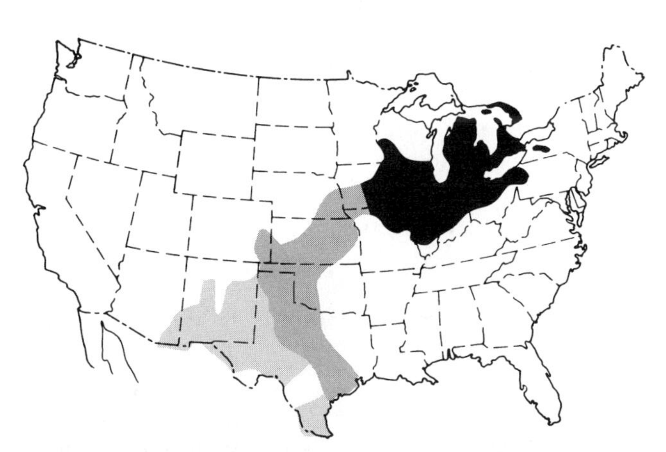

- Desert massasauga, *Sistrurus catenatus edwardsi*
- Western massasauga, *Sistrurus catenatus tergeminus*
- Eastern massasauga, *Sistrurus catenatus catenatus*

FIGURE 176.9. Distribution of the massasauga. Adapted from Ellenhorn MJ, Barceloux DG. Medical Toxicology: Diagnosis of Human Poisoning. 1st Edition, New York: Elsevier, 1988:1125.

rattlesnakes (*Crotalus*). The massasauga averages about 45–75 cm (18–30 in) in length with a slender tail and small rattles.

VENOM

The venom of the southeastern pygmy rattlesnake (*Sistrurus miliarius barbouri*) contains a unique component called barbourin, which inhibits platelet aggregation by binding to the platelet membrane glycoprotein IIb-IIIa receptor and preventing the binding of fibrinogen.[192] Eptifibatide is a cyclic heptapeptide inhibitor of platelet membrane glycoprotein IIb-IIIa receptor derived from barbourin that has undergone clinical trials as an antiplatelet drug for acute coronary syndromes.[193]

CLINICAL RESPONSE

The clinical presentation of envenomations by *Sistrurus* species is similar, but less severe, than envenomation by true rattlesnakes (*Crotalus* spp.). Often bites occur without envenomation and the clinical picture of *Sistrurus* envenomation frequently resembles envenomation by a copperhead rather than *Crotalus* species. Edema is usually mild, but hemorrhagic blebs frequently develop after massasauga envenomation. Some nausea may occur, but other systemic symptoms (vomiting, paresthesias, fasciculations) are unusual. Laboratory abnormalities are rare, but severe thrombocytopenia may develop.[194] Rarely, anuric renal failure may develop in envenomated patients with rhabdomyolysis, myoglobinuria, and coagulation abnormalities.[195]

TREATMENT

There are few clinical data on the use of antivenom for *Sistrurus* envenomations. In general, the management of *Sistrurus* envenomation follows protocols for *Crotalus* envenomations, with gradation of clinical symptoms and administration of CroFab™ or antivenom (Crotalidae) polyvalent as needed.[196] Most envenomations by *Sistrurus* species do not require the administration of antivenom.

Copperhead

Copperheads (*Agkistrodon contortrix*) account for about 25% of venomous snakebites in the United States. The head is solid copper to red brown. The rest of the body is marked with dark cross bands that resemble dumbbells. Highland moccasin is another common name. These snakes possess no rattles.

EXPOSURE

Five subspecies of *Agkistrodon contortrix* (copperhead) range from the eastern United States to southern Texas. The adult copperhead averages about 60–90 cm (24–30 in) in length. The head is copper-colored and the body is thick with reddish brown hourglass designs and no rattles. These snakes inhabit mountains and wooded hillsides as well as streams beds and lowlands in the southern portion of the United States. Copperheads inflict a large portion (about 25–30%) of venomous snakebites, in part, because of their propensity to invade human living areas.[197] Along the Atlantic seaboard, copperhead bites are more common than rattlesnake bites. However, clinical effects from envenomation are usually less severe than other *Crotalus* bites because copperhead venom is the least venomous among North American pit vipers. The incidence of copperhead bite is highest in North Carolina, West Virginia, and Arkansas.

HISTOPATHOLOGY AND PATHOPHYSIOLOGY

In general, the venom of the copperhead contains relatively weak thrombin-like and anticoagulant activity compared with rattlesnakes.[198,199] Like rattlesnake venom, copperhead venom contains enzymes (fibrolase) that incompletely split fibrin and directly inactivate clot-stabilizing factors without activation of plasminogen.[200] Fibrolase is a zinc metalloproteinase that degrades the α-chain of fibrin and the Aα-chain of fibrinogen, but fibrolase does not activate plasminogen.[201] Alfimeprase is a recombinant product derived from a truncated form of fibrolase that possesses direct proteolytic activity against the fibrinogen Aα chain. Copperhead venom also contains a protein C activator. Disintegrins antagonize the effects of integrins, which are membrane-bound molecules that enhance cell movement by providing traction for migrating cells, by assembling extracellular matrices to serve as tracks for migration, and by transmitting guidance signals that direct cells or cell processes to their targets. Contortrostatin is a disintegrin found in the venom of the southern copperhead that has been tested in clinical trials as an agent to reduce the progression of breast cancer by inhibiting angiogenesis.[202] Contortrostatin does not destroy tumor cells.

CLINICAL RESPONSE

Envenomation by copperhead snakes typically produces mild to moderate effects secondary to local tissue reaction that are usually less severe than rattlesnake envenomations. Clinical features of a copperhead envenomation include edema, ecchymosis, and mild pain, and the pain may be out of proportion to local tissue injury. Pain and swelling may continue more than 4 hours after envenomation.[203] Lymphadenopathy occurs commonly, and occasionally with nausea and vomiting develops after envenomation. The clinical presentation of copperhead envenomations are similar between subspecies, although case reports of bites by the Southern copperhead (*A. contortrix contortrix*) suggest that this subspecies produces more severe systemic complaints (vomiting, diarrhea, hematuria) than other subspecies.[204] Systematic symptoms (e.g., fasciculations, paresthesias, coagulopathies, thrombocytopenia) are unusual, and severe envenomations are rare.[205] Most symptoms of copperhead envenomation resolve within one week, although occasionally pain, edema, limb dysfunction, and poor healing may last several months.[206] Death due to copperhead envenomation is rare.[207]

TREATMENT

There are limited data on the treatment of copperhead envenomation, in part, because most copperhead envenomations produce mild to moderate local effects without significant systemic toxicity. Antivenin (Crotalidae) polyvalent is effective against copperhead envenomations, but the administration of antivenom is usually necessary only for the presence of systemic complications in children or the elderly.[208,209] Although there are little clinical data on the use of CroFab™ for the local tissue effects following copperhead envenomation,[210] this antivenom is probably at least as efficacious and much less immunogenic than the Wyeth (Crotalidae) antivenom based on preliminary data.[211] Guidelines for the administration of CroFab™ following copperhead and rattlesnake evenomation are similar pending further data to clarify the benefits and risks associated with the use of CroFab™ for the alleviation of moderate local tissue effects. General wound care, first aid, and antivenom administration are similar to the treatment of *Crotalus* envenomations. The venom of *Agkistrodon* species contains fractions that may prolong neuromuscular blockage with anesthetic agents (e.g., vecuronium).[212]

Cottonmouth

Cottonmouths (water moccasins) account for about 1% of venomous snakebites in the United States. The common name, cottonmouth, reflects the white buccal mucosa characteristic of this species (*Agkistrodon piscivorus*).

EXPOSURE

Three species of cottonmouth are distributed across the semiaquatic environments of the southeastern United States. The usual habitat of the cottonmouth includes bayous, lakes, swamps, streams, drainage ditches, and rice fields with the highest estimated bite rates occurring in Louisiana, Mississippi, Arkansas, Georgia and Florida.[213] Characteristics of this genus include a unique white buccal mucosa on a dark head, a broad heavy body, and the lack of a rattle on the tail. Mature members have a broad head and a dark olive color that provides camouflage in swampy marshes. The cottonmouth shares the features of pit viper characteristics with rattlesnakes. Cottonmouths are capable of envenomating victims while submerged.

CLINICAL RESPONSE

Cottonmouth bites produce an envenomation intermediate in severity between those of rattlesnakes and copperheads. The clinical manifestations should be graded similarly to rattlesnake envenomation.

TREATMENT

Although there are few data on the clinical efficacy of CroFab™ after envenomation by the cottonmouth, this antivenom contains antibodies to cottonmouth venom formed during the production process. The clinical grading and indications for CroFab™ use during cottonmouth envenomation is similar to the use of CroFab™ during rattlesnake envenomation. However, the frequency and dose required to treat cottonmouth envenomation is less than rattlesnake envenomations.

CENTRAL AND SOUTH AMERICA

EXPOSURE

The most important venomous snakes in Central and South America belong to the Crotalinae subfamily.[214,215] The pit vipers are distributed throughout Central and South America including species from the genera *Atropoides*, *Bothriechis*, *Bothrops*, *Cerrophidion*, *Ophryacus*, and *Porthidium*. Central American venomous crotaline snakes include rattlesnakes (*Crotalus* spp.), cantil (*Agkistrodon* spp.), palm pit vipers (*Bothriechis* spp.), forest pit vipers, and lanceheads (*Bothrops* spp.), bushmaster (*Lachesis muta*), and the hog-nosed and montane pit vipers (*Porthidium*, *Atropoides*, and *Cerrophidion* species).[216] *Bothrops atrox* inhabits most of the northern part of South America. The heads of these pit vipers are distinctly wider than the necks. *Bothriechis schlegelii* (palm pit viper) is an arboreal snake in the rain forests, whereas *Bothrops jararaca* (Jararacucu) is a long, wide-bodied, terrestrial snake of the savanna in South America from Brazil to southern Bolivia, Paraguay, and northeastern Argentina. The bushmaster (*Lachesis muta*) is the largest viper in the world with a length up to 3 m (~10 ft). The tail ends in a horny spine rather than the terminal rattles present in crotalids. This timid snake lives in the undergrowth of tropical rain forests from Nicaragua to South America. In South America, the representative of the Crotalinae subfamily is *Crotalus durissus terrificus*, and these crotalines prefer secluded dry, rocky areas. Bite marks of crotalines typically demonstrate isolated single or double puncture wounds, but the presence of these marks does not necessarily indicate envenomation.[217]

VENOM

The venom of pit vipers produces hemorrhage and tissue necrosis from the presence of highly active proteases (e.g., zinc metalloproteinases) and thrombin-like enzymes. High concentrations of hemorrhagic proteases occur in the venoms of pit vipers, particularly *Protobothrops flavoviridis*. These proteinases act on basal membranes causing disruption of the endothelial lining of the small blood vessels. The thrombin-like enzymes split fibrinopeptide A or B from fibrinogen, leading to a consumptive coagulopathy. Some crotaline venoms contain muscle toxins (e.g., crotamine) that produce skeletal muscle damage and rhabdomyolysis.[218,219] Aspercetin is a disulfide-linked heterodimer that is a platelet-aggregating protein isolated from the venom of *Bothrops asper*.[220] Aspercetin-mediated platelet aggregation results from the interaction of von Willebrand factor with platelet glycoprotein receptor Ib (GPIb), and this protein has functional similarities to botrocetin from *Bothrops jararaca* venom. The venom of *Crotalus durissus terrificus* (South American rattlesnake) contains a neurotoxin (crotoxin) that blocks presynaptic neuromuscular transmission via phospholipase A_2. This neurotoxin is similar to the neurotoxin in the Mojave rattlesnake (*Crotalus scutulatus scutulatus*).

CLINICAL RESPONSE

Pit viper envenomations produce the rapid onset of edema and pain that may progress to hemorrhagic vesicles and necrosis along with variable degrees of a coagulopathy. The clinical pattern of envenomation by Central and South American pit vipers is generally similar to North American rattlesnake evenomation, and the severity of envenomation depends on a variety of factors including the potency and amount of venom injected, the health of the victim, the pit viper species, and the size of the snake. Similar to North American rattlesnakes, some bites by South American pit viper do not result in envenomation. In a study of 41 bites by the South American rattlesnake (*Crotalus durissus*), 12% of the victims did not develop signs or symptoms of envenomation.[221]

Particularly after envenomation by lance-headed viper (*Bothrops* spp.), a consumption coagulopathy may develop with hemorrhagic vesicles, local and intraoral bleeding, hematuria, hypotension, hypofibrinogenemia, fibrin-split products, and prolonged coagulation profiles. Necrosis, gangrene, rhabdomyolysis, myoglobinuria, and acute renal failure may complicate pit viper envenomations, particularly bites by the South American rattlesnake (*Crotalus durissus terrificus*),[222] bushmaster (*Lachesis* species),[223] and lance-headed vipers (*Bothrops* spp.).[224] The venom of the *Bothrops jararaca* contains relatively higher concentrations of muscle toxins, and therefore these envenomations commonly cause rhabdomyolysis and myoglobinuria in addition to necrosis and coagulopathy compared with other *Bothrops* species (e.g., *B. neuwiedii*).[225,226] Envenomation by *Lachesis* species (Bushmaster) produces vomiting, profuse diarrhea, diaphoresis, salivation, hypotension, bradycardia and respiratory distress. In contrast to North American crotalids, envenomations by some South American rattlesnakes (e.g., *Crotalus durissus*) also produce neurological symptoms similar to elapid envenomation including myalgias, weakness, ptosis, ophthalmoplegia, dysarthria, and dysphagia that progresses to respiratory failure.[227,228] The prevalence of cerebrovascular complications (intracerebral hemorrhage, subarachnoid bleed, cerebellar hemorrhage) following envenomation by *Bothrops* species is low, but the prognosis for patients with intracranial bleeding is poor in most, but not all patients.[229] Envenomations by *Bothriechis* species (arboreal snakes) and montane pit vipers (*Cerrophidion* spp.) are usually less severe than lanceheaded vipers and the South American rattlesnakes.

TREATMENT

The local treatment of South American pit viper envenomations is similar to the treatment of North American rattlesnakes (see section North America, subsection *True Rattlesnakes*). First aid treatment includes rest, reassurance, immobilization of the bitten extremity at or below the heart, and rapid transport to medical facilities. Moderate to severe envenomations require the intravenous administration of antivenom. Although there are some antigenic similarities between venoms of different species of pit vipers, there is clinically significant variability between venoms from various pit viper species and subspecies.[230,231] Consequently, antivenom developed from the specific area where the envenomation occurred is usually preferable, if available, to polyvalent antivenom. A comparative study of patients receiving Colombian, Ecuadorian, and Brazilian antivenom for pit viper (i.e., *Bothrops* species, *Lachesis muta*) envenomations detected little difference in the efficacy of these three types of antivenom, although the incidence of early anaphylactoid reactions were higher for the Colombian and Brazilian antivenom than for the Ecuadorian antivenom.[232] Crotalidae polyvalent antivenom with antibodies against venom from the tropical rattlesnake (*Crotalus durissus terrificus*) and fer-de-lance (*Bothrops atrox*) has been used successfully for the treatment of envenomation by the hog-nosed viper (*Porthidium nasutum*).[233] Victims with clinically significant envenomations should be evaluated for rhabdomyolysis (serum creatinine, urine myoglobin, serum creatine phosphokinase, serum electrolytes) and coagulation abnormalities (fibrin split products, fibrinogen, complete blood count, prothrombin time, activated partial thromboplastin time). Antibiotics are not usually require unless the wound is heavily contaminated or a secondary infection (e.g., *Aeromonas hydrophilia, Staphylococcus, Streptococcus, enterobacteria*) is suspected.[234,235]

ASIA

EXPOSURE

Table 176.10 lists common Asian pit vipers. The Old World Asian pit vipers are not closely related to the New World pit vipers (*Agkistrodon*), and therefore the Asian pit vipers are no longer classified in the genus *Agkistrodon*. The Asian counterparts of the South American lance-headed vipers (*Bothrops* spp.) are the Asian lance-headed vipers, tree vipers, and bamboo vipers (*Trimeresurus* spp., *Ovophis* spp., *Protobothrops* spp., *Tropidolaemus wagleri*). Well-known *Trimeresurus* species include *T. albolabris* (white-lipped tree viper), *T. flavoviridis* (habu), *T. gramineus* (green tree viper), and *T. stejnegeri* (Chinese green tree viper). Figure 176.10 displays a specimen of the Mangrove pit viper (*Trimeresurus purpureomaculatus* Gray) from India and Southeast Asia. Representatives of the New World *Agkistrodon* species in Asia include the Malayan pit viper (*Calloselasma rhodostoma*), Asian pit vipers (*Gloydius* spp.), and hundred-pace or sharp-nosed pit

TABLE 176.10. Important Venomous Asian Pit Vipers[a]

Species	Common Name	Description	Distribution	Habitat
Calloselasma rhodostoma Boie	Malayan pit viper	Wide-bodied snake up to 1.5 m (~5 ft) with distinctive pit between eyes and nostrils and variable dark triangular markings.	Malayan peninsula and Indonesia	Humid rain forests, rice fields, plantations
Deinagkistrodon acutus Günther	Sharp-nosed pit viper	Heavy-bodied snake with variable coloration, pits between nostrils and eyes, and triangular markings	China	Terrestrial
Gloydius spp.	Mamushi, Himalayan pit viper, Central Asian pit viper	Heavy-bodied snake with variable coloration, pits between nostrils and eyes, and triangular markings	Central Asia/Middle East to Japan	Widespread distribution—dry deserts to cool mountain areas up to 3,000 m (~10,000 ft)
Hypnale hypnale Merrem	Hump-nosed viper	Heavy-bodied snake with variable coloration, pits between nostrils and eyes, and triangular markings	South India, Sri Lanka	Terrestrial
Protobothrops elegans Gray	Japanese lace-headed viper, Sakishima-habu	Monochromatic large yellow eyes with highly variable coloration ranging from green yellow to brown with light markings.	Japan	Arboreal in trees and bushes
Tropidolaemus wagleri Wagler	Wagler's or temple pit viper	Large triangular shaped head with a wide variety of colors, a relatively thin body, and a prehensile tail to aid climbing	Malaysia	Arboreal in trees and bushes
Protobothrops flavoviridis Hallowell	Habu	Large, heavy-bodies snake up to 2 m (~5–6 ft) in length.	Southern Asia	Terrestrial in forests, mangrove areas

[a]Source: Adapted from Ref 2.

FIGURE 176.10. Mangrove Pit Viper (*Trimeresurus purpureomaculatus* Gray). © 1997 Rune Midtgaard. See color insert.

viper (*Deinagkistrodon acutus*). The most common envenomations by snakes in Taiwan involve *Trimeresurus mucrosquamatus* and *Trimeresurus stejnegeri*,[236] whereas the most common pit viper envenomations in South Vietnam involve the green pit viper (*Trimeresurus* spp.) and the Malayan pit viper (*Calloselasma rhodostoma*).[237] Bites by *Deinagkistrodon acutus* occur primarily in southern and eastern Taiwan. These pit vipers have retractable fangs. The hump-nosed viper is one of the smallest venomous snakes in India with a total length below 55 cm (~22 in.). Distinctive features include five large symmetrical plate scales on the triangular head and weakly keeled body scales.

VENOM

Clotting enzymes and hemorrhagic factors are the most important constituents of the venom from Asian pit vipers. The venom of the Malayan pit viper (*Calloselasma rhodostoma*) directly lyses fibrinopeptide A from fibrinogen to form fibrin similar to effect of thrombin. A purified fraction of venom from the Malayan pit viper *Calloselasma rhodostoma* is called ancrod, which contains a serine protease capable of cleaving fibrinopeptide A from fibrinogen that produces a hemorrhagic defibrination coagulopathy. This fraction of the venom acts as cofactor in tPA-induced plasminogen activation, leading to the production of large amounts of fibrinogen and fibrin degradation products (D-dimer).[238] The venom of *Gloydius halys* Pallas contains metalloproteinase-like prothrombin activators similar to the venom of South American fer-de-lance (*Bothrops atrox*).[239]

CLINICAL RESPONSE

Envenomations by Asian pit vipers produce local pain and swelling along with varying degrees of coagulopathy. Typically, envenomation by the humped-nose viper (*Hypnale hypnale*) produces signs of local envenomation manifested by pain, swelling, and induration at the site of the bite that occasionally progresses to hemorrhagic blister formation and regional lymphadenopathy.[240] Signs of systemic envenomation are rare, but severe envenomation (renal failure, coagulation disorder with low fibrinogen concentrations, and increased fibrin degradation products) may occur.[241,242] Platelet counts, prothrombin time, and the partial thromboplastin time are usually normal.[243] Characteristic features of envenomation by *Deinagkistrodon acutus* (hundred-pace viper), *Gloydius* species, *Trimeresurus* species, and *Protobothrops* species include extensive swelling and hemorrhage at the site of envenomation progressing to necrosis along with various degrees of a consumption coagulopathy. Envenomation by arboreal green pit vipers (*Trimeresurus albolabris, T. macrops* Kramer) often causes only painful local edema and mild coagulation abnormalities (thrombocytopenia, prolonged clotting times). Occasionally internal bleeding may result from a defibrination syndrome manifest by low fibrinogen concentrations and elevated fibrin degradation products.[244,245] Bites by the Malayan pit viper cause local tissue damage that often resolves over 1–2 weeks.[246] However, the venom of the Malayan pit viper induces plasmin formation, and the resulting clinical effect following *Calloselasma* envenomation is a defibrination syndrome. Bites by the mamushi pit viper (*Gloydius halys*) usually produce local symptoms without serious systemic complications.[247] However, in addition to local tissue damage, envenomations by *Gloydius halys* may produce blurred vision, ptosis, diplopia, difficulty in opening the mouth, dysphonia, dysphagia, generalized weakness, and in severe cases hemoglobinuria, disseminated intravascular coagulopathy, renal failure, and respiratory failure.[248–250]

TREATMENT

Treatment of Asian pit viper envenomation is similar to crotaline envenomation. Most envenomations require antivenom with the exception of some tree vipers (*Trimeresurus* species). These latter envenomations are usually mild similar to European vipers, and antivenom is not usually necessary unless the envenomation is associated with a significant coagulopathy. Envenomation by the hump-nosed viper (*Hypnale hypnale*) may not require polyvalent antivenom unless the envenomation is associated with systemic complications, and there is no specific polyvalent antivenom for envenomation by this viper.[251] The specificity and efficacy of antivenom varies with different localities and species. The antivenom should be administered intravenously as soon as possible within four to six hours after envenomation.[252] Coagulation profiles and complete blood count should be followed serially until the parameters substantially improve. The presence of dark urine indicates the need to evaluate the patient for myoglobinuria, hematuria, and renal dysfunction. Generous intravenous fluid replacement should be administered to maintain adequate urine flow if rhabdomyolysis occurs.

ELAPIDS

The family **Elapidae** is divided into the following three subfamilies: 1) **Elapinae** (North American coral snakes [*Micrurus*], Asian and African cobras [*Naja*], Asian kraits [*Bungarus*] and coral snakes [*Calliophis*, *Maticora*], African mambas [*Dendroaspis*], and all venomous snakes in Australia (See Table 176.11 and Table 176.12), 2) **Hydrophiinae** (sea snakes), and 3) **Laticaudinae** (sea kraits).[253,254] In contrast to vipers, elapids have more slender bodies, less triangular heads, and fewer body scale rows. Elapid fangs are usually smaller in proportion to body size compared with vipers. Although some Australian elapids possess potent myotoxins and procoagulants, most elapid venoms contain primarily postsynaptic neurotoxins.

NORTH AMERICAN CORAL SNAKES

EXPOSURE

Two genera of poisonous coral snakes (*Micrurus, Micruroides*) represent the cobra family in the United States. Coral snakes are slender snakes ranging between 2 ft and 4 ft (~60–120 cm) with brightly colored, circumferen-

TABLE 176.11. Scientific and Common Names of Venomous Coral Snakes and Cobras from the Family Elapidae[a]

Group Name	Scientific Name	Common Name
Coral Snakes		
	Calliophis spp.	Oriental coral snake
	Micruroides euryxanthus Kennicott	Sonoran or Western coral snake
	Micruroides euryxanthus euryxanthus Kennicott	Arizona coral snake
	Micrurus frontalis Duméril, Bibron & Duméril	Brazilian giant coral snake
	Micrurus fulvius L.	Northern coral snake
	Micrurus tener Baird & Girard	Texas coral snake
Cobras		
	Boulengerina species	Water cobras
	Hemachatus haemachatus Lacépède	Rinkhals cobra
	Naja anchietae Bocage	Anchita's cobra
	Naja atra Cantor	Taiwan or Chinese Cobra
	Naja haje L.	Egyptian or brown cobra
	Naja katiensis Angel	Western brown spitting or Mali cobra
	Naja kaouthia Lesson	Monocled or Thai cobra
	Naja melanoleuca Hallowell	Forest or black, white-lipped cobra
	Naja mossambica Peters	Spitting cobra
	Naja naja L.	Indian cobra or Cobra de Capello
	Naja nigricollis Reinhardt	Black-necked spitting cobra
	Naja nivea L.	Cape or yellow cobra
	Naja oxiana Eichwald	Central Asian cobra
	Naja pallida Boulenger	Red spitting cobra
	Naja philippinensis Taylor	Philippine cobra
	Naja sagittifera Wall	Indian cobra
	Naja siamensis Laurenti	Indochinese spitting cobra
	Naja sputatrix Boie	Malayan cobra
	Naja sumatrana Müller	Equatorial spitting cobra
	Ophiophagus Hannah Cantor	King cobra
	Paranaja multifasciata Werner	Burrowing cobra
	Pseudohaje goldii Boulenger	African tree cobra, Gold's tree cobra
	Pseudohaje nigra Günther	Hoodless cobra
	Walterinnesia aegyptia Lataste	Desert blacksnake or black desert cobra

[a]Source: Adapted from Ref 1.

tial rings on their body. In North America, the coral snake's red and yellow rings are contiguous, whereas nonvenomous North American snakes have a black ring separating the red and yellow rings. The following mnemonic describes the difference between coral snake and nonpoisonous snakes: "Red on yellow, kill a fellow; red on black, good for Jack."

In contrast to pit vipers, coral snakes have round pupils, fixed fangs, small heads, and no facial pits. Generally, coral snakes are docile, preferring to chew rather than bite with their fixed fangs. The classification of coral snake remains controversial. The three species of coral snakes in the United States are *Micruroides euryxanthus* Kennicott (Sonoran or Western coral snake), Texas coral snake (*Micrurus tener* Baird & Girard) and the Northern coral snake (*Micrurus fulvius* L.). Subspecies of the Sonoran coral snake include *Micruroides euryxanthus australis* Zweifel & Norris, *Micruroides euryxanthus euryxanthus* Kennicott (Arizona coral snake), and *Micruroides euryxanthus neglectus* Roze. Subspecies of the Texan coral snake include *Micrurus tener fitzingeri* Jan, *M. tener maculatus* Roze, *M. tener microgalbineus* Brown & Smith, and *M. tener tener* Baird & Girard.

Geographic Distribution

Coral snakes (*Micrurus* spp., *Micruroides* spp.) range from the southeastern United States to central Argentina, and these snakes have adapted to a variety of habitats from desert areas to tropical rain forests. The Northern coral snake (*Micrurus fulvius* L.) inhabits the southeastern United States from Arkansas to Florida, whereas the Texas coral snake (*Micrurus tener* Baird & Girard) lives in Texas, Louisiana, Arkansas, and Northern Mexico. The Western or Sonoran coral snake

TABLE 176.12. Australian, Asian, and Other Elapids[a]

Australian and Pacific Elapids		
	Acanthophis antarcticus Shaw	Common death adder
	Acanthophis praelongus Ramsey	Northern death adder
	Acanthophis pyrrhus Boulenger	Desert death adder
	Acanthophis rugosus Loveridge	Jayan death adder
	Acanthophis wellsi Hoser	Pilbara or black head death adder
	Austrelaps labialis Jan	Pygmy copperhead
	Austrelaps ramsayi Krefft	Highland copperhead
	Austrelaps superbus Günther	Lowland copperhead
	Demansia Psammophis Schlegel	Spotted-headed snake
	Hoplocephalus bitorquatus Jan	Pale headed snake
	Hoplocephalus bungaroides Schlegel	Broad-headed snake
	Hoplocephalus stephensii Krefft	Stephen's banded snake
	Notechis scutatus Peters	Tiger snake
	Oxyuranus microlepidotus McCoy	Inland Taipan
	Oxyuranus scutellatus Peters	Taipan
	Pseudechis australis Gray	King brown or Mulga snake
	Pseudechis butleri Smith	Butler's mulga snake
	Pseudechis colletti Boulenger	Collett's snake
	Pseudechis guttatus De Vis	Spotted or blue bellied black snake
	Pseudechis papuanus Peters & Doria	Papuan black snake
	Pseudechis porphyriacus Shaw	Red-bellied black snake
	Pseudonaja affinis Günther	Dugite
	Pseudonaja guttata Parker	Speckled brown snake
	Pseudonaja inframacula Waite	Peninsula brown snake
	Pseudonaja ingrami Boulenger	Ingram's brown snake
	Pseudonaja nuchalis Günther	Western brown snake
	Pseudonaja textilis Duméril, Bibron & Duméril	Eastern brown snake
	Rhinoplocephalus nigrescens Günther	Eastern small-eyed snake
	Tropidechis carinatus Krefft	Rough-scaled snake
Other Elapids		
	Bungarus caeruleus Schneider	Indian or Blue (common) krait
	Bungarus candidus L.	Malayan krait
	Bungarus fasciatus Schneider	Banded krait
	Bungarus multicinctus Blyth	Formosan or Chinese krait
	Dendroaspis angusticeps Smith	Eastern green mamba
	Dendroaspis jamesoni Traill	Jameson's or green mamba
	Dendroaspis polylepis Günther	Black mamba
	Dendroaspis viridis Hallowell	Western green mamba

[a]Source: Adapted from Ref 1.

(*Micruroides euryxanthus* Kennicott) is found in the southwestern United States (Arizona, New Mexico, Northern Mexico).

Habitats/Characteristics

In contrast to *Crotalus* species, coral snakes possess black snouts and round pupils. There are no facial pits or rattles. North American coral snakes have small round heads and eyes, fixed fangs, and tricolor, slender bodies consisting of black and red bands separated by yellow or white stripes. The black bands on coral snakes are wide and circumferential, whereas black bands on scarlet king snakes are usually narrow and do not encircle the body. Coral snakebites produce a row of teeth marks rather than individual fang marks because the coral snake needs prolonged contact (i.e., approximately 30 s) to infuse their venom into the skin. This shy, nocturnal snake avoids human contact, and most bites result from provocation of the snake. Bites from coral snakes are relatively rare, accounting for $<1\%$ of annual poisonous snakebites in the United States. Coral snakebites are often superficial, leading to a lower rate of envenomation compared with rattlesnake bites. The coloration of coral snakes varies more in Central and South America than in the United States; therefore, the differentiation of coral snakes from nonvenomous snakes is difficult in Central and South America.

VENOM

Coral snake venom is primarily a neurotoxin and causes little local tissue reaction. Because of the more primitive venom apparatus, envenomation is less common following coral snakebites than following *Crotalus* snakebites. All elapids, including coral snakes, have a venom apparatus with proteroglyphous dentition characterized by large recurved, front-fangs and an *almost-closed* venom groove that terminates near the tip of the fang. Injection of venom requires pressure on the venom gland created by surrounding muscle. Effective envenomation by a coral snake requires several seconds to allow the venom to flow down the grooves on the anterior fangs. The teeth of the coral snake are small, averaging about 2 mm (~0.1 in.) in length. The bite marks are usually obvious with minor bleeding from the puncture wounds, but occasionally the teeth marks may be absent.[255] Neurologic toxicity (i.e., blockade of neuromuscular transmission) results from the competitive, nondepolarizing blockade of nicotinic acetylcholine receptors at the neuromuscular junction similar to cobra neurotoxin and α-bungarotoxin (Formosan krait).[256] Elapid venom typically affects peripheral nerves rather than the central nervous system because the neurotoxins do not penetrate the blood–brain barrier. The acetylcholinesterase activity of coral snake venom is limited compared with some other elapid venoms. Envenomations by Eastern and Texas coral snakes produce more severe neurotoxic effects compared with the Arizona coral snake.[257] Elapid venoms contain few proteases compared with viperid venoms.

CLINICAL RESPONSE

Approximately 25–40% of the bites from coral snakes do not cause envenomation, but even following envenomation local signs of envenomation may be absent.[258] Although minor pain may occur at the bite site, local edema, erythema, and pain are usually minimal.[259] Typically, the onset of neurological effects occurs several hours after envenomation with the development of ptosis, but envenomation by the Eastern coral snake may cause neurological symptoms up to 12 hours after envenomation.[258] The clinical features of coral snake envenomation include cranial nerve palsies (dysphagia, drooling, dysarthria, ptosis) followed by limb weakness, loss of deep tendon reflexes, dyspnea, and respiratory paralysis. Euphoria, lightheadedness, and drowsiness are early symptoms of envenomation. Fasciculations, tremor, muscle weakness, increased salivation, nausea, and vomiting also occur. Neurological symptoms may progress to bulbar palsy and respiratory depression within 10 hours after envenomation, manifested by diplopia, dysarthria, drooling, slurred speech, ptosis, dysphagia, and dyspnea. Seizures are uncommon, but convulsions may occur, particularly in small children.[260] Signs of bulbar palsy usually, but not always, precede the development of respiratory failure. Envenomation by the Arizona coral snake causes headache, blurred vision, abdominal pain, and gait disturbances, but these neurological symptoms are not usually as severe as envenomations by the Eastern coral snakes.[261] In a retrospective case series of bites by the Texas coral snake (*Micrurus tener*), systemic symptoms were reported in only 7% of the cases and none of the symptoms were life-threatening.[262] Over one-half of these patients received coral snake antivenom. Following severe envenomations, case reports suggest that muscle weakness and paresthesias may persist for several weeks, particularly in patients without antivenom therapy. Cardiac and hematological effects are usually minimal following envenomation by coral snakes, and laboratory abnormalities are typically none to minimal.

TREATMENT

First-Aid

The patient should be transported to an emergency medical facility as soon as possible. Capture of the snake is helpful to identify the species, provided the snake can be captured safely with minimal risk to the rescuer. *Ineffective* first aid measures include giving ethanol to the victim, the use of tourniquets, and the utilization of incision and drainage. The wound should be washed with soap and water as soon as possible to remove any remaining venom on the skin.

Antivenom

North American coral snake (*Micrurus fulvius* L.) antivenom (Wyeth Pharmaceuticals, Collegeville, PA) is an effective, lyophilized, equine-based antivenom produced by the immunization of healthy horses with Northern coral snake (*Micrurus fulvius* L.) venom. Indications for this antivenom include the following clinical situations: 1) confirmed Texas and Eastern coral snake envenomations (i.e., bite marks or prolonged contact with coral snake) with neurological symptoms, and 2) patients with suspected coral snakebite and early neurological symptoms of envenomation (lethargy, muscle weakness, ptosis, dyspnea). Results of cross-neutralizations tests indicate that this antivenom is not effective for envenomations by the Western or Sonoran coral snake (*Micru-*

roides euryxanthus Kennicott). Although the clinical effects of some coral snakebites produce only local reactions, there are few data on the clinical course of these envenomations without antivenom. The antivenom should be administered before the development of the bulbar palsy (i.e., within 4–8 hours of the bite) because the antivenom does not reverse preexisting neurotoxicity.[263,264] Precautions for the development of hypersensitivity responses include intravenous access, slow initial titration of the antivenom, and the immediate availability of epinephrine and H_1- and H_2-histamine receptor blockers, similar to the administration of other equine antivenom (e.g., antivenin [Crotalidae] polyvalent). The efficacy of skin testing is controversial because of the false-positive and false-negative tests and the delay in administering antivenom. Skin testing involves the intradermal injection of 0.02–0.3 mL of a 1:10 dilution of normal horse serum or antivenom along with an intradermal control test on the opposite extremity with a similar amount of sodium chloride. A positive test occurs when a wheal develops within 5–30 minutes with or without pseudopodia and surrounding erythema.

Approximately 5–10 vials of antivenom should be administered in 250–500 mL of normal saline over 1–2 hours with the actual dose depending on the potential severity of the envenomation. The initial infusion should be slow (3–5 mL/h over the first 5 minutes) with the infusion increased in the absence of adverse effects.[265] If neurological signs worsen, an infusion of an additional 3–5 vials of antivenom should be given. The administration of coral snake antivenom does not rapidly reverse the clinical features of coral snake envenomation; therefore, the patient should be monitored closely for signs of respiratory failure. Urticaria is the most common adverse effect during the administration of coral snake antivenom, and the urticaria usually resolves after slowing of the antivenom infusion and the use of diphenhydramine.[258] The administration of coral snake antivenom to envenomated patients with a positive skin test requires careful evaluation of the potential seriousness of the envenomation and the benefits of early treatment. Severely envenomated patients have survived without antivenom following the use of supportive care with early elective intubation.

Experimental studies suggest that Fab-based ovine antivenom preparations are less immunogenic than the Wyeth antivenom,[266] but the ovine preparation is not commercially available. Rodent studies suggest that the use of Mexican *Micrurus* antivenom (Instituto Bioclon, Mexico City, Mexico) and Tiger Snake (*Notechis scutatus*) antivenom (CSL Limited, Parkville, Victoria, Australia) reduces lethality in mice envenomated by the Northern coral snake (*Micrurus fulvius* L.).[267] However, there are few human data on the clinical effectiveness of these antivenoms for envenomations by US coral snakes. In general, immunochemical assays suggest that antivenoms demonstrate greater reactivity toward venoms from geographically related snakes.[268] There is no commercial antivenom available for the Arizona or Sonoran coral snake (*Micruroides euryxanthus*), but symptoms following envenomation from snakes in this genus are usually mild. In the US, an antivenin index is available from the American Zoo and Aquarium Association (301-562-0777) or the American Association of Poison Control Centers (800-222-1222). Otherwise, treatment is supportive.

Supplemental Care

Respiratory failure is the major life-threatening complication of coral snake envenomation. All victims of coral snake envenomations should be carefully observed for at least 24–48 hours for signs of bulbar palsy (ptosis, diplopia on upward gaze, dysphagia) and impending respiratory failure. Rhabdomyolysis may develop in patients with severe envenomation. Serum sickness may develop 7–14 days after the administration of antivenom similar to that following administration of antivenin (Crotalidae) polyvalent, although the incidence is lower (approximately 10%). Treatment of serum sickness is similar (i.e., corticosteroids) to the regimen outlined for serum sickness following crotalid envenomations.

CENTRAL AND SOUTH AMERICA

The main venomous representatives of the family Elapidae in Central and South America are the coral snakes (*Micrurus* spp., *Micruroides* spp., *Leptomicrurus* spp.). These small snakes are distributed throughout tropical and subtropical Central and South America, but bites are relatively rare because of their reclusive habits. Their wide-ranging locations include seclusion in rocky desert environments and submersion under leaves in forest habitats. Many snake species mimic the red-yellow-black coloration of most coral snakes. However, some coral snakes are bicolored, and differentiation of coral snakes from nonvenomous snakes may be difficult for Central and South American coral snakes. The venom apparatus of coral snakes and other elapids are similar with two fixed, grooved, front fangs up to a few millimeters in length. Bite marks are usually present, but the bite marks from these small fangs frequently are difficult to detect. The venom of coral snakes contains neurotoxins that block neuromuscular transmission at the postsynaptic membrane. These venoms are low in tissue-toxic enzymes (e.g., proteinases, phosphodiesterases, and 5′-nucleotidase activity); consequently, local tissue damage is minimal.

The rapid onset of local pain may develop following coral snake envenomations.[269] Localized swelling usually is present, but necrosis does not usually occur. Neurological symptoms of envenomation are similar to other elapids, manifest by paresthesias, weakness, ptosis, ophthalmoplegia, dysphagia, and dysarthria that progresses to respiratory failure. These symptoms may be delayed up to 12 hours. Other symptoms include metallic taste, myalgias, fasciculations, lightheadedness, nausea, and vomiting. First aid involves rest, reassurance, sitting-up to minimize respiratory difficulty, and rapid transport to medical facilities for the intravenous administration of monovalent antivenom. Treatment is similar to the treatment of envenomation by North American coral snakes. All patients should be observed for at least 24 hours.

MIDDLE AND NEAR EAST/AFRICA

Geographically, the Middle East includes Asian Turkey, Lebanon, Syria, Jordan, Israel, Palestine, Saudi Arabia and the Arabian Peninsula, Iraq, Iran, and the former Southern Asiatic Soviet Republics. This region is predominantly arid, and vipers are the most common cause of serious snakebites in these areas. Elapids are present in restricted ranges, primarily in areas close to northern Africa. Venomous members of the family Elapidae in the Middle East include the Egyptian cobra (*Naja haje*) and the desert black snake or desert cobra (*Walterinnesia aegyptia*). Spitting cobras include *Naja nigricollis*, *N. mossambica*, *N. pallida*, *N. katiensis*, and *Hemachatus haemachatus*. Mambas (*Dendroaspis* spp.) and cobras (*Naja* spp.) are the most important venomous elapid species in Africa (see Table 176.13). Most of these snakes exhibit warning sounds (e.g., cobras lift their head and neck). Although some cobras (*Naja melanoleuca*, *Hemachatus haemachatus*) are nocturnal, most elapids are diurnal and active at dawn. Mambas are predominately arboreal and diurnal. Nonspitting cobras and mambas produce highly potent neurotoxic venoms that act on postsynaptic membranes of the motor nerve endplates similar to curare. The venom of the desert cobra (*Walterinnesia aegyptia*) contains at least three neurotoxins that block neuromuscular transmission postsynaptically.[270] Most elapid venoms contain acetylcholinesterases including the desert cobra.[271] Mamba venom also contains potent neurotoxins called dendrotoxins that inhibit voltage-gated KV1.1, KV1.2, and KV 1.6 potassium channels and block muscarinic receptors.[272] Fasciculin is a peptide that inhibits acetylcholinesterase and produces muscle fasciculations.

Clinical features of envenomation by nonspitting cobras and mambas involve primarily peripheral neuro-

TABLE 176.13. Important African, Middle Eastern, and Near Eastern Venomous Elapids[a]

Species	Common Name	Description	Distribution	Habitat
Naja species	Cobra	Large, heavy bodied up to 1–3 m (~3–10 ft) with characteristic defensive posture of forming round hood from spreading cervical ribs	*N. haje*: Africa to Middle East *N. melanoleuca*: West to South African forests *N. nigricollis*: South of Sahara *N. mossambica*: Southern Africa *N. pallida*: Southern Africa *N. nivea*: Western South Africa *N. katiensis*: West Africa *Hemachatus haemachatus*: South Africa *Walterinnesia aegyptia*: North Africa to Middle East/Iran	Terrestrial habitat of rain forests, swamps, river beds to dry savannah and mountain areas
Dendroaspis species	Mamba	Agile, slender, long snake up to 4 m (~13 ft) that retreats quickly when disturbed	*D. angusticeps*: Coastal Eastern to Southern Africa *D. jamesoni*: Central Africa *D. polylepis*: Eastern and South Africa *D. viridis*: West Africa	Black mambas are primarily terrestrial (wooded savannah, rocky hills, riverine forests), but they may climb trees. Green mambas are almost exclusively arboreal.

[a]Source: Adapted from Ref 2.

logical symptoms including slurred speech, dysphagia, ptosis, and ophthalmoplegia. Symptoms usually develop within 1 hour of envenomation and may progress to paralysis and to respiratory failure. Envenomation by mambas cause the rapid onset of perioral and generalized paresthesias, nausea, vomiting, hypersalivation, and hyperacusis followed by generalized weakness, ptosis, dysphagia, diaphoresis, goose flesh, and drowsiness. Local tissue necrosis is rare, except following envenomation by spitting cobras, such as *Naja nigricollis* (black-necked spitting cobra)[273] and *Naja mossambica* (spitting cobra).[274] Intraocular instillation of spitting cobra venom causes severe conjunctivitis, blepharospasm, local edema, and conjunctival congestion. Complications include corneal ulcers, anterior uveitis, and blindness.[275]

First-aid treatment includes reassurance, rest, and transport to a medical facility as soon as possible. There is no direct scientific data to support the use of compression dressings. Copious irrigation of the eyes is indicated as soon as possible after ocular contact with spitting cobra venom. The eye should be examined by slit lamp for evidence of corneal injury, and treated with topical antibiotics. There are no data to support the intraocular instillation of antivenom. Treatment primarily involves supportive care and the intravenous administration of antivenom as soon as possible after neurological symptoms develop. The erratic and incomplete administration of antivenom following intramuscular injection severely limits the effectiveness of the intramuscular route.[276] The South African polyvalent elapid antivenom covers all African species of elapids.

ASIA

EXPOSURE

Cobras and kraits are the main representatives of venomous snakes in Asia. There are about 10 species of Asian cobras that now represent separate species rather than subspecies of *Naja naja*.[277] These Asian cobras include *Naja atra* Cantor (South China, North Vietnam, Central Taiwan), *Naja kaouthia* Lesson (Thailand, Burma, West Malaysia, Indochina, Bengal), *Naja naja* L. (India, Pakistan, Sri Lanka, Nepal, Bangladesh, South Vietnam), *Naja oxiana* Eichwald (Middle East, North Pakistan, *Naja philippinensis* Taylor (Philippines), *Naja sagittifera* Wall (Andaman Islands), *Naja siamensis* Laurenti (Thailand, Indochina), *Naja sputatrix* Boie (Java, Sunda Islands), *Naja sumatrana* Müller (Malayan Peninsula, Sumatra, Borneo), and *Ophiophagus hannah* Cantor (King Cobra of India, South China, Southeast Asia). These large, wide-bodied snakes grow up to 4m (~13ft) in length (King Cobra), and they spread their cervical ribs, inflate their bodies, and hiss when disturbed. Coloration varies from light-brown to olive-black with most a monochromatic pattern. Several Asian cobras can discharge their venom several meters in the air toward their victims including the *Naja philippinensis*, *Naja siamensis*, *Naja sputatrix*, and *Naja sumatrana*. Figure 176.11 displays a specimen of the Siamese Spitting Cobra (*Naja siamensis* Laurenti). Occasionally, several other cobra species (*N. atra*, *N. samarensis*) display this behavior. Eastern cobras are primarily nocturnal, and evenomation may occur during the night when these snakes are searching for prey (e.g., rats, mice).

Kraits are slender snakes about 1m (~3ft) in length with white or yellow stripes on a black background. Some species are all black. There are approximately 13 species distributed throughout Asia with the exception of the Philippines including *Bungarus caeruleus* Schneider (Indian subcontinent), *B. candidus* L. (Cambodia, Indonesia, Malaysia, Thailand), *B. fasciatus* Schneider (Southeast Asia), and *B. multicinctus* Blyth (Taiwan, South China). Asian kraits are nocturnal, and bites often occur at night without the perception of the bite by the victim after the snake enters a house in search of prey (mice, rats, lizards).[278] They are terrestrial snakes found in open forests and bush as well as agricultural areas (e.g., rice fields).

VENOM

These snakes have anterior fangs that are relatively small and fixed. The venoms of Asian cobras (e.g., cobratoxin from *N. naja atra*) are similar to African cobra venoms consisting of curare-like neurotoxins that block neuromuscular transmission at the postsynaptic nicotine

FIGURE 176.11. Siamese Spitting Cobra (*Naja siamensis* Laurenti). © 1996 Rune Midtgaard. See color insert.

receptors in motor endplates.[279] Both cobra and krait venoms also contain cardiotoxic and cytotoxic components that lyze cell membranes.[280] Asian krait venom also contains curare-like neurotoxins (α-bungarotoxin), which bind postsynaptically to the α-subunit of the nicotinic acetylcholine receptor of the neuromuscular junction, and phospholipase A_2-containing β-bungarotoxin that acts presynaptically.[281] The latter bungarotoxin is the main constituent of krait venom. κ-Bungarotoxins are unique components of krait venom that are structurally similar to α-bungarotoxins and also bind to nicotinic acetylcholine receptors. However, these unique components are minor constituents of krait venom.[282] The exact mechanism of action of bungarotoxins is unclear.[283]

CLINICAL RESPONSE

Cobra envenomations typically produce signs of severe neurotoxicity that may rapidly progress to respiratory failure within 30 minutes of envenomation, particularly following envenomation by the King Cobra (*Ophiophagus hannah*).[284] A case report describes the onset of cranial neuropathy, respiratory failure, and coagulopathy 10 hours after envenomation by an Asiatic cobra (*N. kaouthia*).[285] Neurological effects following cobra envenomation include ptosis, blurred or double vision, dysphagia, drooling, dysarthria, loss of deep tendon reflexes, flaccid paralysis, respiratory failure, and coma. Although most cobra bites cause minimal tissue damage involving the local skin and subcutaneous tissues, bites from some Asian cobras (e.g., *Naja naja, Naja atra*), including spitting cobras, frequently can cause severe local pain, edema, vesiculation, and necrosis in addition to neurological symptoms.[286,287] In a study of 85 patients presenting to provincial hospitals in Thailand with envenomations by *Naja kaouthia* (Thai cobra), 27% of the patients required minor surgical debridement of the wounds and one patient required amputation of his thumb.[288]

Envenomation by Asian kraits (e.g., *Bungarus caeruleus, Bungarus multicinctus*) produces dyspnea, dysphagia, ptosis, blurred vision, generalized neuromuscular weakness, abdominal pain, vomiting, chest pain, myalgias, alteration of consciousness, prolonged respiratory failure, and occasionally a delayed neuropathy.[289] Neurological symptoms typically develop within 1–6 after envenomation, and local reactions (i.e., swelling, erythema) to the bites are generally minimal or mild.[290,291] Typically, the local tissue damage from krait venom is less than Asian cobra venom. Cardiotoxicity is unusual following krait envenomations, but rare case reports associate cardiogenic pulmonary edema with severe krait envenomation.[292] Coagulation profiles usually remain normal following cobra and krait envenomations.

TREATMENT

First-aid treatment involves immediate transport to a medical facility for intravenous administration of antivenom along with rest and reassurance. Although the use of a proximal tourniquet may delay the onset of symptoms, rapid deterioration to respiratory arrest may occur after removal of the tourniquet.[293] Consequently, once applied the tourniquet should be removed only when facilities are available to provide rapid intubation and respiratory support. Most cases of cobra and krait envenomation require the use of antivenom. Because of the high variability in the composition of cobra and krait venom, clinically significant variation occurs in the efficacy of antivenom produced from venom of different species. Although case reports describe the successful use of polyvalent snake venom (e.g., 10 vials, South African Institute of Medical Research [Johannesburg, South Africa] polyvalent snake antivenom) for Asian cobra envenomation, polyvalent cobra antivenom may not effectively treat envenomation of all Asian cobra species. A mono-specific Thai Red Cross Society King Cobra antivenom effectively reverses the neurotoxicity associated with King Cobra envenomation in doses of 15–20 vials.[294] Although current antivenom prevents further deterioration, the polyvalent antivenom does not usually reverse existing respiratory insufficiency or the delayed neuropathy.[295] Cobra antivenom is not efficacious for krait envenomations. High doses of polyvalent antisnake venom (Haffkine Bio-pharmaceutical, Acharya Donde Marg, Parel, Mumbai, India) may be necessary to treat the neurological symptoms associated with krait envenomations.[296] An experimental enzyme-linked immunosorbent assay (ELISA) has been developed to detect Taiwan cobra (*Naja atra*) venom in the serum of snakebite victims.[297] Serum concentrations of this venom remain elevated for at least 24–48 hours after envenomation unless the administration of antivenom neutralizes the venom.

AUSTRALIA

EXPOSURE

Geographical Distribution

All of the venomous snakes in Australia belong to the family Elapidae, and a majority of snakes in Australia are venomous. The distribution of snakes in New Guinea including Papua New Guinea and the Indonesian portion (Irian Jaya) is similar to Australia. The major groups of elapids in Australia include the following species: 1) tiger snakes (*Austrelaps*, *Hoplocephalus*, *Notechis*, *Tropidechis*), 2) brown snakes (*Pseudonaja*), 3) Taipan (*Oxyuranus*), black/mulga snakes (*Pseudechis*), 4) small-eyed snakes (*Micropechis*), and 5) death adders (*Acanthophis*). Death adders are the most widely distributed venomous elapids ranging from continental Australia to the Indonesian islands of Seram and Tanimbar. The common brown snake (*Pseudonaja textilis*) and other brown snakes (*P. inframacula*, *P. nuchalis*, *P. affinis*) are the most common venomous snakes in South Australia. Envenomations by snakes from the genus *Pseudonaja* are the leading cause of mortality from snakebites in Australia.[298] Other Elapidae species associated with Australian fatalities include the tiger snake (*Notechis scutatus*), taipan (*Oxyuranus scutellatus*), and the common death adder (*Acanthophis antarcticus*). Figure 176.12 displays a specimen of the common tiger snake (*Notechis scutatus* Peters). Snakes from the genus *Pseudechis* (e.g., *P. porphyriacus*, red-bellied black snake) are also widely distributed venomous snakes in Australia.[299,300] *Pseudechis australis* causes the largest number of snake envenomation in Australia.[301] In general, these snakes produce little local tissue damage. Figure 176.13 displays a specimen of the King Brown Snake (*Pseudechis australis* Gray). The death adder (*Acanthophis* spp.) inhabits most of Australia and a number of eastern Indonesian islands, whereas the taipan (*Oxyuranus* spp.) lives in the Northern coast and some interior sections.

Habitats/Characteristics

Australian elapids occur in a wide variety of habitats including some areas near population centers. Table 176.14 lists the habitat and physical characteristics of important venomous elapids in Australia.

VENOM

The fangs of Australian elapids are fixed and relatively nonmobile with lengths up to 13 mm (~0.5 in.). The amount of venom ejected with each bite is highly variable, and therefore the degree of envenomation is also highly variable. The Australian small-scaled snake (*Oxyuranus microlepidotus* McCoy) is an Australian Elapidae species that, potentially, is the most poisonous snake in the world on the basis of an LD_{50} in mice of 0.01 mg/kg.[302] Experimental studies in rodents suggest that the venom from this Elapidae species contains a component that relaxes vascular smooth muscle and produces hypotension as well as a component that results in the synthesis of arachidonic acid metabolites.[303] Most venomous Australian elapids (e.g., tiger snakes, taipans, rough-scaled snakes) produce venom that contains neurotoxins that block postsynaptic membranes and the presynaptic release of acetylcholine at the neuromuscular junction.[304,305] The latter neurotoxins (e.g., notexin, taipoxin, textilotoxin) contain phospholipases A_2. Notexin produces rhabdomyolysis as well as paralysis. This neurotoxin causes prolonged paralysis by destroying the terminal bulbs and irreversibly inhibiting the release of acetylcholine.[306]

Venom from the viper-like common death adder (*Acanthophis antarcticus*) contains at least five postsynaptic neurotoxins and four phospholipase A_2 components. Some, but not all species of *Acanthophis* possess myotoxic venom. An *in vitro* study of *Acanthophis* species from Australia and neighboring islands demonstrated myotoxicity from venoms extracted from species

FIGURE 176.12. Common tiger snake (*Notechis scutatus* Peters). © 2002 Rune Midtgaard. See color insert.

FIGURE 176.13. King brown snake (*Pseudechis australis* Gray). © 1997 Rune Midtgaard. See color insert.

TABLE 176.14. Habitat and Characteristics of Important Australian Venomous Elapids[a]

Species	Common Name	Description	Distribution	Habitat
Notechis scutatus Peters	Common tiger snake	Wide-bodied snake up to 1.2m (~4ft) in length with brown to black background and highly variable yellow bands	Southeastern Australia near densely populated areas	River banks, moist forests
Notechis ater Krefft	Black tiger snake	Wide-bodied snake up to 1.2m (~4ft) in length with brown to black background and highly variable black bands	Southeastern Australia near densely populated areas	River banks, moist forests
Acanthophis spp.	Death adder	Broad, triangular head with wide body similar to vipers; background color ranges from light-brown to red and dark-gray with irregular strips	Mainland Australia, New Guinea, Molucca Islands	Sand or soil near shrubs and trees—primarily nocturnal
Pseudechis australis Gray	Mulga snake	Slender snake up to 2–3m (~6–10ft) in length with light brown background	Southern Australia; other species throughout Australia and New Guinea	Wide variety of habitats including deserts and tropical forests
Pseudonaja textilis Duméril, Bibron & Duméril	Brown snake	Slender snake up to 1.5–2m (~5–6ft) in length with brown background; other species vary between yellow, orange, brown, olive, and black backgrounds, occasionally with stripes or spots	Eastern Australia; other species range over all of Australia and Eastern New Guinea	Species inhabit all dry and wet areas
Oxyuranus scutellatus Peters	Taipan	Large, slender snake up to 2m (~6ft) with light-brown, yellow or dark-brown background	North and North East Australia and new Guinea	Wide habitat including tropical rain forests, open savanna, woodlands, and dry areas

[a]Source: Adapted from Ref 2.

of *A. praelongus* Ramsey (northern death adder), *A. rugosus* Loveridge, and *A. wellsi* Hoser (black head death adder), but not from venoms extracted from *A. hawkei* Wells & Wellington (Barkly Tableland death adder), *A. pyrrhus* Boulenger (desert death adder), and regional variants of *A. antarcticus* Shaw.[307] *Pseudechis* venoms contain phospholipase A_2, pre- and postsynaptic neurotoxins, platelet inhibitors, myotoxins, and variety of factors with pro- and anticoagulant, hemorrhagic, and necrotic activities.

Venoms from the brown snake and taipan contain enzymes that interfere with hemostasis and produce clinically similar envenomations. These venoms probably act at the site of the prothrombin complex rather than thrombin (e.g., crotalines) or fibrinogen.[308] The venom of brown snakes (*Pseudonaja* spp.) contains a procoagulant, textarin, which converts prothrombin to thrombin.[309] This procoagulant is a serine protease with a sialic acid component that contributes to the coagulant effects.[310] There is substantial interspecies variability with the venom of some species requiring factor V for the procoagulant activity, whereas other species do not. This venom can cause severe defibrination coagulopathy and secondary renal damage. Brown snake venom rarely causes paralysis and never causes myolysis. The black tiger snake (*Notechis ater*) and Stephen's banded snake (*Hoplocephalus stephensi* Krefft) also produce a procoagulant that causes a defibrination-type coagulopathy after envenomation. This coagulant resembles factor Xa and requires factor V for procoagulant activity.[311] Variation in the composition of the venom, including the procoagulant, during the year adds to the variability in the clinical response to an envenomation.[312] Most Australian elapid venoms do not contain hemorrhagic factors that cause local tissue edema and necrosis. However, the venom of the brown snake (*Pseudonaja textilis*) does contain at least two group 1B phospholipase A_2 enzymes.[313]

HISTOPATHOLOGY AND PATHOPHYSIOLOGY

Postmortem findings following death from elapid snake envenomation are usually unremarkable other than evidence of coagulation disorders.

CLINICAL RESPONSE

Envenomations by Australian elapid snakes produce a wide variation of clinical effects ranging from local swelling (black-bellied swamp snake, *Hemiaspis signata* Jan)[314] to myoglobinuria, paralysis (Rough-scaled snake, *Tropidechis carinatus* Krefft),[315,316] and sudden death (brown snake, *Pseudonaja* spp.).[317] In general, Elapidae venoms produce only minor local damage and swelling at the bite site with the exception of the mulga snake (*Pseudechis australis*).[318] Initial symptoms include lightheadedness, diaphoresis, headache, nausea, vomiting, abdominal pain, and painful regional lymph nodes. Typical neurological symptoms include slurred speech, ophthalmoplegia, ptosis, dysphagia, and paralysis of respiratory muscles. Systemic symptoms following envenomation by the red-bellied black snake (*Pseudechis porphyriacus*) and the spotted black snake (*Pseudechis* guttatus) are typically mild (e.g., headache, blurred vision, nausea, vomiting, diarrhea, upper abdominal cramping, diaphoresis).[319] The brown snake (*Pseudonaja* spp.), black tiger snake (*Notechis ater*), and the Stephen's banded snake (*Hoplocephalus stephensii* Krefft) possess procoagulants in their venoms that cause a defibrination-type coagulopathy and hemorrhage. Serious complications of Australian elapid envenomations include hyperkalemia, intracranial hemorrhage, convulsions, rhabdomyolysis, myalgias, acute renal failure, hemorrhagic shock, cardiovascular failure, and respiratory arrest.[320,321] Envenomation by the death adder produces neurotoxicity (ptosis, ophthalmoplegia, fixed dilated pupils, dysarthria, respiratory failure) with little effect on coagulation or local tissue.[322] The onset of flaccid paralysis after envenomation by death adders can occur 12–18 hours following the bite. Taipan snakes cause a variety of clinical effects including paralysis, complete defibrination coagulopathy, hemorrhage, rhabdomyolysis, and secondary renal failure. Tiger snakes produce a neurotoxin and a myotoxin that causes paralysis, rhabdomyolysis, local edema, and secondary renal failure.[323] The rhabdomyolysis associated with tiger and malga snake envenomations may also be delayed 12–18 hours after the bite. Slight prolongation of the PT and PTT may occur, but these changes usually resolve within 12–18 hours and defibrination does not develop. Some, but not all species of death adders (*Acanthophis* spp.) can produce rhabdomyolysis.[324]

DIAGNOSTIC TESTING

Venom Detection Kit

The venom detection kit issued by Commonwealth Serum Laboratories in Melbourne, Australia is the only commercial diagnostic kit for detection of snake venoms by ELISA methods from urine or from the snakebite site.[325,326] Blood is an unreliable test sample for this kit because of the difficulty interpreting the results.[327] The bite site is the most reliable sample site for the snake venom kits; a urine sample is suitable only when systemic envenomation occurs. This kit detects the venom from the five major venomous snakes in Australia including the common brown snake (*Pseudonaja textilis*), tiger snake (*Notechis scutatus*), death adder (*Acanthophis antarcticus*), king brown snake (*Pseudechis australis*), and taipan (*Oxyuranus scutellatus*). The use of these kits allows the use of specific rather than polyvalent antivenom. The venom detection kits determine the most appropriate antivenom to administer, but this test does not determine the need for antivenom or the recent occurrence of envenomation. Both false-positives (e.g., brown snake) and false-negatives occur, even during severe envenomation.

Abnormalities

Marked defibrination occurs with variable thrombocytopenia following envenomation by brown snakes (*Pseudonaja* spp.), taipans (*Oxyuranus* spp.), or tiger snakes (*Tropidechis* spp., *Hoplocephalus* spp., *Notechis* spp.) that produce procoagulants. These coagulation defects involve the consumption of factor V, factor VIII, fibrinogen, and partial consumption of prothrombin with little effect on other coagulation factors.[328] Fibrin degradation products are markedly elevated with prolongation of the PT and PTT as well as low fibrinogen concentrations. The severity of these coagulation abnormalities is variable with some envenomations resulting in mild changes.[329] In contrast to envenomations by brown or tiger snakes, envenomations by black snakes (e.g., mulga snakes, some *Pseudechis* spp.) produce an anticoagulation syndrome characterized by prolongation of the PT and aPTT times, but normal fibrinogen concentrations and fibrin split products. Rhabdomyolysis may cause hyperkalemia, elevated serum creatinine, and myoglobinuria, as well as increased serum creatine phosphokinase.

TREATMENT

The principal methods of treating Elapidae envenomations is identification of the involved species, determination of the presence of neurological symptoms and signs, close observation of the victim for respiratory failure, administration of antivenom when necessary, and supportive care.

First-Aid

The snakebite victim should be rapidly transported to a medical facility with antivenom resources. In Australia, the Sutherland pressure immobilization first aid technique is the first aid measure of choice despite the lack of clinical trials.[330] This technique involves the pressure immobilization by crepe bandage and splint. A broad bandage is firmly applied over the bite site similar to wrapping a sprained ankle. The bandage covers the entire limb including toes or fingers. The extremity is then immobilized. In Australia, the wound is not cleansed in the field because of interference with the identification of the snake by ELISA testing of dermal traces of venom. Tight tourniquets should be avoided, and the use of constricting bands may increase damage from snake venoms that produce local damage. Once applied, the dressing should not be removed until the patient is hospitalized with intravenous access and antivenom ready to administer. Victims should rest, seated in an upright position with back support to maximize diaphragmatic function. They should be observed closely for respiratory failure. Walking should be minimized to reduce the flow of the venom through the lymphatic system.

Antivenom

Most Australian elapid envenomations with systemic symptoms (e.g., ptosis, ophthalmoplegia, prolonged prothrombin time, rhabdomyolysis, myoglobinuria) require the intravenous administration of antivenom. Identification of the species causing the envenomation is important to the administration of the proper antivenom. The lack of pain and bite marks suggests envenomation by brown snakes. The presence of obvious redness and bruising along with marked swelling within 3 hours suggests the mulga, red-bellied black, or yellow-faced whip snakes, whereas the lack of swelling in the presence of obvious bite marks suggests the tiger or the rough-scaled snakes. Figure 176.14 outlines the use of laboratory data and clinical presentation to determine venomous Australian species of Elapidae snakes, and Table 176.15 lists the appropriate antivenom for these species. The administration of antivenom should be guided by the clinical presentation and laboratory data (i.e., swab of bite site for venom detection kit).[331] Because of the false positive tests associated with the use of the venom detection kits, a positive result on the venom detection kit *alone* is not an indication to administer antivenom and a negative result does not necessarily exclude envenomation.[332] The venom detection kit is designed to help choose the appropriate antivenom rather than confirm snake envenomation. However, *in vitro* studies suggest that there is some cross-neutralization between commercial antivenoms (e.g.,

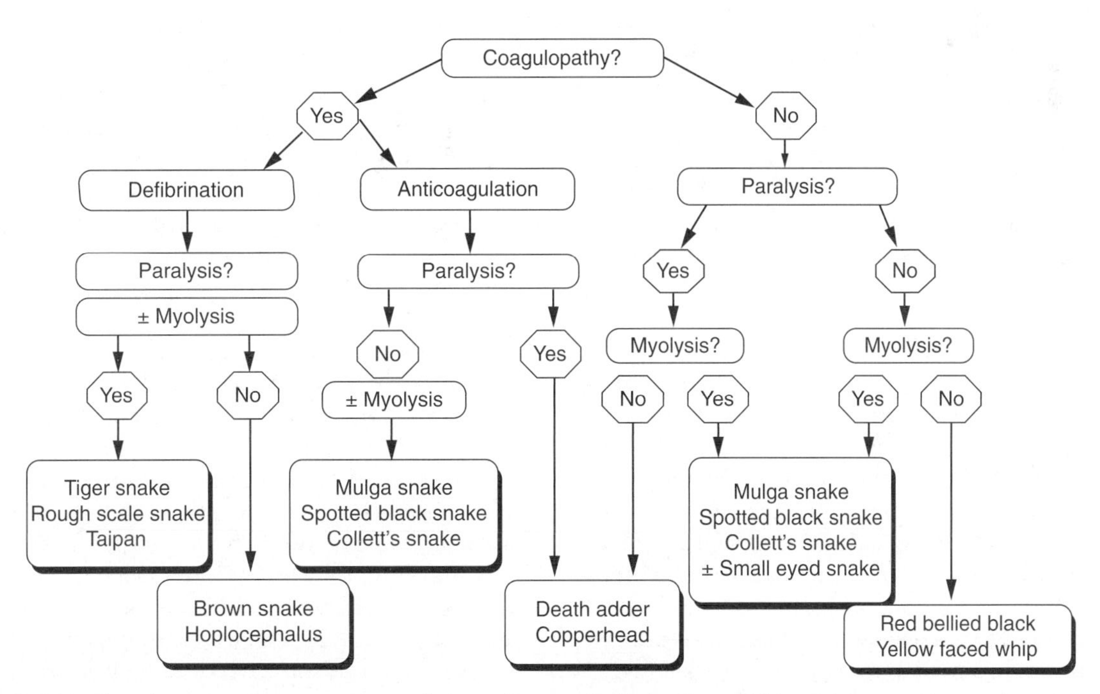

FIGURE 176.14. Flowchart for the assessment of Australian snakebites. From White J. Envenoming and antivenom use in Australia. Toxicon 1998;36:14902. Reprinted with permission from Elsevier.

TABLE 176.15. Major Venomous Snakes of Australia[a]

Group Scientific name	Common Name	Type of Antivenom[b]
Acanthophis antarcticus Shaw	Common death adder	CSL death adder antivenom
Acanthophis pyrrhus Boulenger	Desert death adder	CSL death adder antivenom
Acanthophis praelongus Ramsey	Northern death adder	CSL death adder antivenom
Austrelaps labialis Jan	Pygmy copperhead	CSL tiger snake antivenom
Austrelaps ramsayi Krefft	Highland copperhead	CSL tiger snake antivenom
Austrelaps superbus Günther	Lowland copperhead	CSL tiger snake antivenom
Hoplocephalus bitorquatus Jan	Pale-headed snake	CSL tiger snake antivenom
Hoplocephalus bungaroides Schlegel	Broad headed snake	CSL tiger snake antivenom
Hoplocephalus stephensii Krefft	Stephen's banded snake	CSL tiger snake antivenom
Notechis scutatus Peters	Common tiger snake	CSL tiger snake antivenom
Notechis scutatus subsp. *occidentalis* Glauert	Western tiger snake	CSL tiger snake antivenom
Oxyuranus scutellatus Peters	Common taipan	CSL taipan antivenom
Oxyuranus microlepidotus McCoy	Inland taipan	CSL taipan antivenom
Pseudechis australis Gray	Mulga or King brown snake	CSL black snake antivenom
Pseudechis butleri Smith	Butler's mulga snake	CSL black snake antivenom
Pseudechis colletti Boulenger	Collett's snake	CSL black snake antivenom
Pseudechis guttatus De Vis	spotted or blue bellied black snake	CSL tiger snake antivenom
Pseudechis porphyriacus Shaw	Red bellied black snake	CSL tiger snake antivenom
Pseudonaja affinis Günther	Dugite	CSL brown snake antivenom
Pseudonaja guttata Parker	Speckled brown snake	CSL brown snake antivenom
Pseudonaja inframacula Waite	Peninsula brown snake	CSL brown snake antivenom
Pseudonaja ingrami Boulenger	Ingram's brown snake	CSL brown snake antivenom
Pseudonaja nuchalis Günther	Western brown snake or gwardar	CSL brown snake antivenom
Pseudonaja textilis Duméril, Bibron & Duméril	Eastern brown snake	CSL brown snake antivenom
Rhinoplocephalus nigrescens Günther	Eastern small-eyed snake	CSL tiger snake antivenom
Tropidechis carinatus Krefft	Rough scaled snake	CSL tiger snake antivenom

[a]Source: Adapted from Ref 331.
[b]Antivenom from Commonwealth Serum Laboratories, Ltd (Melbourne, Australia).

brown snake and tiger snake antivenoms).[333] For patients with clear signs of system envenomation, urine may be sampled; however, blood samples are usually unreliable.

The presence of systemic signs of envenomation is the main criteria for the administration of antivenom including evidence of paralysis (ptosis, weakness, ophthalmoplegia), altered consciousness, internal hemorrhage, hypotension, myoglobinuria, or nonclotting blood. The antivenom is diluted 1:10, if possible. Infusion begins slowly to check for an anaphylactic response, and then, in the absence of a response, the dose is infused over 15–20 min. Repeat dosing of antivenom may be necessary to reverse the consumptive coagulopathy and hypofibrinogenemia associated with brown snake envenomations, particularly if there is no improvement in coagulation factors three hours after initial administration of antivenom. In a case series of four brown snake envenomations, at least 9 vials of antivenom was required to reverse the envenomation-induced coagulopathy.[334] In another case series, 90% of severe brown snake envenomations as defined by nondetectable fibrinogen concentrations (<30 mg/dL) responded to between 1–10 ampules with a range up to 23 ampules.[335] The recommend dose of antivenom for severe brown snake envenomation was 10 ampules, but a recent case series on severe brown snake envenomation suggests that even severe envenomations respond to an initial dose of one to two ampules (1000 U/ampule).[336] Repeat dosing with antivenom may be required to completely neutralize all the antivenom including late release from the envenomation site. Regeneration of fibrinogen concentrations requires at least 3–6 hours after adequate neutralization of the venom. The improvement in the PT and aPTT times is a sensitive marker of the complete neutralization of venom by antivenom, and detectable fibrinogen concentrations add little additional information. In a study of Australian snake envenomations, the mean times from initial antivenom dose to recover of PT < 24 seconds and measurable aPTT were 9.2 hours and 5.2 hours, respectively, compared with 8.8 hours for detectable fibrinogen.[337] There are no clinical data to support the use of premedication or sensitivity testing prior to the administration of antivenom.[338] *In vitro* neurotoxicity studies of black snakes (*Pseudechis* spp.) suggest that tiger snake venom neutralizes the neurotoxic effects of envenomations by *Pseudechis* species, particularly *P. porphyriacus*.

Supplemental Care

Envenomated patients must be observed closely for the development of respiratory failure. Generous fluid replacement is important for patients with myoglobinuria, but the role of urinary alkalinization is unclear. These patients should be evaluated for the development of acute renal failure. Supportive care includes monitoring of coagulation disorders (PT, aPTT, fibrinogen level, fibrin or fibrinogen degradation products, and platelet count), as well as kidney function. The use of cryoprecipitate, fresh-frozen plasma, and red blood cell (RBC) transfusions may be necessary for life-threatening internal bleeding. Although fresh-frozen plasma and cryoprecipitate should never be withheld during life-threatening bleeding, neutralizing doses of antivenom should usually be administered as soon as possible before large doses of fresh-frozen plasma and cryoprecipitate.

COLUBRIDS

Despite the large size of the family Colubridae, this family contains only a few venomous species (e.g., bird snake, African boomslang, keelback). These snakes have posterior fangs, and they rarely bite humans. Secretions from the venom gland of colubrid snakes are highly variable, and serious envenomations are unusual. In 1957, a professional herpetologist died after envenomation by a juvenile boomslang (*Dispholidus typus* Smith) that he thought was only mildly venomous.[339] Many of these severe reactions probably involve an allergic response.[340] Potentially dangerous snakes from the Colubridae family include the African boomslang (*Dispholidus typus* Smith), the African bird or twig snake (*Thelotornis kirtlandii* Hallowell, *Thelotornis capensis* Smith), the East Asian yamakagashi (*Rhabdophis tigrinus* Smith), and the South American colubrid (*Philodryas olfersii* Lichtenstein).[341]

EXPOSURE

Geographic Distribution

This enormous family comprises 1,400–1,700 species divided into a variety of subfamilies compared with approximate 500 species in the families Elapidae and Viperidae. Almost all of the snakes in these latter two families contain specialized venom glands. Table 176.16 lists the location and effects of colubrid species commonly associated with envenomations.

There are approximately 38 genera from the family Colubridae in the United States. Case reports suggest that local tissue reaction may result from envenomations by some colubrid snakes in the following colubrid genera from North and Central America: *Alsophis, Conophis*, *Coniophanes*, *Crisantophis*, *Dryophiops*, *Erythrolamprus*, *Leptodeira*, *Oxybelis*, and *Pliocercus*.[342–344]

Most snakebites in South America result from contact with pit vipers (*Bothrops, Crotalus, Lachesis* spp.) or coral snakes (*Micrurus* spp.) rather than colubrid species.[345] Rear-fanged colubrid species associated with colubrid envenomations in South America include snakes from the following genera: *Clelia, Elapomorphus, Helicops, Liophis, Philodryas, Tachymenis*, and *Thamnodynastes*. These envenomations usually produce only mild local symptoms (pain, edema, ecchymosis).[346,347] Snake envenomations in Chile result only from colubrid species because of the lack of front-fanged venomous snakes in this country. Most Colubrid bites in Europe, the Middle East, and Africa produce variable amounts of local edema and pain.[348] However, envenomations by the African tree snakes (*Dispholidus typus, Thelotornis capensis*) may cause serious coagulation disorders.[349,350]

Habitat/Characteristics

Colubrid species have elliptical pupils, and they lack the loreal pit associated with crotaline species. Envenomations by colubrid snakes are uncommon and almost always result from handling the snakes. They are active hunters that move frequently, rarely bite humans, and avoid human contact unless threatened. Bites on young children from the brown tree snake occurred primarily at night on the extremities.[351]

Venom Apparatus

Most snakes from this family have aglyphous dentition, and therefore these snakes lack the specialized venom-injecting apparatus present in venomous species from the families Viperidae and Elapidae. However, about one-third of the species from the Colubridae family have opisthoglyphous dentition characterized by rear fangs that can deliver toxic saliva during chewing (see Figure 176.15). The efficiency of the venom-delivery is relatively poor. In an experimental study of brown tree

TABLE 176.16. Colubrid Species Commonly Associated with Envenomations

Species	Common Name	Location	Effects
Amplorhinus multimaculatus Smith	Cape many-spotted snake	Marshes of southern Africa	Local pain/swelling
Boiga blandingii Hallowell	Tree snake	Central Africa	Local pain/swelling
Boiga irregularis Merrem	Brown tree snake	Coastal Australia, Papua New Guinea, NW Melanesia	Local, nonspecific effects
Coluber ravergieri[a]	Middle Eastern racer	Middle East	Local pain/swelling
Crotaphopeltis hotamboeia Laurenti	Herald snake	Sub-Saharan Africa	Local pain/swelling
Dipsadoboa aulica Günther	Marbled tree snake	Southern Africa	Local pain/swelling
Dispholidus typus Smith	Boomslang	Senegal and Eritrea south to Cape of South Africa	Minimal pain, rarely severe coagulopathy
Malpolon monspessulanus Hermann	Montpellier snake	Southern Europe, Mediterranean	Local pain/swelling
Pliocercus elapoides Cope	Variegated false coral snake	Mexico, Central America	Local pain/swelling
Philodryas olfersii Lichtenstein	Green snake	Brazil	Local pain/swelling with local anticoagulant effects
Psammophis phillipsi Hallowell	Olive grass racer	Southern Africa	Local pain/swelling
Psammophylax rhombeatus L.	Spotted skaapsteker	South Africa	Local pain/swelling
Rhabdophis subminiatus Schlegel	Red-necked keelback	Asia	Minimal pain, rarely severe coagulopathy
Rhabdophis tigrinus Boie	Yamakagashi	Asia	Minimal pain, rarely severe coagulopathy
Thelotornis capensis Smith	Savannah vine snake (African bird snake), Eastern twig snake	Savannas/ forests of Eastern/ Southern Africa	Minimal pain, rarely severe coagulopathy
Stenorrhina freminvillei Duméril, Bibron & Duméril	Blood snake	Central America, Mexico	Local pain/swelling

[a]Ambiguous synonym for *Hemorrhois nummifer* Reuss.

snakes feeding on mice, only about one-half of the contents of the Duvernoy's gland were released during the feeding, and only approximately one-half of the injected material reached the viscera of the prey.[352] The rest of the material remained on the skin.

About 12% of US colubrid species (e.g., hog-nosed snake) and over 100 genera in the Colubridae family have enlarged rear teeth (fangs) that contain lateral or anterior grooves.[353] Both rear-fanged (i.e., opisthoglyphous) and nonfanged (aglyphous) colubrid snakes possess a specialized salivary gland called Duvernoy's glands, which are located above some rear maxillary teeth and below the orbit.[354] The small size of the fangs and the difficulty injecting large amounts of venom limit the toxicity of colubrid species. In addition, the secretions from these glands are highly variable between species with some species secreting only serous material, whereas other species produce proteins and mucous in these glands.[355] Prolonged contact with the snake (>1 minute) enhances the ability of the snake to inject secretions from the Duvernoy's gland into the victim and to increase the risk of a clinically significant envenomation.

PRINCIPAL TOXINS

There are over 700 potentially venomous colubrid species, but the data on the composition of these venoms are limited. In a study of the secretions from 12 colubrid snakes, most of the venoms demonstrated variable amounts of proteolytic activity with little evidence of hyaluronidase, antithrombin activity or kallikrein-like effects.[356] However, venom from the red-necked keelback (*Rhabdophis subminiatus* Schlegel) and several other colubrid species causes serious coagulation abnormalities as a result of prothrombin activation (factor II) and fibrinolysis secondary to intravascular fibrin deposition.[357,358] This venom produces little consumption of antithrombin III despite the extensive formation of thrombin.[359] Variable amounts of phospholipase A_2 (*Trimorphodon biscutatus* Duméril, Bibron & Duméril), proteases, hemorrhagic metalloproteinase (*Dispholidus typus*), myotoxins (*Philodryas olfersii*), and phosphodiesterases are also present in the Duvernoy's gland secretions from some colubrid species.

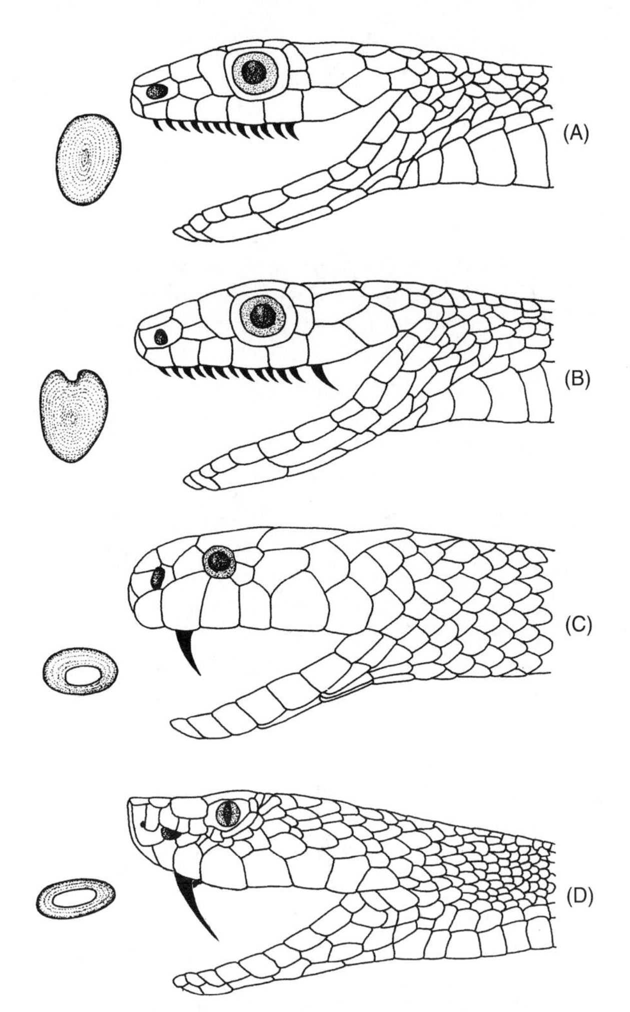

FIGURE 176.15. Dentition of various snake families. (A) Colubridae—aglyphous (ungrooved teeth). (B) Colubridae—opisthoglyphous (posterior grooved teeth). (C) Elapidae—proteroglyphous (front fangs with almost closed groove). (D) Viperidae—solenoglyphous (front, hallow teeth). Adapted from Mebs D. Venomous and Poisonous Animals. A Handbook for Biologists, Toxicologists, Toxinologists, Physicians and Pharmacists. Stuttgart: Medpharm Scientific Publishers, 2002:242.

CLINICAL RESPONSE

Local Effects

Most colubrid bites result only in minor abrasions and anxiety without systemic effects. However, bites by a few colubrid species produce local effects including mild pain, ecchymosis, and localized swelling, particularly when the bite is prolonged (e.g., >30 seconds). The clinical effects of envenomation involve edema, local pain, hemorrhagic vesicles, and enlargement of regional lymph nodes.[360–362] Symptoms usually resolve within 24–72 hours without necrosis.[363,364] Occasionally, the swelling progresses to involve the whole extremity, but systemic symptoms (weakness, nausea, diaphoresis, gastrointestinal distress, vomiting) are rare. Case reports indicate that prolonged contact (1–5 minutes) with large wandering garter snakes (*Thamnophis elegans* subsp. *vagrans* Baird & Girard) produce pain, edema, and localized hemorrhage that involve the region around the bite, but systemic symptoms do not usually develop.[365,366] Bites from a number of Central American Colubridae species produce local pain and swelling including snakes belonging to the following genera: *Conophis, Leptodeira, Leptophis, Oxybelis, Philodryas, Stenorrhina, Tachymenis*, and *Trimorphodon*. In a Brazilian study of 43 patients admitted with the diagnosis of *Philodryas olfersii* envenomation, the most common clinical features were local pain (37.2%), swelling (35%), erythema (19%), and ecchymosis (9%).[367] There was no evidence of abnormal coagulability in this case series. However, *in vitro* studies indicate that Duvernoy's gland secretions from *Philodryas patagoniensis* Girard contain metalloproteinases and serine protease capable of hydrolyzing activity fibrinogen and fibrin.[368] In a series of envenomation by the brown tree snake (*Boiga irregularis*) seen in an Emergency Department in Guam, all bites were associated with localized pain, discoloration, and edema.

Systemic Toxicity

Most envenomations by colubrid snakes do not produce severe toxicity or coagulation disorders. However, several colubrid species can produce severe consumption coagulopathies and hemorrhagic diatheses including bites by the African tree snakes (*Dispholidus typus, Thelotornis capensis*), Japanese yamakagashi (*Rhabdophis trigrinus*), and the South Asian red-neck keelback (*Rhabdophis subminiatus*). Although many bites by these colubrid snakes cause minor symptoms, some envenomations produce severe coagulation abnormalities, particularly following prolonged bites. The clinical presentation of the hematological disorder associated with severe envenomations by these colubrid snakes is similar. Laboratory testing in severe envenomations demonstrates hemolysis and a coagulopathy with depletion of fibrinogen, factors V and VIII, markedly prolonged prothrombin and thromboplastin times, elevated fibrin split products, absent fibrinogen, fragmented erythrocytes, and thrombocytopenia that begins within 2–3 hours after envenomation.[369] Within 24 hours of a severe envenomation, extensive ecchymotic lesions develop, particularly near areas of trauma. Serious internal bleeding does not usually develop until 1–4 days after envenomation, and complications include upper intracranial hemorrhage, gastrointestinal bleeding, massive hematuria, hemoglobinuria, and acute renal failure.[370,371] Hematological abnormalities may remain abnormal for 1–4 weeks after envenomation.[372] The bite wound is not usually painful, but frontal headache, weakness, paresthesias, nausea, persistent vomiting, and severe abdominal cramps may develop within 4 hours. In a series of envenomation by the brown tree snake (*Boiga irregularis*) seen in an Emergency Department in Guam, systemic complaints included lethargy, ptosis, and disorientation. Two infants required respiratory support.

TREATMENT

There are very few data on the management of colubrid envenomation. Management primarily involves local wound care. First-aid involves limiting the duration of bite to minimize the possibility of envenomation by placing the body of the snake on the nearest surface and striking its head. Identification of the envenomating snake is important to eliminate the risk of using inappropriate therapy (i.e., antivenom). The use of tourniquets, incisions, suction, or cryotherapy for colubrid bites is inappropriate. The wound should be cleansed with a disinfectant, and the tetanus immune status updated as necessary. There are no data to support the prophylactic use of antibiotics. The bite wound should be reevaluated in 24–48 hours for the presence of secondary infections. Although rare, hemorrhage may develop following envenomation by the red-necked keelback, African tree snakes (*Dispholidus typus*, *Thelotornis capensis*), and Japanese yamakagashi. Supportive care includes monitoring of coagulation disorders and the use of cryoprecipitate, fresh-frozen plasma, and RBC transfusions when life-threatening internal bleeding occurs. Envenomations by these snakes should be observed for 24–48 hours for the development of coagulation disorders. Heparin is not recommended for the treatment of a DIC-like consumptive coagulopathy. There is no commercial antivenom available to treat colubrid envenomations, and the use of colubrid antivenom is experimental because of the lack of adequate clinical data on efficacy. There are no data to support the use of Crotalidae or Elapidae antivenom following envenomation by colubrid species.

References

1. Russell FE. Snake Venom Poisoning. Great Neck, NY: Scholium International Inc, 1983.
2. Mebs D. Venomous and Poisonous Animals. A Handbook for Biologists, Toxicologists, Toxinologists, Physicians and Pharmacists. Stuttgart: Medpharm Scientific Publishers, 2002.
3. Nelson BK. Snake envenomation incidence, clinical presentation and management. Med Toxicol Adverse Drug Exp 1989;4:17–31.
4. Chajeck T, Rubinger D, Alkan M, Melmed RM, Gunders AE. Anaphylactoid reaction and tissue damage following bite by *Atractaspis engaddensis*. Trans Roy Soc Trop Med Hyg 1974;68:333–337.
5. Hayashi MA, Ligny-Lemaire C, Wollberg Z, Wery M, Galat A, Ogawa T, et al. Long-sarafotoxins: characterization of a new family of endothelin-like peptides. Peptides 2004;25:1243–1251.
6. Kochva E, Bdolah A, Wollberg Z. Sarafotoxins and endothelins: evolution, structure and function. Toxicon 1993; 31:541–568.
7. Trape JF, Pison G, Guyavarch E, Mane Y. High mortality from snakebite in south-eastern Senegal. Trans R Soc Trop Med Hyg 2001;95:420–423.
8. Tilbury CR, Branch WR. Observations on the bite of the southern burrowing asp (*Atractaspis bibronii*) in Natal. S Afr Med J 1989;75:327–331.
9. Kurnik D, Haviv Y, Kochva E. A snake bite by the burrowing asp, *Atractaspis engaddensis*. Toxicon 1999; 37:223–227.
10. White J. Bites and stings from venomous animals: a global overview. Ther Drug Monit 2000;22:65–68.
11. Chippaux J-P. Snake-bites: appraisal of the global situation. Bull World Health Organ 1998;76:515–524.
12. Georgieva DN, Rypniewski W, Gabdoulkhakov A, Genov N, Betzel C. Asp49 phospholipase A(2)-elaidoylamide complex: a new mode of inhibition. Biochem Biophys Res Commun 2004;319:1314–1321.
13. Beer E, Putorti F. Dysphonia, an uncommon symptom of systemic neurotoxic envenomation by *Vipera aspis* bite. Report of two cases. Toxicon 1998;36:697–701.
14. Lee CY, Tsai MC, Chen YM, Ritonja A, Gubensek F. Mode of neuromuscular blocking action of toxic phospholipase A_2 from *Vipera ammodytes* venom. Arch Int Pharmacol Ther 1984;268:313–324.
15. Frangides CY, Koulouras V, Kouni SN, Tzortzatos GV, Nikolaou A, Pneumaticos J, et al. Snake venom poisoning in Greece. Experiences with 147 cases. Eur J Intern Med 2006;17:24–27.
16. Persson H, Irestedt B. A study of 136 cases of adder bite treated in Swedish hospitals during one year. Acta Med Scand 1981;210:433–439.
17. Reading CJ. Incidence, pathology, and treatment of adder (*Vipera berus* L.) bites in man. J Accid Emerg Med 1996;13:346–351.
18. Karlson-Stiber C, Salmonson H, Persson H. A nationwide study of *Vipera berus* bites during one year—epidemiology and morbidity of 231 cases. Clin Toxicol 2005;44:25–30.
19. Petite J. Viper bites: treat or ignore? Review of a series of 99 patients bitten by *Vipera aspis* in an alpine Swiss area. Swiss Med Wkly 2005;135:618–625.

20. Tucker SC, Josty I. Compartment syndrome in the hand following an adder bite. J Hand Surg (Br) 2005:30B:434–435.
21. Vigasio A, Battiston B, De Filippo G, Brunelli G, Calabrese S. Compartmental syndrome due to viper bite. Arch Orthop Trauma Surg 1991;110:175–177.
22. Audebert F, Sorkine M, Robbe-Vincent A, Bon C. Viper bites in France: clinical and biological evaluation; kinetics of envenomations. Hum Exp Toxicol 1994;13;683–688.
23. Audebert F, Sorkine M, Bon C. Envenoming by viper bites in France: clinical gradation and biological quantification by ELISA. Toxicon 1992;30:599–609.
24. Weinelt W, Sattler RW, Mebs D. Persistent paresis of the facialis muscle after European adder (*Vipera berus*) bite on the forehead. Toxicon 2002;40:1627–1629.
25. Gronlund J, Vuori A, Nieminen S. Adder bites. A report of 68 cases. Scand J Surg 2003;92:171–174.
26. Radonic V, Budimir D, Bradaric N. Envenomation by the horned viper (*Vipera ammodytes* L.). Mil Med 1997;162:179–182.
27. Aravanis C, Ioannidis PJ, Ktenas J. Acute myocardial infarction and cerebrovascular accident in a young girl after a viper bite. Br Heart J 1982;47:500–503.
28. Beer E, Musiani R. A case of intestinal infarction following *Vipera aspis* bite. Toxicon 1998;36:729–733.
29. Tiwari I, Johnston WJ. Blood coagulability and viper envenomation. Lancet 1986;1(8481):613–614.
30. Denis D, Lamireau T, Llanas B, Bedry R, Fayon M. Rhabdomyolysis in European viper bite. Acta Paediatr 1998;87:1013–1015.
31. Audebert F, Grosselet O, Sabouraud A, Bon C. Quantitation of venom antigens from European vipers in human serum or urine by ELISA. J Anal Toxicol 1993;17: 236–240.
32. Persson H, Irestedt B. A study of 136 cases of adder bite treated in Swedish hospitals during one year. Acta Med Scand 1981;210:433–439.
33. Bentur Y, Zveibel F, Adler M, Raikhlin B. Delayed administration of *Vipera xanthina palestinae* antivenin. Clin Toxicol 1997;3:257–261.
34. Coppola M, Hogan DE. Venomous snakes of southwest Asia. Am J Emerg Med 1992;10:230–236.
35. Trape JF, Pison G, Gyavarch E, Mane YP. High mortality from snakebite in south-eastern Senegal. Trans Roy Soc Trop Med Hyg 2001;95:420–423.
36. Safer AB, Grace MS. Infrared imaging in vipers: differential responses of crotaline and viperine snakes to paired thermal targets. Behavioural Brain Res 2004;154: 55–61.
37. Marrakchi N, Barbouche R, Guermazi S, Karoui H, Bon C, El Ayeb M. Cerastotin, a serine protease from *Cerastes cerastes* venom, with platelet-aggregating and agglutinating properties. Eur J Biochem 1997;247:121–128.
38. Laraba-Djebari F, Martin-Eauclaire MF, Mauco G, Marchot P. Afaâcytin, an alpha beta-fibrinogenase from *Cerastes cerastes* (horned viper) venom, activates purified factor X and induces serotonin release from human blood platelets. Eur J Biochem 1995;233:756–765.
39. Marrakchi N, Zingali RB, Karoui H, Bon C, el Ayeb M. Cerastocytin, a new thrombin-like platelet activator from the venom of the Tunisian viper *Cerastes cerastes*. Biochim Biophys Acta 1995;1244:147–156.
40. Pirkle H, Theodor I, Miyada D, Simmons G. Thrombin-like enzyme from the venom of *Bitis gabonica*. Purification, properties, and coagulant actions. J Biol Chem 1986;261:8830–8835.
41. Lavonas EJ, Tomaszewski CA, Ford MD, Rouse AM, Kerns WP 2nd. Severe puff adder (*Bitis arietans*) envenomation with coagulopathy. J Toxicol Clin Toxicol 2002;40:911–918.
42. Mebs D, Holada K, Kornalik F, Simak J, Vankova H, Muller D, et al. Severe coagulopathy after a bite of a green bush viper (*Atheris squamiger*): case report and biochemical analysis of the venom. Toxicon 1998;36: 1333–1340.
43. Blaylock RS. Time of onset of clinical envenomation following snakebite. S Afr J Med 1983;64:357–360.
44. Pugh RN, Theakston RD. A clinical study of viper bite poisoning. Ann Trop Med Parasitol 1987;81:135–149.
45. Marsh N, DeRoos F, Touger M. Gaboon viper (*Bitis gabonica*) envenomation resulting from captive specimens—a review of five cases. Clin Toxicol 2007; 45:60–64
46. Warrell DA, Ormerod LD, Davidson NM. Bites by the night adder (*Causus maculatus*) and burrowing vipers (genus *Atractaspis*) in Nigeria. Am J Trop Med Hyg 1976;25:517–524.
47. Lifshitz M, Phillip M, Bernstein T, Sofer S. Snake bite by *Cerastes vipera* in children: report of two cases. Wilderness Environ Med 1995;6:269–272.
48. Schneemann M, Cathomas R, Laidlaw ST, El Nahas AM, Theakston RD, Warrell DA. Life-threatening envenoming by the Saharan horned viper (*Cerastes cerastes*) causing micro-angiopathic haemolysis, coagulopathy and acute renal failure: clinical cases and review. QJM 2004;97:717–727.
49. Lifshitz M, Kastel H, Harman-Boehm I. *Cerastes cerastes* envenomation in an 18-year-old female: a case report. Toxicon 2002;40:1227–1229.
50. Warrell DA, Davidson NMcD, Greenwood BM, Ormerod LD, Pope HM, Watkins BJ, Prentice CR. Poisoning by bites of the saw-scaled or carpet viper (*Echis carinatus*) in Nigeria. Q J Med 1977;46:33–62.
51. Schulchynska-Castel H, Dvilansky A, Keynan A. *Echis colorata* bites: clinical evaluation of 42 patients. A retrospective study. Isr J Med Sci 1986;22:880–884.
52. Karlson-Stiber C, Persson H, Heath A, Smith D, al-Abdulla IH, Sjostrom L. First clinical experiences with specific sheep Fab fragments in snake bite. Report of a multicentre study of *Vipera berus* envenoming. J Intern Med 1997;241:53–58.

53. Bentur Y, Raikhlin-Eisenkraft B, Galperin M. Evaluation of antivenom therapy in *Vipera palaestinae* bites. Toxicon 24;44:53–57.
54. Al-Abdulla I, Garnvwa JM, Rawat S, Smith DS, Landon J, Nasidi A. Formulation of a liquid ovine Fab-based antivenom for the treatment of envenomation by the Nigerian carpet viper (*Echis ocellatus*). Toxicon 2003;42: 399–404.
55. De Silva A, Ranasinghe L. Epidemiology of snake bite in Sri Lanka: a review. Ceylon Med J 1983;28:144–154.
56. Okuda D, Nozaki C, Sekiya F, Morita T. Comparative Biochemistry of Disintegrins Isolated from Snake Venom: Consideration of the Taxonomy and Geographical Distribution of Snakes in the Genus *Echis*. J Biochem 2001;129:615–620.
57. Warrell DA. Snake venoms in science and clinical medicine. 1. Russell's viper: biology, venom and treatment of bites. Trans R Soc Trop Med Hyg 1989;83:732–740.
58. Cockram CS, Chan JC, Chow KY. Bites by the white-lipped pit viper (*Trimeresurus albolabris*) and other species in Hong Kong. A survey of 4 years' experience at the Prince of Wales Hospital. J Trop Med Hyg 1990;93:79–86.
59. Seneviratne U, Dissanayake S. Neurological manifestations of snake bite in Sri Lanka. J Postgrad Med 2002;48:275–279.
60. Tin-Nu-Swe, Tin-Tun, Myint-Lwin, Thein-Than, Tun-Pe, Robertson JI, et al. Renal ischaemia, transient glomerular leak and acute renal tubular damage in patients envenomed by Russell's vipers (*Daboia russelii siamensis*) in Myanmar. Trans R Soc Trop Med Hyg 1993;87: 678–681.
61. Kularatne S. Epidemiology and clinical picture of the Russell's viper (*Daboia russelii russelii*) bite in Anuradhapura, Sri Lanka: a prospective study of 336 patients. Southeast Asian J Trop Med Public Health 2003;34: 855–862.
62. Hung D-Z, Wu M-L, Deng J-F, Lin-Shiau S-Y. Russell's viper snakebite in Taiwan: differences from other Asian countries. Toxicon 2002;40:1291–1298.
63. Theakston RD, Phillips RE, Looareesuwan S, Echeverria P, Makin T, Warrell DA. Bacteriological studies of the venom and mouth cavities of wild Malayan pit vipers (*Calloselasma rhodostoma*) in southern Thailand. Trans R Soc Trop Med Hyg 1990;84:875–879.
64. Phillips RE, Theakston RD, Warrell DA, Galagedera Y, Abeysekera DT, Dissanayaka P, et al. Paralysis, rhabdomyolysis and haemolysis caused by bites of Russell's viper (*Vipera russelii pulchella*) in Sri Lanka: failure of Indian (Haffkine) antivenom. Q J Med 1988;68: 691–715.
65. Ariaratnam CA, Sjostrom L, Raziek Z, Kularatne SA, Arachchi RW, Sheriff MH, et al. An open, randomized comparative trial of two antivenoms for the treatment of envenoming by Sri Lankan Russell's viper (*Daboia russelii russelii*). Trans R Soc Trop Med Hyg 2001;95: 74–80.
66. Hung D-Z, Yu Y-J, Hsu C-L, Lin T-J. Antivenom treatment and renal dysfunction in Russell's viper snakebite in Taiwan: a case series. Roy Soc Trop Med Hyg 2006;100:489–494.
67. Chou T-S, Lin T-J, Kuo M-C, Tsai M-S, Hung D-Z, Tsai J-L. Eight cases of acute renal failure from *Vipera russelii formosensis* venom after administration of antivenom. Vet Hum Toxicol 2002;44:278–282.
68. Jorge MT, Ribeiro LA, O'Connell. Prognostic factors for amputation in the case of envenoming by snakes of the *Bothrops* genus (*Viperidae*). Ann Trop Med Parsitol 1999;93:401–408.
69. Cardoso JLC, Fan HW, Franca Fos, Jorge MT, Leite RP, Nishioka SA, et al. Randomized comparative trial of three antivenoms in the treatment of envenoming by lance-headed vipers (*Bothrops jararaca*) in Sao Paulo, Brazil. Q J Med 1993;86:315–325.
70. Sellahewa KH, Kumararatne MP. Envenomation by the hump-nosed viper (*Hypnale hypnale*). Am J Trop Med Hyg 1994;51:823–825.
71. Joseph JK, Simpson ID, Menon NC, Jose MP, Kulkarni KJ, Raghavendra GB, Warrell DA. First authenticated cases of life-threatening envenoming by the hump-nosed pit viper (*Hypnale hypnale*) in India. Trans R Soc Trop Med Hyg 2007;101:85–90.
72. Gold BS, Dart RC, Barish RA. Bites of venomous snakes. N Engl J Med 2002;347:347–356.
73. Clement JF, Pietrusko RG. Pit viper snakebite in the United States. J Fam Pract 1978;6:269–279.
74. Parrish HM. Incidence of treated snakebites in the United States. Pub Health Rep 1966;81:269–276.
75. Russell FE, Carlson RW, Wainschel J, Osbourne AH. Snake venom poisoning in the United States: Experience with 550 cases. JAMA 1975;233:341–344.
76. Hardy DL. Fatal rattlesnake envenomation in Arizona: 1969–1984. J Toxicol Clin Toxicol 1986;24:1–10.
77. Parrish HM, Goldner JC, Silberg SL. Comparison between snakebites in children and adults. Pediatrics 1965;36:251–256.
78. Butner AN. Rattlesnake bites in Northern California. West J Med 1983;139:179–183.
79. Ennik F. Deaths from bites and stings of venomous animals. West J Med 1980;133:463–468.
80. Langley RL, Morrow WE. Deaths resulting from animal attacks in the United States. Wilderness Environ Med 1997;8:8–16.
81. Gold BS, Wingert WA. Snake venom poisoning in the United States: a review of therapeutic practice. Southern Med J 1994;87:579–589.
82. Plowman DM, Reynolds TL, Joyce SM. Poisonous snakebite in Utah. West J Med 1995;163:547–551.
83. Curry SC, Horning D, Brady P, Requa R, Kunkel DB, Vance MV. The legitimacy of rattlesnake bites in central Arizona. Ann Emerg Med 1989;18:658–663.

84. Downey DJ, Omer GE, Moneim MS. New Mexico rattlesnake bites: demographic review and guidelines for treatment. J Trauma 1991;31:1380–1386.
85. Wingert W, Chan L. Rattlesnake bites in Southern California and rationale for recommended treatment. West J Med 1988;148:37–44.
86. Parrish HM, Wiechmann GH. Rattlesnake bites in the Eastern United States. South Med J 1968;61: 118–126.
87. Norris RL. First report of a bite by the mottled rock rattlesnake (*Crotalus lepidus lepidus*). Toxicon 2005;46: 414–417.
88. Schaper A, de Haro L, Desel H, Ebbecke M, Langer C. Rattlesnake bites in Europe—experiences from Southeastern France and Northern Germany. J Toxicol Clin Toxicol 2004;42:635–641.
89. Bush SP, Cardwell MD. Mojave rattlesnake (*Crotalus scutulatus scutulatus*) identification. Wilderness Environ Med 1999;10:6–9.
90. Klauber LM. Bodily Function in Rattlesnakes. Berkeley, CA: University of California Press; 1982.
91. Zamudio KR, Hardy DL Sr, Martins M, Greene HW. Fang tip spread, puncture distance, and suction for snake bite. Toxicon 2000;38:723–728.
92. Young BA, Kardong KV. Mechanisms controlling venom expulsion in the western diamondback rattlesnake, *Crotalus atrox*. J Exp Zool Part A Ecol Genet Physiol 2007;307:18–27.
93. Minton SA, Weinstein SA. Geographic and ontogenic variation in venom of the western diamondback rattlesnake (*Crotalus atrox*). Toxicon 1986;24:71–80.
94. Wingert W. Management of crotalid envenomations. In: Tu AT, editor. Handbook of Natural Toxins. Vol 5. Reptile venoms and toxins. New York, Marcel Dekker, Inc.; 1991:611–643.
95. Gregory VM, Russell FE, Brewer JR, Zawadski LR. Seasonal variations in rattlesnake venom proteins. Proc West Pharmacol Soc 1984;27:233–236.
96. Lyons WJ. Profound thrombocytopenia associated with *Crotalus ruber ruber* envenomation: a clinical case. Toxicon 1971;9:237–240.
97. Budzynski AZ, Pandya BV, Rubin RN, Brizuela BS, Soszka T, Stewart GJ. Fibrinogenolytic afibrinogenemia after envenomation by Western diamondback rattlesnake (*Crotalus atrox*). Blood 1984;63:1–14.
98. Ownby CL, Reisbeck SL, Allen R. Levels of therapeutic antivenin and venom in a human snakebite victim. South Med J 1996;89:803–806.
99. Amaral CF, Campolina D, Dias MB, Bueno CM, Chavez-Olortegui C, Penaforte CL, et al. Time factor in the detection of circulating whole venom and crotoxin and efficacy of antivenom therapy in patients envenomed by *Crotalus durissus*. Toxicon 1997;35:699–704.
100. Gutierrez J-M, Rucavado A. Snake venom metalloproteinases: their role in the pathogenesis of local tissue damage. Biochimie 2000;82:841–850.
101. Gopalakrishnakone P, Hawgood BJ, Holbrooke SE, Marsh NA, Santana De Sa S, Tu AT. Sites of action of Mojave toxin isolated from the venom of the Mojave rattlesnake. Br J Pharmacol 1980;69:421–431.
102. Fraenkel-Conrat H. Snake venom neurotoxins related to phospholipase A_2. J Toxicol Toxin Rev 1982–1983;1: 205–221.
103. Moura-da-Silva A, Laing G, Paine M. Processing of pro-tumor necrosis factor-alpha by venom metalloproteinases: a hypothesis explaining local tissue damage following snake bite. Eur J Immunol 1996;26: 2000–2005.
104. Harris JB. Toxic constituents of animal venoms arid poisons: 1. Incidence of poisoning, clinical and experimental studies and reptile venoms. Adverse Drug React Acute Poisoning Rev 1982;1:65–92.
105. Ownby C, Cameron D, Tuy A. Isolation of myotoxic component from rattlesnake (*Crotalus viridis viridis*) venom. Am J Pathol 1985;85:149–166.
106. Henschen-Edman AH, Theodor I, Edwards BFP, Pirkle H. Crotalase, a fibrinogen-clotting snake venom enzyme: primary structure and evidence for a fibrinogen recognition exosite different from thrombin. Thromb Haemost 1999;81:81–86.
107. Meier J, Stocker K. Effects of snake venoms on hemostasis. Crit Rev Toxicol 1991;21:171–182.
108. Schmaier AH, Claypool W, Colman RW. Crotalocytins: recognition and purification of a timber rattlesnake platelet aggregating protein. Blood 1980;56:1013–1019.
109. Simon TL, Grace TG. Envenomation coagulopathy in wounds from pit vipers. N Engl J Med 1981;305: 443–447.
110. Abel JH Jr, Nelson AW, Bonilla CA. *Crotalus adamanteus* basic protein toxin: electron microscopic evaluation of myocardial damage. Toxicon 1973;11:59–62.
111. Russell FE. Snake venom poisoning. Vet Hum Toxicol 1991;33:584–586.
112. LoVecchio F, Debus DM. Snakebite envenomation in children: a 10-year retrospective review. Wilderness Environ Med 2001;12:184–189.
113. Troutman WG, Wilson Lee. Topical ophthalmic exposure to rattlesnake venom. Am J Emerg Med 1989;7:307–308.
114. Chiodini RJ, Sundberg JP. Salmonellosis in reptiles: a review. Am J Epidemiol 1981;113:494–499.
115. Waterman SH, Jurrez G, Carr SJ, Kilman L. *Salmonella arizona* infection in Latinos associated with a rattlesnake folk medicine. Am J Public Health 1990;80:286–289.
116. Kelly J, Hopkin R, Rimsza ME. Rattlesnake meat ingestion and *Salmonella arizona* infection in children: case report and review of the literature. Pediat Inf Dis J 1995;14:320–323.
117. Cone LA, Boughton WH, Cone LA, Lehv LH. Rattlesnake capsule-induced *Salmonella arizonae* bacteremia. West J Med 1990;153:315–316.

118. Furlow TG, Brennan LV. Purpura following timber rattlesnake (*Crotalus horridus horridus*) envenomation. Cutis 1985;35:234–236.
119. Bush SP, Siedenburg E. Neurotoxicity associated with suspected southern Pacific rattlesnake (*Crotalus viridis helleri*) envenomation. Wilderness Environ Med 1999;10:247–249.
120. Richardson WH, Goto CS, Gutglass DJ, Williams SR, Clark RF. Rattlesnake envenomation with neurotoxicity refractory to treatment with crotaline Fab antivenom. Clin Toxicol 2007;45:472–475.
121. Brick JF, Gutmann L, Brick J, Apelgren KN, Riggs JE. Timber rattlesnake venom-induced myokymia: evidence for peripheral nerve origin. Neurology 1987;37:1545–1546.
122. LoVecchio F, Pizon AF, Wallace KL, Kunkel DB. Myokymia after snake envenomation in Arizona. Wilderness Environ Med 2005;16:116–117.
123. Morgan DL, Blair HW, Ramsey RP. Suicide attempt by the intravenous injection of rattlesnake venom. South Med J 2006;99:282–284.
124. Hogan DE, Dire DJ. Anaphylactic shock secondary to rattlesnake bite. Ann Emerg Med 1990;19:814–816.
125. Parrish HM, Poland CB. Effects of repeated poisonous snake bites in man. Am J Med Sci 1959;237:277–282.
126. Ellis EF, Smith RT. Systemic anaphylaxis after rattlesnake bite. JAMA 1965;193:151–152.
127. Tu AT, Farid T, Morinaga M. Analyses of snake venom for forensic investigation. J Nat Toxins 2001;10:167–175.
128. Minton SA, Weinstein SA, Wilde CE. An enzyme-linked immunoassay for detection of North American pit viper venom. J Toxicol Clin Toxicol 1984;22:303–316.
129. Holstege CP, Miller MB, Wermuth M, Furbee B, Curry SC. Crotalid snake envenomation. Crit Care Clinics 1997;13:889–921.
130. Tanen DA, Ruha A-M, Graeme KA, Curry SC. Epidemiology and hospital course of rattlesnake envenomations cared for at a tertiary referral center in Central Arizona. Acad Emerg Med 2001;8:177–182.
131. Bond GR, Burkhard KK. Thrombocytopenia following timber rattlesnake envenomation. Ann Emerg Med 1997;30:40–44.
132. Hasiba U, Rosenbach LM, Rockwell D, Lewis JH. DIC-like syndrome after envenomation by the snake, *Crotalus horridus horridus*. N Engl J Med 1975;292:505–507.
133. Gibly RL, Walter FG, Nowlin SW, Berg RA. Intravascular hemolysis associated with North American Crotalid envenomation. Clin Toxicol 1998;36:337–343.
134. Carroll RR, Hall EL, Kitchens CS. Canebrake rattlesnake envenomation. Ann Emerg Med 1997;30:45–48.
135. Moss ST, Bogdan G, Nordt SP, Williams SR, Dart RC, Clark RF. An examination of serial urinalyses in patients with North American Crotalid envenomation. Clin Toxicol 1998;36:329–335.
136. Johnson EK, Kardong KV, Mackessy SP. Electric shocks are ineffective in treatment of lethal effects of rattlesnake envenomation in mice. Toxicon 1987;25:1347–1349.
137. Dart RC, Gustafson RA. Failure of electric shock treatment for rattlesnake envenomation. Ann Emerg Med 1991;20:659–661.
138. McKinney PE. Out-of-hospital and interhospital management of Crotaline snakebite. Ann Emerg Med 2001;37:168–174.
139. Griffen D, Donovan JW. Significant envenomation from a preserved rattlesnake head (in a patient with a history of immediate hypersensitivity to antivenin). Ann Emerg Med 1986;15:955–958.
140. Stewart M, Greenland S, Hoffman J. First aid treatment of poisonous snake bite: are currently recommended procedures justified? Ann Emerg Med 1981;10:331–335.
141. Burgess JL, Dart RC, Egen NB, Mayersohn M. Effects of constriction bands on rattlesnake venom absorption: a pharmacokinetic study. Ann Emerg Med 1992;21:1086–1093.
142. Andrews CE, Dees JE, Edwards RO, Jackson KW, Snyder CC, Moseley T, et al. Venomous snakebite in Florida. J Fla Med Assoc 1968;55:308–316.
143. Alberts MB, Shalit M, LoGalbo F. Suction for venomous snakebite: a study of "mock venom" extraction in a human model. Ann Emerg Med 2004;43:181–186.
144. Bush SP, Hegewald KG, Green SM, Cardwell MD, Hayes WK. Effects of negative pressure venom extraction device (Extractor) on local tissue injury after artificial rattlesnake envenomation in a porcine mode. Wilderness Environ Med 2000;11:180–188.
145. Kerns W II, Tomaszewski C. Airway obstruction following canebrake rattlesnake envenomation. J Emerg Med 2001;20:3444–380.
146. Gerkin R, Sergent KC, Curry SC, Vance M, Nielsen DR, Kazan A. Life-threatening airway obstruction from rattlesnake bite to the tongue. Ann Emerg Med 1987;16:813–816.
147. Hinze JD, Barker JA, Jones TR, Winn RE. Life-threatening upper airway edema caused by a distal rattlesnake bite. Ann Emerg Med 2001;38:79–82.
148. Seger D, Kahn S, Krenzelok EP. Treatment of US Crotalidae bites. Toxicol Rev 2005;24:217–227.
149. Galan JA, Sanchez EE, Rodriguez-Acosta A, Perez JC. Neutralization of venoms from two Southern Pacific rattlesnakes (*Crotalus helleri*) with commercial antivenoms and endothermic animal sera. Toxicon 2004;43:791–799.
150. Sanchez EE, Galan JA, Perez JC, Rodriguez-Acosta A, Chase PB, Perez JC. The efficacy of two antivenoms against the venom of North American snakes. Toxicon 2003;41:357–365.
151. Consroe P, Egen NB, Russell FE, Gerrish K, Smith DC, Sidki A, Landon JT. Comparison of a new ovine antigen binding fragment (Fab) antivenin for United States Crotalidae with the commercial antivenin for protection against venom-induced lethality in mice. Am J Trop Med Hyg 1995;53:507–510.

152. Dart RC, McNally J. Efficacy, safety, and use of snake antivenoms in the United States. Ann Emerg Med 2001;37:181–188.

153. Pizon AF, Riley BD, LoVecchio F, Gill R. Safety and efficacy of Crotalidae Polyvalent Immune Fab in pediatric crotaline envenomation. Acad Emerg Med 2007; 14:373–376.

154. Bush SP, Green SM, Moynihan JA, Hayes WK, Cardwell MD. Crotalidae polyvalent immune Fab (ovine) antivenom is efficacious for envenomations by Southern Pacific rattlesnakes (*Crotalus helleri*). Ann Emerg Med 2002; 40:619–624.

155. Decker WW, Bogdan GM, Garcia RA, Wollen P, Dart RC. Heat and motion stability of polyvalent crotalidae antivenin, ovine Fab. Toxicol 1998;36:377–382.

156. Bebarta V, Dart RC. Effectiveness of delayed use of Crotalidae Polyvalent Immune Fab (ovine) antivenom. J Toxicol Clin Toxicol 2004;42:321–324.

157. Seifert SA, Boyer LV. Recurrence phenomena after immunoglobulin therapy for snake envenomations: Part 1. Pharmacokinetics and pharmacodynamics of immunoglobulin antivenoms and related antibodies. Ann Emerg Med 2001;37:189–195.

158. Boyer LV, Seifert SA, Cain JS. Recurrence phenomena after immunoglobulin therapy for snake envenomations: Part 2. Guidelines for clinical management with crotaline Fab antivenom. Ann Emerg Med 2001;37;196–201.

159. Dart RC, Seifert SA, Carroll L, Clark RF, Hall E, Boyer-Hassen LV, et al. Affinity-purified, mixed monospecific crotalid antivenom ovine Fab for the treatment of crotalid venom poisoning. Ann Emerg Med 1997;30: 33–39.

160. Clark RF, McKinney PE, Chase PB, Walter FG. Immediate and delayed allergic reactions to Crotalidae Polyvalent Immune Fab (ovine) antivenom. Ann Emerg Med 2002;39:671–676.

161. Holstege CP, Wu J, Baer AB. Immediate hypersensitivity reaction associated with the rapid infusion of Crotalidae Polyvalent Immune Fab (ovine). Ann Emerg Med 2002; 39;677–679.

162. Dart RC, Seifert SA, Boyer LV, Clark RF, Hall E, McKinney P, et al. A randomized multicenter trial of Crotalinae Polyvalent Immune Fab (ovine) antivenom for the treatment for crotaline snakebite in the United States. Arch Intern Med 2001;161:2030–2036.

163. Boyer LV, Seifert SA, Clark RF, McNally JT, Williams SR, Nordt SP, Walter FG, Dart RC. Recurrent and persistent coagulopathy following pit viper envenomation. Arch Intern Med 1999;159:706–710.

164. Offerman SR, Barry JD, Schneir A, Clark RF. Biphasic rattlesnake venom-induced thrombocytopenia. J Emerg Med 2003;24:289–293.

165. Bush SP, Wu VH, Corbett SW. Rattlesnake venom-induced thrombocytopenia. Response to antivenin (Crotalidae) polyvalent: a case series. Acad Emerg Med 2000;7:181–185.

166. Dunnihoo DR, Rush BM, Wise RB, Brooks GG, Otterson WN. Snake bite poisoning in pregnancy: a review of the literature. J Reprod Med 1992;37:653–658.

167. Parrish HM, Khan MJ. Snakebite during pregnancy. A report of 4 cases. Obstet Gynecol 1966;27:468–471.

168. Corey EJ. Chemical studies on slow reacting substances/ leukotrienes. Experientia 1982;38:1259–1275.

169. Barach EM, Nowak RM, Lee TG, Tomlanovich MC. Epinephrine for treatment of anaphylactic shock. JAMA 1984;251:2118–2122.

170. Jansen PW, Perkin RM, van Stralen D. Mojave rattlesnake envenomation: prolonged neurotoxicity and rhabdomyolysis. Ann Emerg Med 1992;21:322–325.

171. Bogdan GM, Dart RC, Falbo SC, McNally J, Spaite D. Recurrent coagulopathy after antivenom treatment of crotalid snakebite. Southern Med J 2000;93:562–566.

172. Jurkovich GJ, Luterman A, McCullar K, Ramenofsky ML, Curreri PW. Complications of Crotalidae antivenin therapy. J Trauma 1988;28:1032–1037.

173. Offerman SR, Smith TS, Derlet RW. Does the aggressive use of polyvalent antivenin for rattlesnake bites result in serious acute side effects? West J Med 2001;175: 88–91.

174. LoVecchio F, Klemens J, roundy EB, Klemes A. Serum sickness following administration of antivenin (Crotalidae) polyvalent in 181 cases of presumed rattlesnake envenomation. Wilderness Environ Med 2003;14: 220–221.

175. Wingert WA, Wainschel J. Diagnosis and management of envenomations by poisonous snakes. South Med J 1975;68:1015–1026.

176. Corrigan P, Russell FE, Wainschel J. Clinical reactions to antivenin. Toxicon 1978;16(suppl 1):457–465.

177. Guisto JA. Severe toxicity from crotalid envenomation after early resolution of symptoms. Ann Emerg Med 1995;26:387–389.

178. Hall EL. role of surgical intervention in the management of Crotaline snake envenomation. Ann Emerg Med 2001;37:175–180.

179. Kerrigan KR, Mertz BL, Nelson SJ, Dye JD. Antibiotic prophylaxis for pit viper envenomation: prospective, controlled trial. World J Surg 1997;21:369–373.

180. Clark RF, Selden BS, Furbee B. The incidence of wound infection following crotalid envenomation. J Emerg Med 1993;11:583–586.

181. Parrish HM, Maclaurin AW, Tuttle RL. North American pit vipers: bacterial flora of the mouths and venom glands. Va Med Mon 1956;83:383–385.

182. Burgess JL, Dart RC. Snake venom coagulopathy: use and abuse of blood products in the treatment of pit viper envenomation. Ann Emerg Med 1991;20:795–801.

183. Camilleri C, Offerman S, Gosselin R, Albertson T. Conservative management of delayed, multicomponent coagulopathy following rattlesnake envenomation. Clin Toxicol 2005;43:201–206.

184. Kitchens CS. Hemostatic aspects of envenomation by North American snakes. Hematol Oncol Clin North Am 1992;6:1189–1195.
185. Ruha A-M, Curry SC, Beuhler M, Katz K, Brooks DE, Graeme KA, et al. Initial postmarketing experience with Crotalidae Polyvalent Immune Fab for treatment of rattlesnake envenomation. Ann Emerg Med 2002;39: 609–615.
186. Tanen DA, Danish DC, Grice GA, Riffenburgh RH, Clark RF. Fasciotomy worsens the amount of myonecrosis in a porcine model of Crotaline envenomation. Ann Emerg Med 2004;44:99–104.
187. Curry S, Kraner J, Kunkel D, Ryan P, Vance M, Requa R, Ruggersi S. Non-invasive vascular studies in management of rattlesnake envenomation to extremities. Ann Emerg Med 1985;14:1081–1084.
188. Hardy DL Sr, Zamudio KR. Compartment syndrome, fasciotomy, and neuropathy after a rattlesnake envenomation: aspects of monitoring and diagnosis. Wilderness Environ Med 2006;17:36–40.
189. Shaw BA, Hosalkar HS. Rattlesnake bites in children: antivenin treatment and surgical indications. J Bone Joint Surg Am 2002;84A:1624–1629.
190. Mars M, Hadley G. Failure of pulse oximetry in the assessment of raised limb intracompartmental pressure. Injury 1994;25:379–381.
191. Cowin DJ, Wright T, Cowin JA. Long-term complications of snake bites to the upper extremity. J Southern Ortho Assoc 1998;7:205–211.
192. Scarborough RM, Rose JW, Hsu MA, Phillips DR, Fried VA, Campbell AM, et al. Barbourin. A GPIIb-IIIa-specific integrin antagonist from the venom of *Sistrurus m. barbouri*. J Biol Chem 1991;266:9359–9362.
193. Scarborough RM. Development of eptifibatide. Am Heart J 1999;138:1093–1104.
194. Christiansen J, Fieselmann J. Massasauga rattlesnake bites in Iowa. Iowa Med 1993;83:187–191.
195. Ahlstrom NG, Luginbuhl W, Tisher CC. Acute anuric renal failure after pigmy rattlesnake bite. South Med J 1991;84:783–785.
196. Hankin FM, Smith MD, Penner JA, Louis DS. Eastern massasauga rattlesnake bites. J Pediatr Orthop 1987; 7:201–205.
197. Parrish HM, Carr C. Bites by copperheads (*Agkistrodon contortrix*) in the United States. JAMA 1967;201: 927–932.
198. Wingert WA, Pattabhiraman TR, Cleland R, Meyer P, Pattabhiraman R, Russell FE. Distribution and pathology of copperhead (*Agkistrodon contortrix*) venom. Toxicon 1980;18:591–601.
199. Guan AL,Retzios AD, Henderson GN, Markland FS Jr. Purification and characterization of a fibrinolytic enzyme from venom of the southern copperhead snake (*Agkistrodon contortrix contortrix*). Arch Biochem Biophys 1991;289:197–207.
200. Swenson S, Bush LR, Markland FS. Chimeric derivative of Fibrolase, a fibrinolytic enzyme from southern copperhead venom, possesses inhibitory activity on platelet aggregation. Arch Biochem Biophy 2000;384: 227–237.
201. Retzios AD, Markland FS Jr. A direct-acting fibrinolytic enzyme from the venom of *Agkistrodon contortrix contortrix*: effects on various components of the human blood coagulation and fibrinolysis system. Throm Res 1988;52:541–552.
202. Zhou Q, Sherwin RP, Parrish C, Richters V, Groshen SG, Tsao-Wei D, Markland FS. Contortrostatin, a dimeric disintegrin from *Agkistrodon contortrix contortrix*, inhibits breast cancer progression. Breast Cancer Res Treat 2000;61:249–260.
203. Scharman EJ. Copperhead snakebites: clinical severity of local effects. Ann Emerg Med 2001;38:55–61.
204. Keyler DE, Vandevoort JT. Copperhead envenomations: clinical profiles of three different subspecies. Vet Hum Toxicol 1999;41:149–152.
205. Thorson A, Lavonas EJ, Rouse AM, Kerns WP II. Copperhead envenomations in the Carolinas. J Toxicol Clin Toxicol 2003;41:29–35.
206. Spiller HA, Bosse GM. Prospective study of morbidity associated with snakebite envenomation. J Toxicol Clin Toxicol 2003;41:125–130.
207. Rudolph R, Neal GE, Williams JS, McMahan AP. Snakebite treatment at a Southeastern regional referral center. Am Surgeon 1995;61:767–772.
208. Whitley RE. Conservative treatment of copperhead snakebites without antivenin. J Trauma 1996;41: 219–221.
209. White RR IV, Weber RA. Poisonous snakebite in Central Texas. Ann Surg 1991;213:466–472.
210. Sheibani-Rad S, Young J. Copperhead snake bite in Connecticut: case report and discussion. Connecticut Med 2006;70:301–303.
211. Lavonas EJ, Gerardo CJ, O'Malley G, Arnold TC, Bush SP, Banner WJ Jr, et al. Initial experience with *Crotalidae* polyvalent immune Fab (ovine) antivenom in the treatment of copperhead snakebite. Ann Emerg Med 2004; 43:200–206.
212. Padda GS, Bowen CH. Anesthetic implication of snakebite envenomation. Anesth Analg 1995;81:649–651.
213. Parrish HM, Donnell HD Jr. Bites by cottonmouth (*Agkistrodon piscivorus*) in the United States. South Med J 1968;61:429–434.
214. Sasa M, Vazquez S. Snakebite envenomation in Costa Rica: a revision of incidence in the decade 1990–2000. Toxicon 2003;41:19–22.
215. da Silva CJ, Jorge MT, Ribeiro LA. Epidemiology of snakebite in a central region of Brazil. Toxicon 2003; 41:251–255.
216. Russell FE, Walter FG, Bey TA, Fernandez MC. Snakes and snakebite in Central America. Toxicon 1997;35: 1469–1522.

217. Nishioka S, de A, Silveira PV, Bauab FA. Bite marks are useful for the differential diagnosis of snakebite in Brazil. Wilderness Environ Med 1995;6:183–188.

218. Oguiur N, Camargo ME, Da Silva AR, Horton DS. Quantification of crotamine, a small basic myotoxin, in South American rattlesnake (*Crotalus durissus terrificus*) venom by enzyme-linked immunosorbent assay with parallel-lines analysis. Toxicon 2000;38:443–448.

219. Mebs D, Ownby CL. Myotoxic components of snake venoms: their biochemical and biological activities. Pharmacol Ther 1990;48:223–236.

220. Rucavado A, Soto M, Kamiguti AS, Theakston RD, Fox JW, Escalante T, Gutierrez JM. Characterization of aspercetin, a platelet aggregating component from the venom of the snake *Bothrops asper* which induces thrombocytopenia and potentiates metalloproteinase-induced hemorrhage. Thromb Haemost 2001;85:710–715.

221. de Rezende NA, Torres FM, Dias MB, Campolina D, Chavez-Olortegui C, Amaral CF. South American rattlesnake bite (*Crotalus durissus* SP) without envenoming: insights on diagnosis and treatment. Toxicon 1998;36:2029–2032.

222. Cupo P, Azevedo-Marques MM, Hering SE. Clinical and laboratory features of South American rattlesnake (*Crotalus durissus terrificus*) envenomation in children. Trans R Soc Trop Med Hyg 1988;82:924–929.

223. Jorge MT, Sano-Martins IS, Tomy SC, Castro SC, Ferrari RA, Ribeiro LA, Warrell DA. Snakebite by the bushmaster (*Lachesis muta*) in Brazil: case report and review of the literature. Toxicon 1997;35:545–554.

224. Bucaretchi F, Herrera SR, Hyslop S, Baracat EC, Vieira RJ. Snakebites by *Bothrops* spp in children in Campinas, Sao Paulo, Brazil. Rev Inst Med Trop Sao Paulo 2001;43:329–333.

225. Ribeiro LA, Jorge MT, Lebrao ML. Prognostic factors for local necrosis in *Bothrops jararaca* (Brazilian pit viper) bites. Trans R Soc Trop Med Hyg 2001;95:630–634.

226. Milani Junior R, Jorge MT, de Campos FP, Martins FP, Bousso A, Cardoso JL, et al. Snake bites by the jararacucu (*Bothrops jararacussu*): clinicopathological studies of 29 proven cases in Sao Paulo State, Brazil. QJ Med 1997;90:323–334.

227. Pinho FM, Zanetta DM, Burdmann EA. Acute renal failure after *Crotalus durissus* snakebite: a prospective survey on 100 patients. Kidney Int 2005;67:659–667.

228. Bucaretchi F, Herrera SR, Hyslop S, Baracat EC, Vieira RJ. Snakebites by *Crotalus durissus* ssp. in children in Campinas, Sao Paulo, Brazil. Rev Inst Med Trop Sao Paulo 2002;44:133–138.

229. Santos-Soares PC, Bacellar A, Povoas HP, Brito AF, Santana DL. Stroke and snakebite: case report. Arq Neuropsiquiatr 2007;65:341–344.

230. Francischetti IM, Gombarovits ME, Valenzuela JG, Carlini CR, Guimaraes JA. Intraspecific variation in the venoms of the South American rattlesnake (*Crotalus durissus terrificus*). Comp Biochem Physiol C Toxicol Pharmacol 2000;127:23–36.

231. Saravia P, Rojas E, Arce V, Guevara C, Lopez JC, Chaves E, Velasquez R, Rojas G, Gutierrez JM. Geographic and ontogenic variability in the venom of the neotropical rattlesnake *Crotalus durissus*: pathophysiological and therapeutic implications. Rev Biol Trop 2002;50:337–346.

232. Smalligan R, Cole J, Brito N, Laing GD, Mertz BL, Manock S, et al. Crotaline snake bite in the Ecuadorian Amazon: randomised double blind comparative trial of three South American polyspecific antivenoms. BMJ 2004;329:1129–1133.

233. Schier JG, Wiener SW, Touger M, Nelson LS, Hoffman RS. Efficacy of Crotalidae polyvalent antivenin for the treatment of hognosed viper (*Porthidium nasutum*) envenomation. Ann Emerg Med 2003;41:391–395.

234. Nishioka Sde A, Jorge MT, Silveira PV, Ribeiro LA. South American rattlesnake bite and soft-tissue infection: report of a case. Rev Soc Bras Med Trop 2000; 33:401–402.

235. Jorge MT, Nishioka S, de A, De Oliveira RB, Ribeiro LA, Silveira PV. Aeromonas hydrophila soft-tissue infection as a complication of snake bite: report of three cases. Ann Trop Med Parasitol 1998;92:213–217.

236. Hung D-Z. Taiwan's venomous snakebite: epidemiological, evolution and geographic differences. Trans Roy Soc Trop Med Hyg 2004;97:96–101.

237. Dong LV, Quyen LK, Eng KH, Gopalakrishnakone P. Immunogenicity of venoms from four common snakes in the South of Vietnam and development of ELISA kit for venom detection. J Immunol Methods 2003;282:13–31.

238. Dempfle CE, Argiriou S, Alesci S, Kucher K, Muller-Peltzer H, Rubsamen K, Heene DL. Fibrin formation and proteolysis during ancrod treatment. Evidence for des-A-profibrin formation and thrombin independent factor XIII activity. Ann NY Acad Sci 2001;936: 210–214.

239. Zhang Y, Lee WH, Gao R, Xiong YL, Wang WY, Zhu SW. Effects of Pallas' viper (*Agkistrodon halys pallas*) venom on blood coagulation and characterization of a prothrombin activator. Toxicon 1998;36:143–152.

240. Sellahewa KH, Kumararatne MP. Envenomation by the hump-nosed viper (*Hypnale hypnale*). Am J Trop Med Hyg 1994;51:823–825.

241. Joseph JK, Simpson ID, Menon NC, Jose MP, Kulkarni KJ, Raghavendra GB, Warrell DA. First authenticated cases of life-threatening envenoming by the hump-nosed pit viper (*Hypnale hypnale*) in India. Trans R Soc Trop Med Hyg 2007;101:85–90.

242. de Silva A, Wijekoon AS, Jayasena L, Abeysekera CK, Bao CX, Hutton RA, Warrell DA. Haemostatic dysfunction and acute renal failure following envenoming by Merrem's hump-nosed viper (*Hypnale hypnale*) in Sri Lanka: first authenticated case. Trans R Soc Trop Med Hyg 1994;88:209–212.

243. Premawardena AP, Seneviratne SL, Gunatilake SB, de Silva HJ. Excessive fibrinolysis: the coagulopathy following Merrem's hump-nosed viper (*Hypnale hypnale*) bites. Am J Trop Med Hyg 1998;58:821–823.
244. Wiwanitkit V. A review of the hematologic effects of green pit viper venom. J Toxicol Toxin Rev 2004;23: 105–110.
245. Rojnuckarin P, Intragumtornchai T, Sattapiboon R, Muanpasitporn C, Pakmanee N, Khow O, Swasdikul D. The effects of green pit viper (*Trimeresurus albolabris* and *Trimeresurus macrops*) venom on the fibrinolytic system in human. Toxicon 1999;37:743–755.
246. Vest DK, Kardong KV. Envenomation by a juvenile Malayan pit viper (*Agkistrodon rhodostoma*). Clin Toxicol 1980;16:299–303.
247. Moore TC. Snakebite from the Korean pit viper. Mil Med 1977;142:546–549.
248. Shanghai Vaccine and Serum Institute, Chekiang Medical College, Chekiang Institute of Traditional Chinese Medicine. *Agkistrodon halys* bite treated with specific antivenin observation of 530 cases. Chin Med J 1976;2:59–62.
249. Kimoto T, Suemitsu K, Nakayama H, Komori E, Ohtani M, Ando S. Therapeutic experience of venomous snakebites by the Japanese viper (*Agkistrodon halys blomhoffii*) with low dose of antivenin: report of 43 consecutive cases. Arch Japan Chir 1997;66:71–77.
250. Ari AB. Patient with purely extraocular manifestations from a pit viper snakebite (*Agkistrodon halys brevicaudus*). Mil Med 2001;166:667–669.
251. Sellahewa KH, Gunawardena G, Kumararatne MP. Efficacy of antivenom in the treatment of severe local envenomation by the hump-nosed viper (*Hypnale hypnale*). Am J Trop Med Hyg 1995;53:260–262.
252. Liao W-B, Lee C-W, Tsai Y-S, Liu B-M, Chung K-J. Influential factors affection prognosis of snakebite patients management: Kaohsiung Chang Gung Memorial Hospital experience. Chang Gung Med J 2000;23: 577–583.
253. White J. Snakebite: an Australian perspective. J Wilderness Med 1991;2:219–244.
254. Chippaux J-P. Snake-bites: appraisal of the global situation. Bull World Health Organ 1998;76;515–524.
255. Norris RL, Dart RC. Apparent coral snake envenomation in a patient without visible fang marks. Am J Emerg Med 1989;7:402–405.
256. Pettigrew LC, Glass JP. Neurologic complications of a coral snake bite. Neurol 1985;35:589–592.
257. Walter FG, Bilden EF, Gibly RL. Envenomations. Crit Care Clinics 1999;1:353–386.
258. Kitchens CS, Van Mierop LH. Envenomation by the Eastern coral snake (*Micrurus fulvius fulvius*): a study of 39 victims. JAMA 1987;258:1615–1618.
259. Parrish HM, Khan MS. Bites by coral snakes: report of 11 representative cases. Am J Med Sci 1967;253: 561–568.
260. Andrews CE, Dees JE, Edwards RO, Jackson KW, Snyder CC, Moseley T, et al. Venomous snakebite in Florida. J Fla Med Assoc 1968;55:308–316.
261. Russell FE. Bites by the Sonoran coral snake, *Micruroides euryxanthus*. Toxicon 1967;5:39–42.
262. Morgan DL, Borys DJ, Stanford R, Kjar D, Tobleman W. Texas coral snake (*Micrurus tener*) bites. South Med J 2007;100:152–156.
263. Watt CH. Poisonous snakebite treatment in the United States. JAMA 1978;240:654–656.
264. van Mierop LH. Poisonous snakebite: a review. II. Symptomatology and treatment. J Fla Med Assoc 1976;3:201–210.
265. Gaar GG. Assessment and management of coral and other exotic snake envenomations. J Florida Med Assoc 1996;83:178–182.
266. Rawat S, Laing G, Smith DC, Theakston D, Landon J. A new antivenom to treat Eastern Coral Snake (*Micrurus fulvius fulvius*) envenoming. Toxicon 1994;32:185–190.
267. Wisniewski MS, Hill RE, Havey JM, Bogdan GM, Dart RC. Australian tiger snake (*Notechis scutatus*) and Mexican coral snake (*Micruris* species) antivenoms prevent death from United States coral snake (*Micrurus fulvius fulvius*) venom in a mouse model. J Toxicol Clin Toxicol 2003;41:7–10.
268. de Roodt AR, Paniagua-Solis JF, Dolab JA, Estevez-Ramirez J, Ramos-Cerrillo B, Litwin S, et al. Effectiveness of two common antivenoms for North, Central, and South American *Micrurus* envenomations. J Toxicol Clin Toxicol 2004;42:171–178.
269. Nishioka SA, Silveira PV, Menzes LB. Coral snake bite and severe local pain. Ann Trop Med Parasitol 1993; 87:429–431.
270. Samejima Y, Aoki-Tomomatsu Y, Yanagisawa M, Mebs D. Amino acid sequence of two neurotoxins from the venom of the Egyptian black snake (*Walterinnesia aegyptia*). Toxicon 1997;35:151–157.
271. Duhaiman AS, Alhomida AS, Rabbani N, KamAl MA, al-Jafari AA. Purification and characterization of acetylcholinesterase from desert cobra (*Walterinnesia aegyptia*) venom. Biochimie 1996;78:46–50.
272. Munday SW, Williams SR, Clark RF. Dendrotoxin poisoning in a neurobiochemist. J Toxicol Clin Toxicol 2003;41:163–165.
273. Mebs D. Myotoxic activity of phospholipases A_2 isolated from cobra venoms: neutralization by polyvalent antivenoms. Toxicon 1986;24:1001–1008.
274. Tilbury CR. Observations on the bite of the Mozambique spitting cobra (*Naja mossambica mossambica*). S Afr Med J 1982;61:308–313.
275. Warrell DA, Ormerod LD. Snake venom ophthalmia and blindness caused by the spitting cobra (*Naja nigricollis*) in Nigeria. Am J Trop Med Hyg 1976;25:525–529.
276. Ismail M, Abd-el Salam MA, Al-Ahaidib MS. Pharmacokinetics of ^{125}I-labelled *Walterinnesia aegyptia* venom and its specific antivenins: flash absorption and distribution

of the venom and its toxin versus slow absorption and distribution of IgG, $F(AB')_2$ and F(AB) of the antivenin. Toxicon 1998;36:93–114.

277. Wuster W. Taxonomic changes and toxinology: systematic revisions of the Asiatic cobras (*Naja naja* species complex). Toxicon 1996;34:399–406.
278. Bawaskar HS, Bawaskar PH. Profile of snakebite envenoming in western Maharashtra, India. Trans R Soc Trop Med Hyg 2002;96:79–84.
279. Hsieh HC, Kumar TK, Yu C. Cloning, overexpression, and characterization of cobrotoxin. Biochem Biophys Res Commun 2004;320:1374–1381.
280. Jiang MS, Fletcher JE, Smith LA. Factors influencing the hemolysis of human erythrocytes by cardiotoxins from *Naja naja kaouthia* and *Naja naja atra* venoms and a phospholipase A_2 with cardiotoxin-like activities from *Bungarus fasciatus* venom. Toxicon 1989;27:247–257.
281. Rowan EG. What does beta-bungarotoxin do at the neuromuscular junction? Toxicon 2001;39:107–118.
282. Prasarnpun S, Walsh J, Awad SS, Harris JB. Envenoming bites by kraits: the biological basis of treatment-resistant neuromuscular paralysis. Brain 2005;128:2987–2996.
283. Fathi HB, Rowan EG, Harvey AL. The facilitatory actions of snake venom phospholipase A(2) neurotoxins at the neuromuscular junction are not mediated through voltage-gated K(+) channels. Toxicon 2001;39:1871–1882.
284. Watt G, Padre L, Tuazon L, Theakston RD, Laughlin L. Bites by the Philippine cobra (*Naja naja philippinensis*): prominent neurotoxicity with minimal local signs. Am J Trop Med Hyg 1988;39:306–311.
285. Khandelwal G, Katz KD, Brooks DE, Gonzalez SM, Ulishney CD. *Naja kaouthia*: two cases of Asiatic cobra envenomations. J Emerg Med 2007;32:171–174.
286. Theakston RD, Phillips RE, Warrell DA, Galagedera Y, Abeysekera DT, Dissanayaka P, et al. Envenoming by the common krait (*Bungarus caeruleus*) and Sri Lankan cobra (*Naja naja naja*): efficacy and complications of therapy with Haffkine antivenom. Trans R Soc Trop Med Hyg 1990;84:301–308.
287. Warrell DA. Venomous bites and stings in the tropical world. Med J Aust 1993;159:773–779.
288. Wongtongkam N, Wilde H, Sitthi-Amorn C, Ratanabanangkoon K. A study of Thai cobra (*Naja kaouthia*) bites in Thailand. Mil Med 2005;170:336–341.
289. Pochanugool C, Wilde H, Jitapunkul S, Limthongkul S. Spontaneous recovery from severe neurotoxic envenoming by a Malayan krait *Bungarus candidus* (Linnaeus) in Thailand. Wilderness Environ Med 1997;8:223–225.
290. Pe T, Myint T, Htut A, Htut T, Myint AA, Aung NN. Envenoming by Chinese krait (*Bungarus multicinctus*) and banded krait (*B. fasciatus*) in Myanmar. Trans R Soc Trop Med Hyg 1997;91:686–688.
291. Bawaskar HS, Bawaskar PH. Envenoming by the common krait (*Bungarus caeruleus*) and Asian cobra (*Naja naja*): clinical manifestations and their management in a rural setting. Wilderness Environ Med 2004;15:257–266.
292. Agarwal R, Singh AP, Aggarwal AN. Pulmonary oedema complicating snake bite due to *Bungarus caeruleus*. Singapore Med J 2007;48:e227–e230.
293. Watt G, Padre L, Tuazon ML, Theakston RD, Laughlin LW. Tourniquet application after cobra bite: delay in the onset of neurotoxicity and the dangers of sudden release. Am J Trop Med Hyg 1988;38:618–622.
294. Gold BS, Pyle P. Successful treatment of neurotoxic king cobra envenomation in Myrtle Beach, South Carolina. Ann Emerg Med 1998;32:736–738.
295. Kularatne SAM. Common krait (*Bungarus caeruleus*) bite in Anuradhapura, Sri Lanka: a prospective clinical study, 1996–98. Postgrad Med J 2002;78:276–280.
296. Sharma SK, Koirala S, Dahal G. Krait bite requiring high dose of antivenom: a case report. Southeast Asian J Trop Med Pub Health 33;170–171.
297. Hung D-Z, Liau M-Y, Lin-Shiau S-Y. The clinical significance of venom detection in patients of cobra snakebite. Toxicon 2003;41:409–415.
298. Sutherland SK. Deaths from snake bite in Australia, 1981–1991. Med J Aust 1992;157:740–746.
299. Shine R. Intraspecific variation in thermoregulation, movements and habitat use by Australian Blacksnakes, *Pseudechis porphyriacus* (Elapidae). J Herpetol 1987; 21:165–177.
300. Pearn J, McGuire B, McGuire L, Richardson P. The envenomation syndrome caused by the Australian Red-bellied Black Snake *Pseudechis porphyriacus*. Toxicon 2000;38:1715–1729.
301. Ramasamy S, Fry BG, Hodgson WC. Neurotoxic effects of venoms from seven species of Australasian black snakes (*Pseudechis*): efficacy of black and tiger snake antivenoms. Clin Exp Pharmacol Physiol 2005;32:7–12.
302. Morrison J, Pearn J, Covacevich J, Tanner C, Coulter A. Studies on the venom of *Oxyuranus microlepidotus*. Clin Toxicol 1983–1984;21:373–385.
303. Bell KL, Sutherland SK, Hodgson WC. Some pharmacological studies of venom from the inland taipan (*Oxyuranus microlepidotus*). Toxicology 1998;36:63–74.
304. Harris JB, Grubb BD, Maltin CA, Dixon R. The neurotoxicity of the venom phospholipases A(2), notexin and taipoxin. Exp Neurol 2000;161:517–526.
305. Kim HS, Tamiya N. The amino acid sequence and position of the free thiol group of a short-chain neurotoxin from common-death-adder (*Acanthophis antarcticus*) venom. Biochem J 1981;199:211–218.
306. Jelinek GA, Breheny FX. Ten years of snake bites at Fremantle Hospital. Med J Aust 1990;153:658–661.
307. Wickramaratna JC, Fry BG, Hodgson WC. Species-dependent variations in the in vitro myotoxicity of death adder (*Acanthophis*) venoms. Toxicol Sci 2003;74:352–360.

308. Marshall LR, Herrmann RP. Coagulant and anticoagulant actions of Australian snake venoms. Thromb Haemost 1983;50:707–711.
309. Stocker K, Hauer H, Muller C, Triplett DA. Isolation and characterization of Textarin, a prothrombin activator from eastern brown snake (*Pseudonaja textilis*) venom. Toxicon 1994;32:1227–1236.
310. Williams V, White J, Mirtschin PF. Comparative study on the procoagulant from the venom of Australian brown snakes (Elapidae; *Pseudonaja* spp.) Toxicon 1994;32: 453–459.
311. Weinstein SA, Williams V, White J. Preliminary characteristics of the prothrombin converting enzyme from venom of Stephen's banded snake (*Hoplocephalus stephensii*). Toxicon 2001;39:1937–1939.
312. Williams V, White J. Variation in the composition of the venom from a single specimen of *Pseudonaja textilis* (common brown snake) over one year. Toxicon 1992; 30:202–206.
313. Armugam A, Gong N, Li X, Siew PY, Chai SC, Nair R, Jeyaseelan K. Group IB phospholipase A2 from *Pseudonaja textilis*. Arch Biochem Biophys 2004;421: 10–20.
314. Isbister GK, Dawson AH, Whyte IM. Two cases of bites by the black-bellied swamp snake (*Hemiaspis signata*). Toxicon 2002;40:317–319.
315. Morrison JJ, Tessseraux I, Pearn JH, Harris J, Masci P. Venom of the Australian rough-scaled snake, *Tropidechis carinatus*: lethal potency and electrophysiological actions. Toxicon 1984;22:759–765.
316. Pearn BR, Pearn JH, DeBuse P, Burke J, Covacevich J. Prolonged intensive therapy after snake bite. A probable case of envenomation by the rough-scaled snake. Med J Aust 1985;142:467–469.
317. Sutherland SK, Leonard RL. Snakebite deaths in Australia 1992–1994 and a management update. Med J Aust 1995;163:616–618.
318. Currie BJ. Snakebite in tropical Australia, Papua New Guinea and Irian Jaya. Emerg Med 2000;12:285–294.
319. Jansen M, McLeod M, White J, Isbister GK. Spotted black snake (*Pseudechis guttatus*) envenoming. Med J Aust 2007;186:41–42.
320. Henderson A, Baldwin LN, May C. Fatal brown snake (*Pseudonaja textilis*) envenomation despite the use of antivenom. Med J Aust 1993;158:709–710.
321. White J, Williams V. Severe envenomation with convulsion following multiple bites by a common brown snake, *Pseudonaja textilis*. Aust Paediatr J 1989;25:109–111.
322. Lalloo DG, Trevett AJ, Black J, Mapao J, Saweri A, Naraqi S, et al. Neurotoxicity, anticoagulant activity and evidence of rhabdomyolysis in patients bitten by death adders (*Acanthophis* spp.) in southern Papua New Guinea. Q J Med 1996;89:25–35.
323. Nocera A, Gallagher J, White J. Severe tiger snake envenomation in a wilderness environment. Med J Aust 1998;168:69–71.
324. Lalloo DG, Trevett AJ, Saweri A, Naraqi S, Theakston RD, Warrell DA. The epidemiology of snake bite in Central Province and National Capital District, Papua New Guinea. Trans R Soc Trop Med Hyg 1995;89: 178–182.
325. Selvanayagam ZE, Gopalakrishnakone P. Tests for detection of snake venoms, toxins and venom antibodies: review on recent trends (1987–1997). Toxicon 1999;37: 565–586.
326. Mead HJ, Jelinek GA. Suspected snakebite in children: a study of 156 patients over 10 years. Med J Aust 1996;164;467–470.
327. Hurrell JGR, Chandler HW. Capillary enzyme immunoassay field kits for the detection of snake venom in clinical specimens: a review of two years' use. Med J Aust 1982;2:236–237.
328. Parkin JD, Ibrahim K, Dauer RJ, Braitberg G. Prothrombin activation in eastern tiger snake bite. Pathology 2002;34:162–166.
329. Fatovich DM, Hitchocock T, White J. Mild snake envenomation. Emerg Med 2002;14:85–88.
330. Sutherland SK, Coulter AR, Harris RD. Rationalisation of first-aid measures for elapid snakebite. Lancet 1979;1(8109):183–185.
331. White J. Envenoming and antivenom use in Australia. Toxicon 1998;36:1483–1492.
332. Jelinek GA, Tweed C, Lynch D, Celenza T, Bush B, Michalopoulos N. Cross reactivity between venomous, mildly venomous, and non-venomous snake venoms with the Commonwealth Serum Laboratories Venom Detection Kit. Emerg Med Australas 2004;16:459–464.
333. O'leary MA, Schneider JJ, Krishnan BP, Lavis C, McKendry A, Ong LK, Isbister GK. Cross-neutralisation of Australian brown and tiger snake venoms with commercial antivenoms: Cross-reactivity or antivenom mixtures? Toxicon 2007;50:206–213.
334. Barrett R, Little M. Five years of snake envenoming in far north Queensland. Emerg Med 2003;15:500–510.
335. Yeung JM, Little M, Murray LM, Jelinek GA, Daly FF. Antivenom dosing in 35 patients with severe brown snake (*Pseudonaja*) envenoming in Western Australia over 10 years. Med J Aust 2004;181:703–705.
336. Isbister GK, O'Leary MA, Schneider JJ, Brown SG, Currie BJ, et al. Efficacy of antivenom against the procoagulant effect of Australian brown snake (*Pseudonaja* spp.) venom: *in vivo* and *in vitro* studies. Toxicon 2007;49:57–67.
337. Isbister GK, Williams V, Brown SG, White J, Currie BJ and Australian Snakebite Project Investigators. Clinically applicable laboratory end-points for treating snakebite coagulopathy. Pathology 2006;38:568–572.
338. Tibballs J. Premedication for snake antivenom. Med J Aust 1994;160:4–6.
339. Pope CH. Fatal bite of captive African rear-fanged snake (*Dispholidus*). Copeia 1958:280–282.

340. Chiszar D, Smith HM. Colubrid envenomations in the United States. J Toxicol Toxin Rev 2002;21:85–104.
341. Mittleman MB, Goris RC. Death caused by the bite of the Japanese colubrid snake *Rhabdophis rigrinus* (Boie) (Reptilia, Serpentes, Colubridae). J Herpetol 1978;12: 108–109.
342. Jaume ML, Garrido OH. [Notes on bites of the snake *Alsophis cantherigerus* Bibron (Reptilia-Serpentes colubridae) in Cuba]. Rev Cubana Med Trop 1980;32:145–148. [Spanish]
343. Greding EJ Jr. [Bite and feeding of the Central American snake *Conophis lineatus dunni* Smith]. Rev Biol Trop 1972;20:29–30. [Spanish]
344. Gutierrez JM, Sasa M. Bites and envenomations by colubrid snakes in Mexico and Central America. J Toxicol Toxin Rev 2002;21;105–115.
345. Prado-Franceschi J, Hyslop S. South American colubrid envenomations. J Toxicol Toxin Rev 2002;21: 117–158.
346. Nishioka SA, Silveira PV. *Philodryas patagoniensis* bite and local envenoming. Rev Inst Med Trop Sao Paulo 1994;36:279–281.
347. Ribeiro LA, Puorto G, Jorge MT. Bites by the colubrid snake *Philodryas olfersii*: a clinical and epidemiological study of 43 cases. Toxicon 1999;37:943–948.
348. Kuch U, Mebs D. Envenomations by colubrid snakes in Africa, Europe, and the Middle East. J Toxicol Toxin Rev 2002;21:159–179.
349. du Toit DM. Boomslang (*Dispholidus typus*) bite. A case report and a review of diagnosis and management. S Afr Med J 1980;57:507–510.
350. Atkinson PM, Bradlow BA, White JA, Greig HB, Gaillard MC. Clinical features of twig snake (*Thelotornis capensis*) envenomation. S Afr Med J 1980;58: 1007–1011.
351. Fritts TH, McCoid MJ, Haddock RL. Risks to infants on Guam from bites of the brown tree snake (*Boiga irregularis*). Am J Trop Med Hyg 1990;42:607–611.
352. Hayes WK, Lavin-Murcio P, Kardong KV. Delivery of Duvernoy's secretion into prey by the brown tree snake, *Boiga irregularis* (Serpentes: Colubridae). Toxicon 1993;31:881–887.
353. McKinstry DM. Morphologic evidence of toxic saliva in colubrid snakes: a checklist of world genera. Herpetaol Rev 1983;14:12–15.
354. McKinstry DM. Evidence of toxic saliva in some colubrid snakes of the United States. Toxicon 1978;16:523–534.
355. De Lisle HF. Venomous colubrid snakes. Bull Chicago Herptol Soc 1982;17:1–17.
356. Hill RE, Mackessy SP. Characterization of venom (Duvernoy's secretion) from twelve species of colubrid snakes and partial sequence of four venom proteins. Toxicon 2000;38:1663–1687.
357. Mandell F, Bates J, Mittleman MS, Loy JW. Major coagulopathy and "non-poisonous" snake bites. Pediatrics 1980;65:314–317.
358. Zotz RB, Mebs D, Hirche H, Paar D. Hemostatic changes due to the venom gland extract of the red-necked keelback snake (*Rhabdophis subminiatus*). Toxicon 1991;29: 1501–1508.
359. Hoffmann JJ, Vijgen M, Smeets RE, Melman PG. Haemostatic effects in vivo after snakebite by the red-necked keelback (*Rhabdophis subminiatus*). Blood Coag Fibrin 1992;3:461–464.
360. Cook DG. A case of envenomation by the neotropical colubrid snake, *Stenorrhina freminvillei*. Toxicon 1984;22: 823–827.
361. Ribeiro LA, Puorto G, Jorge MT. Bites by the colubrid snake *Philodryas olfersii*: a clinical and epidemiological study of 43 cases. Toxicon 1999;37:943–948.
362. dos Santos-Costa MC, Outeiral AB, D'Agostini FM, Cappellari LH. Envenomation by the neotropical colubrid *Bioruna maculata* (Boulenger, 1896): a case report. Rev Inst Med Trop 2000;4:283–286.
363. Kirchberg JS, Davidson TM. Envenomation by the colubrid snake *Atractaspis bibronii*: a case report. Toxicon 1991;29:379–381.
364. Morris MA. Envenomation from the bite of *Heterodon nasicus* (Serpentes: Colubridae). Herpetologica 1985;41: 361–363.
365. Vest DK. Envenomation following the bite of a wandering garter snake (*Thamnophis elegans vagrans*). Clin Toxicol 1981;18:573–579.
366. Gomez HF, Davis M, Phillips S, McKinney P, Brent J. Human envenomation from a wandering garter snake. Ann Emerg Med 1994;23:1119–1122.
367. Ribeiro LA, Puorto G, Jorge MT. Bites by the colubrid snake *Philodryas olfersii*: a clinical and epidemiological study of 43 cases. Toxicon 1999;37:943–948.
368. Peichoto ME, Leiva LC, Guaimas Moya LE, Rey L, Acosta O. Duvernoy's gland secretion of *Philodryas patagoniensis* from the northeast of Argentina: its effects on blood coagulation. Toxicon 2005;45:527–534.
369. Gomperts ED, Demetriou D. Laboratory studies and clinical features in a case of boomslang envenomation. S Afr Med J 1977;51:173–175.
370. Seow E, Kuperan P, Goh SK, Gopalakrishnakone P. Morbidity after a bite from a "non-venomous" pet snake. Singapore Med J 2000;41:34–35.
371. Cable D, McGehee W, Wingert WA, Russell FE. Prolonged defibrination after a bite from a "nonvenomous' snake. JAMA 1984;251:925–926.
372. Smeets RE, Melman PG, Hoffmann JJ, Mulder AW. Severe coagulopathy after a bite from a "harmless" snake (*Rhabdophis subminiatus*). J Int Med 1991;230: 351–354.

Chapter 177

SEA SNAKES

IDENTIFYING CHARACTERISTICS

Sea snakes are marine animals that rarely leave the sea. These marine snakes formerly were classified as a separate family (Hydrophiidae), but now these snakes belong to two Elapidae subfamilies, Hydrophiinae (sea snakes) and Laticaudinae (sea kraits).[1] Although sea snakes are all venomous, bites by these sea creatures are rare because of their reclusive disposition. Sea snake envenomations may cause fatalities as a result of neurotoxicity.[2] The common feature of sea snakes is the broad, flat tail that functions as a rudder and propeller. Sea snakes swim in an undulating fashion, powered by a flat, paddle-shaped tail.

Sea snakes are well adapted to the sea, and these snakes probably are the most abundant snakes in the world. There are over 50 species of sea snakes. Table 177.1 lists the common species of sea snakes. Sea kraits (*Laticauda* spp.) evolved independently from the genera of true sea snakes (*Enhydrina, Hydrophis, Lapemis, Pelamis*), probably originating from the Southeast Asian *Calliophis* genus of elapids. True sea snakes (Hydrophiinae) have nostril valves.

EXPOSURE

Geographic Distribution

Sea snakes inhabit tropical and temperate waters in the Indian and Pacific Oceans. Envenomation by sea snakes is most common in Malaysia, Southeast Asia, and the Persian Gulf.[3] There are no reports of sea snakes in the Atlantic Ocean or in the Caribbean Sea. *Pelamis platurus* Smith is the most common species, extending from the west coast of the Americas to the east coast of Africa. The waters off the tip of Africa and Cape Horn are too cool to allow the spread of sea snakes, which require temperate conditions above 20°C (65°F). Several species of sea snakes inhabit freshwater in Philippine lakes.[4]

Behavior

Sea snakes are docile, diurnal creatures, and most human encounters with these snakes do not result in bites or envenomation. These snakes feed primarily on small fish around rocky bottoms and coral reefs as well as fish that inhabit various depths in the water column.[5] Occasionally, sea snakes consume fish eggs or marine invertebrates. Most bites occur in fishermen when they inadvertently handle or step on the snakes after these snakes become entangled in fishing nets. Sea snake bites also occur when sunbathers and swimmers walk in shallow water, particularly near river estuaries. The sea kraits are less aquatic than true sea snakes, seeking rock crevices or land debris during inactive periods. Most sea snake species, except for *Pelamis* species, are bottom feeders that spend little time at the surface of the sea. Sea kraits (i.e., *Laticauda* species) lay their eggs on land, whereas other sea snakes are ovoviviparous and remain in the sea. Sea snakes do possess gills, and they must surface to breath. However, they can remain under the surface of the water for 2 hours.

Medical Toxicology of Natural Substances, by Donald G. Barceloux, MD

TABLE 177.1. Common sea snakes

Species	Common Name
Aipysurus laevis Lacépède	Olive-brown sea snake
Astrotia stokesii Gray	Stokes's sea snake
Enhydrina schistosa Daudin	Beaked or common sea snake
Hydrophis cyanocinctus Daudin	Annulated sea snake
Hydrophis fasciatus Schneider (*Hydrophis gracilis*)	Small-headed sea snake
Hydrophis ornatus Gray	Reef sea snake
Hydrophis spiralis Shaw	Yellow sea snake
Lapemis hardwickii Gray	Hardwicke's sea snake
Laticauda colubrina Schneider	Yellow-lipped sea krait
Pelamis platurus Smith	Pelagic or yellow-bellied sea snake

PRINCIPAL TOXINS

Venom Composition

The venom of sea snakes is similar to the curare-like venom of terrestrial elapids, which inhibit postsynaptic neuromuscular transmission by binding to the α subunits of the nicotinic receptors at the neuromuscular junction.[6] The venom from the yellow-bellied sea snake (*Pelamis platurus*) contains two structurally similar neurotoxins (pelamis toxin a and b) that bind to postsynaptic acetylcholine receptors similar to α-bungarotoxin.[7] The sea snake venoms (e.g., *Enhydrina schistosa* Daudin, *Hydrophis cyanocinctus* Daudin) contain phospholipase A_2 (PLA_2) enzymes. These phospholipases demonstrate close sequence homologies to other elapid PLA_2 enzymes.[8] PLA_2 enzymes catalyze the hydrolysis of the 2-acyl ester bond of the 1,2-diacyl-*sn*-3-phosphoglycerol lipids in the presence of calcium to produce nonesterified fatty acids and lysophospholipids. These types of enzymes demonstrate a wide variety of pharmacological effects including hemolysis, myonecrosis, edema, cardiotoxicity, and neurotoxicity. PLA_2 enzymes within the same venom produce different effects despite similar structure.[9] Potentially, envenomations by sea snakes can cause skeletal muscle damage, myoglobinuria, and hyperkalemia.[10,11] There is substantial geographical variation in the composition of venom from sea snakes, probably as a result of differences in the prey of these snakes.[12]

Venom Apparatus

Sea snakes have proteroglyphous dentition similar to all elapids. The front fangs are small (2–7 mm/~0.1–0.3 in.) and hollow with the open tip communicating with venom glands. Fang marks are characterized by one to four tiny, punctate wounds that appear inconspicuous.[13] Most sea snake bites do not result in envenomation.[14] Multiple bite marks and serrated lacerations suggest serious envenomation when accompanied by symptoms.[14,15]

DOSE RESPONSE

Although the sea snake neurotoxin is more potent than terrestrial elapid venom, the small fangs inject substantially less venom from their crude venom apparatus compared with terrestrial elapids. Consequently, envenomation is less common with sea snakes than terrestrial elapids.

CLINICAL RESPONSE

Most sea snake bites do not result in envenomation,[16] and symptoms of envenomation may not develop even if abrasions or bite marks with surrounding erythema are present.[17] Characteristically, sea snake envenomations produce minimal local pain and swelling in contrast to the stings of other marine animals.[18] Occasionally, regional lymph node enlargement may occur near the site of the bite.[19] Neurologic symptoms typically begin within 30–60 minutes. Symptoms of envenomation include generalized myalgias, malaise, anxiety, euphoria, diaphoresis, and vomiting. Serious signs of envenomation include bulbar palsy (dysphagia, ptosis, external ocular muscle weakness, pupillary dilation, dysarthria), progressive muscle weakness, and respiratory depression. Muscle fasciculations do not usually occur. Clinical evidence of serious envenomations usually occurs within 2 hours unless delayed by the use of tourniquets on the affected extremity. The absence of symptoms within 6–8 hours indicates that envenomation did not occur.

DIAGNOSTIC TESTING

Envenomation by sea snakes does not typically produce significant laboratory abnormalities, such as coagulation disorders, electrolyte imbalance, rhabdomyolysis, disseminated intravascular coagulation, or hepatorenal dysfunction. Laboratory evidence of rhabdomyolysis (dark urine, severe myalgias, myoglobinuria, hyperkalemia, hyperphosphatemia, hypocalcemia) develops occasionally, particularly following envenomation by the common sea snake (*Enhydrina schistosa* Daudin).[20] Myoglobinuria commonly begins about 3–6 hours after envenomation along with evidence of rhabdomyolysis. In addition, skeletal muscle enzyme concentrations are often elevated and a leukocytosis frequently occurs. In serious cases, hyperkalemia and acute renal tubular necrosis may occur within several days of envenomation.

TREATMENT

First Aid

Reassurance, immobilization of the affected limb in a dependent position, and immediate medical attention at a facility that possesses antivenin are the most important first-aid measures. Absorption of sea snake venom is rapid. Animal studies suggest that a lymph-restrictive bandage may be useful at the site of the envenomation if applied immediately (i.e., within the first 5–10 minutes of envenomation).[21] A thick gauze pad placed directly over the bite wound with a wide circumferential bandage provides lymphatic-occlusive pressure. Previously applied tourniquets should not be removed until supportive care (intravenous access, respiratory equipment) is available because of the potential deterioration of the patient following tourniquet removal.[15] The use of tourniquets is not part of the hospital management of snakebites. Incision and drainage even within the first 10 minutes is probably not efficacious.

Stabilization

Respiratory depression and aspiration are the immediate threatening complications. Symptomatic patient should be monitored for hyperkalemia as well as impending respiratory failure (i.e., cyanosis, increasing lethargy, disorientation, hypoxemia, hypercarbia).

Antivenom

Polyvalent equine sea snake antivenin (Commonwealth Laboratories, Melbourne, Australia) is indicated for all symptomatic patients (e.g., moderate myalgias, neurologic signs). Skin testing for horse serum sensitivity is usually recommended prior to the administration of antivenom. This antivenin is produced from the venoms of the common sea snake (*Enhydrina schistosa* Daudin) and the Australian tiger snake (*Notechis scutatus* Peters). *In vitro* studies indicate that this antivenom is effective against the potent postsynaptic neurotoxicity of all Indo-Pacific sea snakes.[22] Although occasionally used when sea snake venom is unavailable, tiger snake (*Notechis scutatus* Peters) antivenom is not as effective *in vitro* against sea snake venom as the specific sea snake antivenom. The sea snake antivenin should be administered as soon as possible (i.e., within 8 hours of envenomation) to all patients demonstrating signs of envenomation.[11] The initial dose is 1–3 vials intravenously, with further vials titrated to the clinical condition of the patient.

Supplemental Care

Patients with bites that do not produce symptoms (e.g., myalgias, muscle stiffness, ptosis, trismus, muscle weakness) within 6–8 hours of the bite can be safely discharged. Patients with serious envenomations should be followed closely for signs of respiratory failure and myoglobinuria. Elevation of serum creatine phosphokinase concentrations above 1000 U/L suggests the need to maintain good diuresis (2–3 mL/kg/h), particularly if myoglobinuria occurs. Envenomated patients should be well hydrated to reduce the risk of myoglobin-induced renal damage. There is limited evidence that alkalinization of the urine (pH>6.5) increases myoglobin clearance. The patient should be monitored at least daily for hyperkalemia and renal failure. Hypocalcemia during rhabdomyolysis does not usually require treatment.

References

1. Mebs D. Venomous and poisonous animals. A handbook for biologists, toxicologists, toxinologists, physicians and pharmacists. Stuttgart: Medpharm Scientific Publishers; 2002.
2. Reid HA. Antivenom in sea snake bit poisoning. Lancet 1975;i:622–623.
3. McGoldrick J, Marx JA. Marine envenomations; part 1: vertebrates. J Emerg Med 1991;9:497–502.
4. Watt G, Theakston DG. Seasnake bites in a freshwater lake. Am J Trop Med Hyg 1985;34:770–773.
5. Lobo AS, Vasudevan K, Pandav B. Tophic ecology of *Lapemis curtus* (Hydrophiinae) along the Western coast of India. Copeia 2005;(3):637–641.
6. Gawade SP, Gaitonde BB. Isolation and characterization of toxic components from the venom of the common Indian sea snake (*Enhydrina schistosa*). Toxicon 1982;20:797–801.
7. Mori N, Ishizaki H, Tu AT. Isolation and characterization of *Pelamis platurus* (yellow-bellied sea snake) postsynaptic isoneurotoxin. J Pharm Pharmacol 1989;41:331–334.
8. Ali SA, Alam JM, Stoeva S, Schutz J, Abbasi A, Zaidi ZH, Voelter W. Sea snake *Hydrophis cyanocinctus* venom. I. Purification, characterization and N-terminal sequence of two phospholipases A_2. Toxicon 1999;37:1505–1520.
9. Ali SA, Alam JM, Abbasi A, Zaidi Z, Stoeva S, Voelter W. Sea snake *Hydrophis cyanocinctus* venom. II. Histopathological changes, induced by a myotoxic phospholipase A_2 (PLA_2-H1). Toxicon 2000;38;687–705.

10. Lind P, Eaker D. Amino acid sequence of a lethal myotoxic phospholipase A2 from the venom of the common sea snake (*Enhydrina schistosa*). Toxicon 1981;19:11–24.
11. Marsden AT, Reid HA. Pathology of sea-snake poisoning. Br Med J 1961;1:1290–1293.
12. Li M, Fry BG, Kini RM. Eggs-only diet: its implications for the toxin profile changes and ecology of the marbled sea snake (*Aipysurus eydouxii*). J Mol Evol 2005;60: 81–89.
13. Heatwole H. Adaptation of marine snakes. Am Sci 1978;66:594–604.
14. Reid HA. Diagnosis, prognosis and treatment of sea-snake bite. Lancet 1961;ii:399–402.
15. Mercer HP, McGill JJ, Ibrahim RA. Envenomation by sea snake in Queensland. Med J Aust 1982;1:130–132.
16. Reid HA. Myoglobinuria and sea snake-bite poisoning. Br Med J 1961;1:1284–1289.
17. Senanayake MP, Ariaratnam CA, Abeywickrema S, Belligaswatte A. Two Sri Lankan cases of identified sea snake bites, without envenoming. Toxicon 2005;45:861–863.
18. Solorzano A. A case of human bite by the pelagic sea snake, *Pelamis platurus* (Serpentes: Hydrophiidae). Rev Biol Trop 1996;43:321–322.
19. Amarasekera N, Jayawardena A, Ariyaratnam A, Hewage UC, De Silva A. Bite of a sea snake (*Hydrophis spiralis*): a case report from Sri Lanka. J Trop Med Hyg 1994;97:195–198.
20. Mercer HP, McGill JJ, Ibrahim RA. Envenomation by sea snake in Queensland. Med J Aust 1981;1:130–132.
21. Anker RL, Straffon WG, Loiselle DS, Anker KM. Retarding the uptake of "mock venom" in humans: comparison of three first-aid treatments. Med J Aust 1982;1: 212–214.
22. Chetty N, Du A, Hodgson WC, Winkel K, Fry BG. The *in vitro* neuromuscular activity of Indo-Pacific sea-snake venoms: efficacy of two commercially available antivenoms. Toxicon 2004;44:193–200.

III Marine Invertebrates

Chapter 178

CONE SHELLS and BLUE-RINGED OCTOPUS (Phylum: Mollusca)

CONE SHELLS

IDENTIFYING CHARACTERISTICS

Conus species comprise a large genus of predatory marine snail, and there are over 500 species of venomous *Conus* species found in all tropical marine habitats. Cone shells are frequently divided into the following groups based on the prey envenomated: worm-hunting (vermivorous), snail-hunting (molluscivorous), and fish-hunting (piscivorous).[1] The fish-hunting cone shells are potentially the most dangerous marine snails to humans.[2]

Common Name: Geographic Cone Shell
Scientific Name: *Conus geographicus* Röding
Family: Conidae
Physical Description: This dangerous, cone-shaped univalve has an attractive shell with a slit-like aperture running the full length of the shell. This species is one of the largest *Conus* species with a shell length up to 15 cm (~6 in.).

EXPOSURE

Clinical drugs from *Conus* species include ziconotide, which is a potent analgesic for opioid-refractory pain as a result of the blockade of *N*-type calcium channels involved in pain pathways. The US Food and Drug Administration (FDA) approved the intrathecal infusion of ziconotide as a treatment for severe chronic or refractory pain.[3]

Geographic Distribution

Cone shells inhabit shallow and deep topical waters of the Indian and Pacific Oceans east to Hawaii, north to Okinawa and south to New Zealand including the whole coast of Australia. A few *Conus* species live in cooler water. The only known species (*Conus ventricosus* Gmelin or *Conus mediterraneus*) in the Mediterranean Sea has not been associated with human envenomation.[4] These organisms are most abundant in shallow waters, particularly among coral reefs where many species inhabit the same reef.

Behavior

Predatory snails from the family Conidae are a diverse group of marine mollusks that use venom to capture prey (fish, mollusks, polychaete worms). These predators feed on worms by paralyzing prey with a detachable radula tooth. During the day, the cone shell seeks shelter under sand or stones. These nocturnal creatures are often at least partially buried in the sand during daylight hours. Most human envenomations occur during the handling of the cones that may wash ashore after storms or dredging operations. Inadvertent contact while digging in sand or coral debris may also occur. Typically, the venom from cone shells that prey on worms and snails are less toxic than other cone shells.

Medical Toxicology of Natural Substances, by Donald G. Barceloux, MD

PRINCIPAL TOXINS

Venom Composition

Conus species produce water soluble toxins that disrupt neuromuscular transmission to paralyze their prey. The venom from each species is distinct, and individual venoms contain over 100 biologically active peptides and proteins.[5] The major biologically active compounds in *Conus* venoms are small (12–50 residues), extensively disulfide-cross-linked, neurotoxic peptides (conotoxins) that target ligand-gated or voltage-gated ion channels, G-protein-coupled receptors, or neurotransmitter transporters.[6] The internal disulfide bonds add stability to the peptide, while altering the conformation to optimize interaction with specific receptors. The venoms of each group of cone shells vary according to the target prey (e.g., piscivorous vs. vermivorous) as well as between individual specimens depending on the specific physiological end-point or "cabal."[7] The "lightning-strike cabal" of fish-hunting *Conus* species rapidly immobilizes prey by injecting peptides that inhibit voltage-gated Na^+ and K^+ channels resulting in massive depolarization of axons in the prey. The venom from these fish-hunting cone shells also contains peptides that inhibit the presynaptic Ca^{++} channels, postsynaptic nicotinic receptors, and the Na^+ channels. This "motor cabal" causes paralysis of the prey by inhibition of neuromuscular transmission.

There are over 5,000 unique peptide sequence present in approximately 500 *Conus* species, and these peptides are arranged into superfamilies (e.g., A, M, O) based on the characteristic arrangement of cysteine residues in the mature peptides.[8] Most families of toxins target a specific, voltage-gated ion channel (Na^+, K^+, Ca^{++}), ligand-gated ion channel (nicotinic acetylcholine, *N*-methyl-D-aspartate, 5-HT-3, α-adrenergic), G-protein-coupled receptors (neurotensin, vasopressin) or neurotransmitter transporters.[9] Different conotoxin families may bind to different sites on the same ion channel (e.g., μ-conotoxins bind to site 1 of the sodium channel whereas δ-conotoxins bind to site 6). Table 178.1 lists some conotoxin families and their respective molecular targets. Peptides of the same superfamily usually share a characteristic sequence of cysteine residues in the mature region of the toxin.[10] Members of the A-superfamily (cysteine framework: CC-C-C-C-C) contain competitive nicotinic acetylcholine receptor antagonists (α-conotoxins, αA-conotoxins) and inhibitors of voltage-gated potassium channels (excitatory κA-conotoxins),[11] whereas the M-superfamily includes voltage-gated sodium channel blockers (μ-conotoxins) and noncompetitive antagonists of the nicotinic acetylcholine receptor (ψ-conotoxins). The M-superfamily

TABLE 178.1. Superfamilies of Disulfide-Rich Conopeptides

Superfamily	Primary Target	Conotoxin Family
O	Sodium channel	δ, μO
	Potassium channel	κ
	Calcium channel	ω
M	Sodium channel	μ
	Potassium channel	κM
	Nicotinic ACh	ψ
A	Potassium channel	K
	Nicotinic ACh	α, αA
	α-Adrenergic receptor	ρ
S	Serotonin receptor	σ
T	Norepinephrine transporter	χ
	Nicotinic receptors	αS
P	Unknown	Unknown
I_1/I_2	Potassium channel	None defined
J	Potassium channel (?)	None defined

conotoxins contain similar cysteine arrangement (CC-C-C-CC) with at least four subfamilies that target different types of ion channels.[12] The O-superfamily (C-C-CC-C-C) is a highly diverse superfamily that contains a variety of active agents including voltage-sensitive calcium channel blockers (ω-conotoxins), delayed inactivators of the voltage-sensitive sodium channel (δ-conotoxins), voltage-gated sodium channel blockers (μO-conotoxins), and voltage-gated potassium channel blockers (κ-conotoxins).[13] The T-superfamily conotoxins are small peptides (9–17 amino acids) that occur in all three major feeding types of cone shells. This superfamily consists of τ-conotoxins (cysteine framework: CC-CC) and χ-conotoxins (cysteine framework: CC-C-C).[14] The other major group of *Conus* peptides contains no or one disulfide linkage. The wide diversity of structures in these superfamilies results in the continual discovery of new superfamilies, such as the V-superfamily (C-C-CC-C-C-C-C).[15] The pharmacological properties of this group of peptides include *N*-methyl-D-aspartate (NMDA)-glutamate receptor antagonists (conantokin), neurotensin-like peptides (contulakin), and vasopressin agonists (conopressin), as well as toxins with poorly defined targets (conorfamide, contryphans).[8]

The I-conotoxin superfamily has four disulfide bonds with the cysteine sequence C-C-CC-CC-C-C; this family was divided into two different gene superfamilies (I_1, I_2) based on different types of signal peptides and distinct functions. Conotoxins from the I_1 superfamily have unique posttranslational modifications at the C-terminus, whereas conotoxins from the I_2 superfamily lack the propeptide region.[16]

Venom Apparatus

All *Conus* species have a venom apparatus that includes a long, highly convoluted venom duct (synthesis, storage of venom), a venom bulb for pumping out the venom, a radula sheath (formation of radula teeth), and harpoon-like teeth along with a long distensible proboscis for delivery of the venom to the prey. Thin-shelled species with wide openings have a more developed venom apparatus; therefore, these species are potentially more harmful.[17] The cone shells have sharp radicular, barbed teeth and a muscular proboscis that propels the hollow tooth (radula tooth) into the prey. Upon touching the prey, the extensible proboscis projects one small (6–8 mm/~0.2–0.3 in.), hollow tooth outward surrounded by the circular muscles at the anterior tip of the proboscis. The venom is then injected into the prey through the hollow tooth.[18] Figure 178.1 displays the venom apparatus of *Conus* species.

Dose Response

In general, the venom of cone shell species is specific to that species; therefore, the potency of venom varies between species.[7] The crude venom from piscivorous cone shells is usually more potent than vermivorous or molluscivorous species. The intraperitoneal LD_{50} of the more potent conotoxins from piscivorous cone shells in rodents is approximately 3–6 μg/kg, similar to ricin. In a study of crude venom from the vermivorous species, *Conus loroisii* Kiener, collected from the Bay of Bengal, the LD_{50} of the crude venom in mice was about 5 mg/kg following intraperitoneal injection.[19] Ziconotide is the synthetic analog of the ω-conotoxin from *Conus magus*.[20] This *N*-type calcium channel blocker inhibits the transmission of nociceptive stimuli, and this drug is approved as nonopioid treatment for chronic pain.

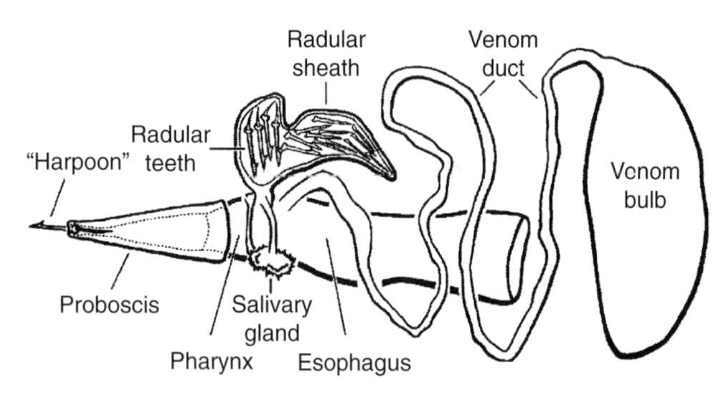

FIGURE 178.1. Venom apparatus of *Conus* species. From Norton RS, Olivera BM. Conotoxins down under. Toxicon 2006;48:782. Reprinted with permission from Elsevier.

HISTOPATHOLOGY AND PATHOPHYSIOLOGY

Every conotoxin has a highly specific ligand that binds to a biologically active receptor, primarily on voltage-gated or ligand-gated ion channels and, in a few cases, on G-protein-linked receptors.[21] These peptides affect the functioning of different voltage-gated ion channels and neurotransmitter-gated receptors. α-Conotoxins are potent nicotinic antagonists similar to cobratoxin and α-bungarotoxin, whereas μ-conotoxins are sodium channel blockers similar to tetrodotoxin and saxitoxin. Voltage-sensitive neuronal calcium channels are the target of ω-conotoxins. The α-conotoxins are competitive antagonists of the nicotinic acetylcholine receptor that can distinguish between muscle- and neuronal-type nicotinic acetylcholine receptors.[22] These nicotinic antagonists target specific receptor subtypes that allow the venom to cause paralysis (muscle-type receptor) and inhibition of the flight and fight response (neural receptors).

CLINICAL RESPONSE

The severity of an envenomation varies substantially depending in part on the cone shell species. Typically, envenomation by a cone shell causes variable pain followed by swelling and numbness at the puncture site. Local effects include immediate pain varying from mild to severe, followed by local edema and erythema that frequently resolve within 1 hour. Hypoesthesia, hyperesthesias, local cyanosis, and ischemia may occur in more serious envenomations, and the swelling and sensitivity may last several weeks. More serious cases involve generalized muscle weakness and neurological signs characterized by ataxia, hoarseness, dysphagia, diplopia, perioral paresthesias, slurred speech, weakness, paralysis, and dyspnea. Intubation may be required as a result of respiratory failure. Gastrointestinal symptoms with the exception of nausea are not usually part of the clinical presentation. Symptoms typically resolve within 6–8 hours.[23] A 29-year-old Filipino male developed altered consciousness, swelling at the site of envenomation, and right-sided facial weakness 1 hour after placing a cone shell snail (*C. geographicus*) next to his extremity while diving.[24] He suddenly developed a respiratory arrest during transfer that did not respond to oxygen, noradrenaline, and artificial ventilation.

DIAGNOSTIC TESTING

The peptides in *Conus* venom are identified by cDNA cloning and polymerase chain reaction (PCR) amplification to determine the amino acid sequence by using primers designed according to the signal sequence of the subfamily or superfamily genes.[25]

TREATMENT

Treatment is supportive with immobilization of the victim, immersion of the affected area in hot water, and careful cleaning of the wound. Hot water immersion (110–115 °F/43–46 °C) for 30 minutes may provide some relief, but there are few clinical data to guide management. After immersion, analgesics may be necessary, although symptoms usually abate by 1 hour. Patients with systemic symptoms should be observed closely for signs of respiratory depression for at least 6 hours. Rarely, ventilatory support is necessary. No antivenin is available.

BLUE-RINGED OCTOPUS

IDENTIFYING CHARACTERISTICS

Three species of blue-ringed octopi (*Hapalochlaena maculosa* Hoyle, *H. lunulata* Quoy and Gaimard, *H. fasciata* Hoyle) are the only species of the 650 species in the class Cephalopoda (e.g., octopus, squid, cuttlefish) that cause serious evenomation in humans.

Common Name: Southern or Lesser Blue-Ringed Octopus

Scientific Name: *Hapalochlaena maculosa* Hoyle

Physical Description: This octopus is about 15–20 cm/~6–8 in. with tentacles extended and weighs <100 g. The lesser blue-ringed octopus has dark brown-yellow bands that cover the body and arms along with iridescent blue rings. These colors darken and the rings fluoresce following stimulation of the octopus.

EXPOSURE

Geographic Distribution

The distribution of the blue-ringed octopus ranges from the Indo–West Pacific north to Okinawa, east to the Philippines, west to India, and south to New Zealand. The southern or lesser blue-ringed octopus (*H. maculosa* Hoyle) inhabits the warm shallow coastal waters of Australia, whereas the range of *Hapalochlaena lunulata* Quoy and Gaimard (greater blue-ringed octopus) includes tropical waters along the northern coastline of Australia near crevices, rock pools, and small caves as well as tropical waters of the western Pacific Ocean. The blue-lined octopus (*Hapalochlaena fasciata* Hoyle) lives in shallow, intertidal rocky shores and coastal waters from southern Queensland to southern New South Wales. All reported cases of envenomation by the blue-ringed octopus have occurred in Australia.

Behavior

The blue-ringed octopus captures prey (crabs, other crustaceans) either by discharging saliva into the vicinity of the prey and waiting for the paralysis of the prey or by covering the prey with the web between its arms to maintain high concentrations of the venom. Aggressive or hungry blue-ringed octopi may dart onto the back of the prey and inject the venom with its beak. Normally, this octopus is nocturnal and shy. Human envenomation by the blue-ringed octopus frequently occurs during inspection of the small, colorful octopus by an unsuspecting victim.

PRINCIPAL TOXINS

Venom Composition

The posterior salivary glands of the blue-ringed octopus contain the neurotoxin, maculotoxin, which is chemically identical to tetrodotoxin.[26,27] The eggs of the Australian blue-ringed octopus also contain tetrodotoxin along with most other soft body parts.[28] In a study of six juvenile blue-ringed octopuses (*Hapalochlaena maculosa*), tetrodotoxin was present in all three body parts (arms > abdomen > cephalothorax) tested.[29] Analysis of the posterior salivary gland and other soft parts from six specimens of *H. maculosus* demonstrated the presence of tetrodotoxin-producing strains of bacteria including strains from the genera *Alteromonas*, *Bacillus*, *Pseudomonas*, and *Vibrio*.[30] Other substances present in the venom or saliva of the octopus include proteolytic enzymes and various amines (e.g., hyaluronidase, hydroxyphenylethanolamine, noradrenaline, serotonin, dopamine, acetyl choline, tryptamine, taurine, tyramine, histamine).[31] The blue-ringed octopus is resistant to the effects of high concentrations of tetrodotoxin and venom extract from their posterior salivary glands, but the mechanism of this resistance is not known.[32] In animal models, tetrodotoxin is several times more toxic than structurally related compounds (e.g., anhydrotetrodotoxin, methoxytetrodotoxin).[33]

Venom Apparatus

Posterior salivary glands are connected to a sharp beak that is located near the central portion at the base of the arms.[34] These glands form compound tubuloacinar exocrine structures that contain muscle fibers in the interstitial regions of the secretory tubules.[35] Contractions of these fibers force secretory material along the salivary ducts into the intestinal blood space. These fibers probably are under voluntary control. The bite usually produces two small, painless puncture wounds that may bleed slightly.

Dose Response

The minimal lethal dose of tetrodotoxin varies substantially between animal species.[33] The oral LD_{50} of tetrodotoxin in mice is approximately 332 µg/kg.[36] Although the hypotensive effects of saxitoxin are weaker than tetrodotoxin, the lethal dose of tetrodotoxin and saxitoxin in animal models is similar.[37]

HISTOPATHOLOGY AND PATHOPHYSIOLOGY

The neurotoxin, maculotoxin, is structurally identical to tetrodotoxin, and presumably the mechanism of toxicity of the venom and tetrodotoxin is similar. Tetrodotoxin binds to a site near the external orifice of the sodium channel and blocks the sodium channel selectively without affecting other types of voltage-activated and transmitter-activated ion channels.[38] The binding and penetration of calcium ions to the sodium channel is probably responsible for the voltage-dependent block. In animal models, the parenteral administration of the extract from posterior salivary glands of *H. maculosa* produces paralysis of the respiratory musculature resulting in anoxia.[39] Rapid administration of this substance also causes severe hypotension secondary to decreased peripheral vascular resistance that typically responds to noradrenaline infusion.

CLINICAL RESPONSE

The bite of the blue-ringed octopus usually is painless, but bleeding may occur at the puncture site. The severity of envenomation ranges from mild to severe.[40] Severe envenomations by the blue-ringed octopus produce the rapid onset of muscle weakness and respiratory depression.[41] Typically, lightheadedness, generalized weakness, nausea, and vomiting begin within 10 minutes of serious envenomations followed by neurological signs (slurred speech, dysphagia, paresthesias, muscle weakness, blurred vision, diplopia). Other neuromuscular signs include myoclonic jerks and muscle spasms. Symptoms usually resolve within 24 hours. Case reports suggest that anaphylactoid, and perhaps anaphylactic, responses may occur after envenomation characterized by pruritus, generalized urticaria, and synovitis.[42]

DIAGNOSTIC TESTING

Methods for the detection of tetrodotoxin include immunoaffinity chromatography and mass spectrometry after conversion of tetrodotoxin to a C9-based derivative. A simple method for the detection of tetrodotoxin in serum and urine samples involves clean up by solid-phase extraction followed by high performance liquid chromatography with post-column derivatization and fluorescence detection.[43] The limits of quantitation for tetrodotoxin in serum and urine samples using this method are 5 ng/mL and 20 ng/mL, respectively. Depending on the dose, detection times for tetrodotoxin in serum and urine samples are less than 1 day and 4–5 days, respectively.

TREATMENT

There are no specific first aid measures. In some locations, the use of compression bandages is recommended, but there are few clinical data to guide management. Because of the potential for respiratory failure following severe envenomations, the victim should be evaluated carefully for respiratory depression with pulse oximeter and evaluation of the adequacy of respirations. Respiratory failure may occur rapidly following severe envenomation, and intubation of the trachea may be necessary to support oxygenation. Otherwise, management is supportive. There are no clinical data to support the use of wound excision and hot water immersion. Patients displaying neurotoxic effects should be observed for 4–8 hours and admitted to an intensive care facility if symptoms progress.

References

1. Terlau H, Shon KJ, Grilley M, Stocker M, Stuhmer W, Olivera BM. Strategy for rapid immobilization of prey by a fish-hunting marine snail. Nature 1996;381:148–151.
2. Norton RS, Olivera BM. Conotoxins down under. Toxicon 2006;48:780–798.
3. Prommer E. Ziconotide: a new option for refractory pain. Drugs Today 2006;42:369–378.
4. Mebs D. Venomous and poisonous animals. A handbook for biologists, toxicologists and toxinologists, physicians and pharmacists. Stuttgart: Medpharm Scientific Publishers; 2002.

5. Zhangsun D, Luo S, Wu Y, Shu X, Hu Y, Xie L. Novel O-superfamily conotoxins identified by cDNA cloning from three vermivorous *Conus* species. Chem Biol Drug Des 2006;68:256–265.
6. Grant MA, Morelli XJ, Rigby AC. Conotoxins and structural biology: a prospective paradigm for drug discovery. Curr Protein Pept Sci 2004;5:235–248.
7. Jakubowski JA, Kelley WP, Sweedler JV, Gilly WF, Schulz JR. Intraspecific variation of venom injected by fish-hunting *Conus* snails. J Exp Biol 2005;208:2873–2883.
8. McIntoxh JM, Jones RM. Cone venom-from accidental stings to deliberate injection. Toxicon 2001;39:1447–1451.
9. Terlau H, Olivera BM. *Conus* venoms: a rich source of novel ion channel-targeted peptides. Physiol Rev 2004;84: 41–68.
10. Olivera BM. *Conus* peptides: biodiversity-based discovery and exogenomics. J Biol Chem 2006;281:31173–31177.
11. Teichert RW, Jacobsen R, Terlau H, Yoshikami D, Olivera BM. Discovery and characterization of the short κA-conotoxins: a novel subfamily of excitatory conotoxins. Toxicon 2007;49:318–328.
12. Han Y-H, Wang Q, Jiang H, Liu L, Xiao C, Yuan D-D, et al. Characterization of novel M-superfamily conotoxins with new disulfide linkage. JEBS J 2006;273:4972–4982.
13. Heinemann SH, Leipold E. Conotoxins of the O-superfamily affecting voltage-gated sodium channels. Cell Mol Life Sci 2007;64:1329–1340.
14. Luo S, Zhangsun D, Wu Y, Shu X, Xie L, Hu Y, et al. Identification and molecular diversity of T-superfamily conotoxins from *Conus lividus* and *Conus litteratus*. Chem Biol Drug Des 2006;68:97–106.
15. Peng C, Liu L, Shao X, Chi C, Wang C. Identification of a novel class of conotoxins defined as V-conotoxins with a unique cysteine pattern and signal peptide sequence. Peptides 2008;29:985–991.
16. Mondal S, Babu RM, Bhavna R, Ramakumar S. I-conotoxin superfamily revisited. J Peptide Sci 2006;12: 679–685.
17. Kohn AJ. Cone shell stings. Recent cases of human injury due to venomous marine snails of the genus *Conus*. Hawaii Med J 1958;17:528–532.
18. Hinegardner RT. The venom apparatus of the cone shell. Hawaii Med J 1953;17:533–536.
19. Saminathan R, Babuji S, Sethupathy S, Viswanathan P, Balasubramanian T, Gopalakrishanakone P. Clinico-toxinological characterization of the acute effects of the venom of the marine snail, *Conus loroisii*. Acta Trop 2006;97:75–87.
20. Prommer E. Ziconotide: a new option for refractory pain. Drugs Today (Barc) 2006;42:369–378.
21. Olivera BM, Cruz LJ. Conotoxins, in retrospect. Toxicon 2001;39:7–14.
22. Arias HR, Blanton MP. α-Conotoxins. Int J Biochem Cell Biol 2000;32:1017–1028.
23. Petrauska LE. A case of cone shell poisoning by bite in Manus Island. Papua New Guinea Med J 1955;1:67–68.
24. Rice RD, Halstead BW. Report of fatal cone shell sting by *Conus geographicus* Linnaeus. Toxicon 1965;5:223–224.
25. Kauferstein S, Melaun C, Mebs D. Direct cDNA cloning of novel conopeptide precursors of the O-superfamily. Peptides 2005;26:361–367.
26. Croft JA, Howden ME. Chemistry of maculotoxin: a potent neurotoxin isolated from *Hapalochlaena maculosa*. Toxicon 1972;10:645–651.
27. Sheumack DD, Howden ME. Maculotoxin: a neurotoxin from the venom glands of the octopus *Hapalochlaena maculosa* identified as tetrodotoxin. Science 1978;199: 188–189.
28. Sheumack DD, Howden ME, Spence I. Occurrence of a tetrodotoxin-like compound in the eggs of the venomous blue-ringed octopus (*Hapalochlaena maculosa*). Toxicon 1984;22:811–812.
29. Yotsu-Yamashita M, Mebs D, Flachsenberger W. Distribution of tetrodotoxin in the body of the blue-ringed octopus (*Hapalochlaena maculosa*). Toxicon 2007;49: 410–412.
30. Hwang DF, Arakawa O, Saito T, Noguchi T, Simidu U, Tsukamoto K, et al. Tetrodotoxin-producing bacteria from the blue-ringed octopus *Octopus maculosus*. Marine Biol 1989;100:327–332.
31. Hartman WJ, Clark WG, Cyr SD, Jordon AL, Leibhold RA. Pharmacologically active amines and their biogenesis in the octopus. Ann N Y Acad Sci 1960;90:637–666.
32. Flachsenberger W, Kerr DIB. Lack of effect of tetrodotoxin and of an extract from the posterior salivary gland of the blue-ringed octopus following injection into the octopus and following application of its brachial nerve. Toxicon 1985;23:997–999.
33. Kao CY. Tetrodotoxin, saxitoxin and their significance in the study of excitation phenomena. Pharmacol Rev 1966;18:997–1049.
34. Sutherland SK, Lane WR. Toxins and mode of envenomation of the common ringed or blue-banded octopus. Med J Aust 1969;1:893–898.
35. Gibbs PJ, Greenaway P. Histological structure of the posterior salivary glands, in the blue ringed octopus *Hapalochlaena maculosa* Hoyle. Toxicon 1978;16:59–70.
36. Sakai F, Sato A, Uraguchi K. [Respiratory paralysis due to tetrodotoxin]. Naunyn Schmiedebergs Arch Exp Pathol Pharmakol 1961;240:313–321. [German]
37. Kao CY. Pharmacology of tetrodotoxin and saxitoxin. Fed Proc 1972;31:1117–1123.
38. Narahashi T, Roy ML, Ginsburg KS. Recent advances in the study of mechanism of action of marine neurotoxins. Neurotoxicology 1994;15:545–554.
39. Flachsenberger WA. Respiratory failure and lethal hypotension due to blue-ringed octopus and tetrodotoxin envenomation observed and counteracted in animal models. Clin Toxicol 1986–87;24:485–502.
40. Walker DG. Survival after severe envenomation by the blue-ringed octopus (*Hapalochlaena maculosa*). Med J Aust 1983;2:663–665.

41. Lane WR, Sutherland SK. The ringed octopus bite: A unique medical emergency. Med J Aust 1967;2:475–476.
42. Edmonds C. A nonfatal case of blue ringed octopus bite. Med J Aust 1969;2:601.
43. O'Leary MA, Schneider JJ, Isbister GK. Use of high performance liquid chromatography to measure tetrodotoxin in serum and urine of poisoned patients. Toxicon 2004;44:549–553.

Chapter 179

JELLYFISH, HYDROIDS, SEA ANEMONES, and CORALS (Phylum: Cnidaria)

Cnidarians (coelenterates) are marine animals with long tentacles that contain multiple venomous stinging cells (nematocysts). This phylum is subdivided into Cubozoa (families: Chirodropidae, Carybdeidae), Hydrozoa (hydroids), Scyphozoa (jellyfish), and Anthozoa (sea anemones, true corals). Cnidarians in this phylum are simple metazoans with radial, biradial, or radio-bilateral symmetry.[1] These delicate creatures are covered with two epithelial layers that surround one internal cavity called the coelenteron. This gastrovascular cavity functions as both a gastrointestinal tract and a circulatory system. The dominant characteristic of this group is the presence of long tentacles surrounding the mouth to this cavity. These tentacles contain stinging organelles called nematocysts that vary in size between 0.005–1 mm (~0.0002–0.04 in.).

Two morphological forms of coelenterates (polyp, medusa) appear alternately during their life cycle. The polyp is the sedentary form that attaches to objects via the caudal end, and polyp feeds through the oral end. The medusa is the mobile form that contains a dome-shaped cavity and a mouth surrounded by tentacles on the concave undersurface. The highest concentrations of nematocysts appear near the mouth and outer surfaces of the tentacles. The cnidarians (coelenterates) cause more marine envenomations than species within any other phylum.

The generic term, jellyfish refers to a multitude of free-floating medusa (jellyfish) belonging to the phylum *Cnidaria*. The fourth class of *Cnidaria*, the anthozoans (e.g., corals, sea anemones), lack the medusa stage and remain attached to the sea bed in the form of a polyp. Figure 179.1 outlines the currently recognized members of the phylum that cause clinically significant envenomations in humans. The specific jellyfish involved in serious envenomations is frequently not identified,[2] and therefore other seriously venomous jellyfish probably exist.

TRUE JELLYFISH and BOX-JELLYFISH

This class of cnidarians is composed of the true medusae or larger jellyfish. Species of true jellyfish (Class: Scyphozoa) are the most conspicuous jellyfish based on the large size and coloration of the medusa stage, such as *Nemopilema nomurai* Kishinouye (Nomura's jellyfish or *Stomolophus nomurai*).[3] These creatures have eight notches along the margin of their dome. The most important venomous species of jellyfish occur in the class Cubozoa, which includes chirodropids (box-jellyfish—*Chironex fleckeri* Southcott, *Chiropsalmus quadrigatus* Haeckel) and carybdeids (Irukandji jellyfish).[4] The carybdeids usually have only one tentacle arising from each corner of the bell, whereas chirodropids usually have many tentacles connected to each

Medical Toxicology of Natural Substances, by Donald G. Barceloux, MD

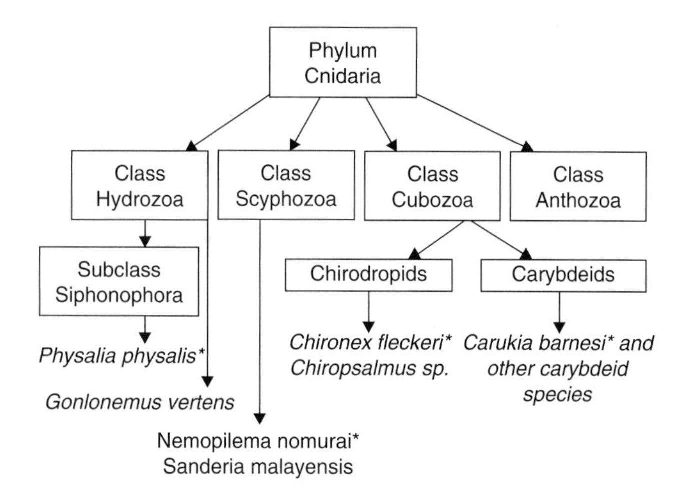

FIGURE 179.1 Overview of medically significant jellyfish with (*) indicating lethal or potentially lethal species. Adapted from Winkel KD, et al. Jellyfish antivenoms: past, present, and future. J Toxicol Toxin Rev 2003;22:117.

corner. Although some species of hydroids (*Physalia physalis* L., *Gonionemus vertens* Agassiz) resemble true jellyfish, these multitentacled hydroids are colonies of hydroids rather than true jellyfish. Fatalities occur as a result of envenomation by species in the three main Cnidarian classes, but the majority of deaths result from contact with chirodropids (i.e., *Chironex fleckeri* Southcott, *Chiropsalmus quadrigatus* Haeckel), primarily in children.[5]

HISTORY

The first reported fatality from envenomation by the North Australian box-jellyfish (*Chironex fleckeri* Southcott) occurred in 1883. In 1944, Southcott observed two different syndromes following jellyfish stings in troops stationed in north Queensland during World War II.[6] In the first syndrome, severe cutaneous pain and markings developed without systemic complaints, whereas the second type caused minor skin marks followed by the delayed onset of severe systemic symptoms. The jellyfish associated with former syndrome (severe cutaneous pain) was discovered in 1955, when Flecker caught several large, box-shaped jellyfish after the death of a young child. Later, Southcott confirmed that this new species causes this syndrome. He named the new species (*Chironex fleckeri* Southcott) after Flecker.[7] In 1952, Flecker named the delayed syndrome, the Irukandji syndrome, after the Aboriginal tribe that inhabited the region where most of the stings occurred.[8] Barnes discovered one of the sources of Irukandji syndrome in 1964 when he and his son developed the syndrome after being stung by a small jellyfish (*Carukia barnesi* Southcott) caught in the local area.[9]

IDENTIFYING CHARACTERISTICS

True jellyfish are the umbrella-shaped, gelatinous medusa stage of a hydroid polyp. The two main classes of true jellyfish are Scyphozoa (true jellyfish) and Cubozoa (box-jellyfish). Scyphozoans are common marine animals throughout the world. Their tentacles appear radially at regular intervals around the bell, and some species possess tentacles arising from inside the bell. True jellyfish associated with human envenomation include the mauve stinger (*Pelagia noctiluca* Forskål), lion's mane jellyfish (*Cyanea capillata* L.), and bluefire jellyfish (*Cyanea lamarckii* Péron & Lesueur). Box-jellyfish (cubozoans) are similar in form to the "true" jellyfish (scyphozoans). However, cubozoans (i.e., box-jellyfish) have a square shape on transverse section along with well-developed eyes and four evenly-spaced groups of one or more tentacles arising from the corners of the bell. The base of each tentacle is distinctively flattened, and the edge of the umbrella turns inward to form a rim (i.e., volarium). Two species of box-jellyfish (*Chironex fleckeri* Southcott, *Chiropsalmus quadrigatus* Haeckel) are indigenous to the southern Pacific Ocean, and these coelenterates are among the most venomous of all sea creatures. Other species of *Chiropsalmus* (e.g., Atlantic species *C. quadrumanus* F. Müller,) produce primarily severe local pain with less significant systemic symptoms,[10] although a fatality occurred in a 4-year old boy soon after envenomation by *C. quadrumanus* F. Müller.[11] The bells are difficult to see, and the tentacles are nearly invisible. Carybdeids are small box-jellyfish species (*Tamoya* spp. [morbakka], *Carybdea rastoni* Haacke [jimble], *Carukia barnesi* Southcott [Irukandji]) that also produce painful stings and an Irukandji-like syndrome. These large, widely distributed carybdeids have single large (i.e., up to 1 m/~3 ft in length) tentacles attached to each corner of a large bell ranging between 4–18 cm (~1.5–7 in.) in diameter.

Common Name: Lion's Mane, Hair Jelly
Scientific Name: *Cyanea capillata* L.
Class: Scyphozoa
Physical Description: *C. capillata* is a large, yellow or reddish-purple jellyfish whereas *C. lamarckii* Péron & Lesueur is blue. These saucer-shaped jellyfish range up to 2 m (~7 ft) in diameter with long, thin tentacles and scalloped edges on the bell.

Common Name: Mauve Stinger, Phosphorescent Floater
Scientific Name: *Pelagia noctiluca* Forskål
Class: Scyphozoa

Physical Description: These yellow to pink jellyfish have four thick oral arms on the bell along with thin tentacles ranging up to 3 m (~10 ft) in length.

Common Name: Sea Nettle
Scientific Name: *Chrysaora quinquecirrha* Desor
Class: Scyphozoa
Physical Description: This clear to light pink jellyfish has a bell containing four central thick oral arms and several thread-like tentacles up to ~3 m (~10 ft) in length.

Common Name: Sea Wasp, Box-Jellyfish, Marine Stinger
Scientific Name: *Chironex fleckeri* Southcott
Class: Cubozoa
Physical Description: These chirodropids have a square shape and more than one tentacle in each corner of the bell. The transparent bell can reach up to 25–30 cm (~10–12) in diameter with 60 tentacles, each 3 m (~10 ft) long. The total maximum potential tentacle length is about 180 m (~600 ft).

Common Name: Box-Jellyfish, Sea Wasp, Fire Medusa, Indringa, Killing Jellyfish
Scientific Name: *Chiropsalmus quadrigatus* Haeckel
Class: Cubozoa
Physical Description: This small jellyfish has a bell up to 7 cm (~3 in.) in diameter, and the appearance is similar to *Chironex fleckeri* except *Chiropsalmus quadrigatus* has fewer and shorter tentacles. These pale blue, transparent, cube-shaped box-jellyfish are often difficult to see. On each of the four corners of the bell, a group of up to 15 tentacles 3 m (~10 ft) long projects into pedaliums (feeding appendages). The outer layer (epidermis) contains the stinging cells (nematocysts). The inner endoderm (gastrodermis) covers the gut. The mesoglea is a jelly-like substance occupying the space between the epidermis and gastrodermis that contains collagen fibers and some cells. A ring of nematocyst-containing tentacles surrounds the mouth.

Common Name: Irukandji
Scientific Name: *Carukia barnesi* Southcott
Class: Cubozoa
Physical Description: This jellyfish has a tiny (<2 cm/~0.8 in.) transparent bell with four small tentacles. These tentacles are attached at the corners of the bell characteristic of box-jellyfish. In contrast to *Chironex* and *Chiropsalmus* box-jellyfish, there is only one tentacle at each corner.

EXPOSURE

Geographic Distribution

The hair jelly (*Cyanea* spp.) inhabits colder areas of the North and Baltic Seas as well as the Atlantic and Pacific Oceans including parts of the Australian coast. The mauve stinger (*Pelagia noctiluca* Forskål) drifts in currents in the Mediterranean Sea, tropical and subtropical Atlantic Ocean, and the English coast. At night, the presence of this jellyfish in the ocean waves produces a rose-red luminescence. The sea nettle (*Chrysaora quinquecirrha* Desor) is a free-floating jellyfish often present in large swarms in the tropical and subtropical Atlantic and Pacific Oceans. The chirodropid-type box-jellyfish appear during the summer months along coastal waters of tropical parts of the northern and southern hemispheres between the Tropic of Capricorn and the Tropic of Cancer. Although stings occur throughout the year, most stings occur in the summer months in shallow coastal waters, particularly during hot, calm days. These chirodropids live in tropical waters along the eastern and western Atlantic Ocean, the east coast and midportion of the Indian Ocean, and the Indo-West Pacific Ocean. Box-jellyfish from the species *Chironex fleckeri* Southcott inhabit coastal areas of northern Australia, adjacent waters of the tropical Indo-West Pacific, and southeastern Asia. These marine animals live in aquatic areas along the shoreline and adjacent estuaries. *Chiropsalmus quadrigatus* Haeckel inhabits the Indian Ocean near the northern Australian coast as well as the Philippine coast. The distribution of Irukandji in Australian tropical waters ranges from north Queensland northwards and westwards to Exmouth, Western Australia. The season for carybdeids is summer, similar to other box-jellyfish. An Irukandji-like syndrome occurred in unprotected divers following stings by jellyfish, when they were swimming over a shallow grassy bottom near the Florida Keys during combat training.[12] However, the etiology of this Irukandji-like syndrome remains unknown because these carybdeids are not known to inhabit this area.

Behavior

Chironex fleckeri Southcott are solitary marine animals that feed on large active prey (e.g., sergestid prawns, plankton, shrimp, small fish) in shallow water near the shoreline. These feeding habits result in incidental contact with humans. The medusa has feeding appendages (pedaliums) that lift food to its mouth. They are strong swimmers that can accelerate in bursts up to 5 ft (~2 m) per second. Their lifespan is measured in months. They reproduce both sexually and asexually. In late summer, they gather in river mouths and estuaries to spawn before dying. Although most (>90%) envenomations occur in Australia from October 1st until June 1st, occasionally *C. fleckeri* envenomations occur in the cooler mid-year, particularly in the Northern Territory of Australia.[13] These stings typically occur in shallow water (≤1 m/<3 ft) on afternoons with low wind speed.

The food source for *Chiropsalmus quadrigatus* Haeckel includes small crustaceans and fish. These creatures tend to congregate near the mouths of creeks and rivers in search of prey, particularly after a rain. In general, these fast swimming jellyfish are not aggressive, and they often trap prey in their tentacles while swimming. Like the other sea wasp (*C. fleckeri* Southcott), reproduction occurs both sexually and asexually. During the sexual phase, the male jelly fertilizes eggs released by the female, and the fertilized eggs develop into larva and then into polyps. The polyps reproduce asexually by budding that releases young polyps to swim freely and to develop into adults. Chirodropids occur in tropical waters in the summer months with the season longest near the equator (e.g., Darwin, Australia) and shorter in areas distant from the equator.

Unlike chirodropids, which live near coastal shorelines, Irukandji stings can occur offshore on the Great Barrier Reef. Most *Chironex* stings occur on the legs, whereas Irukandji stings are more common on the arms.[14] Although Irukandji stings can occur anytime of year, most envenomations in Australian waters occur January through May.[15] Factors that favor Irukandji stings include high, incoming tide, warm ambient temperatures (>28 °C/82 °F), and windy days.

PRINCIPAL TOXINS

Venom Composition

The venoms from *Chiropsalmus quadrigatus* and *Chironex fleckeri* are among the most potent in the world. Animal experiments indicate that these venoms contain dermonecrotic, hemolytic, and lethal factors, but the chemical composition of these factors is not well defined.[16,17] The venoms from coelenterate species contain various enzymes and inflammatory mediators. Constituents include serotonin, ATPase, hyaluronidases, fibrinolysin, peripheral calcium antagonist, and a neurotoxic agent (tetramine), as well as the low-weight-molecular fractions that cause hemolysis, local necrosis, and myocardial depression.[18] Although the composition of the venoms from *Chiropsalmus quadrigatus* and *Chironex fleckeri* is similar, the clinical features of envenomation by the latter are more intense as a result of greater venom injection. Characterization of these thermolabile venoms is complicated by the propensity of the venoms to aggregate and to adhere to equipment surfaces.[19] Additionally, these venoms are relatively unstable when exposed to extreme pH changes and certain chemicals. An *in vivo* study on the stability of venom from *Chironex fleckeri* indicated that the response to this venom in rats was not significantly altered by preparing the venom between pH 5 and pH 9 or between temperatures 4 °C/39 °F to 30 °C/86 °F for $1\frac{1}{2}$ hours.[20] Boiling, but not freeze drying, abolished the cardiovascular effects of this venom. Tentacles may detach from the cnidarians (coelenterates) in rough waters, and the free-floating nematocysts may cause local irritation called sea lice. Even though the nematocysts are detached and the coelenterate is dead, the nematocysts may continue to discharge for a substantial period.

Venom Apparatus

Cnidarians rely primarily on chemical and mechanical perturbations rather than vision to detect and catch prey with their nematocyst-containing tentacles. Each double-walled capsular nematocyst is enclosed by specialized epidermal cells called a cnidoblast (cnidocyst). Some species have a triggering devise (cnidocil), whereas other species possess specialized chemoreceptors that activate the cnidoblast in the presence of prey. The two classes of chemoreceptors include chemoreceptors specific for free and conjugated *N*-acetylated sugars, whereas the other class exhibits broad specificity for low-molecular-weight amino acids. The presence of submicromolar concentrations of known chemical triggers (*N*-acetylated sugars, mucin) sensitize the mechanoreceptors to the frequencies (5–40 Hz) that correspond to the movements of swimming prey.[21] Although there are at least 27 types of nematocysts, most types of nematocysts share the common form of a tightly coiled, open-ended, threadlike tube surrounded by a venom capsule or cnidoblast (see Figure 179.2).[1] On top of this capsule is a lid-like covering or apical cap (i.e., operculum). The shape of the undischarged nematocyst is usually spherical or rod-like. The fluid within the nematocyst is the venom. The harpoon-like, hollow tube contains many spicules, some of which anchor the tube to the skin following discharge. The tube discharges through the operculum when the cnidoblast senses chemical changes or mechanical stimulation. The discharge of the nematocyst requires only a few milliseconds, although the everted thread tube penetrates small dermal capillaries and injects the surrounding venom intravascularly.[22] Absorption of the venom into the lymphatic system occurs from the surrounding dermal tissue following extrusion of venom through the hollow lumen of the thread tube. The severity of the envenomation depends on the ability of the thread tube to penetrate the human

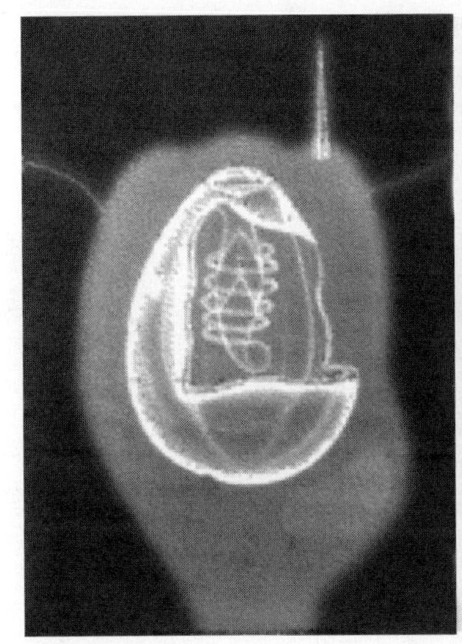
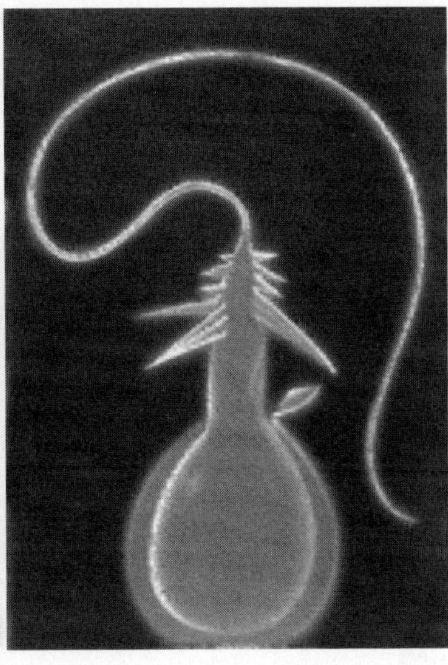

FIGURE 179.2 Uncharged and discharged nematocyst. Adapted from Halstead BW. Coelenterate (Cnidarian) stings and wounds. Clin Dermatol 1987;5:12. Reproduced with permission from Elsevier.

epidermis and to transfer venom to the dermis and capillaries.

Chironex fleckeri tentacles contain at least four types of nematocysts (microbasic mastigophores, atrichous isorhizas, holotrichous isorhizas, microbasic euryteles).[22] The eurytele nematocysts possess a tube with a butt composed of three bulbs and a middle bulb bearing long spines, whereas each holotrichous isorhiza contains a coiled tube with small spines along its length. The atrichous isorhiza nematocysts have spineless tubes that fold loosely in the undischarged condition. The microbasic mastigophore contains a capsule composed of a granular capsular matrix and an inverted, smooth-walled tube that narrows to form a tightly coiled, helical thread. These cigar-shaped structures are the most important type of nematocyst involved with envenomation.

Dose Response

Large coelenterates may discharge hundreds of thousands of nematocysts. The severity of the envenomation depends on the size and quantity of nematocysts, duration of contact, the size (width, length, thickness) of contact with the envenomated skin, presence of protective garments or hair, individual penetrating power of the nematocysts, individual sensitivity of the victim, and the cnidarian (coelenterate) species. Severe envenomations by *Chironex fleckeri* in children typically involve 10–15 tentacles. Envenomation of a 5-year-old boy by *Chironex fleckeri* that involved about 4 m (~13 ft) of tentacle contact was associated with a fatal outcome despite supportive care.[23] The boy arrived at the emergency department in cardiopulmonary arrest within 15 minutes of the envenomation.

HISTOPATHOLOGY AND PATHOPHYSIOLOGY

Mechanism of Toxicity

Death from box-jellyfish venom probably results from respiratory failure or direct cardiotoxicity.[24] These venoms contain cardiotoxic, myotoxic, neurotoxic, cutaneous vasodilating, dermatonecrotic, and hemolytic properties, but these effects are not well-characterized.[25,26] Jellyfish venoms cause nonspecific production of large cation-selective channels or pores that modulate excitable neuromuscular membranes.[27] The venom also produces perturbations in the ionic calcium and/or sodium transport across excitable membranes, but the exact role of pore formation in the development of toxic cardiac and neurological effects remains unclear.[28] The principal clinical reaction of jellyfish envenomation is a toxic rather than immune response because repeat envenomations display similar symptoms and immunity to jellyfish envenomation does not usually develop.[29] However, allergic reactions may play a significant role in a few human envenomations. Specific anti-jellyfish venom IgE levels persisted several years in several envenomated patients with exaggerated responses to sea nettle (*Chrysaora quinquecirrha*), and these antibodies cross-reacted with Portuguese man-of-war (*Physalia*) venom.[30] A patient with a prolonged pruritic, erythematous rash following a jellyfish envenomation developed a positive response to a radioallergosorbent test for jellyfish antigens.[31] The venom from the *Carukia barnesi* (Irukandji) contains a neuronal sodium channel agonist that stimulates the release of massive quantities of catecholamines.[3]

Postmortem Examination

Postmortem examination of a 4-year-old boy, who died soon after a *Chiropsalmus quadrumanus* envenomation, demonstrated frothy white foam in the airways, widespread interstitial and perivascular lymphoid infiltration of the myocardium without necrosis, and focal subendocardial hemorrhage in the septum of the left ventricle.[11] There was marked passive congestion of the internal organs including the lungs. These findings were consistent with a massive catecholamine discharge and acute cardiac dysrhythmia with congestive heart failure. In the postmortem examination of a five-year old boy who died within 40 minutes of an envenomation by *Chironex fleckeri*, dermal changes were minimal.[32] Dermal abnormalities included congestion of superficial and mid-dermal capillaries, occasional melanin-containing melanophages, and mild perivascular edema. There was no evidence of an acute inflammatory infiltrate in the area of the envenomation. Electromicroscopy revealed that the nematocyst tubes penetrated the keratinocytes rather than following intercellular spaces, and the tubes from the nematocysts were present in the deep layers of the epidermis and dermis.

CLINICAL RESPONSE

Jellyfish envenomation (e.g., sea nettle) causes painful, linear, erythematous, vesicular lesions on the skin similar to lesions produced by a whip or a branding iron. Most stings occur on the lower legs or body with tentacles frequently adherent to the victim like a sticky thread. These stings usually produce only localized pain, and frequently the envenomating jellyfish is not identified. The pain typically resolves within 24 hours, although dermal lesions (hemorrhagic vesicles, urticaria) may persist for a month.[33] Systemic symptoms include headache, lightheadedness, malaise, muscle cramps, diaphoresis, and syncope. Recurrent eruptions at the site of envenomation, persistent granulomatous lesions, and activation of herpes labialis after the envenomation are rare sequelae. Atrophy of subcutaneous fat, granuloma annulare, and keloid formation may also occur at the site of injury.[34] Other complications of exposure to jellyfish include the development paralytic ileus after envenomation,[35] and ciguatera poisoning after ingestion of a jellyfish body without the tentacles.[36] Sea bather's eruption is a pruritic, maculopapular or vesicular skin eruption that primarily affects the cutaneous areas underneath swim wear.[37] These lesions develop up to 24 hours after exposure to the thimble jellyfish (*Linuche unguiculata* Schwartz) present in Florida coastal waters, the Gulf of Mexico, and the Mexican Caribbean. Pain and systemic symptoms are usually absent.

Chirodropids (Multitentacled Box-Jellyfish)

Symptoms range from mild discomfort caused by a contact dermatitis to serious systemic complaints. Envenomations by the most venomous, multitentacled box-jellyfish (*Chiropsalmus quadrigatus, Chironex fleckeri*) produce immediate, severely painful, cross-hatched areas of erythema, edema, and vesiculation that may result in necrosis. The majority of *Chironex fleckeri* envenomations are small, and these stings cause severe local cutaneous pain.[14] Most of these stings are not life-threatening.[13]

Severe symptoms include pulmonary edema, respiratory arrest, and cardiovascular collapse. There have been over 60 fatalities from *Chironex fleckeri* envenomations, primarily in children.[5] Cardiorespiratory arrest usually develops within a few minutes of a fatal envenomation; therefore, intervention with antivenom is typically too late to prevent death.[38] Delayed hypersensitivity reactions over the area of the envenomation frequently occur about 7–14 days after envenomation. These reactions involve a pruritic, erythematous maculopapular rash in a pattern over the initial tentacle contact consistent with papular urticaria.[39] Rarely, acute myocardial infarction develops in patients with chest pain after envenomation by jellyfish. A 45-year-old diver without risk factors for coronary artery disease had an acute inferior myocardial infarction manifest by chest pain and diaphoresis after envenomation by an unidentified jellyfish on the left forearm while diving near Qatar.[40] Coronary angiogram 4 days after the envenomation and after receiving streptokinase demonstrated normal coronary arteries.

Irukandji Syndrome (*Carukia barnesi*)

In contrast to the sting of chirodropids, the sting of the Irukandji is usually mild without significant tissue damage. The clinical features of this syndrome range from mild pain to serious cardiac dysfunction. Systemic symptoms typically begin about 30 minutes (range: ~5–40 minutes) after envenomation. Several types of clinical patterns may develop following envenomation by these jellyfish including 1) acute muscular pain in the chest and back, 2) sympathomimetic effects—sweating, headache, tachycardia, nausea, vomiting, hypertension, and 3) cardiorespiratory failure.[3] The initial symptom typically involves dull aching pain in the sacrum followed by the rapid development of diffuse muscle spasms and severe, colicky chest and abdominal pain.

These symptoms usually diminish within 24 hours, but complete resolution of symptoms may require up to one week.[12]

Many of the clinical features of the Irukandji syndrome are similar to the excessive release of adrenaline (e.g., pheochromocytoma) characterized by profuse sweating, piloerection, anxiety, restlessness, severe headache, tremor, pallor, hypertension, and tachycardia. Other serious complications include non-cardiogenic pulmonary edema and/or left ventricular dysfunction.[2,41] Complications of Irukandji envenomation include severe hypertension and fatal intracerebral hemorrhage.[42] Although most envenomations associated with the Irukandji syndrome result from *Carukia barnesi*, other jellyfish (*Tamoya* spp, *Carybdea* spp.) also cause this syndrome based on analysis of residual nematocysts by skin scrapings.[43,44] However, the clinical features of these envenomations are usually less severe than *Chironex* and *Chiropsalmus* box-jellyfish envenomations.[45]

DIAGNOSTIC TESTING

Analytical Methods

Identification of the envenomating jellyfish involves the harvesting of the nematocysts from the skin using either 1) the scraping of the sting site (after appropriate first-aid measures) with a scalpel and preservation of the specimen in 1–4% formalin, or 2) the application of 4–8 cm (~2–3 in.) of sticky-tape to the sting site and placement of tape on a glass slide. Both methods have similar yields of nematocysts following box-jellyfish envenomation, [46] although the sensitivity of these tests for Irukandji syndrome is unknown because of the low nematocyst identification rates.

Abnormalities

Case reports suggest that elevated serum troponin concentrations occur after jellyfish envenomation, even in the absence of significant reduction in cardiac contractility or elevated serum creatine kinase concentrations.[43] Following Irukandji stings, direct cardiac damage may occur, manifest by elevated cardiac troponin I, elevated creatine kinase, reduced cardiac index, nonspecific electrocardiographic abnormalities (i.e., T-wave inversion, ST-segment depression), and global ventricular dysfunction with poor ejection fraction. In a case series of 128 confirmed cases of Irukandji syndrome, 20% of the serum troponin I concentrations were abnormal with a range between 1.0–34 μg/L.[44] A case report associated elevation of serum creatine kinase and serum troponin I with myocarditis.[2]

TREATMENT

First Aid

CHIRODROPIDS (MULTITENTACLED BOX-JELLYFISH)

Most victims of *Chironex* and *Chiropsalmus* box-jellyfish envenomation develop only local pain that responds to symptomatic treatment (cold packs, oral analgesics). However, these envenomations potentially may produce cardiorespiratory arrest; therefore, each victim should be evaluated immediately for the necessity of respiratory and cardiac support. Because of the potential seriousness of these envenomations, bystanders should immediately call for medical help. The presence of impaired circulation, breathing, or consciousness indicates a serious envenomation. The patient should be kept calm with limited motion of the affected extremity.

Although the covering of the contact site with vinegar for 30 seconds deactivates the nematocysts and prevents further envenomation, vinegar does not provide pain relief.[47] Cold packs may help relieve some of the pain, but there are few clinical data to determine the efficacy of cold or heat treatment. There are limited data suggesting that cold packs improve pain relief following envenomation by some jellyfish [*Physalia* spp., *Cyanea* sp. (hair jellyfish), *Tamoya* sp. (Moreton Bay stinger), *Carybdea rastoni*].[48] Other studies suggest that hot packs (45 °C/113 °F) may relieve the pain associated with envenomations by other chirodropids species [*Carybdea alata* (Hawaiian box-jellyfish)].[49,50] There are inadequate clinical data to support the use of pressure immobilization bandaging as a first-aid measure.[51] Although there are few data to confirm the efficacy of compression bandages during jellyfish envenomation, the use of compression bandages is recommended in some parts of Australia for application over the sting site of seriously envenomated victims.[4] If vinegar is not available, the tentacles must be removed by the rescuer prior to the use of compression bandages. These compression bandages are not recommended for minor to moderate jellyfish envenomations. In Australia, CSL *Chironex* antivenom is available for the treatment of serious *Chironex* box-jellyfish envenomations by trained health professionals on the beach. The dose is 1 ampule diluted 1:10 in saline administered intravenously over 10 minutes or 3 ampules intramuscularly. Up to 6 ampules may be administered intravenously during cardiopulmonary resuscitation (CPR).

IRUKANDJI SYNDROME

Most patients with Irukandji syndrome develop only minor symptoms that respond to local care. Occasionally, noncardiogenic pulmonary edema occurs that requires intubation; therefore, victims of the Irukandji syndrome with any respiratory symptoms should be observed for respiratory distress with pulse oximetry, particularly if ongoing pain develops in an area of the body remote to the envenomation. The presence of continuous pain indicates the need for cardiac monitoring and evaluation of cardiac function (electrocardiogram, chest x-ray, cardiac troponin, creatine phosphokinase).

Non-Chirodropids

First-aid measures for jellyfish stings other than chirodropids are not well defined. The tentacles should be removed carefully from the skin. The rescuer may sustain some minor prickling sensations, but significant envenomation does not usually occur in the rescuer. If available, the use of double gloves reduces the risk of envenomation. Ice packs are a traditional part of the first aid that typically involves the application of ice for 5–15 minutes to the area of the sting after the removal of the jellyfish. Reapplication of the ice may be necessary if the pain recurs. However, the superiority of cold packs over the use of hot water for the treatment of pain associated with true jellyfish envenomations has not been well documented. Patients with systemic complaints should be referred for medical care.

Stabilization

The initial treatment of *Chironex* and *Chiropsalmus* box-jellyfish envenomations requires supplemental oxygen, intravenous access, and careful monitoring of cardiorespiratory status with pulse oximeter and cardiac monitor. Victims should be evaluated carefully for the need for endotracheal intubation. Patients with Irukandji syndrome may develop severe hypertension and chest pain that may require the use of intravenous sodium nitroprusside and intravenous nitrates. Intravenous phentolamine (5–10mg IV) is an alternative to sodium nitroprusside for severe hypertension, but there are few clinical data to guide management. These patients should be observed closely for the development of pulmonary edema and myocardial dysfunction. Standard treatment for pulmonary edema includes diuretics (furosemide), morphine, and intravenous sodium nitrates. Testing includes electrocardiogram, cardiac enzymes (creatine phosphokinase with MB fraction, troponin), electrolytes, and measures of hepatorenal function.

Decontamination

The area envenomated by box-jellyfish should be irrigated copiously with 1–2L vinegar for at least 30 seconds to inactivate any remaining nematocysts. After deactivation, the clinging tentacles should be washed off with either fresh or salt water.

Antivenom

The box-jellyfish antivenom is a purified ovine immunoglobulin that is available for the treatment of severe envenomations by *Chironex fleckeri* and the closely related cubozoan, *Chiropsalmus quadrigatus*. Indications include cardiorespiratory arrest, serious cardiac dysrhythmias, clinically significant dyspnea or dysphagia, severe pain unresponsive to narcotics, or extensive skin lesions, particularly in cosmetically important areas.[52] *In vitro* studies suggest that the antivenom is less effective against the neurotoxic and myotoxic effects of the venom.[53] Evidence supporting the efficacy of the antivenom in humans is primarily anecdotal.[52] Early use of the antivenom is important because there is no clear data to indicate that the antivenom can neutralize venom already bound to tissues, and the antivenom is unlikely to relieve pain following administration of the antivenom over 4–6 hours after envenomation.[54] The usual dose is 1 vial (20,000U) for moderately severe stings to 3 vials for life-threatening stings. Adverse reactions to the antivenom are rare.[52] The efficacy of the current box-jellyfish antivenom (CSL Limited, Parkville, Australia) for the treatment of Irukandji syndrome remains unclear. This antivenom is not effective following envenomation by scyphozoan or hydrozoan jellyfish.[3]

Supplemental Care

Most patients with jellyfish stings develop only minor local symptoms, and these patients should be treated symptomatically (local ice packs, oral analgesics) and observed for 2 hours. If systemic symptoms do not occur, these patients may be discharged home with advice to return to the emergency department if systemic symptoms develop. Major envenomations frequently require intravenous narcotic analgesics (e.g., morphine, fentanyl, hydromorphone, meperidine) for pain relief. Following *Chironex* and *Chiropsalmus* box-jellyfish stings, patients with pain unresponsive to cold packs and analgesics are candidates for the administration of 1 ampule of dilute antivenom. Symptomatic patients should be monitored for the development of pulmonary edema and myocardial dysfunction, particularly during the first 24 hours. Delayed hypersensitivity responses to envenomation may develop in 7–14 days, and the pruritic rash usually responds to topical steroids creams and oral antihistamines. Patients with Irukandji syndrome may also develop severe pain that requires narcotic analgesics. The use of intravenous magnesium sulfate is an experimental treatment for the pain and hypertension associated with the Irukandji syndrome based on case reports.[55] Patients with Irukandji syndrome and continuing pain despite adequate narcotic analgesics are at risk of developing cardiac complications.[43] These patients should be observed at least until the patient is pain-free 6 hours after the last dose of analgesics.[56] Observation should include a cardiac evaluation (creatine phosphokinase-MB, troponin I, chest x-ray, electrocardiogram, and echocardiography if evidence of cardiac dysfunction). Measurement of pulmonary artery pressures may be necessary to optimize the administration of vasopressors (dopamine, dobutamine, norepinephrine) for persistent hypotension. Secondary wound infections (*Vibrio vulnificus, Pseudomonas aeruginosa, Aeromonas salmonicida, Staphylococcus aureus, Streptococcus pyogenes*) are rare,[57] and when present usually respond to ciprofloxacin 500–750 mg twice daily.

HYDROIDS

Hydroids (Class: Hydrozoa) commonly occur as plume-like tufts growing on rocks, under corals, in seaweed, and along side underwater structures (pilings, piers). Small medusae project from branching tufts of the polyps. Hydroids are not true jellyfish, and *Physalia* species (Portuguese man-of-war) are colonial hydroids (siphonophores) rather than true jellyfish. The most toxicologically significant species of hydroids are the fire corals (*Millepora* spp.) and the Portuguese man-of-war (*Physalia* spp.) Other less common species of hydroids involved with adverse health effects include the tiny *Gonionemus vertens* Agassiz (orange-striped jellyfish) that has produced systemic symptoms similar to the Irukandji syndrome in Japan. Stings from hydroids produce symptoms ranging from mild, prickling sensations to extremely painful lesions (*Physalia* species) and severe systemic symptoms (*Gonionemus vertens* A. Agassiz).

Orange-Striped Jellyfish

This tiny hydrozoan (*Gonionemus vertens* A. Agassiz) is distributed worldwide, but serious envenomations have been reported only in the Sea of Japan, primarily during the hot summer months.[58] The bell is small (5–25 mm/~0.2–1 in.) with many tentacles attached to the entire edge of the bell. A distinctive symmetrical, right-angled cross is present on the transparent bell. *Gonionemus vertens* (*Gonionemus oshoro* Uchida) inhabits the Sea of Japan near Yamagata and the South Kuril Islands. Symptoms range from severe myalgias, muscle fasciculations, and arthralgias to rhinitis, lacrimation, hoarseness, cough, dyspnea, and mild hypertension. The symptoms last several hours to several days. Neuropsychiatric symptoms (malaise, depression, hallucinations) were associated with exposure to this hydrozoan, but there were no biomarkers to document envenomation by *Gonionemus* species.

Fire Corals

IDENTIFYING CHARACTERISTICS

Fire or stinging corals belong to the order Anthoathecata and Milleporidae family. These marine animals are not true corals. Although superficially similar to "true corals," these marine organisms share physical characteristics (e.g., nematocysts) with jellyfish and anemones. Fire coral (*Millepora* spp.) are a colony of nematocyst-bearing coelenterates with a calcareous skeleton rather than a true coral. These colony-forming polypoid hydrozoans secrete massive calcium carbonate skeletons that are covered with a thin epidermal layer. The skeleton has a smooth surface covered in pores of two different sizes for defensive polyps and feeding polyps. Defensive polyps (dactylozooids) arise from separate openings that encircle the gastrozooids (feeding polyps). *Millepora* species are upright, branching, calcareous growths. Fire coral may grow in a variety of shapes, including column-like, encrusting (flattened), branching, rounded, plate-like or box-shaped. Live colonies are usually yellow-brown with pale edges or branch tips, but some species may be white, yellow, pink, or green.

EXPOSURE

Geographic Distribution

Millepora species inhabit tropical and subtropical coastal areas in tropical and subtropical seas including the Mediterranean Sea, the Red Sea, and the Caribbean Sea. These organisms live in a wide range of coastal reef habitats including the Caribbean coast near Mexico (*M. complanata* Lamarck), Marshall Islands (*M. tenera* Boschma), the Red Sea (*M. platyphylla* Hemprich &

Ehrenberg, *M. dichotoma* Forskal), and the Florida Keys (*M. alcicornis* L.).

Behavior

The polyps live only 20 minutes out of water, and they are easily destroyed by host defenses. Tiny, hair-like dactylozooids may be visible during the day or night.

PRINCIPAL TOXINS

The venom from fire coral contains hemolytic, dermonecrotic, and lethal factors.[59–61] Extracts from the venom stimulate smooth muscle contraction, probably as a result of an increase in Ca^{++} permeability via *L*-type voltage-dependent Ca^{++} channels.[62] Hydroids from the genus *Millepora* possess nematocysts, like other members of the phylum Cnidaria. After chemical or mechanical stimulation, these nematocysts discharge toxins into skin in contact with the hydroid. Envenomation occurs when the nematocyst-containing polyps protrude through pores in the exoskeleton.

CLINICAL RESPONSE

Contact of unprotected extremities with most hydroids produces primarily mechanical injury along with varying degrees of erythema, edema, and pruritus. Rarely, these lesions progress to an indolent ulcer with irregular, necrotic borders. Stings by fire coral usually produce only minor discomfort characterized by small maculopapular lesions that develop about 1–10 hours after contact. Depending on the severity of the envenomation, pain may develop immediately along with an intense pruritus and urticarial wheals 5–30 minutes after envenomation followed by erythematous, edema, and vesiculation within 6 hours. Brief contact of unprotected skin with fire coral can produce more serious stings that classically involve the following four stages: 1) acute urticaria, 2) acute vesiculobullous dermatitis, 3) subacute granulomatous dermatitis, and 4) chronic lichenoid dermatitis.[63] Pruritus may persist during the development of chronic lichenified changes 3 weeks after envenomation. Resolution may require up to 4 months with some residual hyperpigmentation and hypertrophic scarring. Rarely, necrosis of the skin may occur.[64] Systemic symptoms (abdominal pain, diarrhea, fever, chills, muscle spasm, malaise) are unusual, and there are no well-documented fatalities from envenomation by fire corals. Other hydrozoans including sea-fern hydroids and other feather hydroids belonging to the order Leptothecata (*Aglaophenia cupressina* Lamouroux, *Macrorhynchia philippina* Kirchenpauer, *Halecium beanii* Johnston, *Nemalecium lighti* Hargitt) also produce dermal injuries characterized by pruritic, erythematous papular-vesicular lesions and rarely erythema multiforme.[65]

TREATMENT

There are few data to guide the management of fire coral stings. Theoretically, the irrigation of the lesion with vinegar would inactivate any remaining nematocysts. The wound should be inspected and, if necessary, probed carefully to detect any remaining skeletal fragments. Lacerations require thorough exploration and irrigation with copious amounts of saline after local anesthesia. Anecdotally, the antiseptic (nondetergent formulation) povidone-iodine (Betadine®, Purdue Pharma, Stamford, CT) solution may be added in very dilute concentration to the irrigating solution for grossly contaminated wounds. However, there are few clinical data to guide management of these wounds. After thorough cleansing, facial wounds can be closed. Old wounds (i.e., extremity wounds presenting 6–8 hours after injury) may benefit from delayed primary closure after cleaning. If no infection develops by the fourth day after cleansing, the wound can be closed safely. There are no data to support the use of prophylactic antibiotics unless the wound is grossly contaminated. Intradermal steroid injections may be beneficial for chronic, pruritic lichenified lesions that do not respond to oral antihistamines and topical steroids.[66]

Portuguese Man-of-War

IDENTIFYING CHARACTERISTICS

Although there are some structural similarities between *Physalia* species and true jellyfish, the Portuguese man-of-war is a colonial hydroid rather than a true jellyfish. This free-floating hydrozoan colony consists of four separate organisms living in a symbiotic relationship. These blue cnidarians (coelenterates) contain gas-forming glands (carbon monoxide, nitrogen) and form a float with a crest (pneumatophore) that functions as a sail. Gastrozooids enable the organism to digest food. The similarity of the body and long tentacles to a Portuguese admiral's hat accounts for the name Portuguese man-of-war. These tentacles sometimes reach 30 m (100 ft) in length and may contain over one million nematocysts.

Common Name: Atlantic Portuguese Man-of-War

Scientific Name: *Physalia physalis* L.

Physical Description: The clearly visible blue float ranges up to 25 cm (~10 in.) in diameter. The gastric cavity is 5–10 cm (~2–4 in.) in diameter, similar to the Pacific variety, but the highly retractable blue tentacles are longer and more numerous than the bluebottle (*P. utriculus*).

Common Name: Pacific (Portuguese) Man-of-War, Bluebottle

Scientific Name: *Physalia utriculus* La Martiniere

Physical Description: The blue body rarely exceeds 8–10 cm (~3–4 in.) in diameter, and the tentacles seldom reach 30 m (~100 ft).

EXPOSURE

Geographic Distribution

The distribution of *Physalia* species is worldwide in the temperate and tropical zones with more serious stings prevalent in warmer waters. The Atlantic Portuguese man-of-war (*P. utriculus* La Martiniere) inhabits the Atlantic Ocean and Caribbean Sea. The smaller version of the Atlantic variety (*Physalia utriculus* La Martiniere) is distributed across the Pacific and Indian Oceans. Small Portuguese man-of-war specimens appear in the fall, whereas larger specimens usually occur in the spring.

Behavior

This hydrozoan siphonophore forms a drifting polyp colony on the surface of the water with the long tentacles trailing underneath the surface. The bluish float is obscure on the surface of the water; therefore, the risk of contact with swimmers is high. The tentacles stick to fishing nets; consequently, fishermen can sustain injuries while retrieving their nets. These organisms are predators that use several markedly elongated tentacles called fishing tentacles to trap small fish. Exposure to active tentacles may result from the detachment of tentacles during mechanical trauma (e.g., storms) and the presence of these tentacles on the shore. Some sea slugs (*Glaucus* spp.) feed on Portuguese man-of-war tentacles. The undischarged nematocysts pass through the intestinal tract and lodge on the dorsal papillae of the sea slug, where contact with bathers produces a contact dermatitis. The Portuguese man-of-war lacks a free-swimming, bell-shaped medusa stage common to jellyfish. The fishing tentacles contain high concentrations of nematocyst, and contact with the fishing tentacles produces a characteristic pattern of injury that resembles a pearl necklace (i.e., a series of bumps).

PRINCIPAL TOXINS

Venom Composition

The major component of *Physalia physalis* venom is physalitoxin (CAS RN:77108-01-1), which is a potent hemolysin. *In vitro* studies indicate that Portuguese man-of war venom alters plasma membranes and causes calcium influx in several cell types including cardiac cells.[67] The venom-induced calcium influx accompanies a dose-dependent release of intracellular lactate dehydrogenase in an animal model that indicates a loss of plasma membrane integrity and cytolysis. *Physalia* venom contains a neurotoxin that produces intense pain and causes prostaglandin-induced vasodilation *in vivo* and *in vitro*.[34] This venom also releases of histamine and nonspecific endonucleases that cleaves DNA over a wide range of pH.

Venom Apparatus

The two sizes of nematocysts in *Physalia* tentacles average about 10 μm and 23 μm in diameter; *in vitro* studies indicate that the smaller nematocysts contain venom, whereas proteins separated from the larger nematocysts by flow cytometry do not cause toxicity.[68] The sensory trigger receptor for the *Physalia* nematocysts is the cnidocil apparatus.

Dose Response

A *P. physalis* evenomation involving a total length of approximately 9–12 ft (~3–4 m) on the upper extremities was associated with the rapid onset of coma, respiratory arrest, hypotension, and severe bradycardia.[69] The 67-year-old woman died 5 days later of aspiration pneumonia and respiratory failure without regaining consciousness.

HISTOPATHOLOGY AND PATHOPHYSIOLOGY

Postmortem examination of a 67-year-old woman, who died 5 days after developing cardiorespiratory failure immediately after a *P. physalis* evenomation, demonstrated patchy areas of dermal necrosis with neutrophilic and lymphocytic infiltration and extravasation of erythrocytes.[69] There were focal areas of arteriosclerotic cardiovascular disease, but there was no evidence of myocardial necrosis. The victim initially developed a respiratory arrest followed by severe bradycardia and asystole.

CLINICAL RESPONSE

This marine organism is the most dangerous hydroid because of its propensity to cause systemic symptoms. However, most envenomations result only in local symptoms, and fatalities are rare. Stings by the Portuguese man-of-war cause immediate pain along with the early appearance of small, multiple, discontinuous, linear, erythematous papules. This distinctive pattern of the envenomation resembles a beaded, urticarial skin eruption with ladder-like spacing.[70] Localized pain may spread to larger muscle masses or to the whole body. Over several days, the papules can progress to vesicles and pustules that produce hemorrhage, ulcerations, and ultimately hyperpigmentation.[6] Delayed and recurrent skin lesions may develop along the linear area of the prior envenomation weeks to months after the initial envenomation including eczematous, urticarial, or vesiculobullous lesions. Rarely, case reports associate jellyfish envenomation and elevated serum titers against *P. physalis* with acute, localized narrowing of major vessels at the site of the sting and limb-threatening distal ischemia.[71] However, the envenomating jellyfish was not positively identified. Chronic, hypersensitivity-like skin responses following envenomation include erythema nodosum, erythema multiforme, granuloma annulare, postinflammatory hyperpigmentation, and hypertrophic scars.[72] Several case reports suggest that repeat stings may produce a hypersensitivity response that causes arterial spasm, local necrosis, and rarely anaphylaxis.[71,73]

Systemic reactions are more common with the Atlantic variety (*P. physalis*) than the Pacific variety, and the clinical features of envenomation include weakness, fever, malaise, headache, nausea, vomiting, severe muscle spasm and pain in the abdomen and back, paresthesias, and anxiety. These symptoms are similar to the Irukandji syndrome.[74] Rarely, respiratory or myocardial depression, seizures, hemolysis, acute renal failure, and anaphylactic reactions occur. Fatalities may occur as a result of respiratory arrest or cardiac dysthymias.[69,75] Symptoms may persist for 48 hours.

DIAGNOSTIC TESTING

Identification of the envenomating jellyfish involves the harvesting of the nematocysts from the skin using either scalpel-blade scraping of the sting site (after appropriate first-aid measures) or sticky-tape sampling. Both methods have similar yields of nematocysts following box-jellyfish envenomation. There are no specific laboratory abnormalities associated with envenomation by *Physalia* spp.

TREATMENT

The patient should be kept calm to limit the spread of the venom. Systemic symptoms may require cardiorespiratory support, particularly for children swallowing tentacles. First aid is similar to that for most cnidarians (coelenterate) venom. Exposure to vinegar produces varying effects on the discharge of nematocysts from different jellyfish species. Although vinegar inhibits the release of nematocysts from the box-jellyfish species (*Chironex fleckeri, Chiropsalmus quadrigatus*),[76] the use of vinegar may increase the discharge of nematocysts from *Physalia* species in Australian waters.[77] However, other studies suggest that this first-aid measure did not increase the discharges from nematocysts from the Atlantic Portuguese man-of-war.[78] Consequently, the safest decontamination measure is to wash off adherent tentacles with fresh sea water. An open-label, randomized comparison trial suggested that immersion of the sting site from *Physalis* species in 45 °C/113 °F water for 20 minutes reduces the pain associated with these stings, when compared with ice pack treatment.[79] Rarely, envenomated patients develop cardiorespiratory dysfunction; therefore, victims of envenomation require evaluation of the adequacy of ventilation and circulation. Elderly patients with cardiopulmonary disease may develop respiratory depression, particularly after the administration of narcotics. There is no antivenom for *Physalia* evenomation.

ANTHOZOANS

The Class Anthozoa contains sea anemones, hard corals, soft coral, zoanthids, sea pens, and sea fans (gorgonians). Anthozoans exist primarily as the polyp form, and there is no distinct medusa stage in the life cycle. Many anthozoans are colonies of polyps, such as anemone-like zoanthids. In some species, the polyps separate almost completely from each other after budding, whereas other species exist as interconnected polyps that form a continuous mat of tough, rubbery/leathery living tissue.

Sea Anemones

Sea anemones are primarily sedentary marine polyps with a columnar body and highly variable size and color. Sea anemones are present in all oceans ranging from

Arctic to Antarctic waters. These marine organisms develop stationary polyps with tentacles surrounding the mouth on an oral pole. The caudal pole stabilizes the body by forming attachments to an object (i.e., reef inhabitants such as coral). They rely on water currents and the movement of prey to bring the prey in contact with their tentacles. The tentacles contain nematocysts that discharge in response to appropriate chemical and mechanical stimuli. Mechanical stimulus alone is sufficient to cause the nematocyst to discharge. The number of nematocysts discharged depends on the frequency of the mechanical stimulation as well as the stimulation of chemoreceptors. There are at least two types of chemoreceptors that bind *N*-acetylated sugars in the mucins and amino compounds in the hemolymph of prey.[80] These chemical stimuli modulate the frequency and amplitude of the nematocyst discharge to increase the ability of the sea anemone to immobilize prey. Sea anemone venom contains toxins that block currents through voltage-gated potassium channels similar to scorpion toxins.[81] Additionally, some sea anemone venom contains small polypeptides (site-3 toxins) that bind to the extracellular surface of voltage-gated sodium channels.[82] After binding to a site near the S4 segment in domain IV, these toxins disrupt the normal fast inactivation transition resulting in a marked prolongation of the action potentials of excitable tissues of nerves and muscles.

Evenomation by sea anemones varies substantially in intensity depending on the species, but local symptoms are usually mild. The characteristic lesions are erythematous, slightly raised with a flower-like pattern of small, irregularly scattered vesicles. Clinical features of envenomation include pruritus, localized burning, edema, erythema, petechial hemorrhages, ecchymosis that rarely progress to necrosis and ulceration. Systemic symptoms are uncommon. Several case reports temporally associated progressive swelling of an extremity after a sea anemone envenomation with the development of acute renal dysfunction[83] and fulminant hepatic failure,[84] but the species was not specifically identified. Mild systemic symptoms (e.g., headache, weakness, lightheadedness) may accompany envenomations by sea anemones.[85]

Certain sea anemone genera (*Sagartia*, *Anemonia*) contain species that proliferate at the base of sponges and produce sponge fisherman's disease.[86] Secondary infections and abscesses may complicate stings by these creatures. The venom from the Okinawan sea anemone (*Phyllodiscus semoni* Kwietniewski) contains hemolytic polypeptides that cause a relatively more severe dermatitis with edema and ulcerations compared with other sea anemone species.[87] Although the effects of most sea anemone envenomation result in mild, local lesions, stings by the blue-gray to light brown sea anemone (*Actinodendron plumosum* Haddon) may cause extremely painful skin lesions that may persist for weeks. Treatment of sea anemone is similar to other coelenterate envenomations.

Hard Coral

Soft and hard corals (Order: Scleractinia) share a very simple body plan and a polyp structure. This group constitutes the major portion of massive coral reefs, where the calcified exoskeletons of hard corals surround small anemone-like polyps. Besides lacking a hard external skeleton and reef-building capacity, soft corals differ from the hard corals by the presence of eight tentacles rather than the multiples of six tentacles on hard coral polyps. Mechanical trauma is the major complication of exposure to hard coral (order: Scleractinia) because of the sharp edges, introduction of foreign bodies, secondary infection, and poor venom apparatuses. Coral injuries typically begin as mild, superficial wounds that progress to clinically significant cellulitis. Treatment of hard coral injuries requires careful debridement and irrigation because wounds may become indolent, like fire coral envenomations.

Soft Coral, Zoanthids, and Palytoxin

Some members of the order Alcyonacea [soft coral (*Palythoa* spp.), sea fans, sea pansies, sea pens] contain the potent marine toxin, palytoxin, but these organisms are not toxicologically significant unless the toxin accumulates in the food chain. However, toxicity usually results from the ingestion of predators that feed on these coral rather than the soft coral because the soft coral lack a venom apparatus to inject the palytoxin. Zoanthids are tropical and subtropical, anemone-like anthozoans (order: Zoantharia) that encrust rocky surfaces in the crevices of wave-swept tide pools or form mats in sandy back reef flats. These common inhabitants are one of the most abundant organisms in shallow coral reefs and tidal pools, and these organisms are soft corals that have no skeleton. Colonies of zoanthids, (e.g., *Palythoa tuberculoa* Klunzinger) are composed of large, flattened, fused polyps with diameters up to 2 cm (~1 in.).

Palytoxin is a highly potent marine compound first extracted from zoanthids of the genus *Palythoa* that live in tidal rock pools in Maui.[88] Species from this genus are widespread throughout coastal areas of the Western Pacific and Caribbean. Other major sources of the palytoxin include zoanthids in the genus *Zoanthus* and dinoflagellates (*Ostreopsis* spp.).[89] In animal studies, palytoxin is a potent toxin. The LD_{50} of palytoxin in rodents following parental administration (10–100 ng/kg) is about 50 times lower than tetrodotoxin and saxitoxin.[90] The concentration of palytoxin in soft corals varies depending on a variety of factors including species, strain, and season. This toxin is a large (approximately 2,680 kDa), water-soluble compound that probably does not accumulate in fatty tissue. Palytoxin also occurs in marine organisms living in close association with zoanthid colonies, such as sea fans, sponges (phylum: *Porifera*), mussels and crustaceans.[91] Accumulation of palytoxin also occurs in marine animals (e.g., xanthid crabs) that graze on tropical coral reefs and in marine animals (e.g., filefish, triggerfish, mackerels, crown-of-thorn starfish, polychaete worms) that feed on these reef-grazing marine animals.[92] Palytoxin acts through the P-type enzyme, Na^+/K^+-ATPase, to convert the Na^+/K^+ ion pump into nonselective cation channels, possibly by disrupting the normal strict coupling between opening of one access pathway in the P-type Na^+/K^+ ATPase and the closing of the other.[93,94] These actions produce cation movement along ion gradients and the depolarization of excitable cells with release of neurotransmitters. The exact mechanism of the release of neuronal transmitters remains unclear. Palytoxin is a potent hemolysin and histaminic agent. Some species of soft coral (*Lophogorgia* spp.) contain lophotoxin (CAS RN:78697-56-0), which binds irreversibly to the muscle nicotinic acetylcholine receptor by forming a covalent bond with a tyrosine residue in the α-subunit of the receptor.[95] There are few data on the clinical significance of these nicotinic antagonists following exposure to this soft coral.

The ingestion of fish contaminated with palytoxin produces ciguatera-like seafood poisoning that includes rhabdomyolysis, severe muscle spasms and acute respiratory distress.[96,97] A monoclonal antibody-based, enzyme-linked immunoassay (ELISA) has been developed to rapidly detect palytoxin.[98] Patients with palytoxin poisoning should be evaluated for the presence of rhabdomyolysis and renal function. There are few data to guide management, and the treatment of palytoxin poisoning is supportive (analgesics, fluids).

References

1. Halstead BW. Coelenterate (Cnidarian) stings and wounds. Clin Dermatol 1987;5:8–13.
2. Little M, Mulcahy RF, Wenck DJ. Life-threatening cardiac failure in a healthy young female with Irukandji syndrome. Anaesth Intensive Care 2001;29:178–180.
3. Winkel KD, Hawdon GM, Fenner PJ, Gershwin L-A, Collins AG, Tibballs J: Jellyfish antivenoms: past, present, and future. J Toxicol Toxin Rev 2003;22:115–127.
4. Fenner PJ. Dangers in the ocean: the traveler and marine envenomation. I. Jellyfish. J Travel Med 1998;5:135–141.
5. Fenner PJ, Williamson JA. Worldwide deaths and severe envenomation from jellyfish stings. Med J Aust 1996;165: 658–661.
6. Southcott RV. Tropical jellyfish and other marine stingings. Mil Med 1959;124:569–579.
7. Southcott RV. Studies on Australian *cubomedusae*, including a new genus and species apparently harmful to man. Aust J Marine Freshwater Res 1956;7:254–280.
8. Flecker H. "Irukandji" stings to north Queensland bathers without production of wheals but with severe general symptoms. Med J Aust 1952;2:89–91.
9. Barnes JH. Cause and effect in Irukandji stingings. Med J Aust 1964;1:897–904.
10. HadDad V Jr, da Silveira FL, Cardoso JL, Morandini AC. A report of 49 cases of cnidarian envenoming from southeastern Brazilian coastal waters. Toxicon 2002; 40:1445–1450.
11. Bengston K, Nichols MM, Schnadig V, Ellis MD. Sudden death in a child following jellyfish envenomation by *Chiropsalmus quadrumanus*. Case report and autopsy findings. JAMA 1991;266:1404–1406.
12. Grady JD, Burnett JW. Irukandji-like syndrome in South Florida divers. Ann Emerg Med 2003;42:763–766.
13. Currie BJ, Jacups SP. Prospective study of *Chironex fleckeri* and other box jellyfish stings in the "top end" of Australia's Northern Territory. Med J Aust 2005;183: 631–636.
14. Fenner PJ, Harrison SL. Irukandji and *Chironex fleckeri* jellyfish envenomation in tropical Australia. Wilderness Environ Med 2000;11:233–240.
15. Macrokanis CJ, Hall NL, Mein JK. Irukandji syndrome in northern Western Australia: an emerging health problem. Med J Aust 2004;181:699–702.
16. Keen TE, Crone HD. The hemolytic properties of extracts of tentacles from the cnidarian *Chironex fleckeri*. Toxicon 1969;7:55–63.
17. Keen TE, Crone HD. Dermatonecrotic properties of extracts from the tentacles of the cnidarian *Chironex fleckeri*. Toxicon 1969;7:173–180.
18. Burnett JW, Calton GJ. The chemistry and toxicology of some venomous pelagic coelenterates. Toxicon 1977;15: 177–196.
19. Othman I, Burnett JW. Techniques applicable for purifying *Chironex fleckeri* (box-jellyfish) venom. Toxicon 1990;28:821–835.
20. Winter KL, Isbister GK, Seymour JE, Hodgson WC. An *in vivo* examination of the stability of venom from the

Australian box jellyfish *Chironex fleckeri*. Toxicon 2007; 49:804–809.

21. Watson GM, Hessingger DA. Cnidocyte mechanoreceptors are tuned to the movements of swimming prey by chemoreceptors. Science 1989;243:1589–1591.
22. Rifkin J, Endean R. The structure and function of the nematocysts of *Chironex fleckeri* Southcott, 1956. Cell Tissue Res 1983;233:563–577.
23. Lumley J, Williamson JA, Fenner PJ, Burnett JW, Colquhoun DM. Fatal envenomation by *Chironex fleckeri*, the north Australian box jellyfish: the continuing search for lethal mechanism. Med J Aust 1988;148:527–533.
24. Burnett JW, Calton GJ. Venomous pelagic coelenterates: chemistry, toxicology, immunology and treatment of their stings. Toxicon 1987;25:581–602.
25. Baxter EH, Marr AG. Sea wasp (*Chironex fleckeri*) venom: lethal, haemolytic and dermonecrotic properties. Toxicon 1969;7:195–210.
26. Endean R, Monks SA, Cameron AM. Toxins from the box-jellyfish *Chironex fleckeri*. Toxicon 1993;31:397–410.
27. Dubois JM, Tanguy J, Burnett JW. Ionic channels induced by sea nettle toxin in the nodal membrane. Biophys J 1983;42:199–202.
28. Burnett J, Weinrich D, Williamson JA, Fenner PJ, Lutz LL, Bloom DA. Autonomic neurotoxicity of jellyfish and marine animal venoms. Clin Auton Res 1998;8:125–130.
29. Burnett JW, Cobbs CS, Kelman SN, Calton GJ. Studies on the serologic response to jellyfish envenomation. J Am Acad Dermatol 1983;9:229–231.
30. Burnett JW, Calton GJ. Use of IgE antibody determinations in cutaneous coelenterate envenomations. Cutis 1981;27:50–52.
31. Fisher AA. Toxic versus allergic reactions to jellyfish. Cutis 1984;34:450–454.
32. Lumley SG. Cutaneous light microscopic and ultrastructural changes in a fatal case of jellyfish envenomation. J Cutaneous Pathol 1988;15:249–255.
33. Burnett JW, Calton GJ, Burnett HW. Jellyfish envenomation syndromes. J Am Acad Dermatol 1986;14:100–108.
34. Burnett JW, Calton GJ, Morgan RJ. Venomous coelenterates. Cutis 1987;39:191–192.
35. Ponampalam R. An unusual case of paralytic ileus after jellyfish envenomation. Emerg Med J 2002;19:357–358.
36. Zlotnick BA, Hintz S, Park DL, Auerbach PS. Ciguatera poisoning after ingestion of imported jellyfish: diagnostic application of serum immunoassay. Wilderness Environ Med 1995;6:288–294.
37. Segura-Puertas L, Ramos ME, Aramburo C, Heimer De La Cotera EP, Burnett JW. One Linuche mystery solved: all 3 stages of the coronate scyphomedusa *Linuche unguiculata* cause seabather's eruption. J Am Acad Dermatol 2001;44:624–628.
38. Flecker H. Fatal stings to N. Queensland bathers. Med J Aust 1952;1:35–38.
39. O'Reilly GM, Isbister GK, Lawrie PM, Treston GT, Currie BJ. Prospective study of jellyfish stings from tropical Australia, including the major box jellyfish *Chironex fleckeri*. Med J Aust 2001;175:652–655.
40. Salam AM, Albinali HA, Gehani AA, Al Suwaidi J. Acute myocardial infarction in a professional diver after jellyfish sting. Mayo Clin Proc 2003;78:1557–1560.
41. Fenner PJ, Williamson JA, Burnett JW, Colquhoun DM, Godfrey S, Gunawardane K, Murtha W. The "Irukandji syndrome" and acute pulmonary oedema. Med J Aust 1988;149:150–156.
42. Fenner PJ, Hadok MC. Fatal envenomation by jellyfish causing Irukandji syndrome. Med J Aust 2002;177: 362–363.
43. Taylor DM, Pereira P, Seymour J, Winkel KD. A sting from an unknown jellyfish species associated with persistent symptoms and raised troponin I levels. Emerg Med 2002;14:175–180.
44. Huynh TT, Seymour J, Pereira P, Mulcahy R, Cullen P, Carrette T, Little M. Severity of Irukandji syndrome and nematocyst identification from skin scrapings. Med J Aust 2003;178:38–41.
45. Fenner PJ, Fitzpatrick PF, Hartwick RJ, Skinner R. "Morbakka", another cubomedusan. Med J Aust 1985;143:550–551, 554–555.
46. Currie BJ, Wood YK. Identification of *Chironex fleckeri* envenomation by nematocyst recovery from skin. Med J Aust 1995;162:478–480.
47. Beadnell CE, Rider TA, Williamson JA, Fenner PJ. Management of a major box jellyfish (*Chironex fleckeri*) sting lessons from the first minutes and hours. Med J Aust 1992;156:655–658.
48. Exton DR, Fenner PJ, Williamson JA. Cold packs: effective topical analgesia in the treatment of painful stings by *Physalia* and other jellyfish. Med J Aust 1989; 151:625–626.
49. Nomura JT, Sato RL, Ahern RM, Snow JL, Kuwaye TT, Yamamoto LG. A randomized paired comparison trial of cutaneous treatments for acute jellyfish (*Carybdea alata*) stings. Am J Emerg Med 2002;20:624–626.
50. Thomas CS, Scott SA, Galanis DJ, Goto RS. Box jellyfish (*Carybdea alata*) in Waikiki: their influx cycle plus the analgesic effect of hot and cold packs on their stings to swimmers at the beach: a randomized, placebo-controlled, clinical trial. Hawaii Med J 2001;60:100–107.
51. Bailey PM, Little M, Jelineck GA, Wilce JA. Jellyfish envenoming syndromes: unknown toxic mechanism and unproven therapies. Med J Aust 2003;178:34–37.
52. Currie B. Clinical implications of research on the box-jellyfish *Chironex fleckeri*. Toxicon 1994;32:1305–1313.
53. Ramasamy S, Isbister GK, Seymour JE, Hodgson WC. The *in vitro* effects of two chirodropid (*Chironex fleckeri* and *Chiropsalmus* sp.) venoms: efficacy of box jellyfish antivenom. Toxicon 2003;41:703–711.
54. Williamson JA, Le Ray LE, Wohlfahrt M, Fenner PJ. Acute management of serious envenomation by box-jellyfish (*Chironex fleckeri*). Med J Aust 1984;141:851–853.

55. Corkeron M, Pereira P, Makrocanis C. Early experience with magnesium administration in Irukandji syndrome. Anaesth Intensive Care 2004;32:666–669.
56. Little M, Mulcahy RF. A year's experience of Irukandji envenomation in far north Queensland. Med J Aust 1998;169:638–641.
57. Reed KC, Crowell MC, Castro MD, Sloan ML. Skin and soft-tissue infections after injury in the ocean: culture methods and antibiotic therapy for marine bacteria. Mil Med 1999;164:198–201.
58. Pigulevsky SV, Michaleff PV. Poisoning by the medusa *Gonionemus vertens* in the Sea of Japan. Toxicon 1969; 7:145–149.
59. Radwan FFY. Comparative toxinological and immunological studies on the nematocyst venoms of the Red Sea fire corals *Millepora dichotoma* and *M. platyphylla*. Comp Biochem Physiol 2002;131(Part C):232–334.
60. Wittle LW, Scura ED, Middlebrook RE. Stinging coral (*Millepora tenera*) toxin: a comparison of crude extracts with isolated nematocyst extracts. Toxicon 1974;12: 481–486.
61. Wittle LW, Wheeler CA. Toxic and immunological properties of stinging coral toxin. Toxicon 1974;12:487–493.
62. Rojas A,Torres M, Rojas JI, Feregrino A, Heimer-de la Cotera E. Calcium-dependent smooth muscle excitatory effect elicited by the venom of the hydrocoral *Millepora complanata*. Toxicon 2002;40:777–785.
63. Addy JH. Red Sea coral contact dermatitis. Int J Dermatol 1991;30:271–273.
64. Sagi A, Rosenberg L, Ben-Meir P, Hauben DJ. "The fire coral" (*Millepora dichotoma*) as a cause of burns: a case report. Burns 1987;13:325–326.
65. Marques AC, Junior VH, Migotto AE. Envenomation by a benthic Hydrozoa (Cnidaria): the case of *Nemalecium lighti* (Haleciidae). Toxicon 2002;40:213–315.
66. Metzker A, Grunwald Z. Coral injury case report of injury, late effect and therapy. Isr J Med Sci 1985;21:992–994.
67. Edwards L, Hessinger DA. Portuguese man-of-war (*Physalia physalis*) venom induces calcium influx into cells by permeabilizing plasma membranes. Toxicon 2000; 38:105–1028.
68. Burnett JW, Ordonez JV, Calton GJ. Differential toxicity of *Physalia physalis* (Portuguese man-o'war) nematocysts separated by flow cytometry. Toxicon 1986;24:514–518.
69. Stein MR, Marraccini JV, Rothschild NE, Burnett JW. Fatal Portuguese man-o'-war (*Physalia physalis*) envenomation. Ann Emerg Med 1989;18:312–315.
70. Ioannides G, Davis JH. Portuguese man-of-war stinging. Arch Dermatol 1965;91:448–451.
71. Williamson JA, Burnett JW, Fenner PJ, Hach-Wunderle V, Hoe LY, Adiga KM. Acute regional vascular insufficiency after jellyfish envenomation. Med J Aust 1988;149: 698–701.
72. Reed KM, Bronstein BR, Baden HP. Delayed and persistent cutaneous reactions to coelenterates. J Am Acad Dermatol 1984;10:462–466.
73. Togias AG, Burnett JW, Kagey-Sobotka A, Lichtenstein LM. Anaphylaxis after contact with a jellyfish. J Allergy Clin Immunol 1985;75:672–675.
74. Burnett JW, Fenner PJ, Kokelj F, Williamson JA. Serious *Physalia* (Portuguese man-of-war) stings: implications for scuba divers. J Wilderness Med 1994;5:71–76.
75. Burnett JW, Gable WD. A fatal jellyfish envenomation by the Portuguese man-o'war. Toxicon 1989;27:823–824.
76. Hartwick R, Callanan V, Williamson J. Disarming the box-jellyfish: nematocyst inhibition in *Chironex fleckeri*. Med J Aust 1980;1:15–20.
77. Fenner PJ, Williamson JA, Burnett JW, Rifkin J. First aid treatment of jellyfish stings in Australia response to a newly differentiated species. Med J Aust 1993;158: 498–501.
78. Burnett JW, Calton GJ, Rubinstein M. First aid for jellyfish envenomation. South Med J 1983;76:870–872.
79. Loten C, Stokes B, Worsley D, Seymour JE, Jiang S, Isbister GK. A randomized controlled trial of hot water (45 °C) immersion versus ice packs for pain relief in bluebottle stings. Med J Aust 2006;184:329–333.
80. Watson GM, Mire P, Hudson RR. Frequency specificity of vibration dependent discharge of nematocysts in sea anemones. J Exp Zool 1998;281:582–593.
81. Gasparini S, Gilquin B, Menez A. Comparison of sea anemone and scorpion toxins binding to Kv1 channels: an example of convergent evolution. Toxicon 2004;43: 901–908.
82. Hanck DA, Sheets MF. Site-3 toxins and cardiac sodium channels. Toxicon 2007;49:181–193.
83. Mizuno M, Nishikawa K, Yuzawa Y, Kanie T, Mori H, Araki Y, et al. Acute renal failure after a sea anemone sting. Am J Kid Dis 2000;36:E10.
84. Garcia PJ, Schein RMH, Burnett JW. Fulminant hepatic failure from a sea anemone sting. Ann Intern Med 1994;120;665–666.
85. Levy S, Masry D, Halstead BW. Report of stingings by the sea anemone *Triactis producta* Klunzinger from Red Sea. Clin Toxicol 1970;3:638–643.
86. Maretic Z, Russell FE. Stings by the sea anemone *Anemonia sulcata* in the Adriatic Sea. Am J Trop Med Hyg 1983;32:891–896.
87. Nagai H, Oshiro N, Takuwa-Kuroda K, Iwanaga S, Nozaki M, Nakajima T. A new polypeptide toxin from the nematocyst venom of an Okinawan sea anemone *Phyllodiscus semoni* (Japanese name "unbachi-isoginchaku"). Biosci Biotechnol Biochem 2002;66:2621–2625.
88. Moore RE, Scheuer PJ. Palytoxin: a new marine toxin from a coelenterate. Science 1971;172:495–498.
89. Ukena T, Satake M, Usami M, Oshima Y, Naoki H, Fujita T, Kan Y, Yasumoto T. Structure elucidation of ostreocin D, a palytoxin analog isolated from the dinoflagellate *Ostreopsis siamensis*. Biosci Biotechnol Biochem 2001; 65:2585–2588.

90. Vick JA, Wiles JS. The mechanism of action and treatment of palytoxin poisoning. Toxicol Appl Pharmacol 1975;34: 214–223.
91. Gleibs S, Mebs D. Distribution and sequestration of palytoxin in coral reef animals. Toxicon 1999;37:1521–1527.
92. Mebs D. Occurrence and sequestration of toxins in food chains. Toxicon 1998;36:1519–1522.
93. Hilgemann DW. From a pump to a pore: how palytoxin opens the gates. Proc Natl Acad Sci 2003;100:386–388.
94. Artigas P, Gadsby DC. Na+/K+-pump ligands modulate gating of palytoxin-induced ion channels. Proc Natl Acad Sci U S A 2003;100:501–505.
95. Groebe DR, Abramson SN. Lophotoxin is a slow binding irreversible inhibitor of nicotinic acetylcholine receptors. J Biol Chem 1995;270:281–286.
96. Taniyama S, Mahmud Y, Terada M, Takatani T, Arakawa O, Noguchi T. Occurrence of a food poisoning incident by palytoxin from a serranid *Epinephelus* sp. in Japan. J Nat Toxins 2002;11:277–282.
97. Kodama AM, Hokama Y, Yasumoto T, Fukui M, Manea SJ, Sutherland N. Clinical and laboratory findings implicating palytoxin as cause of ciguatera poisoning due to *Decapterus macrosoma* (mackerel). Toxicon 1989;27: 1051–1053.
98. Bignami GS, Raybould TJ, Sachinvala ND, Grothaus PG, Simpson SB, Lazo CB, et al. Monoclonal antibody-based enzyme-linked immunoassays for the measurement of palytoxin in biological samples. Toxicon 1992;30: 687–700.

Chapter 180

SPONGES
(Phylum: Porifera)

IDENTIFYING CHARACTERISTICS

The phylum Porifera (sponges) contains three classes: the largest class, Demospongiae, Calcarea (sponges with limy spicules), and Hexactinellida, (glass sponges).[1] There are at least 5,000 species of sponges, and over 600 species are associated with human toxicity.[2,3] These simple animals are extremely diverse, occurring in many colors, shapes, compressibility, and sizes.

Species associated with human toxicity include the Caribbean and Hawaiian fire sponge (*Tedania ignis*),[4,5] the Western Tropical Atlantic poison bun sponge (*Neofibularia nolitangere* Duchassaing & Michelotti), the Northeastern American red sponge (*Microciona prolifera*), and the Southern Australian stinging sponges (*Neofibularia mordens*, *Lissodendoryx* spp.).[6] Sponges are small multicellular animals with hollow centers, branched tubes, and mineralized skeletons that contain calcium carbonate or silica spicules. Dense masses of spicules are attached with spongin to form the sponge skeleton. Spongin is a fibrillar, intracellular collagen that cements the spicules at points of contact to form the skeleton.

Common Name: Australian Stinging Sponge
Scientific Name: *Neofibularia mordens* Hartman
Physical Description: This sponge is gray to royal blue on the outer surface with an ochre or yellow interior.

Common Name: Red-Bearded Sponge, Northeastern American Red Sponge, Oyster Sponge, Red Moss Sponge
Scientific Name: *Microciona prolifera*
Physical Description: This orange-red sponge forms dense clusters of curved and twisted, finger-like lobes. This sponge reaches 20 cm (~8 in.) in height

Common Name: Hawaiian Fire Sponge, Caribbean Fire Sponge, Scarlet Sponge, Fire Sponge
Scientific Name: *Tedania ignis* Duchassaing & Michelotti
Physical Description: The fire sponge is an orange to bright red sponge with distinctive, volcano-like ostia. The colony reaches 4–12 in. (~10–30 cm) in diameter.

EXPOSURE

Sponges occur worldwide, primarily in shallow waters. Most, but not all, of these sessile, long-lived organisms inhabit marine environments near coastlines. *Neofibularia mordens* Hartman inhabits coastal waters near Australia and New Zealand. The red-bearded sponge (*Microciona prolifera*) grows in small clumps on oyster shells and pilings in shallow, subtidal areas along the Atlantic and Pacific coasts. The range of the red-beard sponge extends from Nova Scotia to Florida and Texas, as well as from Washington to California on the Pacific Coast. The Caribbean fire sponge (*Tedania ignis*) lives in shallow coastal waters and tidal zones of the tropical western Atlantic and Pacific Oceans, particularly on the roots of mangrove trees or on other sponges. Other sponges associated with dermal lesions include *Haliclona* spp. (eastern Australia), *Biemna saucia* Hooper,

Medical Toxicology of Natural Substances, by Donald G. Barceloux, MD

Capon & Hodder (northern Australia), *Ephydatia* spp. (central Australia), *Neofibularia nolitangere* (North America), *Tedania nigrescens* Schmidt, and *Tedania anhelans* Lieberkuhn (New South Wales, Australia).[7]

Sponges have porous exteriors composed of fibrous material (calcium carbonate, silicon dioxide, protein, or spongin). The surface structure of the sponge contains pores or ostia. Through these ostia, water and food enter a system of canals that direct these substances to chambers lined with hairlike flagellum and cells containing sticky, collar-shaped structures. Marine sponges are filter feeders that provide refuge for many small invertebrates and symbiotic microorganisms.[8]

PRINCIPAL TOXINS

Sponges are comprised of a skeleton of silica or calcium carbonate spicules and spongin fibers covered with a surface liquid or slime. Sponges do not possess specialized venom glands, but many species produce crinotoxins (surface liquids or slimes). This slime probably causes a contact dermatitis when the fine, sharp spicules of the sponge break during human contact; however, there are few data on the chemical composition of the slime or any active ingredients.[9] Calcareous spicules may contaminate the wound and contribute to mechanical trauma. These spicules range in size up to about 10 mm with one to four rays (actins).[10] A thin organic sheath covers the spicules. These biominerals have various shapes and consist mainly of magnesium-calcite. Drying reduces the toxicity of stinging sponges, and moisture can reactivate the dermal toxicity of the sponges.[6] *Neofibularia mordens* remains toxic despite 4 years of frozen storage.

The cause of these dermal lesions remains unclear with a variety of potential sources including chemical toxicity, trauma, fungi, viruses, cyanobacteria, and bacteria (*Bacillus* spp., *Pseudomonas* spp.).[11] Hypersensitivity probably does not contribute to the acute phases of the sponge-induced contact dermatitis.[12] The contact dermatitis associated with stinging sponges is distinct from sponge fisherman's disease. This latter entity develops after contact with nematocysts of sea anemones (*Sagartia* spp., *Actinia* spp.), which proliferate at the base of the sponge. Consequently, sponge fisherman's disease occurs when sponge collectors are stung by the nematocyst of the sea anemones that grow near the sponge. Dogger Bank Itch is a severe allergic contact dermatitis that develops in Danish and British trawlermen in the North Sea as a result of repeated exposure to the marine bryozoan, *Alcyonidium gelatinosum* L. This organism contains the allergen, (2-hydroxyethyl) dimethylsulfoxonium ion, which is also present in some sponges (e.g., *Theonella mirabilis* de Laubenfels).[13] However, the clinical significance of this allergen in marine sponges remains unclear.

CLINICAL RESPONSE

Direct contact with these sponges produces a contact dermatitis similar to poison oak or poison ivy (*Rhus*), and systemic reactions do not usually occur following contact with stinging sponges. Dermal injuries to the extremities usually occur when the sponges are handled with unprotected hands after removal from the marine environment. Typically, the victim develops a burning pruritus and erythema within 1 hour of contact. By the following day, mild edema, pain, and local stiffness occur that may spread beyond the area of contact. The intensity of the symptoms associated with these lesions depends, in part, on the species of sponge. The erythematous lesions may progress to vesiculation several days after contact. Pain and local stiffness may persist a few weeks to over several months. Late manifestations include exfoliation or desquamation of the involved skin 1–3 weeks following contact with the stinging sponge. A case report associated the handling of a specimen of Bermuda fire sponge (*Tedania ignis*) with an initial burning, pruritic erythema that resolved over 8 hours. Ten days later, pruritic target lesions appeared on the volar surface of the arms. These lesions became generalized to the face, palms, and soles consistent with erythema multiforme.[14] An anaphylactoid reaction occurred in this patient following patch testing to the Bermuda fire sponge. Secondary infections may develop following contamination of the skin lesion with marine organisms (*Mycobacterium marinum*, *Erysipelothrix rhusiopathiae*, *Erysipelothrix insidiosa*, *Vibrio parahaemolyticus*).

TREATMENT

Contact with sponges causes primarily local reactions. Treatment is supportive. The wound should be cleaned with saline after thorough inspection and removal of all spicules by careful scraping or the use of adhesive tape. Case reports suggest that dilute vinegar soaks (30 mL vinegar per liter of water) or 5% acetic acid soaks several times per day for 5–30 minutes may reduce symptoms of pain and pruritus.[6] Desquamation occurs despite vinegar soaks. Anecdotal experience does not support the use of topical steroids, antihistamines, or prophylactic antibiotics for the treatment of symptoms associated with stinging sponges. The role of topical steroids and antihistamines in delayed reactions remains undefined. Lymphadenopathy may result from the envenomation, and the presence of lymphadenopathy alone does not indicate an infection. Patients should be followed for signs of secondary infection.

References

1. Leys SP, Mackie GO, Reiswig HM. The biology of glass sponges. Adv Mar Biol 2007;52:1–145.
2. Russell FE. Venomous and poisonous marine animal injuries. Vet Hum Toxicol 1991;33:334–337.
3. Halstead BW. Poisonous and Venomous Marine Animals of the World. 2nd rev ed. Princeton, NJ: Darwin Press; 1988.
4. Sims JK, Irei MY. Human Hawaiian marine sponge poisoning. Hawaii Med J 1979;39:263–270.
5. Yaffee HS. Irritation from red sponge. N Engl J Med 1970;282:51.
6. Southcott RV, Coulter JR. The effects of the southern Australian marine stinging sponge: *Neofibularia mordens* and *Lissodendoryx* spp. Med J Aust 1971;2:895–901.
7. Isbister GK, Hooper JN. Clinical effects of stings by sponges of the genus *Tedania* and a review of sponge stings worldwide. Toxicon 2005;46:782–785.
8. Wehrl M, Steinert M, Hentschel U. Bacterial uptake by the marine sponge *Aplysina aerophoba*. Microb Ecol 2007;53:355–365.
9. Flachsenberger W, Holmes NJ, Leigh C, Kerr DI. Properties of the extract and spicules of the dermatitis inducing sponge *Neofibularia mordens* Hartman. Clin Toxicol 1987;25:255–272.
10. Sethmann I, Wörheide G. Structure and composition of calcareous sponge spicules: A review and comparison to structurally related biominerals. Micron 2008;39:209–228.
11. Webster NS. Sponge disease: a global threat? Environ Microbiol 2007;9:1363–1375.
12. Burnett JW, Calton GJ, Morgan RJ. Dermatitis due to stinging sponges. Cutis 1987;39:476.
13. Warabi K, Nakao Y, Matsunaga S, Fukuyama T, Kan T, Yokoshima S, Fusetani N. Dogger Bank Itch revisited: isolation of (2-hydroxyethyl) dimethylsulfoxonium chloride as a cytotoxic constituent from the marine sponge *Theonella* aff. *mirabilis*. Comp Biochem Physiol 2001;128(Part B):27–30.
14. Yaffee HS, Stargardter F. Erythema multiforme from *Tedania ignis*: report of a case and an experimental study of the mechanism of cutaneous irritation from the fire sponge. Arch Dermatol 1963;87:601–604.

Chapter 181

STARFISH, SEA URCHINS, SEA CUCUMBERS, and FIREWORMS
(Phylum: Echinodermata)

STARFISH

IDENTIFYING CHARACTERISTICS

Crown-of-thorns starfish belong to the phylum Echinodermata (Class: Asteroidea), and this organism is the main venomous starfish.

Common Name: Crown-of-Thorns Starfish, Crown-of-Thorns Sea Star

Scientific Name: *Acanthaster planci* L.

Family: Acanthasteridae

Physical Description: Crown-of-thorns is an attractive starfish that possesses dorsal spines covered with a venom-secreting sheath. This organism grows up to 50cm (~20in.) in diameter. The radially arranged body has five arms surrounding a central disk. A green to green-red-purple epidermis covers this starfish that contains mucoid-secreting glands. Figure 181.1 displays a specimen of the crown-of-thorns sea star.

EXPOSURE

Starfish are widely distributed throughout tropical coastal waters. Crown-of-thorns (*Acanthaster planci*) is plentiful in the Great Barrier Reef off Australia. This starfish inhabits coral reefs in the Indo–Pacific Ocean and the Red Sea. The proliferation of this starfish causes considerate destruction of the coral reefs.[1]

PRINCIPAL TOXINS

The needle-sharp dorsal spines of *A. planci* extend 1–5cm (~0.4–2in.) in length with an elongated conical shape. These spines contain a dense calcitic cortex and a less dense medullary pulp that breaks easily. The venom from crown-of-thorns starfish contains myonecrotic, hemorrhagic, capillary permeability, and histamine-releasing factors. The constituents of the venom include phospholipases A_2,[2] thornasteroside A, acanthaglycosides A–F, versicoside A, plancinin (anticoagulant factor), and a lethal factor.[3,4] Although phospholipases are abundant in the venom of snakes and insects, relatively few marine species (i.e., the snail *Conus magus* L., the Mediterranean jellyfish *Rhopilema nomadica* Galil, Spannier & Ferguson) possess this enzyme. As purified by conventional column chromatography, the lethal factor is a basic glycoprotein with a molecular weight of about 20–25kDa.[5] This lethal factor is a mixture of toxins (plancitoxin I and II), which are hepatotoxic and structurally similar to mammalian deoxyribonucleases II (DNase II).[6] Several species of Japanese starfish (*Astropecten scoparius* Valenciennes, *A. latespinosus* Meissner, *A. polyacanthus* Müller & Troschel) contain tetrodotoxin and one Japanese starfish (*Astropecten amurensis* Fisher) contains paralytic shellfish poison.[7,8]

FIGURE 181.1. Crown-of-Thorns sea star (*Acanthaster planci* L.). Photograph courtesy of Paul Auerbach, MD. See color insert.

CLINICAL RESPONSE

Dermal contact with the spines of the crown-of-thorns starfish produces the immediate onset of acute burning pain followed within minutes by bluish discoloration at the puncture site.[9] During the next 4–6 hours, increasing erythema and edema occur around the wound site. The pain usually resolves within a few hours and the edema subsides within 12–24 hours. The wound appears normal by about 48 hours after envenomation. Following severe envenomations, limitation of motion may develop in affected digits. Systemic complaints are extremely rare, although anecdotal reports suggest that muscle weakness, nausea, vomiting, and paresthesias may develop. A chronic synovitis with limitation of motion may result from retained foreign bodies and granuloma formation.[10]

DIAGNOSTIC TESTING

Spines from crown-of-thorns starfish contain calcium carbonate, which is visible on radiography. There are no significant laboratory abnormalities following envenomation.

TREATMENT

Management of dermal contact with sea urchins is supportive. Hot water immersion may reduce pain, but there are few clinical data to guide management. All wounds should be thoroughly explored with magnification, if necessary, for the presence of foreign bodies. Radiography may help determine the presence of foreign bodies because the spines are radiopaque. However, small spines may not be detectable by plain radiographs. Antibiotics (ciprofloxacin 500–750 mg twice daily, doxycycline 100 mg twice daily) are indicated for severely contaminated wounds, but there are no data to support the use of prophylactic antibiotics. Steroid creams are appropriate only for the treatment of persistent lesions.

SEA URCHINS

IDENTIFYING CHARACTERISTICS

Sea urchins are free-living, bottom-dwelling, nocturnal, slow-moving, shy creatures that possess globular bodies covered with calcareous spines of varying shapes and sizes.[11] Although most species of sea urchins do not have venom glands, about 80 of 600 species of sea urchin are venomous.[12] There are 12 orders of the class Echinoidea in the phylum Echinodermata based on morphologic variation in color, size, and shape of the spines.

Common Name: Sea Urchins

Scientific Name: Class: Echinoidea

Physical Description: These ubiquitous creatures resemble globular pincushions surrounded by numerous sharp, detachable, spines. Figure 181.2 displays a specimen of the fire sea urchin (*Asthenosoma varium* Grube). Some species, such as the long-spined sea urchin (*Diadema antillarum* Philippi), contain long spines that easily fracture on contact with human skin.

EXPOSURE

Sea urchins live below rocks or between coral in shallow waters. The hard lime shells of these marine animals are usually globular or egg-shaped with calcareous spines projecting from the shell in many shapes and sizes. Painful injuries result from contact with sharp spines or venomous pedicellariae. These latter organelles are movable, jaw-tipped stalks that protrude between the spines. The pedicellariae contain venom glands and valves that extrude the venom on contact. The pedicellariae of some species (e.g., *Toxopneustes* spp.) are strong enough to penetrate human skin, but many sea urchin species do not envenomate humans. Sea urchin species in the Echinothuriidae and Diadematidae families possess long, slender spines containing venomous pedicellariae. Sea urchins are distributed worldwide with tropical venomous species including sea urchins from the genera *Asthenosoma, Diadema*, and *Echino-*

FIGURE 181.2. Fire sea urchin (*Asthenosoma varium* Grube). Photograph courtesy of Paul Auerbach, MD. See color insert.

thrix. Most injuries result from either stepping on the spines or carelessly handling the sea urchin. Sea urchin roe is a Japanese and Korean delicacy served with sushi, that is a potential food allergen and may cause anaphylaxis in sensitized individuals.[13,14]

PRINCIPAL TOXINS

Venom Composition

Venoms from species in the phylum Echinodermata contain a variety of chemical substances including serotonin, steroid glycosides, quaternary ammonium bases, and acetylcholine-like substances.[15] The venoms from some sea urchin species also include neurotoxins, hemolysins, serotonin, and proteases. The venom of species from the genus *Tripneustes* contains histamine- and bradykinin-like factors that can produce severe pain and cranial nerve palsies following injection by the pedicellariae.[16] However, the pretreatment of rodents with antihistamines does not reduce the toxicity of the venom.[17]

Venom Apparatus

Venomous and nonvenomous sea urchins inflict a painful puncture wound following contact of the spines with unprotected skin. These spines are brittle and may lodge in the wound. An indentation occurs at the base of the spine that fits over a tubercle with muscle fibers extending from the tubercle to a circular ledge (milled ring). The long, sharp spines of some species are hollow and these specialized spines contain venom. Figure 181.3 displays the morphology of sea urchin spines. The nerve ring in the epithelium at the base of the spine stimulates surrounding spines to bend toward the stimulated spine. The complexity of the venom apparatus varies between sea urchin genera. Certain species (e.g., *Asthenosoma*) possess well-developed venom spines, whereas other species (e.g., *Tripneustes*, *Toxopneustes*) have specialized venom organs called pedicellariae. These small, pincer-like organs are modified spines with flexible heads that contain an organelle capable of injecting venom. The pincers open on contact and discharge venom into the wound created by the spine. Tiny barbs with reverse orientation appear on the outer surface of the slender spines. A proteinaceous covering surrounds the spines. All species of sea urchins can inflict mechanical trauma that may produce local tissue reaction from retained debris.

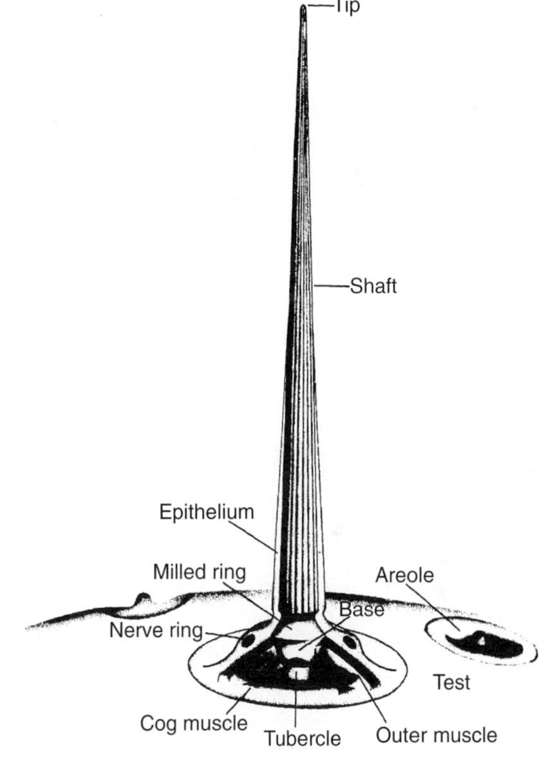

FIGURE 181.3. Morphology of sea urchin spine. From Baden HP. Injuries from sea urchins. Clin Dermatol 1987;5:113. Reprinted with permission from Elsevier.

Dose Response

The toxins in sea urchin venom are not well defined, and there are inadequate data to determine a dose response in humans.

HISTOPATHOLOGY AND PATHOPHYSIOLOGY

Sea-urchin granulomas are chronic granulomatous skin lesions that demonstrate a wide morphologic spectrum; these lesions are not always related to retained foreign bodies. In a histological study of 50 biopsy specimens from 35 patients with sea urchin granulomas, foreign body and sarcoid were the most common patterns associated with the predominant granulomatous inflammatory reaction.[18] Other types of granulomatous reactions were tuberculoid, necrobiotic and suppurative granulomas. The remaining biopsies demonstrated a predominant inflammatory reaction with features of nonspecific chronic inflammation or suppurative dermatitis. Focal necrosis and microabscesses were common, but the identification of foreign materials in the granulomas was uncommon. The delayed reaction associated with sea urchin spines is classified as allergic foreign-body type with the identity of the antigen unknown. Chronic bacterial infection (e.g., *Mycobacterium marinum*) may contribute to the granulomatous reaction.[19]

CLINICAL RESPONSE

Following contact with sea urchins, clinical effects range from mild, local irritation for a few days to granuloma formation, infection, and rarely systemic symptoms. The initial contact with the spines produces little discomfort. However, contact with some sea urchin spines produces intense pain within several minutes of envenomation along with some localized erythema, local black discoloration, and edema. The pain is out of proportion to the extent of the trauma caused by the spines. Most symptoms resolve within 1–2 days, if there are no retained foreign bodies. Colored sea urchins leave multiple punctate areas of discoloration within human skin when the spines fracture. Periungual and subungual nodular lesions may develop from retained spine fragments.[20] In uncomplicated cases, these lesions resolve completely within one month as a result of the resorption of the spines and the extrusion of the spines from the skin.[11] Secondary bacterial infection, abscess formation and chronic foreign body granulomas may complicate the puncture wounds associated with contact with sea urchins.[21] Residual spine fragments can produce pink to bluish, nontender sarcoid-like granulomas up to 1–5 mm (~0.04–0.2 in.) in diameter.[22,23] Spines embedded near nerves and joint spaces may cause an acute synovitis characterized by hyperesthesias, sensory deficits, and painful joints, particularly in the hand.[24] Systemic symptoms associated with sea urchin envenomations include myalgias, nausea, syncope, lightheadedness, paresthesias, muscle spasm, dyspnea, and, rarely, respiratory depression and hypotension.[25]

DIAGNOSTIC TESTING

The calcium carbonate spines of sea urchins are radiopaque, and the use of soft tissue x-rays may help localize the presence of retained spines.[26] Magnetic resonance imaging (MRI) helps define the soft tissue and bony changes associated with synovial arthritis following contact with sea urchins.[12]

TREATMENT

Treatment of sea urchin envenomation includes the careful removal of the protruding spines and immediate immersion of the affected area in hot water (110–115 °F/43–46 °C). Technically, the removal of sea urchin spines is difficult because of the brittle properties of the spines; therefore, aggressive débridement of the area around the puncture wounds is not indicated. The wound should be thoroughly cleaned similar to the treatment of fire coral envenomation. Analgesics may be necessary after hot water immersion. Detached pedicellariae may secrete venom for several hours, and these structures should be removed following envenomation by the pedicellariae-containing species (e.g., *Toxopneustes pileolus* Lamarck). Surgical excision of embedded spines is not indicated routinely because fragmentation of the spines causes additional injury or infection. The presence of skin discoloration at the puncture site does not necessarily indicate the presence of retained spines because residual dye from the spines causes temporary tattooing that usually resolves within 48 hours.[27] Any attempt to remove embedded spines should be performed with good lighting sources and a bloodless field, preferably with radiological direction. Chronic granulomas usually require complete excision when intradermal steroids do not eliminate the chronic granuloma. Anecdotal evidence does not support the use of ammonia, systemic steroids, or prophylactic antibiotics. There are no antidotes. Secondary wound infections (*Vibrio vulnificus, Pseudomonas aeruginosa, Aeromonas salmonicida, Staphylococcus aureus, Streptococcus pyogenes*) are rare,[28] and when present usually respond to ciprofloxacin 500–750 mg twice daily.

SEA CUCUMBERS

IDENTIFYING CHARACTERISTICS

Common Name: Sea Cucumber

Scientific Name: *Bohadschia argus* Jaeger

Family: Holothuriidae (Class: Holothuroidea)

Physical Description: Sea cucumbers are bottom-dwelling scavengers with elongated, sausage-shaped bodies. Tiny tube feet and tentacles surround the mouth.

EXPOSURE

The habitat of sea cucumbers includes both shallow and deep water. Many sea cucumbers feed on jellyfish, and extrude undigested, intact nematocysts through their skin or mouth as a defensive maneuver. During threatening periods, a group of tubules located at the anal end absorb water and form long, sticky threads for protection. Additionally, sea cucumbers may eviscerate their intestinal tract through their mouth as a means to expose threatening predators to toxic nematocyst ingested from other marine animals.[29] Dried sea cucumbers are a delicacy in many eastern Asian countries (e.g., China, Malaysia). Known as trepang or Bêche-de-mer, aphrodisiac effects are attributed to the consumption of this delicacy.

PRINCIPAL TOXINS

Venom Composition

Specialized glands along the ventral and lateral sides of the sea cucumbers produce a toxic, steroidal saponin called holothurin A. This marine toxin is an anionic surfactant and detergent that is an irritant and a hemolysin.[30,31] This compound also impairs transmission across cholinergic neuromuscular junctions. Holothurin activates calmodulin-dependent bovine brain 3′-5′ phosphodiesterase, which is antagonized by classical calmodulin antagonists (e.g., chlordecone).[32]

Venom Apparatus

Certain sea cucumber species eject adherent, white or pink, Cuvierian tubules when mechanically stimulated. These tubules protrude from the cloaca (posterior body). The adhesiveness and high tensile strength allows Cuvierian tubules to entangle and immobilize potential predators.[33] Other species secrete a limy mucous that causes a local inflammatory reaction similar to contact with the Cuvierian tubules.

Dose Response

There are inadequate data on holothurin and other potential toxins in sea cucumbers to determine dose response of these toxins in human.

HISTOPATHOLOGY AND PATHOPHYSIOLOGY

The toxins in sea cucumber are not well described.

CLINICAL RESPONSE

Contact with the venom from sea cucumbers results in a macular-papular eruption. Eye irritation may be severe. Additionally, the sea cucumber secretes undigested nematocysts in self-defense that may cause coelenterate envenomation following direct contact.

DIAGNOSTIC TESTING

There are no commercial biomarkers of envenomation by sea cucumbers.

TREATMENT

Treatment of sea cucumber lesions and other coelenterate envenomation is similar.

FIREWORMS

Fireworms (Class: Polychaeta) are a heterogeneous group of animals that produce a variety of toxin for the acquisition of prey and for defensive purposes. For example, *Paranemertes* species contain pyridine alkaloids, similar to anabasine present in wild tobacco plants.[34] Nereistoxin (4-*N*, *N*-dimethylamino-1,2-dithiolane, CAS RN: 1631-58-9) is a neurotoxin from the salivary glands of the marine annelid worm, *Lumbriconereis heteropoda* Marenzeller, that blocks nicotinic acetylcholine receptors, particularly in insects.[35] Glycerotoxin (CAS RN: 92229-20-4) from the venom of *Glycera convoluta* is a novel 320 kDa protein capable of reversibly stimulating acetylcholine release in the frog, but not at the murine neuromuscular junction because of differences in the subtype of voltage-dependent Ca^{2+} channels. Experimental studies indicate that glycerotoxin acts via Ca(v)2.2 (N-type) channels, but this toxin does not activate Ca(v)2.1 (P/Q-type) channels.[36] Despite the presence of these toxins in worm species, systemic symptoms do not usually occur following contact with these worms. Injuries usually result from the handling of these worms (*Glycera* spp., *Eunice* spp.) and the subsequent contact with barbed bristles. Clinical effects typically include local burning pain, swelling, and erythema. Treatment is supportive with good wound care and special attention directed toward identifying the presence of foreign bodies to prevent the development of granulomas.

References

1. Mebs D. Venomous and poisonous animals. A handbook for biologists, toxicologists and toxinologists, physicians and pharmacists. Stuttgart: Medpharm Scientific Publishers; 2002.
2. Shiomi KA, Kazama A, Shimakura K, Nagashima Y. Purification and properties of phospholipases A_2 from the crown-of-thorns starfish (*Acanthaster planci*) venom. Toxicon 1998;36:589–599.
3. Komori T. Toxin from the starfish *Acanthaster planci* and *Asterina pectinifera*. Toxicon 1997;35:1537–1548.
4. Karasudani I, Koyama T, Nakandakari S, Aniya Y. Purification of anticoagulant factor from the spine venom of the crown-of-thorns starfish, *Acanthaster planci*. Toxicon 1996;34:871–879.
5. Shiomi K, Yamamoto S, Yamanaka H, Kikuchi T. Purification and characterization of a lethal factor in venom from the crown-of-thorns starfish (*Acanthaster planci*). Toxicon 1988;26:1077–1083.
6. Shiomi K, Midorikawa S, Ishida M, Nagashima Y, Nagai H. Plancitoxins, lethal factors from the crown-of-thorns starfish *Acanthaster planci* are deoxyribonucleases II. Toxicon 2004;44:499–506.
7. Lin S-J, Tsai Y-H, Lin H-P, Hwang D-F. Paralytic toxins in Taiwanese starfish *Astropecten scoparius*. Toxicon 1998;36:799–803.
8. Asakawa M, Nishimura F, Miyazawa K, Noguchi T. Occurrence of paralytic shellfish poison in the starfish, *Asterias amurensis* in Kure Bay, Hiroshima Prefecture, Japan. Toxicon 1997;35:1081–1087.
9. Odom CB, Fischermann EA. Crown-of-thorns starfish wounds—some observations on injury sites. Hawaii Med J 1972;31:99–100.
10. Adler M, Kaul A, Jawad ASM. Foreign body synovitis induced by a crown-of-thorns starfish. Rheumatol 2002;41:230–231.
11. Baden HP. Injuries from sea urchins. Clin Dermatol 1987;5:112–117.
12. Liram N, Gomori M, Perouansky M. Sea urchin puncture resulting in PIP joint synovial arthritis: case report and MRI study. J Travel Med 2000;7:43–45.
13. Rodriguez V, Bartolomé B, Armisén M, Vidal C. Food allergy to *Paracentrotus lividus* (sea urchin roe). Ann Allergy Asthma Immunol 2007;98:393–396.
14. Hickey RW. Sea urchin roe (uni) anaphylaxis. Ann Allergy Asthma Immunol 2007;98:493–494.
15. Manowitz NR, Rosenthal RR. Cutaneous systemic reactions to toxins and venoms of common marine organisms. Cutis 1979;23:450–454.
16. Takei M, Nakagawa H, Endo K. Mast cell activation by pedicellarial toxin of sea urchin, *Toxopneustes pileolus*. FEBS Lett 1993;328:59–62.
17. Mebs D. A toxin from the sea urchin *Tripneustes gratilla*. Toxicon 1984;22:306–307.
18. De La Torre C, Toribio J. Sea-urchin granuloma: histologic profile. A pathologic study of 50 biopsies. J Cutan Pathol 2001;28:223–228.
19. De la Torre C, Vega A, Carracedo A, Toribio J. Identification of *Mycobacterium marinum* in sea-urchin granulomas. Br J Dermatol 2002;145:114–116.
20. Haneke E, Tosti A, Piraccini BM. Sea urchin granuloma of the nail apparatus: report of 2 cases. Dermatol 1996;192:140–142.
21. Baden HP, Burnett JW. Injuries from sea urchins. South J Med 1977;70:459–460.
22. O'Neal RL, Halstead BW, Howard LD Jr. Injury to human tissues from sea urchin spines. Calif Med 1964;101: 199–202.
23. Kinmont PD. Sea urchin sarcoidal granuloma. Br J Dermatol 1965;77:335–343.
24. Strauss MB, McDonald RI. Hand injuries from sea urchin spines. Clin Ortho Res 1976;114:216–218.
25. Cracchiolo A 3rd, Goldberg L. Local and systemic reactions to puncture injuries by the sea urchin spine and the date palm thorn. Arthritis Rheum 1977;20:1206–1212.
26. Newmeyer WL 3rd. Management of sea urchin spines in the hand. J Hand Surg 1988;13:455–457.
27. Nassab R, Rayatt S, Peart F. The management of hand injuries caused by sea urchin spines. J Hand Surg (Br) 2005;30B:432–433.
28. Reed KC, Crowell MC, Castro MD, Sloan ML. Skin and soft-tissue infections after injury in the ocean: culture methods and antibiotic therapy for marine bacteria. Mil Med 1999;164:198–201.
29. Byrne M. The morphology of autotomy structures in the sea cucumber *Eupentacta quinquesemita* before and during evisceration. J Exp Biol 2001;204:849–863.
30. Thron CD, Patterson RN, Friess SL. Further biological properties of the sea cucumber toxin holothurin A. Toxicol Appl Pharmacol 1963;5:1–11.
31. Friess SL, Standaert FG, Whitcomb ER, Nigrelli RF, Chanley JD, Sobotka H. Some pharmacologic properties of holothurin A, a glycosidic mixture from the sea cucumber. Ann N Y Acad Sci 1960;90:893–901.

32. Vig PJ, Mehrotra BD, Desaiah D. Holothurin: an activator of bovine brain 3′-5′ phosphodiesterase. Res Commun Chem Pathol Pharmacol 1990;67:419–422.
33. DeMoor S, Waite JH, Jangoux M, Flammang P. Characterization of the adhesive from Cuvierian tubules of the sea cucumber *Holothuria forskali* (Echinodermata, Holothuroidea). Mar Biotechnol (NY) 2003;5:45–57.
34. Kem WR. A study of the occurrence of anabaseine in *Paranemertes* and other nemertines. Toxicon 1971;9: 23–32.
35. Raymond Delpech V, Ihara M, Coddou C, Matsuda K, Sattelle DB. Action of nereistoxin on recombinant neuronal nicotinic acetylcholine receptors expressed in *Xenopus laevis* oocytes. Invert Neurosci 2003;5:29–35.
36. Meunier FA, Feng ZP, Molgo J, Zamponi GW, Schiavo G. Glycerotoxin from *Glycera convoluta* stimulates neurosecretion by up-regulating N-type Ca^{2+} channel activity. EMBO J 2002;21:6733–6743.

IV Eels and Lampreys

Chapter 182

MORAY EEL
(Superclass: Agnatha)

IDENTIFYING CHARACTERISTICS

Moray eels belong to the family Muraenidae and the superclass of jawless fish, Agnatha. Eels are long bony fish with no premaxillary bones. These fish apprehend their prey by biting and then transport prey by extreme protraction and retraction of their pharyngeal jaw apparatus.[1] The elongated, muscular body and the lack of paired fins (pectoral and pelvic) allow the eel to move more easily within the narrow crevices of the reef. The dorsal, anal, and caudal fins are fleshy ridges covered with thickened skin to protect the eel from abrasions. Because the eel lacks paired fins, they swim forward or backward in an S-shaped motion by moving their entire body side to side. These bottom dwelling creatures do not usually have scales. The moray eel is heavier and more compressed than most eel species. The absence of pectoral fins and presence of small round gill openings distinguish the moray eel from other marine eels. These nocturnal creatures reach up to 6ft (~2m) in length.

Common Name: California Moray Eel

Scientific Name: *Gymnothorax mordax* Ayres

Physical Description: This brown to green-brown, scaleless eel reaches up to 5ft (~1.5m) in length. Like other moray eels, this eel has a large mouth, strong teeth, small, round gill openings, and an elongated face with a broad head.

EXPOSURE

Moray eels are bottom dwelling creatures that inhabit tropical and temperate waters along shallow, rocky reefs. Human exposure occurs in these marine environments or in aquariums. The distribution of moray eels is worldwide in tropical waters with a few species present in temperate waters. The California moray lives in coastal waters from Point Conception to Baja California, whereas the spotted moray (*Gymnothorax moringa* Cuvier) lives in Atlantic coastal waters from North Carolina to Rio de Janeiro, Brazil. The blackedge moray (*Gymnothorax nigromarginatus* Girard) inhabits coastal waters from Florida to Yucatan, Mexico. The green moray eel lives in coastal waters from New Jersey to Bermuda as well as the Gulf of Mexico and the Atlantic coast to Brazil. These shy, bottom dwelling creatures rarely attack humans. Most bites result from provocation by humans, such as the blind exploration of marine caves and crevasses with hands. Moray eels emerge from holes and crevices at night to search for inactive, diurnal soft-bodied prey. They have a keen sense of smell due to the presence of lamellae richly covered with stereocilia. Morphology studies indicate that moray eels (e.g., *Echidna* spp.) with large numbers of stereocilia forage for prey primarily by olfaction, whereas other species with less stereocilia search for food visually.[2] *Gymnothorax* species use both methods to hunt prey. Banded sea kraits (*Laticauda colubrina* Schneider)

Medical Toxicology of Natural Substances, by Donald G. Barceloux, MD

prey on eels of the genus *Gymnothorax*, and moray eels from the Pacific have developed resistance to massive doses of sea krait venom as a result of the coevolution of these two species.[3]

HISTOPATHOLOGY AND PATHOPHYSIOLOGY

Moray eels have multiple rows of sharp, jagged teeth that cause ragged, deep wounds following bites. Muscular jaws, tenacious bites, and the sharp teeth of the moray eel produce deep lacerations and crushing injuries that are predisposed to bacterial infections. These animals do not possess a venom apparatus.[4] Therefore, injuries from moray eel bites are traumatic rather than toxic—similar to shark and barracuda bites. The viscera of moray eels from ciguatera endemic areas contain several ciguatoxins (CTX-1, CTX-2, CTX-3).[5] Ciguatoxins are potent sodium channel-activating toxins that share binding sites with brevetoxins. CTX-2 and CTX-3 are less polar and less potent than CTX-1.

CLINICAL RESPONSE

Moray eels can inflict deep, ragged lacerations that cause neurovascular or tendon injuries.[6] Rare case reports associate bites by moray eels with rectal perforation.[7] These wounds are predisposed to bacterial infections. Most marine bacteria are motile, halophilic, Gram-negative rods that are facultative anaerobes.[8] Evaluation of marine environments suggests that *Vibrio* species may contaminate moray eel bites, particularly *Vibrio vulnificus*. These bacteria are ubiquitous marine organisms comprised of two biotypes.[9] Biotype 1 is an opportunistic human pathogen associated with the consumption of raw oysters or with exposure of open wounds to seawater while swimming or handling contaminated fish, particularly in geographical areas with relatively high water temperatures.[10] Biotype 2 is an obligate eel pathogen that rarely infects humans.[11] Potentially, rapidly progressive cellulitis and septicemia may complicate moray eel-induced wounds, particularly in immunocompromised hosts. Soft tissue infections secondary to *V. vulnificus* usually demonstrate bronze discoloration, frequently with the formation of bullous lesions.[12] Case reports associate an indolent, ulcerative infection and regional lymphangitis secondary to *Mycobacterium marinum* following the handling of freshwater eels.[13]

Moray meat is a delicacy, and the consumption of smoked moray eel meat has been associated with an outbreak of *Salmonella blockley*.[14] Portions of the moray eel may contain ciguatoxin, particular the viscera, and the consumption of moray eel may cause ciguatera food poisoning.[15,16]

DIAGNOSTIC TESTING

Analytical methods to identify multiple ciguatoxins in moray eels include reverse-phase high performance liquid chromatography/mass spectrometry.[17]

TREATMENT

Because moray eels inflict deep puncture wounds, bites should be thoroughly evaluated for tendon, nerve, fascia, ligament, and vascular injures. All wounds should be copiously irrigated with normal saline after control of bleeding, evaluation of vital signs, and physical examination. The wound should be explored for foreign bodies (sand, teeth fragments) after radiographic evaluation for retained foreign bodies. Débridement may be necessary for crushed or devitalized tissue depending on the extent of injury and the presence of adjacent neurovascular structures. Wound closure should be loose because of the risk of infection, depending on physician judgment. Abrasions and superficial lacerations in victims with normal host defenses probably do not require prophylactic antibiotics. Although there are few clinical data, anecdotal evidence suggests that victims with deep wounds should receive prophylactic antibiotics, particularly when the wound may contain a retained foreign body. There are few clinical data on the effectiveness of antibiotics for the treatment of infections associated with moray eel bites. Most *Vibrio* species are sensitive to ciprofloxacin, cefuroxime, trimethoprim-sulfamethoxazole, and tetracycline.[18] *Vibrio* species and other Gram-negative bacteria are frequently resistant to penicillin and first-generation cephalosporins. The long, fanged teeth of the moray eel penetrate deep into soft tissue; therefore, victims of moray eel bites should be followed closely for the development of closed-space infections in the affected extremity.

References

1. Mehta RS, Wainwright PC. Functional morphology of the pharyngeal jaw apparatus in moray eels. J Morphol 2008;269:604–619.
2. Fishelson L. Comparative morphology and cytology of the olfactory organs in Moray eels with remarks on their foraging behavior. Anat Rec 1995;243:403–412.
3. Heatwole H, Powell J. Resistance of eels (*Gymnothorax*) to the venom of sea kraits (*Laticauda colubrina*): a test of coevolution. Toxicon 1998;36:619–625.
4. Halstead BW. Poisonous and venomous marine animals of the world. 2nd rev ed. Princeton, NJ: Darwin Press; 1988.
5. Lewis RJ, Sellin M, Poli MA, Norton RS, MacLeod JK, Sheil MM. Purification and characterization of ciguatoxins

from moray eel (*Lycodontis javanicus*, Muranenidae). Toxicon 1991;29:1115–1127.

6. Riordan C, Hussain M, McCann J. Moray eel attack in the tropics: a case report and review of the literature. Wilderness Environ Med 2004;15:194–197.
7. Lo SF, Wong SH, Leung LS, Law IC, Yip AW. Traumatic rectal perforation by an eel. Surg 2004;135: 110–111.
8. Pien FD, Ang KS, Nakashima NT, Evans DG, Grote JA, Hefley ML, Kubota EA. Bacterial flora of marine penetrating injuries. Diagn Microbiol Infect Dis 1983;1: 229–232.
9. Amaro C, Biosca EG, Fouz B, Alcaide E, Esteve C. Evidence that water transmits *Vibrio vulnificus* biotype 2 infection to eels. Appl Environ Microbiol 1995;61: 1133–1137.
10. Dalsgaard A, Frimodt-Moller N, Bruun B, Hoi L, Larsen JL. Clinical manifestations and molecular epidemiology of *Vibrio vulnificus* infections in Denmark. Eur J Clin Microbiol Infect Dis 1996;15:227–232.
11. Veenstra J, Rietra PJ, Stoutenbeek CP, Coster JM, de Gier HH, Dirks-Go S. Infection by an indole-negative variant of *Vibrio vulnificus* transmitted by eels. J Infect Dis 1992;166:209–210.
12. Howard RJ, Burgess GH. Surgical hazards posed by marine and freshwater animals in Florida. Am J Surg 1993;166:563–567.
13. Zeeli T, Samra Z, Pitlik S. Ill from eel? Lancet Infect Dis 2003;3:168.
14. Fell G, Hamouda O, Lindner R, Rehmet S, Liesegang A, Prager R, et al. An outbreak of *Salmonella blockley* infections following smoked eel consumption in Germany. Epidemiol Infect 2000;125:9–12.
15. Yasumoto T, Scheuer PJ. Marine toxins of the pacific—VIII ciguatoxin from moray eel livers. Toxicon 1969;7: 172–176.
16. Nukina M, Koyanagi LM, Scheuer PJ. Two interchangeable forms of ciguatoxin. Toxicon 1984;22:169–176.
17. Lewis RJ, Jones A. Characterization of ciguatoxins and ciguatoxin congeners present in ciguateric fish by gradient reverse-phase high-performance liquid chromatography/mass spectrometry. Toxicon 1997;35:159–168.
18. Erickson T, Vanden Hoek TL, Kuritza A, Leiken JB. The emergency management of moray eel bites. Ann Emerg Med 1992;21:212–216.

Chapter 183

BONY FISH
(Class: Osteichthyes)

LIONFISH, SCORPIONFISH, and STONEFISH

IDENTIFYING CHARACTERISTICS

The Scorpaenidae family contains venomous species of the genera *Pterois* (lionfish), *Scorpaena* (scorpionfish, sculpin), and *Synanceia* (stonefish). This family consists of 70 genera and about 350 species; venomous species from the family Scorpaenidae are some of the most venomous bony fish species. Approximately 80 of the 350 species are venomous. Venomous species of this family include the South Australian cobbler or soldierfish (*Gymnapistes marmoratus* Cuvier), lionfish (*Dendrochirus* species), sculpin (*Scorpaena* species), and fortescue (*Centropogon australis* White). Classification of these fish is based on the morphology of the venom apparatus. The bony plate that covers the facial region from the eyes to the gill helps identify this diverse family (i.e., "mail-cheeked" fish). Stonefish species (*Synanceia* spp.) include the common or estuarine stonefish (*S. horrida* L.), the reef stonefish (*S. verrucosa* Bloch & Schneider), the dwarf stonefish (*Synanceia nana* Eschmeyer & Rama-Rao), the Indian stonefish (*S. horrida* L.), and the pitted stonefish (*Synanceia erosa* Cuvier or *Erosa erosa*). The bullrout (*Notesthes robusta* Günther) is a venomous member of the Tetrarogidae family.

Common Name: Lionfish, Zebrafish, Red Lionfish, Butterfly Cod, Lion Fish, Ornate Butterfly-Cod, Peacock Lionfish, Red Firefish, Scorpion Volitans, Turkey Fish, and Turkeyfish

Scientific Name: *Pterois* spp. (*P. volitans* L., *P. radiata* Cuvier, *P. antennata* Bloch)

Physical Description: This group of colorful, graceful fish has fan-like fins. The length of this beautiful Asian tropical fish is about 30–38 cm (~12–15 in.) with vertical, dark stripes and long slender dorsal and ventral spines. Although coloration is highly variable, these fish are frequently zebra-banded in red and white. There are numerous spiny projections and fleshy tabs on the head. The eyes protrude dorsally over a large mouth that has numerous skin appendages. Figure 183.1 displays a specimen of the lionfish (*Pterois volitans* L.).

Common Name: Sculpin

Scientific Name: *Scorpaena guttata* Girard

Physical Description: The sculpin has a stocky, slightly compressed body with a large head, protruding eyes, and a large mouth and pectoral fins. The color is red to brown with dark patches over the body and fins. Most sculpins are small (<6 in./<15 cm), but the cabezon (*Scorpaenichthys marmoratus* Ayres) reaches 39 in. (~100 cm) in length.

Medical Toxicology of Natural Substances, by Donald G. Barceloux, MD

FIGURE 183.1. Lionfish (*Pterois volitans* L.). Photograph courtesy of Paul Auerbach, MD. See color insert.

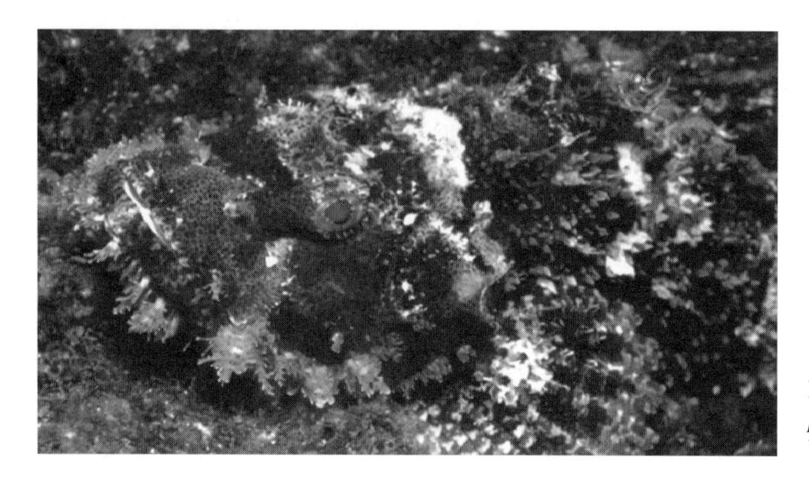

FIGURE 183.2. Spotted scorpionfish (*Scorpaena plumieri* Bloch). Photograph courtesy of Paul Auerbach, MD. See color insert.

Common Name: Spotted Scorpionfish

Scientific Name: *Scorpaena plumieri* Bloch

Physical Description: This common scorpionfish grows up to 45 cm (~18 in.) with fleshy plumes (cirri) over the eyes and characteristic skin flaps on the head. Distinctive features include three bars on the tail and white spots on the dark background of the inside of the pectoral fins. These spots are visible only when threatened or while swimming. Figure 183.2 displays a specimen of the spotted scorpionfish.

Common Name: Stonefish, Australian Stonefish, Estuarine Stonefish

Scientific Name: *Synanceia horrida* L. (*Synanceia trachynis* Richardson)

Physical Description: This well-camouflaged bottom-dwelling fish has a tough, uneven skin that is often covered with slime and wart-like bumps. This bulky, compact fish blends easily in the background with a color that resembles the surrounding environment (e.g., dark, mottled green-brown). The adult fish is about 20–30 cm (~8–12 in.) in length with elevated, dorsally positioned eyes.[1] Figure 183.3 displays a specimen of the estuarine stonefish.

EXPOSURE

Geographic Distribution

Envenomations by species of the Scorpaenidae family are the most common fish envenomations excluding stingrays. Widely distributed, Scorpaenidae species generally inhabit sandy, rocky, or coral coastlines in shallow tropical and temperate waters. Stonefish inhabit the warm temperate and tropical waters of the Indian and

FIGURE 183.3. Australian or estuarine stonefish (*Synanceia horrida* L.). Photograph courtesy of Paul Auerbach, MD. See color insert.

Pacific Oceans north to China, east to Hawaii, west to the African coast, and south to Australia. Sculpin occur throughout marine waters, primarily in tropical and temperate oceans. The native distribution of the lionfish is the coral reefs of the Indo–Pacific Ocean. Lionfish (*Pterois volitans* L.) are frequently found in private aquariums. The bullrout (*Notesthes robusta* Günther) inhabits slow-moving streams and tidal waters along the eastern coast of Australia. Disturbance of this bottom-dwelling fish near rocks and weeds causes the erection of spines and the infliction of painful injuries similar to stonefish.[2]

Behavior

Lionfish

The most commonly encountered species in this genus is the lion fish (*Pterois volitans* L.). These fish respond aggressively to perceived threats by erecting venomous spines. Lionfish are solitary predators that feed on small fish, shrimp, and crab. They inhabit shallow coral and rocky reefs up to depths of 50m (~160ft). These fish prefer shelter under ledges or in caves or crevices during daylight hours. In these areas, the lionfish exhibit almost motionless posture with the head tilted slightly downward. Most lionfish envenomations occur while handling fish in an aquarium rather than in their natural habitat.[3]

Scorpionfish and Sculpin

Sculpin are bottom-dwelling fish that live in shallow waters off the coasts of California and Hawaii. These fish inhabit areas over hard, rocky bottoms at depths ranging from immediately below the surface to 600ft (~180m). Occasionally, these fish are present near sand or mud bottoms. This diverse group of stationary and slow-moving fish has differing habitats and camouflage abilities. Envenomation by the California scorpionfish or sculpin (*Scorpaena guttata* Girard) occurs following contact with the retractable spines during the removal of the fish from a net or a hook. Occasionally, envenomation occurs when humans pursue this slow-moving fish in open seas.

Stonefish

The stonefish (*Synanceia horrida* L.) lives in shallow coastal waters where the fish resembles a stone or dead coral in the crevices of coral reefs. Stonefish are sedentary fish that prefer the shallow waters of coral reefs, kelp beds, and rocky coastline, where inadvertent contact usually occurs when humans step on the camouflaged, bottom-dwelling fish. This fish feeds on small marine animals by sucking the prey into their mouth through rapidly opening and closing their mouth.

PRINCIPAL TOXINS

Venom Composition

Although comparison of venoms from these three families is difficult as a result of methodological differences, the venoms of most species of the Scorpaenidae demonstrate similar cardiovascular and neuromuscular effects.[4,5] The highly complex venom of these fish is heat-labile. The venom retains potency up to a few days after the death of the fish. Although potency between species is similar, the differences in amount and delivery mechanism between species cause a wide variation in symptoms following envenomation.

Scorpionfish

Scorpionfish venom (i.e., *Scorpaena plumieri* Bloch) contains a proteolytic enzyme (protease), which hydrolyzes casein and gelatin.[6]

Stonefish

Venoms from stonefish contain factors that affect the cardiovascular and neuromuscular systems including the development of increased vascular permeability and ventricular fibrillation in animal studies.[7] These venoms demonstrate hemolytic and hyaluronidase activity.[8] Stonustoxin (CAS RN: 137803-80-6) is a toxic protein isolated from the venom of *Synanceia horrida* L. In animal models, the administration of this toxin causes death following the development of endothelium-dependent vasodilation and rapid, irreversible hypotension.[9] This toxin also produces edema, increased vascular permeability, platelet aggregation, and hemolysis. Verrucotoxin is the major component of venom from the stonefish (*Synanceia verrucosa*).[10] *In vitro* studies suggest that verrucotoxin inhibits ATP-sensitive K^+ current through the muscarinic M_3 receptor-PKC pathway.[11]

Venom Apparatus

The venom apparatus of fish typically involves grooved spines containing glandular venom tissue surrounded by an integument or sheath. This apparatus is usually located near the oral pole. These fish have no voluntary control over envenomation other than the angulation of the spines.

Lionfish

The venom apparatus of the lionfish consists of long thin spines involving 12 or 13 dorsal spines, 3 anal spines and 3 pelvic spines associated with paired venom glands at the base of the spines. The pectoral spines do not have venom glands. A thin integumentary sheath covers the venomous spines, but there are no venom ducts. The venom diffuses into the wound when contact with the venomous spines tears the integumentary sheath over the venomous spine.

Scorpionfish

Scorpionfish have variable numbers of long, heavy dorsal, anal, and pelvic spines with moderate-sized venom glands. The dorsal and pelvic spines are covered by thick integumentary sheaths that produce envenomation following discharge into the wound. In contrast to stonefish, scorpionfish and lionfish have a poorly developed venom apparatus that lacks an excretory duct. The spines of scorpionfish are short and strong with the venom gland being more developed than the venom gland in lionfish.

Stonefish

The well-developed venom apparatus of the stonefish consists of short, thick spines, paired venom glands, and very thick sheaths on the dorsal spines. There are 13 dorsal spines, 2 pelvic spines, and 3 anal spines. A fibrous capsule covers the venom glands and divides the glandular tissue into many septa containing numerous nerves and blood vessels.[12] Inadvertent contact with the fish causes release of the venom from the dorsal spines into the wound. Mechanical pressure applied at the base of the venomous spines releases venom from the paired venom glands located in the two lateral grooves at the base of each spine. Regeneration of the venom gland and the venom around the spines requires about 3 weeks.[13]

Dose Response

There are inadequate data in humans to determine the dose response of stonefish venom or stonustoxin. Each spine has two venom glands, which contain about 5–10 mg of crude venom with an average venom yield of approximately 6 mg per spine.[14,15] In animal experiments, the LD_{50} of stonefish venom and snake venom is similar.[16] The estimated intravenous LD_{50} in mice is approximately 22 times greater for stonustoxin than for the crude venom.[17] The amount of venom injected during a stonefish bite is not well defined. The sting of a single stonefish spine on the knee of a 52-year-old woman was associated with intense pain that resolved over 14 hours.[18] She developed no systemic symptoms, and the only sign of envenomation was mild swelling around the area of the sting.

HISTOPATHOLOGY AND PATHOPHYSIOLOGY

These fish contain highly complex venom with enzymatic as well as pronounced cardiac and neuromuscular activity that results from increased intracellular calcium concentrations, possibly via the formation of pores.[19] A nondialyzable, high-molecular-weight, heat-labile protein component produces neuromuscular and cardiac effects including vasodilation, hypotension, muscular weakness, and respiratory paralysis.[20,21] Animal studies suggest that lionfish (*P. volitans*) venom produces cardiovascular effects primarily by acting on muscarinic cholinergic receptors and adrenoceptors.[20] Death from stonefish envenomation is attributed to respiratory arrest secondary to skeletal muscle paralysis.[22] However, animal studies suggest that the venom from stonefish cause airway inflammation and acute lung injury.[23]

CLINICAL RESPONSE

Lionfish and Scorpionfish

The clinical manifestations of scorpion fish (*Scorpaena* spp.) and lionfish (*Pterois* spp.) are similar except the latter produces more severe local symptoms and the former produces more systemic symptoms. Typically, envenomations following contact with the dorsal or pelvic spines produce almost immediate, intense, pulsatile pain that may progress to involve the entire extremity within a few minutes. Symptoms range from erythema and ecchymosis to vesicle formation and local necrosis.[24] Substantial local swelling, hyperesthesias, hypesthesias, paresthesias, erythema, and pain may extend up the involved extremity.[3,25] Frequently, there is bluish discoloration around the puncture site. The maximum intensity of the pain occurs within approximately 1 hour, and the pain usually resolves within 8–12 hours (range: up to 1 day). Hypesthesia may persist along with necrosis when sloughing occurs at the site of envenomation.[26] Multiple stings may produce regional lymphadenopathy and local lymphadenitis. Occasionally, systemic effects of envenomation include nausea, vomiting, abdominal pain, headache, diaphoresis, generalized weakness, muscle cramps, tremulousness, dyspnea, tachycardia, chest pain, and mild hypotension.[27] Systemic symptoms are less common after lionfish envenomations than after sculpin envenomations.

Stonefish

Stonefish envenomation causes immediate, severe local pain that may progress proximally along with local edema, erythema, and bluish discoloration around the puncture wounds.[28] The pain typically peaks 1–2 hours after envenomation and resolves within 12 hours.[29] Vesicles can form in the immediate area of envenomation followed by sloughing of the tissue, cellulitis, and hyperesthesias. Local symptoms usually resolve within 1–2 weeks, but surgical débridement may be necessary for puncture wounds with retained spines or local necrosis.[1] Systemic effects include diaphoresis, nausea, and lightheadedness. Stonefish produce the most severe systemic effects following envenomations by species from the family Scorpaenidae with fatalities occurring rarely as a result of respiratory insufficiency.[30,31] Although uncommon, serious clinical features of stonefish envenomation include bradycardia, syncope, dysrhythmias, hypotension, convulsions, paralysis, and pulmonary edema.[32] Complications include localized necrosis and delayed healing,[33] secondary infection, cutaneous granulomas, and persistent paresthesias and hypesthesias. Extensive tissue necrosis suggests the presence of secondary infection.

DIAGNOSTIC TESTING

Clinically significant laboratory abnormalities do not usually occur following envenomation by these fish. There are no commercial analytical methods for the confirmation of envenomation by these fish.

TREATMENT

First Aid

First aid consists primarily of immersing the envenomated area with hot water, limiting movement, and referral to a health care facility if systemic symptoms develop. There are no specific substances that effectively neutralize the injected venom. Because the venom is heat labile, rapid hot water immersion for 30–60 minutes at 45 °C (113 °F) is usually the most effective first-aid measure.[3] Relief generally occurs soon after immersion, and the soaks may be stopped after complete relief. To prevent thermal injury, the temperature of the water should be tested by immersing the extremity opposite the envenomation into the water bath. After initial immersion of the envenomated extremity in hot water, the extremity should be removed every 15–20 minutes to determine if the pain continues. Narcotic analgesics should be withheld until the hot water immersion is completed to prevent scalding.

Antivenom

Most definite stonefish envenomations require antivenom (Commonwealth Serum Laboratories, Melbourne, Australia), which is an equine $F(ab)_2$ antivenom similar to snake antivenom. Indications for the administration of antivenom include severe pain unresponsive to analgesics, systemic effects (weakness, paralysis), or large stonefish envenomation (i.e., multiple puncture wounds). After skin testing and other precautions necessary for equine products, the usual dose of stonefish antivenom is 1 vial (2,000 Units) intramuscularly for definite clinical evidence of envenomation and the presence of one or two puncture wounds. There are few data on the intravenous administration of this antivenom, but 1 vial is typically diluted in 50–100 mL normal saline and infused slowly. Two milliliters of the antivenom neutralizes about 20 mg of stonefish venom *in vitro*.[14] Two to three vials may be necessary for larger envenomations (i.e., more than 3–4 puncture wounds). Resources for antivenom include local regional poison centers and regional zoos. Currently, there are inadequate data to support the use of this antivenom for other venomous stings. However, envenomations by other members of the family Scorpaenidae do not usually require antivenin.

Supplemental Care

Local wound care includes irrigation, débridement as necessary, and tetanus prophylaxis as indicated. Visible foreign material, integument or spines should be removed with forceps to prevent the subsequent development of granulomas. Imaging studies may be necessary to exclude retained spines. Grossly contaminated wounds may require surgical therapy including the use of operating microscopes. Prophylactic antibiotics are not recommended, but the patient should be warned about signs and symptoms of infection. Magnesium sulfate (Epsom salts) soaks twice a day for several days may provide symptomatic relief for persistent symptoms. Treatment of systemic symptoms is supportive. Intravenous fluids may be necessary to correct hypotension. The administration of antihistamines, epinephrine, or corticosteroids does not alter the duration or intensity of the symptoms associated with envenomation.

WEEVER FISH

IDENTIFYING CHARACTERISTICS

There are four European species of weever fish (Family: Trachinidae) including the greater weever fish (*Trachinus draco* L.), the lesser weever fish (*Echiichthys vipera* Cuvier or *Trachinus vipera*), spotted weever fish (*Trachinus araneus* Cuvier), and the streaked weever fish (*Trachinus radiatus* Cuvier). The stargazer (*Uranoscopus* spp.) is a close relative of the weever fish with a more bulky body. Although this fish has pectoral spines capable of causing painful stings, these spines are not venomous.[34]

Common Name: Lesser Weever Fish

Scientific Name: *Echiichthys vipera* Cuvier (*Trachinus vipera*)

Physical Description: This short (<14cm/6in.), stout fish has a laterally flattened body with yellowish-brown coloration of the dorsal surface. There are dark spots on the head and back, whereas the dorsal fin is entirely black.

Common Name: Greater Weever Fish

Scientific Name: *Trachinus draco* L.

Physical Description: The grayish-yellow body of the greater weever fish is elongated and flattened laterally. The bottom surface of the fish is pale, and there are narrow diagonal stripes on the flanks and large black spots on the dorsal fin. Typically, the greater weever fish exceeds 40cm (~16in.) in length.

EXPOSURE

Geographic Distribution

Weever fish inhabit the Mediterranean Sea, the Atlantic coast of Europe, the Black Sea, and the Pacific coast of Chile, where these creatures are the most venomous fish in temperate waters. The lesser weever fish lives in the Mediterranean (Baltic to North Africa) and eastern Atlantic coastal waters, while the greater weever fish is also found in the North Sea.[35] In France, weever fish are a commercial delicacy. The spotted weever fish (*Trachinus araneus* Cuvier) inhabits shallow bottoms in Mediterranean coastal waters, whereas the streaked weever fish lives in the Mediterranean and eastern Atlantic coastal waters from Morocco to Senegal.

Behavior

The lesser weever fish prefers sand bottoms ranging from 1–50m (~3–160ft) in depth, where these partially submerged fish wait for prey (crab, shrimp). During the day, these fish lie motionless with only their eyes and dorsal fin exposed while waiting for prey. Envenomation usually occurs when bathers step on the partially submerged dorsal spines. Greater weever fish prefer sandy bottoms in water deeper than 100m (325ft). Envenomations primarily occur when fishermen attempt to remove the greater weever fish from their nets.

PRINCIPAL TOXINS

Venom Composition

The venom contains several heat-labile proteins, including a high-molecular-weight protein, a kinin or kininlike substance, and possibly serotonin as well as adrenaline, noradrenaline, and several enzymes.[36] Dracotoxin (CAS RN: 140609-96-7) is a protein isolated from the crude venom of the greater weever fish (*Trachinus draco* L.)

that possesses membrane depolarizing and hemolytic activities in animal models.[37] In animal studies, the crude venom did not possess any proteolytic or histamine-releasing activities, and the lethal activity was unstable at room temperature especially at low protein concentrations.[38] Purification the lethal extract from the lesser weever fish venom (*Echiichthys vipera* Cuvier) suggests that this toxin is comprised of four identical subunits with a total molecular weight of approximately 324kDa.[39]

Venom Apparatus

Spines occur only on the dorsal fin and gill cover. The dorsal spines contain two grooves composed of holocrine glandular tissue. Like those of the lion fish, the spines are covered with a thin integument of stratified columnar epithelium that ruptures during contact with the victim and thus releases the venom.

Dose Response

There are few data on the dose response of active components in weever fish venom. Limited data suggests that there is substantial variation in the sensitivity of animal species to the venom from weever fish. In a study administering venom intraperitoneally or intravenously into a tail vein, guinea pigs and rabbits were 10 times and 25 times, respectively, more sensitive to the venom than mice based on the lethal dose on a weight basis.[40]

HISTOPATHOLOGY AND PATHOPHYSIOLOGY

In animal studies, the intravenous injection of weever fish venom causes necrosis of the ears and tip of the tail along with hair loss.[40] The venom of weever fish contains proteins that have antigenic properties.[41]

CLINICAL RESPONSE

Most weever fish stings cause mild to moderate symptoms without systemic complaints.[42] A burning, intense pain develops immediately after envenomation that may require narcotic analgesics. The pain occasionally radiates up the extremity and into the chest. Local symptoms include erythema, edema, pallor, ecchymosis, paraesthesias, and lymphadenitis that occur within several hours. The pain usually resolves within 24 hours, although slow recovery of a local inflammatory reaction may cause persistent limitation of motion, hyperesthesia, cold sensitivity, and pain.[43] The persistence of an inflammatory reaction requires consideration of the presence of a foreign body. Significant local trauma may occur because of the ability of the spines to penetrate protective clothing including leather shoes. Systemic symptoms are unusual.[35] Rare case reports associate systemic symptoms with weever fish envenomation including headache, delirium, nausea, vomiting, diaphoresis, syncope, and hypotension. Fatalities following weever fish envenomation are not well documented and deaths associated with stings from these fish probably resulted from sepsis.[44] A case report associated the death of an 18-year-old man with suspected envenomation by weever fish off the coast of Majorca (Spain).[45] He swam back to the boat after the envenomation, but was found in full cardiopulmonary arrest 1 hour later.

DIAGNOSTIC TESTING

The specific toxins associated with weever fish envenomation have not been identified, and there are no commercial biomarkers for the confirmation of envenomation by weever fish.

TREATMENT

Treatment is supportive. The best first-aid measure is immediate hot water immersion (110–115°F/~43–46°C) for 30–60 minutes until the pain resolves. The temperature of the water should be tested on the uninvolved extremity prior to hot water immersion. All wounds should be irrigated as soon as possible and all visible foreign bodies (spines, integumentary sheaths) removed. Local wound care requires irrigation, cleansing, and débridement with tetanus immunizations updated as needed. Secondary infections do not usually occur and prophylactic antibiotics are unnecessary.

CATFISH

IDENTIFYING CHARACTERISTICS

Catfish are members of the class Osteichthyes, order Siluriformes with about 1,000 species inhabiting primarily freshwater. North American families of catfish include Ictaluridae (freshwater) and Ariidae (saltwater). The characteristic perioral barbells do not produce envenomation. Table 183.1 outlines families of catfish involved with human envenomations.

TABLE 183.1. Catfish Associated with Human Envenomation

Family	Habitat	Distribution	Examples
Ariidae	Marine	Worldwide	Sea catfish (USA)
Bagridae	Freshwater	Asia	Akaza (Japan)
Clariidae	Freshwater	Indo-Pacific	Ikan Keli (Malaysia)
Doradidae	Freshwater	South America	Bagre (Amazon)
Heteropneustidae	Freshwater	Asia	Pla Cheet (Thailand)
Ictaluridae	Freshwater	North America	Madtom (USA)
Pimelodidae	Freshwater	South America	Bagre (Amazon)
Plotosidae	Marine	Indo-Pacific	Oriental catfish (USA)
Siluridae	Freshwater	Africa, Asia, Europe	European catfish

Source: Adapted from Ref 56.

EXPOSURE

Geographic Distribution

Although most of the 1,000 catfish species are freshwater, bottom-dwelling scavengers, a few species are found in salt water. Freshwater catfish in North America capable of causing envenomation include the brown bullhead (*Ameiurus nebulosus* Lesueur or *Ictalurus nebulosus*), white catfish (*Ameiurus catus* L. *or Ictalurus catus* L.), blue catfish (*I. furcatus* Valenciennes), channel catfish (*Ictalurus punctatus* Rafinesque), and Carolina mudtom (*Noturus furiosus* Jordan & Meek).[46] Stinging catfish in Asia include *Heteropneustes fossilis* Bloch, which are facultative air-breathers that inhabit swamps of Southeast Asia.[47]

Behavior

Freshwater catfish typically inhabit dirty, slow-moving waters, whereas the marine catfish (*Plotosus lineatus* Thunberg) live in kelp beds and travel extensively in large schools. Some unusual characteristics include the suctorial lips of a South American species that allows the catfish to climb cliffs, male catfish that incubate eggs in their mouths, Amazon catfish (e.g., *Plectrochilus diabolicus* Meyer or *Urinophilus diabolicus*) that lodge in the human urethra requiring endoscopic removal, and Egyptian catfish (*Malapterus* spp.) that possess electric plates capable of killing prey. Catfish envenomations occur when fishermen extricate the catfish or when bathers wade in shallow water.[48,49] As a defensive maneuver, catfish can lock the dorsal spines in an extended position.

PRINCIPAL TOXINS

Venom Composition

Animal studies indicate that some catfish venoms contain hemolysins, lethal factors, dermonecrotic fractions, and edema-producing substances, but the clinical relevance of these compounds is unclear.[50,51] The skin secretions of some catfish species also contain toxins. Analysis of crude skin toxin from the Suez Gulf oriental catfish (*Plotosus lineatus* Thunberg) demonstrated a variety of substances including hemolysin, hyaluronidase, and lipase along with smaller amounts of phospholipase A_2, cholinesterase, proteinase, and 5′-nucleotidase.[52] These substances (crinotoxins) may account for the pain associated with direct contact with the skin rather than the spines. Although mudtoms (*Noturus* spp.) appear most toxic to humans, studies in animals suggest that species differences do not entirely account for the wide variation in venom potency.[53]

Venom Apparatus

Catfish have a single spine on the dorsal fin and spines on each of the pectoral spines that are capable of causing trauma to human skin. Figure 183.4 demonstrates the dorsal spine of a typical catfish. The cartilaginous, serrated spines are held erect when the catfish is disturbed. In some catfish species, an integumentary sheath covers the spine. Penetration of the skin by the spine tears the integumentary sheath, exposes the venom gland, and releases venom into the wound. Many of the spines contain sharp, serrated teeth that inhibit removal of the spine from the surrounding area. These spines rest flat against the skin of the catfish until a stimulus causes the fish to erect the spines. The severity of an evenomation ranges from mild following exposure to fork-tailed catfish (family: Ariidae) to potentially serious envenomation by some marine catfish, such as the Oriental or striped catfish (*Plotosus lineatus* Thunberg) or the Arabian Gulf catfish (*Netuma thalassina* Rüppell).[54]

Dose Response

In mice, the intraperitoneal LD_{50} of whole skin crinotoxin from the oriental catfish (*Plotosus lineatus*) was 27.9 μg/g body weight, which is low compared with snake

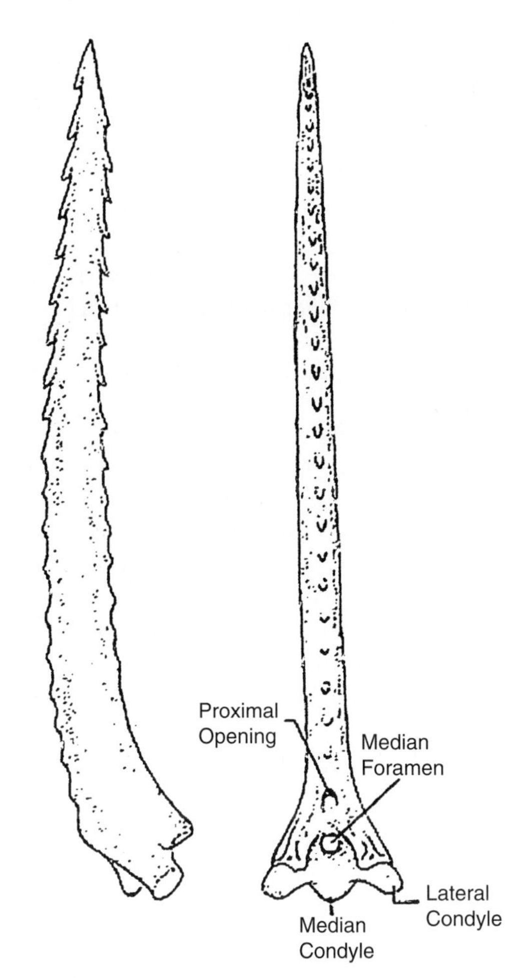

FIGURE 183.4. Dorsal spine of catfish (*Galeichthys felis*). From Blomkalns AL, et al. Catfish spine envenomation: a case report and literature review. Wilderness Environ Med 1999;10:245. Copyright Wilderness Medical Society. Reprinted with permission.

venom.[52] This crude venom has high hyaluronidase and lipase activities.

HISTOPATHOLOGY AND PATHOPHYSIOLOGY

Catfish venom contains varying concentrations of dermonecrotic, inflammatory, hemolytic, and vasoconstrictive factors. Infections (e.g., *Vibrio* spp., *Aeromonas* spp.) may complicate catfish-induced wounds, particularly following retained spines. Chronic, indolent infections may result from the contamination of wounds by *Mycobacterium marinum*.[55]

CLINICAL RESPONSE

Envenomations by the spines of the catfish produce a wide range of symptoms from mild to severe pain.[56] Typically, contact with the venomous spines causes localized pain at the puncture site, occasionally with paresthesias and pain radiating up the extremity.[57] Symptoms usually resolve in 1–2 hours.[58] Significant edema, erythema along with some bleeding at the puncture site may occur following envenomation by some of the more venomous species (e.g., *Plotosus lineatus*). The pain associated with these envenomations may persist 24–48 hours. Edema and erythema may appear along with localized hypesthesias, regional lymphadenopathy, mild fever, and necrosis following more severe envenomations. Local spasm and muscle fasciculations occasionally occur.[49] Although most envenomations resolve without complications, occasionally dermal necrosis may occur that rarely is associated with gangrene.[59] Infection is the most serious common complication of contact with catfish spines, and serious infections (e.g., septic arthritis, acute tenosynovitis, large wound infections) may develop, particularly in immunocompromised patients or patients with chronic illnesses (e.g., liver disease). The microbiology of freshwater and marine catfish differ significantly with the most serious infections resulting from *Vibrio* species (marine) and *Aeromonas* species (freshwater).[56] Other sources of infection include *Klebsiella, Erysipelothrix, Nocardia, Mycobacterium, Actinomyces, Sporothrix, Enterobacteriaceae, Pseudomonas* species, and bacteria flora (*Streptococcus* spp., *Staphylococcus aureus*) of human skin.[60]

DIAGNOSTIC TESTING

Clinically significant laboratory abnormalities do not usually occur after envenomation by catfish. There are no commercial analytical methods to confirm envenomation by catfish.

TREATMENT

Management is supportive and similar to that for Scorpaenidae envenomations. However, catfish venom may not be as heat-labile as venom from other species of the family Scorpaenidae.[50] The wound should be carefully cleansed after the administration of sufficient local anesthetics to allow adequate cleansing of the wound. The affected extremity of symptomatic patients should be immersed in warm water (45 °C/113 °F) for 30–60 minutes depending on the amount of pain relief or the necessity of other procedures (radiography, wound cleansing). Although not all catfish spines are radiopaque, soft-tissue x-rays should be taken if there is a significant possibility of a retained foreign body. These films require careful examination to detect the presence of tiny serrations from catfish spines. If there is radiographic evidence of foreign bodies, the wound should be explored, irrigated, and débrided. Removal of embedded spines may be difficult and require consultation with a surgical specialist. Although there are few data on the use of prophylactic antibiotics, patients with catfish envenomation frequently receive antibiotics (ciprofloxacin), particularly following the removal of foreign bodies, delayed presentation, or evidence of an immunocompromised host. *Aeromonas* species frequently are resistant to penicillin and cephalosporins.[61] Superficial abrasions or lacerations can be treated with local cleansing and topical antibiotics. Victims of catfish envenomation should be followed closely (24–48 hours) for the development of infection, particularly if prophylactic antibiotics are not administered.

RABBITFISH

This narrow, oval-shaped fish (Family: Siganidae) grows to about 30 cm (~1 ft) in length and inhabits coral reefs of the Indo-Pacific and the Mediterranean Sea. They possess numerous spines that contain venom glands along the lateral grooves. After penetration of the skin, the integumentary sheath ruptures and spreads venom in the victim. Immediate pain results from the envenomation, but the pain usually resolves soon after exposure. Rarely, case reports associate infectious complications including necrotizing fascitis with wounds caused by these fish.[62]

DOGFISH

These fish (Squalus spp.) are members of the order Squaliformes, and live throughout the temperate zones (tropics to sub-Antarctic seas). They prefer cool waters and protected bays, where they feed on a variety of vertebrate and invertebrate marine animals. Thick venomous spines protrude from the front of each dorsal fin; envenomation causes painful, erythematous, and edematous puncture wounds. Case reports associate local edema, erythema, and excruciating pain for 6 hours with contact with the anterior spine of a dogfish (*Squalus cubensis* Howell Rivero).[63] The local edema resolved after 2 weeks. Systemic symptoms are rare, but may be serious. Treatment is supportive.

TOADFISH

Toadfish (*Thalassophryne* spp., Family: Batrachoididae) inhabit shallow areas near the Atlantic Coast from Massachusetts to Central America, particularly in river mouths and estuaries. Other toadfish live in the Red Sea and Indo-Pacific (*Allenbatrachus grunniens* L. or *Batrachoides grunniens, Thalassothia cirrhosa* Klunzinger or *Barchatus cirrhosus*) and in the Mediterranean Sea (*Halobatrachus didactylus* Bloch & Schneider or *Batrachoides didactylus*). They are bottom-dwelling fish with large mouths and flat heads; they often hide in sand, rocks, or kelp beds. Two hallow spines with large venom glands at the base project from the dorsal fin, and envenomation by these spines causes local pain and swelling. Systemic symptoms do not usually develop after envenomation. The niquim (*Thalassophryne nattereri* Steindachner) is a venomous toadfish found in estuaries of the western Atlantic from Tobago to San Salvador and Brazil. This marine fish measures up to 14 cm (~5 in.). A substantial number of envenomations in fishermen from northeastern Brazil is attributed to the niquim.[64] Envenomation results from contact with venomous spines located near the opercular and dorsal fins that inject a protease with tissue-kallikrein-like activity.[65,66] The contamination of wounds with skin secretions from toadfish may contribute to the toxic effects of the envenomation.[34] Clinical features of envenomation include intense local pain, mild erythema, severe edema, and necrosis.[67] Complications include local necrosis and wound infections. Systemic symptoms do not usually develop after envenomation. Treatment is supportive.

PORCUPINE FISH

Porcupine fish (*Diodon Hystrix* L.) are widely distributed in marine environments from temperate to tropical waters. In the eastern Pacific, this fish ranges from southern California to Chile including the Galapagos Islands (Ecuador). The porcupine fish is the only member of the *Diodon* genus (order: Tetraodontiformes) found in the Mediterranean Sea. These fish are members of the puffer fish family (Diodontidae). Porcupine fish also live in the western Atlantic from Massachusetts and Bermuda to the northern Gulf of Mexico, south to Brazil. These shy inhabitants of warm coral reefs and shoals range in size between 20–50 cm (~8–20 in.). They are nocturnal, solitary creatures with approximately 20 spines in a row between the snout and dorsal fin that become erect during defensive maneuvers. Otherwise, these spines are kept close to their bodies. The secretions of the skin of the porcupine fish contain tetrodotoxin as well as other thermolabile neurotoxins.[68] Four minor puncture wounds on the left middle finger after contact with a dead porcupine fish caused the onset of paresthesias and numbness within 1½ hours after envenomation. Over the next 2 hours, paresthesias and paresis of the left arm with lightheadedness developed.[69] No respiratory insufficiency occurred and symptoms (numbness, headache) resolved within 3 days. There were no laboratory abnormalities and the patient recovered without sequelae. Treatment is supportive with normal wound care procedures for minor puncture wounds and observation for respiratory insufficiency.

ICHTHYOTOXIC ACUTE RENAL FAILURE

In rural Asian populations, the ingestion of raw freshwater fish gallbladders is a traditional treatment for reduced visual acuity, arthritis, and impotence. Case reports associate the development of acute renal failure with the ingestion of raw gallbladders from freshwater fish. Fish species associated with this syndrome include the common carp (*Cyprinus carpio* L.), grass carp (*Ctenopharyngodon idella* Valenciennes), silver carp (*Hypophthalmichthys molitrix* Valenciennes), black carp (*Mylopharyngodon piceus* Richardson), bighead carp (*Aristichthys nobilis* Richardson), roho labeo (*Labeo rohita* Hamilton), blackshark minnow (*Labeo chrysophekadion* Bleeker), and the bony-lipped barb fish (*Osteochilus vittatus* Valenciennes).[70] Animal studies suggest that the C27 bile salt, 5-α-cyprinol, may contribute to the renal toxicity associated with the consumption of raw freshwater fish gallbladder,[71] but the exact nephrotoxin has not been identified. Renal biopsies from these patients are consistent with acute tubular necrosis. Some hepatotoxicity also occurs, primarily in fulminant cases. Clinical features typically involve the early onset of gastrointestinal distress (abdominal cramps, vomiting, diarrhea) followed in a few days by renal dysfunction. The gastrointestinal symptoms start from ½–12 hours (mean: ~4–5 hours) after ingestion followed by oliguria 48–72 hours after ingestion.[72] In a case series of 17 patients with this illness, the mean time to peak creatinine concentration was 8.6 ± 3.0 days. Rarely, fatal hepatorenal failure occurs.[73] Treatment is supportive with hemodialysis as needed.

References

1. Russell FE. Injuries by venomous animals in the US. JAMA 1961;177:903–907.
2. Patkin M, Freeman D. Bullrout stings. Med J Aust 1969;2: 14–16.
3. Kizer KW, McKinney HE, Auerbach PS. Scorpaenidae envenomation. A five-year poison center experience. JAMA 1985;253:807–810.
4. Carlson RW, Schaeffer RC Jr, Whigham H, Weil MH, Russell FE. Some pharmacological properties of the venom of the scorpionfish *Scorpaena guttata*. II. Toxicon 1973;11:167–180.
5. Church JE, Hodgson WC. The pharmacological activity of fish venoms. Toxicon 2002;40:1083–1093.
6. Carrijo LC, Andrich F, de Lima ME, Cordeiro MN, Richardson M, Figueiredo SG. Biological properties of the venom from the scorpionfish (*Scorpaena plumieri*) and purification of a gelatinolytic protease. Toxicon 2005;45: 843–850.
7. Saunders RP. Venom of the stonefish *Synanceja verrucosa*. Science 1959;129:272–274.
8. Khoo HE. Bioactive proteins from stonefish venom. Clin Exp Pharmacol Physiol 2002;29:802–806.
9. Gwee MC, Gopalakrishnakone P, Yuen R, Khoo HE, Low KS. A review of stonefish venoms and toxins. Pharmacol Ther 1994;64:509–528.
10. Garnier P, Sauviat MP, Goudey-Perriere F, Perriere C. Cardiotoxicity of verrucotoxin, a protein isolated from the venom of *Synanceia verrucosa*. Toxicon 1997;35: 47–55.
11. Wang JW, Yazawa K, Hao LY, Onoue Y, Kameyama M. Verrucotoxin inhibits KATP channels in cardiac myocytes through a muscarinic M3 receptor-PKC pathway. Eur J Pharmacol 2007;563:172–179.
12. Gopalakrishnakone P, Gwee MC. The structure of the venom gland of stonefish *Synanceja horrida*. Toxicon 1993;31:979–988.
13. Saunders PR. Venom of the stone fish *Synanceja verrucosa*. Science 1959;129:272–274.
14. Cooper NK. Stone fish and stingrays—some notes on the injuries that they cause to man. JR Army Med Corps 1991;137:136–140.
15. Wiener S. Observations on the venom of the stone fish (*Synanceja trachynis*). Med J Aust 1959;46:620–627.
16. Khoo HE, Yuen R, Poh CH, Tan CH. Biological activities of *Synanceja horrida* (stonefish) venom. Nat Toxins 1992;1:54–60.
17. Poh CH, Yuen R, Khoo HE, Chung M, Gwee M, Gopalakrishnakone P. Purification and partial characterization of stonustoxin (lethal factor) from *Synanceja horrida* venom. Comp Biochem Physiol B 1991;99:793–798.
18. Brenneke F, Hatz C. Stonefish envenomation—a lucky outcome. Travel Med Infect Dis 2006;4:281–285.

19. Church JE, Moldrich RX, Beart PM, Hodgson WC. Modulation of intracellular Ca^{2+} levels by Scorpaenidae venoms. Toxicon 2003;41:679–689.
20. Church JE, Hodgson WC. Adrenergic and cholinergic activity contributes to the cardiovascular effects of lionfish (*Pterois volitans*) venom. Toxicon 2002;40:787–796.
21. Coats JA, Pattabhiraman TR, Russell FE, Gonzalez H. Some physiopharmacologic properties of scorpionfish venom. Proc West Pharmacol Soc 1980;23:113–115.
22. Austin L, Gillis RG, Youatt G. Stonefish venom: some biochemical and chemical observations. Aust J Exp Biol Med Sci 1965;43:79–90.
23. Boletini-Santos D, Komegae EN, Figueiredo SG, Haddad V Jr, Lopes-Ferreira M, Lima C. Systemic response induced by Scorpaena plumieri fish venom initiates acute lung injury in mice. Toxicon 2008;51:585–596.
24. Vetrano SJ, Lebowitz JB, Marcus S. Lionfish envenomation. J Emerg Med 2002;23:379–382.
25. Aldred B, Erickson T, Lipscomb J. Lionfish envenomations in an urban wilderness. Wilderness Environ Med 1996;7: 291–296.
26. Wasserman GS, Johnston RM. Poisoning from a lionfish sting. Hum Vet Toxicol 1979;21:344–345.
27. Haddad V Jr, Martins IA, Makyama HM. Injuries caused by scorpionfishes (*Scorpaena plumieri* Bloch, 1789 and *Scorpaena brasiliensis* Cuvier, 1829) in the Southwestern Atlantic Ocean (Brazilian coast): epidemiologic, clinic and therapeutic aspects of 23 stings in humans. Toxicon 2003;42:79–83.
28. Grandcolas N, Galéa J, Ananda R, Rakotoson R, D'Andréa C, Harms JD, Staikowsky F. [Stonefish stings: difficult analgesia and notable risk of complications]. Presse Med 2008;37:395–400. [French]
29. Lee JY, Teoh LC, Leo SP. Stonefish envenomations of the hand—a local marine hazard: a series of 8 cases and review of the literature. Ann Acad Med Singapore 2004;33: 515–520,
30. Wiener S. Stone-fish sting and its treatment. Med J Aust 1958;45:218–222.
31. Phoon WO, Alfred ER. A study of stonefish (*Synanceja*) stings in Singapore with a review of the venomous fishes of Malaysia. Singapore Med J 1965;6:158–163.
32. Bagnis R. [Apropos of 51 cases of poisonous stings by the tropical "hogfish" *Synanceja verrucosa* in the Society Islands and the Tuamotu Archipelago]. Med Trop (Mars) 1968;28:612–620. [French]
33. Lyon RM. Stonefish poisoning. Wilderness Environ Med 2004;15:284–288.
34. Mebs D. Venomous and poisonous animals. A handbook for biologists, toxicologists and toxinologists, physicians and pharmacists. Stuttgart: Medpharm Scientific Publishers; 2002.
35. Cain D. Weeverfish sting: An unusual problem. Br Med J 1983:287:406–407.
36. Russell FE. Weeverfish sting: The last word. Br Med J 1983;287:981–982.
37. Chhatwal I, Dreyer F. Isolation and characterization of dracotoxin from the venom of the greater weever fish *Trachinus draco*. Toxicon 1992;30:87–93.
38. Chhatwal I, Dreyer F. Biological properties of a crude venom extract from the greater weever fish *Trachinus draco*. Toxicon 1992;30:77–85.
39. Perriere C, Goudey-Perriere F, Petek F. Purification of a lethal fraction from the venom of the weever fish, *Trachinus vipera* C.V. Toxicon 1988;26:1222–1227.
40. Skeie E. Toxin of the weeverfish (*Trachinus draco*) experimental studies on animals. Acta Pharmacol Toxicol 1962;19:107–120.
41. Skeie E. Weeverfsih toxin some physico-chemical and immunological observations. Acta Pathol Microbiol Scand 1962;56:229–238.
42. Davies RS, Evans RJ. Weever fish stings: a report of two cases presenting to an accident and emergency department. J Accid Emerg Med 1996;13:139–141.
43. Halpern P, Sorkine P, Raskin Y. Envenomation by *Trachinus draco* in the eastern Mediterranean. Eur J Emerg Med 2002;9:274–277.
44. Russell FE, Emery JA. Venom of weevers *Trachinus draco* and *Trachinus vipera*. Ann NY Acad Sci 1960;90: 805–819.
45. Borondo JC, Sanz P, Nogué S, Poncela JL, Garrido P, Valverde JL. Fatal weeverfish sting. Hum Exp Toxicol 2001;20:118–119.
46. McKinstry DM. Catfish stings in the United States: case report and review. J Wilderness Med 1993;4: 293–303.
47. Satora L, Pach J, Targosz D, Szkonicka B. Stinging catfish poisoning. Clin Toxicol 2005;43:893–894.
48. Fredette SR, Derk FF, Nardoza AJ. Catfish spine injury of the foot. J Am Podiatr Med Assoc 1997;87:187–189.
49. Scoggins CH. Catfish stings. JAMA 1975;231:176–177.
50. Calton GJ, Burnett JW. Catfish (*Ictalurus catus*) fin venom. Toxicon 1975;13:399–403.
51. Shiomi K, Takamiya M, Yamanaka H, Kikuchi T, Konno K. Hemolytic, lethal and edema-forming activities of the skin secretion from the oriental catfish (*Plotosus lineatus*). Toxicon 1986;24:1015–1018.
52. Fahim FA, Esmat AY, Mady EA, Ahmed SM, Zaki MA. Biological activities of the crude skin toxin of the Suez Gulf oriental catfish (*Plotosus lineatus*) and its antitumor effect *in vivo* (mice). J Nat Toxins 2002;11:283–295.
53. Birkhead WS. Toxicity of stings of ariid and ictalurid catfishes. Copeia 1972;(4):790–807.
54. Isbister GK. Venomous fish stings in tropical northern Australia. Am J Emerg Med 2001;19:561–565.
55. Ajmal N, Nanney LB, Wolfort SF. Catfish spine envenomation: a case of delayed presentation. Wilderness Environ Med 2003;14:101–105.
56. Blomkalns AL, Otten EJ. Catfish spine envenomation: a case report and literature review. Wilderness Environ Med 1999;10:242–246.

57. Das SK, Johnson MB, Cohly HH. Catfish stings in Mississippi. South Med J 1995;88:809–812.
58. Pacy H. Australian catfish injuries with report of a typical case. Med J Aust 1966;2:63–65.
59. Mann JW 3rd, Werntz JR. Catfish stings to the hand. J Hand Surg [Am] 1991;16:318–321.
60. Murphey DK, Septimus EJ, Waagner DC. Catfish-related injury and infection: report of two cases and review of the literature. Clin Infect Dis 1992;14:689–693.
61. Hanson PG, Standridge J, Jarrett F, Maki DG. Freshwater wound infection due to *Aeromonas hydrophilia*. JAMA 1977;238:1053–1054.
62. Yuen KY, Ma L, Wong SS, Ng WF. Fatal necrotizing fasciitis due to *Vibrio damsela*. Scand J Infect Dis 1993; 25:659–661.
63. Haddad V Jr, Gadig OB. The spiny dogfish (*Squalus cubensis/megalops* group): the envenoming of a fisherman, with taxonomic and toxinological comments on the *Squalus* genus. Toxicon 2005;46:828–830.
64. Lopes-Ferreira M, Barbaro KC, Cardoso DF, Moura-Da-Silva AM, Mota I. *Thalassophryne nattereri* fish venom: biological and biochemical characterization and serum neutralization of its toxic activities. Toxicon 1998;36: 405–410.
65. Lopes-Ferreira M, Emim JA, Oliveira V, Puzer L, Cezari MH, Araujo Mda S, et al. Kininogenase activity of *Thalassophryne nattereri* fish venom. Biochem Pharmacol 2004;68:2151–2157.
66. Faco PE, Havt A, Barbosa PS, Nobre AC, Bezerra GP, Menezes DB, et al. Effects of *Thalassophryne nattereri* fish venom in isolated perfused rat kidney. Toxicon 2003;42:509–514.
67. Haddad V Jr, Pardal PP, Cardoso JL, Martins IA. The venomous toadfish *Thalassophryne nattereri* (niquim or miquim): report of 43 injuries provoked in fishermen of Salinpolis (Para State) and Aracaju (Sergipe State), Brazil. Rev Inst Med Trop Sao Paulo 2003;45:221–223.
68. Malpezzi EL, de Freitas JC, Rantin FT. Occurrence of toxins, other than paralysing type, in the skin of Tetraodontiformes fish. Toxicon 1997;35:57–65.
69. van Gorcum TF, Janse M, Leenders ME, De Vries I, Meulenbelt J. Intoxication following minor stabs from the spines of a porcupine fish. Clin Toxicol 2006;44: 391–393.
70. Xuan BH, thi TX, Nguyen ST, Goldfarb DS, Stokes MB, Rabenou RA. Ichthyotoxic ARF after fish gallbladder ingestion: a large case series from Vietnam. Am J Kid Dis 2003;41:220–224.
71. Hwang DF, Yeh YH, Lai YS, Deng JF. Identification of cyprinol and cyprinol sulfate from grass carp bile and their toxic effects in rats. Toxicon 2001;39:411–414.
72. Sahoo RN, Mohapatra MK, Sahoo B, Das GC. Acute renal failure associated with freshwater fish toxin. Trop Geogr Med 1995;47:94–95.
73. Lin Y-F, Lin S-H. Simultaneous acute renal and hepatic failure after ingestion raw carp gall bladder. Nephrol Dial Transplant 1999;14:2011–2012.

Chapter 184

CARTILAGINOUS FISH
(Class: Chondrichthyes)

STINGRAY

IDENTIFYING CHARACTERISTICS

In contrast to reptiles, fish do not possess venom glands with muscles to inject venom into prey. Some bony and cartilaginous fish have sharp spines connected to a venom gland or spines covered with venom-producing epithelial tissue for defensive purposes rather than acquisition of prey. Relatively few fish species developed these spines covered with a sheath of venom-producing tissue. Most rays including the giant manta rays and devil rays from the Myliobatidae family do not possess venomous spines. However, stingrays have a long tail with venomous spines; these creatures constitute an important group of venomous marine animals. There are several families of marine stingrays and one family of freshwater stingrays.

Common Name: Stingrays

Scientific Name of Marine Families: Dasyatidae (whip rays, stingrays), Gymnuridae (butterfly rays), Myliobatidae (devil rays, bat rays, eagle rays, cow-nose rays), and Urolophidae (round stingrays, stingarees)

Scientific Name of Freshwater Family: Potamotrygonidae (river rays)

Physical Description: The diameter of the stingrays ranges between a few inches to over 5 ft (~1.5 m) with a whip-like tail containing one to four spines. These creatures have flattened dorsal and ventral surfaces with enlarged pectoral fins and the absence of a dorsal fin. Two eyes protrude on the dorsal surface. The most common color of the body is gray.

EXPOSURE

Geographic Distribution

Marine stingrays inhabit shallow waters up to 50 m (~160 ft) in depth in temperate, subtropical, and tropical areas including brackish lagoons and river estuaries. There are at least 11 species in US coastal waters. The marine families of stingrays are distributed worldwide including Dasyatidae (whip rays, stingrays), Urolophidae (round stingrays, stingarees), and Myliobatidae (devil rays, bat rays, eagle rays, cow-nose rays), whereas neotropical freshwater species of the family Potamotrygonidae (river rays) live in the Amazon, the Paraná (South America), the Niger (Africa), and the Mekong (Laos, Vietnam) river basins. Some species in the family Potamotrygonidae prefer the most extreme freshwater conditions of Brazil and the Paraná River basin. The butterfly rays (Gymnuridae) live in American waters as well as other marine waters around the world.

Behavior

The favorite habitats of these passive, reclusive bottom-dwellers are sandy or muddy bottoms of sheltered lagoons and estuaries, where they feed on crustaceans

Medical Toxicology of Natural Substances, by Donald G. Barceloux, MD

and mollusks. Their excellent camouflage and habit of submerging in the bottom cause accidental contact with exposed human extremities. Pressure on the back causes the stingray to envenomate the victim. The weight of the foot stimulates the stingray to thrust its tail forward, driving the spine deep into the victim's foot or lower leg. Incidental contact with the wings of a ray also results in the thrusting of the tip of the tail into the victim. The entire tip of the spine may remain embedded in the victim. In a case series of stingray envenomations, the lower extremity was involved in over 75% of the envenomations.[1] Rarely, serious envenomation occurs on the chest or abdomen while a human is swimming.[2]

PRINCIPAL TOXINS

Venom Composition

Experimental studies indicate that stingray venom potentially contains cardiotoxins (atrioventricular block, QRS prolongation, hypotension, bradycardia, ST-T wave changes), respiratory depressants, and convulsants.[1] Other fractions include phosphodiesterase, 5′-nucleotidase, and serotonin.[5] In rodent studies, the venom from Brazilian *Potamotrygon* freshwater stingray venoms produced mild proteolytic activity, but no hemorrhagic activity.[3] The venom of the smooth back river stingray (*Potamotrygon orbignyi* Castelnau) contains a novel peptide, orpotrin, which is a strong vasoconstrictor in *in vitro* studies.[4]

Venom Apparatus

All stingrays possess one to four bony, reverse-serrated spines located on the dorsal surface of their powerful tail. The arrow-like tip has proximally directed, serrated spines that cause jagged, bleeding lacerations. A thin layer of skin (integumentary sheath) surrounds the elongated spine and covers two ventrolateral grooves that contain venom glands. The spine is usually covered with a film of mucus and venom. When the venom gland on the tail enters the skin, the integumentary sheath surrounding the spine ruptures. Venom then flows down two ventrolateral grooves of the spine into the victim's skin. The powerful tail and serrated spines also cause traumatic wounds, such as penetrating injuries to the abdomen, chest, or throat. Severe trauma (e.g., fatal evisceration wounds to the abdomen or heart) may occur in children or adults.[1] Different families of stingrays possess significant variation in the sophistication of the venom apparatus. For example, the family Gymnuridae (butterfly rays) possesses the least developed system, whereas stingrays from Urolophidae (round stingrays, stingarees) have a well-developed muscular apparatus. The Dasyatidae (devil rays, bat rays) possess a partially developed venom apparatus including the California bat stingray (*Myliobatis californicus* Gill) commonly found off the US coast. Many stingrays lose or tear their integumentary sheaths (45% in one study of 4,000),[1] so that stings often result in trauma without envenomation.

Dose Response

There are inadequate data on stingray venom to determine a dose-response effect for human evenomation. Serious injuries following stingray envenomations usually result from trauma as well as toxicity.

HISTOPATHOLOGY AND PATHOPHYSIOLOGY

The wound by a stingray causes traumatic injuries and envenomation. Consequently, a stingray envenomation can produce a variety of complications including secondary infection, osteomyelitis, intraabdominal trauma and peritonitis, and intrathoracic injuries (e.g., cardiac tamponade, right ventricular perforation).[5,6]

CLINICAL RESPONSE

The clinical features of a marine and fresh water stingray envenomation usually involve only local signs and symptoms (local pain, erythema, edema, ecchymosis, pruritus).[7,8] Intense pain out of proportion to the degree of trauma typically begins within 10 minutes of the envenomation, and the presence of pain disproportional to the injury suggests a stingray envenomation.[9] Edema around the wound is a constant finding, but the extent of the edema is variable. Cyanosis or erythema may surround the wound edges. The wound may be contaminated with debris from the integumentary sheath or surrounding fluids; bleeding is often profuse. If untreated, the pain reaches peak intensity within 60–90 minutes and resolves over 6–48 hours. Systemic symptoms vary considerably in intensity, and these clinical effects include abdominal pain, nausea and vomiting, muscle cramps, diaphoresis, tachycardia, syncope, headache, and muscle fasciculations.[10] Cardiac dysrhythmias, hypotension, and convulsions may occur during severe envenomations, but fatalities are not well documented in the absence of severe trauma. A supraventricular bigeminy developed in a 44-year-old male diver about 30 minutes after being stung by a Japanese red stingray (*Dasyatis akajei* Müller & Henle).[11] His rhythm spontaneously converted to his preexisting incomplete right bundle branch block within a few minutes, and no complications occurred.

Complications of stingray attacks include vascular injury with tissue necrosis and pseudoaneurysm.[12] Case reports of traumatic injuries following stingray envenomations include puncture wounds of the liver, perforated bowel, cardiac tamponade, and exsanguination secondary to femoral artery injury.[2,5,13] Freshwater stingrays inflict painful stings that often develop cutaneous necrosis.[14] Secondary infections may frequently develop. Systemic symptoms include nausea, vomiting, sweating, and agitation, but serious cardiac complications do not usually occur without direct trauma to the heart.

DIAGNOSTIC TESTING

The spines of stingrays are usually radiopaque, and plain films will demonstrate retained foreign bodies from the sting.[15] Most stingray envenomations are not associated with laboratory abnormalities. Deaths typically result from thoracic or abdominal trauma with associated hemorrhage.

TREATMENT

Treatment depends on the severity of the envenomation, the size and depth of the wound, and the extent of systemic symptoms. Lacerations of major arteries or organs may result in life-threatening hemorrhage that requires aggressive fluid resuscitation and blood transfusions.[16] First-aid procedures involve washing the wound with water and removing the integumentary sheath from the wound, if possible. Primary treatment is hot water immersion at 45 °C (113 °F) for 30–60 minutes until the pain resolves. This treatment reduces the acute pain associated with stingray envenomations, but hot water immersion does not alter the skin necrosis associated with freshwater stingray envenomations.[14] After soaking, the wound should be thoroughly débrided and copiously irrigated with saline or a very dilute povidone-iodine (Betadine®, Purdue Pharma, Stamford, CT) solution. If a retained stingray spine is suspected, plain radiographs should be obtained. Local anesthetics (e.g., nerve blocks are preferable) and a bloodless field are usually necessary to adequately remove the remnants of the integumentary sheath and to excise devitalized tissue.[17] Standard prophylactic antibiotics may be necessary because of the common occurrence of wound infections, and tetanus immunization should be updated as required.[18] The wound should be followed closely for signs of infection. Grossly contaminated wounds or wounds older than 6 hours on the extremity may be best managed by delayed primary closure. Large, contaminated wounds probably require prophylactic antibiotics. Although there are few clinical data, choices for antibiotics for suspected *Vibrio* or *Aeromonas* species in these large wounds include a third-generation cephalosporin (e.g., ceftazidime) or the combination of a first-generation cephalosporin or penicillinase-resistant penicillin plus a quinolone (e.g., ciprofloxacin). Elevation of the extremity is strongly recommended. Systemic symptoms should respond to the usual supportive measures. There are no adequate clinical data to support the use of steroids, cryotherapy, potassium permanganate, ammonia, or antihistamines.

RATFISH

Ratfish (chimaeras, Family: Chimaeridae) have large heads and tapering bodies with a long, narrow tails. The spotted ratfish (*Hydrolagus colliei* Lay & Bennett) is a relative of stingrays and sharks that prefers the cool waters of the Pacific Ocean from Alaska to Baja California. The European ratfish (*Chimaera monstrosa* L.) and the small ratfish (*Hydrolagus affinis* Capello) inhabit the northern Atlantic Ocean with the latter also occurring in the Mediterranean Sea. The first dorsal fin contains a strong, serrated spine capable of delivering venom from a glandular epithelium surrounding the slightly grooved spine. This large venomous spine potentially can produce a painful sting, but clinical evidence of envenomation in humans is limited and controversial.[19] Reported clinical effects include localized pain and erythema that usually responds to the treatment measures described for stingray and scorpion fish envenomations. Other types of fish in the class Chondrichthyes capable of producing local, painful puncture wounds include spiny dogfish (*Squalus acanthias* L.) and horned sharks (*Heterodontus* spp.). Hot water immersion at 45 °C (113 °F) for 30–60 minutes, thorough irrigation, and removal of debris are the main elements of supportive care.

References

1. Russell FE, Panos TC, Kang LW, Warner AM, Colket TC 3rd. Studies on the mechanism of death from stingray venom; a report of two fatal cases. Am J Med Sci 1958;235:566–584.
2. Cross TB. An unusual sting ray injury-the skindiver at risk. Med J Aust 1976;2:947–948.
3. Magalhaes KW, Lima C, Piran-Soares AA, Marques EE, Hiruma-Lima CA, Lopes-Ferreira M. Biological and biochemical properties of the Brazilian Potamotrygon stingrays: *Potamotrygon* cf. *scobina* and *Potamotrygon* gr. *orbignyi*. Toxicon 2006;47:575–583.

4. Conceicao K, Konno K, Melo RL, Marques EE, Hiruma-Lima CA, Lima C, et al. Orpotrin: a novel vasoconstrictor peptide from the venom of the Brazilian stingray *Potamotrygon* gr. *orbignyi*. Peptides 2006;27:3039–3046.
5. Liggins JB. An unusual bathing fatality. N Z Med J 1939;38:27–29.
6. Fenner PJ, Williamson JA, Skinner RA. Fatal and non-fatal stingray envenomation. Med J Aust 1989;151: 621–625.
7. Brisset IB, Schaper A, Pommier P, de Haro L. Envenomation by Amazonian freshwater stingray *Potamotrygon motoro*: 2 cases reported in Europe. Toxicon 2006;47: 32–34.
8. Forrester MB. Pattern of stingray injuries reported to Texas poison centers from 1998 to 2004. Hum Exp Toxicol 2005;24;639–642.
9. Rathjen WF, Halstead BW. Report on two fatalities due to stingrays. Toxicon 1969;6:301–302.
10. Evans RJ, Davies RS. Stingray injury. J Accid Emerg Med 1996;13:224–225.
11. Ikeda T. Supraventricular bigeminy following a stingray envenomation: a case report. Hawaii Med J 1989;48: 162–164.
12. Campbell J, Grenon M, You CK. Pseudoaneurysm of the superficial femoral artery resulting from stingray envenomation. Ann Vasc Surg 2003;17:217–20.
13. Cadzow WH. Puncture wound of the liver by stingray spines. Med J Aust 1960;47:936–937.
14. Haddad V Jr, Neto DG, de Paula Neto JB, de Luna Marques FP, Barbaro KC. Freshwater stingrays: a study of epidemiologic, clinic and therapeutic aspects based on 84 envenomings in humans and some enzymatic activities of the venom. Toxicon 2004;43:287–294.
15. Cook MD, Matteucci MJ, Lall R, Ly BT. Stingray envenomation. J Emerg Med 2006;30:345–347.
16. Derr C, O'Connor BJ, MacLeod SL. Laceration of the popliteal artery and compartment syndrome resulting from stingray envenomation. Am J Emerg Med 2007;25: 96–97.
17. Barss P. Wound necrosis caused by the venom of stingrays. Pathological findings and surgical management. Med J Aust 1984;141:854–855.
18. Clark RF, Girard RH, Rao D, Ly BT, Davis DP. Stingray envenomation: a retrospective review of clinical presentation and treatment in 119 cases. J Emerg Med 2007;33: 33–37.
19. Hardy R, Mackie PR. Observations on the chemical composition and toxicity of ratfish (*Chimaera monstrosa*). J Sci Food Agric 1971;22:382–388.

VI Mammals

Chapter 185

MAMMALS

Although there are numerous nonmammalian vertebrates with lethal venoms, few mammalian species are venomous. The best-known venomous mammals belong to the orders Monotremata (platypus, spiny anteaters) and Soricomorpha (shrews). Fossil records indicate that some extinct mammals (North American eutherian, *Bisonalveus browni* Gazin) evolved specialized teeth as salivary venom delivery systems distinctly different from the present-day platypus or shrew.[1] Genome analysis indicates that reptile and platypus venom proteins have been co-opted independently from the same gene families.[2] The monotremes have one opening for their reproductive, excretory, and digestive systems, and in contrast to all other mammals, the female monotremes lay eggs instead of giving live birth. The Australian spiny anteater (echidna) has venomous spines on its hind legs, but there are few data on envenomations by this mammal.

PLATYPUS

Platypuses (*Ornithorhynchus anatinus* Shaw) are furry, dark-brown creatures that inhabit fresh water streams along the eastern Australian coast, Tasmania, and New Guinea where they feed on insect larvae, crustaceans, and mollusks on the bottom of streams. While foraging for food, they dive many times (~75 dives) every hour, spending several seconds between dives on the surface.[3] The mature male platypus weighs about 2kg (~4lbs) and measures about 60cm (~2ft) in length including a 20cm (~8in.) beaver-like tail. The female lays two or three eggs, which hatch after a one-week incubation by the mother. The male Australian duckbilled platypus has venomous, erectile, keratinized spurs on each inner hind leg just above the webbed foot. This shy, stocky creature has venom glands under the dorsal thigh muscles that connect to the spur via a venom duct. During envenomation, the male platypus grabs the victim between his hind legs and then drives the spurs into the victim while squeezing his thighs (and venom gland). Platypus venom is a complex mixture containing several polypeptides and nonproteins including defensins-like peptides,[4] L-to-D-amino-acid-residue isomerases,[5] and C-type natriuretic peptide.[6] The latter compound is a 39-residue peptide that is structurally similar to mammalian C-type natriuretic peptides. These compounds promote mast cell histamine release as well as causing hypotension and vasodilation in animal models via an increase in smooth muscle cGMP.[7] The role of defensins-like peptides in platypus venom is unclear because animal studies indicate these polypeptides do not affect muscles or Na^+-channel currents.[8] Envenomation produces immediate, severe pain along with hyperalgias and edema that may cause functional impairment of the envenomated extremity.[9] Symptoms may persist for 2–3 weeks, and muscle wasting can occur. Systemic effects following envenomation do not usually occur.[10] Treatment is supportive including ice, elevation, splints, and analgesia (regional anesthetic blocks, narcotic as necessary) to relieve the pain. Heat does not usually relieve the pain associated with enven-

Medical Toxicology of Natural Substances, by Donald G. Barceloux, MD

omation. There is no clinical data to support the use of steroids.

SHREWS

Shrews are insectivores that belong to the order Soricomorpha and the Soricidae family; these animals are the smallest mammals on earth. Fossil records of the extinct giant shrews from the Early Pleistocene era indicate that these mammals had a rudimentary envenomation apparatus, probably as a means to increase body mass and hunt larger-sized prey.[11] These animals have short, dense fur and a long, slender snout. They live in a variety of habitats including forests, river banks, deserts, and even human dwellings. Venomous shrews include the American short-tailed shrew (*Blarina brevicauda* Say), the Haitian solenodon (*Solenodon paradoxus* Brandt), the European water shrew (*Neomys fodiens* Pennant), and the Mediterranean shrew (*Neomys anomalus* Cabrera).[12] These mammals possess enlarged and granular submaxillary venom glands that potentially could cause poisoning, but human envenomation is not well described.[13] Some shrews probably do not cause envenomation, such as the Indian shrew (*Suncus murinus* L.).[14]

American Short-Tailed Shrew

Based on limited experimental data, the venom from *Blarina brevicauda* Say is more toxic than other shrews.[12] Blarina toxin is an *N*-linked glycoprotein with a tissue kallikrein-like activity isolated from the submaxillary and sublingual glands of the American short-tailed shrew.[15] This toxin produces neurotoxic effects in animal models including respiratory depression, paralysis, convulsions, and death following intraperitoneal injection in mice.[16] A 32-kDa *N*-glycosylated protease, blarinasin, is a kallikrein-like protease that accounts for 70–75% of the kallikrein-like enzymes in the salivary gland of this shrew based on comparative studies.[17] The effect of envenomation by this shrew is variable, partly as a result of the inefficient venom apparatus.[18] The clinical significance of this toxin to humans is unclear because of the absence of reports of human toxicity following contact with shrews other than reports of minor local effects.[19] These mammals are potential vectors for bacteria and parasites.

Haitian Solenodon

The Haitian insectivore *Solenodon paradoxus* Brandt secretes venom from a mandibular gland that connects via a groove to the second lower incisor. These grooved incisors are missing in the American short-tailed shrew and the European water shrew. *Solenodon paradoxus* Brandt comprises one of only two surviving species of West Indian insectivores.[20] Human envenomations are not well documented, in part because of the limited habitat (i.e., Haiti, Dominican Republic) of this mammal. Mammalian bite produces a variety of infectious complications as well as serious diseases (e.g., rabies), but the mechanism is infectious rather than toxic.

References

1. Fox RC, Scott CS. First evidence of a venom delivery apparatus in extinct mammals. Nature 2005;435:1091–1093.
2. Warren WC, Hillier LW, Marshall Graves JA, Birney E, Ponting CP, Grützner F, et al. Genome analysis of the platypus reveals unique signatures of evolution. Nature 2008;453:175–183.
3. Bethge P, Munks S, Otley H, Nicol S. Diving behaviour, dive cycles and aerobic dive limit in the platypus *Ornithorhynchus anatinus*. Comp Biochem Physiol A Mol Integr Physiol 2003;136:799–809.
4. Torres AM, Tsampazi C, Geraghty DP, Bansal PS, Alewood PF, Kuchel PW. D-Amino acid residue in a defensins-like peptide from platypus venom: effect on structure and chromatographic properties. Biochem J 2005;391:215–220.
5. Torres AM, Tsampazi M, Kennett EC, Belov K, Geraghty DP, Bansal PS, et al. Characterization and isolation of L-to-D-amino-acid-residue isomerase from platypus venom. Amino Acids 2007;32:63–68.
6. Torres AM, Alewood D, Alewood PF, Gallagher CH, Kuchel PW. Conformations of platypus venom C-type natriuretic peptide in aqueous solution and sodium dodecyl sulfate micelles. Toxicon 2002;40:711–719.
7. de Plater GM, Martin RL, Milburn PJ. A C-type natriuretic peptide from the venom of the platypus (*Ornithorhynchus anatinus*): structure and pharmacology. Comp Biochem Physiol 1998;120(Part C):99–110.
8. Torres AM, Kuchel PW. The β-defensin-fold family of polypeptides. Toxicon 2004;44:581–588.
9. Fenner PJ, Williamson JA, Myers D. Platypus envenomation—a painful learning experience. Med J Aust 1992;157:829–832.
10. Tonkin MA, Negrine J. Wild platypus attack in the Antipodes. J Hand Surg (Br) 1994;19B:162–164.

11. Cuenca-Bescós G, Rofes J. First evidence of poisonous shrews with an envenomation apparatus. Naturwissenschaften 2007;94:113–116.
12. Dufton MJ. Venomous mammals. Pharmacol Ther 1992;53:199–315.
13. Pearson OP. The submaxillary glands of shrews. Anat Rec 1950;107:161–169.
14. Akhter MH. Absence of salivary toxin in chhachhundar the Indian shrew, *Suncus murinus*. Indian J Physiol Pharmacol 1974;18:191–194.
15. Kita M, Nakamura Y, Okumura Y, Ohdachi SD, Oba Y, Yoshikuni M, et al. Blarina toxin, a mammalian lethal venom from the short-tailed shrew *Blarina brevicauda*: Isolation and characterization. Proc Natl Acad Sci U S A 2004;101:7542–7547.
16. Ellis S, Krayer O. Properties of a toxin from the salivary gland of the shrew, *Blarina brevicauda*. J Pharmacol Exp Ther 1955;114:127–137.
17. Kita M, Okumura Y, Ohdachi SD, Oba Y, Yoshikuni M, Nakamura Y, et al. Purification and characterization of blarinasin, a new tissue kallikrein-like protease from the short-tailed shrew *Blarina brevicauda*: comparative studies with blarina toxin. Bio Chem 2005;386:177–182.
18. Tomasi TE. Function of venom in the short-tailed shrew, *Blarina brevicauda*. J Mammalogy 1978;59:852–854.
19. Maynard CJ. Singular effects produced by the bite of a short-tailed shrew *Blarina brevicauda*. Cont Sci Newtonwille 1889;1:57–59.
20. Roca AL, Bar-Gal GK, Eizirik E, Helgen KM, Maria R, Springer MS, et al. Mesozoic origin for West Indian insectivores. Nature 2004;429:649–651.

INDEX

Medical Toxicology of Natural Substances, by Donald G. Barceloux, MD